国外优秀数学著作
原 版 系 列

离散与组合数学手册

Handbook of Discrete and Combinatorial Mathematics, 2e

[美] 肯尼斯·H. 罗森 (Kenneth H. Rosen) 主编

（上册·第二版）

（英文）

哈尔滨工业大学出版社
HARBIN INSTITUTE OF TECHNOLOGY PRESS

黑版贸审字 08－2020－201 号

**图书在版编目(CIP)数据**

离散与组合数学手册：第二版＝Handbook of Discrete and Combinatorial Mathematics，2e. 上册：英文/(美)肯尼斯·H. 罗森(Kenneth H. Rosen)主编. —哈尔滨：哈尔滨工业大学出版社，2023.1
ISBN 978-7-5767-0651-2

Ⅰ.①离… Ⅱ.①肯… Ⅲ.①离散数学－英文 ②组合数学－英文 Ⅳ.①O158 ②O157

中国国家版本馆 CIP 数据核字(2023)第 030262 号

LISAN YU ZUHE SHUXUE SHOUCE：DI-ER BAN(SHANGCE)

策划编辑 刘培杰 杜莹雪
责任编辑 刘立娟
封面设计 孙茵艾
出版发行 哈尔滨工业大学出版社
社　　址 哈尔滨市南岗区复华四道街 10 号　邮编 150006
传　　真 0451－86414749
网　　址 http://hitpress.hit.edu.cn
印　　刷 哈尔滨市石桥印务有限公司
开　　本 787 mm×1 092 mm　1/16　印张 102.25　字数 1 959 千字
版　　次 2023 年 1 月第 1 版　2023 年 1 月第 1 次印刷
书　　号 ISBN 978-7-5767-0651-2
定　　价 248.00 元(全 2 册)

(如因印装质量问题影响阅读，我社负责调换)

# *PREFACE*

The importance of discrete and combinatorial mathematics has increased dramatically within the last few decades. This second edition has been written to update all content and to broaden the coverage. We have been gratified by the success of the first edition of the *Handbook*. We hope that the many readers who have asked for a second edition will find it worth the wait.

The purpose of the *Handbook of Discrete and Combinatorial Mathematics* is to provide a comprehensive reference volume for computer scientists, engineers, mathematicians, as well as students, physical and social scientists, and reference librarians, who need information about discrete and combinatorial mathematics.

This first edition of this book was the first resource that presented such information in a ready-reference form designed for all those who use aspects of this subject in their work or studies. This second edition is a major revision of the first edition. It includes extensive additions and updates, summarized later in this preface. The scope of this handbook includes the many areas generally considered to be parts of discrete mathematics, focusing on the information considered essential to its application in computer science, engineering, and other disciplines. Some of the fundamental topic areas covered in this edition include:

| | |
|---|---|
| logic and set theory | graph theory |
| enumeration | trees |
| integer sequences | network flows |
| recurrence relations | combinatorial designs |
| generating functions | computational geometry |
| number theory | coding theory |
| abstract algebra | cryptography |
| linear algebra | discrete optimization |
| discrete probability theory | automata theory |
| data mining | data structures and algorithms |
| discrete bioinformatics | |

### Format

The material in the *Handbook* is presented so that key information can be located and used quickly and easily. Each chapter includes a glossary that provides succinct definitions of the most important terms from that chapter. Individual topics are covered in sections and subsections within chapters, each of which is organized into clearly identifiable parts: definitions, facts, and examples. Lists of facts include:

- information about how material is used and why it is important
- historical information
- key theorems
- the latest results
- the status of open questions

- tables of numerical values, generally not easily computed
- summary tables
- key algorithms in simple pseudocode
- information about algorithms, such as their complexity
- major applications
- pointers to additional resources, both websites and printed material.

Facts are presented concisely and are listed so that they can be easily found and understood. Cross-references linking portions of the *Handbook* are also provided. Readers who wish to study a topic further can consult the resources listed.

The material in the *Handbook* has been chosen for inclusion primarily because it is important and useful. Additional material has been added to ensure comprehensiveness so that readers encountering new terminology and concepts from discrete mathematics in their explorations will be able to get help from this book.

Examples are provided to illustrate some of the key definitions, facts, and algorithms. Some curious and entertaining facts and puzzles that some readers may find intriguing are also included. Readers will also find an extensive collection of biographies after the main chapters, highlighting the lives of many important contributors to discrete mathematics.

Each chapter of the book includes a list of references divided into a list of printed resources and a list of relevant websites.

## How This Book Was Developed

The organization and structure of the first edition of this *Handbook* were developed by a team that included the chief editor, three associate editors, a project editor, and the editor from CRC Press. This team put together a proposed table of contents which was then analyzed by members of a group of advisory editors, each an expert in one or more aspects of discrete mathematics. These advisory editors suggested changes, including the coverage of additional important topics. Once the table of contents was fully developed, the individual sections of the book were prepared by a group of more than 70 contributors from industry and academia who understand how this material is used and why it is important. Contributors worked under the direction of the associate editors and chief editor, with these editors ensuring consistency of style as well as clarity and comprehensiveness in the presentation of material. Material was carefully reviewed by authors and our team of editors to ensure accuracy and consistency of style.

For the second edition, a new team was assembled. The first goal of this team was to put together a new table of contents. This involved identifying opportunities for new chapters and new sections to broaden the scope and appeal of the second edition. With the help of previous and new contributors, additional material was developed and existing material was updated and expanded, following the style and maintaining, or improving, the presentation in the first edition.

## Changes in the Second Edition

The development of the second edition of this book was a major effort, extending over many years. Since the first edition appeared in 1999, many new discoveries have been made and new areas have grown in importance. Important changes include:

- an increase from 17 to 20 chapters and over 360 additional pages

- new chapters on discrete bioinformatics and data mining

- individual chapters on coding theory and on cryptography with expanded coverage (previously covered in a single chapter)

- new sections on many topics, including

| | |
|---|---|
| Algebraic Number Theory | Elliptic Curves |
| Singular Value Decomposition | Hidden Markov Models |
| Probabilistic Method | Perfect Graphs |
| Expander Graphs | Small-World Networks |
| Combinatorial Auctions | Very Large-Scale Neighborhood Search |
| Tabu Search | Quantum Error-Correcting Codes |
| Classical Cryptography | Cryptographic Hashing Functions |
| Cryptographic Mechanisms | Modern Private Key Cryptography |
| Cryptographic Applications | Association Methods |
| Classification | Clustering |
| Outlier Analysis | Sequence Alignment |
| Phylogenetic Trees | Discrete-Time Dynamical Systems |
| RNA Folding | Food Webs and Competition Graphs |
| Neural Codes | Genome Assembly |

- thousands of updates and additions to existing sections, with major changes or new subsections including

| | |
|---|---|
| Primes Numbers | Combinatorial Matrix Theory |
| Cyclic Codes | Public Key Cryptography |
| Partitions | Asymptotics of Sequences |
| Factorization | Distance, Connectivity, Traversability, and Matching |
| Expander Graphs | Graph Colorings, Labelings, and Related Parameters |
| Simulation | Communication Networks |
| Location Theory | Difficult Routing and Assignment Problems |

- more than 30 new biographies

- hundreds of new web resources, which have been accessed to verify their availability in mid-2017

## Acknowledgments

First and foremost, we thank Bob Ross, the CRC editor for this book, for his support and his encouragement in completing this *Handbook*. We would also like to thank Bob Stern, our previous editor, for initiating the project of updating the *Handbook* and for supporting us patiently over many years. We would also like to thank Wayne Yuhasz, our original CRC editor, who commissioned the first edition. Thanks also go to the staff at CRC who helped with the production.

We would also like to thank the many people who were involved with this project. First, we would like to thank the team of advisory editors who helped make this reference relevant, useful, unique, and up-to-date. We want to thank our many contributors for their wonderful contributions and to several who helped edit individual chapters. With the passing of time, some of our advisory editors and contributors are no longer with us. We fondly remember and appreciate these colleagues and friends. (We have used the symbol ⊛ to designate these individuals.)

Finally, as Chief Editor, I would like to express my gratitude to my two associate editors for this edition. Without either of them this new edition would not have been possible. Both were essential in our quest to bring this *Handbook* up to date and to extend its scope. Douglas Shier confronted every problem that got in our way and found a workable solution. He kept the progress on this project going with his commitment and his abilities. Wayne Goddard used his masterful LaTeX skills to overcome all the typesetting challenges that confronted us as we migrated to a new typesetting environment.

MATLAB® is a registered trademark of The MathWorks, Inc. For product information please contact: The MathWorks, Inc., 3 Apple Hill Drive, Natick, MA, 01760-2098 USA. Tel: 508-647-7000, Fax: 508-647-7001, E-mail: info@mathworks.com, Web: www.mathworks.com

# *ADVISORY EDITORIAL BOARD*

# CONTRIBUTORS

Ravindra K. Ahuja
  *University of Florida*

Maria Albareda
  *Universitat Politècnica de Catalunya*

Daniel Aloise
  *Universidade Federal do Rio Grande
  do Norte*

Douglas Altner
  *MITRE*

Stephen F. Altschul
  *National Center for Biotechnology*

Nancy M. Amato
  *Texas A&M University*

George E. Andrews
  *Pennsylvania State University*

Niranjan Balachandran
  *Indian Institute of Technology, Bombay*

R. B. Bapat
  *Indian Statistical Institute*

M. Gisela Bardossy
  *University of Baltimore*

Joseph R. Barr
  *HomeUnion*

Lowell W. Beineke
  *Indiana University – Purdue University
  Fort Wayne*

Edward A. Bender
  *University of California at San Diego*

Karoly Bezdek
  *Cornell University*

Vladimir Boginski
  *University of Florida*

Joel V. Brawley
  *Clemson University*

Graham Brightwell
  *London School of Economics*

Stefan A. Burr
  *City College of New York*

Lisa Carbone
  *Rutgers University*

Jianer Chen
  *Texas A&M University*

Sunil Chopra
  *Northwestern University*

Maria Chudnovsky
  *Princeton University*

Charles J. Colbourn
  *Arizona State University*

Thomas Cormen
  *Dartmouth College*

Margaret Cozzens
  *Rutgers University*

Carina Curto
  *Pennyslvania State University*

Geir Dahl
  *University of Oslo*

Mukesh Dalal
  *i2 Technologies*

Robert W. Day
  *University of Connecticut*

Tamal K. Dey
  *Ohio State University*

Elena Dimitrova
  *Clemson University*

Jeffrey H. Dinitz
*University of Vermont*

Michael Doob
*University of Manitoba*

Thomas A. Dowling
*Ohio State University*

⊛S. E. Elmaghraby
*North Carolina State University*

Susanna S. Epp
*DePaul University*

Joan Feigenbaum
*Yale University*

W. Randolph Franklin
*Rensselaer Polytechnic Institute*

Joseph A. Gallian
*University of Minnesota Duluth*

William Gasarch
*University of Maryland*

Bart E. Goddard
*University of Texas*

Wayne Goddard
*Clemson University*

⊛Charles H. Goldberg
*The College of New Jersey*

Bruce L. Golden
*University of Maryland*

Jon F. Grantham
*Institute for Defense Analyses*

Ralph P. Grimaldi
*Rose-Hulman Institute of Technology*

Jonathan L. Gross
*Columbia University*

Jerrold W. Grossman
*Oakland University*

András Gyárfás
*Hungarian Academy of Sciences*

⊛S. Louis Hakimi
*University of California at Davis*

Richard Hammack
*Virginia Commonwealth University*

Pierre Hansen
*HEC Montréal*

Teresa W. Haynes
*East Tennessee State University*

Qijun He
*Virginia Tech*

Lane A. Hemaspaandra
*University of Rochester*

Michael A. Henning
*University of Johannesburg*

Glenn Hurlbert
*Virginia Commonwealth University*

Wilfried Imrich
*Montanuniversität Leoben*

Vladimir Itskov
*Pennsylvania State University*

Patrick Jaillet
*Massachusetts Institute of Technology*

Andy Jenkins
*University of Georgia*

Shaoquan Jiang
*Mianyang Normal University*

David Joyner
*United States Naval Academy*

Bharat K. Kaku
*Georgetown University*

Sampath Kannan
*University of Pennsylvania*

Victor J. Katz
*University of the District of Columbia*

Khoongming Khoo
*DSO National Laboratories, Singapore*

Jongeun Kim
*University of Florida*

Sandi Klavžar
*University of Ljubljana*

Dina Kravets
*Retired*

Michael Krebs
*California State University, Los Angeles*

Vidyadhar G. Kulkarni
*University of North Carolina*

Manuel Laguna
*University of Colorado*

Charles C. Y. Lam
*California State University, Bakersfield*

Josef Lauri
*University of Malta*

Lawrence M. Leemis
*The College of William and Mary*

Matthew Macauley
*Clemson University*

Bennet Manvel
*Colorado State University*

⊛Carla D. Martin
*National Security Agency*

Stephen B. Maurer
*Swarthmore College*

Alfred J. Menezes
*University of Waterloo*

Mike Mesterton-Gibbons
*Florida State University*

John G. Michaels
*SUNY Brockport*

Milena Mihail
*Georgia Institute of Technology*

Victor S. Miller
*Institute for Defense Analyses*

Esmond Ng
*Lawrence Berkeley National Laboratory*

Beth Novick
*Clemson University*

J. B. Orlin
*Massachusetts Institute of Technology*

James G. Oxley
*Louisiana State University*

János Pach
*EPFL Lausanne and Rényi Institute*

Edward W. Packel
*Lake Forest College*

⊛Uri Peled
*University of Illinois at Chicago*

Barry Peyton
*Dalton State College*

Adolfo Piperno
*Sapienza University of Rome*

Michael D. Plummer
*Vanderbilt University*

Carl Pomerance
*Dartmouth University*

Mihai Pop
*University of Maryland*

Svetlana Poznanović
*Clemson University*

Viera Krnanova Proulx
*Northeastern University*

Robert G. Rieper
*William Patterson University*

David Riley
*University of Wisconsin*

Kenneth H. Rosen
*Monmouth University*

Juanjo Rué
*Universitat Politècnica de Catalunya*

Joseph Rusinko
*Hobart and William Smith Colleges*

Bruce E. Sagan
*Michigan State University*

Aarto Salomaa
*University of Turku*

Edward R. Scheinerman
*Johns Hopkins University*

Richard Scherl
*Monmouth University*

Anthony Shaheen
*California State University, Los Angeles*

Jeff Shalit
*University of Waterloo*

Tony Shaska
*Oakland University*

Douglas R. Shier
*Clemson University*

Andrew V. Sills
*Georgia Southern University*

David Simchi-Levi
*Northwestern University*

Paul K. Stockmeyer
*The College of William and Mary*

Vladimir Stozhkov
*University of Florida*

Ileana Streinu
*Smith College*

Alan C. Tucker
*SUNY Stony Brook*

Peter R. Turner
*United States Naval Academy*

Paul C. van Oorschot
*Entrust Technologies*

Xingyin Wang
*Singapore University of Technology and Design*

Narada Warakagoda
*Telenor*

Lawrence C. Washington
*University of Maryland*

Douglas B. West
*University of Illinois at Urbana–Champaign*

Arthur T. White
*Western Michigan University*

Jay Yellen
*Florida Institute of Technology*

# *CONTENTS*

# 1

## *FOUNDATIONS*

## INTRODUCTION

This chapter covers material usually referred to as the foundations of mathematics, including logic, sets, and functions. In addition to covering these foundational areas, this chapter includes material that shows how these topics are applied to discrete mathematics, computer science, and electrical engineering. For example, this chapter covers methods of proof, program verification, and fuzzy reasoning.

## GLOSSARY

**action**:  a literal or a print command in a production system.

**aleph-null**:  the cardinality $\aleph_0$ of the set $\mathcal{N}$ of natural numbers.

**AND**:  the logical operator for conjunction, also written $\wedge$.

**antecedent**:  in a conditional proposition $p \to q$ ("if $p$ then $q$"), the proposition $p$ ("if-clause") that precedes the arrow.

**antichain**:  a subset of a poset in which no two elements are comparable.

**antisymmetric**:  the property of a binary relation $R$ that if $aRb$ and $bRa$, then $a = b$.

**argument form**:  a sequence of statement forms, each called a *premise* of the argument, followed by a statement form called a *conclusion* of the argument.

**assertion** (or **program assertion**):  a program comment specifying some conditions on the values of the computational variables; these conditions are supposed to hold whenever program flow reaches the location of the assertion.

**asymmetric**:  the property of a binary relation $R$ that if $aRb$, then $b\not{R}a$.

**asymptotic**:  A function $f$ is asymptotic to a function $g$, written $f(x) \sim g(x)$, if $f(x) \neq 0$ for sufficiently large $x$ and $\lim_{x \to \infty} \frac{g(x)}{f(x)} = 1$.

**atom** (or **atomic formula**):  a simplest formula of predicate logic.

**atomic formula**:  See *atom*.

**atomic proposition**:  a proposition that cannot be analyzed into smaller parts and logical operations.

**automated reasoning**:  the process of proving theorems using a computer program that can draw conclusions that follow logically from a set of given facts.

**axiom**:  a statement that is assumed to be true; a postulate.

**axiom of choice**:  the assertion that given any nonempty collection $\mathcal{A}$ of pairwise disjoint sets, there is a set that consists of exactly one element from each of the sets in $\mathcal{A}$.

**axiom** (or **semantic axiom**):  a rule for a programming language construct prescribing the change of values of computational variables when an instruction of that construct-type is executed.

**basis step**:  a proof of the basis premise (first case) in a proof by mathematical induction.

**big-oh notation**:  $f$ is $O(g)$, written $f = O(g)$, if there are constants $C$ and $k$ such that $|f(x)| \leq C|g(x)|$ for all $x > k$.

**bijection** (or **bijective function**): a function that is one-to-one and onto.

**bijective function**: See *bijection*.

**binary relation** (from a set $A$ to a set $B$): any subset of $A \times B$.

**binary relation** (on a set $A$): a binary relation from $A$ to $A$; i.e., a subset of $A \times A$.

**body** (of a clause $A_1, \ldots, A_n \leftarrow B_1, \ldots, B_m$ in a logic program): the literals $B_1, \ldots, B_m$ after $\leftarrow$.

**cardinal number** (or **cardinality**) of a set: for a finite set, the number of elements; for an infinite set, the order of infinity. The cardinal number of $S$ is written $|S|$.

**cardinality**: See *cardinal number*.

**Cartesian product** (of sets $A$ and $B$): the set $A \times B$ of ordered pairs $(a, b)$ with $a \in A$ and $b \in B$ (more generally, the **iterated Cartesian product** $A_1 \times A_2 \times \cdots \times A_n$ is the set of ordered $n$-tuples $(a_1, a_2, \ldots, a_n)$, with $a_i \in A_i$ for each $i$).

**ceiling** (of $x$): the smallest integer that is greater than or equal to $x$, written $\lceil x \rceil$.

**chain**: a subset of a poset in which every pair of elements are comparable.

**characteristic function** (of a set $S$): the function from $S$ to $\{0, 1\}$ whose value at $x$ is 1 if $x \in S$ and 0 if $x \notin S$.

**clause** (in a logic program): a closed formula of the form $\forall x_1 \ldots \forall x_s (A_1 \vee \cdots \vee A_n \leftarrow B_1 \wedge \cdots \wedge B_m)$.

**closed formula**: for a function value $f(x)$, an algebraic expression in $x$.

**closure** (of a relation $R$ with respect to a property $\mathcal{P}$): the relation $S$, if it exists, that has property $\mathcal{P}$ and contains $R$, such that $S$ is a subset of every relation that has property $\mathcal{P}$ and contains $R$.

**codomain** (of a function): the set in which the function values occur.

**comparable** (elements in a poset): elements that are related by the partial order relation.

**complement** (of a relation): given a relation $R$, the relation $\overline{R}$ where $a\overline{R}b$ if and only if $a\cancel{R}b$.

**complement** (of a set): given a set $A$ in a "universal" domain $U$, the set $\overline{A}$ of objects in $U$ that are not in $A$.

**complement operator**: a function $[0, 1] \rightarrow [0, 1]$ used for complementing fuzzy sets.

**complete**: the property of a set of axioms that it is possible to prove all true statements.

**complex number**: a number of the form $a + bi$, where $a$ and $b$ are real numbers, and $i^2 = -1$; the set of all complex numbers is denoted $\mathcal{C}$.

**composite key**: given an $n$-ary relation $R$ on $A_1 \times A_2 \times \cdots \times A_n$, a product of domains $A_{i_1} \times A_{i_2} \times \cdots \times A_{i_m}$ such that for each $m$-tuple $(a_{i_1}, a_{i_2}, \ldots, a_{i_m}) \in A_{i_1} \times A_{i_2} \times \cdots \times A_{i_m}$, there is at most one $n$-tuple in $R$ that matches $(a_{i_1}, a_{i_2}, \ldots, a_{i_m})$ in coordinates $i_1, i_2, \ldots, i_m$.

**composition** (of relations): for $R$ a relation from $A$ to $B$ and $S$ a relation from $B$ to $C$, the relation $S \circ R$ from $A$ to $C$ such that $a(S \circ R)c$ if and only if there exists $b \in B$ such that $aRb$ and $bSc$.

**composition** (of functions): the function $f \circ g$ whose value at $x$ is $f(g(x))$.

**compound proposition**:  a proposition built up from atomic propositions and logical connectives.

**computer-assisted proof**:  a proof that relies on checking the validity of a large number of cases using a special purpose computer program.

**conclusion** (of an argument form):  the last statement of an argument form.

**conclusion** (of a proof):  the last proposition of a proof; the objective of the proof is demonstrating that the conclusion follows from the premises.

**condition**:  the disjunction $A_1 \vee \cdots \vee A_n$ of atomic formulas.

**conditional statement**:  the compound proposition $p \rightarrow q$ ("if $p$ then $q$") that is true except when $p$ is true and $q$ is false.

**conjunction**:  the compound proposition $p \wedge q$ ("$p$ and $q$") that is true only when $p$ and $q$ are both true.

**conjunctive normal form**:  for a proposition in the variables $p_1, p_2, \ldots, p_n$, an equivalent proposition that is the conjunction of disjunctions, with each disjunction of the form $x_{k_1} \vee x_{k_2} \vee \cdots \vee x_{k_m}$, where $x_{k_j}$ is either $p_{k_j}$ or $\neg p_{k_j}$.

**consequent**:  in a conditional proposition $p \rightarrow q$ ("if $p$ then $q$") the proposition $q$ ("then-clause") that follows the arrow.

**consistent**:  the property of a set of axioms that no contradiction can be deduced from the axioms.

**construct** (or **program construct**):  the general form of a programming instruction such as an assignment, a conditional, or a while-loop.

**continuum hypothesis**:  the assertion that the cardinal number of the real numbers is the smallest cardinal number greater than the cardinal number of the natural numbers.

**contradiction**:  a self-contradictory proposition, one that is always false.

**contradiction** (in an indirect proof):  the negation of a premise.

**contrapositive** (of the conditional proposition $p \rightarrow q$):  the  conditional  proposition $\neg q \rightarrow \neg p$.

**converse** (of the conditional proposition $p \rightarrow q$):  the conditional proposition $q \rightarrow p$.

**converse relation**:  another name for the inverse relation.

**corollary**:  a theorem that is derived as an easy consequence of another theorem.

**correct conclusion**:  the conclusion of a valid proof, when all the premises are true.

**countable set**:  a set that is finite or denumerable.

**counterexample**:  a case that makes a statement false.

**definite clause**:  clause with at most one atom in its head.

**denumerable set**:  a set that can be placed in one-to-one correspondence with the natural numbers.

**diagonalization proof**:  any proof that involves something analogous to the diagonal of a list of sequences.

**difference**:  a binary relation $R - S$ such that $a(R - S)b$ if and only if $aRb$ is true and $aSb$ is false.

**difference** (of sets): the set $A - B$ of objects in $A$ that are not in $B$.

**direct proof**: a proof of $p \to q$ that assumes $p$ and shows that $q$ must follow.

**disjoint** (pair of sets): two sets with no members in common.

**disjunction**: the statement $p \vee q$ ("$p$ or $q$") that is true when at least one of the two propositions $p$ and $q$ is true; also called *inclusive or*.

**disjunctive normal form**: for a proposition in the variables $p_1, p_2, \ldots, p_n$, an equivalent proposition that is the disjunction of conjunctions, with each conjunction of the form $x_{k_1} \wedge x_{k_2} \wedge \cdots \wedge x_{k_m}$, where $x_{k_j}$ is either $p_{k_j}$ or $\neg p_{k_j}$.

**disproof**: a proof that a statement is false.

**divisibility lattice**: the lattice consisting of the positive integers under the relation of divisibility.

**domain** (of a function): the set on which a function acts.

**element** (of a set): member of the set; the notation $a \in A$ means that $a$ is an element of $A$.

**elementary projection function**: the function $\pi_i \colon X_1 \times \cdots \times X_n \to X_i$ such that $\pi(x_1, \ldots, x_n) = x_i$.

**empty set**: the set with no elements, written $\emptyset$ or $\{\ \}$.

**epimorphism**: an onto function.

**equality** (of sets): the property that two sets have the same elements.

**equivalence class**: given an equivalence relation on a set $A$ and $a \in A$, the subset of $A$ consisting of all elements related to $a$.

**equivalence relation**: a binary relation that is reflexive, symmetric, and transitive.

**equivalent propositions**: two compound propositions (on the same simple variables) with the same truth table.

**existential quantifier**: the quantifier $\exists x$, read "there is an $x$".

**existentially quantified predicate**: a statement $(\exists x)P(x)$ that there exists a value of $x$ such that $P(x)$ is true.

**exponential function**: any function of the form $b^x$, $b$ a positive constant, $b \neq 1$.

**fact set**: the set of ground atomic formulas.

**factorial** (function): the function $n!$ whose value on the argument $n$ is the product $1 \cdot 2 \cdot 3 \ldots n$; that is, $n! = 1 \cdot 2 \cdot 3 \ldots n$.

**finite**: the property of a set that it is either empty or else can be put in a one-to-one correspondence with a set $\{1, 2, 3, \ldots, n\}$ for some positive integer $n$.

**first-order logic**: See *predicate calculus*.

**floor** (of $x$): the greatest integer less than or equal to $x$, written $\lfloor x \rfloor$.

**formula**: a logical expression constructed from atoms with conjunctions, disjunctions, and negations, possibly with some logical quantifiers.

**full conjunctive normal form**: conjunctive normal form where each disjunction is a disjunction of all variables or their negations.

**full disjunctive normal form**: disjunctive normal form where each conjunction is a conjunction of all variables or their negations.

**fully parenthesized proposition**:  any proposition that can be obtained using the following recursive definition: (1) each variable is fully parenthesized; (2) if $P$ and $Q$ are fully parenthesized, so are $(\neg P)$, $(P \wedge Q)$, $(P \vee Q)$, $(P \rightarrow Q)$, and $(P \leftrightarrow Q)$.

**function** $f \colon A \rightarrow B$:  a rule that assigns to every object $a$ in the domain set $A$ exactly one object $f(a)$ in the codomain set $B$.

**functionally complete set**:  a set of logical connectives from which all other connectives can be derived by composition.

**fuzzy logic**:  a system of logic in which each statement has a truth value in the interval $[0, 1]$.

**fuzzy set**:  a set in which each element is associated with a number in the interval $[0, 1]$ that measures its degree of membership.

**generalized continuum hypothesis**:  the assertion that for every infinite set $S$ there is no cardinal number greater than $|S|$ and less than $|\mathcal{P}(S)|$.

**goal**:  a clause with an empty head.

**graph** (of a function):  given a function $f \colon A \rightarrow B$, the set $\{(a, b) \mid b = f(a)\} \subseteq A \times B$.

**greatest lower bound** (of a subset of a poset):  an element of the poset that is a lower bound of the subset and is greater than or equal to every other lower bound of the subset.

**ground formula**:  a formula without any variables.

**halting function**:  the function that maps computer programs to the set $\{0, 1\}$, with value 1 if the program always halts, regardless of input, and 0 otherwise.

**Hasse diagram**:  a directed graph that represents a poset.

**head** (of a clause $A_1, \ldots, A_n \leftarrow B_1, \ldots, B_m$):  the literals $A_1, \ldots, A_n$ before $\leftarrow$.

**identity function** (on a set):  given a set $A$, the function from $A$ to itself whose value at $x$ is $x$.

**image set** (of a function):  the set of function values as $x$ ranges over all objects of the domain.

**implication**:  formally, the relation $P \Rightarrow Q$ that a proposition $Q$ is true whenever proposition $P$ is true; informally, a synonym for the conditional statement $p \rightarrow q$.

**incomparable**:  two elements in a poset that are not related by the partial order relation.

**independent**:  the property of a set of axioms that none of the axioms can be deduced from the other axioms.

**indirect proof**:  a proof of $p \rightarrow q$ that assumes $\neg q$ is true and proves that $\neg p$ is true.

**induced partition** (on a set under an equivalence relation):  the collection of equivalence classes under the relation.

**induction**:  See *mathematical induction*.

**induction hypothesis**:  in a mathematical induction proof, the statement $P(x_k)$ in the induction step.

**induction step**:  in a mathematical induction proof, a proof of the induction premise "if $P(x_k)$ is true, then $P(x_{k+1})$ is true".

**inductive proof**:  See *mathematical induction*.

*infinite* (set): a set that is not finite.

*injection* (or *injective function*): a one-to-one function.

*instance* (of a formula): formula obtained using a substitution.

*instantiation*: substitution of concrete values for the free variables of a statement or sequence of statements; an instance of a production rule.

*integer*: a whole number, possibly zero or negative; i.e., one of the elements in the set $\mathcal{Z} = \{\ldots, -2, -1, 0, 1, 2, \ldots\}$.

*intersection*: the set $A \cap B$ of objects common to both sets $A$ and $B$.

*intersection relation*: for binary relations $R$ and $S$ on $A$, the relation $R \cap S$ where $a(R \cap S)b$ if and only if $aRb$ and $aSb$.

*interval* (in a poset): given $a \leq b$ in a poset, a subset of the poset consisting of all elements $x$ such that $a \leq x \leq b$.

*inverse function*: for a one-to-one, onto function $f: X \to Y$, the function $f^{-1}: Y \to X$ whose value at $y \in Y$ is the unique $x \in X$ such that $f(x) = y$.

*inverse image* (under $f: X \to Y$ of a subset $T \subseteq Y$): the subset $\{x \in X \mid f(x) \in T\}$, written $f^{-1}(T)$.

*inverse relation*: for a binary relation $R$ from $A$ to $B$, the relation $R^{-1}$ from $B$ to $A$ where $bR^{-1}a$ if and only if $aRb$.

*invertible* (function): a one-to-one and onto function; a function that has an inverse.

*irrational number*: a real number that is not rational.

*irreflexive*: the property of a binary relation $R$ on $A$ that $a\cancel{R}a$, for all $a \in A$.

*lattice*: a poset in which every pair of elements has both a least upper bound and a greatest lower bound.

*least upper bound* (of a subset of a poset): an element of the poset that is an upper bound of the subset and is less than or equal to every other upper bound of the subset.

*lemma*: a theorem that is an intermediate step in the proof of a more important theorem.

*linearly ordered*: the property of a poset that every pair of elements are comparable, also called *totally ordered*.

*literal*: an atom or its negation.

*little-oh notation*: $f$ is $o(g)$ if $\lim_{x \to \infty} \left| \frac{f(x)}{g(x)} \right| = 0$.

*logarithmic function*: a function $\log_b x$ ($b$ a positive constant, $b \neq 1$) defined by the rule $\log_b x = y$ if and only if $b^y = x$.

*logic program*: a finite sequence of definite clauses.

*logically equivalent propositions*: compound propositions that involve the same variables and have the same truth table.

*logically implies*: A compound proposition $P$ logically implies a compound proposition $Q$ if $Q$ is true whenever $P$ is true.

*loop invariant*: an expression that specifies the circumstance under which the loop body will be executed again.

**lower bound** (for a subset of a poset): an element of the poset that is less than or equal to every element of the subset.

**mathematical induction**: a method of proving that every item of a sequence of propositions such as $P(n_0), P(n_0+1), P(n_0+2), \ldots$ is true by showing: (1) $P(n_0)$ is true, and (2) for all $n \geq n_0$, $P(n) \to P(n+1)$ is true.

**maximal element** (in a poset): an element that has no element greater than it.

**maximum element** (in a poset): an element greater than or equal to every element.

**membership function** (in fuzzy logic): a function from elements of a set to $[0,1]$.

**membership table** (for a set expression): a table used to calculate whether an object lies in the set described by the expression, based on its membership in the sets mentioned by the expression.

**minimal element** (in a poset): an element that has no element smaller than it.

**minimum element** (in a poset): an element less than or equal to every element.

**monomorphism**: a one-to-one function.

**multi-valued logic**: a logic system with a set of more than two truth values.

**multiset**: an extension of the set concept, in which each element may occur arbitrarily many times.

**mutually disjoint** (family of sets): See *pairwise disjoint*.

**n-ary predicate**: a statement involving $n$ variables.

**n-ary relation**: any subset of $A_1 \times A_2 \times \cdots \times A_n$.

**naive set theory**: set theory where any collection of objects can be considered to be a valid set, with paradoxes ignored.

**NAND**: the logical connective "not and".

**natural number**: a nonnegative integer (or "counting" number); i.e., an element of $\mathcal{N} = \{0, 1, 2, 3, \ldots\}$. *Note*: Sometimes 0 is not regarded as a natural number.

**negation**: the statement $\neg p$ ("not $p$") that is true if and only if $p$ is not true.

**NOP**: pronounced "no-op", a program instruction that does nothing to alter the values of computational variables or the order of execution.

**NOR**: the logical connective "not or".

**NOT**: the logical connective meaning "not", used in place of $\neg$.

**null set**: the set with no elements, written $\emptyset$ or $\{\ \}$.

**omega notation**: $f$ is $\Omega(g)$ if there are constants $C$ and $k$ such that $|g(x)| \leq C|f(x)|$ for all $x > k$.

**one-to-one** (function): a function $f \colon X \to Y$ that assigns distinct elements of the codomain to distinct elements of the domain; thus, if $x_1 \neq x_2$, then $f(x_1) \neq f(x_2)$.

**onto** (function): a function $f \colon X \to Y$ whose image equals its codomain; i.e., for every $y \in Y$, there is an $x \in X$ such that $f(x) = y$.

**OR**: the logical operator for disjunction, also written $\vee$.

**pairwise disjoint**: the property of a family of sets that each two distinct sets in the family have empty intersection; also called *mutually disjoint*.

**paradox**: a statement that contradicts itself.

**partial function**: a function $f: X \to Y$ that assigns a well-defined object in $Y$ to some (but not necessarily all) the elements of its domain $X$.

**partial order**: a binary relation that is reflexive, antisymmetric, and transitive.

**partially ordered set**: a set with a partial order relation defined on it.

**partition** (of a set $S$): a pairwise disjoint family $\mathcal{P} = \{A_i\}$ of nonempty subsets of $S$ whose union is $S$.

**Peano definition**: a recursive description of the natural numbers that uses the concept of successor.

**Polish prefix notation**: the style of writing compound propositions in prefix notation where sometimes the usual operand symbols are replaced as follows: N for $\neg$, K for $\wedge$, A for $\vee$, C for $\to$, E for $\leftrightarrow$.

**poset**: a partially ordered set.

**postcondition**: an assertion that appears immediately after the executable portion of a program fragment or of a subprogram.

**postfix notation**: the style of writing compound logical propositions where operators are written to the right of the operands.

**power** (of a relation): for a relation $R$ on $A$, the relation $R^n$ on $A$ where $R^0 = I$, $R^1 = R$, and $R^n = R^{n-1} \circ R$ for all $n > 1$.

**power set**: given a set $A$, the set $\mathcal{P}(A)$ of all subsets of $A$.

**precondition**: an assertion that appears immediately before the executable portion of a program fragment or of a subprogram.

**predicate**: a statement involving one or more variables that range over various domains.

**predicate calculus**: the symbolic study of quantified predicate statements.

**prefix notation**: the style of writing compound logical propositions where operators are written to the left of the operands.

**premise**: a proposition taken as the foundation of a proof, from which the conclusion is to be derived.

**prenex normal form**: the form of a well-formed formula in which every quantifier occurs at the beginning and the scope is whatever follows the quantifiers.

**preorder**: a binary relation that is reflexive and transitive.

**primary key**: for an $n$-ary relation on $A_1, A_2, \ldots, A_n$, a coordinate domain $A_j$ such that for each $x \in A_j$ there is at most one $n$-tuple in the relation whose $j$th coordinate is $x$.

**production rule**: a formula of the form $C_1, \ldots, C_n \to A_1, \ldots, A_m$ where each $C_i$ is a condition and each $A_i$ is an action.

**production system**: a set of production rules and a fact set.

**program construct**: See *construct*.

**program fragment**: any sequence of program code, from a single instruction to an entire program.

**program semantics** (or **semantics**): the meaning of an instruction or of a program fragment; i.e., the effect of its execution on the computational variables.

**projection function**:  a function defined on a set of $n$-tuples that selects the elements in certain coordinate positions.

**proof** (of a conclusion from a set of premises):  a sequence of statements (called steps) terminating in the conclusion, such that each step is either a premise or follows from previous steps by a valid argument.

**proof by contradiction**:  a proof that assumes the negation of the statement to be proved and shows that this leads to a contradiction.

**proof done by hand**:  a proof done by a human without the use of a computer.

**proper subset**:  given a set $S$, a subset $T$ of $S$ such that $S$ contains at least one element not in $T$.

**proposition**:  a declarative sentence or statement that is unambiguously either true or false.

**propositional calculus**:  the symbolic study of propositions.

**range** (of a function):  the image set of a function; sometimes used as synonym for codomain.

**rational number**:  the ratio $\frac{a}{b}$ of two integers such that $b \neq 0$; the set of all rational numbers is denoted $\mathcal{Q}$.

**real number**:  a number expressible as a finite (i.e., terminating) or infinite decimal; the set of all real numbers is denoted $\mathcal{R}$.

**recursive definition** (of a function with domain $\mathcal{N}$):  a set of initial values and a rule for computing $f(n)$ in terms of values $f(k)$ for $k < n$.

**recursive definition** (of a set $S$):  a form of specification of membership of $S$, in which some *basis* elements are named individually, and in which a computable rule is given to construct each other element in a finite number of steps.

**refinement of a partition**:  given a partition $\mathcal{P}_1 = \{A_j\}$ on a set $S$, a partition $\mathcal{P}_2 = \{B_i\}$ on the same set $S$ such that every $B_i \in \mathcal{P}_2$ is a subset of some $A_j \in \mathcal{P}_1$.

**reflexive**:  the property of a binary relation $R$ that $aRa$.

**relation** (from set $A$ to set $B$):  a binary relation from $A$ to $B$.

**relation** (on a set $A$):  a binary relation from $A$ to $A$.

**restriction** (of a function):  given $f\colon X \to Y$ and a subset $S \subseteq X$, the function $f|S$ with domain $S$ and codomain $Y$ whose rule is the same as that of $f$.

**reverse Polish notation**:  postfix notation.

**rule of inference**:  a valid argument form.

**satisfiable compound proposition**:  a compound proposition that is true for at least one assignment of truth variables to its variables.

**scope** (of a quantifier):  the predicate to which the quantifier applies.

**semantic axiom**:  See *axiom*.

**semantics**:  See *program semantics*.

**sentence**:  a well-formed formula with no free variables.

**sequence** (in a set):  a list of objects from a set $S$, with repetitions allowed; that is, a function $f\colon \mathcal{N} \to S$ (an infinite sequence, often written $a_0, a_1, a_2, \ldots$) or a function $f\colon \{1, 2, \ldots, n\} \to S$ (a finite sequence, often written $a_1, a_2, \ldots, a_n$).

**set**: a well-defined collection of objects.

**singleton**: a set with one element.

**specification**: in program correctness, a precondition and a postcondition.

**statement form**: a declarative sentence containing some variables and logical symbols which becomes a proposition if concrete values are substituted for all free variables.

**string**: a finite sequence in a set $S$, usually written so that consecutive entries are juxtaposed (i.e., written with no punctuation or extra space between them).

**strongly correct code**: code whose execution terminates in a computational state satisfying the postcondition, whenever the precondition holds before execution.

**subset** (of a set $S$): any set $T$ of objects that are also elements of $S$, written $T \subseteq S$.

**substitution**: a set of pairs of variables and terms.

**surjection** or (**surjective function**): an onto function.

**symmetric**: the property of a binary relation $R$ that if $aRb$ then $bRa$.

**symmetric difference** (of relations): for relations $R$ and $S$ on $A$, the relation $R \oplus S$ where $a(R \oplus S)b$ if and only if exactly one of the following is true: $aRb$, $aSb$.

**symmetric difference** (of sets): for sets $A$ and $B$, the set $A \oplus B$ containing each object that is an element of $A$ or an element of $B$, but not an element of both.

**system of distinct representatives**: given sets $A_1, A_2, \ldots, A_n$ (some of which may be equal), a set $\{a_1, a_2, \ldots, a_n\}$ of $n$ distinct elements with $a_i \in A_i$ for $i = 1, 2, \ldots, n$.

**tautology**: a compound proposition whose form makes it always true, regardless of the truth values of its atomic parts.

**term** (in a domain): either a fixed element of a domain $S$ or an $S$-valued variable.

**theorem**: a statement derived as the conclusion of a valid proof from axioms and definitions.

**theta notation**: $f$ is $\Theta(g)$, written $f = \Theta(g)$, if there are positive constants $C_1, C_2$, and $k$ such that $C_1|g(x)| \leq |f(x)| \leq C_2|g(x)|$ for all $x > k$.

**totally ordered**: the property of a poset that every pair of elements are comparable; also called *linearly ordered*.

**transitive**: the property of a binary relation $R$ that if $aRb$ and $bRc$, then $aRc$.

**transitive closure**: for a relation $R$ on $A$, the smallest transitive relation containing $R$.

**transitive reduction** (of a relation): a relation with the same transitive closure as the original relation and with a minimum number of ordered pairs.

**truth table**: for a compound proposition, a table that gives the truth value of the proposition for each possible combination of truth values of the atomic variables in the proposition.

**two-valued logic**: a logic system where each statement has exactly one of the two values: true or false.

**union**: the set $A \cup B$ of objects in one or both of the sets $A$ and $B$.

**union relation**: for $R$ and $S$ binary relations on $A$, the relation $R \cup S$ where $a(R \cup S)b$ if and only if $aRb$ or $aSb$.

**universal domain**: the collection of all possible objects in the context of the immediate discussion.

**universal quantifier**:  the quantifier $\forall x$, read "for all $x$" or "for every $x$".

**universally quantified predicate**:  a statement $(\forall x)P(x)$ that $P(x)$ is true for every $x$ in its universe of discourse.

**universe of discourse**:  the range of possible values of a variable, within the context of the immediate discussion.

**upper bound** (for a subset of a poset):  an element of the poset that is greater than or equal to every element of the subset.

**valid argument form**:  an argument form such that in any instantiation where all the premises are true, the conclusion is also true.

**Venn diagram**:  a figure composed of possibly overlapping circles or ellipses, used to picture membership in various combinations of the sets.

**verification** (of a program):  a formal argument for the correctness of a program with respect to its specifications.

**weakly correct code**:  code whose execution results in a computational state satisfying the postcondition, whenever the precondition holds before execution and the execution terminates.

**well-formed formula** (**wff**):  a proposition or predicate with quantifiers that bind one or more of its variables.

**well-ordered**:  the property of a set that every nonempty subset has a minimum element.

**well-ordering principle**:  the axiom that every nonempty subset of integers, each greater than a fixed integer, contains a smallest element.

**XOR**:  the logical connective "not or".

**Zermelo-Fraenkel axioms**:  a set of axioms for set theory.

**zero-order logic**:  propositional calculus.

## 1.1   PROPOSITIONAL AND PREDICATE LOGIC

Logic is the basis for distinguishing what may be correctly inferred from a given collection of facts. Propositional logic, where there are no quantifiers (so quantifiers range over nothing) is called zero-order logic. Predicate logic, where quantifiers range over members of a universe, is called first-order logic. Higher-order logic includes second-order logic (where quantifiers can range over relations over the universe), third-order logic (where quantifiers can range over relations over relations), and so on. Logic has many applications in computer science, including circuit design (§5.8.3) and verification of computer program correctness (§1.6). This section defines the meaning of the symbolism and various logical properties that are usually used without explicit mention. See [FlPa95], [Me09], and [Mo76].

Here, only two-valued logic is studied; i.e., each statement is either true or false. Multi-valued logic, in which statements have one of more than two values, is discussed in §1.7.2.

## 1.1.1   PROPOSITIONS AND LOGICAL OPERATIONS

### Definitions:

A **truth value** is either true or false, abbreviated $T$ and $F$, respectively.

A **proposition** (in a natural language such as English) is a declarative sentence that has a well-defined truth value.

A **propositional variable** is a mathematical variable, often denoted by $p$, $q$, or $r$, that represents a proposition.

**Propositional logic** (or **propositional calculus** or **zero-order logic**) is the study of logical propositions and their combinations using logical connectives.

A **logical connective** is an operation used to build more complicated logical expressions out of simpler propositions, whose truth values depend only on the truth values of the simpler propositions.

A proposition is **atomic** or **simple** if it cannot be syntactically analyzed into smaller parts; it is usually represented by a single logical variable.

A proposition is **compound** if it contains one or more logical connectives.

A **truth table** is a table that prescribes the defining rule for a logical operation. That is, for each combination of truth values of the operands, the table gives the truth value of the expression formed by the operation and operands.

The unary connective **negation** (denoted by $\neg$) is defined by the following truth table:

| $p$ | $\neg p$ |
|-----|----------|
| $T$ | $F$      |
| $F$ | $T$      |

*Note*: The negation $\neg p$ is also written $p'$, $\bar{p}$, or $\sim p$.

The common binary connectives are:

| | | |
|---|---|---|
| $p \wedge q$ | **conjunction** | *p and q* |
| $p \vee q$ | **disjunction** | *p or q* |
| $p \rightarrow q$ | **conditional** | *if p then q* |
| $p \leftrightarrow q$ | **biconditional** | *p if and only if q* |
| $p \oplus q$ | **exclusive or** | *p xor q* |
| $p \downarrow q$ | **not or** | *p nor q* |
| $p \mid q$  or  $p \uparrow q$ | **not and** | *p nand q* |

The connective $\mid$ is called the *Sheffer stroke*. The connective $\downarrow$ is called the *Peirce arrow*.

The values of the compound propositions obtained by using the binary connectives are given in the following table:

| $p$ | $q$ | $p \vee q$ | $p \wedge q$ | $p \to q$ | $p \leftrightarrow q$ | $p \oplus q$ | $p \downarrow q$ | $p \mid q$ |
|---|---|---|---|---|---|---|---|---|
| $T$ | $T$ | $T$ | $T$ | $T$ | $T$ | $F$ | $F$ | $F$ |
| $T$ | $F$ | $T$ | $F$ | $F$ | $F$ | $T$ | $F$ | $T$ |
| $F$ | $T$ | $T$ | $F$ | $T$ | $F$ | $T$ | $F$ | $T$ |
| $F$ | $F$ | $F$ | $F$ | $T$ | $T$ | $F$ | $T$ | $T$ |

In the conditional $p \to q$, $p$ is the **antecedent** and $q$ is the **consequent**. The conditional $p \to q$ is often read informally as "$p$ implies $q$".

**Infix notation** is the style of writing compound propositions where binary operators are written between the operands and negation is written to the left of its operand.

**Prefix notation** is the style of writing compound propositions where operators are written to the left of the operands.

**Postfix notation** (or **reverse Polish notation**) is the style of writing compound propositions where operators are written to the right of the operands.

**Polish notation** is the style of writing compound propositions where operators are written using prefix notation and where the usual operand symbols are replaced as follows: N for $\neg$, K for $\wedge$, A for $\vee$, C for $\to$, E for $\leftrightarrow$. (Jan Łukasiewicz, 1878–1956)

A **fully parenthesized proposition** is any proposition that can be obtained using the following recursive definition: (1) each variable is fully parenthesized; (2) if $P$ and $Q$ are fully parenthesized, so are $(\neg P)$, $(P \wedge Q)$, $(P \vee Q)$, $(P \to Q)$, and $(P \leftrightarrow Q)$.

**Facts:**

**1.** The conditional connective $p \to q$ represents the following English constructs:

- if $p$ then $q$
- $p$ only if $q$
- $q$ follows from $p$
- $p$ is a sufficient condition for $q$
- $q$ if $p$
- $p$ implies $q$
- $q$ whenever $p$
- $q$ is a necessary condition for $p$.

**2.** The biconditional connective $p \leftrightarrow q$ represents the following English constructs:

- $p$ if and only if $q$ (often written $p$ iff $q$)
- $p$ and $q$ imply each other
- $p$ is a necessary and sufficient condition for $q$
- $p$ and $q$ are equivalent.

**3.** In computer programming and circuit design, the following notation for logical operators is used: $p$ AND $q$ for $p \wedge q$, $p$ OR $q$ for $p \vee q$, NOT $p$ for $\neg p$, $p$ XOR $q$ for $p \oplus q$, $p$ NOR $q$ for $p \downarrow q$, $p$ NAND $q$ for $p \mid q$.

**4.** *Order of operations*:  In an unparenthesized compound proposition using only the five standard operators $\neg$, $\wedge$, $\vee$, $\to$, and $\leftrightarrow$, the following order of precedence is typically used when evaluating a logical expression, at each level of precedence moving from left to right: first $\neg$, then $\wedge$ and $\vee$, then $\to$, finally $\leftrightarrow$. Parenthesized expressions are evaluated proceeding from the innermost pair of parentheses outward, analogous to the evaluation of an arithmetic expression.

**5.** It is often preferable to use parentheses to show precedence, except for negation operators, rather than to rely on precedence rules.

**6.** No parentheses are needed when a compound proposition is written in either prefix or postfix notation. However, parentheses may be necessary when a compound proposition is written in infix notation.

**7.** The number of nonequivalent logical statements with two variables is 16, because each of the four lines of the truth table has two possible entries, $T$ or $F$. Here are examples of compound propositions that yield each possible combination of truth values. (**T** represents a tautology and **F** a contradiction. See §1.1.2.)

| $p$ | $q$ | **T** | $p \vee q$ | $q \to p$ | $p \to q$ | $p \mid q$ | $p$ | $q$ | $p \leftrightarrow q$ |
|---|---|---|---|---|---|---|---|---|---|
| $T$ | $T$ | $T$ | $T$ | $T$ | $T$ | $F$ | $T$ | $T$ | $T$ |
| $T$ | $F$ | $T$ | $T$ | $T$ | $F$ | $T$ | $T$ | $F$ | $F$ |
| $F$ | $T$ | $T$ | $T$ | $F$ | $T$ | $T$ | $F$ | $T$ | $F$ |
| $F$ | $F$ | $T$ | $F$ | $T$ | $T$ | $T$ | $F$ | $F$ | $T$ |

| $p$ | $q$ | $p \oplus q$ | $\neg q$ | $\neg p$ | $p \wedge q$ | $p \wedge \neg q$ | $\neg p \wedge$ | $p \downarrow q$ | **F** |
|---|---|---|---|---|---|---|---|---|---|
| $T$ | $T$ | $F$ | $F$ | $F$ | $T$ | $F$ | $F$ | $F$ | $F$ |
| $T$ | $F$ | $T$ | $T$ | $F$ | $F$ | $T$ | $F$ | $F$ | $F$ |
| $F$ | $T$ | $T$ | $F$ | $T$ | $F$ | $F$ | $T$ | $F$ | $F$ |
| $F$ | $F$ | $F$ | $T$ | $T$ | $F$ | $F$ | $F$ | $T$ | $F$ |

**8.** The number of different possible logical connectives on $n$ variables is $2^{2^n}$, because there are $2^n$ rows in the truth table.

**9.** The problem of determining whether a compound proposition is satisfiable, known as the *Propositional Satisfiability Problem* (abbreviated as SAT), is important in many practical applications, such as in circuit design and in artificial intelligence, and it is also important in the study of algorithms.

**10.** No efficient (polynomial-time) algorithm has been found for solving SAT. Because it is NP-complete (see Section 16.4.1), if a polynomial-time algorithm could be found for solving this problem, the famous P versus NP problem would be solved in the affirmative.

**Examples:**

**1.** "$1+1 = 3$" and "Romulus and Remus founded New York City" are false propositions.

**2.** "$1 + 1 = 2$" and "The year 1996 was a leap year" are true propositions.

**3.** "Go directly to jail" is not a proposition, because it is imperative, not declarative.

**4.** "$x > 5$" is not a proposition, because its truth value cannot be determined unless the value of $x$ is known.

**5.** "This sentence is false" is not a proposition, because it cannot be given a truth value without creating a contradiction.

**6.** In a truth table evaluation of the compound proposition $p \vee (\neg p \wedge q)$ from the innermost parenthetic expression outward, the steps are to evaluate $\neg p$, next $(\neg p \wedge q)$, and then $p \vee (\neg p \wedge q)$:

| $p$ | $q$ | $\neg p$ | $(\neg p \wedge q)$ | $p \vee (\neg p \wedge q)$ |
|---|---|---|---|---|
| $T$ | $T$ | $F$ | $F$ | $T$ |
| $T$ | $F$ | $F$ | $F$ | $T$ |
| $F$ | $T$ | $T$ | $T$ | $T$ |
| $F$ | $F$ | $T$ | $F$ | $F$ |

**7.** The statements in the left column are evaluated using the order of precedence indicated in the fully parenthesized form in the right column:

$$p \vee q \wedge r \qquad\qquad ((p \vee q) \wedge r)$$
$$p \leftrightarrow q \rightarrow r \qquad\qquad (p \leftrightarrow (q \rightarrow r))$$
$$\neg q \vee \neg r \rightarrow s \wedge t \quad (((\neg q) \vee (\neg r)) \rightarrow (s \wedge t))$$

**8.** The infix statement $p \wedge q$ in prefix notation is $\wedge p\, q$, in postfix notation is $p\, q \wedge$, and in Polish notation is $K\, p\, q$.

**9.** The infix statement $p \rightarrow \neg(q \vee r)$ in prefix notation is $\rightarrow p \neg \vee q\, r$, in postfix notation is $p\, q\, r \vee \neg \rightarrow$, and in Polish notation is $C\, p\, \mathrm{N}\, \mathrm{A}\, q\, r$.

---

## 1.1.2  EQUIVALENCES, IDENTITIES, AND NORMAL FORMS

### Definitions:

A **tautology** is a compound proposition that is always true, regardless of the truth values of its underlying atomic propositions.

A **contradiction** (or **self-contradiction**) is a compound proposition that is always false, regardless of the truth values of its underlying atomic propositions. (The term self-contradiction is used for such a proposition when discussing indirect mathematical arguments, because "contradiction" has another meaning in that context. See §1.5.)

A **contingency** is a compound proposition that is neither a tautology nor a contradiction.

A compound proposition is **satisfiable** if there is at least one assignment of truth values for which it is true.

A compound proposition $P$ **logically implies** a compound proposition $Q$, written $P \Rightarrow Q$, if $Q$ is true whenever $P$ is true. In this case, $P$ is **stronger than** $Q$, and $Q$ is **weaker than** $P$.

Compound propositions $P$ and $Q$ are **logically equivalent**, written $P \equiv Q$, $P \Leftrightarrow Q$, or $P$ iff $Q$, if they have the same truth values for all possible truth values of their variables.

A logical equivalence that is frequently used is sometimes called a **logical identity**.

A collection $\mathcal{C}$ of connectives is **functionally complete** if every compound proposition is equivalent to a compound proposition constructed using only connectives in $\mathcal{C}$.

A **disjunctive normal expression** in the propositions $p_1, p_2, \ldots, p_n$ is a disjunction of one or more propositions, each of the form $x_{k_1} \wedge x_{k_2} \wedge \cdots \wedge x_{k_m}$, where $x_{k_j}$ is either $p_{k_j}$ or $\neg p_{k_j}$.

A **disjunctive normal form** (**DNF**) for a proposition $P$ is a disjunctive normal expression that is logically equivalent to $P$.

A **conjunctive normal expression** in the propositions $p_1, p_2, \ldots, p_n$ is a conjunction of one or more compound propositions, each of the form $x_{k_1} \vee x_{k_2} \vee \cdots \vee x_{k_m}$, where $x_{k_j}$ is either $p_{k_j}$ or $\neg p_{k_j}$.

A **conjunctive normal form** (**CNF**) for a proposition $P$ is a conjunctive normal expression that is logically equivalent to $P$.

A compound proposition $P$ using only the connectives $\neg$, $\wedge$, and $\vee$ has a **logical dual** (denoted $P'$ or $P^d$), obtained by interchanging $\wedge$ and $\vee$, and interchanging the constant **T** (true) and the constant **F** (false).

The **converse** of the conditional proposition $p \to q$ is the proposition $q \to p$.

The **contrapositive** of the conditional proposition $p \to q$ is the proposition $\neg q \to \neg p$.

The **inverse** of the conditional proposition $p \to q$ is the proposition $\neg p \to \neg q$.

**Facts:**

1. $P \Leftrightarrow Q$ is true if and only if $P \Rightarrow Q$ and $Q \Rightarrow P$.

2. $P \Leftrightarrow Q$ is true if and only if $P \leftrightarrow Q$ is a tautology.

3. The following table lists several logical identities.

| name | rule |
|---|---|
| *Commutative laws* | $p \wedge q \Leftrightarrow q \wedge p, \quad p \vee q \Leftrightarrow q \vee p$ |
| *Associative laws* | $p \wedge (q \wedge r) \Leftrightarrow (p \wedge q) \wedge r, \quad p \vee (q \vee r) \Leftrightarrow (p \vee q) \vee r$ |
| *Distributive laws* | $p \wedge (q \vee r) \Leftrightarrow (p \wedge q) \vee (p \wedge r)$ |
| | $p \vee (q \wedge r) \Leftrightarrow (p \vee q) \wedge (p \vee r)$ |
| *DeMorgan's laws* | $\neg(p \wedge q) \Leftrightarrow (\neg p) \vee (\neg q), \quad \neg(p \vee q) \Leftrightarrow (\neg p) \wedge (\neg q)$ |
| *Excluded middle* | $p \vee \neg p \Leftrightarrow \mathbf{T}$ |
| *Contradiction* | $p \wedge \neg p \Leftrightarrow \mathbf{F}$ |
| *Double negation law* | $\neg(\neg p) \Leftrightarrow p$ |
| *Contrapositive law* | $p \to q \Leftrightarrow \neg q \to \neg p$ |
| *Conditional as disjunction* | $p \to q \Leftrightarrow \neg p \vee q$ |
| *Negation of conditional* | $\neg(p \to q) \Leftrightarrow p \wedge \neg q$ |
| *Biconditional as implication* | $(p \leftrightarrow q) \Leftrightarrow (p \to q) \wedge (q \to p)$ |
| *Idempotent laws* | $p \wedge p \Leftrightarrow p, \quad p \vee p \Leftrightarrow p$ |
| *Absorption laws* | $p \wedge (p \vee q) \Leftrightarrow p, \quad p \vee (p \wedge q) \Leftrightarrow p$ |
| *Dominance laws* | $p \vee \mathbf{T} \Leftrightarrow \mathbf{T}, \quad p \wedge \mathbf{F} \Leftrightarrow \mathbf{F}$ |
| *Exportation law* | $p \to (q \to r) \Leftrightarrow (p \wedge q) \to r$ |
| *Identity laws* | $p \wedge \mathbf{T} \Leftrightarrow p, \quad p \vee \mathbf{F} \Leftrightarrow p$ |

4. There are different ways to establish logical identities (equivalences):

   - truth tables (showing that both expressions have the same truth values);
   - using known logical identities and equivalences to establish new ones;
   - taking the dual of a known identity (Fact 7).

5. Logical identities are used in circuit design to simplify circuits. See §5.8.4.

6. Each of the following sets of connectives is functionally complete:

$$\{\wedge, \vee, \neg\}, \quad \{\wedge, \neg\}, \quad \{\vee, \neg\}, \quad \{\,|\,\}, \quad \{\downarrow\}.$$

However, these sets of connectives are not functionally complete:

$$\{\wedge\}, \quad \{\vee\}, \quad \{\wedge, \vee\}.$$

7. If $P \Leftrightarrow Q$ is a logical identity, then so is $P' \Leftrightarrow Q'$, where $P'$ and $Q'$ are the duals of $P$ and $Q$, respectively.

**8.** Every proposition has a disjunctive normal form and a conjunctive normal form, which can be obtained by Algorithms 1 and 2.

---

**Algorithm 1:  Disjunctive normal form of proposition $P$.**

write the truth table for $P$

for each line of the truth table on which $P$ is true, form a "line term"
$x_1 \wedge x_2 \wedge \cdots \wedge x_n$, where $x_i := p_i$ if $p_i$ is true on that line of the truth table
and $x_i := \neg p_i$ if $p_i$ is false on that line

form the disjunction of all these line terms

---

**Algorithm 2:  Conjunctive normal form of proposition $P$.**

write the truth table for $P$

for each line of the truth table on which $P$ is false, form a "line term"
$x_1 \vee x_2 \vee \cdots \vee x_n$, where $x_i := p_i$ if $p_i$ is false on that line of the truth table
and $x_i := \neg p_i$ if $p_i$ is true on that line

form the conjunction of all these line terms

---

**Examples:**

**1.**  The proposition $p \vee \neg p$ is a tautology (the *law of the excluded middle*).

**2.**  The proposition $p \vee \neg p$ is a self-contradiction.

**3.**  The proposition $(p \vee \neg q) \wedge (q \vee \neg r) \wedge (r \vee \neg p)$ is satisfiable because it is true when $p$, $q$, and $r$ are all false. Note, however, that $(p \leftrightarrow q) \wedge (\neg p \leftrightarrow q)$ is unsatisfiable, as it is false for each of the four possible assignments of truth values for $p$ and $q$.

**4.**  A proof that $p \leftrightarrow q$ is logically equivalent to $(p \wedge q) \vee (\neg p \wedge \neg q)$ can be carried out using a truth table:

| $p$ | $q$ | $p \leftrightarrow q$ | $\neg p$ | $\neg q$ | $p \wedge q$ | $\neg p \wedge \neg q$ | $(p \wedge q) \vee (\neg p \wedge \neg q)$ |
|-----|-----|-----------------------|----------|----------|--------------|------------------------|---------------------------------------------|
| $T$ | $T$ | $T$ | $F$ | $F$ | $T$ | $F$ | $T$ |
| $T$ | $F$ | $F$ | $F$ | $T$ | $F$ | $F$ | $F$ |
| $F$ | $T$ | $F$ | $T$ | $F$ | $F$ | $F$ | $F$ |
| $F$ | $F$ | $T$ | $T$ | $T$ | $F$ | $T$ | $T$ |

Since the third and eighth columns of the truth table are identical, the two statements are equivalent.

**5.**  A proof that $p \leftrightarrow q$ is logically equivalent to $(p \wedge q) \vee (\neg p \wedge \neg q)$ can be given by a series of logical equivalences. Reasons are given at the right.

$$
\begin{aligned}
p \leftrightarrow q \quad &\Leftrightarrow (p \rightarrow q) \wedge (q \rightarrow p) &&\text{biconditional as implication} \\
&\Leftrightarrow (\neg p \vee q) \wedge (\neg q \vee p) &&\text{conditional as disjunction} \\
&\Leftrightarrow [(\neg p \vee q) \wedge \neg q] \vee [(\neg p \vee q) \wedge p] &&\text{distributive law} \\
&\Leftrightarrow [(\neg p \wedge \neg q) \vee (q \wedge \neg q)] \vee [(\neg p \wedge p) \vee (q \wedge p)] &&\text{distributive law} \\
&\Leftrightarrow [(\neg p \wedge \neg q) \vee \mathbf{F}] \vee [\mathbf{F} \vee (q \wedge p)] &&\text{contradiction} \\
&\Leftrightarrow [(\neg p \wedge \neg q) \vee \mathbf{F}] \vee [(q \wedge p) \vee \mathbf{F}] &&\text{commutative law} \\
&\Leftrightarrow (\neg p \wedge \neg q) \vee (q \wedge p) &&\text{identity law} \\
&\Leftrightarrow (\neg p \wedge \neg q) \vee (p \wedge q) &&\text{commutative law} \\
&\Leftrightarrow (p \wedge q) \vee (\neg p \wedge \neg q) &&\text{commutative law}
\end{aligned}
$$

**6.** The proposition $p \downarrow q$ is logically equivalent to $\neg(p \vee q)$. Its DNF is $\neg p \wedge \neg q$, and its CNF is $(\neg p \vee \neg q) \wedge (\neg p \vee q) \wedge (p \vee \neg q)$.

**7.** The proposition $p|q$ is logically equivalent to $\neg(p \wedge q)$. Its DNF is $(p \wedge \neg q) \vee (\neg p \wedge q) \vee (\neg p \wedge \neg q)$, and its CNF is $\neg p \vee \neg q$.

**8.** The DNF and CNF for Examples 6 and 7 were obtained by using Algorithm 1 and Algorithm 2 to construct the following table of terms:

| $p$ | $q$ | $p \downarrow q$ | *DNF terms* | *CNF terms* |
|-----|-----|-----|-----|-----|
| $T$ | $T$ | $F$ | | $\neg p \vee \neg q$ |
| $T$ | $F$ | $F$ | | $\neg p \vee q$ |
| $F$ | $T$ | $F$ | | $p \vee \neg q$ |
| $F$ | $F$ | $T$ | $\neg p \wedge \neg q$ | |

| $p$ | $q$ | $p \mid q$ | *DNF terms* | *CNF terms* |
|-----|-----|-----|-----|-----|
| $T$ | $T$ | $F$ | | $\neg p \vee \neg q$ |
| $T$ | $F$ | $T$ | $p \wedge \neg q$ | |
| $F$ | $T$ | $T$ | $\neg p \wedge q$ | |
| $F$ | $F$ | $T$ | $\neg p \wedge \neg q$ | |

**9.** The dual of $p \wedge (q \vee \neg r)$ is $p \vee (q \wedge \neg r)$.

**10.** Let $S$ be the proposition in three propositional variables $p$, $q$, and $r$ that is true when precisely two of the variables are true. Then the disjunctive normal form for $S$ is

$$(p \wedge q \wedge \neg r) \vee (p \wedge \neg q \wedge r) \vee (\neg p \wedge q \wedge r)$$

and the conjunctive normal form for $S$ is

$$(\neg p \vee \neg q \vee \neg r) \wedge (\neg p \vee q \vee r) \wedge (p \vee \neg q \vee r) \wedge (p \vee q \vee \neg r) \wedge (p \vee q \vee r).$$

## 1.1.3   PREDICATE LOGIC

**Definitions:**

A **predicate** is a declarative statement with the symbolic form $P(x)$ or $P(x_1, \ldots, x_n)$ about one or more variables $x$ or $x_1, \ldots, x_n$ whose values are unspecified.

**Predicate logic** (or **predicate calculus** or **first-order logic**) is the study of statements whose variables have quantifiers.

The **universe of discourse** (or **universe** or **domain**) of a variable is the set of possible values of the variable in a predicate.

An **instantiation** of the predicate $P(x)$ is the result of substituting a fixed constant value $c$ from the domain of $x$ for each free occurrence of $x$ in $P(x)$. This is denoted by $P(c)$.

The **existential quantification** of a predicate $P(x)$ whose variable ranges over a domain set $D$ is the proposition $(\exists x \in D)P(x)$ or $(\exists x)P(x)$ that is true if there is at least one $c$ in $D$ such that $P(c)$ is true. The *existential quantifier symbol* $\exists$ is read "there exists" or "there is".

The **universal quantification** of a predicate $P(x)$ whose variable ranges over a domain set $D$ is the proposition $(\forall x \in D)P(x)$ or $(\forall x)P(x)$, which is true if $P(c)$ is true for every element $c$ in $D$. The *universal quantifier symbol* $\forall$ is read "for all", "for each", or "for every".

The **unique existential quantification** of a predicate $P(x)$ whose variable ranges over a domain set $D$ is the proposition $(\exists! x)P(x)$ that is true if $P(c)$ is true for exactly one $c$ in $D$. The *unique existential quantifier symbol* $\exists!$ is read "there is exactly one".

The **scope** of a quantifier is the predicate to which it applies.

A variable $x$ in a predicate $P(x)$ is a **bound variable** if it lies inside the scope of an $x$-quantifier. Otherwise it is a **free variable**.

A **well-formed formula** (**wff**) (or **statement**) is either a proposition or a predicate with quantifiers that bind one or more of its variables.

A **sentence** (**closed wff**) is a well-formed formula with no free variables.

A well-formed formula is in **prenex normal form** if all the quantifiers occur at the beginning and the scope is whatever follows the quantifiers.

A well-formed formula is **atomic** if it does not contain any logical connectives; otherwise the well-formed formula is **compound**.

**Higher-order logic** is the study of statements that allow quantifiers to range over relations over a universe (second-order logic), relations over relations over a universe (third-order logic), etc.

**Facts:**
**1.** If a predicate $P(x)$ is atomic, then the scope of $(\forall x)$ in $(\forall x)P(x)$ is implicitly the entire predicate $P(x)$.

**2.** If a predicate is a compound form, such as $P(x) \wedge Q(x)$, then $(\forall x)[P(x) \wedge Q(x)]$ means that the scope is $P(x) \wedge Q(x)$, whereas $(\forall x)P(x) \wedge Q(x)$ means that the scope is only $P(x)$, in which case the free variable $x$ of the predicate $Q(x)$ has no relationship to the variable $x$ of $P(x)$.

**3.** Universal statements in predicate logic are analogues of conjunctions in propositional logic. If variable $x$ has domain $D = \{x_1, \ldots, x_n\}$, then $(\forall x \in D)P(x)$ is true if and only if $P(x_1) \wedge \cdots \wedge P(x_n)$ is true.

**4.** Existential statements in predicate logic are analogues of disjunctions in propositional logic. If variable $x$ has domain $D = \{x_1, \ldots, x_n\}$, then $(\exists x \in D)P(x)$ is true if and only if $P(x_1) \vee \cdots \vee P(x_n)$ is true.

**5.** Adjacent universal quantifiers [existential quantifiers] can be transposed without changing the meaning of a logical statement:

$$(\forall x)(\forall y)P(x,y) \Leftrightarrow (\forall y)(\forall x)P(x,y)$$
$$(\exists x)(\exists y)P(x,y) \Leftrightarrow (\exists y)(\exists x)P(x,y)$$

**6.** Transposing adjacent logical quantifiers of different types can change the meaning of a statement. (See Example 4.)

**7.** *Rules for negations of quantified statements:*

$$\neg(\forall x)P(x) \Leftrightarrow (\exists x)[\neg P(x)]$$
$$\neg(\exists x)P(x) \Leftrightarrow (\forall x)[\neg P(x)]$$
$$\neg(\exists! x)P(x) \Leftrightarrow \neg(\exists x)P(x) \vee (\exists y)(\exists z)[(y \neq z) \wedge P(y) \wedge P(z)].$$

**8.** Every quantified statement is logically equivalent to some statement in prenex normal form.

**9.** Every statement with a unique existential quantifier is equivalent to a statement that uses only existential and universal quantifiers, according to the rule:

$$(\exists! x)P(x) \Leftrightarrow (\exists x)\big[P(x) \wedge (\forall y)[P(y) \rightarrow (x = y)]\big]$$

where $P(y)$ means that $y$ has been substituted for all free occurrences of $x$ in $P(x)$, and where $y$ is a variable that does not occur in $P(x)$.

**10.** If a statement uses only the connectives $\vee$, $\wedge$, and $\neg$, the following equivalences can be used along with Fact 7 to convert the statement into prenex normal form. The letter $A$ represents a wff without the variable $x$.

$$
\begin{aligned}
(\forall x)P(x) \wedge (\forall x)Q(x) &\Leftrightarrow (\forall x)[P(x) \wedge Q(x)] \\
(\forall x)P(x) \vee (\forall x)Q(x) &\Leftrightarrow (\forall x)(\forall y)[P(x) \vee Q(y)] \\
(\exists x)P(x) \wedge (\exists x)Q(x) &\Leftrightarrow (\exists x)(\exists y)[P(x) \wedge Q(y)] \\
(\exists x)P(x) \vee (\exists x)Q(x) &\Leftrightarrow (\exists x)[P(x) \vee Q(x)] \\
(\forall x)P(x) \wedge (\exists x)Q(x) &\Leftrightarrow (\forall x)(\exists y)[P(x) \wedge Q(y)] \\
(\forall x)P(x) \vee (\exists x)Q(x) &\Leftrightarrow (\forall x)(\exists y)[P(x) \vee Q(y)] \\
A \vee (\forall x)P(x) &\Leftrightarrow (\forall x)[A \vee P(x)] \\
A \vee (\exists x)P(x) &\Leftrightarrow (\exists x)[A \vee P(x)] \\
A \wedge (\forall x)P(x) &\Leftrightarrow (\forall x)[A \wedge P(x)] \\
A \wedge (\exists x)P(x) &\Leftrightarrow (\exists x)[A \wedge P(x)].
\end{aligned}
$$

**Examples:**

**1.** The statement $(\forall x \in \mathcal{R})(\forall y \in \mathcal{R})\,[x + y = y + x]$ is syntactically a predicate preceded by two universal quantifiers. It asserts the commutative law for the addition of real numbers.

**2.** The statement $(\forall x)(\exists y)\,[xy = 1]$ expresses the existence of multiplicative inverses for all numbers in whatever domain is under discussion. Thus, it is true for the positive real numbers, but it is false when the domain is the entire set of reals, since zero has no multiplicative inverse.

**3.** The statement $(\forall x \neq 0)(\exists y)\,[xy = 1]$ asserts the existence of multiplicative inverses for nonzero numbers.

**4.** $(\forall x)(\exists y)\,[x + y = 0]$ expresses the true proposition that every real number has an additive inverse, but $(\exists y)(\forall x)\,[x + y = 0]$ is the false proposition that there is a "universal additive inverse" that when added to any number always yields the sum 0.

**5.** In the statement $(\forall x \in \mathcal{R})\,[x + y = y + x]$, the variable $x$ is bound and the variable $y$ is free.

**6.** "Not all men are mortal" is equivalent to "there exists at least one man who is not mortal". Also, "there does not exist a cow that is blue" is equivalent to the statement "every cow is a color other than blue".

**7.** The statement $(\forall x)\,P(x) \rightarrow (\forall x)\,Q(x)$ is not in prenex form. An equivalent prenex form is $(\forall x)(\exists y)\,[P(y) \rightarrow Q(x)]$.

**8.** The following table illustrates the differences in meaning among the four different ways to quantify a predicate with two variables:

| statement | meaning |
|---|---|
| $(\exists x)(\exists y)\,[x+y=0]$ | There is a pair of numbers whose sum is zero. |
| $(\forall x)(\exists y)\,[x+y=0]$ | Every number has an additive inverse. |
| $(\exists x)(\forall y)\,[x+y=0]$ | There is a universal additive inverse $x$. |
| $(\forall x)(\forall y)\,[x+y=0]$ | The sum of every pair of numbers is zero. |

**9.** The statement $(\forall x)(\exists! y)\,[x+y=0]$ asserts the existence of *unique* additive inverses.

## 1.2  SET THEORY

Sets are used to group objects and to serve as the basic elements for building more complicated objects and structures. Counting elements in sets is an important part of discrete mathematics. General reference books that cover the material of this section are [Cu16], [FlPa95], [Ha11], and [Ka10].

### 1.2.1  SETS

**Definitions:**

A **set** is any well-defined collection of objects, each of which is called a **member** or an **element** of the set. The notation $x \in A$ means that the object $x$ is a member of the set $A$. The notation $x \notin A$ means that $x$ is not a member of $A$.

A **roster** for a finite set specifies the membership of a set $S$ as a list of its elements within braces, i.e., in the form $S = \{a_1, \ldots, a_n\}$. Order of the list is irrelevant, as is the number of occurrences of an object in the list.

A **defining predicate** specifies a set in the form $S = \{x \mid P(x)\}$, where $P(x)$ is a predicate containing the free variable $x$. This means that $S$ is the set of all objects $x$ (in whatever domain is under discussion) such that $P(x)$ is true.

A **recursive description** of a set $S$ gives a roster $B$ of *basic objects* of $S$ and a set of operations for constructing additional objects of $S$ from objects already known to be in $S$. That is, any object that can be constructed by a finite sequence of applications of the given operations to objects in $B$ is also a member of $S$. There may also be a list of axioms that specify when two sequences of operations yield the same result.

The set with no elements is called the **null set** or the **empty set**, denoted $\emptyset$ or $\{\ \}$.

A **singleton** is a set with one element.

The set $\mathcal{N}$ of **natural numbers** is the set $\{0, 1, 2, \ldots\}$. (Sometimes 0 is excluded from the set of natural numbers; when the set of natural numbers is encountered, check to see how it is being defined.)

The set $\mathcal{Z}$ of **integers** is the set $\{\ldots, -2, -1, 0, 1, 2, \ldots\}$.

The set $\mathcal{Q}$ of **rational numbers** is the set of all fractions $\frac{a}{b}$, where $a$ is any integer and $b$ is any nonzero integer.

The set $\mathcal{R}$ of **real numbers** is the set of all numbers that can be written as terminating or nonterminating decimals.

The set $\mathcal{C}$ of **complex numbers** is the set of all numbers of the form $a + bi$, where $a, b \in \mathcal{R}$ and $i = \sqrt{-1}$ $(i^2 = -1)$.

Sets $A$ and $B$ are **equal**, written $A = B$, if they have exactly the same elements:

$$A = B \;\Leftrightarrow\; (\forall x)\left[(x \in A) \leftrightarrow (x \in B)\right].$$

Set $B$ is a **subset** of set $A$, written $B \subseteq A$ or $A \supseteq B$, if each element of $B$ is an element of $A$:

$$B \subseteq A \;\Leftrightarrow\; (\forall x)\left[(x \in B) \rightarrow (x \in A)\right].$$

Set $B$ is a **proper subset** of $A$ if $B$ is a subset of $A$ and $A$ contains at least one element not in $B$. (The notation $B \subset A$ is often used to indicate that $B$ is a proper subset of $A$, but sometimes it is used to mean an arbitrary subset. Sometimes the proper subset relationship is written $B \subsetneq A$, to avoid all possible notational ambiguity.)

A set is **finite** if it is either empty or else can be put in a one-to-one correspondence with the set $\{1, 2, 3, \ldots, n\}$ for some positive integer $n$.

A set is **infinite** if it is not finite.

The **cardinality** $|S|$ of a finite set $S$ is the number of elements in $S$.

A **multiset** is an unordered collection in which elements can occur arbitrarily often, not just once. The number of occurrences of an element is called its **multiplicity**.

An **axiom** (**postulate**) is a statement that is assumed to be true.

A set of axioms is **consistent** if no contradiction can be deduced from the axioms.

A set of axioms is **complete** if it is possible to prove all true statements.

A set of axioms is **independent** if none of the axioms can be deduced from the other axioms.

A **set paradox** is a question in the language of set theory that seems to have no unambiguous answer.

**Naive set theory** is set theory where any collection of objects can be considered to be a valid set, with paradoxes ignored.

**Facts:**
1. The theory of sets was first developed by Georg Cantor (1845–1918).

2. $A = B$ if and only if $A \subseteq B$ and $B \subseteq A$.

3. $\mathcal{N} \subset \mathcal{Z} \subset \mathcal{Q} \subset \mathcal{R} \subset \mathcal{C}$.

4. Every rational number can be written as a decimal that is either terminating or else repeating (i.e., the same block repeats end-to-end forever).

5. Real numbers can be represented as the points on the number line, and include all rational numbers and all irrational numbers (such as $\sqrt{2}$, $\pi$, $e$, etc.).

6. There is no set of axioms for set theory that is both complete and consistent.

7. Naive set theory ignores paradoxes. To avoid such paradoxes, more axioms are needed.

**Examples:**

**1.** The set $\{x \in \mathcal{N} \mid 3 \leq x < 10\}$, described by the defining predicate $3 \leq x < 10$, is equal to the set $\{3, 4, 5, 6, 7, 8, 9\}$, which is described by a roster.

**2.** If $A$ is the set with two objects, one being the number 5 and the other being the set whose elements are the letters $x$, $y$, and $z$, then $A = \{5, \{x, y, z\}\}$. In this example, $5 \in A$, but $x \notin A$, since $x$ is not either member of $A$.

**3.** The set $E$ of even natural numbers can be described recursively as follows:

> Basic objects: $0 \in E$,
> Recursion rule: if $n \in E$, then $n + 2 \in E$.

**4.** *The liar's paradox:* A person says "I am lying". Is the person lying or is the person telling the truth? If the person is lying, then "I am lying" is false, and hence the person is telling the truth. If the person is telling the truth, then "I am lying" is true, and the person is lying. This is also called the *paradox of Epimenides*. This paradox also results from considering the statement "This statement is false".

**5.** *The barber's paradox:* In a small village populated only by men there is exactly one barber. The villagers follow the following rule: the barber shaves a man if and only if the man does not shave himself. Question: does the barber shave himself? If "yes" (i.e., the barber shaves himself), then according to the rule he does not shave himself. If "no" (i.e., the barber does not shave himself), then according to the rule he does shave himself. This paradox illustrates a danger in describing sets by defining predicates.

**6.** *Russell's paradox:* This paradox, named for the British logician Bertrand Russell (1872–1970), shows that the "set of all sets" is an ill-defined concept. If it really were a set, then it would be an example of a set that is a member of itself. Thus, some "sets" would contain themselves as elements and others would not. Let $S$ be the "set" of "sets that are not elements of themselves"; i.e., $S = \{A \mid A \notin A\}$. Question: is $S$ a member of itself? If "yes", then $S$ is not a member of itself, because of the defining membership criterion. If "no", then $S$ is a member of itself, due to the defining membership criterion. One resolution is that the collection of all sets is not a set. (See Chapter 4 of [MiRo91].)

**7.** Paradoxes such as those in Example 6 led Alfred North Whitehead (1861–1947) and Bertrand Russell to develop a version of set theory by categorizing sets based on *set types*: $T_0, T_1, \ldots$. The lowest type $T_0$ consists only of individual elements. For $i > 0$, type $T_i$ consists of sets whose elements come from type $T_{i-1}$. This forces sets to belong to exactly one type. The expression $A \in A$ is always false. In this situation Russell's paradox cannot happen.

---

## 1.2.2  SET OPERATIONS

**Definitions:**

The ***intersection*** of sets $A$ and $B$ is the set $A \cap B = \{x \mid (x \in A) \wedge (x \in B)\}$. More generally, the intersection of any family of sets is the set of objects that are members of every set in the family. The notation

$$\bigcap_{i \in I} A_i = \{x \mid x \in A_i \text{ for all } i \in I\}$$

is used for the intersection of the family of sets $A_i$ indexed by the set $I$.

Two sets $A$ and $B$ are ***disjoint*** if $A \cap B = \emptyset$.

A collection of sets $\{a_i \mid i \in I\}$ is **disjoint** if $\bigcap_{i \in I} A_i = \emptyset$.

A collection of sets is **pairwise disjoint** (or **mutually disjoint**) if every pair of sets in the collection are disjoint.

The **union** of sets $A$ and $B$ is the set $A \cup B = \{x \mid (x \in A) \vee (x \in B)\}$. More generally, the union of a family of sets is the set of objects that are members of at least one set in the family. The notation

$$\bigcup_{i \in I} A_i = \{x \mid x \in A_i \text{ for some } i \in I\}$$

is used for the union of the family of sets $A_i$ indexed by the set $I$.

A **partition** of a set $S$ is a pairwise disjoint family $\mathcal{P} = \{A_i\}$ of nonempty subsets whose union is $S$.

The partition $\mathcal{P}_2 = \{B_i\}$ of a set $S$ is a **refinement** of the partition $\mathcal{P}_1 = \{A_j\}$ of the same set if for every subset $B_i \in \mathcal{P}_2$ there is a subset $A_j \in \mathcal{P}_1$ such that $B_i \subseteq A_j$.

The **complement** of the set $A$ is the set $\overline{A} = U - A = \{x \mid x \notin A\}$ containing every object not in $A$, where the context provides that the objects range over some specific universal domain $U$. (The notation $A'$ or $A^c$ is sometimes used instead of $\overline{A}$.)

The **set difference** is the set $A - B = A \cap \overline{B} = \{x \mid (x \in A) \wedge (x \notin B)\}$. The set difference is sometimes written $A \setminus B$.

The **symmetric difference** of $A$ and $B$ is the set $A \oplus B = \{x \mid (x \in A - B) \vee (x \in B - A)\}$. This is sometimes written $A \triangle B$.

The **Cartesian product** $A \times B$ of two sets $A$ and $B$ is the set $\{(a, b) \mid (a \in A) \wedge (b \in B)\}$, which contains all ordered pairs whose first coordinate is from $A$ and whose second coordinate is from $B$. The Cartesian product of $A_1, \ldots, A_n$ is the set $A_1 \times A_2 \times \cdots \times A_n = \prod_{i=1}^{n} A_i = \{(a_1, a_2, \ldots, a_n) \mid (\forall i)(a_i \in A_i)\}$, which contains all ordered $n$-tuples whose $i$th coordinate is from $A_i$. The Cartesian product $A \times A \times \cdots \times A$ is also written $A^n$. If $S$ is any set, the Cartesian product of the collection of sets $A_s$, where $s \in S$, is the set $\prod_{s \in S} A_s$ of all functions $f \colon S \to \bigcup_{s \in S} A_s$ such that $f(s) \in A_s$ for all $s \in S$.

The **power set** of $A$ is the set $\mathcal{P}(A)$ of all subsets of $A$. The alternative notation $2^A$ for $\mathcal{P}(A)$ emphasizes the fact that the power set has $2^n$ elements if $A$ has $n$ elements.

A **set expression** is any expression built up from sets and set operations.

A **set equation** (or **set identity**) is an equation whose left side and right side are both set expressions.

A **system of distinct representatives** (**SDR**) for a collection of sets $A_1, A_2, \ldots, A_n$ (some of which may be equal) is a set $\{a_1, a_2, \ldots, a_n\}$ of $n$ distinct elements such that $a_i \in A_i$ for $i = 1, 2, \ldots, n$.

A **Venn diagram** is a family of $n$ simple closed curves (typically circles or ellipses) arranged in the plane so that all possible intersections of the interiors are nonempty and connected. (John Venn, 1834–1923.)

A Venn diagram is **simple** if at most two curves intersect at any point of the plane.

A Venn diagram is **reducible** if there is a sequence of curves whose iterative removal leaves a Venn diagram at each step.

A **membership table** is a table used to calculate whether an object lies in the set described by a set expression, based on its membership in the sets mentioned by the expression.

**Facts:**

**1.** If a collection of sets is pairwise disjoint, then the collection is disjoint. The converse is false.

**2.** The following figure illustrates Venn diagrams for two and three sets.

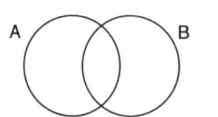 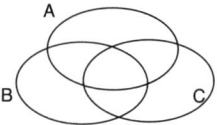

**3.** The following figure gives the Venn diagrams for sets constructed using various set operations.

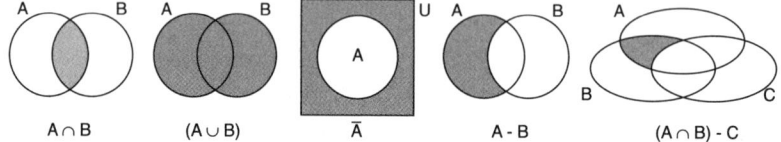

$A \cap B$    $(A \cup B)$    $\overline{A}$    $A - B$    $(A \cap B) - C$

**4.** Intuition regarding set identities can be gleaned from Venn diagrams, but it can be misleading to use Venn diagrams when proving theorems unless great care is taken to make sure that the diagrams are sufficiently general to illustrate all possible cases.

**5.** Venn diagrams are often used as an aid to inclusion/exclusion counting. (See §2.4.)

**6.** Venn gave examples of Venn diagrams with four ellipses and asserted that no Venn diagram could be constructed with five ellipses.

**7.** Peter Hamburger and Raymond Pippert (1996) constructed a simple, reducible Venn diagram with five congruent ellipses. (Two ellipses are *congruent* if they are the exact same size and shape, and differ only by their placement in the plane.)

**8.** Many of the logical identities given in §1.1.2 correspond to set identities, given in the following table.

| name | rule |
|---|---|
| *Commutative laws* | $A \cap B = B \cap A, \quad A \cup B = B \cup A$ |
| *Associative laws* | $A \cap (B \cap C) = (A \cap B) \cap C$ |
|  | $A \cup (B \cup C) = (A \cup B) \cup C$ |
| *Distributive laws* | $A \cap (B \cup C) = (A \cap B) \cup (A \cap C)$ |
|  | $A \cup (B \cap C) = (A \cup B) \cap (A \cup C)$ |
| *DeMorgan's laws* | $\overline{A \cap B} = \overline{A} \cup \overline{B}, \quad \overline{A \cup B} = \overline{A} \cap \overline{B}$ |
| *Complement laws* | $A \cap \overline{A} = \emptyset, \quad A \cup \overline{A} = U$ |
| *Double complement law* | $\overline{\overline{A}} = A$ |
| *Idempotent laws* | $A \cap A = A, \quad A \cup A = A$ |
| *Absorption laws* | $A \cap (A \cup B) = A, \quad A \cup (A \cap B) = A$ |
| *Dominance laws* | $A \cap \emptyset = \emptyset, \quad A \cup U = U$ |
| *Identity laws* | $A \cup \emptyset = A, \quad A \cap U = A$ |

**9.** In a computer, a subset of a relatively small universal domain can be represented by a bit string. Each bit location corresponds to a specific object of the universal domain, and the bit value indicates the presence (1) or absence (0) of that object in the subset.

**10.** In a computer, a subset of a relatively large ordered datatype or universal domain can be represented by a *binary search tree*. (See §18.2.3.)

**11.** For any two finite sets $A$ and $B$, $|A \cup B| = |A| + |B| - |A \cap B|$ (inclusion/exclusion principle). (See §2.4.)

**12.** Set identities can be proved by any of the following approaches:

- a containment proof: show the left side is a subset of the right side, and the right side is a subset of the left side;

- a membership table: construct the analogue of the truth table for each side of the equation;

- using other set identities.

**13.** For all sets $A$, $|A| < |\mathcal{P}(A)|$.

**14.** *Hall's theorem:* A collection of sets $A_1, A_2, \ldots, A_n$ has a system of distinct representatives if and only if for all $k = 1, \ldots, n$ every collection of $k$ subsets $A_{i_1}, A_{i_2}, \ldots, A_{i_k}$ satisfies $|A_{i_1} \cup A_{i_2} \cup \cdots \cup A_{i_k}| \geq k$.

**15.** If a collection of sets $A_1, A_2, \ldots, A_n$ has a system of distinct representatives and if an integer $m$ has the property that $|A_i| \geq m$ for each $i$, then

- if $m \geq n$ there are at least $\frac{m!}{(m-n)!}$ systems of distinct representatives;

- if $m < n$ there are at least $m!$ systems of distinct representatives.

**16.** Systems of distinct representatives can be phrased in terms of 0-1 matrices and graphs. See §6.6, §8.12.2, and §10.4.3.

**Examples:**

**1.** $\{1,2\} \cap \{2,3\} = \{2\}$.

**2.** The collection of sets $\{1,2\}$, $\{4,5\}$, $\{6,7,8\}$ is pairwise disjoint, and hence disjoint.

**3.** The collection of sets $\{1,2\}$, $\{2,3\}$, $\{1,3\}$ is disjoint, but not pairwise disjoint.

**4.** $\{1,2\} \cup \{2,3\} = \{1,2,3\}$.

**5.** Suppose that for every positive integer $n$, $[j \bmod n] = \{k \in \mathcal{Z} \mid k \bmod n = j\}$, for $j = 0, 1, \ldots, n-1$. (See §1.3.1.) Then $\{[0 \bmod 3], [1 \bmod 3], [2 \bmod 3]\}$ is a partition of the integers. Moreover, $\{[0 \bmod 6], [1 \bmod 6], \ldots, [5 \bmod 6]\}$ is a refinement of this partition.

**6.** Within the context of $\mathcal{Z}$ as universal domain, the complement of the set of positive integers is the set consisting of the negative integers and 0.

**7.** $\{1,2\} - \{2,3\} = \{1\}$.

**8.** $\{1,2\} \times \{2,3\} = \{(1,2), (1,3), (2,2), (2,3)\}$.

**9.** $\mathcal{P}(\{1,2\}) = \{\emptyset, \{1\}, \{2\}, \{1,2\}\}$.

**10.** If $L$ is a line in the plane, and if for each $x \in L$, $C_x$ is the circle of radius 1 centered at point $x$, then $\bigcup_{x \in L} C_x$ is an infinite strip of width 2, and $\bigcap_{x \in L} C_x = \emptyset$.

**11.** The five-fold Cartesian product $\{0,1\}^5$ contains 32 different 5-tuples, including, for instance, $(0,0,1,0,1)$.

**12.** The set identity $\overline{A \cap B} = \overline{A} \cup \overline{B}$ is verified by the following membership table. Begin by listing the possibilities for elements being in or not being in the sets $A$ and $B$, using 1 to mean "is an element of" and 0 to mean "is not an element of". Proceed to find the element values for each combination of sets. The two sides of the equation are the same since the columns for $\overline{A \cap B}$ and $\overline{A} \cup \overline{B}$ are identical:

| $A$ | $B$ | $A \cap B$ | $\overline{A \cap B}$ | $\overline{A}$ | $\overline{B}$ | $\overline{A} \cup \overline{B}$ |
|---|---|---|---|---|---|---|
| 1 | 1 | 1 | 0 | 0 | 0 | 0 |
| 1 | 0 | 0 | 1 | 0 | 1 | 1 |
| 0 | 1 | 0 | 1 | 1 | 0 | 1 |
| 0 | 0 | 0 | 1 | 1 | 1 | 1 |

**13.** The collection of sets $A_1 = \{1, 2\}$, $A_2 = \{2, 3\}$, $A_3 = \{1, 3, 4\}$ has systems of distinct representatives, for example $\{1, 2, 3\}$ and $\{2, 3, 4\}$.

**14.** The collection of sets $A_1 = \{1, 2\}$, $A_2 = \{1, 3\}$, $A_3 = \{2, 3\}$, $A_4 = \{1, 2, 3\}$, $A_5 = \{2, 3, 4\}$ does not have a system of distinct representatives since $|A_1 \cup A_2 \cup A_3 \cup A_4| < 4$.

## 1.2.3  INFINITE SETS

### Definitions:
The **Peano definition** for the natural numbers $\mathcal{N}$:

- 0 is a natural number;
- every natural number $n$ has a successor $s(n)$;
- axioms:
  - $\diamond$ 0 is not the successor of any natural number;
  - $\diamond$ two different natural numbers cannot have the same successor;
  - $\diamond$ if $0 \in T$ and if $(\forall n \in \mathcal{N}) \left[ (n \in T) \to (s(n) \in T) \right]$, then $T = \mathcal{N}$.

(This axiomatization is named for Giuseppe Peano, 1858–1932.)

A set is **denumerable** (or **countably infinite**) if it can be put in a one-to-one correspondence with the set of natural numbers $\{0, 1, 2, 3, \ldots\}$. (See §1.3.1.)

A **countable** set is a set that is either finite or denumerable. All other sets are **uncountable**.

The **ordinal numbers** (or **ordinals**) are defined recursively as follows:

- the empty set is the ordinal number 0;
- if $\alpha$ is an ordinal number, then so is the *successor* of $\alpha$, written $\alpha^+$ or $\alpha + 1$, which is the set $\alpha \cup \{\alpha\}$;
- if $\beta$ is any set of ordinals closed under the successor operation, then $\beta$ is an ordinal, called a *limit ordinal*.

The ordinal $\alpha$ is said to be **less than** the ordinal $\beta$, written $\alpha < \beta$, if $\alpha \subseteq \beta$ (which is equivalent to $\alpha \in \beta$).

The **sum** of ordinals $\alpha$ and $\beta$, written $\alpha + \beta$, is the ordinal corresponding to the well-ordered set given by all the elements of $\alpha$ in order, followed by all the elements of $\beta$ (viewed as being disjoint from $\alpha$) in order. (See Fact 26 and §1.4.3.)

The **product** of ordinals $\alpha$ and $\beta$, written $\alpha \cdot \beta$, is the ordinal equal to the Cartesian product $\alpha \times \beta$ with ordering $(a_1, b_1) < (a_2, b_2)$ whenever $b_1 < b_2$, or $b_1 = b_2$ and $a_1 < a_2$ (this is reverse lexicographic order).

Two sets have the **same cardinality** (or are **equinumerous**) if they can be put into one-to-one correspondence (§1.3.1). When the equivalence relation "equinumerous" is used on all sets (§1.4.2), the sets in each equivalence class have the same **cardinal number**. The cardinal number of a set $A$ is written $|A|$. It can also be regarded as the smallest ordinal number among all those ordinal numbers with the same cardinality.

An **order relation** can be defined on cardinal numbers of sets by the rule $|A| \leq B$ if there is a one-to-one function $f \colon A \to B$. If $|A| \leq |B|$ and $|A| \neq |B|$, write $|A| < |B|$.

The **sum** of cardinal numbers **a** and **b**, written $\mathbf{a} + \mathbf{b}$, is the cardinal number of the union of two disjoint sets $A$ and $B$ such that $|A| = \mathbf{a}$ and $|B| = \mathbf{b}$.

The **product** of cardinal numbers **a** and **b**, written **ab**, is the cardinal number of the Cartesian product of two sets $A$ and $B$ such that $|A| = \mathbf{a}$ and $|B| = \mathbf{b}$.

**Exponentiation** of cardinal numbers, written $\mathbf{a}^{\mathbf{b}}$, is the cardinality of the set $A^B$ of all functions from $B$ to $A$, where $|A| = \mathbf{a}$ and $|B| = \mathbf{b}$.

**Facts:**

**1.** Axiom 3 in the Peano definition of the natural numbers is the principle of mathematical induction. (See §1.5.4.)

**2.** The finite cardinal numbers are written $0, 1, 2, 3, \ldots$.

**3.** The cardinal number of any finite set with $n$ elements is $n$.

**4.** The first infinite cardinal numbers are written $\aleph_0, \aleph_1, \aleph_2, \ldots, \aleph_\omega, \ldots$.

**5.** For each ordinal $\alpha$, there is a cardinal number $\aleph_\alpha$.

**6.** The cardinal number of any denumerable set, such as $\mathcal{N}$, $\mathcal{Z}$, and $\mathcal{Q}$, is $\aleph_0$.

**7.** The cardinal number of $\mathcal{P}(\mathcal{N})$, $\mathcal{R}$, and $\mathcal{C}$ is denoted **c** (standing for the *continuum*).

**8.** The set of algebraic numbers (all solutions of polynomials with integer coefficients) is denumerable.

**9.** The set $\mathcal{R}$ is uncountable (proved by Georg Cantor in late 19th century, using a diagonal argument). (See §1.5.5.)

**10.** Every subset of a countable set is countable.

**11.** The countable union of countable sets is countable.

**12.** Every set containing an uncountable subset is uncountable.

**13.** The *continuum problem*, posed by Georg Cantor (1845–1918) and restated by David Hilbert (1862–1943) in 1900, is the problem of determining the cardinality $|\mathcal{R}|$ of the real numbers.

**14.** The *continuum hypothesis* is the assertion that $|\mathcal{R}| = \aleph_1$, the first cardinal number larger than $\aleph_0$. Equivalently, $2^{\aleph_0} = \aleph_1$. (See Fact 35.) Kurt Gödel (1906–1978) proved in 1938 that the continuum hypothesis is consistent with various other axioms of set theory. Paul Cohen (1934–2007) demonstrated in 1963 that the continuum hypothesis cannot be proved from those other axioms; i.e., it is independent of the other axioms of set theory.

**15.** The *generalized continuum hypothesis* is the assertion that $2^{\aleph_\alpha} = \aleph_{\alpha+1}$ for all ordinals $\alpha$. That is, for infinite sets there is no cardinal number strictly between $|S|$ and $|\mathcal{P}(S)|$.

**16.** The generalized continuum hypothesis is consistent with and independent of the usual axioms of set theory.

**17.** There is no largest cardinal number.

**18.** $|A| < |\mathcal{P}(A)|$ for all sets $A$.

**19.** *Schröder-Bernstein theorem*: If $|A| \leq |B|$ and $|B| \leq |A|$, then $|A| = |B|$. (This is also called the Cantor-Schröder-Bernstein theorem.)

**20.** The ordinal number $1 = 0^+ = \{\emptyset\} = \{0\}$, the ordinal number $2 = 1^+ = \{0, 1\}$, etc. In general, for finite ordinals, $n + 1 = n^+ = \{0, 1, 2, \ldots, n\}$.

**21.** The first limit ordinal is $\omega = \{0, 1, 2, \ldots\}$. Then $\omega + 1 = \omega^+ = \omega \cup \{\omega\} = \{0, 1, 2, \ldots, \omega\}$, and so on. The next limit ordinal is $\omega + \omega = \{0, 1, 2, \ldots, \omega, \omega+1, \omega+2, \ldots\}$, also denoted $\omega \cdot 2$. The process never stops, because the next limit ordinal can always be formed as the union of the infinite process that has gone before.

**22.** Limit ordinals have no immediate predecessors.

**23.** The first ordinal that, viewed as a set, is not countable, is denoted $\omega_1$.

**24.** For ordinals the following are equivalent: $\alpha < \beta$, $\alpha \in \beta$, $\alpha \subset \beta$.

**25.** Every set of ordinal numbers has a smallest element; i.e., the ordinals are well-ordered. (See §1.4.3.)

**26.** Ordinal numbers correspond to well-ordered sets (§1.4.3). Two well-ordered sets represent the same ordinal if they can be put into an order-preserving one-to-one correspondence.

**27.** Addition and multiplication of ordinals are associative operations.

**28.** Ordinal addition and multiplication for finite ordinals (those less than $\omega$) are the same as ordinary addition and multiplication on the natural numbers.

**29.** Addition of infinite ordinals is not commutative. (See Example 2.)

**30.** Multiplication of infinite ordinals is not commutative. (See Example 3.)

**31.** The ordinals 0 and 1 are identities for addition and multiplication, respectively.

**32.** Multiplication of ordinals is distributive over addition on the left: $\alpha(\beta + \gamma) = \alpha\beta + \alpha\gamma$. It is not distributive on the right.

**33.** In the definition of the cardinal number $\mathbf{a^b}$, when $\mathbf{a} = 2$, the set $A$ can be taken to be $A = \{0, 1\}$ and an element of $A^B$ can be identified with a subset of $B$ (namely, those elements of $B$ sent to 1 by the function). Thus $2^{|B|} = |\mathcal{P}(B)|$, the cardinality of the power set of $B$.

**34.** If $\mathbf{a}$ and $\mathbf{b}$ are cardinals, at least one of which is infinite, then $\mathbf{a} + \mathbf{b} = \mathbf{a} \cdot \mathbf{b} =$ the larger of $\mathbf{a}$ and $\mathbf{b}$.

**35.** $\mathbf{c}^{\aleph_0} = \aleph_0^{\aleph_0} = 2^{\aleph_0}$.

**36.** The usual rules for finite arithmetic continue to hold for infinite cardinal arithmetic (commutativity, associativity, distributivity, and rules for exponents).

**Examples:**

**1.** $\omega_1 > \omega \cdot 2$, $\omega_1 > \omega^2$, $\omega_1 > \omega^\omega$.

**2.** $1 + \omega = \omega$, but $\omega + 1 > \omega$.

**3.** $2 \cdot \omega = \omega$, but $\omega \cdot 2 > \omega$.

**4.** $\aleph_0 \cdot \aleph_0 = \aleph_0 + \aleph_0 = \aleph_0$.

## 1.2.4   AXIOMS FOR SET THEORY

Set theory can be viewed as an axiomatic system, with undefined terms "set" (the universe of discourse) and "is an element of" (a binary relation denoted $\in$).

**Definitions:**
The **Axiom of choice** (**AC**) states: If $\mathcal{A}$ is any set whose elements are pairwise disjoint nonempty sets, then there exists a set $X$ that has as its elements exactly one element from each set in $\mathcal{A}$.

The **Zermelo-Fraenkel** (**ZF**) **axioms** for set theory: (The axioms are stated informally.)

- *Extensionality* (*equality*):  Two sets with the same elements are equal.
- *Pairing*:  For every $a$ and $b$, the set $\{a, b\}$ exists.
- *Specification* (*subset*):  If $A$ is a set and $P(x)$ is a predicate with free variable $x$, then the subset of $A$ exists that consists of those elements $c \in A$ such that $P(c)$ is true. (The specification axiom guarantees that the intersection of two sets exists.)
- *Union*:  The union of a set (i.e., the set of all the elements of its elements) exists. (The union axiom together with the pairing axiom implies the existence of the union of two sets.)
- *Power set*:  The power set (set of all subsets) of a set exists.
- *Empty set*:  The empty set exists.
- *Regularity* (*foundation*):  Every nonempty set contains a "foundational" element; that is, every nonempty set contains an element that is not an element of any other element in the set. (The regularity axiom prevents anomalies such as a set being an element of itself.)
- *Replacement*:  If $f$ is a function defined on a set $A$, then the collection of images $\{f(a) \mid a \in A\}$ is a set. The replacement axiom (together with the union axiom) allows the formation of large sets by expanding each element of a set into a set.
- *Infinity*:  An infinite set, such as $\omega$ (§1.2.3), exists.

**Facts:**
**1.** The axiom of choice is consistent with and independent of the other axioms of set theory; it can be neither proved nor disproved from the other axioms of set theory.

**2.** The axioms of ZF together with the axiom of choice are denoted ZFC.

**3.** The following propositions are equivalent to the axiom of choice:
- *The well-ordering principle*: Every set can be well-ordered; i.e., for every set $A$ there exists a total ordering on $A$ such that every subset of $A$ contains a smallest element under this ordering.
- *Generalized axiom of choice* (*functional version*): If $\mathcal{A}$ is any collection of nonempty sets, then there is a function $f$ whose domain is $\mathcal{A}$, such that $f(X) \in X$ for all $X \in \mathcal{A}$.
- *Zorn's lemma*: Every nonempty partially ordered set in which every chain (totally ordered subset) contains an upper bound (an element greater than all the other elements in the chain) has a maximal element (an element that is less than no other element). (§1.4.3.)

- *The Hausdorff maximal principle*: Every chain in a partially ordered set is contained in a maximal chain (a chain that is not strictly contained in another chain). (§1.4.3.)
- *Trichotomy*: Given any two sets $A$ and $B$, either there is a one-to-one function from $A$ to $B$, or there is a one-to-one function from $B$ to $A$; i.e., either $|A| \le |B|$ or $|B| \le |A|$.

## 1.3  FUNCTIONS

A function is a rule that associates to each object in one set an object in a second set (these sets are often sets of numbers). For instance, the expected population in future years, based on demographic models, is a function from calendar years to numbers. Encryption is a function from confidential information to apparent nonsense messages, and decryption is a function from apparent nonsense back to confidential information. Computer scientists and mathematicians are often concerned with developing methods to calculate particular functions quickly.

### 1.3.1  BASIC TERMINOLOGY FOR FUNCTIONS

**Definitions:**
A **function** $f$ from a set $A$ to a set $B$, written $f\colon A \to B$, is a rule that assigns to every object $a \in A$ exactly one element $f(a) \in B$. The set $A$ is the **domain** of $f$; the set $B$ is the **codomain** of $f$; the element $f(a)$ is the **image** of $a$ or the **value** of $f$ at $a$. A function $f$ is often identified with its **graph** $\{(a,b) \mid a \in A \text{ and } b = f(a)\} \subseteq A \times B$.

*Note:* The function $f\colon A \to B$ is sometimes represented by the "maps to" notation $x \mapsto f(x)$ or by the variation $x \mapsto expr(x)$, where $expr(x)$ is an expression in $x$. The notation $f(x) = expr(x)$ is a form of the "maps to" notation without the symbol $\mapsto$.

The rule defining a function $f\colon A \to B$ is called **well-defined** since to each $a \in A$ there is associated exactly one element of $B$.

If $f\colon A \to B$ and $S \subseteq A$, the **image** of the subset $S$ under $f$ is the set $f(S) = \{f(x) \mid x \in S\}$.

If $f\colon A \to B$ and $T \subseteq B$, the **pre-image** or **inverse image** of the subset $T$ under $f$ is the set $f^{-1}(T) = \{x \mid f(x) \in T\}$.

The **image** of a function $f\colon A \to B$ is the set $f(A) = \{f(x) \mid x \in A\}$.

The **range** of a function $f\colon A \to B$ is the image set $f(A)$. (Some authors use "range" as a synonym for "codomain".)

A function $f\colon A \to B$ is **one-to-one** (**1–1**, **injective**, or a **monomorphism**) if distinct elements of the domain are mapped to distinct images; i.e., $f(a_1) \neq f(a_2)$ whenever $a_1 \neq a_2$. An **injection** is an injective function.

A function $f\colon A \to B$ is **onto** (**surjective**, or an **epimorphism**) if every element of the codomain $B$ is the image of at least one element of $A$; i.e., if $(\forall b \in B)(\exists a \in A)\,[f(a) = b]$ is true. A **surjection** is a surjective function.

A function $f\colon A \to B$ is **bijective** (or a **one-to-one correspondence**) if it is both injective and surjective; i.e., it is 1–1 and onto. A **bijection** is a bijective function.

If $f\colon A \to B$ and $S \subseteq A$, the **restriction** of $f$ to $S$ is the function $f_S\colon S \to B$ where $f_S(x) = f(x)$ for all $x \in S$. The function $f$ is an **extension** of $f_S$. The restriction of $f$ to $S$ is also written $f|_S$.

A **partial function** on a set $A$ is a rule $f$ that assigns to each element in a subset of $A$ exactly one element of $B$. The subset of $A$ on which $f$ is defined is the **domain of definition** of $f$. In a context that includes partial functions, a rule that applies to all of $A$ is called a **total function**.

Given a 1–1 onto function $f\colon A \to B$, the **inverse function** $f^{-1}\colon B \to A$ has the rule that for each $y \in B$, $f^{-1}(y)$ is the object $x \in A$ such that $f(x) = y$.

If $f\colon A \to B$ and $g\colon B \to C$, then the **composition** is the function $g{\circ}f\colon A \to C$ defined by the rule $(g{\circ}f)(x) = g(f(x))$ for all $x \in A$. The function to the right of the raised circle is applied first.

*Note:* Care must be taken since some sources define the composition $(g{\circ}f)(x) = f(g(x))$ so that the order of application reads left to right.

If $f\colon A \to A$, the **iterated functions** $f^n\colon A \to A$ $(n \geq 2)$ are defined recursively by the rule $f^n(x) = f \circ f^{n-1}(x)$.

A function $f\colon A \to A$ is **idempotent** if $f \circ f = f$.

A function $f\colon A \to A$ is an **involution** if $f \circ f = i_A$. (See Example 1.)

A function whose domain is a Cartesian product $A_1 \times \cdots \times A_n$ is often regarded as a function of $n$ variables (also called a **multivariate** function), and the value of $f$ at $(a_1, \ldots, a_n)$ is usually written $f(a_1, \ldots, a_n)$.

An **(n-ary) operation** on a set $A$ is a function $f\colon A^n \to A$, where $A^n = A \times \cdots \times A$ (with $n$ factors in the product). A 1-ary operation is called **monadic** or **unary**, and a 2-ary operation is called **binary**.

## Facts:

**1.** The graph of a function $f\colon A \to B$ is a binary relation on $A \times B$. (§1.4.1.)

**2.** The graph of a function $f\colon A \to B$ is a subset $S$ of $A \times B$ such that for each $a \in A$ there is exactly one $b \in B$ such that $(a, b) \in S$.

**3.** In general, two or more different objects in the domain of a function might be assigned the same value in the codomain. If this occurs, the function is not 1–1.

**4.** If $f\colon A \to B$ is bijective, then: $f{\circ}f^{-1} = i_B$ (Example 1), $f^{-1}{\circ}f = i_A$, $f^{-1}$ is bijective, and $(f^{-1})^{-1} = f$.

**5.** Function composition is associative: $(f{\circ}g){\circ}h = f{\circ}(g{\circ}h)$, whenever $h\colon A \to B$, $g\colon B \to C$, and $f\colon C \to D$.

**6.** Function composition is not commutative; that is, $f{\circ}g \neq g{\circ}f$ in general. (See Example 12.)

**7.** *Set operations with functions:* If $f\colon A \to B$ with $S_1, S_2 \subseteq A$ and $T_1, T_2 \subseteq B$, then

- $f(S_1 \cup S_2) = f(S_1) \cup f(S_2)$;
- $f(S_1 \cap S_2) \subseteq f(S_1) \cap f(S_2)$, with equality if $f$ is injective;
- $f(\overline{S_1}) \supseteq \overline{f(S_1)}$ (i.e., $f(A - S_1) \supseteq B - f(S_1)$), with equality if $f$ is injective;

- $f^{-1}(T_1 \cup T_2) = f^{-1}(T_1) \cup f^{-1}(T_2)$;
- $f^{-1}(T_1 \cap T_2) = f^{-1}(T_1) \cap f^{-1}(T_2)$;
- $f^{-1}(\overline{T_1}) = \overline{f^{-1}(T_1)}$ (i.e., $f^{-1}(B - T_1) = A - f^{-1}(T_1)$);
- $f^{-1}(f(S_1)) \supseteq S_1$, with equality if $f$ is injective;
- $f(f^{-1}(T_1)) \subseteq T_1$, with equality if $f$ is surjective.

**8.** If $f\colon A \to B$ and $g\colon B \to C$ are both bijective, then $(g \circ f)^{-1} = f^{-1} \circ g^{-1}$.

**9.** If an operation $*$ (such as addition) is defined on a set $B$, then that operation can be extended to the set of all functions from a set $A$ to $B$, by setting $(f * g)(x) = f(x) * g(x)$.

**10.** *Numbers of functions:* If $|A| = m$ and $|B| = n$, the numbers of different types of functions $f\colon A \to B$ are given in the following list:

- all: $n^m$   (§2.2.1)
- one-to-one: $P(n, m) = n(n - 1)(n - 2) \ldots (n - m + 1)$ if $n \ge m$   (§2.2.1)
- onto: $\sum_{j=0}^{n} (-1)^j \binom{n}{j} (n - j)^m$ if $m \ge n$   (§2.4.2)
- partial: $(n + 1)^m$   (§2.3.2)

**Examples:**

**1.** The following are some common functions:

- ***exponential function to base $b$*** (for $b > 0$, $b \ne 1$):   the function $f\colon \mathcal{R} \to \mathcal{R}^+$ where $f(x) = b^x$. (See the following figure.)  ($\mathcal{R}^+$ is the set of positive real numbers.)

- ***logarithm function with base $b$*** (for $b > 0$, $b \ne 1$):   the function $\log_b\colon \mathcal{R}^+ \to \mathcal{R}$ that is the inverse of the exponential function to base $b$; that is,

$$\log_b x = y \text{ if and only if } b^y = x.$$

- ***common logarithm function:***   the function $\log_{10}\colon \mathcal{R}^+ \to \mathcal{R}$ (also written $\log$) that is the inverse of the exponential function to base 10; i.e., $\log_{10} x = y$ when $10^y = x$. (See the following figure.)

- ***binary logarithm function:***   the function $\log_2\colon \mathcal{R}^+ \to \mathcal{R}$ (also denoted $\log$ or $\lg$) that is the inverse of the exponential function to base 2; i.e., $\log_2 x = y$ when $2^y = x$. (See the following figure.)

- ***natural logarithm function:***   the function $\ln\colon \mathcal{R}^+ \to \mathcal{R}$ is the inverse of the exponential function to base $e$; i.e., $\ln(x) = y$ when $e^y = x$, where $e = \lim_{n \to \infty} (1 + \frac{1}{n})^n \approx 2.718281828459$. (See the following figure.)

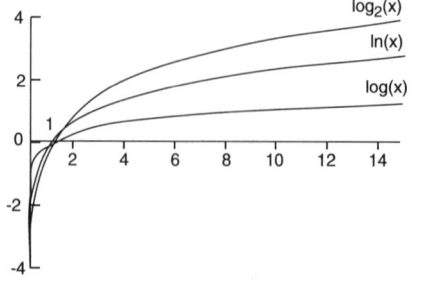

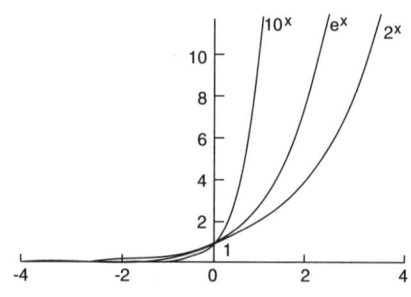

- **iterated logarithm**:   the function $\log^* \colon \mathcal{R}^+ \to \{0, 1, 2, \ldots\}$, where $\log^* x$ is the smallest nonnegative integer $k$ such that $\log^{(k)} x \leq 1$; the function $\log^{(k)}$ is defined recursively by

$$
\log^{(k)} x = \begin{cases}
x & \text{if } k = 0 \\
\log(\log^{(k-1)} x) & \text{if } \log^{(k-1)} x \text{ is defined and positive} \\
\text{undefined} & \text{otherwise.}
\end{cases}
$$

- **mod function**:   for a given positive integer $n$, the function $f \colon \mathcal{Z} \to \mathcal{N}$ defined by the rule $f(k) = k \bmod n$, where $k \bmod n$ is the remainder when the division algorithm is used to divide $k$ by $n$. (See §4.1.2.)
- **identity function** on a set $A$:   the function $i_A \colon A \to A$ such that $i_A(x) = x$ for all $x \in A$.
- **characteristic function** of $S$:   for $S \subseteq A$, the function $\chi_S \colon A \to \{0, 1\}$ given by $\chi_S(x) = 1$ if $x \in S$ and $\chi_S(x) = 0$ if $x \notin S$.
- **projection function**:   the function $\pi_j \colon A_1 \times \cdots \times A_n \to A_j$ $(j = 1, 2, \ldots, n)$ such that $\pi_j(a_1, \ldots, a_n) = a_j$.
- **permutation**:   a function $f \colon A \to A$ that is 1–1 and onto.
- **floor function** (sometimes referred to, especially in number theory, as the **greatest integer function**):   the function $\lfloor\ \rfloor \colon \mathcal{R} \to \mathcal{Z}$ where $\lfloor x \rfloor = $ the greatest integer less than or equal to $x$. The floor of $x$ is also written $[x]$. (See the following figure.) Thus $\lfloor \pi \rfloor = 3$, $\lfloor 6 \rfloor = 6$, and $\lfloor -0.2 \rfloor = -1$.
- **ceiling function**:   the function $\lceil\ \rceil \colon \mathcal{R} \to \mathcal{Z}$ where $\lceil x \rceil = $ the smallest integer greater than or equal to $x$. (See the following figure.) Thus $\lceil \pi \rceil = 4$, $\lceil 6 \rceil = 6$, and $\lceil -0.2 \rceil = 0$.

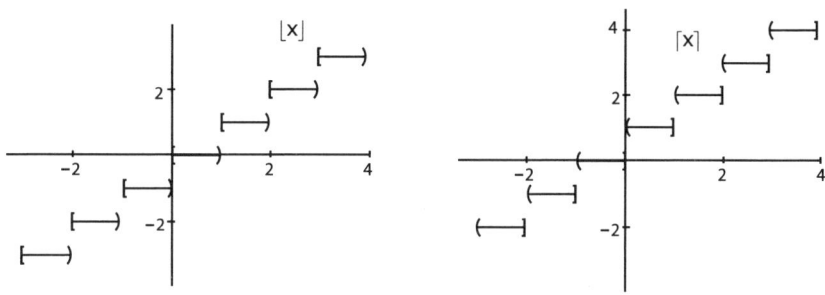

**2.** The floor and ceiling functions are total functions from the reals $\mathcal{R}$ to the integers $\mathcal{Z}$. They are onto, but not one-to-one.

**3.** *Properties of the floor and ceiling functions ($m$ and $n$ represent arbitrary integers):*

- $\lfloor x \rfloor = n$  if and only if  $n \leq x < n + 1$  if and only if  $x - 1 < n \leq x$;
- $\lceil x \rceil = n$  if and only if  $n - 1 < x \leq n$  if and only if  $x \leq n < x + 1$;
- $\lfloor x \rfloor < n$  if and only if  $x < n$;  $\lceil x \rceil \leq n$  if and only if  $x \leq n$;
- $n \leq \lfloor x \rfloor$  if and only if  $n \leq x$;  $n < \lceil x \rceil$  if and only if  $n < x$;
- $x - 1 < \lfloor x \rfloor \leq x \leq \lceil x \rceil < x + 1$;
- $\lfloor x \rfloor = x$  if and only if $x$ is an integer;
- $\lceil x \rceil = x$  if and only if $x$ is an integer;

- $\lfloor -x \rfloor = -\lceil x \rceil$; $\lceil -x \rceil = -\lfloor x \rfloor$;
- $\lfloor x + n \rfloor = \lfloor x \rfloor + n$; $\lceil x + n \rceil = \lceil x \rceil + n$;
- the interval $[x_1, x_2]$ contains $\lfloor x_2 \rfloor - \lceil x_1 \rceil + 1$ integers;
- the interval $[x_1, x_2)$ contains $\lceil x_2 \rceil - \lceil x_1 \rceil$ integers;
- the interval $(x_1, x_2]$ contains $\lfloor x_2 \rfloor - \lfloor x_1 \rfloor$ integers;
- the interval $(x_1, x_2)$ contains $\lceil x_2 \rceil - \lfloor x_1 \rfloor - 1$ integers;
- if $f(x)$ is a continuous, monotonically increasing function, and whenever $f(x)$ is an integer, $x$ is also an integer, then $\lfloor f(x) \rfloor = \lfloor f(\lfloor x \rfloor) \rfloor$ and $\lceil f(x) \rceil = \lceil f(\lceil x \rceil) \rceil$;
- if $n > 0$, then $\lfloor \frac{x+m}{n} \rfloor = \lfloor \frac{\lfloor x \rfloor + m}{n} \rfloor$ and $\lceil \frac{x+m}{n} \rceil = \lceil \frac{\lceil x \rceil + m}{n} \rceil$ (a special case of the preceding fact);
- if $m > 0$, then $\lfloor mx \rfloor = \lfloor x \rfloor + \lfloor x + \frac{1}{m} \rfloor + \cdots + \lfloor x + \frac{m-1}{m} \rfloor$.

**4.** The logarithm function $\log_b x$ is bijective from the positive reals $\mathcal{R}^+$ to the reals $\mathcal{R}$.

**5.** The logarithm function $x \mapsto \log_b x$ is the inverse of the function $x \mapsto b^x$, if the codomain of $x \mapsto b^x$ is the set of positive real numbers. If the domain and codomain are considered to be $\mathcal{R}$, then $x \mapsto \log_b x$ is only a partial function, because the logarithm of a nonpositive number is not defined.

**6.** All logarithm functions are related according to the following change of base formula: $\log_b x = \frac{\log_a x}{\log_a b}$.

**7.** $\log^* 2 = 1$, $\log^* 4 = 2$, $\log^* 16 = 3$, $\log^* 65536 = 4$, $\log^* 2^{65536} = 5$.

**8.** The diagrams in the following figure illustrate a function that is onto but not 1–1 and a function that is 1–1 but not onto.

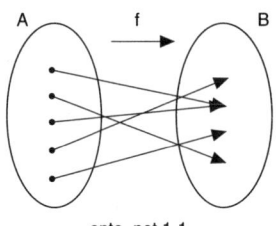
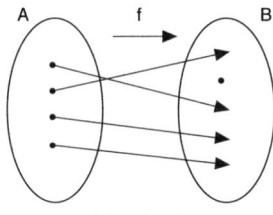

onto, not 1-1                    1-1, not onto

**9.** If the domain and codomain are considered to be the nonnegative reals, then the function $x \mapsto x^2$ is a bijection, and $x \mapsto \sqrt{x}$ is its inverse.

**10.** If the codomain is considered to be the subset of complex numbers with polar coordinate $0 \le \theta < \pi$, then $x \mapsto \sqrt{x}$ can be regarded as a total function.

**11.** Division of real numbers is a multivariate function from $\mathcal{R} \times (\mathcal{R} - \{0\})$ to $\mathcal{R}$, given by the rule $f(x, y) = \frac{x}{y}$. Similarly, addition, subtraction, and multiplication are functions from $\mathcal{R} \times \mathcal{R}$ to $\mathcal{R}$.

**12.** If $f(x) = x^2$ and $g(x) = x + 1$, then $(f \circ g)(x) = (x + 1)^2$ and $(g \circ f)(x) = x^2 + 1$. (Therefore, composition of functions is not commutative.)

**13.** *Collatz conjecture:* If $f : \{1, 2, 3, \ldots\} \to \{1, 2, 3, \ldots\}$ is defined by the rule $f(n) = \frac{n}{2}$ if $n$ is even and $f(n) = 3n + 1$ if $n$ is odd, then for each positive integer $m$ there is a positive integer $k$ such that the iterated function $f^k(m) = 1$. It is not known whether this conjecture is true.

## 1.3.2  COMPUTATIONAL REPRESENTATION

A given function may be described by several different rules. These rules can then be used to evaluate specific values of the function. There is often a large difference in the time required to compute the value of a function using different computational rules. The speed usually depends on the representation of the data as well as on the computational process.

### Definitions:

A (*computational*) *representation* of a function is a way to calculate its values.

A *closed formula* for a function value $f(x)$ is an algebraic expression in the argument $x$.

A *table of values* for a function $f\colon A \to B$ with finite domain $A$ is any explicit representation of the set $\{(a, f(a)) \in A \times B \mid a \in A\}$.

An *infinite sequence* in a set $S$ is a function from the natural numbers $\{0,1,2,\ldots\}$ to the set $S$. It is commonly represented as a list $x_0, x_1, x_2, \ldots$ such that each $x_j \in S$. Sequences are often permitted to start at the index 1 or elsewhere, rather than 0.

A *finite sequence* in a set $S$ is a function from $\{1, 2, \ldots, n\}$ to the set $S$. It is commonly represented as a list $x_1, x_2, \ldots, x_n$ such that each $x_j \in S$. Finite sequences are often permitted to start at the index 0 (or at some other value of the index), rather than at the index 1.

A value of a sequence is also called an *entry*, an *item*, or a *term*.

A *string* is a representation of a sequence as a list in which the successive entries are juxtaposed without intervening punctuation or extra spacing.

A *recursive definition* of a function $f$ with domain $S$ has two parts: there is a set of *base values* (or *initial values*) $B$ on which the value of $f$ is specified, and there is a rule for calculating $f(x)$ for every $x \in S - B$ in terms of previously defined values of $f$.

*Ackermann's function* (Wilhelm Ackermann, 1896–1962) is defined recursively by

$$A(x, y, z) = \begin{cases} x + y & \text{if } z = 0 \\ 0 & \text{if } y = 0,\, z = 1 \\ 1 & \text{if } y = 0,\, z = 2 \\ x & \text{if } y = 0,\, z > 2 \\ A(x, A(x, y - 1, z), z - 1) & \text{if } y,\, z > 0. \end{cases}$$

An alternative version of Ackermann's function, with two variables, is defined recursively by

$$A(m, n) = \begin{cases} n + 1 & \text{if } m = 0 \\ A(m - 1, 1) & \text{if } m > 0,\, n = 0 \\ A(m - 1, A(m, n - 1)) & \text{if } m,\, n > 0. \end{cases}$$

Another alternative version of Ackermann's function is defined recursively by the rule $A(n) = A_n(n)$, where $A_1(n) = 2n$ and $A_m(n) = A_{m-1}^{(n)}(1)$ if $m \geq 2$.

The (input-independent) *halting function* maps computer programs to the set $\{0, 1\}$, with value 1 if the program always halts, regardless of input, and 0 otherwise.

**Facts:**

**1.** If $f\colon \mathcal{N} \to \mathcal{R}$ is recursively defined, the set of base values is frequently the set $\{f(0), f(1), \ldots, f(j)\}$ and there is a rule for calculating $f(n)$ for every $n > j$ in terms of $f(i)$ for one or more $i < n$.

**2.** There are functions whose values cannot be computed. (See Example 5.)

**3.** There are recursively-defined functions that cannot be represented by a closed formula.

**4.** It is possible to find closed formulas for the values of some functions defined recursively. See Chapter 3 for more information.

**5.** Computer software developers often represent a table as a *binary search tree* (§18.2.3).

**6.** In Ackermann's function of three variables $A(x, y, z)$, as the variable $z$ ranges from 0 to 3, $A(x, y, z)$ is the sum of $x$ and $y$, the product of $x$ and $y$, $x$ raised to the exponent $y$, and the iterated exponentiation of $x$ $y$ times. That is, $A(x, y, 0) = x + y$, $A(x, y, 1) = xy$, $A(x, y, 2) = x^y$, $A(x, y, 3) = x^{x^{\cdot^{\cdot^{x}}}}$  ($y$ $x$s in the exponent).

**7.** The version of Ackermann's function with two variables, $A(x, y)$, has the following properties: $A(1, n) = n + 2$, $A(2, n) = 2n + 3$, $A(3, n) = 2^{n+3} - 3$.

**8.** $A(m, n)$ is an example of a well-defined total function that is computable, but not primitive recursive. (See §17.2.1.)

**Examples:**

**1.** The function that maps each month to its ordinal position is represented by the table

$$\{(\mathrm{Jan}, 1), (\mathrm{Feb}, 2), \ldots, (\mathrm{Dec}, 12)\}.$$

**2.** The function defined by the recurrence relation

$$f(0) = 0; \quad f(n) = f(n-1) + 2n - 1 \text{ for } n \geq 1$$

has the closed form $f(x) = x^2$.

**3.** The function defined by the recurrence relation

$$f(0) = 0, \, f(1) = 1; \quad f(n) = f(n-1) + f(n-2) \text{ for } n \geq 2$$

generates the Fibonacci sequence $0, 1, 1, 2, 3, 5, 8, \ldots$ (see §3.1.2) and has the closed form

$$f(n) = \frac{(1 + \sqrt{5})^n - (1 - \sqrt{5})^n}{2^n \sqrt{5}}.$$

**4.** The factorial function $n!$ is recursively defined by the rules

$$0! = 1; \quad n! = n \cdot (n-1)! \text{ for } n \geq 1.$$

It has no known closed formula in terms of elementary functions.

**5.** It is impossible to construct an algorithm to compute the halting function.

**6.** The halting function from the Cartesian product of the set of computer programs and the set of strings to $\{0, 1\}$ whose value is 1 if the program halts when given that string as input and 0 if the program does not halt when given that string as input is noncomputable.

**7.** The following is not a well-defined function $f\colon \{1,2,3,\ldots\} \to \{1,2,3,\ldots\}$

$$f(n) = \begin{cases} 1 & \text{if } n = 1 \\ 1 + f(\frac{n}{2}) & \text{if } n \text{ is even} \\ f(3n-1) & \text{if } n \text{ is odd, } n > 1 \end{cases}$$

since evaluating $f(5)$ leads to the contradiction $f(5) = f(5) + 3$.

**8.** It is not known whether the following is a well-defined function $f\colon \{1,2,3,\ldots\} \to \{1,2,3,\ldots\}$

$$f(n) = \begin{cases} 1 & n = 1 \\ 1 + f(\frac{n}{2}) & n \text{ even} \\ f(3n+1) & n \text{ odd, } n > 1. \end{cases}$$

(See §1.3.1, Example 13.)

---

## 1.3.3  ASYMPTOTIC BEHAVIOR

The asymptotic growth of functions is commonly described with various special pieces of notation and is regularly used in the analysis of computer algorithms to estimate the length of time the algorithms take to run and the amount of computer memory they require.

**Definitions:**

A function $f\colon \mathcal{R} \to \mathcal{R}$ or $f\colon \mathcal{N} \to \mathcal{R}$ is **bounded** if there is a constant $k$ such that $|f(x)| \le k$ for all $x$ in the domain of $f$.

For functions $f, g\colon \mathcal{R} \to \mathcal{R}$ or $f, g\colon \mathcal{N} \to \mathcal{R}$ (sequences of real numbers) the following are used to compare their growth rates:

- $f$ is **big-oh** of $g$ ($g$ **dominates** $f$) if there exist constants $C$ and $k$ such that $|f(x)| \le C|g(x)|$ for all $x > k$.

  Notation: $f$ is $O(g)$, $f(x) \in O(g(x))$, $f \in O(g)$, $f = O(g)$.

- $f$ is **little-oh** of $g$ if $\lim_{x \to \infty} \left| \frac{f(x)}{g(x)} \right| = 0$; i.e., for every $C > 0$ there is a constant $k$ such that $|f(x)| \le C|g(x)|$ for all $x > k$.

  Notation: $f$ is $o(g)$, $f(x) \in o(g(x))$, $f \in o(g)$, $f = o(g)$.

- $f$ is **big omega of** $g$ if there are constants $C$ and $k$ such that $|g(x)| \le C|f(x)|$ for all $x > k$.

  Notation: $f$ is $\Omega(g)$, $f(x) \in \Omega(g(x))$, $f \in \Omega(g)$, $f = \Omega(g)$.

- $f$ is **little omega of** $g$ if $\lim_{x \to \infty} \left| \frac{g(x)}{f(x)} \right| = 0$.

  Notation: $f$ is $\omega(g)$, $f(x) \in \omega(g(x))$, $f \in \omega(g)$, $f = \omega(g)$.

- $f$ is **theta of** $g$ if there are positive constants $C_1$, $C_2$, and $k$ such that $C_1|g(x)| \le |f(x)| \le C_2|g(x)|$ for all $x > k$.

  Notation: $f$ is $\Theta(g)$, $f(x) \in \Theta(g(x))$, $f \in \Theta(g)$, $f = \Theta(g)$, $f \approx g$.

- $f$ is **asymptotic** to $g$ if $\lim_{x \to \infty} \frac{g(x)}{f(x)} = 1$. This relation is sometimes called **asymptotic equality** and is denoted $f \sim g$ or $f(x) \sim g(x)$.

## Facts:

**1.** The notations $O(\ )$, $o(\ )$, $\Omega(\ )$, $\omega(\ )$, and $\Theta(\ )$ all stand for *collections* of functions. Hence the equality sign, as in $f = O(g)$, does not mean equality of functions.

**2.** The symbols $O(g)$, $o(g)$, $\Omega(g)$, $\omega(g)$, and $\Theta(g)$ are frequently used to represent a typical element of the class of functions it represents, as in an expression such as $f(n) = n \log n + o(n)$.

**3.** *Growth rates*:

- $O(g)$: the set of functions that grow no more rapidly than a positive multiple of $g$;
- $o(g)$: the set of functions that grow less rapidly than a positive multiple of $g$;
- $\Omega(g)$: the set of functions that grow at least as rapidly as a positive multiple of $g$;
- $\omega(g)$: the set of functions that grow more rapidly than a positive multiple of $g$;
- $\Theta(g)$: the set of functions that grow at the same rate as a positive multiple of $g$.

**4.** Asymptotic notation can be used to describe the growth of infinite sequences, since infinite sequences are functions from $\{0, 1, 2, \ldots\}$ or $\{1, 2, 3, \ldots\}$ to $\mathcal{R}$ (by considering the term $a_n$ as $a(n)$, the value of the function $a(n)$ at the integer $n$).

**5.** The big-oh notation was introduced in 1892 by Paul Bachmann (1837–1920) in the study of the rates of growth of various functions in number theory.

**6.** The big-oh symbol is often called a *Landau symbol*, after Edmund Landau (1877–1938), who popularized this notation.

**7.** *Properties of big-oh*:

- if $f \in O(g)$ and $c$ is a constant, then $cf \in O(g)$;
- if $f_1, f_2 \in O(g)$, then $f_1 + f_2 \in O(g)$;
- if $f_1 \in O(g_1)$ and $f_2 \in O(g_2)$, then
    - $\diamond$ $(f_1 + f_2) \in O(g_1 + g_2)$
    - $\diamond$ $(f_1 + f_2) \in O(\max(|g_1|, |g_2|))$
    - $\diamond$ $(f_1 f_2) \in O(g_1 g_2)$;
- if $f$ is a polynomial of degree $n$, then $f \in O(x^n)$;
- if $f$ is a polynomial of degree $m$ and $g$ a polynomial of degree $n$, with $m \geq n$, then $\frac{f}{g} \in O(x^{m-n})$;
- if $f$ is a bounded function, then $f \in O(1)$;
- for all $a, b > 1$, $O(\log_a x) = O(\log_b x)$;
- if $f \in O(g)$ and $|h(x)| \geq |g(x)|$ for all $x > k$, then $f \in O(h)$;
- if $f \in O(x^m)$, then $f \in O(x^n)$ for all $n > m$.

**8.** Some of the most commonly used benchmark big-oh classes are: $O(1)$, $O(\log x)$, $O(x)$, $O(x \log x)$, $O(x^2)$, $O(2^x)$, $O(x!)$, and $O(x^x)$. If $f$ is big-oh of any function in this list, then $f$ is also big-oh of each of the following functions in the list:

$$O(1) \subset O(\log x) \subset O(x) \subset O(x \log x) \subset O(x^2) \subset O(2^x) \subset O(x!) \subset O(x^x).$$

The benchmark functions are drawn in the following figure.

**9.** *Properties of little-oh*:

- if $f \in o(g)$, then $cf \in o(g)$ for all nonzero constants $c$;
- if $f_1 \in o(g)$ and $f_2 \in o(g)$, then $f_1 + f_2 \in o(g)$;

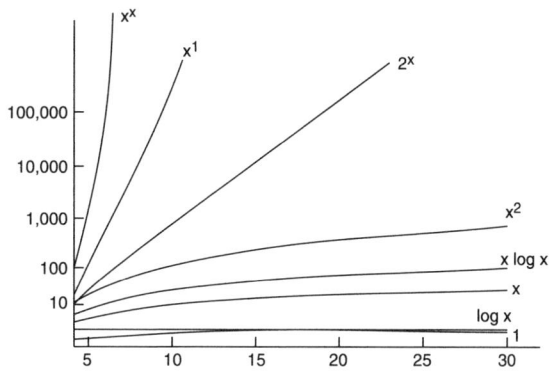

- if $f_1 \in o(g_1)$ and $f_2 \in o(g_2)$, then
  - $\diamond$ $(f_1 + f_2) \in o(g_1 + g_2)$
  - $\diamond$ $(f_1 + f_2) \in o(\max(|g_1|, |g_2|))$
  - $\diamond$ $(f_1 f_2) \in o(g_1 g_2)$;
- if $f$ is a polynomial of degree $m$ and $g$ a polynomial of degree $n$ with $m < n$, then $\frac{f}{g} \in o(1)$;
- the set membership $f(x) \in L + o(1)$ is equivalent to $f(x) \to L$ as $x \to \infty$, where $L$ is a constant.

**10.** If $f \in o(g)$, then $f \in O(g)$; the converse is not true.

**11.** If $f \in O(g)$ and $h \in o(f)$, then $h \in o(g)$.

**12.** If $f \in o(g)$ and $h \in O(f)$, then $h \in O(g)$.

**13.** If $f \in O(g)$ and $h \in O(f)$, then $h \in O(g)$.

**14.** If $f_1 \in o(g_1)$ and $f_2 \in O(g_2)$, then $f_1 f_2 \in o(g_1 g_2)$.

**15.** $f \in O(g)$ if and only if $g \in \Omega(f)$.

**16.** $f \in \Theta(g)$ if and only if $f \in O(g)$ and $g \in O(f)$.

**17.** $f \in \Theta(g)$ if and only if $f \in O(g)$ and $f \in \Omega(g)$.

**18.** If $f(x) = a_n x^n + \cdots + a_1 x + a_0$ $(a_n \neq 0)$, then $f \sim a_n x^n$.

**19.** $f \sim g$ if and only if $\left(\frac{f}{g} - 1\right) \in o(1)$ (provided $g(x) = 0$ only finitely often).

**Examples:**

**1.** $5x^8 + 10^{200}x^5 + 3x + 1 \in O(x^8)$.

**2.** $x^3 \in O(x^4)$, $x^4 \notin O(x^3)$.

**3.** $x^3 \in o(x^4)$, $x^4 \notin o(x^3)$.

**4.** $x^3 \notin o(x^3)$.

**5.** $x^2 \in O(5x^2)$, $x^2 \notin o(5x^2)$.

**6.** $\sin(x) \in O(1)$.

**7.** $\frac{x^7 - 3x}{8x^3 + 5} \in O(x^4)$, $\frac{x^7 - 3x}{8x^3 + 5} \in \Theta(x^4)$.

**8.** $1 + 2 + 3 + \cdots + n \in O(n^2)$.

**9.** $1 + \frac{1}{2} + \frac{1}{3} + \cdots + \frac{1}{n} \in O(\log n)$.

**10.** $\log(n!) \in O(n \log n)$.

**11.** $8x^5 \in \Theta(3x^5)$.

**12.** $x^3 \in \Omega(x^2)$.

**13.** $2^n + o(n^2) \sim 2^n$.

**14.** Sometimes asymptotic equality does not behave like equality: $\ln n \sim \ln(2n)$, but $n \not\sim 2n$ and $\ln n - \ln n \not\sim \ln(2n) - \ln n$.

**15.** $\pi(n) \sim \frac{n}{\ln n}$ where $\pi(n)$ is the number of primes less than or equal to $n$.

**16.** If $p_n$ is the $n$th prime, then $p_n \sim n \ln n$.

**17.** *Stirling's formula:* $n! \sim \sqrt{2\pi n}(\frac{n}{e})^n$.

## 1.4  RELATIONS

Relationships between two sets (or among more than two sets) occur frequently throughout mathematics and its applications. Examples of such relationships include integers and their divisors, real numbers and their logarithms, corporations and their customers, cities and airlines that serve them, people and their relatives. These relationships can be described as subsets of product sets.

Functions are a special type of relation. Equivalence relations can be used to describe similarity among elements of sets and partial order relations describe the relative size of elements of sets.

### 1.4.1  BINARY RELATIONS AND THEIR PROPERTIES

**Definitions:**
A **binary relation** from set $A$ to set $B$ is any subset $R$ of $A \times B$.

An element $a \in A$ **is related to** $b \in B$ in the relation $R$ if $(a, b) \in R$, often written $aRb$. If $(a, b) \notin R$, write $a\not{R}b$.

A **binary relation** (**relation**) on a set $A$ is a binary relation from $A$ to $A$; i.e., a subset of $A \times A$.

A binary relation $R$ on $A$ can have the following properties (to have the property, the relation must satisfy the property for all $a, b, c \in A$):

- **reflexivity**:  $aRa$
- **irreflexivity**:  $a\not{R}a$
- **symmetry**:  if $aRb$, then $bRa$
- **asymmetry**:  if $aRb$, then $b\not{R}a$
- **antisymmetry**:  if $aRb$ and $bRa$, then $a = b$
- **transitivity**:  if $aRb$ and $bRc$, then $aRc$
- **intransitivity**:  if $aRb$ and $bRc$, then $a\not{R}c$

Binary relations $R$ and $S$ from $A$ to $B$ can be combined in the following ways to yield other relations:

- **complement** of $R$:  the relation $\overline{R}$ from $A$ to $B$ where $a\overline{R}b$ if and only if $a\not{R}b$ (i.e., $\neg(aRb)$)

- **difference**:  the binary relation $R - S$ from $A$ to $B$ such that $a(R - S)b$ if and only if $aRb$ and $\neg(aSb)$

- **intersection**:  the relation $R \cap S$ from $A$ to $B$ where $a(R \cap S)b$ if and only if $aRb$ and $aSb$

- **inverse** (**converse**):  the relation $R^{-1}$ from $B$ to $A$ where $bR^{-1}a$ if and only if $aRb$

- **symmetric difference**:  the relation $R \oplus S$ from $A$ to $B$ where $a(R \oplus S)b$ if and only if exactly one of the following is true: $aRb, aSb$

- **union**:  the relation $R \cup S$ from $A$ to $B$ where $a(R \cup S)b$ if and only if $aRb$ or $aSb$.

The **closure** of a relation $R$ with respect to a property $\mathcal{P}$ is the relation $S$, if it exists, that has property $\mathcal{P}$ and contains $R$, such that $S$ is a subset of every relation that has property $\mathcal{P}$ and contains $R$.

A relation $R$ on $A$ is **connected** if for all $a, b \in A$ with $a \neq b$, either $aRb$ or there are $c_1, c_2, \ldots, c_k \in A$ such that $aRc_1, c_1Rc_2, \ldots, c_{k-1}Rc_k, c_kRb$.

If $R$ is a relation on $A$, the **connectivity relation associated with** $R$ is the relation $R'$ where $aR'b$ if and only if $aRb$ or there are $c_1, c_2, \ldots, c_k \in A$ such that $aRc_1, c_1Rc_2, \ldots, c_{k-1}Rc_k, c_kRb$.

If $R$ is a binary relation from $A$ to $B$ and if $S$ is a binary relation from $B$ to $C$, then the **composition** of $R$ and $S$ is the binary relation $S \circ R$ from $A$ to $C$ where $a(S \circ R)c$ if and only if there is an element $b \in B$ such that $aRb$ and $bSc$.

The **$n$th power** ($n$ a nonnegative integer) of a relation $R$ on a set $A$, is the relation $R^n$, where $R^0 = \{(a, a) \mid a \in A\} = I_A$ (see Example 4), $R^1 = R$ and $R^n = R^{n-1} \circ R$ for all integers $n > 1$.

A **transitive reduction** of a relation, if it exists, is a relation with the same transitive closure as the original relation and with a minimal superset of ordered pairs.

### Notation:

**1.** If a relation $R$ is symmetric, $aRb$ is often written $a \sim b$, $a \approx b$, or $a \equiv b$.

**2.** If a relation $R$ is antisymmetric, $aRb$ is often written $a \leq b$, $a < b$, $a \subset b$, $a \subseteq b$, $a \preceq b$, $a \prec b$, or $a \sqsubseteq b$.

### Facts:

**1.** A binary relation $R$ from $A$ to $B$ can be viewed as a function from the Cartesian product $A \times B$ to the boolean domain $\{\text{TRUE}, \text{FALSE}\}$ (often written $\{T, F\}$). The truth value of the pair $(a, b)$ determines whether $a$ is related to $b$.

**2.** Under the *infix convention* for a binary relation, $aRb$ ($a$ *is related to* $b$) means $R(a, b) = \text{TRUE}$; $a\cancel{R}b$ ($a$ *is not related to* $b$) means $R(a, b) = \text{FALSE}$.

**3.** A binary relation $R$ from $A$ to $B$ can be represented in any of the following ways:

- a set $R \subseteq A \times B$, where $(a, b) \in R$ if and only if $aRb$ (this is the definition of $R$);

- a directed graph $D_R$ whose vertices are the elements of $A \cup B$, with an edge from vertex $a$ to vertex $b$ if $aRb$ (§8.3.1);

- a matrix (the adjacency matrix for the directed graph $D_R$): if $A = \{a_1, \ldots, a_m\}$ and $B = \{b_1, \ldots, b_n\}$, the matrix for the relation $R$ is the $m \times n$ matrix $M_R$ with entries $m_{ij}$ where $m_{ij} = 1$ if $a_iRb_j$ and $m_{ij} = 0$ otherwise.

**4.** $R$ is a reflexive relation on $A$ if and only if $\{(a,a) \mid a \in A\} \subseteq R$; i.e., $R$ is a reflexive relation on $A$ if and only if $I_A \subseteq R$.

**5.** $R$ is symmetric if and only if $R = R^{-1}$.

**6.** $R$ is an antisymmetric relation on $A$ if and only if $R \cap R^{-1} \subseteq \{(a,a) \mid a \in A\}$.

**7.** $R$ is transitive if and only if $R \circ R \subseteq R$.

**8.** A relation $R$ can be both symmetric and antisymmetric: for example, the equality relation on a set $A$.

**9.** For a relation $R$ that is both symmetric and antisymmetric: $R$ is reflexive if and only if $R$ is the equality relation on some set; $R$ is irreflexive if and only if $R = \emptyset$.

**10.** The closure of a relation $R$ with respect to a property $\mathcal{P}$ is the intersection of all relations $Q$ with property $\mathcal{P}$ such that $R \subseteq Q$, if there is at least one such relation $Q$.

**11.** The transitive closure of a relation $R$ is the connectivity relation $R'$ associated with $R$, which is equal to the union $\bigcup_{i=1}^{\infty} R^i$ of all the positive powers of the relation.

**12.** A transitive reduction of a relation may contain pairs not in the original relation (Example 8).

**13.** Transitive reductions are not necessarily unique (Example 9).

**14.** If $R$ is a relation on $A$ and $x, y \in A$ with $x \neq y$, then $x$ is related to $y$ in the transitive closure of $R$ if and only if there is a nontrivial directed path from $x$ to $y$ in the directed graph $D_R$ of the relation.

**15.** The following table shows how to obtain various closures of a relation and gives the matrices for the various closures of a relation $R$ with matrix $M_R$ on a set $A$ where $|A| = n$.

| relation | set | matrix |
|---|---|---|
| reflexive closure | $R \cup \{(a,a) \mid a \in A\}$ | $M_R \vee I_n$ |
| symmetric closure | $R \cup R^{-1}$ | $M_R \vee M_{R^{-1}}$ |
| transitive closure | $\bigcup_{i=1}^{n} R^i$ | $M_R \vee M_R^{[2]} \vee \cdots \vee M_R^{[n]}$ |

Here the matrix $I_n$ is the $n \times n$ identity matrix, $M_R^{[i]}$ is the $i$th boolean power of the matrix $M_R$ for the relation $R$, and $\vee$ is the join operator (defined by $0 \vee 0 = 0$ and $0 \vee 1 = 1 \vee 0 = 1 \vee 1 = 1$).

**16.** The following table provides formulas for the number of binary relations with various properties on a set with $n$ elements.

| type of relation | number of relations |
|---|---|
| all relations | $2^{n^2}$ |
| reflexive | $2^{n(n-1)}$ |
| symmetric | $2^{n(n+1)/2}$ |
| transitive | no known simple closed formula (§3.1.8) |
| antisymmetric | $2^n \cdot 3^{n(n-1)/2}$ |
| asymmetric | $3^{n(n-1)/2}$ |
| irreflexive | $2^{n(n-1)}$ |
| equivalence (§1.4.2) | $B_n = $ Bell number $= \sum_{k=1}^{n} \left\{ {n \atop k} \right\}$ where $\left\{ {n \atop k} \right\}$ is a Stirling subset number (§2.5.2) |
| partial order (§1.4.3) | no known simple closed formula (§3.1.8) |

**Algorithm:**

**1.** Warshall's algorithm, also called the Roy-Warshall algorithm (B. Roy and S. Warshall described the algorithm in 1959 and 1960, respectively), Algorithm 1, is an algorithm of order $n^3$ for finding the transitive closure of a relation on a set with $n$ elements. (Stephen Warshall, 1935–2006.)

---

**Algorithm 1**: **Warshall's algorithm.**

input: $M = [m_{ij}]_{n \times n}$ = the matrix representing the binary relation $R$

output: $M$ = the transitive closure of relation $R$

**for** $k := 1$ **to** $n$

   **for** $i := 1$ **to** $n$

      **for** $j := 1$ **to** $n$

         $m_{ij} := m_{ij} \vee (m_{ik} \wedge m_{kj})$

---

**Examples:**

**1.** Some common relations and whether they have certain properties are given in the following table:

| set | relation | reflexive | symmetric | antisymmetric | transitive |
|---|---|---|---|---|---|
| any nonempty set | $=$ | yes | yes | yes | yes |
| any nonempty set | $\neq$ | no | yes | no | no |
| $\mathcal{R}$ | $\leq$ (or $\geq$) | yes | no | yes | yes |
| $\mathcal{R}$ | $<$ (or $>$) | no | no | yes | yes |
| positive integers | is a divisor of | yes | no | yes | yes |
| nonzero integers | is a divisor of | yes | no | no | yes |
| integers | congruence mod $n$ | yes | yes | no | yes |
| any set of sets | $\subseteq$ (or $\supseteq$) | yes | no | yes | yes |
| any set of sets | $\subset$ (or $\supset$) | no | no | yes | yes |

**2.** If $A$ is any set, the *universal relation* is the relation $R$ on $A \times A$ such that $aRb$ for all $a, b \in A$; i.e., $R = A \times A$.

**3.** If $A$ is any set, the *empty relation* is the relation $R$ on $A \times A$ where $aRb$ is never true; i.e., $R = \emptyset$.

**4.** If $A$ is any set, the relation $R$ on $A$ where $aRb$ if and only if $a = b$ is the *identity* (or *diagonal*) *relation* $I = I_A = \{(a,a) \mid a \in A\}$, which is also written $\Delta$ or $\Delta_A$.

**5.** Every function $f: A \to B$ induces a binary relation $R_f$ from $A$ to $B$ under the rule $aR_f b$ if and only if $f(a) = b$.

**6.** For $A = \{2, 3, 4, 6, 12\}$, suppose that $aRb$ means that $a$ is a divisor of $b$. Then $R$ can be represented by the set

$$\{(2,2), (2,4), (2,6), (2,12), (3,3), (3,6), (3,12), (4,4), (4,12), (6,6), (6,12), (12,12)\}.$$

The relation $R$ can also be represented by the digraph with the adjacency matrix

$$\begin{pmatrix} 1 & 0 & 1 & 1 & 1 \\ 0 & 1 & 0 & 1 & 1 \\ 0 & 0 & 1 & 0 & 1 \\ 0 & 0 & 0 & 1 & 1 \\ 0 & 0 & 0 & 0 & 1 \end{pmatrix}.$$

**7.** The transitive closure of the relation $\{(1,3),(2,3),(3,2)\}$ on $\{1,2,3\}$ is the relation $\{(1,2),(1,3),(2,2),(2,3),(3,2),(3,3)\}$.

**8.** The transitive closure of the relation $R = \{(1,2),(2,3),(3,1)\}$ on $\{1,2,3\}$ is the universal relation $\{1,2,3\} \times \{1,2,3\}$. A transitive reduction of $R$ is the relation given by $\{(1,3),(3,2),(2,1)\}$. This shows that a transitive reduction may contain pairs that are not in the original relation.

**9.** If $R = \{(a,b) \mid aRb$ for all $a,b \in \{1,2,3\}\}$, then the relations $\{(1,2),(2,3),(3,1)\}$ and $\{(1,3),(3,2),(2,1)\}$ are both transitive reductions for $R$. Thus, transitive reductions are not unique.

---

## 1.4.2   EQUIVALENCE RELATIONS

Equivalence relations are binary relations that describe various types of similarity or "equality" among elements in a set. The elements that look alike or behave in a similar way are grouped together in equivalence classes, resulting in a partition of the set. Any element chosen from an equivalence class essentially "mirrors" the behavior of all elements in that class.

### Definitions:

An **equivalence relation** on $A$ is a binary relation on $A$ that is reflexive, symmetric, and transitive.

If $R$ is an equivalence relation on $A$, the **equivalence class** of $a \in A$ is the set $R[a] = \{b \in A \mid aRb\}$. When it is clear from context which equivalence relation is intended, the notation for the induced equivalence class can be abbreviated $[a]$.

The **induced partition** on a set $A$ under an equivalence relation $R$ is the set of equivalence classes.

### Facts:

**1.** A nonempty relation $R$ is an equivalence relation if and only if $R \circ R^{-1} = R$.

**2.** The induced partition on a set $A$ actually is a partition of $A$; i.e., the equivalence classes are all nonempty, every element of $A$ lies in some equivalence class, and two classes $[a]$ and $[b]$ are either disjoint or equal.

**3.** There is a one-to-one correspondence between the set of all possible equivalence relations on a set $A$ and the set of all possible partitions of $A$. (Fact 2 shows how to obtain a partition from an equivalence relation. To obtain an equivalence relation from a partition of $A$, define $R$ by the rule $aRb$ if and only if $a$ and $b$ lie in the same element of the partition.)

**4.** For any set $A$, the coarsest partition (with only one set in the partition) of $A$ is induced by the equivalence relation in which every pair of elements are related. The finest partition (with each set in the partition having cardinality 1) of $A$ is induced by the equivalence relation in which no two different elements are related.

**5.** The set of all partitions of a set $A$ is partially ordered under refinement (§1.2.2 and §1.4.3). This partial ordering is a lattice (§5.7).

**6.** To find the smallest equivalence relation containing a given relation, first take the transitive closure of the relation, then take the reflexive closure of that relation, and finally take the symmetric closure.

**Examples:**

**1.** For any function $f\colon A \to B$, define the relation $a_1 R a_2$ to mean that $f(a_1) = f(a_2)$. Then $R$ is an equivalence relation. Each induced equivalence class is the inverse image $f^{-1}(b)$ of some $b \in B$.

**2.** Write $a \equiv b \pmod{n}$ ("$a$ is congruent to $b$ modulo $n$") when $a, b$ and $n > 0$ are integers such that $n \mid b - a$ ($n$ divides $b - a$). Congruence mod $n$ is an equivalence relation on the integers.

**3.** The equivalence relation of congruence modulo $n$ on the integers $\mathcal{Z}$ yields a partition with $n$ equivalence classes: $[0] = \{kn \mid k \in \mathcal{Z}\}$, $[1] = \{1 + kn \mid k \in \mathcal{Z}\}$, $[2] = \{2 + kn \mid k \in \mathcal{Z}\}, \ldots, [n-1] = \{(n-1) + kn \mid k \in \mathcal{Z}\}$.

**4.** The isomorphism relation on any set of groups is an equivalence relation. (The same result holds for rings, fields, etc.) (See Chapter 5.)

**5.** The congruence relation for geometric objects in the plane is an equivalence relation.

**6.** The similarity relation for geometric objects in the plane is an equivalence relation.

### 1.4.3 PARTIALLY ORDERED SETS

Partial orderings extend the relationship of $\leq$ on real numbers and allow a comparison of the relative "size" of elements in various sets. They are developed in greater detail in Chapter 11.

**Definitions:**

A **preorder** on a set $S$ is a binary relation $\leq$ on $S$ that has the following properties for all $a, b, c \in S$:

- reflexive: $a \leq a$
- transitive: if $a \leq b$ and $b \leq c$, then $a \leq c$.

A **partial ordering** (or **partial order**) on a set $S$ is a binary relation $\leq$ on $S$ that has the following properties for all $a, b, c \in S$:

- reflexive: $a \leq a$
- antisymmetric: if $a \leq b$ and $b \leq a$, then $a = b$
- transitive: if $a \leq b$ and $b \leq c$, then $a \leq c$.

*Notation:* The expression $c \geq b$ means that $b \leq c$. The symbols $\preceq$ and $\succeq$ are often used in place of $\leq$ and $\geq$. The expression $a < b$ (or $b > a$) means that $a \leq b$ and $a \neq b$.

A **partially ordered set** (or **poset**) is a set with a partial ordering defined on it.

A **directed ordering** on a set $S$ is a partial ordering that also satisfies the following property: if $a, b \in S$, then there is a $c \in S$ such that $a \leq c$ and $b \leq c$.

*Note:* Some authors do not require that antisymmetry hold in the definition of directed ordering.

Two elements $a$ and $b$ in a poset are **comparable** if either $a \leq b$ or $b \leq a$. Otherwise, they are **incomparable**.

A **totally ordered** (or **linearly ordered**) set is a poset in which every pair of elements are comparable.

A **chain** is a subset of a poset in which every pair of elements are comparable.

An **antichain** is a subset of a poset in which no two distinct elements are comparable.

An **interval** in a poset $(S, \leq)$ is a subset $[a, b] = \{x \mid x \in S, a \leq x \leq b\}$.

An element $b$ in a poset is **minimal** if there exists no element $c$ such that $c < b$.

An element $b$ in a poset is **maximal** if there exists no element $c$ such that $c > b$.

An element $b$ in a poset $S$ is a **maximum element** (or **greatest element**) if every element $c$ satisfies the relation $c \leq b$.

An element $b$ in a poset $S$ is a **minimum element** (or **least element**) if every element $c$ satisfies the relation $c \geq b$.

A **well-ordered** set is a poset $(S, \leq)$ in which every nonempty subset contains a minimum element.

An element $b$ in a poset $S$ is an **upper bound** for a subset $U \subseteq S$ if every element $c$ of $U$ satisfies the relation $c \leq b$.

An element $b$ in a poset $S$ is a **lower bound** for a subset $U \subseteq S$ if every element $c$ of $U$ satisfies the relation $c \geq b$.

A **least upper bound** for a subset $U$ of a poset $S$ is an upper bound $b$ such that if $c$ is any other upper bound for $U$ then $c \geq b$.

A **greatest lower bound** for a subset $U$ of a poset $S$ is a lower bound $b$ such that if $c$ is any other lower bound for $U$ then $c \leq b$.

A **lattice** is a poset in which every pair of elements, $x$ and $y$, have both a least upper bound $\mathrm{lub}(x, y)$ and a greatest lower bound $\mathrm{glb}(x, y)$ (§5.7).

The **Cartesian product** of two posets $(S_1, \leq_1)$ and $(S_2, \leq_2)$ is the poset with domain $S_1 \times S_2$ and relation $\leq_1 \times \leq_2$ given by the rule $(a_1, a_2) \leq_1 \times \leq_2 (b_1, b_2)$ if and only if $a_1 \leq_1 b_1$ and $a_2 \leq_2 b_2$.

The element $c$ **covers** another element $b$ in a poset if $b < c$ and there is no element $d$ such that $b < d < c$.

A **Hasse diagram** (**cover diagram**) for a poset $(S, \leq)$ is a directed graph (§11.1) whose vertices are the elements of $S$ such that there is an arc from $b$ to $c$ if $c$ covers $b$, all arcs are directed upward when drawing the diagram, and arrows on the arcs are omitted.

## Facts:

**1.** $R$ is a partial order on a set $S$ if and only if $R^{-1}$ is a partial order on $S$.

**2.** The only partial order that is also an equivalence relation is the relation of equality.

**3.** The Cartesian product of two posets, each with at least two elements, is not totally ordered.

**4.** In the Hasse diagram for a poset, there is a path from vertex $b$ to vertex $c$ if and only if $b \leq c$. (When $b = c$, it is the path of length 0.)

**5.** Least upper bounds and greatest lower bounds are unique, if they exist.

**Examples:**

**1.** The positive integers are partially ordered under the relation of divisibility, in which $b \leq c$ means that $b$ divides $c$. In fact, they form a lattice (§5.7.1), called the *divisibility lattice*. The least upper bound of two numbers is their least common multiple, and the greatest lower bound is their greatest common divisor.

**2.** The set of all powers of two (or of any other positive integer) forms a chain in the divisibility lattice.

**3.** The set of all primes forms an antichain in the divisibility lattice.

**4.** The set $\mathcal{R}$ of real numbers with the usual definition of $\leq$ is a totally ordered set.

**5.** The set of all logical propositions on a fixed set of logical variables $p, q, r, \ldots$ is partially ordered under inverse implication, so that $B \leq A$ means that $A \rightarrow B$ is a tautology.

**6.** The complex numbers, ordered under magnitude, do *not* form a poset, because they do not satisfy the axiom of antisymmetry.

**7.** The set of all subsets of any set forms a lattice under the relation of subset inclusion. The least upper bound of two subsets is their union, and the greatest lower bound is their intersection. Part (a) in the following figure gives the Hasse diagram for the lattice of all subsets of $\{a, b, c\}$.

**8.** Part (b) of the following figure shows the Hasse diagram for the lattice of all positive integer divisors of 12.

**9.** Part (c) of the following figure shows the Hasse diagram for the set $\{1, 2, 3, 4, 5, 6\}$ under divisibility.

**10.** Part (d) of the following figure shows the Hasse diagram for the set $\{1, 2, 3, 4\}$ with the usual definition of $\leq$.

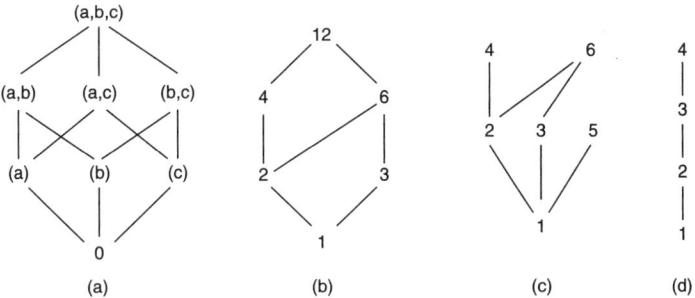

**11.** *Multilevel security policy:*   The flow of information is often restricted by using security clearances. Documents are put into security classes, $(L, C)$, where $L$ is an element of a totally ordered set of authority levels (such as "unclassified", "confidential", "secret", "top secret") and $C$ is a subset (called a "compartment") of a set of subject areas. The subject areas might consist of topics such as agriculture, Eastern Europe, economy, crime, and trade. A document on how trade affects the economic structure of Eastern Europe might be assigned to the compartment {trade, economy, Eastern Europe}. The set of security classes is made into a lattice by the rule: $(L_1, C_1) \leq (L_2, C_2)$ if and only if $L_1 \leq L_2$ and $C_1 \subseteq C_2$. Information is allowed to flow from class $(L_1, C_1)$ to class $(L_2, C_2)$ if and only if $(L_1, C_1) \leq (L_2, C_2)$. For example, a document with security class (secret, {trade, economy}) flows to both (top secret, {trade, economy}) and (secret, {trade, economy, Eastern Europe}), but not vice versa. This set of security classes forms a lattice (§5.7.1).

### 1.4.4  $n$-ARY RELATIONS

**Definitions:**

An **$n$-ary relation** on sets $A_1, A_2, \ldots, A_n$ is any subset $R$ of $A_1 \times A_2 \times \cdots \times A_n$.

The sets $A_i$ are called the **domains** of the relation and the number $n$ is called the **degree** of the relation.

A **primary key** of an $n$-ary relation $R$ on $A_1 \times A_2 \times \cdots \times A_n$ is a domain $A_i$ such that each $a_i \in A_i$ is the $i$th coordinate of at most one $n$-tuple in $R$.

A **composite key** of an $n$-ary relation $R$ on $A_1 \times A_2 \times \cdots \times A_n$ is a product of domains $A_{i_1} \times A_{i_2} \times \cdots \times A_{i_m}$ such that for each $m$-tuple $(a_{i_1}, a_{i_2}, \ldots, a_{i_m}) \in A_{i_1} \times A_{i_2} \times \cdots \times A_{i_m}$, there is at most one $n$-tuple in $R$ that matches $(a_{i_1}, a_{i_2}, \ldots, a_{i_m})$ in coordinates $i_1, i_2, \ldots, i_m$.

The **projection function** $P_{i_1, i_2, \ldots, i_k} : A_1 \times A_2 \times \cdots \times A_n \to A_{i_1} \times A_{i_2} \times \cdots \times A_{i_k}$ is given by the rule

$$P_{i_1, i_2, \ldots, i_k}(a_1, a_2, \ldots, a_n) = (a_{i_1}, a_{i_2}, \ldots, a_{i_k}).$$

That is, $P_{i_1, i_2, \ldots, i_k}$ selects the elements in coordinate positions $i_1, i_2, \ldots, i_k$ from the $n$-tuple $(a_1, a_2, \ldots, a_n)$.

The **join** $J_k(R, S)$ of an $m$-ary relation $R$ and an $n$-ary relation $S$, where $k \le m$ and $k \le n$, is a relation of degree $m + n - k$ such that

$$(a_1, \ldots, a_{m-k}, c_1, \ldots, c_k, b_1, \ldots, b_{n-k}) \in J_k(R, S)$$

if and only if

$$(a_1, \ldots, a_{m-k}, c_1, \ldots, c_k) \in R \text{ and } (c_1, \ldots, c_k, b_1, \ldots, b_{n-k}) \in S.$$

**Facts:**

**1.** An $n$-ary relation on sets $A_1, A_2, \ldots, A_n$ can be regarded as a function $R$ from $A_1 \times A_2 \times \cdots \times A_n$ to the Boolean domain $\{\text{TRUE}, \text{FALSE}\}$, where $(a_1, a_2, \ldots, a_n) \in R$ if and only if $R(a_1, a_2, \ldots, a_n) = \text{TRUE}$.

**2.** $n$-ary relations are essential models in the construction of database systems.

**Examples:**

**1.** Let $A_1$ be the set of all men and $A_2$ the set of all women, in a nonpolygamous society. Let $mRw$ mean that $m$ and $w$ are presently married. Then each of $A_1$ and $A_2$ is a primary key.

**2.** Let $A_1$ be the set of all telephone numbers and $A_2$ the set of all persons. Let $nRp$ mean that telephone number $n$ belongs to person $p$. Then $A_1$ is a primary key if each number is assigned to at most one person, and $A_2$ is a primary key if each person has at most one phone number.

**3.** In a conventional telephone directory, the name and address domains can form a composite key, unless there are two persons with the same name (no distinguishing middle initial or suffix such as "Jr.") at the same address.

**4.** Let $A = B = C = \mathcal{Z}$, and let $R$ be the relation on $A \times B \times C$ such that $(a, b, c) \in R$ if and only if $a + b = c$. The set $A \times B$ is a composite key. There is no primary key.

**5.** Let $A$ = all students at a certain college, $B$ = all student ID numbers being used at the college, $C$ = all major programs at the college. Suppose a relation $R$ is defined on $A \times B \times C$ by the rule $(a, b, c) \in R$ means student $a$ with ID number $b$ has major $c$. If each student has exactly one major and if there is a one-to-one correspondence between students and ID numbers, then $A$ and $B$ are each primary keys.

**6.** Let $A$ = all employee names at a certain corporation, $B$ = all Social Security numbers, $C$ = all departments, $D$ = all job titles, $E$ = all salary amounts, and $F$ = all calendar dates. On $A \times B \times C \times D \times E \times F \times F$ let $R$ be the relation such that $(a, b, c, d, e, f, g) \in R$ means employee named $a$ with Social Security number $b$ works in department $c$, has job title $d$, earns an annual salary $e$, was hired on date $f$, and had the most recent performance review on date $g$. The projection $P_{1,5}$ (projection onto $A \times E$) gives a list of employees and their salaries.

## 1.5  PROOF TECHNIQUES

A proof is a derivation of new facts from old ones. A proof makes possible the derivation of properties of a mathematical model from its definition, or the drawing of scientific inferences based on data that have been gathered. Axioms and postulates capture all basic truths used to develop a theory. Constructing proofs is one of the principal activities of mathematicians.

Furthermore, proofs play an important role in computer science—in such areas as verification of the correctness of computer programs, verification of communications protocols, automatic reasoning systems, and logic programming.

### 1.5.1  RULES OF INFERENCE

**Definitions:**

A *proposition* is a declarative sentence that is unambiguously either true or false. (See §1.1.1.)

A *theorem* is a proposition derived as the conclusion of a valid proof from axioms and definitions.

A *lemma* is a theorem that is an intermediate step in the proof of a more important theorem.

A *corollary* is a theorem that is derived as an easy consequence of another theorem.

A *statement form* is a declarative sentence containing some variables and logical symbols, such that the sentence becomes a proposition if concrete values are substituted for all the free variables.

An *argument form* is a sequence of statement forms.

The final statement form in an argument form is called the *conclusion* (of the argument). The conclusion is often preceded by the word "therefore" (symbolized ∴).

The statement forms preceding the conclusion in an argument form are called *premises* (of the argument).

If concrete values are substituted for the free variables of an argument form, an **argument of that form** is obtained.

An **instantiation** of an argument is the substitution of concrete values into all free variables of the premises and conclusion.

A **valid argument form** is an argument form such that in every instantiation in which all the premises are true, the conclusion is also true.

A **rule of inference** is an alternative name for a valid argument form, which is used when the form is frequently applied.

## Facts:

**1.** *Substitution rule:* Any variable occurring in an argument may be replaced by an expression of the same type without affecting the validity of the argument, as long as the replacement is made everywhere the variable occurs.

**2.** The following table gives rules of inference for arguments with compound statements.

| name | argument form | name | argument form |
|---|---|---|---|
| *Modus ponens* (method of affirming) | $p \rightarrow q$ <br> $p$ <br> $\therefore q$ | *Modus tollens* (method of denying) | $p \rightarrow q$ <br> $\neg q$ <br> $\therefore \neg p$ |
| *Hypothetical syllogism* | $p \rightarrow q$ <br> $q \rightarrow r$ <br> $\therefore p \rightarrow r$ | *Disjunctive syllogism* | $p \vee q$ <br> $\neg p$ <br> $\therefore q$ |
| *Disjunctive addition* | $p$ <br> $\therefore p \vee q$ | *Dilemma by cases* | $p \vee q$ <br> $p \rightarrow r$ <br> $q \rightarrow r$ <br> $\therefore r$ |
| *Constructive dilemma* | $p \vee r$ <br> $p \rightarrow q$ <br> $r \rightarrow s$ <br> $\therefore q \vee s$ | *Destructive dilemma* | $\neg q \vee \neg s$ <br> $p \rightarrow q$ <br> $r \rightarrow s$ <br> $\therefore \neg p \vee \neg r$ |
| *Conjunctive addition* | $p$ <br> $q$ <br> $\therefore p \wedge q$ | *Conditional proof* | $p$ <br> $p \wedge q \rightarrow r$ <br> $\therefore q \rightarrow r$ |
| *Conjunctive simplification* | $p \wedge q$ <br> $\therefore p$ | *Rule of contradiction* | given the contradiction $c$ <br> $\neg p \rightarrow c$ <br> $\therefore p$ |

**3.** The following table gives rules of inference for arguments with quantifiers.

| name | argument form |
|------|---------------|
| *Universal instantiation* | $(\forall x \in D)\, Q(x)$ |
| | $\therefore Q(a)$  (*a* any particular element of *D*) |
| *Generalizing from the* | $Q(a)$  (*a* an arbitrarily chosen element of *D*) |
| *generic particular* | $\therefore (\forall x \in D)\, Q(x)$ |
| *Existential specification* | $(\exists x \in D)\, Q(x)$ |
| | $\therefore Q(a)$  (for at least one $a \in D$) |
| *Existential generalization* | $Q(a)$  (for at least one element $a \in D$) |
| | $\therefore (\exists x \in D)\, Q(x)$ |

**4.** Substituting $R(x) \to S(x)$ in place of $Q(x)$ and $z$ in place of $x$ in generalizing from the generic particular gives the following inferential rule:

*Universal modus*      $R(a) \to S(a)$ for any particular but arbitrarily chosen $a \in D$

*ponens:*      $\therefore (\forall z \in D)\,[R(z) \to S(z)].$

**5.** The rule of generalizing from the generic particular determines the outline of most mathematical proofs.

**6.** The rule of existential specification is used in deductive reasoning to give names to quantities that are known to exist but whose exact values are unknown.

**7.** A useful strategy for determining whether a statement is true is to first try to prove it using a variety of approaches and proof methods. If this is unsuccessful, the next step may be to try to disprove the statement, such as by trying to construct or prove the existence of a counterexample. If this does not work, the next step is to try to prove the statement again, and so on. This is one of the many ways in which many mathematicians attempt to develop new results.

**Examples:**

**1.** Suppose that $D$ is the set of all objects in the physical universe, $P(x)$ is "$x$ is a human being", $Q(x)$ is "$x$ is mortal", and $a$ is the Greek philosopher Socrates.

| *argument form* | *an argument of that form* |
|-----------------|----------------------------|
| $(\forall x \in D)\,[P(x) \to Q(x)]$ | $\forall$ objects $x$, ($x$ is a human being) $\to$ ($x$ is mortal). |
| | (*informally*: All human beings are mortal.) |
| $P(a)$ (for particular $a \in D$) | Socrates is a human being. |
| $\therefore Q(a)$ | $\therefore$Socrates is mortal. |

**2.** The argument form shown below is invalid: there is an argument of this form (shown next to it) that has true premises and a false conclusion.

| *argument form* | *an argument of that form* |
|-----------------|----------------------------|
| $(\forall x \in D)\,[P(x) \to Q(x)]$ | $\forall$ objects $x$, ($x$ is a human being) $\to$ ($x$ is mortal). |
| | (*informally*: All human beings are mortal.) |
| $Q(a)$ (for particular $a \in D$) | My cat Bunbury is mortal. |
| $\therefore P(a)$ | $\therefore$My cat Bunbury is a human being. |

In this example, $D$ is the set of all objects in the physical universe, $P(x)$ is "$x$ is a human being", $Q(x)$ is "$x$ is mortal", and $a$ is my cat Bunbury.

**3.** The distributive law for real numbers, $(\forall a, b, c \in \mathcal{R})[ac + bc = (a + b)c]$, implies that $2\sqrt{2} + 3\sqrt{2} = (2 + 3)\sqrt{2}$ (because 2, 3, and $\sqrt{2}$ are particular real numbers).

**4.** Since 2 is a prime number that is not odd, the rule of existential generalization implies the truth of the statement "$\exists$ a prime number $n$ such that $n$ is not odd".

**5.** To prove that the square of every even integer is even, by the rule of generalizing from the generic particular, begin by supposing that $n$ is any particular but arbitrarily chosen even integer. The job of the proof is to deduce that $n^2$ is even.

**6.** By definition, every even integer equals twice some integer. So if at some stage of a reasoning process there is a particular even integer $n$, it follows from the rule of existential specification that $n = 2k$ for *some* integer $k$ (even though the numerical values of $n$ and $k$ may be unknown).

---

## 1.5.2   PROOFS

### Definitions:

A (*logical*) *proof* of a statement is a finite sequence of statements (called the **steps** of the proof) leading from a set of premises to the given statement. Each step of the proof must either be a premise or follow from some previous steps by a valid rule of inference.

In a **mathematical proof**, the set of premises may contain any item of previously proved or agreed upon mathematical knowledge (definitions, axioms, theorems, etc.) as well as the specific hypotheses of the statement to be proved.

A **direct proof** of a statement of the form $p \rightarrow q$ is a proof that assumes $p$ to be true and then shows that $q$ is true.

An **indirect proof** of a statement of the form $p \rightarrow q$ is a proof that assumes that $\neg q$ is true and then shows that $\neg p$ is true. That is, a proof of this form is a direct proof of the contrapositive $\neg q \rightarrow \neg p$.

A **proof by contradiction** assumes the negation of the statement to be proved and shows that this leads to a contradiction.

### Facts:

**1.** A useful strategy to determine if a statement of the form $(\forall x \in D)[P(x) \rightarrow Q(x)]$ is true or false is to imagine an element $x \in D$ that satisfies $P(x)$ and, using this assumption (and other facts), investigate whether $x$ must also satisfy $Q(x)$. If the answer for all such $x$ is "yes", the given statement is true and the result of the investigation is a direct proof. If it is possible to find an $x \in D$ for which $Q(x)$ is false, the statement is false and this value of $x$ is a counterexample. If the investigation shows that is not possible to find an $x \in D$ for which $Q(x)$ is false, the given statement is true and the result of the investigation is a proof by contradiction.

**2.** There are many types of techniques that can be used to prove theorems. The following table describes how to approach proofs of various types of statements.

| *statement* | *technique of proof* |
|---|---|
| $p \to q$ | *Direct proof*: Assume that $p$ is true. Use rules of inference and previously accepted axioms, definitions, theorems, and facts to deduce that $q$ is true. |
| $(\forall x \in D)P(x)$ | *Direct proof*: Suppose that $x$ is an arbitrary element of $D$. Use rules of inference and previously accepted axioms, definitions, and facts to deduce that $P(x)$ is true. |
| $(\exists x \in D)P(x)$ | *Constructive direct proof*: Use rules of inference and previously accepted axioms, definitions, and facts to actually find an $x \in D$ for which $P(x)$ is true. |
|  | *Nonconstructive direct proof*: Deduce the existence of $x$ from other mathematical facts without a description of how to compute it. |
| $(\forall x \in D)(\exists y \in E)P(x, y)$ | *Constructive direct proof*: Assume that $x$ is an arbitrary element of $D$. Use rules of inference and previously accepted axioms, definitions, and facts to show the existence of a $y \in E$ for which $P(x, y)$ is true, in such a way that $y$ can be computed as a function of $x$. |
|  | *Nonconstructive direct proof*: Assume $x$ is an arbitrary element of $D$. Deduce the existence of $y$ from other mathematical facts without a description of how to compute it. |
| $p \to q$ | *Proof by cases*: Suppose $p \equiv p_1 \vee \cdots \vee p_k$. Prove that each conditional $p_i \to q$ is true. The basis for division into cases is the logical equivalence $[(p_1 \vee \cdots \vee p_k) \to q] \equiv [(p_1 \to q) \wedge \cdots \wedge (p_k \to q)]$. |
| $p \to q$ | *Indirect proof or proof by contraposition*: Assume that $\neg q$ is true (that is, assume that $q$ is false). Use rules of inference and previously accepted axioms, definitions, and facts to show that $\neg p$ is true (that is, $p$ is false). |
| $p \to q$ | *Proof by contradiction*: Assume that $p \to q$ is false (that is, assume that $p$ is true and $q$ is false). Use rules of inference and previously accepted axioms, definitions, and facts to show that a contradiction results. This means that $p \to q$ cannot be false, and hence must be true. |
| $(\exists x \in D)P(x)$ | *Proof by contradiction*: Assume that there is no $x \in D$ for which $P(x)$ is true. Show that a contradiction results. |
| $(\forall x \in D)P(x)$ | *Proof by contradiction*: Assume that there is some $x \in D$ for which $P(x)$ is false. Show that a contradiction results. |
| $p \to (q \vee r)$ | *Proof of a disjunction*: Prove that one of its logical equivalences $(p \wedge \neg q) \to r$ or $(p \wedge \neg r) \to q$ is true. |
| $p_1, \ldots, p_k$ are equivalent | *Proof by cycle of implications*: Prove $p_1 \to p_2$, $p_2 \to p_3$, $\ldots$, $p_{k-1} \to p_k$, $p_k \to p_1$. This is equivalent to proving $(p_1 \to p_2) \wedge (p_2 \to p_3) \wedge \cdots \wedge (p_{k-1} \to p_k) \wedge (p_k \to p_1)$. |

## Examples:

**1.** In the following direct proof (see the table in Fact 2, item 2), the domain $D$ is the set of all pairs of integers, $x$ is $(m, n)$, and the predicate $P(m, n)$ is "if $m$ and $n$ are even, then $m + n$ is even".

*Theorem*: For all integers $m$ and $n$, if $m$ and $n$ are even, then $m + n$ is even.

*Proof*: Suppose $m$ and $n$ are arbitrarily chosen even integers. [$m + n$ *must be shown to be even.*]

1.    $m = 2r$, $n = 2s$ for some integers $r$ and $s$    (by definition of even)
2.    $m + n = 2r + 2s$    (by substitution)
3.    $m + n = 2(r + s)$    (by factoring out the 2)
4.    $r + s$ is an integer    (it is a sum of two integers)
5.  $\therefore m + n$ is even    (by definition of even)

The following partial expansion of the proof shows how some of the steps are justified by rules of inference combined with previous mathematical knowledge:

1.  Every even integer equals twice some integer:
     [$\forall$ even $x \in \mathcal{Z}$  $(x = 2y$ for some $y \in \mathcal{Z})$]
         $m$ is a particular even integer.
     $\therefore m = 2r$ for some integer $r$.

3.  Every integer is a real number: [$\forall n \in \mathcal{Z}$  $(n \in \mathcal{R})$]
         $r$ and $s$ are particular integers.
     $\therefore r$ and $s$ are real numbers.
      The distributive law holds for real numbers:  [$\forall a, b, c \in \mathcal{R}$  $(ab + ac = a(b + c))$]
         2, $r$, and $s$ are particular real numbers.
     $\therefore 2r + 2s = 2(r + s)$.

4.  Any sum of two integers is an integer:  [$\forall m, n \in \mathcal{Z}$  $(m + n \in \mathcal{Z})$]
         $r$ and $s$ are particular integers.
     $\therefore r + s$ is an integer.

5.  Any integer that equals twice some integer is even:
     [$\forall x \in \mathcal{Z}$  (if $x = 2y$ for some $y \in \mathcal{Z}$, then $x$ is even.)]
         $2(r + s)$ equals twice the integer $r + s$.
     $\therefore 2(r + s)$ is even.

2.  *A constructive existence proof*:

   *Theorem*:  Given any integer $n$, there is an integer $m$ with $m > n$.

   *Proof*: Suppose that $n$ is an integer. Let $m = n + 1$. Then $m$ is an integer and $m > n$.

The proof is constructive because it established the existence of the desired integer $m$ by showing that its value can be computed by adding 1 to the value of $n$.

3.  *A nonconstructive existence proof*:

   *Theorem*:  Given a nonnegative integer $n$, there is always a prime number $p$ that is greater than $n$.

   *Proof*: Suppose that $n$ is a nonnegative integer. Consider $n! + 1$. Then $n! + 1$ is divisible by some prime number $p$ because every integer greater than 1 is divisible by a prime number, and $n! + 1 > 1$. Also, $p > n$ because when $n! + 1$ is divided by any positive integer less than or equal to $n$, the remainder is 1 (since any such number is a factor of $n!$).

The proof is a nonconstructive existence proof because it demonstrated the existence of the number $p$, but it offered no computational rule for finding it.

4.  *A proof by cases*:

Theorem: For all odd integers $n$, the number $n^2 - 1$ is divisible by 8.

Proof: Suppose $n$ is an odd integer. When $n$ is divided by 4, the remainder is 0, 1, 2, or 3. Hence $n$ has one of the four forms $4k$, $4k+1$, $4k+2$, or $4k+3$ for some integer $k$. But $n$ is odd. So $n \neq 4k$ and $n \neq 4k+2$. Thus either $n = 4k+1$ or $n = 4k+3$ for some integer $k$.

Case 1 [$n = 4k+1$ for some integer $k$]:  In this case $n^2 - 1 = (4k+1)^2 - 1 = 16k^2 + 8k + 1 - 1 = 16k^2 + 8k = 8(2k^2 + k)$, which is divisible by 8 because $2k^2 + k$ is an integer.

Case 2 [$n = 4k+3$ for some integer $k$]:  In this case $n^2 - 1 = (4k+3)^2 - 1 = 16k^2 + 24k + 9 - 1 = 16k^2 + 24k + 8 = 8(2k^2 + 3k + 1)$, which is divisible by 8 because $2k^2 + 3k + 1$ is an integer.

So in either case $n^2 - 1$ is divisible by 8, and thus the given statement is proved.

5.  *A proof by contraposition*:

Theorem: For all integers $n$, if $n^2$ is even, then $n$ is even.

Proof: Suppose that $n$ is an integer that is not even. Then when $n$ is divided by 2 the remainder is 1, or, equivalently, $n = 2k+1$ for some integer $k$. By substitution, $n^2 = (2k+1)^2 = 4k^2 + 4k + 1 = 2(2k^2 + 2k) + 1$. It follows that when $n^2$ is divided by 2 the remainder is 1 (because $2k^2 + 2k$ is an integer). Thus, $n^2$ is not even.

In this proof by contraposition, a direct proof was given of the contrapositive "if $n$ is not even, then $n^2$ is not even".

6.  *A proof by contradiction*:

Theorem: $\sqrt{2}$ is irrational.

Proof: Suppose not; that is, suppose that $\sqrt{2}$ were a rational number. By definition of rational, there would exist integers $a$ and $b$ such that $\sqrt{2} = \frac{a}{b}$, or, equivalently, $2b^2 = a^2$. Now the prime factorization of the left-hand side of this equation contains an odd number of factors and that of the right-hand side contains an even number of factors (because every prime factor in an integer occurs twice in the prime factorization of the square of that integer). But this is impossible because the prime factorization of every integer is unique. This yields a contradiction, which shows that the original supposition was false. Hence $\sqrt{2}$ is irrational.

7.  *A proof by cycle of implications*:

Theorem: For all positive integers $a$ and $b$, the following statements are equivalent:

(1)  $a$ is a divisor of $b$;

(2)  the greatest common divisor of $a$ and $b$ is $a$;

(3)  $\lfloor \frac{b}{a} \rfloor = \frac{b}{a}$.

Proof : Let $a$ and $b$ be positive integers.

(1) $\rightarrow$ (2): Suppose that $a$ is a divisor of $b$. Since $a$ is also a divisor of $a$, $a$ is a common divisor of $a$ and $b$. But no integer greater than $a$ is a divisor of $a$. So the greatest common divisor of $a$ and $b$ is $a$.

(2) $\rightarrow$ (3): Suppose that the greatest common divisor of $a$ and $b$ is $a$. Then $a$ is a divisor of both $a$ and $b$, so $b = ak$ for some integer $k$. Then $\frac{b}{a} = k$, an integer, and so by definition of floor, $\lfloor \frac{b}{a} \rfloor = k = \frac{b}{a}$.

(3) → (1): Suppose that $\lfloor \frac{b}{a} \rfloor = \frac{b}{a}$. Let $k = \lfloor \frac{b}{a} \rfloor$. Then $k = \lfloor \frac{b}{a} \rfloor = \frac{b}{a}$, and $k$ is an integer by definition of floor. Multiplying the outer parts of the equality by $a$ gives $b = ak$, so by definition of divisibility, $a$ is a divisor of $b$.

8. *A proof of a disjunction*:

   *Theorem*: For all integers $a$ and $p$, if $p$ is prime, then either $p$ is a divisor of $a$, or $a$ and $p$ have no common factor greater than 1.

   *Proof*: Suppose $a$ and $p$ are integers and $p$ is prime, but $p$ is not a divisor of $a$. Since $p$ is prime, its only positive divisors are 1 and $p$. So, since $p$ is not a divisor of $a$, the only possible positive common divisor of $a$ and $p$ is 1. Hence $a$ and $p$ have no common divisor greater than 1.

---

## 1.5.3  DISPROOFS

**Definitions:**

A **disproof** of a statement is a proof that the statement is false.

A **counterexample** to a statement of the form $(\forall x \in D)P(x)$ is an element $b \in D$ for which $P(b)$ is false.

**Facts:**

1. The method of disproof by counterexample is based on the following fact:

$$\neg[(\forall x \in D)\,P(x)] \;\Leftrightarrow\; (\exists x \in D)\,[\neg P(x)].$$

2. The following table describes how to give various types of disproofs.

| statement | technique of disproof |
|---|---|
| $(\forall x \in D)P(x)$ | *Constructive disproof by counterexample*: Exhibit a specific $a \in D$ for which $P(a)$ is false. |
| $(\forall x \in D)P(x)$ | *Existence disproof*: Prove the existence of some $a \in D$ for which $P(a)$ is false. |
| $(\exists x \in D)P(x)$ | Prove that there is no $a \in D$ for which $P(a)$ is true. |
| $(\forall x \in D)\,[P(x) \to Q(x)]$ | Find an element $a \in D$ with $P(a)$ true and $Q(a)$ false. |
| $(\forall x \in D)(\exists y \in E)\,P(x,y)$ | Find an element $a \in D$ with $P(a,y)$ false for every $y \in E$. |
| $(\exists x \in D)(\forall y \in E)\,P(x,y)$ | Prove that there is no $a \in D$ for which $P(a,y)$ is true for every possible $y \in E$. |

**Examples:**

1. The statement $(\forall a,b \in \mathcal{R})\,[a^2 < b^2 \to a < b]$ is disproved by the following counterexample: $a = 2$, $b = -3$. Then $a^2 < b^2$ (because $4 < 9$) but $a \not< b$ (because $2 \not< -3$).

2. The statement "every prime number is odd" is disproved by exhibiting the counterexample $n = 2$, since $n$ is prime and not odd.

## 1.5.4   MATHEMATICAL INDUCTION

### Definitions:

The **principle of mathematical induction (weak form)** is the following rule of inference for proving that all the items in a list $x_0, x_1, x_2, \ldots$ have some property $P(x)$:

| | |
|---|---|
| $P(x_0)$ is true | *basis premise* |
| $(\forall k \geq 0)$ [if $P(x_k)$ is true, then $P(x_{k+1})$ is true] | *induction premise* |
| $\therefore (\forall n \geq 0)[P(x_n)$ is true]. | *conclusion* |

The antecedent $P(x_k)$ in the induction premise "if $P(x_k)$ is true, then $P(x_{k+1})$ is true" is called the **induction hypothesis**.

The **basis step** of a proof by mathematical induction is a proof of the basis premise.

The **induction step** of a proof by mathematical induction is a proof of the induction premise.

The **principle of mathematical induction (strong form)** is the following rule of inference for proving that all the items in a list $x_0, x_1, x_2, \ldots$ have some property $P(x)$:

| | |
|---|---|
| $P(x_0)$ is true | *basis premise* |
| $(\forall k \geq 0)$ [if $P(x_0), P(x_1), \ldots, P(x_k)$ are all true, then $P(x_{k+1})$ is true] | *(strong) induction premise* |
| $\therefore (\forall n \geq 0)[P(x_n)$ is true]. | *conclusion* |

The **well-ordering principle for the integers** is the following axiom:   If $S$ is a nonempty set of integers such that every element of $S$ is greater than some fixed integer, then $S$ contains a least element.

### Facts:

**1.** Typically, the principle of mathematical induction is used to prove that one of the following sequences of statements is true: $P(0), P(1), P(2), \ldots$ or $P(1), P(2), P(3), \ldots$. In these cases the principle of mathematical induction has the form:   if $P(0)$ is true and $P(n) \to P(n+1)$ is true for all $n \geq 0$, then $P(n)$ is true for all $n \geq 0$; or if $P(1)$ is true and $P(n) \to P(n+1)$ is true for all $n \geq 1$, then $P(n)$ is true for all $n \geq 1$.

**2.** If the truth of $P(n+1)$ can be obtained from the previous statement $P(n)$, the weak form of the principle of mathematical induction can be used. If the truth of $P(n+1)$ requires the use of one or more statements $P(k)$ for $k \leq n$, then the strong form should be used.

**3.** Mathematical induction can also be used to prove statements that can be phrased in the form "For all integers $n \geq k$, $P(n)$ is true".

**4.** Mathematical induction can often be used to prove summation formulas and inequalities.

**5.** There are alternative forms of mathematical induction, such as the following:

- if $P(0)$ and $P(1)$ are true, and if $P(n) \to P(n+2)$ is true for all $n \geq 0$, then $P(n)$ is true for all $n \geq 0$;

- if $P(0)$ and $P(1)$ are true, and if $[P(n) \wedge P(n+1)] \to P(n+2)$ is true for all $n \geq 0$, then $P(n)$ is true for all $n \geq 0$.

**6.** The weak form of the principle of mathematical induction, the strong form of the principle of mathematical induction, and the well-ordering principle for the integers are all regarded as axioms for the integers. This is because they cannot be derived from the usual simpler axioms used in the definition of the integers. (See the Peano definition of the natural numbers in §1.2.3.)

**7.** The weak form of the principle of mathematical induction, the strong form of the principle of mathematical induction, and the well-ordering principle for the integers are all equivalent. In other words, each of them can be proved from each of the others.

**8.** The earliest recorded use of mathematical induction occurs in 1575 in the book *Arithmeticorum Libri Duo* by Francesco Maurolico, who used the principle to prove that the sum of the first $n$ odd positive integers is $n^2$.

### Examples:

**1.** *A proof using the weak form of mathematical induction:* (In this proof, $x_0, x_1, x_2, \ldots$ is the sequence $1, 2, 3, \ldots$, and the property $P(x_n)$ is the equation $1+2+\cdots+n = \frac{n(n+1)}{2}$.)

> *Theorem:* For all integers $n \geq 1$, $1 + 2 + \cdots + n = \frac{n(n+1)}{2}$.
>
> *Proof:*
>
> *Basis Step:*   For $n = 1$ the left-hand side of the formula is 1, and the right-hand side is $\frac{1(1+1)}{2}$, which is also equal to 1. Hence $P(1)$ is true.
>
> *Induction Step:*   Let $k$ be an integer, $k \geq 1$, and suppose that $P(k)$ is true. That is, suppose that $1+2+\cdots+k = \frac{k(k+1)}{2}$ holds (the induction hypothesis). It must be shown that $P(k+1)$ is true: $1 + 2 + \cdots + (k+1) = \frac{(k+1)((k+1)+1)}{2}$, or, equivalently, that $1 + 2 + \cdots + (k+1) = \frac{(k+1)(k+2)}{2}$. But, by substitution from the induction hypothesis,
>
> $$1 + 2 + \cdots + (k+1) = (1 + 2 + \cdots + k) + (k+1)$$
> $$= \frac{k(k+1)}{2} + (k+1)$$
> $$= \frac{(k+1)(k+2)}{2}.$$
>
> Thus, $1 + 2 + \cdots + (k+1) = \frac{(k+1)(k+2)}{2}$ and so $P(k+1)$ is true.

**2.** *A proof using the weak form of mathematical induction:*

> *Theorem:* For all integers $n \geq 4$, $2^n < n!$.
>
> *Proof:*
>
> *Basis Step:*   For $n = 4$, $2^4 < 4!$ is true since $16 < 24$.
>
> *Induction Step:*   Let $k$ be an integer, $k \geq 4$, and suppose that $2^k < k!$ is true. The following shows that $2^{k+1} < (k+1)!$ must also be true:
>
> $$2^{k+1} = 2 \cdot 2^k < 2 \cdot k! < (k+1)k! = (k+1)!.$$

**3.** *A proof using the weak form of mathematical induction:*

> *Theorem:* For all integers $n \geq 8$, $n$ cents in postage can be made using only 3-cent and 5-cent stamps.
>
> *Proof:*   Let $P(n)$ be the predicate "$n$ cents postage can be made using only 3-cent and 5-cent stamps".
>
> *Basis Step:*   $P(8)$ is true since 8 cents in postage can be made using one 3-cent stamp and one 5-cent stamp.

*Induction Step*: Let $k$ be an integer, $k \geq 8$, and suppose that $P(k)$ is true. The following shows that $P(k+1)$ must also be true. If the pile of stamps for $k$ cents postage has in it any 5-cent stamps, then remove one 5-cent stamp and replace it with two 3-cent stamps. If the pile for $k$ cents postage has only 3-cent stamps, there must be at least three 3-cent stamps in the pile (since $k \neq 3$ or 6). Remove three 3-cent stamps and replace them with two 5-cent stamps. In either case, a pile of stamps for $k+1$ cents postage results.

4. *A proof using an alternative form of mathematical induction* (Fact 5):

   *Theorem*: For all integers $n \geq 0$, $F_n < 2^n$. ($F_k$ are Fibonacci numbers; see §3.1.2.)

   *Proof*: Let $P(n)$ be the predicate "$F_n < 2^n$".

   *Basis Step*: $P(0)$ and $P(1)$ are both true since $F_0 = 0 < 1 = 2^0$ and $F_1 = 1 < 2 = 2^1$.

   *Induction Step*: Let $k$ be an integer, $k \geq 0$, and suppose that $P(k)$ and $P(k+1)$ are true. Then $P(k+2)$ is also true: $F_{k+2} = F_k + F_{k+1} < 2^k + 2^{k+1} < 2^{k+1} + 2^{k+1} = 2 \cdot 2^{k+1} = 2^{k+2}$.

5. *A proof using the strong form of mathematical induction*:

   *Theorem*: Every integer $n \geq 2$ is divisible by some prime number.

   *Proof*: Let $P(n)$ be the sentence "$n$ is divisible by some prime number".

   *Basis Step*: Since 2 is divisible by 2 and 2 is a prime number, $P(2)$ is true.

   *Induction Step*: Let $k$ be an integer with $k > 2$, and suppose that $P(i)$ is true (the induction hypothesis) for all integers $i$ with $2 \leq i < k$. That is, suppose for all integers $i$ with $2 \leq i < k$ that $i$ is divisible by a prime number. (*It must now be shown that $k$ is divisible by a prime number.*)

   Now either the number $k$ is prime or $k$ is not prime. If $k$ is prime, then $k$ is divisible by a prime number, namely itself. If $k$ is not prime, then $k = a \cdot b$ where $a$ and $b$ are integers, with $2 \leq a < k$ and $2 \leq b < k$. By the induction hypothesis, the number $a$ is divisible by a prime number $p$, and so $k = ab$ is also divisible by that prime $p$. Hence, regardless of whether $k$ is prime or not, $k$ is divisible by a prime number.

6. *A proof using the well-ordering principle*:

   *Theorem*: Every integer $n \geq 2$ is divisible by some prime number.

   *Proof*: Suppose, to the contrary, that there exists an integer $n \geq 2$ that is divisible by no prime number. Thus, the set $S$ of all integers $\geq 2$ that are divisible by no prime number is nonempty. Of course, no number in $S$ is prime, since every number is divisible by itself.

   By the well-ordering principle for the integers, the set $S$ contains a least element $k$. Since $k$ is not prime, there must exist integers $a$ and $b$ with $2 \leq a < k$ and $2 \leq b < k$, such that $k = a \cdot b$. Moreover, since $k$ is the least element of the set $S$ and since both $a$ and $b$ are smaller than $k$, it follows that neither $a$ nor $b$ is in $S$. Hence, the number $a$ (in particular) must be divisible by some prime number $p$. But then, since $a$ is a factor of $k$, the number $k$ is also divisible by $p$, which contradicts the fact that $k$ is in $S$. This contradiction shows that the original supposition is false, or, in other words, that the theorem is true.

7. *A proof using the well-ordering principle*:

*Theorem*: Every decreasing sequence of nonnegative integers is finite.

*Proof*: Suppose $a_1, a_2, \ldots$ is a decreasing sequence of nonnegative integers: $a_1 > a_2 > \cdots$. By the well-ordering principle, the set $\{a_1, a_2, \ldots\}$ contains a least element $a_n$. This number must be the last in the sequence (and hence the sequence is finite). If $a_n$ is not the last term, then $a_{n+1} < a_n$, which contradicts the fact that $a_n$ is the smallest element.

---

### 1.5.5   DIAGONALIZATION ARGUMENTS

**Definitions:**

The **diagonal** of an infinite list of sequences $s_1, s_2, s_3, \ldots$ is the infinite sequence whose $j$th element is the $j$th entry of sequence $s_j$.

A **diagonalization proof** is any proof that involves the diagonal of a list of sequences, or something analogous to this.

**Facts:**

**1.** A diagonalization argument can be used to prove the existence of nonrecursive functions.

**2.** A diagonalization argument can be used to prove that no computer algorithm can ever be developed to determine whether an arbitrary computer program, given as input with a given set of data, will terminate (*the Turing Halting Problem*).

**3.** A diagonalization argument can be used to prove that every mathematical theory (under certain reasonable hypotheses) will contain statements whose truth or falsity is impossible to determine within the theory (*Gödel's Incompleteness Theorem*).

**Example:**

**1.** *A diagonalization proof*:

*Theorem*: The set of real numbers between 0 and 1 is uncountable. (Georg Cantor, 1845–1918).

*Proof*: Suppose, to the contrary, that the set of real numbers between 0 and 1 is countable. The decimal representations of these numbers can be written in a list as follows:

$$0.a_{11}a_{12}a_{13} \ldots a_{1n} \ldots$$
$$0.a_{21}a_{22}a_{23} \ldots a_{2n} \ldots$$
$$0.a_{31}a_{32}a_{33} \ldots a_{3n} \ldots$$
$$\vdots$$
$$0.a_{n1}a_{n2}a_{n3} \ldots a_{nn} \ldots$$

From this list, construct a new decimal number $0.b_1 b_2 b_3 \ldots b_n \ldots$ by specifying that

$$b_i = \begin{cases} 5 & \text{if } a_{ii} \neq 5 \\ 6 & \text{if } a_{ii} = 5. \end{cases}$$

For each integer $i \geq 1$, $0.b_1 b_2 b_3 \ldots b_n \ldots$ differs from the $i$th number in the list in the $i$th decimal place, and hence $0.b_1 b_2 b_3 \ldots b_n \ldots$ is not in the list. Consequently, no such listing of all real numbers between 0 and 1 is possible, and hence, the set of real numbers between 0 and 1 is uncountable.

# 1.6  AXIOMATIC PROGRAM VERIFICATION

Axiomatic program verification is used to prove that a sequence of programming instructions achieves its specified objective. Semantic axioms for the programming language constructs are used in a formal logic argument as rules of inference. Comments called *assertions*, within the sequence of instructions, provide the main details of the argument. The presently high expense of creating verified software can be justified for code that is frequently reused, where the financial benefit is otherwise adequately large, or where human life is concerned, for instance, in airline traffic control. This section presents a representative sample of axioms for typical programming language constructs.

## 1.6.1  ASSERTIONS AND SEMANTIC AXIOMS

The correctness of a program can be argued formally based on a set of semantic axioms that define the behavior of individual programming language constructs [Ap81], [Fl67], [Ho69]. (Some alternative proofs of correctness use denotational semantics [Sc86], [St77] or operational semantics [We72].) In addition, it is possible to synthesize code, using techniques that permit the axioms to guide the selection of appropriate instructions [Di76], [Gr13]. Code specifications and intermediate conditions are expressed in the form of program assertions.

### Definitions:

An **assertion** is a program comment containing a logical statement that constrains the values of the computational variables. These constraints are expected to hold when execution flow reaches the location of the assertion.

A **semantic axiom** for a type of programming instruction is a rule of inference that prescribes the change of value of the variables of computation caused by the execution of that type of instruction.

The assertion **false** represents an inconsistent set of logical conditions. A computer program cannot meet such a specification.

Given two constraints $A$ and $B$ on computational variables, a statement that $B$ follows from $A$ purely for reasons of logic and/or mathematics is called a **logical implication**.

The **postcondition** for an instruction or program fragment is the assertion that immediately follows it in the program.

The **precondition** for an instruction or program fragment is the assertion that immediately precedes it in the program.

The assertion **true** represents the empty set of logical conditions.

### Notation:

**1.** To say that whenever the precondition {Apre} holds, the execution of a program fragment called "Code" will cause the postcondition {Apost} to hold, the following notation styles can be used:

- *Horizontal notation:* {Apre} Code {Apost}

- *Vertical notation:* {Apre}
  Code
  {Apost}
- Flowgraph notation:

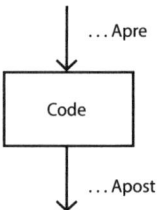

**2.** Curly braces { ... } enclose assertions in generic program code. They do not denote a set.

**3.** Semantic axioms have a finite list of premises and a conclusion. They are represented in the following format:

$$\{\text{Premise 1}\}$$
$$\vdots$$
$$\{\text{Premise } n\}$$
- - - - - - - - - -
$$\{\text{Conclusion}\}$$

**4.** The circumstance that $A$ logically implies $B$ is denoted $A \Rightarrow B$.

---

### 1.6.2    NOP, ASSIGNMENT, AND SEQUENCING AXIOMS

Formal axioms of pure mathematical consequence (no operation, from a computational perspective) and of straight-line sequential flow are used as auxiliaries to verify correctness, even of sequences of simple assignment statements.

**Definitions:**
A **NOP** ("no-op") is a (possibly empty) program fragment whose execution does not alter the state of any computational variables or the sequence of flow.

The **Axiom of NOP** states

$$\{\text{Apre}\} \Rightarrow \{\text{Apost}\} \qquad \text{Premise 1}$$
- - - - - - - - - - - - - -
$$\{\text{Apre}\}\ \text{NOP}\ \{\text{Apost}\} \qquad \text{Conclusion}$$

*Note*: The Axiom of NOP is frequently applied to empty program fragments in order to facilitate a clear logical argument.

An **assignment instruction** $X := E$; means that the variable $X$ is to be assigned the value of the expression $E$.

In a logical assertion $A(X)$ with possible instances of the program variable $X$, the **result of replacing each instance** of $X$ in $A$ by the program expression $E$ is denoted $A(X \leftarrow E)$.

The **Axiom of Assignment** states

$$\{true\} \qquad\qquad\qquad\qquad \text{No premises}$$

- - - - - - - - - - - - -

$$\{A(X \leftarrow E)\}X := E; \{A(X)\} \quad \text{Conclusion}$$

The following **Axiom of Sequence** provides that two consecutive instructions in the program code are executed one immediately after the other:

$$\{\text{Apre}\} \text{ Code1 } \{\text{Amid}\} \qquad \text{Premise 1}$$
$$\{\text{Amid}\} \text{ Code2 } \{\text{Apost}\} \qquad \text{Premise 2}$$

- - - - - - - - - - - - - - -

$$\{\text{Apre}\} \text{ Code1, Code2 } \{\text{Apost}\} \quad \text{Conclusion}$$

(Commas are used as separators in program code.)

**Examples:**

1. *Example of NOP:* Suppose that $X$ is a numeric program variable.

$$\{X = 3\} \Rightarrow \{X > 0\} \qquad \text{mathematical fact}$$

- - - - - - - - - - - - - - -

$$\{X = 3\} \text{ NOP } \{X > 0\} \quad \text{by Axiom of NOP}$$

2. Suppose that $X$ and $Y$ are integer-type program variables. The Axiom of Assignment alone implies correctness of all the following examples:

(a)  $\{X = 4\} \quad X := X * 2; \quad \{X = 8\}$
$A(X)$ is $\{X = 8\}$; $E$ is $X * 2$; $A(X \leftarrow E)$ is $\{X * 2 = 8\}$, which is equivalent to $\{X = 4\}$.

(b)  $\{true\} \quad X := 2; \quad \{X = 2\}$
$A(X)$ is $\{X = 2\}$; $E$ is $2$; $A(X \leftarrow E)$ is $\{2 = 2\}$, which is equivalent to $\{true\}$.

(c)  $\{(-9 < X) \wedge (X < 0)\} \quad Y := X; \quad \{(-9 < Y) \wedge (Y < 0)\}$
$A(Y)$ is $\{(-9 < Y) \wedge (Y < 0)\}$; $E$ is $X$; $A(Y \leftarrow E)$ is $\{(-9 < X) \wedge (X < 0)\}$.

(d)  $\{Y = 1\} \quad X := 0; \quad \{Y = 1\}$
$A(X)$ is $\{Y = 1\}$; $E$ is $0$; $A(X \leftarrow E)$ is $\{Y = 1\}$.

(e)  $\{false\} \quad X := 8; \quad \{X = 2\}$
$A(X)$ is $\{X = 2\}$; $E$ is $8$; $A(X \leftarrow E)$ is $\{8 = 2\}$, which is equivalent to $\{false\}$.

3. *Examples of sequence:*

(a)  $\{X = 1\} X := X + 1; \{X > 0\}$

   i.   $\{X = 1\} \Rightarrow \{X > -1\}$           mathematics

   ii.  $\{X = 1\} \ NOP \ \{X > -1\}$       Axiom of NOP

  iii. $\{X > -1\} X := X + 1; \{X > 0\}$    Axiom of Assignment

  iv. $\{X = 1\} \ NOP, \ X := X + 1; \ \{X > 0\}$    Axiom of Sequence ii, iii

   v.  $\{X = 1\} X := X + 1; \{X > 0\}$      definition of NOP.

(b)  $\{Y = a \wedge X = b\} Z := Y; \ Y := X; \ X := Z; \ \{X = a \wedge Y = b\}$

   i.   $\{Y = a \wedge X = b\} Z := Y; \ \{Z = a \wedge X = b\}$   Axiom of Assignment

   ii.  $\{Z = a \wedge X = b\} Y := X; \ \{Z = a \wedge Y = b\}$   Axiom of Assignment

  iii. $\{Y = a \wedge X = b\} Z := Y, \ Y := X,$          Axiom of Sequence
        $\{Z = a \wedge Y = b\}$                       i, ii

  iv. $\{Z = a \wedge Y = b\} X := Z; \ \{X = a \wedge Y = b\}$   Axiom of Assignment

   v.  $\{Y = a \wedge X = b\} Z := Y, \ Y := X, \ X := Z,$   Axiom of Sequence
        $\{X = a \wedge Y = b\}$                       iii, iv

### 1.6.3   AXIOMS FOR CONDITIONAL EXECUTION CONSTRUCTS

**Definitions:**

A **conditional assignment construct** is any type of program instruction containing a logical condition and an imperative clause such that the imperative clause is to be executed if and only if the logical condition is true. Some types of conditional assignment contain more than one logical condition and more than one imperative clause.

An **if-then** instruction **if IfCond then ThenCode** has one logical condition (which follows the keyword if) and one imperative clause (which follows the keyword then).

The **Axiom of If-then** states

$$\{Apre \wedge IfCond\} \text{ ThenCode } \{Apost\} \qquad \text{Premise 1}$$
$$\{Apre \wedge \neg IfCond\} \Rightarrow \{Apost\} \qquad \text{Premise 2}$$

- - - - - - - - - - - - - - - - - - - - - - - - - - - -

$$\{Apre\} \text{ if IfCond then ThenCode } \{Apost\} \qquad \text{Conclusion}$$

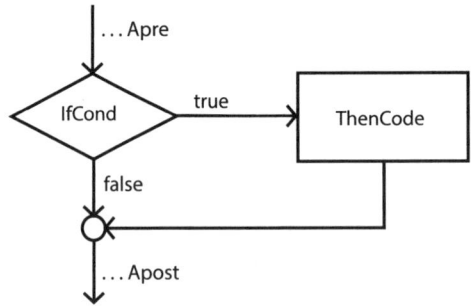

An **if-then-else** instruction **if IfCond then ThenCode else ElseCode** has one logical condition, which follows the keyword if, and two imperative clauses, one after the keyword then, and the other after the keyword else.

The **Axiom of If-then-else** states

$$\{Apre \wedge IfCond\} \text{ ThenCode } \{Apost\} \qquad\qquad \text{Premise 1}$$
$$\{Apre \wedge \neg IfCond\} \text{ ElseCode } \{Apost\} \qquad\qquad \text{Premise 2}$$

- - - - - - - - - - - - - - - - - - - - - - - - - -

$$\{Apre\} \text{ if IfCond then ThenCode else ElseCode } \{Apost\} \qquad \text{Conclusion}$$

**Examples:**

**1.** *If-then:*

$\{true\}$ if $X = 3$ then $Y := X;$ $\{X = 3 \rightarrow Y = 3\}$

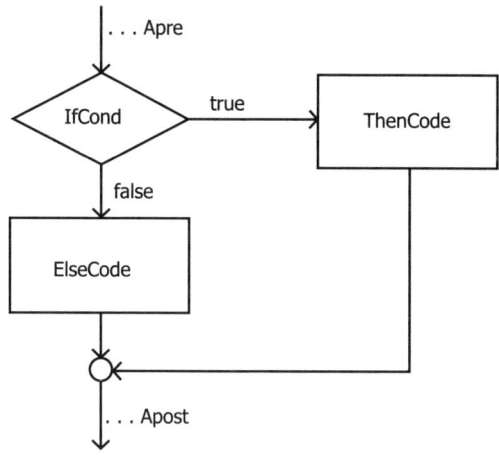

i. $\{X = 3\}\ Y := X;\ \{X = 3\ \wedge\ Y = 3\}$          Axiom of Assignment

ii. $\{X = 3\ \wedge\ Y = 3\}\ NOP\ \{(X = 3) \to (Y = 3)\}$      Axiom of NOP

     (Step ii uses a logic fact: $p \wedge q \Rightarrow p \to q$)

iii. $\{X = 3\}\ Y := X; \{X = 3 \to Y = 3\}$         Axiom of Sequence i, ii

     (Step iii establishes Premise 1 for Axiom of If-then)

iv. $\{\neg(X = 3)\} \Rightarrow \{X = 3 \to Y = 3\}$          Logic fact

     (Step iv establishes Premise 2 for Axiom of If-then)

v. $\{\text{true}\}\ \underline{\text{if }} X = 3 \ \underline{\text{then }} Y := X;\ \{X = 3 \to Y = 3\}$    Axiom of If-then iii, iv.

**2.** *If-then-else*:

$\{X > 0\}$

    $\underline{\text{if }} (X > Y)\ \underline{\text{then }} M := X;\ \underline{\text{else }} M := Y;$

$\{(X > 0) \wedge (X > Y \to M = X) \wedge (X \leq Y \to M = Y)\}$

i. $\{X > 0 \wedge X > Y\}\ M := X;\ \{X > 0 \wedge (X > Y \to M = X) \wedge (X \leq Y \to M = Y)\}$

      by Axiom of Assignment and Axiom of NOP (establishes Premise 1)

ii. $\{X > 0 \wedge \neg(X > Y)\}\ M := Y;\ \{X > 0 \wedge (X > Y \to M = X) \wedge (X \leq Y \to M = Y)\}$

      by Axiom of Assignment and Axiom of NOP (establishes Premise 2)

iii. Conclusion now follows from Axiom of If-then-else.

---

### 1.6.4   AXIOMS FOR LOOP CONSTRUCTS

#### Definitions:

A **while-loop** instruction <u>**while**</u> **WhileCond** <u>**do**</u> **LoopBody** has one logical condition called the **while-condition**, which follows the keyword <u>while</u>, and a sequence of instructions called the **loop-body**. At the outset of execution, the while condition is tested for its truth value. If it is true, then the loop body is executed. This two-step process of test and execute continues until the while condition becomes false, after which the flow of execution passes to whatever program instruction follows the while-loop.

A loop is **weakly correct** if whenever the precondition is satisfied at the outset of execution and the loop is executed to termination, the resulting computational state satisfies the postcondition.

A loop is **strongly correct** if it is weakly correct and if whenever the precondition is satisfied at the outset of execution, the computation terminates.

The **Axiom of While** defines weak correctness of a while-loop (i.e., the axiom ignores the possibility of an infinite loop) in terms of a logical condition called the **loop invariant** denoted "LoopInv" satisfying the following condition:

$\{\text{Apre}\} \Rightarrow \{\text{LoopInv}\}$                                     "Initialization" Premise

$\{\text{LoopInv} \wedge \text{WhileCond}\} \text{ LoopBody } \{\text{LoopInv}\}$                 "Preservation" Premise

$\{\text{LoopInv} \wedge \neg\text{WhileCond}\} \Rightarrow \{\text{Apost}\}$               "Finalization" Premise

- - - - - - - - - - - - - - - - - - - - - - - - - - - - - - - - - - - -

$\{\text{Apre}\} \underline{\text{ while }} \{\text{LoopInv}\} \text{ WhileCond } \underline{\text{do}} \text{ LoopBody } \{\text{Apost}\}$      Conclusion

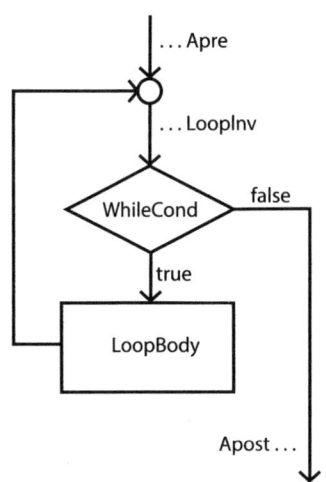

**Example:**

**1.** Suppose that $J$, $N$, and $P$ are integer-type program variables.

$\{\text{Apre} : J = 0 \wedge P = 1 \wedge N \geq 0\}$
$\underline{\text{while }} \{\text{LoopInv} : P = 2^J \wedge J \leq N\} \; (J < N) \; \underline{\text{do}}$
    $P := P * 2;$
    $J := J + 1;$
$\underline{\text{endwhile}}$
$\{\text{Apost} : P = 2^N\}$

i.    $\{\text{Apre} : J = 0 \wedge P = 1 \wedge N \geq 0\} \Rightarrow \{\text{LoopInv} : P = 2^J \wedge J \leq N\}$
        Initialization Premise trivially true by mathematics

ii.    $\{\text{LoopInv} \wedge \text{WhileCond} : (P = 2^J \wedge J \leq N) \wedge (J < N)\}$
        $P := P * 2;$
        $J := J + 1;$
        $\{\text{LoopInv} : P = 2^J \wedge J \leq N\}$
            Preservation Premise proved by using Axiom of Assignment twice
                and Axiom of Sequence

iii.    $\{\text{LoopInv} \wedge \neg\text{WhileCond} : (P = 2^J \wedge J \leq N) \wedge \neg(J < N)\} \Rightarrow \{\text{Apost} : P = 2^N\}$
        Finalization Premise provable by mathematics

iv.    Conclusion now follows from Axiom of While.

**Fact:**

**1.** Proof of termination of a loop is usually achieved by mathematical induction.

---

### 1.6.5 AXIOMS FOR SUBPROGRAM CONSTRUCTS

The parameterless procedure is the simplest subprogram construct. Procedures with parameters and functional subprograms have somewhat more complicated semantic axioms.

**Definitions:**

A **procedure** is a sequence of instructions that lies outside the main sequence of instructions in a program. It consists of a **procedure name**, followed by a **procedure body**.

A **call** instruction **<u>call</u> ProcName** is executed by transferring control to the first executable instruction of the procedure ProcName.

A **return** instruction causes a procedure to transfer control to the executable instruction immediately following the most recently executed call to that procedure. An implicit return is executed after the last instruction in the procedure body is executed. It is good programming style to put a *return* there.

In the following **Axiom of Procedure (parameterless)**, *Apre* and *Apost* are the precondition and postcondition of the instruction <u>call</u> *ProcName*; *ProcPre* and *ProcPost* are the precondition and postcondition of the procedure whose name is *ProcName*.

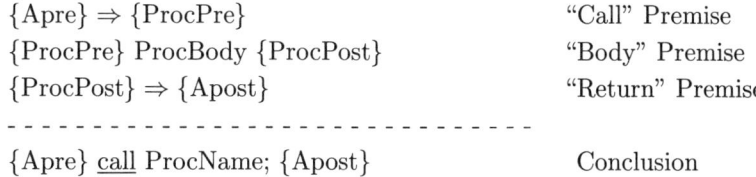

| | |
|---|---|
| $\{Apre\} \Rightarrow \{ProcPre\}$ | "Call" Premise |
| $\{ProcPre\}$ ProcBody $\{ProcPost\}$ | "Body" Premise |
| $\{ProcPost\} \Rightarrow \{Apost\}$ | "Return" Premise |
| - - - - - - - - - - - - - - - - - - - - - - - - - - - - | |
| $\{Apre\}$ <u>call</u> ProcName; $\{Apost\}$ | Conclusion |

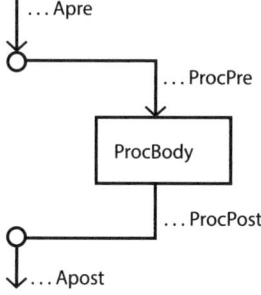

# 1.7   LOGIC-BASED COMPUTER PROGRAMMING PARADIGMS

Mathematical logic is the basis for several different computer software paradigms. These include logic programming, fuzzy reasoning, production systems, artificial intelligence, and expert systems.

## 1.7.1   LOGIC PROGRAMMING

A computer program in the imperative paradigm (familiar in languages like C, BASIC, FORTRAN, and ALGOL) is a list of instructions that describes a precise sequence of actions that a computer should perform. To initiate a computation, one supplies the iterative program plus specific input data to the computer. *Logic programming* provides an alternative paradigm in which a program is a list of "clauses", written in predicate logic, that describe an allowed range of behavior for the computer. To initiate a computation, the computer is supplied with the logic program plus another clause called a "goal". The aim of the computation is to establish that the goal is a logical consequence of the clauses constituting the logic program. The computer simplifies the goal by executing the program repeatedly until the goal becomes empty, or until it cannot be further simplified.

**Definitions:**

A **term** in a domain $S$ is either a fixed element of $S$ or an $S$-valued variable.

An **$n$-ary predicate** on a set $S$ is a function $P\colon S^n \to \{T, F\}$.

An **atomic formula** (or **atom**) is an expression of the form $P(t_1, \ldots, t_n)$, where $n \geq 0$, $P$ is an $n$-ary predicate, and $t_1, \ldots, t_n$ are terms.

A **formula** is a logical expression constructed from atoms with conjunctions, disjunctions, and negations, possibly with some logical quantifiers.

A **substitution** for a formula is a finite set of the form $\{v_1/t_1, \ldots, v_n/t_n\}$, where each $v_i$ is a distinct variable, and each $t_i$ is a term distinct from $v_i$.

The **instance** of a formula $\psi$ using the substitution $\theta = \{v_1/t_1, \ldots, v_n/t_n\}$ is the formula obtained from $\psi$ by simultaneously replacing each occurrence of the variable $v_i$ in $\psi$ by the term $t_i$. The resulting formula is denoted by $\psi\theta$.

A **closed formula** in logic programming is a program without any free variables.

A **ground formula** is a formula without any variables at all.

A **clause** is a formula of the form $\forall x_1 \ldots \forall x_s (A_1 \vee \cdots \vee A_n \leftarrow B_1 \wedge \cdots \wedge B_m)$ with no free variables, where $s, n, m \geq 0$, and $A$'s and $B$'s are atoms. In logic programming, such a clause may be denoted by $A_1, \ldots, A_n \leftarrow B_1, \ldots, B_m$.

The **head** of a clause $A_1, \ldots, A_n \leftarrow B_1, \ldots, B_m$ is the sequence $A_1, \ldots, A_n$.

The **body** of a clause $A_1, \ldots, A_n \leftarrow B_1, \ldots, B_m$ is the sequence $B_1, \ldots, B_m$.

A **definite clause** is a clause of the form $A \leftarrow B_1, \ldots, B_m$ or $\leftarrow B_1, \ldots, B_m$, which contains at most one atom in its head.

An **indefinite clause** is a clause that is not definite.

A *logic program* is a finite sequence of definite clauses.

A *goal* is a definite clause $\leftarrow B_1, \ldots, B_m$ whose head is empty. (Prescribing a goal for a logic program $P$ tells the computer to derive an instance of that goal by manipulating the logical clauses in $P$.)

An *answer* to a goal $G$ for a logic program $P$ is a substitution $\theta$ such that $G\theta$ is a logical consequence of $P$.

A *definite answer* to a goal $G$ for a logic program $P$ is an answer in which every variable is substituted by a constant.

**Facts:**

**1.** A definite clause $A \leftarrow B_1, \ldots, B_m$ represents the following logical constructs:

If every $B_i$ is true, then $A$ is also true;
Statement $A$ can be proved by proving every $B_i$.

**2.** *Definite answer property*: If a goal $G$ for a logic program $P$ has an answer, then it has a definite answer.

**3.** The definite answer property does not hold for indefinite clauses. For example, although $G = \exists x Q(x)$ is a logical consequence of $P = \{Q(a), Q(b) \leftarrow\}$, no ground instance of $G$ is a logical consequence of $P$.

**4.** Logic programming is Turing-complete (§17.5); i.e., any computable function can be represented using a logic program.

**5.** Building on the work of logician J. Alan Robinson in 1965, computer scientists Robert Kowalski and Alain Colmerauer of Imperial College and the University of Marseille-Aix, respectively, in 1972 independently developed the programming language PROLOG (PROgramming in LOGic) based on a special subset of predicate logic.

**6.** The first PROLOG interpreter was implemented in ALGOL-W in 1972 at the University of Marseille-Aix. Since then, several variants of PROLOG have been introduced, implemented, and used in practical applications. The basic paradigm behind all these languages is called Logic Programming.

**7.** In PROLOG, the relation "is" means equality.

**Examples:**

**1.** The following three clauses are definite:

$$P \leftarrow Q, R \qquad P \leftarrow \qquad \leftarrow Q, R.$$

**2.** The clause $P, S \leftarrow Q, R$ is indefinite.

**3.** The substitution $\{X/a, Y/b\}$ for the atom $P(X, Y, Z)$ yields the instance $P(a, b, Z)$.

**4.** The goal $\leftarrow P$ to the program $\{P \leftarrow\}$ has a single answer, given by the empty substitution. This means the goal can be achieved.

**5.** The goal $\leftarrow P$ to the program $\{Q \leftarrow\}$ has no answer. This means it cannot be derived from that program.

**6.** The logic program consisting of the following two definite clauses P1 and P2 computes a complete list of the pairs of vertices in an arbitrary graph that have a path joining them:

P1. $\text{path}(V, V) \leftarrow$
P2. $\text{path}(U, V) \leftarrow \text{path}(U, W), \text{edge}(W, V)$

Definite clauses P3 and P4 comprise a representation of a graph with nodes 1, 2, and 3, and edges $(1, 2)$ and $(2, 3)$:

    P3.  edge(1,2) ←
    P4.  edge(2,3) ←

The goal G represents a query asking for a complete list of the pairs of vertices in an arbitrary graph that have a path joining them:

    G.  ← path$(Y, Z)$

There are three distinct answers of the goal G to the logic program consisting of definite clauses P1 to P4, corresponding to the paths $(1, 2)$, $(1, 2, 3)$, and $(2, 3)$, respectively:

    A1.  $\{Y/1, Z/2\}$
    A2.  $\{Y/1, Z/3\}$
    A3.  $\{Y/2, Z/3\}$

**7.** The following logic program computes the Fibonacci sequence $0, 1, 1, 2, 3, 5, 8, 13, \ldots$, where the predicate $fib(N, X)$ is true if $X$ is the $N$th number in the Fibonacci sequence:

    $fib(0, 0) \leftarrow$
    $fib(1, 1) \leftarrow$
    $fib(N, X + Y) \leftarrow N > 1, fib(N - 1, X), fib(N - 2, Y)$

The goal "← $fib(6, X)$" is answered $\{X/8\}$, the goal "← $fib(X, 8)$" is answered $\{X/6\}$, and the goal "← $fib(N, X)$" has the following infinite sequence of answers:

    $\{N/0, X/0\}$
    $\{N/1, X/1\}$
    $\{N/2, X/1\}$
    $\vdots$

**8.** Consider the problem of finding an assignment of digits (integers $0, 1, \ldots, 9$) to letters such that adding two given words produces the third given word, as in this example:

|   | S | E | N | D |
|---|---|---|---|---|
| + | M | O | R | E |
| M | O | N | E | Y |

One solution to this particular puzzle is given by the following assignment:

$$D = 0, \; E = 0, \; M = 1, \; N = 0, \; O = 0, \; R = 0, \; S = 9, \; Y = 0.$$

The following PROLOG program solves all such puzzles:

    $between(X, X, Z) \leftarrow X < Z.$
    $between(X, Y, Z) \leftarrow between(K, Y, Z), X$ is $K - 1.$
    $val([], 0) \leftarrow.$
    $val([X|Y], A) \leftarrow val(Y, B), between(0, X, 9), A$ is $10 * B + X.$
    $solve(X, Y, Z) \leftarrow val(X, A), val(Y, B), val(Z, C), C$ is $A + B.$

The specific example given above is captured by the following goal:

    ← $solve([D, N, E, S], [E, R, O, M], [Y, E, N, O, M]).$

The predicate $between(X, Y, Z)$ means $X \leq Y \leq Z$. The predicate $val(L, N)$ means that the number $N$ is the value of $L$, where $L$ is the kind of list of letters that occurs on a line of these puzzles. The notation $[X|L]$ means the list obtained by writing list $L$ after item $X$. The predicate $solve(X, Y, Z)$ means that the value of list $Z$ equals the sum of the values of list $X$ and list $Y$.

This example illustrates the ease of writing logic programs for some problems where conventional imperative programs are more difficult to write.

## 1.7.2  FUZZY SETS AND LOGIC

Fuzzy set theory and fuzzy logic are used to model imprecise meanings, such as "tall", that are not easily represented by predicate logic. In particular, instead of assigning either "true" or "false" to the statement "John is tall", fuzzy logic assigns a real number between 0 and 1 that indicates the degree of "tallness" of John. Fuzzy set theory assigns a real number between 0 and 1 to John that indicates the extent to which he is a member of the set of tall people. See [Ka86], [Ka92], [KaLa94], [YaFi94], [YaZa94], [Za65], [Zi01].

### Definitions:

A **fuzzy set** $F = (X, \mu)$ consists of a set $X$ (the **domain**) and a **membership function** $\mu \colon X \to [0, 1]$. Sometimes the set is written $\{ (x, \mu(x)) \mid x \in X \}$ or $\{ \mu(x) \, x \mid x \in X \}$.

The **fuzzy intersection** of fuzzy sets $(A, \mu_A)$ and $(B, \mu_B)$ is the fuzzy set $A \cap B$ with domain $A \cap B$ and membership function $\mu_{A \cap B}(x) = \min(\mu_A(x), \mu_B(x))$.

The **fuzzy union** of fuzzy sets $(A, \mu_A)$ and $(B, \mu_B)$ is the fuzzy set $A \cup B$ with domain $A \cup B$ and membership function $\mu_{A \cup B}(x) = \max(\mu_A(x), \mu_B(x))$.

The **fuzzy complement** of the fuzzy set $(A, \mu)$ is the fuzzy set $\neg A$ or $\overline{A}$ with domain $A$ and membership function $\mu_{\overline{A}}(x) = 1 - \mu(x)$.

The **$n$th constructor** $con(\mu, n)$ of a membership function $\mu$ is the function $\mu^n$. That is, $con(\mu, n)(x) = (\mu(x))^n$.

The **$n$th dilutor** $dil(\mu, n)$ of a membership function $\mu$ is the function $\mu^{1/n}$. That is, $dil(\mu, n)(x) = (\mu(x))^{1/n}$.

A **T-norm operator** is a function $f \colon [0, 1] \times [0, 1] \to [0, 1]$ with the following properties:

- $f(x, y) = f(y, x)$                                         *commutativity*
- $f(f(x, y), z) = f(x, f(y, z))$                             *associativity*
- if $x \leq v$ and $y \leq w$, then $f(x, y) \leq f(v, w)$   *monotonicity*
- $f(a, 1) = a$.                                              *1 is a unit element*

The **fuzzy intersection** $A \cap_f B$ of fuzzy sets $(A, \mu_A)$ and $(B, \mu_B)$ **relative to the T-norm operator** $f$ is the fuzzy set with domain $A \cap B$ and membership function $\mu_{A \cap_f B}(x) = f(\mu_A(x), \mu_B(x))$.

An **S-norm operator** is a function $f \colon [0, 1] \times [0, 1] \to [0, 1]$ with the following properties:

- $f(x, y) = f(y, x)$                                         *commutativity*
- $f(f(x, y), z) = f(x, f(y, z))$                             *associativity*
- if $x \leq v$ and $y \leq w$, then $f(x, y) \leq f(v, w)$   *monotonicity*
- $f(a, 1) = 1$.

The **fuzzy union** $A \cup_f B$ of fuzzy sets $(A, \mu_A)$ and $(B, \mu_B)$ **relative to the S-norm operator** $f$ is the fuzzy set with domain $A \cup B$ and membership function $\mu_{A \cup_f B}(x) = f(\mu_A(x), \mu_B(x))$.

A **complement operator** is a function $f \colon [0,1] \to [0,1]$ with the following properties:

- $f(0) = 1$
- if $x < y$ then $f(x) > f(y)$
- $f(f(x)) = x$.

The **fuzzy complement** $\neg_f A$ of the fuzzy set $(A, \mu)$ **relative to the complement operator** $f$ is the fuzzy set with domain $A$ and membership function $\mu_{\neg_f}(x) = f(\mu(x))$.

A **fuzzy system** consists of a base collection of fuzzy sets, intersections, unions, complements, and implications.

A **hedge** is a monadic operator corresponding to linguistic adjectives such as "very", "about", "somewhat", or "quite" that modify membership functions.

A **two-valued logic** is a logic where each statement has exactly one of the two values: true or false.

A **multi-valued logic** (**n-valued logic**) is a logic with a set of $n$ ($\geq 2$) truth values; i.e., there is a set of $n$ numbers $v_1, v_2, \ldots, v_n \in [0,1]$ such that every statement has exactly one truth value $v_i$.

**Fuzzy logic** is the study of statements where each statement has assigned to it a truth value in the interval $[0,1]$ that indicates the extent to which the statement is true.

If statements $p$ and $q$ have truth values $v_1$ and $v_2$ respectively, the **truth value** of $p \vee q$ is $\max(v_1, v_2)$, the truth value of $p \wedge q$ is $\min(v_1, v_2)$, and the truth value of $\neg p$ is $1 - v_1$.

**Facts:**

**1.** Fuzzy set theory and fuzzy logic were developed by Lofti Zadeh in 1965.

**2.** Fuzzy set theory and fuzzy logic are parallel concepts: given a predicate $P(x)$, the fuzzy truth value of the statement $P(a)$ is the fuzzy set value assigned to $a$ as an element of $\{\, x \mid P(x) \,\}$.

**3.** The usual minimum function $\min(x, y)$ is a T-norm. The usual real maximum function $\max(x, y)$ is an S-norm. The function $c(x) = 1 - x$ is a complement operator.

**4.** Several other kinds of T-norms, S-norms, and complement operators have been defined.

**5.** The words "T-norm" and "S-norm" come from multi-valued logics.

**6.** The only difference between T-norms and S-norms is that the T-norm specifies $f(a, 1) = a$, whereas the S-norm specifies $f(a, 1) = 1$.

**7.** Several standard classes of membership functions have been defined, including step, sigmoid, and bell functions.

**8.** Constructors and dilutors of membership functions are also membership functions.

**9.** The large number of practical applications of fuzzy set theory can generally be divided into three types: machine systems, human-based systems, human-machine systems. Some of these applications are based on fuzzy set theory alone and some on a variety of hybrid configurations involving neurofuzzy approaches, or in combination with neural networks, genetic algorithms, or case-based reasoning.

**10.** The first fuzzy expert system that set a trend in practical fuzzy thinking was the design of a cement kiln called Linkman, produced by Blue Circle Cement and SIRA in Denmark in the early 1980s. The system incorporates the experience of a human operator in a cement production facility.

**11.** The Sendai Subway Automatic Train Operations Controller was designed by Hitachi in Japan. In that system, speed control during cruising, braking control near station zones, and switching of control are determined by fuzzy IF-THEN rules that process sensor measurements and consider factors related to travelers' comfort and safety. In operation since 1986, this most celebrated application encouraged many applications based on fuzzy set controllers in the areas of home appliances (refrigerators, vacuum cleaners, washers, dryers, rice cookers, air conditioners, shavers, blood-pressure measuring devices), video cameras (including fuzzy automatic focusing, automatic exposure, automatic white balancing, image stabilization), automotive (fuzzy cruise control, fuel injection, transmission and brake systems), robotics, and aerospace.

**12.** Applications to finance started with the Yamaichi Fuzzy Fund, which is a fuzzy trading system. This was soon followed by a variety of financial applications world-wide.

**13.** Research activities will soon result in commercial products related to the use of fuzzy set theory in the areas of audio and video data compression (such as HDTV), robotic arm movement control, computer vision, coordination of visual sensors with mechanical motion, aviation (such as unmanned platforms), and telecommunication.

**14.** *Current status*:   Most applications of fuzzy sets and logic are directly related to structured numerical model-free estimators. Presently, most applications are designed with linguistic variables, where proper levels of granularity are being used in the evaluations of those variables, expressing the ambiguity and subjectivity in human thinking. Fuzzy systems capture expert knowledge and through the processing of fuzzy IF-THEN rules are capable of processing knowledge combining the antecedents of each fuzzy rule, calculating the conclusions, and aggregating them to the final decision.

**15.** One way to model *fuzzy implication* $A \to B$ is to define $A \to B$ as $\neg_c A \cup_f B$ relative to some complement operator $c$ and to some S-norm operator $f$. Several other ways have also been considered.

**16.** A fuzzy system is used computationally to control the behavior of an external system.

**17.** Large fuzzy systems have been used in specifying complex real-world control systems. The success of such systems depends crucially on the specific engineering parameters. The correct values of these parameters are usually obtained by trial-and-readjustment.

**18.** A two-valued logic is a logic that assumes the law of the excluded middle: $p \vee \neg p$ is a tautology.

**19.** Every $n$-valued logic is a fuzzy logic.

**Examples:**

**1.** A committee consisting of five people met ten times during the past year. Person $A$ attended 7 meetings, $B$ attended all 10 meetings, $C$ attended 6 meetings, $D$ attended no meetings, and $E$ attended 9 meetings. The set of committee members can be described by the following fuzzy set that reflects the degree to which each of the members attended meetings, using the function $\mu \colon \{A, B, C, D, E\} \to [0, 1]$ with the rule $\mu(x) = \frac{1}{10}$(number of meetings attended):

$$\{(A, 0.7), (B, 1.0), (C, 0.6), (D, 0.0), (E, 0.9)\},$$

which can also be written as

$$\{0.7A, 1.0B, 0.6C, 0.0D, 0.9E\}.$$

Person $B$ would be considered a "full" member and person $D$ a "nonmember".

**2.** Four people are rated on amount of activity in a political party, yielding the fuzzy set

$$P_1 = \{0.8A, 0.45B, 0.1C, 0.75D\},$$

and based on their degree of conservatism in their political beliefs, as

$$P_2 = \{0.6A, 0.85B, 0.7C, 0.35D\}.$$

The fuzzy union of the sets is

$$P_1 \cup P_2 = \{0.8A, 0.85B, 0.7C, 0.75D\},$$

the fuzzy intersection is

$$P_1 \cap P_2 = \{0.6A, 0.45B, 0.1C, 0.35D\}$$

and the fuzzy complement of $P_1$ (measurement of political inactivity) is

$$\overline{P_1} = \{0.2A, 0.55B, 0.9C, 0.25D\}.$$

**3.** In the fuzzy set with domain $T$ and membership function

$$\mu_T(h) = \begin{cases} 0 & \text{if } h \leq 170 \\ \frac{h-170}{20} & \text{if } 170 < h < 190 \\ 1 & \text{otherwise} \end{cases}$$

the number 160 is not a member, the number 195 is a member, and the membership of 182 is 0.6. The graph of $\mu_T$ is given in the following figure.

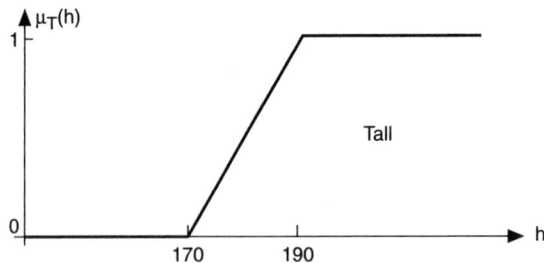

**4.** The fuzzy set $(T, \mu_T)$ of Example 3 can be used to define the fuzzy set "Tall" $= (H, \mu_H)$ of tall people, by the rule $\mu_H(x) = \mu_T(height(x))$ where $height(x)$ is the height of person $x$ calibrated in centimeters.

**5.** The second constructor $con(\mu_H, 2)$ of the fuzzy set "Tall" can be used to define a fuzzy set "Quite tall", whose graph is given in the following figure.

**6.** The second dilutor $dil(\mu_H, 2)$ of the fuzzy set "Tall" defines the fuzzy set "Somewhat tall", whose graph is given in the following figure.

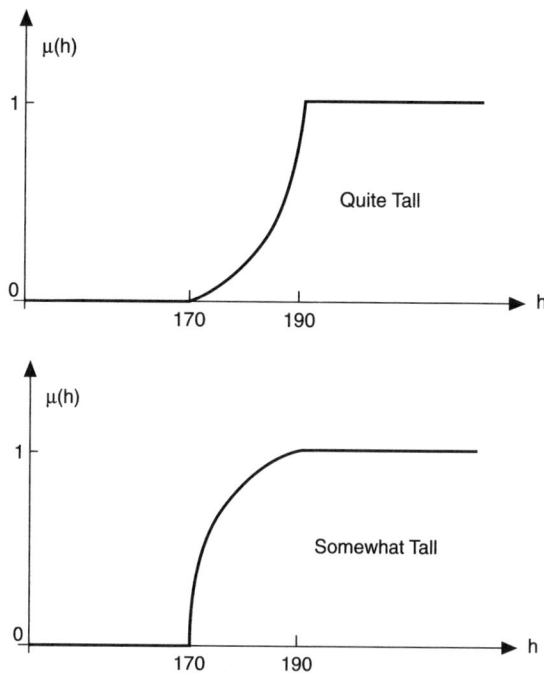

**7.** The concept of "being healthy" can be modeled using fuzzy logic. The truth value 0.95 could be assigned to "Fran is healthy" if Fran is almost always healthy. The truth value 0.4 could be assigned to "Leslie is healthy" if Leslie is healthy somewhat less than half the time. The truth of the statements "Fran and Leslie are healthy" would be 0.4 and "Fran is not healthy" would be 0.05.

**8.** *Behavior closed-loop control systems*: The behavior of some closed-loop control systems can be specified using fuzzy logic. For example, consider an automated heater whose output setting is to be based on the readings of a temperature sensor. A fuzzy set "cold" and the implication "very cold → high" could be used to relate the temperature to the heater settings. The exact behavior of this system is determined by the degree of the constructor used for "very" and by the specific choices of S-norm and complement operators used to define the fuzzy implication—the "engineering parameters" of the system.

### 1.7.3  PRODUCTION SYSTEMS

Production systems are a logic-based computer programming paradigm introduced by Allen Newell and Herbert Simon in 1975. They are commonly used in intelligent systems for representing an expert's knowledge used in solving some real-world task, such as a physician's knowledge of making medical diagnoses.

**Definitions:**

A *fact set* is a set of ground atomic formulas. These formulas represent the information relevant to the system.

A *condition* is a disjunction $A_1 \vee \cdots \vee A_n$, where $n \geq 0$ and each $A_i$ is a literal.

A condition $C$ is *true* in a fact set $S$ if:

- $C$ is empty, or
- $C$ is a positive literal and $C \in S$, or
- $C$ is a negative literal $\neg A$, and $B \notin S$ for each ground instance $B$ of $A$, or
- $C = A_1 \vee \cdots \vee A_n$, and some condition $A_i$ is true in $S$.

A **print command** "print(x)", means that the value of the term $x$ is to be printed.

An **action** is either a literal or a print command.

A **production rule** is of the form $C_1, \ldots, C_n \rightarrow A_1, \ldots, A_m$, where $n, m \geq 1$, each $C_i$ is a condition, each $A_i$ is an action, and each variable in each action appears in some positive literal in some condition.

The **antecedent** of the rule $C_1, \ldots, C_n \rightarrow A_1, \ldots, A_m$ is $C_1, \ldots, C_n$.

The **consequent** of the rule $C_1, \ldots, C_n \rightarrow A_1, \ldots, A_m$ is $A_1, \ldots, A_m$.

An **instantiation** of a production rule is the rule obtained by replacing each variable in each positive literal in each condition of the rule by a constant.

A **production system** consists of a fact set and a set of production rules.

**Facts:**

**1.** Given a fact set $S$, an instantiation $C_1, \ldots, C_n \rightarrow A_1, \ldots, A_m$ of a production rule denotes the following operation:

> if each condition $C_i$ is true in $S$ then
> > for each $A_i$:
> > > if $A_i$ is an atom, add it to $S$
> > > if $A_i$ is a negative literal $\neg B$, then remove $B$ from $S$
> > > if $A_i$ is "print(c)", then print $c$.

**2.** In addition to "print", production systems allow several other system-level commands.

**3.** OPS5 and CLIPS are currently the most popular languages for writing production systems. They are available for most operating systems, including UNIX and DOS.

**4.** To initialize a computation prescribed by a production system, the initial fact set and all the production rules are supplied as input. The command "run1" non-deterministically selects an instantiation of a production rule such that all conditions in the antecedent hold in the fact set, and it "fires" the rule by carrying out the actions in the consequent. The command "run" keeps on selecting and firing rules until no more rule instantiations can be selected.

**5.** Production systems are Turing complete.

**Examples:**

**1.** The fact set $S = \{N(3), 3 > 2, 2 > 1\}$ may represent that "3 is a natural number", that "3 is greater than 2", and that "2 is greater than 1".

**2.** If the fact set $S$ of Example 1 and the production $N(x) \rightarrow \text{print}(x)$ are supplied as input, the command "run" will yield the instantiation $N(3) \rightarrow \text{print}(3)$ and fire it to print 3.

**3.** The production rule $N(x), x > y \rightarrow \neg N(x), N(y)$ has $N(3), 3 > 2 \rightarrow \neg N(3), N(2)$ as an instantiation. If operated on fact set $S$ of Example 1, this rule will change $S$ to $\{3 > 2, 2 > 1, N(2)\}$.

**4.** The production system consisting of the following two production rules can be used to add a set of numbers in a fact set:

$$\neg S(x) \rightarrow S(0)$$
$$S(x), N(y) \rightarrow \neg S(x), \neg N(y), S(x + y).$$

For example, starting with the fact set $\{N(1), N(2), N(3), N(4)\}$, this production system will produce the fact set $\{S(10)\}$.

---

### 1.7.4   AUTOMATED REASONING

Computers have been used to help prove theorems by verifying special cases. But even more, they have been used to carry out reasoning without external intervention. Developing computer programs that can draw conclusions from a given set of facts is the goal of automated reasoning. There are now automated reasoning programs that can prove results that people have not been able to prove. Automated reasoning can help in verifying the correctness of computer programs, verifying protocol design, verifying hardware design, creating software using logic programming, solving puzzles, and proving new theorems.

### Definitions:

**Automated reasoning** is the process of using a computer program to draw conclusions which follow logically from a set of given facts.

**Automated theorem proving** is the use of automated reasoning to prove theorems.

A **proof assistant** is a computer program that attempts to prove a theorem with the help of user input.

A **computer-assisted proof** is a proof that relies on checking the validity of a large number of cases using a special-purpose computer program.

A **proof done by hand** is a proof done by a human without the use of a computer.

### Facts:

**1.** Computer-assisted proofs have been used to settle several well-known conjectures, including the Four Color Theorem (§8.6.4), the nonexistence of a finite projective plane of order 10 (§12.2.3), and the densest packing of unit balls in $\mathcal{R}^3$ (§13.2.1).

**2.** The computer-assisted proofs of both the Four Color Theorem and the nonexistence of a finite projective plane of order 10 rely on having a computer verify certain facts about a large number of cases using special-purpose software.

**3.** Hardware, system software, and special-purpose program errors can invalidate a computer-assisted proof. This makes the verification of computer-assisted proofs important. However, such verification may be impractical.

**4.** Automated reasoning software has been developed for both first-order and higher-order logics. A database of automated reasoning systems can be found at

- http://www-formal.stanford.edu/clt/ARS/systems.html

**5.** Automated theorem provers in first-order logic are computer programs that, when given a statement in predicate logic (a conjecture), attempt to find a proof that the statement is always true (and hence, a theorem) or that is sometimes false. Generally, the search for the proof is automatic, that is, without any intervention by a user.

**6.** Higher-order logic (HOL) extends first-order logic (FOL) by allowing the use of quantifiers over functions and predicates. Computer programs designed to find proofs of statements in HOL are often proof assistants rather than automated theorem provers. These systems rely on user input to help guide their operation.

**7.** The class of problems an automated theorem prover is designed to solve is known as the *problem domain* of this solver. Some automated theorem provers, such as most FOL provers, have a wide problem domain, but other automated theorem provers are built to solve problems in specific domains or of a specific form.

**8.** Proofs by mathematical induction pose particular problems for automated proof systems, because of the difficulty in searching how to apply an induction hypotheses, the problem of identifying all cases required for the basis step of a proof, and complications from the use of recursion. Automated proof systems have been built for mathematical induction with varying degrees of success. (See Example 4.)

**9.** Automated reasoning software has been used to prove new results in many areas, including settling long-standing, well-known open conjectures (such as the Robbins problem described in Example 2).

**10.** Proofs generated by automated reasoning software can usually be checked without using computers or by using software programs that check the validity of proofs.

**11.** Proofs done by humans often use techniques ill-suited for implementation in automated proof software.

**12.** Automated proof systems rely on proof procedures suitable for computer implementation, such as resolution and the semantic tableaux procedure. (See [Fi13] or [Wo96] for details.)

**13.** The effectiveness of automated proof systems depends on following strategies that help programs prove results efficiently.

**14.** Restriction strategies are used to block paths of reasoning that are considered to be unpromising.

**15.** Direction strategies are used to help programs select the approaches to take next.

**16.** Look-ahead strategies let programs draw conclusions before they would ordinarily be drawn following the basic rules of the program.

**17.** Redundancy-control strategies are used to eliminate some of the redundancy in retained information.

**18.** Several efforts have been focused on capturing mathematical knowledge that has been proved by automated reasoning into a database that can be used in automated reasoning systems. (See Examples 6-7.)

**19.** Automated theorem proving and automated reasoning are useful for verifying that computer programs are correct and that hardware incorporates the functions for which it is intended. Automated theorem proving can also be used to establish that network protocols are secure.

**20.** To learn more about automated reasoning and automated theorem proving, consult [Fi13] and [RoVo01].

### Examples:

**1.** The OTTER system (now known as Prover9) is an automated reasoning system based on resolution for first-order logic that was developed at Argonne National Laboratory [Wo96]. It was used to establish many previously unknown results in a wide variety of areas, including algebraic curves, lattices, Boolean algebra, groups, semigroups, and logic. A summary of these results can be found at

- http://www.cs.unm.edu/~mccune/prover9

**2.** The automated reasoning system EQP, developed at Argonne National Laboratory, settled the Robbins problem in 1996. This problem was first proposed in the 1930s by Herbert Robbins and was actively worked on by many mathematicians. The Robbins problem can be stated as follows. Can the equivalence

$$\neg(\neg p) \Leftrightarrow p$$

be derived from the commutative and associative laws for the "or" operator $\vee$ and the identity

$$\neg(\neg(p \vee q) \vee \neg(p \vee \neg q)) \Leftrightarrow p?$$

The EQP system, using some earlier work that established a sufficient condition for the truth of Robbins' problem, found a 15-step proof of the theorem after approximately 8 days of searching on a UNIX workstation when provided with one of several different search strategies.

**3.** Isabelle is an interactive theorem prover developed at the University of Cambridge and Technische Universität München. Isabelle can be used to encode first-order logic, higher-order logic, Zermelo-Fraenkel set theory, and other logics. Isabelle's main proof method is a higher-order version of resolution. Isabelle is interactive and also features efficient automatic reasoning tools. It comes with a large library of formally verified proofs from number theory, analysis, algebra, and set theory. It can be accessed at

- https://isabelle.in.tum.de/index.html

**4.** The Boyer-Mayer theorem prover (NQTHM) has had some notable successes with automating the proofs of theorems using mathematical induction. (See [BoMo79].)

**5.** The WZ computer-assisted and automated theorem prover, developed by H. Wilf and D. Zeilberger, has become a valuable tool for proving combinatorial identities. (See Section 3.7 for details.)

**6.** The goal of the QED Project, active from 1993 until 1996 was to build a repository that represents all important, established mathematical knowledge. It was hoped that it could help mathematicians cope with the explosion of mathematical knowledge and help in developing and verifying computer systems.

**7.** The Mizar system, founded by A. Trybulec, has a formal language for writing mathematical definitions and proofs, a proof assistant for mechanically checking proofs in this language, and a library of formalized theorems, called the Mizar Mathematical Library (MML), that can be used to help prove new results. In 2014 the MML included more than 10,000 formalized definitions and more than 50,000 theorems. This library is available at

- http://mizar.uwb.edu.pl/library

---

## REFERENCES

*Printed Resources*:

[Ap81] K. R. Apt, "Ten years of Hoare's logic: a survey—Part 1", *ACM Transactions on Programming Languages and Systems* 3 (1981), 431–483.

[BoMo79] R. S. Boyer and J. S. Moore, *A Computational Logic*, Academic Press, 1979.

[Cu16] D. W. Cunningham, *Set Theory: A First Course*, Cambridge, 2016.

[Di76] E. W. Dijkstra, *A Discipline of Programming*, Prentice-Hall, 1976.

[DuPr80] D. Dubois and H. Prade, *Fuzzy Sets and Systems—Theory and Applications*, Academic Press, 1980.

[Ep10] S. S. Epp, *Discrete Mathematics with Applications*, 4th ed., Cengage, 2010.

[Fi13] M. Fitting, *First-Order Logic and Automated Theorem Proving*, 2nd ed., Springer, 2013.

[FlPa95] P. Fletcher and C. W. Patty, *Foundations of Higher Mathematics*, Cengage, 1995.

[Fl67] R. W. Floyd, "Assigning meanings to programs", *Proceedings of the American Mathematical Society Symposium in Applied Mathematics* 19 (1967), 19–32.

[Gr13] D. Gries, *The Science of Programming*, Springer, 2013.

[Ha11] P. Halmos, *Naive Set Theory*, Martino Fine Books, 2011 (reprint of Van Nostrand 1960 edition).

[Ho69] C. A. R. Hoare, "An axiomatic basis for computer programming", *Communications of the ACM* 12 (1969), 576–580.

[Ka10] E. Kamke, *Theory of Sets*, translated by F. Bagemihl, Dover, 2010.

[Ka86] A. Kandel, *Fuzzy Mathematical Techniques with Applications*, Addison-Wesley, 1986.

[Ka92] A. Kandel, ed., *Fuzzy Expert Systems*, CRC Press, 1992.

[KaLa94] A. Kandel and G. Langholz, eds., *Fuzzy Control Systems*, CRC Press, 1994.

[Kr12] S. G. Krantz, *The Elements of Advanced Mathematics*, 3rd ed., CRC Press, 2012.

[Ll93] J. W. Lloyd, *Foundations of Logic Programming*, 2nd ed., Springer, 1993.

[Ma09] V. W. Marek, *Introduction to Mathematics of Satisfiability*, CRC Press, 2009

[Me09] E. Mendelson, *Introduction to Mathematical Logic*, 5th ed., CRC Press, 2009.

[MiRo91] J. G. Michaels and K. H. Rosen, eds., *Applications of Discrete Mathematics*, McGraw-Hill, 1991.

[Mo76] J. D. Monk, *Mathematical Logic*, Springer, 1976.

[ReCl90] S. Reeves and M. Clark, *Logic for Computer Science*, Addison-Wesley, 1990.

[RoVo01] J. A. Robinson and A. Voronkov, eds., *Handbook of Automated Reasoning*, Volumes 1 and 2, MIT Press, 2001.

[Ro12] K. H. Rosen, *Discrete Mathematics and Its Applications*, 7th ed., McGraw-Hill, 2012.

[Sc86] D. A. Schmidt, *Denotational Semantics—A Methodology for Language Development*, Allyn & Bacon, 1986.

[St77] J. E. Stoy, *Denotational Semantics: The Scott-Strachey Approach to Programming Language Theory*, MIT Press, 1977.

[WaHa78] D. A. Waterman and F. Hayes-Roth, *Pattern-Directed Inference Systems*, Academic Press, 1978.

[We72] P. Wegner, "The Vienna definition language", *ACM Computing Surveys* 4 (1972), 5–63.

[Wo96] L. Wos, *The Automation of Reasoning: An Experiment's Notebook with OTTER Tutorial*, Academic Press, 1996.

[YaFi94] R. R. Yager and D. P. Filev, *Essentials of Fuzzy Modeling and Control*, Wiley, 1994.

[YaZa94] R. R. Yager and L. A. Zadeh, eds., *Fuzzy Sets, Neural Networks and Soft Computing*, Van Nostrand Reinhold, 1994.

[Za65] L. A. Zadeh, "Fuzzy Sets", *Information and Control* 8 (1965), 338–353.

[Zi01] H-J. Zimmermann, *Fuzzy Set Theory and Its Applications*, 4th ed., Kluwer, 2001.


**Web Resources:**

`http://mizar.uwb.edu.pl/library` (Mizar Mathematical Library.)

`http://plato.stanford.edu/archives/win1997/entries/russell-paradox` (The online Stanford Encyclopedia of Philosophy's discussion of Russell's paradox.)

`http://www.austinlinks.com/Fuzzy` (Fuzzy Logic Archive: a tutorial on fuzzy logic and fuzzy systems, and examples of how fuzzy logic is applied.)

`http://www.cs.unm.edu/~mccune/prover9` (Prover9 automated theorem prover.)

`http://www.cut-the-knot.org/selfreference/russell.shtml` (Russell's paradox.)

`http://www-formal.stanford.edu/clt/ARS/systems.html` (A database of existing mechanized reasoning systems.)

`http://www-groups.dcs.st-and.ac.uk/history/HistTopics/Beginnings_of_set_theory.html` (The beginnings of set theory.)

`http://www.mcs.anl.gov/research/projects/AR/` (A summary of new results in mathematics obtained with Argonne's Automated Deduction Software.)

`http://www.rbjones.com/rbjpub/logic/log025.htm` (Information on logic.)

`https://isabelle.in.tum.de/index.html` (Isabelle proof assistant.)

# 2

# COUNTING METHODS

# INTRODUCTION

Many problems in mathematics, computer science, and engineering involve counting objects with particular properties. Although there are no absolute rules that can be used to solve all counting problems, many counting problems that occur frequently can be solved using a few basic rules together with a few important counting techniques. This chapter provides information on how many standard counting problems are solved.

# GLOSSARY

**binomial coefficient**:  the coefficient $\binom{n}{k}$ of $x^k y^{n-k}$ in the expansion of $(x+y)^n$.

**coloring pattern** (with respect to a set of symmetries of a figure):  a set of mutually equivalent colorings.

**combination** (from a set $S$):  a subset of $S$; any unordered selection from $S$.  A $k$-**combination** from a set is a subset of $k$ elements of the set.

**combination coefficient**:  the number $C(n, k)$ (equal to $\binom{n}{k}$) of ways to make an unordered choice of $k$ items from a set of $n$ items.

**combination-with-replacement** (from a set $S$):  any unordered selection with replacement; a multiset of objects from $S$.

**combination-with-replacement coefficient**:  the number of ways to choose a multiset of $k$ items from a set of $n$ items, written $C^R(n, k)$.

**composition**:  a partition in which the order of the parts is taken into account.

**cycle index**:  for a permutation group $G$, the multivariate polynomial $P_G$ obtained by dividing the sum of the cycle structure representations of all the permutations in $G$ by the number of elements of $G$.

**cycle structure** (of a permutation):  a multivariate monomial whose exponents record the number of cycles of each size.

**derangement**:  a permutation on a set that leaves no element fixed.

**exponential generating function** (for $\{a_k\}_0^\infty$):  the formal sum $\sum_{k=0}^\infty a_k \frac{x^k}{k!}$, or any equivalent closed-form expression.

**falling power**:  the product $x^{\underline{k}} = x(x-1)\ldots(x-k+1)$ of $k$ consecutive factors starting with $x$, each factor decreasing by 1.

**Ferrers diagram**:  a geometric, left-justified, and top-justified array of cells, boxes, dots or nodes representing a partition of an integer, in which each row of dots corresponds to a part of the partition.

**Gaussian binomial coefficient**:  the algebraic expression $\begin{bmatrix} n \\ k \end{bmatrix}$ in the variable $q$ defined for nonnegative integers $n$ and $k$ by $\begin{bmatrix} n \\ k \end{bmatrix} = \frac{q^n-1}{q-1} \cdot \frac{q^{n-1}-1}{q^2-1} \ldots \frac{q^{n+1-k}-1}{q^k-1}$ for $0 < k \le n$ and $\begin{bmatrix} n \\ 0 \end{bmatrix} = 1$.

**generating function** (or **ordinary generating function**) for $\{a_k\}_0^\infty$:  the formal sum $\sum_{k=0}^\infty a_k x^k$, or any equivalent closed-form expression.

**hook** (of a cell in a Ferrers diagram):  the set of cells directly to the right or directly below a given cell, together with the cell itself.

**hooklength** (of a cell in a Ferrers diagram): the number of cells in the hook of that cell.

**Kronecker delta function**: the function $\delta(x, y)$ defined by the rule $\delta(x, y) = 1$ if $x = y$ and 0 otherwise.

**lexicographic order**: the order in which a list of strings would appear in a dictionary.

**Möbius function**: the function $\mu(m)$ where

$$\mu(m) = \begin{cases} 1 & \text{if } m = 1 \\ (-1)^k & \text{if } m \text{ is a product of } k \text{ distinct primes} \\ 0 & \text{if } m \text{ is divisible by the square of a prime,} \end{cases}$$

or a generalization of this function to partially ordered sets.

**multinomial coefficient**: the coefficient $\binom{n}{k_1 \; k_2 \; \dots \; k_m}$ of $x_1^{k_1} x_2^{k_2} \dots x_m^{k_m}$ in the expansion of $(x_1 + x_2 + \dots + x_m)^n$.

**ordered selection** (of $k$ items from a set $S$): a nonrepeating list of $k$ items from $S$.

**ordered selection with replacement** (of $k$ items from a set $S$): a possibly-repeating list of $k$ items from $S$.

**ordinary generating function** (for the sequence $\{a_k\}_0^\infty$): See *generating function*.

**overpartition**: a partition in which at most one occurrence of each integer appearing as a part is distinguished as being overlined.

**partially ordered set** (or **poset**): a set $S$ together with a binary relation $\leq$ that is reflexive, antisymmetric, and transitive, written $(S, \leq)$.

**partition**: an unordered decomposition of an integer into a sum of positive integers.

**Pascal's triangle**: a triangular table with the binomial coefficient $\binom{n}{k}$ appearing in row $n$, column $k$.

**pattern inventory**: a generating function that enumerates the number of coloring patterns.

**permutation**: a one-to-one mapping of a set of elements onto itself, or an arrangement of the set into a list. A **k-permutation** of a set is an ordered nonrepeating sequence of $k$ elements of the set.

**permutation coefficient**: the number of ways to choose a nonrepeating list of $k$ items from a set of $n$ items, written $P(n, k)$.

**permutation group**: a nonempty set $P$ of permutations on a set $S$, such that $P$ is closed under composition and under inversion.

**permutation-with-replacement coefficient**: the number of ways to choose a possibly repeating list of $k$ items from a set of $n$ items, written $P^R(n, k)$.

**poset**: See *partially ordered set*.

**problème des ménages**: the problem of finding the number of ways that married couples can be seated around a circular table so that no men are adjacent, no women are adjacent, and no husband and wife are adjacent.

**problème des rencontres**: given balls 1 through $n$ drawn out of an urn one at a time, the problem of finding the probability that ball $i$ is never the $i$th one drawn.

**Stirling cycle number**: the number $\left[{n \atop k}\right]$ of ways to partition $n$ objects into $k$ nonempty cycles.

**Stirling number of the first kind**: the coefficient $s(n,k)$ of $x^k$ in the polynomial $x(x-1)(x-2)\ldots(x-n+1)$.

**Stirling number of the second kind**: the coefficient $S(n,k)$ of $x^{\underline{k}}$ in the representation $x^n = \sum_k S(n,k)x^{\underline{k}}$ of $x^n$ as a linear combination of falling powers.

**Stirling subset number**: the number $\left\{{n \atop k}\right\}$ of ways to partition $n$ objects into $k$ non-empty subsets.

**symmetry** (of a figure): a spatial motion that maps the figure onto itself.

**tree diagram**: a tree that displays the different alternatives in some counting process.

**unordered selection** (of $k$ items from a set $S$): a subset of $k$ items from $S$.

**unordered selection** (of $k$ items from a set $S$ with replacement): a selection of $k$ objects in which each object in the selection set $S$ can be chosen arbitrarily often and such that the order in which the objects are selected does not matter.

**Young tableau**: an array obtained by replacing each cell of a Ferrers diagram by a positive integer.

## 2.1  SUMMARY OF COUNTING PROBLEMS

Table 1 lists many important counting problems, gives the number of objects being counted, together with a reference to the section of this *Handbook* where details can be found. Table 2 lists several important counting rules and methods, and gives the types of counting problems that can be solved using these rules and methods. Notation used in this table is given at the end of the table.

### Table 1: Counting problems.

| objects | number of objects | reference |
|---|:---:|:---:|
| Arranging objects in a row: | | |
| $n$ distinct objects | $n! = P(n,n) = n(n-1)\ldots 2\cdot 1$ | §2.3.1 |
| $k$ out of $n$ distinct objects | $n^{\underline{k}} = P(n,k) = n(n-1)\ldots(n-k+1)$ | §2.3.1 |
| some of the $n$ objects are identical: $k_1$ of a first kind, $k_2$ of a second kind, $\ldots$, $k_j$ of a $j$th kind, and where $k_1 + k_2 + \cdots + k_j = n$ | $\left({n \atop k_1\ k_2\ \ldots\ k_j}\right) = \frac{n!}{k_1!\,k_2!\ldots k_j!}$ | §2.3.2 |
| none of the $n$ objects remains in its original place (derangements) | $D_n = n!\left(1-\frac{1}{1!}+\cdots+(-1)^n\frac{1}{n!}\right)$ | §2.4.2 |
| Arranging objects in a circle (where rotations, but not reflections, are equivalent): | | |
| $n$ distinct objects | $(n-1)!$ | §2.2.1 |
| $k$ out of $n$ distinct objects | $\frac{P(n,k)}{k}$ | §2.2.1 |
| Choosing $k$ objects from $n$ distinct objects: | | |
| order matters, no repetitions | $P(n,k) = \frac{n!}{(n-k)!} = n^{\underline{k}}$ | §2.3.1 |

| objects | number of objects | reference |
|---|---|---|
| order matters, repetitions allowed | $P^R(n,k) = n^k$ | §2.3.3 |
| order does not matter, no repetitions | $C(n,k) = \binom{n}{k} = \frac{n!}{k!(n-k)!}$ | §2.3.2 |
| order does not matter, repetitions allowed | $C^R(n,k) = \binom{n+k-1}{k}$ | §2.3.3 |
| Subsets: | | |
| of size $k$ from a set of size $n$ | $\binom{n}{k}$ | §2.3.2 |
| of all sizes from a set of size $n$ | $2^n$ | §2.3.4 |
| of $\{1,\ldots,n\}$, without consecutive elements | $F_{n+2}$ | §3.1.2 |
| Placing $n$ objects into $k$ cells: | | |
| distinct objects, distinct cells | $k^n$ | §2.2.1 |
| distinct objects, distinct cells, no cell empty | $\left\{{n \atop k}\right\}k!$ | §2.5.2 |
| distinct objects, identical cells | $\left\{{n \atop 1}\right\}+\left\{{n \atop 2}\right\}+\cdots+\left\{{n \atop k}\right\} = B_n$ | §2.5.2 |
| distinct objects, identical cells, no cell empty | $\left\{{n \atop k}\right\}$ | §2.5.2 |
| distinct objects, distinct cells, with $k_i$ in cell $i$ $(i=1,\ldots,n)$, where $k_1 + k_2 + \cdots + k_j = n$ | $\binom{n}{k_1\,k_2\,\ldots\,k_j}$ | §2.3.2 |
| identical objects, distinct cells | $\binom{n+k-1}{n}$ | §2.3.3 |
| identical objects, distinct cells, no cell empty | $\binom{n-1}{k-1}$ | §2.3.3 |
| identical objects, identical cells | $p_k(n)$ | §2.5.1 |
| identical objects, identical cells, no cell empty | $p_k(n) - p_{k-1}(n)$ | §2.5.1 |
| Placing $n$ distinct objects into $k$ nonempty cycles | $\left[{n \atop k}\right]$ | §2.5.2 |
| Solutions to $x_1 + \cdots + x_n = k$: | | |
| nonnegative integers | $\binom{k+n-1}{k} = \binom{k+n-1}{n-1}$ | §2.3.3 |
| positive integers | $\binom{k-1}{n-1}$ | §2.3.3 |
| integers where $0 \le a_i \le x_i$ for all $i$ | $\binom{k-(a_1+\cdots+a_n)+n-1}{n-1}$ | §2.3.3 |
| integers where $0 \le x_i \le a_i$ for one or more $i$ | inclusion/exclusion principle | §2.4.2 |
| integers with $x_1 \ge \cdots \ge x_n \ge 1$ | $p_n(k) - p_{n-1}(k)$ | §2.5.1 |
| integers with $x_1 \ge \cdots \ge x_n \ge 0$ | $p_n(k)$ | §2.5.1 |
| Solutions to $x_1 + x_2 + \cdots + x_n = n$ in nonnegative integers where $x_1 \ge x_2 \ge \cdots \ge x_n \ge 0$ | $p(n)$ | §2.5.1 |
| Solutions to $x_1 + 2x_2 + 3x_3 + \cdots + nx_n = n$ in nonnegative integers | $p(n)$ | §2.5.1 |

| objects | number of objects | reference |
|---|:---:|:---:|
| Functions from a $k$-element set to an $n$-element set: | | |
| all functions | $n^k$ | §2.2.1 |
| one-to-one functions ($n \geq k$) | $n^{\underline{k}} = \frac{n!}{(n-k)!} = P(n,k)$ | §2.2.1 |
| onto functions ($n \leq k$) | inclusion/exclusion | §2.4.2 |
| partial functions | $\binom{k}{0}+\binom{k}{1}n+\cdots+\binom{k}{k}n^k = (n+1)^k$ | §2.3.2 |
| Bit strings of length $n$: | | |
| all strings | $2^n$ | §2.2.1 |
| with given entries in $k$ positions | $2^{n-k}$ | §2.2.1 |
| with exactly $k$ 0s | $\binom{n}{k}$ | §2.3.2 |
| with at least $k$ 0s | $\binom{n}{k}+\binom{n}{k+1}+\cdots+\binom{n}{n}$ | §2.3.2 |
| with equal numbers of 0s and 1s | $\binom{n}{n/2}$ | §2.3.2 |
| Palindromes | $2^{\lceil n/2 \rceil}$ | §2.2.1 |
| with an even number of 0s | $2^{n-1}$ | §2.3.4 |
| without consecutive 0s | $F_{n+2}$ | §3.1.2 |
| Partitions of a positive integer $n$ into positive summands: | | §2.5.1 |
| total number | $p(n)$ | |
| into at most $k$ parts | $p_k(n)$ | |
| into exactly $k$ parts | $p_k(n) - p_{k-1}(n)$ | |
| into parts each of size $\leq k$ | $p_k(n)$ | |
| Partitions of a set of size $n$: | | |
| all partitions | $B(n)$ | §2.5.2 |
| into $k$ parts | $\left\{ {n \atop k} \right\}$ | §2.5.2 |
| into $k$ parts, each part having at least two elements | $b(n,k)$ | §3.1.8 |
| Paths: | | |
| from $(0,0)$ to $(2n,0)$ made up of line segments from $(i, y_i)$ to $(i+1, y_{i+1})$; integer $y_i \geq 0$, $y_{i+1} = y_i \pm 1$ | $C_n$ | §3.1.3 |
| from $(0,0)$ to $(2n,0)$ made up of line segments from $(i, y_i)$ to $(i+1, y_{i+1})$; integer $y_i > 0$ (for $0 < i < 2n$), $y_{i+1} = y_i \pm 1$ | $C_{n-1}$ | §3.1.3 |
| from $(0,0)$ to $(m,n)$ that move 1 unit up or right at each step | $\binom{m+n}{n}$ | §2.3.2 |
| Permutations of $\{1, \ldots, n\}$: | | |
| all permutations | $n!$ | §2.3.1 |
| with $k$ cycles, all cycles of length $\geq 2$ | $d(n,k)$ | §3.1.8 |
| with $k$ descents | $E(n,k)$ | §3.1.5 |
| with $k$ excedances | $E(n,k)$ | §3.1.5 |
| alternating, $n$ even | $(-1)^{n/2}E_n$ | §3.1.7 |

| objects | number of objects | reference |
|---|---|---|
| alternating, $n$ odd | $T_n$ | §3.1.7 |
| Symmetries of regular figures: | | §2.6 |
| $n$-gon | $2n$ | |
| tetrahedron | 12 | |
| cube | 24 | |
| octahedron | 24 | |
| dodecahedron | 60 | |
| icosahedron | 60 | |
| Coloring regular 2-dimensional & 3-dimensional figures with $\leq k$ colors: | | §2.6 |
| corners of an $n$-gon, allowing rotations and reflections | $\frac{1}{2n}\sum_{d\mid n}\varphi(d)k^{\frac{n}{d}} + \frac{1}{2}k^{\frac{(n+1)}{2}}$, $n$ odd | |
| | $\frac{1}{2n}\sum_{d\mid n}\varphi(d)k^{\frac{n}{d}} + \frac{1}{4}(k^{\frac{n}{2}} + k^{\frac{(n+2)}{2}})$, $n$ even | |
| corners of an $n$-gon, allowing only rotations | $\frac{1}{n}\sum_{d\mid n}\varphi(d)k^{\frac{n}{d}}$ | |
| corners of a triangle, allowing rotations and reflections | $\frac{1}{6}[k^3 + 3k^2 + 2k]$ | |
| corners of a triangle, allowing only rotations | $\frac{1}{3}[k^3 + 2k]$ | |
| corners of a square, allowing rotations and reflections | $\frac{1}{8}[k^4 + 2k^3 + 3k^2 + 2k]$ | |
| corners of a square, allowing only rotations | $\frac{1}{4}[k^4 + k^2 + 2k]$ | |
| corners of a pentagon, allowing rotations and reflections | $\frac{1}{10}[k^5 + 5k^3 + 4k]$ | |
| corners of a pentagon, allowing only rotations | $\frac{1}{5}[k^5 + 4k]$ | |
| corners of a hexagon, allowing rotations and reflections | $\frac{1}{12}[k^6 + 3k^4 + 4k^3 + 2k^2 + 2k]$ | |
| corners of a hexagon, allowing only rotations | $\frac{1}{6}[k^6 + k^3 + 2k^2 + 2k]$ | |
| corners of a tetrahedron | $\frac{1}{12}[k^4 + 11k^2]$ | |
| edges of a tetrahedron | $\frac{1}{12}[k^6 + 3k^4 + 8k^2]$ | |
| faces of a tetrahedron | $\frac{1}{12}[k^4 + 11k^2]$ | |
| corners of a cube | $\frac{1}{24}[k^8 + 17k^4 + 6k^2]$ | |
| edges of a cube | $\frac{1}{24}[k^{12} + 6k^7 + 3k^6 + 8k^4 + 6k^3]$ | |
| faces of a cube | $\frac{1}{24}[k^6 + 3k^4 + 12k^3 + 8k^2]$ | |
| Number of sequences of wins/ losses in $\frac{n+1}{2}$-out-of-$n$ playoff series ($n$ odd) | $2C(n, \frac{n+1}{2})$ | §2.3.2 |
| Sequences $a_1, \ldots, a_{2n}$ having $n$ 1s and $n-1$s; each partial sum $a_1 + \cdots + a_k \geq 0$ | $C_n$ | §3.1.3 |

| objects | number of objects | reference |
|---|---|---|
| Well-formed sequences of parentheses of length $2n$ | $C_n$ | §3.1.3 |
| Well-parenthesized products of $n+1$ variables | $C_n$ | §3.1.3 |
| Triangulations of a convex $(n+2)$-gon | $C_n$ | §3.1.3 |

*Notation:*

$B(n)$ or $B_n$: Bell number

$b(n,k)$: associated Stirling number of the second kind

$C_n = \frac{1}{n+1}\binom{2n}{n}$: Catalan number

$C(n,k) = \binom{n}{k} = \frac{n!}{k!(n-k)!}$: binomial coefficient

$d(n,k)$: associated Stirling number of the first kind

$E_n$: Euler number

$\varphi$: Euler phi-function

$E(n,k)$: Eulerian number

$F_n$: Fibonacci number

$n^{\underline{k}} = n(n-1)\dots(n-k+1) = P(n,k)$: falling power

$P(n,k) = \frac{n!}{(n-k)!}$: $k$-permutation

$p(n)$: number of partitions of $n$

$p_k(n)$: number of partitions of $n$ into at most $k$ summands

$p_k^*(n)$: number of partitions of $n$ into exactly $k$ summands

$\left[\begin{smallmatrix}n\\k\end{smallmatrix}\right]$: Stirling cycle number

$\left\{\begin{smallmatrix}n\\k\end{smallmatrix}\right\}$: Stirling subset number

$T_n$: tangent number

**Table 2: Methods of counting and the problems they solve.**

| statement | technique of proof |
|---|---|
| rule of sum (§2.2.1) | problems that can be broken into disjoint cases, each of which can be handled separately |
| rule of product (§2.2.1) | problems that can be broken into sequences of independent counting problems, each of which can be solved separately |
| rule of quotient (§2.2.1) | problems of counting arrangements, where the arrangements can be divided into collections that are all of the same size |
| pigeonhole principle (§2.2.3) | problems with two sets of objects, where one set of objects needs to be matched with the other |
| inclusion/exclusion principle (§2.4) | problems that involve finding the size of a union of sets, where some or all the sets in the union may have common elements |
| permutations (§2.2.1, §2.3.1, §2.3.3) | problems that require counting the number of selections or arrangements, where order within the selection or arrangement matters |
| combinations (§2.3.2, §2.3.3) | problems that require counting the number of selections or sets of choices, where order within the selection does not matter |

| statement | technique of proof |
|---|---|
| recurrence relations (§2.2.4) | problems that require an answer depending on the integer $n$, where the solution to the problem for a given size $n$ can be related to one or more cases of the problem for smaller sizes |
| generating functions (§2.2.5) | problems that can be solved by finding a closed form for a function that represents the problem and then manipulating the closed form to find a formula for the coefficients |
| Pólya counting (§2.6.5) | problems that require a listing or number of patterns, where the patterns are not to be regarded as different under certain types of motions (such as rotations and reflections) |
| Möbius inversion (§2.7.1) | problems that involve counting certain types of circular permutations |

# 2.2   BASIC COUNTING TECHNIQUES

Most counting methods are based directly or indirectly on the fundamental principles and techniques presented in this section. The rules of sum, product, and quotient are the most basic and are applied more often than any other. The section also includes some applications of the pigeonhole principle, a brief introduction to generating functions, and several examples illustrating the use of tree diagrams and Venn diagrams.

## 2.2.1   RULES OF SUM, PRODUCT, AND QUOTIENT

**Definitions:**
The **rule of sum** states that when there are $m$ cases such that the $i$th case has $n_i$ options, for $i = 1, \ldots, m$, and no two of the cases have any options in common, the total number of options is $n_1 + n_2 + \cdots + n_m$.

The **rule of product** states that when a procedure can be broken down into $m$ steps, such that there are $n_1$ options for Step 1, and such that after the completion of Step $i-1$ ($i = 2, \ldots, m$) there are $n_i$ options for Step $i$, the number of ways of performing the procedure is $n_1 n_2 \ldots n_m$.

The **rule of quotient** states that when a set $S$ is partitioned into equal-sized subsets of $m$ elements each, there are $\frac{|S|}{m}$ subsets.

An **$m$-permutation** of a set $S$ with $n$ elements is a nonrepeating ordered selection of $m$ elements of $S$, that is, a sequence of $m$ distinct elements of $S$. An $n$-permutation is simply called a **permutation** of $S$.

**Facts:**
**1.** The rule of sum can be stated in set-theoretic terms: if sets $S_1, \ldots, S_m$ are finite and pairwise disjoint, then $|S_1 \cup S_2 \cup \cdots \cup S_m| = |S_i| + |S_2| + \cdots + |S_m|$.
**2.** The rule of product can be stated in set-theoretic terms: if sets $S_1, \ldots, S_m$ are finite, then $|S_1 \times S_2 \times \cdots \times S_m| = |S_1| \cdot |S_2| \cdot \cdots \cdot |S_m|$.

**3.** The rule of quotient can be stated in terms of the equivalence classes of an equivalence relation on a finite set $S$: if every class has $m$ elements, then there are $|S|/m$ equivalence classes.

**4.** Venn diagrams (§1.2.2) are often used as an aid in counting the elements of a subset, as an auxiliary to the rule of sum. This generalizes to the principle of inclusion/exclusion (§2.4).

**5.** Counting problems can often be solved by using a combination of counting methods, such as the rule of sum and the rule of product.

### Examples:

**1.** *Counting bit strings:* There are $2^n$ bit strings of length $n$, since such a bit string consists of $n$ bits, each of which is either 0 or 1.

**2.** *Counting bit strings with restrictions:* There are $2^{n-2}$ bit strings of length $n$ $(n \geq 2)$ that begin with two 1s, since forming such a bit string consists of filling in $n-2$ positions with 0s or 1s.

**3.** *Counting palindromes:* A palindrome is a string of symbols that is unchanged if the symbols are written in reverse order, such as *rpnbnpr* or 10011001. There are $k^{\lceil n/2 \rceil}$ palindromes of length $n$ where the symbols are chosen from a set of $k$ symbols.

**4.** *Counting the number of variable names:* Determine the number of variable names, subject to the following rules: a variable name has four or fewer characters, the first character is a letter, the second and third are letters or digits, and the fourth must be X or Y or Z. Partition the names into four sets, $S_1, S_2, S_3, S_4$, containing names of length 1, 2, 3, and 4 respectively. Then $|S_1| = 26$, $|S_2| = 26 \times 36$, $|S_3| = 26 \times 36^2$, and $|S_4| = 26 \times 36^2 \times 3$. Therefore the total number of names equals $|S_1|+|S_2|+|S_3|+|S_4| = 135{,}746$.

**5.** *Counting functions:* There are $n^m$ functions from a set $A = \{a_1, \ldots, a_m\}$ to a set $B = \{b_1, \ldots, b_n\}$. (Construct each function $f : A \to B$ by an $m$-step process, where Step $i$ involves selecting the value $f(a_i)$.)

**6.** *Counting one-to-one functions:* There are $n(n-1)\ldots(n-m+1)$ one-to-one functions from $A = \{a_1, \ldots, a_m\}$ to $B = \{b_1, \ldots, b_n\}$. If values $f(a_1), \ldots, f(a_{i-1})$ have already been selected in set $B$ during the first $i-1$ steps, then there are $n-i+1$ possible values remaining for $f(a_i)$.

**7.** *Counting permutations:* There are $n(n-1)\ldots(n-m+1) = \frac{n!}{(n-m)!}$ $m$-permutations of an $n$-element set. (Each one-to-one function in Example 6 may be viewed as an $m$-permutation of $B$.) (Permutations are discussed in §2.3.)

**8.** *Counting circular permutations:* There are $(n-1)!$ ways to seat $n$ people around a round table (where rotations are regarded as equivalent, but the clockwise/counter-clockwise distinction is maintained). The total number of arrangements is $n!$ and each equivalence class contains $n$ configurations. By the rule of quotient, the number of arrangements is $\frac{n!}{n} = (n-1)!$.

**9.** *Counting restricted circular permutations:* If $n$ women and $n$ men are to be seated around a circular table, with no two of the same sex seated next to each other, the number of possible arrangements is $n(n-1)!^2$.

## 2.2.2   TREE DIAGRAMS

When a counting problem can be decomposed into cases, a tree can be used to make sure that every case is counted, and that no case is counted twice.

**Definitions:**

A **tree diagram** is a line-drawing of a tree, often with its branches and/or nodes labeled. The **root** represents the start of a procedure and the **branches** at each node represent the options for the next step.

**Facts:**

**1.** Tree diagrams can be used as an important auxiliary to the rules of sum and product.

**2.** The objective in a tree-counting approach is often one of the following:

- the number of leaves (endnodes);
- the number of nodes;
- the sum of the path products.

**Examples:**

**1.** There are six possible sequences of wins and losses when the home team (H) plays the visiting team (V) in a best 2-out-of-3 playoff. In the following tree diagram each edge label indicates whether the home team won or lost the corresponding game, and the label at each final node is the outcome of the playoff. The number of different possible sequences equals the number of endnodes, namely 6.

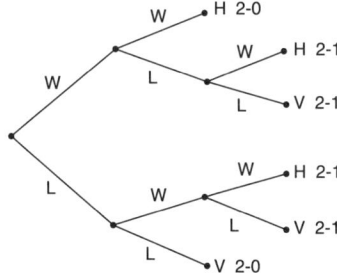

**2.** Suppose that an experimental process begins by tossing two identical dice. If the dice match, the process continues for a second round; if not, the process stops at one round. Thus, an experimental outcome sequence consists of one or two unordered pairs of numbers from 1 to 6. The three paths in the following tree represent the three different kinds of outcome sequences. The total number of possible outcomes is the sum of the path products $6^2 + 6 \cdot 15 + 15 = 141$.

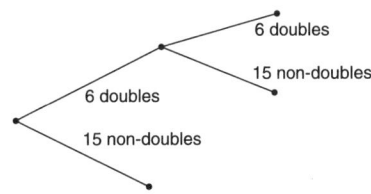

### 2.2.3   PIGEONHOLE PRINCIPLE

**Definitions:**

The **pigeonhole principle** (**Dirichlet drawer principle**) states that if $n + 1$ objects (pigeons) are placed into $n$ boxes (pigeonholes), then some box contains more than one object. (Peter Gustav Lejeune Dirichlet, 1805–1859)

The **generalized pigeonhole principle** states that if $m$ objects are placed into $k$ boxes, then some box contains at least $\lceil \frac{m}{k} \rceil$ objects.

The **set-theoretic form of the pigeonhole principle** states that if $f \colon S \to T$, where $S$ and $T$ are finite and any two of the following conditions hold, then so does the third:

- $f$ is one-to-one;
- $f$ is onto;
- $|S| = |T|$.

**Examples:**

**1.** Among any group of eight people, at least two were born on the same day of the week. This follows since there are seven pigeonholes (the seven days of the week) and more than seven pigeons (the eight people).

**2.** Among any group of 25 people, at least four were born on the same day of the week. This follows from the generalized pigeonhole principle with $m = 25$ and $k = 7$, yielding $\lceil \frac{m}{k} \rceil = \lceil \frac{25}{7} \rceil = 4$.

**3.** Suppose that a dresser drawer contains many black socks and blue socks. If choosing in total darkness, a person must grab at least three socks to be absolutely certain of having a pair of the same color. The two colors are pigeonholes; the pigeonhole principle says that three socks (the pigeons) are enough.

**4.** What is the minimum number of points whose placement in the interior of a $2 \times 2$ square guarantees that at least two of them are less than $\sqrt{2}$ units apart? Four points are not enough, since they could be placed near the respective corners of the $2 \times 2$ square. To see that five is enough, partition the $2 \times 2$ square into four $1 \times 1$ squares. By the pigeonhole principle, one of these $1 \times 1$ squares must contain at least two of the points, and these two must be less than $\sqrt{2}$ units apart.

**5.** In any set of $n + 1$ positive integers, each less than or equal to $2n$, there are at least two such that one is a multiple of the other. To see this, express each of the $n + 1$ numbers in the form $2^k \cdot q$, where $q$ is odd. Since there are only $n$ possible odd values for $q$ between 1 and $2n$, at least two of the $n + 1$ numbers must have the same $q$, and the result follows.

**6.** Let $B_1$ and $B_2$ be any two bit strings, each consisting of five ones and five zeros. Then there is a cyclic shift of bit string $B_2$ so that the resulting string $B_2'$ matches $B_1$ in at least five of its positions. For example, if $B_1 = 1010101010$ and $B_2 = 0001110101$, then $B_2' = 1000111010$ satisfies the condition. Observe that there are 10 possible cyclic shifts of bit string $B_2$. For $i = 1, \ldots, 10$, the $i$th bit of exactly five of these strings will match the $i$th bit of $B_1$. Thus, there is a total of 50 bitmatches over the set of 10 cyclic shifts. The generalized pigeonhole principle implies that there is at least one cyclic shift having $\lceil \frac{50}{10} \rceil = 5$ matching bits.

**7.** Every sequence of $n^2 + 1$ distinct real numbers must have an increasing or decreasing subsequence of length $n + 1$. Given a sequence $a_1, \ldots, a_{n^2+1}$, for each $a_j$ let $d_j$ and $i_j$ be the lengths of the longest decreasing and increasing subsequences beginning with $a_j$. This gives a sequence of $n^2 + 1$ ordered pairs $(d_j, i_j)$. If there were no increasing or decreasing subsequence of length $n + 1$, then there are only $n^2$ possible ordered pairs $(d_j, i_j)$, since $1 \leq d_j \leq n$ and $1 \leq i_j \leq n$. By the pigeonhole principle, at least two ordered pairs must be identical. Hence there are $p$ and $q$ such that $d_p = d_q$ and $i_p = i_q$. If $a_p < a_q$, then the sequence $a_p$ followed by the increasing subsequence starting at $a_q$ gives an increasing subsequence of length greater than $i_q$, a contradiction. A similar contradiction to the choice of $d_p$ follows if $a_q < a_p$. Hence a decreasing or increasing subsequence of length $n + 1$ must exist.

## 2.2.4   SOLVING COUNTING PROBLEMS USING RECURRENCE RELATIONS

Certain types of counting problems can be solved by modeling the problem using a recurrence relation (§3.3) and then working with the recurrence relation.

**Facts:**

**1.** The following general procedure is used for solving a counting problem using a recurrence relation:

- Let $a_n$ be the solution of the counting problem for the parameter $n$;
- Determine a recurrence relation for $a_n$, together with the appropriate number of initial conditions;
- Find the particular value of the sequence that solves the original counting problem by repeated use of the recurrence relation or by finding an explicit formula for $a_n$ and evaluating it at $n$.

**2.** There are many techniques for solving recurrence relations which may be useful in the solution of counting problems. Section 3.3 provides general material on recurrence relations and contains many examples illustrating how counting problems are solved using recurrence relations.

**Examples:**

**1.** *Tower of Hanoi*:   The Tower of Hanoi puzzle consists of three pegs mounted on a board and $n$ disks of different sizes. Initially the disks are on the first peg in order of decreasing size. See the following figure, using four disks. The rules allow disks to be moved one at a time from one peg to another, with no disk ever placed atop a smaller one. The goal of the puzzle is to move the tower of disks to the second peg, with the largest on the bottom. How many moves are needed to solve this puzzle for 64 disks?

Let $a_n$ be the minimum number of moves to solve the Tower of Hanoi puzzle with $n$ disks. Transferring the $n-1$ smallest disks from peg 1 to peg 3 requires $a_{n-1}$ moves. One move is required to transfer the largest disk to peg 2, and transferring the $n-1$ disks now on peg 3 to peg 2, placing them atop the largest disk, requires $a_{n-1}$ moves. Hence the puzzle with $n$ disks can be solved using $2a_{n-1}+1$ moves. The puzzle for $n$ disks cannot be solved in fewer steps, since then the puzzle with $n-1$ disks could be solved using fewer than $a_{n-1}$ moves. Hence $a_n = 2a_{n-1}+1$. The initial condition is $a_1 = 1$. Iterating shows that $a_n = 2a_{n-1}+1 = 2^2 a_{n-2}+2+1 = \cdots = 2^{n-1}a_1 + 2^{n-2} + \cdots + 2^2 + 2 + 1 = 2^n - 1$. So $2^{64} - 1$ moves are required to solve this problem for 64 disks. (Example 3 of §3.3.3 and Example 1 of §3.3.4 provide alternative methods for solving this recurrence relation.)

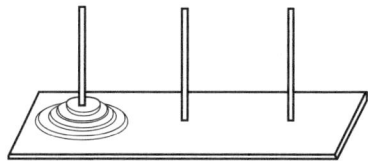

**2.** *Reve's puzzle*:   The Reve's puzzle is the variation of the Tower of Hanoi puzzle that follows the same rules as the Tower of Hanoi puzzle, but uses four pegs.

The minimum number of moves needed to solve the Reve's puzzle for $n$ disks is not known, but it is conjectured that this number is $R(n) = \sum_{i=1}^{k} i 2^{i-1} - \left(\frac{k(k+1)}{2} - n\right)2^{k-1}$ where $k$ is the smallest integer such that $n \le \frac{k(k+1)}{2}$.

The following recursive approach, the *Frame-Stewart algorithm*, gives a method for solving the Reve's puzzle by moving $n$ disks from peg 1 to peg 4 in $R(n)$ moves. If $n = 1$, move the single disk from peg 1 to peg 4. If $n > 1$: recursively move the $n - k$ smallest disks from peg 1 to peg 2 using the Frame-Stewart algorithm; then move the $k$ largest disks from peg 1 to peg 4 using the 3-peg algorithm from Example 1 on pegs 1, 3, and 4; and finally recursively move the $n - k$ smallest disks from peg 2 to peg 4 using the Frame-Stewart algorithm.

**3.** How many strings of four decimal digits contain an even number of 0s? Let $a_n$ be the number of strings of $n$ decimal digits that contain an even number of 0s. To obtain such a string: (1) append a nonzero digit to a string of $n - 1$ decimal digits that has an even number of 0s, which can be done in $9a_{n-1}$ ways; or (2) append a 0 to a string of $n - 1$ decimal digits that has an odd number of 0s, which can be done in $10^{n-1} - a_{n-1}$ ways. Hence $a_n = 9a_{n-1} + (10^{n-1} - a_{n-1}) = 8a_{n-1} + 10^{n-1}$. The initial condition is $a_1 = 9$. It follows that $a_2 = 8a_1 + 10 = 82$, $a_3 = 8a_2 + 100 = 756$, and $a_4 = 8a_3 + 1{,}000 = 7{,}048$.

## 2.2.5  SOLVING COUNTING PROBLEMS USING GENERATING FUNCTIONS

Some counting problems can be solved by finding a closed form for the function that represents the problem and then manipulating the closed form to find the relevant coefficient.

**Facts:**

**1.** Use the following procedure for solving a counting problem by using a generating function:

- Let $a_n$ be the solution of the counting problem for the parameter $n$;
- Find a closed form for the generating function $f(x)$ that has $a_n$ as the coefficient of $x^n$ in its power series;
- Solve the counting problem by computing $a_n$ by expanding the closed form and examining the coefficient of $x^n$.

**2.** Generating functions can be used to solve counting problems that reduce to finding the number of solutions to an equation of the form $x_1 + x_2 + \cdots + x_n = k$, where $k$ is a positive integer and the $x_i$s are integers subject to constraints.

**3.** There are many techniques for manipulating generating functions (§3.2, §3.3.4) which may be useful in the solution of counting problems. Section 3.2 contains examples of counting problems solved using generating functions.

**Examples:**

**1.** How many ways are there to distribute eight identical cookies to three children if each child receives at least two and no more than four cookies. Let $c_n$ be the number of ways to distribute $n$ identical cookies in this way. Then $c_n$ is the coefficient of $x^n$ in $(x^2 + x^3 + x^4)^3$, since a distribution of $n$ cookies to the three children is equivalent to a solution of $x_1 + x_2 + x_3 = 8$ with $2 \le x_i \le 4$ for $i = 1, 2, 3$. Expanding this product shows that $c_8$, the coefficient of $x_8$, is 6. Hence there are six ways to distribute the cookies.

**2.** An urn contains colored balls, where each ball is either red, blue, or black, there are at least ten balls of each color, and balls of the same color are indistinguishable. Find the number of ways to select ten balls from the urn so that an odd number of red balls, an even number of blue balls, and at least five black balls are selected. If $x_1$, $x_2$, and $x_3$ denote the number of red balls, blue balls, and black balls selected, respectively, the answer is provided by the number of nonnegative integer solutions of $x_1 + x_2 + x_3 = 10$ with $x_1$ odd, $x_2$ even, $x_3 \geq 5$. This is the coefficient of $x^{10}$ in the generating function $f(x) = (x + x^3 + x^5 + x^7 + x^9 + \cdots)(1 + x^2 + x^4 + x^6 + x^8 + x^{10} + \cdots)(x^5 + x^6 + x^7 + x^8 + x^9 + x^{10} + \cdots)$. Since the coefficient of $x^{10}$ in the expansion is 6, there are six ways to select the balls as specified.

## 2.3 PERMUTATIONS AND COMBINATIONS

Permutations count the number of arrangements of objects, and combinations count the number of ways to select objects from a set. A permutation coefficient counts the number of ways to arrange a set of objects, whereas a combination coefficient counts the number of ways to select a subset.

### 2.3.1 ORDERED SELECTION: FALLING POWERS

Falling powers mathematically model the process of selecting $k$ items from a collection of $n$ items in circumstances where the ordering of the selection matters and repetition is not allowed.

**Definitions:**
An **ordered selection** of $k$ items from a set $S$ is a nonrepeating list of $k$ items from $S$.

The **falling power** $x^{\underline{k}}$ is the product $x(x-1)\ldots(x-k+1)$ of $k$ decreasing factors starting at the real number $x$.

The number **$n$-factorial**, $n!$ ($n$ a nonnegative integer), is defined by the rule $0! = 1$, $n! = n(n-1)\ldots 3 \cdot 2 \cdot 1$ if $n \geq 1$.

A **permutation of a list** is any rearrangement of the list.

A **permutation of a set** of $n$ items is an arrangement of those items into a list. (Often, such a list and/or the permutation itself is represented by a string whose entries are in the list order.)

A **$k$-permutation of a set** of $n$ items is an ordered selection of $k$ items from that set. A $k$-permutation can be written as a sequence or a string.

The **permutation coefficient** $P(n, k)$ is the number of ways to choose an ordered selection of $k$ items from a set of $n$ items: that is, the number of $k$-permutations.

A **derangement of a list** is a permutation of the entries such that no entry remains in the original position.

**Facts:**
**1.** The falling power $x^{\underline{k}}$ is analogous to the ordinary power $x^k$, which is the product of $k$ constant factors $x$. The underline in the exponent of the falling power is a reminder that consecutive factors drop.

**2.** $P(n, k) = n^{\underline{k}} = \frac{n!}{(n-k)!}$.

3. For any integer $n$, $n^{\underline{n}} = n!$.

4. The numbers $P(n,k) = n^{\underline{k}}$ are given in Table 1.

5. A repetition-free list of length $n$ has approximately $n!/e$ derangements.

**Table 1: Permutation coefficients $P(n,k) = n^{\underline{k}}$.**

| $n \backslash k$ | 0 | 1 | 2 | 3 | 4 | 5 | 6 | 7 | 8 | 9 | 10 |
|---|---|---|---|---|---|---|---|---|---|---|---|
| 0 | 1 | | | | | | | | | | |
| 1 | 1 | 1 | | | | | | | | | |
| 2 | 1 | 2 | 2 | | | | | | | | |
| 3 | 1 | 3 | 6 | 6 | | | | | | | |
| 4 | 1 | 4 | 12 | 24 | 24 | | | | | | |
| 5 | 1 | 5 | 20 | 60 | 120 | 120 | | | | | |
| 6 | 1 | 6 | 30 | 120 | 360 | 720 | 720 | | | | |
| 7 | 1 | 7 | 42 | 210 | 840 | 2,520 | 5,040 | 5,040 | | | |
| 8 | 1 | 8 | 56 | 336 | 1,680 | 6,720 | 20,160 | 40,320 | 40,320 | | |
| 9 | 1 | 9 | 72 | 504 | 3,024 | 15,120 | 60,480 | 181,440 | 362,880 | 362,880 | |
| 10 | 1 | 10 | 90 | 720 | 5,040 | 30,240 | 151,200 | 604,800 | 1,814,400 | 3,628,800 | 3,628,800 |

**Examples:**

1. $(4.2)^{\underline{3}} = 4.2 \cdot 3.2 \cdot 2.2 = 29.568$.

2. *Dealing a row of playing cards*: Suppose that five cards are to be dealt from a deck of 52 cards and placed face up in a row. There are $P(52,5) = 52^{\underline{5}} = 52 \cdot 51 \cdot 50 \cdot 49 \cdot 48 = 311{,}875{,}200$ ways to do this.

3. *Placing distinct balls into distinct bins*:   $k$ differently-colored balls are to be placed into $n$ bins $(n \geq k)$, with at most one ball to a bin. The number of different ways to arrange the balls is $P(n,k) = n^{\underline{k}}$. (Think of the balls as if they were numbered 1 to $k$, so that placing ball $j$ into a bin corresponds to placing that bin into the $j$th position of the list.)

4. *Counting ballots*: Voters are asked to rank their three top choices from the eleven candidates running for office. A first choice vote is worth 3 points, second choice 2 points, and third choice 1 point. Since a completed ballot is an *ordered* selection in this situation, each voter has $P(11,3) = 11^{\underline{3}} = 11 \cdot 10 \cdot 9 = 990$ distinct ways to cast a vote.

5. *License plate combinations*: The license plates in a state have three letters (from the upper-case Roman alphabet of 26 letters) followed by four digits. There are $P(26,3) = 15{,}600$ ways to select the letters and $P(10,4) = 5{,}040$ ways to select the digits. By the rule of product there are $P(26,3)P(10,4) = 15{,}600 \cdot 5{,}040 = 78{,}624{,}000$ strings.

6. *Circular permutations of distinct objects*: See Example 8 of §2.2.1. Also see Example 3 of §2.7.1 for problems that allow identical objects.

7. *Increasing and decreasing subsequences of permutations*: Young tableaux (§2.8) can be used to find the number of permutations of $\{1,2,\ldots,n\}$ with specified lengths of their longest increasing subsequences and longest decreasing subsequences.

---

## 2.3.2   UNORDERED SELECTION: BINOMIAL COEFFICIENTS

Binomial coefficients mathematically model the process of selecting $k$ items from a collection of $n$ items in circumstances where the ordering of the selection does not matter, and repetitions are not allowed.

**Definitions:**

An **unordered selection** of $k$ items from a set $S$ is a subset of $k$ items from $S$.

A **$k$-combination** from a set $S$ is an unordered selection of $k$ items.

The **combination coefficient** $C(n, k)$ is the number of $k$-combinations of $n$ objects.

The **binomial coefficient** $\binom{n}{k}$ is the coefficient of $x^k y^{n-k}$ in the expansion of $(x+y)^n$.

The **extended binomial coefficient** (**generalized binomial coefficient**) $\binom{n}{k}$ is zero whenever $k$ is negative. When $n$ is a negative integer and $k$ is a nonnegative integer, its value is $(-1)^k \binom{k-n-1}{k}$.

The **multicombination coefficient** $C(n\colon k_1, k_2, \ldots, k_m)$, where $n = k_1 + k_2 + \cdots + k_m$, denotes the number of ways to partition $n$ items into subsets of sizes $k_1, k_2, \ldots, k_m$.

The **multinomial coefficient** $\binom{n}{k_1\ k_2\ \ldots\ k_m}$ is the coefficient of $x_1^{k_1} x_2^{k_2} \ldots x_m^{k_m}$ in the expansion of $(x_1 + x_2 + \cdots + x_m)^n$.

The **Gaussian binomial coefficient** is defined for nonnegative integers $n$ and $k$ by

$$\begin{bmatrix} n \\ k \end{bmatrix} = \frac{q^n - 1}{q - 1} \cdot \frac{q^{n-1} - 1}{q^2 - 1} \cdot \frac{q^{n-2} - 1}{q^3 - 1} \cdots \frac{q^{n+1-k} - 1}{q^k - 1} \quad \text{for } 0 < k \le n$$

and $\begin{bmatrix} n \\ 0 \end{bmatrix} = 1$, where $q$ is a variable. (See also §2.5.1.)

**Facts:**

1. $C(n, k) = \frac{P(n,k)}{k!} = \frac{n^{\underline{k}}}{k!} = \frac{n!}{k!(n-k)!} = \binom{n}{k}$.

2. *Pascal's recursion:* $\binom{n}{k} = \binom{n-1}{k-1} + \binom{n-1}{k}$, where $n > 0$ and $k > 0$.

3. *Subsets:* There are $C(n, k)$ subsets of size $k$ that can be chosen from a set of size $n$.

4. The numbers $C(n, k) = \binom{n}{k}$ are given in Table 2.

**Table 2: Combination coefficients (binomial coefficients)** $C(n, k) = \binom{n}{k}$.

| $n \backslash k$ | 0 | 1 | 2 | 3 | 4 | 5 | 6 | 7 | 8 | 9 | 10 | 11 | 12 |
|---|---|---|---|---|---|---|---|---|---|---|---|---|---|
| 0 | 1 | | | | | | | | | | | | |
| 1 | 1 | 1 | | | | | | | | | | | |
| 2 | 1 | 2 | 1 | | | | | | | | | | |
| 3 | 1 | 3 | 3 | 1 | | | | | | | | | |
| 4 | 1 | 4 | 6 | 4 | 1 | | | | | | | | |
| 5 | 1 | 5 | 10 | 10 | 5 | 1 | | | | | | | |
| 6 | 1 | 6 | 15 | 20 | 15 | 6 | 1 | | | | | | |
| 7 | 1 | 7 | 21 | 35 | 35 | 21 | 7 | 1 | | | | | |
| 8 | 1 | 8 | 28 | 56 | 70 | 56 | 28 | 8 | 1 | | | | |
| 9 | 1 | 9 | 36 | 84 | 126 | 126 | 84 | 36 | 9 | 1 | | | |
| 10 | 1 | 10 | 45 | 120 | 210 | 252 | 210 | 120 | 45 | 10 | 1 | | |
| 11 | 1 | 11 | 55 | 165 | 330 | 462 | 462 | 330 | 165 | 155 | 11 | 1 | |
| 12 | 1 | 12 | 66 | 220 | 495 | 792 | 924 | 792 | 495 | 220 | 66 | 12 | 1 |

Sometimes the entries in Table 2 are arranged into the form called *Pascal's triangle*, in which each entry is the sum of the two numbers diagonally above the number (Pascal's recursion, Fact 2).

$$
\begin{array}{ccccccccccccccccccc}
&&&&&&&&& 1 \\
&&&&&&&& 1 && 1 \\
&&&&&&& 1 && 2 && 1 \\
&&&&&& 1 && 3 && 3 && 1 \\
&&&&& 1 && 4 && 6 && 4 && 1 \\
&&&& 1 && 5 && 10 && 10 && 5 && 1 \\
&&& 1 && 6 && 15 && 20 && 15 && 6 && 1 \\
&& 1 && 7 && 21 && 35 && 35 && 21 && 7 && 1 \\
& 1 && 8 && 28 && 56 && 70 && 56 && 28 && 8 && 1 \\
1 && 9 && 36 && 84 && 126 && 126 && 84 && 36 && 9 && 1 \\
\end{array}
$$

$$1 \quad 10 \quad 45 \quad 120 \ 210 \ 252 \ 210 \ 120 \ 45 \quad 10 \quad 1$$

**5.** The extended binomial coefficients satisfy Pascal's recursion. Their definition is constructed precisely to achieve this purpose.

**6.** $C(n\colon k_1, k_2, \ldots, k_m) = \frac{n!}{k_1! k_2! \ldots k_m!} = \binom{n}{k_1 \ k_2 \ \ldots \ k_m}$. The number of strings of length $n$ with $k_i$ objects of type $i$ $(i = 1, 2, \ldots, m)$ is $\frac{n!}{k_1! k_2! \ldots k_m!}$.

**7.** $C(n, k) = C(n\colon k, n - k) = C(n, n - k)$. That is, the number of unordered selections of $k$ objects chosen from $n$ objects is equal to the number of unordered selections of $n - k$ objects chosen from $n$ objects.

**8.** *Gaussian binomial coefficient identities:*

- $\begin{bmatrix} n \\ k \end{bmatrix} = \begin{bmatrix} n \\ n-k \end{bmatrix}$;

- $\begin{bmatrix} n \\ k \end{bmatrix} + \begin{bmatrix} n \\ k-1 \end{bmatrix} q^{n+1-k} = \begin{bmatrix} n+1 \\ k \end{bmatrix}$.

**9.** $(1 + x)(1 + qx)(1 + q^2 x) \ldots (1 + q^{n-1} x) = \sum_{k=0}^{n} \begin{bmatrix} n \\ k \end{bmatrix} q^{k(k-1)/2} x^k$.

**10.** $\lim_{q \to 1} \begin{bmatrix} n \\ k \end{bmatrix} = \binom{n}{k}$.

**11.** $\begin{bmatrix} n \\ k \end{bmatrix} = a_0 + a_1 q + a_2 q^2 + \cdots + a_{k(n-k)} q^{k(n-k)}$ where each $a_i$ is an integer and $\sum_{i=0}^{k(n-k)} a_i = \binom{n}{k}$.

**Examples:**

**1.** *Subsets:* A set with 20 elements has $C(20, 4)$ subsets with four elements. The total number of subsets of a set with 20 elements is equal to $C(20, 0) + C(20, 1) + \cdots + C(20, 20)$, which is equal to $2^{20}$. (See §2.3.4.)

**2.** *Nondistinct balls into distinct bins:* Suppose $k$ identically colored balls are to be placed into $n$ bins $(n \geq k)$, at most one ball to a bin. The number of different ways to do this is $C(n, k) = \frac{n^k}{k!}$. (This amounts to selecting from the $n$ bins the $k$ bins into which the balls are placed.)

**3.** *Counting ballots:* Each voter is asked to identify three choices for trustee from eleven candidates nominated for the position, without specifying any order of preference. Since a completed ballot is an *unordered* selection in this situation, each voter has $C(11, 3) = \frac{11 \cdot 10 \cdot 9}{3!} = 165$ distinct ways to cast a vote.

**4.** *Counting bit strings with exactly $k$ 0s:* There are $\binom{n}{k}$ bit strings of length $n$ with exactly $k$ 0s, since each such bit string is determined by choosing a subset of size $k$ from the $n$ positions; 0s are placed in these $k$ positions, and 1s in the remaining positions.

**5.** *Counting bit strings with at least $k$ 0s:* There are $\binom{n}{k} + \binom{n}{k+1} + \cdots + \binom{n}{n}$ bit strings of length $n$ with at least $k$ 0s, since each such bit string is determined by choosing a subset of size $k, k+1, \ldots,$ or $n$ from the $n$ positions; 0s are placed in these positions, and 1s in the remaining positions.

**6.** *Counting bit strings with equal numbers of 0s and 1s:* For $n$ even, there are $\binom{n}{n/2}$ bit strings of length $n$ with equal numbers of 0s and 1s, since each such bit string is determined by choosing a subset of size $\frac{n}{2}$ from the $n$ positions; 0s are placed in these positions, and 1s in the remaining positions.

**7.** *Counting strings with repeated letters:* The word "MISSISSIPPI" has eleven letters, with "I" and "S" appearing four times each, "P" appearing twice, and "M" once. There are $C(11\colon 4,4,2,1) = \frac{11!}{4!4!2!1!} = 34{,}650$ possible different strings obtainable by permuting the letters. This counting problem is equivalent to partitioning 11 items into subsets of sizes 4, 4, 2, 1.

**8.** *Counting circular strings with repeated letters:* See §2.7.1.

**9.** *Counting paths:* The number of paths in the plane from $(0,0)$ to a point $(m,n)$ $(m, n \geq 0)$ that move one unit upward or one unit to the right at each step is $\binom{m+n}{n}$. Using $U$ for "up" and $R$ for "right", each path can be described by a string of $m$ $R$s and $n$ $U$s.

**10.** *Playoff series:* In a series of playoff games, such as the World Series or Stanley Cup finals, the winner is the first team to win more than half the maximum number of games possible, $n$ (odd). The winner must win $\frac{n+1}{2}$ games. The number of possible win-loss sequences of such a series is $2C(n, \frac{n+1}{2})$. For example, in the World Series between teams $A$ and $B$, any string of length 7 with exactly 4 $A$s represents a winning sequence for $A$. (The string $AABABBA$ means that $A$ won a seven-game series by winning the first, second, fourth, and seventh games; the string $AAAABBB$ means that $A$ won the series by winning the first four games.) There are $C(7,4)$ ways for $A$ to win the World Series, and $C(7,4)$ ways for $B$ to win the World Series.

**11.** *Dealing a hand of playing cards:* A hand of five cards (where order does not matter) can be dealt from a deck of 52 cards in $C(52,5) = \frac{52^{\underline{5}}}{5!} = 2{,}598{,}960$ ways.

**12.** *Poker hands:* Table 3 contains the number of combinations of five cards that form various poker hands (where an ace can be high or low).

**13.** *Counting partial functions:* There are $\binom{k}{0} + \binom{k}{1}n + \binom{k}{2}n^2 + \cdots + \binom{k}{k}n^k$ partial functions $f\colon A \to B$ where $|A| = k$ and $|B| = n$. Each partial function is determined by choosing a domain of definition for the function, which can be done, for each $j = 0, \ldots, n$, in $\binom{k}{j}$ ways. Once a domain of definition is determined, there are $n^j$ ways to define a function on that set. (The sum can be simplified to $(n+1)^k$.)

**14.** $\begin{bmatrix}3\\1\end{bmatrix} = \frac{q^3-1}{q-1} = 1 + q + q^2$.

**15.** $\begin{bmatrix}6\\2\end{bmatrix} = \frac{q^6-1}{q-1} \cdot \frac{q^5-1}{q^2-1} = \frac{q^6-1}{q^2-1} \cdot \frac{q^5-1}{q-1} = (q^4+q^2+1)(q^4+q^3+q^2+q+1) = 1+q+2q^2+ 2q^3 + 3q^4 + 2q^5 + 2q^6 + q^7 + q^8$. The sum of these coefficients is $15 = \binom{6}{2}$, as Fact 11 predicts.

**16.** A particle moves in the plane from $(0,0)$ to $(n-k,k)$ by moving one unit at a time in either the positive $x$ or positive $y$ direction. The number of such paths where the area bounded by the path, the $x$-axis, and the vertical line $x = n - k$ is $i$ units is equal to $a_i$, where $a_i$ is the coefficient of $q^i$ in the expansion of the Gaussian binomial coefficient $\begin{bmatrix}n\\k\end{bmatrix}$ in Fact 11.

### Table 3: Number of poker hands.

| type of hand | formula | explanation |
|---|---|---|
| *royal flush* (ace, king, queen, jack, 10 in same suit) | $4$ | 4 choices for a suit, and 1 royal flush in each suit |
| *straight flush* (5 cards of five consecutive ranks, all in one suit, but not a royal flush) | $\binom{4}{1}9$ | 4 choices for a suit, and in each suit there are 9 ways to get five cards in a row |
| *four of a kind* (4 cards in one rank and a fifth card) | $\binom{13}{1}\binom{48}{1}$ | 13 choices for a rank, only 1 way to select the four cards in that rank, and 48 ways to select a fifth card |
| *full house* (3 cards of one rank, 2 of another rank) | $13\binom{4}{3}12\binom{4}{2}$ | 13 ways to select a rank for the 3-of-a-kind, and $\binom{4}{3}$ ways to choose 3 of this rank; 12 ways to select a rank for the pair, and $\binom{4}{2}$ ways to get a pair of this rank |
| *flush* (5 cards in one suit, but neither royal nor straight flush) | $4\binom{13}{5}-4\cdot10$ | 4 ways to select a suit, $\binom{13}{5}$ ways to choose five cards in that suit; then subtract royal and straight flushes |
| *straight* (5 cards in five consecutive ranks, but not all of the same suit) | $10\cdot4^5-4\cdot10$ | 10 ways to choose five ranks in a row, and 4 ways to choose a card from each rank; then subtract royal and straight flushes |
| *three of a kind* (3 cards of one rank, and 2 cards of two different ranks) | $13\binom{4}{3}\binom{12}{2}4^2$ | 13 ways to select one rank, $\binom{4}{3}$ ways to choose 3 cards of that rank; $\binom{12}{2}$ ways to pick two other ranks; and $4^2$ ways to pick a card of each of those two ranks |
| *two pairs* (2 cards in each of two different ranks, and a fifth card of a third rank) | $\binom{13}{2}\binom{4}{2}\binom{4}{2}44$ | $\binom{13}{2}$ ways to select two ranks, $\binom{4}{2}$ ways to choose 2 cards in each of these ranks, and $\binom{44}{1}$ way to pick a nonmatching fifth card |
| *one pair* (2 cards in one rank, plus 3 cards from three other ranks) | $13\binom{4}{2}\binom{12}{3}4^3$ | 13 ways to select a rank, $\binom{4}{2}$ ways to choose two cards in that rank; $\binom{12}{3}$ ways to pick three other ranks, and $4^3$ ways to pick one card from each of those ranks |

## 2.3.3   SELECTION WITH REPETITION

Some problems concerning counting the number of ways to select $k$ objects from a set of $n$ objects permit choices of objects to be repeated. Some of these situations are also modeled by binomial coefficients.

**Definitions:**

An **ordered selection with replacement** is an ordered selection in which each object in the selection set can be chosen arbitrarily often.

An **ordered selection with specified replacement** fixes the number of times each object is to be chosen.

An **unordered selection with replacement** is a selection in which each object in the selection set can be chosen arbitrarily often.

The **permutation-with-replacement coefficient** $P^R(n, k)$ is the number of ways to choose a possibly repeating list of $k$ items from a set of $n$ items.

The **combination-with-replacement coefficient** $C^R(n, k)$ is the number of ways to choose a multiset of $k$ items from a set of $n$ items.

**Facts:**

**1.** An ordered selection with replacement can be thought of as obtaining an ordered list of names, obtained by selecting an object from a set, writing its name, placing it back in the set, and repeating the process.

**2.** The number of ways to make an ordered selection with replacement of $k$ items from $n$ distinct items (with arbitrary repetition) is $n^k$. Thus $P^R(n, k) = n^k$.

**3.** The number of ways to make an ordered selection of $n$ items from a set of $q$ distinct items, with exactly $k_i$ selections of object $i$, is $\frac{n!}{k_1! k_2! \ldots k_q!}$.

**4.** An unordered selection with replacement can be thought of as obtaining a collection of names, obtained by selecting an object from a set, writing its name, placing it back in the set, and repeating the process. The resulting collection is a multiset (§1.2.1).

**5.** The number of ways to make an unordered selection with replacement of $k$ items from a set of $n$ items is $C(n + k - 1, k)$. Thus $C^R(n, k) = C(n + k - 1, k)$.

*Combinatorial interpretation*: It is sufficient to show that the $k$-multisets that can be chosen from a set of $n$ items are in one-to-one correspondence with the bit strings of length $(n+k-1)$ with $k$ ones. To indicate that $k_j$ copies of item $j$ are selected, for $j = 1, \ldots, n$, write a string of $k_1$ ones, then a "0", then a string of $k_2$ ones, then another "0", then a string of $k_3$ ones, then another "0", and so on, until after the string of $k_{n-1}$ ones and the last "0", there appears the final string of $k_n$ ones. The resulting bit string has length $n + k - 1$ (since it has $k$ ones and $n - 1$ zeros). Every such bit string describes a possible selection. Thus the number of possible selections is $C(n+k-1, k) = C(n+k-1, n-1)$.

**6.** *Integer solutions to the equation* $x_1 + x_2 + \cdots + x_n = k$:

- The number of nonnegative integer solutions is $C(n+k-1, k) = C(n+k-1, n-1)$. [In the combinatorial argument of Fact 5, there are $n$ strings of ones. The first string of ones can be regarded as the value for $x_1$, the second string of ones as the value for $x_2$, etc.]

- The number of positive integer solutions is $C(k - 1, n - 1)$.

- The number of nonnegative integer solutions where $x_i \geq a_i$ for $i = 1, \ldots, n$ is $C(n + k - 1 - (a_1 + \cdots + a_n), n - 1)$ (if $a_1 + \cdots + a_n \leq k$). [Let $x_i = y_i + a_i$ for each $i$, yielding the equation $y_1 + y_2 + \cdots + y_n = k - (a_1 + \cdots + a_n)$ to be solved in nonnegative integers.]

- The number of nonnegative integer solutions where $x_i \leq a_i$ for $i = 1, \ldots, n$ can be obtained using the inclusion/exclusion principle. See §2.4.2.

**Examples:**

**1.** *Distinct balls into distinct bins*: $k$ differently colored balls are to be placed into $n$ bins, with arbitrarily many balls to a bin. The number of different ways to do this is $n^k$. (Apply the rule of product to the number of possible bin choices for each ball.)

**2.** *Binary strings:*   The number of sequences (bit strings) of length $n$ that can be constructed from the symbol set $\{0, 1\}$ is $2^n$.

**3.** *Colored balls into distinct bins with colors repeated:*   $k$ balls are colored so that $k_1$ balls have color 1, $k_2$ have color 2, ..., and $k_q$ have color $q$. The number of ways these $k$ balls can be placed into $n$ distinct bins $(n \geq k)$, at most one per bin, is $\frac{P(n,k)}{k_1!k_2!...k_q!}$.

*Note:*   This is more general than Fact 2, since $n$ can exceed the sum of all the $k_i$s. If $n$ equals this sum, then $P(n, n) = n!$ and the two formulas agree.

**4.**   When three dice are rolled, the "outcome" is the number of times each of the numbers 1 to 6 appears. For instance, two 3s and a 5 is an outcome. The number of different possible outcomes is $C(6 + 3 - 1, 3) = \binom{8}{3} = 56$.

**5.** *Nondistinct balls into distinct bins with multiple balls per bin allowed:*   The number of ways that $k$ identical balls can be placed into $n$ distinct bins, with any number of balls allowed in each bin, is $C(n + k - 1, k)$.

**6.** *Nondistinct balls into distinct bins with no bin allowed to be empty:*   The number of ways that $k$ identical balls can be placed into $n$ distinct bins, with any number of balls allowed in each bin and no bin allowed to remain empty, is $C(k - 1, n - 1)$.

**7.**   How many ways are there to choose one dozen donuts when there are 7 different kinds of donuts, with at least 12 of each type available? Order is not important, so a multiset of size 12 is being constructed from 7 distinct types. Accordingly, there are $C(7 + 12 - 1, 12) = 18{,}564$ ways to choose the dozen donuts.

**8.**   The number of nonnegative integer solutions to the equation $x_1 + x_2 + \cdots + x_7 = 12$ is $C(7 + 12 - 1, 12)$, since this is a rephrasing of Example 7.

**9.**   The number of nonnegative integer solutions to $x_1 + x_2 + \cdots + x_5 = 36$, where $x_1 \geq 4$, $x_3 = 11$, and $x_4 \geq 7$ is $C(17, 3)$. [It is easiest to think of purchasing 36 donuts, where at least 4 of type 1, exactly 11 of type 3, and at least 7 of type 4 must be purchased. Begin with an empty bag, and put in 4 of type 1, 11 of type 3, and 7 of type 4. This leaves 14 donuts to be chosen, and they must be of types 1, 2, 4, or 5, which is equivalent to finding the number of nonnegative integer solutions to $x_1 + x_2 + x_4 + x_5 = 14$.]

## 2.3.4   BINOMIAL COEFFICIENT IDENTITIES

**Facts:**

**1.**   Table 4 lists some identities involving binomial coefficients.

**2.**   Combinatorial identities, such as those in Table 4, can be proved algebraically by using techniques such as substitution, differentiation, or the principle of mathematical induction (see Facts 4 and 5); they can also be proved by using combinatorial proofs. (See Fact 3.)

**Table 4: Binomial coefficient identities.**

| | |
|---|---|
| *Factorial expansion* | $\binom{n}{k} = \frac{n!}{k!(n-k)!}, \ k = 0, 1, 2, \ldots, n$ |
| *Symmetry* | $\binom{n}{k} = \binom{n}{n-k}, \ k = 0, 1, 2, \ldots, n$ |
| *Monotonicity* | $\binom{n}{0} < \binom{n}{1} < \cdots < \binom{n}{\lfloor n/2 \rfloor}, \ n \geq 0$ |
| *Pascal's identity* | $\binom{n}{k} = \binom{n-1}{k-1} + \binom{n-1}{k}, \ k = 0, 1, 2, \ldots, n$ |
| *Binomial theorem* | $(x + y)^n = \sum_{k=0}^{n} \binom{n}{k} x^k y^{n-k}, \ n \geq 0$ |
| *Counting all subsets* | $\sum_{k=0}^{n} \binom{n}{k} = 2^n, \ n \geq 0$ |
| *Even and odd subsets* | $\sum_{k=0}^{n} (-1)^k \binom{n}{k} = 0, \ n \geq 0$ |
| *Sum of squares* | $\sum_{k=0}^{n} \binom{n}{k}^2 = \binom{2n}{n}, \ n \geq 0$ |
| *Square of row sums* | $\left[ \sum_{k=0}^{n} \binom{n}{k} \right]^2 = \sum_{k=0}^{2n} \binom{2n}{k}, \ n \geq 0$ |
| *Absorption/extraction* | $\binom{n}{k} = \frac{n}{k} \binom{n-1}{k-1}, \ k \neq 0$ |
| *Trinomial revision* | $\binom{n}{m}\binom{m}{k} = \binom{n}{k}\binom{n-k}{m-k}, \ 0 \leq k \leq m \leq n$ |
| *Parallel summation* | $\sum_{k=0}^{m} \binom{n+k}{k} = \binom{n+m+1}{m}, \ m, n \geq 0$ |
| *Diagonal summation* | $\sum_{k=0}^{n-m} \binom{m+k}{m} = \binom{n+1}{m+1}, \ n \geq m \geq 0$ |
| *Vandermonde convolution* | $\sum_{k=0}^{r} \binom{m}{k}\binom{n}{r-k} = \binom{m+n}{r}, \ m, n, r \geq 0$ |
| *Diagonal sums in Pascal's triangle* (§2.3.2) | $\sum_{k=0}^{\lfloor n/2 \rfloor} \binom{n-k}{k} = F_{n+1}$ *(Fibonacci numbers)*, $n \geq 0$ |
| *Other common identities* | $\sum_{k=0}^{n} k\binom{n}{k} = n2^{n-1}, \ n \geq 0$ |
| | $\sum_{k=0}^{n} k^2 \binom{n}{k} = n(n+1)2^{n-2}, \ n \geq 0$ |
| | $\sum_{k=0}^{n} (-1)^k k\binom{n}{k} = 0, \ n \geq 0$ |
| | $\sum_{k=0}^{n} \frac{\binom{n}{k}}{k+1} = \frac{2^{n+1}-1}{n+1}, \ n \geq 0$ |
| | $\sum_{k=0}^{n} (-1)^k \frac{\binom{n}{k}}{k+1} = \frac{1}{n+1}, \ n \geq 0$ |
| | $\sum_{k=1}^{n} (-1)^{k-1} \frac{\binom{n}{k}}{k} = 1 + \frac{1}{2} + \frac{1}{3} + \cdots + \frac{1}{n}, \ n > 0$ |
| | $\sum_{k=0}^{n-1} \binom{n}{k}\binom{n}{k+1} = \binom{2n}{n-1}, \ n > 0$ |
| | $\sum_{k=0}^{m} \binom{m}{k}\binom{n}{p+k} = \binom{m+n}{m+p}, \ m, n, p \geq 0, \ n \geq p + m$ |

**3.** The following give combinatorial interpretations of some of the identities involving binomial coefficients in Table 4.

- *Symmetry:* In choosing a subset of $k$ items from a set of $n$ items, the number of ways to select which $k$ items to include must equal the number of ways to select which $n - k$ items to exclude.

- *Pascal's recursion:* In choosing $k$ objects from a list of $n$ distinct objects, the number of ways that include the last object is $\binom{n-1}{k-1}$, and the number of ways that exclude the last object is $\binom{n-1}{k}$. Their sum is then the total number of ways to choose $k$ objects from a set of $n$, namely $\binom{n}{k}$.

- *Binomial theorem:* The coefficient of $x^k y^{n-k}$ in the expansion $(x + y)^n = (x + y)(x + y) \ldots (x + y)$ equals the number of ways to choose $k$ factors from among the $n$ factors $(x + y)$ in which $x$ contributes to the resultant term.

- *Counting all subsets:* Summing the numbers of subsets of all possible sizes yields the total number of different possible subsets.

- *Sum of squares:* Choose a committee of size $n$ from a group of $n$ men and $n$ women. The left side, rewritten as $\binom{n}{k}\binom{n}{n-k}$, describes the process of selecting committees according to the number of men, $k$, and the number of women, $n - k$, on the committee. The right side gives the total number of committees possible.

- *Absorption/extraction*: From a group of $n$ people, choose a committee of size $k$ and a person on the committee to be its chairperson. Equivalently, first select a chairperson from the entire group, and then select the remaining $k-1$ committee members from the remaining $n-1$ people.

- *Trinomial revision*: The left side describes the process of choosing a committee of size $m$ from $n$ people and then a subcommittee of size $k$. The right side describes the process where the subcommittee of size $k$ is first chosen from the $n$ people and then the remaining $m-k$ members of the committee are selected from the remaining $n-k$ people.

- *Vandermonde convolution*:    Given $m$ men and $n$ women, form committees of size $r$. The summands give the numbers of committees broken down by number of men, $k$, and number of women, $r-k$, on the committee; the right side gives the total number of committees.

**4.** The formula for counting all subsets can be obtained from the binomial theorem by substituting 1 for $x$ and 1 for $y$.

**5.** The formula for even and odd subsets can be obtained from the binomial theorem by substituting 1 for $x$ and $-1$ for $y$.

**6.** A set $A$ of size $n$ has $2^{n-1}$ subsets with an even number of elements and $2^{n-1}$ subsets with an odd number of elements. (The *even and odd subsets* identity in Table 4 shows that $\sum \binom{n}{k}$ for $k$ even is equal to $\sum \binom{n}{k}$ for $k$ odd. Since the total number of subsets is $2^n$, each side must equal $2^{n-1}$.)

---

### 2.3.5  GENERATING PERMUTATIONS AND COMBINATIONS

There are various systematic ways to generate permutations and combinations of the set $\{1, \ldots, n\}$.

**Definitions:**
A list of strings from an ordered set is in **lexicographic order** if the strings are sorted as they would appear in a dictionary.

If the elements in the strings are ordered by a relation $<$, string $a_1 a_2 \ldots a_m$ **precedes** $b_1 b_2 \ldots b_n$ if any of the following happens: $a_1 < b_1$; there is a positive integer $k$ such that $a_1 = b_1, \ldots, a_k = b_k$ and $a_{k+1} < b_{k+1}$; or $m < n$ and $a_1 = b_1, \ldots, a_m = b_m$.

**Algorithms:**
Here $k$ is a given positive integer less than or equal to $n$ and $r(k)$ is a randomly chosen integer in the range $\{1, 2, \ldots, k\}$ Algorithms 1, 2, and  5 give ways to generate all permutations, $k$-permutations, and $k$-combinations of $\{1, 2, \ldots, n\}$ in lexicographic order. Algorithms 3, 4, and 6 give ways to randomly generate a permutation, $k$-permutation, and $k$-combination of $\{1, 2, \ldots, n\}$.

---

**Algorithm 1:  Generate permutations of $\{1, \ldots, n\}$ in lexicographic order**

$a_1 a_2 \ldots a_n := 1\, 2 \ldots n$
**while** $a_1 a_2 \ldots a_n \neq n\ n{-}1 \ldots 1$
    $m :=$ the rightmost location such that $a_m$ is followed by a larger number
    $a'_1 a'_2 \ldots a'_{m-1} := a_1 a_2 \ldots a_{m-1}$   {Retain everything to the left of $a_m$}

---

$a'_m :=$ the smallest number larger than $a_m$ to the right of $a_m$

$a'_{m+1}a'_{m+2}\ldots a'_n :=$ everything else, in ascending order

$a_1a_2\ldots a_n := a'_1a'_2\ldots a'_n$

output $a_1a_2\ldots a_n$

---

**Algorithm 2**: **Generate $k$-permutations of $\{1,\ldots,n\}$ in lexicographic order**

$a_1a_2\ldots a_k := 1\,2\,\ldots\,k$

**while** $a_1a_2\ldots a_k \neq n\ n{-}1\,\ldots\,n-(k-1)$

   $m :=$ the rightmost location such that $a_m$ is followed by a larger number

   $a'_1a'_2\ldots a'_{m-1} := a_1a_2\ldots a_{m-1}$   {Retain everything to the left of $a_m$}

   $a'_m :=$ the smallest number larger than $a_m$ to the right of $a_m$

   $a'_{m+1}a'_{m+2}\ldots a'_k :=$ everything else, in ascending order

   $a_1a_2\ldots a_k := a'_1a'_2\ldots a'_k$

   output $a_1a_2\ldots a_k$

---

**Algorithm 3**: **Generate a random permutation of $\{1,\ldots,n\}$.**

$a_1a_2\ldots a_n := 1\,2\,\ldots\,n$

**for** $i := 0$ **to** $n-2$

   interchange $a_{n-i}$ and $a_{r(n-i)}$

output $a_1a_2\ldots a_n$

---

**Algorithm 4**: **Generate a random $k$-permutation of $\{1,\ldots,n\}$.**

$a_1a_2\ldots a_n :=$ a random permutation of $\{1,\ldots,n\}$ generated from Algorithm 3

output $a_1a_2\ldots a_k$

---

**Algorithm 5**: **Generate $k$-combinations of $\{1,\ldots,n\}$ in lexicographic order.**

$a_1a_2\ldots a_k := 1\,2\,\ldots\,k$   {First combination in lexicographic order}

**while** $a_1a_2\ldots a_k \neq n{-}k{+}1\ n{-}k{+}2\ \ldots\ n$

   $m :=$ the rightmost location among $1,\ldots,k$ such that a number larger than

     $a_m$ but smaller than $n$ is not in the combination

   $a'_1a'_2\ldots a'_{m-1} := a_1a_2\ldots a_{m-1}$   {Retain everything to the left of $a_m$}

   $a'_m := a_m + 1$

   $a'_{m+1}a'_{m+2}\ \cdots\ a'_k := a_m{+}2\ a_m{+}3\ldots a_m{+}k{-}m{+}1$   {Continue consecutively}

   output $a_1a_2\ldots a_k := a'_1a'_2\ldots a'_k$

---

**Algorithm 6**: **Generate a random $k$-combination of $\{1,\ldots,n\}$.**

$a_1a_2\ldots a_k :=$ a $k$-permutation of $\{1,\ldots,n\}$ generated by Algorithm 4

output $a_1a_2\ldots a_k$   {Ignore the order in which elements are written}

---

## Examples:

**1.** The lexicographic order for the 3-permutations of $\{1,2,3\}$ is 123, 132, 213, 231, 312, 321.

**2.** The lexicographic order of the $C(5,3) = 10$ 3-combinations of $\{1,2,3,4,5\}$ is 123, 124, 125, 134, 135, 145, 234, 235, 245, 345.

**3.** *Generating permutations*: What permutation follows 3142765 in the lexicographic ordering of the permutations of $\{1,\ldots,7\}$? Step 1 of the while-loop of Algorithm 1 leads to the fourth digit, namely the digit 2, as the first digit from the right that has larger digits following it. Steps 2 and 3 show that the next permutation starts with 3145 since 5 is the smallest digit greater than 2 and following it. Finally, Step 4 yields 2, 6, and 7 (in numerical order) as the digits that follow. Thus, the permutation immediately following 3142765 is 3145267.

**4.** *Generating combinations*: What 5-combination follows 12478 in the lexicographic ordering of 5-combinations of $\{1,\ldots,8\}$? Step 1 of the while-loop of Algorithm 5 leads to the third digit, namely the digit 4, as the first digit from the right that can be safely increased by 1. Step 2 shows that the next permutation starts with 125 since the third digit is increased by 1. Finally, Step 3 yields 6 and 7 as the following digits (add 1 to the newly-listed previous digit until the new selection of $k$ digits is complete). Thus, the combination after 12478 is 12567.

## 2.4   INCLUSION/EXCLUSION

The principle of inclusion/exclusion is used to count the elements in a non-disjoint union of finite sets. Many counting problems can be solved by applying this principle to a well-chosen collection of sets. The techniques involved in this process are best illustrated with examples.

### 2.4.1   PRINCIPLE OF INCLUSION/EXCLUSION

The number of elements in the union of two finite sets $A$ and $B$ is $|A| + |B|$, provided that the sets have no element in common. In the general case, however, some elements in common to both sets have been included in the sum twice. The sum is adjusted to exclude the double-counting of these common elements by subtracting their number:

$$|A \cup B| = |A| + |B| - |A \cap B|.$$

Similarly, the number of elements in the union of three finite sets is

$$|A \cup B \cup C| = |A| + |B| + |C| - |A \cap B| - |A \cap C| - |B \cap C| + |A \cap B \cap C|.$$

The following Venn diagram (§1.2.2) illustrates these two cases. These simple equations generalize to the case of $n$ sets.

**Facts:**

**1.** *Inclusion/exclusion principle*: The number of elements in the union of $n$ finite sets $A_1, A_2, \ldots, A_n$ is

$$|A_1 \cup A_2 \cup \cdots \cup A_n| = \sum_{1 \le i \le n} |A_i| - \sum_{1 \le i < j \le n} |A_i \cap A_j| + \sum_{1 \le i < j < k \le n} |A_i \cap A_j \cap A_k|$$
$$- \cdots + (-1)^{n+1} |A_1 \cap A_2 \cap \cdots \cap A_n|$$

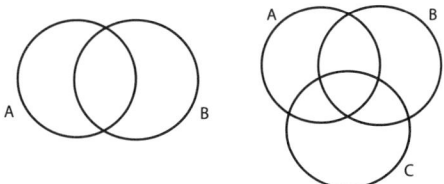

or, alternatively,

$$|A_1 \cup A_2 \cup \cdots \cup A_n| = \sum_{k=1}^{n} (-1)^{k+1} \sum_{1 \le i_1 < \cdots < i_k \le n} |A_{i_1} \cap A_{i_2} \cap \cdots \cap A_{i_k}|.$$

Sometimes the inner sum of the alternative formula is denoted $S_k$.

**2.** The inclusion/exclusion formula for $n$ sets has $2^n - 1$ terms, one for each possible nonempty intersection. The coefficient of a term is $-1$ if the term corresponds to intersections of an even number of sets, and $+1$ otherwise.

**3.** The principle is often applied to the complement of a set. Let $A_i$ be the subset of elements in a universal set $U$ that have property $P_i$. The number of elements that have properties $P_{i_1}, P_{i_2}, \ldots, P_{i_k}$ is often written $N(P_{i_1} P_{i_2} \ldots P_{i_k})$ and the number of elements that have none of these properties is often written $N(P'_{i_1} P'_{i_2} \ldots P'_{i_k})$. The number of elements in $U$ that have none of the properties is

$$N(P'_1 P'_2 \ldots P'_n) = |U| - \sum_{1 \le i \le n} N(P_i) + \sum_{1 \le i < j \le n} N(P_i P_j) - \cdots + (-1)^n N(P_1 P_2 \ldots P_n).$$

**Examples:**

**1.** Of 70 people surveyed, 37 drink coffee, 23 drink tea, and 25 drink neither. Find the number who drink both coffee and tea. Using $C$ to represent the set of coffee drinkers and $T$ to represent the set of tea drinkers, the size of $C \cap T$ must be found. Since $|\overline{T \cup C}| = 25$, the Venn diagram in part (a) of the following figure shows that $|C \cup T| = 45$. According to the inclusion/exclusion principle, $|C \cap T| = |C| + |T| - |C \cup T| = 37 + 23 - 45 = 15$, illustrated in part (b) of the figure.

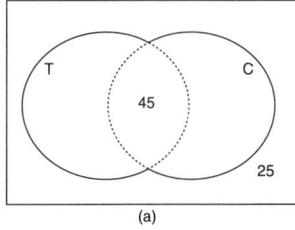
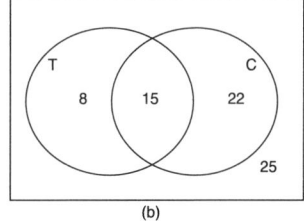

(a)    (b)

**2.** Suppose that 16 high-school juniors enroll in Algebra, 17 in Biology, and 30 in Chemistry; that 5 students enroll in both Algebra and Biology, 4 in both Algebra and Chemistry, and 7 in both Biology and Chemistry; that 3 students enroll in all three; and that every junior takes at least one of these three subjects. Then the total number of students in the junior class is $16 + 17 + 30 - (5 + 4 + 7) + 3 = 50$.

**3.** Each of 11 linguists translates at least one of the languages Amharic and Burmese into English. The numbers who translate only Amharic or Burmese are both odd primes. More linguists translate Burmese than Amharic. How many can translate Amharic?

Based on experimentation or on an analytic approach, the only possible assignment of numbers to regions that fits all these facts leads to 6, as shown in the following figure.

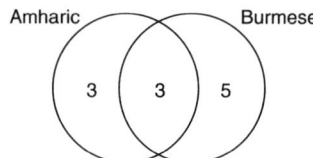

**4.** At a party for 28 people, three kinds of pizza were served: anchovy, broccoli, and cheese. Everyone ate at least one kind. No two of the seven different possible selections of one or more kinds of pizza were eaten by the same number of partygoers. Each of the three possible exclusive selections (one kind of pizza only) was eaten by an odd number of partygoers, and each of the three possible combinations of two kinds of pizza was eaten by an even number of partygoers. If a total of 18 partygoers ate cheese pizza, how many ate both anchovy and broccoli?

The answer is 2. Experimentation or an analytic approach leads to the possible assignments of numbers to regions that fit all these facts, shown in the following figure.

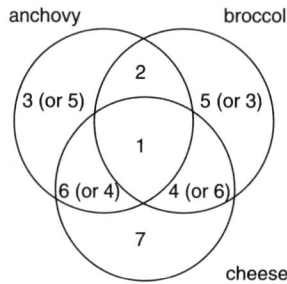

**5.** To count the number of ways to select a 5-card hand from a standard 52-card deck so that the hand contains at least one card from each of the four suits, let $A_1$, $A_2$, $A_3$, and $A_4$ be the subsets of 5-card hands that do not contain a club, diamond, heart, or spade, respectively. Then

$$|A_i| = \binom{52-13}{5} = \binom{39}{5} \quad \text{with } \binom{4}{1} \text{ choices for } i$$
$$|A_i \cap A_j| = \binom{52-26}{5} = \binom{26}{5} \quad \text{with } \binom{4}{2} \text{ choices for } i \text{ and } j$$
$$|A_i \cap A_j \cap A_k| = \binom{52-39}{5} = \binom{13}{5} \quad \text{with } \binom{4}{3} \text{ choices for } i, j, \text{ and } k.$$

There are $\binom{52}{5}$ possible 5-card hands, so by complementation and the principle of inclusion/exclusion, those that contain at least one card from each suit is

$$\binom{52}{5} - \binom{4}{1}\binom{39}{5} + \binom{4}{2}\binom{26}{5} - \binom{4}{3}\binom{13}{5} = 685{,}464.$$

## 2.4.2   APPLYING INCLUSION/EXCLUSION TO COUNTING PROBLEMS

### Definitions:

A **derangement** on a set is a permutation that leaves no element fixed. The number of derangements on a set of cardinality $n$ is denoted $D_n$.

A **rencontre number** $D_{n,k}$ is the number of permutations on a set of $n$ elements that leave exactly $k$ elements fixed.

**Facts:**

**1.** The number of onto functions from an $n$-element set to a $k$-element set $(n \geq k)$ is

$$\sum_{j=0}^{k}(-1)^j \binom{k}{j}(k-j)^n$$

(See Example 3.)

**2.** The following binomial coefficient identities can all be derived by combinatorial arguments using inclusion/exclusion:

- $\sum_{k=0}^{m}(-1)^k \binom{n}{k}\binom{n-k}{m-k} = 0$
- $\sum_{k=m}^{n}(-1)^{k-m}\binom{n}{k} = \binom{n-1}{m-1}$
- $\sum_{k=0}^{n}(-1)^k \binom{n}{k}\left(\frac{n-k+r-1}{r}\right) = \binom{r-1}{n-1}$.

**3.** $D_n = n!(1 - \frac{1}{1!} + \frac{1}{2!} - \cdots + (-1)^n \frac{1}{n!})$. (See Example 8.)

**4.** $\frac{D_n}{n!} \to e^{-1} \approx 0.368$ as $n \to \infty$.

**5.** $D_n = nD_{n-1} + (-1)^n$ for $n \geq 1$.

**6.** $D_n = (n-1)(D_{n-1} + D_{n-2})$ for $n \geq 2$.

**7.** The following table gives some values of $D_n$:

| $n$ | $D_n$ | $n$ | $D_n$ | $n$ | $D_n$ | $n$ | $D_n$ |
|---|---|---|---|---|---|---|---|
| 1 | 0 | 4 | 9 | 7 | 1,854 | 10 | 1,334,961 |
| 2 | 1 | 5 | 44 | 8 | 14,833 | 11 | 14,684,570 |
| 3 | 2 | 6 | 265 | 9 | 133,496 | 12 | 176,214,841 |

**8.** $D_{n,0} = D_n$.

**9.** $D_{n,k} = \binom{n}{k}D_{n-k}$

**10.** The following table gives some values of $D_{n,k}$:

| $n\backslash k$ | 0 | 1 | 2 | 3 | 4 | 5 | 6 | 7 | 8 | 9 | 10 |
|---|---|---|---|---|---|---|---|---|---|---|---|
| 0 | 1 | | | | | | | | | | |
| 1 | 0 | 1 | | | | | | | | | |
| 2 | 1 | 0 | 1 | | | | | | | | |
| 3 | 2 | 3 | 0 | 1 | | | | | | | |
| 4 | 9 | 8 | 6 | 0 | 1 | | | | | | |
| 5 | 44 | 45 | 20 | 10 | 0 | 1 | | | | | |
| 6 | 265 | 264 | 135 | 40 | 15 | 0 | 1 | | | | |
| 7 | 1,854 | 1,855 | 924 | 315 | 70 | 21 | 0 | 1 | | | |
| 8 | 14,833 | 14,832 | 7,420 | 2,464 | 630 | 112 | 28 | 0 | 1 | | |
| 9 | 133,496 | 133,497 | 66,744 | 22,260 | 5,544 | 1,134 | 168 | 36 | 0 | 1 | |
| 10 | 1,334,961 | 1,334,960 | 667,485 | 222,480 | 55,650 | 11,088 | 1,890 | 240 | 45 | 0 | 1 |

**Examples:**

**1.** The inclusion/exclusion principle can be used to establish the binomial coefficient identity

$$\binom{n}{m} = \sum_{k=1}^{m}(-1)^{k+1}\binom{n-k}{m-k}\binom{n}{k}.$$

Let $A_i$ denote the subset of $m$-combinations that contain object $i$. Thus, the $k$-fold intersection $A_{i_1} \cap A_{i_2} \cap \cdots \cap A_{i_k}$ consists of all the $m$-combinations that contain all the objects $i_1, i_2, \ldots, i_k$. Since there are $\binom{n-k}{m-k}$ ways to complete an $m$-combination in this intersection, it follows that $|A_{i_1} \cap A_{i_2} \cap \cdots \cap A_{i_k}| = \binom{n-k}{m-k}$. Since the $k$ objects themselves can be specified in $\binom{n}{k}$ ways, it follows that

$$\sum_{1 \le i_1 < i_2 < \cdots < i_k \le n} |A_{i_1} \cap A_{i_2} \cap \cdots \cap A_{i_k}| = \binom{n-k}{m-k}\binom{n}{k}, \qquad k \le m.$$

Since $A_1 \cup A_2 \cup \cdots \cup A_n$ is the set of all $m$-combinations selected from $1, 2, \ldots, n$ that contain at least one of the objects $1, 2, \ldots, n$, it must be the set of all $m$-combinations.

**2.** *Sieve of Eratosthenes*:   The sieve of Eratosthenes (276–194 BCE) is a method for finding all primes less than or equal to a given positive integer $n$. Begin with the list of integers 2 through $n$, and delete all multiples of the first number in the list, 2, but not including 2. The first integer remaining after 2 is 3; delete all multiples of 3, not including 3. The first integer remaining after 3 is 5; delete all multiples of 5, not including 5. Continue the process. The remaining integers are the primes less than or equal to $n$. (See §4.4.2.)

The inclusion/exclusion principle can be used to obtain the number of primes less than or equal to $n$. (A number $x \le n$ is *prime* if and only if $x$ has a prime factor less than or equal to $\lfloor \sqrt{n} \rfloor$.) Let $P_i$ be the property: a number is greater than the $i$th prime and divisible by the $i$th prime. Then the number of primes less than or equal to $n$ is $N(P_1' P_2' \ldots P_k')$, where there are $k$ primes less than or equal to $\lfloor \sqrt{n} \rfloor$. (§2.4.1, Fact 3.)

For example, the number of primes less than or equal to 100 is $N(P_1' P_2' P_3' P_4') = 99 - \lfloor \frac{100}{2} \rfloor - \lfloor \frac{100}{3} \rfloor - \lfloor \frac{100}{5} \rfloor - \lfloor \frac{100}{7} \rfloor + \lfloor \frac{100}{2 \cdot 3} \rfloor + \lfloor \frac{100}{2 \cdot 5} \rfloor + \lfloor \frac{100}{2 \cdot 7} \rfloor + \lfloor \frac{100}{3 \cdot 5} \rfloor + \lfloor \frac{100}{3 \cdot 7} \rfloor + \lfloor \frac{100}{5 \cdot 7} \rfloor - \lfloor \frac{100}{2 \cdot 3 \cdot 5} \rfloor - \lfloor \frac{100}{2 \cdot 3 \cdot 7} \rfloor - \lfloor \frac{100}{2 \cdot 5 \cdot 7} \rfloor - \lfloor \frac{100}{3 \cdot 5 \cdot 7} \rfloor + \lfloor \frac{100}{2 \cdot 3 \cdot 5 \cdot 7} \rfloor = 99 - 50 - 33 - 20 - 14 + 16 + 10 + 7 + 6 + 4 + 2 - 3 - 2 - 1 - 0 + 0 = 21.$

**3.** *Number of onto functions*:   The number of onto functions from an $n$-element set to a $k$-element set $(n \ge k)$ is $\sum_{j=0}^{k}(-1)^j \binom{k}{j}(k-j)^n$. The number of onto functions from an $n$-element set to a $k$-element set equals the number of ways that $n$ different objects can be distributed among $k$ different boxes with none left empty. Let $A_i$ be the subset of distributions with box $i$ empty. Then

$$|A_i| = (k-1)^n \quad \text{with } \binom{k}{1} \text{ choices for } i$$
$$|A_i \cap A_j| = (k-2)^n \quad \text{with } \binom{k}{2} \text{ choices for } i \text{ and } j$$
$$\vdots$$
$$|A_{i_1} \cap A_{i_2} \cap \cdots \cap A_{i_k}| = (k-k)^n \quad \text{with } \binom{k}{k} \text{ choices for } i_1, i_2, \ldots, i_k.$$

The number of distributions that leave no box empty is then $\sum_{j=0}^{k}(-1)^j \binom{k}{j}(k-j)^n$.

The number of onto functions from an $n$-element set to a $k$-element set for some values of $n$ and $k$ $(n \ge k)$ are given in the following table:

| $n$\\$k$ | 1 | 2 | 3 | 4 | 5 | 6 | 7 | 8 | 9 |
|---|---|---|---|---|---|---|---|---|---|
| 1 | 1 | | | | | | | | |
| 2 | 1 | 2 | | | | | | | |
| 3 | 1 | 6 | 6 | | | | | | |
| 4 | 1 | 14 | 36 | 24 | | | | | |
| 5 | 1 | 30 | 150 | 240 | 120 | | | | |
| 6 | 1 | 62 | 540 | 1560 | 1800 | 720 | | | |
| 7 | 1 | 126 | 1806 | 8400 | 16,800 | 15,120 | 5,040 | | |
| 8 | 1 | 254 | 5796 | 40,824 | 126,000 | 191,520 | 141,120 | 40,320 | |
| 9 | 1 | 510 | 18,150 | 186,480 | 834,120 | 1,905,120 | 2,328,480 | 1,451,520 | 362,880 |

**4.** There are 584 nonnegative integer solutions to $x_1 + x_2 + x_3 + x_4 = 20$ where $x_1 \leq 8$, $x_2 \leq 10$, and $x_3 \leq 5$. [Let $A_1$ be the set of solutions where $x_1 \geq 9$, $A_2$ the set of solutions where $x_2 \geq 11$, and $A_3$ the set of solutions where $x_3 \geq 6$. The final answer, obtained using the inclusion/exclusion principle and the techniques used in the examples of §2.3.3, is equal to $C(23,3) - |A_1 \cup A_2 \cup A_3| = C(23,3) - (C(14,3) + C(12,3) + C(17,3) - C(3,3) - C(8,3) - C(6,3) + 0) = 584.$]

**5.** The permutations $\begin{pmatrix} 1 & 2 & 3 \\ 2 & 3 & 1 \end{pmatrix}$ and $\begin{pmatrix} 1 & 2 & 3 \\ 3 & 1 & 2 \end{pmatrix}$ are derangements of $1, 2, 3$, but the permutations $\begin{pmatrix} 1 & 2 & 3 \\ 1 & 2 & 3 \end{pmatrix}$, $\begin{pmatrix} 1 & 2 & 3 \\ 1 & 3 & 2 \end{pmatrix}$, $\begin{pmatrix} 1 & 2 & 3 \\ 3 & 2 & 1 \end{pmatrix}$, and $\begin{pmatrix} 1 & 2 & 3 \\ 2 & 1 & 3 \end{pmatrix}$ are not.

**6.** *Problème des rencontres*: In the *problème des rencontres* (*matching problem*) an urn contains balls numbered 1 through $n$, and they are drawn out one at a time. A match occurs if ball $i$ is the $i$th ball drawn. The probability that no matches occur when all the balls are drawn is $\frac{D_n}{n!}$. The problem was studied by Pierre-Rémond de Montmort (1678–1719) who studied the card game treize, in which matchings of pairs of cards were counted when two decks of cards were laid out face-up.

**7.** *Problème des ménages*: The *problème des ménages*, first raised by François Lucas (1842–1891), requires that $n$ married couples be seated around a circular table so that no men are adjacent, no women are adjacent, and no husband and wife are adjacent. There are $2n! \sum_{1=0}^{n} (-1)^i (n-i)! \binom{2n-i}{i} \frac{2n}{2n-i}$ ways to seat the people. (There are $2n!$ ways to seat the $n$ women. Regardless of how this is done, by the inclusion/exclusion principle there are $\sum_{1=0}^{n} (-1)^i (n-i)! \binom{2n-i}{i} \frac{2n}{2n-i}$ ways to seat the $n$ men.)

**8.** *Determining the number $D_n$ of derangements of $\{1, \ldots, n\}$*: Let $A_i$ be the subset of permutations that fix object $i$. The permutations in the subset $A_1 \cup A_2 \cup \cdots \cup A_n$ are those that fix at least one object. Then

$$|A_i| = (n-1)! \quad \text{with } \binom{n}{1} \text{ choices for } i$$
$$|A_i \cap A_j| = (n-2)! \quad \text{with } \binom{n}{2} \text{ choices for } i \text{ and } j$$
$$\vdots$$
$$|A_{i_1} \cap A_{i_2} \cap \cdots \cap A_{i_k}| = (n-k)! \quad \text{with } \binom{n}{k} \text{ choices for } i_1, i_2, \ldots, i_k$$

Complementation and inclusion/exclusion now yield the formula in Fact 3:

$$D_n = n! - \sum_{k=1}^{n} (-1)^{k+1} \binom{n}{k} (n-k)! = n! \sum_{k=0}^{n} (-1)^k \frac{1}{k!}.$$

As $n$ becomes large, $\frac{D_n}{n!}$ approaches $e^{-1} \approx 0.368$ very rapidly.

**9.** *Hatcheck problem*: The hatchecker at a restaurant neglects to place claim checks on $n$ hats. Each of the $n$ customers is given a randomly selected hat upon exiting. What is the probability that no one receives the correct hat?

There are $n!$ possible permutations of the $n$ hats, and there are $D_n$ cases in which no one gets the correct hat. Thus, by Example 8, the probability is approximately $e^{-1}$, regardless of the number of diners.

**10.** *Rook polynomials/arrangements of objects with restricted positions*: This describes a family of assignment or matching problems, such as matching applicants to jobs where some applicants cannot be assigned to certain jobs, the problème des ménages, and the problème des rencontres. In terms of matching $n$ applicants to $n$ jobs, set up an $n \times n$ "board of possibilities" where the rows are labeled by the applicants and the columns are labeled by the jobs. Square $(i, j)$ is a *forbidden square* if applicant $i$ cannot perform job $j$; the remaining squares are *allowable squares*. An allowable arrangement is an arrangement where only allowable squares are chosen, with exactly one square chosen in each row and column.

These problems can be rephrased in terms of placing rooks on a chessboard: given a chessboard with some squares forbidden, find the number of ways of placing rooks on the allowable squares of the chessboard so that no rook can capture any other rook. (In chess a rook can move any number of squares vertically or horizontally.) For a given $n \times n$ board $B$, let $A_i$ be the number of ways to place $n$ nontaking rooks on $B$ so that the rook in row $i$ is on a forbidden square. The total number of ways to place $n$ nontaking rooks on allowable squares is

$$n! - |A_1 \cup \cdots \cup A_n| = n! - r_1(B)(n-1)! + r_2(B)(n-2)! - \cdots + (-1)^n r_n(B)0!$$

where the coefficients $r_i(B)$ are the number of ways to place $i$ nontaking rooks on forbidden squares of $B$.

A *rook polynomial* for an $n \times n$ board $B$ is a polynomial of the form

$$R(x, B) = r_0(B) + r_1(B)x + r_2(B)x^2 + \cdots + r_n(B)x^n,$$

where $r_0(B)$ is defined to be 1.

The numbers $r_i(B)$ can sometimes be found more easily by using a combination of the following two reduction techniques:

- $R(x, B) = R(x, B_1) \cdot R(x, B_2)$, if all forbidden squares of $B$ appear in two disjoint sub-boards $B_1$ and $B_2$ (the sub-boards $B_1$ and $B_2$ are disjoint if the row labels of $B$ are partitioned into two parts $S_1$ and $S_2$, the column labels of $B$ are partitioned into two parts $T_1$ and $T_2$, and $B_1$ is obtained from $S_1 \times T_1$ and $B_2$ is obtained from $S_2 \times T_2$).

- $R(x, B) = xR(x, B_1) + R(x, B_2)$, where there is a square $(i, j)$ of $B$, $B_1$ is obtained from $B$ by removing all squares in row $i$ and all squares in column $j$, and $B_2$ is obtained from $B$ by making square $(i, j)$ allowable.

It may be necessary to use these techniques repeatedly to obtain boards that are simple enough that the rook polynomial coefficients can be easily found.

**11.** Rook polynomials can be used to find the number of derangements of $n$ objects. The forbidden squares of the board $B$ are the squares $(i, i)$. The first reduction technique of Example 10 used repeatedly breaks $B$ into $B_1, \ldots, B_n$ where $B_i$ consists only of square $(i, i)$. Then

$$R(x, B) = R(x, B_1)R(x, B_2) \ldots R(x, B_n) = (1+x) \ldots (1+x) = (1+x)^n = \sum_{i=0}^{n} \binom{n}{i} x^i.$$

Therefore, the number of derangements is

$$n! - \left[ \binom{n}{1}(n-1)! - \binom{n}{2}(n-2)! + \cdots + (-1)^{n+1}\binom{n}{n}0! \right] = n!\sum_{k=0}^{n}(-1)^k \frac{1}{k!}.$$

# 2.5  PARTITIONS

Each way to write a positive integer $n$ as a sum of positive integers is called a partition of $n$. Similarly, each way to decompose a set $S$ into a family of mutually disjoint nonempty subsets is called a partition of $S$. In a cyclic partition of a set, the elements of each subset are arranged into cycles, and two cyclic partitions in the same family of subsets are distinct if any of the cycle arrangements are different. The main concerns are counting the number of essentially different partitions of integers and sets, and with counting cyclic partitions of sets.

## 2.5.1  PARTITIONS OF INTEGERS

A positive integer can be decomposed into a sum of positive integers in various ways, taking into account restrictions on the number of parts or on the properties of the parts.

### Definitions:

A **partition** of a positive integer $n$ is a representation of $n$ as the sum of positive integers. The parts are usually written in nonascending order, but order is ignored.

A **Ferrers diagram** or **Young diagram** of a partition is an array of boxes, nodes, or dots into rows of nonincreasing size so that each row represents one part of the partition (also see §2.8 on Young Tableaux).

The **conjugate** of a partition is the partition obtained by transposing the rows and columns of its Ferrers diagram.

Suppose we have two distinguished copies of the positive integers: one ordinary copy and one copy where each integer $k$ is marked with an overline ($\bar{k}$). An **overpartition** of a positive integer $n$ is a representation of $n$ as a sum of positive integers in which at most one occurrence of each integer appearing as a part may be overlined.

A **composition** is a partition in which the order of the parts is taken into account.

A **vector partition** is a decomposition of an $n$-tuple of nonnegative integers into a sum of nonzero $n$-tuples of nonnegative integers, where order is ignored.

A **vector composition** is the same as a vector partition, except that order is taken into account.

### Facts:

**1.** The following table gives the notation for various functions that count partitions:

| function | type of partitions counted |
|---|---|
| $p(n)$ | number of partitions of $n$ |
| $Q(n)$ | number of partitions of $n$ into distinct parts |
| $\mathcal{O}(n)$ | number of partitions of $n$ into odd parts |
| $p_m(n)$ | number of partitions of $n$ with at most $m$ parts |
| $q_m(n)$ | number of partitions of $n$ with no part larger than $m$ |
| $p(N, M, n)$ | number of partitions of $n$ into at most $M$ parts, with each part no larger than $N$ |
| $\bar{p}(n)$ | number of overpartitions of $n$ |

**2.** $p(m, n, n) = q_m(n)$.

**3.** $p(n, m, n) = p_m(n)$.

**4.** $p_m(n) = q_m(n)$.

**5.** $\mathcal{O}(n) = Q(n)$.

**6.** The number of compositions of $n$ into $k$ parts is $\binom{n-1}{n-k} = \binom{n-1}{k-1}$.

**7.** The number of compositions of $n$ is $2^{n-1}$.

**8.** The number of compositions of $n + 1$ into parts greater than $1$ = the number of compositions of $n$ into odd parts = the number of compositions of $n - 1$ into 1s and 2s = $F_n$, where $F_n$ denotes the $n$th Fibonacci number: $F_0 = 0$, $F_1 = 1$, $F_n = F_{n-1} + F_{n-2}$ for $n \geq 2$ (§3.1.2).

**9.** The partition function $p(n)$ satisfies these congruences (see [Kn93] for details):

$$p(5n + 4) \equiv 0 \pmod 5$$
$$p(7n + 5) \equiv 0 \pmod 7$$
$$p(11n + 6) \equiv 0 \pmod{11}.$$

**10.** The partition functions $p(n)$ and $p_m(n)$ satisfy these recurrences:

$$p(n) - p(n - 1) - p(n - 2) + p(n - 5) + p(n - 7) + \cdots$$
$$+ (-1)^k p(n - \frac{k}{2}(3k - 1)) + (-1)^k p(n - \frac{k}{2}(3k + 1)) + \cdots = 0, \ n > 0$$
$$p_m(n) = p_m(n - m) + p_{m-1}(n).$$

**11.** The asymptotic behavior of $p(n)$, $Q(n)$, $p_m(n)$, and $\bar{p}(n)$ is as follows (see Chapters 5 and 6 of [An98], [HaRa18], or [Kn93] for details):

$$p(n) \sim \frac{1}{4n\sqrt{3}} e^{\pi\sqrt{2n/3}} \quad \text{as } n \to \infty,$$

$$Q(n) \sim \frac{1}{4 \cdot 3^{1/4}} n^{-3/4} e^{\pi\sqrt{n/3}} \quad \text{as } n \to \infty,$$

$$\bar{p}(n) \sim \frac{1}{8n} e^{\pi\sqrt{n}} \quad \text{as } n \to \infty,$$

$$p_m(n) \sim \frac{n^{m-1}}{m!(m-1)!} \quad \text{as } n \to \infty, \text{ with } m \text{ fixed.}$$

**12.** The following are generating functions (§3.2) for partition functions:

$$\sum_{n\geq 0} p(n)q^n = \prod_{i=1}^{\infty}(1 + q^i + q^{i+i} + \cdots) = \prod_{i=1}^{\infty}\left(\sum_{m=0}^{\infty} q^{mi}\right) = \prod_{i=1}^{\infty} \frac{1}{1 - q^i}$$

$$= \left(\sum_{k=-\infty}^{\infty} (-1)^k q^{k(3k-1)/2}\right)^{-1} = \frac{e^{\pi i \tau/12}}{\eta(\tau)},$$

where $q = e^{2\pi i \tau}$ and $\eta(\tau)$ is the Dedekind eta function.

$$\sum_{n\geq 0} Q(n)q^n = \prod_{i=1}^{\infty}(1 + q^i).$$

$$\sum_{n\geq 0} p_m(n)q^n = \prod_{i=1}^{m}(1 + q^i + q^{i+i} + \cdots) = \prod_{i=1}^{m}\left(\sum_{m=0}^{\infty} q^{mi}\right) = \prod_{i=1}^{m} \frac{1}{1 - q^i}.$$

$$\sum_{n\geq 0} p(N, M, n)q^n = \prod_{j=1}^{N} \frac{(1 - q^{N+M+1-j})}{1 - q^j} = \frac{\prod_{j=1}^{N+M}(1 - q^j)}{\prod_{j=1}^{N}(1 - q^j)\prod_{j=1}^{M}(1 - q^j)}.$$

$$\sum_{n\geq 0}\bar{p}(n)q^n = \prod_{i=1}^{\infty}\frac{1+q^i}{1-q^i} = \left(\sum_{k=-\infty}^{\infty}(-1)^k q^{k^2}\right)^{-1} = \frac{1}{\vartheta_4(0,q)},$$

where $\vartheta_4(z,q) = \sum_{n=-\infty}^{\infty}(-1)^n q^{n^2} e^{2niz}$ is an elliptic theta function of Jacobi; see Chapter XXI of [WhWa96].

*Note:*   Even though these expressions for $p(N,M,n)$ look like quotients of polynomials they are actually just polynomials of degree $NM$. They are called *Gaussian polynomials* or *q-binomial coefficients*. (See Chapters 1 and 2 of [An98], Chapter 19 of [HaWr08], or [Ma04] for details. Also see §2.3.2.)

**13.** The following are additional generating functions for partition functions (see Chapter 2 of [An98] or Section 8.10 of [GaRa04] for details):

$$\sum_{n=1}^{\infty}p(n)q^n = 1 + \sum_{n=1}^{\infty}\frac{q^n}{(1-q)(1-q^2)\cdots(1-q^n)}$$

$$= 1 + \sum_{n=1}^{\infty}\frac{q^{n^2}}{(1-q)^2(1-q^2)^2\cdots(1-q^n)^2}$$

$$\sum_{n=1}^{\infty}Q(n)q^n = 1 + \sum_{n=1}^{\infty}\frac{q^{n(n+1)/2}}{(1-q)(1-q^2)\cdots(1-q^n)}$$

$$= 1 + q + \sum_{n=2}^{\infty}q^n(1+q)(1+q^2)\cdots(1+q^{n-1})$$

$$\sum_{n=1}^{\infty}p_m(n)q^n = 1 + \sum_{n=1}^{\infty}\frac{(1-q^m)(1-q^{m+1})\cdots(1-q^{m+n-1})}{(1-q)(1-q^2)\cdots(1-q^n)}q^n$$

$$\sum_{n=1}^{\infty}\bar{p}(n)q^n = 1 + \sum_{n=1}^{\infty}\frac{q^{n(n+1)/2}(1+1)(1+q)(1+q^2)\cdots(1+q^{n-1})}{(1-q)^2(1-q^2)^2\cdots(1-q^n)^2}.$$

**14.** See [GrKnPa94] for an algorithm for generating partitions.

**15.** The following table gives some values of $p_m(n)$. More extensive tables appear in [GuGwMi58]:

| $n \backslash m$ | 0 | 1 | 2 | 3 | 4 | 5 | 6 | 7 | 8 | 9 | 10 |
|---|---|---|---|---|---|---|---|---|---|---|---|
| 0 | 1 | 1 | 1 | 1 | 1 | 1 | 1 | 1 | 1 | 1 | 1 |
| 1 | 0 | 1 | 1 | 1 | 1 | 1 | 1 | 1 | 1 | 1 | 1 |
| 2 | 0 | 1 | 2 | 2 | 2 | 2 | 2 | 2 | 2 | 2 | 2 |
| 3 | 0 | 1 | 2 | 3 | 3 | 3 | 3 | 3 | 3 | 3 | 3 |
| 4 | 0 | 1 | 3 | 4 | 5 | 5 | 5 | 5 | 5 | 5 | 5 |
| 5 | 0 | 1 | 3 | 5 | 6 | 7 | 7 | 7 | 7 | 7 | 7 |
| 6 | 0 | 1 | 4 | 7 | 9 | 10 | 11 | 11 | 11 | 11 | 11 |
| 7 | 0 | 1 | 4 | 8 | 11 | 13 | 14 | 15 | 15 | 15 | 15 |
| 8 | 0 | 1 | 5 | 10 | 15 | 18 | 20 | 21 | 22 | 22 | 22 |
| 9 | 0 | 1 | 5 | 12 | 18 | 23 | 26 | 28 | 29 | 30 | 30 |
| 10 | 0 | 1 | 6 | 14 | 23 | 30 | 35 | 38 | 40 | 41 | 42 |

**16.** The following table gives values for $p(n)$ and $Q(n)$:

| $n$ | $p(n)$ | $Q(n)$ | $n$ | $p(n)$ | $Q(n)$ | $n$ | $p(n)$ | $Q(n)$ |
|---|---|---|---|---|---|---|---|---|
| 0 | 1 | 1 | 17 | 297 | 38 | 34 | 12,310 | 512 |
| 1 | 1 | 1 | 18 | 385 | 46 | 35 | 14,883 | 585 |
| 2 | 2 | 1 | 19 | 490 | 54 | 36 | 17,977 | 668 |
| 3 | 3 | 2 | 20 | 627 | 64 | 37 | 21,637 | 760 |
| 4 | 5 | 2 | 21 | 792 | 76 | 38 | 26,015 | 864 |
| 5 | 7 | 3 | 22 | 1,002 | 89 | 39 | 31,185 | 982 |
| 6 | 11 | 4 | 23 | 1,255 | 104 | 40 | 37,338 | 1,113 |
| 7 | 15 | 5 | 24 | 1,575 | 122 | 41 | 44,583 | 1,260 |
| 8 | 22 | 6 | 25 | 1,958 | 142 | 42 | 53,174 | 1,426 |
| 9 | 30 | 8 | 26 | 2,436 | 165 | 43 | 63,261 | 1,610 |
| 10 | 42 | 10 | 27 | 3,010 | 192 | 44 | 75,175 | 1,816 |
| 11 | 56 | 12 | 28 | 3,718 | 222 | 45 | 89,134 | 2,048 |
| 12 | 77 | 15 | 29 | 4,565 | 256 | 46 | 105,558 | 2,304 |
| 13 | 101 | 18 | 30 | 5,604 | 296 | 47 | 124,754 | 2,590 |
| 14 | 135 | 22 | 31 | 6,842 | 340 | 48 | 147,273 | 2,910 |
| 15 | 176 | 27 | 32 | 8,349 | 390 | 49 | 173,525 | 3,264 |
| 16 | 231 | 32 | 33 | 10,143 | 448 | 50 | 204,226 | 3,658 |

**Examples:**

**1.** The number 4 has five partitions:

$$4 \quad 3+1 \quad 2+2 \quad 2+1+1 \quad 1+1+1+1.$$

**2.** The number 4 has eight compositions:

$$4 \quad 1+3 \quad 3+1 \quad 2+2 \quad 2+1+1 \quad 1+2+1 \quad 1+1+2 \quad 1+1+1+1.$$

**3.** The number 4 has fourteen overpartitions:

$$4 \quad \bar{4} \quad 3+1 \quad \bar{3}+1 \quad 3+\bar{1} \quad \bar{3}+\bar{1} \quad 2+2 \quad 2+\bar{2} \quad 2+1+1 \quad \bar{2}+1+1 \quad 2+1+\bar{1} \quad \bar{2}+1+\bar{1}$$

$$1+1+1+1 \quad 1+1+1+\bar{1}$$

**4.** The four vector partitions of $(2, 1)$ are

$$(2, 1) \quad (2, 0) + (0, 1) \quad (1, 0) + (1, 0) + (0, 1) \quad (1, 0) + (1, 1)$$

**5.** The partition $5+4+4+2+1+1+1$ of 18 has the Ferrers diagram in part (a) of the following figure. Its conjugate is the partition $7+4+3+3+1$, with the Ferrers diagram in part (b) of the figure.

**6.** *Identical balls into identical bins:* The number of ways that $n$ identical balls can be placed into $k$ identical bins, with any number of balls allowed in each bin, is given by $p_k(n)$.

**7.** *Identical balls into identical bins with no bin allowed to be empty:* The number of ways that $n$ identical balls can be placed into $k$ identical bins $(n \geq k)$, with any number of balls allowed in each bin and no bin allowed to remain empty, is given by $p_k(n) - p_{k-1}(n)$.

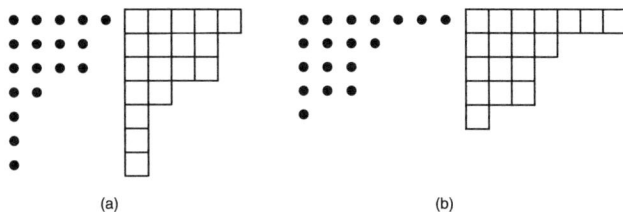

(a)                                    (b)

**8.** *Making change:* Imagine an unlimited supply of silver coins in all possible unit denominations $1, 2, 3, 4, \ldots$, and exactly one gold coin for each unit denomination $1, 2, 3, 4, \ldots$. Within a given denomination, the coins are only distinguishable by metal content. The number of ways to "make change" for $n$ units using only the silver coins is $p(n)$, while the number of ways to make change for $n$ units using the available silver *and* gold coins is $\bar{p}(n)$.

---

### 2.5.2   STIRLING COEFFICIENTS

#### Definitions:

A *cyclic partition* of a set is a partition of the set (into disjoint subsets whose union is the entire set) where the elements of each subset are arranged into cycles. Two cyclic partitions using the same family of subsets are distinct if any of the cycle arrangements are different.

The **Stirling cycle number** $\left[ \begin{smallmatrix} n \\ k \end{smallmatrix} \right]$ is the number of ways to partition $n$ objects into $k$ nonempty cycles.

The **Stirling number of the first kind** $s(n, k)$ is the coefficient of $x^k$ in the polynomial $x(x-1)(x-2)\ldots(x-n+1)$. Thus,

$$\sum_{k=0}^{n} s(n,k)x^k = x(x-1)(x-2)\ldots(x-n+1).$$

The **Stirling subset number** $\left\{ \begin{smallmatrix} n \\ k \end{smallmatrix} \right\}$ is the number of ways to partition a set of $n$ objects into $k$ nonempty subsets.

The **Stirling numbers of the second kind** $S(n, k)$ are defined implicitly by the equation

$$x^n = \sum_{k=0}^{n} S(n,k)x(x-1)(x-2)\ldots(x-k+1).$$

The **Bell number** $B_n$ (§3.1.8) is the number of partitions of a set of $n$ objects. (Eric Temple Bell, 1883–1960)

#### Facts:

**1.** $s(n, k)(-1)^{n-k} = \left[ \begin{smallmatrix} n \\ k \end{smallmatrix} \right]$.

**2.** $S(n, k) = \left\{ \begin{smallmatrix} n \\ k \end{smallmatrix} \right\}$.

**3.** The following table gives Stirling numbers of the first kind, $s(n, k)$:

| $n \backslash k$ | 0 | 1 | 2 | 3 | 4 | 5 | 6 | 7 | 8 | 9 | 10 |
|---|---|---|---|---|---|---|---|---|---|---|---|
| 0 | 1 | | | | | | | | | | |
| 1 | 0 | 1 | | | | | | | | | |
| 2 | 0 | −1 | 1 | | | | | | | | |
| 3 | 0 | 2 | −3 | 1 | | | | | | | |
| 4 | 0 | −6 | 11 | −6 | 1 | | | | | | |
| 5 | 0 | 24 | −50 | 35 | −10 | 1 | | | | | |
| 6 | 0 | −120 | 274 | −225 | 85 | −15 | 1 | | | | |
| 7 | 0 | 720 | −1,764 | 1,624 | −735 | 175 | −21 | 1 | | | |
| 8 | 0 | −5,040 | 13,068 | −13,132 | 6,769 | −1,960 | 322 | −28 | 1 | | |
| 9 | 0 | 40,320 | −109,584 | 118,124 | −67,284 | 22,449 | −4,536 | 546 | −36 | 1 | |
| 10 | 0 | −362,880 | 1,026,576 | −1,172,700 | 723,680 | −269,325 | 63,273 | −9,450 | 870 | −45 | 1 |

**4.** The following table gives Stirling subset numbers of the second kind, $S(n, k) = \left\{ {n \atop k} \right\}$:

| $n \backslash k$ | 0 | 1 | 2 | 3 | 4 | 5 | 6 | 7 | 8 | 9 | 10 |
|---|---|---|---|---|---|---|---|---|---|---|---|
| 0 | 1 | | | | | | | | | | |
| 1 | 0 | 1 | | | | | | | | | |
| 2 | 0 | 1 | 1 | | | | | | | | |
| 3 | 0 | 1 | 3 | 1 | | | | | | | |
| 4 | 0 | 1 | 7 | 6 | 1 | | | | | | |
| 5 | 0 | 1 | 15 | 25 | 10 | 1 | | | | | |
| 6 | 0 | 1 | 31 | 90 | 65 | 15 | 1 | | | | |
| 7 | 0 | 1 | 63 | 301 | 350 | 140 | 21 | 1 | | | |
| 8 | 0 | 1 | 127 | 966 | 1,701 | 1,050 | 266 | 28 | 1 | | |
| 9 | 0 | 1 | 255 | 3,035 | 7,770 | 6,951 | 2,646 | 462 | 36 | 1 | |
| 10 | 0 | 1 | 511 | 9,330 | 34,501 | 42,525 | 22,827 | 5,880 | 750 | 45 | 1 |

**5.** $B_n = \sum_{k=1}^{n} \left\{ {n \atop k} \right\}.$

**6.** The first fifteen Bell numbers are:

$B_1 = 1$          $B_2 = 2$          $B_3 = 5$          $B_4 = 15$
$B_5 = 52$         $B_6 = 203$        $B_7 = 877$        $B_8 = 4,140$
$B_9 = 21,147$     $B_{10} = 115,975$ $B_{11} = 678,570$ $B_{12} = 4,213,597$
$B_{13} = 27,644,437$  $B_{14} = 190,899,322$  $B_{15} = 1,382,958,545.$

**7.** The following table lists some identities involving Stirling numbers:

$$\left[\begin{matrix}n\\k\end{matrix}\right] = (n-1)\left[\begin{matrix}n-1\\k\end{matrix}\right] + \left[\begin{matrix}n-1\\k-1\end{matrix}\right],\ (k>0) \qquad\qquad \textit{cycle number recursion}$$

$$\left[\begin{matrix}n\\0\end{matrix}\right] = \begin{cases} 0, & \text{if } n \neq 0 \\ 1, & \text{if } n = 0 \end{cases}$$

$$\left\{\begin{matrix}n\\k\end{matrix}\right\} = k\left\{\begin{matrix}n-1\\k\end{matrix}\right\} + \left\{\begin{matrix}n-1\\k-1\end{matrix}\right\},\ (k>0) \qquad\qquad \textit{subset number recursion}$$

$$\left\{\begin{matrix}n\\0\end{matrix}\right\} = \begin{cases} 0, & \text{if } n \neq 0 \\ 1, & \text{if } n = 0 \end{cases}$$

$$\sum_k \left[\begin{matrix}n\\k\end{matrix}\right]\left\{\begin{matrix}k\\m\end{matrix}\right\}(-1)^{n-k} = \begin{cases} 0, & \text{if } n \neq m \\ 1, & \text{if } n = m \end{cases} \qquad\qquad \textit{inversion formulas}$$

$$\sum_k \left\{\begin{matrix}n\\k\end{matrix}\right\}\left[\begin{matrix}k\\m\end{matrix}\right](-1)^{n-k} = \begin{cases} 0, & \text{if } n \neq m \\ 1, & \text{if } n = m \end{cases}$$

$$\left\{\begin{matrix}n\\1\end{matrix}\right\} = \left\{\begin{matrix}n\\n\end{matrix}\right\} = 1$$

$$\left\{\begin{matrix}n\\2\end{matrix}\right\} = 2^{n-1} - 1$$

$$\left\{\begin{matrix}n\\k\end{matrix}\right\}k! = \text{the number of onto functions}$$
$$\text{from an } n\text{-set to a } k\text{-set}$$

$$\sum_{k=0}^{n} \left[\begin{matrix}n\\k\end{matrix}\right] = n!$$

$$\sum_{n=0}^{\infty} S(n+k,k)x^n = \frac{1}{(1-x)(1-2x)\dots(1-kx)}$$

$$\sum_{n=0}^{\infty} \frac{s(n,k)x^n}{n!} = \frac{(\log(1+x))^k}{k!}$$

$$\sum_{n=0}^{\infty} \frac{S(n,k)x^n}{n!} = \frac{1}{k!}(e^x - 1)^k$$

**8.** The following give combinatorial interpretations of some of the identities involving Stirling numbers:

- *Stirling cycle number recursion:* When partitioning $n$ objects into $k$ cycles, there are $\left[\begin{matrix}n-1\\k-1\end{matrix}\right]$ ways in which the last object has a cycle to itself. Otherwise, there are $\left[\begin{matrix}n-1\\k\end{matrix}\right]$ ways to partition the other $n-1$ objects into $k$ cycles, and then $n-1$ choices of a location into which the last object can be inserted.

- *Stirling subset number recursion:* When partitioning $n$ objects into $k$ nonempty subsets, there are $\left\{\begin{matrix}n-1\\k-1\end{matrix}\right\}$ ways in which the last object has a subset to itself. Otherwise, there are $\left\{\begin{matrix}n-1\\k\end{matrix}\right\}$ ways to partition the other $n-1$ objects into $k$ subsets, and then $k$ choices of a subset into which the last object can be inserted.

- $\sum_{k=0}^{n} \left[\begin{matrix}n\\k\end{matrix}\right] = n!$: The partitions into cycles are in a one-to-one correspondence with the permutations of $n$ objects, since each permutation can be represented as a composition of disjoint cycles.

- $\left[\begin{matrix}n\\k\end{matrix}\right]$ is the number of ways to seat $n$ individuals around $k$ identical circular tables so that no table is empty.

- $H_n = \frac{1}{n!}\left[\begin{matrix}n+1\\2\end{matrix}\right]$, where $H_n$ is the $n$th Harmonic number (§3.1.7).

**Examples:**

**1.** $x(x-1)(x-2)(x-3) = x^4 - 6x^3 + 11x^2 - 6x$, and hence there are $\left[{4\atop 2}\right] = 11$ permutations of $\{1,2,3,4\}$ with 2 cycles: $(12)(34)$, $(13)(24)$, $(14)(23)$, $(1)(234)$, $(1)(324)$, $(2)(134)$, $(2)(314)$, $(3)(124)$, $(3)(214)$, $(4)(123)$, $(4)(213)$. Also, $s(4,2) = (-1)^{4-2} \cdot 11$.

**2.** $H_3 = 1 + \frac{1}{2} + \frac{1}{3} = \frac{11}{6}$. As seen in Example 1, $\left[{4\atop 2}\right] = 11$. So we have $H_3 = \frac{1}{3!}\left[{4\atop 2}\right]$.

**3.** $x^4 = x(x-1)(x-2)(x-3) + 6x(x-1)(x-2) + 7x(x-1) + x$, and hence there are exactly $\left\{{4\atop 2}\right\} = 7$ set-partitions of $\{1,2,3,4\}$ into two blocks:  $\{1\}$ & $\{2,3,4\}$,  $\{2\}$ & $\{1,3,4\}$, $\{3\}$ & $\{1,2,4\}$,  $\{4\}$ & $\{1,2,3\}$,  $\{1,2\}$ & $\{3,4\}$,  $\{1,3\}$ & $\{2,4\}$,  $\{1,4\}$ & $\{2,3\}$.

# 2.6  BURNSIDE/PÓLYA COUNTING FORMULA

Burnside's lemma and Pólya's formula are used to count the number of "really different" configurations, such as tic-tac-toe patterns and placement of beads on a bracelet, in which various symmetries play a role. One of the scientific applications of Pólya's formula is the enumeration of isomers of a chemical compound. From a mathematical perspective, Burnside/Pólya methods count orbits under a permutation group action. (See §5.3.1.)

## 2.6.1  PERMUTATION GROUPS AND CYCLE INDEX POLYNOMIALS

**Definitions:**

A **permutation on a set** $S$ is a one-to-one mapping of $S$ onto itself. In this context, the elements of $S$ are called **objects**.

A permutation $\pi$ of a finite set $S$ is **cyclic** if there is a subcollection of objects that can be arranged in a cycle $(a_1 a_2 \ldots a_n)$ so that each object $a_j$ is mapped by $\pi$ onto the next object in the cycle and every object of $S$ not in this cycle is fixed by $\pi$, that is, mapped to itself.

The **tabular form** of a permutation $\pi$ on a finite set $S$ is a matrix with two rows. In the first row, each object from $S$ is listed once. Below the object $a$ is its image $\pi(a)$, in this form:

$$\begin{pmatrix} a_1 & a_2 & \cdots & a_n \\ \pi(a_1) & \pi(a_2) & \cdots & \pi(a_n) \end{pmatrix}$$

The **cycle decomposition (form)** of a permutation $\pi$ is a concatenation of cyclic permutations whose object subcollections are disjoint and whose product is $\pi$. (Sometimes the 1-cycles are explicitly written and sometimes they are omitted.)

A set $P$ of permutations of a set $S$ is **closed under composition** if the composition of each pair of permutations in $P$ is also in $P$.

A set $P$ of permutations of a set $S$ is **closed under inversion** if for every permutation $\pi \in P$, $\pi^{-1} \in P$.

A **permutation group** $\mathcal{G} = (P, S)$ is a nonempty set $P$ of permutations on a set $S$ such that $P$ is closed under composition and inversion.

The **cycle structure** of a permutation $\pi$ is an expression (multivariate polynomial) of the form $x_1^{m_1} x_2^{m_2} \ldots x_k^{m_k}$, where $m_j$ is the number of cycles of size $j$ in the cyclic decomposition of $\pi$.

The **cycle index** of a permutation group $\mathcal{G}$ is the multivariate polynomial that is the sum of the cycle structures of all the permutations in $\mathcal{G}$, divided by the number of permutations in $\mathcal{G}$. The cycle index polynomial is written $P_{\mathcal{G}}(x_1, x_2, \ldots, x_n)$. (The notation $P_{\mathcal{G}}$ honors George Pólya (1887–1985) who greatly advanced the application of the cycle index polynomial to counting.)

**Facts:**

**1.** Every permutation has a tabular form.

**2.** The tabular form of a permutation is unique up to the order in which the objects of the permuted set are listed in the first row.

**3.** Every permutation has a cycle decomposition.

**4.** The cycle decomposition of a permutation into a product of disjoint cyclic permutations is unique up to the order of the factors.

**5.** The collection of all permutations on a set $S$ forms a permutation group.

**Examples:**

**1.** The permutation $\begin{pmatrix} a & b & c & d \\ c & d & a & b \end{pmatrix}$ has the cycle decomposition $(ac)(bd)$.

**2.** The symmetric group $\Sigma_3$ of all 6 possible permutations on $\{a, b, c\}$ has the following elements:

$$(a)(b)(c), \ (ab)(c), \ (ac)(b), \ (a)(bc), \ (abc), \ (acb)$$

with respective cycle structures

$$x_1^3, \ x_1 x_2, \ x_1 x_2, \ x_1 x_2, \ x_3, \ x_3.$$

Thus, the cycle index polynomial is

$$P_{\Sigma_3} = \tfrac{1}{6}\left(x_1^3 + 3x_1 x_2 + 2x_3\right).$$

**3.** The group $\Sigma_4$ of all 24 permutations on $\{a, b, c, d\}$ has the following elements:

| | | | | | |
|---|---|---|---|---|---|
| $(a)(b)(c)(d)$ | $(ab)(c)(d)$ | $(ac)(b)(d)$ | $(ad)(b)(c)$ | $(a)(bc)(d)$ | $(a)(bd)(c)$ |
| $(a)(b)(cd)$ | $(abc)(d)$ | $(acb)(d)$ | $(abd)(c)$ | $(adb)(c)$ | $(acd)(b)$ |
| $(adc)(b)$ | $(a)(bcd)$ | $(a)(bdc)$ | $(ab)(cd)$ | $(ac)(bd)$ | $(ad)(bc)$ |
| $(abcd)$ | $(abdc)$ | $(acbd)$ | $(acdb)$ | $(adbc)$ | $(adcb)$ |

The cycle index polynomial is

$$P_{\Sigma_4} = \tfrac{1}{24}\left[x_1^4 + 6x_1^2 x_2 + 8x_1 x_3 + 3x_2^2 + 6x_4\right].$$

## 2.6.2  ORBITS AND SYMMETRIES

**Definitions:**

Given a permutation group $\mathcal{G} = (P, S)$, the **orbit** of $a \in S$ is the set $\{\pi(a) \mid \pi \in P\}$.

A **symmetry of a figure** (or **symmetry motion**) is a spatial motion of the figure onto itself.

**Facts:**

**1.** Given a permutation group $\mathcal{G} = (P, S)$, the relation $R$ defined by

$$aRb \iff \text{ there exists } \pi \in P \text{ such that } \pi(a) = b$$

is an equivalence relation (§1.4.2), and the equivalence classes under it are precisely the orbits.

**2.** The set of all symmetries on a figure forms a group.

**3.** The set of symmetries on a polygon induces a permutation group action on its corner set and a permutation group action on its edge set.

**Examples:**

**1.** Acting on the set $\{a, b, c, d, e\}$ is the following permutation group:

$$(a)(b)(c)(d)(e), (ab)(c)(d)(e), (a)(b)(cd)(e), \text{ and } (ab)(cd)(e).$$

The orbits of this group are $\{a, b\}, \{c, d\}, \{e\}$. The cycle index is

$$\tfrac{1}{4}\left[x_1^5 + 2x_1^3 x_2 + x_1 x_2^2\right].$$

**2.** A square with corners $a, b, c, d$ (in clockwise order) has eight possible symmetries: four rotations in the plane around the center of the square and four reflections (which could also be achieved by $180°$ spatial rotations out of the plane). See the following figure.

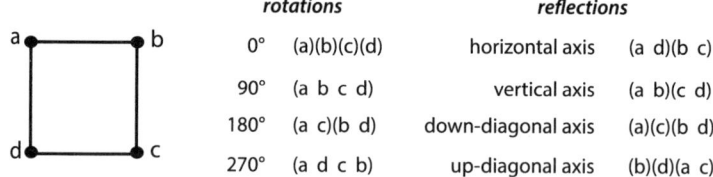

| | rotations | | reflections | |
|---|---|---|---|---|
| | 0° | (a)(b)(c)(d) | horizontal axis | (a d)(b c) |
| | 90° | (a b c d) | vertical axis | (a b)(c d) |
| | 180° | (a c)(b d) | down-diagonal axis | (a)(c)(b d) |
| | 270° | (a d c b) | up-diagonal axis | (b)(d)(a c) |

There is only one orbit, $\{a, b, c, d\}$, and the cycle index for the group of symmetries of a square acting on its corner set (the dihedral group $D_4$) is

$$P_{D_4} = \tfrac{1}{8}\left[x_1^4 + 2x_4 + 3x_2^2 + 2x_1^2 x_2\right].$$

**3.** A pentagon has ten different symmetries. Five are rotations in the plane around the center of the pentagon: $0° = (a)(b)(c)(d)(e), 72° = (abcde), 144° = (acebd), 216° = (adbec),$ and $288° = (aedcb)$; five are reflections (or equivalently, spatial rotations of $180°$ out of the plane) around axis lines through a corner and the middle of an opposite side: $(a)(be)(cd), (b)(ac)(de), (c)(ae)(bd), (d)(ab(ce),$ and $(e)(ad)(bc)$. See the following figure. There is only one orbit, $\{a, b, c, d, e\}$, and the associated cycle index is $\tfrac{1}{10}\left[x_1^5 + 4x_5 + 5x_1 x_2^2\right]$.

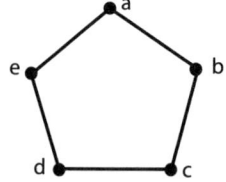

## 2.6.3 COLOR PATTERNS AND INDUCED PERMUTATIONS

### Definitions:

A *coloring* of a set $S$ from a set of $n$ colors is a function from $S$ to the set $\{1,\ldots,n\}$, whose elements are regarded as "colors". The set of all such colorings is denoted $C(S,n)$.

A *corner coloring* of a (polygonal or polyhedral) geometric figure is a coloring of its set of corners.

An *edge coloring* of a geometric figure is a coloring of its set of edges.

Let $c_1$ and $c_2$ be colorings of the set $S$ and let $\pi$ be a permutation of $S$. Write $\pi(c_1) = c_2$ if $c_1(a) = c_2(\pi(a))$ for every $a \in S$. The correspondence $c_1 \mapsto c_1 \circ \pi^{-1}$ is the **map induced by $\pi$ on the colorings of $S$**. (The composition $c_1 \circ \pi^{-1}$ assigns a color to every object $a \in S$, namely the color $c_1(\pi^{-1}(a))$.

Two corner colorings of a figure are **equivalent** if one can be mapped to the other by a symmetry. Similar definitions apply to edge colorings and to face colorings.

Two colorings $c_1$ and $c_2$ of a set $S$ are **equivalent under a group** $\mathcal{G} = (P, S)$ if there is a permutation $\pi \in P$ such that $\pi(c_1) = c_2$.

A **corner coloring pattern of a figure with respect to a set of symmetries** is a set of mutually equivalent colorings of the figure.

### Facts:

**1.** Let $\mathcal{G} = (P, S)$ be a permutation group. Then the induced action of $P$ on the set $C(S, n)$ of colorings with $n$ colors is a permutation group action.

**2.** When $P$ acts on the set $C(S, n)$ of colorings of $S$, the numbers of permuted objects and orbits, and the cycle index polynomial, are different from when $P$ acts on $S$ itself.

**3.** In permuting the set $S$ of corners of a figure, a symmetry of a figure simultaneously induces a permutation of the set of all its corner colorings. An analogous fact holds for edge colorings.

### Examples:

**1.** In Example 2 of §2.6.2, a permutation group of eight elements acts on the four corners of a square. There is only one orbit, and the cycle index is $\frac{1}{8}\left[x_1^4 + 2x_4 + 3x_2^2 + 2x_1^2 x_2\right]$. The following figure shows what happens when the same group acts on the set of black-white colorings. The permuted set has 16 colorings, there are six orbits, and the cycle index polynomial is $\frac{1}{8}\left[x_1^{16} + 2x_1^2 x_2 x_4^3 + 3x_1^4 x_2^6 + 2x_1^8 x_2^4\right]$.

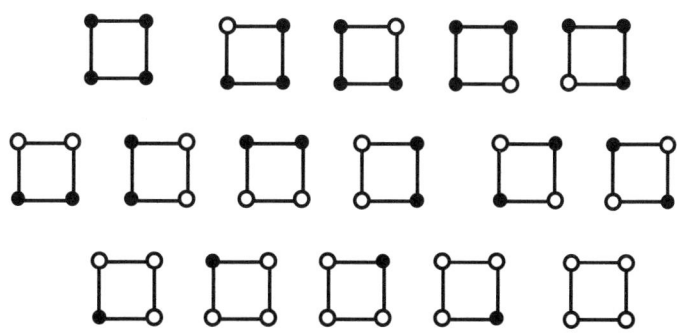

### 2.6.4   FIXED POINTS AND BURNSIDE'S LEMMA

**Definition:**
An element $a \in S$ is a **fixed point** of the permutation $\pi$ if $\pi(a) = a$. The set of all fixed points of $\pi$ is denoted $\mathit{fix}(\pi)$.

**Facts:**
**1.** The number of fixed points of a permutation $\pi$ equals the number of 1-cycles in its cycle decomposition.

**2.** *Burnside's lemma:*  Let $\mathcal{G}$ be a group of permutations acting on a set $S$. Then the number of orbits induced on $S$ is given by

$$\frac{1}{|\mathcal{G}|} \sum_{\pi \in \mathcal{G}} |\mathit{fix}(\pi)|.$$

*Note:*  The theorem commonly called "Burnside's lemma" originated with Georg Frobenius (1848–1917). A widely available book by William Burnside (1852–1927) published in 1911 stated and proved the same result, without mentioning its prior discovery.

**3.** Evaluation of the sum in Burnside's lemma is simplified by using the cycle index polynomial and Fact 1. For each term in the polynomial, multiply the coefficient by the exponent of $x_1$, and then sum these products.

**4.** *Special Burnside's lemma (for colorings):*  Let $\mathcal{G}$ be a group of permutations acting on a set $S$. Then the number of orbits induced on $C(S, n)$ (the set of colorings of $S$ from a set of $n$ colors) is given by substituting $n$ for each variable in the cycle index polynomial.

**5.** The following table gives information on the number of corner coloring patterns of selected figures.

| figure | colors | | | |
|---|---|---|---|---|
| | 2 | 3 | 4 | $m$ |
| triangle | 4 | 11 | 20 | $\frac{1}{6}[m^3 + 3m^2 + 2m]$ |
| square | 6 | 21 | 55 | $\frac{1}{8}[m^4 + 2m^3 + 3m^2 + 2m]$ |
| pentagon | 8 | 39 | 136 | $\frac{1}{10}[m^5 + 5m^3 + 4m]$ |
| hexagon | 13 | 92 | 430 | $\frac{1}{12}[m^6 + 3m^4 + 4m^3 + 2m^2 + 2m]$ |
| heptagon | 18 | 198 | 1,300 | $\frac{1}{14}[m^7 + 7m^4 + 6m]$ |
| octagon | 30 | 498 | 4,183 | $\frac{1}{16}[m^8 + 4m^5 + 5m^4 + 2m^2 + 4m]$ |
| nonagon | 46 | 1,219 | 15,084 | $\frac{1}{18}[m^9 + 9m^5 + 2m^3 + 6m]$ |
| decagon | 78 | 3,210 | 53,764 | $\frac{1}{20}[m^{10} + 5m^6 + 6m^5 + 4m^2 + 4m]$ |
| tetrahedron | 5 | 15 | 36 | $\frac{1}{12}[m^4 + 11m^2]$ |
| cube | 23 | 333 | 2914 | $\frac{1}{24}[m^8 + 17m^4 + 6m^2]$ |

**Examples:**
**1.** In Example 1 of §2.6.2, the permutation group is

$$\{(a)(b)(c)(d)(e),\ (ab)(c)(d)(e),\ (a)(b)(cd)(e),\ (ab)(cd)(e)\}.$$

The cycle index is $\frac{1}{4}\left[x_1^5 + 2x_1^3 x_2 + x_1 x_2^2\right]$. By Burnside's lemma and Fact 3 there are $\frac{1}{4}[1 \cdot 5 + 2 \cdot 3 + 1 \cdot 1] = \frac{12}{4} = 3$ orbits. The orbits are $\{a, b\}, \{c, d\}, \{e\}$.

**2.** Example 1 of §2.6.3 shows 16 colorings of the corners of the square with colors black or white. There are six orbits, and the cycle index for the action on the colorings is $\frac{1}{8}\left[x_1^{16} + 2x_1^2x_2x_4^3 + 3x_1^4x_2^6 + 2x_1^8x_2^4\right]$. By Burnside's lemma and Fact 3, there are $\frac{1}{8}\left[1 \cdot 16 + 2 \cdot 2 + 3 \cdot 4 + 2 \cdot 8\right] = \frac{48}{8} = 6$ orbits.

It is simpler to apply the Special Burnside's lemma to the cycle index for the action on the square (from Example 2 of §2.6.2), $\frac{1}{8}\left[x_1^4 + 2x_4 + 3x_2^2 + 2x_1^2x_2\right]$, which yields $\frac{1}{8}\left[1 \cdot 2^4 + 2 \cdot 2 + 3 \cdot 2^2 + 2 \cdot 2^2 \cdot 2\right] = 6$ orbits.

**3.** (*Continuing Example 3 of §2.6.2*):   The cycle index of the group of symmetries of the pentagon is $\frac{1}{10}\left[x_1^5 + 4x_5 + 5x_1x_2^2\right]$. By the Special Burnside's lemma, the number of $m$-colorings of the corners of an unoriented pentagon is $\frac{1}{10}\left[m^5 + 4m + 5m^3\right]$. For $m = 3$, the formula gives $\frac{1}{10}(243 + 12 + 135) = 39$ 3-coloring patterns of a pentagon.

**4.** A cube has 24 rotational symmetries, which act on the corners. The identity symmetry has cycle structure $x_1^8$. There are three additional classes of symmetries, as follows:

(a) Rotations of $90°, 180°$, or $270°$ about an axis line through the middles of opposite faces, for example, through $abcd$ and $efgh$ in part (a) of the following figure. A $90°$ rotation, such as $(abcd)(efgh)$, has cycle structure $x_4^2$. All $270°$ rotations have that same structure. A $180°$ rotation, such as $(ac)(bd)(eg)(fh)$, has cycle structure $x_2^4$. There are three pairs of opposite faces, and so the total contribution to the cycle index of opposite-face rotations is $6x_4^2 + 3x_2^4$.

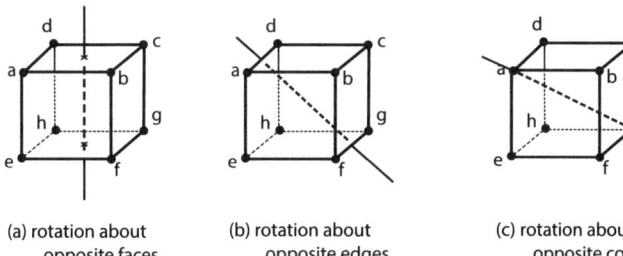

(a) rotation about          (b) rotation about          (c) rotation about
    opposite faces              opposite edges              opposite corners

(b) Rotating $180°$ about an axis line through the middles of opposite edges, for example, through edges $ad$ and $fg$ in part (b) of the figure. This rotation, $(ad)(bh)(ce)(fg)$, has cycle structure $x_2^4$. There are six pairs of opposite edges, and so the total contribution of opposite-edge rotations is $6x_2^4$.

(c) Rotating $120°$ or $240°$ about an axis line through opposite corners, for example, about the line through corners $a$ and $g$ in part (c) of the figure. Any $120°$ rotation, such as $(a)(bde)(chf)(g)$, has cycle structure $x_1^2x_3^2$. A $240°$ rotation has the same structure. There are four pairs of opposite corners, and so the contribution of opposite-corner rotations is $8x_1^2x_3^2$.

Collect terms to obtain the cycle index $\frac{1}{24}\left[x_1^8 + 6x_4^2 + 9x_2^4 + 8x_1^2x_3^2\right]$. Thus, the number of $m$-colorings of the corners of an unoriented cube is $\frac{1}{24}\left[m^8 + 6m^2 + 9m^4 + 8m^4\right]$. For $m = 2$ and 3, the formula gives 23 2-coloring patterns and 333 3-coloring patterns.

### 2.6.5  PÓLYA'S ENUMERATION FORMULA

**Definition:**

A ***pattern inventory*** is a generating function (§3.2) that enumerates the numbers of coloring patterns of a given figure.

**Facts:**

**1.** *Pólya's enumeration formula:*    Let $\mathcal{G} = (P, S)$ be a permutation group and let $\{c_1, \ldots, c_n\}$ be a set of names for $n$ colors for the objects of $S$. Then the pattern inventory with respect to $\mathcal{G}$ for the set of all $n$-colorings of $S$ is given by substituting $(c_1^j + \cdots + c_n^j)$ for $x_j$ in the cycle index $P_{\mathcal{G}}(x_1, \ldots, x_m)$.

*Note*: This theorem was published in 1937. Essentially the same result was derived by H. Redfield in 1927.

**2.** Pólya's enumeration formula has many applications in enumerating various families of graphs. This approach was pioneered by F. Harary [HaPa73].

**3.** Pólya's enumeration formula has many applications in which some practical question is modeled as a graph coloring problem.

**Examples:**

**1.** The pattern inventory of black-white colorings of the corners of a triangle is

$$1b^3 + 1b^2w + 1bw^2 + 1w^3.$$

This means there is one coloring pattern with all three corners black, one with two black corners and one white corner, etc.

**2.** (*Continuing Example 2 of §2.6.4*):  For corner colorings of the square, the cycle index is

$$P_{D_4}(x_1, x_2, x_3, x_4) = \tfrac{1}{8} \left[ x_1^4 + 2x_1^2 x_2 + 3x_2^2 + 2x_4 \right].$$

By Pólya's enumeration formula, the pattern inventory for black-white colorings of the corners of the square is

$$\begin{aligned}
P_{D_4}&[(b+w), (b^2+w^2), (b^3+w^3), (b^4+w^4)] \\
&= \tfrac{1}{8} \left[ (b+w)^4 + 2(b+w)^2(b^2+w^2) + 3(b^2+w^2)^2 + 2(b^4+w^4) \right] \\
&= \tfrac{1}{8} \left[ 8b^4 + 8b^3w + 16b^2w^2 + 8bw^3 + 8w^4 \right] = 1b^4 + 1b^3w + 2b^2w^2 + 1bw^3 + 1w^4.
\end{aligned}$$

This pattern inventory may be confirmed by examining the drawing in Example 1 of §2.6.3.

**3.** (*Continuing Example 3 of §2.6.4*):  For corner colorings of the pentagon, the cycle index is

$$P_{D_5}(x_1, x_2, \ldots, x_5) = \tfrac{1}{10} \left[ x_1^5 + 4x_5 + 5x_1 x_2^2 \right].$$

By Pólya's enumeration formula, the pattern inventory for black-white colorings of the corners of the pentagon (confirmable by drawing pictures) is

$$\begin{aligned}
P_{D_5}&((b+w), (b^2+w^2), (b^3+w^3), (b^4+w^4), (b^5+w^5)) \\
&= \tfrac{1}{10} \left[ (b+w)^5 + 4(b^5+w^5) + 5(b+w)(b^2+w^2)^2 \right] \\
&= \tfrac{1}{10} \left[ 10b^5 + 10b^4w + 20b^3w^2 + 20b^2w^3 + 10bw^4 + 10w^5 \right] \\
&= 1b^5 + 1b^4w + 2b^3w^2 + 2b^2w^3 + 1bw^4 + 1w^5.
\end{aligned}$$

**4.** (*Continuing Example 4 of §2.6.4*):  For corner colorings of the cube, the cycle index is

$$P_{\mathcal{G}}(x_1, \ldots, x_4) = \tfrac{1}{24} \left[ x_1^8 + 6x_4^2 + 9x_2^4 + 8x_1^2 x_3^2 \right].$$

By Pólya's enumeration formula, the pattern inventory for black-white colorings of the corners of the cube is

$$P_{\mathcal{G}}\left((b+w),(b^2+w^2),(b^3+w^3),(b^4+w^4)\right)$$
$$= \tfrac{1}{24}\left[(b+w)^8 + 6(b^4+w^4)^2 + 9(b^2+w^2)^4 + 8(b+w)^2(b^3+w^3)^2\right]$$
$$= b^8 + b^7w + 3b^6w^2 + 3b^5w^3 + 7b^4w^4 + 3b^3w^5 + 3b^2w^6 + bw^7 + w^8.$$

**5.** *Organic chemistry*: Two structurally different compounds with the same chemical formula are called *isomers*. For instance, to two of the six carbons $(C)$ in a ring there might be attached a hydrogen $(H)$, and to each of the four other carbons some other radical $(R)$, thereby yielding the chemical formula $C_6H_2R_4$. The number of different isomers (structurally different arrangements of the radicals) is the same as the number of coloring patterns of a hexagon when two of the corners are "colored" $H$ and four "colored" $R$. The cycle index for the symmetries of a hexagon, in terms of corner permutations, is

$$P_{D_6}(x_1,\ldots,x_6) = \tfrac{1}{12}\left[x_1^6 + 2x_6 + 2x_3^2 + 4x_2^3 + 3x_1^2x_2^2\right].$$

Substituting $(H^j + R^j)$ for $x^j$ yields a pattern inventory listing the number of isomers of $C_6H_iR_{6-i}$:

$$\tfrac{1}{12}\left[(H+R)^6 + 2(H^6+R^6) + 2(H^3+R^3)^2 + 4(H^2+R^2)^3 + 3(H+R)^2(H^2+R^2)^2\right]$$
$$= \tfrac{1}{12}\left[12H^8 + 12H^5R + 36H^4R^2 + 36H^3R^3 + 36H^2R^4 + 12HR^5 + 12R^6\right]$$
$$= 1H^8 + 1H^5R + 3H^4R^2 + 3H^3R^3 + 3H^2R^4 + 1HR^5 + 1R^6.$$

The three possible coloring patterns corresponding to $3H^2R^4$ are shown in the following figure:

---

## 2.7   MÖBIUS INVERSION COUNTING

Möbius inversion is an important tool used to solve a variety of counting problems such as counting how many numbers are relatively prime to some given number (without individually checking each smaller number) and counting certain types of circular arrangements. It generalizes the principle of inclusion/exclusion. (Augustus Ferdinand Möbius, 1790–1868)

## 2.7.1   MÖBIUS INVERSION

**Definitions:**

The **Kronecker delta function** $\delta(x, y)$ is defined by the rule

$$\delta(x, y) = \begin{cases} 1 & \text{if } x = y \\ 0 & \text{otherwise.} \end{cases}$$

The **Möbius function** is the function $\mu$ from the set of positive integers to the set of integers where

$$\mu(m) = \begin{cases} 1 & \text{if } m = 1 \\ (-1)^k & \text{if } m = p_1 p_2 \dots p_k \text{ (the product of } k \text{ distinct primes)} \\ 0 & \text{if } m \text{ is divisible by the square of a prime.} \end{cases}$$

*Note:* See Chapter 11 for Möbius functions defined on partially ordered sets.

For a positive integer $n$, the **Euler phi-function** $\phi(n)$ is the number of positive integers less than $n$ that are relatively prime to $n$.

**Facts:**

1. For the Möbius function $\mu$ defined on the set of positive integers:

   - $\mu$ is *multiplicative*: if $\gcd(m, n) = 1$, then $\mu(mn) = \mu(m)\mu(n)$;
   - $\mu$ is not *completely multiplicative*: $\mu(mn) = \mu(m)\mu(n)$ is not always true;
   - $\sum_{d|n} \mu(d)$, where the sum is taken over all positive divisors of $n$, is 1 if $n = 1$ and 0 if $n > 1$.

2. *Möbius inversion formula:* If $f(n)$ and $g(n)$ are defined for all positive integers and $f(n) = \sum_{d|n} g(d)$, then

$$g(n) = \sum_{d|n} \mu(d) f(n/d).$$

3. For every positive integer $n$, $n = \sum_{d|n} \phi(d)$.

**Examples:**

1. Applying the Möbius inversion formula to the expression given in Fact 3 produces

$$\phi(n) = \sum_{d|n} \mu(d) \frac{n}{d}.$$

As an illustration, $\phi(12) = \sum_{d|12} \mu(d) \frac{12}{d} = [\mu(1) \cdot 12 + \mu(2) \cdot 6 + \mu(3) \cdot 4 + \mu(4) \cdot 3 + \mu(6) \cdot 2 + \mu(12) \cdot 1] = 12 - 6 - 4 + 0 + 2 + 0 = 4$.

2. *Circular permutations with repetitions:* Given an alphabet of $m$ letters, how many circular permutations of length $n$ are possible, if repeated letters are allowed and two permutations are the same if the second can be obtained from the first by rotation? The problem was first solved by Percy A. MacMahon in 1892.

A circular permutation of length $n$ has a period $d$, where $d|n$. (The *period* of a circular permutation, viewed as a circular string, is the length of the shortest substring that repeats end-to-end to give the entire string.) Let $g(d)$ be the number of length $d$ circular permutations that have period $d$. A circular permutation of length $n$ can be constructed from one of length $d$ (where $d|n$) by concatenating it with itself $\frac{n}{d}$ times. For example, the circular permutation $aabbaabb$ (where beginning and end are joined) of period four can be obtained by taking the circular permutation $aabb$ and opening it up at one of four spots between the letters, to obtain any of four linear strings $aabb$, $abba$, $bbaa$, and $baab$. Join one of these to itself, obtaining $aabbaabb$, $abbaabba$, $bbaabbaa$, and $baabbaab$, and then join the beginning and the end to form the circular permutation $aabbaabb$.

For any positive integer $k$, there are $dg(d)$ linear strings of length $k$ obtained by taking $\frac{k}{d}$ repetitions of the linear strings of length $d$ that have period $d$, where $d|k$. Therefore, the total number of linear strings of length $k$ where the objects are chosen from $m$ types is $\sum_{d|k} dg(d) = m^k$. Applying the Möbius inversion formula to $m^k$ and $g$ yields $g(k) = \frac{1}{k} \sum_{d|k} \mu(d)m^{k/d}$. So the total number of circular permutations of length $n$ where the elements are chosen from an alphabet of size $m$ is $\sum_{d|n} g(d)$, which is equal to

$$\sum_{k|n}\left(\tfrac{1}{k}\sum_{d|k}\mu(d)m^{k/d}\right).$$

**3.** *Circular permutations with repetitions and specified numbers of each type of object:* Suppose there are a total of $n$ objects of $t$ types, with $a_i$ of type $i$ $(i = 1, \ldots, t)$, where $a_1 + \cdots + a_t = n$. If $a = \gcd(a_1, \ldots, a_t)$, these circular permutations can be generated as in Example 2 by taking a circular permutation of period $d$ (where $d|a$) with $\frac{a_i d}{n}$ objects of type $i$ $(i = 1, \ldots, t)$, breaking it open, and laying it end-to-end $\frac{n}{d}$ times. Let $g(k)$ be the number of such circular permutations of length $k$ that have period $k$. Then the total number of linear strings of length $n$ with $a_i$ objects of type $i$ is given by $\sum_{d|a} dg(d) = \frac{n!}{a_1! \ldots a_t!}$. By the Möbius inversion formula,

$$g(k) = \tfrac{1}{k}\sum_{d|a}\mu(d)\frac{(k/d)!}{(a_1/d)!(a_2/d)!\ldots(a_t/d)!}.$$

Summing $g(k)$ over all divisors of $a$ gives the desired total number of circular permutations:

$$\sum_{k|a}g(k) = \sum_{k|a}\left(\tfrac{1}{k}\sum_{d|a}\mu(d)\frac{(k/d)!}{(a_1/d)!\ldots(a_t/d)!}\right).$$

# 2.8 YOUNG TABLEAUX

Arrays called Young tableaux were introduced by the Reverend Alfred Young (1873–1940). These arrays are used in combinatorics and the theories of symmetric functions, which are the subject of this section. Young tableaux are also used in the analysis of representations of the symmetric group. They make it possible to approach many results about representation theory from a concrete combinatorial viewpoint.

## 2.8.1  TABLEAUX COUNTING FORMULAS

### Definitions:

The **hook** $H_{i,j}$ of cell $(i,j)$ in the Ferrers diagram for a partition $\lambda$ is the set

$$\{\, (k,j) \in \lambda \mid k \geq i \,\} \ \cup \ \{\, (i,k) \in \lambda \mid k \geq j \,\};$$

that is, the set consisting of the cell $(i,j)$, all cells in its row to its right, and all cells in its column below it.

The **hooklength** $h_{i,j}$ of cell $(i,j)$ is the number $|H_{i,j}|$ of cells in its hook.

A **Young tableau** is an array obtained by replacing each cell of the Ferrers diagram by a positive integer.

The **shape** of a Young tableau is the partition corresponding to the underlying Ferrers diagram. The notation $\lambda \vdash n$ indicates that $\lambda$ partitions the number $n$.

A Young tableau is **semistandard** (an **SSYT**) if the entries in each row are weakly increasing and the entries in each column are strictly increasing.

A semistandard Young tableau of shape $\lambda \vdash n$ is **standard** (an **SYT**) if each number $1, \ldots, n$ occurs exactly once as an entry. The number of SYT of shape $\lambda$ is denoted $f_\lambda$.

If $G$ is a group (see §5.2) then an **involution** is an element $g \in G$ such that $g^2$ is the identity. The number of involutions in the symmetric group $S_n$ (or $\Sigma_n$) (the group of all permutations on the set $\{1, 2, \ldots, n\}$) is denoted $\mathrm{inv}(n)$.

The following table summarizes notation for Young tableaux:

| notation | meaning |
|---|---|
| $\lambda = (\lambda_1, \ldots, \lambda_l)$ | partition (with parts $\lambda_1 \geq \lambda_2 \geq \cdots \geq \lambda_l$) |
| $\lambda \vdash n$ | $\lambda$ partitions the number $n$ |
| $(i, j)$ | cell in a Ferrers diagram |
| $H_{i,j}$ | hook of cell $(i, j)$ |
| $h_{i,j}$ | hooklength of hook $H_{i,j}$ |
| $f_\lambda$ | number of SYT of shape $\lambda$ |
| $\mathrm{inv}(n)$ | number of involutions in $S_n$ |

### Facts:

**1.** *Frame-Robinson-Thrall hook formula* (1954):  The number of SYT of fixed shape $\lambda$ is

$$f_\lambda = \frac{n!}{\prod_{(i,j) \in \lambda} h_{i,j}}.$$

**2.** *Frobenius determinantal formula* (1900):  The number of SYT of fixed shape $\lambda = (\lambda_1, \ldots, \lambda_l)$ is the determinant

$$f_\lambda = n! \left| \frac{1}{(\lambda_i + j - i)!} \right|_{1 \leq i,j \leq l}.$$

**3.** Summations involving the number of SYT:

$$\sum_{\lambda \vdash n} f_\lambda = \mathrm{inv}(n), \qquad \sum_{\lambda \vdash n} f_\lambda^2 = n!.$$

**4.** Young tableaux can be used to find the number of permutations with specified lengths for their longest increasing subsequences and longest decreasing subsequences [Be71].

**Examples:**

**1.** If $\lambda = (3, 2)$ then a complete list of SYT is

$$
\begin{array}{ccc}
1 \; 2 \; 3 & \quad & 1 \; 2 \; 4 & \quad & 1 \; 2 \; 5 & \quad & 1 \; 3 \; 4 & \quad & 1 \; 3 \; 5 \\
4 \; 5 & & 3 \; 5 & & 3 \; 4 & & 2 \; 5 & & 2 \; 4
\end{array}
$$

**2.** If $\lambda = (2, 2)$ then a complete list of SSYT with entries at most 3 is

$$
\begin{array}{cccccc}
1 \; 1 & 1 \; 1 & 2 \; 2 & 1 \; 1 & 1 \; 2 & 1 \; 2 \\
2 \; 2 & 3 \; 3 & 3 \; 3 & 2 \; 3 & 2 \; 3 & 3 \; 3
\end{array}
$$

**3.** For the partition $(3, 2)$, $H_{1,1} = \{(1, 1), (2, 1), (1, 2), (1, 3)\}$. In the following diagram each cell of $(3, 2)$ is replaced with its hooklength.

$$
\begin{array}{ccc}
4 & 3 & 1 \\
2 & 1 &
\end{array}
$$

The hook formula (Fact 1) gives the number of SYT of shape $(3, 2)$: $f_{(3,2)} = \frac{5!}{4 \cdot 3 \cdot 2 \cdot 1^2} = 5$. The determinantal formula (Fact 2) gives the same result:

$$
f_{(3,2)} = 5! \begin{vmatrix} \frac{1}{3!} & \frac{1}{4!} \\ \frac{1}{1!} & \frac{1}{2!} \end{vmatrix} = 5.
$$

**4.** For the partitions of $n = 3$, $f_{(3)} = 1$, $f_{(2,1)} = 2$, $f_{(1,1,1)} = 1$, so the summation formulas become

$$
\sum_{\lambda \vdash 3} f_\lambda = 4 = \mathrm{inv}(3);
$$

$$
\sum_{\lambda \vdash 3} f_\lambda^2 = 6 = 3!.
$$

## 2.8.2  TABLEAUX ALGORITHMS

**Definitions:**

An **inner corner** of a partition $\lambda$ is a cell $(i, j) \in \lambda$ such that $(i + 1, j), (i, j + 1) \notin \lambda$.

An **outer corner** of a partition $\lambda$ is a cell $(i, j) \notin \lambda$ such that $(i - 1, j), (i, j - 1) \in \lambda$.

**Facts:**

**1.** The Greene-Nijenhuis-Wilf algorithm (1979) (Algorithm 1) successively finds inner corners of a tableau of shape $\lambda$, eventually producing a random tableau of the specified shape.

---

**Algorithm 1:  Greene-Nijenhuis-Wilf.**
input: a shape $\lambda_I$ such that $\lambda_I \vdash n$
output: a standard Young tableau of shape $\lambda_I$, uniformly at random
$\lambda := \lambda_I$

---

**while** $\lambda$ is nonempty
   {Find an inner corner $(i,j) \in \lambda$}
   choose (with probability $\frac{1}{|\lambda|}$) any cell $(i,j) \in \lambda$
   **while** the current cell $(i,j)$ is not an inner corner
      choose (with probability $\frac{1}{h_{i,j}}$) a pair $(i',j') \in H_{i,j} - \{(i,j)\}$
      $(i,j) := (i',j')$
   assign label $n$ to inner corner $(i,j)$
   $\lambda := \lambda - \{(i,j)\}$

---

**2.** The Robinson-Schensted algorithm (1938, 1961) (Algorithm 2) constructs a pair $(P,Q)$ of SYT of the same shape, following a specified permutation $\pi$ that guides the insertion of entries into the two tableaux. At each step $i$, the entry $\pi_i$ is inserted into the tableau $P$ whereas $i$ is inserted (at the same position) into $Q$.

---

**Algorithm 2: Robinson-Schensted.**

input: a permutation $\pi \in S_n$ where $\pi = \begin{pmatrix} 1 & 2 & \cdots & n \\ \pi_1 & \pi_2 & \cdots & \pi_n \end{pmatrix}$

output: a pair $(P,Q)$ of standard Young tableaux of the same shape $\lambda \vdash n$
$P_0 := \emptyset$; $Q_0 := \emptyset$
**for** $k := 1$ **to** $n$
   $r := 1$; $c := 1$; $b := \pi_k$; $P_k := P_{k-1}$; $exit :=$ FALSE
   **while** $exit =$ FALSE
      {Find next insertion row $r$ in tableau $P_k$}
      **while** $row_r(P_k) \neq \emptyset$ **and** $\pi_j > \max\{row_r(P_k)\}$
         $r := r + 1$
      {Find next insertion column $c$ in tableau $P_k$}
      $c := 1$
      **while** $P_k[r,c] \neq \emptyset$ **and** $\pi_k < P_k[r,c]$
         $c := c + 1$
      {Insert $b$}
      **if** $P_k[r,c] = \emptyset$ **then**
         $P_k[r,c] := b$; $exit =$ TRUE
      **else**
         $bb := P_k[r,c]$; $P_k[r,c] := b$; $b := bb$
   $Q_k[r,c] := k$
$P := P_n$; $Q := Q_n$

---

**Examples:**

**1.** The diagrams in the following figure illustrate a plausible sequence of current cells chosen as the first step of Algorithm 1, in order to find an inner corner of a tableau of shape $\lambda = (5,5,5,2)$.

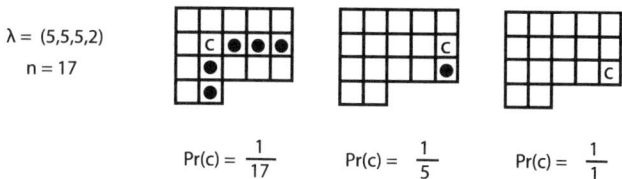

$\lambda = (5,5,5,2)$

$n = 17$

$Pr(c) = \dfrac{1}{17}$    $Pr(c) = \dfrac{1}{5}$    $Pr(c) = \dfrac{1}{1}$

**2.** The permutation $\pi = \begin{pmatrix} 1 & 2 & 3 & 4 & 5 & 6 & 7 \\ 6 & 2 & 3 & 1 & 7 & 5 & 4 \end{pmatrix}$ yields the following sequence of tableaux pairs $(P_k, Q_k)$ during the execution of Algorithm 2.

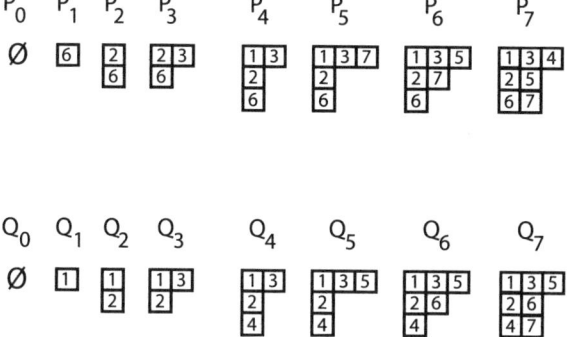

---

# REFERENCES

*Printed Resources*:

[An98] G. E. Andrews, *The Theory of Partitions*, Encyclopedia of Mathematics and its Applications, Vol. 2, Cambridge University Press, 1998.

[BeGo75] E. A. Bender and J. R. Goldman, "On the applications of Möbius inversion in combinatorial analysis", *American Mathematical Monthly* 82 (1975), 789–803. (An expository paper for a general mathematical audience.)

[Be71] C. Berge, *Principles of Combinatorics*, Academic Press, 1971.

[Co79] D. I. A. Cohen, *Basic Techniques of Combinatorial Theory*, Wiley, 1979.

[GaRa04] G. Gasper and M. Rahman, *Basic Hypergeometric Series*, 2nd ed., Cambridge University Press, 2004.

[GrKnPa94] R. L. Graham, D. E. Knuth, and O. Patashnik, *Concrete Mathematics*, 2nd ed., Addison-Wesley, 1994.

[GuGwMi58] H. Gupta, A. E. Gwyther, and J. C. P. Miller, *Tables of Partitions*, Royal Society Mathematical Tables, Vol. 4, 1958.

[Ha98] M. Hall, Jr., *Combinatorial Theory*, 2nd ed., Wiley, 1998. (Section 2.1 of this text contains a brief introduction to Möbius inversion.)

[HaPa73] F. Harary and E. M. Palmer, *Graphical Enumeration*, Academic Press, 1973.

[HaRa18] G. H. Hardy and S. Ramanujan, "Asymptotic formulae in combinatory analysis", *Proceedings of the London Mathematical Society*, Ser. 2, 17 (1918), 75–115. (Reprinted in *Collected Papers of S. Ramanujan*, Chelsea, 2000, 276–309.)

[HaWr08] G. H. Hardy and E. M. Wright, *An Introduction to the Theory of Numbers*, 6th ed., Oxford University Press, 2008.

[JaKe09] G. D. James and A. Kerber, *The Representation Theory of the Symmetric Group*, Encyclopedia of Mathematics and Its Applications, Vol. 16, Cambridge University Press, 2009.

[Kn93] M. I. Knopp, *Modular Functions in Analytic Number Theory*, 2nd ed., Chelsea, 1993.

[Kn97] D. E. Knuth, *The Art of Computer Programming, Vol. 1: Fundamental Algorithms*, 3rd ed., Addison-Wesley, 1997.

[Ma99] I. G. Macdonald, *Symmetric Functions and Hall Polynomials*, 2nd ed., Oxford University Press, 1999.

[Ma04] P. A. MacMahon, *Combinatory Analysis*, Vols. I–II, Cambridge University Press, London, 1916. (Reissued: Dover Publications, 2004.)

[Pa06] E. W. Packel, *The Mathematics of Games and Gambling*, 2nd ed., Mathematical Association of America, 2006.

[PoRe11] G. Pólya and R. C. Read, *Combinatorial Enumeration of Groups, Graphs, and Chemical Compounds*, Springer-Verlag, 2011.

[Ro12] K. H. Rosen, *Discrete Mathematics and Its Applications*, 7th ed., McGraw-Hill, 2012.

[Sa01] B. E. Sagan, *The Symmetric Group: Representations, Combinatorial Algorithms, and Symmetric Functions*, 2nd ed., Springer, 2001.

[St11] R. P. Stanley, *Enumerative Combinatorics*, Vol. I, 2nd ed., Cambridge University Press, 2011. (Chapter 3 of this text contains an introduction to Möbius functions and incidence algebras.)

[Tu12] A. Tucker, *Applied Combinatorics*, 6th ed., Wiley, 2012.

[WhWa96] E. T. Whittaker and G. N. Watson, *A Course of Modern Analysis*, 4th ed., Cambridge University Press, 1996.

[Wi06] H. S. Wilf, *generatingfunctionology*, 3rd ed., A K Peters, 2006.

**Web Resources**:

http://theory.cs.uvic.ca (Combinatorial object server.)

http://theory.cs.uvic.ca/amof/ (AMOF: the Amazing Mathematical Object Factory.)

http://www.cs.sunysb.edu/~algorith/ (The Stony Brook Algorithm Repository; see Section 1.3 on combinatorial problems.)

# 3

# *SEQUENCES*

---

## INTRODUCTION

Sequences of integers occur regularly in combinatorial applications. For example, the solution to a counting problem that depends on a parameter $k$ can be viewed as the $k$th term of a sequence. This chapter provides a guide to particular sequences that arise in applied settings. Such (infinite) sequences can frequently be represented in a finite form. Specifically, sequences can be expressed using generating functions, recurrence relations, or by an explicit formula for the $k$th term of the sequence.

---

## GLOSSARY

**antidifference** (of a function $f$): any function $g$ such that $\Delta g = f$. It is the discrete analogue of antidifferentiation.

**ascent** (in a permutation $\pi$): any index $i$ such that $\pi_i < \pi_{i+1}$.

**asymptotic equality** (of functions): the function $f(n)$ is asymptotic to $g(n)$, written $f(n) \sim g(n)$, if $f(n) \neq 0$ for sufficiently large $n$ and $\lim_{n \to \infty} \frac{g(n)}{f(n)} = 1$.

**Bernoulli numbers**: the numbers $B_n$ produced by the recursive definition $B_0 = 1$, $\sum_{j=0}^{n} \binom{n+1}{j} B_j = 0$, $n \geq 1$.

**Bernoulli polynomials**: the polynomial $B_m(x) = \sum_{k=0}^{m} \binom{m}{k} B_k x^{m-k}$ where $B_k$ is the $k$th Bernoulli number.

**big oh** (of the function $f$): the set of all functions that do not grow faster than some constant multiple of $f$, written $O(f(n))$.

**big omega** (of the function $f$): the set of all functions that grow at least as fast as some constant multiple of $f$, written $\Omega(f(n))$.

**big theta** (of the function $f$): the set of all functions that grow roughly as fast as some constant multiple of $f$, written $\Theta(f(n))$.

**binomial convolution** (of the sequences $\{a_n\}$ and $\{b_n\}$): the sequence whose $r$th term is formed by summing products of the form $\binom{r}{k} a_k b_{r-k}$.

**Catalan number**: the number $C_n = \frac{1}{n+1} \binom{2n}{n}$.

**characteristic equation**: an equation derived from a linear recurrence relation with constant coefficients, whose roots are used to construct solutions to the recurrence relation.

**closed form** (for a sum): an algebraic expression for the value of a sum with variable limits, which has a fixed number of terms; hence the time needed to calculate it does not grow with the size of the set or interval of summation.

**convolution** (of the sequences $\{a_n\}$ and $\{b_n\}$): the sequence whose $r$th term is formed by summing products of the form $a_k b_{r-k}$ where $0 \leq k \leq r$.

**de Bruijn sequence**: a circular ordering of letters from a fixed alphabet with $p$ letters such that each $n$ consecutive letters (wrapping around from the end of the sequence to the beginning, if necessary) forms a different word.

**difference operator**: the operator $\Delta$ where $\Delta f(x) = f(x + 1) - f(x)$ on integer or real-valued functions. It is the discrete analogue of the differentiation operator.

**difference sequence** (for the sequence $A = \{\, a_j \mid j = 0, 1, \ldots \,\}$): the sequence $\Delta A = \{\, a_{j+1} - a_j \mid j = 0, 1, \ldots \,\}$.

**difference table** (for a function $f$): a table whose $k$th row is the $k$th difference sequence for $f$.

**discordant permutation**: a permutation that assigns to every element an image different from those assigned by all other members of a given set of permutations.

**dissimilar hypergeometric terms**: terms in two hypergeometric series such that their ratio is not a rational function.

**divide-and-conquer algorithm**: a recursive procedure that solves a given problem by first breaking it into smaller subproblems (of nearly equal size) and then combining their respective solutions.

**doubly hypergeometric**: a property of function $F(n, k)$ that $\frac{F(n+1,k)}{F(n,k)}$ and $\frac{F(n,k+1)}{F(n,k)}$ are rational functions of $n$ and $k$.

**Eulerian number**: the number of permutations of $\{1, 2, \ldots, n\}$ with exactly $k$ ascents.

**excedance** (of a permutation $\pi$): any index $i$ such that $\pi_i > i$.

**exponential generating function** (for the sequence $a_0, a_1, a_2, \ldots$): the function $f(x) = a_0 + a_1 x + a_2 \frac{x^2}{2!} + \cdots$ or any equivalent closed form expression.

**falling power** (of $x$): the product $x^{\underline{n}} = x(x - 1)(x - 2) \ldots (x - n + 1)$ of $n$ successive descending factors, starting with $x$; the discrete analogue of exponentiation.

**Fibonacci numbers**: the numbers $F_n$ produced by the recursive definition $F_0 = 0$, $F_1 = 1$, $F_n = F_{n-1} + F_{n-2}$ if $n \geq 2$.

**figurate number**: the number of cells in an array of cells bounded by some regular geometrical figure.

**first-order linear recurrence relation with constant coefficients**: an equation of the form $C_0 a_n + C_1 a_{n-1} = f(n)$, $n \geq 1$, with $C_0$ and $C_1$ nonzero real constants.

**generating function** (for the sequence $a_0, a_1, a_2, \ldots$): the function $f(x) = a_0 + a_1 x + a_2 x^2 + \cdots$ or any equivalent closed form expression; sometimes called the *ordinary* generating function for the sequence.

**geometric series**: an infinite series where the ratio between two consecutive terms is a constant.

**Gray code** (of size $n$): a circular ordering of all binary strings of length $n$ in which adjacent strings differ in exactly one bit.

**harmonic number**: the sum $H_n = \sum_{i=1}^{n} \frac{1}{i}$, which is the discrete analogue of the natural logarithm.

**homogeneous recurrence relation**: a recurrence relation satisfied by the identically zero sequence.

**hypergeometric series**: a series where the ratio of two consecutive terms is a rational function.

**indefinite sum** (of the function $f$): the family of all antidifferences of $f$.

**Lah coefficients**: the coefficients resulting from expressing the rising factorial in terms of the falling factorials.

**linear recurrence relation with constant coefficients**: an equation of the form $C_0 a_n + C_1 a_{n-1} + \cdots + C_k a_{n-k} = f(n)$, $n \geq k$, where the $C_i$ are real constants with $C_0$ and $C_k$ nonzero.

**little oh** (of the function $f$): the set of all functions that grow slower than every constant multiple of $f$, written $o(f(n))$.

**little omega** (of the function $f$): the set of all functions that grow faster than every constant multiple of $f$, written $\omega(f(n))$.

**Lucas numbers**: the numbers $L_n$ produced by the recursive definition $L_0 = 2$, $L_1 = 1$, $L_n = L_{n-1} + L_{n-2}$ if $n \geq 2$.

**nonhomogeneous recurrence relation**: a recurrence relation that is not homogeneous.

**polyomino**: a connected configuration of regular polygons (e.g., triangles, squares, or hexagons) in the plane, generalizing a domino.

**power sum**: the sum of the $k$th powers of the integers $1, 2, \ldots, n$.

**radius of convergence** (for the series $\sum a_n x^n$): the number $r$ $(0 \leq r \leq \infty)$ such that the series converges for all $|x| < r$ and diverges for all $|x| > r$.

**Ramsey number**: the number $R(m, n)$ defined as the smallest positive integer $k$ with the following property: if $S$ is a set of size $k$ and the 2-element subsets of $S$ are partitioned into two collections, $C_1$ and $C_2$, then there is a subset of $S$ of size $m$ such that each of its 2-element subsets belong to $C_1$ or there is a subset of $S$ of size $n$ such that each of its 2-element subsets belong to $C_2$.

**recurrence relation**: an equation expressing a term of a sequence as a function of prior terms in the sequence.

**rising power** (of $x$): the product $x^{\overline{n}} = x(x+1)(x+2)\ldots(x+n-1)$ of $n$ successive ascending terms, starting with $x$.

**second-order linear recurrence relation with constant coefficients**: an equation of the form $C_0 a_n + C_1 a_{n-1} + C_2 a_{n-2} = f(n)$, $n \geq 2$, where $C_0, C_1, C_2$ are real constants with $C_0$ and $C_2$ nonzero.

**sequence**: a function from $\{0, 1, 2, \ldots\}$ to the real numbers (often the integers).

**shift operator**: the operator $E$ defined by $Ef(x) = f(x+1)$ on integer or real-valued functions.

**similar hypergeometric terms**: terms in two hypergeometric series such that their ratio is a rational function.

**standardized form** (for a sum): a sum over an integer interval, in which the lower limit of the summation is zero.

**Stirling's approximation formula**: the asymptotic estimate $\sqrt{2\pi n}(n/e)^n$ for $n!$.

**tangent numbers**: numbers generated by the exponential generating function $\tan x$.

## 3.1   SPECIAL SEQUENCES

### 3.1.1   REPRESENTATIONS OF SEQUENCES

A given infinite sequence $a_0, a_1, a_2, \ldots$ can often be represented in a more useful or more compact form. Namely, there may be a closed form expression for $a_n$ as a function of $n$, the terms of the sequence may appear as coefficients in a simple generating function, or the sequence may be specified by a recurrence relation. Each representation has advantages, in either defining the sequence or establishing information about its terms.

**Definitions:**

A **sequence** $\{\, a_n \mid n \geq 0 \,\}$ is a function from the set of nonnegative integers to the real numbers (often the integers). The **terms** of the sequence $\{\, a_n \mid n \geq 0 \,\}$ are the values $a_0, a_1, a_2, \ldots$.

A **closed form** for the sequence $\{a_n\}$ is an algebraic expression for $a_n$ as a function of $n$.

A **recurrence relation** (§3.3) is an equation expressing a term of a sequence as a function of prior terms in the sequence. A **solution** of a recurrence relation is a sequence whose terms satisfy the relation.

The (**ordinary**) **generating function** (§3.2.1) for the sequence $\{a_n\}$ is the function
$$f(x) = \sum_{i=0}^{\infty} a_i x^i$$
or any equivalent closed form expression.

The **exponential generating function** (§3.2.2) for the sequence $\{a_n\}$ is the function
$$g(x) = \sum_{i=0}^{\infty} a_i \frac{x^i}{i!}$$
or any equivalent closed form expression.

**Facts:**

**1.** An important way in which many sequences are represented is by using a recurrence relation. Although not all sequences can be represented by useful recurrence relations, many sequences that arise in the solution of counting problems can be so represented.

**2.** An important way to study a sequence is by using its generating function. Information about terms of the sequence can often be obtained by manipulating the generating function.

**Examples:**

**1.** The Fibonacci numbers $F_n$ (§3.1.2) arise in many applications and are given by the sequence $0, 1, 1, 2, 3, 5, 8, 13, \ldots$. This infinite sequence can be finitely encoded by means of the recurrence relation

$$F_n = F_{n-1} + F_{n-2}, \quad n \geq 2, \quad \text{with } F_0 = 0 \text{ and } F_1 = 1.$$

Alternatively, a closed form expression for this sequence is given by

$$F_n = \tfrac{1}{\sqrt{5}}\left[\left(\tfrac{1+\sqrt{5}}{2}\right)^n - \left(\tfrac{1-\sqrt{5}}{2}\right)^n\right], \quad n \geq 0.$$

The Fibonacci numbers can be represented in a third way, via the generating function $f(x) = \frac{x}{1-x-x^2}$. Namely, when this rational function is expanded in terms of powers of

$x$, the resulting coefficients generate the values $F_n$:

$$\frac{x}{1-x-x^2} = 0x^0 + 1x^1 + 1x^2 + 2x^3 + 3x^4 + 5x^5 + 8x^6 + 13x^7 + \cdots .$$

**2.** Table 1 gives closed form expressions for the generating functions of several combinatorial sequences discussed in this *Handbook*. In this table, $r$ is any real number. Generating functions for other sequences can be found in §3.2.1, Tables 1 and 2.

**Table 1: Generating functions for particular sequences.**

| sequence | notation | reference | closed form |
|---|---|---|---|
| $1, 2, 3, 4, 5, \ldots$ | $\{n\}$ | | $\frac{1}{(1-x)^2}$ |
| $1^2, 2^2, 3^2, 4^2, 5^2, \ldots$ | $\{n^2\}$ | | $\frac{1+x}{(1-x)^3}$ |
| $1^3, 2^3, 3^3, 4^3, 5^3, \ldots$ | $\{n^3\}$ | | $\frac{1+4x+x^2}{(1-x)^4}$ |
| $1, r, r^2, r^3, r^4, \ldots$ | $\{r^n\}$ | | $\frac{1}{1-rx}$ |
| Fibonacci | $F_n$ | §3.1.2 | $\frac{x}{1-x-x^2}$ |
| Lucas | $L_n$ | §3.1.2 | $\frac{2-x}{1-x-x^2}$ |
| Catalan | $C_n$ | §3.1.3 | $\frac{1-\sqrt{1-4x}}{2x}$ |
| Harmonic | $H_n$ | §3.1.7 | $\frac{1}{1-x} \ln \frac{1}{1-x}$ |
| Binomial | $\binom{m}{n}$ | §2.3.2 | $(1+x)^m$ |

**3.** Table 2 gives closed form expressions for the exponential generating functions of several combinatorial sequences discussed in this *Handbook*. Generating functions for other sequences can be found in §3.2.2, Tables 4 and 5.

**Table 2: Exponential generating functions for particular sequences.**

| sequence | notation | reference | closed form |
|---|---|---|---|
| $1, 1, 1, 1, 1, \ldots$ | $\{1\}$ | | $e^x$ |
| $1, r, r^2, r^3, r^4, \ldots$ | $\{r^n\}$ | | $e^{rx}$ |
| Derangements | $D_n$ | §2.4.2 | $\frac{e^{-x}}{1-x}$ |
| Bernoulli | $B_n$ | §3.1.4 | $\frac{x}{e^x-1}$ |
| Tangent | $T_n$ | §3.1.7 | $\tan x$ |
| Euler | $E_n$ | §3.1.7 | $\operatorname{sech} x$ |
| Euler | $|E_n|$ | §3.1.7 | $\sec x$ |
| Stirling cycle number | $\left[ {n \atop k} \right]$ | §2.5.2 | $\frac{1}{k!} \left[ \ln \frac{1}{1-x} \right]^k$ |
| Stirling subset number | $\left\{ {n \atop k} \right\}$ | §2.5.2 | $\frac{1}{k!} \left[ e^x - 1 \right]^k$ |

**4.** Table 3 gives recurrence relations defining particular combinatorial sequences discussed in this *Handbook*.

**Table 3: Recurrence relations for particular sequences.**

| sequence | notation | reference | recurrence relation |
|---|---|---|---|
| Derangements | $D_n$ | §2.4.2 | $D_n = (n-1)(D_{n-1} + D_{n-2})$, $D_0 = 1, D_1 = 0$ |
| Fibonacci | $F_n$ | §3.1.2 | $F_n = F_{n-1} + F_{n-2}, F_0 = 0, F_1 = 1$ |
| Lucas | $L_n$ | §3.1.2 | $L_n = L_{n-1} + L_{n-2}, L_0 = 2, L_1 = 1$ |

| sequence | notation | reference | recurrence relation |
|---|---|---|---|
| Catalan | $C_n$ | §3.1.3 | $C_n = C_0 C_{n-1} + C_1 C_{n-2} + \cdots$ $+ C_{n-1} C_0, \qquad C_0 = 1$ |
| Bernoulli | $B_n$ | §3.1.4 | $\sum_{j=0}^{n} \binom{n+1}{j} B_j = 0, \ \ B_0 = 1$ |
| Eulerian | $E(n, k)$ | §3.1.5 | $E(n, k) = (k+1) E(n-1, k)$ $+ (n-k) E(n-1, k-1),$ $E(n, 0) = 1, \ n \geq 1$ |
| Binomial | $\binom{n}{k}$ | §2.3.2 | $\binom{n}{k} = \binom{n-1}{k} + \binom{n-1}{k-1},$ $\binom{n}{0} = 1, \ \ n \geq 0$ |
| Stirling cycle number | $\left[ {n \atop k} \right]$ | §2.5.2 | $\left[ {n \atop k} \right] = (n-1) \left[ {n-1 \atop k} \right] + \left[ {n-1 \atop k-1} \right],$ $\left[ {0 \atop 0} \right] = 1; \ \left[ {n \atop 0} \right] = 0, n \geq 1$ |
| Stirling subset number | $\left\{ {n \atop k} \right\}$ | §2.5.2 | $\left\{ {n \atop k} \right\} = k \left\{ {n-1 \atop k} \right\} + \left\{ {n-1 \atop k-1} \right\},$ $\left\{ {0 \atop 0} \right\} = 1; \ \left\{ {n \atop 0} \right\} = 0, n \geq 1$ |

## 3.1.2  FIBONACCI NUMBERS

Fibonacci numbers form an important sequence encountered in biology, physics, number theory, computer science, and combinatorics. [BePhHo10], [Gr12], [PhBeHo01], [Va07]

**Definitions:**

The **Fibonacci numbers** $F_0, F_1, F_2, \ldots$ are produced by the recursive definition $F_0 = 0$, $F_1 = 1$, $F_n = F_{n-1} + F_{n-2}$, $n \geq 2$.

A **generalized Fibonacci sequence** is any sequence $G_0, G_1, G_2, \ldots$ such that $G_n = G_{n-1} + G_{n-2}$ for $n \geq 2$.

The **Lucas numbers** $L_0, L_1, L_2, \ldots$ are produced by the recursive definition $L_0 = 2$, $L_1 = 1$, $L_n = L_{n-1} + L_{n-2}$, $n \geq 2$. (François Lucas, 1842–1891)

**Facts:**

**1.** The Fibonacci numbers $F_n$ and Lucas numbers $L_n$ for $n = 0, 1, 2, \ldots, 50$ are shown in the following table.

**2.** The Fibonacci numbers were initially studied by Leonardo of Pisa (c. 1170–1250), who was the son of Bonaccio; consequently these numbers have been called *Fibonacci* numbers after Leonardo, the son of Bonaccio (Filius Bonaccii).

**3.** $\lim\limits_{n \to \infty} \dfrac{F_{n+1}}{F_n} = \lim\limits_{n \to \infty} \dfrac{L_{n+1}}{L_n} = \frac{1}{2}(1 + \sqrt{5}) \approx 1.61803$, the *golden ratio*.

| $n$ | $F_n$ | $L_n$ | $n$ | $F_n$ | $L_n$ | $n$ | $F_n$ | $L_n$ |
|---|---|---|---|---|---|---|---|---|
| 0 | 0 | 2 | 17 | 1,597 | 3,571 | 34 | 5,702,887 | 12,752,043 |
| 1 | 1 | 1 | 18 | 2,584 | 5,778 | 35 | 9,227,465 | 20,633,239 |
| 2 | 1 | 3 | 19 | 4,181 | 9,349 | 36 | 14,930,352 | 33,385,282 |
| 3 | 2 | 4 | 20 | 6,765 | 15,127 | 37 | 24,157,817 | 54,018,521 |
| 4 | 3 | 7 | 21 | 10,946 | 24,476 | 38 | 39,088,169 | 87,403,803 |
| 5 | 5 | 11 | 22 | 17,711 | 39,603 | 39 | 63,245,986 | 141,422,324 |
| 6 | 8 | 18 | 23 | 28,657 | 64,079 | 40 | 102,334,155 | 228,826,127 |
| 7 | 13 | 29 | 24 | 46,368 | 103,682 | 41 | 165,580,141 | 370,248,451 |
| 8 | 21 | 47 | 25 | 75,025 | 167,761 | 42 | 267,914,296 | 599,074,578 |
| 9 | 34 | 76 | 26 | 121,393 | 271,443 | 43 | 433,494,437 | 969,323,029 |
| 10 | 55 | 123 | 27 | 196,418 | 439,204 | 44 | 701,408,733 | 1,568,397,607 |
| 11 | 89 | 199 | 28 | 317,811 | 710,647 | 45 | 1,134,903,170 | 2,537,720,636 |
| 12 | 144 | 322 | 29 | 514,229 | 1,149,851 | 46 | 1,836,311,903 | 4,106,118,243 |
| 13 | 233 | 521 | 30 | 832,040 | 1,860,498 | 47 | 2,971,215,073 | 6,643,838,879 |
| 14 | 377 | 843 | 31 | 1,346,269 | 3,010,349 | 48 | 4,807,526,976 | 10,749,957,122 |
| 15 | 610 | 1,364 | 32 | 2,178,309 | 4,870,847 | 49 | 7,778,742,049 | 17,393,796,001 |
| 16 | 987 | 2,207 | 33 | 3,524,578 | 7,881,196 | 50 | 12,586,269,025 | 28,143,753,123 |

**4.** Fibonacci numbers arise in numerous applications from various areas. For example, they occur in models of population growth of rabbits (Example 3), in modeling plant growth (Example 8), in counting the number of spanning trees of wheel graphs of length $n$ (Example 12), in counting the number of bit strings of length $n$ without consecutive 0s (Example 13), and in many other contexts. See [Gr12] and [Va07] for additional applications of the Fibonacci numbers. The journal *Fibonacci Quarterly* is devoted to the study of the Fibonacci numbers and related topics, a tribute to how widely the Fibonacci numbers arise in mathematics and its applications to other areas. Further readings about the Fibonacci numbers can be found at

- http://www.maths.surrey.ac.uk/hosted-sites/R.Knott/Fibonacci/ fibrefs.html.

**5.** Many properties of the Fibonacci numbers were derived by F. Lucas, who also is responsible for naming them the "Fibonacci" numbers.

**6.** *Binet form* (Jacques Binet, 1786–1856): If $\alpha = \frac{1}{2}(1 + \sqrt{5})$ and $\beta = \frac{1}{2}(1 - \sqrt{5})$ then

$$F_n = \frac{\alpha^n - \beta^n}{\sqrt{5}} = \frac{\alpha^n - \beta^n}{\alpha - \beta}, \qquad F_n \sim \frac{\alpha^n}{\sqrt{5}}.$$

Also,

$$L_n = \alpha^n + \beta^n, \qquad L_n \sim \alpha^n.$$

**7.** $F_n = \frac{1}{2}(F_{n-2} + F_{n+1})$ for all $n \geq 2$. That is, each Fibonacci number is the average of the terms occurring two places before and one place after it in the sequence.

**8.** $L_n = \frac{1}{2}(L_{n-2} + L_{n+1})$ for all $n \geq 2$. That is, each Lucas number is the average of the terms occurring two places before and one place after it in the sequence.

**9.** $F_0 + F_1 + F_2 + \cdots + F_n = F_{n+2} - 1$ for all $n \geq 0$.

**10.** $F_0 - F_1 + F_2 - \cdots + (-1)^n F_n = (-1)^n F_{n-1} - 1$ for all $n \geq 1$.

**11.** $F_1 + F_3 + F_5 + \cdots + F_{2n-1} = F_{2n}$ for all $n \geq 1$.

**12.** $F_0 + F_2 + F_4 + \cdots + F_{2n} = F_{2n+1} - 1$ for all $n \geq 0$.

**13.** $F_0^2 + F_1^2 + F_2^2 + \cdots + F_n^2 = F_n F_{n+1}$ for all $n \geq 0$.

**14.** $F_1 F_2 + F_2 F_3 + F_3 F_4 + \cdots + F_{2n-1} F_{2n} = F_{2n}^2$ for all $n \geq 1$.

**15.** $F_1 F_2 + F_2 F_3 + F_3 F_4 + \cdots + F_{2n} F_{2n+1} = F_{2n+1}^2 - 1$ for all $n \geq 1$.

**16.** If $k \geq 1$ then $F_{n+k} = F_k F_{n+1} + F_{k-1} F_n$ for all $n \geq 0$.

**17.** *Cassini's identity:* $F_{n+1} F_{n-1} - F_n^2 = (-1)^n$ for all $n \geq 1$. (Jean Dominique Cassini, 1625–1712)

**18.** $F_{n+1}^2 + F_n^2 = F_{2n+1}$ for all $n \geq 0$.

**19.** $F_{n+2}^2 - F_{n+1}^2 = F_n F_{n+3}$ for all $n \geq 0$.

**20.** $F_{n+2}^2 - F_n^2 = F_{2n+2}$ for all $n \geq 0$.

**21.** $F_{n+2}^3 + F_{n+1}^3 - F_n^3 = F_{3n+3}$ for all $n \geq 0$.

**22.** $\gcd(F_n, F_m) = F_{\gcd(n,m)}$. This implies that $F_n$ and $F_{n+1}$ are relatively prime, and that $F_k$ divides $F_{nk}$.

**23.** Fibonacci numbers arise as sums of diagonals in Pascal's triangle (§2.3.2):

$$F_{n+1} = \sum_{j=0}^{\lfloor n/2 \rfloor} \binom{n-j}{j} \text{ for all } n \geq 0.$$

**24.** $F_{3n} = \sum_{j=0}^{n} \binom{n}{j} 2^j F_j$ for all $n \geq 0$.

**25.** The Fibonacci sequence $F_0, F_1, F_2, \ldots$ has the generating function $\frac{x}{1-x-x^2}$.

**26.** Fibonacci numbers with negative indices can be defined using the recursive definition $F_{n-2} = F_n - F_{n-1}$. Then $F_{-n} = (-1)^{n-1} F_n$, $n \geq 1$.

**27.** The units digits of the Fibonacci numbers form a sequence that repeats after 60 terms. (Joseph Lagrange, 1736–1813)

**28.** The number of binary strings of length $n$ that contain no consecutive 0s is counted by $F_{n+2}$. (See §3.3.2, Example 12.)

**29.** $L_0 + L_1 + L_2 + \cdots + L_n = L_{n+2} - 1$ for all $n \geq 0$.

**30.** $L_0^2 + L_1^2 + L_2^2 + \cdots + L_n^2 = L_n L_{n+1} + 2$ for all $n \geq 0$.

**31.** $L_n = F_{n-1} + F_{n+1}$, $n \geq 1$. Hence, any formula containing Lucas numbers can be translated into a formula involving Fibonacci numbers.

**32.** The Lucas sequence $L_0, L_1, L_2, \ldots$ has the generating function $\frac{2-x}{1-x-x^2}$.

**33.** Lucas numbers with negative indices can be defined by extending the recursive definition. Then $L_{-n} = (-1)^n L_n$ for all $n \geq 1$.

**34.** $F_n = \frac{L_{n-1} + L_{n+1}}{5}$, $n \geq 1$. Hence, any formula involving Fibonacci numbers can be translated into a formula involving Lucas numbers.

**35.** $2^{n+1} F_{n+1} = \sum_{i=0}^{n} 2^i L_i$ for all $n \geq 0$.

**36.** If $G_0, G_1, \ldots$ is a sequence of generalized Fibonacci numbers, then $F_n = F_{n-1} G_0 + F_n G_1$ for all $n \geq 1$.

**Examples:**

**1.** The Fibonacci number $F_8$ can be computed using the initial values $F_0 = 0$ and $F_1 = 1$ and the recurrence relation $F_n = F_{n-1} + F_{n-2}$ repeatedly: $F_2 = F_1 + F_0 = 1 + 0 = 1$, $F_3 = F_2 + F_1 = 1 + 1 = 2$, $F_4 = F_3 + F_2 = 2 + 1 = 3$, $F_5 = F_4 + F_3 = 3 + 2 = 5$, $F_6 = F_5 + F_4 = 5 + 3 = 8$, $F_7 = F_6 + F_5 = 8 + 5 = 13$, $F_8 = F_7 + F_6 = 13 + 8 = 21$.

**2.** Each male bee (drone) is produced asexually from a female, whereas each female bee is produced from both a male and female. The ancestral tree for a single male bee is shown below. This male has one parent, two grandparents, three great grandparents, and in general $F_{k+2}$ $k$th-order grandparents, $k \geq 0$.

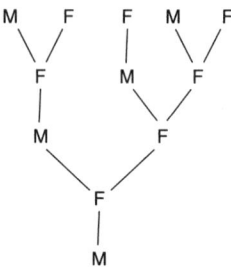

**3.** *Rabbit breeding:* This problem was originally posed by Fibonacci. A single pair of immature rabbits is introduced into a habitat. It takes two months before a pair of rabbits can breed; each month thereafter each pair of breeding rabbits produces another pair. At the start of months 1 and 2, only the original pair $A$ is present. In the third month, $A$ as well as their newly born pair $B$ are present; in the fourth month, $A$, $B$ as well as the new pair $C$ (progeny of $A$) are present; in the fifth month, $A$, $B$, $C$ as well as the new pairs $D$ (progeny of $A$) and $E$ (progeny of $B$) are present. If $P_n$ is the number of pairs present in month $n$, then $P_1 = 1$, $P_2 = 1$, $P_3 = 2$, $P_4 = 3$, $P_5 = 5$. In general, $P_n$ equals the number present in the previous month $P_{n-1}$ plus the number of breeding pairs in the previous month (which is $P_{n-2}$, the number present two months earlier). Thus $P_n = F_n$ for $n \geq 1$.

**4.** Let $S_n$ denote the number of subsets of $\{1, 2, \ldots, n\}$ that do not contain consecutive elements. For example, when $n = 3$ the allowable subsets are $\emptyset, \{1\}, \{2\}, \{3\}, \{1, 3\}$. Therefore, $S_3 = 5$. In general, $S_n = F_{n+2}$ for $n \geq 1$.

**5.** Draw $n$ dots in a line. If each domino can cover exactly two such dots, in how many ways can (nonoverlapping) dominoes be placed? The following figure shows the number of possible solutions for $n = 2, 3, 4$. To find a general expression for $D_n$, the number of possible placements of dominoes with $n$ dots, consider the rightmost dot in any such placement $P$. If this dot is not covered by a domino, then $P$ minus the last dot determines a solution counted by $D_{n-1}$. If the last dot is covered by a domino, then the last two dots in $P$ are covered by this domino. Removing this rightmost domino then gives a solution counted by $D_{n-2}$. Taking into account these two possibilities $D_n = D_{n-1} + D_{n-2}$ for $n \geq 3$ with $D_1 = 1$, $D_2 = 2$. Thus $D_n = F_{n+1}$ for $n \geq 1$.

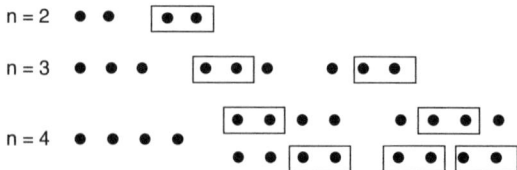

**6.** *Compositions:* Let $T_n$ be the number of ordered compositions (§2.5.1) of the positive integer $n$ into summands that are odd. For example, $4 = 1 + 3 = 3 + 1 = 1 + 1 + 1 + 1$ and $5 = 5 = 1 + 1 + 3 = 1 + 3 + 1 = 3 + 1 + 1 = 1 + 1 + 1 + 1 + 1$. Therefore, $T_4 = 3$ and $T_5 = 5$. In general, $T_n = F_n$ for $n \geq 1$.

**7.** *Compositions:* Let $B_n$ be the number of ordered compositions (§2.5.1) of the positive integer $n$ into summands that are either 1 or 2. For example, $3 = 1 + 2 = 2 + 1 = 1 + 1 + 1$

and $4 = 2 + 2 = 1 + 1 + 2 = 1 + 2 + 1 = 2 + 1 + 1 = 1 + 1 + 1 + 1$. Therefore, $B_3 = 3$ and $B_4 = 5$. In general, $B_n = F_{n+1}$ for $n \geq 1$.

**8.** *Botany*: It has been observed in pine cones (and other botanical structures) that the number of rows of scales winding in one direction is a Fibonacci number while the number of rows of scales winding in the other direction is an adjacent Fibonacci number.

**9.** *Continued fractions*: The continued fraction $1 + \frac{1}{1} = \frac{2}{1}$, the continued fraction $1 + \frac{1}{1+\frac{1}{1}} = \frac{3}{2}$ and the continued fraction $1 + \frac{1}{1+\frac{1}{1+\frac{1}{1}}} = \frac{5}{3}$. In general, a continued fraction composed entirely of 1s equals the ratio of successive Fibonacci numbers.

**10.** *Independent sets on a path*: Consider a path graph on vertices $1, 2, \ldots, n$, with edges joining vertices $i$ and $i + 1$ for $i = 1, 2, \ldots, n - 1$. An independent set of vertices (§8.6.3) consists of vertices no two of which are joined by an edge. By an analysis similar to that in Example 5, the number of independent sets in a path graph on $n$ vertices equals $F_{n+2}$.

**11.** *Independent sets on a cycle*: Consider a cycle graph on vertices $1, 2, \ldots, n$, with edges joining vertices $i$ and $i + 1$ for $i = 1, 2, \ldots, n - 1$ as well as vertices $n$ and 1. Then the number of independent sets (§8.6.3) in a cycle graph on $n$ vertices equals $L_n$.

**12.** *Spanning trees*: The number of spanning trees of the wheel graph $W_n$ (§8.1.3) equals $L_{2n} - 2$.

**13.** If $A$ is the $2 \times 2$ matrix $\begin{pmatrix} 1 & 1 \\ 1 & 0 \end{pmatrix}$, then $A^n = \begin{pmatrix} F_{n+1} & F_n \\ F_n & F_{n-1} \end{pmatrix}$ for $n \geq 1$.

---

### 3.1.3  CATALAN NUMBERS

The sequence of integers called the Catalan numbers arises in counting a variety of combinatorial structures, such as voting sequences, certain types of binary trees, paths in the plane, and triangulations of polygons.

**Definitions:**
The **Catalan numbers** $C_0, C_1, C_2, \ldots$ satisfy the nonlinear recurrence relation $C_n = C_0 C_{n-1} + C_1 C_{n-2} + \cdots + C_{n-1} C_0$, $n \geq 1$, with $C_0 = 1$. (See §3.3.1, Example 9.) (Eugène Catalan, 1814–1894)

**Well-formed (or balanced) sequences of parentheses** of length $2n$ are defined recursively as follows: the empty sequence is well-formed; if sequence $A$ is well-formed so is $(A)$; if sequences $A$ and $B$ are well-formed so is $AB$.

**Well-parenthesized products** of variables are defined recursively as follows: single variables are well-parenthesized; if $A$ and $B$ are well-parenthesized so is $(AB)$.

**Facts:**
**1.** The first 12 Catalan numbers $C_n$ are given in the following table.

| $n$ | 0 | 1 | 2 | 3 | 4 | 5 | 6 | 7 | 8 | 9 | 10 | 11 |
|---|---|---|---|---|---|---|---|---|---|---|---|---|
| $C_n$ | 1 | 1 | 2 | 5 | 14 | 42 | 132 | 429 | 1,430 | 4,862 | 16,796 | 58,786 |

**2.** $\lim\limits_{n \to \infty} \frac{C_{n+1}}{C_n} = 4$.

**3.** Catalan numbers arise in a variety of applications, such as in counting binary trees on $n$ vertices, triangulations of a convex $n$-gon, and well-formed sequences of $n$ left and $n$ right parentheses. See the examples below as well as [Gr12] and [MiRo91].

**4.** $C_n = \frac{1}{n+1}\binom{2n}{n}$ for all $n \geq 0$.

**5.** The Catalan numbers $C_0, C_1, C_2, \ldots$ have the generating function $\frac{1-\sqrt{1-4x}}{2x}$.

**6.** $C_n \sim \frac{4^n}{\sqrt{\pi n^3}}$.

**7.** $C_n = \binom{2n}{n} - \binom{2n}{n-1} = \binom{2n-1}{n} - \binom{2n-1}{n+1}$ for all $n \geq 1$.

**8.** $C_{n+1} = \frac{2(2n+1)}{n+2}C_n$ for all $n \geq 0$.

**Examples:**

**1.** The number of binary trees (§9.1.2) on $n$ vertices is $C_n$.

**2.** The number of left-right binary trees (§9.3.3) on $2n+1$ vertices is $C_n$.

**3.** The number of ordered trees (§9.1.2) on $n$ vertices is $C_{n-1}$.

**4.** Suppose that a coin is tossed $2n$ times, coming up heads exactly $n$ times and tails exactly $n$ times. The number of sequences of tosses in which the cumulative number of heads is always at least as large as the cumulative number of tails is $C_n$. For example, when $n = 3$ there are $C_3 = 5$ such sequences of six tosses: HTHTHT, HTHHTT, HHTTHT, HHTHTT, HHHTTT.

**5.** In Example 4, the number of sequences of tosses in which the cumulative number of heads always exceeds the cumulative number of tails (until the very last toss) is $C_{n-1}$. For example, when $n = 3$ there are $C_2 = 2$ such sequences of six tosses: HHTHTT, HHHTTT.

**6.** *Triangulations:* Let $T_n$ be the number of triangulations of a convex $n$-gon, using $n-3$ nonintersecting diagonals. For instance, the following figure shows the $T_5 = 5$ triangulations of a pentagon. In general, $T_n = C_{n-2}$ for $n \geq 3$.

**7.** Suppose $2n$ points are placed in fixed positions, evenly distributed on the circumference of a circle. Then there are $C_n$ ways to join $n$ pairs of the points so that the resulting chords do not intersect. The following figure shows the $C_3 = 5$ solutions for $n = 3$.

**8.** *Well-formed sequences of parentheses:* The sequence of parentheses (()()) involving three left and three right parentheses is well-formed, whereas the sequence ( ) ) ( ( ) is not syntactically meaningful. There are five such well-formed sequences in this case:

$$()()(),\ ()(()),\ (())(),\ (()()),\ ((()))。$$

Notice that if each left parenthesis is replaced by an H and each right parenthesis by a T, then these five balanced sequences correspond exactly to the five coin tossing sequences listed in Example 4. In general, the number of balanced sequences involving $n$ left and $n$ right parentheses is $C_n$.

**9.** Consider the following procedure composed of $n$ nested **for** loops:

```
count := 0
for i_1 := 1 to 1
```

> **for** $i_2 := 1$ **to** $i_1 + 1$
>> **for** $i_3 := 1$ **to** $i_2 + 1$
>>> $\vdots$
>>>> **for** $i_n := 1$ **to** $i_{n-1} + 1$
>>>> $count := count + 1$

Then the value of *count* upon exit from this procedure is $C_n$.

**10.** *Well-parenthesized products*: The product $x_1 x_2 x_3$ (relative to some binary "multiplication" operation) can be evaluated as either $(x_1 x_2)x_3$ or $x_1(x_2 x_3)$. In the former, $x_1$ and $x_2$ are first combined and then the result is combined with $x_3$. In the latter, $x_2$ and $x_3$ are first combined and then the result is combined with $x_1$. Let $P_n$ indicate the number of ways to evaluate the product $x_1 x_2 \ldots x_n$ of $n$ variables, using a binary operation. Note that $P_3 = 2$. In general, $P_n = C_{n-1}$. This was the problem originally studied by Catalan. (See §3.3.1, Example 9.)

**11.** The numbers $1, 2, \ldots, 2n$ are to be placed in the $2n$ positions of an $2 \times n$ array $A = (a_{ij})$. Such an arrangement is *monotone* if the values increase within each row and within each column. Then there are $C_n$ ways to form a monotone $2 \times n$ array containing the entries $1, 2, \ldots, 2n$. For instance, the following is one of the $C_4 = 14$ monotone $2 \times 4$ arrays:

$$A = \begin{pmatrix} 1 & 3 & 5 & 6 \\ 2 & 4 & 7 & 8 \end{pmatrix}.$$

---

### 3.1.4  BERNOULLI NUMBERS AND POLYNOMIALS

The Bernoulli numbers are important in obtaining closed form expressions for the sums of powers of integers. These numbers also arise in expansions involving other combinatorial sequences.

**Definitions:**

The **Bernoulli numbers** $B_n$ satisfy the recurrence relation $\sum_{j=0}^{n} \binom{n+1}{j} B_j = 0$ for all $n \geq 1$, with $B_0 = 1$. (Jakob Bernoulli, 1654–1705)

The **Bernoulli polynomials** $B_m(x)$ are given by $B_m(x) = \sum_{k=0}^{m} \binom{m}{k} B_k x^{m-k}$.

**Facts:**

**1.** The first 14 Bernoulli numbers $B_n$ are shown in the following table.

| $n$ | 0 | 1 | 2 | 3 | 4 | 5 | 6 | 7 | 8 | 9 | 10 | 11 | 12 | 13 |
|---|---|---|---|---|---|---|---|---|---|---|---|---|---|---|
| $B_n$ | 1 | $-\frac{1}{2}$ | $\frac{1}{6}$ | 0 | $-\frac{1}{30}$ | 0 | $\frac{1}{42}$ | 0 | $-\frac{1}{30}$ | 0 | $\frac{5}{66}$ | 0 | $-\frac{691}{2{,}730}$ | 0 |

**2.** $B_{2k+1} = 0$ for all $k \geq 1$.

**3.** The nonzero Bernoulli numbers alternate in sign.

**4.** $B_n = B_n(0)$.

**5.** The Bernoulli numbers have the exponential generating function $\sum_{n=0}^{\infty} B_n \frac{x^n}{n!} = \frac{x}{e^x - 1}$.

**6.** The Bernoulli numbers can be expressed in terms of the Stirling subset numbers (§2.5.2): $B_n = \sum_{j=0}^{n} (-1)^j \left\{{n \atop j}\right\} \frac{j!}{j+1}$ for all $n \geq 0$.

**7.** The Bernoulli numbers appear as coefficients in the Maclaurin expansion of $\tan x$, $\cot x$, $\csc x$, $\tanh x$, $\coth x$, and $\operatorname{csch} x$.

**8.** The Bernoulli polynomials can be used to obtain closed form expressions for the sum of powers of the first $n$ positive integers. (See §3.5.4.)

**9.** The first 14 Bernoulli polynomials $B_m(x)$ are shown in the following table.

| $n$ | $B_n(x)$ |
|---|---|
| 0 | $1$ |
| 1 | $x - \frac{1}{2}$ |
| 2 | $x^2 - x + \frac{1}{6}$ |
| 3 | $x^3 - \frac{3}{2}x^2 + \frac{1}{2}x$ |
| 4 | $x^4 - 2x^3 + x^2 - \frac{1}{30}$ |
| 5 | $x^5 - \frac{5}{2}x^4 + \frac{5}{3}x^3 - \frac{1}{6}x$ |
| 6 | $x^6 - 3x^5 + \frac{5}{2}x^4 - \frac{1}{2}x^2 + \frac{1}{42}$ |
| 7 | $x^7 - \frac{7}{2}x^6 + \frac{7}{2}x^5 - \frac{7}{6}x^3 + \frac{1}{6}x$ |
| 8 | $x^8 - 4x^7 + \frac{14}{3}x^6 - \frac{7}{3}x^4 + \frac{2}{3}x^2 - \frac{1}{30}$ |
| 9 | $x^9 - \frac{9}{2}x^8 + 6x^7 - \frac{21}{5}x^5 + 2x^3 - \frac{3}{10}x$ |
| 10 | $x^{10} - 5x^9 + \frac{15}{2}x^8 - 7x^6 + 5x^4 - \frac{3}{2}x^2 + \frac{5}{66}$ |
| 11 | $x^{11} - \frac{11}{2}x^{10} + \frac{55}{6}x^9 - 11x^7 + 11x^5 - \frac{11}{2}x^3 + \frac{5}{6}x$ |
| 12 | $x^{12} - 6x^{11} + 11x^{10} - \frac{33}{2}x^8 + 22x^6 - \frac{33}{2}x^4 + x^2 - \frac{691}{2,730}$ |
| 13 | $x^{13} - \frac{13}{2}x^{12} + 13x^{11} - \frac{143}{6}x^9 + \frac{286}{7}x^7 - \frac{429}{10}x^5 + \frac{65}{3}x^3 - \frac{691}{210}x$ |

**10.** $\int_0^1 B_m(x)\, dx = 0$ for all $m \geq 1$.

**11.** $\frac{dB_m(x)}{dx} = mB_{m-1}(x)$ for all $m \geq 1$.

**12.** $B_{m+1}(x+1) - B_{m+1}(x) = (m+1)x^m$ for all $m \geq 0$.

**13.** The Bernoulli polynomials have the exponential generating function given by $\sum_{m=0}^{\infty} B_m(x) \frac{t^m}{m!} = \frac{te^{xt}}{e^t - 1}$.

---

## 3.1.5   EULERIAN NUMBERS

Eulerian numbers are important in counting numbers of permutations with certain numbers of increases and decreases.

**Definitions:**
Let $\pi = (\pi_1, \pi_2, \ldots, \pi_n)$ be a permutation of $\{1, 2, \ldots, n\}$.

An **ascent** of the permutation $\pi$ is any index $i$ ($1 \leq i < n$) such that $\pi_i < \pi_{i+1}$. A **descent** of the permutation $\pi$ is any index $i$ ($1 \leq i < n$) such that $\pi_i > \pi_{i+1}$.

An **excedance** of the permutation $\pi$ is any index $i$ ($1 \leq i \leq n$) such that $\pi_i > i$. A **weak excedance** of the permutation $\pi$ is any index $i$ ($1 \leq i \leq n$) such that $\pi_i \geq i$.

The **Eulerian number** $E(n, k)$, also written $\left\langle {n \atop k} \right\rangle$, is the number of permutations of $\{1, 2, \ldots, n\}$ with exactly $k$ ascents.

## Facts:

1. $E(n, k)$ is the number of permutations of $\{1, 2, \ldots, n\}$ with exactly $k$ descents.

2. $E(n, k)$ is the number of permutations of $\{1, 2, \ldots, n\}$ with exactly $k$ excedances.

3. $E(n, k)$ is the number of permutations of $\{1, 2, \ldots, n\}$ with exactly $k + 1$ weak excedances.

4. The Eulerian numbers can be used to obtain closed form expressions for the sum of powers of the first $n$ positive integers (§3.5.4).

5. Eulerian numbers $E(n, k)$ ($1 \leq n \leq 10$, $0 \leq k \leq 8$) are given in the following table.

| $n \backslash k$ | 0 | 1 | 2 | 3 | 4 | 5 | 6 | 7 | 8 |
|---|---|---|---|---|---|---|---|---|---|
| 1 | 1 | | | | | | | | |
| 2 | 1 | 1 | | | | | | | |
| 3 | 1 | 4 | 1 | | | | | | |
| 4 | 1 | 11 | 11 | 1 | | | | | |
| 5 | 1 | 26 | 66 | 26 | 1 | | | | |
| 6 | 1 | 57 | 302 | 302 | 57 | 1 | | | |
| 7 | 1 | 120 | 1,191 | 2,416 | 1,191 | 120 | 1 | | |
| 8 | 1 | 247 | 4,293 | 15,619 | 15,619 | 4,293 | 247 | 1 | |
| 9 | 1 | 502 | 14,608 | 88,234 | 156,190 | 88,234 | 14,608 | 502 | 1 |
| 10 | 1 | 1,013 | 47,840 | 455,192 | 1,310,354 | 1,310,354 | 455,192 | 47,840 | 1,013 |

6. $E(n, 0) = E(n, n - 1) = 1$ for all $n \geq 1$.

7. *Symmetry:*  $E(n, k) = E(n, n - 1 - k)$ for all $n \geq 1$.

8. $E(n, k) = (k + 1)E(n - 1, k) + (n - k)E(n - 1, k - 1)$ for all $n \geq 2$.

9. $\sum\limits_{k=0}^{n-1} E(n, k) = n!$ for all $n \geq 1$.

10. *Worpitzky's identity:*   $x^n = \sum\limits_{k=0}^{n-1} E(n, k)\binom{x+k}{n}$ for all $n \geq 1$. (Julius Worpitzky, 1835–1895)

11. $E(n, k) = \sum\limits_{j=0}^{k} (-1)^j \binom{n+1}{j}(k + 1 - j)^n$ for all $n \geq 1$.

12. The Bernoulli numbers (§3.1.4) can be expressed as alternating sums of Eulerian numbers: $B_m = \frac{m}{2^m(2^m - 1)} \sum\limits_{k=0}^{m-2} (-1)^k E(m - 1, k)$ for $m \geq 2$.

13. The Stirling subset numbers (§2.5.2) can be expressed in terms of the Eulerian numbers: $\left\{ {n \atop m} \right\} = \frac{1}{m!} \sum\limits_{k=0}^{n-1} E(n, k)\binom{k}{n-m}$ for $n \geq m$ and $n \geq 1$.

14. The Eulerian numbers have the following (bivariate) generating function in variables $x, t$: $\sum\limits_{m=0}^{\infty} \sum\limits_{n=0}^{\infty} E(n, m)x^m \frac{t^n}{n!} = \frac{1 - x}{e^{(x-1)t} - x}$.

## Examples:

1. The permutation $\pi = (\pi_1, \pi_2, \pi_3, \pi_4) = (1, 2, 3, 4)$ has three ascents since $1 < 2 < 3 < 4$ and it is the only permutation in $S_4$ with three ascents; note that $E(4, 3) = 1$. There are $E(4, 1) = 11$ permutations in $S_4$ with one ascent: $(1, 4, 3, 2)$, $(2, 1, 4, 3)$, $(2, 4, 3, 1)$, $(3, 1, 4, 2)$, $(3, 2, 1, 4)$, $(3, 2, 4, 1)$, $(3, 4, 2, 1)$, $(4, 1, 3, 2)$, $(4, 2, 1, 3)$, $(4, 2, 3, 1)$, and $(4, 3, 1, 2)$.

**2.** The permutation $\pi = (2, 4, 3, 1)$ has two excedances since $2 > 1$ and $4 > 2$. There are $E(4, 2) = 11$ such permutations in $S_4$.

**3.** The permutation $\pi = (1, 3, 2)$ has two weak excedances since $1 \geq 1$ and $3 \geq 2$. There are $E(3, 1) = 4$ such permutations in $S_3$: $(1, 3, 2)$, $(2, 1, 3)$, $(2, 3, 1)$, and $(3, 2, 1)$.

**4.** When $n = 3$, Worpitzky's identity (Fact 10) states that

$$x^3 = E(3, 0)\binom{x}{3} + E(3, 1)\binom{x+1}{3} + E(3, 2)\binom{x+2}{3} = \binom{x}{3} + 4\binom{x+1}{3} + \binom{x+2}{3}.$$

This is verified algebraically since $\binom{x}{3} + 4\binom{x+1}{3} + \binom{x+2}{3} = \frac{1}{6}(x(x-1)(x-2) + 4(x+1)x(x-1) + (x+2)(x+1)x) = \frac{x}{6}(x^2 - 3x + 2 + 4x^2 - 4 + x^2 + 3x + 2) = \frac{x}{6}(6x^2) = x^3$.

---

## 3.1.6   RAMSEY NUMBERS

The Ramsey numbers arise from the work of Frank P. Ramsey (1903–1930), who in 1930 published a paper [Ra30] dealing with set theory that generalized the pigeonhole principle. (Also see §7.11.1, §8.11.2.) [GrRoSp13], [MiRo91], [RoTe09]

**Definitions:**

The **Ramsey number** $R(m, n)$ is the smallest positive integer $k$ with the following property: if $S$ is a set of size $k$ and the 2-element subsets of $S$ are partitioned into 2 collections, $C_1$ and $C_2$, then there is a subset of $S$ of size $m$ such that each of its 2-element subsets belong to $C_1$ or there is a subset of $S$ of size $n$ such that each of its 2-element sets belong to $C_2$.

The **Ramsey number** $R(m_1, \ldots, m_n; r)$ is the smallest positive integer $k$ with the following property: if $S$ is a set of size $k$ and the $r$-element subsets of $S$ are partitioned into $n$ collections $C_1, \ldots, C_n$, then for some $j$ there is a subset of $S$ of size $m_j$ such that each of its $r$-element subsets belong to $C_j$.

The **Schur number** $S(n)$ is the smallest integer $k$ with the following property: if $\{1, \ldots, k\}$ is partitioned into $n$ subsets $A_1, \ldots, A_n$, then there is a subset $A_i$ such that the equation $x + y = z$ has a solution where $x, y, z \in A_i$. (Issai Schur, 1875–1941)

**Facts:**

**1.** *Ramsey's theorem:*   The Ramsey numbers $R(m, n)$ and $R(m_1, \ldots, m_n; r)$ are well defined for all $m, n \geq 1$ and for all $m_1, \ldots, m_n \geq 1$, $r \geq 1$.

**2.** Ramsey numbers $R(m, n)$ can be phrased in terms of coloring edges of the complete graphs $K_n$: the Ramsey number $R(m, n)$ is the smallest positive integer $k$ such that, if each edge of $K_k$ is colored red or blue, then either the red subgraph contains a copy of $K_m$ or else the blue subgraph contains a copy of $K_n$. (See §8.11.2.)

**3.** *Symmetry:*   $R(m, n) = R(n, m)$.

**4.** $R(m, 1) = R(1, m) = 1$ for every $m \geq 1$.

**5.** $R(m, 2) = R(2, m) = m$ for every $m \geq 1$.

**6.** The values of few Ramsey numbers are known. What is currently known about Ramsey numbers $R(m, n)$, for $3 \leq m \leq 10$ and $3 \leq n \leq 10$, and bounds on other Ramsey numbers are displayed in Table 4. The entries in the body of this table are $R(m, n)$ $(m, n \leq 10)$ when known, or the best known range $r_1 \leq R(m, n) \leq r_2$ when not known. The Ramsey numbers $R(3, 3)$, $R(3, 4)$, $R(3, 5)$, and $R(4, 4)$ were found by A. M. Gleason and R. E. Greenwood in 1955; $R(3, 6)$ was found by J. G. Kalbfleisch in 1966; $R(3, 7)$

was found by J. E. Graver and J. Yackel in 1968; $R(3,8)$ was found by B. McKay and Z. Ke Min; $R(3,9)$ was found by C. M. Grinstead and S. M. Roberts in 1982; $R(4,5)$ was found by B. McKay and S. Radziszowski in 1993.

**Table 4: Some classical Ramsey numbers.**

| $m \backslash n$ | 3 | 4 | 5 | 6 | 7 | 8 | 9 | 10 |
|---|---|---|---|---|---|---|---|---|
| 3 | 6 | 9 | 14 | 18 | 23 | 28 | 36 | 40-42 |
| 4 | – | 18 | 25 | 36-41 | 49-61 | 58-84 | 73-115 | 92-149 |
| 5 | – | – | 43-49 | 58-87 | 80-143 | 101-216 | 126-316 | 144-442 |
| 6 | – | – | – | 102-165 | 113-298 | 132-495 | 169-780 | 179-1,171 |
| 7 | – | – | – | – | 205-540 | 217-1,031 | 241-1,713 | 289-2,826 |
| 8 | – | – | – | – | – | 282-1,870 | 317-3,583 | $\leq 6,090$ |
| 9 | – | – | – | – | – | – | 565-6,588 | 581-12,677 |
| 10 | – | – | – | – | – | – | – | 798-23,556 |

Bounds for $R(m,n)$ for $m = 3$ and $4$, with $11 \leq n \leq 15$:

| | |
|---|---|
| $47 \leq R(3,11) \leq 50$ | $98 \leq R(4,11) \leq 191$ |
| $52 \leq R(3,12) \leq 59$ | $128 \leq R(4,12) \leq 238$ |
| $59 \leq R(3,13) \leq 68$ | $133 \leq R(4,13) \leq 291$ |
| $66 \leq R(3,14) \leq 77$ | $141 \leq R(4,14) \leq 349$ |
| $73 \leq R(3,15) \leq 87$ | $153 \leq R(4,15) \leq 417$ |

**7.** If $m_1 \leq m_2$ and $n_1 \leq n_2$, then $R(m_1,n_1) \leq R(m_2,n_2)$.

**8.** $R(m,n) \leq R(m,n-1) + R(n-1,m)$ for all $m, n \geq 2$.

**9.** If $m \geq 3$, $n \geq 3$, and if $R(m,n-1)$ and $R(m-1,n)$ are even, then $R(m,n) \leq R(m,n-1) + R(m-1,n) - 1$.

**10.** $R(m,n) \leq \binom{m+n-2}{m-1}$. (Erdős and Szekeres, 1935)

**11.** The Ramsey numbers $R(m,n)$ satisfy the following asymptotic relationship:
$$\frac{\sqrt{2}}{e}(1 + o(1))m 2^{m/2} \leq R(m,m) \leq \binom{2m+2}{m+1} \cdot O((\log m)^{-1}).$$

**12.** There exist constants $c_1$ and $c_2$ such that $c_1 m \ln m \leq R(3,m) \leq c_2 m \ln m$.

**13.** The problem of finding the Ramsey numbers $R(m_1, \ldots, m_n; 2)$ can be phrased in terms of coloring edges of the complete graphs $K_n$. $R(m_1, \ldots, m_n; 2)$ is equal to the smallest positive integer $k$ with the following property: no matter how the edges of $K_k$ are colored with the $n$ colors $1, 2, \ldots, n$, there is some $j$ such that $K_k$ has a subgraph $K_{m_j}$ of color $j$. (The edges of $K_k$ are the 2-element subsets; $C_j$ is the set of edges of color $j$.)

**14.** $R(m_1, m_2; 2) = R(m_1, m_2)$.

**15.** Very little is also known about the numbers $R(m_1, \ldots, m_n; 2)$ if $n \geq 3$.

**16.** $R(2, \ldots, 2; 2) = 2$.

**17.** If each $m_i \geq 3$, the only Ramsey number whose value is known is $R(3,3,3;2) = 17$.

**18.** $R(m, r, r, \ldots, r; r) = m$ if $m \geq r$.

**19.** $R(m_1, \ldots m_n; 1) = m_1 + \cdots + m_n - (n-1)$.

**20.** Ramsey theory is a generalization of the pigeonhole principle. In the terminology of Ramsey numbers, the fact that $R(2, \ldots, 2; 1) = n + 1$ means that $n + 1$ is the smallest positive integer with the property that if $S$ has size $n + 1$ and the subsets of $S$ are partitioned into $n$ sets $C_1, \ldots, C_n$, then for some $j$ there is a subset of $S$ of size 2 such that each of its elements belong to $C_j$. Hence, some $C_j$ has at least two elements. If $S$ is a set of $n + 1$ pigeons and the subset $C_j$ $(j = 1, \ldots, n)$ is the set of pigeons roosting in pigeonhole $j$, then some pigeonhole must have at least two pigeons in it. The Ramsey numbers $R(2, \ldots, 2; 1)$ give the smallest number of pigeons that force at least two to roost in the same pigeonhole.

**21.** *Schur's theorem:*  $S(k) \leq R(3, \ldots, 3; 2)$ (where there are $k$ 3s in the notation for the Ramsey number).

**22.** The following Schur numbers are known: $S(1) = 2$, $S(2) = 5$, $S(3) = 14$.

**23.** The equation $x + y = z$ in the definition of Schur numbers has been generalized to equations of the form $x_1 + \cdots + x_{n-1} = x_n$, $n \geq 4$. [BeBr82].

**24.** *Convex sets:*  Ramsey numbers play a role in constructing convex polygons. Suppose $m$ is a positive integer and there are $n$ given points, no three of which are collinear. If $n \geq R(m, 5; 4)$, then a convex $m$-gon can be obtained from $m$ of the $n$ points [ErSz35]. This paper provided the impetus for the study of Ramsey numbers and suggested the possibility of its wide applicability in mathematics.

**25.** It remains an unsolved problem to find the smallest integer $x$ (which depends on $m$) such that if $n \geq x$, then a convex $m$-gon can be obtained from $m$ of the $n$ points.

**26.** Extensive information on Ramsey number theory, including bounds on Ramsey numbers, can be found at S. Radziszowski's web site:

- `http://www.cs.rit.edu/~spr/ElJC/eline.html`

**Examples:**

**1.** If six people are at a party, then either three of these six are mutual friends or three are mutual strangers. If six is replaced by five, the result is not true. These facts follow since $R(3, 3) = 6$. (See Fact 2. The six people can be regarded as vertices, with a red edge joining friends and a blue edge joining strangers.)

**2.** If the set $\{1, \ldots, k\}$ is partitioned into two subsets $A_1$ and $A_2$, then the equation $x + y = z$ may or may not have a solution where $x, y, z \in A_1$ or $x, y, z \in A_2$. If $k \geq 5$, a solution is guaranteed since $S(2) = 5$. If $k < 5$, no solution is guaranteed—take $A_1 = \{1, 4\}$ and $A_2 = \{2, 3\}$.

---

### 3.1.7  OTHER SEQUENCES

Additional sequences that regularly arise in discrete mathematics are now described.

---

#### ▷ Euler Polynomials

**Definition:**

The **Euler polynomials** $E_n(x)$ have the following exponential generating function:

$$\sum_{n=0}^{\infty} E_n(x)\frac{t^n}{n!} = \frac{2e^{xt}}{e^t + 1}.$$

**Facts:**

**1.**  $E_n(x + 1) + E_n(x) = 2x^n$ for all $n \geq 0$.

**2.** The Euler polynomials can be expressed in terms of the Bernoulli numbers (§3.1.4):

$$E_{n-1}(x) = \frac{1}{n} \sum_{k=1}^{n} (2 - 2^{k+1})\binom{n}{k} B_k x^{n-k} \text{ for all } n \geq 1.$$

**3.** The alternating sum of powers of the first $n$ integers can be expressed in terms of the Euler polynomials: $\sum_{j=1}^{n}(-1)^{n-j}j^k = \frac{1}{2}\big[E_k(n+1)+(-1)^nE_k(0)\big]$.

**4.** The first 14 Euler polynomials $E_n(x)$ are shown in the following table.

| $n$ | $E_n(x)$ |
|---|---|
| 0 | 1 |
| 1 | $x - \frac{1}{2}$ |
| 2 | $x^2 - x$ |
| 3 | $x^3 - \frac{3}{2}x^2 + \frac{1}{4}$ |
| 4 | $x^4 - 2x^3 + x$ |
| 5 | $x^5 - \frac{5}{2}x^4 + \frac{5}{2}x^2 - \frac{1}{2}$ |
| 6 | $x^6 - 3x^5 + 5x^3 - 3x$ |
| 7 | $x^7 - \frac{7}{2}x^6 + \frac{35}{4}x^4 - \frac{21}{2}x^2 + \frac{17}{8}$ |
| 8 | $x^8 - 4x^7 + 14x^5 - 28x^3 + 17x$ |
| 9 | $x^9 - \frac{9}{2}x^8 + 21x^6 - 63x^4 + \frac{153}{2}x^2 - \frac{31}{2}$ |
| 10 | $x^{10} - 5x^9 + 30x^7 - 126x^5 + 255x^3 - 155x$ |
| 11 | $x^{11} - \frac{11}{2}x^{10} + \frac{165}{4}x^8 - 231x^6 + \frac{2,805}{4}x^4 - \frac{1,705}{2}x^2 + \frac{691}{4}$ |
| 12 | $x^{12} - 6x^{11} + 55x^9 - 396x^7 + 1,683x^5 - 3,410x^3 + 2,073x$ |
| 13 | $x^{13} - \frac{13}{2}x^{12} + \frac{143}{2}x^{10} - \frac{1,287}{2}x^8 + \frac{7,293}{2}x^6 - \frac{22,165}{2}x^4 + \frac{26,949}{2}x^2 - \frac{5,461}{2}$ |

---

### ▷ Euler and Tangent Numbers

**Definitions:**

The **Euler numbers** $E_n$ are given by $E_n = 2^n E_n(\frac{1}{2})$, where $E_n(x)$ is an Euler polynomial.

The **tangent numbers** $T_n$ have the exponential generating function $\tan x$: $\sum_{n=0}^{\infty} T_n \frac{x^n}{n!} = \tan x$.

**Facts:**

**1.** The first twelve Euler numbers $E_n$ and tangent numbers $T_n$ are shown in the following table.

| $n$ | 0 | 1 | 2 | 3 | 4 | 5 | 6 | 7 | 8 | 9 | 10 | 11 |
|---|---|---|---|---|---|---|---|---|---|---|---|---|
| $E_n$ | 1 | 0 | $-1$ | 0 | 5 | 0 | $-61$ | 0 | 1,385 | 0 | $-50,521$ | 0 |
| $T_n$ | 0 | 1 | 0 | 2 | 0 | 16 | 0 | 272 | 0 | 7,936 | 0 | 353,792 |

**2.** $E_{2k+1} = T_{2k} = 0$ for all $k \geq 0$.

**3.** The nonzero Euler numbers alternate in sign.

**4.** The Euler numbers have the exponential generating function $\frac{2}{e^t+e^{-t}} = \operatorname{sech} t$.

**5.** The exponential generating function for $|E_n|$ is $\sum_{n=0}^{\infty}|E_n|\frac{t^n}{n!} = \sec t$.

**6.** The tangent numbers can be expressed in terms of the Bernoulli numbers (§3.1.4): $T_{2n-1} = (-1)^{n-1}\frac{4^n(4^n-1)}{2n}B_{2n}$ for all $n \geq 1$.

**7.** The tangent numbers can be expressed as an alternating sum of Eulerian numbers (§3.1.5): $T_{2n+1} = \sum_{k=0}^{2n}(-1)^{n-k}E(2n+1,k)$ for all $n \geq 0$.

**8.** $(-1)^n E_{2n}$ counts the number of *alternating* permutations in $S_{2n}$: that is, the number of permutations $\pi = (\pi_1, \pi_2, \ldots, \pi_{2n})$ on $\{1, 2, \ldots, 2n\}$ with $\pi_1 > \pi_2 < \pi_3 > \pi_4 < \cdots > \pi_{2n}$.

**9.** $T_{2n+1}$ counts the number of alternating permutations in $S_{2n+1}$.

**Examples:**

**1.** The permutation $\pi = (\pi_1, \pi_2, \pi_3, \pi_4) = (2, 1, 4, 3)$ is alternating since $2 > 1 < 4 > 3$. In all there are $(-1)^2 E_4 = 5$ alternating permutations in $S_4$: $(2, 1, 4, 3)$, $(3, 1, 4, 2)$, $(3, 2, 4, 1)$, $(4, 1, 3, 2)$, $(4, 2, 3, 1)$.

**2.** The permutation $\pi = (\pi_1, \pi_2, \pi_3, \pi_4, \pi_5) = (4, 1, 3, 2, 5)$ is alternating since $4 > 1 < 3 > 2 < 5$. In all there are $T_5 = 16$ alternating permutations in $S_5$.

---

▷ **Harmonic Numbers**

**Definition:**

The **harmonic numbers** $H_n$ are given by $H_n = \sum_{i=1}^{n} \frac{1}{i}$ for $n \geq 0$, with $H_0 = 0$.

**Facts:**

**1.** $H_n$ is the discrete analogue of the natural logarithm (§3.4.1).

**2.** The first twelve harmonic numbers $H_n$ are shown in the following table.

| $n$ | 0 | 1 | 2 | 3 | 4 | 5 | 6 | 7 | 8 | 9 | 10 | 11 |
|---|---|---|---|---|---|---|---|---|---|---|---|---|
| $H_n$ | 0 | 1 | $\frac{3}{2}$ | $\frac{11}{6}$ | $\frac{25}{12}$ | $\frac{137}{60}$ | $\frac{49}{20}$ | $\frac{363}{140}$ | $\frac{761}{280}$ | $\frac{7,129}{2,520}$ | $\frac{7,381}{2,520}$ | $\frac{83,711}{27,720}$ |

**3.** The harmonic numbers can be expressed in terms of the Stirling cycle numbers (§2.5.2): $H_n = \frac{1}{n!}\left[\begin{smallmatrix} n+1 \\ 2 \end{smallmatrix}\right]$, $n \geq 1$.

**4.** $\sum_{i=1}^{n} H_i = (n+1)\left[H_{n+1} - 1\right]$ for all $n \geq 1$.

**5.** $\sum_{i=1}^{n} iH_i = \binom{n+1}{2}\left[H_{n+1} - \frac{1}{2}\right]$ for all $n \geq 1$.

**6.** $\sum_{i=1}^{n} \binom{i}{k} H_i = \binom{n+1}{k+1}\left[H_{n+1} - \frac{1}{k+1}\right]$ for all $n \geq 1$.

**7.** $H_n \to \infty$ as $n \to \infty$.

**8.** $H_n \sim \ln n + \gamma + \frac{1}{2n} - \frac{1}{12n^2} + \frac{1}{120n^4}$, where $\gamma \approx 0.577215664901533$ denotes Euler's constant.

**9.** The harmonic numbers have the generating function $\frac{1}{1-x} \ln \frac{1}{1-x}$.

**Example:**

**1.** Fact 8 yields the approximation $H_{10} \approx 2.928968257896$. The actual value is $H_{10} = 2.928968253968\ldots$, so the approximation is accurate to 9 significant digits. The approximation $H_{20} \approx 3.597739657206$ is accurate to 10 digits, and the approximation $H_{40} \approx 4.27854303893$ is accurate to 12 digits.

---

▷ **Gray Codes**

**Definition:**

A **Gray code** of size $n$ is an ordering $G^n = (g_1, g_2, \ldots, g_{2^n})$ of the $2^n$ binary strings of length $n$ such that $g_k$ and $g_{k+1}$ differ in exactly one bit, for $1 \leq k < 2^n$. Usually it is required that $g_{2^n}$ and $g_1$ also differ in exactly one bit.

**Facts:**

**1.** Gray codes exist for all $n \geq 1$. Sample Gray codes $G^n$ are shown in the following table.

| $n$ | $G^n$ | | | | | | | |
|---|---|---|---|---|---|---|---|---|
| 1 | 0 | 1 | | | | | | |
| 2 | 00 | 10 | 11 | 01 | | | | |
| 3 | 000 | 100 | 110 | 010 | 011 | 111 | 101 | 001 |
| 4 | 0000 | 1000 | 1100 | 0100 | 0110 | 1110 | 1010 | 0010 | 0011 |
| | 1011 | 1111 | 0111 | 0101 | 1101 | 1001 | 0001 | | |
| 5 | 00000 | 10000 | 11000 | 01000 | 01100 | 11100 | 10100 | 00100 | 00110 |
| | 10110 | 11110 | 01110 | 01010 | 11010 | 10010 | 00010 | 00011 | 10011 |
| | 11011 | 01011 | 01111 | 11111 | 10111 | 00111 | 00101 | 10101 | 11101 |
| | 01101 | 01001 | 11001 | 10001 | 00001 | | | | |

**2.** A Gray code of size $n \geq 2$ corresponds to a Hamilton cycle in the $n$-cube (§8.4.4).

**3.** Gray codes correspond to an ordering of all subsets of $\{1, 2, \ldots, n\}$ such that adjacent subsets differ by the insertion or deletion of exactly one element. Each subset $A$ corresponds to a binary string $a_1 a_2 \ldots a_n$ where $a_i = 1$ if $i \in A$, $a_i = 0$ if $i \notin A$.

**4.** A Gray code $G^n$ can be recursively obtained in the following way:

- *first half of $G^n$*: Add a 0 to the end of each string in $G^{n-1}$.
- *second half of $G^n$*: Add a 1 to the end of each string in the reversal of $G^{n-1}$.

---

▷ **de Bruijn Sequences**

**Definitions:**

A **$(p, n)$ de Bruijn sequence** on the alphabet $\Sigma = \{0, 1, \ldots, p - 1\}$ is a sequence $(s_0, s_1, \ldots, s_{L-1})$ of $L = p^n$ elements $s_i \in \Sigma$ such that each consecutive subsequence $(s_i, s_{i+1}, \ldots, s_{i+n-1})$ of length $n$ is distinct. Here the addition of subscripts is done modulo $L$ so that the sequence is considered as a circular ordering. (Nicolaas G. de Bruijn, 1918–2012)

The **de Bruijn diagram** $D_{p,n}$ is a directed graph whose vertices correspond to all possible strings $s_1 s_2 \ldots s_{n-1}$ of $n-1$ symbols from $\Sigma$. There are $p$ arcs leaving the vertex $s_1 s_2 \ldots s_{n-1}$, each labeled with a distinct symbol $\alpha \in \Sigma$ and leading to the adjacent vertex $s_2 s_3 \ldots s_{n-1} \alpha$.

**Facts:**

**1.** The de Bruijn diagram $D_{p,n}$ has $p^{n-1}$ vertices and $p^n$ arcs.

**2.** $D_{p,n}$ is a strongly connected digraph (§8.3.2).

**3.** $D_{p,n}$ is an Eulerian digraph (§8.4.3).

**4.** There are $(p!)^{p^{n-1}} p^{-n}$ distinct $(p, n)$ de Bruijn sequences.

**5.** Any Euler circuit in $D_{p,n}$ produces a $(p, n)$ de Bruijn sequence.

**6.** de Bruijn sequences exist for all $p$ (with $n \geq 1$). Sample de Bruijn sequences are shown in the following table.

| $(p, n)$ | de Bruijn sequence |
|---|---|
| $(2, 1)$ | 01 |
| $(2, 2)$ | 0110 |
| $(2, 3)$ | 01110100 |
| $(2, 4)$ | 0101001101111000 |
| $(3, 2)$ | 012202110 |
| $(3, 3)$ | 012001110100022212202112102 |
| $(4, 2)$ | 0113102212033230 |

**7.** A de Bruijn sequence can be generated from an alphabet $\Sigma = \{0, 1, \ldots, p - 1\}$ of $p$ symbols using Algorithm 1.

---

**Algorithm 1**:  **Generating a $(p, n)$ de Bruijn sequence.**

1. Start with the sequence $S$ containing $n$ zeros.
2. Append the largest symbol from $\Sigma$ to $S$ so that the newly formed sequence $S'$ of $n$ symbols does not already appear as a subsequence of $S$. Let $S = S'$.
3. Repeat Step 2 as long as possible.
4. When Step 2 cannot be applied, remove the last $n - 1$ symbols from $S$.

---

**Examples:**

**1.** The de Bruijn diagram $D_{2,3}$ is shown in the following figure. An Eulerian circuit is obtained by visiting in order the vertices $11, 10, 01, 10, 00, 00, 01, 11, 11$. The de Bruijn sequence 01000111 is obtained by reading off the edge labels $\alpha$ as this circuit is traversed.

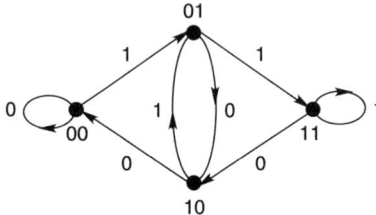

**2.** When $p = 2$ and $n = 3$, the formula in Fact 4 gives

$$\frac{(2!)^{2^{3-1}}}{2^3} = \frac{2^4}{2^3} = 2$$

distinct (2,3) de Bruijn sequences. These are 01000111 and its reversal 11100010.

---

▷ **Self-generating Sequences**

**Definition:**

Some unusual sequences defined by simple recurrence relations or rules are informally called **self-generating sequences**.

**Examples:**

**1.** *Hofstadter G-sequence*: This sequence is defined by $a(n) = n - a(a(n-1))$, with initial condition $a(0) = 0$. The initial terms of this sequence are 0, 1, 1, 2, 3, 3, 4, 4, 5, 6, 7, 8, 8, 9, 9, 10, .... It is easy to show this sequence is well defined. A formula for the $n$th term of this sequence is $a(n) = \lfloor (n+1)\mu \rfloor$, where $\mu = (-1 + \sqrt{5})/2$. [Ho79]

**2.** *Variations of the Hofstadter G-sequence about which little is known*: These include the sequence defined by $a(n) = n - a(a(a(n-1)))$ with $a(0) = 1$, whose initial terms are $0, 1, 1, 2, 3, 4, 4, 5, 5, 6, 7, 7, 8, 9, 10, 10, 11, 12, 13, \ldots$ and the sequence defined by $a(n) = n - a(a(a(a(n-1))))$ with $a(0) = 1$, whose initial terms are 0, 1, 1, 2, 3, 4, 5, 5, 6, 6, 7, 8, 8, 9, 10, 11, 11, 12, 13, 14, ....

**3.** The sequence $a(n) = a(n - a(n-1)) + a(n - a(n-2))$, with $a(0) = a(1) = 1$, was also defined by Hofstadter. The initial terms of this sequence are 1, 1, 2, 3, 3, 4, 5, 5, 6, 6, 6, 8, 8, 8, 10, 10, 10, 12, ....

**4.** The intertwined sequence $F(n)$ and $M(n)$ are defined by $F(n) = n - F(M(n-1))$ and $M(n) = n - M(F(n-1))$, with initial conditions $F(0) = 1$ and $M(0) = 0$. The initial terms of the sequence $F(n)$ (sometimes called the "female" sequence of the pair) begins with the terms $1, 1, 2, 2, 3, 3, 4, 5, 5, 6, 6, 7, 8, 8, 9, 9, 10, \ldots$ and the initial terms of the sequence $M(n)$ (sometimes called the "male" sequence of the pair) begins with the terms $0, 0, 1, 2, 2, 3, 4, 4, 5, 6, 6, 7, 7, 8, 9, 9, 10, \ldots$.

**5.** *Golomb's self-generating sequence*: This sequence is the unique nondecreasing sequence $a_1, a_2, a_3, \ldots$ with the property that it contains exactly $a_k$ occurrences of the integer $k$ for each integer $k$. The initial terms of this sequence are 1, 2, 2, 3, 3, 4, 4, 4, 5, 5, 5, 6, 6, 6, 6, ....

**6.** If $f(n)$ is the largest integer $m$ such that $a_m = n$ where $a_k$ is the $k$th term of Golomb's self-generating sequence, then $f(n) = \sum_{k=1}^{n} a_k$ and $f(f(n)) = \sum_{k=1}^{n} k a_k$.

---

### 3.1.8  MINIGUIDE TO SEQUENCES

This section lists the numerical values of various integer sequences, classified according to the type of combinatorial structure that produces the terms. This listing supplements many of the tables presented in this *Handbook*. A tabulation of over 5,400 integer sequences is provided in [SlPl95], and over 250,000 entries are available through an online database. (See Fact 4.)

**Definitions:**

The **power sum** $S^k(n) = \sum_{j=1}^{n} j^k$ is the sum of the $k$th powers of the first $n$ positive integers. The sum of the $k$th powers of the first $n$ odd integers is denoted $O^k(n) = \sum_{j=1}^{n} (2j-1)^k$.

The **associated Stirling number of the first kind** $d(n, k)$ is the number of $k$-cycle permutations of an $n$-element set with all cycles of length $\geq 2$.

The **associated Stirling number of the second kind** $b(n, k)$ is the number of $k$-block partitions of an $n$-element set with all blocks of size $\geq 2$.

The **double factorial** $n!!$ is the product $n(n-2)\ldots 6\cdot 4\cdot 2$ if $n$ is an even positive integer and $n(n-2)\ldots 5\cdot 3\cdot 1$ if $n$ is an odd positive integer.

The **Lah coefficients** $L(n,k)$ are the coefficients of $x^{\underline{k}}$ (§3.4.2) resulting from the expansion of $x^{\overline{n}}$ (§3.4.2):

$$x^{\overline{n}} = \sum_{k=1}^{n} L(n,k)x^{\underline{k}}.$$

A permutation $\pi$ is **discordant** from a set $A$ of permutations when $\pi(i) \neq \alpha(i)$ for all $i$ and all $\alpha \in A$. Usually $A$ consists of the identity permutation $\iota$ and powers of the $n$-cycle $\sigma_n = (1\ 2\ \ldots\ n)$ (see §5.3.1).

A **necklace** with $n$ beads in $c$ colors corresponds to an equivalence class of functions from an $n$-set to a $c$-set, under cyclic or dihedral equivalence.

A **figurate number** is the number of cells in an array of cells bounded by some regular geometrical figure.

A **polyomino** with $p$ polygons (cells) is a connected configuration of $p$ regular polygons in the plane. The polygons usually considered are either triangles, squares, or hexagons.

**Facts:**

**1.** A database of integer sequences was initiated by Neal J. A. Sloane in 1964, while he was exploring the sequence 1, 8, 78, 944, ... that arose in determining the average height of vertices in rooted trees.

**2.** Sloane published the book [Sl73], which contained over 2,300 integer sequences.

**3.** Together with Simon Plouffe, Sloane published a revised and expanded version of the book in 1995 [SlPl95]; it contained over 5,400 integer sequences.

**4.** *World Wide Web page:* Sequences can be accessed and identified using a web page maintained by The OEIS Foundation:

- http://oeis.org

Sequences are arranged in lexicographic order, after ignoring any initial 0 or 1 terms as well as the signs of terms, and are uniquely identified as A plus six digits.

**5.** Each entry in the following miniguide lists initial terms of the sequence, provides a brief description, and gives the reference number used in the online database of Fact 4.

**Examples:**

**1.** The initial five terms $1, 2, 6, 20, 70, 252$ of an unknown sequence were entered at the site listed in Fact 4. Several matching sequences were identified, the first of which (A000984) corresponds to the central binomial coefficients $\binom{2n}{n}$. Entering the terms above, separated by commas, identifies sequences in which the specified numbers occur consecutively in the order given.

**2.** Entering the initial terms 1 1 2 3 5 8 13 21 34 into the OEIS site produced over 10,000 matching sequences. In this case, spaces were used rather than commas to delimit the terms and so sequences matching the given terms in any order are returned. The first sequence produced was the Fibonacci sequence (A000045); another was the sequence $a_n = \left\lceil e^{\frac{n-1}{2}} \right\rceil$ (A005181).

▷ **Miniguide to Sequences from Discrete Mathematics**

The following miniguide contains a selection of important sequences, grouped by functional problem area (such as graph theory, algebra, number theory). The sequences are listed in a logical, rather than lexicographic, order within each identifiable grouping. This listing supplements existing tables within the *Handbook*. References to appropriate sections of the *Handbook* are also provided. The notation "A$xxxxxx$" is the reference number used in the OEIS database (Fact 4).

**Powers of Integers** (§3.1.1, §3.5.4)

1, 2, 4, 8, 16, 32, 64, 128, 256, 512, 1024, 2048, 4096, 8192, 16384, 32768, 65536, 131072
$$2^n \text{ [A000079]}$$

1, 3, 9, 27, 81, 243, 729, 2187, 6561, 19683, 59049, 177147, 531441, 1594323, 4782969
$$3^n \text{ [A000244]}$$

1, 4, 16, 64, 256, 1024, 4096, 16384, 65536, 262144, 1048576, 4194304, 16777216, 67108864
$$4^n \text{ [A000302]}$$

1, 5, 25, 125, 625, 3125, 15625, 78125, 390625, 1953125, 9765625, 48828125, 244140625
$$5^n \text{ [A000351]}$$

1, 6, 36, 216, 1296, 7776, 46656, 279936, 1679616, 10077696, 60466176, 362797056
$$6^n \text{ [A000400]}$$

1, 7, 49, 343, 2401, 16807, 117649, 823543, 5764801, 40353607, 282475249, 1977326743
$$7^n \text{ [A000420]}$$

1, 8, 64, 512, 4096, 32768, 262144, 2097152, 16777216, 134217728, 1073741824, 8589934592
$$8^n \text{ [A001018]}$$

1, 9, 81, 729, 6561, 59049, 531441, 4782969, 43046721, 387420489, 3486784401, 31381059609
$$9^n \text{ [A001019]}$$

1, 4, 9, 16, 25, 36, 49, 64, 81, 100, 121, 144, 169, 196, 225, 256, 289, 324, 361, 400, 441, 484
$$n^2 \text{ [A000290]}$$

1, 8, 27, 64, 125, 216, 343, 512, 729, 1000, 1331, 1728, 2197, 2744, 3375, 4096, 4913, 5832
$$n^3 \text{ [A000578]}$$

1, 16, 81, 256, 625, 1296, 2401, 4096, 6561, 1000014641, 20736, 28561, 38416, 50625, 65536
$$n^4 \text{ [A000583]}$$

1, 32, 243, 1024, 3125, 7776, 16807, 32768, 59049, 100000, 161051, 248832, 371293, 537824
$$n^5 \text{ [A000584]}$$

1, 64, 729, 4096, 15625, 46656, 117649, 262144, 531441, 1000000, 1771561, 2985984
$$n^6 \text{ [A001014]}$$

1, 128, 2187, 16384, 78125, 279936, 823543, 2097152, 4782969, 10000000, 19487171
$$n^7 \text{ [A001015]}$$

1, 256, 6561, 65536, 390625, 1679616, 5764801, 16777216, 43046721, 100000000, 214358881
$$n^8 \text{ [A001016]}$$

1, 512, 19683, 262144, 1953125, 10077696, 40353607, 134217728, 387420489, 1000000000
$$n^9 \text{ [A001017]}$$

1, 3, 6, 10, 15, 21, 28, 36, 45, 55, 66, 78, 91, 105, 120, 136, 153, 171, 190, 210, 231, 253, 276, 300
$$S^1(n) \text{ [A000217]}$$

1, 5, 14, 30, 55, 91, 140, 204, 285, 385, 506, 650, 819, 1015, 1240, 1496, 1785, 2109, 2470, 2870
$$S^2(n) \text{ [A000330]}$$

1, 9, 36, 100, 225, 441, 784, 1296, 2025, 3025, 4356, 6084, 8281, 11025, 14400, 18496, 23409
$$S^3(n) \text{ [A000537]}$$

1, 17, 98, 354, 979, 2275, 4676, 8772, 15333, 25333, 39974, 60710, 89271, 127687, 178312
$$S^4(n) \text{ [A000538]}$$

1, 33, 276, 1300, 4425, 12201, 29008, 61776, 120825, 220825, 381876, 630708, 1002001
$$S^5(n) \text{ [A000539]}$$

1, 65, 794, 4890, 20515, 67171, 184820, 446964, 978405, 1978405, 3749966, 6735950
$$S^6(n) \text{ [A000540]}$$

1, 129, 2316, 18700, 96825, 376761, 1200304, 3297456, 8080425, 18080425, 37567596
$$S^7(n) \text{ [A000541]}$$

1, 257, 6818, 72354, 462979, 2142595, 7907396, 24684612, 67731333, 167731333, 382090214
$$S^8(n) \text{ [A000542]}$$

1, 513, 20196, 282340, 2235465, 12313161, 52666768, 186884496, 574304985, 1574304985
$$S^9(n) \text{ [A007487]}$$

3, 6, 14, 36, 98, 276, 794, 2316, 6818, 20196, 60074, 179196, 535538, 1602516, 4799354
$$S^n(3) \text{ [A001550]}$$

4, 10, 30, 100, 354, 1300, 4890, 18700, 72354, 282340, 1108650, 4373500, 17312754
$$S^n(4) \text{ [A001551]}$$

5, 15, 55, 225, 979, 4425, 20515, 96825, 462979, 2235465, 10874275, 53201625, 261453379
$$S^n(5) \text{ [A001552]}$$

6, 21, 91, 441, 2275, 12201, 67171, 376761, 2142595, 12313161, 71340451, 415998681
$$S^n(6) \text{ [A001553]}$$

7, 28, 140, 784, 4676, 29008, 184820, 1200304, 7907396, 52666768, 353815700, 2393325424
$$S^n(7) \text{ [A001554]}$$

8, 36, 204, 1296, 8772, 61776, 446964, 3297456, 24684612, 186884496, 1427557524
$$S^n(8) \text{ [A001555]}$$

9, 45, 285, 2025, 15333, 120825, 978405, 8080425, 67731333, 574304985, 4914341925
$$S^n(9) \text{ [A001556]}$$

1, 5, 32, 288, 3413, 50069, 873612, 17650828, 405071317, 10405071317, 295716741928
$$S^n(n) \text{ [A001923]}$$

1, 28, 153, 496, 1225, 2556, 4753, 8128, 13041, 19900, 29161, 41328, 56953, 76636, 101025
$$O^3(n) \text{ [A002593]}$$

1, 82, 707, 3108, 9669, 24310, 52871, 103496, 187017, 317338, 511819, 791660, 1182285
$$O^4(n) \text{ [A002309]}$$

1, 244, 3369, 20176, 79225, 240276, 611569, 1370944, 2790801, 5266900, 9351001, 15787344
$$O^5(n) \text{ [A002594]}$$

## Factorial Numbers

1, 1, 2, 6, 24, 120, 720, 5040, 40320, 362880, 3628800, 39916800, 479001600, 6227020800
$$n! \text{ [A000142]}$$

1, 4, 36, 576, 14400, 518400, 25401600, 1625702400, 131681894400, 13168189440000
$$(n!)^2 \text{ [A001044]}$$

$2, 3, 8, 30, 144, 840, 5760, 45360, 403200, 3991680, 43545600, 518918400, 6706022400$
$$n! + (n-1)! \text{ [A001048]}$$

$1, 2, 8, 48, 384, 3840, 46080, 645120, 10321920, 185794560, 3715891200, 81749606400$
$$n!!, n \text{ even [A000165]}$$

$1, 1, 3, 15, 105, 945, 10395, 135135, 2027025, 34459425, 654729075, 13749310575$
$$n!!, n \text{ odd [A001147]}$$

$1, 1, 2, 12, 288, 34560, 24883200, 125411328000, 5056584744960000$
$$\text{product of } n \text{ factorials [A000178]}$$

$1, 2, 6, 30, 210, 2310, 30030, 510510, 9699690, 223092870, 6469693230, 200560490130$
$$\text{product of first } n \text{ primes [A002110]}$$

**Binomial Coefficients** (§2.3.2)
$1, 3, 6, 10, 15, 21, 28, 36, 45, 55, 66, 78, 91, 105, 120, 136, 153, 171, 190, 210, 231, 253, 276$
$$\binom{n}{2} \text{ [A000217]}$$

$1, 4, 10, 20, 35, 56, 84, 120, 165, 220, 286, 364, 455, 560, 680, 816, 969, 1140, 1330, 1540, 1771$
$$\binom{n}{3} \text{ [A000292]}$$

$1, 5, 15, 35, 70, 126, 210, 330, 495, 715, 1001, 1365, 1820, 2380, 3060, 3876, 4845, 5985, 7315$
$$\binom{n}{4} \text{ [A000332]}$$

$1, 6, 21, 56, 126, 252, 462, 792, 1287, 2002, 3003, 4368, 6188, 8568, 11628, 15504, 20349$
$$\binom{n}{5} \text{ [A000389]}$$

$1, 7, 28, 84, 210, 462, 924, 1716, 3003, 5005, 8008, 12376, 18564, 27132, 38760, 54264, 74613$
$$\binom{n}{6} \text{ [A000579]}$$

$1, 8, 36, 120, 330, 792, 1716, 3432, 6435, 11440, 19448, 31824, 50388, 77520, 116280, 170544$
$$\binom{n}{7} \text{ [A000580]}$$

$1, 9, 45, 165, 495, 1287, 3003, 6435, 12870, 24310, 43758, 75582, 125970, 203490, 319770$
$$\binom{n}{8} \text{ [A000581]}$$

$1, 10, 55, 220, 715, 2002, 5005, 11440, 24310, 48620, 92378, 167960, 293930, 497420, 817190$
$$\binom{n}{9} \text{ [A000582]}$$

$1, 11, 66, 286, 1001, 3003, 8008, 19448, 43758, 92378, 184756, 352716, 646646, 1144066$
$$\binom{n}{10} \text{ [A001287]}$$

$1, 2, 3, 6, 10, 20, 35, 70, 126, 252, 462, 924, 1716, 3432, 6435, 12870, 24310, 48620, 92378$
$$\text{central binomial coefficients } \binom{n}{\lfloor n/2 \rfloor} \text{ [A001405]}$$

$1, 2, 6, 20, 70, 252, 924, 3432, 12870, 48620, 184756, 705432, 2704156, 10400600, 40116600$
$$\text{central binomial coefficients } \binom{2n}{n} \text{ [A000984]}$$

$1, 3, 10, 35, 126, 462, 1716, 6435, 24310, 92378, 352716, 1352078, 5200300, 20058300$
$$\binom{2n+1}{n} \text{ [A001700]}$$

**Stirling Cycle Numbers/Stirling Numbers of the First Kind** (§2.5.2)
$1, 3, 11, 50, 274, 1764, 13068, 109584, 1026576, 10628640, 120543840, 1486442880$
$$\begin{bmatrix} n \\ 2 \end{bmatrix} \text{ [A000254]}$$

$1, 6, 35, 225, 1624, 13132, 118124, 1172700, 12753576, 150917976, 1931559552, 26596717056$
$$\begin{bmatrix} n \\ 3 \end{bmatrix} \text{ [A000399]}$$

$1, 10, 85, 735, 6769, 67284, 723680, 8409500, 105258076, 1414014888, 20313753096$
$$\begin{bmatrix} n \\ 4 \end{bmatrix} \text{ [A000454]}$$

$1, 15, 175, 1960, 22449, 269325, 3416930, 45995730, 657206836, 9957703756, 159721605680$
$$\begin{bmatrix} n \\ 5 \end{bmatrix} \text{ [A000482]}$$

1, 21, 322, 4536, 63273, 902055, 13339535, 206070150, 3336118786, 56663366760
$$\begin{bmatrix} n \\ 6 \end{bmatrix} \text{ [A001233]}$$

1, 28, 546, 9450, 157773, 2637558, 44990231, 790943153, 14409322928, 272803210680
$$\begin{bmatrix} n \\ 7 \end{bmatrix} \text{ [A001234]}$$

2, 11, 35, 85, 175, 322, 546, 870, 1320, 1925, 2717, 3731, 5005, 6580, 8500, 10812, 13566
$$\begin{bmatrix} n \\ n-2 \end{bmatrix} \text{ [A000914]}$$

6, 50, 225, 735, 1960, 4536, 9450, 18150, 32670, 55770, 91091, 143325, 218400, 323680
$$\begin{bmatrix} n \\ n-3 \end{bmatrix} \text{ [A001303]}$$

24, 274, 1624, 6769, 22449, 63273, 157773, 357423, 749463, 1474473, 2749747, 4899622
$$\begin{bmatrix} n \\ n-4 \end{bmatrix} \text{ [A000915]}$$

## Stirling Subset Numbers/Stirling Numbers of the Second Kind (§2.5.2)

1, 6, 25, 90, 301, 966, 3025, 9330, 28501, 86526, 261625, 788970, 2375101, 7141686
$$\left\{ \begin{matrix} n \\ 3 \end{matrix} \right\} \text{ [A000392]}$$

1, 10, 65, 350, 1701, 7770, 34105, 145750, 611501, 2532530, 10391745, 42355950, 171798901
$$\left\{ \begin{matrix} n \\ 4 \end{matrix} \right\} \text{ [A000453]}$$

1, 15, 140, 1050, 6951, 42525, 246730, 1379400, 7508501, 40075035, 210766920, 1096190550
$$\left\{ \begin{matrix} n \\ 5 \end{matrix} \right\} \text{ [A000481]}$$

1, 21, 266, 2646, 22827, 179487, 1323652, 9321312, 63436373, 420693273, 2734926558
$$\left\{ \begin{matrix} n \\ 6 \end{matrix} \right\} \text{ [A000770]}$$

1, 28, 462, 5880, 63987, 627396, 5715424, 49329280, 408741333, 3281882604, 25708104786
$$\left\{ \begin{matrix} n \\ 7 \end{matrix} \right\} \text{ [A000771]}$$

1, 7, 25, 65, 140, 266, 462, 750, 1155, 1705, 2431, 3367, 4550, 6020, 7820, 9996, 12597, 15675
$$\left\{ \begin{matrix} n \\ n-2 \end{matrix} \right\} \text{ [A001296]}$$

1, 15, 90, 350, 1050, 2646, 5880, 11880, 22275, 39325, 66066, 106470, 165620, 249900
$$\left\{ \begin{matrix} n \\ n-3 \end{matrix} \right\} \text{ [A001297]}$$

1, 31, 301, 1701, 6951, 22827, 63987, 159027, 359502, 752752, 1479478, 2757118, 4910178
$$\left\{ \begin{matrix} n \\ n-4 \end{matrix} \right\} \text{ [A001298]}$$

1, 1, 3, 7, 25, 90, 350, 1701, 7770, 42525, 246730, 1379400, 9321312, 63436373, 420693273
$$\max_k \left\{ \begin{matrix} n \\ k \end{matrix} \right\} \text{ [A002870]}$$

## Associated Stirling Numbers of the First Kind (§3.1.8)

3, 20, 130, 924, 7308, 64224, 623376, 6636960, 76998240, 967524480, 13096736640
$$d(n, 2) \text{ [A000276]}$$

15, 210, 2380, 26432, 303660, 3678840, 47324376, 647536032, 9418945536, 145410580224
$$d(n, 3) \text{ [A000483]}$$

2, 20, 210, 2520, 34650, 540540, 9459450, 183783600, 3928374450, 91662070500
$$d(n, n-3) \text{ [A000906]}$$

6, 130, 2380, 44100, 866250, 18288270, 416215800, 10199989800, 268438920750
$$d(n, n-4) \text{ [A000907]}$$

1, 120, 7308, 303660, 11098780, 389449060, 13642629000, 486591585480, 17856935296200
$$d(2n, n-2) \text{ [A001785]}$$

1, 24, 924, 26432, 705320, 18858840, 520059540, 14980405440, 453247114320
$$d(2n+1, n-1) \text{ [A001784]}$$

**Associated Stirling Numbers of the Second Kind** (§3.1.8)

$3, 10, 25, 56, 119, 246, 501, 1012, 2035, 4082, 8177, 16368, 32751, 65518, 131053, 262124$
$$b(n, 2) \text{ [A000247]}$$

$15, 105, 490, 1918, 6825, 22935, 74316, 235092, 731731, 2252341, 6879678, 20900922$
$$b(n, 3) \text{ [A000478]}$$

$1, 25, 490, 9450, 190575, 4099095, 94594500, 2343240900, 62199262125, 1764494857125$
$$b(2n, n - 1) \text{ [A000497]}$$

$1, 56, 1918, 56980, 1636635, 47507460, 1422280860, 44346982680, 1446733012725$
$$b(2n + 1, n - 1) \text{ [A000504]}$$

**Lah Coefficients** (§3.1.8)

$1, 6, 36, 240, 1800, 15120, 141120, 1451520, 16329600, 199584000, 2634508800$
$$L(n, 2) \text{ [A001286]}$$

$1, 12, 120, 1200, 12600, 141120, 1693440, 21772800, 299376000, 4390848000, 68497228800$
$$L(n, 3) \text{ [A001754]}$$

$1, 20, 300, 4200, 58800, 846720, 12700800, 199584000, 3293136000, 57081024000$
$$L(n, 4) \text{ [A001755]}$$

$1, 30, 630, 11760, 211680, 3810240, 69854400, 1317254400, 25686460800, 519437318400$
$$L(n, 5) \text{ [A001777]}$$

$1, 42, 1176, 28224, 635040, 13970880, 307359360, 6849722880, 155831195520$
$$L(n, 6) \text{ [A001778]}$$

**Eulerian Numbers** (§3.1.5)

$1, 4, 11, 26, 57, 120, 247, 502, 1013, 2036, 4083, 8178, 16369, 32752, 65519, 131054, 262125$
$$E(n, 1) \text{ [A000295]}$$

$1, 11, 66, 302, 1191, 4293, 14608, 47840, 152637, 478271, 1479726, 4537314, 13824739$
$$E(n, 2) \text{ [A000460]}$$

$1, 26, 302, 2416, 15619, 88234, 455192, 2203488, 10187685, 45533450, 198410786$
$$E(n, 3) \text{ [A000498]}$$

$1, 57, 1191, 15619, 156190, 1310354, 9738114, 66318474, 423281535, 2571742175$
$$E(n, 4) \text{ [A000505]}$$

$1, 120, 4293, 88234, 1310354, 15724248, 162512286, 1505621508, 12843262863$
$$E(n, 5) \text{ [A000514]}$$

$1, 247, 14608, 455192, 9738114, 162512286, 2275172004, 27971176092, 311387598411$
$$E(n, 6) \text{ [A001243]}$$

$1, 502, 47840, 2203488, 66318474, 1505621508, 27971176092, 447538817472, 6382798925475$
$$E(n, 7) \text{ [A001244]}$$

**Other Special Sequences** (§3.1)

$1, 1, 2, 3, 5, 8, 13, 21, 34, 55, 89, 144, 233, 377, 610, 987, 1597, 2584, 4181, 6795, 10946, 17711$
$$\text{Fibonacci numbers, } n \geq 1 \text{ [A000045]}$$

$1, 3, 4, 7, 11, 18, 29, 47, 76, 123, 199, 322, 521, 843, 1364, 2207, 3571, 5778, 9349, 15127$
$$\text{Lucas numbers, } n \geq 1 \text{ [A000204]}$$

$1, 1, 2, 5, 14, 42, 132, 429, 1430, 4862, 16796, 58786, 208012, 742900, 2674440, 9694845$
$$\text{Catalan numbers, } n \geq 0 \text{ [A000108]}$$

$1, 3, 11, 25, 137, 49, 363, 761, 7129, 7381, 83711, 86021, 1145993, 1171733, 1195757$
$$\text{numerators of harmonic numbers, } n \geq 1 \text{ [A001008]}$$

1, 2, 6, 12, 60, 20, 140, 280, 2520, 2520, 27720, 27720, 360360, 360360, 360360, 720720
$$\text{denominators of harmonic numbers, } n \geq 1 \text{ [A002805]}$$

1, 1, 1, 1, 1, 5, 691, 7, 3617, 43867, 174611, 854513, 236364091, 8553103, 23749461029
$$\text{numerators of Bernoulli numbers } |B_{2n}|, \; n \geq 0 \text{ [A000367]}$$

1, 6, 30, 42, 30, 66, 2730, 6, 510, 798, 330, 138, 2730, 6, 870, 14322, 510, 6, 1919190, 6, 13530
$$\text{denominators of Bernoulli numbers } |B_{2n}|, \; n \geq 0 \text{ [A002445]}$$

1, 1, 5, 61, 1385, 50521, 2702765, 199360981, 19391512145, 2404879675441
$$\text{Euler numbers } |E_{2n}|, \; n \geq 0 \text{ [A000364]}$$

1, 2, 16, 272, 7936, 353792, 22368256, 1903757312, 209865342976, 29088885112832
$$\text{tangent numbers } T_{2n+1}, \; n \geq 0 \text{ [A000182]}$$

1, 1, 2, 5, 15, 52, 203, 877, 4140, 21147, 115975, 678570, 4213597, 27644437, 190899322
$$\text{Bell numbers, } n \geq 0 \text{ [A000110]}$$

## Numbers of Certain Algebraic Structures (§1.4, §5.2)

1, 1, 1, 2, 1, 1, 1, 3, 2, 1, 1, 2, 1, 1, 1, 5, 1, 2, 1, 2, 1, 1, 1, 3, 2, 1, 3, 2, 1, 1, 1, 7, 1, 1, 1, 4, 1, 1, 1, 3
$$\text{abelian groups of order } n \text{ [A000688]}$$

1, 1, 1, 2, 1, 2, 1, 5, 2, 2, 1, 5, 1, 2, 1, 14, 1, 5, 1, 5, 2, 2, 1, 15, 2, 2, 5, 4, 1, 4, 1, 51, 1, 2, 1, 14, 1, 2
$$\text{groups of order } n \text{ [A000001]}$$

2, 3, 5, 7, 11, 13, 17, 19, 23, 29, 31, 37, 41, 43, 47, 53, 59, 60, 61, 67, 71, 73, 79, 83, 89, 97, 101
$$\text{orders of simple groups [A005180]}$$

60, 168, 360, 504, 660, 1092, 2448, 2520, 3420, 4080, 5616, 6048, 6072, 7800, 7920, 9828
$$\text{orders of noncyclic simple groups [A001034]}$$

1, 1, 2, 5, 16, 63, 318, 2045, 16999, 183231, 2567284, 46749427, 1104891746, 33823827452
$$\text{partially ordered sets on } n \text{ elements [A000112]}$$

1, 2, 13, 171, 3994, 154303, 9415189, 878222530, 122207703623, 24890747921947
$$\text{transitive relations on } n \text{ elements [A006905]}$$

1, 5, 52, 1522, 145984, 48464496, 56141454464, 229148550030864, 3333310786076963968
$$\text{relations on } n \text{ unlabeled points [A001173]}$$

1, 2, 1, 2, 3, 6, 9, 18, 30, 56, 99, 186, 335, 630, 1161, 2182, 4080, 7710, 14532, 27594, 52377
$$\text{binary irreducible polynomials of degree } n \text{ [A001037]}$$

## Permutations (§5.3.1)

*by cycles*

1, 1, 1, 3, 15, 75, 435, 3045, 24465, 220185, 2200905, 24209955, 290529855, 3776888115
$$\text{no 2-cycles [A000266]}$$

1, 1, 2, 4, 16, 80, 520, 3640, 29120, 259840, 2598400, 28582400, 343235200, 4462057600
$$\text{no 3-cycles [A000090]}$$

1, 1, 2, 6, 18, 90, 540, 3780, 31500, 283500, 2835000, 31185000, 372972600, 4848643800
$$\text{no 4-cycles [A000138]}$$

0, 1, 1, 3, 9, 45, 225, 1575, 11025, 99225, 893025, 9823275, 108056025, 1404728325
$$\text{no even length cycles [A000246]}$$

*discordant* (§2.4.2, §3.1.8)

1, 0, 1, 2, 9, 44, 265, 1854, 14833, 133496, 1334961, 14684570, 176214841, 2290792932
$$\text{derangements, discordant for } \iota \text{ [A000166]}$$

1, 1, 0, 1, 2, 13, 80, 579, 4738, 43387, 439792, 4890741, 59216642, 775596313, 10927434464
$$\text{menage numbers, discordant for } \iota \text{ and } \sigma_n \text{ [A000179]}$$

$0, 1, 2, 20, 144, 1265, 12072, 126565, 1445100, 17875140, 238282730, 3407118041$
$$\textbf{discordant for } \iota, \sigma_n, \sigma_n^2 \text{ [A000183]}$$

*by order*
$1, 2, 3, 4, 6, 6, 12, 15, 20, 30, 30, 60, 60, 84, 105, 140, 210, 210, 420, 420, 420, 420, 840, 840$
$$\textbf{max order } \text{[A000793]}$$

$1, 2, 3, 4, 6, 12, 15, 20, 30, 60, 84, 105, 140, 210, 420, 840, 1260, 1540, 2310, 2520, 4620, 5460$
$$\textbf{max order, increasing } \text{[A002809]}$$

$1, 2, 4, 16, 56, 256, 1072, 11264, 78976, 672256, 4653056, 49810432, 433429504, 4448608256$
$$\textbf{order a power of 2 } \text{[A005388]}$$

$0, 1, 3, 9, 25, 75, 231, 763, 2619, 9495, 35695, 140151, 568503, 2390479, 10349535, 46206735$
$$\textbf{order 2 } \text{[A001189]}$$

$0, 0, 2, 8, 20, 80, 350, 1232, 5768, 31040, 142010, 776600, 4874012, 27027728, 168369110$
$$\textbf{order 3 } \text{[A001471]}$$

$0, 0, 0, 6, 30, 180, 840, 5460, 30996, 209160, 1290960, 9753480, 69618120, 571627056$
$$\textbf{order 4 } \text{[A001473]}$$

$0, 0, 1, 3, 6, 10, 30, 126, 448, 1296, 4140, 17380, 76296, 296088, 1126216, 4940040, 23904000$
$$\textbf{odd, order 2 } \text{[A001465]}$$

**Necklaces** (§2.6)
$1, 2, 3, 4, 6, 8, 14, 20, 36, 60, 108, 188, 352, 632, 1182, 2192, 4116, 7712, 14602, 27596, 52488$
$$\textbf{2 colors, } n \textbf{ beads } \text{[A000031]}$$

$1, 3, 6, 11, 24, 51, 130, 315, 834, 2195, 5934, 16107, 44368, 122643, 341802, 956635, 2690844$
$$\textbf{3 colors, } n \textbf{ beads } \text{[A001867]}$$

$1, 4, 10, 24, 70, 208, 700, 2344, 8230, 29144, 104968, 381304, 1398500, 5162224, 19175140$
$$\textbf{4 colors, } n \textbf{ beads } \text{[A001868]}$$

$1, 5, 15, 45, 165, 629, 2635, 11165, 48915, 217045, 976887, 4438925, 20346485, 93900245$
$$\textbf{5 colors, } n \textbf{ beads } \text{[A001869]}$$

**Number Theory** (§4.4, §4.6)
$2, 3, 5, 7, 11, 13, 17, 19, 23, 29, 31, 37, 41, 43, 47, 53, 59, 61, 67, 71, 73, 79, 83, 89, 97, 101, 103$
$$\textbf{primes } \text{[A000040]}$$

$0, 1, 2, 2, 3, 3, 4, 4, 4, 4, 5, 5, 6, 6, 6, 6, 7, 7, 8, 8, 8, 8, 9, 9, 9, 9, 9, 9, 10, 10, 11, 11, 11, 11, 11, 11$
$$\textbf{number of primes} \le n \text{ [A000720]}$$

$1, 1, 1, 1, 1, 2, 1, 1, 1, 2, 1, 2, 1, 2, 2, 1, 1, 2, 1, 2, 2, 2, 1, 2, 1, 2, 1, 2, 1, 3, 1, 1, 2, 2, 2, 2, 1, 2, 2, 2$
$$\textbf{number of distinct primes dividing } n \text{ [A001221]}$$

$2, 3, 5, 7, 13, 17, 19, 31, 61, 89, 107, 127, 521, 607, 1279, 2203, 2281, 3217, 4253, 4423, 9689$
$$\textbf{Mersenne primes } \text{[A000043]}$$

$1, 1, 1, 2, 1, 2, 1, 3, 2, 2, 1, 4, 1, 2, 2, 5, 1, 4, 1, 4, 2, 2, 1, 7, 2, 2, 3, 4, 1, 5, 1, 7, 2, 2, 2, 9, 1, 2, 2, 7$
$$\textbf{number of ways of factoring } n \text{ [A001055]}$$

$1, 1, 2, 2, 4, 2, 6, 4, 6, 4, 10, 4, 12, 6, 8, 8, 16, 6, 18, 8, 12, 10, 22, 8, 20, 12, 18, 12, 28, 8, 30, 16$
$$\textbf{Euler totient function } \text{[A000010]}$$

$561, 1105, 1729, 2465, 2821, 6601, 8911, 10585, 15841, 29341, 41041, 46657, 52633, 62745$
$$\textbf{Carmichael numbers } \text{[A002997]}$$

$1, 2, 2, 3, 2, 4, 2, 4, 3, 4, 2, 6, 2, 4, 4, 5, 2, 6, 2, 6, 4, 4, 2, 8, 3, 4, 4, 6, 2, 8, 2, 6, 4, 4, 4, 9, 2, 4, 4, 8$
$$\textbf{number of divisors of } n \text{ [A000005]}$$

$1, 3, 4, 7, 6, 12, 8, 15, 13, 18, 12, 28, 14, 24, 24, 31, 18, 39, 20, 42, 32, 36, 24, 60, 31, 42, 40, 56$
**sum of divisors of** $n$ [A000203]

$6, 28, 496, 8128, 33550336, 8589869056, 137438691328, 2305843008139952128$
**perfect numbers** [A000396]

**Partitions** (§2.5.1)

$1, 1, 2, 3, 5, 7, 11, 15, 22, 30, 42, 56, 77, 101, 135, 176, 231, 297, 385, 490, 627, 792, 1002, 1255$
**partitions of** $n$ [A000041]

$1, 1, 1, 2, 2, 3, 4, 5, 6, 8, 10, 12, 15, 18, 22, 27, 32, 38, 46, 54, 64, 76, 89, 104, 122, 142, 165, 192$
**partitions of** $n$ **into distinct parts** [A000009]

$1, 3, 6, 13, 24, 48, 86, 160, 282, 500, 859, 1479, 2485, 4167, 6879, 11297, 18334, 29601, 47330$
**planar partitions of** $n$ [A000219]

**Figurate Numbers** (§3.1.8)
*polygonal*

$1, 3, 6, 10, 15, 21, 28, 36, 45, 55, 66, 78, 91, 105, 120, 136, 153, 171, 190, 210, 231, 253, 276$
**triangular** [A000217]

$1, 5, 12, 22, 35, 51, 70, 92, 117, 145, 176, 210, 247, 287, 330, 376, 425, 477, 532, 590, 651, 715$
**pentagonal** [A000326]

$1, 6, 15, 28, 45, 66, 91, 120, 153, 190, 231, 276, 325, 378, 435, 496, 561, 630, 703, 780, 861, 946$
**hexagonal** [A000384]

$1, 7, 18, 34, 55, 81, 112, 148, 189, 235, 286, 342, 403, 469, 540, 616, 697, 783, 874, 970, 1071$
**heptagonal** [A000566]

$1, 8, 21, 40, 65, 96, 133, 176, 225, 280, 341, 408, 481, 560, 645, 736, 833, 936, 1045, 1160, 1281$
**octagonal** [A000567]

*pyramidal*

$1, 4, 10, 20, 35, 56, 84, 120, 165, 220, 286, 364, 455, 560, 680, 816, 969, 1140, 1330, 1540, 1771$
**3-dimensional triangular, height** $n$ [A000292]

$1, 5, 14, 30, 55, 91, 140, 204, 285, 385, 506, 650, 819, 1015, 1240, 1496, 1785, 2109, 2470, 2870$
**3-dimensional square, height** $n$ [A000330]

$1, 6, 18, 40, 75, 126, 196, 288, 405, 550, 726, 936, 1183, 1470, 1800, 2176, 2601, 3078, 3610$
**3-dimensional pentagonal, height** $n$ [A002411]

$1, 7, 22, 50, 95, 161, 252, 372, 525, 715, 946, 1222, 1547, 1925, 2360, 2856, 3417, 4047, 4750$
**3-dimensional hexagonal, height** $n$ [A002412]

$1, 8, 26, 60, 115, 196, 308, 456, 645, 880, 1166, 1508, 1911, 2380, 2920, 3536, 4233, 5016, 5890$
**3-dimensional heptagonal, height** $n$ [A002413]

$1, 5, 15, 35, 70, 126, 210, 330, 495, 715, 1001, 1365, 1820, 2380, 3060, 3876, 4845, 5985, 7315$
**4-dimensional triangular, height** $n$ [A000332]

$1, 6, 20, 50, 105, 196, 336, 540, 825, 1210, 1716, 2366, 3185, 4200, 5440, 6936, 8721, 10830$
**4-dimensional square, height** $n$ [A002415]

$1, 7, 25, 65, 140, 266, 462, 750, 1155, 1705, 2431, 3367, 4550, 6020, 7820, 9996, 12597, 15675$
**4-dimensional pentagonal, height** $n$ [A001296]

$1, 8, 30, 80, 175, 336, 588, 960, 1485, 2200, 3146, 4368, 5915, 7840, 10200, 13056, 16473$
**4-dimensional hexagonal, height** $n$ [A002417]

$1, 9, 35, 95, 210, 406, 714, 1170, 1815, 2695, 3861, 5369, 7280, 9660, 12580, 16116, 20349$
**4-dimensional heptagonal, height $n$** [A002418]

**Polyominoes** (§3.1.8)
$1, 1, 2, 5, 12, 35, 108, 369, 1285, 4655, 17073, 63600, 238591, 901971, 3426576, 13079255$
**squares, $n$ cells** [A000105]

$1, 1, 1, 3, 4, 12, 24, 66, 160, 448, 1186, 3334, 9235, 26166, 73983, 211297, 604107, 1736328$
**triangles, $n$ cells** [A000577]

$1, 1, 3, 7, 22, 82, 333, 1448, 6572, 30490, 143552, 683101, 3274826, 15796897, 76581875$
**hexagons, $n$ cells** [A000228]

$1, 1, 2, 8, 29, 166, 1023, 6922, 48311, 346543, 2522572, 18598427, 138462649, 1039496297$
**cubes, $n$ cells** [A000162]

**Trees** (§9.3)
$1, 1, 1, 2, 3, 6, 11, 23, 47, 106, 235, 551, 1301, 3159, 7741, 19320, 48629, 123867, 317955$
**$n$ unlabeled vertices** [A000055]

$1, 1, 2, 4, 9, 20, 48, 115, 286, 719, 1842, 4766, 12486, 32973, 87811, 235381, 634847, 1721159$
**rooted, $n$ unlabeled vertices** [A000081]

$1, 1, 3, 16, 125, 1296, 16807, 262144, 4782969, 100000000, 2357947691, 61917364224$
**$n$ labeled vertices** [A000272]

$1, 2, 9, 64, 625, 7776, 117649, 2097152, 43046721, 1000000000, 25937424601, 743008370688$
**rooted, $n$ labeled vertices** [A000169]

*by diameter*
$1, 2, 5, 8, 14, 21, 32, 45, 65, 88, 121, 161, 215, 280, 367, 471, 607, 771, 980, 1232, 1551, 1933$
**diameter 4, $n \geq 5$ vertices** [A000094]

$1, 2, 7, 14, 32, 58, 110, 187, 322, 519, 839, 1302, 2015, 3032, 4542, 6668, 9738, 14006, 20036$
**diameter 5, $n \geq 6$ vertices** [A000147]

$1, 3, 11, 29, 74, 167, 367, 755, 1515, 2931, 5551, 10263, 18677, 33409, 59024, 102984, 177915$
**diameter 6, $n \geq 7$ vertices** [A000251]

$1, 3, 14, 42, 128, 334, 850, 2010, 4625, 10201, 21990, 46108, 94912, 191562, 380933, 746338$
**diameter 7, $n \geq 8$ vertices** [A000550]

$1, 4, 19, 66, 219, 645, 1813, 4802, 12265, 30198, 72396, 169231, 387707, 871989, 1930868$
**diameter 8, $n \geq 9$ vertices** [A000306]

*by height*
$1, 3, 8, 18, 38, 76, 147, 277, 509, 924, 1648, 2912, 5088, 8823, 15170, 25935, 44042, 74427$
**height 3, $n \geq 4$ vertices** [A000235]

$1, 4, 13, 36, 93, 225, 528, 1198, 2666, 5815, 12517, 26587, 55933, 116564, 241151, 495417$
**height 4, $n \geq 5$ vertices** [A000299]

$1, 5, 19, 61, 180, 498, 1323, 3405, 8557, 21103, 51248, 122898, 291579, 685562, 1599209$
**height 5, $n \geq 6$ vertices** [A000342]

*series-reduced*
$1, 1, 0, 1, 1, 2, 2, 4, 5, 10, 14, 26, 42, 78, 132, 249, 445, 842, 1561, 2988, 5671, 10981, 21209$
**$n$ vertices** [A000014]

$1, 1, 0, 2, 4, 6, 12, 20, 39, 71, 137, 261, 511, 995, 1974, 3915, 7841, 15749, 31835, 64540$
**rooted, $n$ vertices** [A001679]

0, 1, 0, 1, 1, 2, 3, 6, 10, 19, 35, 67, 127, 248, 482, 952, 1885, 3765, 7546, 15221, 30802, 62620
**planted, $n$ vertices** [A001678]

**Graphs** (§8.1, §8.3, §8.4, §8.9)
1, 2, 4, 11, 34, 156, 1044, 12346, 274668, 12005168, 1018997864, 165091172592
**$n$ vertices** [A000088]

*chromatic number*
4, 6, 7, 7, 8, 9, 9, 10, 10, 10, 11, 11, 12, 12, 12, 13, 13, 13, 13, 14, 14, 14, 15, 15, 15, 15, 16, 16
**surface, connectivity $n \geq 1$** [A000703]
4, 7, 8, 9, 10, 11, 12, 12, 13, 13, 14, 15, 15, 16, 16, 16, 17, 17, 18, 18, 19, 19, 19, 20, 20, 20, 21
**surface, genus $n \geq 0$** [A000934]

*genus*
0, 0, 0, 0, 1, 1, 1, 2, 3, 4, 5, 6, 8, 10, 11, 13, 16, 18, 20, 23, 26, 29, 32, 35, 39, 43, 46, 50, 55, 59, 63
**complete graphs, $n$ vertices** [A00933]

*connected*
1, 1, 2, 6, 21, 112, 853, 11117, 261080, 11716571, 1006700565, 164059830476, 50335907869219
**$n$ vertices** [A001349]
1, 1, 0, 2, 5, 32, 234, 3638, 106147, 6039504, 633754161, 120131932774, 41036773627286
**series-reduced, $n$ vertices** [A005636]
1, 1, 3, 5, 12, 30, 79, 227, 710, 2322, 8071, 29503, 112822, 450141, 1867871, 8037472
**$n$ edges** [A002905]
1, 1, 4, 38, 728, 26704, 1866256, 251548592, 66296291072, 34496488594816
**$n$ labeled vertices** [A001187]

*directed*
1, 3, 16, 218, 9608, 1540944, 882033440, 1793359192848, 13027956824399552
**$n$ vertices** [A000273]
1, 3, 9, 33, 139, 718, 4535, 35979, 363083, 4717687, 79501654, 1744252509, 49872339897
**transitive, $n$ vertices** [A001930]
1, 1, 2, 4, 12, 56, 456, 6880, 191536, 9733056, 903753248, 154108311168, 48542114686912
**tournaments, $n$ vertices** [A000568]
1, 4, 29, 355, 6942, 209527, 9535241, 642779354, 63260289423, 8977053873043
**transitive, $n$ labeled vertices** [A000798]

*various*
1, 2, 2, 4, 3, 8, 4, 14, 9, 22, 8, 74, 14, 56, 48, 286, 36, 380, 60, 1214, 240, 816, 188, 15506, 464
**transitive, $n$ vertices** [A006799]
1, 1, 2, 3, 7, 16, 54, 243, 2038, 33120, 1182004, 87723296, 12886193064, 3633057074584
**all degrees even, $n$ vertices** [A002854]
1, 0, 1, 1, 4, 8, 37, 184, 1782, 31026, 1148626, 86539128, 12798435868, 3620169692289
**Eulerian, $n$ vertices** [A003049]
1, 0, 1, 3, 8, 48, 383, 6196, 177083, 9305118, 883156024, 152522187830
**Hamiltonian, $n$ vertices** [A003216]
1, 2, 2, 4, 3, 8, 6, 22, 26, 176, 546, 19002, 389454, 50314870, 2942198546, 1698517037030
**regular, $n$ vertices** [A005176]
0, 1, 1, 3, 10, 56, 468, 7123, 194066, 9743542, 900969091, 153620333545, 48432939150704
**nonseparable, $n$ vertices** [A002218]

$1, 2, 4, 11, 33, 142, 822, 6966, 79853, 1140916, 18681008, 333312451$
**planar, $n$ vertices** [A005470]

## 3.2   GENERATING FUNCTIONS

Generating functions express an infinite sequence as coefficients arising from a power series in an auxiliary variable. The closed form of a generating function is a concise way to represent such an infinite sequence. Properties of the sequence can be explored by analyzing the closed form of an associated generating function. Two types of generating functions are discussed in this section—ordinary generating functions and exponential generating functions. The former arise when counting configurations in which order is not important, while the latter are appropriate when order matters.

### 3.2.1   ORDINARY GENERATING FUNCTIONS

**Definitions:**

The (**ordinary**) **generating function** for the sequence $a_0, a_1, a_2, \ldots$ of real numbers is the formal power series $f(x) = a_0 + a_1 x + a_2 x^2 + \cdots = \sum_{i=0}^{\infty} a_i x^i$ or any equivalent closed form expression.

The **convolution** of the sequence $a_0, a_1, a_2, \ldots$ and the sequence $b_0, b_1, b_2, \ldots$ is the sequence $c_0, c_1, c_2, \ldots$ in which $c_t = a_0 b_t + a_1 b_{t-1} + a_2 b_{t-2} + \cdots + a_t b_0 = \sum_{k=0}^{t} a_k b_{t-k}$.

**Facts:**

**1.** Generating functions are considered as algebraic forms and can be manipulated as such, without regard to actual convergence of the power series.

**2.** A rational form (the ratio of two polynomials) is a concise expression for the generating function of the sequence obtained by carrying out long division on the polynomials. (See Example 1.)

**3.** Generating functions are often useful for constructing and verifying identities involving binomial coefficients and other special sequences. (See Example 10.)

**4.** Generating functions can be used to derive formulas for the sums of powers of integers. (See Example 17.)

**5.** Generating functions can be used to solve recurrence relations. (See §3.3.4.)

**6.** Each sequence $\{a_n\}$ defines a *unique* generating function $f(x)$, and conversely.

**7.** *Related generating functions:*   Suppose $f(x) = \sum_{k=0}^{\infty} a_k x^k$ and $g(x) = \sum_{k=0}^{\infty} b_k x^k$ are generating functions for the sequences $a_0, a_1, a_2, \ldots$ and $b_0, b_1, b_2, \ldots$, respectively. Table 1 gives some related generating functions.

**Table 1: Related generating functions.**

| generating function | sequence |
|---|---|
| $x^n f(x)$ | $\underbrace{0, 0, 0, \ldots, 0}_{n}, a_0, a_1, a_2, \ldots$ |

| generating function | sequence |
|---|---|
| $f(x) - a_n x^n$ | $a_0, a_1, \ldots, a_{n-1}, 0, a_{n+1}, \ldots$ |
| $a_0 + a_1 x + \cdots + a_n x^n$ | $a_0, a_1, \ldots, a_n, 0, 0, \ldots$ |
| $f(x^2)$ | $a_0, 0, a_1, 0, a_2, 0, a_3, \ldots$ |
| $\frac{f(x) - a_0}{x}$ | $a_1, a_2, a_3, \ldots$ |
| $\frac{f(x) - a_0 - a_1 x - \cdots - a_n x^n}{x^{n+1}}$ | $a_{n+1}, a_{n+2}, a_{n+3}, \ldots$ |
| $f'(x)$ | $a_1, 2a_2, 3a_3, \ldots, ka_k, \ldots$ |
| $x f'(x)$ | $0, a_1, 2a_2, 3a_3, \ldots, ka_k, \ldots$ |
| $\int_0^x f(t)\, dt$ | $0, a_0, \frac{a_1}{2}, \frac{a_2}{3}, \ldots, \frac{a_k}{k+1}, \ldots$ |
| $\frac{f(x)}{1-x}$ | $a_0, a_0 + a_1, a_0 + a_1 + a_2, a_0 + a_1 + a_2 + a_3, \ldots$ |
| $(1+x)f(x)$ | $a_0, a_0 + a_1, a_1 + a_2, \ldots, a_k + a_{k+1}, \ldots$ |
| $(1 + x + x^2)f(x)$ | $a_0, a_0 + a_1, a_0 + a_1 + a_2, a_1 + a_2 + a_3, \ldots$ |
| $rf(x) + sg(x)$ | $ra_0 + sb_0, ra_1 + sb_1, ra_2 + sb_2, \ldots$ |
| $f(x)g(x)$ | $a_0 b_0, a_0 b_1 + a_1 b_0, a_0 b_2 + a_1 b_1 + a_2 b_0, \ldots$ |
| | (the convolution of $\{a_n\}$ and $\{b_n\}$) |

## Examples:

**1.** The sequence $0, 1, 4, 9, 16, \ldots$ of squares of the nonnegative integers has the generating function $0 + x + 4x^2 + 9x^3 + 16x^4 + \cdots$. However, this generating function has a concise closed form expression, namely $\frac{x + x^2}{1 - 3x + 3x^2 - x^3}$. Verification is obtained by carrying out long division on the indicated polynomials. This concise form can be used to deduce properties involving the sequence, such as an explicit algebraic expression for the sum of squares of the first $n$ positive integers. (See Example 17.)

**2.** The generating function for the sequence $1, 1, 1, 1, 1, \ldots$ is $1 + x + x^2 + x^3 + x^4 + \cdots = \frac{1}{1-x}$. Differentiating both sides of this expression produces $1 + 2x + 3x^2 + 4x^3 + \cdots = \frac{1}{(1-x)^2}$. Thus, $\frac{1}{(1-x)^2}$ is a closed form expression for the generating function of the sequence $1, 2, 3, 4, \ldots$. (See Table 2.)

**3.** Table 2 gives closed form expressions for the generating functions of particular sequences. In this table, $r$ is an arbitrary real number, $F_n$ is the $n$th Fibonacci number (§3.1.2), $L_n$ is the $n$th Lucas number (§3.1.2), $C_n$ is the $n$th Catalan number (§3.1.3), and $H_n$ is the $n$th harmonic number (§3.1.7).

### Table 2: Generating functions for particular sequences.

| sequence | closed form |
|---|---|
| $1, 1, 1, 1, 1, \ldots$ | $\frac{1}{1-x}$ |
| $1, 1, \ldots, 1, 0, 0, \ldots$ ($n$ 1s) | $\frac{1 - x^n}{1 - x}$ |
| $1, 1, \ldots, 1, 1, 0, 1, 1, \ldots$ (0 following $n$ 1s) | $\frac{1}{1-x} - x^n$ |
| $1, -1, 1, -1, 1, -1, \ldots$ | $\frac{1}{1+x}$ |
| $1, 0, 1, 0, 1, \ldots$ | $\frac{1}{1-x^2}$ |
| $1, 2, 3, 4, 5, \ldots$ | $\frac{1}{(1-x)^2}$ |
| $1, 4, 9, 16, 25, \ldots$ | $\frac{1+x}{(1-x)^3}$ |
| $1, r, r^2, r^3, r^4, \ldots$ | $\frac{1}{1 - rx}$ |
| $0, r, 2r^2, 3r^3, 4r^4, \ldots$ | $\frac{rx}{(1 - rx)^2}$ |

| sequence | closed form |
|---|---|
| $0, 1, \frac{1}{2}, \frac{1}{3}, \frac{1}{4}, \frac{1}{5}, \ldots$ | $\ln \frac{1}{1-x}$ |
| $\frac{1}{0!}, \frac{1}{1!}, \frac{1}{2!}, \frac{1}{3!}, \frac{1}{4!}, \ldots$ | $e^x$ |
| $0, 1, -\frac{1}{2}, \frac{1}{3}, -\frac{1}{4}, \frac{1}{5}, \ldots$ | $\ln(1+x)$ |
| $F_0, F_1, F_2, F_3, F_4, \ldots$ | $\frac{x}{1-x-x^2}$ |
| $L_0, L_1, L_2, L_3, L_4, \ldots$ | $\frac{2-x}{1-x-x^2}$ |
| $C_0, C_1, C_2, C_3, C_4, \ldots$ | $\frac{1-\sqrt{1-4x}}{2x}$ |
| $H_0, H_1, H_2, H_3, H_4, \ldots$ | $\frac{1}{1-x} \ln \frac{1}{1-x}$ |

**4.** For every positive integer $n$, the binomial theorem (§2.3.4) states that

$$(1+x)^n = \binom{n}{0} + \binom{n}{1}x + \binom{n}{2}x^2 + \cdots + \binom{n}{n}x^n = \sum_{k=0}^{n} \binom{n}{k}x^k,$$

so $(1+x)^n$ is a closed form for the generating function of $\binom{n}{0}, \binom{n}{1}, \binom{n}{2}, \ldots, \binom{n}{n}, 0, 0, 0, \ldots$.

**5.** For every positive integer $n$, the Maclaurin series expansion for $(1+x)^{-n}$ is

$$(1+x)^{-n} = 1 + (-n)x + \frac{(-n)(-n-1)x^2}{2!} + \cdots$$

$$= 1 + \sum_{k=1}^{\infty} \frac{(-n)(-n-1)(-n-2)\ldots(-n-k+1)}{k!}x^k.$$

Consequently, $(1+x)^{-n}$ is the generating function for the sequence $\binom{-n}{0}, \binom{-n}{1}, \binom{-n}{2}, \ldots$, where $\binom{-n}{k} = (-1)^k \binom{n+k-1}{k}$ is an extended binomial coefficient (§2.3.2).

**6.** Using Example 5, the expansion of $f(x) = (1-3x)^{-8}$ is

$$(1-3x)^{-8} = (1+y)^{-8} = \sum_{k=0}^{\infty} \binom{-8}{k}y^k = \sum_{k=0}^{\infty} \binom{-8}{k}(-3x)^k.$$

So the coefficient of $x^4$ in $f(x)$ is $\binom{-8}{4}(-3)^4 = (-1)^4\binom{8+4-1}{4}(81) = \binom{11}{4}(81) = 26{,}730$.

**7.** Table 3 gives additional examples of generating functions related to binomial expansions. In this table, $m$ and $n$ are positive integers, and $r$ is any real number.

**Table 3: Examples of binomial-type generating functions.**

| generating function | expansion |
|---|---|
| $(1+x)^n$ | $\binom{n}{0} + \binom{n}{1}x + \binom{n}{2}x^2 + \cdots + \binom{n}{n}x^n = \sum_{k=0}^{n}\binom{n}{k}x^k$ |
| $(1+rx)^n$ | $\binom{n}{0} + \binom{n}{1}rx + \binom{n}{2}r^2x^2 + \cdots + \binom{n}{n}r^nx^n = \sum_{k=0}^{n}\binom{n}{k}r^kx^k$ |
| $(1+x^m)^n$ | $\binom{n}{0} + \binom{n}{1}x^m + \binom{n}{2}x^{2m} + \cdots + \binom{n}{n}x^{nm} = \sum_{k=0}^{n}\binom{n}{k}x^{km}$ |
| $(1+x)^{-n}$ | $\binom{-n}{0} + \binom{-n}{1}x + \binom{-n}{2}x^2 + \cdots = \sum_{k=0}^{\infty}(-1)^k\binom{n+k-1}{k}x^k$ |
| $(1+rx)^{-n}$ | $\binom{-n}{0} + \binom{-n}{1}rx + \binom{-n}{2}r^2x^2 + \cdots = \sum_{k=0}^{\infty}(-1)^k\binom{n+k-1}{k}r^kx^k$ |
| $(1-x)^{-n}$ | $\binom{-n}{0} + \binom{-n}{1}(-x) + \binom{-n}{2}(-x)^2 + \cdots = \sum_{k=0}^{\infty}\binom{n+k-1}{k}x^k$ |
| $(1-rx)^{-n}$ | $\binom{-n}{0} + \binom{-n}{1}(-rx) + \binom{-n}{2}(-rx)^2 + \cdots = \sum_{k=0}^{\infty}\binom{n+k-1}{k}r^kx^k$ |
| $\dfrac{x^n}{(1-x)^{n+1}}$ | $\binom{n}{n}x^n + \binom{n+1}{n}x^{n+1} + \binom{n+2}{n}x^{n+2} + \cdots = \sum_{k=n}^{\infty}\binom{k}{n}x^k$ |

**8.** For any real number $r$, the Maclaurin series expansion for $(1+x)^r$ is

$$(1+x)^r = \binom{r}{0}1 + \binom{r}{1}x + \binom{r}{2}x^2 + \cdots,$$

where $\binom{r}{k} = \frac{r(r-1)(r-2)\ldots(r-k+1)}{k!}$ if $k > 0$ and $\binom{r}{0} = 1$.

**9.** Using Example 8, the expansion of $f(x) = \sqrt{1+x}$ is

$$\sqrt{1+x} = (1+x)^{1/2} = \binom{1/2}{0}1 + \binom{1/2}{1}x + \binom{1/2}{2}x^2 + \cdots$$

$$= 1 + \tfrac{1}{2}x + \frac{\frac{1}{2}\cdot\frac{-1}{2}}{2!}x^2 + \frac{\frac{1}{2}\cdot\frac{-1}{2}\cdot\frac{-3}{2}}{3!}x^3 + \frac{\frac{1}{2}\cdot\frac{-1}{2}\cdot\frac{-3}{2}\cdot\frac{-5}{2}}{4!}x^4 + \cdots$$

$$= 1 + \tfrac{1}{2}x - \tfrac{1}{8}x^2 + \tfrac{1}{16}x^3 - \tfrac{5}{128}x^4 + \cdots.$$

Thus $\sqrt{1+x}$ is the generating function for the sequence $1, \tfrac{1}{2}, -\tfrac{1}{8}, \tfrac{1}{16}, -\tfrac{5}{128}, \ldots$.

**10.** Vandermonde's convolution identity (§2.3.4) can be obtained from the generating functions $f(x) = (1+x)^m$ and $g(x) = (1+x)^n$. First, $(1+x)^m(1+x)^n = (1+x)^{m+n}$. Equating coefficients of $x^r$ on both sides of this equation and using Table 3 produces

$$\sum_{k=0}^{m} \binom{m}{k}\binom{n}{r-k} = \binom{m+n}{r}.$$

**11.** Twenty identical computer terminals are to be distributed into five distinct rooms so each room receives at least two terminals. The number of such distributions is the coefficient of $x^{20}$ in the expansion of $f(x) = (x^2+x^3+x^4+\cdots)^5 = x^{10}(1+x+x^2+\cdots)^5 = \frac{x^{10}}{(1-x)^5}$. Thus the coefficient of $x^{20}$ in $f(x)$ is the coefficient of $x^{10}$ in $(1-x)^{-5}$, which from Table 3 is $\binom{5+10-1}{10} = \binom{14}{10} = 1{,}001$.

**12.** Suppose in Example 11 that each room can accommodate at most seven terminals. Now the generating function is $g(x) = (x^2+x^3+x^4+x^5+x^6+x^7)^5 = x^{10}(1+x+x^2+x^3+x^4+x^5)^5 = x^{10}\left(\frac{1-x^6}{1-x}\right)^5$. Consequently, the number of allowable distributions is the coefficient of $x^{10}$ in $\left(\frac{1-x^6}{1-x}\right)^5 = (1-x^6)^5(1-x)^{-5} = \left[1 - \binom{5}{1}x^6 + \binom{5}{2}x^{12} - \cdots - x^{30}\right]\left[\binom{-5}{0} + \binom{-5}{1}(-x) + \binom{-5}{2}(-x)^2 + \cdots\right]$. This coefficient is $\left[\binom{-5}{10}(-1)^{10} - \binom{5}{1}\binom{-5}{4}(-1)^4\right] = \binom{14}{10} - \binom{5}{1}\binom{8}{4} = 651$.

**13.** *Unordered selections with replacement:* $k$ objects are selected from $n$ distinct objects, with repetition allowed. For each of the $n$ distinct objects, the power series $1 + x + x^2 + \cdots$ represents the possible choices (namely none, one, two, ...) for that object. The generating function for all $n$ objects is then

$$f(x) = (1+x+x^2+\cdots)^n = \left(\tfrac{1}{1-x}\right)^n = (1-x)^{-n} = \sum_{k=0}^{\infty}\binom{n+k-1}{k}x^k.$$

The number of selections with replacement is the coefficient of $x^k$ in $f(x)$, namely $\binom{n+k-1}{k}$.

**14.** Suppose there are $p$ types of objects, with $n_i$ indistinguishable objects of type $i$. The number of ways to pick a total of $k$ objects (where the number of selected objects of type $i$ is at most $n_i$) is the coefficient of $x^k$ in the generating function

$$\prod_{i=1}^{p}(1+x+x^2+\cdots+x^{n_i}).$$

**15.** *Partitions:* Generating functions can be found for $p(n)$, the number of partitions of the positive integer $n$ (§2.5.1). The number of 1s that appear as summands in a partition of $n$ is 0 or 1 or 2 or $\ldots$, recorded as the terms in the power series $1 + x + x^2 + x^3 + \cdots$. The power series $1 + x^2 + x^4 + x^6 + \cdots$ records the number of 2s that can appear in a partition of $n$, and so forth. For example, $p(12)$ is the coefficient of $x^{12}$ in

$$(1+x+x^2+\cdots)(1+x^2+x^4+\cdots)\ldots(1+x^{12}+x^{24}+\cdots) = \prod_{i=1}^{12}\tfrac{1}{1-x^i},$$

or in $(1+x+x^2+\cdots+x^{12})(1+x^2+x^4+\cdots+x^{12})\ldots(1+x^{12})$. In general, the function $P(x) = \prod_{i=1}^{\infty}\frac{1}{1-x^i}$ is the generating function for the sequence $p(0), p(1), p(2), \ldots$, where $p(0)$ is defined as 1.

**16.** The function $P_d(x) = (1+x)(1+x^2)(1+x^3)\ldots = \prod_{i=1}^{\infty}(1+x^i)$ generates $Q(n)$, the number of partitions of $n$ into distinct summands (see §2.5.1). The function $P_o(x) = \frac{1}{1-x} \cdot \frac{1}{1-x^3} \cdot \frac{1}{1-x^5} \cdots = \prod_{j=0}^{\infty}(1-x^{2j+1})^{-1}$ is the generating function for $\mathcal{O}(n)$, the number of partitions of $n$ with all summands odd (see §2.5.1). Then

$$P_d(x) = (1+x)(1+x^2)(1+x^3)(1+x^4)\ldots$$
$$= \frac{1-x^2}{1-x} \cdot \frac{1-x^4}{1-x^2} \cdot \frac{1-x^6}{1-x^3} \cdot \frac{1-x^8}{1-x^4} \cdots = \frac{1}{1-x} \cdot \frac{1}{1-x^3} \cdots = P_o(x),$$

so $Q(n) = \mathcal{O}(n)$ for every nonnegative integer $n$.

**17.** *Summation formulas:* Generating functions can be used to produce the formula $1^2+2^2+\cdots+n^2 = \frac{1}{6}n(n+1)(2n+1)$. (See §3.5.4 for an extensive tabulation of summation formulas.) Manipulating the expansion $(1-x)^{-1} = 1+x+x^2+x^3+\cdots$ produces

$$x\frac{d}{dx}\left[x\frac{d}{dx}(1-x)^{-1}\right] = \frac{x(1+x)}{(1-x)^3} = x + 2^2x^2 + 3^2x^3 + \cdots.$$

So $\frac{x(1+x)}{(1-x)^3}$ is the generating function for the sequence $0^2, 1^2, 2^2, 3^2, \ldots$ and, by Fact 7, $\frac{x(1+x)}{(1-x)^4}$ generates the sequence $0^2, 0^2+1^2, 0^2+1^2+2^2, 0^2+1^2+2^2+3^2, \ldots$. Consequently, $\sum_{i=0}^{n} i^2$ is the coefficient of $x^n$ in

$$(x+x^2)(1-x)^{-4} = (x+x^2)\left[\binom{-4}{0} + \binom{-4}{1}(-x) + \binom{-4}{2}(-x)^2 + \cdots\right].$$

The answer is then $\binom{-4}{n-1}(-1)^{n-1} + \binom{-4}{n-2}(-1)^{n-2} = \binom{n+2}{n-1} + \binom{n+1}{n-2} = \frac{1}{6}n(n+1)(2n+1)$.

**18.** *Catalan numbers:* The Catalan numbers (§3.1.3) $C_0, C_1, C_2, \ldots$ satisfy the recurrence relation $C_n = C_0C_{n-1} + C_1C_{n-2} + \cdots + C_{n-1}C_0$, $n \geq 1$, with $C_0 = 1$. (See §3.3.1.) Hence their generating function $f(x) = \sum_{k=0}^{\infty} C_k x^k$ satisfies $xf^2(x) = f(x) - 1$, yielding $f(x) = \frac{1}{2x}(1 - \sqrt{1-4x}) = \frac{1}{2x}(1 - (1-4x)^{1/2})$. (The negative square root is chosen since the numbers $C_i$ cannot be negative.) Applying Example 8 to $(1-4x)^{1/2}$ yields $f(x) = \frac{1}{2x}\left[1 - \sum_{k=0}^{\infty}\binom{1/2}{k}(-4)^k x^k\right] = \frac{1}{2x}\left[1 - \sum_{k=0}^{\infty}\frac{-1}{2k-1}\binom{2k}{k}x^k\right] = \sum_{k=0}^{\infty}\frac{1}{k+1}\binom{2k}{k}x^k$. Thus $C_n = \frac{1}{n+1}\binom{2n}{n}$.

## 3.2.2 EXPONENTIAL GENERATING FUNCTIONS

Encoding the terms of a sequence as coefficients of $\frac{x^k}{k!}$ is often helpful in obtaining information about a sequence, such as in counting permutations of objects (where the order of listing objects is important). The functions that result are called exponential generating functions.

**Definitions:**

The **exponential generating function** for the sequence $a_0, a_1, a_2, \ldots$ of real numbers is the formal power series $f(x) = a_0 + a_1x + a_2\frac{x^2}{2!} + \cdots = \sum_{i=0}^{\infty} a_i\frac{x^i}{i!}$ or any equivalent closed form expression.

The **binomial convolution** of the sequence $a_0, a_1, a_2, \ldots$ and the sequence $b_0, b_1, b_2, \ldots$ is the sequence $c_0, c_1, c_2, \ldots$ in which $c_t = \binom{t}{0}a_0b_t + \binom{t}{1}a_1b_{t-1} + \binom{t}{2}a_2b_{t-2} + \cdots + \binom{t}{t}a_tb_0 = \sum_{k=0}^{t}\binom{t}{k}a_kb_{t-k}$, the coefficient of $\frac{x^t}{t!}$.

**Facts:**

**1.** Each sequence $\{a_n\}$ defines a *unique* exponential generating function $f(x)$, and conversely.

**2.** *Related exponential generating functions*: Suppose $f(x) = \sum_{k=0}^{\infty} a_k \frac{x^k}{k!}$ and $g(x) = \sum_{k=0}^{\infty} b_k \frac{x^k}{k!}$ are exponential generating functions for the sequences $a_0, a_1, a_2, \ldots$ and $b_0, b_1, b_2, \ldots$, respectively. Table 4 gives some related exponential generating functions. [Here $P(n,k) = \binom{n}{k}k! = \frac{n!}{(n-k)!}$ is the number of $k$-permutations of a set with $n$ distinct objects. (See §2.3.1.)]

**Table 4: Related exponential generating functions.**

| generating function | closed form |
|---|---|
| $xf(x)$ | $0, a_0, 2a_1, 3a_2, \ldots, (k+1)a_k, \ldots$ |
| $x^n f(x)$ | $\underbrace{0,0,0,\ldots,0}_{n}, P(n,n)a_0, P(n+1,n)a_1, \ldots, P(n+k,n)a_k, \ldots$ |
| $f'(x)$ | $a_1, a_2, a_3, \ldots, a_k, \ldots$ |
| $\int_0^x f(t)dt$ | $0, a_0, a_1, a_2, \ldots$ |
| $rf(x) + sg(x)$ | $ra_0 + sb_0, ra_1 + sb_1, ra_2 + sb_2, \ldots$ |
| $f(x)g(x)$ | $\binom{0}{0}a_0 b_0, \binom{1}{0}a_0 b_1 + \binom{1}{1}a_1 b_0, \binom{2}{0}a_0 b_2 + \binom{2}{1}a_1 b_1 + \binom{2}{2}a_2 b_0, \ldots$ (binomial convolution of $\{a_k\}$ and $\{b_k\}$) |

**Examples:**

**1.** The binomial theorem (§2.3.4) gives

$$(1+x)^n = \binom{n}{0} + \binom{n}{1}x + \binom{n}{2}x^2 + \binom{n}{3}x^3 + \cdots + \binom{n}{n}x^n$$
$$= P(n,0) + P(n,1)x + P(n,2)\frac{x^2}{2!} + P(n,3)\frac{x^3}{3!} + \cdots + P(n,n)\frac{x^n}{n!}.$$

Hence $(1+x)^n$ is the exponential generating function for the sequence $P(n,0)$, $P(n,1)$, $P(n,2)$, $P(n,3)$, $\ldots, P(n,n), 0, 0, 0, \ldots$.

**2.** The Maclaurin series expansion for $e^x$ is $e^x = 1 + x + \frac{x^2}{2!} + \frac{x^3}{3!} + \cdots$, so the function $e^x$ is the exponential generating function for the sequence $1, 1, 1, 1, \ldots$. The function $e^{-x} = 1 - x + \frac{x^2}{2!} - \frac{x^3}{3!} + \cdots$ is the exponential generating function for the sequence $1, -1, 1, -1, \ldots$. Consequently,

$$\tfrac{1}{2}(e^x + e^{-x}) = 1 + \tfrac{x^2}{2!} + \tfrac{x^4}{4!} + \cdots$$

is the exponential generating function for $1, 0, 1, 0, 1, 0, \ldots$, while

$$\tfrac{1}{2}(e^x - e^{-x}) = x + \tfrac{x^3}{3!} + \tfrac{x^5}{5!} + \cdots$$

is the exponential generating function for $0, 1, 0, 1, 0, 1, \ldots$.

**3.** The function $f(x) = \frac{1}{1-x} = \sum_{i=0}^{\infty} x^i = \sum_{i=0}^{\infty} i! \frac{x^i}{i!}$ is the exponential generating function for the sequence $0!, 1!, 2!, 3!, \ldots$.

**4.** Table 5 gives closed form expressions for the exponential generating functions of particular sequences. In this table, $\left[{n \atop k}\right]$ is a Stirling cycle number, $\left\{{n \atop k}\right\}$ is a Stirling subset number, $B_n$ is the $n$th Bell number (§2.5.2), and $D_n$ is the number of derangements of $n$ objects (§2.4.2).

**Table 5: Exponential generating functions for particular sequences.**

| sequence | closed form |
|:---:|:---:|
| $1,1,1,1,1,\ldots$ | $e^x$ |
| $1,-1,1,-1,1,\ldots$ | $e^{-x}$ |
| $1,0,1,0,1,\ldots$ | $\frac{1}{2}(e^x + e^{-x})$ |
| $0,1,0,1,0,\ldots$ | $\frac{1}{2}(e^x - e^{-x})$ |
| $0,1,2,3,4,\ldots$ | $xe^x$ |
| $P(n,0), P(n,1),\ldots, P(n,n), 0, 0, \ldots$ | $(1+x)^n$ |
| $0,\ldots,0, \begin{bmatrix}n\\n\end{bmatrix}, \begin{bmatrix}n+1\\n\end{bmatrix}, \ldots$ | $\frac{1}{n!}\left[\ln\frac{1}{(1-x)}\right]^n$ |
| $0,\ldots,0, \left\{{n\atop n}\right\}, \left\{{n+1\atop n}\right\}, \ldots$ | $\frac{1}{n!}[e^x - 1]^n$ |
| $B_0, B_1, B_2, B_3, B_4, \ldots$ | $e^{e^x - 1}$ |
| $D_0, D_1, D_2, D_3, D_4, \ldots$ | $\frac{e^{-x}}{1-x}$ |

**5.** The number of ways to permute five of the eight letters in TERMINAL is found using the exponential generating function $f(x) = (1+x)^8$. Here each of the eight letters in TERMINAL is accounted for by the factor $(1+x)$, where $1 (= x^0)$ indicates the letter does not occur in the permutation and $x (= x^1)$ indicates that it does. The coefficient of $\frac{x^5}{5!}$ in $f(x)$ is $\binom{8}{5}5! = P(8,5) = 6{,}720$.

**6.** The number of ways to permute five of the letters in TRANSPORTATION is found as the coefficient of $\frac{x^5}{5!}$ in the exponential generating function $f(x) = (1+x+\frac{x^2}{2!}+\frac{x^3}{3!})(1+x+\frac{x^2}{2!})^4(1+x)^3$. Here the factor $1+x+\frac{x^2}{2!}+\frac{x^3}{3!}$ accounts for the letter T which can be used 0, 1, 2, or 3 times. The factor $1+x+\frac{x^2}{2!}$ occurs four times—for each of R, A, N, and O. Each of the letters S, P, and I produces the factor $(1+x)$. The coefficient of $x^5$ in $f(x)$ is found to be $\frac{487}{3}$, so the answer is $(\frac{487}{3})5! = 19{,}480$.

**7.** The number of ternary sequences (made up of 0s, 1s, and 2s) of length 10 with at least one 0 and an odd number of 1s can be found using the exponential generating function

$$f(x) = (x + \tfrac{x^2}{2!} + \tfrac{x^3}{3!} + \cdots)(x + \tfrac{x^3}{3!} + \tfrac{x^5}{5!} + \cdots)(1 + x + \tfrac{x^2}{2!} + \tfrac{x^3}{3!} + \cdots)$$
$$= (e^x - 1)\tfrac{1}{2}(e^x - e^{-x})e^x = \tfrac{1}{2}(e^{3x} - e^{2x} - e^x + 1)$$
$$= \tfrac{1}{2}\left(\sum_{i=0}^{\infty}\frac{(3x)^i}{i!} - \sum_{i=0}^{\infty}\frac{(2x)^i}{i!} - \sum_{i=0}^{\infty}\frac{x^i}{i!} + 1\right).$$

The answer is the coefficient of $\frac{x^{10}}{10!}$ in $f(x)$, which is $\frac{1}{2}(3^{10} - 2^{10} - 1^{10}) = 29{,}012$.

**8.** Suppose in Example 7 that no symbol may occur exactly two times. The exponential generating function is then $f(x) = (1+x+\frac{x^3}{3!}+\frac{x^4}{4!}+\cdots)^3 = (e^x - \frac{x^2}{2})^3 = e^{3x} - \frac{3}{2}x^2 e^{2x} + \frac{3}{4}x^4 e^x - \frac{1}{8}x^6$. The number of ternary sequences is the coefficient of $\frac{x^{10}}{10!}$ in $f(x)$, namely $3^{10} - \frac{3}{2}(10)(9)2^8 + \frac{3}{4}(10)(9)(8)(7)1^6 = 28{,}269$.

**9.** Exponential generating functions can be used to count the number of onto functions $\varphi\colon A \to B$ where $|A| = m$ and $|B| = n$. Each such function is specified by the sequence of $m$ values $\varphi(a_1), \varphi(a_2), \ldots, \varphi(a_m)$, where each element $b \in B$ occurs at least once in this sequence. Element $b$ contributes a factor $(x + \frac{x^2}{2!} + \frac{x^3}{3!} + \cdots) = (e^x - 1)$ to the exponential generating function $f(x) = (e^x - 1)^n$. The number of onto functions is the coefficient of $\frac{x^m}{m!}$ in $f(x)$, or $n!$ times the coefficient of $\frac{x^m}{m!}$ in $\frac{(e^x-1)^n}{n!}$. From Table 5, the answer is then $n!\left\{{m\atop n}\right\}$. (Also see §2.5.2.)

## 3.3 RECURRENCE RELATIONS

In a number of counting problems, it may be difficult to find the solution directly. However, it is frequently possible to express the solution to a problem of a given size in terms of solutions to problems of smaller size. This interdependence of solutions produces a recurrence relation. Although there is no practical systematic way to solve all recurrence relations, this section contains methods for solving certain types of recurrence relations, thereby providing an explicit formula for the original counting problem. The topic of recurrence relations provides the discrete counterpart to concepts in the study of ordinary differential equations.

### 3.3.1 BASIC CONCEPTS

**Definitions:**

A **recurrence relation** for the sequence $a_0, a_1, a_2, \ldots$ is an equation relating the term $a_n$ to certain of the preceding terms $a_i$, $i < n$, for each $n \geq n_0$.

The recurrence relation is **linear** if it expresses $a_n$ as a linear function of a fixed number of preceding terms. Otherwise the relation is **nonlinear**.

The recurrence relation is **$k$th-order** if $a_n$ can be expressed in terms of $a_{n-1}, a_{n-2}, \ldots,$ $a_{n-k}$.

The recurrence relation is **homogeneous** if the zero sequence $a_0 = a_1 = \cdots = 0$ satisfies the relation. Otherwise the relation is **nonhomogeneous**.

A $k$th-order linear homogeneous recurrence relation **with constant coefficients** is an equation of the form $C_0 a_n + C_1 a_{n-1} + \cdots + C_k a_{n-k} = 0$, $n \geq k$, where the $C_i$ are real constants with $C_0 \neq 0$, $C_k \neq 0$. **Initial conditions** for this recurrence relation specify particular values for $k$ of the $a_i$ (typically $a_0, a_1, \ldots, a_{k-1}$).

**Facts:**

**1.** A $k$th-order linear homogeneous recurrence relation with constant coefficients can also be written $C_0 a_{n+k} + C_1 a_{n+k-1} + \cdots + C_k a_n = 0$, $n \geq 0$.

**2.** There are in general an infinite number of solution sequences $\{a_n\}$ to a $k$th-order linear homogeneous recurrence relation (with constant coefficients).

**3.** A $k$th-order linear homogeneous recurrence relation with constant coefficients together with $k$ initial conditions on consecutive terms $a_0, a_1, \ldots, a_{k-1}$ uniquely determines the sequence $\{a_n\}$. This is not necessarily the case for nonlinear relations (see Example 2) or when nonconsecutive initial conditions are specified (see Example 3).

**4.** The same recurrence relation can be written in different forms by adjusting the subscripts. For example, the recurrence relation $a_n = 3a_{n-1}$, $n \geq 1$, can be written as $a_{n+1} = 3a_n$, $n \geq 0$.

**Examples:**

**1.** The relation $a_n - a_{n-1}^2 + 2a_{n-2} = 0$, $n \geq 2$, is a nonlinear homogeneous recurrence relation with constant coefficients. If the initial conditions $a_0 = 0$, $a_1 = 1$ are imposed, this defines a unique sequence $\{a_n\}$ whose first few terms are $0, 1, 1, -1, -1, 3, 11, 115, \ldots$.

**2.** The first-order (constant coefficient) recurrence relation $a_{n+1}^2 - a_n = 3$, $a_0 = 1$ is nonhomogeneous and nonlinear. Even though one initial condition is specified, this does not uniquely specify a solution sequence. Namely, the two sequences $1, -2, 1, 2, \ldots$ and $1, -2, -1, \sqrt{2}, \ldots$ satisfy the recurrence relation and the given initial condition.

**3.** The second-order relation $a_{n+2} - a_n = 0$, $n \geq 0$, with nonconsecutive initial conditions $a_1 = a_3 = 0$ does not uniquely specify a solution sequence. Both $a_n = (-1)^n + 1$ and $a_n = 2(-1)^n + 2$ satisfy the recurrence and the given initial conditions.

**4.** *Compound interest:* If an initial investment of $P$ dollars is made at a rate of $r$ percent compounded annually, then the amount $a_n$ after $n$ years is given by the recurrence relation $a_n = a_{n-1}(1 + \frac{r}{100})$, where $a_0 = P$. [The amount at the end of the $n$th year is equal to the amount $a_{n-1}$ at the end of the $(n-1)$st year plus the interest on $a_{n-1}$, namely $\frac{r}{100} a_{n-1}$.]

**5.** *Fibonacci sequence:* The Fibonacci numbers satisfy the second-order linear homogeneous recurrence relation $a_n - a_{n-1} - a_{n-2} = 0$.

**6.** *Bit strings:* Let $a_n$ be the number of bit strings of length $n$. Then $a_0 = 1$ (the empty string) and $a_n = 2a_{n-1}$ if $n > 0$. [Every bit string of length $n - 1$ gives rise to two bit strings of length $n$, by placing a 0 or a 1 at the end of the string of length $n - 1$.]

**7.** *Bit strings with no consecutive 0s:* See Example 12 of §3.3.2.

**8.** *Permutations:* Let $a_n$ denote the number of permutations of $\{1, 2, \ldots, n\}$. Then $a_n$ satisfies the first-order linear homogeneous recurrence relation (with nonconstant coefficients) $a_{n+1} = (n+1)a_n$, $n \geq 1$, $a_1 = 1$. This follows since any $n$-permutation $\pi$ can be transformed into an $(n+1)$-permutation by inserting the element $n+1$ into any of the $n+1$ available positions—either at the beginning or end of $\pi$, or between two adjacent elements of $\pi$. To solve for $a_n$, repeatedly apply the recurrence relation and its initial condition: $a_n = na_{n-1} = n(n-1)a_{n-2} = n(n-1)(n-2)a_{n-3} = \cdots = n(n-1)(n-2)\ldots 2a_1 = n!$.

**9.** *Catalan numbers:* The Catalan numbers (§3.1.3, §3.3.1) satisfy the nonlinear homogeneous recurrence relation $C_n - C_0 C_{n-1} - C_1 C_{n-2} - \cdots - C_{n-1} C_0 = 0$, $n \geq 1$, with initial condition $C_0 = 1$. Given the product of $n + 1$ variables $x_1 x_2 \ldots x_{n+1}$, let $C_n$ be the number of ways in which the multiplications can be carried out. For example, there are five ways to form the product $x_1 x_2 x_3 x_4$: $((x_1 x_2)x_3)x_4$, $(x_1(x_2 x_3))x_4$, $(x_1 x_2)(x_3 x_4)$, $x_1((x_2 x_3)x_4)$, and $x_1(x_2(x_3 x_4))$. No matter how the multiplications are performed, there will be an outermost product of the form $(x_1 x_2 \ldots x_i)(x_{i+1} \ldots x_{n+1})$. The number of ways in which the product $x_1 x_2 \ldots x_i$ can be formed is $C_{i-1}$ and the number of ways in which the product $x_{i+1} \ldots x_{n+1}$ can be formed is $C_{n-i}$. Thus, $(x_1 x_2 \ldots x_i)(x_{i+1} \ldots x_{n+1})$ can be obtained in $C_{i-1} C_{n-i}$ ways. Summing these over the values $i = 1, 2, \ldots, n$ yields the recurrence relation.

**10.** *Tower of Hanoi:* See Example 1 of §2.2.4.

**11.** *Onto functions:* The number of onto functions $\varphi: A \to B$ can be found by developing a nonhomogeneous linear recurrence relation based on the size of $B$. Let $|A| = m$ and let $a_n$ be the number of onto functions from $A$ to a set with $n$ elements. Then $a_n = n^m - \binom{n}{1}a_1 - \binom{n}{2}a_2 - \cdots - \binom{n}{n-1}a_{n-1}$, $n \geq 2$, $a_1 = 1$. This follows since the total number of functions from $A$ to $B$ is $n^m$ and the number of functions that map $A$ onto a proper subset of $B$ with exactly $j$ elements is $\binom{n}{j}a_j$.

For example, if $m = 7$ and $n = 4$, applying this recursion gives $a_2 = 2^7 - 2(1) = 126$, $a_3 = 3^7 - 3(1) - 3(126) = 1{,}806$, $a_4 = 4^7 - 4(1) - 6(126) - 4(1{,}806) = 8{,}400$. Thus there are 8,400 onto functions in this case.

## 3.3.2    HOMOGENEOUS RECURRENCE RELATIONS

It is assumed throughout this subsection that the recurrence relations are linear with constant coefficients.

**Definitions:**

A *geometric progression* is a sequence $a_0, a_1, a_2, \ldots$ for which $\frac{a_1}{a_0} = \frac{a_2}{a_1} = \cdots = \frac{a_{n+1}}{a_n} = \cdots = r$, the *common ratio*.

The *characteristic equation* for the $k$th-order recurrence relation $C_0 a_n + C_1 a_{n-1} + \cdots + C_k a_{n-k} = 0$, $n \geq k$, is the equation $C_0 r^k + C_1 r^{k-1} + \cdots + C_k = 0$. The *characteristic roots* are the roots of this equation.

The sequences $\{a_n^{(1)}\}, \{a_n^{(2)}\}, \ldots, \{a_n^{(k)}\}$ are *linearly dependent* if there exist constants $t_1, t_2, \ldots, t_k$, not all zero, such that $\sum_{i=1}^{k} t_i a_n^{(i)} = 0$ for all $n \geq 0$. Otherwise, they are *linearly independent*.

**Facts:**

**1.** *General method for solving a linear homogeneous recurrence relation with constant coefficients:* First find the general solution. Then use the initial conditions to find the particular solution.

**2.** If the $k$ characteristic roots $r_1, r_2, \ldots, r_k$ are distinct, then $r_1^n, r_2^n, \ldots, r_k^n$ are linearly independent solutions of the homogeneous recurrence relation. The general solution is $a_n = c_1 r_1^n + c_2 r_2^n + \cdots + c_k r_k^n$, where $c_1, c_2, \ldots, c_k$ are arbitrary constants.

**3.** If a characteristic root $r$ has multiplicity $m$, then $r^n, nr^n, \ldots, n^{m-1}r^n$ are linearly independent solutions of the homogeneous recurrence relation. The linear combination $c_1 r^n + c_2 n r^n + \cdots + c_m n^{m-1} r^n$ is also a solution, where $c_1, c_2, \ldots, c_m$ are arbitrary constants.

**4.** Facts 2 and 3 can be used together. If there are $k$ characteristic roots $r_1, r_2, \ldots, r_k$, with respective multiplicities $m_1, m_2, \ldots, m_k$ (where some of the $m_i$ can equal 1), the the general solution is a sum of sums, each of the form appearing in Fact 3.

**5.** *de Moivre's theorem:* For any positive integer $n$, $(\cos\theta + i\sin\theta)^n = \cos n\theta + i\sin n\theta$. This result is used to find solutions of recurrence relations when the characteristic roots are complex numbers. (See Example 10.)

**6.** *Solving first-order recurrence relations:* The solution of the homogeneous recurrence relation $a_{n+1} = d a_n$, $n \geq 0$, with initial condition $a_0 = A$, is $a_n = A d^n$, $n \geq 0$.

**7.** *Solving second-order recurrence relations:* Let $r_1, r_2$ be the characteristic roots associated with the second-order homogeneous relation $C_0 a_n + C_1 a_{n-1} + C_2 a_{n-2} = 0$. There are three possibilities:

- $r_1, r_2$ are distinct real numbers: $r_1^n$ and $r_2^n$ are linearly independent solutions of the recurrence relation. The general solution has the form

$$a_n = c_1 r_1^n + c_2 r_2^n,$$

  where the constants $c_1, c_2$ are found from the values of $a_n$ for two distinct values of $n$ (often $n = 0, 1$).

- $r_1, r_2$ form a complex conjugate pair $a \pm bi$: The general solution is

$$a_n = c_1(a + bi)^n + c_2(a - bi)^n = (\sqrt{a^2 + b^2})^n (k_1 \cos n\theta + k_2 \sin n\theta),$$

with $\theta = \arctan(b/a)$. Here $(\sqrt{a^2 + b^2})^n \cos n\theta$ and $(\sqrt{a^2 + b^2})^n \sin n\theta$ are linearly independent solutions.

- $r_1, r_2$ are real and equal: $r_1^n$ and $nr_1^n$ are linearly independent solutions of the recurrence relation. The general solution is

$$a_n = c_1 r_1^n + c_2 n r_1^n.$$

## Examples:

**1.** The geometric progression $7, 21, 63, 189, \ldots$, with common ratio 3, satisfies the first-order homogeneous recurrence relation $a_{n+1} - 3a_n = 0$ for all $n \geq 0$.

**2.** The first-order homogeneous recurrence relation $a_{n+1} = 3a_n$, $n \geq 0$, does not determine a *unique* geometric progression. Any geometric sequence with ratio 3 is a solution; for example the geometric progression in Example 1 (with $a_0 = 7$), as well as the geometric progression $5, 15, 45, 135, \ldots$ (with $a_0 = 5$).

**3.** The first-order recurrence relation $a_{n+1} = 3a_n$, $n \geq 0$, $a_0 = 7$ is easily solved using Fact 6. The general solution is $a_n = 7(3^n)$ for all $n \geq 0$.

**4.** *Compound interest:* If interest is compounded quarterly, how long does it take for an investment of \$500 to double when the annual interest rate is 8%? If $a_n$ denotes the value of the investment after $n$ quarters have passed, then $a_{n+1} = a_n + 0.02a_n = (1.02)a_n$, $n \geq 0$, $a_0 = 500$. [Here the quarterly rate is $0.08/4 = 0.02 = 2\%$.] By Fact 6, the solution is $a_n = 500(1.02)^n$, $n \geq 0$. The investment doubles when $1000 = 500(1.02)^n$, so $n = \frac{\log 2}{\log 1.02} \approx 35.003$. Consequently, after 36 quarters (or 9 years) the initial investment of \$500 (more than) doubles.

**5.** *Population growth:* The number of bacteria in a culture (approximately) triples in size every hour. If there are (approximately) 100,000 bacteria in a culture after six hours, how many were there at the start? Define $p_n$ to be the number of bacteria in the culture after $n$ hours have elapsed. Then $p_{n+1} = 3p_n$ for $n \geq 0$. From Fact 6, $p_n = p_0(3^n)$. So $100,000 = p_0(3^6)$ and $p_0 \approx 137$.

**6.** *Fibonacci sequence:* The Fibonacci sequence $0, 1, 1, 2, 3, 5, 8, 13, \ldots$ arises in varied applications (§3.1.2). Its terms satisfy the second-order homogeneous recurrence relation $F_n = F_{n-1} + F_{n-2}$, $n \geq 2$, with initial conditions $F_0 = 0$, $F_1 = 1$. An explicit formula can be obtained for $F_n$ using Fact 7. The characteristic equation is $r^2 - r - 1 = 0$, with distinct real roots $\frac{1 \pm \sqrt{5}}{2}$. Thus the general solution is

$$F_n = c_1 \left(\frac{1+\sqrt{5}}{2}\right)^n + c_2 \left(\frac{1-\sqrt{5}}{2}\right)^n.$$

Using the initial conditions $F_0 = 0$, $F_1 = 1$ gives $c_1 = \frac{1}{\sqrt{5}}$, $c_2 = -\frac{1}{\sqrt{5}}$ and the explicit formula

$$F_n = \frac{1}{\sqrt{5}}\left[\left(\frac{1+\sqrt{5}}{2}\right)^n - \left(\frac{1-\sqrt{5}}{2}\right)^n\right], \quad n \geq 0.$$

**7.** *Lucas sequence:* Related to the sequence of Fibonacci numbers is the sequence of Lucas numbers $2, 1, 3, 4, 7, 11, 18, \ldots$ (see §3.1.2). The terms of this sequence satisfy the same second-order homogeneous recurrence relation $L_n = L_{n-1} + L_{n-2}$, $n \geq 2$, but with the different initial conditions $L_0 = 2$, $L_1 = 1$. The formula for $L_n$ is

$$L_n = \left(\frac{1+\sqrt{5}}{2}\right)^n + \left(\frac{1-\sqrt{5}}{2}\right)^n, \quad n \geq 0.$$

**8.** *Random walk:* A particle undergoes a random walk in one dimension, along the $x$-axis. Barriers are placed at positions $x = 0$ and $x = T$. At any instant, the particle moves with probability $p$ one unit to the right; with probability $q = 1-p$ it moves one unit to the left. Let $a_n$ denote the probability that the particle, starting at position $x = n$, reaches the barrier $x = T$ before it reaches the barrier $x = 0$. It can be reasoned that $a_n$ satisfies the second-order recurrence relation $a_n = pa_{n+1} + qa_{n-1}$ or $pa_{n+1} - a_n + qa_{n-1} = 0$. In this case the two initial conditions are $a_0 = 0$ and $a_T = 1$. The characteristic equation $pr^2 - r + q = (pr - q)(r - 1) = 0$ has roots $1$, $\frac{q}{p}$. When $p \neq q$, the roots are distinct and the first case of Fact 7 can be used to determine $a_n$; when $p = q$, the third case of Fact 7 must be used. (Explicit solutions are given in §7.5.2, Fact 10.)

**9.** The second-order relation $a_n + 4a_{n-1} - 21a_{n-2} = 0$, $n \geq 2$, has the characteristic equation $r^2 + 4r - 21 = 0$, with distinct real roots $3$ and $-7$. The general solution to the recurrence relation is

$$a_n = c_1(3^n) + c_2(-7)^n, \quad n \geq 0,$$

where $c_1$, $c_2$ are arbitrary constants. If the initial conditions specify $a_0 = 1$ and $a_1 = 1$, then solving the equations $1 = a_0 = c_1 + c_2$, $1 = a_1 = 3c_1 - 7c_2$ gives $c_1 = \frac{4}{5}$, $c_2 = \frac{1}{5}$. In this case, the unique solution is

$$a_n = \tfrac{4}{5}(3^n) + \tfrac{1}{5}(-7)^n, \quad n \geq 0.$$

**10.** The second-order relation $a_n - 6a_{n-1} + 58a_{n-2} = 0$, $n \geq 2$, has the characteristic equation $r^2 - 6r + 58 = 0$, with complex conjugate roots $r = 3 \pm 7i$. The general solution is

$$a_n = c_1(3 + 7i)^n + c_2(3 - 7i)^n, \quad n \geq 0.$$

Using Fact 5, $(3 + 7i)^n = [\sqrt{3^2 + 7^2}(\cos\theta + i\sin\theta)]^n = (\sqrt{58})^n(\cos n\theta + i\sin n\theta)$, where $\theta = \arctan\frac{7}{3}$. Likewise $(3 - 7i)^n = (\sqrt{58})^n(\cos n\theta - i\sin n\theta)$. This gives the general solution

$$a_n = (\sqrt{58})^n[(c_1 + c_2)\cos n\theta + (c_1 - c_2)i\sin n\theta] = (\sqrt{58})^n(k_1\cos n\theta + k_2\sin n\theta).$$

If the initial conditions $a_0 = 1$ and $a_1 = 1$ are specified, then $1 = a_0 = k_1$, $1 = a_1 = \sqrt{58}(\cos\theta + k_2\sin\theta)$, yielding $k_1 = 1$, $k_2 = -\frac{2}{7}$. Thus

$$a_n = (\sqrt{58})^n(\cos n\theta - \tfrac{2}{7}\sin n\theta), \quad n \geq 0.$$

**11.** The second-order relation $a_{n+2} - 6a_{n+1} + 9a_n = 0$, $n \geq 0$, has the characteristic equation $r^2 - 6r + 9 = (r - 3)^2 = 0$, with the repeated roots $3$, $3$. The general solution to this recurrence is

$$a_n = c_1(3^n) + c_2n3^n, \quad n \geq 0.$$

If the initial conditions are $a_0 = 2$ and $a_1 = 4$, then $2 = a_0 = c_1$, $4 = 2(3) + c_2(1)(3)$, giving $c_1 = 2$, $c_2 = -\frac{2}{3}$. Thus

$$a_n = 2(3^n) - \tfrac{2}{3}n3^n = 2(3^n - n3^{n-1}), \quad n \geq 0.$$

**12.** For $n \geq 1$, let $a_n$ count the number of binary strings of length $n$ that contain no consecutive 0s. Here $a_1 = 2$ (for the two strings 0 and 1) and $a_2 = 3$ (for the strings 01, 10, 11). For $n \geq 3$, a string counted in $a_n$ ends in either 1 or 0. If the $n$th bit is 1, then the preceding $n - 1$ bits provide a string counted in $a_{n-1}$; if the $n$th bit is 0 then the last two bits are 10, and the preceding $n - 2$ bits give a string counted in $a_{n-2}$. Thus $a_n = a_{n-1} + a_{n-2}$, $n \geq 3$, with $a_1 = 2$ and $a_2 = 3$. The solution to this relation is simply $a_n = F_{n+2}$, the Fibonacci sequence shifted two places. An explicit formula for $a_n$ is obtained using the result in Example 6.

**13.** The third-order recurrence relation $a_{n+3} - a_{n+2} - 4a_{n+1} + 4a_n = 0$, $n \geq 0$, has the characteristic equation $r^3 - r^2 - 4r + 4 = (r-2)(r+2)(r-1) = 0$, with characteristic roots 2, $-2$, and 1. The general solution is given by

$$a_n = c_1 2^n + c_2(-2)^n + c_3 1^n = c_1 2^n + c_2(-2)^n + c_3, \quad n \geq 0.$$

**14.** The general solution of the third-order recurrence relation $a_{n+3} - 3a_{n+2} - 3a_{n+1} + a_n = 0$, $n \geq 0$, is

$$a_n = c_1 1^n + c_2 n 1^n + c_3 n^2 1^n = c_1 + c_2 n + c_3 n^2, \quad n \geq 0.$$

Here the characteristic roots are 1, 1, 1.

**15.** The fourth-order relation $a_{n+4} + 2a_{n+2} + a_n = 0$, $n \geq 0$, has the characteristic equation $r^4 + 2r^2 + 1 = (r^2 + 1)^2 = 0$. Since the characteristic roots are $\pm i$, $\pm i$, the general solution is

$$a_n = c_1 i^n + c_2(-i)^n + c_3 n i^n + c_4 n(-i)^n$$
$$= k_1 \cos \tfrac{n\pi}{2} + k_2 \sin \tfrac{n\pi}{2} + k_3 n \cos \tfrac{n\pi}{2} + k_4 n \sin \tfrac{n\pi}{2}, \quad n \geq 0.$$

### 3.3.3  NONHOMOGENEOUS RECURRENCE RELATIONS

It is assumed throughout this subsection that the recurrence relations are linear with constant coefficients.

**Definition:**
The $k$th-order **nonhomogeneous** recurrence relation has the form $C_0 a_n + C_1 a_{n-1} + \cdots + C_k a_{n-k} = f(n)$, $n \geq k$, where $C_0 \neq 0$, $C_k \neq 0$, and $f(n) \neq 0$ for at least one value of $n$.

**Facts:**
**1.** *General solution:* The general solution of the nonhomogeneous $k$th-order recurrence relation has the form
$$a_n = a_n^{(h)} + a_n^{(p)},$$
where $a_n^{(h)}$ is the general solution of the homogeneous relation $C_0 a_n + C_1 a_{n-1} + \cdots + C_k a_{n-k} = 0$, $n \geq k$, and $a_n^{(p)}$ is a particular solution for the given relation $C_0 a_n + C_1 a_{n-1} + \cdots + C_k a_{n-k} = f(n)$, $n \geq k$.

**2.** Given the nonhomogeneous first-order relation $C_0 a_n + C_1 a_{n-1} = kr^n$, $n \geq 1$, where $r$ and $k$ are nonzero constants:

- If $r^n$ is *not* a solution of the associated homogeneous relation, then $a_n^{(p)} = Ar^n$ for $A$ a constant.
- If $r^n$ is a solution of the associated homogeneous relation, then $a_n^{(p)} = Bnr^n$ for $B$ a constant.

**3.** Given the nonhomogeneous second-order relation $C_0 a_n + C_1 a_{n-1} + C_2 a_{n-2} = kr^n$, $n \geq 2$, where $r$ and $k$ are nonzero constants:

- If $r^n$ is *not* a solution of the associated homogeneous relation, then $a_n^{(p)} = Ar^n$ for $A$ a constant.
- If $a_n^{(h)} = c_1 r^n + c_2 r_1^n$, for $r \neq r_1$, then $a_n^{(p)} = Bnr^n$ for $B$ a constant.

- If $a_n^{(h)} = c_1 r^n + c_2 n r^n$, then $a_n^{(p)} = C n^2 r^n$ for $C$ a constant.

**4.** Consider the $k$th-order nonhomogeneous recurrence relation $C_0 a_n + C_1 a_{n-1} + \cdots + C_k a_{n-k} = f(n)$. If $f(n)$ is a constant multiple of one of the forms in the first column of Table 1, then the associated *trial solution* $t(n)$ is the corresponding entry in the second column of the table. [Here $A, B, A_0, A_1, \ldots, A_t, r, \alpha$ are real constants.]

- If no summand of $t(n)$ solves the associated homogeneous relation, then $a_n^{(p)} = t(n)$ is a particular solution.

- If a summand of $t(n)$ solves the associated homogeneous relation, then multiply $t(n)$ by the *smallest* (positive integer) power of $n$—say $n^s$—so that no summand of the adjusted trial solution $n^s t(n)$ solves the associated homogeneous relation. Then $a_n^{(p)} = n^s t(n)$ is a particular solution.

- If $f(n)$ is a sum of constant multiples of the forms in the first column of Table 1, then (adjusted) trial solutions are formed for each summand using the first two parts of Fact 4. Adding the resulting trial solutions then provides a particular solution of the nonhomogeneous relation.

**Table 1: Trial particular solutions for $C_0 a_n + C_1 a_{n-1} + \cdots + C_k a_{n-k} = f(n)$.**

| $f(n)$ | $t(n)$ |
|---|---|
| $c$, a constant | $A$ |
| $n^t$ ($t$ a positive integer) | $A_t n^t + A_{t-1} n^{t-1} + \cdots + A_1 n + A_0$ |
| $r^n$ | $A r^n$ |
| $\sin n\alpha$ | $A \sin n\alpha + B \cos n\alpha$ |
| $\cos n\alpha$ | $A \sin n\alpha + B \cos n\alpha$ |
| $n^t r^n$ | $r^n (A_t n^t + A_{t-1} n^{t-1} + \cdots + A_1 n + A_0)$ |
| $r^n \sin n\alpha$ | $r^n (A \sin n\alpha + B \cos n\alpha)$ |
| $r^n \cos n\alpha$ | $r^n (A \sin n\alpha + B \cos n\alpha)$ |

**Examples:**

**1.** Consider the nonhomogeneous relation $a_n + 4a_{n-1} - 21a_{n-2} = 5(4^n)$, $n \geq 2$. The solution is $a_n = a_n^{(h)} + a_n^{(p)}$, where $a_n^{(h)}$ is the solution of $a_n + 4a_{n-1} - 21a_{n-2} = 0$, $n \geq 2$. So

$$a_n^{(h)} = c_1(3^n) + c_2(-7)^n, \quad n \geq 0.$$

From the third entry in Table 1 $a_n^{(p)} = A(4^n)$ for some constant $A$. Substituting this into the given nonhomogeneous relation yields $A(4^n) + 4A(4^{n-1}) - 21A(4^{n-2}) = 5(4^n)$. Dividing through by $4^{n-2}$ gives $16A + 16A - 21A = 80$, or $A = 80/11$. Consequently,

$$a_n = c_1(3^n) + c_2(-7)^n + \tfrac{80}{11}(4^n), \quad n \geq 0.$$

If the initial conditions are $a_0 = 1$ and $a_1 = 2$, then $c_1$ and $c_2$ are found using $1 = c_1 + c_2 + \frac{80}{11}$, $2 = 3c_1 - 7c_2 + \frac{320}{11}$, yielding

$$a_n = -\tfrac{71}{10}(3^n) + \tfrac{91}{110}(-7)^n + \tfrac{80}{11}(4^n), \quad n \geq 0.$$

**2.** Suppose the given recurrence relation is $a_n + 4a_{n-1} - 21a_{n-2} = 8(3^n)$, $n \geq 2$. Then it is still true that

$$a_n^{(h)} = c_1(3^n) + c_2(-7)^n, \quad n \geq 0,$$

where $c_1$ and $c_2$ are arbitrary constants. By the second part of Fact 3, a particular solution is $a_n^{(p)} = An3^n$. Substituting $a_n^{(p)}$ gives $An3^n + 4A(n-1)3^{n-1} - 21A(n-2)3^{n-2} = 8(3^n)$. Dividing by $3^{n-2}$ produces $9An + 12A(n-1) - 21A(n-2) = 72$, so $A = \frac{12}{5}$. Thus

$$a_n = c_1(3^n) + c_2(-7)^n + \tfrac{12}{5}n3^n, \quad n \geq 0.$$

**3.** *Tower of Hanoi:* (See Example 1 of §2.2.4.) If $a_n$ is the minimum number of moves needed to transfer the $n$ disks, then $a_n$ satisfies the first-order nonhomogeneous relation

$$a_n - 2a_{n-1} = 1, \quad n \geq 1,$$

where $a_0 = 0$. Here $a_n^{(h)} = c(2^n)$ for an arbitrary constant $c$, and $a_n^{(p)} = A$, using the first entry of Table 1. So $A = 2A + 1$ or $A = -1$. Hence $a_n = c(2^n) - 1$ and $0 = a_0 = c(2^0) - 1$ implies $c = 1$, giving

$$a_n = 2^n - 1, \quad n \geq 0.$$

**4.** How many regions are formed if $n$ lines are drawn in the plane, in general position (no two parallel and no three intersecting at a point)? If $a_n$ denotes the number of regions thus formed, then $a_1 = 2$, $a_2 = 4$, and $a_3 = 7$ are easily determined. A general formula can be found by developing a recurrence relation for $a_n$. Namely, if line $n + 1$ is added to the diagram with $a_n$ regions formed by $n$ lines, this new line intersects all the other $n$ lines. These intersection points partition line $n + 1$ into $n + 1$ segments, each of which splits an existing region in two. As a result, $a_{n+1} = a_n + (n + 1)$, $n \geq 1$, a first-order nonhomogeneous recurrence relation. Solving this relation with the initial condition $a_1 = 1$ produces $a_n = \frac{1}{2}(n^2 + n + 2)$.

---

### 3.3.4   METHOD OF GENERATING FUNCTIONS

Generating functions (see §3.2.1) can be used to solve individual recurrence relations as well as simultaneous systems of recurrence relations. This technique is analogous to the use of Laplace transforms in solving systems of differential equations.

**Facts:**

**1.** To solve the $k$th-order recurrence relation $C_0 a_{n+k} + \cdots + C_k a_n = f(n)$, $n \geq 0$, carry out the following steps:

- Multiply both sides of the recurrence equation by $x^{n+k}$ and sum the result.
- Take this new equation, rewrite it in terms of the generating function $f(x) = \sum_{n=0}^{\infty} a_n x^n$ and solve for $f(x)$.
- Expand the expression found for $f(x)$ in terms of powers of $x$ so that the coefficient $a_n$ can be identified.

**2.** To solve a system of $k$th-order recurrence relations, carry out the following steps:

- Multiply both sides of each recurrence equation by $x^{n+k}$ and sum the results.
- Rewrite the system of equations in terms of the generating functions $f(x)$, $g(x)$, ... for $a_n, b_n, \ldots$, and solve for these generating functions.
- Expand the expressions found for each generating function in terms of powers of $x$ so that the coefficients $a_n, b_n, \ldots$ can be identified.

**Examples:**

**1.** The nonhomogeneous first-order relation $a_{n+1} - 2a_n = 1$, $n \geq 0$, $a_0 = 0$, arises in the *Tower of Hanoi* problem (§3.3.3, Example 3). Begin by applying the first step of Fact 1:

$$a_{n+1}x^{n+1} - 2a_n x^{n+1} = x^{n+1},$$

$$\sum_{n=0}^{\infty} a_{n+1}x^{n+1} - 2\sum_{n=0}^{\infty} a_n x^{n+1} = \sum_{n=0}^{\infty} x^{n+1}.$$

Then apply the second step of Fact 1:

$$\sum_{n=0}^{\infty} a_{n+1}x^{n+1} - 2x\sum_{n=0}^{\infty} a_n x^n = x\sum_{n=0}^{\infty} x^n,$$

$$(f(x) - a_0) - 2xf(x) = \frac{x}{1-x},$$

$$(f(x) - 0) - 2xf(x) = \frac{x}{1-x}.$$

Solving for $f(x)$ gives

$$f(x) = \frac{x}{(1-x)(1-2x)} = \frac{1}{1-2x} - \frac{1}{1-x} = \sum_{n=0}^{\infty} (2x)^n - \sum_{n=0}^{\infty} x^n = \sum_{n=0}^{\infty} (2^n - 1)x^n.$$

Since $a_n$ is the coefficient of $x^n$ in $f(x)$, $a_n = 2^n - 1$, $n \geq 0$.

**2.** To solve the nonhomogeneous second-order relation $a_{n+2} - 2a_{n+1} + a_n = 2^n$, $n \geq 0$, $a_0 = 1$, $a_1 = 2$, apply the first step of Fact 1:

$$a_{n+2}x^{n+2} - 2a_{n+1}x^{n+2} + a_n x^{n+2} = 2^n x^{n+2},$$

$$\sum_{n=0}^{\infty} a_{n+2}x^{n+2} - 2\sum_{n=0}^{\infty} a_{n+1}x^{n+2} + \sum_{n=0}^{\infty} a_n x^{n+2} = \sum_{n=0}^{\infty} 2^n x^{n+2}.$$

The second step of Fact 1 produces

$$\sum_{n=0}^{\infty} a_{n+2}x^{n+2} - 2x\sum_{n=0}^{\infty} a_{n+1}x^{n+1} + x^2\sum_{n=0}^{\infty} a_n x^n = x^2\sum_{n=0}^{\infty} (2x)^n,$$

$$[f(x) - a_0 - a_1 x] - 2x[f(x) - a_0] + x^2 f(x) = \frac{x^2}{1-2x},$$

$$[f(x) - 1 - 2x] - 2x[f(x) - 1] + x^2 f(x) = \frac{x^2}{1-2x}.$$

Solving for $f(x)$ gives

$$f(x) = \frac{1}{1-2x} = \sum_{n=0}^{\infty} (2x)^n = \sum_{n=0}^{\infty} 2^n x^n.$$

Thus $a_n = 2^n$, $n \geq 0$, is the solution of the given recurrence relation.

**3.** Fact 2 can be used to solve the system of recurrence relations

$$a_{n+1} = 2a_n - b_n + 2$$
$$b_{n+1} = -a_n + 2b_n - 1$$

for $n \geq 0$, with $a_0 = 0$ and $b_0 = 1$. Multiplying by $x^{n+1}$ and summing yields

$$\sum_{n=0}^{\infty} a_{n+1}x^{n+1} = 2x\sum_{n=0}^{\infty} a_n x^n - x\sum_{n=0}^{\infty} b_n x^n + 2x\sum_{n=0}^{\infty} x^n$$

$$\sum_{n=0}^{\infty} b_{n+1}x^{n+1} = -x\sum_{n=0}^{\infty} a_n x^n + 2x\sum_{n=0}^{\infty} b_n x^n - x\sum_{n=0}^{\infty} x^n.$$

These two equations can be rewritten in terms of the generating functions $f(x) = \sum_{n=0}^{\infty} a_n x^n$ and $g(x) = \sum_{n=0}^{\infty} b_n x^n$ as

$$f(x) - a_0 = 2xf(x) - xg(x) + 2x\frac{1}{1-x}$$
$$g(x) - b_0 = -xf(x) + 2xg(x) - x\frac{1}{1-x}.$$

Solving this system (with $a_0 = 0$, $b_0 = 1$) produces

$$f(x) = \frac{x(1-2x)}{(1-x)^2(1-3x)} = -\frac{3}{4}\frac{1}{1-x} + \frac{1}{2}\frac{1}{(1-x)^2} + \frac{1}{4}\frac{1}{(1-3x)}$$

$$= -\frac{3}{4}\sum_{n=0}^{\infty} x^n + \frac{1}{2}\sum_{n=0}^{\infty}\binom{-2}{n}x^n + \frac{1}{4}\sum_{n=0}^{\infty}(3x)^n$$

$$= -\frac{3}{4}\sum_{n=0}^{\infty} x^n + \frac{1}{2}\sum_{n=0}^{\infty}\binom{n+1}{n}x^n + \frac{1}{4}\sum_{n=0}^{\infty} 3^n x^n$$

and

$$g(x) = \frac{1-4x+2x^2}{(1-x)^2(1-3x)} = \frac{3}{4}\frac{1}{1-x} + \frac{1}{2}\frac{1}{(1-x)^2} - \frac{1}{4}\frac{1}{(1-3x)}$$

$$= \frac{3}{4}\sum_{n=0}^{\infty} x^n + \frac{1}{2}\sum_{n=0}^{\infty}\binom{n+1}{n}x^n - \frac{1}{4}\sum_{n=0}^{\infty} 3^n x^n.$$

It then follows that

$$a_n = -\frac{3}{4} + \frac{1}{2}(n+1) + \frac{1}{4}(3^n), \quad n \geq 0$$
$$b_n = \frac{3}{4} + \frac{1}{2}(n+1) - \frac{1}{4}(3^n), \quad n \geq 0.$$

## 3.3.5   DIVIDE-AND-CONQUER RELATIONS

Certain algorithms proceed by breaking up a given problem into subproblems of nearly equal size; solutions to these subproblems are then combined to produce a solution to the original problem. Analysis of such "divide-and-conquer" algorithms results in special types of recurrence relations that can be solved both exactly and asymptotically.

### Definitions:
The ***time-complexity function*** $f(n)$ for an algorithm gives the (maximum) number of operations required to solve any instance of size $n$. The function $f(n)$ is ***monotone increasing*** if $m < n \Rightarrow f(m) \leq f(n)$ where $m$ and $n$ are positive integers.

A ***recursive divide-and-conquer algorithm*** splits a given problem of size $n = b^k$ into $a$ subproblems of size $\frac{n}{b}$ each. It requires (at most) $h(n)$ operations to create the subproblems and subsequently combine their solutions.

Let $S = S_b$ be the set of integers $\{1, b, b^2, \ldots\}$ and let $\mathcal{Z}^+$ be the set of positive integers.

If $f(n)$ and $g(n)$ are functions on $\mathcal{Z}^+$, then $g$ ***dominates*** $f$ ***on*** $S$, written $f \in O(g)$ on $S$, if there are positive constants $A \in \mathcal{R}$, $k \in \mathcal{Z}^+$ such that $|f(n)| \leq A|g(n)|$ holds for all $n \in S$ with $n \geq k$.

### Facts:
**1.** The time-complexity function $f(n)$ of a recursive divide-and-conquer algorithm is defined for $n \in S$ and satisfies the recurrence relation

$$f(1) = c,$$
$$f(n) = af(n/b) + h(n), \quad \text{for } n = b^k, \ k \geq 1,$$

where $a, b, c \in \mathcal{Z}^+$ and $b \geq 2$.

2. *Solving* $f(n) = af(n/b) + c$, $f(1) = c$:

- If $a = 1$: $f(n) = c(\log_b n + 1)$ for $n \in S$. Thus $f \in O(\log_b n)$ on $S$. If, in addition, $f(n)$ is monotone increasing, then $f \in O(\log_b n)$ on $\mathcal{Z}^+$.

- If $a \geq 2$: $f(n) = c(an^{\log_b a} - 1)/(a-1)$ for $n \in S$. Thus $f \in O(n^{\log_b a})$ on $S$. If, in addition, $f(n)$ is monotone increasing, then $f \in O(n^{\log_b a})$ on $\mathcal{Z}^+$.

3. Let $f(n)$ be any function satisfying the inequality relations

$$f(1) \leq c,$$
$$f(n) \leq af(n/b) + c, \quad \text{for} \ \ n = b^k, \ k \geq 1,$$

where $a, b, c \in \mathcal{Z}^+$ and $b \geq 2$.

- If $a = 1$:  $f \in O(\log_b n)$ on $S$. If, in addition, $f(n)$ is monotone increasing, then $f \in O(\log_b n)$ on $\mathcal{Z}^+$.

- If $a \geq 2$:  $f \in O(n^{\log_b a})$ on $S$. If, in addition, $f(n)$ is monotone increasing, then $f \in O(n^{\log_b a})$ on $\mathcal{Z}^+$.

4. *Solving for a monotone increasing* $f(n)$ *where* $f(n) = af(n/b) + rn^d$ ($n = b^k$, $k \geq 1$), $f(1) = c$, *where* $a, b, c, d \in \mathcal{Z}^+$, $b \geq 2$, *and* $r$ *is a positive real number:*

- If $a < b^d$: $f \in O(n^d)$ on $\mathcal{Z}^+$.

- If $a = b^d$: $f \in O(n^d \log_b n)$ on $\mathcal{Z}^+$.

- If $a > b^d$: $f \in O(n^{\log_b a})$ on $\mathcal{Z}^+$.

The same asymptotic results hold if inequalities $\leq$ replace equalities in the given recurrence relation.

**Examples:**

1. If $f(n)$ satisfies the recurrence relation $f(n) = f(\frac{n}{2}) + 3$, $n \in S_2$, $f(1) = 3$, then Fact 2 gives $f(n) = 3(\log_2 n + 1)$. Thus $f \in O(\log_2 n)$ on $S_2$.

2. If $f(n)$ satisfies the recurrence relation $f(n) = 4f(\frac{n}{3}) + 7$, $n \in S_3$, $f(1) = 7$, then Fact 2 gives $f(n) = 7(4n^{\log_3 4} - 1)/3$. Thus $f \in O(n^{\log_3 4})$ on $S_3$.

3. *Binary search:* The binary search algorithm (§18.2.3) is a recursive procedure to search for a specified value in an ordered list of $n$ items. Its complexity function satisfies $f(n) = f(\frac{n}{2}) + 2$, $n \in S_2$, $f(1) = 2$. Since the complexity function $f(n)$ is monotone increasing in the list size $n$, Fact 2 shows that $f \in O(\log_2 n)$.

4. *Merge sort:* The merge sort algorithm (§18.3.4) is a recursive procedure for sorting the $n$ elements of a list. It repeatedly divides a given list into two nearly equal sublists, sorts those sublists, and combines the sorted sublists. Its complexity function satisfies $f(n) = 2f(\frac{n}{2}) + (n-1)$, $n \in S_2$, $f(1) = 0$. Since $f(n)$ is monotone increasing and satisfies the inequality relation $f(n) \leq 2f(\frac{n}{2}) + n$, Fact 4 gives $f \in O(n \log_2 n)$.

5. *Matrix multiplication:* The Strassen algorithm is a recursive procedure for multiplying two $n \times n$ matrices (see §6.3.3). One version of this algorithm requires seven multiplications of $\frac{n}{2} \times \frac{n}{2}$ matrices and 15 additions of $\frac{n}{2} \times \frac{n}{2}$ matrices. Consequently, its complexity function satisfies $f(n) = 7f(\frac{n}{2}) + 15n^2/4$, $n \in S_2$, $f(1) = 1$. From the third part of Fact 4, $f \in O(n^{\log_2 7})$ on $\mathcal{Z}^+$. This algorithm requires approximately $O(n^{2.81})$ operations to multiply $n \times n$ matrices, compared to $O(n^3)$ for the standard method.

## 3.4  FINITE DIFFERENCES

The difference and antidifference operators are the discrete analogues of ordinary differentiation and antidifferentiation. Difference methods can be used for curve-fitting and for solving recurrence relations.

### 3.4.1  THE DIFFERENCE OPERATOR

The difference operator plays a role in combinatorial modeling analogous to that of the derivative operator in continuous analysis.

**Definitions:**

Let $f: \mathcal{N} \to \mathcal{R}$.

The **difference operator** $\Delta f(x) = f(x+1) - f(x)$ is the discrete analogue of the differentiation operator.

The **kth difference** of $f$ is the operator $\Delta^k f(x) = \Delta^{k-1} f(x+1) - \Delta^{k-1} f(x)$, for $k \geq 1$, with $\Delta^0 f = f$.

The **shift operator** $E$ is defined by $Ef(x) = f(x+1)$.

The **harmonic sum** $H_n = \sum_{i=1}^{n} \frac{1}{i}$ is the discrete analogue of the natural logarithm (§3.1.7).

*Note:* Most of the results stated in this subsection are also valid for functions on non-discrete domains. The functional notation that is used for most of this subsection, instead of the more usual subscript notation for sequences, makes the results easier to read and helps underscore the parallels between discrete and ordinary calculus.

**Facts:**

1. *Linearity:* $\Delta(\alpha f + \beta g) = \alpha \Delta f + \beta \Delta g$, for all constants $\alpha$ and $\beta$.

2. *Product rule:* $\Delta(f(x)g(x)) = (Ef(x))\Delta g(x) + (\Delta f(x))g(x)$. This is analogous to the derivative formula for the product of functions.

3. $\Delta^m x^n = 0$, for $m > n$, and $\Delta^n x^n = n!$.

4. $\Delta^n f(x) = \sum_{k=0}^{n} (-1)^k \binom{n}{k} f(x+n-k)$.

5. $f(x+n) = \sum_{k=0}^{n} \binom{n}{k} \Delta^k f(x)$.

6. *Leibniz's theorem:* $\Delta^n(f(x)g(x)) = \sum_{k=0}^{n} \binom{n}{k} \Delta^k f(x) \Delta^{n-k} g(x+k)$.

7. *Quotient rule:* $\Delta\left(\frac{f(x)}{g(x)}\right) = \frac{g(x)\Delta f(x) - f(x)\Delta g(x)}{g(x)g(x+1)}$.

8. The shift operator $E$ satisfies $\Delta f = Ef - f$, written equivalently as $E = 1 + \Delta$.

9. $E^n f(x) = f(x+n)$.

10. The equation $\Delta C(x) = 0$ implies that $C$ is periodic with period 1. Moreover, if the domain is restricted to the integers (e.g., if $C(n)$ is a sequence), then $C$ is constant.

**Examples:**

**1.** If $f(x) = x^3$ then $\Delta f(x) = (x+1)^3 - x^3 = 3x^2 + 3x + 1$.

**2.** The following table gives formulas for the differences of some important functions. In this table, the notation $x^{\underline{n}}$ refers to the $n$th falling power of $x$ (§3.4.2).

| $f(x)$ | $\Delta f(x)$ |
|---|---|
| $\binom{x}{n}$ | $\binom{x}{n-1}$ |
| $(x+a)^{\underline{n}}$ | $n(x+a)^{\underline{n-1}}$ |
| $x^n$ | $\binom{n}{1}x^{n-1} + \binom{n}{2}x^{n-2} + \cdots + 1$ |
| $a^x$ | $(a-1)a^x$ |
| $H_x$ | $x^{\underline{-1}} = \frac{1}{x+1}$ |
| $\sin x$ | $2\sin(\frac{1}{2})\cos(x+\frac{1}{2})$ |
| $\cos x$ | $-2\sin(\frac{1}{2})\sin(x+\frac{1}{2})$ |

**3.** $\Delta^2 f(x) = f(x+2) - 2f(x+1) + f(x)$, from Fact 4.

**4.** $f(x+3) = f(x) + 3\Delta f(x) + 3\Delta^2 f(x) + \Delta^3 f(x)$, from Fact 5.

**5.** The shift operator can be used to find the exponential generating function (§3.2.2) for the sequence $\{a_k\}$, where $a_k$ is a polynomial in variable $k$ of degree $n$.

$$\sum_{k=0}^{\infty} \frac{a_k x^k}{k!} = \sum_{k=0}^{\infty} \frac{E^k(a_0)x^k}{k!} = \left(\sum_{k=0}^{\infty} \frac{x^k E^k}{k!}\right) a_0$$

$$= e^{xE} a_0 = e^{x(1+\Delta)} a_0 = e^x e^{x\Delta} a_0$$

$$= e^x \left(a_0 + \frac{x\Delta a_0}{1!} + \frac{x^2 \Delta^2 a_0}{2!} + \cdots + \frac{x^n \Delta^n a_0}{n!}\right).$$

For example, if $a_k = k^2 + 1$ then $\sum_{k=0}^{\infty} \frac{(k^2+1)x^k}{k!} = e^x(1+x+x^2)$.

## 3.4.2   CALCULUS OF DIFFERENCES: FALLING AND RISING POWERS

Falling powers provide a natural analogue between the calculus of finite sums and differences and the calculus of integrals and derivatives. Stirling numbers provide a means of expressing ordinary powers in terms of falling powers and vice versa.

**Definitions:**

The **$n$th falling power** of $x$, written $x^{\underline{n}}$, is the discrete analogue of exponentiation and is defined by

$$x^{\underline{n}} = x(x-1)(x-2)\ldots(x-n+1)$$

$$x^{\underline{-n}} = \frac{1}{(x+1)(x+2)\ldots(x+n)}$$

$$x^{\underline{0}} = 1.$$

The **$n$th rising power** of $x$, written $x^{\overline{n}}$, is defined by

$$x^{\overline{n}} = x(x+1)(x+2)\ldots(x+n-1),$$

$$x^{\overline{-n}} = \frac{1}{(x-n)(x-n+1)\ldots(x-1)},$$

$$x^{\overline{0}} = 1.$$

**Facts:**

**1.** *Conversion between falling and rising powers:*

$$x^{\underline{n}} = (-1)^n(-x)^{\overline{n}} = (x - n + 1)^{\overline{n}} = \frac{1}{(x+1)^{\overline{-n}}} ,$$

$$x^{\overline{n}} = (-1)^n(-x)^{\underline{n}} = (x + n - 1)^{\underline{n}} = \frac{1}{(x-1)^{\underline{-n}}} ,$$

$$x^{\underline{-n}} = \frac{1}{(x+1)^{\overline{n}}} ,$$

$$x^{\overline{-n}} = \frac{1}{(x-1)^{\underline{n}}} .$$

**2.** *Laws of exponents:*

$$x^{\underline{m+n}} = x^{\underline{m}}(x - m)^{\underline{n}} ,$$

$$x^{\overline{m+n}} = x^{\overline{m}}(x + m)^{\overline{n}} .$$

**3.** *Binomial theorem:* $(x + y)^{\underline{n}} = \binom{n}{0}x^{\underline{n}} + \binom{n}{1}x^{\underline{n-1}}y^{\underline{1}} + \cdots + \binom{n}{n}y^{\underline{n}} .$

**4.** The action of the difference operator on falling powers is analogous to the action of the derivative on ordinary powers: $\Delta x^{\underline{n}} = nx^{\underline{n-1}} .$

**5.** There is no chain rule for differences, but the binomial theorem implies the rule

$$\Delta(x + a)^{\underline{n}} = n(x + a)^{\underline{n-1}} .$$

**6.** *Newton's theorem:* If $f(x)$ is a polynomial of degree $n$, then

$$f(x) = \sum_{k=0}^{n} \frac{\Delta^k f(0)}{k!} x^{\underline{k}} .$$

This is an analogue of *Maclaurin's theorem.*

**7.** If $f(x) = x^n$ then $\Delta^k f(0) = \left\{\begin{smallmatrix} n \\ k \end{smallmatrix}\right\} \cdot k!$, where $\left\{\begin{smallmatrix} n \\ k \end{smallmatrix}\right\}$ is a Stirling subset number (§2.5.2).

**8.** Falling powers can be expressed in terms of ordinary powers using Stirling cycle numbers (§2.5.2):

$$x^{\underline{n}} = \sum_{k=1}^{n} \left[\begin{smallmatrix} n \\ k \end{smallmatrix}\right](-1)^{n-k} x^k.$$

**9.** Rising powers can be expressed in terms of ordinary powers using Stirling cycle numbers (§2.5.2):

$$x^{\overline{n}} = \sum_{k=1}^{n} \left[\begin{smallmatrix} n \\ k \end{smallmatrix}\right] x^k.$$

**10.** Ordinary powers can be expressed in terms of falling or rising powers using Stirling subset numbers (§2.5.2):

$$x^n = \sum_{k=1}^{n} \left\{\begin{smallmatrix} n \\ k \end{smallmatrix}\right\} x^{\underline{k}} = \sum_{k=1}^{n} \left\{\begin{smallmatrix} n \\ k \end{smallmatrix}\right\}(-1)^{n-k} x^{\overline{k}}.$$

**Examples:**

**1.** Fact 8, together with Fact 3 of §2.5.2, gives

$$x^{\underline{0}} = x^0,$$

$$x^{\underline{1}} = x^1,$$

$$x^{\underline{2}} = x^2 - x^1,$$

$$x^{\underline{3}} = x^3 - 3x^2 + 2x^1,$$

$$x^{\underline{4}} = x^4 - 6x^3 + 11x^2 - 6x^1.$$

**2.** Fact 10, together with Fact 4 of §2.5.2, gives

$$x^0 = x^{\underline{0}},$$
$$x^1 = x^{\underline{1}},$$
$$x^2 = x^{\underline{2}} + x^{\underline{1}},$$
$$x^3 = x^{\underline{3}} + 3x^{\underline{2}} + x^{\underline{1}},$$
$$x^4 = x^{\underline{4}} + 6x^{\underline{3}} + 7x^{\underline{2}} + x^{\underline{1}}.$$

### 3.4.3  DIFFERENCE SEQUENCES AND DIFFERENCE TABLES

New sequences can be obtained from a given sequence by repeatedly applying the difference operator.

**Definitions:**
The **difference sequence** for the sequence $A = \{\, a_j \mid j = 0, 1, \ldots \}$ is the sequence $\Delta A = \{\, a_{j+1} - a_j \mid j = 0, 1, \ldots \}$.

The **$k$th difference sequence** for $f: \mathcal{N} \to \mathcal{R}$ is given by $\Delta^k f(0), \Delta^k f(1), \Delta^k f(2), \ldots$.

The **difference table** for $f: \mathcal{N} \to \mathcal{R}$ is the table $T_f$ whose $k$th row is the $k$th difference sequence for $f$. That is, $T_f[k, l] = \Delta^k f(l) = \Delta^{k-1} f(l+1) - \Delta^{k-1} f(l)$.

**Facts:**
**1.** The leftmost column of a difference table completely determines the entire table, via Newton's theorem (Fact 6, §3.4.2).

**2.** The difference table of an $n$th degree polynomial consists of $n + 1$ nonzero rows followed by all zero rows.

**Examples:**
**1.** If $A = 0, 1, 4, 9, 16, 25, \ldots$ is the sequence of squares of integers, then its difference sequence is $\Delta A = 1, 3, 5, 7, 9, \ldots$. Observe that $\Delta(x^2) = 2x + 1$.

**2.** The difference table for $x^{\underline{3}}$ is given by

|  | 0 | 1 | 2 | 3 | 4 | 5 | $\cdots$ |
|---|---|---|---|---|---|---|---|
| $\Delta^0 x^{\underline{3}} = x^{\underline{3}}$ | 0 | 0 | 0 | 6 | 24 | 60 | $\cdots$ |
| $\Delta^1 x^{\underline{3}} = 3x^{\underline{2}}$ | 0 | 0 | 6 | 18 | 36 | | $\cdots$ |
| $\Delta^2 x^{\underline{3}} = 6x^{\underline{1}}$ | 0 | 6 | 12 | 18 | | $\cdots$ | |
| $\Delta^3 x^{\underline{3}} = 6$ | 6 | 6 | 6 | $\cdots$ | | | |
| $\Delta^4 x^{\underline{3}} = 0$ | 0 | 0 | $\cdots$ | | | | |

**3.** The difference table for $x^3$ is given by

|  | 0 | 1 | 2 | 3 | 4 | 5 | $\cdots$ |
|---|---|---|---|---|---|---|---|
| $\Delta^0 x^3 = x^3$ | 0 | 1 | 8 | 27 | 64 | 125 | $\cdots$ |
| $\Delta^1 x^3 = 3x^2 + 3x + 1$ | 1 | 7 | 19 | 37 | 61 | | $\cdots$ |
| $\Delta^2 x^3 = 6x + 6$ | 6 | 12 | 18 | 24 | | $\cdots$ | |
| $\Delta^3 x^3 = 6$ | 6 | 6 | 6 | $\cdots$ | | | |
| $\Delta^4 x^3 = 0$ | 0 | 0 | $\cdots$ | | | | |

**4.** The difference table for $3^x$ is given by

|  | 0 | 1 | 2 | 3 | 4 | 5 | $\cdots$ |
|---|---|---|---|---|---|---|---|
| $\Delta^0 3^x = 3^x$ | 1 | 3 | 9 | 27 | 81 | 243 | $\cdots$ |
| $\Delta^1 3^x = 2 \cdot 3^x$ | 2 | 6 | 18 | 54 | 162 | $\cdots$ | |
| $\Delta^2 3^x = 4 \cdot 3^x$ | 4 | 12 | 36 | 108 | $\cdots$ | | |
| $\Delta^3 3^x = 8 \cdot 3^x$ | 8 | 24 | 72 | $\cdots$ | | | |
| $\Delta^4 3^x = 16 \cdot 3^x$ | 16 | 48 | $\cdots$ | | | | |
| $\vdots$ | $\vdots$ | | | | | | |
| $\Delta^k 3^x = 2^k 3^x$ | | | | | | | |

**5.** *Application to curve-fitting:* Find the polynomial $p(x)$ of smallest degree that passes through the points $(0,5), (1,5), (2,3), (3,5), (4,17), (5,45)$. The difference table for the sequence $5, 5, 3, 5, 17, 45$ is

| 5 | 5 | 3 | 5 | 17 | 45 | $\cdots$ |
|---|---|---|---|---|---|---|
| 0 | -2 | 2 | 12 | 28 | $\cdots$ | |
| -2 | 4 | 10 | 16 | $\cdots$ | | |
| 6 | 6 | 6 | $\cdots$ | | | |
| 0 | 0 | $\cdots$ | | | | |

Newton's theorem shows that the polynomial of smallest degree is $p(x) = 5 - x^{\underline{2}} + x^{\underline{3}} = x^3 - 4x^2 + 3x + 5$.

---

### 3.4.4   DIFFERENCE EQUATIONS

Difference equations are analogous to differential equations and many of the techniques are as fully developed. Difference equations provide a way to solve recurrence relations.

**Definitions:**

A **difference equation** is an equation involving the difference operator and/or higher-order differences of an unknown function.

An **antidifference** of the function $f$ is any function $g$ such that $\Delta g = f$. The notation $\Delta^{-1} f$ denotes any such function.

**Facts:**

**1.** Any recurrence relation (§3.3) can be expressed as a difference equation, and vice versa, by using Facts 4 and 5 of §3.4.1.

**2.** The solution to a recurrence relation can sometimes be easily obtained by converting it to a difference equation and applying difference methods.

**Examples:**

**1.** To find an antidifference of $10 \cdot 3^x$, refer to the table given in Example 2 of §3.4.1: $\Delta^{-1}(10 \cdot 3^x) = 5\Delta^{-1}(2 \cdot 3^x) = 5 \cdot 3^x + C$. (Also see Fact 5 of §3.5.3.)

**2.** To find an antidifference of $3x$, first express $x$ as $x^{\underline{1}}$ and then use the table found in Example 2 of §3.4.1: $\Delta^{-1} 3x = 3\Delta^{-1} x^{\underline{1}} = \frac{3}{2} x^{\underline{2}} + C = \frac{3}{2} x(x-1) + C$.

**3.** To find an antidifference of $x^2$, express $x^2$ as $x^{\underline{2}} + x^{\underline{1}}$ and then use the table found in Example 2 of §3.4.1: $\Delta^{-1}x^2 = \Delta^{-1}(x^{\underline{2}} + x^{\underline{1}}) = \Delta^{-1}x^{\underline{2}} + \Delta^{-1}x^{\underline{1}} = \frac{1}{3}x^{\underline{3}} + \frac{1}{2}x^{\underline{2}} + C = \frac{1}{3}x(x-1)(x-2) + \frac{1}{2}x(x-1) + C$.

**4.** The following are examples of difference equations:

$$\Delta^3 f(x) + x^4 \Delta^2 f(x) - f(x) = 0,$$
$$\Delta^3 f(x) + f(x) = x^2.$$

**5.** To solve the recurrence relation $a_{n+1} = a_n + 5^n$, $n \geq 0$, with $a_0 = 2$, first note that $\Delta a_n = 5^n$. Thus $a_n = \Delta^{-1}5^n = \frac{1}{4}(5^n) + C$. The initial condition $a_0 = 2$ now implies that $a_n = \frac{1}{4}(5^n + 7)$.

**6.** To solve the equation $a_{n+1} = (na_n + n)/(n+1)$, $n \geq 1$, the recurrence relation is first rewritten as $(n+1)a_{n+1} - na_n = n$, which is equivalent to $\Delta(na_n) = n$. Thus $na_n = \Delta^{-1}n = \frac{1}{2}n^{\underline{2}} + C$, which implies that $a_n = \frac{1}{2}(n-1) + C(\frac{1}{n})$.

**7.** To solve $a_n = 2a_{n-1} - a_{n-2} + 2^{n-2} + n - 2$, $n \geq 2$, with $a_0 = 4$, $a_1 = 5$, the recurrence relation is rewritten as $a_{n+2} - 2a_{n+1} + a_n = 2^n + n$, $n \geq 0$. Now, by applying Fact 4 of §3.4.1, the left-hand side may be replaced by $\Delta^2 a_n$. If the antidifference operator is applied twice to the resulting difference equation and the initial conditions are substituted, the solution obtained is

$$a_n = 2^n + \tfrac{1}{6}n^{\underline{3}} + c_1 n + c_2 = 2^n + \tfrac{1}{6}n(n-1)(n-2) + 3.$$

## 3.5   FINITE SUMS AND SUMMATION

Finite sums arise frequently in combinatorial mathematics and in the analysis of running times of algorithms. There are a few basic rules for transforming sums into possibly more tractable equivalent forms, and there is a calculus for evaluating these standard forms.

### 3.5.1   SIGMA NOTATION

A complex form of symbolic representation of discrete sums using the uppercase Greek letter $\Sigma$ (sigma) was introduced by Joseph Fourier in 1820 and has evolved into several variations.

**Definitions:**
The **sigma expression** $\sum_{i=a}^{b} f(i)$ has the value $f(a) + f(a+1) + \cdots + f(b-1) + f(b)$ if $a \leq b$ ($a, b \in \mathcal{Z}$), and 0 otherwise. In this expression, $i$ is the **index of summation** or **summation variable**, which ranges from the **lower limit** $a$ to the **upper limit** $b$. The interval $[a, b]$ is the **interval of summation**, and $f(i)$ is a **term** or **summand** of the summation.

A sigma expression $S_n = \sum_{i=0}^{n} f(i)$ is in **standardized form** if the lower limit is zero and the upper limit is an integer-valued expression.

A **sigma expression** $\sum_{k \in K} g(k)$ **over the set** $K$ has as its value the sum of all the values $g(k)$, where $k \in K$.

A *closed form* for a sigma expression with an indefinite number of terms is an algebraic expression with a fixed number of terms, whose value equals the sum.

A *partial sum* of the (standardized) sigma expression $S_n = \sum_{i=0}^{n} f(i)$ is the sigma expression $S_k = \sum_{i=0}^{k} f(i)$, where $0 \le k \le n$.

An *iterated sum* or *multiple sum* is an expression with two or more sigmas, as exemplified by the double sum $\sum_{i=c}^{d} \sum_{j=a}^{b} f(i,j)$. Evaluation proceeds from the innermost sigma outward.

A lower or upper limit for an inner sum of an iterated sum is *dependent* if it depends on an outer variable. Otherwise, that limit is *independent*.

### Examples:

**1.** The sum $f(1) + f(2) + f(3) + f(4) + f(5)$ may be represented as $\sum_{i=1}^{5} f(i)$.

**2.** Sometimes the summand is written as an expression, such as $\sum_{n=1}^{50}(n^2 + n)$, which means the same as $\sum_{n=1}^{50} f(n)$, where $f(n) = n^2 + n$. Brackets or parentheses can be used to distinguish what is in the summand of such an "anonymous function" from whatever is written to the immediate right of the sigma expression. They may be omitted when such a summand is very simple.

**3.** Sometimes the property defining the indexing set is written as part of the $\Sigma$, as in the expressions $\sum_{1 \le k \le n} a_k$ or $\sum_{k \in K} b_k$.

**4.** The right-hand side of the equation $\sum_{j=0}^{n} x^j = \frac{x^{n+1}-1}{x-1}$ is a closed form for the sigma expression on the left-hand side.

**5.** The operational meaning of the multiple sum with independent limits $\sum_{i=1}^{3} \sum_{j=2}^{4} \frac{i}{j}$ is first to expand the inner sum, obtaining the single sum $\sum_{i=1}^{3} \left[ \frac{i}{2} + \frac{i}{3} + \frac{i}{4} \right]$. Expansion of the outer sum then yields $\left[ \frac{1}{2} + \frac{1}{3} + \frac{1}{4} \right] + \left[ \frac{2}{2} + \frac{2}{3} + \frac{2}{4} \right] + \left[ \frac{3}{2} + \frac{3}{3} + \frac{3}{4} \right] = \frac{13}{2}$.

**6.** The multiple sum with dependent limits $\sum_{i=1}^{3} \sum_{j=i}^{4} \frac{i}{j}$ is evaluated by first expanding the inner sum, obtaining $\left[ \frac{1}{1} + \frac{1}{2} + \frac{1}{3} + \frac{1}{4} \right] + \left[ \frac{2}{2} + \frac{2}{3} + \frac{2}{4} \right] + \left[ \frac{3}{3} + \frac{3}{4} \right] = 6$.

---

## 3.5.2 ELEMENTARY TRANSFORMATION RULES FOR SUMS

Sums can be transformed using a few simple rules. A well-chosen sequence of transformations often simplifies evaluation.

### Facts:

**1.** *Distributivity rule:* $\sum_{k \in K} c a_k = c \sum_{k \in K} a_k$, for any constant $c$.

**2.** *Associativity rule:* $\sum_{k \in K} (a_k + b_k) = \sum_{k \in K} a_k + \sum_{k \in K} b_k$.

**3.** *Rearrangement rule:* $\sum_{k \in K} a_k = \sum_{k \in K} a_{\rho(k)}$, if $\rho$ is a permutation of the integers in $K$.

**4.** *Telescoping for sequences:* For any sequence $\{ a_j \mid j = 0, 1, \dots \}$, $\sum_{i=m}^{n} (a_{i+1} - a_i) = a_{n+1} - a_m$.

**5.** *Telescoping for functions:* For any function $f \colon \mathcal{N} \to \mathcal{R}$, $\sum_{i=m}^{n} \Delta f(i) = f(n+1) - f(m)$.

**6.** *Perturbation method:* Given a standardized sum $S_n = \sum_{i=0}^{n} f(i)$, form the equation

$$\sum_{i=0}^{n} f(i) + f(n+1) = f(0) + \sum_{i=1}^{n+1} f(i) = f(0) + \sum_{i=0}^{n} f(i+1).$$

Algebraic manipulation often leads to a closed form for $S_n$.

**7.** *Interchanging independent indices of a double sum:* When the lower and upper limits of the inner variable of a double sum are independent of the outer variable, the order of summation can be changed, simply by swapping the inner sigma, limits and all, with the outer sigma. That is,

$$\sum_{i=c}^{d} \sum_{j=a}^{b} f(i,j) = \sum_{j=a}^{b} \sum_{i=c}^{d} f(i,j).$$

**8.** *Interchanging dependent indices of a double sum:* When either the lower or upper limit of the inner variable $j$ of a double sum of an expression $f(i,j)$ is dependent on the outer variable $i$, the order of summation can still be changed by swapping the inner sum with the outer sum. However, the limits of the new inner variable $i$ must be written as functions of the new outer variable $j$ so that the entire set of pairs $(i,j)$ over which $f(i,j)$ is summed is the same as before. One particular case of interest is the interchange

$$\sum_{i=1}^{n} \sum_{j=i}^{n} f(i,j) = \sum_{j=1}^{n} \sum_{i=1}^{j} f(i,j).$$

**Examples:**

**1.** The following summation can be evaluated using Fact 4 (telescoping for sequences):

$$\sum_{i=1}^{n} \frac{1}{i(i+1)} = -\sum_{i=1}^{n}\left(\frac{1}{i+1} - \frac{1}{i}\right) = 1 - \frac{1}{n+1}.$$

**2.** Evaluate $S_n = \sum_{i=0}^{n} x^i$, using the perturbation method.

$$\sum_{i=0}^{n} x^i + x^{n+1} = x^0 + \sum_{i=1}^{n+1} x^i = 1 + x \sum_{i=1}^{n+1} x^{i-1},$$

$$S_n + x^{n+1} = 1 + x \sum_{i=0}^{n} x^i = 1 + xS_n,$$

giving $S_n = \frac{x^{n+1}-1}{x-1}$.

**3.** Evaluate $S_n = \sum_{i=0}^{n} i2^i$, using the perturbation method.

$$\sum_{i=0}^{n} i2^i + (n+1)2^{n+1} = 0 \cdot 2^0 + \sum_{i=1}^{n+1} i2^i = \sum_{i=0}^{n} (i+1)2^{i+1},$$

$$S_n + (n+1)2^{n+1} = 2\sum_{i=0}^{n} i2^i + 2\sum_{i=0}^{n} 2^i = 2S_n + 2(2^{n+1} - 1),$$

giving $S_n = (n+1)2^{n+1} - 2(2^{n+1} - 1) = (n-1)2^{n+1} + 2.$

**4.** Interchange independent indices of a double sum:

$$\sum_{i=1}^{3} \sum_{j=2}^{4} \frac{i}{j} = \sum_{j=2}^{4} \sum_{i=1}^{3} \frac{i}{j} = \sum_{j=2}^{4}\left[\frac{1}{j} + \frac{2}{j} + \frac{3}{j}\right] = \sum_{j=2}^{4} \frac{6}{j} = 6\sum_{j=2}^{4} \frac{1}{j} = 6\left[\frac{1}{2} + \frac{1}{3} + \frac{1}{4}\right] = \frac{13}{2}.$$

**5.** Interchange dependent indices of a double sum:

$$\sum_{i=1}^{3}\sum_{j=i}^{3}\frac{i}{j} = \sum_{j=1}^{3}\sum_{i=1}^{j}\frac{i}{j} = \sum_{j=1}^{3}\frac{1}{j}\sum_{i=1}^{j}i = \frac{1}{1}\cdot 1 + \frac{1}{2}\cdot 3 + \frac{1}{3}\cdot 6 = \frac{9}{2}.$$

### 3.5.3   ANTIDIFFERENCES AND SUMMATION FORMULAS

Some standard combinatorial functions analogous to polynomials and exponential functions facilitate the development of a calculus of finite differences, analogous to the differential calculus of continuous mathematics. The *fundamental theorem of discrete calculus* is useful in deriving a number of summation formulas.

**Definitions:**

An **antidifference** of the function $f$ is any function $g$ such that $\Delta g = f$, where $\Delta$ is the difference operator (§3.4.1). The notation $\Delta^{-1}f$ denotes any such function.

The **indefinite sum** of the function $f$ is the infinite family of all antidifferences of $f$. The notation $\sum f(x)\delta x + c$ is sometimes used for the indefinite sum to emphasize the analogy with integration.

**Facts:**

**1.** *Fundamental theorem of discrete calculus:*

$$\sum_{k=a}^{b} f(k) = \Delta^{-1}f(k)\Big|_{a}^{b+1} = \Delta^{-1}f(b+1) - \Delta^{-1}f(a).$$

*Note:* The upper evaluation point is one more than the upper limit of the sum.

**2.** *Linearity:* $\Delta^{-1}(\alpha f + \beta g) = \alpha\Delta^{-1}f + \beta\Delta^{-1}g$, for any constants $\alpha$ and $\beta$.

**3.** *Summation by parts:*

$$\sum_{i=a}^{b} f(i)\Delta g(i) = f(b+1)g(b+1) - f(a)g(a) - \sum_{i=a}^{b} g(i+1)\Delta f(i).$$

This result, which generalizes Fact 5 of §3.5.2, is a direct analogue of integration by parts in continuous analysis.

**4.** *Abel's transformation:*

$$\sum_{k=1}^{n} f(k)g(k) = f(n+1)\sum_{k=1}^{n} g(k) - \sum_{k=1}^{n}\left(\Delta f(k)\sum_{r=1}^{k} g(r)\right).$$

**5.** The following table gives the antidifferences of selected functions. In this table, $H_x$ indicates the harmonic sum (§3.4.1), $x^{\underline{n}}$ is the $n$th falling power of $x$ (§3.4.2), and $\left\{{n\atop k}\right\}$ is a Stirling subset number (§2.5.2).

| $f(x)$ | $\Delta^{-1}f(x)$ | $f(x)$ | $\Delta^{-1}f(x)$ |
|---|---|---|---|
| $\binom{x}{n}$ | $\binom{x}{n+1}$ | $(x+a)^{\underline{n}}$ | $\frac{(x+a)^{\underline{n+1}}}{n+1}$, $n \neq -1$ |
| $(x+a)^{\underline{-1}}$ | $H_{x+a}$ | $a^x$ | $\frac{a^x}{(a-1)}$, $a \neq 1$ |
| $a^x$ | $\frac{a^x}{(a-1)}$, $a \neq 1$ | $xa^x$ | $\frac{a^x}{(a-1)}\left(x - \frac{a}{a-1}\right)$, $a \neq 1$ |
| $x^n$ | $\sum_{k=1}^{n}\frac{\left\{{n\atop k}\right\}}{k+1}x^{\underline{k+1}}$ | $(-1)^x$ | $\frac{1}{2}(-1)^{x+1}$ |
| $\sin x$ | $\frac{-1}{2\sin(\frac{1}{2})}\cos(x - \frac{1}{2})$ | $\cos x$ | $\frac{1}{2\sin(\frac{1}{2})}\sin(x - \frac{1}{2})$ |

**6.** The following table gives finite sums of selected functions.

| summation | formula | summation | formula |
|---|---|---|---|
| $\sum_{k=1}^{n} k^{\underline{m}}$ | $\frac{(n+1)^{\underline{m+1}}}{m+1}$, $m \neq -1$ | $\sum_{k=1}^{n} k^m$ | $\sum_{j=1}^{m} \frac{\{{}_j^m\}(n+1)^{\underline{j+1}}}{j+1}$ |
| $\sum_{k=0}^{n} a^k$ | $\frac{a^{n+1}-1}{a-1}$, $a \neq 1$ | $\sum_{k=1}^{n} ka^k$ | $\frac{(a-1)(n+1)a^{n+1}-a^{n+2}+a}{(a-1)^2}$, $a \neq 1$ |
| $\sum_{k=1}^{n} \sin k$ | $\frac{\sin(\frac{n+1}{2})\sin(\frac{n}{2})}{\sin(\frac{1}{2})}$ | $\sum_{k=1}^{n} \cos k$ | $\frac{\cos(\frac{n+1}{2})\sin(\frac{n}{2})}{\sin(\frac{1}{2})}$ |

**Examples:**

**1.** $\sum_{k=1}^{n} k^3 = \sum_{k=1}^{n}(k^{\underline{1}} + 3k^{\underline{2}} + k^{\underline{3}}) = \left(\frac{k^{\underline{2}}}{2} + k^{\underline{3}} + \frac{k^{\underline{4}}}{4}\right)\Big|_1^{n+1} = \frac{n^2(n+1)^2}{4}$.

**2.** To evaluate $\sum_{k=1}^{n} k(k+2)(k+3)$, first rewrite its summand:

$$\sum_{k=1}^{n} k(k+2)(k+3) = \Delta^{-1}[(k+1-1)(k+2)(k+3)]\Big|_1^{n+1}$$

$$= \left[\Delta^{-1}(k+3)^{\underline{3}} - \Delta^{-1}(k+3)^{\underline{2}}\right]\Big|_1^{n+1}$$

$$= \left[\frac{(k+3)^{\underline{4}}}{4} - \frac{(k+3)^{\underline{3}}}{3}\right]\Big|_1^{n+1}$$

$$= \frac{(n+4)^{\underline{4}}}{4} - \frac{(n+4)^{\underline{3}}}{3} + 2$$

$$= \frac{(n+4)(n+3)(n+2)(3n-1)+24}{12}.$$

**3.** $\sum_{k=1}^{n} k3^k = \Delta^{-1}(k3^k)\Big|_1^{n+1} = 3^k\left[\frac{k}{2} - \frac{3}{4}\right]\Big|_1^{n+1} = \frac{(2n-1)3^{n+1}+3}{4}$.

**4.** Summation by parts can be used to calculate $\sum_{j=0}^{n} jx^j$, using $f(j) = j$ and $\Delta g(j) = x^j$. Thus $g(j) = x^j/(x-1)$, and Fact 3 yields

$$\sum_{j=0}^{n} jx^j = \frac{(n+1)x^{n+1}}{(x-1)} - 0 - \sum_{j=0}^{n}\frac{x^{j+1}}{(x-1)} = \frac{(n+1)x^{n+1}}{(x-1)} - \frac{x}{x-1}\sum_{j=0}^{n} x^j$$

$$= \frac{(n+1)x^{n+1}}{(x-1)} - \frac{x}{x-1}\frac{x^{n+1}-1}{(x-1)} = \frac{(n+1)(x-1)x^{n+1}-x^{n+2}+x}{(x-1)^2}.$$

**5.** Summation by parts also yields an antiderivative of $x3^x$:

$$\Delta^{-1}(x3^x) = \Delta^{-1}\left(x\Delta(\tfrac{1}{2}\cdot 3^x)\right) = \tfrac{1}{2}x3^x - \Delta^{-1}(\tfrac{1}{2}\cdot 3^{x+1}\cdot 1) = 3^x\left(\tfrac{x}{2} - \tfrac{3}{4}\right).$$

---

### 3.5.4  STANDARD SUMS

Many useful summation formulas are derivable by combinations of elementary manipulation and finite calculus. Such sums can be expressed in various ways, using different combinatorial coefficients. (See §3.1.8.)

**Definition:**
The **power sum** $S^k(n) = \sum_{j=1}^{n} j^k = 1^k + 2^k + 3^k + \cdots + n^k$ is the sum of the $k$th powers of the first $n$ positive integers.

**Facts:**

**1.** $S^k(n)$ is a polynomial in $n$ of degree $k+1$ with leading coefficient $1\frac{1}{k+1}$. The continuous analogue of this fact is the familiar $\int_a^b x^k dx = \frac{1}{k+1}(b^{k+1} - a^{k+1})$.

**2.** The power sum $S^k(n)$ can be expressed using the Bernoulli polynomials (§3.1.4) as

$$S^k(n) = \frac{1}{k+1}[B_{k+1}(n+1) - B_{k+1}(0)].$$

**3.** When $S^k(n)$ is expressed in terms of binomial coefficients with the second entry fixed at $k+1$, the coefficients are the Eulerian numbers (§3.1.5).

$$S^k(n) = \sum_{i=0}^{k-1} E(k,i)\binom{n+i+1}{k+1}.$$

**4.** When $S^k(n)$ is expressed in terms of binomial coefficients with the first entry fixed at $n+1$, the coefficients are products of factorials and Stirling subset numbers (§2.5.2).

$$S^k(n) = \sum_{i=1}^{k} i!\left\{{k \atop i}\right\}\binom{n+1}{i+1}.$$

**5.** Formulas for the power sums described in Facts 1, 3, and 4 are given in the next three tables, for small values of $k$.

| summation | formula |
|-----------|---------|
| $\sum_{j=1}^n j$ | $\frac{1}{2}n(n+1)$ |
| $\sum_{j=1}^n j^2$ | $\frac{1}{6}n(n+1)(2n+1)$ |
| $\sum_{j=1}^n j^3$ | $\frac{1}{4}n^2(n+1)^2$ |
| $\sum_{j=1}^n j^4$ | $\frac{1}{30}n(n+1)(2n+1)(3n^2+3n-1)$ |
| $\sum_{j=1}^n j^5$ | $\frac{1}{12}n^2(n+1)^2(2n^2+2n-1)$ |
| $\sum_{j=1}^n j^6$ | $\frac{1}{42}n(n+1)(2n+1)(3n^4+6n^3-n^2-3n+1)$ |
| $\sum_{j=1}^n j^7$ | $\frac{1}{24}n^2(n+1)^2(3n^4+6n^3-n^2-4n+2)$ |
| $\sum_{j=1}^n j^8$ | $\frac{1}{90}n(n+1)(2n+1)(5n^6+15n^5+5n^4-15n^3-n^2+9n-3)$ |
| $\sum_{j=1}^n j^9$ | $\frac{1}{20}n^2(n+1)^2(2n^6+6n^5+n^4-8n^3+n^2+6n-3)$ |

| summation | formula |
|-----------|---------|
| $\sum_{j=1}^n j$ | $\binom{n+1}{2}$ |
| $\sum_{j=1}^n j^2$ | $\binom{n+1}{3} + \binom{n+2}{3}$ |
| $\sum_{j=1}^n j^3$ | $\binom{n+1}{4} + 4\binom{n+2}{4} + \binom{n+3}{4}$ |
| $\sum_{j=1}^n j^4$ | $\binom{n+1}{5} + 11\binom{n+2}{5} + 11\binom{n+3}{5} + \binom{n+4}{5}$ |
| $\sum_{j=1}^n j^5$ | $\binom{n+1}{6} + 26\binom{n+2}{6} + 66\binom{n+3}{6} + 26\binom{n+4}{6} + \binom{n+5}{6}$ |

| summation | formula |
|-----------|---------|
| $\sum_{j=1}^n j$ | $\binom{n+1}{2}$ |
| $\sum_{j=1}^n j^2$ | $\binom{n+1}{2} + 2\binom{n+1}{3}$ |
| $\sum_{j=1}^n j^3$ | $\binom{n+1}{2} + 6\binom{n+1}{3} + 6\binom{n+1}{4}$ |
| $\sum_{j=1}^n j^4$ | $\binom{n+1}{2} + 14\binom{n+1}{3} + 36\binom{n+1}{4} + 24\binom{n+1}{5}$ |
| $\sum_{j=1}^n j^5$ | $\binom{n+1}{2} + 30\binom{n+1}{3} + 150\binom{n+1}{4} + 240\binom{n+1}{5} + 120\binom{n+1}{6}$ |

**Examples:**

**1.** To find the third power sum $S^3(n) = \sum_{j=1}^{n} j^3$ via Fact 2, use the Bernoulli polynomial $B_4(x) = x^4 - 2x^3 + x^2 - \frac{1}{30}$ from the table found in §3.1.4. Thus

$$S^3(n) = \frac{1}{4}\left[B_4(x)\right]\Big|_0^{n+1} = \frac{(n+1)^4 - 2(n+1)^3 + (n+1)^2}{4} = \frac{n^2(n+1)^2}{4}.$$

**2.** Power sums can be found using antidifferences and Stirling numbers of both types. For example, to find $S^3(n) = \sum_{x=1}^{n} x^3$ first compute

$$\Delta^{-1}x^3 = \Delta^{-1}\left(\left\{^3_1\right\}x^{\underline{1}} + \left\{^3_2\right\}x^{\underline{2}} + \left\{^3_3\right\}x^{\underline{3}}\right) = \frac{x^{\underline{2}}}{2} + x^{\underline{3}} + \frac{x^{\underline{4}}}{4}.$$

Each term $x^{\underline{m}}$ is then expressed in terms of ordinary powers of $x$

$$x^{\underline{2}} = \left[^2_2\right]x^2 - \left[^2_1\right]x^1 = x^2 - x,$$
$$x^{\underline{3}} = \left[^3_3\right]x^3 - \left[^3_2\right]x^2 + \left[^3_1\right]x^1 = x^3 - 3x^2 + 2x,$$
$$x^{\underline{4}} = \left[^4_4\right]x^4 - \left[^4_3\right]x^3 + \left[^4_2\right]x^2 - \left[^4_1\right]x^1 = x^4 - 6x^3 + 11x^2 - 6x,$$

so $\Delta^{-1}x^3 = \frac{1}{2}(x^2 - x) + (x^3 - 3x^2 + 2x) + \frac{1}{4}(x^4 - 6x^3 + 11x^2 - 6x) = \frac{1}{4}(x^4 - 2x^3 + x^2)$. Evaluating this antidifference between the limits $x = 1$ and $x = n + 1$ gives $S^3(n) = \frac{1}{4}n^2(n+1)^2$. See §3.5.3, Fact 1.

# 3.6   ASYMPTOTICS OF SEQUENCES

An exact formula for the terms of a sequence may be unwieldy. For example, it is difficult to estimate the magnitude of the central binomial coefficient $\binom{2n}{n} = \frac{(2n)!}{(n!)^2}$ from the definition of the factorial function alone. On the other hand, Stirling's approximation formula (§3.6.2) leads to the asymptotic estimate $\frac{4^n}{\sqrt{\pi n}}$. In applying asymptotic analysis, various "rules of thumb" help bypass tedious derivations. In practice, these rules almost always lead to correct results that can be proved by more rigorous methods. In the following discussions of asymptotic properties, the parameter tending to infinity is denoted by $n$. Both the subscripted notation $a_n$ and the functional notation $f(n)$ are used to denote a sequence. The notation $f(n) \sim g(n)$ ($f$ is *asymptotic* to $g$) means that $f(n) \neq 0$ for sufficiently large $n$ and $\lim_{n\to\infty} \frac{g(n)}{f(n)} = 1$.

## 3.6.1   APPROXIMATE SOLUTIONS TO RECURRENCES

Although recurrences are a natural source of sequences, they often yield only crude asymptotic information. As a general rule, it helps to derive a summation or a generating function from the recurrence before obtaining asymptotic estimates.

**Facts:**

**1.** *Rule of thumb:* Suppose a recurrence for some sequence $a_n$ can be transformed into a recurrence for a related sequence $b_n$, so that the transformed sequence is approximately homogeneous and linear with constant coefficients (§3.3). Let $\rho$ be the largest positive root of the characteristic equation for the homogeneous constant coefficient recurrence. Then it is probably true that $\frac{b_{n+1}}{b_n} \sim \rho$; i.e., $b_n$ grows roughly like $\rho^n$.

**2.** Nonlinear recurrences are *not* covered by Fact 1.

**3.** Recurrences without fixed degree such as divide-and conquer recurrences (§3.3.5), in which the difference between the largest and smallest subscripts is unbounded, are *not* covered by Fact 1. See [GrKn07] for appropriate techniques.

**Examples:**

**1.** Consider the recurrence $D_{n+1} = n(D_n + D_{n-1})$ for $n \geq 1$, and define $d_n = \frac{D_n}{n!}$. Then $d_{n+1} = \frac{n}{n+1}d_n + \frac{1}{n+1}d_{n-1}$, which is quite close to the constant coefficient recurrence $\hat{d}_{n+1} = \hat{d}_n$. Since the characteristic root for this latter approximate recurrence is $\rho = 1$, Fact 1 suggests that $\frac{d_{n+1}}{d_n} \sim 1$, which implies that $d_n$ is close to constant. Thus, we expect the original variable $D_n$ to grow like $n!$. Indeed, if the initial conditions are $D_0 = D_1 = 1$, then $D_n = n!$. With initial conditions $D_0 = 1$, $D_1 = 0$, then $D_n$ is the number of *derangements* of $n$ objects (§2.4.2), in which case $D_n$ is the closest integer to $\frac{n!}{e}$ for $n \geq 1$.

**2.** The accuracy of Example 1 is unusual. By way of contrast, the number $I_n$ of *involutions* of an $n$-set (§2.8.1) satisfies the recurrence $I_{n+1} = I_n + nI_{n-1}$ for $n \geq 1$ with $I_0 = I_1 = 1$. By defining $i_n = I_n/(n!)^{1/2}$, then

$$i_{n+1} = \frac{i_n}{(n+1)^{1/2}} + \frac{i_{n-1}}{(1+1/n)^{1/2}},$$

which is nearly the same as the constant coefficient recurrence $\hat{i}_{n+1} = \hat{i}_{n-1}$. The characteristic equation $\rho^2 = 1$ has roots $\pm 1$, so Fact 1 suggests that $i_n$ is nearly constant and hence that $I_n$ grows like $\sqrt{n!}$. The approximation in this case is not so good, because $I_n/\sqrt{n!} \sim e^{\sqrt{n}}/(8\pi en)^{1/4}$, which is not a constant.

## 3.6.2  ANALYTIC METHODS FOR DERIVING ASYMPTOTIC ESTIMATES

Concepts and methods from continuous mathematics can be useful in analyzing the asymptotic behavior of sequences. The notion of a generating function (§3.2) will play an important role here.

**Definitions:**

The **radius of convergence** of the series $\sum a_n x^n$ is the number $r$ such that the series converges for all $|x| < r$ and diverges for all $|x| > r$, where $0 \leq r \leq \infty$.

Given a series $A(x) = \sum a_n x^n$, we write $[x^n]A(x) = a_n$.

The **gamma function** is the function $\Gamma(x) = \int_0^\infty t^{x-1}e^{-t}\,dt$.

**Facts:**

**1.** Stirling's approximation:  $n! \sim \sqrt{2\pi n}(\frac{n}{e})^n$.

**2.** $\Gamma(x+1) = x\Gamma(x)$, $\Gamma(n+1) = n!$, and $\Gamma(\frac{1}{2}) = \sqrt{\pi}$.

**3.** The radius of convergence of $\sum a_n x^n$ is given by $\frac{1}{r} = \limsup_{n\to\infty} |a_n|^{1/n}$.

**4.** From Fact 3, it follows that $|a_n|$ tends to behave like $r^{-n}$. Most analytic methods are refinements of this idea: while the absolute value of the radius of convergence gives the exponential growth of the coefficients, the nature of the point where the series ceases to be analytic gives the subexponential growth of the coefficients. See Fact 7.

**5.** The behavior of $f(z)$ near singularities on its circle of convergence determines the dominant asymptotic behavior of the coefficients of $f$. Estimates are often based on Cauchy's integral formula: $a_n = \oint f(z)z^{-n-1}\,dz$.

**6.** *Rule of thumb:* Consider the set of values of $x$ for which $f(x) = \sum a_n x^n$ is either infinite or undefined, or involves computing a nonintegral power of 0. The absolute value of the least such $x$ is normally the radius of convergence of $f(x)$. If there is no such $x$, then $r = \infty$.

**7.** *Rule of thumb:* As seen in Fact 4, the *modulus* of the radius of convergence of $f(x) = \sum a_n x^n$ determines the exponential growth of $a_n$. Additionally, the *nature* of the function at the radius of convergence determines the subexponential growth of the coefficients.

**8.** *Rule of thumb:* A very general example of Fact 7 is the following. Suppose that $0 < r < \infty$ is the radius of convergence of $f(x)$, that $g(x)$ has a larger radius of convergence, and that

$$f(x) - g(x) \sim C\left(-\ln(1-\tfrac{x}{r})\right)^b \left(1-\tfrac{x}{r}\right)^c \quad \text{as } x \to r^-$$

for some constants $C$, $b$, and $c$, where it is not the case that both $b = 0$ and $c$ is a nonnegative integer. (Often $g(x) = 0$.) Then it is probably true that

$$[x^n]f(x) \sim \begin{cases} C\binom{n-c-1}{n}(\ln n)^b r^{-n}, & \text{if } c \neq 0, \\ Cb\,\frac{(\ln n)^{b-1}}{n}\,r^{-n}, & \text{if } c = 0. \end{cases}$$

**9.** *Rule of thumb:* Let $a(x) = \frac{d\ln f(x)}{d\ln x}$ and $b(x) = \frac{da(x)}{d\ln x}$. Suppose that $a(r_n) = n$ has a solution with $0 < r_n < r$ and that $b(r_n) \in o(n^2)$. Then it is probably true that

$$a_n \sim \frac{f(r_n)r_n^{-n}}{\sqrt{2\pi b(r_n)}}.$$

**Examples:**

**1.** The number $D_n$ of derangements (§2.4.2) has the exponential generating function $f(x) = \sum D_n \frac{x^n}{n!} = \frac{e^{-x}}{1-x}$. Since evaluation for $x = 1$ involves division by 0, it follows that $r = 1$. Since $\frac{e^{-x}}{1-x} \sim \frac{e^{-1}}{1-x}$ as $x \to 1^-$, take $g(x) = 0$, $C = e^{-1}$, $b = 0$, and $c = -1$. Fact 8 suggests that $D_n \sim \frac{n!}{e}$, which is correct.

**2.** The number $b_n$ of left-right binary $n$-leaved trees has the generating function $f(x) = \frac{1}{2}(1 - \sqrt{1-4x})$. (See §9.3.3.) In this case $r = \frac{1}{4}$ since $f(\frac{1}{4})$ requires computing a fractional power of 0. Take $g(x) = \frac{1}{2}$, $C = \frac{1}{2}$, $b = 0$, and $c = \frac{1}{2}$ to suspect from Fact 8 that

$$b_n \sim -\frac{1}{2}\binom{n-\frac{3}{2}}{n}4^n = \frac{-\Gamma(n-\frac{1}{2})4^n}{2\Gamma(n+1)\Gamma(-\frac{1}{2})} \sim \frac{4^{n-1}}{\sqrt{\pi n^3}},$$

which is valid. (Facts 1 and 2 have also been used.) This estimate converges rather rapidly—by the time $n = 40$, the estimate is less than 0.1% below $b_{40}$.

**3.** Since $\sum \frac{x^n}{n!} = e^x$, $n!$ can be estimated by taking $a(x) = b(x) = x$ and $r_n = n$ in Fact 9. This gives $\frac{1}{n!} \sim \frac{e^n n^{-n}}{\sqrt{2\pi n}}$, which is Stirling's asymptotic formula.

**4.** The number $B_n$ of partitions of an $n$-set (§2.5.2) satisfies $\sum B_n \frac{x^n}{n!} = \exp(e^x - 1)$. In this case, $r = \infty$. Since $a(x) = xe^x$ and $b(x) = x(x+1)e^x$, it follows that $r_n$ is the solution to $r_n \exp(r_n) = n$ and that $b(r_n) = (r_n+1)n \sim nr_n \in o(n^2)$. Fact 9 suggests

$$B_n \sim \frac{n!\exp(e^{r_n} - 1)}{r_n^n\sqrt{2\pi nr_n}} = \frac{n!\exp(n/r_n - 1)}{r_n^n\sqrt{2\pi nr_n}}.$$

This estimate is correct, though the estimate converges quite slowly, as shown in the following table.

| $n$ | 10 | 20 | 100 | 200 |
|---|---|---|---|---|
| estimate | $1.49 \times 10^5$ | $6.33 \times 10^{13}$ | $5.44 \times 10^{115}$ | $7.01 \times 10^{275}$ |
| $B_n$ | $1.16 \times 10^5$ | $5.17 \times 10^{13}$ | $4.76 \times 10^{115}$ | $6.25 \times 10^{275}$ |
| ratio | 1.29 | 1.22 | 1.14 | 1.12 |

Improved asymptotic estimates exist.

**5.** Analytic methods can sometimes be used to obtain asymptotic results when only a functional equation is available (see also §3.6.3.) For example, if $a_n$ is the number of $n$-leaved rooted trees in which each non-leaf node has exactly two children (with left and right not distinguished), the generating function for $a_n$ satisfies $f(x) = x + (f(x)^2 + f(x^2))/2$, from which it can be deduced that $a_n \sim Cn^{-3/2}r^{-n}$, where $r = 0.4026975\ldots$ and $C = 0.31877\ldots$ can easily be computed to any desired degree of accuracy. See [BeWi06] for more information.

## 3.6.3   GENERATING FUNCTION SCHEMES

Specific general expressions defining a given generating function can be applied to automatically give asymptotic results for their coefficients.

**Facts:**
**1.** *Rule of thumb:* Assume that the generating function $A(x) = \sum a_n x^n$ satisfies the equation $A(x) = x\phi(A(x))$, where $\phi(t)$ is a nonlinear function, analytic at 0 with $\phi(0) \neq 0$ and $[t^n]\phi(t) \geq 0$ for all values of $n > 0$. This is the so-called *Lagrange scheme*. Then, by means of the *Lagrange inversion formula*, we have

$$a_n = \frac{1}{n}[t^{n-1}]\phi(t)^n.$$

Sometimes this exact formula is useless, and one can exploit analytic methods to provide asymptotic estimates for $a_n$. More precisely, if $R$ is the radius of convergence of $\phi(t)$ and there exists a solution $\tau$ to the equation $\phi(\tau) = \tau\phi'(\tau)$ with $0 < \tau < R$, then

$$a_n \sim \sqrt{\frac{\phi(\tau)}{\phi''(\tau)}} \frac{1}{\sqrt{2\pi n^3}} \left(\frac{\phi(\tau)}{\tau}\right)^n.$$

**2.** *Rule of thumb:* Systems of functional equations have the same asymptotic form as in Fact 1. Let $A_1(x), \ldots, A_k(x)$ satisfy a functional system of equations of the form $A_1(x) = F_1(x, A_1(x), \ldots, A_k(x))$, $\ldots$, $A_k(x) = F_k(x, A_1(x), \ldots, A_k(x))$, where the functions $F_i$ are nonlinear. Additionally, the system is *irreducible*, namely one cannot simplify it by eliminating equations. Then, under certain technical but general conditions for the functions $F_i$ it is probably true that, for some constants $C_i$, the exponential growth of the coefficients is of the form

$$[x^n]A_i(x) \sim C_i \frac{1}{\sqrt{n^3}} \rho^{-n}.$$

The constant $\rho$ can be computed analytically as a solution of a system of equations. See [Dr97] for details.

**Examples:**

**1.** The Catalan numbers $C_n$ (see §3.1.3) are encoded by the generating function $C(x)$, which satisfies the functional equation $C(x) = 1 + xC(x)^2$. Writing $C(x) - 1 = U(x)$ we obtain $U(x) = x(1 + U(x))^2$. Then $\phi(t) = (1+t)^2$, and Fact 1 applies directly. This yields $\tau = 1$ and $\phi(\tau)/\tau = 4$. Hence, by Fact 1

$$[x^n]U(x) = [x^n]C(x) = \frac{1}{n+1}\binom{2n}{n} \sim \frac{1}{\sqrt{\pi n^3}}4^n.$$

Compare this result with Example 2 of §3.6.2. Observe that the asymptotic estimate can be obtained either by applying Stirling's formula to the Catalan number (expressed using a binomial coefficient), or directly by using Fact 1.

**2.** Consider the family of rooted trees where each internal vertex has outdegree equal to 2, 3 or 4. Then the generating function $T(x)$ associated with this family satisfies the equation $T(x) = x(1 + T(x)^2 + T(x)^3 + T(x)^4)$, where $x$ encodes vertices (see [FlSe09]). In this case, $\phi(t) = (1 + t^2 + t^3 + t^4)$ and the positive solution to the equation $\phi(\tau) = \tau\phi'(\tau)$ is given by $\tau = 0.56773\ldots$. Finally, the growth constant is $\phi(\tau)/\tau = 2.83443\ldots$.

---

### 3.6.4  ASYMPTOTIC ESTIMATES OF MULTIPLY-INDEXED SEQUENCES

Asymptotic estimates for multiply-indexed sequences are considerably more difficult to obtain. To begin with, the meaning of a formula such as

$$\binom{n}{k} \sim \frac{2^n}{\sqrt{\pi n/2}}\exp\left(-\frac{(n-2k)^2}{2n}\right)$$

must be carefully stated, because both $n$ and $k$ are tending to $\infty$, and the formula is valid only when this happens in such a way that $|2k - n| \in o(n^{3/4})$.

**Facts:**

**1.** Most estimates of multiple summations are based on summing over one index at a time.

**2.** Some analytic results are available in the research literature [Od95]. In particular, Drmota [Dr97] shows that Fact 2 in §3.6.3 can be extended in some situations in order to deal with multiply-indexed sequences.

**3.** Very recently, some tools arising from multidimensional complex analysis have been exploited to study such questions [PeWi13].

**Example:**

**1.** Consider the number of rooted plane trees, expressed in terms of the number of vertices $n$. This sequence is given by the Catalan numbers. We consider also the number of leaves $r$. Denote by $t_{n,r}$ the number of rooted plane trees with $n$ vertices and $r$ leaves, and write $T(x, u) = \sum t_{n,r} x^n u^r$. Then it can be shown that $T(x, u)$ satisfies the functional equation

$$T(x, u) = xu + \frac{xT(x, u)}{1 - T(x, u)}.$$

Application of the Lagrange inversion formula (see Fact 1 in §3.6.3) gives

$$t_{n,r} = \frac{1}{n-1}\binom{n-1}{n-r}\binom{n-1}{r}.$$

Using Stirling's approximation formula we get

$$t_{n,r} \sim \frac{1}{\pi n^2} 2^{2n-1} \exp\left(-\frac{(n-2k)^2}{n}\right).$$

Such an estimate can be obtained as well by means of the tools of [Dr97] using complex analytic techniques.

## 3.7 MECHANICAL SUMMATION PROCEDURES

This section describes mechanical procedures that have been developed to evaluate sums of terms involving binomial coefficients and related factors. These procedures can not only be used to find explicit formulas for many sums, but can also be used to show that no simple closed formulas exist for certain sums. The invention of these mechanical procedures has been a surprising development in combinatorics. The material presented here is mostly adapted from [PeWiZe96], a comprehensive source for material on this topic.

### 3.7.1 HYPERGEOMETRIC SERIES

**Definitions:**

A **geometric series** is a series of the form $\sum_{k=0}^{\infty} a_k$ where the ratio between two consecutive terms is a constant, i.e., where the ratio $\frac{a_{k+1}}{a_k}$ is a constant for all $k = 0, 1, 2, \ldots$.

A **hypergeometric series** is a series of the form $\sum_{k=0}^{\infty} t_k$ where $t_0 = 1$ and the ratio of two consecutive terms is a rational function of the summation index $k$, i.e., the ratio $\frac{t_{k+1}}{t_k} = \frac{P(k)}{Q(k)}$ where $P(k)$ and $Q(k)$ are polynomials in the integer $k$. The terms of a hypergeometric series are called **hypergeometric terms**.

When the numerator $P(k)$ and denominator $Q(k)$ of this ratio are completely factored to give

$$\frac{P(k)}{Q(k)} = \frac{(k+a_1)(k+a_2)\ldots(k+a_p)}{(k+b_1)(k+b_2)\ldots(k+b_q)(k+1)},$$

where $x$ is a constant, this hypergeometric series is denoted by

$$_pF_q = \begin{bmatrix} a_1 & a_2 & \ldots & a_p \\ b_1 & b_2 & \ldots & b_q \end{bmatrix}; x \end{bmatrix}.$$

**Note:** If there is no factor $k+1$ in the denominator $Q(k)$ when it is factored, by convention the factor $k+1$ is added to both the numerator $P(k)$ and denominator $Q(k)$. Also, a horizontal dash is used to indicate the absence of factors in the numerator or in the denominator.

The hypergeometric terms $s_n$ and $t_n$ are **similar**, denoted $s_n \sim t_n$, if their ratio $s_n/t_n$ is a rational function of $n$. Otherwise, these terms are called **dissimilar**.

**Facts:**

**1.** A geometric series is also a hypergeometric series.

**2.** If $s_n$ is a hypergeometric term, then $\frac{1}{s_n}$ is also a hypergeometric term. (Equivalently, if $\sum_{k=0}^{\infty} s_n$ is a hypergeometric series, then $\sum_{k=0}^{\infty} \frac{1}{s_n}$ also is.)

**3.** In common usage, instead of stating that the series $\sum_{k=0}^{\infty} s_n$ is a hypergeometric series, it is stated that $s_n$ is a hypergeometric term. This means exactly the same thing.

**4.** If $s_n$ and $t_n$ are hypergeometric terms, then $s_n t_n$ is a hypergeometric term. (Equivalently, if $\sum_{k=0}^{\infty} s_n$ and $\sum_{k=0}^{\infty} t_n$ are hypergeometric series, then $\sum_{k=0}^{\infty} s_n t_n$ is a hypergeometric series.)

**5.** If $s_n$ is a hypergeometric term and $s_n$ is not a constant, then $s_{n+1} - s_n$ is a hypergeometric term similar to $s_n$.

**6.** If $s_n$ and $t_n$ are hypergeometric terms and $s_n + t_n \neq 0$ for all $n$, then $s_n + t_n$ is hypergeometric if and only if $s_n$ and $t_n$ are similar.

**7.** If $t_n^{(1)}, t_n^{(2)}, \ldots, t_n^{(k)}$ are hypergeometric terms with $\sum_{i=1}^{k} t_n^{(i)} = 0$, then $t_n^{(i)} \sim t_n^{(j)}$ for some $i$ and $j$ with $1 \leq i < j \leq k$.

**8.** A sum of a fixed number of hypergeometric terms can be expressed as a sum of pairwise dissimilar hypergeometric terms.

**9.** The terms of a hypergeometric series can be expressed using rising powers $a^{\overline{n}}$ (also known as rising factorials and denoted by $(a)_n$) (see §3.4.2) as follows:

$$
{}_pF_q = \begin{bmatrix} a_1 & a_2 & \cdots & a_p \\ b_1 & b_2 & \cdots & b_q \end{bmatrix} = \sum_{k=0}^{\infty} \frac{(a_1)^{\overline{k}}(a_2)^{\overline{k}} \cdots (a_p)^{\overline{k}}}{(b_1)^{\overline{k}}(b_2)^{\overline{k}} \cdots (b_q)^{\overline{k}}} \frac{x^k}{k!}.
$$

**10.** There are a large number of well-known hypergeometric identities (see Facts 12–17, for example) that can be used as a starting point when a closed form for a sum of hypergeometric terms is sought.

**11.** There are many rules that transform a hypergeometric series with one parameter set into a different hypergeometric series with a second parameter set. Such transformation rules can be helpful in constructing closed forms for sums of hypergeometric terms.

**12.** ${}_1F_1 \begin{bmatrix} 1 \\ 1 \end{bmatrix} = e^x$.

**13.** ${}_1F_0 \begin{bmatrix} a \\ - \end{bmatrix} = \dfrac{1}{(1-x)^a}$.

**14.** *Gauss's ${}_2F_1$ identity:* If $b$ is zero or a negative integer or the real part of $c - a - b$ is positive, then

$$
{}_2F_1 \begin{bmatrix} a & b \\ c \end{bmatrix} = \frac{\Gamma(c-a-b)\Gamma(c)}{\Gamma(c-a)\Gamma(c-b)}
$$

where $\Gamma$ is the gamma function (so $\Gamma(n) = (n-1)!$ when $n$ is a positive integer).

**15.** *Kummer's ${}_2F_1$ identity:* If $a - b + c = 1$, then

$$
{}_2F_1 \begin{bmatrix} a & b \\ c \end{bmatrix} = \frac{\Gamma(\frac{b}{2}+1)\Gamma(b-a+1)}{\Gamma(b+1)\Gamma(\frac{b}{2}-a+1)}
$$

and when $b$ is a negative integer, this can be expressed as

$$
{}_2F_1 \begin{bmatrix} a & b \\ c \end{bmatrix} = 2\cos(\tfrac{\pi b}{2}) \frac{\Gamma(|b|)\Gamma(b-a+1)}{\Gamma(\frac{|b|}{2})\Gamma(\frac{b}{2}-a+1)}.
$$

**16.** *Saalschütz's* $_3F_2$ *identity*: If $d + e = a + b + c + 1$ and $c$ is a negative integer, then

$$_3F_2\begin{bmatrix} a & b & c \\ d & e \end{bmatrix};1\end{bmatrix} = \frac{(d-a)^{\overline{|c|}}(d-b)^{\overline{|c|}}}{d^{\overline{|c|}}(d-a-b)^{\overline{|c|}}}.$$

**17.** *Dixon's identity*: If $1 + \frac{a}{2} - b - c > 0$, $d = a - b + 1$, and $e = a - c + 1$, then

$$_3F_2\begin{bmatrix} a & b & c \\ d & e \end{bmatrix};1\end{bmatrix} = \frac{(\frac{a}{2})!(a-b)!(a-c)!(\frac{a}{2}-b-c)!}{a!(\frac{a}{2}-b)!(\frac{a}{2}-c)!(a-b-c)!}.$$

The more familiar form of this identity reads

$$\sum_k (-1)^k \binom{a+b}{a+k}\binom{a+c}{c+k}\binom{b+c}{b+k} = \frac{(a+b+c)!}{a!b!c!}.$$

**18.** *Clausen's* $_4F_3$ *identity*: If $d$ is a negative integer or zero and $a + b + c - d = \frac{1}{2}$, $e = a + b + \frac{1}{2}$, and $a + f = d + 1 = b + g$, then

$$_4F_3\begin{bmatrix} a & b & c & d \\ e & f & g \end{bmatrix};1\end{bmatrix} = \frac{(2a)^{\overline{|d|}}(a+b)^{\overline{|d|}}(2b)^{\overline{|d|}}}{(2a+2b)^{\overline{|d|}}(a)^{\overline{|d|}}(b)^{\overline{|d|}}}.$$

**Examples:**

**1.** The series $\sum_{k=0}^{\infty} 3 \cdot (-5)^k$ is a geometric series. The series $\sum_{k=0}^{\infty} n2^n$ is not a geometric series.

**2.** The series $\sum_{k=0}^{\infty} t_k$ is a hypergeometric series when $t_k$ equals $2^k$, $(k+1)^2$, $\frac{1}{2k+3}$, or $\frac{1}{(2k+1)(k+3)!}$, but is not hypergeometric when $t_k = 2^k + 1$.

**3.** The series $\sum_{k=0}^{\infty} \frac{3^k}{k!^4}$ equals $_0F_3\begin{bmatrix} - \\ 1 \ 1 \ 1 \end{bmatrix};3\end{bmatrix}$ since the ratio of the $(k+1)$st and $k$th terms is $\frac{3}{(k+1)^4}$.

**4.** A closed form for $S_n = \sum_{k=0}^{\infty}(-1)^k\binom{2n}{k}^2$ can be found by first noting that $S_n = {}_2F_1\begin{bmatrix} -2n & -2n \\ 1 \end{bmatrix};-1\end{bmatrix}$ since the ratio between successive terms of the sum is $\frac{-(k-2n)^2}{(k+1)^2}$. This shows that Kummer's $_2F_1$ identity can be invoked with $a = -2n$, $b = -2n$, and $c = 1$, producing the equality $S_n = \frac{2(-1)^n(2n-1)!}{n!(n-1)!} = (-1)^n\binom{2n}{n}$.

**5.** An example of a transformation rule for hypergeometric functions is provided by

$$_2F_1\begin{bmatrix} a & b \\ c \end{bmatrix};x\end{bmatrix} = (1-x)^{c-a-b}{}_2F_1\begin{bmatrix} c-a & c-b \\ c \end{bmatrix};x\end{bmatrix}.$$

---

## 3.7.2   ALGORITHMS TO PRODUCE CLOSED FORMS FOR SUMS OF HYPERGEOMETRIC TERMS

**Definitions:**

A function $F(n,k)$ is called **doubly hypergeometric** if both $\frac{F(n+1,k)}{F(n,k)}$ and $\frac{F(n,k+1)}{F(n,k)}$ are rational functions of $n$ and $k$.

A function $F(n, k)$ is a **proper hypergeometric term** if it can be expressed as

$$F(n, k) = P(n, k)\frac{\prod_{i=1}^{G}(a_i n + b_i k + c_i)!}{\prod_{i=1}^{H}(u_i n + v_i k + w_i)!}x^k$$

where $x$ is a variable, $P(n, k)$ is a polynomial in $n$ and $k$, $G$ and $H$ are nonnegative integers, and all the coefficients $a_i$, $b_i$, $u_i$, and $v_i$ are integers.

A function $F(n, k)$ of the form

$$F(n, k) = P(n, k)\frac{\displaystyle\prod_{i=1}^{G}(a_i n + b_i k + c_i)!}{\displaystyle\prod_{i=1}^{H}(u_i n + v_i k + w_i)!}x^k$$

is said to be **well-defined** at $(n, k)$ if none of the terms $(a_i n + b_i k + c_i)$ in the product is a negative integer. The function $F(n, k)$ is defined to have the value 0 if $F$ is well-defined at $(n, k)$ and there is a term $(u_i n + v_i k + w_i)$ in the product that is a negative integer or $P(n, k) = 0$.

**Facts:**

**1.** If $F(n, k)$ is a proper hypergeometric term, then there exist positive integers $L$ and $M$ and polynomials $a_{i,j}(n)$ for $i = 0, 1, \ldots, L$ and $j = 0, 1, \ldots, M$, not all zero, such that

$$\sum_{i=0}^{L}\sum_{j=0}^{M} a_{i,j}(n)F(n - j, k - i) = 0$$

for all pairs $(n, k)$ with $F(n, k) \neq 0$ and all the values of $F(n, k)$ in this double sum are well-defined. Moreover, there is such a recurrence with $M$ equal to $M' = \sum_s |b_s| + \sum_t |v_t|$ and $L$ equal to $L' = \deg(P) + 1 + M'(-1 + \sum_s |a_s| + \sum_t |u_t|)$, where the $a_i$, $b_i$, $u_i$, $v_i$ and $P$ come from an expression of $F(n, k)$ as a hypergeometric term as specified in the definition.

**2.** *Sister Celine's algorithm:* This algorithm, developed in 1945 by Sister Mary Celine Fasenmyer (1906–1996), can be used to find recurrence relations for sums of the form $f(n) = \sum_k F(n, k)$ where $F$ is a doubly hypergeometric function. The algorithm finds a recurrence of the form $\sum_{i=0}^{L}\sum_{j=0}^{M} a_{i,j}(n)F(n - j, k - i) = 0$ by proceeding as follows:

- Start with trial values of $L$ and $M$, such as $L = 1$, $M = 1$.
- Assume that a recurrence relation of the type sought exists with these values of $L$ and $M$, with the coefficients $a_{i,j}(n)$ to be determined, if possible.
- Divide each term in the sum of the recurrence by $F(n, k)$, then reduce each fraction $F(n - j, k - i)/F(n, k)$, simplifying the ratios of factorials so only rational functions of $n$ and $k$ are left.
- Combine the terms in the sum using a common denominator, collecting the numerator into a single polynomial in $k$.
- Solve the system of linear equations for the $a_{i,j}(n)$ that results when the coefficients of each power of $k$ in the numerator polynomial are equated to zero.
- If these steps fail, repeat the procedure with larger values of $L$ and $M$; by Fact 2, this procedure is guaranteed to eventually work.

**3.** *Gosper's algorithm*: This algorithm, developed by R. W. Gosper, Jr., can be used to determine, given a hypergeometric term $t_n$, whether there is a hypergeometric term $z_n$ such that $z_{n+1} - z_n = t_n$. When there is such a hypergeometric term $z_n$, the algorithm also produces such a term.

**4.** Gosper's algorithm takes a hypergeometric term $t_n$ as input and performs the following general steps (for details see [PeWiZe96]):

- Let $r(n) = t_{n+1}/t_n$; this is a rational function of $n$ since $t$ is hypergeometric.
- Find polynomials $a(n)$, $b(n)$, and $c(n)$ such that $\gcd(a(n), b(n+h)) = 1$ whenever $h$ is a nonnegative integer; this is done using the following steps:
  - ◇ Let $r(n) = K \cdot \frac{f(n)}{g(n)}$ where $f(n)$ and $g(n)$ are monic relatively prime polynomials and $K$ is a constant, let $R(h)$ be the resultant of $f(n)$ and $g(n+h)$ (which is the product of the zeros of $g(n+h)$ at the zeros of $f(n)$), and let $S = \{h_1, h_2, \ldots, h_N\}$ be the set of nonnegative integer zeros of $R(h)$ where $0 \le h_1 < h_2 < \cdots < h_N$.
  - ◇ Let $p_0(n) = f(n)$ and $q_0(n) = g(n)$; then for $j = 1, 2, \ldots, N$ carry out the following steps:
    $$s_j(n) := \gcd(p_{j-1}(n), q_{j-1}(n + h_j));$$
    $$p_j(n) := p_{j-1}(n)/s_j(n);$$
    $$q_j(n) := q_{j-1}(n)/s_j(n - h_j).$$
- Take $a(n) := Kp_N(n)$; $b(n) := q_N(n)$; $c(n) := \prod_{i=1}^{N} \prod_{j=1}^{h_i} s_i(n - j)$.
- Find a nonzero polynomial $x(n)$ such that $a(n)x(n+1) - b(n-1)x(n) = c(n)$ if one exists; such a polynomial can be found using the method of undetermined coefficients to find a nonzero polynomial of degree $d$ or less, where the degree $d$ depends on the polynomials $a(n)$, $b(n)$, and $c(n)$. If no such polynomial exists, then the algorithm fails. The degree $d$ is determined by the following rules:
  - ◇ When $\deg a(n) \ne \deg b(n)$ or $\deg a(n) = \deg b(n)$ but the leading coefficients of $a(n)$ and $b(n)$ differ, then $d = \deg c(n) - \max(\deg a(n), \deg b(n))$.
  - ◇ When $\deg a(n) = \deg b(n)$ and the leading coefficients of $a(n)$ and $b(n)$ agree, $d = \max(\deg c(n) - \deg a(n) + 1, (B - A)/L)$ where $a(n) = Ln^k + An^{k-1} + \cdots$ and $b(n-1) = Ln^k + Bn^{k-1} + \cdots$; if this $d$ is negative, then no such polynomial $x(n)$ exists.
- Let $z_n = t_n \cdot b(n-1)x(n)/c(n)$; it follows that $z_{n+1} - z_n = t_n$.

**5.** When Gosper's algorithm fails, this shows that a sum of hypergeometric terms cannot be expressed as a hypergeometric term plus a constant.

**6.** Programs in both *Maple* and *Mathematica* implementing algorithms described in this section can be found at the following sites:

- http://www.math.upenn.edu/~wilf/AeqB.html
- http://www.math.rutgers.edu/~zeilberg/programs.html

### Examples:

**1.** The function $F(n, k) = \frac{1}{5n+2k+2}$ is a proper hypergeometric term since $F(n, k)$ can be expressed as $F(n, k) = \frac{(5n+2k+1)!}{(5n+2k+2)!}$.

**2.** The function $F(n, k) = \frac{1}{n^2+k^3+5}$ is not a proper hypergeometric term.

**3.** Sister Celine's algorithm can be used to find a recurrence relation satisfied by the function $f(n) = \sum_k F(n, k)$ where $F(n, k) = k\binom{n}{k}$ for $n = 0, 1, 2, \ldots$. The algorithm proceeds by finding a recurrence relation of the form $a(n)F(n, k) + b(n)F(n+1, k) +$

$c(n)F(n, k+1) + d(n)F(n+1, k+1) = 0$. Since $F(n, k) = k\binom{n}{k}$, this recurrence relation simplifies to $a(n) + b(n) \cdot \frac{n+1}{n+1-k} + c(n) \cdot \frac{n-k}{k} + d(n) \cdot \frac{n+1}{k} = 0$. Putting the left-hand side of this equation over a common denominator and expressing it as a polynomial in $k$, four equations in the unknowns $a(n)$, $b(n)$, $c(n)$, and $d(n)$ are produced. These equations have the following solutions: $a(n) = t(-1 - \frac{1}{n})$, $b(n) = 0$, $c(n) = t(-1 - \frac{1}{n})$, $d = t$, where $t$ is a constant. This produces the recurrence relation $(-1 - \frac{1}{n})F(n, k) + (-1 - \frac{1}{n})F(n, k+1) + F(n+1, k+1) = 0$, which can be summed over all integers $k$ and simplified to produce the recurrence relation $f(n+1) = 2 \cdot \frac{n+1}{n} f(n)$, with $f(1) = 1$. From this it follows that $f(n) = n2^{n-1}$.

**4.** As shown in [PeWiZe96], Sister Celine's algorithm can be used to find an identity for $f(n) = \sum_k F(n, k)$ where $F(n, k) = \binom{n}{k}\binom{2n}{k}(-2)^{n-k}$. A recurrence for $F(n, k)$ can be found using her techniques (which can be carried out using either *Maple* or *Mathematica* software, for example). An identity that can be found this way is: $-8(n-1)F(n-2, k-1) - 2(2n-1)F(n-1, k-1) + 4(n-1)F(n-2, k) + 2(2n-1)F(n-1, k) + nF(n, k) = 0$. When this is summed over all integers $k$, the recurrence relation $nf(n) - 4(n-1)f(n-2) = 0$ is obtained. From the definition of $f$ it follows that $f(0) = 1$ and $f(1) = 0$. From the initial conditions and the recurrence relation for $f(n)$, it follows that $f(n) = 0$ when $n$ is odd and $f(n) = \binom{n}{n/2}$ when $n$ is even. (This is known as the *Reed-Dawson identity*.)

**5.** Gosper's algorithm can be used to find a closed form for $S_n = \sum_{k=1}^{n} k\,k!$. Let $t_n = n\,n!$. Following Gosper's algorithm gives $r(n) = \frac{t_{n+1}}{t_n} = \frac{(n+1)^2}{n}$, $a(n) = n + 1$, $b(n) = 1$, and $c(n) = n$. The polynomial $x(n)$ must satisfy $(n+1)x(n+1) - x(n) = n$; the polynomial $x(n) = 1$ is such a solution. It follows that $z_n = n!$ satisfies $z_{n+1} - z_n = t_n$. Hence $s_n = z_n - z_1 = n! - 1$ and $S_n = s_{n+1} = (n+1)! - 1$.

**6.** Gosper's algorithm can be used to show that $S_n = \sum_{k=0}^{n} k!$ cannot be expressed as a hypergeometric term plus a constant. Let $t_n = n!$. Following Gosper's algorithm gives $r(n) = \frac{t_{n+1}}{t_n} = n + 1$, $a(n) = n + 1$, $b(n) = 1$, $c(n) = 1$. The polynomial $x(n)$ must satisfy $(n+1)x(n+1) - x(n) = 1$ and must have a degree less than zero. It follows that there is no closed form for $\sum_{k=0}^{n} k!$ of the type specified.

---

### 3.7.3   CERTIFYING THE TRUTH OF COMBINATORIAL IDENTITIES

**Definitions:**

A pair of functions $(F, G)$ is called a **WZ pair** (after Wilf and Zeilberger) if $F(n+1, k) - F(n, k) = G(n, k+1) - G(n, k)$. If $(F, G)$ is a WZ pair, then $F$ is called the **WZ mate** of $G$ and vice versa.

A **WZ certificate** $R(n, k)$ is a function that can be used to verify the hypergeometric identity $\sum_k f(n, k) = r(n)$ by creating a WZ pair $(F, G)$ with $F(n, k) = \frac{f(n,k)}{r(n)}$ when $r(n) \neq 0$ and $F(n, k) = f(n, k)$ when $r(n) = 0$ and $G(n, k) = R(n, k)F(n, k)$. When a hypergeometric identity is proved using a a WZ certificate, this proof is called a **WZ proof**.

**Facts:**

**1.** If $(F, G)$ is a WZ pair such that for each integer $n \geq 0$, $\lim_{k \to \pm\infty} G(n, k) = 0$, then $\sum_k F(n, k)$ is a constant for $n = 0, 1, 2, \ldots$.

**2.** If $(F, G)$ is a WZ pair such that for each integer $k$, the limit $f_k = \lim_{n \to \infty} F(n, k)$ exists and is finite, for every nonnegative integer $n$ it is the case that $\lim_{k \to \pm\infty} G(n, k) = 0$, and $\lim_{L \to \infty} \sum_{n \geq 0} G(n, -L) = 0$, then $\sum_{n \geq 0} G(n, k) = \sum_{j \leq k-1}(f_j - F(0, j))$.

**3.** An identity $\sum_k f(n,k) = r(n)$ can be verified using its WZ certificate $R(n,k)$ as follows:

- If $r(n) \neq 0$, define $F(n,k)$ by $F(n,k) = \frac{f(n,k)}{r(n)}$, else define $F(n,k) = f(n,k)$; define $G(n,k)$ by $G(n,k) = R(n,k)F(n,k)$.
- Confirm that $(F,G)$ is a WZ pair, i.e., that $F(n+1,k) - F(n,k) = G(n,k+1) - G(n,k)$, by dividing the factorials out and verifying the polynomial identity that results.
- Verify that the original identity holds for a particular value of $n$.

**4.** The WZ certificate of an identity $\sum_k f(n,k) = r(n)$ can be found using the following steps:

- If $r(n) \neq 0$, define $F(n,k)$ to be $F(n,k) = \frac{f(n,k)}{r(n)}$, else define $F(n,k)$ to be $F(n,k) = f(n,k)$.
- Let $f(k) = F(n+1,k) - F(n,k)$; provide $f(k)$ as input to Gosper's algorithm.
- If Gosper's algorithm produces $G(n,k)$ as output, it is the WZ mate of $F$ and the function $R(n,k) = \frac{G(n,k)}{F(n,k)}$ is the WZ certificate of the identity $\sum_k F(n,k) = C$ where $C$ is a constant.

If Gosper's algorithm fails, this algorithm also fails.

**Examples:**

**1.** To prove the identity $f(n) = \sum_k \binom{n}{k}^2 = \binom{2n}{n}$, express it in the form $\sum_k F(n,k) = 1$ where $F(n,k) = \binom{n}{k}^2 / \binom{2n}{n}$. The identity can be proved by taking the function $R(n,k) = \frac{k^2(3n-2k+3)}{2(2n+1)(n-k+1)^2}$ as its WZ certificate. (This certificate can be obtained using Gosper's algorithm.)

**2.** To prove Gauss's $_2F_1$ identity via a WZ proof, express it in the form $\sum_k F(n,k) = 1$ where $F(n,k) = \frac{(n+k)!(b+k)!(c-n-1)!(c-b-1)!}{(c+k)!(n-1)!(c-n-b-1)!(k+1)!(b-1)!}$. The identity can then be proved by taking the function $R(n,k) = \frac{(k+1)(k+c)}{n(n+1-c)}$ as its WZ certificate. (This certificate can be obtained using Gosper's algorithm.)

## REFERENCES

*Printed Resources*:

[AbSt65] M. Abramowitz and I. A. Stegun, eds., *Handbook of Mathematical Functions*, reprinted by Dover, 1965. (An invaluable and general reference for dealing with functions, with a chapter on sums in closed form.)

[BeWi06] E. A. Bender and S. G. Williamson, *Foundations of Combinatorics with Applications*, Dover, 2006. (Section 12.4 contains further discussion of rules of thumb for asymptotic estimation.)

[BePhHo10] G. E. Bergum, A. N. Philippou, and A. F. Horadam, eds., *Applications of Fibonacci Numbers*, Kluwer, 2010.

[BeBr82] A. Beutelspacher and W. Brestovansky, "Generalized Schur numbers", *Combinatorial Theory*, Lecture Notes in Mathematics 969, Springer-Verlag, 1982, 30–38.

[Br09] R. Brualdi, *Introductory Combinatorics*, 5th ed., Pearson, 2009.

[De10] N. G. de Bruijn, *Asymptotic Methods in Analysis*, North-Holland, 1958. Third edition reprinted by Dover Publications, 2010. (This monograph covers a variety of topics in asymptotics from the viewpoint of an analyst.)

[Dr97] M. Drmota, "Systems of functional equations", *Random Structures & Algorithms* 10 (1997), 103–124.

[Dr09] M. Drmota, *Random Trees: An Interplay Between Combinatorics and Probability*, Springer-Verlag, 2009. (This book covers the study of asymptotic estimates with detailed asymptotic results for a wide variety of families of trees.)

[ErSz35] P. Erdős and G. Szekeres, "A combinatorial problem in geometry", *Compositio Mathematica* 2 (1935), 463–470.

[FlSaZi91] P. Flajolet, B. Salvy, and P. Zimmermann, "Automatic average-case analysis of algorithms", *Theoretical Computer Science* 79 (1991), 37–109. (Describes computer software to automate asymptotic average time analysis of algorithms.)

[FlSe09] P. Flajolet and R. Sedgewick, *Analytic Combinatorics*, Cambridge University Press, 2009. (Reference for the use of complex analytic techniques in order to analyze generating functions.)

[GrKnPa94] R. L. Graham, D. E. Knuth, and O. Patashnik, *Concrete Mathematics*, 2nd ed., Addison-Wesley, 1994. (A superb compendium of special sequences, their properties and analytical techniques.)

[GrRoSp13] R. Graham, B. Rothschild, and J. Spencer, *Ramsey Theory*, 2nd ed., Wiley, 2013.

[GrKn07] D. H. Greene and D. E. Knuth, *Mathematics for the Analysis of Algorithms*, 3rd ed., Birkhäuser, 2007. (Parts of this text discuss various asymptotic methods.)

[Gr04] R. P. Grimaldi, *Discrete and Combinatorial Mathematics*, 5th ed., Addison-Wesley, 2004.

[Gr12] R. P. Grimaldi, *Fibonacci and Catalan Numbers*, Wiley, 2012.

[Ha75] E. R. Hansen, *A Table of Series and Products*, Prentice-Hall, 1975. (A reference giving many summations and products in closed form.)

[Ho79] D. R. Hofstadter, *Gödel, Escher, Bach*, Basic Books, 1979.

[Ko01] T. Koshy, *Fibonacci and Lucas Numbers with Applications*, Wiley, 2001.

[Ko09] T. Koshy, *Catalan Numbers with Applications*, Oxford University Press, 2009.

[MiRo91] J. G. Michaels and K. H. Rosen, eds., *Applications of Discrete Mathematics*, McGraw-Hill, 1991. (Chapters 6, 7, and 8 discuss Stirling numbers, Catalan numbers, and Ramsey numbers.)

[Mi90] R. E. Mickens, *Difference Equations: Theory and Applications*, 2nd ed., Chapman and Hall/CRC, 1990.

[NiZuMo91] I. Niven, H. S. Zuckerman, and H. L. Montgomery, *An Introduction to the Theory of Numbers*, 5th ed., Wiley, 1991.

[Od95] A. M. Odlyzko, "Asymptotic Enumeration Methods", in *Handbook of Combinatorics*, R. L. Graham, M. Grötschel, and L. Lovász (eds.), North-Holland, 1995, 1063–1229. (This is an encyclopedic presentation of methods with a 400$^+$ item bibliography.)

[PeWi13] R. Pemantle and M. Wilson, *Analytic Combinatorics in Several Variables*, Cambridge University Press, 2013.

[PeWiZe96] M. Petkovšek, H. S. Wilf, and D. Zeilberger, *A=B*, A K Peters/CRC Press, 1996.

[PhBeHo01] A. N. Philippou, G. E. Bergum, and A. F. Horadam, eds., *Fibonacci Numbers and Their Applications*, Mathematics and Its Applications, Vol. 28, Springer, 2001.

[Ra30] F. Ramsey, "On a problem of formal logic", *Proceedings of the London Mathematical Society* 30 (1930), 264–286.

[Ri02] J. Riordan, *An Introduction to Combinatorial Analysis*, Dover, 2002.

[RoTe09] F. S. Roberts and B. Tesman, *Applied Combinatorics*, 2nd ed., Chapman and Hall/CRC Press, 2009.

[Ro12] K. H. Rosen, *Discrete Mathematics and Its Applications*, 7th ed., McGraw-Hill, 2012.

[Sl73] N. J. A. Sloane, *A Handbook of Integer Sequences*, Academic Press, 1973. (The original compilation of integer sequences.)

[SlPl95] N. J. A. Sloane and S. Plouffe, *The Encyclopedia of Integer Sequences*, Academic Press, 1995. (The definitive reference work on integer sequences.)

[StMc77] D. F. Stanat and D. F. McAllister, *Discrete Mathematics in Computer Science*, Prentice-Hall, 1977.

[St11] R. P. Stanley, *Enumerative Combinatorics*, Vol. I, 2nd ed., Cambridge University Press, 2011.

[St01] R. P. Stanley, *Enumerative Combinatorics*, Vol. II, Cambridge University Press, 2001.

[To85] I. Tomescu, *Problems in Combinatorics and Graph Theory*, translated by R. Melter, Wiley, 1985.

[Va07] S. Vajda, *Fibonacci and Lucas Numbers, and the Golden Section*, Dover, 2007.

[Wi05] H. S. Wilf, *generatingfunctionology*, 3rd ed., CRC Academic Press, 2005.

**Web Resources**:

`http://oeis.org` (On-Line Encyclopedia of Integer Sequences.)

`http://www.cs.rit.edu/~spr/ElJC/eline.html` (Contains information on Ramsey numbers.)

`http://www.math.rutgers.edu/~zeilberg/programs.html` (Contains programs in *Maple* implementing algorithms described in §3.7.2.)

`http://www.maths.surrey.ac.uk/hosted-sites/R.Knott/Fibonacci/fibrefs.html` (Contains links to sites dealing with Fibonacci numbers as well as a list of relevant books and articles.)

`http://www.math.upenn.edu/~wilf/AeqB.html` (Contains programs in *Maple* and *Mathematica* implementing algorithms described in §3.7.2.)

# 4

**NUMBER THEORY**

## INTRODUCTION

This chapter covers the basics of number theory. Number theory, a subject with a long and rich history, has become increasingly important because of its applications to computer science and cryptography. The core topics of number theory, such as divisibility, radix representations, greatest common divisors, primes, factorization, congruences, diophantine equations, and continued fractions, are covered here. Algorithms for finding greatest common divisors, large primes, and factorizations of integers are described.

There are many famous problems in number theory, including some that have been solved only recently, such as Fermat's last theorem, and others that have eluded resolution, such as the Goldbach conjecture. The status of such problems is described in this chapter. New discoveries in number theory, such as new large primes, are being made at an increasingly fast pace. This chapter describes the current state of knowledge and provides pointers to Internet sources where the latest facts can be found.

## GLOSSARY

**algebraic integer**:  a root of a monic polynomial with integer coefficients.

**algebraic number**:  a root of a polynomial with integer coefficients.

**arithmetic function**:  a function defined for all positive integers.

**Bachet's equation**:  a diophantine equation of the form $y^2 = x^3 + k$, where $k$ is a given integer.

**base**:  the positive integer $b$, with $b > 1$, in the expansion $n = a_k b^k + a_{k-1} b^{k-1} + \cdots + a_1 b + a_0$, where $0 \le a_i \le b-1$ for $i = 0, 1, 2, \ldots, k$.

**binary coded decimal expansion**:  the expansion produced by replacing each decimal digit of an integer by the four-bit binary expansion of that digit.

**binary representation of an integer**:  the base two expansion of this integer.

**Carmichael number**:  a positive integer that is a pseudoprime to all bases.

**Catalan's equation**: the diophantine equation $x^m - y^n = 1$, where solutions in integers greater than 1 are sought for $x$, $y$, $m$, and $n$.

**Chinese remainder theorem**: the theorem that states that given a set of congruences $x \equiv a_i \pmod{m_i}$ for $i = 1, 2, \ldots, n$, where the integers $m_i$, $i = 1, 2, \ldots, n$, are pairwise relatively prime, there is a unique simultaneous solution of these congruences modulo $M = m_1 m_2 \ldots m_n$.

**complete system of residues modulo $m$**: a set of integers such that every integer is congruent modulo $m$ to exactly one integer in the set.

**composite**: a positive integer that has a factor other than 1 and itself.

**congruence class of $a$ modulo $m$**: the set of integers congruent to $a$ modulo $m$.

**congruent integers modulo $m$**: two integers with a difference divisible by $m$.

**continued fraction**: a finite or infinite expression of the form $a_0 + 1/(a_1 + 1/(a_2 + \cdots$; usually abbreviated $[a_0, a_1, a_2, \ldots]$.

**convergent**: a rational fraction obtained by truncating a continued fraction.

**coprime** (integers): integers that have no positive common divisor other than 1; see *relatively prime*.

**cyclotomic field**: a field obtained by adjoining a primitive root of unity to the field of rational numbers.

**diophantine approximation**: the approximation of a number by numbers belonging to a specified set, often the set of rational numbers.

**diophantine equation**: an equation together with the restriction that the only solutions of the equation of interest are those belonging to a specified set, often the set of integers or the set of rational numbers.

**Dirichlet's theorem** (on primes in arithmetic progressions): the theorem that states that there are infinitely many primes in each arithmetic progression of the form $an + b$, where $a$ and $b$ are relatively prime positive integers.

**discrete logarithm of $a$ to the base $r$ modulo $m$**: the integer $x$ such that $r^x \equiv a \pmod{m}$, where $r$ is a primitive root of $m$ and $\gcd(a, m) = 1$.

**divides**: the integer $a$ divides the integer $b$, written $a \mid b$, if there is an integer $c$ such that $b = ac$.

**divisor**: (1) an integer $d$ such that $d$ divides $a$ for a given integer $a$, or (2) the positive integer $d$ that is divided into the integer $a$ to yield $a = dq + r$ with $0 \leq r < d$.

**elliptic curve**: for prime $p > 3$, the set of solutions $(x, y)$ to the congruence $y^2 \equiv x^3 + ax + b \pmod{p}$, where $4a^3 + 27b^2 \not\equiv 0 \pmod{p}$, together with a special point $\mathcal{O}$, called the *point at infinity*.

**elliptic curve method (ECM)**: a factoring technique invented by Lenstra that is based on the theory of elliptic curves.

**Euler phi-function**: the function $\phi(n)$ whose value at the positive integer $n$ is the number of positive integers not exceeding $n$ relatively prime to $n$.

**Euler's theorem**: the theorem that states that if $n$ is a positive integer and $a$ is an integer with $\gcd(a, n) = 1$, then $a^{\phi(n)} \equiv 1 \pmod{n}$, where $\phi(n)$ is the value of the Euler phi-function at $n$.

**exactly divides**: if $p$ is a prime and $n$ is a positive integer, $p^r$ exactly divides $n$, written $p^r \| n$, if $p^r$ divides $n$, but $p^{r+1}$ does not divide $n$.

**factor** (of an integer $n$):  an integer that divides $n$.

**factorization algorithm**:  an algorithm whose input is a positive integer and whose output is the prime factorization of this integer.

**Farey series** (of order $n$):  the set of fractions $\frac{h}{k}$, where $h$ and $k$ are relatively prime nonnegative integers with $0 \le h \le k \le n$ and $k \ne 0$.

**Fermat equation**:  the diophantine equation $x^n + y^n = z^n$, where $n$ is an integer greater than 2 and $x$, $y$, $z$ are nonzero integers.

**Fermat number**:  a number of the form $2^{2^n} + 1$, where $n$ is a nonnegative integer.

**Fermat prime**:  a prime Fermat number.

**Fermat's last theorem**:  the theorem that states that if $n$ is a positive integer greater than 2, then the equation $x^n + y^n = z^n$ has no solutions in integers with $xyz \ne 0$.

**Fermat's little theorem**:  the theorem that states that if $p$ is prime and $a$ is an integer, then $a^p \equiv a \pmod{p}$.

**Fibonacci numbers**:  the sequence of numbers defined by $F_0 = 0$, $F_1 = 1$, and $F_n = F_{n-1} + F_{n-2}$ for $n = 2, 3, 4, \dots$.

**fundamental theorem of arithmetic**:  the theorem that states that every positive integer has a unique representation as the product of primes written in nondecreasing order.

**Gaussian integers**:  the set of numbers of the form $a + bi$, where $a$ and $b$ are integers and $i$ is $\sqrt{-1}$.

**greatest common divisor gcd** (of a set of integers):  the largest integer that divides all integers in the set; the greatest common divisor of the integers $a_1, a_2, \dots, a_n$ is denoted $\gcd(a_1, a_2, \dots, a_n)$.

**hexadecimal representation** (of an integer):  the base sixteen representation of this integer.

**index of $a$ to the base $r$ modulo $m$**:  the smallest nonnegative integer $x$, denoted $\text{ind}_r a$, such that $r^x \equiv a \pmod{m}$, where $r$ is a primitive root of $m$ and $\gcd(a, m) = 1$.

**integer of an algebraic number field**:  an algebraic integer that belongs to this field.

**inverse of an integer $a$ modulo $m$**:  an integer $\bar{a}$ such that $a\bar{a} \equiv 1 \pmod{m}$; here $\gcd(a, m) = 1$.

**irrational number**:  a real number that is not the ratio of two integers.

**Jacobi symbol**:  a generalization of the Legendre symbol.

**Kronecker symbol**:  a generalization of the Legendre and Jacobi symbols.

**least common multiple** (of a set of integers):  the smallest positive integer that is divisible by all integers in the set.

**least nonnegative residue of $a$ modulo $m$**:  the remainder when $a$ is divided by $m$; it is the smallest nonnegative integer congruent to $a$ modulo $m$, written $a \bmod m$.

**Legendre symbol**:  the symbol $\left(\frac{a}{p}\right)$ that has the value 1 if $a$ is a square modulo $p$ and $-1$ if $a$ is not a square modulo $p$; here $p$ is a prime and $a$ is an integer not divisible by $p$.

**linear congruential method**:  a method for generating a sequence of pseudo-random numbers based on a congruence of the form $x_{n+1} \equiv ax_n + c \pmod{m}$.

**Mersenne prime**: a prime of the form $2^p - 1$, where $p$ is a prime.

**Möbius function**: the arithmetic function $\mu(n)$, where $\mu(n) = 1$ if $n = 1$, $\mu(n) = 0$ if $n$ has a square factor larger than 1, and $\mu(n) = (-1)^s$ if $n$ is square free and is the product of $s$ different primes.

**modulus**: the integer $m$ in a congruence $a \equiv b \,(\text{mod } m)$.

**multiple of an integer $a$**: an integer $b$ such that $a$ divides $b$.

**multiplicative function**: a function $f$ such that $f(mn) = f(m)f(n)$ whenever $m$ and $n$ are relatively prime positive integers.

**mutually relatively prime** (set of integers): integers with no common factor greater than 1.

**number field**: a field that is a finite degree extension of $\mathcal{Q}$.

**number field sieve**: a factoring algorithm, currently the best one known for large numbers with no small prime factors.

**octal representation of an integer**: the base eight representation of this integer.

**one's complement expansion**: an $n$ bit representation of an integer $x$ with $|x| < 2^{n-1}$, for a specified positive integer $n$, in which the leftmost bit is 0 if $x \geq 0$ and 1 if $x < 0$, and the remaining $n - 1$ bits are those of the binary expansion of $x$ if $x \geq 0$, and the complements of the bits in the expansion of $|x|$ if $x < 0$.

**order of an integer $a$ modulo $m$**: the least positive integer $t$, denoted by $\text{ord}_m a$, such that $a^t \equiv 1 \pmod{m}$; here $\gcd(a, m) = 1$.

**pairwise relatively prime**: integers with the property that each two of them are relatively prime.

**palindrome**: a finite sequence that reads the same forward and backward.

**partial quotient**: a term $a_i$ of a continued fraction.

**Pell's equation**: the diophantine equation $x^2 - dy^2 = 1$, where $d$ is a positive integer that is not a perfect square.

**perfect number**: a positive integer whose sum of positive divisors, other than the integer itself, equals this integer.

**periodic base $b$ expansion**: a base $b$ expansion in which the terms beyond a certain point are repetitions of the same block of integers.

**powerful integer**: an integer $n$ with the property that $p^2$ divides $n$ whenever $p$ is a prime that divides $n$.

**primality test**: an algorithm that determines whether a positive integer is prime.

**prime**: a positive integer greater than 1 that has exactly two factors, 1 and itself.

**prime factorization**: the factorization of an integer into primes.

**prime number theorem**: the theorem that states that the number of primes not exceeding a positive real number $x$ is asymptotic to $\frac{x}{\log x}$ (where $\log x$ denotes the natural logarithm of $x$).

**prime-power factorization**: the factorization of an integer into powers of distinct primes.

**primitive root of an integer $n$**: an integer $r$ such that the least positive residues of the powers of $r$ run through all positive integers relatively prime to $n$ and less than $n$.

***probabilistic primality test***: an algorithm that determines whether an integer is prime with a small probability of a false positive result.

***pseudoprime to the base b***: a composite positive integer $n$ such that $b^n \equiv b \pmod{n}$.

***pseudo-random number generator***: a deterministic method to generate numbers that share many properties with numbers really chosen randomly.

***Pythagorean triple***: positive integers $x$, $y$, and $z$ such that $x^2 + y^2 = z^2$.

***quadratic field***: the set of numbers $Q(\sqrt{d}) = \{a + b\sqrt{d} \mid a, b \text{ rational}\}$ for some square-free integer $d$.

***quadratic irrational***: an irrational number that is the root of a quadratic polynomial with integer coefficients.

***quadratic nonresidue*** (of $m$): an integer that is not a perfect square modulo $m$.

***quadratic reciprocity***: the law that states that given two odd primes $p$ and $q$, if at least one of them is of the form $4n + 1$, then $p$ is a quadratic residue of $q$ if and only if $q$ is a quadratic residue of $p$ and if both primes are of the form $4n + 3$, then $p$ is a quadratic residue of $q$ if and only if $q$ is a quadratic nonresidue of $p$.

***quadratic residue*** (of $m$): an integer that is a perfect square modulo $m$.

***quadratic sieve***: a factoring algorithm invented by Pomerance in 1981.

***rational number***: a real number that is the ratio of two integers. The set of rational numbers is denoted $\mathcal{Q}$.

***reduced system of residues modulo m***: pairwise incongruent integers modulo $m$ such that each integer in the set is relatively prime to $m$ and every integer relatively prime to $m$ is congruent to an integer in the set.

***relatively prime*** (integers): two integers with no common divisor greater than 1; see *coprime*.

***remainder*** (of the integer $a$ when divided by the positive integer $d$): the integer $r$ in the equation $a = dq + r$ with $0 \le r < d$, written $r = a \bmod d$.

***root*** (of a function $f$ modulo $m$): an integer $r$ such that $f(r) \equiv 0 \pmod{m}$.

***sieve of Eratosthenes***: a procedure for finding all primes less than a specified integer.

***smooth number***: an integer all of whose prime divisors are small.

***square root*** (of $a$ modulo $m$): an integer $r$ whose square is congruent to $a$ modulo $m$.

***square-free integer***: an integer not divisible by any perfect square other than 1.

***ten most wanted numbers***: a collection of large integers on a list, maintained by a group of researchers, whose currently unknown factorizations are actively sought; these integers are somewhat beyond the realm of numbers that can be factored using known techniques.

***terminating base-b expansion***: a base-$b$ expansion with only a finite number of nonzero coefficients.

***totient function***: the Euler phi-function.

***transcendental number***: a complex number that cannot be expressed as the root of an algebraic equation with integer coefficients.

***trial division***: a factorization technique that proceeds by dividing an integer by successive primes.

***twin primes***:  a pair of primes that differ by two.

***two's complement expansion***:  an $n$ bit representation of an integer $x$, with $-2^{n-1} \leq x \leq 2^{n-1} - 1$, for a specified positive integer $n$, in which the leftmost bit is 0 if $x \geq 0$ and 1 if $x < 0$, and the remaining $n - 1$ bits are those from the binary expansion of $x$ if $x \geq 0$ and are those of the binary expansion of $2^n - |x|$ if $x < 0$.

***ultimately periodic***:  a sequence (typically a base-$k$ expansion or continued fraction) $(a_i)_{i \geq 0}$ that eventually repeats; that is, there exist $k$ and $N$ such that $a_{n+k} = a_n$ for all $n \geq N$.

***unit*** (of a number field):  an algebraic integer that divides 1 in this field.

***Waring's problem***:  the problem of determining the smallest number $g(k)$ such that every integer is the sum of $g(k)$ $k$th powers of integers.

## 4.1  BASIC CONCEPTS

The basic concepts of number theory include the classification of numbers into different sets of special importance, the notion of divisibility, and the representation of integers. For more information about these basic concepts, see introductory number theory texts, such as [Ro10].

Unless otherwise specified, log will indicate the natural logarithm function.

### 4.1.1  NUMBERS

**Definitions:**

The ***integers*** are the elements of the set $\mathcal{Z} = \{\ldots, -3, -2, -1, 0, 1, 2, 3, \ldots\}$.

The ***natural numbers*** are the integers in the set $\mathcal{N} = \{0, 1, 2, 3, \ldots\}$.

The ***rational numbers*** are real numbers that can be written as $a/b$ where $a$ and $b$ are integers with $b \neq 0$. Numbers that are not rational are called ***irrational***. The set of rational numbers is denoted by $\mathcal{Q}$.

The ***algebraic numbers*** are real numbers that are solutions of equations of the form $a_n x^n + \cdots + a_1 x + a_0 = 0$, where $a_i$ is an integer for $i = 0, 1, \ldots, n$. Real numbers that are not algebraic are called ***transcendental***.

**Facts:**

**1.** The following table summarizes information and notation about some important types of numbers.

| name | definition | examples |
|------|-----------|----------|
| natural numbers $\mathcal{N}$ | $\{0, 1, 2, \ldots\}$ | $0, 43$ |
| integers $\mathcal{Z}$ | $\{\ldots, -2, -1, 0, 1, 2, \ldots\}$ | $0, 43, -314$ |
| Gaussian integers $\mathcal{Z}[i]$ | $\{a + bi \mid a, b \in \mathcal{Z}\}$ | $3, 4 + 3i, 7i$ |
| rational numbers $\mathcal{Q}$ | $\{\frac{a}{b} \mid a, b \in \mathcal{Z}; b \neq 0\}$ | $0, \frac{22}{7}$ |
| quadratic irrationals | irrational root of quadratic equation $a_2 x^2 + a_1 x + a_0 = 0$; all $a_i \in \mathcal{Q}$ | $\sqrt{2}, \frac{2+\sqrt{5}}{3}$ |
| irrational numbers | $\mathcal{R} - \mathcal{Q}$ | $\sqrt{2}, \pi, e$ |
| algebraic numbers $\overline{\mathcal{Q}}$ | root of algebraic equation $a_n x^n + \cdots + a_0 = 0, n \geq 1$, $a_0, \ldots, a_n \in \mathcal{Z}$ | $i, \sqrt{2}, \sqrt[3]{\frac{3}{2}}$ |
| algebraic integers $\mathcal{A}$ | root of *monic* algebraic equation $x^n + a_{n-1} x^{n-1} + \cdots + a_0 = 0$, $n \geq 1, a_0, a_1, \ldots, a_{n-1} \in \mathcal{Z}$ | $i, \sqrt{2}, \frac{1+\sqrt{5}}{2}$ |
| transcendental numbers | $\mathcal{C} - \overline{\mathcal{Q}}$ | $\pi, e, i \log 2$ |
| real numbers $\mathcal{R}$ | completion of $\mathcal{Q}$ | $0, \frac{1}{3}, \sqrt{2}, \pi$ |
| complex numbers $\mathcal{C}$ | $\overline{\mathcal{R}}$ or $\mathcal{R}[i]$ | $3 + 2i, e + i\pi$ |

**2.** Some sources do not include 0 in the set of natural numbers. So, always check the definition of natural numbers used in your sources.

**3.** A real number is rational if and only if its decimal expansion terminates or is periodic. (See §4.1.3).

**4.** The number $N^{1/m}$ is irrational where $N$ and $m$ are positive integers, unless $N$ is the $m$th power of an integer $n$.

**5.** The number $\log_b a$ is irrational, where $a$ and $b$ are positive integers greater than 1, if there is a prime that divides exactly one of $a$ and $b$.

**6.** If $x$ is a root of an equation $x^m + a_{m-1} x^{m-1} + \cdots + a_0 = 0$ in which the coefficients $a_i$ $(i = 0, 1, \ldots, m-1)$ are integers, then $x$ is either integer or irrational.

**7.** The set of algebraic numbers is countable (§1.2.3). Hence, almost all real numbers are transcendental. (However, showing a particular number of interest is transcendental is usually difficult.)

**8.** Both $e$ and $\pi$ are transcendental. The transcendence of $e$ was proven by Hermite in 1873, and $\pi$ was proven transcendental by Lindemann in 1882. Proofs of the transcendence of $e$ and $\pi$ can be found in [HaWr08].

**9.** *Gelfond-Schneider theorem*: If $\alpha$ and $\beta$ are algebraic numbers with $\alpha$ not equal to 0 or 1 and $\beta$ irrational, then $\alpha^\beta$ is transcendental. (For a proof see [Ba90].)

**10.** *Baker's linear forms in logarithms*: If $\alpha_1, \ldots, \alpha_n$ are nonzero algebraic numbers and $\log \alpha_1, \ldots, \log \alpha_n$ are linearly independent over $\mathcal{Q}$, then $1, \log \alpha_1, \ldots, \log \alpha_n$ are linearly independent over $\overline{\mathcal{Q}}$, where $\overline{\mathcal{Q}}$ is the closure of $\mathcal{Q}$. (Consult [Ba90] for a proof and applications of this theorem.)

**Examples:**

**1.** The numbers $\frac{11}{17}, -\frac{3345}{7}, -1, \frac{578}{579}$, and 0 are rational.

**2.** The number $\log_2 10$ is irrational.

**3.** The numbers $\sqrt{2}, 1 + \sqrt{2}$, and $\frac{1+\sqrt{2}}{5}$ are irrational.

**4.** The number $x = 0.10100100010000\ldots$, with a decimal expansion consisting of blocks where the $n$th block is a 1 followed by $n$ 0s, is irrational, since this decimal expansion does not terminate and is not periodic.

**5.** The decimal expansion of $\frac{22}{7}$ is periodic, since $\frac{22}{7} = 3.\overline{142857}$. However, the decimal expansion of $\pi$ neither terminates, nor is periodic, with $\pi = 3.141592653589793\ldots$.

**6.** It is not known whether Euler's constant $\gamma = \lim_{n\to\infty}\left(\sum_{k=1}^{n}\frac{1}{k} - \log n\right)$ (where $\log x$ denotes the natural logarithm of $x$) is rational or irrational.

**7.** The numbers $2$, $\frac{1}{2}$, $\sqrt{17}$, $\sqrt[3]{5}$, and $1 + \sqrt[6]{2}$ are algebraic.

**8.** By the Gelfond-Schneider theorem (Fact 9), $\sqrt{2}^{\sqrt{2}}$ is transcendental.

**9.** By Baker's linear forms in logarithms theorem (Fact 10), since $\log_2 10$ is irrational, it is transcendental.

## 4.1.2  DIVISIBILITY

The notion of the divisibility of one integer by another is the most basic concept in number theory. Introductory number theory texts, such as [HaWr08], [NiZuMo91], and [Ro10], are good references for this material.

**Definitions:**
Suppose $a$ and $d$ are integers with $d > 0$. Then in the equation $a = dq + r$ with $0 \le r < d$, $a$ is the **dividend**, $d$ is the **divisor**, $q$ is the **quotient**, and $r$ is the **remainder**.

Let $m$ and $n$ be integers with $m \ge 1$ and $n = dm + r$ with $0 \le r < m$. Then $n \bmod m$, the value of the **mod** $m$ function at $n$, is $r$, the remainder when $n$ is divided by $m$.

If $a$ and $b$ are integers and $a \ne 0$, then $a$ **divides** $b$, written $a|b$, if there is an integer $c$ such that $b = ac$. If $a$ divides $b$, then $a$ is a **factor** or **divisor** of $b$, and $b$ is a **multiple** of $a$. If $a$ is a positive divisor of $b$ that does not equal $b$, then $a$ is a **proper divisor** of $b$. The notation $a \nmid b$ means that $a$ does not divide $b$.

A **prime** is a positive integer divisible by exactly two distinct positive integers, 1 and itself. A positive integer, other than 1, that is not prime is called **composite**.

An integer is **square free** if it is not divisible by any perfect square other than 1.

An integer $n$ is **powerful** if whenever a prime $p$ divides $n$, $p^2$ divides $n$.

If $p$ is prime and $n$ is a positive integer, then $p^r$ **exactly divides** $n$, written $p^r||n$, if $p^r$ divides $n$, but $p^{r+1}$ does not divide $n$.

**Facts:**
**1.** If $a$ is a nonzero integer, then $a|0$.

**2.** If $a$ is an integer, then $1|a$.

**3.** If $a$ and $b$ are positive integers and $a|b$, then the following statements are true:
- $a \le b$;
- $\frac{b}{a}$ divides $b$;
- $a^k$ divides $b^k$ for every positive integer $k$;
- $a$ divides $bc$ for every integer $c$.

**4.**  If $a$, $b$, and $c$ are integers such that $a|b$ and $b|c$, then $a|c$.

**5.**  If $a$, $b$, and $c$ are integers such that $a|b$ and $a|c$, then $a|bm + cn$ for all integers $m$ and $n$.

**6.**  If $a$ and $b$ are integers such that $a|b$ and $b|a$, then $a = \pm b$.

**7.**  If $a$ and $b$ are integers and $m$ is a nonzero integer, then $a|b$ if and only if $ma|mb$.

**8.**  *Division algorithm*:  If $a$ and $d$ are integers with $d$ positive, then there are unique integers $q$ and $r$ such that $a = dq + r$ with $0 \le r < d$. (*Note*: The division algorithm is not an algorithm, in spite of its name.)

**9.**  The quotient $q$ and remainder $r$ when the integer $a$ is divided by the positive integer $d$ are given by $q = \lfloor \frac{a}{d} \rfloor$ and $r = a - d \lfloor \frac{a}{d} \rfloor$, respectively.

**10.**  If $a$ and $d$ are positive integers, then there are unique integers $q$, $r$, and $e$ such that $a = dq + er$ where $e = \pm 1$ and $-\frac{d}{2} < r \le \frac{d}{2}$.

**11.**  There are several divisibility tests that are easily performed using the decimal expansion of an integer. These include:

- An integer is divisible by 2 if and only if its last digit is even. It is divisible by 4 if and only if the integer made up of its last two digits is divisible by four. More generally, it is divisible by $2^j$ if and only if the integer made up of the last $j$ decimal digits of $n$ is divisible by $2^j$.

- An integer is divisible by 5 if and only if its last digit is divisible by 5 (which means it is either 0 or 5). It is divisible by 25 if and only if the integer made up of the last two digits is divisible by 25. More generally, it is divisible by $5^j$ if and only if the integer made up of the last $j$ digits of $n$ is divisible by $5^j$.

- An integer is divisible by 3, or by 9, if and only if the sum of the decimal digits of $n$ is divisible by 3, or by 9, respectively.

- An integer is divisible by 11 if and only if the integer formed by alternately adding and subtracting the decimal digits of the integer is divisible by 11.

- An integer is divisible by 7, 11, or 13 if and only if the integer formed by successively adding and subtracting the three-digit integers formed from successive blocks of three decimal digits of the original number, where digits are grouped starting with the rightmost digit, is divisible by 7, 11, or 13, respectively.

**12.**  If $d|b - 1$, then $n = (a_k \ldots a_1 a_0)_b$ (this notation is defined in §4.1.3) is divisible by $d$ if and only if the sum of the base $b$ digits of $n$, $a_k + \cdots + a_1 + a_0$, is divisible by $d$.

**13.**  If $d|b + 1$, then $n = (a_k \ldots a_1 a_0)_b$ is divisible by $d$ if and only if the alternating sum of the base $b$ digits of $n$, $(-1)^k a_k + \cdots - a_1 + a_0$, is divisible by $d$.

**14.**  If $p^r || a$ and $p^s || b$, where $p$ is a prime and $a$ and $b$ are positive integers, then $p^{r+s} || ab$.

**15.**  If $p^r || a$ and $p^s || b$, where $p$ is a prime and $a$ and $b$ are positive integers, then $p^{\min(r,s)} || a + b$.

**16.**  There are infinitely many primes. (See §4.4.2.)

**17.**  There are efficient algorithms that can produce large integers that have an extremely high probability of being prime. (See §4.4.4.)

**18.**  *Fundamental theorem of arithmetic*:  Every positive integer can be written as the product of primes in exactly one way, where the primes occur in nondecreasing order in the factorization.

**19.**  Many different algorithms have been devised to find the factorization of a positive integer into primes. Using some recently invented algorithms and the powerful computer systems available today, it is feasible to factor integers with over 100 digits. (See §4.5.1.)

**20.** The relative ease of producing large primes compared with the apparent difficulty of factoring large integers is the basis for an important cryptosystem called RSA. (See Chapter 15.)

**Examples:**

**1.** The integers 0, 3, $-12$, 21, 342, and $-1{,}113$ are divisible by 3; the integers $-1$, 7, 29, and $-1{,}111$ are not divisible by 3.

**2.** The quotient and remainder when 214 is divided by 6 are 35 and 4, respectively, since $214 = 35 \cdot 6 + 4$.

**3.** The quotient and remainder when $-114$ is divided by 7 are $-17$ and 5, respectively, since $-114 = -17 \cdot 7 + 5$.

**4.** With $a = 214$ and $d = 6$, the expansion of Fact 10 is $214 = 36 \cdot 6 - 2$ (so that $e = -1$ and $r = 2$).

**5.** 11 **mod** 4 = 3, 100 **mod** 7 = 2, and $-22$ **mod** 5 = 3.

**6.** The following are primes: 2, 3, 17, 101, 641. The following are composites: 4, 9, 91, 111, 1001.

**7.** The integers 15, 105, and 210 are square free; the integers 12, 99, and 270 are not.

**8.** The integer 72 is powerful since 2 and 3 are the only primes that divide 72 and $2^2 = 4$ and $3^2 = 9$ both divide 72, but 180 is not powerful since 5 divides 180, but $5^2$ does not.

**9.** The integer 32,688,048 is divisible by 2, 4, 8, and 16 since 2|8, 4|48, 8|048, and 16|8,048, but it is not divisible by 32 since 32 does not divide 88,048.

**10.** The integer 723,160,823 is divisible by 11 since the alternating sum of its digits, $3 - 2 + 8 - 0 + 6 - 1 + 3 - 2 + 7 = 22$, is divisible by 11.

**11.** Since $3^3|216$, but $3^4 \nmid 216$, it follows that $3^3\|216$.

---

### 4.1.3  RADIX REPRESENTATIONS

The representation of numbers in different bases has been important in the development of mathematics from its earliest days and is extremely important in computer arithmetic. For further details on this topic, see [Kn97], [Ko93], and [Sc85].

**Definitions:**

The **base $b$ expansion** of a positive integer $n$, where $b$ is an integer greater than 1, is the unique expansion of $n$ as $n = a_k b^k + a_{k-1} b^{k-1} + \cdots + a_1 b + a_0$; here $k$ is a nonnegative integer, $a_j$ is a nonnegative integer less than $b$ for $j = 0, 1, \ldots, k$, and the initial coefficient $a_k \neq 0$. This expansion is written as $(a_k a_{k-1} \ldots a_1 a_0)_b$.

The integer $b$ in the base $b$ expansion of an integer is called the **base** or **radix** of the expansion.

The coefficients $a_j$ in the base $b$ expansion of an integer are called the base $b$ **digits** of the expansion.

Base 10 expansions are called **decimal** expansions. The digits are called **decimal digits**.

Base 2 expansions are called **binary** expansions. The digits are called **binary digits** or **bits**.

Base 8 expansions are called **octal** expansions.

Base 16 expansions are called **hexadecimal** expansions. The 16 hexadecimal digits are $0, 1, 2, 3, 4, 5, 6, 7, 8, 9, A, B, C, D, E, F$ (where $A, B, C, D, E, F$ correspond to the decimal numbers $10, 11, 12, 13, 14, 15$, respectively).

The **binary coded decimal** expansion of an integer is the bit string formed by replacing each digit in the decimal expansion of the integer by the four bit binary expansion of that digit.

The **one's complement** expansion of an integer $x$ with $|x| < 2^{n-1}$, for a specified positive integer $n$, uses $n$ bits, where the leftmost bit is $0$ if $x \geq 0$ and $1$ if $x < 0$, and the remaining $n-1$ bits are those from the binary expansion of $x$ if $x \geq 0$ and are the complements of the bits in the binary expansion of $|x|$ if $x < 0$. (*Note:* The one's complement representation $11 \ldots 1$, consisting of $n$ 1s, is usually considered to be the negative representation of the number $0$.)

The **two's complement** expansion of an integer $x$ with $-2^{n-1} \leq x \leq 2^{n-1}-1$, for a specified positive integer $n$, uses $n$ bits, where the leftmost bit is $0$ if $x \geq 0$ and $1$ if $x < 0$, and the remaining $n-1$ bits are those from the binary expansion of $x$ if $x \geq 0$ and are those of the binary expansion of $2^n - |x|$ if $x < 0$.

The **base $b$ expansion** (where $b$ is an integer greater than 1) of a real number $x$ with $0 \leq x < 1$ is the unique expansion of $x$ as $x = \sum_{j=1}^{\infty} \frac{c_j}{b^j}$ where $c_j$ is a nonnegative integer less than $b$ for $j = 1, 2, \ldots$ and for every integer $N$ there is a coefficient $c_n \neq b-1$ for some $n > N$. This expansion is written as $(.c_1 c_2 c_3 \ldots)_b$.

A base $b$ expansion $(.c_1 c_2 c_3 \ldots)_b$ **terminates** if there is a positive integer $n$ such that $c_n = c_{n+1} = c_{n+2} = \cdots = 0$.

A base $b$ expansion $(.c_1 c_2 c_3 \ldots)_b$ is **periodic** if there are positive integers $N$ and $k$ such that $c_{n+k} = c_n$ for all $n \geq N$.

The **periodic base $b$ expansion** $(.c_1 c_2 \ldots c_{N-1} c_N \ldots c_{N+k-1} c_N \ldots c_{N+k-1} c_N \ldots)_b$ is denoted by $(.c_1 c_2 \ldots c_{N-1} \overline{c_N \ldots c_{N+k-1}})_b$. The part of the periodic base $b$ expansion preceding the periodic part is the **pre-period** and the periodic part is the **period**, where the period and pre-period are taken to have minimal possible length.

## Facts:

**1.** If $b$ is a positive integer greater than 1, then every positive integer $n$ has a unique base $b$ expansion.

**2.** *Converting from base 10 to base $b$:* Take the positive integer $n$ and divide it by $b$ to obtain $n = bq_0 + a_0$, $0 \leq a_0 < b$. Then divide $q_0$ by $b$ to obtain $q_0 = bq_1 + a_1$, $0 \leq a_1 < b$. Continue this process, successively dividing the quotients by $b$, until a quotient of zero is obtained, after $k$ steps. The base $b$ expansion of $n$ is then $(a_{k-1} \ldots a_1 a_0)_b$. (See Algorithm 1.)

---

**Algorithm 1:  Constructing base $b$ expansions.**

**procedure** *base $b$ expansion*($n$: positive integer)

$q := n$

$k := 0$

**while** $q \neq 0$

    $a_k := q \bmod b$

    $q := \lfloor \frac{q}{b} \rfloor$

---

$$k := k + 1$$
{The base $b$ expansion of $n$ is $(a_{k-1} \ldots a_1 a_0)_b$}

---

**3.** *Converting from base 2 to base $2^k$:* Group the bits in the base 2 expansion into blocks of $k$ bits, starting from the right, and then convert each block of $k$ bits into a base $2^k$ digit. For example, converting from binary (base 2) to octal (base 8) is done by grouping the bits of the binary expansion into blocks of three bits starting from the right and converting each block into an octal digit. Similarly, converting from binary to hexadecimal (base 16) is done by grouping the bits of the binary expansion into blocks of four bits starting from the right and converting each block into a hex digit.

**4.** *Converting from base $2^k$ to binary (base 2):* convert each base $2^k$ digit into a block of $k$ bits and string together these bits in the order the original digits appear. For example, to convert from hexadecimal to binary, convert each hex digit into the block of four bits that represent this hex digit and then string together these blocks of four bits in the correct order.

**5.** Every positive integer can be expressed uniquely as the sum of distinct powers of 2. This follows since every positive integer has a unique base 2 expansion, with the digits either 0 or 1.

**6.** There are $\lfloor \log_b n \rfloor + 1$ decimal digits in the base $b$ expansion of the positive integer $n$.

**7.** The number $x$ with one's complement representation $(a_{n-1} a_{n-2} \ldots a_1 a_0)$ can be found using the equation

$$x = -a_{n-1}(2^{n-1} - 1) + \sum_{i=0}^{n-2} a_i 2^i.$$

**8.** The number $x$ with two's complement representation $(a_{n-1} a_{n-2} \ldots a_1 a_0)$ can be found using the equation

$$x = -a_{n-1} \cdot 2^{n-1} + \sum_{i=0}^{n-2} a_i 2^i.$$

**9.** Two's complement representations of integers are often used by computers because addition and subtraction of integers, where these integers may be either positive or negative, can be performed easily using these representations.

**10.** Define a function $\operatorname{Lg} n$ by the rule

$$\operatorname{Lg} n = \begin{cases} 1 & \text{if } n = 0; \\ 1 + \lfloor \log_2 |n| \rfloor & \text{if } n \neq 0. \end{cases}$$

Then $\operatorname{Lg} n$ is the number of bits in the base 2 expansion of $n$, not counting the sign bit. (Compare with Fact 6.)

**11.** The bit operations for the basic operations are given in the following table, adapted from [BaSh96]. This table displays the number of bit operations used by the standard, naive algorithms, doing things bit by bit (addition with carries, subtraction with borrows, standard multiplication by each bit and shifting and adding, and standard division), and a big-oh estimate for the number of bits required to do the operations using the algorithm with the currently best known computational complexity. (The function Lg is defined in Fact 10; the function $\mu(m, n)$ is defined by the rule $\mu(m, n) = m(\operatorname{Lg} n)(\operatorname{Lg} \operatorname{Lg} n)$ if $m \geq n$ and $\mu(m, n) = n(\operatorname{Lg} m)(\operatorname{Lg} \operatorname{Lg} m)$ otherwise.)

| operations | number of bits for operation (following naive algorithm) | best known complexity (sophisticated algorithm) |
|---|---|---|
| $a \pm b$ | $\operatorname{Lg} a + \operatorname{Lg} b$ | $O(\operatorname{Lg} a + \operatorname{Lg} b)$ |
| $a \cdot b$ | $\operatorname{Lg} a \cdot \operatorname{Lg} b$ | $O(\mu(\operatorname{Lg} a, \operatorname{Lg} b))$ |
| $a = qb + r$ | $\operatorname{Lg} q \cdot \operatorname{Lg} b$ | $O(\mu(\operatorname{Lg} q, \operatorname{Lg} b))$ |

**12.** If $b$ is a positive integer greater than 1 and $x$ is a real number with $0 \le x < 1$, then $x$ can be uniquely written as $x = \sum_{j=1}^{\infty} \frac{c_j}{b^j}$ where $c_j$ is a nonnegative integer less than $b$ for all $j$, with the restriction that for every positive integer $N$ there is an integer $n$ with $n > N$ and $c_n \ne b - 1$ (in other words, it is not the case that from some point on, all the coefficients are $b - 1$).

**13.** A periodic or terminating base $b$ expansion, where $b$ is a positive integer, represents a rational number.

**14.** The base $b$ expansion of a rational number, where $b$ is a positive integer, either terminates or is periodic.

**15.** If $0 < x < 1$, $x = \frac{r}{s}$ where $r$ and $s$ are relatively prime positive integers, and $s = TU$ where every prime factor of $T$ divides $b$ and $\gcd(U, b) = 1$, then the period length of the base $b$ expansion of $x$ is $\operatorname{ord}_U b$ (defined in §4.7.1) and the pre-period length is the smallest positive integer $N$ such that $T$ divides $b^N$.

**16.** The period length of the base $b$ expansion of $\frac{1}{m}$ ($b$ and $m$ positive integers greater than 1) is $m - 1$ if and only if $m$ is prime and $b$ is a primitive root of $m$. (See §4.7.1.)

## Examples:

**1.** The binary (base 2), octal (base 8), and hexadecimal (base 16) expansions of the integer 2001 are $(11111010001)_2$, $(3721)_8$, and $(7D1)_{16}$, respectively. The octal and hexadecimal expansions can be obtained from the binary expansion by grouping together, from the right, the bits of the binary expansion into groups of three bits and four bits, respectively.

**2.** The hexadecimal expansion $2FB3$ can be converted to a binary expansion by replacing each hex digit by a block of four bits to give 10111110110011. (The initial two 0s in the four-bit expansion of the initial hex digit 2 are omitted.)

**3.** The binary coded decimal expansion of 729 is 011100101001.

**4.** The nine-bit one's complement expansions of 214 and $-113$ (taking $n = 9$ in the definition) are 011010110 and 110001110.

**5.** The nine-bit two's complement expansions of 214 and $-113$ (taking $n = 9$ in the definition) are 011010110 and 110001111.

**6.** By Fact 7 the integer with a nine-bit one's complement representation of 101110111 equals $-1(256 - 1) + 119 = -136$.

**7.** By Fact 8 the integer with a nine-bit two's complement representation of 101110111 equals $-256 + 119 = -137$.

**8.** By Fact 15 the pre-period of the decimal expansion of $\frac{5}{28}$ has length 2 and the period has length 6 since $28 = 4 \cdot 7$ and $\operatorname{ord}_7 10 = 6$. This is verified by noting that $\frac{5}{28} = (.17\overline{857142})_{10}$.

## 4.2  GREATEST COMMON DIVISORS

The concept of the greatest common divisor of two integers plays an important role in number theory. The Euclidean algorithm, an algorithm for computing greatest common divisors, was known in ancient times and was one of the first algorithms that was studied for what is now called its computational complexity. The Euclidean algorithm and its extensions are used extensively in number theory and its applications, including those to cryptography. For more information about the contents of this section consult [HaWr08], [NiZuMo91], or [Ro10].

### 4.2.1  INTRODUCTION

**Definitions:**

The **greatest common divisor** of the integers $a$ and $b$, not both zero, written $\gcd(a, b)$, is the largest integer that divides both $a$ and $b$.

The integers $a$ and $b$ are **relatively prime** (or **coprime**) if they have no positive divisors in common other than 1, i.e., if $\gcd(a, b) = 1$.

The **greatest common divisor** of the integers $a_i$, $i = 1, 2, \ldots, k$, not all zero, written $\gcd(a_1, a_2, \ldots, a_k)$, is the largest integer that divides all the integers $a_i$.

The integers $a_1, a_2, \ldots, a_k$ are **pairwise relatively prime** if $\gcd(a_i, a_j) = 1$ for $i \neq j$.

The integers $a_1, a_2, \ldots, a_k$ are **mutually relatively prime** if $\gcd(a_1, a_2, \ldots, a_k) = 1$.

The **least common multiple** of nonzero integers $a$ and $b$, written $\operatorname{lcm}(a, b)$, is the smallest positive integer that is a multiple of both $a$ and $b$.

The **least common multiple** of nonzero integers $a_1, \ldots, a_k$, written $\operatorname{lcm}(a_1, \ldots, a_k)$, is the smallest positive integer that is a multiple of all the integers $a_i$, $i = 1, 2, \ldots, k$.

The **Farey series of order $n$** is the set of fractions $\frac{h}{k}$ in which $h$ and $k$ are integers, $0 \leq h \leq k \leq n$, $k \neq 0$, and $\gcd(h, k) = 1$, in ascending order, with 0 and 1 included in the forms $\frac{0}{1}$ and $\frac{1}{1}$, respectively.

**Facts:**

1. If $d|a$ and $d|b$, then $d|\gcd(a, b)$.
2. If $a|m$ and $b|m$, then $\operatorname{lcm}(a, b)|m$.
3. If $a$ is a positive integer, then $\gcd(0, a) = a$.
4. If $a$ and $b$ are positive integers with $a < b$, then $\gcd(a, b) = \gcd(b \bmod a, a)$.
5. If $a$ and $b$ are integers with $\gcd(a, b) = d$, then $\gcd(\frac{a}{d}, \frac{b}{d}) = 1$.
6. If $a$, $b$, and $c$ are integers, then $\gcd(a + cb, b) = \gcd(a, b)$.
7. If $a$, $b$, and $c$ are integers with not both $a$ and $b$ zero and $c \neq 0$, then $\gcd(ac, bc) = |c|\gcd(a, b)$.
8. If $a$ and $b$ are integers with $\gcd(a, b) = 1$, then $\gcd(a + b, a - b) = 1$ or 2. (This greatest common divisor is 2 when both $a$ and $b$ are odd.)
9. If $a$, $b$, and $c$ are integers with $\gcd(a, b) = \gcd(a, c) = 1$, then $\gcd(a, bc) = 1$.

**10.** If $a$, $b$, and $c$ are mutually relatively prime nonzero integers, then $\gcd(a, bc) = \gcd(a, b) \cdot \gcd(a, c)$.

**11.** If $a$ and $b$ are integers, not both zero, then $\gcd(a, b)$ is the least positive integer of the form $ma + nb$ where $m$ and $n$ are integers.

**12.** The probability that two randomly selected integers are relatively prime is $\frac{6}{\pi^2}$. More precisely, if $R(n)$ equals the number of pairs of integers $a, b$ with $1 \leq a \leq n$, $1 \leq b \leq n$, and $\gcd(a, b) = 1$, then $\frac{R(n)}{n^2} = \frac{6}{\pi^2} + O(\frac{\log n}{n})$.

**13.** If $a$ and $b$ are positive integers, then $\gcd(2^a - 1, 2^b - 1) = 2^{(a,b)} - 1$.

**14.** If $a$, $b$, and $c$ are integers and $a|bc$ and $\gcd(a, b) = 1$, then $a|c$.

**15.** If $a$, $b$, and $c$ are integers, $a|c$, $b|c$ and $\gcd(a, b) = 1$, then $ab|c$.

**16.** If $a_1, a_2, \ldots, a_k$ are integers, not all zero, then $\gcd(a_1, \ldots, a_k)$ is the least positive integer that is a linear combination with integer coefficients of $a_1, \ldots, a_k$.

**17.** If $a_1, a_2, \ldots, a_k$ are integers, not all zero, and $d|a_i$ holds for $i = 1, 2, \ldots, k$, then $d| \gcd(a_1, a_2, \ldots, a_k)$.

**18.** If $a_1, \ldots, a_n$ are integers, not all zero, then the greatest common divisor of these $n$ integers is the same as the greatest common divisor of the set of $n - 1$ integers made up of the first $n - 2$ integers and the greatest common divisor of the last two. That is, $\gcd(a_1, \ldots, a_n) = \gcd(a_1, \ldots, a_{n-2}, \gcd(a_{n-1}, a_n))$.

**19.** If $a$ and $b$ are nonzero integers and $m$ is a positive integer, then $\operatorname{lcm}(ma, mb) = m \cdot \operatorname{lcm}(a, b)$

**20.** If $b$ is a common multiple of the integers $a_1, a_2, \ldots, a_k$, then $b$ is a multiple of $\operatorname{lcm}(a_1, \ldots, a_k)$.

**21.** The common multiples of the integers $a_1, \ldots, a_k$ are the integers $0$, $\operatorname{lcm}(a_1, \ldots, a_k)$, $2 \cdot \operatorname{lcm}(a_1, \ldots, a_k), \ldots$.

**22.** If $a_1, a_2, \ldots, a_n$ are pairwise relatively prime integers, then $\operatorname{lcm}(a_1, \ldots, a_n) = a_1 a_2 \ldots a_n$.

**23.** If $a_1, a_2, \ldots, a_n$ are integers, not all zero, then $\operatorname{lcm}(a_1, a_2, \ldots, a_{n-1}, a_n) = \operatorname{lcm}(\operatorname{lcm}(a_1, a_2, \ldots, a_{n-1}), a_n)$.

**24.** If $a = p_1^{a_1} p_2^{a_2} \cdots p_n^{a_n}$ and $b = p_1^{b_1} p_2^{a_2} \cdots p_n^{b_n}$, where the $p_i$ are distinct primes for $i = 1, \ldots, n$, and each exponent is a nonnegative integer, then

$$\gcd(a, b) = p_1^{\min(a_1, b_1)} p_2^{\min(a_2, b_2)} \cdots p_n^{\min(a_n, b_n)},$$

where $\min(x, y)$ denotes the minimum of $x$ and $y$, and

$$\operatorname{lcm}(a, b) = p_1^{\max(a_1, b_1)} p_2^{\max(a_2, b_2)} \cdots p_n^{\max(a_n, b_n)},$$

where $\max(x, y)$ denotes the maximum of $x$ and $y$.

**25.** If $a$ and $b$ are positive integers, then $ab = \gcd(a, b) \cdot \operatorname{lcm}(a, b)$.

**26.** If $a$, $b$, and $c$ are positive integers, then $\operatorname{lcm}(a, b, c) = \dfrac{abc \cdot \gcd(a, b, c)}{\gcd(a, b) \cdot \gcd(a, c) \cdot \gcd(b, c)}$.

**27.** If $a$, $b$, and $c$ are positive integers, then $\gcd(\operatorname{lcm}(a, b), \operatorname{lcm}(a, c)) = \operatorname{lcm}(a, \gcd(b, c))$ and $\operatorname{lcm}(\gcd(a, b), \gcd(a, c)) = \gcd(a, \operatorname{lcm}(b, c))$.

**28.** If $\frac{a}{b}$, $\frac{c}{d}$, and $\frac{e}{f}$ are successive terms of a Farey series, then $\frac{c}{d} = \frac{a+e}{b+f}$.

**29.** If $\frac{a}{b}$ and $\frac{c}{d}$ are successive terms of a Farey series, then $ad - bc = -1$.

**30.** If $\frac{a}{b}$ and $\frac{c}{d}$ are successive terms of a Farey series of order $n$, then $b + d > n$.

**31.** Farey series are named after an English geologist who published a note describing their properties in the *Philosophical Magazine* in 1816. The eminent French mathematician Cauchy supplied proofs of the properties stated, but not proved, by Farey. Also, according to [Di71], these properties had been stated and proved by Haros in 1802.

**Examples:**

**1.** $\gcd(12, 15) = 3$, $\gcd(14, 25) = 1$, $\gcd(0, 100) = 100$, and $\gcd(3, 39) = 3$.

**2.** $\gcd(2^7 3^3 5^4 7^2 11^3 17^3, 2^4 3^5 5^2 7^2 11^2 13^3) = 2^4 3^3 5^2 7^2 11^2$.

**3.** $\operatorname{lcm}(2^7 3^3 5^4 7^2 11^3 17^3, 2^4 3^5 5^2 7^2 11^2 13^3) = 2^7 3^5 5^4 7^2 11^3 13^3 17^3$.

**4.** $\gcd(18, 24, 36) = 6$ and $\gcd(10, 25, 35, 245) = 5$.

**5.** The integers 15, 21, and 35 are mutually relatively prime since $\gcd(15, 21, 35) = 1$. However, they are not pairwise relatively prime since $\gcd(15, 35) = 5$.

**6.** The integers 6, 35, and 143 are both mutually relatively prime and pairwise relatively prime.

**7.** The Farey series of order 5 is $\frac{0}{1}, \frac{1}{5}, \frac{1}{4}, \frac{1}{3}, \frac{2}{5}, \frac{1}{2}, \frac{3}{5}, \frac{2}{3}, \frac{3}{4}, \frac{4}{5}, \frac{1}{1}$.

---

### 4.2.2 THE EUCLIDEAN ALGORITHM

Finding the greatest common divisor of two integers is one of the most common problems in number theory and its applications. An algorithm for this task was known in ancient times by Euclid. His algorithm and its extensions are among the most commonly used algorithms. For more information about these algorithms see [BaSh96] and [Kn97].

**Definition:**

The **Euclidean algorithm** is an algorithm that computes the greatest common divisor of two integers $a$ and $b$ with $a \le b$, by replacing them with $a$ and $b \bmod a$, and repeating this step until one of the integers reached is zero.

**Facts:**

**1.** *The Euclidean algorithm:* The greatest common divisor of two positive integers can be computed using the recurrence given in §4.2.1, Fact 4 together with §4.1.2, Fact 3. The resulting algorithm proceeds by successively replacing a pair of positive integers with a new pair of integers formed from the smaller of the two integers and the remainder when the larger is divided by the smaller, stopping once a zero remainder is reached. The last nonzero remainder is the greatest common divisor of the original two integers. (See Algorithm 1.)

---

**Algorithm 1:** The Euclidean algorithm.

**procedure** $gcd(a, b$: positive integers)

  $r_0 := a$; $r_1 := b$

  $i := 1$

  **while** $r_i \ne 0$

    $r_{i+1} := r_{i-1} \bmod r_i$

    $i := i + 1$

  $\{\gcd(a, b) \text{ is } r_{i-1}\}$

**2.** *Lamé's theorem:*    The number of divisions needed to find the greatest common divisor of two positive integers using the Euclidean algorithm does not exceed five times the number of decimal digits in the smaller of the two integers. (This was proved by Gabriel Lamé (1795–1870).) (See [BaSh96] or [Ro10] for a proof.)

**3.** The Euclidean algorithm finds the greatest common divisor of the Fibonacci numbers (§3.1.2) $F_{n+1}$ and $F_{n+2}$ (where $n$ is a positive integer) using exactly $n$ division steps. If the Euclidean algorithm uses exactly $n$ division steps to find the greatest common divisor of the positive integers $a$ and $b$ (with $a < b$), then $a \geq F_{n+1}$ and $b \geq F_{n+2}$.

**4.** The Euclidean algorithm uses $O((\log b)^3)$ bit operations to find the greatest common divisor of two integers $a$ and $b$ with $a < b$.

**5.** The Euclidean algorithm uses $O(\text{Lg } a \cdot \text{Lg } b)$ bit operations to find the greatest common divisor of two integers $a$ and $b$.

**6.** *Least remainder Euclidean algorithm:*    The greatest common divisor of two integers $a$ and $b$ (with $a < b$) can be found by replacing $a$ and $b$ with $a$ and the least remainder of $b$ when divided by $a$. (The *least remainder* of $b$ when divided by $a$ is the integer of smallest absolute value congruent to $b$ modulo $a$. It equals $b \bmod a$ if $b \bmod a \leq \frac{a}{2}$, and $(b \bmod a) - a$ if $b \bmod a > \frac{a}{2}$.) Repeating this procedure until a remainder of zero is reached produces the greatest common divisor of $a$ and $b$ as the last nonzero remainder.

**7.** The number of divisions used by the least remainder Euclidean algorithm to find the greatest common divisor of two integers is less than or equal to the number of divisions used by the Euclidean algorithm to find this greatest common divisor.

**8.** *Binary greatest common divisor algorithm:*    The greatest common divisor of two integers $a$ and $b$ can also be found using an algorithm known as the *binary greatest common divisor algorithm*. It is based on the following reductions: if $a$ and $b$ are both even, then $\gcd(a,b) = 2\gcd(\frac{a}{2}, \frac{b}{2})$; if $a$ is even and $b$ is odd, then $\gcd(a,b) = \gcd(\frac{a}{2}, b)$ (and if $a$ is odd and $b$ is even, switch them); and if $a$ and $b$ are both odd, then $\gcd(a,b) = \gcd(\frac{|a-b|}{2}, b)$. To stop, the algorithm uses the rule that $\gcd(a,a) = a$.

**9.** *Extended Euclidean algorithm:*    The extended Euclidean algorithm finds $\gcd(a,b)$ and expresses it in the form $\gcd(a,b) = ma + nb$ for some integers $m$ and $n$. The two-pass version proceeds by first working through the steps of the Euclidean algorithm to find $\gcd(a,b)$, and then working backwards through the steps to express $\gcd(a,b)$ as a linear combination of each pair of successive remainders until the original integers $a$ and $b$ are reached. The one-pass version of this algorithm keeps track of how each successive remainder can be expressed as a linear combination of successive remainders. When the last step is reached both $\gcd(a,b)$ and integers $m$ and $n$ with $\gcd(a,b) = ma + nb$ are produced. The one-pass version is displayed as Algorithm 2.

---

**Algorithm 2:    The extended Euclidean algorithm.**

**procedure** $gcdex(a, b$: positive integers)

$\quad r_0 := a; \; r_1 := b$

$\quad m_0 := 1; \; m_1 = 0$

$\quad n_0 := 0; \; n_1 := 1$

$\quad i := 1$

$\quad$ **while** $r_i \neq 0$

$\quad\quad r_{i+1} := r_{i-1} \bmod r_i$

$\quad\quad m_{i+1} := m_{i-1} - \left\lfloor \frac{r_{i-1}}{r_i} \right\rfloor m_i$

$$n_{i+1} := n_{i-1} - \left\lfloor \frac{r_{i-1}}{r_i} \right\rfloor n_i$$
$$i := i + 1$$
$$\{\gcd(a,b) \text{ is } r_{i-1} \text{ and } \gcd(a,b) = m_{i-1}a + n_{i-1}b\}$$

**Examples:**

**1.** When the Euclidean algorithm is used to find $\gcd(53, 77)$, the following steps result:

$$77 = 1 \cdot 53 + 24,$$
$$53 = 2 \cdot 24 + 5,$$
$$24 = 4 \cdot 5 + 4,$$
$$5 = 1 \cdot 4 + 1,$$
$$4 = 4 \cdot 1.$$

This shows that $\gcd(53, 77) = 1$. Working backwards through these steps to perform the two-pass version of the Euclidean algorithm gives

$$1 = 5 - 1 \cdot 4$$
$$= 5 - 1 \cdot (24 - 4 \cdot 5) \;=\; 5 \cdot 5 - 1 \cdot 24$$
$$= 5 \cdot (53 - 2 \cdot 24) - 1 \cdot 24 \;=\; 5 \cdot 53 - 11 \cdot 24$$
$$= 5 \cdot 53 - 11 \cdot (77 - 1 \cdot 53) \;=\; 16 \cdot 53 - 11 \cdot 77.$$

**2.** The steps of the least-remainder algorithm when used to compute $\gcd(57, 93)$ are

$$\gcd(57, 93) = \gcd(57, 21) = \gcd(21, 6) = \gcd(6, 3) = 3.$$

**3.** The steps of the binary GCD algorithm when used to compute $\gcd(108, 194)$ are

$$\gcd(108, 194) = 2 \cdot \gcd(54, 97) = 2 \cdot \gcd(27, 97) = 2 \cdot \gcd(27, 35)$$
$$= 2 \cdot \gcd(4, 35) = 2 \cdot \gcd(2, 35) = 2 \cdot \gcd(1, 35) = 2.$$

## 4.3   CONGRUENCES

### 4.3.1   INTRODUCTION

**Definitions:**

If $m$ is a positive integer and $a$ and $b$ are integers, then **$a$ *is congruent to $b$ modulo $m$***, written $a \equiv b \pmod{m}$, if $m$ divides $a - b$. If $m$ does not divide $a - b$, $a$ and $b$ are ***incongruent*** modulo $m$, written $a \not\equiv b \pmod{m}$.

A ***complete system of residues modulo $m$*** is a set of integers such that every integer is congruent modulo $m$ to exactly one of the integers in the set.

If $m$ is a positive integer and $a$ is an integer with $a = bm + r$, where $0 \le r \le m - 1$, then $r$ is the ***least nonnegative residue of $a$ modulo $m$***. When $a$ is not divisible by $m$, $r$ is the ***least positive residue of $a$ modulo $m$***.

The **congruence class of a modulo m** is the set of integers congruent to $a$ modulo $m$ and is written $[a]_m$. Any integer in $[a]_m$ is called a **representative** of this class.

If $m$ is a positive integer and $a$ is an integer relatively prime to $m$, then $\bar{a}$ is an **inverse of a modulo m** if $a\bar{a} \equiv 1 \pmod{m}$. An inverse of $a$ modulo $m$ is also written $a^{-1} \bmod m$.

If $m$ is a positive integer, then a **reduced residue system modulo m** is a set of integers such that every integer relatively prime to $m$ is congruent modulo $m$ to exactly one integer in the set.

If $m$ is a positive integer, the set of congruence classes modulo $m$ is written $\mathcal{Z}_m$. (See §5.2.1.)

If $m$ is a positive integer greater than 1, the set of congruence classes of elements relatively prime to $m$ is written $\mathcal{Z}_m^{\star}$; that is, $\mathcal{Z}_m^{\star} = \{[a]_m \in \mathcal{Z}_m \mid \gcd(a, n) = 1\}$. (See §5.2.1.)

**Facts:**

1. If $m$ is a positive integer and $a$, $b$, and $c$ are integers, then

   - $a \equiv a \pmod{m}$;
   - $a \equiv b \pmod{m}$ if and only if $b \equiv a \pmod{m}$;
   - if $a \equiv b \pmod{m}$ and $b \equiv c \pmod{m}$, then $a \equiv c \pmod{m}$.

Consequently, congruence modulo $m$ is an equivalence relation. (See §1.4.2 and §5.2.1.)

2. If $m$ is a positive integer and $a$ is an integer, then $m$ divides $a$ if and only if $a \equiv 0 \pmod{m}$.

3. If $m$ is a positive integer and $a$ and $b$ are integers with $a \equiv b \pmod{m}$, then $\gcd(a, m) = \gcd(b, m)$.

4. If $a$, $b$, $c$, and $m$ are integers with $m$ positive and $a \equiv b \pmod{m}$, then $a + c \equiv b + c \pmod{m}$, $a - c \equiv b - c \pmod{m}$, and $ac \equiv bc \pmod{m}$.

5. If $m$ is a positive integer and $a$, $b$, $c$, and $d$ are integers with $a \equiv b \pmod{m}$ and $c \equiv d \pmod{m}$, then $ac \equiv bd \pmod{m}$.

6. If $a$, $b$, $c$, and $m$ are integers, $m$ is positive, $d = \gcd(c, m)$, and $ac \equiv bc \pmod{m}$, then $a \equiv b \pmod{\frac{m}{d}}$.

7. If $a$, $b$, $c$, and $m$ are integers, $m$ is positive, and $c$ and $m$ are relatively prime, and $ac \equiv bc \pmod{m}$, then $a \equiv b \pmod{m}$.

8. If $a$, $b$, $k$, and $m$ are integers with $k$ and $m$ positive and $a \equiv b \pmod{m}$, then $a^k \equiv b^k \pmod{m}$.

9. Suppose $a$, $b$, and $m$ are integers with $a \equiv b \pmod{m}$. If $c$ is an integer, it does not necessarily follow that $c^a \equiv c^b \pmod{m}$.

10. If $f(x_1, \ldots, x_n)$ is a polynomial with integer coefficients and $a_1, \ldots, a_n, b_1, \ldots, b_n$ are integers with $a_i \equiv b_i \pmod{m}$ for all $i$, then $f(a_1, \ldots, a_n) \equiv f(b_1, \ldots, b_n) \pmod{m}$.

11. If $a$, $b$, and $m_i$ are integers with $m_i$ positive and $a \equiv b \pmod{m_i}$ for $i = 1, 2, \ldots, k$, then $a \equiv b \pmod{\operatorname{lcm}(m_1, m_2, \ldots, m_k)}$.

12. If $a$ and $b$ are integers, $m_i$ $(i = 1, 2, \ldots, k)$ are pairwise relatively prime positive integers, and $a \equiv b \pmod{m_i}$ for $i = 1, 2, \ldots, k$, then $a \equiv b \pmod{m_1 m_2 \ldots m_k}$.

13. The congruence class $[a]_m$ is the set of integers $\{a, a \pm m, a \pm 2m, \ldots\}$. If $a \equiv b \pmod{m}$, then $[a]_m = [b]_m$. The congruence classes modulo $m$ are the equivalence classes of the congruence modulo $m$ equivalence relation. (See §5.2.1.)

**14.** Addition, subtraction, and multiplication of congruence classes modulo $m$, where $m$ is a positive integer, are defined by $[a]_m + [b]_m = [a+b]_m$, $[a]_m - [b]_m = [a-b]_m$, and $[a]_m[b]_m = [ab]_m$. Each of these operations is well defined, in the sense that using representatives of the congruence classes other than $a$ and $b$ does not change the resulting congruence class.

**15.** If $m$ is a positive integer, then $(\mathcal{Z}_n, +)$, where $+$ is the operation of addition of congruence classes defined in Fact 14 and in §5.2.1, is an abelian group. The identity element in this group is $[0]_m$ and the inverse of $[a]_m$ is $[-a]_m = [m-a]_m$.

**16.** If $m$ is a positive integer greater than 1 and $a$ is relatively prime to $m$, then $a$ has an inverse modulo $m$.

**17.** An inverse of $a$ modulo $m$, where $m$ is a positive integer and $\gcd(a, m) = 1$, may be found by using the extended Euclidean algorithm to find integers $x$ and $y$ such that $ax + my = 1$, which implies that $x$ is an inverse of $a$ modulo $m$.

**18.** If $m$ is a positive integer, then $(\mathcal{Z}_m^\star, \cdot)$, where $\cdot$ is the multiplication operation on congruence classes, is an abelian group. (See §5.2.1.) The identity element of this group is $[1]_m$ and the inverse of the class $[a]_m$ is the class $[\bar{a}]_m$, where $\bar{a}$ is an inverse of $a$ modulo $m$.

**19.** If $a_i$ $(i = 1, \ldots, m)$ is a complete residue system modulo $m$, where $m$ is a positive integer, and $r$ and $s$ are integers with $\gcd(m, r) = 1$, then $ra_i + s$ is a complete system of residues modulo $m$.

**20.** If $a$ and $b$ are integers and $m$ is a positive integer with $0 \le a < m$ and $0 \le b < m$, then $(a+b) \bmod m = a+b$ if $a+b < m$, and $(a+b) \bmod m = a+b-m$ if $a+b \ge m$.

**21.** Computing the least positive residue modulo $m$ of powers of integers is important in cryptology (see Chapter 15). An efficient algorithm for computing $b^n \bmod m$ where $n$ is a positive integer with binary expansion $n = (a_{k-1} \ldots a_1 a_0)_2$ is to find the least positive residues of $b, b^2, b^4, \ldots, b^{2^{k-1}}$ modulo $m$ by successively squaring and reducing modulo $m$, multiplying together the least positive residues modulo $m$ of $b^{2^j}$ for those $j$ with $a_j = 1$, reducing modulo $m$ after each multiplication.

**22.** *Wilson's theorem:* If $p$ is prime, then $(p-1)! \equiv -1 \pmod p$.

**23.** If $n$ is a positive integer greater than 1 such that $(n-1)! \equiv -1 \pmod n$ then $n$ is prime.

**24.** *Fermat's little theorem:* If $p$ is a prime and $a$ is an integer not divisible by $p$ then $a^{p-1} \equiv 1 \pmod p$.

**25.** *Euler's theorem:* If $m$ is a positive integer and $a$ is an integer relatively prime to $m$, then $a^{\phi(m)} \equiv 1 \pmod m$, where $\phi(m)$ is the number of positive integers not exceeding $m$ that are relatively prime to $m$.

**26.** If $a$ is an integer and $p$ is a prime that does not divide $a$, then from Fermat's little theorem it follows that $a^{p-2}$ is an inverse of $a$ modulo $p$.

**27.** If $a$ and $m$ are relatively prime integers with $m > 1$, then $a^{\phi(m)-1}$ is an inverse of $a$ modulo $m$. This follows directly from Euler's theorem.

**28.** *Linear congruential method:* One of the most commonly used methods for generating pseudo-random numbers is the *linear congruential method*. It starts with integers $m$, $a$, $c$, and $x_0$ where $2 \le a < m$, $0 \le c < m$, and $0 \le x_0 \le m$. The sequence of pseudo-random numbers is defined recursively by

$$x_{n+1} = (ax_n + c) \bmod m, \quad n = 0, 1, 2, 3, \ldots.$$

Here $m$ is the *modulus*, $a$ is the *multiplier*, $c$ is the *increment*, and $x_0$ is the *seed* of the generator.

**29.** Big-oh estimates for the number of bit operations required to do modular addition, modular subtraction, modular multiplication, modular inversion, and modular exponentiation are summarized in the following table.

| name | operation | number of bit operations |
|---|---|---|
| modular addition | $(a+b) \bmod m$ | $O(\log m)$ |
| modular subtraction | $(a-b) \bmod m$ | $O(\log m)$ |
| modular multiplication | $(a \cdot b) \bmod m$ | $O(\log^2 m)$ |
| modular inversion | $(a^{-1}) \bmod m$ | $O(\log^2 m)$ |
| modular exponentiation | $a^k \bmod m, \; k < m$ | $O(\log^3 m)$ |

**Examples:**

**1.** $23 \equiv 5 \pmod 9$, $-17 \equiv 13 \pmod{15}$, and $99 \equiv 0 \pmod{11}$, but $11 \not\equiv 3 \pmod 5$, $-3 \not\equiv 8 \pmod 6$, and $44 \not\equiv 0 \pmod 7$.

**2.** To find an inverse of 53 modulo 77, use the extended Euclidean algorithm to obtain $16 \cdot 53 - 11 \cdot 77 = 1$ (see Example 1 of §4.2.2). This implies that 16 is an inverse of 53 modulo 77.

**3.** Since 11 is prime, by Wilson's theorem it follows that $10! \equiv -1 \pmod{11}$.

**4.** $5! \equiv 0 \pmod 6$, which provides an impractical verification that 6 is not prime.

**5.** To find the least positive residue of $3^{201}$ modulo 11, note that by Fermat's little theorem $3^{10} \equiv 1 \pmod{11}$. Hence $3^{201} = (3^{10})^{20} \cdot 3 \equiv 3 \pmod{11}$.

**6.** *Zeller's congruence:*   A congruence can be used to determine the day of the week of any date in the Gregorian calendar, the calendar used in most of the world. Let $w$ represent the day of the week, with $w = 0, 1, 2, 3, 4, 5, 6$ for Sunday, Monday, Tuesday, Wednesday, Thursday, Friday, Saturday, respectively. Let $k$ represent the day of the month. Let $m$ represent the month with $m = 11, 12, 1, 2, 3, 4, 5, 6, 7, 8, 9, 10$ for January, February, March, April, May, June, July, August, September, October, November, December, respectively. Let $N$ represent the previous year if the month is January or February or the current year otherwise, with $C$ the century of $N$ and $Y$ the particular year of the century of $N$ so that $N = 100Y + C$. Then the day of the week can be found using the congruence

$$w \equiv k + \lfloor 2.6m - 0.2 \rfloor - 2C + Y + \lfloor \tfrac{Y}{4} \rfloor + \lfloor \tfrac{C}{4} \rfloor \pmod 7.$$

**7.** January 1, 1900 was a Monday. This follows by Zeller's congruence with $C = 18$, $Y = 99$, $m = 11$, and $k = 1$, noting that to apply this congruence January is considered the eleventh month of the preceding year.

## 4.3.2  LINEAR AND POLYNOMIAL CONGRUENCES

**Definitions:**

A *linear congruence in one variable* is a congruence of the form $ax \equiv b \pmod m$, where $a$, $b$, and $m$ are integers, $m$ is positive, and $x$ is an unknown.

If $f$ is a polynomial with integer coefficients, an integer $r$ is a *solution* of the congruence $f(x) \equiv 0 \pmod m$, or a *root* of $f(x)$ modulo $m$, if $f(r) \equiv 0 \pmod m$.

## Facts:

**1.** If $a$, $b$, and $m$ are integers, $m$ is positive, and $\gcd(a, m) = d$, then the congruence $ax \equiv b \pmod{m}$ has exactly $d$ incongruent solutions modulo $m$ if $d|b$, and no solutions if $d \nmid b$.

**2.** If $a$, $b$, and $m$ are integers, $m$ is positive, and $\gcd(a, m) = 1$, then the solutions of $ax \equiv b \pmod{m}$ are all integers $x$ with $x \equiv \bar{a}b \pmod{m}$.

**3.** If $a$ and $b$ are positive integers and $p$ is a prime that does not divide $a$, then the solutions of $ax \equiv b \pmod{p}$ are the integers $x$ with $x \equiv a^{p-2}b \pmod{p}$.

**4.** *Thue's lemma:*  If $p$ is a prime and $a$ is an integer not divisible by $p$, then the congruence $ax \equiv y \pmod{p}$ has a solution $x_0, y_0$ with $0 < |x_0| < \sqrt{p}$, $0 < |y_0| < \sqrt{p}$.

**5.** *Chinese remainder theorem:*  If $m_i$, $i = 1, 2, \ldots, r$, are pairwise relatively prime positive integers, then the system of simultaneous congruences $x \equiv a_i \pmod{m_i}$, $i = 1, 2, \ldots, r$, has a unique solution modulo $M = m_1 m_2 \ldots m_r$ which is given by $x \equiv a_1 M_1 y_1 + a_2 M_2 y_2 + \cdots + a_r M_r y_r$ where $M_k = \frac{M}{m_k}$ and $y_k$ is an inverse of $M_k$ modulo $m_k$, $k = 1, 2, \ldots, r$.

**6.** Problems involving the solution of a system of simultaneous congruences arose in the writing of ancient mathematicians, including the Chinese mathematician Sun-Tsu, and in other works by Indian and Greek mathematicians. (See [Di71] for details.)

**7.** The system of simultaneous congruences $x \equiv a_i \pmod{m_i}$, $i = 1, 2, \ldots, r$, has a solution if and only if $\gcd(m_i, m_j)$ divides $a_i - a_j$ for all pairs of integers $(i, j)$ with $1 \leq i < j \leq r$. If a solution exists, it is unique modulo $\operatorname{lcm}(m_1, m_2, \ldots, m_r)$.

**8.** If $a, b, c, d, e, f$, and $m$ are integers with $m$ positive such that $\gcd(ad - bc, m) = 1$, then the system of congruences $ax + by \equiv e \pmod{m}$, $cx + dy \equiv f \pmod{m}$ has a unique solution given by $x \equiv g(de - bf) \pmod{m}$, $y \equiv g(af - ce) \pmod{m}$ where $g$ is an inverse of $ad - bc$ modulo $m$.

**9.** *Lagrange's theorem:*  If $p$ is prime, then the polynomial $f(x) = a_n x^n + \cdots + a_1 x + a_0$ with $a_n \not\equiv 0 \pmod{p}$ has at most $n$ roots modulo $p$.

**10.** If $f(x) = a_n x^n + \cdots + a_1 x + a_0$, where $a_i$ $(i = 1, \ldots, n)$ is an integer and $p$ is prime, has more than $n$ roots modulo $p$, then $p$ divides $a_i$ for all $i = 1, \ldots, n$.

**11.** If $m_1, m_2, \ldots, m_r$ are pairwise relatively prime positive integers with product $m = m_1 m_2 \ldots m_r$, and $f$ is a polynomial with integer coefficients, then $f(x)$ has a root modulo $m$ if and only if $f(x)$ has a root modulo $m_i$, for all $i = 1, 2, \ldots, r$. Furthermore, if $f(x)$ has $n_i$ incongruent roots modulo $m_i$ and $n$ incongruent roots modulo $m$, then $n = n_1 n_2 \ldots n_r$.

**12.** If $p$ is prime, $k$ is a positive integer, and $s$ is a root of $f(x)$ modulo $p^k$, then

- if $p \nmid f'(s)$, then there is a unique root $t$ of $f(x)$ modulo $p^{k+1}$ with $t \equiv s \pmod{p^k}$, namely $t = s + p^k u$ where $u$ is the unique solution of $f'(s)u \equiv -f(s)/p^k \pmod{p}$;

- if $p|f'(s)$ and $p^{k+1}|f(s)$, then there are exactly $p$ incongruent roots of $f(x)$ modulo $p^{k+1}$ congruent to $s$ modulo $p$, given by $s + p^k i$, $i = 0, 1, \ldots, p-1$;

- if $p|f'(s)$ and $p^{k+1} \nmid f(s)$, then there are no roots of $f(x)$ modulo $p^{k+1}$ that are congruent to $s$ modulo $p^k$.

**13.** *Finding roots of a polynomial modulo $m$, where $m$ is a positive integer:*  First find roots of the polynomial modulo $p^r$ for each prime power in the prime-power factorization of $m$ (Fact 14) and then use the Chinese remainder theorem (Fact 5) to find solutions modulo $m$.

**14.** Finding solutions modulo $p^r$ reduces to first finding solutions modulo $p$. In particular, if there are no roots of $f(x)$ modulo $p$, there are no roots of $f(x)$ modulo $p^r$. If $f(x)$

has roots modulo $p$, choose one, say $r$ with $0 \le r < p$. By Fact 12, corresponding to $r$ there are 0, 1, or $p$ roots of $f(x)$ modulo $p^2$.

**Examples:**

1.  There are three incongruent solutions of $6x \equiv 9 \pmod{15}$ since $\gcd(6, 15) = 3$ and $3|9$. The solutions are those integers $x$ with $x \equiv 4, 9$, or 14 (mod 15).

2.  The linear congruence $2x \equiv 7 \pmod 6$ has no solutions since $\gcd(2, 6) = 2$ and $2 \nmid 7$.

3.  The solutions of the linear congruence $3x \equiv 5 \pmod{11}$ are those integers $x$ with $x \equiv \bar{3} \cdot 5 \equiv 4 \cdot 5 \equiv 9 \pmod{11}$.

4.  It follows from the Chinese remainder theorem (Fact 5) that the solutions of the systems of simultaneous congruences $x \equiv 1 \pmod 3$, $x \equiv 2 \pmod 4$, and $x \equiv 3 \pmod 5$ are all integers $x$ with $x \equiv 1 \cdot 20 \cdot 2 + 2 \cdot 15 \cdot 3 + 3 \cdot 12 \cdot 3 \equiv 58 \pmod{60}$.

5.  The simultaneous congruences $x \equiv 4 \pmod 9$ and $x \equiv 7 \pmod{15}$ can be solved by noting that the first congruence implies that $x - 4 = 9t$ for some integer $t$, so that $x = 9t + 4$. Inserting this expression for $x$ into the second congruence gives $9t + 4 \equiv 7 \pmod{15}$. This implies that $3t \equiv 1 \pmod 5$, so that $t \equiv 2 \pmod 5$ and $t = 5u + 2$ for some integer $u$. Hence $x = 45u + 22$ for some integer $u$. The solutions of the two simultaneous congruences are those integers $x$ with $x \equiv 22 \pmod{45}$.

## 4.4   PRIME NUMBERS

One of the most powerful tools in number theory is the fact that each composite integer can be decomposed into a product of primes. Primes may be thought of as the building blocks of the integers in the sense that they can be decomposed only in trivial ways, for example, $3 = 1 \times 3$. Prime numbers, once of only theoretical interest, now are important in many applications, especially in the area of cryptography where large primes play a crucial role in the area of public key cryptosystems (see Chapter 15). From ancient to modern times, mathematicians have devoted long hours to the study of primes and their properties. Even so, many questions about primes have only partially been answered or remain complete mysteries, including questions that ask whether there are infinitely many primes of certain forms. There have been many recent discoveries concerning prime numbers, such as the discovery of new Mersenne primes or the proof that infinitely often there are pairs of primes at most a bounded distance apart.. The current state of knowledge on some of these questions and the latest discoveries are described in this section.

Additional information about primes can be found in [CrPo05] and [Ri96], as well as the *Prime Pages* website `http://www.utm.edu/research/primes/`.

### 4.4.1   BASIC CONCEPTS

**Definitions:**

A *prime* is a natural number greater than 1 that is exactly divisible only by 1 and itself.

A *composite* is a natural number greater than 1 that is not a prime. That is, a composite may be factored into the product of two natural numbers both smaller than itself.

**Facts:**

**1.** The number 1 is not considered to be prime.

**2.** The following table lists the primes less than 10,000. The prime number $p_{10n+k}$ is found by looking at the row beginning with $n..$ and at the column beginning with $..k$.

| | ..0 | ..1 | ..2 | ..3 | ..4 | ..5 | ..6 | ..7 | ..8 | ..9 |
|---|---|---|---|---|---|---|---|---|---|---|
| | | 2 | 3 | 5 | 7 | 11 | 13 | 17 | 19 | 23 |
| 1.. | 29 | 31 | 37 | 41 | 43 | 47 | 53 | 59 | 61 | 67 |
| 2.. | 71 | 73 | 79 | 83 | 89 | 97 | 101 | 103 | 107 | 109 |
| 3.. | 113 | 127 | 131 | 137 | 139 | 149 | 151 | 157 | 163 | 167 |
| 4.. | 173 | 179 | 181 | 191 | 193 | 197 | 199 | 211 | 223 | 227 |
| 5.. | 229 | 233 | 239 | 241 | 251 | 257 | 263 | 269 | 271 | 277 |
| | | | | | | | | | | |
| 6.. | 281 | 283 | 293 | 307 | 311 | 313 | 317 | 331 | 337 | 347 |
| 7.. | 349 | 353 | 359 | 367 | 373 | 379 | 383 | 389 | 397 | 401 |
| 8.. | 409 | 419 | 421 | 431 | 433 | 439 | 443 | 449 | 457 | 461 |
| 9.. | 463 | 467 | 479 | 487 | 491 | 499 | 503 | 509 | 521 | 523 |
| 10.. | 541 | 547 | 557 | 563 | 569 | 571 | 577 | 587 | 593 | 599 |
| | | | | | | | | | | |
| 11.. | 601 | 607 | 613 | 617 | 619 | 631 | 641 | 643 | 647 | 653 |
| 12.. | 659 | 661 | 673 | 677 | 683 | 691 | 701 | 709 | 719 | 727 |
| 13.. | 733 | 739 | 743 | 751 | 757 | 761 | 769 | 773 | 787 | 797 |
| 14.. | 809 | 811 | 821 | 823 | 827 | 829 | 839 | 853 | 857 | 859 |
| 15.. | 863 | 877 | 881 | 883 | 887 | 907 | 911 | 919 | 929 | 937 |
| | | | | | | | | | | |
| 16.. | 941 | 947 | 953 | 967 | 971 | 977 | 983 | 991 | 997 | 1009 |
| 17.. | 1013 | 1019 | 1021 | 1031 | 1033 | 1039 | 1049 | 1051 | 1061 | 1063 |
| 18.. | 1069 | 1087 | 1091 | 1093 | 1097 | 1103 | 1109 | 1117 | 1123 | 1129 |
| 19.. | 1151 | 1153 | 1163 | 1171 | 1181 | 1187 | 1193 | 1201 | 1213 | 1217 |
| 20.. | 1223 | 1229 | 1231 | 1237 | 1249 | 1259 | 1277 | 1279 | 1283 | 1289 |
| | | | | | | | | | | |
| 21.. | 1291 | 1297 | 1301 | 1303 | 1307 | 1319 | 1321 | 1327 | 1361 | 1367 |
| 22.. | 1373 | 1381 | 1399 | 1409 | 1423 | 1427 | 1429 | 1433 | 1439 | 1447 |
| 23.. | 1451 | 1453 | 1459 | 1471 | 1481 | 1483 | 1487 | 1489 | 1493 | 1499 |
| 24.. | 1511 | 1523 | 1531 | 1543 | 1549 | 1553 | 1559 | 1567 | 1571 | 1579 |
| 25.. | 1583 | 1597 | 1601 | 1607 | 1609 | 1613 | 1619 | 1621 | 1627 | 1637 |
| | | | | | | | | | | |
| 26.. | 1657 | 1663 | 1667 | 1669 | 1693 | 1697 | 1699 | 1709 | 1721 | 1723 |
| 27.. | 1733 | 1741 | 1747 | 1753 | 1759 | 1777 | 1783 | 1787 | 1789 | 1801 |
| 28.. | 1811 | 1823 | 1831 | 1847 | 1861 | 1867 | 1871 | 1873 | 1877 | 1879 |
| 29.. | 1889 | 1901 | 1907 | 1913 | 1931 | 1933 | 1949 | 1951 | 1973 | 1979 |
| 30.. | 1987 | 1993 | 1997 | 1999 | 2003 | 2011 | 2017 | 2027 | 2029 | 2039 |
| | | | | | | | | | | |
| 31.. | 2053 | 2063 | 2069 | 2081 | 2083 | 2087 | 2089 | 2099 | 2111 | 2113 |
| 32.. | 2129 | 2131 | 2137 | 2141 | 2143 | 2153 | 2161 | 2179 | 2203 | 2207 |

|       | ..0  | ..1  | ..2  | ..3  | ..4  | ..5  | ..6  | ..7  | ..8  | ..9  |
|-------|------|------|------|------|------|------|------|------|------|------|
| 33..  | 2213 | 2221 | 2237 | 2239 | 2243 | 2251 | 2267 | 2269 | 2273 | 2281 |
| 34..  | 2287 | 2293 | 2297 | 2309 | 2311 | 2333 | 2339 | 2341 | 2347 | 2351 |
| 35..  | 2357 | 2371 | 2377 | 2381 | 2383 | 2389 | 2393 | 2399 | 2411 | 2417 |
|       |      |      |      |      |      |      |      |      |      |      |
| 36..  | 2423 | 2437 | 2441 | 2447 | 2459 | 2467 | 2473 | 2477 | 2503 | 2521 |
| 37..  | 2531 | 2539 | 2543 | 2549 | 2551 | 2557 | 2579 | 2591 | 2593 | 2609 |
| 38..  | 2617 | 2621 | 2633 | 2647 | 2657 | 2659 | 2663 | 2671 | 2677 | 2683 |
| 39..  | 2687 | 2689 | 2693 | 2699 | 2707 | 2711 | 2713 | 2719 | 2729 | 2731 |
| 40..  | 2741 | 2749 | 2753 | 2767 | 2777 | 2789 | 2791 | 2797 | 2801 | 2803 |
|       |      |      |      |      |      |      |      |      |      |      |
| 41..  | 2819 | 2833 | 2837 | 2843 | 2851 | 2857 | 2861 | 2879 | 2887 | 2897 |
| 42..  | 2903 | 2909 | 2917 | 2927 | 2939 | 2953 | 2957 | 2963 | 2969 | 2971 |
| 43..  | 2999 | 3001 | 3011 | 3019 | 3023 | 3037 | 3041 | 3049 | 3061 | 3067 |
| 44..  | 3079 | 3083 | 3089 | 3109 | 3119 | 3121 | 3137 | 3163 | 3167 | 3169 |
| 45..  | 3181 | 3187 | 3191 | 3203 | 3209 | 3217 | 3221 | 3229 | 3251 | 3253 |
|       |      |      |      |      |      |      |      |      |      |      |
| 46..  | 3257 | 3259 | 3271 | 3299 | 3301 | 3307 | 3313 | 3319 | 3323 | 3329 |
| 47..  | 3331 | 3343 | 3347 | 3359 | 3361 | 3371 | 3373 | 3389 | 3391 | 3407 |
| 48..  | 3413 | 3433 | 3449 | 3457 | 3461 | 3463 | 3467 | 3469 | 3491 | 3499 |
| 49..  | 3511 | 3517 | 3527 | 3529 | 3533 | 3539 | 3541 | 3547 | 3557 | 3559 |
| 50..  | 3571 | 3581 | 3583 | 3593 | 3607 | 3613 | 3617 | 3623 | 3631 | 3637 |
|       |      |      |      |      |      |      |      |      |      |      |
| 51..  | 3643 | 3659 | 3671 | 3673 | 3677 | 3691 | 3697 | 3701 | 3709 | 3719 |
| 52..  | 3727 | 3733 | 3739 | 3761 | 3767 | 3769 | 3779 | 3793 | 3797 | 3803 |
| 53..  | 3821 | 3823 | 3833 | 3847 | 3851 | 3853 | 3863 | 3877 | 3881 | 3889 |
| 54..  | 3907 | 3911 | 3917 | 3919 | 3923 | 3929 | 3931 | 3943 | 3947 | 3967 |
| 55..  | 3989 | 4001 | 4003 | 4007 | 4013 | 4019 | 4021 | 4027 | 4049 | 4051 |
|       |      |      |      |      |      |      |      |      |      |      |
| 56..  | 4057 | 4073 | 4079 | 4091 | 4093 | 4099 | 4111 | 4127 | 4129 | 4133 |
| 57..  | 4139 | 4153 | 4157 | 4159 | 4177 | 4201 | 4211 | 4217 | 4219 | 4229 |
| 58..  | 4231 | 4241 | 4243 | 4253 | 4259 | 4261 | 4271 | 4273 | 4283 | 4289 |
| 59..  | 4297 | 4327 | 4337 | 4339 | 4349 | 4357 | 4363 | 4373 | 4391 | 4397 |
| 60..  | 4409 | 4421 | 4423 | 4441 | 4447 | 4451 | 4457 | 4463 | 4481 | 4483 |
|       |      |      |      |      |      |      |      |      |      |      |
| 61..  | 4493 | 4507 | 4513 | 4517 | 4519 | 4523 | 4547 | 4549 | 4561 | 4567 |
| 62..  | 4583 | 4591 | 4597 | 4603 | 4621 | 4637 | 4639 | 4643 | 4649 | 4651 |
| 63..  | 4657 | 4663 | 4673 | 4679 | 4691 | 4703 | 4721 | 4723 | 4729 | 4733 |
| 64..  | 4751 | 4759 | 4783 | 4787 | 4789 | 4793 | 4799 | 4801 | 4813 | 4817 |
| 65..  | 4831 | 4861 | 4871 | 4877 | 4889 | 4903 | 4909 | 4919 | 4931 | 4933 |
|       |      |      |      |      |      |      |      |      |      |      |
| 66..  | 4937 | 4943 | 4951 | 4957 | 4967 | 4969 | 4973 | 4987 | 4993 | 4999 |
| 67..  | 5003 | 5009 | 5011 | 5021 | 5023 | 5039 | 5051 | 5059 | 5077 | 5081 |
| 68..  | 5087 | 5099 | 5101 | 5107 | 5113 | 5119 | 5147 | 5153 | 5167 | 5171 |

|       | ..0  | ..1  | ..2  | ..3  | ..4  | ..5  | ..6  | ..7  | ..8  | ..9  |
|-------|------|------|------|------|------|------|------|------|------|------|
| 69..  | 5179 | 5189 | 5197 | 5209 | 5227 | 5231 | 5233 | 5237 | 5261 | 5273 |
| 70..  | 5279 | 5281 | 5297 | 5303 | 5309 | 5323 | 5333 | 5347 | 5351 | 5381 |
|       |      |      |      |      |      |      |      |      |      |      |
| 71..  | 5387 | 5393 | 5399 | 5407 | 5413 | 5417 | 5419 | 5431 | 5437 | 5441 |
| 72..  | 5443 | 5449 | 5471 | 5477 | 5479 | 5483 | 5501 | 5503 | 5507 | 5519 |
| 73..  | 5521 | 5527 | 5531 | 5557 | 5563 | 5569 | 5573 | 5581 | 5591 | 5623 |
| 74..  | 5639 | 5641 | 5647 | 5651 | 5653 | 5657 | 5659 | 5669 | 5683 | 5689 |
| 75..  | 5693 | 5701 | 5711 | 5717 | 5737 | 5741 | 5743 | 5749 | 5779 | 5783 |
|       |      |      |      |      |      |      |      |      |      |      |
| 76..  | 5791 | 5801 | 5807 | 5813 | 5821 | 5827 | 5839 | 5843 | 5849 | 5851 |
| 77..  | 5857 | 5861 | 5867 | 5869 | 5879 | 5881 | 5897 | 5903 | 5923 | 5927 |
| 78..  | 5939 | 5953 | 5981 | 5987 | 6007 | 6011 | 6029 | 6037 | 6043 | 6047 |
| 79..  | 6053 | 6067 | 6073 | 6079 | 6089 | 6091 | 6101 | 6113 | 6121 | 6131 |
| 80..  | 6133 | 6143 | 6151 | 6163 | 6173 | 6197 | 6199 | 6203 | 6211 | 6217 |
|       |      |      |      |      |      |      |      |      |      |      |
| 81..  | 6221 | 6229 | 6247 | 6257 | 6263 | 6269 | 6271 | 6277 | 6287 | 6299 |
| 82..  | 6301 | 6311 | 6317 | 6323 | 6329 | 6337 | 6343 | 6353 | 6359 | 6361 |
| 83..  | 6367 | 6373 | 6379 | 6389 | 6397 | 6421 | 6427 | 6449 | 6451 | 6469 |
| 84..  | 6473 | 6481 | 6491 | 6521 | 6529 | 6547 | 6551 | 6553 | 6563 | 6569 |
| 85..  | 6571 | 6577 | 6581 | 6599 | 6607 | 6619 | 6637 | 6653 | 6659 | 6661 |
|       |      |      |      |      |      |      |      |      |      |      |
| 86..  | 6673 | 6679 | 6689 | 6691 | 6701 | 6703 | 6709 | 6719 | 6733 | 6737 |
| 87..  | 6761 | 6763 | 6779 | 6781 | 6791 | 6793 | 6803 | 6823 | 6827 | 6829 |
| 88..  | 6833 | 6841 | 6857 | 6863 | 6869 | 6871 | 6883 | 6899 | 6907 | 6911 |
| 89..  | 6917 | 6947 | 6949 | 6959 | 6961 | 6967 | 6971 | 6977 | 6983 | 6991 |
| 90..  | 6997 | 7001 | 7013 | 7019 | 7027 | 7039 | 7043 | 7057 | 7069 | 7079 |
|       |      |      |      |      |      |      |      |      |      |      |
| 91..  | 7103 | 7109 | 7121 | 7127 | 7129 | 7151 | 7159 | 7177 | 7187 | 7193 |
| 92..  | 7207 | 7211 | 7213 | 7219 | 7229 | 7237 | 7243 | 7247 | 7253 | 7283 |
| 93..  | 7297 | 7307 | 7309 | 7321 | 7331 | 7333 | 7349 | 7351 | 7369 | 7393 |
| 94..  | 7411 | 7417 | 7433 | 7451 | 7457 | 7459 | 7477 | 7481 | 7487 | 7489 |
| 95..  | 7499 | 7507 | 7517 | 7523 | 7529 | 7537 | 7541 | 7547 | 7549 | 7559 |
|       |      |      |      |      |      |      |      |      |      |      |
| 96..  | 7561 | 7573 | 7577 | 7583 | 7589 | 7591 | 7603 | 7607 | 7621 | 7639 |
| 97..  | 7643 | 7649 | 7669 | 7673 | 7681 | 7687 | 7691 | 7699 | 7703 | 7717 |
| 98..  | 7723 | 7727 | 7741 | 7753 | 7757 | 7759 | 7789 | 7793 | 7817 | 7823 |
| 99..  | 7829 | 7841 | 7853 | 7867 | 7873 | 7877 | 7879 | 7883 | 7901 | 7907 |
| 100.. | 7919 | 7927 | 7933 | 7937 | 7949 | 7951 | 7963 | 7993 | 8009 | 8011 |
|       |      |      |      |      |      |      |      |      |      |      |
| 101.. | 8017 | 8039 | 8053 | 8059 | 8069 | 8081 | 8087 | 8089 | 8093 | 8101 |
| 102.. | 8111 | 8117 | 8123 | 8147 | 8161 | 8167 | 8171 | 8179 | 8191 | 8209 |
| 103.. | 8219 | 8221 | 8231 | 8233 | 8237 | 8243 | 8263 | 8269 | 8273 | 8287 |
| 104.. | 8291 | 8293 | 8297 | 8311 | 8317 | 8329 | 8353 | 8363 | 8369 | 8377 |

|       | ..0  | ..1  | ..2  | ..3  | ..4  | ..5  | ..6  | ..7  | ..8  | ..9  |
|-------|------|------|------|------|------|------|------|------|------|------|
| 105.. | 8387 | 8389 | 8419 | 8423 | 8429 | 8431 | 8443 | 8447 | 8461 | 8467 |
| 106.. | 8501 | 8513 | 8521 | 8527 | 8537 | 8539 | 8543 | 8563 | 8573 | 8581 |
| 107.. | 8597 | 8599 | 8609 | 8623 | 8627 | 8629 | 8641 | 8647 | 8663 | 8669 |
| 108.. | 8677 | 8681 | 8689 | 8693 | 8699 | 8707 | 8713 | 8719 | 8731 | 8737 |
| 109.. | 8741 | 8747 | 8753 | 8761 | 8779 | 8783 | 8803 | 8807 | 8819 | 8821 |
| 110.. | 8831 | 8837 | 8839 | 8849 | 8861 | 8863 | 8867 | 8887 | 8893 | 8923 |
| 111.. | 8929 | 8933 | 8941 | 8951 | 8963 | 8969 | 8971 | 8999 | 9001 | 9007 |
| 112.. | 9011 | 9013 | 9029 | 9041 | 9043 | 9049 | 9059 | 9067 | 9091 | 9103 |
| 113.. | 9109 | 9127 | 9133 | 9137 | 9151 | 9157 | 9161 | 9173 | 9181 | 9187 |
| 114.. | 9199 | 9203 | 9209 | 9221 | 9227 | 9239 | 9241 | 9257 | 9277 | 9281 |
| 115.. | 9283 | 9293 | 9311 | 9319 | 9323 | 9337 | 9341 | 9343 | 9349 | 9371 |
| 116.. | 9377 | 9391 | 9397 | 9403 | 9413 | 9419 | 9421 | 9431 | 9433 | 9437 |
| 117.. | 9439 | 9461 | 9463 | 9467 | 9473 | 9479 | 9491 | 9497 | 9511 | 9521 |
| 118.. | 9533 | 9539 | 9547 | 9551 | 9587 | 9601 | 9613 | 9619 | 9623 | 9629 |
| 119.. | 9631 | 9643 | 9649 | 9661 | 9677 | 9679 | 9689 | 9697 | 9719 | 9721 |
| 120.. | 9733 | 9739 | 9743 | 9749 | 9767 | 9769 | 9781 | 9787 | 9791 | 9803 |
| 121.. | 9811 | 9817 | 9829 | 9833 | 9839 | 9851 | 9857 | 9859 | 9871 | 9883 |
| 122.. | 9887 | 9901 | 9907 | 9923 | 9929 | 9931 | 9941 | 9949 | 9967 | 9973 |

**3.** *Fundamental theorem of arithmetic:* Every natural number greater than 1 is either prime or can be written as a product of prime factors in a unique way, up to the order of the prime factors. That is, every composite $n$ can be expressed uniquely as $n = p_1 p_2 \ldots p_k$, where $p_1 \leq p_2 \leq \cdots \leq p_k$ are primes. This is sometimes also known as the *unique factorization theorem.*

**4.** The unique factorization of a positive integer $n$ formed by grouping together equal prime factors produces the unique *prime-power factorization* $n = p_1^{a_1} p_2^{a_2} \ldots p_k^{a_k}$.

**5.** The following table lists the prime-power factorization of all positive integers below 2,500. Numbers appearing in boldface are prime.

|    | 0 | 1 | 2 | 3 | 4 | 5 | 6 | 7 | 8 | 9 |
|----|---|---|---|---|---|---|---|---|---|---|
| 0  |            |        | **2**      | **3**  | $2^2$      | **5**  | $2 \cdot 3$     | **7**  | $2^3$        | $3^2$         |
| 1  | $2 \cdot 5$     | **11** | $2^2 \cdot 3$   | **13** | $2 \cdot 7$     | $3 \cdot 5$ | $2^4$       | **17** | $2 \cdot 3^2$     | **19**        |
| 2  | $2^2 \cdot 5$   | $3 \cdot 7$ | $2 \cdot 11$    | **23** | $2^3 \cdot 3$   | $5^2$  | $2 \cdot 13$    | $3^3$  | $2^2 \cdot 7$     | **29**        |
| 3  | $2 \cdot 3 \cdot 5$  | **31** | $2^5$      | $3 \cdot 11$ | $2 \cdot 17$  | $5 \cdot 7$ | $2^2 \cdot 3^2$ | **37** | $2 \cdot 19$     | $3 \cdot 13$       |
| 4  | $2^3 \cdot 5$   | **41** | $2 \cdot 3 \cdot 7$  | **43** | $2^2 \cdot 11$  | $3^2 \cdot 5$ | $2 \cdot 23$    | **47** | $2^4 \cdot 3$     | $7^2$         |
| 5  | $2 \cdot 5^2$   | $3 \cdot 17$ | $2^2 \cdot 13$  | **53** | $2 \cdot 3^3$   | $5 \cdot 11$ | $2^3 \cdot 7$   | $3 \cdot 19$ | $2 \cdot 29$     | **59**        |
| 6  | $2^2 \cdot 3 \cdot 5$ | **61** | $2 \cdot 31$    | $3^2 \cdot 7$ | $2^6$      | $5 \cdot 13$ | $2 \cdot 3 \cdot 11$ | **67** | $2^2 \cdot 17$    | $3 \cdot 23$       |
| 7  | $2 \cdot 5 \cdot 7$  | **71** | $2^3 \cdot 3^2$ | **73** | $2 \cdot 37$    | $3 \cdot 5^2$ | $2^2 \cdot 19$  | $7 \cdot 11$ | $2 \cdot 3 \cdot 13$  | **79**        |
| 8  | $2^4 \cdot 5$   | $3^4$  | $2 \cdot 41$    | **83** | $2^2 \cdot 3 \cdot 7$ | $5 \cdot 17$ | $2 \cdot 43$    | $3 \cdot 29$ | $2^3 \cdot 11$    | **89**        |
| 9  | $2 \cdot 3^2 \cdot 5$ | $7 \cdot 13$ | $2^2 \cdot 23$  | $3 \cdot 31$ | $2 \cdot 47$    | $5 \cdot 19$ | $2^5 \cdot 3$   | **97** | $2 \cdot 7^2$     | $3^2 \cdot 11$     |
| 10 | $2^2 \cdot 5^2$ | **101** | $2 \cdot 3 \cdot 17$ | **103** | $2^3 \cdot 13$  | $3 \cdot 5 \cdot 7$ | $2 \cdot 53$    | **107** | $2^2 \cdot 3^3$   | **109**       |

| | 0 | 1 | 2 | 3 | 4 | 5 | 6 | 7 | 8 | 9 |
|---|---|---|---|---|---|---|---|---|---|---|
| 11 | $2 \cdot 5 \cdot 11$ | $3 \cdot 37$ | $2^4 \cdot 7$ | **113** | $2 \cdot 3 \cdot 19$ | $5 \cdot 23$ | $2^2 \cdot 29$ | $3^2 \cdot 13$ | $2 \cdot 59$ | $7 \cdot 17$ |
| 12 | $2^3 \cdot 3 \cdot 5$ | $11^2$ | $2 \cdot 61$ | $3 \cdot 41$ | $2^2 \cdot 31$ | $5^3$ | $2 \cdot 3^2 \cdot 7$ | **127** | $2^7$ | $3 \cdot 43$ |
| 13 | $2 \cdot 5 \cdot 13$ | **131** | $2^2 \cdot 3 \cdot 11$ | $7 \cdot 19$ | $2 \cdot 67$ | $3^3 \cdot 5$ | $2^3 \cdot 17$ | **137** | $2 \cdot 3 \cdot 23$ | **139** |
| 14 | $2^2 \cdot 5 \cdot 7$ | $3 \cdot 47$ | $2 \cdot 71$ | $11 \cdot 13$ | $2^4 \cdot 3^2$ | $5 \cdot 29$ | $2 \cdot 73$ | $3 \cdot 7^2$ | $2^2 \cdot 37$ | **149** |
| 15 | $2 \cdot 3 \cdot 5^2$ | **151** | $2^3 \cdot 19$ | $3^2 \cdot 17$ | $2 \cdot 7 \cdot 11$ | $5 \cdot 31$ | $2^2 \cdot 3 \cdot 13$ | **157** | $2 \cdot 79$ | $3 \cdot 53$ |
| 16 | $2^5 \cdot 5$ | $7 \cdot 23$ | $2 \cdot 3^4$ | **163** | $2^2 \cdot 41$ | $3 \cdot 5 \cdot 11$ | $2 \cdot 83$ | **167** | $2^3 \cdot 3 \cdot 7$ | $13^2$ |
| 17 | $2 \cdot 5 \cdot 17$ | $3^2 \cdot 19$ | $2^2 \cdot 43$ | **173** | $2 \cdot 3 \cdot 29$ | $5^2 \cdot 7$ | $2^4 \cdot 11$ | $3 \cdot 59$ | $2 \cdot 89$ | **179** |
| 18 | $2^2 \cdot 3^2 \cdot 5$ | **181** | $2 \cdot 7 \cdot 13$ | $3 \cdot 61$ | $2^3 \cdot 23$ | $5 \cdot 37$ | $2 \cdot 3 \cdot 31$ | $11 \cdot 17$ | $2^2 \cdot 47$ | $3^3 \cdot 7$ |
| 19 | $2 \cdot 5 \cdot 19$ | **191** | $2^6 \cdot 3$ | **193** | $2 \cdot 97$ | $3 \cdot 5 \cdot 13$ | $2^2 \cdot 7^2$ | **197** | $2 \cdot 3^2 \cdot 11$ | **199** |
| 20 | $2^3 \cdot 5^2$ | $3 \cdot 67$ | $2 \cdot 101$ | $7 \cdot 29$ | $2^2 \cdot 3 \cdot 17$ | $5 \cdot 41$ | $2 \cdot 103$ | $3^2 \cdot 23$ | $2^4 \cdot 13$ | $11 \cdot 19$ |
| 21 | $2 \cdot 3 \cdot 5 \cdot 7$ | **211** | $2^2 \cdot 53$ | $3 \cdot 71$ | $2 \cdot 107$ | $5 \cdot 43$ | $2^3 \cdot 3^3$ | $7 \cdot 31$ | $2 \cdot 109$ | $3 \cdot 73$ |
| 22 | $2^2 \cdot 5 \cdot 11$ | $13 \cdot 17$ | $2 \cdot 3 \cdot 37$ | **223** | $2^5 \cdot 7$ | $3^2 \cdot 5^2$ | $2 \cdot 113$ | **227** | $2^2 \cdot 3 \cdot 19$ | **229** |
| 23 | $2 \cdot 5 \cdot 23$ | $3 \cdot 7 \cdot 11$ | $2^3 \cdot 29$ | **233** | $2 \cdot 3^2 \cdot 13$ | $5 \cdot 47$ | $2^2 \cdot 59$ | $3 \cdot 79$ | $2 \cdot 7 \cdot 17$ | **239** |
| 24 | $2^4 \cdot 3 \cdot 5$ | **241** | $2 \cdot 11^2$ | $3^5$ | $2^2 \cdot 61$ | $5 \cdot 7^2$ | $2 \cdot 3 \cdot 41$ | $13 \cdot 19$ | $2^3 \cdot 31$ | $3 \cdot 83$ |
| 25 | $2 \cdot 5^3$ | **251** | $2^2 \cdot 3^2 \cdot 7$ | $11 \cdot 23$ | $2 \cdot 127$ | $3 \cdot 5 \cdot 17$ | $2^8$ | **257** | $2 \cdot 3 \cdot 43$ | $7 \cdot 37$ |
| 26 | $2^2 \cdot 5 \cdot 13$ | $3^2 \cdot 29$ | $2 \cdot 131$ | **263** | $2^3 \cdot 3 \cdot 11$ | $5 \cdot 53$ | $2 \cdot 7 \cdot 19$ | $3 \cdot 89$ | $2^2 \cdot 67$ | **269** |
| 27 | $2 \cdot 3^3 \cdot 5$ | **271** | $2^4 \cdot 17$ | $3 \cdot 7 \cdot 13$ | $2 \cdot 137$ | $5^2 \cdot 11$ | $2^2 \cdot 3 \cdot 23$ | **277** | $2 \cdot 139$ | $3^2 \cdot 31$ |
| 28 | $2^3 \cdot 5 \cdot 7$ | **281** | $2 \cdot 3 \cdot 47$ | **283** | $2^2 \cdot 71$ | $3 \cdot 5 \cdot 19$ | $2 \cdot 11 \cdot 13$ | $7 \cdot 41$ | $2^5 \cdot 3^2$ | $17^2$ |
| 29 | $2 \cdot 5 \cdot 29$ | $3 \cdot 97$ | $2^2 \cdot 73$ | **293** | $2 \cdot 3 \cdot 7^2$ | $5 \cdot 59$ | $2^3 \cdot 37$ | $3^3 \cdot 11$ | $2 \cdot 149$ | $13 \cdot 23$ |
| 30 | $2^2 \cdot 3 \cdot 5^2$ | $7 \cdot 43$ | $2 \cdot 151$ | $3 \cdot 101$ | $2^4 \cdot 19$ | $5 \cdot 61$ | $2 \cdot 3^2 \cdot 17$ | **307** | $2^2 \cdot 7 \cdot 11$ | $3 \cdot 103$ |
| 31 | $2 \cdot 5 \cdot 31$ | **311** | $2^3 \cdot 3 \cdot 13$ | **313** | $2 \cdot 157$ | $3^2 \cdot 5 \cdot 7$ | $2^2 \cdot 79$ | **317** | $2 \cdot 3 \cdot 53$ | $11 \cdot 29$ |
| 32 | $2^6 \cdot 5$ | $3 \cdot 107$ | $2 \cdot 7 \cdot 23$ | $17 \cdot 19$ | $2^2 \cdot 3^4$ | $5^2 \cdot 13$ | $2 \cdot 163$ | $3 \cdot 109$ | $2^3 \cdot 41$ | $7 \cdot 47$ |
| 33 | $2 \cdot 3 \cdot 5 \cdot 11$ | **331** | $2^2 \cdot 83$ | $3^2 \cdot 37$ | $2 \cdot 167$ | $5 \cdot 67$ | $2^4 \cdot 3 \cdot 7$ | **337** | $2 \cdot 13^2$ | $3 \cdot 113$ |
| 34 | $2^2 \cdot 5 \cdot 17$ | $11 \cdot 31$ | $2 \cdot 3^2 \cdot 19$ | $7^3$ | $2^3 \cdot 43$ | $3 \cdot 5 \cdot 23$ | $2 \cdot 173$ | **347** | $2^2 \cdot 3 \cdot 29$ | **349** |
| 35 | $2 \cdot 5^2 \cdot 7$ | $3^3 \cdot 13$ | $2^5 \cdot 11$ | **353** | $2 \cdot 3 \cdot 59$ | $5 \cdot 71$ | $2^2 \cdot 89$ | $3 \cdot 7 \cdot 17$ | $2 \cdot 179$ | **359** |
| 36 | $2^3 \cdot 3^2 \cdot 5$ | $19^2$ | $2 \cdot 181$ | $3 \cdot 11^2$ | $2^2 \cdot 7 \cdot 13$ | $5 \cdot 73$ | $2 \cdot 3 \cdot 61$ | **367** | $2^4 \cdot 23$ | $3^2 \cdot 41$ |
| 37 | $2 \cdot 5 \cdot 37$ | $7 \cdot 53$ | $2^2 \cdot 3 \cdot 31$ | **373** | $2 \cdot 11 \cdot 17$ | $3 \cdot 5^3$ | $2^3 \cdot 47$ | $13 \cdot 29$ | $2 \cdot 3^3 \cdot 7$ | **379** |
| 38 | $2^2 \cdot 5 \cdot 19$ | $3 \cdot 127$ | $2 \cdot 191$ | **383** | $2^7 \cdot 3$ | $5 \cdot 7 \cdot 11$ | $2 \cdot 193$ | $3^2 \cdot 43$ | $2^2 \cdot 97$ | **389** |
| 39 | $2 \cdot 3 \cdot 5 \cdot 13$ | $17 \cdot 23$ | $2^3 \cdot 7^2$ | $3 \cdot 131$ | $2 \cdot 197$ | $5 \cdot 79$ | $2^2 \cdot 3^2 \cdot 11$ | **397** | $2 \cdot 199$ | $3 \cdot 7 \cdot 19$ |
| 40 | $2^4 \cdot 5^2$ | **401** | $2 \cdot 3 \cdot 67$ | $13 \cdot 31$ | $2^2 \cdot 101$ | $3^4 \cdot 5$ | $2 \cdot 7 \cdot 29$ | $11 \cdot 37$ | $2^3 \cdot 3 \cdot 17$ | **409** |
| 41 | $2 \cdot 5 \cdot 41$ | $3 \cdot 137$ | $2^2 \cdot 103$ | $7 \cdot 59$ | $2 \cdot 3^2 \cdot 23$ | $5 \cdot 83$ | $2^5 \cdot 13$ | $3 \cdot 139$ | $2 \cdot 11 \cdot 19$ | **419** |
| 42 | $2^2 \cdot 3 \cdot 5 \cdot 7$ | **421** | $2 \cdot 211$ | $3^2 \cdot 47$ | $2^3 \cdot 53$ | $5^2 \cdot 17$ | $2 \cdot 3 \cdot 71$ | $7 \cdot 61$ | $2^2 \cdot 107$ | $3 \cdot 11 \cdot 13$ |
| 43 | $2 \cdot 5 \cdot 43$ | **431** | $2^4 \cdot 3^3$ | **433** | $2 \cdot 7 \cdot 31$ | $3 \cdot 5 \cdot 29$ | $2^2 \cdot 109$ | $19 \cdot 23$ | $2 \cdot 3 \cdot 73$ | **439** |
| 44 | $2^3 \cdot 5 \cdot 11$ | $3^2 \cdot 7^2$ | $2 \cdot 13 \cdot 17$ | **443** | $2^2 \cdot 3 \cdot 37$ | $5 \cdot 89$ | $2 \cdot 223$ | $3 \cdot 149$ | $2^6 \cdot 7$ | **449** |
| 45 | $2 \cdot 3^2 \cdot 5^2$ | $11 \cdot 41$ | $2^2 \cdot 113$ | $3 \cdot 151$ | $2 \cdot 227$ | $5 \cdot 7 \cdot 13$ | $2^3 \cdot 3 \cdot 19$ | **457** | $2 \cdot 229$ | $3^3 \cdot 17$ |
| 46 | $2^2 \cdot 5 \cdot 23$ | **461** | $2 \cdot 3 \cdot 7 \cdot 11$ | **463** | $2^4 \cdot 29$ | $3 \cdot 5 \cdot 31$ | $2 \cdot 233$ | **467** | $2^2 \cdot 3^2 \cdot 13$ | $7 \cdot 67$ |
| 47 | $2 \cdot 5 \cdot 47$ | $3 \cdot 157$ | $2^3 \cdot 59$ | $11 \cdot 43$ | $2 \cdot 3 \cdot 79$ | $5^2 \cdot 19$ | $2^2 \cdot 7 \cdot 17$ | $3^2 \cdot 53$ | $2 \cdot 239$ | **479** |
| 48 | $2^5 \cdot 3 \cdot 5$ | $13 \cdot 37$ | $2 \cdot 241$ | $3 \cdot 7 \cdot 23$ | $2^2 \cdot 11^2$ | $5 \cdot 97$ | $2 \cdot 3^5$ | **487** | $2^3 \cdot 61$ | $3 \cdot 163$ |
| 49 | $2 \cdot 5 \cdot 7^2$ | **491** | $2^2 \cdot 3 \cdot 41$ | $17 \cdot 29$ | $2 \cdot 13 \cdot 19$ | $3^2 \cdot 5 \cdot 11$ | $2^4 \cdot 31$ | $7 \cdot 71$ | $2 \cdot 3 \cdot 83$ | **499** |
| 50 | $2^2 \cdot 5^3$ | $3 \cdot 167$ | $2 \cdot 251$ | **503** | $2^3 \cdot 3^2 \cdot 7$ | $5 \cdot 101$ | $2 \cdot 11 \cdot 23$ | $3 \cdot 13^2$ | $2^2 \cdot 127$ | **509** |
| 51 | $2 \cdot 3 \cdot 5 \cdot 17$ | $7 \cdot 73$ | $2^9$ | $3^3 \cdot 19$ | $2 \cdot 257$ | $5 \cdot 103$ | $2^2 \cdot 3 \cdot 43$ | $11 \cdot 47$ | $2 \cdot 7 \cdot 37$ | $3 \cdot 173$ |
| 52 | $2^3 \cdot 5 \cdot 13$ | **521** | $2 \cdot 3^2 \cdot 29$ | **523** | $2^2 \cdot 131$ | $3 \cdot 5^2 \cdot 7$ | $2 \cdot 263$ | $17 \cdot 31$ | $2^4 \cdot 3 \cdot 11$ | $23^2$ |
| 53 | $2 \cdot 5 \cdot 53$ | $3^2 \cdot 59$ | $2^2 \cdot 7 \cdot 19$ | $13 \cdot 41$ | $2 \cdot 3 \cdot 89$ | $5 \cdot 107$ | $2^3 \cdot 67$ | $3 \cdot 179$ | $2 \cdot 269$ | $7^2 \cdot 11$ |
| 54 | $2^2 \cdot 3^3 \cdot 5$ | **541** | $2 \cdot 271$ | $3 \cdot 181$ | $2^5 \cdot 17$ | $5 \cdot 109$ | $2 \cdot 3 \cdot 7 \cdot 13$ | **547** | $2^2 \cdot 137$ | $3^2 \cdot 61$ |
| 55 | $2 \cdot 5^2 \cdot 11$ | $19 \cdot 29$ | $2^3 \cdot 3 \cdot 23$ | $7 \cdot 79$ | $2 \cdot 277$ | $3 \cdot 5 \cdot 37$ | $2^2 \cdot 139$ | **557** | $2 \cdot 3^2 \cdot 31$ | $13 \cdot 43$ |
| 56 | $2^4 \cdot 5 \cdot 7$ | $3 \cdot 11 \cdot 17$ | $2 \cdot 281$ | **563** | $2^2 \cdot 3 \cdot 47$ | $5 \cdot 113$ | $2 \cdot 283$ | $3^4 \cdot 7$ | $2^3 \cdot 71$ | **569** |
| 57 | $2 \cdot 3 \cdot 5 \cdot 19$ | **571** | $2^2 \cdot 11 \cdot 13$ | $3 \cdot 191$ | $2 \cdot 7 \cdot 41$ | $5^2 \cdot 23$ | $2^6 \cdot 3^2$ | **577** | $2 \cdot 17^2$ | $3 \cdot 193$ |

| | 0 | 1 | 2 | 3 | 4 | 5 | 6 | 7 | 8 | 9 |
|---|---|---|---|---|---|---|---|---|---|---|
| 58 | $2^2 \cdot 5 \cdot 29$ | $7 \cdot 83$ | $2 \cdot 3 \cdot 97$ | $11 \cdot 53$ | $2^3 \cdot 73$ | $3^2 \cdot 5 \cdot 13$ | $2 \cdot 293$ | **587** | $2^2 \cdot 3 \cdot 7^2$ | $19 \cdot 31$ |
| 59 | $2 \cdot 5 \cdot 59$ | $3 \cdot 197$ | $2^4 \cdot 37$ | **593** | $2 \cdot 3^3 \cdot 11$ | $5 \cdot 7 \cdot 17$ | $2^2 \cdot 149$ | $3 \cdot 199$ | $2 \cdot 13 \cdot 23$ | **599** |
| 60 | $2^3 \cdot 3 \cdot 5^2$ | **601** | $2 \cdot 7 \cdot 43$ | $3^2 \cdot 67$ | $2^2 \cdot 151$ | $5 \cdot 11^2$ | $2 \cdot 3 \cdot 101$ | **607** | $2^5 \cdot 19$ | $3 \cdot 7 \cdot 29$ |
| 61 | $2 \cdot 5 \cdot 61$ | $13 \cdot 47$ | $2^2 \cdot 3^2 \cdot 17$ | **613** | $2 \cdot 307$ | $3 \cdot 5 \cdot 41$ | $2^3 \cdot 7 \cdot 11$ | **617** | $2 \cdot 3 \cdot 103$ | **619** |
| 62 | $2^2 \cdot 5 \cdot 31$ | $3^3 \cdot 23$ | $2 \cdot 311$ | $7 \cdot 89$ | $2^4 \cdot 3 \cdot 13$ | $5^4$ | $2 \cdot 313$ | $3 \cdot 11 \cdot 19$ | $2^2 \cdot 157$ | $17 \cdot 37$ |
| 63 | $2 \cdot 3^2 \cdot 5 \cdot 7$ | **631** | $2^3 \cdot 79$ | $3 \cdot 211$ | $2 \cdot 317$ | $5 \cdot 127$ | $2^2 \cdot 3 \cdot 53$ | $7^2 \cdot 13$ | $2 \cdot 11 \cdot 29$ | $3^2 \cdot 71$ |
| 64 | $2^7 \cdot 5$ | **641** | $2 \cdot 3 \cdot 107$ | **643** | $2^2 \cdot 7 \cdot 23$ | $3 \cdot 5 \cdot 43$ | $2 \cdot 17 \cdot 19$ | **647** | $2^3 \cdot 3^4$ | $11 \cdot 59$ |
| 65 | $2 \cdot 5^2 \cdot 13$ | $3 \cdot 7 \cdot 31$ | $2^2 \cdot 163$ | **653** | $2 \cdot 3 \cdot 109$ | $5 \cdot 131$ | $2^4 \cdot 41$ | $3^2 \cdot 73$ | $2 \cdot 7 \cdot 47$ | **659** |
| 66 | $2^2 \cdot 3 \cdot 5 \cdot 11$ | **661** | $2 \cdot 331$ | $3 \cdot 13 \cdot 17$ | $2^3 \cdot 83$ | $5 \cdot 7 \cdot 19$ | $2 \cdot 3^2 \cdot 37$ | $23 \cdot 29$ | $2^2 \cdot 167$ | $3 \cdot 223$ |
| 67 | $2 \cdot 5 \cdot 67$ | $11 \cdot 61$ | $2^5 \cdot 3 \cdot 7$ | **673** | $2 \cdot 337$ | $3^3 \cdot 5^2$ | $2^2 \cdot 13^2$ | **677** | $2 \cdot 3 \cdot 113$ | $7 \cdot 97$ |
| 68 | $2^3 \cdot 5 \cdot 17$ | $3 \cdot 227$ | $2 \cdot 11 \cdot 31$ | **683** | $2^2 \cdot 3^2 \cdot 19$ | $5 \cdot 137$ | $2 \cdot 7^3$ | $3 \cdot 229$ | $2^4 \cdot 43$ | $13 \cdot 53$ |
| 69 | $2 \cdot 3 \cdot 5 \cdot 23$ | **691** | $2^2 \cdot 173$ | $3^2 \cdot 7 \cdot 11$ | $2 \cdot 347$ | $5 \cdot 139$ | $2^3 \cdot 3 \cdot 29$ | $17 \cdot 41$ | $2 \cdot 349$ | $3 \cdot 233$ |
| 70 | $2^2 \cdot 5^2 \cdot 7$ | **701** | $2 \cdot 3^3 \cdot 13$ | $19 \cdot 37$ | $2^6 \cdot 11$ | $3 \cdot 5 \cdot 47$ | $2 \cdot 353$ | $7 \cdot 101$ | $2^2 \cdot 3 \cdot 59$ | **709** |
| 71 | $2 \cdot 5 \cdot 71$ | $3^2 \cdot 79$ | $2^3 \cdot 89$ | $23 \cdot 31$ | $2 \cdot 3 \cdot 7 \cdot 17$ | $5 \cdot 11 \cdot 13$ | $2^2 \cdot 179$ | $3 \cdot 239$ | $2 \cdot 359$ | **719** |
| 72 | $2^4 \cdot 3^2 \cdot 5$ | $7 \cdot 103$ | $2 \cdot 19^2$ | $3 \cdot 241$ | $2^2 \cdot 181$ | $5^2 \cdot 29$ | $2 \cdot 3 \cdot 11^2$ | **727** | $2^3 \cdot 7 \cdot 13$ | $3^6$ |
| 73 | $2 \cdot 5 \cdot 73$ | $17 \cdot 43$ | $2^2 \cdot 3 \cdot 61$ | **733** | $2 \cdot 367$ | $3 \cdot 5 \cdot 7^2$ | $2^5 \cdot 23$ | $11 \cdot 67$ | $2 \cdot 3^2 \cdot 41$ | **739** |
| 74 | $2^2 \cdot 5 \cdot 37$ | $3 \cdot 13 \cdot 19$ | $2 \cdot 7 \cdot 53$ | **743** | $2^3 \cdot 3 \cdot 31$ | $5 \cdot 149$ | $2 \cdot 373$ | $3^2 \cdot 83$ | $2^2 \cdot 11 \cdot 17$ | $7 \cdot 107$ |
| 75 | $2 \cdot 3 \cdot 5^3$ | **751** | $2^4 \cdot 47$ | $3 \cdot 251$ | $2 \cdot 13 \cdot 29$ | $5 \cdot 151$ | $2^2 \cdot 3^3 \cdot 7$ | **757** | $2 \cdot 379$ | $3 \cdot 11 \cdot 23$ |
| 76 | $2^3 \cdot 5 \cdot 19$ | **761** | $2 \cdot 3 \cdot 127$ | $7 \cdot 109$ | $2^2 \cdot 191$ | $3^2 \cdot 5 \cdot 17$ | $2 \cdot 383$ | $13 \cdot 59$ | $2^8 \cdot 3$ | **769** |
| 77 | $2 \cdot 5 \cdot 7 \cdot 11$ | $3 \cdot 257$ | $2^2 \cdot 193$ | **773** | $2 \cdot 3^2 \cdot 43$ | $5^2 \cdot 31$ | $2^3 \cdot 97$ | $3 \cdot 7 \cdot 37$ | $2 \cdot 389$ | $19 \cdot 41$ |
| 78 | $2^2 \cdot 3 \cdot 5 \cdot 13$ | $11 \cdot 71$ | $2 \cdot 17 \cdot 23$ | $3^3 \cdot 29$ | $2^4 \cdot 7^2$ | $5 \cdot 157$ | $2 \cdot 3 \cdot 131$ | **787** | $2^2 \cdot 197$ | $3 \cdot 263$ |
| 79 | $2 \cdot 5 \cdot 79$ | $7 \cdot 113$ | $2^3 \cdot 3^2 \cdot 11$ | $13 \cdot 61$ | $2 \cdot 397$ | $3 \cdot 5 \cdot 53$ | $2^2 \cdot 199$ | **797** | $2 \cdot 3 \cdot 7 \cdot 19$ | $17 \cdot 47$ |
| 80 | $2^5 \cdot 5^2$ | $3^2 \cdot 89$ | $2 \cdot 401$ | $11 \cdot 73$ | $2^2 \cdot 3 \cdot 67$ | $5 \cdot 7 \cdot 23$ | $2 \cdot 13 \cdot 31$ | $3 \cdot 269$ | $2^3 \cdot 101$ | **809** |
| 81 | $2 \cdot 3^4 \cdot 5$ | **811** | $2^2 \cdot 7 \cdot 29$ | $3 \cdot 271$ | $2 \cdot 11 \cdot 37$ | $5 \cdot 163$ | $2^4 \cdot 3 \cdot 17$ | $19 \cdot 43$ | $2 \cdot 409$ | $3^2 \cdot 7 \cdot 13$ |
| 82 | $2^2 \cdot 5 \cdot 41$ | **821** | $2 \cdot 3 \cdot 137$ | **823** | $2^3 \cdot 103$ | $3 \cdot 5^2 \cdot 11$ | $2 \cdot 7 \cdot 59$ | **827** | $2^2 \cdot 3^2 \cdot 23$ | **829** |
| 83 | $2 \cdot 5 \cdot 83$ | $3 \cdot 277$ | $2^6 \cdot 13$ | $7^2 \cdot 17$ | $2 \cdot 3 \cdot 139$ | $5 \cdot 167$ | $2^2 \cdot 11 \cdot 19$ | $3^3 \cdot 31$ | $2 \cdot 419$ | **839** |
| 84 | $2^3 \cdot 3 \cdot 5 \cdot 7$ | $29^2$ | $2 \cdot 421$ | $3 \cdot 281$ | $2^2 \cdot 211$ | $5 \cdot 13^2$ | $2 \cdot 3^2 \cdot 47$ | $7 \cdot 11^2$ | $2^4 \cdot 53$ | $3 \cdot 283$ |
| 85 | $2 \cdot 5^2 \cdot 17$ | $23 \cdot 37$ | $2^2 \cdot 3 \cdot 71$ | **853** | $2 \cdot 7 \cdot 61$ | $3^2 \cdot 5 \cdot 19$ | $2^3 \cdot 107$ | **857** | $2 \cdot 3 \cdot 11 \cdot 13$ | **859** |
| 86 | $2^2 \cdot 5 \cdot 43$ | $3 \cdot 7 \cdot 41$ | $2 \cdot 431$ | **863** | $2^5 \cdot 3^3$ | $5 \cdot 173$ | $2 \cdot 433$ | $3 \cdot 17^2$ | $2^2 \cdot 7 \cdot 31$ | $11 \cdot 79$ |
| 87 | $2 \cdot 3 \cdot 5 \cdot 29$ | $13 \cdot 67$ | $2^3 \cdot 109$ | $3^2 \cdot 97$ | $2 \cdot 19 \cdot 23$ | $5^3 \cdot 7$ | $2^2 \cdot 3 \cdot 73$ | **877** | $2 \cdot 439$ | $3 \cdot 293$ |
| 88 | $2^4 \cdot 5 \cdot 11$ | **881** | $2 \cdot 3^2 \cdot 7^2$ | **883** | $2^2 \cdot 13 \cdot 17$ | $3 \cdot 5 \cdot 59$ | $2 \cdot 443$ | **887** | $2^3 \cdot 3 \cdot 37$ | $7 \cdot 127$ |
| 89 | $2 \cdot 5 \cdot 89$ | $3^4 \cdot 11$ | $2^2 \cdot 223$ | $19 \cdot 47$ | $2 \cdot 3 \cdot 149$ | $5 \cdot 179$ | $2^7 \cdot 7$ | $3 \cdot 13 \cdot 23$ | $2 \cdot 449$ | $29 \cdot 31$ |
| 90 | $2^2 \cdot 3^2 \cdot 5^2$ | $17 \cdot 53$ | $2 \cdot 11 \cdot 41$ | $3 \cdot 7 \cdot 43$ | $2^3 \cdot 113$ | $5 \cdot 181$ | $2 \cdot 3 \cdot 151$ | **907** | $2^2 \cdot 227$ | $3^2 \cdot 101$ |
| 91 | $2 \cdot 5 \cdot 7 \cdot 13$ | **911** | $2^4 \cdot 3 \cdot 19$ | $11 \cdot 83$ | $2 \cdot 457$ | $3 \cdot 5 \cdot 61$ | $2^2 \cdot 229$ | $7 \cdot 131$ | $2 \cdot 3^3 \cdot 17$ | **919** |
| 92 | $2^3 \cdot 5 \cdot 23$ | $3 \cdot 307$ | $2 \cdot 461$ | $13 \cdot 71$ | $2^2 \cdot 3 \cdot 7 \cdot 11$ | $5^2 \cdot 37$ | $2 \cdot 463$ | $3^2 \cdot 103$ | $2^5 \cdot 29$ | **929** |
| 93 | $2 \cdot 3 \cdot 5 \cdot 31$ | $7^2 \cdot 19$ | $2^2 \cdot 233$ | $3 \cdot 311$ | $2 \cdot 467$ | $5 \cdot 11 \cdot 17$ | $2^3 \cdot 3^2 \cdot 13$ | **937** | $2 \cdot 7 \cdot 67$ | $3 \cdot 313$ |
| 94 | $2^2 \cdot 5 \cdot 47$ | **941** | $2 \cdot 3 \cdot 157$ | $23 \cdot 41$ | $2^4 \cdot 59$ | $3^3 \cdot 5 \cdot 7$ | $2 \cdot 11 \cdot 43$ | **947** | $2^2 \cdot 3 \cdot 79$ | $13 \cdot 73$ |
| 95 | $2 \cdot 5^2 \cdot 19$ | $3 \cdot 317$ | $2^3 \cdot 7 \cdot 17$ | **953** | $2 \cdot 3^2 \cdot 53$ | $5 \cdot 191$ | $2^2 \cdot 239$ | $3 \cdot 11 \cdot 29$ | $2 \cdot 479$ | $7 \cdot 137$ |
| 96 | $2^6 \cdot 3 \cdot 5$ | $31^2$ | $2 \cdot 13 \cdot 37$ | $3^2 \cdot 107$ | $2^2 \cdot 241$ | $5 \cdot 193$ | $2 \cdot 3 \cdot 7 \cdot 23$ | **967** | $2^3 \cdot 11^2$ | $3 \cdot 17 \cdot 19$ |
| 97 | $2 \cdot 5 \cdot 97$ | **971** | $2^2 \cdot 3^5$ | $7 \cdot 139$ | $2 \cdot 487$ | $3 \cdot 5^2 \cdot 13$ | $2^4 \cdot 61$ | **977** | $2 \cdot 3 \cdot 163$ | $11 \cdot 89$ |
| 98 | $2^2 \cdot 5 \cdot 7^2$ | $3^2 \cdot 109$ | $2 \cdot 491$ | **983** | $2^3 \cdot 3 \cdot 41$ | $5 \cdot 197$ | $2 \cdot 17 \cdot 29$ | $3 \cdot 7 \cdot 47$ | $2^2 \cdot 13 \cdot 19$ | $23 \cdot 43$ |
| 99 | $2 \cdot 3^2 \cdot 5 \cdot 11$ | **991** | $2^5 \cdot 31$ | $3 \cdot 331$ | $2 \cdot 7 \cdot 71$ | $5 \cdot 199$ | $2^2 \cdot 3 \cdot 83$ | **997** | $2 \cdot 499$ | $3^3 \cdot 37$ |
| 100 | $2^3 \cdot 5^3$ | $7 \cdot 11 \cdot 13$ | $2 \cdot 3 \cdot 167$ | $17 \cdot 59$ | $2^2 \cdot 251$ | $3 \cdot 5 \cdot 67$ | $2 \cdot 503$ | $19 \cdot 53$ | $2^4 \cdot 3^2 \cdot 7$ | **1009** |
| 101 | $2 \cdot 5 \cdot 101$ | $3 \cdot 337$ | $2^2 \cdot 11 \cdot 23$ | **1013** | $2 \cdot 3 \cdot 13^2$ | $5 \cdot 7 \cdot 29$ | $2^3 \cdot 127$ | $3^2 \cdot 113$ | $2 \cdot 509$ | **1019** |
| 102 | $2^2 \cdot 3 \cdot 5 \cdot 17$ | **1021** | $2 \cdot 7 \cdot 73$ | $3 \cdot 11 \cdot 31$ | $2^{10}$ | $5^2 \cdot 41$ | $2 \cdot 3^3 \cdot 19$ | $13 \cdot 79$ | $2^2 \cdot 257$ | $3 \cdot 7^3$ |
| 103 | $2 \cdot 5 \cdot 103$ | **1031** | $2^3 \cdot 3 \cdot 43$ | **1033** | $2 \cdot 11 \cdot 47$ | $3^2 \cdot 5 \cdot 23$ | $2^2 \cdot 7 \cdot 37$ | $17 \cdot 61$ | $2 \cdot 3 \cdot 173$ | **1039** |
| 104 | $2^4 \cdot 5 \cdot 13$ | $3 \cdot 347$ | $2 \cdot 521$ | $7 \cdot 149$ | $2^2 \cdot 3^2 \cdot 29$ | $5 \cdot 11 \cdot 19$ | $2 \cdot 523$ | $3 \cdot 349$ | $2^3 \cdot 131$ | **1049** |

|  | 0 | 1 | 2 | 3 | 4 | 5 | 6 | 7 | 8 | 9 |
|---|---|---|---|---|---|---|---|---|---|---|
| 105 | $2 \cdot 3 \cdot 5^2 \cdot 7$ | **1051** | $2^2 \cdot 263$ | $3^4 \cdot 13$ | $2 \cdot 17 \cdot 31$ | $5 \cdot 211$ | $2^5 \cdot 3 \cdot 11$ | $7 \cdot 151$ | $2 \cdot 23^2$ | $3 \cdot 353$ |
| 106 | $2^2 \cdot 5 \cdot 53$ | **1061** | $2 \cdot 3^2 \cdot 59$ | **1063** | $2^3 \cdot 7 \cdot 19$ | $3 \cdot 5 \cdot 71$ | $2 \cdot 13 \cdot 41$ | $11 \cdot 97$ | $2^2 \cdot 3 \cdot 89$ | **1069** |
| 107 | $2 \cdot 5 \cdot 107$ | $3^2 \cdot 7 \cdot 17$ | $2^4 \cdot 67$ | $29 \cdot 37$ | $2 \cdot 3 \cdot 179$ | $5^2 \cdot 43$ | $2^2 \cdot 269$ | $3 \cdot 359$ | $2 \cdot 7^2 \cdot 11$ | $13 \cdot 83$ |
| 108 | $2^3 \cdot 3^3 \cdot 5$ | $23 \cdot 47$ | $2 \cdot 541$ | $3 \cdot 19^2$ | $2^2 \cdot 271$ | $5 \cdot 7 \cdot 31$ | $2 \cdot 3 \cdot 181$ | **1087** | $2^6 \cdot 17$ | $3^2 \cdot 11^2$ |
| 109 | $2 \cdot 5 \cdot 109$ | **1091** | $2^2 \cdot 3 \cdot 7 \cdot 13$ | **1093** | $2 \cdot 547$ | $3 \cdot 5 \cdot 73$ | $2^3 \cdot 137$ | **1097** | $2 \cdot 3^2 \cdot 61$ | $7 \cdot 157$ |
| 110 | $2^2 \cdot 5^2 \cdot 11$ | $3 \cdot 367$ | $2 \cdot 19 \cdot 29$ | **1103** | $2^4 \cdot 3 \cdot 23$ | $5 \cdot 13 \cdot 17$ | $2 \cdot 7 \cdot 79$ | $3^3 \cdot 41$ | $2^2 \cdot 277$ | **1109** |
| 111 | $2 \cdot 3 \cdot 5 \cdot 37$ | $11 \cdot 101$ | $2^3 \cdot 139$ | $3 \cdot 7 \cdot 53$ | $2 \cdot 557$ | $5 \cdot 223$ | $2^2 \cdot 3^2 \cdot 31$ | **1117** | $2 \cdot 13 \cdot 43$ | $3 \cdot 373$ |
| 112 | $2^5 \cdot 5 \cdot 7$ | $19 \cdot 59$ | $2 \cdot 3 \cdot 11 \cdot 17$ | **1123** | $2^2 \cdot 281$ | $3^2 \cdot 5^3$ | $2 \cdot 563$ | $7^2 \cdot 23$ | $2^3 \cdot 3 \cdot 47$ | **1129** |
| 113 | $2 \cdot 5 \cdot 113$ | $3 \cdot 13 \cdot 29$ | $2^2 \cdot 283$ | $11 \cdot 103$ | $2 \cdot 3^4 \cdot 7$ | $5 \cdot 227$ | $2^4 \cdot 71$ | $3 \cdot 379$ | $2 \cdot 569$ | $17 \cdot 67$ |
| 114 | $2^2 \cdot 3 \cdot 5 \cdot 19$ | $7 \cdot 163$ | $2 \cdot 571$ | $3^2 \cdot 127$ | $2^3 \cdot 11 \cdot 13$ | $5 \cdot 229$ | $2 \cdot 3 \cdot 191$ | $31 \cdot 37$ | $2^2 \cdot 7 \cdot 41$ | $3 \cdot 383$ |
| 115 | $2 \cdot 5^2 \cdot 23$ | **1151** | $2^7 \cdot 3^2$ | **1153** | $2 \cdot 577$ | $3 \cdot 5 \cdot 7 \cdot 11$ | $2^2 \cdot 17^2$ | $13 \cdot 89$ | $2 \cdot 3 \cdot 193$ | $19 \cdot 61$ |
| 116 | $2^3 \cdot 5 \cdot 29$ | $3^3 \cdot 43$ | $2 \cdot 7 \cdot 83$ | **1163** | $2^2 \cdot 3 \cdot 97$ | $5 \cdot 233$ | $2 \cdot 11 \cdot 53$ | $3 \cdot 389$ | $2^4 \cdot 73$ | $7 \cdot 167$ |
| 117 | $2 \cdot 3^2 \cdot 5 \cdot 13$ | **1171** | $2^2 \cdot 293$ | $3 \cdot 17 \cdot 23$ | $2 \cdot 587$ | $5^2 \cdot 47$ | $2^3 \cdot 3 \cdot 7^2$ | $11 \cdot 107$ | $2 \cdot 19 \cdot 31$ | $3^2 \cdot 131$ |
| 118 | $2^2 \cdot 5 \cdot 59$ | **1181** | $2 \cdot 3 \cdot 197$ | $7 \cdot 13^2$ | $2^5 \cdot 37$ | $3 \cdot 5 \cdot 79$ | $2 \cdot 593$ | **1187** | $2^2 \cdot 3^3 \cdot 11$ | $29 \cdot 41$ |
| 119 | $2 \cdot 5 \cdot 7 \cdot 17$ | $3 \cdot 397$ | $2^3 \cdot 149$ | **1193** | $2 \cdot 3 \cdot 199$ | $5 \cdot 239$ | $2^2 \cdot 13 \cdot 23$ | $3^2 \cdot 7 \cdot 19$ | $2 \cdot 599$ | $11 \cdot 109$ |
| 120 | $2^4 \cdot 3 \cdot 5^2$ | **1201** | $2 \cdot 601$ | $3 \cdot 401$ | $2^2 \cdot 7 \cdot 43$ | $5 \cdot 241$ | $2 \cdot 3^2 \cdot 67$ | $17 \cdot 71$ | $2^3 \cdot 151$ | $3 \cdot 13 \cdot 31$ |
| 121 | $2 \cdot 5 \cdot 11^2$ | $7 \cdot 173$ | $2^2 \cdot 3 \cdot 101$ | **1213** | $2 \cdot 607$ | $3^5 \cdot 5$ | $2^6 \cdot 19$ | **1217** | $2 \cdot 3 \cdot 7 \cdot 29$ | $23 \cdot 53$ |
| 122 | $2^2 \cdot 5 \cdot 61$ | $3 \cdot 11 \cdot 37$ | $2 \cdot 13 \cdot 47$ | **1223** | $2^3 \cdot 3^2 \cdot 17$ | $5^2 \cdot 7^2$ | $2 \cdot 613$ | $3 \cdot 409$ | $2^2 \cdot 307$ | **1229** |
| 123 | $2 \cdot 3 \cdot 5 \cdot 41$ | **1231** | $2^4 \cdot 7 \cdot 11$ | $3^2 \cdot 137$ | $2 \cdot 617$ | $5 \cdot 13 \cdot 19$ | $2^2 \cdot 3 \cdot 103$ | **1237** | $2 \cdot 619$ | $3 \cdot 7 \cdot 59$ |
| 124 | $2^3 \cdot 5 \cdot 31$ | $17 \cdot 73$ | $2 \cdot 3^3 \cdot 23$ | $11 \cdot 113$ | $2^2 \cdot 311$ | $3 \cdot 5 \cdot 83$ | $2 \cdot 7 \cdot 89$ | $29 \cdot 43$ | $2^5 \cdot 3 \cdot 13$ | **1249** |
| 125 | $2 \cdot 5^4$ | $3^2 \cdot 139$ | $2^2 \cdot 313$ | $7 \cdot 179$ | $2 \cdot 3 \cdot 11 \cdot 19$ | $5 \cdot 251$ | $2^3 \cdot 157$ | $3 \cdot 419$ | $2 \cdot 17 \cdot 37$ | **1259** |
| 126 | $2^2 \cdot 3^2 \cdot 5 \cdot 7$ | $13 \cdot 97$ | $2 \cdot 631$ | $3 \cdot 421$ | $2^4 \cdot 79$ | $5 \cdot 11 \cdot 23$ | $2 \cdot 3 \cdot 211$ | $7 \cdot 181$ | $2^2 \cdot 317$ | $3^3 \cdot 47$ |
| 127 | $2 \cdot 5 \cdot 127$ | $31 \cdot 41$ | $2^3 \cdot 3 \cdot 53$ | $19 \cdot 67$ | $2 \cdot 7^2 \cdot 13$ | $3 \cdot 5^2 \cdot 17$ | $2^2 \cdot 11 \cdot 29$ | **1277** | $2 \cdot 3^2 \cdot 71$ | **1279** |
| 128 | $2^8 \cdot 5$ | $3 \cdot 7 \cdot 61$ | $2 \cdot 641$ | **1283** | $2^2 \cdot 3 \cdot 107$ | $5 \cdot 257$ | $2 \cdot 643$ | $3^2 \cdot 11 \cdot 13$ | $2^3 \cdot 7 \cdot 23$ | **1289** |
| 129 | $2 \cdot 3 \cdot 5 \cdot 43$ | **1291** | $2^2 \cdot 17 \cdot 19$ | $3 \cdot 431$ | $2 \cdot 647$ | $5 \cdot 7 \cdot 37$ | $2^4 \cdot 3^4$ | **1297** | $2 \cdot 11 \cdot 59$ | $3 \cdot 433$ |
| 130 | $2^2 \cdot 5^2 \cdot 13$ | **1301** | $2 \cdot 3 \cdot 7 \cdot 31$ | **1303** | $2^3 \cdot 163$ | $3^2 \cdot 5 \cdot 29$ | $2 \cdot 653$ | **1307** | $2^2 \cdot 3 \cdot 109$ | $7 \cdot 11 \cdot 17$ |
| 131 | $2 \cdot 5 \cdot 131$ | $3 \cdot 19 \cdot 23$ | $2^5 \cdot 41$ | $13 \cdot 101$ | $2 \cdot 3^2 \cdot 73$ | $5 \cdot 263$ | $2^2 \cdot 7 \cdot 47$ | $3 \cdot 439$ | $2 \cdot 659$ | **1319** |
| 132 | $2^3 \cdot 3 \cdot 5 \cdot 11$ | **1321** | $2 \cdot 661$ | $3^3 \cdot 7^2$ | $2^2 \cdot 331$ | $5^2 \cdot 53$ | $2 \cdot 3 \cdot 13 \cdot 17$ | **1327** | $2^4 \cdot 83$ | $3 \cdot 443$ |
| 133 | $2 \cdot 5 \cdot 7 \cdot 19$ | $11^3$ | $2^2 \cdot 3^2 \cdot 37$ | $31 \cdot 43$ | $2 \cdot 23 \cdot 29$ | $3 \cdot 5 \cdot 89$ | $2^3 \cdot 167$ | $7 \cdot 191$ | $2 \cdot 3 \cdot 223$ | $13 \cdot 103$ |
| 134 | $2^2 \cdot 5 \cdot 67$ | $3^2 \cdot 149$ | $2 \cdot 11 \cdot 61$ | $17 \cdot 79$ | $2^6 \cdot 3 \cdot 7$ | $5 \cdot 269$ | $2 \cdot 673$ | $3 \cdot 449$ | $2^2 \cdot 337$ | $19 \cdot 71$ |
| 135 | $2 \cdot 3^3 \cdot 5^2$ | $7 \cdot 193$ | $2^3 \cdot 13^2$ | $3 \cdot 11 \cdot 41$ | $2 \cdot 677$ | $5 \cdot 271$ | $2^2 \cdot 3 \cdot 113$ | $23 \cdot 59$ | $2 \cdot 7 \cdot 97$ | $3^2 \cdot 151$ |
| 136 | $2^4 \cdot 5 \cdot 17$ | **1361** | $2 \cdot 3 \cdot 227$ | $29 \cdot 47$ | $2^2 \cdot 11 \cdot 31$ | $3 \cdot 5 \cdot 7 \cdot 13$ | $2 \cdot 683$ | **1367** | $2^3 \cdot 3^2 \cdot 19$ | $37^2$ |
| 137 | $2 \cdot 5 \cdot 137$ | $3 \cdot 457$ | $2^2 \cdot 7^3$ | **1373** | $2 \cdot 3 \cdot 229$ | $5^3 \cdot 11$ | $2^5 \cdot 43$ | $3^4 \cdot 17$ | $2 \cdot 13 \cdot 53$ | $7 \cdot 197$ |
| 138 | $2^2 \cdot 3 \cdot 5 \cdot 23$ | **1381** | $2 \cdot 691$ | $3 \cdot 461$ | $2^3 \cdot 173$ | $5 \cdot 277$ | $2 \cdot 3^2 \cdot 7 \cdot 11$ | $19 \cdot 73$ | $2^2 \cdot 347$ | $3 \cdot 463$ |
| 139 | $2 \cdot 5 \cdot 139$ | $13 \cdot 107$ | $2^4 \cdot 3 \cdot 29$ | $7 \cdot 199$ | $2 \cdot 17 \cdot 41$ | $3^2 \cdot 5 \cdot 31$ | $2^2 \cdot 349$ | $11 \cdot 127$ | $2 \cdot 3 \cdot 233$ | **1399** |
| 140 | $2^3 \cdot 5^2 \cdot 7$ | $3 \cdot 467$ | $2 \cdot 701$ | $23 \cdot 61$ | $2^2 \cdot 3^3 \cdot 13$ | $5 \cdot 281$ | $2 \cdot 19 \cdot 37$ | $3 \cdot 7 \cdot 67$ | $2^7 \cdot 11$ | **1409** |
| 141 | $2 \cdot 3 \cdot 5 \cdot 47$ | $17 \cdot 83$ | $2^2 \cdot 353$ | $3^2 \cdot 157$ | $2 \cdot 7 \cdot 101$ | $5 \cdot 283$ | $2^3 \cdot 3 \cdot 59$ | $13 \cdot 109$ | $2 \cdot 709$ | $3 \cdot 11 \cdot 43$ |
| 142 | $2^2 \cdot 5 \cdot 71$ | $7^2 \cdot 29$ | $2 \cdot 3^2 \cdot 79$ | **1423** | $2^4 \cdot 89$ | $3 \cdot 5^2 \cdot 19$ | $2 \cdot 23 \cdot 31$ | **1427** | $2^2 \cdot 3 \cdot 7 \cdot 17$ | **1429** |
| 143 | $2 \cdot 5 \cdot 11 \cdot 13$ | $3^3 \cdot 53$ | $2^3 \cdot 179$ | **1433** | $2 \cdot 3 \cdot 239$ | $5 \cdot 7 \cdot 41$ | $2^2 \cdot 359$ | $3 \cdot 479$ | $2 \cdot 719$ | **1439** |
| 144 | $2^5 \cdot 3^2 \cdot 5$ | $11 \cdot 131$ | $2 \cdot 7 \cdot 103$ | $3 \cdot 13 \cdot 37$ | $2^2 \cdot 19^2$ | $5 \cdot 17^2$ | $2 \cdot 3 \cdot 241$ | **1447** | $2^3 \cdot 181$ | $3^2 \cdot 7 \cdot 23$ |
| 145 | $2 \cdot 5^2 \cdot 29$ | **1451** | $2^2 \cdot 3 \cdot 11^2$ | **1453** | $2 \cdot 727$ | $3 \cdot 5 \cdot 97$ | $2^4 \cdot 7 \cdot 13$ | $31 \cdot 47$ | $2 \cdot 3^6$ | **1459** |
| 146 | $2^2 \cdot 5 \cdot 73$ | $3 \cdot 487$ | $2 \cdot 17 \cdot 43$ | $7 \cdot 11 \cdot 19$ | $2^3 \cdot 3 \cdot 61$ | $5 \cdot 293$ | $2 \cdot 733$ | $3^2 \cdot 163$ | $2^2 \cdot 367$ | $13 \cdot 113$ |
| 147 | $2 \cdot 3 \cdot 5 \cdot 7^2$ | **1471** | $2^6 \cdot 23$ | $3 \cdot 491$ | $2 \cdot 11 \cdot 67$ | $5^2 \cdot 59$ | $2^2 \cdot 3^2 \cdot 41$ | $7 \cdot 211$ | $2 \cdot 739$ | $3 \cdot 17 \cdot 29$ |
| 148 | $2^3 \cdot 5 \cdot 37$ | **1481** | $2 \cdot 3 \cdot 13 \cdot 19$ | **1483** | $2^2 \cdot 7 \cdot 53$ | $3^3 \cdot 5 \cdot 11$ | $2 \cdot 743$ | **1487** | $2^4 \cdot 3 \cdot 31$ | **1489** |
| 149 | $2 \cdot 5 \cdot 149$ | $3 \cdot 7 \cdot 71$ | $2^2 \cdot 373$ | **1493** | $2 \cdot 3^2 \cdot 83$ | $5 \cdot 13 \cdot 23$ | $2^3 \cdot 11 \cdot 17$ | $3 \cdot 499$ | $2 \cdot 7 \cdot 107$ | **1499** |
| 150 | $2^2 \cdot 3 \cdot 5^3$ | $19 \cdot 79$ | $2 \cdot 751$ | $3^2 \cdot 167$ | $2^5 \cdot 47$ | $5 \cdot 7 \cdot 43$ | $2 \cdot 3 \cdot 251$ | $11 \cdot 137$ | $2^2 \cdot 13 \cdot 29$ | $3 \cdot 503$ |
| 151 | $2 \cdot 5 \cdot 151$ | **1511** | $2^3 \cdot 3^3 \cdot 7$ | $17 \cdot 89$ | $2 \cdot 757$ | $3 \cdot 5 \cdot 101$ | $2^2 \cdot 379$ | $37 \cdot 41$ | $2 \cdot 3 \cdot 11 \cdot 23$ | $7^2 \cdot 31$ |

| | 0 | 1 | 2 | 3 | 4 | 5 | 6 | 7 | 8 | 9 |
|---|---|---|---|---|---|---|---|---|---|---|
| 152 | $2^4 \cdot 5 \cdot 19$ | $3^2 \cdot 13^2$ | $2 \cdot 761$ | **1523** | $2^2 \cdot 3 \cdot 127$ | $5^2 \cdot 61$ | $2 \cdot 7 \cdot 109$ | $3 \cdot 509$ | $2^3 \cdot 191$ | $11 \cdot 139$ |
| 153 | $2 \cdot 3^2 \cdot 5 \cdot 17$ | **1531** | $2^2 \cdot 383$ | $3 \cdot 7 \cdot 73$ | $2 \cdot 13 \cdot 59$ | $5 \cdot 307$ | $2^9 \cdot 3$ | $29 \cdot 53$ | $2 \cdot 769$ | $3^4 \cdot 19$ |
| 154 | $2^2 \cdot 5 \cdot 7 \cdot 11$ | $23 \cdot 67$ | $2 \cdot 3 \cdot 257$ | **1543** | $2^3 \cdot 193$ | $3 \cdot 5 \cdot 103$ | $2 \cdot 773$ | $7 \cdot 13 \cdot 17$ | $2^2 \cdot 3^2 \cdot 43$ | **1549** |
| 155 | $2 \cdot 5^2 \cdot 31$ | $3 \cdot 11 \cdot 47$ | $2^4 \cdot 97$ | **1553** | $2 \cdot 3 \cdot 7 \cdot 37$ | $5 \cdot 311$ | $2^2 \cdot 389$ | $3^2 \cdot 173$ | $2 \cdot 19 \cdot 41$ | **1559** |
| 156 | $2^3 \cdot 3 \cdot 5 \cdot 13$ | $7 \cdot 223$ | $2 \cdot 11 \cdot 71$ | $3 \cdot 521$ | $2^2 \cdot 17 \cdot 23$ | $5 \cdot 313$ | $2 \cdot 3^3 \cdot 29$ | **1567** | $2^5 \cdot 7^2$ | $3 \cdot 523$ |
| 157 | $2 \cdot 5 \cdot 157$ | **1571** | $2^2 \cdot 3 \cdot 131$ | $11^2 \cdot 13$ | $2 \cdot 787$ | $3^2 \cdot 5^2 \cdot 7$ | $2^3 \cdot 197$ | $19 \cdot 83$ | $2 \cdot 3 \cdot 263$ | **1579** |
| 158 | $2^2 \cdot 5 \cdot 79$ | $3 \cdot 17 \cdot 31$ | $2 \cdot 7 \cdot 113$ | **1583** | $2^4 \cdot 3^2 \cdot 11$ | $5 \cdot 317$ | $2 \cdot 13 \cdot 61$ | $3 \cdot 23^2$ | $2^2 \cdot 397$ | $7 \cdot 227$ |
| 159 | $2 \cdot 3 \cdot 5 \cdot 53$ | $37 \cdot 43$ | $2^3 \cdot 199$ | $3^3 \cdot 59$ | $2 \cdot 797$ | $5 \cdot 11 \cdot 29$ | $2^2 \cdot 3 \cdot 7 \cdot 19$ | **1597** | $2 \cdot 17 \cdot 47$ | $3 \cdot 13 \cdot 41$ |
| 160 | $2^6 \cdot 5^2$ | **1601** | $2 \cdot 3^2 \cdot 89$ | $7 \cdot 229$ | $2^2 \cdot 401$ | $3 \cdot 5 \cdot 107$ | $2 \cdot 11 \cdot 73$ | **1607** | $2^3 \cdot 3 \cdot 67$ | **1609** |
| 161 | $2 \cdot 5 \cdot 7 \cdot 23$ | $3^2 \cdot 179$ | $2^2 \cdot 13 \cdot 31$ | **1613** | $2 \cdot 3 \cdot 269$ | $5 \cdot 17 \cdot 19$ | $2^4 \cdot 101$ | $3 \cdot 7^2 \cdot 11$ | $2 \cdot 809$ | **1619** |
| 162 | $2^2 \cdot 3^4 \cdot 5$ | **1621** | $2 \cdot 811$ | $3 \cdot 541$ | $2^3 \cdot 7 \cdot 29$ | $5^3 \cdot 13$ | $2 \cdot 3 \cdot 271$ | **1627** | $2^2 \cdot 11 \cdot 37$ | $3^2 \cdot 181$ |
| 163 | $2 \cdot 5 \cdot 163$ | $7 \cdot 233$ | $2^5 \cdot 3 \cdot 17$ | $23 \cdot 71$ | $2 \cdot 19 \cdot 43$ | $3 \cdot 5 \cdot 109$ | $2^2 \cdot 409$ | **1637** | $2 \cdot 3^2 \cdot 7 \cdot 13$ | $11 \cdot 149$ |
| 164 | $2^3 \cdot 5 \cdot 41$ | $3 \cdot 547$ | $2 \cdot 821$ | $31 \cdot 53$ | $2^2 \cdot 3 \cdot 137$ | $5 \cdot 7 \cdot 47$ | $2 \cdot 823$ | $3^3 \cdot 61$ | $2^4 \cdot 103$ | $17 \cdot 97$ |
| 165 | $2 \cdot 3 \cdot 5^2 \cdot 11$ | $13 \cdot 127$ | $2^2 \cdot 7 \cdot 59$ | $3 \cdot 19 \cdot 29$ | $2 \cdot 827$ | $5 \cdot 331$ | $2^3 \cdot 3^2 \cdot 23$ | **1657** | $2 \cdot 829$ | $3 \cdot 7 \cdot 79$ |
| 166 | $2^2 \cdot 5 \cdot 83$ | $11 \cdot 151$ | $2 \cdot 3 \cdot 277$ | **1663** | $2^7 \cdot 13$ | $3^2 \cdot 5 \cdot 37$ | $2 \cdot 7^2 \cdot 17$ | **1667** | $2^2 \cdot 3 \cdot 139$ | **1669** |
| 167 | $2 \cdot 5 \cdot 167$ | $3 \cdot 557$ | $2^3 \cdot 11 \cdot 19$ | $7 \cdot 239$ | $2 \cdot 3^3 \cdot 31$ | $5^2 \cdot 67$ | $2^2 \cdot 419$ | $3 \cdot 13 \cdot 43$ | $2 \cdot 839$ | $23 \cdot 73$ |
| 168 | $2^4 \cdot 3 \cdot 5 \cdot 7$ | $41^2$ | $2 \cdot 29^2$ | $3^2 \cdot 11 \cdot 17$ | $2^2 \cdot 421$ | $5 \cdot 337$ | $2 \cdot 3 \cdot 281$ | $7 \cdot 241$ | $2^3 \cdot 211$ | $3 \cdot 563$ |
| 169 | $2 \cdot 5 \cdot 13^2$ | $19 \cdot 89$ | $2^2 \cdot 3^2 \cdot 47$ | **1693** | $2 \cdot 7 \cdot 11^2$ | $3 \cdot 5 \cdot 113$ | $2^5 \cdot 53$ | **1697** | $2 \cdot 3 \cdot 283$ | **1699** |
| 170 | $2^2 \cdot 5^2 \cdot 17$ | $3^5 \cdot 7$ | $2 \cdot 23 \cdot 37$ | $13 \cdot 131$ | $2^3 \cdot 3 \cdot 71$ | $5 \cdot 11 \cdot 31$ | $2 \cdot 853$ | $3 \cdot 569$ | $2^2 \cdot 7 \cdot 61$ | **1709** |
| 171 | $2 \cdot 3^2 \cdot 5 \cdot 19$ | $29 \cdot 59$ | $2^4 \cdot 107$ | $3 \cdot 571$ | $2 \cdot 857$ | $5 \cdot 7^3$ | $2^2 \cdot 3 \cdot 11 \cdot 13$ | $17 \cdot 101$ | $2 \cdot 859$ | $3^2 \cdot 191$ |
| 172 | $2^3 \cdot 5 \cdot 43$ | **1721** | $2 \cdot 3 \cdot 7 \cdot 41$ | **1723** | $2^2 \cdot 431$ | $3 \cdot 5^2 \cdot 23$ | $2 \cdot 863$ | $11 \cdot 157$ | $2^6 \cdot 3^3$ | $7 \cdot 13 \cdot 19$ |
| 173 | $2 \cdot 5 \cdot 173$ | $3 \cdot 577$ | $2^2 \cdot 433$ | **1733** | $2 \cdot 3 \cdot 17^2$ | $5 \cdot 347$ | $2^3 \cdot 7 \cdot 31$ | $3^2 \cdot 193$ | $2 \cdot 11 \cdot 79$ | $37 \cdot 47$ |
| 174 | $2^2 \cdot 3 \cdot 5 \cdot 29$ | **1741** | $2 \cdot 13 \cdot 67$ | $3 \cdot 7 \cdot 83$ | $2^4 \cdot 109$ | $5 \cdot 349$ | $2 \cdot 3^2 \cdot 97$ | **1747** | $2^2 \cdot 19 \cdot 23$ | $3 \cdot 11 \cdot 53$ |
| 175 | $2 \cdot 5^3 \cdot 7$ | $17 \cdot 103$ | $2^3 \cdot 3 \cdot 73$ | **1753** | $2 \cdot 877$ | $3^3 \cdot 5 \cdot 13$ | $2^2 \cdot 439$ | $7 \cdot 251$ | $2 \cdot 3 \cdot 293$ | **1759** |
| 176 | $2^5 \cdot 5 \cdot 11$ | $3 \cdot 587$ | $2 \cdot 881$ | $41 \cdot 43$ | $2^2 \cdot 3^2 \cdot 7^2$ | $5 \cdot 353$ | $2 \cdot 883$ | $3 \cdot 19 \cdot 31$ | $2^3 \cdot 13 \cdot 17$ | $29 \cdot 61$ |
| 177 | $2 \cdot 3 \cdot 5 \cdot 59$ | $7 \cdot 11 \cdot 23$ | $2^2 \cdot 443$ | $3^2 \cdot 197$ | $2 \cdot 887$ | $5^2 \cdot 71$ | $2^4 \cdot 3 \cdot 37$ | **1777** | $2 \cdot 7 \cdot 127$ | $3 \cdot 593$ |
| 178 | $2^2 \cdot 5 \cdot 89$ | $13 \cdot 137$ | $2 \cdot 3^4 \cdot 11$ | **1783** | $2^3 \cdot 223$ | $3 \cdot 5 \cdot 7 \cdot 17$ | $2 \cdot 19 \cdot 47$ | **1787** | $2^2 \cdot 3 \cdot 149$ | **1789** |
| 179 | $2 \cdot 5 \cdot 179$ | $3^2 \cdot 199$ | $2^8 \cdot 7$ | $11 \cdot 163$ | $2 \cdot 3 \cdot 13 \cdot 23$ | $5 \cdot 359$ | $2^2 \cdot 449$ | $3 \cdot 599$ | $2 \cdot 29 \cdot 31$ | $7 \cdot 257$ |
| 180 | $2^3 \cdot 3^2 \cdot 5^2$ | **1801** | $2 \cdot 17 \cdot 53$ | $3 \cdot 601$ | $2^2 \cdot 11 \cdot 41$ | $5 \cdot 19^2$ | $2 \cdot 3 \cdot 7 \cdot 43$ | $13 \cdot 139$ | $2^4 \cdot 113$ | $3^3 \cdot 67$ |
| 181 | $2 \cdot 5 \cdot 181$ | **1811** | $2^2 \cdot 3 \cdot 151$ | $7^2 \cdot 37$ | $2 \cdot 907$ | $3 \cdot 5 \cdot 11^2$ | $2^3 \cdot 227$ | $23 \cdot 79$ | $2 \cdot 3^2 \cdot 101$ | $17 \cdot 107$ |
| 182 | $2^2 \cdot 5 \cdot 7 \cdot 13$ | $3 \cdot 607$ | $2 \cdot 911$ | **1823** | $2^5 \cdot 3 \cdot 19$ | $5^2 \cdot 73$ | $2 \cdot 11 \cdot 83$ | $3^2 \cdot 7 \cdot 29$ | $2^2 \cdot 457$ | $31 \cdot 59$ |
| 183 | $2 \cdot 3 \cdot 5 \cdot 61$ | **1831** | $2^3 \cdot 229$ | $3 \cdot 13 \cdot 47$ | $2 \cdot 7 \cdot 131$ | $5 \cdot 367$ | $2^2 \cdot 3^3 \cdot 17$ | $11 \cdot 167$ | $2 \cdot 919$ | $3 \cdot 613$ |
| 184 | $2^4 \cdot 5 \cdot 23$ | $7 \cdot 263$ | $2 \cdot 3 \cdot 307$ | $19 \cdot 97$ | $2^2 \cdot 461$ | $3^2 \cdot 5 \cdot 41$ | $2 \cdot 13 \cdot 71$ | **1847** | $2^3 \cdot 3 \cdot 7 \cdot 11$ | $43^2$ |
| 185 | $2 \cdot 5^2 \cdot 37$ | $3 \cdot 617$ | $2^2 \cdot 463$ | $17 \cdot 109$ | $2 \cdot 3^2 \cdot 103$ | $5 \cdot 7 \cdot 53$ | $2^6 \cdot 29$ | $3 \cdot 619$ | $2 \cdot 929$ | $11 \cdot 13^2$ |
| 186 | $2^2 \cdot 3 \cdot 5 \cdot 31$ | **1861** | $2 \cdot 7^2 \cdot 19$ | $3^4 \cdot 23$ | $2^3 \cdot 233$ | $5 \cdot 373$ | $2 \cdot 3 \cdot 311$ | **1867** | $2^2 \cdot 467$ | $3 \cdot 7 \cdot 89$ |
| 187 | $2 \cdot 5 \cdot 11 \cdot 17$ | **1871** | $2^4 \cdot 3^2 \cdot 13$ | **1873** | $2 \cdot 937$ | $3 \cdot 5^4$ | $2^2 \cdot 7 \cdot 67$ | **1877** | $2 \cdot 3 \cdot 313$ | **1879** |
| 188 | $2^3 \cdot 5 \cdot 47$ | $3^2 \cdot 11 \cdot 19$ | $2 \cdot 941$ | $7 \cdot 269$ | $2^2 \cdot 3 \cdot 157$ | $5 \cdot 13 \cdot 29$ | $2 \cdot 23 \cdot 41$ | $3 \cdot 17 \cdot 37$ | $2^5 \cdot 59$ | **1889** |
| 189 | $2 \cdot 3^3 \cdot 5 \cdot 7$ | $31 \cdot 61$ | $2^2 \cdot 11 \cdot 43$ | $3 \cdot 631$ | $2 \cdot 947$ | $5 \cdot 379$ | $2^3 \cdot 3 \cdot 79$ | $7 \cdot 271$ | $2 \cdot 13 \cdot 73$ | $3^2 \cdot 211$ |
| 190 | $2^2 \cdot 5^2 \cdot 19$ | **1901** | $2 \cdot 3 \cdot 317$ | $11 \cdot 173$ | $2^4 \cdot 7 \cdot 17$ | $3 \cdot 5 \cdot 127$ | $2 \cdot 953$ | **1907** | $2^2 \cdot 3^2 \cdot 53$ | $23 \cdot 83$ |
| 191 | $2 \cdot 5 \cdot 191$ | $3 \cdot 7^2 \cdot 13$ | $2^3 \cdot 239$ | **1913** | $2 \cdot 3 \cdot 11 \cdot 29$ | $5 \cdot 383$ | $2^2 \cdot 479$ | $3^3 \cdot 71$ | $2 \cdot 7 \cdot 137$ | $19 \cdot 101$ |
| 192 | $2^7 \cdot 3 \cdot 5$ | $17 \cdot 113$ | $2 \cdot 31^2$ | $3 \cdot 641$ | $2^2 \cdot 13 \cdot 37$ | $5^2 \cdot 7 \cdot 11$ | $2 \cdot 3^2 \cdot 107$ | $41 \cdot 47$ | $2^3 \cdot 241$ | $3 \cdot 643$ |
| 193 | $2 \cdot 5 \cdot 193$ | **1931** | $2^2 \cdot 3 \cdot 7 \cdot 23$ | **1933** | $2 \cdot 967$ | $3^2 \cdot 5 \cdot 43$ | $2^4 \cdot 11^2$ | $13 \cdot 149$ | $2 \cdot 3 \cdot 17 \cdot 19$ | $7 \cdot 277$ |
| 194 | $2^2 \cdot 5 \cdot 97$ | $3 \cdot 647$ | $2 \cdot 971$ | $29 \cdot 67$ | $2^3 \cdot 3^5$ | $5 \cdot 389$ | $2 \cdot 7 \cdot 139$ | $3 \cdot 11 \cdot 59$ | $2^2 \cdot 487$ | **1949** |
| 195 | $2 \cdot 3 \cdot 5^2 \cdot 13$ | **1951** | $2^5 \cdot 61$ | $3^2 \cdot 7 \cdot 31$ | $2 \cdot 977$ | $5 \cdot 17 \cdot 23$ | $2^2 \cdot 3 \cdot 163$ | $19 \cdot 103$ | $2 \cdot 11 \cdot 89$ | $3 \cdot 653$ |
| 196 | $2^3 \cdot 5 \cdot 7^2$ | $37 \cdot 53$ | $2 \cdot 3^2 \cdot 109$ | $13 \cdot 151$ | $2^2 \cdot 491$ | $3 \cdot 5 \cdot 131$ | $2 \cdot 983$ | $7 \cdot 281$ | $2^4 \cdot 3 \cdot 41$ | $11 \cdot 179$ |
| 197 | $2 \cdot 5 \cdot 197$ | $3^3 \cdot 73$ | $2^2 \cdot 17 \cdot 29$ | **1973** | $2 \cdot 3 \cdot 7 \cdot 47$ | $5^2 \cdot 79$ | $2^3 \cdot 13 \cdot 19$ | $3 \cdot 659$ | $2 \cdot 23 \cdot 43$ | **1979** |
| 198 | $2^2 \cdot 3^2 \cdot 5 \cdot 11$ | $7 \cdot 283$ | $2 \cdot 991$ | $3 \cdot 661$ | $2^6 \cdot 31$ | $5 \cdot 397$ | $2 \cdot 3 \cdot 331$ | **1987** | $2^2 \cdot 7 \cdot 71$ | $3^2 \cdot 13 \cdot 17$ |

| | 0 | 1 | 2 | 3 | 4 | 5 | 6 | 7 | 8 | 9 |
|---|---|---|---|---|---|---|---|---|---|---|
| 199 | $2 \cdot 5 \cdot 199$ | $11 \cdot 181$ | $2^3 \cdot 3 \cdot 83$ | **1993** | $2 \cdot 997$ | $3 \cdot 5 \cdot 7 \cdot 19$ | $2^2 \cdot 499$ | **1997** | $2 \cdot 3^3 \cdot 37$ | **1999** |
| 200 | $2^4 \cdot 5^3$ | $3 \cdot 23 \cdot 29$ | $2 \cdot 7 \cdot 11 \cdot 13$ | **2003** | $2^2 \cdot 3 \cdot 167$ | $5 \cdot 401$ | $2 \cdot 17 \cdot 59$ | $3^2 \cdot 223$ | $2^3 \cdot 251$ | $7^2 \cdot 41$ |
| 201 | $2 \cdot 3 \cdot 5 \cdot 67$ | **2011** | $2^2 \cdot 503$ | $3 \cdot 11 \cdot 61$ | $2 \cdot 19 \cdot 53$ | $5 \cdot 13 \cdot 31$ | $2^5 \cdot 3^2 \cdot 7$ | **2017** | $2 \cdot 1009$ | $3 \cdot 673$ |
| 202 | $2^2 \cdot 5 \cdot 101$ | $43 \cdot 47$ | $2 \cdot 3 \cdot 337$ | $7 \cdot 17^2$ | $2^3 \cdot 11 \cdot 23$ | $3^4 \cdot 5^2$ | $2 \cdot 1013$ | **2027** | $2^2 \cdot 3 \cdot 13^2$ | **2029** |
| 203 | $2 \cdot 5 \cdot 7 \cdot 29$ | $3 \cdot 677$ | $2^4 \cdot 127$ | $19 \cdot 107$ | $2 \cdot 3^2 \cdot 113$ | $5 \cdot 11 \cdot 37$ | $2^2 \cdot 509$ | $3 \cdot 7 \cdot 97$ | $2 \cdot 1019$ | **2039** |
| 204 | $2^3 \cdot 3 \cdot 5 \cdot 17$ | $13 \cdot 157$ | $2 \cdot 1021$ | $3^2 \cdot 227$ | $2^2 \cdot 7 \cdot 73$ | $5 \cdot 409$ | $2 \cdot 3 \cdot 11 \cdot 31$ | $23 \cdot 89$ | $2^{11}$ | $3 \cdot 683$ |
| 205 | $2 \cdot 5^2 \cdot 41$ | $7 \cdot 293$ | $2^2 \cdot 3^3 \cdot 19$ | **2053** | $2 \cdot 13 \cdot 79$ | $3 \cdot 5 \cdot 137$ | $2^3 \cdot 257$ | $11^2 \cdot 17$ | $2 \cdot 3 \cdot 7^3$ | $29 \cdot 71$ |
| 206 | $2^2 \cdot 5 \cdot 103$ | $3^2 \cdot 229$ | $2 \cdot 1031$ | **2063** | $2^4 \cdot 3 \cdot 43$ | $5 \cdot 7 \cdot 59$ | $2 \cdot 1033$ | $3 \cdot 13 \cdot 53$ | $2^2 \cdot 11 \cdot 47$ | **2069** |
| 207 | $2 \cdot 3^2 \cdot 5 \cdot 23$ | $19 \cdot 109$ | $2^3 \cdot 7 \cdot 37$ | $3 \cdot 691$ | $2 \cdot 17 \cdot 61$ | $5^2 \cdot 83$ | $2^2 \cdot 3 \cdot 173$ | $31 \cdot 67$ | $2 \cdot 1039$ | $3^3 \cdot 7 \cdot 11$ |
| 208 | $2^5 \cdot 5 \cdot 13$ | **2081** | $2 \cdot 3 \cdot 347$ | **2083** | $2^2 \cdot 521$ | $3 \cdot 5 \cdot 139$ | $2 \cdot 7 \cdot 149$ | **2087** | $2^3 \cdot 3^2 \cdot 29$ | **2089** |
| 209 | $2 \cdot 5 \cdot 11 \cdot 19$ | $3 \cdot 17 \cdot 41$ | $2^2 \cdot 523$ | $7 \cdot 13 \cdot 23$ | $2 \cdot 3 \cdot 349$ | $5 \cdot 419$ | $2^4 \cdot 131$ | $3^2 \cdot 233$ | $2 \cdot 1049$ | **2099** |
| 210 | $2^2 \cdot 3 \cdot 5^2 \cdot 7$ | $11 \cdot 191$ | $2 \cdot 1051$ | $3 \cdot 701$ | $2^3 \cdot 263$ | $5 \cdot 421$ | $2 \cdot 3^4 \cdot 13$ | $7^2 \cdot 43$ | $2^2 \cdot 17 \cdot 31$ | $3 \cdot 19 \cdot 37$ |
| 211 | $2 \cdot 5 \cdot 211$ | **2111** | $2^6 \cdot 3 \cdot 11$ | **2113** | $2 \cdot 7 \cdot 151$ | $3^2 \cdot 5 \cdot 47$ | $2^2 \cdot 23^2$ | $29 \cdot 73$ | $2 \cdot 3 \cdot 353$ | $13 \cdot 163$ |
| 212 | $2^3 \cdot 5 \cdot 53$ | $3 \cdot 7 \cdot 101$ | $2 \cdot 1061$ | $11 \cdot 193$ | $2^2 \cdot 3^2 \cdot 59$ | $5^3 \cdot 17$ | $2 \cdot 1063$ | $3 \cdot 709$ | $2^4 \cdot 7 \cdot 19$ | **2129** |
| 213 | $2 \cdot 3 \cdot 5 \cdot 71$ | **2131** | $2^2 \cdot 13 \cdot 41$ | $3^3 \cdot 79$ | $2 \cdot 11 \cdot 97$ | $5 \cdot 7 \cdot 61$ | $2^3 \cdot 3 \cdot 89$ | **2137** | $2 \cdot 1069$ | $3 \cdot 23 \cdot 31$ |
| 214 | $2^2 \cdot 5 \cdot 107$ | **2141** | $2 \cdot 3^2 \cdot 7 \cdot 17$ | **2143** | $2^5 \cdot 67$ | $3 \cdot 5 \cdot 11 \cdot 13$ | $2 \cdot 29 \cdot 37$ | $19 \cdot 113$ | $2^2 \cdot 3 \cdot 179$ | $7 \cdot 307$ |
| 215 | $2 \cdot 5^2 \cdot 43$ | $3^2 \cdot 239$ | $2^3 \cdot 269$ | **2153** | $2 \cdot 3 \cdot 359$ | $5 \cdot 431$ | $2^2 \cdot 7^2 \cdot 11$ | $3 \cdot 719$ | $2 \cdot 13 \cdot 83$ | $17 \cdot 127$ |
| 216 | $2^4 \cdot 3^3 \cdot 5$ | **2161** | $2 \cdot 23 \cdot 47$ | $3 \cdot 7 \cdot 103$ | $2^2 \cdot 541$ | $5 \cdot 433$ | $2 \cdot 3 \cdot 19^2$ | $11 \cdot 197$ | $2^3 \cdot 271$ | $3^2 \cdot 241$ |
| 217 | $2 \cdot 5 \cdot 7 \cdot 31$ | $13 \cdot 167$ | $2^2 \cdot 3 \cdot 181$ | $41 \cdot 53$ | $2 \cdot 1087$ | $3 \cdot 5^2 \cdot 29$ | $2^7 \cdot 17$ | $7 \cdot 311$ | $2 \cdot 3^2 \cdot 11^2$ | **2179** |
| 218 | $2^2 \cdot 5 \cdot 109$ | $3 \cdot 727$ | $2 \cdot 1091$ | $37 \cdot 59$ | $2^3 \cdot 3 \cdot 7 \cdot 13$ | $5 \cdot 19 \cdot 23$ | $2 \cdot 1093$ | $3^7$ | $2^2 \cdot 547$ | $11 \cdot 199$ |
| 219 | $2 \cdot 3 \cdot 5 \cdot 73$ | $7 \cdot 313$ | $2^4 \cdot 137$ | $3 \cdot 17 \cdot 43$ | $2 \cdot 1097$ | $5 \cdot 439$ | $2^2 \cdot 3^2 \cdot 61$ | $13^3$ | $2 \cdot 7 \cdot 157$ | $3 \cdot 733$ |
| 220 | $2^3 \cdot 5^2 \cdot 11$ | $31 \cdot 71$ | $2 \cdot 3 \cdot 367$ | **2203** | $2^2 \cdot 19 \cdot 29$ | $3^2 \cdot 5 \cdot 7^2$ | $2 \cdot 1103$ | **2207** | $2^5 \cdot 3 \cdot 23$ | $47^2$ |
| 221 | $2 \cdot 5 \cdot 13 \cdot 17$ | $3 \cdot 11 \cdot 67$ | $2^2 \cdot 7 \cdot 79$ | **2213** | $2 \cdot 3^3 \cdot 41$ | $5 \cdot 443$ | $2^3 \cdot 277$ | $3 \cdot 739$ | $2 \cdot 1109$ | $7 \cdot 317$ |
| 222 | $2^2 \cdot 3 \cdot 5 \cdot 37$ | **2221** | $2 \cdot 11 \cdot 101$ | $3^2 \cdot 13 \cdot 19$ | $2^4 \cdot 139$ | $5^2 \cdot 89$ | $2 \cdot 3 \cdot 7 \cdot 53$ | $17 \cdot 131$ | $2^2 \cdot 557$ | $3 \cdot 743$ |
| 223 | $2 \cdot 5 \cdot 223$ | $23 \cdot 97$ | $2^3 \cdot 3^2 \cdot 31$ | $7 \cdot 11 \cdot 29$ | $2 \cdot 1117$ | $3 \cdot 5 \cdot 149$ | $2^2 \cdot 13 \cdot 43$ | **2237** | $2 \cdot 3 \cdot 373$ | **2239** |
| 224 | $2^6 \cdot 5 \cdot 7$ | $3^3 \cdot 83$ | $2 \cdot 19 \cdot 59$ | **2243** | $2^2 3 \cdot 11 \cdot 17$ | $5 \cdot 449$ | $2 \cdot 1123$ | $3 \cdot 7 \cdot 107$ | $2^3 \cdot 281$ | $13 \cdot 173$ |
| 225 | $2 \cdot 3^2 \cdot 5^3$ | **2251** | $2^2 \cdot 563$ | $3 \cdot 751$ | $2 \cdot 7^2 \cdot 23$ | $5 \cdot 11 \cdot 41$ | $2^4 \cdot 3 \cdot 47$ | $37 \cdot 61$ | $2 \cdot 1129$ | $3^2 \cdot 251$ |
| 226 | $2^2 \cdot 5 \cdot 113$ | $7 \cdot 17 \cdot 19$ | $2 \cdot 3 \cdot 13 \cdot 29$ | $31 \cdot 73$ | $2^3 \cdot 283$ | $3 \cdot 5 \cdot 151$ | $2 \cdot 11 \cdot 103$ | **2267** | $2^2 \cdot 3^4 \cdot 7$ | **2269** |
| 227 | $2 \cdot 5 \cdot 227$ | $3 \cdot 757$ | $2^5 \cdot 71$ | **2273** | $2 \cdot 3 \cdot 379$ | $5^2 \cdot 7 \cdot 13$ | $2^2 \cdot 569$ | $3^2 \cdot 11 \cdot 23$ | $2 \cdot 17 \cdot 67$ | $43 \cdot 53$ |
| 228 | $2^3 \cdot 3 \cdot 5 \cdot 19$ | **2281** | $2 \cdot 7 \cdot 163$ | $3 \cdot 761$ | $2^2 \cdot 571$ | $5 \cdot 457$ | $2 \cdot 3^2 \cdot 127$ | **2287** | $2^4 \cdot 11 \cdot 13$ | $3 \cdot 7 \cdot 109$ |
| 229 | $2 \cdot 5 \cdot 229$ | $29 \cdot 79$ | $2^2 \cdot 3 \cdot 191$ | **2293** | $2 \cdot 31 \cdot 37$ | $3^3 \cdot 5 \cdot 17$ | $2^3 \cdot 7 \cdot 41$ | **2297** | $2 \cdot 3 \cdot 383$ | $11^2 \cdot 19$ |
| 230 | $2^2 \cdot 5^2 \cdot 23$ | $3 \cdot 13 \cdot 59$ | $2 \cdot 1151$ | $7^2 \cdot 47$ | $2^8 \cdot 3^2$ | $5 \cdot 461$ | $2 \cdot 1153$ | $3 \cdot 769$ | $2^2 \cdot 577$ | **2309** |
| 231 | $2 \cdot 3 \cdot 5 \cdot 7 \cdot 11$ | **2311** | $2^3 \cdot 17^2$ | $3^2 \cdot 257$ | $2 \cdot 13 \cdot 89$ | $5 \cdot 463$ | $2^2 \cdot 3 \cdot 193$ | $7 \cdot 331$ | $2 \cdot 19 \cdot 61$ | $3 \cdot 773$ |
| 232 | $2^4 \cdot 5 \cdot 29$ | $11 \cdot 211$ | $2 \cdot 3^3 \cdot 43$ | $23 \cdot 101$ | $2^2 \cdot 7 \cdot 83$ | $3 \cdot 5^2 \cdot 31$ | $2 \cdot 1163$ | $13 \cdot 179$ | $2^3 \cdot 3 \cdot 97$ | $17 \cdot 137$ |
| 233 | $2 \cdot 5 \cdot 233$ | $3^2 \cdot 7 \cdot 37$ | $2^2 \cdot 11 \cdot 53$ | **2333** | $2 \cdot 3 \cdot 389$ | $5 \cdot 467$ | $2^5 \cdot 73$ | $3 \cdot 19 \cdot 41$ | $2 \cdot 7 \cdot 167$ | **2339** |
| 234 | $2^2 \cdot 3^2 \cdot 5 \cdot 13$ | **2341** | $2 \cdot 1171$ | $3 \cdot 11 \cdot 71$ | $2^3 \cdot 293$ | $5 \cdot 7 \cdot 67$ | $2 \cdot 3 \cdot 17 \cdot 23$ | **2347** | $2^2 \cdot 587$ | $3^4 \cdot 29$ |
| 235 | $2 \cdot 5^2 \cdot 47$ | **2351** | $2^4 \cdot 3 \cdot 7^2$ | $13 \cdot 181$ | $2 \cdot 11 \cdot 107$ | $3 \cdot 5 \cdot 157$ | $2^2 \cdot 19 \cdot 31$ | **2357** | $2 \cdot 3^2 \cdot 131$ | $7 \cdot 337$ |
| 236 | $2^3 \cdot 5 \cdot 59$ | $3 \cdot 787$ | $2 \cdot 1181$ | $17 \cdot 139$ | $2^2 \cdot 3 \cdot 197$ | $5 \cdot 11 \cdot 43$ | $2 \cdot 7 \cdot 13^2$ | $3^2 \cdot 263$ | $2^6 \cdot 37$ | $23 \cdot 103$ |
| 237 | $2 \cdot 3 \cdot 5 \cdot 79$ | **2371** | $2^2 \cdot 593$ | $3 \cdot 7 \cdot 113$ | $2 \cdot 1187$ | $5^3 \cdot 19$ | $2^3 \cdot 3^3 \cdot 11$ | **2377** | $2 \cdot 29 \cdot 41$ | $3 \cdot 13 \cdot 61$ |
| 238 | $2^2 \cdot 5 \cdot 7 \cdot 17$ | **2381** | $2 \cdot 3 \cdot 397$ | **2383** | $2^4 \cdot 149$ | $3^2 \cdot 5 \cdot 53$ | $2 \cdot 1193$ | $7 \cdot 11 \cdot 31$ | $2^2 \cdot 3 \cdot 199$ | **2389** |
| 239 | $2 \cdot 5 \cdot 239$ | $3 \cdot 797$ | $2^3 \cdot 13 \cdot 23$ | **2393** | $2 \cdot 3^2 \cdot 7 \cdot 19$ | $5 \cdot 479$ | $2^2 \cdot 599$ | $3 \cdot 17 \cdot 47$ | $2 \cdot 11 \cdot 109$ | **2399** |
| 240 | $2^5 \cdot 3 \cdot 5^2$ | $7^4$ | $2 \cdot 1201$ | $3^3 \cdot 89$ | $2^2 \cdot 601$ | $5 \cdot 13 \cdot 37$ | $2 \cdot 3 \cdot 401$ | $29 \cdot 83$ | $2^3 \cdot 7 \cdot 43$ | $3 \cdot 11 \cdot 73$ |
| 241 | $2 \cdot 5 \cdot 241$ | **2411** | $2^2 \cdot 3^2 \cdot 67$ | $19 \cdot 127$ | $2 \cdot 17 \cdot 71$ | $3 \cdot 5 \cdot 7 \cdot 23$ | $2^4 \cdot 151$ | **2417** | $2 \cdot 3 \cdot 13 \cdot 31$ | $41 \cdot 59$ |
| 242 | $2^2 \cdot 5 \cdot 11^2$ | $3^2 \cdot 269$ | $2 \cdot 7 \cdot 173$ | **2423** | $2^3 \cdot 3 \cdot 101$ | $5^2 \cdot 97$ | $2 \cdot 1213$ | $3 \cdot 809$ | $2^2 \cdot 607$ | $7 \cdot 347$ |
| 243 | $2 \cdot 3^5 \cdot 5$ | $11 \cdot 13 \cdot 17$ | $2^7 \cdot 19$ | $3 \cdot 811$ | $2 \cdot 1217$ | $5 \cdot 487$ | $2^2 \cdot 3 \cdot 7 \cdot 29$ | **2437** | $2 \cdot 23 \cdot 53$ | $3^2 \cdot 271$ |
| 244 | $2^3 \cdot 5 \cdot 61$ | **2441** | $2 \cdot 3 \cdot 11 \cdot 37$ | $7 \cdot 349$ | $2^2 \cdot 13 \cdot 47$ | $3 \cdot 5 \cdot 163$ | $2 \cdot 1223$ | **2447** | $2^4 \cdot 3^2 \cdot 17$ | $31 \cdot 79$ |
| 245 | $2 \cdot 5^2 \cdot 7^2$ | $3 \cdot 19 \cdot 43$ | $2^2 \cdot 613$ | $11 \cdot 223$ | $2 \cdot 3 \cdot 409$ | $5 \cdot 491$ | $2^3 \cdot 307$ | $3^3 \cdot 7 \cdot 13$ | $2 \cdot 1229$ | **2459** |

| | 0 | 1 | 2 | 3 | 4 | 5 | 6 | 7 | 8 | 9 |
|---|---|---|---|---|---|---|---|---|---|---|
| 246 | $2^2 \cdot 3 \cdot 5 \cdot 41$ | $23 \cdot 107$ | $2 \cdot 1231$ | $3 \cdot 821$ | $2^5 \cdot 7 \cdot 11$ | $5 \cdot 17 \cdot 29$ | $2 \cdot 3^2 \cdot 137$ | **2467** | $2^2 \cdot 617$ | $3 \cdot 823$ |
| 247 | $2 \cdot 5 \cdot 13 \cdot 19$ | $7 \cdot 353$ | $2^3 \cdot 3 \cdot 103$ | **2473** | $2 \cdot 1237$ | $3^2 \cdot 5^2 \cdot 11$ | $2^2 \cdot 619$ | **2477** | $2 \cdot 3 \cdot 7 \cdot 59$ | $37 \cdot 67$ |
| 248 | $2^4 \cdot 5 \cdot 31$ | $3 \cdot 827$ | $2 \cdot 17 \cdot 73$ | $13 \cdot 191$ | $2^2 \cdot 3^3 \cdot 23$ | $5 \cdot 7 \cdot 71$ | $2 \cdot 11 \cdot 113$ | $3 \cdot 829$ | $2^3 \cdot 311$ | $19 \cdot 131$ |
| 249 | $2 \cdot 3 \cdot 5 \cdot 83$ | $47 \cdot 53$ | $2^2 \cdot 7 \cdot 89$ | $3^2 \cdot 277$ | $2 \cdot 29 \cdot 43$ | $5 \cdot 499$ | $2^6 \cdot 3 \cdot 13$ | $11 \cdot 227$ | $2 \cdot 1249$ | $3 \cdot 7^2 \cdot 17$ |

**Examples:**

1. $6 = 2 \times 3$.

2. $245 = 5 \times 7^2$.

3. $10! = 2^8 \times 3^4 \times 5^2 \times 7$.

4. $68{,}718{,}821{,}377 = (2^{17} - 1) \cdot (2^{19} - 1)$ (both factors are Mersenne primes; see §4.4.3).

5. The largest prime known is $2^{74,207,281} - 1$, which has 22,338,618 decimal digits. It is a Mersenne prime (see Table 2).

---

### 4.4.2 COUNTING PRIMES

**Definitions:**

The value of the **prime counting function** $\pi(x)$ at $x$, where $x$ is a positive real number, equals the number of primes less than or equal to $x$.

The **li function** is defined by $\mathrm{li}(x) = \int_0^x \frac{dt}{\ln t}$, for $x \geq 2$. (The principal value is taken for the integral at the singularity $t = 1$.)

**Twin primes** are primes that differ by exactly 2.

**Facts:**

1. Euclid (ca. 300 BCE) proved that there are infinitely many primes. He observed that the product of a finite list of primes, plus one, must be divisible by a prime not on that list.

2. Leonhard Euler (1707–1783) showed that the sum of the reciprocals of the primes up to $n$ tends toward infinity as $n$ tends toward infinity, which also implies that there are infinitely many primes. (There are many other proofs as well.)

3. There is no useful, exact formula known which will produce the $n$th prime, given $n$. It is relatively easy to construct a useless (that is, impractical) one. For example, let $\alpha = \sum_{n=1}^{\infty} p_n / 2^{2^n}$, where $p_n$ is the $n$th prime. Then the $n$th prime is $\lfloor 2^{2^n} \alpha \rfloor - 2^{2^{n-1}} \lfloor 2^{2^{n-1}} \alpha \rfloor$, where $\lfloor x \rfloor$ is the greatest integer less than or equal to $x$.

4. If $f(x)$ is a polynomial with integer coefficients that is not constant, then there are infinitely many integers $n$ for which $|f(n)|$ is not prime.

5. There are polynomials with integer coefficients with the property that the set of positive values taken by each of these polynomials as the variables range over the set of nonnegative integers is the set of prime numbers. The existence of such polynomials has essentially no practical value for constructing primes. For example, there are polynomials in 26 variables of degree 25, in 42 variables of degree 5, and in 12 variables of degree 13,697 that have this property. (See [Ri96].)

6. $\frac{p_n}{n \log n} \to 1$ as $n \to \infty$. (This follows from the prime number theorem, Fact 10.)

7. An inexact and rough formula for the $n$th prime is $n \log n$.

**8.** $p_n > n \log n$ for all $n$. (J. B. Rosser)

**9.** *The sieve of Eratosthenes*:   Eratosthenes (3rd century BCE) developed a method (Algorithm 1) for listing all prime numbers less than a fixed bound $N$. This method starts from the prime 2 and successively eliminates all of its multiple from the list. It continues through the list and successively eliminates multiples of the next (prime) value. Only trial values up to the square root of $N$ need to be tested.

---

**Algorithm 1**:   **Sieve of Eratosthenes.**

make a list of the numbers from 2 to $N$

$i := 1$

**while** $i \leq \sqrt{N}$

   $i := i + 1$

   **if** $i$ is not already crossed out **then** cross out all proper multiples of $i$ that
      are less than or equal to $N$

{The numbers *not* crossed out then comprise the primes up to $N$}

---

**10.** *Prime number theorem*:    $\pi(x)$, when divided by $\frac{x}{\log x}$, tends to 1 as $x$ tends to infinity. That is, $\pi(x)$ is asymptotic to $\frac{x}{\log x}$ as $x \to \infty$.

**11.** The prime number theorem was first conjectured by Carl Friedrich Gauss (1777–1855) in 1792, and was first proved in 1896 independently by Charles de la Vallée Poussin (1866–1962) and Jacques Hadamard (1865–1963). They proved it in the stronger form $|\pi(x) - \mathrm{li}(x)| < c_1 x e^{-c_2 \sqrt{\log x}}$, where $c_1$ and $c_2$ are positive constants. Their proofs used functions of a complex variable. The first elementary proofs (not using complex variables) of the prime number theorem were supplied in 1949 by Paul Erdős (1913–1996) and Atle Selberg (1917–2007).

**12.** Integration by parts shows that $\mathrm{li}(x)$ is asymptotic to $\frac{x}{\log x}$ as $x \to \infty$.

**13.** $|\pi(x) - \mathrm{li}(x)| < c_3 x e^{-c_4 (\log x)^{3/5} (\log \log x)^{-1/5}}$ for certain positive constants $c_3$ and $c_4$. (I. M. Vinogradov and Nikolai Korobov, 1958.)

**14.** If the Riemann hypothesis (see Open Problem 1) is true, $|\pi(x) - \mathrm{li}(x)|$ is bounded by $\sqrt{x} \log x$ for all $x \geq 2$. Note that conversely, if $|\pi(x) - \mathrm{li}(x)| < \sqrt{x} \log x$ for all $x \geq 2$, then the Riemann hypothesis is true.

**15.** J. E. Littlewood (1885–1977) showed that $\pi(x) - \mathrm{li}(x)$ changes sign infinitely often. However, no explicit number $x$ with $\pi(x) - \mathrm{li}(x) > 0$ is known. Carter Bays and Richard H. Hudson have shown that such a number $x$ exists below $1.4 \cdot 10^{316}$. It is also known (D. Platt and T. Trudgian) that $\pi(x) < \mathrm{li}(x)$ for all $x < 1.39 \cdot 10^{17}$.

**16.** The largest exactly computed value of $\pi(x)$ is $\pi(10^{26})$. This value, computed by D. B. Staple in 2015, is about $1.56 \cdot 10^{11}$ below $\mathrm{li}(10^{26})$. (See the following table.)

| $n$ | $\pi(10^n)$ | $\approx \pi(10^n) - \text{li}(10^n)$ |
|---|---|---|
| 1 | 4 | $-2$ |
| 2 | 25 | $-5$ |
| 3 | 168 | $-10$ |
| 4 | 1,229 | $-17$ |
| 5 | 9,592 | $-38$ |
| 6 | 78,498 | $-130$ |
| 7 | 664,579 | $-339$ |
| 8 | 5,761,455 | $-754$ |
| 9 | 50,847,534 | $-1,701$ |
| 10 | 455,052,511 | $-3,104$ |
| 11 | 4,118,054,813 | $-11,588$ |
| 12 | 37,607,912,018 | $-38,263$ |
| 13 | 346,065,536,839 | $-108,971$ |
| 14 | 3,204,941,750,802 | $-314,890$ |
| 15 | 29,844,570,422,669 | $-1,052,619$ |
| 16 | 279,238,341,033,925 | $-3,214,632$ |
| 17 | 2,623,557,157,654,233 | $-7,956,589$ |
| 18 | 24,739,954,287,740,860 | $-21,949,555$ |
| 19 | 234,057,667,276,344,607 | $-99,877,775$ |
| 20 | 2,220,819,602,560,918,840 | $-222,744,644$ |
| 21 | 21,127,269.486,018,731,928 | $-597,394,254$ |
| 22 | 201,467,286,689,315,906,290 | $-1,932,355,208$ |
| 23 | 1,925,320,391,606,803,968,923 | $-7,250,186,216$ |
| 24 | 18,435,599,767,349,200,867,866 | $-17,146,907,278$ |
| 25 | 176,846,309,399,143,769,411,680 | $-55,160,980,939$ |
| 26 | 1,699,246,750,872,437,141,327,603 | $-155,891,678,121$ |

**17.** *Dirichlet's theorem on primes in arithmetic progressions:*    Given coprime integers $a, b$ with $b$ positive, there are infinitely many primes $p \equiv a \pmod{b}$. G. L. Dirichlet proved this in 1837.

**18.** The number of primes $p$ less than $x$ such that $p \equiv a \pmod{b}$ is asymptotic to $\frac{1}{\phi(b)}\pi(x)$ as $x \to \infty$, if $a$ and $b$ are coprime and $b$ is positive. (Here $\phi$ is the Euler phi-function; see §4.6.2.)

**Open Problems:**

**1.** *Riemann hypothesis:*    The *Riemann hypothesis* (RH), posed in 1859 by Bernhard Riemann (1826–1866), is a conjecture about the location of zeros of the *Riemann zeta function*, the function of the complex variable $s$ defined by the series $\zeta(s) = \sum_{n=1}^{\infty} n^{-s}$ when the real part of $s$ is $> 1$, and defined by the formula

$$\zeta(s) = \frac{s}{s-1} - s \int_1^{\infty}(x - \lfloor x \rfloor)x^{-s-1}\,dx$$

in the larger region when the real part of $s$ is $> 0$, except for the single point $s = 1$, where it remains undefined. The Riemann hypothesis asserts that all of the solutions to $\zeta(s) = 0$ in this larger region lie on the vertical line in the complex number plane with real part $\frac{1}{2}$. Its proof would imply a better error estimate for the prime number theorem. While believed to be true, it has not been proved. (See Fact 14.)

**2.** *Extended Riemann hypothesis*:    There is a generalized form of the Riemann hypothesis known as the *extended Riemann hypothesis (ERH)* or the *generalized Riemann hypothesis (GRH)*, which also has important consequences in number theory. (For example, see §4.4.4.)

**3.** *Hypothesis H*: The *hypothesis H* of Andrzej Schinzel and Wacław Sierpiński (1882–1969) asserts that for every collection of irreducible nonconstant polynomials $f_1(x), \ldots, f_k(x)$ with integral coefficients and positive leading coefficients, if there is no fixed integer greater than 1 dividing the product $f_1(m) \ldots f_k(m)$ for all integers $m$, then there are infinitely many integers $m$ such that each of the numbers $f_1(m), \ldots, f_k(m)$ is prime. The case when each of the polynomials is linear was previously conjectured by L. E. Dickson, and is known as the *prime $k$-tuples conjecture*. The only case of Hypothesis H that has been proved is the case of a single linear polynomial; this is Dirichlet's theorem (Fact 17). The case of the two linear polynomials $x$ and $x + 2$ corresponds to the twin prime conjecture (Open Problem 4). Among the many consequences of Hypothesis H is the assertion that there are infinitely many primes of the form $m^2 + 1$.

**4.** *Twin primes*:    It has been conjectured that there are infinitely many twin primes, that is, pairs of primes that differ by 2.

Let $d_n$ denote the difference between the $(n+1)$st prime and the $n$th prime. The sequence $d_n$ is unbounded. The prime number theorem implies that on average $d_n$ is about $\log n$. The twin prime conjecture asks whether $d_n$ is 2 infinitely often. Although no proof of the infinitude of twin primes has been found, some results have recently been proved about the infinitude of primes that differ by a particular positive integer greater than two. In 2013 Yitang Zhang surprised the mathematical community when he proved that there is an integer $N$ with $N < 70{,}000{,}000$ such that infinitely many pairs of primes differ by $N$. In later work by the Polymath Project, a collaborative effort to further mathematical research, Zhang's bound was reduced to 246. (Under the assumption that certain hypotheses hold, this bound can be reduced to as little as 6.) The hope is that this work may eventually lead to a proof of the twin prime conjecture.

**5.** It is conjectured that $d_n$ can be as big as $\log^2 n$ infinitely often, but not much bigger. Roger Baker and Glyn Harman have recently shown that $d_n < n^{.535}$ for all large numbers $n$. For the other direction, in 2014 Ford, Green, Konyagin, and Tao showed that $d_n > c \log n (\log \log n)(\log \log \log \log n)/(\log \log \log n)$ infinitely often. Several improvements have been made on the constant $c$, but this ungainly expression has stubbornly resisted improvement.

**6.** Christian Goldbach (1690–1764) conjectured that every integer greater than 5 is the sum of three primes.

**7.** *Goldbach's conjecture*:    Christian Goldbach (1690–1764) conjectured that every even integer greater than 2 is a sum of two primes.

- In 2013 the Goldbach conjecture was verified up to $4 \cdot 10^{18}$ by Olveira e Silva, Herzog, and Pardi.

- Goldbach also conjectured that all odd integers greater than 7 are the sum of three primes. This was proved by Harald Helfgott in 2013 and involved a verification of this conjecture for all odd integers $n$ with $7 \leq n < 10^{27}$ done by David Platt.

- In 1966 J. R. Chen proved that every sufficiently large even number is either the sum of two primes or the sum of a prime and a number that is the product of two primes.

**8.** In 2004 B. Green and T. Tao proved the conjecture made in 1770 that the sequence of prime numbers contains arbitrarily long arithmetic progressions.

**Examples:**

**1.** A method for showing that there are infinitely many primes is to note that the integer $n! + 1$ must have a prime factor greater than $n$, so there is no largest prime. Note that $n! + 1$ is prime for $n = 1, 2, 3, 11, 27, 37, 41, 73, 77, 116, 154, 320, 340, 399$, and $427$, but is composite for all positive integers less than $427$ not listed.

**2.** For $p$ a prime, let $Q(p)$ equal one more than the product of the primes not exceeding $p$. For example $Q(5) = 2 \cdot 3 \cdot 5 + 1 = 31$. Then $Q(p)$ is prime for $p = 2, 3, 5, 7, 11, 31, 379, 1019, 1021, 2657, 3229, 4547, 4787, 11549, 13649$; it is composite for all $p < 11213$ not in this list. For example, $Q(13) = 2 \cdot 3 \cdot 5 \cdot 7 \cdot 11 \cdot 13 + 1$ is composite.

**3.** There are six primes not exceeding 16, namely $2, 3, 5, 7, 11, 13$. Hence $\pi(16) = 6$.

**4.** The expression $n^2 + 1$ is prime for $n = 1, 2, 4, 6, 10, \ldots$, but it is unknown whether there are infinitely many primes of this form when $n$ is an integer. (See Open Problem 3.)

**5.** The polynomial $f(n) = n^2 + n + 41$ takes on prime values for $n = 0, 1, 2, \ldots, 39$, but $f(40) = 1681 = 41^2$.

**6.** Applying Dirichlet's theorem with $a = 123$ and $b = 1,000$, there are infinitely many primes that end in the digits 123. The first such prime is 1,123.

**7.** The pairs 17, 19 and 191, 193 are twin primes. As of the publication of this book, the largest known twin primes, $2{,}996{,}863{,}034{,}895 \cdot 2^{1,290,000} \pm 1$, have 388,342 decimal digits. They were found in 2016 by the Twin Prime Search, a dedicated effort for finding twin primes begun in 2006, in conjunction with PrimeGrid, a distributed computing project for finding world record prime numbers of various types:

- https://www.primegrid.com

### 4.4.3  NUMBERS OF SPECIAL FORM

Numbers of the form $b^n \pm 1$, for $b$ a small number, are often easier to factor or test for primality than other numbers of the same size. They also have a colorful history.

**Definitions:**

A **Cunningham number** is a number of the form $b^n \pm 1$, where $b$ and $n$ are natural numbers, and $b$ is "small" — 2, 3, 5, 6, 7, 10, 11, or 12. They are named after Allan Cunningham, who, along with H. J. Woodall, published in 1925 a table of factorizations of many of these numbers.

A **Fermat number** $F_m$ is a Cunningham number of the form $2^{2^m} + 1$. (See Table 1.)

A **Fermat prime** is a Fermat number that is prime.

A **Mersenne number** $M_n$ is a Cunningham number of the form $2^n - 1$.

A **Mersenne prime** is a Mersenne number that is prime. (See Table 2.)

The **cyclotomic polynomials** $\Phi_k(x)$ are defined recursively by the equation $x^n - 1 = \prod_{d|n} \Phi_d(x)$.

A **perfect number** is a positive integer that is equal to the sum of all its proper divisors.

**Facts:**

**1.** If $M_n$ is prime, then $n$ is prime, but the converse is not true.

**2.** If $b > 2$ or $n$ is composite, then a nontrivial factorization of $b^n - 1$ is given by $b^n - 1 = \prod_{d|n} \Phi_d(b)$, though the factors $\Phi_d(b)$ are not necessarily primes.

**3.** The number $b^n + 1$ can be factored as the product of $\Phi_d(b)$, where $d$ runs over the divisors of $2n$ that are not divisors of $n$. When $n$ is not a power of 2 and $b \geq 2$, this factorization is nontrivial.

**4.** Some numbers of the form $b^n \pm 1$ also have so-called *Aurifeuillian factorizations*, named after A. Aurifeuille. For more details, see [BrEtal88].

**5.** The only primes of the form $b^n - 1$ (with $n > 1$) are Mersenne primes.

**6.** The only primes of the form $2^n + 1$ are Fermat primes.

**7.** Fermat numbers are named after Pierre de Fermat (1601–1695), who observed that $F_0$, $F_1$, $F_2$, $F_3$, and $F_4$ are prime and stated (incorrectly) that all such numbers are prime. Euler proved this was false, by showing that $F_5 = 2^{32} + 1 = 641 \cdot 6{,}700{,}417$.

**8.** $F_4$ is the largest known Fermat prime. It is conjectured that all larger Fermat numbers are composite.

**9.** The smallest Fermat number that has not yet been completely factored is $F_{12} = 2^{2^{12}} + 1$, which has a 1187-digit composite factor.

**10.** As of 2016, we know that $F_m$ is composite for all $m$ with $5 \leq m \leq 32$. In all, we know 288 values of $m$ for which $F_m$ is composite. The largest such $F_m$, with $m = 3{,}329{,}780$, was established as composite in 2014 when it was shown to be divisible by $193 \cdot 2^{3,329,782} + 1$.

**11.** At least one prime factor is known for the $F_m$ in Fact 10 except for $F_{20}$ and $F_{24}$. The complete factorization of $F_m$ is only known for $0 \leq m \leq 11$.

**12.** Table 1 provides known factorizations of $F_m$ for $0 \leq m \leq 24$. Here $p_k$ indicates a $k$-digit prime, and $c_k$ indicates a $k$-digit composite. All other numbers appearing in the right column have been proved prime.

**13.** For up-to-date information about the factorization of Fermat numbers, see

  • http://www.maths.dur.ac.uk/users/dzmitry.badziahin/Fermat
    %20factoring%20status.html

**14.** *Pepin's criterion:* For $m \geq 1$, $F_m$ is prime if and only if $3^{(F_m - 1)/2} \equiv -1 \pmod{F_m}$.

**15.** For $m \geq 2$, every factor of $F_m$ is of the form $2^{m+2}k + 1$.

**16.** Mersenne numbers are named after Marin Mersenne (1588–1648), who made a list of what he thought were all the Mersenne primes $M_p$ with $p \leq 257$. (The status of $M_p$ for the primes up to 23 was known prior to Mersenne.) His list consisted of the primes $p = 2, 3, 5, 7, 13, 17, 19, 31, 67, 127,$ and 257. However, it was later shown that $M_{67}$ and $M_{257}$ are composite, while $M_{61}$, $M_{89}$, and $M_{107}$, missing from the list, are prime.

**17.** It is not known whether there are infinitely many Mersenne primes, nor whether infinitely many Mersenne numbers with a prime exponent are composite, though it is conjectured that both are true.

**18.** Euclid showed that the product of a Mersenne prime $2^p - 1$ with $2^{p-1}$ is perfect. Euler showed that every even perfect number is of this form.

**Table 1: Fermat numbers.**

| $m$ | factorization of $F_m$ |
|---|---|
| 0 | 3 |
| 1 | 5 |
| 2 | 17 |
| 3 | 257 |
| 4 | 65,537 |
| 5 | $641 \times p_7$ |
| 6 | $274,177 \times p_{14}$ |
| 7 | $59,649,589,127,497,217 \times p_{22}$ |
| 8 | $1,238,926,361,552,897 \times p_{62}$ |
| 9 | $2,424,833 \times 7,455,602,825,647,884,208,337,395,736,200,454,918,783,366,342,657$ $\times p_{99}$ |
| 10 | $45,592,577 \times 6,487,031,809$ $\times 4,659,775,785,220,018,543,264,560,743,076,778,192,897 \times p_{252}$ |
| 11 | $319,489 \times 974,849 \times 167,988,556,341,760,475,137$ $\times 3,560,841,906,445,833,920,513 \times p_{564}$ |
| 12 | $114,689 \times 26,017,793 \times 63,766,529 \times 190,274,191,361$ $\times 1,256,132,134,125,569$ $\times 568,630,647,535,356,955,169,033,410,940,867,804,839,360,742,060,818,433 \times c_{1,133}$ |
| 13 | $2,710,954,639,361 \times 2,663,848,877,152,141,313$ $\times 3,603,109,844,542,291,969 \times 319,546,020,820,551,643,220,672,513 \times c_{2,391}$ |
| 14 | $116,928,085,873,074,369,829,035,993,834,596,371,340,386,703,423,373,313 \times c_{4880}$ |
| 15 | $1,214,251,009 \times 2,327,042,503,868,417$ $\times 168,768,817,029,516,972,383,024,127,016,961 \times c_{9,808}$ |
| 16 | $825,753,601 \times 188,981,757,975,021,318,420,037,633 \times c_{19,694}$ |
| 17 | $31,065,037,602,817$ $\times 7,751,061,099,802,522,589,358,967,058,392,886,922,693,580,423,169 \times c_{39,395}$ |
| 18 | $13,631,489 \times 81,274,690,703.860,512,587,777 \times c_{78,884}$ |
| 19 | $70,525,124,609 \times 646,730,219,521 \times 37,590,055,514,133,754,286,524,446,080,499,713$ $\times c_{157,770}$ |
| 20 | $c_{315,653}$ |
| 21 | $4,485,296,422,913 \times c_{631,294}$ |
| 22 | $64,658,705,994,591,851,009,055,774,868,504,577 \times c_{1,262,577}$ |
| 23 | $167,772,161 \times c_{2,525,215}$ |
| 24 | $c_{5,050,446}$ |

**19.** The *Lucas-Lehmer test* can be used to determine whether a given Mersenne number is prime or composite. (See Algorithm 2.)

---

**Algorithm 2:   Lucas-Lehmer test.**

$p :=$ an odd prime; $u := 4$; $i := 0$
**while** $i \leq p - 2$
$\quad i := i + 1$
$\quad u := u^2 - 2 \bmod 2^p - 1$
{If $u = 0$ then $2^p - 1$ is prime, else $2^p - 1$ is composite}

---

**20.** Table 2 lists all known Mersenne primes. An asterisk indicates that a discovery was made as part of the GIMPS program (see Fact 21). As of 2016, the largest known Mersenne prime is $2^{74,207,281} - 1$. As of 2016 we know that there are no additional Mersenne primes less than the 44th prime listed, but there may be one or more Mersenne primes between the 44th Mersenne prime and the largest prime on this list. When a new Mersenne prime is found by computer, there may be other numbers of the form $M_p$ less than this prime not yet checked for primality. It can take months, or even years, to do this checking. A new Mersenne prime may even be found this way, as was the case for the 29th.

### Table 2: Mersenne primes.

| $n$ | exponent | decimal digits | year discovered | discoverer (computer used) |
|---|---|---|---|---|
| 1 | 2 | 1 | ancient times | |
| 2 | 3 | 1 | ancient times | |
| 3 | 5 | 2 | ancient times | |
| 4 | 7 | 3 | ancient times | |
| 5 | 13 | 4 | 1461 | anonymous |
| 6 | 17 | 6 | 1588 | Cataldi |
| 7 | 19 | 6 | 1588 | Cataldi |
| 8 | 31 | 10 | 1750 | Euler |
| 9 | 61 | 19 | 1883 | Pervushin |
| 10 | 89 | 27 | 1911 | Powers |
| 11 | 107 | 33 | 1913 | Fauquembergue |
| 12 | 127 | 39 | 1876 | Lucas |
| 13 | 521 | 157 | 1952 | Robinson (SWAC) |
| 14 | 607 | 183 | 1952 | Robinson (SWAC) |
| 15 | 1,279 | 386 | 1952 | Robinson (SWAC) |
| 16 | 2,203 | 664 | 1952 | Robinson (SWAC) |
| 17 | 2,281 | 687 | 1952 | Robinson (SWAC) |
| 18 | 3,217 | 969 | 1957 | Riesel (BESK) |
| 19 | 4,253 | 1,281 | 1961 | Hurwitz (IBM 7090) |
| 20 | 4,423 | 1,332 | 1961 | Hurwitz (IBM 7090) |
| 21 | 9,689 | 2,917 | 1963 | Gillies (ILLIAC 2) |
| 22 | 9,941 | 2,993 | 1963 | Gillies (ILLIAC 2) |
| 23 | 11,213 | 3,376 | 1963 | Gillies (ILLIAC 2) |
| 24 | 19,937 | 6,002 | 1971 | Tuckerman (IBM 360/91) |
| 25 | 21,701 | 6,533 | 1978 | Noll and Nickel (Cyber 174) |
| 26 | 23,209 | 6,987 | 1979 | Noll (Cyber 174) |
| 27 | 44,497 | 13,395 | 1979 | Nelson and Slowinski (Cray 1) |
| 28 | 86,243 | 25,962 | 1982 | Slowinski (Cray 1) |
| 29 | 110,503 | 33,265 | 1988 | Colquitt and Welsh (NEC SX-W) |
| 30 | 132,049 | 39,751 | 1983 | Slowinski (Cray X-MP) |
| 31 | 216,091 | 65,050 | 1985 | Slowinski (Cray X-MP) |
| 32 | 756,839 | 227,832 | 1992 | Slowinski and Gage (Cray 2) |
| 33 | 859,433 | 258,716 | 1994 | Slowinski and Gage (Cray 2) |
| 34 | 1,257,787 | 378,632 | 1996 | Slowinski and Gage (Cray T94) |
| 35 | 1,398,269 | 420,921 | 1996 | Armengaud* (90 MHz Pentium) |
| 36 | 2,976,221 | 895,932 | 1997 | Spence* (100 MHz Pentium) |

| $n$ | exponent | decimal digits | year discovered | discoverer (computer used) |
|---|---|---|---|---|
| 37 | 3,021,377 | 909,526 | 1998 | Clarkson* (200 MHz Pentium) |
| 38 | 6,972,593 | 2,098,960 | 1999 | Hajratwala* (350 MHz Pentium) |
| 39 | 13,466,917 | 4,053,946 | 2001 | Cameron* (800 MHz Athlon) |
| 40 | 20,996,011 | 6,320,430 | 2003 | Schafer* (2 GHz Dell Dimension) |
| 41 | 24,036,583 | 7,235,733 | 2004 | Findley* (2.4 GHz Pentium) |
| 42 | 25,964,951 | 7,816,230 | 2005 | Nowak* (2.4 GHz Pentium) |
| 43 | 30,402,457 | 9,152,052 | 2005 | (Cooper and Boone)* (2 GHz Pentium) |
| 44 | 32,582,657 | 9,808,358 | 2006 | (Cooper,and Boone)* (3 GHz Pentium) |
| 45 | 37,156,667 | 11,185,272 | 2008 | Elvenich* (2.83 GHz Core 2 Duo) |
| 46 | 42,643,801 | 12,837,064 | 2009 | Strindmo* (3 GHz Core 2 Duo) |
| 47 | 43,112,609 | 12,978,189 | 2008 | Smith* (Dell Optiplex) |
| 48 | 57,885,161 | 17,425,170 | 2013 | Cooper* (3 GHz Core 2 Duo) |
| 49 | 74,207,281 | 22,338,618 | 2016 | Cooper* (3.6 GHz Intel i7-4790) |

**21.** George Woltman launched the Great Internet Mersenne Prime Search (GIMPS) in 1996. GIMPS provides free software for PCs. GIMPS has played a role in discovering the last 15 Mersenne primes. Thousands of people participate in GIMPS over PrimeNet, a virtual supercomputer of distributed PCs, together running more than 250 Teraflops in the quest for Mersenne primes. Consult the GIMPS website at http://www.mersenne.org for more information about this quest and how to join it.

**22.** As of 2016, the two smallest composite Mersenne numbers not completely factored were $2^{671} - 1$ and $2^{683} - 1$.

**23.** It is not known whether any odd perfect numbers exist. Odd perfect numbers, if they exist, must satisfy many different conditions that have been proved over the last 130 years. As of 2016, we know that there are none below $10^{300}$, a result of R. P. Brent, G. L. Cohen, and H. J. J. te Riele in 1991. Among other known conditions are that an odd perfect number must have at least 75 prime factors (proved by K. G. Hare in 2005) and at least nine distinct prime factors (proved by P. P. Neilsen in 2006).

**24.** The best reference for the history of the factorization of Cunningham numbers is [BrEtal88].

**25.** The current version of the Cunningham table, maintained by Sam Wagstaff, can be found at https://homes.cerias.purdue.edu/~ssw/cun/index.html.

**Examples:**

**1.** The Mersenne number $M_{11} = 2^{11} - 1$ is not prime since $M_{11} = 23 \cdot 89$.

**2.** To factor $342 = 7^3 - 1$ note that $7^3 - 1 = (7 - 1) \cdot (7^2 + 7 + 1) = 6 \cdot 57$.

**3.** To factor $3^7 + 1$ note that $3^7 + 1 = \Phi_2(3)\Phi_{14}(3) = 4 \cdot 547$.

**4.** An example of an Aurifeuillian factorization is given by $2^{4k-2} + 1 = (2^{2k-1} - 2^k + 1) \cdot (2^{2k-1} + 2^k + 1)$.

**5.** $\Phi_1(x) = x - 1$ and $x^3 - 1 = \Phi_1(x)\Phi_3(x)$, so $\Phi_3(x) = (x^3 - 1)/\Phi_1(x) = x^2 + x + 1$.

## 4.4.4   PSEUDOPRIMES AND PRIMALITY TESTING

Finding efficient ways to determine whether a positive integer is prime has fascinated mathematicians from ancient to modern times. (See [CrPo05] and [Po10].) Many of the most suggested approaches begin by studying the congruence in Fermat's little theorem when the modulus is not prime. This leads to the notions of a pseudoprime and to probabilistic primality tests. These tests are applied to produce large primes that are used in public key cryptography (see Chapter 15).

**Definitions:**

A **pseudoprime to the base b** is a composite number $n$ such that $b^n \equiv b \pmod{n}$.

A **pseudoprime** is a pseudoprime to the base 2.

A **Carmichael number** is a pseudoprime to all bases.

A **strong pseudoprime to the base b** is an odd composite number $n = 2^s d + 1$, with $d$ odd, and either $b^d \equiv 1 \pmod{n}$ or $b^{2^r d} \equiv -1 \pmod{n}$ for some integer $r$, $0 \le r < s$.

A **witness** for an odd composite number $n$ is a base $b$, with $1 < b < n$, to which $n$ is not a strong pseudoprime. Thus, $b$ is a "witness" to $n$ being composite.

A **primality proof** is an irrefutable verification that an integer is prime.

**Facts:**

**1.** By Fermat's little theorem (§4.3.1), $b^{p-1} \equiv 1 \pmod{p}$ for all primes $p$ and all integers $b$ that are not multiples of $p$. Thus, the only numbers $n > 1$ with $b^{n-1} \equiv 1 \pmod{n}$ are primes and pseudoprimes to the base $b$ (which are coprime to $b$). Similarly, the numbers $n$ which satisfy the strong pseudoprime congruence conditions are the odd primes not dividing $b$ and the strong pseudoprimes to the base $b$.

**2.** The smallest pseudoprime is 341.

**3.** There are infinitely many pseudoprimes; however, Paul Erdős has proved that pseudoprimes are rare compared to primes. The same results are true for pseudoprimes to any fixed base $b$. (See [CrPo05] or [Ri96] for details.)

**4.** In 1910, Robert D. Carmichael gave the first examples of Carmichael numbers. The first 16 Carmichael numbers are

$$561 = 3 \cdot 11 \cdot 17 \qquad 1{,}105 = 5 \cdot 13 \cdot 17 \qquad 1{,}729 = 7 \cdot 13 \cdot 19$$
$$2{,}465 = 5 \cdot 17 \cdot 29 \qquad 2{,}821 = 7 \cdot 13 \cdot 31 \qquad 6{,}601 = 7 \cdot 23 \cdot 41$$
$$8{,}911 = 7 \cdot 19 \cdot 67 \qquad 10{,}585 = 5 \cdot 29 \cdot 73 \qquad 15{,}841 = 7 \cdot 31 \cdot 73$$
$$29{,}341 = 13 \cdot 37 \cdot 61 \qquad 41{,}041 = 7 \cdot 11 \cdot 13 \cdot 41 \qquad 46{,}657 = 13 \cdot 37 \cdot 97$$
$$52{,}633 = 7 \cdot 73 \cdot 103 \qquad 62{,}745 = 3 \cdot 5 \cdot 47 \cdot 89 \qquad 63{,}973 = 7 \cdot 13 \cdot 19 \cdot 37$$
$$75{,}361 = 11 \cdot 17 \cdot 31$$

**5.** If $n$ is a Carmichael number, then $n$ is the product of at least three distinct odd primes with the property that if $q$ is any one of these primes, then $q-1$ divides $n-1$.

**6.** There are a finite number of Carmichael numbers that are the product of exactly $r$ primes with the first $r-2$ primes specified.

**7.** If $m$ is a positive integer such that $6m+1$, $12m+1$, and $18m+1$ are all primes, then $(6m+1)(12m+1)(18m+1)$ is a Carmichael number.

**8.** In 1994, W. R. Alford (1937–2001), Andrew Granville (born 1962), and Carl Pomerance (born 1944) showed that there are infinitely many Carmichael numbers.

**9.** There are infinitely many numbers that are simultaneously strong pseudoprimes to each base in any given finite set. Each odd composite $n$, however, can be a strong pseudoprime to at most one-fourth of the bases $b$ with $1 \le b \le n - 1$.

**10.** Michael Rabin, building upon the notion of a strong pseudoprime introduced by J. L. Selfridge (born 1927), suggested Algorithm 3 (often referred to as the *Miller-Rabin test*).

---

**Algorithm 3**:  **Strong probable prime test (to a random base).**

input: positive numbers $n$, $d$, $s$, with $d$ odd and $n = 2^s d + 1$.

$b :=$ a random integer such that $1 < b < n$

$c := b^d \bmod n$

**if** $c = 1$ or $c = n - 1$, **then** declare $n$ a *probable prime* and stop

compute sequentially $c^2 \bmod n$, $c^4 \bmod n, \ldots, c^{2^{s-1}} \bmod n$

**if** one of these is $n - 1$, **then** declare $n$ a *probable prime* and stop

**else** declare $n$ composite and stop

---

**11.** A "probable prime" is not necessarily a prime, but the chances are good. The probability that an odd composite is not declared composite by Algorithm 3 is at most $\frac{1}{4}$, so the probability it passes $k$ independent iterations is at most $4^{-k}$. Suppose this test is applied to random odd inputs $n$ with the hope of finding a prime. That is, random odd numbers $n$ (chosen between two consecutive powers of 2) are tested until one is found that passes each of $k$ independent iterations of the test. Ronald Burthe showed in 1995 that the probability that the output of this procedure is composite is less than $4^{-k}$.

**12.** Gary Miller proved in 1976 that if the extended Riemann hypothesis (§4.4.2) is true, then every odd composite $n$ has a witness less than $c \log^2 n$, for some constant $c$. Eric Bach showed in 1985 that one may take $c = 2$. Therefore, if an odd number $n > 1$ passes the strong probable prime test for every base $b$ less than $2 \log^2 n$, and if the extended Riemann hypothesis is true, then $n$ is prime.

**13.** In practice, one can test whether numbers under $2.5 \cdot 10^{10}$ are prime by a small number of strong probable prime tests. Pomerance, Selfridge, and Wagstaff have verified (1980) that there are no numbers less than this bound that are simultaneously strong pseudoprimes to the bases 2, 3, 5, 7, and 11. Thus, any number less than $2.5 \cdot 10^{10}$ that passes those strong pseudoprime tests is a prime.

**14.** Gerhard Jaeschke showed in 1993 that the test described in Fact 13 works far beyond $2.5 \cdot 10^{10}$; the first number for which it fails is 2,152,302,898,747.

**15.** Only primes pass the strong pseudoprime tests to all the bases 2, 3, 5, 7, 11, 13, and 17 until the composite number 341,550,071,728,321 is reached.

**16.** While pseudoprimality tests are usually quite efficient at recognizing composites, the task of *proving* that a number is prime can be more difficult.

**17.** In 1983, Leonard Adleman, Carl Pomerance, and Robert Rumely developed the *APR algorithm*, which can prove that a number $n$ is prime in time proportional to $(\log n)^{c \log \log \log n}$, where $c$ is a positive constant. See [Co93] and [CrPo05] for details.

**18.** Recently, Oliver Atkin and François Morain developed an algorithm to prove primality. It is difficult to predict in advance how long it will take, but in practice it has been fast. One advantage of their algorithm is that, unlike APR, it produces a polynomial-time primality proof, though the running time to find the proof may be a bit longer. An implementation called ECPP (*elliptic curve primality proving*) is available from `http://www.lix.polytechnique.fr/Labo/Francois.Morain/Prgms/getecpp.english.html`.

**19.** In 1987, Carl Pomerance showed that every prime $p$ has a primality proof whose verification involves just $c \log p$ multiplications with integers the size of $p$. It may be difficult, however, to find such a short primality proof.

**20.** In 2002, Manindra Agrawal, Neeraj Kayal, and Nitin Saxena introduced a test for determining whether a positive integer is prime that was a major breakthrough because it works in polynomial time (in terms of the number of digits of the number tested), it works for all positive integers (unlike some other tests that only work for integers of a special form), it is deterministic (that is, it always determines whether a number is prime unlike probabilistic primality tests), and its correctness does not depend on any unproven hypotheses.

**21.** The original AKS test determines whether a positive integer $n$ is prime in about $O(\log^{7.5} n)$ time. Improvements in the test made by Carl Pomerance and H. W. Lenstra reduce this time complexity to about $O(\log^6 n)$.

**22.** The importance of the AKS test is more theoretical than practical at the present time. It establishes that there is a polynomial-time algorithm for determining whether a positive integer is prime. However, not even the faster versions of this test run fast enough to be used to find large primes, unlike probabilistic primality tests. (See Fact 18.)

## 4.5   FACTORIZATION

Determining the prime factorization of positive integers is a question that has been studied for many years (see [CrPo05] for example). Furthermore, in the past four decades, this question has become relevant for an extremely important application, the security of public key cryptosystems. The question of exactly how to decompose a composite number into the product of its prime factors is a difficult one that continues to be the subject of much research.

### 4.5.1   FACTORIZATION ALGORITHMS

**Definition:**
A *smooth number* is an integer all of whose prime divisors are small.

**Facts:**
**1.** The simplest algorithm for factoring an integer is *trial division*, Algorithm 1. While simple, this algorithm is useful only for numbers that have a fairly small prime factor. It can be modified so that after $j = 3$, the number $j$ is incremented by 2, and there are other improvements of this kind.

---

**Algorithm 1**:   **Trial division.**

input: an integer $n \geq 2$

output: $j$ (smallest prime factor of $n$) or statement that $n$ is prime

$j := 2$

---

> **while** $j \leq \sqrt{n}$
>     **if** $j|n$ **then** print that $j$ is a prime factor of $n$ and stop  {$n$ is not prime}
>     $j := j + 1$
>   **if** no factor is found **then** declare $n$ prime

**2.** Many factoring algorithms have been developed in the last half century. Coupled with the development of modern computers, increasing larger and larger integers can be factored. Among these factoring algorithms are the quadratic sieve (QS), Pollard $p - 1$ factorization, the number field sieve (NFS), the elliptic curve method (ECM), and the continued fraction method (CFRAC).

**3.** Currently, one of the fastest algorithms for numbers that are feasible to factor but do not have a small prime factor is the *quadratic sieve* (QS), Algorithm 2, invented by Carl Pomerance in 1981. (For numbers at the far range of feasibility, the *number field sieve* is faster; see Fact 10.)

---

**Algorithm 2**:   **Quadratic sieve.**
  input: $n$ (an odd composite number that is not a power)
  output: $g$ (a nontrivial factor of $n$)
  find $a_1, \ldots, a_k$ such that each $a_i^2 - n$ is smooth
  find a subset of the numbers $a_i^2 - n$ whose product is a square, say $x^2$
  reduce $x$ modulo $n$
  $y :=$ the product of the $a_i$ used to form the square
  reduce $y$ modulo $n$
  {This gives a congruence $x^2 \equiv y^2$ (mod $n$); equivalently $n|(x^2 - y^2)$}
  $g := \gcd(x - y, n)$
  **if** $g$ is not a nontrivial factor **then** find new $x$ and $y$ (if necessary, find more $a_i$)

---

**4.** The greatest common divisor calculation may be quickly done via the Euclidean algorithm. If $x \not\equiv \pm y$ (mod $n$), then $g$ will be a nontrivial factor of $n$. (Among all solutions to the congruence $x^2 \equiv y^2$ (mod $n$) with $xy$ coprime to $n$, at least half of them lead to a nontrivial factorization of $n$.) Finding the $a_i$s is at the heart of the algorithm and is accomplished using a sieve not unlike the sieve of Eratosthenes, but applied to the consecutive values of the quadratic polynomial $a^2 - n$. If $a$ is chosen near $\sqrt{n}$, then $a^2 - n$ will be relatively small, and thus more likely to be smooth. So one sieves the polynomial $a^2 - n$, where $a$ runs over integers near $\sqrt{n}$, for values that are smooth. When enough smooth values are collected, the subset with product a square may be found via a linear algebra subroutine applied to a matrix formed out of the exponents in the prime factorizations of the smooth values. The linear algebra may be done modulo 2.

**5.** The current formulation of QS involves many improvements, the most notable of them is the *multiple polynomial variation* of James Davis and Peter Montgomery.

**6.** In 1994, QS was used to factor a 129-digit composite that was the product of a 64-digit prime and a 65-digit prime. This number had been proposed as a challenge to those who would try to crack the famous RSA cryptosystem.

**7.** In 1985, Hendrik W. Lenstra, Jr. (born 1949) invented the *elliptic curve method* (ECM), which has the advantage that, like trial division, the running time is based on the size of the smallest prime factor. Thus, it can be used to find comparatively small factors of numbers whose size would be prohibitively large for the quadratic sieve. It can be best understood by first examining the $p-1$ *method* of John Pollard, Algorithm 3.

**8.** The Pollard algorithm (Algorithm 3) is successful and efficient if $p-1$ happens to be smooth for some prime $p|n$. If the prime factors $p$ of $n$ have the property that $p-1$ is not smooth, Algorithm 3 will eventually be successful if a high enough bound $B$ is chosen, but in this case it will not be any more efficient than trial division, Algorithm 1. ECM gets around this restriction on the numbers that can be efficiently factored by randomly searching through various mathematical objects called *elliptic curve groups*, each of which has $p+1-a$ elements, where $|a| < 2\sqrt{p}$ and $a$ depends on the curve. ECM is successful when a group is encountered such that $p+1-a$ is a smooth number.

---

**Algorithm 3:  $p-1$ factorization method.**

input: $n$ (composite number), $B$ (a bound)

output: a nontrivial factor of $n$

$b := 2$

{Loop on $b$}

**if** $b|n$ **then** stop  {$b$ is a prime factor of $n$}

$M := 1$

**while** $M \leq B$

$\quad g := \gcd(b^{\text{lcm}(1,2,...,M)} - 1, n)$

$\quad$ **if** $n > g > 1$ **then** output $g$ and stop  {$g$ is a nontrivial factor of $n$}

$\quad$ **else if** $g = n$ **then** choose first prime larger than $b$ and go to beginning of

$\qquad\qquad$ the $b$-loop

$\quad$ **else** $M := M + 1$

---

**9.** As of 2016, prime factors as large as 83 digits have been found using ECM. (After such a factor is discovered it may turn out that the remaining part of the number is a prime and the factorization is now complete. This last prime may be very large, as with the tenth and eleventh Fermat numbers; see Table 1 of §4.4.3. In such cases the success of ECM is measured by the *second* largest prime factor in the prime factorization, though in some sense the method has discovered the largest prime factor as well.)

**10.** The *number field sieve* (NFS), originally suggested by Pollard for numbers of special form, and developed for general composite numbers by Buhler, Lenstra, and Pomerance, is currently the fastest factoring algorithm for very large numbers with no small prime factors.

**11.** The number field sieve is similar to QS in that one attempts to assemble two squares $x^2$ and $y^2$ whose difference is a multiple of $n$, and this is done via a sieve and linear algebra modulo 2. However, NFS is much more complicated than QS. Although faster for very large numbers, the complexity of the method makes it unsuitable for numbers much smaller than 100 digits. The exact crossover with QS depends a great deal on the implementations and the hardware employed. The two are roughly within an order of magnitude of each other for numbers between 100 and 150 digits, with QS having the edge at the lower end and NFS the edge at the upper end.

**12.** Part of the NFS algorithm requires expressing a small multiple of the number to be factored by a polynomial of moderate degree. The running time depends, in part, on the size of the coefficients of this polynomial. For Cunningham numbers, this polynomial can be easy to find. (For example, in the notation of §4.4.3, $8F_9 = 8(2^{2^9}+1) = f(2^{103})$, where $f(x) = x^5 + 8$.) This version is called the *special number field sieve* (SNFS). The version for general numbers, the *general number field sieve* (GNFS), has somewhat greater complexity. The greatest success of SNFS has been the factorization of a 180-digit Cunningham number, while the greatest success of GNFS has been the factorization of a 232-digit number of no special form and with no small prime factor.

**13.** See [Co93], [CrPo05], [Po90], and [Po94] for fuller descriptions of the factoring algorithms described here, as well as others, including the *continued fraction* (CFRAC) method. Until the advent of QS, this had been the fastest known practical algorithm.

**14.** The factorization algorithms QS, ECM, SNFS, and GNFS are fast in practice, but analyses of their running times depend on heuristic arguments and unproved hypotheses. The fastest algorithm whose running time has been rigorously analyzed is the *class group relations method* (CGRM). It, however, is not practical. It is a probabilistic algorithm whose expected running time is bounded by $e^{c\sqrt{\log n \log\log n}}$, where $c$ tends to 1 as $n$ tends to infinity through the odd composite numbers that are not powers. This result was proved in 1992 by Lenstra and Pomerance.

**15.** These algorithms are summarized in Table 1. $L(a, b)$ means that the running time to factor $n$ is bounded by $e^{c(\log n)^a (\log\log n)^{1-a}}$, where $c$ tends to $b$ as $n$ tends to infinity through the odd composite non-powers. Running times are measured in the number of arithmetic steps with integers at most the size of $n$.

**Table 1: Comparison of various factoring methods.**

| algorithm | year introduced | greatest success | running time | rigorously analyzed |
|---|---|---|---|---|
| trial division | antiquity | – | $\sqrt{n}$ | yes |
| CFRAC | 1970 | 63-digit number | $L\left(\frac{1}{2}, \sqrt{\frac{3}{2}}\right)$ | no |
| $p-1$ | 1974 | 58-digit factor | – | yes |
| QS | 1981 | 135-digit number | $L(\frac{1}{2}, 1)$ | no |
| ECM | 1985 | 83-digit factor | $L(\frac{1}{2}, 1)$ | no |
| SNFS | 1988 | 355-digit number | $L\left(\frac{1}{3}, \sqrt[3]{\frac{32}{9}}\right)$ | no |
| CGRM | 1992 | – | $L(\frac{1}{2}, 1)$ | yes |
| GNFS | 1993 | 221-digit number | $L\left(\frac{1}{3}, \sqrt[3]{\frac{64}{9}}\right)$ | no |

**16.** The running time for trial division in Table 1 is a worst-case estimate, achieved when $n$ is prime or the product of two primes of the same magnitude. When $n$ is composite, trial division will discover the least prime factor $p$ of $n$ in roughly $p$ steps. The record for the largest prime factor discovered via trial division is not known, nor is the largest number proved prime by this method, though the feat of Euler of proving that the Mersenne number $2^{31} - 1$ is prime, using only trial division and hand calculations, should certainly be noted. (Euler surely knew, though, that any prime factor of $2^{31} - 1$ is 1 mod 31, so only 1 out of every 31 trial divisors needed to be tested.)

**17.** The running time of the $p-1$ method is about $B$, where $B$ is the least number such that for some prime factor $p$ of $n$, $p - 1$ divides $\text{lcm}(1, 2, \ldots, B)$.

**18.** There are variants of CFRAC and GNFS that have smaller heuristic complexity estimates, but the ones in Table 1 are for the fastest practical version.

**19.** The running time bound for ECM is a worst-case estimate. It is more appropriate to measure ECM as a function of the least prime factor $p$ of $n$. This heuristic complexity bound is $e^{c\sqrt{\log p \log \log p}}$, where $c$ tends to $\sqrt{2}$ as $p$ tends to infinity.

**20.** The following table lists the largest hard number factored as a function of time. It was compiled with the assistance of Sam Wagstaff. It should be remarked that there is no firm definition of a "hard number". What is meant here is that the number was factored by an algorithm that is not sensitive to any particular form the number may have, nor sensitive to the size of the prime factors.

| *year* | *method* | *digits* |
|---|---|---|
| 1970 | CFRAC | 39 |
| 1979 | CFRAC | 46 |
| 1982 | CFRAC | 54 |
| 1983 | QS | 67 |
| 1986 | QS | 87 |
| 1988 | QS | 102 |
| 1990 | QS | 116 |
| 1994 | QS | 129 |
| 1995 | GNFS | 130 |
| 1999 | GNFS | 155 |
| 2003 | GNFS | 174 |
| 2005 | GNFS | 200 |
| 2009 | GNFS | 232 |

**21.** It is unknown whether there is a polynomial-time factorization algorithm. Whether there are any factorization algorithms that surpass the quadratic sieve, the elliptic curve method, and the number field sieve in their respective regions of superiority is an area of much current research.

**22.** A subjective measurement of progress in factorization can be gained by looking at the "ten most wanted numbers" to be factored. The list is maintained by Sam Wagstaff and can be found at `http://homes.cerias.purdue.edu/~ssw/cun//want131`. As of May 2016, "number one" on this list is $2^{1207} - 1$.

**23.** In 1994 Peter Shor invented an algorithm for factoring integers on a quantum computer that runs in polynomial time. More precisely, this algorithm factors a positive integer $n$ in $O(\log^2 n \log \log n \log \log \log n)$ time. At the present time, only rudimentary quantum computers have been built. Many technological challenges must be solved before useful quantum computing is feasible. As of 2016, the largest integer that has been factored using a quantum computer is 56,153; this was done using an algorithm different from Shor's algorithm.

## 4.6 ARITHMETIC FUNCTIONS

Functions whose domains are the set of positive integers play an important role in number theory. Such functions are called arithmetic functions and are the subject of this section. The information presented here includes definitions and properties of many important arithmetic functions, asymptotic estimates on the growth of these functions, and algebraic properties of sets of certain arithmetic functions. For more information on the topics covered in this section, see [Ap76].

## 4.6.1   MULTIPLICATIVE AND ADDITIVE FUNCTIONS

### Definitions:

An **arithmetic function** is a function that is defined for all positive integers.

An arithmetic function is **multiplicative** if $f(mn) = f(m)f(n)$ whenever $m$ and $n$ are relatively prime positive integers.

An arithmetic function is **completely multiplicative** if $f(mn) = f(m)f(n)$ for all positive integers $m$ and $n$.

If $f$ is an arithmetic function, then $\sum_{d|n} f(d)$, the value of the **summatory function** of $f$ at $n$, is the sum of $f(d)$ over all positive integers $d$ that divide $n$.

An arithmetic function $f$ is **additive** if $f(mn) = f(m) + f(n)$ whenever $m$ and $n$ are relatively prime positive integers.

An arithmetic function $f$ is **completely additive** if $f(m,n) = f(m) + f(n)$ whenever $m$ and $n$ are positive integers.

### Facts:

**1.** If $f$ is a multiplicative function and $n = p_1^{a_1} p_2^{a_2} \ldots p_s^{a_s}$ is the prime-power factorization of $n$, then $f(n) = f(p_1^{a_1}) f(p_2^{a_2}) \ldots f(p_s^{a_s})$.

**2.** If $f$ is multiplicative, then $f(1) = 1$.

**3.** If $f$ is a completely multiplicative function and $n = p_1^{a_1} p_2^{a_2} \ldots p_s^{a_s}$, then $f(n) = f(p_1)^{a_1} f(p_2)^{a_2} \ldots f(p_s)^{a_s}$.

**4.** If $f$ is multiplicative, then the arithmetic function $F(n) = \sum_{d|n} f(d)$ is multiplicative.

**5.** If $f$ is an additive function, then $f(1) = 0$.

**6.** If $f$ is an additive function and $a$ is a positive real number, then $F(n) = a^{f(n)}$ is multiplicative.

**7.** If $f$ is a completely additive function and $a$ is a positive real number, then $F(n) = a^{f(n)}$ is completely multiplicative.

### Examples:

**1.** The function $f(n) = n^2$ is multiplicative. Even more, it is completely multiplicative.

**2.** The function $I(n) = \lfloor \frac{1}{n} \rfloor$ (so that $I(1) = 1$ and $I(n) = 0$ if $n$ is a positive integer greater than 1) is completely multiplicative.

**3.** The Euler phi-function, the number of divisors function, the sum of divisors function, and the Möbius function (defined in subsequent sections) are all multiplicative. None of these functions is completely multiplicative.

## 4.6.2   EULER'S PHI-FUNCTION

### Definition:

If $n$ is a positive integer then $\phi(n)$, the value of the **Euler phi-function** at $n$, is the number of positive integers not exceeding $n$ that are relatively prime to $n$. The Euler phi-function is also known as the **totient function**.

**Facts:**

1. The Euler $\phi$ function is multiplicative, but not completely multiplicative.

2. If $p$ is a prime, then $\phi(p) = p - 1$.

3. If $p$ is a positive integer with $\phi(p) = p - 1$, then $p$ is prime.

4. If $p$ is a prime and $a$ is a positive integer, then $\phi(p^a) = p^a - p^{a-1}$.

5. If $n$ is a positive integer with prime-power factorization $n = p_1^{a_1} p_2^{a_2} \ldots p_k^{a_k}$, then $\phi(n) = n \prod_{j=1}^{k} (1 - \frac{1}{p_j})$.

6. If $n$ is a positive integer greater than 2, then $\phi(n)$ is even.

7. If $n$ has $r$ distinct odd prime factors, then $2^r$ divides $\phi(n)$.

8. If $m$ and $n$ are positive integers and $\gcd(m, n) = d$, then $\phi(mn) = \frac{\phi(m)\phi(n)d}{\phi(d)}$.

9. If $m$ and $n$ are positive integers and $m|n$, then $\phi(m)|\phi(n)$.

10. If $n$ is a positive integer, then $\sum_{d|n} \phi(d) = \sum_{d|n} \phi(\frac{n}{d}) = n$.

11. If $n$ is a positive integer with $n \geq 5$, then $\phi(n) > \frac{n}{6 \log \log n}$.

12. $\sum_{k=1}^{n} \phi(k) = \frac{3n^2}{\pi^2} + O(n \log n)$

13. $\sum_{k=1}^{n} \frac{\phi(k)}{k} = \frac{6n}{\pi^2} + O(n \log n)$

**Examples:**

1. Table 1 in §4.6.3 displays the values of $\phi(n)$ for $1 \leq n \leq 1{,}000$.

2. To see that $\phi(10) = 4$, note that the positive integers not exceeding 10 relatively prime to 10 are 1, 3, 7, and 9.

3. To find $\phi(720)$, note that $\phi(720) = \phi(2^4 3^2 5) = 720(1 - \frac{1}{2})(1 - \frac{1}{3})(1 - \frac{1}{5}) = 192$.

---

## 4.6.3   SUM AND NUMBER OF DIVISORS FUNCTIONS

**Definitions:**

If $n$ is a positive integer, then $\sigma(n)$, the value of the **sum of divisors function** at $n$, is the sum of the positive integer divisors of $n$.

A positive integer $n$ is **perfect** if and only if it equals the sum of its proper divisors (or equivalently, if $\sigma(n) = 2n$).

A positive integer $n$ is **abundant** if the sum of the proper divisors of $n$ exceeds $n$ (or equivalently, if $\sigma(n) > 2n$).

A positive integer $n$ is **deficient** if the sum of the proper divisors of $n$ is less than $n$ (or equivalently, if $\sigma(n) < 2n$).

The positive integers $m$ and $n$ are **amicable** if $\sigma(m) = \sigma(n) = m + n$.

If $n$ is a positive integer, then $\tau(n)$, the value of the **number of divisors function** at $n$, is the number of positive integer divisors of $n$.

**Facts:**

1. The number of divisors function is multiplicative, but not completely multiplicative.

2. The number of divisors function is the summatory function of $f(n) = 1$; that is, $\tau(n) = \sum_{d|n} 1$.

**3.** The sum of divisors function is multiplicative, but not completely multiplicative.

**4.** The sum of divisors function is the summatory function of $f(n) = n$; that is, $\sigma(n) = \sum_{d|n} d$.

**5.** If $n$ is a positive integer with prime-power factorization $n = p_1^{a_1} p_2^{a_2} \ldots p_k^{a_k}$, then $\sigma(n) = \prod_{j=1}^{k} (p_j^{a_j+1} - 1)/(p_j - 1)$.

**6.** If $n$ is a positive integer with prime-power factorization $n = p_1^{a_1} p_2^{a_2} \ldots p_k^{a_k}$, then $\tau(n) = \prod_{j=1}^{k} (a_j + 1)$.

**7.** If $n$ is a positive integer, then $\tau(n)$ is odd if and only if $n$ is a perfect square.

**8.** If $k$ is an integer greater than 1, then the equation $\tau(n) = k$ has infinitely many solutions.

**9.** If $n$ is a positive integer, then $(\sum_{d|n} \tau(d))^2 = \sum_{d|n} \tau(d)^3$.

**10.** A positive integer $n$ is an even perfect number if and only if $n = 2^{m-1}(2^m - 1)$ where $m$ is an integer, $m \geq 2$, and $2^m - 1$ is prime (so that it is a Mersenne prime (§4.4.3)). Hence, the number of known even perfect numbers equals the number of known Mersenne primes.

**11.** It is unknown whether there are any odd perfect numbers. However, it is known that there are no odd perfect numbers less than $10^{300}$ and that any odd perfect number must have at least eight different prime factors.

**12.** $\sum_{k=1}^{n} \sigma(k) = \frac{\pi^2 n^2}{12} + O(n \log n)$.

**13.** $\sum_{k=1}^{n} \tau(k) = n \log n + (2\gamma - 1)n + O(\sqrt{n})$, where $\gamma$ is Euler's constant.

**14.** If $m$ and $n$ are amicable, then $m$ is the sum of the proper divisors of $n$, and vice versa.

### Examples:

**1.** Table 1 lists the values of $\sigma(n)$ and $\tau(n)$ for $1 \leq n \leq 1,000$.

**2.** To find $\tau(720)$, note that $\tau(720) = \tau(2^4 \cdot 3^2 \cdot 5) = (4+1)(2+1)(1+1) = 30$.

**3.** To find $\sigma(200)$, note that $\sigma(200) = \sigma(2^3 \cdot 5^2) = \frac{2^4-1}{2-1} \cdot \frac{5^3-1}{5-1} = 15 \cdot 31 = 465$.

**4.** The integers 6 and 28 are perfect; the integers 9 and 16 are deficient; the integers 12 and 945 are abundant.

**5.** The integers 220 and 284 form the smallest pair of amicable numbers.

### Table 1: Values of $\phi(n)$, $\sigma(n)$, $\tau(n)$, and $\mu(n)$ for $1 \leq n \leq 1000$.

| $n$ | $\phi$ | $\sigma$ | $\tau$ | $\mu$ | $n$ | $\phi$ | $\sigma$ | $\tau$ | $\mu$ | $n$ | $\phi$ | $\sigma$ | $\tau$ | $\mu$ | $n$ | $\phi$ | $\sigma$ | $\tau$ | $\mu$ | $n$ | $\phi$ | $\sigma$ | $\tau$ | $\mu$ |
|---|---|---|---|---|---|---|---|---|---|---|---|---|---|---|---|---|---|---|---|---|---|---|---|---|
| 1 | 1 | 1 | 1 | 1 | 2 | 1 | 3 | 2 | −1 | 3 | 2 | 4 | 2 | −1 | 4 | 2 | 7 | 3 | 0 | 5 | 4 | 6 | 2 | −1 |
| 6 | 2 | 12 | 4 | 1 | 7 | 6 | 8 | 2 | −1 | 8 | 4 | 15 | 4 | 0 | 9 | 6 | 13 | 3 | 0 | 10 | 4 | 18 | 4 | 1 |
| 11 | 10 | 12 | 2 | −1 | 12 | 4 | 28 | 6 | 0 | 13 | 12 | 14 | 2 | −1 | 14 | 6 | 24 | 4 | 1 | 15 | 8 | 24 | 4 | 1 |
| 16 | 8 | 31 | 5 | 0 | 17 | 16 | 18 | 2 | −1 | 18 | 6 | 39 | 6 | 0 | 19 | 18 | 20 | 2 | −1 | 20 | 8 | 42 | 6 | 0 |
| 21 | 12 | 32 | 4 | 1 | 22 | 10 | 36 | 4 | 1 | 23 | 22 | 24 | 2 | −1 | 24 | 8 | 60 | 8 | 0 | 25 | 20 | 31 | 3 | 0 |
| 26 | 12 | 42 | 4 | 1 | 27 | 18 | 40 | 4 | 0 | 28 | 12 | 56 | 6 | 0 | 29 | 28 | 30 | 2 | −1 | 30 | 8 | 72 | 8 | −1 |
| 31 | 30 | 32 | 2 | −1 | 32 | 16 | 63 | 6 | 0 | 33 | 20 | 48 | 4 | 1 | 34 | 16 | 54 | 4 | 1 | 35 | 24 | 48 | 4 | 1 |
| 36 | 12 | 91 | 9 | 0 | 37 | 36 | 38 | 2 | −1 | 38 | 18 | 60 | 4 | 1 | 39 | 24 | 56 | 4 | 1 | 40 | 16 | 90 | 8 | 0 |
| 41 | 40 | 42 | 2 | −1 | 42 | 12 | 96 | 8 | −1 | 43 | 42 | 44 | 2 | −1 | 44 | 20 | 84 | 6 | 0 | 45 | 24 | 78 | 6 | 0 |
| 46 | 22 | 72 | 4 | 1 | 47 | 46 | 48 | 2 | −1 | 48 | 16 | 124 | 10 | 0 | 49 | 42 | 57 | 3 | 0 | 50 | 20 | 93 | 6 | 0 |
| 51 | 32 | 72 | 4 | 1 | 52 | 24 | 98 | 6 | 0 | 53 | 52 | 54 | 2 | −1 | 54 | 18 | 120 | 8 | 0 | 55 | 40 | 72 | 4 | 1 |
| 56 | 24 | 120 | 8 | 0 | 57 | 36 | 80 | 4 | 1 | 58 | 28 | 90 | 4 | 1 | 59 | 58 | 60 | 2 | −1 | 60 | 16 | 168 | 12 | 0 |
| 61 | 60 | 62 | 2 | −1 | 62 | 30 | 96 | 4 | 1 | 63 | 36 | 104 | 6 | 0 | 64 | 32 | 127 | 7 | 0 | 65 | 48 | 84 | 4 | 1 |

| $n$ | $\phi$ | $\sigma$ | $\tau$ | $\mu$ | $n$ | $\phi$ | $\sigma$ | $\tau$ | $\mu$ | $n$ | $\phi$ | $\sigma$ | $\tau$ | $\mu$ | $n$ | $\phi$ | $\sigma$ | $\tau$ | $\mu$ | $n$ | $\phi$ | $\sigma$ | $\tau$ | $\mu$ |
|---|---|---|---|---|---|---|---|---|---|---|---|---|---|---|---|---|---|---|---|---|---|---|---|---|
| 66 | 20 | 144 | 8 | −1 | 67 | 66 | 68 | 2 | −1 | 68 | 32 | 126 | 6 | 0 | 69 | 44 | 96 | 4 | 1 | 70 | 24 | 144 | 8 | −1 |
| 71 | 70 | 72 | 2 | −1 | 72 | 24 | 195 | 12 | 0 | 73 | 72 | 74 | 2 | −1 | 74 | 36 | 114 | 4 | 1 | 75 | 40 | 124 | 6 | 0 |
| 76 | 36 | 140 | 6 | 0 | 77 | 60 | 96 | 4 | 1 | 78 | 24 | 168 | 8 | −1 | 79 | 78 | 80 | 2 | −1 | 80 | 32 | 186 | 10 | 0 |
| 81 | 54 | 121 | 5 | 0 | 82 | 40 | 126 | 4 | 1 | 83 | 82 | 84 | 2 | −1 | 84 | 24 | 224 | 12 | 0 | 85 | 64 | 108 | 4 | 1 |
| 86 | 42 | 132 | 4 | 1 | 87 | 56 | 120 | 4 | 1 | 88 | 40 | 180 | 8 | 0 | 89 | 88 | 90 | 2 | −1 | 90 | 24 | 234 | 12 | 0 |
| 91 | 72 | 112 | 4 | 1 | 92 | 44 | 168 | 6 | 0 | 93 | 60 | 128 | 4 | 1 | 94 | 46 | 144 | 4 | 1 | 95 | 72 | 120 | 4 | 1 |
| 96 | 32 | 252 | 12 | 0 | 97 | 96 | 98 | 2 | −1 | 98 | 42 | 171 | 6 | 0 | 99 | 60 | 156 | 6 | 0 | 100 | 40 | 217 | 9 | 0 |
| 101 | 100 | 102 | 2 | −1 | 102 | 32 | 216 | 8 | −1 | 103 | 102 | 104 | 2 | −1 | 104 | 48 | 210 | 8 | 0 | 105 | 48 | 192 | 8 | −1 |
| 106 | 52 | 162 | 4 | 1 | 107 | 106 | 108 | 2 | −1 | 108 | 36 | 280 | 12 | 0 | 109 | 108 | 110 | 2 | −1 | 110 | 40 | 216 | 8 | −1 |
| 111 | 72 | 152 | 4 | 1 | 112 | 48 | 248 | 10 | 0 | 113 | 112 | 114 | 2 | −1 | 114 | 36 | 240 | 8 | −1 | 115 | 88 | 144 | 4 | 1 |
| 116 | 56 | 210 | 6 | 0 | 117 | 72 | 182 | 6 | 0 | 118 | 58 | 180 | 4 | 1 | 119 | 96 | 144 | 4 | 1 | 120 | 32 | 360 | 16 | 0 |
| 121 | 110 | 133 | 3 | 0 | 122 | 60 | 186 | 4 | 1 | 123 | 80 | 168 | 4 | 1 | 124 | 60 | 224 | 6 | 0 | 125 | 100 | 156 | 4 | 0 |
| 126 | 36 | 312 | 12 | 0 | 127 | 126 | 128 | 2 | −1 | 128 | 64 | 255 | 8 | 0 | 129 | 84 | 176 | 4 | 1 | 130 | 48 | 252 | 8 | −1 |
| 131 | 130 | 132 | 2 | −1 | 132 | 40 | 336 | 12 | 0 | 133 | 108 | 160 | 4 | 1 | 134 | 66 | 204 | 4 | 1 | 135 | 72 | 240 | 8 | 0 |
| 136 | 64 | 270 | 8 | 0 | 137 | 136 | 138 | 2 | −1 | 138 | 44 | 288 | 8 | −1 | 139 | 138 | 140 | 2 | −1 | 140 | 48 | 336 | 12 | 0 |
| 141 | 92 | 192 | 4 | 1 | 142 | 70 | 216 | 4 | 1 | 143 | 120 | 168 | 4 | 1 | 144 | 48 | 403 | 15 | 0 | 145 | 112 | 180 | 4 | 1 |
| 146 | 72 | 222 | 4 | 1 | 147 | 84 | 228 | 6 | 0 | 148 | 72 | 266 | 6 | 0 | 149 | 148 | 150 | 2 | −1 | 150 | 40 | 372 | 12 | 0 |
| 151 | 150 | 152 | 2 | −1 | 152 | 72 | 300 | 8 | 0 | 153 | 96 | 234 | 6 | 0 | 154 | 60 | 288 | 8 | −1 | 155 | 120 | 192 | 4 | 1 |
| 156 | 48 | 392 | 12 | 0 | 157 | 156 | 158 | 2 | −1 | 158 | 78 | 240 | 4 | 1 | 159 | 104 | 216 | 4 | 1 | 160 | 64 | 378 | 12 | 0 |
| 161 | 132 | 192 | 4 | 1 | 162 | 54 | 363 | 10 | 0 | 163 | 162 | 164 | 2 | −1 | 164 | 80 | 294 | 6 | 0 | 165 | 80 | 288 | 8 | −1 |
| 166 | 82 | 252 | 4 | 1 | 167 | 166 | 168 | 2 | −1 | 168 | 48 | 480 | 16 | 0 | 169 | 156 | 183 | 3 | 0 | 170 | 64 | 324 | 8 | −1 |
| 171 | 108 | 260 | 6 | 0 | 172 | 84 | 308 | 6 | 0 | 173 | 172 | 174 | 2 | −1 | 174 | 56 | 360 | 8 | −1 | 175 | 120 | 248 | 6 | 0 |
| 176 | 80 | 372 | 10 | 0 | 177 | 116 | 240 | 4 | 1 | 178 | 88 | 270 | 4 | 1 | 179 | 178 | 180 | 2 | −1 | 180 | 48 | 546 | 18 | 0 |
| 181 | 180 | 182 | 2 | −1 | 182 | 72 | 336 | 8 | −1 | 183 | 120 | 248 | 4 | 1 | 184 | 88 | 360 | 8 | 0 | 185 | 144 | 228 | 4 | 1 |
| 186 | 60 | 384 | 8 | −1 | 187 | 160 | 216 | 4 | 1 | 188 | 92 | 336 | 6 | 0 | 189 | 108 | 320 | 8 | 0 | 190 | 72 | 360 | 8 | −1 |
| 191 | 190 | 192 | 2 | −1 | 192 | 64 | 508 | 14 | 0 | 193 | 192 | 194 | 2 | −1 | 194 | 96 | 294 | 4 | 1 | 195 | 96 | 336 | 8 | −1 |
| 196 | 84 | 399 | 9 | 0 | 197 | 196 | 198 | 2 | −1 | 198 | 60 | 468 | 12 | 0 | 199 | 198 | 200 | 2 | −1 | 200 | 80 | 465 | 12 | 0 |
| 201 | 132 | 272 | 4 | 1 | 202 | 100 | 306 | 4 | 1 | 203 | 168 | 240 | 4 | 1 | 204 | 64 | 504 | 12 | 0 | 205 | 160 | 252 | 4 | 1 |
| 206 | 102 | 312 | 4 | 1 | 207 | 132 | 312 | 6 | 0 | 208 | 96 | 434 | 10 | 0 | 209 | 180 | 240 | 4 | 1 | 210 | 48 | 576 | 16 | 1 |
| 211 | 210 | 212 | 2 | −1 | 212 | 104 | 378 | 6 | 0 | 213 | 140 | 288 | 4 | 1 | 214 | 106 | 324 | 4 | 1 | 215 | 168 | 264 | 4 | 1 |
| 216 | 72 | 600 | 16 | 0 | 217 | 180 | 256 | 4 | 1 | 218 | 108 | 330 | 4 | 1 | 219 | 144 | 296 | 4 | 1 | 220 | 80 | 504 | 12 | 0 |
| 221 | 192 | 252 | 4 | 1 | 222 | 72 | 456 | 8 | −1 | 223 | 222 | 224 | 2 | −1 | 224 | 96 | 504 | 12 | 0 | 225 | 120 | 403 | 9 | 0 |
| 226 | 112 | 342 | 4 | 1 | 227 | 226 | 228 | 2 | −1 | 228 | 72 | 560 | 12 | 0 | 229 | 228 | 230 | 2 | −1 | 230 | 88 | 432 | 8 | −1 |
| 231 | 120 | 384 | 8 | −1 | 232 | 112 | 450 | 8 | 0 | 233 | 232 | 234 | 2 | −1 | 234 | 72 | 546 | 12 | 0 | 235 | 184 | 288 | 4 | 1 |
| 236 | 116 | 420 | 6 | 0 | 237 | 156 | 320 | 4 | 1 | 238 | 96 | 432 | 8 | −1 | 239 | 238 | 240 | 2 | −1 | 240 | 64 | 744 | 20 | 0 |
| 241 | 240 | 242 | 2 | −1 | 242 | 110 | 399 | 6 | 0 | 243 | 162 | 364 | 6 | 0 | 244 | 120 | 434 | 6 | 0 | 245 | 168 | 342 | 6 | 0 |
| 246 | 80 | 504 | 8 | −1 | 247 | 216 | 280 | 4 | 1 | 248 | 120 | 480 | 8 | 0 | 249 | 164 | 336 | 4 | 1 | 250 | 100 | 468 | 8 | 0 |
| 251 | 250 | 252 | 2 | −1 | 252 | 72 | 728 | 18 | 0 | 253 | 220 | 288 | 4 | 1 | 254 | 126 | 384 | 4 | 1 | 255 | 128 | 432 | 8 | −1 |
| 256 | 128 | 511 | 9 | 0 | 257 | 256 | 258 | 2 | −1 | 258 | 84 | 528 | 8 | −1 | 259 | 216 | 304 | 4 | 1 | 260 | 96 | 588 | 12 | 0 |
| 261 | 168 | 390 | 6 | 0 | 262 | 130 | 396 | 4 | 1 | 263 | 262 | 264 | 2 | −1 | 264 | 80 | 720 | 16 | 0 | 265 | 208 | 324 | 4 | 1 |
| 266 | 108 | 480 | 8 | −1 | 267 | 176 | 360 | 4 | 1 | 268 | 132 | 476 | 6 | 0 | 269 | 268 | 270 | 2 | −1 | 270 | 72 | 720 | 16 | 0 |
| 271 | 270 | 272 | 2 | −1 | 272 | 128 | 558 | 10 | 0 | 273 | 144 | 448 | 8 | −1 | 274 | 136 | 414 | 4 | 1 | 275 | 200 | 372 | 6 | 0 |
| 276 | 88 | 672 | 12 | 0 | 277 | 276 | 278 | 2 | −1 | 278 | 138 | 420 | 4 | 1 | 279 | 180 | 416 | 6 | 0 | 280 | 96 | 720 | 16 | 0 |
| 281 | 280 | 282 | 2 | −1 | 282 | 92 | 576 | 8 | −1 | 283 | 282 | 284 | 2 | −1 | 284 | 140 | 504 | 6 | 0 | 285 | 144 | 480 | 8 | −1 |
| 286 | 120 | 504 | 8 | −1 | 287 | 240 | 336 | 4 | 1 | 288 | 96 | 819 | 18 | 0 | 289 | 272 | 307 | 3 | 0 | 290 | 112 | 540 | 8 | −1 |
| 291 | 192 | 392 | 4 | 1 | 292 | 144 | 518 | 6 | 0 | 293 | 292 | 294 | 2 | −1 | 294 | 84 | 684 | 12 | 0 | 295 | 232 | 360 | 4 | 1 |
| 296 | 144 | 570 | 8 | 0 | 297 | 180 | 480 | 8 | 0 | 298 | 148 | 450 | 4 | 1 | 299 | 264 | 336 | 4 | 1 | 300 | 80 | 868 | 18 | 0 |
| 301 | 252 | 352 | 4 | 1 | 302 | 150 | 456 | 4 | 1 | 303 | 200 | 408 | 4 | 1 | 304 | 144 | 620 | 10 | 0 | 305 | 240 | 372 | 4 | 1 |
| 306 | 96 | 702 | 12 | 0 | 307 | 306 | 308 | 2 | −1 | 308 | 120 | 672 | 12 | 0 | 309 | 204 | 416 | 4 | 1 | 310 | 120 | 576 | 8 | −1 |
| 311 | 310 | 312 | 2 | −1 | 312 | 96 | 840 | 16 | 0 | 313 | 312 | 314 | 2 | −1 | 314 | 156 | 474 | 4 | 1 | 315 | 144 | 624 | 12 | 0 |
| 316 | 156 | 560 | 6 | 0 | 317 | 316 | 318 | 2 | −1 | 318 | 104 | 648 | 8 | −1 | 319 | 280 | 360 | 4 | 1 | 320 | 128 | 762 | 14 | 0 |
| 321 | 212 | 432 | 4 | 1 | 322 | 132 | 576 | 8 | −1 | 323 | 288 | 360 | 4 | 1 | 324 | 108 | 847 | 15 | 0 | 325 | 240 | 434 | 6 | 0 |
| 326 | 162 | 492 | 4 | 1 | 327 | 216 | 440 | 4 | 1 | 328 | 160 | 630 | 8 | 0 | 329 | 276 | 384 | 4 | 1 | 330 | 80 | 864 | 16 | 1 |
| 331 | 330 | 332 | 2 | −1 | 332 | 164 | 588 | 6 | 0 | 333 | 216 | 494 | 6 | 0 | 334 | 166 | 504 | 4 | 1 | 335 | 264 | 408 | 4 | 1 |
| 336 | 96 | 992 | 20 | 0 | 337 | 336 | 338 | 2 | −1 | 338 | 156 | 549 | 6 | 0 | 339 | 224 | 456 | 4 | 1 | 340 | 128 | 756 | 12 | 0 |

| n | φ | σ | τ | μ | n | φ | σ | τ | μ | n | φ | σ | τ | μ | n | φ | σ | τ | μ | n | φ | σ | τ | μ |
|---|---|---|---|---|---|---|---|---|---|---|---|---|---|---|---|---|---|---|---|---|---|---|---|---|
| 341 | 300 | 384 | 4 | 1 | 342 | 108 | 780 | 12 | 0 | 343 | 294 | 400 | 4 | 0 | 344 | 168 | 660 | 8 | 0 | 345 | 176 | 576 | 8 | -1 |
| 346 | 172 | 522 | 4 | 1 | 347 | 346 | 348 | 2 | -1 | 348 | 112 | 840 | 12 | 0 | 349 | 348 | 350 | 2 | -1 | 350 | 120 | 744 | 12 | 0 |
| 351 | 216 | 560 | 8 | 0 | 352 | 160 | 756 | 12 | 0 | 353 | 352 | 354 | 2 | -1 | 354 | 116 | 720 | 8 | -1 | 355 | 280 | 432 | 4 | 1 |
| 356 | 176 | 630 | 6 | 0 | 357 | 192 | 576 | 8 | -1 | 358 | 178 | 540 | 4 | 1 | 359 | 358 | 360 | 2 | -1 | 360 | 96 | 1170 | 24 | 0 |
| 361 | 342 | 381 | 3 | 0 | 362 | 180 | 546 | 4 | 1 | 363 | 220 | 532 | 6 | 0 | 364 | 144 | 784 | 12 | 0 | 365 | 288 | 444 | 4 | 1 |
| 366 | 120 | 744 | 8 | -1 | 367 | 366 | 368 | 2 | -1 | 368 | 176 | 744 | 10 | 0 | 369 | 240 | 546 | 6 | 0 | 370 | 144 | 684 | 8 | -1 |
| 371 | 312 | 432 | 4 | 1 | 372 | 120 | 896 | 12 | 0 | 373 | 372 | 374 | 2 | -1 | 374 | 160 | 648 | 8 | -1 | 375 | 200 | 624 | 8 | 0 |
| 376 | 184 | 720 | 8 | 0 | 377 | 336 | 420 | 4 | 1 | 378 | 108 | 960 | 16 | 0 | 379 | 378 | 380 | 2 | -1 | 380 | 144 | 840 | 12 | 0 |
| 381 | 252 | 512 | 4 | 1 | 382 | 190 | 576 | 4 | 1 | 383 | 382 | 384 | 2 | -1 | 384 | 128 | 1020 | 16 | 0 | 385 | 240 | 576 | 8 | -1 |
| 386 | 192 | 582 | 4 | 1 | 387 | 252 | 572 | 6 | 0 | 388 | 192 | 686 | 6 | 0 | 389 | 388 | 390 | 2 | -1 | 390 | 96 | 1008 | 16 | 1 |
| 391 | 352 | 432 | 4 | 1 | 392 | 168 | 855 | 12 | 0 | 393 | 260 | 528 | 4 | 1 | 394 | 196 | 594 | 4 | 1 | 395 | 312 | 480 | 4 | 1 |
| 396 | 120 | 1092 | 18 | 0 | 397 | 396 | 398 | 2 | -1 | 398 | 198 | 600 | 4 | 1 | 399 | 216 | 640 | 8 | -1 | 400 | 160 | 961 | 15 | 0 |
| 401 | 400 | 402 | 2 | -1 | 402 | 132 | 816 | 8 | -1 | 403 | 360 | 448 | 4 | 1 | 404 | 200 | 714 | 6 | 0 | 405 | 216 | 726 | 10 | 0 |
| 406 | 168 | 720 | 8 | -1 | 407 | 360 | 456 | 4 | 1 | 408 | 128 | 1080 | 16 | 0 | 409 | 408 | 410 | 2 | -1 | 410 | 160 | 756 | 8 | -1 |
| 411 | 272 | 552 | 4 | 1 | 412 | 204 | 728 | 6 | 0 | 413 | 348 | 480 | 4 | 1 | 414 | 132 | 936 | 12 | 0 | 415 | 328 | 504 | 4 | 1 |
| 416 | 192 | 882 | 12 | 0 | 417 | 276 | 560 | 4 | 1 | 418 | 180 | 720 | 8 | -1 | 419 | 418 | 420 | 2 | -1 | 420 | 96 | 1344 | 24 | 0 |
| 421 | 420 | 422 | 2 | -1 | 422 | 210 | 636 | 4 | 1 | 423 | 276 | 624 | 6 | 0 | 424 | 208 | 810 | 8 | 0 | 425 | 320 | 558 | 6 | 0 |
| 426 | 140 | 864 | 8 | -1 | 427 | 360 | 496 | 4 | 1 | 428 | 212 | 756 | 6 | 0 | 429 | 240 | 672 | 8 | -1 | 430 | 168 | 792 | 8 | -1 |
| 431 | 430 | 432 | 2 | -1 | 432 | 144 | 1240 | 20 | 0 | 433 | 432 | 434 | 2 | -1 | 434 | 180 | 768 | 8 | -1 | 435 | 224 | 720 | 8 | -1 |
| 436 | 216 | 770 | 6 | 0 | 437 | 396 | 480 | 4 | 1 | 438 | 144 | 888 | 8 | -1 | 439 | 438 | 440 | 2 | -1 | 440 | 160 | 1080 | 16 | 0 |
| 441 | 252 | 741 | 9 | 0 | 442 | 192 | 756 | 8 | -1 | 443 | 442 | 444 | 2 | -1 | 444 | 144 | 1064 | 12 | 0 | 445 | 352 | 540 | 4 | 1 |
| 446 | 222 | 672 | 4 | 1 | 447 | 296 | 600 | 4 | 1 | 448 | 192 | 1016 | 14 | 0 | 449 | 448 | 450 | 2 | -1 | 450 | 120 | 1209 | 18 | 0 |
| 451 | 400 | 504 | 4 | 1 | 452 | 224 | 798 | 6 | 0 | 453 | 300 | 608 | 4 | 1 | 454 | 226 | 684 | 4 | 1 | 455 | 288 | 672 | 8 | -1 |
| 456 | 144 | 1200 | 16 | 0 | 457 | 456 | 458 | 2 | -1 | 458 | 228 | 690 | 4 | 1 | 459 | 288 | 720 | 8 | 0 | 460 | 176 | 1008 | 12 | 0 |
| 461 | 460 | 462 | 2 | -1 | 462 | 120 | 1152 | 16 | 1 | 463 | 462 | 464 | 2 | -1 | 464 | 224 | 930 | 10 | 0 | 465 | 240 | 768 | 8 | -1 |
| 466 | 232 | 702 | 4 | 1 | 467 | 466 | 468 | 2 | -1 | 468 | 144 | 1274 | 18 | 0 | 469 | 396 | 544 | 4 | 1 | 470 | 184 | 864 | 8 | -1 |
| 471 | 312 | 632 | 4 | 1 | 472 | 232 | 900 | 8 | 0 | 473 | 420 | 528 | 4 | 1 | 474 | 156 | 960 | 8 | -1 | 475 | 360 | 620 | 6 | 0 |
| 476 | 192 | 1008 | 12 | 0 | 477 | 312 | 702 | 6 | 0 | 478 | 238 | 720 | 4 | 1 | 479 | 478 | 480 | 2 | -1 | 480 | 128 | 1512 | 24 | 0 |
| 481 | 432 | 532 | 4 | 1 | 482 | 240 | 726 | 4 | 1 | 483 | 264 | 768 | 8 | -1 | 484 | 220 | 931 | 9 | 0 | 485 | 384 | 588 | 4 | 1 |
| 486 | 162 | 1092 | 12 | 0 | 487 | 486 | 488 | 2 | -1 | 488 | 240 | 930 | 8 | 0 | 489 | 324 | 656 | 4 | 1 | 490 | 168 | 1026 | 12 | 0 |
| 491 | 490 | 492 | 2 | -1 | 492 | 160 | 1176 | 12 | 0 | 493 | 448 | 540 | 4 | 1 | 494 | 216 | 840 | 8 | -1 | 495 | 240 | 936 | 12 | 0 |
| 496 | 240 | 992 | 10 | 0 | 497 | 420 | 576 | 4 | 1 | 498 | 164 | 1008 | 8 | -1 | 499 | 498 | 500 | 2 | -1 | 500 | 200 | 1092 | 12 | 0 |
| 501 | 332 | 672 | 4 | 1 | 502 | 250 | 756 | 4 | 1 | 503 | 502 | 504 | 2 | -1 | 504 | 144 | 1560 | 24 | 0 | 505 | 400 | 612 | 4 | 1 |
| 506 | 220 | 864 | 8 | -1 | 507 | 312 | 732 | 6 | 0 | 508 | 252 | 896 | 6 | 0 | 509 | 508 | 510 | 2 | -1 | 510 | 128 | 1296 | 16 | 1 |
| 511 | 432 | 592 | 4 | 1 | 512 | 256 | 1023 | 10 | 0 | 513 | 324 | 800 | 8 | 0 | 514 | 256 | 774 | 4 | 1 | 515 | 408 | 624 | 4 | 1 |
| 516 | 168 | 1232 | 12 | 0 | 517 | 460 | 576 | 4 | 1 | 518 | 216 | 912 | 8 | -1 | 519 | 344 | 696 | 4 | 1 | 520 | 192 | 1260 | 16 | 0 |
| 521 | 520 | 522 | 2 | -1 | 522 | 168 | 1170 | 12 | 0 | 523 | 522 | 524 | 2 | -1 | 524 | 260 | 924 | 6 | 0 | 525 | 240 | 992 | 12 | 0 |
| 526 | 262 | 792 | 4 | 1 | 527 | 480 | 576 | 4 | 1 | 528 | 160 | 1488 | 20 | 0 | 529 | 506 | 553 | 3 | 0 | 530 | 208 | 972 | 8 | -1 |
| 531 | 348 | 780 | 6 | 0 | 532 | 216 | 1120 | 12 | 0 | 533 | 480 | 588 | 4 | 1 | 534 | 176 | 1080 | 8 | -1 | 535 | 424 | 648 | 4 | 1 |
| 536 | 264 | 1020 | 8 | 0 | 537 | 356 | 720 | 4 | 1 | 538 | 268 | 810 | 4 | 1 | 539 | 420 | 684 | 6 | 0 | 540 | 144 | 1680 | 24 | 0 |
| 541 | 540 | 542 | 2 | -1 | 542 | 270 | 816 | 4 | 1 | 543 | 360 | 728 | 4 | 1 | 544 | 256 | 1134 | 12 | 0 | 545 | 432 | 660 | 4 | 1 |
| 546 | 144 | 1344 | 16 | 1 | 547 | 546 | 548 | 2 | -1 | 548 | 272 | 966 | 6 | 0 | 549 | 360 | 806 | 6 | 0 | 550 | 200 | 1116 | 12 | 0 |
| 551 | 504 | 600 | 4 | 1 | 552 | 176 | 1440 | 16 | 0 | 553 | 468 | 640 | 4 | 1 | 554 | 276 | 834 | 4 | 1 | 555 | 288 | 912 | 8 | -1 |
| 556 | 276 | 980 | 6 | 0 | 557 | 556 | 558 | 2 | -1 | 558 | 180 | 1248 | 12 | 0 | 559 | 504 | 616 | 4 | 1 | 560 | 192 | 1488 | 20 | 0 |
| 561 | 320 | 864 | 8 | -1 | 562 | 280 | 846 | 4 | 1 | 563 | 562 | 564 | 2 | -1 | 564 | 184 | 1344 | 12 | 0 | 565 | 448 | 684 | 4 | 1 |
| 566 | 282 | 852 | 4 | 1 | 567 | 324 | 968 | 10 | 0 | 568 | 280 | 1080 | 8 | 0 | 569 | 568 | 570 | 2 | -1 | 570 | 144 | 1440 | 16 | 1 |
| 571 | 570 | 572 | 2 | -1 | 572 | 240 | 1176 | 12 | 0 | 573 | 380 | 768 | 4 | 1 | 574 | 240 | 1008 | 8 | -1 | 575 | 440 | 744 | 6 | 0 |
| 576 | 192 | 1651 | 21 | 0 | 577 | 576 | 578 | 2 | -1 | 578 | 272 | 921 | 6 | 0 | 579 | 384 | 776 | 4 | 1 | 580 | 224 | 1260 | 12 | 0 |
| 581 | 492 | 672 | 4 | 1 | 582 | 192 | 1176 | 8 | -1 | 583 | 520 | 648 | 4 | 1 | 584 | 288 | 1110 | 8 | 0 | 585 | 288 | 1092 | 12 | 0 |
| 586 | 292 | 882 | 4 | 1 | 587 | 586 | 588 | 2 | -1 | 588 | 168 | 1596 | 18 | 0 | 589 | 540 | 640 | 4 | 1 | 590 | 232 | 1080 | 8 | -1 |
| 591 | 392 | 792 | 4 | 1 | 592 | 288 | 1178 | 10 | 0 | 593 | 592 | 594 | 2 | -1 | 594 | 180 | 1440 | 16 | 0 | 595 | 384 | 864 | 8 | -1 |
| 596 | 296 | 1050 | 6 | 0 | 597 | 396 | 800 | 4 | 1 | 598 | 264 | 1008 | 8 | -1 | 599 | 598 | 600 | 2 | -1 | 600 | 160 | 1860 | 24 | 0 |
| 601 | 600 | 602 | 2 | -1 | 602 | 252 | 1056 | 8 | -1 | 603 | 396 | 884 | 6 | 0 | 604 | 300 | 1064 | 6 | 0 | 605 | 440 | 798 | 6 | 0 |
| 606 | 200 | 1224 | 8 | -1 | 607 | 606 | 608 | 2 | -1 | 608 | 288 | 1260 | 12 | 0 | 609 | 336 | 960 | 8 | -1 | 610 | 240 | 1116 | 8 | -1 |
| 611 | 552 | 672 | 4 | 1 | 612 | 192 | 1638 | 18 | 0 | 613 | 612 | 614 | 2 | -1 | 614 | 306 | 924 | 4 | 1 | 615 | 320 | 1008 | 8 | -1 |

| n | φ | σ | τ | μ | n | φ | σ | τ | μ | n | φ | σ | τ | μ | n | φ | σ | τ | μ | n | φ | σ | τ | μ |
|---|---|---|---|---|---|---|---|---|---|---|---|---|---|---|---|---|---|---|---|---|---|---|---|---|
| 616 | 240 | 1440 | 16 | 0 | 617 | 616 | 618 | 2 | −1 | 618 | 204 | 1248 | 8 | −1 | 619 | 618 | 620 | 2 | −1 | 620 | 240 | 1344 | 12 | 0 |
| 621 | 396 | 960 | 8 | 0 | 622 | 310 | 936 | 4 | 1 | 623 | 528 | 720 | 4 | 1 | 624 | 192 | 1736 | 20 | 0 | 625 | 500 | 781 | 5 | 0 |
| 626 | 312 | 942 | 4 | 1 | 627 | 360 | 960 | 8 | −1 | 628 | 312 | 1106 | 6 | 0 | 629 | 576 | 684 | 4 | 1 | 630 | 144 | 1872 | 24 | 0 |
| 631 | 630 | 632 | 2 | −1 | 632 | 312 | 1200 | 8 | 0 | 633 | 420 | 848 | 4 | 1 | 634 | 316 | 954 | 4 | 1 | 635 | 504 | 768 | 4 | 1 |
| 636 | 208 | 1512 | 12 | 0 | 637 | 504 | 798 | 6 | 0 | 638 | 280 | 1080 | 8 | −1 | 639 | 420 | 936 | 6 | 0 | 640 | 256 | 1530 | 16 | 0 |
| 641 | 640 | 642 | 2 | −1 | 642 | 212 | 1296 | 8 | −1 | 643 | 642 | 644 | 2 | −1 | 644 | 264 | 1344 | 12 | 0 | 645 | 336 | 1056 | 8 | −1 |
| 646 | 288 | 1080 | 8 | −1 | 647 | 646 | 648 | 2 | −1 | 648 | 216 | 1815 | 20 | 0 | 649 | 580 | 720 | 4 | 1 | 650 | 240 | 1302 | 12 | 0 |
| 651 | 360 | 1024 | 8 | −1 | 652 | 324 | 1148 | 6 | 0 | 653 | 652 | 654 | 2 | −1 | 654 | 216 | 1320 | 8 | −1 | 655 | 520 | 792 | 4 | 1 |
| 656 | 320 | 1302 | 10 | 0 | 657 | 432 | 962 | 6 | 0 | 658 | 276 | 1152 | 8 | −1 | 659 | 658 | 660 | 2 | −1 | 660 | 160 | 2016 | 24 | 0 |
| 661 | 660 | 662 | 2 | −1 | 662 | 330 | 996 | 4 | 1 | 663 | 384 | 1008 | 8 | −1 | 664 | 328 | 1260 | 8 | 0 | 665 | 432 | 960 | 8 | −1 |
| 666 | 216 | 1482 | 12 | 0 | 667 | 616 | 720 | 4 | 1 | 668 | 332 | 1176 | 6 | 0 | 669 | 444 | 896 | 4 | 1 | 670 | 264 | 1224 | 8 | −1 |
| 671 | 600 | 744 | 4 | 1 | 672 | 192 | 2016 | 24 | 0 | 673 | 672 | 674 | 2 | −1 | 674 | 336 | 1014 | 4 | 1 | 675 | 360 | 1240 | 12 | 0 |
| 676 | 312 | 1281 | 9 | 0 | 677 | 676 | 678 | 2 | −1 | 678 | 224 | 1368 | 8 | −1 | 679 | 576 | 784 | 4 | 1 | 680 | 256 | 1620 | 16 | 0 |
| 681 | 452 | 912 | 4 | 1 | 682 | 300 | 1152 | 8 | −1 | 683 | 682 | 684 | 2 | −1 | 684 | 216 | 1820 | 18 | 0 | 685 | 544 | 828 | 4 | 1 |
| 686 | 294 | 1200 | 8 | 0 | 687 | 456 | 920 | 4 | 1 | 688 | 336 | 1364 | 10 | 0 | 689 | 624 | 756 | 4 | 1 | 690 | 176 | 1728 | 16 | 1 |
| 691 | 690 | 692 | 2 | −1 | 692 | 344 | 1218 | 6 | 0 | 693 | 360 | 1248 | 12 | 0 | 694 | 346 | 1044 | 4 | 1 | 695 | 552 | 840 | 4 | 1 |
| 696 | 224 | 1800 | 16 | 0 | 697 | 640 | 756 | 4 | 1 | 698 | 348 | 1050 | 4 | 1 | 699 | 464 | 936 | 4 | 1 | 700 | 240 | 1736 | 18 | 0 |
| 701 | 700 | 702 | 2 | −1 | 702 | 216 | 1680 | 16 | 0 | 703 | 648 | 760 | 4 | 1 | 704 | 320 | 1524 | 14 | 0 | 705 | 368 | 1152 | 8 | −1 |
| 706 | 352 | 1062 | 4 | 1 | 707 | 600 | 816 | 4 | 1 | 708 | 232 | 1680 | 12 | 0 | 709 | 708 | 710 | 2 | −1 | 710 | 280 | 1296 | 8 | −1 |
| 711 | 468 | 1040 | 6 | 0 | 712 | 352 | 1350 | 8 | 0 | 713 | 660 | 768 | 4 | 1 | 714 | 192 | 1728 | 16 | 1 | 715 | 480 | 1008 | 8 | −1 |
| 716 | 356 | 1260 | 6 | 0 | 717 | 476 | 960 | 4 | 1 | 718 | 358 | 1080 | 4 | 1 | 719 | 718 | 720 | 2 | −1 | 720 | 192 | 2418 | 30 | 0 |
| 721 | 612 | 832 | 4 | 1 | 722 | 342 | 1143 | 6 | 0 | 723 | 480 | 968 | 4 | 1 | 724 | 360 | 1274 | 6 | 0 | 725 | 560 | 930 | 6 | 0 |
| 726 | 220 | 1596 | 12 | 0 | 727 | 726 | 728 | 2 | −1 | 728 | 288 | 1680 | 16 | 0 | 729 | 486 | 1093 | 7 | 0 | 730 | 288 | 1332 | 8 | −1 |
| 731 | 672 | 792 | 4 | 1 | 732 | 240 | 1736 | 12 | 0 | 733 | 732 | 734 | 2 | −1 | 734 | 366 | 1104 | 4 | 1 | 735 | 336 | 1368 | 12 | 0 |
| 736 | 352 | 1512 | 12 | 0 | 737 | 660 | 816 | 4 | 1 | 738 | 240 | 1638 | 12 | 0 | 739 | 738 | 740 | 2 | −1 | 740 | 288 | 1596 | 12 | 0 |
| 741 | 432 | 1120 | 8 | −1 | 742 | 312 | 1296 | 8 | −1 | 743 | 742 | 744 | 2 | −1 | 744 | 240 | 1920 | 16 | 0 | 745 | 592 | 900 | 4 | 1 |
| 746 | 372 | 1122 | 4 | 1 | 747 | 492 | 1092 | 6 | 0 | 748 | 320 | 1512 | 12 | 0 | 749 | 636 | 864 | 4 | 1 | 750 | 200 | 1872 | 16 | 0 |
| 751 | 750 | 752 | 2 | −1 | 752 | 368 | 1488 | 10 | 0 | 753 | 500 | 1008 | 4 | 1 | 754 | 336 | 1260 | 8 | −1 | 755 | 600 | 912 | 4 | 1 |
| 756 | 216 | 2240 | 24 | 0 | 757 | 756 | 758 | 2 | −1 | 758 | 378 | 1140 | 4 | 1 | 759 | 440 | 1152 | 8 | −1 | 760 | 288 | 1800 | 16 | 0 |
| 761 | 760 | 762 | 2 | −1 | 762 | 252 | 1536 | 8 | −1 | 763 | 648 | 880 | 4 | 1 | 764 | 380 | 1344 | 6 | 0 | 765 | 384 | 1404 | 12 | 0 |
| 766 | 382 | 1152 | 4 | 1 | 767 | 696 | 840 | 4 | 1 | 768 | 256 | 2044 | 18 | 0 | 769 | 768 | 770 | 2 | −1 | 770 | 240 | 1728 | 16 | 1 |
| 771 | 512 | 1032 | 4 | 1 | 772 | 384 | 1358 | 6 | 0 | 773 | 772 | 774 | 2 | −1 | 774 | 252 | 1716 | 12 | 0 | 775 | 600 | 992 | 6 | 0 |
| 776 | 384 | 1470 | 8 | 0 | 777 | 432 | 1216 | 8 | −1 | 778 | 388 | 1170 | 4 | 1 | 779 | 720 | 840 | 4 | 1 | 780 | 192 | 2352 | 24 | 0 |
| 781 | 700 | 864 | 4 | 1 | 782 | 352 | 1296 | 8 | −1 | 783 | 504 | 1200 | 8 | 0 | 784 | 336 | 1767 | 15 | 0 | 785 | 624 | 948 | 4 | 1 |
| 786 | 260 | 1584 | 8 | −1 | 787 | 786 | 788 | 2 | −1 | 788 | 392 | 1386 | 6 | 0 | 789 | 524 | 1056 | 4 | 1 | 790 | 312 | 1440 | 8 | −1 |
| 791 | 672 | 912 | 4 | 1 | 792 | 240 | 2340 | 24 | 0 | 793 | 720 | 868 | 4 | 1 | 794 | 396 | 1194 | 4 | 1 | 795 | 416 | 1296 | 8 | −1 |
| 796 | 396 | 1400 | 6 | 0 | 797 | 796 | 798 | 2 | −1 | 798 | 216 | 1920 | 16 | 1 | 799 | 736 | 864 | 4 | 1 | 800 | 320 | 1953 | 18 | 0 |
| 801 | 528 | 1170 | 6 | 0 | 802 | 400 | 1206 | 4 | 1 | 803 | 720 | 888 | 4 | 1 | 804 | 264 | 1904 | 12 | 0 | 805 | 528 | 1152 | 8 | −1 |
| 806 | 360 | 1344 | 8 | −1 | 807 | 536 | 1080 | 4 | 1 | 808 | 400 | 1530 | 8 | 0 | 809 | 808 | 810 | 2 | −1 | 810 | 216 | 2178 | 20 | 0 |
| 811 | 810 | 812 | 2 | −1 | 812 | 336 | 1680 | 12 | 0 | 813 | 540 | 1088 | 4 | 1 | 814 | 360 | 1368 | 8 | −1 | 815 | 648 | 984 | 4 | 1 |
| 816 | 256 | 2232 | 20 | 0 | 817 | 756 | 880 | 4 | 1 | 818 | 408 | 1230 | 4 | 1 | 819 | 432 | 1456 | 12 | 0 | 820 | 320 | 1764 | 12 | 0 |
| 821 | 820 | 822 | 2 | −1 | 822 | 272 | 1656 | 8 | −1 | 823 | 822 | 824 | 2 | −1 | 824 | 408 | 1560 | 8 | 0 | 825 | 400 | 1488 | 12 | 0 |
| 826 | 348 | 1440 | 8 | −1 | 827 | 826 | 828 | 2 | −1 | 828 | 264 | 2184 | 18 | 0 | 829 | 828 | 830 | 2 | −1 | 830 | 328 | 1512 | 8 | −1 |
| 831 | 552 | 1112 | 4 | 1 | 832 | 384 | 1778 | 14 | 0 | 833 | 672 | 1026 | 6 | 0 | 834 | 276 | 1680 | 8 | −1 | 835 | 664 | 1008 | 4 | 1 |
| 836 | 360 | 1680 | 12 | 0 | 837 | 540 | 1280 | 8 | 0 | 838 | 418 | 1260 | 4 | 1 | 839 | 838 | 840 | 2 | −1 | 840 | 192 | 2880 | 32 | 0 |
| 841 | 812 | 871 | 3 | 0 | 842 | 420 | 1266 | 4 | 1 | 843 | 560 | 1128 | 4 | 1 | 844 | 420 | 1484 | 6 | 0 | 845 | 624 | 1098 | 6 | 0 |
| 846 | 276 | 1872 | 12 | 0 | 847 | 660 | 1064 | 6 | 0 | 848 | 416 | 1674 | 10 | 0 | 849 | 564 | 1136 | 4 | 1 | 850 | 320 | 1674 | 12 | 0 |
| 851 | 792 | 912 | 4 | 1 | 852 | 280 | 2016 | 12 | 0 | 853 | 852 | 854 | 2 | −1 | 854 | 360 | 1488 | 8 | −1 | 855 | 432 | 1560 | 12 | 0 |
| 856 | 424 | 1620 | 8 | 0 | 857 | 856 | 858 | 2 | −1 | 858 | 240 | 2016 | 16 | 1 | 859 | 858 | 860 | 2 | −1 | 860 | 336 | 1848 | 12 | 0 |
| 861 | 480 | 1344 | 8 | −1 | 862 | 430 | 1296 | 4 | 1 | 863 | 862 | 864 | 2 | −1 | 864 | 288 | 2520 | 24 | 0 | 865 | 688 | 1044 | 4 | 1 |
| 866 | 432 | 1302 | 4 | 1 | 867 | 544 | 1228 | 6 | 0 | 868 | 360 | 1792 | 12 | 0 | 869 | 780 | 960 | 4 | 1 | 870 | 224 | 2160 | 16 | 1 |
| 871 | 792 | 952 | 4 | 1 | 872 | 432 | 1650 | 8 | 0 | 873 | 576 | 1274 | 6 | 0 | 874 | 396 | 1440 | 8 | −1 | 875 | 600 | 1248 | 8 | 0 |
| 876 | 288 | 2072 | 12 | 0 | 877 | 876 | 878 | 2 | −1 | 878 | 438 | 1320 | 4 | 1 | 879 | 584 | 1176 | 4 | 1 | 880 | 320 | 2232 | 20 | 0 |
| 881 | 880 | 882 | 2 | −1 | 882 | 252 | 2223 | 18 | 0 | 883 | 882 | 884 | 2 | −1 | 884 | 384 | 1764 | 12 | 0 | 885 | 464 | 1440 | 8 | −1 |
| 886 | 442 | 1332 | 4 | 1 | 887 | 886 | 888 | 2 | −1 | 888 | 288 | 2280 | 16 | 0 | 889 | 756 | 1024 | 4 | 1 | 890 | 352 | 1620 | 8 | −1 |

| $n$ | $\phi$ | $\sigma$ | $\tau$ | $\mu$ | $n$ | $\phi$ | $\sigma$ | $\tau$ | $\mu$ | $n$ | $\phi$ | $\sigma$ | $\tau$ | $\mu$ | $n$ | $\phi$ | $\sigma$ | $\tau$ | $\mu$ | $n$ | $\phi$ | $\sigma$ | $\tau$ | $\mu$ |
|---|---|---|---|---|---|---|---|---|---|---|---|---|---|---|---|---|---|---|---|---|---|---|---|---|
| 891 | 540 | 1452 | 10 | 0 | 892 | 444 | 1568 | 6 | 0 | 893 | 828 | 960 | 4 | 1 | 894 | 296 | 1800 | 8 | −1 | 895 | 712 | 1080 | 4 | 1 |
| 896 | 384 | 2040 | 16 | 0 | 897 | 528 | 1344 | 8 | −1 | 898 | 448 | 1350 | 4 | 1 | 899 | 840 | 960 | 4 | 1 | 900 | 240 | 2821 | 27 | 0 |
| 901 | 832 | 972 | 4 | 1 | 902 | 400 | 1512 | 8 | −1 | 903 | 504 | 1408 | 8 | −1 | 904 | 448 | 1710 | 8 | 0 | 905 | 720 | 1092 | 4 | 1 |
| 906 | 300 | 1824 | 8 | −1 | 907 | 906 | 908 | 2 | −1 | 908 | 452 | 1596 | 6 | 0 | 909 | 600 | 1326 | 6 | 0 | 910 | 288 | 2016 | 16 | 1 |
| 911 | 910 | 912 | 2 | −1 | 912 | 288 | 2480 | 20 | 0 | 913 | 820 | 1008 | 4 | 1 | 914 | 456 | 1374 | 4 | 1 | 915 | 480 | 1488 | 8 | −1 |
| 916 | 456 | 1610 | 6 | 0 | 917 | 780 | 1056 | 4 | 1 | 918 | 288 | 2160 | 16 | 0 | 919 | 918 | 920 | 2 | −1 | 920 | 352 | 2160 | 16 | 0 |
| 921 | 612 | 1232 | 4 | 1 | 922 | 460 | 1386 | 4 | 1 | 923 | 840 | 1008 | 4 | 1 | 924 | 240 | 2688 | 24 | 0 | 925 | 720 | 1178 | 6 | 0 |
| 926 | 462 | 1392 | 4 | 1 | 927 | 612 | 1352 | 6 | 0 | 928 | 448 | 1890 | 12 | 0 | 929 | 928 | 930 | 2 | −1 | 930 | 240 | 2304 | 16 | 1 |
| 931 | 756 | 1140 | 6 | 0 | 932 | 464 | 1638 | 6 | 0 | 933 | 620 | 1248 | 4 | 1 | 934 | 466 | 1404 | 4 | 1 | 935 | 640 | 1296 | 8 | −1 |
| 936 | 288 | 2730 | 24 | 0 | 937 | 936 | 938 | 2 | −1 | 938 | 396 | 1632 | 8 | −1 | 939 | 624 | 1256 | 4 | 1 | 940 | 368 | 2016 | 12 | 0 |
| 941 | 940 | 942 | 2 | −1 | 942 | 312 | 1896 | 8 | −1 | 943 | 880 | 1008 | 4 | 1 | 944 | 464 | 1860 | 10 | 0 | 945 | 432 | 1920 | 16 | 0 |
| 946 | 420 | 1584 | 8 | −1 | 947 | 946 | 948 | 2 | −1 | 948 | 312 | 2240 | 12 | 0 | 949 | 864 | 1036 | 4 | 1 | 950 | 360 | 1860 | 12 | 0 |
| 951 | 632 | 1272 | 4 | 1 | 952 | 384 | 2160 | 16 | 0 | 953 | 952 | 954 | 2 | −1 | 954 | 312 | 2106 | 12 | 0 | 955 | 760 | 1152 | 4 | 1 |
| 956 | 476 | 1680 | 6 | 0 | 957 | 560 | 1440 | 8 | −1 | 958 | 478 | 1440 | 4 | 1 | 959 | 816 | 1104 | 4 | 1 | 960 | 256 | 3048 | 28 | 0 |
| 961 | 930 | 993 | 3 | 0 | 962 | 432 | 1596 | 8 | −1 | 963 | 636 | 1404 | 6 | 0 | 964 | 480 | 1694 | 6 | 0 | 965 | 768 | 1164 | 4 | 1 |
| 966 | 264 | 2304 | 16 | 1 | 967 | 966 | 968 | 2 | −1 | 968 | 440 | 1995 | 12 | 0 | 969 | 576 | 1440 | 8 | −1 | 970 | 384 | 1764 | 8 | −1 |
| 971 | 970 | 972 | 2 | −1 | 972 | 324 | 2548 | 18 | 0 | 973 | 828 | 1120 | 4 | 1 | 974 | 486 | 1464 | 4 | 1 | 975 | 480 | 1736 | 12 | 0 |
| 976 | 480 | 1922 | 10 | 0 | 977 | 976 | 978 | 2 | −1 | 978 | 324 | 1968 | 8 | −1 | 979 | 880 | 1080 | 4 | 1 | 980 | 336 | 2394 | 18 | 0 |
| 981 | 648 | 1430 | 6 | 0 | 982 | 490 | 1476 | 4 | 1 | 983 | 982 | 984 | 2 | −1 | 984 | 320 | 2520 | 16 | 0 | 985 | 784 | 1188 | 4 | 1 |
| 986 | 448 | 1620 | 8 | −1 | 987 | 552 | 1536 | 8 | −1 | 988 | 432 | 1960 | 12 | 0 | 989 | 924 | 1056 | 4 | 1 | 990 | 240 | 2808 | 24 | 0 |
| 991 | 990 | 992 | 2 | −1 | 992 | 480 | 2016 | 12 | 0 | 993 | 660 | 1328 | 4 | 1 | 994 | 420 | 1728 | 8 | −1 | 995 | 792 | 1200 | 4 | 1 |
| 996 | 328 | 2352 | 12 | 0 | 997 | 996 | 998 | 2 | −1 | 998 | 498 | 1500 | 4 | 1 | 999 | 648 | 1520 | 8 | 0 | 1000 | 400 | 2340 | 16 | 0 |

## 4.6.4  THE MÖBIUS FUNCTION AND OTHER IMPORTANT ARITHMETIC FUNCTIONS

### Definitions:

If $n$ is a positive integer, $\mu(n)$, the value of the **Möbius function**, is defined by

$$\mu(n) = \begin{cases} 1, & \text{if } n = 1 \\ 0, & \text{if } n \text{ has a square factor larger than 1} \\ (-1)^s, & \text{if } n \text{ is square free and is the product of } s \text{ different primes.} \end{cases}$$

If $n > 1$ is a positive integer, with prime-power factorization $p_1^{a_1} p_2^{a_2} \ldots p_m^{a_m}$, then $\lambda(n)$, the value of **Liouville's function** at $n$, is given by $\lambda(n) = (-1)^{a_1 + a_2 + \cdots + a_m}$, with $\lambda(1) = 1$.

If $n$ is a positive integer with prime-power factorization $n = p_1^{a_1} p_2^{a_2} \ldots p_m^{a_m}$, then the arithmetic functions $\Omega$ and $\omega$ are defined by $\Omega(1) = \omega(1) = 0$ and for $n > 1$, $\Omega(n) = \sum_{i=1}^{m} a_i$ and $\omega(n) = m$. That is, $\Omega(n)$ is the sum of exponents in the prime-power factorization of $n$ and $\omega(n)$ is the number of distinct primes in the prime-power factorization of $n$.

### Facts:

1. The Möbius function is multiplicative, but not completely multiplicative.

2. *Möbius inversion formula:* If $f$ is an arithmetic function and $F(n) = \sum_{d|n} f(d)$, then $f(n) = \sum_{d|n} \mu(d) F(\frac{n}{d})$.

3. If $n$ is a positive integer, then $\phi(n) = \sum_{d|n} \mu(d) \frac{n}{d}$.

4. If $f$ is multiplicative, then $\sum_{d|n} \mu(d) f(d) = \prod_{p|n} (1 - f(p))$.

5. If $f$ is multiplicative, then $\sum_{d|n} \mu(d)^2 f(d) = \prod_{p|n} (1 + f(p))$.

**6.** If $n$ is a positive integer, then $\sum_{d|n} \mu(d) = \begin{cases} 1 & \text{if } n = 1; \\ 0 & \text{if } n > 1. \end{cases}$

**7.** If $n$ is a positive integer, then $\sum_{d|n} \lambda(d) = \begin{cases} 1 & \text{if } n \text{ is a perfect square}; \\ 0 & \text{if } n \text{ is not a perfect square}. \end{cases}$

**8.** In 1897 Mertens showed that $|\sum_{k=1}^{n} \mu(k)| < \sqrt{n}$ for all positive integers $n$ not exceeding 10,000 and conjectured that this inequality holds for all positive integers $n$. However, in 1985 Odlyzko and teRiele disproved this conjecture, which went by the name *Mertens' conjecture*, without giving an explicit integer $n$ for which the conjecture fails. In 1987 Pintz showed that there is at least one counterexample $n$ with $n \leq 10^{65}$, again without giving an explicit counterexample $n$. Finding such an integer $n$ requires more computing power than is currently available.

**9.** Liouville's function is completely multiplicative.

**10.** The function $\omega$ is additive and the function $\Omega$ is completely additive.

**Examples:**

**1.** Table 1 lists the values of $\mu(n)$ for $1 \leq n \leq 1,000$.

**2.** $\mu(12) = 0$ since $2^2 | 12$ and $\mu(105) = \mu(3 \cdot 5 \cdot 7) = (-1)^3 = -1$.

**3.** $\lambda(720) = \lambda(2^4 \cdot 3^2 \cdot 5) = (-1)^{4+2+1} = (-1)^7 = -1$.

**4.** $\Omega(720) = \Omega(2^4 \cdot 3^2 \cdot 5) = 4 + 2 + 1 = 7$ and $\omega(720) = \omega(2^4 \cdot 3^2 \cdot 5) = 3$.

---

## 4.6.5  DIRICHLET PRODUCTS

**Definitions:**

If $f$ and $g$ are arithmetic functions, then the **Dirichlet product** of $f$ and $g$ is the function $f \star g$ defined by $(f \star g)(n) = \sum_{d|n} f(d) g(\frac{n}{d})$.

If $f$ and $g$ are arithmetic functions such that $f \star g = g \star f = I$, where $I(n) = \lfloor \frac{1}{n} \rfloor$, then $g$ is the **Dirichlet inverse** of $f$.

**Facts:**

**1.** If $f$ and $g$ are arithmetic functions, then $f \star g = g \star f$.

**2.** If $f$, $g$, and $h$ are arithmetic functions, then $(f \star g) \star h = f \star (g \star h)$.

**3.** If $f$, $g$, and $h$ are arithmetic functions, then $f \star (g + h) = (f \star g) + (f \star h)$.

**4.** Because of Facts 1–3, the set of arithmetic functions with the operations of Dirichlet product and ordinary addition of functions forms a ring. (See Chapter 5.)

**5.** If $f$ is an arithmetic function with $f(1) \neq 0$, then there is a unique Dirichlet inverse of $f$, which is written as $f^{-1}$. Furthermore, $f^{-1}$ is given by the recursive formulas $f^{-1}(1) = \frac{1}{f(1)}$ and $f^{-1}(n) = -\frac{1}{f(1)} \sum_{\substack{d|n \\ d<n}} f(\frac{n}{d}) f^{-1}(d)$ for $n > 1$.

**6.** The set of all arithmetic functions $f$ with $f(1) \neq 0$ forms an abelian group with respect to the operation $\star$, where the identity element is the function $I$.

**7.** If $f$ and $g$ are arithmetic functions with $f(1) \neq 0$ and $g(1) \neq 0$, then $(f * g)^{-1} = f^{-1} \star g^{-1}$.

**8.** If $u$ is the arithmetic function with $u(n) = 1$ for all positive integers $n$, then $\mu \star u = I$, so $u = \mu^{-1}$ and $\mu = u^{-1}$.

**9.** If $f$ is a multiplicative function, then $f$ is completely multiplicative if and only if $f^{-1}(n) = \mu(n)f(n)$ for all positive integers $n$.

**10.** If $f$ and $g$ are multiplicative functions, then $f \star g$ is also multiplicative.

**11.** If $f$ and $g$ are arithmetic functions and both $f$ and $f \star g$ are multiplicative, then $g$ is also multiplicative.

**12.** If $f$ is multiplicative, then $f^{-1}$ exists and is multiplicative.

**Examples:**

**1.** The identity $\phi(n) = \sum_{d|n} \mu(d)\frac{n}{d}$ (§4.6.4, Fact 3) implies that $\phi = \mu \star N$ where $N$ is the multiplicative function $N(n) = n$.

**2.** Since the function $N$ is completely multiplicative, $N^{-1} = \mu N$ by Fact 9.

**3.** From Example 1 and Facts 7 and 8, it follows that $\phi^{-1} = \mu^{-1} \star \mu N = \mu \star \mu N$. Hence $\phi^{-1}(n) = \sum_{d|n} d\mu(d)$.

# 4.7   PRIMITIVE ROOTS AND QUADRATIC RESIDUES

A primitive root of an integer, when it exists, is an integer whose powers run through a complete system of residues modulo this integer. When a primitive root exists, it is possible to use the theory of indices to solve certain congruences. This section provides the information needed to understand and employ primitive roots.

The question of which integers are perfect squares modulo a prime is one that has been studied extensively. An integer that is a perfect square modulo $n$ is called a quadratic residue of $n$. The law of quadratic reciprocity provides a surprising link between the answer to the question of whether a prime $p$ is a perfect square modulo a prime $q$ and the answer to the question of whether $q$ is a perfect square modulo $p$. This section provides information that helps determine whether an integer is a quadratic residue modulo a given integer $n$.

There are important applications of the topics covered in this section, including applications to public key cryptography and authentication schemes. (See Chapter 15.)

## 4.7.1   PRIMITIVE ROOTS

**Definitions:**

If $a$ and $m$ are relatively prime positive integers, then the **order of $a$ modulo $m$**, denoted $\text{ord}_m a$, is the least positive integer $x$ such that $a^x \equiv 1 \pmod{m}$.

If $r$ and $n$ are relatively prime integers and $n$ is positive, then $r$ is a **primitive root modulo $m$** if $\text{ord}_n r = \phi(n)$. A primitive root modulo $m$ is also said to be a primitive root of $m$ and $m$ is said to have a primitive root.

If $m$ is a positive integer, then the **minimum universal exponent modulo $m$** is the smallest positive integer $\lambda(m)$ for which $a^{\lambda(m)} \equiv 1 \pmod{m}$ for all integers $a$ relatively prime to $m$.

**Facts:**

**1.** The positive integer $n$, with $n > 1$, has a primitive root if and only if $n = 2$, $4$, $p^t$, or $2p^t$, where $p$ is an odd prime and $t$ is a positive integer.

**2.** There are $\phi(d)$ incongruent integers modulo $p$ if $p$ is prime and $d$ is a positive divisor of $p - 1$.

**3.** There are $\phi(p - 1)$ primitive roots of $p$ if $p$ is a prime.

**4.** If the positive integer $m$ has a primitive root, then it has a total of $\phi(\phi(m))$ incongruent primitive roots.

**5.** If $r$ is a primitive root of the odd prime $p$, then either $r$ or $r + p$ is a primitive root modulo $p^2$.

**6.** If $r$ is a primitive root of $p^2$, where $p$ is prime, then $r$ is a primitive root of $p^k$ for all positive integers $k$.

**7.** It is an unsettled conjecture (stated by E. Artin) whether $2$ is a primitive root of infinitely many primes. More generally, given any prime $p$ it is unknown whether $p$ is a primitive root of infinitely many primes.

**8.** It is known that given any three primes, at least one of these primes is a primitive root of infinitely many primes [GuMu84].

**9.** Given a set of $n$ primes, $p_1, p_2, \ldots, p_n$, there are $\prod_{k=1}^{n} \phi(p_k - 1)$ integers $x$ with $1 < x \le \prod_{k=1}^{n} p_k$ such that $x$ is a primitive root of $p_k$ for $k = 1, 2, \ldots, n$. Such an integer $x$ is a called a *common primitive root* of the primes $p_1, \ldots, p_n$.

**10.** Let $g_p$ denote the smallest positive integer that is a primitive root modulo $p$, where $p$ is a prime. It is known that $g_p$ is not always small; in particular it has been shown by Fridlender and Salié that there is a positive constant $C$ such that $g_p > C \log p$ for infinitely many primes $p$ [Ri96].

**11.** Burgess has shown that $g_p$ does not grow too rapidly; in particular he showed that $g_p \le Cp^{\frac{1}{4}+\epsilon}$ for $\epsilon > 0$, $C$ a constant, and $p$ sufficiently large [Ri96].

**12.** The minimum universal exponent modulo the powers of $2$ are $\lambda(2) = 1$, $\lambda(2^2) = 2$, and $\lambda(2^k) = 2^{k-2}$ for $k = 3, 4, \ldots$.

**13.** If $m$ is a positive integer with prime-power factorization $2^k q_1^{a_1} \ldots q_r^{a_r}$, where $k$ is a nonnegative integer, then the least universal exponent of $m$ is given by $\lambda(m) = \text{lcm}(\lambda(2^k), \phi(q_1^{a_1}), \ldots, \phi(q_r^{a_r}))$.

**14.** For every positive integer $m$, there is an integer $a$ such that $\text{ord}_m a = \lambda(m)$.

**15.** There are six positive integers $m$ with $\lambda(m) = 2$: $m = 3, 4, 6, 8, 12, 24$.

**16.** The following table displays the least primitive root $\omega$ of each prime less than 10,000.

| $p$ | $\omega$ | $p$ | $\omega$ | $p$ | $\omega$ | $p$ | $\omega$ | $p$ | $\omega$ | $p$ | $\omega$ | $p$ | $\omega$ | $p$ | $\omega$ | $p$ | $\omega$ |
|---|---|---|---|---|---|---|---|---|---|---|---|---|---|---|---|---|---|
| 3 | 2 | 5 | 2 | 7 | 3 | 11 | 2 | 13 | 2 | 17 | 3 | 19 | 2 | 23 | 5 | 29 | 2 |
| 31 | 3 | 37 | 2 | 41 | 6 | 43 | 3 | 47 | 5 | 53 | 2 | 59 | 2 | 61 | 2 | 67 | 2 |
| 71 | 7 | 73 | 5 | 79 | 3 | 83 | 2 | 89 | 3 | 97 | 5 | 101 | 2 | 103 | 5 | 107 | 2 |
| 109 | 6 | 113 | 3 | 127 | 3 | 131 | 2 | 137 | 3 | 139 | 2 | 149 | 2 | 151 | 6 | 157 | 5 |
| 163 | 2 | 167 | 5 | 173 | 2 | 179 | 2 | 181 | 2 | 191 | 19 | 193 | 5 | 197 | 2 | 199 | 3 |
| 211 | 2 | 223 | 3 | 227 | 2 | 229 | 6 | 233 | 3 | 239 | 7 | 241 | 7 | 251 | 6 | 257 | 3 |
| 263 | 5 | 269 | 2 | 271 | 6 | 277 | 5 | 281 | 3 | 283 | 3 | 293 | 2 | 307 | 5 | 311 | 17 |
| 313 | 10 | 317 | 2 | 331 | 3 | 337 | 10 | 347 | 2 | 349 | 2 | 353 | 3 | 359 | 7 | 367 | 6 |

| $p$ | $\omega$ | $p$ | $\omega$ | $p$ | $\omega$ | $p$ | $\omega$ | $p$ | $\omega$ | $p$ | $\omega$ | $p$ | $\omega$ | $p$ | $\omega$ | $p$ | $\omega$ |
|---|---|---|---|---|---|---|---|---|---|---|---|---|---|---|---|---|---|
| 373 | 2 | 379 | 2 | 383 | 5 | 389 | 2 | 397 | 5 | 401 | 3 | 409 | 21 | 419 | 2 | 421 | 2 |
| 431 | 7 | 433 | 5 | 439 | 15 | 443 | 2 | 449 | 3 | 457 | 13 | 461 | 2 | 463 | 3 | 467 | 2 |
| 479 | 13 | 487 | 3 | 491 | 2 | 499 | 7 | 503 | 5 | 509 | 2 | 521 | 3 | 523 | 2 | 541 | 2 |
| 547 | 2 | 557 | 2 | 563 | 2 | 569 | 3 | 571 | 3 | 577 | 5 | 587 | 2 | 593 | 3 | 599 | 7 |
| 601 | 7 | 607 | 3 | 613 | 2 | 617 | 3 | 619 | 2 | 631 | 3 | 641 | 3 | 643 | 11 | 647 | 5 |
| 653 | 2 | 659 | 2 | 661 | 2 | 673 | 5 | 677 | 2 | 683 | 5 | 691 | 3 | 701 | 2 | 709 | 2 |
| 719 | 11 | 727 | 5 | 733 | 6 | 739 | 3 | 743 | 5 | 751 | 3 | 757 | 2 | 761 | 6 | 769 | 11 |
| 773 | 2 | 787 | 2 | 797 | 2 | 809 | 3 | 811 | 3 | 821 | 2 | 823 | 3 | 827 | 2 | 829 | 2 |
| 839 | 11 | 853 | 2 | 857 | 3 | 859 | 2 | 863 | 5 | 877 | 2 | 881 | 3 | 883 | 2 | 887 | 5 |
| 907 | 2 | 911 | 17 | 919 | 7 | 929 | 3 | 937 | 5 | 941 | 2 | 947 | 2 | 953 | 3 | 967 | 5 |
| 971 | 6 | 977 | 3 | 983 | 5 | 991 | 6 | 997 | 7 | 1009 | 11 | 1013 | 3 | 1019 | 2 | 1021 | 10 |
| 1031 | 14 | 1033 | 5 | 1039 | 3 | 1049 | 3 | 1051 | 7 | 1061 | 2 | 1063 | 3 | 1069 | 6 | 1087 | 3 |
| 1091 | 2 | 1093 | 5 | 1097 | 3 | 1103 | 5 | 1109 | 2 | 1117 | 2 | 1123 | 2 | 1129 | 11 | 1151 | 17 |
| 1153 | 5 | 1163 | 5 | 1171 | 2 | 1181 | 7 | 1187 | 2 | 1193 | 3 | 1201 | 11 | 1213 | 2 | 1217 | 3 |
| 1223 | 5 | 1229 | 2 | 1231 | 3 | 1237 | 2 | 1249 | 7 | 1259 | 2 | 1277 | 2 | 1279 | 3 | 1283 | 2 |
| 1289 | 6 | 1291 | 2 | 1297 | 10 | 1301 | 2 | 1303 | 6 | 1307 | 2 | 1319 | 13 | 1321 | 13 | 1327 | 3 |
| 1361 | 3 | 1367 | 5 | 1373 | 2 | 1381 | 2 | 1399 | 13 | 1409 | 3 | 1423 | 3 | 1427 | 2 | 1429 | 6 |
| 1433 | 3 | 1439 | 7 | 1447 | 3 | 1451 | 2 | 1453 | 2 | 1459 | 3 | 1471 | 6 | 1481 | 3 | 1483 | 2 |
| 1487 | 5 | 1489 | 14 | 1493 | 2 | 1499 | 2 | 1511 | 11 | 1523 | 2 | 1531 | 2 | 1543 | 5 | 1549 | 2 |
| 1553 | 3 | 1559 | 19 | 1567 | 3 | 1571 | 2 | 1579 | 3 | 1583 | 5 | 1597 | 11 | 1601 | 3 | 1607 | 5 |
| 1609 | 7 | 1613 | 3 | 1619 | 2 | 1621 | 2 | 1627 | 3 | 1637 | 2 | 1657 | 11 | 1663 | 3 | 1667 | 2 |
| 1669 | 2 | 1693 | 2 | 1697 | 3 | 1699 | 3 | 1709 | 3 | 1721 | 3 | 1723 | 3 | 1733 | 2 | 1741 | 2 |
| 1747 | 2 | 1753 | 7 | 1759 | 6 | 1777 | 5 | 1783 | 10 | 1787 | 2 | 1789 | 6 | 1801 | 11 | 1811 | 6 |
| 1823 | 5 | 1831 | 3 | 1847 | 5 | 1861 | 2 | 1867 | 2 | 1871 | 14 | 1873 | 10 | 1877 | 2 | 1879 | 6 |
| 1889 | 3 | 1901 | 2 | 1907 | 2 | 1913 | 3 | 1931 | 2 | 1933 | 5 | 1949 | 2 | 1951 | 3 | 1973 | 2 |
| 1979 | 2 | 1987 | 2 | 1993 | 5 | 1997 | 2 | 1999 | 3 | 2003 | 5 | 2011 | 3 | 2017 | 5 | 2027 | 2 |
| 2029 | 2 | 2039 | 7 | 2053 | 2 | 2063 | 5 | 2069 | 2 | 2081 | 3 | 2083 | 2 | 2087 | 5 | 2089 | 7 |
| 2099 | 2 | 2111 | 7 | 2113 | 5 | 2129 | 3 | 2131 | 2 | 2137 | 10 | 2141 | 2 | 2143 | 3 | 2153 | 3 |
| 2161 | 23 | 2179 | 7 | 2203 | 5 | 2207 | 5 | 2213 | 2 | 2221 | 2 | 2237 | 2 | 2239 | 3 | 2243 | 2 |
| 2251 | 7 | 2267 | 2 | 2269 | 2 | 2273 | 3 | 2281 | 7 | 2287 | 19 | 2293 | 2 | 2297 | 5 | 2309 | 2 |
| 2311 | 3 | 2333 | 2 | 2339 | 2 | 2341 | 7 | 2347 | 3 | 2351 | 13 | 2357 | 2 | 2371 | 2 | 2377 | 5 |
| 2381 | 3 | 2383 | 5 | 2389 | 2 | 2393 | 3 | 2399 | 11 | 2411 | 6 | 2417 | 3 | 2423 | 5 | 2437 | 2 |
| 2441 | 6 | 2447 | 5 | 2459 | 2 | 2467 | 2 | 2473 | 5 | 2477 | 2 | 2503 | 3 | 2521 | 17 | 2531 | 2 |
| 2539 | 2 | 2543 | 5 | 2549 | 2 | 2551 | 6 | 2557 | 2 | 2579 | 2 | 2591 | 7 | 2593 | 7 | 2609 | 3 |
| 2617 | 5 | 2621 | 2 | 2633 | 3 | 2647 | 3 | 2657 | 3 | 2659 | 2 | 2663 | 5 | 2671 | 7 | 2677 | 2 |
| 2683 | 2 | 2687 | 5 | 2689 | 19 | 2693 | 2 | 2699 | 2 | 2707 | 2 | 2711 | 7 | 2713 | 5 | 2719 | 3 |
| 2729 | 3 | 2731 | 3 | 2741 | 2 | 2749 | 6 | 2753 | 3 | 2767 | 3 | 2777 | 3 | 2789 | 2 | 2791 | 6 |
| 2797 | 2 | 2801 | 3 | 2803 | 2 | 2819 | 2 | 2833 | 5 | 2837 | 2 | 2843 | 2 | 2851 | 2 | 2857 | 11 |
| 2861 | 2 | 2879 | 7 | 2887 | 5 | 2897 | 3 | 2903 | 5 | 2909 | 2 | 2917 | 5 | 2927 | 5 | 2939 | 2 |
| 2953 | 13 | 2957 | 2 | 2963 | 2 | 2969 | 3 | 2971 | 10 | 2999 | 17 | 3001 | 14 | 3011 | 2 | 3019 | 2 |
| 3023 | 5 | 3037 | 2 | 3041 | 3 | 3049 | 11 | 3061 | 6 | 3067 | 2 | 3079 | 6 | 3083 | 2 | 3089 | 3 |
| 3109 | 6 | 3119 | 7 | 3121 | 7 | 3137 | 3 | 3163 | 3 | 3167 | 5 | 3169 | 7 | 3181 | 7 | 3187 | 2 |
| 3191 | 11 | 3203 | 2 | 3209 | 3 | 3217 | 5 | 3221 | 10 | 3229 | 6 | 3251 | 6 | 3253 | 2 | 3257 | 3 |

| $p$ | $\omega$ | $p$ | $\omega$ | $p$ | $\omega$ | $p$ | $\omega$ | $p$ | $\omega$ | $p$ | $\omega$ | $p$ | $\omega$ | $p$ | $\omega$ | $p$ | $\omega$ |
|---|---|---|---|---|---|---|---|---|---|---|---|---|---|---|---|---|---|
| 3259 | 3 | 3271 | 3 | 3299 | 2 | 3301 | 6 | 3307 | 2 | 3313 | 10 | 3319 | 6 | 3323 | 2 | 3329 | 3 |
| 3331 | 3 | 3343 | 5 | 3347 | 2 | 3359 | 11 | 3361 | 22 | 3371 | 2 | 3373 | 5 | 3389 | 3 | 3391 | 3 |
| 3407 | 5 | 3413 | 2 | 3433 | 5 | 3449 | 3 | 3457 | 7 | 3461 | 2 | 3463 | 3 | 3467 | 2 | 3469 | 2 |
| 3491 | 2 | 3499 | 2 | 3511 | 7 | 3517 | 2 | 3527 | 5 | 3529 | 17 | 3533 | 2 | 3539 | 2 | 3541 | 7 |
| 3547 | 2 | 3557 | 2 | 3559 | 3 | 3571 | 2 | 3581 | 2 | 3583 | 3 | 3593 | 3 | 3607 | 5 | 3613 | 2 |
| 3617 | 3 | 3623 | 5 | 3631 | 15 | 3637 | 2 | 3643 | 2 | 3659 | 2 | 3671 | 13 | 3673 | 5 | 3677 | 2 |
| 3691 | 2 | 3697 | 5 | 3701 | 2 | 3709 | 2 | 3719 | 7 | 3727 | 3 | 3733 | 2 | 3739 | 7 | 3761 | 3 |
| 3767 | 5 | 3769 | 7 | 3779 | 2 | 3793 | 5 | 3797 | 2 | 3803 | 2 | 3821 | 3 | 3823 | 3 | 3833 | 3 |
| 3847 | 5 | 3851 | 2 | 3853 | 2 | 3863 | 5 | 3877 | 2 | 3881 | 13 | 3889 | 11 | 3907 | 2 | 3911 | 13 |
| 3917 | 2 | 3919 | 3 | 3923 | 2 | 3929 | 3 | 3931 | 2 | 3943 | 3 | 3947 | 2 | 3967 | 6 | 3989 | 2 |
| 4001 | 3 | 4003 | 2 | 4007 | 5 | 4013 | 2 | 4019 | 2 | 4021 | 2 | 4027 | 3 | 4049 | 3 | 4051 | 10 |
| 4057 | 5 | 4073 | 3 | 4079 | 11 | 4091 | 2 | 4093 | 2 | 4099 | 2 | 4111 | 12 | 4127 | 5 | 4129 | 13 |
| 4133 | 2 | 4139 | 2 | 4153 | 5 | 4157 | 2 | 4159 | 3 | 4177 | 5 | 4201 | 11 | 4211 | 6 | 4217 | 3 |
| 4219 | 2 | 4229 | 2 | 4231 | 3 | 4241 | 3 | 4243 | 2 | 4253 | 2 | 4259 | 2 | 4261 | 2 | 4271 | 7 |
| 4273 | 5 | 4283 | 2 | 4289 | 3 | 4297 | 5 | 4327 | 3 | 4337 | 3 | 4339 | 10 | 4349 | 2 | 4357 | 2 |
| 4363 | 2 | 4373 | 2 | 4391 | 14 | 4397 | 2 | 4409 | 3 | 4421 | 3 | 4423 | 3 | 4441 | 21 | 4447 | 3 |
| 4451 | 2 | 4457 | 3 | 4463 | 5 | 4481 | 3 | 4483 | 2 | 4493 | 2 | 4507 | 2 | 4513 | 7 | 4517 | 2 |
| 4519 | 3 | 4523 | 5 | 4547 | 2 | 4549 | 6 | 4561 | 11 | 4567 | 3 | 4583 | 5 | 4591 | 11 | 4597 | 5 |
| 4603 | 2 | 4621 | 2 | 4637 | 2 | 4639 | 3 | 4643 | 5 | 4649 | 3 | 4651 | 3 | 4657 | 15 | 4663 | 3 |
| 4673 | 3 | 4679 | 11 | 4691 | 2 | 4703 | 5 | 4721 | 6 | 4723 | 2 | 4729 | 17 | 4733 | 5 | 4751 | 19 |
| 4759 | 3 | 4783 | 6 | 4787 | 2 | 4789 | 2 | 4793 | 3 | 4799 | 7 | 4801 | 7 | 4813 | 2 | 4817 | 3 |
| 4831 | 3 | 4861 | 11 | 4871 | 11 | 4877 | 2 | 4889 | 3 | 4903 | 3 | 4909 | 6 | 4919 | 13 | 4931 | 6 |
| 4933 | 2 | 4937 | 3 | 4943 | 7 | 4951 | 6 | 4957 | 2 | 4967 | 5 | 4969 | 11 | 4973 | 2 | 4987 | 2 |
| 4993 | 5 | 4999 | 3 | 5003 | 2 | 5009 | 3 | 5011 | 2 | 5021 | 3 | 5023 | 3 | 5039 | 11 | 5051 | 2 |
| 5059 | 2 | 5077 | 2 | 5081 | 3 | 5087 | 5 | 5099 | 2 | 5101 | 6 | 5107 | 2 | 5113 | 19 | 5119 | 3 |
| 5147 | 2 | 5153 | 5 | 5167 | 6 | 5171 | 2 | 5179 | 2 | 5189 | 2 | 5197 | 7 | 5209 | 17 | 5227 | 2 |
| 5231 | 7 | 5233 | 10 | 5237 | 3 | 5261 | 2 | 5273 | 3 | 5279 | 7 | 5281 | 7 | 5297 | 3 | 5303 | 5 |
| 5309 | 2 | 5323 | 5 | 5333 | 2 | 5347 | 3 | 5351 | 11 | 5381 | 3 | 5387 | 2 | 5393 | 3 | 5399 | 7 |
| 5407 | 3 | 5413 | 5 | 5417 | 3 | 5419 | 3 | 5431 | 3 | 5437 | 5 | 5441 | 3 | 5443 | 2 | 5449 | 7 |
| 5471 | 7 | 5477 | 2 | 5479 | 3 | 5483 | 2 | 5501 | 2 | 5503 | 3 | 5507 | 2 | 5519 | 13 | 5521 | 11 |
| 5527 | 5 | 5531 | 10 | 5557 | 2 | 5563 | 2 | 5569 | 13 | 5573 | 2 | 5581 | 6 | 5591 | 11 | 5623 | 5 |
| 5639 | 7 | 5641 | 14 | 5647 | 3 | 5651 | 2 | 5653 | 5 | 5657 | 3 | 5659 | 2 | 5669 | 3 | 5683 | 2 |
| 5689 | 11 | 5693 | 2 | 5701 | 2 | 5711 | 19 | 5717 | 2 | 5737 | 5 | 5741 | 2 | 5743 | 10 | 5749 | 2 |
| 5779 | 2 | 5783 | 7 | 5791 | 6 | 5801 | 3 | 5807 | 5 | 5813 | 2 | 5821 | 6 | 5827 | 2 | 5839 | 6 |
| 5843 | 2 | 5849 | 3 | 5851 | 2 | 5857 | 7 | 5861 | 3 | 5867 | 5 | 5869 | 2 | 5879 | 11 | 5881 | 31 |
| 5897 | 3 | 5903 | 5 | 5923 | 2 | 5927 | 5 | 5939 | 2 | 5953 | 7 | 5981 | 3 | 5987 | 2 | 6007 | 3 |
| 6011 | 2 | 6029 | 2 | 6037 | 5 | 6043 | 5 | 6047 | 5 | 6053 | 2 | 6067 | 2 | 6073 | 10 | 6079 | 17 |
| 6089 | 3 | 6091 | 7 | 6101 | 2 | 6113 | 3 | 6121 | 7 | 6131 | 2 | 6133 | 5 | 6143 | 5 | 6151 | 3 |
| 6163 | 3 | 6173 | 2 | 6197 | 2 | 6199 | 3 | 6203 | 2 | 6211 | 2 | 6217 | 5 | 6221 | 3 | 6229 | 2 |
| 6247 | 5 | 6257 | 3 | 6263 | 5 | 6269 | 2 | 6271 | 11 | 6277 | 2 | 6287 | 7 | 6299 | 2 | 6301 | 10 |
| 6311 | 7 | 6317 | 2 | 6323 | 2 | 6329 | 3 | 6337 | 10 | 6343 | 3 | 6353 | 3 | 6359 | 13 | 6361 | 19 |
| 6367 | 3 | 6373 | 2 | 6379 | 2 | 6389 | 2 | 6397 | 2 | 6421 | 6 | 6427 | 3 | 6449 | 3 | 6451 | 3 |
| 6469 | 2 | 6473 | 3 | 6481 | 7 | 6491 | 2 | 6521 | 6 | 6529 | 7 | 6547 | 2 | 6551 | 17 | 6553 | 10 |

| $p$ | $\omega$ | $p$ | $\omega$ | $p$ | $\omega$ | $p$ | $\omega$ | $p$ | $\omega$ | $p$ | $\omega$ | $p$ | $\omega$ | $p$ | $\omega$ | $p$ | $\omega$ |
|---|---|---|---|---|---|---|---|---|---|---|---|---|---|---|---|---|---|
| 6563 | 5 | 6569 | 3 | 6571 | 3 | 6577 | 5 | 6581 | 14 | 6599 | 13 | 6607 | 3 | 6619 | 2 | 6637 | 2 |
| 6653 | 2 | 6659 | 2 | 6661 | 6 | 6673 | 5 | 6679 | 7 | 6689 | 3 | 6691 | 2 | 6701 | 2 | 6703 | 5 |
| 6709 | 2 | 6719 | 11 | 6733 | 2 | 6737 | 3 | 6761 | 3 | 6763 | 2 | 6779 | 2 | 6781 | 2 | 6791 | 7 |
| 6793 | 10 | 6803 | 2 | 6823 | 3 | 6827 | 2 | 6829 | 2 | 6833 | 3 | 6841 | 22 | 6857 | 3 | 6863 | 5 |
| 6869 | 2 | 6871 | 3 | 6883 | 2 | 6899 | 2 | 6907 | 2 | 6911 | 7 | 6917 | 2 | 6947 | 2 | 6949 | 2 |
| 6959 | 7 | 6961 | 13 | 6967 | 5 | 6971 | 2 | 6977 | 3 | 6983 | 5 | 6991 | 6 | 6997 | 5 | 7001 | 3 |
| 7013 | 2 | 7019 | 2 | 7027 | 2 | 7039 | 3 | 7043 | 2 | 7057 | 5 | 7069 | 2 | 7079 | 7 | 7103 | 5 |
| 7109 | 2 | 7121 | 3 | 7127 | 5 | 7129 | 7 | 7151 | 7 | 7159 | 3 | 7177 | 10 | 7187 | 2 | 7193 | 3 |
| 7207 | 3 | 7211 | 2 | 7213 | 5 | 7219 | 2 | 7229 | 2 | 7237 | 2 | 7243 | 2 | 7247 | 5 | 7253 | 2 |
| 7283 | 2 | 7297 | 5 | 7307 | 2 | 7309 | 6 | 7321 | 7 | 7331 | 2 | 7333 | 6 | 7349 | 2 | 7351 | 6 |
| 7369 | 7 | 7393 | 5 | 7411 | 2 | 7417 | 5 | 7433 | 3 | 7451 | 2 | 7457 | 3 | 7459 | 2 | 7477 | 2 |
| 7481 | 6 | 7487 | 5 | 7489 | 7 | 7499 | 2 | 7507 | 2 | 7517 | 2 | 7523 | 2 | 7529 | 3 | 7537 | 7 |
| 7541 | 2 | 7547 | 2 | 7549 | 2 | 7559 | 13 | 7561 | 13 | 7573 | 2 | 7577 | 3 | 7583 | 5 | 7589 | 2 |
| 7591 | 6 | 7603 | 2 | 7607 | 5 | 7621 | 2 | 7639 | 7 | 7643 | 2 | 7649 | 3 | 7669 | 2 | 7673 | 3 |
| 7681 | 17 | 7687 | 6 | 7691 | 2 | 7699 | 3 | 7703 | 5 | 7717 | 2 | 7723 | 3 | 7727 | 5 | 7741 | 7 |
| 7753 | 10 | 7757 | 2 | 7759 | 3 | 7789 | 2 | 7793 | 3 | 7817 | 3 | 7823 | 5 | 7829 | 2 | 7841 | 12 |
| 7853 | 2 | 7867 | 3 | 7873 | 5 | 7877 | 2 | 7879 | 3 | 7883 | 2 | 7901 | 2 | 7907 | 2 | 7919 | 7 |
| 7927 | 3 | 7933 | 2 | 7937 | 3 | 7949 | 2 | 7951 | 6 | 7963 | 5 | 7993 | 5 | 8009 | 3 | 8011 | 14 |
| 8017 | 5 | 8039 | 11 | 8053 | 2 | 8059 | 3 | 8069 | 2 | 8081 | 3 | 8087 | 5 | 8089 | 17 | 8093 | 2 |
| 8101 | 6 | 8111 | 11 | 8117 | 2 | 8123 | 2 | 8147 | 2 | 8161 | 7 | 8167 | 3 | 8171 | 2 | 8179 | 2 |
| 8191 | 17 | 8209 | 7 | 8219 | 2 | 8221 | 2 | 8231 | 11 | 8233 | 10 | 8237 | 2 | 8243 | 2 | 8263 | 3 |
| 8269 | 2 | 8273 | 3 | 8287 | 3 | 8291 | 2 | 8293 | 2 | 8297 | 3 | 8311 | 3 | 8317 | 6 | 8329 | 7 |
| 8353 | 5 | 8363 | 2 | 8369 | 3 | 8377 | 5 | 8387 | 2 | 8389 | 6 | 8419 | 3 | 8423 | 5 | 8429 | 2 |
| 8431 | 3 | 8443 | 2 | 8447 | 5 | 8461 | 6 | 8467 | 2 | 8501 | 7 | 8513 | 5 | 8521 | 13 | 8527 | 5 |
| 8537 | 3 | 8539 | 2 | 8543 | 5 | 8563 | 2 | 8573 | 2 | 8581 | 6 | 8597 | 2 | 8599 | 3 | 8609 | 3 |
| 8623 | 3 | 8627 | 2 | 8629 | 6 | 8641 | 17 | 8647 | 3 | 8663 | 5 | 8669 | 2 | 8677 | 2 | 8681 | 15 |
| 8689 | 13 | 8693 | 2 | 8699 | 2 | 8707 | 5 | 8713 | 5 | 8719 | 3 | 8731 | 2 | 8737 | 5 | 8741 | 2 |
| 8747 | 2 | 8753 | 3 | 8761 | 23 | 8779 | 11 | 8783 | 5 | 8803 | 2 | 8807 | 5 | 8819 | 2 | 8821 | 2 |
| 8831 | 7 | 8837 | 2 | 8839 | 3 | 8849 | 3 | 8861 | 2 | 8863 | 3 | 8867 | 2 | 8887 | 3 | 8893 | 5 |
| 8923 | 2 | 8929 | 11 | 8933 | 2 | 8941 | 6 | 8951 | 13 | 8963 | 2 | 8969 | 3 | 8971 | 2 | 8999 | 7 |
| 9001 | 7 | 9007 | 3 | 9011 | 2 | 9013 | 5 | 9029 | 2 | 9041 | 3 | 9043 | 3 | 9049 | 7 | 9059 | 2 |
| 9067 | 3 | 9091 | 3 | 9103 | 6 | 9109 | 10 | 9127 | 3 | 9133 | 6 | 9137 | 3 | 9151 | 3 | 9157 | 6 |
| 9161 | 3 | 9173 | 2 | 9181 | 2 | 9187 | 3 | 9199 | 3 | 9203 | 2 | 9209 | 3 | 9221 | 2 | 9227 | 2 |
| 9239 | 19 | 9241 | 13 | 9257 | 3 | 9277 | 5 | 9281 | 3 | 9283 | 2 | 9293 | 2 | 9311 | 7 | 9319 | 3 |
| 9323 | 2 | 9337 | 5 | 9341 | 2 | 9343 | 5 | 9349 | 2 | 9371 | 2 | 9377 | 3 | 9391 | 3 | 9397 | 2 |
| 9403 | 3 | 9413 | 3 | 9419 | 2 | 9421 | 2 | 9431 | 7 | 9433 | 5 | 9437 | 2 | 9439 | 22 | 9461 | 3 |
| 9463 | 3 | 9467 | 2 | 9473 | 3 | 9479 | 7 | 9491 | 2 | 9497 | 3 | 9511 | 3 | 9521 | 3 | 9533 | 2 |
| 9539 | 2 | 9547 | 2 | 9551 | 11 | 9587 | 2 | 9601 | 13 | 9613 | 2 | 9619 | 2 | 9623 | 5 | 9629 | 2 |
| 9631 | 3 | 9643 | 2 | 9649 | 7 | 9661 | 2 | 9677 | 2 | 9679 | 3 | 9689 | 3 | 9697 | 10 | 9719 | 17 |
| 9721 | 7 | 9733 | 2 | 9739 | 3 | 9743 | 5 | 9749 | 2 | 9767 | 5 | 9769 | 13 | 9781 | 6 | 9787 | 3 |
| 9791 | 11 | 9803 | 2 | 9811 | 3 | 9817 | 5 | 9829 | 10 | 9833 | 3 | 9839 | 7 | 9851 | 2 | 9857 | 5 |
| 9859 | 2 | 9871 | 3 | 9883 | 2 | 9887 | 5 | 9901 | 2 | 9907 | 2 | 9923 | 2 | 9929 | 3 | 9931 | 10 |
| 9941 | 2 | 9949 | 2 | 9967 | 3 | 9973 | 11 | | | | | | | | | | | | |

**Examples:**

**1.** Since $2^1 \equiv 2$, $2^2 \equiv 4$, and $2^3 \equiv 1 \pmod 7$, it follows that $\mathrm{ord}_7 2 = 3$.

**2.** The integers 2, 6, 7, 8 form a complete set of incongruent primitive roots modulo 11.

**3.** The integer 10 is a primitive root of 487, but it is not a primitive root of $487^2$.

**4.** There are $\phi(6)\phi(10) = 2 \cdot 4 = 8$ common primitive roots of 7 and 11 between 1 and $7 \cdot 11 = 77$. They are the integers 17, 19, 24, 40, 52, 61, 68, 73.

**5.** From Facts 12 and 13 it follows that the minimum universal exponent of 1,200 is $\lambda(7{,}200) = \lambda(2^5 \cdot 3^2 \cdot 5^2) = \mathrm{lcm}(2^3, \phi(3^2), \phi(5^2)) = \mathrm{lcm}(8, 6, 20) = 120$.

## 4.7.2  INDEX ARITHMETIC

**Definition:**

If $m$ is a positive integer with primitive root $r$ and $a$ is an integer relatively prime to $m$, then the unique nonnegative integer $x$ not exceeding $\phi(m)$ with $r^x \equiv a \pmod m$ is the **index of $a$ to the base $r$ modulo $m$**, or the **discrete logarithm of $a$ to the base $r$ modulo $m$**.

The index is denoted $\mathrm{ind}_r a$ (where the modulus $m$ is fixed).

**Facts:**

**1.** Table 1 displays, for each prime $p < 100$, the indices of all numbers not exceeding $p$ using the least primitive root $g$ of $p$ as the base. If $g^x = y$, the table on the left has a $y$ in position $x$, while the one on the right has an $x$ in position $y$.

### Table 1: Indices for primes less than 100.

3:

| N | 0 | 1 | 2 | 3 | 4 | 5 | 6 | 7 | 8 | 9 |
|---|---|---|---|---|---|---|---|---|---|---|
| 0 |   | 2 | 1 |   |   |   |   |   |   |   |

| I | 0 | 1 | 2 | 3 | 4 | 5 | 6 | 7 | 8 | 9 |
|---|---|---|---|---|---|---|---|---|---|---|
| 0 | 1 | 2 | 1 |   |   |   |   |   |   |   |

5:

| N | 0 | 1 | 2 | 3 | 4 | 5 | 6 | 7 | 8 | 9 |
|---|---|---|---|---|---|---|---|---|---|---|
| 0 |   | 4 | 1 | 3 | 2 |   |   |   |   |   |

| I | 0 | 1 | 2 | 3 | 4 | 5 | 6 | 7 | 8 | 9 |
|---|---|---|---|---|---|---|---|---|---|---|
| 0 | 1 | 2 | 4 | 3 | 1 |   |   |   |   |   |

7:

| N | 0 | 1 | 2 | 3 | 4 | 5 | 6 | 7 | 8 | 9 |
|---|---|---|---|---|---|---|---|---|---|---|
| 0 |   | 6 | 2 | 1 | 4 | 5 | 3 |   |   |   |

| I | 0 | 1 | 2 | 3 | 4 | 5 | 6 | 7 | 8 | 9 |
|---|---|---|---|---|---|---|---|---|---|---|
| 0 | 1 | 3 | 2 | 6 | 4 | 5 | 1 |   |   |   |

11:

| N | 0 | 1 | 2 | 3 | 4 | 5 | 6 | 7 | 8 | 9 |
|---|---|---|---|---|---|---|---|---|---|---|
| 0 |   | 10 | 1 | 8 | 2 | 4 | 9 | 7 | 3 | 6 |
| 1 | 5 |   |   |   |   |   |   |   |   |   |

| I | 0 | 1 | 2 | 3 | 4 | 5 | 6 | 7 | 8 | 9 |
|---|---|---|---|---|---|---|---|---|---|---|
| 0 | 1 | 2 | 4 | 8 | 5 | 10 | 9 | 7 | 3 | 6 |
| 1 | 1 |   |   |   |   |   |   |   |   |   |

13:

| N | 0 | 1 | 2 | 3 | 4 | 5 | 6 | 7 | 8 | 9 |
|---|---|---|---|---|---|---|---|---|---|---|
| 0 |   | 12 | 1 | 4 | 2 | 9 | 5 | 11 | 3 | 8 |
| 1 | 10 | 7 | 6 |   |   |   |   |   |   |   |

| I | 0 | 1 | 2 | 3 | 4 | 5 | 6 | 7 | 8 | 9 |
|---|---|---|---|---|---|---|---|---|---|---|
| 0 | 1 | 2 | 4 | 8 | 3 | 6 | 12 | 11 | 9 | 5 |
| 1 | 10 | 7 | 1 |   |   |   |   |   |   |   |

17:

| N | 0 | 1 | 2 | 3 | 4 | 5 | 6 | 7 | 8 | 9 |
|---|---|---|---|---|---|---|---|---|---|---|
| 0 |   | 16 | 14 | 1 | 12 | 5 | 15 | 11 | 10 | 2 |
| 1 | 3 | 7 | 13 | 4 | 9 | 6 | 8 |   |   |   |

| I | 0 | 1 | 2 | 3 | 4 | 5 | 6 | 7 | 8 | 9 |
|---|---|---|---|---|---|---|---|---|---|---|
| 0 | 1 | 3 | 9 | 10 | 13 | 5 | 15 | 11 | 16 | 14 |
| 1 | 8 | 7 | 4 | 12 | 2 | 6 | 1 |   |   |   |

19:

| N | 0 | 1 | 2 | 3 | 4 | 5 | 6 | 7 | 8 | 9 |
|---|---|---|---|---|---|---|---|---|---|---|
| 0 |   | 18 | 1 | 13 | 2 | 16 | 14 | 6 | 3 | 8 |
| 1 | 17 | 12 | 15 | 5 | 7 | 11 | 4 | 10 | 9 |  |

| I | 0 | 1 | 2 | 3 | 4 | 5 | 6 | 7 | 8 | 9 |
|---|---|---|---|---|---|---|---|---|---|---|
| 0 | 1 | 2 | 4 | 8 | 16 | 13 | 7 | 14 | 9 | 18 |
| 1 | 17 | 15 | 11 | 3 | 6 | 12 | 5 | 10 | 1 |  |

23:

| N | 0 | 1 | 2 | 3 | 4 | 5 | 6 | 7 | 8 | 9 |
|---|---|---|---|---|---|---|---|---|---|---|
| 0 |   | 22 | 2 | 16 | 4 | 1 | 18 | 19 | 6 | 10 |
| 1 | 3 | 9 | 20 | 14 | 21 | 17 | 8 | 7 | 12 | 15 |
| 2 | 5 | 13 | 11 |  |  |  |  |  |  |  |

| I | 0 | 1 | 2 | 3 | 4 | 5 | 6 | 7 | 8 | 9 |
|---|---|---|---|---|---|---|---|---|---|---|
| 0 | 1 | 5 | 2 | 10 | 4 | 20 | 8 | 17 | 16 | 11 |
| 1 | 9 | 22 | 18 | 21 | 13 | 19 | 3 | 15 | 6 | 7 |
| 2 | 12 | 14 | 1 |  |  |  |  |  |  |  |

29:

| N | 0 | 1 | 2 | 3 | 4 | 5 | 6 | 7 | 8 | 9 |
|---|---|---|---|---|---|---|---|---|---|---|
| 0 |   | 28 | 1 | 5 | 2 | 22 | 6 | 12 | 3 | 10 |
| 1 | 23 | 25 | 7 | 18 | 13 | 27 | 4 | 21 | 11 | 9 |
| 2 | 24 | 17 | 26 | 20 | 8 | 16 | 19 | 15 | 14 |  |

| I | 0 | 1 | 2 | 3 | 4 | 5 | 6 | 7 | 8 | 9 |
|---|---|---|---|---|---|---|---|---|---|---|
| 0 | 1 | 2 | 4 | 8 | 16 | 3 | 6 | 12 | 24 | 19 |
| 1 | 9 | 18 | 7 | 14 | 28 | 27 | 25 | 21 | 13 | 26 |
| 2 | 23 | 17 | 5 | 10 | 20 | 11 | 22 | 15 | 1 |  |

31:

| N | 0 | 1 | 2 | 3 | 4 | 5 | 6 | 7 | 8 | 9 |
|---|---|---|---|---|---|---|---|---|---|---|
| 0 |   | 30 | 24 | 1 | 18 | 20 | 25 | 28 | 12 | 2 |
| 1 | 14 | 23 | 19 | 11 | 22 | 21 | 6 | 7 | 26 | 4 |
| 2 | 8 | 29 | 17 | 27 | 13 | 10 | 5 | 3 | 16 | 9 |
| 3 | 15 |  |  |  |  |  |  |  |  |  |

| I | 0 | 1 | 2 | 3 | 4 | 5 | 6 | 7 | 8 | 9 |
|---|---|---|---|---|---|---|---|---|---|---|
| 0 | 1 | 3 | 9 | 27 | 19 | 26 | 16 | 17 | 20 | 29 |
| 1 | 25 | 13 | 8 | 24 | 10 | 30 | 28 | 22 | 4 | 12 |
| 2 | 5 | 15 | 14 | 11 | 2 | 6 | 18 | 23 | 7 | 21 |
| 3 | 1 |  |  |  |  |  |  |  |  |  |

37:

| N | 0 | 1 | 2 | 3 | 4 | 5 | 6 | 7 | 8 | 9 |
|---|---|---|---|---|---|---|---|---|---|---|
| 0 |   | 36 | 1 | 26 | 2 | 23 | 27 | 32 | 3 | 16 |
| 1 | 24 | 30 | 28 | 11 | 33 | 13 | 4 | 7 | 17 | 35 |
| 2 | 25 | 22 | 31 | 15 | 29 | 10 | 12 | 6 | 34 | 21 |
| 3 | 14 | 9 | 5 | 20 | 8 | 19 | 18 |  |  |  |

| I | 0 | 1 | 2 | 3 | 4 | 5 | 6 | 7 | 8 | 9 |
|---|---|---|---|---|---|---|---|---|---|---|
| 0 | 1 | 2 | 4 | 8 | 16 | 32 | 27 | 17 | 34 | 31 |
| 1 | 25 | 13 | 26 | 15 | 30 | 23 | 9 | 18 | 36 | 35 |
| 2 | 33 | 29 | 21 | 5 | 10 | 20 | 3 | 6 | 12 | 26 |
| 3 | 11 | 22 | 7 | 14 | 28 | 19 | 1 |  |  |  |

41:

| N | 0 | 1 | 2 | 3 | 4 | 5 | 6 | 7 | 8 | 9 |
|---|---|---|---|---|---|---|---|---|---|---|
| 0 |   | 40 | 26 | 15 | 12 | 22 | 1 | 39 | 38 | 30 |
| 1 | 8 | 3 | 27 | 31 | 25 | 37 | 24 | 33 | 16 | 9 |
| 2 | 34 | 14 | 29 | 36 | 13 | 4 | 17 | 5 | 11 | 7 |
| 3 | 23 | 28 | 10 | 18 | 19 | 21 | 2 | 32 | 35 | 6 |
| 4 | 20 |  |  |  |  |  |  |  |  |  |

| I | 0 | 1 | 2 | 3 | 4 | 5 | 6 | 7 | 8 | 9 |
|---|---|---|---|---|---|---|---|---|---|---|
| 0 | 1 | 6 | 36 | 11 | 25 | 27 | 39 | 29 | 10 | 19 |
| 1 | 32 | 28 | 4 | 24 | 21 | 3 | 18 | 26 | 33 | 34 |
| 2 | 40 | 35 | 5 | 30 | 16 | 14 | 2 | 12 | 31 | 22 |
| 3 | 9 | 13 | 37 | 17 | 20 | 38 | 23 | 15 | 8 | 7 |
| 4 | 1 |  |  |  |  |  |  |  |  |  |

43:

| N | 0 | 1 | 2 | 3 | 4 | 5 | 6 | 7 | 8 | 9 |
|---|---|---|---|---|---|---|---|---|---|---|
| 0 |   | 42 | 27 | 1 | 12 | 25 | 28 | 35 | 39 | 2 |
| 1 | 10 | 30 | 13 | 32 | 20 | 26 | 24 | 38 | 29 | 19 |
| 2 | 37 | 36 | 15 | 16 | 40 | 8 | 17 | 3 | 5 | 41 |
| 3 | 11 | 34 | 9 | 31 | 23 | 18 | 14 | 7 | 4 | 33 |
| 4 | 22 | 6 | 21 |  |  |  |  |  |  |  |

| I | 0 | 1 | 2 | 3 | 4 | 5 | 6 | 7 | 8 | 9 |
|---|---|---|---|---|---|---|---|---|---|---|
| 0 | 1 | 3 | 9 | 27 | 38 | 28 | 41 | 37 | 25 | 32 |
| 1 | 10 | 30 | 4 | 12 | 36 | 22 | 23 | 26 | 35 | 19 |
| 2 | 14 | 42 | 40 | 34 | 16 | 5 | 15 | 2 | 6 | 18 |
| 3 | 11 | 33 | 13 | 39 | 31 | 7 | 21 | 20 | 17 | 8 |
| 4 | 24 | 29 | 1 |  |  |  |  |  |  |  |

47:

| N | 0 | 1 | 2 | 3 | 4 | 5 | 6 | 7 | 8 | 9 |
|---|---|---|---|---|---|---|---|---|---|---|
| 0 |   | 46 | 18 | 20 | 36 | 1 | 38 | 32 | 8 | 40 |
| 1 | 19 | 7 | 10 | 11 | 4 | 21 | 26 | 16 | 12 | 45 |
| 2 | 37 | 6 | 25 | 5 | 28 | 2 | 29 | 14 | 22 | 35 |
| 3 | 39 | 3 | 44 | 27 | 34 | 33 | 30 | 42 | 17 | 31 |
| 4 | 9 | 15 | 24 | 13 | 43 | 41 | 23 |  |  |  |

| I | 0 | 1 | 2 | 3 | 4 | 5 | 6 | 7 | 8 | 9 |
|---|---|---|---|---|---|---|---|---|---|---|
| 0 | 1 | 5 | 25 | 31 | 14 | 23 | 21 | 11 | 8 | 40 |
| 1 | 12 | 13 | 18 | 43 | 27 | 41 | 17 | 38 | 2 | 10 |
| 2 | 3 | 15 | 28 | 46 | 42 | 22 | 16 | 33 | 24 | 26 |
| 3 | 36 | 39 | 7 | 35 | 34 | 29 | 4 | 20 | 6 | 30 |
| 4 | 9 | 45 | 37 | 44 | 32 | 19 | 1 |  |  |  |

**53:**

| N | 0 | 1 | 2 | 3 | 4 | 5 | 6 | 7 | 8 | 9 |
|---|---|---|---|---|---|---|---|---|---|---|
| 0 |   | 52 | 1 | 17 | 2 | 47 | 18 | 14 | 3 | 34 |
| 1 | 48 | 6 | 19 | 24 | 15 | 12 | 4 | 10 | 35 | 37 |
| 2 | 49 | 31 | 7 | 39 | 20 | 42 | 25 | 51 | 16 | 46 |
| 3 | 13 | 33 | 5 | 23 | 11 | 9 | 36 | 30 | 38 | 41 |
| 4 | 50 | 45 | 32 | 22 | 8 | 29 | 40 | 44 | 21 | 28 |
| 5 | 43 | 27 | 26 |   |   |   |   |   |   |   |

| I | 0 | 1 | 2 | 3 | 4 | 5 | 6 | 7 | 8 | 9 |
|---|---|---|---|---|---|---|---|---|---|---|
| 0 | 1 | 2 | 4 | 8 | 16 | 32 | 11 | 22 | 44 | 35 |
| 1 | 17 | 34 | 15 | 30 | 7 | 14 | 28 | 3 | 6 | 12 |
| 2 | 24 | 48 | 43 | 33 | 13 | 26 | 52 | 51 | 49 | 45 |
| 3 | 37 | 21 | 42 | 31 | 9 | 18 | 36 | 19 | 38 | 23 |
| 4 | 46 | 39 | 25 | 50 | 47 | 41 | 29 | 5 | 10 | 20 |
| 5 | 40 | 27 | 1 |   |   |   |   |   |   |   |

**59:**

| N | 0 | 1 | 2 | 3 | 4 | 5 | 6 | 7 | 8 | 9 |
|---|---|---|---|---|---|---|---|---|---|---|
| 0 |   | 58 | 1 | 50 | 2 | 6 | 51 | 18 | 3 | 42 |
| 1 | 7 | 25 | 52 | 45 | 19 | 56 | 4 | 40 | 43 | 38 |
| 2 | 8 | 10 | 26 | 15 | 53 | 12 | 46 | 34 | 20 | 28 |
| 3 | 57 | 49 | 5 | 17 | 41 | 24 | 44 | 55 | 39 | 37 |
| 4 | 9 | 14 | 11 | 33 | 27 | 48 | 16 | 23 | 54 | 36 |
| 5 | 13 | 32 | 47 | 22 | 35 | 31 | 21 | 30 | 29 |   |

| I | 0 | 1 | 2 | 3 | 4 | 5 | 6 | 7 | 8 | 9 |
|---|---|---|---|---|---|---|---|---|---|---|
| 0 | 1 | 2 | 4 | 8 | 16 | 32 | 5 | 10 | 20 | 40 |
| 1 | 21 | 42 | 25 | 50 | 41 | 23 | 46 | 33 | 7 | 14 |
| 2 | 28 | 56 | 53 | 47 | 35 | 11 | 22 | 44 | 29 | 58 |
| 3 | 57 | 55 | 51 | 43 | 27 | 54 | 49 | 39 | 19 | 38 |
| 4 | 17 | 34 | 9 | 18 | 36 | 13 | 26 | 52 | 45 | 31 |
| 5 | 3 | 6 | 12 | 24 | 48 | 37 | 15 | 30 | 1 |   |

**61:**

| N | 0 | 1 | 2 | 3 | 4 | 5 | 6 | 7 | 8 | 9 |
|---|---|---|---|---|---|---|---|---|---|---|
| 0 |   | 60 | 1 | 6 | 2 | 22 | 7 | 49 | 3 | 12 |
| 1 | 23 | 15 | 8 | 40 | 50 | 28 | 4 | 47 | 13 | 26 |
| 2 | 24 | 55 | 16 | 57 | 9 | 44 | 41 | 18 | 51 | 35 |
| 3 | 29 | 59 | 5 | 21 | 48 | 11 | 14 | 39 | 27 | 46 |
| 4 | 25 | 54 | 56 | 43 | 17 | 34 | 58 | 20 | 10 | 38 |
| 5 | 45 | 53 | 42 | 33 | 19 | 37 | 52 | 32 | 36 | 31 |
| 6 | 30 |   |   |   |   |   |   |   |   |   |

| I | 0 | 1 | 2 | 3 | 4 | 5 | 6 | 7 | 8 | 9 |
|---|---|---|---|---|---|---|---|---|---|---|
| 0 | 1 | 2 | 4 | 8 | 16 | 32 | 3 | 6 | 12 | 24 |
| 1 | 48 | 35 | 9 | 18 | 36 | 11 | 22 | 44 | 27 | 54 |
| 2 | 47 | 33 | 5 | 10 | 20 | 40 | 19 | 38 | 15 | 30 |
| 3 | 60 | 59 | 57 | 53 | 45 | 29 | 58 | 55 | 49 | 37 |
| 4 | 13 | 26 | 52 | 43 | 25 | 50 | 39 | 17 | 34 | 7 |
| 5 | 14 | 28 | 56 | 51 | 41 | 21 | 42 | 23 | 46 | 31 |
| 6 | 1 |   |   |   |   |   |   |   |   |   |

**67:**

| N | 0 | 1 | 2 | 3 | 4 | 5 | 6 | 7 | 8 | 9 |
|---|---|---|---|---|---|---|---|---|---|---|
| 0 |   | 66 | 1 | 39 | 2 | 15 | 40 | 23 | 3 | 12 |
| 1 | 16 | 59 | 41 | 19 | 24 | 54 | 4 | 64 | 13 | 10 |
| 2 | 17 | 62 | 60 | 28 | 42 | 30 | 20 | 51 | 25 | 44 |
| 3 | 55 | 47 | 5 | 32 | 65 | 38 | 14 | 22 | 11 | 58 |
| 4 | 18 | 53 | 63 | 9 | 61 | 27 | 29 | 50 | 43 | 46 |
| 5 | 31 | 37 | 21 | 57 | 52 | 8 | 26 | 49 | 45 | 36 |
| 6 | 56 | 7 | 48 | 35 | 6 | 34 | 33 |   |   |   |

| I | 0 | 1 | 2 | 3 | 4 | 5 | 6 | 7 | 8 | 9 |
|---|---|---|---|---|---|---|---|---|---|---|
| 0 | 1 | 2 | 4 | 8 | 16 | 32 | 64 | 61 | 55 | 43 |
| 1 | 19 | 38 | 9 | 18 | 36 | 5 | 10 | 20 | 40 | 13 |
| 2 | 26 | 52 | 37 | 7 | 14 | 28 | 56 | 45 | 23 | 46 |
| 3 | 25 | 50 | 33 | 66 | 65 | 63 | 59 | 51 | 35 | 3 |
| 4 | 6 | 12 | 24 | 48 | 29 | 58 | 49 | 31 | 62 | 57 |
| 5 | 47 | 27 | 54 | 41 | 15 | 30 | 60 | 53 | 39 | 11 |
| 6 | 22 | 44 | 21 | 42 | 17 | 34 | 1 |   |   |   |

**71:**

| N | 0 | 1 | 2 | 3 | 4 | 5 | 6 | 7 | 8 | 9 |
|---|---|---|---|---|---|---|---|---|---|---|
| 0 |   | 70 | 6 | 26 | 12 | 28 | 32 | 1 | 18 | 52 |
| 1 | 34 | 31 | 38 | 39 | 7 | 54 | 24 | 49 | 58 | 16 |
| 2 | 40 | 27 | 37 | 15 | 44 | 56 | 45 | 8 | 13 | 68 |
| 3 | 60 | 11 | 30 | 57 | 55 | 29 | 64 | 20 | 22 | 65 |
| 4 | 46 | 25 | 33 | 48 | 43 | 10 | 21 | 9 | 50 | 2 |
| 5 | 62 | 5 | 51 | 23 | 14 | 59 | 19 | 42 | 4 | 3 |
| 6 | 66 | 69 | 17 | 53 | 36 | 67 | 63 | 47 | 61 | 41 |
| 7 | 35 |   |   |   |   |   |   |   |   |   |

| I | 0 | 1 | 2 | 3 | 4 | 5 | 6 | 7 | 8 | 9 |
|---|---|---|---|---|---|---|---|---|---|---|
| 0 | 1 | 7 | 49 | 59 | 58 | 51 | 2 | 14 | 27 | 47 |
| 1 | 45 | 31 | 4 | 28 | 54 | 23 | 19 | 62 | 8 | 56 |
| 2 | 37 | 46 | 38 | 53 | 16 | 41 | 3 | 21 | 5 | 35 |
| 3 | 32 | 11 | 6 | 42 | 10 | 70 | 64 | 22 | 12 | 13 |
| 4 | 20 | 69 | 57 | 44 | 24 | 26 | 40 | 67 | 43 | 17 |
| 5 | 48 | 52 | 9 | 63 | 15 | 34 | 25 | 33 | 18 | 55 |
| 6 | 30 | 68 | 50 | 66 | 36 | 39 | 60 | 65 | 29 | 61 |
| 7 | 1 |   |   |   |   |   |   |   |   |   |

73:

| N | 0 | 1 | 2 | 3 | 4 | 5 | 6 | 7 | 8 | 9 |
|---|---|---|---|---|---|---|---|---|---|---|
| 0 |   | 72 | 8 | 6 | 16 | 1 | 14 | 33 | 24 | 12 |
| 1 | 9 | 55 | 22 | 59 | 41 | 7 | 32 | 21 | 20 | 62 |
| 2 | 17 | 39 | 63 | 46 | 30 | 2 | 67 | 18 | 49 | 35 |
| 3 | 15 | 11 | 40 | 61 | 29 | 34 | 28 | 64 | 70 | 65 |
| 4 | 25 | 4 | 47 | 51 | 71 | 13 | 54 | 31 | 38 | 66 |
| 5 | 10 | 27 | 3 | 53 | 26 | 56 | 57 | 68 | 43 | 5 |
| 6 | 23 | 58 | 19 | 45 | 48 | 60 | 69 | 50 | 37 | 52 |
| 7 | 42 | 44 | 36 | | | | | | | |

| I | 0 | 1 | 2 | 3 | 4 | 5 | 6 | 7 | 8 | 9 |
|---|---|---|---|---|---|---|---|---|---|---|
| 0 | 1 | 5 | 25 | 52 | 41 | 59 | 3 | 15 | 2 | 10 |
| 1 | 50 | 31 | 9 | 45 | 6 | 30 | 4 | 20 | 27 | 62 |
| 2 | 18 | 17 | 12 | 60 | 8 | 40 | 54 | 51 | 36 | 34 |
| 3 | 24 | 47 | 16 | 7 | 35 | 29 | 72 | 68 | 48 | 21 |
| 4 | 32 | 14 | 70 | 58 | 71 | 63 | 23 | 42 | 64 | 28 |
| 5 | 67 | 43 | 69 | 53 | 46 | 11 | 55 | 56 | 61 | 13 |
| 6 | 65 | 33 | 19 | 22 | 37 | 39 | 49 | 26 | 57 | 66 |
| 7 | 38 | 44 | 1 | | | | | | | |

79:

| N | 0 | 1 | 2 | 3 | 4 | 5 | 6 | 7 | 8 | 9 |
|---|---|---|---|---|---|---|---|---|---|---|
| 0 |   | 78 | 4 | 1 | 8 | 62 | 5 | 53 | 12 | 2 |
| 1 | 66 | 68 | 9 | 34 | 57 | 63 | 16 | 21 | 6 | 32 |
| 2 | 70 | 54 | 72 | 26 | 13 | 46 | 38 | 3 | 61 | 11 |
| 3 | 67 | 56 | 20 | 69 | 25 | 37 | 10 | 19 | 36 | 35 |
| 4 | 74 | 75 | 58 | 49 | 76 | 64 | 30 | 59 | 17 | 28 |
| 5 | 50 | 22 | 42 | 77 | 7 | 52 | 65 | 33 | 15 | 31 |
| 6 | 71 | 45 | 60 | 55 | 24 | 18 | 73 | 48 | 29 | 27 |
| 7 | 41 | 51 | 14 | 44 | 23 | 47 | 40 | 43 | 39 | |

| I | 0 | 1 | 2 | 3 | 4 | 5 | 6 | 7 | 8 | 9 |
|---|---|---|---|---|---|---|---|---|---|---|
| 0 | 1 | 3 | 9 | 27 | 2 | 6 | 18 | 54 | 4 | 12 |
| 1 | 36 | 29 | 8 | 24 | 72 | 58 | 16 | 48 | 65 | 37 |
| 2 | 32 | 17 | 51 | 74 | 64 | 34 | 23 | 69 | 49 | 68 |
| 3 | 46 | 59 | 19 | 57 | 13 | 39 | 38 | 35 | 26 | 78 |
| 4 | 76 | 70 | 52 | 77 | 73 | 61 | 25 | 75 | 67 | 43 |
| 5 | 50 | 71 | 55 | 7 | 21 | 63 | 31 | 14 | 42 | 47 |
| 6 | 62 | 28 | 5 | 15 | 45 | 56 | 10 | 30 | 11 | 33 |
| 7 | 20 | 60 | 22 | 66 | 40 | 41 | 44 | 53 | 1 | |

83:

| N | 0 | 1 | 2 | 3 | 4 | 5 | 6 | 7 | 8 | 9 |
|---|---|---|---|---|---|---|---|---|---|---|
| 0 |   | 82 | 1 | 72 | 2 | 27 | 73 | 8 | 3 | 62 |
| 1 | 28 | 24 | 74 | 77 | 9 | 17 | 4 | 56 | 63 | 47 |
| 2 | 29 | 80 | 25 | 60 | 75 | 54 | 78 | 52 | 10 | 12 |
| 3 | 18 | 38 | 5 | 14 | 57 | 35 | 64 | 20 | 48 | 67 |
| 4 | 30 | 40 | 81 | 71 | 26 | 7 | 61 | 23 | 76 | 16 |
| 5 | 55 | 46 | 79 | 59 | 53 | 51 | 11 | 37 | 13 | 34 |
| 6 | 19 | 66 | 39 | 70 | 6 | 22 | 15 | 45 | 58 | 50 |
| 7 | 36 | 33 | 65 | 69 | 21 | 44 | 49 | 32 | 68 | 43 |
| 8 | 31 | 42 | 41 | | | | | | | |

| I | 0 | 1 | 2 | 3 | 4 | 5 | 6 | 7 | 8 | 9 |
|---|---|---|---|---|---|---|---|---|---|---|
| 0 | 1 | 2 | 4 | 8 | 16 | 32 | 64 | 45 | 7 | 14 |
| 1 | 28 | 56 | 29 | 58 | 33 | 66 | 49 | 15 | 30 | 60 |
| 2 | 37 | 74 | 65 | 47 | 11 | 22 | 44 | 5 | 10 | 20 |
| 3 | 40 | 80 | 77 | 71 | 59 | 35 | 70 | 57 | 31 | 62 |
| 4 | 41 | 82 | 81 | 79 | 75 | 67 | 51 | 19 | 38 | 76 |
| 5 | 69 | 55 | 27 | 54 | 25 | 50 | 17 | 34 | 68 | 53 |
| 6 | 23 | 46 | 9 | 18 | 36 | 72 | 61 | 39 | 78 | 73 |
| 7 | 63 | 43 | 3 | 6 | 12 | 24 | 48 | 13 | 26 | 52 |
| 8 | 21 | 42 | 1 | | | | | | | |

89:

| N | 0 | 1 | 2 | 3 | 4 | 5 | 6 | 7 | 8 | 9 |
|---|---|---|---|---|---|---|---|---|---|---|
| 0 |   | 88 | 16 | 1 | 32 | 70 | 17 | 81 | 48 | 2 |
| 1 | 86 | 84 | 33 | 23 | 9 | 71 | 64 | 6 | 18 | 35 |
| 2 | 14 | 82 | 12 | 57 | 49 | 52 | 39 | 3 | 25 | 59 |
| 3 | 87 | 31 | 80 | 85 | 22 | 63 | 34 | 11 | 51 | 24 |
| 4 | 30 | 21 | 10 | 29 | 28 | 72 | 73 | 54 | 65 | 74 |
| 5 | 68 | 7 | 55 | 78 | 19 | 66 | 41 | 36 | 75 | 43 |
| 6 | 15 | 69 | 47 | 83 | 8 | 5 | 13 | 56 | 38 | 58 |
| 7 | 79 | 62 | 50 | 20 | 27 | 53 | 67 | 77 | 40 | 42 |
| 8 | 46 | 4 | 37 | 61 | 26 | 76 | 45 | 60 | 44 | |

| I | 0 | 1 | 2 | 3 | 4 | 5 | 6 | 7 | 8 | 9 |
|---|---|---|---|---|---|---|---|---|---|---|
| 0 | 1 | 3 | 9 | 27 | 81 | 65 | 17 | 51 | 64 | 14 |
| 1 | 42 | 37 | 22 | 66 | 20 | 60 | 2 | 6 | 18 | 54 |
| 2 | 73 | 41 | 34 | 13 | 39 | 28 | 84 | 74 | 44 | 43 |
| 3 | 40 | 31 | 4 | 12 | 36 | 19 | 57 | 82 | 68 | 26 |
| 4 | 78 | 56 | 79 | 59 | 88 | 86 | 80 | 62 | 8 | 24 |
| 5 | 72 | 38 | 25 | 75 | 47 | 52 | 67 | 23 | 69 | 29 |
| 6 | 87 | 83 | 71 | 35 | 16 | 48 | 55 | 76 | 50 | 61 |
| 7 | 5 | 15 | 45 | 46 | 49 | 58 | 85 | 77 | 53 | 70 |
| 8 | 32 | 7 | 21 | 63 | 11 | 33 | 10 | 30 | 1 | |

97:

| N | 0 | 1 | 2 | 3 | 4 | 5 | 6 | 7 | 8 | 9 |
|---|---|---|---|---|---|---|---|---|---|---|
| 0 |    | 96 | 34 | 70 | 68 | 1 | 8 | 31 | 6 | 44 |
| 1 | 35 | 86 | 42 | 25 | 65 | 71 | 40 | 89 | 78 | 81 |
| 2 | 69 | 5 | 24 | 77 | 76 | 2 | 59 | 18 | 3 | 13 |
| 3 | 9 | 46 | 74 | 60 | 27 | 32 | 16 | 91 | 19 | 95 |
| 4 | 7 | 85 | 39 | 4 | 58 | 45 | 15 | 84 | 14 | 62 |
| 5 | 36 | 63 | 93 | 10 | 52 | 87 | 37 | 55 | 47 | 67 |
| 6 | 43 | 64 | 80 | 75 | 12 | 26 | 94 | 57 | 61 | 51 |
| 7 | 66 | 11 | 50 | 28 | 29 | 72 | 53 | 21 | 33 | 30 |
| 8 | 41 | 88 | 23 | 17 | 73 | 90 | 38 | 83 | 92 | 54 |
| 9 | 79 | 56 | 49 | 20 | 22 | 82 | 48 |  |  |  |

| I | 0 | 1 | 2 | 3 | 4 | 5 | 6 | 7 | 8 | 9 |
|---|---|---|---|---|---|---|---|---|---|---|
| 0 | 1 | 5 | 25 | 28 | 43 | 21 | 8 | 40 | 6 | 30 |
| 1 | 53 | 71 | 64 | 29 | 48 | 46 | 36 | 83 | 27 | 38 |
| 2 | 93 | 77 | 94 | 82 | 22 | 13 | 65 | 34 | 73 | 74 |
| 3 | 79 | 7 | 35 | 78 | 2 | 10 | 50 | 56 | 86 | 42 |
| 4 | 16 | 80 | 12 | 60 | 9 | 45 | 31 | 58 | 96 | 92 |
| 5 | 72 | 69 | 54 | 76 | 89 | 57 | 91 | 67 | 44 | 26 |
| 6 | 33 | 68 | 49 | 51 | 61 | 14 | 70 | 59 | 4 | 20 |
| 7 | 3 | 15 | 75 | 84 | 32 | 63 | 24 | 23 | 18 | 90 |
| 8 | 62 | 19 | 95 | 87 | 47 | 41 | 11 | 55 | 81 | 17 |
| 9 | 85 | 37 | 88 | 52 | 66 | 39 | 1 |  |  |  |

**2.** If $m$ is a positive integer with primitive root $r$ and $a$ is a positive integer relatively prime to $m$, then $a \equiv r^{\text{ind}_r a} \pmod{m}$.

**3.** If $m$ is a positive integer with primitive root $r$, then $\text{ind}_r 1 = 0$ and $\text{ind}_r r = 1$.

**4.** If $m > 2$ is an integer with primitive root $r$, then $\text{ind}_r(-1) = \frac{\phi(m)}{2}$.

**5.** If $m$ is a positive integer with primitive root $r$, and $a$ and $b$ are integers relatively prime to $m$, then

- $\text{ind}_r 1 \equiv 0 \pmod{\phi(m)}$;
- $\text{ind}_r(ab) \equiv \text{ind}_r a + \text{ind}_r b \pmod{\phi(m)}$;
- $\text{ind}_r a^k \equiv k \cdot \text{ind}_r a \pmod{\phi(m)}$ if $k$ is a positive integer.

**6.** If $m$ is a positive integer, and if $r$ and $s$ are both primitive roots modulo $m$, then $\text{ind}_r a \equiv \text{ind}_s a \cdot \text{ind}_r s \pmod{\phi(m)}$.

**7.** If $m$ is a positive integer with primitive root $r$, and $a$ and $b$ are integers both relatively prime to $m$, then the exponential congruence $a^x \equiv b \pmod{m}$ has a solution if and only if $\gcd(\text{ind}_r a, \phi(m)) \mid \text{ind}_r b$. Furthermore, if there is a solution to this exponential congruence, then there are exactly $\gcd(\text{ind}_r a, \phi(m))$ incongruent solutions.

**8.** There are a wide variety of algorithms for computing discrete logarithms, including those known as the baby-step, giant-step algorithm, the Pollard rho algorithm, the Pollig-Hellman algorithm, and the index-calculus algorithm. (See [MeVoVa96].)

**9.** The fastest algorithms known for computing discrete logarithms, relative to a fixed primitive root, of a given prime $p$ are index-calculus algorithms, which have subexponential computational complexity. In particular, there is an algorithm based on the number field sieve that runs using $L_p(\frac{1}{3}, 1.923) = O(\exp((1.923 + o(1))(\log p)^{\frac{1}{3}}(\log \log p)^{\frac{2}{3}}))$ bit operations. (See [MeVoVa96].)

**10.** Many cryptographic methods rely on the intractability of finding discrete logarithms of integers relative to a fixed primitive root $r$ of a fixed prime $p$.

**Examples:**

**1.** To solve $3x^{30} \equiv 4 \pmod{37}$ take indices to the base 2 (2 is the smallest primitive root of 37) to obtain $\text{ind}_2(3x^{30}) \equiv \text{ind}_2 4 = 2 \pmod{36}$. Since $\text{ind}_2(3x^{30}) \equiv \text{ind}_2 3 + 30 \cdot \text{ind}_2 x = 26 + 30 \cdot \text{ind}_2 x \pmod{36}$, it follows that $30 \cdot \text{ind}_2 x \equiv 12 \pmod{36}$. The solutions to this congruence are those $x$ such that $\text{ind}_2(x) \equiv 4, 10, 16, 22, 28, 34 \pmod{36}$. From the table of indices (Table 1), the solutions are those $x$ with $x \equiv 16, 25, 9, 21, 12, 28 \pmod{37}$.

**2.** To solve $7^x \equiv 6 \pmod{17}$ take indices to the base 3 (3 is the smallest primitive root of 17) to obtain   $\text{ind}_3(7^x) \equiv \text{ind}_3 6 = 15 \pmod{16}$. Since $\text{ind}_3(7^x) \equiv x \cdot \text{ind}_3 7 \equiv$

$11x \pmod{16}$, it follows that $11x \equiv 15 \pmod{16}$. Since all the steps in this computation are reversible, it follows that the solutions of the original congruence are the solutions of this linear congruence, namely those $x$ with $x \equiv 13 \pmod{16}$.

---

### 4.7.3   QUADRATIC RESIDUES

**Definitions:**

If $m$ and $k$ are positive integers and $a$ is an integer relatively prime to $m$, then $a$ is a **$k$th power residue** of $m$ if the congruence $x^k \equiv a \pmod{m}$ has a solution.

If $a$ and $m$ are relatively prime integers and $m$ is positive, then $a$ is a **quadratic residue** of $m$ if the congruence $x^2 \equiv a \pmod{m}$ has a solution. If $x^2 \equiv a \pmod{m}$ has no solution, then $a$ is a **quadratic nonresidue** of $m$.

If $p$ is an odd prime and $p$ does not divide $a$, then the **Legendre symbol** $\left(\frac{a}{p}\right)$ is 1 if $a$ is a quadratic residue of $p$ and $-1$ if $a$ is a quadratic nonresidue of $p$. This symbol is named after the French mathematician Adrien-Marie Legendre (1752–1833).

If $n$ is an odd positive integer with prime-power factorization $n = p_1^{t_1} p_2^{t_2} \ldots p_m^{t_m}$ and $a$ is an integer relatively prime to $n$, then the **Jacobi symbol** $\left(\frac{a}{n}\right)$ is defined by

$$\left(\tfrac{a}{n}\right) = \prod_{i=1}^{m} \left(\tfrac{a}{p_i}\right)^{t_i},$$

where the symbols on the right-hand side of the equality are Legendre symbols. This symbol is named after the German mathematician Carl Gustav Jacob Jacobi (1804–1851).

Let $a$ be a positive integer that is not a perfect square and such that $a \equiv 0$ or $1 \pmod{4}$. The **Kronecker symbol** (named after the German mathematician Leopold Kronecker (1823–1891)), which is a generalization of the Legendre symbol, is defined as

- $\left(\tfrac{a}{2}\right) = \begin{cases} 1 & \text{if } a \equiv 1 \pmod{8} \\ -1 & \text{if } a \equiv 5 \pmod{8} \end{cases}$

- $\left(\tfrac{a}{p}\right) =$ the Legendre symbol $\left(\tfrac{a}{p}\right)$ if $p$ is an odd prime such that $p$ does not divide $a$

- $\left(\tfrac{a}{n}\right) = \prod_{j=1}^{r} \left(\tfrac{a}{p_j}\right)^{t_j}$ if $\gcd(a, n) = 1$ and $n = \prod_{j=1}^{r} p_j^{t_j}$ is the prime factorization of $n$.

**Facts:**

**1.** If $p$ is an odd prime, then there are an equal number of quadratic residues modulo $p$ and quadratic nonresidues modulo $p$ among the integers $1, 2, \ldots, p-1$. In particular, there are $\frac{p-1}{2}$ integers of each type in this set.

**2.** *Euler's criterion:* If $p$ is an odd prime and $a$ is a positive integer not divisible by $p$, then $\left(\tfrac{a}{p}\right) \equiv a^{(p-1)/2} \pmod{p}$.

**3.** If $p$ is an odd prime, and $a$ and $b$ are integers not divisible by $p$ with $a \equiv b \pmod{p}$, then $\left(\tfrac{a}{p}\right) = \left(\tfrac{b}{p}\right)$.

**4.** If $p$ is an odd prime, and $a$ and $b$ are integers not divisible by $p$, then $\left(\tfrac{a}{p}\right)\left(\tfrac{b}{p}\right) = \left(\tfrac{ab}{p}\right)$.

**5.** If $p$ is an odd prime, and $a$ and $b$ are integers not divisible by $p$, then $\left(\tfrac{a^2}{p}\right) = 1$.

**6.** If $p$ is an odd prime, then $\left(\frac{-1}{p}\right) = \begin{cases} 1 & \text{if } p \equiv 1 \, (\text{mod } 4) \\ -1 & \text{if } p \equiv -1 \, (\text{mod } 4). \end{cases}$

**7.** If $p$ is an odd prime, then $-1$ is a quadratic residue of $p$ if $p \equiv 1 \, (\text{mod } 4)$ and a quadratic nonresidue of $p$ if $p \equiv -1 \, (\text{mod } 4)$. (This is a direct consequence of Fact 6.)

**8.** *Gauss' lemma:* If $p$ is an odd prime, $a$ is an integer with $\gcd(a, p) = 1$, and $s$ is the number of least positive residues of $a, 2a, \ldots, \frac{p-1}{2}a$ greater than $\frac{p}{2}$, then $\left(\frac{a}{p}\right) = (-1)^s$.

**9.** If $p$ is an odd prime, then $\left(\frac{2}{p}\right) = (-1)^{(p^2-1)/8}$.

**10.** The integer 2 is a quadratic residue of all primes $p$ with $p \equiv \pm 1 \, (\text{mod } 8)$ and a quadratic nonresidue of all primes $p \equiv \pm 3 \, (\text{mod } 8)$. (This follows from Fact 9.)

**11.** *Law of quadratic reciprocity:* If $p$ and $q$ are odd primes, then

$$\left(\frac{p}{q}\right)\left(\frac{q}{p}\right) = (-1)^{\frac{p-1}{2} \cdot \frac{q-1}{2}}.$$

This law was first proved by Carl Friedrich Gauss (1777–1855).

**12.** Many different proofs of the law of quadratic reciprocity have been discovered. By one count, there are more than 150 different proofs. Gauss published eight different proofs himself.

**13.** The law of quadratic reciprocity implies that if $p$ and $q$ are odd primes, then $\left(\frac{p}{q}\right) = \left(\frac{q}{p}\right)$ if either $p \equiv 1 \, (\text{mod } 4)$ or $q \equiv 1 \, (\text{mod } 4)$, and $\left(\frac{p}{q}\right) = -\left(\frac{q}{p}\right)$ if $p \equiv q \equiv 3 \, (\text{mod } 4)$.

**14.** If $m$ is an odd positive integer, and $a$ and $b$ are integers relatively prime to $m$ with $a \equiv b \, (\text{mod } m)$, then $\left(\frac{a}{m}\right) = \left(\frac{b}{m}\right)$.

**15.** If $m$ is an odd positive integer, and $a$ and $b$ are integers relatively prime to $m$, then $\left(\frac{ab}{m}\right) = \left(\frac{a}{m}\right)\left(\frac{b}{m}\right)$.

**16.** If $m$ is an odd positive integer and $a$ is an integer relatively prime to $m$, then $\left(\frac{a^2}{m}\right) = 1$.

**17.** If $m$ and $n$ are relatively prime odd positive integers, and $a$ is an integer relatively prime to $m$ and $n$, then $\left(\frac{a}{mn}\right) = \left(\frac{a}{m}\right)\left(\frac{a}{n}\right)$.

**18.** If $m$ is an odd positive integer, then the value of the Jacobi symbol $\left(\frac{a}{m}\right)$ does not determine whether $a$ is a perfect square modulo $m$.

**19.** If $m$ is an odd positive integer, then $\left(\frac{-1}{m}\right) = (-1)^{\frac{m-1}{2}}$.

**20.** If $m$ is an odd positive integer, then $\left(\frac{2}{m}\right) = (-1)^{\frac{m^2-1}{8}}$.

**21.** *Reciprocity law for Jacobi symbols:* If $m$ and $n$ are relatively prime odd positive integers, then

$$\left(\frac{m}{n}\right)\left(\frac{n}{m}\right) = (-1)^{\frac{m-1}{2} \frac{n-1}{2}}.$$

**22.** The number of integers in a reduced set of residues modulo $n$ with $\left(\frac{k}{n}\right) = 1$ equals the number with $\left(\frac{k}{n}\right) = -1$.

**23.** The Legendre symbol $\left(\frac{a}{p}\right)$, where $p$ is prime and $0 \leq a < p$, can be evaluated using $O((\log_2 p)^2)$ bit operations.

**24.** The Jacobi symbol $\left(\frac{a}{n}\right)$, where $n$ is a positive integer and $0 \leq a < n$, can be evaluated using $O((\log_2 n)^2)$ bit operations.

**25.** Let $p$ be an odd prime. Even though half the integers $x$ with $1 \leq x < p$ are quadratic nonresidues of $p$, there is no known polynomial-time deterministic algorithm for finding such an integer. However, picking integers at random produces a probabilistic algorithm that has 2 as the expected number of iterations done before a nonresidue is found.

**26.** Let $m$ be a positive integer with a primitive root. If $k$ is a positive integer and $a$ is an integer relatively prime to $m$, then $a$ is a $k$th power residue of $m$ if and only if $a^{\phi(m)/d} \equiv 1 \,(\mathrm{mod}\ m)$ where $d = \gcd(k, \phi(m))$. Moreover, if $a$ is a $k$th power residue of $m$, then there are exactly $d$ incongruent solutions modulo $m$ of the congruence $x^k \equiv a \,(\mathrm{mod}\ m)$.

**27.** If $p$ is a prime, $k$ is a positive integer, and $a$ is an integer with $\gcd(a, p) = 1$, then $a$ is a $k$th power residue of $p$ if and only if $a^{(p-1)/d} \equiv 1 \,(\mathrm{mod}\ p)$, where $d = \gcd(k, p-1)$.

**28.** The $k$th roots of a $k$th power residue modulo $p$, where $p$ is a prime, can be computed using a primitive root and indices to this primitive root. This is only practical for small primes $p$. (See §4.7.1.)

**Examples:**

**1.** The integers 1, 3, 4, 5, and 9 are quadratic residues of 11; the integers 2, 6, 7, 8, and 10 are quadratic nonresidues of 11. Hence $\left(\frac{1}{11}\right) = \left(\frac{3}{11}\right) = \left(\frac{4}{11}\right) = \left(\frac{5}{11}\right) = \left(\frac{9}{11}\right) = 1$ and $\left(\frac{2}{11}\right) = \left(\frac{6}{11}\right) = \left(\frac{7}{11}\right) = \left(\frac{8}{11}\right) = \left(\frac{10}{11}\right) = -1$.

**2.** To determine whether 11 is a quadratic residue of 19, note that using the law of quadratic reciprocity (Fact 11) and Facts 3, 4, and 10 it follows that $\left(\frac{11}{19}\right) = -\left(\frac{19}{11}\right) = -\left(\frac{8}{11}\right) = -\left(\frac{2}{11}\right)^3 = -(-1)^3 = 1$.

**3.** To evaluate the Jacobi symbol $\left(\frac{2}{45}\right)$, note that $\left(\frac{2}{45}\right) = \left(\frac{2}{3^2 \cdot 5}\right) = \left(\frac{2}{3}\right)^2 \cdot \left(\frac{2}{5}\right) = (-1)^2(-1) = -1$.

**4.** The Jacobi symbol $\left(\frac{5}{21}\right) = 1$, but 5 is not a quadratic residue of 21.

**5.** The integer 6 is a fifth power residue of 101 since $6^{(101-1)/5} = 6^{20} \equiv 1 \,(\mathrm{mod}\ 101)$.

**6.** From Example 5 it follows that 6 is a fifth power residue of 101. The solutions of the congruence $x^5 \equiv 6 \,(\mathrm{mod}\ 101)$, the fifth roots of 6, can be found by taking indices to the primitive root 2 modulo 101. Since $\mathrm{ind}_2 6 = 70$, this gives $\mathrm{ind}_2 x^5 = 5 \cdot \mathrm{ind}_2 x \equiv 70 \,(\mathrm{mod}\ 100)$. The solutions of this congruence are the integers $x$ with $\mathrm{ind}_2 x \equiv 14 \,(\mathrm{mod}\ 20)$. This implies that the fifth roots of 6 are the integers with $\mathrm{ind}_2 x \equiv 14, 34, 54, 74,$ and $94$. These are the integers $x$ with $x \equiv 22, 70, 85, 96,$ and $30 \,(\mathrm{mod}\ 101)$.

**7.** The integer 5 is not a sixth power residue of 17 since $5^{\frac{16}{\gcd(6,16)}} = 5^8 \equiv -1 \,(\mathrm{mod}\ 17)$.

## 4.7.4   MODULAR SQUARE ROOTS

**Definition:**

If $m$ is a positive integer and $a$ is an integer, then $r$ is a **square root of $a$ modulo $m$** if $r^2 \equiv a \,(\mathrm{mod}\ m)$.

**Facts:**

**1.** If $p$ is a prime of the form $4n + 3$ and $a$ is a perfect square modulo $p$, then the two square roots of $a$ modulo $p$ are $\pm a^{(p+1)/4}$.

**2.** If $p$ is a prime of the form $8n + 5$ and $a$ is a perfect square modulo $p$, then the two square roots of $a$ modulo $p$ are $x \equiv \pm a^{(p+3)/8}(\mathrm{mod}\ p)$ if $a^{(p-1)/4} \equiv 1 \,(\mathrm{mod}\ p)$ and $x \equiv \pm 2^{(p-1)/4}a^{(p+3)/8}(\mathrm{mod}\ p)$ if $a^{(p-1)/4} \equiv -1 \,(\mathrm{mod}\ p)$.

**3.** If $n$ is a positive integer that is the product of two distinct primes $p$ and $q$, and $a$ is a perfect square modulo $n$, then there are four distinct square roots of $a$ modulo $n$. These square roots can be found by finding the two square roots of $a$ modulo $p$ and the two square roots of $a$ modulo $q$ and then using the Chinese remainder theorem to find the four square roots of $a$ modulo $n$.

**4.** A square root of an integer $a$ that is a square modulo $p$, where $p$ is an odd prime, can be found by an algorithm that uses an average of $O((\log_2 p)^3)$ bit operations. (See [MeVoVa96].)

**5.** If $n$ is an odd integer with $r$ distinct prime factors, $a$ is a perfect square modulo $n$, and $\gcd(a, n) = 1$, then $a$ has exactly $2^r$ incongruent square roots modulo $n$.

**Examples:**

**1.** Using Legendre symbols it can be shown that 11 is a perfect square modulo 19. Using Fact 1 it follows that the square roots of 11 modulo 19 are given by $x \equiv \pm 11^{(19+1)/4} = \pm 11^5 \equiv \pm 7 \pmod{19}$.

**2.** There are four incongruent square roots of 860 modulo $11021 = 103 \cdot 107$. To find these solutions, first note that $x^2 \equiv 860 = 36 \pmod{103}$ so that $x \equiv \pm 6 \pmod{103}$ and $x^2 \equiv 860 = 4 \pmod{107}$ so that $x \equiv \pm 2 \pmod{107}$. The Chinese remainder theorem can be used to find these square roots. They are $x \equiv -212, -109, 109, 212 \pmod{11021}$.

**3.** The square roots of 121 modulo 315 are 11, 74, 101, 151, 164, 214, 241, and 304. As predicted by Fact 5, $121 = 11^2$ has $2^3 = 8$ square roots modulo $315 = 3^2 \cdot 5 \cdot 7$.

# 4.8  DIOPHANTINE EQUATIONS

An important area of number theory is devoted to finding solutions of equations where the solutions are restricted to belong to the set of integers, or some other specified set, such as the set of rational numbers. An equation with the added proviso that the solutions must be integers (or must belong to some other specified countable set, such as the set of rational numbers) is called a *diophantine equation*. This name comes from the ancient Greek mathematician Diophantus (ca. 250 A.D.), who wrote extensively on such equations.

Diophantine equations have both practical and theoretical importance. Their practical importance arises when variables in an equation represent quantities of objects, for example. Fermat's last theorem, which states that there are no nontrivial solutions in integers $n > 2$, $x$, $y$, and $z$ to the diophantine equation $x^n + y^n = z^n$ has long interested mathematicians and non-mathematicians alike. This theorem was proved only in the mid-1990s, even though many brilliant scholars sought a proof during the last three centuries.

More information about diophantine equations can be found in [Di71], [Gu94], and [Mo69].

## 4.8.1  LINEAR DIOPHANTINE EQUATIONS

**Definition:**

A *linear diophantine equation* is an equation of the form $a_1 x_1 + a_2 x_2 + \cdots + a_n x_n = c$, where $c, a_1, \ldots, a_n$ are integers and where integer solutions are sought for the unknowns $x_1, x_2, \ldots, x_n$.

## Facts:

**1.** Let $a$ and $b$ be integers with $\gcd(a,b) = d$. The linear diophantine equation $ax+by = c$ has no solutions if $d \nmid c$. If $d|c$, then there are infinitely many solutions in integers. Moreover, if $x = x_0$, $y = y_0$ is a particular solution, then all solutions are given by $x = x_0 + \frac{b}{d}n$, $y = y_0 - \frac{a}{d}n$, where $n$ is an integer.

**2.** A linear diophantine equation $a_1x_1 + a_2x_2 + \cdots + a_nx_n = c$ has solutions in integers if and only if $\gcd(a_1, a_2, \ldots, a_n)|c$. In that case, there are infinitely many solutions.

**3.** A solution $(x_0, y_0)$ of the linear diophantine equation $ax + by = c$ where $\gcd(a,b)|c$ can be found by first expressing $\gcd(a,b)$ as a linear combination of $a$ and $b$, and then multiplying by $c/\gcd(a,b)$. (See §4.2.1.)

**4.** A linear diophantine equation $a_1x_1 + a_2x_2 + \cdots + a_nx_n = c$ in $n$ variables can be solved by a reduction method. To find a particular solution, first let $b = \gcd(a_2, \ldots, a_n)$ and let $(x_1, y)$ be a solution of the diophantine equation $a_1x_1 + by = c$. Iterate this procedure on the diophantine equation in $n-1$ variables, $a_2x_2 + a_3x_3 + \cdots + a_nx_n = y$ until an equation in two variables is obtained.

**5.** The solution to a system of $r$ linear diophantine equations in $n$ variables is obtained by using Gaussian elimination (§6.4.2) to reduce to a single diophantine equation in two or more variables.

**6.** If $a$ and $b$ are relatively prime positive integers and $n$ is a positive integer, then the diophantine equation $ax+by = n$ has a *nonnegative* integer solution if $n \geq (a-1)(b-1)$.

**7.** If $a$ and $b$ are relatively prime positive integers, then there are exactly $(a-1)(b-1)/2$ nonnegative integers $n$ less than $ab - a - b$ such that the equation $ax + by = n$ has a nonnegative solution.

**8.** If $a$ and $b$ are relatively prime positive integers, then there are no nonnegative solutions of $ax + by = ab - a - b$.

## Examples:

**1.** To solve the linear diophantine equation $17x + 13y = 100$, express $\gcd(17,13) = 1$ as a linear combination of 17 and 13. Using the steps of the Euclidean algorithm, it follows that $4 \cdot 13 - 3 \cdot 17 = 1$. Multiplying by 100 yields $100 = 400 \cdot 13 - 300 \cdot 17$. Since a particular solution is $x = -300$, $y = 400$, all solutions are then given by $x = -300 + 13n$, $y = 400 - 17n$, where $n$ ranges over the set of integers.

**2.** A traveller has exactly \$510 in travelers checks where each check is either a \$20 or a \$50 check. How many checks of each denomination can there be?

The answer to this question is given by the set of solutions in nonnegative integers to the linear diophantine equation $20x + 50y = 510$. There are infinitely many solutions in integers, which can be shown to be given by $x = -102 + 5n$, $y = 51 - 2n$. Since both $x$ and $y$ must be nonnegative, it follows that $n = 21, 22, 23, 24$, or 25. Therefore there could be three \$20 checks and nine \$50 checks, eight \$20 checks and seven \$50 checks, thirteen \$20 checks and five \$50 checks, eighteen \$20 checks and three \$50 checks, or twenty-three \$20 checks and one \$50 check.

**3.** The linear diophantine equation $12x_1 + 21x_2 + 9x_3 + 15x_4 = 9$ has infinitely many solutions since $\gcd(12, 21, 9, 15) = 3$, which divides 9. To find a particular solution, first divide both sides of the equation by 3 to get $4x_1 + 7x_2 + 3x_3 + 5x_4 = 3$. Now $1 = \gcd(7,3,5)$, so solve $4x_1 + 1y = 3$, as in Example 1, to get $x_1 = 1$, $y = -1$. Next we solve $7x_2 + 3x_3 + 5x_4 = -1$. Since $1 = \gcd(3,5)$, solve $7x_2 + 1z = -1$ to get $x_2 = 1$, $z = -8$. Finally, solve $3x_3 + 5x_4 = -8$ to get $x_3 = -1$, $x_4 = -1$. This gives the particular solution $x = (1, 1, -1, -1)$.

**4.** To solve the following system of linear diophantine equations in integers

$$x + y + z + w = 100$$
$$x + 2y + 3z + 4w = 300$$
$$x + 4y + 9z + 16w = 1000,$$

first reduce the system by elimination to

$$x + y + z + w = 100$$
$$y + 2z + 3w = 200$$
$$2z + 6w = 300.$$

The solution to the last equation is $z = 150 + 3t$, $w = -t$, where $t$ is an integer. Back-substitution gives

$$y = 200 - 2(150 + 3t) - 3(-t) = -100 - 3t$$
$$x = 100 - (-100 - 3t) - (150 + 3t) - (-t) = 50 + t.$$

## 4.8.2  PYTHAGOREAN TRIPLES

**Definitions:**
A **Pythagorean triple** is a solution $(x, y, z)$ of the equation $x^2 + y^2 = z^2$, where $x$, $y$, and $z$ are positive integers.

A Pythagorean triple is **primitive** if $\gcd(x, y, z) = 1$.

**Facts:**
1. Pythagorean triples represent the lengths of sides of right triangles.
2. All primitive Pythagorean triples are given by

$$x = 2mn, \ y = m^2 - n^2, \ z = m^2 + n^2$$

where $m$ and $n$ are relatively prime positive integers of opposite parity with $m > n$.
3. All Pythagorean triples can be found by taking

$$x = 2mnt, \ y = (m^2 - n^2)t, \ z = (m^2 + n^2)t$$

where $t$ is a positive integer and $m$ and $n$ are as in Fact 2.
4. Given a Pythagorean triple $(x, y, z)$ with $y$ odd, then $m$ and $n$ from Fact 2 can be found by taking $m = \sqrt{\frac{z+y}{2}}$ and $n = \sqrt{\frac{z-y}{2}}$.
5. The following table lists all Pythagorean triples with $z \le 100$.

| $m$ | $n$ | $x = 2mn$ | $y = m^2 - n^2$ | $z = m^2 + n^2$ |
|---|---|---|---|---|
| 2 | 1 | 4 | 3 | 5 |
| 3 | 1 | 6 | 8 | 10 |
| 3 | 2 | 12 | 5 | 13 |
| 4 | 1 | 8 | 15 | 17 |
| 4 | 2 | 16 | 12 | 20 |

| $m$ | $n$ | $x = 2mn$ | $y = m^2 - n^2$ | $z = m^2 + n^2$ |
|---|---|---|---|---|
| 4 | 3 | 24 | 7 | 25 |
| 5 | 1 | 10 | 24 | 26 |
| 5 | 2 | 20 | 21 | 29 |
| 5 | 3 | 30 | 16 | 34 |
| 5 | 4 | 40 | 9 | 41 |
| 6 | 1 | 12 | 35 | 37 |
| 6 | 2 | 24 | 32 | 40 |
| 6 | 3 | 36 | 27 | 45 |
| 6 | 4 | 48 | 20 | 52 |
| 6 | 5 | 60 | 11 | 61 |
| 7 | 1 | 14 | 48 | 50 |
| 7 | 2 | 28 | 45 | 53 |
| 7 | 3 | 42 | 40 | 58 |
| 7 | 4 | 56 | 33 | 65 |
| 7 | 5 | 70 | 24 | 74 |
| 7 | 6 | 84 | 13 | 85 |
| 8 | 1 | 16 | 63 | 65 |
| 8 | 2 | 32 | 60 | 68 |
| 8 | 3 | 48 | 55 | 73 |
| 8 | 4 | 64 | 48 | 80 |
| 8 | 5 | 80 | 39 | 89 |
| 8 | 6 | 96 | 28 | 100 |
| 9 | 1 | 18 | 80 | 82 |
| 9 | 2 | 36 | 77 | 85 |
| 9 | 3 | 54 | 72 | 90 |
| 9 | 4 | 72 | 65 | 97 |

**6.** The solutions of the diophantine equation $x^2 + y^2 = 2z^2$ can be obtained by transforming this equation into $\left(\frac{x+y}{2}\right)^2 + \left(\frac{x-y}{2}\right)^2 = z^2$, which shows that $(\frac{x+y}{2}, \frac{x-y}{2}, z)$ is a Pythagorean triple. All solutions are given by $x = (m^2 - n^2 + 2mn)t$, $y = (m^2 - n^2 - 2mn)t$, $z = (m^2 + n^2)t$ where $m$, $n$, and $t$ are integers.

**7.** The solutions of the diophantine equation $x^2 + 2y^2 = z^2$ are given by $x = (m^2 - 2n^2)t$, $y = 2mnt$, $z = m^2 + 2n^2$ where $m$, $n$, and $t$ are positive integers.

**8.** The solutions of the diophantine equation $x^2 + y^2 + z^2 = w^2$ where $y$ and $z$ are even are given by $x = \frac{m^2 + n^2 - r^2}{r}$, $y = 2m$, $z = 2n$, $w = \frac{m^2 + n^2 + r^2}{r}$, where $m$ and $n$ are positive integers and $r$ runs through the divisors of $m^2 + n^2$ less than $(m^2 + n^2)^{1/2}$.

**9.** The solutions of the diophantine equation $x^2 + y^2 = z^2 + w^2$, with $x > z$, are given by $x = \frac{ms+nr}{2}$, $y = \frac{ns-mr}{2}$, $z = \frac{ms-nr}{2}$, $w = \frac{ns+mr}{2}$, where if $m$ and $n$ are both odd, then $r$ and $s$ are either both odd or both even.

### 4.8.3  FERMAT'S LAST THEOREM

**Definitions:**

The **Fermat equation** is the diophantine equation $x^n + y^n = z^n$, where $x$, $y$, $z$ are integers and $n$ is a positive integer greater than 2.

A **nontrivial solution** to the Fermat equation $x^n + y^n = z^n$ is a solution in integers $x$, $y$, and $z$ in which none of $x$, $y$, and $z$ are zero.

Let $p$ be an odd prime and let $\mathcal{K} = \mathcal{Q}(\omega)$ be the degree-$p$ cyclotomic extension of the rational numbers (§5.6.2). If $p$ does not divide the class number of $\mathcal{K}$ (see [Co93]), then $p$ is said to be **regular**. Otherwise $p$ is **irregular**.

**Facts:**

**1.** *Fermat's last theorem:*  The statement that the diophantine equation $x^n + y^n = z^n$ has no nontrivial solutions in the positive integers for $n \geq 3$, is called *Fermat's last theorem*. The statement was made more than 300 years ago by Pierre de Fermat (1601–1665) and resisted proof until recently.

**2.** Fermat wrote in the margin of his copy of the works of Diophantus, next to the discussion of the equation $x^2 + y^2 = z^2$, the following: "However, it is impossible to write a cube as the sum of two cubes, a fourth power as the sum of two fourth powers and in general any power the sum of two similar powers. For this I have discovered a truly wonderful proof, but the margin is too small to contain it." In spite of this quotation, no proof was found of this statement until 1994, even though many mathematicians actively worked on finding such a proof. Most mathematicians would find it shocking if Fermat actually had found a proof.

**3.** Fermat's last theorem was finally proved in 1994 by Andrew Wiles [Wi95]. Wiles collected the Wolfskehl Prize, worth approximately \$50,000 in 1997 for this proof.

**4.** That there are no nontrivial solutions of the Fermat equation for $n = 4$ was demonstrated by Fermat with an elementary proof using the *method of infinite descent*. This method proceeds by showing that for every solution in positive integers, there is a solution such that the values of each of the integers $x$, $y$, and $z$ is smaller, contradicting the well-ordering property of the set of integers.

**5.** The method of infinite descent invented by Fermat can be used to show that the more general diophantine equation $x^4 + y^4 = z^2$ has no nontrivial solutions in integers $x$, $y$, and $z$.

**6.** The diophantine equation $x^4 - y^4 = z^2$ has no nontrivial solutions, as can be shown using the method of infinite descent.

**7.** The sum of two cubes may equal the sum of two other cubes. That is, there are nontrivial solutions of the diophantine equation $x^3 + y^3 = z^3 + w^3$. The smallest solution is $x = 1$, $y = 12$, $z = 9$, $w = 10$.

**8.** The sum of three cubes may also be a cube. In fact, the solutions of $x^3 + y^3 + z^3 = w^3$ are given by $x = 3a^2 + 5b(a-b)$, $y = 4a(a-b) + 6b^2$, $z = 5a(a-b) - 3b^2$, $w = 6a^2 - 4b(a+b)$ where $a$ and $b$ are integers.

**9.** Euler conjectured that there were four fourth powers of positive integers whose sum is also the fourth power of an integer. In other words, he conjectured that there are nontrivial solutions to the diophantine equation $v^4 + w^4 + x^4 + y^4 = z^4$. The first such example was found in 1911 when it was discovered (by R. Norrie) that $30^4 + 120^4 + 272^4 + 315^4 = 353^4$.

**10.** Euler also conjectured that the sum of the fourth powers of three positive integers can never be the fourth power of an integer and that the sum of fifth powers of four positive integers can never be the fifth power of an integer, and so on. In other words, he conjectured that there were no nontrivial solutions to the diophantine equations $w^4 + x^4 + y^4 = z^4$, $v^5 + w^5 + x^5 + y^5 = z^5$, and so on. He was mistaken. The smallest counterexamples known are $95{,}800^4 + 217{,}519^4 + 414{,}560^4 = 422{,}481^4$ and $27^5 + 84^5 + 110^5 + 133^5 = 144^5$.

**11.** If $n = mp$ for some integer $m$ and $p$ is prime, then the Fermat equation can be rewritten as $(x^m)^p + (y^m)^p = (z^m)^p$. Since the only positive integers greater than 2 without an odd prime factor are powers of 2 and $x^4 + y^4 = z^4$ has no nontrivial solutions in integers, Fermat's last theorem can be demonstrated by showing that $x^p + y^p = z^p$ has no nontrivial solutions in integers $x$, $y$, and $z$ when $p$ is an odd prime.

**12.** An odd prime $p$ is regular if and only if it does not divide the numerator of any of the numbers $B_2, B_4, \dots, B_{p-3}$, where $B_k$ is the $k$th Bernoulli number. (See §3.1.4.)

**13.** There is a relatively simple proof of Fermat's last theorem for exponents that are regular primes.

**14.** The smallest irregular primes are 37, 59, 67, 101, 103, 149, and 157.

**15.** Wiles' proof of Fermat's last theorem is based on the theory of elliptic curves. The proof is based on relating to integers $a$, $b$, $c$, and $n$ that supposedly satisfy the Fermat equation $a^n + b^n = c^n$ the elliptic curve $y^2 = x(x + a^n)(x - b^n)$ (called the *associated Frey curve*) and deriving a contradiction using sophisticated results from the theory of elliptic curves.

For more details, see Wiles' original proof [Wi95], the popular account [Si97], and the following two sites:

- http://www.scienceandreason.net/flt/flt01.htm
- http://www.pbs.org/wgbh/nova/proof

---

## 4.8.4    PELL'S, BACHET'S, AND CATALAN'S EQUATIONS

### Definitions:
***Pell's equation*** is a diophantine equation of the form $x^2 - dy^2 = 1$, where $d$ is a square-free positive integer. This diophantine equation is named after John Pell (1611–1685).

***Bachet's equation*** is a diophantine equation of the form $y^2 = x^3 + k$. This diophantine equation is named after Claude Gaspar Bachet (1587–1638).

***Catalan's equation*** is the diophantine equation $x^m - y^n = 1$, where a solution is sought with integers $x > 0$, $y > 0$, $m > 1$, and $n > 1$. This diophantine equation is named after Eugène Charles Catalan (1814–1894).

### Facts:
**1.** If $x, y$ is a solution to the diophantine equation $x^2 - dy^2 = n$ with $d$ square free and $n^2 < d$, then the rational number $\frac{x}{y}$ is a convergent of the simple continued fraction for $\sqrt{d}$. (See §4.9.2.)

**2.** An equation of the form $ax'^2 + bx' + c = y'^2$ can be transformed by means of the relations $x = 2ax' + b$ and $y = 2y'$ into an equation of the form $x^2 - dy^2 = n$, where $n = b^2 - 4ac$ and $d = a$.

**3.** It is ironic that John Pell apparently had little to do with finding the solutions to the diophantine equation $x^2 - dy^2 = 1$. Euler gave this equation its name following a mistaken reference. Fermat conjectured an infinite number of solutions to this equation in 1657; this was eventually proved by Lagrange in 1768.

**4.** Let $x, y$ be the least positive solution to $x^2 - dy^2 = 1$, with $d$ square free. Then every positive solution is given by

$$x_k + y_k\sqrt{d} = (x + y\sqrt{d})^k,$$

where $k$ ranges over the positive integers.

**5.** The following table gives the smallest positive solutions to Pell's equation $x^2 - dy^2 = 1$ with $d$ a square-free positive integer less than 100.

| $d$ | $x$ | $y$ | $d$ | $x$ | $y$ |
|---|---|---|---|---|---|
| 2 | 3 | 2 | 51 | 50 | 7 |
| 3 | 2 | 1 | 53 | 66,249 | 9,100 |
| 5 | 9 | 4 | 55 | 89 | 12 |
| 6 | 5 | 2 | 57 | 151 | 20 |
| 7 | 8 | 3 | 58 | 19,603 | 2,574 |
| 10 | 19 | 6 | 59 | 530 | 69 |
| 11 | 10 | 3 | 61 | 1,766,319,049 | 226,153,980 |
| 13 | 649 | 180 | 62 | 63 | 8 |
| 14 | 15 | 4 | 65 | 129 | 16 |
| 15 | 4 | 1 | 66 | 65 | 8 |
| 17 | 33 | 8 | 67 | 48,842 | 5,967 |
| 19 | 170 | 39 | 69 | 7,775 | 936 |
| 21 | 55 | 12 | 70 | 251 | 30 |
| 22 | 197 | 42 | 71 | 3,480 | 413 |
| 23 | 24 | 5 | 73 | 2,281,249 | 267,000 |
| 26 | 51 | 10 | 74 | 3,699 | 430 |
| 29 | 9,801 | 1,820 | 77 | 351 | 40 |
| 30 | 11 | 2 | 78 | 53 | 6 |
| 31 | 1,520 | 273 | 79 | 80 | 9 |
| 33 | 23 | 4 | 82 | 163 | 18 |
| 34 | 35 | 6 | 83 | 82 | 9 |
| 35 | 6 | 1 | 85 | 285,769 | 30,996 |
| 37 | 73 | 12 | 86 | 10,405 | 1,122 |
| 38 | 37 | 6 | 87 | 28 | 3 |
| 39 | 25 | 4 | 89 | 500,001 | 53,000 |
| 41 | 2,049 | 320 | 91 | 1,574 | 165 |
| 42 | 13 | 2 | 93 | 12,151 | 1,260 |
| 43 | 3,482 | 531 | 94 | 2,143,295 | 221,064 |
| 46 | 24,335 | 3,588 | 95 | 39 | 4 |
| 47 | 48 | 7 | 97 | 62,809,633 | 6,377,352 |

**6.** If $k = 0$, then the formulae $x = t^2, y = t^3$ give an infinite number of solutions to the Bachet equation $y^2 = x^3 + k$.

**7.** There are no solutions to Bachet's equation for the following values of $k$: $-144, -105,$ $-78, -69, -42, -34, -33, -31, -24, -14, -5, 7, 11, 23, 34, 45, 58, 70.$

**8.** The following table lists solutions to Bachet's equation for various values of $k$.

| $k$ | $x$ |
|---|---|
| 0 | $t^2$ ($t$ any integer) |
| 1 | $0, -1, 2$ |
| 17 | $-1, -2, 2, 4, 8, 43, 52, 5334$ |
| $-2$ | 3 |
| $-4$ | $2, 5$ |
| $-7$ | $2, 32$ |
| $-15$ | 1 |

**9.** If $k < 0$, $k$ is square free, $k \equiv 2$ or $3 \pmod 4$, and the class number of the field $\mathcal{Q}(\sqrt{-k})$ is not a multiple of 3, then the only solution of the Bachet equation $y^2 = x^3 + k$ for $x$ is given by whichever of $-(4k \pm 1)/3$ is an integer. The first few values of such $k$ are 1, 2, 5, 6, 10, 13, 14, 17, 21, and 22.

**10.** Solutions to the Catalan equation give consecutive integers that are powers of integers.

**11.** The Catalan equation has the solution $x = 3$, $y = 2$, $m = 2$, $n = 3$, so $8 = 2^3$ and $9 = 3^2$ are consecutive powers of integers. The *Catalan conjecture* is that this is the only solution.

**12.** Levi ben Gerson showed in the 14th century that 8 and 9 are the only consecutive powers of 2 and 3, so that the only solution in positive integers of $3^m - 2^n = \pm 1$ is $m = 2$ and $n = 3$.

**13.** Euler proved that the only solution in positive integers of $x^3 - y^2 = \pm 1$ is $x = 2$ and $y = 3$.

**14.** Lebesgue showed in 1850 that $x^m - y^2 = 1$ has no solutions in positive integers when $m$ is an integer greater than 3.

**15.** The diophantine equations $x^3 - y^n = 1$ and $x^m - y^3 = 1$ with $m > 2$ were shown to have no solutions in positive integers in 1921, and in 1964 it was shown that $x^2 - y^n = 1$ has no solutions in positive integers.

**16.** R. Tijdeman showed in 1976 that there are only finitely many solutions in integers to the Catalan equation $x^m - y^n = 1$ by showing that there is a computable constant $C$ such that for every solution, $x^m < C$ and $y^n < C$. However, the enormous size of the constant $C$ makes it infeasible to establish the Catalan conjecture using this result.

**17.** In 2002 Catalan's conjecture was proved by Preda Mihăilescu using the theories of cyclotomic fields and Galois modules. Consequently, the statement that Catalan's equation has exactly one solution (given in Fact 11) is now called Mihăilescu's theorem.

**Examples:**

**1.** To solve the diophantine equation $x^2 - 13y^2 = 1$, note that the simple continued fraction for $\sqrt{13}$ is $[3; \overline{1, 1, 1, 1, 6}]$, with convergents $3, 4, \frac{7}{2}, \frac{11}{3}, \frac{18}{5}, \frac{119}{33}, \frac{137}{38}, \frac{256}{71}, \frac{393}{109}, \frac{649}{180}, \ldots$. The smallest positive solution to the equation is $x = 649$, $y = 180$. A second solution is given by $(649 + 180\sqrt{13})^2 = 842{,}401 + 233{,}640\sqrt{13}$, that is, $x = 842{,}401$, $y = 233{,}640$.

**2.** Congruence considerations can be used to show that there are no solutions of Bachet's equation for $k = 7$. Modulo 8, every square is congruent to 0, 1, or 4; therefore if $x$ is even, then $y^2 \equiv 7 \pmod 8$, a contradiction. Likewise if $x \equiv 3 \pmod 4$, then $y^2 \equiv 2 \pmod 8$, also impossible. So assume that $x \equiv 1 \pmod 4$. Add one to both sides and factor to get $y^2 + 1 = x^3 + 8 = (x+2)(x^2 - 2x + 4)$. Now $x^2 - 2x + 4 \equiv 3 \pmod 4$, so it must have a prime divisor $p \equiv 3 \pmod 4$. Then $y^2 \equiv -1 \pmod p$, which implies that $-1$ is a quadratic residue modulo $p$. (See §4.7.3.) But $p \equiv 3 \pmod 4$, so $-1$ cannot be a quadratic residue modulo $p$. Therefore, there are no solutions when $k = 7$.

## 4.8.5 SUMS OF SQUARES AND WARING'S PROBLEM

### Definitions:

If $k$ is a positive integer, then $g(k)$ is the smallest positive integer such that every positive integer can be written as a sum of $g(k)$ $k$th powers.

If $k$ is a positive integer, then $G(k)$ is the smallest positive integer such that every *sufficiently large* positive integer can be written as a sum of $G(k)$ $k$th powers.

The determination of $g(k)$ is called **Waring's problem**. (Edward Waring, 1741–1793)

### Facts:

**1.** A positive integer $n$ is the sum of two squares if and only if each prime factor of $n$ of the form $4k + 3$ appears to an even power in the prime factorization of $n$.

**2.** If $m = a^2 + b^2$ and $n = c^2 + d^2$, then the number $mn$ can be expressed as the sum of two squares as follows: $mn = (ac + bd)^2 + (ad - bc)^2$.

**3.** If $n$ is representable as the sum of two squares, then it is representable in $4(d_1 - d_2)$ ways (where the order of the squares and their signs matter), where $d_1$ is the number of divisors of $n$ of the form $4k + 1$ and $d_2$ is the number of divisors of $n$ of the form $4k + 3$.

**4.** An integer $n$ is the sum of three squares if and only if $n$ is not of the form $4^m(8k+7)$, where $m$ is a nonnegative integer.

**5.** The positive integers less than 100 that are not the sum of three squares are $7, 15, 23, 28, 31, 39, 47, 55, 60, 63, 71, 79, 87, 92$, and $95$.

**6.** *Lagrange's four-square theorem:* Every positive integer is the sum of 4 squares, some of which may be zero. (Joseph Lagrange, 1736–1813)

**7.** A useful lemma due to Lagrange is the following. If $m = a^2 + b^2 + c^2 + d^2$ and $n = e^2 + f^2 + g^2 + h^2$, then $mn$ can be expressed as the sum of four squares as follows: $mn = (ae + bf + cg + dh)^2 + (af - be + ch - dg)^2 + (ag - ce + df - bh)^2 + (ah - de + bg - cf)^2$.

**8.** The number of ways $n$ can be written as the sum of four squares is $8(s - s_4)$, where $s$ is the sum of the divisors of $n$ and $s_4$ is the sum of the divisors of $n$ that are divisible by 4.

**9.** It is known that $g(k)$ always exists.

**10.** For $6 \le k \le 471{,}600{,}000$ the following formula holds except possibly for a finite number of positive integers $k$: $g(k) = \lfloor (\frac{3}{2})^k \rfloor + 2^k - 2$ where $\lfloor x \rfloor$ represents the floor (greatest integer) function.

**11.** The exact value of $G(k)$ is known only for two values of $k$, $G(2) = 4$ and $G(4) = 16$.

**12.** From Facts 6 and 11, it follows that $G(2) = g(2) = 4$.

**13.** If $k$ is an integer with $k \ge 2$, then $G(k) \le g(k)$.

**14.** If $k$ is an integer with $k \ge 2$, then $G(k) \ge k + 1$.

**15.** Hardy and Littlewood showed that $G(k) \le (k-2)2^{k-1} + 5$ and conjectured that $G(k) < 2k+1$ when $k$ is not a power of 2 and $G(k) < 4k$ when $k$ is a power of 2.

**16.** The best upper bound known for $G(k)$ is $G(k) < ck \log k$ for some constant $c$.

**17.** The known values and established estimates for $g(k)$ and $G(k)$ for $2 \le k \le 8$ are given in the following table.

| | |
|---|---|
| $g(2) = 4$ | $G(2) = 4$ |
| $g(3) = 9$ | $4 \le G(3) \le 7$ |
| $g(4) = 19$ | $G(4) = 16$ |
| $g(5) = 37$ | $6 \le G(5) \le 17$ |
| $g(6) = 73$ | $9 \le G(6) \le 24$ |
| $143 \le g(7) \le 3,806$ | $8 \le G(7) \le 33$ |
| $279 \le g(8) \le 36,119$ | $32 \le G(8) \le 42$ |

**18.** There are many related diophantine equations concerning sums and differences of powers. For instance $x = 1$, $y = 12$, $z = 9$, and $w = 10$ is the smallest solution to $x^3 + y^3 = z^3 + w^3$.

# 4.9   DIOPHANTINE APPROXIMATION

Diophantine approximation is the study of how closely a number $\theta$ can be approximated by numbers of some particular kind. Usually $\theta$ is an irrational (real) number, and the goal is to approximate $\theta$ using rational numbers $\frac{p}{q}$. Standard references are [Ca57] and [Sc80].

## 4.9.1   CONTINUED FRACTIONS

**Definitions:**
A *continued fraction* is a (finite or infinite) expression of the form

$$a_0 + \cfrac{1}{a_1 + \cfrac{1}{a_2 + \cfrac{1}{a_3 + \cfrac{1}{\ddots}}}}$$

The terms $a_0, a_1, \ldots$ are called the **partial quotients**. If the partial quotients are all integers, and $a_i \ge 1$ for $i \ge 1$, then the continued fraction is said to be **simple**. For convenience, the above expression is usually abbreviated as $[a_0, a_1, a_2, a_3, \ldots]$.

A continued fraction that has an expansion with a block that repeats after some point is called **ultimately periodic**. The ultimately periodic continued fraction expansion $[a_0, a_1, \ldots,\ a_N, a_{N+1}, \ldots, a_{N+k}, a_{N+1}, \ldots, a_{N+k}, a_{N+1}, \ldots]$ is often abbreviated as $[a_0, a_1, \ldots, a_N, \overline{a_{N+1}, \ldots, a_{N+k}}]$. The terms $a_0, a_1, \ldots, a_N$ are called the **pre-period** and the terms $a_{N+1}, a_{N+2}, \ldots, a_{N+k}$ are called the **period**.

## Facts:

**1.** Every irrational number has a unique expansion as a simple continued fraction.

**2.** Every rational number has exactly two simple continued fraction expansions, one with an odd number of terms and one with an even number of terms. Of these, the one with the larger number of terms ends with 1.

**3.** The simple continued fraction for a real number $r$ is finite if and only if $r$ is rational.

**4.** The simple continued fraction for a real number $r$ is infinite and ultimately periodic if and only if $r$ is a quadratic irrational.

**5.** The simple continued fraction for $\sqrt{d}$, where $d$ a positive integer that is not a square, is as follows: $\sqrt{d} = [a_0, \overline{a_1, a_2, \ldots, a_n, 2a_0}]$, where the sequence $(a_1, a_2, \ldots, a_n)$ is a palindrome.

**6.** The following table illustrates the three types of continued fractions.

| type | kind of number | example |
|---|---|---|
| finite | rational | $\frac{355}{113} = [3, 7, 16]$ |
| ultimately periodic | quadratic irrational | $\sqrt{2} = [1, 2, 2, 2, \ldots]$ |
| infinite, but not ultimately periodic | neither rational nor quadratic irrational | $\pi = [3, 7, 15, 1, 292, \ldots]$ |

**7.** The continued fraction for a real number can be computed by Algorithm 1.

---

**Algorithm 1:  The continued fraction algorithm.**

**procedure** $CFA(x$: real number$)$

$i := 0$

$x_0 := x$

$a_0 := \lfloor x_0 \rfloor$

output$(a_0)$

**while** $(x_i \neq a_i)$

$\quad x_{i+1} := \frac{1}{x_i - a_i}$

$\quad i := i + 1$

$\quad a_i := \lfloor x_i \rfloor$

$\quad$ output$(a_i)$

{Return the finite or infinite sequence $(a_0, a_1, \ldots)$}

---

**8.** The following table gives continued fractions for $\sqrt{d}$, with $d$ square free and $d \leq 100$.

| $d$ | $\sqrt{d}$ | $d$ | $\sqrt{d}$ |
|---|---|---|---|
| 2 | $[1, \overline{2}]$ | 53 | $[7, \overline{3, 1, 1, 3, 14}]$ |
| 3 | $[1, \overline{1, 2}]$ | 54 | $[7, \overline{2, 1, 6, 1, 2, 14}]$ |
| 5 | $[2, \overline{4}]$ | 55 | $[7, \overline{2, 2, 2, 14}]$ |
| 6 | $[2, \overline{2, 4}]$ | 56 | $[7, \overline{2, 14}]$ |
| 7 | $[2, \overline{1, 1, 1, 4}]$ | 57 | $[7, \overline{1, 1, 4, 1, 1, 14}]$ |
| 8 | $[2, \overline{1, 4}]$ | 58 | $[7, \overline{1, 1, 1, 1, 1, 1, 14}]$ |
| 10 | $[3, \overline{6}]$ | 59 | $[7, \overline{1, 2, 7, 2, 1, 14}]$ |
| 11 | $[3, \overline{3, 6}]$ | 60 | $[7, \overline{1, 2, 1, 14}]$ |

| $d$ | $\sqrt{d}$ | $d$ | $\sqrt{d}$ |
|---|---|---|---|
| 12 | $[3, \overline{2,6}]$ | 61 | $[7, \overline{1,4,3,1,2,2,1,3,4,1,14}]$ |
| 13 | $[3, \overline{1,1,1,1,6}]$ | 62 | $[7, \overline{1,6,1,14}]$ |
| 14 | $[3, \overline{1,2,1,6}]$ | 63 | $[7, \overline{1,14}]$ |
| 15 | $[3, \overline{1,6}]$ | 65 | $[8, \overline{16}]$ |
| 17 | $[4, \overline{8}]$ | 66 | $[8, \overline{8,16}]$ |
| 18 | $[4, \overline{4,8}]$ | 67 | $[8, \overline{5,2,1,1,7,1,1,2,5,16}]$ |
| 19 | $[4, \overline{2,1,3,1,2,8}]$ | 68 | $[8, \overline{4,16}]$ |
| 20 | $[4, \overline{2,8}]$ | 69 | $[8, \overline{3,3,1,4,1,3,3,16}]$ |
| 21 | $[4, \overline{1,1,2,1,1,8}]$ | 70 | $[8, \overline{2,1,2,1,2,16}]$ |
| 22 | $[4, \overline{1,2,4,2,1,8}]$ | 71 | $[8, \overline{2,2,1,7,1,2,2,16}]$ |
| 23 | $[4, \overline{1,3,1,8}]$ | 72 | $[8, \overline{2,16}]$ |
| 24 | $[4, \overline{1,8}]$ | 73 | $[8, \overline{1,1,5,5,1,1,16}]$ |
| 26 | $[5, \overline{10}]$ | 74 | $[8, \overline{1,1,1,1,16}]$ |
| 27 | $[5, \overline{5,10}]$ | 75 | $[8, \overline{1,1,1,16}]$ |
| 28 | $[5, \overline{3,2,3,10}]$ | 76 | $[8, \overline{1,2,1,1,5,4,5,1,1,2,1,16}]$ |
| 29 | $[5, \overline{2,1,1,2,10}]$ | 77 | $[8, \overline{1,3,2,3,1,16}]$ |
| 30 | $[5, \overline{2,10}]$ | 78 | $[8, \overline{1,4,1,16}]$ |
| 31 | $[5, \overline{1,1,3,5,3,1,1,10}]$ | 79 | $[8, \overline{1,7,1,16}]$ |
| 32 | $[5, \overline{1,1,1,10}]$ | 80 | $[8, \overline{1,16}]$ |
| 33 | $[5, \overline{1,2,1,10}]$ | 82 | $[9, \overline{18}]$ |
| 34 | $[5, \overline{1,4,1,10}v]$ | 83 | $[9, \overline{9,18}v]$ |
| 35 | $[5, \overline{1,10}]$ | 84 | $[9, \overline{6,18}]$ |
| 37 | $[6, \overline{12}]$ | 85 | $[9, \overline{4,1,1,4,18}]$ |
| 38 | $[6, \overline{6,12}]$ | 86 | $[9, \overline{3,1,1,1,8,1,1,1,3,18}]$ |
| 39 | $[6, \overline{4,12}]$ | 87 | $[9, \overline{3,18}]$ |
| 40 | $[6, \overline{3,12}]$ | 88 | $[9, \overline{2,1,1,1,2,18}]$ |
| 41 | $[6, \overline{2,2,12}]$ | 89 | $[9, \overline{2,3,3,2,18}]$ |
| 42 | $[6, \overline{2,12}]$ | 90 | $[9, \overline{2,18}]$ |
| 43 | $[6, \overline{1,1,3,1,5,1,3,1,1,12}]$ | 91 | $[9, \overline{1,1,5,1,5,1,1,18}]$ |
| 44 | $[6, \overline{1,1,1,2,1,1,1,12}]$ | 92 | $[9, \overline{1,1,2,4,2,1,1,18}]$ |
| 45 | $[6, \overline{1,2,2,2,1,12}]$ | 93 | $[9, \overline{1,1,1,4,6,4,1,1,1,18}]$ |
| 46 | $[6, \overline{1,3,1,1,2,6,2,1,1,3,1,12}]$ | 94 | $[9, \overline{1,2,3,1,1,5,1,8,1,5,1,1,3,2,1,18}]$ |
| 47 | $[6, \overline{1,5,1,12}]$ | 95 | $[9, \overline{1,2,1,18}]$ |
| 48 | $[6, \overline{1,12}]$ | 96 | $[9, \overline{1,3,1,18}]$ |
| 50 | $[7, \overline{14}]$ | 97 | $[9, \overline{1,5,1,1,1,1,1,1,5,1,18}]$ |
| 51 | $[7, \overline{7,14}]$ | 98 | $[9, \overline{1,8,1,18}]$ |
| 52 | $[7, \overline{4,1,2,1,4,14}]$ | 99 | $[9, \overline{1,18}]$ |

**9.** Continued fraction expansions for certain quadratic irrationals are given in the following table.

| $d$ | continued fraction expansion for $\sqrt{d}$ |
|---|---|
| $\sqrt{n^2-1}$ | $[\,n-1,\overline{1,2n-2}\,]$ |
| $\sqrt{n^2-2}$ | $[\,n-1,\overline{1,n-2,1,2n-2}\,]$ |
| $\sqrt{n^2+1}$ | $[\,n,\overline{2n}\,]$ |
| $\sqrt{n^2+2}$ | $[\,n,\overline{n,2n}\,]$ |
| $\sqrt{n^2-n}$ | $[\,n-1,\overline{2,2n-2}\,]$ |
| $\sqrt{n^2+n}$ | $[\,n,\overline{2,2n}\,]$ |
| $\sqrt{4n^2+4}$ | $[\,2n,\overline{n,4n}\,]$ |
| $\sqrt{4n^2-n}$ | $[\,2n-1,\overline{1,2,1,4n-2}\,]$ |
| $\sqrt{4n^2+n}$ | $[\,2n,\overline{4,4n}\,]$ |
| $\sqrt{9n^2+2n}$ | $[\,3n,\overline{3,6n}\,]$ |

**10.** Continued fraction expansions for some famous numbers are given in the following table.

| number | continued fraction expansion |
|---|---|
| $\pi$ | $[\,3,7,15,1,292,1,1,1,2,1,3,1,14,2,1,1,2,2,2,2,1,84,2,1,1,15,3,\ldots]$ |
| $\gamma$ | $[\,0,1,1,2,1,2,1,4,3,13,5,1,1,8,1,2,4,1,1,40,1,11,3,7,1,7,1,1,5,\ldots]$ |
| $\sqrt[3]{2}$ | $[\,1,3,1,5,1,1,4,1,1,8,1,14,1,10,2,1,4,12,2,3,2,1,3,4,1,1,2,14,\ldots]$ |
| $\log 2$ | $[\,0,1,2,3,1,6,3,1,1,2,1,1,1,1,3,10,1,1,1,2,1,1,1,1,3,2,3,1,13,7,\ldots]$ |
| $e$ | $[\,2,1,2,1,1,4,1,1,6,1,1,8,1,1,10,1,1,12,\ldots]$ |
| $e^{\frac{1}{n}}$ | $[\,1,\ \overline{n-1,\ 1,1,\ 3n-1,\ 1,1,\ 5n-1,\ 1,1,\ 7n-1},\ldots]$ |
| $e^{\frac{2}{2n+1}}$ | $[\,1,\ \overline{(6n+3)k+n,\ (24n+12)k+12n+6,\ (6n+3)k+5n+2,\ 1,\ 1}_{k\geq 0}\,]$ |
| $\tanh\frac{1}{n}$ | $[\,0,\ n,\ 3n,\ 5n,\ 7n,\ldots]$ |
| $\tan\frac{1}{n}$ | $[\,0,n-1,1,3n-2,1,5n-2,1,7n-2,1,9n-2,\ldots]$ |
| $\frac{1+\sqrt{5}}{2}$ | $[\,1,1,1,1,\ldots]$ |

**Examples:**

**1.** To find the continued fraction representation of $\frac{62}{23}$, apply Algorithm 1 to obtain

$$\tfrac{62}{23} = 2 + \cfrac{1}{\frac{23}{16}}, \quad \tfrac{23}{16} = 1 + \cfrac{1}{\frac{16}{7}}, \quad \tfrac{16}{7} = 2 + \cfrac{1}{\frac{7}{2}}, \quad \tfrac{7}{2} = 3 + \cfrac{1}{2}.$$

Combining these equations shows that $\frac{62}{23} = [2,1,2,3,2]$. Since $2 = 1 + \frac{1}{1}$, it also follows that $\frac{62}{23} = [2,1,2,3,1,1]$.

**2.** Applying Algorithm 1 to find the continued fraction of $\sqrt{6}$, it follows that

$$a_0 = \lfloor\sqrt{6}\rfloor = 2, \quad a_1 = \left\lfloor\tfrac{\sqrt{6}+2}{2}\right\rfloor = 2, \quad a_2 = \lfloor\sqrt{6}+2\rfloor = 4, \quad a_3 = a_1, \quad a_4 = a_2,\ldots .$$

Hence $\sqrt{6} = [\,2,\overline{2,4}\,]$.

**3.** The continued fraction expansion of $e$ is $e = [\,2,1,2,1,1,4,1,1,6,\ldots]$. This expansion is often abbreviated as $[\,2,\overline{1,2k,1}_{k\geq 1}\,]$. (See [Pe54].)

## 4.9.2  CONVERGENTS

**Definitions:**

Let $p_{-2} = 0$, $q_{-2} = 1$, $p_{-1} = 1$, $q_{-1} = 0$, and define $p_n = a_n p_{n-1} + p_{n-2}$ and $q_n = a_n q_{n-1} + q_{n-2}$ for $n \geq 0$. Then $\frac{p_n}{q_n} = [a_0, a_1, \ldots, a_n]$. The fraction $\frac{p_n}{q_n}$ is called the $n$th *convergent*.

**Facts:**

1. $p_n q_{n-1} - p_{n-1} q_n = (-1)^{n+1}$ for $n \geq 0$.

2. Let $\theta = [a_0, a_1, a_2, \ldots]$ be an irrational number. Then $\left| \theta - \frac{p_n}{q_n} \right| < \frac{1}{a_{n+1} q_n^2}$.

3. If $n > 1$, $0 < q \leq q_n$, and $\frac{p}{q} \neq \frac{p_n}{q_n}$, then $\left| \theta - \frac{p}{q} \right| > \left| \theta - \frac{p_n}{q_n} \right|$.

4. $[\ldots, a, b, 0, c, d, \ldots] = [\ldots, a, b + c, d, \ldots]$.

5. Almost all real numbers have unbounded partial quotients.

6. For almost all real numbers, the frequency with which the partial quotient $k$ occurs is $\log_2 \left( 1 + \frac{1}{k(k+2)} \right)$. Hence, the partial quotient 1 occurs about 41.5% of the time, the partial quotient 2 occurs about 17.0% of the time, etc.

7. For almost all real numbers,

$$\lim_{n \to \infty} (a_1 a_2 \ldots a_n)^{\frac{1}{n}} = K \approx 2.68545.$$

$K$ is called *Khintchine's constant*.

8. *Lévy's law:* For almost all real numbers,

$$\lim_{n \to \infty} (p_n)^{\frac{1}{n}} = \lim_{n \to \infty} (q_n)^{\frac{1}{n}} = e^{\frac{\pi^2}{12 \log 2}}.$$

**Examples:**

1. We compute the first eight convergents to $\pi$ as follows:

| $n =$ | $-2$ | $-1$ | 0 | 1 | 2 | 3 | 4 | 5 | 6 | 7 | 8 |
|---|---|---|---|---|---|---|---|---|---|---|---|
| $a_n =$ | | | 3 | 7 | 15 | 1 | 292 | 1 | 1 | 1 | 2 |
| $p_n =$ | 0 | 1 | 3 | 22 | 333 | 355 | 103,993 | 104,348 | 208,341 | 312,689 | 833,719 |
| $q_n =$ | 1 | 0 | 1 | 7 | 106 | 113 | 33,102 | 33,215 | 66,317 | 99,532 | 265,381 |

2. In order to find a rational fraction $\frac{p}{q}$ in lowest terms that approximates $e$ to within $10^{-6}$, we compute the convergents $q_n$ until $a_{n+1}(q_n)^2 < 10^{-6}$:

| $n =$ | $-2$ | $-1$ | 0 | 1 | 2 | 3 | 4 | 5 | 6 | 7 | 8 | 9 | 10 |
|---|---|---|---|---|---|---|---|---|---|---|---|---|---|
| $a_n =$ | | | 2 | 1 | 2 | 1 | 1 | 4 | 1 | 1 | 6 | 1 | 1 |
| $p_n =$ | 0 | 1 | 2 | 3 | 8 | 11 | 19 | 87 | 106 | 193 | 1,264 | 1,457 | 2,721 |
| $q_n =$ | 1 | 0 | 1 | 1 | 3 | 4 | 7 | 32 | 39 | 71 | 465 | 536 | 1,001 |

Hence, $\frac{2721}{1001} \approx 2.71828171$ is the desired fraction.

### 4.9.3  APPROXIMATION THEOREMS

**Facts:**

**1.** *Dirichlet's theorem:*  If $\theta$ is irrational, then

$$\left|\theta - \tfrac{p}{q}\right| < \tfrac{1}{q^2}$$

for infinitely many $p, q$.

**2.** *Dirichlet's theorem in d dimensions:*  If $\theta_1, \theta_2, \ldots, \theta_d$ are real numbers with at least one $\theta_i$ irrational, then

$$\left|\theta_i - \tfrac{p_i}{q}\right| < \tfrac{1}{q^{1+\frac{1}{d}}}$$

for infinitely many $p_1, p_2, \ldots, p_d, q$.

**3.** *Hurwitz's theorem:*  If $\theta$ is an irrational number, then

$$\left|\theta - \tfrac{p}{q}\right| < \tfrac{1}{\sqrt{5}q^2}$$

for infinitely many $p, q$. The constant $\sqrt{5}$ is best possible.

**4.** *Liouville's theorem:*  Let $\theta$ be an irrational algebraic number of degree $n$. Then there exists a constant $c$ (depending on $\theta$) such that

$$\left|\theta - \tfrac{p}{q}\right| > \tfrac{c}{q^n}$$

for all rationals $\frac{p}{q}$ with $q > 0$. The number $\theta$ is called a *Liouville number* if $\left|\theta - \tfrac{p}{q}\right| < q^{-n}$ has a solution for all $n \geq 0$. An example of a Liouville number is $\sum_{k \geq 1} 2^{-k!}$.

**5.** *Roth's theorem:*  Let $\theta$ be an irrational algebraic number, and let $\epsilon$ be any positive number. Then

$$\left|\theta - \tfrac{p}{q}\right| > \tfrac{1}{q^{2+\epsilon}}$$

for all but finitely many rationals $\frac{p}{q}$ with $q > 0$.

### 4.9.4  IRRATIONALITY MEASURES

**Definitions:**

Let $\theta$ be a real irrational number. Then the real number $\mu$ is said to be an **irrationality measure** for $\theta$ if for every $\epsilon > 0$ there exists a positive real $q_0 = q_0(\epsilon)$ such that $\left|\theta - \tfrac{p}{q}\right| > q^{-(\mu+\epsilon)}$ for all integers $p$, $q$ with $q > q_0$.

**Fact:**

**1.** The following table shows the best irrationality measures known for some important numbers.

| number $\theta$ | measure $\mu$ | discoverer |
|:---:|:---:|:---:|
| $\pi$ | 7.6064 | Salikhov (2008) |
| $\pi^2$ | 5.4413 | Rhin and Viola (1996) |
| $\zeta(3)$ | 5.5139 | Rhin and Viola (2001) |
| $\log 2$ | 3.5746 | Marcovecchio (2009) |
| $\frac{\pi}{\sqrt{3}}$ | 4.2304 | Androsenko (2015) |

# 4.10  ALGEBRAIC NUMBER THEORY

When working on questions involving integers or rational numbers, one is often led to look at extensions of the rational numbers. These yield rings of algebraic integers, which have many properties similar to the usual integers. A notable exception is the frequent lack of unique factorization into primes.

Many classical questions can be rephrased in terms of algebraic number theory, for example, factorization of polynomials modulo primes. The number field sieve, which is based on algebraic number theory, is the strongest general purpose factorization algorithm currently being used. Algebraic number theory is an active subject of research and remains a basic tool in diophantine equations and in many other areas of mathematics.

Standard references are [La00], [IrRo98], [Mu11], [StTa16], and [Wa97].

## 4.10.1  BASIC CONCEPTS

### Definitions:

A complex number $\alpha$ is an **algebraic number** if it is a root of a monic polynomial with rational coefficients. (A **monic polynomial** is a polynomial with leading coefficient equal to 1.)

An algebraic number $\alpha$ is an **algebraic integer** if it is a root of a monic polynomial with integer coefficients.

If $\alpha$ is an algebraic number with minimal polynomial $f(x)$ of degree $n$, then the $n-1$ other roots of $f(x)$ are called the **conjugates** of $\alpha$.

If $\alpha$ and $\beta$ are algebraic integers and there is an algebraic integer $\gamma$ such that $\alpha\gamma = \beta$, then $\alpha$ **divides** $\beta$, written $\alpha|\beta$.

An algebraic number $\alpha$ is of **degree** $n$ if it is a root of a monic polynomial with rational coefficients of degree $n$ but is not a root of any monic polynomial with rational coefficients of degree less than $n$.

The **integers** of an algebraic number field are the algebraic integers that belong to this field.

A **number field** is a field that is a finite degree extension of $\mathcal{Q}$.

An algebraic integer $\epsilon$ is a **unit** if there exists an algebraic integer $\gamma$ such that $\epsilon\gamma = 1$.

### Facts:

**1.** The integers are the only rational numbers that are algebraic integers.

**2.** Sums and products of algebraic integers are algebraic integers, so the algebraic integers in a field form a ring.

**3.** Each element of a number field is an algebraic number.

### Examples:

**1.** Two of the most common examples of number fields are quadratic fields and cyclotomic fields. (See §4.10.2 and §4.10.6.)

**2.** The numbers $2 + \sqrt{3}$ and $2 - \sqrt{3}$ are algebraic integers because they are roots of the polynomial $X^2 - 4X + 1$. They are conjugates of each other. They are units because $(2 + \sqrt{3})(2 - \sqrt{3}) = 1$.

**3.** $1/\sqrt{2}$ is an algebraic number but is not an algebraic integer. It is a root of the polynomial is $X^2 - (1/2)$, which does not have integer coefficients, and there is no monic polynomial with integer coefficients that has $1/\sqrt{2}$ as a root.

**4.** The minimal polynomial of $\sqrt[3]{2}$ is $X^3 - 2$. The other roots of this polynomial are $\omega\sqrt[3]{2}$ and $\omega^2\sqrt[3]{2}$, where $\omega$ is a primitive cube root of unity, so these are the conjugates of $\sqrt[3]{2}$.

**5.** Let $i = \sqrt{-1}$. The field $\mathcal{Q}(i) = \{a + bi \mid a, b \in \mathcal{Q}\}$ is a number field of degree 2.

**6.** The field $\mathcal{Q}(\sqrt[3]{2}) = \{a + b\sqrt[3]{2} + c\sqrt[3]{4} \mid a, b, c \in \mathcal{Q}\}$ is a number field of degree 3.

**7.** The algebraic integers in $\mathcal{Q}(\sqrt[3]{2})$ are $\mathcal{Z}[\sqrt[3]{2}] = \{a + b\sqrt[3]{2} + c\sqrt[3]{4} \mid a, b, c \in \mathcal{Z}\}$.

**8.** Algebraic numbers arise naturally when solving diophantine equations involving integers. (See §4.10.4, Example 6.)

---

## 4.10.2    QUADRATIC FIELDS

### Definitions:

If $d \neq 1$ is a square-free integer, then $\mathcal{Q}(\sqrt{d}) = \{a + b\sqrt{d} \mid a \text{ and } b \text{ are rational numbers}\}$ is called a **quadratic field**. If $d > 1$, then $\mathcal{Q}(\sqrt{d})$ is called a **real quadratic field**; if $d < 0$, then $\mathcal{Q}(\sqrt{d})$ is called an **imaginary quadratic field**.

A number $\alpha$ in $\mathcal{Q}(\sqrt{d})$ is a **quadratic integer** (or an **integer** when the context is clear) if $\alpha$ is an algebraic integer.

The integers of $\mathcal{Q}(\sqrt{-1})$ are called the **Gaussian integers**. (These are the numbers in $\mathcal{Z}[i] = \{a + bi \mid a, b \text{ are integers}\}$. See §5.4.2.)

If $\alpha = a + b\sqrt{d}$ belongs to $\mathcal{Q}(\sqrt{d})$, then its **conjugate**, denoted by $\overline{\alpha}$, is the number $a - b\sqrt{d}$.

If $\alpha$ belongs to $\mathcal{Q}(\sqrt{d})$, then the **norm** of $\alpha$ is the number $N(\alpha) = \alpha\overline{\alpha}$.

### Facts:

**1.** The integers of the field $\mathcal{Q}(\sqrt{d})$, where $d$ is a square-free integer, are the numbers $a + b\sqrt{d}$ when $d \equiv 2$ or $3 \pmod{4}$. When $d \equiv 1 \pmod{4}$, the integers of the field are the numbers $\frac{a + b\sqrt{d}}{2}$, where $a$ and $b$ are integers which are either both even or both odd.

**2.** If $d < 0$, $d \neq -1$, $d \neq -3$, then there are exactly two units, $\pm 1$, in $\mathcal{Q}(\sqrt{d})$. There are exactly four units in $Q(\sqrt{-1})$, namely $\pm 1$ and $\pm i$. There are exactly six units in $\mathcal{Q}(\sqrt{-3})$: $\pm 1, \pm\frac{-1+\sqrt{-3}}{2}, \pm\frac{-1-\sqrt{-3}}{2}$.

**3.** If $d > 1$, there are infinitely many units in $\mathcal{Q}(\sqrt{d})$. Furthermore, there is a unit $\epsilon_0 > 1$, called the *fundamental unit* of $\mathcal{Q}(\sqrt{d})$, such that all units are of the form $\pm\epsilon_0^n$ where $n$ is an integer.

### Examples:

**1.** The conjugate of $-2 + 3i$ in the ring of Gaussian integers is $-2 - 3i$. Consequently, $N(-2 + 3i) = (-2 - 3i)(-2 + 3i) = 13$.

**2.** The number $1 + \sqrt{2}$ is a fundamental unit of $\mathcal{Q}(\sqrt{2})$. Therefore, all units are of the form $\pm(1 + \sqrt{2})^n$ where $n = 0, \pm 1, \pm 2, \ldots$.

## 4.10.3   PRIMES AND UNIQUE FACTORIZATION IN QUADRATIC FIELDS

### Definitions:

An integer $\pi$ in $\mathcal{Q}(\sqrt{d})$, not zero or a unit, is **irreducible** in $\mathcal{Q}(\sqrt{d})$ if, whenever $\pi = \alpha\beta$ where $\alpha$ and $\beta$ are integers in $\mathcal{Q}(\sqrt{d})$, either $\alpha$ or $\beta$ is a unit.

An integer $\pi$ in $\mathcal{Q}(\sqrt{d})$, not zero or a unit, is **prime** in $\mathcal{Q}(\sqrt{d})$ if, whenever $\pi \mid \alpha\beta$ where $\alpha$ and $\beta$ are integers in $\mathcal{Q}(\sqrt{d})$, either $\pi \mid \alpha$ or $\pi \mid \beta$.

If $\alpha$ and $\beta$ are nonzero integers in $\mathcal{Q}(\sqrt{d})$ and $\alpha = \beta\epsilon$ where $\epsilon$ is a unit, then $\beta$ is called an **associate** of $\alpha$.

A quadratic field $\mathcal{Q}(\sqrt{d})$ is a **Euclidean field** with respect to the norm if, given integers $\alpha$ and $\beta$ in $\mathcal{Q}(\sqrt{d})$ where $\beta$ is not zero, there are integers $\delta$ and $\gamma$ in $\mathcal{Q}(\sqrt{d})$ such that $\alpha = \gamma\beta + \delta$ and $|N(\delta)| < |N(\beta)|$.

A quadratic field $\mathcal{Q}(\sqrt{d})$ has the **unique factorization property** if whenever $\alpha \neq 0$ is a non-unit integer in $\mathcal{Q}(\sqrt{d})$ with two factorizations $\alpha = \epsilon\pi_1\pi_2\ldots\pi_r = \epsilon'\pi_1'\pi_2'\ldots\pi_s'$ where $\epsilon$ and $\epsilon'$ are units and each $\pi_i$ and $\pi_j'$ is prime, then $r = s$ and the primes $\pi_i$ and $\pi_j'$ can be paired off into pairs of associates.

### Facts:

**1.** If $\alpha$ is an integer in $\mathcal{Q}(\sqrt{d})$ and $N(\alpha)$ is an integer that is prime, then $\alpha$ is irreducible.

**2.** The integers of $\mathcal{Q}(\sqrt{d})$ are a unique factorization domain if and only if every irreducible is prime.

**3.** A Euclidean quadratic field has the unique factorization property.

**4.** The quadratic field $\mathcal{Q}(\sqrt{d})$ is Euclidean if and only if $d$ is one of the following integers: $-11, -7, -3, -2, -1, 2, 3, 5, 6, 7, 11, 13, 17, 19, 21, 29, 33, 37, 41, 57, 73$.

**5.** If $d < 0$, then the imaginary quadratic field $\mathcal{Q}(\sqrt{d})$ has the unique factorization property if and only if $d = -1, -2, -3, -7, -11, -19, -43, -67$, or $-163$. This theorem was stated as a conjecture by Gauss in the 19th century and was proved in the 1960s by Harold Stark and Alan Baker independently.

**6.** In 1952, the engineer Kurt Heegner (1893–1965) had published a proof of the result of Baker and Stark, but it was not accepted because it relied on some defective statements of Heinrich Weber. After the proofs by Baker and Stark, the gaps in Heegner's proof were filled in by Stark, Max Deuring, and Bryan Birch, independently.

**7.** It is unknown whether infinitely many real quadratic fields $\mathcal{Q}(\sqrt{d})$ have the unique factorization property.

**8.** Of the 60 real quadratic fields $\mathcal{Q}(\sqrt{d})$ with $2 \leq d \leq 100$, exactly 38 have the unique factorization property, namely those with $d = 2, 3, 5, 6, 7, 11, 13$ 14, 17, 19, 21, 22, 23, 29, 31, 33, 37, 38, 41, 43, 46, 47, 53, 57, 59, 61, 62, 67, 69, 71, 73, 77, 83, 86, 89, 93, 94, and 97.

**9.** The *Cohen-Lenstra heuristics* [CoLe84] predict that, as $p$ runs through the primes $p \equiv 1 \pmod 4$, approximately 75.446% of the fields $\mathcal{Q}(\sqrt{p})$ have the unique factorization property.

## Examples:

**1.** The number $2 + i$ is an irreducible Gaussian integer. This follows since its norm $N(2 + i) = (2 + i)(2 - i) = 5$ is a prime integer. Its associates are itself and the three Gaussian integers $(-1)(2 + i) = -2 - i$, $i(2 + i) = -1 + 2i$, and $-i(2 + i) = 1 - 2i$.

**2.** The integers of $\mathcal{Q}(\sqrt{-5})$ are the numbers of the form $a + b\sqrt{-5}$, where $a$ and $b$ are integers. The field $\mathcal{Q}(\sqrt{-5})$ does not have the unique factorization property. To see this, note that $6 = 2 \cdot 3 = (1 + \sqrt{-5})(1 - \sqrt{-5})$ and each of 2, 3, $1 + \sqrt{-5}$, and $1 - \sqrt{-5}$ are irreducibles in this quadratic field. For example, to see that $1 + \sqrt{-5}$ is irreducible, suppose that $1 + \sqrt{-5} = (a + b\sqrt{-5})(c + d\sqrt{-5})$. This implies that $6 = (a^2 + 5b^2)(c^2 + 5d^2)$, which is impossible unless $a = \pm 1$, $b = 0$ or $c = \pm 1$, $d = 0$. Consequently, one of the factors must be a unit.

---

## 4.10.4   GENERAL NUMBER FIELDS

### Definitions:

If $I$ and $J$ are ideals of the ring of algebraic integers of a number field, $I$ **divides** $J$ if there exists an ideal $I'$ of this ring such that $II' = J$.

Let $K$ be a number field of degree $n$ over $\mathcal{Q}$. Choose algebraic integers $\beta_1, \ldots, \beta_n$ in $K$ such that every algebraic integer in $K$ is a linear combination of $\beta_1, \ldots, \beta_n$ with integer coefficients. Form the $n \times n$ matrix $(a_{ij})$, where the $ij$th entry is the trace from $K$ to $\mathcal{Q}$ of $\beta_i \beta_j$. The **discriminant** of $K$, denoted $D_K$, is the determinant of the matrix $(a_{ij})$.

Two nonzero ideals $I_1$ and $I_2$ of the ring of algebraic integers of a number field $K$ are in the same **ideal class** if there is an element $x$ in the field such that $I_1 = xI_2$, where the latter is the set $\{xi \mid i \in I_2\}$.

The set of ideal classes forms a group, called the **ideal class group** of $K$.

The order of the ideal class group is called the **class number** of $K$, and is often denoted $h$ or $h_K$.

If $\alpha$ and $\beta$ are algebraic integers in a number field $K$ and $I$ is an ideal of the ring of integers of $K$, then $\alpha$ is **congruent** to $\beta$ mod $I$, written $\alpha \equiv \beta \pmod{I}$, if $\alpha - \beta \in I$.

Let $K$ be a number field and let $\mathcal{O}_K$ be the ring of algebraic integers in $K$. Let $I$ be a nonzero ideal of $\mathcal{O}_K$. The **norm** of $I$, denoted $N(I)$, is the order of $\mathcal{O}_K / I$.

Let $K$ be a number field of degree $n$ over $\mathcal{Q}$. Write $K = \mathcal{Q}(\alpha)$ for some $\alpha$, and let $m(X)$ be the minimal polynomial of $\alpha$. This polynomial has $r_1$ real roots and $r_2$ pairs of complex roots. The numbers $r_1$ and $r_2$ are called the **number of real embeddings** of $K$ and the **number of pairs of complex embeddings**, respectively.

### Facts:

**1.** Algebraic number theory arose from attempts to generalize the law of quadratic reciprocity (see §4.7.3) to higher power reciprocity laws.

**2.** Cubic reciprocity is best stated using the Eisenstein integers, namely the algebraic integers $\mathcal{Z}[\omega]$, where $\omega$ is a primitive cube root of unity. Let $\pi_1$ and $\pi_2$ be two primes of $\mathcal{Z}[\omega]$ that are relatively prime to each other and do not divide 3. The *cubic residue symbol* $\left(\dfrac{\pi_1}{\pi_2}\right)_3$ is the unique power of $\omega$ satisfying

$$\pi_1^{(N(\pi_2)-1)/3} \equiv \left(\frac{\pi_1}{\pi_2}\right)_3 \pmod{\pi_2}.$$

**3.** *Cubic reciprocity* says that if $\pi_1$ and $\pi_2$ are congruent to 2 modulo 3, then

$$\left(\frac{\pi_1}{\pi_2}\right)_3 = \left(\frac{\pi_2}{\pi_1}\right)_3.$$

This was proved by Gotthold Eisenstein (1823–1852), Gauss, and others.

**4.** Gauss proved a quartic (or biquadratic) reciprocity law in the context of the Gaussian integers $\mathcal{Z}[i]$, where $i = \sqrt{-1}$.

**5.** Ernst Kummer (1810–1893) studied $p$th power reciprocity laws for odd primes $p$ in the context of the $p$th cyclotomic field $\mathcal{Q}(\zeta)$ (see §4.10.6), where $\zeta$ is a primitive $p$th root of unity.

**6.** Kummer realized that unique factorization can fail in cyclotomic fields and was thus led to his theory of ideal numbers (in modern terminology, these are essentially ring homomorphisms from the ring of integers of a cyclotomic field to a finite field; see [Ma77]) and ideal class groups.

**7.** Richard Dedekind (1831–1916) reinterpreted Kummer's ideal numbers as ideals in the ring of algebraic integers.

**8.** In modern language, Kummer showed that every nonzero ideal is uniquely a product of prime ideals.

**9.** The ideal class group is a finite abelian group.

**10.** The class number of a number field $K$ is 1 if and only if the ring of algebraic integers in $K$ is a unique factorization domain.

**11.** The concept of the class number of a number field can be useful for finding solutions to diophantine equations in integers. See Example 7 and §4.10.6, Example 3.

**12.** Let $K = \mathcal{Q}(\sqrt{d})$, where $d \neq 1$ is a square-free integer that can be chosen positive or negative. The discriminant $D_K$ is $4d$ if $d \equiv 2,3 \pmod 4$, and $d$ if $d \equiv 1 \pmod 4$.

**13.** If $I$ and $J$ are nonzero ideals of the ring of algebraic integers of a number field, then $I$ divides $J$ if and only if $J \subseteq I$.

**14.** Let $K$ be a number field, let $r_1$ be the number of real embeddings of $K$, and let $r_2$ be the number of pairs of complex embeddings. Let $W_K$ be the group of roots of unity contained in $K$. The *Dirichlet unit theorem* says that the group of units of the ring of algebraic integers in $K$ is the direct sum of $W_K$ and a free abelian group of rank $r_1 + r_2 - 1$.

**15.** The *Minkowski bound* says that if $K$ is a number field of degree $n$ over $\mathcal{Q}$ with $r_2$ pairs of complex embeddings, then each ideal class contains an ideal $I$ with

$$1 \leq N(I) \leq \tfrac{n!}{n^n} \left(\tfrac{4}{\pi}\right)^{r_2} \sqrt{|D_K|},$$

where $D_K$ is the discriminant of $K$.

**16.** The *Cohen-Lenstra heuristics* [CoLe84] predict that if $p$ is an odd prime, then $p$ divides the class number of a certain fraction of imaginary quadratic fields. This fraction is

$$1 - \prod_{m=1}^{\infty}(1 - p^m).$$

For $p = 3$, this is $0.43987\ldots$. For real quadratic fields, the predicted fraction is

$$1 - \prod_{m=2}^{\infty}(1 - p^m).$$

For $p = 3$, this is $0.15981\ldots$. Both of these predictions agree well with computed data.

**Examples:**

**1.** If $K = \mathcal{Q}(\sqrt{-d})$ is an imaginary quadratic field, $r_1 = 0$ and $r_2 = 1$. Therefore, $r_1 + r_2 - 1 = 0$, so the group of units consists of only the roots of unity in $K$.

**2.** If $K = \mathcal{Q}(\sqrt{d})$ is a real quadratic field, $r_1 = 2$ and $r_2 = 0$. Therefore, $r_1 + r_2 - 1 = 1$, so the group of units is the direct sum of $\pm 1$ and an infinite cyclic group. The generator $\epsilon_0 > 1$ for this infinite cyclic group is called the *fundamental unit* of $K$.

**3.** Let $a$ be an integer such that $a^2 + 3a + 9$ is square free. Let $\epsilon_1$ and $\epsilon_2$ be two of the roots of $X^2 - aX - (a+3)X - 1$. Then the group of units of the ring of algebraic integers in $\mathcal{Q}(\epsilon_1)$ is the direct sum of $\pm 1$ and the rank 2 free abelian group generated by $\epsilon_1$ and $\epsilon_2$. The fields of this form are known as the *simplest cubic fields*.

**4.** The ring of algebraic integers in $\mathcal{Q}(\sqrt{-5})$ is the set of numbers of the form $a + b\sqrt{-5}$, where $a, b$ are integers. The ideal $I$ generated by 2 and $1 + \sqrt{-5}$ and the ideal $J$ generated by 3 and $1 - \sqrt{-5}$ are not principal. Let $x = (1 - \sqrt{-5})/2$. Then $xI = J$, so $I$ and $J$ are in the same ideal class. The product $IJ$ is by definition the ideal generated by the products of the generators of $I$ and $J$, so it is the ideal generated by 6, $2 - 2\sqrt{-5}$, $3 + 3\sqrt{-5}$, and 6. Since $6 - (2 - 2\sqrt{-5}) - (3 + \sqrt{-5}) = 1 - \sqrt{-5}$ is in $IJ$, and all of these generators of $IJ$ are multiples of $1 - \sqrt{-5}$, we find that $IJ$ is the principal ideal generated by $1 - \sqrt{-5}$.

**5.** Let $K = \mathcal{Q}(\sqrt{-5})$. In the definition of $D_K$, take $\beta_1 = 1$ and $\beta_2 = \sqrt{-5}$. Then the matrix $(a_{ij})$ is

$$\begin{pmatrix} 2 & 0 \\ 0 & -10 \end{pmatrix},$$

which has determinant $D_K = -20$. The Minkowski bound is $4\sqrt{5}/\pi < 3$, so each ideal class contains an ideal of norm 1 or 2. Only the full ring of algebraic integers has norm 1, and the ideal $I$ in the previous example is the only ideal of norm 2, so there are only two ideal classes. Therefore, the class number is 2.

**6.** Suppose we want to find all integer solutions of $x^2 + 1 = y^3$. Factor the left side as $(x + i)(x - i)$. It can be shown that $x + i$ and $x - i$ are relatively prime in $\mathcal{Z}[i]$, which has the unique factorization property. Since their product is a cube, each factor is a unit times a cube in $\mathcal{Z}[i]$. In particular, $x + i = (a + bi)^3$. The imaginary parts yield $1 = b(3a^2 - b^2)$, which implies that $b = \pm 1 = (3a^2 - b^2)$. This yields $a = 0$, hence $x = 0$. The only integer solution to $x^2 + 1 = y^3$ is therefore $(x, y) = (0, 1)$.

**7.** Suppose we want to find all solutions of $x^2 + 13 = y^3$. Factor the left side as $(x + \sqrt{-13})(x - \sqrt{-13})$. It can be shown that the ideals $I_1 = (x + \sqrt{-13})\mathcal{Z}[\sqrt{-13}]$ and $I_2 = (x - \sqrt{-13})\mathcal{Z}[\sqrt{-13}]$ are relatively prime, in the sense that they have no common prime ideal factors. The product $I_1 I_2$ equals the cube of the ideal $y\mathcal{Z}[\sqrt{-13}]$. Unique factorization into prime ideals implies that both $I_1$ and $I_2$ must be cubes of ideals. Therefore, $J^3 = I_1$ for some ideal $J$ of $\mathcal{Z}[\sqrt{-13}]$. But $I_1$ is a principal ideal, so $J$ has order 1 or 3 in the ideal class group. The class number of $\mathcal{Q}(\sqrt{-13})$ is 2, so $J^2$ must also be principal. It follows that $J$ is principal: $J = (a + b\sqrt{-13})\mathcal{Z}[\sqrt{-13}]$. Since $J^3 = I_1$, and the units of $\mathcal{Z}[\sqrt{-13}]$ are $\pm 1$, which are cubes, we must have $(a + b\sqrt{-13})^3 = x + \sqrt{-13}$. The coefficients of $\sqrt{-13}$ yield $1 = b(3a^2 - 13b^2)$, so $b = \pm 1 = 3a^2 - 13b^2$. This yields $a = \pm 4$, and then $x = \pm 70$, $y = 17$.

**8.** Let $\omega$ be a primitive cube root of unity. Let $\pi_1 = -3\omega - 1$ and $\pi_2 = -6\omega - 1$. Then $\pi_1 \equiv \pi_2 \equiv 2 \,(\mathrm{mod}\ 3)$. Since $N(\pi_2) = 31$, the cubic residue symbol is

$$\left(\frac{\pi_1}{\pi_2}\right)_3 \equiv \pi_1^{(31-1)/3} = -3725 - 18357\omega = 1 + (-6\omega - 1)(2952 - 129\omega) \equiv 1 \,(\mathrm{mod}\ \pi_2).$$

Since $\pi_2 = 1 + 2\pi_1 \equiv 1 \pmod{\pi_1}$, the cubic residue symbol is $\left(\frac{\pi_2}{\pi_1}\right)_3 = 1$.

**9.** In Example 8, cubic reciprocity plus the fact that $\pi_2 \equiv 1 \pmod{\pi_1}$ could be used to calculate $\left(\frac{\pi_1}{\pi_2}\right)_3$:

$$\left(\frac{\pi_1}{\pi_2}\right)_3 = \left(\frac{\pi_2}{\pi_1}\right)_3 = \left(\frac{1}{\pi_1}\right)_3 = 1.$$

This implies that $\pi_1$ is congruent to a cube mod $\pi_2$. A calculation yields $(12 + w)^3 \equiv \pi_1 \pmod{\pi_2}$.

---

## 4.10.5  PROPERTIES OF NUMBER FIELDS

### Definitions:

The ring of algebraic integers in a number field is called **monogenic** if the ring is of the form

$$\mathcal{Z}[\alpha] = \{a_0 + a_1\alpha + a_2\alpha^2 + \cdots + a_m\alpha^m \mid m \geq 0, a_i \in \mathcal{Z} \text{ for } 0 \leq i \leq m\}.$$

Let $p$ be a prime number and let $K$ be a number field of degree $n$ over $\mathcal{Q}$. Let $\mathcal{O}_K$ be the ring of algebraic integers in $K$. Factor the ideal $p\mathcal{O}_K$ into prime ideals in $\mathcal{O}_K$ as $P_1^{e_1} P_2^{e_2} \cdots P_g^{e_g}$, where $P_1, \ldots, P_g$ are distinct prime ideals of $\mathcal{O}_K$ and $e_1, \ldots, e_g$ are positive integers. If some $e_i > 1$ then $p$ **ramifies** in $K$. If $g = n$ then $p$ is said to **split completely** in $K$. If $g = 1$ and $e_1 = 1$, then $p$ is **inert** in $K$.

The **Dedekind zeta function** of a number field $K$ is defined to be

$$\zeta_K(s) = \sum_I N(I)^{-s},$$

where $s$ is a complex number with $\mathcal{R}(s) > 1$ and the sum is over the nonzero ideals $I$ of $\mathcal{O}_K$.

### Facts:

**1.** A prime number $p$ ramifies in a number field $K$ if and only if $p$ divides the discriminant $D_K$.

**2.** If $K \neq \mathcal{Q}$, then $|D_K| > 1$, so there is always at least one prime number that ramifies in $K$.

**3.** Let $K$ be a number field and suppose its ring of integers is $\mathcal{Z}[\alpha]$. Let $f(X)$ be the minimal polynomial of $\alpha$. Let $p$ be a prime and suppose $f(X) \equiv h_1(X)^{e_1} h_2(X)^{e_2} \cdots h_g(X)^{e_g}$ $\pmod{p}$ is the factorization of $f(X)$ mod $p$ into monic polynomials that are distinct mod $p$ and irreducible mod $p$. Then the ideal generated by $p$ factors into prime ideals as $P_1^{e_1} P_2^{e_2} \cdots P_g^{e_g}$ in the ring of integers of $K$. The prime ideal $P_i$ is generated by $p$ and $h_i(\alpha)$.

**4.** The ring of algebraic integers in $\mathcal{Q}(w, \sqrt[3]{2})$, where $w$ is a primitive cube root of unity, is $\mathcal{Z}[\alpha]$, where $\alpha = (1 - \sqrt[3]{2} + w\sqrt[3]{4})/(1 - w)$.

**5.** Let $w$ be a primitive cube root of unity. The ring of algebraic integers of the field $\mathcal{Q}(w, \sqrt[3]{d})$ is not monogenic for any cubefree integers $d \neq \pm 2, \pm 4$. (See [Ch02].)

**6.** Let $K$ be a Galois extension of $\mathcal{Q}$ of odd prime degree $p$ over $\mathcal{Q}$. If $2p+1$ is composite, then the ring of algebraic integers in $K$ is not monogenic. If $2p + 1 = \ell$ is prime, then the ring is monogenic if and only if $K = \mathcal{Q}(\cos(2\pi/\ell))$, in which case the ring is $\mathcal{Z}[2\cos(2\pi/\ell)]$. (See [Gr86].)

**7.** The *number field sieve* is a method that uses algebraic numbers to produce relations $x^2 \equiv y^2 \,(\text{mod } n)$ in integers. It is the basis for a very powerful factorization algorithm that has been used to factor integers of around 200 decimal digits. (See §4.5.1.)

**8.** The Dedekind zeta function of a number field, although defined initially only for $\mathcal{R}(s) > 1$, has a natural extension to a function analytic in the whole complex plane, except for a simple pole at 1.

**9.** The *Generalized Riemann Hypothesis* predicts that each zero of the Dedekind zeta function of a number field with real part between 0 and 1 must have real part equal to $1/2$.

**10.** The Generalized Riemann Hypothesis implies the truth of Artin's conjecture (see §4.7.1) that every integer other than $-1$ and perfect squares is a primitive root for infinitely many primes.

**Examples:**

**1.** If $p \equiv 1 \,(\text{mod } 4)$, then $p$ can be written as a sum of two squares: $p = a^2 + b^2$. Therefore, $p = (a + bi)(a - bi)$ in $\mathcal{Q}(i)$, which means that $p$ splits completely in $\mathcal{Q}(i)$. If $p \equiv 3 \,(\text{mod } 4)$, then $p\mathcal{Z}[i]$ is a prime ideal, so $p$ is inert. The relation $2 = -i(1 + i)^2$ written in terms of ideals becomes $(2) = (1 + i)^2$, so 2 ramifies.

**2.** The polynomial $X^2 + 5$ factors mod 3 as $(X + 1)(X + 2)$. Therefore (Fact 3), the ideal generated by 3 in $\mathcal{Z}[\sqrt{-5}]$ factors as $P_1 P_2$, where $P_1$ is generated by $1 + \sqrt{-5}$ and 3, and $P_2$ is generated by $2 + \sqrt{-5}$ and 3.

**3.** The polynomial $X^2 + 5$ is irreducible mod 11. Therefore (Fact 3), the ideal generated by 11 in $\mathcal{Z}[\sqrt{-5}]$ is a prime ideal.

**4.** The polynomial $X^2 + 5$ factors mod 2 as $(X + 1)^2$. Therefore (Fact 3), the ideal generated by 2 in $\mathcal{Z}[\sqrt{-5}]$ factors as $P^2$, where $P$ is the ideal generated by 2 and $1 + \sqrt{-5}$.

## 4.10.6  CYCLOTOMIC FIELDS

**Definitions:**

The $n$th **cyclotomic field** is the field obtained by adjoining a primitive $n$th root of unity to the field of rational numbers, that is, the field $\mathcal{Q}(\zeta)$, where $\zeta = e^{2\pi i/n}$.

A prime $p$ is called **regular** if $p$ does not divide the class number of the $p$th cyclotomic field. If $p$ divides this class number, $p$ is called **irregular**.

**Facts:**

**1.** Cyclotomic fields arose in the work of Gauss on constructibility of regular polygons by compass and straightedge.

**2.** Cyclotomic fields also arose as a natural setting for generalizations of quadratic reciprocity. (See §4.10.4.)

**3.** Kummer developed much of the theory of cyclotomic fields in the mid-1800s.

**4.** The classical proof of Fermat's last theorem for regular primes by Kummer relied on his introduction of ideal numbers and ideal class groups to establish a replacement for the unique factorization property in cyclotomic fields. See Example 4.

**5.** Kummer's work superseded incorrect proofs of Lamé and Cauchy, who had incorrectly and implicitly assumed that the $p$th cyclotomic field always has unique factorization.

**6.** Extensions of Kummer's work led to much progress on Fermat's last theorem through the 1900s, even though the eventual proof of the theorem was by different techniques.

**7.** The $n$th cyclotomic field has degree $\phi(n)$ over $\mathcal{Q}$, where $\phi$ is Euler's function.

**8.** The ring of algebraic integers in the $n$th cyclotomic field is $\mathcal{Z}[\zeta] = \{a_0 + a_1\zeta + \cdots + a_{\phi(n)-1}\zeta^{\phi(n)-1} \mid a_i \in \mathcal{Z}\}$, where $\zeta$ is a primitive $n$th root of unity. Therefore, this ring is monogenic.

**9.** The *Kronecker-Weber theorem* says that if $K$ is a number field such that $K/\mathcal{Q}$ is a Galois extension with abelian Galois group, then $K$ is contained in a cyclotomic field.

**10.** Let $n \geq 3$ be an integer. The $n$th cyclotomic field contains the field $\mathcal{Q}(\cos(2\pi/n))$, and the class number $h_n^+$ of this latter field divides the class number $h_n$ of the $n$th cyclotomic field.

**11.** The quotient $h_n^- = h_n/h_n^+$ can be computed fairly quickly. The computation of $h_n^+$ is difficult.

**12.** The largest prime $p$ for which $h_p^+$ has been computed is $p = 151$ (and $h_{151}^+ = 1$).

**13.** *Vandiver's conjecture* predicts that if $p$ is prime then $p$ does not divide $h_p^+$. This has been verified for all $p < 163 \cdot 10^6$.

**14.** There are infinitely many irregular primes. It has not been proved that there are infinitely many regular primes. Heuristic models predict that about 61% $(= e^{-1/2})$ of primes are regular. This prediction agrees well with computational data.

**15.** Let $p$ be prime. The numbers $\epsilon_a = \sin(2\pi a/p)/\sin(2\pi/p)$ for $2 \leq a \leq (p-1)/2$ are units of the ring of algebraic integers of the $p$th cyclotomic field. They, along with $-1$ and the $p$th roots of unity, generate a subgroup, called the *cyclotomic units*, of index $h_p^+$ in the group of units of the ring of algebraic integers of this field.

**16.** Preda Mihăilescu's proof of the Catalan conjecture (§4.8.4) relied heavily on properties of cyclotomic fields.

**Examples:**

**1.** Let $p$ be an odd prime. The class number $h_p$ of the $p$th cyclotomic field is 1 if and only if $p \leq 19$. The class number of the 23rd cyclotomic field is 3.

**2.** The class number of the 37th cyclotomic field is $h_{37} = 37$. This factors as $h_{37}^- = 37$ and $h_{37}^+ = 1$. The class number of the 59th cyclotomic field is 41241, which factors as $h_{59}^- = 41241 = 3 \cdot 59 \cdot 233$ and $h_{59}^+ = 1$. Both 37 and 59 are irregular primes.

**3.** The smallest prime $p$ with $h_p^+ > 1$ is $p = 163$, for which $h_{163}^+ = 4$.

**4.** *First case of Fermat's last theorem for regular primes:* Suppose $x, y, z$ are nonzero integers not divisible by the odd, regular prime $p$, and $x^p + y^p = z^p$. Let $\zeta$ be a primitive $p$th root of unity. Then $\prod_{j=1}^{p}(x + \zeta^j y) = z^p$. If $\gcd(x,y,z) = 1$ then the factors $x + \zeta^j y$ are pairwise relatively prime. Since their product is a $p$th power, the ideal generated by each factor is the $p$th power of an ideal: $(x + \zeta^j y) = I_j^p$. If the class number of the $p$th cyclotomic field is not a multiple of $p$, then this implies that $I_j$ must be principal, so $x + \zeta^j y$ is a unit times the $p$th power of an element of the field. This is the same as can be obtained when there is unique factorization in the $p$th cyclotomic field. The remainder of the argument is fairly straightforward. See [Wa97].

# 4.11   ELLIPTIC CURVES

Elliptic curves have been studied since the time of Diophantus and continue to be a very active area of research. The most important fact is that the points on an elliptic curve have a group structure. Elliptic curves arise as diophantine equations, and they played a crucial role in the proof of Fermat's last theorem. Elliptic curves over finite fields are a basic tool in cryptography, mostly because of the difficulty of the Discrete Log Problem.

## 4.11.1   BASIC CONCEPTS

### Definitions:

If $F$ is a field of characteristic not 2 or 3, an ***elliptic curve E defined over F*** is a curve that can be given by the **Weierstrass equation**

$$y^2 = x^3 + Ax + B,$$

where $A, B \in F$. It is required that the cubic polynomial $x^3 + Ax + B$ have no repeated roots, which is equivalent to $4A^3 + 27B^2 \neq 0$.

If $K$ is a field containing $F$, then we denote

$$E(K) = \{(x, y) \mid (x, y) \text{ lies on } E \text{ and } x, y \in K\} \cup \{\infty\},$$

where $\infty$ is the point at infinity. (See Fact 10.)

If $E$ is given by $y^2 = x^3 + Ax + B$, then the ***j-invariant*** is defined as

$$j(E) = (1728)4A^3/(4A^3 + 27B^2).$$

Given points $P$ and $Q$ on an elliptic curve $E$ in Weierstrass form, the **group law** (or **addition law**) is defined as follows. Draw the line through $P$ and $Q$ (the tangent line to $E$ if $P = Q$). This line intersects $E$ at a third point $R'$ (if the line is vertical, $R' = \infty$; see Facts 10 and 11). Reflect $R'$ across the $x$-axis to get a point $R$. This operation is denoted $P + Q = R$ (it is not simply coordinate addition).

If $n$ is a positive integer, then the ***multiple n of point p*** is the sum of $n$ copies of $P$: i.e., $nP = P + P + \cdots + P$.

A ***torsion point*** is a point $P$ on an elliptic curve $E$ such that $nP = \infty$ for some integer $n > 0$, where $\infty$ is the point at infinity (see Fact 10).

### Facts:

**1.** Comprehensive coverage of elliptic curves can be found in [Si09], [SiTa15], and [Wa08].

**2.** Although elliptic curves can be defined over arbitrary fields, the most important examples are over finite fields, the rational numbers, and the complex numbers.

**3.** Elliptic curves are not ellipses. They received their name because of their close relation with elliptic integrals, which occur in many physical problems, including finding the arc length of an ellipse.

**4.** The addition of points satisfies the associative law: $(P + Q) + R = P + (Q + R)$.

**5.** If $K$ is a field containing the coefficients of $E$, the addition law makes $E(K)$ into an abelian group, where the point at infinity is the identity element.

**6.** The inverse of a point $P = (x, y)$ is $-P = (x, -y)$. Points can be subtracted: $Q - P = Q + (-P)$.

**7.** *Addition of points:* Let $E$ be an elliptic curve defined by $y^2 = x^3 + Ax + B$. Let $P_1 = (x_1, y_1)$ and $P_2 = (x_2, y_2)$ be points on $E$ with $P_1, P_2 \neq \infty$. Let $P_1 + P_2 = P_3 = (x_3, y_3)$.

- If $x_1 \neq x_2$, then

$$x_3 = m^2 - x_1 - x_2, \qquad y_3 = m(x_1 - x_3) - y_1, \qquad \text{where } m = \frac{y_2 - y_1}{x_2 - x_1}.$$

- If $x_1 = x_2$ but $y_1 \neq y_2$, then $P_1 + P_2 = \infty$.
- If $P_1 = P_2$ and $y_1 \neq 0$, then

$$x_3 = m^2 - 2x_1, \qquad y_3 = m(x_1 - x_3) - y_1, \qquad \text{where } m = \frac{3x_1^2 + A}{2y_1}.$$

- If $P_1 = P_2$ and $y_1 = 0$, then $P_1 + P_2 = \infty$.
- $P_1 + \infty = P_1$.

**8.** The following figure gives a picture of how addition works for elliptic curves over the real numbers. For most fields, this picture cannot be drawn, but it is helpful for geometric intuition.

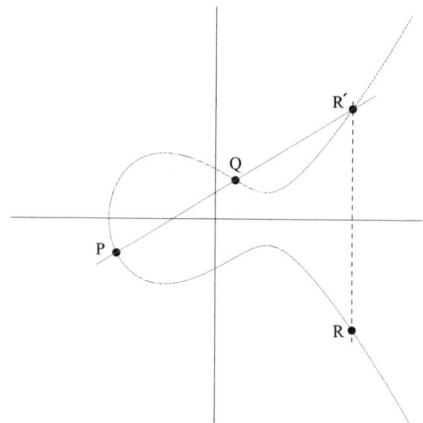

**9.** A curve $y^2 = x^3 + Ax + B$ can be put into homogeneous form $y^2 z = x^3 + Axz^2 + Bz^3$ and the points on the curve are points $(x : y : z)$ in projective space. If $z \neq 0$, such a point can be rescaled to $(x/z : y/z : 1)$, which is a point on the original nonhomogeneous curve.

**10.** In homogeneous coordinates, the curve in projective space contains the point $(x : y : z) = (0 : 1 : 0)$, which is called the *point at infinity*. In most situations, it is easiest to regard the point at infinity as a formal symbol $\infty$ that acts as the identity of the group law. The point at infinity is usually denoted $\infty$ or $\mathcal{O}$.

**11.** Vertical lines ($x = \text{constant}$) in the plane extend in projective space to contain the point at infinity.

**12.** When the characteristic of the field is 2 or 3, the *generalized Weierstrass form* is used:

$$y^2 + a_1 xy + a_3 y = x^3 + a_2 x^2 + a_4 x + a_6,$$

with $a_1, a_2, a_3, a_4, a_6 \in F$. The reflection across the $x$-axis in the group law is replaced by the map $(x, y) \mapsto (x, -y - a_1 x - a_3)$. See [Si09].

**13.** When the characteristic of the field is not 2 or 3, there are situations where the generalized Weierstrass form $y^2 + a_1 xy + a_3 y = x^3 + a_2 x^2 + a_4 x + a_6$ is useful, but a change of variables can be used to put the curve into Weierstrass form.

**14.** Elliptic curves played a crucial role in the proof of Fermat's last theorem (§4.8.3). Suppose $a^n + b^n = c^n$, where $a, b, c, n$ are positive integers and $n \geq 3$. Form the elliptic curve

$$y^2 = x(x - a^n)(x + b^n).$$

The fact that the cubic polynomial has discriminant $(abc)^{2n}$, which is a $2n$th power, forces the elliptic curve to have properties so restrictive that it can be shown that such a curve cannot exist. Therefore, $a, b, c$ cannot exist.

**15.** In the 1980s, Hendrik Lenstra developed a method for factorization of integers using elliptic curves. It is similar to the Pollard $p-1$ method (see §4.5.1), but has the advantage that it can use several elliptic curves in parallel. If a set of curves does not produce the factorization, another set can be used, an aspect that has no analogue for the Pollard $p-1$ algorithm. The method is effective for factoring integers of around 50 decimal digits.

**16.** Shafi Goldwasser and Joe Kilian invented a primality test based on elliptic curves. It was refined by Oliver Atkin and François Morain and has been used to prove the primality of integers with more than 20,000 decimal digits. (See §4.4.4, Fact 18.)

**17.** Suppose we have a curve defined over a field of characteristic not 2 by an equation $v^2 = au^4 + bu^3 + cu^2 + du + e$ and suppose we have a point $(p, q)$ lying on the curve. There is a change of variables that transforms the equation to Weierstrass form. See [Co92].

**18.** A nonsingular plane cubic curve can be changed to a curve in Weierstrass form; a sketch of the procedure is given in [SiTa15]. See Example 4 for a simple example.

**19.** Let $E$ be an elliptic curve over $F$, let $n$ be a positive integer not divisible by the characteristic of $F$, and let $E[n]$ be the set of points $P \in E(\overline{F})$ (where $\overline{F}$ is the algebraic closure of $F$) that satisfy $nP = \infty$. Then $E[n] \simeq \mathcal{Z}/n\mathcal{Z} \oplus \mathcal{Z}/n\mathcal{Z}$.

**20.** Let $E$ be defined over a field $F$ of characteristic $p > 0$, let $m > 0$, and let $E[p^m]$ be the set of points $P \in E(\overline{F})$ that satisfy $p^m P = \infty$. Then $E[p^m] \simeq$ either $\mathcal{Z}/p^m \mathcal{Z}$ or 0. The second case happens if and only if $E$ is supersingular (see §4.11.3).

**21.** Let $E$ be an elliptic curve over $F$, let $n$ be a positive integer not divisible by the characteristic of $F$, and let $E(F)[n]$ be the set of points $P \in E(F)$ that satisfy $nP = \infty$. If $E(F)[n] \simeq \mathcal{Z}/n\mathcal{Z} \oplus \mathcal{Z}/n\mathcal{Z}$, then $F$ contains the $n$th roots of unity. When $F = GF(q)$, this means that $n \mid q - 1$.

**Examples:**

**1.** The elliptic curve $y^2 = x^3 + 73$ contains the points $(2, 9)$ and $(3, 10)$. The line through these two points is $y = x + 7$, which intersects the elliptic curve in $(2, 9)$, $(3, 10)$, and $(-4, 3)$. The reflection across the $x$-axis of the third point is $(-4, -3)$, so $(2, 9) + (3, 10) = (-4, -3)$.

**2.** The line $y = (2/3)(x - 2) + 9$ is tangent to the elliptic curve $y^2 = x^3 + 73$ at the point $(2, 9)$. The line also intersects the curve at $(-32/9, 143/27)$. The reflection across the $x$-axis is $(-32/9, -143/27)$, so $2(2, 9) = (2, 9) + (2, 9) = (-32/9, -143/27)$.

**3.** The point $P = (2,3)$ is a torsion point on the elliptic curve $y^2 = x^3 + 1$. The multiples of this point are

$$2P = (0,1), \quad 3P = (-1,0), \quad 4P = (0,-1), \quad 5P = (2,-3), \quad 6P = \infty.$$

**4.** The change of variables $x = 12c/(a+b)$, $y = 36(a-b)/(a+b)$ changes the elliptic curve $y^2 = x^3 - 432$ into the Fermat equation of exponent 3, namely, $a^3 + b^3 = c^3$.

**5.** The *cannonball problem*, which goes back at least to Edouard Lucas in 1875, asks if there is a set of cannonballs that can be put into a square pyramid and also can be arranged in a square array. If the square pyramid has $x$ layers and the square array is $y \times y$, the problem asks for integer solutions to $y^2 = x(x+1)(2x+1)/6$, an equation that can be transformed into Weierstrass form $y_1^2 = x_1^2 - 36x_1$ with the change of variables $x = (x_1 - 6)/12$, $y = y_1/72$. Besides the trivial solutions $x = 0$ and $x = 1$, the only other nonnegative integral solution is $x = 24$, corresponding to 4,900 cannonballs.

**6.** The *congruent number problem* asks which positive integers occur as areas of right triangles with rational sides. Such an integer was called a *congruent number* by Fibonacci. The problem of finding congruent numbers appears in an Arab manuscript before 1000 CE. For example, 5 occurs as the area of the right triangle with sides $3/2$, $20/3$, $41/6$.

**7.** Fermat used his method of infinite descent to show that 1 is not a congruent number.

**8.** Under the assumption of the truth of the conjecture of Birch and Swinnerton-Dyer (see §4.11.4, Fact 5) for the elliptic curve $y^2 = x^3 - n^2 x$, Jerrold Tunnell gave a simple criterion for determining whether the integer $n$ is a congruent number.

**9.** The congruent numbers less than 40 are 5, 6, 7, 13, 14, 15, 20, 21, 22, 23, 24, 28, 29, 30, 31, 34, 37, 38, 39.

**10.** Finding a rational right triangle of area $n$ is equivalent to finding an $x$ such that $x - n$, $x$, and $x + n$ are all squares of rational numbers.

**11.** Nathan Fine [Fi76] showed, implicitly using the theory of elliptic curves, that every positive rational occurs as the area of some triangle (not necessarily a right triangle) with rational sides.

**12.** Several applications of elliptic curves to recreational mathematics are given in [Ma16].

---

## 4.11.2  ISOMORPHISMS, ENDOMORPHISMS, AND ISOGENIES

**Definitions:**

Let $E_1, E_2$ be elliptic curves defined over $F$, where $E_i$ is given by $y^2 = x^3 + A_i x + B_i$ for $i = 1, 2$. A map $\phi : E_1(\overline{F}) \to E_2(\overline{F})$ is said to be **given by rational functions** (where $\overline{F}$ is the algebraic closure of $F$) if there are rational functions $f(x,y)$ and $g(x,y)$ such that $\phi(x,y) = (f(x,y), g(x,y))$.

A map given by rational functions is **defined over a field** $K$ if the coefficients of the rational functions $f$ and $g$ can be chosen to be elements of $K$.

An **endomorphism** of an elliptic curve is a homomorphism, given by rational functions, from an elliptic curve to itself. The zero endomorphism maps every point to the point $\infty$.

An elliptic curve $E$ has **complex multiplication** (**CM**) if there are endomorphisms of $E$ that are not given by multiplication by an integer.

Let $E_1, E_2$ be elliptic curves defined over $F$. An **isogeny** is a homomorphism, given by rational functions, from $E_1(\overline{F})$ to $E_2(\overline{F})$, where $\overline{F}$ is the algebraic closure of $F$.

Two elliptic curves $E_i$, $i = 1, 2$, given in Weierstrass form by $y_i^2 = x_i^3 + A_i x_i + B_i$, for $i = 1, 2$, are **isomorphic** over a field $K$ (containing all $A_i, B_i$) if there is a constant $u \in K$ such that the change of variables

$$x_2 = u^2 x_1, \quad y_2 = u^3 y_1$$

maps the points on $E_1$ with coordinates in the algebraic closure of $K$ to those on $E_2$.

Two elliptic curves $E_1$ and $E_2$ defined over a field $F$ are said to be **twists** of each other if they are isomorphic over the algebraic closure of $F$.

**Facts:**

**1.** If a map is given by rational functions $f$ and $g$, as in the definitions, then $g^2 = f^3 + A_2 f + B_2$ for all $x, y \in \overline{F}$ satisfying $y^2 = x^3 + A_1 x + B_1$.

**2.** Let $E$ be an elliptic curve and let $n$ be an integer. The map $P \mapsto nP$ is given by rational functions (for $n = 2$, see the addition formula (§4.11.1) applied to the case $P_1 = P_2$).

**3.** Two elliptic curves defined over a field $F$ are isomorphic over the algebraic closure of $F$ if and only if they have the same $j$-invariant.

**4.** Let $E$ be defined over $F$. The endomorphisms of $E$ defined over the algebraic closure of $F$ form a ring (under addition and composition of functions).

**5.** If the characteristic of $F$ is 0, this ring of endomorphisms is isomorphic either to $\mathcal{Z}$ or to a subring of the ring of algebraic integers (see §4.10.2) in an imaginary quadratic field.

**6.** If $F$ is a finite field of characteristic $p$, this ring of endomorphisms is isomorphic either to a subring of the ring of algebraic integers (see §4.10.2) in an imaginary quadratic field or to a noncommutative ring (in technical terms: a maximal order in a quaternion algebra over $\mathcal{Q}$ that is ramified only at $p$ and $\infty$).

**7.** A noncommutative endomorphism ring occurs if and only if $E$ is supersingular.

**8.** The most common situation where $E_1$ and $E_2$ are twists of each other is when $E_i$ is defined by $y^2 = x^3 + A_i x + B_i$ and there is a nonzero $d \in F$ such that $A_2 = d^2 A_1$ and $B_2 = d^3 B_1$, in which case the isomorphism is given by $(x, y) \mapsto (dx, d^{3/2} y)$.

**Examples:**

**1.** The elliptic curves $y_1^2 = x_1^3 + 3x_1 + 7$ and $y_2^2 = x_2^3 + 12x_2 + 56$ are twists of each other. They both have the $j$-invariant $6912/53$. The change of variables $x_2 = 2x_1$, $y_2 = 2^{3/2} y_1$ transforms one curve to the other.

**2.** Let $E$ be the elliptic curve $y^2 = x^3 + x$ defined over $\mathcal{C}$. The map $\phi : (x, y) \mapsto (-x, iy)$ gives an endomorphism of $E$ that is not multiplication by an integer, so $E$ has complex multiplication. It is easy to see that $\phi^4$ is the identity. The ring $\mathcal{Z}[i]$, with $i$ corresponding to $\phi$, is the endomorphism ring of $E$.

**3.** Let $E_1 : y_1^2 = x_1^3 + ax_1^2 + bx_1$ be an elliptic curve over a field of characteristic not 2. We require $b \neq 0$ and $a^2 - 4b \neq 0$ in order to have $4A^3 + 27B^2 \neq 0$ when the curve is changed to Weierstrass form. Let $E_2$ be the elliptic curve $y_2^2 = x_2^3 - 2ax_2^2 + (a^2 - 4b)x_2$. Define $\alpha$ by

$$(x_2, y_2) = \alpha(x_1, y_1) = \left( \frac{y_1^2}{x_1^2}, \frac{y_1(x_1^2 - b)}{x_1^2} \right).$$

Then $\alpha$ is an isogeny from $E_1$ to $E_2$. The kernel of $\alpha$ is $\{\infty, (0,0)\}$.

---

## 4.11.3  ELLIPTIC CURVES OVER FINITE FIELDS

### Definitions:

Given a point $P$ on an elliptic curve and multiples $nP$ and $mP$, where $m, n$ are integers that are not revealed, the **computational Diffie-Hellman problem** is that of finding $mnP$.

Given two points $P$ and $Q$ on an elliptic curve such that $Q = nP$ for some integer $n$, the **discrete logarithm problem** is that of finding an integer $k$ such that $Q = kP$.

If $E$ is an elliptic curve defined over a finite field $GF(q)$, the **Frobenius map** $\phi_q$ maps a point $(x, y)$ (with $x, y$ in the algebraic closure of $GF(q)$) to the point $(x^q, y^q)$.

A **supersingular elliptic curve** is an elliptic curve defined over a field $F$ of characteristic $p$ that has no points of order $p$ with coordinates in the algebraic closure of $F$.

Let $A$, $B$, $C$, be groups, and let $*$ denote their group operations. A **bilinear pairing** is a map

$$A \times B \to C, \quad (a, b) \mapsto c = \langle a, b \rangle,$$

such that

$$\langle a_1 * a_2, b \rangle = \langle a_1, b \rangle * \langle a_2, b \rangle, \qquad \langle a, b_1 * b_2 \rangle = \langle a, b_1 \rangle * \langle a, b_2 \rangle$$

for all $a, a_1, a_2 \in A$ and all $b, b_1, b_2 \in B$. Two common pairings are the Tate-Lichtenbaum pairing and the Weil pairing (see Fact 7).

### Facts:

**1.** The set of points $E(GF(q))$ forms a finite group. It has the structure

$$\mathcal{Z}/m\mathcal{Z} \quad \text{or} \quad \mathcal{Z}/m\mathcal{Z} \times \mathcal{Z}/n\mathcal{Z}$$

for some integers $m, n$.

**2.** *Hasse's theorem:* Let $E$ be an elliptic curve defined over $GF(q)$ and let $N$ be the number of points on $E$ with coordinates in $GF(q)$. Then

$$q + 1 - 2\sqrt{q} \leq N \leq q + 1 + 2\sqrt{q}.$$

**3.** Let $E$ be an elliptic curve defined over $GF(q)$ and let $a = \#E(GF(q))$. Write

$$X^2 - aX + q = (X - \alpha)(X - \beta).$$

Then the number of points on $E$ with coordinates in $GF(q^n)$ is

$$q^n + 1 - \alpha^n - \beta^n.$$

**4.** The Frobenius map $\phi_q$ of an elliptic curve $E$ defined over $GF(q)$ is an endomorphism of $E$, and $\phi_q^2 - a\phi_q + q$ is the zero endomorphism, where $a = q + 1 - \#E(GF(q))$.

**5.** There is an efficient algorithm due to Schoof-Elkies-Atkin that counts the number of points on an elliptic curve over a finite field.

**6.** An elliptic curve over a finite field $F$ of characteristic $p$ is supersingular if and only if the order of $E(F)$ is congruent to 1 modulo $p$.

**7.** Let $E$ be an elliptic curve defined over a finite field $F = GF(q)$ and let $n$ be a positive integer such that $q-1$ is a multiple of $n$. Let $E[n]$ be the set of points $P$ with coordinates in the algebraic closure of $F$ such that $nP = \infty$. The *Tate-Lichtenbaum pairing* is a nondegenerate bilinear pairing

$$\tau_n : E[n] \times (E(F)/nE(F)) \to \mu_n,$$

where $\mu_n$ is the group of $n$th roots of unity in $F$. The *Weil pairing* is a nondegenerate bilinear pairing

$$e_n : E[n] \times E[n] \to \mu_n.$$

**8.** The Weil pairing and the Tate-Lichtenbaum pairing can be computed efficiently by Miller's algorithm [Mi04].

**9.** The Tate-Lichtenbaum pairing (and the Weil pairing) can be used to attack the discrete logarithm problem. Let $E$ be defined over $GF(q)$. Suppose that $P, Q$ are points of order $\ell$ on $E$ and $kP = Q$. Let $q^n$ be chosen so that $\ell \mid q^n - 1$ (the smallest such $n > 0$ is called the *embedding degree*). Choose a random point $R \in E(GF(q^n))$ and compute $\alpha = \tau_\ell(P, R)$ and $\beta = \tau_\ell(Q, R)$. Then $\beta = \alpha^k$, so we have a discrete logarithm problem in $GF(q^n)$. If $n$ is small, it is easier to attack this discrete logarithm problem (for fields) than to use methods that work on arbitrary elliptic curves.

**Examples:**

**1.** Let $E$ be the elliptic curve $y^2 = x^3 + 4x + 1$ defined over $GF(5)$. The points in $E(GF(5))$ are $\infty, (0,1), (0,4), (1,1), (1,4), (3,0), (4,1), (4,4)$. The addition law yields $(0,1) + (1,4) = (3,0)$ and $2(1,1) = (4,1)$. There are eight points, so $a = 5 + 1 - 8 = -2$. The polynomial $X^2 + 2X + 5$ factors as $(X + 1 - 2i)(X + 1 + 2i)$. Therefore (see Fact 3), the number of points in $E(GF(5^n))$ is $5^n + 1 - (-1 + 2i)^n - (-1 - 2i)^n$. The group $E(GF(5))$ is cyclic of order 8.

**2.** Let $E$ be the elliptic curve $y^2 = x^3 + 4x + 1$ defined over $GF(5)$. An example of the discrete logarithm problem is to find $k$ such that $k(0,1) = (1,1)$. The answer in this case is $k = 5$.

**3.** There is a list of 15 elliptic curves recommended by NIST (see [FIPS186-3]).

**4.** The elliptic curve $E$ defined by $y^2 \equiv x^3 + 1 \pmod 5$ has six points with coordinates in $\mathbb{Z}/5\mathbb{Z}$. Therefore, it is supersingular. Let $\omega$ be a primitive cube root of unity in the algebraic closure of $GF(5)$. Then $\alpha : (x, y) \mapsto (\omega x, y)$ is an endomorphism of $E$. Let $\phi = \phi_5$ be the Frobenius map. Then $\phi\alpha(x, y) = (\omega^2 x^5, y^5)$, but $\alpha\phi(x, y) = (\omega x^5, y^5)$. Therefore, the ring of endomorphisms of $E$ is noncommutative.

---

## 4.11.4  ELLIPTIC CURVES OVER $\mathcal{Q}$

**Definition:**

Let $E$ be an elliptic curve given in generalized Weierstrass form with integer coefficients. For each prime $p$, look at the equation for $E$ modulo $p$. For all primes not in a finite set $S_E$, this yields a nonsingular curve modulo $p$. Let $a_p = p + 1 - \#E(GF(p))$. The **L-series** for $E$ is

$$L(E, s) = \prod_{p \in S_E} (L_p(s))^{-1} \prod_{p \notin S_E} \left(1 - a_p p^{-s} + p^{1-2s}\right)^{-1} = \sum_{n=1}^{\infty} a_n n^{-s},$$

where the finitely many factors $L_p(s)$ depend on the properties of $E$ modulo $p$, and where $s$ is a complex variable with $\mathcal{R}(s) > 3/2$.

**Facts:**

**1.** *Mordell-Weil theorem*: If $F$ is a finite extension of $\mathcal{Q}$ and $E$ is defined over $F$, then $E(F)$ is a finitely generated abelian group, which means that $E(F) \simeq \mathcal{Z}^r \oplus T$, where $T$ is a finite group. The integer $r$ is called the *Mordell-Weil rank* of $E(F)$.

**2.** In 2006, Noam Elkies discovered an elliptic curve defined over $\mathcal{Q}$ whose Mordell-Weil rank is at least 28. It has been conjectured that there are elliptic curves of arbitrarily large Mordell-Weil rank. On the other hand, Andrew Granville and Bjorn Poonen have each produced heuristic arguments that indicate that there should be an upper bound on the rank of elliptic curves.

**3.** Manjul Barghava and Arul Shankar have proved that the average rank of elliptic curves defined over $\mathcal{Q}$ is at most $7/6$.

**4.** If $E$ is an elliptic curve defined over $\mathcal{Q}$, the L-series of $E$ has a continuation to an analytic function defined in the whole complex plane. It has a functional equation relating $L(E, s)$ and $L(E, 2 - s)$. This was conjectured by Taniyama, Shimura, and Weil and was proved by Wiles, Taylor, Breuil, Conrad, and Diamond.

**5.** *Conjecture of Birch and Swinnerton-Dyer*: Let $E$ be defined over $\mathcal{Q}$. The *analytic rank* of $E$ is the order of vanishing of $L(E, s)$ at $s = 1$ (that is, the degree of the first nonzero term in the Taylor series at $s = 1$). It is conjectured that the analytic rank of $E$ equals the Mordell-Weil rank of $E(\mathcal{Q})$. This is known to be true when the analytic rank is 0 or 1 by work of Victor Kolyvagin and others.

**6.** *Lutz-Nagell theorem*: Let $E$ be defined over $\mathcal{Q}$ by the equation $y^2 = x^3 + Ax + B$ with $A, B \in \mathcal{Z}$. Let $\infty \neq (x, y) \in E(\mathcal{Q})$ be a torsion point. Then $x, y \in \mathcal{Z}$, and either $y = 0$ or $y^2 \mid 4A^3 + 27B^2$.

**7.** *Mazur's theorem*: Let $E$ be defined over $\mathcal{Q}$. The torsion subgroup of $E(\mathcal{Q})$ has one of the following forms:

$$\mathcal{Z}/n\mathcal{Z} \text{ for } 1 \leq n \leq 10 \text{ or } n = 12; \quad \mathcal{Z}/n\mathcal{Z} \times \mathcal{Z}/2\mathcal{Z} \text{ for } 1 \leq n \leq 4.$$

**Examples:**

**1.** Let $E$ be defined by $y^2 = x^3 + 1$. Then $4A^3 + 27B^2 = 27$. Suppose $(x, y)$ is a torsion point with $y \neq 0$. The Lutz-Nagell theorem says that $y^2 \mid 27$, so $y = \pm 1$ or $\pm 3$. This yields the points $(0, \pm 1)$ and $(2, \pm 3)$. It can be checked that $3(0, -1) = 3(0, 1) = \infty$ and $6(2, 3) = 6(2, -3) = \infty$, so these are torsion points. The only point with $y = 0$ is $(-1, 0)$. Therefore, the torsion subgroup of $E(\mathcal{Q})$ has order 6 and is

$$\{\infty, (-1, 0), (0, -1), (0, 1), (2, -3), (2, 3)\}.$$

**2.** Let $E$ be defined by $y^2 = x^3 - 16x + 16$. The Lutz-Nagell theorem (Fact 6) implies that if $(x, y)$ is a torsion point with $y \neq 0$, then $y^2 \mid 2^8 \cdot 37$, so $y = \pm 2^j$ with $0 \leq j \leq 4$. These do not yield torsion points, and neither does $y = 0$, so the torsion subgroup of $E(\mathcal{Q})$ is trivial. Therefore, $E(\mathcal{Q}) \simeq \mathcal{Z}^r$ for some $r$. In fact, $r = 1$ and the points with rational coordinates are the multiples of the point $P = (0, 4)$. The first few multiples of $P$ are

$$2P = (4, 4), \quad 3P = (-4, -4), \quad 4P = (8, -20), \quad 5P = (1, -1),$$
$$6P = (24, 116), \quad 7P = (-20/9, 172/27), \quad 8P = (84/25, -52/125).$$

## 4.11.5  ELLIPTIC CURVES OVER $\mathcal{C}$

**Definition:**
Given a lattice $L = \mathcal{Z}\omega_1 + \mathcal{Z}\omega_2$ in $\mathcal{C}$, the **Weierstrass $\wp$-function** is given by

$$\wp(z) = \wp(z; L) = \tfrac{1}{z^2} + \sum_{\substack{\omega \in L \\ \omega \neq 0}} \left( \tfrac{1}{(z-\omega)^2} - \tfrac{1}{\omega^2} \right).$$

**Facts:**
**1.** If $E$ is defined over the complex numbers $\mathcal{C}$, then $E(\mathcal{C})$ is isomorphic to $\mathcal{C}/L$, where $L$ is a lattice in $\mathcal{C}$ (that is, $L$ is an additive subgroup of $\mathcal{C}$ of the form $\mathcal{Z}\omega_1 + \mathcal{Z}\omega_2$, where $\omega_1, \omega_2$ are complex numbers that are linearly independent over the real numbers).

**2.** The Weierstrass $\wp$-function is meromorphic in $\mathcal{C}$, has a double pole at each $\omega \in L$, and is doubly periodic: $\wp(z + \omega_1) = \wp(z + \omega_2) = \wp(z)$ for all $z \in \mathcal{C}$.

**3.** Given a lattice $L$ in $\mathcal{C}$, there are constants $g_2, g_3$ such that the associated Weierstrass $\wp$-function satisfies

$$\wp'(z)^2 = 4\wp(z)^3 - g_2\wp(z) - g_3$$

for all $z \in \mathcal{C}$. This means that there is a map

$$z \mapsto (\wp(z), \tfrac{1}{2}\wp'(z))$$

from $\mathcal{C}/L$ to $E(\mathcal{C})$, where $E$ is the elliptic curve defined over $\mathcal{C}$ by $y^2 = x^3 - (g_2/4)x - (g_3/4)$. This is an isomorphism of groups.

**4.** When an elliptic curve with complex multiplication is represented over the complex numbers as $\mathcal{C}/L$ for a lattice $L$, then the complex multiplication is given by multiplication by (nonreal) complex numbers $\alpha$ such that $\alpha L \subseteq L$.

## REFERENCES

*Printed Resources*:

[AlWi04] S. Alaca and K. S. Williams, *Introductory Algebraic Number Theory*, Cambridge University Press, 2004.

[An98] G. E. Andrews, *The Theory of Partitions*, Encyclopedia of Mathematics and Its Applications, Vol. 2, Cambridge University Press, 1998.

[Ap76] T. M. Apostol, *Introduction to Analytic Number Theory*, Springer-Verlag, 1976.

[BaSh96] E. Bach and J. Shallit, *Algorithmic Number Theory, Volume 1, Efficient Algorithms*, MIT Press, 1996.

[Ba90] A. Baker, *Transcendental Number Theory*, Cambridge University Press, 1990.

[Br89] D. M. Bressoud, *Factorization and Primality Testing*, Springer-Verlag, 1989.

[BrEtal88] J. Brillhart, D. H. Lehmer, J. L. Selfridge, B. Tuckerman, and S. S. Wagstaff, Jr., *Factorizations of $b^n \pm 1$, $b = 2, 3, 5, 6, 7, 10, 11, 12$ up to High Powers*, 2nd ed., American Mathematical Society, 1988.

[Ca57] J. W. S. Cassels, *An Introduction to Diophantine Approximation*, Cambridge University Press, 1957.

[Ch02] M-L. Chang, "Non-monogeneity in a family of sextic fields", *Journal of Number Theory* 97 (2002), 252–268.

[Co93] H. Cohen, *A Course in Computational Algebraic Number Theory*, Springer-Verlag, 1993.

[CoLe84] H. Cohen and H. W. Lenstra, Jr., "Heuristics on class groups of number fields", *Number theory, Noordwijkerhout 1983*, Lecture Notes in Mathematics 1068, Springer, 1984, 33–62.

[Co92] I. Connell, "Addendum to a paper of Harada and Lang", *Journal of Algebra* 145 (1992), 463–467.

[CrPo05] R. E. Crandall and C. Pomerance, *Primes: A Computational Perspective*, 2nd ed., Springer-Verlag, 2005.

[Da13] A. Das, *Computational Number Theory*, CRC Press, 2013.

[Di71] L. E. Dickson, *History of the Theory of Numbers*, Chelsea Publishing Company, 1971.

[Fi76] N. Fine, "On rational triangles", *American Mathematical Monthly* 83 (1976), 517–521.

[FIPS186-3] "Digital Signature Standard (DSS)", Federal Information Processing Standards Publication 186-3, U.S. Department of Commerce/National Institute of Standards and Technology, June 2009.

[Gr86] M-N. Gras, "Non monogénéité de l'anneau des entiers des extensions cycliques de Q de degré premier $\ell \geq 5$", *Journal of Number Theory* 23 (1986), 347–353.

[GuMu84] R. Gupta and M. R. Murty, "A remark on Artin's Conjecture", *Inventiones Mathematicae* 78 (1984), 127–130.

[Gu94] R. K. Guy, *Unsolved Problems in Number Theory*, 2nd ed., Springer-Verlag, 1994.

[HaWr08] G. H. Hardy and E. M. Wright, *An Introduction to the Theory of Numbers*, 6th ed., Oxford University Press, 2008.

[IrRo98] K. Ireland and M. Rosen, *A Classical Introduction to Modern Number Theory*, 2nd ed., Springer, 1998.

[Kn97] D. E. Knuth, *The Art of Computer Programming, Volume 2: Seminumerical Algorithms*, 3rd ed., Addison-Wesley, 1997.

[Ko93] I. Koren, *Computer Arithmetic Algorithms*, Prentice Hall, 1993.

[La00] S. Lang *Algebraic Number Theory*, 2nd ed., Springer, 2000.

[Ma16] A. J. MacLeod, "Elliptic curves in recreational number theory", 2016; available at https://arxiv.org/pdf/1610.03430v1.pdf.

[Ma77] B. Mazur, "Book Review: Ernst Edward Kummer, Collected Papers", *Bulletin of the American Mathematical Society* 83 (1977), 976–988.

[MeVoVa96] A. J. Menezes, P. C. van Oorschot, and S. A Vanstone, *Handbook of Applied Cryptography*, CRC Press, 1996.

[Mi04] V. Miller, "The Weil pairing and its efficient calculation", *Journal of Cryptology* 17 (2004), 235–161.

[Mo69] L. J. Mordell, *Diophantine Equations*, Academic Press, 1969.

[Mu11] R. A. Mullin, *Algebraic Number Theory*, 2nd ed., CRC Press, 2011.

[NiZuMo91] I. Niven, H. S. Zuckerman, and H. L. Montgomery, *An Introduction to the Theory of Numbers*, 5th ed., Wiley, 1991.

[Pe54] O. Perron, *Die Lehre von den Kettenbrüchen*, 3rd ed., Teubner Verlagsgesellschaft, 1954.

[Po90] C. Pomerance, ed., *Cryptology and Computational Number Theory*, Proceedings of Symposia in Applied Mathematics 42, American Mathematical Society, 1990.

[Po94] C. Pomerance, "The number field sieve", in *Mathematics of Computation 1943–1993: A Half-Century of Computational Mathematics*, W. Gautschi (ed.), Proceedings of Symposia in Applied Mathematics 48, American Mathematical Society, 1994, 465–480.

[Po10] C. Pomerance, "Primality testing: variations on a theme of Lucas", *Congressus Numerantium* 201 (2010), 301-312.

[Ri96] P. Ribenboim, *The New Book of Prime Number Records*, Springer-Verlag, 1996.

[Ro10] K. H. Rosen, *Elementary Number Theory and Its Applications*, 6th ed., Pearson, 2010.

[Sc80] W. M. Schmidt, *Diophantine Approximation*, Lecture Notes in Mathematics 785, Springer, 1980.

[Sc85] N. R. Scott, *Computer Number Systems and Arithmetic*, Prentice Hall, 1985.

[Si09] J. H. Silverman, *The Arithmetic of Elliptic Curves*, 2nd ed., Springer, 2009.

[SiTa15] J. H. Silverman and J. T. Tate, *Rational Points on Elliptic Curves*, 2nd ed., Springer, 2015.

[Si97] S. Singh, *The Quest to Solve the World's Greatest Mathematical Problem*, Walker & Co., 1997.

[StTa16] I. Stewart and D. Tall, *Algebraic Number Theory and Fermat's Last Theorem*, 4th ed., CRC Press, 2016.

[Wa13] S. S. Wagstaff, Jr., *Joy of Factoring*, American Mathematical Society, 2013.

[Wa97] L. Washington, *Introduction to Cyclotomic Fields*, 2nd ed., Springer, 1997.

[Wa08] L. Washington, *Elliptic Curves: Number Theory and Cryptography*, 2nd ed., Chapman & Hall/CRC Press, 2008.

[Wi95] A. J. Wiles, "Modular elliptic curves and Fermat's last theorem", *Annals of Mathematics* 141 (1995), 443–551.

**Web Resources**:

http://www-groups.dcs.st-and.ac.uk/~history/HistTopics/Fermat's_last_theorem.html (History of Fermat's last theorem.)

`http://www.jmilne.org/math/CourseNotes/ant.html` (J. S. Milne, "Algebraic number theory" course notes.)

`http://www.mersenne.org` (The Great Internet Mersenne Prime Search webpage.)

`http://www.numbertheory.org/ntw/number_theory.html` (The Number Theory webpage.)

`http://www.pbs.org/wgbh/nova/proof` (NOVA Online: Proof of Fermat's last theorem.)

`http://www.scienceandreason.net/flt/flt01.htm` (The mathematics of Fermat's last theorem.)

`http://www.utm.edu/research/primes` (Prime number research, records, and resources.)

`https://homes.cerias.purdue.edu/~ssw/cun/index.html` (The Cunningham Project website.)

# 5

# ALGEBRAIC STRUCTURES

*John G. Michaels*

## INTRODUCTION

Many of the most common mathematical systems, including the integers, the rational numbers, and the real numbers, have an underlying algebraic structure. This chapter examines the structure and properties of various types of algebraic objects. These objects arise in a variety of settings and occur in many different applications, including counting techniques, coding theory, information theory, engineering, and circuit design.

## GLOSSARY

**abelian group**:  a group in which $a \star b = b \star a$ for all $a, b$ in the group.

**absorption laws**:  in a lattice $a \vee (a \wedge b) = a$ and $a \wedge (a \vee b) = a$.

**algebraic element** (over a field):  given a field $F$, an element $\alpha \in K$ (extension of $F$) such that there exists $p(x) \in F[x]$ $(p(x) \neq 0)$ such that $p(\alpha) = 0$. Otherwise $\alpha$ is **transcendental** over $F$.

**algebraic extension** (of a field):  given a field $F$, a field $K$ such that $F$ is a subfield of $K$ and all elements of $K$ are algebraic over $F$.

**algebraic integer**:  an algebraic number that is a zero of a monic polynomial with coefficients in $\mathcal{Z}$.

**algebraic number**:  a complex number that is algebraic over $\mathcal{Q}$.

**algebraic structure**:  $(S, \star_1, \star_2, \ldots, \star_n)$ where $S$ is a nonempty set and $\star_1, \ldots, \star_n$ are binary or monadic operations defined on $S$.

**alternating group** (on $n$ elements):  the subgroup $A_n$ of all even permutations in $S_n$.

**associative property**:  the property of a binary operator $\star$ that $(a \star b) \star c = a \star (b \star c)$.

**atom**:  an element $a$ in a bounded lattice such that $0 < a$ and there is no element $b$ such that $0 < b < a$.

**automorphism**:  an isomorphism of an algebraic structure onto itself.

**automorphism** $\varphi$ **fixes set** $S$ **elementwise**:  $\varphi(a) = a$ for all $a \in S$.

**binary operation** (on a set $S$):  a function $\star \colon S \times S \to S$.

**Boolean algebra**:  a bounded, distributive, complemented lattice. Equivalent definition: $(B, +, \cdot, ', 0, 1)$ where $B$ is a set with two binary operations, $+$ (addition) and $\cdot$ (multiplication), one monadic operation, $'$ (complement), and two distinct elements, $0$ and $1$, that satisfy the commutative laws $(a + b = b + a, \; a \cdot b = b \cdot a)$, distributive laws $(a \cdot (b + c) = (a \cdot b) + (a \cdot c), \; a + (b \cdot c) = (a + b) \cdot (a + c))$, identity laws $(a + 0 = a, \; a \cdot 1 = a)$, and complement laws $(a + a' = 1, \; a \cdot a' = 0)$.

**Boolean function of degree** $n$:  a function $f \colon \{0,1\}^n = \{0,1\} \times \cdots \times \{0,1\} \to \{0,1\}$.

**bounded lattice**:  a lattice having elements $0$ (**lower bound**) and $1$ (**upper bound**) such that $0 \le a$ and $a \le 1$ for all $a$.

**cancellation properties**:  if $ab = ac$ and $a \neq 0$, then $b = c$ (**left cancellation property**); if $ba = ca$ and $a \neq 0$, then $b = c$ (**right cancellation property**).

**characteristic** (of a field):  the smallest positive integer $n$ such that $1 + 1 + \cdots + 1 = 0$ ($n$ summands). If no such $n$ exists, the field has characteristic $0$ (or characteristic $\infty$).

**closure property**: a set $S$ is closed under an operation $\star$ if the range of $\star$ is a subset of $S$.

**commutative property**: the property of an operation $\star$ that $a \star b = b \star a$.

**commutative ring**: a ring in which multiplication is commutative.

**complemented lattice**: a bounded lattice such that for each element $a$ there is an element $b$ such that $a \vee b = 1$ and $a \wedge b = 0$.

**conjunctive normal form (CNF)** (of a Boolean function): a Boolean function written as a product of maxterms.

**coset**: For a subgroup $H$ of group $G$ and $a \in G$, a **left coset** is $aH = \{ah \mid h \in H\}$; a **right coset** is $Ha = \{ha \mid h \in H\}$.

**cycle of length $n$**: a permutation on a set $S$ that moves elements only in a single orbit of size $n$.

**cyclic group**: a group $G$ with an element $a \in G$ such that $G = \{a^n \mid n \in \mathcal{Z}\}$.

**cyclic subgroup** (generated by $a$): $\{a^n \mid n \in \mathcal{Z}\} = \{\ldots, a^{-2}, a^{-1}, e, a, a^2, \ldots\}$, often written $(a)$, $\langle a \rangle$, or $[a]$. The element $a$ is a **generator** of the subgroup.

**degree** (of field $K$ over field $F$): $[K \colon F] = $ the dimension of $K$ as a vector space over $F$.

**degree** (of a permutation group): the size of the set on which the permutations are defined.

**dihedral group**: the group $D_n$ of symmetries (rotations and reflections) of a regular $n$-gon.

**disjunctive normal form (DNF)** (of a Boolean function): a Boolean function written as a sum of minterms.

**distributive lattice**: a lattice that satisfies $a \wedge (b \vee c) = (a \wedge b) \vee (a \wedge c)$ and $a \vee (b \wedge c) = (a \vee b) \wedge (a \vee c)$ for all $a, b, c$ in the lattice.

**division ring**: a nontrivial ring in which every nonzero element is a unit.

**dual** (of an expression in a Boolean algebra): the expression obtained by interchanging the operations $+$ and $\cdot$, and interchanging the elements 0 and 1, in the original expression.

**duality principle**: the principle stating that an identity between Boolean expressions remains valid when the duals of the expressions are taken.

**Euclidean domain**: an integral domain with a Euclidean norm defined on it.

**Euclidean norm** (on an integral domain): given an integral domain $I$, a function $\delta \colon I - \{0\} \to \mathcal{N}$ such that for all $a, b \in I$, $\delta(a) \leq \delta(ab)$; and for all $a, d \in I$ $(d \neq 0)$ there are $q, r \in I$ such that $a = dq + r$, where either $r = 0$ or $\delta(r) < \delta(d)$.

**even permutation**: a permutation that can be written as a product of an even number of transpositions.

**extension field** (of field $F$): a field $K$ such that $F$ is a subfield of $K$.

**field**: an algebraic structure $(F, +, \cdot)$ where $F$ is a set closed under two binary operations $+$ and $\cdot$, $(F, +)$ is an abelian group, the nonzero elements form an abelian group under multiplication, and the distributive law $a \cdot (b + c) = a \cdot b + a \cdot c$ holds.

**finite field**: a field with a finite number of elements.

**finitely generated group**: a group with a finite set of generators.

**fixed field** (of a set of automorphisms of a field):  given a set $\Phi$ of automorphisms of a field $F$, the set $\{\, a \in F \mid a\varphi = a \text{ for all } \varphi \in \Phi \,\}$.

**free monoid** (generated by a set):  given a set $S$, the monoid consisting of all words on $S$ under concatenation.

**functionally complete**:  property of a set of operators in a Boolean algebra that every Boolean function can be written using only these operators.

**Galois extension** (of a field $F$):  a field $K$ that is a normal, separable extension of $F$.

**Galois field**:  $GF(p^n)$ = the algebraic extension $\mathcal{Z}_p[x]/(f(x))$ of the finite field $\mathcal{Z}_p$ where $p$ is a prime and $f(x)$ is an irreducible polynomial over $\mathcal{Z}_p$ of degree $n$.

**Galois group** (of $K$ over $F$):  the group of automorphisms $G(K/F)$ of field $K$ that fix field $F$ elementwise.

**group**:  an algebraic structure $(G, \star)$, where $G$ is a set closed under the binary operation $\star$, the operation $\star$ is associative, $G$ has an identity element, and every element of $G$ has an inverse in $G$.

**homomorphism of groups**:  a function $\varphi \colon S \to T$, where $(S, \star_1)$ and $(T, \star_2)$ are groups, such that $\varphi(a \star_1 b) = \varphi(a) \star_2 \varphi(b)$ for all $a, b \in S$.

**homomorphism of rings**:  a function $\varphi \colon S \to T$, where $(S, +_1, \cdot_1)$ and $(T, +_2, \cdot_2)$ are rings such that $\varphi(a +_1 b) = \varphi(a) +_2 \varphi(b)$ and $\varphi(a \cdot_1 b) = \varphi(a) \cdot_2 \varphi(b)$ for all $a, b \in S$.

**ideal**:  a subring of a ring that is closed under left and right multiplication by elements of the ring.

**identity**:  an element $e$ in an algebraic structure $S$ such that $e \star a = a \star e = a$ for all $a \in S$.

**improper subgroups** (of $G$):  the subgroups $G$ and $\{e\}$.

**index of $H$ in $G$**:  the number of left (or right) cosets of $H$ in $G$.

**integral domain**:  a commutative ring with unity that has no zero divisors.

**inverse of an element** $a$:  an element $a'$ such that $a \star a' = a' \star a = e$.

**involution**:  a function that is the identity when it is composed with itself.

**irreducible element in a ring**:  a noninvertible element that cannot be written as the product of noninvertible elements.

**irreducible polynomial**:  a polynomial $p(x)$ of degree $n > 0$ over a field that cannot be written as $p_1(x) \cdot p_2(x)$ where $p_1(x)$ and $p_2(x)$ are polynomials of smaller degrees. Otherwise $p(x)$ is **reducible**.

**isomorphic**:  property of algebraic structures of the same type, $G$ and $H$, that there is an isomorphism from $G$ onto $H$, written $G \cong H$.

**isomorphism**:  a one-to-one and onto function between two algebraic structures that preserves the operations on the structures.

**isomorphism of groups**:  for groups $(G_1, \star_1)$ and $(G_2, \star_2)$, a function $\varphi \colon G_1 \to G_2$ that is one-to-one, onto $G_2$, and satisfies the property $\varphi(a \star_1 b) = \varphi(a) \star_2 \varphi(b)$.

**isomorphism of permutation groups**:  for permutation groups $(G, X)$ and $(H, Y)$, a pair of functions $(\alpha \colon G \to H, f \colon Y \to Y)$ such that $\alpha$ is a group isomorphism and $f$ is a bijection.

**isomorphism of rings**: for rings $(R_1, +_1, \cdot_1)$ and $(R_2, +_2, \cdot_2)$, a function $\varphi \colon R_1 \to R_2$ that is one-to-one, onto $R_2$, and satisfies the properties $\varphi(a +_1 b) = \varphi(a) +_2 \varphi(b)$ and $\varphi(a \cdot_1 b) = \varphi(a) \cdot_2 \varphi(b)$.

**kernel** (of a group homomorphism): given a group homomorphism $\varphi$, the set $\varphi^{-1}(e) = \{ x \mid \varphi(x) = e \}$, where $e$ is the group identity.

**kernel** (of a ring homomorphism): given a ring homomorphism $\varphi$, the set $\varphi^{-1}(0) = \{ x \mid \varphi(x) = 0 \}$.

**Klein four-group**: the group under composition of the four rigid motions of a rectangle that leave the rectangle in its original location.

**lattice**: a nonempty partially ordered set in which $\inf\{a, b\}$ and $\sup\{a, b\}$ exist for all $a, b$. $(a \vee b = \sup\{a, b\}, a \wedge b = \inf\{a, b\}.)$ Equivalently, a nonempty set closed under two binary operations $\vee$ and $\wedge$ that satisfy the associative laws, the commutative laws, and the absorption laws $(a \vee (a \wedge b) = a, a \wedge (a \vee b) = a)$.

**left divisor of zero**: $a \neq 0$ with $b \neq 0$ such that $ab = 0$.

**literal**: a Boolean variable or its complement.

**maximal ideal**: an ideal in a ring $R$ that is not properly contained in any ideal of $R$ except $R$ itself.

**maxterm of the Boolean variables** $x_1, \ldots, x_n$: a sum of the form $y_1 + \cdots + y_n$ where for each $i$, $y_i$ is equal to $x_i$ or $x_i'$.

**minimal polynomial** (of an element with respect to a field): given a field $F$ and $\alpha \in F$, the monic irreducible polynomial $f(x) \in F[x]$ of smallest degree with $f(\alpha) = 0$.

**minterm of the Boolean variables** $x_1, \ldots, x_n$: a product of the form $y_1 \ldots y_n$ where for each $i$, $y_i$ is equal to $x_i$ or $x_i'$.

**monadic operation**: a function from a set into itself.

**monoid**: an algebraic structure $(S, \star)$ such that $\star$ is associative and $S$ has an identity.

**normal extension of** $F$: a field $K$ such that $K/F$ is algebraic and every irreducible polynomial in $F[x]$ with a root in $K$ has all its roots in $K$ (**splits in $K$**).

**normal subgroup** (of a group): given a group $G$, a subgroup $H \subseteq G$ such that $aH = Ha$ for all $a \in G$.

**octic group**: See *dihedral group*.

**odd permutation**: a permutation that can be written as a product of an odd number of transpositions.

**orbit** (of an object $a \in S$ under permutation $\sigma$): $\{\ldots, a\sigma^{-2}, a\sigma^{-1}, a, a\sigma, a\sigma^2, \ldots\}$.

**order** (of an algebraic structure): the number of elements in the underlying set.

**order** (of a group element): for an element $a \in G$, the smallest positive integer $n$ such that $a^n = e$ ($na = 0$ if $G$ is written additively). If there is no such integer, then $a$ has **infinite** order.

**p-group**: for prime $p$, a group such that every element has a power of $p$ as its order.

**permutation**: a one-to-one and onto function $\sigma \colon S \to S$, where $S$ is any nonempty set.

**permutation group**: a collection of permutations on a set of objects that form a group under composition.

**polynomial** (in the variable $x$ over a ring): an expression of the form $p(x) = a_n x^n + a_{n-1} x^{n-1} + \cdots + a_1 x^1 + a_0 x^0$ where $a_n, \ldots, a_0$ are elements of the ring. For a polynomial $p(x)$, the largest integer $k$ such that $a_k \neq 0$ is the **degree** of $p(x)$. The **constant polynomial** $p(x) = a_0$ has degree 0, if $a_0 \neq 0$. If $p(x) = 0$ (the **zero polynomial**), the degree of $p(x)$ is undefined (or $-\infty$).

**polynomial ring** (over a ring $R$): $R[x] = \{p(x) \mid p(x) \text{ is a polynomial in } x \text{ over } R\}$ with the usual definitions of addition and multiplication.

**prime ideal** (of a ring $R$): an ideal $I \neq R$ with property that $ab \in I$ implies that $a \in I$ or $b \in I$.

**proper subgroup** (of a group $G$): any subgroup of $G$ except $G$ and $\{e\}$.

**quotient group (factor group)**: for a normal subgroup $H$ of $G$, the group $G/H = \{aH \mid a \in G\}$, where $aH \cdot bH = (ab)H$.

**quotient ring**: for $I$ an ideal in a ring $R$, the ring $R/I = \{a + I \mid a \in R\}$, where $(a + I) + (b + I) = (a + b) + I$ and $(a + I) \cdot (b + I) = (ab) + I$.

**reducible** (polynomial): a polynomial that is not irreducible.

**right divisor of zero**: $b \neq 0$ with $a \neq 0$ such that $ab = 0$.

**ring**: an algebraic structure $(R, +, \cdot)$ where $R$ is a set closed under two binary operations $+$ and $\cdot$, $(R, +)$ is an abelian group, $R$ satisfies the associative law for multiplication, and $R$ satisfies the left and right distributive laws for multiplication over addition.

**ring with unity**: a ring with an identity for multiplication.

**root field**: a splitting field.

**semigroup**: an algebraic structure $(S, \star)$ where $S$ is a nonempty set that is closed under the associative binary operation $\star$.

**separable extension** (of field $F$): a field $K$ such that every element of $K$ is the root of a separable polynomial in $F[x]$.

**separable polynomial**: a polynomial $p(x) \in F[x]$ of degree $n$ that has $n$ distinct roots in its splitting field.

**sign** (of a permutation): the value $+1$ if the permutation has an even number of transpositions when the permutation is written as a product of transpositions, and $-1$ otherwise.

**simple group**: a group whose only normal subgroups are $\{e\}$ and $G$.

**skew field**: a division ring.

**splitting field** (for nonconstant $p(x) \in F[x]$): the field $K = F(\alpha_1, \ldots, \alpha_n)$ where $p(x) = \alpha(x - \alpha_1) \ldots (x - \alpha_n)$, $\alpha \in F$.

**subfield** (of a field $K$): a subset $F \subseteq K$ that is a field using the same operations used in $K$.

**subgroup** (of a group $G$): a subset $H \subseteq G$ such that $H$ is a group using the same group operation used in $G$.

**subgroup generated by** $\{a_i \mid i \in S\}$: for a given group $G$ where $a_i \in G$ for all $i$ in $S$, the smallest subgroup of $G$ containing $\{a_i \mid i \in S\}$.

**subring** (of a ring $R$): a subset $S \subseteq R$ that is a ring using the same operations used in $R$.

**Sylow *p*-subgroup** (of $G$):  a subgroup of $G$ that is a $p$-group and is not properly contained in any $p$-group of $G$.

**symmetric group**:  the group of all permutations on $\{1, 2, \ldots, n\}$ under the operation of composition.

**transcendental element** (over a field $F$):  given a field $F$ and an extension field $K$, an element of $K$ that is not a root of any nonzero polynomial in $F[x]$.

**transposition**:  a cycle of length 2.

**unary operation**:  See *monadic operation*.

**unit** (in a ring):  an element with a multiplicative inverse in the ring.

**unity** (in a ring):  a multiplicative identity not equal to 0.

**word** (on a set):  a finite sequence of elements of the set.

**zero** (of a polynomial $f$):  an element $a$ such that $f(a) = 0$.

# 5.1  ALGEBRAIC MODELS

## 5.1.1  DOMAINS AND OPERATIONS

**Definitions:**

An **$n$-ary operation** on a set $S$ is a function $\star\colon S \times S \times \cdots \times S \to S$, where the domain is the product of $n$ factors.

A **binary operation** on a set $S$ is a function $\star\colon S \times S \to S$.

A **monadic operation** (or **unary operation**) on a set $S$ is a function $\star\colon S \to S$.

An **algebraic structure** $(S, \star_1, \star_2, \ldots, \star_n)$ consists of a nonempty set $S$ (the **domain**) with one or more $n$-ary operations $\star_i$ defined on $S$.

A binary operation can have some of the following properties:

- **associative property**:  $a \star (b \star c) = (a \star b) \star c$ for all $a$, $b$, $c \in S$;
- **existence of an identity element**:  there is an element $e \in S$ such that $e \star a = a \star e = a$ for all $a \in S$  ($e$ is an *identity* for $S$);
- **existence of inverses**:  for each element $a \in S$ there is an element $a' \in S$ such that $a' \star a = a \star a' = e$  ($a'$ is an *inverse* of $a$);
- **commutative property**:  $a \star b = b \star a$ for all $a, b \in S$.

**Example:**

**1.** The most important types of algebraic structures with one binary operation are listed in the following table. A checkmark means that the property holds.

| | closed | associative | commutative | existence of identity | existence of inverses |
|---|---|---|---|---|---|
| semigroup | ✓ | ✓ | | | |
| monoid | ✓ | ✓ | | | |
| group | ✓ | ✓ | | ✓ | ✓ |
| abelian group | ✓ | ✓ | ✓ | ✓ | ✓ |

## 5.1.2   SEMIGROUPS AND MONOIDS

**Definitions:**

A **semigroup** $(S, \star)$ consists of a nonempty set $S$ and an associative binary operation $\star$ on $S$.

A **monoid** $(S, \star)$ consists of a nonempty set $S$ and an associative binary operation $\star$ on $S$ such that $S$ has an identity.

A nonempty subset $T$ of a semigroup $(S, \star)$ is a **subsemigroup** of $S$ if $T$ is closed under $\star$.

A subset $T$ of a monoid $(S, \star)$ with identity $e$ is a **submonoid** of $S$ if $T$ is closed under $\star$ and $e \in T$.

Two semigroups [monoids] $(S_1, \star_1)$ and $(S_2, \star_2)$ are **isomorphic** if there is a function $\varphi \colon S_1 \to S_2$ that is one-to-one, onto $S_2$, and such that $\varphi(a \star_1 b) = \varphi(a) \star_2 \varphi(b)$ for all $a, b \in S_1$.

A **word** on a set $S$ (the *alphabet*) is a finite sequence of elements of $S$.

The **free monoid [free semigroup] generated by** $S$ is the monoid [semigroup] $(S^*, \star)$ where $S^*$ is the set of all words on a set $S$ and the operation $\star$ is defined on $S^*$ by concatenation: $x_1 x_2 \ldots x_m \star y_1 y_2 \ldots y_n = x_1 x_2 \ldots x_m y_1 y_2 \ldots y_n$. $(S^*, \star)$ is also called the **free monoid [free semigroup] on** $S^*$.

**Facts:**

**1.** Every monoid is a semigroup.

**2.** Every semigroup $(S, \star)$ is isomorphic to a subsemigroup of some semigroup of transformations on some set. Hence, every semigroup can be regarded as a semigroup of transformations. An analogous result is true for monoids.

**Examples:**

**1.** *Free semigroups and monoids:* The free monoid generated by $S$ is a monoid with the empty word $e = \lambda$ (the sequence consisting of zero elements) as the identity.

**2.** The possible input tapes to a computer form a free monoid on the set of symbols (such as the ASCII symbols) in the computer alphabet.

**3.** *Semigroup and monoid of transformations on a set $S$:* Let $S$ be a nonempty set and let $\mathcal{F}$ be the set of all functions $f \colon S \to S$. With the operation $\star$ defined by composition, $(f \star g)(x) = f(g(x))$, $(\mathcal{F}, \star)$ is the *semigroup [monoid] of transformations on $S$*. The identity of $\mathcal{F}$ is the identity transformation $e \colon S \to S$ where $e(x) = x$ for all $x \in S$.

**4.** The set of closed walks based at a fixed vertex $v$ in a graph forms a monoid under the operation of concatenation. The null walk is the identity. (§8.4.1.)

**5.** For a fixed positive integer $n$, the set of all $n \times n$ matrices with elements in any ring with unity (§5.4.1) where $\star$ is matrix multiplication (using the operations in the ring) is a semigroup and a monoid. The identity is the identity matrix.

**6.** The sets

$\mathcal{N} = \{0, 1, 2, 3, \ldots\}$ (natural numbers),
$\mathcal{Z} = \{\ldots, -2, -1, 0, 1, 2, \ldots\}$ (integers),
$\mathcal{Q}$ (the set of rational numbers),
$\mathcal{R}$ (the set of real numbers),
$\mathcal{C}$ (the set of complex numbers),

where $\star$ is either addition or multiplication, are all semigroups and monoids. Using either addition or multiplication, each semigroup is a subsemigroup of each of those following it in this list. Likewise, using either addition or multiplication, each monoid is a submonoid of each of those following it in this list. For example, $(\mathcal{Q}, +)$ is a subsemigroup and submonoid of $(\mathcal{R}, +)$ and $(\mathcal{C}, +)$. Under addition, $e = 0$; under multiplication, $e = 1$.

# 5.2  GROUPS

## 5.2.1  BASIC CONCEPTS

**Definitions:**

A **group** $(G, \star)$ consists of a set $G$ with a binary operator $\star$ defined on $G$ such that $\star$ has the following properties:

- **associative property**: $a \star (b \star c) = (a \star b) \star c$ for all $a$, $b$, $c \in G$;
- **identity property**: $G$ has an element $e$ (*identity of $G$*) that satisfies $e \star a = a \star e = a$ for all $a \in G$;
- **inverse property**: for each element $a \in G$ there is an element $a^{-1} \in G$ (*inverse of $a$*) such that $a^{-1} \star a = a \star a^{-1} = e$.

If $a \star b = b \star a$ for all $a, b \in G$, the group $G$ is **commutative** or **abelian**. (Niels H. Abel, 1802–1829)

The **order** of a finite group $G$, denoted $|G|$, is the number of elements in the group.

The (**external**) **direct product** of groups $(G_1, \star_1)$ and $(G_2, \star_2)$ is the group $G_1 \times G_2 = \{(a_1, a_2) \mid a_1 \in G_1, a_2 \in G_2\}$ where multiplication $\star$ is defined by the rule $(a_1, a_2) \star (b_1, b_2) = (a_1 \star_1 b_1, a_2 \star_2 b_2)$. The direct product can be extended to $n$ groups: $G_1 \times G_2 \times \cdots \times G_n$. The direct product is also called the **direct sum** and written $G_1 \oplus G_2 \oplus \cdots \oplus G_n$, especially if the groups are abelian. If $G_i = G$ for all $i$, the direct product can be written $G^n$.

The group $G$ is **finitely generated** if there are $a_1, a_2, \ldots, a_n \in G$ such that every element of $G$ can be written as $a_{k_1}^{\epsilon_1} a_{k_2}^{\epsilon_2} \ldots a_{k_j}^{\epsilon_j}$ where $k_i \in \{1, \ldots, n\}$ and $\epsilon_i \in \{1, -1\}$, for some $j \geq 0$; where the empty product is defined to be $e$.

*Note:* Frequently the operation $\star$ is multiplication or addition. If the operation is addition, the group $(G, +)$ is an *additive group*. If the operation is multiplication, the group $(G, \cdot)$ is a *multiplicative group*.

|  | operation $*$ | identity $e$ | inverse $a^{-1}$ |
|---|---|---|---|
| *additive group* | $a+b$ | $0$ | $-a$ |
| *multiplicative group* | $a \cdot b$ or $ab$ | $1$ or $e$ | $a^{-1}$ |

**Facts:**

**1.** Every group has exactly one identity element.

**2.** In every group every element has exactly one inverse.

**3.** *Cancellation laws:*  In all groups,

   - if $ab = ac$ then $b = c$ *(left cancellation law)*;
   - if $ba = ca$, then $b = c$ *(right cancellation law)*.

**4.** $(a^{-1})^{-1} = a$.

**5.** $(ab)^{-1} = b^{-1}a^{-1}$. More generally, $(a_1 a_2 \ldots a_k)^{-1} = a_k^{-1} a_{k-1}^{-1} \ldots a_1^{-1}$.

**6.** If $a$ and $b$ are elements of a group $G$, the equations $ax = b$ and $xa = b$ have unique solutions in $G$. The solutions are $x = a^{-1}b$ and $x = ba^{-1}$, respectively.

**7.** The direct product $G_1 \times \cdots \times G_n$ is abelian when each group $G_i$ is abelian.

**8.** $|G_1 \times \cdots \times G_n| = |G_1| \cdot \cdots \cdot |G_n|$.

**9.** The identity for $G_1 \times \cdots \times G_n$ is $(e_1, \ldots, e_n)$ where $e_i$ is the identity of $G_i$. The inverse of $(a_1, \ldots, a_n)$ is $(a_1, \ldots, a_n)^{-1} = (a_1^{-1}, \ldots, a_n^{-1})$.

**10.** The structure of a group can be determined by a single rule (see Example 2) or by a group table listing all products (see Examples 2 and 3).

**Examples:**

**1.** Table 1 displays information on several common groups. All groups listed have infinite order, except for the following: the group of complex $n$th roots of unity has order $n$, the group of all bijections $f \colon S \to S$ where $|S| = n$ has order $n!$, $\mathcal{Z}_n$ has order $n$, $\mathcal{Z}_n^*$ has order $\phi(n)$ (Euler phi-function), $S_n$ has order $n!$, $A_n$ has order $n!/2$, $D_n$ has order $2n$, and the quaternion group has order 8.

All groups listed in the table are abelian except for: the group of bijections, $GL(n, \mathcal{R})$, $S_n$, $A_n$, $D_n$, and $\mathbf{Q}$.

**Table 1: Examples of groups.**

| set | operation | identity | inverses |
|---|---|---|---|
| $\mathcal{Z}, \mathcal{Q}, \mathcal{R}, \mathcal{C}$ | addition | $0$ | $-a$ |
| $\mathcal{Z}^n$, $n$ a positive integer (also $\mathcal{Q}^n, \mathcal{R}^n, \mathcal{C}^n$) | coordinatewise addition | $(0, \ldots, 0)$ | $-(a_1, \ldots, a_n) = (-a_1, \ldots, -a_n)$ |
| the set of all complex numbers of modulus $1 = \{e^{i\theta} = \cos\theta + i\sin\theta \mid 0 \le \theta < 2\pi\}$ | multiplication | $e^{i0} = 1$ | $(e^{i\theta})^{-1} = e^{-i\theta}$ |
| the *complex $n$th roots of unity* (solutions to $z^n = 1$) $\{e^{2\pi i k/n} \mid k = 0, 1, \ldots, n-1\}$ | multiplication | $1$ | $(e^{2\pi i k/n})^{-1} = e^{2\pi i (n-k)/n}$ |
| $\mathcal{R}-\{0\}, \mathcal{Q}-\{0\}, \mathcal{C}-\{0\}$ | multiplication | $1$ | $1/a$ |
| $\mathcal{R}^*$ (positive real numbers) | multiplication | $1$ | $1/a$ |

| set | operation | identity | inverses |
|---|---|---|---|
| all rotations of the plane around the origin; $r_\alpha =$ counterclockwise rotation through an angle of $\alpha°$: $r_\alpha(x,y) = (x\cos\alpha - y\sin\alpha,\ x\sin\alpha + y\sin\alpha)$ | composition: $r_{\alpha_2} \circ r_{\alpha_1} = r_{\alpha_1 + \alpha_2}$ | $r_0$ (the $0°$ rotation) | $r_\alpha^{-1} = r_{-\alpha}$ |
| all 1–1, onto functions (bijections) $f\colon S \to S$ where $S$ is any nonempty set | composition of functions | $i\colon S \to S$ where $i(x) = x$ for all $x \in S$ | $f^{-1}(y) = x$ if and only if $f(x) = y$ |
| $\mathcal{M}_{m\times n} =$ all $m \times n$ matrices with entries in $\mathcal{R}$ | matrix addition | $O_{m\times n}$ (zero matrix) | $-A$ |
| $GL(n, \mathcal{R}) =$ all $n \times n$ invertible, or nonsingular, matrices with entries in $\mathcal{R}$; (the general linear group) | matrix multiplication | $I_n$ (identity matrix) | $A^{-1}$ |
| $\mathcal{Z}_n = \{0, 1, \ldots, n-1\}$ | $(a+b) \bmod n$ | 0 | $n - a\ (a \neq 0)$ $-0 = 0$ |
| $\mathcal{Z}_n^* = \{k \mid k \in \mathcal{Z}_n,\ k$ relatively prime to $n\},\ n > 1$ | $ab \bmod n$ | 1 | see Example 2 |
| $S_n =$ all permutations of $\{1, 2, \ldots, n\}$; (symmetric group) (See §5.3.) | composition of permutations | identity permutation | inverse permutation |
| $A_n =$ all even permutations of $\{1, 2, \ldots, n\}$; (alternating group) (See §5.3.) | composition of permutations | identity permutation | inverse permutation |
| $D_n =$ symmetries (rotations and reflections) of a regular $n$-gon; (dihedral group) | composition of functions | rotation through $0°$ | $r_\alpha^{-1} = r_{-\alpha}$; reflections are their own inverses |
| $\mathbf{Q} =$ quaternion group (see Example 3) | | | |

**2.** The groups $\mathcal{Z}_n$ and $\mathcal{Z}_n^*$ (see Table 1): In the groups $\mathcal{Z}_n$ and $\mathcal{Z}_n^*$ an element $a$ can be viewed as the equivalence class $\{b \in \mathcal{Z} \mid b \bmod n = a \bmod n\}$, which can be written $\bar{a}$ or $[a]$. To find the inverse $a^{-1}$ of $a \in \mathcal{Z}_n^*$, use the extended Euclidean algorithm to find integers $a^{-1}$ and $k$ such that $aa^{-1} + nk = \gcd(a, n) = 1$. The following are the group tables for $\mathcal{Z}_2 = \{0, 1\}$ and $\mathcal{Z}_3 = \{0, 1, 2\}$:

| + | 0 | 1 |
|---|---|---|
| 0 | 0 | 1 |
| 1 | 1 | 0 |

| + | 0 | 1 | 2 |
|---|---|---|---|
| 0 | 0 | 1 | 2 |
| 1 | 1 | 2 | 0 |
| 2 | 2 | 0 | 1 |

**3.** Quaternion group: $\mathbf{Q} = \{1, -1, i, -i, j, -j, k, -k\}$ where multiplication is defined by the following relations:

$$i^2 = j^2 = k^2 = -1, \qquad ij = -ji = k, \qquad jk = -kj = i, \qquad ki = -ik = j$$

where 1 is the identity. These relations yield the multiplication table shown next. Here, $1^{-1} = 1$, $(-1)^{-1} = -1$, $x$ and $-x$ are inverses for $x = i, j, k$. The group is nonabelian.

| $\cdot$ | 1 | $-1$ | $i$ | $-i$ | $j$ | $-j$ | $k$ | $-k$ |
|---|---|---|---|---|---|---|---|---|
| 1 | 1 | $-1$ | $i$ | $-i$ | $j$ | $-j$ | $k$ | $-k$ |
| $-1$ | $-1$ | 1 | $-i$ | $i$ | $-j$ | $j$ | $-k$ | $k$ |
| $i$ | $i$ | $-i$ | $-1$ | 1 | $k$ | $-k$ | $-j$ | $j$ |
| $-i$ | $-i$ | $i$ | 1 | $-1$ | $-k$ | $k$ | $j$ | $-j$ |
| $j$ | $j$ | $-j$ | $-k$ | $k$ | $-1$ | 1 | $i$ | $-i$ |
| $-j$ | $-j$ | $j$ | $k$ | $-k$ | 1 | $-1$ | $-i$ | $i$ |
| $k$ | $k$ | $-k$ | $j$ | $-j$ | $-i$ | $i$ | $-1$ | 1 |
| $-k$ | $-k$ | $k$ | $-j$ | $j$ | $i$ | $-i$ | 1 | $-1$ |

The quaternion group **Q** can also be defined as the following group of 8 matrices, where $i$ is the complex number such that $i^2 = -1$ and the group operation is matrix multiplication.

$$\begin{pmatrix} 1 & 0 \\ 0 & 1 \end{pmatrix}, \begin{pmatrix} -1 & 0 \\ 0 & -1 \end{pmatrix}, \begin{pmatrix} -i & 0 \\ 0 & i \end{pmatrix}, \begin{pmatrix} i & 0 \\ 0 & -i \end{pmatrix},$$

$$\begin{pmatrix} 0 & 1 \\ -1 & 0 \end{pmatrix}, \begin{pmatrix} 0 & -1 \\ 1 & 0 \end{pmatrix}, \begin{pmatrix} 0 & i \\ i & 0 \end{pmatrix}, \begin{pmatrix} 0 & -i \\ -i & 0 \end{pmatrix}$$

**4.** The set $\{a, b, c, d\}$ with either of the following multiplication tables is not a group. In the first case there is an identity, $a$, and each element has an inverse, but the associative law fails: $(bc)d \neq b(cd)$. In the second case there is no identity (hence inverses are not defined) and the associative law fails.

| $\cdot$ | $a$ | $b$ | $c$ | $d$ |
|---|---|---|---|---|
| $a$ | $a$ | $b$ | $c$ | $d$ |
| $b$ | $b$ | $d$ | $a$ | $c$ |
| $c$ | $c$ | $a$ | $b$ | $d$ |
| $d$ | $d$ | $c$ | $b$ | $a$ |

| $\cdot$ | $a$ | $b$ | $c$ | $d$ |
|---|---|---|---|---|
| $a$ | $a$ | $c$ | $b$ | $d$ |
| $b$ | $d$ | $b$ | $a$ | $c$ |
| $c$ | $b$ | $d$ | $c$ | $a$ |
| $d$ | $c$ | $a$ | $d$ | $b$ |

## 5.2.2   GROUP ISOMORPHISM AND HOMOMORPHISM

**Definitions:**

For groups $G$ and $H$, a function $\varphi : G \to H$ such that $\varphi(ab) = \varphi(a)\varphi(b)$ for all $a, b \in G$ is a **homomorphism**. The notation $a\varphi$ is sometimes used instead of $\varphi(a)$.

For groups $G$ and $H$, a function $\varphi : G \to H$ is an **isomorphism** from $G$ to $H$ if $\varphi$ is a homomorphism that is 1–1 and onto $H$. In this case $G$ is **isomorphic** to $H$, written $G \cong H$.

An isomorphism $\varphi : G \to G$ is an **automorphism**.

The **kernel** of $\varphi$ is the set $\{g \in G \mid \varphi(g) = e\}$, where $e$ is the identity of the group $G$.

**Facts:**

**1.** If $\varphi$ is an isomorphism, $\varphi^{-1}$ is an isomorphism.

**2.** Isomorphism is an equivalence relation: $G \cong G$ (reflexive); if $G \cong H$, then $H \cong G$ (symmetric); if $G \cong H$ and $H \cong K$, then $G \cong K$ (transitive).

**3.** If $\varphi\colon G \to H$ is a homomorphism, then $\varphi(G)$ is a group (a subgroup of $H$).

**4.** If $\varphi\colon G \to H$ is a homomorphism, then the kernel of $\varphi$ is a group (a subgroup of $G$).

**5.** If $p$ is prime there is only one group of order $p$ (up to isomorphism), the group $(\mathcal{Z}_p, +)$.

**6.** *Cayley's theorem:* If $G$ is a finite group of order $n$, then $G$ is isomorphic to a subgroup of the group $S_n$ of permutations on $n$ objects. (Arthur Cayley, 1821–1895) The isomorphism is obtained by associating with each $a \in G$ the map $\pi_a\colon G \to G$ given by the rule $\pi_a(g) = ga$ for all $g \in G$.

**7.** $\mathcal{Z}_m \times \mathcal{Z}_n$ is isomorphic to $\mathcal{Z}_{mn}$ if and only if $m$ and $n$ are relatively prime.

**8.** If $n = n_1 n_2 \dots n_k$ where the $n_i$ are powers of distinct primes, then $\mathcal{Z}_n$ is isomorphic to $\mathcal{Z}_{n_1} \times \mathcal{Z}_{n_2} \times \cdots \times \mathcal{Z}_{n_k}$.

**9.** *Fundamental theorem of finite abelian groups:* Every finite abelian group $G$ (order $\geq 2$) is isomorphic to a direct product of cyclic groups where each cyclic group has order a power of a prime. That is, $G$ is isomorphic to $\mathcal{Z}_{n_1} \times \mathcal{Z}_{n_2} \times \cdots \times \mathcal{Z}_{n_k}$ where each cyclic order $n_i$ is a power of some prime. In addition, the set $\{n_1, \dots, n_k\}$ is unique.

**10.** Every finite abelian group is isomorphic to a subgroup of $\mathcal{Z}_n^*$ for some $n$.

**11.** *Fundamental theorem of finitely generated abelian groups:* If $G$ is a finitely generated abelian group, then there are unique integers $n \geq 0$, $n_1, n_2, \dots, n_k \geq 2$ where $n_{i+1} \mid n_i$ for $i = 1, 2, \dots, k-1$ such that $G$ is isomorphic to $\mathcal{Z}^n \times \mathcal{Z}_{n_1} \times \mathcal{Z}_{n_2} \times \cdots \times \mathcal{Z}_{n_k}$.

**Examples:**

**1.** The following table lists the number of nonisomorphic groups and abelian groups of all orders from 1 to 60.

| order | 1 | 2 | 3 | 4 | 5 | 6 | 7 | 8 | 9 | 10 | 11 | 12 | 13 | 14 | 15 | 16 | 17 | 18 | 19 | 20 |
|---|---|---|---|---|---|---|---|---|---|---|---|---|---|---|---|---|---|---|---|---|
| groups | 1 | 1 | 1 | 2 | 1 | 2 | 1 | 5 | 2 | 2 | 1 | 5 | 1 | 2 | 1 | 14 | 1 | 5 | 1 | 5 |
| abelian | 1 | 1 | 1 | 2 | 1 | 1 | 1 | 3 | 2 | 1 | 1 | 2 | 1 | 1 | 1 | 5 | 1 | 2 | 1 | 2 |
| order | 21 | 22 | 23 | 24 | 25 | 26 | 27 | 28 | 29 | 30 | 31 | 32 | 33 | 34 | 35 | 36 | 37 | 38 | 39 | 40 |
| groups | 2 | 2 | 1 | 15 | 2 | 2 | 5 | 4 | 1 | 4 | 1 | 51 | 1 | 2 | 1 | 14 | 1 | 2 | 2 | 14 |
| abelian | 1 | 1 | 1 | 3 | 2 | 1 | 3 | 2 | 1 | 1 | 1 | 7 | 1 | 1 | 1 | 4 | 1 | 1 | 1 | 3 |
| order | 41 | 42 | 43 | 44 | 45 | 46 | 47 | 48 | 49 | 50 | 51 | 52 | 53 | 54 | 55 | 56 | 57 | 58 | 59 | 60 |
| groups | 1 | 6 | 1 | 4 | 2 | 2 | 1 | 52 | 2 | 5 | 1 | 5 | 1 | 15 | 2 | 13 | 2 | 2 | 1 | 13 |
| abelian | 1 | 1 | 1 | 2 | 2 | 1 | 1 | 5 | 2 | 2 | 1 | 2 | 1 | 3 | 1 | 3 | 1 | 1 | 1 | 2 |

**2.** All groups of order 12 or less are listed by order in the following table.

| order | groups |
|---|---|
| 1 | $\{e\}$ |
| 2 | $\mathcal{Z}_2$ |
| 3 | $\mathcal{Z}_3$ |
| 4 | $\mathcal{Z}_4$, if there is an element of order 4 (group is cyclic) <br> $\mathcal{Z}_2 \times \mathcal{Z}_2 \cong$ Klein four-group, if no element has order 4 (§5.3.2) |

| order | groups |
|---|---|
| 5 | $\mathcal{Z}_5$ |
| 6 | $\mathcal{Z}_6$, if there is an element of order 6 (group is cyclic) |
|   | $S_3 \cong D_3$, if there is no element of order 6 (§5.3.1, §5.3.2) |
| 7 | $\mathcal{Z}_7$ |
| 8 | $\mathcal{Z}_8$, if there is an element of order 8 (group is cyclic) |
|   | $\mathcal{Z}_2 \times \mathcal{Z}_4$, if there is an element $a$ of order 4, but none of order 8, and |
|   |     if there is an element $b \notin (a)$ such that $ab = ba$ and $b^2 = e$ |
|   | $\mathcal{Z}_2 \times \mathcal{Z}_2 \times \mathcal{Z}_2$, if every element has order 1 or 2 |
|   | $D_4$, if there is an element $a$ of order 4, but none of order 8, and if |
|   |     there is an element $b \notin (a)$ such that $ba = a^3 b$ and $b^2 = e$ |
|   | Quaternion group, if there is an element $a$ of order 4, none of order 8, |
|   |     and an element $b \notin (a)$ such that $ba = a^3 b$ and $b^2 = a^2$ (§5.2.1) |
| 9 | $\mathcal{Z}_9$, if there is an element of order 9 (group is cyclic) |
|   | $\mathcal{Z}_3 \times \mathcal{Z}_3$, if there is no element of order 9 |
| 10 | $\mathcal{Z}_{10}$, if there is an element of order 10 (group is cyclic) |
|   | $D_5$, if there is no element of order 10 |
| 11 | $\mathcal{Z}_{11}$ |
| 12 | $\mathcal{Z}_{12} \cong \mathcal{Z}_3 \times \mathcal{Z}_4$, if there is an element of order 12 (group is cyclic) |
|   | $\mathcal{Z}_2 \times \mathcal{Z}_6 \cong \mathcal{Z}_2 \times \mathcal{Z}_2 \times \mathcal{Z}_3$, if group is abelian but noncyclic |
|   | $D_6$, if group is nonabelian and has an element of order 6 but none of |
|   |     order 4 |
|   | $A_4$, if group is nonabelian and has no element of order 6 |
|   | The group generated by $a$ and $b$, where $a$ has order 4, $b$ has order 3, |
|   |     and $ab = b^2 a$ |

## 5.2.3  SUBGROUPS

### Definitions:

A **subgroup** of a group $(G, \star)$ is a subset $H \subseteq G$ such that $(H, \star)$ is a group (with the same group operation as in $G$). Write $H \leq G$ if $H$ is a subgroup of $G$.

If $a \in G$, the set $(a) = \{\ldots, a^{-2} = (a^{-1})^2, a^{-1}, a^0 = e, a, a^2, \ldots\} = \{a^n \mid n \in \mathcal{Z}\}$ is the **cyclic subgroup** generated by $a$. The element $a$ is a **generator** of $G$.

$G$ and $\{e\}$ are **improper subgroups** of $G$. All other subgroups of $G$ are **proper subgroups** of $G$.

### Facts:

1. If $G$ is a group, then $\{e\}$ and $G$ are subgroups of $G$.

2. If $G$ is a group and $a \in G$, the set $(a)$ is a subgroup of $G$.

3. Every subgroup of an abelian group is abelian.

**4.** If $H$ is a subgroup of a group $G$, then the identity element of $H$ is the identity element of $G$; the inverse (in the subgroup $H$) of an element $a$ in $H$ is the inverse (in the group $G$) of $a$.

**5.** *Lagrange's theorem*: Let $G$ be a finite group. If $H$ is any subgroup of $G$, then the order of $H$ is a divisor of the order of $G$. (Joseph-Louis Lagrange, 1736–1813)

**6.** If $d$ is a divisor of the order of a group $G$, there may be no subgroup of order $d$. (The group $A_4$, of order 12, has no subgroup of order 6. See §5.3.2.)

**7.** If $G$ is a finite abelian group, then the converse of Lagrange's theorem is true for $G$.

**8.** If $G$ is finite (not necessarily abelian) and $p$ is a prime that divides the order of $G$, then $G$ has a subgroup of order $p$.

**9.** If $G$ has order $p^m n$ where $p$ is prime and $p$ does not divide $n$, then $G$ has a subgroup of order $p^m$, called a *Sylow subgroup* or *Sylow p-subgroup*. See §5.2.6.

**10.** A subset $H$ of a group $G$ is a subgroup of $G$ if and only if the following are all true: $H \neq \emptyset$; $a, b \in H$ implies $ab \in H$; and $a \in H$ implies $a^{-1} \in H$.

**11.** A subset $H$ of a group $G$ is a subgroup of $G$ if and only if $H \neq \emptyset$ and $a, b \in H$ implies that $ab^{-1} \in H$.

**12.** If $H$ is a nonempty finite subset of a group $G$ with the property that $a, b \in H$ implies that $ab \in H$, then $H$ is a subgroup of $G$.

**13.** The intersection of any collection of subgroups of a group $G$ is a subgroup of $G$.

**14.** The union of subgroups is not necessarily a subgroup. See Example 12.

**Examples:**

**1.** *Additive subgroups*: Each of the following can be viewed as a subgroup of all the groups listed after it: $(\mathcal{Z}, +)$, $(\mathcal{Q}, +)$, $(\mathcal{R}, +)$, $(\mathcal{C}, +)$.

**2.** For $n$ any positive integer, the set $n\mathcal{Z} = \{nz \mid z \in \mathcal{Z}\}$ is a subgroup of $\mathcal{Z}$.

**3.** $\mathcal{Z}_2$ is not a subgroup of $\mathcal{Z}_4$ (the group operations are not the same).

**4.** The set of odd integers under addition is not a subgroup of $(\mathcal{Z}, +)$ (the set of odd integers is not closed under addition).

**5.** $(\mathcal{N}, +)$ is not a subgroup of $(\mathcal{Z}, +)$ ($\mathcal{N}$ does not contain its inverses).

**6.** The group $\mathcal{Z}_6$ has the following four subgroups: $\{0\}$, $\{0, 3\}$, $\{0, 2, 4\}$, $\mathcal{Z}_6$.

**7.** *Multiplicative subgroups*: Each of the following can be viewed as a subgroup of all the groups listed after it: $(\mathcal{Q} - \{0\}, \cdot)$, $(\mathcal{R} - \{0\}, \cdot)$, $(\mathcal{C} - \{0\}, \cdot)$.

**8.** The set of $n$ complex $n$th roots of unity can be viewed as a subgroup of the set of all complex numbers of modulus 1 under multiplication, which is a subgroup of $(\mathcal{C} - \{0\}, \cdot)$.

**9.** If $nd = 360$ ($n$ and $d$ positive integers) and $r_k$ is the counterclockwise rotation of the plane about the origin through an angle of $k°$, then $\{r_k \mid k = 0, d, 2d, 3d, \ldots, (n-1)d\}$ is a subgroup of the group of all rotations of the plane around the origin.

**10.** The set of all $n \times n$ nonsingular diagonal matrices is a subgroup of the set of all $n \times n$ nonsingular matrices under multiplication.

**11.** If $n = mk$, then $\{0, m, 2m, \ldots, (k-1)m\}$ is a subgroup of $(\mathcal{Z}_n, +)$ isomorphic to $\mathcal{Z}_k$.

**12.** The union of subgroups need not be a subgroup: $\{2n \mid n \in \mathcal{Z}\}$ and $\{3n \mid n \in \mathcal{Z}\}$ are subgroups of $\mathcal{Z}$, but their union is not a subgroup of $\mathcal{Z}$ since $2 + 3 = 5 \notin \{2n \mid n \in \mathcal{Z}\} \cup \{3n \mid n \in \mathcal{Z}\}$.

### 5.2.4   COSETS AND QUOTIENT GROUPS

**Definitions:**

If $H$ is a subgroup of a group $G$ and $a \in G$, then the set $aH = \{ah \mid h \in H\}$ is a **left coset** of $H$ in $G$. The set $Ha = \{ha \mid h \in H\}$ is a **right coset** of $H$ in $G$. (If $G$ is written additively, the cosets are written $a + H$ and $H + a$.)

The **index** of a subgroup $H$ in a group $G$, written $(G : H)$ or $[G : H]$, is the number of left (or right) cosets of $H$ in $G$.

A **normal** subgroup of a group $G$ is a subgroup $H$ of $G$ such that $aH = Ha$ for all $a \in G$. The notation $H \triangleleft G$ means that $H$ is a normal subgroup of $G$.

If $H$ is a normal subgroup of $G$, the **quotient group** (or **factor group of $G$ modulo $H$**) is the group $G/H = \{aH \mid a \in G\}$, where $aH \cdot bH = (ab)H$.

If $G$ is a group and $a \in G$, an element $b \in G$ is a **conjugate** of $a$ if $b = gag^{-1}$ for some $g \in G$.

If $G$ is a group and $a \in G$, the set $\{x \mid x \in G,\ ax = xa\}$ is the **centralizer** (or **normalizer**) of $a$.

If $G$ is a group, the set $\{x \mid x \in G,\ gx = xg \text{ for all } g \in G\}$ is the **center** of $G$.

If $H$ is a subgroup of group $G$, the set $\{x \mid x \in G,\ xHx^{-1} = H\}$ is the **normalizer** of $H$.

**Facts:**

1. If $H$ is a subgroup of a group $G$, then the following are equivalent:

   - $H$ is a normal subgroup of $G$;
   - $aHa^{-1} = a^{-1}Ha = H$ for all $a \in G$;
   - $a^{-1}ha \in H$ for all $a \in G$, $h \in H$;
   - for all $a \in G$ and $h_1 \in H$, there is $h_2 \in H$ such that $ah_1 = h_2a$.

2. If group $G$ is abelian, then every subgroup $H$ of $G$ is normal. If $G$ is not abelian, it may happen that $H$ is not normal.

3. If group $G$ is finite, then $(G : H) = |G|/|H|$.

4. $\{e\}$ and $G$ are normal subgroups of group $G$.

5. In the group $G/H$, the identity is $eH = H$ and the inverse of $aH$ is $a^{-1}H$.

6. *Fundamental homomorphism theorem:* If $\varphi : G \to H$ is a homomorphism and has kernel $K$, then $K$ is a normal subgroup of $G$ and $G/K$ is isomorphic to $\varphi(G)$.

7. If $H$ is a normal subgroup of a group $G$ and $\varphi : G \to G/H$ is defined by $\varphi(g) = gH$, then $\varphi$ is a homomorphism onto $G/H$ with kernel $H$.

8. If $H$ is a normal subgroup of a finite group $G$, then $G/H$ has $|G|/|H|$ cosets.

9. If $H$ and $K$ are normal subgroups of a group $G$, then $H \cap K$ is a normal subgroup of $G$.

10. For all $a \in G$, the centralizer of $a$ is a subgroup of $G$.

11. The center of a group is a subgroup of the group.

12. The normalizer of a subgroup of group $G$ is a subgroup of $G$.

**13.** The index of the centralizer of $a \in G$ is equal to the number of distinct conjugates of $a$ in $G$.

**14.** If a group $G$ contains normal subgroups $H$ and $K$ such that $H \cap K = \{e\}$ and $\{hk \mid h \in H, k \in K\} = G$, then $G$ is isomorphic to $H \times K$.

**15.** If $G$ is a group such that $|G| = ab$ where $a$ and $b$ are relatively prime, and if $G$ contains normal subgroups $H$ of order $a$ and $K$ of order $b$, then $G$ is isomorphic to $H \times K$.

**Examples:**

**1.** $\mathcal{Z}/n\mathcal{Z}$ is isomorphic to $\mathcal{Z}_n$, since $\varphi \colon \mathcal{Z} \to \mathcal{Z}_n$ defined by $\varphi(g) = g \bmod n$ has kernel $n\mathcal{Z}$.

**2.** The left cosets of the subgroup $H = \{0, 4\}$ in $\mathcal{Z}_8$ are $H + 0 = \{0, 4\}$, $H + 1 = \{1, 5\}$, $H + 2 = \{2, 6\}$, $H + 3 = \{3, 7\}$. The index of $H$ in $\mathcal{Z}_8$ is $(\mathcal{Z}_8, H) = 4$.

**3.** $\{(1), (12)\}$ is not a normal subgroup of the symmetric group $S_3$ (§5.3.1).

---

### 5.2.5   CYCLIC GROUPS AND ORDER

**Definitions:**

A group $(G, \cdot)$ is **cyclic** if there is $a \in G$ such that $G = \{a^n \mid n \in \mathcal{Z}\}$, where $a^0 = e$ and $a^{-n} = (a^{-1})^n$ for all positive integers $n$. If $G$ is written additively, $G = \{na \mid n \in \mathcal{Z}\}$, where $0a = 0$ and if $n > 0$, $na = a + a + a + \cdots + a$ ($n$ terms) and $-na = (-a) + (-a) + \cdots + (-a)$ ($n$ terms).

The element $a$ is called a **generator** of $G$ and the group $(G, \cdot)$ is written $((a), \cdot)$, $(a)$, or $\langle a \rangle$.

The **order of an element** $a \in G$, written $|(a)|$ or $\mathrm{ord}(a)$, is the smallest positive integer $n$ such that $a^n = e$ ($na = 0$ if $G$ is written additively). If there is no such integer, then $a$ has **infinite** order.

A subgroup $H$ of a group $(G, \cdot)$ is a **cyclic subgroup** if there exists $a \in H$ such that $H = \{a^n \mid n \in \mathcal{Z}\}$.

**Facts:**

**1.** The order of an element $a$ is equal to the number of elements in $(a)$.

**2.** Every group of prime order is cyclic.

**3.** Every cyclic group is abelian. However, not every abelian group is cyclic: for example, $(\mathcal{R}, +)$ and the Klein four-group.

**4.** If $G$ is an infinite cyclic group, then $G \cong (\mathcal{Z}, +)$.

**5.** If $G$ is a finite cyclic group of order $n$, then $G \cong (\mathcal{Z}_n, +)$.

**6.** If $G$ is a group of order $n$, then the order of every element of $G$ is a divisor of $n$.

**7.** *Cauchy's theorem*: If $G$ is a group of order $n$ and $p$ is a prime that divides $n$, then $G$ contains an element of order $p$. (Augustin-Louis Cauchy, 1789–1857)

**8.** If $G$ is a cyclic group of order $n$ generated by $a$, then $G = \{a, a^2, a^3, \ldots, a^n\}$ and $a^n = e$. If $k$ and $n$ are relatively prime, then $a^k$ is also a generator of $G$, and conversely.

**9.** If $G$ is a group and $a \in G$, then $(a)$ is a cyclic subgroup of $G$.

**10.** Every subgroup of a cyclic group is cyclic.

**11.** If $G$ is a group of order $n$ and there is an element $a \in G$ of order $n$, then $G$ is cyclic and $G = (a)$.

**Examples:**

**1.** $(\mathcal{Z}, +)$ is cyclic and is generated by each of 1 and $-1$.

**2.** $(\mathcal{Z}_n, +)$ is cyclic and is generated by each element of $\mathcal{Z}_n$ that is relatively prime to $n$. If $a \in \mathcal{Z}_n$, then $a$ has order $n/\gcd(a, n)$.

**3.** $(\mathcal{Z}_p, +)$, $p$ prime, is a cyclic group generated by each of the elements $1, 2, \ldots, p - 1$. If $a \neq 0$, $a$ has order $p$.

**4.** $(\mathcal{Z}_n^*, \cdot)$ is cyclic if and only if $n = 2$, $4$, $p^k$, or $2p^k$, where $k \geq 1$ and $p$ is an odd prime.

## 5.2.6   SYLOW THEORY

The Sylow theorems are used to help classify the nonisomorphic groups of a given order by guaranteeing the existence of subgroups of certain orders. (Peter Ludvig Mejdell Sylow, 1832–1918)

**Definitions:**

For prime $p$, a group $G$ is a ***p-group*** if every element of $G$ has order $p^n$ for some $n \geq 0$.

For prime $p$, a **Sylow *p*-subgroup** (**Sylow subgroup**) of $G$ is a subgroup of $G$ that is a $p$-group and is not properly contained in any $p$-group in $G$.

**Facts:**

**1.** *Sylow's theorem*:   If $G$ is a group of order $p^m \cdot q$ where $p$ is a prime, $m \geq 1$, and $p$ does not divide $q$, then

- $G$ contains subgroups of orders $p, p^2, \ldots, p^m$ (hence, if prime $p$ divides the order of a finite group $G$, then $G$ contains an element of order $p$);

- if $H$ and $K$ are Sylow $p$-subgroups of $G$, there is $g \in G$ such that $K = gHg^{-1}$ ($K$ is *conjugate* to $H$);

- the number of Sylow $p$-subgroups of $G$ is $kp + 1$ for some integer $k$ such that $(kp + 1) \mid q$.

**2.** If $G$ is a group of order $pq$ where $p$ and $q$ are primes and $p < q$, then $G$ contains a normal subgroup of order $q$.

**3.** If $G$ is a group of order $pq$ where $p$ and $q$ are primes, $p < q$, and $p$ does not divide $q - 1$, then $G$ is cyclic.

**Examples:**

**1.** Every group of order 15 is cyclic (by Fact 3).

**2.** Every group of order 21 contains a normal subgroup of order 7 (by Fact 2).

## 5.2.7   SIMPLE GROUPS

Simple groups arise as a fundamental part of the study of finite groups and the structure of their subgroups. An extensive, lengthy search by many mathematicians for all finite simple groups ended in 1980 when, as the result of hundreds of articles written by over one hundred mathematicians, the classification of all finite simple groups was completed. See [As00] and [Go82] for details.

**Definitions:**

A group $G \neq \{e\}$ is **simple** if its only normal subgroups are $\{e\}$ and $G$.

A **composition series** for a group $G$ is a finite sequence of subgroups $H_1 = G$, $H_2, \ldots$, $H_{n-1}$, $H_n = \{e\}$ such that $H_{i+1}$ is a normal subgroup of $H_i$ and $H_i/H_{i+1}$ is simple, for $i = 1, \ldots, n-1$.

A finite group $G$ is **solvable** if it has a sequence of subgroups $H_1 = G$, $H_2, \ldots$, $H_{n-1}$, $H_n = \{e\}$ such that $H_{i+1}$ is a normal subgroup of $H_i$ and $H_i/H_{i+1}$ is abelian, for $i = 1, \ldots, n-1$.

A **sporadic** group is one of 26 nonabelian finite simple groups that is not an alternating group or a group of Lie type [Go82].

**Facts:**

**1.** Every finite group has a composition series. Thus, simple groups (the quotient groups in the series) can be regarded as the building blocks of finite groups.

**2.** Some infinite groups, such as $(\mathcal{Z}, +)$, do not have composition series.

**3.** Every abelian group is solvable.

**4.** An abelian group $G$ is simple if and only if $G \cong \mathcal{Z}_p$ where $p$ is prime.

**5.** If $G$ is a nonabelian solvable group, then $G$ is not simple.

**6.** Every group of prime order is simple.

**7.** Every group of order $p^n$ ($p$ prime) is solvable.

**8.** Every group of order $p^n q^m$ ($p$, $q$ primes) is solvable.

**9.** If $G$ is a solvable, simple finite group, then $G$ is either $\{e\}$ or $\mathcal{Z}_p$ ($p$ prime).

**10.** If $G$ is a simple group of odd order, then $G \cong \mathcal{Z}_p$ for some prime $p$.

**11.** There is no infinite simple, solvable group.

**12.** *Burnside's conjecture/Feit-Thompson theorem*:  In 1911 William Burnside conjectured that all groups of odd order are solvable. This conjecture was proved in 1963 by Walter Feit and John Thompson. (See Fact 13.)

**13.** Every nonabelian simple group has even order. (This follows from the Feit-Thompson theorem.)

**14.** The proof of the Burnside conjecture provided the impetus for a massive program to classify all finite simple groups. This program, organized by Daniel Gorenstein, led to hundreds of journal articles and concluded in 1980 when the classification problem was finally solved (Fact 15). [GoLySo02]

**15.** *Classification theorem for finite simple groups*:  Every finite simple group is of one of the following types:

- abelian: $\mathcal{Z}_p$ where $p$ is prime (§5.2.1);
- nonabelian:
  - ⋄ alternating groups $A_n$ ($n \neq 4$) (§5.3.2);
  - ⋄ groups of *Lie* type, which fall into 6 classes of classical groups and 10 classes of exceptional simple groups [Ca89];
  - ⋄ *sporadic* groups. There are 26 sporadic groups, listed here from smallest to largest order. The letters in the names of the groups reflect the names of some of the people who conjectured the existence of the groups or proved the groups simple. $M_{11}$ (order 7,920), $M_{12}$, $M_{22}$, $M_{23}$, $M_{24}$, $J_1$, $J_2$, $J_3$, $J_4$, $HS$, $Mc$, $Suz$, $Ru$, $He$, $Ly$, $ON$, $.1$, $.2$, $.3$, $M(22)$, $M(23)$, $M(24)'$, $F_5$, $F_3$, $F_2$, $F_1$ (the *monster* or *Fischer-Griess* group of order $\approx 10^{54}$).

## 5.2.8  GROUP PRESENTATIONS

**Definitions:**

The **balanced alphabet** on the set $X = \{x_1, \ldots, x_n\}$ is the set $\{x_1, x_1^{-1}, \ldots, x_n, x_n^{-1}\}$, whose elements are often called **symbols**.

Symbols $x_j$ and $x_j^{-1}$ of a balanced alphabet are **inverses** of each other. A double inverse $(x_j^{-1})^{-1}$ is understood as the identity operator.

A **word** in $X$ is a string $s_1 s_2 \ldots s_n$ of symbols from the balanced alphabet on $X$.

The **inverse of a word** $s = s_1 s_2 \ldots s_n$ is the word $s^{-1} = s_n^{-1} \ldots s_2^{-1} s_1^{-1}$.

The **free semigroup** $W(X)$ has the set of words in $X$ as its domain and string concatenation as its product operation.

A **trivial relator** in the set $X = \{x_1, \ldots, x_n\}$ is a word of the form $x_j x_j^{-1}$ or $x_j^{-1} x_j$.

A word $u$ is **freely equivalent** to a word $v$, denoted $u \sim v$, if $v$ can be obtained from $u$ by iteratively inserting and deleting trivial relators, in the usual sense of those string operations. This is an equivalence relation, whose classes are called **free equivalence classes**.

A **reduced word** is a word containing no instances of a trivial relator as a substring.

The **free group** $F[X]$ has the set of free equivalence classes of words in $X$ as its domain and class concatenation as its product operation.

A **group presentation** is a pair $(X : R)$, where $X$ is an alphabet and $R$ is a set of words in $X$ called **relators**. A group presentation is **finite** if $X$ and $R$ are both finite.

A word $u$ is **R-equivalent** to a word $v$ under the group presentation $(X : R)$, denoted $u \sim_R v$, if $v$ can be obtained from $u$ by iteratively inserting and deleting relators from $R$ or trivial relators. This is an equivalence relation, whose classes are called **R-equivalence classes**.

The group $\mathcal{G}(X : R)$ **presented** by the group presentation $(X : R)$ has the set of $R$-equivalence classes as its domain and class concatenation as its product operation. Moreover, any group $G$ isomorphic to $\mathcal{G}(X : R)$ is said to be **presented** by the group presentation $(X : R)$.

The group $G$ is **finitely presentable** if it has a presentation whose alphabet and relator set are both finite.

The **commutator** of the words $u$ and $v$ is the word $u^{-1} v^{-1} u v$. Any word of this form is called a commutator.

A **conjugate** of the word $v$ is any word of the form $u^{-1} v u$.

**Facts:**

**1.** Max Dehn (1911) formulated three fundamental decision problems for finite presentations:

- *word problem:* Given an arbitrary presentation $(X : R)$ and an arbitrary word $w$, decide whether $w$ is equivalent to the empty word (i.e., the group identity).

- *conjugacy problem:* Given an arbitrary presentation $(X : R)$ and two arbitrary words $w_1$ and $w_2$, decide whether $w_1$ is equivalent to a conjugate of $w_2$.

- *isomorphism problem:* Given two arbitrary presentations $(X : R)$ and $(Y : S)$, decide whether they present isomorphic groups.

**2.** W. W. Boone (1955) and P. S. Novikov (1955) constructed presentations in which the word problem is recursively unsolvable. This implies that there is no single finite procedure that works for all finite presentations, thereby negatively solving Dehn's word problem and conjugacy problem.

**3.** M. O. Rabin (1958) proved that it is impossible to decide even whether a presentation presents the trivial group, which immediately implies that Dehn's isomorphism problem is recursively unsolvable.

**4.** The word problem is recursively solvable in various special classes of group presentations, including the following: presentations with no relators (i.e., free groups), presentations with only one relator, presentations in which the relator set includes the commutator of each pair of generators (i.e., abelian groups).

**5.** The group presentation $\mathcal{G}(X : R)$ is the quotient of the free group $F[X]$ by the normalizer of the relator set $R$.

**6.** More information on group presentations can be found in [CoMo80], [CrFo08], and [MaKaSo04].

**Examples:**

**1.** The cyclic group $\mathcal{Z}_k$ has the presentation $(x : x^k)$.

**2.** The direct sum $\mathcal{Z}_r \oplus \mathcal{Z}_s$ has the presentation $(x, y : x^r, y^s, x^{-1}y^{-1}xy)$.

**3.** The dihedral group $\mathcal{D}_q$ has the presentation $(x, y : x^q, y^2, y^{-1}xyx)$.

# 5.3   PERMUTATION GROUPS

Permutations, as arrangements, are important tools used extensively in combinatorics (§2.3 and §2.7). The set of permutations on a given set forms a group, and it is this algebraic structure that is examined in this section.

## 5.3.1   BASIC CONCEPTS

**Definitions:**

A **permutation** is a one-to-one and onto function $\sigma \colon S \to S$, where $S$ is any nonempty set. If $S = \{a_1, a_2, \ldots, a_n\}$, a permutation $\sigma$ is sometimes written as the $2 \times n$ matrix

$$\sigma = \begin{pmatrix} a_1 & a_2 & \cdots & a_n \\ a_1\sigma & a_2\sigma & \cdots & a_n\sigma \end{pmatrix}$$

where $a_i\sigma$ means $\sigma(a_i)$.

A permutation $\sigma \colon S \to S$ is a **cycle of length $n$** if there is a subset of $S$ of size $n$, $\{a_1, a_2, \ldots, a_n\}$, such that $a_1\sigma = a_2, a_2\sigma = a_3, \ldots, a_n\sigma = a_1$, and $a\sigma = a$ for all other elements of $S$. Write $\sigma = (a_1 \ a_2 \ \cdots \ a_n)$. A **transposition** is a cycle of length 2.

A **permutation group** $(G, X)$ is a collection $G$ of permutations on a nonempty set $X$ (whose elements are called *objects*) such that these permutations form a group under composition. That is, if $\sigma$ and $\tau$ are permutations in $G$, $\sigma\tau$ is the permutation in $G$ defined by the rule $a(\sigma\tau) = (a\sigma)\tau$. The **order** of the permutation group is $|G|$. The **degree** of the permutation group is $|X|$.

The **symmetric group on $n$ elements** is the group $S_n$ of all permutations on the set $\{1, 2, \ldots, n\}$ under composition. (See Fact 1.)

An **isomorphism** from a permutation group $(G, X)$ to a permutation group $(H, Y)$ is a pair of functions $(\alpha\colon G{\to}H,\ f\colon X{\to}Y)$ such that $\alpha$ is a group isomorphism and $f$ is one-to-one and onto $Y$.

If $\sigma_1 = (a_{i_1}\ a_{i_2}\ \ldots\ a_{i_m})$ and $\sigma_2 = (a_{j_1}\ a_{j_2}\ \ldots\ a_{j_n})$ are cycles on $S$, then $\sigma_1$ and $\sigma_2$ are **disjoint cycles** if the sets $\{a_{i_1}, a_{i_2}, \ldots, a_{i_m}\}$ and $\{a_{j_1}, a_{j_2}, \ldots, a_{j_n}\}$ are disjoint.

An **even permutation** [**odd permutation**] is a permutation that can be written as a product of an even [odd] number of transpositions.

The **sign** of a permutation (where the permutation is written as a product of transpositions) is $+1$ if it has an even number of transpositions and $-1$ if it has an odd number of transpositions.

The **identity permutation** on $S$ is the permutation $\iota\colon S \to S$ such that $x\iota = x$ for all $x \in S$.

An **involution** is a permutation $\sigma$ such that $\sigma^2 = \iota$ (the identity permutation).

The **orbit** of $a \in S$ under $\sigma$ is the set $\{\ldots, a\sigma^{-2}, a\sigma^{-1}, a, a\sigma, a\sigma^2, \ldots\}$.

**Facts:**

**1.** *Symmetric group of degree $n$:*   The set of permutations on a nonempty set $X$ is a group, where the group operation is composition of permutations: $\sigma_1\sigma_2$ is defined by $x(\sigma_1\sigma_2) = (x\sigma_1)\sigma_2$. The identity is the identity permutation $\iota$. The inverse of $\sigma$ is the permutation $\sigma^{-1}$, where $x\sigma^{-1} = y$ if and only if $y\sigma = x$. If $|X| = n$, the group of permutations is written $S_n$, the *symmetric group of degree $n$*.

**2.** Multiplication of permutations is not commutative. (See Examples 1 and 4.)

**3.** A permutation $\pi$ is an involution if and only if $\pi = \pi^{-1}$.

**4.** The number of involutions in $S_n$, denoted $\mathrm{inv}(n)$, is equal to the number of Young tableaux that can be formed from the set $\{1, 2, \ldots, n\}$. (See §2.8.)

**5.** Permutations can be used to find determinants of matrices. (See §6.3.)

**6.** Every permutation on a finite set can be written as a product of disjoint cycles.

**7.** Cycle notation is not unique: for example, $(1\ 4\ 7\ 5) = (4\ 7\ 5\ 1) = (7\ 5\ 1\ 4) = (5\ 1\ 4\ 7)$.

**8.** Every permutation is either even or odd, and no permutation is both even and odd. Hence, every permutation has a unique sign.

**9.** Each cycle of length $k$ can be written as a product of $k - 1$ transpositions:

$$(x_1\ x_2\ x_3\ \ldots\ x_k) = (x_1\ x_2)(x_1\ x_3)(x_1\ x_4)\ldots(x_1\ x_k).$$

**10.** $S_n$ has order $n!$.

**11.** $S_n$ is not abelian for $n \geq 3$. For example, $(1\ 2)(1\ 3) \neq (1\ 3)(1\ 2)$.

**12.** The order of a permutation that is a single cycle is the length of the cycle. For example, $(1\ 5\ 4)$ has order 3.

**13.** The order of a permutation that is written as a product of disjoint cycles is equal to the least common multiple of the lengths of the cycles.

**14.** *Cayley's theorem:* If $G$ is a finite group of order $n$, then $G$ is isomorphic to a subgroup of $S_n$. (See §5.2.2.)

**15.** Let $G$ be a group of permutations on a set $X$ (such a group is said to *act on* $X$). Then $G$ induces an equivalence relation $R$ on the set $X$ by the following rule: for $a, b \in X$, $aRb$ if and only if there is a permutation $\sigma \in G$ such that $a\sigma = b$.

**Examples:**

**1.** If $\sigma = \begin{pmatrix} 1 & 2 & 3 & 4 & 5 \\ 5 & 1 & 2 & 4 & 3 \end{pmatrix}$, $\tau = \begin{pmatrix} 1 & 2 & 3 & 4 & 5 \\ 4 & 5 & 1 & 3 & 2 \end{pmatrix}$, then $\sigma\tau = \begin{pmatrix} 1 & 2 & 3 & 4 & 5 \\ 2 & 4 & 5 & 3 & 1 \end{pmatrix}$ and

$\tau\sigma = \begin{pmatrix} 1 & 2 & 3 & 4 & 5 \\ 4 & 3 & 5 & 2 & 1 \end{pmatrix}$. Note that $\sigma\tau \neq \tau\sigma$.

**2.** All elements of $S_n$ can be written in *cycle notation*. For example,

$$\sigma = \begin{pmatrix} 1 & 2 & 3 & 4 & 5 & 6 & 7 \\ 4 & 6 & 3 & 7 & 1 & 2 & 5 \end{pmatrix} = (1\ 4\ 7\ 5)(2\ 6)(3).$$

Each cycle describes the orbit of the elements in that cycle. For example, $(1\ 4\ 7\ 5)$ is a cycle of length 4, and indicates that $1\sigma = 4$, $4\sigma = 7$, $7\sigma = 5$, and $5\sigma = 1$. The cycle $(3)$ indicates that $3\sigma = 3$. If a cycle has length 1, that cycle can be omitted when a permutation is written as a product of cycles: $(1\ 4\ 7\ 5)(2\ 6)(3) = (1\ 4\ 7\ 5)(2\ 6)$.

**3.** Multiplication of permutations written in cycle notation can be performed easily. For example: if $\sigma = (1\ 5\ 3\ 2)$ and $\tau = (1\ 4\ 3)(2\ 5)$, then $\sigma\tau = (1\ 5\ 3\ 2)(1\ 4\ 3)(2\ 5) = (1\ 2\ 4\ 3\ 5)$. (Moving from left to right through the product of cycles, trace the orbit of each element. For example, $3\sigma = 2$ and $2\tau = 5$; therefore $3\sigma\tau = 5$.)

**4.** Multiplication of cycles need not be commutative. For example, $(1\ 2)(1\ 3) = (1\ 2\ 3)$, $(1\ 3)(1\ 2) = (1\ 3\ 2)$, but $(1\ 2\ 3) \neq (1\ 3\ 2)$. However, disjoint cycles commute.

**5.** If the group of permutations $G = \{\iota, (1\ 2), (3\ 5)\}$ acts on the set $S = \{1, 2, 3, 4, 5\}$, then the partition of $S$ resulting from the equivalence relation induced by $G$ is $\{\{1, 2\}, \{3, 5\}, \{4\}\}$. (See Fact 15.)

**6.** Let group $G = \{\iota, (1\ 2)\}$ act on $X = \{1, 2\}$ and group $H = \{\iota, (1\ 2)(3)\}$ act on $Y = \{1, 2, 3\}$. The permutation groups $(G, X)$ and $(H, Y)$ are not isomorphic since there is no bijection between $X$ and $Y$ (even though $G$ and $H$ are isomorphic groups).

---

## 5.3.2 EXAMPLES OF PERMUTATION GROUPS

**Definitions:**

The **alternating group** on $n$ elements ($n \geq 2$) is the subgroup $A_n$ of $S_n$ consisting of all even permutations.

The **dihedral group** (**octic group**) $D_n$ is the group of rigid motions (rotations and reflections) of a regular polygon with $n$ sides under composition.

The **Klein four-group** (or **Viergruppe** or the **group of the rectangle**) is the group under composition of the four rigid motions of a rectangle that leave the rectangle in its original location. (Felix Klein, 1849–1925)

Given a permutation $\sigma \colon S \to S$, the **induced pair permutation** is the permutation $\sigma^{(2)}$ on unordered pairs of elements of $S$ given by the rule $\sigma^{(2)}(\{x, y\}) = \{\sigma(x), \sigma(y)\}$.

Given a permutation group $G$ acting on a set $S$, the **induced pair-action group** $G^{(2)}$ is the group of induced pair-permutations $\{\sigma^{(2)} \mid \sigma \in G\}$ under composition.

Given a permutation $\sigma \colon S \to S$, the **ordered pair-permutation** is the permutation $\sigma^{[2]}$ on the set $S \times S$ given by the rule $\sigma^{[2]}((x, y)) = (\sigma(x), \sigma(y))$.

Given a permutation group $G$ acting on a set $S$, the **ordered pair-action group** $G^{[2]}$ is the group of ordered pair-permutations $\{\sigma^{[2]} \mid \sigma \in G\}$ under composition.

**Facts:**

**1.** Some common subgroups of $S_n$ are listed in the following table.

| subgroup | order | description |
|---|---|---|
| symmetric group $S_n$ | $n!$ | all permutations of $\{1, 2, \ldots, n\}$ |
| alternating group $A_n$ | $n!/2$ | all even permutations of $\{1, 2, \ldots, n\}$ |
| dihedral group $D_n$ | $2n$ | rigid motions of regular $n$-gon in 3-dimensional space (Example 2) |
| Klein 4-group (subgroup of $S_4$) | 4 | rigid motions of rectangle in 3-dimensional space (Example 3) |
| identity | 1 | consists only of identity permutation |

**2.** The group $A_n$ is abelian if $n = 2$ or $3$, and is nonabelian if $n \geq 4$.

**3.** The group $D_n$ has order $2n$. The elements consist of the $n$ rotations and $n$ reflections of a regular polygon with $n$ sides. The $n$ rotations are the counterclockwise rotations about the center through angles of $\frac{360k}{n}$ degrees $(k = 0, 1, \ldots, n-1)$. (Clockwise rotations can be written in terms of counterclockwise rotations.) If $n$ is odd, the $n$ reflections are reflections in lines through a vertex and the center; if $n$ is even, the reflections are reflections in lines joining opposite vertices and in lines joining midpoints of opposite sides.

**4.** The elements of $D_n$ can be written as permutations of $\{1, 2, \ldots, n\}$. See the following figure for the rigid motions in $D_4$ (the rigid motions of the square) and the following table for the group multiplication table for $D_4$.

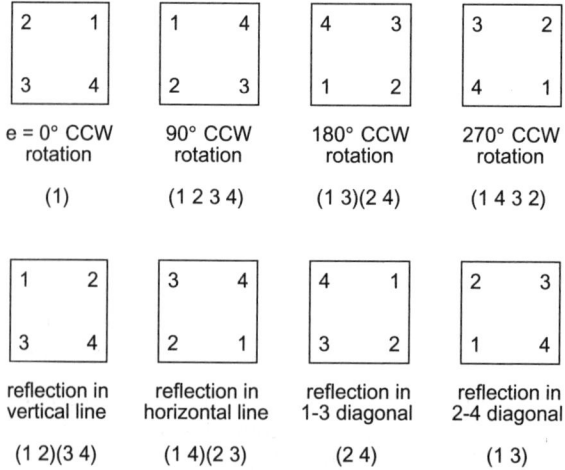

| · | (1) | (1234) | (13)(24) | (1432) | (12)(34) | (14)(23) | (24) | (13) |
|---|---|---|---|---|---|---|---|---|
| (1) | (1) | (1234) | (13)(24) | (1432) | (12)(34) | (14)(23) | (24) | (13) |
| (1234) | (1234) | (13)(24) | (1432) | (1) | (24) | (13) | (14)(23) | (12)(34) |
| (13)(24) | (13)(24) | (1432) | (1) | (1234) | (14)(23) | (12)(34) | (13) | (24) |
| (1432) | (1432) | (1) | (1234) | (13)(24) | (13) | (24) | (12)(34) | (14)(23) |
| (12)(34) | (12)(34) | (13) | (14)(23) | (24) | (1) | (13)(24) | (1432) | (1234) |
| (14)(23) | (14)(23) | (24) | (12)(34) | (13) | (13)(24) | (1) | (1234) | (1432) |
| (24) | (24) | (21)(34) | (13) | (14)(23) | (1234) | (1432) | (1) | (13)(24) |
| (13) | (13) | (14)(23) | (24) | (12)(34) | (1432) | (1234) | (13)(24) | (1) |

**5.** The Klein four-group consists of the following four rigid motions of a rectangle: the rotations about the center through $0°$ or $180°$, and reflections through the horizontal or vertical lines through its center, as illustrated in the following figure. The following table is the multiplication table for the Klein four-group.

| 2 | 1 | 4 | 3 | 1 | 2 | 3 | 4 |
| 3 | 4 | 1 | 2 | 4 | 3 | 2 | 1 |

| e = 0° CCW rotation | 180° CCW rotation | reflection in vertical line | reflection in horizontal line |
|---|---|---|---|
| (1) | (1 3)(2 4) | (1 2)(3 4) | (1 4)(2 3) |

| · | (1) | (13)(24) | (12)(34) | (14)(23) |
|---|---|---|---|---|
| (1) | (1) | (13)(24) | (12)(34) | (14)(23) |
| (13)(24) | (13)(24) | (1) | (14)(23) | (12)(34) |
| (12)(34) | (12)(34) | (14)(23) | (1) | (13)(24) |
| (14)(23) | (14)(23) | (12)(34) | (13)(24) | (1) |

**6.** The Klein four-group is isomorphic to $\mathcal{Z}_8^*$.

**7.** The induced permutation group $S_n^{(2)}$ and the ordered-pair-action group $S_n^{[2]}$ are used in enumerative graph theory. (See §8.9.2.)

**8.** The induced permutation group $S_n^{(2)}$ has $\binom{n}{2}$ objects and $n!$ permutations.

**9.** The ordered-pair-action permutation group $S_n^{[2]}$ has $n^2$ objects and $n!$ permutations.

## 5.4   RINGS

### 5.4.1   BASIC CONCEPTS

**Definitions:**

A **ring** $(R, +, \cdot)$ consists of a set $R$ closed under binary operations $+$ and $\cdot$ such that

- $(R, +)$ is an abelian group; i.e., $(R, +)$ satisfies

⋄ associative property: $a + (b + c) = (a + b) + c$ for all $a, b, c \in R$;

⋄ identity property: $R$ has an *identity element* 0 that satisfies $0+a = a+0 = a$ for all $a \in R$;

⋄ inverse property: for each $a \in R$ there is an *additive inverse element* $-a \in R$ (the *negative* of $a$) such that $-a + a = a + (-a) = 0$;

⋄ commutative law: $a + b = b + a$ for all $a, b \in R$;

• the operation $\cdot$ is associative: $a \cdot (b \cdot c) = (a \cdot b) \cdot c$ for all $a, b, c \in R$;

• The *distributive properties* for multiplication over addition hold for all $a, b, c \in R$:

⋄ left distributive property: $a \cdot (b + c) = a \cdot b + a \cdot c$;

⋄ right distributive property: $(a + b) \cdot c = a \cdot c + b \cdot c$.

A ring $R$ is **commutative** if the multiplication operation is commutative: $a \cdot b = b \cdot a$ for all $a, b \in R$.

A ring $R$ is a **ring with unity** if there is an identity 1 ($\neq 0$) for multiplication; i.e., $1 \cdot a = a \cdot 1 = a$ for all $a \in R$. The multiplicative identity is the **unity** of $R$.

An element $x$ in a ring $R$ with unity is a **unit** if $x$ has a multiplicative inverse; i.e., there is $x^{-1} \in R$ such that $x \cdot x^{-1} = x^{-1} \cdot x = 1$.

**Subtraction** in a ring is defined by the rule $a - b = a + (-b)$.

**Facts:**

**1.** Multiplication $a \cdot b$ is often written $ab$ or $a \times b$.

**2.** The order of precedence of operations in a ring follows that for real numbers: multiplication is to be done before addition. That is, $a + bc$ means $a + (bc)$ rather than $(a+b)c$.

**3.** In all rings, $a0 = 0a = 0$.

**4.** *Properties of subtraction:*

$$-(-a) = a \qquad (-a)(-b) = ab \qquad a(b-c) = ab - ac \qquad (a-b)c = ac - bc$$

$$a(-b) = (-a)b = -(ab) \qquad (-1)a = -a \text{ (if the ring has unity).}$$

**5.** The set of all units of a ring is a group under the multiplication defined on the ring.

**Examples:**

**1.** The following table gives several examples of rings, together with their associated identity and unity elements.

| set and addition and multiplication operations | 0 | 1 |
|---|---|---|
| $\{0\}$, usual addition and multiplication;  (*trivial ring*) | 0 | none |
| $\mathcal{Z}, \mathcal{Q}, \mathcal{R}, \mathcal{C}$, with usual $+$ and $\cdot$ | 0 | 1 |
| $\mathcal{Z}_n = \{0, 1, \ldots, n-1\}$ ($n$ a positive integer), $a+b = (a+b) \bmod n$, $a \cdot b = (ab) \bmod n$; (*modular ring*) | 0 | 1 |
| $\mathcal{Z}[\sqrt{2}] = \{a+b\sqrt{2} \mid a, b \in \mathcal{Z}\}$, $(a+b\sqrt{2})+(c+d\sqrt{2}) = (a+c)+(b+d)\sqrt{2}$, $(a+b\sqrt{2})\cdot(c+d\sqrt{2}) = (ac+2bd) + (ad+bc)\sqrt{2}$ [Similar rings can be constructed using $\sqrt{n}$ ($n$ an integer) if $\sqrt{n}$ not an integer.] | $0+0\sqrt{2}$ | $1+0\sqrt{2}$ |
| $\mathcal{Z}[i] = \{a + bi \mid a, b \in \mathcal{Z}\}$; (*Gaussian integers;* see §5.4.2, Example 2.) | $0+0i$ | $1+0i$ |

| set and addition and multiplication operations | 0 | 1 |
|---|---|---|
| $\mathcal{M}_{n \times n}(R)$ = all $n \times n$ matrices with entries in a ring $R$ with unity, matrix addition and multiplication; (*matrix ring*) | $O_n$ (zero matrix) | $I_n$ (identity matrix) |
| $R = \{f \mid f : A \to B\}$ (*A* any nonempty set and *B* any ring), $(f+g)(x) = f(x)+g(x)$, $(f \cdot g)(x) = f(x) \cdot g(x)$; (*ring of functions*) | $f$ such that $f(x)=0$ for all $x \in A$ | $f$ such that $f(x)=1$ for all $x \in A$ (if $B$ has unity) |
| $\mathcal{P}(S)$ = all subsets of a set $S$, $A+B = A \Delta B = (A \cup B) - (A \cap B)$ (*symmetric difference*), $A \cdot B = A \cap B$; (*Boolean ring*) | $\emptyset$ | $S$ |
| $\{a+bi+cj+dk \mid a, b, c, d \in \mathcal{R}\}$, $i, j, k$ in quaternion group, elements are added and multiplied like polynomials using $ij = k$, etc.; (*ring of real quaternions*, §5.2.1) | $0+0i+0j+0k$ | $1+0i+0l+0k$ |

2. *Polynomial rings*: For a ring $R$, the set

$$R[x] = \{a_n x^n + \cdots + a_1 x + a_0 \mid a_0, a_1, \ldots, a_n \in R\}$$

forms a ring, where the elements are added and multiplied using the "usual" rules for addition and multiplication of polynomials. The additive identity 0 is the constant polynomial $p(x) = 0$; the unity is the constant polynomial $p(x) = 1$ if $R$ has a unity 1. (See §5.5.)

3. *Product rings*: For rings $R$ and $S$, the set $R \times S = \{(r, s) \mid r \in R, s \in S\}$ forms a ring, where

$$(r_1, s_1) + (r_2, s_2) = (r_1 + r_2, s_1 + s_2);$$
$$(r_1, s_1) \cdot (r_2, s_2) = (r_1 r_2, s_1 s_2).$$

The additive identity is $(0,0)$. Unity is $(1,1)$ if $R$ and $S$ each have unity 1. Product rings can have more than two factors: $R_1 \times R_2 \times \cdots \times R_k$ or $R^n = R \times \cdots \times R$.

---

## 5.4.2 SUBRINGS AND IDEALS

**Definitions:**

A subset $S$ of a ring $(R, +, \cdot)$ is a **subring** of $R$ if $(S, +, \cdot)$ is a ring using the same operations $+$ and $\cdot$ that are used in $R$.

A subset $I$ of a ring $(R, +, \cdot)$ is an **ideal** of $R$ if

- $(I, +, \cdot)$ is a subring of $(R, +, \cdot)$;
- $I$ is closed under left and right multiplication by elements of $R$: if $x \in I$ and $r \in R$, then $rx \in I$ and $xr \in I$.

In a commutative ring $R$, an ideal $I$ is **principal** if there is $r \in R$ such that $I = Rr = \{xr \mid x \in R\}$. $I$ is the **principal ideal generated by** $r$, written $I = (r)$.

In a commutative ring $R$, an ideal $I \neq R$ is **maximal** if the only ideal properly containing $I$ is $R$.

In a commutative ring $R$, an ideal $I \neq R$ is **prime** if $ab \in I$ implies that $a \in I$ or $b \in I$.

**Facts:**

**1.** If $S$ is a nonempty subset of a ring $(R, +, \cdot)$, then $S$ is a subring of $R$ if and only if $S$ is closed under subtraction and multiplication.

**2.** An ideal in a ring $(R, +, \cdot)$ is a subgroup of the group $(R, +)$, but not necessarily conversely.

**3.** The intersection of ideals in a ring is an ideal.

**4.** If $R$ is any ring, then $R$ and $\{0\}$ are ideals, called *trivial* ideals.

**5.** In a commutative ring with unity, every maximal ideal is a prime ideal.

**6.** Every ideal $I$ in the ring $\mathcal{Z}$ is a principal ideal. $I = (r)$ where $r$ is the smallest positive integer in $I$.

**7.** If $R$ is a commutative ring with unity, then $R$ is a field (see §5.6) if and only if the only ideals of $R$ are $R$ and $\{0\}$.

**8.** An ideal in a ring is the analogue of a normal subgroup in a group.

**9.** The second condition in the definition of ideal can be stated as $rI \subseteq I$ ($I$ is a *left ideal*) and $Ir \subseteq I$ ($I$ is a *right ideal*). (If $A$ is a subset of a ring $R$ and $r \in R$, then $rA = \{ra \mid a \in A\}$ and $Ar = \{ar \mid a \in A\}$.)

**Examples:**

**1.** With the usual definitions of $+$ and $\cdot$, each of the following rings can be viewed as a subring of all the rings listed after it: $\mathcal{Z}$, $\mathcal{Q}$, $\mathcal{R}$, $\mathcal{C}$.

**2.** *Gaussian integers:* $\mathcal{Z}[i] = \{a + bi \mid a, b \in \mathcal{Z}\}$ using the addition and multiplication of $\mathcal{C}$ is a subring of the ring of complex numbers.

**3.** The ring $\mathcal{Z}$ is a subring of $\mathcal{Z}[\sqrt{2}]$ and $\mathcal{Z}[\sqrt{2}]$ is a subring of $\mathcal{R}$.

**4.** Each set $n\mathcal{Z}$ ($n$ an integer) is a principal ideal in the ring $\mathcal{Z}$.

---

### 5.4.3   RING HOMOMORPHISM AND ISOMORPHISM

**Definitions:**

If $R$ and $S$ are rings, a function $\varphi \colon R \to S$ is a **ring homomorphism** if for all $a, b \in R$

- $\varphi(a + b) = \varphi(a) + \varphi(b)$     ($\varphi$ preserves addition)
- $\varphi(ab) = \varphi(a)\varphi(b)$.     ($\varphi$ preserves multiplication)

*Note:* $\varphi(a)$ is sometimes written $a\varphi$.

If a ring homomorphism $\varphi$ is also one-to-one and onto $S$, then $\varphi$ is a **ring isomorphism** and $R$ and $S$ are **isomorphic**, written $R \cong S$.

A **ring endomorphism** is a ring homomorphism $\varphi \colon R \to R$.

A **ring automorphism** is a ring isomorphism $\varphi \colon R \to R$.

The **kernel** of a ring homomorphism $\varphi \colon R \to S$ is $\varphi^{-1}(0) = \{x \in R \mid \varphi(x) = 0\}$.

**Facts:**

**1.** If $\varphi$ is a ring isomorphism, then $\varphi^{-1}$ is a ring isomorphism.

**2.** The kernel of a ring homomorphism from $R$ to $S$ is an ideal of the ring $R$.

**3.** If $\varphi\colon R \to S$ is a ring homomorphism, then $\varphi(R)$ is a subring of $S$.

**4.** If $\varphi\colon R \to S$ is a ring homomorphism and $R$ has unity, then either $\varphi(1) = 0$ or $\varphi(1)$ is unity for $\varphi(R)$.

**5.** If $\varphi$ is a ring homomorphism, then $\varphi(0) = 0$ and $\varphi(-a) = -\varphi(a)$.

**6.** A ring homomorphism is a ring isomorphism between $R$ and $\varphi(R)$ if and only if the kernel of $\varphi$ is $\{0\}$.

**7.** *Homomorphisms preserve subrings:*   Let $\varphi\colon R \to S$ be a ring homomorphism. If $A$ is a subring of $R$, then $\varphi(A)$ is a subring of $S$. If $B$ is a subring of $S$, then $\varphi^{-1}(B)$ is a subring of $R$.

**8.** *Homomorphisms preserve ideals:*   Let $\varphi\colon R \to S$ be a ring homomorphism. If $A$ is an ideal of $R$, then $\varphi(A)$ is an ideal of $S$. If $B$ is an ideal of $S$, then $\varphi^{-1}(B)$ is an ideal of $R$.

**Examples:**

**1.** The function $\varphi\colon \mathcal{Z} \to \mathcal{Z}_n$ defined by the rule $\varphi(a) = a \bmod n$ is a ring homomorphism.

**2.** If $R$ and $S$ are rings, then the function $\varphi\colon R \to S$ defined by the rule $\varphi(a) = 0$ for all $a \in R$ is a ring homomorphism.

**3.** The function $\varphi\colon \mathcal{Z} \to R$ ($R$ any ring with unity) defined by the rule $\varphi(x) = x \cdot 1$ is a ring homomorphism. The kernel of $\varphi$ is the subring $n\mathcal{Z}$ for some nonnegative integer $n$, called the *characteristic* of $R$.

**4.** Let $\mathcal{P}(S)$ be the ring of all subsets of a set $S$ (see §5.4.1). If $|S| = 1$, then $\mathcal{P}(S) \cong \mathcal{Z}_2$ with the ring isomorphism $\varphi$ where $\varphi(\emptyset) = 0$ and $\varphi(S) = 1$. More generally, if $|S| = n$, then $\mathcal{P}(S) \cong \mathcal{Z}_2^n = \mathcal{Z}_2 \times \cdots \times \mathcal{Z}_2$.

**5.** $\mathcal{Z}_n \cong \mathcal{Z}/(n)$ for all positive integers $n$. (See §5.4.4.)

**6.** $\mathcal{Z}_m \times \mathcal{Z}_n \cong \mathcal{Z}_{mn}$, if $m$ and $n$ are relatively prime.

## 5.4.4   QUOTIENT RINGS

**Definitions:**

If $I$ is an ideal in a ring $R$ and $a \in R$, then the set $a + I = \{a + x \mid x \in I\}$ is a **coset** of $I$ in $R$.

The set of all cosets, $R/I = \{a + I \mid a \in R\}$, is a ring, called the **quotient ring**, where addition and multiplication are defined by the following rules:

- $(a + I) + (b + I) = (a + b) + I$;
- $(a + I) \cdot (b + I) = (ab) + I$.

**Facts:**

**1.** If $R$ is commutative, then $R/I$ is commutative.

**2.** If $R$ has unity 1, then $R/I$ has the coset $1 + I$ as unity.

**3.** If $I$ is an ideal in ring $R$, the function $\varphi\colon R \to R/I$ defined by the rule $\varphi(x) = x + I$ is a ring homomorphism, called the *natural map*. The kernel of $\varphi$ is $I$.

**4.** *Fundamental homomorphism theorem!for rings*: If $\varphi$ is a ring homomorphism and $K$ is the kernel of $\varphi$, then $\varphi(R) \cong R/K$.

**5.** If $R$ is a commutative ring with unity and $I$ is an ideal in $R$, then $I$ is a maximal ideal if and only if $R/I$ is a field (see §5.6).

**Examples:**

**1.** For each integer $n$, $\mathcal{Z}/n\mathcal{Z}$ is a quotient ring, isomorphic to $\mathcal{Z}_n$.

**2.** See §5.6.1 for Galois rings.

---

## 5.4.5   RINGS WITH ADDITIONAL PROPERTIES

Beginning with rings, as additional requirements are added, the following hierarchy of sets of algebraic structures is obtained:

$$\text{rings} \;\supset\; \begin{array}{c}\text{commutative}\\ \text{rings with}\\ \text{unity}\end{array} \;\supset\; \begin{array}{c}\text{integral}\\ \text{domains}\end{array} \;\supset\; \begin{array}{c}\text{Euclidean}\\ \text{domains}\end{array} \;\supset\; \begin{array}{c}\text{principal}\\ \text{ideal}\\ \text{domains}\end{array}$$

**Definitions:**

The **cancellation properties** in a ring $R$ state that for all $a, b, c \in R$

> if $ab = ac$ and $a \neq 0$, then $b = c$   (*left cancellation property*)
> if $ba = ca$ and $a \neq 0$, then $b = c$   (*right cancellation property*).

Let $R$ be a ring and let $a, b \in R$ where $a \neq 0, b \neq 0$. If $ab = 0$, then $a$ is a **left divisor of zero** and $b$ is a **right divisor of zero**.

An **integral domain** is a commutative ring with unity that has no zero divisors.

A **principal ideal domain** (**PID**) is an integral domain in which every ideal is a principal ideal.

A **division ring** is a ring with unity in which every nonzero element is a unit (i.e., every nonzero element has a multiplicative inverse).

A **field** is a commutative ring with unity such that each nonzero element has a multiplicative inverse. (See §5.6.)

A **Euclidean norm** on an integral domain $R$ is a function $\delta\colon R - \{0\} \to \{0, 1, 2, \ldots\}$ such that

- $\delta(a) \leq \delta(ab)$ for all $a, b \in R - \{0\}$;
- the following generalization of the division algorithm for integers holds: for all $a, d \in R$ where $d \neq 0$, there are elements $q, r \in R$ such that $a = dq + r$, where either $r = 0$ or $\delta(r) < \delta(d)$.

A **Euclidean domain** is an integral domain with a Euclidean norm defined on it.

**Facts:**

**1.** The cancellation properties hold in an integral domain.

**2.** Every finite integral domain is a field.

**3.** Every integral domain can be imbedded in a field. Given an integral domain $R$, there is a field $F$ and a ring homomorphism $\varphi \colon R \to F$ such that $\varphi(1) = 1$.

**4.** A ring with unity is a division ring if and only if the nonzero elements form a group under the multiplication defined on the ring.

**5.** *Wedderburn's theorem:* Every finite division ring is a field. (J. H. M. Wedderburn, 1882–1948)

**6.** Every commutative division ring is a field.

**7.** In a Euclidean domain, if $b \neq 0$ is not a unit, then $\delta(ab) > \delta(a)$ for all $a \neq 0$. For $b \neq 0$, $b$ is a unit in $R$ if and only if $\delta(b) = \delta(1)$.

**8.** In every Euclidean domain, a Euclidean algorithm for finding the gcd can be carried out.

**Examples:**

**1.** Some common Euclidean domains are given in the following table.

| set | Euclidean norm |
|---|---|
| $\mathcal{Z}$ | $\delta(a) = \lvert a \rvert$ |
| $\mathcal{Z}[i]$   (Gaussian integers) | $\delta(a + bi) = a^2 + b^2$ |
| $F$ (any field) | $\delta(a) = 1$ |
| polynomial ring $F[x]$   ($F$ any field) | $\delta(p(x)) = $ degree of $p(x)$ |

**2.** The following table gives examples of rings with additional properties.

| ring | commuta- tive ring with unity | integral domain | principal ideal domain | Euclidean domain | division ring | field |
|---|---|---|---|---|---|---|
| $\mathcal{Z}$ | yes | yes | yes | yes | no | no |
| $\mathcal{Q}, \mathcal{R}, \mathcal{C}$ | yes | yes | yes | yes | yes | yes |
| $\mathcal{Z}_p$ ($p$ prime) | yes | yes | yes | yes | yes | yes |
| $\mathcal{Z}_n$ ($n$ composite) | yes | no | no | no | no | no |
| real quaternions | no | no | no | no | yes | no |
| $\mathcal{Z}[x]$ | yes | no | no | no | no | no |
| $\mathcal{M}_{n \times n}$ | no | no | no | no | no | no |

# 5.5   POLYNOMIAL RINGS

## 5.5.1   BASIC CONCEPTS

**Definitions:**

A *polynomial in the variable $x$ over a ring $R$* is an expression of the form

$$f(x) = a_n x^n + a_{n-1} x^{n-1} + \cdots + a_1 x^1 + a_0 x^0$$

where $a_n, \ldots, a_0 \in R$.

For a polynomial $f(x) \neq 0$, the largest integer $k$ such that $a_k \neq 0$ is the **degree** of $f(x)$, written $\deg f(x)$.

A **constant polynomial** is a polynomial $f(x) = a_0$. If $a_0 \neq 0$, $f(x)$ has degree 0. If $f(x) = 0$ (the **zero polynomial**), the degree of $f(x)$ is undefined. (The degree of the zero polynomial is also said to be $-\infty$.)

The **polynomial ring** (in one variable $x$) over a ring $R$ consists of the set

$$R[x] = \{f(x) \mid f(x) \text{ is a polynomial over } R \text{ in the variable } x\}$$

with addition and multiplication defined by the following rules:

- $(a_n x^n + \cdots + a_1 x^1 + a_0 x^0) + (b_m x^m + \cdots + b_1 x^1 + b_0 x^0)$
  $= a_n x^n + \cdots + a_{m+1} x^{m+1} + (a_n + b_n) x^n + \cdots + (a_1 + b_1) x^1 + (a_0 + b_0) x^0$
  if $n \geq m$, and
- $(a_n x^n + \cdots + a_1 x^1 + a_0 x^0)(b_m x^m + \cdots + b_1 x^1 + b_0 x^0)$
  $= c_{n+m} x^{n+m} + \cdots + c_1 x^1 + c_0 x^0$
  where $c_i = a_0 b_i + a_1 b_{i-1} + \cdots + a_i b_0$ for $i = 0, 1, \ldots, m+n$.

A polynomial $f(x) \in R[x]$ of degree $n$ is **monic** if $a_n = 1$.

The **value** of a polynomial $f(x) = a_n x^n + a_{n-1} x^{n-1} + \cdots + a_1 x^1 + a_0 x^0$ at $c \in R$ is the element $f(c) = a_n c^n + a_{n-1} c^{n-1} + \cdots + a_1 c + a_0 \in R$.

An element $c \in R$ is a **zero** of the polynomial $f(x)$ if $f(c) = 0$.

If $R$ is a subring of a commutative ring $S$, an element $a \in S$ is **algebraic** over $R$ if there is a nonzero $f(x) \in R[x]$ such that $f(a) = 0$.

If $p(x)$ is not algebraic over $R$, then $p(x)$ is **transcendental** over $R$.

A polynomial $f(x) \in R[x]$ of degree $n$ is **irreducible** over $R$ if $f(x)$ cannot be written as $f_1(x) f_2(x)$ (**factors** of $f(x)$) where $f_1(x)$ and $f_2(x)$ are polynomials over $R$ of degrees less than $n$. Otherwise $f(x)$ is **reducible** over $R$.

The **polynomial ring** (in the variables $x_1, x_2, \ldots, x_n$ with $n > 1$) over a ring $R$ is defined by the rule $R[x_1, x_2, \ldots, x_n] = (R[x_1, x_2, \ldots, x_{n-1}])[x_n]$.

**Facts:**

**1.** Polynomials over an arbitrary ring $R$ generalize polynomials with coefficients in $\mathcal{R}$ or $\mathcal{C}$. The notation and terminology follow the usual conventions for polynomials with real (or complex) coefficients:

- the elements $a_n, \ldots, a_0$ are *coefficients*;
- subtraction notation can be used: $a_i x^i + (-a_j) x^j = a_i x^i - a_j x^j$;
- the term $1 x^i$ can be written as $x^i$;
- the term $x^1$ can be written $x$;
- the term $x^0$ can be written 1;
- terms $0 x^i$ can be omitted.

**2.** There is a distinction between a polynomial $f(x) \in R[x]$ and the function it defines using the rule $f(c) = a_n c^n + a_{n-1} c^{n-1} + \cdots + a_1 c + a_0$ for $c \in R$. The same function might be defined by infinitely many polynomials. For example, the polynomials $f_1(x) = x \in \mathcal{Z}_2[x]$ and $f_2(x) = x^2 \in \mathcal{Z}_2[x]$ define the same function: $f_1(0) = f_2(0) = 0$ and $f_1(1) = f_2(1) = 1$.

**3.** If $R$ is a ring, $R[x]$ is a ring.

**4.** If $R$ is a commutative ring, then $R[x]$ is a commutative ring.

**5.** If $R$ is a ring with unity, then $R[x]$ has the constant polynomial $f(x) = 1$ as unity.

**6.** If $R$ is an integral domain, then $R[x]$ is an integral domain. If $f_1(x)$ has degree $m$ and $f_2(x)$ has degree $n$, then the degree of $f_1(x)f_2(x)$ is $m + n$.

**7.** If ring $R$ is not an integral domain, then $R[x]$ is not an integral domain. If $f_1(x)$ has degree $m$ and $f_2(x)$ has degree $n$, then the degree of $f_1(x)f_2(x)$ can be smaller than $m + n$. (For example, in $\mathcal{Z}_6[x]$, $(3x^2)(2x^3) = 0$.)

**8.** *Factor theorem:* If $R$ is a commutative ring with unity and $f(x) \in R[x]$ has degree $\geq 1$, then $f(a) = 0$ if and only if $x - a$ is a factor of $f(x)$.

**9.** If $R$ is an integral domain and $p(x) \in R[x]$ has degree $n$, then $p(x)$ has at most $n$ zeros in $R$. If $R$ is not an integral domain, then a polynomial may have more zeros than its degree; for example, $x^2 + x \in \mathcal{Z}_6[x]$ has four zeros — $0, 2, 3, 5$.

## 5.5.2  POLYNOMIALS OVER A FIELD

### Facts:

**1.** Even though $F$ is a field (§5.6.1), $F[x]$ is never a field. (The polynomial $f(x) = x$ has no multiplicative inverse in $F[x]$.)

**2.** If $f(x)$ has degree $n$, then $f(x)$ has at most $n$ distinct zeros.

**3.** *Irreducibility over a finite field:* If $F$ is a finite field and $n$ is a positive integer, then there is an irreducible polynomial over $F$ of degree $n$.

**4.** *Unique factorization theorem:* If $f(x)$ is a polynomial over a field $F$ and is not the zero polynomial, then $f(x)$ can be uniquely factored (ignoring the order in which the factors are written) as $af_1(x) \cdots f_k(x)$ where $a \in F$ and each $f_i(x)$ is a monic polynomial that is irreducible over $F$.

**5.** *Eisenstein's irreducibility criterion:* If $f(x) \in \mathcal{Z}[x]$ has degree $n > 0$, if there is a prime $p$ such that $p$ divides every coefficient of $f(x)$ except $a_n$, and if $p^2$ does not divide $a_0$, then $f(x)$ is irreducible over $\mathcal{Q}$. (Gotthold Eisenstein, 1823–1852)

**6.** *Division algorithm for polynomials:* If $F$ is a field with $a(x)$, $d(x) \in F[x]$ and $d(x)$ is not the zero polynomial, then there are unique polynomials $q(x)$ (quotient) and $r(x)$ (remainder) in $F[x]$ such that $a(x) = d(x)q(x) + r(x)$ where $\deg r(x) < \deg d(x)$ or $r(x) = 0$. If $d(x)$ is monic, then the division algorithm for polynomials can be extended to all rings with unity.

**7.** *Irreducibility over the real numbers $\mathcal{R}$:* If $f(x) \in \mathcal{R}[x]$ has degree at least 3, then $f(x)$ is reducible. The only irreducible polynomials in $\mathcal{R}[x]$ are of degree 1 or 2; for example $x^2 + 1$ is irreducible over $\mathcal{R}$.

**8.** *Fundamental theorem of algebra (irreducibility over the complex numbers $\mathcal{C}$):* If $f(x) \in \mathcal{C}[x]$ has degree $n \geq 1$, then $f(x)$ can be completely factored:

$$f(x) = c(x - c_1)(x - c_2) \ldots (x - c_n)$$

where $c, c_1, \ldots, c_n \in \mathcal{C}$.

**9.** If $F$ is a field and $f(x) \in F[x]$ has degree 1 (i.e., $f(x)$ is linear), then $f(x)$ is irreducible.

**10.** If $F$ is a field and $f(x) \in F[x]$ has degree $\geq 2$ and has a zero, then $f(x)$ is reducible. (If $f(x)$ has $a$ as a zero, then $f(x)$ can be written as $(x - a)f_1(x)$ where $\deg f_1(x) = \deg f(x)-1$.) The converse is false: a polynomial may have no zeros, but still be reducible. (See Example 2.)

**11.** If $F$ is a field and $f(x) \in F[x]$ has degree 2 or 3, then $f(x)$ is irreducible if and only if $f(x)$ has no zeros.

**Examples:**

**1.** In $\mathcal{Z}_5[x]$, if $a(x) = 3x^4 + 2x^3 + 2x + 1$ and $d(x) = x^2 + 2$, then $q(x) = 3x^2 + 2x + 4$ and $r(x) = 3x + 3$. To obtain $q(x)$ and $r(x)$, use the same format as for long division of natural numbers, with arithmetic operations carried out in $\mathcal{Z}_5$:

$$
\begin{array}{r}
3x^2 + 2x + 4 \\
x^2 + 2 \overline{)\,3x^4 + 2x^3 + 0x^2 + 2x + 1} \\
\end{array}
$$

$$
\begin{array}{ll}
3x^4 \qquad + x^2 & \\
\quad 2x^3 + 4x^2 & [-x^2 = 4x^2 \text{ over } \mathcal{Z}_5] \\
\quad 2x^3 \qquad + 4x & \\
\qquad 4x^2 + 3x & [2x - 4x = -2x = 3x \text{ over } \mathcal{Z}_5] \\
\qquad 4x^2 \qquad + 3 & \\
\qquad\qquad 3x + 3 &
\end{array}
$$

**2.** Polynomials can have no zeros, but be reducible. The polynomial $f(x) = x^4 + x^2 + 1 \in Z_2[x]$ has no zeros (since $f(0) = f(1) = 1$), but $f(x)$ can be factored as $(x^2 + x + 1)^2$. Similarly, $x^4 + 2x^2 + 1 = (x^2 + 1)^2 \in \mathcal{R}[x]$.

## 5.6   FIELDS

### 5.6.1   BASIC CONCEPTS

**Definitions:**

A **field** $(F, +, \cdot)$ consists of a set $F$ together with two binary operations, $+$ and $\cdot$, such that

- $(F, +, \cdot)$ is a ring;
- $(F - \{0\}, \cdot)$ is a commutative group.

A **subfield** $F$ of field $(K, +, \cdot)$ is a subset of $K$ that is a field using the same operations as those in $K$.

If $F$ is a subfield of $K$, then $K$ is called an **extension field** of $F$. Write $K/F$ to indicate that $K$ is an extension field of $F$.

For $K$ an extension field of $F$, the **degree of $K$ over $F$** is $[K : F] = $ the dimension of $K$ as a vector space over $F$. (See §6.1.3.)

A **field isomorphism** is a function $\varphi \colon F_1 \to F_2$, where $F_1$ and $F_2$ are fields, such that $\varphi$ is one-to-one, onto $F_2$, and satisfies the following for all $a, b \in F_1$:

- $\varphi(a + b) = \varphi(a) + \varphi(b)$;

- $\varphi(ab) = \varphi(a)\varphi(b)$.

A **field automorphism** is an isomorphism $\varphi\colon F \to F$, where $F$ is a field. The set of all automorphisms of $F$ is denoted $\mathrm{Aut}(F)$.

The **characteristic** of a field $F$ is the smallest positive integer $n$ such that $1+\cdots+1 = 0$, where there are $n$ summands. If there is no such integer, $F$ has characteristic 0 (also called characteristic $\infty$).

### Facts:

**1.** Every field is a commutative ring with unity. A field satisfies all properties of a commutative ring with unity, and has the additional property that every nonzero element has a multiplicative inverse.

**2.** Every finite integral domain is a field.

**3.** A field is a commutative division ring.

**4.** If $F$ is a field and $a$, $b \in F$ where $a \neq 0$, then $ax + b = 0$ has a unique solution in $F$.

**5.** If $F$ is a field, every ideal in $F[x]$ is a principal ideal.

**6.** If $p$ is a prime and $n$ is any positive integer, then there is exactly one field (up to isomorphism) with $p^n$ elements, the *Galois field* $GF(p^n)$. (§5.6.1)

**7.** If $\varphi\colon F \to F$ is a field automorphism, then

- $-\varphi(a) = \varphi(-a)$
- $\varphi(a^{-1}) = \varphi(a)^{-1}$

for all $a \neq 0$.

**8.** The intersection of all subfields of a field $F$ is a field, called the *prime field* of $F$.

**9.** If $F$ is a field, $\mathrm{Aut}(F)$ is a group under composition of functions.

**10.** The characteristic of a field is either 0 or prime.

**11.** Every field of characteristic 0 is isomorphic to a field that is an extension of $\mathcal{Q}$ and has $\mathcal{Q}$ as its prime field.

**12.** Every field of characteristic $p > 0$ is isomorphic to a field that is an extension of $\mathcal{Z}_p$ and has $\mathcal{Z}_p$ as its prime field.

**13.** If field $F$ has characteristic $p > 0$, then $(a+b)^p = a^p + b^p$ for all $a, b \in F$.

**14.** If field $F$ has characteristic $p > 0$, $f(x) \in \mathcal{Z}_p[x]$, and $\alpha \in F$ is a zero of $f(x)$, then $\alpha^p, \alpha^{p^2}, \alpha^{p^3}, \ldots$ are also zeros of $f(x)$.

**15.** If $p$ is not a prime, then $\mathcal{Z}_p$ is not a field since $\mathcal{Z}_p - \{0\}$ will fail to be closed under multiplication. For example, $\mathcal{Z}_6$ is not a field since $2 \in \mathcal{Z}_6 - \{0\}$ and $3 \in \mathcal{Z}_6 - \{0\}$, but $2 \cdot 3 = 0 \notin \mathcal{Z}_6 - \{0\}$.

### Examples:

**1.** The following table gives several examples of fields.

| set and operations | $-a$ | $a^{-1}$ | characteristic | order |
|---|---|---|---|---|
| $Q$, $R$, $C$, with usual addition and multiplication | $-a$ | $1/a$ | $0$ | infinite |
| $Z_p = \{0, 1, \ldots, p-1\}$ ($p$ prime), addition and multiplication **mod** $p$ | $p - a$ $(-0 = 0)$ | $a^{-1} = b$, where $ab$ **mod** $p = 1$ | $p$ | $p$ |
| $F[x]/(f(x))$, $f(x)$ irreducible over field $F$, coset addition and multiplication (Example 2) | $-[a+(f(x))]=$ $-a+(f(x))$ | $[a+(f(x))]^{-1}=$ $a^{-1}+(f(x))$ | varies | varies |
| $GF(p^n)=Z_p[x]/(f(x))$, $f(x)$ of degree $n$ irreducible over $Z_p$ ($p$ prime), addition and multiplication of cosets (*Galois field*) | $-[a+(f(x))]=$ $-a+(f(x))$ | $[a+(f(x))]^{-1}=$ $a^{-1}+(f(x))$ | $p$ | $p^n$ |

**2.** The field $F[x]/(f(x))$: If $F$ is any field and $f(x) \in F[x]$ of degree $n$ is irreducible over $F$, the quotient ring structure $F[x]/(f(x))$ is a field. The elements of $F[x]/(f(x))$ are cosets of polynomials in $F[x]$ modulo $f(x)$, where $(f(x))$ is the principal ideal generated by $f(x)$. Polynomials $f_1(x)$ and $f_2(x)$ lie in the same coset if and only if $f(x)$ is a factor of $f_1(x) - f_2(x)$.

Using the division algorithm for polynomials, any polynomial $g(x) \in F[x]$ can be written as $g(x) = f(x)q(x) + r(x)$ where $q(x)$ and $r(x)$ are unique polynomials in $F[x]$ and $r(x)$ has degree $< n$. The equivalence class $g(x) + (f(x))$ can be identified with the polynomial $r(x)$, and thus $F[x]/(f(x))$ can be regarded as the field of all polynomials in $F[x]$ of degree $< n$.

## 5.6.2  EXTENSION FIELDS AND GALOIS THEORY

Throughout this subsection assume that field $K$ is an extension of field $F$.

**Definitions:**

For $\alpha \in K$, $F(\alpha)$ is the smallest field containing $\alpha$ and $F$, called the **field extension** of $F$ by $\alpha$.

For $\alpha_1, \ldots, \alpha_n \in K$, $F(\alpha_1, \ldots, \alpha_n)$ is the smallest field containing $\alpha_1, \ldots, \alpha_n$ and $F$, called the **field extension** of $F$ by $\alpha_1, \ldots, \alpha_n$.

If $K$ is an extension field of $F$ and $\alpha \in K$, then $\alpha$ is **algebraic** over $F$ if $\alpha$ is a root of a nonzero polynomial in $F[x]$. If $\alpha$ is not the root of any nonzero polynomial in $F[x]$, then $\alpha$ is **transcendental** over $F$.

A complex number is an **algebraic number** if it is algebraic over $Q$.

An **algebraic integer** is an algebraic number $\alpha$ that is a zero of a polynomial of the form $x^n + a_{n-1}x^{n-1} + \cdots + a_1 x + a_0$ where each $a_i \in Z$.

An extension field $K$ of $F$ is an **algebraic extension** of $F$ if every element of $K$ is algebraic over $F$. Otherwise $K$ is a **transcendental extension** of $F$.

An extension field $K$ of $F$ is a **finite extension** of $F$ if $K$ is finite-dimensional as a vector space over $F$ (see Fact 11). The dimension of $K$ over $F$ is written $[K : F]$.

Let $\alpha$ be algebraic over a field $F$. The **minimal polynomial of $\alpha$ with respect to $F$** is the monic irreducible polynomial $f(x) \in F[x]$ of smallest degree such that $f(\alpha) = 0$.

A polynomial $f(x) \in F[x]$ **splits** over $K$ if $f(x) = \alpha(x - \alpha_1)\ldots(x - \alpha_n)$ where $\alpha, \alpha_1, \ldots, \alpha_n \in K$.

$K$ is a **splitting field** (**root field**) of a nonconstant $f(x) \in F[x]$ if $f(x)$ splits over $K$ and $K$ is the smallest field with this property.

A polynomial $f(x) \in F[x]$ of degree $n$ is **separable** if $f(x)$ has $n$ distinct roots in its splitting field.

$K$ is a **separable extension** of $F$ if every element of $K$ is the root of a separable polynomial in $F[x]$.

$K$ is a **normal extension** of $F$ if $K/F$ is algebraic and every irreducible polynomial in $F[x]$ with a root in $K$ has all its roots in $K$ (i.e., splits in $K$).

$K$ is a **Galois extension** of $F$ if $K$ is a normal, separable extension of $F$.

A field automorphism $\varphi$ **fixes set $S$ elementwise** if $\varphi(x) = x$ for all $x \in S$.

The **fixed field** of a subset $A \subseteq \mathrm{Aut}(F)$ is $F_A = \{x \in F \mid \varphi(x) = x \text{ for all } \varphi \in A\}$.

The **Galois group** of $K$ over $F$ is the group of automorphisms $G(K/F)$ of $K$ that fix $F$ elementwise. If $K$ is a splitting field of $f(x) \in F[x]$, $G(K/F)$ is also known as the **Galois group** of $f(x)$. (Évariste Galois, 1811–1832)

**Facts:**

**1.**  The elements of $K$ that are algebraic over $F$ form a subfield of $K$.

**2.**  The algebraic numbers in $\mathcal{C}$ form a field; the algebraic integers form a subring of $\mathcal{C}$, called the *ring of algebraic integers*.

**3.**  Every nonconstant polynomial has a unique splitting field, up to isomorphism.

**4.**  If $f(x) \in F[x]$ splits as $\alpha(x - \alpha_1)\ldots(x - \alpha_n)$, then the splitting field for $f(x)$ is $F(\alpha_1, \ldots, \alpha_n)$.

**5.**  If $F$ is a field and $p(x) \in F[x]$ is a nonconstant polynomial, then there is an extension field $K$ of $F$ and $\alpha \in K$ such that $p(\alpha) = 0$.

**6.**  If $f(x)$ is irreducible over $F$, then the ring $F[x]/(f(x))$ is an algebraic extension of $F$ and contains a root of $f(x)$.

**7.**  The field $F$ is isomorphic to a subfield of any algebraic extension $F[x]/(f(x))$. The element $0 \in F$ corresponds to the coset of the zero polynomial; all other elements of $F$ appear in $F[x]/(f(x))$ as cosets of the constant polynomials.

**8.**  Every minimal polynomial is irreducible.

**9.**  If $K$ is a field extension of $F$ and $\alpha \in K$ is a root of an irreducible polynomial $f(x) \in F[x]$ of degree $n \geq 1$, then $F(\alpha) = \{c_{n-1}\alpha^{n-1} + \cdots + c_1\alpha + c_0 \mid c_i \in F \text{ for all } i\}$.

**10.**  If $K$ is an extension field of $F$ and $\alpha \in K$ is algebraic over $F$, then

- there is a unique monic irreducible polynomial $f(x) \in F[x]$ of smallest degree (the *minimum polynomial*) such that $f(\alpha) = 0$;
- $F(\alpha) \cong F[x]/(f(x))$;

- if the degree of $\alpha$ over $F$ is $n$, then $K = \{a_0 + a_1\alpha + a_2\alpha^2 + \cdots + a_{n-1}\alpha^{n-1} \mid a_0, a_1, \ldots, a_{n-1} \in F\}$; in fact, $K$ is an $n$-dimensional vector space over $F$, with basis $1, \alpha, \alpha^2, \ldots, \alpha^{n-1}$.

**11.** If $K$ is an extension field of $F$ and $x \in K$ is transcendental over $F$, then $F(\alpha) \cong$ the field of all fractions $f(x)/g(x)$ where $f(x)$, $g(x) \in F[x]$ and $g(x)$ is not the zero polynomial.

**12.** $K$ is a splitting field of some polynomial $f(x) \in F[x]$ if and only if $K$ is a Galois extension of $F$.

**13.** If $K$ is a splitting field for separable $f(x) \in F[x]$ of degree $n$, then $G(K, F)$ is isomorphic to a subgroup of the symmetric group $S_n$.

**14.** If $K$ is a splitting field of $f(x) \in F[x]$, then

- every element of $G(K/F)$ permutes the roots of $f(x)$ and is completely determined by its effect on the roots of $f(x)$;
- $G(K/F)$ is isomorphic to a group of permutations of the roots of $f(x)$.

**15.** If $K$ is a splitting field for separable $f(x) \in F[x]$, then $|G(K/F)| = [K : F]$.

**16.** For $[K : F]$ finite, $K$ is a normal extension of $F$ if and only if $K$ is a splitting field of some polynomial in $F[x]$.

**17.** *The fundamental theorem of Galois theory*: If $K$ is a normal extension of $F$, where $F$ is either finite or has characteristic 0, then there is a one-to-one correspondence $\Phi$ between the lattice of all fields $K'$, where $F \subseteq K' \subseteq K$, and the lattice of all subgroups $H$ of the Galois group $G(K/F)$:

$$\Phi(K') = G(K/K') \quad \text{and} \quad \Phi^{-1}(H) = K_H.$$

The correspondence $\Phi$ has the following properties:

- for fields $K'$ and $K''$ where $F \subseteq K' \subseteq K$ and $F \subseteq K'' \subseteq K$

$$K' \subseteq K'' \longleftrightarrow \Phi(K'') \subseteq \Phi(K').$$

  That is, $G(K/K'') \subseteq G(K/K')$.

- $\Phi$ interchanges the operations meet and join for the lattice of subfields and the lattice of subgroups:

$$\Phi(K' \wedge K'') = G(K/K') \vee G(K/K''),$$
$$\Phi(K' \vee K'') = G(K/K') \wedge G(K/K'').$$

  (Note: In the lattice of fields [groups], $A \wedge B = A \cap B$ and $A \vee B$ is the smallest field [group] containing $A$ and $B$.)

- $K'$ is a normal extension of $F$ if and only if $G(K/K')$ is a normal subgroup of $G(K/F)$.

**18.** *Formulas for solving polynomial equations of degrees 2, 3, or 4*:

- *second-degree (quadratic) equation* $ax^2 + bx + c = 0$: the quadratic formula gives the solutions $\dfrac{-b \pm \sqrt{b^2 - 4ac}}{2a}$;
- *third-degree (cubic) equation* $a_3x^3 + a_2x^2 + a_1x + a_0 = 0$:

  (1)  divide by $a_3$ to obtain $x^3 + b_2x^2 + b_1x + b_0 = 0$,

(2) make the substitution $x = y - \frac{b_2}{3}$ to obtain an equation of the form $y^3 + cy + d = 0$, with solutions $y = \sqrt[3]{\frac{-d}{2} + \sqrt{\frac{d^2}{4} + \frac{c^3}{27}}} + \sqrt[3]{\frac{-d}{2} - \sqrt{\frac{d^2}{4} + \frac{c^3}{27}}}$,

(3) use the substitution $x = y - \frac{b_2}{3}$ to obtain the solutions to the original equation;

- *fourth-degree (quartic) equation* $a_4 x^4 + a_3 x^3 + a_2 x^2 + a_1 x + a_0 = 0$:

  (1) divide by $a_4$ to obtain $x^4 + ax^3 + bx^2 + cx + d = 0$,

  (2) solve the *resolvent* equation $y^3 - by^s + (ac - 4d)y + (-a^2 d + 4bd - c^2) = 0$ to obtain a root $z$,

  (3) solve the pair of quadratic equations:

$$x^2 + \tfrac{a}{2}x + \tfrac{z}{2} = \pm\sqrt{\left(\tfrac{a^2}{4} - b + z\right)x^2 + \left(\tfrac{a}{2}z - c\right)x + \left(\tfrac{z^2}{4} - d\right)}$$

to obtain the solutions to the original equation.

**19.** A general method for solving cubic equations algebraically was given by Nicolo Fontana (1500–1557), also called Tartaglia. The method is often referred to as Cardano's method because Girolamo Cardano (1501–1576) published the method. Ludovico Ferrari (1522–1565), a student of Cardano, discovered a general method for solving quartic equations algebraically.

**20.** *Equations of degree 5 or more:* In 1824 Abel proved that the general quintic polynomial equation $a_5 x^5 + \cdots + a_1 x + a_0 = 0$ (and those of higher degree) are not solvable by radicals; that is, there can be no formula for writing the roots of such equations using only the basic arithmetic operations and the taking of $n$th roots. Évariste Galois (1811–1832) demonstrated the existence of such equations that are not solvable by radicals and related solvability by radicals of polynomial equations to determining whether the associated permutation group (the Galois group) of roots is solvable. (See Application 1.)

**Examples:**

**1.** $\mathcal{C}$ *as an algebraic extension of* $\mathcal{R}$: Let $f(x) = x^2 + 1 \in \mathcal{R}[x]$ and $\alpha = x + (x^2 + 1) \in \mathcal{R}[x]/(x^2 + 1)$. Then $\alpha^2 = -1$. Thus, $\alpha$ behaves like $i$ (since $i^2 = -1$). Hence $\mathcal{R}[x]/(x^2 + 1) = \{c_1\alpha + c_0 \mid c_1, c_0 \in \mathcal{R}\} \cong \{c_1 i + c_0 \mid c_0, c_1 \in \mathcal{R}\} = \mathcal{C}$.

**2.** *Algebraic extensions of* $\mathcal{Z}_p$: If $f(x) \in \mathcal{Z}_p$ is an irreducible polynomial of degree $n$, then the algebraic extension $\mathcal{Z}_p[x]/(f(x))$ is a *Galois field*.

**3.** If $f(x) = x^4 - 2x^2 - 3 \in \mathcal{Q}[x]$, its splitting field is

$$\mathcal{Q}(\sqrt{3}, i) = \{a + b\sqrt{3} + ci + di\sqrt{3} \mid a, b, c, d \in \mathcal{Q}\}.$$

There are three intermediate fields: $\mathcal{Q}(\sqrt{3})$, $\mathcal{Q}(i)$, and $\mathcal{Q}(i\sqrt{3})$, as illustrated below. The Galois group $G(\mathcal{Q}(\sqrt{3}, i)/\mathcal{Q}) = \{e, \phi_1, \phi_2, \phi_3\}$ where

$$\phi_1(a + b\sqrt{3} + ci + di\sqrt{3}) = a + b\sqrt{3} - ci - di\sqrt{3},$$
$$\phi_2(a + b\sqrt{3} + ci + di\sqrt{3}) = a - b\sqrt{3} + ci - di\sqrt{3},$$
$$\phi_3(a + b\sqrt{3} + ci + di\sqrt{3}) = a - b\sqrt{3} - ci + di\sqrt{3} = \phi_2\phi_1 = \phi_1\phi_2,$$
$$e(a + b\sqrt{3} + ci + di\sqrt{3}) = a + b\sqrt{3} + ci + di\sqrt{3}.$$

$G(\mathcal{Q}(\sqrt{3}, i), \mathcal{Q})$ has the following subgroups:

$$G = G(\mathcal{Q}(\sqrt{3}, i), \mathcal{Q}) = \{e, \phi_1, \phi_2, \phi_3\},$$
$$H_1 = G(\mathcal{Q}(\sqrt{3}, i), \mathcal{Q}(\sqrt{3})) = \{e, \phi_1\},$$
$$H_2 = G(\mathcal{Q}(\sqrt{3}, i), \mathcal{Q}(i)) = \{e, \phi_2\},$$
$$H_3 = G(\mathcal{Q}(\sqrt{3}, i), \mathcal{Q}(i\sqrt{3})) = \{e, \phi_3\},$$
$$\{e\} = G(\mathcal{Q}(\sqrt{3}, i), \mathcal{Q}(\sqrt{3}, i)).$$

The correspondence between fields and Galois groups is shown in the following table and figure.

| field | Galois group |
|---|---|
| $\mathcal{Q}(\sqrt{3}, i)$ | $\{e\}$ |
| $\mathcal{Q}(\sqrt{3})$ | $H_1$ |
| $\mathcal{Q}(i\sqrt{3})$ | $H_3$ |
| $\mathcal{Q}(i)$ | $H_2$ |
| $\mathcal{Q}$ | $G$ |

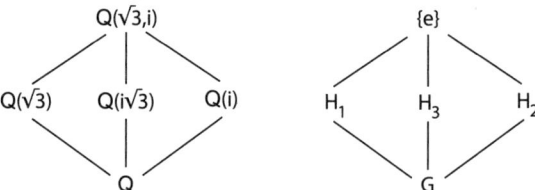

**4.** *Cyclotomic extensions:*   The $n$th roots of unity are the solutions to $x^n - 1 = 0$: $1, \omega, \omega^2, \ldots, \omega^{n-1}$, where $\omega = e^{2\pi i/n}$. The extension field $\mathcal{Q}(\omega)$ is a *cyclotomic extension* of $\mathcal{Q}$. If $p > 2$ is prime, then $G(\mathcal{Q}(\omega), \mathcal{Q})$ is a cyclic group of order $p-1$ and is isomorphic to $\mathcal{Z}_p^*$ (the multiplicative group of nonzero elements of $\mathcal{Z}_p$).

## Applications:

**1.** *Solvability by radicals:*   A polynomial equation $f(x) = 0$ is *solvable by radicals* if each root can be expressed in terms of the coefficients of the polynomial, using only the operations of addition, subtraction, multiplication, division, and the taking of $n$th roots.

If $F$ is a field of characteristic 0 and $f(x) \in F[x]$ has $K$ as splitting field, then $f(x) = 0$ is solvable by radicals if and only if $G(K, F)$ is a solvable group. Since there are polynomials whose Galois groups are not solvable, there are polynomials whose roots cannot be found by elementary algebraic methods. For example, the polynomial $x^5 - 36x + 2$ has the symmetric group $S_5$ as its Galois group, which is not solvable. Hence, the roots of $x^5 - 36x + 2 = 0$ cannot be found by elementary algebraic methods. This example shows that there can be no algebraic formula for solving all fifth-degree equations.

**2.** *Straightedge and compass constructibility:*   Using only a straightedge (unmarked ruler) and a compass, there is no general method for

- trisecting angles (given an angle whose measure is $\alpha$, to construct an angle with measure $\frac{\alpha}{3}$);
- duplicating the cube (given the side of a cube $C_1$, to construct the side of a cube $C_2$ that has double the volume of $C_1$);

- squaring the circle (given a circle of area $A$, to construct a square with area $A$);
- constructing a regular $n$-gon for all $n \geq 3$.

Straightedge and compass constructions yield only lengths that can be obtained by addition, subtraction, multiplication, division, and taking square roots. Beginning with lengths that are rational numbers, each of these operations yields field extensions $\mathcal{Q}(a)$ and $\mathcal{Q}(b)$ where $a$ and $b$ are coordinates of a point constructed from points in $\mathcal{Q} \times \mathcal{Q}$. These operations force $[\mathcal{Q}(a) \colon \mathcal{Q}]$ and $[\mathcal{Q}(b) \colon \mathcal{Q}]$ to be powers of 2. However, trisecting angles, duplicating cubes, and squaring circles all yield extensions of $\mathcal{Q}$ such that the degrees of the extensions are not powers of 2. Hence these three types of constructions are not possible with straightedge and compass.

## 5.6.3  FINITE FIELDS

Finite fields have a wide range of applications in various areas of computer science and engineering: coding theory, combinatorics, computer algebra, cryptology, generation of pseudorandom numbers, switching circuit theory, and symbolic computation.

Throughout this subsection assume that $F$ is a finite field.

**Definitions:**

A **finite field** is a field with a finite number of elements.

The **Galois field** $GF(p^n)$ is the algebraic extension $\mathcal{Z}_p[x]/(f(x))$ of the finite field $\mathcal{Z}_p$ where $p$ is a prime and $f(x)$ is an irreducible polynomial over $\mathcal{Z}_p$ of degree $n$. (See Fact 1.)

A **primitive element** of $GF(p^n)$ is a generator of the cyclic group of nonzero elements of $GF(p^n)$ under multiplication.

Let $\alpha$ be a primitive element of $GF(p^n)$. The **discrete exponential function** (with base $\alpha$) is the function $\exp_\alpha \colon \{0, 1, 2, \ldots, p^n - 2\} \to GF(p^n)^*$ defined by the rule $\exp_\alpha k = \alpha^k$.

Let $\alpha$ be a primitive element of $GF(p^n)$. The **discrete logarithm** or **index function** (with base $\alpha$) is the function $\mathrm{ind}_\alpha \colon GF(p^n)^* \to \{0, 1, 2, \ldots, p^n - 2\}$ where $\mathrm{ind}_\alpha(x) = k$ if and only if $x = \alpha^k$.

Let $\alpha$ be a primitive element of $GF(p^n)$. The **Zech logarithm** (**Jacobi logarithm**) is the function $Z \colon \{1, \ldots, p^n - 1\} \to \{0, \ldots, p^n - 2\}$ such that $\alpha^{Z(k)} = 1 + \alpha^k$; if $1 + \alpha^k = 0$, then $Z(k) = 0$.

**Facts:**

**1.** *Existence of finite fields:*  For each prime $p$ and positive integer $n$ there is exactly one field (up to isomorphism) with $p^n$ elements—the field $GF(p^n)$, also written $F_{p^n}$.

**2.** *Construction of finite fields:*  Given an irreducible polynomial $f(x) \in \mathcal{Z}_p[x]$ of degree $n$ and a zero $\alpha$ of $f(x)$,

$$GF(p^n) \cong \mathcal{Z}_p[x]/(f(x)) \cong \{c_{n-1}\alpha^{n-1} + \cdots + c_1\alpha + c_0 \mid c_i \in \mathcal{Z}_p \text{ for all } i\}.$$

**3.** If $F$ is a finite field, then

- $F$ has $p^n$ elements for some prime $p$ and positive integer $n$;
- $F$ has characteristic $p$ for some prime $p$;

- $F$ is an extension of $\mathcal{Z}_p$.

**4.** $[GF(p^n)\colon \mathcal{Z}_p] = n.$

**5.** $GF(p^n) = $ the field of the $p^n$ roots of $x^{p^n} - x \in \mathcal{Z}_p[x].$

**6.** The minimal polynomial of $\alpha \in GF(p^n)$ with respect to $\mathcal{Z}_p$ is

$$f(x) = (x - \alpha)(x - \alpha^p)(x - \alpha^{p^2})\ldots(x - \alpha^{p^i})$$

where $i$ is the smallest positive integer such that $\alpha^{p^{i+1}} = \alpha.$

**7.** If a field $F$ has order $p^n$, then every subfield of $F$ has order $p^k$ for some $k$ that divides $n$.

**8.** The multiplicative group of nonzero elements of a finite field $F$ is a cyclic group.

**9.** If a field $F$ has $m$ elements, then the multiplicative order of each nonzero element of $F$ is a divisor of $m - 1$.

**10.** If a field $F$ has $m$ elements and $d$ is a divisor of $m - 1$, then there is an element of $F$ of order $d$.

**11.** Each discrete logarithm function has the following properties:

$\operatorname{ind}_\alpha(xy) \equiv \operatorname{ind}_\alpha x + \operatorname{ind}_\alpha y \pmod{p^n - 1};$
$\operatorname{ind}_\alpha(xy^{-1}) \equiv \operatorname{ind}_\alpha x - \operatorname{ind}_\alpha y \pmod{p^n - 1};$
$\operatorname{ind}_\alpha(x^k) \equiv k\operatorname{ind}_\alpha x \pmod{p^n - 1}.$

**12.** The discrete logarithm function $\operatorname{ind}_\alpha$ is the inverse of the discrete exponential function $\exp_\alpha$. That is, $\operatorname{ind}_\alpha x = y$ if and only if $\exp_\alpha y = x.$

**13.** A discrete logarithm function can be used to facilitate multiplication and division of elements of $GF(p^n).$

**14.** The Zech logarithm facilitates the addition of elements $\alpha^i$ and $\alpha^j$ $(i > j)$ in $GF(p^n)$, since $\alpha^i + \alpha^j = \alpha^j(\alpha^{i-j} + 1) = \alpha^j \cdot \alpha^{Z(i-j)} = \alpha^{j+Z(i-j)}.$ (Note that the values of the Zech logarithm function depend on the primitive element used.)

**15.** There are $\frac{1}{k}\sum_{d|k}\mu(\frac{k}{d})p^{nd}$ irreducible polynomials of degree $k$ over $GF(p^n)$, where $\mu$ is the Möbius function (§2.7).

**Examples:**

**1.** If $p$ is prime, $\mathcal{Z}_p$ is a finite field and $\mathcal{Z}_p \cong GF(p).$

**2.** The field $\mathcal{Z}_2 = F_2$:

| + | 0 | 1 |
|---|---|---|
| 0 | 0 | 1 |
| 1 | 1 | 0 |

| · | 0 | 1 |
|---|---|---|
| 0 | 0 | 0 |
| 1 | 0 | 1 |

**3.** The field $\mathcal{Z}_3 = F_3$:

| + | 0 | 1 | 2 |
|---|---|---|---|
| 0 | 0 | 1 | 2 |
| 1 | 1 | 2 | 0 |
| 2 | 2 | 0 | 1 |

| · | 0 | 1 | 2 |
|---|---|---|---|
| 0 | 0 | 0 | 0 |
| 1 | 0 | 1 | 2 |
| 2 | 0 | 2 | 1 |

**4.** Construction of $GF(2^2) = F_4$:

$$GF(2^2) = \mathcal{Z}_2[x]/(x^2 + x + 1) = \{c_1\alpha + c_0 \mid c_1, c_0 \in \mathcal{Z}_2\} = \{0, 1, \alpha, \alpha + 1\}$$

where $\alpha$ is a zero of $x^2 + x + 1$; i.e., $\alpha^2 + \alpha + 1 = 0$. The nonzero elements of $GF(p^n)$ can also be written as powers of $\alpha$ as $\alpha$, $\alpha^2 = -\alpha - 1 = \alpha + 1$, $\alpha^3 = \alpha \cdot \alpha^2 = \alpha(\alpha + 1) = \alpha^2 + \alpha = (\alpha + 1) + \alpha = 2\alpha + 1 = 1$.

Thus, $GF(2^2) = \{0, 1, \alpha, \alpha^2\}$ has the following addition and multiplication tables:

| + | 0 | 1 | $\alpha$ | $\alpha^2$ |
|---|---|---|---|---|
| 0 | 0 | 1 | $\alpha$ | $\alpha^2$ |
| 1 | 1 | 0 | $\alpha^2$ | $\alpha$ |
| $\alpha$ | $\alpha$ | $\alpha^2$ | 0 | 1 |
| $\alpha^2$ | $\alpha^2$ | $\alpha$ | 1 | 0 |

| $\cdot$ | 0 | 1 | $\alpha$ | $\alpha^2$ |
|---|---|---|---|---|
| 0 | 0 | 0 | 0 | 0 |
| 1 | 0 | 1 | $\alpha$ | $\alpha^2$ |
| $\alpha$ | 0 | $\alpha$ | $\alpha^2$ | 1 |
| $\alpha^2$ | 0 | $\alpha^2$ | 1 | $\alpha$ |

5. *Construction of $GF(2^3) = F_8$:*   Let $f(x) = x^3 + x + 1 \in \mathcal{Z}_2[x]$ and let $\alpha$ be a root of $f(x)$. Then $GF(2^3) = \{c_2\alpha^2 + c_1\alpha + c_0 \mid c_0, c_1, c_2 \in \mathcal{Z}_2\}$ where $\alpha^3 + \alpha + 1 = 0$.

The elements of $GF(2^3)$ (using $\alpha$ as generator) are

$$0, \qquad \alpha, \qquad \alpha^2, \qquad \alpha^3 = \alpha + 1$$
$$\alpha^4 = \alpha^2 + \alpha, \quad \alpha^5 = \alpha^2 + \alpha + 1, \quad \alpha^6 = \alpha^2 + 1, \quad 1 \,(= \alpha^7).$$

Multiplication is carried out using the ordinary rules of exponents and the fact that $\alpha^7 = 1$. The following Zech logarithm values can be used to construct the table for addition: $Z(1) = 3$, $Z(2) = 6$, $Z(3) = 1$, $Z(4) = 5$, $Z(5) = 4$, $Z(6) = 2$, $Z(7) = 0$. For example $\alpha^3 + \alpha^5 = \alpha^3 \cdot \alpha^{Z(5-3)} = \alpha^3 \cdot \alpha^6 = \alpha^9 = \alpha^2$.

Using strings of 0s and 1s to represent the elements, $0 = 000$, $1 = 001$, $\alpha = 010$, $\alpha + 1 = 011$, $\alpha^2 = 100$, $\alpha^2 + \alpha = 110$, $\alpha^2 + 1 = 101$, $\alpha^2 + \alpha + 1 = 111$, yields the following tables for addition and multiplication:

| + | 000 | 001 | 010 | 011 | 100 | 101 | 110 | 111 |
|---|---|---|---|---|---|---|---|---|
| 000 | 000 | 001 | 010 | 011 | 100 | 101 | 110 | 111 |
| 001 | 001 | 000 | 011 | 010 | 101 | 100 | 111 | 110 |
| 010 | 010 | 011 | 000 | 001 | 110 | 111 | 100 | 101 |
| 011 | 011 | 010 | 001 | 000 | 111 | 110 | 101 | 100 |
| 100 | 100 | 101 | 110 | 111 | 000 | 001 | 010 | 011 |
| 101 | 101 | 100 | 111 | 110 | 001 | 000 | 011 | 010 |
| 110 | 110 | 111 | 100 | 101 | 010 | 011 | 000 | 001 |
| 111 | 111 | 110 | 101 | 100 | 011 | 010 | 001 | 000 |

| $\cdot$ | 000 | 001 | 010 | 011 | 100 | 101 | 110 | 111 |
|---|---|---|---|---|---|---|---|---|
| 000 | 000 | 000 | 000 | 000 | 000 | 000 | 000 | 000 |
| 001 | 000 | 001 | 010 | 011 | 100 | 101 | 110 | 111 |
| 010 | 000 | 010 | 100 | 110 | 011 | 001 | 111 | 101 |
| 011 | 000 | 011 | 110 | 101 | 111 | 100 | 001 | 010 |
| 100 | 000 | 100 | 011 | 111 | 110 | 010 | 101 | 001 |
| 101 | 000 | 101 | 001 | 100 | 010 | 111 | 011 | 110 |
| 110 | 000 | 110 | 111 | 001 | 101 | 011 | 010 | 100 |
| 111 | 000 | 111 | 101 | 010 | 001 | 110 | 100 | 011 |

The same field can be constructed using $g(x) = x^3 + x^2 + 1$ instead of $f(x) = x^3 + x + 1$ and $\beta$ as a root of $g(x)$ $(\beta^3 + \beta^2 + 1 = 0)$. The elements (using $\beta$ as generator) are $0$, $\beta$, $\beta^2$, $\beta^3 = \beta^2 + 1$, $\beta^4 = \beta^2 + \beta + 1$, $\beta^5 = \beta + 1$, $\beta^6 = \beta^2 + \beta$, $1 \,(= \beta^7)$.

The polynomial $g(x)$ yields the following Zech logarithm values, which can be used to construct the table for addition: $Z(1) = 5$, $Z(2) = 3$, $Z(3) = 2$, $Z(4) = 6$, $Z(5) = 1$, $Z(6) = 4$, $Z(7) = 0$. This field is isomorphic to the field defined using $f(x) = x^3 + x + 1$.

**6.** The following table lists the irreducible polynomials of degree at most 8 in $\mathcal{Z}_2[x]$. Each polynomial is represented by the string of its coefficients, beginning with the highest power. For example, $x^3 + x + 1$ is represented by 1011. For more extensive tables of irreducible polynomials over certain finite fields, see [LiNi94].

| | | | | | | |
|---|---|---|---|---|---|---|
| degree 1: | 10 | 11 | | | | |
| degree 2: | 111 | | | | | |
| degree 3: | 1011 | 1101 | | | | |
| degree 4: | 10011 | 11001 | 11111 | | | |
| degree 5: | 100101 | 101001 | 101111 | 110111 | 111011 | 111101 |
| degree 6: | 1000011 | 1001001 | 1010111 | 1011011 | 1100001 | 1100111 |
| | 1101101 | 1110011 | 1110101 | | | |
| degree 7: | 10000011 | 10001001 | 10001111 | 10010001 | 10011101 | 10100111 |
| | 10101011 | 10111001 | 10111111 | 11000001 | 11001011 | 11010011 |
| | 11010101 | 11100101 | 11101111 | 11110001 | 11110111 | 11111101 |
| degree 8: | 100011011 | 100011101 | 100101011 | 100101101 | 100111001 | 100111111 |
| | 101001101 | 101011111 | 101100011 | 101100101 | 101101001 | 101110001 |
| | 101110111 | 101111011 | 110000111 | 110001011 | 110001101 | 110011111 |
| | 110100011 | 110101001 | 110110001 | 110111101 | 111000011 | 111001111 |
| | 111010111 | 111011101 | 111100111 | 111110011 | 111110101 | 111111001 |

# 5.7   LATTICES

## 5.7.1   BASIC CONCEPTS

**Definitions:**

A *lattice* $(L, \vee, \wedge)$ is a nonempty set $L$ closed under two binary operations $\vee$ (*join*) and $\wedge$ (*meet*) such that the following laws are satisfied for all $a, b, c \in L$:

- *associative laws:* $a \vee (b \vee c) = (a \vee b) \vee c$ and $a \wedge (b \wedge c) = (a \wedge b) \wedge c$;
- *commutative laws:* $a \vee b = b \vee a$ and $a \wedge b = b \wedge a$;
- *absorption laws:* $a \vee (a \wedge b) = a$ and $a \wedge (a \vee b) = a$.

Lattices $L_1$ and $L_2$ are **isomorphic** (as lattices) if there is a function $\varphi: L_1 \to L_2$ that is one-to-one, onto $L_2$ and preserves $\vee$ and $\wedge$: $\varphi(a \vee b) = \varphi(a) \vee \varphi(b)$ and $\varphi(a \wedge b) = \varphi(a) \wedge \varphi(b)$ for all $a, b \in L_1$.

$L_1$ is a **sublattice** of lattice $L$ if $L_1 \subseteq L$ and $L_1$ is a lattice using the same operations as those used in $L$.

The **dual** of a statement in a lattice is the statement obtained by interchanging the operations $\vee$ and $\wedge$, and interchanging the elements 0 (lower bound) and 1 (upper bound). (See §5.7.2.)

An **order relation** $\leq$ can be defined on a lattice so that $a \leq b$ means that $a \vee b = b$, or, equivalently, that $a \wedge b = a$. Write $a < b$ if $a \leq b$ and $a \neq b$.

**Facts:**

**1.** If $L$ is a lattice and $a, b \in L$, then $a \wedge b$ and $a \vee b$ are unique.

**2.** *Lattices as partially ordered sets:* Every lattice is a partially ordered set using the order relation $\leq$. (See §1.4.3; also see Chapter 11 for extended coverage.)

**3.** Every partially ordered set $L$ in which glb $\{a, b\}$ and lub $\{a, b\}$ exist for all $a, b \in L$ can be regarded as a lattice by defining $a \vee b = $ lub $\{a, b\}$ and $a \wedge b = $ glb $\{a, b\}$.

**4.** *The duality principle holds in all lattices:* If a theorem is the consequence of the definition of lattice, then the dual of the statement is also a theorem.

**5.** *Lattice diagrams:* Every finite lattice can be pictured in a poset diagram (Hasse diagram), called a *lattice diagram*.

**6.** *Idempotent laws:* $a \vee a = a$ and $a \wedge a = a$ for all $a \in L$.

**Example:**

**1.** The following table gives examples of lattices.

| set | $\vee$ (join) | $\wedge$ (meet) |
|---|---|---|
| $\mathcal{N}$ | $a \vee b = \text{lcm}\{a, b\}$ | $a \wedge b = \gcd\{a, b\}$ |
| $\mathcal{N}$ | $a \vee b = \max\{a, b\}$ | $a \wedge b = \min\{a, b\}$ |
| $\mathcal{Z}_2^n$ | $(a_1, \ldots, a_n) \vee (b_1, \ldots, b_n) = (\max(a_1, b_1), \ldots, \max(a_n, b_n))$ | $(a_1, \ldots, a_n) \vee (b_1, \ldots, b_n) = (\min(a_1, b_1), \ldots, \min(a_n, b_n))$ |
| all subgroups of a group $G$ | $H_1 \vee H_2 = $ the intersection of all subgroups of $G$ containing $H_1$ and $H_2$ | $H_1 \wedge H_2 = H_1 \cap H_2$ |
| all subsets of set $S$ | $A_1 \vee A_2 = A_1 \cup A_2$ | $A_1 \wedge A_2 = A_1 \cap A_2$ |

## 5.7.2  SPECIALIZED LATTICES

**Definitions:**

A lattice $L$ is **distributive** if the following are true for all $a, b, c \in L$:

- $a \wedge (b \vee c) = (a \wedge b) \vee (a \wedge c)$;
- $a \vee (b \wedge c) = (a \vee b) \wedge (a \vee c)$.

A **lower bound** (**smallest element**, **least element**) in a lattice $L$ is an element $0 \in L$ such that $0 \wedge a = 0$ (equivalently, $0 \leq a$) for all $a \in L$.

An **upper bound** (**largest element**, **greatest element**) in a lattice $L$ is an element $1 \in L$ such that $1 \vee a = 1$ (equivalently, $a \leq 1$) for all $a \in L$.

A lattice $L$ is **bounded** if $L$ contains a lower bound 0 and an upper bound 1.

A lattice $L$ is **complemented** if

- $L$ is bounded;
- for each $a \in L$ there is an element $b \in L$ (called a **complement** of $a$) such that $a \vee b = 1$ and $a \wedge b = 0$.

An element $a$ in a bounded lattice $L$ is an **atom** if $0 < a$ and there is no element $b \in L$ such that $0 < b < a$.

## Facts:

**1.** Each of the distributive properties in a lattice implies the other.

**2.** Not every lattice is distributive. (See Example 1.)

**3.** If a lattice is not distributive, it must contain a sublattice isomorphic to one of the two lattices in the following figure.

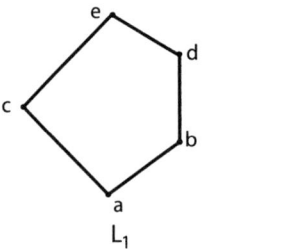 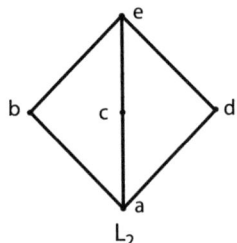

**4.** Every finite lattice is bounded: if $L = \{a_1, \ldots, a_n\}$, then

$$1 = a_1 \vee \cdots \vee a_n \qquad \text{and} \qquad 0 = a_1 \wedge \cdots \wedge a_n.$$

**5.** Some infinite lattices are bounded, while others are not. (See Examples 2 and 3.)

**6.** In a complemented lattice, complements are not necessarily unique. See the lattice in Example 4.

**7.** If $L$ is a finite, complemented, distributive lattice and $a \in L$, then there is exactly one set of atoms $\{a_1, \ldots, a_k\}$ such that $a = a_1 \vee \cdots \vee a_k$.

## Examples:

**1.** Neither lattice in Fact 3 is distributive. For example, in lattice $L_1$, $d \vee (b \wedge c) = d$, but $(d \vee b) \wedge (d \vee c) = b$. And in $L_2$, $d \vee (b \wedge c) = d$, but $(d \vee b) \wedge (d \vee c) = a$.

**2.** The lattice $(\mathcal{N}, \vee, \wedge)$ where $a \vee b = \max(a, b)$ and $a \wedge b = \min(a, b)$ is not bounded; there is a lower bound (the integer 0), but there is no upper bound.

**3.** The following infinite lattice is bounded. The element 1 is an upper bound and the element 0 is a lower bound.

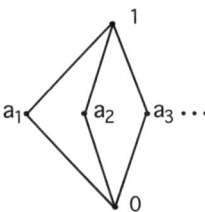

**4.** The lattice in Example 3 is complemented, but complements are not unique in that lattice. For example, the element $a_1$ has $a_2, a_3, \ldots$ as complements.

**5.** In lattice $L_1$ of Fact 3, $b$ and $c$ are atoms. In the lattice of all subsets of a set $S$ (see Example 1), the atoms are the subsets of $S$ of size 1.

# 5.8  BOOLEAN ALGEBRAS

Boolean algebra is a generalization of the algebra of sets and the algebra of logical propositions. It forms an abstract model of the design of circuits.

## 5.8.1  BASIC CONCEPTS

### Definitions:

A **Boolean algebra** $(B, +, \cdot, ', 0, 1)$ consists of a set $B$ closed under two binary operations, $+$ (*addition*) and $\cdot$ (*multiplication*), and one monadic operation, $'$ (*complementation*), and having two distinct elements, 0 and 1, such that the following laws are true for all $a, b, c \in B$:

- *commutative laws:* $a + b = b + a$ and $a \cdot b = b \cdot a$;
- *distributive laws:* $a \cdot (b + c) = (a \cdot b) + (a \cdot c)$ and $a + (b \cdot c) = (a + b) \cdot (a + c)$;
- *identity laws:* $a + 0 = a$ and $a \cdot 1 = a$;
- *complement laws:* $a + a' = 1$ and $a \cdot a' = 0$.

(George Boole, 1813–1864)

*Notation:*   It is common practice to omit the "$\cdot$" symbol in a Boolean algebra, writing $ab$ instead of $a \cdot b$. The complement operation is also written using an overline: $x' = \overline{x}$. By convention, complementation is done first, then multiplication, and finally addition. For example, $a + bc'$ means $a + (b(c'))$.

The **dual** of a statement in a Boolean algebra is the statement obtained by interchanging the operations $+$ and $\cdot$, and interchanging the elements 0 and 1 in the original statement.

Boolean algebras $B_1$ and $B_2$ are **isomorphic** (as Boolean algebras) if there is a function $\varphi \colon B_1 \to B_2$ that is one-to-one and onto $B_2$ such that for all $a, b \in B_1$

- $\varphi(a + b) = \varphi(a) + \varphi(b)$;
- $\varphi(ab) = \varphi(a)\varphi(b)$;
- $\varphi(a') = \varphi(a)'$.

An element $a \neq 0$ in a Boolean algebra is an **atom** if the following holds: if $xa = x$, then either $x = 0$ or $x = a$; that is, if $x \leq a$, then either $x = 0$ or $x = a$ (see Fact 1).

The binary operation NAND, written $|$, is defined by $a \mid b = (ab)'$.

The binary operation NOR, written $\downarrow$, is defined by $a \downarrow b = (a + b)'$.

The binary operation XOR, written $\oplus$, is defined by $a \oplus b = ab' + a'b$.

**Facts:**

**1.** Every Boolean algebra is a bounded, distributive, complemented lattice where $a \vee b = a + b$ and $a \wedge b = ab$. Hence, every Boolean algebra is a partially ordered set (where $a \leq b$ if and only if $a + b = b$, or, equivalently, $ab = a$ or $a' + b = 1$ or $ab' = 0$).

**2.** The *duality principle* holds in all Boolean algebras: if a theorem is the consequence of the definition of Boolean algebra, then the dual of the theorem is also a theorem.

**3.** *Structure of Boolean algebras:* Every finite Boolean algebra is isomorphic to $\{0, 1\}^n$ for some positive integer $n$. Hence every finite Boolean algebra has $2^n$ elements. The atoms are the $n$ $n$-tuples of 0s and 1s with a 1 in exactly one position.

**4.** If $B$ is a finite Boolean algebra and $b \in B$ ($b \neq 0$), there is exactly one set of atoms $a_1, \ldots, a_k$ such that $b = a_1 + \cdots + a_k$.

**5.** If a Boolean algebra $B$ has $n$ atoms, then $B$ has $2^n$ elements.

**6.** The following laws hold in the Boolean algebra $B$, for all $a, b, c \in B$:

- *associative laws:* $a + (b + c) = (a + b) + c$ and $a(bc) = (ab)c$;
  (Hence there is no ambiguity in writing $a + b + c$ and $abc$.)
- *idempotent laws:* $a + a = a$, $aa = a$;
- *absorption laws:* $a(a + b) = a$, $a + ab = a$;
- *domination (boundedness) laws:* $a + 1 = 1$ and $a0 = 0$;
- *double complement (involution) law:* $(a')' = a$;
- *De Morgan's laws:* $(a + b)' = a'b'$ and $(ab)' = a' + b'$;
- *uniqueness of complement:* if $a + b = 1$ and $ab = 0$, then $b = a'$.

**7.** Since every Boolean algebra is a lattice, every finite Boolean algebra can be pictured using a partially ordered set diagram. (§11.1)

**Examples:**

**1.** $\{0, 1\}$ is a Boolean algebra, where addition, multiplication, and complementation are defined in the following tables:

| + | 0 | 1 |
|---|---|---|
| 0 | 0 | 1 |
| 1 | 1 | 1 |

| · | 0 | 1 |
|---|---|---|
| 0 | 0 | 0 |
| 1 | 0 | 1 |

| $x$ | $x'$ |
|---|---|
| 0 | 1 |
| 1 | 0 |

**2.** If $S$ is any set, then $\mathcal{P}(S)$ (the set of all subsets of $S$) is a Boolean algebra where

$$A_1 + A_2 = A_1 \cup A_2, \quad A_1 \cdot A_2 = A_1 \cap A_2, \quad A' = \overline{A}$$

and $0 = \emptyset$ and $1 = S$.

**3.** Given $n$ variables, the set of all compound propositions in these variables (identified with their truth tables) is a Boolean algebra where

$$p + q = p \vee q \quad p \cdot q = p \wedge q \quad \overline{p} = \neg p$$

and 0 is a contradiction (the truth table with only values $F$) and 1 is a tautology (the truth table with only values $T$).

**4.** If $B$ is any Boolean algebra, then $B^n = \{(a_1, \ldots, a_n) \mid a_i \in B \text{ for all } i\}$ is a Boolean algebra, where the operations are performed coordinatewise:

$$(a_1, \ldots, a_n) + (b_1, \ldots, b_n) = (a_1 + b_1, \ldots, a_n + b_n);$$
$$(a_1, \ldots, a_n) \cdot (b_1, \ldots, b_n) = (a_1 \cdot b_1, \ldots, a_n \cdot b_n);$$
$$(a_1, \ldots, a_n)' = (a_1', \ldots, a_n').$$

In this Boolean algebra $0 = (0, \ldots, 0)$ and $1 = (1, \ldots, 1)$.

**5.** The statements in each of the following pairs are duals of each other:

$$a + b = cd, \quad ab = c + d;$$
$$a + (b + c) = (a + b) + c, \quad a(bc) = (ab)c;$$
$$a + 1 = 1, \ a0 = 0.$$

## 5.8.2   BOOLEAN FUNCTIONS

**Definitions:**

A **Boolean expression** in the variables $x_1, \ldots, x_n$ is an expression defined recursively by

- 0, 1, and all variables $x_i$ are Boolean expressions in $x_1, \ldots, x_n$;
- if $E$ and $F$ are Boolean expressions in the variables $x_1, \ldots, x_n$, then $(EF)$, $(E+F)$, and $E'$ are Boolean expressions in the variables $x_1, \ldots, x_n$.

A **Boolean function** of degree $n$ is a function $f \colon \{0, 1\}^n \to \{0, 1\}$.

A **literal** is a Boolean variable or its complement.

A **minterm** of the Boolean variables $x_1, \ldots, x_n$ is a product of the form $y_1 \ldots y_n$ where for each $i$, $y_i$ is equal to $x_i$ or $x_i'$.

A **maxterm** of the Boolean variables $x_1, \ldots, x_n$ is a sum of the form $y_1 + \cdots + y_n$ where for each $i$, $y_i$ is equal to $x_i$ or $x_i'$.

A Boolean function of degree $n$ is in **disjunctive normal form** (**DNF**) (or **sum-of-products expansion**) if it is written as a sum of distinct minterms in the variables $x_1, \ldots, x_n$. (*Note:* disjunctive normal form is sometimes called *full* disjunctive normal form.)

A Boolean function is in **conjunctive normal form** (**CNF**) (or **product-of-sums expansion**) if it is written as a product of distinct maxterms.

A set of operators in a Boolean algebra is **functionally complete** if every Boolean function can be written using only these operators.

**Facts:**

**1.** Every Boolean function can be written as a Boolean expression.

**2.** There are $2^{2^n}$ Boolean functions of degree $n$. Examples of the 16 different Boolean functions with two variables, $x$ and $y$, are given in the following table.

| $x$ | $y$ | $1$ | $x+y$ | $x+y'$ | $x'+y$ | $x\|y$ | $x$ | $y$ | $x \oplus y$ | $(x \oplus y)'$ | $y'$ | $x'$ | $xy$ | $xy'$ | $x'y$ | $x \downarrow y$ | $0$ |
|---|---|---|---|---|---|---|---|---|---|---|---|---|---|---|---|---|---|
| 1 | 1 | 1 | 1 | 1 | 1 | 0 | 1 | 1 | 0 | 1 | 0 | 0 | 1 | 0 | 0 | 0 | 0 |
| 1 | 0 | 1 | 1 | 1 | 0 | 1 | 1 | 0 | 1 | 0 | 1 | 0 | 0 | 1 | 0 | 0 | 0 |
| 0 | 1 | 1 | 1 | 0 | 1 | 1 | 0 | 1 | 1 | 0 | 0 | 1 | 0 | 0 | 1 | 0 | 0 |
| 0 | 0 | 1 | 0 | 1 | 1 | 1 | 0 | 0 | 0 | 1 | 1 | 1 | 0 | 0 | 0 | 1 | 0 |

**3.** Every Boolean function (not identically 0) can be written in disjunctive normal form. Either of the following two methods can be used:

(a) Rewrite the expression for the function so that no parentheses remain. For each term that does not have a literal for a variable $x_i$, multiply that term by $x_i + x_i'$. Multiply out so that no parentheses remain. Use the idempotent law to remove any duplicate terms or duplicate factors.

(b) Make a table of values for the function. For each row where the function has the value 1, form a minterm that yields 1 in only that row. Form the sum of these minterms.

**4.** Every Boolean function (not identically 1) can be written in conjunctive normal form. Any of the following three methods can be used:

(a) Write the negation of the expression in disjunctive normal form. Use De Morgan's laws to take the negation of this expression.

(b) Make a table of values for the function. For each row where the function has the value 0, form a minterm that yields 1 in only that row. Form the sum of these minterms. Use De Morgan's laws to take the complement of this sum.

(c) Make a table of values for the function. For each row where the function has the value 0, form a maxterm that yields 0 in only that row. Form the product of these maxterms.

**5.** The following are examples of functionally complete sets, with explanations showing how any Boolean function can be written using only these operations:

- $\{+, \cdot, '\}$   disjunctive normal form uses only the operators $+$, $\cdot$, and $'$;

- $\{+, '\}$   De Morgan's law $(a \cdot b)' = a' + b'$ allows the replacement of any occurrence of $a \cdot b$ with an expression that does not use $\cdot$;

- $\{\cdot, '\}$   De Morgan's law $a + b = (a' \cdot b')'$ allows the replacement of any occurrence of $a + b$ with an expression that does not use $+$;

- $\{|\}$   write the expression for any function in DNF; use $a' = a \,|\, a$, $a+b = (a \,|\, a) \,|\, (b \,|\, b)$, and $a \cdot b = (a \,|\, b) \,|\, (a \,|\, b)$ to replace each occurrence of $'$, $+$, and $\cdot$ with $|$;

- $\{\downarrow\}$   write the expression for any function in DNF; use $a' = a \downarrow a$, $a+b = (a \downarrow b) \downarrow (a \downarrow b)$, and $a \cdot b = (a \downarrow a) \downarrow (b \downarrow b)$ to replace each occurrence of $'$, $+$, and $\cdot$ with $\downarrow$.

**6.** The set $\{+, \cdot\}$ is not functionally complete.

**Examples:**

**1.** The function $f \colon \{0, 1\}^3 \to \{0, 1\}$ defined by $f(x, y, z) = x(z' + y'z) + x'$ is a Boolean function in the Boolean variables $x, y, z$. Multiplying out the expression for this function yields $f(x, y, z) = xz' + xy'z + x'$. In this form the second term, $xy'z$, is a minterm in the three variables $x, y, z$. The first and third terms are not minterms: the first term, $xz'$, does not use a literal for $y$, and the third term, $x'$, does not use literals for $y$ and $z$.

**2.** *Writing a Boolean function in disjunctive normal form:* To write the function $f$ from Example 1 in DNF using Fact 3(a), replace the terms $xz'$ and $x'$ with equivalent minterms by multiplying these terms by $1 (= a + a')$ for each missing variable $a$:

$$xz' = xz' \cdot 1 = xz'(y + y') = xyz' + xy'z';$$
$$x' = x' \cdot 1 \cdot 1 = x'(y + y')(z + z') = x'yz + x'yz' + x'y'z + x'y'z'.$$

Therefore,

$$f(x, y, z) = x(z' + y'z) + x'$$
$$= xz' + xy'z + x'$$
$$= xyz' + xy'z' + xy'z + x'yz + x'yz' + x'y'z + x'y'z'.$$

Alternatively, using Fact 3(b), the table of values for $f$ yields 1 in all rows except the row in which $x = y = z = 1$. Therefore minterms are obtained for the other rows, yielding the same sum of seven minterms.

**3.** *Writing a Boolean function in conjunctive normal form:* Using Fact 4(a) to write the function $f(x, y) = xy' + x'y$ in CNF, first rewrite the negation of $f$ in DNF, obtaining $f'(x, y) = xy + x'y'$. The negation of $f'$ is $f''(x, y) = f(x, y) = (x' + y')(x + y)$.

Alternatively, using Fact 4(c), the function $f$ has value 0 only when $x = y = 1$ and $x = y = 0$. The maxterms that yield 0 in exactly one of these rows are $x' + y'$ and $x + y$. This yields the CNF $f(x, y) = (x' + y')(x + y)$.

---

### 5.8.3   LOGIC GATES

Boolean algebra can be used to model circuitry, with 0s and 1s as inputs and outputs. The elements of these circuits are *gates* that implement the Boolean operations.

**Facts:**

**1.** The following figure gives representations for the three standard Boolean operators, $+$, $\cdot$, and $'$, together with representations for three related operators. (For example, the AND gate takes two inputs, $x$ and $y$, and produces one output, $xy$.)

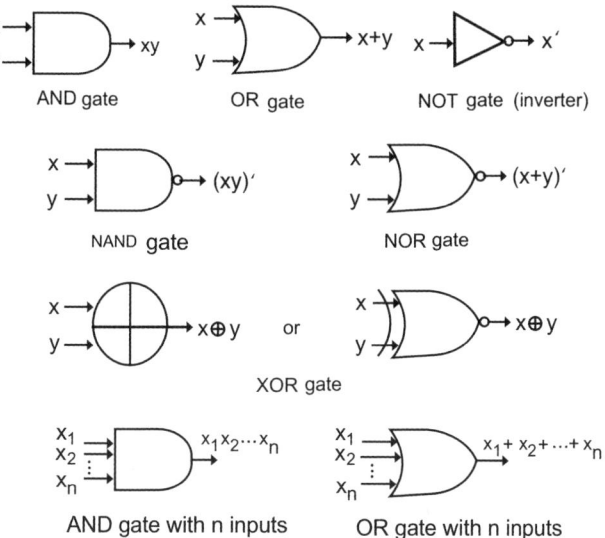

**2.** Gates can be extended to include cases where there are more than two inputs. The figure of Fact 1 also shows an AND gate and an OR gate with multiple inputs. These correspond to $x_1 x_2 \ldots x_n$ and $x_1 + x_2 + \cdots + x_n$. (Since both operations satisfy the associative laws, no parentheses are needed.)

**Examples:**

**1.** *The gate diagram for a half-adder:*   A *half-adder* is a Boolean circuit that adds two bits, $x$ and $y$, producing two outputs:

> a *sum bit* $s = (x+y)(xy)'$   ($s = 0$ if $x = y = 0$ or $x = y = 1$; $s = 1$ otherwise);
>
> a *carry bit* $c = xy$   ($c = 1$ if and only if $x = y = 1$).

The gate diagram for a half-adder is given in the following figure. This circuit is an example of a *multiple output circuit* since there is more than one output.

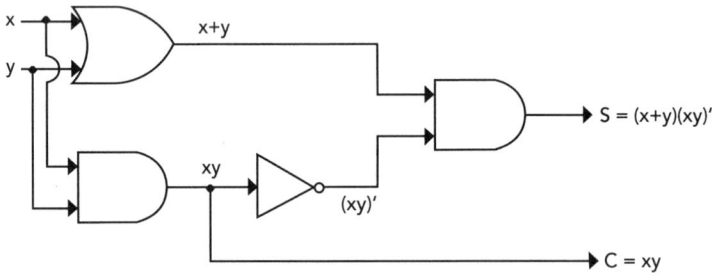

**2.**   *The gate diagram for a full-adder:*   A *full-adder* is a Boolean circuit that adds three bits ($x$, $y$, and a carry bit $c$) and produces two outputs (a sum bit $s$ and a carry bit $c'$). The full-adder gate diagram is given in the following figure.

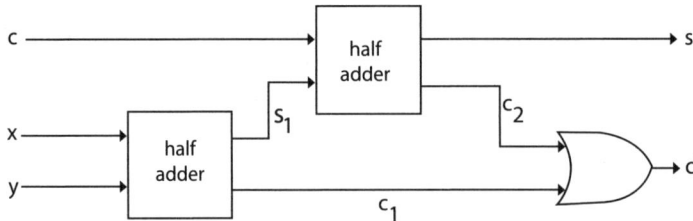

## 5.8.4   MINIMIZATION OF CIRCUITS

Boolean expressions that appear to be different can yield the same combinatorial circuit. For example, $xyz + xyz' + x'y$ and $y$ (as functions of $x$ and $y$) have the same table of values and hence yield the same circuit. (The first expression can be simplified to give the second: $xyz+xyz'+x'y = xy(z+z')+x'y = xy \cdot 1+x'y = xy+x'y = (x+x')y = 1 \cdot y = y$.)

**Definitions:**

A Boolean expression is **minimal** (as a sum-of-products) if among all equivalent sum-of-products expressions it has the fewest number of summands, and among all sum-of-products expressions with that number of summands it uses the smallest number of literals in the products.

A **Karnaugh map** for a Boolean expression written in disjunctive normal form is a diagram (constructed using the following algorithm) that displays the minterms in the Boolean expression.

**Facts:**

**1.** Minimization of circuits is an NP-hard problem.

**2.** *Don't care conditions*: In some circuits, it may be known that some elements of the input set for the Boolean function will never be used. Consequently, the values of the expression for these elements is irrelevant. The values of the circuit function for these unused elements of the input set are called *don't care conditions*, and the values can be arbitrarily chosen to be 0 or 1. The blocks in the Karnaugh map where the function values are irrelevant are marked with $d$. In the simplification process of the Karnaugh map, 1s can be substituted for any or all of the $d$s in order to cover larger blocks of boxes and achieve a simpler equivalent expression.

**Algorithm:**

There is an algorithm for minimizing Boolean expressions by systematically grouping terms together. When carried out visually, the method uses a Karnaugh map (Maurice Karnaugh, born 1924). When carried out numerically using bit strings, the method is called the Quine-McCluskey method (Willard Quine, 1908–2000; Edward McCluskey, 1929–2016).

**1.** *Karnaugh map method*:

(a) Write the Boolean expression in disjunctive normal form.

(b) Obtain the Karnaugh map for this Boolean expression. The layout of the table depends on the number of variables under consideration.

The grids for Boolean expressions with two variables ($x$ and $y$), three variables ($x$, $y$, and $z$), and four variables ($w$, $x$, $y$, and $z$) are shown in the following figure. Each square in each grid corresponds to exactly one minterm—the product of the row heading and the column heading. For example, the upper right box in the grid of part (a) of the figure represents the minterm $xy'$; the lower right box in the grid of part (c) of the figure represents $w'xyz'$.

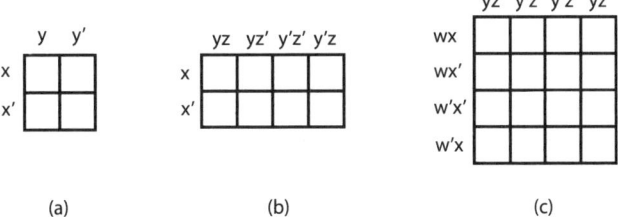

(a)                    (b)                    (c)

The headings are placed in a certain order—adjacent squares in any row (or column) differ in exactly one literal in their row headings (or column headings). The first and last squares in any row (or column) are to be regarded as adjacent. (The variable names can be permuted; for example, in part (b) of the figure, the row headings can be $y$ and $y'$ and the column headings can be $xz$, $xz'$, $x'z'$, and $x'z$. The column headings could also have been written in order as $yz$, $y'z$, $y'z'$, $yz'$ or $y'z$, $y'z'$, $yz'$, $yz$.)

The Karnaugh map for the Boolean expression is obtained by placing a checkmark in each square corresponding to a minterm in the expression.

(c) Find the best covering. A geometric version of the distributive law is used to "cover" groups of the adjacent marked squares, with every marked square covered at least once and each group covered being as large as possible. The possible ways of covering squares depends on the number of variables.

(For example, working with two variables and using the distributive law, $x'y + x'y' = x'(y + y') = x'1 = x'$. This corresponds to covering the two boxes in the bottom row of the first $2 \times 2$ grid in the following figure and noting that the only literal common to both boxes is $x'$.

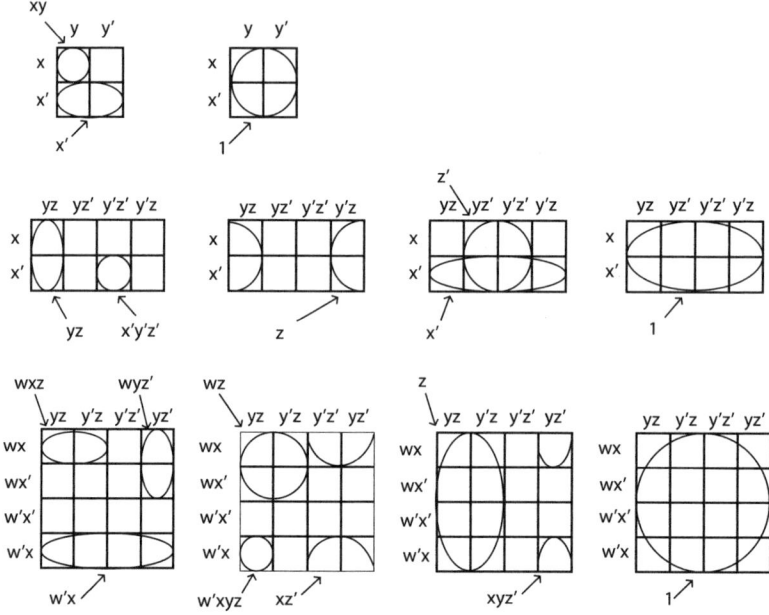

Similarly, working with three variables, $xyz' + xy'z' + x'yz' + x'y'z' = xz'(y + y') + x'z'(y + y') = xz' + x'z' = (x + x')z' = z'$. This corresponds to covering the four boxes in the second and third columns of the third $2 \times 4$ grid in the second row of the figure and noting that $z'$ is the only common literal.)

The following table shows what groups of boxes can be covered, for expressions with 2, 3, and 4 variables. These are the combinations whose expressions can be simplified to a single minterm. Examples for 2, 3, and 4 variables are shown in the previous figure. (The method is awkward to use when there are more than 4 variables.)

| # variables | groups of boxes that can be covered |
|:---:|:---|
| 2 | $1\times1$, $1\times2$, $2\times1$, $2\times2$ |
| 3 | $1\times1$, $1\times2$, $1\times4$, $2\times1$, $2\times2$, $2\times4$ |
| 4 | $1\times1$, $1\times2$, $1\times4$, $2\times1$, $2\times2$, $2\times4$, $4\times1$, $4\times2$, $4\times4$ |

To obtain the minimization, cover boxes according to the following rules:
- cover all marked boxes at least once;
- cover the largest possible blocks of marked boxes;
- do not cover any unmarked box;
- use the fewest blocks possible.

(d) Find the product of common literals for each of the blocks and form the sum of these products to obtain the minimization.

**2.** *Quine-McCluskey method:*

(a) Write the Boolean expression in disjunctive normal form, and in each summand list the variables in alphabetical order. Identify with each term a bit string, using a 1 if the literal is not a complement and 0 if the literal is a complement. (For example, $v'wx'yz$ is represented by 01011.)

(b) Form a table with the following columns:

   *column 1*:  Make a numbered list of the terms and their bit strings, beginning with the terms with the largest number of uncomplemented variables. (For example, $wxy'z$ precedes $wx'yz'$.)

   *column 2*:  Make a list of pairs of terms from column 1 where the literals in the two terms differ in exactly one position. Use a distributive law to add and simplify the two terms and write the numbers of these terms and the sum of the terms in the second column, along with its bit string, using "−" in place of the variable that no longer appears in the sum. (For example, $xyz'$ and $xy'z'$ can be combined to yield $xz'$ with bit string $1-0$.)

   *columns 3, 4, etc.*:  To obtain column 3 combine the terms in column 2 in pairs according to the same procedure as that used to construct column 2. Repeat this process until no more terms can be combined.

(c) Form a table with a row for each of the terms that cannot be used to form terms with fewer variables and a column for each of the original terms in the disjunctive normal form of the original expression. Mark the square in the $ij$-position if the minterm in column $j$ could be a summand for the term in row $i$.

(d) Find a set of rows, with as few rows as possible, such that every column has been marked at least once in at least one row. The sum of the products labeling these rows minimizes the original expression.

### Examples:

**1.** Simplify $w'x'y + w'z(xy + x'y') + w'x'z' + w'xyz' + wx'y'z'$ (an expression in four variables) using a Karnaugh map.

First write the expression in disjunctive normal form:

$$w'x'y + w'z(xy + x'y') + w'x'z' + w'xyz' + wx'y'z'$$
$$= w'x'yz + w'x'yz' + w'xyz + w'x'y'z + w'x'yz' + w'x'y'z' + w'xyz' + wx'y'z'$$
$$= w'x'yz + w'x'yz' + w'xyz + w'x'y'z + w'x'y'z' + w'xyz' + wx'y'z'.$$

Next, draw its Karnaugh map. See part (a) of the following figure. A covering is given in part (b) of the figure. Note that in order to use larger blocks, some squares have been covered more than once. Also note that $w'x'yz$, $w'xyz$, $w'x'yz'$, and $w'xyz'$ are covered with one $2 \times 2$ block rather than with two $1 \times 2$ blocks. In the three blocks the common literals are $w'x'$, $w'y$, and $x'y'z'$.

Finally, form the sum of these products: $w'x' + w'y + x'y'z'$.

**2.** Minimize $w'xy'z + wxyz' + wx'yz' + w'x'yz + wxyz + w'x'y'z + w'xyz$ (an expression in four variables) using the Quine-McCluskey method.

Step (b) of the Quine-McCluskey method yields the table that follows.

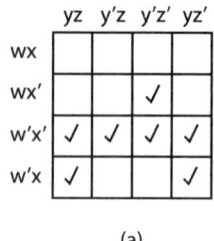

(a)

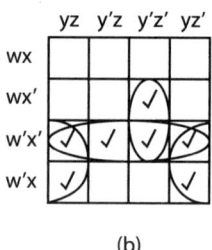

(b)

| 1 | $wxyz$ | 1111 | 1,2 | $wxy$ | 111– | 3,5,6,7 | $w'z$ | 0--1 |
|---|--------|------|-----|-------|------|---------|-------|------|
| 2 | $wxyz'$ | 1110 | 1,3 | $xyz$ | –111 | | | |
| 3 | $w'xyz$ | 0111 | 2,4 | $wyz'$ | 1–10 | | | |
| 4 | $wx'yz'$ | 1010 | 3,5 | $w'yz$ | 0–11 | | | |
| 5 | $w'x'yz$ | 0011 | 3,6 | $w'xz$ | 01–1 | | | |
| 6 | $w'xy'z$ | 0101 | 5,7 | $w'x'z$ | 00–1 | | | |
| 7 | $w'x'y'z$ | 0001 | 6,7 | $w'y'z$ | 0–01 | | | |

The four terms $w'z$, $wxy$, $xyz$, $wyz'$ were not used in combining terms, so they become the names of the rows in the next table:

| | $wxyz$ | $wxyz'$ | $w'xyz$ | $wx'yz'$ | $w'x'yz$ | $w'xy'z$ | $w'x'y'z$ |
|---|--------|---------|---------|----------|----------|----------|-----------|
| $w'x$ | | | ✓ | | ✓ | ✓ | ✓ |
| $wxy$ | ✓ | ✓ | | | | | |
| $xyz$ | ✓ | | ✓ | | | | |
| $wyz'$ | | ✓ | | ✓ | | | |

There are two ways to cover the seven minterms:

$$w'x, wxy, wyz' \quad \text{or} \quad w'x, xyz, wyz'.$$

This yields two ways to minimize the original expression:

$$w'x + wxy + wyz' \quad \text{and} \quad w'z + xyz + wyz'.$$

---

# REFERENCES

***Printed Resources:***

[Ar10] M. Artin, *Algebra*, 2nd ed., Pearson, 2010.

[As00] M. Aschbacher, *Finite Group Theory*, 2nd ed., Cambridge University Press, 2000.

[BiBa70] G. Birkhoff and T. C. Bartee, *Modern Applied Algebra*, McGraw-Hill, 1970.

[BiMa98] G. Birkhoff and S. Mac Lane, *A Survey of Modern Algebra*, A K Peters/CRC Press, 1998.

[Bl87] N. J. Bloch, *Abstract Algebra with Applications*, Prentice-Hall, 1987.

[Ca89] R. W. Carter, *Simple Groups of Lie Type*, Wiley, 1989.

[Ch08] L. Childs, *A Concrete Introduction to Higher Algebra*, 3rd ed., Springer-Verlag, 2008.

[CoMo80] H. S. M. Coxeter and W. O. J. Moser, *Generators and Relations for Discrete Groups*, 4th ed., Springer-Verlag, 1980.

[CrFo08] R. H. Crowell and R. H. Fox, *Introduction to Knot Theory*, Dover, 2008.

[Fr02] J. B. Fraleigh, *A First Course in Abstract Algebra*, 7th ed., Pearson, 2002.

[Go82] D. Gorenstein, *Finite Simple Groups: An Introduction to Their Classification*, Springer, 1982.

[GoLySo02] D. Gorenstein, R. Lyons, and R. Solomon, *The Classification of the Finite Simple Groups*, American Mathematical Society, 2002.

[Ha99] M. Hall, Jr., *The Theory of Groups*, 2nd ed., American Mathematical Society, 1999.

[He75] I. N. Herstein, *Topics in Algebra*, 2nd ed., Wiley, 1975.

[Hu74] T. W. Hungerford, *Algebra*, Springer, 1974.

[Ka72] I. Kaplansky, *Fields and Rings*, 2nd ed., University of Chicago Press, 1972.

[LiNi94] R. Lidl and H. Niederreiter, *Introduction to Finite Fields and Their Applications*, revised edition, Cambridge University Press, 1994.

[LiPi98] R. Lidl and G. Pilz, *Applied Abstract Algebra*, 2nd ed., Springer-Verlag, 1998.

[MaBi99] S. Mac Lane and G. Birkhoff, *Algebra*, 3rd ed., American Mathematical Society, 1999.

[MaKaSo04] W. Magnus, A. Karrass, and D. Solitar, *Combinatorial Group Theory*, 2nd ed., Dover, 2004.

[Mc73] N. H. McCoy, *The Theory of Rings*, Chelsea, 1973.

[McBe77] N. H. McCoy and T. Berger, *Algebra: Groups, Rings and Other Topics*, Allyn & Bacon, 1977.

[McFr12] N. H. McCoy and P. Franklin, *Rings and Ideals* (Carus Monograph No. 8), Literary Licensing, LLC, 2012.

[Mc03] R. J. McEliece, *Finite Fields for Computer Scientists and Engineers*, Kluwer Academic Publishers, 2003.

[MeEtal10] A. Menezes, I. Blake, X. Gao, R. Mullin, S. Vanstone, and T. Yaghoobian, *Applications of Finite Fields*, Kluwer Academic Publishers, 2010.

[Ro99] J. J. Rotman, *An Introduction to the Theory of Groups*, 4th ed., Springer, 1999.

[ThFe63] J. G. Thompson and W. Feit, "Solvability of groups of odd order," *Pacific Journal of Mathematics* 13 (1963), 775–1029.

[va91] B. L. van der Waerden, *Modern Algebra* (2 volumes), Springer, 1991.

**Web Resources**:

http://magma.maths.usyd.edu.au/magma/ (The Magma Computer Algebra System, successor to CAYLEY, developed by the Computational Algebra Group at the University of Sydney.)

http://www.gap-system.org (Home page for GAP – Groups, Algorithms and Programming, a system for computational discrete algebra.)

# 6

# *LINEAR ALGEBRA*

## INTRODUCTION

Concepts from linear algebra play an important role in various applications of discrete mathematics, as in coding theory, computer graphics, marketing, generation of pseudo-random numbers, graph theory, and combinatorial designs. This chapter discusses fundamental concepts of linear algebra, computational aspects, and various applications.

## GLOSSARY

***access*** (of a class):   The class $C_i$ of vertices has access to class $C_j$ if either $i = j$ or there is a path from a vertex in $C_i$ to a vertex in $C_j$.

***adjacency matrix*** (of a graph):   a square $(0, 1)$-matrix whose $(i, j)$ entry is 1 if and only if vertices $i$ and $j$ are adjacent.

***adjoint***:   See *Hermitian adjoint*.

***algebraic multiplicity***:   given an eigenvalue, the multiplicity of the eigenvalue as a root of the characteristic equation.

***augmented matrix*** (of a linear system):   the matrix obtained by appending the right-hand side vector to the coefficient matrix as its rightmost column.

***back substitution***:   a procedure for solving an upper triangular linear system.

***basic class*** (of a matrix):   a class such that the Perron root of the corresponding principal submatrix equals that of the entire matrix.

***basis***:   an independent spanning set of vectors in a vector space.

***characteristic equation***:   for a square matrix $A$, the equation $p_A(\lambda) = 0$, where $p_A(\lambda)$ is the characteristic polynomial of $A$.

***characteristic polynomial***:   for a square matrix $A$, the polynomial (in the indefinite symbol $\lambda$) given by $p_A(\lambda) = \det(\lambda I - A)$.

***Cholesky decomposition***:   expressing a matrix $A$ as $A = LL^T$, where $L$ is lower triangular and every entry on the main diagonal of $L$ is positive.

***circulant***:   a matrix in which every row is obtained by a single cyclic shift of the previous row.

***class*** (of a matrix):   a maximal set of row indices such that the corresponding vertices have mutual access in the directed graph of the matrix.

***complete pivoting***:   an implementation of Gaussian elimination in which a pivot of largest magnitude is selected at each step.

***completion*** (of a partially specified matrix):   a matrix in which values are chosen for each unspecified entry.

***condition number***:   given a matrix $A$, the number $\kappa(A) = \|A\| \, \|A^{-1}\|$.

***conjugate sequence*** (of a sequence):   the sequence whose $n$th term is the number of terms not less than $n$ in the given sequence.

**dependent set**: a set of vectors in a vector space that are not independent.

**determinant**: given an $n \times n$ matrix $A$, $\det A = \sum_{\sigma \in S_n} \operatorname{sgn}(\sigma) a_{1\sigma(1)} a_{2\sigma(2)} \cdots a_{n\sigma(n)}$, where $S_n$ is the symmetric group on $n$ elements and the coefficient $\operatorname{sgn}(\sigma)$ is the sign of the permutation $\sigma$: 1 if $\sigma$ is an even permutation and $-1$ if $\sigma$ is an odd permutation.

**diagonal matrix**: a square matrix with nonzero elements only on the main diagonal.

**diagonalizable matrix**: a square matrix that is *similar* to a diagonal matrix.

**difference** (of matrices of the same dimensions): the matrix each of whose elements is the difference between corresponding elements of the original matrices.

**dimension**: for a vector space $V$, the number of vectors in any basis for $V$.

**directed graph** (of a matrix $A$): the graph $G(A)$ with vertices corresponding to the rows of $A$ and an edge from $i$ to $j$ whenever $a_{ij}$ is nonzero.

**direct sum** (of subspaces): given subspaces $U$ and $W$, the sum of subspaces in which $U$ and $W$ have only the zero vector in common.

**distance** (between vectors): given vectors $v$ and $w$, the length of the vector $v - w$.

**dominant eigenvalue**: given a matrix, an eigenvalue of the matrix of maximum modulus.

**dot product** (of real vectors): given real vectors $x = (x_1, \ldots, x_n)$ and $y = (y_1, \ldots, y_n)$, the number $x \cdot y = \sum_{i=1}^{n} x_i y_i$.

**doubly stochastic matrix**: a matrix with all entries nonnegative and with all row and column sums equal to 1.

**eigenvalue**: given a square matrix $A$, a scalar $\lambda$ such that $Ax = \lambda x$ for some nonzero vector $x$.

**eigenvector**: given a square matrix $A$, a nonzero vector $x$ such that the vector $Ax$ is a scalar multiple of $x$.

**eigenspace**: given a square matrix $A$, the vector space $\{\, x \mid Ax = \lambda x \,\}$ for some scalar $\lambda$.

**exponent** (of a matrix): given a matrix $A$, the least positive integer $m$, if it exists, such that $A^m$ has all positive entries.

**fill**: in Gaussian elimination, those nonzero entries created in the triangular factors of a matrix corresponding to zero entries in the original matrix.

**final class**: given a matrix, a class of the matrix with access to no other class.

**flop**: a multiply-add operation involving a single multiplication followed by a single addition.

**forward substitution**: a procedure for solving a lower triangular linear system.

**fully indecomposable matrix**: a matrix that is not partly decomposable.

**Gaussian elimination**: a solution procedure that at each step uses one equation to eliminate one variable from the system of equations.

**geometric multiplicity**: the dimension of the eigenspace.

**Geršgorin discs**: regions in the complex plane that collectively are guaranteed to contain all the eigenvalues of a given matrix.

**growth factor**:  a ratio that measures how large the entries of a matrix become during Gaussian elimination.

**Hermitian adjoint**:  given a matrix $A$, the matrix $A^*$ obtained from the transpose $A^T$ by replacing each entry by its complex conjugate.

**Hermitian matrix**:  a complex matrix whose transpose is its (elementwise) complex conjugate.

**idempotent matrix**:  a matrix $A$ such that $A^2 = A$.

**identity matrix**:  a diagonal matrix in which each diagonal element is 1.

**ill-conditioned system**:  a linear system $Ax = b$ whose solution $x$ is extremely sensitive to errors in the data $A$ and $b$.

**independent set**:  a set of vectors in a vector space that is not dependent.

**index of cyclicity**:  for a matrix, the number of eigenvalues with maximum modulus.

**inner product**:  a field-valued function of two vector variables used to define a notion of orthogonality (that is, perpendicularity).  In real or complex vector spaces it is also used to introduce length, distance, and convergence.

**inverse**:  given a square matrix $A$, the square matrix $A^{-1}$ whose product with the original matrix is the identity matrix.

**invertible matrix**:  a matrix that has an inverse.

**irreducible matrix**:  a matrix that is not reducible.

**isomorphic** (vector spaces):  vector spaces that are structurally identical.

**kernel** (of a linear transformation):  the set of all vectors that are mapped to the zero vector by the linear transformation.

**Laplacian matrix** (of a graph):  the difference between the diagonal matrix associated with the graph (having vertex degrees on the diagonal) and the adjacency matrix of the graph.

**length** (of a vector):  the square root of the inner product of the vector with itself.

**linear combination** (of vectors):  given vectors $v_1, v_2, \ldots, v_t$, a vector of the form $a_1 v_1 + a_2 v_2 + \cdots + a_t v_t$, where the $a_i$ are scalars.

**linear operator**:  a linear transformation from a vector space to itself.

**linear system**:  a set of $m$ linear equations in $n$ variables $x$, represented by $Ax = b$; here $A$ is the coefficient matrix and $b$ is the right-hand side vector.

**linear transformation**:  a function $T$ from one vector space over $F$ to another vector space over $F$ satisfying $T(au+v) = aT(u) + T(v)$ for all vectors $u$, $v$ and all scalars $a$.

**lower triangular matrix**:  a matrix in which all nonzero elements occur either on or below the diagonal.

**LU decomposition**:  expressing a matrix $A$ as the product $A = LU$, where $L$ is unit lower triangular and $U$ is upper triangular.

**Markowitz pivoting**:  a simple greedy strategy for reducing the number of nonzero entries introduced during the $LU$ decomposition of a sparse matrix.

**matrix** (of a linear transformation):  given a linear transformation $T$, a matrix associated with $T$ that represents $T$ with respect to a fixed basis.

**minimal polynomial**: for a matrix $A$, the monic polynomial $q(\cdot)$ of minimum degree such that $q(A) = 0$.

**minimum degree algorithm**: a version of the Markowitz pivoting strategy for symmetric coefficient matrices.

**minor**: the determinant of a square submatrix of a given matrix.

**modulus**: the absolute value of a complex number.

**nilpotent matrix**: a matrix $A$ such that $A^k = 0$ for some positive integer $k$.

**nonnegative matrix**: a matrix with each entry nonnegative.

**nonsingular matrix**: a matrix that has an inverse.

**normal matrix**: a matrix $A$ such that $AA^* = A^*A$ ($A^*$ is the Hermitian adjoint of $A$).

**nullity** (of a linear transformation): the dimension of the kernel of the linear transformation.

**nullity** (of a matrix): the dimension of the null space of the matrix.

**null space** (of a matrix $A$): the set of all vectors $x$ for which $Ax = 0$.

**numerically stable algorithm**: an algorithm whose accuracy is not greatly harmed by roundoff errors.

**numerically unstable algorithm**: an algorithm that can return an inaccurate solution even when the solution is relatively insensitive to errors in the data.

**orthogonal matrix**: a real square matrix whose inverse is its transpose.

**orthogonal set** (of vectors): a set of vectors in which any two distinct vectors have inner product 0.

**orthonormal set** (of vectors): a set of unit length orthogonal vectors.

**partial pivoting**: an implementation of Gaussian elimination which at step $k$ selects the pivot of largest magnitude in column $k$.

**partly decomposable** (matrix): an $n \times n$ matrix containing a zero submatrix of size $k \times (n-k)$ for some $1 \leq k \leq n-1$.

**permanent** (of an $n \times n$ matrix $A$): $\mathrm{per}(A) = \sum_{\sigma \in S_n} a_{1\sigma(1)} a_{2\sigma(2)} \cdots a_{n\sigma(n)}$, where $S_n$ is the symmetric group on $n$ elements.

**permutation matrix**: a square $(0,1)$-matrix in which the entry 1 occurs exactly once in each row and exactly once in each column.

**Perron root**: the spectral radius of a nonnegative matrix.

**pivot**: the coefficient of the eliminated variable in the equation used to eliminate it.

**positive definite matrix**: a Hermitian matrix $A$ such that $x^*Ax > 0$ for all $x \neq 0$.

**positive matrix**: a matrix with each entry positive.

**positive semidefinite matrix**: a Hermitian matrix $A$ such that $x^*Ax \geq 0$ for all $x$.

**power** (of a square matrix): the square matrix obtained by multiplying the matrix by itself the required number of times.

**primitive matrix**: a matrix with a finite exponent.

**principal minor** (of a matrix): the determinant of a principal submatrix of the matrix.

**principal submatrix** (of a matrix $A$):  the matrix obtained from $A$ by deleting all but a specified set of rows and the same set of columns.

**product** (of matrices):  for an $m \times n$ matrix $A$ and an $n \times p$ matrix $B$, the $m \times p$ matrix $AB$ whose $ij$-entry is the scalar product of row $i$ of $A$ and column $j$ of $B$.

**range** (of a linear transformation $T$):  the set of all vectors $w$ for which $T(v) = w$ has a solution.

**rank** (of a linear transformation $T$):  the dimension of the range of $T$.

**rank** (of a matrix):  the maximum number of linearly independent rows (or columns) in the matrix.

**reducible matrix**:  a matrix $A$ with $a_{ij} = 0$ for all $i \in S$, $j \notin S$, for some set $S$.

**roundoff errors**:  the errors associated with storing and computing numbers in finite precision arithmetic on a digital computer.

**row stochastic matrix**:  a matrix with all entries nonnegative and row sums 1.

**scalar**:  an element of a field.

**scalar multiple** (of a matrix):  the matrix obtained by multiplying each element of the original matrix by the scalar.

**scalar product**:  See *dot product*.

**sign pattern** (of a matrix):  the matrix obtained by replacing each entry of the given matrix by its sign (0, 1, or $-1$).

**similar matrices**:  square matrices $A$ and $B$ satisfying the equation $P^{-1}BP = A$ for some invertible matrix $P$.

**singular matrix**:  a matrix that has no inverse.

**singular value decomposition** (of a matrix $A$):  the representation $A = U\Sigma V^{T}$, where $U, V$ are orthogonal matrices and $\Sigma$ is a diagonal matrix.

**singular values** (of a matrix $A$):  the positive square roots of the eigenvalues of $AA^{*}$, where $A^{*}$ is the Hermitian adjoint of $A$.

**skew-Hermitian matrix**:  a matrix equal to the negative of its Hermitian adjoint.

**skew-symmetric matrix**:  a matrix equal to the negative of its transpose.

**span** (of a set of vectors):  all vectors obtainable as linear combinations of the given vectors.

**spanning set**:  a set of vectors in a vector space $V$ whose span equals $V$.

**sparse matrix**:  a matrix that has relatively few nonzero entries.

**spectral radius** (of a matrix):  the maximum modulus of an eigenvalue of the matrix.

**square matrix**:  a matrix having the same number of rows and columns.

**strictly diagonally dominant matrix**:  a square matrix each of whose diagonal elements exceeds in modulus the sum of the moduli of all other elements in that row.

**strictly totally positive matrix**:  a matrix with all minors positive.

**submatrix** (of a matrix $A$):  the matrix obtained from $A$ by deleting all but a certain set of rows and a certain set of columns.

**subspace**:  a vector space within a vector space.

**sum** (of matrices): for two matrices of the same dimensions, the matrix each of whose elements is the sum of the corresponding elements of the original matrices.

**sum** (of subspaces): given subspaces $U$ and $W$, the subspace consisting of all possible sums $u + w$ where $u \in U$ and $w \in W$.

**symmetric matrix**: a matrix that equals its transpose.

**term rank** (of a $(0,1)$-matrix): the maximum number of 1s such that no two are in the same row or column.

**tournament matrix**: a $(0,1)$-matrix $A = (a_{ij})$ with zero diagonal entries and whose off-diagonal entries satisfy $a_{ij} + a_{ji} = 1$.

**trace**: given a square matrix, the sum of the diagonal elements of the matrix.

**transpose** (of a matrix): for a matrix $A$, the matrix $A^T$ whose columns are the rows of the original matrix.

**tridiagonal matrix**: a matrix whose nonzero entries are either on the main diagonal or immediately above or below the main diagonal.

**unitary matrix**: a square matrix whose inverse is its Hermitian adjoint.

**unit triangular matrix**: a (lower or upper) triangular matrix having all diagonal entries 1.

**upper triangular matrix**: a matrix in which all nonzero elements occur either on or above the main diagonal.

**vector**: an individual object of a vector space.

**vector space**: a collection of objects that can be added and multiplied by scalars, always yielding another object in the collection.

**well-conditioned system**: a linear system $Ax = b$ whose solution $x$ is relatively insensitive to errors in the data $A$ and $b$.

**$(0,1)$-matrix**: a matrix with each entry either 0 or 1.

# 6.1   VECTOR SPACES

The concept of a "vector" comes initially from the physical world, where a vector is a quantity having both magnitude and direction (for example, force and velocity). The mathematical concept of a vector space generalizes these ideas, with applications in coding theory, finite geometry, cryptography, and other areas of discrete mathematics.

## 6.1.1   BASIC CONCEPTS

### Definitions:

A **vector space** over a field $F$ (§5.6.1) is a triple $(V, \oplus, \cdot)$ consisting of a set $V$ and two operations, $\oplus$ (vector addition) and $\cdot$ (scalar multiplication), such that

- $(V, \oplus)$ is an abelian group (§5.2.1); i.e., $\oplus$ is a function $(u, v) \to u \oplus v$ from $V \times V$ to $V$ satisfying

⋄ $(u \oplus v) \oplus w = u \oplus (v \oplus w)$  for all $u, v, w \in V$;

⋄ there is a vector 0 such that $v \oplus 0 = v$  for all $v \in V$;

⋄ for each $v \in V$ there is $-v \in V$ such that $v \oplus (-v) = 0$;

⋄ $u \oplus v = v \oplus u$  for all $u, v \in V$;

• the operation $\cdot$ is a function $(a, v) \to a \cdot v$ from $F \times V$ to $V$ such that for all $a, b \in F$ and $u, v \in V$ the following properties hold:

⋄ $a \cdot (b \cdot v) = (ab) \cdot v$;

⋄ $(a + b) \cdot v = (a \cdot v) \oplus (b \cdot v)$;

⋄ $a \cdot (u \oplus v) = (a \cdot u) \oplus (a \cdot v)$;

⋄ $1 \cdot v = v$.

Here, $ab$ and $a + b$ represent multiplication and addition of elements $a, b \in F$.

The **scalars** are the elements of $F$, the **vectors** are the elements of $V$, and the set $V$ itself is often also called the **vector space**.

The **difference** of two vectors $u$ and $v$ is the vector $u - v = u \oplus (-v)$ where $-v$ is the negative of $v$ in the abelian group $(V, \oplus)$.

*Notation:* While vector addition $\oplus$ and field addition $+$ can be quite different, it is customary to use the same notation $+$ for both. It is also customary to write $av$ instead of $a \cdot v$, and to use the symbol 0 for the additive identities of the vector space $V$ and the field $F$.

## Facts:
Assume that $V$ is a vector space over $F$.

1. $a0 = 0$ and $0v = 0$ for all $a \in F$ and $v \in V$.

2. $(-1)v = -v$ for all $v \in V$.

3. If $av = 0$ for $a \in F$ and $v \in V$, then either $a = 0$ or $v = 0$.

4. *Cancellation property:* For all $u, v, w \in V$, if $u + v = w + v$, then $u = w$.

5. $a(u - v) = au - av$ for all $a \in F$ and $u, v \in V$.

## Examples:
1. *Force vectors:* Forces in the plane can be represented by geometric vectors such as $F_1$ and $F_2$ in part (a) of the following figure; addition of these vectors is carried out using the so-called parallelogram law. By introducing a coordinate system and locating the initial point of each directed line segment at the origin $(0, 0)$, each geometric vector can be named by its terminal point. Thus, a vector in the plane becomes a pair $(x, y) \in \mathcal{R}^2$ of real numbers. The parallelogram law of addition translates into componentwise addition (part (c) of the figure), while stretching (respectively, shrinking, negating) translates to componentwise multiplication by a real number $r > 1$ (respectively, $0 < r < 1$, $r = -1$). Three-dimensional force vectors are similarly represented using triples $(x, y, z) \in \mathcal{R}^3$.

2. *Euclidean space:* Generalizing Example 1, $n$-dimensional Euclidean space consists of all $n$-tuples of real numbers $\mathcal{R}^n = \{ (x_1, x_2, \ldots, x_n) \mid x_i \in \mathcal{R} \}$.

3. If $F$ is any field, then $F^n = \{ (x_1, x_2, \ldots, x_n) \mid x_i \in F \}$ is a vector space, where addition and scalar multiplication are componentwise:

$$(x_1, x_2, \ldots, x_n) + (y_1, y_2, \ldots, y_n) = (x_1 + y_1, x_2 + y_2, \ldots, x_n + y_n)$$
$$a(x_1, x_2, \ldots, x_n) = (ax_1, ax_2, \ldots, ax_n),$$

where $a \in F$. When $F = \mathcal{R}$, these are the vectors mentioned in Examples 1 and 2.

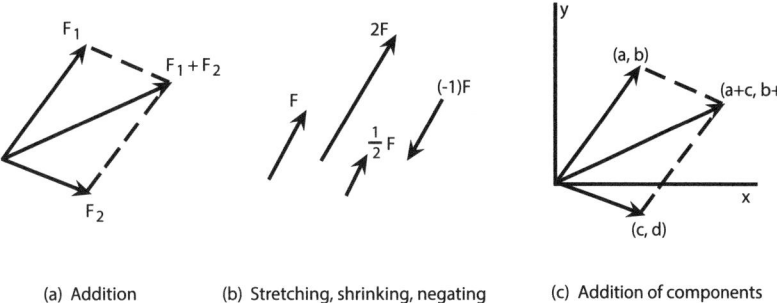

(a) Addition          (b) Stretching, shrinking, negating          (c) Addition of components

**4.** *A vector space over $\mathcal{Z}_2$:*  $V$ consists of the 128 subsets of the set $\{1, 2, \ldots, 7\}$ as represented by binary 7-tuples; for example, the subset $\{1, 4, 5, 7\}$ corresponds to $(1, 0, 0, 1, 1, 0, 1)$ and the subset $\{1, 2, 3, 4\}$ to $(1, 1, 1, 1, 0, 0, 0)$. The operations on $V$ are componentwise addition and scalar multiplication mod 2. In this vector space, the sum of two members of $V$ corresponds to the symmetric difference (§1.2.2) of the associated sets. (This example is a special case of Example 3.)

**5.** *A finite affine plane over $\mathcal{Z}_5$:*  $V$ consists of all pairs $(x, y)$ where $x, y \in \mathcal{Z}_5$ and where addition and scalar multiplication are componentwise modulo 5. This special case of Example 3 arises in finite geometry where the 25 members of $V$ are thought of as "points" and the sets of solutions to equations of the form $ax + by = c$ (where $a, b, c \in \mathcal{Z}_5$ with one of $a$ or $b \neq 0$) are viewed as "lines".

**6.** *Infinite binary sequences:*  $V$ consists of all infinite binary sequences $\{\, (s_1, s_2, \ldots) \mid s_i \in \mathcal{Z}_2 \,\}$ where addition and multiplication are componentwise mod 2. As in Example 4, each $s \in V$ may be viewed as a subset of the positive integers, but each $s$ may also be viewed as a potential "message" or "data" stream; for example, each group of 7 consecutive members of $s$ could represent a letter in the 7-bit ASCII code.

**7.** $V = F^{m \times n}$, the set of all $m \times n$ matrices over $F$, is a vector space, where vector addition is the usual matrix addition and scalar multiplication is the usual scalar-by-matrix multiplication (§6.3.2). When $m = 1$, this reduces to Example 3.

**8.** Let $V = E$ be a field and $F$ a subfield. Then $V$ is a vector space over $F$ where vector addition and scalar multiplication are the addition and multiplication of $E$. In particular, the finite field $F_q$ of prime power order $q = p^n$ is a vector space over the subfield $F_p$.

**9.** Let $V = F[x]$, the set of all polynomials (§5.5.2) over $F$ in an indeterminate $x$. Then $V$ is a vector space over $F$, where addition is ordinary polynomial addition and scalar multiplication is the usual scalar-by-polynomial multiplication.

**10.** For a nonempty set $X$ and a given vector space $U$ over $F$, let $V$ denote the set of all functions from $X$ to $U$. The sum $f + g$ of two vectors (functions) $f, g \in V$ is defined by $(f + g)(x) = f(x) + g(x)$ for all $x \in X$ and the scalar multiplication $af$ of $a \in F$ by $f \in V$ is defined by $(af)(x) = af(x)$. (For specific cases of this general vector space, see §6.1.2, Examples 13–15.)

## 6.1.2  SUBSPACES

### Definitions:

A **subspace** of a vector space $V$ is a nonempty subset $W$ of $V$ that is a vector space under the addition and scalar multiplication operations inherited from $V$.

The **sum** of two subspaces $U, W \subseteq V$ is the set $\{\, u+w \mid u \in U, \; w \in W \,\}$. If $U \cap W = \{0\}$, their sum is called the **direct sum**, denoted $U \oplus W$.

If $A$ is an $m \times n$ matrix over $F$, the **null space** $NS(A)$ of $A$ is $\{\, x \in F^{n \times 1} \mid Ax = 0 \,\}$. The null space of $A$ is also called the **right null space** when contrasted with the **left null space** $LNS(A)$ defined by $\{\, y \in F^{1 \times m} \mid yA = 0 \,\}$.

### Facts:

Assume that $V$ is a vector space over $F$.

**1.**  $W \subseteq V$ is a subspace of $V$ if and only if $W \neq \emptyset$ and for all $a, b \in F$ and $u, v \in W$, $au + bv \in W$.

**2.**  $W \subseteq V$ is a subspace of $V$ if and only if $W \neq \emptyset$ and for all $a \in F$ and $u, v \in W$, $u + v \in W$ and $au \in W$.

**3.**  Every subspace of $V$ contains 0, the zero vector.

**4.**  The sets $\{0\}$ and $V$ are subspaces of $V$.

**5.**  The intersection of any collection of subspaces of $V$ is a subspace of $V$.

**6.**  The sum of any collection of subspaces of $V$ is a subspace of $V$.

**7.**  Each member of $U \oplus W$ can be expressed as a sum $u + w$ for a unique $u \in U$ and a unique $w \in W$.

**8.**  The set of solutions to a homogeneous linear equation in the unknowns $x_1, x_2, \ldots, x_n$ is a subspace of $F^n$. Namely, for any fixed $(a_1, a_2, \ldots, a_n) \in F^n$, the set $W = \{\, x \in F^n \mid a_1 x_1 + a_2 x_2 + \cdots + a_n x_n = 0 \,\}$ is a subspace of $F^n$.

**9.**  The set of solutions to any collection of homogeneous linear equations in the unknowns $x_1, x_2, \ldots, x_n$ is a subspace of $F^n$. In particular, if $W$ is a subspace of $F^n$ then the set of all $x = (x_1, x_2, \ldots, x_n) \in F^n$ satisfying $a_1 x_1 + a_2 x_2 + \cdots + a_n x_n = 0$ for all $(a_1, a_2, \ldots, a_n) \in W$ is a subspace of $V$ called the *orthogonal complement* of $W$ and denoted $W^{\perp}$.

**10.**  The null space $NS(A)$ of an $m \times n$ matrix $A$ over $F$ is a subspace of $F^{n \times 1}$.

**11.**  The left null space $LNS(A)$ of an $m \times n$ matrix $A$ is a subspace of $F^{1 \times m}$ and equals $(NS(A^T))^T$ where $T$ denotes transpose.

### Examples:

**1.**  The set of all 3-tuples of real numbers of the form $(a, b, 2a + 3b)$ where $a, b \in \mathcal{R}$ is a subspace of $\mathcal{R}^3$. This subspace can also be described as the set of solutions $(x, y, z)$ to the homogeneous linear equation $2x + 3y - z = 0$.

**2.**  The set of all 4-tuples of real numbers of the form $(a, -a, 0, b)$ where $a, b \in \mathcal{R}$ is a subspace of $\mathcal{R}^4$. This subspace can also be described as the set of solutions $(x_1, x_2, x_3, x_4)$ to the pair of equations $x_1 + x_2 = 0$ and $x_3 = 0$.

**3.**  For $V = \mathcal{Z}_5^2$, the set of all solutions to the equation $x + 2y = 0$ forms a subspace. It consists of the finite set $\{(0,0), (3,1), (1,2), (4,3), (2,4)\}$ and can also be described as the set of all pairs in $V$ of the form $(3a, a)$. The set $S$ of solutions to $x + 2y = 1$, namely $\{(1,0), (4,1), (2,2), (0,3), (3,4)\}$, is not a subspace of $V$ since for example $(1,0) + (4,1) = (0,1) \notin S$. However $S$ is a "line" in the affine plane described in Example 5 of §6.1.1.

**4.**  In the vector space $V = \mathcal{Z}_2^7$, the set of 7-tuples with an even number of 1s is a subspace. This subspace can also be described as the collection of all members of $V$ whose components sum to 0.

**5.** *Coding theory*: In the vector space $F^n$ over the finite field $F = GF(q)$, a linear code (§14.2) is simply any subspace of $F^n$. In particular, an $(n, k)$-code is a $k$-dimensional subspace of $F^n$.

**6.** *Binary codes*: A linear binary code is any subspace of the vector space $F^n$ where $F$ is the finite field on two elements, $GF(2)$. Generalizing Example 4, the set of all binary $n$-tuples with an even number of 1s is a subspace of $F^n$ and so is a linear binary code.

**7.** Consider the undirected graph (§8.1) in the following figure, where the edges have been labeled with the integers $\{1, 2, \ldots, 7\}$. Associate with this graph the vector space $V = \mathcal{Z}_2^7$ where, as in Example 4 (§6.1.1), each binary 7-tuple is identified with a subset of edges. One subspace $W$ of $V$, called the *cycle space* of the graph, corresponds to the (edge-disjoint) union of cycles in the graph. For example, $(1, 1, 0, 1, 0, 1, 1) \in W$ as it corresponds to the cycle $1, 2, 6, 7, 4$, and so is $(1, 1, 1, 0, 1, 1, 1)$ which corresponds to the edge-disjoint union of cycles $1, 2, 3$ and $5, 6, 7$. The sum of these two members of $W$ is $(0, 0, 1, 1, 1, 0, 0)$ which corresponds to the cycle $3, 4, 5$.

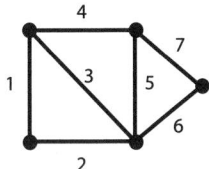

**8.** The set of $n \times n$ symmetric matrices (§6.3.1) over a field $F$ is a subspace of $F^{n \times n}$, and so is the set of $n \times n$ upper triangular matrices (§6.3.1) over $F$.

**9.** For an $m \times m$ matrix $A$ over $F$ and $\lambda \in F$, the set $W = \{\, X \in F^{m \times n} \mid AX = \lambda X \,\}$ is a subspace of $F^{m \times n}$. (This space is related to the *eigenspaces* of $A$ discussed in §6.5.2.)

**10.** For a given $n \times n$ matrix $A$ over $F$, the set $W = \{\, X \in F^{n \times n} \mid XA = AX \,\}$ is a subspace of $F^{n \times n}$. (This is the space of matrices that commute with $A$.)

**11.** Let field $E$ be a vector space over subfield $F$, and let $K$ denote the set of all elements $\alpha \in E$ that satisfy a polynomial equation of the form $f(\alpha) = 0$ for some nonzero $f(x) \in F[x]$. Then $K$ is a subfield of $E$ containing $F$ (the field of *algebraic elements* of $E$ over $F$) and consequently is a subspace of $E$ over $F$. (See §5.6.2.)

**12.** For each fixed $n \geq 1$, the set of all polynomials of degree $\leq n$ is a subspace of $F[x]$. (See §6.1.1, Example 9.)

**13.** In Example 10 of §6.1.1, take $X = [a, b]$ where $a, b \in \mathcal{R}$ with $a < b$, and take $U = \mathcal{R}$ as a vector space over itself. The resulting $V$, the set of all real-valued functions on $[a, b]$, is a vector space. The set $C[a, b]$ of continuous real-valued functions on $[a, b]$ is a subspace of $V$.

**14.** In Example 10 of §6.1.1, take $X = \{1, 2, \ldots, 7\}$ and take $U = \mathcal{Z}_2$ as a vector space over itself. The resulting $V$, the set of all functions from $\{1, 2, \ldots, 7\}$ to $\mathcal{Z}_2$, can be thought of as the vector space of binary 7-tuples $V = \mathcal{Z}_2^7$.

**15.** In Example 10 of §6.1.1, take both $X$ and $U$ to be vector spaces over $F$. Then $V$ is the vector space of all functions from $X$ to $U$. The collection of those $T \in V$ satisfying $T(a\alpha + b\beta) = aT(\alpha) + bT(\beta)$ for all $a, b \in F$ and $\alpha, \beta \in X$ is a subspace of $V$. (This space is the space of *linear transformations* considered in §6.2.)

### 6.1.3  LINEAR COMBINATIONS, INDEPENDENCE, BASIS, AND DIMENSION

**Definitions:**

If $v_1, v_2, \ldots, v_t$ are vectors from a vector space $V$ over $F$, then a vector $w \in V$ is a **linear combination** of $v_1, v_2, \ldots, v_t$ if $w = a_1 v_1 + a_2 v_2 + \cdots + a_t v_t$ for some scalars $a_i \in F$. The zero vector is considered a linear combination of $\emptyset$.

For $S \subseteq V$, the **span** of $S$, denoted $\mathrm{Span}(S)$, is the set of all (finite) linear combinations of members of $S$; that is, $\mathrm{Span}(S)$ consists of all finite sums $a_1 v_1 + a_2 v_2 + \cdots + a_t v_t$ where $v_i \in S$ and $a_i \in F$. (The span of the empty set is taken to be $\{0\}$.) $\mathrm{Span}(S)$ is also called the space **generated** or **spanned** by $S$. (See Fact 1.)

The **row space** $RS(A)$ of an $m \times n$ matrix $A$ over $F$ (§6.3.1) is $\mathrm{Span}(R_1, R_2, \ldots, R_m)$, where $R_1, R_2, \ldots, R_m$ are the rows of $A$ viewed as vectors in $F^{1 \times n}$.

The **column space** $CS(A)$ of $A$ is $\mathrm{Span}(C_1, C_2, \ldots, C_n)$, where $C_1, C_2, \ldots, C_n$ are the columns of $A$.

A subset $S \subseteq V$ is called a **spanning set** for $V$ if $\mathrm{Span}(S) = V$.

A subset $S \subseteq V$ is (**linearly**) **independent** if every finite subset $\{v_1, v_2, \ldots, v_t\}$ of $S$ has the property that the only scalars $a_1, a_2, \ldots, a_t$ satisfying $a_1 v_1 + a_2 v_2 + \cdots + a_t v_t = 0$ are $a_1 = a_2 = \cdots = a_t = 0$.

A subset $S \subseteq V$ is (**linearly**) **dependent** if it is not independent.

A **basis** for $V$ is an independent spanning set.

A vector space $V$ is **finite dimensional** if it has a finite basis; otherwise, $V$ is **infinite dimensional**.

The **dimension** $\dim V$ of a vector space $V$ is the cardinality of any basis for $V$. (See Fact 8.)

If $\mathcal{B} = (v_1, v_2, \ldots, v_n)$ is an ordered basis for $V$, then the **coordinates** of $v$ with respect to $\mathcal{B}$ are the scalars $a_1, a_2, \ldots, a_n$ such that $v = a_1 v_1 + a_2 v_2 + \cdots + a_n v_n$. (See Fact 14.) The **coordinate vector** $[v]_\mathcal{B}$ of $v$ with respect to $\mathcal{B}$ (written as a column) is $[v]_\mathcal{B} = (a_1, a_2, \ldots, a_n)^T$ where $T$ denotes transpose (§6.3.1).

*Note:* Some writers distinguish between the coordinates written as a row and as a column, calling the row $(a_1, a_2, \ldots, a_n)$ the *coordinate vector* of $v$ with respect to $\mathcal{B}$ and the column $(a_1, a_2, \ldots, a_n)^T$ the *coordinate matrix* of $v$ with respect to $\mathcal{B}$.

The **row rank** of a matrix $A$ over $F$ is $\dim RS(A)$, and the **column rank** of $A$ is $\dim CS(A)$. The **rank** of $A$ is the size of the largest square submatrix of $A$ with nonzero determinant (§6.3.4); that is, rank $A = r$ if there exists an $r \times r$ submatrix of $A$ whose determinant is nonzero, and every $t \times t$ submatrix of $A$ with $t > r$ has zero determinant.

The **nullity** of a matrix $A$ is $\dim NS(A)$.

Two vector spaces $V$ and $U$ over the same field $F$ are **isomorphic** if there exists a bijective mapping $T \colon V \to U$ such that $T(v + w) = T(v) + T(w)$ and $T(av) = aT(v)$ for all $v, w \in V$ and $a \in F$. The mapping $T$ is called an **isomorphism**.

**Facts:**

**1.** Span$(S)$ is a subspace of $V$. In particular, $RS(A)$ is a subspace of $F^{1\times n}$ and $CS(A)$ is a subspace of $F^{m\times 1}$.

**2.** Span$(S)$ is the intersection of all subspaces of $V$ that contain $S$; thus, Span$(S)$ is the smallest subspace of $V$ containing $S$ in that it lies inside every subspace of $V$ containing $S$.

**3.** A set $\{v\}$ consisting of a single vector from $V$ is dependent if and only if $v = 0$.

**4.** A set of two or more vectors is dependent if and only if some vector in the set is a linear combination of the remaining vectors in the set.

**5.** Any superset of a dependent set is dependent, and any subset of an independent set is independent. (The empty set is independent.)

**6.** If $V$ has a basis of $n$ elements, then every subset of $V$ with more than $n$ elements is dependent.

**7.** If $W$ is a subspace of $V$ then $\dim W \le \dim V$.

**8.** Every vector space $V$ has a basis, and every two bases for $V$ have the same number of elements (cardinality). For infinite-dimensional vector spaces, this fact relies on the *axiom of choice* (§1.2.4).

**9.** Every independent subset of $V$ can be extended to a basis for $V$. More generally, if $S$ is an independent set, then every maximal independent set containing $S$ is a basis for $V$ containing $S$. For infinite-dimensional vector spaces, this fact relies on the axiom of choice. (An independent set is *maximal* if every set properly containing it is dependent.)

**10.** Every spanning set contains a basis for $V$. More generally, if $S$ is a spanning set, then every minimal spanning subset of $S$ is a basis for $V$. For infinite-dimensional vector spaces, this fact relies on the axiom of choice. (A spanning set is *minimal* if it contains no proper subset that spans $V$.)

**11.** *Rank-nullity theorem*: If $A$ is an $m \times n$ matrix over $F$ then

- $\dim RS(A) + \dim NS(A) = n$;
- $\dim CS(A) + \dim NS(A) = n$;
- $\dim RS(A) + \dim LNS(A) = m$;
- $\dim CS(A) + \dim LNS(A) = m$.

**12.** For every matrix $A$, row rank $A$ = column rank $A$ = rank $A$. Thus, the (maximum) number of independent rows of $A$ equals the (maximum) number of independent columns.

**13.** The set of solutions to the $m$ homogeneous linear equations $\sum_{j=1}^{n} a_{ij}x_j = 0$ in $n$ unknowns has dimension $n - r$, where $r$ is the rank of the $m \times n$ coefficient matrix $A = (a_{ij})$.

**14.** If $\mathcal{B}$ is a basis for a vector space $V$ (finite or infinite), then each $v \in V$ can be expressed as $v = a_1 v_1 + a_2 v_2 + \cdots + a_t v_t$, where $a_i \in F$ and $v_i \in \mathcal{B}$. If $v = b_1 v_1 + b_2 v_2 + \cdots + b_t v_t$ is another expression for $v$ in terms of elements of $\mathcal{B}$ (where possibly some zero coefficients have been inserted to make the two expressions have equal length), then $a_i = b_i$ for $i = 1, 2, \ldots, t$. (If $\mathcal{B}$ is finite, this justifies the definition of the coordinate vector $[v]_{\mathcal{B}}$.)

**15.** If $\mathcal{B} = (v_1, v_2, \ldots, v_n)$ is an ordered basis for $V$, then the function $T\colon V \to F^{n\times 1}$ defined by $T(v) = [v]_{\mathcal{B}}$ is an isomorphism, so $V$ is isomorphic to $F^{n\times 1}$.

**16.** Two vector spaces over $F$ are isomorphic if and only if they have the same dimension.

**Examples:**

**1.** The vector space $F^n$ has dimension $n$. The *standard basis* is the ordered basis $(e_1, e_2, \ldots, e_n)$ where $e_i$ is the vector with 1 in position $i$ and 0s elsewhere. (The spaces $F^n$, $F^{1 \times n}$, and $F^{n \times 1}$ are isomorphic and are often identified and used interchangeably.)

**2.** The vector space $F^{m \times n}$ of $m \times n$ matrices over $F$ has dimension $mn$; the *standard basis* is $\{ E_{ij} \mid 1 \le i \le m,\ 1 \le j \le n \}$ where $E_{ij}$ is the $m \times n$ matrix with a 1 in position $(i, j)$ and 0s elsewhere. It is isomorphic to $F^{mn}$.

**3.** The subspace of $\mathcal{R}^3$ containing all 3-tuples of the form $(a, b, 2a + 3b)$ has dimension 2. One basis for this subspace is $\mathcal{B}_1 = ((1, 0, 2), (0, 1, 3))$ and another is $\mathcal{B}_2 = ((1, 1, 5), (1, -1, -1))$. The vector $w = (5, -1, 7)$ is in the subspace since $w = 5(1, 0, 2) + (-1)(0, 1, 3) = 2(1, 1, 5) + 3(1, -1, -1)$. The coordinate vector of $w$ with respect to $\mathcal{B}_1$ is $(5, -1)^T$ and the coordinate vector of $w$ with respect to $\mathcal{B}_2$ is $(2, 3)^T$.

**4.** If $W$ is the subspace of $V = \mathcal{Z}_2^5$ containing all members of $V$ whose components sum to 0, then $W$ has dimension 4. In fact $W = \{ (a, b, c, d, a + b + c + d) \mid a, b, c, d \in \mathcal{Z}_2 \}$. One ordered basis for this space is $((1, 0, 0, 0, 1), (0, 1, 0, 0, 1), (0, 0, 1, 0, 1), (0, 0, 0, 1, 1))$.

**5.** *Binary codes:* More generally, consider the set of all binary $n$-tuples with an even number of 1s; this is the linear binary code mentioned in Example 6, §6.1.2. These vectors form a subspace $W$ of $V = \mathcal{Z}_2^n$ of dimension $n - 1$. A basis for $W$ consists of the following $n - 1$ vectors, each of which has exactly two 1s: $(1, 0, \ldots, 1), (0, 1, \ldots, 1), \ldots, (0, 0, \ldots, 1, 1)$. Consequently there are $2^{n-1}$ vectors in the code $W$.

**6.** The field $\mathcal{C}$ of complex numbers is two-dimensional as a vector space over $\mathcal{R}$; it has the ordered basis $(1, i)$, where $i = \sqrt{-1}$. Any two complex numbers, neither of which is a real multiple of the other, form a basis.

**7.** Both $\mathcal{C}$ and $\mathcal{R}$ are infinite-dimensional vector spaces over the rational field $\mathcal{Q}$.

**8.** The vector space $F[x]$ is an infinite-dimensional space over $F$; $(1, x, x^2, x^3, \ldots)$ is an ordered basis. The subspace of all polynomials of degree $\le n$ has dimension $n + 1$; $(1, x, x^2, \ldots, x^n)$ is an ordered basis.

---

## 6.1.4   INNER PRODUCTS, LENGTH, AND ORTHOGONALITY

By imposing additional structure on real and complex vector spaces, the concepts of length, distance, and orthogonality can be introduced. These concepts are motivated by the corresponding geometric notions for physical vectors. Also, for real vector spaces the geometric idea of angle can be formulated analytically.

**Definitions:**

An **inner product** on a vector space $V$ over $\mathcal{R}$ is a function $\langle \cdot, \cdot \rangle \colon V \times V \to \mathcal{R}$ such that for all $u, v, w \in V$ and $a, b \in \mathcal{R}$ the following hold:

- $\langle u, v \rangle = \langle v, u \rangle$;
- $\langle u, u \rangle \ge 0$ with equality if and only if $u = 0$;
- $\langle au + bv, w \rangle = a \langle u, w \rangle + b \langle v, w \rangle$.

An **inner product** on a vector space $V$ over $\mathcal{C}$ is a function $\langle \cdot, \cdot \rangle \colon V \times V \to \mathcal{C}$ such that for all $u, v, w \in V$ and $a, b \in \mathcal{C}$ the following hold:

- $\langle u, v \rangle = \overline{\langle v, u \rangle}$ (where bar denotes complex conjugation);
- $\langle u, u \rangle \ge 0$ with equality if and only if $u = 0$;

- $\langle au + bv, w \rangle = a \langle u, w \rangle + b \langle v, w \rangle$.

*Note*: The first property implies that $\langle u, u \rangle$ is real, so the second property makes sense.

An **inner product space** is a vector space over $\mathcal{R}$ or $\mathcal{C}$ on which an inner product is defined. Such a space is called a real or complex inner product space, depending on its scalar field.

The **norm** (**length**) of a vector $v \in V$ is $\|v\| = \sqrt{\langle v, v \rangle}$.

A vector $v \in V$ is a **unit vector** if and only if $\|v\| = 1$.

The **distance** $d(v, w)$ from $v$ to $w$ is $d(v, w) = \|v - w\|$.

In a real inner product space, the **angle** between nonzero vectors $v$ and $w$ is the real number $\theta$, $0 \le \theta \le \pi$, such that $\cos \theta = \dfrac{\langle v, w \rangle}{\|v\| \cdot \|w\|}$.

Two vectors $v$ and $w$ are **orthogonal** if and only if $\langle v, w \rangle = 0$.

A subset $S \subseteq V$ is an **orthogonal set** if $\langle v, w \rangle = 0$ for all $v, w \in S$ with $v \ne w$.

A subset $S \subseteq V$ is an **orthonormal set** if $S$ is an orthogonal set and $\|v\| = 1$ for all $v \in S$.

If $W$ is a subspace of an inner product space $V$, then the **orthogonal complement** $W^\perp = \{ v \in V \mid \langle v, w \rangle = 0 \text{ for all } w \in W \}$.

**Facts:**

**1.** *Standard inner product on* $\mathcal{R}^n$: The real-valued function defined by $\langle x, y \rangle = x_1 y_1 + x_2 y_2 + \cdots + x_n y_n$ is an inner product on $V = \mathcal{R}^n$.

**2.** *Standard inner product on* $\mathcal{C}^n$: The complex-valued function defined by $\langle x, y \rangle = x_1 \bar{y}_1 + x_2 \bar{y}_2 + \cdots + x_n \bar{y}_n$ is an inner product on $V = \mathcal{C}^n$.

**3.** If $A$ is an $n \times n$ real positive definite matrix (§6.3.2), then the function defined by $\langle x, y \rangle = x^T A y$ is an inner product on $\mathcal{R}^n$. (Here $x^T$ denotes the transpose of $x$.)

**4.** If $H$ is an $n \times n$ complex positive definite matrix (§6.3.2), then the function defined by $\langle x, y \rangle = y^* H x$ is an inner product on $\mathcal{C}^n$. ($y^*$ is the conjugate-transpose of $y$.)

**5.** The function $\langle f, g \rangle = \int_a^b f(x) g(x) \, dx$ is an inner product on the vector space $C[a, b]$ of continuous real-valued functions on the interval $[a, b]$.

**6.** The inner product $\langle \cdot, \cdot \rangle$ on an inner product space $V$ is an inner product on any subspace $W$ of $V$.

**7.** If $W$ is a subspace of an inner product space $V$, then the orthogonal complement $W^\perp$ is a subspace of $V$ and $V = W \oplus W^\perp$.

**8.** The norm function satisfies the following properties for all scalars $a$ and all vectors $v, w \in V$:

- $\|v\| \ge 0$ with equality if and only if $v = 0$;
- $\|av\| = |a| \cdot \|v\|$, where $|a|$ denotes the absolute value of $a$;
- $|\langle v, w \rangle| \le \|v\| \cdot \|w\|$  (*Cauchy-Schwarz inequality*);
- $\|v + w\| \le \|v\| + \|w\|$  (*triangle inequality*);
- if $v \ne 0$, then $\dfrac{1}{\|v\|} v$ is a unit vector (the *normalization* of $v$).

**9.** The distance function on a vector space $V$ satisfies the following properties for all $v, w, z \in V$:

- $d(v, w) \geq 0$ with equality if and only if $v = w$;
- $d(v, w) = d(w, v)$;
- $d(v, z) \leq d(v, w) + d(w, z)$  (*triangle inequality*).

**10.** For real inner product spaces, two nonzero vectors are orthogonal if and only if the angle between them is $\theta = \frac{\pi}{2}$.

**11.** An orthogonal set $S$ of nonzero vectors can be converted to an orthonormal set by normalizing each vector in $S$.

**12.** An orthogonal set of nonzero vectors is independent. An orthonormal set is independent.

**13.** If $V$ is an $n$-dimensional inner product space, any orthonormal set contains at most $n$ vectors, and any orthonormal set of $n$ vectors is a basis for $V$.

**14.** Every subspace $W$ of an $n$-dimensional space $V$ has an orthonormal (orthogonal) basis.

**15.** *Gram-Schmidt orthogonalization:*   From any ordered basis $(w_1, w_2, \ldots, w_m)$ for a subspace $W$, an orthonormal basis $(u_1, u_2, \ldots, u_m)$ for $W$ can be constructed using Algorithm 1. (Jörgen Gram, 1850–1916; Erhardt Schmidt, 1876–1959)

---

**Algorithm 1**:   **Gram-Schmidt orthogonalization process.**

input: an ordered basis $(w_1, w_2, \ldots, w_m)$

output: an orthonormal basis $(u_1, u_2, \ldots, u_m)$

$u_1 := \dfrac{1}{a_1} w_1$, where $a_1 := \|w_1\|$

**for** $j := 2$ **to** $m$

$$a_j := \left\| w_j - \sum_{i=1}^{j-1} \langle w_j, u_i \rangle u_i \right\|$$

$$u_j := \frac{1}{a_j} \left( w_j - \sum_{i=1}^{j-1} \langle w_j, u_i \rangle u_i \right)$$

---

**16.** The standard basis is orthonormal with respect to the standard inner product.

**17.** If $(u_1, u_2, \ldots, u_m)$ is an orthonormal basis for a subspace $W$ of $V$ and $w \in W$, then $w = \langle w, u_1 \rangle u_1 + \langle w, u_2 \rangle u_2 + \cdots + \langle w, u_m \rangle u_m$.

**18.** *Projection vector:*   Let $W$ be a subspace of a vector space $V$ and let $v$ be a vector in $V$.

- There is a unique vector $p \in W$ nearest to $v$; that is, the vector $p$ minimizes $\|v - w\|$ over all $w \in W$. This vector $p$ is called the *projection* of $v$ onto $W$, written $p = \text{proj}_W(v)$.

- If $(u_1, u_2, \ldots, u_m)$ is any orthonormal basis for $W$, then the projection of $v$ onto $W$ is given by $\text{proj}_W(v) = \langle v, u_1 \rangle u_1 + \langle v, u_2 \rangle u_2 + \cdots + \langle v, u_m \rangle u_m$.

- The vector $\text{proj}_W(v)$ is the unique vector $w \in W$ such that $v - w$ is orthogonal to every vector in $W$.

**19.** *Projection matrix:* If $V = \mathcal{R}^n$ is equipped with the standard inner product and $(u_1, u_2, \ldots, u_m)$ is an orthonormal basis for a subspace $W$, then the projection of each $x \in \mathcal{R}^n$ onto $W$ is given by $\text{proj}_W(x) = Ax$, where $A = GG^T$ with $G = (u_1, u_2, \ldots, u_m)$ being the $n \times m$ matrix with the $u_i$ as columns.

**20.** The projection matrix $A$ is symmetric and satisfies $A^2 = A$.

**Examples:**

Consider the vector space $\mathcal{R}^4$ with the standard inner product $\langle x, y \rangle = x^T y$, and let $W$ be the subspace spanned by the three vectors $w_1 = (1,1,1,1)^T$, $w_2 = (3,1,3,1)^T$, $w_3 = (3,1,1,1)^T$.

1. $\langle w_1, w_2 \rangle = 8$ and $\|w_1\| = 2$.

2. The angle $\theta$ between $w_1$ and $w_2$ satisfies $\cos\theta = \frac{8}{2\sqrt{20}} = \frac{2}{\sqrt{5}}$ (so $\theta \approx 0.4636$ radians).

3. The distance from $w_1$ to $w_2$ is $d(w_1, w_2) = \|w_1 - w_2\| = \|(-2,0,-2,0)^T\| = 2\sqrt{2}$.

4. The orthogonal complement $W^\perp$ of $W$ is the set of vectors of the form $(0, a, 0, -a)$.

5. The Gram-Schmidt process applied to $(w_1, w_2, w_3)$ yields

$$u_1 = \tfrac{1}{a_1} w_1 = (\tfrac{1}{2}, \tfrac{1}{2}, \tfrac{1}{2}, \tfrac{1}{2})^T, \text{where } a_1 = \|w_1\| = 2;$$

$$u_2 = \tfrac{1}{a_2}(w_2 - \langle w_2, u_1 \rangle u_1) = \tfrac{1}{a_2}((3,1,3,1)^T - 4(\tfrac{1}{2}, \tfrac{1}{2}, \tfrac{1}{2}, \tfrac{1}{2})^T)$$
$$= \tfrac{1}{a_2}(1,-1,1,-1)^T = (\tfrac{1}{2}, -\tfrac{1}{2}, \tfrac{1}{2}, -\tfrac{1}{2})^T, \text{where } a_2 = \|(1,-1,1,-1)^T\| = 2;$$

$$u_3 = \tfrac{1}{a_3}(w_3 - \langle w_3, u_1 \rangle u_1 - \langle w_3, u_2 \rangle u_2)$$
$$= \tfrac{1}{a_3}\left((3,1,1,1)^T - 3(\tfrac{1}{2}, \tfrac{1}{2}, \tfrac{1}{2}, \tfrac{1}{2})^T - 1(\tfrac{1}{2}, -\tfrac{1}{2}, \tfrac{1}{2}, -\tfrac{1}{2})^T\right)$$
$$= \tfrac{1}{a_3}(1,0,-1,0)^T = (\tfrac{1}{\sqrt{2}}, 0, -\tfrac{1}{\sqrt{2}}, 0)^T, \text{where } a_3 = \|(1,0,-1,0)^T\| = \sqrt{2}.$$

6. The vector in $W$ that is nearest to $v = (3,6,3,4)^T$ is $p = \text{proj}_W(v) = \langle v, u_1 \rangle u_1 + \langle v, u_2 \rangle u_2 + \langle v, u_3 \rangle u_3 = 8u_1 + (-2)u_2 + 0u_3 = (3,5,3,5)^T$. Further, $v - p = (0,1,0,-1)^T$ is orthogonal to every vector in $W$; if $u_4 = (0, \tfrac{1}{\sqrt{2}}, 0, -\tfrac{1}{\sqrt{2}})^T$ is the normalization of $v - p$ then $(u_1, u_2, u_3, u_4)$ is an orthonormal basis for $\mathcal{R}^4$.

7. The projection of any $x \in \mathcal{R}^4$ onto $W$ is given by $\text{proj}_W(x) = Ax$, where

$$A = GG^T = (u_1, u_2, u_3)(u_1, u_2, u_3)^T = \begin{pmatrix} 1 & 0 & 0 & 0 \\ 0 & \tfrac{1}{2} & 0 & \tfrac{1}{2} \\ 0 & 0 & 1 & 0 \\ 0 & \tfrac{1}{2} & 0 & \tfrac{1}{2} \end{pmatrix}.$$

Thus, if $x = (3,6,3,4)^T$, its projection onto $W$ is computed as $Ax = (3,5,3,5)^T$, consistent with the answer found in Example 6.

## 6.2  LINEAR TRANSFORMATIONS

Linear transformations are special types of functions that map one vector space to another. They are called "linear" because of their effect on the lines of a vector space, where by a "line" is meant a set of vectors $w$ of the form $w = au + v$ where $u \neq 0$ and $v$ are fixed vectors in the space and $a$ varies over all values in the scalar field. Linear transformations carry lines in one vector space to lines or points in the other.

## 6.2.1   LINEAR TRANSFORMATIONS, RANGE, AND KERNEL

### Definitions:

Let $V$ and $W$ be vector spaces over the same field $F$. A **linear transformation** is a function $T: V \to W$ satisfying $T(au + v) = aT(u) + T(v)$ for all $u, v \in V$ and $a \in F$.

The **range** $R_T$ of a linear transformation $T$ is $R_T = \{\, T(v) \mid v \in V \,\}$.

The **kernel** $\ker T$ of a linear transformation $T$ is $\ker T = \{\, v \in V \mid T(v) = 0 \,\}$.

The **rank** of $T$ is the dimension of $R_T$. ($R_T$ is a subspace of $W$ by Fact 5.)

The **nullity** of $T$ is the dimension of $\ker T$. ($\ker T$ is a subspace of $V$ by Fact 5.)

A **linear operator** on $V$ is a linear transformation from $V$ to $V$.

### Facts:

**1.** For any vector spaces $V$ and $W$ over $F$, the zero function $Z: V \to W$ defined by $Z(v) = 0$ for all $v \in V$ is a linear transformation from $V$ to $W$.

**2.** For any vector space $V$ over $F$, the identity function $I: V \to V$ defined by $I(v) = v$ for all $v \in V$ is a linear operator on $V$.

**3.** The following four statements are equivalent for a function $T: V \to W$:

- $T$ is a linear transformation;
- $T(u + v) = T(u) + T(v)$ and $T(au) = aT(u)$ for all $u, v \in V$ and $a \in F$;
- $T(au + bv) = aT(u) + bT(v)$ for all $u, v \in V$ and $a, b \in F$;
- $T(\sum_{i=1}^{t} a_i v_i) = \sum_{i=1}^{t} a_i T(v_i)$ for all finite subsets $\{v_1, v_2, \ldots, v_t\} \subseteq V$ and scalars $a_i \in F$.

**4.** If $T: V \to W$ is a linear transformation, then

- $T(0) = 0$;
- $T(-v) = -T(v)$ for all $v \in V$;
- $T(u - v) = T(u) - T(v)$ for all $u, v \in V$.

**5.** If $T: V \to W$ is a linear transformation, then $R_T$ is a subspace of $W$ and $\ker T$ is a subspace of $V$.

**6.** If $T: V \to W$ is a linear transformation, then the rank of $T$ plus the nullity of $T$ equals the dimension of its domain: $\dim R_T + \dim(\ker T) = \dim V$.

**7.** If $T: V \to W$ is a linear transformation and if the vectors $\{v_1, v_2, \ldots, v_n\}$ span $V$, then $\{T(v_1), T(v_2), \ldots, T(v_n)\}$ span $R_T$.

**8.** If $T: V \to W$ is a linear transformation, then $T$ is completely determined by its action on a basis for $V$. That is, if $\mathcal{B}$ is a basis for $V$ and $f$ is any function from $\mathcal{B}$ to $W$, then there exists a unique linear transformation $T$ such that $T(v) = f(v)$ for all $v \in \mathcal{B}$.

**9.** A linear transformation $T: V \to W$ is one-to-one if and only if $\ker T = \{0\}$.

**10.** A linear transformation $T: V \to W$ is onto if and only if for *every* basis $\mathcal{B}$ of $V$, the set $\{\, T(v) \mid v \in \mathcal{B} \,\}$ spans $W$.

**11.** A linear transformation $T: V \to W$ is onto if and only if for *some* basis $\mathcal{B}$ of $V$, the set $\{T(v) \mid v \in \mathcal{B}\}$ spans $W$.

**12.** If $T: V \to W$ is a bijective linear transformation, then its inverse $T^{-1}: W \to V$ is also a bijective linear transformation.

**13.** For each fixed $m \times n$ matrix $A$ over $F$, the function $T: F^{n \times 1} \to F^{m \times 1}$ defined by $T(x) = Ax$ is a linear transformation.

**14.** Every linear transformation $T: F^{n \times 1} \to F^{m \times 1}$ has the form $T(x) = Ax$ for some unique $m \times n$ matrix $A$ over $F$.

**15.** The range $R_T$ of the linear transformation $T(x) = Ax$ is equal to the column space of $A$, and $\ker T$ is equal to the null space of $A$. (See §6.1.2, §6.1.3.)

**16.** If $T$ is a linear transformation from $V$ to $W$ and if $T(v_0) = w_0 \in R_T$, then the solution set $S$ to the equation $T(v) = w_0$ is $S = \{v_0 + u \mid u \in \ker T\}$.

**Examples:**

**1.** The function $T: \mathcal{R}^{2 \times 1} \to \mathcal{R}^{2 \times 1}$ given by $T\begin{pmatrix} x_1 \\ x_2 \end{pmatrix} = \begin{pmatrix} x_1 - 3x_2 \\ -2x_1 + 6x_2 \end{pmatrix}$ is a linear trans-

formation. It has the form $T(x) = Ax$, where $A = \begin{pmatrix} 1 & -3 \\ -2 & 6 \end{pmatrix}$. The kernel of $T$ is

$\{(3a, a)^T \mid a \in \mathcal{R}\}$ and the range of $T$ is $\{(b, -2b)^T \mid b \in \mathcal{R}\}$.

**2.** For each fixed matrix $A \in F^{n \times n}$, the function $T: F^{n \times n} \to F^{n \times n}$ defined by $T(X) = AX - XA$ is a linear transformation whose kernel is the set of matrices commuting

with $A$. Specifically, let $n = 2$, $F = \mathcal{R}$, and $A = \begin{pmatrix} 1 & -3 \\ -2 & 6 \end{pmatrix}$. Then $\dim \mathcal{R}^{2 \times 2} =$

$4$, and by computation $T\left[\begin{pmatrix} 1 & 0 \\ 0 & 0 \end{pmatrix}\right] = \begin{pmatrix} 0 & 3 \\ -2 & 0 \end{pmatrix}$, $T\left[\begin{pmatrix} 0 & 1 \\ 0 & 0 \end{pmatrix}\right] = \begin{pmatrix} 2 & -5 \\ 0 & -2 \end{pmatrix}$. Thus,

$\dim R_T \geq 2$. Since both the identity matrix $I$ and $A$ itself are in $\ker T$, $\dim(\ker T) \geq 2$. By Fact 6, it follows that $\dim R_T = 2$ and $\dim(\ker T) = 2$. Therefore $(I, A)$ forms a

basis for $\ker T$, and the matrices $\begin{pmatrix} 0 & 3 \\ -2 & 0 \end{pmatrix}$ and $\begin{pmatrix} 2 & -5 \\ 0 & -2 \end{pmatrix}$ are a basis for $R_T$. From

Fact 16, the solutions to $T(x) = \begin{pmatrix} 0 & 3 \\ -2 & 0 \end{pmatrix}$ are precisely the set of matrices of the form

$\begin{pmatrix} 1 & 0 \\ 0 & 0 \end{pmatrix} + a \begin{pmatrix} 1 & 0 \\ 0 & 1 \end{pmatrix} + b \begin{pmatrix} 1 & -3 \\ -2 & 6 \end{pmatrix}$ with $a, b \in \mathcal{R}$.

**3.** The function $E(x_1, x_2, x_3, x_4) = (x_1, x_2, x_3, x_4, x_1 + x_3 + x_4, x_1 + x_2 + x_4, x_1 + x_2 + x_3)$, where $x_i \in \mathcal{Z}_2$, is a linear transformation important in coding theory. It represents an encoding of 4-bit binary vectors into 7-bit binary vectors ("codewords") before being sent over a noisy channel (§14.2). The kernel of the transformation consists of only the zero vector $0 = (0, 0, 0, 0)$, and so the transformation is one-to-one. The collection of codewords (that is, the range of $E$), is a 16-member, 4-dimensional subspace of $\mathcal{Z}_2^7$ having the special property that any two of its distinct members differ in at least three components. This means that if, during transmission of a codeword, an error is made in any single one of its components, then the error can be detected and corrected as there will be a unique codeword that differs from the received vector in a single component.

**4.** Continuing with Example 3, the linear transformation $D(z_1, z_2, z_3, z_4, z_5, z_6, z_7) = (z_1 + z_3 + z_4 + z_5, z_1 + z_2 + z_4 + z_6, z_1 + z_2 + z_3 + z_7)$ is used in decoding the (binary)

received vector $z$. This transformation has the special property that its kernel is precisely the set of codewords defined in Example 3. Thus, if $D(z) \neq 0$, then a transmission error has been made.

**5.** For $\mathcal{C}$ as a vector space over $\mathcal{R}$ and any $z_0 \in \mathcal{C}$, the function $T: \mathcal{C} \to \mathcal{C}$ defined by $T(z) = z_0 z$ is a linear operator; in particular, if $z_0 = \cos\theta + i\sin\theta$, then $T$ is a *rotation* by the angle $\theta$. ($T(z)$ is also a linear operator on $\mathcal{C}$ as a vector space over itself.)

**6.** For any fixed real-valued continuous function $g$ on the interval $[a, b]$, the function $T$ from the space $C[a, b]$ of continuous functions on $[a, b]$ to the space $D[a, b]$ of continuously differentiable functions on $[a, b]$ given by $T(f)(x) = \int_a^x g(t)f(t)dt$ is a linear transformation.

**7.** For the vector space $V$ of functions $p: \mathcal{R} \to \mathcal{R}$ with continuous derivatives of all orders, the mapping $T: V \to V$ defined by $T(p) = p'' - 3p' + 2p$ (where $p'$ and $p''$ are the first and second derivatives of $p$) is a linear transformation. Its kernel is the solution set to the homogeneous differential equation $p'' - 3p' + 2p = 0$: namely, $p(x) = Ae^x + Be^{2x}$, where $A, B \in \mathcal{R}$. Since $T(x^2) = 2 - 6x + 2x^2$, the set of all solutions to $T(p) = 2 - 6x + 2x^2$ is $x^2 + Ae^x + Be^{2x}$ (by Fact 16).

**8.** If $v_0$ is a fixed vector in a real inner product space $V$, then $T: V \to \mathcal{R}$ given by $T(v) = \langle v, v_0 \rangle$ is a linear transformation.

**9.** For $W$ a subspace of the inner product space $V$, the projection $\text{proj}_W$ of $V$ onto $W$ is a linear transformation. (See §6.1.4.)

---

## 6.2.2   VECTOR SPACES OF LINEAR TRANSFORMATIONS

### Definitions:

If $S$ and $T$ are linear transformations from $V$ to $W$, the **sum** (**addition**) of $S$ and $T$ is the function $S + T$ defined by $(S + T)(v) = S(v) + T(v)$ for all $v \in V$.

If $T$ is a linear transformation from $V$ to $W$, the **scalar product** (**scalar multiplication**) of $a \in F$ by $T$ is the function $aT$ defined by $(aT)(v) = aT(v)$ for all $v \in V$.

If $T: V \to W$ and $S: W \to U$ are linear transformations, then the **product** (**multiplication**, **composition**) of $S$ and $T$ is the function $S \circ T$ defined by $(S \circ T)(v) = S(T(v))$.

*Note:* Some writers use the notation $vT$ to denote the image of $v$ under the transformation $T$, in which case $T \circ S$ is used instead of $S \circ T$ to denote the product; that is, $v(T \circ S) = (vT)S$.

### Facts:

**1.** The sum of two linear transformations from $V$ to $W$ is a linear transformation from $V$ to $W$.

**2.** The product of a scalar and a linear transformation is a linear transformation.

**3.** If $T: V \to W$ and $S: W \to U$ are linear transformations, then their product $S \circ T$ is a linear transformation from $V$ to $U$.

**4.** The set of linear transformations from $V$ to $W$ with the operations of addition and scalar multiplication forms a vector space over $F$. This vector space is denoted $L(V, W)$.

**5.** The set $L(V, V)$ of linear operators on $V$ with the operations of addition, scalar multiplication, and multiplication forms an *algebra* with identity over $F$. Namely, $L(V, V)$ is a vector space over $F$ and is a ring with identity under the addition and multiplication operations. In addition, $a(S \circ T) = (aS) \circ T = S \circ (aT)$ holds for all scalars $a \in F$ and all $S, T \in L(V, V)$. The identity mapping is the multiplicative identity of the algebra.

**6.** If $\dim V = n$ and $\dim W = m$, then $\dim L(V, W) = nm$.

### Examples:

**1.** Consider $L(F^{n \times 1}, F^{m \times 1})$. If $T$ and $S$ are in $L(F^{n \times 1}, F^{m \times 1})$, then $T(x) = Ax$ and $S(x) = Bx$ for unique $m \times n$ matrices $A$ and $B$ over $F$. Then $(T + S)(x) = (A + B)x$, $(aT)(x) = aAx$, and in case $m = n$, $(T \circ S)(x) = ABx$.

**2.** Let $V = C[a, b]$ be the space of real-valued continuous functions on the interval $[a, b]$, and let $T$ and $S$ be linear operators defined by $T(f)(x) = \int_a^x e^{-t} f(t) \, dt$ and $S(f)(x) = \int_a^x e^t f(t) \, dt$. Then $(T + S)(f)(x) = \int_a^x (e^{-t} + e^t) f(t) \, dt$, $(cT)(f)(x) = \int_a^x ce^{-t} f(t) \, dt$, and $(T \circ S)(f)(x) = \int_a^x \int_a^t e^{s-t} f(s) \, ds \, dt$.

**3.** Let $V$ be the real vector space of all functions $p: \mathcal{R} \to \mathcal{R}$ with continuous derivatives of all orders, and let $D$ be the derivative function. Then $D: V \to V$ is a linear operator on $V$ and so is a function such as $T = D^2 - 3D + 2I$ where $D^2 = D \circ D$ and $I$ is the identity operator on $V$. The action of $T$ on $p \in V$ is given by $T(p) = p'' - 3p' + 2p$.

---

## 6.2.3   MATRICES OF LINEAR TRANSFORMATIONS

### Definitions:

If $T: V \to W$ is a linear transformation where $\dim V = n$, $\dim W = m$, and if $\mathcal{B} = (v_1, v_2, \ldots, v_n)$ and $\mathcal{B}' = (v_1', v_2', \ldots, v_m')$ are ordered bases for $V$ and $W$, respectively, then the **matrix of $T$ with respect to $\mathcal{B}$ and $\mathcal{B}'$** is the $m \times n$ matrix $[T]_{\mathcal{B}, \mathcal{B}'}$ whose $j$th column is $[T(v_j)]_{\mathcal{B}'}$, the coordinate vector (§6.1.3) of $T(v_j)$ with respect to $\mathcal{B}'$.

If $T: V \to V$ is a linear operator on $V$, then the **matrix of $T$ with respect to $\mathcal{B}$** is the $n \times n$ matrix $[T]_{\mathcal{B}, \mathcal{B}}$ denoted simply as $[T]_{\mathcal{B}}$.

### Facts:

Assume that $T$ and $S$ are linear transformations from $V$ to $W$, $\mathcal{B}$ and $\mathcal{B}'$ are respective bases for $V$ and $W$, and $A$ and $B$ are the matrices defined by $A = [T]_{\mathcal{B}, \mathcal{B}'}$ and $B = [S]_{\mathcal{B}, \mathcal{B}'}$.

**1.** $[T(v)]_{\mathcal{B}'} = [T]_{\mathcal{B}, \mathcal{B}'} [v]_{\mathcal{B}}$ for all $v \in V$; that is, if $y = [T(v)]_{\mathcal{B}'}$ and $x = [v]_{\mathcal{B}}$, then $y = Ax$.

**2.** $\ker T = \{ x_1 v_1 + x_2 v_2 + \cdots + x_n v_n \mid (x_1, x_2, \ldots, x_n)^T \in NS(A) \}$, where $\mathcal{B} = (v_1, v_2, \ldots, v_n)$.

**3.** $T$ is one-to-one if and only if $NS(A) = \{0\}$.

**4.** $R_T = \{ y_1 v_1' + y_2 v_2' + \cdots + y_m v_m' \mid (y_1, y_2, \ldots, y_m)^T \in CS(A) \}$, where $\mathcal{B}' = (v_1', v_2', \ldots, v_m')$.

**5.** $T$ is onto if and only if $CS(A) = F^{m \times 1}$.

**6.** $T$ is bijective if and only if $m = n$ and $A$ is invertible. In this case, $[T^{-1}]_{\mathcal{B}', \mathcal{B}} = A^{-1}$.

**7.** $[T + S]_{\mathcal{B}, \mathcal{B}'} = A + B$, $[aT]_{\mathcal{B}, \mathcal{B}'} = aA$ for all $a \in F$, and the mapping $f$ from $L(V, W)$ to $F^{m \times n}$ defined by $f(T) = [T]_{\mathcal{B}, \mathcal{B}'}$ is an isomorphism.

**8.** If $U$ is a vector space over $F$, $\mathcal{B}''$ is a basis for $U$, and $R: W \to U$ is a linear transformation, then $[R \circ T]_{\mathcal{B}, \mathcal{B}''} = CA$ where $C = [R]_{\mathcal{B}', \mathcal{B}''}$; that is, $[R \circ T]_{\mathcal{B}, \mathcal{B}''} = [R]_{\mathcal{B}', \mathcal{B}''} [T]_{\mathcal{B}, \mathcal{B}'}$.

**9.** The algebra $L(V, V)$ is isomorphic to the matrix algebra $F^{n \times n}$.

**10.** If $I: V \to V$ is the identity mapping, then $[I]_{\mathcal{B},\mathcal{B}} = [I]_{\mathcal{B}}$ equals the identity matrix for any basis $\mathcal{B}$.

**11.** If $A$ is an $m \times n$ matrix over $F$ with $\mathcal{B}$ and $\mathcal{B}'$ being arbitrary bases for $V$ and $W$, respectively, then there exists a unique linear transformation $T: V \to W$ such that $A = [T]_{\mathcal{B},\mathcal{B}'}$.

**12.** Linear transformations are used extensively in computer graphics. (See Example 5.) Further information can be found in [PoGe89].

### Examples:

**1.** Consider $T: \mathcal{R}^{2 \times 1} \to \mathcal{R}^{2 \times 1}$ given by $T\begin{pmatrix} x_1 \\ x_2 \end{pmatrix} = \begin{pmatrix} x_1 - 3x_2 \\ -2x_1 + 6x_2 \end{pmatrix}$ and the bases $\mathcal{B} = (v_1, v_2)$ and $\mathcal{B}' = (v_1', v_2')$, where $v_1 = (1,0)^T$, $v_2 = (0,1)^T$ and $v_1' = (1,1)^T$, $v_2' = (2,1)^T$. Since

$$T(v_1) = (1,-2)^T = (-5)v_1' + 3v_2',$$
$$T(v_2) = (-3,6)^T = 15v_1' + (-9)v_2',$$

it follows that $[T(v_1)]_{\mathcal{B}'} = (-5,3)^T$ and $[T(v_2)]_{\mathcal{B}'} = (15,-9)^T$; hence, the matrix of $T$ relative to $\mathcal{B}$ and $\mathcal{B}'$ is $[T]_{\mathcal{B},\mathcal{B}'} = \begin{pmatrix} -5 & 15 \\ 3 & -9 \end{pmatrix}$. Similarly, $[T]_{\mathcal{B},\mathcal{B}} = [T]_{\mathcal{B}} = \begin{pmatrix} 1 & -3 \\ -2 & 6 \end{pmatrix}$, and $[T]_{\mathcal{B}',\mathcal{B}'} = [T]_{\mathcal{B}'} = \begin{pmatrix} 10 & 5 \\ -6 & -3 \end{pmatrix}$.

**2.** Consider $T$ of Example 1 where $A = [T]_{\mathcal{B},\mathcal{B}'} = \begin{pmatrix} -5 & 15 \\ 3 & -9 \end{pmatrix}$. Since $NS(A) = \{(3a, a)^T \mid a \in \mathcal{R}\}$ and $CS(A) = \{(-5b, 3b)^T \mid b \in \mathcal{R}\}$, Fact 2 gives $\ker T = \{3av_1 + av_2 = (3a, a)^T \mid a \in \mathcal{R}\}$ and Fact 4 gives $R_T = \{(-5b)v_1' + 3bv_2' = (b, -2b)^T \mid b \in \mathcal{R}\}$. $T$ is not one-to-one since $NS(A) \neq \{0\}$ and is not onto since $CS(A) \neq \mathcal{R}^{2 \times 1}$. (Any one of the three matrices found in Example 1 could have been used to determine $\ker T$ and $R_T$ and to reach these same conclusions.)

**3.** Consider the linear operator on $\mathcal{R}^{2 \times 2}$ defined by $T(X) = AX - XA$ where $A = \begin{pmatrix} 1 & -3 \\ -2 & 6 \end{pmatrix}$, and let $\mathcal{B} = (E_{11}, E_{12}, E_{21}, E_{22})$ be the standard basis. (Here, $E_{ij}$ has a 1 in position $(i,j)$ and 0s elsewhere.) Then

$$T(E_{11}) = AE_{11} - E_{11}A = \begin{pmatrix} 0 & 3 \\ -2 & 0 \end{pmatrix} = 0E_{11} + 3E_{12} + (-2)E_{21} + 0E_{22},$$

so $(0, 3, -2, 0)^T$ is the first column of $[T]_{\mathcal{B}}$. Similar calculations yield

$$[T]_{\mathcal{B}} = \begin{pmatrix} 0 & 2 & -3 & 0 \\ 3 & -5 & 0 & -3 \\ -2 & 0 & 5 & 2 \\ 0 & -2 & 3 & 0 \end{pmatrix}.$$

The null space of this $4 \times 4$ matrix is $\{(5a + b, 3a, 2a, b)^T \mid a, b \in \mathcal{R}\}$, so that those matrices $X$ commuting with $A$ (that is, in $\ker T$) have the form $X = \begin{pmatrix} 5a + b & 3a \\ 2a & b \end{pmatrix}$.

**4.** Consider $\mathcal{C}$ as a vector space over $\mathcal{R}$ and the rotation operator in Example 5 of §6.2.1; namely, $T(z) = z_0 z$ where $z_0 = \cos\theta + i\sin\theta$. If $\mathcal{B}$ is the standard basis, $\mathcal{B} = (1, i)$, then the matrix of $T$ relative to $\mathcal{B}$ is $[T]_{\mathcal{B}} = \begin{pmatrix} \cos\theta & -\sin\theta \\ \sin\theta & \cos\theta \end{pmatrix}$.

**5.** *Computer graphics*: The polygon in part (a) of the following figure can be rotated by applying the transformation $T$ in Example 4 to its vertices $(-2, -2)$, $(1, -1)$, $(2, 1)$, $(-1, 3)$. The matrix of vertex coordinates is

$$X = \begin{pmatrix} -2 & 1 & 2 & -1 \\ -2 & -1 & 1 & 3 \end{pmatrix}.$$

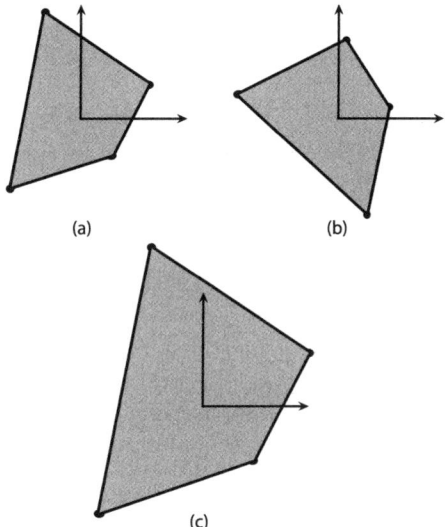

(a)                    (b)

(c)

For a rotation of $\frac{\pi}{3}$, the matrix of $T$ is

$$A = \begin{pmatrix} \frac{1}{2} & -\frac{\sqrt{3}}{2} \\ \frac{\sqrt{3}}{2} & \frac{1}{2} \end{pmatrix}$$

and

$$AX \approx \begin{pmatrix} 0.732 & 1.366 & 0.134 & -3.098 \\ -2.732 & 0.366 & 2.232 & 0.634 \end{pmatrix},$$

producing the rotated polygon shown in part (b) of the figure. To perform a "zoom in" operation, the original polygon can be rescaled by 50% by applying the transformation $S\begin{pmatrix} x \\ y \end{pmatrix} = \begin{pmatrix} 1.5x \\ 1.5y \end{pmatrix}$. As the matrix for $S$ relative to the standard basis is $D = \begin{pmatrix} 1.5 & 0 \\ 0 & 1.5 \end{pmatrix}$, the vertex coordinates $X$ are transformed into $DX = \begin{pmatrix} -3 & 1.5 & 3 & -1.5 \\ -3 & -1.5 & 1.5 & 4.5 \end{pmatrix}$; see part (c) of the figure. Reflection through the $x$-axis would involve the transformation $R\begin{pmatrix} x \\ y \end{pmatrix} = \begin{pmatrix} x \\ -y \end{pmatrix}$, represented by the diagonal matrix $C = \begin{pmatrix} 1 & 0 \\ 0 & -1 \end{pmatrix}$. In computer graphics, the vertices of an object are actually given $(x, y, z)$ coordinates and three-dimensional versions of the above transformations can be applied to move and reshape the object as well as render the scene when the user's viewpoint is changed.

## 6.2.4  CHANGE OF BASIS

### Definitions:

Let $\mathcal{B} = (v_1, v_2, \ldots, v_n)$ and $\mathcal{B}' = (v_1', v_2', \ldots, v_n')$ be two ordered bases for $V$, and let $I$ denote the identity mapping from $V$ to $V$. The matrix $P = [I]_{\mathcal{B}, \mathcal{B}'}$ is the **transition matrix** from $\mathcal{B}$ to $\mathcal{B}'$. It is also called the **change of basis matrix** from basis $\mathcal{B}$ to basis $\mathcal{B}'$.

If $A$ and $B$ are two $n \times n$ matrices over a field $F$, then $B$ is **similar** to $A$ if there exists an invertible $n \times n$ matrix $P$ over $F$ such that $P^{-1}BP = A$.

### Facts:

**1.** The transition matrix $P = [I]_{\mathcal{B}, \mathcal{B}'}$ is invertible; its inverse is $P^{-1} = [I]_{\mathcal{B}', \mathcal{B}}$.

**2.** If $x = [v]_{\mathcal{B}}$ and $y = [v]_{\mathcal{B}'}$, then $y = Px$ where $P = [I]_{\mathcal{B}, \mathcal{B}'}$.

**3.** When $\mathcal{B} = \mathcal{B}'$, the transition matrix $P = [I]_{\mathcal{B}, \mathcal{B}} = [I]_{\mathcal{B}}$ is the $n \times n$ identity matrix.

**4.** If $T$ is a linear operator on $V$ with $A$ and $B$ the matrices of $T$ relative to bases $\mathcal{B}$ and $\mathcal{B}'$, respectively, then $B$ is similar to $A$. Specifically, $P^{-1}BP = A$ where $P = [I]_{\mathcal{B}, \mathcal{B}'}$.

**5.** If $A$ and $B$ are similar $n \times n$ matrices, then $A$ and $B$ represent the same linear operator $T$ relative to suitably chosen bases. More specifically, suppose $P^{-1}BP = A$, $\mathcal{B} = (v_1, v_2, \ldots, v_n)$ is any basis for $V$, and $T$ is the unique linear transformation with $A = [T]_{\mathcal{B}}$. Then $B = [T]_{\mathcal{B}'}$ where $\mathcal{B}' = (v_1', v_2', \ldots, v_n')$ is the basis for $V$ given by $v_j' = \sum_{i=1}^{n} p_{ij}^{-1} v_i$.

### Examples:

**1.** Consider the $\mathcal{R}^{2 \times 1}$ bases $\mathcal{B} = (v_1, v_2)$ and $\mathcal{B}' = (v_1', v_2')$, where $v_1 = (1, 0)^T$, $v_2 = (0, 1)^T$ and $v_1' = (1, 1)^T$, $v_2' = (2, 1)^T$. Since $v_1 = (-1)v_1' + v_2'$ and $v_2 = 2v_1' + (-1)v_2'$, the transition matrix from $\mathcal{B}$ to $\mathcal{B}'$ is $P = [I]_{\mathcal{B}, \mathcal{B}'} = \begin{pmatrix} -1 & 2 \\ 1 & -1 \end{pmatrix}$, and its inverse $P^{-1} = \begin{pmatrix} 1 & 2 \\ 1 & 1 \end{pmatrix}$ is the transition matrix $[I]_{\mathcal{B}', \mathcal{B}}$. If $v = x_1 v_1 + x_2 v_2$ where $x_i \in \mathcal{R}$, then by Fact 2, $v = y_1 v_1' + y_2 v_2'$ where $y_1 = (-1)x_1 + 2x_2$ and $y_2 = x_1 + (-1)x_2$.

**2.** Consider $T \colon \mathcal{R}^{2 \times 1} \to \mathcal{R}^{2 \times 1}$ given by $T \begin{pmatrix} x_1 \\ x_2 \end{pmatrix} = \begin{pmatrix} x_1 - 3x_2 \\ -2x_1 + 6x_2 \end{pmatrix}$, and the same bases $\mathcal{B}$ and $\mathcal{B}'$ specified in Example 1. The matrix of $T$ with respect to $\mathcal{B}$ is $[T]_{\mathcal{B}} = A = \begin{pmatrix} 1 & -3 \\ -2 & 6 \end{pmatrix}$ and the matrix of $T$ with respect to $\mathcal{B}'$ is $[T]_{\mathcal{B}'} = B = \begin{pmatrix} 10 & 5 \\ -6 & -3 \end{pmatrix}$. Moreover, $A$ and $B$ are similar; indeed, as Fact 4 shows, $A = P^{-1}BP$ where $P = \begin{pmatrix} -1 & 2 \\ 1 & -1 \end{pmatrix}$ is determined in Example 1.

# 6.3   MATRIX ALGEBRA

Matrices naturally arise in the analysis of linear systems and in representing discrete structures. This section studies important types of matrices, their properties, and methods for efficient matrix computation.

## 6.3.1   BASIC CONCEPTS AND SPECIAL MATRICES

**Definitions:**

The $m \times n$ matrix $A = (a_{ij})$ is a rectangular array of $mn$ real or complex numbers $a_{ij}$, arranged into $m$ **rows** and $n$ **columns**.

The **transpose** of the $m \times n$ matrix $A = (a_{ij})$ is the $n \times m$ matrix $A^T = (b_{ij})$ in which $b_{ij} = a_{ji}$.

The $i$th row of $A$, denoted $A(i,:)$, is the array $a_{i1}\ a_{i2}\ \dots\ a_{in}$. The elements in the $i$th row can be regarded as a **row vector** $(a_{i1}, a_{i2}, \dots, a_{in})$ in $\mathcal{R}^n$ or $\mathcal{C}^n$. The $j$th column of $A$, denoted $A(:,j)$, is the array

$$a_{1j}$$
$$a_{2j}$$
$$\vdots$$
$$a_{mj}$$

which can be identified with the **column vector** $(a_{1j}, a_{2j}, \dots, a_{mj})^T$.

A matrix is **sparse** if it has relatively few nonzero entries.

A **submatrix** of the matrix $A$ contains the elements occurring in rows $i_1 < i_2 < \cdots < i_k$ and columns $j_1 < j_2 < \cdots < j_r$ of $A$. A **principal submatrix** of the matrix $A$ contains the elements occurring in rows $i_1 < i_2 < \cdots < i_k$ and columns $i_1 < i_2 < \cdots < i_k$ of $A$. This principal submatrix has **order** $k$ and is written $A[i_1, i_2, \dots, i_k]$.

Two matrices $A$ and $B$ are **equal** if they are both $m \times n$ matrices with $a_{ij} = b_{ij}$ for all $i = 1, 2, \dots, m$ and $j = 1, 2, \dots, n$.

The **Hermitian adjoint** of the $m \times n$ matrix $A = (a_{ij})$ is the $n \times m$ matrix $A^* = (b_{ij})$ in which $b_{ij}$ is the complex conjugate of $a_{ji}$.

If $m = n$, the matrix $A = (a_{ij})$ is **square** with **diagonal** elements $a_{11}, a_{22}, \dots, a_{nn}$. The **main diagonal** contains the diagonal elements of $A$. An **off-diagonal** element is any $a_{ij}$ with $i \neq j$. The **trace** of $A$, tr $A$, is the sum of the diagonal elements of $A$.

The following table defines special types of square matrices.

| matrix | definition |
|---|---|
| identity | $I_n = (e_{ij})$ where $e_{ij} = \begin{cases} 1 & \text{if } i = j \\ 0 & \text{if } i \neq j \end{cases}$ ($n \times n$ matrix; each diagonal entry is 1; each off-diagonal entry is 0) |
| diagonal | $D = (d_{ij})$ where $d_{ij} = 0$ if $i \neq j$ (nonzero entries occur only |

| matrix | definition |
|---|---|
|  | on the main diagonal) |
| lower triangular | $L = (l_{ij})$ where $l_{ij} = 0$ if $j > i$ (nonzero entries occur only on or below the diagonal) |
| upper triangular | $U = (u_{ij})$ where $u_{ij} = 0$ if $j < i$ (nonzero entries occur only on or above the diagonal) |
| unit triangular | triangular matrix with all diagonal entries 1 |
| tridiagonal | $A = (a_{ij})$ where $a_{ij} = 0$ if $|i - j| > 1$ (nonzero entries occur only on or immediately above or below the diagonal) |
| symmetric | real matrix $A$ for which $A = A^T$ |
| skew-symmetric | real matrix $A$ for which $A = -A^T$ |
| Hermitian | complex matrix $A$ for which $A = A^*$ |
| skew-Hermitian | complex matrix $A$ for which $A = -A^*$ |

**Facts:**

**1.** Triangular matrices arise in the solution of systems of linear equations (§6.4).

**2.** A tridiagonal matrix can be represented as follows, where the diagonal lines represent the (possibly) nonzero entries.

**3.** Tridiagonal matrices are particular types of sparse matrices. Such matrices arise in discretized versions of continuous problems, the solution of difference equations (§3.3, §3.4.4), and the solution of eigenvalue problems (§6.5).

**4.** Sparse matrices frequently arise in the solution of large systems of linear equations (§6.4), since in many physical models a given variable typically interacts with relatively few others. Linear systems derived from sparse matrices require less storage space and can be solved more efficiently than those derived from a "dense" matrix.

**5.** Forming the transpose of a square matrix corresponds to "reflecting" the matrix elements with respect to the main diagonal.

**6.** Any skew-symmetric matrix $A$ must have $a_{ii} = 0$ for all $i$.

**7.** Any Hermitian matrix $A$ must have $a_{ii}$ real for all $i$.

**8.** If $A$ is real then $A^* = A^T$.

**9.** The columns of the identity matrix $I_n$ are the standard basis vectors for $\mathcal{R}^n$ (§6.1.3).

**10.** Viewed as a linear transformation (§6.2), the identity matrix represents the identity transformation; that is, it leaves all vectors unchanged.

**11.** Viewed as linear transformations, diagonal matrices with positive diagonal entries leave the directions of the basis vectors unchanged, but alter the relative scale of the basis vectors.

**Examples:**

**1.** The $2 \times 2$ and $3 \times 3$ identity matrices are $I_2 = \begin{pmatrix} 1 & 0 \\ 0 & 1 \end{pmatrix}$ and $I_3 = \begin{pmatrix} 1 & 0 & 0 \\ 0 & 1 & 0 \\ 0 & 0 & 1 \end{pmatrix}$.

**2.** The matrix $A = \begin{pmatrix} 6 & 0 & 1 \\ 0 & 2 & 4 \\ 1 & 4 & 3 \end{pmatrix}$ is symmetric.

**3.** The matrix $A = \begin{pmatrix} 1 & 2 - 3i \\ 2 + 3i & -4 \end{pmatrix}$ is Hermitian.

**4.** A $2 \times 2$ diagonal matrix transforms the unit square in $\mathcal{R}^2$ into a rectangle with sides parallel to the coordinate axes. The following figure shows the effect of the diagonal matrix $\begin{pmatrix} 3 & 0 \\ 0 & 2 \end{pmatrix}$ on certain vectors and on the unit square in $\mathcal{R}^2$. The standard basis vectors $\{(1,0)^T, (0,1)^T\}$ have been transformed to $\{(3,0)^T, (0,2)^T\}$.

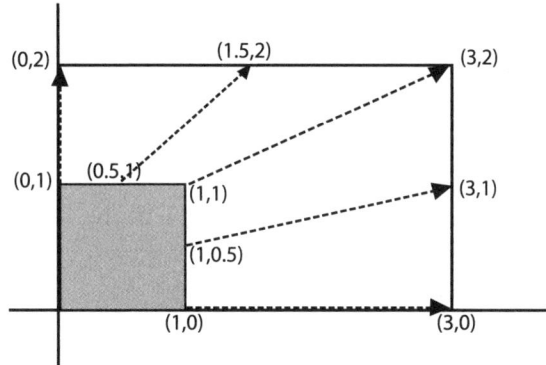

**5.** A $3 \times 3$ diagonal matrix transforms the unit cube into a rectangular parallelepiped.

**6.** The standard basis vectors are all eigenvectors of a diagonal matrix with the corresponding diagonal elements as their associated eigenvalues (§6.5).

### 6.3.2 OPERATIONS OF MATRIX ALGEBRA

**Definitions:**

The **scalar product** (**dot product**) of real vectors $x = (x_1, x_2, \ldots, x_n)$ and $y = (y_1, y_2, \ldots, y_n)$ is the number $x \cdot y = \sum_{i=1}^{n} x_i y_i$.

The $n \times n$ matrix $A$ is **nonsingular** (**invertible**) if there exists an $n \times n$ matrix $A^{-1}$ such that $AA^{-1} = A^{-1}A = I$. Any such matrix $A^{-1}$ is an **inverse** of $A$.

An **orthogonal** matrix is a real square matrix $A$ such that $A^T A = I$.

A **unitary** matrix is a complex square matrix $A$ such that $A^* A = I$, where $A^*$ is the Hermitian adjoint of $A$ (§6.3.1).

A **positive definite** matrix is a real symmetric (or complex Hermitian) matrix $A$ such that $x^* A x > 0$ for all $x \neq 0$.

The nonnegative **powers** of a square matrix $A$ are given by $A^0 = I$, $A^n = AA^{n-1}$. If $A$ is nonsingular then $A^{-n} = (A^{-1})^n$.

The following table defines various operations defined on matrices $A = (a_{ij})$ and $B = (b_{ij})$. (See Facts 1, 2, 5, 6 for restrictions on the sizes of the matrices.)

| operation | definition |
|---|---|
| sum $A + B$ | $A + B = (c_{ij})$ where $c_{ij} = a_{ij} + b_{ij}$ |
| difference $A - B$ | $A - B = (c_{ij})$ where $c_{ij} = a_{ij} - b_{ij}$ |
| scalar multiple $\alpha A$ | $\alpha A = (c_{ij})$ where $c_{ij} = \alpha a_{ij}$ |
| product $AB$ | $AB = (c_{ij})$ where $c_{ij} = \sum_k a_{ik} b_{kj}$ |

## Facts:

1. Matrices of different dimensions cannot be added or subtracted.

2. Square matrices of the same dimension can be multiplied.

3. Real or complex matrix addition satisfies the following properties:

   - commutative:  $A + B = B + A$;
   - associative:  $A + (B + C) = (A + B) + C$,    $A(BC) = (AB)C$;
   - distributive:  $A(B + C) = AB + AC$,   $(A + B)C = AC + BC$;
   - $\alpha(A + B) = \alpha A + \alpha B$, $\alpha(AB) = (\alpha A)B = A(\alpha B)$ for all scalars $\alpha$.

4. Matrix multiplication is not, in general, commutative—even when both products are defined. (See Example 3.)

5. The product $AB$ is defined if and only if the number of columns of $A$ equals the number of rows of $B$. That is, $A$ must be an $m \times n$ matrix and $B$ must be an $n \times p$ matrix.

6. The $ij$th element of the product $C = AB$ is the scalar product of row $i$ of $A$ and column $j$ of $B$:

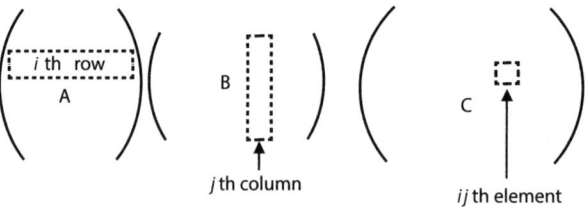

7. Multiplication by identity matrices of the appropriate dimension leaves a matrix unchanged: if $A$ is $m \times n$, then $I_m A = A I_n = A$.

8. Multiplication by diagonal matrices has the effect of scaling the rows or columns of a matrix. Pre-multiplication by a diagonal matrix scales the rows:

$$
\begin{pmatrix}
d_{11} & 0 & \cdots & 0 \\
0 & d_{22} & \cdots & 0 \\
\vdots & \vdots & & \vdots \\
0 & 0 & \cdots & d_{nn}
\end{pmatrix}
\begin{pmatrix}
a_{11} & \cdots & a_{1p} \\
a_{21} & \cdots & a_{2p} \\
\vdots & & \vdots \\
a_{n1} & \cdots & a_{np}
\end{pmatrix}
=
\begin{pmatrix}
d_{11}a_{11} & \cdots & d_{11}a_{1p} \\
d_{22}a_{21} & \cdots & d_{22}a_{2p} \\
\vdots & & \vdots \\
d_{nn}a_{n1} & \cdots & d_{nn}a_{np}
\end{pmatrix}.
$$

Post-multiplication by a diagonal matrix scales the columns:

$$
\begin{pmatrix}
a_{11} & \cdots & a_{1n} \\
a_{21} & \cdots & a_{2n} \\
\vdots & & \vdots \\
a_{m1} & \cdots & a_{mn}
\end{pmatrix}
\begin{pmatrix}
d_{11} & 0 & \cdots & 0 \\
0 & d_{22} & \cdots & 0 \\
\vdots & \vdots & & \vdots \\
0 & 0 & \cdots & d_{nn}
\end{pmatrix}
=
\begin{pmatrix}
d_{11}a_{11} & \cdots & d_{nn}a_{1n} \\
d_{11}a_{21} & \cdots & d_{nn}a_{2n} \\
\vdots & & \vdots \\
d_{11}a_{m1} & \cdots & d_{nn}a_{mn}
\end{pmatrix}.
$$

**9.** Any Hermitian matrix can be expressed as $A + iB$ where $A$ is symmetric and $B$ is skew-symmetric.

**10.** The inverse of a (nonsingular) matrix is unique.

**11.** If $A$ is nonsingular, the solution of the system of linear equations (§6.4) $Ax = b$ is given by (but almost never computed by) $x = A^{-1}b$.

**12.** The product of nonsingular matrices $A$ and $B$ is nonsingular, with $(AB)^{-1} = B^{-1}A^{-1}$. Conversely, if $A$ and $B$ are square matrices with $AB$ nonsingular, then $A$ and $B$ are nonsingular.

**13.** For a nonsingular matrix regarded as a linear transformation (§6.2), the inverse matrix represents the inverse transformation.

**14.** Sums of lower (upper) triangular matrices are lower (upper) triangular.

**15.** Products of lower (upper) triangular matrices are lower (upper) triangular.

**16.** A triangular matrix $A$ is nonsingular if and only if $a_{ii} \neq 0$ for all $i$.

**17.** If a lower (upper) triangular matrix is nonsingular then its inverse is lower (upper) triangular.

**18.** *Properties of transpose:*

- $(A^T)^T = A$;
- $(A + B)^T = A^T + B^T$;
- $(AB)^T = B^T A^T$;
- $AA^T$ and $A^T A$ are symmetric;
- if $A$ is nonsingular then so is $A^T$; moreover $(A^T)^{-1} = (A^{-1})^T$.

**19.** *Properties of Hermitian adjoint:*

- $(A^*)^* = A$;
- $(A + B)^* = A^* + B^*$;
- $(AB)^* = B^* A^*$;
- $AA^*$ and $A^* A$ are Hermitian;
- if $A$ is nonsingular, then so is $A^*$; moreover $(A^*)^{-1} = (A^{-1})^*$.

**20.** If $A$ is orthogonal, then $A$ is nonsingular and $A^{-1} = A^T$.

**21.** The rows (columns) of an orthogonal matrix are orthonormal with respect to the standard inner product on $\mathcal{R}^n$ (§6.1.4).

**22.** Products of orthogonal matrices are orthogonal.

**23.** If $A$ is unitary, then $A$ is nonsingular and $A^{-1} = A^*$.

**24.** The rows (columns) of a unitary matrix are orthonormal with respect to the standard inner product on $\mathcal{C}^n$ (§6.1.4).

**25.** Products of unitary matrices are unitary.

**26.** Positive definite matrices are nonsingular.

**27.** All eigenvalues (§6.5) of a positive definite matrix are positive.

**28.** Powers of a positive definite matrix are positive definite.

**29.** If $A$ is skew-symmetric, then $I + A$ is positive definite.

**30.** If $A$ is nonsingular, then $A^T A$ is positive definite.

**Examples:**

**1.** Let $A = \begin{pmatrix} 1 & 2 & 3 \\ 4 & 5 & 6 \end{pmatrix}$ and $B = \begin{pmatrix} 7 & 8 & 9 \\ 0 & 1 & 2 \end{pmatrix}$. Then $A + B = \begin{pmatrix} 8 & 10 & 12 \\ 4 & 6 & 8 \end{pmatrix}$ and $A - B = \begin{pmatrix} -6 & -6 & -6 \\ 4 & 4 & 4 \end{pmatrix}$.

**2.** The scalar product of the vectors $a = (1, 0, -1)$ and $b = (4, 3, 2)$ is $a \cdot b = (1)(4) + (0)(3) + (-1)(2) = 2$.

**3.** Let $A = \begin{pmatrix} 1 & 0 \\ 1 & 1 \end{pmatrix}$, $B = \begin{pmatrix} 2 & 3 \\ 4 & 1 \end{pmatrix}$, and $C = \begin{pmatrix} 1 & 1 & 2 \\ 0 & 2 & 3 \end{pmatrix}$. Then $AB$ and $BA$ are both defined with $AB = \begin{pmatrix} 1 & 0 \\ 1 & 1 \end{pmatrix}\begin{pmatrix} 2 & 3 \\ 4 & 1 \end{pmatrix} = \begin{pmatrix} 2 & 3 \\ 6 & 4 \end{pmatrix}$, whereas $BA = \begin{pmatrix} 2 & 3 \\ 4 & 1 \end{pmatrix}\begin{pmatrix} 1 & 0 \\ 1 & 1 \end{pmatrix} = \begin{pmatrix} 5 & 3 \\ 5 & 1 \end{pmatrix}$. Also, $AC$ is defined but $CA$ is not defined.

**4.** The matrices $A, B$ of Example 1 cannot be multiplied since $A$ has 3 columns and $B$ has 2 rows; see Fact 5. However, all the products $A^T B$, $AB^T$, $B^T A$, $BA^T$ exist: $A^T B = \begin{pmatrix} 1 & 4 \\ 2 & 5 \\ 3 & 6 \end{pmatrix}\begin{pmatrix} 7 & 8 & 9 \\ 0 & 1 & 2 \end{pmatrix} = \begin{pmatrix} 7 & 12 & 17 \\ 14 & 21 & 28 \\ 21 & 30 & 39 \end{pmatrix}$, $AB^T = \begin{pmatrix} 1 & 2 & 3 \\ 4 & 5 & 6 \end{pmatrix}\begin{pmatrix} 7 & 0 \\ 8 & 1 \\ 9 & 2 \end{pmatrix} = \begin{pmatrix} 50 & 8 \\ 122 & 17 \end{pmatrix}$, $B^T A = \begin{pmatrix} 7 & 14 & 21 \\ 12 & 21 & 30 \\ 17 & 28 & 39 \end{pmatrix}$, $BA^T = \begin{pmatrix} 50 & 122 \\ 8 & 17 \end{pmatrix}$. Note that $(B^T A)^T = A^T B$, as guaranteed by Fact 18.

**5.** *Multiplication by a diagonal matrix:* $\begin{pmatrix} 3 & 0 \\ 0 & 2 \end{pmatrix}\begin{pmatrix} 1 & 2 & 3 \\ 4 & 5 & 6 \end{pmatrix} = \begin{pmatrix} 3 & 6 & 9 \\ 8 & 10 & 12 \end{pmatrix}$ and $\begin{pmatrix} 1 & 2 & 3 \\ 4 & 5 & 6 \end{pmatrix}\begin{pmatrix} 2 & 0 & 0 \\ 0 & 3 & 0 \\ 0 & 0 & 1 \end{pmatrix} = \begin{pmatrix} 2 & 6 & 3 \\ 8 & 15 & 6 \end{pmatrix}$.

**6.** The $2 \times 2$ matrix $A = \begin{pmatrix} a & b \\ c & d \end{pmatrix}$ is nonsingular if $\Delta = ad - bc \neq 0$; in this case $A^{-1} = \dfrac{1}{\Delta}\begin{pmatrix} d & -b \\ -c & a \end{pmatrix}$.

**7.** The matrix $A = \frac{1}{9}\begin{pmatrix} 4 & 8 & 1 \\ 7 & -4 & 4 \\ -4 & 1 & 8 \end{pmatrix}$ is orthogonal.

**8.** If $A = \frac{1}{2} \begin{pmatrix} 1 & -i & -1+i \\ i & 1 & 1+i \\ 1+i & -1+i & 0 \end{pmatrix}$ then $A^* = \frac{1}{2} \begin{pmatrix} 1 & -i & 1-i \\ i & 1 & -1-i \\ -1-i & 1-i & 0 \end{pmatrix}$.

Since $A^*A = I$ the matrix $A$ is unitary.

**9.** Every $2 \times 2$ orthogonal matrix $Q$ can be written as $Q = \begin{pmatrix} \cos\theta & -\sin\theta \\ \sin\theta & \cos\theta \end{pmatrix}$ for some real $\theta$. Geometrically, the matrix $Q$ effects a counterclockwise rotation by the angle $\theta$.

**10.** For the matrix $Q$ in Example 9, $Q^2 = \begin{pmatrix} \cos^2\theta - \sin^2\theta & -2\sin\theta\cos\theta \\ 2\sin\theta\cos\theta & \cos^2\theta - \sin^2\theta \end{pmatrix}$. Since this must be the same as a rotation by an angle of $2\theta$, then $Q^2 = \begin{pmatrix} \cos 2\theta & -\sin 2\theta \\ \sin 2\theta & \cos 2\theta \end{pmatrix}$.

Equating these two expressions for $Q^2$ gives the double angle formulas of trigonometry.

**11.** The matrix $\begin{pmatrix} 4 & 2i & -3+i \\ -2i & -8 & 6+3i \\ -3-i & 6-3i & 5 \end{pmatrix}$ is Hermitian. It can be written (see Fact 9)

as $A + Bi = \begin{pmatrix} 4 & 0 & -3 \\ 0 & -8 & 6 \\ -3 & 6 & 5 \end{pmatrix} + \begin{pmatrix} 0 & 2i & i \\ -2i & 0 & 3i \\ -i & -3i & 0 \end{pmatrix}$ with $A$ symmetric and $B$ skew-symmetric.

---

### 6.3.3   FAST MULTIPLICATION OF MATRICES

A variety of methods have been devised to multiply matrices more efficiently than by simply using the definition in §6.3.2. This section presents alternative methods for carrying out matrix multiplication.

**Definitions:**
The **shift left operation** $\text{shL}(A(i,:), k)$ rotates elements of row $i$ in matrix $A$ exactly $k$ places to the left, where data shifted off the left side of the matrix are wrapped around to the right side.

The **shift up operation** $\text{shU}(B(:,j), k)$ rotates elements of column $j$ in matrix $B$ exactly $k$ places up, where data shifted off the top of the matrix are wrapped around to the bottom.

These operations can also be applied simultaneously to every row of $A$ or every column of $B$, denoted $\text{shL}(A, k)$ and $\text{shU}(B, k)$ respectively.

**Facts:**
**1.** The basic definition given in §6.3.2 can be used to multiply the $m \times n$ matrix $A$ and the $n \times p$ matrix $B$. The associated algorithm (Algorithm 1) requires $O(mnp)$ operations (additions and multiplications of individual elements).

---

**Algorithm 1**:   **Basic matrix multiplication.**
input: $m \times n$ matrix $A$, $n \times p$ matrix $B$

output: $m \times p$ matrix $C = AB$
for $i := 1$ to $m$
  for $j := 1$ to $p$
   $C(i,j) := 0$
   for $k := 1$ to $n$
    $C(i,j) := C(i,j) + A(i,k)B(k,j)$

**2.** *Matrix multiplication in scalar product form*: Algorithm 1 can be rewritten in terms of the scalar product operation, giving Algorithm 2.

**Algorithm 2**: **Scalar product form of matrix multiplication.**
input: $m \times n$ matrix $A$, $n \times p$ matrix $B$
output: $m \times p$ matrix $C = AB$
for $i := 1$ to $m$
  for $j := 1$ to $p$
   $C(i,j) := A(i,:) \cdot B(:,j)$

**3.** Algorithm 2 is well suited for fast multiplication on computers designed for efficient scalar product operations. It requires $O(mp)$ scalar products.

**4.** *Matrix multiplication in linear combination form*: Algorithm 3 carries out matrix multiplication by taking a linear combination of columns of $A$ to obtain each column of the product.

**Algorithm 3**: **Column linear combination form of matrix multiplication.**
input: $m \times n$ matrix $A$, $n \times p$ matrix $B$
output: $m \times p$ matrix $C = AB$
for $j := 1$ to $p$
  $C(:,j) := 0$
  for $k := 1$ to $n$
   $C(:,j) := C(:,j) + B(k,j)A(:,k)$

**5.** The inner loop of Algorithm 3 performs a "vector + (scalar × vector)" operation, well suited to a vector computer using efficiently pipelined arithmetic processing.

**6.** Algorithm 3 is often used for fast general matrix multiplication on vector machines since it is based on a natural vector operation. If these vector operations can be performed on all elements simultaneously, then $O(np)$ vector operations are needed.

**7.** Access to matrix elements in Algorithm 3 is by column. There are other rearrangements of the algorithm which access matrix information by row.

**8.** *Fast multiplication on array processors*: Algorithm 4 multiplies two $n \times n$ (or smaller dimension) matrices on a computer with an $n \times n$ array of processors. It uses various shift operations on the arrays and the array-multiplication operation ($*$) of elementwise multiplication.

---

**Algorithm 4**:  **Array processor matrix multiplication.**

input: $n \times n$ matrices $A, B$

output: $n \times n$ matrix $C = AB$

{Preshift the matrix arrays}

**for** $i := 1$ **to** $n$

  shL$(A(i,:), i-1)$       {Shift $i$th row $i-1$ places left}

  shU$(B(:,i), i-1)$       {Shift $i$th column $i-1$ places up}

$C := 0$      {Initialize product array}

**for** $k := 1$ **to** $n$

  $C := C + A * B$

  shL$(A, 1)$

  shU$(B, 1)$

---

**9.** At each step Algorithm 4 shifts $A$ one place to the left and shifts $B$ one place up so that components of the array product are correct new terms for the corresponding elements of $C = AB$. Each matrix is preshifted so the first step complies with this requirement.

**10.** Two $n \times n$ matrices can be multiplied in $O(n)$ time using Algorithm 4 on an array processor.

**11.** *Strassen's algorithm:* Algorithm 5 recursively carries out matrix multiplication for $n \times n$ matrices $A$ and $B$ where $n = 2^k$. The basis of Strassen's algorithm is partitioning the two factors into square blocks with dimension half that of the original matrices.

---

**Algorithm 5**:  **Strassen's algorithm for $2^k \times 2^k$ matrices.**

**procedure** *Strassen*$(A, B)$

input: $2^k \times 2^k$ matrices $A, B$

output: $2^k \times 2^k$ matrix $C = AB$

**if** $k = 1$ **then** use Algorithm 6

**else**

  partition $A, B$ into four $2^{k-1} \times 2^{k-1}$ blocks $A = \begin{pmatrix} A_{11} & A_{12} \\ A_{21} & A_{22} \end{pmatrix}, B = \begin{pmatrix} B_{11} & B_{12} \\ B_{21} & B_{22} \end{pmatrix}$

  $P := Strassen((A_{11} + A_{22}), (B_{11} + B_{22}))$

  $Q := Strassen((A_{21} + A_{22}), B_{11}); \quad R := Strassen(A_{11}, (B_{12} - B_{22}))$

  $S := Strassen(A_{22}, (B_{21} - B_{11})); \quad T := Strassen((A_{11} + A_{12}), B_{22})$

  $U := Strassen((A_{21} - A_{11}), (B_{11} + B_{12}))$

  $V := Strassen((A_{12} - A_{22}), (B_{21} + B_{22}))$

  $C_{11} := P + S - T + V; \quad C_{12} := R + T; \quad C_{21} := Q + S; \quad C_{22} := P - Q + R + U$

$C := \begin{pmatrix} C_{11} & C_{12} \\ C_{21} & C_{22} \end{pmatrix}$

---

**12.** Strassen's algorithm ultimately requires the fast multiplication of $2 \times 2$ matrices (Algorithm 6).

---

**Algorithm 6**:   **Strassen's algorithm for $2 \times 2$ matrices.**

input: $2 \times 2$ matrices $A, B$

output: $2 \times 2$ matrix $C = AB$

$p := (a_{11} + a_{22})(b_{11} + b_{22}); \; q := (a_{21} + a_{22})b_{11}; \; r := a_{11}(b_{12} - b_{22})$

$s := a_{22}(b_{21} - b_{11}); \; t := (a_{11} + a_{12})b_{22}; \; u := (a_{21} - a_{11})(b_{11} + b_{12})$

$v := (a_{12} - a_{22})(b_{21} + b_{22})$

$c_{11} := p + s - t + v; \; c_{12} := r + t; \; c_{21} := q + s; \; c_{22} := p - q + r + u$

---

**13.** Algorithm 6 multiplies two $2 \times 2$ matrices using only 7 multiplications and 18 additions instead of the normal 8 multiplications and 4 additions. For most modern computers saving one multiplication at the cost of 14 extra additions would not represent a gain.

**14.** Strassen's algorithm can be extended to $n \times n$ matrices where $n$ is not a power of 2. The general algorithm requires $O(n^{\log_2 7}) \approx O(n^{2.807})$ multiplications. Details of this algorithm and its efficiency can be found in [GoVL13].

**Examples:**

**1.** This example illustrates Algorithm 4 for $4 \times 4$ array matrix multiplication. The preshift and the first array multiplication yield the arrays

| $a_{11}$ | $a_{12}$ | $a_{13}$ | $a_{14}$ | $b_{11}$ | $b_{22}$ | $b_{33}$ | $b_{44}$ | $a_{11}b_{11}$ | $a_{12}b_{22}$ | $a_{13}b_{33}$ | $a_{14}b_{44}$ |
|---|---|---|---|---|---|---|---|---|---|---|---|
| $a_{22}$ | $a_{23}$ | $a_{24}$ | $a_{21}$ | $b_{21}$ | $b_{32}$ | $b_{43}$ | $b_{14}$ | $a_{22}b_{21}$ | $a_{23}b_{32}$ | $a_{24}b_{43}$ | $a_{21}b_{14}$ |
| $a_{33}$ | $a_{34}$ | $a_{31}$ | $a_{32}$ | $b_{31}$ | $b_{42}$ | $b_{13}$ | $b_{24}$ | $a_{33}b_{31}$ | $a_{34}b_{42}$ | $a_{31}b_{13}$ | $a_{32}b_{24}$ |
| $a_{44}$ | $a_{41}$ | $a_{42}$ | $a_{43}$ | $b_{41}$ | $b_{12}$ | $b_{23}$ | $b_{34}$ | $a_{44}b_{41}$ | $a_{41}b_{12}$ | $a_{42}b_{23}$ | $a_{43}b_{34}$ |

The next shifts and multiply-accumulate operation produce

| $a_{12}$ | $a_{13}$ | $a_{14}$ | $a_{11}$ | $b_{21}$ | $b_{32}$ | $b_{43}$ | $b_{14}$ |
|---|---|---|---|---|---|---|---|
| $a_{23}$ | $a_{24}$ | $a_{21}$ | $a_{22}$ | $b_{31}$ | $b_{42}$ | $b_{13}$ | $b_{24}$ |
| $a_{34}$ | $a_{31}$ | $a_{32}$ | $a_{33}$ | $b_{41}$ | $b_{12}$ | $b_{23}$ | $b_{34}$ |
| $a_{41}$ | $a_{42}$ | $a_{43}$ | $a_{44}$ | $b_{11}$ | $b_{22}$ | $b_{33}$ | $b_{44}$ |

| $a_{11}b_{11} + a_{12}b_{21}$ | $a_{12}b_{22} + a_{13}b_{32}$ | $a_{13}b_{33} + a_{14}b_{43}$ | $a_{14}b_{44} + a_{11}b_{14}$ |
|---|---|---|---|
| $a_{22}b_{21} + a_{23}b_{31}$ | $a_{23}b_{32} + a_{24}b_{42}$ | $a_{24}b_{43} + a_{21}b_{13}$ | $a_{21}b_{14} + a_{22}b_{24}$ |
| $a_{33}b_{31} + a_{34}b_{41}$ | $a_{34}b_{42} + a_{31}b_{12}$ | $a_{31}b_{13} + a_{32}b_{23}$ | $a_{32}b_{24} + a_{33}b_{34}$ |
| $a_{44}b_{41} + a_{41}b_{11}$ | $a_{41}b_{12} + a_{42}b_{22}$ | $a_{42}b_{23} + a_{43}b_{33}$ | $a_{43}b_{34} + a_{44}b_{44}$ |

At subsequent stages the remaining terms get added in to the appropriate elements of the product matrix. The total cost of matrix multiplication is therefore reduced to $n$ parallel multiply-accumulate operations plus some communication costs, which for a typical distributed memory array processor are generally small.

**2.** Algorithm 6 is illustrated using the matrices $A = \begin{pmatrix} 3 & 4 \\ -1 & 2 \end{pmatrix}$, $B = \begin{pmatrix} 7 & 3 \\ 1 & -3 \end{pmatrix}$. Then

$p = 5 \cdot 4 = 20, \qquad q = 1 \cdot 7 = 7, \qquad r = 3 \cdot 6 = 18, \qquad s = 3 \cdot (-6) = -12,$

$t = 7 \cdot (-3) = -21, \quad u = (-4) \cdot 10 = -40, \quad v = 2 \cdot (-2) = -4,$

giving the following elements of $C = AB$: $c_{11} = 20 - 12 + 21 - 4 = 25$, $c_{12} = 18 - 21 = -3$, $c_{21} = 7 - 12 = -5$, $c_{22} = 20 - 7 + 18 - 40 = -9$.

## 6.3.4  DETERMINANTS

**Definitions:**

For an $n \times n$ matrix $A$ with $n > 1$, $A_{ij}$ denotes the $(n-1) \times (n-1)$ matrix obtained by deleting row $i$ and column $j$ from $A$.

The **determinant** $\det A$ of an $n \times n$ matrix $A$ can be defined recursively:

- if $A = (a)$ is a $1 \times 1$ matrix, then $\det A = a$;
- if $n > 1$, then $\det A = \sum_{j=1}^{n}(-1)^{j+1}a_{1j}\det A_{1j}$.

A **minor** of a matrix is the determinant of a square submatrix of the given matrix. A **principal minor** is the determinant of a principal submatrix.

*Notation:*  The determinant of $A = (a_{ij})$ is commonly written using vertical bars:

$$\det A = |A| = \begin{vmatrix} a_{11} & a_{12} & \cdots & a_{1n} \\ a_{21} & a_{22} & \cdots & a_{2n} \\ \vdots & \vdots & & \vdots \\ a_{n1} & a_{n2} & \cdots & a_{nn} \end{vmatrix}.$$

**Facts:**

1. *Laplace expansion:*  For any $r$,

$$\det A = \sum_{j=1}^{n}(-1)^{r+j}a_{rj}\det A_{rj} = \sum_{i=1}^{n}(-1)^{i+r}a_{ir}\det A_{ir}.$$

2. If $A = (a_{ij})$ is $n \times n$, then $\det A = \sum_{\sigma \in S_n} \text{sgn}(\sigma)\, a_{1\sigma(1)}a_{2\sigma(2)}\cdots a_{n\sigma(n)}$. Here $S_n$ is the set of all permutations on $\{1, 2, \ldots, n\}$, and $\text{sgn}(\sigma)$ equals 1 if $\sigma$ is even and $-1$ if $\sigma$ is odd (§5.3.1).

3. $\det \begin{pmatrix} a & b \\ c & d \end{pmatrix} = \begin{vmatrix} a & b \\ c & d \end{vmatrix} = ad - bc.$

4. $\det \begin{pmatrix} a & b & c \\ d & e & f \\ g & h & i \end{pmatrix} = \begin{vmatrix} a & b & c \\ d & e & f \\ g & h & i \end{vmatrix} = aei + bfg + cdh - afh - bdi - ceg.$

5. $\det AB = \det A \det B = \det BA$ for all $n \times n$ matrices $A$, $B$.

6. $\det A^T = \det A$ for all $n \times n$ matrices $A$.

7. $\det \alpha A = \alpha^n \det A$ for all $n \times n$ matrices $A$ and all scalars $\alpha$.

8. $\det I = 1$.

9. If $A$ has two identical rows (or two identical columns), then $\det A = 0$.

10. Interchanging two rows (or two columns) of a matrix changes the sign of the determinant.

11. Multiplying one row (or column) of a matrix by a scalar multiplies its determinant by that same scalar.

12. Adding a multiple of one row (column) to another row (column) leaves the value of the determinant unchanged.

**13.** If $D = (d_{ij})$ is an $n \times n$ diagonal matrix, then $\det D = d_{11}d_{22}\ldots d_{nn}$.

**14.** If $T = (t_{ij})$ is an $n \times n$ triangular matrix, then $\det T = t_{11}t_{22}\ldots t_{nn}$.

**15.** If $A$ and $D$ are square matrices, then $\det \begin{pmatrix} A & B \\ 0 & D \end{pmatrix} = \det A \det D = \det \begin{pmatrix} A & 0 \\ C & D \end{pmatrix}$.

**16.** $A$ is nonsingular if and only if $\det A \neq 0$.

**17.** If $A$ is nonsingular then $\det(A^{-1}) = \dfrac{1}{\det A}$.

**18.** If $A$ and $D$ are nonsingular, then $\det \begin{pmatrix} A & B \\ C & D \end{pmatrix} = \det A \det(D - CA^{-1}B) =$ $\det D \det(A - BD^{-1}C)$.

**19.** The determinant of a Hermitian matrix (§6.3.1) is real.

**20.** The determinant of a skew-symmetric matrix (§6.3.1) of odd size is zero.

**21.** The determinant of an orthogonal matrix (§6.3.2) is $\pm 1$.

**22.** The $n \times n$ symmetric (or Hermitian) matrix $A$ is positive definite if and only if all its leading principal submatrices $A[1], A[1,2], \ldots, A[1,2,\ldots,n]$ have a positive determinant.

**23.** The $n \times n$ *Vandermonde matrix*

$$\begin{pmatrix} 1 & x_1 & \cdots & x_1^{n-1} \\ 1 & x_2 & \cdots & x_2^{n-1} \\ \vdots & \vdots & & \vdots \\ 1 & x_n & \cdots & x_n^{n-1} \end{pmatrix}$$

has determinant $\prod_{i<j}(x_j - x_i)$.

**24.** If the $n \times n$ matrix $A = (a_{ij})$ has diagonal elements $a_{ii} = x$ and off-diagonal elements $a_{ij} = y$, then $\det A = (x - y)^{n-1}(x - y + ny)$.

**25.** The equation of the straight line through points $(a_1, b_1)$ and $(a_2, b_2)$ is given by

$$\begin{vmatrix} x & y & 1 \\ a_1 & b_1 & 1 \\ a_2 & b_2 & 1 \end{vmatrix} = 0.$$

**26.** The equation of the circle through points $(a_1, b_1)$, $(a_2, b_2)$, $(a_3, b_3)$ is given by

$$\begin{vmatrix} x^2 + y^2 & x & y & 1 \\ a_1^2 + b_1^2 & a_1 & b_1 & 1 \\ a_2^2 + b_2^2 & a_2 & b_2 & 1 \\ a_3^2 + b_3^2 & a_3 & b_3 & 1 \end{vmatrix} = 0.$$

**27.** If the three points $(a_1, b_1)$, $(a_2, b_2)$, $(a_3, b_3)$ are listed in counterclockwise order, then the area of the triangle they form is given by

$$\frac{1}{2} \begin{vmatrix} a_1 & b_1 & 1 \\ a_2 & b_2 & 1 \\ a_3 & b_3 & 1 \end{vmatrix}.$$

**28.** The parallelepiped $P = \{\alpha_1 a_1 + \alpha_2 a_2 + \cdots + \alpha_n a_n \mid 0 \leq \alpha_i \leq 1\}$ spanned by the vectors $a_1, a_2, \ldots, a_n$ has volume $|\det A|$, where $A$ has columns $a_1, a_2, \ldots, a_n$.

**29.** *Computation*: The determinant is (almost) never computed from the definition or from Fact 2. Instead it is calculated using Facts 12 and 14. (See Example 1.)

**Examples:**

**1.** Determinants can be calculated by using row operations to create a triangular matrix, and then applying Fact 14:

$$\det \begin{pmatrix} -1 & 2 & 1 \\ 0 & 5 & 2 \\ 3 & 4 & 3 \end{pmatrix} = \det \begin{pmatrix} -1 & 2 & 1 \\ 0 & 5 & 2 \\ 0 & 10 & 6 \end{pmatrix} = \det \begin{pmatrix} -1 & 2 & 1 \\ 0 & 5 & 2 \\ 0 & 0 & 2 \end{pmatrix} = -10.$$

Here the second matrix is obtained from the first by adding 3 times row 1 to row 3; the third matrix is obtained from the second by adding $-2$ times row 2 to row 3.

**2.** Determinants can be calculated by using row and column interchanges to obtain a form with exploitable zeros:

$$\det \begin{pmatrix} 4 & 5 & 1 & 6 \\ 0 & 6 & 0 & 3 \\ 0 & 5 & 0 & 2 \\ 3 & 3 & 2 & 4 \end{pmatrix} = -\det \begin{pmatrix} 4 & 1 & 5 & 6 \\ 0 & 0 & 6 & 3 \\ 0 & 0 & 5 & 2 \\ 3 & 2 & 3 & 4 \end{pmatrix} = \det \begin{pmatrix} 4 & 1 & 5 & 6 \\ 3 & 2 & 3 & 4 \\ 0 & 0 & 5 & 2 \\ 0 & 0 & 6 & 3 \end{pmatrix}.$$

Here the second matrix is obtained from the first by interchanging columns 2 and 3; the third matrix is obtained from the second by interchanging rows 2 and 4. The third matrix has block triangular form, with diagonal blocks $A = \begin{pmatrix} 4 & 1 \\ 3 & 2 \end{pmatrix}$ and $D = \begin{pmatrix} 5 & 2 \\ 6 & 3 \end{pmatrix}$.
By Fact 15, the original determinant equals $\det A \det D = 5 \cdot 3 = 15$.

**3.** The symmetric matrix $A = \begin{pmatrix} 3 & 1 & 0 \\ 1 & 5 & 3 \\ 0 & 3 & 4 \end{pmatrix}$ is positive definite, since its leading principal

minors (Fact 22) are positive: $\det(3) = 3 > 0$, $\det \begin{pmatrix} 3 & 1 \\ 1 & 5 \end{pmatrix} = 14 > 0$, and (by Fact 1)

$$\det A = 3 \det \begin{pmatrix} 5 & 3 \\ 3 & 4 \end{pmatrix} - \det \begin{pmatrix} 1 & 3 \\ 0 & 4 \end{pmatrix} = 3 \cdot 11 - 4 = 29 > 0.$$

**4.** The equation of the line through points $(1,3)$ and $(4,5)$ can be found using Fact 25:

$$\begin{vmatrix} x & y & 1 \\ 1 & 3 & 1 \\ 4 & 5 & 1 \end{vmatrix} = \begin{vmatrix} x & y & 1 \\ 1 & 3 & 1 \\ 0 & -7 & -3 \end{vmatrix} = \begin{vmatrix} x & y-\frac{7}{3} & 1 \\ 1 & \frac{2}{3} & 1 \\ 0 & 0 & -3 \end{vmatrix} = (\frac{2}{3}x - y + \frac{7}{3})(-3) = 0,$$

giving $\frac{2}{3}x - y + \frac{7}{3} = 0$ or $y = \frac{2}{3}x + \frac{7}{3}$.

**5.** By Fact 27, the area of the triangle formed by the points $(0,0)$, $(1,3)$, and $(4,5)$ is

$$\frac{1}{2} \begin{vmatrix} 0 & 0 & 1 \\ 4 & 5 & 1 \\ 1 & 3 & 1 \end{vmatrix} = \frac{1}{2} \begin{vmatrix} 4 & 5 \\ 1 & 3 \end{vmatrix} = \frac{7}{2}.$$

**6.** *Cayley's formula:* The determinant of the $(n-1) \times (n-1)$ matrix

$$T_n = \begin{pmatrix} n-1 & -1 & \cdots & -1 \\ -1 & n-1 & \cdots & -1 \\ \vdots & \vdots & & \vdots \\ -1 & -1 & \cdots & n-1 \end{pmatrix}$$

counts the number of spanning trees of a complete graph. (See §9.2.2.) Using Fact 24, $\det T_n = n^{n-2}[n - (n-1)] = n^{n-2}$.

---

### 6.3.5   RANK

**Definition:**

The **rank** of an $m \times n$ matrix $A$, written rank $A$, is the size of the largest square nonsingular submatrix of $A$.

**Facts:**

**1.** rank $A =$ rank $A^T$.

**2.** The rank of $A$ equals the maximum number of linearly independent rows or linearly independent columns in $A$.

**3.** rank$(A + B) \leq$ rank $A +$ rank $B$.

**4.** rank $AB \leq \min\{$rank $A,$ rank $B\}$.

**5.** If $A$ is nonsingular then rank $AB =$ rank $B$ and rank $CA =$ rank $C$.

**6.** rank $A = \dim CS(A)$, where $CS(A)$ is the column space of $A$ and $\dim V$ denotes the dimension of the vector space $V$. (See §6.1.3.)

**7.** rank $A = \dim RS(A)$, where $RS(A)$ is the row space of $A$. (See §6.1.3.)

**8.** An $n \times n$ matrix $A$ is nonsingular if and only if rank $A = n$.

**9.** Every matrix of rank $r$ can be written as a sum of $r$ matrices of rank 1.

**10.** If $a$ and $b$ are nonzero $n \times 1$ vectors, then $ab^T$ is an $n \times n$ matrix of rank 1.

**11.** The rank of a matrix is not always easy to compute. In the absence of severe roundoff errors, it can be obtained by counting the number of nonzero rows at the end of the Gaussian elimination procedure (§6.4.2).

**12.** *System of linear equations:* Consider the system $Ax = b$, where $A$ is $m \times n$. Let $A_b = (A : b)$ denote the $m \times (n+1)$ matrix whose $(n+1)$st column is the vector $b$. Then the system $Ax = b$ has

- a unique solution $\Leftrightarrow$ rank $A =$ rank $A_b = n$;
- infinitely many solutions $\Leftrightarrow$ rank $A =$ rank $A_b < n$;
- no solution $\Leftrightarrow$ rank $A <$ rank $A_b$.

**Examples:**

**1.** The matrix $A = \begin{pmatrix} 1 & -1 & 2 \\ 3 & 4 & -1 \\ 5 & 2 & 3 \end{pmatrix}$ is singular since $\det A = 0$. However, the submatrix

$A[1,2] = \begin{pmatrix} 1 & -1 \\ 3 & 4 \end{pmatrix}$ has determinant 7 and so is nonsingular, showing that rank $A = 2$.

The matrix $A$ has two linearly independent rows: row $3 = 2 \times$ (row 1) + (row 2). Likewise, it has two linearly independent columns: column $3 =$ (column 1) $-$ (column 2). This again confirms (by Fact 2) that rank $A = 2$.

**2.** Consider the system of equations $Ax = b$, where $A$ is the matrix in Example 1 and $b = (0, 7, 7)^T$. Since rank $A =$ rank $A_b = 2 < 3$, this system has infinitely many solutions $x$. In fact, the set of solutions is given by $\{(1 - \alpha, 1 + \alpha, \alpha)^T \mid \alpha \in \mathcal{R}\}$.

**3.** The matrix $A = \begin{pmatrix} 1 & x & x^2 \\ x & x^2 & x^3 \\ x^2 & x^3 & x^4 \end{pmatrix}$ can be expressed as the product $aa^T$ where $a$ is the column vector $(1, x, x^2)^T$. By Fact 10, $A$ has rank 1.

---

### 6.3.6  IDENTITIES OF MATRIX ALGEBRA

**Facts:**

**1.** *Cauchy-Binet formula:* If $C$ is $m \times m$ and $C = AB$ where $A$ is $m \times n$ and $B$ is $n \times m$, then the determinant of $C$ is given by the sum of all products of order $m$ minors of $A$ and the corresponding order $m$ minors of $B$:

$$\det C = \sum_{1 \le s_1 < s_2 < \cdots < s_m \le n} \begin{vmatrix} a_{1s_1} & a_{1s_2} & \cdots & a_{1s_m} \\ a_{2s_1} & a_{2s_2} & \cdots & a_{2s_m} \\ \vdots & \vdots & & \vdots \\ a_{ms_1} & a_{ms_2} & \cdots & a_{ms_m} \end{vmatrix} \cdot \begin{vmatrix} b_{s_11} & b_{s_12} & \cdots & b_{s_1m} \\ b_{s_21} & b_{s_22} & \cdots & b_{s_2m} \\ \vdots & \vdots & & \vdots \\ b_{s_m1} & b_{s_m2} & \cdots & b_{s_mm} \end{vmatrix}.$$

- If $m = n$ there is only one possible selection; the Cauchy-Binet formula for this case reduces to $\det C = \det A \det B$; (see Fact 5, §6.3.4).

- If $m > n$ no possible selections exist (the sum is empty), so $\det C = 0$.

**2.** *Courant-Fischer minimax identity:* If the eigenvalues (§6.5) of an $n \times n$ Hermitian matrix $A$ are ordered so that $\lambda_1 \ge \lambda_2 \ge \cdots \ge \lambda_n$, then $\lambda_k = \max_{\dim V = k} \min_{0 \neq x \in V} \dfrac{x^T A x}{x^T x}$ where $V$ is a linear subspace of $\mathcal{C}^n$.

**3.** *Hadamard's inequality:* This gives an upper bound for the determinant of an $n \times n$ matrix $A$ in terms of the $l_2$ norms (§6.4.5) of its rows (or columns):

- in terms of rows: $\quad (\det A)^2 \le \prod_{i=1}^{n} \left( \sum_{j=1}^{n} |a_{ij}|^2 \right) = \prod_{i=1}^{n} \|A(i, :)\|^2;$

- in terms of columns: $\quad (\det A)^2 \le \prod_{j=1}^{n} \left( \sum_{i=1}^{n} |a_{ij}|^2 \right) = \prod_{j=1}^{n} \|A(:, j)\|^2.$

**4.** *Sherman-Morrison identity:* If $A$ is a nonsingular $n \times n$ matrix and $u, v \in \mathcal{R}^n$, then

$$(A - uv^T)^{-1} = A^{-1} + \frac{1}{1 - v^T A^{-1} u} (A^{-1} uv^T A^{-1}).$$

**5.** *Woodbury's identity:* If $A$ is nonsingular, then

$$(A - UV^T)^{-1} = A^{-1} + A^{-1} U (I - V^T A^{-1} U)^{-1} V^T A^{-1}.$$

**6.** Suppose $A$ is a nonsingular $n \times n$ matrix, with $S$ a set of $k$ indices $i_1 < i_2 < \cdots < i_k$ and $\overline{S}$ the set of remaining indices in $\{1, 2, \ldots, n\}$. Then the principal minors of $A^{-1}$ are related to the principal minors of $A$ via

$$\det A^{-1}[S] = \frac{1}{\det A} \det A[\overline{S}].$$

**7.** *Jacobi's identity:*   If the $n \times n$ system of linear differential equations $\dfrac{dx}{dt} = P(t)x$ has the linearly independent family of solutions $X(:, j)(t)$ for $j = 1, 2, \ldots, n$, then the determinant of the (variable) matrix $X(t)$ whose columns are the $X(:, j)(t)$ is given by

$$\det X(t_1) = c \exp \left( \int_{t_0}^{t_1} \operatorname{tr} P(t)\, dt \right),$$

where $c$ is a constant and $\operatorname{tr} P(t) = p_{11}(t) + p_{22}(t) + \cdots + p_{nn}(t)$ is the trace of the matrix $P(t)$.

## 6.4   LINEAR SYSTEMS

The need to find solutions of linear systems arises in numerous branches of science and engineering (physics, biology, chemistry, structural engineering, electrical engineering, civil engineering) as well as in statistics and applied mathematics. This section discusses various techniques for the efficient solution of such systems, especially important when these systems are large and sparse.

### 6.4.1   BASIC CONCEPTS

This subsection is concerned with representing and solving a system of $m$ linear equations in $n$ unknowns. Throughout, the focus will be on systems whose data are real numbers. The extension to linear systems with complex data is straightforward.

**Definitions:**

A **linear equation** in unknowns $x_1, x_2, \ldots, x_n$ is an equation of the form $\sum_{j=1}^{n} a_j x_j = b$, where the **coefficients** $a_j \in \mathcal{R}$ and the **right-hand side** $b \in \mathcal{R}$. A **solution** of this equation is any set of values $x_1, x_2, \ldots, x_n$ satisfying the given equation.

A **system of linear equations** in unknowns $x_1, x_2, \ldots, x_n$ is a collection of $m$ equations $\sum_{j=1}^{n} a_{ij} x_j = b_i$, $i = 1, 2, \ldots, m$ where all $a_{ij} \in \mathcal{R}$ and all $b_i \in \mathcal{R}$. A **solution** of this system is any set of values $x_1, x_2, \ldots, x_n$ satisfying (simultaneously) the $m$ given equations. The **coefficient matrix** of this system is the $m \times n$ matrix $A = (a_{ij})$, and the **augmented matrix** is the $m \times (n+1)$ matrix $A_b = (A : b)$.

A **homogeneous** system has right-hand sides $b_i = 0$ for all $i = 1, 2, \ldots, m$; otherwise the system is **nonhomogeneous**.

**Back substitution** is a simple and efficient iterative procedure for solving an upper triangular linear system $Ux = b$, one unknown at a time.

**Forward substitution** is a simple and efficient iterative procedure for solving a lower triangular linear system $Lx = b$, one unknown at a time.

**Facts:**

**1.** A system of $m$ linear equations in the unknowns $x_1, x_2, \ldots, x_n$ can be represented by the linear system $Ax = b$ where $A$ is the $m \times n$ coefficient matrix, $x = (x_1, x_2, \ldots, x_n)^T$ is the column vector of unknowns, and $b = (b_1, b_2, \ldots, b_m)^T$ is the column vector of right-hand sides.

**2.** Given the linear system $Ax = b$, where $A$ is $m \times n$

- the system has no solution when $\operatorname{rank} A < \operatorname{rank} A_b$;
- the system has a unique solution when $\operatorname{rank} A = \operatorname{rank} A_b = n$;
- the system has infinitely many solutions when $\operatorname{rank} A = \operatorname{rank} A_b < n$; in this case the set of solutions is an affine subspace of dimension $n - \operatorname{rank} A$ (§6.1.3).

**3.** If the square matrix $A$ is nonsingular (§6.3.2), then $Ax = b$ has the unique solution vector $x = A^{-1}b$.

**4.** If the square matrix $L = (l_{ij})$ is lower triangular, then $Lx = b$ has a unique solution whenever $l_{ii} \neq 0$ for all $i$. In this case the solution can be found using *forward substitution* (Algorithm 1).

---

**Algorithm 1:   Forward substitution.**

input: $n \times n$ nonsingular matrix $L$, $n \times 1$ vector $b$

output: $n \times 1$ vector $x = L^{-1}b$

$$x_1 := \frac{b_1}{l_{11}}$$

**for** $i := 2$ **to** $n$

$$x_i := \frac{1}{l_{ii}}\left(b_i - \sum_{j=1}^{i-1} l_{ij}x_j\right)$$

---

**5.** If the square matrix $U = (u_{ij})$ is upper triangular, then $Ux = b$ has a unique solution whenever $u_{ii} \neq 0$ for all $i$. In this case the solution can be found using *back substitution* (Algorithm 2).

---

**Algorithm 2:   Back substitution.**

input: $n \times n$ nonsingular matrix $U$, $n \times 1$ vector $b$

output: $n \times 1$ vector $x = U^{-1}b$

$$x_n := \frac{b_n}{u_{nn}}$$

**for** $i := n - 1$ **down to** 1

$$x_i := \frac{1}{u_{ii}}\left(b_i - \sum_{j=i+1}^{n} u_{ij}x_j\right)$$

---

**Examples:**

**1.** The system of linear equations

$$x_1 + 3x_2 + 4x_3 = 1$$
$$3x_1 + 5x_2 + 0x_3 = 7$$

corresponds to the linear system $Ax = b$, where $A = \begin{pmatrix} 1 & 3 & 4 \\ 3 & 5 & 0 \end{pmatrix}$ and $b = \begin{pmatrix} 1 \\ 7 \end{pmatrix}$. Since $\operatorname{rank} A = \operatorname{rank} A_b = 2 < 3$ the system has an infinite number of solutions. In fact the set of solutions can be expressed as $\{(4 + 5a, -1 - 3a, a)^T \mid a \in \mathcal{R}\}$. Equivalently, it can be expressed as the affine subspace $\{(4, -1, 0)^T + a(5, -3, 1)^T \mid a \in \mathcal{R}\}$ of dimension $n - \operatorname{rank} A = 3 - 2 = 1$.

**2.** The system of linear equations

$$5x_1 - 3x_2 + 4x_3 = 4$$
$$-x_2 + 5x_3 = 7$$
$$3x_3 = 6$$

has the upper triangular coefficient matrix $U = \begin{pmatrix} 5 & -3 & 4 \\ 0 & -1 & 5 \\ 0 & 0 & 3 \end{pmatrix}$. Using Algorithm 2, the

unique solution is $x_3 = \frac{6}{3} = 2$, $x_2 = \frac{1}{-1}(7 - 5 \cdot 2) = 3$, $x_1 = \frac{1}{5}(4 - (-3) \cdot 3 - 4 \cdot 2) = 1$.

---

### 6.4.2  GAUSSIAN ELIMINATION

Solving a system of linear equations via Gaussian elimination is one of the most common computations performed by scientists and engineers. Gaussian elimination successively eliminates variables from the original system, creating a triangular system that is easily solved (§6.4.1).

*Note:* This subsection deals only with linear systems $Ax = b$, where $A$ is a nonsingular $n \times n$ real matrix and $b \in \mathcal{R}^n$.

#### Definitions:

**Gaussian elimination** is a method for solving systems of linear equations; at each step, one equation is used to eliminate one variable from the rest of the equations. The coefficient of the eliminated variable in the eliminated equation is the **pivot**.

A **flop** is a multiply-add operation of the form $t = s + ab$, especially when performed in floating point arithmetic on a digital computer.

**Roundoff errors** are the errors associated with storing and computing numbers in finite precision arithmetic on a digital computer.

A **numerically stable algorithm** is a method whose accuracy is not greatly harmed by roundoff errors.

A **numerically unstable algorithm** is a method that may return an inaccurate solution even when the solution is relatively insensitive to errors in the data.

#### Facts:

**1.** Gaussian elimination is easily extended to linear systems for which the data $A$ and $b$ are complex. Extension to rectangular $m \times n$ linear systems is more involved, but not difficult [GoVL13].

**2.** In Gaussian elimination, the coefficient of a variable in one of the equations can be used as a pivot if and only if its value is nonzero.

**3.** In practice, careful choice of pivots is needed to ensure accuracy or improve efficiency or both. (See Example 2.)

**4.** Assume freedom at each step to choose any available nonzero pivot. Then Gaussian elimination succeeds (using exact arithmetic) if and only if $A$ is nonsingular.

**5.** Gaussian elimination transforms the initial linear system into a second linear system such that

- the solutions of the two systems are identical;

- the solution of the second system is easily obtained by back substitution.

**6.** $Ax = b$ can be solved by Gaussian elimination and back substitution (Algorithm 3), assuming that at each step there is a nonzero pivot on the main diagonal.

---

**Algorithm 3**: **Gaussian elimination and back substitution.**

input: $n \times n$ nonsingular matrix $A$, $n \times 1$ vector $b$

output: $n \times 1$ vector $x = A^{-1}b$

{Gaussian elimination}

**for** $j := 1$ **to** $n - 1$

    {$a_{jj}$ is the pivot}

    **for** $i := j + 1$ **to** $n$

        {Eliminate $x_j$ from equation $i$}

        compute multiplier $m := -a_{ij}/a_{jj}$

        add $m \times$ row $j$ to row $i$

{Back substitution}

use Algorithm 2 on the resulting upper triangular matrix to obtain the values

    $x_n, x_{n-1}, \ldots, x_1$

---

**7.** Algorithm 3, implemented to take advantage of created 0s, requires $\frac{1}{3}n^3 + O(n^2)$ flops.

**8.** To solve $Ax = b$ by computing the inverse and then forming the product $A^{-1}b$ requires $n^3 + O(n^2)$ flops.

**9.** *Cramer's rule:* This method for solving $Ax = b$ expresses each entry of the solution $x = (x_1, x_2, \ldots, x_n)^T$ as the ratio of two determinants:

$$x_1 = \frac{\det A_1}{\det A}, \quad x_2 = \frac{\det A_2}{\det A}, \ldots, \quad x_n = \frac{\det A_n}{\det A},$$

where $A_i$ is obtained from $A$ by substituting column vector $b$ for the $i$th column of $A$. (Gabriel Cramer, 1704–1752)

**10.** Cramer's rule is of extremely limited use numerically because

- it requires far more flops than Gaussian elimination and back substitution;

- it is numerically unstable.

**Examples:**

**1.** The following system is solved by first applying Gaussian elimination:

$$x_1 + x_2 + 2x_3 + x_4 = 1$$
$$2x_1 + 4x_2 + 5x_3 + 4x_4 = 5 \quad (-\tfrac{2}{1} \times \text{equation 1})$$
$$x_1 + 7x_2 + 7x_3 + 6x_4 = 6 \quad (-\tfrac{1}{1} \times \text{equation 1})$$
$$2x_1 + 4x_2 + 9x_3 + 5x_4 = 3 \quad (-\tfrac{2}{1} \times \text{equation 1})$$

$$x_1 + x_2 + 2x_3 + x_4 = 1$$
$$2x_2 + x_3 + 2x_4 = 3$$
$$6x_2 + 5x_3 + 5x_4 = 5 \quad (-\tfrac{6}{2} \times \text{equation 2})$$
$$2x_2 + 5x_3 + 3x_4 = 1 \quad (-\tfrac{2}{2} \times \text{equation 2})$$

$$x_1 + x_2 + 2x_3 + x_4 = 1$$
$$2x_2 + x_3 + 2x_4 = 3$$
$$2x_3 - x_4 = -4$$
$$4x_3 + x_4 = -2 \quad (-\tfrac{4}{2} \times \text{equation 3})$$

$$x_1 + x_2 + 2x_3 + x_4 = 1$$
$$2x_2 + x_3 + 2x_4 = 3$$
$$2x_3 - x_4 = -4$$
$$3x_4 = 6$$

The solution is then obtained by back substitution:

$$x_4 = 6/3 = 2,$$
$$x_3 = [-4 + 1 \cdot 2]/2 = -1,$$
$$x_2 = [3 - 1 \cdot (-1) - 2 \cdot 2]/2 = 0,$$
$$x_1 = [1 - 1 \cdot 0 - 2 \cdot (-1) - 1 \cdot 2]/1 = 1.$$

**2.** Suppose the following system is solved, rounding all results to three significant digits:

$$0.0001x_1 + x_2 = 1$$
$$0.5x_1 + 0.5x_2 = 1 \quad (-\tfrac{0.5}{0.0001} \times \text{equation 1})$$

$$0.0001x_1 + x_2 = 1$$
$$-5000x_2 = -5000$$

Using back substitution produces $x_2 = 1$ and $x_1 = 0$. However, the correct solution to this simple linear system is $x_1 = \frac{10000}{9999}$ and $x_2 = \frac{9998}{9999}$, which to three significant digits becomes $x_1 = 1$ and $x_2 = 1$. Consequently, simply choosing any nonzero pivot can produce inaccurate results.

---

**6.4.3** *LU DECOMPOSITION*

Gaussian elimination can be formulated as *LU* decomposition of the coefficient matrix.

**Definitions:**

An **LU decomposition** of a square matrix $A$ expresses $A = LU$, where $L = (l_{ij})$ is unit lower triangular and $U = (u_{ij})$ is upper triangular.

A **permutation matrix** is a square matrix with entries 0 or 1, where the entry 1 occurs precisely once in each row and once in each column.

**Facts:**

**1.** A square matrix has an $LU$ decomposition if and only if every principal submatrix (§6.3.1) is nonsingular.

**2.** If $P$ is a permutation matrix, then the product $PA$ rearranges the rows of $A$ and the product $AP$ rearranges the columns of $A$.

**3.** The matrix $A$ is nonsingular if and only if there exists a permutation matrix $P$ such that $PA$ has an $LU$ decomposition. The $LU$ decomposition of $PA$ is unique.

**4.** It may be necessary to rearrange the rows of $A$ to avoid a zero pivot.

**5.** Assume $A$ has an $LU$ decomposition, and consider Gaussian elimination applied to $Ax = b$ with pivots on the main diagonal. The following statements express $LU$ decomposition as a reformulation of Gaussian elimination:

- the entry $u_{ij}$ $(1 \leq i \leq j \leq n)$ is the coefficient of $x_j$ in equation $i$ after Gaussian elimination has been completed;

- to eliminate $x_j$ from equation $i$, $i > j$, Gaussian elimination adds $-l_{ij} \times$ equation $j$ to equation $i$.

**6.** If $A$ has an $LU$ decomposition, then the linear system $Ax = b$ can be solved as follows (see Algorithm 4):

- compute the decomposition $A = LU$;

- solve $Ly = b$; that is, perform forward substitution;

- solve $Ux = y$; that is, perform back substitution.

**7.** It is inefficient to solve a nontrivial sequence of linear systems $Ax_1 = b_1$, $Ax_2 = b_2$, ..., $Ax_p = b_p$ by repeating Gaussian elimination for each system. Only one $LU$ decomposition is needed, followed by $p$ forward substitution steps and $p$ back substitution steps.

**8.** An $LU$ decomposition of an $n \times n$ matrix requires $n^2 + O(n)$ storage locations and $\frac{1}{3}n^3 + O(n^2)$ flops.

---

**Algorithm 4:**   **LU decomposition with forward and back substitution.**

input: $n \times n$ nonsingular matrix $A$, $n \times 1$ vector $b$

output: $n \times 1$ vector $x = A^{-1}b$

{Compute $A = LU$}

**for** $k := 1$ **to** $n - 1$

$\quad u_{kk} := a_{kk}$

$\quad$ **for** $i := k + 1$ **to** $n$

$\quad\quad l_{ik} := a_{ik}/a_{kk};\ u_{ki} := a_{ki}$

$\quad\quad$ **for** $j := k + 1$ **to** $n$

$\quad\quad\quad$ **for** $i := k + 1$ **to** $n$

$\quad\quad\quad\quad a_{ij} := a_{ij} - l_{ik}u_{kj}$

{Solve $Ly = b$; that is, perform forward substitution}

for $i := 1$ to $n$

$$y_i := b_i - \sum_{j=1}^{i-1} l_{ij} y_j$$

{Solve $Ux = y$; that is, perform back substitution}

for $i := n$ down to 1

$$x_i := \frac{1}{u_{ii}} \left( y_i - \sum_{j=i+1}^{n} u_{ij} x_j \right)$$

**Examples:**

**1.** The following matrix $A$ has no $LU$ decomposition because $a_{11} = 0$ (see Fact 1):

$$A = \begin{pmatrix} 0 & 1 \\ 2 & 3 \end{pmatrix}.$$

However, rearranging the rows of $A$ (Fact 4) produces

$$PA = \begin{pmatrix} 0 & 1 \\ 1 & 0 \end{pmatrix} \begin{pmatrix} 0 & 1 \\ 2 & 3 \end{pmatrix} = \begin{pmatrix} 2 & 3 \\ 0 & 1 \end{pmatrix} = \begin{pmatrix} 1 & 0 \\ 0 & 1 \end{pmatrix} \begin{pmatrix} 2 & 3 \\ 0 & 1 \end{pmatrix} = LU.$$

**2.** The unique $LU$ decomposition of the matrix $A$ in Example 1, §6.4.2 is

$$\begin{pmatrix} 1 & 1 & 2 & 1 \\ 2 & 4 & 5 & 4 \\ 1 & 7 & 7 & 6 \\ 2 & 4 & 9 & 5 \end{pmatrix} = \begin{pmatrix} 1 & 0 & 0 & 0 \\ 2 & 1 & 0 & 0 \\ 1 & 3 & 1 & 0 \\ 2 & 1 & 2 & 1 \end{pmatrix} \begin{pmatrix} 1 & 1 & 2 & 1 \\ 0 & 2 & 1 & 2 \\ 0 & 0 & 2 & -1 \\ 0 & 0 & 0 & 3 \end{pmatrix}.$$

## 6.4.4   CHOLESKY DECOMPOSITION

For symmetric positive definite linear systems, Cholesky decomposition (which exploits symmetry in the coefficient matrix) is roughly twice as efficient as $LU$ decomposition.

**Definition:**

A **Cholesky decomposition** of $A$ expresses $A = LL^T$, where $L$ is lower triangular and every entry on the main diagonal of $L$ is positive.

**Facts:**

**1.** A matrix has a Cholesky decomposition if and only if it is symmetric and positive definite.

**2.** When $A$ is symmetric and positive definite, the linear system $Ax = b$ can be solved as follows:

- compute a Cholesky decomposition $A = LL^T$;
- solve $Ly = b$; i.e., perform forward substitution;
- solve $L^T x = y$; i.e., perform back substitution.

**3.** A simple symmetric variant of the standard $LU$ decomposition algorithm is used to compute Cholesky decomposition [GoVL13, St88].

**4.** Cholesky decomposition requires $\frac{1}{2}n^2+O(n)$ storage locations and $\frac{1}{6}n^3+O(n^2)$ flops, in contrast to the $n^2 + O(n)$ storage locations and $\frac{1}{3}n^3 + O(n^2)$ flops required by $LU$ decomposition.

**Example:**

**1.** The matrix $A = \begin{pmatrix} 1 & -1 & 3 \\ -1 & 2 & -1 \\ 3 & -1 & 14 \end{pmatrix}$ is clearly symmetric. It is positive definite since

its principal submatrices $(1)$, $\begin{pmatrix} 1 & -1 \\ -1 & 2 \end{pmatrix}$, and $A$ have positive determinants. (See §6.3.4, Fact 22.) Matrix $A$ can be written as $A = LL^T$, where $L$ is the lower triangular matrix $\begin{pmatrix} 1 & 0 & 0 \\ -1 & 1 & 0 \\ 3 & 2 & 1 \end{pmatrix}$. To solve the linear system $Ax = b$ with $b = (1,0,6)^T$, first solve the lower triangular system $Ly = b$, yielding $y = (1,1,1)^T$. Then solve the upper triangular system $L^T x = y$, yielding $x = (-3,-1,1)^T$.

## 6.4.5  CONDITIONING OF LINEAR SYSTEMS

Errors in the data $A$ and $b$ lead to errors in the solution $x$. The *condition number* of $A$ can be used to bound relative error in the solution in terms of relative errors in the data.

**Definitions:**
A (***generalized***) ***vector norm*** on $\mathcal{R}^n$ is a real-valued function $\|\cdot\|$ satisfying the following properties for all real scalars $a$ and all vectors $x, y \in \mathcal{R}^n$:

- $\|x\| \geq 0$ with equality if and only if $x = 0$;
- $\|ax\| = |a| \cdot \|x\|$, where $|a|$ denotes the absolute value of $a$;
- $\|x + y\| \leq \|x\| + \|y\|$.

The ***matrix norm*** induced by the vector norm $\|\cdot\|$ is defined by $\|A\| = \max_{\|x\|=1} \|Ax\|$.

The ***condition number*** of a nonsingular matrix $A$ is the number $\kappa(A) = \|A\| \, \|A^{-1}\|$. The larger the condition number of a matrix, the more ***ill conditioned*** it is; the smaller the condition number of a matrix, the more ***well conditioned*** it is.

**Facts:**
**1.** The definition of a vector norm given here generalizes that of a vector norm derived from an inner product space (§6.1.4).

**2.** The matrix norm induced by a vector norm satisfies

- $\|X\| \geq 0$ with equality if and only if $X = 0$;
- $\|aX\| = |a| \cdot \|X\|$, where $|a|$ denotes the absolute value of $a$;
- $\|X + Y\| \leq \|X\| + \|Y\|$;
- $\|XY\| \leq \|X\|\|Y\|$.

**3.** $\kappa(A) \geq 1$.

**4.** Consider the linear system $Ax = b \neq 0$, where $A$ is nonsingular. Suppose that changing from $A$ to $A + \Delta A$ and $b$ to $b + \Delta b$ changes the solution from $x$ to $x + \Delta x$. If $\|(\Delta A)\| \, \|A^{-1}\| < 1$, the relative error in $x$ can be bounded in terms of relative errors in the data:

$$\frac{\|\Delta x\|}{\|x\|} \leq \left(\frac{\kappa(A)}{1 - \|(\Delta A)\| \, \|A^{-1}\|}\right)\left(\frac{\|\Delta b\|}{\|b\|} + \frac{\|\Delta A\|}{\|A\|}\right).$$

**5.** The following are consequences of Fact 4:

- For an ill-conditioned linear system $Ax = b$, *some* small errors in $A$ or $b$ can potentially be amplified into large errors in $x$;

- For a well-conditioned linear system $Ax = b$, where $1 - \|(\Delta A)\| \, \|A^{-1}\|$ is not approximately zero, *all* small errors in $A$ or $b$ result in no more than modest errors in $x$.

**6.** Assume $A$ is nonsingular, let $Ax = b \neq 0$, and view $\hat{x}$ as an approximation to the solution $x$. Then the residual $r = A\hat{x} - b$ and the error $\hat{x} - x$ satisfy

$$\frac{\|\hat{x} - x\|}{\|x\|} \leq \kappa(A)\frac{\|r\|}{\|b\|}.$$

**7.** Whenever $A$ is ill conditioned, a small relative residual $\|r\|/\|b\|$ may not imply a small relative error $\|\hat{x} - x\|/\|x\|$.

**Examples:**

**1.** The standard Euclidean norm (§6.1.4) on $\mathcal{R}^n$ defined by $\|x\|_2 = (\sum_{i=1}^{n} x_i^2)^{1/2}$ is a (generalized) vector norm.

**2.** The $l_1$ norm on $\mathcal{R}^n$ defined by $\|x\|_1 = \sum_{i=1}^{n} |x_i|$ is a (generalized) vector norm.

**3.** In coding theory (§14.1), the *Hamming distance* between two codewords $x, y \in \mathcal{Z}_2^n$ is just $\|x - y\|_1$.

**4.** The $l_\infty$ norm on $\mathcal{R}^n$ defined by $\|x\|_\infty = \max_{1 \leq i \leq n} |x_i|$ is a (generalized) vector norm.

**5.** The matrix norm induced by $\|x\|_1$ is given by $\|A\|_1 = \max_{1 \leq j \leq n} \sum_{i=1}^{n} |a_{ij}|$.

**6.** The matrix norm induced by $\|x\|_\infty$ is given by $\|A\|_\infty = \max_{1 \leq i \leq n} \sum_{j=1}^{n} |a_{ij}|$.

**7.** The matrix norm induced by $\|x\|_2$, also called the *spectral norm*, is given by $\|A\|_2 = \max\{\sqrt{\lambda} \mid \lambda \text{ an eigenvalue of } A^T A\}$.

**8.** Consider the linear system $Ax = b$, where $A = \begin{pmatrix} 1 & 2 \\ 2.001 & 4 \end{pmatrix}$ and $b = \begin{pmatrix} 2 \\ 4 \end{pmatrix}$. Then $\|A\|_\infty = 6.001$ and $\|A^{-1}\|_\infty = 3{,}000$, so that $\kappa(A) = 18{,}003$. The solution of $Ax = b$ is $x = (0, 1)^T$ whereas the solution of the slightly perturbed system with $\hat{b} = \begin{pmatrix} 2 \\ 4.001 \end{pmatrix}$ is $x = (1, 0.5)^T$. Even though the change in the right-hand side is small, the large condition number allows for radical changes in the solution vector, as seen here.

## 6.4.6  PIVOTING FOR STABILITY

Gaussian elimination can be numerically unstable. Numerical stability can be vastly improved by the addition of pivoting strategies that select large pivots.

**Definitions:**

Let $a_{ij}^{(k)}$ denote the $ij$-entry of the current matrix after step $k$ of Gaussian elimination (or $LU$ decomposition). The **growth factor** is defined by $\dfrac{\max\limits_{i,j,k} |a_{ij}^{(k)}|}{\max\limits_{i,j} |a_{ij}|}$.

**Partial pivoting** is a solution strategy which at step $k$ of Gaussian elimination exchanges row $k$ with the row $i \geq k$ having the entry of largest magnitude in column $k$.

**Complete pivoting** is a solution strategy which at step $k$ of Gaussian elimination exchanges row $k$ and column $k$ with, respectively, the row $i \geq k$ and the column $j \geq k$ containing the entry of largest magnitude.

**Facts:**

**1.** For general coefficient matrices, Gaussian elimination (that is, $LU$ decomposition) without pivoting is numerically unstable.

**2.** To improve the numerical stability of Gaussian elimination, it suffices to introduce a pivoting strategy that keeps the growth factor small.

**3.** For Gaussian elimination with complete pivoting the growth factor is bounded above by

$$n^{1/2}\big(2^1 3^{1/2} 4^{1/3} \ldots n^{1/(n-1)}\big)^{1/2},$$

which is a relatively slow-growing function of $n$; hence, Gaussian elimination with complete pivoting is numerically stable.

**4.** For Gaussian elimination with partial pivoting, the growth factor is bounded above by $2^{n-1}$, and moreover there are contrived examples for which the growth factor is $2^{n-1}$. Hence, Gaussian elimination with partial pivoting can be numerically unstable.

**5.** In practice, partial pivoting is preferred over complete pivoting for the following two reasons:

- Despite contrived examples having an exponential growth factor, partial pivoting limits the growth factor in practice almost as well as complete pivoting;

- Partial pivoting is significantly more efficient than complete pivoting; it compares $\frac{1}{2}n^2 + O(n)$ pairs of potential pivots, while complete pivoting compares $\frac{1}{3}n^3 + O(n^2)$ pairs.

**Example:**

**1.** $LU$ decomposition applied to the following matrix shows that partial pivoting can produce a growth factor of $2^{n-1}$ (see Fact 4). Observe that $\max_{i,j} |a_{ij}| = 1$ and $\max_{i,j,k} |a_{ij}^{(k)}| = u_{nn} = 2^{n-1}$; hence the growth factor is $2^{n-1}$.

$$\begin{pmatrix} 1 & 0 & 0 & \cdots & 0 & 1 \\ -1 & 1 & 0 & \cdots & 0 & 1 \\ -1 & -1 & 1 & \cdots & 0 & 1 \\ \vdots & \vdots & \vdots & & \vdots & \vdots \\ -1 & -1 & -1 & \cdots & 0 & 1 \\ -1 & -1 & -1 & \cdots & 1 & 1 \\ -1 & -1 & -1 & \cdots & -1 & 1 \end{pmatrix}$$

## 6.4.7   PIVOTING TO PRESERVE SPARSITY

Many, if not most, linear systems that arise in practice have relatively few nonzero entries in the coefficient matrix. Some pivoting strategies aim to preserve many zero entries in the triangular factors; the $LU$ decomposition algorithm can then save time and space by omitting zero entries from the computation.

### Definitions:

A matrix is **sparse** if it has relatively few nonzero entries. The number of nonzero entries of matrix $A$ is denoted $|A|$. The $i$th row of $A$ is denoted $A(i,:)$ and the $j$th column of $A$ is denoted $A(:,j)$. (See §6.3.1.)

**Fill** refers to nonzero entries in the triangular factors whose corresponding positions in the coefficient matrix are occupied by zeros.

The **upper bandwidth** and **lower bandwidth** of a matrix $A$ are given respectively by $\mathrm{ub}(A) = \max\{\,(j-i) \mid a_{ij} \neq 0,\, i < j\,\}$, $\mathrm{lb}(A) = \max\{\,(i-j) \mid a_{ij} \neq 0,\, i > j\,\}$.

A **banded** $LU$ decomposition algorithm stores and computes all entries of $L$ and $U$ within the band defined by $\mathrm{lb}(A)$ and $\mathrm{ub}(A)$.

A **general sparse** $LU$ decomposition algorithm stores and computes only the nonzero entries in the triangular factors, irrespective of the banded structure.

The **Markowitz pivoting strategy** for Gaussian elimination chooses at step $k$ from among all available pivots one that minimizes the product $(|L(:,k)| - 1)(|U(k,:)| - 1)$.

The **minimum degree algorithm** is a restricted version of the Markowitz pivoting strategy; it assumes (and preserves) symmetry in the coefficient matrix. At step $k$ of Gaussian elimination, this algorithm chooses from among the entries on the main diagonal a pivot that minimizes $|L(:,k)|$.

*Note:* The realistic "no-cancellation" assumption will be made throughout. Namely, once an entry becomes nonzero during a triangular decomposition, it will be nonzero upon termination.

### Facts:

**1.** The amount of fill in triangular factors often varies greatly with the choice of pivots.

**2.** Under the no-cancellation assumption, bandwidth reduction and fill reduction become combinatorial optimization problems.

**3.** The following problems are provably intractable (i.e., NP-hard; see §17.5):

- for a symmetric matrix $A$, find a permutation matrix $P$ that minimizes the bandwidth $\mathrm{lb}(PAP^T)$;

- for a nonsingular matrix $A$, find permutation matrices $P$ and $Q$ such that the $LU$ decomposition $PAQ = LU$ exists and $|L| + |U|$ is minimum;

- for a symmetric positive definite matrix $A$, find a permutation matrix $P$ that minimizes $|L|$, where $L$ is the Cholesky factor of $PAP^T$.

**4.** In view of Fact 3, various heuristics are used to reduce bandwidth or to reduce fill.

**5.** Assume that $A$ has an $LU$ decomposition. Then $\text{lb}(L) = \text{lb}(A)$ and $\text{ub}(U) = \text{ub}(A)$.

**6.** The chief advantage of a banded $LU$ decomposition algorithm over a general sparse $LU$ decomposition algorithm is its simplicity. The same advantage holds for *profile* and *skyline* methods, both of which are generalizations of the banded approach [GeLi81].

**7.** For most problems encountered in practice, a banded $LU$ decomposition algorithm, even if $A$ has been permuted so that $\text{lb}(A)$ and $\text{ub}(A)$ are minimum, requires much more space and work than a general sparse $LU$ decomposition algorithm coupled with the Markowitz pivoting strategy. The same comment applies to profile and skyline methods.

**8.** Let $A$ be a symmetric positive definite matrix, and let $P$ be a permutation matrix with the same number of rows and columns.

- the Cholesky decomposition of $PAP^T$ exists and is numerically stable;

- the undirected graph (§8.1) $G$ of the Cholesky factor of $PAP^T$ is a *chordal graph* and $P$ defines a *perfect elimination ordering* of $G$ [GeLi81].

**9.** General sparse Cholesky decomposition can be handled in a clean, modular fashion:

- using only the positions of nonzeros in $A$ as input, compute a permutation $P$ to reduce fill in the Cholesky factor of $PAP^T$ (using, for example, the minimum degree algorithm);

- construct data structures to contain the nonzeros of the Cholesky factor;

- after putting the nonzero entries of $PAP^T$ into the data structures, compute the Cholesky factor of $PAP^T$ in the provided data structures;

- perform forward and back substitutions to solve the linear system.

**10.** For symmetric positive definite matrices arising from two-dimensional and three-dimensional partial differential equations, the *nested dissection* algorithm often computes a more effective fill-reducing permutation than does the minimum degree algorithm [GeLi81].

**11.** The interplay between pivoting for stability and pivoting for sparsity complicates general sparse $LU$ factorization. The best approach is not yet certain.

**12.** A number of robust and well-tested software packages for solving linear systems are available at

- http://www.mathworks.com
- http://www.nag.com
- http://www.netlib.org
- http://www.roguewave.com

**Examples:**

**1.** For any "arrowhead" matrix there is a pivot sequence that completely fills the matrix and another that creates no fill, making it the canonical example used to illustrate Fact 1. The following is a $4 \times 4$ arrowhead matrix that fills in completely. (Here $\star$ occupies a position that is nonzero in $A$, $\bullet$ is a fill entry in $L$ or $U$, and a space is a zero.)

$$A = \begin{pmatrix} \star & \star & \star & \star \\ \star & \star & & \\ \star & & \star & \\ \star & & & \star \end{pmatrix}, \qquad A = LU = \begin{pmatrix} \star & & & \\ \star & \star & & \\ \star & \bullet & \star & \\ \star & \bullet & \bullet & \star \end{pmatrix} \begin{pmatrix} \star & \star & \star & \star \\ & \star & \bullet & \bullet \\ & & \star & \bullet \\ & & & \star \end{pmatrix}.$$

Reversing the pivot sequence, however, results in no fill:

$$\widehat{A} = PAP^T = \begin{pmatrix} & & & 1 \\ & & 1 & \\ & 1 & & \\ 1 & & & \end{pmatrix} \begin{pmatrix} \star & \star & \star & \star \\ \star & \star & & \\ \star & & \star & \\ \star & & & \star \end{pmatrix} \begin{pmatrix} & & & 1 \\ & & 1 & \\ & 1 & & \\ 1 & & & \end{pmatrix} = \begin{pmatrix} \star & & & \star \\ & \star & & \star \\ & & \star & \star \\ \star & \star & \star & \star \end{pmatrix},$$

$$\widehat{A} = \widehat{L}\widehat{U} = \begin{pmatrix} \star & & & \\ & \star & & \\ & & \star & \\ \star & \star & \star & \star \end{pmatrix} \begin{pmatrix} \star & & & \star \\ & \star & & \star \\ & & \star & \star \\ & & & \star \end{pmatrix}.$$

**2.** The following table illustrates how Fact 7 typically manifests itself in practice. The four problems arise in finite element modeling of actual structures. The table records data for two distinct methods:

- a profile-reducing permutation from the reverse Cuthill-McKee algorithm [GeLi81] in tandem with a profile factorization algorithm;
- a fill-reducing permutation from the minimum degree algorithm [GeLi81] in tandem with a general sparse factorization algorithm.

Recorded for each method are the number of nonzero entries in the Cholesky factor (expressed in millions) and the number of flops needed to compute the factor (expressed in millions).

| problem | $n$ | $|A|$ | $|L|\,(\times 10^{-6})$ | | No. flops $(\times 10^{-6})$ | |
|---|---|---|---|---|---|---|
| | | | profile reduction | general sparse | profile reduction | general sparse |
| coliseum | 1,806 | 63,454 | 0.190 | 0.112 | 11.803 | 4.952 |
| winter sports arena | 3,562 | 159,910 | 0.538 | 0.279 | 44.245 | 16.352 |
| nuclear power station | 11,948 | 149,090 | 5.908 | 0.663 | 2,135.2 | 70.779 |
| 76 story skyscraper | 15,439 | 252,241 | 2.637 | 1.417 | 232.79 | 142.57 |

## 6.5   EIGENANALYSIS

Identifying the eigenvalues and eigenvectors of a matrix facilitates the study of complicated systems and the analysis of their behavior over time. A basis consisting of eigenvectors yields a particularly simple representation of a linear transformation (§6.2). Eigenvalues can also provide useful information about discrete structures (§8.10.2).

### 6.5.1   EIGENVALUES AND CHARACTERISTIC POLYNOMIALS

**Definitions:**

A complex number $\lambda$ is an **eigenvalue** of the $n \times n$ complex matrix $A$ if there exists a nonzero vector $x \in C^n$ (an **eigenvector** of $A$ corresponding to $\lambda$) such that $Ax = \lambda x$.

The **characteristic polynomial** of the square matrix $A$ is the polynomial $p_A(\lambda) = \det(\lambda I - A)$.

The **characteristic equation** of $A$ is the equation $p_A(\lambda) = 0$.

A **nilpotent** matrix is a square matrix $A$ such that $A^k = 0$ for some positive integer $k$.

An **idempotent** matrix is a square matrix $A$ such that $A^2 = A$.

Let $S_k(A)$ denote the sum of all order $k$ principal minors of the matrix $A$.

**Facts:**

**1.** The characteristic polynomial $p_A(\lambda)$ of an $n \times n$ matrix $A$ is a monic polynomial of degree $n$ in $\lambda$.

**2.** The coefficient of $\lambda^{n-1}$ in $p_A(\lambda)$ is $-\operatorname{tr} A$.

**3.** The constant term in $p_A(\lambda)$ is $(-1)^n \det A$.

**4.** $p_A(\lambda) = \sum_{k=0}^{n} (-1)^k S_k(A) \lambda^{n-k}$.

**5.** Similar matrices (§6.2.4) have the same characteristic polynomial.

**6.** The roots of the characteristic equation are the eigenvalues of $A$.

**7.** *Cayley-Hamilton theorem:* If $p_A(\cdot)$ is the characteristic polynomial of $A$ then $p_A(A)$ is the zero matrix.

**8.** An $n \times n$ matrix has $n$ (not necessarily distinct) eigenvalues.

**9.** The matrix $A$ is singular if and only if 0 is an eigenvalue of $A$.

**10.** The characteristic equation of $A = \begin{pmatrix} a & b \\ c & d \end{pmatrix}$ is $p_A(\lambda) = \lambda^2 - (a+d)\lambda + (ad - bc)$.

**11.** The eigenvalues of $A = \begin{pmatrix} a & b \\ c & d \end{pmatrix}$ are given by $\dfrac{(a+d) \pm \sqrt{(a-d)^2 + 4bc}}{2}$.

**12.** If the $n \times n$ matrix $A$ has eigenvalues $\lambda_1, \lambda_2, \ldots, \lambda_n$ then

- $\sum_{i=1}^{n} \lambda_i = \operatorname{tr} A$;
- $\prod_{i=1}^{n} \lambda_i = \det A$;
- the $k$th elementary symmetric function $\sum_{i_1 < \cdots < i_k} \lambda_{i_1} \ldots \lambda_{i_k}$ equals $S_k(A)$.

**13.** The eigenvalues are continuous functions of the entries of a matrix. More precisely, given an $n \times n$ matrix $A$ with eigenvalues $\lambda_1, \lambda_2, \ldots, \lambda_n$ and $\epsilon > 0$, there exists $\delta > 0$ such that for any $n \times n$ matrix $B$, with eigenvalues $\mu_1, \mu_2, \ldots, \mu_n$ and satisfying $\max_{i,j} |a_{ij} - b_{ij}| < \delta$, there exists a permutation $\tau$ of $1, 2, \ldots, n$ such that $|\lambda_i - \mu_{\tau(i)}| < \epsilon$, $i = 1, 2, \ldots, n$.

**14.** The following table gives the eigenvalues of certain specialized matrices $A$, whose eigenvalues are $\lambda_1, \lambda_2, \ldots, \lambda_n$. In this table $k$ is any positive integer.

| matrix | eigenvalues |
|---|---|
| diagonal matrix | diagonal elements |
| upper (or lower) triangular matrix | diagonal elements |
| $A^T$ | eigenvalues of $A$ |
| $A^*$ | complex conjugates of the eigenvalues of $A$ |
| $A^k$ | $\lambda_1^k, \ldots, \lambda_n^k$ |
| $A^{-k}$, $A$ nonsingular | $\lambda_1^{-k}, \ldots, \lambda_n^{-k}$ |
| $q(A)$, where $q(\cdot)$ is a polynomial | $q(\lambda_1), \ldots, q(\lambda_n)$ |
| $SAS^{-1}$, $S$ nonsingular | eigenvalues of $A$ |
| $AB$, where $A$ is $m \times n$, $B$ is $n \times m$, $m \geq n$ | eigenvalues of $BA$; and 0 ($m - n$ times) |
| $(a - b)I_n + bJ_n$, where $J_n$ is the $n \times n$ matrix of all 1s | $a + (n - 1)b$; and $a - b$ ($n - 1$ times) |
| $A$ $n \times n$ nilpotent | 0 ($n$ times) |
| $A$ $n \times n$ idempotent of rank $r$ | 1 ($r$ times); and 0 ($n - r$ times) |

**Examples:**

**1.** The characteristic polynomial for $A = \begin{pmatrix} 1 & 4 \\ 2 & 3 \end{pmatrix}$ is $p_A(\lambda) = \begin{vmatrix} \lambda - 1 & -4 \\ -2 & \lambda - 3 \end{vmatrix} = \lambda^2 - 4\lambda - 5 = (\lambda + 1)(\lambda - 5)$, so the eigenvalues are $\lambda = -1$ and $\lambda = 5$. The vector $x = (2, -1)^T$ is an eigenvector for $\lambda = -1$ since $Ax = (-2, 1)^T = -x$. The vector $x = (1, 1)^T$ is an eigenvector for $\lambda = 5$ since $Ax = (5, 5)^T = 5x$.

**2.** For the matrix in Example 1,

$$p_A(A) = A^2 - 4A - 5I = \begin{pmatrix} 9 & 16 \\ 8 & 17 \end{pmatrix} - \begin{pmatrix} 4 & 16 \\ 8 & 12 \end{pmatrix} - \begin{pmatrix} 5 & 0 \\ 0 & 5 \end{pmatrix} = \begin{pmatrix} 0 & 0 \\ 0 & 0 \end{pmatrix},$$

as required by Fact 7.

**3.** The characteristic polynomial of the matrix $A = \begin{pmatrix} 3 & 0 & 2 \\ 4 & 1 & 4 \\ 2 & 0 & 3 \end{pmatrix}$ can be calculated by

using Facts 1–4. Since $\mathrm{tr}\, A = 7$, $\det A = 5$, and $S_2(A) = \begin{vmatrix} 1 & 4 \\ 0 & 3 \end{vmatrix} + \begin{vmatrix} 3 & 2 \\ 2 & 3 \end{vmatrix} + \begin{vmatrix} 3 & 0 \\ 4 & 1 \end{vmatrix} = 11$,

it follows that $p_A(\lambda) = \lambda^3 - 7\lambda^2 + 11\lambda - 5$. Thus $p_A(5) = 0$, showing that $\lambda = 5$ is an eigenvalue of $A$. An eigenvector corresponding to $\lambda = 5$ is $x = (1, 2, 1)^T$ since $Ax = (5, 10, 5)^T = 5x$.

**4.** The matrix $A$ in Example 3 is nonsingular since $p_A(0) \neq 0$, so 0 is not an eigenvalue of $A$ (Fact 9). The inverse of $A$ can be calculated using the Cayley-Hamilton theorem: $A^3 - 7A^2 + 11A - 5I = 0$, so $5I = A^3 - 7A^2 + 11A = A(A^2 - 7A + 11I)$ and $I =$

$$A[\tfrac{1}{5}(A^2 - 7A + 11I)]. \text{ Consequently, } A^{-1} = \tfrac{1}{5}(A^2 - 7A + 11I) = \tfrac{1}{5}\begin{pmatrix} 3 & 0 & -2 \\ -4 & 5 & -4 \\ -2 & 0 & 3 \end{pmatrix}.$$

## 6.5.2   EIGENVECTORS AND DIAGONALIZATION

### Definitions:

Let $\lambda$ be an eigenvalue of the $n \times n$ (complex) matrix $A$. The **algebraic multiplicity** of $\lambda$ is its multiplicity as a root of the characteristic polynomial.

The **eigenspace** of $A$ corresponding to $\lambda$ is the vector space $\{\, x \in C^n \mid Ax = \lambda x \,\}$.

The **geometric multiplicity** of $\lambda$ is the dimension of the eigenspace of $A$ corresponding to $\lambda$.

The square matrix $A$ is **diagonalizable** if there exists a nonsingular matrix $P$ such that $P^{-1}AP$ is a diagonal matrix.

The **minimal polynomial** of the square matrix $A$ is the monic polynomial $q(\cdot)$ of minimum degree such that $q(A) = 0$.

The square matrix $A$ is **normal** if $AA^* = A^*A$.

The **singular values** of an $n \times n$ matrix $A$ are the (positive) square roots of the eigenvalues of $AA^*$, written $\sigma_1(A) \leq \sigma_2(A) \leq \cdots \leq \sigma_n(A)$.

A **row stochastic** matrix is a matrix with all entries nonnegative and row sums 1.

### Facts:

**1.** The eigenspace corresponding to $\lambda$ is a subspace of the vector space $C^n$. Specifically, it is the null space (§6.1.2) of the matrix $A - \lambda I$.

**2.** Eigenvectors corresponding to distinct eigenvalues are linearly independent.

**3.** If $\lambda, \mu$ are distinct eigenvalues of $A$ and if $Ax = \lambda x$ and $A^*y = \mu y$, then $x, y$ are orthogonal.

**4.** The algebraic multiplicity is never less than the geometric multiplicity, but sometimes it is greater. (See Example 3.)

**5.** The minimal polynomial is unique.

**6.** If $A$ can be diagonalized to a diagonal matrix $D$, then the eigenvalues of $A$ appear along the diagonal of $D$.

**7.** The following conditions are equivalent for an $n \times n$ matrix $A$:

- $A$ is diagonalizable;
- $A$ has $n$ linearly independent eigenvectors;
- the minimal polynomial of $A$ has distinct linear factors;
- the algebraic multiplicity of each eigenvalue of $A$ equals its geometric multiplicity.

**8.** If the $n \times n$ matrix $A$ has $n$ distinct eigenvalues then $A$ is diagonalizable.

**9.** If the $n \times n$ matrix $A$ has $n$ linearly independent eigenvectors $v_1, v_2, \ldots, v_n$ then $A$ is diagonalizable using the matrix $P$ whose columns are the vectors $v_1, v_2, \ldots, v_n$.

**10.** Hermitian, skew-Hermitian, and unitary matrices are normal matrices.

**11.** *Spectral theorem for normal matrices:* If $A$ is an $n \times n$ normal matrix, then it can be diagonalized by a unitary matrix. That is, there exists an $n \times n$ unitary matrix $U$ such that $U^*AU = \text{diag}(\lambda_1, \lambda_2, \ldots, \lambda_n)$, the diagonal matrix with the eigenvalues $\lambda_1, \lambda_2, \ldots, \lambda_n$ of $A$ along its diagonal.

**12.** If $A$ is normal, then it has a *spectral decomposition* $A = \sum_{i=1}^{n} \lambda_i u_i u_i^*$, where $\{u_1, u_2, \ldots, u_n\}$ is an orthonormal basis for $\mathcal{C}^n$.

**13.** Diagonalization results for special types of normal matrices are given in the following table.

| matrix $A$ | eigenvalues | diagonalization result |
|---|---|---|
| Hermitian | real | Fact 11 |
| real symmetric | real | there exists a real orthogonal $P$ such that $P^T AP$ is diagonal |
| skew-Hermitian | purely imaginary | Fact 11 |
| real skew-symmetric | purely imaginary | there exists a real orthogonal $Q$ such that $Q^T AQ$ is a direct sum of matrices, each of which is a $2 \times 2$ real skew-symmetric or null matrix |
| unitary | all with modulus 1 | Fact 11 |

**14.** If $A, B$ are normal and commute, they can be simultaneously diagonalized. Namely, there exists a unitary $U$ such that $U^*AU$ and $U^*BU$ are both diagonal.

**15.** For any square matrix $A$, the rank of $A$ is never less than the number of nonzero eigenvalues (counting multiplicities) of $A$.

**16.** If $A$ is normal then its rank equals the number of nonzero eigenvalues.

**17.** *Schur's triangularization theorem:* Suppose $A$ is a square matrix. Then there exists a unitary $U$ such that $U^*AU$ is upper triangular with the eigenvalues of $A$ on its diagonal.

**18.** If $A, B$ are square matrices that commute then there exists a unitary $U$ such that $U^*AU$ and $U^*BU$ are both upper triangular.

**19.** *Jordan canonical form:* Let $A$ be an $n \times n$ matrix with the distinct eigenvalues $\lambda_1, \lambda_2, \ldots, \lambda_k$ having (algebraic) multiplicities $r_1, r_2, \ldots, r_k$ respectively. Then there exists a nonsingular matrix $P$ such that $P^{-1}AP = \text{diag}(\Lambda_1, \Lambda_2, \ldots, \Lambda_k)$, where

$$\Lambda_i = \begin{pmatrix} \lambda_i & * & 0 & \cdots & 0 & 0 \\ 0 & \lambda_i & * & \cdots & 0 & 0 \\ \vdots & \vdots & & & \vdots & \vdots \\ 0 & 0 & 0 & \cdots & \lambda_i & * \\ 0 & 0 & 0 & \cdots & 0 & \lambda_i \end{pmatrix}$$

is an $r_i \times r_i$ matrix and each $*$ is either 0 or 1. Furthermore, the number of 1s is $r_i$ minus the geometric multiplicity of $\lambda_i$.

**20.** The rank of a square matrix equals the number of nonzero singular values.

**21.** If $A$ is a square matrix and if $U$ and $V$ are unitary, then $A$ and $UAV$ have the same singular values.

**22.** *Singular value decomposition*: If $A$ is an $n \times n$ matrix then there exist $n \times n$ unitary matrices $U, V$ such that $UAV$ is diagonal with $\sigma_1(A), \sigma_2(A), \ldots, \sigma_n(A)$ on the diagonal.

**23.** *QR factorization*: If $A$ is an $n \times n$ matrix then there exists a unitary matrix $Q$ and an upper triangular matrix $R$ such that $A = QR$.

**24.** The $QR$ factorization of a matrix can be calculated using Gram-Schmidt orthogonalization (§6.1.4).

**Examples:**

**1.** Let $x, y$ be vectors of size $n \times 1$ and let $A = xy^T$. Then the eigenvalues of $A$ are given by $y^T x$ and 0, the latter with multiplicity $n - 1$.

**2.** The matrix given in Example 3, §6.5.1 has the characteristic polynomial $p_A(\lambda) = \lambda^3 - 7\lambda^2 + 11\lambda - 5 = (\lambda-1)^2(\lambda-5)$. The eigenvalues are $\lambda = 1$ with algebraic multiplicity 2 and $\lambda = 5$ with algebraic multiplicity 1.

For $\lambda = 1$ the eigenspace is the null space of $A - \lambda I = \begin{pmatrix} 2 & 0 & 2 \\ 4 & 0 & 4 \\ 2 & 0 & 2 \end{pmatrix}$. It consists of all

vectors of the form $(a, b, -a)^T$ and so is spanned by the linearly independent eigenvectors $(1, 0, -1)^T$ and $(0, 1, 0)^T$. Thus the geometric multiplicity of $\lambda = 1$ is 2, the same as its algebraic multiplicity 1.

As seen in Example 3 of §6.5.1, the eigenvalue $\lambda = 5$ has the eigenvector $(1, 2, 1)^T$, linearly

independent of the previous two eigenvectors (Fact 2). If $P = \begin{pmatrix} 1 & 0 & 1 \\ 0 & 1 & 2 \\ -1 & 0 & 1 \end{pmatrix}$ is the

matrix containing these eigenvectors then $P^{-1}AP = \mathrm{diag}(1, 1, 5)$, thereby diagonalizing $A$ (Fact 9).

**3.** By using *Maple*, the characteristic polynomial of the matrix

$$A = \begin{pmatrix} 7 & 2 & 4 & 0 & 3 \\ 0 & 6 & 0 & 0 & 0 \\ 0 & -2 & 4 & 0 & 0 \\ 3 & 2 & 4 & 4 & 3 \\ 3 & 0 & 2 & 0 & 7 \end{pmatrix}$$

is found to be $\lambda^5 - 28\lambda^4 + 300\lambda^3 - 1552\lambda^2 + 3904\lambda - 3840 = (\lambda - 4)^3(\lambda - 6)(\lambda - 10)$, so the eigenvalue $\lambda = 4$ of $A$ has algebraic multiplicity 3. The eigenspace for $\lambda = 4$ is the null space of

$$A - \lambda I = \begin{pmatrix} 3 & 2 & 4 & 0 & 3 \\ 0 & 2 & 0 & 0 & 0 \\ 0 & -2 & 0 & 0 & 0 \\ 3 & 2 & 4 & 0 & 3 \\ 3 & 0 & 2 & 0 & 3 \end{pmatrix},$$

which is spanned by $(1, 0, 0, -1)^T$ and $(0, 0, 0, 1, 0)^T$. So $\lambda = 4$ has geometric multiplicity 2. By Fact 7, $A$ is not diagonalizable. The minimal polynomial of $A$ is $(\lambda - 4)^2(\lambda - 6)(\lambda - 10)$, which has the repeated linear factor $\lambda - 4$.

**4.** The conclusion of Fact 16 need not hold if $A$ is not normal. For example, the matrix $\begin{pmatrix} 0 & 1 \\ 0 & 0 \end{pmatrix}$ has rank 1 but has no nonzero eigenvalues.

**5.** *Matrix powers:*  The matrix $A = \begin{pmatrix} \frac{1}{2} & \frac{1}{2} \\ \frac{3}{4} & \frac{1}{4} \end{pmatrix}$ is a row stochastic matrix and the powers $A^n$ of such matrices are important in the analysis of Markov chains (§7.7). The eigenvalues of $A$ are $\lambda = 1$ and $\lambda = -\frac{1}{4}$, with corresponding eigenvectors $(1,1)^T$ and $(2,-3)^T$. Thus $P^{-1}AP = D = \text{diag}(1, -\frac{1}{4})$, where $P = \begin{pmatrix} 1 & 2 \\ 1 & -3 \end{pmatrix}$. Consequently $A = PDP^{-1}$, $A^2 = PDP^{-1}PDP^{-1} = PD^2P^{-1}$, and in general $A^n = PD^nP^{-1}$. Since $D^n = \text{diag}(1^n, (-\frac{1}{4})^n) = \text{diag}(1, \alpha^n)$, the $n$th power of $A$ can be computed as $A^n = \frac{1}{5}\begin{pmatrix} 3 + 2\alpha^n & 2 - 2\alpha^n \\ 3 - 3\alpha^n & 2 + 3\alpha^n \end{pmatrix}$. Since $|\alpha| < 1$, $A^n \to \begin{pmatrix} \frac{3}{5} & \frac{2}{5} \\ \frac{3}{5} & \frac{2}{5} \end{pmatrix}$ as $n \to \infty$.

---

### 6.5.3  LOCALIZATION

Since analytic computation of eigenvalues can be complicated, there are several simple methods available for (geometrically) estimating the eigenvalues of a matrix. These methods can be informative in cases when only the approximate location of eigenvalues is needed.

**Definitions:**

The **spectral radius** of $A$, $\rho(A)$, is the maximum modulus of an eigenvalue of $A$.

Let $A$ be an $n \times n$ matrix and let $\alpha_i = \sum_{j \neq i} |a_{ij}|$, $i = 1, 2, \ldots, n$.

The **Geršgorin discs** associated with $A$ are the discs

$$\{\, z \in \mathcal{C} \mid |z - a_{ii}| \leq \alpha_i \,\}, \quad i = 1, 2, \ldots, n.$$

The **ovals of Cassini** associated with $A$ are the ellipses

$$\{\, z \in \mathcal{C} \mid |z - a_{ii}||z - a_{jj}| \leq \alpha_i \alpha_j \,\}, \quad i \neq j.$$

A **strictly diagonally dominant** matrix is a square matrix $A$ satisfying $|a_{ii}| > \alpha_i$ for $i = 1, 2, \ldots, n$.

**Facts:**

**1.** $\rho(A)$ is the radius of the smallest disc, centered at the origin of the complex plane, enclosing all of the eigenvalues of $A$.

**2.** $\rho(A) \leq \min \{ \max_i \sum_j |a_{ij}|, \max_j \sum_i |a_{ij}| \}$.

**3.** The spectral radius of a row stochastic matrix is 1.

**4.** All the eigenvalues of $A$ are contained in the union of the associated Geršgorin discs.

**5.** A connected region formed by precisely $k \leq n$ Geršgorin discs contains exactly $k$ eigenvalues of $A$.

**6.** All the eigenvalues of $A$ are contained in the union of the $\frac{n(n-1)}{2}$ ovals of Cassini associated with $A$.

**Examples:**

**1.** By Fact 2, the spectral radius of the symmetric matrix

$$A = \begin{pmatrix} 8 & -2 & 1 & 1 \\ -2 & -8 & -2 & -1 \\ 1 & -2 & 7 & 1 \\ 1 & -1 & 1 & 8 \end{pmatrix}$$

is bounded by the maximum absolute row (column) sum 13. Since the eigenvalues of a real symmetric matrix are real, the spectral radius bound gives the interval $[-13, 13]$ enclosing all eigenvalues. The Geršgorin discs are the intervals

$$8 \pm 4 = [4, 12], \ -8 \pm 5 = [-13, -3], \ 7 \pm 4 = [3, 11], \ 8 \pm 3 = [5, 11].$$

The second interval is disjoint from the others, so one eigenvalue is localized in the interval $[-13, -3]$ while the other three are in the interval $[3, 12]$. The actual eigenvalues of $A$ are (approximately) $-8.51, 6.31, 7.03, 10.2$, consistent with the above intervals. Also, 0 is not in any of the four Geršgorin discs so 0 is not an eigenvalue and $A$ is nonsingular. Since the eigenvalues of $A^{-1}$ are the reciprocals of the eigenvalues of $A$ (see §6.5.1), it follows that the eigenvalues of the symmetric matrix $A^{-1}$ are localized to the intervals $[-\frac{1}{3}, -\frac{1}{13}]$ and $[\frac{1}{12}, \frac{1}{3}]$.

**2.** Using Fact 4, the eigenvalues of the matrix $A = \begin{pmatrix} 2 & 1 & -1 \\ 0 & 6 & 2 \\ 1 & -1 & 8 \end{pmatrix}$ are located in the union of the discs

$$D_1 = \{ z \mid |z - 2| \le 2 \}, \ D_2 = \{ z \mid |z - 6| \le 2 \}, \ D_3 = \{ z \mid |z - 8| \le 2 \}.$$

Since $A$ and $A^T$ have the same eigenvalues, an alternative set of disks can be formed based on the absolute *column* sums of $A$: namely

$$\widehat{D}_1 = \{ z \mid |z - 2| \le 1 \}, \ \widehat{D}_2 = \{ z \mid |z - 6| \le 2 \}, \ \widehat{D}_3 = \{ z \mid |z - 8| \le 3 \}.$$

Here $\widehat{D}_1$ is disjoint from both $\widehat{D}_2$ and $\widehat{D}_3$, and so one eigenvalue of $A$ is localized to $\widehat{D}_1$, and the other two to $\widehat{D}_2 \cup \widehat{D}_3$. In fact, the eigenvalues of $A$ are $2.24$ and $6.88 \pm 0.91i$, approximately.

**3.** The row stochastic matrix $A = \begin{pmatrix} \frac{1}{2} & \frac{1}{2} \\ \frac{3}{4} & \frac{1}{4} \end{pmatrix}$ has Geršgorin discs

$$D_1 = \{ z \mid |z - \frac{1}{2}| \le \frac{1}{2} \} \text{ and } D_2 = \{ z \mid |z - \frac{1}{4}| \le \frac{3}{4} \}.$$

Since $D_1 \subseteq D_2$ all eigenvalues must lie in $D_2$. As seen in Example 5 of §6.5.2, the eigenvalues of $A$ are 1 and $-\frac{1}{4}$.

**4.** Suppose $A$ is strictly diagonally dominant. Then all Geršgorin discs for $A$ reside in the positive right-half plane so all the eigenvalues must have positive real part. In particular, 0 is not an eigenvalue and $A$ must be nonsingular.

**5.** If the $n \times n$ matrix $A$ satisfies $a_{ii}a_{jj} > \alpha_i \alpha_j$ for all $i \ne j$ then $A$ must be nonsingular, since by Fact 6 zero is not an eigenvalue of $A$. The matrix of Example 2 satisfies this condition since $a_{ii}a_{jj} \ge 12 > 4 = \alpha_i \alpha_j$ for all $i \ne j$, and so it must be nonsingular.

## 6.5.4   COMPUTATION OF EIGENVALUES

The eigenvalues of a matrix can be obtained, in theory, by forming the characteristic equation and finding its roots. Since this is not a practical solution method for problems of realistic size, a variety of iterative techniques have been developed.

**Definitions:**

A *dominant* eigenvalue of a matrix is an eigenvalue with the maximum modulus.

Let $U(\theta; i, j)$ be the $n \times n$ matrix obtained by replacing the $2 \times 2$ principal submatrix of the identity matrix, corresponding to rows $i$ and $j$, with the rotation matrix

$$\begin{pmatrix} \cos\theta & \sin\theta \\ -\sin\theta & \cos\theta \end{pmatrix}.$$

**Facts:**

**1.** *Power method*:   The power method (Algorithm 1) is a simple technique for finding the dominant eigenvalue and an associated eigenvector of a nonsingular matrix $A$ having a unique dominant eigenvalue.

---

**Algorithm 1**:   **Power method.**

input: $n \times n$ nonsingular matrix $A$

output: approximations $x_k$ to an eigenvector of $A$

{Initialization}

choose any vector $x_0 \in \mathcal{C}^n$ with $\|x_0\| = 1$

{Iterative step}

**for** $k := 1$ **to** ...

$$x_k := \frac{A x_{k-1}}{\|A x_{k-1}\|}$$

---

**2.** In Algorithm 1, the $k$th estimate $x_k = \dfrac{A^k x_0}{\|A^k x_0\|}$.

**3.** The sequence $x_k$ converges to an eigenvector of $A$.

**4.** The sequence $\|A x_k\|$ approaches the dominant eigenvalue.

**5.** The power method is best suited for large sparse matrices.

**6.** The rate of convergence of the power method is dictated by the ratio of the largest to the second largest (in modulus) eigenvalue of $A$. The larger this ratio (the more separated these two eigenvalues in modulus), the faster the convergence of the method.

**7.** *QR method*:   This method (Algorithm 2) calculates the eigenvalues of a given $n \times n$ matrix $A$.

---

**Algorithm 2**:   $QR$ **method.**

input: $n \times n$ matrix $A$

output: $n \times n$ matrices $A_k$

{Initialization}

---

$A := Q_0 R_0$ (a $QR$ factorization of $A$)

{Iterative step}

**for** $k := 1$ **to** ...

$\quad A_k := R_{k-1} Q_{k-1}$

$\quad$ obtain a $QR$ factorization $A_k = Q_k R_k$

**8.** The $QR$ factorization in Algorithm 2 produces a unitary matrix $Q_k$ and an upper triangular matrix $R_k$. (See §6.5.2, Fact 23.)

**9.** Under certain conditions (for example, if the eigenvalues of $A$ have distinct moduli) the sequence $A_k$ in Algorithm 2 converges to an upper triangular matrix whose diagonal entries are the eigenvalues of $A$.

**10.** If $A$ is real then its $QR$ factors are real and can be calculated using real arithmetic. In this case, if $A$ has nonreal eigenvalues then under certain conditions, the limiting matrix is block triangular with $1 \times 1$ and $2 \times 2$ diagonal blocks.

**11.** The $QR$ method is not well suited for large sparse matrices since the factors $Q, R$ can quickly fill with nonzeros.

**12.** Often as a preparatory step for the $QR$ method the matrix is first reduced to *Hessenberg form* (upper triangular form in which there may be one nonzero diagonal below the main diagonal) by using Householder transformations [Da95].

**13.** The convergence of the $QR$ method can be very slow if the matrix has two eigenvalues that are close in moduli.

**14.** More effective versions of the $QR$ method are available which make use of certain *shift strategies* [GoVL13].

**15.** *Jacobi method*: This method (Algorithm 3) finds the eigenvalues of a real symmetric $n \times n$ matrix $A$ having at least one nonzero off-diagonal entry.

---

**Algorithm 3**:  **Jacobi method.**

input: $n \times n$ real symmetric matrix $A$

output: $n \times n$ matrices $A_k$

{Initialization}

$A_1 = (a_{ij}^{(1)}) := A$

{Iterative step}

**for** $k := 1$ **to** ...

$\quad$ choose $r, s$ $(r < s)$ with $|a_{rs}^{(k)}|$ as large as possible

$\quad$ define $\theta$ by $\cot 2\theta = \dfrac{a_{rr}^{(k)} - a_{ss}^{(k)}}{2a_{rs}^{(k)}}$

$\quad A_{k+1} = (a_{ij}^{(k+1)}) := U(\theta; r, s)^T A_k \, U(\theta; r, s)$

---

**16.** The sequence $A_k$ in Algorithm 3 converges to a real diagonal matrix with the eigenvalues of $A$ on the diagonal.

**17.** The orthogonal matrix $U(\theta; r, s)$ represents a (clockwise) plane rotation by the angle $\theta$.

**18.** The Jacobi method is particularly appropriate when $A$ is nearly diagonal, although in general the $QR$ method exhibits faster convergence.

**19.** A variant of the Jacobi method, the *serial Jacobi method*, uses plane rotation pairs cyclically—for example, $(1, 2), (1, 3), \ldots, (1, n), (2, 3), \ldots, (2, n), \ldots$.

**20.** For further information on numerical computation of eigenvalues, see [GoVL13, Da95].

**21.** A number of robust and well-tested software packages for carrying out eigensystem analysis are available at

- http://www.mathworks.com
- http://www.nag.com
- http://www.netlib.org
- http://www.roguewave.com

**Examples:**

**1.** The power method, when applied to the matrix in Example 3 of §6.5.1, produces the following sequence of vectors $x_k$ and scalars $\|Ax_k\|$:

| $k$ | 0 | 1 | 2 | 3 | 4 | 5 |
|---|---|---|---|---|---|---|
| $x_k$ | $\begin{pmatrix} 1 \\ 0 \\ 0 \end{pmatrix}$ | $\begin{pmatrix} 0.557 \\ 0.743 \\ 0.371 \end{pmatrix}$ | $\begin{pmatrix} 0.436 \\ 0.805 \\ 0.403 \end{pmatrix}$ | $\begin{pmatrix} 0.414 \\ 0.814 \\ 0.407 \end{pmatrix}$ | $\begin{pmatrix} 0.409 \\ 0.816 \\ 0.408 \end{pmatrix}$ | $\begin{pmatrix} 0.409 \\ 0.816 \\ 0.408 \end{pmatrix}$ |
| $\|Ax_k\|$ | 5.385 | 5.537 | 5.107 | 5.021 | 5.004 | 5.001 |

The scalars $\|Ax_k\|$ approach the dominant eigenvalue 5 and the vectors $x_k$ approach a multiple of the eigenvector $(1, 2, 1)^T$.

**2.** The eigenvalues of the matrix $A = \begin{pmatrix} 1 & 4 \\ 2 & 3 \end{pmatrix}$ can be approximated using the $QR$ method. $A = Q_0 R_0$ with $Q_0 = \begin{pmatrix} -0.447 & -0.894 \\ -0.894 & 0.447 \end{pmatrix}$ and $R_0 = \begin{pmatrix} -2.236 & -4.472 \\ 0 & -2.236 \end{pmatrix}$.

Then $A_1 = R_0 Q_0 = \begin{pmatrix} 5 & 0 \\ 2 & -1 \end{pmatrix}$. Continuing this process produces

$$A_2 = \begin{pmatrix} 4.862 & 2.345 \\ 0.345 & -0.862 \end{pmatrix}, \quad A_3 = \begin{pmatrix} 5.023 & -1.927 \\ 0.073 & -1.023 \end{pmatrix}, \quad A_4 = \begin{pmatrix} 4.995 & 2.014 \\ 0.014 & -0.995 \end{pmatrix}.$$

The sequence $A_k$ approaches an upper triangular matrix with the eigenvalues 5 and $-1$ on its diagonal.

**3.** The eigenvalues of the matrix $A = \begin{pmatrix} 0 & 1 & 1 \\ 1 & 4 & -3 \\ 1 & -3 & 4 \end{pmatrix}$ can be approximated using the

Jacobi method. The largest off-diagonal $|a_{rs}|$ of $A_1 = A$ occurs for $r = 2$, $s = 3$ giving $\theta = \frac{\pi}{4} = 0.7854$. Applying the matrix $U(\theta; 2, 3) = \begin{pmatrix} 1 & 0 & 0 \\ 0 & 0.7071 & 0.7071 \\ 0 & -0.7071 & 0.7071 \end{pmatrix}$ produces $A_2 =$

$$U(\theta; 2, 3)^T A_1 \, U(\theta; 2, 3) = \begin{pmatrix} 0 & 0 & 1.4142 \\ 0 & 7 & 0 \\ 1.4142 & 0 & 1 \end{pmatrix}.$$ The largest magnitude off-diagonal

entry of $A_2$ is $|a_{13}|$, giving $\theta = 0.6155$, $U(\theta; 1, 3) = \begin{pmatrix} 0.8165 & 0 & 0.5774 \\ 0 & 1 & 0 \\ -0.5774 & 0 & 0.8165 \end{pmatrix}$, and $A_3 =$

$$U(\theta; 1, 3)^T A_2 \, U(\theta; 1, 3) = \begin{pmatrix} -1 & 0 & 0 \\ 0 & 7 & 0 \\ 0 & 0 & 2 \end{pmatrix}.$$ So the eigenvalues of $A$ are $-1, 2, 7$.

---

### 6.5.5  SPECIAL CLASSES

This section discusses eigenvalues and eigenvectors of specially structured matrices, such as Hermitian, positive definite, nonnegative, totally positive, and circulant matrices.

**Definitions:**

If $x, y \in \mathcal{R}^n$, then $x$ **majorizes** $y$ if $\sum_{i=1}^n x_i = \sum_{i=1}^n y_i$ and for $k = 1, 2, \ldots, n-1$ the sum of the $k$ largest components of $x$ is at least as large as the sum of the $k$ largest components of $y$. A similar definition holds for infinite sequences with finitely many nonzero terms.

A Hermitian $n \times n$ matrix $A$ is **positive definite** if $x^* A x > 0$ for all nonzero $x \in \mathcal{C}^n$. It is **positive semidefinite** if $x^* A x \geq 0$ for all $x \in \mathcal{C}^n$.

If $A$ and $B$ are $n \times n$ Hermitian matrices then $A$ **dominates** $B$ in **Löwner order** if $A - B$ is positive semidefinite, written $A \succeq B$.

A matrix is **nonnegative** (**positive**) if each of its entries is nonnegative (positive).

The $n \times n$ matrix $A$ is **reducible** if either it is the $1 \times 1$ zero matrix or there exists a permutation matrix $P$ such that $PAP^T$ is of the form $\begin{pmatrix} B & 0 \\ C & D \end{pmatrix}$, where $B$ and $D$ are square. A matrix is **irreducible** if it is not reducible.

A **strictly totally positive** matrix has all of its minors positive.

A **circulant** matrix has the form

$$\begin{pmatrix} a_0 & a_1 & a_2 & \cdots & a_n \\ a_n & a_0 & a_1 & \cdots & a_{n-1} \\ a_{n-1} & a_n & a_0 & \cdots & a_{n-2} \\ \vdots & \vdots & \vdots & & \vdots \\ a_1 & a_2 & a_3 & \cdots & a_0 \end{pmatrix}.$$

*Notation:* Let $\lambda_1(A) \leq \lambda_2(A) \leq \cdots \leq \lambda_n(A)$ be the eigenvalues of an $n \times n$ Hermitian matrix $A$.

**Facts:**

**1.** *Cauchy's interlacing theorem:* Let $A$ be an $n \times n$ Hermitian matrix and let $B$ be a principal submatrix of $A$ of order $n-1$. Then

$$\lambda_i(A) \leq \lambda_i(B) \leq \lambda_{i+1}(A), \quad i = 1, 2, \ldots, n-1.$$

**2.** *Weyl's theorem:*   Let $A, B$ be $n \times n$ Hermitian matrices and let $j, k$ be integers satisfying $1 \leq j, k \leq n$.

- If $j + k \geq n + 1$, then $\lambda_{j+k-n}(A + B) \leq \lambda_j(A) + \lambda_k(B)$;
- If $j + k \leq n + 1$, then $\lambda_j(A) + \lambda_k(B) \leq \lambda_{j+k-1}(A + B)$.

**3.** Interpretations of the $k$th smallest eigenvalue of a Hermitian matrix are given in the following table.

| eigenvalue | variational characterization |
|---|---|
| $\lambda_1(A)$ | $\min(x^* A x)$, minimum over all unit vectors $x$ |
| $\lambda_n(A)$ | $\max(x^* A x)$, maximum over all unit vectors $x$ |
| $\lambda_k(A)$, $k = 2, \ldots, n$ | $\min(x^* A x)$, minimum over all unit vectors $x$ orthogonal to the eigenspaces of $\lambda_1, \ldots, \lambda_{k-1}$ |
| $\lambda_{n-k}(A)$, $k = 1, \ldots, n-1$ | $\max(x^* A x)$, maximum over all unit vectors $x$ orthogonal to the eigenspaces of $\lambda_{n-k+1}, \ldots, \lambda_n$ |

**4.** *Schur's majorization theorem:*   If $A$ is an $n \times n$ Hermitian matrix, then $(\lambda_1(A), \lambda_2(A), \ldots, \lambda_n(A))$ majorizes $(a_{11}, a_{22}, \ldots, a_{nn})$. Specifically, if $a_{11} \geq a_{22} \geq \cdots \geq a_{nn}$ then

$$\sum_{i=1}^{k} \lambda_{n-i+1}(A) \geq \sum_{i=1}^{k} a_{ii}, \quad k = 1, 2, \ldots, n.$$

**5.** *Hoffman-Wielandt theorem:*   If $A, B$ are $n \times n$ Hermitian matrices, then

$$\sum_{i=1}^{n} (\lambda_i(A + B) - \lambda_i(A))^2 \leq \sum_{i,j=1}^{n} |b_{ij}|^2.$$

**6.** *Sylvester's law of inertia:*   If $A$ is an $n \times n$ Hermitian matrix and if $X$ is a nonsingular $n \times n$ matrix, then $A$ and $X^T A X$ have the same number of positive eigenvalues as well as the same number of negative eigenvalues.

**7.** A Hermitian matrix is positive definite (positive semidefinite) if and only if all its eigenvalues are positive (nonnegative).

**8.** If $A, B$ are $n \times n$ positive semidefinite matrices and $A \succeq B$, then $\lambda_i(A) \geq \lambda_i(B)$, $i = 1, 2, \ldots, n$.

**9.** If $A, B$ are $n \times n$ positive semidefinite matrices, then $\lambda_{i+j-n}(AB) \leq \lambda_i(A)\lambda_j(B)$ holds for $1 \leq i, j \leq n$ and $i + j \geq n + 1$.

**10.** If $A, B$ are $n \times n$ positive semidefinite matrices, then

$$\prod_{i=1}^{k} \lambda_i(AB) \geq \prod_{i=1}^{k} \lambda_i(A)\lambda_i(B), \quad k = 1, 2, \ldots, n.$$

**11.** *Kantorovich inequality:*   If $A$ is an $n \times n$ positive definite matrix and if $x \in C^n$ is a unit vector, then

$$(x^* A x)(x^* A^{-1} x) \leq \frac{(\lambda_1(A) + \lambda_n(A))^2}{4\lambda_1(A)\lambda_n(A)}.$$

**12.** *Perron-Frobenius theorem:*   If $A$ is an irreducible nonnegative square matrix, then the spectral radius of $A$ (the *Perron root* of $A$) is an eigenvalue of $A$ with algebraic multiplicity 1 and it has an associated positive eigenvector. If $A$ is positive then the spectral radius exceeds the modulus of any other eigenvalue.

**13.** If $A$ is a nonnegative square matrix, then the spectral radius of $A$ is an eigenvalue of $A$ and it has an associated nonnegative eigenvector.

**14.** Let $A$ be an $n \times n$ strictly totally positive matrix. Then the eigenvalues of $A$ are distinct and positive: $\lambda_1(A) < \lambda_2(A) < \cdots < \lambda_n(A)$. The real eigenvector corresponding to $\lambda_{n-k}$ has exactly $k$ variations in sign.

**15.** If $A$ is an $n \times n$ strictly totally positive matrix, then $(\lambda_1(A), \lambda_2(A), \ldots, \lambda_n(A))$ majorizes $(a_{11}, a_{22}, \ldots, a_{nn})$.

**16.** An $(n+1) \times (n+1)$ circulant matrix has eigenvalues $\lambda_j = a_0 + a_1\omega^j + a_2\omega^{2j} + \cdots + a_n\omega^{nj}$, $j = 0, 1, \ldots, n$ with $(1, \omega^j, \omega^{2j}, \ldots, \omega^{nj})$, $j = 0, 1, \ldots, n$ the corresponding eigenvectors, where $\omega = e^{\frac{2\pi i}{n+1}}$.

**Examples:**

**1.** The matrix $A = \begin{pmatrix} 0 & 1 & 1 \\ 1 & 4 & -3 \\ 1 & -3 & 4 \end{pmatrix}$ has eigenvalues $-1$, $2$, and $7$ (§6.5.4, Example 3).

The principal submatrix $A[1,2] = \begin{pmatrix} 0 & 1 \\ 1 & 4 \end{pmatrix}$ has eigenvalues $2 \pm \sqrt{5}$, which are approximately equal to $-0.2361$ and $4.2361$. As required by Fact 1, these latter two eigenvalues interlace those of $A$: $-1 \leq -0.2361 \leq 2 \leq 4.2361 \leq 7$. Similarly, the principal submatrix

$A[2,3] = \begin{pmatrix} 4 & -3 \\ -3 & 4 \end{pmatrix}$ has eigenvalues $1$ and $7$, which interlace those of $A$.

**2.** The matrix in Example 1 has the eigenvalue sequence $(-1, 2, 7)$. This sequence majorizes (see Fact 4) the sequence $(0, 4, 4)$ of diagonal elements: $7 \geq 4$, $7 + 2 \geq 4 + 4$, and $7 + 2 - 1 \geq 4 + 4 + 0$.

**3.** The irreducible matrix $A$ in Example 3 of §6.5.3 is positive with eigenvalues $1$ and $-\frac{1}{4}$. Thus $\rho(A) = 1$ and it exceeds the modulus of any other eigenvalue. As required by Fact 12, there is a positive eigenvector associated with $\lambda = 1$, namely $(1, 1)^T$.

**4.** The matrix $A = \begin{pmatrix} 2 & 0 & 3 \\ 1 & 4 & 5 \\ 2 & 0 & 1 \end{pmatrix}$ is nonnegative with eigenvalues $-1$ (algebraic multiplicity 1) and $4$ (algebraic multiplicity 2). So the spectral radius is $4$ and (see Fact 13) $\lambda = 4$ must be an eigenvalue. In addition, there is a nonnegative eigenvector associated with $\lambda = 4$, namely $(0, 1, 0)^T$.

# 6.6   COMBINATORIAL MATRIX THEORY

Matrices and graphs represent two different ways of viewing certain discrete structures. At times a matrix perspective can lend insight into graphical or combinatorial structures. At other times the graph associated with a matrix can provide useful information about matrix properties.

## 6.6.1   MATRICES OF ZEROS AND ONES

### Definitions:

A **(0, 1)-matrix** is a matrix with each entry either 0 or 1.

The **term rank** $\rho_t(A)$ of a $(0,1)$-matrix $A$ is the maximum number of 1s such that no two are in the same row or column.

An $n \times n$ $(0,1)$-matrix is **partly decomposable** if it has a $k \times (n-k)$ zero submatrix for some $1 \le k \le n-1$; otherwise $A$ is **fully indecomposable**.

For vectors $x = (x_1, x_2, \ldots, x_n)$ and $y = (y_1, y_2, \ldots, y_n)$ we say that $x$ is **majorized** by $y$ (§6.5.5) if the sum of the $k$ largest components in $x$ is no more than the sum of the $k$ largest components in $y$ for $k \le n$, with equality for $k = n$. This is denoted by $x \preceq y$.

A vector $x = (x_1, x_2, \ldots, x_n)$ is **monotone** if $x_1 \ge x_2 \ge \cdots \ge x_n$. If $R = (r_1, r_2, \ldots, r_m)$ is a vector of nonnegative integers with $r_i \le n$, then its **conjugate** vector is $R^* = (r_1^*, r_2^*, \ldots, r_n^*)$ where $r_i^*$ is the number of $r_j$ with value no less than $i$. Thus $R^*$ is monotone.

For nonnegative integral vectors $R = (r_1, r_2, \ldots, r_m)$ and $S = (s_1, s_2, \ldots, s_n)$ define $\mathcal{A}(R, S)$ as the class of $(0,1)$-matrices $A = (a_{ij})$ with row sum vector $R$ and column sum vector $S$, i.e., $\sum_{j=1}^{n} a_{ij} = r_i$ $(i \le m)$ and $\sum_{i=1}^{m} a_{ij} = s_j$ $(j \le n)$. We may assume $\sum_i r_i = \sum_j s_j$. Using permutations of rows and columns one may assume that $R$ and $S$ are monotone. Then the **structure matrix** of $\mathcal{A}(R, S)$ is the $(m+1) \times (n+1)$ matrix $T$ with entries $t_{kl} = kl - \sum_{j=1}^{l} s_j + \sum_{i=k+1}^{m} r_i$ $(0 \le k \le m,\ 0 \le l \le n)$.

An $n \times n$ $(0,1)$-matrix $A = (a_{ij})$ satisfying $a_{ii} = 0$ for $i \le n$ and $a_{ij} + a_{ji} = 1$ for $1 \le i < j \le n$ is called a **tournament** matrix. Its row sum vector is called the **score vector**, and it records the number of wins of each player in a round-robin tournament where $a_{ij} = 1$ means that Player $i$ won against Player $j$.

### Facts:

**1.** *König's theorem:*  The term rank $\rho_t(A)$ of a $(0,1)$-matrix $A$ equals the minimum number of rows and columns required to cover all 1s in the matrix.

**2.** $\text{rank}(A) \le \rho_t(A)$.

**3.** *Frobenius-König theorem:*  Let $A$ be an $n \times n$ $(0,1)$-matrix. Then the term rank of $A$ is less than $n$ if and only if $A$ has a zero submatrix of size $r \times s$ with $r + s = n + 1$.

**4.** Let $A$ be an $n \times n$ $(0,1)$-matrix each of whose row sums and column sums is $k$. Then $A$ can be expressed as a sum of $k$ permutation matrices (§6.4.3).

**5.** Let $A$ be a square $(0,1)$-matrix and let $B$ be the matrix obtained from $A$ by replacing each 0 entry on the main diagonal of $A$ by 1. Then $A$ is irreducible (§6.5.5) if and only if $B$ is fully indecomposable.

**6.** Let $A, B$ be $n \times n$ fully indecomposable matrices. Then the matrix obtained by replacing every nonzero entry in $AB$ by 1 is fully indecomposable.

**7.** *Gale-Ryser theorem:* Let $R = (r_1, r_2, \ldots, r_m)$ and $S = (s_1, s_2, \ldots, s_n)$ be nonnegative integral vectors with $r_i \le n$ and $s_j \le m$. Then the class $\mathcal{A}(R, S)$ is nonempty (i.e., there is an $m \times n$ $(0,1)$-matrix $A = (a_{ij})$ with row sum vector $R$ and column sum vector $S$) if and only if $S \preceq R^*$.

**8.** The class $\mathcal{A}(R, S)$ contains a unique matrix if and only if $S = R^*$.

**9.** Let $T = (t_{kl})$ be the structure matrix of a nonempty class $\mathcal{A}(R, S)$ and let $A \in \mathcal{A}(R, S)$. Let $A_1$ be the leading $k \times l$ submatrix of $A$, and let $A_2$ be the submatrix corresponding to rows $k+1, \ldots, m$ and columns $l+1, \ldots, n$. Then $t_{kl}$ equals the number of zeros in $A_1$ plus the number of ones in $A_2$.

**10.** *Ford-Fulkerson theorem:* Suppose $R$ and $S$ are as in Gale-Ryser, with $R$ and $S$ monotone. Then the class $\mathcal{A}(R, S)$ is nonempty if and only if the structure matrix $T$ is nonnegative.

**11.** *Ryser's theorem:* If $A$ and $B$ are distinct matrices in $\mathcal{A}(R, S)$, then $B$ can be obtained from $A$ by a sequence of interchanges; each such interchange replaces a $2 \times 2$ identity matrix $I_2$ (a submatrix of $A$) by the matrix obtained by complementing zeros and ones, or vice versa.

**12.** *Landau's theorem:* Let $R = (r_1, r_2, \ldots, r_n)$ be a monotone nonnegative integral vector. Then there exists a tournament matrix with score vector $R$ if and only if $\sum_{i=1}^{k} r_i \geq \binom{k}{2}$ $(1 \leq k \leq n)$, with equality for $k = n$.

**13.** *Hardy-Littlewood-Pólya theorem:* If $x, y \in \mathcal{R}^n$, then $x \preceq y$ if and only if there is a doubly stochastic matrix (§6.6.3) $A$ with $x = Ay$.

**Examples:**

**1.** The following matrix contains a $2 \times 4$ zero submatrix, occurring in rows 1, 3 and columns 1, 2, 4, 5. By Fact 3, this means that the matrix must have term rank less than 5. In fact, the matrix has term rank 3. Namely, the starred entries represent a set of 3 entries, no two of which are in the same row or column, and 3 is the largest number of nonzero entries with this property. Rows $2, 4$ and column 3 cover all the 1s in the matrix, and no smaller number suffices, as guaranteed by Fact 1, and $\rho_t(A) = 3$.

$$\begin{pmatrix} 0 & 0 & 1^* & 0 & 0 \\ 1^* & 1 & 1 & 0 & 1 \\ 0 & 0 & 1 & 0 & 0 \\ 1 & 0 & 1 & 1 & 1^* \\ 0 & 0 & 1 & 0 & 0 \end{pmatrix}.$$

**2.** The matrix

$$A = \begin{pmatrix} 0 & 1 & 1 & 1 \\ 1 & 0 & 1 & 1 \\ 1 & 1 & 0 & 1 \\ 1 & 1 & 1 & 0 \end{pmatrix}$$

has all row and column sums equal to 3. By Fact 4, it can be expressed as the sum of three permutation matrices. For example

$$A = \begin{pmatrix} 0 & 0 & 1 & 0 \\ 0 & 0 & 0 & 1 \\ 1 & 0 & 0 & 0 \\ 0 & 1 & 0 & 0 \end{pmatrix} + \begin{pmatrix} 0 & 0 & 0 & 1 \\ 0 & 0 & 1 & 0 \\ 0 & 1 & 0 & 0 \\ 1 & 0 & 0 & 0 \end{pmatrix} + \begin{pmatrix} 0 & 1 & 0 & 0 \\ 1 & 0 & 0 & 0 \\ 0 & 0 & 0 & 1 \\ 0 & 0 & 1 & 0 \end{pmatrix}.$$

**3.** *Assignment problem:* There are $n$ applicants for $n$ vacant jobs. Each applicant is qualified for exactly $k \geq 1$ jobs and for each job there are exactly $k$ qualified applicants. Is it possible to assign each applicant to a (distinct) job for which the applicant is qualified? To answer this question form the $(0, 1)$-matrix $A$ where $a_{ij} = 1$ if applicant $i$ is qualified for job $j$, otherwise $a_{ij} = 0$. All row and column sums of $A$ equal $k$, so (by Fact 4) $A$

can be expressed as the sum of $k \geq 1$ permutation matrices. Select any one of these permutation matrices and use it to define an assignment of applicants to jobs. Thus it *is* possible in this case to fill each job with a different qualified applicant.

**4.** Let

$$
\bar{A} = \begin{pmatrix} 1 & 1 & 1 & 1 & 1 & 1 & 0 & 0 \\ 1 & 1 & 1 & 1 & 1 & 0 & 0 & 0 \\ 1 & 1 & 1 & 1 & 1 & 0 & 0 & 0 \\ 1 & 1 & 1 & 1 & 0 & 0 & 0 & 0 \\ 1 & 1 & 1 & 1 & 0 & 0 & 0 & 0 \end{pmatrix}, \quad A = \begin{pmatrix} 1 & 1 & 1 & 0 & 1 & 0 & 1 & 1 \\ 1 & 1 & 0 & 1 & 1 & 0 & 1 & 0 \\ 1 & 1 & 0 & 1 & 0 & 1 & 0 & 1 \\ 1 & 0 & 1 & 1 & 0 & 1 & 0 & 0 \\ 0 & 1 & 1 & 0 & 1 & 1 & 0 & 0 \end{pmatrix}.
$$

Both these matrices have row sum vector $R = (6, 5, 5, 4, 4)$. The column sum vector of $\bar{A}$ is $S(\bar{A}) = (5, 5, 5, 5, 3, 1, 0, 0)$ which is equal to the conjugate $R^*$ of $R$. The column sum vector of $A$ is $S = (4, 4, 3, 3, 3, 3, 2, 2)$, so $A \in \mathcal{A}(R, S)$. The majorization $S \preceq R^*$ holds (Fact 7). $\bar{A}$ is the unique matrix in the class $\mathcal{A}(R, R^*)$.

**5.** The matrix

$$
A = \begin{pmatrix} 0 & 1 & 1 & 1 \\ 0 & 0 & 1 & 0 \\ 0 & 0 & 0 & 1 \\ 0 & 1 & 0 & 0 \end{pmatrix}
$$

is a tournament matrix with score vector $(3, 1, 1, 1)$. It may represent a round-robin tournament where Player 1 wins against all the others while, e.g., Player 2 only defeats Player 3.

---

### 6.6.2   MATRICES AND GRAPHS

Different matrices may be used to represent graphs and directed graphs. This makes it possible to use linear algebra and related fields, such as optimization, to investigate graph properties and problems formulated in terms of graphs.

**Definitions:**
Let $G$ be a simple graph $G$ with vertices $1, 2, \ldots, n$ and edges $e_1, e_2, \ldots, e_m$. The **adjacency matrix** of $G$ is the square $(0, 1)$-matrix $A_G$ of size $n \times n$ where $A_G(i, j)$ is 1 if vertices $i$ and $j$ are adjacent, and 0 otherwise. The **incidence matrix** of $G$ is the $(0, 1)$-matrix $M_{I,G}$ of size $n \times m$ where $M_{I,G}(i, j)$ is 1 if vertex $i$ is incident with edge $e_j$, and is 0 otherwise.

The **Laplacian matrix** $L_G$ of $G$ is defined as $L_G = D_G - A_G$ where $D_G$ is the diagonal matrix whose diagonal entry $d_i$ is the degree of vertex $i$, and $A_G$ is the adjacency matrix of $G$.

Let $D$ be a directed graph with vertices $1, 2, \ldots, n$ and arcs $e_1, e_2, \ldots, e_m$; we assume no parallel edges and no loops. The **adjacency matrix** of $D$ is the square $(0, 1)$-matrix $A_D$ where $A_D(i, j)$ is 1 if there is an arc with tail $i$ and head $j$, and 0 otherwise. The **incidence matrix** of $D$ is the $n \times m$ $(0, \pm 1)$-matrix $M_{I,D}$ where $M_{I,D}(i, j)$ is 1 (resp. $-1$) if vertex $i$ is the head (resp. tail) of arc $e_j$, and is 0 otherwise.

An **oriented incidence matrix** of a simple graph $G$ is the incidence matrix of the directed graph $D$ obtained by some orientation of each of its edges.

A matrix $A$ is called **totally unimodular** if the determinant of every square submatrix of $A$ (i.e., a minor) is 0 or $\pm 1$.

A **polyhedron** is a set of the form $\{x \in \mathcal{R}^n \mid Ax \leq b\}$ for some real $m \times n$ matrix $A$ and $b \in \mathcal{R}^m$, i.e., it is the solution set of a linear inequality system.

A **polytope** $P$ is the convex hull of a finite set of vectors, say $x_1, x_2, \ldots, x_k$ in $\mathcal{R}^n$, so it consists of all convex combinations $\sum_{i=1}^{k} c_i x_i$ where $c_i \geq 0$ $(1 \leq i \leq k)$, $\sum_{i=1}^{k} c_i = 1$. An **extreme point** of $P$ is a point in $P$ which cannot be written as a convex combination of other points in $P$.

The **graph** $G(A)$ of a symmetric $n \times n$ matrix $A = (a_{ij})$ is the graph with vertices $1, 2, \ldots, n$ and an edge $\{i, j\}$ whenever $a_{ij} \neq 0$.

**Facts:**

**1.** If $D$ is a directed graph, $A_D^k = A_D \cdots A_D$ ($k$ times) is the matrix whose $(i, j)$th entry equals the number of directed walks of length $k$ from vertex $i$ to vertex $j$ in the graph $G$.

**2.** Two graphs $G$ and $G'$ on $n$ vertices are isomorphic if and only if there is a permutation matrix $P$ such that $A_G = PA_{G'}P^T$.

**3.** For a graph $G$, $M_{I,G}M_{I,G}^T = D_G + A_G$.

**4.** For a directed graph $D$, $M_{I,D}M_{I,D}^T = D_G - A_G$.

**5.** If $G$ is a graph and $M_{I,D}$ is an oriented incidence matrix of $G$, then $M_{I,D}M_{I,D}^T = D_G - A_G = L_G$.

**6.** For every graph $G$ the matrix $L_G$ is symmetric and positive semidefinite, and the multiplicity of 0 as an eigenvalue of $L_G$ equals the number of connected components of $G$.

**7.** If $\mu_G$ is the vector of the eigenvalues of $L_G$ ($n$ eigenvalues, if $G$ has $n$ vertices), then the following majorizations hold: $d_G \preceq \mu_G \preceq d_G^*$, where $d_G$ is the degree vector of $G$ and $d_G^*$ is its conjugate vector (§6.6.1).

**8.** If a graph $G$ with $n$ vertices is connected then (any) oriented incidence matrix has rank $n - 1$. More generally, if $G$ has $k$ connected components, then the rank is $n - k$.

**9.** *Matrix-tree theorem* (§8.10.2): For any graph $G$ the absolute value of the determinant of any $(n - 1) \times (n - 1)$ submatrix of the Laplacian matrix $L_G$ equals the number of spanning trees in $G$.

**10.** If $D$ is a directed graph, then its incidence matrix $M_{I,D}$ is totally unimodular.

**11.** If $G$ is a simple graph, then its incidence matrix $M_{I,G}$ is totally unimodular if and only if $G$ is bipartite.

**12.** Every polyhedron is a convex set (so it is closed under taking convex combinations).

**13.** *Farkas-Minkowski-Weyl theorem*: A set $P$ is a polytope if and only if $P$ is a bounded polyhedron.

**14.** *Integrality theorem*: If $A$ is totally unimodular, and $b, b', c, c'$ are integral vectors with $b \leq b'$ and $c \leq c'$, then the polytope $\{x \mid b \leq Ax \leq b', \ c \leq c \leq c'\}$ has only integral extreme points.

**15.** If $G$ is a bipartite graph, then the matching polytope, defined as the convex hull of the incidence vectors of matchings (§10.2.1) in $G$, equals $\{x \mid M_{I,G}x \leq 1, \ 0 \leq x \leq 1\}$, where 0 and 1 are the zero vector and the all ones vector, respectively. This follows from the integrality theorem.

**16.** The minimum rank of a symmetric matrix $A$ satisfying $G(A) = G$, where $G$ is a given matrix, can be computed when $G$ is a tree.

**Examples:**

**1.** Let $G$ be the graph with vertices $1, 2, 3, 4$ and edges $\{1,2\}$, $\{2,3\}$, $\{3,4\}$, $\{1,4\}$, $\{2,4\}$. Its adjacency matrix $A_G$, incidence matrix $M_{I,G}$ (with this edge ordering), and Laplacian matrix $L_G$ are

$$A_G = \begin{pmatrix} 0 & 1 & 0 & 1 \\ 1 & 0 & 1 & 1 \\ 0 & 1 & 0 & 1 \\ 1 & 1 & 1 & 0 \end{pmatrix}, \quad M_{I,G} = \begin{pmatrix} 1 & 0 & 0 & 1 & 0 \\ 1 & 1 & 0 & 0 & 1 \\ 0 & 1 & 1 & 0 & 0 \\ 0 & 0 & 1 & 1 & 1 \end{pmatrix}, \quad L_G = \begin{pmatrix} 2 & -1 & 0 & -1 \\ -1 & 3 & -1 & -1 \\ 0 & -1 & 2 & -1 \\ -1 & -1 & -1 & 3 \end{pmatrix}.$$

The determinant of each $3 \times 3$ submatrix of $L_G$ is $\pm 8$, so the number of spanning trees of $G$ is 8.

**2.** An oriented incidence matrix for the graph $G$ above is

$$M_{I,D} = \begin{pmatrix} 1 & 0 & 0 & 1 & 0 \\ -1 & 1 & 0 & 0 & 1 \\ 0 & -1 & -1 & 0 & 0 \\ 0 & 0 & 1 & -1 & -1 \end{pmatrix}.$$

$M_{I,D}$ is totally unimodular while $M_{I,G}$ is not. As $G$ contains an odd cycle, the submatrix of $M_{I,G}$ consisting of rows $2, 3, 4$ and columns $2, 3, 5$ has determinant 2 (Fact 11).

**3.** For the graph $G$ in Example 1, the degree vector is $d_G = (2, 3, 2, 3)$, its conjugate is $d_G^* = (4, 4, 2, 0)$, and the vector of eigenvalues of the Laplacian matrix is $\mu_G = (4, 4, 2, 0)$. So, in this case $\mu_G = d_G^*$. This equality always holds for the class of *threshold graphs*; such graphs are constructed by starting with the empty graph and successively adding either isolated vertices or a vertex connected to all previous vertices. The graph $G$ in Example 1 is a threshold graph.

**4.** Let $D$ be a directed graph with $n$ vertices and $m$ edges. Also, let $b \in \mathcal{R}^n$ with $\sum_{i=1}^{n} b_i = 0$ and let $d \in \mathcal{R}^m$ be nonnegative. Any vector in the polyhedron $P_{D,b} := \{x \mid M_{I,D} x = b,\ 0 \le x \le d\}$ is called a *network flow*. The variable $x_e$ represents the flow from tail to head of the arc $e$, and the linear equations assure that the difference between the flow into a vertex and the flow out of the same vertex is constant, and equal to the corresponding entry of $b$. The vector $d$ represents arc capacities. If $b$ and $d$ are integral, then the extreme points of $P_{D,b}$ are integral. The linear optimization problem $\min\{c^T x \mid x \in P_{D,b}\}$ is known as the *minimum cost network flow problem* (§10.5.1) and it has many important applications.

## 6.6.3   NONNEGATIVE MATRICES

This subsection discusses nonnegative matrices and special classes of nonnegative matrices such as primitive and doubly stochastic matrices. Certain results highlight the relationship between the Perron root (§6.5.5, Fact 12) and the directed graph of a matrix.

**Definitions:**

The **directed graph** $G(A)$ of an $n \times n$ matrix $A$ consists of $n$ vertices $1, 2, \ldots, n$ with an arc from $i$ to $j$ if and only if $a_{ij} \ne 0$.

Vertices $i$ and $j$ of $G(A)$ are **equivalent** if $i = j$ or if there is a path in $G(A)$ from $i$ to $j$ and a path from $j$ to $i$. The corresponding equivalence classes (§1.4.2) of this relation are the **classes** of $A$.

Class $C_i$ has **access** to class $C_j$ if $i = j$ or if there is a path in $G(A)$ from a vertex in $C_i$ to a vertex in $C_j$. A class is **final** if it has access to no other class. Class $C_i$ is **basic** if $\rho(A[C_i]) = \rho(A)$, where $\rho(\cdot)$ is the spectral radius (§6.5.3) and $A[C_i]$ is the principal submatrix of $A$ defined by indices in class $C_i$.

Let $A$ be an $n \times n$ nonnegative irreducible matrix. The number $h$ of eigenvalues of $A$ of modulus $\rho(A)$ is called the **index of cyclicity** of $A$. The matrix $A$ is **primitive** if $h = 1$.

The **exponent** of $A$, written $\exp(A)$, is the least positive integer $m$ with $A^m > 0$.

A square matrix is **doubly stochastic** if it is nonnegative and all row and column sums are 1.

If $A$ is an $n \times n$ matrix and $\sigma \in S_n$, the symmetric group on $n$ elements (§5.3.1), then the set $\{a_{1\sigma(1)}, a_{2\sigma(2)}, \ldots, a_{n\sigma(n)}\}$ is the **diagonal** of $A$ corresponding to $\sigma$.

A diagonal of $A$ is **positive** if each entry in it is positive.

Matrices $A$ and $B$ of the same size have the same **pattern** if the following condition holds: $a_{ij} = 0$ if and only if $b_{ij} = 0$.

The matrix $A$ has a **doubly stochastic pattern** if there exists a doubly stochastic matrix $B$ such that $A$ and $B$ have the same pattern.

An $n \times n$ matrix is called **completely positive** if $A = BB^T$ for some nonnegative $n \times k$ matrix.

**Facts:**

**1.** The matrix $A$ is irreducible if and only if $G(A)$ is strongly connected (§8.3.2).

**2.** *Frobenius normal form:* If the $n \times n$ matrix $A$ has $k$ classes, then there exists a permutation matrix $P$ such that

$$PAP^T = \begin{pmatrix} A_{11} & 0 & \cdots & 0 \\ A_{21} & A_{22} & \cdots & 0 \\ \vdots & \vdots & & \vdots \\ A_{k1} & A_{k2} & \cdots & A_{kk} \end{pmatrix}$$

where each $A_{ii}$, $1 \leq i \leq k$, is either irreducible or a $1 \times 1$ zero matrix.

**3.** The classes of a nonnegative $n \times n$ matrix $A$ are in one-to-one correspondence with the strong components (§8.3.2) of $G(A)$ and hence can be found in linear time.

**4.** Let $A$ be an $n \times n$ nonnegative matrix. There is a positive eigenvector corresponding to $\rho(A)$ if and only if the basic classes of $A$ are the same as its final classes.

**5.** Let $A$ be an $n \times n$ nonnegative matrix with eigenvalue $\lambda$. There exists a nonnegative eigenvector for $\lambda$ if and only if there exists a class $C_i$ satisfying both of the following:

- $\rho(A[C_i]) = \lambda$;
- if $C_j$ $(j \neq i)$ is any class that has access to $C_i$, then $\rho(A[C_j]) < \rho(A[C_i])$.

**6.** The $n \times n$ nonnegative matrix $A$ is primitive if and only if $\exp(A) < \infty$.

**7.** A nonnegative irreducible matrix with positive trace is primitive.

**8.** Suppose $A$ is an $n \times n$ nonnegative irreducible matrix. Let $S_i$ be the set of all the lengths of cycles in $G(A)$ passing through vertex $i$, and let $h_i$ be the greatest common divisor of all the elements of $S_i$. Then $h_1 = h_2 = \cdots = h_n$ and this common value equals the index of cyclicity of $A$.

**9.** Let $A$ be a nonnegative irreducible $n \times n$ matrix with $p \geq 1$ nonzero elements on the main diagonal. Then $A$ is primitive and $\exp(A) \leq 2n - p - 1$.

**10.** Let $A$ be a primitive $n \times n$ matrix, and let $s$ be the smallest length of a directed cycle in $G(A)$. Then $\exp(A) \leq n + s(n - 2)$.

**11.** Let $A$ be an $n \times n$ primitive $(0, 1)$-matrix, $n \geq 2$. Then $\exp(A) \leq (n - 1)^2 + 1$. Equality holds if and only if there exists a permutation matrix $P$ such that

$$
PAP^T = \begin{pmatrix}
0 & 1 & 0 & \cdots & 0 \\
0 & 0 & 1 & \cdots & 0 \\
\vdots & \vdots & \vdots & & \vdots \\
1 & 0 & 0 & \cdots & 1 \\
1 & 0 & 0 & \cdots & 0
\end{pmatrix}.
$$

**12.** The set $\Omega_n$ of $n \times n$ doubly stochastic matrices is a compact convex set.

**13.** *Birkhoff-von Neumann theorem:*    Every $A \in \Omega_n$ can be expressed as a convex combination of $n \times n$ permutation matrices: namely, $A = c_1 P_1 + c_2 P_2 + \cdots + c_t P_t$ for some permutation matrices $P_1, P_2, \ldots, P_t$ and some positive real numbers $c_1, c_2, \ldots, c_t$ with $c_1 + c_2 + \cdots + c_t = 1$.

**14.** The following conditions are equivalent for an $n \times n$ matrix $A$:

- $A$ has doubly stochastic pattern;
- There exist permutation matrices $P, Q$ such that $PAQ$ is a direct sum of fully indecomposable matrices;
- Every nonzero entry of $A$ is contained in a positive diagonal.

**15.** Let $A$ be an $n \times n$ nonnegative idempotent matrix of rank $k$. Then there exists a permutation matrix $P$ such that

$$
PAP^T = \begin{pmatrix}
J & JU & 0 & 0 \\
0 & 0 & 0 & 0 \\
VJ & VJU & 0 & 0 \\
0 & 0 & 0 & 0
\end{pmatrix},
$$

where $J$ is a direct sum of $k$ positive idempotent matrices of rank 1, and where $U$ and $V$ are arbitrary nonnegative matrices of appropriate sizes.

**16.** A nonnegative symmetric matrix $A$ of rank $k$ is idempotent if and only if there exists a permutation matrix $P$ such that

$$
PAP^T = \begin{pmatrix}
J & 0 \\
0 & 0
\end{pmatrix},
$$

where $J$ is a direct sum of $k$ positive symmetric rank one idempotent matrices.

**17.** Each completely positive $n \times n$ matrix is nonnegative, symmetric, and positive semidefinite. For $n \leq 4$ the converse is also true.

**18.**  The set of $n \times n$ completely positive matrices is a closed convex cone and its extreme rays are generated by the rank one matrices $xx^T$ where $x$ is a nonnegative vector in $\mathcal{R}^n$.

**19.**  A graph $G$ is completely positive (meaning that every nonnegative positive semidefinite matrix with graph $G$ is completely positive) if and only if $G$ does not have an odd cycle of length greater than 4.

**Examples:**

**1.**  The following nonnegative matrix is in Frobenius normal form

$$\left(\begin{array}{c|cc|cc|cc}
5 & 0 & 0 & 0 & 0 & 0 & 0 \\
\hline
0 & 1 & 1 & 0 & 0 & 0 & 0 \\
0 & 2 & 0 & 0 & 0 & 0 & 0 \\
\hline
2 & 4 & 1 & 0 & 3 & 0 & 0 \\
1 & 2 & 0 & 2 & 1 & 0 & 0 \\
\hline
0 & 1 & 1 & 0 & 0 & 3 & 2 \\
0 & 2 & 1 & 0 & 0 & 2 & 3
\end{array}\right)$$

with four classes $C_1 = \{1\}$, $C_2 = \{2,3\}$, $C_3 = \{4,5\}$ and $C_4 = \{6,7\}$. Class $C_3$ has access to $C_1$ and $C_2$ while class $C_4$ has access to $C_2$. Classes $C_1$ and $C_2$ are final, since they have access to no other classes. The eigenvalues of $A$ are $-2$, $-1$, $1$, $2$, $3$, $5$, $5$ so $\rho(A) = 5$. Classes $C_1$ and $C_4$ are basic since $\rho(A[C_1]) = \rho(A[C_4]) = 5$. Since no class has access to $C_3$, Fact 5 shows there is a nonnegative eigenvector of $A$ for the eigenvalue $\rho(A[C_3]) = 3$, namely $(0,0,0,1,1,0,0)^T$. However there is no nonnegative eigenvector of $A$ for $\rho(A[C_2]) = 2$ since class $C_3$ has access to class $C_2$ and $\rho(A[C_3]) \geq \rho(A[C_2])$.

**2.**  The directed graph of the matrix

$$A = \begin{pmatrix} 0 & 1 & 7 & 0 & 0 \\ 0 & 0 & 0 & 0 & 1 \\ 0 & 0 & 0 & 1 & 0 \\ 1 & 0 & 0 & 0 & 0 \\ 1 & 0 & 0 & 0 & 0 \end{pmatrix}$$

is the union of the cycles $1, 2, 5, 1$ and $1, 3, 4, 1$. The greatest common divisor of the lengths of all cycles passing through vertex 1 is 3, which by Fact 8 must be the index of cyclicity. In fact, $A$ has eigenvalues $2$, $-1 \pm i\sqrt{3}$, $0$, $0$ and thus there are 3 eigenvalues with modulus $\rho(A) = 2$.

**3.**  By Fact 13 every doubly stochastic matrix is a convex combination of permutation matrices. For example, the doubly stochastic matrix

$$\begin{pmatrix} .4 & .3 & .3 & 0 \\ .5 & 0 & .4 & .1 \\ .1 & .6 & 0 & .3 \\ 0 & .1 & .3 & .6 \end{pmatrix}$$

can be expressed as

$$.2\begin{pmatrix} 0 & 0 & 1 & 0 \\ 1 & 0 & 0 & 0 \\ 0 & 1 & 0 & 0 \\ 0 & 0 & 0 & 1 \end{pmatrix} + .3\begin{pmatrix} 0 & 1 & 0 & 0 \\ 1 & 0 & 0 & 0 \\ 0 & 0 & 0 & 1 \\ 0 & 0 & 1 & 0 \end{pmatrix} + .1\begin{pmatrix} 0 & 0 & 1 & 0 \\ 0 & 0 & 0 & 1 \\ 1 & 0 & 0 & 0 \\ 0 & 1 & 0 & 0 \end{pmatrix} + .4\begin{pmatrix} 1 & 0 & 0 & 0 \\ 0 & 0 & 1 & 0 \\ 0 & 1 & 0 & 0 \\ 0 & 0 & 0 & 1 \end{pmatrix}.$$

### 6.6.4  PERMANENTS

The permanent of a matrix is defined as a sum of terms, each corresponding to the product of elements along a diagonal of the matrix. Permanents arise in the study of systems of distinct representatives and in other combinatorial problems.

**Definition:**

The **permanent** of the $n \times n$ matrix $A$ is $\text{per}(A) = \sum_{\sigma \in S_n} a_{1\sigma(1)} a_{2\sigma(2)} \cdots a_{n\sigma(n)}$, where $S_n$ is the symmetric group on $n$ elements. (See §5.3.1.)

**Facts:**

**1.** The permanent of $A$ is an unsigned version of the determinant of $A$. (See §6.3.4, Fact 2.)

**2.** Computing the permanent of a square $(0,1)$-matrix is #P-complete.

**3.** *Laplace expansion:*  Suppose $A$ is an $n \times n$ matrix and $A_{ij}$ is the submatrix of $A$ obtained by deleting the $i$th row and the $j$th column. Then for $i = 1, 2, \ldots, n$

$$\text{per}(A) = \sum_{j=1}^{n} a_{ij} \, \text{per}(A_{ij}).$$

A similar expansion holds with respect to any column.

**4.** $\text{per}(A^T) = \text{per}(A)$.

**5.** Interchanging two rows (or two columns) of $A$ does not change $\text{per}(A)$.

**6.** Multiplying any row (or column) of $A$ by the scalar $\alpha$ multiplies $\text{per}(A)$ by $\alpha$.

**7.** Unlike the determinant, the permanent is not multiplicative with respect to matrix multiplication. (See Example 2.)

**8.** The permanent of a triangular matrix (§6.3.1) is equal to the product of its diagonal entries.

**9.** The permanent of a block diagonal matrix is equal to the product of the permanents of its diagonal blocks.

**10.** For each positive integer $n$ the permanent of the $n \times n$ matrix

$$\begin{pmatrix} 0 & 1 & \cdots & 1 \\ 1 & 0 & \cdots & 1 \\ \vdots & \vdots & & \vdots \\ 1 & 1 & \cdots & 0 \end{pmatrix}$$

is $n! \sum_{r=0}^{n} (-1)^r \frac{1}{r!}$ and it represents the number of derangements (§2.4.2) of order $n$.

**11.** The permanent of an $n \times n$ $(0,1)$-matrix $A$ counts the number of assignments ($n \times n$ permutation submatrices) consistent with the 1 entries of $A$.

**12.** *Minc-Brégman inequality:*  Let $A$ be an $n \times n$ $(0,1)$-matrix with row sums $r_1, r_2, \ldots, r_n$. Then

$$\text{per}(A) \le \prod_{i=1}^{n} (r_i!)^{1/r_i}.$$

**13.** If $A$ is a nonnegative $n \times n$ matrix with row sums $r_1, r_2, \ldots, r_n$ then $\text{per}(A) \le r_1 r_2 \ldots r_n$.

**14.** Let $A$ be a fully indecomposable nonnegative integral $n \times n$ matrix and let $s(A)$ denote the sum of the entries in $A$. Then

$$s(A) - 2n + 2 \le \text{per}(A) \le 2^{s(A)-2n} + 1.$$

**15.** *Alexandroff inequality*: Let $A$ be a nonnegative $n \times n$ matrix and let $A_i$ be the $i$th column of $A$, $i = 1, 2, \ldots, n$. Then

$$(\text{per}(A))^2 \ge \text{per}(A_1, \ldots, A_{n-2}, A_{n-1}, A_{n-1}) \, \text{per}(A_1, \ldots, A_{n-2}, A_n, A_n).$$

**16.** The definition of the permanent can be extended to $m \times n$ matrices with $m \le n$ by summing over all permutations in $S_m$.

**17.** If $A$ is an $m \times n$ $(0,1)$-matrix, then $\text{per}(A) > 0$ if and only if $A$ has term rank $m$.

**18.** *van der Waerden-Egorychev-Falikman inequality*: If $A$ is a doubly stochastic $n \times n$ matrix then $\text{per}(A) \ge \frac{n!}{n^n}$, and equality holds if and only if $A = J_n$, the matrix with each entry $\frac{1}{n}$.

*Note*: This result was first conjectured by B. L. van der Waerden in 1926. Despite repeated attempts to prove it, the conjecture remained unresolved until finally established in 1980 by G. P. Egorychev. The conjecture was also proved independently by D. I. Falikman in 1981, apart from establishing the uniqueness of the minimizing matrix $A$. A self-contained exposition of Egorychev's proof is given in [Kn81].

**19.** Let $A$ be the $m \times n$ incidence matrix of $m$ subsets of a given $n$-set $X$: namely, $a_{ij} = 1$ if $j \in X_i$ and $a_{ij} = 0$ otherwise. Then $\text{per}(A)$ counts the number of SDRs (systems of distinct representatives, §1.2.2) selected from the sets $X_1, X_2, \ldots, X_m$.

**Examples:**

**1.** For the matrix

$$A = \begin{pmatrix} 1 & 0 & 2 \\ 3 & 1 & 1 \\ 1 & 5 & 2 \end{pmatrix},$$

evaluation of $\text{per}(A)$ by the definition gives $\text{per}(A) = 1 \cdot 1 \cdot 2 + 2 \cdot 3 \cdot 5 + 2 \cdot 1 \cdot 1 + 1 \cdot 1 \cdot 5 = 39$. Using the Laplace expansion on row 1 gives $\text{per}(A) = 1 \cdot \text{per} \begin{pmatrix} 1 & 1 \\ 5 & 2 \end{pmatrix} + 2 \cdot \text{per} \begin{pmatrix} 3 & 1 \\ 1 & 5 \end{pmatrix} = 1 \cdot 7 + 2 \cdot 16 = 39$.

**2.** If $A = \begin{pmatrix} 1 & 1 \\ 0 & 1 \end{pmatrix}$ and $B = \begin{pmatrix} 1 & 0 \\ 1 & 1 \end{pmatrix}$, then $C = AB = \begin{pmatrix} 2 & 1 \\ 1 & 1 \end{pmatrix}$. Notice that $\text{per}(AB) = 3 \ne 1 \cdot 1 = \text{per}(A) \, \text{per}(B)$.

**3.** *Assignments*: Suppose there are 4 applicants for 4 jobs, where the qualifications of each applicant $i$ for each job $j$ can be specified by the $(0,1)$-matrix

$$A = \begin{pmatrix} 0 & 1 & 0 & 1 \\ 1 & 1 & 0 & 1 \\ 0 & 0 & 1 & 1 \\ 1 & 1 & 1 & 0 \end{pmatrix}.$$

Then the number of different assignments of jobs to qualified applicants (see §6.6.1, Example 3) equals $\text{per}(A) = 4$. In fact, these are given by those permutations $\sigma$ where $\{(\sigma(1), \sigma(2), \sigma(3), \sigma(4))\} = \{(2,1,4,3), (2,4,3,1), (4,1,3,2), (4,2,3,1)\}$.

**4.** *Ménage problem:*  Suppose that 5 wives are seated around a circular table, leaving one vacant space between consecutive women. Find the number of ways to seat in these vacant spots their 5 husbands so that no man is seated next to his wife. Suppose that the wives occupy positions $W_1, W_2, \ldots, W_5$ listed in a clockwise fashion around the table and that $X_i$ is the vacant position to the right of $W_i$. Let $A$ be the $5 \times 5$ $(0, 1)$-matrix where $a_{ij} = 1$ if and only if husband $H_i$ can be assigned to position $X_i$ without violating the requirements of the problem. Then

$$A = \begin{pmatrix} 0 & 0 & 1 & 1 & 1 \\ 1 & 0 & 0 & 1 & 1 \\ 1 & 1 & 0 & 0 & 1 \\ 1 & 1 & 1 & 0 & 0 \\ 0 & 1 & 1 & 1 & 0 \end{pmatrix}.$$

By Fact 11, the number of possible assignments for each fixed placement of wives is $\mathrm{per}(A) = 13$. (Also see §2.4.2, Example 7.)

**5.** Count the number of nontaking rooks on a chessboard with restricted positions (§2.4.2). Specifically, suppose that positions $(1, 1), (2, 3), (3, 1), (4, 2), (4, 3)$ of a $4 \times 4$ chessboard cannot be occupied by rooks. In the remaining positions, 4 rooks are to be placed so they are nontaking: no two are in the same row or in the same column. This can be solved (see Fact 11) by finding all permutations consistent with the 1s in the matrix

$$A = \begin{pmatrix} 0 & 1 & 1 & 1 \\ 1 & 1 & 0 & 1 \\ 0 & 1 & 1 & 1 \\ 1 & 0 & 0 & 1 \end{pmatrix}.$$

Here $\mathrm{per}(A) = 6$ is easily found using the Laplace expansion on the first column of $A$, so there are 6 placements of nontaking rooks.

## 6.6.5   MATRIX COMPLETION AND SIGN-SOLVABILITY

Matrix completion deals with filling in missing entries in a partial matrix to obtain a matrix with certain properties. Applications of this concept can be found in data analysis and control, for example. The linear algebra of sign-solvability is motivated by qualitative analysis in economics.

### Definitions:
A real symmetric $n \times n$ matrix $A$ is called **positive semidefinite** if $x^T A x \geq 0$ for all $x \in \mathcal{R}^n$, i.e., the quadratic form $Q(x) = x^T A x = \sum_{i,j} a_{ij} x_i x_j$ is never negative.

A real symmetric $n \times n$ matrix $A$ is called **copositive** if $x^T A x \geq 0$ for all nonnegative $x \in \mathcal{R}^n$.

A **partial matrix** is a rectangular array in which some entries are specified and others are not. If values are chosen for each unspecified entry, one obtains a **completion** of the partial matrix.

The **sign pattern** of a matrix $A$ is the matrix obtained from $A$ by replacing each entry by its sign $(0, 1, \text{ or } -1)$. The **qualitative class** $\mathcal{Q}(A)$ of $A$ consists of all matrices with the same sign pattern as $A$.

A system of linear equations $Ax = b$ is **sign-solvable** if for each $\tilde{A} \in \mathcal{Q}(A)$ and $\tilde{b} \in \mathcal{Q}(b)$, the system $\tilde{A}x = \tilde{b}$ has a solution and all solutions obtained in this way lie in the same qualitative class. Thus, changing numerical values, but not the sign, of the data in the linear system does not change the sign of the solutions.

Matrix $A$ is termed an **L-matrix** if every matrix in the qualitative class of $A$ has linearly independent rows. A square $L$-matrix is called a **sign-nonsingular (SNS) matrix**.

**Facts:**

**1.** The set of $n \times n$ copositive matrices is a convex cone which is the dual of the cone of $n \times n$ completely positive matrices.

**2.** Let $A$ be a partial matrix whose fully specified principal submatrices are positive semidefinite, and with positive diagonal elements. Then $A$ can be completed to a positive semidefinite matrix if and only the graph of $A$ (corresponding to specified entries) is chordal, i.e., every cycle of length at least 4 has a chord.

**3.** Every partial copositive matrix has a completion that is copositive.

**4.** The linear system $Ax = 0$ is sign-solvable if and only if $A^T$ is an $L$-matrix.

**5.** If the linear system $Ax = b$ is sign-solvable, then $A^T$ is an $L$-matrix.

**6.** A matrix $A$ is an $L$-matrix if and only if for every diagonal matrix $D$ with diagonal elements $0, \pm 1$ the matrix $DA$ contains a nonzero column which is unsigned (i.e., it does not contain both a positive and a negative entry).

**7.** The recognition problem for $L$-matrices is NP-complete.

**8.** A square matrix $A$ is an SNS-matrix if and only if $\det A \neq 0$ and the determinant of every matrix in $\mathcal{Q}(A)$ has the same sign.

**9.** $Ax = b$ is sign-solvable if and only if $A$ is an SNS-matrix and each matrix obtained from $A$ by replacing a column by $b$ is either an SNS-matrix or has an identically zero determinant (that is, each of the $n!$ terms in the determinant expression is zero).

**Examples:**

**1.** The following matrix $A$ is a partial copositive matrix and $B$ is a copositive completion of $A$.

$$ A = \begin{pmatrix} 8 & ? & -1 \\ ? & 2 & -1 \\ -1 & -1 & 1 \end{pmatrix}, \quad B = \begin{pmatrix} 8 & 4 & -1 \\ 4 & 2 & -1 \\ -1 & -1 & 1 \end{pmatrix}. $$

**2.** The linear system $Ax = b$ whose sign patterns are

$$ A' = \begin{pmatrix} 1 & 1 \\ 1 & -1 \end{pmatrix}, \quad b' = \begin{pmatrix} 1 \\ 0 \end{pmatrix} $$

is sign-solvable. The determinant of every matrix in $\mathcal{Q}(A)$ is negative, so $A$ is an SNS-matrix. Moreover, when $\tilde{A} \in \mathcal{Q}(A)$ and $\tilde{b} \in \mathcal{Q}(b)$, every solution $x = \begin{pmatrix} x_1 \\ x_2 \end{pmatrix}$ of $\tilde{A}x = \tilde{b}$ satisfies that $x_1$ and $x_2$ have the same sign (by the second equation) and actually $x_1, x_2 > 0$ (by the first equation). The notion of sign-solvability is motivated by qualitative problems in economics; see [BrSh09] and [Sa83].

**3.** A matrix $A$ whose sign pattern is

$$ A' = \begin{pmatrix} 1 & 0 & 0 & * \\ 0 & -1 & 0 & * \\ 0 & 0 & 1 & * \end{pmatrix} $$

where $*$ denotes any sign, is an $L$-matrix. This is seen from the definition or by Fact 6.

## 6.7   SINGULAR VALUE DECOMPOSITION

The singular value decomposition (SVD) is a matrix factorization that is used widely in many applications. Its importance in data analysis as well as in theory has changed the field of linear algebra over the last three decades. In recent years, the SVD and higher-order equivalents have become even more prominent due to increased computational memory and speed, as well as the multidimensional nature of many datasets.

### 6.7.1   BASIC CONCEPTS

**Definitions:**

A *singular value decomposition (SVD)* of a matrix $A \in \mathcal{R}^{m \times n}$ expresses $A = U\Sigma V^T$, where $U \in \mathcal{R}^{m \times m}$, $V \in \mathcal{R}^{n \times n}$ are orthogonal matrices and $\Sigma \in \mathcal{R}^{m \times n}$ is diagonal. The diagonal entries $\sigma_i$ of $\Sigma$ also satisfy $\sigma_1 \geq \sigma_2 \geq \cdots \geq \sigma_p \geq 0$, where $p = \min(m, n)$.

The *singular values* of $A$ are the diagonal entries of $\Sigma$.

The *left singular vectors* of $A$ are the $m \times 1$ column vectors of $U$.

The *right singular vectors* of $A$ are the $n \times 1$ column vectors of $V$.

If the SVD of an $m \times n$ matrix $A$ is given by $A = U\Sigma V^T$ then the **thin (or reduced) SVD** of $A$ $(m \geq n)$ expresses $A = U_1\Sigma_1 V^T$, where $U_1$ is $m \times n$ and contains the first $n$ columns of $U$ and $\Sigma_1$ is an $n \times n$ diagonal matrix containing the $n$ singular values of $A$.

The *matrix p-norm* for $1 \leq p \leq \infty$ of a matrix $A \in \mathcal{R}^{m \times n}$ corresponding to the vector $p$-norm is the real number

$$\|A\|_p = \sup\left\{ \tfrac{\|Ax\|_p}{\|x\|_p} \mid x \neq 0,\ x \in \mathcal{R}^n \right\}.$$

The *Frobenius norm* of a matrix $A \in \mathcal{R}^{m \times n}$, denoted $\|A\|_F$, is the real number

$$\|A\|_F = \sqrt{\sum_{i=1}^{m} \sum_{j=1}^{n} |a_{ij}|^2}.$$

**Facts:**

Most of these facts can be found in an advanced linear algebra or matrix theory book. See [GoVL13, §2.4], [Ma13], [TrBa97, Lectures 4–5] for more details and proofs.

**1.** Any $m \times n$ matrix $A$ has an SVD factorization. If $A \in \mathcal{C}^{m \times n}$ then the SVD is written as $A = U\Sigma V^*$, where $U$ and $V$ are unitary, but $\Sigma \in \mathcal{R}^{m \times n}$.

**2.** The singular values of $A$ are uniquely determined.

**3.** If $A$ is square and the singular values are distinct, then the left and right singular vectors are uniquely determined up to (complex) signs.

**4.** Suppose $A = U\Sigma V^T$ is the SVD of $A$ and there are $r$ nonzero singular values. Denote $u_i$ as the $i$th column of $U$ and $v_j$ as the $j$th column of $V$. Then

- $\text{rank}(A) = r$;
- $\text{null}(A) = \text{span}\{v_{r+1}, \dots, v_n\}$;
- $\text{range}(A) = \text{span}\{u_1, \dots, u_r\}$;
- $\text{null}(A^T) = \text{span}\{u_{r+1}, \dots, u_m\}$;
- $\text{range}(A^T) = \text{span}\{v_1, \dots, v_r\}$.

**5.** If the SVD of $A$ is given by $A = U\Sigma V^T$ then

- the nonzero singular values of $A$ are the square roots of the nonzero eigenvalues of both $AA^T$ and $A^T A$;
- the columns of $U$ are the orthonormal eigenvectors of $AA^T$;
- the columns of $V$ are the orthonormal eigenvectors of $A^T A$.

**6.** If $A$ is symmetric then the singular values of $A$ are the absolute values of the eigenvalues of $A$.

**7.** The image of the unit sphere in $\mathcal{R}^n$ under the map $A = U\Sigma V^T$ is a hyperellipse in $\mathcal{R}^m$. The nonzero singular values are the lengths of the principal semi-axes of the hyperellipse and the left singular vectors are the unit vectors in the direction of the principal semi-axes.

**8.** The SVD can be written as a sum of rank-1 matrices:

$$A = \sum_{i=1}^{r} \sigma_i u_i v_i^T, \tag{1}$$

where $r = \text{rank}(A)$, $\sigma_i$ is the $i$th singular value, and $u_i$, $v_i$ are the $i$th left and right singular vectors, respectively.

**9.** If $A$ is $m \times n$ and has nonzero singular values $\sigma_1, \dots, \sigma_r$ then

- $||A||_2 = \sigma_1$;
- $||A||_F = \sqrt{\sigma_1^2 + \sigma_2^2 + \cdots + \sigma_r^2}$;
- $\min_{x \neq 0} \dfrac{||Ax||_2}{||x||_2} = \sigma_n$ if $m \geq n$.

**10.** If $k < r = \text{rank}(A)$ and $A_k = \sum_{i=1}^{k} \sigma_i u_i v_i^T$ then

- $\min_{\text{rank}(B)=k} ||A - B||_2 = ||A - A_k||_2 = \sigma_{k+1}$;
- $\min_{\text{rank}(B)=k} ||A - B||_F = ||A - A_k||_F = \sqrt{\sigma_{k+1}^2 + \cdots + \sigma_r^2}$.

**11.** If $A$ is an $n \times n$ square matrix then $|\det(A)| = \prod_{i=1}^{n} \sigma_i$.

**12.** If $A$ is square then it can be factored into its *polar decomposition* $A = QP$, where $Q$ is orthogonal and $P$ is symmetric and positive definite. If $A = U\Sigma V^T$ is the SVD of $A$ then the polar decomposition of $A$ is given exactly by $A = (UV^T)(V\Sigma V^T)$.

## Examples:

**1.** Let $A = \begin{bmatrix} 3 & 2 \\ 2 & 3 \\ 2 & -2 \end{bmatrix}$. The SVD of $A = U \Sigma V^T$ is

$$U = \begin{bmatrix} -\sqrt{2}/2 & 1/\sqrt{18} & -2/3 \\ -\sqrt{2}/2 & -1/\sqrt{18} & 2/3 \\ 0 & 4/\sqrt{18} & 1/3 \end{bmatrix}, \quad V = \begin{bmatrix} -\sqrt{2}/2 & \sqrt{2}/2 \\ -\sqrt{2}/2 & -\sqrt{2}/2 \end{bmatrix}, \quad \Sigma = \begin{bmatrix} 5 & 0 \\ 0 & 3 \\ 0 & 0 \end{bmatrix}.$$

- $\text{rank}(A) = 2$ since there are two nonzero singular values.
- $||A||_2 = 5$ and $||A||_F = \sqrt{5^2 + 3^2} = \sqrt{34}$.
- $A$ can be written as a sum of two rank-1 (outer product) matrices:

$$A = 5 \begin{bmatrix} -\sqrt{2}/2 \\ -\sqrt{2}/2 \\ 0 \end{bmatrix} \begin{bmatrix} -\sqrt{2}/2 \\ -\sqrt{2}/2 \end{bmatrix}^T + 3 \begin{bmatrix} 1/\sqrt{18} \\ -1/\sqrt{18} \\ 4/\sqrt{18} \end{bmatrix} \begin{bmatrix} \sqrt{2}/2 \\ -\sqrt{2}/2 \end{bmatrix}^T.$$

- The eigenvalues of $A^T A$ are $\lambda_1 = 25$ and $\lambda_2 = 9$, which are the squares of the singular values.

**2.** Let $A = \begin{bmatrix} -1 & 2 & 2 \\ 2 & -1 & 2 \\ 2 & 2 & -1 \end{bmatrix}$. The singular values of $A$ are $\sigma_1 = 3, \sigma_2 = 3, \sigma_3 = 3$.

Thus the determinant is the product of the singular values: $\det(A) = 27$.

---

## 6.7.2  COMPUTATION OF THE SVD

Mathematically, the SVD of $A$ can be computed using the eigenvalue decomposition of $A^T A$. This algorithm is not preferred in practice for two reasons. First, the algorithm is numerically unstable since forming $A^T A$ can result in a loss of information [GoVL13, §5.3.2]. Second, if $A$ is large, forming $A^T A$ may be troublesome. Even if $A^T A$ can be formed, computing its eigenvalue decomposition may be impractical. In practice, the main SVD algorithm is iterative and similar to the QR algorithm used to compute eigenvalues (§6.5.4) without explicit formation of $A^T A$. Here we present an overview of the common SVD algorithms and refer to [GoVL13], [ClDh13] for details and references.

### Definitions:
A **Householder matrix** is an $m \times m$ matrix $P$ such that

$$P = I - \beta v v^T, \qquad \beta = \tfrac{2}{v^T v},$$

where $v \in \mathcal{R}^m$ is nonzero. A Householder matrix represents a linear transformation that reflects a vector in the hyperplane $\text{span}\{v\}^\perp$.

An upper (lower) **bidiagonal** matrix $B = (b_{ij})$ is a matrix in which the only nonzero $b_{ij}$ occur when $i = j$ or $i = j - 1$ ($i = j + 1$).

**Facts:**

**1.** *SVD with Golub-Kahan bidiagonalization:* This method (Algorithm 1) computes the SVD of a matrix $A$ using a two-phase approach; a bidiagonalization phase followed by an iterative procedure to diagonalize the bidiagonal matrix.

---

**Algorithm 1: SVD with Golub-Kahan bidiagonalization.**

input: $m \times n$ matrix $A$, $m \geq n$

output: matrices $U, V, \Sigma$ such that $A = U\Sigma V^T$ gives the SVD of $A$.

1. *bidiagonalization.* $A$ is brought to upper bidiagonal form (see [GoKa65])

{Initialization}

$A_1 := A$

{Iterative step}

**for** $k := 1$ **to** $n$

   $A_{k+1} := U_k^T A_k$

   **if** $k \leq n - 2$

      $A_{k+1} := A_{k+1} V_k$

$B := A_{n+1}$

$\tilde{U}_1 := U_1 U_2 \cdots U_n$

$\tilde{V}_1 := V_1 V_2 \cdots V_{n-2}$

2. *diagonalization.* Upper bidiagonal $B = \tilde{U}_1^T A \tilde{V}_1$ is diagonalized using a variant of the QR algorithm (see [GoVL13, §8.6.3])

{Iterative step}

**for** $k := 1$ **to** $n - 1$

   $B := B V_k(\theta_1; k, k+1)$

   $B := U_k(\theta_2; k, k+1)^T B$

$\tilde{U}_2 := U_1 U_2 \cdots U_{n-1}$

$\tilde{V}_2 := V_1 V_2 \cdots V_{n-1}$

$U := \tilde{U}_1 \tilde{U}_2$

$V := \tilde{V}_1 \tilde{V}_2$

$\Sigma := B$

---

**2.** The bidiagonalization step of Algorithm 1 finds $\tilde{U}_1$ and $\tilde{V}_1$ simultaneously by *implicitly* applying the QR method for the symmetric eigenvalue problem to $A^T A$ [GoVL13, §8.6.3].

**3.** In the bidiagonalization step of Algorithm 1, the orthogonal matrix $U_k$ is a Householder matrix that reflects the vector $A_k([k, k+1, \ldots, n], k)$ onto a multiple of $e_1$. Similarly, the orthogonal matrix $V_k$ is a Householder matrix that reflects the vector $A_k(k, [k+1, k+2, \ldots, n])$ onto a multiple of $e_1$.

**4.** In the diagonalization step of Algorithm 1, the orthogonal matrix $V_k(\theta_1; k, k+1)$ is a rotation matrix defined by $\theta_1$ such that

$$y^T V_k(\theta_1; k, k+1) = \alpha \begin{bmatrix} 1 & 0 \end{bmatrix},$$

where $\alpha \in \mathcal{R}$ and

$$y^T = \begin{cases} \begin{bmatrix} t_{11} - \lambda & t_{12} \end{bmatrix} & \text{for } k = 1 \\ \begin{bmatrix} b_{k,k+1} & b_{k,k+2} \end{bmatrix} & \text{for } 1 < k < n - 1. \end{cases}$$

In the above, $\lambda$ is the eigenvalue of the trailing $2 \times 2$ submatrix of $T$ that is closest to $t_{nn}$ for $T = B^T B$.

**5.** In the diagonalization step of Algorithm 1, the orthogonal matrix $U_k(\theta_2; k, k+1)$ is a rotation matrix defined by $\theta_2$ such that

$$U_k(\theta_2; k, k+1)^T \begin{bmatrix} b_{kk} \\ b_{k+1,k} \end{bmatrix} = \beta \begin{bmatrix} 1 \\ 0 \end{bmatrix},$$

where $\beta \in \mathcal{R}$.

**6.** Various improvements to the bidiagonalization step have been proposed and there is a long history of work related to improving SVD algorithms in general. Specialized SVD algorithms for structured matrices have also been developed [GoVL13, §8.6], [ClDh13].

**7.** *Two-Sided Jacobi SVD:* This method (Algorithm 2) computes the SVD of a matrix by computing the SVD of $2 \times 2$ submatrices.

---

**Algorithm 2:   Two-Sided Jacobi SVD.**

input: $m \times n$ matrix $A$

output: matrices $U, V, \Sigma$ such that $A = U\Sigma V^T$ gives the SVD of $A$.

{Initialization}

$A_1 = (a_{ij}^{(1)}) := A$

{Iterative step}

**for** $k := 1$ **to** $\ldots$ until convergence criteria met

  choose $r, s$ $(r < s)$ with $|a_{rs}^{(k)}|$ as large as possible

  define $\phi$ by $\tan \phi = \dfrac{a_{rs}^{(k)} - a_{sr}^{(k)}}{a_{rr}^{(k)} + a_{ss}^{(k)}}$

  $B = (b_{ij}) := Q_k(\phi; r, s)^T A_k$

  define $\theta$ by $\cot 2\theta = \dfrac{b_{rr} - b_{ss}}{2b_{rs}}$

  $A_{k+1} = (a_{ij}^{(k+1)}) := J_k(\theta; r, s)^T B J_k(\theta; r, s)$

$\Sigma := A_{k+1}$

$U := Q_1 J_1 Q_2 J_2 \cdots Q_k J_k$

$V := J_1 J_2 \cdots J_k$

---

**8.** The sequence $A_k$ in Algorithm 2 converges to a matrix with maximum sums of squares of its diagonal elements. Choosing the largest $|a_{rs}|$ works to ensure linear convergence so that the sums of squares of the off-diagonal elements decrease at each step. However, in practice the asymptotic convergence rate of Algorithm 2 is quadratic [GoVL13, §8.5]. The convergence criteria is met when the sums of squares of the off-diagonals are less than some prescribed tolerance.

**9.** The orthogonal matrix $Q(\phi; r, s)$ in Algorithm 2 represents a rotation by the angle $\phi$, chosen so that the $2 \times 2$ $(r, s)$ submatrix of $B = Q(\phi; r, s)A_k$ is symmetric.

**10.** The orthogonal matrix $J(\theta; r, s)$ in Algorithm 2 is the same as $U(\theta; r, s)$ in Algorithm 3 of §6.5.4, since the $2 \times 2$ $(r, s)$ submatrix of $B$ in the Two-Sided Jacobi Algorithm is symmetric.

### 6.7.3   SOLVING LINEAR SYSTEMS

The SVD is ubiquitous in real-world problems. Its widespread appeal is due to its stability in computation and ability to provide meaningful solutions of linear systems that may be ill-conditioned (i.e., close to singular). The SVD is the factorization of choice used to find the best solution to a non-square linear system or systems that are not full rank. It is also the most precise and numerically stable method to compute the rank of a matrix.

**Definitions:**

A **rank-deficient** matrix $A \in \mathcal{R}^{m \times n}$ has $\text{rank}(A) < \min(m, n)$.

An **underdetermined** system of linear equations is a system $Ax = b$, with $A \in \mathcal{R}^{m \times n}$, $x \in \mathcal{R}^n$, $b \in \mathcal{R}^m$, where $m < n$.

An **overdetermined** system of linear equations is a system $Ax = b$, with $A \in \mathcal{R}^{m \times n}$, $x \in \mathcal{R}^n$, $b \in \mathcal{R}^m$, where $m > n$.

A **least squares** problem finds a minimum norm solution to an overdetermined system of linear equations. That is, given $A \in \mathcal{R}^{m \times n}$, $b \in \mathcal{R}^m$, with $m > n$, solve

$$\min_{x \in \mathcal{R}^n} \|Ax - b\|_2.$$

The **Moore-Penrose pseudo-inverse** of $A \in \mathcal{R}^{m \times n}$ is the unique matrix $A^+ \in \mathcal{R}^{n \times m}$ that satisfies

$$
\begin{aligned}
AA^+A &= A, & A^+AA^+ &= A^+, \\
(AA^+)^T &= AA^+, & (A^+A)^T &= A^+A.
\end{aligned}
$$

The **numerical rank** of a matrix $A \in \mathcal{R}^{m \times n}$ for some (small) tolerance $\delta$ is the number $\hat{r}$ such that the computed singular values $(\sigma_i)$ of $A$ satisfy

$$\sigma_1 \geq \sigma_2 \geq \cdots \geq \sigma_{\hat{r}} > \delta > \sigma_{\hat{r}+1} \geq \cdots \geq \sigma_p, \quad p = \min(m, n).$$

Usually $\delta = \epsilon \|A\|_\infty$ where $\epsilon$ is the machine epsilon.

**Facts:**

**1.** The most precise and numerically stable method to compute the rank of a matrix is to compute its numerical rank using the SVD.

**2.** Let the SVD of $A \in \mathcal{R}^{m \times n}$ be given by $A = U\Sigma V^T$, where $r = \text{rank}(A)$, $\sigma_i$ is the $i$th singular value, $u_i$ and $v_i$ are the $i$th left and right singular vectors, respectively.

- The least squares solution $\tilde{x}$ to the overdetermined system $Ax = b$ is

$$\tilde{x} = \sum_{i=1}^{r} \frac{u_i^T b}{\sigma_i} v_i.$$

- The smallest 2-norm residual is given by

$$\|A\tilde{x} - b\|_2^2 = \sum_{i=r+1}^{m} (u_i^T b)^2.$$

- The smallest 2-norm solution above is also valid if the system is underdetermined.

**3.** The Moore-Penrose pseudo-inverse $A^+ \in \mathcal{R}^{n \times m}$ of $A \in \mathcal{R}^{m \times n}$ is computed using the SVD of $A$. Specifically, if $A = U\Sigma V^T$ is the SVD of $A$, then $A^+ = V\Sigma^+ U^T$ where $\Sigma^+ \in \mathcal{R}^{n \times m}$ has diagonal elements $1/\sigma_i$ for $i = 1, \ldots, r$ followed by $n - r$ zeros ($r = \operatorname{rank}(A)$).

- The least squares solution $\tilde{x}$ to the overdetermined system $Ax = b$ can also be written as $\tilde{x} = A^+ b$.

- The pseudo-inverse $A^+ \in \mathcal{R}^{n \times m}$ of $A \in \mathcal{R}^{m \times n}$ is the unique Frobenius norm solution to the problem

$$\min_{X \in \mathcal{R}^{n \times m}} \|AX - I_m\|_F.$$

- If $A$ is square and nonsingular, then $A^+ = A^{-1}$.

**Examples:**

1. Suppose $A = \begin{bmatrix} 3 & 2 \\ 2 & 3 \\ 2 & -2 \end{bmatrix}$ and $b = \begin{bmatrix} 1 \\ 2 \\ 2 \end{bmatrix}$.

   - The Moore-Penrose pseudo-inverse of $A$ is $A^+ = \frac{1}{45} \begin{bmatrix} 7 & 2 & 10 \\ 2 & 7 & -10 \end{bmatrix}$.

   - The least squares solution to $Ax = b$ is $\tilde{x} = A^+ b = \frac{1}{45} \begin{bmatrix} 31 \\ -4 \end{bmatrix}$.

2. Note that $(AB)^+ \neq B^+ A^+$ in general, even if $A, B$ are square. For example, let

$$A = \begin{bmatrix} 1 & 0 \\ 0 & 0 \end{bmatrix}, \qquad B = \begin{bmatrix} 1 & 1 \\ 0 & 1 \end{bmatrix}.$$

Then $(AB)^+ = \begin{bmatrix} \frac{1}{2} & 0 \\ \frac{1}{2} & 0 \end{bmatrix}$; however $B^+ A^+ = \begin{bmatrix} 1 & 0 \\ 0 & 0 \end{bmatrix}$.

## 6.7.4   APPLICATIONS

The SVD is utilized in many applications. It plays a major role in statistics where it is used in Principal Component Analysis and multivariate data analysis. Here we list some main application areas of the SVD with the understanding that the list is not exhaustive, but rather it gives a flavor for the types of problems that are well suited for the SVD.

**Definitions:**

**Dimension reduction** is a method that reduces the number of dimensions of a dataset while maintaining as much information about the dataset as possible.

Let $A \in \mathcal{R}^{m \times n}$ comprise a matrix whose $i$th column represents $m$ measurements of the $i$th variable centered about zero. Then $\frac{1}{m-1} A^T A$ is a **covariance matrix**, in which the $(i, j)$-element measures the covariance between column $i$ and column $j$ of $A$.

**Alternating least squares** (**ALS**) is a method that solves a multidimensional least squares problem such that holding all variables constant but one reduces the problem to a linear least squares problem (§6.7.3) in that one variable. The multidimensional problem is solved by alternating which set of variables is held constant with the goal to eventually converge to the multidimensional least squares solution, though convergence is not guaranteed.

## Examples:

**1.** *Principal Component Analysis* (PCA): PCA is a dimension reduction technique that performs an orthogonal transformation of a dataset $A \in \mathcal{R}^{m \times n}$ in order to reduce its dimensionality while retaining as much of the variation as possible [Jo02]. PCA computes the eigenvalue decomposition of the covariance matrix of $A$. The eigenvectors (*principal components*) corresponding to the dominant eigenvalues contain most of the variation of the data and are uncorrelated. PCA is equivalent to computing the SVD of the data matrix due to Fact 5, §6.7.1.

**2.** *Latent Semantic Indexing* (LSI): Given a set of $n$ documents and a set of $m$ terms appearing in those documents, set $A$ to be the $m \times n$ *term-document matrix* where the $(i, j)$-element of $A$ is the number of times term $i$ appears in document $j$. LSI uses the SVD of $A$ in order to correlate related terms in a set of documents [De88], [BeDuOB95], [Me00, §5.12]. This is often used to find relevant documents based on a user query of a set of terms. Due to the typically large size of $A$ in real-world applications, Fact 10 in Section §6.7.1 is used to compute the nearest rank-$k$ approximation to $A$ where $k \ll \text{rank}(A)$.

**3.** *Eigenfaces:* Given a set of $N$ grayscaled facial images all of size $m \times n$, set $A$ to be the $mn \times N$ matrix where the $j$th column is the vectorized matrix of pixels of image $j$. Eigenfaces is a facial recognition algorithm that uses PCA to project an $m \times n$ query image onto the dominant eigenspaces of the covariance matrix of $A$. This is equivalent to computing the SVD of $A$. See [MuMaHe04], [TuPe91a], [TuPe91b] for more information.

**4.** *Collaborative Filtering:* Collaborative filtering is one approach to a class of methods called *recommender systems* that work to give suggestions on an item $i$ to a user $u$. Recommender systems have become extremely popular recently due to their substantial improvement to the Netflix recommender system publicized by the winner of the Netflix prize [KoBeVo09]. They are now in use in many internet sites, including Amazon.com, YouTube, Netflix, Yahoo, and more. Collaborative filtering is used when explicit feedback of user ratings towards items is not available and user behavior ("click history") is used instead.

Suppose $A$ is the $m \times n$ user-item matrix where the $(u, i)$-entry is a rating $r_{ui}$ which indicates the preference by user $u$ of item $i$ (we use the $r_{ui}$ terminology to remain consistent with the collaborative filtering literature). The general model is based on the SVD and in its most basic form involves finding (not necessarily orthogonal) matrices $X, Y$ that solve

$$\min\{||A^T - Y^T X||_2^2 \mid X \in \mathcal{R}^{k \times m}, Y \in \mathcal{R}^{k \times n}\},$$

where $k \ll \text{rank}(A)$. There are many extensions to this basic model, including regularization and addition of bias terms. Since $A$ is typically very large, the SVD cannot be explicitly computed and ALS is used instead to find the best $X, Y$. The $i$th column of $Y$ measures the extent to which item $i$ possesses the $k$ factors, and the $u$th column of $X$ measures the extent of interest user $u$ has in items high in the $k$ factors. Therefore, $r_{ui}$ is viewed as an inner product of column $i$ of $Y$ and column $u$ of $X$. See [KoBe11] for a detailed treatment of collaborative filtering.

**5.** *Data Mining:* The SVD is used in a variety of problems related to clustering and multivariate analysis of large datasets (see for example [MaKeHo05]). PCA, LSI, and

Eigenfaces are three examples of data mining techniques, but there are more. The SVD is also used in graph partitioning for the ratio cut problem and is used in the HITS Algorithm for web searching and queries [Kl99]. The SVD appears in chemical physics to obtain approximate solutions to the coupled-cluster equations and is a popular tool for electronic structure calculations [Ki03]. It is useful in genetics in a PCA context to understand the dynamics of gene expression and classify genes into groups of similar regulation and function [AlBrBo00], [HoEtal00]. The SVD has found its way into political science where it was used to investigate the ideologies of members of Congress [MaPo12], [PoEtal05], [PoEtal07]. It has also been applied in robotics [BeKu02], financial mathematics [FeEtal11], compressed sensing [XuLi10], and geophysics [Sm94].

**6.** *Cross-Modal Factor Analysis* (CFA): CFA is a technique to rotate both optimally and rigidly one $m$-element data matrix into another. The rotation is rigid in the sense that both distances and angles between the elements in the initial data matrix are preserved. The rotation is optimal in the sense that the sum of the $m$ squared (Euclidean) distances between the corresponding initial and target data point pairs is minimized. CFA is also known as the *orthogonal Procrustes problem* [GoVL13]. In particular, to map an $m \times n$ data matrix $B$ optimally into an $m \times n$ data matrix $A$ solve

$$\min_{Q \in \mathcal{R}^{n \times n}} \|A - BQ\|_F^2, \quad \text{subject to } Q^T Q = I_n.$$

The optimal $Q$ is $Q = UV^T$ where $U, V$ are found from the SVD $B^T A = U\Sigma V^T$. Note that if $B$ is an $m \times n$ data matrix and $A = BQ$ for some unknown orthogonal transformation $Q$, then the minimum squared Frobenius norm is 0 and CFA may be applied to recover the matrix $Q$.

**7.** *Canonical Correlation Analysis* (CCA): CCA is a way of measuring correlations between two groups of random variables, $x = (x_1, \ldots, x_m)$ and $y = (y_1, \ldots, y_n)$ [Ho36], [Mu82], [BjGo73]. The goal is to find linear combinations of the first group and linear combinations of the second group that are maximally correlated. The solution is found using the SVD of the *cross-covariance matrix*, in which the $(i, j)$-element is the covariance between variables $x_i$ and $y_j$.

**8.** *Geometric Applications:* Fact 7, §6.7.1 is used in geology to measure the grain sizes of irregularly shaped crystals. This is done by finding ellipsoids that approximate the size and shape of each grain [MaPo12], [Hi06], [Ma98]. The SVD is also used in shape matching problems; see for example [RuGa06].

**9.** *Compression:* The SVD is used in many areas to obtain low-rank approximations to a dataset using Fact 10, §6.7.1. It is used in signal processing as a noise reduction method, where a low-rank approximation represents a filtered signal with less noise [Me00, §5.12]. Using the SVD as a noise filter is also popular in image deblurring [HaNaOL06]. Similarly, the SVD is used in image processing to obtain a low-rank approximation to an image as a way to compress the image signal [Ka96].

**10.** *Other Applications:* The SVD is used in entanglement, a quantum form of correlation, where it is used to diagonalize the one-particle reduced density matrix to obtain the natural orbitals (i.e., the singular vectors) and their occupation numbers (i.e., the singular values) [PaYo01], [ScLoMa01], [MaPo12]. It is also important for theoretical endeavors, such as path-following methods for computing curves of equilibria in dynamical systems [DiGaPa06].

## 6.7.5  Higher Dimensions

More and more applications involve data that are best represented in higher dimensions. In particular, data compression and PCA are two significant motivating areas that have prompted higher-dimensional equivalents of the matrix SVD. Thus, information is stored in multidimensional (or multiway) arrays, i.e., tensors, instead of as matrices. Nearly all the fields mentioned in §6.7.4 and many others use higher-order tensors for multidimensional data analysis.

**Definitions:**
An **order-$N$ tensor** is a multiway array with $N$ indices:

$$\mathcal{A} = (a_{i_1 i_2 \ldots i_N}) \in \mathcal{R}^{I_1 \times I_2 \times \cdots \times I_N}.$$

The **outer product**, denoted by "o," of a set of $N$ vectors $x_1 \in \mathcal{R}^{I_1}, x_2 \in \mathcal{R}^{I_2}, \ldots, x_N \in \mathcal{R}^{I_N}$ is the order-$N$ tensor

$$\mathcal{A} = x_1 \circ x_2 \circ \cdots \circ x_N \in \mathcal{R}^{I_1 \times I_2 \times \cdots \times I_N},$$

where each element $a_{i_1 i_2 \ldots i_N}$ of $\mathcal{A}$ is the product of the corresponding vector elements:

$$a_{i_1 i_2 \ldots i_N} = x_1(i_1) x_2(i_2) \ldots x_N(i_N) \quad \text{for all } 1 \le i_j \le I_j.$$

Here $x_j(i_j)$ refers to the $i_j$th element of the vector $x_j$ for $j = 1, \ldots, N$.

If an order-$N$ tensor $\mathcal{A}$ can be expressed as an outer product of $N$ vectors, then $\mathcal{A}$ is a **rank-one tensor**.

The **rank** of an order-$N$ tensor $\mathcal{A}$ is the minimum number of rank-one tensors that sum to $\mathcal{A}$. For $\mathcal{A} \in \mathcal{R}^{I \times J \times K}$ (illustrated for notational simplicity), its rank decomposition is of the form

$$\mathcal{A} = \sum_{i=1}^{r} x_i \circ y_i \circ z_i, \tag{2}$$

where $r$ is minimal, $x_i \in \mathcal{R}^I$, $y_i \in \mathcal{R}^J$, and $z_i \in \mathcal{R}^K$ for $i = 1, \ldots, r$.

The **maximum rank** of a tensor is the largest attainable rank for that given order and sized tensor.

The **typical rank** of a tensor is any rank that occurs with probability greater than zero (i.e., on a set with positive Lebesgue measure).

Let $\mathcal{A}$ be an order-$N$ tensor. The **Higher-Order Singular Value Decomposition (HOSVD)** is a sum of rank-one tensors where the corresponding vectors are orthonormal. If $N = 3$ and $\mathcal{A} \in \mathcal{R}^{I \times J \times K}$ this is usually written as

$$\mathcal{A} = \sum_{i=1}^{I} \sum_{j=1}^{J} \sum_{k=1}^{K} \sigma_{ijk} (u_i \circ v_j \circ w_k), \tag{3}$$

where $u_i, v_j, w_k$ are the $i$th, $j$th, $k$th columns of orthogonal matrices $U, V, W$, respectively and $\sigma_{ijk}$ are elements from the tensor $\Sigma \in \mathcal{R}^{I \times J \times K}$.

**Facts:**

**1.** A first-order tensor is a vector, a second-order tensor is a matrix, a third-order tensor is a "cube," and so on.

**2.** Suppose the SVD of $A \in \mathcal{R}^{m \times n}$ is given by $A = U \Sigma V^T$. Using the outer product notation, equation (1) can be re-written as

$$A = \sum_{i=1}^{r} \sigma_i (u_i \circ v_i). \tag{4}$$

**3.** The definition of rank for matrices and order-2 tensors coincide. Indeed, the minimal $r$ such that a matrix $A$ is a sum of rank-one matrices is given by its SVD.

**4.** The HOSVD, illustrated for order-3 tensors in (3), is not a rank-revealing decomposition for tensors of order $N \geq 3$.

**5.** Extending (4) for higher-order tensors defines the rank of a higher-order tensor. The extension is easily seen by comparing (2) and (4). In contrast with matrices, the set of vectors in (2) is not necessarily orthogonal, whereas a rank decomposition for matrices can be chosen to be the SVD in which case the columns of $U$ and $V$ in (4) are orthonormal. Furthermore, it can be shown that an orthogonal rank decomposition of the form (2) may not exist.

**6.** The HOSVD can be computed directly by computing the SVDs of matrix "flattenings" in each dimension. Therefore, the HOSVD exists for any tensor $\mathcal{A}$ [GoVL13, §12.5], [KoBa09], [DeDeVa00].

**7.** The maximum rank and typical rank for order-2 tensors (matrices) are identical. If $A \in \mathcal{R}^{m \times n}$ then the maximum (typical) rank of $A$ is $\min(m, n)$.

**8.** Suppose $\mathcal{A}$ is an order-$N$ real-valued tensor, where $N \geq 3$. There are several differences between tensor rank and matrix rank:

- rank($\mathcal{A}$) may be different when considered over $\mathcal{R}$ and $\mathcal{C}$. See [Ma11], [KoBa09] for examples and explanations. Contrast this with matrices where the rank (and SVD) is the same when a real-valued matrix is considered over $\mathcal{R}$ and $\mathcal{C}$.

- Except in special cases, there is no direct algorithm to determine rank($\mathcal{A}$), and in fact the problem is NP-hard [Ha90]. In practice the rank of a tensor is determined numerically by iterative algorithms; see [KoBa09].

- $\mathcal{A}$ may have a different maximum and typical rank.

- $\mathcal{A}$ may have more than one typical rank, whereas a tensor over $\mathcal{C}$ always has one typical rank.

- If $\mathcal{A} \in \mathcal{R}^{I \times J \times K}$, only the weak upper bound on its maximum rank is known:

$$\text{rank}(\mathcal{A}) \leq \min\{IJ, IK, JK\}.$$

- If $\mathcal{A} \in \mathcal{R}^{2 \times 2 \times 2}$, then the maximum rank of $\mathcal{A}$ is 3 over $\mathcal{R}$ and 2 over $\mathcal{C}$. The typical ranks of $\mathcal{A}$ are 2 and 3 (when considered over $\mathcal{R}$).

- If $\mathcal{A} \in \mathcal{R}^{n \times n \times 2}$, then the maximum rank of $\mathcal{A}$ is $3n/2$ over $\mathcal{R}$ and $n$ over $\mathcal{C}$. The typical ranks for $\mathcal{A}$ are either $n$ or $n + 1$. See [KoBa09], [CoEtal09] for more special cases of typical rank.

**9.** Fact 10, §6.7.1 does not extend to tensors of order $N \geq 3$. The best low-rank approximation to a higher-order tensor may not exist.

**10.** The ordering of the dimensions of the tensor does not affect the rank.

**Examples:**

**1.** The set of $n \times n \times n$ real-valued rank-deficient tensors $\mathcal{A}$ over $\mathcal{R}$ has nonzero measure. If $\mathcal{A} \in \mathcal{R}^{2 \times 2 \times 2}$, Monte Carlo simulations show that a rank-2 tensor is generated 79% of the time and a rank-3 tensor is generated 21% of the time.

**2.** Suppose $\mathcal{A} \in \mathcal{R}^{2 \times 2 \times 2}$ where the front and back slices of $\mathcal{A}$ are defined respectively by the matrices

$$\begin{bmatrix} 1 & 0 \\ 0 & 1 \end{bmatrix} \quad \text{and} \quad \begin{bmatrix} 0 & 1 \\ -1 & 0 \end{bmatrix}.$$

Then $\text{rank}(\mathcal{A}) = 3$ over $\mathcal{R}$:

$$\mathcal{A} = \left( \begin{bmatrix} 1 \\ 0 \end{bmatrix} \circ \begin{bmatrix} 1 \\ 0 \end{bmatrix} \circ \begin{bmatrix} 1 \\ -1 \end{bmatrix} \right) + \left( \begin{bmatrix} 0 \\ 1 \end{bmatrix} \circ \begin{bmatrix} 0 \\ 1 \end{bmatrix} \circ \begin{bmatrix} 1 \\ 1 \end{bmatrix} \right) + \left( \begin{bmatrix} 1 \\ -1 \end{bmatrix} \circ \begin{bmatrix} 1 \\ 1 \end{bmatrix} \circ \begin{bmatrix} 0 \\ 1 \end{bmatrix} \right).$$

However, $\text{rank}(\mathcal{A}) = 2$ over $\mathcal{C}$:

$$\mathcal{A} = \left( \begin{bmatrix} 1/\sqrt{2} \\ -i/\sqrt{2} \end{bmatrix} \circ \begin{bmatrix} 1/\sqrt{2} \\ i/\sqrt{2} \end{bmatrix} \circ \begin{bmatrix} 1 \\ i \end{bmatrix} \right) + \left( \begin{bmatrix} 1/\sqrt{2} \\ i/\sqrt{2} \end{bmatrix} \circ \begin{bmatrix} 1/\sqrt{2} \\ -i/\sqrt{2} \end{bmatrix} \circ \begin{bmatrix} 1 \\ -i \end{bmatrix} \right).$$

**3.** The HOSVD is a form of higher-order PCA (see §6.7.4) where the matrices $U, V, W$ in (3) are thought of as the principal components in each dimension. The tensor $\Sigma = (\sigma_{ijk})$ in (3) is called the *core tensor* and its entries give the interaction between the different components. However, the higher-order PCA is used as a dimension reduction technique and therefore usually has the following form, illustrated for $\mathcal{A} \in \mathcal{R}^{I \times J \times K}$

$$\mathcal{A} \approx \sum_{i=1}^{P} \sum_{j=1}^{Q} \sum_{k=1}^{R} \sigma_{ijk} (u_i \circ v_j \circ w_k), \tag{5}$$

where $P < I, Q < J, R < K$ and $u_i, v_j, w_k$ are the $i$th, $j$th, $k$th columns of matrices $U \in \mathcal{R}^{I \times P}$, $V \in \mathcal{R}^{J \times Q}$, $W \in \mathcal{R}^{K \times R}$, respectively where $U, V, W$ (usually) have orthonormal columns.

**4.** The Eigenfaces algorithm (Example 3, §6.7.4) has been generalized for higher-order tensors as *TensorFaces* [VaTe02].

# REFERENCES

*Printed Resources*:

[AhMaOr93] R. K. Ahuja, T. L. Magnanti, and J. B. Orlin, *Network Flows: Theory, Algorithms, and Applications*, Prentice-Hall, 1993.

[AlBrBo00] O. Alter, P. O. Brown, and D. Botstein, "Singular value decomposition for genome-wide expression data processing and modeling", *Proceedings of the National Academy of Sciences USA* 97 (2000), 10101–10106.

[An87] T. Ando, "Totally positive matrices", *Linear Algebra and Its Applications* 90 (1987), 165–219. (Discusses basic properties of totally positive matrices.)

[An14] H. Anton, *Elementary Linear Algebra*, 11th ed., Wiley, 2014.

[Ba14] R. B. Bapat, *Graphs and Matrices*, 2nd ed., Springer, 2014. (Contains an introduction to combinatorial matrix theory and spectral graph theory.)

[BaRa97] R. B. Bapat and T. E. S. Raghavan, *Nonnegative Matrices and Applications*, Encyclopedia of Mathematical Sciences, No. 64, Cambridge University Press, 1997.

[Be97] R. Bellman, *Introduction to Matrix Analysis*, 2nd ed., Society for Industrial and Applied Mathematics, 1997.

[BeKu02] C. Belta and V. Kumar, "An SVD-based projection method for interpolation on $SE(3)$", *IEEE Transactions on Robotics and Automation* 18 (2002), 334–345.

[BePl94] A. Berman and R. J. Plemmons, *Nonnegative Matrices in the Mathematical Sciences*, Society for Industrial and Applied Mathematics, 1994. (Second revised edition of an authentic modern introduction.)

[BeDuOB95] M. W. Berry, S. T. Dumais, and G. W. O'Brien, "Using linear algebra for intelligent information retrieval", *SIAM Review* 37 (1995), 573–595.

[BjGo73] A. Björk and G. H. Golub, "Numerical methods for computing the angles between linear subspaces", *Mathematics of Computation* 27 (1973), 579–594.

[Br06] R. A. Brualdi, *Combinatorial Matrix Classes*, Cambridge University Press, 2006. (Discusses matrix classes such as $\mathcal{A}(R, S)$ and doubly stochastic matrices in detail.)

[BrRy91] R. A. Brualdi and H. J. Ryser, *Combinatorial Matrix Theory*, Cambridge University Press, 1991. (Discusses combinatorial results pertaining to matrices and graphs, as well as permanents.)

[BrSh09] R. A. Brualdi and B. L. Shader, *Matrices of Sign-Solvable Linear Systems*, Cambridge University Press, 2009. (Treats the concept of sign-solvability in detail.)

[ClDh13] A. K. Cline and I. S. Dhillon, "Computation of the singular value decomposition", in *Handbook of Linear Algebra*, 2nd ed., L. Hogben (ed.), Chapman & Hall/CRC Press, 2013, Chapter 58.

[CoEtal09] P. Comon, J. M. F. Ten Berge, L. De Lathauwer, and J. Castaing, "Generic and typical ranks of multi-way arrays", *Linear Algebra and Its Applications* 430 (2009), 2997–3007.

[Da10] B. N. Datta, *Numerical Linear Algebra and Applications*, 2nd ed., Society for Industrial and Applied Mathematics, 2010.

[De88] S. Deerwester, "Improving information retrieval with latent semantic indexing", *Proceedings of the 51st Annual Meeting of the American Society for Information Science* 25 (1988), 36–40.

[DeDeVa00] L. De Lathauwer, B. De Moor, and J. Vandewalle, "A multilinear singular value decomposition", *SIAM Journal on Matrix Analysis and Applications* 21 (2000), 1253–1278.

[DiGaPa06] L. Dieci, M. G. Gasparo, and A. Papini, "Path following by SVD", in *Computational Science – ICCS 2006*, Lecture Notes in Computer Science 3994, V. N. Alexandrov, G. D. van Albada, P. M. A. Sloot, and J. Dongarra (eds.), Springer, 2006, 677–684.

[DuErRe89] I. S. Duff, A. M. Erisman, and J. K. Reid, *Direct Methods for Sparse Matrices*, 2nd ed., Oxford University Press, 1989.

[FeEtal11] D. J. Fenn, M. A. Porter, S. Williams, M. McDonald, N. F. Johnson, and N. S. Jones, "Temporal evolution of financial market correlations", *Physical Review E* 84 (2011), 026109.

[Ga59] F. R. Gantmacher, *The Theory of Matrices*, Volumes I and II, Chelsea, 1959. (The first volume contains an introduction to matrix theory; the second discusses special topics such as nonnegative matrices and totally positive matrices.)

[GeLi81] A. George and J. W-H. Liu, *Computer Solution of Large Sparse Positive Definite Systems*, Prentice-Hall, 1981.

[GoKa65] G. H. Golub and W. Kahan, "Calculating the singular values and pseudo-inverse of a matrix", *SIAM Journal on Numerical Analysis* 2 (1965), 205–224.

[GoVL13] G. H. Golub and C. Van Loan, *Matrix Computations*, 4th ed., Johns Hopkins University Press, 2013. (Discusses elementary and advanced matrix computation problems; a remarkable blend of theoretical aspects and numerical issues.)

[HaNaOL06] P. C. Hansen, J. G. Nagy, and D. P. O'Leary, *Deblurring Images: Matrices, Spectra, and Filtering*, Society for Industrial and Applied Mathematics, 2006.

[Ha90] J. Håstad, "Tensor rank is NP-complete", *Journal of Algorithms* 11 (1990), 644–654.

[Hi06] M. D. Higgins, *Quantitative Textural Measurements in Igneous and Metamorphic Petrology*, Cambridge University Press, 2006.

[Ho13] L. Hogben, *Handbook of Linear Algebra*, 2nd ed., Chapman & Hall/CRC Press, 2013. (This handbook gives an excellent overview of linear algebra and treats combinatorial matrix theory in detail.)

[HoEtal00] N. S. Holter, M. Mitra, A. Maritan, M. Cieplak, J. R. Banavar, and N. V. Fedoroff, "Fundamental patterns underlying gene expression profiles: simplicity from complexity", *Proceedings of the National Academy of Sciences USA* 97 (2000), 8409–8414.

[HoJo90] R. A. Horn and C. R. Johnson, *Matrix Analysis*, Cambridge University Press, 1990. (A comprehensive, modern introduction to matrix theory.)

[Ho36] H. Hotelling, "Relations between two sets of variables", *Biometrika* 28 (1936), 321–377.

[Jo02] I. Joliffe, *Principal Component Analysis*, 2nd ed., Springer Series in Statistics, 2002.

[Ka96] D. Kalman, "A singularly valuable decomposition: the SVD of a matrix", *The College Mathematics Journal* 27 (1996), 2–23.

[Ki03] T. Kinoshita, "Singular value decomposition approach for the approximate coupled-cluster method", *Journal of Chemical Physics* 119 (2003), 7756–7762.

[Kl99] J. Kleinberg, "Authoritative sources in a hyperlinked environment", *Journal of the ACM* 46 (1999), 604–632.

[Kn81] D. E. Knuth, "A permanent inequality", *American Mathematical Monthly* 88 (1981), 731–740.

[KoBa09] T. Kolda and B. Bader, "Tensor decompositions and applications", *SIAM Review* 51 (2009), 455–500.

[KoBe11] Y. Koren and R. M. Bell, "Advances in collaborative filtering", in *Recommender Systems Handbook*, F. Ricci, L. Rokach, B. Shapira, and P. Kantor (eds.), Springer, 2011, 145–186.

[KoBeVo09] Y. Koren, R. M. Bell, and C. Volinsky, "Matrix factorization techniques for recommender systems", *IEEE Computer* 42 (2009), 30–37.

[MaKeHo05] K. Marida, J. T. Kent, and S. Holmes, *Multivariate Analysis*, Academic Press, New York, 2005.

[Ma98] A. D. Marsh, "On the interpretation of crystal size distributions in magmatic systems", *Journal of Petrology* 39 (1998), 553–599.

[MaOlAr11] A. W. Marshall, I. Olkin, and B. C. Arnold, *Inequalities: Theory of Majorization and Its Applications*, 2nd ed., Springer, 2011. (A comprehensive introduction to majorization theory and its many applications.)

[Ma11] C. D. Martin, "The rank of a $2 \times 2 \times 2$ tensor", *Linear and Multilinear Algebra* 59 (2011), 943–950.

[MaPo12] C. D. Martin and M. A. Porter, "The extraordinary SVD", *MAA Monthly* 119 (2012), 838–851.

[Ma13] R. Mathias, "Singular values and singular value inequalities", in *Handbook of Linear Algebra*, 2nd ed., L. Hogben (ed.), Chapman & Hall/CRC Press, 2013, Chapter 24.

[Me00] C. D. Meyer, *Matrix Analysis and Applied Linear Algebra*, Society for Industrial and Applied Mathematics, 2000.

[Mi84] H. Minc, *Permanents*, Cambridge University Press, 1984.

[Mi88] H. Minc, *Nonnegative Matrices*, Wiley, 1988.

[Mu82] K. E. Muller, "Understanding canonical correlation through the general linear model and principal components", *American Statistician* 36 (1982), 342–354.

[MuMaHe04] N. Muller, L. Magaia, and B. M. Herbst, "Singular value decomposition, eigenfaces, and 3D reconstructions", *SIAM Review* 46 (2004), 518–545.

[PaYo01] R. Paškauskas and L. You, "Quantum correlations in two-boson wave functions", *Physical Review A* 64 (2001), 042310.

[PoGe89] C. K. Pokorny and C. F. Gerald, *Computer Graphics: The Principles Behind the Art and Science*, Franklin, Beedle & Associates, 1989.

[PoEtal07] M. A. Porter, P. J. Mucha, M. E. J. Newman, and A. J. Friend, "Community structure in the United States House of Representatives", *Physica A* 386 (2007), 414–438.

[PoEtal05] M. A. Porter, P. J. Mucha, M. E. J. Newman, and C. M. Warmbrand, "A network analysis of committees in the United States House of Representatives", *Proceedings of the National Academy of Sciences USA* 102 (2005), 7057–7062.

[RuGa06] M. Rudek and P. R. Gardel Kurka, "3-D measurement from images using a range box", *ABCM Symposium Series in Mechatronics* 2 (2006), 629–636.

[Sa83] P. A. Samuelson, *Foundations of Economic Analysis*, Harvard University Press, 1983.

[ScLoMa01] J. Schliemann, D. Loss, and A. H. MacDonald, "Double-occupancy errors, adiabaticity, and entanglement of spin qubits in quantum dots", *Physical Review B* 63 (2001), 085311.

[Sc86] H. Schneider, "The influence of the marked reduced graph of a nonnegative matrix on the Jordan form and on related properties: a survey", *Linear Algebra and Its Applications* 84 (1986), 161–189.

[Sc98] A. Schrijver, *Theory of Linear and Integer Programming*, Wiley, 1998.

[Sc03] A. Schrijver, *Combinatorial Optimization – Polyhedra and Efficiency*, Springer, 2003. (A main reference on combinatorial optimization and polyhedral combinatorics.)

[Sm94] C. Small, "A global analysis of midocean ridge axial topography", *Geophysical Journal International* 116 (1994), 64–84.

[St05] G. Strang, *Linear Algebra and Its Applications*, 4th ed., Cengage Learning, 2005.

[Te91] J. M. F. Ten Berge, "Kruskal's polynomial for $2 \times 2 \times 2$ arrays and a generalization to $2 \times n \times n$ arrays", *Psychometrika* 56 (1991), 631–636.

[TrBa97] L. N. Trefethen and D. Bau, III, *Numerical Linear Algebra*, Society for Industrial and Applied Mathematics, 1997.

[TuPe91a] M. A. Turk and A. P. Pentland, "Eigenfaces for recognition", *Journal of Cognitive Neuroscience* 3 (1991), 71–86.

[TuPe91b] M. A. Turk and A. P. Pentland, "Face recognition using Eigenfaces", *Proceedings of Computer Vision and Pattern Recognition* 3 (1991), 586–591.

[VaTe02] M. A. O. Vasilescu and D. Terzopoulos, "Multilinear image analysis for facial recognition", in *ICPR 2002: Proceedings of the 16th International Conference on Pattern Recognition*, 2002, 511–514.

[Wi88] J. H. Wilkinson, *The Algebraic Eigenvalue Problem*, Oxford University Press, 1988. (An excellent reference on eigenvalues and matrix computations in general.)

[XuLi10] L. Xu and Q. Liang, "Computation of the singular value decomposition", in *Wireless Algorithms, Systems, and Applications*, Lecture Notes in Computer Science 6221, G. Pandurangan, V. S. A. Kumar, G. Ming, Y. Liu, and Y. Li (eds.), Springer, 2010, 338–342.

**Web Resources**:

`http://www.mathworks.com` (Home page for MATLAB, a high-level language for dense and sparse systems, eigenanalysis of dense and sparse matrices, and calculation of characteristic polynomials.)

`http://www.nag.com` (Contains Fortran, and C libraries for dense and sparse systems and eigenanalysis of dense and sparse matrices.)

`http://www.netlib.org` (Contains LINPACK, a collection of Fortran routines for relatively small dense systems; also contains EISPACK, a collection of Fortran routines for analyzing eigenvalues and eigenvectors of several classes of matrices.)

`http://www.netlib.org` (Contains LAPACK/CLAPACK, which supersedes LINPACK and hosts a collection of Fortran and C routines for dense and banded problems, ideal for shared-memory vector and parallel processors.)

`http://www.roguewave.com/products-services/imsl-numerical-libraries` (Contains IMSL, Fortran, and C libraries for dense and sparse systems and for eigenanalysis of dense and banded problems.)

`http://www.theory.caltech.edu/~preskill/ph219/chap2_13.pdf/` (Lecture notes for Physics 219: Quantum Information and Computation by J. Preskill, California Institute of Technology.)

# 7

# DISCRETE PROBABILITY

## INTRODUCTION

This chapter discusses aspects of discrete probability that are relevant to mathematics, computer science, engineering, and other disciplines. Topics covered include random variables, important discrete probability distributions, random walks, Markov models, queues, simulation, and the probabilistic method. Various applications to genetics, telephone network performance, reliability, average-case algorithm analysis, and combinatorics are presented.

## GLOSSARY

**absorbing boundary**: a boundary that stops the motion of a random walk whose trajectory comes into contact with it.

**all-terminal reliability**: the probability that a given network is connected.

**antithetic variates**: a variance reduction technique, based on negatively correlated variates, used in the simulation analysis of a given system.

**aperiodic state**: a state of a Markov chain that is not periodic.

**arrival process**: the statistical description of the time between successive arrivals to a queueing system.

**average-case complexity** (of an algorithm): the average number of operations required by the algorithm, taken over all problem instances of a given size.

**Bernoulli random variable**: the discrete random variable $X \in \{0, 1\}$ with probability distribution $Pr(X = 0) = 1 - p$ and $Pr(X = 1) = p$, for some $0 < p < 1$.

**binomial random variable**: the discrete random variable $X \in \{0, 1, \ldots, n\}$ with probability distribution $Pr(X = k) = \binom{n}{k} p^k (1 - p)^{n-k}$, for some $0 < p < 1$.

**Bose-Einstein model**: a probability model in which $k$ indistinguishable balls are randomly placed into $n$ distinguishable urns; several balls are allowed to occupy the same urn.

**boundary**: a point or set of points restricting the trajectory of a random walk.

**branching process**: a special type of Markov chain used to model the growth, and possible extinction, of populations.

**closed class**: a communicating class of states of a Markov chain in which transitions from these states never lead to states outside the class.

***coherent system***: a system of components for which increasing the number of operating components will not degrade the performance of the system.

***common random numbers***: a variance reduction technique in which alternative system configurations are analyzed using the same set of random numbers.

***communicating class***: a maximal set of states in a Markov chain that are reachable from one another by a finite number of transitions.

***conditional probability***: the probability $Pr(A|B)$ that the event $A$ occurs, given that another event $B$ has occurred.

***cutset***: a minimal set of edges in a graph the removal of which disconnects the graph.

***decoding problem*** (for an HMM): the problem of determining the most likely sequence of states that could have generated the observed data of an HMM with specified parameters.

***density function***: a nonnegative real-valued function $f(x)$ that determines the distribution of a continuous random variable $X$ via $Pr(a < X < b) = \int_a^b f(x)\,dx$.

***dependent events***: events that are not independent.

***discrete-event simulation***: a simulation of a time-evolving stochastic process in which changes to the state of the system can only occur at discrete instants.

***discrete-time Markov chain***: a probabilistic model of a randomly evolving system whose future is independent of the past if the present state is known.

***distribution*** (of a random variable): a probability measure associated with the values attained by the random variable.

***elastic boundary***: a boundary that could be absorbing or reflecting, usually depending on some given probability.

***evaluation problem*** (for an HMM): the problem of determining the probability of obtaining the observed data produced by an HMM with specified parameters.

***event***: a subset of the sample space.

***expected value*** (of a random variable): the average value taken on by the random variable.

***experiment***: any physically or mentally conceivable action having a measurable result.

***extinction probability***: in a branching process, the probability that the population eventually dies out.

***Fermi-Dirac model***: a probability model in which $k$ indistinguishable balls are randomly placed into $n$ distinguishable urns; at most one ball can occupy each urn.

***first passage time***: the time to first visit a given set of states in a Markov chain.

***floating-point arithmetic***: the "real number" arithmetic of computers.

***flop***: a unit for floating-point computations that is useful in assessing the complexity of an algorithm.

***gambler's ruin***: a one-dimensional random walk in which a gambler wins or loses one unit at each play of a game, with the game terminating whenever the gambler amasses a known amount or loses the entire initial stake.

***geometric random variable***: the discrete random variable $X \in \{1, 2, \ldots\}$ with probability distribution $Pr(X = k) = (1 - p)^{k-1}p$, for some $0 < p < 1$.

**hidden Markov model (HMM)**:  a stochastic model in which (hidden) states follow a Markov chain, yet one only observes a sequence of state-dependent outcomes.

**hypergeometric random variable**:  the discrete random variable that counts the number of red balls obtained when randomly selecting a fixed number of balls from an urn containing a specified number of red and black balls.

**independent events**:  events in which knowledge of whether one of the events did or did not occur does not alter the probability of occurrence of any of the other events.

**independent random variables**:  random variables whose joint distribution is the product of their individual distributions.

**irreducible chain**:  a Markov chain that can visit any state from any other state in a finite number of steps.

**irrelevant edge**:  an edge of a two-terminal network not appearing on any simple path joining the two terminals of the network.

**K-cutset**:  a minimal set of edges in a graph, the removal of which disconnects some pair of vertices in $K$.

**K-tree**:  a minimal set of edges in a graph that connects all vertices in $K$.

**learning problem** (for an HMM):  the problem of finding parameters of an HMM that are most likely to have produced the observed data.

**machine unit**:  a measure of the precision of floating-point arithmetic.

**Maxwell-Boltzmann model**:  a probability model in which $k$ distinguishable balls are randomly placed into $n$ distinguishable urns; several balls can occupy the same urn.

**mincut**:  a minimal set of components in a coherent system such that the system fails whenever these specified components fail.

**minpath**:  a minimal set of components in a coherent system such that the system operates whenever these specified components operate.

**Monte Carlo simulation**:  a simulation used to study both deterministic and stochastic phenomenon in which the passage of time is not material.

**overflow**:  the result of a floating-point arithmetic operation that exceeds the available range of numbers.

**parallel system**:  a system of components that fails only when all components fail.

**periodic state**:  a state of a Markov chain that can only be revisited at multiples of a certain number $d > 1$ (the *period* of the state).

**Poisson random variable**:  the discrete random variable $X \in \{0, 1, \ldots\}$ with probability distribution $Pr(X = k) = \frac{e^{-\lambda}\lambda^k}{k!}$, for some $\lambda > 0$.

**probability**:  a numerical value between 0 and 1 measuring the likelihood of occurrence of an event; the larger the number, the more likely the event.

**pseudo-random numbers**:  numbers generated in a predictable fashion, but that appear to behave like independent and identically distributed random numbers.

**purely multiplicative linear congruential generator**:  a widely used method of producing a stream of pseudo-random numbers.

**queue capacity**:  the maximum number of customers allowed at any time in a queueing system, either waiting or being served.

**queue discipline**: the protocol according to which customers are selected for service from among those waiting for service.

**queueing system**: a stochastic process in which customers arrive, await service, and are served.

**queue-length process**: a stochastic process describing the number of customers in the queueing system.

**Ramsey number $R(k, k)$**: the smallest integer $N$ such that any coloring of the edges of the complete graph on $n \geq N$ vertices possesses a complete subgraph on $k$ vertices with all edges colored the same.

**random numbers**: real numbers generated uniformly over the interval $(0, 1)$.

**random variable**: a function that assigns a real number to each outcome in the sample space.

**random walk**: a stochastic process based on the problem of determining the probable location of a point subject to random motions.

**recurrent state**: a state of a Markov chain from which the probability of return to itself is 1.

**recurrent walk**: a random walk that returns to its starting location with probability 1.

**reflecting boundary**: a boundary that redirects the motion of a random walk whose trajectory comes into contact with it.

**relative error**: the (percent) error in a computation relative to the true value.

**reliability**: the probability that a given system functions at a random instant of time.

**roundoff error**: the error resulting from abbreviating a number to the precision of the machine.

**sample size**: the number of possible outcomes of an experiment.

**sample space**: the set of all possible outcomes of an experiment.

**series system**: a system of components that operates only when all components operate.

**service-time distribution**: the statistical distribution of time required to serve a customer in a queueing system.

**simple path**: a path containing no repeated vertices.

**simulation**: a technique for studying numerically the behavior of complex stochastic systems and estimating their performance.

**single-station queueing system**: a system in which customers arrive at a service facility, wait for service, and depart after service completion.

**s-t cutset**: a minimal set of edges in a graph the removal of which leaves no *s-t* path.

**stability condition**: the set of parameter values for which the queue-length process (or the waiting-time process) has a steady-state distribution.

**steady-state distribution**: in a queueing system, the limiting probability distribution of the number of customers in the system.

**stochastic process**: a collection of random variables, typically indexed by time (discrete or continuous).

**structure function**:  a binary-valued function defined on all subsets of components; its value indicates whether or not the system operates when the specified components all operate.

**traffic intensity**:  in a queueing system, the ratio of the maximum arrival rate to the maximum service rate.

**trajectory**:  the successive positions traced out by a particle undergoing a random walk.

**transient state**:  a state in a Markov chain from which the probability of return to itself is less than 1.

**transient walk**:  a random walk that is not recurrent.

**transition probability**:  the probability of reaching a specified state in a Markov chain by a single transition (step) from a given state.

**transition probability matrix**:  the matrix of one-step transition probabilities for a Markov chain.

**two-terminal network**:  a network in which two vertices (or terminals) are specified.

**two-terminal reliability**:  the probability that the specified vertices of a two-terminal network are connected by a path of operating edges.

**underflow**:  the result of a floating-point operation that is smaller than the smallest representable number.

**uniform random variable**:  the continuous random variable $X \in (\alpha, \beta)$ with density function $f(x) = \frac{1}{\beta - \alpha}$.

**variance** (of a random variable):  a measure of dispersion of the random variable, equal to the average square of the deviation of the random variable from its expected value.

**variance reduction techniques**:  methods for obtaining greater precision for a fixed amount of sampling.

**waiting-time process**:  a stochastic process describing the time spent in the system by the customers.

---

# 7.1  FUNDAMENTAL CONCEPTS

### Definitions:

An **experiment** is any physically or mentally conceivable undertaking that results in a measurable outcome.

The **sample space** is the set $\Omega$ of all possible outcomes of an experiment.

The **sample size** of an experiment is the number of possible outcomes of the experiment.

An **event** in the sample space $\Omega$ is a subset of $\Omega$.

For a family of events $\{A_j \mid j \in J\}$, the **union** $\bigcup_{j \in J} A_j$ is the set of outcomes belonging to at least one $A_j$; the **intersection** $\bigcap_{j \in J} A_j$ is the set of all outcomes belonging to every $A_j$.

The **complement** $\overline{A}$ of an event $A$ is the set of outcomes in the sample space not belonging to $A$.

The events $A$ and $B$ are **disjoint** if $A \cap B = \emptyset$. The events $A_1, A_2, A_3, \ldots$ are **pairwise disjoint** if every pair $A_i, A_j$ of distinct events is disjoint.

A **probability measure** on the sample space $\Omega$ is a function $Pr$ from the set of subsets of $\Omega$ into the interval $[0, 1]$ satisfying

- $Pr(\Omega) = 1;$

- $Pr(\bigcup\limits_{k=1}^{\infty} A_k) = \sum\limits_{k=1}^{\infty} Pr(A_k)$, if the events $\{A_k\}$ are pairwise disjoint.

A **fair (unbiased) coin** is a coin that is just as likely to land Heads $(H)$ as it is to land Tails $(T)$.

A **red/blue spinner** is a disk consisting of two sectors, one red with area $r$ and one blue with area $b$.

**Facts:**

**1.** $Pr(\emptyset) = 0$.

**2.** $Pr(A)$ has the interpretation as the long-run proportion of time that the event $A$ occurs in repeated trials of the experiment.

**3.** $Pr(\bigcup\limits_{k=1}^{n} A_k) = \sum\limits_{k=1}^{n} Pr(A_k)$, if the $n$ events $\{A_k\}$ are pairwise disjoint.

**4.** If all outcomes are equally likely and the sample space has $k$ elements, where $k$ is a positive integer, then the probability of event $A$ is the number of elements of $A$ divided by the size of the sample space; that is, $Pr(A) = \dfrac{|A|}{k}$.

**5.** *Principle of inclusion-exclusion (simple form):* For events $A$ and $B$,

$$Pr(A \cup B) = Pr(A) + Pr(B) - Pr(A \cap B).$$

**6.** *Principle of inclusion-exclusion (general form):* For any events $A_1, A_2, \ldots, A_n$,

$$Pr(\bigcup\limits_{r=1}^{n} A_r) = \sum\limits_{i} Pr(A_i) - \sum\limits_{i<j} Pr(A_i \cap A_j) + \sum\limits_{i<j<k} Pr(A_i \cap A_j \cap A_k) -$$
$$\cdots + (-1)^{n+1} Pr(A_1 \cap A_2 \cap \cdots \cap A_n).$$

**7.** *Sieve principle:* If $A_1, A_2, \ldots, A_n$ are events, then

$$Pr(\text{exactly } k \text{ of the } A_j \text{ occur}) = \sum\limits_{r=k}^{n} (-1)^{r+k} \binom{r}{k} \sum\limits_{j_1<j_2<\cdots<j_r} Pr(A_{j_1} \cap A_{j_2} \cap \cdots \cap A_{j_r}).$$

**8.** *Boole's inequality:* If $A_1, A_2, \ldots, A_n$ are events, then

$$Pr(\bigcup\limits_{k=1}^{n} A_k) \leq \sum\limits_{k=1}^{n} Pr(A_k).$$

If $A_1, A_2, \ldots$ is an infinite sequence of events, then

$$Pr(\bigcup\limits_{k=1}^{\infty} A_k) \leq \sum\limits_{k=1}^{\infty} Pr(A_k).$$

(George Boole, 1815–1864.)

**9.** *Bonferroni's inequality:* If $A_1, A_2, \ldots, A_n$ are events, then

$$Pr(\bigcup_{k=1}^{n} A_k) \geq \sum_{k=1}^{n} Pr(A_k) - \sum_{k<j} Pr(A_k \cap A_j).$$

**10.** $Pr(\overline{A}) = 1 - Pr(A)$.

**11.** *Monotonicity:* If $A \subseteq B$, then $Pr(A) \leq Pr(B)$.

**12.** If $A_1 \subseteq A_2 \subseteq A_3 \subseteq \cdots$ is an increasing sequence of events, then

$$\lim_{n \to \infty} Pr(A_n) = Pr(\bigcup_{n=1}^{\infty} A_n).$$

**13.** If $A_1 \supseteq A_2 \supseteq A_3 \supseteq \cdots$ is a decreasing sequence of events, then

$$\lim_{n \to \infty} Pr(A_n) = Pr(\bigcap_{n=1}^{\infty} A_n).$$

**Examples:**

**1.** The following table gives examples of specific experiments, their sample spaces, and the corresponding sample size.

| experiment | sample space | sample size |
|---|---|---|
| toss a coin | $\{H, T\}$ | 2 |
| toss a coin $n$ times | $\{(\omega_1, \ldots, \omega_n) \mid \omega_i \text{ is } H \text{ or } T\}$ | $2^n$ |
| roll a die | $\{1, 2, 3, 4, 5, 6\}$ | 6 |
| roll a pair of dice | $\{(1,1), (1,2), \ldots, (6,5), (6,6)\}$ | 36 |
| draw a card from a standard deck | $\{2\clubsuit, 2\diamondsuit, \ldots, A\heartsuit, A\spadesuit\}$ | 52 |
| spin a red/blue spinner | $\{\text{red, blue}\}$ | 2 |

**2.** The following are various events defined for the experiment of rolling a pair of dice (see the table of Example 1).

sum of dice is 9:  $A = \{(3,6), (4,5), (5,4), (6,3)\}$
both dice are multiples of 3:  $B = \{(3,3), (3,6), (6,3), (6,6)\}$
sum of dice $\leq 4$:  $C = \{(1,1), (1,2), (1,3), (2,1), (2,2), (3,1)\}$.

**3.** The events $A = \{\text{sum of dice} = 9\}$ and $C = \{\text{sum of dice} \leq 4\}$ are disjoint. The events $A_i = \{\text{sum of the dice is } i\}$, $2 \leq i \leq 12$, are pairwise disjoint.

**4.** *Random selection of an integer:* Let $\Omega = \{1, 2, 3, \ldots, n\}$ be the sample space corresponding to the experiment of randomly selecting an integer between 1 and $n$, and define $Pr(j) = Pr(\{j\}) = \frac{1}{n}$. By Fact 4, $Pr(3 \leq j \leq n) = Pr(\{3 \leq j \leq n\}) = \frac{n-2}{n}$.

**5.** For a red/blue spinner, the sample space is $\Omega = \{\text{red, blue}\}$. If the spinner is equally likely to land at any location, then $Pr(\text{red}) = \frac{r}{r+b}$ and $Pr(\text{blue}) = \frac{b}{r+b}$.

**6.** Toss a fair coin $n$ times and interpret Heads as 1 and Tails as 0. The sample space $\Omega = \{(\omega_1, \omega_2, \ldots, \omega_n) \mid \omega_j \in \{0, 1\}\}$ consists of all possible 0-1 sequences of length $n$. Since $|\Omega| = 2^n$, each probability $Pr((\omega_1, \omega_2, \ldots, \omega_n))$ is assigned the value $\frac{1}{2^n}$. By Fact 4, $Pr(A) = \frac{|A|}{2^n}$ holds for all $A \subseteq \Omega$.

For example, the probability of no tails appearing in four coin tosses is the probability of event $A = \{(1,1,1,1)\}$, so $Pr(A) = \frac{1}{16}$. The probability of exactly one tail is the probability of event $B = \{(0,1,1,1), (1,0,1,1), (1,1,0,1), (1,1,1,0)\}$; hence $Pr(B) = \frac{4}{16} = \frac{1}{4}$. The probability of at least two tails is, using Fact 10, $1 - Pr(A) - Pr(B) = 1 - \frac{5}{16} = \frac{11}{16}$.

**7.** *Derangements:*   Let $D_n$ be the set of derangements (§2.4.2) on the $n$ elements $\{1, 2, \ldots, n\}$ and define $A_j$ to be the set of all permutations fixing $j$. For any permutation $\sigma$, $Pr(\sigma) = \frac{1}{n!}$. Also, for $j_1 < j_2 < \cdots < j_k$, $Pr(A_{j_1} \cap A_{j_2} \cap \cdots \cap A_{j_k}) = \frac{(n-k)!}{n!}$ and $\sum_{j_1 < j_2 < \cdots < j_k} Pr(A_{j_1} \cap A_{j_2} \cap \cdots \cap A_{j_k}) = \binom{n}{k} \frac{(n-k)!}{n!} = \frac{1}{k!}$. By Fact 6, $Pr(\bigcup_{r=1}^{n} A_r) = \sum_i Pr(A_i) - \sum_{i<j} Pr(A_i \cap A_j) + \sum_{i<j<k} Pr(A_i \cap A_j \cap A_k) - \cdots + (-1)^{n+1} Pr(A_1 \cap A_2 \cap \cdots \cap A_n) = 1 - \frac{1}{2!} + \frac{1}{3!} - \cdots + (-1)^{n+1} \frac{1}{n!}$. Hence, $Pr(D_n) = 1 - 1 + \frac{1}{2!} - \frac{1}{3!} - \cdots + \frac{(-1)^n}{n!} \approx e^{-1} \approx 0.36788$.

**8.** *5-card stud poker:*   Five cards are drawn from a well-shuffled deck of 52 playing cards. The sample space consists of the $\binom{52}{5} = 2{,}598{,}960$ possible five-card hands. The approximate probabilities of various events are displayed in the following table. (See §2.3.2, Example 12 for further details.) As seen from the probabilities given in the table, obtaining a five-card hand containing three of a kind is approximately ten times more likely than obtaining a five-card hand containing a flush, which in turn is approximately ten times more likely than obtaining a five-card hand containing four of a kind.

| type of hand | example | hand enumeration | probability |
|---|---|---|---|
| one pair | $7\heartsuit, 7\diamondsuit, K\clubsuit, J\spadesuit, 2\heartsuit$ | $\binom{13}{1}\binom{4}{2}\binom{12}{3}4^3$ | 0.42 |
| two pairs | $7\spadesuit, 7\heartsuit, K\diamondsuit, K\spadesuit, 3\clubsuit$ | $\binom{13}{2}\binom{4}{2}\binom{4}{2}44$ | 0.048 |
| three of a kind | $7\clubsuit, 7\heartsuit, 7\diamondsuit, 3\diamondsuit, 5\spadesuit$ | $\binom{13}{1}\binom{4}{3}\binom{12}{2}4^2$ | 0.021 |
| straight | $7\clubsuit, 8\spadesuit, 9\diamondsuit, 10\clubsuit, J\heartsuit$ | $10(4^5 - 4)$ | 0.0039 |
| flush | $3\diamondsuit, 6\diamondsuit, 7\diamondsuit, J\diamondsuit, K\diamondsuit$ | $4(\binom{13}{5} - 10)$ | 0.0020 |
| full house | $3\heartsuit, 3\diamondsuit, 3\spadesuit, 7\clubsuit, 7\heartsuit$ | $13 \cdot 12 \cdot \binom{4}{3}\binom{4}{2}$ | 0.0014 |
| four of a kind | $A\clubsuit, A\diamondsuit, A\heartsuit, A\spadesuit, 7\spadesuit$ | $13 \cdot 48$ | 0.00024 |
| straight flush | $7\diamondsuit, 8\diamondsuit, 9\diamondsuit, 10\diamondsuit, J\diamondsuit$ | $\binom{4}{1}9$ | 0.000014 |
| royal flush | $10\heartsuit, J\heartsuit, Q\heartsuit, K\heartsuit, A\heartsuit$ | 4 | 0.0000015 |

# 7.2   INDEPENDENCE AND DEPENDENCE

Sequences of independent events are often encountered when an experiment is repeated (without changes). Independent events correspond, intuitively, to events that do not affect the outcome of one another. The treatment of dependent events requires conditional probabilities.

## 7.2.1   BASIC CONCEPTS

**Definitions:**

Two events $A$ and $B$ are **independent** if $Pr(A \cap B) = Pr(A) \, Pr(B)$.

The $n$ events $A_1, A_2, \ldots, A_n$ are **independent** if for all $k$ ($2 \le k \le n$) and $j_1, j_2, \ldots, j_k$ ($1 \le j_1 < j_2 < \cdots < j_k \le n$),

$$Pr(A_{j_1} \cap A_{j_2} \cap \cdots \cap A_{j_k}) = Pr(A_{j_1}) \, Pr(A_{j_2}) \ldots Pr(A_{j_k}).$$

The infinite collection of events $\{A_n \mid n \geq 1\}$ is **independent** if for all finite $k \geq 2$ the events $A_1, A_2, \ldots, A_k$ are independent.

Let $B$ be an event with $Pr(B) > 0$. The **conditional probability of $A$ given $B$** is
$$Pr(A|B) = \frac{Pr(A \cap B)}{Pr(B)}.$$

**Facts:**

**1.** If events $A$ and $B$ are independent, then so are $A$ and $\overline{B}$, $\overline{A}$ and $B$, and $\overline{A}$ and $\overline{B}$.

**2.** If $A_1, A_2, \ldots$ are independent events, then $Pr(\bigcap\limits_{k=1}^{\infty} A_k) = \prod\limits_{k=1}^{\infty} Pr(A_k)$.

**3.** The function $\phi_B: A \to Pr(A|B)$ is a probability measure (§7.1).

**4.** $Pr(A \cap B) = Pr(A|B)\, Pr(B)$.

**5.** If $A$ and $B$ are independent, then $Pr(A|B) = Pr(A)$. This equation captures the notion that for independent events $A$ and $B$ the knowledge that one of the events has occurred does not affect the probability of the other occurring.

**6.** Pairwise independence of a collection of events does not necessarily imply that all events are independent (see Example 3).

**7.** *Law of total probability:*  For any event $A$ and any partition of $\Omega$ into the events $B_1, B_2, \ldots, B_n$,
$$Pr(A) = \sum_{i=1}^{n} Pr(A \cap B_i) = \sum_{i=1}^{n} Pr(A|B_i)\, Pr(B_i).$$

**8.** *Bayes' formula:*  For any event $A$ and any partition of $\Omega$ into the events $B_1, B_2, \ldots, B_n$,
$$Pr(B_1|A) = \frac{Pr(B_1 \cap A)}{Pr(A)} = \frac{Pr(A|B_1)\, Pr(B_1)}{\sum\limits_{i=1}^{n} Pr(A|B_i)\, Pr(B_i)}.$$

(Thomas Bayes, 1702–1761.)

**9.** *Chain rule:*  For any events $A_1, A_2, \ldots, A_n$ satisfying $Pr(\bigcap\limits_{k=1}^{n-1} A_k) > 0$,

$$Pr(A_1 \cap A_2 \cap \cdots \cap A_n) = Pr(A_1)\, Pr(A_2|A_1)\, Pr(A_3|A_1 \cap A_2) \ldots Pr(A_n \mid \bigcap_{k=1}^{n-1} A_k).$$

**Examples:**

**1.** *Tossing two fair coins:*  The sample space for this experiment consists of the four outcomes $HH, HT, TH, TT$. For example, the outcome $HT$ means that the first coin turns up Heads and the second Tails. Because both coins are fair, all four outcomes are equally likely and in particular $Pr(HT) = \frac{1}{4}$. Since $Pr(H) = Pr(T) = \frac{1}{2}$, $Pr(HT) = \frac{1}{4} = \frac{1}{2} \cdot \frac{1}{2} = Pr(H)\, Pr(T)$. Thus, the events "Heads on the first coin" and "Tails on the second coin" are independent.

**2.** *Tossing a fair coin $n$ times:*  As in Example 6 of §7.1, let 1 stand for Heads and 0 for Tails. For each $1 \leq i \leq n$ select $\epsilon_i \in \{0, 1\}$ and define $A_i = \{\epsilon_i$ occurs on the $i$th toss$\}$. Since all outcomes are equally likely, $Pr(A_1 \cap A_2 \cap \cdots \cap A_n) = (\frac{1}{2})^n = \frac{1}{2} \times \frac{1}{2} \times \cdots \times \frac{1}{2} = Pr(A_1)\, Pr(A_2) \ldots Pr(A_n)$. Also, for all $j_1, j_2, \ldots, j_k$ $(2 \leq k \leq n)$, $Pr(A_{j_1} \cap A_{j_2} \cap \cdots \cap A_{j_k}) = (\frac{1}{2})^k = \frac{1}{2} \times \frac{1}{2} \times \cdots \times \frac{1}{2} = Pr(A_{j_1})\, Pr(A_{j_2}) \ldots Pr(A_{j_k})$. Therefore the events $A_1, A_2, \ldots, A_n$ are independent.

**3.** Let $\Omega = \{a, b, c, d\}$ be a sample space with equiprobable outcomes. Let $A = \{a, b\}$, $B = \{a, c\}$, and $C = \{a, d\}$. Here $Pr(A) = Pr(B) = Pr(C) = \frac{1}{2}$. Also $Pr(A \cap B) = \frac{1}{4} = \frac{1}{2} \cdot \frac{1}{2} = Pr(A)\,Pr(B)$, $Pr(A \cap C) = \frac{1}{4} = \frac{1}{2} \cdot \frac{1}{2} = Pr(A)\,Pr(C)$, and $Pr(B \cap C) = \frac{1}{4} = \frac{1}{2} \cdot \frac{1}{2} = Pr(B)\,Pr(C)$. Yet $Pr(A \cap B \cap C) = \frac{1}{4} \neq \frac{1}{2} \cdot \frac{1}{2} \cdot \frac{1}{2} = Pr(A)\,Pr(B)\,Pr(C)$. In this example, any two of the events are independent, but all three are not.

**4.** *Gambler's fallacy:*  Suppose that a fair coin is tossed five times, turning up Heads on all five tosses. What is the probability that the next (sixth) toss turns up Tails? A common fallacy is to believe that a Tail is more likely to turn up next, since in the long run 50% of the coins should turn up Tails (and 50% Heads).

The appropriate sample space consists of $2^6 = 64$ equiprobable outcomes, representing any sequence of six Heads and/or Tails. The required probability is $Pr(A|B)$, where $A = \{(H, H, H, H, H, T)\}$ and $B = \{(H, H, H, H, H, H), (H, H, H, H, H, T)\}$. Then $Pr(A|B) = \frac{Pr(A \cap B)}{Pr(B)} = \frac{Pr(A)}{Pr(B)} = \frac{1}{2}$. Consequently, a Tail turning up next is just as likely as a Head.

**5.** An urn contains seven blue marbles and five red marbles. An experiment consists of drawing (without replacement) a marble at random, observing its color, and then drawing a second marble at random. Let $B_i$ be the event "the $i$th marble drawn is blue" and let $R_i$ be the event "the $i$th marble drawn is red", where $i \in \{1, 2\}$. Then $Pr(B_1) = \frac{7}{12}$, $Pr(R_2|B_1) = \frac{5}{11}$, $Pr(B_1 \cap R_2) = Pr(R_2|B_1)\,Pr(B_1) = \frac{5}{11} \cdot \frac{7}{12} = \frac{35}{132}$. By Fact 7,

$$Pr(R_2) = Pr(R_2|R_1)\,Pr(R_1) + Pr(R_2|B_1)\,Pr(B_1) = \frac{4}{11} \cdot \frac{5}{12} + \frac{5}{11} \cdot \frac{7}{12} = \frac{55}{132}.$$

By Fact 8,

$$Pr(B_1|R_2) = \frac{Pr(R_2|B_1)\,Pr(B_1)}{Pr(R_2|R_1)\,Pr(R_1) + Pr(R_2|B_1)\,Pr(B_1)} = \frac{\frac{5}{11} \cdot \frac{7}{12}}{\frac{4}{11} \cdot \frac{5}{12} + \frac{5}{11} \cdot \frac{7}{12}} = \frac{7}{11}.$$

**6.** A particular family is known to have two children (one 9 years old, the other 10 years old). When a census taker comes to the house, a girl answers the doorbell. What is the probability that the other child is also a girl?

To answer this question, construct the sample space $\Omega = \{(b, b), (b, g), (g, b), (g, g)\}$, where, for example, the ordered pair $(b, g)$ means that the younger child is a boy and the older child is a girl. Assume that all four outcomes in the sample space are equiprobable. The required probability is $Pr(A|B)$, where $A = \{(g, g)\}$ and $B = \{(b, g), (g, b), (g, g)\}$. Then

$$Pr(A|B) = \frac{Pr(A \cap B)}{Pr(B)} = \frac{Pr(\{(g, g)\})}{Pr(\{(b, g), (g, b), (g, g)\})} = \frac{\frac{1}{4}}{\frac{3}{4}} = \frac{1}{3}.$$

**7.** *Genetics:*  Genes are responsible for physical traits of all living things. Each gene is composed of two alleles. Dominant alleles are represented with capital letters and recessive alleles with lower case letters. The basic discoveries concerning genetics were made by Gregor Mendel (1822–1884). One of the genes that is responsible for eye color exhibits two alleles—a dominant one $B$, for brown eyes, and a recessive one $b$, for blue eyes. In a certain population the genotype probabilities are

$Pr(\text{an individual has genotype } BB) = 0.2$

$Pr(\text{an individual has genotype } Bb) = 0.5$

$Pr(\text{an individual has genotype } bb) = 0.3.$

Let $E_b$ be the event that an offspring receives a $b$ allele from its mother, and let $F_b$ be the event that it receives a $b$ allele from its father. Conditioning on the genotype $(BB, Bb, bb)$ of the offspring produces

$$Pr(E_b) = Pr(E_b|BB)\,Pr(BB) + Pr(E_b|Bb)\,Pr(Bb) + Pr(E_b|bb)\,Pr(bb)$$
$$= 0 \times 0.2 \,+\, 0.5 \times 0.5 \,+\, 1 \times 0.3 = 0.55;$$

similarly $Pr(F_b) = 0.55$.

Let $C$ be the event that the offspring has blue eyes (that is, has genotype $bb$). By independence,

$$Pr(C) = Pr(E_b \cap F_b) = Pr(E_b)\,Pr(F_b) = (0.55)^2 = 0.3025.$$

**8.** In Example 7, let $A$ be the event that the father has blue eyes (has genotype $bb$). If the father has blue eyes, then the offspring will have blue eyes (event $C$) if and only if it receives a $b$ allele from its mother, giving $Pr(C|A) = Pr(E_b) = 0.55$. We also have $Pr(C|\text{father is } Bb) = Pr(C \text{ and mother is } Bb|\text{father is } Bb) + Pr(C \text{ and mother is } bb| \text{father is } Bb) = 0.25 \times 0.5 \,+\, 0.5 \times 0.3 = 0.275$ and $Pr(C|\text{father is } BB) = 0$.

The conditional probability that the father has blue eyes if the offspring has blue eyes is obtained from Fact 8 (interchanging phenotype with genotype when convenient) as

$$Pr(A|C) = \frac{Pr(C|\text{father has } bb)\,Pr(\text{father has } bb)}{Pr(C|bb)\,Pr(bb) + Pr(C|Bb)\,Pr(Bb) + Pr(C|BB)\,Pr(BB)}$$
$$= \frac{0.55 \times 0.3}{0.55 \times 0.3 \,+\, 0.275 \times 0.5 \,+\, 0} \approx 0.545.$$

**9.** *Let's Make a Deal*: A game show contestant is told there is a fabulous prize hidden behind one of three doors ($A$, $B$, or $C$). The contestant guesses that the prize is behind door $A$. At this point the game show host (who knows what is behind each door, and in particular knows that the prize is not behind door $B$) opens door $B$, revealing that the prize is not there. The contestant is then offered the opportunity to change her guess. Should she? Intuition might suggest that nothing is to be gained by changing the guess (the prize, it is argued, is now equally likely to be behind either door $A$ or door $C$). Using conditional probabilities, however, shows that it is definitely worthwhile to now guess that the prize is behind door $C$, assuming that the host is known to always open a door with no prize and to choose randomly if both remaining doors do not hide the prize.

It is reasonable to assume that the prize is equally likely to be hidden behind each of the doors. Thus, if $H_X$ denotes the event in which the prize is hidden behind door $X$, then $Pr(H_A) = Pr(H_B) = Pr(H_C) = \frac{1}{3}$. If $O_X$ denotes the event that door $X$ is opened, then $Pr(O_B|H_A) = Pr(O_C|H_A) = \frac{1}{2}$, whereas $Pr(O_B|H_B) = 0$ and $Pr(O_B|H_C) = 1$. By Fact 8,

$$Pr(H_A|O_B) = \frac{Pr(H_A \cap O_B)}{Pr(O_B)}$$
$$= \frac{Pr(H_A)\,Pr(O_B|H_A)}{Pr(H_A)\,Pr(O_B|H_A) + Pr(H_B)\,Pr(O_B|H_B) + Pr(H_C)\,Pr(O_B|H_C)}$$
$$= \frac{\frac{1}{3} \cdot \frac{1}{2}}{\frac{1}{3} \cdot \frac{1}{2} \,+\, 0 \,+\, \frac{1}{3} \cdot 1} = \frac{1}{3}.$$

Similarly $Pr(H_C|O_B) = \frac{2}{3}$, so it is twice as likely for the prize to be hidden behind door $C$ as behind door $A$, given that door $B$ is shown to contain no prize. A web-based simulation of this situation, in which prizes are randomly hidden behind doors, enables one to verify experimentally this conclusion; see for example

- http://math.ucsd.edu/~crypto/Monty/monty.html

## 7.2.2  URN MODELS

Several applications can be viewed as the result of placing balls into urns.

**Definitions:**

In the following models, $k$ balls are randomly placed in $n$ distinguishable urns labeled 1 through $n$.

- *Model 1* (**Maxwell-Boltzmann**):  The balls are distinguishable and multiple occupancy is permitted.
- *Model 2*:  The balls are distinguishable and multiple occupancy is not permitted.
- *Model 3* (**Fermi-Dirac**):  The balls are indistinguishable and multiple occupancy is not permitted.
- *Model 4* (**Bose-Einstein**):  The balls are indistinguishable and multiple occupancy is permitted.
- *Model 5*:  The balls are distinguishable, no urn is allowed to remain empty, and multiple occupancy is permitted.

**Facts:**

**1.** The following table shows, for different urn models, the probability of the event $(k_1, k_2, \ldots, k_n)$, in which $k_1$ balls are in urn 1, $k_2$ balls are in urn 2, $\ldots$, $k_n$ balls are in urn $n$, with the restrictions $\sum_{j=1}^{n} k_j = k$, $k_j \geq 0$. In Model 2, $n^{\underline{k}}$ is a falling power (see §3.4.2). In Models 2 and 3, every $k_j \in \{0, 1\}$ and the models are meaningful only if $k \leq n$. In Model 5, every $k_j \geq 1$, $k \geq n$, and $\left\{{k \atop n}\right\}$ is a Stirling subset number (§2.5.2).

| model | sample size | enumeration of $(k_1, \ldots, k_n)$ | probability of $(k_1, \ldots, k_n)$ |
|---|---|---|---|
| 1 | $n^k$ | $\binom{k}{k_1\ k_2\ \ldots\ k_n}$ | $\binom{k}{k_1\ k_2\ \ldots\ k_n} n^{-k}$ |
| 2 | $n^{\underline{k}}$ | $k!$ | $\binom{n}{k}^{-1}$ |
| 3 | $\binom{n}{k}$ | $1$ | $\binom{n}{k}^{-1}$ |
| 4 | $\binom{n+k-1}{k}$ | $1$ | $\binom{n+k-1}{k}^{-1}$ |
| 5 | $n!\left\{{k \atop n}\right\}$ | $\binom{k}{k_1\ k_2\ \ldots\ k_n}$ | $\binom{k}{k_1\ k_2\ \ldots\ k_n}/\left(n!\left\{{k \atop n}\right\}\right)$ |

**2.** The Maxwell-Boltzmann model was originally proposed to explain the distribution of $k$ subatomic particles into $n$ different energy states. It has been replaced by the Bose-Einstein model (appropriate for particles with integer "spin", such as photons and pi mesons) and by the Fermi-Dirac model (appropriate for particles with half-integer "spin", such as protons and neutrons).

**3.** *Pólya's urn scheme* (George Pólya, 1887–1985):  In this model, an urn contains $b$ black and $r$ red balls. At each step one ball is drawn and replaced, and $c$ additional balls of the same color are placed in that urn. This scheme models the spread of a contagious disease where an infected person infects $c$ other persons.

**4.** The case $c = 0$ in Pólya's urn scheme corresponds to sampling balls with replacement.

**5.** The case $c = -1$ in Pólya's urn scheme corresponds to sampling balls without replacement.

**6.** The following table shows how to calculate several types of probabilities using Pólya's urn scheme, where $b$ is the number of black balls, $r$ is the number of red balls, and $c$ is the number of additional balls added each time.

| event | probability |
|---|---|
| drawing a black | $\frac{b}{b+r}$ |
| drawing a black then red | $\frac{br}{(b+r)(b+r+c)}$ |
| drawing in order black, red, black | $\frac{br(b+c)}{(b+r)(b+r+c)(b+r+2c)}$ |
| drawing $k$ black and $n-k$ red balls in a prescribed order | $\frac{b(b+c)...(b+(k-1)c)r(r+c)...(r+(n-k-1)c)}{(b+r)(b+r+c)(b+r+2c)...(b+r+(n-1)c)}$ |
| drawing $k$ black balls in $n$ drawings; the order of drawing does not matter | $\frac{\binom{-b/c}{k}\binom{-r/c}{n-k}}{\binom{-(b+r)/c}{n}}$ |

**Examples:**

**1.** *Partial derivatives:*   For analytic functions $f$, the order in which derivatives are taken does not matter. As an example, the mixed second partial derivatives $f_{xy}$ and $f_{yx}$ are equal, as are $f_{xxy}$ and $f_{xyx}$. Consequently, the number of different third-order partial derivatives of a function of $n$ variables is the number of ways to distribute $k = 3$ indistinguishable balls into $n$ urns (variables). Each such distribution corresponds to selecting the number of times each variable occurs in forming the partial derivative. Using the entry for Model 4 in the table for Fact 1, there are $\binom{n+2}{3}$ third-order partial derivatives of $f$. When $n = 3$ this gives $\binom{5}{3} = 10$ different third-order partial derivatives of $f(x, y, z)$: namely, $f_{xxx}, f_{yyy}, f_{zzz}, f_{xxy}, f_{xxz}, f_{xyy}, f_{yyz}, f_{xzz}, f_{yzz}, f_{xyz}$. In general, there are $\binom{n+k-1}{k}$ different $k$th-order partial derivatives of $f$.

**2.** Model 3 provides a model for the occurrence of misprints on the pages of a book. Here the $n$ urns correspond to the $n$ symbols printed sequentially in the book and $k$ is the number of misprints. Each symbol is either correct or a misprint, so multiple occupancy does not occur.

Also, assuming that the misprints are not generated in a systematic fashion, the $k$ balls can be considered indistinguishable, with misprints equally likely to occur at any location on the page.

**3.** *Lottery odds:*   A lottery is conducted by selecting five different numbers from $1, 2, \ldots, 9$. This can be viewed using urn Model 3, in which the five selected numbers correspond to $k = 5$ identical balls placed into $n = 9$ distinguished urns. The number of such selections, by the table of Fact 1, is $\binom{9}{5} = 126$. Only one of these 126 selections matches all the five winning numbers, so $Pr(\text{match } 5) = \frac{1}{126}$.

To match exactly four of the five winning numbers, select the matching numbers in $\binom{5}{4} = 5$ ways and select the (single) nonmatching number in $\binom{4}{1} = 4$ ways, giving $Pr(\text{match } 4) = \frac{5 \cdot 4}{126} = \frac{20}{126}$.

To match exactly three of the winning numbers, select the matching numbers in $\binom{5}{3} = 10$ ways and select the two nonmatching numbers in $\binom{4}{2} = 6$ ways, giving $Pr(\text{match } 3) = \frac{10 \cdot 6}{126} = \frac{60}{126}$.

**4.** In a number of state lotteries, $k = 6$ numbers are drawn from $1, 2, \ldots, n$. The following table gives the probability of matching exactly six, exactly five, and exactly four of the six winning numbers, for values of $n = 35, \ldots, 60$.

| $n$ | match 6 | match 5 | match 4 |
|---|---|---|---|
| 35 | 1/1,623,160 | 87/811,580 | 87/23,188 |
| 36 | 1/1,947,792 | 15/162,316 | 2,175/649,264 |
| 37 | 1/2,324,784 | 31/387,464 | 2,325/774,928 |
| 38 | 1/2,760,681 | 64/920,227 | 2,480/920,227 |
| 39 | 1/3,262,623 | 66/1,087,541 | 2,640/1,087,541 |
| 40 | 1/3,838,380 | 17/319,865 | 561/255,892 |
| 41 | 1/4,496,388 | 35/749,398 | 2,975/1,498,796 |
| 42 | 1/5,245,786 | 108/2,622,893 | 675/374,699 |
| 43 | 1/6,096,454 | 111/3,048,227 | 4,995/3,048,227 |
| 44 | 1/7,059,052 | 57/1,764,763 | 10,545/7,059,052 |
| 45 | 1/8,145,060 | 39/1,357,510 | 741/543,004 |
| 46 | 1/9,366,819 | 80/3,122,273 | 3,900/3,122,273 |
| 47 | 1/10,737,573 | 82/3,579,191 | 4,100/3,579,191 |
| 48 | 1/12,271,512 | 21/1,022,626 | 4,305/4,090,504 |
| 49 | 1/13,983,816 | 43/2,330,636 | 645/665,896 |
| 50 | 1/15,890,700 | 22/1,324,225 | 473/529,690 |
| 51 | 1/18,009,460 | 27/1,800,946 | 1,485/1,800,946 |
| 52 | 1/20,358,520 | 69/5,089,630 | 3,105/4,071,704 |
| 53 | 1/22,957,480 | 141/11,478,740 | 3,243/4,591,496 |
| 54 | 1/25,827,165 | 32/2,869,685 | 376/573,937 |
| 55 | 1/28,989,675 | 98/9,663,225 | 392/644,215 |
| 56 | 1/32,468,436 | 25/2,705,703 | 875/1,546,116 |
| 57 | 1/36,288,252 | 17/2,016,014 | 2,125/4,032,028 |
| 58 | 1/40,475,358 | 52/6,745,893 | 1,105/2,248,631 |
| 59 | 1/45,057,474 | 53/7,509,579 | 3,445/7,509,579 |
| 60 | 1/50,063,860 | 81/12,515,965 | 4,293/10,012,772 |

**5.** Let an urn contain $c = 1$ red ball and $b = 9$ black balls. In Pólya's urn scheme with $c = 1$, the probability of obtaining the sequence $RRB$ (two red balls and then a black ball) is found using conditional probabilities as

$$Pr(RRB) = Pr(B|RR)\,Pr(R|R)\,Pr(R) = \tfrac{9}{12} \cdot \tfrac{2}{11} \cdot \tfrac{1}{10}.$$

Likewise,

$$Pr(BRR) = Pr(R|BR)\,Pr(R|B)\,Pr(B) = \tfrac{2}{12} \cdot \tfrac{1}{11} \cdot \tfrac{9}{10},$$

and

$$Pr(RBR) = Pr(R|RB)\,Pr(B|R)\,Pr(R) = \tfrac{2}{12} \cdot \tfrac{9}{11} \cdot \tfrac{1}{10}.$$

Thus,

$$Pr(RRB) = Pr(BRR) = Pr(RBR) = \tfrac{3}{220},$$

agreeing with the value obtained using the table of Fact 6, with $k = 1$ and $n = 3$.

The probability of obtaining two red balls and one black ball in some order is then $Pr(RRB) + Pr(BRR) + Pr(RBR) = \frac{9}{220}$. Using the extended binomial coefficients (§2.3.2), the corresponding entry in the table of Fact 6 can be verified for this case, where $k = 1$ and $n = 3$:

$$Pr(\text{ exactly one black ball }) = \frac{\binom{-9}{1}\binom{-1}{2}}{\binom{-10}{3}} = \frac{(-1)^1\binom{9}{1}(-1)^2\binom{2}{2}}{(-1)^3\binom{12}{3}} = \frac{9}{220},$$

agreeing with the value already found.

# 7.3   RANDOM VARIABLES

## 7.3.1   DISTRIBUTIONS

### Definitions:

A *random variable* $X$ is a real-valued function on a probability space $\Omega$.

The random variable $X: \Omega \to \mathcal{R}$ is *discrete* if the range of $X$ is finite or countable.

The real-valued function $f: \mathcal{R} \to \mathcal{R}$ is a *density function* if

- $f(x) \geq 0$ for all $x \in \mathcal{R}$;
- $\int_{-\infty}^{\infty} f(x)\, dx = 1$.

The random variable $X$ is (*absolutely*) *continuous* if there exists a density function $f$ such that $Pr(a < X < b) = \int_a^b f(x)\, dx$ for all $a < b$.

The *distribution* $\mu_X$ of the random variable $X$ is given by $\mu_X(B) = Pr(X \in B)$ for every interval $B$.

The *cumulative distribution function* of a random variable $X$ is given by $F(x) = Pr(X \leq x)$.

A *random vector* is a function $X = (X_1, \ldots, X_k): \Omega \to \mathcal{R}^k$.

The *joint distribution* $\mu_{X_1, \ldots, X_k}$ of the random vector $(X_1, \ldots, X_k)$ is defined by $\mu_{X_1, \ldots, X_k}(B_1, \ldots, B_k) = Pr(X_1 \in B_1, \ldots, X_k \in B_k)$ for any $k$ intervals $B_1, \ldots, B_k$.

The random variables $X_1, \ldots, X_n$ are *independent* if for any intervals $B_1, \ldots, B_n$ $Pr(X_1 \in B_1, \ldots, X_n \in B_n) = Pr(X_1 \in B_1) \ldots Pr(X_n \in B_n)$.

### Facts:

1. The cumulative distribution function $F(x)$ is a nondecreasing function of $x$.
2. $\lim_{x \to \infty} F(x) = 1$, $\lim_{x \to -\infty} F(x) = 0$.
3. $Pr(a < X \leq b) = F(b) - F(a)$ for $a < b$.
4. If $X$ is a discrete random variable, then $\sum_k Pr(X = k) = 1$.
5. If $X$ is a continuous random variable, then $\frac{d}{dx} F(x) = f(x)$.

**6.** Some important discrete random variables are described in Table 1. Here $q = 1 - p$.

**Table 1: Discrete random variables.**

| distribution | description of event $(X = k)$ | range of $X$ | $Pr(X = k)$ |
|---|---|---|---|
| Bernoulli $B(1, p)$ | $k = 0$ indicates a failure, $k = 1$ indicates a success | $0, 1$ | $q$ $p$ |
| binomial $B(n, p)$ | $k$ successes in $n$ trials, each with probability $p$ of success | $0, 1, \ldots, n$ | $\binom{n}{k} p^k q^{n-k}$ |
| Poisson $P(\lambda)$ | $k$ arrivals to a counter over a unit period of time, at average rate $\lambda$ | $0, 1, 2, \ldots$ | $\frac{e^{-\lambda} \lambda^k}{k!}$ |
| geometric $G(p)$ | $k$ trials before first success occurs | $1, 2, \ldots$ | $q^{k-1} p$ |
| Pascal $NB(r, p)$ | $k$ trials before $r$th success occurs | $r, r+1, \ldots$ | $\binom{k-1}{r-1} q^{k-r} p^r$ |
| hypergeometric $(N, m, n)$ | sample $n$ items from $N$ items, where $m$ are defective and $N - m$ are not; $k$ = number of defectives selected | $0, 1, \ldots, n$ | $\frac{\binom{m}{k}\binom{N-m}{n-k}}{\binom{N}{n}}$ |

**7.** Some important continuous random variables are described in Table 2. Here it is understood that the density function $f(x) = 0$ outside the specified range.

**8.** If $X_1$ and $X_2$ are independent binomial random variables (see Table 1) with parameters $n_1, p$ and $n_2, p$, respectively, then $X_1 + X_2$ is also a binomial random variable with parameters $n_1 + n_2, p$.

**9.** If $X_1$ and $X_2$ are independent Poisson random variables (see Table 1) with parameters $\lambda_1$ and $\lambda_2$, respectively, then $X_1 + X_2$ is also a Poisson random variable with parameter $\lambda_1 + \lambda_2$.

**10.** If $X_1$ and $X_2$ are independent normal random variables (see Table 2) with parameters $\mu_1, \sigma_1^2$ and $\mu_2, \sigma_2^2$, respectively, then $X_1 + X_2$ is also a normal random variable with parameters $\mu_1 + \mu_2, \sigma_1^2 + \sigma_2^2$.

**Table 2: Continuous random variables.**

| distribution | range of $X$ | density function $f(x)$ |
|---|---|---|
| uniform $(\alpha, \beta)$ | $(\alpha, \beta)$ | $\frac{1}{\beta - \alpha}$, $\alpha < \beta$ |
| exponential $(\lambda)$ | $[0, \infty)$ | $\lambda e^{-\lambda x}$, $\lambda > 0$ |
| standard normal $(0, 1)$ | $(-\infty, \infty)$ | $\frac{1}{\sqrt{2\pi}} e^{-x^2/2}$ |
| normal $(\mu, \sigma^2)$ | $(-\infty, \infty)$ | $\frac{1}{\sigma\sqrt{2\pi}} e^{-\frac{1}{2}(\frac{x-\mu}{\sigma})^2}$, $\sigma > 0$ |
| gamma $\Gamma(n, \lambda)$ | $[0, \infty)$ | $\frac{\lambda^n x^{n-1} e^{-\lambda x}}{\Gamma(n)}$, $\Gamma(n) = \int_0^\infty t^{n-1} e^{-t}\, dt$ |
| Cauchy $(\alpha)$ | $(-\infty, \infty)$ | $\frac{\alpha}{\pi(\alpha^2 + x^2)}$, $\alpha > 0$ |
| beta $(p, q)$ | $[0, 1]$ | $\frac{\Gamma(p+q)}{\Gamma(p)\Gamma(q)} x^{p-1}(1-x)^{q-1}$, $p, q > 0$ |
| chi square $\chi^2(r)$ | $[0, \infty)$ | $\frac{1}{2^{r/2}\Gamma(r/2)} x^{\frac{r}{2}-1} e^{-\frac{x}{2}}$, $r > 0$ |
| F-distribution $F_{m,n}$ | $[0, \infty)$ | $\frac{\Gamma((m+n)/2)}{\Gamma(m/2)\Gamma(n/2)} \left(\frac{m}{n}\right)^{m/2} \frac{x^{(m-2)/2}}{(1+(m/n)x)^{(m+n)/2}}$ |
| t-distribution $t_k$ | $(-\infty, \infty)$ | $\frac{\Gamma((k+1)/2)}{\Gamma(k/2)} \frac{1}{\sqrt{k\pi}} \frac{1}{(1+x^2/k)^{(k+1)/2}}$ |
| Rayleigh $R(\sigma)$ | $[0, \infty)$ | $\frac{x e^{-x^2/2\sigma^2}}{\sigma^2}$, $\sigma > 0$ |

**Examples:**

**1.** A spinner has three sectors—red, white, and blue—with sector areas 0.2, 0.7, and 0.1, respectively. Define a random variable $X$ according to the rule $X = 1$ if the spinner points on red, $X = 2$ if it points on white, and $X = 3$ if it points on blue. The distribution of the discrete random variable $X$ is displayed in the following table.

| event | $i$ | $Pr(X = i)$ |
|-------|-----|-------------|
| red   | 1   | 0.2         |
| white | 2   | 0.7         |
| blue  | 3   | 0.1         |

**2.** *Bernoulli random variable:*  Let $A \subset \Omega$ be a fixed subset, with $Pr(A) = p$ for some $0 < p < 1$. Define the random variable $X$ by $X(\omega) = 1$ for $\omega \in A$ and $X(\omega) = 0$ for $\omega \notin A$. Often it is said that a *success* occurs whenever $\omega \in A$ and a *failure* occurs otherwise. Then $X$ is a Bernoulli random variable with $Pr(X = 1) = p$ and $Pr(X = 0) = 1 - p$. (Jakob Bernoulli, 1654–1705.)

**3.** *Binomial random variable:*  Suppose that a die is thrown and that the occurrence of either a one or a six results in a "success". A single roll of the die constitutes a Bernoulli trial (Example 2) with probability of success $p = \frac{1}{3}$. The number of successes in 10 successive independent trials is a binomial random variable $X$ with parameters $n = 10$ and $p = \frac{1}{3}$. In general, the number of successes $X$ is a discrete random variable with possible values $0, 1, 2, \ldots, n$ and its distribution is given in Table 1.

**4.** A dart is thrown at a circular target of radius 1. Assume that the target is never missed and that any point on the target is as equally likely to be hit as any other point. Let $X$ be the dart's distance from the center of the target. Since $X$ can assume any value between 0 and 1, $X$ is a continuous random variable. For $0 \leq a < b \leq 1$, $Pr(a < X < b) = Pr$(the dart lands in the annulus with radii $a$ and $b$)$= \frac{1}{\pi} \times$(the area of the annulus with radii $a$ and $b$)$= \frac{1}{\pi}(\pi b^2 - \pi a^2) = b^2 - a^2$.

**5.** *Hypergeometric random variable:*  A total of $n$ balls are selected from an urn containing $N$ balls, of which $m$ are red and $N - m$ are black. Let $X$ be the number of red balls selected. Then $X$ is a discrete random variable having the distribution

$$Pr(X = k) = \frac{\binom{m}{k}\binom{N-m}{n-k}}{\binom{N}{n}}, \quad 0 \leq k \leq m.$$

**6.** *Multinomial random variable:*  Cast $n$ identical balls into $N$ labeled boxes in such a way that the probability that a ball ends up in box $j$ is $p_j$, where $\sum_{j=1}^{N} p_j = 1$. Let $X_j$ denote the number of balls in box $j$ $(1 \leq j \leq N)$. For a vector $(x_1, x_2, \ldots, x_N)$ with $\sum_{j=1}^{N} x_j = n$, the probability that box 1 contains $x_1$ balls, box 2 contains $x_2$ balls,..., box $N$ contains $x_N$ balls is given by

$$Pr(X_1 = x_1, X_2 = x_2, \ldots, X_N = x_N) = \frac{n!}{x_1! x_2! \ldots x_N!} p_1^{x_1} p_2^{x_2} \cdots p_N^{x_N}$$

$$= \binom{n}{x_1 \ x_2 \ \cdots \ x_N} p_1^{x_1} p_2^{x_2} \cdots p_N^{x_N},$$

expressed using the multinomial coefficients (§2.3.2).

**7.** *Joint distribution:*  Two fair coins are tossed once, resulting in four equally likely outcomes $\{(T, T), (T, H), (H, T), (H, H)\}$. Let the random variable $X$ be the total number of heads observed, and let the random variable $Y$ be the number of heads on the first coin minus the number of heads on the second coin. The joint probability distribution $\mu_{X,Y}$ is given by $Pr(X = 0, Y = 0) = Pr(X = 1, Y = -1) = Pr(X = 1, Y = 1) = Pr(X = 2, Y = 0) = \frac{1}{4}$. Thus $Pr(X = 0) = Pr(X = 2) = \frac{1}{4}$, $Pr(X = 1) = \frac{1}{2}$ and $Pr(Y = 1) = Pr(Y = -1) = \frac{1}{4}$, $Pr(Y = 0) = \frac{1}{2}$. Since $Pr(X = 0, Y = 0) = \frac{1}{4} \neq \frac{1}{8} = Pr(X = 0) Pr(Y = 0)$, the variables $X$ and $Y$ are not independent.

## 7.3.2   MEAN, VARIANCE, AND HIGHER MOMENTS

### Definitions:

The **mean (expected value)** $EX$ of a discrete random variable $X$ is given by $EX = \sum_k k \, Pr(X = k)$.

The **mean** of a continuous random variable $X$ with density function $f$ is given by $EX = \int_{-\infty}^{\infty} x f(x) \, dx$.

The **variance** $\text{Var}(X)$ of a random variable $X$ is $\text{Var}(X) = E((X - EX)^2)$.

The **standard deviation** of $X$ is $\sqrt{\text{Var}(X)}$.

The **covariance** $\text{Cov}(X,Y)$ of two random variables $X$ and $Y$ is given by $\text{Cov}(X,Y) = E((X - EX)(Y - EY))$.

The **correlation** $\rho_{X,Y}$ of two random variables $X$ and $Y$ is $\rho_{X,Y} = \dfrac{\text{Cov}(X,Y)}{\sqrt{\text{Var}(X)\,\text{Var}(Y)}}$.

The **kth moment** of a random variable $X$ is $E(X^k)$.

### Facts:

**1.** The expected value $EX$ of a random variable $X$ measures the "weighted average" of $X$ or the "center of gravity" of its distribution.

**2.** $E(X + Y) = EX + EY$.

**3.** $E(cX) = cEX$ for all constants $c$.

**4.** $E(c) = c$ for all constants $c$.

**5.** If $X$ is a nonnegative integer random variable, then $EX = \sum_{n=0}^{\infty} Pr(X > n)$.

**6.** If $X$ is a nonnegative continuous random variable, then $EX = \int_0^{\infty} Pr(X > x) \, dx$.

**7.** If $X$ and $Y$ are independent, then $E(XY) = (EX)(EY)$.

**8.** If $g$ is a real-valued function and $X$ is a discrete random variable, then $E(g(X)) = \sum_k g(k) \, Pr(X = k)$.

**9.** If $g$ is an integrable real-valued function and $X$ is continuous with density $f(x)$, then $E(g(X)) = \int g(t) f(t) \, dt$.

**10.** The variance $\text{Var}(X)$ of a random variable $X$ measures the "dispersion" of $X$ about its expected value $EX$.

**11.** $\text{Var}(X) \geq 0$; $\text{Var}(X) = 0$ if and only if for some constant $c$, $Pr(X = c) = 1$.

**12.** $\text{Var}(cX) = c^2 \text{Var}(X)$ for all constants $c$.

**13.**  $\mathrm{Var}(X+Y) = \mathrm{Var}(X) + \mathrm{Var}(Y) + 2\mathrm{Cov}(X,Y)$.

**14.**  $\mathrm{Var}(X+Y) = \mathrm{Var}(X) + \mathrm{Var}(Y)$ if $X$ and $Y$ are independent.

**15.**  $\mathrm{Var}(X) = E(X^2) - (EX)^2$.

**16.**  $\mathrm{Cov}(X,Y) = 0$ if $X$ and $Y$ are independent. The converse is false. (Example 4.)

**17.**  $\mathrm{Cov}(X,Y) = E(XY) - (EX)(EY)$.

**18.**  The correlation $\rho_{X,Y}$ is a scale-invariant measure of the degree of linear relationship between two random variables $X$ and $Y$. Specifically, $\rho_{X,Y} = 1$ only when $Y = aX + b$ for some constants $a > 0$ and $b$. Similarly, $\rho_{X,Y} = -1$ only when $Y = aX + b$ for some constants $a < 0$ and $b$.

**19.**  $|\rho_{X,Y}| \leq 1$.

**20.**  *Bienaymé-Chebyshev's inequality*:  $Pr(|X - EX| \geq t) \leq \frac{\mathrm{Var}(X)}{t^2}$, for any value $t > 0$. (Irénée-Jules Bienaymé, 1796–1878 and Pafnuty Lvovich Chebyshev, 1821–1894.)

**21.**  *Kolmogorov's inequality*:  Suppose $X_1, X_2, \ldots, X_n$ are independent random variables, and let $S_k = X_1 + X_2 + \cdots + X_k$ for $1 \leq k \leq n$. Then for any value $t > 0$ the probability that $|S_k - ES_k| < t$ holds for all $k = 1, 2, \ldots, n$ is at least $1 - \frac{\mathrm{Var}(S_n)}{t^2}$. (Andrey Nikolayevich Kolmogorov, 1903–1987.)

## Examples:

**1.**  The random variable $X$ is the number of heads obtained in three tosses of a fair coin. It follows a binomial distribution (Table 1, §7.3.1), with $n = 3$ and $p = \frac{1}{2}$. Thus $Pr(X = 0) = \frac{1}{8}$, $Pr(X = 1) = \frac{3}{8}$, $Pr(X = 2) = \frac{3}{8}$, and $Pr(X = 3) = \frac{1}{8}$. Using the definition of expected value, $EX = \sum_k k\, Pr(X = k) = 0 \cdot \frac{1}{8} + 1 \cdot \frac{3}{8} + 2 \cdot \frac{3}{8} + 3 \cdot \frac{1}{8} = \frac{3}{2}$. In general, the mean of a binomial distribution with parameters $n$ and $p$ is $np$; see the corresponding entry in Table 3 of §7.3.3.

**2.**  The variance of the discrete random variable $X$ in Example 1 can be found using $\mathrm{Var}(X) = E((X - EX)^2) = E((X - \frac{3}{2})^2) = (0 - \frac{3}{2})^2 \cdot \frac{1}{8} + (1 - \frac{3}{2})^2 \cdot \frac{3}{8} + (2 - \frac{3}{2})^2 \cdot \frac{3}{8} + (3 - \frac{3}{2})^2 \cdot \frac{1}{8} = \frac{3}{4}$. In general, the variance of a binomial distribution with parameters $n$ and $p$ is $np(1 - p)$; see the corresponding entry in Table 3 of §7.3.3.

**3.**  Suppose $X$ is a Bernoulli random variable with parameter $p$, so $Pr(X = 0) = 1 - p$ and $Pr(X = 1) = p$. Then $EX = 0 \cdot (1 - p) + 1 \cdot p = p$ and $\mathrm{Var}(X) = E((X - p)^2) = (0 - p)^2 \cdot (1 - p) + (1 - p)^2 \cdot p = p^2(1 - p) + p(1 - p)^2 = p(1 - p)$. Also, $E(X^2) = 0^2 \cdot (1 - p) + 1^2 \cdot p = p$ and using Fact 15 $\mathrm{Var}(X) = E(X^2) - (EX)^2 = p - p^2 = p(1 - p)$, as before.

**4.**  *Covariance and independence*:  In Example 7 of §7.3.1, $EX = 0 \cdot \frac{1}{4} + 1 \cdot \frac{1}{2} + 2 \cdot \frac{1}{4} = 1$ and $EY = -1 \cdot \frac{1}{4} + 0 \cdot \frac{1}{2} + 1 \cdot \frac{1}{4} = 0$. Also, $E(XY) = -1 \cdot \frac{1}{4} + 0 \cdot \frac{1}{2} + 1 \cdot \frac{1}{4} = 0$. By Fact 17, $\mathrm{Cov}(X,Y) = E(XY) - (EX)(EY) = 0 - 1 \cdot 0 = 0$. In this example, variables $X$ and $Y$ have zero covariance (and zero correlation); however (see §7.3.1, Example 7) they are not independent random variables.

**5.**  The moments of the normal random variable $X$ with parameters $\mu = 0$ and $\sigma = 1$ are $E(X^{2k}) = 1 \cdot 3 \cdots (2k - 1)$ and $E(x^{2k-1}) = 0$ for $k \geq 1$.

**6.**  A manufacturing plant produces ball bearings with an average diameter of 50 mm and a variance of 11 mm$^2$. Without any further information about the shape of the distribution of the diameters $X$, Fact 20 shows that the probability $Pr(|X - 50| \geq 8)$ of exceeding the nominal diameter by more than 8 mm is no more than $\frac{\mathrm{Var}(X)}{8^2} = \frac{11}{64} = 0.172$. Thus, no more than 17.2% of the ball bearings produced can exceed the stated tolerance.

**7.**  *Average-case algorithm analysis*:  A simple algorithm for locating an item in an (unordered) list $A = [a_1, a_2, \ldots, a_n]$ is called a *linear search*; it sequentially examines each entry of list $A$ and compares the given item *key* with each $a_k$ until a match is found, or until the entire list is searched, in which case *key* is known not to be in the list. To obtain the average case complexity of this algorithm, suppose that *key* is known to occur in $A$ and that it is equally likely (with probability $\frac{1}{n}$) to be at each of the $n$ positions of $A$. If *key* is in fact located at position $k$ of $A$, then $k$ comparisons are required by the algorithm. The expected number of comparisons needed is thus $EX = \sum_{k=1}^{n} k\, Pr(X = k) = \sum_{k=1}^{n} k \cdot \frac{1}{n} = \frac{1}{n} \sum_{k=1}^{n} k = \frac{1}{n} \frac{n(n+1)}{2} = \frac{(n+1)}{2}$. Consequently, the average-case complexity of linear search is $O(n)$; see §1.3.3.

---

### 7.3.3   GENERATING FUNCTIONS

**Definitions:**

The **probability generating function** of a discrete random variable $X$ is the function $\phi(t) = E(t^X) = \sum_k t^k\, Pr(X = k)$, defined for $|t| \leq 1$.

The **moment generating function** of a discrete random variable $X$ is the function $\psi(t) = E(e^{tX}) = \sum_k e^{tk}\, Pr(X = k)$, defined for all $t$ such that $\psi(t)$ converges.

The **moment generating function** of a continuous random variable $X$ with density $f$ is the function $\psi(t) = E(e^{tX}) = \int e^{tx} f(x)\, dx$, defined for all $t$ such that $\psi(t)$ converges.

The **characteristic function** (Fourier transform) of a discrete random variable $X$ is $\chi(t) = E(e^{itX}) = \sum_k e^{itk}\, Pr(X = k)$, defined for all $t \in \mathcal{R}$.

The **characteristic function** of a continuous random variable $X$ with density $f$ is $\chi(t) = E(e^{itX}) = \int e^{itx} f(x)\, dx$, defined for all $t \in \mathcal{R}$.

**Facts:**

**1.** The expected value of a random variable $X$ can be expressed in terms of the first derivative of its generating function: $EX = \phi'(1) = \psi'(0) = -i\chi'(0)$.

**2.** The variance of a random variable $X$ can be expressed in terms of the first and second derivatives of its generating function: $\mathrm{Var}(X) = \phi''(1) + \phi'(1) - [\phi'(1)]^2 = \psi''(0) - [\psi'(0)]^2 = [\chi'(0)]^2 - \chi''(0)$.

**3.** $\frac{d^k}{dt^k} \phi(1) = E(X(X-1)(X-2)\ldots(X-k+1))$.

**4.** $\frac{d^k}{dt^k} \psi(0) = E(X^k)$.

**5.** $\frac{d^k}{dt^k} \chi(0) = i^k E(X^k)$.

**6.** For any of the three types of generating functions defined, the generating function of the sum of independent random variables is the product of their respective generating functions.

**Examples:**

**1.** The binomial random variable with parameters $n$ and $p$ is the sum of $n$ independent Bernoulli random variables with parameter $p$. The probability generating function for a Bernoulli random variable is $\phi(t) = E(t^X) = t^0(1-p) + t^1 p = q + pt$, where $q = 1 - p$. By Fact 6 the probability generating function for a binomial random variable is $[\phi(t)]^n = (q + pt)^n$.

**2.** Table 3 shows the mean, variance, probability generating function, moment generating function, and characteristic function of several important discrete distributions. Here $q = 1 - p$. An asterisk ($*$) signifies that the entry is not available in simple form.

**Table 3: Moments and generating functions for discrete distributions.**

| distribution | mean | variance | $\phi(t)$ | $\psi(t)$ | $\chi(t)$ |
|---|---|---|---|---|---|
| Bernoulli $B(1,p)$ | $p$ | $pq$ | $q+pt$ | $q+pe^t$ | $q+pe^{it}$ |
| Binomial $B(n,p)$ | $np$ | $npq$ | $(q+pt)^n$ | $(q+pe^t)^n$ | $(q+pe^{it})^n$ |
| Poisson $P(\lambda)$ | $\lambda$ | $\lambda$ | $e^{\lambda(t-1)}$ | $e^{\lambda(e^t-1)}$ | $e^{\lambda(e^{it}-1)}$ |
| geometric $G(p)$ | $\frac{1}{p}$ | $\frac{q}{p^2}$ | $\frac{pt}{1-qt}$ | $\frac{pe^t}{1-qe^t}$ | $\frac{pe^{it}}{1-qe^{it}}$ |
| Pascal $NB(r,p)$ | $\frac{r}{p}$ | $\frac{rq}{p^2}$ | $(\frac{pt}{1-qt})^r$ | $(\frac{pe^t}{1-qe^t})^r$ | $(\frac{pe^{it}}{1-qe^{it}})^r$ |
| hypergeometric $(N,m,n)$ | $\frac{mn}{N}$ | $\frac{m(N-m)n(N-n)}{N^2(N-1)}$ | * | * | * |

**3.** Table 4 shows the mean, variance, moment generating function, and characteristic function of several important continuous distributions. An asterisk ($*$) indicates that the entry is not available.

**Table 4: Moments and generating functions for continuous distributions.**

| distribution | mean | variance | $\phi(t)$ | $\chi(t)$ |
|---|---|---|---|---|
| uniform $(\alpha,\beta)$ | $\frac{\alpha+\beta}{2}$ | $\frac{(\beta-\alpha)^2}{12}$ | $\frac{e^{\beta t}-e^{\alpha t}}{t(\beta-\alpha)}$ | $\frac{e^{i\beta t}-e^{i\alpha t}}{it(\beta-\alpha)}$ |
| exponential $(\lambda)$ | $\frac{1}{\lambda}$ | $\frac{1}{\lambda^2}$ | $\frac{\lambda}{\lambda-t}$ | $\frac{\lambda}{\lambda-it}$ |
| standard normal $(0,1)$ | $0$ | $1$ | $e^{t^2/2}$ | $e^{-t^2/2}$ |
| normal $(\mu,\sigma^2)$ | $\mu$ | $\sigma^2$ | $e^{\mu t+\frac{\sigma^2t^2}{2}}$ | $e^{i\mu t-\frac{\sigma^2t^2}{2}}$ |
| gamma $\Gamma(n,\lambda)$ | $\frac{n}{\lambda}$ | $\frac{n}{\lambda^2}$ | $(\frac{\lambda}{\lambda-t})^n$ | $(\frac{\lambda}{\lambda-it})^n$ |
| Cauchy $(\alpha)$ | $\infty$ | $\infty$ | $\infty$ | $\frac{1}{\alpha^2}e^{-\alpha|t|}$ |
| beta $(p,q)$ | $\frac{p}{p+q}$ | $\frac{pq}{(p+q)^2(p+q+1)}$ | * | * |
| chi square $\chi^2(r)$ | $r$ | $2r$ | $(1-2t)^{-r/2}$ | $(1-2it)^{-r/2}$ |
| F-distribution $F_{m,n}$ | $\frac{n}{n-2}$ | $\frac{2n^2(m+n-2)}{m(n-2)^2(n-4)}$ | * | * |
| t-distribution $t_k$ | $0$ | $\frac{k}{k-2}$ | * | * |
| Rayleigh $R(\sigma)$ | $\sqrt{\frac{\pi}{2}}\sigma$ | $2(1-\frac{\pi}{4})\sigma^2$ | * | * |

**4.** The moments of a binomial random variable $X$ can be found from its moment generating function $\psi(t)=(q+pe^t)^n$. For example, $\psi'(t)=n(q+pe^t)^{n-1}(pe^t)$ and using Fact 1 produces $EX=\psi'(0)=n(q+p)^{n-1}p=np$.

**5.** From Table 4 the moment generating function for the exponential distribution with parameter $\lambda$ is $\psi(t)=\frac{\lambda}{\lambda-t}$. Then $\psi'(t)=\frac{\lambda}{(\lambda-t)^2}$ and $\psi''(t)=\frac{2\lambda}{(\lambda-t)^3}$, giving $\psi'(0)=\frac{\lambda}{\lambda^2}=\frac{1}{\lambda}$ and $\psi''(0)=\frac{2\lambda}{\lambda^3}=\frac{2}{\lambda^2}$. By Facts 1 and 2, $EX=\psi'(0)=\frac{1}{\lambda}$ and $\mathrm{Var}(X)=\psi''(0)-[\psi'(0)]^2=\frac{2}{\lambda^2}-(\frac{1}{\lambda})^2=\frac{1}{\lambda^2}$.

## 7.4  DISCRETE PROBABILITY COMPUTATIONS

Many discrete probability computations are much less straightforward than may at first be imagined because of difficulties arising from the finiteness of computer arithmetic systems. Good algorithm design can usually avoid such problems.

## 7.4.1   INTEGER COMPUTATIONS

Enumeration of combinatorial objects, such as permutations and combinations, requires the computation of integer factorials. In practice, these factorials can only be computed as integers for small values. Since in most cases the factorials are not themselves the primary objective of the computation, potential numerical difficulties can be overcome by carefully designed recursive algorithms.

### Definitions:

The $N$-bit **binary representation** of the positive integer $n$ is $(b_{N-1}b_{N-2}\ldots b_1b_0)_2$ where $b_i \in \{0,1\}$ and $n = \sum_{i=0}^{N-1} b_i 2^i$. (See §4.1.3.) Each $b_i$ is a **binary digit** (**bit**).

The **two's complement** representation of the signed integer $n$ is $(b_{N-1}b_{N-2}\ldots b_1b_0)_{2'}$ where $n = -b_{N-1}2^{N-1} + \sum_{i=0}^{N-2} b_i 2^i$.

Integer **wraparound** is the phenomenon of adding 1 to the largest representable integer and obtaining the smallest representable integer.

### Facts:

**1.** Signed integers are usually represented in a computer as two's complement binary words of a fixed wordlength; commonly 8, 16, or 32 bits are used.

**2.** A two's complement integer using $N$-bit words is interpreted by treating the most significant bit as a coefficient of $2^{N-1}$.

**3.** The range of representable integers in $N$-bit two's complement is from $-2^{N-1} = (10\ldots00)_{2'}$ to $2^{N-1} - 1 = (01\ldots11)_{2'}$. Arithmetic operations can generate no carries beyond this range.

**4.** Integer wraparound is a consequence of Fact 3 since (in regular binary arithmetic) $(01\ldots11)_2 + (00\ldots01)_2 = (10\ldots00)_2$. Some systems have integer range checking available to avoid the effect of wraparound.

**5.** Permutations and combinations (§2.3) are usually expressed in terms of integer factorials.

**6.** Binomial coefficients (§2.3.2) can be computed using integer arithmetic provided the result is within the range being used. Algorithm 1 breaks the computation of a binomial coefficient into a recursive loop using $\binom{n}{k} = \frac{n!}{k!(n-k)!} = \frac{n(n-1)\ldots(n-k+1)}{k(k-1)\ldots 1} = \left(\frac{n}{1}\right)\left(\frac{n-1}{2}\right)\ldots\left(\frac{n+1-k}{k}\right)$ and the result $\binom{n}{k} = \binom{n}{n-k}$. (See §2.3.2, Fact 7.)

---

**Algorithm 1**:   **Integer computation of binomial coefficients.**

input: positive integers $n, k$

output: $b = \binom{n}{k}$

if $2k > n$ then $k := n - k$

$b := 1$

for $i := 1$ to $k$

$\quad b := [b \cdot (n + 1 - i)]$ **div** $i$

---

**7.** By doing the multiplication before the integer division in Algorithm 1, the numerator necessarily has the appropriate factors to ensure an exact integer result.

**Examples:**

**1.** For $N = 8$, 16, and 32 bits, the two's complement binary integer ranges are given in the following table.

| $N$ | minimum value | maximum value |
|---|---|---|
| 8 | $-128$ | 127 |
| 16 | $-32768$ | 32767 |
| 32 | $-2147483648$ | 2147483647 |

**2.** For $N = 8$, the integer 86 has the two's complement representation $(01010110)_{2'}$ and the integer $-86$ has the two's complement representation $-128 + 42 = (10101010)_{2'}$.

**3.** For 8-bit integers, the effect of integer wraparound is shown by $127 + 1 = -128$. Similarly, we would have $64 \times 2 = -128$.

**4.** For 8-bit two's complement integers, only $1!, 2!, 3!, 4!, 5!$ can be computed correctly. Subsequent factorials would generate integer answers—but wrong ones. In particular, 6! would evaluate to $-48$. Namely, the 8-bit two's complement representation of $5! = 120$ is 01111000 and $6 = 00000110$ so that, with no carries to the left of the 8th bit, $6! = 6 \times 5!$ is represented by the sum of 11100000 and 11110000 (which are respectively 01111000 shifted 2 and 1 places left.) This sum (again without carries to the left of the leading bit) is 11010000, which represents $(-128) + 64 + 16 = -48$.

**5.** Using 16-bit integers with wraparound, the binomial coefficient $\binom{12}{8}$ cannot be computed directly from its definition since neither 12! nor 8! can be computed correctly. Thus Algorithm 1 finds instead $\binom{12}{4} = \binom{12}{8}$ since $2 \times 8 > 12$. This is computed as $\binom{12}{1}\binom{11}{2}\binom{10}{3}\binom{9}{4}$, with each multiplication being performed before its associated division: $(12/1)$ is multiplied by 11, divided by 2, multiplied by 10, divided by 3, multiplied by 9, and divided by 4. This produces the intermediate results 12, 132, 66, 660, 220, 1980, 495, so that the correct final result is obtained without any intermediate computation exceeding the integer range.

---

## 7.4.2  FLOATING-POINT COMPUTATIONS

To compute discrete probabilities (e.g., binomial probabilities), careful attention must be given to the underlying floating-point computation model and its properties.

**Definitions:**

Let $F$ be the set of numbers representable in a particular floating-point system, with $\Omega$ the largest positive number in $F$ and $\omega$ the smallest positive number in $F$.

The floating-point arithmetic operations in $F$ are denoted by $\oplus, \ominus, \otimes, \oslash$ when it is necessary to distinguish them from their real counterparts $+, -, \times, /$.

The number $x$ is represented in the computer in binary **floating-point** form by the approximation $x \approx \pm f \times 2^E$ where the **fraction** or **mantissa** $f$ is a binary fraction of fixed length and the **exponent** $E$ is an integer within a fixed range. Usually the floating point representation is **normalized** so that $f \in [1, 2)$.

Floating-point arithmetic is subject to **roundoff error**, the error introduced by abbreviating the representation of a number or an arithmetic result to a finite wordlength.

The usual measure of error for floating-point computation is **relative error**, which is given for an approximation $x^*$ to a quantity $x$ by $\dfrac{|x^* - x|}{|x|} \approx \dfrac{|x^* - x|}{|x^*|}$.

A **floating-point operation** (or **flop**) is any arithmetic operation performed using floating-point arithmetic.

**Overflow** results from a floating-point operation where the magnitude of the result is too large for the available range of the floating-point system being used.

**Underflow** results from a floating-point operation where the magnitude of the result is too small for the available range of the floating-point system being used.

The **machine unit** $\mu$ of a floating-point system is the smallest positive number that can be added to 1 and produce a result recognized in the machine as greater than 1: namely, $\mu = \min\{x \in F \mid 1 \oplus x > 1\}$.

**Facts:**

**1.** Roundoff errors are propagated in subsequent computations.

**2.** The two expressions given for relative error are often used interchangeably.

**3.** Overflow and underflow result from the finite range of available exponents. The limits of these ranges and the details of the implementation vary with both the hardware and software being used. See [IE08] for the most common implementations.

**4.** Usually an overflow condition terminates a program, while underflow results are normally replaced by 0.

**5.** Because of the finite mantissa length (and independent of the rounding rule), most axioms of the real number system fail for floating-point arithmetic [St74]. The following table summarizes similarities and differences between the real numbers $\mathcal{R}$ and the floating-point system $F$. The second column of the table describes the property, assuming $a, b, c \in \mathcal{R}$. If the property fails in $F$, a brief reason for the failure is also given.

| property | description in $\mathcal{R}$ | valid in $F$? |
|---|---|---|
| closure $+$ | $a + b \in \mathcal{R}$ | NO: overflow |
| closure $\times$ | $a \times b \in \mathcal{R}$ | NO: overflow |
| commutativity | $a + b = b + a,\ a \times b = b \times a$ | YES |
| associativity $+$ | $(a + b) + c = a + (b + c)$ | NO: $a = 1,\ b = c = \frac{\mu}{2}$ |
| special case | $(a + b) - a = b$ | NO: $a = 1,\ b = \frac{\mu}{2}$ |
| associativity $\times$ | $(a \times b) \times c = a \times (b \times c)$ | NO: roundoff, overflow, or underflow |
| distributive law | $a \times (b + c) = (a \times b) + (a \times c)$ | NO: roundoff, overflow, or underflow |
| existence of zero | $(\exists\, 0)\ a + 0 = a$ | YES |
| unique negative | $\exists!\, (-a)\ a + (-a) = 0$ | NO: $[-(1 \oplus \mu) \otimes a] \oplus a$ $= 0$ if $\mu \times a < \omega$ |
| existence of one | $(\exists\, 1)\ a \times 1 = a$ | YES |
| zero divisors | $a \times b = 0 \Rightarrow a = 0$ or $b = 0$ | NO: $a \otimes b = 0 \Rightarrow$ $a < \sqrt{\omega}$ or $b < \sqrt{\omega}$ |
| total ordering | $a < b$ or $a = b$ or $a > b$ | YES |
| order-preservation | $a > b \Rightarrow a + c > b + c$ | NO: roundoff |
| special case | $x > 0 \Rightarrow 1 + x > 1$ | NO: $x < \mu$ |

**6.** In the previous table, most of the properties that fail in $F$ hold approximately—at least for arguments of the same sign. These failures are not critical to most computations, but they can be important for computations such as summing sets of numbers and evaluating binomial probabilities.

**7.** The existence of the machine unit $\mu$ ensures that some of the order properties of $\mathcal{R}$ will not carry over to $F$.

**8.** The machine unit $\mu$ is not the same as the smallest representable positive number $\omega$ in $F$.

**9.** The relative error in subtraction is essentially unbounded due to cancellation.

**10.** IEEE arithmetic is required to deliver the same result as if rounding were performed on the infinite precision computation assuming that the data are exact.

**11.** A sum of terms of the same sign should generally be summed from smallest to largest.

**12.** Improved accuracy in computing a summation is possible by regarding the partial sums as members of a (reduced) list of summands and always adding the two smallest terms in the current list. However, the overhead would be prohibitive in most cases.

**13.** Special care must be taken in computing a summation if its terms are computed recursively, since the smallest term can underflow.

**14.** There is no completely reliable method for summing terms of mixed sign.

**15.** For alternating series, special transformations such as Euler's method can be used [BuTu92, Chapter 1].

**16.** Algorithm 2 computes the cumulative sum of binomial probabilities (§7.3.1) using the definition $B(N,p\,;k) = \sum_{i=0}^{k} \binom{N}{i} p^i (1-p)^{N-i}$.

---

**Algorithm 2**:   **Recursive computation of binomial probabilities.**

   input: positive integers $N, k$; real number $p$
   output: $s = B(N, p\,; k)$
   $q := 1 - p$
   $t := q^N$
   $s := t$
   **for** $i := 1$ **to** $k$
      $t := t * \left(\dfrac{p}{q}\right) * \dfrac{(N+1-i)}{i}$
      $s := s + t$

---

**17.** Algorithm 2 will only work for small values of $N$.

**18.** If $k$ is not too large, Algorithm 2 does in fact sum terms from smallest to largest.

**19.** To compute $B(N,p\,;k)$ for large values of $k$, use the fact that $B(N,p\,;k) = 1 - B(N,1-p\,;N-k-1)$ and then apply Algorithm 2.

**20.** If $q^N$ underflows to 0, then Algorithm 2 returns 0 for all values of $k$.

**21.** Algorithm 3 gives an alternative way to calculate the individual binomial probability term $\binom{N}{r} p^r (1-p)^{N-r}$. It computes the logarithm of each factor recursively and then exponentiates this at the end.

---

**Algorithm 3**:   **Logarithmic computation of binomial probability terms.**

input: positive integers $N, r$; real number $p$

output: $b = \binom{N}{r} p^r (1-p)^{N-r}$

$q := 1 - p$

$t := r * \ln p + (N - r) \ln q$

**for** $i := 1$ **to** $r$

$\quad t := t + \ln(N + 1 - i) - \ln i$

$b := e^t$

---

**22.** Algorithm 3 must be safeguarded to ensure that $e^t$ underflows to 0 for large negative arguments $t$.

**23.** Using logarithms is a frequently applied technique for computing products of many factors with widely varying magnitudes. It is one step along the way toward using the symmetric level-index scheme for number representation and arithmetic [ClOlTu89].

**Examples:**

**1.** *Summations*:  If the first $2^{24}$ terms of the harmonic series are summed using IEEE single precision floating-point arithmetic, both forward and backward, then the sums differ by approximately 11%. Specifically, $\left( \cdots \left( \left( 1 \oplus \frac{1}{2} \right) \oplus \frac{1}{3} \right) \oplus \cdots \oplus 2^{-24} \right) \approx 15.40$, while summing the same terms from right-to-left yields $17.23$.

**2.** *Binomial probabilities*:   The computation of binomial probabilities is thoroughly discussed in Section 2.6 of [St74] with reference to the specific case where $N = 2000$, $k = 200$, and $p = 0.1$. Using Algorithm 2 in this case gives the initial value $t = 0$ and therefore the final result is $s = 0$. The true value of the final probability is approximately $0.5$.

**3.** If the final term in Example 2 is computed in IEEE single precision, the binomial coefficient itself would overflow. It is certainly greater than $10^{200}$. Also, both $(0.1)^{200}$ and $(0.9)^{1800}$ would underflow. However the true value of this term is around $0.03$.

---

# 7.5   RANDOM WALKS

Random walks are special stochastic processes whose applications include various models of particle motion, covering topics as diverse as crystallography, gambling, stock markets, biology, genetics, astronomy, and statistical sampling of large graphs. This section examines an important special class of random walks, sometimes referred to as lattice random walks, whose trajectories are generated by the summation of independent and identically distributed discrete random variables.

---

## 7.5.1   GENERAL CONCEPTS

**Definitions:**

A **stochastic process** is a collection of random variables, typically indexed by time (discrete or continuous).

A **d-dimensional random walk** is a stochastic process on the integer lattice $\mathcal{Z}^d$ whose **trajectories** are defined by an initial position $S_0 = a$ and the sequence of sums $S_n = a + X_1 + X_2 + \cdots + X_n$, $n \geq 1$, where the displacements $X_1, X_2, \ldots$ are independent and identically distributed random variables on $\mathcal{Z}^d$.

A random walk is **simple** if the values $X_i$ are restricted to the $2d$ points of $\mathcal{Z}^d$ of unit Euclidean distance from the origin. (That is, the random walk proceeds at each time step to a point one unit from the current point along some coordinate axis.) A **symmetric** random walk is a simple walk in which the $2d$ values of $X_i$ have the same probability.

Random walks that return to the initial position with probability 1 are **recurrent**; otherwise they are **transient**.

An **absorbing boundary** is a point or a set of points on the lattice that stops the motion of a random walk whose trajectory comes into contact with it. A **reflecting boundary** is a point or a set of points that redirects the motion of a random walk. Both are special cases of an **elastic** boundary, which stops or redirects the motion depending on some given probability.

The **gambler's ruin problem** is a simple one-dimensional random walk with absorbing boundaries at values 0 and $b$. It colorfully illustrates the fortunes of a gambler, who starts with $a$ dollars and who at each play of a game has a fixed probability of winning one dollar. The game ends once the gambler has either amassed the amount $S_n = b$ or goes broke $S_n = 0$.

For $k \in \mathcal{Z}^d$, the **first passage time** $T_k$ into point $k$ is the first time at which the random walk reaches the point $k$: namely, $T_k = \min\{i \geq 1 \mid S_i = k\}$. More generally, the **hitting time** $T_A$ for entering set $A \subseteq \mathcal{Z}^d$ is the first time at which the random walk reaches some point in set $A$: namely, $T_A = \min\{i \geq 1 \mid S_i \in A\}$.

The following is the **basic initial problem** of random walks.

- For $k \in \mathcal{Z}^d$, find $Pr(S_n = k)$, the probability that a "particle", executing the random walk and starting at point $a$ at time 0, will be at point $k$ at time $n$.

The following are **first passage time problems**.

- Find the probability $Pr(T_k = n)$ that, starting at point $a$ at time 0, the first visit to point $k$ occurs at time $n$.

- Find the probability $Pr(T_A = n)$ that, starting at point $a$ at time 0, the first visit to $A$ occurs at time $n$; characterize $S_{T_A}$, the point at which $A$ is first visited.

Other classical problems in random walks include

- *Range problem*: Find or approximate the probability distribution and/or the mean of the number of distinct points visited by a random walk up to time $n$.

- *Occupancy problem*: Find or approximate the probability distribution and/or the mean of the number of times a given point or a set of points has been visited up to time $n$.

- *Boundary problem*: Address all previous problems under absorbing, reflecting, and/or elastic boundary conditions.

**Examples:**

**1.** *Coin tossing*:   Tossing a coin $n$ times can be viewed as a one-dimensional random walk $(d = 1)$ on the integers $\mathcal{Z}$. This walk begins at the origin $(a = 0)$ with $X_i = 1$ if the result of the $i$th toss is a Head and $X_i = -1$ if the result of the $i$th toss is a Tail. Since each step $X_i$ is of unit length, this is a *simple* one-dimensional random walk. If the tosses are independent events, then $Pr(X_i = 1) = p$ and $Pr(X_i = -1) = 1 - p$ holds for all $i$, where $0 < p < 1$. The random variable $S_n$ is the cumulative number of Heads minus the cumulative number of Tails in $n$ tosses. The walk is symmetric if $p = \frac{1}{2}$. A return to the origin means that $S_n = 0$: that is, the number of Heads and Tails have equalized after $n$ tosses.

**2.** *Gambler's ruin*:   A gambler repeatedly plays a game of chance, in which a dollar is won at each turn with probability $p$ and a dollar is lost with probability $1 - p$. For example, suppose the gambler starts with 90 dollars, and stops whenever his current fortune is 0 (a ruin) or 100 (a positive net gain of 10 dollars). What is the gambler's ultimate probability of being ruined? Of success? On average how many expected plays does it take for the game to be over? What is the expected net gain for the gambler? If $p = 0.5$ the answers are 0.1, 0.9, 900, and 0 respectively. If $p = 0.45$ they are 0.866, 0.134, 765.6, and $-76.6$ respectively. (See §7.5.2, Example 4.)

---

### 7.5.2   ONE-DIMENSIONAL SIMPLE RANDOM WALKS

A number of results are known for random walks in one dimension that take a succession of unit steps (in either the positive or negative direction).

**Definitions:**

The **one-dimensional simple random walk** (see §7.5.1) corresponds to a particle moving randomly on the set $\mathcal{Z}$ of integers. It begins at the origin at time 0 and at each time $1, 2, \ldots$ thereafter, moves either one step up (right) with probability $p$, or one step down (left) with probability $1 - p$. This random walk is **symmetric** when $p = \frac{1}{2}$.

The **trajectory** of a one-dimensional simple random walk is described by $S_0 = 0$ and $S_n = X_1 + X_2 + \cdots + X_n$, $n \geq 1$, where the $X_i$ are independent and have a Bernoulli distribution (§7.3.1), with $Pr(X_i = 1) = p$ and $Pr(X_i = -1) = q = 1 - p$ for $p \in (0, 1)$.

Suppose a trajectory is graphically represented by plotting $S_n$ as a function of $n$, so that the point $(n, k)$ corresponds to $S_n = k$. Linking successive points with straight lines produces a **path** between points. Define $N(n, k)$ to be the number of paths from $(0, 0)$ to $(n, k)$.

In a random walk starting at $S_0 = a > 0$ with absorbing boundaries at 0 and $b > a$:

- $q_a$ is the probability that the random walk will be absorbed at 0;
- $p_a$ is the probability that the random walk will be absorbed at $b$;
- $D_a$ is the time until absorption.

**Facts:**

**1.** *Reflection principle*:   Let $n_2 > n_1 \geq 0$, $k_1 > 0$, $k_2 > 0$. The number of paths from $(n_1, k_1)$ to $(n_2, k_2)$ that touch or cross the $x$-axis equals the number of paths from $(n_1, -k_1)$ to $(n_2, k_2)$.

**2.** $N(n, k) = \binom{n}{(n+k)/2}$, if $\frac{n+k}{2}$ is an integer in $\{0, 1, \ldots, n\}$; $N(n, k) = 0$ otherwise.

**3.** If $n \geq 1$ is fixed and $-n \leq k \leq n$, then $Pr(S_n = k) = N(n, k)p^{\frac{1}{2}(n+k)}q^{\frac{1}{2}(n-k)}$.

**4.** *Ballot theorem:*  For $k > 0$, the number of paths from $(0, 0)$ to $(n, k)$ that do not return to or cross the $x$-axis is $\frac{k}{n}N(n, k)$.

**5.** For $n \geq 1$, the first return time $T_0$ to the origin satisfies

- $Pr(T_0 > n) = E(\frac{|S_n|}{n})$;
- $Pr(T_0 = 2n) = \frac{1}{2n-1}\binom{2n}{n}p^n q^n$;
- $Pr(T_0 > 2n) = Pr(S_{2n} = 0) = \binom{2n}{n}2^{-2n}$, if the walk is symmetric.

**6.** *Recurrent walks:*  $Pr(T_0 < \infty) = 1$ (the walk is recurrent) if and only if $p = q = \frac{1}{2}$. In this case $E(T_0) = \infty$.

**7.** For $k \neq 0$ and $n > 0$, $Pr(T_k = n) = \frac{|k|}{n}Pr(S_n = k)$.

**8.** For $k > 0$ and $n > 0$, the maximum value $M_n = \max\{S_0, S_1, \ldots, S_n\}$ satisfies

- $M_n \geq k$ if and only if $T_k \leq n$;
- $Pr(M_n \geq k) = Pr(S_n = k) + \sum_{i \geq k+1}[1 + (\frac{q}{p})^{i-k}]Pr(S_n = i)$;
- $Pr(M_n = k) = Pr(S_n = k) + Pr(S_n = k + 1)$, if the walk is symmetric.

**9.** *Arc sine laws:*  Let $W_n$ be the number of times among $\{0, 1, \ldots, n\}$ at which a random walk is positive and let $L_n$ be the time of the last visit to $0$ up to time $n$. For a symmetric random walk,

- $Pr(W_{2n} = 2k) = Pr(L_{2n} = 2k) = Pr(S_{2k} = 0)\,Pr(S_{2n-2k} = 0)$;
- as $n \to \infty$, $Pr(\frac{W_{2n}}{2n} \leq x) \approx \frac{2}{\pi}\arcsin\sqrt{x}$, for $x \in [0, 1]$.

**10.** *Gambler's ruin problem:*  In this random walk with absorbing boundaries (§7.5.1), $q_a$ is the probability of the gambler (having an initial capital of $a$) being ruined and $p_a$ is the probability of eventually winning (achieving a total of $b$). Facts 11–17 refer to the gambler's ruin problem.

**11.** If $p \neq q$, $q_a = \frac{(q/p)^b - (q/p)^a}{(q/p)^b - 1}$ and $p_a = 1 - q_a$.

**12.** If $p = q = \frac{1}{2}$, $q_a = 1 - \frac{a}{b}$ and $p_a = \frac{a}{b}$.

**13.** The expected gain in the gambler's ruin problem is $b(1 - q_a) - a$, which is $0$ if and only if $p = q = \frac{1}{2}$.

**14.** $Pr(D_a = n) = b^{-1}2^n p^{(n-a)/2}q^{(n+a)/2}\sum_{k=1}^{b-1}\cos^{n-1}\frac{\pi k}{b}\sin\frac{\pi k}{b}\sin\frac{\pi ak}{b}$.

**15.** If $p \neq q$, $E(D_a) = \frac{a}{q-p} - \frac{b}{q-p}\frac{1-(q/p)^a}{1-(q/p)^b}$.

**16.** If $p = q = \frac{1}{2}$, $E(D_a) = a(b - a)$.

**17.** *Limiting case of the gambler's ruin problem:*  When $b = \infty$,

- $q_a = (\frac{q}{p})^a$ if $p > q$, and $q_a = 1$ otherwise;
- $Pr(D_a = n) = 2^n p^{(n-a)/2}q^{(n+a)/2}\int_0^1 \cos^{n-1}\pi x \sin\pi x \sin\pi ax\,dx$
  $$= \frac{a}{n}\binom{n}{(n+a)/2}p^{\frac{1}{2}(n-a)}q^{\frac{1}{2}(n+a)};$$
- if $p < q$, $E(D_a) = \frac{a}{q-p}$;
- if $p = q = \frac{1}{2}$, $E(D_a) = \infty$.

**18.** *Random walks with one reflecting boundary:*  Consider a random walk starting at $S_0 = a \geq 0$ with a reflecting boundary at $0$.

- The position at time $n \geq 1$ is given by $S_n = \max\{0, S_{n-1} + X_n\}$.
- When $p < q$ and as $n \to \infty$, there is a stationary distribution for the random walk, coinciding with the distribution of $M = \sup_{i \geq 0} S_i$, and given by $Pr(M = k) = (1 - \frac{p}{q})(\frac{p}{q})^k$ for all $k \geq 0$.

## Examples:

**1.** A graphical representation of the trajectory for a one-dimensional simple random walk is shown in the following figure. Here $T_0 = 2$ and $T_2 = 4$; $M_3 = 1$ and $M_4 = 2$; $W_7 = 4$; and $L_7 = 6$.

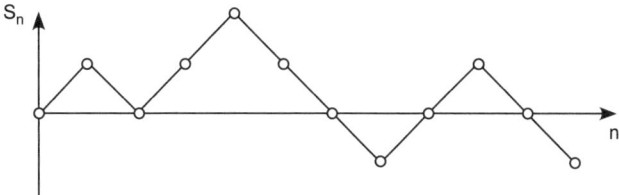

**2.** The ballot theorem takes its name from the following problem. Suppose that, in a ballot, candidate A scores $x$ votes and candidate B scores $y$ votes, $x > y$. What is the probability that, during the ballot, A is always ahead of B? By Fact 4, the answer is $\frac{x-y}{x+y}$. As an illustration, if $\frac{x}{x+y} = 0.52$, this probability is 0.04.

**3.** How much time does a symmetric random walk spend to the left of the origin? Contrary to intuition, with non-negligible probability, the fraction of time spent to the left (or to the right) of the origin is near 0 or 1, but not near $\frac{1}{2}$.

For example, when $n$ is large, Fact 9 shows that the probability a symmetric random walk spends at least 97.6% of the time to the left of the origin is approximately $0.1 = \frac{2}{\pi} \arcsin \sqrt{0.024}$. Symmetrically, there is a 0.1 probability that it spends at least 97.6% of the time to the right of the origin. Altogether, with probability 0.2 a symmetric random walk spends at least 97.6% of the time entirely on one side of the origin.

**4.** *Gambler's ruin:* Suppose that the probability of winning one dollar is $p = 0.45$, so $q = 0.55$. A gambler begins with an initial stake of $a = 90$ and will quit whenever the current winnings reach $b = 100$. Using Fact 11, the probability of ruin (i.e., losing the entire original stake) is

$$q_a = \frac{(11/9)^{100} - (11/9)^{90}}{(11/9)^{100} - 1} \approx 0.866.$$

The expected net gain is, by Fact 13, $b(1 - q_a) - a = 100(0.134) - 90 = -76.6$. The average duration (number of plays of the game) is found from Fact 15 to be 765.6 plays. Surprisingly, even though the probability $p$ of winning is only slightly less than 0.5 and the gambler starts within close reach of the desired goal, the gambler can expect to be ruined with high probability and the average number of plays of the game is large.

**5.** The average duration of a fair game in the gambler's ruin problem is considerably longer than would be naively expected. When one player has one dollar and the adversary 1,000 dollars, Fact 16 shows that the average duration is 1,000 trials.

## 7.5.3   GENERALIZED RANDOM WALKS

Two generalizations of one-dimensional random walks are covered here. In the first case, a one-dimensional walk is now allowed to be based on an arbitrary (as opposed to

Bernoulli) distribution. In the second case, a symmetric random walk is considered in higher dimensions.

**Definitions:**

In a one-dimensional random walk on $\mathcal{Z}$ starting at $S_0 = a > 0$ with a given boundary $b > a$, let $u_a$ be the probability that the particle will arrive at some position which is less than or equal to 0 before reaching any position greater than or equal to $b$.

Let $R_n$ be the number of distinct points visited by a random walk up to time $n$.

**Facts:**

**1.** If $X_1, X_2, \ldots$ are arbitrarily distributed independent random variables, many basic qualitative laws are preserved for one-dimensional walks.

- In the case of two absorbing boundaries, the particle will reach one of them with probability 1.

- In the case of a single absorbing boundary at 0, if $E(X_i) \le 0$ then the particle will reach 0 with probability 1.

- An unrestricted walk with $E(X_i) = 0$ and $\mathrm{Var}(X_i) < \infty$ will return to its initial position with probability 1, and the expected return time is infinite.

**2.** *General ruin problem:* Assume that at each step the particle has probability $p_k$ to move from any point $i$ to $i + k$, where $k \in \mathcal{Z}$. The particle starts from position $a$.

- $u_a = 1$ if $a \le 0$, and $u_a = 0$ if $a \ge b$.

- For $0 < a < b$, $u_a = \sum\limits_{i \in \mathcal{Z}} u_i p_{i-a}$. This corresponds to a system of $b - 1$ linear equations in $b - 1$ unknowns that has a unique solution.

**3.** *Local central limit theorem:* For a $d$-dimensional symmetric (simple) random walk,

- $\left| Pr(S_n = k) - 2(\frac{d}{2\pi n})^{d/2} e^{-\frac{d|k|^2}{2n}} \right| \le O(n^{-(d+2)/2})$;

- $\left| Pr(S_n = k) - 2(\frac{d}{2\pi n})^{d/2} e^{-\frac{d|k|^2}{2n}} \right| \le |k|^{-2} O(n^{-d/2})$.

**4.** *Pólya's theorem:* For the symmetric random walks in 1 or 2 dimensions, there is probability 1 that the walk will eventually return to its initial position (recurrent random walk). In dimension $d \ge 3$ this probability is strictly less than 1. (George Pólya, 1887–1985.)

**5.** For symmetric random walks in $d = 3$ dimensions, the probability of an eventual return to the initial position is approximately 0.34 and the expected number of returns is approximately 0.53. The following table gives the approximate return probabilities $Pr(T_0 < \infty)$ for dimensions $d \le 10$.

| $d$ | $Pr(T_0 < \infty)$ |
|-----|-----|
| 3 | 0.341 |
| 4 | 0.193 |
| 5 | 0.135 |
| 6 | 0.105 |
| 7 | 0.0858 |
| 8 | 0.0729 |
| 9 | 0.0634 |
| 10 | 0.0562 |

**6.** *Range problems:*  As $n \to \infty$,

- if $d = 1$, $E(R_n) \approx (\frac{8n}{\pi})^{1/2}$;

- if $d = 2$, $E(R_n) \approx \frac{\pi n}{\log n}$;

- if $d \geq 3$, $E(R_n) \approx c_d n$ for some constant $c_d$.

**Example:**

**1.** *Absorbing boundaries:*  A particle starts at position $a \in \{0, 1, 2, 3, 4\}$ on a line and with equal probabilities moves one or two positions, either to the right or left. Upon reaching a position $x \leq 0$ or $x \geq 4$, the particle is stopped. This is a form of the general ruin problem with $p_{-2} = p_{-1} = p_1 = p_2 = \frac{1}{4}$. We are interested in the probability $u_a$ of absorption at position $x \leq 0$, given that the particle starts at position $a$. Using Fact 2, $u_0 = 1$, $u_4 = 0$ and $u_1, u_2, u_3$ satisfy the following equations:

$$u_1 = \tfrac{1}{2} + \tfrac{1}{4}u_2 + \tfrac{1}{4}u_3$$
$$u_2 = \tfrac{1}{4} + \tfrac{1}{4}u_1 + \tfrac{1}{4}u_3$$
$$u_3 = \tfrac{1}{4}u_1 + \tfrac{1}{4}u_2.$$

Solving this linear system produces the unique solution $u_1 = \frac{7}{10}$, $u_2 = \frac{1}{2}$, $u_3 = \frac{3}{10}$. Intuitively, these are reasonable values since starting at the middle position $a = 2$ it should be equally likely for the particle to be absorbed at either boundary; starting at position $a = 1$ there should be a greater chance of absorption at the left boundary (probability $\frac{7}{10}$) than at the right boundary (probability $1 - \frac{7}{10} = \frac{3}{10}$).

## 7.5.4  APPLICATIONS OF RANDOM WALKS

Random walk methodology is central to a number of diverse problem settings. This section describes several important applications. Additional examples are found in [BaNi70], [Be93], and [We94].

**Examples:**

**1.** *Biological migration:*  The name "random walk" first appears in a query sent by Karl Pearson (1857–1936) to the journal *Nature* in 1905. Pearson's problem refers to a walk in the plane, with successive steps of length $l_1, l_2, \ldots$ at angles $\Theta_1, \Theta_2, \ldots$ with respect to the $x$-axis, the $\Theta_i$ being chosen randomly. The problem is to find, after some fixed time, the probability distribution of the distance from the initial position. The question was motivated by a theory of biological migration which Pearson developed at that time, but soon discarded. Nevertheless, Pearson's random walk was born and it has since been applied to many biological models.

**2.** *Biology:*  Other, more recent, examples of random walk applications include DNA sequencing in genetics, bacterial migration in porous media, and molecular diffusion. In the latter example, diffusion of molecules occurs as a result of the thermal energy of the molecules. The motion of the molecules, perturbed through interactions with other molecules, is then modeled as a random walk. See [Be93] for further details.

**3.** *Physical sciences:*  There are many applications in the physical sciences, including the classical Scher-Montroll model of electrical transport in amorphous semiconductors (a continuous-time random walk model), models of diffusion on tenuously connected structures such as percolation clusters, inference of molecular structure from data collected in x-ray scattering experiments, configurational statistics of polymers, and reaction kinetics in confined geometries. Details can be found in [BaNi70] and [We94].

**4.** *Sequential sampling*:  A major application of random walks in statistics is in connection with Wald's theory of sequential sampling. In this context, the $X_i$ represent certain characteristics of samples or observations. Measurements are taken as long as the random walk remains within a given region. Termination with acceptance or rejection of the appropriate hypothesis occurs depending on which part of the boundary is reached.

**5.** *Stock prices*:  One of the early applications of computers in economics was to analyze economic time series and, in particular, the behavior of stock market prices over time. It first came as a surprise when Kendall found in 1953 that he could not identify any predictable patterns in stock prices. However, it soon became apparent that random price movements indicated a well-functioning or efficient market and that a random walk could be used as a model for the underlying market. See [Ma90]. In fact, at the beginning of this century, Bachelier had already developed a diffusion model for the stock market. Other macroeconomic time series have also been modeled using random walks.

**6.** *Astronomy*:  The problem of the mean motion of a planet in the presence of perturbations due to other planets has a very long history (Lagrange). Statistical properties of perturbed orbits can be analyzed with the help of Pearson's random walk model. The escape of comets from the solar system has also been modeled as a random walk among energy states. Details are provided in [BaNi70].

## 7.6   SYSTEM RELIABILITY

System reliability involves the study of the overall performance of systems of interconnected components. Examples of such systems are communication, transportation, and electrical power distribution systems, as well as computer networks.

### 7.6.1   GENERAL CONCEPTS

**Definitions:**

Suppose a given system is composed of a set $N = \{1, 2, \ldots, n\}$ of failure-prone **components**. At any instant of time, each component is found in one of two **states**: either **operating** or **failed**.

The **reliability** of component $i$ is the probability $p_i$ that component $i$ is operating at a given instant of time. The **unreliability** (or failure probability) of component $i$ is $q_i = 1 - p_i$.

At any instant of time, the system is found in one of two **states**: **operating** or **failed**.

The **structure function** $\phi$ is a binary-valued function defined on all subsets $S \subseteq N$. Specifically, $\phi(S) = 1$ if the system operates when all components in $S$ operate and all components of $N - S$ fail; otherwise $\phi(S) = 0$.

The structure function $\phi$ is **monotone** if $S \subseteq T \Rightarrow \phi(S) \leq \phi(T)$. In words, monotonicity means that the addition of more operating components to an already functioning system cannot result in system failure.

The structure function $\phi$ is **nontrivial** if $\phi(\emptyset) = 0$ and $\phi(N) = 1$.

A **coherent** system $(N, \phi)$ has a structure function $\phi$ that is monotone and nontrivial.

The **dual** of the system $(N, \phi)$ is the system $(N, \phi^D)$, defined by $\phi^D(S) = 1 - \phi(N - S)$.

The **reliability** $R_{N,\phi}$ of the system $(N, \phi)$ is the probability that the system functions at a random instant of time: $R_{N,\phi} = Pr(\phi(S) = 1)$.

The **unreliability** $U_{N,\phi}$ of the system is given by $U_{N,\phi} = Pr(\phi(S) = 0) = 1 - R_{N,\phi}$.

**Facts:**

**1.** System reliability depends not only on the reliability of its individual components, but also on the manner in which they are interconnected.

**2.** If the state of any component is statistically independent of the state of any other component, then the probability that $S \subseteq N$ is precisely the set of operating components is given by $\text{prob}(S) = \prod\limits_{i \in S} p_i \prod\limits_{j \notin S} q_j$.

**3.** The dual of the dual of a system $(N, \phi)$ is the original system $(N, \phi)$.

**Examples:**

**1.** Consider a system built from the set of components $N = \{1, 2, 3, 4\}$. Associate with each component a known weight: $w_1 = 5$, $w_2 = 7$, $w_3 = 4$, $w_4 = 8$. If $S \subseteq N$ is the set of operating components, then the system is considered to operate if $\sum_{i \in S} w_i > 12$. Thus $\phi(S) = 1$ for precisely the following sets $S$:

$$\{1, 4\}, \ \{2, 4\}, \ \{1, 2, 3\}, \ \{1, 2, 4\}, \ \{1, 3, 4\}, \ \{2, 3, 4\}, \ \{1, 2, 3, 4\}.$$

In all other cases $\phi(S) = 0$. This structure function is nontrivial and monotone, so that $(N, \phi)$ is a coherent system. The reliability of the system is then

$$R_{N,\phi} = p_1 q_2 q_3 p_4 + q_1 p_2 q_3 p_4 + p_1 p_2 p_3 q_4 + p_1 p_2 q_3 p_4 + p_1 q_2 p_3 p_4 + q_1 p_2 p_3 p_4 + p_1 p_2 p_3 p_4.$$

The dual system $(N, \phi^D)$ has the structure function $\phi^D$, where $\phi^D(T) = 0$ for precisely the following component sets $T$:

$$\emptyset, \{1\}, \ \{2\}, \ \{3\}, \ \{4\}, \ \{1, 3\}, \ \{2, 3\}.$$

In all other cases $\phi^D(T) = 1$. For example, $\phi^D(\{1, 3\}) = 1 - \phi(\{2, 4\}) = 1 - 1 = 0$.

**2.** For critical financial transactions, calculations are carried out simultaneously by three separate microprocessors. The three results are compared and the result is accepted if any two of the processors agree (or all three agree). Here the system has components $\{1, 2, 3\}$ corresponding to the three microprocessors. A component fails if it gives the wrong answer, and the system fails if this "majority rule" produces an incorrect (or inconclusive) answer. Thus, $\phi(S) = 1$ if and only if $S$ is $\{1, 2\}$, $\{1, 3\}$, $\{2, 3\}$, or $\{1, 2, 3\}$.

This structure function is nontrivial and monotone, so the system is coherent. If the microprocessors are identical and each operates independently with probability $p$, then $Pr(\text{exactly two components work}) = 3p^2(1 - p)$, $Pr(\text{all components work}) = p^3$, and

$$R_{N,\phi} = 3p^2(1 - p) + p^3 = 3p^2 - 2p^3.$$

For example, if $p = 0.95$ then $R_{N,\phi} = 0.99275$. In this case, even though any single microprocessor has a 5% failure rate, the system as a whole has only a 0.7% failure rate.

**3.** *Telephone network*: The components of this system are individual communication links (or trunk lines) joining nearby locations. Any telephone call placed between two distant locations in this system needs to be routed along available communication links.

However, as a result of hardware or software malfunctions, or as a result of overloaded circuits, certain links may be unavailable at a given instant of time. Thus, a telephone network can be modeled as a system whose components are subject to failure at random times. The reliability of the entire system is the probability that the system functions at a random instant of time; that is, that at any random point in time users can successfully complete their calls.

## 7.6.2  COHERENT SYSTEMS

It is assumed throughout this subsection that the system $(N, \phi)$ is coherent.

**Definitions:**
A **minpath** $P$ is a minimal set of components such that $\phi(P) = 1$: i.e., $\phi(P) = 1$ and $\phi(S) = 0$ for all proper subsets $S \subset P$. The collection of all minpaths for $(N, \phi)$ is denoted $\mathcal{P}$.

A **mincut** $C$ is a minimal set of components such that $\phi(N - C) = 0$: i.e., $\phi(N - C) = 0$ and $\phi(N - S) = 1$ for all proper subsets $S \subset C$. The collection of all mincuts for $(N, \phi)$ is denoted $\mathcal{C}$.

Let $G = (V, E)$ be an undirected graph with vertex set $V$ and edge set $E$ (§8.1.1).

- A **simple path** in $G$ is a path that contains no repeated vertices.
- An **s-t cutset** of $G$ is a minimal set of edges, the removal of which leaves no $s$-$t$ path in $G$.
- A **cutset** of $G$ is a minimal set of edges, the removal of which disconnects $G$.

If $K \subseteq V$, a **K-tree** is a minimal set $F$ of edges in $G$ such that every two vertices of $K$ are joined by a path in $F$.

If $K \subseteq V$, a **K-cutset** is a minimal set of edges in $G$, the removal of which disconnects some pair of vertices in $K$.

**Facts:**
**1.** The dual of a coherent system is itself coherent.

**2.** The structure function of a coherent system $(N, \phi)$ can be completely described using its minpaths $\mathcal{P}$ or using its mincuts $\mathcal{C}$. Specifically
- $\phi(S) = 1$ if and only if $S$ contains some minpath $P$;
- $\phi(S) = 0$ if and only if $N - S$ contains some mincut $C$.

**3.** The minpaths of $(N, \phi)$ are the mincuts of the dual $(N, \phi^D)$, and conversely.

**4.** The mincuts of $(N, \phi)$ are the minpaths of the dual $(N, \phi^D)$, and conversely.

**5.** Every minpath of $(N, \phi)$ and every mincut of $(N, \phi)$ have nonempty intersection.

**6.** If $P$ is a minimal set of components that has nonempty intersection with every mincut of $(N, \phi)$, then $P$ is a minpath of $(N, \phi)$.

**7.** If $C$ is a minimal set of components that has nonempty intersection with every minpath of $(N, \phi)$, then $C$ is a mincut of $(N, \phi)$.

**8.** A $K$-tree has the topology of a tree (§9.1.1) whose leaf vertices are in $K$.

## Examples:

**1.** *Series system:* This system $(N, \phi)$ operates only when all components of $N$ operate. See the following figure. General characteristics are listed in Table 1.

$$\boxed{1}\!-\!\boxed{2}\!-\!\boxed{3}\!-\cdots-\!\boxed{n}$$

### Table 1: Characteristics of series and parallel systems.

|  | series | parallel |
|---|---|---|
| structure function | $\phi(S) = 0$, if $S \subset N$ <br> $\phi(N) = 1$ | $\phi(S) = 1$, if $S \neq \emptyset$ <br> $\phi(\emptyset) = 0$ |
| minpaths | $P_1 = N$ | $P_1 = \{1\}, P_2 = \{2\}, \ldots, P_n = \{n\}$ |
| mincuts | $C_1 = \{1\}, C_2 = \{2\}, \ldots, C_n = \{n\}$ | $C_1 = N$ |
| reliability | $p_1 p_2 \ldots p_n$ | $1 - (1 - p_1)(1 - p_2) \ldots (1 - p_n)$ |
| unreliability | $1 - p_1 p_2 \ldots p_n$ | $(1 - p_1)(1 - p_2) \ldots (1 - p_n)$ |
| dual | parallel, $n$ components | series, $n$ components |

**2.** *Parallel system:* This system $(N, \phi)$ fails only when all components of $N$ fail. See the following figure. General characteristics are listed in Table 1.

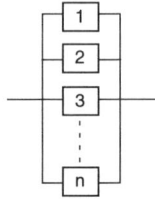

**3.** *k-out-of-n success system:* This system $(N, \phi)$ operates only when at least $k$ out of the $n$ components operate. The following figure illustrates a 2-out-of-3 success system. General characteristics are listed in Table 2. The special case $k = 1$ gives a parallel system; $k = n$ gives a series system.

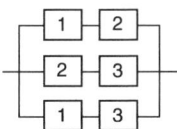

**4.** *k-out-of-n failure system:* This system $(N, \phi)$ fails only when at least $k$ out of the $n$ components fail. This is the same as an $(n - k + 1)$-out-of-$n$ success system. General characteristics are listed in Table 2. The special case $k = 1$ gives a series system; $k = n$ gives a parallel system.

**5.** *Two-terminal network:* Two vertices $s, t$ of an undirected graph $G = (V, E)$ are specified and a message is to be sent from vertex $s$ to vertex $t$. Assume that only the edges are failure-prone, so $N = E$. The system operates when there exists some path of operating edges joining $s$ to $t$ in the graph.

- *structure function:* $\phi(S) = 1$ if there exists a path from $s$ to $t$ in the subgraph defined by edges $S$; $\phi(S) = 0$ otherwise;

- *minpaths*: all simple $s$-$t$ paths of $G$;
- *mincuts*: all $s$-$t$ cutsets of $G$;
- *reliability*: $R_{N,\phi} =$ the probability that a message sent from $s$ will arrive at $t = \sum\{\text{prob}(S) \mid S \text{ contains some simple } s\text{-}t \text{ path}\}$;
- *unreliability*: $U_{N,\phi} = \sum\{\text{prob}(N - S) \mid S \text{ contains some } s\text{-}t \text{ cutset}\}$.

**Table 2: Characteristics of $k$-out-of-$n$ systems.**

|  | $k$-out-of-$n$ success | $k$-out-of-$n$ failure |
|---|---|---|
| structure function | $\phi(S) = 1$ if $\|S\| \geq k$ <br> $\phi(S) = 0$ if $\|S\| < k$ | $\phi(S) = 0$ if $\|S\| \leq n - k$ <br> $\phi(S) = 1$ if $\|S\| > n - k$ |
| minpaths | $S \subseteq N$ with $\|S\| = k$ | $S \subseteq N$ with $\|S\| = n - k + 1$ |
| mincuts | $S \subseteq N$ with $\|S\| = n - k + 1$ | $S \subseteq N$ with $\|S\| = k$ |
| reliability | $\sum\{\text{prob}(S) \mid \|S\| \geq k\}$ | $\sum\{\text{prob}(N - S) \mid \|S\| < k\}$ |
| unreliability | $\sum\{\text{prob}(S) \mid \|S\| < k\}$ | $\sum\{\text{prob}(N - S) \mid \|S\| \geq k\}$ |
| dual | $k$-out-of-$n$ failure | $k$-out-of-$n$ success |

**6.** *All-terminal network*:   A message is to be disseminated among all vertices $V$ in the undirected graph $G = (V, E)$. The system operates when the operating edges in the graph allow all vertices to mutually communicate.

- *structure function*: $\phi(S) = 1$ if the subgraph defined by vertices $V$ and edges $S$ is connected; $\phi(S) = 0$ otherwise;
- *minpaths*: all spanning trees (§9.2) of $G$;
- *mincuts*: all cutsets of $G$;
- *reliability*: $R_{N,\phi} =$ probability that $G$ is connected $= \sum\{\text{prob}(S) \mid S \text{ contains}$ some spanning tree of $G\}$;
- *unreliability*: $U_{N,\phi} = \sum\{\text{prob}(N - S) \mid S \text{ contains some cutset of } G\}$.

**7.** *All-terminal network, equal operating probabilities*:   Suppose the undirected graph $G = (V, E)$ has $n = |V|$ vertices and $m = |E|$ edges, and suppose each edge has the same operating probability $p$. Let $q = 1 - p$ be the common edge failure probability. Then the all-terminal reliability $R(G, p)$ can be calculated as

$$R(G, p) = \sum_{i=0}^{m} F_i p^{m-i} q^i,$$

where $F_i$ counts the number of operating system states in which exactly $m - i$ edges are functioning.

- $0 \leq F_i \leq \binom{m}{i}$ for $i = 0, 1, \ldots, m$;
- $F_i = 0$ for $i > m - n + 1$;
- $F_i = \binom{m}{i}$ for $i < c$, where $c$ is the minimum cardinality of a cutset of $G$;
- Sperner's bound: $iF_i \leq (m - i + 1)F_{i-1}$.

**8.** *$K$-terminal network*:   A message is to be disseminated among a fixed subset $K$ of vertices in the undirected graph $G = (V, E)$. The system operates when the operating edges of the graph allow all vertices in $K$ to mutually communicate.

- *structure function*: $\phi(S) = 1$ if the subgraph defined by vertices $K$ and edges $S$ is connected; $\phi(S) = 0$ otherwise;

- *minpaths*: all $K$-trees of $G$;
- *mincuts*: all $K$-cutsets of $G$;
- *reliability*: $R_{N,\phi}$ = probability that $K$ is connected = $\sum\{\text{prob}(S) \mid S$ contains some $K$-tree of $G\}$;
- *unreliability*: $U_{N,\phi} = \sum\{\text{prob}(N - S) \mid S$ contains some $K$-cutset of $G\}$;
- *special cases*: $K = \{s, t\}$ gives the two-terminal network problem; $K = V$ gives the all-terminal network problem.

**9.** Examples 5–8 are defined in terms of undirected networks. The two-terminal, all-terminal, and $K$-terminal reliability problems described in these examples can also be defined for directed networks.

**10.** Consider the coherent system $(N, \phi)$ on components $N = \{1, 2, 3, 4, 5\}$ with min-paths $P_1 = \{1, 2\}, P_2 = \{1, 5\}, P_3 = \{3, 5\}, P_4 = \{2, 3, 4\}$. To illustrate Fact 2, notice that $\phi(\{3, 5\}) = 1$ and $\phi(\{2, 3, 5\}) = 1$ since $P_3$ is a minpath. Also, $\phi(\{2, 5\}) = 0$ since $\{2, 5\}$ contains no minpath. By Fact 7, $C_1 = \{1, 3\}$ is a mincut for this system since $C_1$ has nonempty intersection with each of $P_1, P_2, \ldots, P_4$ and since neither $\{1\}$ nor $\{3\}$ has this property. Likewise, $C_2 = \{2, 5\}$ and $C_3 = \{1, 4, 5\}$ are mincuts for this system. Fact 4 shows that the dual $(N, \phi^D)$ has as its minpaths the mincuts of $(N, \phi)$: namely, $\{1, 3\}$, $\{2, 5\}$, and $\{1, 4, 5\}$. This means $\phi^D(\{1, 3, 4\}) = 1$ since $\{1, 3, 4\}$ contains the minpath $\{1, 3\}$. Alternatively, from the definition $\phi^D(\{1, 3, 4\}) = 1 - \phi(\{2, 5\}) = 1 - 0 = 1$.

---

### 7.6.3 CALCULATING SYSTEM RELIABILITY

Four general approaches can be used to calculate the reliability $R_{N,\phi}$ of a coherent system $(N, \phi)$. These are state-space enumeration, inclusion-exclusion, disjoint products, and factoring.

**Notation:**
Let $\mathcal{P} = \{P_1, P_2, \ldots, P_m\}$ be the minpaths and let $\mathcal{C} = \{C_1, C_2, \ldots, C_r\}$ be the mincuts of the coherent system $(N, \phi)$.

- $E_i$ is the event that all components of minpath $P_i$ operate (with no stipulation as to the states of the other components);
- $F_i$ is the event that all components of $C_i$ fail (with no stipulation as to the states of the other components);
- $E_i E_j$ denotes the event that both $E_i$ and $E_j$ occur;
- $(N, \phi_{+i})$ is the system derived from $(N, \phi)$ in which component $i$ always works;
- $(N, \phi_{-i})$ is the system derived from $(N, \phi)$ in which component $i$ always fails.

**Facts:**
**1.** Calculation of the reliability $R_{N,\phi}$ is in general quite difficult; specifically, it is a #P-complete problem (see §17.5.3).
**2.** $R_{N,\phi} = \sum\{\text{prob}(S) \mid \phi(S) = 1\}$.
**3.** $R_{N,\phi} = \sum\{\text{prob}(S) \mid S$ contains some $P \in \mathcal{P}\}$.
**4.** $R_{N,\phi} = 1 - U_{N,\phi} = 1 - \sum\{\text{prob}(N - S) \mid S$ contains some $C \in \mathcal{C}\}$.
**5.** *State-space enumeration*: System reliability can be found by enumerating all operating (or all failed) states of the system, using Facts 2–4.

**6.** $R_{N,\phi} = Pr(E_1 \cup E_2 \cup \cdots \cup E_m)$.

**7.** $U_{N,\phi} = Pr(F_1 \cup F_2 \cup \cdots \cup F_r)$.

**8.** Applying the inclusion-exclusion principle (§2.4) to Fact 6 produces

$$R_{N,\phi} = \sum_i Pr(E_i) - \sum_{i<j} Pr(E_i E_j) + \cdots + (-1)^{m+1} Pr(E_1 E_2 \ldots E_m).$$

**9.** Applying the inclusion-exclusion principle (§2.4) to Fact 7 produces

$$U_{N,\phi} = \sum_i Pr(F_i) - \sum_{i<j} Pr(F_i F_j) + \cdots + (-1)^{r+1} Pr(F_1 F_2 \ldots F_r).$$

**10.** *Inclusion-exclusion:* This approach calculates system reliability using Facts 8–9.

**11.** $R_{N,\phi} = Pr(E_1) + Pr(\overline{E}_1 E_2) + \cdots + Pr(\overline{E}_1 \overline{E}_2 \ldots \overline{E}_{m-1} E_m)$.

**12.** $U_{N,\phi} = Pr(F_1) + Pr(\overline{F}_1 F_2) + \cdots + Pr(\overline{F}_1 \overline{F}_2 \ldots \overline{F}_{r-1} F_r)$.

**13.** *Disjoint products:* This approach calculates system reliability using the law of total probability (§7.2.1). (See Facts 11–12.)

**14.** $R_{N,\phi} = p_i R_{N,\phi+i} + (1 - p_i) R_{N,\phi-i}$.

**15.** *Factoring:* Rather than requiring an enumeration of the minpaths or mincuts of the system, this method (based on Fact 14) concentrates on the state of an individual component $i$: it is either operating (with probability $p_i$) or failed (with probability $q_i = 1 - p_i$).

**16.** The factoring method is applied most productively when the system $(N, \phi)$ has additional structure. For example, this approach can be used to determine the reliability of $k$-out-of-$n$ systems and two-terminal networks. (See §7.6.4.)

**Examples:**

**1.** *State-space enumeration:* Consider the coherent system $(N, \phi)$ with $N = \{1, 2, 3, 4\}$ and minpaths $P_1 = \{2, 3\}$, $P_2 = \{1, 2, 4\}$, $P_3 = \{1, 3, 4\}$. By Fact 2 of §7.6.2, the operating states of the system are $\{2, 3\}, \{1, 2, 3\}, \{1, 2, 4\}, \{1, 3, 4\}, \{2, 3, 4\}$, and $\{1, 2, 3, 4\}$. By Fact 3, $R_{N,\phi} = q_1 p_2 p_3 q_4 + p_1 p_2 p_3 q_4 + p_1 p_2 q_3 p_4 + p_1 q_2 p_3 p_4 + q_1 p_2 p_3 p_4 + p_1 p_2 p_3 p_4$.

**2.** *Inclusion-exclusion:* Consider the coherent system on $N = \{1, 2, 3, 4\}$ with minpaths $P_1 = \{1, 2\}$, $P_2 = \{2, 4\}$, $P_3 = \{1, 3, 4\}$. Event $E_1$ has probability $p_1 p_2$, event $E_1 E_2$ has probability $p_1 p_2 p_4$, etc. Fact 8 gives

$$R_{N,\phi} = p_1 p_2 + p_2 p_4 + p_1 p_3 p_4 - p_1 p_2 p_4 - p_1 p_2 p_3 p_4 - p_1 p_2 p_3 p_4 + p_1 p_2 p_3 p_4$$
$$= p_1 p_2 + p_2 p_4 + p_1 p_3 p_4 - p_1 p_2 p_4 - p_1 p_2 p_3 p_4.$$

**3.** *Disjoint products:* Fact 11 is applied to the coherent system on $N = \{1, 2, 3, 4, 5, 6\}$ with minpaths $P_1 = \{1, 5\}$, $P_2 = \{1, 3, 6\}$, $P_3 = \{2, 4, 5\}$, $P_4 = \{2, 6\}$. For simplicity of notation, let the event $\{e \text{ operates}\}$ be denoted by $e$, and let the event $\{e \text{ fails}\}$ be denoted by $\bar{e}$. Identities of set theory (see §1.2.2) can then be used to obtain

$$Pr(E_1) = p_1 p_5;$$

$$Pr(\overline{E}_1 E_2) = Pr((\bar{1} \cup \bar{5})136) = Pr(\bar{5}136) = p_1 p_3 q_5 p_6;$$

$$Pr(\overline{E}_1 \overline{E}_2 E_3) = Pr((\bar{1} \cup \bar{5})(\bar{1} \cup \bar{3} \cup \bar{6})245) = Pr(\bar{1}(\bar{1} \cup \bar{3} \cup \bar{6})245)$$
$$= Pr(\bar{1}245) = q_1 p_2 p_4 p_5;$$

$$Pr(\overline{E}_1 \overline{E}_2 \overline{E}_3 E_4) = Pr((\bar{1} \cup \bar{5})(\bar{1} \cup \bar{3} \cup \bar{6})(\bar{2} \cup \bar{4} \cup \bar{5})26)$$
$$= Pr((\bar{1} \cup \bar{5})(\bar{1} \cup \bar{3})(\bar{4} \cup \bar{5})26)$$
$$= Pr((\bar{1} \cup \bar{3}\bar{5})(\bar{4} \cup \bar{5})26)$$
$$= Pr((\bar{1}\bar{4} \cup \bar{1}\bar{5} \cup \bar{3}\bar{5})26).$$

Since the events $\overline{14}$, $\overline{15}$, and $\overline{35}$ above are not disjoint, Fact 11 can be reapplied to this new union of events, yielding

$$Pr(\overline{E}_1\overline{E}_2\overline{E}_3 E_4) = Pr((\overline{14} \cup 4\overline{15} \cup 1\overline{35})26)$$
$$= Pr(\overline{1}\overline{4}26 \cup 4\overline{1}\overline{5}26 \cup 1\overline{3}\overline{5}26)$$
$$= q_1 p_2 q_4 p_6 + q_1 p_2 p_4 q_5 p_6 + p_1 p_2 q_3 q_5 p_6.$$

The final expression for system reliability is then

$$R_{N,\phi} = p_1 p_5 + p_1 p_3 q_5 p_6 + q_1 p_2 p_4 p_5 + q_1 p_2 q_4 p_6 + q_1 p_2 p_4 q_5 p_6 + p_1 p_2 q_3 q_5 p_6.$$

## 7.6.4  SPECIALIZED ALGORITHMS FOR CALCULATING RELIABILITY

The general methods for calculating system reliability discussed in §7.6.3 can often be streamlined when the system has a special structure. This section describes algorithms for calculating the reliability of series-parallel, $k$-out-of-$n$, and certain network systems.

**Definitions:**

The **two-terminal reliability** of a network $G$ is the probability $R_{st}(G)$ that vertices $s$ and $t$ are connected in $G$.

The **all-terminal reliability** of a network $G$ is the probability $R_V(G)$ that $G$ is connected.

If $G$ is a two-terminal network with specified vertices $s$ and $t$, then an **irrelevant** edge is one not appearing in any simple $s$-$t$ path of $G$.

Let $G_S = (V_S, E_S)$ be the subgraph of a directed graph $G = (V, E)$ induced by edges $S \subseteq E$. If $G_S$ is acyclic and has no irrelevant edges, the **domination** of $G_S$ is $d_S = (-1)^{|E_S|-|V_S|+1}$; in all other cases, define $d_S = 0$.

Define $f_k(n)$ to be the reliability of a $k$-out-of-$n$ success system having the components $N = \{1, 2, \ldots, n\}$ and corresponding reliabilities $p_1, p_2, \ldots, p_n$.

**Facts:**

1. A parallel system has reliability $f_1(n) = 1 - (1 - p_1)(1 - p_2)\ldots(1 - p_n)$.

2. A series system has reliability $f_n(n) = p_1 p_2 \ldots p_n$.

3. *Series-parallel system:* If a system is constructed from series and parallel subsystems (with no component appearing in more than one subsystem), then the reliability of the overall system is calculated by successively applying Facts 1 and 2.

4. Applying the factoring approach of §7.6.3 to a $k$-out-of-$n$ success system gives $f_k(n) = p_n f_{k-1}(n-1) + (1 - p_n)f_k(n-1)$.

5. *$k$-out-of-$n$ system:* Repeated application of Facts 1, 2, and 4 produces the reliability of any $k$-out-of-$n$ success system. Since a $k$-out-of-$n$ failure system is the same as an $(n - k + 1)$-out-of-$n$ success system, the reliability of any $k$-out-of-$n$ failure system is found in a similar way.

6. For a two-terminal undirected network $G$, the system $(N, \phi_{-e})$ corresponds to the two-terminal network $G - e$ with edge $e$ deleted.

7. For a two-terminal undirected network $G$, the system $(N, \phi_{+e})$ corresponds to the two-terminal network $G/e$ in which edge $e$ is contracted.

**8.** *Two-terminal undirected network*:  Algorithm 1 is a recursive procedure that calculates $R_{st}(G)$.  Based on the factoring approach of §7.6.3, it splits the initial reliability calculation for $G$ into calculations for the smaller networks $G - e$ and $G/e$.

---

**Algorithm 1**:   **Two-terminal reliability for undirected networks.**

**procedure** $R_{st}(G)$

perform series and parallel reductions on $G$, producing network $H$

**if** $H$ consists of the single edge $(s,t)$ **then** return the reliability of $(s,t)$

**else**

    select an edge $e$ from $H$

    let $p_e$ be the reliability of $e$ in $H$

    return  $p_e\, R_{st}(H/e) + (1 - p_e)\, R_{st}(H - e)$

---

**9.** For the sake of efficiency, Algorithm 1 carries out any applicable series and parallel reductions before selecting an edge on which to factor.

**10.** To avoid redundant calculations in Algorithm 1, edge $e$ should be chosen from $H$ so that $H/e$ and $H - e$ do not contain irrelevant edges.

**11.** If $G$ is a two-terminal directed network, then $R_{st}(G) = \sum_S d_S \prod_{i \in S} p_i$.

**12.** The expression in Fact 11 is obtained by using the inclusion-exclusion expansion (§7.6.3, Fact 8) applied to the simple $s$-$t$ paths of $G$.  Remarkably, a number of terms in this expansion cancel one another and the remaining coefficients are either $+1$ or $-1$.

**13.** In an undirected network $G = (V, E)$, the minpaths for the all-terminal problem are the spanning trees of $G$.

**14.** *All-terminal reliability*:  Algorithm 2 calculates $R_V(G)$ using the disjoint-products expansion (§7.6.3, Fact 11), applied to the spanning trees in lexicographic (dictionary) order.  Here each term of the expansion reduces to a single product involving $p_i$ and $q_i = 1 - p_i$.

---

**Algorithm 2**:   **All-terminal reliability for undirected networks.**

input: undirected network $G$

output: $R_V(G)$

let $T_1, T_2, \ldots, T_m$ be the spanning trees of $G$, listed in lexicographic order

**for** $k := 1$ **to** $m$

    $S_k := T_k$

    $F_k := \emptyset$

    **for** $j := 1$ **to** $k - 1$

        $F_k := F_k \cup \min\{r \mid r \in T_j - T_k\}$

    $g_k := \prod_{i \in S_k} p_i \prod_{i \in F_k} q_i$

    $R_V(G) := \sum_{k=1}^{m} g_k$

---

**Examples:**

**1.** A system on four components is built up from series and parallel subsystems.  Subsystem A has components 1 and 2 in series and subsystem B consists of component 3.  Subsystem C has these two subsystems in parallel and its reliability is $1 - (1 - p_1 p_2)(1 - p_3) =$

$p_1p_2 + p_3 - p_1p_2p_3$. The entire system is constructed from subsystem D (component 4 alone) in series with subsystem C, so it has reliability $p_4(p_1p_2 + p_3 - p_1p_2p_3) = p_1p_2p_4 + p_3p_4 - p_1p_2p_3p_4$.

**2.** To calculate the reliability of a 2-out-of-3 success system with components $1, 2, 3$, Facts 1 and 2 are first used to obtain

$$f_1(2) = 1 - (1 - p_1)(1 - p_2) = p_1 + p_2 - p_1p_2, \quad f_2(2) = p_1p_2.$$

Fact 4 then gives the system reliability

$$R_{N,\phi} = f_2(3) = p_3f_1(2) + (1 - p_3)f_2(2) = p_1p_2 + p_1p_3 + p_2p_3 - 2p_1p_2p_3.$$

**3.** The two-terminal *bridge* network $G$ is shown in the following figure, with $s = a$ and $t = d$.

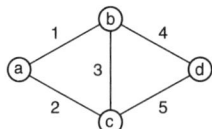

No series or parallel reductions can be performed on $G$, so factoring with respect to edge $e = 3$ produces the networks $G_1 = G/e$ and $G_2 = G - e$ shown in the next figure.

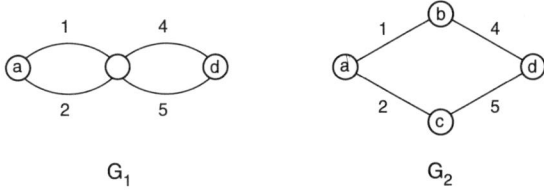

$$G_1 \qquad\qquad G_2$$

Since both $G_1$ and $G_2$ are series-parallel networks,

$$\begin{aligned}
R_{st}(G_1) &= [1 - (1 - p_1)(1 - p_2)][1 - (1 - p_4)(1 - p_5)] \\
&= (p_1 + p_2 - p_1p_2)(p_4 + p_5 - p_4p_5), \\
R_{st}(G_2) &= [1 - (1 - p_1p_4)(1 - p_2p_5)] = p_1p_4 + p_2p_5 - p_1p_2p_4p_5.
\end{aligned}$$

Algorithm 1 then produces

$$\begin{aligned}
R_{st}(G) &= p_3(p_1 + p_2 - p_1p_2)(p_4 + p_5 - p_4p_5) + (1 - p_3)(p_1p_4 + p_2p_5 - p_1p_2p_4p_5) \\
&= p_1p_4 + p_2p_5 + p_1p_3p_5 + p_2p_3p_4 - p_1p_3p_4p_5 - p_2p_3p_4p_5 - p_1p_2p_3p_4 \\
&\quad - p_1p_2p_3p_5 - p_1p_2p_4p_5 + 2p_1p_2p_3p_4p_5.
\end{aligned}$$

**4.** Consider a directed version of the two-terminal network in the figure of Example 3, in which there are oppositely directed edges $3 = (b, c)$ and $6 = (c, b)$. Edges 1 and 2 are directed out of $s = a$, while edges 4 and 5 are directed into $t = d$. The cyclic subgraph defined by edges $S = \{1, 2, 3, 4, 5, 6\}$ has $d_S = 0$. Also the subgraph defined by $S = \{1, 2, 3, 4\}$ has the irrelevant edges 2 and 3, so that $d_S = 0$. On the other hand, $S = \{1, 2, 3, 5\}$ defines an acyclic network without irrelevant edges, giving $d_S = (-1)^{4-4+1} = -1$ and the term $-p_1p_2p_3p_5$. Similarly, $S = \{1, 4\}$ produces $d_S = (-1)^{2-3+1} = +1$ and the term $+p_1p_4$.

After generating all acyclic networks without irrelevant edges, Fact 11 is applied to obtain

$$R_{st}(G) = p_1p_4 + p_2p_5 + p_1p_3p_5 + p_2p_4p_6 - p_1p_2p_4p_5 - p_1p_3p_4p_5$$
$$- p_1p_2p_4p_6 - p_1p_2p_3p_5 - p_2p_4p_5p_6 + p_1p_2p_3p_4p_5 + p_1p_2p_4p_5p_6.$$

**5.** The bridge network $G$ in Example 3 has eight spanning trees, given in lexicographic order by $T_1 = \{1, 2, 4\}$, $T_2 = \{1, 2, 5\}$, $T_3 = \{1, 3, 4\}$, $T_4 = \{1, 3, 5\}$, $T_5 = \{1, 4, 5\}$, $T_6 = \{2, 3, 4\}$, $T_7 = \{2, 3, 5\}$, $T_8 = \{2, 4, 5\}$. Applying Algorithm 2 gives

$$\begin{array}{lll}
S_1 = \{1, 2, 4\}, & F_1 = \emptyset, & g_1 = p_1p_2p_4 \\
S_2 = \{1, 2, 5\}, & F_2 = \{4\}, & g_2 = p_1p_2q_4p_5 \\
S_3 = \{1, 3, 4\}, & F_3 = \{2\}, & g_3 = p_1q_2p_3p_4 \\
S_4 = \{1, 3, 5\}, & F_4 = \{2, 4\}, & g_4 = p_1q_2p_3q_4p_5 \\
\vdots & \vdots & \vdots \\
S_8 = \{2, 4, 5\}, & F_8 = \{1, 3\}, & g_8 = q_1p_2q_3p_4p_5
\end{array}$$

Summing these eight terms then yields

$$R_V(G) = p_1p_2p_4 + p_1p_2q_4p_5 + p_1q_2p_3p_4 + p_1q_2p_3q_4p_5 + \cdots + q_1p_2q_3p_4p_5.$$

**6.** The all-terminal reliability of the bridge network $G$ given in Example 3 can also be calculated by determining the coefficients $F_i$ discussed in Example 7 of §7.6.2. Here we assume that all edges operate with probability $p$ and fail with probability $q = 1 - p$. Notice that $c = 2$ is the minimum cardinality of a cutset in $G$. Therefore, $F_0 = \binom{5}{0} = 1$ and $F_1 = \binom{5}{1} = 5$. At the other extreme, $F_i = 0$ for $i > m - n + 1 = 2$. The only remaining coefficient is $F_2$, obtained by noting that $G$ remains connected whenever any three of the five edges operate, except for $\{1, 2, 3\}$ and $\{3, 4, 5\}$, so that $F_2 = \binom{5}{2} - 2 = 8$. This gives $R(G, p) = p^5 + 5p^4q + 8p^3q^2$.

# 7.7   DISCRETE-TIME MARKOV CHAINS

Many physical systems evolve randomly in time, e.g., the population of a country, the value of a company's stock, the number of customers waiting at a checkout counter, and the functional state of a machine subject to failures and repairs. A discrete-time Markov chain can be used to model such situations when the set of possible states of the system is finite (or countable) and the system changes state at discrete time points. Such Markov chain models find applications in diverse fields, such as biology, computer science, inventory, production, queueing systems, and demography. In addition, many recursive algorithms can be viewed as a manifestation of an underlying discrete-time Markov chain.

## 7.7.1   MARKOV CHAINS

**Definitions:**

A sequence of random variables $\{X_n \mid n \geq 0\}$ is a (**discrete-time**) **Markov chain** (**DTMC**) on a (countable) state space $S$ if $X_n \in S$ for all $n \geq 0$ and $X_{n+1}$ depends (probabilistically) on the previous states of the system only via $X_n$:

$$Pr(X_{n+1} = j \mid X_n = i, X_{n-1} = i_{n-1}, ..., X_0 = i_0) = Pr(X_{n+1} = j \mid X_n = i),$$

for all $i_0, i_1, ..., i_{n-1}, i, j \in S$.

A Markov chain $\{X_n \mid n \geq 0\}$ is **time-homogeneous** if

$$Pr(X_{n+1} = j \mid X_n = i) = p_{ij}, \text{ for all } n \geq 0.$$

*Note:* Only time-homogeneous discrete-time Markov chains will be considered in this and later sections.

The matrix $P = (p_{ij})$ is the (**one-step**) **transition probability matrix** of the discrete-time Markov chain.

The **initial distribution** for a DTMC is the vector $a = (a_i)$, where $a_i = Pr(X_0 = i)$ for $i \in S$.

The **transition diagram** of a DTMC is the directed graph (§8.3.1) $G = (V, E)$, where $V = S$ is the state space and $E = \{(i, j) \in S \times S \mid p_{ij} > 0\}$.

A **stochastic** matrix $M = (m_{ij})$ has $m_{ij} \geq 0$ for all $i, j$ and $\sum_j m_{ij} = 1$ for all $i$.

**Facts:**

**1.** The first systematic study of Markov chains was carried out by Andrei Andreevich Markov (1856–1922); this work initiated the study of stochastic processes (sequences of random variables).

**2.** A DTMC on state space $S$ is completely described by the initial distribution $a$ and the transition probability matrix $P$.

**3.** The transition probability matrix $P$ is a stochastic matrix.

**Examples:**

**1.** Consider a DTMC on the set $S = \{1, 2, 3, 4, 5, 6\}$ with the following transition probability matrix

$$P = \begin{pmatrix} 0.4 & 0.6 & 0 & 0 & 0 & 0 \\ 0.7 & 0.3 & 0 & 0 & 0 & 0 \\ 0 & 0 & 0 & 1 & 0 & 0 \\ 0 & 0 & 1 & 0 & 0 & 0 \\ 0 & 0 & 0 & 0 & 1 & 0 \\ 0.1 & 0.1 & 0.1 & 0.1 & 0.1 & 0.5 \end{pmatrix}$$

To completely describe this DTMC, it is also necessary to specify the initial distribution. For example, $a = (0, 0, 0, 0, 0, 1)$ means that the system starts off in state 6.

**2.** *Simple random walk with absorbing states:* This is a DTMC on $S = \{0, 1, 2, ..., N\}$ with transition probabilities $p_{i,i+1} = p$, $p_{i,i-1} = 1 - p = q$, $1 \leq i \leq N - 1$, where $0 \leq p \leq 1$ is a given number. Also, $p_{0,0} = p_{N,N} = 1$, meaning that states 0 and $N$ are absorbing: once the DTMC visits these states it cannot leave them. This Markov model is also (more colorfully) known as the *gambler's ruin problem* (§7.5.1, Example 2). The transition diagram of this DTMC is given in the following figure.

**3.** *Simple random walk with reflecting states:* This is a variant of Example 2, in which the boundary states 0 and $N$ are reflecting: namely, $p_{0,1} = p_{N,N-1} = 1$. The transition diagram of this DTMC is shown next.

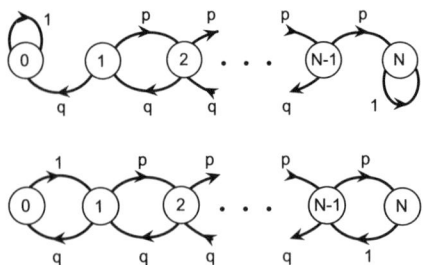

**4.** *Weather:*   A simplified model of the daily weather results in a DTMC. Suppose that each day is either sunny (0) or rainy (1) and that tomorrow's weather depends only on today's weather. Specifically, suppose that a rainy day follows a sunny day with probability 0.3 and a sunny day follows a rainy day with probability 0.4. This is a DTMC with state space $S = \{0, 1\}$ and transition probability matrix

$$P = \begin{pmatrix} 0.7 & 0.3 \\ 0.4 & 0.6 \end{pmatrix}.$$

**5.** *Urns:*   Urn B contains nine black and one white ball, while Urn R contains six red and four white balls. Balls are successively drawn with replacement from an urn. If the ball drawn is colored, the drawing continues from the same urn. If the ball drawn is white, the drawing continues from the other urn. Define the state of the system to be the urn being sampled, so $S = \{B, R\}$. This is a DTMC with transition probabilities $p_{BB} = 0.9$, $p_{BR} = 0.1$, $p_{RR} = 0.6$, $p_{RB} = 0.4$.

**6.** *Ehrenfest diffusion model:*   Suppose that there are $M$ molecules in a vessel, separated into two chambers by a membrane, across which molecules can pass. A state of the system at any instant is given by $(k_1, k_2)$, where there are $k_1$ molecules in the first chamber and $k_2 = M - k_1$ in the second chamber. Transitions from the current state $(k_1, k_2)$ occur by the movement of a single molecule from the first chamber to the second, resulting in state $(k_1 - 1, k_2 + 1)$, or from the second chamber to the first, resulting in state $(k_1 + 1, k_2 - 1)$.

In the Ehrenfest model of this process, the probability of transition from $(k_1, k_2)$ to $(k_1 - 1, k_2 + 1)$ is given by $\frac{k_1}{M}$, whereas the probability of transition to $(k_1 + 1, k_2 - 1)$ is $\frac{k_2}{M} = 1 - \frac{k_1}{M}$. This quantifies the idea that if more molecules are present in (say) chamber 1, then it is more likely for some molecule to transfer next from chamber 1 to chamber 2. This is a DTMC with state space $S = \{(0, M), (1, M - 1), \ldots, (M, 0)\}$ and the transition probabilities specified.

---

## 7.7.2   TRANSIENT ANALYSIS

Transient analysis of a DTMC involves the computation of $Pr(X_n = j)$, the probability of the Markov chain being in state $j$ after $n$ steps.

**Definitions:**

For $i, j \in S$ the **$n$-step transition probability** $p_{ij}^{(n)}$ is the probability of being in state $j$ after $n \geq 0$ steps, if the Markov chain starts in state $i$: $p_{ij}^{(n)} = Pr(X_n = j \mid X_0 = i)$.

The **$n$-step transition probability matrix** is given by $P^{(n)} = (p_{ij}^{(n)})$.

**Facts:**

1.  $Pr(X_n = j) = \sum_{i \in S} Pr(X_0 = i) \, Pr(X_n = j \mid X_0 = i) = \sum_{i \in S} a_i p_{ij}^{(n)}$.

2.  *Chapman-Kolmogorov equations:*  $P^{(n+m)} = P^{(n)} P^{(m)} = P^{(m)} P^{(n)}$ for all $m, n \geq 0$.

3.  If $P^n$ denotes the $n$th power of $P$, then $P^{(n)} = P^n$, $n \geq 0$.

4.  If $a$ is the initial distribution of a DTMC, the (absolute) probabilities $Pr(X_n = j)$ are the entries of the vector $aP^n$.

**Examples:**

1.  For Example 4 of §7.7.1, the two-step transition probability matrix is

$$P^{(2)} = P^2 = \begin{pmatrix} 0.7 & 0.3 \\ 0.4 & 0.6 \end{pmatrix} \begin{pmatrix} 0.7 & 0.3 \\ 0.4 & 0.6 \end{pmatrix} = \begin{pmatrix} 0.61 & 0.39 \\ 0.52 & 0.48 \end{pmatrix}.$$

Note that $P^{(2)}$ is again a stochastic matrix. To illustrate, if Friday is sunny then the conditional probability that the following Sunday is sunny is given by $p_{00}^{(2)} = 0.61$.

2.  A general two-state DTMC on $S = \{0, 1\}$ can be represented by the stochastic transition probability matrix

$$P = \begin{pmatrix} 1 - p & p \\ q & 1 - q \end{pmatrix}.$$

Direct calculation gives the two-step transition probability matrix

$$P^{(2)} = \begin{pmatrix} (1-p)^2 + pq & p(2 - p - q) \\ q(2 - p - q) & (1-q)^2 + pq \end{pmatrix},$$

which can be rewritten as

$$P^{(2)} = \frac{1}{p+q} \begin{pmatrix} q + p(1 - p - q)^2 & p - p(1 - p - q)^2 \\ q - q(1 - p - q)^2 & p + q(1 - p - q)^2 \end{pmatrix}.$$

In general,

$$P^{(n)} = \frac{1}{p+q} \begin{pmatrix} q + p(1 - p - q)^n & p - p(1 - p - q)^n \\ q - q(1 - p - q)^n & p + q(1 - p - q)^n \end{pmatrix}.$$

3.  *Limiting probabilities:*  Suppose in Example 2 that $0 < p < 1$ and $0 < q < 1$, so that $|1 - p - q| < 1$. From the final expression obtained in Example 2, it is seen that $P^{(n)}$ tends to the limiting matrix

$$\frac{1}{p+q} \begin{pmatrix} q & p \\ q & p \end{pmatrix}.$$

Consequently, if $a = (a_1, a_2)$ is any initial distribution, then the limiting probabilities $aP^n$ approach

$$\frac{1}{p+q} \begin{pmatrix} a_1 & a_2 \end{pmatrix} \begin{pmatrix} q & p \\ q & p \end{pmatrix} = \frac{1}{p+q} \begin{pmatrix} q & p \\ q & p \end{pmatrix}$$

since $a_1 + a_2 = 1$. For example, the limiting probability of being in state 0 is $\frac{q}{p+q}$, independent of the initial state of the DTMC. (See §7.7.4.)

## 7.7.3   CLASSIFICATION OF STATES

### Definitions:

State $j \in S$ is **accessible** from state $i \in S$ (written $i \to j$) if it is possible to make a sequence of transitions leading from state $i$ to state $j$: that is, $p_{ij}^{(n)} > 0$ for some $n \geq 0$.

States $i, j \in S$ **communicate** (written $i \leftrightarrow j$) if they are mutually accessible from one another: $i \to j$ and $j \to i$.

Set $C \subseteq S$ is a (maximal) **communicating class** if

- $i, j \in C \Rightarrow i \leftrightarrow j$;
- $i \in C, i \leftrightarrow j \Rightarrow j \in C$.

A communicating class $C$ is **closed** if transitions from the states of $C$ never lead to states outside $C$: $i \in C, j \notin C \Rightarrow j$ is not accessible from $i$.

A DTMC is **irreducible** if $i \leftrightarrow j$ for all $i, j \in S$; otherwise it is **reducible**.

For $j \in S$ define

- $T_j = \min\{n > 0 \mid X_n = j\}$;
- $f_j = Pr(T_j < \infty \mid X_0 = j)$;
- $f_j(n) = Pr(T_j = n \mid X_0 = j)$;
- $m_j = E(T_j \mid X_0 = j)$.

State $j \in S$ is **recurrent** if return to that state is certain: $f_j = 1$; if $f_j < 1$ then state $j$ is **transient**. A recurrent state $j \in S$ is **positive recurrent** if $m_j < \infty$ and **null recurrent** if $m_j = \infty$.

A recurrent state $j$ has **period** $d$ if $d$ is the largest integer satisfying $\sum_{n=0}^{\infty} f_j(nd) = 1$. If $d = 1$ then state $j$ is **aperiodic**.

### Facts:

**1.** Generally, all classes that are not closed can be lumped into a single set $T$ of transient states. Thus, the state space of a DTMC can be partitioned into closed classes $C_1, C_2, \ldots, C_K$ and the set $T$.

**2.** State $j$ is accessible from state $i$ if and only if there is a directed path from vertex $i$ to vertex $j$ in the transition diagram of the DTMC.

**3.** A set of states $C$ is a communicating class if and only if the corresponding set of vertices forms a strongly connected component (§8.3.2) in the transition diagram.

**4.** Tarjan [Ta72] describes an algorithm to find the strongly connected components, which runs in time linear in the number of arcs in the transition diagram (i.e., the number of nonzero entries of $P$).

**5.** Transience, positive recurrence, and null recurrence are class properties:

- if $i$ is transient and $i \leftrightarrow j$ then $j$ is transient;
- if $i$ is positive recurrent and $i \leftrightarrow j$ then $j$ is positive recurrent;
- if $i$ is null recurrent and $i \leftrightarrow j$ then $j$ is null recurrent.

In other words, states in a communicating class are all simultaneously transient or null recurrent or positive recurrent.

**6.** By Fact 5, a communicating class or an irreducible DTMC can be termed positive recurrent, null recurrent, or transient if all of its states are positive recurrent, null recurrent, or transient.

**7.** A finite communicating class is positive recurrent if it is closed, and transient otherwise.

**8.** A finite state irreducible DTMC is positive recurrent.

**9.** Null recurrent states do not occur in a finite state DTMC.

**10.** Establishing recurrence or transience in an infinite state DTMC is a difficult task and has to be done on a case-by-case basis.

**11.** We have not defined period for a transient state since for such a state the concept is not needed. Some references do however define period for all states.

**12.** The period of state $j$ is the greatest common divisor of all integers $n \geq 0$ such that $p_{jj}^{(n)} > 0$.

**13.** The period of state $j$ is the greatest common divisor of all the lengths of the directed cycles in the transition diagram that contain state $j$.

**14.** Periodicity is a class property: if $i$ has period $d$ and $i \leftrightarrow j$, then $j$ has period $d$.

**15.** The period of a state in a finite irreducible DTMC is at most equal to the number of states in the DTMC.

**16.** By Fact 14, a recurrent communicating class or a recurrent irreducible DTMC can be termed periodic if all states in it are periodic with $d > 1$, else it is termed aperiodic.

**Examples:**
**1.** For the DTMC in Example 1 of §7.7.1, it is seen that $1 \to 2$, $2 \to 1$, and $1 \leftrightarrow 2$. However, 3 is not accessible from 1. The communicating classes are $C_1 = \{1, 2\}$, $C_2 = \{3, 4\}$, $C_3 = \{5\}$, $C_4 = \{6\}$. This DTMC is reducible. Classes $C_1, C_2, C_3$ are closed, but $C_4$ is not. States $1, 2, 3, 4, 5$ are positive recurrent and state 6 is transient. Classes $C_1, C_2, C_3$ are positive recurrent.

**2.** Consider the random walk in Example 2 of §7.7.1 with $0 < p < 1$. There are three communicating classes: $C_1 = \{0\}$, $C_2 = \{1, 2, ..., N - 1\}$, and $C_3 = \{N\}$. This DTMC is reducible. Here $C_1$ and $C_3$ are closed, while $C_2$ is not. States 0 and $N$ are positive recurrent. The rest are transient.

**3.** For the DTMC in Example 1 of §7.7.1, states 3 and 4 have period 2; states 1, 2, and 5 are aperiodic. A period is not associated with state 6 since it is transient. Classes $\{1, 2\}$ and $\{5\}$ are aperiodic, while the class $\{3, 4\}$ is periodic with period 2.

**4.** For the DTMC in Example 2 of §7.7.1, states 0 and $N$ are aperiodic. Period is not defined for the rest of the states as they are transient.

**5.** Example 3 of §7.7.1 is an irreducible chain. All states are positive recurrent and have period 2.

## 7.7.4  LIMITING BEHAVIOR

To establish possible equilibrium configurations of DTMCs, it is necessary to study the behavior of the $n$-step transition probabilities $Pr(X_n = j \mid X_0 = i)$ as $n \to \infty$.

**Facts:**

**1.** Let $\{X_n \mid n \geq 0\}$ be an irreducible DTMC with transition probability matrix $P$ and finite state space $S$. Then there exists a unique solution $\pi = (\pi_j)$ to the equations

$$\pi = \pi P, \quad \sum_{j \in S} \pi_j = 1.$$

**2.** The long run fraction of the visits to state $j$ is given by $\pi_j$, regardless of the initial state. Also, $m_j$, the expected time between two consecutive visits to state $j$, is $\frac{1}{\pi_j}$.

**3.** If the DTMC is aperiodic, then $\lim_{n \to \infty} Pr(X_n = j \mid X_0 = i) = \pi_j$ for all $i \in S$.

**4.** Let $\{X_n \mid n \geq 0\}$ be a finite state reducible DTMC with $K$ closed communicating classes $C_1, C_2, \ldots, C_K$ and the set of transient states $T$. Then $\lim_{n \to \infty} Pr(X_n = j \mid X_0 = i) = \pi_{ij}$, where

    (a) if $j \in T$, then $\pi_{ij} = 0$;

    (b) if $i$ and $j$ belong to different closed classes, then $\pi_{ij} = 0$;

    (c) if $i$ and $j$ belong to the same closed class $C_r$, then $\pi_{ij} = \pi_j$, where $\{\pi_j\}$ are the limiting probabilities calculated by using Fact 1 for the irreducible DTMC formed by the states in $C_r$;

    (d) if $i \in T$ and $j \in C_r$, then $\pi_{ij} = \alpha_{ir}\pi_j$, where $\{\pi_j\}$ are as in (c) and $\alpha_{ir}$ is the probability that the DTMC eventually visits the class $C_r$ starting from state $i$.

**5.** In Fact 4(c), if $C_r$ is periodic then limiting probabilities do not exist and $\pi_j$ is interpreted as the long run fraction of the time the DTMC spends in state $j$ starting from state $i$.

**6.** The $\{\alpha_{ir}\}$ in Fact 4(d) are given by the unique solution to

$$\alpha_{ir} = \sum_{j \in C_r} p_{ij} + \sum_{j \in T} p_{ij}\alpha_{jr}.$$

**Examples:**

**1.** The DTMC in Example 3 of §7.7.1 is irreducible and periodic with $d = 2$. Using Fact 1, its limiting behavior is described by the equations

$$\pi_0 = q\pi_1$$
$$\pi_1 = \pi_0 + q\pi_2$$
$$\pi_i = p\pi_{i-1} + q\pi_{i+1} \quad \text{for } 2 \leq i \leq N - 2$$
$$\pi_{N-1} = p\pi_{N-2} + \pi_N$$
$$\pi_N = p\pi_{N-1}$$

and $\sum_{j=0}^{N} \pi_j = 1$. Solving these equations gives $\pi_j = \frac{\rho_j}{c}$, where

$$\rho_0 = 1$$
$$\rho_j = \frac{(\frac{p}{q})^j}{p} \quad \text{for } 1 \leq j \leq N - 1$$
$$\rho_N = \left(\frac{p}{q}\right)^{N-1}$$

and the normalizing constant is $c = \sum_{j=0}^{N} \rho_j$. This DTMC is periodic and hence these $\pi_j$ represent the long run fraction of the time the DTMC spends in state $j$. Here $\lim_{n \to \infty} Pr(X_n = j \mid X_0 = i)$ does not exist since the probabilities under question keep oscillating with period 2.

**2.** For the DTMC in Example 1 of §7.7.1, $C_1 = \{1,2\}$, $C_2 = \{3,4\}$, $C_3 = \{5\}$, and $T = \{6\}$. Therefore, $\pi_1 = \frac{7}{13}$, $\pi_2 = \frac{6}{13}$, $\pi_3 = \frac{1}{2}$, $\pi_4 = \frac{1}{2}$, $\pi_5 = 1$, $\alpha_{61} = \frac{2}{5}$, $\alpha_{62} = \frac{2}{5}$, $\alpha_{63} = \frac{1}{5}$. By Fact 4 the limiting matrix $(\pi_{ij})$ is given by

$$
\begin{pmatrix}
\frac{7}{13} & \frac{6}{13} & 0 & 0 & 0 & 0 \\
\frac{7}{13} & \frac{6}{13} & 0 & 0 & 0 & 0 \\
0 & 0 & \frac{1}{2} & \frac{1}{2} & 0 & 0 \\
0 & 0 & \frac{1}{2} & \frac{1}{2} & 0 & 0 \\
0 & 0 & 0 & 0 & 1 & 0 \\
\frac{14}{65} & \frac{12}{65} & \frac{1}{5} & \frac{1}{5} & \frac{1}{5} & 0
\end{pmatrix}.
$$

States 3 and 4 are periodic, and hence the third and fourth columns need to be interpreted as the long run fraction of the time the discrete-time Markov chain spends in those states.

**3.** The Ehrenfest diffusion model (§7.7.1, Example 6) is an irreducible DTMC with period 2. The solution $\pi$ (Fact 1) is given by $\pi_j = \binom{M}{j}2^{-M}$ $(0 \le j \le M)$. The binomial distribution (§7.3.1) describes the long run fraction of time the system spends in each state.

## 7.7.5 FIRST PASSAGE TIMES

### Definitions:

Let $\{X_n \mid n \ge 0\}$ be a DTMC on state space $S$ with transition probability matrix $P$. Let $A \subseteq S$ be a given subset of states.

The **first passage time** $T_A$ into set $A$ is the first time at which the Markov chain reaches some state in set $A$; i.e., $T_A = \min\{n \ge 0 \mid X_n \in A\}$.

For $i \in S$, let $\alpha_i = Pr(T_A < \infty \mid X_0 = i)$ and let $\tau_i = E(T_A \mid X_0 = i)$.

### Facts:

**1.** The $\{\alpha_i\}$ are given by the unique solution to

$$
\alpha_i = \sum_{j \in S} p_{ij}\alpha_j
$$

with the boundary conditions $\alpha_i = 1$ if $i \in A$ and $\alpha_i = 0$ if no state in $A$ is accessible from $i$.

**2.** If $\alpha_i = 1$ for all $i \in S$, then $\{\tau_i\}$ are given by the unique solution to

$$
\tau_i = 1 + \sum_{j \in S} p_{ij}\tau_j
$$

with the boundary condition $\tau_i = 0$ if $i \in A$.

### Examples:

**1.** Consider the DTMC in §7.7.1, Example 2 with $A = \{0\}$. The equations of Fact 1 are $\alpha_i = q\alpha_{i-1} + p\alpha_{i+1}$, $1 \le i \le N-1$ with the boundary conditions $\alpha_0 = 1$ and $\alpha_N = 0$. If $q \ne p$, the solution is given by

$$
\alpha_i = \frac{(\frac{q}{p})^i - (\frac{q}{p})^N}{1 - (\frac{q}{p})^N}, \quad 0 \le i \le N.
$$

If $q = p$, the solution is $\alpha_i = 1 - \frac{i}{N}$.

**2.** Consider the DTMC in §7.7.1, Example 2 with $A = \{0, N\}$. In this case $\alpha_i = 1$ for all $i$. The equations of Fact 2 are $\tau_i = 1 + q\tau_{i-1} + p\tau_{i+1}$, $1 \le i \le N-1$, with the boundary conditions $\tau_0 = 0$ and $\tau_N = 0$. If $q \ne p$, the solution is given by

$$\tau_i = \frac{i}{q-p} - \frac{N}{q-p}\frac{1-(\frac{q}{p})^i}{1-(\frac{q}{p})^N}, \quad 0 \le i \le N.$$

If $q = p$, the solution is given by $\tau_i = i(N-i)$.

---

## 7.7.6  BRANCHING PROCESSES

Branching processes are a special type of Markov chain used to study the growth (and possible extinction) of populations in biology and sociology as well as particles in physics.

### Definitions:
Suppose $\{Y_{ni} \mid n, i \ge 1\}$ are independent and identically distributed random variables having common probability distribution function $p_k = Pr(Y_{ni} = k)$, $k \ge 0$, with mean $m$ and variance $\sigma^2$. Then the DTMC $\{X_n \mid n \ge 0\}$ is a **branching process** if $X_0 = 1$,

$$X_{n+1} = \sum_{i=1}^{X_n} Y_{ni}.$$

A branching process is **stable** if $m < 1$, **critical** if $m = 1$, and **unstable** if $m > 1$.

The **extinction probability** of a branching process is the probability that the population becomes extinct, where $X_0 = 1$.

### Facts:
**1.** It is convenient to think of the random variable $X_n$ as the number of individuals in the $n$th generation and the random variable $Y_{ni}$ as the number of offspring of the $i$th individual in the $n$th generation.

**2.** The transient behavior of the branching process is given by

$$E(X_n) = m^n,$$

$$\text{Var}(X_n) = \begin{cases} n\sigma^2, & \text{if } m = 1 \\ \sigma^2 m^{n-1}\dfrac{m^n - 1}{m - 1}, & \text{if } m \ne 1. \end{cases}$$

**3.** State 0 is absorbing for a branching process. Absorption in state 0 is certain if and only if $m \le 1$, while the expected time until extinction (i.e., absorption in state 0) is finite if and only if $m < 1$.

**4.** The probability of extinction $\rho$ in an unstable branching process is given by the unique solution in $(0, 1)$ to the equation

$$\rho = \sum_{n=0}^{\infty} p_n \rho^n.$$

**5.** The expected total number of individuals ever born in a stable branching process until it becomes extinct is given by

$$E\left(\sum_{n=0}^{\infty} X_n\right) = \frac{1}{1 - m}.$$

**6.** There is no simple expression for the expected time until absorption for a general stable branching process.

**Examples:**

**1.** The branching process with $p_0 = \frac{1}{2}$, $p_1 = \frac{1}{8}$, $p_2 = \frac{3}{8}$ has mean $m = \frac{7}{8} < 1$ and is stable. With probability 1, the population will die out.

**2.** The branching process with $p_0 = \frac{1}{4}$, $p_1 = \frac{1}{4}$, $p_2 = \frac{1}{2}$ has mean $m = \frac{5}{4} > 1$ and is unstable. The probability of extinction $\rho_0$ is found as the smallest positive root of the equation $\rho = \frac{1}{4} + \frac{1}{4}\rho + \frac{1}{2}\rho^2$. The roots of this equation are $\frac{1}{2}$ and 1, so the probability of extinction is $\rho_0 = \frac{1}{2}$. If the initial population is $X_0 = 10$ instead of $X_0 = 1$, then the probability that the initial population eventually becomes extinct is $\rho_0^{10} = \frac{1}{1024}$.

## 7.8   HIDDEN MARKOV MODELS

A Hidden Markov Model (HMM) is a statistical tool used in modeling discrete-time stochastic systems. The theory of HMMs assumes that the system under consideration has only a finite number of states. Still the HMM is a very powerful tool because of its dual stochastic nature (i.e., the probabilistic modeling of both inter-state transitions and intra-state processes). This property makes it suitable for modeling systems that change their statistical properties over time. HMMs are widely used in speech recognition and have found applications in several other areas such as bioinformatics, gene sequence modeling, economic/financial analysis, speech synthesis, and web usage analysis.

### 7.8.1   BASIC CONCEPTS

**Definitions:**

A **Hidden Markov Model** (**HMM**) consists of a finite set $S = \{1, 2, \ldots, N\}$ of **states**, where each state $j \in S$ is associated with a (generally multidimensional) probability distribution $b_j(o)$. In a particular state, at a given time $t \geq 1$, an outcome or **observation** $o_t$ can be generated, according to the associated probability distribution.

Transitions among the states are governed by a set of **transition probabilities** $a_{ij} = Pr(q_{t+1} = j | q_t = i)$, where $q_t \in S$ is a random variable denoting the state at time $t$. The matrix $A = (a_{ij})$ is the **transition probability matrix** of the HMM.

The **initial state distribution** of the HMM is the vector $\pi = (\pi_i)$, where $\pi_i = Pr(q_1 = i)$, $i \in S$.

While a transition from $q_t$ to the next state $q_{t+1}$ can in general depend on all past states, the model is simplified by truncating the dependency to only the past $k$ states. This gives a **$k$th-order HMM** in which $Pr(q_{t+1} = j | q_t = i_t, q_{t-1} = i_{t-1}, \ldots, q_{t-k+1} = i_{t-k+1}) = Pr(q_{t+1} = j | q_t = i_t, q_{t-1} = i_{t-1}, \ldots, q_1 = i_1)$, where $i_1, i_2, \ldots, i_t, j \in S$. If $k = 1$, this HMM has the **Markov property**:

$$a_{ij} = Pr(q_{t+1} = j | q_t = i) = Pr(q_{t+1} = j | q_t = i_t, q_{t-1} = i_{t-1}, \ldots, q_1 = i_1),$$

for all $i_1, i_2, \ldots, i_t, j \in S$.

The HMM is **stationary** or **time homogeneous** if $Pr(q_{s+1} = j | q_s = i) = Pr(q_{t+1} = j | q_t = i)$ for all $s, t \geq 1$.

The HMM has the **conditional independence property** if $Pr(O|q_1, q_2, \ldots, q_T, \lambda) = \prod_{t=1}^{T} Pr(o_t|q_t, \lambda)$, where $O = (o_1, o_2, \ldots, o_T)$ is a sequence of observations, $q_t \in S$, and $\lambda$ denotes the parameter set of the HMM.

The HMM is a **Discrete Hidden Markov Model** (**DHMM**) if all the observations it can generate can be drawn from a finite set. That is, $o_t \in \{\nu_1, \nu_2, \ldots, \nu_M\}$ for all $t \geq 1$, where $\nu_k$, $k \in \mathcal{M} = \{1, 2, \ldots, M\}$, denotes the $k$th observation symbol in the alphabet. Then the probability distribution associated with each state can be represented by the finite set of probabilities $b_j(o) = b_{jk} = Pr(o = \nu_k|q_t = j)$, where $j \in S$, $k \in \mathcal{M}$, and $o$ is the current observation. The matrix $B = (b_{jk})$ is termed the **observation probability matrix** of the DHMM.

The HMM is a **Continuous Density Hidden Markov Model** (**CDHMM**) if it can generate all the observations within a given (finite or infinite) interval. In this case the observation probability distribution can be modeled analytically using a density function such as a multidimensional Gaussian distribution

$$N(\mu, \Sigma, o) = (2\pi)^{-\frac{D}{2}} |\Sigma|^{-\frac{1}{2}} \exp\left(-\tfrac{1}{2}(o - \mu)^T \Sigma^{-1} (o - \mu)\right)$$

where $D$ is the dimension of the observation vector, $\mu$ is the distribution mean, and $\Sigma$ is the distribution covariance matrix.

**Facts:**

**1.** A discrete-time Markov chain (DTMC) [see §7.7] can be extended to an HMM by associating a probability distribution with each state of the DTMC.

**2.** Unlike a DTMC, in the case of an HMM, only the outcome $o_t$, not the state $q_t$, is visible to an external observer and therefore states are "hidden" to the outside. This accounts for the name *Hidden Markov Model*.

**3.** A DHMM can be completely defined by the 3-tuple $\lambda = (A, B, \pi)$, where $A$, $B$, and $\pi$ are the transition probability matrix, the observation probability matrix, and the initial state distribution, respectively.

**4.** For a CDHMM a weighted sum of a finite number $M$ of Gaussians is often used to model the distributions $b_j(o)$ as

$$b_j(o) = \sum_{m=1}^{M} c_{jm} N(\mu_{jm}, \Sigma_{jm}, o)$$

for all $j \in S$, where $c_{jm}$, $m \in \mathcal{M}$, is the weighting coefficient.

**5.** A CDHMM can be completely defined by the 5-tuple $\lambda = (A, C, \mu^*, \Sigma^*, \pi)$, where $A$ and $\pi$ are as in Fact 3, and the weighting coefficient matrix, the set of mean vectors, and the set of covariance matrices are given by $C = (c_{jm})$, $\mu^* = \{\mu_{jm}|j \in S, m \in \mathcal{M}\}$, and $\Sigma^* = \{\Sigma_{jm}|j \in S, m \in \mathcal{M}\}$.

**6.** The initial state probabilities satisfy $\sum_{i=1}^{N} \pi_i = 1$.

**7.** The transition probabilities satisfy $a_{ij} \geq 0$, $i, j \in S$ and $\sum_{j=1}^{N} a_{ij} = 1$, $i \in S$.

**8.** The observation probabilities for a DHMM satisfy $b_{jk} \geq 0$, $j \in S$, $k \in \mathcal{M}$ and $\sum_{k=1}^{M} b_{jk} = 1$, $j \in S$.

**9.** The weighting coefficients for a CDHMM satisfy $c_{jm} \geq 0$, $j \in S$, $m \in M$ and $\sum_{m=1}^{M} c_{jm} = 1$, $j \in S$.

**Examples:**

**1.** Suppose that an experimenter tosses one of three (possibly biased) coins and reports to the observer only the outcome Heads ($H$) or Tails ($T$). This situation can be modeled as a DHMM in which the state $j$ refers to coin $j$, which has probability $p_j$ of producing Heads. The experimenter successively chooses coins according to a DTMC with transition probabilities $a_{ij}$, where $i, j \in \{1, 2, 3\}$. For coin $j$ the observation probabilities for $H$ and $T$ are $b_j(H) = p_j$ and $b_j(T) = 1 - p_j$, respectively. This DHMM has $N = 3$ states and $M = 2$ observation symbols.

**2.** Suppose that there are $N$ urns, each containing some mix of $M$ colored balls. At each time step, a ball is randomly selected from one of the urns and its color is revealed to the observer. The ball is then returned to that urn. However, the observer does not know what urn was sampled.

This situation can be modeled as a DHMM in which the state corresponds to the urn sampled, with state transition probabilities indicating the next urn sampled. For a given urn, the mixture of colors dictates the observation probabilities for the $M$ possible colors. Notice that the states (urns) are hidden from view, and one only is allowed to observe the color drawn.

**3.** Suppose that we are interested in determining overall annual temperatures (Hot or Cold) at a specific location at some point in the distant past. However, we do not have records of the temperatures (say over $T$ years) but do have samples of trees harvested (and preserved) over that time period. In particular, we can observe the size of tree rings $o_1, o_2, \ldots, o_T$ over that time period, where each $o_t \in \{\text{small, medium, large}\}$. It is reasonable that temperature (Hot, Cold) may follow a 2-state DTMC, with transition probabilities indicating that hot years are more likely to follow hot years, and similarly for cold years. Likewise, the size of tree rings may be presumed to be correlated with overall annual temperature. This situation can be modeled as a DHMM, since we only observe the size of tree rings, but not the state of the system (temperature).

As a specific example, the following figure shows the state transition diagram for the Markov chain on states Hot and Cold. Below this are the corresponding columns of the matrix $B = (b_{jk})$. The observations $o_1, o_2, o_3$ indicate respectively the ring sizes (small, medium, large).

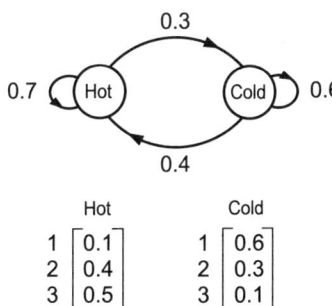

**4.** A variety of additional applications, together with reference sources, are given in the following table.

| application | references |
|---|---|
| biological sequencing | [DuEtal98] |
| computer vision | [BuCa01] |
| economic analysis | [BhHa04] |
| handwriting recognition | [GuScDe97] |
| power systems | [SaGu93] |
| protein folding | [StEtal11] |
| speech recognition | [GaYo07] |
| speech synthesis | [ToEtal13] |
| web usage | [SaTa12] |

## 7.8.2  THE EVALUATION PROBLEM

This subsection discusses the problem of determining the likelihood of obtaining the observed data, given the specification of the parameters of some given HMM. This provides a measure for scoring how well a particular model matches the observation sequence. For simplicity, the focus here is on DHMMs, with specified parameter set $\lambda = (A, B, \pi)$.

**Definitions:**

Assume that a given sequence of observations $O = (o_1, o_2, \ldots, o_T)$ has been generated by a given HMM with parameter set $\lambda$. The **evaluation problem** consists of determining the probability $Pr(O|\lambda)$.

A **state** $Q$ of the system is given by $Q = (q_1, q_2, \ldots q_T)$ and the set of all system states is denoted $\Omega$. **Direct evaluation** of $Pr(O|\lambda)$ is obtained by enumerating all states of the system: $Pr(O|\lambda) = \sum_{Q \in \Omega} Pr(O, Q|\lambda) = \sum_{Q \in \Omega} Pr(O|Q, \lambda) \, Pr(Q|\lambda)$.

The **forward algorithm** for computing $Pr(O|\lambda)$ makes use of the **forward variables** $\alpha_t(i) = Pr(o_1, o_2, \ldots, o_t, q_t = i|\lambda)$, which can be interpreted as the probability of the partial observation sequence $o_1, o_2, \ldots, o_t$, when it terminates at the state $i$.

The **backward algorithm** for computing $Pr(O|\lambda)$ makes use of the **backward variables** $\beta_t(i) = Pr(o_{t+1}, o_{t+2}, \ldots, o_T|q_t = i, \lambda)$, which can be interpreted as the probability of the partial observation sequence $o_{t+1}, o_{t+2}, \ldots, o_T$, given that the current state is $i$.

**Facts:**

1. In the direct evaluation method, $Pr(Q|\lambda)$ can be calculated using $\pi$ and $A$.

2. In the direct evaluation method, $Pr(O|Q, \lambda)$ can be calculated using $B$.

3. Since the state space $\Omega$ has size $O(N^T)$, direct evaluation of $Pr(O|\lambda)$ has $O(TN^T)$ computational complexity and so is feasible only if $N$ and $T$ are very small.

4. The forward variables $\alpha_t(j)$ satisfy the recursive relationship

$$\alpha_{t+1}(j) = b_j(o_{t+1}) \sum_{i=1}^{N} \alpha_t(i) a_{ij}, \; j \in S, \; t \in \{1, 2, \ldots, T-1\},$$

with $\alpha_1(j) = \pi_j b_j(o_1), \; j \in S$.

**5.** The conditional probability of the observation sequence given the HMM is related to the forward variables through

$$Pr(O|\lambda) = \sum_{i=1}^{N} Pr(o_1, o_2, \ldots, o_T, q_T = i|\lambda) = \sum_{i=1}^{N} \alpha_T(i).$$

**6.** The forward algorithm, shown as Algorithm 1, is based on Facts 4 and 5.

---

**Algorithm 1:  Forward algorithm.**

input: $A, B, \pi$

output: $Pr(O|\lambda)$

**for** $j := 1$ **to** $N$

   $\alpha_1(j) := \pi_j b_j(o_1)$

**for** $t := 1$ **to** $T - 1$

   **for** $j := 1$ **to** $N$

      $\alpha_{t+1}(j) := b_j(o_{t+1}) \sum_{i=1}^{N} \alpha_t(i) a_{ij}$

$Pr(O|\lambda) := \sum_{i=1}^{N} \alpha_T(i)$

---

**7.** The computational complexity of the forward algorithm is $O(N^2 T)$, which is linear in $T$ compared to the exponential complexity of direct evaluation (Fact 3).

**8.** The backward variables $\beta_t(i)$ satisfy the recursive relationship

$$\beta_t(i) = \sum_{j=1}^{N} \beta_{t+1}(j) a_{ij} b_j(o_{t+1}), \ i \in S, \ t \in \{T - 1, T - 2, \ldots, 1\},$$

with $\beta_T(i) = 1$, $i \in S$.

**9.** The conditional probability of the observation sequence given the HMM is related to the backward variables through

$$Pr(O|\lambda) = \sum_{i=1}^{N} \pi_i b_i(o_1) \beta_1(i)$$

since

$$\begin{aligned} \pi_i b_i(o_1) \beta_1(i) &= Pr(q_1 = i) \, Pr(o_1|q_1 = i) \, Pr(o_2, o_3, \ldots, o_T|q_1 = i, \lambda) \\ &= Pr(o_1, o_2, \ldots, o_T, q_1 = i|\lambda). \end{aligned}$$

**10.** The backward algorithm, shown as Algorithm 2, is based on Facts 8 and 9.

---

**Algorithm 2:  Backward algorithm.**

input: $A, B, \pi$

output: $Pr(O|\lambda)$

**for** $i := 1$ **to** $N$

   $\beta_T(i) := 1$

**for** $t := T - 1$ **to** $1$

   **for** $i := 1$ **to** $N$

      $\beta_t(i) := \sum_{j=1}^{N} \beta_{t+1}(j) a_{ij} b_j(o_{t+1})$

$Pr(O|\lambda) := \sum_{i=1}^{N} \pi_i b_i(o_1) \beta_1(i)$

---

**11.** The computational complexity of the backward algorithm is also $O(N^2T)$.

**12.** Using the definitions of $\alpha_t(i)$ and $\beta_t(i)$ together with the conditional independence assumption, it can be shown that

$$\alpha_t(i)\beta_t(i) = Pr(O, q_t = i|\lambda), \ i \in S, \ t \in \{1, 2, \ldots, T\}.$$

**13.** Fact 11 gives another way to calculate $Pr(O|\lambda)$, using both forward and backward variables:

$$Pr(O|\lambda) = \sum_{i=1}^{N} Pr(O, q_t = i|\lambda) = \sum_{i=1}^{N} \alpha_t(i)\beta_t(i), \ t \in \{1, 2, \ldots, T\}.$$

**Examples:**

**1.** We use the direct evaluation approach to calculate the probability of observing the sequence (medium, large, small) for Example 3 in §7.8.1. For ease of representation, the observations are encoded using small = 1, medium = 2, and large = 3, so the observed sequence is $2, 3, 1$. Also, we encode the states as Hot = 1 and Cold = 2. In this problem, the initial state probabilities $\pi = (0.8, 0.2)$ are used.

To illustrate, for state $Q = $ Hot Hot Cold = 112, we calculate $Pr(Q|\lambda) = Pr(112|\lambda) = (0.8)(0.7)(0.3) = 0.168$ and $Pr(O|Q, \lambda) = Pr(231|Q, \lambda) = (0.4)(0.5)(0.6) = 0.12$. Consequently, $Pr(O, Q|\lambda) = Pr(O|Q, \lambda) Pr(Q|\lambda) = (0.12)(0.168) = 0.02016$. Similar calculations are shown in the following table. Adding the entries in the last column gives $Pr(O|\lambda) = 0.03628$.

| $Q$ | $Pr(Q|\lambda)$ | $Pr(O|Q, \lambda)$ | $Pr(O, Q|\lambda)$ |
|-----|-----------------|--------------------|--------------------|
| 111 | 0.392 | 0.02 | 0.00784 |
| 112 | 0.168 | 0.12 | 0.02016 |
| 121 | 0.096 | 0.004 | 0.000384 |
| 122 | 0.144 | 0.024 | 0.003456 |
| 211 | 0.056 | 0.015 | 0.00084 |
| 212 | 0.024 | 0.09 | 0.00216 |
| 221 | 0.048 | 0.003 | 0.000144 |
| 222 | 0.072 | 0.018 | 0.001296 |

**2.** We illustrate the application of the forward algorithm to the same problem studied in Example 1.

$\alpha_1(1) = \pi_1 b_1(2) = (0.8)(0.4) = 0.32, \ \alpha_1(2) = \pi_2 b_2(2) = (0.2)(0.3) = 0.06;$
$\alpha_2(1) = b_1(3)[\alpha_1(1)a_{11} + \alpha_1(2)a_{21}] = 0.5[(0.32)(0.7) + (0.06)(0.4)] = 0.124;$
$\alpha_2(2) = b_2(3)[\alpha_1(1)a_{12} + \alpha_1(2)a_{22}] = 0.1[(0.32)(0.3) + (0.06)(0.6)] = 0.0132;$
$\alpha_3(1) = b_1(1)[\alpha_2(1)a_{11} + \alpha_2(2)a_{21}] = 0.1[(0.124)(0.7) + (0.0132)(0.4)] = 0.009208;$
$\alpha_3(2) = b_2(1)[\alpha_2(1)a_{12} + \alpha_2(2)a_{22}] = 0.6[(0.124)(0.3) + (0.0132)(0.6)] = 0.027072.$

As a result, $Pr(O|\lambda) = 0.009208 + 0.027072 = 0.03628.$

**3.** We illustrate the application of the backward algorithm to the same problem studied in Example 1.

$\beta_3(1) = 1, \ \beta_3(2) = 1;$
$\beta_2(1) = \beta_3(1)a_{11}b_1(o_3) + \beta_3(2)a_{12}b_2(o_3) = 1(0.7)(0.1) + 1(0.3)(0.6) = 0.25;$
$\beta_2(2) = \beta_3(1)a_{21}b_1(o_3) + \beta_3(2)a_{22}b_2(o_3) = 1(0.4)(0.1) + 1(0.6)(0.6) = 0.4;$

$$\beta_1(1) = \beta_2(1)a_{11}b_1(o_2) + \beta_2(2)a_{12}b_2(o_2) = (0.25)(0.7)(0.5) + (0.4)(0.3)(0.1) = 0.0995;$$
$$\beta_1(2) = \beta_2(1)a_{21}b_1(o_2) + \beta_2(2)a_{22}b_2(o_2) = (0.25)(0.4)(0.5) + (0.4)(0.6)(0.1) = 0.074.$$

Then $Pr(O|\lambda) = \pi_1 b_1(o_1)\beta_1(1) + \pi_2 b_2(o_1)\beta_1(2) = (0.8)(0.4)(0.0995) + (0.2)(0.3)(0.074) = 0.03628.$

---

### 7.8.3  THE DECODING PROBLEM

This subsection discusses the problem of finding a state sequence that best explains the observed data, given the specification of the parameters of an HMM. The focus here is on DHMMs, with parameter set $\lambda = (A, B, \pi)$.

**Definitions:**
Given an HMM with parameter set $\lambda$ and a sequence of observations $O = (o_1, o_2, \ldots, o_T)$, the **decoding problem** deals with finding a state sequence $Q = (q_1.q_2, \ldots, q_T)$ that maximizes the conditional probability $Pr(Q|O, \lambda)$.

The **direct approach** to the decoding problem proceeds by calculating $Pr(Q|O, \lambda)$ for all possible state sequences and picking one that achieves the highest value.

The **Viterbi algorithm** for solving the decoding problem makes use of the auxiliary variable

$$\delta_t(i) = \max_{q_1, q_2, \ldots, q_{t-1}} Pr(q_1, q_2, \ldots, q_{t-1}, q_t = i, o_1, o_2, \ldots, o_t|\lambda),$$

which is the largest joint probability that the partial observation sequence and the state sequence can have up to time $t$, when the current state is $i$.

**Facts:**
**1.** Since the observation sequence $O$ is given, the decoding problem can be solved by finding a state sequence $Q$ that maximizes the joint probability $Pr(Q, O|\lambda)$.
**2.** Since the state space $\Omega$ has size $O(N^T)$, the direct approach has exponential complexity.
**3.** From the Markov property and the conditional independence property of the HMM, the following recursion holds:

$$\delta_{t+1}(j) = b_j(o_{t+1})[\max_{i=1,\ldots,N} \delta_t(i)a_{ij}], \; j \in S, \; t \in \{1, 2, \ldots, T-1\},$$

with $\delta_1(j) = \pi_j b_j(o_1)$, $j \in S$.
**4.** The Viterbi algorithm (Algorithm 3), based on Facts 1 and 3, produces a state sequence $Q^*$ that maximizes the joint probability $Pr(Q, O|\lambda)$.

---

**Algorithm 3**: Viterbi algorithm.
input: $A, B, \pi, O$
output: state $Q^* = (q_1^*, q_2^*, \ldots, q_T^*)$ that maximizes $Pr(Q, O|\lambda)$, and the maximum joint probability $p^*$.
**for** $j := 1$ **to** $N$
$\quad \delta_1(j) := \pi_j b_j(o_1)$
$\quad \psi_1(j) := 0$

$$\boxed{\begin{aligned}
&\textbf{for } t := 1 \textbf{ to } T-1 \\
&\quad \textbf{for } j := 1 \textbf{ to } N \\
&\qquad \delta_{t+1}(j) := b_j(o_{t+1})[\max_{i=1,\ldots,N} \delta_t(i)a_{ij}] \\
&\qquad \psi_{t+1}(j) := \operatorname{argmax}_{i=1,\ldots,N} \delta_t(i)a_{ij} \\
&p^* := \max_{j=1,\ldots,N} \delta_T(j) \\
&q_T^* := \operatorname{argmax}_{j=1,\ldots,N} \delta_T(j) \\
&\textbf{for } t := T \textbf{ to } 2 \quad \{\text{Backtrack to find the state } Q^*\} \\
&\quad q_{t-1}^* := \psi_t(q_t^*) \\
&Q^* = (q_1^*, q_2^*, \ldots, q_T^*)
\end{aligned}}$$

**5.** Algorithm 3 uses the predecessor array $\psi$ in order to retrieve a state sequence that achieves the maximum joint probability $p^*$.

**6.** Algorithm 3 has $O(N^2 T)$ complexity.

**7.** Calculations in the Viterbi algorithm are quite similar to those in the forward algorithm, except that the *max* operator is applied within the main loop as opposed to the *summation* operator.

**8.** Algorithm 3 can be interpreted as a search in an acyclic network whose nodes are formed by the states of the HMM at each time instant $t \in \{1, 2, \ldots, T\}$.

**Examples:**

**1.** We use the direct evaluation approach to calculate a state sequence $Q^*$ that maximizes the joint probability $Pr(Q, O|\lambda)$, where $O = 231$. The last column of the table shown in Example 1 of §7.8.2 displays the joint probabilities $Pr(Q, O|\lambda)$. The maximum such probability is then 0.02016, achieved by the state sequence $Q^* = 112$, which corresponds to the sequence of temperatures Hot Hot Cold.

**2.** We illustrate the application of the Viterbi algorithm to the same problem studied in Example 1.

$$\begin{aligned}
&\delta_1(1) = \pi_1 b_1(2) = (0.8)(0.4) = 0.32, \; \delta_1(2) = \pi_2 b_2(2) = (0.2)(0.3) = 0.06; \\
&\delta_2(1) = b_1(3) \max[\delta_1(1)a_{11}, \delta_1(2)a_{21}] = 0.5 \max[0.224, 0.024] = 0.112, \; \psi_2(1) = 1; \\
&\delta_2(2) = b_2(3) \max[\delta_1(1)a_{12}, \delta_1(2)a_{22}] = 0.1 \max[0.096, 0.036] = 0.0096, \; \psi_2(2) = 1; \\
&\delta_3(1) = b_1(1) \max[\delta_2(1)a_{11}, \delta_2(2)a_{21}] = 0.1 \max[0.0784, 0.00384] = 0.00784, \\
&\qquad \psi_3(1) = 1; \\
&\delta_3(2) = b_2(1) \max[\delta_2(1)a_{12}, \delta_2(2)a_{22}] = 0.6 \max[0.0336, 0.00576] = 0.02016, \\
&\qquad \psi_3(2) = 1.
\end{aligned}$$

So the maximum joint probability is $p^* = \max[\delta_3(1), \delta_3(2)] = \max[0.00784, 0.02016)] = 0.02016$. Also $q_3^* = 2$, $q_2^* = \psi_3(q_3^*) = \psi_3(2) = 1$, $q_1^* = \psi_2(q_2^*) = \psi_2(1) = 1$, giving $Q^* = 112$.

## 7.8.4  THE LEARNING PROBLEM

This subsection discusses the problem of optimizing the model parameters $\lambda = (A, B, \pi)$ of an HMM to best describe a given sequence of observations. This problem arises in "training" an HMM in order to mimic the observed behavior of a system.

## Definitions:

Let the HMM have parameter set $\lambda = (A, B, \pi)$ and suppose the sequence of observations $O = (o_1, o_2, \ldots, o_T)$ is given. The **learning problem** deals with adjusting the model parameters $\lambda$ in order to maximize $L = Pr(O|\lambda)$.

While there is no known way to analytically solve this problem, there are iterative procedures that adjust the model parameters so that $L$ is locally maximized. The **Baum-Welch algorithm** is such a procedure which makes use of the auxiliary variables $\xi_t(i, j) = Pr(q_t = i, q_{t+1} = j|O, \lambda)$ and $\gamma_t(i) = Pr(q_t = i|O, \lambda)$.

Here $\xi_t(i, j)$ is the probability of being in state $i$ at time $t$ and in state $j$ at time $t + 1$, whereas $\gamma_t(i)$ is the a posteriori probability of being in state $i$ at time $t$, given the observation sequence and the model.

## Facts:

**1.** The Baum-Welch algorithm, also known as the *forward-backward* algorithm, can be derived using arguments of occurrence counting between $q_t$ and $q_{t+1}$.

**2.** The Baum-Welch algorithm is a special case of the well-known *Expectation Maximization (EM)* method. It is guaranteed to converge to at least a local maximum.

**3.** The following relationship holds:

$$\xi_t(i, j) = \frac{Pr(q_t = i, q_{t+1} = j, O|\lambda)}{Pr(O|\lambda)} = \frac{\alpha_t(i)a_{ij}\beta_{t+1}(j)b_j(o_{t+1})}{\sum_{\ell=1}^{N} \sum_{m=1}^{N} \alpha_t(\ell)a_{\ell m}\beta_{t+1}(m)b_m(o_{t+1})},$$

where $i, j \in S$ and $t \in \{1, 2, \ldots, T - 1\}$.

**4.** The relationship between $\gamma_t(i)$ and $\xi_t(i, j)$ is given by

$$\gamma_t(i) = \sum_{j=1}^{N} \xi_t(i, j), \ i \in S, \ t \in \{1, 2, \ldots, T - 1\}.$$

**5.** Using Facts 3 and 4, the Baum-Welch algorithm can be constructed as shown in Algorithm 4.

---

**Algorithm 4: Baum-Welch algorithm.**

input: $\lambda = (A, B, \pi)$, $O$

output: $\overline{\lambda} = (\overline{A}, \overline{B}, \overline{\pi})$ that maximizes $Pr(O|\overline{\lambda})$

initialize $\lambda_0$ to a set of random values

**while** $Pr(O|\lambda) - Pr(O|\lambda_0) > \epsilon$

   run the forward algorithm to find $\alpha_t(i)$ for all $i \in S$ and $t \in \{1, 2, \ldots, T\}$

   run the backward algorithm to find $\beta_t(i)$ for all $i \in S$ and $t \in \{1, 2, \ldots, T\}$

   **for** $i := 1$ **to** $N$

      **for** $j := 1$ **to** $N$

         **for** $t := 1$ **to** $T - 1$

$$\xi_t(i, j) := \frac{\alpha_t(i)a_{ij}\beta_{t+1}(j)b_j(o_{t+1})}{\sum_{\ell=1}^{N} \sum_{m=1}^{N} \alpha_t(\ell)a_{\ell m}\beta_{t+1}(m)b_m(o_{t+1})}$$

   **for** $i := 1$ **to** $N$

      **for** $t := 1$ **to** $T - 1$

         $\gamma_t(i) := \sum_{j=1}^{N} \xi_t(i, j)$

$$\text{\textbf{for} } i := 1 \text{ \textbf{to} } N$$
$$\bar{\pi}_i := \gamma_1(i)$$
$$\text{\textbf{for} } i := 1 \text{ \textbf{to} } N$$
$$\quad \text{\textbf{for} } j := 1 \text{ \textbf{to} } N$$
$$\quad \bar{a}_{ij} := \frac{\sum \{\xi_t(i,j) : t = 1, \ldots, T-1\}}{\sum \{\gamma_t(i) : t = 1, \ldots, T-1\}}$$
$$\text{\textbf{for} } j := 1 \text{ \textbf{to} } N$$
$$\quad \text{\textbf{for} } k := 1 \text{ \textbf{to} } M$$
$$\quad \bar{b}_j(k) := \frac{\sum \{\gamma_t(j) : t = 1, \ldots, T-1; O_t = \nu_k\}}{\sum \{\gamma_t(j) : t = 1, \ldots, T-1\}}$$
$$\bar{A} := (\bar{a}_{ij}), \ \bar{B} := (\bar{b}_j(k)), \ \bar{\pi} := (\bar{\pi}_i)$$
$$\bar{\lambda} := (\bar{A}, \bar{B}, \bar{\pi}), \ \lambda_0 := \lambda$$
$$\lambda := \bar{\lambda}$$
$$\text{\textbf{return} } \bar{\lambda}$$

## 7.9  QUEUEING THEORY

Queueing theory provides a set of tools for the analysis of systems in which customers arrive at a service facility. It has its origins in the works of A. K. Erlang (starting in 1908) in telephony. Since then it has found many applications in diverse areas such as manufacturing, inventory systems, computer science, analysis of algorithms, and telecommunications. Although queueing theory uses the terminology of servers providing service to customers, in actual applications the customers may be people, jobs, computational steps, or messages, and the servers may be human beings, machines, telephone circuits, communication channels, or computers.

### 7.9.1  SINGLE-STATION QUEUES

The simplest queueing system is a single-station queue in which customers arrive, wait for service, and depart after service completion. In this and subsequent sections we restrict ourselves to single-station queues.

**Definitions:**

A **queueing system** consists of a set of **customers**, who arrive at a service facility according to a specified **arrival process**. If a server is available then the customer is served immediately, with the length of time required to carry out the service determined by a **service-time distribution**. If a server is not free, the customer joins the **queue** and is later served according to a **service discipline**, which specifies the order in which customers are selected for service from the queue. Throughout, the service discipline is assumed to be First-Come-First-Served (FCFS). Alternative service disciplines include Last-Come-First-Served, randomly, or according to a tiered priority scheme.

The **queue capacity** is the maximum number of customers allowed in the system, either being served or awaiting service. Unless otherwise specified, the queue capacity is assumed to be infinite.

In a **single-station** queueing system, customers arrive, wait for service, and depart after service completion.

An **exponential distribution** with parameter $\lambda$ is a density function (§7.3.1) having the form $f(x) = \lambda e^{-\lambda x}$ for $x \geq 0$.

An arrival process is **Poisson** if the interarrival times (times between successive arrivals) are independent and identically distributed exponential random variables.

A random variable has an **Erlang distribution** with phase parameter $k$ if it is the sum of $k \geq 1$ independent and identically distributed exponential random variables.

**Facts:**

**1.** If the random variable $X$ has an exponential distribution with parameter $\lambda$, then $E(X) = \frac{1}{\lambda}$ and $\text{Var}(X) = \frac{1}{\lambda^2}$.

**2.** If the arrival process is Poisson with parameter $\lambda$, then the number of customers arriving in an interval of time of length $x$ is a Poisson random variable (§7.3.1) with parameter $\lambda x$.

**3.** The Erlang distribution is a special type of gamma distribution (§7.3.1).

**4.** *Kendall's notation*: A single-station queueing system can be described by the 5-tuple interarrival-time distribution/service-time distribution/number of servers/waiting room capacity/service discipline.

**5.** The following symbols are standard in describing queueing systems according to the scheme in Fact 4.

- $M$ — exponential ($M$ for Memoryless);
- $E_k$ — Erlang with phase parameter $k$;
- $D$ — deterministic (constant);
- $G$ — general.

**6.** More complicated queueing systems can consist of networks of queues, multiple types of customers, and priority schemes.

**7.** A list of over 300 books on queueing theory can be found at

- http://web2.uwindsor.ca/math/hlynka/qbook.html

**8.** A compilation of queueing theory software can be found at

- http://web2.uwindsor.ca/math/hlynka/qsoft.html

**Examples:**

**1.** A single-station queueing system is depicted by the schematic diagram in the following figure. Here customers randomly join the system (according to the arrival process), wait for service in the waiting room, are served (which takes a random amount of time), and then depart from the system.

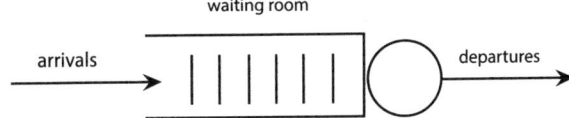

**2.** An $M/G/3/10/\text{LCFS}$ system has Poisson arrivals (exponential interarrival times), general service times, three servers, room for ten customers (including those in service), and Last-Come-First-Served service discipline.

**3.** An $M/M/1$ queue has Poisson arrivals, exponential service times, a single server, infinite waiting room, and FCFS service discipline.

**4.** *Airplane landings*: The landing of aircraft at an airport can be viewed as a queueing system in which the aircraft are the customers and the runways are the servers. Aircraft arrive according to a certain stochastic arrival process, and the length of time to land follows a certain service-time distribution. Those aircraft that are unable to land must join the queue of circling aircraft, awaiting service (normally according to a FCFS discipline, except in the case of an emergency landing, which would be a type of priority scheme).

**5.** *Communication network*: Messages arrive according to a Poisson process at rate $\lambda$ per second and are to be transmitted over a particular data link. The time required to transmit a single message is exponentially distributed, with average duration $\frac{1}{\mu}$ seconds. Messages waiting to be sent are stored in an input buffer. If the buffer has infinite capacity, then this system is an $M/M/1$ queue. If the input buffer has finite capacity $c$, then this system is an $M/M/1/c$ queue.

**6.** *Banking*: Customers arriving at a bank form a single common queue and are served by the $s$ available tellers in FCFS order. If arrivals are Poisson and the length of time to service a customer is exponential, this system can be modeled as an $M/M/s$ queue.

**7.** *Call center*: A call center has $c$ operators (servers) available to provide technical assistance to customers calling a central telephone number. Calls are routed to an available agent if one is free; otherwise the caller receives a busy signal. If arrivals are Poisson and the length of time spent on a call follows an arbitrary distribution, then this is an $M/G/c/c$ queue. It is also known as a *loss system*, since any calls to the system receive a busy signal when all servers are occupied, and hence these calls are "lost" to the system.

---

### 7.9.2  GENERAL SYSTEMS

This section presents results applicable to single-station queues with general arrival patterns, service-time distributions, and queue disciplines.

**Definitions:**

For a single-station queueing system, define

- $A_n = $ the arrival time of the $n$th customer to the system;
- $S_n = $ the service time of the $n$th customer;
- $D_n = $ the departure time of the $n$th customer;
- $A(t) = $ the total number of arrivals up to and including time $t$;
- $D(t) = $ the total number of departures up to and including time $t$;
- $X(t) = $ the total number of customers waiting in the system at time $t$.

The stochastic process $\{X(t) \mid t \geq 0\}$ is the **queue-length process**.

The state distribution following an arbitrary departure is $\pi_j = \lim_{n \to \infty} Pr(X(D_n^+) = j)$, for $j \geq 0$.

The state distribution prior to an arbitrary arrival is $\pi_j^* = \lim_{n\to\infty} Pr(X(A_n^-) = j)$, for $j \geq 0$.

The state distribution at an arbitrary time point, or **steady-state distribution**, is $p_j = \lim_{t\to\infty} Pr(X(t) = j)$, for $j \geq 0$.

The queue-length process (or the queueing system) is **stable** if the steady-state distribution $\{p_j \mid j \geq 0\}$ exists and $\sum_{j=0}^{\infty} p_j = 1$.

The **waiting time** of the $n$th customer is $W_n = D_n - A_n$; it includes the service time.

The steady-state **expected waiting time** is $W = \lim_{n\to\infty} \frac{1}{n} \sum_{k=1}^{n} W_k$, if the limit exists.

The **long-run arrival rate** is $\lambda = \lim_{n\to\infty} \frac{n}{A_n}$, if the limit exists.

The steady-state **expected number** in the system is $L = \lim_{t\to\infty} \frac{1}{t} \int_0^t X(u)\, du$, if the limit exists.

### Facts:

**1.** The number of customers in the system at any time equals the total number of arrivals up to that time minus the number of departures up to that time: that is, $X(t) = A(t) - D(t)$ for $t \geq 0$.

**2.** A sample path of the queue-length process $\{X(t) \mid t \geq 0\}$ is piecewise constant, with upward jumps at points of arrival and downward jumps at points of departure.

**3.** Suppose all the jumps in the sample paths of $\{X(t) \mid t \geq 0\}$ are of size $\pm 1$, with probability 1. If either $\pi_j$ or $\pi_j^*$ exists, then $\pi_j = \pi_j^*$ for all $j \geq 0$.

**4.** *PASTA (Poisson Arrivals See Time Averages)*: If $\{A(t) \mid t \geq 0\}$ is Poisson and, for every $s \geq 0$, $\{A(t) \mid t \geq s\}$ is independent of $\{X(u) \mid 0 \leq u < s\}$, then $p_j = \pi_j^*$ for all $j \geq 0$.

**5.** *Little's law* (J. D. C. Little, 1961):   $L = \lambda W$.

### Examples:

**1.** Suppose arrivals to a system occur deterministically, every three minutes, and service times are deterministic, each taking two minutes. Since every arriving customer is served immediately, either $X(t) = 0$ (no customers) or $X(t) = 1$ (a single customer). Every arrival finds an empty system and every departure leaves an empty system: $\pi_0 = 1 = \pi_0^*$, $\pi_1 = 0 = \pi_1^*$, as required by Fact 3. (The steady-state distribution does not exist.)

**2.** On average $\lambda = 24$ customers per hour arrive at a copy shop. Typically, there are $L = 9$ customers in the store at any time. Using Little's law, $W = \frac{L}{\lambda} = 0.375$ hour so that each customer spends on average 0.375 hours (or 22.5 minutes) in the shop.

**3.** The steady-state queue length or waiting time in a queueing system can be reduced by increasing the service rate. Suppose the long-run arrival rate doubles, but the service rate is increased so that the steady-state expected waiting time remains the same. Then by Little's law the steady-state expected queue length will also double.

**4.** *Machine repair*: A single machine is either working or being repaired. Suppose that the average time between breakdowns is exponentially distributed with mean $\frac{1}{\lambda}$ and the time to repair the machine is is exponentially distributed with mean $\frac{1}{\mu}$. This is then an $M/M/1/1$ queueing system with a single customer, corresponding to a broken down machine. $X(t) = 0$ signifies that the machine is working, and $X(t) = 1$ signifies that the machine is being repaired. A sample path of this system is shown in the following figure, with the machine initially working. Over a long period of time, after $N$ breakdowns and subsequent repairs, the machine is working for $N(\frac{1}{\lambda})$ units of time and is being repaired for $N(\frac{1}{\mu})$ units of time. The long run proportion of time the machine is working is then

$$\frac{N(\frac{1}{\lambda})}{N(\frac{1}{\lambda}) + N(\frac{1}{\mu})} = \frac{\mu}{\lambda + \mu}.$$

This value also turns out to be the steady-state probability of finding the system in state 0, with the machine working.

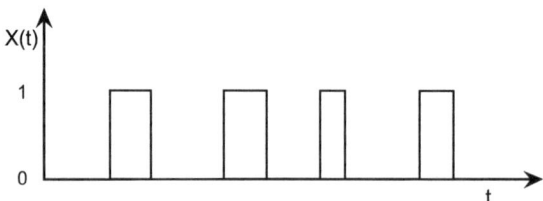

### 7.9.3  SPECIAL QUEUEING SYSTEMS

This section summarizes analytical results about special types of single-station queues.

**Notation:**

- $\frac{1}{\lambda}$ = expected interarrival time;
- $\tilde{A}(\cdot)$ = Laplace transform of the interarrival-time density;
- $\frac{1}{\mu}$ = expected service time;
- $\tilde{B}(\cdot)$ = Laplace transform of the service-time density;
- $\sigma^2$ = variance of the service time;
- $s$ = number of servers;
- $\rho = \frac{\lambda}{s\mu}$ = traffic intensity.

The probability generating function (§7.3.3) for the steady-state distribution $\{p_j\}$ of a queueing system is $\phi(z) = \sum\limits_{j=0}^{\infty} p_j z^j$, $|z| < 1$.

The Laplace transform for the waiting-time distribution $f(w)$ of a queueing system is $\psi(s) = \int_0^\infty e^{-sw} f(w)\, dw$, $Re(s) > 0$.

**Facts:**

**1.** The $M/M/1$ queue is stable if $\rho < 1$. The following results hold when the queue is stable.

$$p_j = (1 - \rho)\rho^j = \pi_j = \pi_j^*$$
$$L = \frac{\rho}{1-\rho}$$
$$W = \frac{1}{\mu - \lambda}.$$

**2.** The $M/M/1/K$ queue is always stable. Assume $\rho \neq 1$.

$$p_j = \tfrac{1-\rho}{1-\rho^{K+1}}\rho^j, \quad 0 \leq j \leq K$$
$$\pi_j^* = \tfrac{p_j}{1-p_K} = \pi_j, \quad 0 \leq j \leq K-1$$
$$L = \tfrac{\rho}{1-\rho}\big(\tfrac{1-\rho^K}{1-\rho^{K+1}} - Kp_K\big)$$
$$W = \tfrac{1}{\mu-\lambda}\big(\tfrac{1-\rho^K}{1-\rho^{K+1}} - Kp_K\big).$$

If $\rho = 1$, the above formulas reduce to

$$p_j = \tfrac{1}{K+1}, \quad 0 \leq j \leq K$$
$$\pi_j^* = \pi_j = \tfrac{1}{K}, \quad 0 \leq j \leq K-1$$
$$L = \tfrac{K}{2}$$
$$W = \tfrac{K}{2\mu}.$$

**3.** The $M/M/s$ queue is stable if $\rho < 1$. The following results hold when the queue is stable.

$$p_0 = \left( \sum_{n=0}^{s-1} \tfrac{(s\rho)^n}{n!} + \tfrac{s^s}{s!}\tfrac{\rho^s}{1-\rho} \right)^{-1}$$
$$p_j = \begin{cases} \tfrac{(s\rho)^j}{j!}p_0, & 0 \leq j < s \\ \tfrac{s^s}{s!}\rho^j p_0, & j \geq s \end{cases}$$
$$\pi_j = \pi_j^* = p_j, \quad j \geq 0$$
$$L = \rho\big(s + \tfrac{p_s}{(1-\rho)^2}\big)$$
$$W = \tfrac{1}{s\mu}\big(s + \tfrac{p_s}{(1-\rho)^2}\big)$$

$E(\text{number of busy servers}) = \rho.$

**4.** The $M/M/\infty$ queue is always stable.

$$p_j = e^{-(\lambda/\mu)}\tfrac{(\lambda/\mu)^j}{j!}, \quad j \geq 0$$
$$\pi_j = \pi_j^* = p_j, \quad j \geq 0$$
$$L = \tfrac{\lambda}{\mu}$$
$$W = \tfrac{1}{\mu}.$$

**5.** The $M/G/1$ queue is stable if $\rho < 1$. The following results hold when the queue is stable.

$$p_0 = 1 - \rho,$$
$$\pi_j = \pi_j^* = p_j$$
$$\phi(z) = (1-\rho)\tfrac{(1-z)\tilde{B}(\lambda(1-z))}{\tilde{B}(\lambda(1-z))-z}$$
$$\psi(s) = (1-\rho)\tfrac{s\tilde{B}(s)}{s-\lambda(1-\tilde{B}(s))}$$
$$L = \rho + \tfrac{\rho^2+\lambda^2\sigma^2}{2(1-\rho)}$$
$$W = \tfrac{1}{\mu} + \tfrac{\lambda((1/\mu)^2+\sigma^2)}{2(1-\rho)}.$$

The last four equations are the various forms of the *Pollaczek-Khintchine formula*. No closed form results are available for $M/G/c$ queues for $2 \leq c < \infty$.

**6.** The $M/G/c/c$ queue, also called a *loss system*, is always stable. The main result is

$$p_j = \frac{\rho^j/j!}{\sum\limits_{n=0}^{c} \rho^n/n!}, \quad 0 \le j \le c.$$

**7.** The $M/G/\infty$ queue is always stable.

$$p_j = e^{-(\lambda/\mu)} \frac{(\lambda/\mu)^j}{j!}, \quad j \ge 0$$
$$\pi_j = \pi_j^* = p_j, \quad j \ge 0$$
$$L = \tfrac{\lambda}{\mu}$$
$$W = \tfrac{1}{\mu}.$$

**8.** The $G/M/1$ queue is stable if $\rho < 1$. When the queue is stable there is a unique solution $\alpha \in (0,1)$ to $\alpha = \tilde{A}(\mu(1-\alpha))$. The following results hold when the queue is stable.

$$\pi_j^* = (1-\alpha)\alpha^j = \pi_j, \quad j \ge 0$$
$$p_0 = 1 - \rho$$
$$p_j = \rho\pi_{j-1}^*, \quad j \ge 1$$
$$L = \tfrac{\rho}{1-\alpha}$$
$$W = \tfrac{1}{\mu}\tfrac{1}{(1-\alpha)}.$$

The $G/M/c$ queue can be analytically solved, but the results are complicated.

**9.** The $G/M/\infty$ queue is always stable.

$$L = \tfrac{\lambda}{\mu}$$

$$\text{variance of number in system } = \tfrac{\lambda}{\mu}(1 - \tfrac{\lambda}{\mu}) + \tfrac{\lambda/\mu}{2}\tfrac{\tilde{A}(\mu)}{1-\tilde{A}(\mu)}.$$

**Examples:**

**1.** At a drop-in legal clinic, the lawyer sees four clients during a typical (eight hour) day. Each client's case requires on average 1.5 hours of the lawyer's time. If arrivals are Poisson and service times are exponentially distributed, then this is an $M/M/1$ queue with $\lambda = \frac{4}{8} = \frac{1}{2}$ customers per hour and $\frac{1}{\mu} = \frac{3}{2}$. Here $\rho = \frac{\lambda}{\mu} = \frac{3}{4} < 1$, so the queue is stable. Using Fact 1, $p_0 = 1 - \rho = \frac{1}{4}$, so there is probability $\frac{1}{4}$ that the lawyer is idle. The expected number of clients in the clinic is $L = \frac{\rho}{1-\rho} = 3$ and the average wait of a client is $W = \frac{1}{\mu - \lambda} = 6$ hours.

**2.** Customers arrive at a service station according to a Poisson process with rate 10 per hour. The manager has two options: (a) employ a single fast server who can service a customer in five minutes on average, or (b) employ two slow servers each taking ten minutes on average to serve a customer. Assume that the service times are exponential. Which option should the manager implement to minimize the expected waiting time in steady state?

Under (a) the system is an $M/M/1$ queue with $\lambda = 10$, $\mu = 12$. Since $\rho = \frac{10}{12} < 1$, the system is stable. By Fact 1, $W = \frac{1}{12-10} = 0.5$ hours. Under (b) the system is an $M/M/2$ queue with $\lambda = 10$ and $\mu = 6$. The system is stable since $\rho = \frac{10}{2\cdot6} < 1$. From Fact 3, $p_0 = \frac{1}{11}$, $p_2 = \frac{25}{198}$, and $W = \frac{6}{11} = 0.55$ hours. Thus option (a) is better. In general, it is better to employ a few fast servers than many slow servers with the same overall service capacity.

**3.** A system manager has the option of using one of three possible servers in a single-server system. The service times under the first server are exponential with mean of six minutes. Under the second server they are uniformly distributed over $[4, 8]$ minutes. Under the third they are constant, equal to six minutes. The customers arrive according to a Poisson process with rate 8 per hour. Which server should be chosen to minimize the expected waiting time in steady state?

The mean service time is six minutes; i.e., $\mu = 10$ per hour, for all three servers. However, the variances $\sigma^2$ are different. This $M/G/1$ system is stable under all three servers since $\rho = \frac{8}{10} < 1$. For server one, $\sigma^2 = 0.01$ (hours)$^2$ and $W = 0.5$ hours. For the second server, $\sigma^2 = \frac{1}{2700} = 0.000370$ (hours)$^2$ and $W = \frac{83}{270} = 0.31$ hours. For the third server, $\sigma^2 = 0.0$ (hours)$^2$ and $W = \frac{3}{10} = 0.3$ hours. Thus, it is best to use the server with constant service times. In general, reducing the variance of the service times has a beneficial effect on the system.

**4.** A small business wants to install a telephone system with multiple lines, though without any capacity for call queueing. This is to be done to ensure that 95% of the calls made to the business get answered. Suppose that the arrival process is Poisson with $\lambda = 10$ calls per hour, and that the average call lasts five minutes. This is an $M/G/c/c$ loss system (Fact 6), and it is necessary to find the smallest value of $c$ such that $p_c \leq 0.05$. Using Fact 6 with $\rho = \frac{10}{12} = \frac{5}{6}$ and $c = 1$ gives $p_1 = \frac{\rho}{1+\rho} = 0.45$. Similar calculations give $p_2 = 0.16$ for $c = 2$ and $p_3 = 0.042$ for $c = 3$. Consequently, three lines are needed to ensure the stipulated grade-of-service requirement.

## 7.10  SIMULATION

Simulation is a technique for numerically estimating the performance of a complex stochastic system when analytic solution is not feasible. This section discusses both *discrete-event* and *Monte Carlo* simulation. In discrete-event simulation models, the passage of time plays a key role, as changes to the state of the system occur only at certain points in simulated time. For example, queueing and inventory systems can be studied by discrete-event simulation models. Monte Carlo simulation models do not, however, require the passage of time. Such models are useful in estimating eigenvalues, estimating $\pi$, and estimating the quantiles of a mathematically intractable test statistic in hypothesis testing. Simulation has been described [BrFoSc87] as "driving a model of a system with suitable inputs and observing the corresponding outputs." Accordingly, the following three subsections discuss input modeling, output analysis, and simulation programming languages.

### 7.10.1  INPUT MODELING

This subsection addresses three key issues in constructing a simulation model:

- determining a source of randomness to drive the probabilistic aspects of the model;

- input model selection to determine the appropriate probabilistic models to drive the simulation;

- random variate generation algorithms that transform random numbers to random variates.

## Definitions:

**Random numbers** are real numbers generated uniformly over the interval $(0, 1)$.

A **random number generator** is any mechanism or algorithm for generating random numbers.

**Pseudo-random numbers** are values generated deterministically, but that appear to behave like independent and identically distributed random numbers.

Let $m$ be a large prime integer. A **purely multiplicative linear congruential generator** (§4.3.1) produces a stream of pseudo-random numbers $\{\frac{x_i}{m} \mid i \geq 1\}$ based on the recursive relationship $x_{i+1} = ax_i \pmod{m}$, where $a$ is an integer **multiplier** between 1 and $m - 1$, and $x_0$ is an integer **seed** between 1 and $m - 1$.

An **input model** characterizes the stochastic elements of a discrete-event simulation model.

A **trace-driven** input model generates a process that is identical to the collected data values without relying on a parametric model.

A **random variate** is a realization of a random variable.

## Facts:

**1.** Stochastic simulations typically derive their source of randomness from random numbers. That is, inputs to the simulated system need to be generated according to a specified probability model, a task that can be accomplished by suitably transforming (uniform) random numbers.

**2.** Desirable properties for random number generators include: uniformity, independence, speed, minimal memory requirements, ease of implementation, portability across various computer systems, reproducibility, and multiple stream capability.

**3.** Although numerous methods have been proposed for generating random numbers, multiplicative linear congruential generators are typically used to produce a stream of pseudo-random numbers.

**4.** Due to the prevalence of 32-bit computer architecture, $m$ is often chosen to be $2^{31} - 1$, which is prime.

**5.** A *full period* generator, which cycles through all $m - 1$ possible $x_i$ values prior to repeating, is obtained by selecting $a$ to be a primitive root modulo $m$. (See §4.7.1.)

**6.** Software for pseudo-random number generators can be found at

- http://random.mat.sbg.ac.at/software
- http://www.taygeta.com/random.html

**7.** Additional information on the theoretical and empirical performance of a variety of pseudo-random number generators is available at

- http://random.mat.sbg.ac.at/generators
- http://random.mat.sbg.ac.at/tests

**8.** If the multiplier $a$ is chosen so that it is "modulus-compatible" with $m$, then potential overflow can be averted for large values of $m$. Two values of $a$ that are often used with $m = 2^{31} - 1$ are $a = 7^5 = 16{,}807$ and $a = 48{,}271$ [PaMi88].

**9.** More detail concerning random number generation is given by [LE94] and [LE12].

**10.** Successful input modeling for a discrete-event simulation requires a close match between the input model and the true underlying probabilistic mechanism associated with the system.

**11.** One of the first steps in determining an appropriate input model for an element of a discrete-event simulation is to assess whether the observations are independent and identically distributed.

**12.** An input model can be specified in several ways: e.g., using a cumulative distribution function, joint probability density function, hazard function, intensity function, or variate-generation algorithm.

**13.** Potential input models for simulation are described in references [BaNe03], [FoEtal11], [JoKoBa94] and at

- `http://www.math.wm.edu/~leemis/chart/UDR/UDR.html`

**14.** Many input models rely on parametric probabilistic models such as the binomial, normal, and Weibull distributions. Maximum likelihood is typically used to estimate parameters of these models.

**15.** Bézier curves [FlWi93] offer a unique combination of the parametric and nonparametric approaches. After an initial distribution is fitted to the dataset, the modeler decides whether differences between the empirical and fitted models represent sampling variability (chance variation) or an aspect of the distribution that should be included in the input model.

**16.** Multivariate distributions (e.g., the multivariate normal distribution with mean $\mu$ and variance-covariance matrix $\Sigma$) are considered by [Jo87].

**17.** Once an input model has been chosen, random variate generation algorithms are used to transform random numbers to random variates from the input model.

**18.** Devroye [De86] gives algorithms for converting random numbers to random variates associated with input models chosen to drive the simulation.

**19.** Techniques commonly used for generating random variates from univariate probability distributions are: inverse transformation, composition, acceptance/rejection, and special properties.

**20.** Algorithm 1, which shows the inversion method for generating a univariate random variate $X$, is based on the probability integral transformation. It is assumed that the cumulative distribution function $F(x)$ for the input model of interest has the inverse $F^{-1}(U)$.

---

**Algorithm 1**:   **Inverse transformation method.**

input: cumulative distribution function $F$

output: a random variate $X$ from this distribution

generate $U$ uniformly over $(0, 1)$

$X := F^{-1}(U)$

---

**21.** Other topics in random variate generation include table methods, generating from multivariate distributions, random sampling, estimating integrals, and generating processes correlated in time.

**Examples:**

**1.** Suppose that a sequence of arrival times (e.g., of customers at a bank) is collected over a 24-hour time period. A *trace-driven* input model for the arrival process is generated by having arrivals occur at the same times as the observed values.

**2.** Let $t_1, t_2, \ldots, t_n$ be the arrival times to a queue collected on the time interval $(0, c]$. If the times between arrivals are independent and identically distributed, a parametric or nonparametric model can be fitted to the data. In the former case, parameters are often estimated by maximizing the *likelihood function* [La15]

$$L(\boldsymbol{\theta}) = \prod_{i=1}^{n} f(x_i, \boldsymbol{\theta}),$$

where $x_i = t_i - t_{i-1}$ for $i = 1, 2, \ldots, n$, $t_0 \equiv 0$, $\boldsymbol{\theta} = (\theta_1, \theta_2, \ldots, \theta_p)$ is a vector of unknown parameters, and $f(x_i, \boldsymbol{\theta})$ is the probability density function of the interarrival times.

**3.** If the interarrival times to a queue (as in Example 2) are not independent and identically distributed, then a nonstationary point process might be considered, such as a nonhomogeneous Poisson process, where the arrival rate $\lambda(t; \theta)$ varies over time. One parametric model is the *power law* process, with intensity function $\lambda(t; \lambda, \kappa) = \lambda^{\kappa} \kappa t^{\kappa-1}$ for $t > 0$, where $\lambda$ and $\kappa$ are positive parameters. The likelihood function for the single realization on $(0, c]$ is $L(\lambda, \kappa) = \left( \prod_{i=1}^{n} \lambda(t_i; \lambda, \kappa) \right) \exp\left( - \int_0^c \lambda(t; \lambda, \kappa)\, dt \right)$. Maximum likelihood estimators can be determined by maximizing $L(\lambda, \kappa)$ or its logarithm with respect to $\lambda$ and $\kappa$. Confidence regions for the unknown parameters can be found by using asymptotic properties of the likelihood ratio statistic or the observed *information matrix* [La15]. As with all statistical modeling, goodness-of-fit tests should be performed in order to assess the model adequacy.

**4.** *Weibull distribution:* The *Weibull distribution* has cumulative distribution function

$$F(x) = 1 - e^{-(\lambda x)^{\kappa}} \text{ for } x > 0.$$

The inverse cumulative distribution function is $F^{-1}(y) = \dfrac{[-\ln(1-y)]^{1/\kappa}}{\lambda}$ for $0 < y < 1$. Algorithm 1 can be used to generate a Weibull variate according to

$$X := \frac{[-\ln(1-U)]^{1/\kappa}}{\lambda}.$$

**5.** *M/M/1 queue:* To simulate the operation of a single-server queue (§7.9.1) having Poisson arrivals at arrival rate $\lambda$ and exponential service times with service rate $\mu$, exponentially generated variates are needed for the interarrival times $\{I_n\}$ and the service times $\{S_n\}$. These are available as a special case of the Weibull distribution (Example 4) with $\kappa = 1$ and can be generated using $I_n = \dfrac{-\ln(1-U_n)}{\lambda}$ and $S_n = \dfrac{-\ln(1-V_n)}{\mu}$, with the $\{U_n\}$, $\{V_n\}$ generated uniformly over $(0, 1)$.

A concrete example is provided in the following table, which shows one simulated run of an $M/M/1$ queue with $\lambda = 0.5$ and $\mu = 0.7$. The table shows, in successive columns, the following values for each customer $n$: the interarrival time $I_n$, the arrival time $A_n$, the service time $S_n$, the beginning time of service $B_n$, the departure time $D_n$, and the sojourn time $W_n = D_n - A_n$. Notice that customers $1, 4, 7, 8, 9$ are served immediately and incur no waiting time in the queue.

| customer | $I_n$ | $A_n$ | $S_n$ | $B_n$ | $D_n$ | $W_n$ |
|----------|-------|-------|-------|-------|-------|-------|
| 1 | 5.44 | 5.44 | 0.78 | 5.44 | 6.22 | 0.78 |
| 2 | 0.61 | 6.05 | 2.77 | 6.22 | 8.99 | 2.94 |
| 3 | 0.35 | 6.40 | 0.96 | 8.99 | 9.95 | 3.55 |
| 4 | 4.12 | 10.52 | 2.42 | 10.52 | 12.94 | 2.42 |
| 5 | 0.54 | 11.06 | 0.88 | 12.94 | 13.82 | 2.76 |
| 6 | 2.07 | 13.13 | 0.87 | 13.82 | 14.69 | 1.56 |
| 7 | 6.82 | 19.95 | 0.86 | 19.95 | 20.81 | 0.86 |
| 8 | 2.19 | 22.14 | 0.76 | 22.14 | 22.90 | 0.76 |
| 9 | 4.09 | 26.23 | 3.31 | 26.23 | 29.54 | 3.31 |
| 10 | 0.02 | 26.25 | 0.01 | 29.54 | 29.55 | 3.30 |

## 7.10.2  OUTPUT ANALYSIS

Once a verified and validated simulation model has been developed, a modeler typically wants to estimate measures of performance associated with outputs of the model. Although there are often several performance measures of interest, a single measure of performance $\theta$ (e.g., the mean waiting time in a queue) is studied here. This subsection discusses using point estimation to compute an estimate for $\theta$, determining a confidence interval for the point estimate, and applying variance reduction techniques to obtain more precise point estimates.

### Definitions:

Suppose $\{Y_i\}$ is the **output stochastic process**. If the output stochastic process consists of independent observations obtained from a population with cumulative distribution function $F_Y$, the **pth quantile** of $F_Y$ is the value $y_p$ such that $F_Y(y_p) = p$. The **median** of $F_Y$ corresponds to $p = 0.5$.

The **sample mean** of the observations $Y_1, Y_2, \ldots, Y_n$ is given by $\overline{Y} = \frac{1}{n} \sum_{i=1}^{n} Y_i$.

If the values $Y_1, Y_2, \ldots, Y_n$ are rearranged so that $Y_{(1)} \leq Y_{(2)} \leq \cdots \leq Y_{(n)}$ then $Y_{(i)}$ is the $i$th **order statistic**.

The **mean** $\mu_Y$ of the process is the asymptotic mean of the output process $\{Y_i\}$.

The **variance** $\sigma_Y^2$ of the process is the asymptotic mean of the output process $\{(Y_i - \overline{Y})^2\}$.

The **probability** $Pr(A)$ of event $A$ is the asymptotic mean of the output process $\{I(A)\}$, where $I$ is the 0-1 indicator function for event $A$.

The output process $Y_1, Y_2, \ldots$ is **covariance stationary** if, for finite mean $\mu$ and finite variance $\sigma^2 > 0$, $E(Y_i) = \mu$, $i = 1, 2, \ldots$, $\mathrm{Var}(Y_i) = \sigma^2$, $i = 1, 2, \ldots$, and $\mathrm{Cov}(Y_i, Y_{i+j})$ is independent of $i$, for $j = 1, 2, \ldots$.

**Variance reduction techniques** are strategies for obtaining greater precision for a fixed amount of sampling.

**Time-persistent statistics** are often calculated in discrete-event simulations. These statistics fundamentally differ from statistics that are based on observations $Y_1, Y_2, \ldots, Y_n$. Time-persistent statistics account for the value of a performance measure of interest (for example, the number of customers in a waiting line) and also the length of time that the value was held.

## Facts:

**1.** The three most common measures of performance to be estimated in Monte Carlo simulation and discrete-event simulation are means, probabilities, and quantiles.

**2.** Point estimates for $\mu_Y$, $\sigma_Y^2$, $Pr(A)$ are typically given by the associated sample statistics, that is, the sample mean, the sample variance, and the fraction of occurrences of the event $A$.

**3.** A simple estimator of $y_p = F_Y^{-1}(p)$ is $Y_{(s)}$, where $s = \lfloor p(n+1) \rfloor$. This estimator can be improved (with respect to bias) by estimating $F_Y^{-1}(p)$ with the linear combination $(1-\alpha)Y_{(s)} + \alpha Y_{(s+1)}$, where $\alpha = p(n+1) - \lfloor p(n+1) \rfloor$.

**4.** Sample means, variances, and quantiles can also be estimated for time-persistent statistics [Ne13].

**5.** *Replication:* This is one of the simplest methods of interval estimation, in which several runs of a simulation model are used. Classical confidence intervals based on the central limit theorem for the measures of interest are often applied to the output.

**6.** The presence of autocorrelation among observations (e.g., the waiting times of adjacent customers in a queue) significantly complicates the statistical analysis of simulation output from a single run.

**7.** To analyze a single simulation run with autocorrelation present, techniques have been developed for determining interval estimates whose actual coverage is close to the stated coverage. For many of these techniques, the output is assumed to be covariance stationary. These techniques include batch means, overlapping batch means, standardized time series, regeneration, spectral analysis, and autoregression [La15].

**8.** *Common random numbers:* This is a variance reduction technique in which two or more alternative system configurations are analyzed using the same set of random numbers for particular purposes (e.g., generating service times). Using common random numbers ensures that the output differences are due to the configurations rather than the sampling variability in the random numbers.

**9.** *Antithetic variates:* This is a second variance reduction technique, applicable to the analysis of a single system. If the random numbers $\{U_i\}$ are used for a particular purpose in one simulation run, then using $\{1 - U_i\}$ in a second run typically induces a negative correlation between the outputs of the two runs. Thus, the average of the output measures from the two runs will have a reduced variance.

**10.** There are a variety of variance reduction techniques. See Wilson [Wi84] for a detailed discussion.

**11.** Other topics in output analysis include initialization bias detection, ranking and selection, comparing alternative system designs, experimental design, and optimization.

## Examples:

**1.** *Confidence intervals for expected waiting times:* Let $X_1, X_2, \ldots, X_n$ be the averages of the waiting times of customers in a single-server queue from $n$ independent replications of a discrete-event simulation model. A $100(1-\alpha)\%$ confidence interval for $\mu$, the steady-state mean waiting time, is

$$\overline{X} - t_{\alpha/2, n-1} \, \tfrac{s}{\sqrt{n}} < \mu < \overline{X} + t_{\alpha/2, n-1} \, \tfrac{s}{\sqrt{n}} \,,$$

where $\overline{X}$ is the sample mean, $s$ is the sample standard deviation, and $t_{\alpha/2, n-1}$ is the $\frac{1}{1-\alpha/2}$ quantile of the $t$ distribution with $n-1$ degrees of freedom. Each replication must be "warmed up" to avoid initialization bias. The asymptotic normality of $X_1, X_2, \ldots, X_n$ is assured by the central limit theorem and independence is based on the use of independent random number streams.

**2.** *M/M/1 queue:*   The simulation in Example 5 of §7.10.1 was executed so that the first 200 customer wait times were collected. The measure of performance $\theta$ for the system is the steady-state expected customer wait time. The initial conditions for each replication are an empty system and an idle server. The stopping time for each replication is when the 200th customer departs. Running this simulation experiment for $n = 100$ replications gave $\overline{X} = 4.72$ and for $n = 500$ replications gave $\overline{X} = 4.76$. For this simple queueing system, the steady-state analytical solution is $W = \frac{1}{\mu - \lambda} = 5.0$ (§7.9.3, Fact 1). These averages are biased low because the early waiting times have a lower expected value than the subsequent waiting times as a result of the initial conditions. To improve these point estimates, the system was permitted to warm up for the first 100 customers and the average waiting time was then calculated for the last 100 customers. In this case, rerunning the simulation gave the improved estimates $\overline{X} = 5.20$ for $n = 100$ and $\overline{X} = 4.93$ for $n = 500$.

**3.** *Common random numbers:*   Law [La15, pp. 596–597] compares the $M/M/1$ and the $M/M/2$ queueing models with a utilization of $\rho = 0.9$ using the waiting times in the queue of the first 100 customers. With $n = 100$ independent replications of each system, the two models are compared in four ways:

- independent runs $(I)$;
- arrival streams using common random numbers $(A)$;
- service times using common random numbers $(S)$;
- arrival streams and service times using common random numbers $(A \& S)$.

Common random numbers is a variance reduction technique that feeds identical interarrival and/or service times into the two different queueing models to increase the likelihood that observed differences in the waiting times are due to the system configurations ($M/M/1$ versus $M/M/2$) rather than sampling error. The mean half-widths of the confidence intervals ($\alpha = 0.10$) reported for this example are $0.70(I), 0.49(A), 0.49(S)$, and $0.04(A \& S)$.

---

### 7.10.3   SIMULATION LANGUAGES

This subsection considers the history and features of simulation programming languages developed over the years.

**Facts:**

**1.** The use of a general-purpose simulation programming language (SPL) expedites model development, input modeling, output analysis, and animation. In addition, SPLs have accelerated the use of simulation as an analysis tool by bringing down the cost of developing a simulation model.

**2.** In a history of the development of SPLs from 1955 to 1986, Nance [Na93] defines six requirements that an SPL must meet:

- random number generation;
- random variate generation;
- list processing capabilities so that objects can be created, altered, and deleted;
- statistical analysis routines;

- summary report generators;
- a timing executive or event calendar to model the passage of time.

3. SPLs may take the form of

- a set of subprograms in a general purpose language (GPL) such as Fortran or C that can be called to meet these six requirements;
- a preprocessor that converts statements or symbols to lines of code in a GPL;
- a conventional programming language.

4. The following table shows a division of the historical record into five distinct periods, including the names of several languages that came into existence in each period.

| period | characteristics | languages |
|--------|-----------------|-----------|
| 1955–1960 | period of search | GSP |
| 1961–1965 | the advent | CLP, CSL, DYNAMO, GASP, GPSS, MILITRAN, OPS, QUIKSCRIPT, SIMSCRIPT, SIMULA, SOL |
| 1966–1970 | formative period | AS, BOSS, Q-GERT, SLANG, SPL |
| 1971–1978 | expansion period | DRAFT, HOCUS, PBQ, SIMPL |
| 1979–1986 | consolidation and regeneration | INS, SIMAN, SLAM |

5. The General Purpose System Simulator (GPSS) was first developed on various IBM computers in the early 1960s. Algol-based SIMULA was also developed in the 1960s and had features that were ahead of its time. These included abstract data types, inheritance, the co-routine concept, and quasi-parallel execution.

6. SIMSCRIPT was developed by the RAND Corporation with the purpose of decreasing model and program development times. SIMSCRIPT models are described in terms of entities, attributes, and sets. The syntax and program organization were influenced by Fortran.

7. The Control and Simulation Language (CSL) takes an "activity scanning" approach to language design, where the activity is the basic descriptive unit.

8. The General Activity Simulation Program (GASP), in common with several other languages, used flow-chart symbols to bridge the gap between personnel unfamiliar with programming and programmers unfamiliar with the application area. Although originally written in Algol, GASP provided Fortran subroutines for list-processing capabilities (e.g., queue insertion).

9. GASP was a forerunner to both the Simulation Language for Alternative Modeling (SLAM) and SIMulation ANalysis (SIMAN) languages.

10. SLAM was the first language to include three modeling perspectives in one language: network (process orientation), discrete-event, and continuous (state variables).

11. SIMAN was the first major SPL executable on an IBM PC.

12. Simulation software has mushroomed, with numerous packages and languages available both for general purpose and application-specific simulations. Special purpose and integrated packages are widespread and available on desktop computers. The 2015 survey [Sw15] compares 55 products, having a wide range of features and capabilities.

13. A recent trend has been the addition of animation to intelligently view simulation output. Several web-based simulation tools can be found at

- http://www.dcs.ed.ac.uk/home/hase/simjava/applets/index.html
- http://www.vissim.com
- http://www.anylogic.com
- http://www.um.es/fem/EjsWiki/index.php/Main/WhatIsEJS

**14.** Software for carrying out Monte Carlo simulation can be found at the site

- http://random.mat.sbg.ac.at/software

**15.** A site that writes code for random variate generation for various probability distributions can be found at

- http://statistik.wu-wien.ac.at/projects/anuran/index.html

**16.** Winter Simulation Conference proceeding papers on simulation languages, modeling, analysis, applications, etc. are available at

- http://www.informs-sim.org

# 7.11   THE PROBABILISTIC METHOD

The Probabilistic Method is a powerful technique for proving the existence of combinatorial structures with certain desired properties. This method seems ubiquitous because it provides a technique to deal with the "local-global" problem. More specifically, when one is confronted with the task of constructing a finite set structure that satisfies a certain combinatorial condition at every element, the total number of choices often grows exponentially, so it is very difficult to decide which choices are good. The probabilistic paradigm provides what one may call conditions for building a global patch from all the local data. The basic premise is the following: Define a suitable probability space and then prove that the probability of making no "bad choice" is positive. In particular, its power is most visible when it is used to prove deterministic statements. We describe here briefly some of the main aspects of the probabilistic method. There are numerous books, monographs, and lecture notes on the subject including [AlSp08], [MoRe00], [Sp87].

## 7.11.1   THE BASIC METHOD

**Definitions:**

Suppose $(\Omega, Pr)$ is a probability on a finite set $\Omega$ and let $X : \Omega \to \mathcal{R}$ be a random variable on $\Omega$. The **expectation** of $X$ is given by $E(X) = \sum_{\omega \in \Omega} X(\omega) \, Pr(\omega)$.

The **Ramsey number** $R(k, k)$ is the least integer $N$ such that for every two-coloring of the edges of the complete graph $K_n$ with $n \geq N$, there is a monochrome complete subgraph on $k$ vertices.

A hypergraph is said to be **$n$-uniform** if each edge size is $n$. We say that a hypergraph $\mathcal{H} = (\mathcal{V}, \mathcal{E})$ has **Property B** (after Felix Bernstein, 1878–1956) if each vertex in $\mathcal{V}$ can be assigned one of two colors so that no edge $E \in \mathcal{E}$ is monochromatic, i.e., for no edge $E \in \mathcal{E}$ is it the case that every vertex of $E$ has been assigned the same color.

A set $B \subset \mathcal{R}$ is called **sum-free** if the sum of any two elements in $B$ does not lie in $B$.

The **girth** of a graph is the size of a smallest cycle in the graph. If the graph is acyclic then its girth is defined to be $\infty$.

**Facts:**

**1.** Suppose $(\Omega, Pr)$ is a probability on the finite set $\Omega$ and $X, Y : \Omega \to \mathcal{R}$ are random variables on $\Omega$. Then

$$
\begin{aligned}
E(X + Y) &= E(X) + E(Y), \\
E(\lambda X) &= \lambda E(X), \text{ for real } \lambda.
\end{aligned}
$$

This is expressed by saying that expectation is linear.

**2.** If $X$ is a random variable defined on a probability space $(\Omega, Pr)$, then $Pr(X \geq E(X)) > 0$ and $Pr(X \leq E(X)) > 0$. In particular, with positive probability, there exist $\omega, \omega' \in \Omega$ such that $X(\omega) \geq E(X)$ and $X(\omega') \leq E(\omega')$.

**3.** *Markov's inequality:* If $X : \Omega \to \mathcal{R}^+$ is a nonnegative valued random variable, then for any $a > 0$

$$
Pr(X > a) \leq \frac{E(X)}{a}.
$$

**Examples:**

We include several examples that illustrate how the method basically works. These examples span several areas of discrete mathematics, including additive combinatorics, extremal combinatorics, combinatorial geometry, and graph theory.

**1.** *Lower bound for $R(k,k)$:* We shall show that $R(k,k) > 2^{k/2}$ [Er47]. To do so, color each edge of $K_n$ red or blue independently, and with equal probability. Then the probability that there exists a monochrome $K_k$ under this coloring is at most $\binom{n}{k} 2^{-\binom{k}{2}+1}$ since there are $\binom{n}{k}$ distinct $k$ subsets of $\{1, 2, \ldots, n\}$ and for each fixed $k$ subset, the probability that all the edges incident with those vertices bear the same color is $2^{-\binom{k}{2}+1}$. Now, a routine calculation shows that for $n = 2^{k/2}$ this expression is less than one under a random two-coloring. In particular, with positive probability, no $k$-subset of the vertex set induces a monochromatic $K_k$.

**2.** *Colored hats and a guessing game:* There are $n$ friends standing in a circle so that everyone can see everybody else. On each person's head a randomly chosen hat— either black or white—is placed. After they have had a look at each other, they must simultaneously shout, "my hat is black", "my hat is white", or "I don't know the color of my hat". The friends may make any strategy prior to the placement of the hats but subsequent to that, they are not allowed to communicate in any manner. They are awarded a grand prize if at least one person gets the color of her hat correct, and *no one gets her hat color wrong*. In the latter case, they are all punished. The friends need to devise a strategy that increases the probability of their getting the reward.

One strategy for the beating the 50-50 odds is the following. Denoting white and black by 1 and 0 respectively, any configuration of hats on their heads is a point in $\{0, 1\}^n$. Suppose there exists a set $L \subset \{0, 1\}^n$ such that for every element $(x_1, x_2, \ldots, x_n) \in W = \{0, 1\}^n \setminus L$ there is an element $(y_1, y_2, \ldots, y_n) \in L$ such that the set $\{i \mid x_i \neq y_i\}$ has size 1. In this case we say that the set $L$ is *desirable*. Here is a strategy for the $n$ friends: Person $i$ knows $x_j$ for all $j \neq i$, so if there is a unique value of $x_i$ so that $(x_1, x_2, \ldots, x_n) \in W$ then person $i$ declares that her hat color is $x_i$. It then follows that the probability of getting the award by following this strategy is $|W|/2^n$. Thus to maximize this probability, they need a "small" desirable set $L$. This can be obtained by making a random choice. Pick a set $X$ by choosing each element of $\{0, 1\}^n$ independently with probability $p$ (to be determined).

If $Y$ is the set of elements which differ from the chosen elements in at least two coordinates then $E(|Y|) = 2^n(1-p)^{n+1}$, so the set $L = X \cup Y$ satisfies the required condition. Furthermore, by linearity of expectation $E(|L|) = E(|X| + |Y|) = 2^n(p + (1-p)^{n+1})$. Minimizing this over $p \in [0,1]$ we get $E(|L|) \leq 2^n O\left(\frac{\log n}{n}\right)$. In particular, the $n$ friends can achieve a success rate of $1 - O\left(\frac{\log n}{n}\right)$. One can also prove similar results when the friends are assigned hats that may take any one of $q$ different colors, for any $q \geq 2$. See [Al08] for other related results.

**3.  *Property B in uniform hypergraphs*:**  Denote by $m(n)$ the least number of edges in an $n$-uniform hypergraph which does *not* have Property B. Then we have [Er63], [Er64]

$$2^{n-1} \leq m(n) \leq O(n^2 2^n).$$

For the lower bound, we need to show that any $n$-uniform hypergraph with fewer than $2^{n-1}$ edges is 2-colorable. To see that, given such a hypergraph $\mathcal{H}$, randomly and uniformly and independently color each vertex red or blue. Then if $N$ denotes the number of monochrome edges, then $E(N) = \sum_{E \in \mathcal{E}} Pr(E \text{ is monochrome}) < 1$ since the probability that any given edge is monochrome is precisely $1/2^{n-1}$. Thus there is some choice of vertex coloring such that no edge is monochrome.

For the upper bound, we pick a random hypergraph as follows. Let $V$ be the set $\{1, 2, \ldots, v\}$ and choose $m$ edges from the set of all $n$-subsets of $V$, independently at random: specifically, pick an edge $E$ with probability $1/\binom{v}{n}$, and repeat this process over $m$ rounds. Here $v, m$ will be chosen later. Now, fix a partition $\Pi$ of $V$ into two parts of sizes $a, b$. Then the probability that an edge that is not monochrome for $\Pi$ is selected for $\mathcal{H}$ equals

$$\frac{\binom{a}{n} + \binom{b}{n}}{\binom{v}{n}} \geq \frac{2\binom{v/2}{n}}{\binom{v}{n}},$$

so the probability that there exists a partition $\Pi$ such that the corresponding coloring for $\mathcal{H}$ has no monochrome edges is at most

$$(\# \text{ of 2-colorings of } V) \cdot (1 - P)^m \leq 2^v e^{-Pm},$$

where $P$ denotes the quantity $\frac{2\binom{v/2}{n}}{\binom{v}{n}}$. Some calculation and optimizing for $m$ shows that for $v = n^2/2$ and $n$ large enough we have $m \leq \frac{n^2}{2}\ln(2) \cdot 2^{n-1} \cdot e^{1+o(1)}$. It is conjectured that $m(n)/2^n = \Theta(n)$. The best known result is $m(n)/2^n = \Omega(\sqrt{\frac{n}{\log n}})$. [RaSr00]

**4.  *Sum-free subsets of a given set*:**  Given a set $B \subset \mathcal{N}$ of size $n$, how large a sum-free subset can it contain? We will show that $B$ contains a sum-free subset of size greater than $|B|/3$. [Er65], [AlKl90]

Pick a prime $p$ of the form $3k+2$ which is larger than twice the maximum absolute value of elements in $B$. By our choice of $p$, all of the elements in $B$ are distinct mod $p$. Now, consider the sets $xB = \{xb \mid b \in B\}$ in $\mathcal{Z}/p\mathcal{Z}$, and let $N(x) = |[k+1, 2k+1] \cap xB|$. Now note in order to find a sum-free subset of $B$ of the desired size, it suffices to find some $x \in \mathcal{Z}/p\mathcal{Z}$ such that $N(x) > n/3$ since the set $\{k+1, k+2, \ldots, 2k+1\}$ in $\mathcal{Z}/p\mathcal{Z}$ is sum-free. To find such an $x$, pick $x \in (\mathcal{Z}/p\mathcal{Z}) \setminus \{0\}$ uniformly at random. Then

$$E(N(x)) = \sum_{b \in B} Pr\left(x \cdot b \in [k+1, 2k+1]\right) = n \cdot \frac{k+1}{3k+1} > n/3.$$

This again implies that there exists some $x$ such that $N(x) > n/3$ and that completes the proof.

Another way to state this result is as follows. The best constant $c$ for which every set of integers of size $n$ contains a sum-free subset of size at least $cn$ is at least as large as $1/3$. It has been shown that in fact $1/3$ is the best possible constant [EbGrMa14].

**Alterations:**

Sometimes the probabilistic method may not produce the desired object itself but can prove the existence of an object which is very "close" to what is desired.

**1.** *Improvement on $R(k,k)$:* Color the edges of the complete graph $K_n$ red or blue with equal probability, and independently for distinct edges. Then the expected number of monochrome copies of $K_k$ is $m = \binom{n}{k} 2^{-\binom{k}{2}+1}$. Thus there is a coloring of the edges in which there are at most $m$ monochrome copies of $K_k$. Now, from each such monochrome copy, delete a vertex; then the resulting graph on $n - m$ vertices has no monochrome $K_k$. Thus we get $R(k,k) > n - \binom{n}{k} 2^{-\binom{k}{2}+1}$. Optimizing this for $n$ (for fixed $k$) we get $R(k,k) > (\frac{1}{e} + o(1))k2^{k/2}$.

**2.** *A conjecture of Danzer-Grünbaum in combinatorial geometry:* Ludwig Danzer and Branko Grünbaum [DaGr62] conjectured the following: "Any configuration of $2n$ points in $\mathcal{R}^n$ contains some three points that form a non-acute angle". In fact, they managed a proof of the same for dimensions $2, 3$. Following [ErFu83], we give a proof that refutes this conjecture for all $n \geq 40$. To do so, we try to pick a subset of vertices of the $n$-dimensional hypercube, and try to avoid right angles. Since the vertices of the hypercube are in one to one correspondence with the set of subsets of $\{1, 2 \ldots, n\}$, we can denote the vertices picked instead by subsets of $\{1, 2, \ldots, n\}$. Now pick each subset of $\{1, 2, \ldots, n\}$ independently with probability $p$ for some $p$ to be determined. Let $X = \#\{\{P, Q, R\} \mid \angle PQR = \frac{\pi}{2}\}$. Suppose the sets $P, Q, R$ form a right angle. If we denote by $V_P$ the vector in $\mathcal{R}^n$ corresponding to the set (vertex) $P$ of the cube, then $\langle V_P - V_Q, V_R - V_Q \rangle = 0$ where $\langle \cdot, \cdot \rangle$ is the usual dot product in $\mathcal{R}^n$. This implies that $\emptyset = \bar{P} \cap Q \cap \bar{R}$ and $\emptyset = P \cap \bar{Q} \cap R$. Therefore $\{P, Q, R\}$ is a right angle triple if and only if $P \cap R \subset Q \subset P \cup R$. Hence if the set of points picked is $S$, then $E(|S|) = 2^n p$ and it is not difficult to see that $E(|X|) = \frac{6^n p^3}{2}$. Now if we remove one of the sets from each triple of $X$ that appear among the sets chosen in $S$ then the resulting set $S'$ has no right angles either, so there is a subset $S'$ of expected size at least $2^n p - \frac{6^n p^3}{2}$ with no right angles. Optimizing for $p$, we get $E(|S'|) \geq \frac{1}{3}\sqrt{\frac{2}{3}} \left(\frac{2}{\sqrt{3}}\right)^n$, so there is a set of size $\Omega((\frac{2}{\sqrt{3}})^n)$ in the $n$-dimensional hypercube in which no three points form a right angle. For $n \geq 40$, this exceeds $2n$.

For more recent and better bounds (including deterministic constructions) for large sets with only acute angles, and other generalizations, see [Be06].

**3.** *Graphs with arbitrarily large girth and large chromatic number:* Given $k, g \geq 3$ there exist graphs $G$ with $\chi(G) > k$ and girth greater than $g$. The sketch of the proof of this result, due to Erdős [Er59], is as follows. Take the vertex set $V = \{1, 2, \ldots, n\}$ and for each pair $i, j$ select the corresponding edge with probability $p = n^{-1+\theta}$ for some $0 < \theta < 1/g$, independently for each pair. By the choice of $p$ it follows that the probability that this graph has an independent set of size greater than $O(n^{1-\theta} \log n)$ is $o(1)$. Also, if $X$ denotes the number of cycles in $G$ of size at most $g$ then $E(X) = o(n)$, so with high probability (probability greater than $\frac{1}{2}$) $X < n/2$. Hence with positive probability there exists a graph for which the number of cycles of size at most $g$ is less than $n/2$ *and* the size of maximum independent set in $G$ is at most $O(n^{1-\theta} \log n)$. Fix such a graph and remove a vertex from each small cycle (i.e., a cycle of size at most $g$). The resulting graph has no small cycles, at least $n/2$ vertices, and $\chi(G) > \Omega\left(\frac{n^\theta}{\log n}\right)$.

## 7.11.2  DEPENDENT RANDOM CHOICE

For some instances of proving the existence of a combinatorial structure with certain desired properties, it is better to choose a different object randomly and then pick a relevant object associated with the randomly chosen object to be our desired object. Since we do not choose our actual objects of interest randomly but rather in some *dependent manner*, this method is referred to as the method of Dependent Random Choice.

**Definitions:**
For a graph $H$, $ex(n; H)$ denotes the maximum number of edges in a graph on $n$ vertices without a copy of $H$.

For a graph $H$, the **Ramsey number** $R(H)$ denotes the minimum number of vertices $N$ such that every two coloring of the edges of $K_N$ admits a monochrome copy of $H$.

The $n$-dimensional **hypercube** $Q_n$ is the graph whose vertices are all the binary sequences of length $n$; two vertices are adjacent if and only if they differ in exactly one coordinate.

**Examples:**
**1.** Suppose $H$ is bipartite with vertex partition $A \cup B$ and let $|A| = a$, $|B| = b$. In [AlKrSu03], the authors prove that if every vertex of $B$ has degree at most $r$, then $ex(n; H) = O(n^{2-\frac{1}{r}})$. To see this, we shall try to embed each vertex of $H$ into any graph $G$ with $Cn^{2-1/r}$ edges for some absolute constant $C$. First we show that $G$ contains a subset of vertices $A_0$ of size at least $a + b$ with the property that every $r$-subset of $A_0$ has at least $a + b$ common neighbors in $G$. To achieve this, pick a subset $T \subset V(G)$ over $t$ rounds, at random (with repetition, and uniformly over each round). Consider the set of common neighbors of $T$ which we denote by $N^*(T)$. If $d$ denotes the average degree of the vertices of $G$, then

$$E(|N^*(T)|) = \sum_{v \in V} Pr(v \in N^*(T)) = \sum_{v \in V} \left(\frac{d(v)}{n}\right)^t \geq \frac{d^t}{n^{t-1}}.$$

Let $Y$ denote the number of $r$-subsets $U$ of $N^*(T)$ such that $U$ has fewer than $a + b$ common neighbors. If $U$ is a subset of size $r$ with fewer than $a + b$ neighbors, the event $U \subset N^*(T)$ occurs only if every element of $T$ was picked from among the common neighbors of $U$, so $Pr(U \subset N^*(T)) \leq (\frac{a+b}{n})^t$, and hence by linearity of expectation, $E(Y) \leq \binom{n}{r}(\frac{a+b}{n})^t$. So, $E(|N^*(T)| - Y) \geq \frac{d^t}{n^{t-1}} - \binom{n}{r}(\frac{a+b}{n})^t$. Thus it follows that there exists a set $A_0 \subset N^*(T)$ of size $\geq \frac{d^t}{n^{t-1}} - \binom{n}{r}(\frac{a+b}{n})^t$ such that every $r$-subset of $A_0$ has at least $a + b$ common neighbors in $G$. Now since $V(H) = A \cup B$, we embed the vertices of $A$ into $A_0$ arbitrarily. To embed the vertices of $B$, let $B = \{v_1, v_2, \ldots, v_b\}$ and suppose that we have already embedded $v_1, \ldots, v_{i-1}$ into $A_0$. Let $W$ denote the set of neighbors of $v_i$ in $A$ (in the graph $H$). Since $A$ has been embedded into $A_0$, this gives a set $U \subset V(G)$ of size at most $r$ corresponding to the set of neighbors of $v_i$. But every $U$ of size at most $r$ has at least $a + b$ neighbors in $G$, of which at most $a + i - 1$ (with $i < b$) vertices are already assigned to some vertex of $A \cup B$. Hence, $v_i$ can be embedded into $A_0$ as well. To complete the proof, some routine calculations show that for $C > \frac{1}{2}(a + \frac{(a+b)^r}{r!})^{\frac{1}{r}}$ we have $|A_0| \geq a + b$.

**2.** *The Ramsey number $R(Q_n)$*: Burr and Erdős [BuEr75] conjectured that $R(Q_n) = O(2^n)$. This conjecture is yet unsettled, but we show here (following [FoSu11]) that $R(Q_n) \leq 2^{3n}$. Set $N = 2^{3n}$ and consider the denser of the two colors that appear in a two coloring of the edges of $K_N$. There are at least $\frac{1}{2}\binom{N}{2}$ edges of this color, so the average degree of the graph $G$ induced by edges of this color is at least $2^{-4/3}N$. Set $t = 3n/2, m = 2^n$ and $a = 2^{n-1}$. Then

$$\frac{d^t}{N^{t-1}} - \binom{N}{n}\left(\frac{m}{N}\right)^t \geq 2^{-4/3}N - N^{n-t}m^t/n! \geq 2^{n-1}.$$

As in the previous proof, it follows that the graph $G$ contains a set $U$ of size $2^{n-1}$ with the property that every $n$ vertices of $U$ have at least $2^n$ common neighbors. Since $Q_n$ is an $n$-regular bipartite graph with parts of size $2^{n-1}$, it follows that $G$ contains $Q_n$.

**3.** For additional examples, see the survey paper [FoSu11].

## 7.11.3   THE SECOND MOMENT

**Definitions:**

The **variance** of the random variable $X$ is $\mathrm{Var}(X) = E(X - E(X))^2$.

A set of positive integers $\{x_1, x_2, \ldots, x_k\}$ has **distinct sums** if the quantities $\sum_{i \in S} x_i$ are all distinct for all subsets $S \subseteq \{1, 2, \ldots, k\}$.

**Facts:**

**1.** *Chebyshev's inequality*: For a random variable $X : \Omega \to \mathcal{R}$ and any $\lambda > 0$

$$Pr(|X - E(X)| > \lambda) \leq \frac{\mathrm{Var}(X)}{\lambda^2}.$$

**2.** If $X$ is a nonnegative integer-valued random variable, then

$$Pr(X = 0) \leq \frac{\mathrm{Var}(X)}{E(X)^2}.$$

**3.** If $\mathrm{Var}(X) = o(E(X)^2)$, then asymptotically almost surely $X \sim E(X)$.

**Examples:**

**1.** Erdős posed the question of estimating the maximum size $f(n)$ of a set $\{x_1, x_2, \ldots, x_k\}$ with distinct sums and $x_k \leq n$. We shall show that $f(n) \leq \log_2 n + \frac{1}{2}\log_2\log_2 n + O(1)$. Suppose $X = \{x_1, x_2, \ldots, x_k\}$ is a maximum sized set with distinct sums. Pick a random subset $S$ of $\{1, 2, \ldots, k\}$ by selecting each element independently with probability $1/2$. This subset yields the random sum $X_S = \sum_{i \in S} x_i$ with $E(X_S) = \frac{1}{2}(x_1 + x_2 + \cdots + x_k)$. Also, $\mathrm{Var}(X_S) = \frac{1}{4}(x_1^2 + x_2^2 + \cdots + x_k^2) \leq \frac{n^2 k}{4}$, so by Chebyshev's inequality we have $Pr(|X_S - E(X_S)| < \lambda) \geq 1 - \frac{n^2 k}{4\lambda^2}$. Since $X$ has distinct sums and there are $2^k$ distinct subsets of $\{x_1, x_2, \ldots, x_k\}$, it follows that for each integer $r$ we have that $Pr(X_S = r)$ is either $\frac{1}{2^k}$ or $0$, depending on whether $r$ appears as a sum of some subset of $X$ or not. This observation coupled with Chebyshev's inequality gives

$$1 - \frac{n^2 k}{4\lambda^2} \leq Pr(|X_S - E(X_S)| < \lambda) \leq \frac{2\lambda + 1}{2^k}.$$

Optimizing for $\lambda$ gives the desired result.

**2.** Given a set $X$ of at least $4k^2$ distinct residue classes modulo a prime $p$, there exists an integer $a$ such that $\{ax \pmod{p} \mid x \in X\}$ intersects every cyclic interval in $\{0, 1, \ldots, p-1\}$ of length at least $p/k$ (see [AlKrNe95], [AlPe92]). To see this, we first note that if some shift $aX + b$ satisfies this property, then so does the set $aX$.

We start with a simple observation. Let $I_1, I_2, \ldots, I_{2k}$ be a (fixed) covering of $\{0, 1, \ldots, p-1\}$ by $2k$ intervals of length $\lceil p/2k \rceil$ each. Suppose that a set $Y$ intersects each of these intervals $I_j$; then in particular, since every cyclic interval $J$ of size at least $p/k$ must contain some $I_j$ it follows that $Y \cap J \neq \emptyset$. Select $a \in \{1, 2, \ldots, p-1\}$ and $b \in \{0, 1, \ldots, p-1\}$ uniformly and independently, and let $X_{a,b} = \{ax+b \mid x \in X\}$. Here the arithmetic is modulo $p$. Fix an interval $I_j$ and consider the random variables $X_x^j$ (for $x \in X$) which equal one if $ax + b \in I_j$ and zero otherwise. Setting $N_j = \sum_{x \in X} X_x^j = |X_{a,b} \cap I_j|$ we see that $E(N_j) = \sum_{x \in X} Pr(ax + b \in I_j) \geq 2k$. Furthermore, $\text{Var}(N_j) = \sum_{x \in X} \text{Var}(X_x) \leq 2k$. The equality in the above holds because the random variables $X_x^j$ (as $x$ varies over $X$) are pairwise independent, though they are not all mutually independent. Hence, by Chebyshev's inequality, $Pr(N_j = 0) < 1/2k$. In particular, the probability that $N_j = 0$ for some $j$ is at most $\sum_j Pr(N_j = 0)$, and this is strictly less than 1. Consequently, there exist $a, b$ such that $X_{a,b}$ intersects every $I_j$, and this completes the proof.

## 7.11.4   OCCURRENCE OF RARE EVENTS: THE LOCAL LEMMA

In all the examples seen so far, the probability of the "bad" event(s) is in fact negligible. But in order to prove that a "good" event occurs, we only need to show that the probability of the bad event is strictly less than one.

**Definitions:**

Suppose $E_1, E_2, \ldots, E_N$ are events in a probability space $(\Omega, Pr)$. A graph $\mathcal{D} = (\mathcal{V}, \mathcal{E})$ with vertex set $\mathcal{V} = \{E_i \mid i = 1, \ldots, N\}$ is called a **dependency digraph** of the events $E_i$ if each $E_i$ is independent of the set $\{E_j \mid (E_i, E_j) \notin \mathcal{E}\}$. In words, $E_i$ is independent of all the events $E_j$ whenever $E_i$ and $E_j$ are nonadjacent in $\mathcal{D}$. We write $i \sim j$ (resp. $i \not\sim j$) to denote $(E_i, E_j) \in \mathcal{D}$ (resp. $(E_i, E_j) \notin \mathcal{D}$).

Suppose graph $G$ admits a partition $V = V_1 \cup V_2 \cup \cdots \cup V_r$ of its vertex set. We say a set $\mathcal{I}$ is an **independent transversal** for this partition if $\mathcal{I}$ contains exactly one element from each $V_i$ and is independent in $G$.

Fix $k \geq 2$. For a $k$ coloring $c : \mathcal{R} \to \{1, 2, \ldots, k\}$ of $\mathcal{R}$ and a fixed subset $T \subset \mathcal{R}$ we say that $T$ is **multicolored with respect to $c$** if every color appears in the coloring induced on $T$.

A **proper $k$-coloring** of (the vertices of) a graph $G = (V, E)$ is a map $C : V \to \{1, 2, \ldots, k\}$ such that for adjacent vertices $x$ and $y$ we have $C(x) \neq C(y)$. For any integer $\beta > 1$ a proper vertex coloring $C$ of $G$ is **$\beta$-frugal** if for any vertex $v$ and any color $c$ there are at most $\beta$ neighbors of $v$ that are colored $c$.

**Facts:**

**1.** *Lovász Local Lemma - symmetric case:*   Suppose $E_1, E_2, \ldots, E_N$ are events in a probability space $(\Omega, Pr)$ and $\mathcal{D}$ is a dependency graph for these events. Let $d$ denote the maximum vertex degree in $\mathcal{D}$, and let $0 < p < 1$ be such that $Pr(E_i) \leq p$ for all $i$. If $ep(d+1) \leq 1$, then
$$Pr\left(\overline{E_1} \wedge \overline{E_2} \wedge \cdots \wedge \overline{E_N}\right) > 0.$$

In particular, with positive probability, none of the events $E_i$ occur. Here, $e \approx 2.71828$ denotes the Euler constant. [ErLo75]

**2.** *Lovász Local Lemma - general case:*   Suppose $E_1, E_2, \ldots, E_N$ are events in a probability space $(\Omega, Pr)$ and $\mathcal{D}$ is a dependency graph for these events. Suppose there are real numbers $x_i, i = 1, \ldots, N$ satisfying $0 \le x_i < 1$ so that for each event $E_i$ we have

$$Pr(E_i) \le x_i \prod_{j \not\sim i}(1 - x_j).$$

Then

$$Pr\left(\overline{E_1} \wedge \overline{E_2} \wedge \cdots \wedge \overline{E_N}\right) \ge \prod_{i=1}^{N}(1 - x_i) > 0.$$

Again, with positive probability, none of the events $E_i$ occur. [ErLo75]

**Examples:**

**1.** *Independent transversals:* Suppose the graph $G$ admits a vertex partition $V = V_1 \cup V_2 \cup \cdots \cup V_r$ and the maximum vertex degree in $G$ equals $d$. Suppose further that each set $V_i$ has size at least $2ed$. Then there exists an independent transversal in $G$ for the partition $(V_1, V_2, \ldots, V_r)$. [Al88]

To see this, pick $v_i \in V_i$ independently and uniformly for each $i$. For each edge $(x, y)$, let $E_{xy}$ denote the event that both vertices $x$ and $y$ are selected. Our chosen set is an independent transversal if and only if for each edge $(x, y) \in E(G)$, the event $E_{xy}$ does not occur. Consider the graph $\mathcal{D}$ whose vertex set consists of the edges of $G$ in which two edges are adjacent in $\mathcal{D}$ if and only if they intersect. Then $\mathcal{D}$ is a dependency graph for the events $E_{xy}$ for each $(x, y) \in E(G)$. To use the local lemma, note that the probability that both vertices of any fixed edge $e \in E(G)$ are included in the transversal is at most $1/(2ed)^2$, and since each vertex in $G$ has degree at most $d$, the maximum degree of $\mathcal{D}$ is at most $2d$. Hence by the local lemma, it follows that with positive probability, none of the events $E_{xy}$ occur, since $\frac{e(2d+1)}{(2ed)^2} \le 1$.

**2.** *A question of Straus:* Does there exist $m(k) > 0$ such that for any given fixed $S$ of size $m$, there exists a $k$-coloring of $\mathcal{R}$ such that every translate $x + S$, $x \in \mathcal{R}$ is multicolored? It turns out that $m(k) = (3 + o(1))k \log k$ suffices. [ErLo75]

To see this, first let $X$ be a finite set of real numbers and let $\mathcal{X} = \bigcup_{x \in X}(x + S)$. Now color every element of $\mathcal{X}$ from $\{1, 2, \ldots, k\}$ independently and uniformly. We claim that $Pr(\wedge_{x \in X}((x + S) \text{ is not multicolored})) > 0$. Fix $x$ and denote the event that $x + S$ is not multicolored by $\mathcal{E}_x$. Then

$$Pr(\mathcal{E}_x) \le k \left(1 - \tfrac{1}{k}\right)^m.$$

Moreover, $\mathcal{E}_x$ is independent of all $\mathcal{E}_y$ as long as $(x + S) \cap (y + S) = \emptyset$. Hence by the local lemma, we are done provided $ekm^2 \left(1 - \frac{1}{k}\right)^m \le 1$. It is a routine calculation to see that this holds if $m > (3 + o(1))k \log k$. Hence it follows that for every finite set $X$, there exists a color choice for the elements of $\mathcal{X}$ such that each translate $x + S$ is multicolored. This coupled with a compactness argument shows that the same result holds for $X = \mathcal{R}$.

**3.** *Frugal colorings:* Suppose $G$ is a regular graph of degree $\Delta$, i.e., every vertex of $G$ has degree $\Delta$. Then for $\Delta$ sufficiently large, there is a proper $\beta$-frugal coloring of $G$ using at most $O(\Delta^{1+\frac{1}{\beta}})$ colors [HiMoRe97]. To see this, as always, color each vertex uniformly and independently from the set of colors $\{1, 2, \ldots, N\}$ where $N = m\Delta^{1+\frac{1}{\beta}}$ for some suitable constant $m$ that shall be determined later. For each edge $uv$ we consider the event $A_{uv}$ that both $u$ and $v$ have been assigned the same color. Likewise, for each independent set $v_1, v_2, \ldots, v_{\beta+1}$ of vertices that are also neighbors of some vertex $x$ in $G$, we consider the event $B(v_1, v_2, \ldots, v_{\beta+1})$ that all the vertices $v_1, \ldots, v_{\beta+1}$ have been assigned the same color. This random coloring describes a frugal coloring if none of the events $A_{uv}$ or $B(v_1, v_2, \ldots, v_{\beta+1})$ occur. Note that each event $A_{uv}$ is independent of all other $A_{wz}$ as long as $\{u, v\} \cap \{w, z\} = \emptyset$ and independent of

$B(v_1, v_2, \ldots, v_{\beta+1})$ if $\{v_1, v_2, \ldots, v_{\beta+1}\} \cap \{u, v\} = \emptyset$. Similarly, $B(v_1, v_2, \ldots, v_{\beta+1})$ is independent of $B(w_1, w_2, \ldots, w_{\beta+1})$ if $\{v_1, v_2, \ldots, v_{\beta+1}\} \cap \{w_1, w_2, \ldots, w_{\beta+1}\} = \emptyset$.

Let $x = 2/N$ and $y = 2/N^\beta$. Following a few routine calculations, it follows that

$$Pr(A_{uv}) \le x(1-x)^{2\Delta-2}(1-y)^{2\Delta\binom{\Delta}{\beta}}$$

and

$$Pr(B(v_1, v_2, \ldots, v_{\beta+1})) \le y(1-x)^{(\beta+1)\Delta}(1-y)^{(\beta+1)\Delta\binom{\Delta}{\beta}},$$

for $m = 12$, so by the general version of the local lemma the proof follows.

To see how good the above bound is, consider the following bipartite graphs $G_\beta$, for any integer $\beta > 1$. The vertex set consists of the 1-dimensional subspaces (*points*) and $(\beta+1)$-dimensional subspaces ($\beta + 1$ *flats*) of a $(\beta+2)$-dimensional vector space over a finite field of $q$ elements. A point $p$ and a $\beta + 1$ flat $\Pi$ are adjacent in $G_\beta$ if $p$ is a subspace of $\Pi$. Note that no color can be used more than $\beta$ times in a $\beta$-frugal coloring. Since any $\beta + 1$ distinct points together lie in a $(\beta+1)$-dimensional flat, the number of colors needed is at least $(q^{\beta+1} + q^\beta + \cdots + 1)/\beta$. Since the maximum degree in $G_\beta$ equals $q^\beta + q^{\beta-1} + \cdots + q + 1$, the number of colors needed in a $\beta$-frugal coloring is at least $\Omega(\Delta^{1+1/\beta})$. Hence the result above is tight up to the constant factor $C$.

## 7.11.5   BASIC CONCENTRATION INEQUALITIES

**Definitions:**

For a hypergraph $\mathcal{H}$, a 2-coloring $\chi$ of the vertices of $\mathcal{H}$, and an edge $e \in \mathcal{E}(\mathcal{H})$ the **discrepancy** $D(e)$ of the edge $e$ is the absolute value of the difference between the number of vertices in the two color classes. The **discrepancy of $\mathcal{H}$ with respect to $\chi$** denoted $D(\mathcal{H}, \chi)$ is defined as the maximum $D(e)$ as $e$ ranges over the edges in $\mathcal{H}$. The **discrepancy of the hypergraph $\mathcal{H}$** is $D(\mathcal{H}) = \min_\chi D(\mathcal{H}, \chi)$ where $\chi$ ranges over all 2-colorings of $\mathcal{H}$.

For graphs $G_1 = (V_1, E_1)$ and $G_2 = (V_2, E_2)$ the **Cartesian product** of $G_1$ and $G_2$ denoted $G_1 \square G_2$ is the graph with vertex set $V_1 \times V_2$ where vertices $(u_1, u_2)$ and $(v_1, v_2)$ are adjacent if either $u_1 = v_1$ and $(u_2, v_2) \in E_2$ or $(u_1, v_1) \in E_1$ and $u_2 = v_2$. The Cartesian product of graphs is associative and commutative.

For vertices $u, v$ in a graph $G$, the **distance** $d_G(u, v)$ between $u$ and $v$ in $G$ is the length of a shortest $u$-$v$ path in $G$. If $u$ and $v$ lie in different connected components, then we write $d_G(u, v) = \infty$. For a subset of vertices $U$, $d_G(x, U) = \min_{u \in U} d_G(x, u)$.

**Facts:**

1. *Chernoff bounds* [Ch52]: Suppose $X$ is a binomial random variable $B(n, p)$, and $\lambda = E(X) = np$. Then for $t \ge 0$

$$Pr(X \ge E(X) + t) \le \exp\left(-\frac{t^2}{2(\lambda+t/3)}\right),$$

$$Pr(X \ge E(X) - t) \le \exp\left(-\frac{t^2}{2\lambda}\right).$$

2. *Chernoff bounds - general case*: If $X_i \sim Ber(p_i)$, $i = 1, \ldots, n$ and $X_i$ are mutually independent Bernoulli random variables defined on $\Omega$ with parameters $p_i$, then the same inequalities as above hold for $X = \sum_{i=1}^n X_i$ with $\lambda = E(X) = \sum_{i=1}^n p_i$.

**3.** *Simple concentration bound:* Suppose $X_1, X_2, \ldots, X_n$ are independent random variables in $(\Omega, Pr)$ and suppose $f(x_1, x_2, \ldots, x_n) : \mathcal{R}^n \to \mathcal{R}$ is a real-valued function that satisfies the *Lipschitz* condition, i.e., $|f(x_1, x_2, \ldots, x_i, \ldots, x_n) - f(x_1, x_2, \ldots, x_i', \ldots, x_n)| \leq c_i$ for each $i$ and for some $c_i > 0$. If $X = f(X_1, X_2, \ldots, X_n)$, then

$$Pr(|X - E(X)| > t) \leq 2 \exp\left(-\frac{t^2}{2\sum_{i=1}^n c_i^2}\right).$$

**4.** *Azuma's inequality* [Az67]: Let $X_0, X_1, \ldots, X_n$ be a martingale such that $|X_{i+1} - X_i| \leq c_i$ for each $i$ and some reals $c_i > 0$. Then for $\lambda > 0$

$$Pr(X_n - X_0 > \lambda) \leq \exp\left(-\frac{\lambda^2}{2\sum_{i=1}^n c_i^2}\right),$$

$$Pr(X_n - X_0 < -\lambda) \leq \exp\left(-\frac{\lambda^2}{2\sum_{i=1}^n c_i^2}\right).$$

**Examples:**

**1.** *Discrepancy:* Let $\mathcal{H}$ be a $k$-uniform hypergraph with $k$ edges. Color each vertex uniformly and independently using one of the two colors. We shall show that with positive probability, this random coloring $\chi$ satisfies $D(\mathcal{H}, \chi) \leq \sqrt{8k \log k}$, so that $D(\mathcal{H}) \leq \sqrt{8k \log k}$. For each edge $e$, we have by the Chernoff bounds that $Pr(D(e) > 2t) \leq 2e^{-t^2/(3k/2)} < \frac{1}{k}$ for $t = \sqrt{k \log k}$. Hence the probability that $D(e)$ exceeds $2t$ for some edge $e \in \mathcal{H}$ is less than 1, so that with positive probability, $\chi$ has the desired property.

**2.** *A result of Erdős, Silverman, and Stein* [ErSiSt83]: Let $\Pi_n$ be a projective plane of order $n$. There exists a subset $S$ of points, and absolute constants $k, K$ such that $k \log n \leq |L \cap S| \leq K \log n$ for every line $L$ in $\Pi_n$.

To see this, choose $S$ at random, with each point $x$ placed in $S$ with probability $p = \frac{f(n)}{n+1}$, for some $f(n)$ to be determined later. Fix a line $L$, and let $S_L = |S \cap L|$. Note that $E(S_L) = (n+1)p = f(n)$. By the Chernoff bound, $Pr[|S_L - f(n)| > f(n)] < 2e^{-f(n)/3}$. Since $\Pi_n$ contains $n^2 + n + 1$ lines,

$$Pr \text{ (there exists } L \text{ such that } |S_l - f(n)| > f(n)) < 4e^{-f(n)/3}n^2.$$

Therefore, if $e^{f(n)/3} > 4n^2$, the desired $S$ exists. It is easy to check that $f(n) = 3 \log 4n^2$ works for the aforementioned argument.

**3.** Consider the graph $G_{r,n} = K_r^{\square n}$, the $r$-fold Cartesian product of the complete graph $K_r$. For $\epsilon > 0$, let $W \subset G_{r,n}$ be a subset of size at least $\epsilon r^n$ and let $U$ be the set of vertices of $G_{r,n}$ whose distance from $W$ is greater than $2\sqrt{n \log n}$. We claim that $|U| \leq \frac{Cr^n}{n^2}$ for some absolute constant $C$ that depends only on $\epsilon$.

Suppose $x$ is a vertex of $G_{r,n}$ obtained by choosing $v_1, v_2, \ldots, v_n$ uniformly and independently among the vertices of $K_r$. Consider the *Doob martingale process* $\{Y_i\}_{i=0,1,\ldots,n}$ of exposing one $v_i$ at a time, i.e., consider the random variables $Y_i = E(d(x, W)|v_1, v_2, \ldots, v_i)$ for $i = 0, 1, \ldots, n$ (here by convention, $Y_0 = E(d(x, W))$) where we write $d(x, W) = d_G(x, W)$ for simplicity. $\{Y_i\}$ is a martingale with $c_i = 1$ for each $i$ since altering $v_i$ changes the value of $Y_{i-1}$ by at most one. By Azuma's inequality, we have $Pr(d(x, W) - E(d(x, W)) < -c\sqrt{n}) \leq e^{-c^2/2}$. Setting $c < \sqrt{2 \log(1/\epsilon)}$ we see that $Pr(d(x, W) < E(d(x, W) - c\sqrt{n}) < \epsilon$. However, note that since $|W| \geq \epsilon r^n$, we have $Pr(d(x, W) = 0) \geq \epsilon$, so in particular, $E(d(x, W) < \sqrt{2 \log(1/\epsilon)}n$. On the other hand, we also have $Pr(d(x, W) - E(d(x, W)) > c\sqrt{n}) \leq e^{-c^2/2}$, so setting $c = 2\sqrt{\log n}$ we get $Pr(d(x, W) > 2\sqrt{n \log n}) \leq O(1/n^2)$. This completes the proof.

## 7.11.6  CORRELATION INEQUALITIES

**Definitions:**

A family of subsets $\mathcal{A}$ of $\{1, 2, \ldots, n\}$ is called an **upset** if whenever $A \in \mathcal{A}$ and $A \subset A'$ then $A' \in \mathcal{A}$ as well. A family of subsets $\mathcal{A}$ of $\{1, 2, \ldots, n\}$ is called a **downset** if whenever $A \in \mathcal{A}$ and $A' \subset A$ then $A' \in \mathcal{A}$ as well.

Suppose $L$ is a finite distributive lattice, i.e., a lattice in which for any $x, y, z \in L$ we have $(x \wedge y) \vee z = (x \vee z) \wedge (y \vee z)$. A function $\mu : L \to \mathcal{R}^+$ is called **log-supermodular** if $\mu(x)\mu(y) \leq \mu(x \vee y)\mu(x \wedge y)$.

A real-valued function $f$ defined on a finite distributive lattice is called **increasing** if $f(x) \leq f(y)$ whenever $x \leq y$. Similarly, a function $f$ is called **decreasing** if $f(x) \geq f(y)$ whenever $x \leq y$.

**Facts:**

**1.** *Four function theorem* [AhDa78]:   Suppose $\alpha, \beta, \gamma, \delta$ are nonnegative real-valued functions defined on the set of all subsets of $\{1, 2, \ldots, n\}$ and suppose for any sets $A, B \subset \{1, 2, \ldots, n\}$

$$\alpha(A)\beta(B) \leq \gamma(A \cup B)\delta(A \cap B)$$

holds. Then for any families of sets $\mathcal{A}, \mathcal{B}$

$$\alpha(\mathcal{A})\beta(\mathcal{B}) \leq \gamma(\mathcal{A} \cup \mathcal{B})\delta(\mathcal{A} \cap \mathcal{B}),$$

where for any function $\phi$, $\phi(\mathcal{A})$ denotes the sum $\sum_{A \in \mathcal{A}} \phi(A)$.

**2.** *Four function theorem - lattice version:*  Suppose $\alpha, \beta, \gamma, \delta$ are nonnegative real-valued functions defined on the distributive lattice $L$ and suppose that for any $a, b \in L$

$$\alpha(a)\beta(b) \leq \gamma(a \vee b)\delta(a \wedge b).$$

Then for any subsets $A, B \subset L$

$$\alpha(A)\beta(B) \leq \gamma(A \vee B)\delta(A \wedge B),$$

where $A \vee B = \{a \vee b \mid a \in A,\ b \in B\}$ and $A \wedge B = \{a \wedge b \mid a \in A,\ b \in B\}$.

**3.** *FKG inequality* [FoKaGi71]: Let $L$ be a finite distributive lattice and let $\mu : L \to \mathcal{R}^+$ be a log-supermodular function. Then for any two increasing functions $f, g$

$$\left( \textstyle\sum_{x \in L} \mu(x)f(x) \right) \left( \textstyle\sum_{x \in L} g(x)\mu(x) \right) \leq \left( \textstyle\sum_{x \in L} \mu(x)f(x)g(x) \right) \left( \textstyle\sum_{x \in L} \mu(x) \right).$$

**4.** *Correlation inequalities:*  Suppose $\mathcal{A}, \mathcal{B}, \mathcal{C}, \mathcal{D}$ are families of subsets of $\{1, 2, \ldots, n\}$ with $\mathcal{A}, \mathcal{B}$ being upsets and $\mathcal{C}, \mathcal{D}$ being downsets. Pick a random subset $R$ of $\{1, 2, \ldots, n\}$ by picking each $i \in \{1, 2, \ldots, n\}$ independently with probability $0 \leq p_i < 1$. Then

$$
\begin{aligned}
Pr(R \in \mathcal{A} \cap \mathcal{B}) &\geq Pr(R \in \mathcal{A}) \cdot Pr(A \in \mathcal{B}), \\
Pr(R \in \mathcal{C} \cap \mathcal{D}) &\geq Pr(R \in \mathcal{C}) \cdot Pr(R \in \mathcal{D}), \\
Pr(R \in \mathcal{A} \cap \mathcal{C}) &\leq Pr(R \in \mathcal{A}) \cdot Pr(R \in \mathcal{C}).
\end{aligned}
$$

**Examples:**

**1.** Suppose $0 < \alpha_1 \le \alpha_2 \le \cdots \le \alpha_n$ and $0 < \beta_1 \le \beta_2 \le \cdots \le \beta_n$ are increasing sequences of positive real numbers. Then

$$n \sum_i \alpha_i \beta_{n+1-i} \le \left( \sum_i \alpha_i \right) \left( \sum_i \beta_i \right) \le n \sum_i \alpha_i \beta_i.$$

To prove the first inequality, we use induction on $n$; if the inequality holds for $n-1$, then applying the inequality for the sequences $\mathfrak{a}_i = (\alpha_1, \alpha_2, \ldots, \alpha_{i-1}, \alpha_{i+1}, \ldots, \alpha_n)$ and $\mathfrak{b}_i = (\beta_1, \beta_2, \ldots, \beta_{n-i}, \beta_{n-i+2}, \ldots, \beta_n)$ and summing these inequalities as $i$ ranges over $1, 2, \ldots, n$, we get

$$\sum_i (A - \alpha_i)(B - \beta_{n+1-i}) \ge (n-1)^2 \sum_i \alpha_i \beta_{n+1-i},$$

where $A = \sum_i \alpha_i$, $B = \sum_i \beta_i$. This inequality simplifies to $(n-2)AB + \sum_i \alpha_i \beta_{n+1-i} \ge (n-1)^2 \sum_i \alpha_i \beta_{n+1-i}$, which establishes the inductive hypothesis. For the second inequality, consider the distributive lattice $L = \{1, 2, \ldots, n\}$ with $i \vee j = \max\{i, j\}$, $i \wedge j = \min\{i, j\}$. It is easy to check that $\alpha(i) = \alpha_i, \beta(i) = \beta_i, \gamma(k) = \max\{\alpha_i \beta_j \mid i \vee j = k\}$, $\delta(i) = 1$ satisfy the hypotheses of the four function theorem. Furthermore, since the sequences $\alpha_i, \beta_i$ are nonnegative and increasing we have $\gamma(i) \le \alpha_i \beta_i$. Now the four function theorem gives $AB \le \sum_i n\gamma(i) \le n \sum_i \alpha_i \beta_i$ which is what we seek.

**2.** Suppose $\mathcal{A}$ is a family of subsets of $\{1, 2, \ldots, n\}$ such that for any $A, B \in \mathcal{A}$ both $A \cap B \ne \emptyset$ and $A \cup B \ne \{1, 2, \ldots, n\}$ hold. Then $|\mathcal{A}| \le 2^{n-2}$. To see why, let $\mathcal{C}$ be a maximal intersecting family containing $\mathcal{A}$, and let $\mathcal{D}$ be a maximal family containing $\mathcal{A}$ with no two sets having $\{1, 2, \ldots, n\}$ as their union. Note that the maximality of $\mathcal{C}$ and $\mathcal{D}$ implies that $\mathcal{C}$ is an upset and $\mathcal{D}$ is a downset. Since $\mathcal{A} \subset \mathcal{C} \cap \mathcal{D}$, by the correlation inequalities

$$|\mathcal{A}| \le |\mathcal{C} \cap \mathcal{D}| \le \frac{|\mathcal{C}| \cdot |\mathcal{D}|}{2^n} = 2^{n-2}.$$

This is also clearly best possible, e.g., consider the family $\mathcal{A} = \{A \mid 1 \in A,\ n \notin A\}$.

**3.** Suppose $\mathcal{A}_1, \mathcal{A}_2, \ldots, \mathcal{A}_k$ (for $k \ge 1$) are intersecting families on $\{1, 2, \ldots, n\}$. Then how large can $\left| \bigcup_{i=1}^k \mathcal{A}_i \right|$ be? Clearly, if we set $\mathcal{A}_i$ to be the intersecting family of all subsets of $\{1, 2, \ldots, n\}$ containing $i$, then $\bigcup_{i=1}^k \mathcal{A}_i$ consists of all those sets that contain at least one of the elements of $\{1, \ldots, k\}$, so clearly $\left| \bigcup_{i=1}^k \mathcal{A}_i \right| = 2^n - 2^{n-k}$. We claim that this is indeed best possible. For $k = 1$ this follows since for any set $A \subset \{1, 2, \ldots, n\}$ we must have $|\mathcal{A} \cap \{A, \overline{A}\}| \le 1$, so an intersecting family of $\{1, 2, \ldots, n\}$ contains at most $2^{n-1}$ sets. In general, let $\mathcal{B} = \bigcup_{i=1}^{k-1} \mathcal{A}_i$. Since each $\mathcal{A}_i$ is an intersecting family, we may assume that each $\mathcal{A}_i$ is an upset since adding larger sets will not violate the property that $\mathcal{A}_i$ is intersecting. Since each $\mathcal{A}_i$ is an upset, it follows that $\mathcal{B}$ is an upset as well. Also, without loss of generality, assume that $\mathcal{A} := \mathcal{A}_k$ is a maximal intersecting family, so $|\mathcal{A}| = 2^{n-1}$. Let us pick a set $T$ uniformly at random from $\{1, 2, \ldots, n\}$. Then

$$
\begin{aligned}
Pr(T \in \mathcal{A} \cup \mathcal{B}) &= Pr(T \in \mathcal{A}) + Pr(T \in \mathcal{B}) - Pr(T \in \mathcal{A} \cap \mathcal{B}) \\
&\le \tfrac{1}{2} + Pr(T \in \mathcal{B}) - \tfrac{1}{2} Pr(T \in \mathcal{B}) \\
&= \tfrac{1}{2} + \tfrac{1}{2} Pr(T \in \mathcal{B}) \\
&\le \tfrac{1}{2} + \tfrac{1}{2} \left( 1 - \tfrac{1}{2^{k-1}} \right) = 1 - \tfrac{1}{2^k},
\end{aligned}
$$

where the first inequality is a consequence of the correlation inequality applied to the upsets $\mathcal{A}, \mathcal{B}$, and the last inequality follows by induction on $k$. But this translates as $\left| \bigcup \mathcal{A}_i \right| \le 2^n \left( 1 - \tfrac{1}{2^k} \right) = 2^n - 2^{n-k}$.

# REFERENCES

***Printed Resources***:

[AhDa78] R. Ahlswede and D. E. Daykin, "An inequality for the weights of two families of sets, their unions and intersections", *Zeitschrift für Wahrscheinlichkeitstheorie verw. Gebiete* 43 (1978), 183–185.

[AlFi02] D. Aldous and J. A. Fill, *Reversible Markov Chains and Random Walks on Graphs*, unfinished monograph, 2002.

[Al88] N. Alon, "The linear arboricity of graphs", *Israel Journal of Mathematics* 62 (1988), 311–325.

[Al08] N. Alon, "Problems and results in extremal combinatorics II", *Discrete Mathematics* 308 (2008), 4460–4472.

[AlKl90] N. Alon and D. J. Kleitman, "Sum-free subsets", in *A Tribute to Paul Erdős*, A. Baker, B. Bollobás, and A. Hajnal (eds.), Cambridge University Press, 13–26, 1990.

[AlKrSu03] N. Alon, M. Krivelevich, and B. Sudakov, "Turán numbers of bipartite graphs and related Ramsey-type questions", *Combinatorics, Probability and Computing* 12 (2003), 477–494.

[AlKrNe95] N. Alon, I. Kriz, and J. Nešetřil, "How to color shift hypergraphs", *Studia Scientiarum Mathematicarum Hungarica* 30 (1995), 1–11.

[AlPe92] N. Alon and Y. Peres, "Uniform dilations", *Geometric and Functional Analysis* 2 (1992), 1–28.

[AlSp08] N. Alon and J. Spencer, *The Probabilistic Method*, 3rd ed., Wiley, 2008.

[Az67] K. Azuma, "Weighted sums of certain dependent variables", *Tohoku Mathematical Journal* 3 (1967), 357–367.

[BaNe03] N. Balakrishnan and V. B. Nevzorov, *A Primer on Statistical Distributions*, Wiley, 2003.

[BaNi70] M. N. Barber and B. W. Ninham, *Random and Restricted Walks: Theory and Applications*, Gordon and Breach, 1970.

[BaPr87] R. E. Barlow and F. Proschan, *Mathematical Theory of Reliability*, Society for Industrial and Applied Mathematics, 1987.

[Be93] H. C. Berg, *Random Walks in Biology*, Princeton University Press, 1993.

[Be06] D. Bevan, "Sets of points determining only acute angles and some related colouring problems", *Electronic Journal of Combinatorics* 13 (2006), # R12.

[BhHa04] R. Bhar and S. Hamori, *Hidden Markov Models: Applications to Financial Economics*, Springer, 2004.

[Bo01] B. Bollobás, *Random Graphs*, 2nd ed., Vol. 73, Cambridge Studies in Advanced Mathematics, Cambridge University Press, 2001.

[BrFoSc87] P. Bratley, B. L. Fox, and L. E. Schrage, *A Guide to Simulation*, 2nd ed., Springer-Verlag, 1987.

[BuTu92] J. L. Buchanan and P. R. Turner, *Numerical Methods and Analysis*, McGraw-Hill, 1992.

[BuCa01] H. Bunke and T. Caelli, *Hidden Markov Models: Applications in Computer Vision*, World Scientific Publishing, 2001.

[BuEr75] S. A. Burr and P. Erdős, "On the magnitude of generalized Ramsey numbers for graphs", *Infinite and Finite Sets I*, Colloq. Math. Soc. Janos Bolyai 10, North-Holland, 214–240, 1975.

[Ch52] H. Chernoff, "A measure of asymptotic efficiency for tests of a hypothesis based on the sum of observations", *Annals of Mathematical Statistics* 23 (1952), 493–507.

[ClOlTu89] C. W. Clenshaw, F. W. J. Olver, and P. R. Turner, "Level-index arithmetic: an introductory survey", in *Numerical Analysis and Parallel Processing, Lecture Notes in Mathematics* 1397 (1989), Springer-Verlag, 95–168.

[Co87] C. J. Colbourn, *The Combinatorics of Network Reliability*, Oxford University Press, 1987.

[Co81] R. B. Cooper, *Introduction to Queueing Theory*, North-Holland, 1981.

[DaGr62] L. Danzer and B. Grünbaum, "Über zwei probleme bezüglich konvexer Körper von P. Erdős und von V. L. Klee", *Mathematische Zeitschrift* 79 (1962), 95–99.

[De86] L. Devroye, *Non-Uniform Random Variate Generation*, Springer-Verlag, 1986.

[DoSn84] P. G. Doyle and J. L. Snell, *Random Walks and Electric Networks*, Mathematical Association of America, 1984. (An excellent treatment of the connection between random walks and electrical networks.)

[DuEtal98] R. Durbin, S. Eddy, A. Krogh, and G. Mitchison, *Biological Sequence Analysis: Probabilistic Models of Proteins and Nucleic Acids*, Cambridge University Press, 1998.

[EbGrMa14] S. Eberhard, B. Green, and F. Manners, "Sets of integers with no large sum-free subset", *Annals of Mathematics* 180 (2014), 621–652.

[Er47] P. Erdős, "Some remarks on the theory of graphs", *Bulletin of the American Mathematical Society* 53 (1947), 292–294.

[Er59] P. Erdős, "Graph theory and probability", *Canadian Journal of Mathematics* 11 (1959), 34–38.

[Er63] P. Erdős, "On a combinatorial problem", *Nordisk Matematisk Tidskrift* 11 (1963), 220–223.

[Er64] P. Erdős, "On a combinatorial problem II", *Acta Mathematica Academiae Scientiarum Hungaricae* 15 (1964), 445–447.

[Er65] P. Erdős, "Extremal problems in number theory", *Proceedings of Symposia in Pure Mathematics (AMS)* Vol. VIII (1965), 181–189.

[ErFu83] P. Erdős and Z. Füredi, "The greatest angle among $n$ points in the $d$-dimensional Euclidean space", *Annals of Discrete Mathematics* 17 (1983), 275–283.

[ErLo75] P. Erdős and L. Lovász, "Problems and results on 3-chromatic hypergraphs and some related questions", in *Infinite and Finite Sets*, A. Hajnal, R. Rado, and V. T. Sos (eds.), North-Holland, 609–628, 1975.

[ErSiSt83] P. Erdős, R. Silverman, and A. Stein, "Intersection properties of families containing sets of nearly the same size", *Ars Combinatoria* 15 (1983), 247–259.

[Fe68] W. Feller, *An Introduction to Probability Theory and Its Applications*, Vol. I, 3rd ed., Wiley, 1968. (A classic text with extensive coverage of probability theory, the combinatorics of simple random walks, and Markov chains.)

[Fe71] W. Feller, *An Introduction to Probability Theory and Its Applications*, Vol. II, 2nd ed., Wiley, 1971. (A companion to the first volume with a treatment of random walks on continuous space and more advanced topics.)

[Fi03] G. S. Fishman, *Monte Carlo: Concepts, Algorithms, and Applications*, Springer-Verlag, 2003. (A comprehensive and integrated treatment of Monte Carlo methods and their applications.)

[FlWi93] M. Flanigan-Wagner and J. R. Wilson, "Using univariate Bézier distributions to model simulation input processes", in *Proceedings of the 1993 Winter Simulation Conference*, 1993, 365–373.

[FoEtal11] C. Forbes, M. Evans, N. Hastings, and B. Peacock, *Statistical Distributions*, 4th ed., Wiley, 2011.

[FoKaGi71] C. M. Fortuin, P. W. Kastelyen, and J. Ginibre, "Correlation inequalities on some partially ordered sets", *Communications in Mathematical Physics* 22 (1971), 89–103.

[FoSu11] J. Fox and B. Sudakov, "Dependent random choice", *Random Structures & Algorithms* 38 (2011), 68–99.

[GaYo07] M. Gales and S. Young, "The application of hidden Markov models in speech recognition", *Foundations and Trends in Signal Processing* 1 (2007), 195–304.

[Ga12] N. Gautam, *Analysis of Queues: Methods and Applications*, CRC Press, 2012.

[GuScDe97] I. Guyon, M. Schenkel, and J. Denker, "Overview and synthesis of on-line cursive handwriting recognition techniques", in *Handbook of Character Recognition and Document Image Analysis*, H. Bunke and P. S. P. Wang (eds.), World Scientific Publishing, 1997, Chapter 7.

[Ha13] M. Harchol-Balter, *Performance Modeling and Design of Computer Systems: Queueing Theory in Action*, Cambridge University Press, 2013.

[GrHa85] D. Gross and C. M. Harris, *Fundamentals of Queueing Theory*, 2nd ed., Wiley, 1985.

[HeSo82] D. P. Heyman and M. J. Sobel, *Stochastic Models in Operations Research*, Vol. 1, McGraw-Hill, 1982.

[HiMoRe97] H. Hind, M. Molloy, and B. Reed, "Colouring a graph frugally", *Combinatorica* 17 (1997), 469–482.

[Ho63] W. Hoeffding, "Probability inequalities for sums of bounded random variables", *Journal of the American Statistical Association* 58 (1963) 13–30.

[IE08] IEEE, *Binary Floating-Point Arithmetic*, IEEE Standard 754-2008, IEEE, 2008.

[JaLuRu00] S. Janson, T. Łuczak, and A. Ruciński, *Random Graphs*, Wiley, 2000.

[Jo87] M. E. Johnson, *Multivariate Statistical Simulation*, Wiley, 1987.

[JoKoBa94] N. L. Johnson, S. Kotz, and N. Balakrishnan, *Continuous Univariate Distributions*, Volume 1, 2nd ed., Wiley, 1994.

[KeSn60] J. G. Kemeny and L. J. Snell, *Finite Markov Chains*, Van Nostrand, 1960.

[Kl75] L. Kleinrock, *Queueing Systems, Vol. I: Theory*, Wiley, 1975.

[Kl66] D. J. Kleitman, "Families of non-disjoint subsets", *Journal of Combinatorial Theory A* 1 (1966), 153–155.

[Ku11] V. G. Kulkarni, *Introduction to Modeling and Analysis of Stochastic Systems*, 2nd ed., Springer, 2011.

[Ku16] V. G. Kulkarni, *Modeling and Analysis of Stochastic Systems*, 3rd ed., Chapman-Hall, 2016.

[La15] A. M. Law, *Simulation Modeling and Analysis*, 5th ed., McGraw-Hill, 2015.

[La91] G. Lawler, *Intersections of Random Walks*, Birkhäuser, 1991. (A monograph on the mathematical analysis of problems dealing with the non-intersection of paths of random walks. Sophisticated mathematical methods.)

[LE94] P. L'Ecuyer, "Uniform random number generation", *Annals of Operations Research* 53 (1994), 77–120.

[LE12] P. L'Ecuyer, "Random number generation" in *Handbook of Computational Statistics*, 2nd ed., J. E. Gentle, W. Haerdle, and Y. Mori (eds.), Springer-Verlag, 35–71, 2012.

[Lo93] L. Lovász, "Random walks on graphs: a survey", in *Combinatorics, Paul Erdös is Eighty*, Bolyai Society, Mathematical Studies, Vol. 2 (1993), 1–46.

[Ma90] B. G. Malkiel, *A Random Walk Down Wall Street*, W. W. Norton, 1990.

[MoRe00] M. Molloy and B. Reed, *Graph Colouring and the Probabilistic Method*, Springer-Verlag, 2000.

[Na93] R. E. Nance, "A history of discrete event simulation programming languages", *ACM SIGPLAN Notices* 28 (1993), 149–175.

[Ne13] B. L. Nelson, *Foundations and Methods of Stochastic Simulation: A First Course*, Springer, 2013.

[PaMi88] S. K. Park and K. W. Miller, "Random number generators: good ones are hard to find", *Communications of the ACM* 31 (1988), 1192–1201.

[RaSr00] J. Radhakrishnan and A. Srinivasan, "Improved bounds and algorithms for hypergraph two-coloring", *Random Structures & Algorithms* 16 (2000), 4–32.

[Ro14] S. M. Ross, *Introduction to Probability Models*, 11th ed., Academic Press, 2014.

[SaTa12] N. Sain and S. Tamrakar, "Web usage mining & pre-fetching based on hidden Markov model and fuzzy clustering", *International Journal of Computer Science and Information Technologies* 3 (2012), 4874–4877.

[SaGu93] L. Satish and B. I. Gururaj, "Use of hidden Markov models for partial discharge pattern classification", *IEEE Transactions on Dielectrics and Electrical Insulation* 28 (1993) 172–182.

[Sc90] B. Schmeiser, "Simulation experiments", in *Handbooks in OR & MS*, D. P. Heyman and M. J. Sobel (eds.), Elsevier, 1990, 296–330.

[Sh91] D. R. Shier, *Network Reliability and Algebraic Structures*, Clarendon Press, 1991.

[Sp87] J. H. Spencer, *Ten Lectures on the Probabilistic Method*, SIAM, 1987.

[Sp76] F. Spitzer, *Principles of Random Walks*, 2nd ed., Springer-Verlag, 1976. (A classic monograph on mathematical properties of lattice random walks.)

[St74] P. H. Sterbenz, *Floating-Point Computation*, Prentice-Hall, 1974.

[StEtal11] J. Stigler, F. Ziegler, A. Gieseke, J. C. M. Gebhardt, and M. Rief, "The complex folding network of single calmodulin molecules", *Science* 334 (2011), 512–516.

[Sw15] J. J. Swain, "Simulated worlds", *OR/MS Today* 42/6 (2015), 36–49.

[Ta72] R. E. Tarjan, "Depth-first search and linear graph algorithms", *SIAM Journal on Computing* 1 (1972), 146–160.

[ToEtal13] K. Tokuda, Y. Nankaku, T. Toda, H. Zen, J. Yamagishi, and K. Oura, "Speech synthesis based on hidden Markov models", *Proceedings of the IEEE* 101 (2013), 1234–1252.

[We94] G. H. Weiss, *Aspects and Applications of the Random Walk*, North Holland, 1994. (A treatment of random walks and their applications with a physical scientist's perspective.)

[Wi84] J. R. Wilson, "Variance reduction techniques for digital simulation", *American Journal of Mathematical and Management Sciences* 4 (1984), 277–312.

[Wo89] R. W. Wolff, *Stochastic Modeling and the Theory of Queues*, Prentice-Hall, 1989.

**Web Resources**:

https://math.dartmouth.edu/~doyle/docs/walks/walks.pdf (A 2006 online version of the Doyle and Snell book on random walks and electrical networks.)

http://math.ucsd.edu/~crypto/Monty/monty.html (Simulation of the *Let's Make a Deal* game show, in which prizes are randomly hidden behind doors.)

http://random.mat.sbg.ac.at/generators/ (Information on a variety of pseudo-random number generators.)

http://random.mat.sbg.ac.at/software (Software for Monte Carlo simulation and for pseudo-random number generators.)

http://random.mat.sbg.ac.at/tests (Tests for uniform pseudo-random number generators.)

http://statistik.wu-wien.ac.at/projects/anuran/index.html (Code generation for non-uniform random variates.)

http://web2.uwindsor.ca/math/hlynka/qbook.html (Provides an extensive list of books on queueing theory.)

http://web2.uwindsor.ca/math/hlynka/qsoft.html (Compilation of software on queueing theory.)

http://ws3.atv.tuwien.ac.at/eurosim/ (Lists a number of commercial and freeware/shareware simulation packages.)

http://www.anylogic.com (Integrated modeling and simulation development environment.)

http://www.dcs.ed.ac.uk/home/hase/simjava/applets/index.html (Java-based discrete event simulation applets.)

http://www.informs-sim.org (Archive of Winter Simulation Conference proceedings, an excellent source for simulation information.)

http://www.math.wm.edu/~leemis/chart/UDR/UDR.html (Interactive chart containing univariate probability distributions.)

http://www.stat.berkeley.edu/~aldous/RWG/book.html (Recompiled monograph of Aldous and Fill on reversible Markov chains and graphs.)

`http://www.um.es/fem/EjsWiki/index.php/Main/WhatIsEJS` (Visual simulation environment.)

`http://www.taygeta.com/random.html` (Software for pseudo-random number generators.)

`http://www.vissim.com` (Graphical language for simulation development.)

# 8

# GRAPH THEORY

## INTRODUCTION

A graph is conceptually a set of points and a set of lines (possibly curved) joining one point to another (or to itself). Graph theory has its origins in many disciplines. Graphs are natural mathematical models of physical situations in which the points represent either objects or locations and the lines represent connections. Graphs are also used to model sociological and abstract situations in which each line represents a relationship between the entities represented by the points. Applications of graphs are wide-ranging—in areas such as circuit design, communications networks, ecology, engineering, operations research, counting, probability, set theory, information theory, and sociology.

This chapter contains an extensive treatment of the various properties of graphs. Further topics in graph theory are covered in Chapter 9 (Trees) and in Chapter 10 (Networks and Flows).

## GLOSSARY

**acyclic digraph**:  a digraph containing no directed cycles.

**acyclic graph**:  a graph containing no cycles.

**adjacency matrix** (of a digraph):  the square matrix $A$ with $A[i,j] =$ the number of edges from vertex $v_i$ to vertex $v_j$.

**adjacency matrix** (of a graph): the square matrix $A$ with $A[i,j] = $ the number of edges between vertices $v_i$ and $v_j$.

**adjacent edges**: two edges with a common endpoint.

**adjacent vertex** (in a digraph) from [to] a vertex $u$: a vertex $v$ such that there is an arc from $u$ to $v$ [to $u$ from $v$].

**adjacent vertices**: two vertices that are endpoints of the same edge.

**admittance matrix**: See *Laplacian*.

**algebraic specification** (of a graph): a specification that uses group elements in the vertex and edge names and uses the group operation in the incidence rule.

**almost every** (**a.e.**) **graph has property** ***P***: the statement that the probability that a random $n$-vertex graph has property $P$ approaches 1 as $n \to \infty$.

**antichain**: a hypergraph in which no edge contains any other edge.

**antihole**: an induced subgraph whose complement is isomorphic to the cycle $C_k$ with $k \geq 4$.

**antimagic graph**: a graph whose $q$ edges can be labeled with distinct integers from 1 to $q$ such that the sums of the labels of the edges incident to each vertex are distinct.

**arc**: another name for a directed edge of a graph.

**articulation point**: synonym for cutpoint.

**attachment of a bridge of a subgraph**: for a bridge $B$ of a subgraph $H$, a vertex of $B \cap H$.

**attribute** (of the edge-set or vertex-set): any additional feature, such as length, cost, or color, that enables a graph to model a real problem.

**automorphism**: for a graph or digraph, an isomorphism from the graph or digraph to itself.

**automorphism group**: the collection $\mathcal{A}ut(G)$ of all automorphisms of a graph or digraph $G$ under the operation of composition.

**basis** (for a digraph): a set of vertices $V'$ of the digraph such that every vertex not in $V'$ is reachable from $V'$ and no proper subset of $V'$ has this property.

**Berge graph**: a graph that has no odd hole and no odd antihole.

**bipartite**: property of a graph that its vertices can be partitioned into two subsets, called "parts", so that no two vertices within the same part are adjacent.

**block**: in a graph, a maximal nonseparable subgraph.

**bond**: a minimal disconnecting set of edges.

**boundary** (of a region of a graph imbedded in a surface): for region $R$, the subgraph containing all vertices and edges incident on $R$; denoted by $\partial R$.

**boundary** (of a set of vertices): for $F$ a subset of vertices in a graph, the set of edges that have one endpoint inside $F$ and one endpoint outside $F$; denoted by $\partial F$.

**bouquet**: a graph $B_n$ with one vertex and $n$ self-loops.

**bridge** (**edge**): a cut-edge.

**cactus**: a connected graph in which every block is either an edge or a cycle.

**Cartesian product**:  for graphs $G$ and $H$, the graph $G \square H$ that has vertex set $V_G \times V_H$ and edge set $\{(g,h)(g',h') \mid gg' \in E_G \text{ and } h = h', \text{ or, } g = g' \text{ and } hh' \in E_H\}$.

**caterpillar**:  a tree that contains a path such that every edge has one or both endpoints in that path.

**Cayley graph** (**or digraph**):  a graph that depicts a group with a prescribed set of generators; the vertices represent group elements, and the edges or arcs (often "colored" by the generators) represent the product rule.

**cellular imbedding**:  an imbedding such that every region is equivalent to the interior of a (unit) disk.

**center**:  in a connected graph, the set of vertices of minimum eccentricity.

**chain**:  a simple hypergraph in which, given any pair of edges, one edge contains the other.

**characteristic polynomial** (of a graph):  the characteristic polynomial of its adjacency matrix.

**chromatic index** (of a graph or hypergraph):  See *edge chromatic number*.

**chromatic number** (of a graph):  the minimum number $\chi(G)$ of colors needed to color the vertices of a graph $G$ so that no vertex is adjacent to a vertex of the same color.

**chromatic number** (of a hypergraph):  the smallest number $\chi(H)$ of independent sets required to partition the vertex set of $H$.

**chromatic number** (of a map):  the minimum number $\chi(M)$ of colors needed to color the regions of the map $M$ so that no color meets itself across an edge.

**chromatic number** (of a surface):  the largest map chromatic number $\chi(S)$ taken over all maps on the surface $S$.

**chromatically $n$-critical graph**:  an $n$-chromatic graph $G$ such that $\chi(G-e) = n-1$ no matter what edge $e$ is removed.

**circuit**:  synonym for a closed walk, a closed trail, or a cycle, depending on the context.

**clique** (in a graph):  a complete subgraph.

**clique** (in a hypergraph):  a simple hypergraph such that every pair of edges has nonempty intersection.

**clique number** (of a graph):  the number $\omega(G)$ of vertices of a largest clique in the graph $G$.

**clique number** (of a hypergraph):  the largest number $\omega(H)$ of edges of any partial clique in the hypergraph $H$.

**clique partition number**:  for a hypergraph $H$, the smallest number $cp(H)$ of cliques required to partition the edge set.

**closed walk** (**trail or path**):  a walk, trail, or path whose origin and terminus are the same.

**$n$-colorable graph**:  a graph having a vertex coloring using at most $n$ colors.

**$n$-colorable map**:  a map having a coloring using at most $n$ colors.

**comparability graph**:  a graph that admits a transitive orientation.

**complement** (of a graph):  the graph $\overline{G}$ with the same vertex set as $G$, but in which two vertices are adjacent if and only if they are not adjacent in $G$.

**complete bipartite graph**: a bipartite graph $K_{r,s}$ whose vertex set has two parts, of sizes $r$ and $s$, respectively, such that every vertex in one part is adjacent to every vertex in the other part.

**complete graph**: the simple graph $K_n$ with $n$ vertices in which every pair of vertices is adjacent.

**complete hypergraph**: the simple $n$-vertex hypergraph $K_n^*$ in which every subset of vertices is an edge.

**complete multipartite** (or **$k$-partite**) **graph**: a $k$-partite simple graph such that every pair of vertices from different parts is joined by an edge; denoted by $K_{n_1,\ldots,n_k}$, where $n_1,\ldots,n_k$ are the sizes of the parts.

**complete $r$-uniform hypergraph**: the simple $n$-vertex hypergraph $K_n^r$ in which every $r$-element subset is an edge.

**complete set of invariants**: a set of invariants that determine a graph or digraph up to isomorphism.

**component**: for a graph, a maximal connected subgraph.

**connected**: property of a graph that each pair of vertices is joined by a path.

**connectivity**: See *vertex connectivity*.

**contraction**: for a graph, the result of a sequence of elementary contractions.

**contraction**, **elementary** (of a graph): the operation of shrinking an edge to a point, so that its endpoints are merged, without otherwise changing the graph.

**contraction**, **elementary** (of a simple graph): replacing two adjacent vertices $u$ and $v$ by one vertex adjacent to all other vertices to which $u$ or $v$ were adjacent.

**converse**: for a digraph, the digraph obtained by reversing the direction of every arc.

**corona** (of a graph): for a graph $H$, the graph obtained by adding for each vertex $v$ of $H$ a new vertex $v'$ and the edge $vv'$; denoted by $H \circ K_1$.

**crosscap**: a subportion of a surface that forms a Möbius band.

**crosscap number** (of a nonorientable surface): for a nonorientable surface $S$, the maximum number $\overline{\gamma}(S)$ of disjoint crosscaps one can find on the surface. The nonorientable surface of crosscap number $k$ is denoted $N_k$.

**crossing number**: for a graph $G$, the minimum number $\nu(G)$ of edge-crossings taken over all normalized planar drawings of $G$.

**cube graph**: See *hypercube graph*.

**cut-edge**: for a graph $G$, an edge $e$ such that $G - e$ has more components than $G$.

**cut-vertex** (or **cutpoint**): for a graph $G$, a vertex $v$ such that $G - v$ has more components than $G$.

**cycle**: a closed path of positive length. See also *$k$-cycle*.

**cycle**, **directed**: a closed directed walk in which all the vertices except the first and last are distinct.

**cycle graph**: a graph $C_n$ with $n$ vertices that is 2-regular and connected.

**cycle rank**: for a connected graph $G$, the number $\beta_1(G)$ of edges in the complement of a spanning tree for $G$; that is, $|E_G| - |V_G| + 1$.

**DAG**: an acronym for directed acyclic graph.

**degree** (of a vertex in a graph):  for a vertex $v$, the number $\deg(v)$ of instances of $v$ as an endpoint; that is, the number of proper edges incident on $v$ plus twice the number of loops at $v$.

**degree** (of a hypergraph vertex):  for a vertex $x$, the number $\deg(x)$ of hypergraph edges containing $x$.

**degree sequence of a graph**:  the sequence of the degrees of its vertices, usually sorted.

**deleting an edge from a graph**:  given a graph $G$ and an edge $e$ of $G$, the operation that results in the subgraph $G - e$, which contains all the vertices of $G$ and all edges except $e$.

**deleting a vertex from a graph**:  given a graph $G$ and a vertex $v$ of $G$, the operation that results in the subgraph $G - v$, which contains all vertices of $G$ except $v$ and all the edges of $G$ except those incident with $v$.

**diameter**:  for a connected graph, the maximum distance between two of its vertices.

**diconnected digraph**:  See *strongly connected digraph*.

**digraph** (or **directed graph**):  a graph in which every edge is directed.

**dipole**:  the graph $D_n$ with two vertices and a multi-edge of multiplicity $n$ joining them.

**direct product**:  for graphs $G$ and $H$, the graph $G \times H$ that has vertex set $V_G \times V_H$ and edge set $\{(g, h)(g', h') \mid gg' \in E_G \text{ and } hh' \in E_H\}$.

**directed cycle**, **path**, **trail**, **walk**:  See *cycle*, *path*, etc.

**directed graph**:  See *digraph*.

**direction** (on an edge):  a sense of forward progression from one end to the other, usually marked by an arrowhead.

**disconnected** (digraph):  a digraph whose underlying graph is disconnected.

**disconnecting set of edges** (in a connected graph):  a set whose removal yields a nonconnected graph.

**disconnecting set of vertices** (in a connected graph):  a set whose removal yields a nonconnected graph.

**distance** (in a connected graph):  for two vertices $v$ and $w$, the length $d(v, w)$ of a shortest path between them.

**distance** (in a connected digraph):  for two vertices $v$ and $w$, the length $d(v, w)$ of a shortest directed path between them.

**dodecahedral graph**:  the 1-skeleton of the dodecahedron, which is a 3-dimensional polyhedron whose 12 faces are all pentagons; this graph has 20 vertices, each of degree 3, and 30 edges.

**dominating set** (of a graph):  a set $S$ of vertices such that every vertex not in $S$ is adjacent to a vertex in $S$.

**domination number** (of a graph):  the minimum cardinality $\gamma(G)$ of a dominating set in graph $G$.

**downset**:  a simple hypergraph in which every subset of every edge is also an edge of the hypergraph.

**dual graph imbedding**:  a new graph imbedding obtained by placing a dual vertex in the interior of each existing ("primal") region and by drawing a dual edge through each existing ("primal") edge connecting the dual vertices on its opposite sides.

**dual** (of a hypergraph):  for a hypergraph $H$, the hypergraph $H^*$ whose incidence matrix is the transpose of the incidence matrix $M(H)$.

**eccentricity** (of a vertex):  for a vertex $v$ in a connected graph, the maximum distance from $v$ to another vertex.

**edge**:  a line, either joining one *vertex* to another or joining a *vertex* to itself; an element of the second constituent set of a *graph*.

**edge chromatic number** (of a graph):  for a graph $G$, the smallest number $n$ such that $G$ is $n$-edge colorable, written $\chi_1(G)$.

**edge chromatic number** (of a hypergraph):  for a hypergraph $H$, the smallest number $q(H)$ of matchings required to partition the edge set of $H$.

**$n$-edge colorable**:  property of a graph that it has an edge coloring using at most $n$ colors.

**edge coloring**:  an assignment of colors to the edges of a graph so that adjacent edges receive different colors. See also *$n$-edge colorable*.

**edge connectivity**:  the cardinality $\kappa'(G)$ of a smallest disconnecting set of edges in graph $G$. See also *k-edge-connected*.

**edge cut**:  See *disconnecting set*.

**edge independence number**:  the cardinality $\alpha_1(G)$ of a largest independent set of edges in graph $G$.

**edge-complement**:  See *complement*.

**edge-deleted subgraph**:  any subgraph obtained from a graph by removing a single edge.

**edge-reconstructible graph**:  a graph that is uniquely determined by its collection of edge-deleted subgraphs.

**edge-reconstructible invariant**:  an invariant that is uniquely determined by the collection of edge-deleted subgraphs of a graph.

**eigenvalue** (of a graph):  a number $\lambda$ such that $Ax = \lambda x$ for some nonzero vector $x$, where $A$ is the adjacency matrix.

**eigenvector** (of a graph):  a nonzero vector $x$ such that $Ax = \lambda x$, where $A$ is the adjacency matrix.

**embedding**:  See *imbedding*.

**empty graph**:  sometimes, a graph with no edges; other times, a graph with no vertices or edges. See *null graph*.

**endpoints** (of an edge):  the *vertices* that are joined by the edge.

**Euler characteristic**:  for a surface $S$, the invariant $\chi(S)$ given by $2 - 2g$ for the orientable surface of genus $g$, and $2 - k$ for the nonorientable surface of crosscap number $k$.

**Euler tour**:  a closed Euler trail.

**Euler trail**:  a trail that contains all the edges of the graph.

**Eulerian graph**:  a graph that has an Euler tour.

**expander family**:  a sequence of graphs, all of the same regularity and whose orders grow to infinity, for which there exists $\epsilon > 0$ such that the isoperimetric constant is at least $\epsilon$ for all the graphs.

**exterior region**:  in a planar graph drawing, the region that extends to infinity.

**extremal graph**:  for a set $\mathcal{G}$ of graphs and an integer $n$, an $n$-vertex graph with $ex(\mathcal{G};n)$ edges that contains no member of $\mathcal{G}$.

**extremal number**:  for a set $\mathcal{G}$ of graphs, the greatest number $ex(\mathcal{G};n)$ of edges in any $n$-vertex simple graph that does *not* contain some member of $\mathcal{G}$ as a subgraph.

**face**:  for an imbedding of a graph in a surface, a region plus its boundary.

**1-factorization**:  a partition of the edges of a graph into 1-factors.

**forest**:  a graph without a cycle.

**four color theorem**:  the fact that every planar map can be properly colored with at most four colors, proved in 1976.

**general graph**:  another name for a graph that might have loops.

**generating set** (for a group):  a subset of group elements such that every group element is a product of generators.

**genus** (of a graph):  the minimum genus of a surface in which the graph has a cellular imbedding.

**genus** (of an orientable surface):  for a surface $S$, the maximum number $\gamma(S)$ of disjoint handles one can find on the surface; the orientable surface of genus $g$ is denoted $S_g$.

**girth**:  for a graph, the number of edges in a shortest cycle, if there is at least one cycle; undefined if the graph has no cycles.

**graceful graph**:  a graph for which there exists an injection $f$ from the vertices to the set $\{0, 1, \ldots, q\}$, where $q$ is the number of edges, such that when each edge $xy$ is assigned the label $|f(x) - f(y)|$, the resulting edge labels are distinct.

**graph**:  a set $V$ of vertices and a set $E$ of edges such that all the endpoints of edges in $E$ are contained in $V$, written $G = (V, E)$, $(V_G, E_G)$, or $(V(G), E(G))$.

**graph model**:  any configuration with underlying graph structure, and possibly some additional attributes on its edges and/or vertices, such as length, direction, or cost.

**graph sum**:  for graphs $G$ and $H$, the graph $G + H$ whose vertex set and edge set are, respectively, the disjoint union of the vertex sets and the edge sets of $G$ and $H$.

**graphical sequence**:  a sequence of nonnegative integers such that there is a *simple graph* for which it is the *degree sequence*.

**Gray code**:  a cyclic ordering of all $2^k$ bitstrings of length $k$, such that each bitstring differs from the next in exactly one bit entry.

**Hamilton cycle**:  a spanning cycle; that is, a cycle including each vertex of a graph exactly once.

**Hamilton path**:  a path that includes all the vertices of a graph.

**Hamiltonian graph**:  a graph that contains a Hamiltonian cycle.

**harmonious graph**:  a graph for which there is an injection $f$ from the vertices to $\mathcal{Z}_q$, where $q$ is the number of edges, such that when each edge $xy$ is assigned the label $f(x) + f(y) \pmod{q}$, the resulting edge labels are distinct; when the graph is a tree, exactly one label may be used on two vertices.

**head** (of an arc):  the vertex the arc goes to.

**Hoffman polynomial** (of a graph): a polynomial $p(x)$ of minimum degree such that $p(A) = J$, where $A$ is the adjacency matrix and $J$ is the matrix with every entry equal to 1.

**hole**: an induced subgraph that is isomorphic to the cycle $C_k$ with $k \geq 4$.

**homeomorphic graphs**: two graphs that can both be obtained from the same graph by a sequence of edge subdivisions.

**hypercube graph**: a graph $Q_d$ whose $2^d$ vertices can be labeled with the bitstrings of length $d$, so that two vertices are adjacent if and only if their labels differ in exactly one bit.

**hypergraph**: a finite set $V$ of "vertices" together with a finite collection $E$ of "edges" (sometimes, "hyperedges"), which are arbitrary subsets of $V$, written $H = (V, E)$.

**icosahedral graph**: the 1-skeleton of the icosahedron, which is a 3-dimensional polyhedron whose 20 faces are triangles; this graph has 12 vertices, each of degree 5, and 30 edges.

**imbedding** (of a graph in a surface): a drawing so that there are no edge-crossings; also *embedding*.

**incidence matrix** (of a digraph with no self-loops): the matrix $M_I$ with $M_I[i, j] = 0$ if vertex $v_i$ is not an endpoint of arc $e_j$, $M_I[i, j] = 1$ if $v_i$ is the head of arc $e_j$, and $M_I[i, j] = -1$ if $v_i$ is the tail of arc $e_j$.

**incidence matrix** (of a graph): the matrix $M_I$ with $M_I[i, j] = 0$ if vertex $v_i$ is not an endpoint of edge $e_j$, $M_I[i, j] = 1$ if $e_j$ is a proper edge with endpoint $v_i$, and $M_I[i, j] = 2$ if $e_j$ is a loop at $v_i$.

**incidence matrix** (of a hypergraph): for a hypergraph, the matrix $[m_{i,j}]$ where $m_{i,j} = 1$ if vertex $x_j$ is in edge $e_i$, and $m_{i,j} = 0$ otherwise.

**incidence rule**: a rule specifying the endpoints of every edge of a graph.

**incident edge** (from [to] a digraph vertex): for a vertex $u$ in a digraph, an arc $e$ such that $u$ is the tail [head] of $e$.

**incident edge** (in a graph): for a vertex $u$ in a graph, an edge $e$ such that $u$ is an endpoint of $e$.

**incident-edge table** (for a graph): a table that lists, for each vertex, the edges having that vertex as an endpoint.

**in-degree**: for a vertex $v$, the number of arcs with head $v$.

**independent set** (in a graph): for a graph $G$, a subset of either $V(G)$ or $E(G)$ such that no two elements are adjacent in $G$.

**independent set** (of hypergraph vertices): a set of vertices that does not (completely) contain any edge of the hypergraph.

**independence number** (of a graph): the number $\alpha(G)$ of vertices in the largest independent subset in $G$.

**independence number** (of a hypergraph): the maximum number $\alpha(H)$ of vertices that form an independent set in $H$.

**induced subgraph** (on a vertex subset): the subgraph of a graph $G$ containing every edge of $G$ that joins two vertices of the prescribed vertex subset.

***intersection graph*** (for a family of subsets):  for a family $\mathcal{F} = \{S_j\}$ of sets, the graph with vertex set $\mathcal{F}$ such that there is an edge between each pair of subsets $S_i$ and $S_j$ whose intersection is nonempty.

***intersection graph*** (of a hypergraph):  for a hypergraph $H$, the simple graph $I(H)$ whose vertices are the edges of $H$, such that two vertices of $I(H)$ are adjacent if and only if the corresponding edges of $H$ have nonempty intersection.

***interval graph***:  the *intersection graph* of a family of subintervals of $[0,1]$.

***invariant***:  a parameter or property of graphs that is preserved by isomorphisms.

***irreducible tournament***:  a tournament with no bipartition $V_1, V_2$ of the vertices such that all arcs between $V_1$ and $V_2$ go from $V_1$ to $V_2$.

***isolated point***:  a vertex of a graph that is not the endpoint of any edge.

***isomorphic*** (pair of graphs):  a pair of graphs with identical mathematical structure; formally, a pair of graphs such that there is an isomorphism from one to the other.

***isomorphism*** (of digraphs):  an isomorphism of the underlying graphs of two digraphs such that the edge-correspondence preserves direction.

***isomorphism*** (of graphs):  for graphs $G$ and $H$, a pair of bijections $f_V \colon V_G \to V_H$ and $f_E \colon E_G \to E_H$ such that for every edge $e \in E_G$, the endpoints of $e$ are mapped onto the endpoints of $f_E(e)$.

***isomorphism*** (of simple graphs):  a bijection between the vertices of two graphs such that a pair of vertices is adjacent in one graph if and only if the corresponding pair of vertices is adjacent in the other graph.

***isomorphism type***:  for a graph [digraph] $G$, the class of all graphs [digraphs] isomorphic to $G$.

***isoperimetric constant***:  for graph $G$ with vertex set $V_G$, the minimum ratio $|\partial F|/|F|$ taken over all $F \subseteq V_G$ such that $0 < |F| \le |V_G|/2$; denoted $h(G)$.

***join***:  for graphs $G$ and $H$, the graph $G * H$ obtained by adding to the disjoint union $G + H$ an edge from each vertex in $G$ to each vertex in $H$.

***k-connected***:  property of a graph $G$ that the smallest size of a disconnecting set of vertices is at least $k$; that is, $\kappa(G) \ge k$.

***k-cycle***:  a cycle of length $k$.

***k-edge-connected***:  property of a graph $G$ that $\kappa'(G) \ge k$.

***k-partite graph***:  a graph whose vertex set can be partitioned into at most $k$ parts in such a way that each edge joins different parts. Equivalent to a *k-colorable graph*.

***k-regular***:  property of a graph or hypergraph that all its vertices have degree $k$.

***king***:  a vertex in a digraph that can reach all other vertices by paths of length 1 or 2.

***Kuratowski graphs***:  the complete graph $K_5$ and the complete bipartite graph $K_{3,3}$.

***labeled graph***:  in applied graph theory, any graph in which the vertices and/or edges have been assigned labels; in pure graph theory, a graph in which standard labels $v_1, v_2, \ldots, v_n$ have been assigned to the vertices.

***$\lambda$ of a graph***:  for a connected $d$-regular graph, the maximum $|\lambda_i|$ taken over all eigenvalues $\lambda_i$ that are not $\pm d$.

***Laplacian*** (of a graph $G$):  the matrix $D - A$ where $D$ is the diagonal matrix with the degree sequence of $G$ on the diagonal and where $A$ is the adjacency matrix.

**length** (of a walk): the number of edge-steps in the sequence that specifies the walk.

**lexicographic product**: for graphs $G$ and $H$, the graph $G \circ H$ that has vertex set $V_G \times V_H$ and edge set $\{(g,h)(g',h') \mid (gg' \in E_G)$, or $(g = g'$ and $hh' \in E_H)\}$.

**line**: synonym for edge, or refers to what is modeled by an edge.

**line graph**: for a graph $G$, the graph $L(G)$ whose vertices correspond to the edges of $G$, with two vertices being adjacent in $L(G)$ whenever the corresponding edges have a common endpoint in $G$.

**linear extension ordering**: a consecutive labeling $v_1, v_2, \ldots, v_n$ of the vertices of a digraph such that, if there is an arc from $v_i$ to $v_j$, then $i < j$.

**link**: See *proper edge*.

**loop** (or **self-loop**): an edge joining a vertex to itself.

**magic graph**: a connected graph whose edges can be labeled with distinct positive integers such that for each vertex $v$ the sum of the labels of all edges incident with $v$ is the same.

**map**: an imbedding of a graph on a surface.

**map chromatic number**: See *chromatic number of a map*.

**map coloring**: an assignment of colors to the regions of a map so that adjacent regions receive different colors.

**matching**: a set of pairwise disjoint edges in a graph or hypergraph.

**matching number**: in a graph, the maximum number of pairwise disjoint edges of the graph; in a hypergraph $H$, the maximum number $\nu(H)$ of pairwise disjoint edges of $H$; that is, the cardinality of the largest partial of $H$ that forms a matching.

**minor**: for a graph $G$, any graph that can be obtained from $G$ by a sequence of edge deletions and contractions.

**Möbius band**: the surface obtained from a rectangular sheet by pasting the left side to the right with a half-twist.

**multi-arc**: two or more arcs, all of which have the same head and the same tail.

**multi-edge**: a set of at least two edges, all of which have the same endpoints.

**multigraph**: a graph with multi-edges.

**neighbor**: for a vertex, any vertex adjacent to it.

**node**: a vertex, or refers to what is modeled by a vertex.

**nonorientable surface**: a surface such that some subportion forms a Möbius band.

**nonorientable surface of crosscap number $k$**: the surface $N_k$ obtained by adding $k$ crosscaps to a sphere.

**nonplanar**: property of a graph that it cannot be drawn in the plane without crossings.

**nonseparable**: property of a connected graph that it has no cut-vertex.

**normal**: property of a hypergraph $H$ that $q(H) = \Delta(H)$.

**normalized drawing**: the usual way a graph is drawn, avoiding pathological contrivances such as overloaded crossings (i.e., more than two edges).

**null graph**: (usually) a graph with no vertices or edges.

**obstruction to n-coloring**: synonym for a chromatically $(n+1)$-critical graph, since a chromatically $(n+1)$-critical subgraph prevents $n$-chromaticity.

**octahedral graph**: the 1-skeleton of the 3-dimensional octahedron, or sometimes, a generalization of this graph.

**1-skeleton** (of a polyhedron): the graph whose vertices and edges are, respectively, the vertices and edges of that polyhedron.

**open**: property of a walk, trail, or path that its final vertex is different from its initial vertex.

**order** (of a graph): for a graph $G$, the cardinality $|V_G|$ of the vertex set.

**order** (of a hypergraph edge): the number of vertices in the edge.

**orientable surface**: any surface obtainable from a sphere by adding handles.

**orientable surface of genus g**: the surface $S_g$ obtained by attaching $g$ handles to a sphere.

**orientation**: an assignment of a direction to every edge of a graph, making it a digraph.

**origin** (of a walk): the initial vertex of the walk.

**out-degree**: for a vertex $v$, the number of arcs with tail $v$.

**partial**: for a hypergraph $H = (V, E)$, a hypergraph $H' = (V, E')$ such that $E' \subseteq E$.

**path**: a trail in which all of its vertices are different, except that the initial and final vertices may be the same. See also $u,v$-path.

**path, directed**: a directed trail in which no vertex is repeated.

**pebbling number** (of a graph): the maximum value $\pi(G)$ of the rooted pebbling number over all roots of the graph $G$.

**pebbling step from u to v**: for an edge $uv$, the removal of two pebbles from $u$ and placement of one pebble on $v$.

**perfect graph**: a graph such that every induced subgraph has vertex chromatic number equal to its clique number.

**perfect matching** (or **1-factor**): a matching that covers all the vertices of a graph.

**Petersen graph**: a 3-regular 10-vertex graph that looks like a 5-cycle joined by its vertices to the vertices of a 5-pointed star drawn in its interior.

**planar**: property of a graph that it can be drawn in the plane without crossings.

**Platonic graph**: the 1-skeleton of a Platonic solid.

**Platonic solid**: any of five 3-dimensional polyhedra whose sides are all identical regular polygons.

**polyhedron**: a generalization of a polygon to higher dimensions; usually a solid 3-dimensional figure subtended by planes.

**proper edge** (or **link**): an edge with two distinct endpoints.

**pseudograph**: synonym for a graph with loops.

**r-partite hypergraph**: an $r$-uniform hypergraph whose vertex set can be partitioned into $r$ blocks so that each edge intersects each block in exactly one vertex.

**r-uniform**: property of a uniform hypergraph that $r$ is the common edge-order.

**radius**: for a connected graph $G$, the minimum eccentricity among the vertices of $G$.

**Ramanujan graph**: A $d$-regular graph $G$ such that $\lambda(G) \le 2\sqrt{d-1}$.

**Ramsey number, classical**: the number $r(m, n)$, which is the smallest positive integer $k$ such that every simple graph with $k$ vertices either contains $K_m$ as a subgraph or has a set of $n$ independent vertices.

**Ramsey number**: the number $R(G, H)$, which is the smallest positive integer $k$ such that, if the edges of $K_k$ are bipartitioned into red and blue classes, then either the red subgraph contains a copy of $G$ or the blue subgraph contains a copy of $H$.

**random graph on $n$ vertices**: an $n$-vertex graph generated by a probability distribution, in which each edge is as likely to occur as any of the others.

**reachable vertex** (from vertex $u$): a vertex $v$ such that there is a $u, v$-path.

**reconstructible**: property of a graph that it is uniquely determined by its collection of vertex-deleted subgraphs.

**reconstructible invariant**: an invariant that is uniquely determined by the collection of vertex-deleted subgraphs of a graph.

**reducible**: property of a digraph that its vertex set can be partitioned into a disjoint union $V_1 \cup V_2$ so that all arcs joining $V_1$ and $V_2$ go from $V_1$ to $V_2$.

**region**: for a graph imbedded in a surface, a maximal expanse of surface containing no vertex and no part of any edge of the graph.

**regular**: property of a graph or hypergraph that all its vertices have the same degree. See also *k-regular*.

**representation** (of a graph): a graph description, such as a drawing, from which a formal specification can be constructed and labeled with the vertex names and edge names, so as to obtain a graph that conforms to the incidence rule for the graph.

**rooted pebbling number** (of a graph): for a connected graph with root vertex $r$, the minimum number $t$ so that from every starting configuration of $t$ pebbles it is possible to move a pebble to $r$ via pebbling steps.

**rotation system** (of an imbedding): a list of the cyclic orderings of the incidence of edges at each vertex.

**Schreier graph**: a graph that depicts the cosets of some subgroup of a group with some set of generators; the vertices represent cosets, and the edges (often "color-coded" for the generators) represent the product rule.

**self-complementary**: property of a graph that it is isomorphic to its complement.

**self-loop**: an edge that joins a vertex to itself; see *loop*.

**simple digraph**: a digraph that has no self-loops and no pair of arcs with the same tail and head.

**simple graph**: a graph with no loops or multi-edges.

**simple hypergraph**: a hypergraph with no repeated edges.

**sink**: a digraph vertex with out-degree zero.

**source**: a digraph vertex with in-degree zero.

**spanning subgraph**: a subgraph of a graph $G$ that includes all vertices of $G$.

**specification** (of a graph): a list of its vertices and a list of its edges, with the incidence rule for determining the endpoints of every edge.

**spectrum** (of a graph): the multiset of its eigenvalues.

**strong component**: in a digraph, a maximal subdigraph that is strongly connected.

**strong orientation**: for a graph, an assignment of a direction to every edge making it a strongly connected digraph.

**strong product**: for graphs $G$ and $H$, the graph $G \boxtimes H$ that has vertex set $V_G \times V_H$ and edge set $\{(g', h)(g', h') \mid (g = g' \text{ or } gg' \in E_G) \text{ and } (h = h' \text{ or } hh' \in E_H)\}$.

**strong tournament**: a tournament in which there is a directed path from every vertex to every other vertex.

**strongly connected**: property of a digraph that every vertex is reachable from every other vertex.

**strongly regular graph** (with parameters $(n,\ k,\ r,\ s)$): an $n$-vertex, $k$-regular graph in which every adjacent pair of vertices is mutually adjacent to $r$ other vertices, and in which every pair of nonadjacent vertices is mutually adjacent to $s$ other vertices.

**subdivision** (of an edge): the operation of inserting a new vertex into the interior of the edge, thereby splitting it into two edges.

**subdivision**: given a graph, any new graph obtained by subdividing one or more edges one or more times.

**subgraph**: given a graph $G$, a graph whose vertices and edges are all in $G$.

**supermagic graph**: a connected graph whose edges can be labeled with consecutive positive integers such that for each vertex $v$ the sum of the labels of all edges incident with $v$ is the same for all $v$.

**tail** (of an arc): the vertex the arc goes from.

**terminus** (of a walk): the last vertex of the walk.

**tetrahedral graph**: another name for the complete graph $K_4$, resulting from the fact that it is equivalent to the 1-skeleton of the 4-sided Platonic solid called a tetrahedron.

**thickness**: for a graph $G$, the minimum number $\theta(G)$ of planar subgraphs whose union is $G$.

**topological sort** (or **topsort**): an algorithm that assigns a linear extension ordering to a DAG.

**tough graph**: a connected graph $G$ such that for every nonempty set $S$ of vertices, the number of components of the graph $G - S$ does not exceed $|S|$.

**total dominating set** (of a graph): a set $S$ of vertices such that every vertex is adjacent to a vertex in $S$.

**total domination number**: the minimum cardinality $\gamma_t(G)$ of a total dominating set in graph $G$.

**tournament**: a digraph with exactly one arc between each pair of distinct vertices.

**trail**: a walk in which no edge occurs more than once.

**trail, directed**: a directed walk in which no arc is repeated.

**transitive**: property of a digraph that whenever it contains an arc from $u$ to $v$ and an arc from $v$ to $w$, it also contains an arc from $u$ to $w$.

***transitive orientation***: for a graph, an assignment of a direction to every edge, making it a transitive digraph.

***transmitter***: in a digraph, a vertex that has an arc to every other vertex.

***transversal***: in a hypergraph, a set of vertices that has nonempty intersection with every edge of the hypergraph.

***transversal number***: the minimum number $\tau(H)$ of vertices taken over all transversals of $H$.

***tree***: a connected graph without a cycle.

***trivial graph***: the graph with one vertex and no edge.

***Turán graph***: the $n$-vertex $k$-partite simple graph $T_k(n)$ with the maximum number of edges.

***u,v-path***: a path whose origin is the vertex $u$ and whose terminus is the vertex $v$.

***underlying graph***: for a digraph, the graph obtained from the digraph by stripping the directions off all the arcs.

***uniform***: property of a hypergraph that all edges have the same number of vertices. See also *r-uniform*.

***unilaterally connected*** (or ***unilateral***): property of a digraph that for every pair of vertices $u, v$, there is either a $uv$-path or a $vu$-path.

***upset***: a simple hypergraph in which every superset of every edge is also an edge of the hypergraph.

***valence***: a synonym for *degree* (adapted from molecular bonds in chemistry).

***vertex***: a point; an element of the first constituent set of a graph.

***vertex coloring***: an assignment of colors to the vertices of a graph so that adjacent vertices receive different colors.

**(vertex) *connectivity***: the smallest number $\kappa(G)$ of vertices whose removal disconnects the graph; by convention, $\kappa(K_n) = n - 1$.

***vertex cut***: See *disconnecting set*.

***vertex-deleted subgraph***: any subgraph obtained from a graph by removing a single vertex and all of its incident edges.

***vertex invariant***: a property at a vertex that is preserved by every isomorphism.

***walk***: an alternating sequence $v_0, e_1, v_1, \ldots, e_r, v_r$ of vertices and edges where consecutive edges are adjacent, so that each edge $e_i$ joins vertices $v_{i-1}$ and $v_i$.

***walk, directed***: an alternating sequence of vertices and arcs $v_0, e_1, v_1, e_2, \ldots, e_n, v_n$ where the arcs align head to tail, so that each vertex is the head of the preceding arc and the tail of the subsequent arc.

***weakly connected*** (or ***weak***) ***digraph***: a digraph whose underlying graph is connected.

***weighted graph***: a graph model in which each edge is assigned a number called the weight or the cost.

***wheel graph***: an $(n + 1)$-vertex graph $W_n$ that "looks like" a wheel whose rim is an $n$-cycle and whose hub vertex is joined by spokes to all the vertices on the rim.

# 8.1    INTRODUCTION TO GRAPHS

Graphs are highly adaptable mathematical structures, and can be represented on a computer so that as new applications arise, existing algorithms can be reused without rewriting. This section provides some of the basic terminology and operations needed for the study of graphs and describes several useful families of graphs.

## 8.1.1    VARIETIES OF GRAPHS AND GRAPH MODELS

Due to the vast breadth of the usefulness of graphs, the terminology varies widely, not only from one type of graph to another, but also from one application to another. The table in Fact 1 gives synonyms for several terms. Definitions for undirected graphs are given first, followed by definitions for directed graphs. Because much of the terminology is similar for the two types of graph, only terms that are different for directed graphs are given.

**Definitions:**

A **graph** $G = (V, E)$ is a set $V$ of *vertices* and a set $E$ of *edges* (both sets are finite and $V$ is nonempty unless specified otherwise) such that each edge is associated with either two vertices or one vertex twice. It is sometimes denoted $(V_G, E_G)$ or $(V(G), E(G))$.

A **vertex** is usually conceptualized as a point. Abstractly, it is a member of the first of the two sets that form a *graph*.

An **edge** is usually conceptualized as a line segment or curve, either joining one vertex to another or joining a vertex to itself. Abstractly, it is a member of the second of the two sets that form a *graph*.

The number of vertices of a graph is often called its **order**. The number of edges of a graph is sometimes called its **size**.

A **proper edge** (or **link**) is an edge that joins one vertex to another.

A **loop** (or **self-loop**) is an edge that joins a vertex to itself.

The **endpoints** of an edge are the vertices that the edge joins. A loop has only one endpoint.

An edge $e$ is **incident with** a vertex $v$ if $v$ is an endpoint of $e$.

An **incidence rule** specifies the endpoints of the edges.

A **simple graph** is a graph $G$ that has no loops and in which no two edges have the same endpoints. In a simple graph, an edge with endpoints $v$ and $w$ can be regarded as the pair $\{v, w\}$ and is often denoted simply as $vw$.

A graph with no edges is called a **null graph**. If further, $V$ is empty, then the result is the **empty graph**. (Note that some researchers invert the meanings of null graph and empty graph.)

The **trivial graph** has just one vertex and no edges.

Vertices $v$ and $w$ are **adjacent** if there is an edge whose endpoints are $v$ and $w$.

Two edges are **adjacent** if they have a common endpoint.

A **neighbor** of a vertex is any vertex to which it is adjacent.

An **attribute** of the edge set or vertex set of a graph is a feature such as length, cost, or color sometimes attached to graphs.

A **graph model** is a graph which may have attributes on its edges or vertices. The vertices and edges of the model may represent arbitrary objects and relationships from the context of the application.

A **weighted graph** is a graph in which each edge is assigned a number called the weight or cost.

A **node** is sometimes a synonym for a vertex and sometimes refers to whatever is modeled by a vertex in a graph model.

A **line** is sometimes a synonym for an edge and sometimes refers to whatever is modeled by an edge in a graph model.

If at least two edges have the same endpoints, the set of all edges with these endpoints is called a **multi-edge**. If $r$ is the number of these edges, this is a sometimes called an **edge of multiplicity** $r$. A graph with a multi-edge is sometimes said to have **multiple edges** or **parallel edges**.

A **multigraph** is another name for a graph with multi-edges but no loops, used for emphasis when the context is largely restricted to simple graphs.

A **pseudograph** (or **general graph**) is another name for a graph in which loops and multi-edges are permitted, used for emphasis when the context is largely restricted to loop-free graphs.

The **degree** of vertex $v$, $\deg(v)$, is the number of proper edges plus *twice* the number of loops incident with $v$. Thus, in a drawing, it is the number of edge-endings at $v$.

The **valence** of a vertex is a synonym for *degree* adapted from terminology in chemistry.

A vertex of degree 0 is called an **isolated vertex**.

A vertex of degree 1 is called a **leaf** or an **end vertex**.

A **regular** graph is a graph in which all vertices have the same degree. It is called **k-regular** if that degree is $k$.

The **degree sequence** of a graph is the sequence of the degrees of its vertices, usually given in increasing or decreasing order.

A **graphical sequence** is a sequence of nonnegative integers that is the degree sequence of some simple graph.

A **digraph** or **directed graph** $D = (V, A)$ is a set $V$ of *vertices* and a set $A$ of *arcs* (both sets are finite and $V$ is nonempty unless specified otherwise) such that each arc is associated with an ordered pair of vertices (which may be the same). It is sometimes denoted $(V_D, A_D)$ or $(V(D), A(D))$.

A **direction** on an edge is an ordering for its endpoints so that the edge goes *from* one endpoint and *to* the other. Any edge, including a loop, can be directed by giving it a sense of forward progression; for example, in a graph drawing, by placing an arrow on the edge.

An **arc** is a directed edge.

The **tail** of an arc is the vertex at which the arc originates. The **head** is the vertex at which the arc terminates.

Two arcs are **parallel** if they have the same head and the same tail. They are **opposite** if the tail of each is the head of the other.

A **strict digraph** has no parallel arcs. In set theory terms, it is the graph of a relation.

A **simple digraph** has no loops and no parallel arcs. In set theory terms, it is the graph of an irreflexive relation.

The **out-degree** of vertex $v$, denoted $\text{od}(v)$ or $\deg^+(v)$, is the number of arcs having $v$ as tail. The **in-degree** of vertex $v$, denoted $\text{id}(v)$ or $\deg^-(v)$, is the number of arcs having $v$ as head.

A vertex with in-degree 0 and positive out-degree is called a **source** or **transmitter**. A vertex with out-degree 0 and positive in-degree is called a **sink** or **receiver**.

The **degree pair** of vertex $v$ is the ordered pair $(\text{od}(v), \text{id}(v))$.

The **degree sequence** of a digraph is the sequence of the degree pairs of its vertices, usually given in increasing or decreasing order of the out-degree.

A **digraphical sequence** is a sequence of ordered pairs of nonnegative integers that is the degree sequence of some simple digraph.

**Facts:**

1. The following table lists some graph theory synonyms.

| |
|---|
| vertex: point, node |
| edge: line, link |
| loop: self-loop |
| neighbor: adjacent vertex |
| arc: directed edge |
| degree: valence |
| number of vertices: order |
| number of edges: size |
| nonsimple graph: pseudograph, general graph |
| loop-free nonsimple graph: multigraph |

2. The following table lists some of the varieties of graphs and digraphs.

| graph variety | loops allowed? | multi-edges allowed? |
|---|---|---|
| simple graph | NO | NO |
| multigraph | NO | YES |
| general graph | YES | YES |
| pseudograph | YES | YES |
| digraph | YES | YES |
| strict digraph | YES | NO* |
| simple digraph | NO | NO* |

*at most one arc in each direction between two vertices

**3.** In a drawing of a graph, the degree of a vertex $v$ equals the number of edge-ends at $v$. The degree of $v$ need *not* equal the number of edges at $v$, since each loop contributes twice toward the degree.

**4.** In every graph, the sum of the degrees equals twice the number of edges. From this it follows that the sum of the degrees of all vertices is even.

**5.** In every graph the number of vertices of odd degree is even.

**6.** The name *handshaking lemma* is commonly applied to various elementary results about the degrees of simple graphs, especially Facts 4 and 5.

**7.** In every simple graph with at least two vertices, there is a pair of vertices with the same degree.

**8.** *Havel's theorem:* A sequence of nonnegative integers is graphical if and only if the sequence obtained by deleting the largest entry $d$ and subtracting 1 from each of the $d$ next largest entries is graphical. (V. Havel, 1955)

**9.** A nonincreasing sequence of nonnegative integers $d_1, d_2, \ldots, d_n$ is graphical if and only if its sum is even, and for $k = 1, 2, \ldots, n$,

$$\sum_{i=1}^{k} d_i \leq k(k-1) + \sum_{i=k+1}^{n} \min\{k, d_i\}.$$

(P. Erdős and T. Gallai, 1960)

**10.** In every digraph, the sum of the in-degrees equals the sum of the out-degrees, and this is equal to the number of arcs.

**11.** A sequence of ordered pairs of nonnegative integers is digraphical if and only if the sequence obtained by replacing the largest first entry $d$ by 0 and subtracting 1 from each of the $d$ largest second entries among the other pairs is digraphical.

**12.** A sequence of ordered pairs of nonnegative integers at most $n-1$, say $(a_1, b_1), (a_2, b_2)$, $\ldots, (a_n, b_n)$, with the $a_i$ nonincreasing is digraphical if and only if $\sum_{i=1}^{n} a_i = \sum_{i=1}^{n} b_i$, and for $k = 1, 2, \ldots, n$,

$$\sum_{i=1}^{k} a_i \leq \sum_{i=1}^{k} \min\{k-1, b_i\} + \sum_{i=k+1}^{n} \min\{k, b_i\}.$$

(D. L. Fulkerson, 1960)

**Examples:**

**1.** The following figure gives examples of the various varieties of graphs. (See the table in Fact 2.)

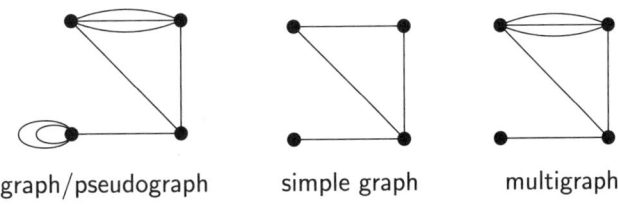

graph/pseudograph       simple graph       multigraph

**2.** The following figure gives examples of the various varieties of digraphs. (See the table in Fact 2.)

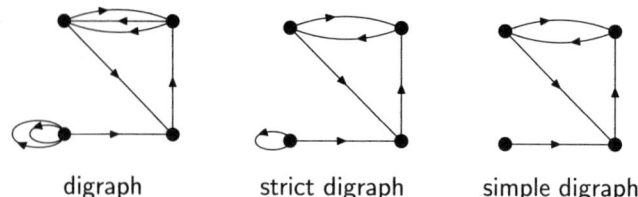

digraph          strict digraph          simple digraph

**3.** *Computer programming flowchart* (always a digraph):  Each vertex represents some programmed operation or decision, and each arc represents the flow of control to the next operation or decision.

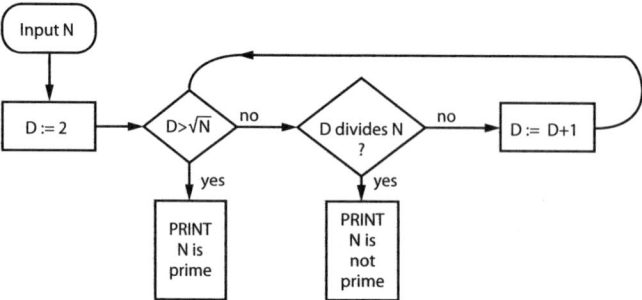

**4.** *Model for social networks*:  Each vertex represents a person in the network, and each edge represents a form of interaction between the persons represented by its endpoints. This is illustrated by the following graph.

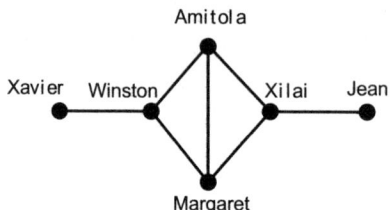

**5.** *Model for road networks* (most edges undirected):  Each vertex represents either an intersection of two roads or the end of a dead-end street.  The absence of an endpoint in the illustration indicates that the road continues beyond what is shown.  Direction on an edge may be used to indicate a one-way road, with undirected edges being two-way roads.

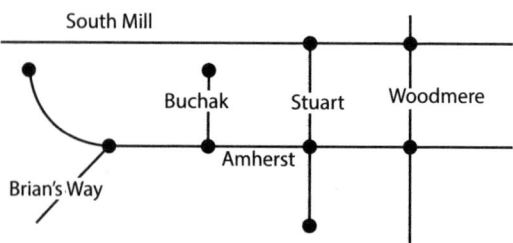

**6.** In the following graph with vertex set $V = \{v_1, v_2, v_3, v_4, v_5\}$ and edge set $E = \{e_1, e_2, e_3, e_4, e_5, e_6, e_7\}$, the vertex $v_5$ is an isolated vertex, and the degree sequence is $(0, 3, 3, 4, 4)$. The edge $e_7$ is a loop, and the three edges $e_4$, $e_5$, and $e_6$ form a multi-edge.

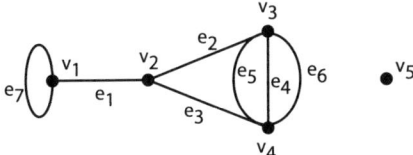

**7.** Deleting the isolated vertex and the self-loop in Example 6 and then reducing the multi-edge to a single edge yields the following simple graph, whose degree sequence is $(1, 2, 2, 3)$.

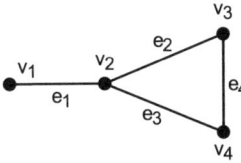

**8.** One possible choice of edge directions for the graph of Example 6 yields this digraph.

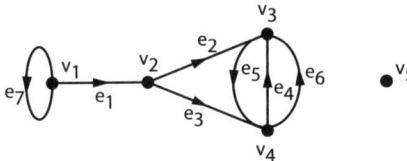

**9.** The sequence $(1, 2, 2, 3, 4, 5)$ is not graphical, by Fact 4, because its sum is odd.

**10.** Havel's reduction (Fact 8) of the sequence $(2, 2, 2, 3, 4, 5)$ is $(1, 1, 1, 2, 3)$. Havel's reduction of that sequence is $(0, 0, 1, 1)$. Since $(0, 0, 1, 1)$ is the degree sequence of a graph with four vertices, two of which are isolated and two of which are joined by an edge, it follows from Havel's theorem that the sequence $(2, 2, 2, 3, 4, 5)$ is graphical.

**11.** The sequence $(2, 1), (2, 0), (1, 1), (1, 3)$, is not digraphical since the sum of the first entries does not equal the sum of the second entries.

**12.** The reduction of the sequence $(3, 1), (1, 2), (1, 2), (1, 1)$ using Fact 11 is the sequence $(0, 1), (1, 1), (1, 1), (1, 0)$. Since the second sequence is digraphical, being the degree sequence of the directed path of length 3, the first sequence is also digraphical.

---

## 8.1.2 GRAPH OPERATIONS

**Definitions:**

A graph $H = (V_H, E_H)$ is a **subgraph** of a graph $G = (V_G, E_G)$ if its vertex set and edge set are subsets of $V_G$ and $E_G$, respectively.

A **spanning subgraph** of a graph $G$ is a subgraph that contains all the vertices of $G$.

The **induced subgraph** on a nonempty set $S$ of vertices in a graph $G$ is the subgraph whose vertex set is $S$ and whose edge set consists of all edges of $G$ having both endpoints in $S$. It may be denoted by $\langle S \rangle$ or $G[S]$ and is the maximal subgraph of $G$ with vertex set $S$.

An **induced subgraph** of $G$ is a subgraph $H$ such that every edge of $G$ that joins two vertices of $H$ is also an edge of $H$.

**Deleting an edge** $e$ from a graph $G$ results in the subgraph $G - e$ that contains all the vertices of $G$ and all the edges of $G$ except for $e$.

**Deleting a set** $Y$ **of edges** from a graph $G$ results in the subgraph $G - Y$ that contains all the vertices of $G$ and all the edges of $G$ except for those in $Y$.

**Deleting a vertex** $v$ from a graph $G$ results in the subgraph $G - v$ that contains all the vertices of $G$ except $v$ and all the edges of $G$ except those incident with $v$. It is the induced subgraph $\langle V_G - \{v\}\rangle$.

**Deleting a set** $S$ **of vertices** from a graph $G$ results in the subgraph $G - S$ that contains all the vertices of $G$ except those in $S$ and all the edges of $G$ except those incident with vertices in $S$. It is the induced subgraph $\langle V_G - S\rangle$.

**Contracting an edge** $e$ in a graph $G$ means shrinking the edge to a point, so that its endpoints are merged, without changing the rest of the graph. The resulting graph is denoted $G/e$ (or $G \cdot e$ or $G \downarrow e$). To construct $G/e$ from $G$, delete the edge $e$ from the edge set and replace all instances of its endpoints in the vertex set and incidence rule by a new vertex.

A **minor** of a graph $G$ is any graph that can be obtained from $G$ by a sequence of edge deletions and contractions and vertex deletions.

The **graph union** $G \cup H$ has as its vertices and edges those vertices and edges that are in $G$ or $H$.

The **graph intersection** $G \cap H$ has as its vertices and edges those vertices and edges that are both in $G$ and in $H$.

The **graph sum** (or **disjoint union**) $G + H$ consists of the disjoint unions of the vertex sets and of the edge sets of the graphs $G$ and $H$.

The **iterated graph sum** $nG$ is the sum of $n$ disjoint copies of $G$.

The **join** $G * H$ is obtained by adding to $G + H$ an edge from each vertex in $G$ to each vertex in $H$.

The **Cartesian product** $G \square H$ (or sometimes $G \times H$) has as its vertices the product $V_G \times V_H$ and as its edges $(V_G \times E_H) \cup (E_G \times V_H)$. The endpoints of the edge $(u, d)$ are the vertices $(u, x)$ and $(u, y)$, where $x$ and $y$ are the endpoints of $d$ in $H$, and those of the edge $(e, w)$ are $(u, w)$ and $(v, w)$, where $u$ and $v$ are the endpoints of $e$.

An **isomorphism** $f \colon G \to H$ (of graphs) establishes their structural equivalence. It is given by a pair of bijections $f_V \colon V_G \to V_H$ and $f_E \colon E_G \to E_H$ such that if $u$ and $v$ are the endpoints of edge $e$ in graph $G$, then $f_V(u)$ and $f_V(v)$ are the endpoints of $f_E(e)$ in graph $H$. The vertex function and the edge function can both be denoted $f$ without the subscript. (See §8.5.)

Two graphs are **isomorphic** if there is an isomorphism between them. This means that they are essentially the same graph except for the names of their vertices and edges.

A **graph mapping** $f \colon G \to H$ (of graphs) is a pair of functions $f_V \colon V_G \to V_H$ and $f_E \colon E_G \to E_H$ such that if $u$ and $v$ are the endpoints of edge $e$ in $G$, then $f(u)$ and $f(v)$ are the endpoints of $f(e)$ in $H$. Such a pair of functions is said to *preserve incidence*. The vertex function and the edge function can both be denoted $f$ without the subscript.

An **automorphism** of a graph $G$ is an isomorphism of $G$ to itself.

The **automorphism group** $Aut(G)$ is the group of all automorphisms of graph $G$.

**Subdividing an edge** $e$ is the operation of inserting a new vertex in the interior of an edge. Combinatorially, this is achieved by deleting $e$ and adding a new vertex adjacent to the endpoints of $e$.

Two graphs are **homeomorphic** if there is a graph from which they can both be obtained by a sequence of edge subdivisions.

The **complement** $\overline{G}$ of a simple graph $G$ has the same vertex set as $G$, with two vertices being adjacent in $\overline{G}$ if and only if they are not adjacent in $G$.

A **self-complementary** graph is a graph that is isomorphic to its complement.

The **line graph** $L(G)$ of a graph $G$ has vertices corresponding to the edges of $G$, with two vertices in $L(G)$ being adjacent whenever the corresponding edges are adjacent in $G$.

**Facts:**

**1.** If a graph $J$ is isomorphic to a subgraph of a graph $G$, then it is commonly said that $J$ "is" a subgraph of $G$, even though $V_J$ and $E_J$ might not be subsets of $V_G$ and $E_G$.

**2.** A graph is a subgraph of its union with any other graph.

**3.** The intersection of two graphs is a subgraph of each of them.

**4.** A graph mapping is the combinatorial counterpart of what is topologically a continuous function from one graph to the other.

**5.** A graph isomorphism is a graph mapping for which both the vertex function and the edge function are bijections.

**6.** There is a self-complementary graph of order $n$ if and only if $n \equiv 0$ or $1$ (mod 4).

**7.** The automorphism group $Aut(G)$ of any simple graph is isomorphic to the automorphism group $Aut(\overline{G})$ of its complement.

**8.** A connected graph $G$ is isomorphic to its line graph if and only if $G$ is a cycle (§8.1.3).

**9.** If two connected graphs have isomorphic line graphs, then either they are isomorphic to each other or they are $K_3$ and $K_{1,3}$ (see §8.1.3).

**10.** $Aut(K_n)$ is isomorphic to the symmetric group $S_n$ (see §5.3.1).

**11.** $Aut(C_n)$ is isomorphic to the dihedral group $D_n$ (see §5.3.2).

**Examples:**

**1.** The dark subgraph spans the following graph, because it contains every vertex of the given graph.

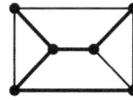

**2.** The Cartesian product $C_4 \,\square\, K_2$ (see §8.1.3) is illustrated as follows:

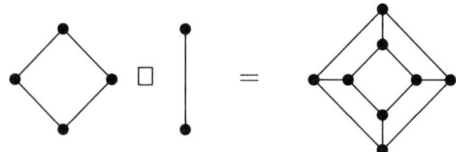

3. The join $\overline{K_2} * P_3$ (see §8.1.3) is illustrated as follows:

4. The following two graphs are homeomorphic, but not isomorphic.

5. The graphs $K_{3,3}$ and $2K_3$ (see §8.1.3) are complements of each other.
6. The path $P_4$ and the cycle $C_5$ (see §8.1.3) are self-complementary.
7. The line graph $L(K_4)$ (see §8.1.3) is isomorphic to the octahedral graph $K_{2,2,2}$.

### 8.1.3   SPECIAL GRAPHS AND GRAPH FAMILIES

**Definitions:**

*Note:* Many of the following graphs are drawn in Figures 1 and 2.

The **bouquet** $B_n$ is the graph with one vertex and $n$ loops.

The **dipole** $D_n$ is the graph with two vertices and an edge of multiplicity $n$ joining them.

The **complete graph** $K_n$ is the simple graph with $n$ vertices in which every pair of vertices is adjacent.

The **$n$-path** $P_n$ consists of a sequence of $n$ vertices $v_1, v_2 \ldots, v_n$ and the $n-1$ edges joining successive vertices in the sequence; that is, the $n-1$ edges are $v_1v_2, v_2v_3, \ldots, v_{n-1}v_n$.

A **path** is a graph that is an $n$-path for some $n \geq 1$.

The **$n$-cycle** $C_n$ consists of a sequence of $n$ vertices $v_1, v_2, \ldots, v_n$ and the $n$ edges joining successive vertices cyclically; that is, the $n$ edges are $v_1v_2, v_2v_3, \ldots, v_{n-1}v_n, v_nv_1$.

A **cycle** is a graph which is an $n$-cycle for some $n > 0$.

The **$n$-wheel** $W_n$ is the join of $K_1$ and the $n$-cycle $C_n$.

A graph is **bipartite** if its vertices can be partitioned into two subsets (the **parts**, or **partite sets**) so that no two vertices in the same part are adjacent.

The **complete bipartite graph** $K_{r,s}$ is the simple bipartite graph in which the two parts have cardinalities $r$ and $s$, such that every vertex in one part is adjacent to every vertex in the other part.

The **complete $r$-partite graph** $K_{n_1,n_2,...,n_r}$ has $r$ disjoint subsets of vertices of orders $n_1, n_2, \ldots, n_r$, with two vertices adjacent if and only if they lie in different subsets. If the $r$ sets all have $t$ vertices, this graph is sometimes denoted $K_{r(t)}$.

A graph $G$ is **connected** if for each pair of vertices in $G$, there is a path in $G$ from one to the other.

A **tree** is a connected graph without any cycles as subgraphs. (See Chapter 9.) A **forest** is a graph without any cycles as subgraphs.

The **Kuratowski graphs** are the graphs $K_5$ and $K_{3,3}$.

The **Petersen graph** is the graph constructed from two disjoint 5-cycles on the vertices $v_1, v_2, v_3, v_4, v_5$ and $w_1, w_2, w_3, w_4, w_5$ by adding the edges $v_1w_2, v_2w_4, v_3w_1, v_4w_3, v_5w_5$. (See Figure 2.)

A **Platonic solid** is a regular 3-dimensional polyhedron.

The **1-skeleton** of a polyhedron is the graph that has as its vertices and edges those of the polyhedron.

The **tetrahedral graph** is the 1-skeleton of the 4-sided Platonic solid called a tetrahedron (its faces are triangles). It has 4 vertices, each of degree 3, and 6 edges.

The **cube graph** $Q_3$ is the 1-skeleton of the 6-sided Platonic solid called a cube (its faces are squares). It has 8 vertices, each of degree 3, and 12 edges.

The **$d$-dimensional hypercube** graph $Q_d$ is a graph with $2^d$ vertices that can be labeled with the $2^d$ bitstrings of length $d$ so that two vertices are adjacent if and only if their labels differ in exactly one bit.

The **octahedral graph** $\mathcal{O}_3$ is the 1-skeleton of the 8-sided Platonic solid called an octahedron (its faces are triangles). It has 6 vertices, each of degree 4, and 12 edges.

The **generalized octahedral graph** $\mathcal{O}_n$ is the graph that can be obtained from the complete graph $K_{2n}$ by removing $n$ mutually nonadjacent edges. It is isomorphic to $K_{n(2)}$.

The **dodecahedral graph** is the 1-skeleton of the 12-sided Platonic solid called a dodecahedron (its faces are pentagons). It has 20 vertices, each of degree 3, and 30 edges.

The **icosahedral graph** is the 1-skeleton of the 20-sided Platonic solid called an icosahedron (its faces are triangles). It has 12 vertices, each of degree 5, and 30 edges.

The **intersection graph** of a finite collection $F = \{S_1, S_2, \ldots, S_k\}$ of subsets of some set has as its vertices the subsets themselves, with an edge between each pair of subsets whose intersection is nonempty.

An **interval graph** is any graph isomorphic to the intersection graph of some collection of intervals on the real line.

**Facts:**
**1.** In a computer program, the trivial graph is often used as the initial value of a graph-valued variable, similar to how an integer-valued variable is initialized to zero. As the program runs, the graph-valued variable can be modified by adding vertices and edges.

**2.** Bouquets and dipoles are fundamental building blocks for graphs constructed by topological techniques.

**3.** Every path is a tree.

**4.** A graph is bipartite if and only if it has no cycles of odd length.

**5.** Every tree is bipartite.

**6.** The hypercube graphs $Q_n$ can be defined recursively as follows: $Q_0 = K_1$, $Q_n = K_2 \square Q_{n-1}$ for $n > 0$.

**7.** The hypercube graph $Q_n$ is bipartite and is isomorphic to the lattice of subsets of a set of $n$ elements. (See §5.7.1.)

**8.** The octahedral graphs $\mathcal{O}_n$ can be defined recursively as follows: $\mathcal{O}_0 = \overline{K_2}$, $\mathcal{O}_n = \overline{K_2} * \mathcal{O}_{n-1}$ for $n > 0$.

**9.** There are exactly five Platonic graphs: the tetrahedral graph $K_4$, the cube $Q_3$, the octahedral graph $\mathcal{O}_3$, the dodecahedral graph, and the icosahedral graph.

**Examples:**

**1.** Figure 1 shows some of the classes of graphs that occur most often in general constructions.

**2.** Figure 2 shows some of the graphs that occur most often as special examples.

## 8.1.4  GRAPH REPRESENTATION AND COMPUTATION

To apply a computer to graph-theoretic computations, it is necessary to specify the underlying graph completely and without ambiguity. The representations are also useful in their own right.

**Definitions:**

A **specification** of a graph is a list of its vertices, a list of its edges, and the incidence rule for determining the endpoints of every edge.

An **endpoint table** for a graph is a tabular description of the incidence rule, that gives the endpoints of every edge. In a digraph or partially directed graph, the tail and head of each arc are distinguished.

An **incident-edge table** for a graph is a tabular description of the incidence rule, that gives for each vertex $v$, a list of the edges having $v$ as an endpoint. If the graph is directed, this list is partitioned into two sublists, according to whether $v$ is tail or head.

A **representation** of a graph $G$ is a graph description, such as a drawing, from which a formal specification could be constructed and labeled with the vertex names and edge names from $G$, so as to obtain a graph that conforms to the incidence rule for $G$.

The **incidence matrix** of a graph (without loops) $G$ with vertices $v_1, v_2, \ldots, v_n$ and edges $e_1, e_2, \ldots, e_m$ is the $n \times m$ matrix $M_I$ with

$$M_I[i,j] = \begin{cases} 0 & \text{if } v_i \text{ is not an endpoint of } e_j \\ 1 & \text{if } v_i \text{ is an endpoint of } e_j. \end{cases}$$

(In case loops are present, if $e_j$ is a loop at $v_i$, then usually $M_I[i,j] = 2$, but sometimes $M_I[i,j] = 1$ even though this violates the properties that every column-sum equals 2 and every row-sum equals the degree of the corresponding vertex.)

Figure 1: Some fundamental infinite classes of graphs.

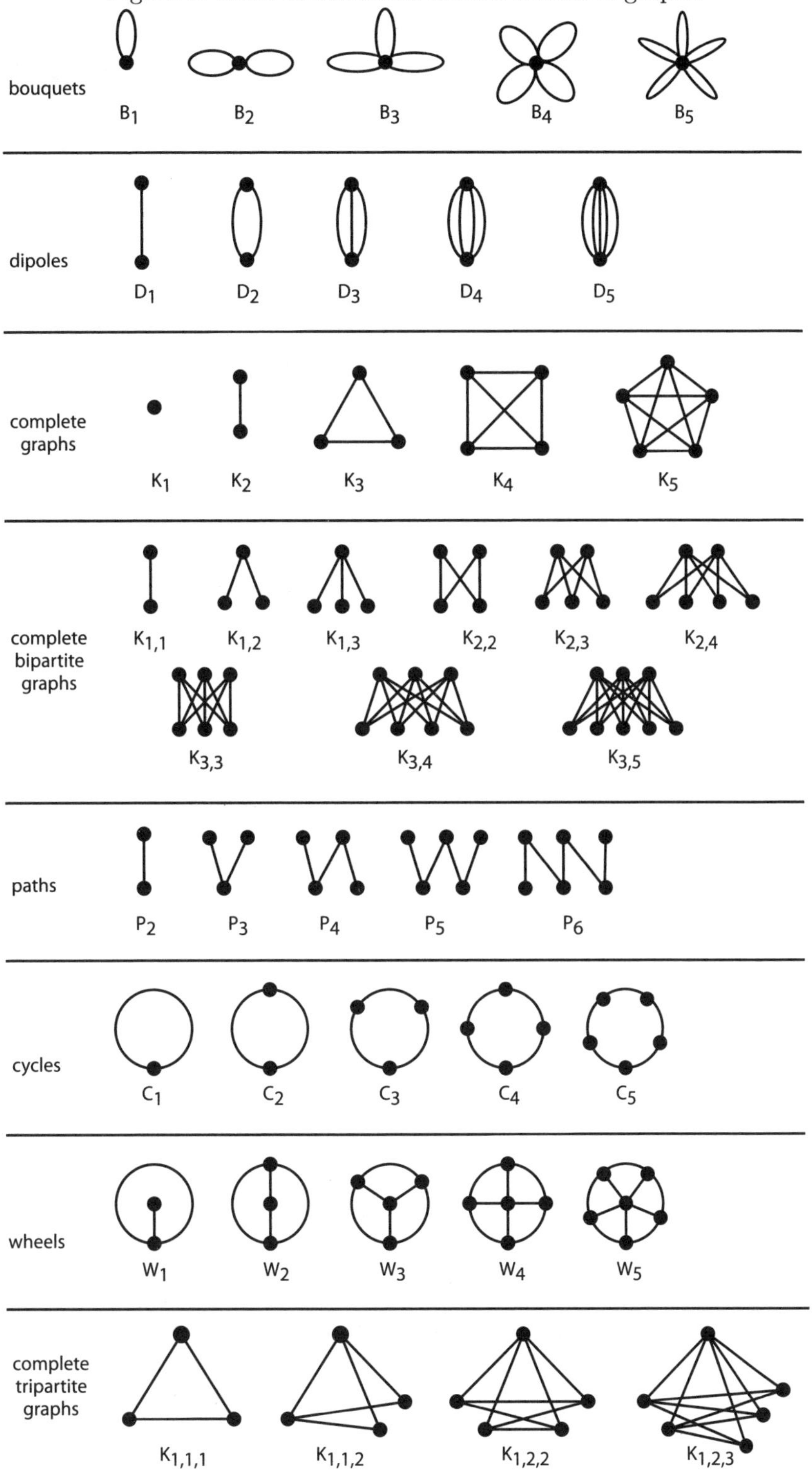

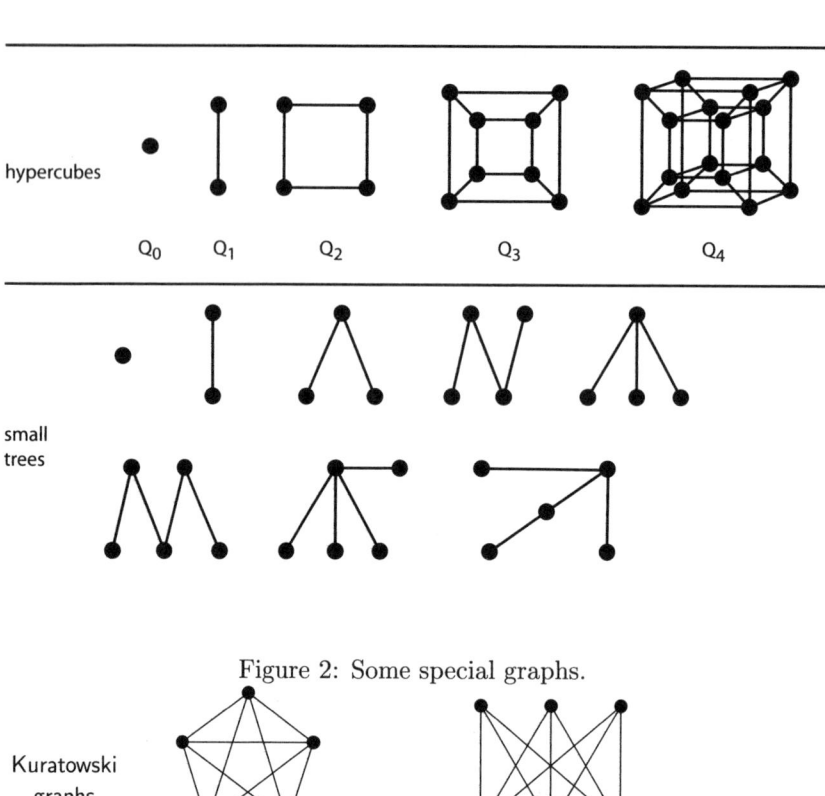

hypercubes

$Q_0$    $Q_1$    $Q_2$    $Q_3$    $Q_4$

small
trees

Figure 2: Some special graphs.

Kuratowski
graphs

$K_5$    $K_{3,3}$

Petersen
graph

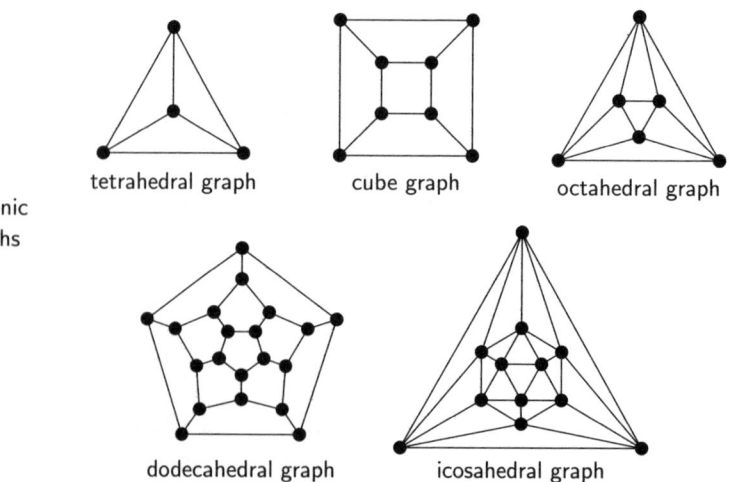

Platonic
graphs

tetrahedral graph    cube graph    octahedral graph

dodecahedral graph    icosahedral graph

The **adjacency matrix** of a loop-free graph $G$ with vertices $v_1, v_2, \ldots, v_n$ is the $n \times n$ matrix $A$ with

$$A[i,j] = \text{the number of edges between } v_i \text{ and } v_j \text{ if } i \neq j.$$

If there are loops, then $A[i,i]$ is usually defined to be the number of loops at $v_i$.

A **normalized drawing** of a graph represents each vertex as a distinct point in the plane and each edge as a possibly curved line between endpoints, such that

- the interior of an edge does not contain any vertex;
- at most two edges cross at any point of the plane;
- two edges cross each other at most once;
- each edge crossing is normal, not a tangency.

A **complete set of operations** on graphs is a set from which all other operations can be constructed. The operations in a complete set are **primitive** if none can be derived from the other operations.

A **graph computation package** is a computer software system that represents graphs and includes a complete set of operations.

**Facts:**

**1.** Despite the redundancy, an incident-edge table is often used with an endpoint table in computer software, since it facilitates fast searching at the cost of relatively little space.

**2.** If a graph is simple, then its edges can be represented as endpoint pairs $vw$. Thus, the graph can be specified as a list of endpoint pairs and a list of isolated vertices.

**3.** If a graph is simple, then its incident-edge table can be represented as a table that gives the list of neighbors of every vertex.

**4.** If $A$ is the adjacency matrix of graph $G$, then the $(i,j)$-entry of $A^k$ is the number of walks (see §8.4.1) of length $k$ from $v_i$ to $v_j$ in $G$.

**5.** *Matrix-tree theorem:* Let $G$ be a graph, and let $A$ be its adjacency matrix and $D$ the diagonal matrix of the degrees of its vertices. Then the value of every cofactor of $D - A$ equals the number of spanning trees of $G$. (G. R. Kirchhoff, 1847)

**6.** Given the incidence matrix of $G$, it is possible to obtain the incidence matrix for a subgraph $H$ of $G$ by deleting all rows and columns corresponding to vertices and edges that are not in $H$.

**7.** The most commonly used complete set of operations is adding a vertex, deleting a vertex, adding an edge, and deleting an edge.

**8.** Graph computation packages are built into mathematical computation systems such as Maple and Mathematica. They usually include display operations.

**Examples:**

**1.** The following normalized drawing, endpoint table, incident-edge table, incidence matrix and adjacency matrix all specify the same graph $G$.

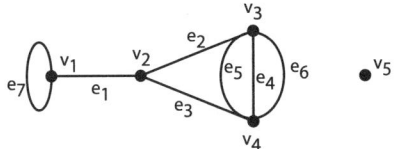

incident-edge table

endpoint table

| $e_1$ | $e_2$ | $e_3$ | $e_4$ | $e_5$ | $e_6$ | $e_7$ |
|---|---|---|---|---|---|---|
| $v_1$ | $v_2$ | $v_2$ | $v_3$ | $v_3$ | $v_3$ | $v_1$ |
| $v_2$ | $v_3$ | $v_4$ | $v_4$ | $v_4$ | $v_4$ | |

| | | |
|---|---|---|
| $v_1$ | $e_1$ | $e_7$ |
| $v_2$ | $e_1$ | $e_2$ | $e_3$ |
| $v_3$ | $e_2$ | $e_4$ | $e_5$ | $e_6$ |
| $v_4$ | $e_3$ | $e_4$ | $e_5$ | $e_6$ |
| $v_5$ | | |

$$M_I = \begin{array}{c|ccccccc} & e_1 & e_2 & e_3 & e_4 & e_5 & e_6 & e_7 \\ \hline v_1 & 1 & 0 & 0 & 0 & 0 & 0 & 2 \\ v_2 & 1 & 1 & 1 & 0 & 0 & 0 & 0 \\ v_3 & 0 & 1 & 0 & 1 & 1 & 1 & 0 \\ v_4 & 0 & 0 & 1 & 1 & 1 & 1 & 0 \\ v_5 & 0 & 0 & 0 & 0 & 0 & 0 & 0 \end{array}$$

$$A = \begin{array}{c|ccccc} & v_1 & v_2 & v_3 & v_4 & v_5 \\ \hline v_1 & 1 & 1 & 0 & 0 & 0 \\ v_2 & 1 & 0 & 1 & 1 & 0 \\ v_3 & 0 & 1 & 0 & 3 & 0 \\ v_4 & 0 & 1 & 3 & 0 & 0 \\ v_5 & 0 & 0 & 0 & 0 & 0 \end{array}$$

**2.** The following normalized drawing, list of endpoint pairs, lists-of-neighbors table, incidence matrix and adjacency matrix all specify the same simple graph $H$. It is a spanning subgraph of the graph $G$ of Example 1, but it is *not* an induced subgraph.

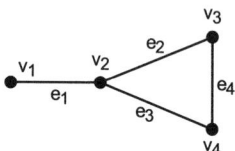

lists-of-neighbors

| | | | |
|---|---|---|---|
| $v_1:$ | $v_2$ | | |
| $v_2:$ | $v_1$ | $v_3$ | $v_4$ |
| $v_3:$ | $v_2$ | $v_4$ | |
| $v_4:$ | $v_2$ | $v_3$ | |

endpoint pairs:

$$v_1v_2, v_2v_3, v_2v_4, v_3v_4$$

$$M_I = \begin{array}{c|cccc} & e_1 & e_2 & e_3 & e_4 \\ \hline v_1 & 1 & 0 & 0 & 0 \\ v_2 & 1 & 1 & 1 & 0 \\ v_3 & 0 & 1 & 0 & 1 \\ v_4 & 0 & 0 & 1 & 1 \end{array}$$

$$A = \begin{array}{c|cccc} & v_1 & v_2 & v_3 & v_4 \\ \hline v_1 & 0 & 1 & 0 & 0 \\ v_2 & 1 & 0 & 1 & 1 \\ v_3 & 0 & 1 & 0 & 1 \\ v_4 & 0 & 1 & 1 & 0 \end{array}$$

**3.** Squaring and cubing the adjacency matrix of Example 2 provides an illustration of Fact 4.

$$A^2 = \begin{array}{c|cccc} & v_1 & v_2 & v_3 & v_4 \\ \hline v_1 & 1 & 0 & 1 & 1 \\ v_2 & 0 & 3 & 1 & 1 \\ v_3 & 1 & 1 & 2 & 1 \\ v_4 & 1 & 1 & 1 & 2 \end{array}$$

$$A^3 = \begin{array}{c|cccc} & v_1 & v_2 & v_3 & v_4 \\ \hline v_1 & 0 & 3 & 1 & 1 \\ v_2 & 3 & 2 & 4 & 4 \\ v_3 & 1 & 4 & 2 & 3 \\ v_4 & 1 & 4 & 3 & 2 \end{array}$$

For instance, the three walks of length 3 from $v_4$ to $v_3$ are as follows:

$$\langle v_4, e_3, v_2, e_3, v_4, e_4, v_3 \rangle, \quad \langle v_4, e_4, v_3, e_4, v_4, e_4, v_3 \rangle, \quad \langle v_4, e_4, v_3, e_2, v_2, e_2, v_3 \rangle.$$

**4.** As an illustration of the Kirchhoff matrix-tree theorem of Fact 5, observe that the graph of Example 2 has the following three spanning trees.

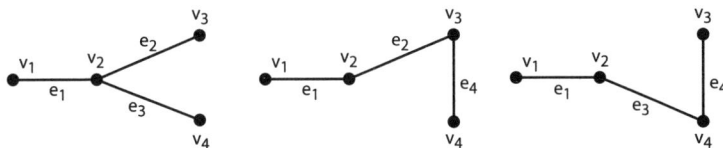

The value of the $(2,2)$-cofactor of the matrix $D - A$ is also equal to 3:

$$D - A = \begin{array}{c} \\ v_1 \\ v_2 \\ v_3 \\ v_4 \end{array} \begin{array}{cccc} v_1 & v_2 & v_3 & v_4 \\ \left( \begin{array}{cccc} 1 & -1 & 0 & 0 \\ -1 & 3 & -1 & -1 \\ 0 & -1 & 2 & -1 \\ 0 & -1 & -1 & 2 \end{array} \right) \end{array} \qquad \text{cofactor} = \begin{vmatrix} 1 & 0 & 0 \\ 0 & 2 & -1 \\ 0 & -1 & 2 \end{vmatrix} = 3.$$

## 8.2   GRAPH MODELS

Modeling with graphs is one of the main ways in which discrete mathematics has been applied to real world problems. This section gives a list of some of the ways in which graphs are used as mathematical models. Further information can be found in for example [ArGr06] and [Ro76].

### 8.2.1   ATTRIBUTES OF A GRAPH MODEL

**Definitions:**

A **mathematical representation** of a physical or behavioral phenomenon is a correspondence between the parts and processes of that phenomenon and a mathematical system of objects and functions.

A **model** of a physical or behavioral phenomenon is the mathematical object or function assigned to that phenomenon under a mathematical representation.

**Modeling** is the mathematical activity of designing models and comprehensive mathematical representations of physical and behavioral phenomena.

A **graph model** is a mathematical representation that involves a graph.

**Examples:**

Table 1 gives many examples of graph models. Each example states what the vertices and edges (or arcs) represent and where in the *Handbook* details can be found.

**Table 1: Directory of graph models.**

| subject area and application | vertex attributes and meaning edge/arc attributes and meaning | reference |
|---|---|---|
| computer programming flowcharts | vertex labels are program steps edge directions show flow | §8.1.1 |
| social organization social networks | vertices are persons edges represent interactions | §8.1.1 |

| subject area and application | vertex attributes and meaning edge/arc attributes and meaning | reference |
|---|---|---|
| civil engineering<br>  road networks | vertices are road intersections<br>  edges are roads | §8.1.1,<br>§8.3.1 |
| operations research<br>  scheduling | vertices are activities<br>  arcs show operational precedence | §8.3.1 |
| sociology<br>  hierarchical dominance | vertices are individuals<br>  arcs show who reports to whom | §8.3.1 |
| computer programming<br>  subprogram-calling diagram | vertices are subprograms<br>  arcs show calling direction | §8.3.1 |
| ecology<br>  food webs | vertices are species<br>  arcs show who eats whom | §8.3.1 |
| operations research<br>  scheduling | vertices are activities<br>  edges are activity conflicts | §8.3.1,<br>§8.6.1 |
| genealogy<br>  family trees | vertices are family members<br>  arcs show parenthood | §8.3.1 |
| set theory<br>  binary relations | vertices are elements<br>  arcs show relatedness | §8.3.1 |
| probabilistic analysis<br>  Markov models | vertices are process states<br>  edges are state transitions | §8.3.2 |
| traffic control<br>  assigning one-way streets | vertices are intersections<br>  edges are streets | §8.3.3 |
| partially ordered sets<br>  Hasse diagrams | vertices are elements<br>  arcs show covering relation | §8.3.4 |
| computer engineering<br>  communications networks | vertices are computational nodes<br>  arcs are communications links | §8.4.2 |
| operations research<br>  transportation networks | vertices are supply and demand nodes<br>  arcs are supply lines | §8.4.2 |
| walking tours<br>  Seven Bridges of Königsberg | vertices are land masses<br>  edges are bridges | §8.4.3 |
| postal delivery routing<br>  Chinese Postman Problem | vertices are street intersections<br>  edges are streets | §8.4.3 |
| information theory<br>  Gray codes | vertices are binary strings<br>  edges are single-bit changes | §8.4.4 |
| radio broadcasting<br>  assignment of frequencies | vertices are broadcast stations<br>  edges are potential interference | §8.6.1 |
| chemistry<br>  preventing explosions | vertices are chemicals<br>  edges are co-combustibility | §8.6.1 |
| cartography<br>  map-coloring | regions are countries<br>  edges are shared borders | §8.6.4 |

| subject area and application | vertex attributes and meaning edge/arc attributes and meaning | reference |
|---|---|---|
| highway construction   avoiding overcrossings | vertices are road intersections   edges are roads | §8.7.1 |
| electrical network boards   avoiding insulation | vertices are circuit components   edges are wires | §8.7.1 |
| VLSI computer chips   minimizing layering | vertices are circuit components   edges are wires | §8.7.4 |
| information management   binary search trees | vertices are data records   edges are decisions | §18.1.4 |
| computer operating systems   priority trees | vertices are jobs   edges are priority relations | §18.1.5 |
| physical chemistry   counting isomers | vertices are atoms   edges are molecular bonds | §9.3.2 |
| network optimization   min-cost spanning trees | edges are connections   edge-labels are costs | §10.1.1 |
| bipartite matching   personnel assignment | parts are people and jobs   edges are job-capabilities | §10.2.2 |
| network optimization   shortest path | vertices are locations   edge-labels are distances | §10.3.1 |
| traveling salesman routing   shortest complete tour | vertices are locations   edge-labels are distances | §10.7.1 |

## 8.3   DIRECTED GRAPHS

Assigning directions to the edges of a graph is useful in modeling, and is natural whenever order is important, e.g., in a hierarchical structure or a one-way road system. Also, any graph may be viewed as a digraph, by replacing each edge with two directed edges, one in each direction. Many graph problems are best solved as special cases of digraph problems, for instance, finding shortest paths, maximum flows, and connectivity.

### 8.3.1   DIGRAPH MODELS AND REPRESENTATIONS

Most terminology applies equally well to graphs and digraphs. The definitions below are special to digraphs or take on a somewhat different meaning for digraphs. Further, where it is clear that only digraphs are being discussed, "directedness" is often an implicit attribute of an "edge", "path", and other terms.

**Definitions:**

A *directed graph*, or *digraph*, consists of a set $V$ of *vertices* and a set $E$ of *directed edges* or *arcs*, and an *incidence function* that assigns to each edge a *tail* and a *head*.

The **tail** of an arc is the vertex it leaves, and the **head** is the vertex it enters. An arc with tail $u$ and head $v$ is often denoted by $uv$.

A **simple digraph** has no self-loops or multi-arcs.

The **underlying graph of a digraph** is the graph obtained from the digraph by replacing every directed edge by an undirected edge.

The **out-degree** of vertex $v$, denoted $\delta^+(v)$, is the number of arcs with tail at $v$.

The **in-degree** of vertex $v$, denoted $\delta^-(v)$, is the number of arcs with head at $v$.

A digraph is **transitive** if whenever it contains an arc from $u$ to $v$ and an arc from $v$ to $w$, it also contains an arc from $u$ to $w$.

The **adjacency matrix** of a digraph is the matrix $A = [a_{ij}]$ where $a_{ij}$ is the number of arcs from $v_i$ to $v_j$.

The **incidence matrix** of a digraph with no self-loops is the matrix $M = [b_{ij}]$, where $b_{ij} = +1$ if $v_i$ is the tail of $e_j$, $b_{ij} = -1$ if $v_i$ is the head of $e_j$, and $b_{ij} = 0$ otherwise. (There is no standard convention for self-loops.)

## Facts:

**1.** *Simple-digraph terminology*: In a context focusing on simple digraphs, there is often a different convention:

- "digraph" refers to a simple digraph;
- a directed graph with multi-arcs is called a *multidigraph*;
- a directed graph with self-loops is called a *pseudodigraph*;

**2.** *Alternative "path" terminology*: There is an alternative convention in which a (directed) "path" may use vertices and arcs more than once, but an "elementary path" does not repeat arcs, and a "simple path" does not repeat vertices (and, hence, does not repeat arcs either). See §8.3.2.

**3.** A digraph is frequently represented by an *arc list*, in which each arc is represented by an ordered pair $uv$, where $u$ is its tail and $v$ is its head. There is a separate entry for each arc, so that $uv$ occurs as often as the number of such arcs. A list of the isolated vertices plus such an arc list completely specifies a digraph.

**4.** Another common specification of a digraph is the *lists-of-neighbors representation*. For each vertex $u$, there is a list, which contains the head of each arc whose tail is $u$. Thus a vertex $v$ occurs in that list as many times as there are arcs from $u$ to $v$.

**5.** The incidence matrix is another way to represent a digraph. Since all but one or two of the entries in every column are zero, the incidence matrix is an inefficient representation.

**6.** The adjacency matrix is a way to specify a digraph when there is no reason to identify the arcs by name.

**7.** A digraph can be represented by a $2 \times |E|$ *incidence table* in which the tail and head of arc $e$ appear in column $e$. Direction on an arc can be indicated by a convention as to whether tail or head appears in the first row.

**8.** The row-sum in a directed adjacency matrix equals the out-degree of the corresponding vertex. The column-sum equals the in-degree.

**9.** In a digraph, the sum of the in-degrees, the sum of the out-degrees, and the number of arcs are all equal to each other; that is, $\sum_{v \in V} \delta^-(v) = \sum_{v \in V} \delta^+(v) = |E|$.

**Examples:**

**1.** The following arc list, incidence table, list-of-neighbors, and adjacency matrix all represent the digraph $G$.

incidence table:

arc list:

$uv, \ vv, \ vw, \ xw, \ xw, \ ux, \ xu$

| | $e_1$ | $e_2$ | $e_3$ | $e_4$ | $e_5$ | $e_6$ | $e_7$ |
|---|---|---|---|---|---|---|---|
| | $u$ | $v$ | $v$ | $x$ | $x$ | $u$ | $x$ |
| | $v$ | $v$ | $w$ | $w$ | $w$ | $x$ | $u$ |

lists-of-neighbors:

$u : v, x$
$v : v, w$
$w : \emptyset$
$x : w, w, u$

adjacency matrix:

$$\begin{array}{c}\\ u \\ v \\ w \\ x \end{array}\begin{array}{cccc} u & v & w & x \\ \begin{pmatrix} 0 & 1 & 0 & 1 \\ 0 & 1 & 1 & 0 \\ 0 & 0 & 0 & 0 \\ 1 & 0 & 2 & 0 \end{pmatrix} \end{array}$$

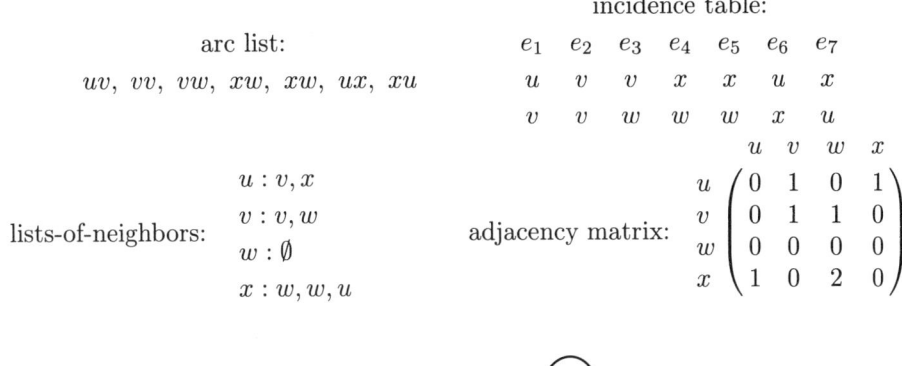

**2.** *Civil Engineering*:  A road network in which at least some of the roads are one-way can be modeled by a digraph. The vertices are road junctures; each two-way road is represented by a pair of arcs, one in each direction. Loops are allowed, and they may represent "circles" that occur in housing developments and in industrial parks. Similarly, multi-arcs may occur.

**3.** *Operations Research*:  A large project consists of many smaller tasks with a *precedence relation*—some tasks must be completed before certain others can begin. The vertices represent tasks, and there is an arc from $u$ to $v$ if task $u$ must be completed before $v$ can begin. For instance, in the following figure it is necessary both that food is loaded and the cabin is cleaned before passengers are loaded.

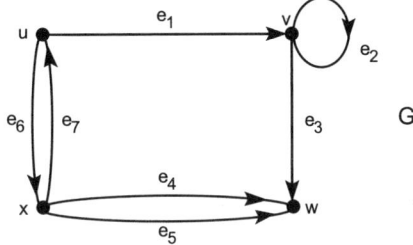

**4.** *Sociology and Sociobiology*:  A business (or army, or society, or ant colony) has a *hierarchical dominance* structure. The vertices are the employees (soldiers, citizens, ants) and there is an arc from $u$ to $v$ if $u$ dominates $v$. If the chain of command is unique, with a single leader, and if only arcs representing immediate authority are included, then the result is a *rooted tree*. (See §9.1.2.)

**5.** *Computer Software Design*:  A large program consists of many subprograms, some of which can invoke others. Let the vertices of $D$ be the subprograms, and let there be an arc from $u$ to $v$ if subprogram $u$ can invoke subprogram $v$. Then the *call graph* $D$ encapsulates all possible ways control can flow within the program. Directed cycles represent indirect recursion, and serve as a warning to the designer to ensure against infinite loops. See the following figure, where subprogram 2 can call itself indirectly.

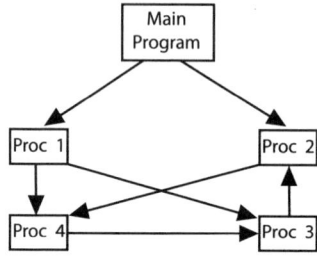

**6.** *Ecology*: A *food web* is a simple digraph in which vertices represent species and in which there is an arc from $u$ to $v$ if species $u$ eats species $v$. The following figure shows a small food web.

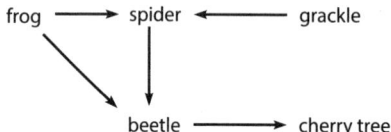

**7.** *Operations Research*: A sequence of books must be printed and bound, using one press and one binding machine. Suppose that book $i$ requires time $p_i$ for printing and time $b_i$ for binding. It is desired to print the books in such an order that the binding machine is never idle: when it finishes one book, the next book should already be printed. The vertices of a digraph $D$ can represent the books. There is an arc from book $i$ to book $j$ if $p_j \le b_i$. Then any path through all the vertices corresponds to a permissible ordering.

**8.** *Genealogy*: A "family tree" is a digraph where the orientation is traditionally given not by arrows but by the direction down for later generations. Despite the name, a family tree is usually not a tree, since people commonly marry distant cousins, knowingly or unknowingly.

**9.** *Binary Relations*: To any binary relation $R$ on a set $V$ (see §1.4.1) a digraph $D(V, R)$ can be associated: the vertices are the elements of $V$, and there is an arc from $u$ to $v$ if $(u, v) \in R$. Conversely, every digraph without multiple arcs defines a binary relation on its vertices. The relation $R$ is transitive if and only if the digraph $D(V, R)$ is transitive.

---

## 8.3.2  PATHS, CYCLES, AND CONNECTEDNESS

### Definitions:

A **directed walk** is a sequence of arcs such that the head of one arc is the tail of the next arc.

The **length** of a directed walk is the number of arcs in the sequence.

A **closed directed walk** is a directed walk that begins and ends at the same vertex.

A **directed trail** is a directed walk in which no arc is repeated.

A **directed path** is a directed trail in which no vertex is repeated.

A **directed cycle** is a closed directed trail in which no vertices are repeated, except the initial and final vertex.

Vertex $v$ is **reachable** from vertex $u$ if there is a directed path from $u$ to $v$.

A **basis for a digraph** is a set of vertices $V'$ such that every vertex not in $V'$ is reachable from $V'$ and such that no proper subset of $V'$ has this property.

The **distance** from a vertex $u$ to a vertex $v$ in a digraph is the length of a shortest directed path from $u$ to $v$.

A digraph is **strongly connected** (or **diconnected**, or **strong**) if every vertex is reachable from every other vertex.

A digraph is **unilaterally connected** (or **unilateral**) if for every pair of vertices $u$ and $v$, there is either a $uv$-path or a $vu$-path.

A digraph is **weakly connected** (or **weak**) if the underlying graph is connected.

The digraph is **disconnected** if the underlying graph is disconnected.

A **strong component** of a digraph is a maximal subgraph that is strongly connected.

A digraph is **reducible** if the vertex set can be partitioned into a disjoint union $V_1 \cup V_2$ so that all arcs joining $V_1$ and $V_2$ go from $V_1$ to $V_2$.

The **condensation** $D^*$ of a digraph $D$ is the simple digraph whose vertices are the strong components $\{V_1, V_2, \ldots, V_k\}$ of $D$, with an arc $V_i V_j \in E_{D^*}$ if and only if there is an arc $vv'$ in $D$ such that $v \in V_i$ and $v' \in V_j$.

The **converse of a digraph** $D$ is obtained by reversing the directions of all the arcs of $D$.

The **directional dual** of a theorem about digraphs is the statement obtained by replacing each property in the theorem statement by its converse.

## Facts:

**1.** Using a pencil on a drawing of a digraph, a directed walk can be traversed by following the arrows without lifting the pencil from the graph.

**2.** Distance in digraphs need not be symmetric. That is, the distance from $u$ to $v$ might be different from the distance from $v$ to $u$.

**3.** If $A$ is the adjacency matrix of $D$, then the $ij$ entry of $A^n$ is the number of $n$-arc walks from $v_i$ to $v_j$.

**4.** Let $\delta^+$ be the smallest out-degree of a simple digraph $D$. If $\delta^+ > 0$, then $D$ has a cycle of length at least $\delta^+ + 1$.

**5.** Let $\delta^-$ be the smallest in-degree of a simple digraph $D$. If $\delta^- > 0$, then $D$ has a cycle of length at least $\delta^- + 1$.

**6.** The directional dual of a theorem about digraphs is a theorem about digraphs.

**7.** Fact 5 is the directional dual of Fact 4.

**8.** A digraph $D$ is Eulerian (see §8.4.3) if and only if the underlying graph is connected and in-degree equals out-degree at every vertex.

**9.** A digraph $D$ has an Euler $uv$-trail (where $u \neq v$) if the following conditions hold:

- the out-degree of vertex $u$ exceeds the in-degree by one; that is, $d^+(u) = d^-(u)+1$;
- the in-degree of $v$ exceeds the out-degree by one; that is, $d^-(v) = d^+(v) + 1$;
- at every other vertex, the in-degree equals the out-degree; that is, $d^-(w) = d^+(w)$ for all $w \neq u, v$.

Other Euler-type results for graphs generalize to digraphs as well.

**10.** Let $\delta$ be the minimum of all in- and out-degrees of simple digraph $D$. If $\delta \geq |V|/2 > 1$, then $D$ contains a Hamilton cycle (see §8.4.4).

**11.** The strong components of a digraph $D$ partition the vertices of $D$, but not the arcs. However, the maximal unilateral subgraphs do partition the arcs. If $V_1, V_2$ are the vertex sets of two strong components of $D$, then all arcs between $V_1$ and $V_2$ face the same way—either from $V_1$ or to $V_1$. See the following figure.

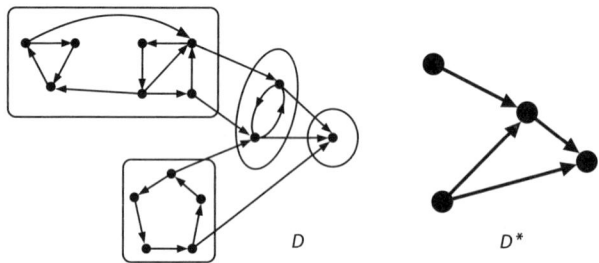

$D$          $D^*$

**12.** The condensation of any digraph is an acyclic digraph (see §8.3.4). See the figure for the previous fact.

**13.** A digraph is reducible if and only if its condensation has at least two vertices.

**14.** A digraph is unilateral if and only if its condensation is a path.

**15.** A set $V'$ is a basis of a digraph $D$ if and only if $V'$ consists of one vertex from each strong component of $D$ that has in-degree 0 in $D^*$. Thus, every basis of a digraph has the same number of vertices.

**16.** The eigenvalues of a digraph $D$ are the union (counting multiplicities) of the eigenvalues of its strong components. (See §8.10.2.)

**Examples:**

**1.** Let $u, v$ be vertices of an $n$-vertex digraph $D$ with adjacency matrix $A$. If $v$ is reachable from $u$, then some $uv$-path has length $\leq n - 1$. Thus, $D$ is strong if and only if every entry of $\sum_{k=0}^{n-1} A^k$ is positive. There are more computationally efficient tests for diconnectivity: the Floyd-Warshall algorithm (see §10.3.3) and directed depth-first search (see §9.2.1).

**2.** Let $M$ be an arbitrary square matrix. Computation of the eigenvalues of $M$ can sometimes be sped up as follows. Create matrix $A$ by replacing each nonzero entry of $M$ by a '1', and then let $D$ be the digraph with adjacency matrix $A$. The eigenvalues of $M$ are the union of the eigenvalues of the minors of $M$ indexed by the strong components of $D$. If $M$ is *sparse* (few nonzeros), then digraph $D$ will usually have many small components and this approach will be efficient.

**3.** *Markov models:* Let $V$ represent a set of states and $E$ the possible transitions of a Markov process (see §7.7). Then walks through $D$ represent "histories" that the process can follow.

## 8.3.3   ORIENTATIONS

There are many natural questions concerning when the edges of an undirected graph could be assigned directions so as to obtain a certain sort of digraph. For instance, when can a graph be oriented to obtain a strong digraph? An application of this last question is to determine when a set of roads could all be made one-way, while keeping all points reachable from all others.

**Definitions:**

An **orientation** of a graph is an assignment of directions to its edges, thereby making it a digraph.

An orientation of a graph is **strong** if, for each pair of vertices $u, v$, there is a directed path from $u$ to $v$ and a directed path from $v$ to $u$.

An orientation of a graph is **transitive** if, whenever there is an arc from $u$ to $v$ and an arc from $v$ to $w$, there is also an arc from $u$ to $w$.

A graph that admits a transitive orientation is called a **comparability graph**.

A **cut-edge** (or **bridge**) of a graph is an edge whose removal would increase the number of components (see §8.4.1).

A **2-edge-connected** graph $G$ is connected and has no cut-edge.

A **generalized circuit** in a graph is a closed walk (see §8.4.1) that uses each edge at most once in each direction.

A **triangular chord** for a closed walk (see §8.4.1) $u_1, u_2, \ldots, u_k, u_1$ is a proper edge that joins two vertices exactly two apart on the walk.

**Facts:**

**1.** Let $\chi(G)$ be the chromatic number (see §8.6.1) of graph $G$. Then every orientation of $G$ has a path of length at least $\chi(G) - 1$.

**2.** A graph $G$ has a strong orientation if and only if $G$ is 2-edge-connected. (H. Robbins, 1939)

**3.** A graph $G$ is a comparability graph if and only if every generalized circuit of $G$ of odd length $> 3$ has a triangular chord.

**4.** Algorithms 1 and 2 give ways of creating a strong orientation in a 2-edge-connected graph.

---

**Algorithm 1**:   **Naive algorithm for creating a strong orientation.**

{This algorithm is good to use by hand for small graphs}

input: a 2-edge-connected graph $G$

output: a strong orientation of $G$

$H :=$ any cycle in $G$

direct $H$

**while** some vertex of $G$ is not in directed subgraph $H$

$\quad v :=$ a vertex not in $H$

$\quad$ find two edge-disjoint paths from $v$ to $H$

{Two such paths exist because $G$ is 2-edge-connected}
direct one path from $v$ to $H$ and the other from $H$ to $v$
$H := H$ with these two subgraphs added
orient any remaining edges arbitrarily

---

**Algorithm 2**:   **Better algorithm for creating a strong orientation.**
{A good algorithm for large graphs or for computer implementation}
input: a 2-edge-connected graph
output: a strong orientation

select an arbitrary vertex as root
construct the depth-first search spanning tree from that root  {See §9.2.1.}
orient the tree edges downward from the root
orient all back edges upward toward the root
orient all cross edges arbitrarily

---

## Examples:

**1.** In the figure below the digraph $D$ is a weak transitive orientation of the graph $G$ and $D'$ is a strong nontransitive orientation.

  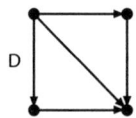  

**2.** The following graph is not transitively orientable, and $x, u, v, y, v, w, z, w, u, x$ are the vertices of a generalized circuit without a triangular chord.

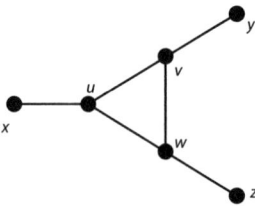

**3.** *Traffic control*: The flow of traffic on crowded city streets can sometimes be improved by making streets one-way. When this is done, it is necessary that a car can travel legally between any two locations. Assigning directions to the edges of the graph representing the street grid is an orientation of this graph, and cars can travel legally between any two points if and only this graph has a strong orientation. Consequently, by Robbins' theorem (Fact 2), to make all the streets one-way without losing mutual accessibility of locations, it is necessary and sufficient that the grid of streets be 2-edge-connected.

## 8.3.4  DIRECTED ACYCLIC GRAPHS

**Definitions:**

A digraph is **acyclic** if it has no *directed* cycles. A directed acyclic graph is sometimes called a DAG.

A **source** of a digraph is a vertex of in-degree zero.

A **sink** of a digraph is a vertex of out-degree zero.

A **linear extension ordering** of the $n$ vertices of a digraph is a consecutive labeling $v_1, v_2, \ldots, v_n$ so that, if there is an arc from $v_i$ to $v_j$, then $i < j$. (See also §11.2.5.)

A **topological sort**, or **topsort**, is an algorithm that assigns a linear extension ordering to a DAG.

**Facts:**

**1.** Every DAG has at least one source, and by duality, at least one sink.

**2.** Every DAG has a unique basis (see §8.3.2), namely, the set of all its sources.

**3.** Topsort yields a linear ordering for the vertices that makes the adjacency matrix of a DAG upper-triangular.

**4.** Doing a preliminary topsort permits some optimization problems about paths to be solved subsequently by a single algorithmic pass through the vertices in the topsort order; see, for example, §16.4.1 (critical paths).

**5.** Algorithm 3 provides a simple algorithm for topological sort.

---

**Algorithm 3**:  **Naive topological sort.**
{Construct a linear extension ordering for a DAG}
input: a DAG $D$
output: a numbering of the vertices in a topsort order

$H := D$; $k := 1$
**while** $V_H \neq \emptyset$
    $v_k :=$ a vertex of $H$ of in-degree 0  {This exists; see Fact 1}
    $H := H - v_k$  {Remaining graph is still a DAG}
    $k := k + 1$

---

**Examples:**

**1.** In the following digraph vertex $w$ is a source and vertex $z$ is a sink. It is a DAG, even though the underling graph has cycles. Labeling the vertices either in the order $w, x, y, z$ or $w, y, x, z$ is a linear extension ordering.

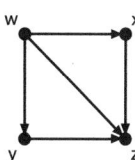

**2.** Consider any digraph whose vertices represent instantaneous events, and whose arcs go from earlier events to later events. Any such digraph is acyclic. Conversely, any digraph whose vertices represent procedural steps and whose arcs represent required precedence can be scheduled (using a topological sort) so that arcs do in fact go forward in time.

**3.** The Hasse diagram of a poset (see §11.1.1) is a DAG, as is the entire graph of a poset (arc from $u$ to $v$ if and only if $u \geq v$).

## 8.3.5   TOURNAMENTS

### Definitions:

A **tournament** is a digraph with exactly one arc between each pair of distinct vertices. An **$n$-tournament** has $n$ vertices.

The **score vector** of a tournament is the sequence of out-degrees of the vertices (number of arcs leaving each vertex), usually in ascending order.

A tournament is **regular** if every vertex has the same out-degree.

A tournament is **strong** if there is a directed path between each pair of vertices in both directions.

A tournament is **transitive** if, whenever there is an arc from $u$ to $v$ and from $v$ to $w$, there is also an arc from $u$ to $w$.

A tournament is **irreducible** if there is no bipartition $V_1, V_2$ of the vertices such that all arcs between $V_1$ and $V_2$ go from $V_1$ to $V_2$.

Vertex $u$ of a tournament **dominates** vertex $v$ if there is an arc from $u$ to $v$.

A **transmitter** in a digraph is a vertex that has an arc to every other vertex.

A **king** in a digraph is a vertex from which there is a path of length 1 or 2 to all other vertices.

A **single-elimination competition** is a contest from which a competitor is eliminated after the first loss.

### Facts:

**1.** Every tournament has a Hamilton path (see §8.4.4), in fact an odd number of them.

**2.** The following statements are equivalent for any $n$-tournament $T$:

- $T$ is strong;
- $T$ is irreducible;
- $T$ has a Hamilton cycle;
- $T$ has cycles of all lengths $3, 4, \ldots, n$;
- Every vertex of $T$ is on cycles of all lengths $3, 4, \ldots, n$.

**3.** Almost all tournaments are strong, in the sense that, as $n \to \infty$, the fraction of labeled $n$-tournaments that are strong approaches 1.

**4.** The following are equivalent for a tournament:

- the tournament is transitive;
- the tournament contains no cycles;

- the tournament contains no 3-cycles;
- the tournament is a total (i.e., linear) order;
- the tournament has a unique Hamilton path.

**5.** Every tournament has a king.

**6.** The king of a tournament is unique if and only if it is a transmitter. Otherwise, there are at least three kings.

**7.** In a tournament, almost every vertex is a king, for as $n \to \infty$, the fraction of $n$-tournaments in which every vertex is a king approaches 1.

**8.** *Score vector characterizations:* A nondecreasing sequence $S$ of nonnegative integers $s_1, s_2, \ldots, s_n$ is the score vector of an $n$-tournament if and only if

$$\sum_{i=1}^{k} s_i \geq \binom{k}{2}, \quad \text{for } k = 1, 2, \ldots, n-1, \quad \text{and} \quad \sum_{i=1}^{n} s_i = \binom{n}{2}.$$

Or equivalently, if and only if the sequence $S'$ obtained by deleting any one $s_i$ and reducing the largest remaining $n - s_i - 1$ terms by 1 is a score vector of an $(n-1)$-tournament.

**9.** The second characterization of Fact 8 leads to a recursive algorithm to construct a tournament having a specified score vector. See Example 4.

**10.** A nonnegative integer sequence $s_1 \leq s_2 \leq \cdots \leq s_n$ is the score vector of a *strong* $n$-tournament if and only if

$$\sum_{i=1}^{k} s_i > \binom{k}{2}, \quad \text{for } k = 1, 2, \ldots, n-1, \quad \text{and} \quad \sum_{i=1}^{n} s_i = \binom{n}{2}.$$

**11.** The two tournaments in the following figure are isomorphic as unlabeled tournaments, but distinct as labeled tournaments.

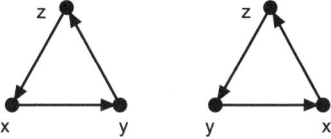

There are $2^{\binom{n}{2}}$ distinct *labeled* tournaments, because for each pair of vertices $\{u, v\}$, there are two choices for which way to direct the edge. If $c_n$ is the number of distinct *unlabeled* $n$-tournaments, then

$$c_n > \frac{2^{\binom{n}{2}}}{n!} \quad \text{and} \quad \lim_{n \to \infty} \frac{c_n}{2^{\binom{n}{2}}} / n! = 1.$$

See §8.9.3 and §8.9.4.

**12.** When a tournament is acyclic, it corresponds to a unique total ordering (Fact 4), so the ranking is clear. However, almost all tournaments are strong (Fact 3). Moreover, in a large tournament, almost every vertex is a king (Fact 7). These are reasons why it is difficult to give a satisfactory general method to rank tournaments.

**13.** *Scheduling tournaments:* To speed up the play of an $n$-tournament, games can be scheduled in parallel. If $n$ is even, then at most $n/2$ of the $\binom{n}{2}$ games can be played simultaneously, so at least $n - 1$ rounds are needed. However, if $n$ is odd, then only $(n - 1)/2$ games can be played simultaneously, so at least $n$ rounds are needed. In fact, this minimum number of rounds can be obtained, and several methods of scheduling tournaments, subject to various additional conditions, have been devised. See [Mo15].

**Examples:**

**1.** A round-robin sports tournament in which there are no ties is a tournament in the mathematical sense defined above. However, a single-elimination competition (e.g., most tennis tournaments) is not a tournament as defined above.

**2.** It has been observed that in every small flock of hens, *every* pair of hens establish a dominance relation—the weaker of the two allows the stronger to peck her. Thus, this pecking order is a tournament.

**3.** In a "paired comparison experiment", a subject is asked to state a preference in each pair chosen from $n$ items. This amounts to a tournament, where there is an arc $ij$ if item $i$ is preferred to item $j$.

**4.** Is there a tournament on vertices $(a, b, c, d, e)$ with respective scores $(1, 2, 2, 2, 3)$? Deleting $e$ according to the second part of Fact 8 leaves vertices $(a, b, c, d)$ with scores $(1, 2, 2, 1)$. Next deleting $d$ leaves $(a, b, c)$ with scores $(1, 1, 1)$. The obvious tournament with such a score vector is a 3-cycle. Next reinsert vertex $d$, making it dominate vertex $a$ only. Then reinsert vertex $e$, making it dominate $a, b, c$. This 5-tournament has the specified score vector $(1, 2, 2, 2, 3)$.

**5.** *Ranking real tournaments*: Ranking teams by their order along a Hamilton path (see Fact 1) is rarely satisfactory, because that order is unique only for transitive tournaments (Fact 4); in most cases, there are a great many Hamilton paths. Ranking by score vector usually creates ties, and a team with few wins may deserve a better rank if those teams it beats have many wins. So one may consider the *second-order score vector*, where each team's score is the sum of the out-degrees of the teams it beats. This can be continued to $n$th-order score vectors. There is a limit ranking obtained this way (often quite satisfactory), related to the eigenvalues of the digraph. See [Mo15] for more detail and references.

# 8.4   DISTANCE, CONNECTIVITY, TRAVERSABILITY, & MATCHINGS

Graphs often serve as models for transportation and communication network problems. Movement from one node to another in the network corresponds to the graph-theoretic notion of a walk, while the connectivity of a graph is a measure of resistance to a communications cutoff.

## 8.4.1   WALKS, DISTANCE, AND CYCLE RANK

**Definitions:**

A **walk** in a graph is an alternating sequence $v_0, e_1, v_1, \ldots, e_r, v_r$ of vertices and edges in which each edge $e_i$ joins vertices $v_{i-1}$ and $v_i$. Such a walk is also called a $v_0, v_r$-*walk*.

The **length** of a walk is the number of occurrences of edges in it. An edge that occurs more than once is counted each time it occurs.

A **trail** is a walk in which all of the edges are different.

A walk or trail is **open** if its final vertex is different from its initial vertex.

A walk or trail is **closed** if its final vertex is the same as its initial vertex.

A **path** is a trail in which all the vertices are different. A path from $v_0$ to $v_r$ is called a $v_0, v_r$-*path*. (The word "path" also refers to a type of graph; see §8.1.3.)

A **cycle** is a closed trail of positive length in which all the vertices are different except for the initial and final vertex. (The word "cycle" also refers to a type of graph; see §8.1.3.)

A graph is **connected** if each pair of vertices is joined by a path.

A **component** of a graph is a maximal connected subgraph of the graph.

The vertex $v$ is **reachable** from vertex $u$ in a graph if there is a $u, v$-path in the graph.

An **isolated vertex** of a graph is a vertex with no incident edges.

The **distance** $d(v, w)$ between two vertices $v$ and $w$ of a graph is the length of a shortest path between them, with $d(v, v) = 0$, and $d(v, w) = \infty$ if there is no path between $v$ and $w$.

The **diameter** of a connected graph is the maximum distance between two of its vertices.

The **eccentricity** of a vertex $v$ of a connected graph is the greatest distance from $v$ to another vertex.

The **radius** of a connected graph is the minimum eccentricity among all the vertices of the graph.

The **center** of a connected graph is the set of vertices of minimum eccentricity.

The **cycle rank** (or **first Betti number**), denoted by $\beta_1(G)$, of a connected graph $G = (V, E)$ is $|E| - |V| + 1$.

**Facts:**

**1.** *Alternative terminology*: Sometimes "circuit" is used to mean what is here called a closed trail.

**2.** In a simple graph, a walk may be represented as a string of vertices $v_0 v_1 \ldots v_r$, without mentioning the edges.

**3.** The distance function on the vertex set of any connected graph $G$ is a metric; i.e., the following rules hold for all vertices $u, v$, and $w$ in $G$:

- $d(v, w) \geq 0$, with equality if and only if $v = w$;
- $d(w, v) = d(v, w)$;
- $d(u, w) \leq d(u, v) + d(v, w)$, with equality if and only if $v$ is on a shortest path from $u$ to $w$.

**4.** There are polynomial-time algorithms for finding a shortest path between vertices. (See §10.3.2.)

**5.** A graph is connected if and only if it has a spanning tree.

**6.** A graph is disconnected if and only if there is a partition of its vertex set into nonempty sets $A$ and $B$ so that no edge has one end in $A$ and the other in $B$.

**7.** The relation "is reachable from" is an equivalence relation on the vertex set. The equivalence classes of this relation induce the components.

**8.** A graph is connected if every vertex is reachable from every other vertex.

**9.** In a simple graph, the minimum length of a cycle is at least 3. In a general graph, a self-loop is a 1-cycle, and a 2-cycle is formed by a pair of vertices joined by a pair of parallel edges.

**10.** The cycle rank $\beta_1(G)$ of a connected graph $G$ equals the number of edges in the complement of a spanning tree for $G$.

**11.** The cycle rank $\beta_1(G)$ of a connected graph $G$ is equal to the rank of a vector space over $\mathcal{Z}_2$ whose domain is the set of cycles of $G$.

**12.** The cycle rank of a connected planar graph $G$ equals the number of regions in a plane drawing of $G$, *minus* the exterior region.

**13.** The following table gives the cycle rank of some families of graphs.

| graph | cycle rank |
|---|---|
| bouquet $B_n$ | $n$ |
| dipole $D_n$ | $n-1$ |
| complete graph $K_n$ | $\frac{(n-2)(n-1)}{2}$ |
| complete bipartite graph $K_{m,n}$ | $(m-1)(n-1)$ |
| cycle graph $C_n$ | $1$ |
| wheel $W_n$ | $n$ |
| hypercube $Q_n$ | $(n-2)2^n + 1$ |
| any tree | $0$ |

**Example:**

**1.** The following connected graph has diameter 3 and radius 2. The vertices in its center are indicated by solid dots.

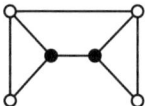

**2.** The cycle rank of the following connected graph is 3. Observe that there are three edges in the complement of the indicated spanning tree.

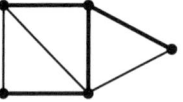

**3.** The following disconnected graph has three components, one of which is an isolated vertex.

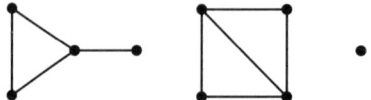

## 8.4.2 CONNECTIVITY

**Definitions:**

A *cut-vertex* of a graph $G$ (or *cut-point* or *articulation point*) is a vertex $v$ such that $G - v$ has more components than $G$. (In topological analysis of nonsimple graphs, sometimes a vertex attached to a self-loop is also considered to be a cut-vertex.)

A *nonseparable* graph is a connected graph with no cut-vertices.

A *block* of a graph is a maximal nonseparable subgraph.

An *cut-edge* of a graph $G$ is an edge $e$ such that $G - e$ has more components than $G$ (in which case there is just one more).

A *disconnecting set of vertices* (or *cutset*) in a connected graph is a set of vertices whose removal yields a disconnected graph.

A *disconnecting set of edges* in a connected graph is a set of edges whose removal yields a disconnected graph.

The *zeroth Betti number* $\beta_0(G)$ of a graph $G$ is the number of components in $G$. Elsewhere this is sometimes denoted by $c(G)$ or $\omega(G)$.

The (*vertex*) *connectivity* $\kappa(G)$ is the number of vertices in a smallest disconnecting set of vertices. By convention, $\kappa(K_n) = n - 1$.

The *edge connectivity* $\kappa'(G)$ is the number of edges in a smallest disconnecting set of edges.

A graph is *k-connected* if $\kappa(G) \geq k$.

A graph is *k-edge-connected* if $\kappa'(G) \geq k$.

**Facts:**

**1.** A vertex is a cut-vertex if and only if it lies on all paths between two other vertices.

**2.** Every nontrivial graph has at least two vertices that are not cut-vertices.

**3.** An edge is a cut-edge if and only if it is not contained in any cycle.

**4.** For any edge $e$ of a graph $G$, $\beta_0(G) + 1 \geq \beta_0(G - e) \geq \beta_0(G)$.

**5.** For any vertex $v$ of a graph $G$, $\beta_0(G - v) \geq \beta_0(G)$; however, $\beta_0(G - v)$ may be arbitrarily greater than $\beta_0(G)$.

**6.** Let $G$ be a 2-connected graph. Then for any two vertices, there is a cycle containing those vertices.

**7.** Let $G$ be a 2-connected graph. Then for any two edges, there is a cycle containing those edges.

**8.** The following statements are equivalent for a connected graph $G$ with at least three vertices:

- $G$ is nonseparable;
- every pair of vertices lies on a cycle;
- every pair of edges lies on a cycle;
- given any three vertices $u, v$, and $w$, there is a path from $u$ to $w$ containing $v$;
- given any three vertices $u, v$, and $w$, there is a path from $u$ to $w$ not containing $v$.

**9.** *Menger's theorem (for vertex connectivity)*: A graph with at least $k + 1$ vertices is $k$-connected if and only if every pair of vertices is joined by $k$ paths which are internally disjoint (i.e., disjoint except for their origin and terminus). (Menger, 1927)

**10.** *Menger's theorem (for edge connectivity)*: A graph is $k$-edge-connected if and only if every pair of vertices is joined by $k$ edge-disjoint paths. (Ford and Fulkerson, 1956; also Elias, Feinstein, and Shannon, 1956)

**11.** For any graph $G$, the vertex connectivity is no more than the edge connectivity, and the edge connectivity is no more than the minimum degree. That is, $\kappa(G) \le \kappa'(G) \le \delta(G)$, where $\delta(G)$ denotes the minimum degree.

**12.** Furthermore, for any positive integers $a \ge b \ge c$, there exists a simple graph $G$ for which $\delta(G) = a$, $\kappa'(G) = b$, and $\kappa(G) = c$. (Chartrand and Harary, 1968)

**13.** The following table gives the vertex connectivity and edge connectivity of some families of graphs.

| graph | $\kappa$ | $\kappa'$ |
|---|---|---|
| complete graph $K_n$ | $n - 1$ | $n - 1$ |
| complete bipartite graph $K_{m,n}$ | $\min(m, n)$ | $\min(m, n)$ |
| cycle graph $C_n$ | 2 | 2 |
| wheel $W_n$ | 3 | 3 |
| hypercube $Q_n$ | $n$ | $n$ |

**Examples:**

**1.** The following graph $G$ has cut-vertices $u$ and $v$. The blocks are illustrated at the right.

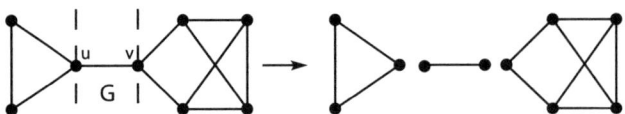

**2.** In the following graph, vertices $u$ and $v$ form a disconnecting set.

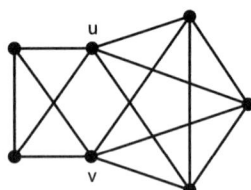

**3.** *Communication networks*: A communication network can be modeled as a graph with vertices representing the nodes and with undirected edges representing direct two-way communications links between nodes. In order that all pairs of nodes be in communication, the graph must be connected. Vertex connectivity and edge connectivity are measures of network reliability.

**4.** *Transportation networks*: Low connectivity in transportation networks results in "bottlenecks", in which many different shipments must all pass through a small number of nodes. High connectivity implies (by Menger's theorem) several alternative routes between nodes.

**5.** Menger's theorem implies that a 2-connected graph has two disjoint paths between each pair of vertices. It does *not* imply that for any path between two vertices, there must be a second such path disjoint from the first, as indicated in the following graph. There are two disjoint paths from the leftmost vertex to the rightmost, but there is no such path disjoint from the one indicated by thick edges.

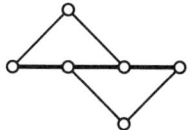

**6.** The following shows a graph $G$ with $\kappa(G) = 2$ and $\kappa'(G) = 3$. On the left there are two internally-disjoint paths between the upper-left vertex and the lower-right vertex, and on the right there are three edge-disjoint paths.

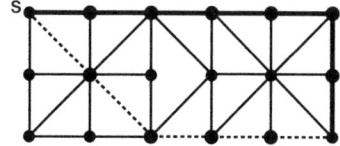 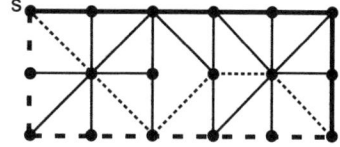

**7.** The following graph illustrates Fact 12, with $\kappa = 2$, $\kappa' = 3$, $\delta = 4$.

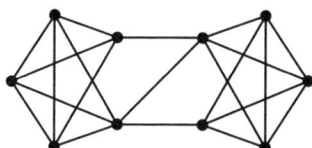

---

**8.4.3   EULER TRAILS AND TOURS**

**Definitions:**

An **Euler trail** in a graph [digraph] is a trail that contains all the edges [arcs] of the graph.

An **Euler tour** or **Euler circuit** in a graph or digraph is a closed Euler trail.

A graph or digraph is **Eulerian** if it has an Euler tour.

**Facts:**

**1.** *Seven bridges of Königsberg problem:*  In Kaliningrad, Russia, two branches of the River Pregel meet and flow past an island into the Baltic Sea. In 1736, when this was the town of Königsberg in East Prussia, there were seven bridges joining the banks of the river, the headland, and the island, as illustrated at the left in the following picture. The celebrated Swiss mathematician Leonhard Euler (1707–1783) was asked whether it was possible to cross all seven bridges without recrossing any of them. In the earliest known paper on graph theory, Euler proved it is impossible, because the graph at the right has no Euler trail.

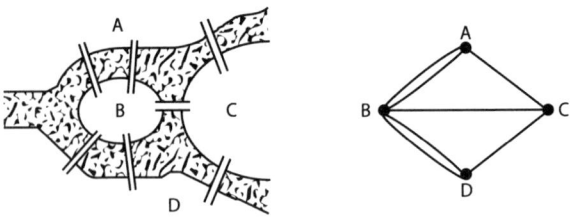

**2.** Euler's work on the seven bridges of Köningsberg problem is commonly described as the founding of graph theory and also as the founding of topology.

**3.** A connected graph is Eulerian if and only if every vertex has even degree. The tour may begin/end at any vertex.

**4.** A connected digraph has a directed Euler tour if and only if the in-degree of every vertex $v$ equals its out-degree.

**5.** A connected graph has an Euler trail between distinct vertices $u$ and $v$ if and only if $u$ and $v$ are the only vertices of odd degree.

**6.** A connected graph (digraph) is Eulerian if and only if there exists a collection of cycles (directed cycles) whose edges partition the edge set of the graph.

**7.** A connected planar (see §8.7.1) graph is Eulerian if and only if its dual (see §8.8.2) is bipartite.

**8.** A graph $G$ can be oriented to have a directed Euler tour if and only if it is an Eulerian graph. (Traversing an Euler tour provides an orientation.)

**9.** The following table indicates which members of several families of graphs are Eulerian.

| graph | Eulerian? |
|---|---|
| bouquet $B_n$ | for all $n$ |
| dipole $D_n$ | for even $n$ |
| complete graph $K_n$ | for odd $n$ |
| complete bipartite graph $K_{m,n}$ | for $m$ and $n$ both even |
| cycle graph $C_n$ | for all $n$ |
| wheel $W_n$ | never |
| hypercube $Q_n$ | for even $n$ |
| tree | only if trivial |

**10.** Algorithm 1 gives a recursive method for finding an Eulerian tour on an Eulerian graph.

---

**Algorithm 1:  Recursive algorithm for finding an Eulerian tour.**

input: a connected graph $G$, all of whose vertices have even degree

output: an Euler tour of $G$

$C :=$ a cycle in the graph $G$; place $C$ on the cycle-queue $\mathcal{Q}$

partition the edge-complement $G - E(C)$ into components $H_1, H_2, \ldots, H_k$

recursively run this algorithm on each component $H_i$

{So far, $E_G$ has been completely partitioned into the cycles on $\mathcal{Q}$}

merge the elements of $\mathcal{Q}$ into an Euler tour for $G$, by traversing the cycle $C$

    and splicing in the tours found for the components $H_i$ whenever possible

---

**11.** Fleury's algorithm for finding an Euler tour or trail is given in Algorithm 2.

---

**Algorithm 2:  Fleury's algorithm for finding an Euler tour/trail.**

input: a connected graph $G$, an initial vertex $v$, and a final vertex $w$; if $v \neq w$, then every vertex except $v$ and $w$ must have even degree (if $v = w$, then all degrees must be even)

output: an Euler trail whose origin is $v$ and whose terminus is $w$

{Find trail edge with origin $v$}

**if** $\deg(v) > 1$ **then** $e :=$ any edge incident at $v$ which is not a cut-edge

**else**  { $\deg(v) = 1$ }

   $e :=$ the unique edge incident at $v$

   $u :=$ the other endpoint of $e$

recursively find an Euler trail from $u$ to $w$ in $G - e$

prepend the edge $e$ to the trail found in the recursive step

{This yields the required Euler trail of $G$}

---

**Examples:**

**1.** The following is an Eulerian graph and one of its Euler tours.

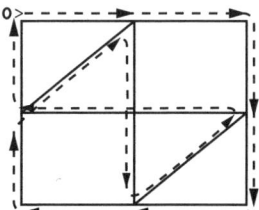

**2.** *Chinese postman problem* (due to Guan Meigu, 1962):  A letter carrier begins at the post office, traverses every street in his territory at least once, and then returns to the post office. His objective is to walk as little as possible. Each edge of a graph representing the street configuration is labeled with the length of the corresponding block. If the graph is Eulerian, then an Euler tour gives an optimal solution. Otherwise, some edges must be retraced. Polynomial-time algorithms to solve this problem are known. See §10.2.3.

**3.** For every letter in an arbitrary $n$-letter alphabet $\mathcal{A}$, there is a string starting and ending with that letter, in which every possible substring of two letters appears consecutively exactly once. To see this, consider the digraph $D$ with the letters of $\mathcal{A}$ as vertices and one arc for each ordered pair. (This is called a de Bruijn digraph.) The digraph $D$ is connected and, at each of the $n$ vertices, in-degree = out-degree = $n$, which implies that it is Eulerian. Thus, the sequence of vertices encountered on a closed Euler tour from any vertex to itself yields the specified string.

In the following figure, $e_1, e_2, \ldots, e_9$ are the arcs of an Euler cycle and the associated string is the sequence of vertices, *aabbccacba*. This result generalizes to substrings of any fixed length, also using Euler tours.

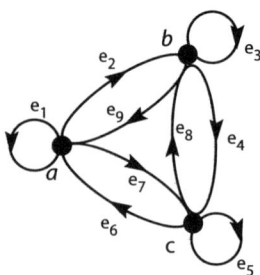

## 8.4.4  HAMILTON CYCLES AND PATHS

### Definitions:

A **Hamilton cycle** in a graph [digraph] is a cycle [directed cycle] that includes all vertices of the graph.

A graph or digraph is **Hamiltonian** if it contains a Hamilton cycle.

A **Hamilton path** in a graph [digraph] is a path [directed path] that includes all vertices of the graph.

A **theta graph** is a subdivision of the complete bipartite graph $K_{2,3}$. Thus, the graph comprises three internally disjoint paths joining the two degree-3 vertices.

A **Gray code** is a cyclic ordering of all $2^k$ length-$k$ bitstrings, such that each bitstring differs from the next in exactly one bit entry.

A **1-tough** (or simply **tough**) graph is a connected graph $G$ such that no matter what nonempty, proper vertex subset $S$ is removed, the resulting number of components of $G - S$ is no more than $|S|$.

### Facts:

**1.** The concept of a Hamilton cycle first arose in a puzzle within the Icosian Game, invented by Sir William Rowan Hamilton (1805–1865), an Irish mathematician. This puzzle involved a dodecahedron whose 20 vertices were labeled with world capitals. It required finding a complete tour of these 20 capitals.

**2.** The recognition problem for Hamiltonian graphs is NP-complete. Thus, unlike the case with Eulerian graphs, there is no easy test to decide whether a graph is Hamiltonian (unless P = NP). However, many of the following facts provide criteria that are often helpful in deciding.

**3.** A Hamiltonian graph has no cut-point. (Thus, it is 2-connected.)

**4.** The previous fact has this generalization: Let $G$ be a Hamiltonian graph and let $S \subseteq V_G$. Then the graph $G - S$ has at most $|S|$ components. That is, Hamiltonian graphs are tough.

**5.** Bipartite Hamiltonian graphs have an equal number of vertices in the two parts of the bipartition.

**6.** If a simple graph has $n \geq 3$ vertices and minimum degree at least $n/2$, then it is Hamiltonian. (Dirac, 1952)

**7.** If a simple graph has $n \geq 3$ vertices, and if every pair of nonadjacent vertices $u$ and $v$ satisfies the inequality $\deg(u) + \deg(v) \geq n$, then it is Hamiltonian. (Ore, 1960)

**8.** Suppose that a simple graph with $n \geq 3$ vertices has degree sequence $d_1 \leq d_2 \leq \cdots \leq d_n$, and that for every $i$ with $1 \leq i \leq n/2$ either $d_i > i$ or $d_{n-i} \geq n - i$. Then the graph is Hamiltonian.

**9.** Every simple graph with $n \geq 3$ vertices and at least $(n^2 - 3n + 6)/2$ edges is Hamiltonian.

**10.** Every graph with at least three vertices whose connectivity ($\kappa$) is at least as large as its independence number ($\alpha$) is a Hamiltonian graph.

**11.** Every 4-connected planar graph is Hamiltonian.

**12.** If the edges of the complete graph $K_n$ are assigned directions, then the resulting digraph always has a Hamilton directed path.

**13.** The edges of the complete graph $K_{2n+1}$ can be partitioned into $n$ Hamilton cycles.

**14.** Theta graphs are non-Hamiltonian and every non-Hamiltonian graph contains a theta subgraph.

**15.** A graph $G$ is non-Hamiltonian if there is an independent set containing more than half the vertices of $G$.

**16.** "Almost all" graphs are Hamiltonian. That is, of the exactly $2^{n(n-1)/2}$ simple graphs on $n$ (labeled) vertices, the proportion that are Hamiltonian tends to 1 as $n \to \infty$.

**17.** Suppose that a simple graph is constructed by the following process: start with $n$ vertices and no edges; until the minimum degree is 2, a possible edge is chosen uniformly at random from among the edges not already in the graph, and added to the graph. With probability tending to 1 as $n \to \infty$, the resulting graph is Hamiltonian.

**18.** The following table indicates which members of several families of graphs are Hamiltonian.

| graph | Hamiltonian? |
|---|---|
| bouquet $B_n$ | for all $n \geq 1$ |
| dipole $D_n$ | for all $n \geq 2$ |
| complete graph $K_n$ | for all $n \geq 3$ |
| complete bipartite graph $K_{m,n}$ | when $m = n$ |
| cycle graph $C_n$ | for all $n \geq 1$ |
| wheel $W_n$ | for all $n \geq 2$ |
| hypercube $Q_n$ | for all $n \geq 2$ |

**Examples:**

**1.** Finding a Hamilton cycle in the dodecahedral graph (see §8.1.3), as illustrated below, is equivalent to solving Hamilton's Icosian Game puzzle. An example of a Hamilton cycle in this graph is: $RSTVWXHJKLMNPCDFGBZQR$.

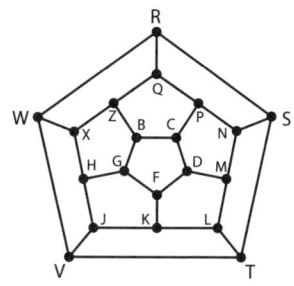

**2.** The following graph has the Hamilton cycle *acefdba*.

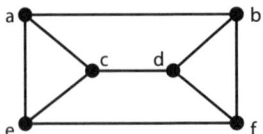

**3.** The following graph is non-Hamiltonian, by Fact 15, since the vertices $u$, $v$, $w$, and $x$ form an independent set.

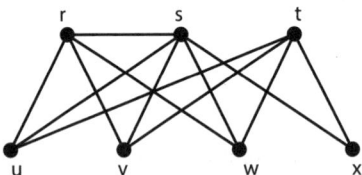

**4.** The 10-cycle $C_{10}$ (§8.1.3) is an example of a graph that satisfies none of the sufficient conditions in Facts 6–10 above for Hamiltonicity, but is nonetheless Hamiltonian.

**5.** The traveling salesman problem (see §10.7.1) is to find a minimum-cost Hamilton cycle in a complete graph whose edges are labeled with costs.

**6.** *Information theory — Gray codes*: In information theory, a cyclic ordering of the $2^n$ length-$n$ bitstrings such that each bitstring differs from its predecessor in exactly one bit is called a *Gray code*. This corresponds to a Hamilton cycle in the $n$-dimensional hypercube $Q_n$.

The following figure shows a Hamilton cycle in the 3-cube giving the Gray code $000 \rightarrow 001 \rightarrow 011 \rightarrow 111 \rightarrow 101 \rightarrow 100 \rightarrow 110 \rightarrow 010 \rightarrow 000$.

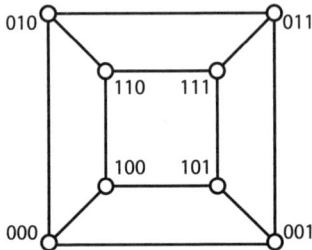

**7.** The Petersen graph (see §8.1.3) is tough but not Hamiltonian.

**8.** The following graph is tough but not Hamiltonian.

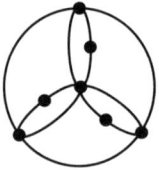

## 8.4.5  MATCHINGS AND FACTORS

**Definitions:**

A **matching** in a graph is a set of disjoint edges.

A **perfect** matching (or **1-factor**) is a matching which covers all the vertices of the graph.

A graph $H$ is a **factor** of a graph $G$ if $H$ is a spanning subgraph of $G$.

A factor which is $k$-regular is called a **k-factor**.

If $f$ is a function on $V(G)$, then a spanning subgraph $H$ is called an **f-factor** if $\deg_H(v) = f(v)$, for all $v \in V(G)$. Of course, if $f(v) = k$ for every vertex, then $f$ is just a $k$-factor.

An **[a, b]-factor** of a graph $G$ is an $f$-factor such that $a \le f(v) \le b$, for all $v \in V(G)$.

If $f$ and $g$ are functions on $V(G)$, then a **(g, f)-factor** is a spanning subgraph $H$ such that $g(v) \le \deg_H(v) \le f(v)$, for all $v \in V(G)$.

A **path factor** is a spanning subgraph where each component is a path.

An **F-factor** is a spanning subgraph where each component is either a $K_2$ or an odd cycle.

A **1-factorization** of a graph is a partition of the edges into edge-disjoint 1-factors.

A graph has the **odd-cycle property** if every pair of odd cycles either has a common vertex or is joined by an edge.

A graph $G$ is **k-tough** if $c(G - S) \le |S|/k$ for all cutsets $S \subseteq V(G)$, where $c(G - S)$ denotes the number of components of $G - S$.

**Facts:**

**1.** *Petersen's theorem*: A 3-regular graph with at most one cut-edge contains a 1-factor and hence also a 2-factor. (J. Petersen, 1891)

**2.** If $\nu(G)$ is the size of a largest matching in $G$ and $\tau(G)$ is the minimum number of vertices necessary to cover all the edges of $G$, then $\nu(G) \le \tau(G)$. *König's theorem* says that if $G$ is bipartite, $\nu(G) = \tau(G)$. (D. König, 1931)

**3.** *Hall's theorem*: Let $G$ be a bipartite graph with vertex partition $V(G) = A \cup B$. Then $G$ has a matching of $A$ onto $B$ if and only if (i) $|A| = |B|$, and (ii) $|N(X)| \ge |X|$ for all $X \subseteq A$, where $N(X)$ denotes the neighborhood of set $X$. (In fact, Hall's and König's theorems are equivalent.) (P. Hall, 1935)

**4.** *Tutte's 1-factor theorem*: A graph $G$ has a 1-factor if and only if for each $S \subseteq V(G)$, it holds that $c_o(G - S) \le |S|$, where $c_o(G - S)$ denotes the number of odd components of $G - S$. (W. Tutte, 1947)

**5.** If a graph of even order is $r$-regular and has the odd-cycle property, then it has a 1-factor.

**6.** If $G$ is $k$-connected and contains a 1-factor and if $|V(G)|$ is sufficiently large, then $G$ has at least $k!$ 1-factors.

**7.** If $G$ is connected and has a unique 1-factor, then $G$ has a cut-edge belonging to this 1-factor.

**8.** If $G$ is an $r$-regular $(r-1)$-edge-connected graph of even order, then $G$ has a 1-factor not containing any $r-1$ prescribed edges. (And hence $G$ has a 1-factor containing any prescribed edge.)

**9.** If $G$ is a connected graph of even order whose automorphism group acts transitively on $V(G)$, then $G$ has a 1-factor containing any given edge.

**10.** A 3-regular graph has exponentially many 1-factors.

**11.** Counting the number of perfect matchings has applications in crystal physics and organic chemistry.

**12.** The first polynomial-time algorithm for matching in an arbitrary graph was formulated by Edmonds [Ed65] and is now called the *Blossom Algorithm*. (See §10.2.3.)

**13.** Currently the fastest algorithm for matching in an arbitrary graph with $n$ vertices and $m$ edges is $O(m\sqrt{n})$.

**14.** If a graph $G$ of order $n$ is $k$-tough and $kn$ is even, then $G$ has a $k$-factor.

**15.** Let $G$ be a connected graph of order $n$ and let $k$ be an integer $\geq 3$ such that $kn$ is even and $n \geq 4k-3$. If $d(u) \geq k$ for all vertices, and $\max\{d(u), d(v)\} \geq n/2$ for all pairs of non-adjacent vertices $u$ and $v$, then $G$ has a $k$-factor.

**16.** Let $G$ be a graph with the odd-cycle property and with a $k$-factor, and let $r$ be an integer, $1 \leq r \leq k$. Then $G$ has an $r$-factor if $r|V(G)|$ is even.

**17.** If $G$ is a graph and $G-v$ has a $k$-factor for all $v \in V(G)$, then $G$ also has a $k$-factor.

**18.** Tutte's $f$-factor theorem: Graph $G$ has an $f$-factor if and only if the following two conditions hold for all disjoint sets $D, S \subseteq V(G)$:

(i) $f(D) - f(S) + d_{G-D}(S) - q_G(D, S, f) \geq 0$, where $q_G(D, S, f)$ denotes the number of components $C$ of $G - (D \cup S)$ such that $e_G(V(C), S) + f(V(C))$ is odd;

(ii) $f(D) - f(S) + d_{G-D}(S) - q_G(D, S, f) \equiv f(V(G)) \pmod{2}$.

(Here $e_G(A, B)$ denotes the number of edges in $G$ joining vertex sets $A$ and $B$.)

**19.** If $G$ is $r$-regular, then $G$ has a $[k, k+1]$-factor, for all $k, 0 \leq k \leq r$.

**20.** The graph $G$ has a $(g, f)$-factor if and only if both the following conditions hold:

(i) $e_G(V(C), S) + f(V(C))$ is odd;

(ii) $f(D) - g(S) + \deg_{G-D}(S) - \hat{q}_G(D, S, g, f) \geq 0$, for all pairs of disjoint sets $D, S \subseteq V(G)$, where $\hat{q}_G(D, S, g, f)$ denotes the number of components $C$ of $G - (D \cup S)$ having $g(v) = f(v)$ for all $v \in V(C)$.

**21.** If $G$ is a simple graph, then $G$ has a path factor if and only if $i(G-S) \leq |2S|$, for all $S \subseteq V(G)$. Here $i(G-S)$ denotes the number of isolated vertices in $G - S$.

**22.** Let $FACT(H)$ denote the problem of finding a factor where each component is isomorphic to $H$. If $H = K_2$, then $FACT(H)$ is just the problem of finding a perfect matching and hence is polynomial-time. However, if $H$ has a component with at least three vertices, then $FACT(H)$ becomes *NP*-complete.

**23.** A graph $G$ has an $F$-factor if and only if $|N(S)| \geq |S|$, for every independent set $S \subseteq V(G)$. (This result can be viewed as a generalization of Hall's Theorem to the non-bipartite case.)

**24.** There is a polynomial-time algorithm for finding an $F$-factor or showing that none exists.

**25.** *1-factorization conjecture:* Every $r$-regular graph of even order $n$ with $r \geq n/2$ admits a 1-factorization. (Or stated differently, all such graphs have *edge chromatic number* $r$.)

**26.** The 1-factorization conjecture is "asymptotically true" in that every $r$-regular graph of even order $n$ with $r \geq (\frac{1}{2} + o(1))n$ has a 1-factorization.

# 8.5   GRAPH ISOMORPHISM AND RECONSTRUCTION

Deciding whether two graph descriptions actually specify structurally identical graphs is called isomorphism testing. Polynomial-time algorithms for isomorphism testing are known only for certain special classes of graphs. However, there are heuristic algorithms to test isomorphism of reasonable-sized graphs. The related problem of reconstructing a graph from its vertex-deleted subgraphs is also still unsettled.

## 8.5.1   ISOMORPHISM INVARIANTS

### Definitions:

An **isomorphism** between two *simple graphs* $G$ and $H$ is a bijection $f: V_G \to V_H$ between their vertex sets that preserves adjacency: $(u, v) \in E_G$ if and only if $(f(u), f(v)) \in E_H$, for any pair of vertices $u, v \in V_G$.

In full generality, a **graph isomorphism** between two graphs $G$ and $H$ is a pair of bijections $f_V: V_G \to V_H$ and $f_E: E_G \to E_H$ such that for every edge $e \in E_G$, the endvertices of $e$ are mapped by $f_V$ onto the endvertices of $f_E(e)$. *Note:* Except when confusion will result, the same notation $f$ can be used for both $f_V$ and $f_E$.

A **digraph isomorphism** is an isomorphism that preserves all arcs.

Two graphs are **isomorphic** if there is an isomorphism from one to the other.

The **isomorphism type** of a graph (digraph) $G$ is the class of all graphs (digraphs) isomorphic to $G$.

A **graph invariant** is a property of graphs such that two isomorphic graphs have the same value with regard to it.

A **vertex invariant** is a property of a vertex that is preserved by isomorphism: if $f: V_G \to V_H$ is an isomorphism, $v \in V_G$ and $p$ is a property of $v$, then the vertex $f(v) \in V_H$ has the same value as $v$ with regard to $p$.

An **automorphism** is an isomorphism from a graph to itself.

The set of all automorphisms of a graph $G$ forms a group under composition called the **automorphism group** $Aut(G)$. (See §5.2.2.)

### Facts:

**1.** Two graphs $G$ and $H$ are isomorphic if there is a bijection $f: V_G \to V_H$ such that for every vertex pair $u, v \in V_G$ the number of edges joining $u$ and $v$ equals the number joining their images $f(u), f(v) \in V_H$.

**2.** Graph invariants are used to distinguish between nonisomorphic graphs.

**3.** Most graph invariants are either too hard to compute or not strong enough at distinguishing similar but nonisomorphic graphs.

**4.** Vertex invariants are often used to partition the vertices of a graph into equivalence classes under graph automorphism, in order to discover the automorphism group of the graph.

**5.** Graphs have many isomorphism invariants, including: elementary invariants (such as the degree sequence), structural invariants (such as the girth), topological invariants (such as genus), chromatic invariants (such as the edge chromatic number), and algebraic invariants (such as eigenvalues).

**Examples:**

**1.** The following figure illustrates an isomorphism $f$ of simple graphs.

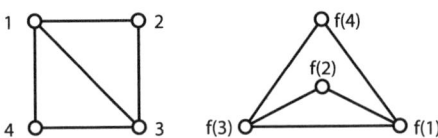

**2.** The following figure illustrates an isomorphism $f$ of nonsimple graphs.

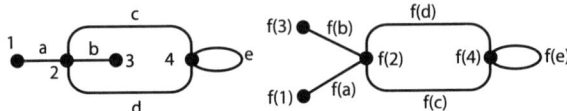

**3.** The following two digraphs are not isomorphic. Even though there are six different isomorphisms of their underlying graphs, none of them preserves the direction of all the edges.

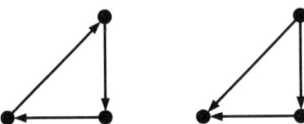

**4.** The next three graph drawings all look different. The following table shows some of their isomorphism invariants, from which it may be concluded that graph B cannot be isomorphic to either graph A or graph C, but that A and C might be isomorphic.

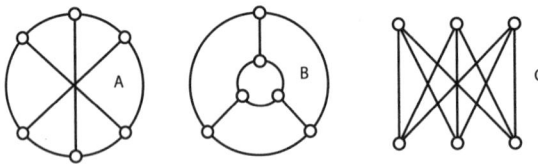

|  | $A$ | $B$ | $C$ |
|---|---|---|---|
| # vertices | 6 | 6 | 6 |
| # edges | 9 | 9 | 9 |
| degree seq. | $3, 3, 3, 3, 3, 3$ | $3, 3, 3, 3, 3, 3$ | $3, 3, 3, 3, 3, 3$ |
| connectivity | 3 | 3 | 3 |
| girth | 4 | 3 | 4 |
| genus | 1 | 0 | 1 |
| chromatic # | 2 | 3 | 2 |

To construct an isomorphism between graphs A and C, assign labels 0, 1, 2, 3, 4, 5 cyclically to the vertices of A. Then assign labels 0, 2, and 4 to the top three vertices of C, and 1, 3, and 5 to the bottom three.

## 8.5.2   ISOMORPHISM TESTING

### Definitions:

An **isomorphism test for graphs** is an algorithm that accepts two graphs as input and outputs "yes" to indicate the decision that they are isomorphic or "no" to indicate that they are not isomorphic. Unless the context explicitly mentions the possibility of error, it is implicitly understood that the decision is correct.

The **eigenvalues** of a graph are the eigenvalues of its adjacency matrix. (See §8.10.2 and §6.5.)

The **average genus** of a graph is the average genus of the imbedding surface, taken over all cellular imbeddings of that graph. (See §8.8.3.)

An **equitable partition** for a graph $G$ is a partition $V_1, \ldots, V_n$ of its vertex set and a set of numbers $\{d_{i,j} \mid 1 \leq i, j \leq n\}$ such that every vertex in $V_i$ is adjacent to exactly $d_{i,j}$ vertices in $V_j$.

A **devil's pair** for an isomorphism-testing approach is a pair of nonisomorphic graphs that the approach fails to distinguish.

### Facts:

**1.** No polynomial-time isomorphism testing algorithm is known. Moreover, it is not known whether isomorphism testing is an NP-complete problem.

**2.** Any two randomly selected graphs are almost always not isomorphic.

**3.** An algorithm that calculates the order of the automorphism group of a graph can be used as a subprogram to test isomorphism of two graphs $G$ and $H$, as follows:

> **if** $|Aut(G)| \neq |Aut(H)|$ **then** $G$ and $H$ are not isomorphic
> **else**
>> **if** $|Aut(G \cup H)| = 2|Aut(G)|^2$ **then** $G$ and $H$ are isomorphic
>> **else** $G$ and $H$ are not isomorphic

Note that the converse is true. Thus the problem of graph isomorphism and the problem of computing the order of the automorphism group of a graph are computationally equivalent.

**4.** Another algebraic approach to isomorphism testing is based on eigenvalues. A devil's pair for simply comparing eigenvalues appears in Example 4.

**5.** One topological approach to isomorphism testing dissects each graph into planar components (§8.7) and combines known efficient tests for isomorphism of planar graphs with careful study of possible interconnections.

**6.** Another topological approach to isomorphism testing is based on the genus distribution (§8.8.4), taken over all cellular imbeddings. Although calculating the genus distribution by brute force would be tedious, one can estimate it by random sampling. Any pair of trees is a trivial devil's pair, but trees are easily tested by another isomorphism algorithm.

**7.** The best known practical isomorphism algorithm is "NAUTY" (an acronym for No AUTomorphisms, Yes?) created by B. D. McKay. This backtrack algorithm repeatedly refines an initial vertex partition. At each stage of the refinement, a cell of size greater than 1 is broken into two parts, one a single vertex, and the coarsest equitable partition is found. The discrete partitions generated in this way correspond to labelings of the graph, organized so as to determine the automorphism group. An isomorphism invariant criterion is used to pick one of these labelings as the "canonical" one used for isomorphism testing. This algorithm can very quickly check isomorphism for most graphs.

**8.** To certify that two graphs are isomorphic, one can give a vertex bijection that realizes the isomorphism. Deciding whether a bijection between the vertex sets of two graphs is the vertex function of a graph isomorphism can be achieved in polynomial time.

**9.** If graph isomorphism is NP-complete, then the complexity hierarchy between P and NP collapses, which is considered unlikely. (See §17.5.)

**10.** The following table shows the best known time bounds for checking isomorphism in various classes of graphs. Almost all of these bounds have been achieved using an algebraic approach.

| class of graphs (on $n$ vertices) | time bound |
|:---:|:---:|
| graphs | $\exp O((\log n)^c)$ |
| trees | $O(n)$ |
| planar graphs | $O(n)$ |
| graphs of genus $g$ | $n^{O(g)}$ |
| cubic graphs | $O(n^3 \log n)$ |
| graphs with max degree $\leq d$ | $n^{O(d)}$ |
| tournaments | $n^{O(\log n)}$ |

**11.** Algorithm 1 gives a naive $O(n^2\,n!)$ method for testing whether or not two graphs are isomorphic.

---

**Algorithm 1:   Naive graph isomorphism-testing algorithm.**

input: simple graphs $G, H$

**if** $|V_G| \neq |V_H|$ or $|E_G| \neq |E_H|$ **then** print "NO" and stop
**for** each bijection $f\colon V_G \to V_H$
   **for** each pair $u, v \in V_G$
      **if** $u, v$ adjacent and $f(u), f(v)$ not adjacent **then** print "NO" and stop
      **if** $u, v$ not adjacent and $f(u), f(v)$ adjacent **then** print "NO" and stop
print "YES" and stop

---

**Examples:**

**1.** The labeling of the following two isomorphic graphs indicates the correspondence between vertices. This is the Petersen graph.

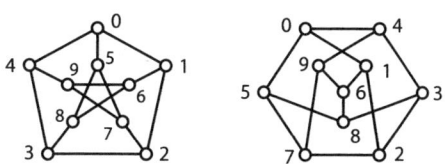

**2.** One devil's pair for isomorphism testing by degree sequence is

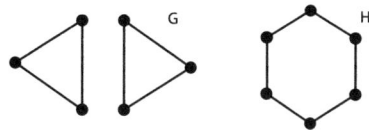

**3.** Another devil's pair for isomorphism testing by degree sequence is

**4.** The following graphs are a devil's pair for isomorphism testing by simple comparison of eigenvalues. They both have characteristic polynomial $\lambda^6 - 7\lambda^4 - 4\lambda^3 + 7\lambda^2 + 4\lambda - 1$. Yet they cannot be isomorphic because their degree sequences are different.

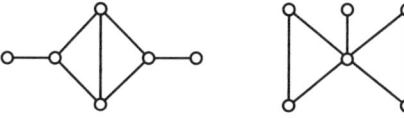

**5.** A 3-connected devil's pair for simply comparing average genus is shown below. Both graphs shown have 8 imbeddings of genus 0, 536 of genus 1, 3416 of genus 2, and 1224 of genus 3.

## 8.5.3  GRAPH RECONSTRUCTION

The question of whether a graph is reconstructible from its subgraphs is one of the most beguiling unsolved problems in graph theory.

### Definitions:

A **vertex-deleted subgraph** of a graph $G$ is a subgraph $G - v$ obtained by removing a single vertex $v$ and all of its incident edges.

An **edge-deleted subgraph** of a graph $G$ is a subgraph $G - e$ obtained by removing a single edge $e$.

The **vertex-deleted subgraph collection** of a graph $G$ is the multi-set of all vertex-deleted subgraphs $G - v$. The number of times a graph appears in the collection equals the number of different vertices whose removal yields that graph. Thus, the cardinality of the collection equals the number of vertices of $G$.

The **edge-deleted subgraph collection** of a graph $G$ is the multi-set of all edge-deleted subgraphs $G - e$. The number of times a graph appears in the collection equals the number of different edges whose removal yields that graph. Thus, the cardinality of the collection equals the number of edges of $G$.

A **reconstructible** graph is a graph $G$ such that no other graph has the same vertex-deleted subgraph collection as $G$.

An **edge-reconstructible** graph is a graph $G$ such that no other graph has the same edge-deleted subgraph collection as $G$.

A **reconstructible invariant** is a graph invariant such that all graphs with the same vertex-deleted subgraph collection have the same value with respect to this invariant.

An **edge-reconstructible invariant** is a graph invariant such that all graphs with the same edge-deleted subgraph collection have the same value with respect to this invariant.

## Conjectures:

**1.** *The graph reconstruction conjecture* (P. Kelly and S. Ulam, 1941): Every graph with more than two vertices is reconstructible.

**2.** *The edge reconstruction conjecture*: Every graph with at least four edges is edge reconstructible.

**3.** *Halin's conjecture*: If two (possibly infinite) graphs with more than two vertices have the same vertex-deleted subgraph collection, then each graph is a subgraph of the other.

**4.** *Ramachandran's conjecture*: Every digraph $D$ is reconstructible from the vertex-deleted subdigraphs $D - v$ together with the additional information consisting of the in- and out-degree of $v$.

## Facts:

**1.** The graph reconstruction conjecture implies the edge reconstruction conjecture, and both are implied by Halin's conjecture.

**2.** The graph reconstruction conjecture does not hold for graphs on two vertices, because $K_2$ and $\overline{K_2}$ have identical sets of deleted subgraphs.

**3.** The edge reconstruction conjecture does not hold for graphs on four edges, because $K_3 + K_1$ (disjoint union) and $K_{1,3}$ have identical collections of edge-deleted subgraphs.

**4.** Computer search has verified the reconstruction conjecture for graphs with eleven or fewer vertices.

**5.** The following table lists some invariants and types of graphs which are known to be reconstructible.

| both edge-reconstructible and reconstructible | | other edge-reconstructible |
|---|---|---|
| *invariants* | *graphs* | *graphs* |
| number of vertices | regular | more edges than non-edges |
| number of edges | disconnected | only two vertex degrees |
| degree sequence | trees | no induced $K_{1,3}$ subgraph |
| connectivity | outerplanar | large with Hamilton path |
| characteristic polynomial | cacti | $2\log_2(2 \text{ max deg}) \leq \text{avg deg}$ |
| | maximal planar | |

**6.** If graph $F$ has fewer vertices than graph $G$, then the number of subgraphs of $G$ isomorphic to $F$ is reconstructible.

**7.** The reconstruction conjecture is not true for directed graphs in general, because nonreconstructible tournaments of arbitrarily large size are known, but Ramachandran's conjecture holds for all known nonreconstructible tournaments.

**8.** Infinite graphs are not reconstructible in general, but Halin's conjecture holds for all known nonreconstructible infinite pairs.

**9.** Almost every graph is uniquely determined by any three vertex-deleted subgraphs.

**Example:**

**1.** The following figure shows a graph (at the left) and its collection of vertex-deleted subgraphs.

# 8.6   GRAPH COLORINGS, LABELINGS, & RELATED PARAMETERS

The vertex set of a graph can be colored so that adjacent vertices are colored differently. Similarly, the edges of a graph without self-loops can be colored so that adjacent edges are colored differently. If a graph is imbedded in a surface so that there are no self-adjacent regions, then the regions can be colored so that adjacent regions receive different colors. These entertaining concepts have many important applications, including assignment and scheduling problems. We consider these and related questions here.

## 8.6.1   VERTEX COLORINGS

**Definitions:**

A (***proper***) ***vertex k-coloring*** (or ***k-coloring***) of a simple graph $G$ is a function $f \colon V_G \to \{1, \dots, k\}$ such that adjacent vertices are assigned different numbers. Often, the set $\{1, \dots, k\}$ is regarded as a set of colors.

A ***coloring*** of a graph is a $k$-coloring for some integer $k$.

An ***improper coloring*** of a graph permits two adjacent vertices to be colored the same.

A graph is ***k-colorable*** (or ***k-vertex colorable***) if it has a $k$-coloring.

The ***chromatic number*** or (***vertex chromatic number***) $\chi(G)$ of a graph $G$ is the minimum number $k$ such that $G$ is $k$-colorable; that is, $\chi(G)$ is the smallest number of colors needed to color the vertices of $G$ so that no adjacent vertices have the same color.

A graph $G$ is ***k-chromatic*** if $\chi(G) = k$.

A graph $G$ is **chromatically k-critical** if $G$ is $k$-chromatic and if $\chi(G - e) = k - 1$ for each edge $e$ of $G$.

An **elementary contraction** of a simple graph $G$ on the edge $e$, denoted $G \cdot e$, is obtained by replacing the edge $e$ and its two endpoints by one vertex adjacent to all the other vertices to which the endpoints were adjacent.

The **chromatic polynomial** of the graph $G$ is the function $\pi_G(t)$ whose value at the integer $t$ is the number of different functions $V_G \to \{1, \ldots, t\}$ that are proper colorings of $G$.

A **list assignment** $L$ of a graph is an assignment of a set $L_v$ of allowable colors at each vertex $v$.

A graph is **k-choosable** if, for every list assignment $L$ with $|L_v| \geq k$ for all $v$, it is possible to choose for each vertex $v$ an element of $L_v$ such that the result is a proper coloring.

The **choice number** $ch(G)$ is the minimum $k$ such that graph $G$ is $k$-choosable.

**Facts:**

**1.** A direct way to calculate the chromatic number of a reasonably small graph is in two steps. First derive an upper bound for the number of colors needed, either by finding a coloring by trial and error or by using the greedy coloring algorithm. Then prove that one fewer colors would be insufficient. This could be achieved by an exponential-time exhaustion algorithm, or by finding an insightful proof for the particular graph.

**2.** $\chi(G) = 2$ if and only if the graph is bipartite and its edgeset is nonempty.

**3.** The odd cycles are the 3-critical graphs.

**4.** The odd wheels $W_{2n+1}$, $n \geq 1$, are 4-critical.

**5.** $\chi(G) \leq 1 + \Delta(G)$, where $\Delta$ denotes maximum degree.

**6.** *Brooks' theorem:*  If $G$ is a connected graph which is neither an odd cycle nor a complete graph, then $\chi(G) \leq \Delta(G)$.

**7.** $\chi(G) \leq 1 + \max \delta(G')$, where $\delta$ denotes minimum degree, and where the maximum is taken over all induced subgraphs $G'$ of $G$.

**8.** $\chi(G) \leq m(G)$, where $m(G)$ is the maximum length of a path in $G$.

**9.** *The Four Color Theorem* (Appel and Haken, 1976):  If $G$ is planar, then $\chi(G) \leq 4$. That is, every planar graph has a proper coloring of its vertices with 4 or fewer colors.

**10.** *Hadwiger's conjecture:*  If $G$ is a connected graph with $\chi(G) = n$, then $K_n$ is a minor of $G$; it is known to be true for $n \leq 5$.

**11.** *Nordhaus-Gaddum inequalities:*  If $G$ is a graph with $|V(G)| = p$ and $\overline{G}$ is its edge-complement, then

- $2\sqrt{p} \leq \chi(G) + \chi(\overline{G}) \leq p + 1$;
- $p \leq \chi(G) \cdot \chi(\overline{G}) \leq (p+1)^2/4$.

**12.** The greedy coloring algorithm (Algorithm 1) produces a vertex coloring of a graph $G$ whose vertices are ordered. (It is called "greedy" because once a color is assigned, it is never changed.)

---

**Algorithm 1**:   **Greedy coloring algorithm.**

input: a graph $G$ with vertex list $v_1, v_2, \ldots, v_n$

$c := 0$ {Initialize color at "color 0"}

**while** some vertex still has no color

  $c := c + 1$  {Get the next unused color}

  **for** $i := 1$ **to** $n$  {Assign the new color to as many vertices as possible}

    **if** $v_i$ is uncolored and no neighbor of $v_i$ has color $c$ **then** assign color $c$ to $v_i$

---

**13.**  The number of colors used by the greedy coloring algorithm depends on the ordering of the vertices of $G$. At least one of the orderings of the vertices of $G$ yields $\chi(G)$, but the number of colors used can exceed $\chi(G)$ arbitrarily.

**14.**  There is no known polynomial-time algorithm for finding $\chi(G)$ exactly. Deciding whether a graph has at most a particular chromatic number is NP-complete, if that number is at least 3.

**15.**  The following table gives the chromatic and edge chromatic numbers (see §8.6.2) of the graphs in some common families.

| graph $G$ | $\chi(G)$ | $\chi_1(G)$ |
|---|---|---|
| path $P_n$, $n \geq 3$ | 2 | 2 |
| cycle $C_n$, $n$ even, $n \geq 2$ | 2 | 2 |
| cycle $C_n$, $n$ odd, $n \geq 3$ | 3 | 3 |
| wheel $W_n$, $n$ even, $n \geq 4$ | 3 | $n$ |
| wheel $W_n$, $n$ odd, $n \geq 3$ | 4 | $n$ |
| complete graph $K_n$, $n$ even, $n \geq 2$ | $n$ | $n-1$ |
| complete graph $K_n$, $n$ odd, $n \geq 3$ | $n$ | $n$ |
| complete bipartite graph $K_{m,n}$, $m, n \geq 1$ | 2 | $\max\{m, n\}$ |
| bipartite $G$, at least one edge | 2 | $\Delta(G)$ |
| Petersen graph | 3 | 4 |
| complete $k$-partite $K_{m_1, \ldots, m_k}$, $m_i \geq 1$ | $k$ | $\max\{m_1, \ldots, m_k\}$ |

**16.**  The choice number is at least the chromatic number.

**17.**  If $G$ is connected but neither complete nor an odd cycle, $ch(G) \leq \Delta(G)$.

**18.**  The choice number of $K_{m,m}$ is $(1 + o(1)) \log_2 m$.

**19.**  Every planar graph is 5-choosable.

**20.**  For every edge $e$ of a simple graph $G$, $\pi_G(t) = \pi_{G-e}(t) - \pi_{G \cdot e}(t)$.

**21.**  The chromatic polynomial $\pi_G(t)$ of a graph with $n$ vertices and $m$ edges is a polynomial in $t$ of degree $n$, whose leading term is $t^n$, whose next term is $-mt^{n-1}$, and whose constant term is 0.

**22.**  The following table gives the chromatic polynomials of some graphs.

| graph | $\pi_G(t)$ |
|---|---|
| $n$-vertex tree | $t(t-1)^{n-1}$ |
| cycle graph $C_n$ | $(t-1)^n + (-1)^n(t-1)$ |
| wheel $W_n$ | $t(t-2)^{n-1} + (-1)^{n-1}t(t-2)$ |
| complete graph $K_n$ | $t^{\underline{n}} = t(t-1)(t-2) \ldots (t-n+1)$ |

**Examples:**

**1.** *Time scheduling:*   Let classes at a school be modeled by the vertices of a simple graph $G$, with two vertices adjacent if and only if there is at least one student in both of the corresponding classes. Then $\chi(G)$ gives the minimum number of time periods for scheduling the classes so as to accommodate all the students.

**2.** *Assignment of radio frequencies:*   If the vertices of a graph $G$ represent radio stations, with two stations adjacent precisely when their broadcast areas overlap, then $\chi(G)$ determines the minimum number of transmission frequencies required to avoid broadcast interference.

**3.** *Separating combustible chemical combinations:*   Let the vertices of graph $G$ represent different kinds of chemicals needed in some manufacturing process. An edge joins each pair of chemicals that might explode if they are combined. The chromatic number of this graph is the number of different storage areas required so that no two chemicals that mix explosively are stored together.

**4.**   Proceeding in the direct way, as described in Fact 1, to color the graph in the following figure quickly yields its chromatic number. Applying the greedy coloring algorithm, with the vertices considered in cyclic order around the 8-cycle, yields a 3-coloring. Since this graph contains an odd cycle (a 5-cycle), it cannot be 2-colored. Thus, $\chi = 3$.

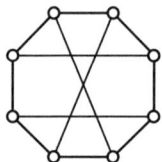

**5.**   In the following figure vertex colorings are indicated for the cycle graphs $C_3$, $C_4$, and $C_5$; in each case three colors are used. Note that $\chi(C_3) = \chi(C_5) = 3$, whereas $\chi(C_4) = 2$ (since the vertex colored "3" could have been colored "1").

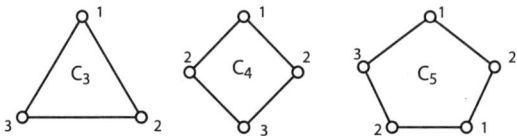

**6.**   The following figure shows three chromatically 4-critical graphs.

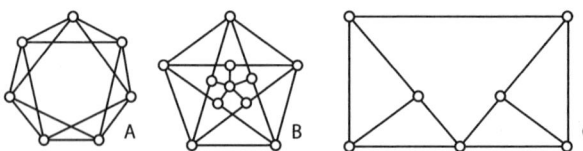

**7.**   A 3-coloring of graph $A$ in the figure of Example 6 would necessarily give some color to three different vertices. Two of these vertices would have to be adjacent (because the edge-complement contains no 3-cycle). Thus, a 3-coloring could not be proper, and hence $\chi = 4$.

**8.**   A 3-coloring of graph $B$ in the figure of Example 6 would need three different colors on the outer 5-cycle. These would force the use of three different colors on the points of the central 5-star. This would force the use of a fourth color on the central vertex. Thus, $\chi = 4$.

**9.** The choice number of $K_{3,3}$ is 3. To see that $ch(K_{3,3}) > 2$, consider the assignment $\{a, b\}$, $\{a, c\}$, and $\{b, c\}$ to the vertices in one partite set, and the same in the other partite set.

---

## 8.6.2 EDGE COLORINGS

**Definitions:**

An *edge coloring* of a graph is an assignment of colors to its edges such that adjacent edges receive different colors.

A graph $G$ is *k-edge colorable* if there is an edge coloring of $G$ using at most $k$ colors.

The *edge chromatic number* $\chi_1(G)$ (or $\chi'(G)$) of a graph $G$ is the minimum $k$ such that $G$ is $k$-edge colorable. If $\chi_1(G) = k$, then $G$ is *edge k-chromatic*.

*Chromatic index* is a synonym for edge chromatic number.

A graph is *edge-chromatically k-critical* if it is edge $k$-chromatic and $\chi_1(G - e) = \chi_1(G) - 1$ for every edge $e$ of $G$.

For a graph $G$, the *line graph* $L(G)$ has as vertices the edges of $G$, with two vertices adjacent in $L(G)$ if and only if the corresponding edges are adjacent in $G$.

**Facts:**

**1.** Every edge coloring of a graph $G$ can be interpreted as a vertex coloring of the associated line graph $L(G)$. Thus, $\chi_1(G) = \chi(L(G))$.

**2.** $\Delta(G) \leq \chi_1(G)$.

**3.** *Vizing's theorem:* If $G$ is a simple graph, then $\chi_1(G) \leq \Delta(G) + 1$.

**4.** *Vizing's general theorem:* If $G$ is a general graph whose maximum edge multiplicity is $\mu$, then $\chi_1(G) \leq \Delta(G) + \mu$.

**5.** Either $\chi_1(G) = \Delta(G)$ ($G$ is of *class one*) or $\chi_1(G) = \Delta(G) + 1$ ($G$ is of *class two*).

**6.** $\chi_1(K_{m,n}) = \chi(L(K_{m,n})) = \chi(K_m \,\square\, K_n) = \max\{m, n\}$, if $m, n \geq 1$.

**7.** If $G$ is bipartite, then $\chi_1(G) = \Delta(G)$.

**8.** $\chi_1(K_n) = n$ if $n$ is odd ($n \neq 1$); $\chi_1(K_n) = n - 1$ if $n$ is even.

**9.** If $G$ is planar and $\Delta(G) \geq 8$, then $\chi_1(G) = \Delta(G)$.

**10.** If $G$ is 3-regular and Hamiltonian, then $\chi_1(G) = \Delta(G)$.

**11.** If $G$ is regular with $|V_G|$ odd and $|E_G| > 0$, then $\chi_1(G) = \Delta(G) + 1$.

**12.** The greedy edge-coloring algorithm (Algorithm 2) produces an edge-coloring of a graph $G$, whose vertices are ordered. The number of colors it assigns depends on the vertex ordering, and it is not necessarily the minimum possible. (It is equivalent to applying the greedy vertex-coloring algorithm to the line graph.)

---

**Algorithm 2: Greedy edge-coloring algorithm.**

input: a graph $G$ with edge list $e_1, e_2, \ldots, e_n$

$c := 0$ {Initialize color at "color 0"}

**while** some edge still has no color

$c := c + 1$  {Get the next unused color}
**for** $i := 1$ **to** $n$
  {Assign the new color to as many edges as possible}
  **if** $e_i$ is uncolored and no neighbor of $e_i$ has color $c$ **then** assign color $c$ to $e_i$

**Examples:**

**1.** The following three graphs are all edge 3-chromatic. None of them is edge-chromatically 3-critical. Since each graph has a vertex of degree three, no 2-edge-coloring is possible.

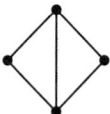  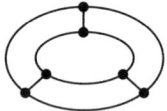

**2.** The following graph is 5-edge-chromatic. Since there are 14 edges, a 4-edge-coloring would have to give the same color to four of them. For this edge-coloring to be proper, these four edges would have to have no endpoints in common. That is impossible, because the graph has only seven vertices.

**3.** The Petersen graph is edge-chromatically 4-critical.

**4.** *Exam scheduling*:   Suppose that each student at a university is to be examined orally by each of their professors at the end of the term. Then the minimum number of examination periods required is the edge chromatic number of the bipartite graph with vertices representing students and professors, and edges connecting students with their professors.

**5.** *Wiring electrical network boards*:  A number of relays, switches, and other electronic devices $D_1, D_2, \ldots, D_n$ on a relay panel are to be connected into a network. The connecting wires are twisted into a cable, with those connected to $D_1$ emerging at one point, those connected to $D_2$ at another, and so forth. The wires emerging from the same point must be colored differently, so that they can be distinguished. The least number of colors required to color the wires is the edge chromatic number of the associated network.

**6.** The following nonsimple graph illustrates Vizing's general theorem. Its highest edge multiplicity is 3, its maximum degree is 6, and its edge chromatic number is 9.

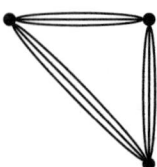

### 8.6.3   CLIQUES AND INDEPENDENCE

**Definitions:**

A **clique** of a graph $G$ is a complete subgraph of $G$. It is **maximal** if it is contained in no larger complete subgraph of $G$.

The **clique number** $\omega(G)$ of a graph $G$ is the number of vertices of a largest clique of $G$.

A subset $W$ of $V(G)$ (or $D$ of $E(G)$) is **independent** if no two elements of $W$ (respectively $D$) are adjacent.

The **vertex independence number** $\alpha(G)$ of $G$ is the size of a largest independent set of vertices in $G$.

The **edge independence number** $\alpha_1(G)$ of a graph $G$ is the size of a largest independent set of edges in $G$. It is also called the **matching number**.

**Facts:**

**1.** The independence number of a graph is equal to the clique number of its edge-complement, and vice versa. That is, $\alpha(G) = \omega(\overline{G})$ and $\omega(G) = \alpha(\overline{G})$.

**2.** The chromatic number of a graph is at least as large as the clique number: $\chi(G) \geq \omega(G)$.

**3.** For each positive integer $n$, there is a graph $G$ with chromatic number $n$ and clique number equal to 2; that is, $G$ contains no triangles.

**4.** If no induced subgraph of a graph is isomorphic to $P_4$, then its chromatic number equals its clique number and the greedy algorithm (§8.6.1, Algorithm 1) always produces a coloring with the minimum number of colors.

**5.** $\dfrac{|V(G)|}{\alpha(G)} \leq \chi(G) \leq |V(G)| + 1 - \alpha(G)$.

**6.** If $|E(G)| > \Delta(G) \cdot \alpha_1(G)$, then $\chi_1(G) = \Delta(G) + 1$.

**Examples:**

**1.** The following graph has three maximal cliques—of sizes 2, 3, and 4. Thus, its clique number is 4.

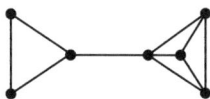

**2.** If $1 \leq m \leq n$, then $\omega(K_{m,n}) = 2$, $\alpha(K_{m,n}) = n$, and $\alpha_1(K_{m,n}) = m$.

**3.** Define $K_{n(m)}$ to be the graph whose edge complement is $nK_m$, the disjoint union of $n$ copies of $K_m$. Then $\omega(K_{n(m)}) = n$, $\alpha(K_{n(m)}) = m$, and $\alpha_1(K_{n(m)}) = \left\lfloor \frac{mn}{2} \right\rfloor$.

### 8.6.4   MAP COLORINGS

**Definitions:**

An **orientable surface** $S$ is a surface homeomorphic to a sphere with $g \geq 0$ handles attached and is denoted by $S_g$.

A **nonorientable surface** $S$ is a surface homeomorphic to a sphere with $k \geq 1$ crosscaps attached and is denoted by $N_k$. (See §8.8.1.)

The **Euler characteristic** of a surface $S$ is $2 - 2g$ if $S$ is homeomorphic to $S_g$, and $2 - k$ if $S$ is homeomorphic to $N_k$. The most usual notation for Euler characteristic in mathematics is $\chi(S)$. However, since that is used for the chromatic number, ad hoc notation such as $eu(S)$ is sometimes used in chromatic graph theory.

A **map** on a surface is an imbedding of a graph on that surface.

A **map coloring** is an assignment of colors to the regions of a map so that adjacent regions (those sharing a one-dimensional boundary portion) receive different colors.

A map $M$ is **n-colorable** if there is a map coloring of $M$ using at most $n$ colors.

The **chromatic number** $\chi(M)$ of a map $M$ is the minimum $n$ such that $M$ is $n$-colorable.

The **chromatic number** $\chi(S)$ of a surface $S$ is the largest chromatic number $\chi(M)$ for all maps $M$ on $S$.

The (**empire**) **chromatic number** $\chi(S, c)$ for a surface $S$ is the largest $\chi(M)$ for all maps $M$ on $S$, where now a country has at most $c \geq 1$ components (regions) and all components of a fixed country are colored alike, but adjacent components of different countries must receive different colors. (Thus $\chi(S) = \chi(S, 1)$.)

**Facts:**

**1.** A region coloring can be regarded as a vertex coloring of the dual graph (see §8.8.2). From this perspective, $\chi$ is the largest value of $\chi(G)$ for all graphs $G$ imbeddable on $S$.

**2.** By stereographic projection (see §8.7.5), $\chi(S_0)$ gives the chromatic number of the plane.

**3.** Let $G$ be a planar cubic block; then $\chi_1(G) = 3$.

**4.** Let $M$ be a plane map whose graph $G$ is connected and bridgeless. Then $\chi(M) = 2$ if and only if $G$ is Eulerian.

**5.** Let $M$ be a plane map for a cubic connected bridgeless graph $G$; then $\chi(M) = 3$ if and only if the dual graph is Eulerian.

**6.** If $G$ is a plane graph without triangles, then $\chi(G) = 3$. (Grötzsch, 1958)

**7.** *The Four Color Theorem* (Appel and Haken, 1976): $\chi(S_0) = 4$. That is, every map on a sphere or plane can be colored with 4 or fewer colors.

**8.** *The Heawood map coloring theorem* (Ringel and Youngs, 1968): For $g > 0$,

$$\chi(S_g) = \left\lfloor \frac{7 + \sqrt{1 + 48g}}{2} \right\rfloor .$$

**9.** *The nonorientable Heawood map coloring theorem* (Ringel, 1954): For $k > 0$,

$$\chi(N_k) = \left\lfloor \frac{7 + \sqrt{1 + 24k}}{2} \right\rfloor$$

except that $\chi(N_2) = 6$.

**10.** $\chi(S, c) \leq \left\lfloor \dfrac{6c + 1 + \sqrt{(6c + 1)^2 - 24\,eu(S)}}{2} \right\rfloor .$

**11.** $\chi(S_0, c) = 6c$ for $c \geq 2$; $\chi(S_1, c) = 6c + 1$ for $c \geq 1$; and $\chi(N_1, c) = 6c$ for $c \geq 1$.

**12.** *History of the four color problem*:   In 1852, Francis Guthrie asked whether four colors suffice to color every planar map. Arthur Cayley in 1878 was first to mention the problem in print. In 1879, A. B. Kempe, a London barrister, published a "proof" of the four color conjecture: every planar map is 4-colorable. In 1890 Percy Heawood found an error in Kempe's argument. A correct proof was established by Kenneth Appel and Wolfgang Haken in 1976.

**13.** *Concepts in the Haken-Appel proof of the four color theorem*:   Appel and Haken found an "unavoidable" set with 1476 graphs, which means that at least one of these graphs must be a subgraph of any minimum counterexample to the four color conjecture. A method called "discharging" is used to find an unavoidable set. Using a computer, they proved that each of these graphs is "reducible", which means that it cannot be a subgraph of a minimum counterexample.

**14.**   A simplified proof of the four color theorem can be found at the website: `http://www.math.gatech.edu/~thomas/FC/ftpinfo.html`

**Examples:**

**1.**   Let $M$ be the tetrahedral map, i.e., an imbedding of $K_4$ in $S_0$. By Fact 4, $\chi(M) \neq 2$, since $K_4$ is not Eulerian. By Fact 5, $\chi(M) \neq 3$, since the dual graph (isomorphic to $K_4$) is also not Eulerian. Thus, $\chi(M) = 4 = \chi(S_0)$.

**2.**   *Cartography*:   If countries on Earth are allowed two components, but no more, then by Fact 11 a map might require twelve colors, but no more.

**3.**   By Fact 8, $\chi(S_1) = 7$. The dual map of the following figure imbeds $K_7$ in the torus. To obtain the torus, paste the left side of the rectangular sheet directly to the right side, and then paste the top to the bottom with a $\frac{2}{7}$ twist.

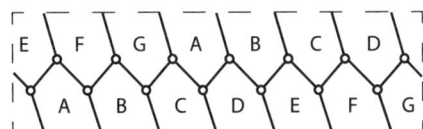

---

## 8.6.5   MORE GRAPH COLORINGS

This subsection deals with the proper coloring of the vertices of a graph where each vertex is assigned more than one color (or label), or colorings of the vertices and edges simultaneously.

**Definitions:**

A (***proper***) ***k-tuple*** coloring of a graph is an assignment of a set of $k$ distinct colors to each vertex of a graph so that adjacent vertices have disjoints sets.

The ***k-tuple chromatic number*** $\chi_k(G)$ of a graph $G$ is the smallest number of colors such that $G$ has a $k$-tuple coloring.

The ***fractional chromatic number*** $\chi_f(G)$ of a graph $G$ is the minimum of the ratio $\chi_k(G)/k$ taken over all $k$.

A ***total coloring*** of a graph is an assignment of colors to both the vertices and edges such that two elements receive different colors if they are adjacent or incident.

The ***total chromatic number*** $\chi_T(G)$ is the minimum number of colors needed for a total coloring of graph $G$.

**Facts:**

1. $\chi_k(G) \le k\chi(G)$.

2. If the clique number $\omega(G)$ (see §8.6.3) of $G$ is equal to $\chi(G)$, then $\chi_k(G) = k\chi(G)$.

3. $\chi_k(K_n) = nk$.

4. If $G$ is bipartite (with at least one edge), then $\chi_k(G) = 2k$.

5. $\chi_f(G) \le \chi(G)$.

6. $\chi_T(K_n) = n$ if $n$ is even, and $\chi_T(K_n) = n + 1$ if $n$ is odd.

7. The *total coloring conjecture* is that $\chi_T(G) \le \Delta(G) + 2$ for all graphs $G$.

**Examples:**

1. *Multiple channel assignment*: Several cities each need to have four broadcast frequencies assigned to them (a generalization of §8.6.1, Example 2). The 4-tuple chromatic number $\chi_4(G)$ is the minimum number of frequencies needed so that there is no broadcast interference.

2. $\chi_2(C_5) = 5$, as illustrated in the following figure. In fact it can be shown that $\chi_f(C_5) = 5/2$.

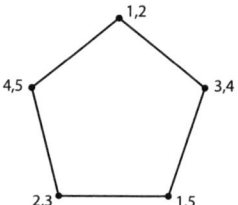

3. *Exam scheduling*: Each final exam at a school is given in two parts, with each part requiring one final exam period. If a graph $G$ is constructed by using the courses as vertices, with an edge joining $v$ and $w$ if there is a student taking courses $v$ and $w$, then $\chi_2(G)$ gives the minimum number of periods required to schedule all the exams so no student has a conflict.

---

### 8.6.6   DOMINATION IN GRAPHS

Domination has applications in areas such as network design, social sciences, optimization, and bioinformatics. For more information see the books [HaHeSl98a], [HaHeSl98b], and [HeYe13].

**Definitions:**

A **dominating set** of a graph is a set $S$ of vertices such that every vertex not in $S$ is adjacent to a vertex in $S$.

A **minimal dominating set** is a dominating set that contains no dominating set as a proper subset.

The **domination number** $\gamma(G)$ is the minimum cardinality of a (minimal) dominating set in a graph $G$, and the **upper domination number** $\Gamma(G)$ is the maximum cardinality of a minimal dominating set.

A **total dominating set** of a graph is a set $S$ of vertices such that every vertex is adjacent to a vertex in $S$. Thus, a total dominating set $S$ is a dominating set with the additional property that the subgraph induced by the set $S$ contains no isolated vertex.

A total dominating set $S$ is **minimal** if it contains no total dominating set as a proper subset.

The **total domination number** $\gamma_t(G)$ of a graph $G$ is the minimum cardinality of a total dominating set, and the **upper total domination number** $\Gamma_t(G)$ of $G$ is the maximum cardinality of a minimal total dominating set.

An **independent dominating set** is a set that is both a dominating set and an independent set.

The **independent domination number**, $i(G)$, and the **(vertex) independence number**, $\alpha(G)$, are the minimum and maximum cardinalities, respectively, of a maximal independent set in $G$.

Let $S$ be a set of vertices in a graph $G$. A vertex $v \in S$ is **$S$-irredundant** if $v$ has no neighbor in $S$ or there exists a vertex $w \notin S$ that is adjacent to $v$ but to no other vertex of $S$. The set $S$ is an **irredundant set** if every vertex in it is $S$-irredundant.

The **irredundance number** of $G$, $ir(G)$, is the minimum cardinality of a maximal irredundant set, while the **upper irredundance number** of $G$, $IR(G)$, is the maximum cardinality of an irredundant set.

The **corona** $H \circ K_1$ of a graph $H$ is the graph obtained from $H$ by adding for each vertex $v$ of $H$ a new vertex $v'$ and the pendant edge $vv'$.

**Facts:**

**1.** If $S$ is a dominating set, then so too is every superset of $S$. However, not every subset of $S$ is necessarily a dominating set.

**2.** Every maximal independent set is a minimal dominating set, and every minimal dominating set is a maximal irredundant set. Thus $ir(G) \leq \gamma(G) \leq i(G) \leq \alpha(G) \leq \Gamma(G) \leq IR(G)$.

**3.** Determining whether a graph has domination number at most $k$ is NP-complete.

**4.** If $G$ is a graph of order $n$, then $1 \leq \gamma(G) \leq n$. Equality of the lower bound is attained if and only if the maximum degree of $G$ is $n-1$, and equality holds for the upper bound if and only if $G$ is the empty graph, that is, a set of isolated vertices.

**5.** If $G = (V, E)$ is a graph with no isolated vertex and $D$ is a minimal dominating set of $G$, then $V \setminus D$ is a dominating set of $G$.

**6.** *Ore's bound.* If $G$ is a graph of order $n$ with no isolated vertex, then $\gamma(G) \leq n/2$.

**7.** If $G$ is a graph of order $n$ with no isolated vertex, then $\gamma(G) = n/2$ if and only if the components of $G$ are the cycle $C_4$ or the corona $H \circ K_1$ for any connected graph $H$.

**8.** For every regular graph $G$ of order $n$ with no isolated vertex, $\Gamma(G) \leq n/2$.

**9.** If graph $G$ has $n$ vertices and $m$ edges and does not contain isolated vertices, then $\gamma(G) \leq (1-b)n/2 + bm$ for all $b \geq 0$. For example, taking $b = 0$, we have Ore's upper bound of $\gamma(G) \leq n/2$. Taking $b = 1/3$, we have that $\gamma(G) \leq (n+m)/3$.

**10.** If $G$ is a connected graph of order $n \geq 8$ with $\delta(G) \geq 2$, then $\gamma(G) \leq 2n/5$, and this bound is tight. (Here $\delta(G)$ denotes the minimum degree.)

**11.** If $G$ is a graph of order $n$ and $\delta(G) \geq 3$, then $\gamma(G) \leq 3n/8$, and this bound is tight.

**12.** If $G$ is a connected graph of order $n \geq 14$ with $\delta(G) \geq 2$ and no induced 4-cycle and no induced 5-cycle, then $\gamma(G) \leq 3n/8$.

**13.** *Vizing's conjecture.* For any graphs $G$ and $H$, $\gamma(G \square H) \geq \gamma(G)\gamma(H)$. (Here $\square$ denotes Cartesian product.)

**14.** For any graphs $G$ and $H$, $\Gamma(G \square H) \geq \Gamma(G)\Gamma(H)$.

**15.** If $G$ is a connected graph of order $n \geq 3$, then $\gamma_t(G) \leq 2n/3$.

**16.** If $G$ is a connected graph of order $n \geq 11$ with $\delta(G) \geq 2$, then $\gamma_t(G) \leq 4n/7$.

**17.** If $G$ is a graph of order $n$ with $\delta(G) \geq 3$, then $\gamma_t(G) \leq n/2$.

**18.** If $G$ is a graph of order $n$ with $\delta(G) \geq 4$, then $\gamma_t(G) \leq 3n/7$.

**19.** If $G$ is a graph of order $n$ with $\delta(G) \geq 5$, then $\gamma_t(G) \leq \left(\frac{4}{11} + \frac{1}{72}\right)n$. The conjectured bound is $\gamma_t(G) \leq 4n/11$.

**20.** If $G$ is a graph of order $n$ with $\delta(G) \geq 1$, then $\gamma_t(G) \leq \left(\frac{1+\ln \delta}{\delta}\right)n$.

**21.** If $G$ is a connected graph of order $n$, girth $g \geq 3$ with $\delta(G) \geq 2$, then

$$\gamma_t(G) \leq \frac{n}{2} + \max\left(1, \frac{n}{2(g+1)}\right),$$

and this bound is sharp.

**22.** For every $k$-regular graph $G$ of order $n$ with no isolates, $\Gamma_t(G) \leq n/(2 - 1/k)$.

**23.** For any isolate-free graphs $G$ and $H$, $2\gamma_t(G \square H) \geq \gamma_t(G)\gamma_t(H)$.

**24.** For any isolate-free graphs $G$ and $H$, $2\Gamma_t(G \square H) \geq \Gamma_t(G)\Gamma_t(H)$.

**25.** For every $0 < \epsilon' < \epsilon$ and $p = (1 + \epsilon')\sqrt{\frac{1}{n}(2 \ln n)}$, almost every graph $G \in \mathcal{G}(n, p)$ has

$$\left(\frac{1}{2\sqrt{2}} - \epsilon\right)\sqrt{n \ln(n)} < \gamma(G) \leq \gamma_t(G) < \left(\frac{1}{\sqrt{2}} + \epsilon\right)\sqrt{n \ln(n)}.$$

**Examples:**

**1.** For the Petersen graph $G$ below, the sets $S_1 = \{2, 5, 7\}$, $S_2 = \{1, 3, 6, 10\}$, and $S_3 = \{1, 2, 3, 4, 5\}$ are all minimal dominating sets. Hence, the Petersen graph contains minimal dominating sets of cardinalities 3, 4, and 5. No set of two vertices dominates all ten vertices in the graph, and so the set $S_1$ is a dominating set of minimum cardinality and $\gamma(G) = 3$. Every minimal dominating set in the Petersen graph has cardinality at most 5, implying that $\Gamma(G) = 5$.

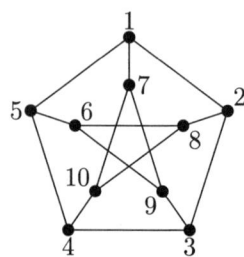

**2.** For the Petersen graph $G$ above, it holds that $\gamma_t(G) = 4$ and the closed neighborhoods $N[v]$ corresponding to the ten vertices $v$ in the Petersen graph form the ten minimum total dominating sets of $G$. Further, $\Gamma_t(G) = 6$ and the set $\{1, 3, 5, 8, 9, 10\}$ is an example of a minimal total dominating set of $G$ of size 6.

## 8.6.7   GRAPH PEBBLING

**Definitions:**

A **configuration** $C$ is an assignment of pebbles to the vertices of a connected graph $G$. The value $C(v)$ denotes the number of pebbles at vertex $v$. The **size** of $C$ is defined as $|C| = \sum_{v \in V(G)} C(v)$.

For an edge $uv$, if $u$ has at least two pebbles on it, then a **pebbling step from $u$ to $v$** is to remove two pebbles from $u$ and place one pebble on $v$. That is, if $C$ is the original configuration, then the resulting configuration $C'$ has $C'(u) = C(u) - 2$, $C'(v) = C(v) + 1$, and $C'(x) = C(x)$ for all $x \in V(G) - \{u, v\}$.

A configuration $C$ is **$r$-solvable** if it is possible from $C$ to place a pebble on vertex $r$ via pebbling steps; it is **$r$-unsolvable** otherwise. The configuration $C$ is **solvable** if it is $r$-solvable for every vertex $r$.

For a connected graph $G$ and a **root** vertex $r$, the **rooted pebbling number** $\pi(G, r)$ is defined to be the minimum number $t$ so that every configuration $C$ of size $t$ is $r$-solvable. The **pebbling number** $\pi(G)$ is the maximum value of $\pi(G, r)$ over all $r$.

A graph $G$ is of **Class 0** if $\pi(G)$ equals its order.

A sequence of paths $\mathcal{P} = (P[1], \ldots, P[h])$ is a **maximum $r$-path partition** of a rooted tree $(T, r)$ if $\mathcal{P}$ forms a partition of $E(T)$, $r$ is a leaf of $P[1]$ and, for all $1 \leq i \leq h$, $P[i]$ is a maximum-length path in $T - \bigcup_{j=1}^{i-1} E(P[j])$ among all such paths with one endpoint in $\bigcup_{j=1}^{i-1} P[j]$.

For $m \geq 2t + 1$, the **Kneser graph** $K(m, t)$ has as vertices all $t$-element subsets of $\{1, 2, \ldots, m\}$ and edges between every pair of disjoint sets.

**Facts:**

**1.** Graph pebbling arose in combinatorial number theory. It has since produced the following more general result. If $g_1, \ldots, g_n$ is a sequence of elements of an abelian group $\Gamma$ of size $n$, then there is a nonempty subsequence $(g_k)_{k \in K}$ such that $\sum_{k \in K} a_k = 0_\Gamma$ and $\sum_{k \in K} 1/|g_k| \leq 1$, where $|g|$ denotes the order of the element $g$ in $\Gamma$ and $0_\Gamma$ is the identity element in $\Gamma$.

**2.** Every rooted graph $(G, r)$ has $\pi(G, r) \geq \max\{n(G), 2^{\mathrm{ecc}(r)}\}$, where $\mathrm{ecc}(r)$ is the maximum distance of a vertex from $r$, and $\pi(G, r) \leq 2^{n(G)-1}$.

**3.** The path on $n$ vertices has pebbling number $\pi(P_n) = 2^{n-1}$.

**4.** The cycle has pebbling number $\pi(C_{2k}) = 2^k$ and $\pi(C_{2k+1}) = \lceil (2^{k+2} - 1)/3 \rceil$ for all $k \geq 1$.

**5.** If $(T, r)$ is a rooted tree, then $\pi(T, r) = \sum_{i=1}^{h} 2^{l_i} - h + 1$, where $l_i$ is the length of path $P[i]$ in a maximum $r$-path partition $\mathcal{P}$ of $T$.

**6.** Complete graphs, complete bipartite graphs other than stars, cubes, the Petersen graph, and split graphs with minimum degree at least 3 are all Class 0.

**7.** For any constant $c > 0$ there is an integer $t_0$ such that the Kneser graph $K(2t + s, t)$ is Class 0 when $t > t_0$ and $s \geq c(t/\log t)^{1/2}$.

**8.** If $G$ is a graph with $n$ vertices and $e$ edges, then $G$ is Class 0 when $e \geq \binom{n-1}{2} + 2$, while $G$ being Class 0 implies that $\kappa(G) \geq 2$, $e \geq \lfloor 3n/2 \rfloor$, and the girth of $G$ is at most $1 + 2\log n$.

**9.** If $\text{diam}(G) = 2$, then $\pi(G) \le n(G) + 1$. If also $\kappa(G) \ge 3$, then $G$ is Class 0.

**10.** If $\text{diam}(G) = d$ and $\kappa(G) \ge 2^{2d+3}$, then $G$ is Class 0.

**11.** There exists a Class 0 graph $G$ on at most $n \ge 3$ vertices with girth at least $\lfloor \sqrt{(\log n)/2 + 1/4} - 1/2 \rfloor$.

**12.** Let $G_{n,p}$ be the random graph on $n$ vertices with edge probability $p$ (see §8.11.3, §10.8.2). If $p \gg (n \log n)^{1/d}/n$ for fixed $d$, then almost surely $G_{n,p}$ is Class 0.

**13.** *Graham's Conjecture* states that every pair of connected graphs $G$ and $H$ satisfy $\pi(G \,\square\, H) \le \pi(G)\pi(H)$. This is known to be true when $G$ and $H$ are both complete graphs, both trees, both cycles, both complete bipartite graphs with at least 15 vertices per part, or both connected graphs on $n$ vertices with minimum degree at least $k \ge 2^{12n/k+15}$.

**14.** Deciding whether a configuration on a graph is solvable is NP-complete, even when restricted to the classes of diameter-two graphs or planar graphs, but is in P when restricted to the classes of complete graphs, trees, diameter-two planar graphs, or outer-planar graphs.

**15.** Deciding whether $\pi(G) \le k$ is $\Pi_2^P$-complete (see §17.5.1).

**16.** $\pi(G)$ can be calculated in polynomial time when $G$ is a tree, a diameter-two graph, a split graph, or a 2-path.

### Examples:

**1.** Two $r$-unsolvable configurations (of maximum size, right) on the path $P_7$ are shown below.

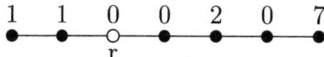     

**2.** A maximum-sized $r$-unsolvable configuration on a tree is shown below.

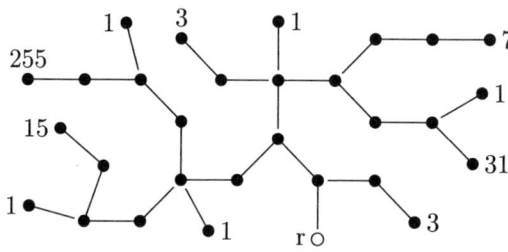

**3.** An $r$-solvable (and solvable) configuration is shown below.

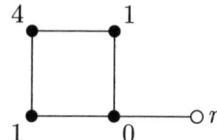

## 8.6.8   GRAPH LABELINGS

### Definitions:

A graph with $q$ edges is **graceful** if there exists an injection $f$ from the vertices to the set $\{0, 1, \ldots, q\}$ such that, when each edge $xy$ is assigned the label $|f(x) - f(y)|$, the resulting edge labels are distinct.

A graph with $q$ edges is **harmonious** if there is an injection $f$ from the vertices to the group of integers modulo $q$ such that when each edge $xy$ is assigned the label $f(x) + f(y)$ (mod $q$), the resulting edge labels are distinct. When a graph is a tree, exactly one label may be used on two vertices.

A connected graph is called **magic** if there is a labeling of the edges with distinct positive integers such that for each vertex $v$ the sum of the labels of all edges incident with $v$ is the same for all $v$. A magic labeling is called **supermagic** if the set of edge labels consists of consecutive positive integers.

A graph with $q$ edges is called **antimagic** if its edges can be labeled with distinct integers from 1 to $q$ such that the sums of the labels of the edges incident to each vertex are distinct.

### Remarks:

**1.** In 1967 A. Rosa introduced graceful labelings as tools for decomposing the complete graph into isomorphic subgraphs.

**2.** In 1980 R. Graham and N. Sloane introduced harmonious labelings as a modular version of additive bases problems stemming from error-correcting codes.

**3.** Determining whether a graph has a harmonious labeling is NP-complete.

**4.** "Almost all" graphs are not graceful or harmonious.

**5.** The conjecture made in 1968 by Ringel and Kotzig that all trees are graceful is still open.

**6.** The conjecture made in 1980 by Graham and Sloane that all trees are harmonious is still open.

**7.** J. Sedláček introduced magic labelings of graphs in 1963.

**8.** B. M. Stewart introduced supermagic labelings of graphs in 1967 as an extension of the classic concept of an $n \times n$ magic square in number theory, which corresponds to a supermagic labeling of $K_{n,n}$.

**9.** In 1990 Hartsfield and Ringel introduced antimagic graphs. Their conjectures that every tree except $P_2$ is magic and every connected graph except $P_2$ is magic are still open.

**10.** There are scores of variations on graceful, harmonious, magic, and antimagic labelings.

### Examples:

**1.** Here are graceful and harmonious labelings of the wheel $W_5$.

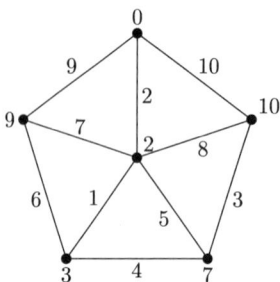
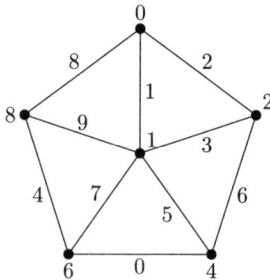

**2.** The following table specifies conditions for which some common graphs are graceful and harmonious.

| Graph | Graceful | Harmonious |
|---|---|---|
| trees | at most 35 vertices | at most 31 vertices |
| caterpilars | all | all |
| $C_n$ | iff $n \equiv 0, 3 \pmod 4$ | iff $n$ odd |
| wheels $W_n$ | all | all |
| grids $P_m \,\square\, P_n$ | all | iff $(m, n) \neq (2, 2)$ |
| $K_n$ | iff $n \leq 4$ | iff $n \leq 4$ |
| $K_{m,n}$ | all | iff $m$ or $n = 1$ |
| $C_m \,\square\, P_n$ | $n = 2$ or $m$ even | $n$ odd; $n = 2, m \neq 4$; $m = 4, n \geq 3$ |
| $C_m \,\square\, C_n$ | $m \equiv 0 \pmod 4$, $n$ even | $m = 4$, $n > 1$ |
| $n$-cube | all | iff $n \geq 4$ |

**3.** Here are magic and antimagic labelings of the wheel $W_5$.

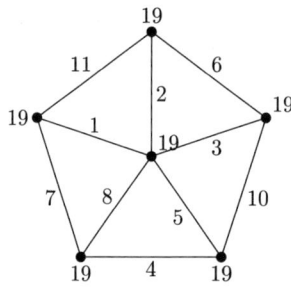
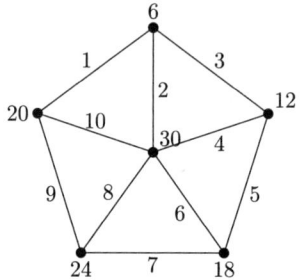

**4.** The following table specifies conditions for which some common magic and antimagic graphs. An M indicates the graph is magic; SPM indicates it is supermagic.

| Graph | Magic | Antimagic |
|---|---|---|
| $P_n$ | M iff $n = 2$ | $n \geq 3$ |
| $C_n$ | M none | all |
| $K_n$ | M if $n = 2$ or $n \geq 5$ | $n \geq 3$ |
| wheels $W_n$ | M $n \geq 4$ | all |
| $K_{m,n}$ | M if $m = n \geq 3$ | all except $K_{1,1}$ |
| $C_m \,\square\, C_n$ | SPM if $m = n$ or $m, n$ even | all |
| $n$-cube | SPM $n = 1$ or $n > 2$ even | all |

# 8.7  PLANAR DRAWINGS

Planarity is an important consideration in physical networks of any kind, because it is usually less expensive to fabricate a planar network. For instance, overpasses are a costly feature in highway design. Moreover, it is less complicated to manufacture a planar electrical network than a nonplanar network.

## 8.7.1  CHARACTERIZING PLANAR GRAPHS

A graph cannot be drawn without edge-crossings in the plane if it "contains" either the complete graph $K_5$ or the complete bipartite graph $K_{3,3}$. Conversely, every graph that "contains" neither of those two graphs can be drawn without crossings.

### Definitions:

A graph **imbedding** (or **embedding**) is a drawing with no crossings at all.

A graph is **planar** if it has an imbedding in the plane.

A graph is **nonplanar** if no imbedding in the plane is possible.

A drawing of a graph is **normalized** if there are no crossings, or if each crossing is a point where the interior of one edge crosses the interior of one other edge. (Edges may be drawn either straight or with curves.)

The graphs $K_5$ and $K_{3,3}$ are called the **Kuratowski graphs**.

### Facts:

**1.** The graphs $K_5$ and $K_{3,3}$ are both nonplanar. See Examples 4 and 5 for proofs that they are not planar.

**2.** *Kuratowski planarity theorem:* A graph is planar if and only if it has no subgraph homeomorphic (see §8.1.2) to $K_5$ or to $K_{3,3}$.

### Examples:

**1.** The drawings of $Q_3$, $K_5$, and $K_{3,3}$ in the following figure all have crossings. However, the graph $Q_3$ is planar, because it can be redrawn without any crossings.

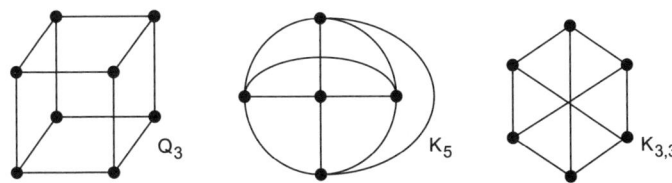

**2.** The drawings of $Q_3$ and $K_5$ in the figure of Example 1 are normalized, but the drawing of $K_{3,3}$ is not normalized, because three lines go through the same point.

**3.** The Petersen graph does not contain $K_{3,3}$ itself as a subgraph. However, if the two edges depicted by broken lines in the following figure are discarded, then the resulting graph is homeomorphic to $K_{3,3}$, so the Petersen graph is not planar.

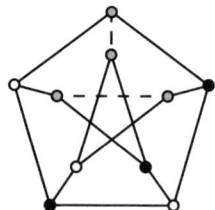

**4.** *Nonplanarity of $K_5$:*   To draw the complete graph on the vertices $v_1, v_2, v_3, v_4, v_5$ in the plane, one might as well start by drawing the 4-cycle $v_1, v_2, v_3, v_4$, which (by the Jordan curve theorem) separates the plane. Next draw the edges between $v_1$ and $v_3$ and between $v_2$ and $v_4$. To avoid crossing each other, one of these edges must go inside the 4-cycle and the other outside, as shown in the following figure. The net result so far is that there are four 3-sided regions, each with three vertices on its boundary. Thus, no matter which region is to contain the vertex $v_5$, that vertex cannot be joined to more than three other vertices without crossing the boundary.

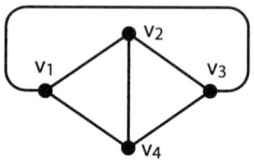

**5.** *Nonplanarity of $K_{3,3}$:*   To form a planar drawing of the complete bipartite graph on the parts $\{v_1, v_3, v_5\}$ and $\{v_2, v_4, v_6\}$, one might as well start by drawing the 6-cycle $v_1, v_2, v_3, v_4, v_5, v_6$, which separates the plane. Next draw the edges between $v_1$ and $v_4$ and between $v_2$ and $v_5$. To avoid crossing each other, one of these edges must go inside the 6-cycle and the other outside. The net result so far is shown in the following figure. It is now clear that $v_3$ and $v_6$ cannot be joined without crossing some other edge.

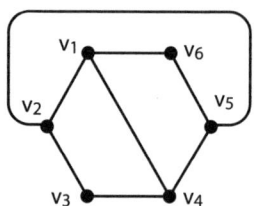

**6.** *Civil engineering:*   Suppose that a number of towns are to be joined by a network of highways. If the network is planar, then the cost of bridges for underpasses and overpasses can be avoided.

**7.** *Electrical networks:*   A planar electrical network with bare wires joining the nodes can be placed directly onto a flat board. Otherwise, insulation is needed to prevent short circuits at wire crossings.

## 8.7.2  NUMERICAL PLANARITY CRITERIA

Certain numerical relationships are true of all planar graphs. One way to show that a graph is nonplanar is to show that it does not satisfy one of these relations.

### Definitions:

A **region** of an imbedded graph is, informally, a piece of what results when the surface is cut open along all the edges. That is, a maximal subsurface containing no vertex and no part of any edge of the graph.

The **boundary of a region** $R$ of an imbedded graph is the subgraph containing all vertices and edges incident on $R$. It is denoted $\partial R$.

A **face** of an imbedded graph is a region plus its boundary.

The **exterior region** of a planar graph drawing is the region that extends to infinity.

The **girth** of a graph is the number of edges in a shortest cycle. The girth is undefined if the graph has no cycles.

### Facts:

**1.** *Euler polyhedral equation*:  Let $G = (V, E)$ be a connected graph imbedded in the plane with face set $F$. Then $|V| - |E| + |F| = 2$.

**2.** *Edge-face inequality*:  Let $G = (V, E)$ be a simple, connected graph imbedded in a surface with face set $F$. Then $2|E| \geq 3|F|$.

**3.** *Edge-face inequality (strong version)*:  Let $G = (V, E)$ be a connected graph, but not a tree, imbedded in a surface with face set $F$. Then $2|E| \geq \text{girth}(G) \cdot |F|$.

**4.** Let $G = (V, E)$ be a simple, connected planar graph. If $G$ is planar then $3|V| - |E| \geq 6$.

**5.** Let $G = (V, E)$ be a connected planar graph that is not a tree. Then $(|V| - 2) \cdot \text{girth}(G) \geq |E| \cdot (\text{girth}(G) - 2)$.

**6.** Let $G = (V, E)$ be a simple, connected, bipartite graph that is not a tree. If $G$ is planar then $|E| \leq 2 \cdot |V| - 4$.

### Examples:

**1.** In the planar imbedding of the following figure, $|V| = 4$, $|E| = 6$, and $|F| = 4$. Thus, $|V| - |E| + |F| = 4 - 6 + 4 = 2$. (The "exterior" region counts as a face.)

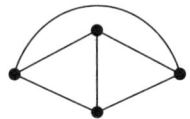

**2.** Fact 4 implies that $K_5$ is nonplanar.

**3.** Fact 5 implies that the Petersen graph, whose girth equals 5, is nonplanar.

**4.** Fact 6 implies that $K_{3,3}$ is nonplanar.

---

## 8.7.3  PLANARITY ALGORITHM

### Definitions:

A **bridge $B$ of a subgraph $H$** in a graph $G$ is a maximal connected subgraph $B$ of the edge-complement $G - H$ such that for any two edges $d$ and $e$ of $B$, there is a path starting with $d$ and terminating in $e$, none of whose internal vertices lies in $B \cap H$.

An **attachment of a bridge** is a vertex both in the bridge and the subgraph. (That is, an attachment is a vertex in which the bridge meets the rest of the graph.)

## Facts:

**1.** Call two edges in the complement of a subgraph $H$ of a graph $G$ "related" if they are both contained in a path in $G$ that has no vertices of $H$ in its interior. Then the bridges of $H$ are the induced subgraphs on the equivalence classes of edges under this relation.

**2.** Informally, a bridge is a subgraph obtained from one of the "pieces" that result by deleting $H$ from $G$ by reattaching the endpoints to the edges that attach to $H$. See Example 1.

**3.** The time needed to test planarity by searching directly for subdivided copies of $K_5$ and $K_{3,3}$ is an exponential function of the number of vertices.

**4.** J. Hopcroft and R. Tarjan developed a planarity-testing algorithm that can be executed in time proportional to the number of vertices ("linear time").

**5.** None of the linear-time planarity algorithms is easy to describe and implement. However, Algorithm 1 below is easily implemented, and its running time is satisfactory for reasonably large graphs.

**6.** Algorithm 1 can be implemented to run in time approximately proportional to the square of the number of vertices ("quadratic time").

---

**Algorithm 1**: **Easy planarity-testing for graph $G$.**

input: a simple, connected graph $G$

$G_0 :=$ an arbitrary cycle in $G$; draw $G_0$ in the plane; $j := 0$

{*Grow a sequence of nested subgraphs $G_0, G_1, \ldots$ until all of $G$ has been drawn in the plane; if this does not happen, then $G$ is nonplanar*}

**while** $G_j \neq G$ {*This possible exit implies $G$ is planar*} **and** $\big(\forall B \in \text{bridges}\,(G_j)\big)$ $\big(\forall v \in \text{attachments}\,(B)\big)\big(\exists \text{ region } R \text{ of } G_j \text{ in plane}\big)\ v \in \partial R$ {*This possible exit implies $G$ is nonplanar*} **do**

{*While-loop body says how to grow subgraph $G_{j+1}$*}

**if** $\big(\exists B \in \text{bridges}\,(G_j)\big)\big(\forall v \in \text{attachments}\,(B)\big)\big(\exists! \text{ region } R \text{ of } G_j\big)\ \big[v \in \partial R\big]$

**then** {*Case 1 — a forced move exists*}

select a path $P$ between two attachments of $B$

obtain subgraph $G_{j+1}$ by drawing path $P$ in region $R$

**else** {*Case 2 — no forced move exists*}

select any bridge, and find two regions for its attachments

select any path between two attachments of that bridge

draw that path into either region to obtain $G_{j+1}$

$j := j + 1$

---

## Example:

**1.** The following figure shows a subgraph and its three bridges $B_1 B_2, B_3$. The subgraph $H$ is the dark cycle. Attachments of the bridges are the vertices along the dark cycle.

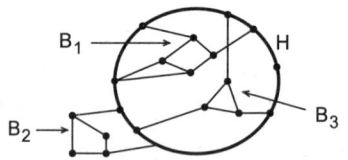

**2.** Suppose that the figure in Example 1 occurred in the execution of Algorithm 1. At the next iteration of the while-loop body, suppose that bridge $B_2$ is selected, and suppose that a path in $B_2$ is drawn outside the dark cycle. Then, on the following iteration of the while-loop body, bridge $B_3$ would be a forced choice, and a path from $B_3$ would have to be drawn inside the dark cycle. Eventually, bridge $B_1$ would have to be drawn outside the dark cycle, thereby yielding a planar drawing of the entire graph.

---

### 8.7.4  CROSSING NUMBER AND THICKNESS

**Definitions:**

The **crossing number** of graph $G$, denoted $\nu(G)$, is the minimum number of edge-crossings possible in a normalized drawing of $G$ in the plane.

The **thickness** of graph $G$, denoted $\theta(G)$, is the minimum number of planar graphs whose union is $G$.

**Facts:**

**1.** $\nu(K_n) \leq \frac{1}{4} \cdot \lfloor \frac{n}{2} \rfloor \cdot \lfloor \frac{n-1}{2} \rfloor \cdot \lfloor \frac{n-2}{2} \rfloor \cdot \lfloor \frac{n-3}{2} \rfloor$.

**2.** For all integers $n \leq 12$, $\nu(K_n) = \frac{1}{4} \cdot \lfloor \frac{n}{2} \rfloor \cdot \lfloor \frac{n-1}{2} \rfloor \cdot \lfloor \frac{n-2}{2} \rfloor \cdot \lfloor \frac{n-3}{2} \rfloor$.

**3.** It is conjectured that equality in Fact 2 holds for all positive integers.

**4.** $\nu(K_{m,n}) \leq \lfloor \frac{m}{2} \rfloor \cdot \lfloor \frac{m-1}{2} \rfloor \cdot \lfloor \frac{n}{2} \rfloor \cdot \lfloor \frac{n-1}{2} \rfloor$.

**5.** For all integers $m$ and $n$ such that $\min(m,n) \leq 6$, $\nu(K_{m,n}) = \lfloor \frac{m}{2} \rfloor \cdot \lfloor \frac{m-1}{2} \rfloor \cdot \lfloor \frac{n}{2} \rfloor \cdot \lfloor \frac{n-1}{2} \rfloor$.

**6.** *Zarankiewicz's conjecture:* The equation of Fact 5 holds for all positive integers $m$ and $n$.

**7.** $\theta(K_n) = \lfloor \frac{n+7}{6} \rfloor$ except that $\theta(K_9) = \theta(K_{10}) = 3$.

**8.** $\theta(Q_n) = \lceil \frac{n+1}{4} \rceil$.

**9.** $\theta(G) \geq \lceil \frac{|E|}{3|V|-6} \rceil$ for all simple graphs.

**Examples:**

**1.** Fact 1 implies that $\nu(K_6) \leq 3$. Thus, it is possible to draw $K_6$ with at most three crossings.

**2.** *Computer engineering:* Facts 7 and 8 yield lower bounds for the minimum number of layers needed for a multi-layer layout of an electronic interconnection network whose architecture is a complete graph or a hypercube graph, respectively.

---

### 8.7.5  STEREOGRAPHIC PROJECTION

**Definitions:**

A continuous one-to-one function from one subset of Euclidean space onto another is a **topological equivalence** if its inverse is continuous. (Informally, this means that either subset could be reshaped into the other without tearing, but only by compressing, stretching, and twisting.)

The **stereographic projection** adds a single point to a plane and thereby closes the "hole at infinity" and converts it into a sphere, as follows. Start with a sphere in 3-space, tangent at its south pole $S$ to the plane $z = 0$ at the origin $(0, 0, 0)$, as shown in the following figure; and through each point $x$ of the sphere draw a ray from the north pole $N$, extending to the point $f(x)$ at which it meets the plane.

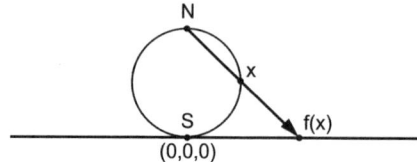

**Facts:**

**1.** The correspondence $x \to f(x)$ from the sphere minus its north pole onto the plane is a topological equivalence. In other words, the sphere minus a point could be stretched apart at the missing point and flattened out so that it covers the plane.

**2.** Any planar imbedding can be transformed into an imbedding in the sphere, which is a closed surface, by using the inverse of stereographic projection and closing up the pinhole. This eliminates the inelegant nuisance of having one "special" region with a hole.

---

## 8.7.6   GEOMETRIC DRAWINGS

Geometric drawing of graphs is a topic in computational geometry. Unlike ordinarily planarity and topological graph theory, its concerns include the exact coordinates in the plane of the images of the vertices and the edges.

**Definitions:**

A **straight-line drawing** of a graph is a drawing in which each edge is represented by a single straight line segment.

An **orthogonal drawing** of a graph is a drawing in which each edge is represented by a chain of horizontal and vertical line segments.

A **polyline drawing** of a graph is a drawing in which each edge is represented by a polygonal path, that is, by a chain of line segments with arbitrary slope.

A **bend** in a polyline drawing is a junction point of two line segments belonging to the same edge.

A **grid drawing** of a graph is a polyline drawing in which vertices, crossings, and bends have integer coordinates.

The **area of a graph drawing** is the area of the convex hull of the drawing.

A **distance-ranked partition** of a graph $G$ with respect to a nonempty vertex subset $S$ has cells $C_j$ for $j = 0, 1, \ldots$. Vertex $v$ is in cell $C_j$ if and only if its shortest path to every vertex of $S$ has length $j$.

A **distance-ranked drawing** of a graph $G$ with respect to a nonempty vertex subset $S$ has the cells of its distance-ranked partition organized into columns from left to right according to ascending distance from $S$.

**Facts:**

**1.** Straight-line and orthogonal drawings are special cases of polyline drawings.

**2.** Polyline drawings can approximate drawings with curved edges.

**3.** Computer systems that support general polyline drawings are more complicated than systems that support only straight-line drawings.

**4.** Many graph drawing problems involve a trade-off between competing objectives, such as the desire to minimize both the area and the number of edge-crossings.

**5.** The area required for a planar polyline grid drawing of an $n$-vertex planar graph is $O(n^2)$.

**6.** The area required for a planar orthogonal grid drawing of an $n$-vertex planar graph is $O(n^2)$.

**7.** The area required for a planar straight line grid drawing of an $n$-vertex planar graph is $O(n^2)$.

**8.** Every planar graph of maximum degree 4 has an orthogonal planar drawing whose total number of bends is at most $2n + 2$.

**9.** Every planar graph of maximum degree 4 has an orthogonal planar drawing such that the maximum number of bends in an edge is at most 2.

**Examples:**

**1.** The following figure shows a nonplanar straight line drawing of the octahedron $K_{2,2,2}$ and a planar polyline drawing of that same graph.

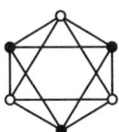 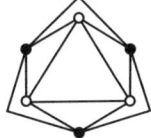

**2.** The following figure shows two orthogonal grid drawings of the octahedron $K_{2,2,2}$. Whereas the lefthand drawing has two edges with three bends, the maximum number of bends in any edge of the middle drawing is two. The righthand drawing has the smallest total number of bends and the smallest area of the three drawings.

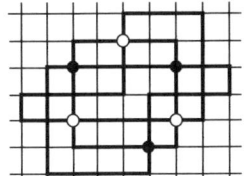 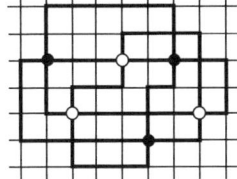 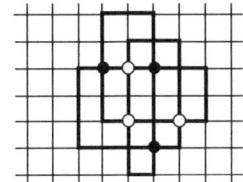

**3.** The following figure shows a distance-ranked drawing of the cube graph $Q_3$ with respect to the vertex 000.

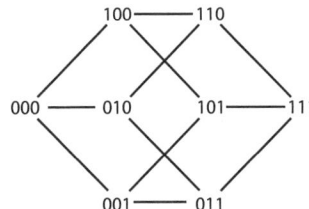

# 8.8 TOPOLOGICAL GRAPH THEORY

Topological graph theory mainly involves placing graphs on closed surfaces. Special emphasis is given to placements that are minimum with respect to some kind of cost or that are highly symmetric. Minimization helps to control the cost of manufacturing networks, and symmetry facilitates the task of routing information through a network.

## 8.8.1 CLOSED SURFACES

Holes in a surface can be closed off by operations like stereographic projection (see §8.7.5). This enables topological graph theory to focus on drawings in closed surfaces.

**Definitions:**

*Adding a handle* to a surface is accomplished in two steps (illustrated in Example 1): (a) punch two disk-like holes into the surface; and (b) reclose the surface by installing a tube that runs from one hole to the other.

An *orientable surface* is defined recursively to be either the sphere $S_0$, or a surface that is obtained from an orientable surface by adding a handle. (See Example 2 for the construction.)

The *genus of an orientable surface* is the number of handles one must add to the sphere to obtain it. Thus, the surface obtained by adding $g$ handles to $S_0$ has genus $g$. It is denoted $S_g$.

The *torus* is the surface $S_1$ of genus 1.

A *Möbius band* is the surface obtained by pasting the left side of a rectangular sheet to the right with a half-twist. A paper ring with a half-twist is a commonplace model of the Möbius band. (See Example 3.)

*Adding a crosscap* to a surface is accomplished by the following two steps: (a) punch one disk-like hole into the surface; and (b) reclose the hole by matching its boundary to the boundary of a Möbius band.

The *nonorientable surface* $N_k$ is obtained by adding $k$ crosscaps to the sphere. The sphere is sometimes regarded as the "surface with crosscap number 0" and denoted $N_0$, even though it is orientable. (See Example 4.)

The subscript $k$ is called the *crosscap number* of the surface $N_k$.

The surfaces $N_1$ and $N_2$ are called the *projective plane* and the *Klein bottle*, respectively.

**Facts:**

**1.** *Classification of closed surfaces*: Every closed surface is equivalent to exactly one of the surfaces $S_g$ $(g \geq 0)$ or $N_k$ $(k \geq 1)$.

**2.** Adding a handle to the nonorientable surface $N_k$ is equivalent to adding two crosscaps. That is, the resulting surface is $N_{k+2}$.

**3.** If a loop is drawn around each handle of $S_g$ and if these $g$ loops are then cut open, the result is a (non-closed) surface that can be stretched and flattened out into a subset of the plane.

**4.** The subscript $g$ equals the maximum number of closed curves on $S_g$ that can be cut open without disconnecting that surface.

**5.** The subscript $k$ equals the maximum number of closed curves on $N_k$ that can be cut open without disconnecting that surface.

**6.** No closed nonorientable surface can be imbedded in $\mathcal{R}^3$.

**7.** *Network layouts:*  The surfaces actually used for computer interconnection network layouts and other practical purposes rarely have graceful curved shapes, because among other reasons, that would obstruct miniaturization and ease of manufacture. Moreover, such surfaces usually have holes. However, the classification theorem and the closing of holes reduce the topology of the layout problems to placing graphs on closed surfaces.

### Examples:

**1.**  Adding a handle is achieved by punching two holes and connecting them with a tube, as illustrated.

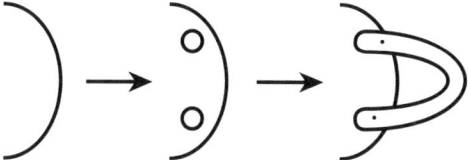

**2.**  To construct the sequence of all orientable surfaces from the sphere $S_0$, each successive handle is added at the right of the previous surface.

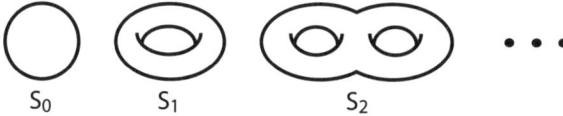

**3.**  The Möbius band is constructed by giving a half-twist to a rectangular strip and then pasting the ends together.

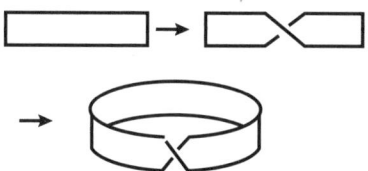

**4.**  To construct the sequence of all nonorientable surfaces from the projective plane $N_1$, each successive crosscap is added by cutting a hole in the previous surface and capping it with a Möbius band.

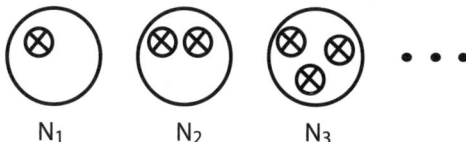

## 8.8.2   DRAWING GRAPHS ON SURFACES

### Definitions:

A **flat-polygon representation of a surface** $S$ is a drawing of a flat polygon with markings to match the sides into pairs such that when the sides are pasted together as the markings indicate, the resulting surface $S$ is obtained. (Certain special flat-polygon representations are called **fundamental polygon representations**.) (See Example 1.)

An **imbedding** (or **embedding**) of a graph is a drawing with no edge-crossings.

A **face** of an imbedding means a region plus its boundary. The set of all faces is denoted $F$.

The **Euler characteristic of an imbedding** of a graph $G = (V, E)$ is the number $|V| - |E| + |F|$.

A **flat-polygon drawing of a graph** on a surface has some graph edges drawn in two or more segments, so that one segment runs from one endpoint of an edge to a side of the flat polygon and another segment runs from the other endpoint to the corresponding position on the matched side of the flat polygon. Sometimes there are also some interior edge segments running between polygon sides. (Flat-polygon drawings are best used for small graphs.)

**Imbedding modification** (or "surgery") on a surface means adding handles and cross-caps to the surface and then drawing one or more edges that traverse the new handles and crosscaps.

The (Poincarè) **duality construction** (see Example 3) is as follows:

- insert into the interior of each (primal) face $f$ a single dual vertex $f^*$;
- through each primal edge $e$ draw a dual edge $e^*$; if edge $e$ lies on the intersection of two primal faces $f$ and $f'$ (possibly $f = f'$), then the dual edge $e^*$ joins the dual vertices $f^*$ and $f'^*$;
- the **dual graph** is the graph $G^* = (\{ f^* \mid f \in F \}, \{ e^* \mid e \in E \})$;
- the **dual imbedding** is the resulting imbedding $G^* \to S$.

### Facts:

**1.** Every closed surface has a flat-polygon representation. This makes it possible to draw a picture in the plane of any graph imbedding in any surface.

**2.** *Euler polyhedral equation for orientable surfaces:* Let $G = (V, E)$ be a connected graph, cellularly imbedded (see §8.8.3) into the surface $S_g$ with face set $F$. Then

$$|V| - |E| + |F| = 2 - 2g = \chi(S_g).$$

**3.** *Euler polyhedral equation for nonorientable surfaces:* Let $G = (V, E)$ be a connected graph, cellularly imbedded into the surface $N_k$ with face set $F$. Then

$$|V| - |E| + |F| = 2 - k = \chi(N_k).$$

**4.** *Edge-face inequality:* Let $G = (V, E)$ be a simple, connected graph imbedded in a surface with face set $F$. Then

$$2|E| \geq 3|F|.$$

**5.** *Edge-face inequality, strong version:* Let $G = (V, E)$ be a connected graph, but not a tree, imbedded in a surface with face set $F$. Then

$$2|E| \geq \text{girth}(G) \cdot |F|.$$

**6.** The most frequently used method for constructing the imbeddings of a recursively constructable sequence of graphs is with voltage graphs. (See §8.10.3.)

**7.** For each surface $S_g$, there is a finite list $\mathcal{L}_g$ of graphs analogous to the Kuratowski graphs $K_5$ and $K_{3,3}$. That is, a graph can be an imbedding in $S_g$ if and only if it contains no subgraph homeomorphic to a graph in $\mathcal{L}_g$. (Robertson and Seymour, 1990)

**Examples:**

**1.** Flat-polygon representations of the double-torus and the Klein bottle are illustrated in the following diagrams.

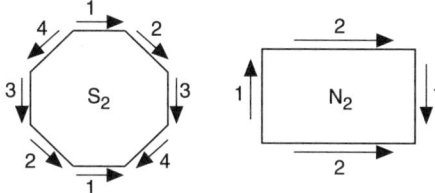

**2.** In the imbedding $K_5 \rightarrow S_1$ illustrated below, edges $c$ and $d$ cross through flat-polygon sides 2 and 1, respectively. The "outer region" is actually 8-sided, with boundary circuit $\langle a, d, b, c, f, d, e, c \rangle$. Two pairs of sides of this region are pasted together. The appearance of this single region as four subregions at the corners of the flat polygon is a side-effect of the particular representation and not a true feature of the imbedding.

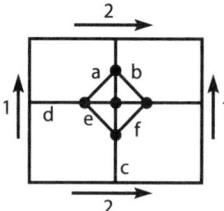

**3.** The Poincare duality construction is illustrated in the following diagram.

## 8.8.3   COMBINATORIAL REPRESENTATION OF GRAPH IMBEDDINGS

### Definitions:

The **rotation** (**in "edge-format"**) at $v$ is obtained from a flat-polygon drawing of a graph by the following sequence of steps: (a) label one end of each edge $+$ and the other end $-$, or put an arrow on each edge so that the head faces toward the $+$ end; and (b) at each vertex, traverse a small circle centered at that vertex, and record the cyclically ordered list of edge-ends encountered; this list is the rotation.

The **vertex-format** of a rotation is obtained by replacing each edge-end in the edge-format by the vertex at the other end of that edge. The vertex format is used only for simple graphs.

A **rotation system** is a complete list of rotations, that is, one for every vertex. If the surface is orientable, it is assumed that the traversals of the small circles around the vertices are in a consistent direction, that is, all clockwise or else all counterclockwise.

An imbedding is **cellular** (or a "2-cell imbedding") if every region is planar and has connected boundary.

### Facts:

**1.** Two cellular imbeddings of a graph are equivalent if and only if they have the same rotation system.

**2.** If a cellular graph imbedding is represented as a rotation system, then the regions can be reconstructed algorithmically.

### Example:

**1.** An imbedding $K_4 \to S_1$ and both formats of its rotation system are shown next.

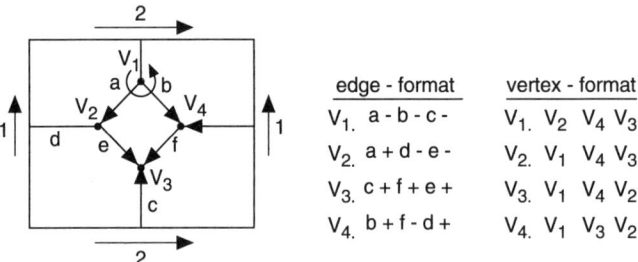

| edge - format | vertex - format |
|---|---|
| $V_1$. a - b - c - | $V_1$. $V_2$ $V_4$ $V_3$ |
| $V_2$. a + d - e - | $V_2$. $V_1$ $V_4$ $V_3$ |
| $V_3$. c + f + e + | $V_3$. $V_1$ $V_4$ $V_2$ |
| $V_4$. b + f - d + | $V_4$. $V_1$ $V_3$ $V_2$ |

## 8.8.4   GENUS AND CROSSCAP NUMBER

### Definitions:

The **minimum genus** (often, simply **genus**) $\gamma_{\min}(G)$ of a connected graph $G$ is the minimum integer $g$ such that there is an imbedding of $G$ into the surface $S_g$.

The **maximum genus** $\gamma_{\max}(G)$ of a connected graph $G$ is the maximum integer $g$ such that there is a cellular imbedding of $G$ into the surface $S_g$.

The **minimum crosscap number** (often, simply **crosscap number**) $\overline{\gamma}_{\min}(G)$ is the minimum integer $k$ such that there is an imbedding of $G$ into $N_k$. Thus, a planar graph has minimum crosscap number zero.

The **maximum crosscap number** $\overline{\gamma}_{\max}(G)$ is the maximum integer $k$ such that there is a cellular imbedding of $G$ into $N_k$. Thus, a planar graph has maximum crosscap number zero.

**Facts:**

**1.** The genus of any planar graph is 0.

**2.** $\gamma_{\min}(G) \geq \dfrac{|E| - 3|V| + 6}{6}$  if $G$ is simple.

**3.** $\gamma_{\min}(G) \geq \dfrac{|E| - 2|V| + 4}{4}$  if $G$ is simple and bipartite.

**4.** $\gamma_{\min}(K_n) = \left\lceil \dfrac{(n-3)(n-4)}{12} \right\rceil$.   (Ringel and Youngs, 1968)

**5.** $\gamma_{\min}(K_{m,n}) = \left\lceil \dfrac{(m-2)(n-2)}{4} \right\rceil$.   (Ringel, 1965)

**6.** $\gamma_{\min}(Q_n) = \left\lceil \dfrac{(m-2)(n-2)}{4} \right\rceil$.   (Ringel, 1955)

**7.** $\overline{\gamma}_{\min}(G) \geq \left\lceil \dfrac{|E| - 3|V| + 6}{3} \right\rceil$  for every simple graph $G$.

**8.** $\overline{\gamma}_{\min}(G) \geq \left\lceil \dfrac{|E| - 2|V| + 4}{2} \right\rceil$  for every simple bipartite graph $G$.

**9.** $\overline{\gamma}_{\min}(K_n) = \left\lceil \dfrac{(n-3)(n-4)}{6} \right\rceil$,   except that $\overline{\gamma}_{\min}(K_7) = 3$.  (Ringel, 1959)

**10.** Many genus and crosscap number formulas can be derived by using *voltage graphs* or *current graphs* (see §8.10.4). [GrTu12]

**11.** For any genus $g$ such that $\gamma_{\min}(G) \leq g \leq \gamma_{\max}(G)$, there is at least one imbedding of the graph $G$ in the surface $S_g$.

**12.** For any crosscap number $k$ such that $\overline{\gamma}_{\min}(G) \leq k \leq \overline{\gamma}_{\max}(G)$, there is at least one imbedding of the graph $G$ in the surface $N_k$.

**13.** The problem of calculating the minimum genus of a graph is NP-hard. (Thomassen, 1989)

**14.** The problem of calculating the maximum genus of a graph is solvable in polynomial time. (Furst, Gross, and McGeoch, 1988)

## 8.9   ENUMERATING GRAPHS

It is often important to know how many graphs there are with some desired property; for example, this is used in computer science to analyze the requirements of algorithms, and in chemistry to catalogue chemical molecules with various shapes. Many of the techniques for counting graphs are based on the master theorem in the historic 1937 work of George Pólya [PóRe87]. Frank Harary and others exploited this master theorem in counting graphs, multigraphs, digraphs, and similar graphical structures. An exhaustive survey of results in graph enumeration can be found in [HaPa73]. Alternatively, if you know the first few terms of a graph-counting sequence, you can likely find more terms, references, and further information in the *On-Line Encyclopedia of Integer Sequences* [OEIS].

### 8.9.1  COUNTING LABELED GRAPHS AND MULTIGRAPHS

When counting graphs, it is important to distinguish between the enumeration of labeled graphs and that of unlabeled graphs. Labeled graphs are relatively easy to count, usually requiring only factorials, exponentials, and binomial coefficients.

**Definitions:**
A *labeled graph* or *labeled multigraph* is a graph or multigraph with standard labels (commonly $v_1, v_2, \ldots, v_n$) assigned to the vertices. Two labeled graphs or multigraphs with the same set of labels are considered the same only if there is an isomorphism from one to the other that preserves the labels.

**Examples:**
**1.** The following figure shows the three isomorphically distinct graphs with 4 vertices and 3 edges. There are 4 essentially different ways to label each of the first two and 12 ways to label the third. Thus there are 20 different labeled graphs with 4 vertices and 3 edges.

**2.** The next figure shows the three isomorphically distinct loopless multigraphs that together with the graphs in the previous figure form the six different multigraphs with 4 vertices and 3 edges. There are 6 essentially different ways to label the first and third graphs in this figure and 24 ways to label the middle graph. Thus the multigraphs in these two figures represent the total of 56 labeled loopless multigraphs with 4 vertices and 3 edges.

**Facts:**
**1.** The number of labeled graphs with $n$ vertices and $m$ edges is the binomial coefficient $\binom{\binom{n}{2}}{m}$. These numbers form sequence A084546 in [OEIS]. See Table 1.

**Table 1: Labeled graphs with $n$ vertices and $m$ edges.**

| $m \backslash n$ | 1 | 2 | 3 | 4 | 5 | 6 | 7 | 8 |
|---|---|---|---|---|---|---|---|---|
| 0 | 1 | 1 | 1 | 1 | 1 | 1 | 1 | 1 |
| 1 | | 1 | 3 | 6 | 10 | 15 | 21 | 28 |
| 2 | | | 3 | 15 | 45 | 105 | 210 | 378 |
| 3 | | | 1 | 20 | 120 | 455 | 1,330 | 3,276 |
| 4 | | | | 15 | 210 | 1,365 | 5,985 | 20,475 |
| 5 | | | | 6 | 252 | 3,003 | 20,349 | 98,280 |
| 6 | | | | 1 | 210 | 5,005 | 54,264 | 376,740 |
| 7 | | | | | 120 | 6,435 | 116,280 | 1,184,040 |
| 8 | | | | | 45 | 6,435 | 203,490 | 3,108,105 |
| 9 | | | | | 10 | 5,005 | 293,930 | 6,906,900 |
| 10 | | | | | 1 | 3,003 | 352,716 | 13,123,110 |
| 11 | | | | | | 1,365 | 352,716 | 21,474,180 |
| 12 | | | | | | 455 | 293,930 | 30,421,755 |
| 13 | | | | | | 105 | 203,490 | 37,442,160 |
| 14 | | | | | | 15 | 116,280 | 40,116,600 |
| total | 1 | 2 | 8 | 64 | 1,024 | 32,768 | 2,097,152 | 268,435,456 |

**2.** For $m > \binom{n}{2}/2$, the number of labeled graphs with $n$ vertices and $m$ edges is the same as the number of labeled graphs with $n$ vertices and $\binom{n}{2} - m$ edges.

**3.** The total number of labeled graphs with $n$ vertices is $2^{\binom{n}{2}}$. This is sequence A006125 in [OEIS]. See Table 1.

**4.** The number $C_n$ of connected labeled graphs with $n$ vertices can be determined from the following recurrence system.

$$C_1 = 1, \quad \text{and} \quad C_n = 2^{\binom{n}{2}} - \frac{1}{n}\sum_{k=1}^{n-1} k\binom{n}{k}2^{\binom{n-k}{2}}C_k \quad \text{for } n > 1.$$

This is sequence A001187 in [OEIS]. See Table 2.

**Table 2: Connected labeled graphs with $n$ vertices.**

| $n$ | 1 | 2 | 3 | 4 | 5 | 6 | 7 | 8 |
|---|---|---|---|---|---|---|---|---|
| $C_n$ | 1 | 1 | 4 | 38 | 728 | 26,704 | 1,866,256 | 251,548,592 |

**5.** Asymptotically, most labeled graphs are connected. Thus the sequence $C_n$ satisfies

$$C_n \sim 2^{\binom{n}{2}}.$$

**6.** The number of labeled loopless multigraphs with $n$ vertices and $m$ edges is the binomial coefficient $\binom{m+\binom{n}{2}-1}{m}$. When $n = 1$, this expression should be interpreted as 1 when $m = 0$ and 0 otherwise. See Table 3. These numbers form the sequence A098568 in [OEIS].

**Table 3: Labeled loopless multigraphs with $n$ vertices and $m$ edges.**

| $m \backslash n$ | 1 | 2 | 3 | 4 | 5 | 6 | 7 | 8 |
|---|---|---|---|---|---|---|---|---|
| 0 | 1 | 1 | 1 | 1 | 1 | 1 | 1 | 1 |
| 1 | | 1 | 3 | 6 | 10 | 15 | 21 | 28 |
| 2 | | 1 | 6 | 21 | 55 | 120 | 231 | 406 |
| 3 | | 1 | 10 | 56 | 220 | 680 | 1,771 | 4,060 |
| 4 | | 1 | 15 | 126 | 715 | 3,060 | 10,626 | 31,465 |
| 5 | | 1 | 21 | 252 | 2,002 | 11,628 | 53,130 | 201,376 |
| 6 | | 1 | 28 | 462 | 5,005 | 38,760 | 230,230 | 1,107,568 |

## 8.9.2   COUNTING UNLABELED GRAPHS AND MULTIGRAPHS

Unlike the enumeration of labeled graphs, the enumeration of unlabeled graphs requires rather sophisticated counting techniques, often utilizing permutation group theory and generating functions.

**Definitions:**

The **symmetric group** $S_n$ is the group of all $n!$ permutations $\gamma$ acting on the set $X_n = \{1, 2, \ldots, n\}$.

The **order** of a permutation group is the number of permutations it contains. The **degree** of a permutation group is the number of objects being permuted. The symmetric group $S_n$ has order $n!$ and degree $n$.

The **cycle index** $Z(G)$ of a permutation group $G$ of order $m$ and degree $d$ is a polynomial in variables $a_1, a_2, \ldots, a_d$ given by the formula

$$Z(G) = \frac{1}{m} \sum_{\gamma \in G} \prod_{k=1}^{d} a_k^{j_k(\gamma)},$$

where $j_k(\gamma)$ is the number of cycles of length $k$ in the permutation $\gamma$. For example, for $G = S_3 = \{(1)(2)(3), (123), (132), (1)(23), (2)(13), (3)(12)\}$, the symmetric group of order 6 and degree 3, the cycle index is $Z(G_3) = \frac{1}{6}\left(a_1^3 + 2a_3 + 3a_1 a_2\right)$.

The **pair permutation** $\gamma^{(2)}$ induced by the permutation $\gamma$ acting on the set $X_n$ is the permutation acting on unordered pairs of distinct elements of $X_n$ defined by the rule

$$\gamma^{(2)}\left(\{x_1, x_2\}\right) = \{\gamma(x_1), \gamma(x_2)\}.$$

The **symmetric pair group** $S_n^{(2)}$ induced by the symmetric group $S_n$ is the permutation group $\{\gamma^{(2)} \mid \gamma \in S_n\}$. This group has order $n!$ and degree $n(n-1)/2$.

**Facts:**

**1.** The cycle index $Z(S_n^{(2)})$ of the symmetric pair group, used in counting graphs with $n$ vertices, is

$$Z\left(S_n^{(2)}\right) = \frac{1}{n!} \sum_{(j)} \frac{n!}{\prod_k k^{j_k} j_k!} \prod_k a_k^{k\binom{j_k}{2}} \left(a_k a_{2k}^{k-1}\right)^{j_{2k}} a_{2k+1}^{k j_{2k+1}} \prod_{r<s} a_{\operatorname{lcm}(r,s)}^{\gcd(r,s) j_r j_s},$$

where $\mathrm{lcm}(r, s)$ and $\gcd(r, s)$ are the least common multiple and greatest common divisor of $r$ and $s$, respectively. The sum is taken over all partitions $(j) = j_1, j_2, \ldots, j_n$ of the integer $n$ as an unordered sum of parts, where $j_k$ is the number of parts of size $k$. For example, in the partition of 7 as $2 + 2 + 3$ we have $j_2 = 2$, $j_3 = 1$, and $j_1 = j_4 = j_5 = j_6 = j_7 = 0$. Explicit formulas for $Z(S_n^{(2)})$ for small values of $n$ are

$$Z\!\left(S_1^{(2)}\right) = 1$$

$$Z\!\left(S_2^{(2)}\right) = a_1$$

$$Z\!\left(S_3^{(2)}\right) = \frac{1}{3!}\left(a_1^3 + 3a_1 a_2 + 2a_3\right)$$

$$Z\!\left(S_4^{(2)}\right) = \frac{1}{4!}\left(a_1^6 + 9a_1^2 a_2^2 + 8a_3^2 + 6a_2 a_4\right)$$

$$Z\!\left(S_5^{(2)}\right) = \frac{1}{5!}\left(a_1^{10} + 10a_1^4 a_2^3 + 20a_1 a_3^3 + 15a_1^2 a_2^4 + 30a_2 a_4^2 + 20a_1 a_3 a_6 + 24a_5^2\right)$$

$$Z\!\left(S_6^{(2)}\right) = \frac{1}{6!}\left(a_1^{15} + 15a_1^7 a_2^4 + 40a_1^3 a_3^4 + 60a_1^3 a_2^6 + 180a_1 a_2 a_4^3 + 120a_1 a_2 a_3^2 a_6\right.$$
$$\left. + 144a_5^3 + 40a_3^5 + 120a_3 a_6^2\right)$$

**2.** Let $G_{n,m}$ denote the number of graphs with $n$ vertices and $m$ edges, and let $g_n(x)$ be the generating function for $n$-vertex graphs, so that

$$g_n(x) = \sum_{m=0}^{\binom{n}{2}} G_{n,m} x^m.$$

Pólya's enumeration theorem states that this generating function $g_n(x)$ can be obtained from the cycle index $Z(S_n^{(2)})$ by replacing each variable $a_i$ with $1 + x^i$. See Table 4. These numbers form sequence A008406 in [OEIS].

**3.** For $m > \binom{n}{2}/2$, the number of graphs with $n$ vertices and $m$ edges is the same as the number of graphs with $n$ vertices and $\binom{n}{2} - m$ edges.

**4.** The total number $G_n$ of graphs with $n$ vertices is obtained from the cycle index $Z(S_n^{(2)})$ by replacing each variable $a_i$ with the number 2. See Table 4. This is sequence A000088 in [OEIS].

**5.** Asymptotically, the sequence $G_n$ satisfies $G_n \sim 2^{\binom{n}{2}}/n!$.

**Table 4: Graphs with $n$ vertices and $m$ edges.**

| $m \backslash n$ | 1 | 2 | 3 | 4 | 5 | 6 | 7 | 8 |
|---|---|---|---|---|---|---|---|---|
| 0 | 1 | 1 | 1 | 1 | 1 | 1 | 1 | 1 |
| 1 | | 1 | 1 | 1 | 1 | 1 | 1 | 1 |
| 2 | | | 1 | 2 | 2 | 2 | 2 | 2 |
| 3 | | | 1 | 3 | 4 | 5 | 5 | 5 |
| 4 | | | | 2 | 6 | 9 | 10 | 11 |
| 5 | | | | 1 | 6 | 15 | 21 | 24 |
| 6 | | | | 1 | 6 | 21 | 41 | 56 |
| 7 | | | | | 4 | 24 | 65 | 115 |

| $m \backslash n$ | 1 | 2 | 3 | 4 | 5 | 6 | 7 | 8 |
|---|---|---|---|---|---|---|---|---|
| 8 | | | | | 2 | 24 | 97 | 221 |
| 9 | | | | | 1 | 21 | 131 | 402 |
| 10 | | | | | 1 | 15 | 148 | 663 |
| 11 | | | | | | 9 | 148 | 980 |
| 12 | | | | | | 5 | 131 | 1,312 |
| 13 | | | | | | 2 | 97 | 1,557 |
| 14 | | | | | | 1 | 65 | 1,646 |
| Total | 1 | 2 | 4 | 11 | 34 | 156 | 1,044 | 12,346 |

**6.** The enumeration of connected graphs requires an auxiliary sequence $A_n$ defined recursively by

$$A_1 = 1, \quad \text{and} \quad A_n = nG_n - \sum_{k=1}^{n-1} A_k \cdot G_{n-k} \quad \text{for } n > 1.$$

This sequence 1, 3 ,7, 27, 106, 681, 5972, 88963, ... is sequence A003083 in [OEIS]. The number $K_n$ of connected graphs with $n$ vertices can then be computed as

$$K_n = \frac{1}{n} \sum_{d|n} \mu(d) A_{n/d},$$

where the sum is over all divisors of $n$ and $\mu$ is the Möbius function defined by

$$\mu(n) = \begin{cases} 1 & \text{if } n = 0 \\ 0 & \text{if } m^2 | n \text{ for some } m > 1 \\ (-1)^k & \text{if } n \text{ is the product of } k \text{ distinct primes.} \end{cases}$$

See Table 5. The sequence $K_n$ is sequence A001349 in [OEIS].

**Table 5: Connected graphs with $n$ vertices.**

| $n$ | 1 | 2 | 3 | 4 | 5 | 6 | 7 | 8 |
|---|---|---|---|---|---|---|---|---|
| $K_n$ | 1 | 1 | 2 | 6 | 21 | 112 | 853 | 11,117 |

**7.** Asymptotically, most graphs are connected. Thus the sequence $K_n$ satisfies $K_n \sim 2^{\binom{n}{2}}/n!$.

**8.** Let $M_{n,k}$ denote the number of loopless multigraphs with $n$ vertices and $k$ edges, and let $m_n(x)$ be the generating function for $n$-vertex loopless multigraphs, so that

$$m_n(x) = \sum_{m=0}^{\binom{n}{2}} M_{n,k} x^k.$$

Pólya's enumeration theorem states that this generating function $m_n(x)$ can be obtained from the cycle index $Z(S_n^{(2)})$ by replacing each variable $a_i$ with the infinite series $1 + x^i + x^{2i} + x^{3i} + \cdots$. See Table 6. These numbers form sequence A192517 in [OEIS]. (Also, column $n = 3$ is sequence A001399, column $n = 4$ is sequence A003082, column $n = 5$ is sequence A014395, and column $n = 6$ is sequence A014396.)

**Table 6: Loopless multigraphs with $n$ vertices and $m$ edges.**

| $m \backslash n$ | 1 | 2 | 3 | 4 | 5 | 6 |
|---|---|---|---|---|---|---|
| 0 | 1 | 1 | 1 | 1 | 1 | 1 |
| 1 |  | 1 | 1 | 1 | 1 | 1 |
| 2 |  | 1 | 2 | 3 | 3 | 3 |
| 3 |  | 1 | 3 | 6 | 7 | 8 |
| 4 |  | 1 | 4 | 11 | 17 | 21 |
| 5 |  | 1 | 5 | 18 | 35 | 52 |
| 6 |  | 1 | 7 | 32 | 76 | 132 |
| 7 |  | 1 | 8 | 48 | 149 | 313 |
| 8 |  | 1 | 10 | 75 | 291 | 741 |
| 9 |  | 1 | 12 | 111 | 539 | 1,684 |
| 10 |  | 1 | 14 | 160 | 974 | 3,711 |

## 8.9.3  COUNTING LABELED DIGRAPHS AND TOURNAMENTS

### Definitions:

A **labeled digraph** is a digraph with distinct labels, typically $v_1, v_2, \ldots, v_n$, assigned to its vertices. Two labeled digraphs with the same set of labels are considered the same only if there is an isomorphism from one to the other that preserves the labels.

A **tournament** (or **round-robin tournament**) is a digraph in which, for each pair $u$, $v$ of distinct vertices, either there exists an arc from $u$ to $v$ or an arc from $v$ to $u$ but not both.

A digraph is **strong** (or **strongly connected**) if for each pair $u$, $v$ of vertices, there exist directed paths from $u$ to $v$ and from $v$ to $u$. A strong tournament is also called an **irreducible** tournament.

### Examples:

**1.** The following figure shows the four isomorphically distinct digraphs with 3 vertices and 3 arcs. The last two are tournaments. There are 6 essentially different ways to label each of the first three digraphs and 2 ways to label the fourth. Thus there are 20 different labeled digraphs with 3 vertices and 3 arcs. Only the last digraph is strong—an irreducible tournament.

**2.** The next figure shows the four isomorphically distinct tournaments with 4 vertices. There are 24 essentially different ways to label the first and last tournaments, and 8 ways to label each of the middle two. Thus there are 64 different labeled tournaments with 4 vertices. Only the last tournament is strong.

**Facts:**

**1.** The number of labeled digraphs with $n$ vertices and $m$ arcs is the binomial coefficient $\binom{n(n-1)}{m}$. See Table 7. These numbers form sequence A123554 in [OEIS].

**Table 7: Labeled digraphs with $n$ vertices and $m$ arcs.**

| $m \backslash n$ | 1 | 2 | 3 | 4 | 5 |
|---|---|---|---|---|---|
| 0 | 1 | 1 | 1 | 1 | 1 |
| 1 | | 2 | 6 | 12 | 20 |
| 2 | | 1 | 15 | 66 | 190 |
| 3 | | | 20 | 220 | 1,140 |
| 4 | | | 15 | 495 | 4,845 |
| 5 | | | 6 | 792 | 15,504 |
| 6 | | | 1 | 924 | 38,760 |
| 7 | | | | 792 | 77,520 |
| 8 | | | | 495 | 125,970 |
| 9 | | | | 220 | 167,960 |
| 10 | | | | 66 | 184,756 |
| Total | 1 | 4 | 64 | 4,096 | 1,048,576 |

**2.** For $m > n(n-1)/2$, the number of labeled digraphs with $n$ vertices and $m$ arcs is the same as the number of labeled digraphs with $n$ vertices and $n(n-1) - m$ arcs.

**3.** The total number of labeled digraphs with $n$ vertices is $2^{n(n-1)}$. See Table 7. This is sequence A053763 in [OEIS].

**4.** The number of labeled tournaments with $n$ vertices is $2^{\binom{n}{2}}$, the same as the number of labeled graphs with $n$ vertices. See Table 8. This is sequence A006125 in [OEIS].

**5.** The number $\widehat{S_n}$ of strong labeled tournaments with $n$ vertices can be computed from the recursive formula

$$\widehat{S_1} = 1, \quad \text{and} \quad \widehat{S_n} = 2^{\binom{n}{2}} - \sum_{k=1}^{n-1} \binom{n}{k} 2^{\binom{n-k}{2}} \widehat{S_k} \quad \text{for } n > 1.$$

See Table 8. This is sequence A054946 in [OEIS].

**Table 8: Labeled tournaments and strong tournaments with $n$ vertices.**

| $n$ | Labeled Tournaments | Strong Labeled Tournaments |
|---|---|---|
| 1 | 1 | 1 |
| 2 | 2 | 0 |
| 3 | 8 | 2 |
| 4 | 64 | 24 |
| 5 | 1,024 | 544 |
| 6 | 32,768 | 22,320 |
| 7 | 2,097,152 | 1,677,488 |
| 8 | 268,435,456 | 236,522,496 |

**6.** Asymptotically, most labeled tournaments are strong. Thus the sequence $\widehat{S_n}$ counting strong labeled tournaments satisfies $\widehat{S_n} \sim 2^{\binom{n}{2}}$.

## 8.9.4  COUNTING UNLABELED DIGRAPHS AND TOURNAMENTS

As with unlabeled graphs, the enumeration of unlabeled digraphs and tournaments requires rather sophisticated counting techniques, often utilizing permutation group theory and generating functions. (See §8.9.2.)

**Definitions:**
The **ordered pair permutation** $\gamma^{[2]}$ induced by the permutation $\gamma$ acting on the set $X_n = \{1, 2, \ldots, n\}$ is the permutation acting on ordered pairs of distinct elements of $X_n$ defined by the rule

$$\gamma^{[2]} \left( (x_1, x_2) \right) = (\gamma(x_1), \gamma(x_2)) .$$

The **reduced ordered pair group** $S_n^{[2]}$ induced by the symmetric group $S_n$ is the permutation group $\{\gamma^{[2]} \mid \gamma \in S_n\}$. This group has order $n!$ and degree $n(n-1)$.

**Facts:**
**1.** The cycle index $Z(S_n^{[2]})$ of the reduced ordered pair group, used in counting digraphs with $n$ vertices, is

$$Z(S_n^{[2]}) = \frac{1}{n!} \sum_{(j)} \frac{n!}{\prod_k k^{j_k} j_k!} \prod_k a_k^{(k-1)j_k + 2k\binom{j_k}{2}} \prod_{r<s} a_{\mathrm{lcm}(r,s)}^{2\gcd(r,s)j_r j_s} ,$$

where $\mathrm{lcm}(r, s)$ and $\gcd(r, s)$ are the least common multiple and greatest common divisor of $r$ and $s$, respectively. The sum is taken over all partitions $(j) = j_1, j_2, \ldots, j_n$ of the integer $n$ as an unordered sum of parts, where $j_k$ is the number of parts of size $k$. (See §2.5 for a discussion of partitions.) Explicit formulas for $Z(S_n^{[2]})$ for small values of $n$ are

$$Z\big(S_1^{[2]}\big) = 1$$

$$Z\big(S_2^{[2]}\big) = \frac{1}{2!}\big(a_1^2 + a_2\big)$$

$$Z\big(S_3^{[2]}\big) = \frac{1}{3!}\big(a_1^6 + 3a_2^3 + 2a_3^2\big)$$

$$Z\big(S_4^{[2]}\big) = \frac{1}{4!}\big(a_1^{12} + 6a_1^2 a_2^5 + 8a_3^4 + 3a_2^6 + 6a_4^3\big)$$

$$Z\big(S_5^{[2]}\big) = \frac{1}{5!}\big(a_1^{20} + 10a_1^6 a_2^7 + 20a_1^2 a_3^6 + 15a_2^{10} + 30a_4^5 + 20a_2 a_3^2 a_6^2 + 24a_5^4\big)$$

$$Z\big(S_6^{[2]}\big) = \frac{1}{6!}\big(a_1^{30} + 15a_1^{12}a_2^9 + 40a_1^6 a_3^8 + 45a_1^2 a_2^{14} + 90a_1^2 a_4^7 + 120a_2^3 a_3^4 a_6^2$$
$$+ 144a_5^6 + 15a_2^{15} + 90a_2 a_4^7 + 40a_3^{10} + 120a_6^5\big).$$

**2.** Let $D_{n,m}$ denote the number of digraphs with $n$ vertices and $m$ arcs, and let $d_n(x)$ be the generating function for $n$-vertex digraphs, so that

$$d_n(x) = \sum_{m=0}^{n(n-1)} D_{n,m} x^m.$$

Pölya's enumeration theorem states that this generating function $d_n(x)$ can be obtained from the cycle index $Z(S_n^{[2]})$ by replacing each variable $a_i$ with $1+x^i$. See Table 9. These numbers form sequence A052283 in [OEIS].

**Table 9: Digraphs with $n$ vertices and $m$ arcs.**

| $m \backslash n$ | 1 | 2 | 3 | 4 | 5 |
|---|---|---|---|---|---|
| 0 | 1 | 1 | 1 | 1 | 1 |
| 1 |  | 1 | 1 | 1 | 1 |
| 2 |  | 1 | 4 | 5 | 5 |
| 3 |  |  | 4 | 13 | 16 |
| 4 |  |  | 4 | 27 | 61 |
| 5 |  |  | 1 | 38 | 154 |
| 6 |  |  | 1 | 48 | 379 |
| 7 |  |  |  | 38 | 707 |
| 8 |  |  |  | 27 | 1,155 |
| 9 |  |  |  | 13 | 1,490 |
| 10 |  |  |  | 5 | 1,670 |
| Total | 1 | 3 | 16 | 218 | 9,608 |

**3.** For $m > n(n-1)/2$, the number of digraphs with $n$ vertices and $m$ arcs is the same as the number of digraphs with $n$ vertices and $n(n-1) - m$ arcs.

**4.** The total number $D_n$ of digraphs with $n$ vertices is obtained from the cycle index $Z(S_n^{[2]})$ by replacing each variable $a_i$ with the number 2. See Table 9. This is sequence A000273 in [OEIS].

**5.** Asymptotically, the sequence $D_n$ satisfies $D_n \sim 2^{n(n-1)}/n!$.

**6.** The number $T_n$ of tournaments with $n$ vertices is given by the formula

$$T_n = \frac{1}{n!} \sum_{(j)}' \frac{n!}{\prod_k k^{j_k} j_k!} 2^{D(j)},$$

where the sum is over all partitions $(j)$ of $n$ into odd size parts, and where

$$D(j) = \frac{1}{2}\left(\sum_{r=1}^n \sum_{s=1}^n \gcd(r,s)j_r j_s - \sum_{k=1}^n j_k\right).$$

See Table 10. This is sequence A000568 in [OEIS].

**7.** The number $S_n$ of strong tournaments with $n$ vertices can be determined by the recurrence relation

$$S_1 = 1, \quad \text{and} \quad S_n = T_n - \sum_{k=1}^{n-1} T_{n-k}S_k \quad \text{for } n > 1,$$

where $T_n$ is the number of tournaments from Fact 6 above. See Table 10. Note that there are no strong tournaments with exactly two vertices. This is sequence A051337 in [OEIS].

**8.** The sequences $T_n$ and $S_n$ satisfy $T_n \sim S_n \sim 2^{\binom{n}{2}}/n!$.

**Table 10: Tournaments and strong tournaments with $n$ vertices.**

| $n$ | Tournaments | Strong Tournaments |
|---|---|---|
| 1 | 1 | 1 |
| 2 | 1 | 0 |
| 3 | 2 | 1 |
| 4 | 4 | 1 |
| 5 | 12 | 6 |
| 6 | 56 | 35 |
| 7 | 456 | 353 |
| 8 | 6,880 | 6,008 |
| 9 | 191,536 | 178,133 |
| 10 | 9,733,056 | 9,355,949 |
| 11 | 903,753,248 | 884,464,590 |
| 12 | 154,108,311,168 | 152,310,149,735 |

# 8.10   GRAPH FAMILIES

This section discusses some graph families and the theory behind them: perfect graphs, eigenvalues, algebraically-defined graphs, expander graphs, and product graphs.

## 8.10.1   PERFECT GRAPHS

The class of perfect graphs was introduced by Claude Berge in 1961. Around the same time Berge made two conjectures about this class of graphs that motivated a great deal of research in the field. Both conjectures have by now been solved, and are known as the Weak and Strong Perfect Graph Theorems. Below is a brief discussion of perfect graphs and of some recent developments on the topic. For more information see the appropriate section in [GrYeZh15].

**Definitions:**

An ***induced subgraph*** of $G$ is a subgraph $H$ such that every edge of $G$ that joins two vertices of $H$ is also an edge of $H$.

A graph $G$ is called ***perfect*** if for every induced subgraph $H$ of $G$, $\chi(H) = \omega(H)$ (where $\chi$ denotes the chromatic number and $\omega$ the clique number; see §8.6.3).

A ***hole*** in a graph is an induced subgraph that is isomorphic to the cycle $C_k$ with $k \geq 4$, and $k$ is the ***length*** of the hole. A hole is ***odd*** if $k$ is odd, and ***even*** otherwise.

An ***antihole*** in a graph is an induced subgraph whose complement is isomorphic to $C_k$ with $k \geq 4$, and $k$ is the ***length*** of the antihole. An antihole is ***odd*** if $k$ is odd, and ***even*** otherwise.

A graph is called ***Berge*** if it has no odd hole and no odd antihole.

A Berge graph $G$ is called **basic** if: (a) $G$ or its complement is bipartite; (b) $G$ or its complement is the line graph of a bipartite graph; or (c) $G$ is a double split graph.

An **even pair** is a pair of vertices $\{u, v\}$ such that every induced path from $u$ to $v$ has even length (in particular, $v$ and $v$ are nonadjacent). The operation of **contracting even pair** $\{u, v\}$ means replacing $\{u, v\}$ by a vertex $w$ such that $w$ is adjacent precisely to all vertices that were adjacent to at least one of $u$ or $v$.

A graph $G$ is called **even contractile** if one can repeatedly contract an even pair and end up with the complete graph on $\omega(G)$ vertices.

A **prism** is an induced subgraph consisting of two triangles $\{a_1, a_2, a_3\}$ and $\{b_1, b_2, b_3\}$ and three disjoint paths $P_1, P_2, P_3$, where $P_i$ is from $a_i$ to $b_i$, and for $1 \leq i < j \leq 3$ the only edges between $V(P_i)$ and $V(P_j)$ are $a_i a_j$ and $b_i b_j$. A prism is **odd** if $P_1, P_2, P_3$ all have odd length, and **even** if $P_1, P_2, P_3$ all have even length.

**Results:**

**1.** An odd hole has $\omega = 2$ and $\chi = 3$; an odd antihole of length $k$ has $\omega = (k-1)/2$ and $\chi = (k+1)/2$. Therefore a perfect graph cannot contain either.

**2.** *Weak Perfect Graph Theorem*: A graph is perfect if and only if its complement is perfect. [Lo72]

**3.** *Strong Perfect Graph Theorem*: A graph is perfect if and only if it has no odd hole and no odd antihole. [ChEtal06]

**4.** The strong theorem implies the weak.

**5.** There is a polynomial-time algorithm to test if an input graph is perfect.

**6.** If a graph is restricted to being perfect, then some algorithmic problems, such as calculating the size of the largest clique or finding an optimal coloring, that are NP-complete in general, can be solved in polynomial time.

**7.** If $G$ is Berge and $G'$ is formed by contracting an even pair, then $G'$ is Berge and $\omega(G') = \omega(G)$.

**8.** If a prism is Berge, then either it is an odd prism or an even prism.

**9.** It is conjectured that if a Berge graph has no odd prism and no antihole of length at least six, then it is even contractile.

**10.** It is known that if a Berge graph has no prism and no antihole of length at least six, then it is even contractile. Also, if a Berge graph has no odd prism and no hole of length four, then it has an even pair.

## 8.10.2  SPECTRAL GRAPH THEORY

In this subsection we consider only undirected graphs.

**Definitions:**

The **characteristic polynomial of a graph** $G$ is the characteristic polynomial $p(x)$ of its **adjacency matrix** $A_G$; that is, $p(x) = \det(xI - A_G)$.

An **eigenvalue** (or **characteristic value**) **of a matrix** $A$ is a number $\lambda$ such that $Ax = \lambda x$, for some nonzero vector $x$; the vector $x$ is an **eigenvector** (or **characteristic vector**).

An **eigenvector of a graph** is an eigenvector of its adjacency matrix, and an **eigenvalue of a graph** an eigenvalue of its adjacency matrix.

The **spectrum of a graph** is the spectrum of its adjacency matrix; that is, the multiset of eigenvalues.

The **Laplacian** (or **admittance matrix**) **of a graph** $G$ is the matrix $D_G - A_G$, where $D_G$ is the diagonal matrix with the degree sequence of $G$ on the diagonal and $A_G$ is the adjacency matrix.

A connected graph $G$ is **strongly regular with parameters** $(n, k, r, s)$ if

- $|V_G| = n$;
- $G$ is $k$-regular, with $k > 0$;
- every pair of adjacent vertices is mutually adjacent to $r$ other vertices;
- every pair of nonadjacent vertices is mutually adjacent to $s$ other vertices.

The **Hoffman polynomial of a graph** is a polynomial $p(x)$ of minimum degree such that $p(A_G) = J$, where $A_G$ is the adjacency matrix and $J$ is the square matrix with every entry equal to 1.

A **cospectral pair of graphs** is a pair of nonisomorphic graphs that have the same spectrum.

## Facts:

**1.** The eigenvalues of a graph are independent of the particular labeling of the vertices; thus, two isomorphic graphs have the same spectrum.

**2.** All the eigenvalues of a graph are real. This is a special case of the well-known linear algebra result that the eigenvalues of any Hermitian matrix are real.

**3.** From linear algebra, it follows that the characteristic polynomial of a graph $G$ satisfies the equation $p(x) = \prod_{i=1}^{n}(x - \lambda_i)$, where $\lambda_1, \ldots, \lambda_n$ are the eigenvalues of $G$.

**4.** If a graph is connected, then its largest eigenvalue has multiplicity 1. This eigenvalue has a corresponding eigenvector with all positive entries, which is the only such eigenvector.

**5.** If $\lambda$ is the largest eigenvalue of a graph and $\mu$ is another eigenvalue, then $\lambda \geq |\mu|$; moreover, $-\lambda$ is an eigenvalue if and only if the graph is bipartite.

**6.** A graph is bipartite if and only if its spectrum is symmetric with respect to 0; that is, $\lambda$ is an eigenvalue if and only if $-\lambda$ is also an eigenvalue.

**7.** The largest eigenvalue of a $k$-regular graph is $k$, and it has multiplicity equal to the number of connected components. The sum of the coordinates of an eigenvector corresponding to any other eigenvalue is 0.

**8.** The $(i, j)$th entry of the $k$th power $A_G^k$ of the adjacency matrix of a graph $G$ is the number of walks of length $k$ starting at vertex $v_i$ and terminating at $v_j$.

**9.** If $\lambda_1, \lambda_2, \ldots, \lambda_n$ are the eigenvalues of a graph $G$, then $\sum_{i=1}^{n} \lambda_i^2 = 2|E_G|$ where $E_G$ is the edge-set of $G$. Also, $\sum_{i=1}^{n} \lambda_i^3 = 6T$ where $T$ is the number of triangles in $G$.

**10.** If $p(x) = x^n + a_{n-1}x^{n-1} + a_{n-2}x^{n-2} + \cdots + a_1 x + a_0$ is the characteristic polynomial of a graph $G$, then $a_{n-1} = 0$, $-a_{n-2}$ is the number of edges, and $-a_{n-3}$ is twice the number of triangles.

**11.** The set of eigenvalues of the disjoint sum $G + H$ is the union of the sets of eigenvalues of $G$ and $H$. The multiplicity of $\lambda$ as an eigenvalue of $G + H$ is the sum of the multiplicity of $\lambda$ as an eigenvalue of $G$ and the multiplicity of $\lambda$ as an eigenvalue of $H$.

**12.** The eigenvalues of the Cartesian product $G \square H$ are $\{\lambda_i + \lambda_j \mid \lambda_i$ an eigenvalue of $G$ and $\lambda_j$ an eigenvalue of $H\}$. The multiplicity of $\lambda_i + \lambda_j$ as an eigenvalue of $G \square H$ is the product of the multiplicity of $\lambda_i$ as an eigenvalue of $G$ and $\lambda_j$ as an eigenvalue of $H$.

**13.** If $G$ is a $k$-regular graph and $\overline{G}$ its complement, then $\lambda < k$ is an eigenvalue of $G$ if and only if $-\lambda - 1$ is an eigenvalue of $\overline{G}$. In this case $\lambda$ and $-\lambda - 1$ have the same multiplicities.

**14.** If $\lambda$ is an eigenvalue of $G$ with multiplicity $m$, then $-\lambda - 1$ is an eigenvalue of $\overline{G}$ with multiplicity $m - 1$, $m$, or $m + 1$.

**15.** If $G$ has $n$ vertices and $\lambda_1 \geq \lambda_2 \geq \cdots \geq \lambda_n$ as eigenvalues, and $H$ is an induced subgraph with $n - 1$ vertices and eigenvalues $\mu_1 \geq \mu_2 \geq \cdots \geq \mu_{n-1}$, then $\lambda_1 \geq \mu_1 \geq \lambda_2 \geq \mu_2 \geq \cdots \geq \mu_{n-1} \geq \lambda_n$. (See §6.5.5.)

**16.** The eigenvalues of a line graph $L(G)$ are greater than or equal to $-2$. Equality is attained unless every connected component of $G$ is a tree or has exactly one circuit, that circuit being odd.

**17.** If $G$ is a $k$-regular graph with $\lambda > -k$ as an eigenvalue, then $\lambda + k - 2$ is an eigenvalue of $L(G)$.

**18.** A graph has a Hoffman polynomial if and only if it is regular and connected.

**19.** A regular connected graph has exactly three distinct eigenvalues if and only if it is strongly regular.

**20.** *Matrix-tree theorem:* If $M_i$ is formed by deleting the $i$th row and column from the Laplacian of $G$, then $\det(M_i)$ is independent of the choice of $i$ and is equal to the number of spanning trees of $G$.

### Examples:

**1.** The edgeless graph on $n$ vertices has one eigenvalue, namely $0$ with multiplicity $n$.

**2.** The eigenvalues of the complete graph $K_n$ are $n - 1$ and $-1$ with respective multiplicities of $1$ and $n - 1$. For instance, the characteristic polynomial of $K_4$ is

$$\begin{vmatrix} x & -1 & -1 & -1 \\ -1 & x & -1 & -1 \\ -1 & -1 & x & -1 \\ -1 & -1 & -1 & x \end{vmatrix} = (x + 1)^3 (x - 3).$$

**3.** The eigenvalues of the complete bipartite graph $K_{m,n}$ are $\sqrt{mn}$, $0$, and $-\sqrt{mn}$, with respective multiplicities of $1$, $mn - 2$, and $1$.

**4.** The eigenvalues of the Petersen graph are $3$, $1$, and $-2$, with respective multiplicities $1$, $5$, and $4$.

**5.** The eigenvalues of the $n$-path $P_n$ are $\{2 \cos \frac{k\pi}{n+1} \mid k = 1, 2, \ldots, n\}$, each with multiplicity $1$.

**6.** The eigenvalues of the $n$-cycle $C_n$, are $\{2 \cos \frac{2k\pi}{n} \mid k = 1, 2, \ldots, n\}$. The eigenvalue $2$, and the eigenvalue $-2$ when $n$ is even, have multiplicity $1$; all other eigenvalues have multiplicity $2$.

**7.** The eigenvalues of the hypercube $Q_d$ are $d, d-2, d-4, \ldots, -d+2, -d$, with respective multiplicities $\binom{d}{0}, \binom{d}{1}, \binom{d}{2}, \ldots, \binom{d}{d-1}, \binom{d}{d}$.

**8.** The eigenvalues of the line graph $L(K_n)$ are $2n - 4$, $n - 4$, and $-2$, with respective multiplicities $1$, $n - 1$, and $\frac{n(n-3)}{2}$.

**9.** The eigenvalues of the line graph $L(K_{m,n})$ are $m+n-2$, $m-2$, $n-2$, and $-2$, with respective multiplicities $1$, $n-1$, $m-1$, and $(m-1)(n-1)$.

**10.** If $G$ is strongly regular with parameters $(n,k,r,s)$, then its eigenvalues are $k$ and $\frac{1}{2}\left(r-s\pm\sqrt{(r-s)^2-4(s-k)}\right)$.

**11.** The smallest pair of cospectral graphs is $K_{1,4}$ and $C_4+K_1$, both with spectrum $\{-2,0,0,0,2\}$. See the following figure. Observe that $K_{1,4}$ is connected and that $C_4+K_1$ is not, and that the two graphs have different degree sequences. This implies that connectedness and degree sequences cannot be determined from spectral properties alone.

## 8.10.3   ALGEBRAIC GRAPHS AND AUTOMORPHISMS

### Definitions:

A *generating subset for a group* $\Gamma$ is a subset $\Sigma$ of group elements such that every group element is a product of elements of $\Sigma$. (*Note:* The group identity is the empty product.)

The *Cayley digraph for group* $\Gamma$ *and generating set* $\Sigma$ has as vertices the elements of $\Gamma$, with an arc $\sigma_\gamma$ from the vertex $\gamma$ to the vertex $\gamma'$ if and only if $\gamma\sigma=\gamma'$.

If $\Gamma$ is a group and $\Sigma$ a symmetric set (meaning $\gamma\in\Gamma$ if and only if $\gamma^{-1}\in\Gamma$), then the undirected *Cayley graph* for $\Gamma$ and $\Sigma$ is the graph obtained by removing all arc directions from the Cayley digraph, and by collapsing each opposite pairs of arcs to a single edge.

The *Schreier graph* [*Schreier digraph*] for a group $\mathcal{A}$ with generating set $x_1,\ldots,x_r$ and subgroup $\mathcal{B}$ has as its vertices the cosets of $\mathcal{B}$ in $\mathcal{A}$. For each coset $\mathcal{B}b$ and each generator $x_j$, there is an edge between vertices $\mathcal{B}b$ and $\mathcal{B}bx_j$ [from $\mathcal{B}b$ to $\mathcal{B}bx_j$].

An *algebraic specification* of a graph is a generalization of Cayley graphs and Schreier graphs. It uses elements of a group as all or part of the names of the vertices and edges, and the group operation in the incidence rule.

A *voltage graph* is a form of algebraic specification in which the vertices and edges are specified as a set of one or more symbols with subscripts, ranging over group elements. Its usual form is a digraph with vertex labels and edge labels.

The *automorphism group* $Aut(G)$ of a graph $G$ is the set of all automorphisms of graph $G$, under the operation of functional composition. (See §8.5.1.)

A graph is *vertex-transitive* if for all vertices $u$ and $v$ there is an automorphism that maps $u$ to $v$.

### Facts:

**1.** Every Cayley graph is regular and vertex-transitive.

**2.** Every Cayley graph [digraph] is a Schreier graph [digraph].

**3.** A simple graph $G$ and its edge-complement $\overline{G}$ have the same automorphism group.

**4.** An automorphism $\varphi$ of a graph $G$ induces an automorphism $\overline{\varphi}$ on the line graph $L(G)$.

**5.** If $G$ is a connected simple graph with at least 4 vertices, then $G$ and its line graph $L(G)$ have isomorphic automorphism groups.

**6.** If the $G$ graph has adjacency matrix $A$, and if the permutation $\varphi$ of $V_G$ has permutation matrix $P$, then $\varphi$ is the vertex map of an automorphism of $G$ if and only if $PA = AP$.

**7.** If all eigenvalues of a graph $G$ have multiplicity 1, then every automorphism has order at most 2.

**8.** *Frucht's theorem*: Let $\Gamma$ be any finite group. Then there exists a graph $G$ whose automorphism group is isomorphic to $\Gamma$. It can be constructed by modifying a Cayley digraph for $\Gamma$.

**9.** Every regular graph of even degree can be specified as a Schreier graph. (J. Gross, 1977)

**10.** The graph specified by a voltage graph is topologically a covering space of the voltage graph. (J. Gross, 1974)

### Examples:

**1.** The $n$-vertex edgeless graph and the complete graph $K_n$ have the symmetric group $S_n$ as their automorphism group. They are the only $n$-vertex graphs with this automorphism group.

**2.** The automorphism group of the complete bipartite graph $K_{m,n}$ is $S_n \times S_m$ if $n \neq m$ and is the wreath product [Ro99] $S_n \wr S_2$ if $m = n$.

**3.** The automorphism group of the $n$-path $P_n$ (with $n > 1$) is $S_2$.

**4.** The automorphism group of the $n$-cycle $C_n$ is the dihedral group $D_n$ of order $2n$. For instance, the 4-cycle $C_4$ with vertices $a$, $b$, $c$, and $d$ (in cyclic order), has the following vertex automorphisms:

$$(a)(b)(c)(d) \quad (a\ b\ c\ d) \quad (a\ c)(b\ d) \quad (a\ d\ c\ b)$$
$$(a\ b)(c\ d) \quad (a\ d)(b\ c) \quad (a)(c)(b\ d) \quad (b)(d)(a\ c).$$

**5.** The Cayley digraph of the group $S_3$ with generating set $\{(1\ 2\ 3), (1\ 2)\}$ is illustrated in the following figure.

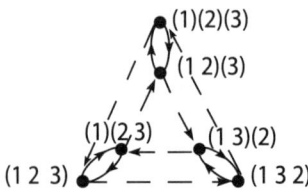

**6.** The Petersen graph is an example of a graph that is vertex-transitive but not a Cayley graph. It has the following algebraic specification:

$$V = \{v_j, w_j \mid j = 0, 1, 2, 3, 4\};$$
$$E = \{x_j\ (u_j \to u_{(j+1)\ \text{mod}\ 5}),\ y_j\ (v_j \to v_{(j+2)\ \text{mod}\ 5}),\ z_j\ (u_j \to v_j) \mid j = 0, 1, 2, 3, 4\}.$$

With appropriate labeling, it corresponds to the voltage graph on the right.

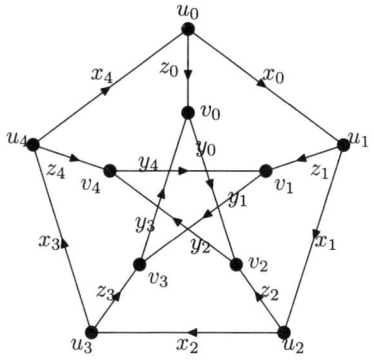
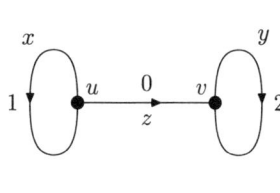

## 8.10.4   EXPANDER GRAPHS

Expander graphs are sparse graphs where every small set of vertices is well-connected to the rest of the graph. For more information on expander graphs, and for proofs of facts stated below, see [HoLiWi06], [KrSh11], [Lu94], and the references contained within. All graphs in this subsection are finite and undirected.

### Definitions:

If $F$ is a subset of vertices in a graph, the **boundary** of $F$, denoted by $\partial F$, is the set of edges that have one endpoint inside $F$ and one endpoint outside $F$.

The **isoperimetric constant** of graph $X$ with vertex set $V$ is defined as

$$h(X) = \min\left\{ \frac{|\partial F|}{|F|} \ \middle| \ F \subseteq V,\ 0 < |F| \le |V|/2 \right\}.$$

It is also known as the Cheeger constant or expansion constant of $X$.

An **expander family** is a sequence $(X_n)$ of graphs, all of the same regularity and whose orders grow to infinity, for which there exists $\epsilon > 0$ such that $h(X_n) \ge \epsilon$ for all $n$.

The **eigenvalues** of graph $G$ are the eigenvalues of its adjacency matrix $A_G$. (See §8.10.2.) Since $A_G$ is symmetric, all its eigenvalues are real. We use the convention that $\lambda_1(X) \ge \lambda_2(X) \ge \cdots \ge \lambda_n(X)$.

If $X$ is a $d$-regular graph, then the **spectral gap** of $X$ is $d - \lambda_2(X)$.

If $X$ is a regular graph of order $n$, we define $\lambda(X)$ as $\max\{|\lambda_2(X)|, |\lambda_n(X)|\}$ if $X$ is not bipartite, and $\max\{|\lambda_2(X)|, |\lambda_{n-1}(X)|\}$ if $X$ is bipartite.

A $d$-regular graph $X$ is **Ramanujan** if $\lambda(X) \le 2\sqrt{d-1}$.

Let $G$ be a finite group and let $\Gamma$ be a subset of $G$. Given a finite-dimensional unitary representation $\pi\colon G \to GL(V)$ of $G$, define $\kappa(G, \Gamma, \pi)$ to be the minimum, over all unit vectors $v \in V$ and all $\gamma \in \Gamma$, of the quantity $\|\pi(\gamma)v - v\|$. The **Kazhdan constant** of the pair $(G, \Gamma)$, denoted by $\kappa(G, \Gamma)$, is defined to be the minimum, over all nontrivial irreducible unitary representations $\pi$ of $G$, of $\kappa(G, \Gamma, \pi)$.

### Facts:

**1.** Expander graphs are used in the design of algorithms, computer networks, and error-correcting codes, as well as in cryptography. They have also been used inside proofs of complexity theory results.

**2.** If $X$ is a $d$-regular graph, then

$$\frac{d - \lambda_2(X)}{2} \le h(X) \le \sqrt{(d + \lambda_2(X))(d - \lambda_2(X))}.$$

**3.** Let $(X_n)$ be a sequence of $d$-regular graphs with $|X_n| \to \infty$. Then $(X_n)$ is an expander family if and only if there exists $\epsilon > 0$ such that the spectral gap is at least $\epsilon$ for all $n$.

**4.** Let $(X_n)$ be a sequence of $d$-regular Ramanujan graphs with $d \ge 3$ and $|X_n| \to \infty$. Then $(X_n)$ is an expander family.

**5.** There exist infinite families of regular bipartite Ramanujan graphs of every degree bigger than 2. (Marcus, Spielman, Srivastava, 2015).

**6.** Let $(X_n)$ be a sequence of $d$-regular graphs with $|X_n| \to \infty$. Then for all $\epsilon > 0$ there is some $X_n$ with $\lambda(X_n) > 2\sqrt{d-1} - \epsilon$. That is, the bound given in the definition of Ramanujan graphs cannot be lowered.

**7.** Let $d$ be a positive integer and let $(G_n)$ be a sequence of finite groups with $|G_n| \to \infty$. For each $n$, let $X_n$ be an (undirected) Cayley graph for $G_n$ with generating set $\Gamma_n$ chosen so that $X_n$ is $d$-regular. (See §8.10.3.) Then $(X_n)$ forms an expander family if and only if there exists $\epsilon > 0$ such that $\kappa(G_n, \Gamma_n) \ge \epsilon$ for all $n$. (It is known that this cannot hold if all the groups are abelian.)

**8.** Let $d \ge 3$ and let $0 < \alpha < 1$ be a real number such that $2^{4/d} < (1-\alpha)^{1-\alpha}(1+\alpha)^{1+\alpha}$. In a certain random graph model, as $n \to \infty$, the probability that $h(X) \ge (1-\alpha)d/2$ for a randomly chosen $d$-regular graph $X$ of order $n$ goes to 1.

**9.** Let $d$ be a positive integer and $\epsilon > 0$. In a certain random graph model, as $n \to \infty$, the probability that $\lambda_2(X) \le 2\sqrt{d-1} + \epsilon$ for a randomly chosen $d$-regular graph $X$ of order $n$ goes to 1.

**Examples:**
**1.** Let $X = (V_X, E_X)$ and $Y = (V_Y, E_Y)$ be regular graphs, where the degree of $X$ equals the order of $Y$. In this example, multiple edges and loops are allowed, and multiple edges are treated as distinct elements. For each $v$ from $V_X$, define the set $E_v$ consisting of all edges $e \in E_X$ incident to $v$, and choose some bijection $L_v \colon V_Y \to E_v$. Define the **zig-zag product** $X \textcircled{z} Y$ to be the graph with vertex set $V_X \times V_Y$ where the number of edges between $(x_1, y_1)$ and $(x_2, y_2)$ is equal to the number of ordered pairs $(z_1, z_2) \in E_Y \times E_Y$ such that $y_1$ is an endpoint of $z_1$, $y_2$ is an endpoint of $z_2$, and $L_{x_1}(z_1(y_1)) = L_{x_2}(z_2(y_2))$.

**2.** If $G$ is a graph with adjacency matrix $A$, we denote by $G^2$ the graph with adjacency matrix $A^2$. Let $W$ be a nonbipartite $d$-regular graph of order $d^4$ with $\lambda(W)/d \le 1/5$. (It can be shown that such a graph $W$ exists.) Define the sequence $W_n$ by $W_1 = W^2$ and $W_{n+1} = W_n^2 \textcircled{z} W$. Then the sequence of graphs $(W_n)$ forms an expander family.

**3.** Construct a family $(X_n)$ of 8-regular graphs as follows. Let the vertex set of $X_n$ be $\mathbb{Z}_n \times \mathbb{Z}_n$. Let the neighbors of $(x, y)$ be the vertices $(x, y+2x)$, $(x+2y, y)$, $(x, y+2x+1)$, $(x + 2y + 1, y)$, $(x, y - 2x)$, $(x - 2y, y)$, $(x, y - 2x - 1)$, and $(x - 2y - 1, y)$ (arithmetic modulo $n$). Then $\lambda(X_n) \le 5\sqrt{2} < 8$. Hence, $(X_n)$ is an expander family by Fact 3.

## 8.10.5   PRODUCT GRAPHS

Graph products provide a convenient language for the description of structures, but they are also algebraic objects with deep results and challenging open problems. Here we describe only the four standard products, although there are many other products that are useful for special purposes, e.g., in the construction of expanders. For more information consult the handbook [HaImKl11].

## Definitions:

Given graphs $G$ and $H$, the **Cartesian product** $G \square H$, the **direct product** $G \times H$, the **strong product** $G \boxtimes H$, and the **lexicographic product** $G \circ H$ have as vertex sets the Cartesian product $V(G) \times V(H)$. Edges are as follows:

$$E(G \square H) = \{(g,h)(g',h') \mid gg' \in E(G) \text{ and } h = h', \text{ or, } g = g' \text{ and } hh' \in E(H)\},$$
$$E(G \times H) = \{(g,h)(g',h') \mid gg' \in E(G) \text{ and } hh' \in E(H)\},$$
$$E(G \boxtimes H) = E(G \square H) \cup E(G \times H),$$
$$E(G \circ H) = \{(g,h)(g',h') \mid gg' \in E(G), \text{ or } g = g' \text{ and } hh' \in E(H)\}.$$

If $\bullet \in \{\square, \boxtimes, \times, \circ\}$ and $A = G \bullet H$, then $G$ and $H$ are **factors** of $A$ with respect to $\bullet$.

If a graph cannot be represented as a product $G \bullet H$ of two graphs on at least two vertices each, then it is **prime** with respect to $\bullet$.

## Facts:

**1.** The standard products are associative in the sense that the map $((g,h),k) \mapsto (g,(h,k))$ is an isomorphism $(G \bullet H) \bullet K \to G \bullet (H \bullet K)$.

**2.** The Cartesian, direct, and strong products are commutative. In general $G \circ H$ and $H \circ G$ are not isomorphic.

**3.** If $G \circ H$ and $H \circ G$ are isomorphic, then $G$ and $H$ are either both complete, both edgeless, or both lexicographic powers of the same graph $X$.

**4.** If $+$ represents disjoint union, then $G \bullet (H + K) = G \bullet H + G \bullet K$ for each of the standard products.

**5.** The one-vertex graph $K_1$ is a unit for the Cartesian, the strong and the lexicographic products; that is, $G = G \square K_1 = G \boxtimes K_1 = G \circ K_1$ for any graph $G$.

**6.** The direct product has no unit in the class of simple graphs. If one allows loops, then its unit is the one-vertex graph with a loop.

**7.** The Cartesian and the strong products of two graphs are connected if and only if both factors are connected.

**8.** The direct product of two graphs is connected if and only if both factors are connected and at most one is bipartite.

**9.** If $G$ is nontrivial, then $G \circ H$ is connected if and only if $G$ is connected. (But if $G = K_1$, then $G \circ H$ is connected if and only if $H$ is connected.)

**10.** Every nontrivial connected graph has a unique representation as a Cartesian product of prime graphs, up to isomorphisms and the order of the factors. The same result holds for the strong product.

**11.** Unique prime factorization over the direct product involves the class $\Gamma_0$ of finite graphs that admit loops. Every connected nonbipartite graph on more than one vertex factors uniquely (up to isomorphism and the order of the factors) into primes with respect to the direct product in $\Gamma_0$.

**12.** Although connected bipartite graphs do not necessarily factor uniquely into primes, such a factoring always contains exactly one prime bipartite factor, which is unique. (R. Hammack and O. Puffenberger, 2015)

**13.** Despite the fact that prime factorization over the direct product need not be unique for bipartite graphs, we have the following cancellation laws:
(i) If $G \times H$ is isomorphic to $G' \times H$ and $G, G', H$ are bipartite, then $G$ is isomorphic to $G'$.
(ii) If $G \times H$ is isomorphic to $G' \times H$ and $H$ has an odd cycle, then $G$ is isomorphic to $G'$.
In general, $G \times H$ being isomorphic to $G' \times H$ requires that $V(G') = V(G)$ and $E(G') = \{x\alpha(y) \mid xy \in E(G)\}$, where $\alpha$ is a permutation of $V(G)$ satisfying $\alpha(x)\alpha^{-1}(y) \in E(G)$ for all $xy \in E(G)$.

**14.** A Cartesian, strong, direct, or lexicographic product has a transitive automorphism group if and only if every factor has a transitive automorphism group.

**15.** Given a connected graph of order $n$ and size $m$, its prime factors over the Cartesian product can be computed in time $O(m)$.

**16.** Given a connected graph, its prime factors over the strong product can be computed in time $O(ma\Delta)$, where $m$ is the size, $a$ the arboricity, and $\Delta$ the maximum degree.

**17.** Over the direct product, the prime factors of a connected nonbipartite graph of order $n$, size $m$, and maximum degree $\Delta$ can be computed in $O(m \min(n^2, \Delta^3))$ time.

**18.** The problem of finding a nontrivial factoring of a graph over the lexicographic product is polynomially equivalent to the graph isomorphism problem.

**19.** The automorphism group of a connected graph is isomorphic to the automorphism group of the disjoint union of its prime factors over the Cartesian product. The same holds: (i) For the strong product, if in addition no two vertices of the graph have identical closed neighborhoods. (ii) For the direct product (in the class $\Gamma_0$ of finite graphs that admit loops), if the graph is nonbipartite and no two vertices have identical open neighborhoods.

**20.** If $G$ and $H$ are graphs on at least two vertices, then the connectivity $\kappa(G \square H)$ is $\min\{\kappa(G)|V(H)|, \kappa(H)|V(G)|, \delta(G) + \delta(H)\}$.

**Open problems:**

**1.** *Hedetniemi's conjecture:*   $\chi(G \times H) = \min\{\chi(G), \chi(H)\}$, where $\chi$ is the chromatic number. This is only known to be true if $G$ or $H$ has chromatic number up to 4. (See §8.6.1)

**2.** *Vizing's conjecture:*   $\gamma(G \square H) \geq \gamma(G)\gamma(H)$, where $\gamma$ is the domination number. (See §8.6.6.)

**3.** *Graham's conjecture:*   $p(G \square H) \leq p(G)p(H)$, where $p$ is the pebbling number. (See §8.6.7.)

**Examples:**

**1.** The figure below shows the four products whose factors are the five-cycle and the path of length two.

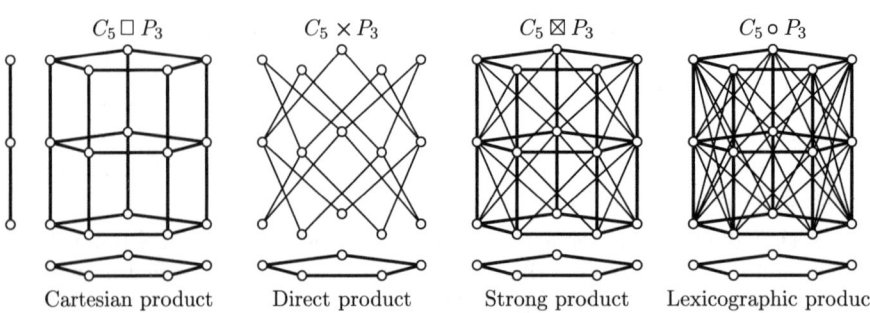

| $C_5 \square P_3$ | $C_5 \times P_3$ | $C_5 \boxtimes P_3$ | $C_5 \circ P_3$ |
|:---:|:---:|:---:|:---:|
| Cartesian product | Direct product | Strong product | Lexicographic product |

**2.** *Hamming graphs* are Cartesian products of complete graphs. Thus hypercubes are Hamming graphs where every factor is a $K_2$.

**3.** Some direct products can be represented as Cartesian products. For example, $K_2 \times K_4 = Q_3$, $C_{2k+1} \times C_{2k+1} = C_{2k+1} \square C_{2k+1}$, and $K_m \times K_n$ is the complement of $K_m \square K_n$.

**4.** A *median graph* is a graph with the property that for any three vertices $u, v, w$, there is exactly one vertex $z$ that is on a shortest $u,v$-path, a shortest $v,w$-path, and a shortest $w,u$-path. These graphs play a role in genetics. Every median graph is the **retract** of an hypercube: that is, the image of an adjacency-preserving mapping $\alpha$ of some hypercube into itself, where $\alpha^2 = \alpha$.

**5.** Large powers of (small) graphs with respect to the direct product can be used to model the internet graph and similar networks.

**6.** The *Shannon capacity* of a graph $G$ is the limit as $n \to \infty$ of the $n$th root of the independence number of the strong product of $n$ copies of $G$. It measures the amount of information involving an alphabet $V(G)$ that can be accurately transmitted across a channel, where $E(G)$ is the set of pairs of symbols that are easily "confoundable" in transmission.

**7.** If $G = \Gamma(A, S)$ and $G' = \Gamma(A', S')$ are Cayley graphs, where $A$ and $A'$ are groups with generating sets $S$ and $S'$, then $G \circ G'$ is the Cayley graph $\Gamma\big(A \times A', (\{1\} \times S') \cup (S \times A')\big)$.

# 8.11   ANALYTIC GRAPH THEORY

Analytic graph theory involves three different perspectives on the properties of graphs that are sufficiently "dense". One analysis is what must happen in a simple graph when the number of edges is sufficiently large. A second analysis is what must happen in at least one of the parts of a partition of the edges of a graph. The third analysis is what happens with a high probability when a graph is randomly chosen according to some distribution.

## 8.11.1   EXTREMAL GRAPH THEORY

We focus on the part of extremal graph theory that investigates the number of edges an $n$-vertex simple graph must have in order to guarantee that it contains a certain graph or type of graph.

**Definitions:**

The **extremal number** $ex(\mathcal{G}; n)$ for a set $\mathcal{G}$ of graphs is the greatest number of edges in any simple graph with $n$ vertices that does *not* contain some member of $\mathcal{G}$ as a subgraph. Also, the notation $ex(G; n)$ is used when $\mathcal{G}$ consists of just one graph $G$.

An **extremal graph** for a set $\mathcal{G}$ of graphs and an integer $n$ is a graph with $n$ vertices and $ex(\mathcal{G}; n)$ edges that contains no member of $\mathcal{G}$.

The **Turán graph** $T_k(n)$ is the $n$-vertex $k$-partite simple graph with the maximum number of edges.

The **Turán number** $t_k(n)$ is the number of edges in the Turán graph $T_k(n)$.

**Facts:**

**1.** If $ex(\mathcal{G}; n) = \binom{n}{2}$, then no graph with $n$ vertices contains any member of $\mathcal{G}$.

**2.** The Turán graph $T_k(n)$ is the unique complete $k$-partite graph with the property that the numbers of vertices in any two of its parts differ by at most 1. In the special case $k = 2$, $T_2(n) = K_{\lfloor n/2 \rfloor, \lceil n/2 \rceil}$. More generally, if $n = tk + r$, where $0 \le r < k$, then there are $r$ parts of size $t + 1$ and $k - r$ parts of size $t$.

**3.** The Turán number $t_k(n)$ equals $\binom{n}{2} - \frac{t(n-k+r)}{2}$, where $n = tk + r$, with $0 \le r < k$. If $k = 2$, this simplifies: $t_2(n) = \lfloor \frac{n^2}{4} \rfloor$.

**4.** *Turán's theorem:* $ex(K_k; n) = t_{k-1}(n)$; furthermore, $T_{k-1}(n)$ is the only extremal graph for $K_k$ and $n$.

**5.** Let $\chi = \chi(G)$ (chromatic number of $G$, §8.6.1) and $p = |G|$. Then $ex(G; n)$ is given by $\left(1 - \frac{1}{\chi - 1}\right)\binom{n}{2} + O(n^{2 - 1/p})$. Furthermore, all the extremal graphs differ from the Turán graph $T_{\chi-1}(n)$ by adding and deleting $O(n^{2-1/p})$ edges, and the minimum degree of all such graphs is $(1 - \frac{1}{\chi - 1})n + O(n^{2-1/p})$. (Erdős, Simonovits)

**6.** Fact 5 is also true for $ex(\mathcal{G}; n)$, where $\chi$ is the smallest chromatic number among the members of $\mathcal{G}$, and $p$ is the smallest order among these members.

**7.** $ex(\mathcal{G}; n) = O(n)$ if and only if $\mathcal{G}$ contains a (tree or) forest.

**8.** There exists a number $t_0$ such that, for $t > t_0$, every tree $T$ of order $t$ satisfies the inequality $ex(T; n) \le \frac{n(t-2)}{2}$ for every $n \ge t + 1$.

**9.** $ex(C_4; n) = \frac{1}{2}(n^{3/2}) + O(n^{4/3})$. (The exponent $\frac{4}{3}$ can be slightly improved.)

**10.** $ex(C_{2m}; n) = O(n^{1 + 1/m})$. This is known to be sharp only for $2m = 4, 6, 10$, but is conjectured to be sharp for all $m$.

**11.** The ratio $\frac{ex(\mathcal{G}; n)}{\binom{n}{2}}$ is monotone nonincreasing; that is, for every set $\mathcal{G}$ and for all $m \le n$, $\frac{ex(\mathcal{G}; m)}{\binom{m}{2}} \ge \frac{ex(\mathcal{G}; n)}{\binom{n}{2}}$.

**12.** The following table summarizes some facts that apply as the number of edges grows.

| # edges | what must occur, but not for smaller # edges | what must occur if $n$ is large enough |
|---|---|---|
| $n$ | some cycle | |
| $\lfloor \frac{3n-1}{2} \rfloor$ | some even cycle | |
| $3n - 5$ | two disjoint cycles | |
| $t_2(n) + 1 = \lfloor \frac{n^2}{4} \rfloor + 1$ | some odd cycle (i.e., $\chi \ge 3$), $C_3, \ldots, C_{\lfloor (n+3)/2 \rfloor}$ | $K_{s,s} + e$ for fixed $s$ |
| $t_2(n) + m$, $m$ fixed | | $m\lfloor \frac{n}{2} \rfloor$ copies of $C_3$; for fixed $s$ $K_{s,s}$ plus $m$ extra edges |
| $t_k(n) + 1$ | $K_k$; also, $\chi \ge k$ | |
| $t_k(n) + m$, $m$ fixed | | for fixed $s$, $K_{s,n}$ plus $m$ extra edges |
| $\binom{n}{2} - n + 3$ | a Hamilton cycle | |

**Examples:**

**1.** $ex(K_2; n) = 0$. The extremal graph is the edgeless graph.

**2.** $ex(P_2; n) = \lfloor \frac{n}{2} \rfloor$. The extremal graph is a perfect matching.

**3.** $ex(K_{1,r}; n) = \lfloor \frac{(r-1)n}{2} \rfloor$. If $(r-1)n$ is even, then any $(r-1)$-regular graph is an extremal graph. If $(r-1)n$ is odd, then any graph with one vertex of degree $r-2$ and all the others of degree $r-1$ is extremal.

**4.** $ex(K_3; n) = \lfloor \frac{n^2}{4} \rfloor$. The Turán graph $T_2(n)$ is the only extremal graph.

**5.** The Turán graph $T_3(10)$ is the 3-partite graph $K_{3,3,4}$. It has 33 edges, which is more than any other 3-partite graph on 10 vertices. Thus, $ex(K_4; 10) = 33$.

## 8.11.2   RAMSEY THEORY FOR GRAPHS

If the edges of a "dense" graph are partitioned into two parts, then at least one of the parts must still be fairly dense. Ramsey theory, which can also be studied in connection with many mathematical objects other than graphs, relies on this idea. (Also see §3.1.6.)

**Definitions:**

The (***classical***) ***Ramsey number*** $r(m, n)$ is the smallest positive integer $k$ such that every $k$-vertex graph contains either the complete graph $K_m$ or $n$ mutually nonadjacent vertices.

The ***Ramsey number*** $r(G, H)$ is the smallest positive integer $k$ such that, if the edges of $K_k$ are partitioned into red and blue classes, then either the red subgraph contains a copy of $G$ or else the blue subgraph contains a copy of $H$. Sometimes $r(G)$ denotes $r(G, G)$.

The ***Ramsey number*** $r(G_1, \ldots, G_s)$ is the smallest number $k$ such that in any $s$-fold partition of the edgeset of $K_k$, there is an index $j$ such that the $j$th part contains the graph $G_j$.

A ***$k$-canonical coloring*** of a complete graph is an edge-coloring in which the vertices can be partitioned into $k$ or fewer parts, such that the color of each edge depends only on the two parts to which its endpoints belong.

The ***arrows notation*** $F \rightarrow (G, H)$ ("*$F$ arrows $(G, H)$*") means that if the edges of the graph $F$ are partitioned into red edges and blue edges, then either the red subgraph contains a copy of $G$ or else the blue subgraph contains a copy of $H$. When $G = H$, the notation $F \rightarrow G$ is used. The notation $F \rightarrow (G_1, \ldots, G_k)$ means that $k$ edge colors are involved.

**Facts:**

**1.** $r(K_m, K_n) = r(m, n)$.

**2.** $r(G, H) = r(H, G)$. That is, Ramsey numbers are symmetric.

**3.** $r(K_n, K_1) = r(K_1, K_n) = 1$ for every $n \geq 1$.

**4.** $r(K_n, K_2) = r(K_2, K_n) = n$ for every $n \geq 1$.

**5.** $r(K_m, K_n) \leq r(K_m, K_{n-1}) + r(K_{m-1}, K_n)$ for all $m, n \geq 2$.

**6.** $r(K_m, K_n) \leq \binom{m+n-2}{m-1}$. (Erdős and Szekeres, 1935)

**7.** If $n \geq 3$, then $2^{n/2} \leq r(K_n, K_n) \leq \binom{2n-2}{n-1} < 4^n$.

**8.** The previous fact can be improved: $O(n2^{n/2}) \leq r(K_n, K_n) \leq \binom{2n-2}{n-1} \cdot O(1/\log n)$.

**9.** There exist constants $c_1$ and $c_2$ such that $c_1 n \ln n \leq r(K_3, K_n) \leq c_2 n \ln n$.

**10.** A 1-canonical coloring assigns every edge the same color.

**11.** A 2-canonical coloring consists of two complete edge-monochromatic subgraphs, such that all edges joining them are of the same color.

**12.** If $\chi(G) = \chi$ and $|V_H| = n$, then $r(G, H) \geq (\chi - 1)(n - 1) + 1$. This fact is based on a $(\chi - 1)$-canonical coloring.

**13.** If $T$ is an $n$-vertex tree, then $r(K_m, T) = (m - 1)(n - 1) + 1$. In other words, the lower bound in the immediately preceding fact determines the Ramsey number.

**14.** Except for $r(C_3, C_3) = r(C_4, C_4) = 6$, the Ramsey numbers $r(C_m, C_n)$ and $r(P_m, C_n)$ are determined by the best possible 2-canonical colorings, which are easy to find.

**15.** For every choice of graphs $G_1, G_2, \ldots, G_k$, there exists a graph $F$ for which $F \to (G_1, \ldots, G_k)$. In particular, the Ramsey number $r(G_1, \ldots, G_k)$ is well defined.

**16.** If $m, n \geq 3$, the values of only nine classical Ramsey numbers are known:

$$
\begin{array}{lll}
r(3,3) = 6 & r(3,4) = 9 & r(3,5) = 14 \\
r(3,6) = 18 & r(3,7) = 23 & r(3,8) = 28 \\
r(3,9) = 36 & r(4,4) = 18 & r(4,5) = 25.
\end{array}
$$

**17.** Estimates on some other Ramsey numbers are given in §3.1.6. In addition to the nine exact results, only one other nontrivial Ramsey number for complete graphs is known:

$$
r(K_3, K_3, K_3) = 17.
$$

## 8.11.3    PROBABILISTIC GRAPH THEORY

Probabilistic graph theory takes two basic directions. It studies random graphs for themselves, and it uses random graphs in deriving graph-theoretical results that are not themselves probabilistic.

**Definitions:**

In **Model 1**, the **random graph** $G_{n,p}$ has $n$ labeled vertices $v_1, \ldots, v_n$, and the probability of any pair of vertices being joined by an edge is $p$, where all these edge probabilities are mutually independent.

In **Model 2**, the **random graph** $G_{n,e}$ has $n$ labeled vertices $v_1, \ldots, v_n$, and exactly $e$ edges, and each such labeled graphs occurs with the same probability $1/\binom{N}{e}$, where $N = \binom{n}{2}$.

We say **almost every (a.e.) graph** has some property $P$ under either Model 1 or Model 2, if the probability that a random graph has property $P$ approaches 1 as $n \to \infty$; where the probability $p$ stays constant under Model 1, but how $e$ varies with $n$ under Model 2 must be specified. If no model is explicitly specified, then Model 1 with $p = 1/2$ is implicit.

**Facts:**

**1.** The number of labeled graphs in the probability space for Model 1 is $2^{\binom{n}{2}}$.

**2.** While Model 2 is sometimes considered more natural, it is in practice usually easier to work with Model 1. Fortunately, Model 1 with $p = \frac{e}{N}$ behaves very similarly to Model 2 in most cases, so that facts about Model 1 usually lead easily to facts about Model 2 as well.

**3.** In Model 1, a graph with $e$ edges occurs with probability $p^e(1-p)^{\binom{n}{2}-e}$. If $p = 1/2$, then every labeled graph on $n$ vertices has the same probability $2^{-\binom{n}{2}}$.

**4.** Random graphs can be used to prove theorems about graphs, especially existence theorems. (See Example 1.)

**5.** Let $b = \frac{1}{p}$ and $d = 2\log_b \frac{en}{2\log_b n} = 1$, where $e = 2.718\ldots$, not the number of edges. Then for every positive $\epsilon < \frac{1}{2}$ the clique number of a.e. graph is either $\lfloor d-\epsilon \rfloor$ or $\lfloor d+\epsilon \rfloor$, where these two values are usually the same when $\epsilon$ is small. This means that the clique number is determined for a.e. graph, unless $d$ is close to an integer, in which case there are two possible values.

**6.** If $p \gg 1/n$ (for example if $p$ is constant), the chromatic number of $G(n,p)$ is

$$\left(\tfrac{1}{2} + o(1)\right) \frac{n}{\log_2(np)} \log_2\left(\tfrac{1}{1-p}\right).$$

**7.** Moreover, if $p = n^{-\alpha}$ for fixed $\alpha > 1/2$, then in Model 1, there exists an $f(n,p)$ so that for almost every graph, $f(n,p) \le \chi \le f(n,p)+1$. That is, the chromatic number $\chi$ takes on one of two consecutive values.

**8.** Almost every graph in Model 1 has its connectivity and its edge connectivity equal to its minimum degree. Furthermore, the common value of these three parameters is $pn - (2p(1-p)n\log n)^{\frac{1}{2}} + o(n\log n)^{\frac{1}{2}}$.

**9.** Generating a random graph $G_{n,p}$ under Model 1 is straightforward, as indicated by Algorithm 1.

---

**Algorithm 1:   Generate random graph $G_{n,p}$ (per Model 1).**

initialize graph with vertex list $v_1, v_2, \ldots, v_n$

**for** $i := 1$ **to** $n-1$

    **for** $j := i+1$ **to** $n$

        join vertices $v_i$ and $v_j$ with probability $p$

---

**10.** To generate a random graph $G_{n,e}$ under Model 2, the possible edges are placed in bijective correspondence with the integers $1, \ldots, \binom{n}{2}$ according to the rule $f(i,j) = \binom{n}{2} - \binom{n-i+1}{2} + j$. Also, the $e$-combinations of the integers $1, \ldots, \binom{n}{2}$ are placed in bijective correspondence with the integers $1, \ldots, \binom{\binom{n}{2}}{e}$ according to the lexicographic ordering of those $e$-combinations (see §2.3.5). This yields Algorithm 2.

---

**Algorithm 2:   Generate random graph $G_{n,e}$ (per Model 2).**

initialize graph with vertex list $v_1, v_2, \ldots, v_n$

generate random integer $r \in \left\{1, \ldots, \left(\binom{n}{2} \atop e\right)\right\}$

convert $r$ to an $e$-combination $C$ in $\left\{1, \ldots, \binom{n}{2}\right\}$

convert $e$-combination $C$ to $e$ edges in $G_{n,e}$

---

**11.** For every fixed $s$, almost every graph contains the complete graph $K_s$. Moreover, for every fixed graph $H$, almost every graph contains $H$.

**12.** Here is a table of some properties of almost every $n$-vertex graph:

| $p$ under Model 1 | $e$ under Model 2 | property of almost every graph |
|---|---|---|
| $o(\frac{1}{n})$ | $o(n)$ | no cycles |
| $\frac{2c}{n},\ 0 < c < \frac{1}{2}$ | $cn,\ 0 < c < \frac{1}{2}$ | cycles are possible, and the largest component has order $\approx \ln n$ |
| $\frac{1}{n}$ | $\frac{n}{2}$ | some cycle exists, and the largest component has order $\Theta(n^{2/3})$ |
| $\frac{2c}{n},\ c > \frac{1}{2}$ | $cn,\ c > \frac{1}{2}$ | the largest component has order $c'n$ |
| $\frac{c \ln n}{n},\ c < 1$ | $\frac{c}{2} n \ln n,\ c < 1$ | the graph is disconnected |
| $\frac{c \ln n}{n},\ c > 1$ | $\frac{c}{2} n \ln n,\ c > 1$ | the graph is connected and Hamiltonian |

**Examples:**

**1.** *Using random graphs to prove theorems:*  Here is a proof that the Ramsey number $r(K_n, K_n)$ (see §8.11.2) is greater than $2^{n/2}$ for all $n \geq 3$. Consider a random red-blue edge-coloring of $K_N$ for some $N > n$ with $p(red) = 1/2$. The probability that any given $K_n$ occurring within this 2-colored $K_N$ is entirely red is $2^{-\binom{n}{2}}$. Of course, the probability that it is colored blue is the same. Thus, the probability that the given subgraph $K_n$ is monochromatic in either color is $2^{1-\binom{n}{2}}$. Since there are $\binom{N}{n}$ different copies of $K_n$ in the colored $K_N$, the expected number of monochromatic $K_n$ is $\binom{N}{n} \cdot 2^{1-\binom{n}{2}}$.

With the choice of $N = \lfloor 2^{n/2} \rfloor$, this expectation is $\binom{N}{n} \cdot 2^{1-\binom{n}{2}} < \frac{2^{1+n/2}}{n!} \cdot \frac{N^n}{2^{n^2/2}} < 1$, i.e., less than 1. Therefore there must be some coloring with no monochromatic $K_n$ at all.

**2.** Other examples in which random graphs are analyzed can be found in the section on the probabilistic method (see §7.11).

# 8.12   HYPERGRAPHS

In ordinary graph theory, an edge of a simple graph can be regarded as a pair of vertices. In hypergraph theory, an "edge" can be regarded as an arbitrary subset of vertices. In this sense, hypergraphs are a natural generalization of graphs. Their systematic study was initiated by C. Berge. They have evolved into a unifying combinatorial concept.

## 8.12.1   HYPERGRAPHS AS A GENERALIZATION OF GRAPHS

**Definitions:**

A **hypergraph** $H = (V, E)$ is a finite set $V$ of "vertices" together with a finite multiset $E$ of "edges" (sometimes, "hyperedges"), which are arbitrary subsets of $V$.

The **order** of a hypergraph edge is its cardinality.

A **partial hypergraph** (or simply a **partial**) of the hypergraph $H = (V, E)$ is a hypergraph $H' = (V, E')$ such that $E' \subseteq E$. This generalizes a spanning subgraph.

A hypergraph $H = (V, E)$ is **simple** if $E$ has no repeated edges.

The **incidence matrix** of a hypergraph $H = (V, E)$ with $E = \{e_1, e_2, \ldots, e_m\}$ and $V = \{x_1, x_2, \ldots, x_n\}$ is the $m \times n$ matrix $M(H) = [m_{i,j}]$ with $m_{i,j} = 1$ if $x_j \in e_i$ and $0$ otherwise.

The **dual hypergraph** of the hypergraph $H$ is the hypergraph $H^*$ whose incidence matrix is the transpose of the incidence matrix $M(H)$. This concept of duality from block design theory differs from the dual of a graph embedded on a surface.

The **degree** $\deg(x)$ of a hypergraph vertex $x$ is the number of hypergraph edges containing $x$.

A hypergraph is **regular** if all vertices have the same degree. If $t$ is the common value of the degrees, then the hypergraph is **$t$-regular**.

A hypergraph is **uniform** if all edges have the same number of vertices. If $r$ is the common value, then the hypergraph is **$r$-uniform**.

The **complete hypergraph** $K_n^*$ has all subsets of $n$ vertices as edges, so that it has $2^n$ edges.

The **complete $r$-uniform hypergraph** $K_n^r$ is the simple hypergraph of order $n$ with all $r$-element subsets as edges, so that it has $\binom{n}{r}$ edges.

The **intersection graph** $I(H)$ of the hypergraph $H$ is a simple graph whose vertices are the edges of $H$. Two vertices of $I(H)$ are adjacent if and only if the corresponding edges of $H$ have nonempty intersection.

An **independent set** of vertices in a hypergraph is a set of vertices that does not (completely) contain any edge of the hypergraph.

**Facts:**

**1.** *How to draw a hypergraph*: First draw the vertices and the hyperedges of order 2, as if they were vertices and edges, respectively, of a graph. Then shade triangular regions corresponding to hyperedges of order 3. Higher-order hyperedges and hyperedges of order 1 can be indicated by drawing enclosures around their vertices.

**2.** Every hypergraph satisfies the generalized Euler equation for degree-sum:

$$\sum_{x \in V} \deg(x) = \sum_{e \in E} |e|.$$

**3.** Every simple graph is a 2-uniform simple hypergraph.

**4.** The intersection graph of a hypergraph generalizes the line graph $L(G)$ of a graph $G$. (See §8.1.2.)

**5.** Every graph is the intersection graph of some hypergraph.

**6.** Every graph of order $n$ is isomorphic to the intersection graph of a hypergraph of order at most $\lfloor n^2/4 \rfloor$.

**7.** When a graph $G$ is regarded as a hypergraph, its dual is a hypergraph whose intersection graph is $G$.

**Examples:**

**1.** The hypergraph $H = (V, E)$ with $V = \{a, b, c, d\}$ and $E = \{ab, bc, bd, acd, c\}$ can be illustrated as follows:

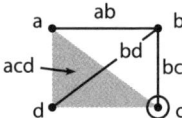

2. The hypergraph of Example 1 has the following incidence matrix:

|     | a | b | c | d |
|-----|---|---|---|---|
| ab  | 1 | 1 | 0 | 0 |
| acd | 1 | 0 | 1 | 1 |
| bc  | 0 | 1 | 1 | 0 |
| bd  | 0 | 1 | 0 | 1 |
| c   | 0 | 0 | 1 | 0 |

3. The dual of the hypergraph of Example 1 has the following incidence matrix:

|       | $v(ab)$ | $v(acd)$ | $v(bc)$ | $v(bd)$ | $v(c)$ |
|-------|---------|----------|---------|---------|--------|
| $e(a)$ | 1       | 1        | 0       | 0       | 0      |
| $e(b)$ | 1       | 0        | 1       | 1       | 0      |
| $e(c)$ | 0       | 1        | 1       | 0       | 1      |
| $e(d)$ | 0       | 1        | 0       | 1       | 0      |

This dual hypergraph may be illustrated as follows:

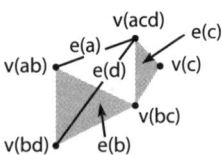

4. The hypergraph of Example 1 has the following intersection graph:

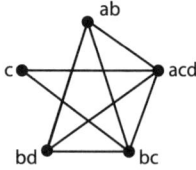

---

## 8.12.2   HYPERGRAPHS AS GENERAL COMBINATORIAL STRUCTURES

### Definitions:

A **transversal** (or **cover** or **blocking set**) in a hypergraph is a set of vertices that has nonempty intersection with every edge of the hypergraph.

A **system of distinct representatives** (**SDR**) in a hypergraph $H = (V, E)$ with $E = \{e_1, e_2, \ldots, e_m\}$ is a transversal of $m$ distinct vertices $x_1, x_2, \ldots, x_m$ such that $x_i \in e_i$ for $i = 1, \ldots, m$.

***Hall's condition*** on a hypergraph is that, for each $t = 1, \ldots, m$ the union of every subset of $t$ edges have at least $t$ vertices. Thus, each partial must have at least as many vertices as edges.

A ***matching*** in a hypergraph is a set of pairwise disjoint edges.

An ***antichain*** is a hypergraph in which no edge contains any other edge.

A ***chain*** is a simple hypergraph in which, given any pair of edges, one edge contains the other.

A ***symmetric chain*** in an $n$-vertex hypergraph $H$ is a chain with edges of order $\frac{n}{2} - t, \ldots, \frac{n}{2} + t$ for some $t \geq 0$.

A ***downset*** (or ***ideal***) is a simple hypergraph in which every subset of every edge is also an edge of the hypergraph.

An ***upset*** (or ***filter***) is a simple hypergraph in which every superset of every edge is also an edge of the hypergraph.

A ***hypergraph clique*** is a simple hypergraph such that every pair of edges has nonempty intersection.

An ***r-partite*** hypergraph is an $r$-uniform hypergraph whose vertex set can be partitioned into $r$ blocks such that each edge intersects each block in exactly one vertex.

A hypergraph is ***unimodular*** if the determinant of every square submatrix of its incidence matrix is equal to 0, 1, or $-1$.

A hypergraph is an ***interval hypergraph*** if its $n$ vertices can be labeled $1, 2, \ldots, n$ so that each edge is labeled by consecutive numbers.

**Facts:**

**1.** Hall's condition is necessary and sufficient for the existence of an SDR in a hypergraph.

**2.** *Sperner's lemma:* If the hypergraph $H$ with $n$ vertices and $m$ edges is an antichain, then $m \leq \binom{n}{\lfloor n/2 \rfloor}$.

**3.** If the hypergraph $H$ with $m_i$ edges of order $i$ for $i = 1, \ldots, n$ is an antichain, then $\sum_{i=0}^{n} m_i \binom{n}{i}^{-1} \leq 1$.

**4.** The complete hypergraph $K_n^*$ can be partitioned into symmetric chains.

**5.** *Kleitman's lemma:* Let $D$ and $U$ be hypergraphs on the same $n$ vertices. Let $D$ be a $d$-edge downset and $U$ a $u$-edge upset. And let $D$ and $U$ have $m$ common edges. Then $du \geq 2^n m$.

**6.** An $n$-vertex hypergraph clique has at most $2^{n-1}$ edges.

**7.** An $r$-uniform $n$-vertex hypergraph clique $n$ has at most $\binom{n-1}{r-1}$ edges if $n \geq 2r$.

**8.** In any $r$-uniform hypergraph $H$, the maximum size $r$-partite partial hypergraph contains at least $\frac{r!}{r^r}$ of the edges of $H$.

**9.** Let $H$ be an $n$-vertex, $m$-edge hypergraph clique, such that each pair of distinct edges intersect in exactly one vertex. Then $m \leq n$. (de Bruijn and Erdős)

**10.** *Fisher's inequality:* Let $H$ be an $n$-vertex, $m$-edge hypergraph clique such that each pair of edges intersects in $\lambda$ vertices. Then $m \leq n$.

**11.** *Modular intersection theorem:* Let $L$ be a set of $s$ integers, and let $p$ be a prime number. Let $H$ be an $r$-uniform hypergraph such that $r \notin L \mod p$ and that the intersection size for each pair of distinct edges is in $L \mod p$. Then $m \leq \binom{n}{s}$.

**Examples:**

1. The Fano plane (see §12.1.1) is the hypergraph with (using mod 7 arithmetic):

$$V = \{1, 2, \ldots, 7\} \text{ and } E = \{ \{1+i, 2+i, 4+i\} \mid 1 \leq i \leq 7 \}.$$

2. A block design is a regular, uniform hypergraph such that each pair of vertices is contained in precisely $\lambda$ edges. Block designs often provide extremal examples in problems of hypergraph theory.

3. A matroid (see §12.4.1) can be regarded as a hypergraph such that under every nonnegative weighting of the vertices, a greedy algorithm would find an edge of maximum weight.

---

## 8.12.3   NUMERICAL INVARIANTS OF HYPERGRAPHS

Calculating formulas for the values of some standard numerical invariants of hypergraphs tends to be quite difficult, even for complete hypergraphs. Two famous examples are Lovász's proof of the Kneser conjecture and Baranyai's proof of the factorization theorem.

**Definitions:**

The **maxdegree** $\Delta(H)$ is the largest degree of any vertex in the hypergraph $H$.

The **chromatic number** $\chi(H)$ is the smallest number of independent sets required to partition the vertex set of $H$. To ensure the existence of such partitions it is assumed that $H$ does not contain any edges with just one vertex.

The **independence number** $\alpha(H)$ is the maximum number of vertices which form an independent set in $H$.

The **chromatic index** $q(H)$ is the smallest number of matchings required to partition the edges of $H$.

A hypergraph $H$ is **normal** if $q(H) = \Delta(H)$.

The **transversal number** $\tau(H)$ is the minimum cardinality (i.e., number of vertices), taken over all transversals of $H$.

The **matching number** $\nu(H)$ is the maximum number of pairwise disjoint edges of $H$, i.e., the cardinality of the largest partial of $H$ which forms a matching.

The **clique partition number** $cp(H)$ is the smallest number of cliques required to partition the edge set of $H$.

The **clique number** $w(H)$ is the largest number of edges of any partial clique in the hypergraph $H$.

**Facts:**

1. Many hypergraph invariants are representable as graph invariants. In particular,

$$w(H) = w(I(H)), \quad \nu(H) = \alpha(I(H)), \quad q(H) = \chi(I(H)), \quad cp(H) = \chi(\overline{I(H)})$$

where $\overline{G}$ denotes the edge-complement of a graph $G$.

2. Every hypergraph $H$ satisfies the following two *min* $\geq$ *max* relations:

$$q(H) \geq cp(H) \geq \Delta(H), \qquad \tau(H) \geq cp(H) \geq \nu(H).$$

**3.** A hypergraph $H$ is normal if and only if $\tau(H') = \nu(H')$ for all partials $H'$ of $H$. (Lovász, 1972)

**4.** The following relations hold in every $n$-vertex hypergraph $H$:

$$\tau(H) = n - \alpha(H), \qquad \chi(H) \geq \frac{n}{\alpha(H)}, \qquad \chi(H) + \alpha(H) \leq n + 1.$$

**5.** The parameters $\chi(H)$ and $\tau(H)$ can be approximated by greedy algorithms.

**6.** The *Kneser conjecture* that $cp(K^r_{2r+k}) = k + 2$ was proved by topological methods. (Lovász and Bárány, 1978)

**7.** The *factorization theorem* that $q(K^r_{kr}) = \binom{kr-1}{r-1}$ was proved by using network flows. (Baranyai, 1975)

**8.** Hypergraphs in the following classes are known to be bicolorable (i.e., $\chi(H) = 2$): normal hypergraphs (including unimodular hypergraphs), $r$-uniform hypergraphs with size at most $2^{r-1}$, $r$-uniform hypergraphs in which each edge intersects at most $2^{r-3}$ other edges (proved by probabilistic methods), finite planes of order at least three.

**Examples:**

**1.** Consider the hypergraph $H$ of §8.12.1, Example 1 with $V = \{a, b, c, d\}$ and $E = \{ab, bc, bd, acd, c\}$. The maximum degree $\Delta(H)$ is 3, since vertex $c$ has degree 3. The chromatic number $\chi(H)$ is 4, since every pair of vertices lies in some edge, so all four vertices must get different colors. The independence number $\alpha(H)$ is 1, since every pair of vertices lies in some edge. The chromatic index $q(H)$ is 4, using the matching $c, ab$.

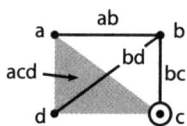

The hypergraph $H$ is not normal, since $q(H) = 4$, but $\Delta(H) = 3$. The transversal number $\tau(H)$ is 2, using the transversal $b, c$. The matching number $\nu(H)$ is 2, using the matching $ab, c$.

**2.** The Fano plane (§8.12.2, Example 1) has the following parameters: $\omega = q = 7$, $\Delta = \tau = \chi = 3$, $\alpha = 4$, $\nu = cp = 1$.

---

# REFERENCES

*Printed Resources*:

[AgGr06] G. Agnarsson and R. Greenlaw, *Graph Theory: Modeling, Applications, and Algorithms*, Pearson, 2006.

[BeWi04] L. W. Beineke and R. J. Wilson, *Topics in Algebraic Graph Theory*, Cambridge University Press, 2004.

[BeWi15] L. W. Beineke and R. J. Wilson, *Topics in Chromatic Graph Theory*, Cambridge University Press, 2015.

[Bi94] N. Biggs, *Algebraic Graph Theory*, 2nd ed., Cambridge University Press, 1994.

[Bo85] B. Bollobás, *Random Graphs*, Academic Press, 1985.

[BoHe77] J. A. Bondy and R. L. Hemminger, "Graph reconstruction — a survey", *Journal of Graph Theory* 1 (1977), 227–268.

[BoMu08] J. A. Bondy and U. S. R. Murty, *Graph Theory*, Springer, 2008.

[Br13] A. Bretto, *Hypergraph Theory: An Introduction*, Springer, 2013.

[ChLeZh15] G. Chartrand, L. Lesniak, and P. Zhang, *Graphs & Digraphs*, 6th ed., Chapman and Hall/CRC, 2015.

[ChEtal06] M. Chudnovsky, N. Robertson, P. Seymour, and R. Thomas, "The strong perfect graph theorem", *Annals of Mathematics* 164 (2006), 51–229.

[CvRoSi10] D. M. Cvetković, P. Rowlinson, and S. Simić, *An Introduction to the Theory of Graph Spectra*, Cambridge University Press, 2010.

[Ed65] J. Edmonds, "Paths, trees and flowers", *Canadian Journal of Mathematics* 17 (1965), 449–467.

[Eu36] L. Euler, "Solutio problematis ad geometriam situs pertinentis", *Commentarii academiae scientiarum Petropolitanae* 8 (1736), 128–140.

[GaJo79] M. R. Garey and D. S. Johnson, *Computers and Intractability: A Guide to the Theory of NP-completeness*, W. H. Freeman, 1979.

[GrTu12] J. L. Gross and T. W. Tucker, *Topological Graph Theory*, Dover, 2012.

[GrYe05] J. L. Gross and J. Yellen, *Graph Theory and Its Applications*, Chapman and Hall/CRC, 2005.

[GrYeZh13] J. L. Gross, J. Yellen, and P. Zhang, *Handbook of Graph Theory*, 2nd ed., Chapman and Hall/CRC, 2013.

[Ha35] P. Hall, "On representatives of subsets", *Journal of the London Mathematical Society* 10 (1935), 26–30.

[HaImKl11] R. Hammack, W. Imrich, and S. Klavžar, *Handbook of Product Graphs*, 2nd ed., CRC Press, 2011.

[HaPa73] F. Harary and E. M. Palmer, *Graphical Enumeration*, Academic Press, 1973.

[HaHeSl98a] T. W. Haynes, S. T. Hedetniemi, and P. J. Slater, *Fundamentals of Domination in Graphs*, Marcel Dekker, 1998.

[HaHeSl98b] T. W. Haynes, S. T. Hedetniemi, and P. J. Slater, *Domination in Graphs: Advanced Topics*, Marcel Dekker, 1998.

[HeYe13] M. A. Henning and A. Yeo, *Total Domination in Graphs*, Springer, 2013.

[HoLiWi06] S. Hoory, N. Linial, and A. Wigderson, "Expander graphs and their applications", *Bulletin of the American Mathematical Society* 43 (2006), 439–561.

[ImKlRa08] W. Imrich, S. Klavžar, and D. F. Rall, *Topics in Graph Theory: Graphs and Their Cartesian Product*, A K Peters, 2008.

[KöScTo93] J. Köbler, U. Schöning, and J. Torán, *The Graph Isomorphism Problem: Its Structural Complexity*, Birkhäuser, 1993.

[KrSh11] M. Krebs and A. Shaheen, *Expander Families and Cayley Graphs: A Beginner's Guide*, Oxford University Press, 2011.

[Lo72] L. Lovász, "Normal hypergraphs and the perfect graph conjecture", *Discrete Mathematics* 2 (1972), 253–267.

[Lu94] A. Lubotzky, *Discrete Groups, Expanding Graphs, and Invariant Measures*, Birkhäuser, 1994.

[McPi14] B. D. McKay and A. Piperno, "Practical graph isomorphism, II", *Journal of Symbolic Computation* 60 (2014), 94–112.

[Mo15] J. W. Moon, *Topics on Tournaments*, Dover, 2015.

[Na78] C. St. J. A. Nash-Williams, "The reconstruction problem", in *Selected Topics in Graph Theory*, L. W. Beineke and R. J. Wilson (eds.), Academic Press, 1978, 205–236.

[PeSk03] S. Pemmaraju and S. Skiena, *Implementing Discrete Mathematics*, Cambridge University Press, 2003.

[PóRe87] G. Pólya and R. C. Read, *Combinatorial Enumeration of Groups, Graphs, and Chemical Compounds*, Springer-Verlag, 1987.

[Re98] R. C. Read, *Atlas of Graphs*, Oxford University Press (Clarendon Press), 1998.

[Ro76] F. S. Roberts, *Discrete Mathematical Models with Applications to Social, Biological, and Environmental Problems*, Prentice-Hall, 1976.

[Ro11] K. H. Rosen, *Discrete Mathematics and Its Applications*, 7th ed., McGraw-Hill, 2011.

[Ro99] J. J. Rotman, *An Introduction to the Theory of Groups*, 4th ed., Springer, 1999

[Ta97] R. Tamassia, "Graph drawing", in *Handbook of Discrete and Computational Geometry*, J. E. Goodman and J. O'Rourke (eds.), CRC Press, 1997, 815–832.

[Th98] R. Thomas, "An update of the four-color theorem", *Notices of the American Mathematical Society* 45 (1998), 848–859.

[Tu47] W. T. Tutte, The factorization of linear graphs, *Journal of the London Mathematical Society* 22 (1947), 107–111.

[We01] D. B. West, *Introduction to Graph Theory*, 2nd ed., Prentice-Hall, 2001.

**Web Resources**:

[OEIS] oeis.org (The On-Line Encyclopedia of Integer Sequences)

www.cs.stonybrook.edu/~algorith/ (The Stony Brook Algorithm Repository)

www.math.gatech.edu/~thomas/FC (Information on the Four Color Theorem.)

Graphviz.org (graph theory software.)

# 9

# *TREES*

## INTRODUCTION

A tree is a connected graph containing no cycles. Trees have applications in a wide variety of disciplines, particularly computer science. For example, they can be used to construct searching algorithms for finding a particular item in a list, to store data, to model decisions and their outcomes, or to design networks.

## GLOSSARY

**ancestor** (of a vertex $v$ in a rooted tree): any vertex on a path to $v$ from the root.

**$m$-ary tree**: a rooted tree in which every internal vertex has at most $m$ children.

**backtrack**: a pair of successive edges in a walk where the second edge is the same as the first, but traversed in the opposite direction.

**balanced tree**: a rooted $m$-ary tree of height $h$ such that all leaves of the tree have height $h$ or $h-1$.

**bihomogeneous tree**: a tree (usually infinite) in which there are exactly two values for the vertex degrees.

**binary search tree** (BST): a type of binary tree used to represent a table of data, which is efficiently accessed by storage and retrieval algorithms.

**binary tree**: an ordered rooted tree in which each vertex has at most two children, that is, a possible "left child" and a possible "right child"; an only child must be designated either as a left child or a right child (this usage is standard for computer science); an $m$-ary tree in which $m = 2$ (in pure graph theory).

**bounded tree**: a (possibly infinite) tree of finite diameter.

**breadth-first search**: a method for visiting all the vertices of a graph in a sequence, in order of their proximity to a designated starting vertex.

**caterpillar**: a tree that contains a path such that every edge has one or both endpoints in that path.

**center** (of a tree): the set of vertices of minimum eccentricity.

**child** (of a vertex $v$ in a rooted tree): a vertex such that $v$ is its immediate ancestor.

**chord**: for a graph $G$ with a spanning tree $T$, an edge $e$ of $G$ such that $e \notin T$.

**complete binary tree**: a binary tree where every parent has two children and all leaves are at the same depth.

**decision tree**: a rooted tree in which every internal vertex represents a decision and each path from the root to a leaf represents a sequence of decisions.

**depth** (of a vertex in a rooted tree): the number of edges in the unique path from the root to that vertex.

**depth-first search**: a method for visiting every vertex of a graph by progressing as far as possible from the most recently visited vertex, before doing any backtracking.

**descendant** (of a vertex $v$ in a rooted tree): a vertex that follows $v$ on a path away from the root.

**diameter** (of a tree): the maximum distance between two vertices in the tree.

**distance** (between two vertices in a tree): the number of edges in the unique (simple) path between these vertices.

**eccentricity** (of a vertex in a tree): the length of the longest simple path beginning at that vertex.

**finite tree**: a tree with a finite number of vertices and edges.

**forest**: a graph with no cycles.

**full *m*-ary tree**: a rooted tree in which every internal vertex has exactly *m* children.

**fundamental cycle**: in a connected graph $G$, the unique cycle created by adding the edge $e \in E_G$ not in $T$ to a spanning tree $T$.

**fundamental edge-cut**: in a connected graph $G$, the partition-cut $\langle X_1, X_2 \rangle$ where $X_1$ and $X_2$ are the vertex sets of the two components of $T - e$ for some edge $e$ in a spanning tree $T$ for $G$.

**fundamental system of cycles**: in a connected graph $G$, the set of fundamental cycles corresponding to the various edges of $G - T$, where $T$ is a spanning tree for $G$.

**fundamental system of edge-cuts**: in a connected graph $G$, the set of fundamental edge-cuts that result from removal of an edge from a spanning tree $T$ for $G$.

**height** (of a rooted tree): the maximum of the levels of its vertices.

**homogeneous**: property of an (infinite) tree that every vertex has the same degree.

**n-homogeneous**: property of an (infinite) tree that every vertex has degree $n$.

**infinite tree**: a tree with an infinite number of vertices and edges.

**inorder traversal** (of an ordered rooted tree): a recursive listing of all vertices starting with the vertices of the first subtree of the root, next the root vertex itself, and then the vertices of the other subtrees as they occur from left to right.

**internal vertex** (of a rooted tree): a vertex with children.

**isomorphism** (of trees): for trees $X$ and $Y$, a pair of bijections $f_V \colon V_X \to V_Y$ and $f_E \colon E_X \to E_Y$ such that if $u$ and $v$ are the endpoints of an edge $e$ in the tree $X$, then $f_V(u)$ and $f_V(v)$ are the endpoints of the edge $f_E(e)$ in the tree $Y$ (see §8.1).

**isomorphism** (of rooted trees): for rooted trees $(T_1, r_1)$ and $(T_2, r_2)$, a tree isomorphism $f \colon T_1 \to T_2$ that takes $r_1$ to $r_2$.

**labeled tree**: a tree with labels such as $v_1, v_2, \ldots, v_n$ assigned to its vertices.

**leaf**: in a rooted tree, a vertex that has no children.

**left child** (of a vertex in an ordered, rooted binary tree): the first child of that vertex.

**left-right tree**: Synonym for *full binary tree*.

**left subtree** (of an ordered, rooted binary tree): the subtree rooted at a left child.

**level** (of a vertex in a rooted tree): the length of the unique path from the root to this vertex.

**locally finite tree**: a tree in which the degree of every vertex is finite.

**nth level** (of a rooted tree): the set of all vertices at depth $n$.

**ordered tree**: a rooted tree in which the children of each internal vertex are linearly ordered.

**parent** (of a vertex $v$, other than the root, in a rooted tree): a vertex that is the immediate predecessor of $v$ on the unique path away from the root to $v$.

**partition-cut of a graph**: given a partition of the set of vertices of $G$ into $X_1$ and $X_2$, the set $\langle X_1, X_2 \rangle$ of edges of $G$ that have one endpoint in $X_1$ and the other in $X_2$.

**postorder traversal** (of an ordered rooted tree): a recursive listing of vertices starting with the vertices of subtrees as they occur from left to right, followed by the root.

**preorder traversal** (of an ordered rooted tree): a recursive listing of vertices starting with the root, then the vertices of the first subtree, followed by the vertices of other subtrees as they occur from left to right.

**reduced tree**: a tree with no vertices of degree 2.

**reduced walk**: a walk in a graph without backtracking.

**regular**: Synonym for *homogeneous*.

**right child** (of a vertex in an ordered rooted binary tree): the second child of that vertex.

**right subtree** (of an ordered, rooted binary tree): the subtree rooted at the right child.

**rooted tree**: a tree in which one vertex is designated as the "root".

**siblings** (in a rooted tree): vertices with the same parent.

**spanning tree** (of a connected graph): a subgraph that is a tree and contains all the vertices of the graph.

**subtree**: a subgraph of a tree that is also a tree.

**tree**: a connected graph with no cycles.

**tree edge**: for a graph $G$ with a spanning tree $T$, an edge $e$ of $G$ such that $e \in T$.

**tree traversal**: a walk that visits all the vertices of a tree.

# 9.1   CHARACTERIZATIONS AND TYPES OF TREES

## 9.1.1   PROPERTIES OF TREES

For trees, as with other graphs, there is a wide variety of terminology in use from one application or specialty to another.

**Definitions:**

A graph is **acyclic** if it contains no subgraph isomorphic to a cycle $C_n$ (see §8.1.3).

A **forest** is an acyclic graph.

A **tree** is an acyclic connected graph. (*Note:* Unless stated otherwise, all trees are assumed to be finite, i.e., to have a finite number of vertices.)

A **walk** is an alternating sequence $v_0, e_1, v_1, \ldots, e_r, v_r$ of vertices and edges in which each edge $e_i$ joins vertices $v_{i-1}$ and $v_i$.

A **path** (sometimes called a **simple path**) is a walk where all the vertices are different.

The **eccentricity** of a vertex is the length of the longest (simple) path beginning at that vertex.

The **center** of a tree contains all vertices with minimum eccentricity.

An **end vertex** of a tree is a vertex of degree 1.

A **caterpillar** is a tree that contains a path such that every edge has one or both endpoints in that path.

### Facts:
1. A (finite) tree with at least two vertices has at least two end vertices.
2. A connected graph with $n$ vertices is a tree if and only if has exactly $n - 1$ edges.
3. A graph is a tree if and only if there is a unique (simple) path between any two vertices.
4. A graph is a forest if and only if every edge is a cut-edge. (See §8.4.2.)
5. Trees are bipartite. Hence, every tree can be colored using two colors.
6. The center of a tree consists of either only one vertex or two adjacent vertices.

### Examples:
1. The following tree has 9 vertices and $9 - 1 = 8$ edges.

2. The following graph is a forest with three components.

3. The center of the following tree consists of the two adjacent vertices $a$ and $b$.

4. The following tree is a caterpillar.

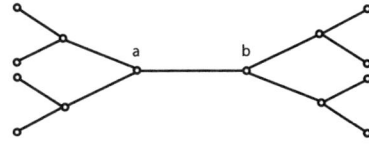

**5.** Neither of the following graphs is a tree. One contains a 3-cycle, and the other contains a 1-cycle (i.e., a self-loop).

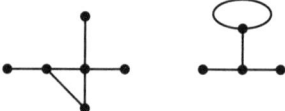

## 9.1.2   ROOTS AND ORDERINGS

Adding some extra structure to trees adapts them to applications in many disciplines, especially computer science.

**Definitions:**

A **rooted tree** $(T, r)$ is a tree $T$ with a distinguished vertex $r$ (the **root**), in which all edges are implicitly directed away from the root.

Two rooted trees $(T_1, r_1)$ and $(T_2, r_2)$ are **isomorphic as rooted trees** if there is an isomorphism $f \colon T_1 \to T_2$ (see §8.1.2) that takes $r_1$ to $r_2$.

A **child** of a vertex $v$ in a rooted tree is a vertex that is the immediate successor of $v$ on a path away from the root.

A **descendant** of a vertex $v$ in a rooted tree is $v$ itself or any vertex that is a successor of $v$ on a path away from the root. A **proper descendant** of a vertex $v$ in a rooted tree is any descendant except $v$ itself.

The **parent** of a vertex $v$ in a rooted tree is a vertex that is the immediate predecessor of $v$ on the path to $v$ away from the root.

The **parent function** of a rooted tree $T$ maps the root of $T$ to the empty set and maps every other vertex to its parent.

An **ancestor** of a vertex $v$ in a rooted tree is $v$ itself or any vertex on the path to $v$ away from the root. A **proper ancestor** of a vertex $v$ in a rooted tree is any ancestor except $v$ itself.

**Siblings** in a rooted tree are vertices with the same parent.

An **internal vertex** in a rooted tree is a vertex with children.

A **leaf** in a rooted tree is a vertex that has no children.

The **depth** of a vertex in a rooted tree is the number of edges in the unique path from the root to that vertex.

The **nth level** in a rooted tree is the set of all vertices at depth $n$.

The **height** of a rooted tree is the maximum depth over all vertices.

An **ordered tree** is a rooted tree in which the children of each internal vertex are linearly ordered.

A **left sibling** of a vertex $v$ in an ordered tree is a sibling that precedes $v$ in the ordering of $v$ and its siblings. A **right sibling** of a vertex $v$ in an ordered tree is a sibling that follows $v$ in the ordering of $v$ and its siblings.

A *plane tree* is a geometric realization (drawing) of an ordered tree such that the left-to-right order of the children of each vertex in the drawing is consistent with the linear ordering of the corresponding vertices in the tree.

In the *level ordering* of the vertices of an ordered tree, $u$ precedes $v$ under any of these circumstances:

- if the depth of $u$ is less than the depth of $v$;
- if $u$ is a left sibling of $v$;
- if the parent of $u$ precedes the parent of $v$.

Two ordered trees $(T_1, r_1)$ and $(T_2, r_2)$ are *isomorphic as ordered trees* if there is a rooted tree isomorphism $f: T_1 \to T_2$ that preserves the ordering at every vertex.

An *m-ary tree* is a rooted tree such that every internal vertex has at most $m$ children.

A *full m-ary tree* is a rooted tree such that every internal vertex has exactly $m$ children.

A *(pure) binary tree* is a rooted tree such that every internal vertex has at most two children. This meaning of "binary tree" occurs commonly in graph theory.

A *binary tree* is a 2-ary tree such that every child, even an only child, is distinguished as *left child* or *right child*. This meaning of "binary tree" occurs commonly in computer science and in permutation groups.

The *principal subtree* at a vertex $v$ of a rooted tree comprises all descendants of $v$ and all edges incident to these descendants. It is itself a rooted tree with $v$ designated as its root.

The *left subtree* of a vertex $v$ in a binary tree is the principal subtree at the left child. The *right subtree* of $v$ is the principal subtree at the right child.

A *balanced tree* of height $h$ is a rooted $m$-ary tree in which all leaves are of height $h$ or $h-1$.

A *complete binary tree* is a binary tree in which every parent has two children and all leaves are at the same depth.

A *complete m-ary tree* is an $m$-ary tree in which every parent has $m$ children and all leaves are at the same depth.

A *decision tree* is a rooted tree in which every internal vertex represents a decision and each path from the root to a leaf represents a sequence of decisions.

A *prefix code* for a finite set $X = \{x_1, \ldots, x_n\}$ is a set $\{c_1, \ldots, c_n\}$ of binary strings in $X$ (called *codewords*) such that no codeword is a prefix of any other codeword.

A *Huffman code* for a set $X$ with a probability measure $Pr$ (see §7.1) is a prefix code $\{c_1, \ldots, c_n\}$ such that $\sum_{j=1}^{n} len(c_j) Pr(x_j)$ is minimum among all prefix codes, where $len(c_j)$ measures the length of $c_j$ in bits.

A *Huffman tree* for a set $X$ with a probability measure $Pr$ is a tree constructed by Huffman's algorithm to produce a Huffman code for $(X, Pr)$.

**Facts:**

**1.** Plane trees are usually drawn so that vertices of the same level in the corresponding ordered tree are at the same vertical position in the plane.

**2.** A rooted tree can be represented by its vertex set plus its parent function.

**3.** The concept of finite binary tree also has the following recursive definition: (basis) an ordered tree with only one vertex is a binary tree; (recursion) an ordered tree with more than one vertex is a binary tree if the root has at most two children and if both its principal subtrees are binary trees.

**4.** A full $m$-ary tree with $k$ internal vertices has $mk+1$ vertices and $(m-1)k+1$ leaves.

**5.** A full $m$-ary tree with $k$ vertices has $(k-1)/m$ internal vertices and $((m-1)k+1)/m$ leaves.

**6.** There are at most $m^h$ leaves in any $m$-ary tree of height $h$.

**7.** A *binary search tree* is a special kind of binary tree used to implement a *random access table* with $O(n)$ maintenance and retrieval algorithms. (See §18.2.3.)

**8.** A balanced binary tree can be used to implement a *priority queue* with $O(n)$ enqueue and dequeue algorithms. (See §18.2.4.)

**9.** Algorithm 1, due to D. Huffman in 1951, constructs a Huffman tree.

---

**Algorithm 1:   Find a Huffman code.**

input: the probabilities $Pr(x_1), \ldots, Pr(x_n)$ on a set $X$

output: a Huffman code for $(X, Pr)$

initialize $F$ to be a forest of isolated vertices, labeled $x_1, \ldots, x_n$, each to be
    regarded as a rooted tree

assign weight $Pr(x_j)$ to the rooted tree $x_j$, for $j = 1, \ldots, n$

**repeat until** forest $F$ is a single tree

    choose two rooted trees, $T$ and $T'$, of smallest weights in forest $F$

    replace trees $T$ and $T'$ in forest $F$ by a tree with a new root whose left subtree
        is $T$ and whose right subtree is $T'$

    label the new edge to $T$ with a 0 and the new edge to $T'$ with a 1

    assign weight $w(T) + w(T')$ to the new tree

**return** tree $F$

{The Huffman code word for $x_i$ is the concatenation of the labels on the unique
    path from the root to $x_i$}

---

**Examples:**

**1.** The following is a rooted tree $(T, d)$ with root $d$.

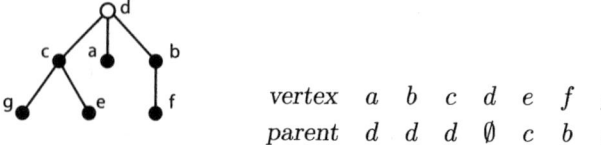

$$
\begin{array}{c|ccccccc}
vertex & a & b & c & d & e & f & g \\
parent & d & d & d & \emptyset & c & b & c \\
\end{array}
$$

**2.** The following is a 2-ary tree of height 4.

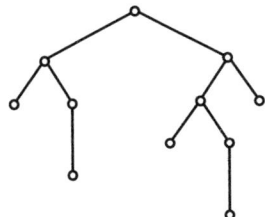

**3.** A balanced binary tree is shown next.

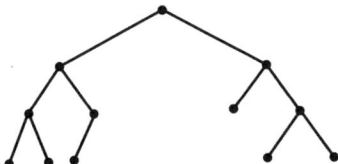

**4.** The following tree is rooted at vertex $r$. Vertices $d$ and $e$ are children of vertex $b$. Vertex $f$ is a descendant of $f$, $d$, $b$, and $r$, but $f$ is not a descendant of vertex $a$. Vertex $a$ is the parent of $c$, which is the only proper descendant of vertex $a$. Vertices $d$ and $e$ are siblings, but $c$ is not a sibling of $d$ or of $e$. The leaves are $c, e, f$; the internal vertices are $a, b, r, d$.

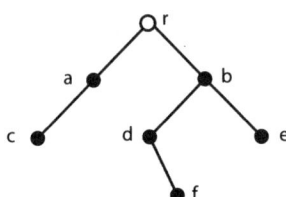

**5.** The following two rooted trees are isomorphic as graphs, but they are considered to be different as rooted trees, because there is no graph isomorphism from one to the other that maps root to root.

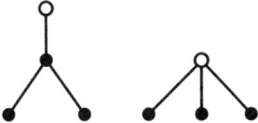

**6.** The following two plane trees are isomorphic as rooted trees, but they are not isomorphic as ordered rooted trees, because there is no rooted tree isomorphism from one to the other that preserves the child ordering at every vertex.

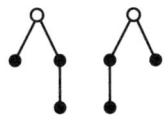

**7.** A complete binary tree of height 2 and a complete 3-ary tree of height 2 are shown next.

**8.** The iterative construction of a *Huffman tree* for the set $X = \{a, b, c, d, e, f\}$ with respective probabilities $\{0.08, 0.10, 0.12, 0.15, 0.20, 0.35\}$ would proceed as follows:

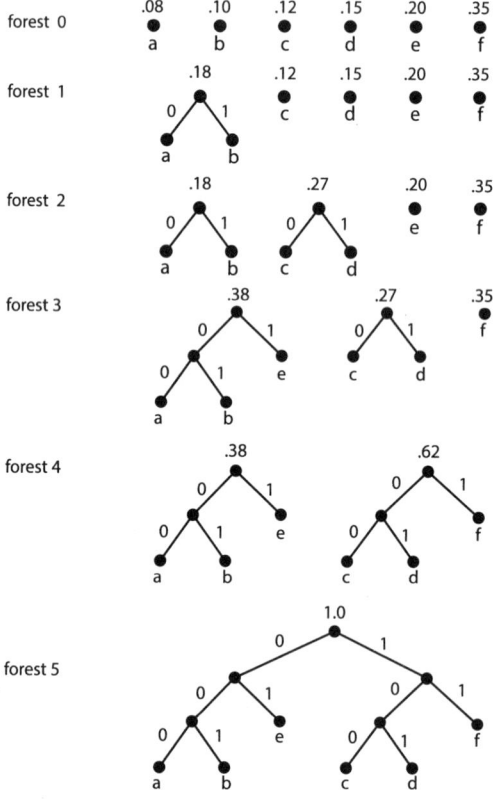

The resulting codes are 000 for $a$, 001 for $b$, 100 for $c$, 101 for $d$, 01 for $e$, and 11 for $f$. So the most frequently used objects in $X$ are represented by the shortest codes.

### 9.1.3  TREE TRAVERSAL

Ordered rooted trees can be used to store data or arithmetic expressions involving numbers, variables, and operations. A tree traversal algorithm gives a systematic method for accessing the information stored in the tree.

**Definitions:**

A **boundary walk** of a plane tree is a walk around the boundary of the single region of the plane imbedding of the tree, starting at the root.

A **backtrack** along a walk in a graph is an instance $\ldots, u, e, v, e, u, \ldots$ of two consecutive edge-steps in which an edge-step traverses the same edge as its predecessor, but in the opposite direction.

A **preorder traversal** of an ordered rooted tree $T$ lists the vertices of $T$ (or their labels) so that each vertex $v$ is followed by all the vertices, in preorder, in its principal subtrees, respecting their left-to-right order.

A **postorder traversal** of an ordered rooted tree $T$ lists the vertices of $T$ (or their labels) so that each vertex $v$ is preceded by all the vertices, in postorder, in its principal subtrees, respecting their left-to-right order.

An **inorder traversal** of an ordered rooted tree $T$ lists the vertices of $T$ (or their labels) so that each vertex $v$ is preceded by all the vertices, in inorder, in its first principal subtree and so that $v$ is followed by the vertices, in inorder, of its other principal subtrees, respecting their left-to-right order.

**Prefix (or Polish) notation** is the form of an arithmetic expression obtained from a preorder traversal of a binary tree representing this expression.

**Postfix (or reverse Polish) notation** is the form of an arithmetic expression obtained from a postorder traversal of a binary tree representing this expression.

**Infix notation** is the form of an arithmetic expression obtained from an inorder traversal of a binary tree representing this expression. A left parenthesis is written immediately before writing the left principal subtree of each vertex, and a right parenthesis is written immediately after writing the right principal subtree.

The **universal address system** of an ordered rooted tree is a labeling in which the root is labeled 0 and for each vertex with label $x$, its $m$ children are labeled $x.1, x.2, \ldots, x.m$, from left to right.

In the **level order** of the vertices of an ordered tree $T$, vertex $u$ precedes vertex $v$ if $u$ is nearer the root, or if $u$ and $v$ are at the same level and $u$ and $v$ have ancestors $u'$ and $v'$ that are siblings and $u'$ precedes $v'$ in the ordering of $T$.

A bijective assignment of labels from an ordered set (such as alphabetic strings or the integers) to the vertices of an ordered tree is **sorted** if the level order of these labels is either ascending or descending.

## Facts:

**1.** The preorder traversal of a plane tree is obtained by a counterclockwise traversal of the boundary walk of the plane region, that is, starting downward toward the left. As each vertex of the tree is encountered for the first time along this walk, it is recorded in the preorder.

**2.** The postorder traversal of a plane tree is obtained by a counterclockwise traversal of the boundary walk of the plane region, that is, starting downward toward the left. As each vertex of the tree is encountered for the last time along this walk, it is recorded in the postorder. Algorithm 2 uses this to find the parents of all vertices.

**3.** The inorder traversal of a plane tree is obtained by a counterclockwise traversal of the boundary walk of the plane region, that is, starting downward toward the left. As each interior vertex of the tree is encountered for the second time along this walk, it is recorded in the inorder. An end vertex is recorded whenever it is encountered for the only time.

**4.** Two nonisomorphic ordered trees with sorted vertex labels can have the same preorder but not the same postorder.

---

**Algorithm 2**:  **Parent-finder for the postorder of a plane tree.**

input: the postorder $v_{p(1)}, \ldots, v_{p(n)}$ of a plane tree with sorted vertex labels and
        a vertex $v_j$

output: the parent of $v_j$

scan the postorder until $v_j$ is encountered

continue scanning until some vertex $v_i$ is encountered such that $i < j$

return $(v_i)$

---

**Examples:**

**1.** The following plane tree has preorder $a\,b\,e\,f\,h\,c\,d\,g\,i\,j\,k$, postorder $e\,h\,f\,b\,c\,i\,j\,k\,g\,d\,a$, and inorder $e\,b\,h\,f\,a\,c\,i\,g\,j\,k\,d$.

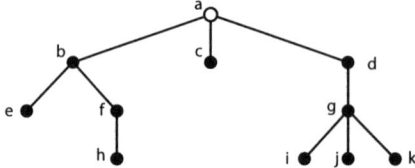

**2.** The following binary tree represents the arithmetic expression that has infix form $(x + y)/(x - 2)$, prefix form $/ + x\,y - x\,2$, and postfix form $x\,y + x\,2 - /$.

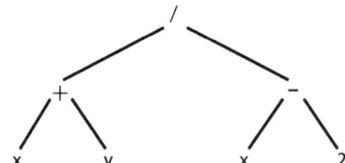

---

## 9.1.4  INFINITE TREES

**Definitions:**

An **infinite tree** is a tree with an infinite number of vertices and edges.

The **diameter** of a tree is the maximum distance between two distinct vertices in the tree.

A **bounded tree** is a tree of finite diameter.

The **degree** of a vertex is the number of edges incident with it.

A **locally finite tree** is a tree in which the degree of every vertex is finite.

A **homogeneous tree** is a tree in which every vertex has the same degree.

An **$n$-homogeneous tree** is a tree in which every vertex has degree $n$.

A **bihomogeneous tree** is a nonhomogeneous tree with a partition of the vertices into two subsets, such that all vertices in the same subset have the same degree.

**Facts:**

**1.** *König's lemma:* An infinite tree that is locally finite contains an infinitely long path.

**Examples:**

**1.** Suppose that two finite bitstrings are considered adjacent if one bitstring can be obtained from the other by appending a 0 or a 1 at the right. The resulting graph is the infinite bihomogeneous tree, in which the empty string $\lambda$ has degree 2 and all other finite bitstrings have degree 3.

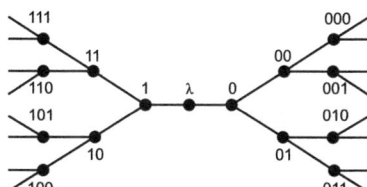

**2.** Consider the set of all finite strings on the alphabet $\{a, a^{-1}, b, b^{-1}\}$ containing no instances of the substrings $aa^{-1}$, $a^{-1}a$, $bb^{-1}$, or $b^{-1}b$. Suppose that two such strings are considered to be adjacent if and only if one of them can be obtained from the other by appending one of the alphabet symbols at the right. Then the resulting graph is a 4-homogeneous tree.

**3.** Consider as vertices the set of infinite bitstrings with at most two 1s. Suppose two such bitstrings are regarded as adjacent if they differ in only one bit and that bit is a rightmost 1 for one of the two bitstrings. This graph is a bounded tree of diameter four.

# 9.2   SPANNING TREES

A spanning tree of a graph $G$ is a subgraph of $G$ that is a tree and contains every vertex of $G$. Spanning trees are very useful in searching the vertices of a graph and in communicating from any given vertex to the other vertices. Minimum spanning trees are covered in §10.1.

## 9.2.1   DEPTH-FIRST AND BREADTH-FIRST SPANNING TREES

**Definitions:**

A *spanning tree* of a graph $G$ is a tree that is a subgraph of $G$ and that contains every vertex of $G$.

A *tree edge* of a graph $G$ with a spanning tree $T$ is an edge $e$ such that $e \in T$.

A *chord* of a graph $G$ with a spanning tree $T$ is an edge $e$ such that $e \notin T$.

A *back edge* of a digraph $G$ with a spanning tree $T$ is a chord that joins one of its endpoints to an ancestor in $T$. A *forward edge* is a chord that joins one of its endpoints to a descendant in $T$. A *cross edge* is a chord that is neither a back edge nor a forward edge.

The **fundamental cycle** of a chord $e$ with respect to a given spanning tree $T$ of a graph $G$ consists of the edge $e$ and the unique path in $T$ joining the endpoints of $e$.

A **depth-first search** (**DFS**) of a graph $G$ is a way to traverse every vertex of a connected graph by constructing a spanning tree, rooted at a given vertex $r$. Each stage of the DFS traversal seeks to move to an unvisited neighbor of the most recently visited vertex, and backtracks only if there is none available. See Algorithm 1.

A **depth-first search tree** is the spanning tree constructed during a depth-first search.

**Backtracking** during a depth-first search means retreating from a vertex with no unvisited neighbors back to its parent in the DFS-tree.

A **breadth-first search** (**BFS**) of a graph $G$ is a way to traverse every vertex of a connected graph by constructing a spanning tree, rooted at a given vertex $r$. After the BFS traversal visits a vertex $v$, all of the previously unvisited neighbors of $v$ are enqueued, and then the traversal removes from the queue whatever vertex is at the front of the queue, and visits that vertex. See Algorithm 2.

A **breadth-first search tree** is the spanning tree constructed during a breadth-first search.

The **fundamental system of cycles** of a connected graph $G$ associated with a spanning tree $T$ is the set of fundamental cycles corresponding to the various edges of $G - T$.

Given two vertex sets $X_1$ and $X_2$ that partition the vertex set of a graph $G$, the **partition-cut** $\langle X_1, X_2 \rangle$ is the set of edges of $G$ that have one endpoint in $X_1$ and the other in $X_2$.

The **fundamental edge-cut** of a connected graph $G$ associated with removal of an edge $e$ from a spanning tree $T$ is the partition-cut $\langle X_1, X_2 \rangle$ where $X_1$ and $X_2$ are the vertex sets of the two components of $T - e$.

The **fundamental system of edge-cuts** of a connected graph $G$ associated with a spanning tree $T$ is the set of fundamental edge-cuts that result from removal of an edge from the tree $T$.

---

**Algorithm 1**:  **Depth-first search spanning tree.**

input: a connected $n$-vertex graph $G$ and a starting vertex $r$

output: the edge set $E_T$ of a spanning tree and an array $X[1..n]$ listing $V_G$ in
    DFS order

initialize all vertices as *unvisited* and all edges as *unused*

$E_T := \emptyset;\ loc := 1$

$dfs(r)$

**procedure** $dfs(u)$

mark $u$ as *visited*

$X[loc] := u$

$loc := loc + 1$

**while** vertex $u$ has any *unused* edges

    $e := $ next *unused* edge at $u$

mark $e$ as *used*

$w :=$ the other endpoint of edge $e$

**if** $w$ is *unvisited* **then**

    add $e$ to $E_T$

    *dfs*$(w)$

---

**Algorithm 2**:  **Breadth-first search spanning tree.**

input: a connected $n$-vertex graph $G$ and a starting vertex $r$.

output: the edge set $E_T$ of a spanning tree and an array $X[1..n]$ listing $V_G$ in
    BFS order

initialize all vertices as *unvisited* and all edges as *unused*

$E_T := \emptyset$; $loc := 1$; $Q := r$  {$Q$ is a queue}

**while** $Q \neq \emptyset$

    $x := front(Q)$

    remove $x$ from $Q$

    *bfs*$(x)$

**procedure** *bfs*$(u)$

mark $u$ as *visited*

$X[loc] := u$

$loc := loc + 1$

**while** vertex $u$ has any *unused* edges

    $e :=$ next *unused* edge at $u$

    mark $e$ as *used*

    $w :=$ the other endpoint of edge $e$

    **if** $w$ is *unvisited* **then**

        add $e$ to $E_T$

        add $w$ to the back of $Q$

**Facts:**

**1.** Every connected graph has at least one spanning tree.

**2.** A connected graph $G$ has $k$ edge-disjoint spanning trees if and only if for every partition of $V_G$ into $m$ nonempty subsets, there are at least $k(m-1)$ edges connecting vertices in different subsets.

**3.** Let $T$ and $T'$ be spanning trees of a graph $G$ and $e \in T - T'$. Then there exists an edge $e' \in T' - T$ such that both $T - e \cup \{e'\}$ and $T' - e' \cup \{e\}$ are spanning trees of $G$.

**4.** In the column vector space of the incidence matrix of $G$ over $Z_2$, every edge set can be represented as a sum of column vectors. Let $T$ be a spanning tree of $G$. Then each cycle $C$ can be written in a unique way as a linear combination of the fundamental cycles of whatever chords of $T$ occur in $C$.

**5.** Depth-first search on an $n$-vertex, $m$-edge graph runs in $O(m)$ time.

**6.** DFS-trees are used to find the components, cutpoints, blocks, and cut-edges of a graph.

**7.** The unique path in the BFS-tree $T$ of a graph $G$ from its root $r$ to a vertex $v$ is a shortest path in $G$ from $r$ to $v$.

**8.** Breadth-first search on an $n$-vertex, $m$-edge graph runs in $O(m)$ time.

**9.** A BFS-tree in a simple graph has no back edges.

**10.** Dijkstra's algorithm (see §10.3.2) constructs a spanning tree $T$ in an edge-weighted graph such that for each vertex $v$, the unique path in $T$ from a specified root $r$ to $v$ is a minimum-cost path in the graph from $r$ to $v$. When all edges have unit weights, Dijkstra's algorithm produces the BFS tree.

**11.** The level order of the vertices of an ordered tree is the order in which they would be traversed in a breadth-first search of the tree.

**12.** The fundamental cycle of an edge $e$ with respect to a spanning tree $T$ such that $e \notin T$ consists of edge $e$ and those edges of $T$ whose fundamental edge-cuts contain $e$.

**13.** The fundamental edge-cut with respect to removal of edge $e$ from a spanning tree $T$ consists of edge $e$ and those edges of $E_G - E_T$ whose fundamental cycles contain $e$.

### Examples:

**1.** Consider the following graph and spanning tree and a digraph on the same vertex and edge set. The tree edges are $a, b, c, e, f, h, k, l$, and the chords are $d, g, i, j$. Chord $d$ is a forward edge, chord $i$ is a back edge, and chords $g$ and $j$ are cross edges.

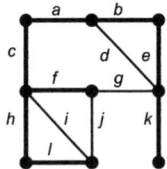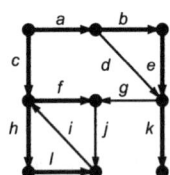

**2.** In the graph of Example 1, the fundamental cycles of the chords $d$, $g$, $i$, and $j$ are $\{d, b, e\}$, $\{g, f, c, a, b, e\}$, $\{i, h, l\}$, and $\{j, f, h, l\}$, respectively. The non-fundamental cycle $\{a, d, g, c, f\}$ is the sum (mod 2) of the fundamental cycles of chords $d$ and $g$.

**3.** A spanning tree and its fundamental system of cycles are shown next.

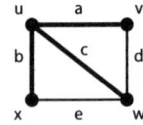

**4.** A spanning tree and its fundamental system of edge-cuts are shown next.

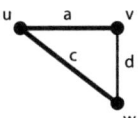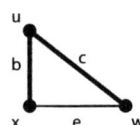

**5.** In the graph of Example 1 suppose that the local order of adjacencies at each vertex is the alphabetic order of the edge labels. Then the construction of the DFS-tree is shown in the following figure.

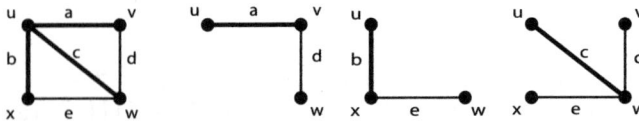

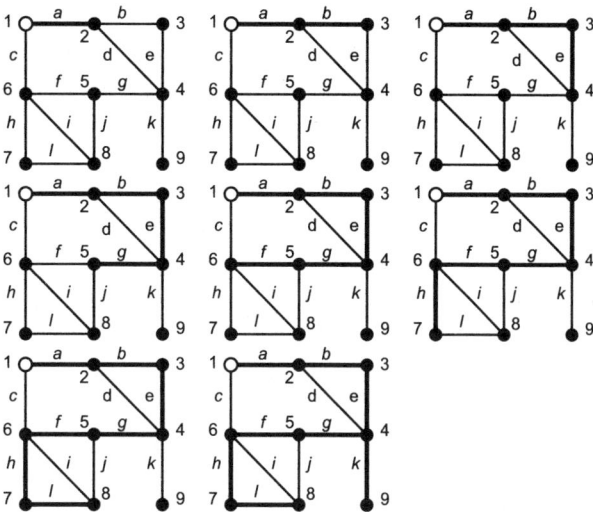

**6.** In the graph of Example 1 suppose that the local order of adjacencies at each vertex is the alphabetic order of the edge labels. Then the construction of the BFS-tree is shown in the following figure.

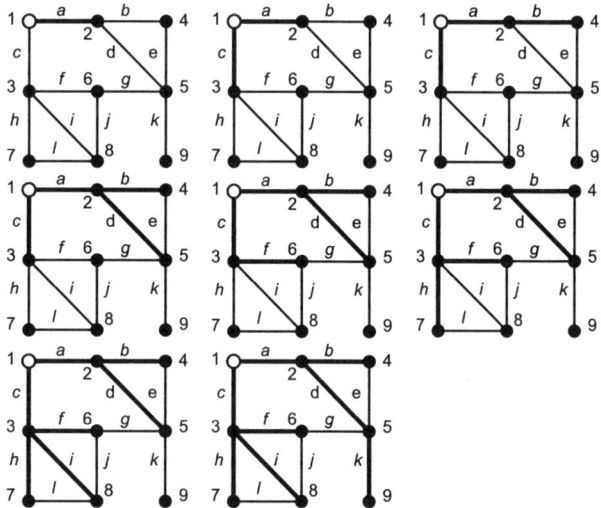

---

## 9.2.2   ENUMERATION OF SPANNING TREES

### Definitions:

The **number of spanning trees** $\tau(G)$ of a graph $G$ counts two spanning trees $T_1$ and $T_2$ as different if their edge sets are different, even if there is an automorphism of $G$ mapping $T_1$ onto $T_2$.

The **degree matrix** $D(G)$ of an $n$-vertex graph $G$ with vertex degree sequence $d_1, \ldots, d_n$ is the $n \times n$ diagonal matrix in which the elements of the main diagonal are the degrees $d_1, \ldots, d_n$ (and the off-diagonal elements are 0s).

**Facts:**

**1.** *Cayley's formula:* $\tau(K_n) = n^{n-2}$, where $K_n$ is the complete graph.

**2.** $\tau(K_{m,n}) = m^{n-1}n^{m-1}$, where $K_{m,n}$ is the complete bipartite graph.

**3.** $\tau(I_s * K_{n-s}) = n^{n-2}(1 - s/n)^{s-1}$, where $I_s$ is the edgeless graph on $n$ vertices and "$*$" denotes the join (see §8.1.2).

**4.** $\tau(W_n) = \left(\frac{3+\sqrt{5}}{2}\right)^n + \left(\frac{3-\sqrt{5}}{2}\right)^n - 2$, where $W_n$ denotes the wheel with $n$ rim vertices.

**5.** *Matrix-tree theorem:* For each $s$ and $t$, the value $\tau(G)$ equals $(-1)^{s+t}$ times the determinant of the matrix obtained by deleting row $s$ and column $t$ from $D(G) - A(G)$, where $A(G)$ is the adjacency matrix for $G$.

**6.** For each edge $e$ of a graph $G$, $\tau(G) = \tau(G - e) + \tau(G/e)$, where "$-e$" denotes edge deletion and "$/e$" denotes edge contraction.

**7.** The number of spanning trees of $K_n$ with degrees $d_1, \ldots, d_n$ equals $\binom{n-2}{d_1-1,\ldots,d_n-1}$ (see §2.3.2). In this formula, the vertices are distinguishable (labeled) and are given their degrees in advance, and the only question is how to realize them with edges.

**Examples:**

**1.** $\tau(K_3) = 3^{3-2} = 3$. Each of the three spanning trees is a path on two edges, as illustrated below. Also, $\tau(K_4) = 4^{4-2} = 16$.

**2.** $\tau(K_{2,n}) = n2^{n-1}$. To confirm this, let $X = \{x_1, x_2\}$ and $|Y| = n$. The spanning tree contains a path of length 2 joining $x_1$ to $x_2$, whose middle vertex in $Y$ can be chosen in $n$ ways. For each of the remaining $n - 1$ vertices of $Y$, there is a choice as to which of $x_1$ and $x_2$ is its neighbor (not both, since that would create a cycle).

**3.** $\tau(I_3 + K_2) = 5^3 \left(1 - \frac{3}{5}\right)^2 = 20$.

**4.** $\tau(W_4) = \left(\frac{3+\sqrt{5}}{2}\right)^4 + \left(\frac{3-\sqrt{5}}{2}\right)^4 - 2 = 45$.

**5.** To illustrate the matrix-tree theorem, consider the following graph $G$. Then

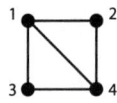

$$D(G) - A(G) = \begin{pmatrix} 3 & -1 & -1 & -1 \\ -1 & 2 & 0 & -1 \\ -1 & 0 & 2 & -1 \\ -1 & -1 & -1 & 3 \end{pmatrix}.$$

Deleting row 2 and column 3, for example, yields

$$\tau(G) = (-1)^{2+3} \begin{vmatrix} 3 & -1 & -1 \\ -1 & 0 & -1 \\ -1 & -1 & 3 \end{vmatrix} = 8.$$

The eight spanning trees of $G$ are

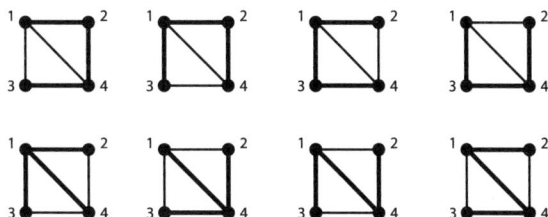

**6.** The recursive formula $\tau(G) = \tau(G-e) + \tau(G/e)$ is illustrated with the same graph $G$ of the previous example and with $e = v_1 v_4$. In the computation $G$ is drawn instead of writing $\tau(G)$, and similarly with the other graphs. This yields

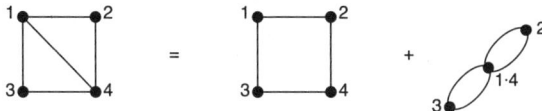

**7.** Let the vertices of $K_5$ be $v_0, v_1, v_2, v_3, v_4$. The number of spanning trees of $K_5$ in which the degrees of $v_0, v_1, v_2, v_3, v_4$ are $3, 2, 1, 1, 1$, respectively, is given by the multinomial coefficient $\binom{5-2}{3-1, 2-1, 1-1, 1-1, 1-1} = \frac{3!}{2! \cdot 1! \cdot 0! \cdot 0! \cdot 0!} = \frac{6}{2 \cdot 1 \cdot 1 \cdot 1} = 3$. The three trees in question are

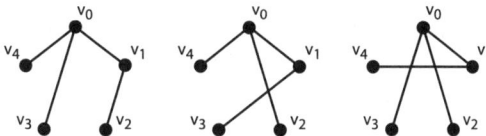

## 9.3  ENUMERATING TREES

Tree counting began with Arthur Cayley in 1889, who wanted to enumerate the saturated hydrocarbons. George Pólya developed an extensive theory in 1937 for counting families of organic chemicals, which was used by Richard Otter in 1948 in his solution of the specific problem of counting saturated hydrocarbons. Tree counting formulas are used in computer science to estimate running times in the design of algorithms.

### 9.3.1  COUNTING GENERIC TREES

When counting generic trees, we must be careful to distinguish among labeled trees, rooted trees, unlabeled trees, and various other objects. While labeled trees can be counted easily, unlabeled trees, both rooted and unrooted, are typically counted using generating functions.

**Definitions:**
A **labeled tree** is a tree in which distinct labels, typically $v_1, v_2, \ldots, v_n$, have been assigned to the vertices. Two labeled trees with the same set of labels are considered the same only if there is an isomorphism from one to the other that preserves the labels.

A **rooted tree** is a tree in which one vertex, the root, is distinguished. Two rooted trees are considered the same only if there is an isomorphism from one to the other that maps the root of the first to the root of the second.

A **rooted labeled tree** is a labeled tree in which one vertex, the root, is distinguished. Two rooted labeled trees with the same set of labels are considered the same only if there is an isomorphism from one to the other that preserves the labels and maps the root of the first to the root of the second.

A **reduced tree** (or **irreducible tree**) is a tree with no vertices of degree 2.

**Examples:**
**1.** The figure below shows the three isomorphically distinct trees with five vertices. There are 60 essentially different ways to label each of the first two and five essentially different ways to label the third. Thus there are 125 different labeled trees with five vertices.

**2.** There are three essentially different ways to root the first tree in the figure of Example 1, four essentially different ways to root the second, and two essentially different ways to root the third. Thus there are nine rooted (unlabeled) trees with five vertices.

**3.** Each of the 125 labeled trees discussed in Example 1 can be rooted at any of its five vertices, yielding 625 possible rooted labeled trees.

**4.** The third tree in the figure of Example 1 is the only reduced tree with five vertices.

**Facts:**
**1.** The number of labeled trees with $n$ vertices is $n^{n-2}$. See Table 1. This is sequence A000272 in [OEIS].

**2.** The number of rooted labeled trees with $n$ vertices is $n^{n-1}$. See Table 1. This is sequence A000169 in [OEIS].

**Table 1: Labeled trees and rooted labeled trees with $n$ vertices.**

| $n$ | Labeled Trees | Rooted Labeled Trees |
|---|---|---|
| 1 | 1 | 1 |
| 2 | 1 | 2 |
| 3 | 3 | 9 |
| 4 | 16 | 64 |
| 5 | 125 | 625 |
| 6 | 1,296 | 7,776 |
| 7 | 16,807 | 117,649 |
| 8 | 262,144 | 2,097,152 |
| 9 | 4,782,969 | 43,046,721 |
| 10 | 100,000,000 | 1,000,000,000 |
| 11 | 2,357,947,691 | 25,937,424,601 |
| 12 | 61,917,364,224 | 743,008,370,688 |

| $n$ | Labeled Trees | Rooted Labeled Trees |
|----|----|----|
| 13 | 1,792,160,394,037 | 23,298,085,122,481 |
| 14 | 56,693,912,375,296 | 793,714,773,254,144 |
| 15 | 1,946,195,068,359,375 | 29,192,926,025,390,625 |
| 16 | 72,057,594,037,927,936 | 1,152,921,504,606,846,980 |

**3.** Let $R_n$ denote the number of (unlabeled) rooted trees with $n$ vertices, and let $r(x)$ be the generating function for rooted trees, so that

$$r(x) = \sum_{n=1}^{\infty} R_n x^n = x + x^2 + 2x^3 + 4x^4 + 9x^5 + 20x^6 + \cdots .$$

The coefficients $R_n$ of this generating function can be determined by means of the recurrence relation

$$r(x) = x \prod_{k=1}^{\infty} (1 - x^k)^{-R_k}.$$

See Table 2. This is sequence A000081 in [OEIS].

**Table 2: Rooted trees, trees, and reduced trees with $n$ vertices.**

| $n$ | Rooted Trees | Trees | Reduced Trees |
|----|----|----|----|
| 1 | 1 | 1 | 1 |
| 2 | 1 | 1 | 1 |
| 3 | 2 | 1 | 0 |
| 4 | 4 | 2 | 1 |
| 5 | 9 | 3 | 1 |
| 6 | 20 | 6 | 2 |
| 7 | 48 | 11 | 2 |
| 8 | 115 | 23 | 4 |
| 9 | 286 | 47 | 5 |
| 10 | 719 | 106 | 10 |
| 11 | 1,842 | 235 | 14 |
| 12 | 4,766 | 551 | 26 |
| 13 | 12,486 | 1,301 | 42 |
| 14 | 32,973 | 3,159 | 78 |
| 15 | 87,811 | 7,741 | 132 |
| 16 | 235,381 | 19,320 | 249 |
| 17 | 634,847 | 48,629 | 445 |
| 18 | 1,721,159 | 123,867 | 842 |
| 19 | 4,688,676 | 317,955 | 1,561 |
| 20 | 12,826,228 | 823,065 | 2,988 |
| 21 | 35,221,832 | 2,144,505 | 5,671 |
| 22 | 97,055,181 | 5,623,756 | 10,981 |
| 23 | 268,282,855 | 14,828,074 | 21,209 |
| 24 | 743,724,984 | 39,299,897 | 41,472 |

| $n$ | Rooted Trees | Trees | Reduced Trees |
|---|---|---|---|
| 25 | 2,067,174,645 | 104,636,890 | 81,181 |
| 26 | 5,759,636,510 | 279,793,450 | 160,176 |
| 27 | 16,083,734,329 | 751,065,460 | 316,749 |
| 28 | 45,007,066,269 | 2,023,443,032 | 629,933 |
| 29 | 126,186,554,308 | 5,469,566,585 | 1,256,070 |
| 30 | 354,426,847,597 | 14,830,871,802 | 2,515,169 |

**4.** Let $T_n$ denote the number of trees with $n$ vertices, and let $t(x)$ be the generating function for trees, so that

$$t(x) = \sum_{n=1}^{\infty} T_n x^n = x + x^2 + x^3 + 2x^4 + 3x^5 + 6x^6 + \cdots .$$

The coefficients $T_n$ of this generating function $t(x)$ can be determined from the generating function $r(x)$ for rooted trees in Fact 3 above by using the formula

$$t(x) = r(x) - \tfrac{1}{2}\left( r^2(x) - r(x^2) \right) .$$

See Table 2. This is sequence A000055 in [OEIS].

**5.** Counting reduced trees requires an auxiliary sequence $Q_n$ with generating function $q(x)$, so that

$$q(x) = \sum_{k=1}^{\infty} Q_k x^k = x + x^3 + x^4 + 2x^5 + 3x^6 + 6x^7 + 10x^8 + \cdots .$$

The coefficients $Q_i$ of this generating function can be determined from the recurrence relation

$$q(x) = \frac{x}{1+x} \prod_{k=1}^{\infty} (1 - x^k)^{-Q_k} .$$

This is sequence A001678 in [OEIS].

**6.** Let $H_n$ denote the number of reduced trees with $n$ vertices, and let $h(x)$ be the generating function for reduced trees, so that

$$h(x) = \sum_{n=1}^{\infty} H_n x^n = x + x^2 + x^4 + x^5 + 2x^6 + 2x^7 + 4x^8 + \cdots .$$

The coefficients $H_n$ of this generating function $h(x)$ can be determined from the auxiliary function $q(x)$ in Fact 5 above by using the formula

$$h(x) = (1+x)q(x) - \left(\frac{1+x}{2}\right) q^2(x) + \left(\frac{1-x}{2}\right) q(x^2).$$

See Table 2. Note that there are no reduced trees with exactly 3 vertices. This is sequence A000014 in [OEIS].

## 9.3.2   COUNTING TREES IN CHEMISTRY

### Definitions:

A *1-4 tree* is a tree in which each vertex has degree 1 or 4.

A *1-rooted 1-4 tree* is a 1-4 tree rooted at a vertex of degree 1.

### Remarks:

**1.** The 1-4 trees model many types of organic chemical molecules such as saturated hydrocarbons or alkanes. These molecules have the chemical formula $C_nH_{2n+2}$ and consist of $n$ carbon atoms of valence 4 and $2n + 2$ hydrogen atoms of valence 1.

**2.** The 1-rooted 1-4 trees model the monosubstituted hydrocarbons such as the alcohols with the chemical formula $C_nH_{2n+1}OH$ and consisting of $n$ carbon atoms, $2n+1$ hydrogen atoms, and an OH group.

**3.** It is convenient when counting alcohols to include the water molecule HOH as an honorary alcohol.

### Examples:

**1.** The figure below shows the three different 1-4 trees with five vertices of degree 4 and 12 vertices of degree 1.

**2.** The first 1-4 tree in Example 1 can be rooted at a vertex of degree 1 in three essentially different ways, the second in four essentially different ways, and the third in essentially only one way. Thus there are eight different 1-rooted 1-4 trees with five vertices of degree 4.

### Facts:

**1.** A 1-4 tree with $n$ vertices of degree 4 always has $2n + 2$ vertices of degree 1.

**2.** Let $A_n$ denote the number of 1-rooted 1-4 trees with $n$ vertices of degree 4, and let $a(x)$ be the generating function for the number of 1-rooted 1-4 trees, so that

$$a(x) = \sum_{n=0}^{\infty} A_n x^n = 1 + x + x^2 + 2x^3 + 4x^4 + 8x^5 + 17x^6 + \cdots .$$

The coefficients $A_n$ of this generating function $a(x)$ can be determined from the recurrence relation

$$a(x) = 1 + \tfrac{x}{6} \left( a^3(x) + 3a(x)a(x^2) + 2a(x^3) \right).$$

See Table 3. This is sequence A000598 in [OEIS].

**Table 3: 1-Rooted 1-4 trees and 1-4 trees with $n$ vertices of degree 4.**

| $n$ | 1-Rooted 1-4 Trees | 1-4 trees |
|---|---|---|
| 1 | 1 | 1 |
| 2 | 1 | 1 |
| 3 | 2 | 1 |
| 4 | 4 | 2 |
| 5 | 8 | 3 |
| 6 | 17 | 5 |
| 7 | 39 | 9 |
| 8 | 89 | 18 |
| 9 | 211 | 35 |
| 10 | 507 | 75 |
| 11 | 1,238 | 159 |
| 12 | 3,057 | 355 |
| 13 | 7,639 | 802 |
| 14 | 19,241 | 1,858 |
| 15 | 48,865 | 4,347 |
| 16 | 124,906 | 10,359 |
| 17 | 321,198 | 24,894 |
| 18 | 830,219 | 60,523 |
| 19 | 2,156,010 | 148,284 |
| 20 | 5,622,109 | 366,319 |
| 21 | 14,715,813 | 910,726 |
| 22 | 38,649,152 | 2,278,658 |
| 23 | 101,821,927 | 5,731,580 |
| 24 | 269,010,485 | 14,490,245 |
| 25 | 712,566,567 | 36,797,588 |
| 26 | 1,891,993,344 | 93,839,412 |
| 27 | 5,034,704,828 | 240,215,803 |
| 28 | 13,425,117,806 | 617,105,614 |
| 29 | 35,866,550,869 | 1,590,507,121 |
| 30 | 95,991,365,288 | 4,111,846,763 |

**3.** Counting (unrooted) 1-4 trees requires first counting 1-4 trees rooted at a vertex of degree 4. Let $Z_n$ be the number of 4-rooted 1-4 trees with $n$ vertices of degree 4, and let $z(x)$ be the generating function for the number of 4-rooted 1-4 trees, so that

$$z(x) = \sum_{n=1}^{\infty} Z_n x^n = x + x^2 + 2x^3 + 4x^4 + 9x^5 + 18x^6 + \cdots .$$

The coefficients $Z_n$ of this generating function $z(x)$ can be obtained by using the formula

$$z(x) = \tfrac{x}{24} \left( a^4(x) + 6a^2(x)a(x^2) + 8a(x)a(x^3) + 3a^2(x^2) + 6a(x^4) \right),$$

where $a(x)$ is the generating function for 1-rooted 1-4 trees from Fact 2. This is sequence A000678 in [OEIS].

**4.** Let $B_n$ denote the number of (unrooted) 1-4 trees with $n$ vertices of degree 4, and let $b(x)$ be the generating function for 1-4 trees, so that

$$b(x) = \sum_{n=0}^{\infty} B_n x^n = 1 + x + x^2 + x^3 + 2x^4 + 3x^5 + 5x^6 + \cdots .$$

The coefficients $B_n$ of this generating function $b(x)$ can be determined from the functions $a(x)$ and $z(x)$ from Facts 2 and 3, respectively, by using the formula

$$b(x) = z(x) + a(x) - \tfrac{1}{2}\left(a^2(x) - a(x^2)\right).$$

See Table 3. This is sequence A000602 in [OEIS].

---

### 9.3.3 COUNTING TREES IN COMPUTER SCIENCE

**Definitions:**

A **binary tree** consists of a root vertex and at most two principal subtrees that are themselves binary trees. Each principal subtree must be specified as either the left subtree or the right subtree. The root vertex of a binary tree is joined by an edge to the root of each principal subtree.

An **ordered tree** consists of a root vertex and a sequence $t_1, t_2, \ldots, t_m$ of $m \geq 0$ principal subtrees that are themselves ordered trees.

The **children** of the root vertex of an ordered tree or a binary tree are the roots of the principal subtrees.

A **full binary tree** (sometimes called a **left-right tree**) is a binary tree in which each vertex has either 0 or 2 children.

**Examples:**

**1.** The following figure shows the five binary trees with three vertices.

**2.** The next figure shows the five ordered trees with four vertices.

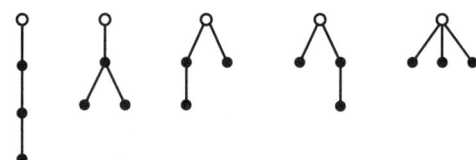

**3.** The following figure shows the five full binary trees with seven vertices.

**Remarks:**

**1.** In computer science, trees are usually drawn with the root at the top.

**2.** Binary trees are easily represented in a computer. Other types of trees are often converted into binary trees for computer representation.

**3.** Ordered trees are used to represent structures such as family trees, showing all descendants of a person represented by the root. The roots of the principal subtrees represent the children of the root person, in order of birth.

**4.** Full binary trees are frequently used to represent arithmetic expressions, in which the leaves correspond to numbers and the other vertices represent binary operations such as $+$, $-$, $\times$, or $\div$.

**Facts:**

**1.** The number of binary trees with $n$ vertices is the Catalan number $C_n$ (see §3.1.3) given by the formula

$$C_n = \frac{1}{n+1}\binom{2n}{n} = \frac{(2n)!}{(n+1)!\,n!} \quad \text{for } n \geq 0.$$

See Table 4. This is sequence A000108 in [OEIS].

**Table 4: The Catalan numbers.**

| $n$ | $C_n$ | $n$ | $C_n$ |
|---|---|---|---|
| 1 | 1 | 17 | 129,644,790 |
| 2 | 2 | 18 | 477,638,700 |
| 3 | 5 | 19 | 1,767,263,190 |
| 4 | 14 | 20 | 6,564,120,420 |
| 5 | 42 | 21 | 24,466,267,020 |
| 6 | 132 | 22 | 91,482,563,640 |
| 7 | 429 | 23 | 343,059,613,650 |
| 8 | 1,430 | 24 | 1,289,904,147,324 |
| 9 | 4,862 | 25 | 4,861,946,401,452 |
| 10 | 16,796 | 26 | 18,367,353,072,152 |
| 11 | 58,786 | 27 | 69,533,550,916,004 |
| 12 | 208,012 | 28 | 263,747,951,750,360 |
| 13 | 742,900 | 29 | 1,002,242,216,651,368 |
| 14 | 2,674,440 | 30 | 3,814,986,502,092,304 |
| 15 | 9,694,845 | 31 | 14,544,636,039,226,909 |
| 16 | 35,357,670 | 32 | 55,534,064,877,048,198 |

**2.** The number of ordered trees with $n$ vertices is the Catalan number $C_{n-1}$. See Table 4.

**3.** The number of full binary trees with $2n+1$ vertices is also $C_n$. See Table 4.

# REFERENCES

*Printed Resources*:

[Ca89] A. Cayley, A theorem on trees, *Quarterly Journal of Pure and Applied Mathematics* 23 (1889), 376–378.

[GrYeZh13] J. Gross, J. Yellen, and P. Zhang, *Handbook of Graph Theory*, 2nd ed., Chapman and Hall/CRC, 2013.

[HaPa73] F. Harary and E. M. Palmer, *Graphical Enumeration*, Academic Press, 1973.

[PóRe87] G. Pólya and R. C. Read, *Combinatorial Enumeration of Groups, Graphs, and Chemical Compounds*, Springer-Verlag, 1987.

[Ro11] K. H. Rosen, *Discrete Mathematics and Its Applications*, 7th ed., McGraw-Hill, 2011.

[Se80] J. P. Serre, *Trees*, Springer-Verlag, 1980.

[Ta83] R. E. Tarjan, *Data Structures and Network Algorithms*, SIAM, 1983.

[We01] D. B. West, *Introduction to Graph Theory*, 2nd ed., Prentice-Hall, 2001.

*Web Resources*:

[OEIS] `oeis.org`  (The On-Line Encyclopedia of Integer Sequences.)

`www.cs.stonybrook.edu/~algorith/`  (The Stony Brook Algorithm Repository.)

# 10

# NETWORKS AND FLOWS

# INTRODUCTION

The vertices and edges of a graph often have quantitative information associated with them, such as supplies and demands (for vertices), and distance, length, capacity, and cost (for edges). Relative to such networks, a number of discrete optimization problems arise in a variety of disciplines: statistics, electrical engineering, operations research, combinatorics, and computer science. Typical applications include designing least cost telecommunication systems, maximizing throughput in a manufacturing system, finding a minimum cost route or set of routes for delivery vehicles, and distributing electricity from a set of supply points to meet customer demands at minimum cost. In this chapter, a number of classical network optimization problems are studied and algorithms are described for their exact or approximate solution. Models and analysis of small-world networks are also presented.

# GLOSSARY

**adjacency matrix**: a 0-1 matrix whose $(i, j)$ entry indicates the absence or presence, respectively, of an arc joining vertex $i$ to vertex $j$ in a graph.

**adjacency set**: the set of arcs emanating from a specified vertex.

**alternating path** (in a matching): a path with edges that are alternately free and matched.

**arc list**: a list of the arcs of a graph, presented in no particular order.

**assignment** (from a set $S$ to a set $T$): a bijective function from $S$ onto $T$.

**augmenting path** (in a flow network): a directed path between two specified vertices in which each arc has a positive residual capacity.

**augmenting path** (in a matching): an alternating path between two free vertices.

**backbone network**: a collection of devices that interconnect vertices at which message exchanges occur in a communication network.

**Barabási-Albert model**: a model of network formation in which a new vertex is (preferentially) joined to existing vertices based on their current degree in the network.

**blossom**: an odd length cycle formed by adding an edge joining two even vertices on an alternating path.

**blocking pair** (relative to a matching): a currently unmatched pair of individuals, each of which prefers one another to their current mate in the matching.

***capacitated concentrator location problem***: an optimization problem in which a minimum cost configuration of concentrators and their connections to terminals are sought so that each concentrator's total capacity is not exceeded.

***capacitated minimum spanning tree***: a minimum cost collection of subtrees joined to a specified root vertex, in which the total amount of demand generated by each subtree is bounded above by a constant.

***capacitated network***: a network in which each arc is assigned a capacity.

***capacity*** (of an arc): the maximum amount of material that can flow along the arc.

***capacity*** (of a cut $[S, \overline{S}]$): the sum of the capacities of arcs $(i, j)$ with $i \in S$ and $j \in \overline{S}$.

***capacity*** (of a path): the smallest capacity of any arc on the path.

***capacity assignment problem***: an optimization problem in which links of different capacities are to be installed at minimum cost to support a number of point-to-point communication demands.

***clustering coefficient*** (of a graph): a measure of the probability that two random neighbors of a vertex are neighbors of each other.

***complete matching***: in a bipartite graph $G = (X \cup Y, E)$, a matching $M$ in which each vertex of $X$ is incident with an edge of $M$.

***composite (hybrid) method***: a heuristic algorithm that combines elements of both construction methods and improvement methods.

***configuration model***: a model that generates graphs by randomly connecting half-edges in a graph having a given degree sequence.

***connected facility location problem***: an optimization problem in which certain facilities are to be opened and customers assigned to them at minimum total cost.

***construction method***: a heuristic algorithm that builds a feasible solution, starting with a trivial configuration.

***cost*** (of a flow): $\sum_{(i,j)} c_{ij} x_{ij}$, where $c_{ij}$ is the cost and $x_{ij}$ is the flow on arc $(i, j)$.

***cut*** (in a graph): the set of edges $[S, \overline{S}]$ in the graph joining vertices in $S$ to vertices in the complementary set $\overline{S}$.

***degree distribution*** (of a graph): the proportion of vertices having each degree in the graph.

***directed network***: a vertex set $V$ and an arc set $E$, where each directed arc has an associated cost, length, weight, or capacity.

***directed out-tree***: a tree rooted at vertex $s$ such that the unique path in the tree from vertex $s$ to every other vertex is a directed path.

***distance label***: an estimate (in particular, an upper bound) on the shortest path length from the source vertex to each network vertex.

***Dorogovtsev-Mendes model***: a small-world network model that constructs a regular (low-dimensional) lattice and then adds an extra vertex, randomly connected to existing vertices in the lattice.

***even vertex*** (in an alternating path $P$): a vertex on $P$ that is reached using an even number of edges of $P$, starting from the origin vertex of $P$.

***exact algorithm***: a procedure that produces a verifiable optimal solution to every problem instance.

***flow***: a feasible assignment of material that satisfies flow conservation and arc capacity restrictions.

***forward star***: a compact representation of a graph in which information about arcs leaving a vertex is stored using consecutive locations of an array.

***free edge*** (in a matching): an edge that does not appear in the matching.

***free vertex*** (in a matching): a vertex that is not incident with any matched edge.

***generalized random graph model***: a model that generates graphs by adding edges with probability proportional to the product of the given vertex degrees.

***heuristic algorithm***: a procedure that produces a feasible, though not necessarily optimal, solution to every problem instance.

***improvement method***: a heuristic algorithm that starts with a suboptimal solution (often randomly generated) and attempts to improve it.

***Jackson-Rogers model***: a two-step model of network formation in which a new vertex is joined randomly to a set of parental vertices and then is joined to some of the neighbors of these parental vertices.

***Kleinberg small-world model***: a modified version of the Newman-Watts model in which edges are randomly added, based on the network distance between their associated vertices.

***Klemm-Eguíluz model***: a model of network formation in which a new vertex is joined either to an existing vertex or to a dynamically changing set of active vertices.

***length*** (of a path): the sum of all costs appearing on the arcs of the path.

***linear assignment problem*** (***LAP***): an optimization problem in which an assignment is sought that minimizes an appropriate set-up cost.

***link capacity***: an upper bound on the amount of traffic that a communication link can carry at any one time.

***linked adjacency list***: a collection of singly-linked lists used to represent a graph.

***local access network***: a network used to transfer traffic between the backbone network and the end users.

***matched edge*** (in a matching): an edge that appears in the matching.

***matched vertex*** (in a matching): a vertex that is incident with a matched edge.

***matching*** (in a graph): a set of pairwise nonadjacent edges in the graph.

***mate*** (of a matched vertex): in a matching, the other endpoint of the matched edge incident with the given vertex.

***maximum flow*** (in a network): a flow in the network having maximum value.

***maximum size matching***: a matching having the largest size.

***maximum spanning tree*** (of a network): a spanning tree of the network with maximum cost.

***maximum weight matching***: a matching having the largest weight.

***metaheuristic***: a general-purpose heuristic procedure (such as tabu search, simulated annealing, genetic algorithms, or neural networks) for solving difficult optimization problems.

***minimum cost flow*** (in a network): a flow in the network having minimum cost.

***minimum cut*** (in a network): a cut in the network having minimum capacity.

***minimum spanning tree*** (of a network): a spanning tree of the network with minimum cost.

***negative cycle***: a directed cycle of negative cost (or length).

***network design problem***: an optimization problem in which a minimum cost assignment of different capacity cables to edges is sought that satisfies given point-to-point demands and that respects all edge capacities.

***Newman-Watts model***: a small-world network model that constructs a regular (low-dimensional) lattice and then adds a random number of extra edges.

***odd vertex*** (in an alternating path $P$): a vertex on $P$ that is reached using an odd number of edges of the path $P$, starting from the origin vertex of $P$.

***perfect matching***: a matching in a graph in which each vertex of the graph is incident with exactly one edge of the matching.

***power-law distribution***: a distribution in which the proportion $p_k$ of vertices having degree $k$ is given by $p_k = Ck^{-\beta}$ for some fixed parameter $\beta$.

***predecessor***: relative to a rooted tree, the vertex preceding a given vertex on the unique path from the root to the given vertex.

***preflow***: a relaxation of flow where inflow into a vertex can be greater than its outflow.

***pseudoflow***: a relaxation of flow where inflow into a vertex need not be equal to its outflow.

***quadratic assignment problem*** (**QAP**): an optimization problem in which an assignment is sought that minimizes the sum of set-up and interaction costs.

***random graph***: a graph obtained by randomly adding edges between a fixed number of vertices.

***reduced cost*** of arc $(i, j)$: relative to given vertex potentials $\pi$, the quantity $c_{ij}^{\pi} = c_{ij} - \pi(i) + \pi(j)$.

***regenerator location problem***: an optimization problem in which a minimum cost placement of regenerators is sought so that all terminal vertices can communicate with one another without loss of signal quality.

***regular lattice***: a graph obtained by placing regularly-spaced vertices and then joining each vertex to its closest neighbors.

***residual capacity*** (of an arc): the maximum additional flow (with respect to a given flow) that can be sent on an arc.

***residual network***: a network consisting of arcs with positive residual capacity.

***s-t cut***: a cut $[S, \overline{S}]$ in which $s \in S$ and $t \in \overline{S}$.

***savings***: the reduction in cost from joining two vertices directly compared to joining both to a central vertex.

***shortest path***: a directed path between specified vertices having minimum total cost (or length).

*size* (of a matching): the number of edges in the matching.

***stable matching***: a matching that contains no blocking pairs.

***stable partners***: individuals that are matched in some stable matching.

**survivable network**:  a network that can survive failures in some of its vertices or edges and still transfer a prespecified amount of traffic.

**traveling salesman problem (TSP)**:  an optimization problem in which a fixed set of cities must be visited in some order at minimum total cost.

**two-phase method**:  a heuristic algorithm that implements a cluster first/route second philosophy.

**undirected network**:  a vertex set $V$ and an edge set $E$, where each undirected edge has an associated cost, length, weight, or capacity.

**value** of a flow:  the total flow leaving the source vertex.

**vehicle routing problem (VRP)**:  an optimization problem in which a given set of customers must be serviced at minimum total cost, using a fleet of vehicles having fixed capacity.

**vertex potential**:  a quantity $\pi(i)$ associated with each vertex $i$ of a network.

**Watts-Strogatz model**:  a small-world network model that constructs a regular (low-dimensional) lattice and then repositions a random number of its edges.

**weight** (of a matching):  the sum of the weights of edges in the matching.

# 10.1   MINIMUM SPANNING TREES

In an undirected network, the minimum spanning tree problem is the problem of identifying a spanning tree of the network that has the smallest possible sum of edge costs. This problem arises in a number of applications, both as a stand-alone problem and as a subproblem in more complex problem settings. It is assumed throughout this section that the network is connected.

## 10.1.1   BASIC CONCEPTS

**Definitions:**

An **undirected network** is a weighted graph (§8.1.1) $G = (V, E)$, where $V$ is the set of vertices, $E$ is the set of undirected edges, and each edge $(i, j) \in E$ has an associated **cost** (or **weight, length**) $c_{ij}$. Let $n = |V|$ and $m = |E|$.

If $T = (V, F)$ is a spanning tree (§9.2) of $G = (V, E)$, then every edge in $F \subseteq E$ is a **tree edge** and every edge in $E - F$ is a **nontree edge** (or **chord**).

A **minimum spanning tree (MST)** of $G$ is a spanning tree of $G$ for which the sum of the edge costs is minimum.

A **maximum spanning tree** of $G$ is a spanning tree of $G$ for which the sum of the edge costs is maximum.

A **cut** of $G = (V, E)$ is a partition of the vertex set $V$ into two parts, $S$ and $\overline{S} = V - S$. Each cut defines the set of edges $[S, \overline{S}] \subseteq E$ having one endpoint in $S$ and the other endpoint in $\overline{S}$.

**Facts:**

**1.** Every spanning tree $T$ of a network $G$ with $n$ vertices contains exactly $n-1$ edges, and every two vertices of $T$ are connected by a unique path.

**2.** Adding an edge to a spanning tree of $G$ produces a unique cycle, called a *fundamental cycle* (§9.2.1).

**3.** Every cut $[S, \overline{S}]$ is a disconnecting set of edges (§8.4.2). However, not every disconnecting set of edges can be represented as a cut $[S, \overline{S}]$; see Example 4.

**4.** Removing an edge from a spanning tree of $G$ produces two subtrees, on vertex sets $S$ and $\overline{S}$, respectively. The associated cut $[S, \overline{S}]$ is called a *fundamental cut*.

**5.** *Path optimality conditions*: A spanning tree $T^*$ is a minimum spanning tree of $G$ if and only if for each nontree edge $(k, l)$ of $G$, $c_{ij} \leq c_{kl}$ holds for all tree edges $(i, j)$ in the fundamental cycle determined by edge $(k, l)$.

**6.** *Cut optimality conditions*: A spanning tree $T^*$ is a minimum spanning tree of $G$ if and only if for each tree edge $(i, j) \in T^*$, $c_{ij} \leq c_{kl}$ holds for all nontree edges $(k, l)$ in the fundamental cut determined by edge $(i, j)$.

**7.** If all edge costs are different, then the minimum spanning tree is unique.

**8.** The minimum spanning tree can be unique even if some of the edge costs are equal; see Example 3.

**9.** Adding a constant to all edge costs of an undirected network does not change the minimum spanning tree(s) of the network. Thus, it is sufficient to have an algorithm that works when all edge costs are positive.

**10.** Multiplying each edge cost of an undirected network by $-1$ converts a maximum spanning tree into a minimum spanning tree, and vice versa. Thus, it is sufficient to have algorithms to find a minimum spanning tree.

**Examples:**

**1.** Part (a) of the following figure shows an undirected network $G$, with costs indicated on each edge. Part (b) shows a spanning tree $T^*$ of $G$.

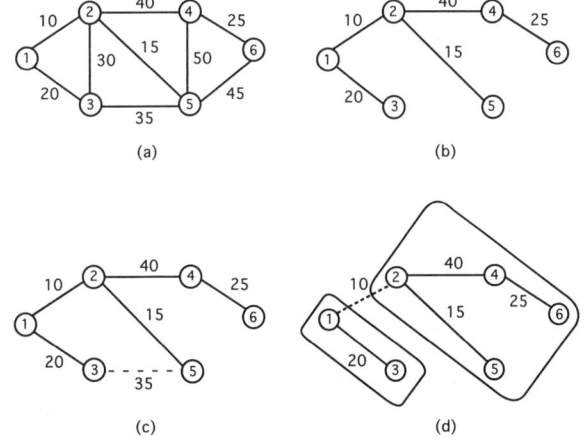

Adding the nontree edge $(3, 5)$ to $T^*$ produces the fundamental cycle $[3, 1, 2, 5, 3]$; see part (c). Since each tree edge in this cycle has cost no more than that of the nontree edge $(3, 5)$, the path optimality condition is satisfied by edge $(3, 5)$. Similarly, it can be verified that the other nontree edges, namely $(2, 3)$, $(4, 5)$, and $(5, 6)$, satisfy the path optimality conditions, establishing by Fact 5 that $T^*$ is a minimum spanning tree. By Fact 7 this is the unique minimum spanning tree.

**2.** For the tree edge $(1,2)$ in part (b) of the figure of Example 1, the fundamental cut $[S, \overline{S}]$ formed by deleting edge $(1,2)$ has $S = \{1,3\}$ and $\overline{S} = \{2,4,5,6\}$; see part (d) of the figure. This cut contains two nontree edges, $(2,3)$ and $(3,5)$. Since each such nontree edge has cost greater than or equal to that of the tree edge $(1,2)$, the cut optimality condition is satisfied for edge $(1,2)$. Similarly, it can be verified that the other tree edges, namely $(1,3)$, $(2,4)$, $(2,5)$, and $(4,6)$, satisfy the cut optimality conditions, establishing by Fact 6 that $T^*$ is a minimum spanning tree.

**3.** The undirected network in part (a) of the following figure has four vertices and five edges, with the edge cost shown beside each edge. This network contains eight spanning trees, which are listed in the table in part (b) of the figure. The spanning tree $T_5$ achieves the minimum cost 7 among all the spanning trees and so is a minimum spanning tree. In fact, $T_5$ is the unique minimum spanning tree, even though the edge costs are not all distinct. See Fact 8.

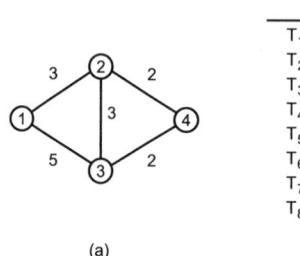

|       | Spanning Tree Edges | Total Cost |
|-------|---------------------|------------|
| $T_1$ | (1,2), (1,3) (2,4)  | 10         |
| $T_2$ | (1,2), (1,3) (3,4)  | 10         |
| $T_3$ | (1,2), (2,3) (2,4)  | 8          |
| $T_4$ | (1,2), (2,3) (3,4)  | 8          |
| $T_5$ | (1,2), (2,4) (3,4)  | 7          |
| $T_6$ | (1,3), (2,3) (2,4)  | 10         |
| $T_7$ | (1,3), (2,3) (3,4)  | 10         |
| $T_8$ | (1,3), (2,4) (3,4)  | 9          |

(a)                                              (b)

**4.** The set of edges $F = \{(1,2), (2,4), (3,4)\}$ is a disconnecting set in the network $G$ of Example 3, since removal of these edges disconnects $G$. However, there is no partition of the vertex set of $G$ into nonempty sets $S$ and $\overline{S}$ for which $F = [S, \overline{S}]$.

## 10.1.2  ALGORITHMS FOR MINIMUM SPANNING TREES

There are several greedy algorithms for constructing minimum spanning trees, based on the optimality conditions in §10.1.1, Facts 5 and 6. Each of these algorithms myopically (greedily) adds an edge to the current configuration based on only local information; nonetheless, these procedures are guaranteed to produce a minimum spanning tree.

**Definitions:**

The **nearest neighbor** operation takes as input a tree $T^*$ having vertex set $S$ and produces a minimum cost edge $(i,j)$ in the cut $[S, \overline{S}]$; i.e., $c_{ij} = \min\{c_{ab} \mid a \in S, \, b \notin S\}$.

The **merge** operation takes as input an edge $(i,j)$ whose two endpoints $i$ and $j$ belong to disjoint trees $T_i$ and $T_j$ and combines the trees into $T_i \cup T_j \cup \{(i,j)\}$.

The graph $G = (V, E)$ is assumed connected, with $n$ vertices and $m$ edges.

**Facts:**

**1.** *Kruskal's algorithm:* This greedy algorithm (Algorithm 1) is based on the path optimality conditions (§10.1.1) and builds a minimum spanning tree by examining edges of $E$ one by one in nondecreasing order of their costs. The edge being examined is added to the current forest if its addition does not create a cycle. (J. B. Kruskal, 1928–2010)

---

**Algorithm 1**: **Kruskal's algorithm.**

input: connected undirected network $G$

output: minimum spanning tree $T^*$

order the edges $(i_1, j_1), (i_2, j_2), \ldots, (i_m, j_m)$ so that $c_{i_1 j_1} \leq c_{i_2 j_2} \leq \cdots \leq c_{i_m j_m}$

$T^* := \emptyset$

**for** $k := 1$ **to** $m$

    **if** $T^* \cup \{(i_k, j_k)\}$ does not contain a cycle **then** $T^* := T^* \cup \{(i_k, j_k)\}$

---

**2.** Kruskal's algorithm can be terminated once $n - 1$ edges have been added to $T^*$.

**3.** Kruskal's algorithm can be implemented in $O(m \log n)$ time. The bottleneck operation is sorting of the edge costs.

**4.** Algorithm 1 was independently discovered by Kruskal (1956) and by H. Loberman and A. Weinberger (1957).

**5.** *Prim's algorithm*: This algorithm (Algorithm 2) is based on the cut optimality conditions (§10.1.1). It maintains a single tree $T^*$, which initially consists of an arbitrary vertex $i_0$. At each iteration, the algorithm adds the least cost edge emanating from $T^*$ until a spanning tree is obtained. (R. C. Prim, born 1921)

---

**Algorithm 2**: **Prim's algorithm.**

input: connected undirected network $G$, vertex $i_0$

output: minimum spanning tree $T^*$

$T^* :=$ the tree consisting of vertex $i_0$

**while** $|T^*| < n - 1$

    $(i, j) := nearest\_neighbor(T^*)$

    $T^* := T^* \cup \{(i, j)\}$

---

**6.** Algorithm 2 was first proposed in 1930 by V. Jarník. Later it was independently discovered by Prim (1957) and by E. W. Dijkstra (1959).

**7.** Running times of several implementations of Prim's algorithm are shown in the following table. See [AhMaOr93] for a discussion of these implementations.

| data structure | running time |
|---|---|
| binary heap | $O(m \log n)$ |
| $d$-heap | $O(m \log_d n)$, with $d = \max\{2, \lceil \frac{m}{n} \rceil\}$ |
| Fibonacci heap | $O(m + n \log n)$ |

**8.** *Sollin's algorithm*: This greedy algorithm (Algorithm 3) is also based on the cut optimality conditions (§10.1.1). It starts with a forest of $n$ trees, each consisting of a single vertex, and builds a minimum spanning tree by repeatedly adding edges to the current forest. At each iteration a least cost edge emanating from each tree is added, leading to the merging of certain trees.

---

**Algorithm 3**:  Sollin's algorithm.

input: connected undirected network $G$

output: minimum spanning tree $T^*$

$T^* :=$ forest of all vertices of $G$, but no edges

**while** $|T^*| < n - 1$

    let $T_1, T_2, \ldots, T_p$ be the trees in the forest $T^*$

    **for** $k := 1$ **to** $p$

        $(i_k, j_k) := nearest\_neighbor(T_k)$

    **for** $k := 1$ **to** $p$

        **if** $i_k$ and $j_k$ belong to different trees **then**

            $merge(i_k, j_k)$

            $T^* := T^* \cup \{(i_k, j_k)\}$

---

**9.**  Each iteration of Algorithm 3 reduces the number of trees in the forest $T^*$ by at least half.

**10.**  Sollin's algorithm performs $O(\log n)$ iterations and can be implemented to run in $O(m \log n)$ time; see [AhMaOr93].

**11.**  A variation of Sollin's algorithm that runs in time $O(m \log \log n)$ can be found in [Ya75].

**12.**  The origins of Algorithm 3 can be traced to O. Borůvka (1926), who first formulated the minimum spanning tree problem in the context of electric power networks. This algorithm was independently proposed in 1938 by G. Choquet for points in a metric space and by G. Sollin in 1961 for arbitrary networks.

**13.**  Sollin's algorithm can be easily parallelized in EREW (exclusive-read, exclusive-write) PRAM (parallel random access machine). (See §17.1.4.) This algorithm assigns a processor to each edge and each vertex of the network [KiLe88].

**14.**  Chong, Han, and Lam [ChHaLa01] developed an algorithm that runs in $O(\log n)$ time using $m$ parallel processors. This running time is best possible.

**15.**  Computational studies have found that the Prim and Sollin algorithms consistently outperform Kruskal's algorithm. Prim's algorithm is faster when the network is dense, whereas Sollin's algorithm is faster when the network is sparse.

**16.**  An excellent discussion of the history of the minimum spanning tree problem is provided in [GrHe85].

**17.**  The fastest deterministic algorithms for the minimum spanning tree problem are due to Chazelle [Ch00] and to Pettie and Ramachandran [PeRa02]. Chazelle's algorithm runs in $O(m\alpha(n, m))$ time where $\alpha(n, m)$ is the inverse Ackermann function, which for all practical purposes is less than 5. Pettie and Ramachandran show how to find a minimum spanning tree in $O(T^*(n, m))$ time, where $T^*(n, m)$ is the minimum number of edge-weight comparisons needed to determine the solution.

**18.**  Karger, Klein, and Tarjan [KaKlTa95] developed a randomized algorithm for the minimum spanning tree problem that runs in $O(m)$ time.

**19.**  Computer codes that implement spanning tree algorithms can be found at

    • http://www3.cs.stonybrook.edu/~algorith/files/minimum-spanning-tree. shtml

**20.** An important variant of the minimum spanning tree problem places constraints on the number of edges incident with a vertex in a candidate spanning tree. Such *degree-constrained* minimum spanning trees are investigated in [GlKl75] and [Vo89].

**21.** Another variant is the *capacitated* minimum spanning tree problem, which arises in the design of local access telecommunication networks. In this problem, a feasible spanning tree is one rooted at a specified central vertex such that the total traffic (number of calls) generated by each subtree connected to the central vertex does not exceed a known capacity. A feasible spanning tree having minimum total cost is then sought. (See §10.6.1 for further details.)

**Examples:**

**1.** For the network shown in part (a) of the following figure, ordering edges by nonde-creasing cost produces the following sequence of edges: $(2,4)$, $(3,5)$, $(3,4)$, $(2,3)$, $(4,5)$, $(1,2)$, $(1,3)$. Kruskal's algorithm adds the edges $(2,4)$, $(3,5)$, $(3,4)$ to $T^*$; discards the edges $(2,3)$ and $(4,5)$; then adds the edge $(1,2)$ to $T^*$ and terminates. Part (b) of the figure shows the resulting minimum spanning tree, having total cost 80.

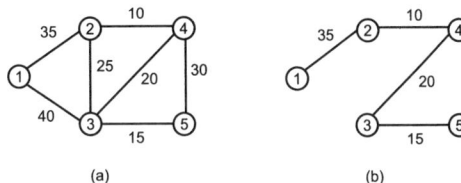

(a)                                    (b)

**2.** Prim's algorithm (Algorithm 2) is applied to the network in part (a) of the figure for Example 1, starting with the initial vertex $i_0 = 3$. The minimum cost edge out of vertex 3 is $(3,5)$, so $T^* = \{(3,5)\}$. Next, the minimum cost edge emanating from $T^*$ is $(3,4)$, giving $T^* = \{(3,5),(3,4)\}$. Subsequent iterations add the edges $(2,4)$ and $(1,2)$, producing the minimum spanning tree $T^* = \{(3,5),(3,4),(2,4),(1,2)\}$. Starting from any other initial vertex $i_0$ would give the same result.

**3.** To apply Sollin's algorithm (Algorithm 3) to the network in part (a) of the figure for Example 1, begin with a forest containing five trees, each consisting of a single vertex. Part (a) of the following figure shows the least cost edge emanating from each of these trees. One iteration of Algorithm 3 produces the two trees shown in part (b) of the following figure. The least cost edge emanating from either of these two trees is $(3,4)$. Adding this edge completes the minimum spanning tree shown in part (c) of the figure.

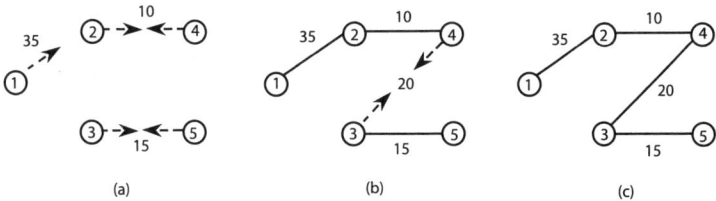

(a)                          (b)                          (c)

---

### 10.1.3   APPLICATIONS

Minimum spanning tree problems arise both directly and indirectly. For direct applications, the points in a given set are to be connected using the least cost collection of edges. For indirect applications, creative modeling of the original problem recasts it as a minimum spanning tree problem.

## Applications:

**1.** *Designing physical systems*: A minimum cost network is to be designed to connect geographically dispersed system components. Each component is represented by a vertex, with potential network connections between vertices represented by edges. A cost is associated with each edge.

**2.** Examples of Application 1 occur in the following:

- Connect terminals in cabling the panels of electrical equipment in order to use the least total cost of wire.

- Construct a pipeline network to connect a number of towns using the smallest possible total cost of pipeline.

- Link isolated villages in a remote region, which are connected by roads but not yet by telephone service. The problem is to determine along which stretches of roads to place telephone lines to link every pair of villages, using the minimum total miles of installed lines.

- Construct a digital computer system, composed of high-frequency circuitry, when it is important to minimize the length of wires between different components to reduce both capacitance and delay line effects.

- Connect a number of computer sites by high-speed lines. Each line is available for leasing at a certain monthly cost, and a configuration is required that connects all the sites at minimum overall cost.

- Design a backbone network of high-capacity links that connect switching devices to support internet traffic. A minimum cost backbone network that maintains acceptable throughput is required.

**3.** *Clustering*: Objects having $k$ measurable characteristics are to be clustered into groups of "similar" objects. First, construct an undirected network, where each object is represented by a vertex and every two distinct vertices are joined by an edge. The cost of edge $(i, j)$ is the distance (in $k$-dimensional space) between the $k$-vectors for objects $i$ and $j$. Applying Kruskal's algorithm to this network then yields a hierarchy of partitions of the vertex set; each partition is defined by the trees comprising the forest obtained at each iteration of Kruskal's algorithm. Such a hierarchy is then used to define clusters of the original objects. This application has been used to represent multidimensional gene expression data [XuOlXu02].

**4.** Computation of minimum spanning trees sometimes arises as a subproblem in a larger optimization problem. For example, one heuristic approach to the *traveling salesman problem* (§10.7.1) involves the calculation of minimum spanning trees.

**5.** *Optimal message passing*: An intelligence service has agents operating in a non-friendly country. Each agent knows some of the other agents and has in place procedures for arranging a rendezvous with someone he knows. For each such possible rendezvous, say between agent $i$ and agent $j$, any message passed between these agents will fall into hostile hands with a certain probability $p_{ij}$. The group leader wants to transmit a confidential message among all the agents while maximizing the probability that no message is intercepted.

If the agents are represented by vertices and each possible rendezvous by an edge, then in the resulting graph $G$ a spanning tree $T$ is required that maximizes the probability that no message is intercepted, given by $\Pi_{(i,j)\in T}(1 - p_{ij})$. Such a tree can be found by defining the cost of each edge $(i, j)$ as $\log(1 - p_{ij})$ and solving a *maximum* spanning tree problem.

**6.** *All-pairs minimax path problem:*   In this variant of the shortest path problem (see §10.3.1), the value of a path $P$ is the maximum cost of an edge in $P$. The all-pairs minimax path problem is to determine a minimum value path between every pair of vertices in a network $G$. It can be shown that if $T^*$ is a minimum spanning tree of $G$, then the unique path in $T^*$ between any pair of vertices is also a minimax path between that pair of vertices.

**7.** Examples of Application 6 arise in the following contexts:

- Determine the trajectory of a spacecraft that keeps the maximum temperature of the surface as small as possible.

- When traveling through a desert, select a route that minimizes the length of the longest stretch between rest areas.

- A person traveling in a wheelchair desires a route that minimizes the maximum ascent along the path segments of the route.

**8.** *Measuring homogeneity of bimetallic objects:* In this application minimum spanning trees are used to determine the degree to which a bimetallic object is homogeneous in composition. First, the composition of the bimetallic object is measured at a set of sample points. A network is then constructed with vertices corresponding to the sample points and with an edge connecting physically adjacent sample points. The cost of edge $(i, j)$ is the product of the physical (Euclidean) distance between sample points $i$ and $j$, and a homogeneity factor between 0 and 1. The homogeneity factor is 0 if the composition of the corresponding samples is identical, and is 1 if the composition is very different. This cost structure gives greater weight to two points if they have different compositions and are far apart. Then the cost of the minimum spanning tree provides an overall measure of the homogeneity of the object.

**9.** Additional applications, with reference sources, are given in the following table.

| application | references |
|---|---|
| two-dimensional storage schemes | [AhMaOr93], [AhEtal95] |
| chemical physics | [AhMaOr93] |
| manufacturing | [EvMi92] |
| network design | [AhMaOr93] |
| network reliability | [AhEtal95] |
| pattern classification | [AhMaOr93], [GrHe85] |
| image segmentation | [FeHu04] |
| automatic speech recognition | [GrHe85] |
| numerical taxonomy | [GrHe85] |
| handwriting recognition | [TaRo04] |

# 10.2   MATCHINGS

In an undirected network, the maximum matching problem is to find a set of nonadjacent edges that has the largest total size or weight. This discrete optimization problem arises in a number of applications, often involving the optimal pairing of a set of objects.

## 10.2.1  BASIC CONCEPTS

**Definitions:**

Let $G = (V, E)$ be an undirected network with vertex set $V$ and edge set $E$ (see §10.1.1). Assume that $G$ contains neither loops nor multiple edges. Each edge $e = (i, j) \in E$ has an associated **weight** $w_e = w_{ij}$. Let $n = |V|$ and $m = |E|$.

The **degree** of vertex $v \in V$ in $G$ is the number of edges in $G$ that are incident with $v$, written $deg(v)$. (See §8.1.1.)

A **matching** in $G = (V, E)$ is a set $M \subseteq E$ of pairwise nonadjacent edges (§8.1.1).

A **perfect matching** in $G = (V, E)$ is a matching $M$ in which each vertex of $V$ is incident with exactly one edge of $M$.

A **vertex cover** in $G = (V, E)$ is a set $S$ of vertices such that every edge of $G$ is incident with some vertex in $S$.

The **size** (**cardinality**) of a matching $M$ is the number of edges in $M$, written $|M|$.

The **weight** of a matching $M$ is $wt(M) = \sum_{e \in M} w_e$.

A **maximum size** matching of $G$ is a matching $M$ having the largest size $|M|$.

A **maximum weight** matching of $G$ is a matching $M$ having the largest weight $wt(M)$.

Relative to a matching $M$ in $G = (V, E)$, edges $e \in M$ are **matched** edges, while edges $e \in E - M$ are **free** edges. Vertex $v$ is **matched** if it is incident with a matched edge; otherwise vertex $v$ is **free**.

Every matched vertex $v$ has a **mate**, the other endpoint of the matched edge incident with $v$.

With respect to a matching $M$, the **weight** $wt(P)$ of path $P$ is the sum of the weights of the free edges in $P$ minus the sum of the weights of the matched edges in $P$.

An **alternating** path has edges that are alternately free and matched. An **augmenting** path is an alternating path that starts and ends at a free vertex.

**Facts:**

**1.** Matchings are useful in a wide variety of applications, such as assigning personnel to jobs, target tracking, crew scheduling, snowplowing streets, scheduling on parallel machines, among others (see §10.2.5).

**2.** In a matching $M$, each vertex of $G$ has degree 0 or 1 relative to the edges in $M$. In a perfect matching $M$, each vertex of $G$ has degree 1 relative to the edges in $M$.

**3.** Any matching $M$ of $G$ contains $2|M|$ matched vertices and $n - 2|M|$ free vertices.

**4.** If $M$ is any matching in $G$, then $|M| \leq \lfloor \frac{n}{2} \rfloor$.

**5.** *Weak duality theorem*: The size of any vertex cover for $G$ is an upper bound on the size of any matching in $G$. As a result, the minimum size of a vertex cover is at least as large as the size of a maximum matching.

**6.** Every augmenting path has an odd number of edges.

**7.** If $M$ is a matching and $P$ is an augmenting path with respect to $M$, then the symmetric difference (§1.2.2) $M \Delta P$ is a matching of size $|M| + 1$.

**8.** If $M$ is a matching and $P$ is an augmenting path with respect to $M$, then $wt(M\Delta P)$ $= wt(M) + wt(P)$.

**9.** *Augmenting path theorem:* $M$ is a maximum size matching if and only if there is no augmenting path with respect to $M$.

**10.** Fact 9 was obtained independently by C. Berge (1957) and also by R. Z. Norman and M. O. Rabin (1959). This result was also recognized in an 1891 paper of J. Petersen.

**11.** Suppose $M$ is a matching having maximum weight among all matchings of a fixed size $k$. If $P$ is an augmenting path of maximum weight, then $M\Delta P$ is a matching having maximum weight among all matchings of size $k + 1$.

**12.** Suppose paths $P_1, P_2, \ldots, P_k$ are obtained as in Fact 11 by augmenting along a maximum weight path. Then $wt(P_1) \geq wt(P_2) \geq \cdots \geq wt(P_k)$.

**13.** *Weighted augmenting path theorem:* $M$ is a maximum weight matching if and only if there is no augmenting path with respect to $M$ that has positive weight.

**14.** The number of perfect matchings of the complete graph (§8.1.3) $K_{2n}$ on $2n$ vertices is $(2n - 1)!! = 1 \cdot 3 \cdot 5 \ldots (2n - 1)$.

**15.** An historical perspective on the theory of matchings is found in [P192].

### Examples:

**1.** The graph $G$ in the following figure has six vertices and eight edges. The highlighted edges display a matching $M$ of size 2. By Fact 4, the size of any matching in $G$ is at most 3. Since the set $S = \{1, 6\}$ of vertices form a vertex cover for $G$ of size 2, Fact 5 shows that $M$ is actually a maximum size matching and $S$ is a minimum size vertex cover.

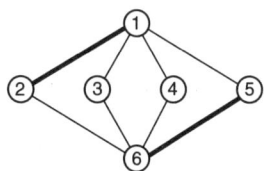

**2.** Part (a) of the following figure displays a network $G$ with the weight $w_e$ shown next to each edge $e$.

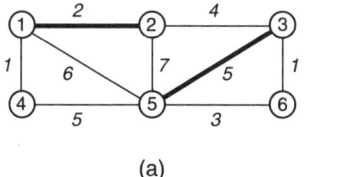     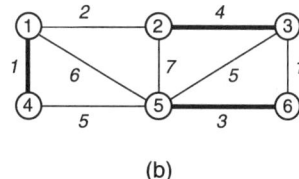

                (a)                             (b)

The matching $M_1 = \{(1, 2), (3, 5)\}$ of size 2 is also shown, with the matched edges highlighted. The mate of vertex 1 is vertex 2, and the mate of vertex 5 is vertex 3. The weight of $M_1$ is $wt(M_1) = 7$. Relative to the matching $M_1$, vertices 4 and 6 are free vertices, and an augmenting path $P$ from 4 to 6 is given by the set of edges $P = \{(1, 4), (1, 2), (2, 3), (3, 5), (5, 6)\}$. Here $wt(P) = 1 + 4 + 3 - 2 - 5 = 1$ and (as guaranteed by Fact 6) path $P$ has an odd number of edges. The matching $M_2 = M_1\Delta P = \{(1, 2), (3, 5)\} \Delta \{(1, 4), (1, 2), (2, 3), (3, 5), (5, 6)\} = \{(1, 4), (2, 3), (5, 6)\}$ is a perfect matching and is highlighted in part (b) of the figure. There are no free vertices relative to matching $M_2$ and no augmenting paths, so $M_2$ is a maximum size matching of $G$. There are other maximum size matchings, such as $\{(1, 4), (2, 5), (3, 6)\}$ and $\{(1, 2), (4, 5), (3, 6)\}$.

**3.** Part (a) of the following figure shows a matching $M_1$ of size 1, with $wt(M_1) = 7$. Since edge $(2,5)$ has maximum weight among all edges, $M_1$ is a maximum weight matching of size 1. Relative to $M_1$ the augmenting path $P_1 = \{(1,5), (2,5), (2,3)\}$ has weight $6 + 4 - 7 = 3$, whereas the augmenting path $\{(3,6)\}$ has weight 1. It can be verified that $P_1$ is a maximum weight augmenting path relative to $M_1$. By Fact 11, $M_2 = M_1 \Delta P_1 = \{(1,5), (2,3)\}$ is a maximum weight matching of size 2 in the network, with $wt(M_2) = wt(M_1) + wt(P_1) = 7 + 3 = 10$; see part (b) of the figure.

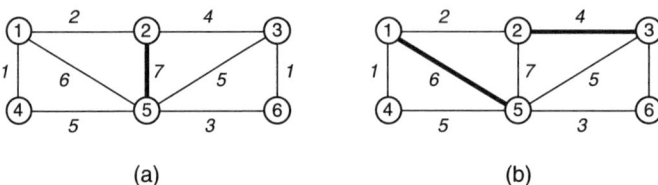

(a)                                      (b)

Relative to $M_2$ there are several augmenting paths between the free vertices 4 and 6:

$$Q_1 = \{(1,4), (1,5), (5,6)\}, \qquad wt(Q_1) = 1 + 3 - 6 = -2,$$
$$Q_2 = \{(1,4), (1,5), (2,5), (2,3), (3,6)\}, \qquad wt(Q_2) = 1 + 7 + 1 - 6 - 4 = -1,$$
$$Q_3 = \{(4,5), (1,5), (1,2), (2,3), (3,6)\}, \qquad wt(Q_3) = 5 + 2 + 1 - 6 - 4 = -2.$$

The maximum weight augmenting path is $Q_2$ and so (by Fact 11) $M_3 = M_2 \Delta Q_2 = \{(1,4), (2,5), (3,6)\}$ is a maximum weight matching of size 3 in the network with $wt(M_3) = 9$. Overall, the maximum weight matching in $G$ is $M_2$ as expected, since all augmenting paths relative to $M_2$ have negative weight (see Fact 12).

**4.** *Paired kidney donations:* To address the severe shortage of organs for use in kidney transplants, some hospitals have organized a paired kidney donation program. While a relative of a patient would gladly donate a kidney to that patient, it may not be possible because of incompatibilities in blood type or unfavorable antibody reactions.

This situation can be illustrated using the following graph $G$. Each of the eight vertices in $G$ represents a pair of incompatible donor-recipients. Two vertices are connected by an edge if the donors $D$ of each pair are compatible matches for the recipients $R$ of the other pair. For example, the edge $(1,2)$ indicates that $D_1$ can donate a kidney to $R_2$ and that $D_2$ can donate a kidney to $R_1$. A largest number of compatible kidney exchanges then corresponds to a maximum size matching in $G$. By Fact 4, there can be at most four compatible matches. However, $S = \{1, 3, 6\}$ is a vertex cover for $G$, meaning a maximum size matching can have size at most 3 (Fact 5). Consequently, at most three compatible matches are possible in this situation—for example, the pairs $(1,2), (3,7), (6,8)$.

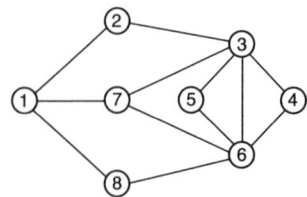

## 10.2.2   MATCHINGS IN BIPARTITE NETWORKS

In this section, algorithms are described for finding maximum size and maximum weight matchings in bipartite networks (§8.1.3). Bipartite networks arise in a number of applications, such as in assigning personnel to jobs or tracking objects over time. Moreover, the algorithms developed for the case of bipartite networks are considerably simpler than those needed for the case of general networks (§10.2.3).

**Definitions:**

Let $G = (X \cup Y, E)$ be a bipartite network with $n$ vertices and $m$ edges, and edge weights $w_{xy}$.

If $S \subseteq X$ then $\Gamma(S) = \{y \in Y \mid (x, y) \in E$ for some $x \in X\}$ is the set of vertices in $Y$ adjacent to some vertex of $X$.

A **complete matching** from $X$ to $Y$ in $G = (X \cup Y, E)$ is a matching $M$ in which each vertex of $X$ is incident with an edge of $M$.

The **directed two-terminal flow network** $G'$ associated with $G = (X \cup Y, E)$ is defined by adding new vertices $s$ and $t$, as well as arcs $(s, x)$ for each $x \in X$ and arcs $(y, t)$ for each $y \in Y$. All other arcs $(x, y)$ of $G'$ correspond to edges $(x, y)$ of $G$ where $x \in X$ and $y \in Y$. Every arc of $G'$ has capacity 1.

**Facts:**

**1.** *Hall's theorem:*  $G = (X \cup Y, E)$ has a complete matching from $X$ to $Y$ if and only if $|\Gamma(S)| \geq |S|$ holds for every $S \subseteq X$. In words, a complete matching exists precisely when every set of vertices in $X$ is adjacent to at least an equal number of vertices in $Y$. (Philip Hall, 1904–1982.)

**2.** *König's theorem:*  For a bipartite network $G$, the maximum size of a matching in $G$ equals the minimum size of a vertex cover for $G$. (Dénes König, 1884–1944.)

**3.** *Sufficient condition for a complete matching:*  Suppose there exists some $k$ such that $deg(x) \geq k \geq deg(y)$ holds in $G = (X \cup Y, E)$ for all $x \in X$ and $y \in Y$. Then $G$ has a complete matching from $X$ to $Y$.

**4.** There is a one-to-one correspondence between matchings of size $k$ in $G$ and integral flows (§10.4.1) of value $k$ in the associated two-terminal flow network $G'$.

**5.** A maximum flow in $G'$, and thereby a maximum size matching of $G$, can be found in $O(m\sqrt{n})$ time.

**6.** Suppose that costs are added to the two-terminal flow network $G'$, using $c_{ij} = 0$ if $i = s$ or $j = t$, and $c_{ij} = -w_{ij}$ otherwise. By starting with the flow (§10.5.1) $x = 0$, the successive shortest path algorithm (§10.5.2) can be applied to $G'$ to produce a minimum cost flow. This will identify (via Fact 4) a matching with maximum weight.

**7.** *Bipartite matching algorithm:*  This method (Algorithm 1), which is based on Fact 9 of §10.2.1, produces a maximum size matching of the bipartite network $G = (X \cup Y, E)$. Each iteration involves a modified breadth-first search of $G$ (see §9.2.1), starting with the free vertices in the set $X$. All vertices of $G$ are structured into levels that alternate between free and matched edges.

---

**Algorithm 1**:  **Bipartite matching algorithm.**
input: undirected bipartite network $G = (X \cup Y, E)$
output: maximum size matching $M$
$M := \emptyset$
**while** true
    let $S_1$ consist of all free vertices of $X$
    mark all vertices of $S_1$ as seen, the remaining vertices as unseen
    **while** there are unseen vertices of $G$
        $S_2 := \{y \mid (x, y) \in E, \ x \in S_1, \ y \text{ unseen}\}$
        **if** some $y \in S_2$ is free **then**
            an augmenting path $P$ to $y$ has been found
            mark all remaining vertices of $G$ as seen
        **else** mark all vertices of $S_2$ as seen
            $S_1 := \{x \mid (y, x) \in M, \ y \in S_2, \ x \text{ unseen}\}$
            mark all vertices of $S_1$ as seen
    **if** an augmenting path $P$ has been found **then** $M := M \Delta P$
    **else** terminate with matching $M$

---

**8.** Algorithm 1 can be implemented to run in $O(nm)$ time.

**9.** *Bipartite weighted matching algorithm*:  This method (Algorithm 2), which is based on Fact 13 of §10.2.1, produces a maximum weight matching of $G = (X \cup Y, E)$. Each iteration develops a longest path tree in $G$, rooted at the set of free vertices in $X$. The tentative largest weight of a path from a free vertex in $X$ to vertex $j$ is maintained in the label $d(j)$.

---

**Algorithm 2**:  **Bipartite weighted matching algorithm.**
input: undirected bipartite network $G = (X \cup Y, E)$, weights $w_e$
output: maximum weight matching $M$
$M := \emptyset$
**while** true
    let $S_1$ consist of all free vertices of $X$
    $d(j) := 0$ for $j \in S_1$, $d(j) := -\infty$ otherwise
    **while** $S_1 \neq \emptyset$
        $S_2 := \emptyset$
        **for** $(x, y) \in E - M$ with $x \in S_1$
            **if** $d(x) + w_{xy} > d(y)$ **then**
                $d(y) := d(x) + w_{xy}$, $S_2 := S_2 \cup \{y\}$
        $S_1 := \emptyset$
        **for** $(y, x) \in M$ with $y \in S_2$
            **if** $d(y) - w_{yx} > d(x)$ **then**
                $d(x) := d(y) - w_{yx}$, $S_1 := S_1 \cup \{x\}$
    let $y$ be a free vertex with maximum label $d(y)$
    let $P$ be the associated augmenting path to $y$

> **if** $d(y) > 0$ **then** $M := M\Delta P$
>
> **else** terminate with matching $M$

**10.** Algorithm 2 can be implemented to run in $O(nm)$ time.

### Examples:

**1.** *Drug testing*:   A drug company wishes to test $n$ antibiotics using $n$ volunteer patients. Some of the patients have known allergic reactions to certain of these antibiotics. To determine whether there is a feasible assignment of the $n$ different antibiotics to $n$ different patients, construct the bipartite network $G = (X \cup Y, E)$, where $X$ is the set of antibiotics and $Y$ is the set of patients. An edge $(i, j) \in E$ exists when patient $j$ is not allergic to antibiotic $i$. A complete matching of $G$ is then sought.

**2.** Part (a) of the following figure shows a bipartite graph $G$ with $X = \{1, 2, 3, 4\}$ and $Y = \{a, b, c, d\}$.

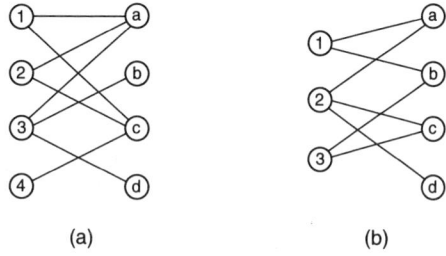

(a)          (b)

Using Fact 1, there cannot be a complete matching from $X$ to $Y$: if $S = \{1, 2, 4\}$ then $\Gamma(S) = \{a, c\}$ and $|\Gamma(S)| < |S|$. There is, however, a matching of size 3: for example, $\{(1, a), (3, b), (4, c)\}$. This matching has maximum size, since by Fact 2 there is a vertex cover $\{3, a, c\}$ of size 3.

**3.** Part (b) of the figure in Example 2 shows a bipartite graph $G$ with $X = \{1, 2, 3\}$ and $Y = \{a, b, c, d\}$. Since $\deg(x) \geq 2 \geq \deg(y)$ holds for all $x \in X$ and $y \in Y$, there must exist a complete matching from $X$ to $Y$. One such complete matching is given by $\{(1, a), (2, c), (3, b)\}$.

**4.** Algorithm 1 can be used to find a maximum size matching in the bipartite graph shown in part (a) of the following figure.

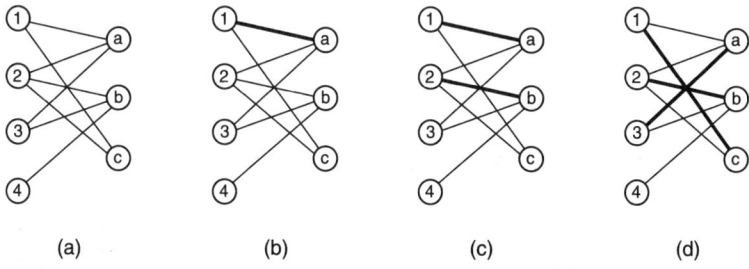

(a)          (b)          (c)          (d)

Relative to the initial empty matching, all vertices of $X$ are free so $S_1 = \{1, 2, 3, 4\}$, giving $S_2 = \{a, b, c\}$. In particular, vertex $a \in S_2$ is free and an augmenting path to $a$ is $P = \{(1, a)\}$. The resulting matching is $M = \{(1, a)\}$, shown in part (b) of the figure.

The second iteration of Algorithm 1 starts with $S_1 = \{2, 3, 4\}$, giving $S_2 = \{a, b, c\}$. An augmenting path to the free vertex $b$ is $P = \{(2, b)\}$, resulting in $M = \{(1, a), (2, b)\}$; see part (c) of the figure.

At the next iteration, $S_1 = \{3, 4\}$ and $S_2 = \{a, b\}$. Since both vertices of $S_2$ are now matched, the algorithm continues with $S_1 = \{1, 2\}$ and $S_2 = \{c\}$. Since $c \in S_2$ is free, with augmenting path $P = \{(3, a), (a, 1), (1, c)\}$, the new matching produced is $M = \{(1, c), (2, b), (3, a)\}$; see part (d) of the figure.

The fourth iteration produces $S_1 = \{4\}$, $S_2 = \{b\}$; $S_1 = \{2\}$, $S_2 = \{a, c\}$; and finally $S_1 = \{1, 3\}$, $S_2 = \emptyset$. No further augmenting paths are found, and the algorithm terminates with the maximum size matching $M = \{(1, c), (2, b), (3, a)\}$.

**5.** Algorithm 2 can be used to find a maximum weight matching in the bipartite network shown in part (a) of the following figure.

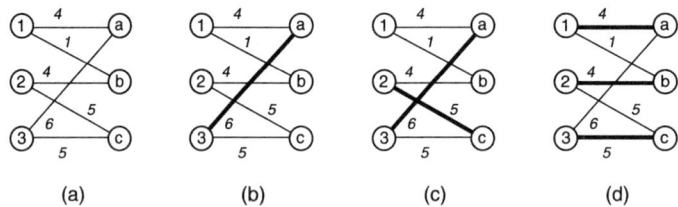

(a)                (b)                (c)                (d)

Relative to the initial empty matching, all vertices of $X$ are free so $S_1 = \{1, 2, 3\}$, with $d(1) = d(2) = d(3) = 0$. The labels on vertices $a, b, c$ are updated to $d(a) = 6$, $d(b) = 4$, $d(c) = 5$, giving $S_2 = \{a, b, c\}$. Since $M = \emptyset$ no further updates occur. The free vertex $a$ has maximum label, and the associated path $P_1 = \{(3, a)\}$ has $wt(P_1) = 6$. The resulting matching $M = \{(3, a)\}$ is shown in part (b) of the figure; it represents the largest weight matching of size 1, with $wt(M) = 6$.

The second iteration starts with $S_1 = \{1, 2\}$. The current labels on vertices $a, b, c$ are then updated to $d(a) = 4$, $d(b) = 4$, $d(c) = 5$, so $S_2 = \{a, b, c\}$. Using the matched edge $(a, 3)$, vertex 3 has its label updated to $d(3) = -2$ and $S_1 = \{3\}$. No further updates occur, and free vertex $c$ with maximum label $d(c) = 5$ is selected. This label corresponds to the augmenting path $P_2 = \{(2, c)\}$, with $wt(P_2) = 5$. The new matching is $M = \{(2, c), (3, a)\}$, with $wt(M) = 11$; see part (c) of the figure.

At the third iteration, $S_1 = \{1\}$ and vertices $a, b$ receive updated labels $d(a) = 4$, $d(b) = 1$. Subsequent updates produce $d(3) = -2$, $d(c) = 3$, $d(2) = -2$, $d(b) = 2$.

Finally, the free vertex $b$ is selected with $d(b) = 2$, corresponding to the augmenting path $P_3 = \{(1, a), (a, 3), (3, c), (c, 2), (2, b)\}$ with $wt(P_3) = 2$. This produces the maximum weight matching $M = \{(1, a), (2, b), (3, c)\}$, with $wt(M) = 13$; see part (d) of the figure. As predicted by Fact 12 of §10.2.1, the weights of the augmenting paths are nonincreasing: $wt(P_1) \geq wt(P_2) \geq wt(P_3)$.

---

## 10.2.3  MATCHINGS IN NONBIPARTITE NETWORKS

This section covers matchings in more general (nonbipartite) networks. Algorithms for constructing maximum size and maximum weight matchings are considerably more intricate than for bipartite networks. The important new concept is that of a "blossom" in a network.

### Definitions:

Suppose $P$ is an alternating path from a free vertex $s$ in network $G = (V, E)$. Then a vertex $v$ on $P$ is **even** (**outer**) if the subpath $P_{sv}$ of $P$ joining $s$ to $v$ has even length. Vertex $v$ on $P$ is **odd** (**inner**) if $P_{sv}$ has odd length.

Suppose $P$ is an alternating path from a free vertex $s$ to an even vertex $v$ and edge $(v, w) \in E$ joins $v$ to another even vertex $w$ on $P$. Then $P \cup \{(v, w)\}$ contains a unique cycle, called a **blossom**.

A **shrunken blossom** results when a blossom $B$ is collapsed into a single vertex $b$, whereby every edge $(x, y)$ with $x \notin B$ and $y \in B$ is transformed into the edge $(x, b)$. The reverse of this process gives an **expanded blossom**.

**Facts:**

**1.** Every blossom $B$ has odd length $2k + 1$ and contains $k$ matched edges, for some $k \geq 1$.

**2.** A bipartite network contains no blossoms.

**3.** *Edmonds' theorem*: Suppose network $G^B$ is formed from $G$ by collapsing blossom $B$. Then $G^B$ contains an augmenting path if and only if $G$ does. (J. Edmonds, born 1965.)

**4.** *General matching algorithm*: This method (Algorithm 3), based on Fact 9 of §10.2.1, produces a maximum size matching of $G$. At each iteration, a forest of trees is grown, rooted at the free vertices of $G$, to find an augmenting path. As encountered, blossoms $B$ are shrunk, with the search continued in the resulting network $G^B$.

---

**Algorithm 3**:  **General matching algorithm.**

input: undirected network $G = (V, E)$

output: maximum size matching $M$

$M := \emptyset$

{Start iteration}

mark all free vertices as even

mark all matched vertices as unreached

mark all free edges as unexamined

**while** there are unexamined edges $(v, w)$ and no augmenting path is found

    mark $(v, w)$ as examined

    {Case 1}

    **if** $v$ is even and $w$ is unreached **then**

        mark $w$ as odd and its mate $z$ as even

        extend the forest by $(v, w)$ and the matched edge $(w, z)$

    {Case 2}

    **if** $v, w$ are even and they belong to different subtrees **then**

        an augmenting path has been found

    {Case 3}

    **if** $v, w$ are even and they belong to the same subtree **then**

        a blossom $B$ has been found

        shrink $B$ to an even vertex $b$

**if** an augmenting path $P$ has been found **then**

    $M := M \Delta P$

    go to {Start iteration}

**else** terminate with matching $M$

---

**5.** Algorithm 3 was initially proposed by Edmonds [Ed65a] with a time bound of $O(n^4)$.

**6.** An improved implementation of Algorithm 3 runs in $O(nm)$ time.

**7.** There are other algorithms for maximum size matchings in nonbipartite networks:

- an algorithm of Gabow [Ga76], which runs in time $O(n^3)$;
- an algorithm of Micali and Vazirani [MiVa80], which runs in $O(m\sqrt{n})$ time.

Computer codes for these algorithms can be found at

- ftp://dimacs.rutgers.edu/pub/netflow/matching/

**8.** *General weighted matching algorithms*:  More complicated algorithms are required for solving weighted matching problems. The first such algorithm, also involving blossoms, was developed by Edmonds [Ed65b] and has a time bound of $O(n^4)$.

**9.** Improved algorithms exist for the weighted matching problem, with running times $O(n^3)$, $O(nm \log n)$, and $O(nm + n^2 \log n)$ respectively. Computer code for the first of these algorithms can be found at

- ftp://dimacs.rutgers.edu/pub/netflow/matching/

**Examples:**

**1.** In part (a) of the following figure, $P = \{(1,2),(2,3),(3,4),(4,5)\}$ is an alternating but not augmenting path, relative to the matching $M = \{(2,3),(4,5)\}$.

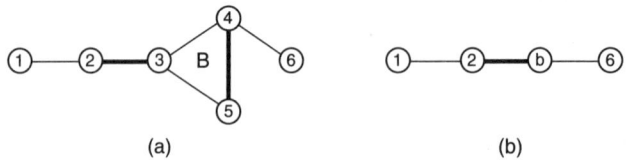

(a)                                          (b)

Relative to this path, vertices 1, 3, 5 are even while vertices 2, 4 are odd. Since $(3,5)$ is an edge joining two even vertices on $P$, the blossom $B = \{(3,4),(4,5),(5,3)\}$ is formed. On the other hand, $Q = \{(1,2),(2,3),(3,5),(5,4),(4,6)\}$ is an augmenting path relative to $M$ so that $M \Delta Q = \{(1,2),(3,5),(4,6)\}$ is a matching of larger size—in fact a matching of maximum size. Notice that relative to path $Q$, vertices 1, 3, 4 are even while vertices 2, 5, 6 are odd.

**2.** Shrinking the blossom $B$ relative to path $P$ in part (a) of the figure of Example 1 produces the network $G^B$ shown in part (b) of that figure; note that vertices $1, b$ are even while vertices 2, 6 are odd. The path $P^B = \{(1,2),(2,b),(b,6)\}$ is now augmenting in $G^B$. By expanding $P^B$ so that $(2,3)$ remains matched and $(4,6)$ remains free, the augmenting path $Q = \{(1,2),(2,3),(3,5),(5,4),(4,6)\}$ in $G$ is obtained.

**3.** Algorithm 3 is applied to the nonbipartite network shown in part (a) of the following figure. Suppose the matching $M = \{(3,4),(6,8)\}$ of size 2 is already available.

*Iteration 1*: The free vertices 1, 2, 5, 7 are marked as even, and the matched vertices 3, 4, 6, 8 are marked as unreached. The initial forest consists of the isolated vertices 1, 2, 5, 7.

- If the free edge $(2,3)$ is examined, then Case 1 applies, so vertex 3 is marked odd and vertex 4 even; the free edge $(2,3)$ and the matched edge $(3,4)$ are added to the forest.

- If the free edge $(4,7)$ is examined, then Case 2 applies, and the augmenting path $P = \{(2,3),(3,4),(4,7)\}$ is found. Using $P$ the new matching $M = \{(2,3),(4,7),(6,8)\}$ of size 3 is obtained; see part (b) of the figure.

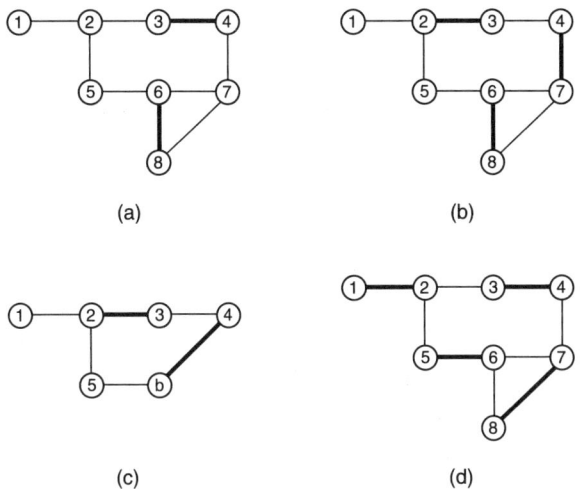

(a)                          (b)

(c)                          (d)

*Iteration 2*: The forest is initialized with the free (even) vertices 1, 5.

- If the free edge $(1, 2)$ is examined, then Case 1 applies, so vertex 2 is marked odd and vertex 3 even; edges $(1, 2)$ and $(2, 3)$ are added to the forest.

- Examining in turn the free edges $(3, 4)$ and $(7, 6)$ makes 4, 6 odd vertices and 7, 8 even. Edges $(3, 4), (4, 7), (7, 6), (6, 8)$ are then added to the subtree rooted at 1.

- If edge $(7, 8)$ is examined, Case 3 applies so the blossom $B = \{(7, 6), (6, 8), (8, 7)\}$ is detected and shrunk to the even vertex $b$; part (c) of the figure shows the resulting $G^B$. The current subtree rooted at 1 now becomes $\{(1, 2), (2, 3), (3, 4), (4, b)\}$.

- If the free edge $(b, 5)$ is examined, then Case 2 applies and the augmenting path $\{(1, 2), (2, 3), (3, 4), (4, b), (b, 5)\}$ is found in $G^B$. The corresponding augmenting path in $G$ is $P = \{(1, 2), (2, 3), (3, 4), (4, 7), (7, 8), (8, 6), (6, 5)\}$. Forming $M \triangle P$ produces the new matching $\{(1, 2), (3, 4), (7, 8), (5, 6)\}$, a maximum size matching; see part (d) of the figure.

## 10.2.4   STABLE MATCHINGS

The stable matching problem, formulated on a bipartite network $G = (X \cup Y, E)$, incorporates preferences of the items/individuals in $X$ for those in $Y$ and likewise the preferences of the items/individuals in $Y$ for those in $X$. In a stable matching there is no incentive for an unmatched pair $(x, y)$ to both find it desirable to switch their existing mates for one another. This is an important model for assignment problems with preferences (e.g., the assignment of medical residents to hospitals, students to schools, and organ donors to patients).

**Definitions:**

Let $G$ be a **complete bipartite graph** (§8.1.3) on the vertex set $X \cup Y$, where $X$ and $Y$ are disjoint sets with $|X| = |Y| = n$ and $(x, y)$ is an edge for every $x \in X$ and $y \in Y$. For each $x \in X$, let $\succ_{\mathbf{x}}$ be an ordering of the elements of $Y$; for each $y \in Y$, let $\succ_{\mathbf{y}}$ be an ordering of the elements of $X$.

Suppose $M$ is a perfect matching in $G = (X \cup Y, E)$. Then $w$ is the **mate** $M(v)$ of $v \in X \cup Y$ if $(v, w) \in M$. The pair $(x, y) \in X \times Y$ is a **blocking pair** or **instability** if

$(x, y) \notin M$, yet $y \succ_x M(x)$ and $x \succ_y M(y)$. In other words, each of $x$ and $y$ prefer one another to their current mate.

A perfect matching $M$ of $G$ is **stable** if there are no blocking pairs relative to $M$.

Individuals $x$ and $y$ are **stable partners** if $(x, y) \in M$ for *some* stable matching $M$.

**Facts:**

**1.** A stable matching in $G = (X \cup Y, E)$ always exists and can be found by using Algorithm 4, called the Gale-Shapley algorithm [GaSh62].

---

**Algorithm 4:   Gale-Shapley algorithm.**

input: complete bipartite graph $G$ on $X \cup Y$, with $|X| = |Y|$

output: stable matching $M$ of $G$

let all $x \in X$ and $y \in Y$ be free

**while** there is some free $x$

   select such an $x$

   let $y$ be the greatest element of $Y$ (according to $\succ_x$) that has not rejected $x$

     **if** $y$ is free, tentatively match $y$ with $x$

        add $(x, y)$ to $M$

   **else**

       **if** $x \succ_y M(y)$ then

          make $M(y)$ free and tentatively match $y$ with $x$

       **else**

          permanently record that $y$ has now rejected $x$

---

**2.** Each iteration of Algorithm 4 involves a free vertex $x$ "proposing" to its most-preferred choice $y \in Y$ that has not yet rejected it. Once a proposal has been rejected, it is never reconsidered. However, the individuals in $Y$ can always "trade up" if a proposal is made from a more preferred partner.

**3.** Surprisingly, Algorithm 4 always terminates and will do so with a stable matching. Moreover, the outcome is the same regardless of the order in which the free vertices in $X$ are processed.

**4.** Algorithm 4 terminates after at most $n^2$ iterations and can be implemented to run in $O(n^2)$ time.

**5.** Algorithm 4 finds the best stable matching for $X$. That is, every $x \in X$ is matched with its *most preferred* stable partner.

**6.** Algorithm 4 finds the worst stable matching for $Y$. That is, every $y \in Y$ is matched with its *least preferred* stable partner.

**7.** By reversing the roles of $X$ and $Y$ in Algorithm 4, we obtain a best stable matching for $Y$.

**8.** While formulated for a complete bipartite graph $G$ with $|X| = |Y|$, the stable matching problem can be generalized to an arbitrary bipartite graph. In this case, edges of $G$ indicate acceptable partners. Stability of $M$ now means that no $x, y$ that are not matched with each other prefer one another to their current partners $M(x), M(y)$. A stable matching may no longer be perfect.

**9.** A number of applications (e.g., assigning students to schools) require the possibility of multiple partners. For example, in the school assignment context, each student would be assigned to a single school, but schools can admit many students. This generalization allows each $y \in Y$ to have an integer quota $q_y \geq 1$. In fact, this situation can be reduced to a standard matching problem by making $q_y$ copies of $y$, and by letting each copy have the same preference ordering over $X$ as the original $y$.

**10.** Since the stable matching produced by the Gale-Shapley algorithm greatly advantages the individuals in $X$ and greatly disadvantages the individuals in $Y$ (see Facts 5–6), work has proceeded to find more equitable stable matchings [GuIr89].

**11.** The 2012 Nobel Prize in Economics was awarded to Lloyd S. Shapley and Alvin E. Roth for their pioneering work on stable matchings and applications to important societal problems.

**Examples:**

**1.** The following table shows the preferences (in descending order) of the individuals in $X = \{1, 2, 3, 4\}$ for those in $Y = \{a, b, c, d\}$ and likewise the preferences of the individuals in $Y$ for those in $X$.

| 1 | $d$ | $b$ | $c$ | $a$ | | $a$ | 1 | 3 | 2 | 4 |
|---|---|---|---|---|---|---|---|---|---|---|
| 2 | $a$ | $d$ | $b$ | $c$ | | $b$ | 4 | 2 | 1 | 3 |
| 3 | $a$ | $c$ | $b$ | $d$ | | $c$ | 2 | 1 | 3 | 4 |
| 4 | $c$ | $b$ | $d$ | $a$ | | $d$ | 4 | 2 | 1 | 3 |

The matching $\{(1,d), (2,b), (3,c), (4,a)\}$ is shown in bold in the following figure. This matching is not stable since $(4, b)$ (shown as dashed) is a blocking pair for the matching. Namely, $b \succ_4 a$ and $4 \succ_b 2$.

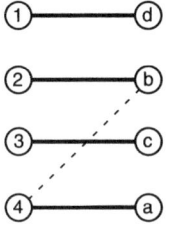

**2.** The matching $\{(1,b), (2,d), (3,a), (4,c)\}$ is stable. To verify this, observe that 3 and 4 are matched with their top choice, so they cannot be part of a blocking pair. Suppose 1 was matched with a higher choice than $b$, namely $d$; however, $d$ prefers the current mate 2 to 1 so $(1, d)$ is not a blocking pair. Similarly, the only possible blocking pair involving 2 is $(2, a)$, but $a$ prefers the current mate 3 to 2.

**3.** The matching $\{(1,c), (2,d), (3,a), (4,b)\}$ is also stable. This is the only other stable matching for the given set of preferences.

**4.** Applying Algorithm 4 to the instance in Example 1 results in the following. First, 1 proposes to $d$ and is tentatively accepted; 2 proposes to $a$ and is tentatively accepted. Next, 3 proposes to $a$; since 3 is more favored by $a$ than its current partner 2, $a$ accepts the proposal and rejects 2. Then 4 proposes to $c$ and is tentatively accepted. The current matching is now $\{(1,d), (3,a), (4,c)\}$. Now 2 proposes to $d$, the most-preferred of the remaining candidates, who accepts the proposal and rejects 1. Then 1 proposes to $b$, the most-preferred of the remaining candidates, who accepts the proposal.

At this point no member of $X$ is free and so the algorithm terminates with the matching $\{(1, b), (2, d), (3, a), (4, c)\}$. This is the stable matching identified in Example 2. Notice that this matching pairs 3 and 4 with their first choices, while 1 and 2 are paired with their second choices.

**5.** As indicated in Example 3, the only other stable matching for the given preferences is the matching $\{(1, c), (2, d), (3, a), (4, b)\}$. Notice that this matching pairs 3 with its first choice, 2 and 4 with their second choice, and 1 with its third choice. This verifies Fact 5, showing that the stable matching found by Algorithm 4 matches each member of $X$ with its most preferred stable partner.

**6.** By reversing the roles of $X$ and $Y$ in Algorithm 4, we obtain the second stable matching $\{(1, c), (2, d), (3, a), (4, b)\}$. This matching, which pairs $b$ with its first choice, and $a, c, d$ with their second choice, is the best stable matching for $Y$. (See Fact 6.)

## 10.2.5   APPLICATIONS

Matching problems, in both bipartite and nonbipartite networks, are useful models in a number of applied areas. This section presents some representative applications of matchings.

### Applications:

**1.** *Linear assignment problem*:   There are $n$ applicants to be assigned to $n$ jobs, with each job being filled with exactly one applicant. The weight $w_{ij}$ measures the suitability of applicant $i$ for job $j$. Finding a valid assignment with the best overall weight is a weighted matching problem on the bipartite network $G = (X \cup Y, E)$, where $X$ is the set of applicants and $Y$ is the set of jobs.

**2.** *Personnel assignment*:   Pairs of pilots are to be assigned to a fleet of aircraft serving international routes. Pilots $i$ and $j$ are considered compatible if they are fluent in a common language and have comparable flight training. Form the network $G$ whose vertices represent the pilots and with edges placed between compatible pairs of pilots. The problem of flying the largest number of aircraft with compatible pilots can then be solved as a maximum size matching problem on $G$.

**3.** Other examples of Application 2 occur in assigning police officers sharing beats, matching pairs of compatible roommates, and assigning pairs of employees with complementary skills to specific projects.

**4.** *Pruned chessboards*:   Several squares ($2k$ in all) are removed from an $n \times n$ chessboard, yielding the pruned chessboard $\mathcal{P}$. Is it then possible to cover the squares of $\mathcal{P}$ using nonoverlapping dominoes, with no squares left uncovered? This can be formulated as a matching problem on the bipartite network $G = (R \cup B, E)$, where $R$ is the set of red squares and $B$ is the set of black squares in $\mathcal{P}$. An edge joins $r \in R$ to $b \in B$ if squares $r$ and $b$ share a common side. Each set of nonoverlapping dominoes on $\mathcal{P}$ corresponds to a matching in $G$.

All squares of $\mathcal{P}$ can be covered using nonoverlapping dominoes if and only if the maximum size matching in $G$ has size $\frac{n^2}{2} - k$. More generally, a maximum size matching in $G$ explicitly provides a way to cover the maximum number of squares of $\mathcal{P}$ using nonoverlapping dominoes.

**5.** *Target tracking*:   The movements of $n$ objects (such as submarines or missiles) are to be followed over time. The locations of the set of objects are known at two distinct times, though without identification of the individual objects. Suppose $X = \{x_1, x_2, \ldots, x_n\}$ and $Y = \{y_1, y_2, \ldots, y_n\}$ represent the spatial coordinates of the objects detected at times $t$ and $t + \Delta t$. If $\Delta t$ is sufficiently small, then the Euclidean distance between a given object's position at these two times should be relatively small. To aid in identifying the objects (as well as their velocities and directions of travel), a pairing between set $X$ and set $Y$ is desired that minimizes the sum of Euclidean distances.

This can be formulated as a maximum weight matching problem on the complete bipartite network $G = (X \cup Y, E)$, where the edge $(i, j)$ indicates pairing position $x_i$ with position $y_j$. The weight of this edge is the negative of the Euclidean distance between $x_i$ and $y_j$. A maximum weight matching of size $n$ in $G$ then provides an optimal (minimum distance) pairing of all observations at the two times $t$ and $t + \Delta t$.

**6.** *Crew scheduling*: Bus drivers are hired to work two four-hour shifts each day. Union rules require a certain minimum amount of time between the shifts that a driver can work. There are also costs associated with getting the driver between the ending location of the first shift and the starting location of the second shift.

The problem of optimally combining pairs of shifts that satisfy union regulations and incur minimum total cost can be formulated as a maximum weight matching problem. Namely, define the network $G$ with vertices representing each shift that must be covered and edges between pairs of compatible shifts (satisfying union regulations). The weight of edge $(i, j)$ is the negative of the cost of assigning a single driver to shifts $i$ and $j$. It is convenient also to add edges $(i, i)$ to $G$ to represent the possibility of needing a part-time driver to cover a single shift; edge $(i, i)$ is given a sufficiently large negative weight to discourage single-shift assignments unless absolutely necessary. A maximum weight matching in the network $G$ then provides a minimum cost pairing of shifts for the bus drivers.

**7.** *Snowplowing streets*: The streets of an area of a city are to be plowed by a single snowplow. Let $G$ be the network representing the street system of the city, with vertices representing street intersections and edges representing streets. Associated with each street $(i, j)$ is its length $c_{ij}$.

If all vertices of $G$ have even degree, then $G$ is an Eulerian graph (§8.4.3) and a circuit that traverses each edge (street) exactly once can be found using the algorithms in §8.4.3. Otherwise, a closed walk of $G$ that covers each street at least once is needed, and one with minimum total length $\sum c_{ij}$ is desired. Let $N$ be the set of vertices of $G$ having odd degree; by Fact 4 of §8.1.1, $|N|$ is an even integer $2k$. Form the complete network $H = (N, E)$ in which the weight of edge $(i, j)$ is the negative of the shortest path distance (§10.3.1) between vertices $i$ and $j$ in $G$. Determine a maximum weight (perfect) matching $M$ of size $k$ in $H$. For each $(i, j)$ in $M$, add the edges of the shortest path between $i$ and $j$ to the network $G$, forming the network $G'$. Every vertex of $G'$ now has even degree, and an Euler circuit of $G'$ provides a minimum cost traversal of the city streets.

This problem is known as the (undirected) *Chinese postman problem*. A directed version of the problem is discussed in §10.5.3, Application 3.

**8.** *Sparse matrices*: In solving large sparse systems of linear equations, it is advantageous to first reorder the rows and columns of the given $n \times n$ coefficient matrix $A$. Specifically, it is desired to place as many nonzero entries on the diagonal of the permuted matrix [Du81]. This can be viewed as a maximum size matching problem on a bipartite graph $G = (X \cup Y, E)$ associated with the coefficient matrix $A$. Here $X$ contains the $n$ row indices, $Y$ contains the $n$ column indices, and edge $(i, j)$ exists whenever $a_{ij} \neq 0$.

**9.** *Stable matchings*: A celebrated application of the stable matching model can be found in the National Resident Matching Program (NRMP), which assigns graduating medical students (residents) to hospitals. The algorithm developed for this centralized matching program (in use from 1952) was very similar in operation to the Gale-Shapley algorithm and produced hospital-optimal stable matchings. Roth and Peranson [RoPe99] helped to redesign the NRMP matching algorithm, now a variant of the resident-optimal Gale-Shapley algorithm.

**10.**  Additional applications, with reference sources, are given in the following table.

| application | references |
|---|---|
| medical residents assignment | [AhMaOr93] |
| school bus driver assignment | [AhMaOr93] |
| oil well drilling | [AhMaOr93], [Ge95], [LoPu86] |
| chemical bonds | [AhMaOr93] |
| inventory depletion | [AhMaOr93] |
| scheduling on machines | [AhMaOr93], [LoPu86] |
| ranks of matrices | [AhMaOr93] |
| doubly stochastic matrices | [LoPu86] |
| nonnegative matrices | [LoPu86] |
| basketball conference scheduling | [EvMi92] |
| major league umpire scheduling | [EvMi92] |
| project scheduling | [Ge95] |
| plotting street maps | [Ge95] |

# 10.3  SHORTEST PATHS

The shortest path problem requires finding paths of minimum cost (or length) from a specified source vertex to every other vertex in a directed network. Shortest path problems lie at the heart of network flows (§10.4–§10.5). They are important both to researchers and to practitioners because

- they arise frequently in application settings where material is to be sent between specified points as quickly, as cheaply, or as reliably as possible;

- they arise as subproblems when solving many combinatorial and network optimization problems;

- they can be solved very efficiently.

## 10.3.1  BASIC CONCEPTS

### Definitions:

A **directed network** is a weighted graph $G = (V, E)$, where $V$ is the set of vertices and $E$ is the set of arcs (directed edges). Each arc $(i, j) \in E$ has an associated **cost** (or **weight, length**) $c_{ij}$. It is possible that certain of the $c_{ij}$ are negative. Let $n = |V|$ and $m = |E|$.

The **adjacency set** $A(i)$ for vertex $i$ is the set of all arcs incident from $i$, written $A(i) = \{(i,j) \mid (i,j) \in E\}$.

A directed path (§8.3.2) $P$ has **length** $\Sigma_{(i,j)\in P}\, c_{ij}$.

A directed cycle (§8.3.2) $W$ for which $\Sigma_{(i,j)\in W}\, c_{ij} < 0$ is called a **negative cycle**.

A **shortest path** from vertex $s$ to vertex $j$ is a directed $s$-$j$ path having minimum length.

A **directed out-tree** is a tree rooted at vertex $s$ in which all arcs are directed away from the root $s$.

A **shortest path tree** is an out-tree $T^*$ rooted at vertex $s$ with the property that the directed path in $T^*$ from $s$ to any other vertex $j$ is a shortest $s$-$j$ path.

A vector $d(\cdot)$ is called a vector of **distance labels** if for every vertex $j \in V$, $d(j)$ is the length of some directed path from the source vertex $s$ to vertex $j$, with $d(s) = 0$. If these labels are the lengths of shortest $s$-$j$ paths, they are called **shortest path distances**.

The directed path $P = [i_0, i_1, \ldots, i_r]$ from vertex $i_0$ to vertex $i_r$ can be represented using **predecessor indices**: $pred(i_1) = i_0, pred(i_2) = i_1, \ldots, pred(i_r) = i_{r-1}$.

**Facts:**

**1.** Shortest paths are useful in a wide variety of applications, such as in efficient routing of messages and distribution of goods, developing optimal investment strategies, scheduling personnel, and approximating piecewise linear functions (see §10.3.5).

**2.** If $P = [s, i_1, \ldots, i_r]$ is a shortest path from $s$ to $i_r$ then $Q = [s, i_1, \ldots, i_k]$ is a shortest path from $s$ to $i_k$ for each $1 \le k < r$.

**3.** *Shortest path optimality conditions:* The vector $d(\cdot)$ of distance labels represents shortest path distances if and only if $d(j) \le d(i) + c_{ij}$ for all $(i,j) \in E$.

**4.** If the network contains a negative cycle accessible from vertex $s$, then distance labels satisfying the conditions in Fact 3 do not exist.

**5.** If the network does not contain any negative cycle, then (unique) distance labels satisfying the conditions in Fact 3 always exist. Furthermore, there is a shortest path tree $T^*$ realizing these shortest path distances.

**Examples:**

**1.** In the directed network of the following figure, arc costs are shown along each arc. Part (b) lists the nine paths from vertex 1 to vertex 6, together with their lengths. Path $P_4$, with length 10, is the (unique) shortest path joining these two vertices. This path can be represented using the predecessor indices: $pred(6) = 4$, $pred(4) = 3$, $pred(3) = 2$, $pred(2) = 1$. By Fact 2, the subpath $Q = [1, 2, 3, 4]$ of $P_4$ is a shortest path from vertex 1 to vertex 4.

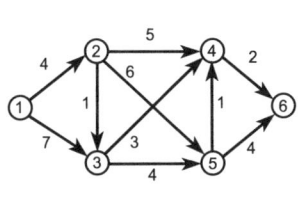

|  | Path | Length |
|---|---|---|
| $P_1$ | [1, 2, 4, 6] | 11 |
| $P_2$ | [1, 2, 5, 6] | 14 |
| $P_3$ | [1, 2, 5, 4, 6] | 13 |
| $P_4$ | [1, 2, 3, 4, 6] | 10 |
| $P_5$ | [1, 2, 3, 5, 6] | 13 |
| $P_6$ | [1, 2, 3, 5, 4, 6] | 12 |
| $P_7$ | [1, 3, 4, 6] | 12 |
| $P_8$ | [1, 3, 5, 4, 6] | 14 |
| $P_9$ | [1, 3, 5, 6] | 15 |

(a)                              (b)

**2.** In the directed network of the following figure, arc costs are shown along each arc and a set of distance labels is shown at each vertex. Part (b) gives paths from vertex $s = 1$ whose lengths equal the corresponding distance labels. These distance labels do not satisfy the optimality conditions of Fact 3 because for the arc $(3,5)$, $d(5) > d(3) + c_{35}$. The out-tree $T$ in this figure defined by predecessor indices $pred(2) = 5$, $pred(3) = 1$, $pred(4) = 2, pred(5) = 3$ has distance labels $d = (0, 5, 5, 25, 0)$. It is a shortest path tree rooted at vertex 1 since the optimality conditions of Fact 3 are satisfied: namely

$$5 \le 0 + 10 \text{ for arc } (1,2), \quad 5 \le 5 + 10 \text{ for arc } (2,3), \quad 0 \le 25 + 15 \text{ for arc } (4,5).$$

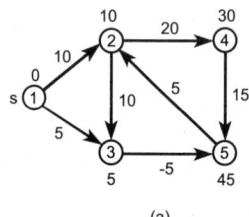

| j | d(j) | path |
|---|------|------|
| 2 | 10 | [1,2] |
| 3 | 5 | [1,3] |
| 4 | 30 | [1,2,4] |
| 5 | 45 | [1,2,4,5] |

(a)　　　　(b)

## 10.3.2   ALGORITHMS FOR SINGLE-SOURCE SHORTEST PATHS

This subsection discusses algorithms for finding shortest path trees from a given source vertex $s$ in a directed network $G$ with $n$ vertices and $m$ arcs.

**Facts:**

**1.** *Label-correcting algorithm:*   A general label-correcting algorithm (Algorithm 1) is based on the shortest path optimality conditions (§10.3.1, Fact 3) and is a very popular algorithm to solve shortest path problems with arbitrary arc costs (L. R. Ford, 1956 and R. E. Bellman, 1958).

---

**Algorithm 1:  Label-correcting algorithm.**
input: directed network $G$, source vertex $s$
output: shortest path tree $T^*$ rooted at $s$
$d(s) := 0$
$pred(s) := 0$
$d(j) := \infty$ for all $j \in V - \{s\}$
LIST $:= \{s\}$
**while** LIST $\ne \emptyset$
　remove a vertex $i$ from LIST
　**for** each $(i,j) \in A(i)$
　　**if** $d(j) > d(i) + c_{ij}$ **then**
　　　$d(j) := d(i) + c_{ij}$
　　　$pred(j) := i$
　　　**if** $j \notin$ LIST **then** add $j$ to LIST

---

**2.** Algorithm 1 maintains a list, LIST, of vertices with the property that if an arc $(i,j)$ violates the optimality condition, then LIST must contain vertex $i$. If LIST is empty, then the current distance labels are optimal. Otherwise some vertex $i$ is removed from LIST and the arcs of $A(i)$ are scanned. If an arc $(i, j) \in A(i)$ violates the optimality condition, then $d(j)$ is updated appropriately.

**3.** When Algorithm 1 terminates, the nonzero predecessor indices define a shortest path tree $T^*$ rooted at the source vertex: namely, $T^* = \{(pred(i), i) \mid i \in V - \{s\}\}$.

**4.** *Convergence*: In Algorithm 1, vertices in LIST can be selected in any order and the algorithm still converges finitely. If all arc costs are integers whose magnitudes are bounded by a constant $C$, then the algorithm performs $O(n^2 C)$ iterations and can be implemented to run in $O(nmC)$ time, regardless of the order in which vertices from LIST are selected.

**5.** *Queue implementation*: Suppose in Algorithm 1 that LIST is maintained as a queue (§18.1.2); that is, vertices in LIST are examined in a first-in-first-out (FIFO) order. This specific implementation examines no vertex more than $n - 1$ times and runs in $O(nm)$ time. This is the best strongly polynomial-time algorithm to solve the shortest path problem with arbitrary arc costs.

**6.** *Deque implementation*: Suppose in Algorithm 1 that LIST is maintained as a deque or two-way list (§18.1.2). Specifically, vertices are removed from the front of the deque, but vertices are added either at the front or at the rear. If the vertex has been in LIST earlier, the algorithm adds it to the front; otherwise, it adds the vertex to the rear. Empirical studies have found that the deque implementation is one of the most efficient algorithms to solve the shortest path problem in practice even though it is not a polynomial-time algorithm.

**7.** *Negative cycle detection*: The queue implementation (Fact 5) of the label-correcting algorithm can be used to detect the presence of a negative cycle. To do so, record the number of times that the algorithm examines each vertex. If the algorithm examines a vertex more than $n - 1$ times, there must exist a negative cycle. In this case, the subgraph formed by the arcs $(pred(i), i)$ will contain a negative cycle.

**8.** Goldberg [Go95] developed a scaling algorithm that determines a negative cost cycle in $O(\sqrt{n}m \log C)$ time. It is the fastest running time for a large range of values of $C$.

**9.** Cherkassky et al. [ChGoRa96], [ChEtal09] have carried out careful computational analyses of algorithms for single-source shortest path problems.

**10.** *Dijkstra's algorithm (1959)*: Dijkstra's algorithm (Algorithm 2) is a popular algorithm for solving shortest path problems with nonnegative arc costs (E. W. Dijkstra, 1930–2002).

---

**Algorithm 2**: **Dijkstra's algorithm.**

   input: directed network $G$ with $c_{ij} \geq 0$, source vertex $s$

   output: shortest path tree $T^*$ rooted at $s$

   $d(s) := 0$

   $pred(s) := 0$

   $d(j) := \infty$ for all $j \in V - \{s\}$

   LIST $:= V$

   **while** LIST $\neq \emptyset$

      {Vertex selection}

      let $i \in$ LIST be a vertex for which $d(i) = \min\{d(j) \mid j \in$ LIST$\}$

      remove vertex $i$ from LIST

{Distance update}
  **for** each $(i,j) \in A(i)$
    **if** $d(j) > d(i) + c_{ij}$ **then**
      $d(j) := d(i) + c_{ij}$
      $pred(j) := i$

**11.** Algorithm 2 performs two steps repeatedly: *vertex selection* and *distance update*. The vertex selection step chooses a vertex $i$ with smallest distance label in LIST for examination. The distance update step scans each arc $(i,j) \in A(i)$ and updates the distance label $d(j)$, if necessary, to restore the optimality condition for arc $(i,j)$.

**12.** Whenever a vertex is selected for examination in Algorithm 2, its distance label is the shortest path distance from $s$; consequently, each vertex is examined only once.

**13.** Using a simple array or linked list representation of LIST, vertex selections take a total of $O(n^2)$ time and distance updates take a total of $O(m)$ time. This implementation of Algorithm 2 runs in $O(n^2)$ time.

**14.** By using more sophisticated data structures, the efficiency of Dijkstra's algorithm can be improved. Currently, two of the best implementations use

- Fibonacci heaps, giving $O(m + n \log n)$ running time [FrTa84];
- radix heaps, giving $O(m + n(\log C)^{1/2})$ running time [AhEtal90].

**15.** Thorup [Th99] developed an algorithm that solves the shortest path problem on undirected networks in $O(m)$ time.

**16.** Goldberg [Go01] gave an implementation of Dijkstra's algorithm that solves the shortest path problem in $O(m)$ expected time if the arc lengths are positive and uniformly distributed.

**17.** A very fast implementation of Dijkstra's algorithm is due to R. Dial [Di69], and it runs in $O(m + nC)$ time.

**18.** A comprehensive discussion of several implementations of Dijkstra's algorithm and the label-correcting algorithm is presented in [AhMaOr93].

**19.** Computer codes that implement algorithms for single-source shortest paths can be found at

- http://www.avglab.com/andrew/soft.html
- http://www.dis.uniroma1.it/challenge9/

**20.** A useful extension of the shortest path problem involves finding the *k shortest paths* in a network. The case $k = 1$ corresponds to a shortest path. More generally, the $k$th shortest path is one having the $k$th smallest length among all paths from $s$ to $t$. Several algorithms for solving the problem of finding the $k$ shortest paths are discussed in [EvMi92].

**Examples:**

**1.** The following figure illustrates three iterations of the label-correcting algorithm applied to Example 2 of §10.3.1. LIST is maintained as a queue.

In the first iteration, the source vertex $s = 1$ is examined, and the distance labels of vertices 2 and 3 are decreased to 10 and 5, respectively. At this point, LIST $= [2, 3]$.

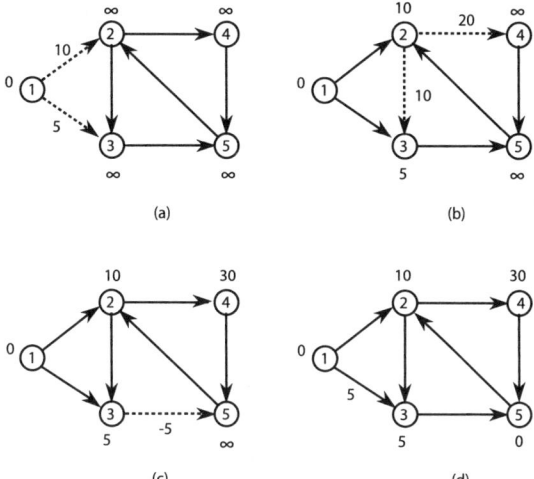

(a)

(b)

(c)

(d)

In the second iteration, vertex 2 is removed from LIST and examined. The distance label of vertex 4 decreases to 30, while the distance label of vertex 3 remains unchanged, giving LIST $= [3, 4]$.

Next, vertex 3 is removed from LIST and examined, triggering a reduction of the distance label of vertex 5 to 0. At this point the current out-tree, defined by the predecessor indices $pred(\cdot)$, consists of arcs $(1, 2)$, $(2, 4)$, $(1, 3)$, and $(3, 5)$.

**2.** Part (b) of the following figure shows the application of Dijkstra's algorithm to the directed network in part (a) with nonnegative arc costs and $s = 1$. Shown at each iteration are the current distance labels, the vertex selected, and the resulting distance updates. Upon termination, the shortest path lengths $d = (0, 6, 4, 9, 7)$ are realized by the optimal tree $T^*$ having arcs $(1, 3)$, $(3, 2)$, $(2, 4)$, and $(2, 5)$.

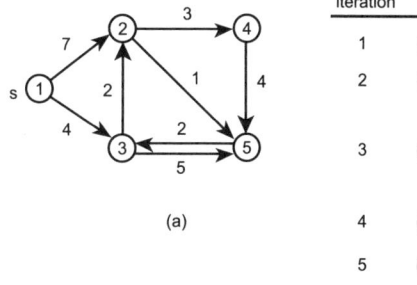

(a)

| iteration | labels | select | updates |
|---|---|---|---|
| 1 | $(0, \infty, \infty, \infty, \infty)$ | 1 | $d(2) = 7$,  $d(3) = 4$ |
| 2 | $(0, 7, 4, \infty, \infty)$ | 3 | $d(2) = \min\{7, 6\} = 6$ |
| | | | $d(5) = 9$ |
| 3 | $(0, 6, 4, \infty, 9)$ | 2 | $d(4) = 9$ |
| | | | $d(5) = \min\{9, 7\} = 7$ |
| 4 | $(0, 6, 4, 9, 7)$ | 5 | $d(3) = \min\{4, 9\} = 4$ |
| 5 | $(0, 6, 4, 9, 7)$ | 4 | $d(5) = \min\{7, 13\} = 7$ |

(b)

## 10.3.3  ALGORITHMS FOR ALL-PAIRS SHORTEST PATHS

This section discusses algorithms for finding shortest path distances between every pair of vertices in a directed network with $n$ vertices and $m$ arcs.

**Definitions:**

Suppose $G = (V, E)$ is a directed network with vertex set $V$ and arc set $E$, and let $c_{ij}$ be the cost of arc $(i, j) \in E$.

The $n \times n$ **arc length matrix** $U = (u_{ij})$ is defined as follows:

$$u_{ij} = \begin{cases} 0 & \text{if } i = j \\ c_{ij} & \text{if } (i, j) \in E \\ \infty & \text{if } i \neq j \text{ and } (i, j) \notin E. \end{cases}$$

Let $d_{ij}$ be the length of a shortest path from vertex $i$ to vertex $j$, with $d_{ii} = 0$.

Define the $n \times n$ matrix $D^k = (d_{ij}^k)$, where $d_{ij}^k$ is the length of a shortest path from vertex $i$ to vertex $j$ subject to the condition that the path contains no more than $k$ arcs.

Define **minsum matrix multiplication** $C = A \otimes B$ by $c_{ij} = \min_{1 \leq p \leq n}\{a_{ip} + b_{pj}\}$. Also, define $A^{\otimes k} = A \otimes A \otimes \cdots \otimes A$ ($k$ times).

In the directed path $[i_0, i_1, \ldots, i_r]$ from $i_0$ to $i_r$, the vertices $i_1, i_2, \ldots, i_{r-1}$ are called **internal vertices**. Let $d^k[i, j]$ be the length of a shortest path from vertex $i$ to vertex $j$ subject to the condition that this path uses only $1, 2, \ldots, k - 1$ as internal vertices. The $n \times n$ matrix $D^{[k]}$ contains the entries $d^k[i, j]$.

**Facts:**

**1.** The length of a shortest path containing at most $k$ arcs can be expressed in terms of shortest path lengths involving at most $k - 1$ arcs. Namely, for all vertices $i$ and $j$

- $d_{ij}^1 = u_{ij}$;
- $d_{ij}^k = \min_{1 \leq p \leq n}\{d_{ip}^{k-1} + u_{pj}\}$ for $2 \leq k \leq n - 1$;
- if there is no negative cycle, then $d_{ij}^{n-1} = d_{ij}$.

**2.** $D^k = U^{\otimes k}$ for all $1 \leq k \leq n - 1$.

**3.** For any pair of vertices $i$ and $j$, the following conditions hold:

- $d^1[i, j] = u_{ij}$;
- $d^{k+1}[i, j] = \min\{d^k[i, j], d^k[i, k] + d^k[k, j]\}$, $1 \leq k \leq n$;
- if there is no negative cycle then $d^{n+1}[i, j] = d_{ij}$.

**4.** The all-pairs shortest path problem can be solved by applying $n$ times either Algorithm 1 or Algorithm 2 of §10.3.2, considering each vertex once as a source.

**5.** Specialized algorithms are available to solve the all-pairs shortest path problem: the matrix multiplication algorithm (Fact 6) and the Floyd-Warshall algorithm (Fact 8).

**6.** *Matrix multiplication algorithm:* This algorithm (Algorithm 3), based on Facts 1 and 2, computes the shortest path distances between all vertex pairs by multiplying two matrices repeatedly, using minsum matrix multiplication.

---

**Algorithm 3:  Matrix multiplication algorithm.**

input: directed network $G$ on $n$ vertices

output: shortest distance matrix $D = (d_{ij})$

form the $n \times n$ arc length matrix $U$

compute $D := U^{\otimes(n-1)}$

---

**7.** If there is no negative cycle, then Algorithm 3 finds all shortest path distances using $O(\log n)$ matrix multiplications, each of which takes $O(n^3)$ time. Hence this algorithm runs in $O(n^3 \log n)$ time and requires $O(n^2)$ space. Zwick [Zw02] showed how the matrix algorithm could be sped up using fast matrix multiplication.

**8.** *Floyd-Warshall algorithm*:   This approach (Algorithm 4) calculates all-pairs shortest path distances in a directed network $G$ and is based on computing conditional shortest path lengths $d[i,j]$.

---

**Algorithm 4:   Floyd-Warshall algorithm.**

input: directed network $G$ on $n$ vertices

output: shortest distance matrix $D = (d[i,j])$

**for** all $(i,j) \in V \times V$   $d[i,j] := \infty$

**for** all $i \in V$   $d[i,i] := 0$

**for** all $(i,j) \in E$   $d[i,j] := c_{ij}$

**for** $k := 1$ **to** $n$

   **for** $(i,j) \in V \times V$

      **if** $d[i,j] > d[i,k] + d[k,j]$ **then** $d[i,j] := d[i,k] + d[k,j]$

---

**9.** If there is no negative cycle, the Floyd-Warshall algorithm correctly computes the matrix of shortest path distances. A single $n \times n$ array $D$ is used to implement the algorithm.

**10.** Algorithm 4 can be used to detect (and identify) negative cycles by monitoring whenever $d[i,i] < 0$ occurs for some vertex $i$.

**11.** Algorithm 4 runs in $O(n^3)$ time and requires $O(n^2)$ space.

**12.** If the underlying network is dense, that is, $m = \Omega(n^2)$, then the $O(n^3)$ time bound for Algorithm 4 is as good as any other discussed in §10.3.2 or §10.3.3.

**13.** Algorithm 4 was first discovered by B. Roy in 1959 in the context of determining the transitive closure of a graph; this same algorithm was independently discovered by S. Warshall in 1962. The method was generalized to computing all-pairs shortest paths by R. W. Floyd, also in 1962.

### Examples:

**1.** The matrix multiplication algorithm is applied to the directed network in the following figure.

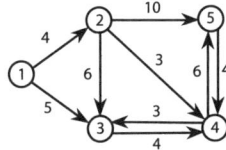

By Facts 1 and 2, the matrix $D^4$ is the matrix of shortest path distances.

$$D^1 = \begin{pmatrix} 0 & 4 & 5 & \infty & \infty \\ \infty & 0 & 6 & 3 & 10 \\ \infty & \infty & 0 & 4 & \infty \\ \infty & \infty & 3 & 0 & 6 \\ \infty & \infty & \infty & 4 & 0 \end{pmatrix} \qquad D^2 = \begin{pmatrix} 0 & 4 & 5 & 7 & 14 \\ \infty & 0 & 6 & 3 & 9 \\ \infty & \infty & 0 & 4 & 10 \\ \infty & \infty & 3 & 0 & 6 \\ \infty & \infty & 7 & 4 & 0 \end{pmatrix}$$

$$D^4 = \begin{pmatrix} 0 & 4 & 5 & 7 & 13 \\ \infty & 0 & 6 & 3 & 9 \\ \infty & \infty & 0 & 4 & 10 \\ \infty & \infty & 3 & 0 & 6 \\ \infty & \infty & 7 & 4 & 0 \end{pmatrix}.$$

2. Algorithm 4 is illustrated using the network shown in Example 1.

$$D^{[1]} = \begin{pmatrix} 0 & 4 & 5 & \infty & \infty \\ \infty & 0 & 6 & 3 & 10 \\ \infty & \infty & 0 & 4 & \infty \\ \infty & \infty & 3 & 0 & 6 \\ \infty & \infty & \infty & 4 & 0 \end{pmatrix} \qquad D^{[3]} = \begin{pmatrix} 0 & 4 & 5 & 7 & 14 \\ \infty & 0 & 6 & 3 & 10 \\ \infty & \infty & 0 & 4 & \infty \\ \infty & \infty & 3 & 0 & 6 \\ \infty & \infty & \infty & 4 & 0 \end{pmatrix}$$

$$D^{[5]} = \begin{pmatrix} 0 & 4 & 5 & 7 & 13 \\ \infty & 0 & 6 & 3 & 9 \\ \infty & \infty & 0 & 4 & 10 \\ \infty & \infty & 3 & 0 & 6 \\ \infty & \infty & 7 & 4 & 0 \end{pmatrix}.$$

It can be verified that $D^{[2]} = D^{[1]}$, $D^{[4]} = D^{[3]}$, and $D^{[6]} = D^{[5]}$. Consequently, the matrix $D^{[5]}$ above gives all shortest path distances.

## 10.3.4  PARALLEL ALGORITHMS

Parallel implementations of certain shortest path algorithms are described here relative to an EREW (exclusive-read, exclusive-write) PRAM (parallel random-access machine). For details of EREW PRAM, see §17.1.4.

**Facts:**

1. *Label-correcting algorithm:* The parallel implementation of the label-correcting algorithm (§10.3.2) associates a processor with each arc and with each vertex of the network. This algorithm maintains a distance label for each vertex, appropriately initialized. Suppose the distance labels are $d(i)$ at the beginning of an iteration. During the iteration, the processor attached to each arc $(i, j)$ computes a temporary label $d'(i, j) = d(i) + c_{ij}$ in $O(1)$ time. Then the processor associated with vertex $j$ examines incoming arcs at vertex $j$ and sets $d(j) := \min\{d'(i, j) \mid 1 \le i \le n\}$.

2. Using a parallel prefix operation, the distance labels can be updated in $O(\log n)$ time. The label-correcting algorithm performs $O(n)$ iterations and so its running time is $O(n \log n)$ using $O(m)$ processors.

3. *Matrix multiplication algorithm:* The matrix multiplication algorithm of §10.3.3 solves the all-pairs shortest path problem by performing $O(\log n)$ matrix multiplications.

4. Unlike a sequential computer, where matrix multiplication takes $O(n^3)$ time, a parallel computer can perform matrix multiplication in $O(\log n)$ time using $O(n^3)$ processors [Le92]. Consequently, this all-pairs shortest path algorithm runs in $O(\log^2 n)$ time using $O(n^3)$ processors.

## 10.3.5 APPLICATIONS

Shortest path problems arise in a variety of applications, both as stand-alone models and as subproblems in more complex problem settings. Shortest path problems also arise in surprising ways that on the surface might not appear to involve networks at all. This subsection presents several models based on determining shortest paths.

### Applications:

**1.** *Distribution*: Material needs to be shipped by truck from a central warehouse to various retailers at minimum cost. The underlying network is an undirected road network, with edges representing the roads joining various cities (vertices). The cost of an edge is the per unit shipping cost. Solving the single-source shortest path problem provides a least-cost shipping pattern for the material.

**2.** *Telephone routing*: A call is to be routed from a specified origin to a specified destination. Here the underlying network is the telephone system, with vertices representing individual users (or switching centers). Since a direct connection between the origin vertex $s$ and the destination vertex $t$ may not be available, one practice is to route the call along a path having the minimum number of arcs (i.e., involving the smallest number of switching centers). This means finding a shortest path with unit lengths on all arcs. Alternatively, each arc can be provided with a measure of delay, and routing can take place along a timewise shortest path from $s$ to $t$.

**3.** *Salesperson routing*: A salesperson is to travel by air from city A to city B. The commission obtained by visiting each city along the way can be estimated. An optimal itinerary can be found by solving a shortest path problem on the underlying airline network, represented as a directed network of nonstop routes (arcs) connecting cities (vertices). Each arc $(i, j)$ is given the net cost $c_{ij} = f_{ij} - r_j$, where $f_{ij}$ is the cost of the flight from city $i$ to city $j$ and $r_j$ is the commission obtained by visiting city $j$. A shortest path from A to B identifies an optimal itinerary.

**4.** *Investment strategy*: An investor has a fixed amount to invest at the beginning of the year. A variety of different financial opportunities are available for investing during the year, with each such opportunity assumed to be available only at the start of each month. Construct the directed network having a vertex for each month as well as a final vertex $t = 13$. The arc $(i, j)$ corresponds to an investment opportunity beginning in month $i$ and maturing at the start of month $j$, with its weight $c_{ij}$ being the negative of the profit earned for the duration of the investment. An optimal investment strategy is identified by a shortest path from vertex 1 to vertex $t$.

**5.** *Equipment replacement*: A job shop must periodically replace its capital equipment because of machine wear. As the machine ages, it breaks down more frequently and so becomes more expensive to operate. Also, as a machine ages its salvage value decreases. Let $c_{ij}$ denote the cost of buying a particularly important machine at the beginning of period $i$, plus the cost of operating the machine over the periods $i, i+1, \ldots, j-1$, minus the salvage cost of the machine at the beginning of period $j$. The problem is to design a replacement plan that minimizes the cost of buying, selling, and operating the machine over a planning horizon of $n$ years, assuming that the job shop must have exactly one machine in service at all times.

This problem can be formulated as a shortest path problem on a network $G$ with vertices $i = 1, 2, \ldots, n+1$; $G$ contains an arc $(i, j)$ with cost $c_{ij}$ for all $i < j$. There is a one-to-one correspondence between directed paths in $G$ from vertex 1 to vertex $n+1$ and equipment replacement plans. The following figure gives a sample network with $n = 5$. The path $[1, 3, 6]$ corresponds to buying the equipment at the beginning of periods 1 and 3. A shortest path from vertex 1 to vertex $n + 1$ identifies an optimal replacement plan.

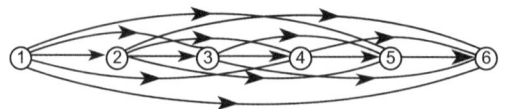

**6.** *Paragraph problem*:   The document processing program TeX uses an optimization procedure to decompose a paragraph into several lines so that when lines are left- and right-justified, the appearance of the paragraph will be the most attractive. Suppose that a paragraph consists of words $i = 1, 2, \ldots, n$. Let $c_{ij}$ denote the attractiveness of a line if it begins with the word $i$ and ends with the word $j - 1$. The program TeX uses formulas to compute the value of each $c_{ij}$. Given the $c_{ij}$, the decision problem is to decompose the paragraph into several lines of text in order to maximize the total attractiveness (of all lines). This problem can be formulated as a shortest path problem in a manner similar to Application 5.

**7.** *Tramp steamer problem*:   A ship travels from port to port carrying cargo and passengers. A voyage of the steamer from port $i$ to port $j$ earns $p_{ij}$ units of profit and requires $t_{ij} \geq 0$ units of time. Here it is assumed that $\sum_{(i,j)\in W} t_{ij} > 0$ for every directed cycle $W$ in $G$. The captain of the ship would like to know whether there exists a tour (directed cycle) $W$ for which the daily profit is greater than a specified threshold $\mu_0$; that is, $\sum_{(i,j)\in W} p_{ij} / \sum_{(i,j)\in W} t_{ij} > \mu_0$. By writing this inequality as $\sum_{(i,j)\in W}(\mu_0 t_{ij} - p_{ij}) < 0$, it is seen that there is a tour $W$ with mean daily profit exceeding $\mu_0$ if and only if $G$ contains a negative cost directed cycle $W$. The shortest path label-correcting algorithm can be used to detect the presence (or absence) of negative cycles (see §10.3.2, Fact 7).

**8.** *System of difference constraints*:   In some linear programming applications (§16.1) with constraints of the form $Ax \leq b$, the $m \times n$ constraint matrix $A$ contains one $+1$ and one $-1$ in each row, with all other entries being zero. Suppose that the $k$th row has a $+1$ entry in column $j_k$ and a $-1$ entry in column $i_k$; entries in the vector $b$ have arbitrary signs. This linear program defines the following set of $m$ *difference constraints* in $n$ variables $x = (x(1), x(2), \ldots, x(n))$: $x(j_k) - x(i_k) \leq b(k)$ for each $k = 1, 2, \ldots, m$. The problem is to determine whether this system of difference constraints has a feasible solution, and if so, to obtain one.

Associate a graph $G$ with this system of difference constraints; $G$ has $n$ vertices corresponding to the $n$ variables, and the arc $(i_k, j_k)$ of length $b(k)$ results from the constraint $x(j_k) - x(i_k) \leq b(k)$. These constraints are identical with the optimality conditions for the shortest path problem in $G$, and they can be satisfied if and only if $G$ contains no negative cycle. In this case the shortest path distances give a solution $x$ satisfying the constraints.

**9.** Examples of Application 8 are also found in telephone operator scheduling, just-in-time scheduling, analyzing the consistency of measurements, and the scaling of data.

**10.** *Maximin paths*:   In a network with capacities (that is, upper bounds on the amount of material that can be sent on each arc), the capacity of a path is the smallest capacity on any of its constituent arcs. A common problem in such networks is to find a path from vertex $s$ to vertex $t$ having the maximum capacity. This represents a path along which the maximum amount of material can flow. Such a *maximin path* can be found efficiently by adapting Dijkstra's shortest path algorithm.

**11.** Additional applications, with reference sources, are given in the following table.

| application | references |
|---|---|
| approximating piecewise linear functions | [AhMaOr93], [AhEtal95] |
| DNA sequence alignment | [AhEtal95] |
| molecular confirmation | [AhMaOr93], [AhEtal95] |
| robot design | [AhMaOr93], [AhEtal95] |
| scaling of matrices | [AhMaOr93], [AhEtal95] |
| knapsack problems | [AhMaOr93], [AhEtal95], [EvMi92] |
| personnel planning | [AhMaOr93] |
| production lot sizing | [EvMi92] |
| transportation planning | [EvMi92] |
| single-crew scheduling | [AhMaOr93] |
| dynamic facility location | [AhMaOr93] |
| UAV path planning | [KiHe03] |
| social networks | [Ne01] |

## 10.4  MAXIMUM FLOWS

The maximum flow problem involves sending the maximum amount of material from a specified source vertex $s$ to another specified sink vertex $t$, subject to capacity restrictions on the amount of material that can flow along each arc. A closely related problem is the minimum cut problem, which is to find a set of arcs with smallest total capacity whose removal separates $s$ and $t$.

### 10.4.1  BASIC CONCEPTS

**Definitions:**

Let $G = (V, E)$ be a directed network with vertex set $V$ and arc set $E$ (see §10.3.1). Each arc $(i,j) \in E$ has an associated **capacity** $u_{ij} \geq 0$. Such a network is called a **capacitated** network. Let $n = |V|$ and $m = |E|$.

Suppose $s$ is a specified **source** vertex and $t$ is a specified **sink** vertex. Then a (**feasible**) **flow** is a function $x = (x_{ij})$ defined on arcs $(i,j) \in E$ satisfying

- *mass balance constraints:* $\sum_{\{j|(i,j)\in E\}} x_{ij} = \sum_{\{j|(j,i)\in E\}} x_{ji}$ for all $i \in V - \{s,t\}$;
- *capacity constraints:* $0 \leq x_{ij} \leq u_{ij}$ for all $(i,j) \in E$.

The arc $(i,j)$ is **saturated** in flow $x$ if $x_{ij} = u_{ij}$.

The **value** of flow $x$ is $v = \sum_{\{j|(s,j)\in E\}} x_{sj}$, the total flow leaving the source vertex.

A **maximum flow** is a flow having maximum value.

A **cut** $[S, \overline{S}]$ partitions the vertex set $V$ into two subsets $S$ and $\overline{S} = V - S$, and consists of all arcs with one endpoint in $S$ and the other in $\overline{S}$. Arcs directed from $S$ to $\overline{S}$ are **forward arcs**, and the set of forward arcs is denoted by $(S, \overline{S})$. Arcs directed from $\overline{S}$ to $S$ are **backward arcs**, and the set of backward arcs is denoted by $(\overline{S}, S)$.

The cut $[S, \overline{S}]$ is an **s-t cut** if $s \in S$ and $t \in \overline{S}$. The **capacity** of the $s$-$t$ cut $[S, \overline{S}]$ is
$$u[S, \overline{S}] = \sum_{(i,j) \in (S, \overline{S})} u_{ij}.$$

A **minimum cut** is an $s$-$t$ cut having minimum capacity.

## Facts:

**1.** The flow $x_{ij}$ on arc $(i, j)$ can represent the number of cars (per hour) traveling along a highway segment, the rate at which oil is pumped through a section of pipe in a distribution system, or the number of messages per unit time that can be sent along a data link in a communication system.

**2.** The mass balance constraints ensure that for all vertices $i$ (other than the source or sink), the total flow out of $i$ equals the total flow into $i$.

**3.** The capacity constraints ensure that the flow on an arc does not exceed its stated capacity.

**4.** Maximum flows arise in a variety of practical problems involving the flow of goods, vehicles, and messages in a network. Maximum flows can also be used to study the connectivity of graphs, the covering of chessboards, the selection of representatives, winning records in tournaments, matrix rounding, and staff scheduling (see §10.4.3).

**5.** For any $s$-$t$ flow $x$, the flow out of $s$ equals the flow into $t$; that is,
$$\sum_{\{j | (s,j) \in E\}} x_{sj} = v = \sum_{\{j | (j,t) \in E\}} x_{jt}.$$

**6.** Removal of the arcs in the $s$-$t$ cut $Z = [S, \overline{S}]$ from $G$ separates vertex $s$ from vertex $t$: namely, there is no $s$-$t$ path in $G - Z$.

**7.** Let $[S, \overline{S}]$ be any $s$-$t$ cut in the network. Then the value of the flow $x$ is given by
$$v = \sum_{(i,j) \in (S, \overline{S})} x_{ij} - \sum_{(j,i) \in (\overline{S}, S)} x_{ji}.$$

That is, the net flow across each $s$-$t$ cut is the same and equals $v$.

**8.** *Weak duality theorem:* The value of every $s$-$t$ flow is less than or equal to the capacity of every $s$-$t$ cut in the network.

**9.** If $x$ is some $s$-$t$ flow whose value equals the capacity of some $s$-$t$ cut $[S, \overline{S}]$, then $x$ is a maximum flow and $[S, \overline{S}]$ is a minimum cut.

**10.** *Max-flow min-cut theorem:* The maximum value of the flow from vertex $s$ to vertex $t$ in a capacitated network equals the minimum capacity among all $s$-$t$ cuts. (L. R. Ford and D. R. Fulkerson, 1956.)

**11.** A systematic study of flows in networks was first carried out by Ford and Fulkerson [FoFu62]. They developed their max-flow min-cut theorem when working on a classified project for Frank S. Ross and Ted Harris. The goal of the project was to evaluate interdiction strategies on the capacity of the Eastern European rail network to support a conventional war. Their 1955 classified report "Fundamentals of a method for evaluating rail net capacities" was declassified in 1999.

## Examples:

**1.** Part (a) of the following figure shows a flow network with $s = 1$ and $t = 6$; capacities are indicated along each arc. The function $x$ given in part (b) satisfies the mass balance constraints and the capacity constraints, and hence is a feasible flow. Relative to this flow, arc $(1,2)$ is not saturated since $x_{12} = 6 < 7 = u_{12}$; on the other hand, arc $(3,5)$ is saturated since $x_{35} = 3 = u_{35}$. The flow has value $v = x_{12}+x_{13} = 6+2 = 8$. Here the flow into vertex 6 is $x_{46}+x_{56} = 5+3 = 8 = v$, as guaranteed by Fact 5. The flow value across the *s-t* cut $[S,\overline{S}]$ with $S = \{1,3\}$ is $x_{12} + x_{34} + x_{35} - x_{23} = 6+3+3-4 = 8$. Similarly, the flow value across the *s-t* cut $[S,\overline{S}]$ with $S = \{1,2,3,5\}$ is $x_{24} + x_{34} + x_{56} - x_{45} = 2+3+3-0 = 8$. (See Fact 7.) This flow is not, however, a maximum flow.

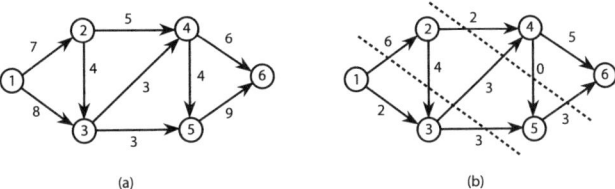

(a)                    (b)

**2.** In Example 1, the *s-t* cut $[S,\overline{S}]$ with $S = \{1,2,3,5\}$ has capacity $u_{24} + u_{34} + u_{56} = 5+3+9 = 17$. Thus the value of any flow in the network is bounded above (see Fact 8) by 17. The *s-t* cut $[S,\overline{S}]$ with $S = \{1,3\}$ has capacity $u_{12}+u_{34}+u_{35} = 7+3+3 = 13$. This cut capacity 13 provides an improved upper bound on the value of a flow. In particular, the flow defined in part (b) of the previous figure has value $v = 8 \leq 13$.

**3.** The following figure shows another feasible flow $x'$ in the network of Example 1. For $x'$, the flow value across the *s-t* cut $[S,\overline{S}]$ with $S = \{1,2,3\}$ is $v = x_{24} + x_{34} + x_{35} = 5+3+3 = 11$, which equals the *s-t* cut capacity $u[S,\overline{S}] = u_{24}+u_{34}+u_{35} = 5+3+3 = 11$. By Fact 9, $x'$ is a maximum flow and $S = \{1,2,3\}$ defines a minimum cut $[S,\overline{S}]$.

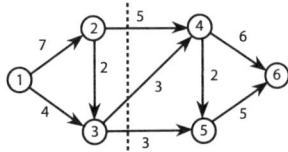

## 10.4.2   ALGORITHMS FOR MAXIMUM FLOWS

There are two main classes of maximum flow algorithms: augmenting path algorithms and preflow-push algorithms. Both types of algorithms work on an auxiliary network (called the residual network) associated with the current solution.

**Definitions:**
Let $G = (V, E)$ be a directed network with $n = |V|$ and $m = |E|$. Let $s$ and $t$ be the specified source and sink, and let $U$ be the largest of the arc capacities $u_{ij}$ in $G$.

Let $x = (x_{ij})$ be a function defined on the arcs $(i,j)$ of $G$. Relative to $x$, the **outflow** from vertex $i$ and **inflow** to vertex $i$ are given, respectively, by

$$ out(i) = \sum_{\{j|(i,j)\in E\}} x_{ij}, \qquad in(i) = \sum_{\{j|(j,i)\in E\}} x_{ji}. $$

The **excess** of vertex $i$ is $e(i) = in(i) - out(i)$.

A **preflow** is any $x = (x_{ij})$ satisfying

- *relaxed mass balance constraints*: $e(i) \geq 0$ for all $i \in V - \{s,t\}$;
- *capacity constraints*: $0 \leq x_{ij} \leq u_{ij}$ for all $(i,j) \in E$.

Vertex $i$ is **active** if $e(i) > 0$.

Given a flow (or a preflow) $x$, the **residual capacity** $r_{ij}$ of the arc $(i,j) \in E$ is the maximum additional flow that can be sent from $i$ to $j$ using arcs $(i,j)$ and $(j,i)$.

The **residual network** $G(x)$ with respect to flow $x$ consists of those arcs of $G$ having positive residual capacity.

An **augmenting path** is a directed path from vertex $s$ to vertex $t$ in $G(x)$.

The **capacity** of a directed path in $G(x)$ is the minimum residual arc capacity appearing on the path.

A set of **distance labels** with respect to a preflow (or flow) $x$ is a function $d: V \to \{0,1,2,\ldots\}$ satisfying

- $d(t) = 0$;
- $d(i) \leq d(j) + 1$ for every arc $(i,j)$ in the residual network $G(x)$.

An arc $(i,j)$ in the residual network $G(x)$ is **admissible** with respect to the distance labels $d(\cdot)$ if $d(i) = d(j) + 1$.

**Facts:**

**1.** The maximum flow problem on an undirected network can be converted to a maximum flow problem on a directed network. Namely, replace every undirected edge $(i,j)$ of capacity $u_{ij}$ by two oppositely directed arcs $(i,j)$ and $(j,i)$, each with capacity $u_{ij}$.

**2.** The residual capacity $r_{ij} = (u_{ij} - x_{ij}) + x_{ji}$. The first term $u_{ij} - x_{ij}$ represents the unused capacity of arc $(i,j)$; the second term $x_{ji}$ represents the amount of flow on arc $(j,i)$ that can be canceled to increase flow from vertex $i$ to vertex $j$.

**3.** The capacity of an augmenting path is always positive.

**4.** *Augmenting path property*:  A flow $x$ is a maximum flow if and only if the residual network $G(x)$ contains no augmenting path.

**5.** *Augmenting path algorithm*:  A general augmenting path algorithm (Algorithm 1) is based on Fact 4. It identifies augmenting paths and sends flows on these paths until the residual network contains no such path.

---

**Algorithm 1**:  **Augmenting path algorithm.**

input: directed network $G$, source vertex $s$, sink vertex $t$

output: maximum flow $x$

$x := 0$

**while** $G(x)$ contains a directed path from $s$ to $t$

    identify an augmenting path $P$ in $G(x)$

    $\delta := \min\{r_{ij} \mid (i,j) \in P\}$

    augment $\delta$ units of flow along $P$ and update $G(x)$

recover an optimal flow $x$ from the final residual network $G(x)$

---

**6.** *Integrality property*:  For networks with integer capacities, Algorithm 1 starts with the zero flow and augments by an integral flow at each iteration. Hence the maximum flow problem with integral capacities always has an optimal integer flow.

**7.** An augmenting path in $G(x)$ can be identified by any search procedure that starts at vertex $s$ and identifies all vertices reachable from $s$ by directed paths (§8.3.2).

**8.** Augmenting the flow along $P$ by $\delta$ decreases the residual capacities of arcs in $P$ by $\delta$ and increases the residual capacities of the reversals of arcs in $P$ by $\delta$.

**9.** At the last iteration of Algorithm 1, let $S$ be the set of vertices reachable from $s$. Then $t \in \overline{S}$ and $[S, \overline{S}]$ is a minimum cut.

**10.** Upon termination of Algorithm 1, an optimal flow $x$ can be reconstructed from the final $G(x)$ using Fact 2. Specifically, let $(i, j) \in E$. If $u_{ij} - r_{ij} \geq 0$ then set $x_{ij} = u_{ij} - r_{ij}$ and $x_{ji} = 0$; otherwise, set $x_{ji} = r_{ij} - u_{ij}$ and $x_{ij} = 0$.

**11.** Algorithm 1 was independently discovered by L. R. Ford and D. R. Fulkerson (1956) and by P. Elias, A. Feinstein, and C. E. Shannon (1956).

**12.** The distance label $d(i)$ is a lower bound on the length (number of arcs) of a shortest (directed) path from vertex $i$ to vertex $t$ in the residual network.

**13.** If some vertex $j$ satisfies $d(j) \geq n$, then vertex $j$ is separated from the sink vertex in the residual network.

**14.** Algorithm 1 runs in pseudopolynomial time $O(nmU)$ for networks with integer arc capacities. The algorithm may not terminate finitely for networks with irrational capacities.

**15.** Two specific implementations of Algorithm 1 run in polynomial time:

- By augmenting flow along a shortest path, the number of augmentations is reduced to $O(nm)$, and using very sophisticated data structures this algorithm can be implemented to run in $O(nm \log n)$ time;

- By augmenting flow along a path with maximum residual capacity, the number of augmentations is $O(m \log U)$ and this algorithm can be implemented to run in $O(nm \log U)$ time.

**16.** *Preflow-push algorithm:* The preflow-push algorithm (Algorithm 2) maintains a preflow at every step and pushes flow on individual arcs instead of along augmenting paths. The basic operation is to select an active vertex and try to remove its excess by pushing flow to neighbors that are "closer" to the sink.

---

**Algorithm 2: Preflow-push algorithm.**

input: directed network $G$, source vertex $s$, sink vertex $t$
output: maximum flow $x$
compute the shortest path lengths $d(\cdot)$ to vertex $t$
$d(s) := n$; $x := 0$; $x_{sj} := u_{sj}$ for all arcs $(s, j) \in E$
**while** the network contains an active vertex
    select an active vertex $i$ and *push_relabel(i)*
recover an optimal flow $x$ from the final residual network $G(x)$

**procedure** *push_relabel(i)*
    **if** the network contains an admissible arc $(i, j)$ **then**
        push $\delta := \min\{e(i), r_{ij}\}$ units of flow from $i$ to $j$
    **else** $d(i) := \min\{d(j) + 1 \mid (i, j) \in E$ and $r_{ij} > 0\}$

---

**17.** The shortest path lengths calculated in Algorithm 2 represent the minimum number of arcs in a path to vertex $t$ and can be efficiently found by carrying out a breadth-first search relative to $t$ (§9.2.1).

**18.** In Algorithm 2, if the active vertex currently being examined has an admissible arc $(i, j)$, then increasing the flow on $(i, j)$ by $\delta$ decreases $r_{ij}$ by $\delta$ and increases $r_{ji}$ by $\delta$. Also, $e(i)$ is decreased by $\delta$ and $e(j)$ is increased by $\delta$.

**19.** In Algorithm 2, if the active vertex currently being examined has no admissible arc, then after its distance label is increased, at least one admissible arc is created.

**20.** The preflow-push algorithm can be implemented to run in $O(n^2 m)$ time. Variations of this algorithm with improved worst-case complexity are described in [AhOrTa89].

**21.** The highest-label preflow-push algorithm [GoTa86] is a specific implementation of Algorithm 2 that always examines vertices with the largest distance label. This $O(n^2 \sqrt{m})$ implementation is currently one of the fastest algorithms to solve the maximum flow problem in practice.

**22.** Algorithm 2 can be implemented to run in $O(nm \log(n^2/m))$ time using a dynamic tree data structure.

**23.** Hochbaum [Ho98] introduced the *pseudoflow algorithm*, which is a variant of the preflow-push algorithm. Implementations of the pseudoflow algorithm are among the fastest for solving the maximum flow problem.

**24.** King, Rao, and Tarjan [KiRaTa94] developed a preflow-push algorithm that runs in $O(nm)$ time whenever $m > n^{1+\epsilon}$ for some fixed $\epsilon > 0$.

**25.** Goldberg and Rao [GoRa98] developed an augmenting path algorithm that has the best time bound if $U$ is not too large relative to $n$. The running time when $m \leq n^{4/3}$ is $O(m^{3/2} \log n \log U)$. The running time when $m > n^{4/3}$ is $O(n^{2/3} m \log(n^2/m) \log U)$.

**26.** Cheriyan and Mehlhorn [ChMe99] developed a preflow-push algorithm that runs in $O(n^3/\log n)$ time. This is the best running time if $m$ is proportional to $n^2$.

**27.** Orlin [Or13] developed an augmenting path algorithm that runs in $O(nm)$ time when $m < n^{1.06}$. When combined with the King, Rao, and Tarjan algorithm, this establishes that a maximum flow can be obtained in $O(nm)$ time. Orlin also showed how to solve the maximum flow problem in $O(n^2/\log n)$ time when $m$ is proportional to $n$.

**28.** A series of papers starting with Christiano et al. [ChEtal11] and leading to a paper by Kelner et al. [KeEtal14] developed $\epsilon$-optimal solutions for the maximum flow problem on undirected graphs. The algorithms rely on the theory of electrical flows. The best of these algorithms runs in nearly linear time.

**29.** The books [AhMaOr93] and [CoEtal01] discuss additional versions of augmenting and preflow-push algorithms, as well as specializations of these algorithms to unit capacity networks, bipartite networks, and planar networks.

**30.** Goldberg and Tarjan [GoTa14] provide a survey of algorithms for the maximum flow problem and discuss the history of running time improvements.

**31.** Computer codes for solving maximum flow/minimum cut problems can be found at

- `ftp://dimacs.rutgers.edu/pub/netflow/maxflow/`
- `http://www.avglab.com/andrew/soft.html`
- `http://riot.ieor.berkeley.edu/Applications/Pseudoflow/maxflow.html`

**Examples:**

**1.** Part (a) of the following figure illustrates a network $G$ with capacities shown next to each arc. A feasible flow from vertex $s = 1$ to vertex $t = 4$ is displayed in part (b); this flow has value $v = x_{12} + x_{13} = 3$. Every path in $G$ from $s$ to $t$ contains a saturated arc: paths $[1, 2, 4]$ and $[1, 2, 3, 4]$ have the saturated arc $(1, 2)$, and path $[1, 3, 4]$ has the saturated arc $(3, 4)$. Consequently, no additional flow can be pushed in the "forward" direction from $s$ to $t$. Yet, the current flow $x$ is not a maximum flow.

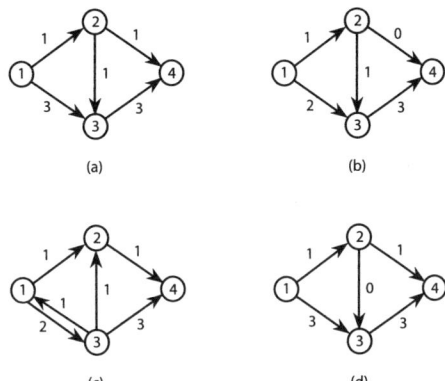

To find additional flow from $s$ to $t$, the residual network $G(x)$ is constructed; see part (c) of the figure. An augmenting path in part (c) is $P = [1, 3, 2, 4]$ with (residual) capacity $\delta = 1$. Adding the flow on $P$ to that in part (b) produces the new flow $x'$ in part (d); notice that the flow on arc $(2, 3)$ in $x$ has been canceled in this process. The resulting flow $x'$ has flow value $v = 4$. Since the $s$-$t$ cut $[S, \overline{S}]$ with $S = \{1, 2, 3\}$ has capacity $u_{24} + u_{34} = 4 = v$, the flow $x'$ is a maximum flow and $S = \{1, 2, 3\}$ defines a cut having minimum capacity.

**2.** The following figure illustrates three iterations of the augmenting path algorithm (Algorithm 1). Part (a) of the figure shows a network with capacities indicated on each arc. Here $s = 1$ and $t = 6$. Initially the flow $x = 0$, so the residual network is identical to the original network with $r_{ij} = u_{ij}$ for every arc $(i, j)$.

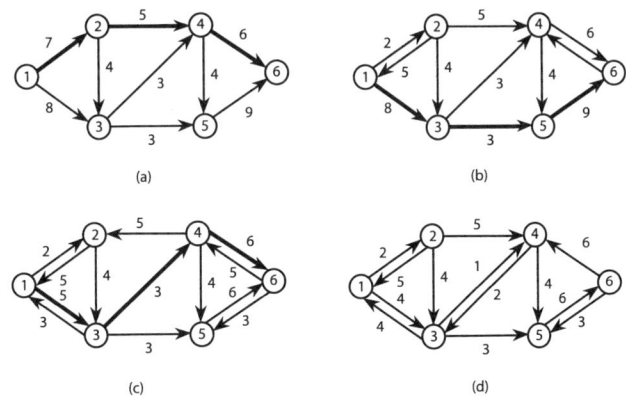

Suppose that the algorithm identifies path $P^1 = [1, 2, 4, 6]$ as the augmenting path. The algorithm augments $\delta = \min\{r_{12}, r_{24}, r_{46}\} = \min\{7, 5, 6\} = 5$ units of flow along $P^1$. This augmentation changes the residual capacities only of arcs in $P^1$ (or their reverse arcs), yielding the new residual network in part (b).

In the second iteration, suppose the algorithm identifies path $P^2 = [1, 3, 5, 6]$ as the next augmenting path. Then flow is increased by $\delta = \min\{8, 3, 9\} = 3$ units along $P^2$; part (c) shows the residual network after the second augmentation.

A third augmentation with $\delta = 1$ occurs along path $P^3 = [1, 3, 4, 6]$ in part (c), giving the residual network shown in part (d).

**3.** The following figure illustrates three iterations of the preflow-push algorithm on the flow network with capacities given in part (a). Here $s = 1$ and $t = 4$; in addition, the pair $(e(i), d(i))$ is shown beside each vertex $i$. Part (b) of the figure gives $G(x)$ corresponding to the initial preflow $x$ with $x_{12} = 2$ and $x_{13} = 4$.

Suppose that the algorithm selects vertex 2 for the push/relabel operation. Then arc $(2, 4)$ is the only admissible arc and the algorithm pushes $\delta = \min\{e(2), r_{24}\} = \min\{2, 1\} = 1$ unit along this arc; part (c) gives the residual network at this stage.

Suppose that the algorithm again selects vertex 2. Since no admissible arc emanates from this vertex, the algorithm performs a relabel operation and gives vertex 2 a new distance label: $d(2) = \min\{d(3) + 1, d(1) + 1\} = \min\{2, 5\} = 2$. The new residual network is the same as the one shown in part (c) except that $d(2) = 2$ instead of 1.

In the third iteration, suppose that vertex 3 is selected. Then $\delta = \min\{e(3), r_{34}\} = \min\{4, 5\} = 4$ units are pushed along the arc $(3, 4)$; part (d) gives the residual network at the end of this iteration.

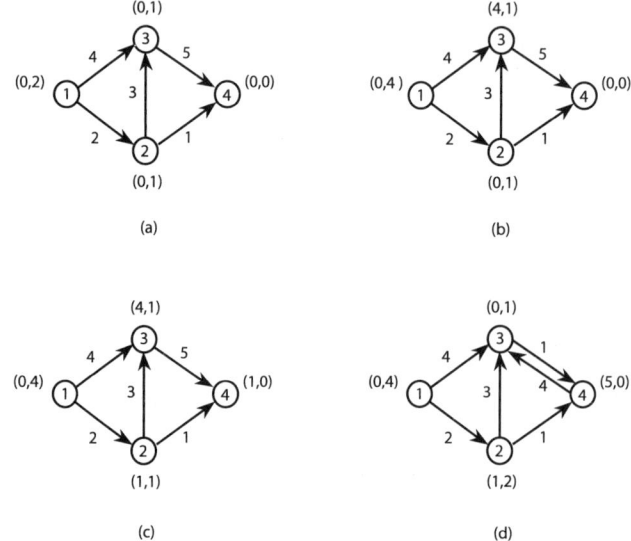

## 10.4.3  APPLICATIONS

A variety of applied problems can be modeled using maximum flows or minimum cuts. The max-flow min-cut theorem (§10.4.1, Fact 10) can also be used to deduce a number of max-min duality results in combinatorial theory. This section discusses a number of such applications.

**Applications:**

**1.** *Distribution network:* Oil needs to be shipped from a refinery to a storage facility using the pipelines of an underlying distribution network. Here the refinery corresponds to a particular vertex $s$ in the distribution network, and the storage facility corresponds to another vertex $t$. The capacity of each arc is the maximum amount of oil per unit time that can flow along it. The maximum flow rate from the source vertex $s$ to the sink vertex $t$ is determined by the value of a maximum $s$-$t$ flow.

**2.** Other examples of Application 1 occur in transportation networks, electrical power networks, and telecommunication networks.

**3.** *System of distinct representatives*: Given is a collection of sets $X_1, X_2, \ldots, X_m$ which are subsets of a given $n$-set $X$. A system of distinct representatives (§1.2.2) for the collection is sought, if one exists.

To solve this problem, set up the bipartite network $(V_1 \cup V_2, E)$ in which there is a vertex of $V_1$ for each set $X_i$ and a vertex of $V_2$ for each element of $X$. An arc $(i, j)$ of infinite capacity joins $i \in V_1$ to $j \in V_2$ if $j \in X_i$. Add a source vertex $s$ joined by arcs of unit capacity to each $i \in V_1$, and a sink vertex $t$ with arcs of unit capacity joining each $j \in V_2$ to $t$. Then a system of distinct representatives exists if and only if the maximum flow in this constructed network has value $m$. In this case, those arcs $(i, j)$, with $i \in V_1$ and $j \in V_2$, having flow $x_{ij} = 1$ identify a system of distinct representatives selected from the $m$ sets.

**4.** *Feasible flow problem*: This problem involves finding a flow $x$ in $G = (V, E)$ so that the net flow at each vertex is a specified value $b(i)$, where $\sum_{i \in V} b(i) = 0$. That is, a flow $x$ on the arcs of network $G$ is required, satisfying

- *mass balance constraints*: $\displaystyle\sum_{\{j \mid (i,j) \in E\}} x_{ij} - \sum_{\{j \mid (j,i) \in E\}} x_{ji} = b(i)$ for all $i \in V$;

- *capacity constraints*: $0 \le x_{ij} \le u_{ij}$ for all $(i, j) \in E$.

This can be modeled as a maximum flow problem. Construct the augmented network $G'$ by adding a source vertex $s$ and a sink vertex $t$ to $G$. For each vertex $i$ with $b(i) > 0$, an arc $(s, i)$ is added to $E$ with capacity $b(i)$; for each vertex $i$ with $b(i) < 0$, an arc $(i, t)$ is added to $E$ with capacity $-b(i)$. Then solve a maximum flow problem from vertex $s$ to vertex $t$ in $G'$. It can be proved that the feasible flow problem for $G$ has a solution if and only if the maximum flow in $G'$ saturates all arcs emanating from vertex $s$ in $G'$.

**5.** Application 4 frequently arises in distribution problems. For example, a known amount of merchandise is available at certain ports and is required at other ports in known quantities. Also the maximum quantity of merchandise that can be shipped on a particular sea route is specified. Determining whether it is possible to satisfy all of the demands by using the available supplies is a feasible flow problem.

**6.** *Graph connectivity*: In a directed graph $G$, the arc connectivity $\kappa'_{ij}$ of vertices $i$ and $j$ is the minimum number of arcs whose removal from $G$ leaves no directed path from $i$ to $j$. The arc connectivity $\kappa'(G)$ is the minimum number of arcs whose removal from $G$ separates some pair of vertices (see §8.4.2). The arc connectivity of a graph is an important measure of the graph's reliability or stability. Since $\kappa'(G) = \min\{\kappa'_{ij} \mid (i, j) \in V \times V, i \ne j\}$, the arc connectivity of a graph can be computed by determining the arc connectivity of $n(n-1)$ pairs of vertices. As a matter of fact, the arc connectivity of $G$ can be found by determining only $n - 1$ arc connectivities.

The arc connectivity $\kappa'_{ij}$ can be found by applying the max-flow min-cut theorem (§10.4.1) to the network obtained from $G$ by setting the capacity of each arc $(i, j)$ to 1. In such a unit capacity network, the maximum $i$-$j$ flow value equals the maximum number of arc-disjoint paths from vertex $i$ to vertex $j$, and the minimum $i$-$j$ cut capacity equals the minimum number of arcs required to separate vertex $i$ and vertex $j$. This shows that the maximum number of arc-disjoint paths from vertex $i$ to vertex $j$ equals the minimum number of arcs whose removal disconnects all paths from vertex $i$ to vertex $j$. (This result is a variation of Menger's theorem in §8.4.2; it was independently discovered by

Ford and Fulkerson and by Elias, Feinstein, and Shannon.) Consequently, $\kappa'_{ij}$ equals the maximum $i$-$j$ flow value in the network, and the arc connectivity $\kappa'(G)$ can be determined by solving $n-1$ maximum flow problems in a unit capacity network.

**7.** *Tournaments:* Consider a round-robin tournament between $n$ teams, assuming each team plays against every other team $c$ times and no game ends in a draw. It is claimed that $\alpha_i$ for $1 \le i \le n$ is the number of victories accrued by the $i$th team at the end of the tournament. Verifying whether the nonnegative integers $\alpha_1, \alpha_2, \ldots, \alpha_n$ are possible winning records for the $n$ teams can be modeled as a feasible flow problem.

Define a directed network $G = (V, E)$ with vertex set $V = \{1, 2, \ldots, n\}$ and arc set $E = \{(i,j) \in V \times V \mid i < j\}$. Let $x_{ij}$, $i < j$, represent the number of times team $i$ defeats team $j$. The total number of times team $i$ defeats teams $i+1, i+2, \ldots, n$ is $\sum_{\{j|(i,j)\in E\}} x_{ij}$. Since the number of times team $i$ defeats a team $j < i$ is $c - x_{ji}$, it follows that the total number of times that team $i$ defeats teams $1, 2, \ldots, i-1$ is $(i-1)c - \sum_{\{j|(j,i)\in E\}} x_{ji}$. However, there are two constraints:

- The total number of wins $\alpha_i$ of team $i$ must equal the total number of times it defeats teams $1, 2, \ldots, n$, giving

$$\sum_{\{j|(i,j)\in E\}} x_{ij} - \sum_{\{j|(j,i)\in E\}} x_{ji} = \alpha_i - (i-1)c \text{ for all } i \in V;$$

- A possible winning record must also satisfy $0 \le x_{ij} \le c$ for all $(i,j) \in E$.

Consequently, $\{\alpha_i\}$ define a possible winning record if these two constraints have a feasible solution $x$. Let $b(i) = \alpha_i - (i-1)c$. Since $\sum_{i\in V} \alpha_i$ and $\sum_{i\in V}(i-1)c$ are both equal to $\frac{cn(n-1)}{2}$, the total number of games played, it follows that $\sum_{i\in V} b(i) = 0$. The problem of finding a feasible solution to the two constraints is then a feasible flow problem.

**8.** *Matchings and covers:* The max-flow min-cut theorem can also be used to prove a max-min result concerning matchings and covers in a directed bipartite graph $G = (V_1 \cup V_2, E)$. (See §8.1.3.) The subset $E' \subseteq E$ is a *matching* (§10.2.1) if no two arcs in $E'$ are incident with the same vertex. The subset $V' \subseteq V_1 \cup V_2$ is a *vertex cover* if every arc in $E$ is incident to at least one vertex in $V'$. Create the network $G'$ from $G$ by adding vertices $s$ and $t$, as well as arcs $(s,i)$ with capacity 1 for all $i \in V_1$ and arcs $(j,t)$ with capacity 1 for all $j \in V_2$. All other arcs of $G'$ correspond to arcs of $G$ and have infinite capacity. Then each matching of cardinality $v$ defines a flow of value $v$ in $G'$, and each $s$-$t$ cut of capacity $v$ induces a corresponding vertex cover with $v$ vertices. Application of the max-flow min-cut theorem establishes the desired result. Namely, in a bipartite graph $G = (V_1 \cup V_2, E)$, the maximum cardinality of any matching equals the minimum cardinality of any vertex cover of $G$.

**9.** *0-1 matrices:* Suppose $A = (a_{ij})$ is a 0-1 matrix. Associate with $A$ the directed bipartite graph $G = (V_1 \cup V_2, E)$, where $V_1$ is the set of row indices and $V_2$ is the set of column indices. Place an arc $(i,j) \in E$ whenever $a_{ij} = 1$. A matching in $G$ now corresponds to a set of "independent" 1s in the matrix $A$: i.e., no two of these 1s are in the same row or the same column. Also, a vertex cover of $G$ corresponds to a set of rows and columns in $A$ that collectively cover all the 1s in the matrix. Applying the result in Application 8 shows that the maximum number of independent 1s in $A$ equals the minimum number of lines (rows and/or columns) needed to cover all the 1s in $A$. This result is known as König's theorem (§6.6.1).

**10.** Additional applications, with reference sources, are given in the following table.

| application | references |
|---|---|
| matrix rounding | [AhMaOr93], [AhEtal95] |
| distributed computing | [AhMaOr93], [AhEtal95] |
| network reliability | [AhMaOr93], [AhEtal95] |
| open pit mining | [AhMaOr93], [AhEtal95] |
| building evacuation | [AhMaOr93] |
| covering sports events | [AhEtal95] |
| nurse staff scheduling | [AhMaOr93], [AhEtal95] |
| bus scheduling | [AhEtal95] |
| machine scheduling | [AhMaOr93], [AhEtal95] |
| tanker scheduling | [AhMaOr93], [AhEtal95] |
| bottleneck assignment | [FoFu62] |
| selecting freight-handling terminals | [AhEtal95] |
| site selection | [EvMi92] |
| material-handling systems | [EvMi92] |
| decompositions of partial orders | [FoFu62] |
| matrices with prescribed row/column sums | [FoFu62] |
| computer vision | [Ro99] |

# 10.5   MINIMUM COST FLOWS

The minimum cost flow problem involves determining the least cost shipment of a commodity through a capacitated network in order to satisfy demands at certain vertices using supplies available at other vertices. This problem generalizes both the shortest path problem (§10.3) and the maximum flow problem (§10.4).

## 10.5.1   BASIC CONCEPTS

**Definitions:**

Let $G = (V, E)$ be a directed network with vertex set $V$ and arc set $E$ (see §10.3.1). Each arc $(i, j) \in E$ has an associated **cost** $c_{ij}$ and a **capacity** $u_{ij} \geq 0$. Let $n = |V|$ and $m = |E|$.

Each vertex $i \in V$ has an associated supply/demand $b(i)$. If $b(i) > 0$, then vertex $i$ is a **supply vertex**; if $b(i) < 0$, then vertex $i$ is a **demand vertex**.

A (**feasible**) **flow** is a function $x = (x_{ij})$ defined on arcs $(i, j) \in E$ satisfying

- *mass balance constraints:* $\displaystyle\sum_{\{j \mid (i,j) \in E\}} x_{ij} - \sum_{\{j \mid (j,i) \in E\}} x_{ji} = b(i)$ for all $i \in V$,
- *capacity constraints:* $0 \leq x_{ij} \leq u_{ij}$ for all $(i, j) \in E$,

where $\displaystyle\sum_{i \in V} b(i) = 0$.

The **cost** of flow $x$ is $\displaystyle\sum_{(i,j) \in E} c_{ij} x_{ij}$.

A *minimum cost flow* is a flow having minimum cost.

A *pseudoflow* is a function $x = (x_{ij})$ satisfying the arc capacity constraints; it may violate the mass balance constraints.

The *residual network* $G(x)$ corresponding to a flow (or pseudoflow) $x$ is defined in the following manner. Replace each arc $(i, j) \in E$ by two arcs $(i, j)$ and $(j, i)$. Arc $(i, j)$ has cost $c_{ij}$ and *residual capacity* $r_{ij} = u_{ij} - x_{ij}$, and arc $(j, i)$ has cost $-c_{ij}$ and *residual capacity* $r_{ji} = x_{ij}$. The residual network consists only of arcs with positive residual capacity.

The *potential* of vertex $i$ is a quantity $\pi(i)$ associated with the mass balance constraint at vertex $i$. With respect to a given set of vertex potentials, the *reduced cost* of an arc $(i, j)$ in the residual network $G(x)$ is $c_{ij}^{\pi} = c_{ij} - \pi(i) + \pi(j)$.

The *cost* of path $P$ in $G(x)$ is $c(P) = \sum_{(i,j) \in P} c_{ij}$; its *reduced cost* is $c^{\pi}(P) = \sum_{(i,j) \in P} c_{ij}^{\pi}$.

A *negative cycle* is a directed cycle $W$ in $G(x)$ for which $c(W) < 0$.

**Facts:**

**1.** The mass balance constraints ensure that the net flow out of each vertex $i$ is equal to $b(i)$. Thus, if there is excess flow out of vertex $i$, then $b(i) > 0$ and $i$ is a supply vertex. If $b(i) < 0$, then more flow enters $i$ than leaves $i$, meaning that vertex $i$ is a demand vertex.

**2.** Minimum cost flows arise in practical problems involving the least cost routing of goods, vehicles, and messages in a network. Minimum cost flows can also be used in models of warehouse layout, production and inventory problems, scheduling of personnel, automatic classification of chromosomes, and racial balancing of schools. (See §10.5.3.)

**3.** Let $\{\pi(i) \mid i \in V\}$ be any set of vertex potentials.

- If $P$ is a path from $i$ to $j$ in $G(x)$, then $c^{\pi}(P) = c(P) - \pi(i) + \pi(j)$.
- If $W$ is a cycle in $G(x)$, then $c^{\pi}(W) = c(W)$.

**4.** *Negative cycle optimality conditions:* A feasible flow $x$ is a minimum cost flow if and only if the residual network $G(x)$ contains no negative cycle.

**5.** *Reduced cost optimality conditions:* A feasible flow $x$ is a minimum cost flow if and only if some set of vertex potentials $\pi$ satisfies $c_{ij}^{\pi} \geq 0$ for every arc $(i, j)$ in $G(x)$.

**6.** *Complementary slackness optimality conditions:* A feasible flow $x$ is a minimum cost flow if and only if there exist vertex potentials $\pi$ such that for every arc $(i, j) \in E$

- if $c_{ij}^{\pi} > 0$, then $x_{ij} = 0$;
- if $c_{ij}^{\pi} < 0$, then $x_{ij} = u_{ij}$;
- if $0 < x_{ij} < u_{ij}$, then $c_{ij}^{\pi} = 0$.

**Examples:**

**1.** In the flow network of part (a) of the following figure, $b(i)$ is shown next to each vertex $i$ and $(c_{ij}, u_{ij})$ is shown next to each arc $(i, j)$.

The function $x = (x_{ij})$ given in part (b) satisfies the mass balance constraints for each vertex. For example, the flow out of vertex 2 is $x_{24} = 6$ and the flow into vertex 2 is $x_{12} + x_{32} = 5$, so that flow out minus flow in equals $6 - 5 = 1 = b(2)$. Also the capacity constraints for all arcs are satisfied: e.g., $x_{12} = 4 \leq 5 = u_{12}$. Thus $x$ is a feasible flow, with cost 163. The residual network $G(x)$ corresponding to the flow $x$ is shown in part (c). Selected arcs of $G(x)$ are labeled with their cost and residual capacity. The directed cycle $W = [1, 2, 3, 1]$ in $G(x)$ has cost $11 - 9 - 10 = -8$ and so $W$ is a negative cycle. By Fact 4, this flow $x$ is not a minimum cost flow.

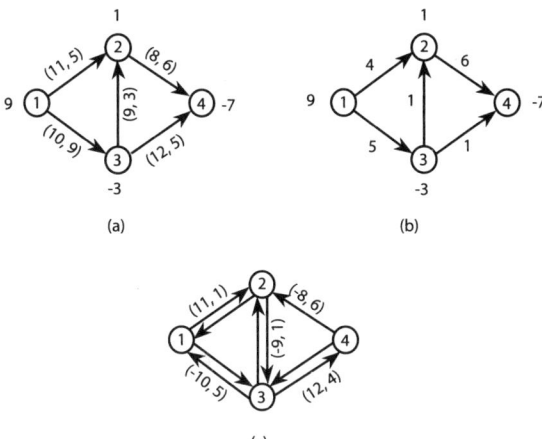

(a)

(b)

(c)

**2.** Part (a) of the following figure shows another feasible flow $x'$ for the network in Example 1, with cost 155.

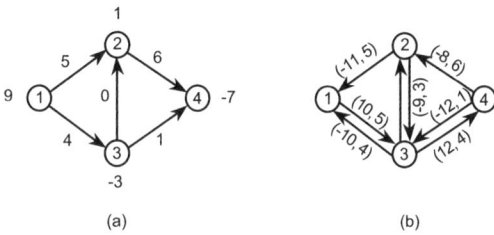

(a)

(b)

The corresponding residual network $G(x')$ is given in part (b), in which each arc is labeled with its cost and its residual capacity. Using the vertex potentials $\pi = (0, -14, -10, -22)$, the reduced cost of arc $(2, 1)$ in the residual network is $c_{21}^{\pi} = -11 - (-14) + 0 = 3$; likewise $c_{32}^{\pi} = 9 - (-10) - 14 = 5$. The remaining reduced costs are found to be zero, so $c_{ij}^{\pi} \geq 0$ for all arcs $(i, j)$ in $G(x)$. By Fact 5, $x'$ is a minimum cost flow for the given network.

**3.** Alternatively, the optimality of the flow $x'$ in part (a) of the figure for Example 2 can be verified using Fact 6. As in Example 2, let $\pi = (0, -14, -10, -22)$. Arc $(3, 2)$ of the original network $G$ in part (a) of the figure of Example 1 has positive reduced cost $c_{32}^{\pi} = 9 - (-10) - 14 = 5$ and $x_{32}' = 0$. Arc $(1, 2)$ has $c_{12}^{\pi} = 11 - 0 - 14 = -3 < 0$ and $x_{12}' = u_{12}$. The remaining arcs $(1, 3), (2, 4), (3, 4)$ have zero reduced cost. Consequently, the complementary slackness optimality conditions are satisfied and the flow $x'$ achieves the minimum cost.

## 10.5.2 ALGORITHMS FOR MINIMUM COST FLOWS

A variety of algorithms are available to solve the minimum cost flow problem. Three algorithms are described in this section: the cycle-canceling algorithm, the successive shortest path algorithm, and the network simplex algorithm.

### Definitions:

Let $G = (V, E)$ be a directed network with $n = |V|$ and $m = |E|$; let $U$ denote the largest arc capacity and let $C$ denote the largest arc cost (in absolute value) in $G$.

For a given pseudoflow $x = (x_{ij})$, the **imbalance** of vertex $i \in V$ is $e(i) = b(i) + \sum_{\{j|(j,i)\in E\}} x_{ji} - \sum_{\{j|(i,j)\in E\}} x_{ij}$.

An **excess vertex** is one with a positive imbalance, and a **deficit vertex** is one with a negative imbalance.

A **spanning tree solution** $x = (x_{ij})$ consists of a spanning tree $T$ of $G = (V, E)$ in which each nontree arc $(i, j)$ has either $x_{ij} = 0$ or $x_{ij} = u_{ij}$.

A spanning tree solution is **feasible** if the mass balance constraints and capacity constraints are satisfied.

**Facts:**

1. *Cycle-canceling algorithm*: The cycle-canceling algorithm (Algorithm 1) is based on the negative cycle optimality conditions (§10.5.1, Fact 4). It starts with a feasible flow and successively augments flow along negative cycles in the residual network until there is no negative cycle.

---

**Algorithm 1:    Cycle-canceling algorithm.**

input: directed network $G$

output: minimum cost flow $x$

establish a feasible flow $x$ in the network

**while** $G(x)$ contains a negative cycle

identify a negative cycle $W$

$\delta := \min\{r_{ij} \mid (i,j) \in W\}$

augment $\delta$ units of flow along $W$ and update $G(x)$

recover an optimal flow $x$ from the final residual network $G(x)$

---

2. As shown in §10.4.3, an initial feasible flow can be found by solving a maximum flow problem.

3. *Integrality property*: For problems with integer arc capacities and integer vertex supplies/demands, Algorithm 1 starts with an integer flow, at each iteration augments by an integral amount of flow, and thus produces an optimal flow that is integer. Thus any minimum cost flow problem with integer supplies, demands, and capacities always has an optimal solution that is integer.

4. A negative cycle $W$ in the residual network can be identified in $O(nm)$ time by using a queue implementation of the label-correcting algorithm (§10.3.2, Fact 5).

5. Augmenting the flow along $W$ by $\delta$ decreases the residual capacities of arcs in $W$ by $\delta$ and increases the residual capacities of the reversals of arcs in $W$ by $\delta$.

6. Upon termination of Algorithm 1, an optimal flow $x$ can be reconstructed from the final $G(x)$; see §10.4.2, Fact 10.

7. For problems with integer supplies, demands, and arc capacities, the cycle-canceling algorithm runs in pseudopolynomial time $O(nm^2 CU)$.

8. If flow is augmented along a negative cycle $W$ in $G(x)$ that minimizes the ratio $\frac{1}{|W|} \sum_{(i,j)\in W} c_{ij}$ among all directed cycles in $G(x)$, then this implementation runs in polynomial time [GoTa88].

**9.** *Successive shortest path algorithm*:   The successive shortest path algorithm (Algorithm 2) starts with the pseudoflow $x = 0$. It proceeds by selecting an excess vertex $k$ and a deficit vertex $l$, and then augmenting flow along a minimum cost path from vertex $k$ to vertex $l$ in $G(x)$.

---

**Algorithm 2**:   **Successive shortest path algorithm.**

input: directed network $G$

output: minimum cost flow $x$

$x := 0$

$e(i) := b(i)$ for all $i \in V$

initialize $V^+ := \{i \mid e(i) > 0\}$ and $V^- := \{i \mid e(i) < 0\}$

**while** $V^+ \neq \emptyset$

  select a vertex $k \in V^+$ and a vertex $l \in V^-$

  identify a shortest path $P$ in $G(x)$ from vertex $k$ to vertex $l$

  $\delta := \min\{e(k), -e(l), \min\{r_{ij} \mid (i,j) \in P\}\}$

  augment $\delta$ units of flow along $P$

  update $e$, $G(x)$, $V^+$, and $V^-$

recover an optimal flow $x$ from the final residual network $G(x)$

---

**10.**   If in Algorithm 2 reduced costs $c_{ij}^{\pi}$ are used instead of arc costs $c_{ij}$, then Dijkstra's algorithm (§10.3.2, Algorithm 2) can be applied to determine a shortest path $P$ in the residual network.

**11.**   Augmenting the flow along $P$ by $\delta$ decreases the residual capacities of arcs in $P$ by $\delta$ and increases the residual capacities of the reversals of arcs in $P$ by $\delta$. It also decreases $e(k)$ by $\delta$ and increases $e(l)$ by $\delta$.

**12.**   The solution maintained by the successive shortest path algorithm always satisfies the reduced cost optimality conditions (§10.5.1, Fact 5). The final solution is in addition feasible, and so is an optimal solution of the minimum cost flow problem.

**13.**   For problems with integer supplies, demands, and arc capacities, the shortest augmenting path algorithm runs in pseudopolynomial time.

**14.**   Several implementations of the shortest augmenting path algorithm run in polynomial or even strongly polynomial time. Orlin [Or88] describes an implementation running in $O(m \log n(m+n \log n))$ time, currently the fastest strongly polynomial-time algorithm to solve the minimum cost flow problem.

**15.**   If a minimum cost flow problem has an optimal solution, then it has an optimal spanning tree solution.

**16.**   Given a spanning tree solution $x$, with flows on nontree arcs $(i,j)$ specified (at either $0$ or $u_{ij}$), the flows on the tree arcs are uniquely determined by the mass balance constraints.

**17.**   Given a spanning tree solution $x$, vertex potentials $\pi$ can be determined such that

- $\pi(1) = 0$;
- $c_{ij}^{\pi} = 0$ for all tree arcs $(i,j)$.

**18.**   *Complementary slackness optimality conditions*:   Suppose $x$ is a feasible spanning tree solution with vertex potentials determined as in Fact 17. Then $x$ is a minimum cost flow if

- $c_{ij}^{\pi} \geq 0$ for all nontree arcs $(i, j)$ with $x_{ij} = 0$;
- $c_{ij}^{\pi} \leq 0$ for all nontree arcs $(i, j)$ with $x_{ij} = u_{ij}$.

**19.** *Network simplex algorithm:* The network simplex algorithm (Algorithm 3) is a specialized version of the well-known linear programming simplex method (§16.1.3). It maintains a spanning tree solution and at each iteration transforms the current spanning tree solution into an improved spanning tree solution until optimality is reached.

---

**Algorithm 3**:  **Network simplex algorithm.**

input: directed network $G$

output: minimum cost flow $x$

determine an initial spanning tree solution with associated tree $T$

let $x$ be the flow and $\pi$ the corresponding vertex potentials

**while** a nontree arc violates the complementary slackness optimality conditions

    select an entering arc $(k, l)$ violating its optimality condition

    add arc $(k, l)$ to $T$, augment the maximum possible flow in the cycle thus

        created, and determine the leaving arc $(p, q)$

    update the tree $T$, the flow $x$, and the vertex potentials $\pi$

---

**20.** Using appropriate data structures, the network simplex algorithm can be implemented very efficiently. The network simplex algorithm is one of the fastest algorithms to solve the minimum cost flow problem in practice.

**21.** The network simplex algorithm has a exponential worst-case time bound. Orlin [Or97] provided the first polynomial-time implementations of the (generic) network simplex algorithm.

**22.** Detailed descriptions of Algorithms 1–3, as well as several other algorithms for finding minimum cost flows, can be found in [AhMaOr93].

**23.** Computer codes for solving the minimum cost flow problem can be found at

- ftp://dimacs.rutgers.edu/pub/netflow/mincost/
- http://www.avglab.com/andrew/soft.html
- http://lemon.cs.elte.hu/trac/lemon

**Examples:**

**1.** The following figure illustrates the cycle-canceling algorithm. Part (a) of this figure depicts the given flow network, with $b(i)$ shown for each vertex $i$ and $(c_{ij}, u_{ij})$ for each arc $(i, j)$. Part (b) shows the residual network corresponding to the flow $x_{12} = x_{24} = 3$ and $x_{13} = x_{34} = 1$.

In the first iteration, suppose the algorithm selects the negative cycle $[2, 3, 4, 2]$ with cost $-1$. Then $\delta = \min\{r_{23}, r_{34}, r_{42}\} = \min\{2, 4, 3\} = 2$ units of flow are augmented along this cycle. Part (c) shows the modified residual network.

In the next iteration, the algorithm selects the cycle $[1, 3, 4, 2, 1]$ with cost $-2$ and augments $\delta = 1$ unit of flow. Part (d) depicts the updated residual network which contains no negative cycle, so the algorithm terminates. From part (d), an optimal flow pattern is deduced: $x_{12} = x_{13} = x_{23} = 2$ and $x_{34} = 4$.

**2.** The successive shortest path algorithm is illustrated using the flow network in part (a) of the figure for Example 1. The initial residual network $G(x)$ for $x = 0$ is the same as

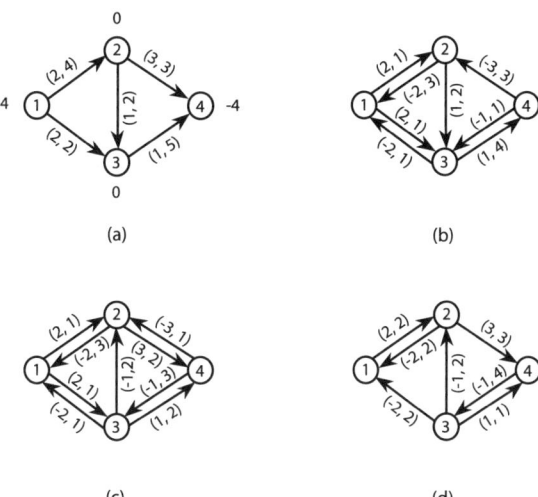

(a)                                    (b)

(c)                                    (d)

that of part (a). Initially, the imbalances are $e = (4, 0, 0, -4)$, so that $V^+ = \{1\}$ and $V^- = \{4\}$, giving $k = 1$ and $l = 4$. The shortest path from vertex 1 to 4 in $G(x)$ is $[1, 3, 4]$, and the algorithm augments $\delta = 2$ units of flow along this path. The following figure shows the residual network after this augmentation, as well as the updated imbalance at each vertex.

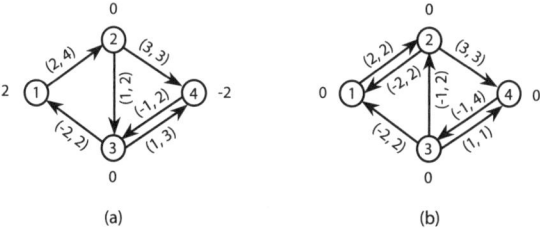

(a)                                    (b)

The sets $V^+$ and $V^-$ do not change, so again $k = 1$ and $l = 4$. The shortest path from vertex 1 to vertex 4 is now $[1, 2, 3, 4]$, and the algorithm augments $\delta = 2$ units of flow along this path. Part (b) of the figure shows the resulting residual network. Now $V^+ = V^- = \emptyset$ and the algorithm terminates.

**3.** The following figure illustrates the network simplex algorithm. Part (a) of this figure depicts the given flow network, with $b(i)$ shown for each vertex $i$ and $(c_{ij}, u_{ij})$ for each arc $(i, j)$.

A feasible spanning tree solution is shown in part (b) of the figure; each nontree arc (dashed line) has flow at either its lower or upper bound. The unique flows on the tree arcs (solid lines) are determined by the mass balance constraints. A set of vertex potentials (obtained using Fact 17) are also shown in part (b). Relative to these potentials $\pi$, the reduced costs for the nontree arcs are given by $c_{23}^\pi = 2 - (-3) - 2 = 3$, $c_{35}^\pi = 4 - (-2) - 5 = 1$, $c_{54}^\pi = 5 - (-5) - 8 = 2$, and $c_{46}^\pi = 3 - (-8) - 9 = 2$. Since arc $(3, 5)$, with flow at its upper bound, violates the optimality conditions of Fact 18, it is added to the current tree producing the cycle $[1, 2, 5, 3, 1]$. The maximum flow that can be sent along this cycle without violating the capacity constraints is 1 unit, which forces the flow on arc $(2, 5)$ to its upper bound. Arc $(2, 5)$ is then removed from the current tree and arc $(3, 5)$ is added to the current tree.

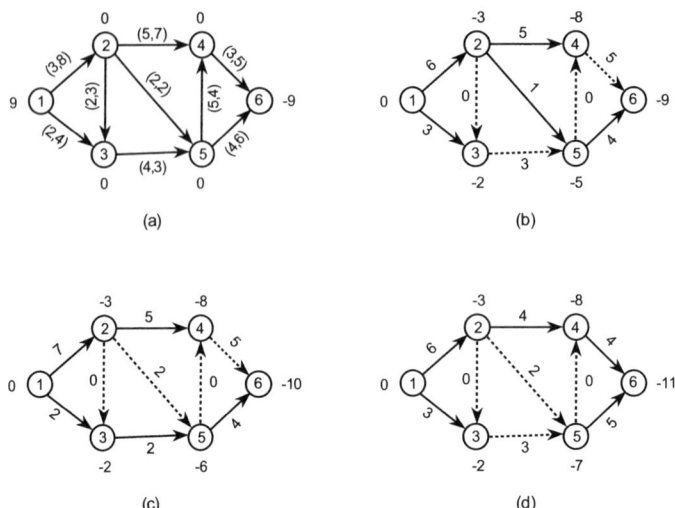

(a)          (b)

(c)          (d)

Part (c) of the figure gives the new flow as well as the new vertex potentials. Since $c_{46}^{\pi} = 3 - (-8) - 10 = 1$, arc $(4, 6)$ is added to the spanning tree, forming the cycle $[1, 3, 5, 6, 4, 2, 1]$. The maximum flow that can be sent along this cycle without violating the capacity constraints is 1 unit, which forces arc $(3, 5)$ out of the tree.

Part (d) gives the new flow as well as the new vertex potentials. Since the complementary slackness optimality conditions are satisfied, the current flow is optimal.

## 10.5.3  APPLICATIONS

Minimum cost flow problems arise in many industrial settings and scientific domains, often in the form of distribution or routing problems. The minimum cost flow problem also has less transparent applications, several of which are presented in this section.

**Applications:**

**1.** *Distribution:* A common application of the minimum cost flow problem involves the distribution at minimum cost of a product from manufacturing plants (with known supplies) to warehouses (with known demands). A similar scenario applies to the distribution of goods from warehouses to retailers as well as the flow of raw materials and intermediate goods through various machining stations in a production line.

**2.** *Routing:* The routing of cars through an urban street network and the routing of calls through a telephone system can be modeled using minimum cost flows. In either case, the items (cars, calls) must be sent from certain specified origins to other specified destinations, with capacity constraints on the total flow on each arc (road, communication link). This is done to minimize total (or average) delay in the system.

**3.** *Directed Chinese postman problem:* Leaving from the post office, a mail carrier needs to visit all houses on a postal route, delivering and collecting letters, and then return to the post office. The carrier would like to cover this route by traveling the minimum possible distance. (See also §8.4.3.) In this variation, known as the *directed Chinese postman problem*, each street is assumed to be directed, so the problem is defined on a directed network $G = (V, E)$ whose arcs $(i, j)$ have an associated nonnegative length $c_{ij}$. It is desired to find a directed walk (§8.3.2) of minimum length that starts at some vertex (the post office), visits each arc of the network at least once, and returns to the starting vertex. In an optimal walk, some arcs may be traversed more than once. If $x_{ij}$ represents the number of times arc $(i, j)$ is traversed, then this problem can be formulated as

$$\text{minimize:} \quad \sum_{(i,j) \in E} c_{ij} x_{ij}$$

$$\text{subject to:} \quad \sum_{\{j | (i,j) \in E\}} x_{ij} - \sum_{\{j | (j,i) \in E\}} x_{ji} = 0 \text{ for all } i \in V,$$

$$x_{ij} \geq 1 \text{ for all } (i, j) \in E.$$

This problem is a minor variant of the minimum cost flow problem where each arc has a lower bound of one unit of flow. From an optimal flow $x^*$ for this problem, an optimal tour can be constructed in the following manner. First, replace each arc $(i, j)$ with $x_{ij}^*$ copies of the arc, each carrying a unit flow. Next, decompose the resulting network into a set of directed cycles. Finally, connect the directed cycles to form a closed walk.

**4.** *Optimal loading of a hopping airplane:*  A small commuter airline uses a plane with capacity of at most $p$ passengers on a "hopping flight", as shown in part (a) of the following figure. The flight visits the cities $1, 2, 3, \ldots, n$ in a fixed sequence. The plane can pick up passengers at any city and drop them off at any other city. Let $b_{ij}$ denote the number of passengers available at city $i$ who want to go to city $j$, and let $f_{ij}$ denote the fare per passenger from city $i$ to city $j$. The airline would like to determine the number of passengers that the plane should carry between various origins and destinations in order to maximize the total fare per trip while never exceeding the capacity of the plane.

Part (b) of the following figure shows a minimum cost flow formulation of this hopping plane flight problem. The network displays data only for those arcs with nonzero cost or finite capacity. Any arc without a displayed cost has zero cost; any arc without a displayed capacity has infinite capacity. For example, three types of passengers are available at vertex 1: those whose destination is vertex 2, vertex 3, or vertex 4. These three types of passengers are represented by the vertices 1-2, 1-3, and 1-4 with supplies $b_{12}$, $b_{13}$, and $b_{14}$. A passenger available at any such vertex, say 1-3, either boards the plane at its origin vertex by flowing through the arc $(1\text{-}3, 1)$, thus incurring a cost of $-f_{13}$ units, or never boards the plane, represented by flowing through the arc $(1\text{-}3, 3)$.

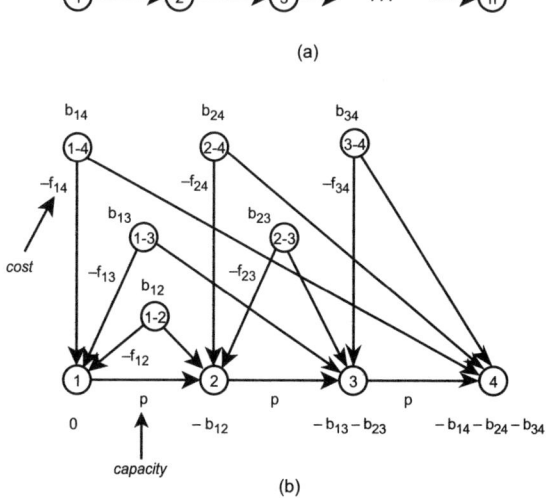

(a)

(b)

**5.** *Leveling mountainous terrain*:   In building road networks through hilly or mountainous terrain, civil engineers must determine how to distribute earth from high points to low points of the terrain to produce a leveled roadbed. To model this, construct a *terrain graph*, an undirected graph $G$ whose vertices represent locations with a demand for earth (low points) or locations with a supply of earth (high points). An edge of $G$ indicates an available route for distributing the earth, and the cost of this edge is the cost per truckload of moving earth between the corresponding two locations. The following figure shows a portion of a sample terrain graph. A leveling plan for a terrain graph is a flow (set of truckloads) that meets the demands at vertices (levels the low points) by the available supplies (earth obtained from high points) at minimum trucking cost. This can be solved as a minimum cost flow problem on the terrain graph.

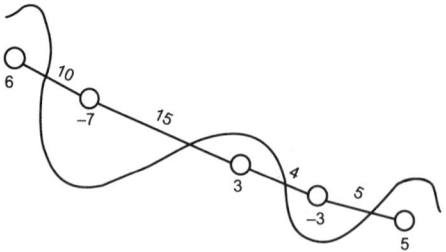

**6.** Additional applications, with reference sources, are given in the following table.

| application | references |
|---|---|
| medical tomography | [AhMaOr93], [AhEtal95] |
| automatic chromosome classification | [AhMaOr93], [AhEtal95] |
| racial balancing of schools | [AhMaOr93], [AhEtal95] |
| controlled matrix rounding | [AhMaOr93], [AhEtal95] |
| building evacuation | [AhMaOr93], [AhEtal95] |
| just-in-time scheduling | [AhMaOr93], [AhEtal95] |
| telephone operator scheduling | [AhMaOr93], [AhEtal95] |
| nurse staff scheduling | [AhMaOr93] |
| machine scheduling | [AhMaOr93] |
| production scheduling | [EvMi92] |
| equipment replacement | [AhMaOr93] |
| microdata file merging | [AhEtal95], [EvMi92] |
| warehouse layout | [AhMaOr93], [AhEtal95] |
| facility location | [AhMaOr93], [AhEtal95] |
| determining service districts | [AhMaOr93], [AhEtal95], [EvMi92] |
| capacity expansion | [AhMaOr93] |
| vehicle fleet planning | [AhMaOr93] |

# 10.6  COMMUNICATION NETWORKS

Modern communication networks consist of two main components. Using high-capacity links, the backbone network interconnects switching centers and gateway vertices that carry and direct traffic through the communication system. Local access networks transfer traffic between the backbone network and the end users. This section presents several optimization models used in the design of such communication networks.

## 10.6.1  CAPACITATED MINIMUM SPANNING TREE PROBLEM

The capacitated minimum spanning tree problem arises in the design of local access tree networks in which end users generate and retrieve data from other sources, always through a specified control center (e.g., a communication switch of the backbone network). In this problem, user sites are to be interconnected at minimum cost by means of subtrees, which are in turn connected to the control center. The total traffic in each subtree is limited by a capacity constraint.

**Definitions:**

Let $N = \{1, 2, \ldots, n\}$ be a set of **terminals** and let $0$ denote a specified **control center**. The complete undirected graph $G = (V, E)$ has vertex set $V = N \cup \{0\}$ and contains all possible edges between distinct vertices of $V$ (§8.1.3).

The **cost** of connecting distinct vertices $i, j \in V$ is $c_{ij} = c_e$, where $e = (i, j) \in E$.

The **demand** $w_i$ at vertex $i \in N$ is the amount of traffic to be transmitted to the control center.

Relative to a spanning tree $T$ (§9.2.1) of $G$, vertex $j$ is a **root vertex** if it is adjacent to vertex $0$. Vertex $i$ is **assigned** to root vertex $j$ if $j$ is on the unique path in $T$ joining $i$ to the control center. The set of all vertices assigned to $j$ defines the **subtree** $T_j$ of $T$. This subtree has **demand** $D(T_j) = \sum_{i \in T_j} w_i$.

A **capacitated minimum spanning tree** (**CMST**) is a spanning tree $T$ of $G$ composed of subtrees $T_{j_1}, T_{j_2}, \ldots, T_{j_r}$ such that

- $\sum_{e \in T} c_e$ is minimum;
- the demand in each $T_j$ is at most $Q$, a specified capacity.

For $j \in N$, let $y_j \in \{0, 1\}$ indicate whether vertex $j$ is a root vertex.

For $i, j \in N$, let $x_{ij} \in \{0, 1\}$ indicate whether vertex $i$ is assigned to root vertex $j$, and let $z_{ij} \in \{0, 1\}$ indicate whether edge $(i, j)$ belongs to the spanning tree $T$.

Given a vector $z$, the subgraph $G(N, E_z)$ of $G$ **induced by** $z$ has vertex set $N$ and edges $e \in E_z$ if $z_e > 0$. Similarly, given a vector $(z, y)$, the subgraph **induced by** $(z, y)$, written $G(V, E_{zy})$, has vertex set $V$; it contains every edge of $E_z$ plus each edge from $j$ to $0$ where $y_j > 0$.

Relative to a given vector $z$, $C(i, j)$ denotes the set of all **i-j cuts** $[S, \overline{S}]$ in the graph $G(N, E_z)$. (See §10.4.1.)

If $S \subseteq V$, then $E(S) = \{(i,j) \in E \mid i,j \in S\}$ contains all edges between vertices of $S$.

For $I \subset N$, let $b(I)$ be the minimum number of subtrees needed to pack all terminals in $I$. That is, $b(I)$ is the optimal solution to the **bin packing problem** (§16.3.2) with bins of capacity $Q$ and items of size $w_i$ for every $i \in I$.

A set $S \subset N$ is a **cover** if $\sum_{i \in S} w_i > Q$. If also $(\sum_{i \in S} w_i) - w_k \leq Q$ for all $k \in S$, then $S$ is a **minimal cover**.

**Facts:**

**1.** The CMST problem has the following 0-1 integer linear programming formulation (§16.1.8):

$$\text{minimize:} \quad \sum_{i=1}^{n-1} \sum_{j=i+1}^{n} c_{ij} z_{ij} + \sum_{j=1}^{n} c_{0j} y_j$$

$$\text{subject to:} \quad \sum_{j=1}^{n} x_{ij} = 1, \text{ for all } i \in \{1,2,\ldots,n\}$$

$$\sum_{i=1}^{n} w_i x_{ij} \leq Q y_j, \text{ for all } j \in \{1,2,\ldots,n\}$$

$$x_{ij} \leq y_j, \text{ for all } i,j \in \{1,2,\ldots,n\}$$

$$x_{ij} \leq \sum_{e \in K} z_e, \text{ for all } i,j \in \{1,2,\ldots,n\} \ (i \neq j) \text{ and for all } K \in C(i,j)$$

$$\sum_{e} z_e + \sum_{j} y_j = n$$

$$x_{ij} \in \{0,1\}, \text{ for all } i,j \in \{1,2,\ldots,n\}$$

$$y_j \in \{0,1\}, \text{ for all } j \in \{1,2,\ldots,n\}$$

$$z_{ij} \in \{0,1\}, \text{ for all } i,j \in \{1,2,\ldots,n\} \ (i \neq j).$$

**2.** In Fact 1,

- the first set of constraints ensures that each vertex is assigned to a root vertex;
- the second set of constraints ensures that the flow through any root vertex is no more than the capacity $Q$;
- the third set of constraints ensures that vertex $i$ can be assigned to vertex $j$ only if $j$ is a root vertex;
- the fourth set of constraints ensures that if vertex $i$ is assigned to root vertex $j$, then there must be a path between $i$ and $j$;
- the fifth set of constraints guarantees that $G(V, E_{zy})$ is a tree.

**3.** *Savings heuristic:* This greedy heuristic (Algorithm 1) begins with $n$ components, each a single vertex, and successively merges pairs of components to reduce the total cost by the largest amount.

**4.** The quantity $s_{uv}$ computed in Algorithm 1 represents the *savings* in joining subtrees $T_u$ and $T_v$ to one another, compared to joining both to vertex 0.

**5.** The savings heuristic, developed by Esau and Williams [EsWi66], was one of the first heuristics developed for the CMST problem.

**6.** The savings heuristic, surprisingly effective in practice, has attracted significant attention. Relatively simple adjustments have been proposed to improve its performance while maintaining its simplicity.

- Dai and Fujino [DaFu00] use component-oriented savings computations in order to increase its efficiency;

- Bruno and Laporte [BrLa02] propose enhancements to the procedure to connect subtrees;
- Öncan and Altinel [ÖnAl09] introduce parametric improvements which control edge selections and combine distance and demand information in its savings criterion.

---

**Algorithm 1: Savings heuristic.**

input: undirected network $G$, control center 0, capacity limit $Q$

output: an approximate capacitated minimum spanning tree $T^*$

$U := \{1, 2, \ldots, n\}$

$T_u := \{u\}$ for $u \in U$

**while** true

    **for** $u \in U$

        compute $f_u$, the minimum cost of connecting the control center 0 to component $T_u$

    $S := \emptyset$

    **for** $u, v \in U$ $(u \neq v)$

        **if** $D(T_u \cup T_v) \leq Q$ **then**

            compute $s_{uv}$, the difference between $\max\{f_u, f_v\}$ and the minimum cost of connecting $T_u$ to $T_v$

            **if** $s_{uv} > 0$ **then** $S := S \cup \{(u, v)\}$

    **if** $S = \emptyset$ **then** return

    **else**

        choose $u_0, v_0$ such that $s_{u_0 v_0} = \max\{s_{uv} \mid (u, v) \in S\}$

        merge $T_{u_0}$ and $T_{v_0}$, creating a new subtree indexed by $\min\{u_0, v_0\}$, and update $U$ appropriately

---

**7.** *Optimal tour partitioning heuristic:* This heuristic (Algorithm 2), developed by Altinkemer and Gavish [AlGa88], is based on finding a traveling salesman tour (§10.7.1) in a certain derived graph.

---

**Algorithm 2: Optimal tour partitioning heuristic.**

input: undirected network $G$, control center 0, capacity limit $Q$

output: an approximate capacitated minimum spanning tree $T^*$

find a traveling salesman tour on the vertex set $V = N \cup \{0\}$

let $0 = x^{(0)}, x^{(1)}, \ldots, x^{(n)}$ be an ordering of the vertices on the tour

construct the directed graph $H$ with vertex set $V$ and arc costs $C_{jk}$:

    **if** $j < k$ and $\sum_{i=j+1}^{k} w_{x^{(i)}} \leq Q$ **then** $C_{jk} := c_{x^{(0)}, x^{(j+1)}} + \sum_{i=j+1}^{k-1} c_{x^{(i)}, x^{(i+1)}}$

    **else** $C_{jk} := \infty$

find a shortest path $P$ from $x^{(0)}$ to $x^{(n)}$ in $H$

use $P = [x^{(0)}, x^{(u)}, x^{(v)}, \ldots, x^{(t)}, x^{(n)}]$ to define $T^*$ via the subtrees

    $\{x^{(0)}, x^{(1)}, x^{(2)}, \ldots, x^{(u)}\}$, $\{x^{(0)}, x^{(u+1)}, x^{(u+2)}, \ldots, x^{(v)}\}$, $\ldots$,

        $\{x^{(0)}, x^{(t+1)}, x^{(t+2)}, \ldots, x^{(n)}\}$

**8.** In Algorithm 2, every path from $x^{(0)}$ to $x^{(n)}$ in the directed graph $H$ generates a collection of subtrees satisfying the capacity restriction.

**9.** The performance of Algorithm 2 depends on the initial traveling salesman tour chosen. If an optimal traveling salesman tour is used, then the worst-case relative error bound of the algorithm is $4 - \frac{4}{Q}$. That is, $\widehat{Z}/Z^* \leq 4 - \frac{4}{Q}$, where $\widehat{Z}$ is the cost of the heuristic solution generated and $Z^*$ is the cost of the optimal design.

**10.** *Exact algorithms:*  A number of exact algorithms are based on mathematical programming approaches:

- Gavish [Ga85] develops a Lagrangian relaxation based algorithm and uses it to solve problems with homogeneous (unit) demands;

- Araque, Hall, and Magnanti [ArHaMa90] derive valid inequalities and facets for the CMST problem;

- Hall [Ha96] and Bienstock, Deng, and Simchi-Levi [BiDeSi94] develop additional valid inequalities and facets and use them in a branch-and-cut algorithm;

- Gouveia and Martins [GoMa99], [GoMa00], [GoMa05] propose branch-and-cut algorithms over hop-indexed formulations with tight lower bounds for unitary demand problems;

- Uchoa et al. [UcEtal08] present a robust branch-cut-and-price algorithm over an arc formulation strengthened by an exponential number of constraints on $q$-arb structures. Various powerful cuts are implemented without increasing the complexity of the problem and which improve upon previous formulations.

**11.** The CMST formulation given in Fact 1 can be improved by adding the various inequalities listed in Facts 12–15.

**12.** *Knapsack inequalities:*  Let $S$ be a minimal cover. For every $l \in N$, the inequality

$$\sum_{i \in S} x_{il} \leq (|S| - 1)y_l$$

is valid for the CMST problem.

**13.** *Subtour elimination inequalities:*  For any $I \subset N$, let $\mathcal{P} = \{S_1, S_2, \ldots, S_{|I|}\}$ be a partition of $N - I$ into $|I|$ subsets, some of which may be empty. For every $i \in I$, let $S_i$ be the unique subset from $\mathcal{P}$ associated with it. Then

$$\sum_{e \in E(I)} z_e + \sum_{j \in I} y_j + \sum_{i \in I} \sum_{j \in S_i} x_{ij} \leq \sum_{i \in I} \sum_{j \in N} x_{ij}$$

is valid for the CMST problem.

**14.** *Generalized subtour elimination inequalities:*  For any $I \subset N$, the inequality

$$\sum_{e \in E(I)} z_e \leq |I| - b(I)$$

is valid for the CMST problem.

**15.** *Cluster inequalities:*  Consider $p$ sets of vertices $S_1, S_2, \ldots, S_p \subset N$ with $p \geq 3$. If the conditions

- $S_0 = \bigcap_{i=1}^{p} S_i \neq \emptyset$, and $S_i - S_0 \neq \emptyset$  for $i = 1, 2, \ldots, p$

- $\sum_{i \in S_k \cup S_l} w_i > Q$  for all $1 \leq k < l \leq p$

are satisfied, then

$$\sum_{i=1}^{p} \sum_{e \in E(S_i)} z_e \leq \sum_{i=1}^{p} |S_i| - 2p + 1$$

is valid for the CMST problem.

**16.** *Metaheuristics*: Since 1995 there has been growing interest in metaheuristics with the aim of solving larger, more difficult instances. The methods proposed include neighborhood exploration [AhOrSh01], [AhOrSh03], [AmDoVo96], [SoDuRi03], tabu search [ShEtal97], ant colony optimization [ReLa06], filter-and-fan algorithms [ReMa11], and genetic algorithms [RuEtal15].

**Examples:**

**1.** The following figure presents data for a problem involving $n = 5$ terminals and a control center 0. Part (a) gives the cost $c_{ij}$ of constructing each edge $(i, j)$ as well as the demand $w_i$ at each vertex $i$. The objective is to construct a minimum cost set of subtrees connected to vertex 0, in which the demand generated by any subtree is at most $Q = 150$. Part (b) shows a feasible capacitated spanning tree $T$, which contains two root vertices (at 2 and 3). The total demand in subtree $T_2$ is $w_2 + w_4 + w_5 = 150 \leq Q$ and the total demand in subtree $T_3$ is $w_1 + w_3 = 95 \leq Q$. The spanning tree $T$ has total cost 21, the sum of the displayed edge costs $c_{ij}$.

| | | $c_{ij}$ | | | | |
|---|---|---|---|---|---|---|
| i \ j | 2 | 3 | 4 | 5 | $c_{0i}$ | $w_i$ |
| 1 | 2 | 4 | 3 | 4 | 8 | 40 |
| 2 | | 5 | 2 | 3 | 5 | 35 |
| 3 | | | 6 | 5 | 7 | 55 |
| 4 | | | | 4 | 6 | 50 |
| 5 | | | | | 6 | 65 |

(a)

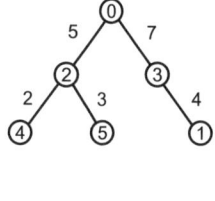

(b)

**2.** Algorithm 1 is applied to the problem data in the figure of Example 1. To begin, five subtrees are selected, each a single vertex and each joined to the control center 0. Thus, $f_1 = 8$, $f_2 = 5$, $f_3 = 7$, $f_4 = 6$, and $f_5 = 6$. Then $s_{12} = 8 - 2 = 6$, $s_{13} = 8 - 4 = 4, \ldots$, $s_{35} = 7 - 5 = 2$, and $s_{45} = 6 - 4 = 2$. The largest savings occurs for $(1, 2)$ so $T_1$ and $T_2$ are merged, giving the new tree $T_1$ with root vertex 2 and the single edge $(1, 2)$. Next, the $f_u$ and $s_{uv}$ are updated. For example, $f_1 = \min\{8, 5\} = 5$, $f_3 = 7$, and $s_{13} = 7 - \min\{c_{13}, c_{23}\} = 7 - 4 = 3$. The largest savings is found to be $s_{14}$, so $T_1$ and $T_4$ are merged, giving the new tree $T_1$ with root vertex 2 and edges $(1, 2)$ and $(2, 4)$. At the next stage $T_3$ and $T_5$ are merged, giving the new tree $T_3$ with root vertex 5 and the single edge $(3, 5)$. Since no further merging can take place (without violating the capacity constraint), the savings heuristic terminates with the spanning tree shown in the following figure (see next page), having total cost 20.

## 10.6.2  CAPACITATED CONCENTRATOR LOCATION PROBLEM

The capacitated concentrator location problem is frequently used to locate concentrators in local access networks and switching centers in the backbone network. In either case,

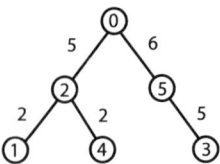

concentrators of fixed capacity are to be located at a subset of possible sites. Each given terminal of the network is to be connected to exactly one concentrator, so that the concentrator's capacity is not exceeded. A feasible configuration having minimum total cost is then sought.

**Definitions:**

Let $N = \{1, 2, \ldots, n\}$ be a specified set of **terminals**, where terminal $i$ uses $w_i$ units of **capacity**. Let $M = \{1, 2, \ldots, m\}$ be a set of possible **sites** for concentrators, each of fixed capacity $Q$.

The bipartite graph $G = (V, E)$ has vertex set $V = N \cup M$ and edges $(i, j)$ for all possible assignments connecting terminal $i \in N$ to a concentrator at site $j \in M$.

If a concentrator is located at site $j$, the **set-up cost** is $v_j$, for $j \in M$. The **connection cost** of connecting terminal $i$ to a concentrator at site $j$ is $c_{ij}$, for $i \in N$ and $j \in M$.

The **capacitated concentrator location problem** (**CCLP**) involves finding locations for concentrators and an assignment of terminals to concentrators such that

- the sum of set-up and connection costs is minimum;
- the total capacity required by the terminals assigned to each concentrator is at most $Q$.

For $j \in M$, let $y_j \in \{0, 1\}$ indicate whether a concentrator is located at site $j$.

For $i \in N$ and $j \in M$, let $x_{ij} \in \{0, 1\}$ indicate whether terminal $i$ is connected to a concentrator at site $j$.

**Facts:**

**1.** There are two variants of capacity constraints, based on (a) each concentrator's processing capacity, and (b) the number of connection ports. Han and Raja [HaRa03] study an extended version of the problem with both types of constraints.

**2.** The classical capacitated concentrator location problem with only processing capacity constraints (i.e., variant (a)) is also known as the *single source capacitated facility location problem*.

**3.** The classical CCLP has the following 0-1 integer linear programming formulation (§16.1.8):

$$\text{minimize:} \quad \sum_{i=1}^{n} \sum_{j=1}^{m} c_{ij} x_{ij} + \sum_{j=1}^{m} v_j y_j$$

$$\text{subject to:} \quad \sum_{j=1}^{m} x_{ij} = 1, \text{ for all } i \in N$$

$$\sum_{i=1}^{n} w_i x_{ij} \leq Q y_j, \text{ for all } j \in M$$

$$x_{ij} \leq y_j, \text{ for all } i \in N, \ j \in M$$

$$x_{ij} \in \{0, 1\}, \text{ for all } i \in N, \ j \in M$$

$$y_j \in \{0, 1\}, \text{ for all } j \in M.$$

**4.** In Fact 3,

- the first set of constraints ensures that each terminal is connected to exactly one concentrator;

- the second set of constraints ensures that the concentrator's capacity is not exceeded;

- third set of constraints ensures that terminal $i$ can be assigned to a concentrator at site $j$ only if a concentrator is located at site $j$.

**5.** In variant (b), $w_i = 1$, for all $i \in N$ and $Q$ is the number of connection ports.

**6.** A number of algorithms have been proposed for this problem, most of which are based on a Lagrangian relaxation approach, while some are based on polyhedral analysis [NeWo99].

**7.** Yang, Chu, and Chen [YaChCh12] present a cut-and-solve algorithm which they apply to large sized instances.

**8.** Guastaroba and Speranza [GuSp14] extend the kernel search framework with variants based on variable fixing to binary integer linear programming models and apply it to this problem with high-quality results. The heuristic is based on optimally solving a sequence of subproblems, where each subproblem is restricted to a subset of the decision variables. The subsets of decision variables are constructed starting from the optimal values of the linear relaxation.

**Examples:**

**1.** The following table shows the data for a problem with four terminals $i$ and three possible sites $j$ for locating concentrators. Let the capacity of any concentrator be $Q = 30$ for variant (a) and $Q = 2$ for variant (b).

$c_{ij}$

| $i$ \ $j$ | 1 | 2 | 3 | $w_i$ |
|---|---|---|---|---|
| 1 | 8 | 6 | 4 | 11 |
| 2 | 2 | 4 | 3 | 12 |
| 3 | 5 | 5 | 3 | 9 |
| 4 | 3 | 2 | 6 | 8 |
| $v_j$ | 9 | 7 | 8 | |

**2.** For variant (a), one feasible configuration is to connect terminals $2, 3, 4$ to a concentrator at site 2, and to connect terminal 1 to a concentrator at site 3. See part (a) of the following figure.

The concentrator at site 2 has total capacity $w_2 + w_3 + w_4 = 29 \le 30$ and the concentrator at site 3 has total capacity $w_1 = 11 \le 30$. The connection cost is $c_{22} + c_{32} + c_{42} + c_{13} = 15$ and the set-up cost is $v_2 + v_3 = 15$, giving a total cost of 30. Another feasible configuration, shown in part (b), is to connect terminals 2 and 4 to a concentrator at site 1, and to connect terminals 1 and 3 to a concentrator at site 3. The connection cost is 12 and the set-up cost is 17, giving the smaller total cost 29.

**3.** For variant (b) only the second configuration is feasible.

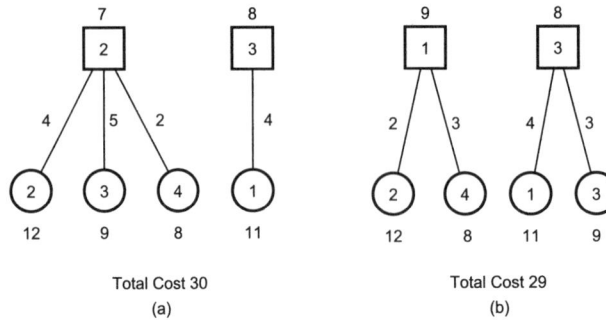

Total Cost 30                    Total Cost 29
(a)                                    (b)

### 10.6.3  NETWORK DESIGN PROBLEM

Fiber-optic and opto-electronic cable technologies, together with traditional copper cables, provide many possible choices for link capacities and offer economies of scale. In the
capacity assignment problem, a point-to-point communication demand is given between
various pairs of vertices of the (typically, backbone) network. The objective is to install
links of several types (capacities) to transfer all communication demand without violating link capacities and to do so at minimum total cost. The special case involving two
types of transmission media, also referred to in the literature as the two-facility network
design problem or the network loading problem, is discussed here.

#### Definitions:

Let $G = (V, E)$ be an undirected graph with vertex set $V$ and edge set $E$.

Each communication demand is represented by a **commodity** $k \in K$, where $K$ is the
set of commodities. Commodity $k \in K$ has a required flow in $G$ of $d_k$ units between its
**origin vertex** $O(k)$ and its **destination vertex** $D(k)$.

Two types of cables can be installed: **low capacity** cables have capacity $L$, and **high
capacity** cables have capacity $H$. Let $a_e$ $(b_e)$ denote the **installation cost** for each low
capacity (high capacity) cable on edge $e \in E$.

The **network design problem** (**NDP**) involves finding a mix of low and high capacity
cables for each edge of $G$ such that

- the total installation cost is minimum;
- all communication demands $d_k$ are met;
- the flow on each edge does not exceed its installed capacity.

Let $x_e = x_{ij}$ $(y_e = y_{ij})$ be the number of low capacity (high capacity) cables installed on
edge $e = (i, j)$.

Let $f_{ij}^k$ be the amount of commodity $k$ that flows from $i$ to $j$ on edge $(i, j)$.

#### Facts:

**1.** The NDP has the following mixed integer linear programming formulation (§16.1.8):

minimize:   $\displaystyle\sum_{e\in E}(a_e x_e + b_e y_e)$

subject to:   $\displaystyle\sum_{j\in V} f^k_{j,O(k)} - \sum_{j\in V} f^k_{O(k),j} = -d_k,$ for all $k \in K$

$\displaystyle\sum_{j\in V} f^k_{j,D(k)} - \sum_{j\in V} f^k_{D(k),j} = d_k,$ for all $k \in K$

$\displaystyle\sum_{j\in V} f^k_{ji} - \sum_{j\in V} f^k_{ij} = 0,$ for all $k \in K$ and for all $i \in V \setminus \{O(k), D(k)\}$

$\displaystyle\sum_{k\in K}(f^k_{ij} + f^k_{ji}) \le Lx_{ij} + Hy_{ij},$ for all $(i,j) \in E$

$x_e, y_e \ge 0$ integer, for all $e \in E$

$f^k_{ij}, f^k_{ji} \ge 0,$ for all $(i,j) \in E$ and for all $k \in K.$

2. In Fact 1,

- the first three sets of constraints are the standard mass balance constraints (see §10.4.1);
- the next set of constraints enforces the capacity constraint on the total flow through edge $e = (i, j)$.

3. The NDP becomes harder to solve when more than one type of facility is involved, due to the complexity of the cost structure. The two-facility NDP is strongly NP-hard as it contains the fixed-charge NDP and the Steiner tree problem as special cases.

4. Various polyhedral approaches for network design problems are discussed in [BiGu95] and [MaMiVa95]. Hamid and Agarwal [HaAg15] consider 3-partitions of the original graph and derive a family of facets that significantly strengthen the linear formulation and reduce the integrality gap.

**Example:**

1. In the network $G$ of the following figure, the costs $(a_e, b_e)$ are shown for each edge $e$; here $L = 2$ and $H = 5$. There are $k = 3$ communication demands (commodities): $d_1 = 12$ between vertices 1 and 4, $d_2 = 10$ between vertices 2 and 5, and $d_3 = 9$ between vertices 1 and 5. A feasible assignment of flows and capacities to edges is displayed in part (b) of the figure. For instance, edge $(1,3)$ carries 7 units of commodity 1 and 9 units of commodity 3, for a total flow of 16 units. There are three high capacity cables and one low capacity cable installed on this edge giving a total capacity of $3H + L = 17$, at a cost of $3 \cdot 5 + 1 \cdot 3 = 18$. The total installation cost for this assignment is 114.

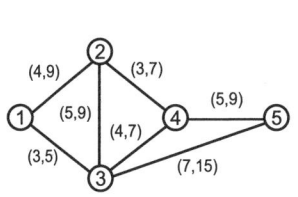

$f^k_{ij}$

| e | 1 | 2 | 3 | $x_e$ | $y_e$ |
|---|---|---|---|---|---|
| (1,2) | 5 | | | 0 | 1 |
| (1,3) | 7 | | 9 | 1 | 3 |
| (2,3) | | 4 | | 0 | 1 |
| (2,4) | 5 | 6 | | 3 | 1 |
| (3,4) | 7 | | 3 | 0 | 2 |
| (3,5) | | 4 | 6 | 0 | 2 |
| (4,5) | | 6 | 3 | 0 | 2 |

(a)                (b)

## 10.6.4   MODELS FOR SURVIVABLE NETWORKS

The introduction of fiber-optic technology has provided high capacity links and makes it possible to design communication networks with low-cost sparse topologies. Unfortunately, sparse networks are very vulnerable; a failure in one edge or vertex can disconnect many users from the rest of the network. This is the prime motivation for studying the design of survivable networks.

**Definitions:**

Let $G = (V, E)$ be an undirected graph with vertex set $V$ and edge set $E$.

The **cost** of establishing edge $e \in E$ is given by $c_e$. The cost of a subnetwork $H = (V, F)$ of $G$ is $\sum_{e \in F} c_e$.

Associated with every vertex $s \in V$ is a corresponding number $r_s$, indicating a desired level of **redundancy**.

A spanning subnetwork $H = (V, F)$ of $G$ is said to satisfy the **edge (vertex) connectivity requirement** if for every distinct pair $s, t \in V$ there are at least $r_{st} = \min\{r_s, r_t\}$ edge-disjoint (vertex-disjoint) paths between $s$ and $t$ in $H$.

Define $x_e$, for $e = (i, j)$, to be the number of edges connecting vertex $i$ to vertex $j$.

**Facts:**

**1.** The problem of designing a minimum cost subnetwork that satisfies all edge connectivity requirements has the following integer programming formulation (§16.1.8):

$$\text{minimize:} \quad \sum_{e \in E} c_e x_e$$

$$\text{subject to:} \quad \sum_{e \in [S, \overline{S}]} x_e \geq \max_{(i,j) \in [S, \overline{S}]} r_{ij}, \text{ for all } S \subset V,\, S \neq \emptyset$$

$$x_e \geq 0 \text{ integer, for all } e \in E.$$

**2.** The model in Fact 1, analyzed by Goemans and Bertsimas [GoBe93], allows multiple edges connecting the same two vertices.

**3.** Grötschel, Monma, and Stoer [GrMoSt92] analyze a related survivability model in which multiple edges are forbidden. In this case, $x_e$ is restricted to be 0 or 1 in the formulation of Fact 1.

**Examples:**

**1.** Part (a) of the following figure shows a network $G$ having four vertices and six edges; the cost $c_e$ of each edge $e$ is also displayed.

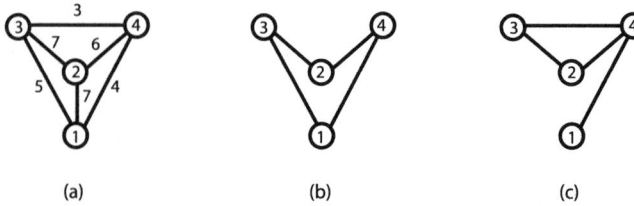

(a)                    (b)                    (c)

Suppose that the specified redundancies are $r_1 = 1$ and $r_2 = r_3 = r_4 = 2$. The spanning subnetwork $H$ shown in part (b), with cost 22, satisfies the vertex connectivity requirement. For example, there are $\min\{r_3, r_4\} = 2$ vertex-disjoint paths joining vertices 3 and 4 in $H$: namely, $[3,2,4]$ and $[3,1,4]$. Also, there are $\min\{r_2, r_4\} = 2$ vertex-disjoint paths joining vertices 2 and 4: $[2,4]$ and $[2,3,1,4]$.

**2.** Part (c) of the figure for Example 1 shows another spanning subnetwork $H'$ with cost 20 that satisfies the stated vertex connectivity requirement. For example, there are $\min\{r_2, r_4\} = 2$ vertex-disjoint paths joining vertices 2 and 4 in $H'$: $[2,4]$ and $[2,3,4]$. Notice that there is $\min\{r_1, r_3\} = 1$ path joining 1 and 3 in $H'$, but not two such vertex-disjoint paths.

## 10.6.5  CONNECTED FACILITY LOCATION PROBLEM

In designing the last mile of telecommunication networks, the fiber-to-the-curb strategy can be modeled as the connected facility location problem. Namely, fiber optic cables run to a cabinet serving a neighborhood; end users connect to this cabinet using copper connections; and expensive switching devices are installed in these cabinets. The relevant problem is to minimize the overall cost of the network by determining positions of cabinets, deciding which customers to connect to them, and how to reconnect cabinets among each other and to the backbone.

**Definitions:**

Let $G = (V, E)$ be an undirected graph with vertex set $V$ and edge set $E$.

The vertex set $V$ is partitioned into three disjoint sets $D, F, S$: $D$ is the set of **demand vertices**, $F$ is the set of potential **facility vertices**, and $S$ is the set of potential **Steiner vertices**. Thus, $V = D \cup F \cup S$.

**Facility opening costs** $c_i \geq 0$ are incurred for each open facility $i \in F$. **Assignment costs** $a_{ij} \geq 0$ are incurred for assigning customer $j \in D$ to a facility $i \in F$. **Core edge costs** $b_{ij} \geq 0$ are incurred for any edge $(i,j) \in E(F \cup S)$ if it is used by the **Steiner tree** $T$ connecting open facilities.

The vertices in $S$ may be viewed as pure Steiner vertices and can only be used in the tree $T$ as Steiner vertices, while the vertices in $F$ may be used as Steiner vertices in the tree $T$ with no facility opening cost incurred if no customers are assigned to them.

The final network cost is given by $\sum_{i \in Z} c_i + \sum_{(i,j) \in E(T)} b_{ij} + \sum_{j \in D} a_{i(j)j}$, where $i(j)$ is the facility serving demand vertex $j$, $Z$ is the set of open facilities, and $T$ is a Steiner tree connecting the open facilities.

The **connected facility location** (**ConFL**) **problem** involves finding a subset of open facilities such that

- each customer is assigned to exactly one open facility;
- a Steiner tree connects all open facilities;
- the total network cost is minimized.

Let $x_e = x_{ij} \in \{0,1\}$ ($y_e = y_{ij} \in \{0,1\}$) indicate whether an assignment edge (core edge) is installed on edge $(i,j)$.

Let $z_i \in \{0,1\}$ indicate whether there is an open facility at vertex $i$.

Specify vertex $r \in D$ as the root vertex for the Steiner tree and designate a commodity $k$ for each demand vertex that has origin $r$ and destination $k \in D$.

Let $f_{ij}^k$ be the amount of commodity $k$ that flows from $i$ to $j$ on edge $(i,j)$ for demand vertex $k \in D$.

**Facts:**

1. Problem ConFL has the following mixed integer programming formulation (§16.1.8):

$$\begin{aligned}
\text{minimize:} \quad & \sum_{i \in F} c_i z_i + \sum_{(i,j) \in E(S \cup F)} b_{ij} y_{ij} + \sum_{i \in F, j \in D} a_{ij} x_{ij} \\
\text{subject to:} \quad & \sum_{i \in F} x_{ij} = 1, \text{ for all } j \in D \\
& x_{ij} \leq z_i, \text{ for all } i \in F \text{ and for all } j \in D \\
& \sum_{j \in V} f_{ji}^k - \sum_{j \in V} f_{i,j}^k = -1, \text{ for } i = r \text{ and for all } k \in D \\
& \sum_{j \in V} f_{ji}^k - \sum_{j \in V} f_{ij}^k = 1, \text{ for } i = k \text{ and for all } k \in D \\
& \sum_{j \in V} f_{ji}^k - \sum_{j \in V} f_{ij}^k = 0, \text{ for } i \in V \setminus \{r, k\} \text{ and for all } k \in D \\
& f_{ij}^k \leq y_{ij}, \text{ for all } (i,j) \in E(S \cup F) \text{ and for all } k \in D \\
& f_{ij}^k \leq x_{ij}, \text{ for all } i \in F, \text{ for all } j \in D, \text{ and for all } k \in D \\
& x_{ij} \in \{0,1\}, \text{ for all } (i,j) \in E \\
& y_{ij} \in \{0,1\}, \text{ for all } (i,j) \in E(S \cup F) \\
& z_l \in \{0,1\}, \text{ for all } l \in S \cup F \\
& f_{ij}^k \geq 0, \text{ for all } (i,j) \in E \text{ and for all } k \in D.
\end{aligned}$$

2. In Fact 1,
   - the first two sets of constraints enforce that each customer is assigned to an open facility;
   - the next set of constraints are the standard mass balance constraints (§10.4.1);
   - the last two sets of constraints ensure that flow is only routed through edges on the tree.

3. The ConFL problem encompasses a family of network problems that combine facility location and network design; namely, the rent-or-buy problem, the Steiner tree star problem, and the general Steiner tree star problem.

4. Various approximation algorithms, heuristics and exact methods have been proposed.

5. Bardossy and Raghavan [BaRa10] propose a dual-based local heuristic to obtain both a high-quality heuristic solution and a tight lower bound. In the first stage the problem is modeled as a directed Steiner tree problem with a unit degree constraint at the root vertex. Then a dual-ascent procedure is applied to obtain a lower bound and a feasible solution to the problem. In the second stage a local search procedure is used to improve the feasible solution.

6. Gollowitzer and Ljubic [GoLj11] present various mixed integer programming formulations and compare the models in terms of their bounds and computational time.

**Example:**

1. Consider the network $G$ with vertices $D = \{1,2,3,4,5,6,7\}$, $F = \{8,9,10,11\}$, and $S = \{12,13\}$, as seen in part (a) of the following figure. Edges of $G$ are defined by a complete bipartite graph between facilities and demand vertices, and a complete graph between facilities and potential Steiner nodes. The assignment costs $a_{ij}$ and the core edge costs $b_{ij}$ are shown in part (c) and part (d), respectively. The facility opening cost is $c_i = 1$ for all facility vertices. A feasible assignment of customers to facilities and the connection of open facilities is displayed in part (b) of the figure. The overall cost for

this solution is 22. The total assignment cost is 10, the total core edge cost for $T$ is 9, and the total facility opening cost is 3. Vertex 9 plays the role of a Steiner vertex and so does not incur an opening cost. This solution turns out to be optimal.

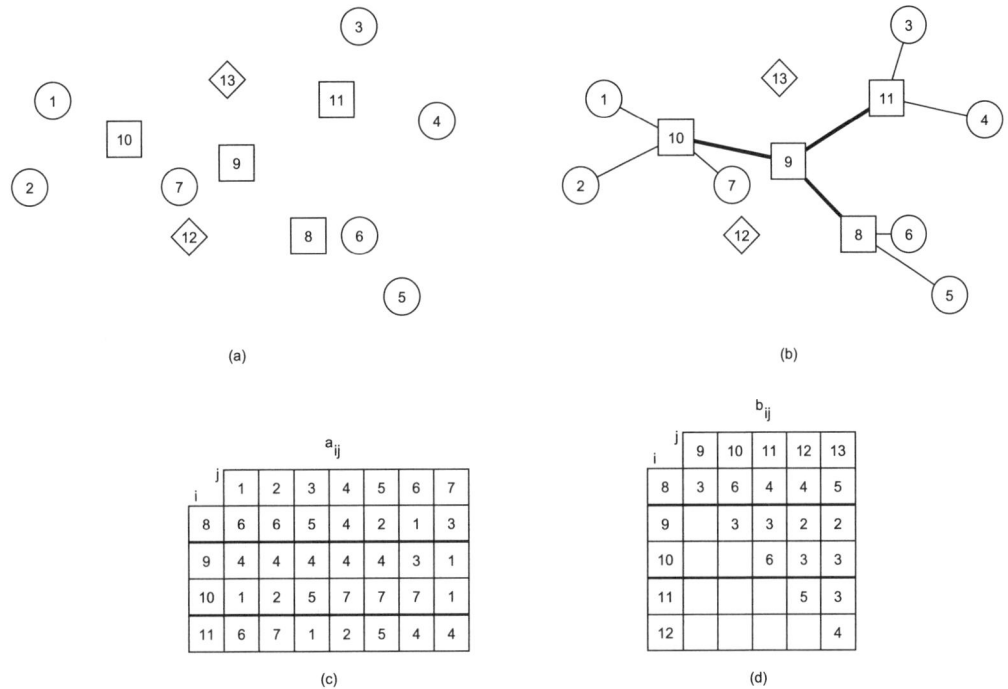

(a)                                                                                      (b)

$a_{ij}$

| j | 1 | 2 | 3 | 4 | 5 | 6 | 7 |
|---|---|---|---|---|---|---|---|
| i |   |   |   |   |   |   |   |
| 8 | 6 | 6 | 5 | 4 | 2 | 1 | 3 |
| 9 | 4 | 4 | 4 | 4 | 4 | 3 | 1 |
| 10 | 1 | 2 | 5 | 7 | 7 | 7 | 1 |
| 11 | 6 | 7 | 1 | 2 | 5 | 4 | 4 |

(c)

$b_{ij}$

| j | 9 | 10 | 11 | 12 | 13 |
|---|---|----|----|----|----|
| i |   |    |    |    |    |
| 8 | 3 | 6  | 4  | 4  | 5  |
| 9 |   | 3  | 3  | 2  | 2  |
| 10 |  |    | 6  | 3  | 3  |
| 11 |  |    |    | 5  | 3  |
| 12 |  |    |    |    | 4  |

(d)

## 10.6.6   REGENERATOR LOCATION PROBLEM

In optical telecommunication networks, an optical signal can only travel a maximum distance before its quality deteriorates to the point that it must be regenerated by a regenerator on the network. As the cost of a regenerator is high, the goal is to deploy as few regenerators as possible in the network, while ensuring that all vertices can communicate with one another.

### Definitions:

Let $G = (N, E)$ be a **communication graph** with vertex set $N$ and edge set $E$.

Let $S \subseteq N$ be the set of **candidate locations** where regenerators can be placed, and let $T \subseteq N$ be the set of **terminal vertices** that must communicate with one another.

Each edge $e = (i, j) \in E$ represents a **simple path** between $i$ and $j$. In other words, the length of the path does not exceed the maximum distance that can be traveled by the optical signal such that no regenerator is necessary between them.

The **installation cost** for placing a regenerator at location $i \in S$ is $c_i$.

The **regenerator location problem** (**RLP**) involves locating regenerators in the candidate locations $S$ such that

- the total regenerator installation cost is minimum;
- all terminal vertices can communicate with one another.

Let $x_i \in \{0,1\}$ indicate whether a regenerator is located at vertex $i$.

Specify a vertex $r \in T$ as the source vertex for all communications. Define a commodity $k$ for each terminal vertex $k \in T \setminus \{r\}$ that has origin $r$ and destination $k$.

Let $y_{rj} \in \{0,1\}$ indicate whether there is any flow from $r$ to $j$ on edge $(r,j)$.

Let $f_{ij}^k$ be the amount of commodity $k$ that flows from $i$ to $j$ on edge $(i,j)$ for terminal vertex $k$.

**Facts:**

**1.** The RLP has the following mixed integer linear programming formulation (§16.1.8):

$$\text{minimize:} \quad \sum_{i \in S} c_i x_i$$

$$\text{subject to:} \quad \sum_{j \in N} f_{ji}^k - \sum_{j \in N} f_{ij}^k = -1, \text{ for } i = r \text{ and for all } k \in T \setminus \{r\}$$

$$\sum_{j \in N} f_{ji}^k - \sum_{j \in N} f_{ij}^k = 1, \text{ for } i = k \text{ and for all } k \in T \setminus \{r\}$$

$$\sum_{j \in N} f_{ji}^k - \sum_{j \in N} f_{ij}^k = 0, \text{ for } i \in N \setminus \{r, k\} \text{ and for all } k \in T \setminus \{r\}$$

$$f_{ij}^k \leq x_j, \text{ for all } (i,j) \in E \text{ and for all } k \in T$$

$$f_{rj}^k \leq y_{rj}, \text{ for all } j \in N \setminus \{r\} \text{ and for all } k \in T$$

$$\sum_{(r,j) \in E} y_{rj} \leq 1 + M x_r$$

$$x_i \in \{0,1\}, \text{ for all } i \in S$$

$$y_{rj} \in \{0,1\}, \text{ for all } (r,j) \in E$$

$$f_{ij}^k \geq 0, \text{ for all } (i,j) \in E \text{ and for all } k \in T.$$

**2.** In Fact 1,

- the first three sets of constraints are the standard mass balance constraints (see §10.4.1);
- the next set of constraints enforces the installation of a regenerator for passing communications;
- the following sets of constraints ensure that the source vertex $r$ has unit degree or a regenerator installed. Here $M$ is a sufficiently large constant with $M \geq |T| - 2$.

**3.** The regenerator location problem is NP-hard.

**4.** Chen, Ljubic, and Raghavan [ChLjRa15] establish that the RLP is equivalent to the vertex-weighted directed Steiner forest problem. Using this fact, they derive various integer and mixed integer programming formulations. They also propose construction heuristics combined with local search.

**5.** When $N = S = T$, Chen, Ljubic, and Raghavan [ChLjRa10] show that the RLP is equivalent to a maximum leaf spanning tree problem.

**Example:**

**1.** In the communication graph $G$ of the following figure, $S = \{1, 3, 6\}$, $T = \{1, 2, 4, 5\}$, and the regenerator installation cost is 1 for all locations. A feasible installation of regenerators that ensures communication between all terminal vertices is displayed in part (b) of the figure, where # indicates the location of regenerators. The total regenerator installation cost is 2.

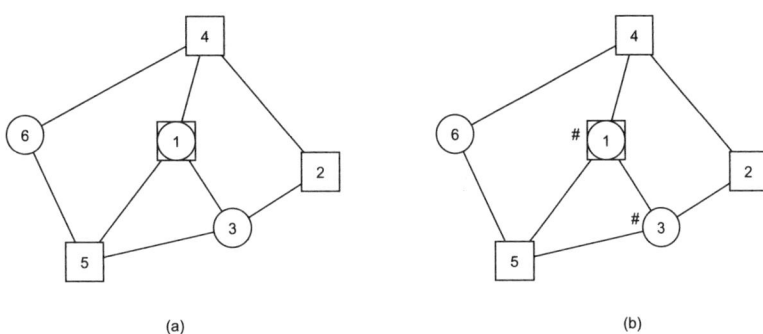

(a)                                                      (b)

## 10.7   DIFFICULT ROUTING AND ASSIGNMENT PROBLEMS

An *exact* algorithm for a combinatorial optimization problem is a procedure that produces a verifiable optimal solution to every instance of this problem. A *heuristic* algorithm produces a feasible (although not necessarily optimal) solution to each problem instance. This section discusses exact and heuristic approaches to three classical combinatorial optimization problems: the traveling salesman problem, the vehicle routing problem, and the quadratic assignment problem. These three problems have in common the goal of minimizing the cost of movement or travel, generally of people or of materials.

### 10.7.1   TRAVELING SALESMAN PROBLEM

In the traveling salesman problem, a salesman starts out from a home city and is to visit in some order a specified set of cities, returning home at the end. This journey is to be designed to incur the minimum total cost (or distance). While the traveling salesman problem has attracted the attention of many mathematicians and computer scientists, it has resisted attempts to develop an efficient solution algorithm.

**Definitions:**

Let $G = (V, E)$ be a complete graph (§8.1.3) with $V = \{1, 2, \ldots, n\}$ the set of vertices and $E$ the set of all edges joining pairs of distinct vertices.

Each edge $(i, j) \in E$ has an associated **cost** or **distance** $c_{ij}$.

The **distance** between set $S \subseteq V$ and vertex $j \notin S$ is $d_S(j) = \min\{c_{ij} \mid i \in S\}$.

A **Hamilton cycle** or **tour** in $G$ is a cycle passing through each vertex $i \in V$ exactly once. (See §8.4.4.)

The **cost** of a cycle $C$ is $\sum\limits_{(i,j) \in C} c_{ij}$.

The **traveling salesman problem** (**TSP**) requires finding a Hamilton cycle in $G$ of minimum total cost.

The costs (distances) satisfy the **triangle inequality** if $c_{ij} \leq c_{ik} + c_{kj}$ holds for all distinct $i, j, k \in V$.

In a **Euclidean** TSP, each vertex $i$ corresponds to a point $x_i$ in $\mathcal{R}^2$ and $c_{ij}$ is the distance between $x_i$ and $x_j$, relative to the standard real inner product (§6.1.4).

**Tour construction** procedures generate an approximately optimal TSP tour from the costs $c_{ij}$.

**Tour improvement** procedures attempt to find a smaller cost tour, given an initial (often random) tour.

**Composite** (**hybrid**) procedures construct a starting tour from one of the tour construction procedures and then attempt to find a smaller cost tour using one or more of the tour improvement procedures.

A **k-change** or **k-opt exchange** of a given tour is obtained by deleting $k$ edges from the tour and adding $k$ other edges to form a new tour. A tour is **k-optimal** (**k-opt**) if it is not possible to improve the tour via a $k$-change.

**Metaheuristics** are general-purpose procedures (such as tabu search, simulated annealing, genetic algorithms, or neural networks) for heuristically solving difficult optimization problems; these general methodologies for searching complex solution spaces can be specialized to handle specific types of optimization problems.

**Facts:**

**1.** The TSP is possibly the most well-known network optimization problem, and it serves as a prototype for difficult combinatorial optimization problems in the theory of algorithmic complexity (§17.5.2).

**2.** The first use of the term "traveling salesman problem" in a mathematical context appears to have occurred in 1931–1932.

**3.** There are numerous applications of the TSP: drilling of printed circuit boards, cluster analysis, sequencing of jobs, x-ray crystallography, archaeology, cutting stock problems, robotics, and order-picking in a warehouse (see Examples 7–11).

**4.** There are $\frac{(n-1)!}{2}$ different Hamilton cycles in the complete graph $G$. This means that brute force enumeration of all Hamilton cycles to solve the TSP is not practical. (See Example 1.)

**5.** The TSP is an NP-hard optimization problem (§17.5.2). This remains true even when the distances satisfy the triangle inequality or represent Euclidean distances.

**6.** If certain edges $(i, j)$ of $G$ are missing, then $c_{ij}$ can be assigned a sufficiently large value $M$; e.g., one can set $M$ to be greater than the sum of the $n$ largest edge costs. The TSP can then be solved on the complete graph $G$. If the (exact) solution obtained has any edges with cost $M$, then there is no Hamilton cycle in the original graph.

**7.** *Asymmetric traveling salesman problem:* Certain applications require finding a minimum cost directed Hamilton cycle in a directed network $H$; here it is not required that $c_{ij} = c_{ji}$ holds for all arcs $(i, j)$ of $H$. This asymmetric (directed) TSP can be transformed into a TSP problem on an undirected network [JüReRi95].

**8.** A seminal paper of G. B. Dantzig, D. R. Fulkerson, and S. M. Johnson (1954) solved a 49-city TSP to optimality by adding cutting planes (§16.1.8) to a linear programming relaxation of the problem.

**9.** Although ingenious exact algorithms for the TSP have been proposed by numerous authors, most encounter problems with storage and/or running time for cases with more than five hundred vertices.

**10.** Exact approaches to the TSP are computationally intensive, especially for large networks. Thus a large number of heuristic approaches have been developed to produce useful, but not necessarily optimal, solutions to the TSP.

**11.** The wealth of TSP heuristics can be categorized into four broad classes: tour construction procedures, tour improvement procedures, composite procedures, and meta-heuristics.

**12.** *Nearest neighbor heuristic:* This construction method (Algorithm 1) builds up a tour by successively adding new vertices that are closest to a growing path.

---

**Algorithm 1:  Nearest neighbor heuristic.**

input: undirected network $G = (V, E)$, costs $c_{ij}$

output: a traveling salesman tour

$i_0 :=$ any vertex of $G$ {the starting vertex}

$W := V - \{i_0\}$

$P := \emptyset$

$v := i_0$

**while** $W \neq \emptyset$

    let $k \in W$ be such that $c_{vk} = \min\{c_{vj} \mid j \in W\}$

    add $(v, k)$ to $P$

    $W := W - \{k\}$

    $v := k$

add $(k, i_0)$ to the path $P$ to produce a tour

---

**13.** Using appropriate data structures, Algorithm 1 can be implemented to run in $O(n^2)$ time.

**14.** Suppose $z_{\text{NN}}$ is the cost of a tour constructed by the nearest neighbor heuristic and $z_{\text{OPT}}$ is the cost of an optimal TSP tour. Then there are examples for which $\frac{z_{\text{NN}}}{z_{\text{OPT}}}$ is $\Theta(\log n)$. This means that the cost of the tour produced by Algorithm 1 cannot be bounded above by a constant times the cost of an optimal TSP tour.

**15.** *Nearest insertion heuristic:* This construction method (Algorithm 2) builds up a tour from smaller cycles by successively adding a vertex that is closest to the current cycle $C$. The new vertex is inserted between two successive vertices in the cycle, in the best possible way.

---

**Algorithm 2:  Nearest insertion heuristic.**

input: undirected network $G = (V, E)$, costs $c_{ij}$

output: a traveling salesman tour

$i :=$ any vertex of $G$ {the starting vertex}

$j :=$ subscript such that $c_{ij} = \min\{c_{ir} \mid r \in V - \{i\}\}$

$S := \{i, j\}$

$C := \{(i, j), (j, i)\}$

**while** $S \neq V$

    let $k$ be such that $d_S(k) = \min\{d_S(r) \mid r \in V - S\}$

    $S := S \cup \{k\}$

find an edge $(u, v) \in C$ so $c_{uk} + c_{kv} - c_{uv} = \min\{c_{xk} + c_{ky} - c_{xy} \mid (x, y) \in C\}$
add $(u, k)$ and $(k, v)$ to $C$, and remove $(u, v)$ from $C$

**16.** Using appropriate data structures, Algorithm 2 can be implemented to run in $O(n^2)$ time.

**17.** Suppose $z_{\mathrm{NI}}$ is the cost of a tour constructed by the nearest insertion heuristic and $z_{\mathrm{OPT}}$ is the cost of an optimal TSP tour. If the values $c_{ij}$ satisfy the triangle inequality, then $\frac{z_{\mathrm{NI}}}{z_{\mathrm{OPT}}} \leq 2$ holds for all TSP instances.

**18.** *Clarke and Wright savings heuristic:*    This construction method (Algorithm 3) builds up a tour by successively adding an edge $(i, j)$ having the largest *savings* $s_{ij}$, the benefit from directly connecting vertices $i$ and $j$ compared with joining each directly to a central vertex.

**Algorithm 3**:    **Clarke and Wright savings heuristic.**
input: undirected network $G = (V, E)$, costs $c_{ij}$
output: a traveling salesman tour
select any vertex (for example, 1) as the starting vertex
compute $s_{ij} = c_{1i} + c_{1j} - c_{ij}$ for distinct $i, j \in V - \{1\}$
order the savings $s_{i_1 j_1} \geq s_{i_2 j_2} \geq \cdots \geq s_{i_t j_t}$
$P := \emptyset$
$k := 0$
**while** $|P| < n - 2$
   $k := k + 1$
   **if** $P \cup \{(i_k, j_k)\}$ is a vertex-disjoint union of paths **then** add $(i_k, j_k)$ to $P$
connect the endpoints of $P$ to vertex 1, forming a tour

**19.** Using appropriate data structures, Algorithm 3 can be implemented to run in $O(n^2 \log n)$ time.

**20.** *Christofides' heuristic:*    This construction method (Algorithm 4) builds up a tour from a minimum spanning tree to which are added certain other small cost edges. It is assumed that the costs satisfy the triangle inequality.

**Algorithm 4**:    **Christofides' heuristic.**
input: undirected network $G$, costs $c_{ij}$
output: a traveling salesman tour
$T :=$ minimum spanning tree of $G$ with respect to the costs $c_{ij}$ (see §10.1)
let $S$ contain all odd-degree vertices in $T$
find a minimum cost perfect matching $M$ (§10.2) relative to vertices $S$ of $G$ and
   using the costs $c_{ij}$
obtain a closed trail $C$ by adding $M$ to the edges of $T$
remove all edges but two incident with vertices of degree greater than 2 by exploiting the triangle inequality, transforming $C$ into a tour

**21.** Using appropriate data structures, Algorithm 4 can be implemented to run in $O(n^3)$ time.

**22.** Suppose $z_C$ is the cost of a tour constructed by Christofides' heuristic and $z_{OPT}$ is the cost of an optimal TSP tour. If the $c_{ij}$ satisfy the triangle inequality, then $\frac{z_C}{z_{OPT}} \le \frac{3}{2}$ holds for all TSP instances.

**23.** The following table [JüReRi95] compares several of the most popular tour construction procedures on a set of 30 Euclidean TSPs from the literature with known optimal solutions. These problems range in size from 105 to 2,392 vertices. Surprisingly, the savings heuristic is the best tour construction heuristic of those tested. These results are consistent with those of other studies.

| heuristic | average percent above optimality |
|---|---|
| nearest neighbor | 24.2 |
| nearest insertion | 20.0 |
| Christofides | 19.5 |
| modified nearest neighbor | 18.6 |
| cheapest insertion | 16.8 |
| random insertion | 11.1 |
| farthest insertion | 9.9 |
| savings | 9.8 |
| modified savings | 9.6 |

**24.** The best known tour improvement heuristics for the TSP involve edge exchanges (Algorithm 5). Often the initial tour is chosen randomly from the set of all possible tours.

---

**Algorithm 5**: **General edge-exchange heuristic.**

input: undirected network $G$, initial tour

output: a traveling salesman tour

**repeat** improve the tour using an allowable edge exchange

**until** no additional improvement can be made

---

**25.** Specialized versions of Algorithm 5 typically use 2-opt exchanges, 3-opt exchanges, and more complicated *Lin-Kernighan* [JüReRi95] edge exchanges. Such exchange techniques have been used to generate excellent solutions to large-scale TSPs in a reasonable amount of time.

**26.** Edge-exchange procedures are typically more expensive computationally than tour construction procedures.

**27.** Tour improvement procedures typically require a "downhill move" (i.e., a strict reduction in cost) in order for edge exchanges to be made. As a result, they terminate with a local minimum solution.

**28.** Since the 2-opt exchange procedure is weaker than the 3-opt procedure, Algorithm 5 will generally terminate at an inferior local optimum using 2-opt exchanges instead of 3-opt exchanges. The Lin-Kernighan procedure will generally terminate with a better local optimum than will a 3-opt exchange procedure.

**29.** In practice, it often makes sense to apply a composite procedure. The strategy is to get a good initial solution rapidly (by tour construction), which is then improved by an edge-exchange procedure.

**30.** The following table [JüReRi95] compares several composite procedures on the same sample problems described in Fact 23. In each case, the initial tour is constructed using the nearest neighbor heuristic (Algorithm 1). The improvement procedures include 2-opt, 3-opt, two variants of Lin-Kernighan, and iterated Lin-Kernighan. Iterated Lin-Kernighan is the most computationally burdensome of the edge-exchange procedures, but it consistently obtains results that are within 1% of optimality.

| heuristic | average percent above optimality |
|---|---|
| 2-opt | 8.3 |
| 3-opt | 3.8 |
| Lin-Kernighan (variant 1) | 1.9 |
| Lin-Kernighan (variant 2) | 1.5 |
| Iterated Lin-Kernighan | 0.6 |

**31.** *Metaheuristics:*  Unlike Algorithm 5 (which permits only downhill moves), meta-heuristics [OsKe95] allow the possibility of nonimproving moves. For example, uphill moves can be accepted either randomly (simulated annealing) or based upon determinis-tic rules (threshold accepting). Memory can be incorporated in order to prevent revisiting local minima already evaluated and to encourage discovering new ones (tabu search, see §16.9).

Other metaheuristics such as evolutionary strategies, genetic algorithms, and neural net-works have also been applied to the TSP. To date, neural networks and tabu search have been less successful than the other approaches.

**32.** For a detailed history of the TSP, see the first chapter of [LaEtal85] as well as the very accessible exposition in [Co12].

**33.** The article [La10] provides a concise guide to the most effective exact and heuristic approaches for the TSP.

**34.** Software for the traveling salesman problem can be found in [LoPu07].

**35.** A library of sample problems, with their best known solutions, is available at

- http://comopt.ifi.uni-heidelberg.de/software/TSPLIB95

**36.** To date, the largest non-trivial instance solved to optimality has 85,900 cities (solved in 2006).

**37.** For exact algorithms, we refer interested readers to [ApEtal07].

**Examples:**

**1.** *Brute force enumeration:*  Suppose that a TSP solution is required for the complete graph $G$ on $n = 25$ cities. By Fact 4, there are $\frac{24!}{2} \approx 3.1 \times 10^{23}$ Hamilton tours in the graph $G$. Even with a supercomputer that is capable of finding and evaluating each such tour in one nanosecond ($10^{-9}$ seconds), it would take over 9.8 million years of uninterrupted computations to determine an optimal TSP tour. This example illustrates how quickly brute force enumeration of Hamilton tours becomes impractical.

**2.** Part (a) of the following figure shows the costs $c_{ij}$ for a five city TSP. An initial tour can be constructed using the nearest neighbor heuristic (Algorithm 1). Let the initial vertex be $i_0 = 1$, so $W = \{2, 3, 4, 5\}$. The closest vertex of $W$ to 1 is 2, with $c_{12} = 1$, so edge $(1, 2)$ is added to the current path. A closest vertex of $W = \{3, 4, 5\}$ to 2 is 5, so edge $(2, 5)$ is added to the path. Continuing in this way, edges $(5, 3)$ and edge $(3, 4)$ are added, giving the path $P = [1, 2, 5, 3, 4]$ and the tour $[1, 2, 5, 3, 4, 1]$ with total cost $1 + 3 + 2 + 3 + 5 = 14$. This tour is displayed in part (b).

$c_{ij}$

| $i$ \ $j$ | 2 | 3 | 4 | 5 |
|---|---|---|---|---|
| 1 | 1 | 2 | 5 | 4 |
| 2 |  | 3 | 4 | 3 |
| 3 |  |  | 3 | 2 |
| 4 |  |  |  | 3 |

(a)

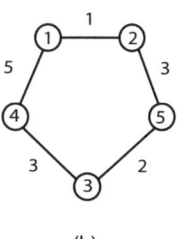

(b)

**3.** Suppose that the nearest insertion heuristic (Algorithm 2) is applied to the problem data in part (a) of the figure for Example 2, starting with the initial vertex $i = 1$. The nearest vertex to $i$ is $j = 2$, giving the initial cycle $C = \{(1,2), (2,1)\}$. The closest vertex to this cycle is $k = 3$, producing the new cycle $C = \{(1,2), (2,3), (3,1)\}$. Relative to $S = \{1,2,3\}$, $d_S(4) = 3$ and $d_S(5) = 2$, so vertex 5 will next be added to the cycle. Since $c_{15} + c_{52} - c_{12} = 6$, $c_{25} + c_{53} - c_{23} = 2$, and $c_{15} + c_{53} - c_{13} = 4$, vertex 5 is inserted between vertices 2 and 3 in the current cycle, giving $C = \{(1,2), (2,5), (5,3), (3,1)\}$. Finally, vertex 4 is added between vertices 2 and 5, producing the tour $C = \{(1,2), (2,4), (4,5), (5,3), (3,1)\}$ with total cost 12.

**4.** The savings heuristic (Algorithm 3) can alternatively be applied to the problem specified in part (a) of the figure for Example 2. The savings $s_{23} = c_{12} + c_{13} - c_{23} = 1 + 2 - 3 = 0$. Similarly, $s_{24} = 2$, $s_{25} = 2$, $s_{34} = 4$, $s_{35} = 4$, and $s_{45} = 6$. This produces the ordered list of edges $[(4,5), (3,4), (3,5), (2,4), (2,5), (2,3)]$. Considering edges in turn from this list gives the path $P = [3,4,5,2]$. Adding edges from the endpoints of $P$ to vertex 1 produces the tour $[1,3,4,5,2,1]$ with total cost 12.

**5.** Christofides' heuristic (Algorithm 4) is now applied to the problem given in part (a) of the figure for Example 2. A minimum spanning tree $T$ consists of the following edges: $(1,2)$, $(1,3)$, $(3,5)$, $(3,4)$; see part (a) of the following figure. Vertices $2,3,4,5$ have odd degree and $\{(2,4), (3,5)\}$ constitutes a minimum cost perfect matching on these vertices. Adding these edges to those of $T$ produces the multi-graph in part (b) of the following figure. Replacing edges $(4,3)$ and $(3,5)$ having aggregate cost 5 by the single edge $(4,5)$ of cost 3 produces the tour in part (c), having total cost 12.

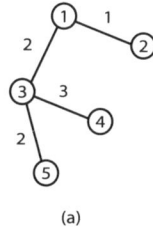
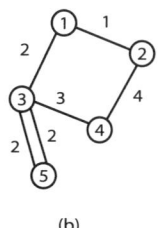
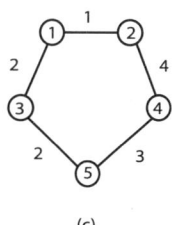

(a)             (b)             (c)

**6.** To illustrate edge exchanges, consider the tour of cost 14 in part (b) of the figure for Example 2. Removal of edges $(1,4)$ and $(3,5)$ disconnects the cycle into two disjoint paths. Join the endpoints of one path to the endpoints of the other with edges $(1,3)$ and $(4,5)$ to create a new tour $[1,2,5,4,3,1]$ of smaller cost 12. No further pairwise exchanges reduce the cost of this tour, so this tour is a 2-opt local minimum solution.

**7.** *Delivery routes:* A delivery truck must visit a set of customers in a city and then return to the central garage after completing the route. Determining an optimal (i.e., shortest time) delivery route can be modeled as a traveling salesman problem on a city street network. Here the vertices represent the customer locations and the cost $c_{ij}$ of edge $(i, j)$ is the driving time between locations $i$ and $j$.

**8.** *Printed circuit boards*:   One application of the TSP occurs in fabricating printed circuit boards. Holes at a number of fixed locations have to be drilled through the board. The objective is to minimize the total time needed to move the drilling head from position to position. Here the vertices $i$ correspond to the locations of the holes as well as the starting position of the drill. The cost $c_{ij}$ represents the time required to move the drilling head from $i$ and reposition it at $j$. A minimum cost traveling salesman tour gives an optimal way of sequencing the drilling of the holes.

**9.** *Order-picking*:   In a warehouse, a customer order requires a certain subset of the items stored there. A vehicle must be sent to pick up these items and then return to the central dispatch location. Here the vertices are the locations of the items as well as the central dispatch location. The costs are the times needed to move the vehicle from one location to the other. A minimum cost traveling salesman tour then gives an optimal order in which to retrieve items from the warehouse.

**10.** *Job sequencing*:   In a factory, materials must be processed by a series of operations on a machine. The set-up time between operations varies depending on the order in which the operations are scheduled. Determining an optimal ordering that minimizes the total set-up time can be formulated as a traveling salesman problem.

**11.**   Additional applications, with reference sources, are given in the following table.

| application | references |
|---|---|
| dating archaeological finds | [AhEtal95] |
| DNA mapping | [AhEtal95] |
| x-ray crystallography | [JüReRi95] |
| engine design | [AhEtal95] |
| robotics | [JüReRi95] |
| clustering | [AhEtal95], [LaEtal85] |
| cutting stock problems | [HoPaRi13] |
| aircraft route assignment | [HoPaRi13] |
| computer wiring | [EvMi92], [JüReRi95], [LaEtal85], |
| genome sequence mapping | [AgEtal00] |
| fuel-optimal imaging strategies | [BaMcBe00] |
| power cables | [SlMaKa97] |

## 10.7.2   VEHICLE ROUTING PROBLEM

Private firms and public organizations that distribute goods or provide services to customer locations rely on a fleet of vehicles. Given demands for service at numerous points in a transportation network, the vehicle routing problem requires determining which customers are to be serviced by each vehicle and the order in which customers on a route are to be visited.

**Definitions:**

Let $G = (V, E)$ be a complete graph with $V = \{1, 2, \ldots, n\}$ the set of vertices and $E$ the set of all edges joining pairs of distinct vertices (§8.1.3).

Vertex 1 is the **central depot**, whereas the other vertices represent **customer locations**. Customer location $i$ has a known **demand** $w_i$.

Each edge $(i, j) \in E$ has an associated **distance** or **cost** $c_{ij}$.

There are also available a number of **vehicles**, each having the same **capacity** $Q$.

A **route** is sequence of customers visited by a vehicle that starts and ends at the central depot.

The **vehicle routing problem** (**VRP**) requires partitioning the set of customers into a set of delivery routes such that

- the total distance traveled by all vehicles is minimum;
- the total demand generated by the customers assigned to each route is $\leq Q$.

In a **construction** heuristic for the VRP, subtours are joined as long as the resulting subtour does not violate the vehicle capacity.

An **improvement** heuristic employs successive edge exchanges that reduce the total distance without violating any vehicle capacity constraint.

A **two-phase** heuristic implements a cluster first/route second philosophy, in which customers are first partitioned into groups $G_k$ with $\sum\limits_{i \in G_k} w_i \leq Q$, after which a minimum distance sequencing of customers is found within each group.

**Facts:**

**1.** The TSP (§10.7.1) is a special case of the VRP in which there is a single vehicle with unlimited capacity.

**2.** The VRP is an NP-hard optimization problem (§17.5.2).

**3.** VRPs with more than 100 vertices are difficult to solve to optimality. The state-of-the-art exact method is branch-and-cut-and-price.

**4.** Most solution strategies for large VRPs are heuristic in nature, involving construction, improvement, and two-phase methods as well as metaheuristics.

**5.** In 1959 G. B. Dantzig and J. H. Ramser first formulated the general vehicle routing problem and developed a heuristic solution procedure. This solution technique was applied to a problem involving the delivery of gasoline to service stations.

**6.** The Clarke and Wright savings heuristic (§10.7.1) is a construction approach, originally proposed for the VRP. Algorithm 6 outlines this heuristic, which begins with each customer served by a different vehicle and successively combines routes in order of non-increasing savings $s_{ij} = c_{1i} + c_{1j} - c_{ij}$ to form a smaller set of feasible routes.

---

**Algorithm 6**: **Clarke and Wright savings heuristic.**

input: undirected network $G = (V, E)$, costs $c_{ij}$, capacity limit $Q$

output: a set of delivery routes

$R_i :=$ route consisting of edges $(1, i)$ and $(i, 1)$ for $i \in V - \{1\}$

compute $s_{ij} = c_{1i} + c_{1j} - c_{ij}$ for distinct $i, j \in V - \{1\}$

order the savings $s_{i_1 j_1} \geq s_{i_2 j_2} \geq \cdots \geq s_{i_t j_t}$

**for** $k := 1$ **to** $t$

    **if** $R_{i_k}$ and $R_{j_k}$ have combined demand at most $Q$ **then** merge $R_{i_k}$ and $R_{j_k}$

**7.** In two-phase methods, a minimum distance ordering of customers within each specified cluster of vertices can be found by solving a TSP (§10.7.1).

**8.** In the past two decades, metaheuristics such as simulated annealing, tabu search, genetic algorithms, record-to-record travel, and iterated neighborhood search have been applied quite successfully to VRPs.

**9.** [LaRoVi14] compared the performances of a selected set of successful metaheuristics. In terms of objective function values, the parallel algorithm that combined a heuristic local search improvement with integer programming [GrGoWa11] performed best on both CMT (50 to 200 customers) and GWKC (240 to 483 customers) benchmark instances.

**10.** Extensions to the basic VRP include modifications for asymmetric distances ($c_{ij}$ need not equal $c_{ji}$), heterogeneous fleets, constraints on the distance of every route, multiple depots, time windows for deliveries, split deliveries, arc routing, inventory routing, prize collecting routing, pick-up and delivery, and multi-objective routing.

**11.** A survey of 25 commercial software products for vehicle routing problems is available in [HaPa16]. This survey discusses interfaces with mapping systems, cloud-based solutions, computer platforms supported, extensions to the basic VRP that are incorporated, and significant installations of the product for industrial customers.

**12.** A list of free vehicle routing software can be found in [WaEtal16].

**13.** Datasets and best-known solutions for VRPs are available at

- http://neo.lcc.uma.es/vrp/

**Examples:**

**1.** The following table gives the data for a VRP involving six customers in which vehicle capacity is 820. The route $[1, 2, 4, 6, 1]$ is not feasible since the total demand of customers on this route is $w_2 + w_4 + w_6 = 486 + 326 + 24 = 836 > 820$. Since $\sum_{i=2}^{7} w_i = 1967$, at least $\lceil 1967/820 \rceil = 3$ routes will be needed to service all demands. The routes $[1, 5, 2, 6, 1]$, $[1, 3, 1]$, and $[1, 4, 7, 1]$ constitute a feasible set of routes with (respective) demands 800, 541, and 626. In this feasible solution, the total distance traveled by the first vehicle is $c_{15} + c_{52} + c_{26} + c_{61} = 131$, by the second is $c_{13} + c_{31} = 114$, and by the third is $c_{14} + c_{47} + c_{71} = 181$, for a total distance of 426.

| customer | 2 | 3 | 4 | 5 | 6 | 7 |
|----------|-----|-----|-----|-----|-----|-----|
| demand | 486 | 541 | 326 | 290 | 24 | 300 |

| $c_{ij}$ | 2 | 3 | 4 | 5 | 6 | 7 |
|----------|-----|-----|-----|-----|-----|-----|
| 1 | 19 | 57 | 51 | 49 | 4 | 92 |
| 2 | | 51 | 10 | 53 | 25 | 53 |
| 3 | | | 49 | 18 | 30 | 47 |
| 4 | | | | 50 | 11 | 38 |
| 5 | | | | | 68 | 9 |
| 6 | | | | | | 94 |

**2.** The Clarke and Wright heuristic is applied to the problem specified in the table of Example 1. For instance, $s_{35} = c_{13} + c_{15} - c_{35} = 57 + 49 - 18 = 88$. The largest savings occurs for $s_{57} = 132$ and $w_5 + w_7 = 590 \le 820$, so the initial routes $[1, 5, 1]$ and $[1, 7, 1]$ are merged to produce the feasible route $[1, 5, 7, 1]$ with distance 150. The next largest savings occur for $s_{47} = 105$, $s_{37} = 102$, and $s_{35} = 88$; however, neither customer 3 nor customer 4 can be inserted into the route $[1, 5, 7, 1]$ without exceeding the vehicle capacity. The next largest savings is $s_{24} = 60$, giving the new feasible route $[1, 2, 4, 1]$ with demand 812 and distance 80. Continuing in this fashion eventually finds $s_{36} = 31$ and constructs the route $[1, 3, 6, 1]$ with demand 565 and distance 91. This feasible set of three routes has total distance $150 + 80 + 91 = 321$, smaller than that for the feasible solution given in Example 1.

## 10.7.3  QUADRATIC ASSIGNMENT PROBLEM

The quadratic assignment problem deals with the relative location of facilities that interact with one another in some manner. The objective is to minimize the total cost of interactions between facilities, with distance often used as a surrogate for measures such as dollar cost, fatigue, or inconvenience.

### Definitions:

There are $n$ **facilities** to be assigned to $n$ predefined **locations**, where each location can accommodate any one facility.

The **fixed cost** $c_{ij}$ is the cost of assigning facility $i$ to location $j$.

The **flow** $f_{ij}$ is the level of interaction between facilities $i$ and $j$.

The **distance** $d_{ij}$ between locations $i$ and $j$ is the per unit cost of interaction between the two locations. Typically, it is measured using the rectilinear or Euclidean distance between the locations.

An **assignment** is a bijection $\rho$ from the set of facilities onto the set of locations.

The **linear assignment problem (LAP)** is the problem of finding an assignment $\rho$ that minimizes $\sum_i c_{i,\rho(i)}$.

The **quadratic assignment problem (QAP)** is the problem of finding an assignment $\rho$ that gives the minimum value $z_{\text{QAP}}$ of $\sum_i c_{i,\rho(i)} + \sum_{i,p} f_{ip} d_{\rho(i),\rho(p)}$.

In some partial assignment for the QAP, let $\mathcal{F}$ be the set of facilities (possibly empty) that have already been assigned and $\mathcal{L}$ be the set of locations having assigned facilities.

### Facts:

**1.** The following table gives a variety of situations that can be formulated using the QAP model.

| facilities | interaction |
|---|---|
| departments in a manufacturing plant | flow of materials |
| departments in an office building | flow of information, movement of people |
| departments in a hospital | movement of patients and medical staff |
| buildings on a campus | movement of students and staff |
| electronic component boards | connections |
| computer keyboard keys | movement of fingers |
| runners in a relay team | transfer time of a baton |
| airport terminals and gates | flow of passengers with connecting flights |
| turbine blades | slight variations in blade weights, that need to be balanced |

**2.** The interdependence of facilities due to interactions between them leads to the quadratic nature of the objective function in the QAP.

**3.** If the facilities are independent of each other (there are no interactions between them), the QAP reduces to the LAP, which can be solved in polynomial time (§10.2.2).

**4.** The TSP is a special case of the QAP (see Example 4).

**5.** The QAP is an NP-hard optimization problem (§17.5.2).

**6.** Exact solution of the QAP is limited to fairly small problems, generally of size 30 or smaller.

**7.** A lower bound on completions of a partial assignment for the QAP is given by

$$\min \sum_{i \in \mathcal{F}} c_{i,\rho(i)} + \sum_{i \in \mathcal{F}} \sum_{p \in \mathcal{F}} f_{ip} d_{\rho(i),\rho(p)}$$
$$+ \sum_{i \in \mathcal{F}} \sum_{p \notin \mathcal{F}} \left( f_{ip} d_{\rho(i),\rho(p)} + f_{pi} d_{\rho(p),\rho(i)} \right)$$
$$+ \sum_{i \notin \mathcal{F}} c_{i,\rho(i)} + \sum_{i \notin \mathcal{F}} \sum_{p \notin \mathcal{F}} f_{ip} d_{\rho(i),\rho(p)}.$$

The first two terms in this expression are the known fixed and interaction costs of assignments already made; the third term captures the interaction costs between assigned facilities and those yet to be assigned; and the last two terms represent the fixed and interaction costs of assignments not yet made.

**8.** A minimum value $z^*$ can be calculated for the last three terms in the lower bound expression of Fact 7 by solving a LAP such that each cost term is a lower bound on the incremental costs that would be incurred if facility $i \notin \mathcal{F}$ is assigned to location $j \notin \mathcal{L}$.

**9.** *Gilmore-Lawler lower bound:* This lower bound for $z_{QAP}$ is given by

$$\sum_{i \in \mathcal{F}} c_{i,\rho(i)} + \sum_{i \in \mathcal{F}} \sum_{p \in \mathcal{F}} f_{ip} d_{\rho(i),\rho(p)} + z^*,$$

where $z^*$ is found as in Fact 8.

**10.** The Gilmore-Lawler lower bound allows the QAP to be solved using a branch-and-bound (implicit enumeration) technique (§16.1.8).

**11.** The following table provides alternative tighter lower bounds for exact solution of the QAP. Here SDP refers to Semidefinite Programming and RLT refers to the Reformulation-Linearization Technique.

| lower bounds | references |
| --- | --- |
| linear programming based | [ReRaDr95] |
| quadratic programming based | [AnBr01] |
| SDP based | [Ro04] |
| lift-and-project SDP | [BuVa06] |
| level-1 RLT | [HaGr98], [ShAd99] |
| level-2 RLT | [AdEtal07] |
| level-3 RLT | [HaEtal12] |
| level-2 RLT interior point | [RaEtal02] |
| bundle method | [ReSo07] |

**12.** There are several ways to linearize the QAP by defining additional variables and constraints. However, none of the linearizations proposed so far has proved to be computationally effective.

**13.** Heuristic methods for solving the QAP can be classified as limited enumeration, construction methods, improvement methods, hybrid methods, and metaheuristics. A survey of exact and heuristic solution methods for the QAP is found in [KuHe87]; experimental comparisons of heuristic approaches appear in [BuSt78] and [Li81].

**14.** *Limited enumeration*: There are two distinct approaches for limiting the search for an optimal QAP solution using a branch-and-bound approach:

- The search can be curtailed by placing a limit on the computation time or the number of subproblems examined. Since an optimal solution is often found fairly early in a branch-and-bound procedure, especially if a good branching rule is available, this approach may find an optimal (or a near-optimal) solution while saving on the significant cost of proving optimality.

- The gap between the lower and upper bound is largest at higher levels of a branch-and-bound tree. Thus a relatively large gap can be used to fathom subproblems at higher levels, and this gap can be decreased gradually as the search reaches lower levels of the tree.

**15.** *Construction methods*: These heuristics start with an empty assignment and add assignments one at a time until a complete solution is obtained. The rule used to choose the next assignment can employ various viewpoints:

- a *local view*: select a facility having the maximum interaction with a facility already assigned; locate it to minimize the cost of interaction between facilities;

- a *global view*: take into account assignments already made as well as future assignments to be made.

**16.** Suppose that $k$ assignments have already been made. Using statistical properties, the expected value for the completion of the partial assignment is given by the following expression, whose terms are analogous to those in Fact 7:

$$EV = \sum_{i \in \mathcal{F}} c_{i,\rho(i)} + \sum_{i \in \mathcal{F}} \sum_{p \in \mathcal{F}} f_{ip} d_{\rho(i),\rho(p)}$$

$$+ \frac{\sum_{i \in \mathcal{F}} \sum_{p \notin \mathcal{F}} \sum_{j \notin \mathcal{L}} \left( f_{ip} d_{\rho(i),j} + f_{pi} d_{j,\rho(i)} \right)}{n - k}$$

$$+ \frac{\sum_{i \notin \mathcal{F}} \sum_{j \notin \mathcal{L}} c_{ij}}{n - k} + \frac{\sum_{i,p \notin \mathcal{F}} f_{ip} \left( \sum_{j,q \notin \mathcal{L}} d_{jq} \right)}{(n - k)(n - k - 1)}.$$

The low computational requirements of computing $EV$ make this a good choice to guide a construction heuristic [GrWh70].

**17.** *Improvement methods*: These heuristics start with a suboptimal solution (often randomly generated) and attempt to improve it through partial changes in the assignments. Several important issues arise in designing an improvement heuristic:

- *type of exchange*: The choices are pairwise, triple, or higher-order exchanges. The use of pairwise exchanges has been found to be the most effective in terms of solution quality and computational burden. Higher-order exchanges can be beneficial but are generally used in a limited way because of the significant increase in computation time.

- *scope of exchange*: The procedure can use a local approach that considers only the exchange of adjacent facilities, or a global approach that considers all possible exchanges. Current computing capabilities allow the use of a global approach, which has been found to be more effective.

- *choice of exchange*: The procedure can effect an exchange as soon as an improving move is found, or can evaluate all possible exchanges and choose the best. The first improvement option is more common.

- *order of evaluation*: The possible exchanges can be evaluated in a random or some predetermined order. This is relevant only if the "first improvement" approach is used, as is often the case. One simple but effective solution is to consider facilities in the fixed order of decreasing total interactions, so that exchanges with potentially large savings are evaluated first.

**18.** *Hybrid methods*: Unlike improvement procedures, which tend to get trapped at local minima, hybrid methods use multiple restarts from a set of diversified solutions. Hybrid procedures combine the power of improvement routines with diversified solutions obtained through construction methods.

**19.** *Metaheuristics*: In recent years, metaheuristics such as simulated annealing, tabu search, and genetic algorithms have been developed to help improvement procedures avoid the trap of local minima and have been applied with success to the QAP. Metaheuristics have been able to find the best known solutions for the commonly used benchmark problems in the literature and remain an active area of research on the QAP.

**20.** A recent survey of the QAP is provided in [Dr15]. The most successful metaheuristic so far for solving the QAP is the hybrid genetic algorithm.

**21.** Research papers, software, datasets, test instances, and solutions can be found in QAPLIB at

- http://anjos.mgi.polymtl.ca/qaplib/

**Examples:**

**1.** The following table gives the data $c_{ij}$, $f_{ij}$ for a QAP with four facilities and four locations.

| $c_{ij}$ | 1 | 2 | 3 | 4 |
|---|---|---|---|---|
| 1 | 1 | 3 | 2 | 1 |
| 2 | 2 | 1 | 4 | 3 |
| 3 | 4 | 2 | 4 | 4 |
| 4 | 3 | 1 | 2 | 2 |

| $f_{ij}$ | 1 | 2 | 3 | 4 |
|---|---|---|---|---|
| 1 | 0 | 1 | 3 | 4 |
| 2 | 1 | 0 | 2 | 1 |
| 3 | 3 | 2 | 0 | 3 |
| 4 | 4 | 1 | 3 | 0 |

The fixed locations $1, 2, 3, 4$ occur at equally-spaced points along a line, with unit distances between successive points, so that $d_{ij} = |i - j|$. For the assignment $\rho$ specified by $\rho(1) = 1, \rho(2) = 4, \rho(3) = 2, \rho(4) = 3$ the fixed cost is $c_{11} + c_{24} + c_{32} + c_{43} = 8$. Because the flows and distances are symmetric, the interaction cost is $2(f_{12}d_{14} + f_{13}d_{12} + f_{14}d_{13} + f_{23}d_{42} + f_{24}d_{43} + f_{34}d_{23}) = 44$. The total cost of assignment $\rho$ is then $8 + 44 = 52$.

**2.** The assignment in Example 1 can be improved by a pairwise exchange. Namely, instead of assigning facilities 1 and 2 (respectively) to locations 1 and 4, they are assigned to the interchanged locations 4 and 1. This gives $\sigma(1) = 4, \sigma(2) = 1, \sigma(3) = 2, \sigma(4) = 3$. Then the fixed cost incurred is $c_{14} + c_{21} + c_{32} + c_{43} = 7$ and the interaction cost is $2(f_{12}d_{41} + f_{13}d_{42} + f_{14}d_{43} + f_{23}d_{12} + f_{24}d_{13} + f_{34}d_{23}) = 40$. The total cost 47 is lower than that for the assignment $\rho$ in Example 1. In fact $\sigma$ is an optimal QAP assignment.

**3.** The QAP arises in designing the layout of a manufacturing facility. A number of products are to be made in this facility and different products require different operations in given sequences for completion. These operations are performed by $n$ departments: e.g., turning, milling, drilling, heat treatment, and assembly. Knowing the sequence of operations and the volume of each product to be produced, it is possible to calculate the flow $f_{ij}$ from any department $i$ to another department $j$. There are $n$ physical locations, with distance $d_{ij}$ between locations $i$ and $j$. The fixed cost of assigning department $i$ to location $j$ is $c_{ij}$, representing the cost of building foundations and installing support equipment (cables, pipes) for the machines. Then the objective is to assign departments to locations in order to minimize the sum of fixed and interaction costs.

**4.** The TSP (§10.7.1) can be formulated as a special case of the QAP, where the $n$ cities correspond to locations and a position number (facility) in the tour is to be associated with each city. Let $f_{12} = f_{23} = \cdots = f_{n1} = 1$ and $f_{ij} = 0$ otherwise. The distance $d_{ij}$ represents the cost of traveling between cities $i$ and $j$, and let all fixed costs $c_{ij}$ be zero. Then a solution to this QAP gives an optimal labeling of cities with their positions in an optimal TSP tour.

# 10.8 SMALL-WORLD NETWORKS

The concept of a "small-world network" is well known in both the scientific literature and in popular culture. It is often associated with the term "six degrees of separation", which indicates the relatively small number of steps in a chain linking random individuals. Small-world networks are neither random, nor perfectly ordered. Rather they display the important characteristics of small average distance between vertices, small diameter, and a high degree of clustering. Many social networks exhibit the small-world property, as well as a number of other naturally occurring networks (the structure of the Internet, ecological food webs, transportation networks, and biological networks). This section presents an introduction to the structure and dynamics of small-world networks.

## 10.8.1 BASIC CONCEPTS

This subsection provides important graph definitions and concepts useful in quantitatively characterizing network structures that arise in nature and society. It allows one to compare and contrast random, structured, and small-world networks.

**Definitions:**

Let $G = (V, E)$ be a simple undirected graph, with $n = |V|$ and $m = |E|$. The vertex subset $S \subseteq V$ forms an **induced subgraph** $G[S] = (S, E[S])$ if $E[S] = \{(i, j) \mid i, j \in S, (i, j) \in E\}$. The largest connected component of $G$ and the corresponding induced subgraph are indicated by $S \subseteq V$ and $G[S]$, respectively.

The **open neighborhood** of vertex $i \in V$ is denoted $N(i) = \{j \mid (i, j) \in E\}$.

The **degree** of vertex $i$ is $d_i = |N(i)|$ and $\bar{d} = \frac{1}{n} \sum_{i \in V} d_i$ is the **average degree** of $G$. A **$d$-regular graph** is a graph in which the degree of each vertex is equal to $d$.

The **triangle degree** of vertex $i$ is $t_i = |\{(j, k) \mid j, k \in N(i), (j, k) \in E\}|$: namely, the number of triangles containing vertex $i$.

The **distance** between vertices $i, j \in V$ is the length $l(i, j)$ of a shortest path between them. The **diameter** of $G$ is the largest distance between any two vertices in $G$: $\mathrm{diam}(G) = \max_{i,j \in V} l(i, j)$.

*Note*: If $G$ is not connected, we consider the diameter of the largest component $\mathrm{diam}(G[S])$ instead.

The **average distance** in $G$ is defined as $\bar{l} = \bar{l}(G) = \sum\limits_{i,j \in V, i<j} l(i,j) / \binom{n}{2}$.

The **clustering coefficient** $C$ of a graph is

$$C = \frac{3 \times (\text{number of triangles})}{\text{number of connected triples}},$$

where a **connected triple** consists of three vertices connected by two or more edges.

The **local clustering coefficient** $C_i$ of vertex $i$ is

$$C_i = \frac{\text{number of triangles connected to vertex } i}{\text{number of connected triples centered on vertex } i}$$

and the **Watts-Strogatz clustering coefficient** $C_{\mathrm{WS}}$ is the average of the local clustering coefficients:

$$C_{\mathrm{WS}} = \frac{1}{n} \sum_{i \in V} C_i.$$

**Facts:**

**1.** Many empirical studies of practical small-world networks imply that distances in the network are considered to be "small" when they are orders of magnitude smaller than the order $n$ of the network [Ba16]. Hence, network distances can be considered "small" when $\bar{l}(G) = O(\ln n)$ and $\mathrm{diam}(G) = O(\ln n)$.

**2.** The number of distinct triangles in $G$ is equal to $\frac{1}{3} \sum_{i \in V} t_i$.

**3.** The clustering coefficient $C$ is a measure of *transitivity*: namely, the probability that two random neighbors of a vertex are neighbors of each other [Ne10].

**4.** The factor 3 in the numerator of $C$ arises since each triangle produces three closed triples.

**5.** Equivalently, $C$ can be defined in terms of $t_i$ and $d_i$:

$$C = \frac{\sum_{i \in V} t_i}{\sum_{i \in V} \binom{d_i}{2}}.$$

**6.** The alternative clustering coefficient $C_{\mathrm{WS}}$ was proposed by Watts and Strogatz [WaSt98].

**7.** Since the local clustering coefficient can be alternatively represented as $C_i = t_i / \binom{d_i}{2}$, $C_{\mathrm{WS}}$ can also be equivalently expressed in terms of $t_i$ and $d_i$:

$$C_{\mathrm{WS}} = \frac{1}{n} \sum_{i \in V} \frac{t_i}{\binom{d_i}{2}}.$$

**8.** The clustering coefficients satisfy $0 \leq C \leq 1$ and $0 \leq C_{\mathrm{WS}} \leq 1$.

**9.** The clustering coefficients $C$ and $C_{\mathrm{WS}}$ are both good measures of network transitivity. Random networks display low transitivity, and thereby both $C$ and $C_{\mathrm{WS}}$ for such networks usually converge to zero as $n \to \infty$. In contrast, these clustering coefficients for a network with high transitivity usually fall within the interval $[0.1, 1]$.

**10.** Whereas random networks tend to have a small average distance and a small clustering coefficient, small-world networks have a small average distance and a large clustering coefficient.

**Examples:**

**1.** The following graph on five vertices has the degree sequence $d_1 = 2$, $d_2 = d_3 = d_4 = 3$, $d_5 = 1$ and the triangle degrees $t_1 = 1$, $t_2 = t_3 = 2$, $t_4 = 1$, $t_5 = 0$.

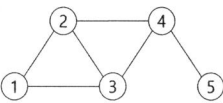

By Fact 5, the clustering coefficient $C$ is computed as

$$C = \frac{\sum_{i=1}^{5} t_i}{\sum_{i=1}^{5} \binom{d_i}{2}} = \frac{1+2+2+1+0}{1+3+3+3+0} = 0.6.$$

By Fact 7, the local clustering coefficients are $C_1 = 1$, $C_2 = C_3 = \frac{2}{3}$, $C_4 = \frac{1}{3}$, $C_5 = 0$ and so $C_{\mathrm{WS}}$ is computed as

$$C_{\mathrm{WS}} = \tfrac{1}{5} \sum_{i=1}^{5} C_i = \tfrac{1}{5}\left(1 + \tfrac{2}{3} + \tfrac{2}{3} + \tfrac{1}{3} + 0\right) = \tfrac{8}{15} \approx 0.53.$$

Notice that $C > C_{\mathrm{WS}}$ holds in this example.

**2.** In general, there is no order relation between these two definitions of the clustering coefficient ($C$ and $C_{\mathrm{WS}}$). For instance, suppose we delete vertex 5 from the graph in Example 1 and consider the graph $\widetilde{G} = G[V \backslash \{5\}]$. Then $C(\widetilde{G}) = 3/4 = 0.75$ and $C_{\mathrm{WS}}(\widetilde{G}) = 5/6 \approx 0.83$, yielding $C < C_{\mathrm{WS}}$ for the graph $\widetilde{G}$.

**3.** Although $C$ need not be equal to $C_{\mathrm{WS}}$, there are types of graphs where the equality $C = C_{\mathrm{WS}}$ holds. For example, every regular graph satisfies $C = C_{\mathrm{WS}}$. Namely, for any $d$-regular graph $G$ with $t_G$ triangles in total

$$C = C_{\mathrm{WS}} = \frac{3t_G}{n\binom{d}{2}}.$$

There are graphs that are not regular, but still satisfy $C = C_{\mathrm{WS}}$, as occurs in the following two graphs. In the first graph $C = C_{\mathrm{WS}} = 1/4$, while in the second graph $C = C_{\mathrm{WS}} = 1/12$. Thus, regularity is a sufficient but not a necessary condition for $C = C_{\mathrm{WS}}$.

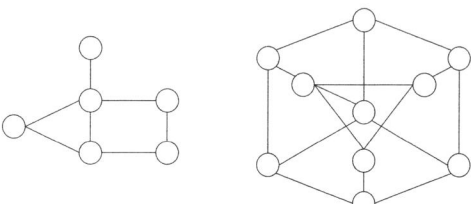

---

## 10.8.2   RANDOM GRAPH MODELS

The classical random graph model was mostly motivated by an observation that social connections are formed with omnipresent randomness. Although this random graph model is not sufficient to explain the properties of social and other real-world networks, it still plays a significant role in studying the small-world phenomena.

**Definitions:**

In the $G_{n,m}$ *random model*, $n$ vertices are connected by $m$ randomly placed edges.

In the $G_{n,p}$ *random model*, each pair of vertices has an edge connecting them with a fixed probability $p$.

Let $n_k$ be the number of vertices having degree $k$ in a simple graph $G = (V, E)$. The *degree distribution* of $G$ is given by $p_k = n_k/n$ for $k = 0, 1, \ldots, n - 1$.

A discrete random variable $\xi$ follows a *binomial distribution* $B(N, p)$ with parameters $N$ and $p$ if

$$Pr(\xi = k) = \binom{N}{k} p^k (1 - p)^{N-k} \text{ for } k = 0, 1, \ldots, N.$$

A discrete random variable $\eta$ follows a *Poisson distribution* $P(\lambda)$ with parameter $\lambda$ if

$$Pr(\eta = k) = \frac{\lambda^k}{k!} e^{-\lambda} \text{ for } k = 0, 1, 2, \ldots.$$

**Facts:**

**1.** Erdős and Rényi [ErRé60] proposed the $G_{n,m}$ random model, while Gilbert [Gi59] proposed the $G_{n,p}$ random model. Both $G_{n,m}$ and $G_{n,p}$ are commonly referred as to *Erdős-Rényi random graphs* or *uniform random graphs*.

**2.** The $G_{n,m}$ and $G_{n,p}$ models behave identically for $p = m/\binom{n}{2}$ under several conditions. While this subsection focuses on the more popular $G_{n,p}$ random graph model, most of the results obtained for $G_{n,p}$ carry over to $G_{n,m}$ [Ne03].

**3.** Since $p_k$ is the fraction of vertices with degree $k$ in $G$, the degree distribution of $G$ reflects the probability that a randomly selected vertex of a graph has a specified degree.

**4.** *Binomial degree distribution:* The degree distribution of $G_{n,p}$ is binomial with the parameters $n - 1$ and $p$ [Ne10]: namely,

$$p_k = \binom{n - 1}{k} p^k (1 - p)^{n-k-1} \text{ for } k = 0, 1, \ldots, n - 1.$$

Since $E[B(N, p)] = Np$ (see §7.3.3), the average degree $\bar{d}$ is well estimated by $(n - 1)p$.

**5.** *Poisson degree distribution:* For large sparse random graphs, the degree distribution of $G_{n,p}$ is well approximated by a Poisson distribution with mean $\bar{d}$. Specifically, if $n$ tends to infinity and $\bar{d} \ll n$, then it can be shown [Ba16] that for $k \ll n$

$$p_k \approx \frac{\bar{d}^k}{k!} e^{-\bar{d}} .$$

Hence $G_{n,p}$ graphs are also referred to as *Poisson random networks*. Although $\bar{d} \ll n$, $\bar{d}$ still can be sufficiently large and can tend to infinity (but at a much slower rate than $n$).

**6.** *Size of a random network:* While the number of edges of $G_{n,m}$ is constant and equal to $m$, the number of edges of $G_{n,p}$ is random but densely concentrated around its expectation $E[m(G_{n,p})] = p\binom{n}{2}$, due to the properties of the binomial distribution.

**7.** *Approximate equivalence of the models:* In the limit of large $n$, models $G_{n,m}$ and $G_{n,p}$ exhibit very similar properties when $m$ is close to $p\binom{n}{2}$. Particularly, if $p(1 - p)n^2 \to \infty$ and a certain property $Q$ holds for $G_{n,m}$ for all consecutive values of $m$ in the range $p\binom{n}{2} - \sqrt{p(1 - p)n} < m < p\binom{n}{2} + \sqrt{p(1 - p)n}$, then $Q$ holds for $G_{n,p}$ as well. The converse holds for a rich class of convex properties [Bo98].

**8.** *Order of the giant component*:  The giant connected component emerges in a uniform random graph after an average degree passes the critical point $\bar{d} = 1$. Poisson networks with $1 < \bar{d} < \ln n/n$ are said to be in a *supercritical regime*. They have a single giant component containing a fraction $u = n_S/n$ of all vertices, which can be approximated as

$$u \approx -\frac{W(-\bar{d}e^{-\bar{d}})}{\bar{d}},$$

where $W$ is the principal branch of *the Lambert W-function* [Ne10]. For random graphs with $\bar{d} > \ln n/n$ (*connected regime*) the giant connected component contains all vertices making $S = V$ and $u = 1$.

**9.** *Diameter*:  The following results hold for *connected* Erdős-Rényi (Poisson) networks.

- Simple calculations can yield a good approximation for the diameter of a uniform random graph [Ba16], [Ne10]:

$$\mathrm{diam}(G_{n,p}) = \Theta\left(\frac{\ln n}{\ln \bar{d}}\right).$$

- If $np \to \infty$ for an Erdős-Rényi graph $G_{n,p}$, then the following result holds asymptotically almost surely [ChLu04]:

$$\mathrm{diam}(G_{n,p}) = (1 + o(1))\frac{\ln n}{\ln np}.$$

- For a Poisson random graph $G_{n,p}$ with $\bar{d} = \lambda > 1$ (thus, $p \approx \lambda/n$), the equality

$$\mathrm{diam}(G_{n,p}) = \frac{\ln n}{\ln \lambda} + 2\frac{\ln n}{\ln \lambda_\star} + O_\mathrm{p}(1)$$

  is valid, where $\lambda_\star$ is a "dual"-parameter of $\lambda$ given by the formula $\lambda_\star = -W(-\lambda e^{-\lambda})$ and $O_\mathrm{p}(1)$ denotes a sequence of random variables $X_n$ *bounded in probability* [JaLuRu00]. This result is a stronger form of the result obtained by Fernholz and Ramachandran [FeRa07] for Poisson random graphs.

**10.** *Clustering coefficient*:  The clustering coefficient of a uniform random graph is $C = p$ or, equivalently, $C = O(1/n)$. For large sparse networks, $C$ converges to 0 as $n \to \infty$. However, Albert and Barabási [AlBa00] argued that clustering coefficients in real-world networks tend not to follow the results obtained for uniform random graphs.

**11.** Although uniform random graphs exhibit small distances (logarithmic in the order of the network), their clustering coefficient tends to zero as $n$ increases. The latter fact is the main flaw of this model in the context of small-world networks. This flaw is addressed by the models presented in the next section.

## 10.8.3   WATTS-STROGATZ MODEL

The Watts-Strogatz model [WaSt98] combines the small diameter and high clustering properties. This subsection summarizes the main characteristics of this model and its extensions.

**Definitions:**

A *lattice* (or **grid**) is a graph whose vertices correspond to integer coordinate points in an $n$-dimensional Euclidean space and whose edges connect every vertex to its $c$ closest vertices.

The *Watts-Strogatz model* (W-S model, $G_{c,p}^{\text{WS}}$) generates networks using the following steps:

1. Define the number of vertices $n$, the lattice parameter $c$, and the rewiring parameter $p \in [0, 1]$, where $n \gg c \gg \ln n \gg 1$ and $c$ must be even.

2. Place $n$ vertices in a "ring" and connect each vertex to its $c$ closest vertices. Note that the resulting graph is a one-dimensional lattice with periodic boundary conditions.

3. "Rewire" each edge $(i, j)$ with probability $p$. This means that we shift one of its ends $j$ to a new position randomly selected from the rest of the vertices in the ring (ensuring that no two vertices are joined by more than one edge and no vertex has a self-loop), whereas the other end $i$ remains at the same position.

The *Newman-Watts model* (N-W model, $G_{c,p}^{\text{NW}}$) was proposed by Newman and Watts [NeWa99] in order to analyze the W-S model theoretically. The N-W model is a modification of the W-S model, obtained by slightly revising the W-S network generation algorithm. Instead of rewiring, the generation algorithm for the N-W model adds extra edges to the underlying lattice. Namely, the algorithm generates a new edge with probability $p$ for each edge in the initial lattice. The endpoints of each newly generated edge are chosen uniformly at random.

The *Dorogovtsev-Mendes model* (D-M model, $G_{n,p}^{\text{DM}}$) was proposed by Dorogovtsev and Mendes [DoMe00] as an alternative small-world network model. The D-M model is built by adding an extra vertex as a "central point" and connecting it with each vertex in the underlying lattice with probability $p$. The edges emanating from the central vertex have weights $1/2$ as opposed to unit weights of the edges on the initial ring.

**Facts:**

**1.** The assumptions stated in the W-S model include sparsity ($n \gg c$), connectivity ($c \gg \ln n$), and a proper size ($\ln n \gg 1$) of the network.

**2.** Consider the W-S model. If $p \to 0$, then $\bar{l}(G_{c,p}^{\text{WS}}) \sim n/(2c) \gg 1$ and $C(G_{c,p}^{\text{WS}}) \approx C_{\text{WS}}(G_{c,p}^{\text{WS}}) \sim (3(c-2))/(4(c-1))$, which is approximately $3/4$ for large values of $c$. If $p \to 1$, then $\bar{l}(G_{c,p}^{\text{WS}}) \approx \bar{l}(G_{n,p}) \sim \ln n / \ln c$ and $C(G_{c,p}^{\text{WS}}) \approx C_{\text{WS}}(G_{c,p}^{\text{WS}}) \approx C(G_{n,p}) \sim c/n$. In particular, small average distance and high clustering are observed simultaneously (i.e., the network becomes "small-world") for small values of $p$, usually $0.001 < p < 0.01$.

**3.** The N-W model is equivalent to the W-S model if $p$ is small, which is the case when the W-S model becomes a small-world network.

**4.** In the N-W model, as well as in the W-S model, the average distance increases linearly with the network size $n$ for $p = 0$, whereas it increases logarithmically with $n$ for $p = 1$. Newman et al. [NeMoWa00] employed a mean-field approximation to show that the transition between these growth rates occurs in the region where the average number of added shortcuts $s = ncp/2$ is about one. More specifically, as $n \to \infty$, $s \gg 1$ and $c, p$ are fixed the average distance between vertices increases logarithmically with $n$, since

$$\bar{l}(G_{c,p}^{\text{NW}}) \approx \frac{\ln(ncp)}{c^2 p} = \Theta\left(\ln n\right).$$

Thus, the N-W model shows that the addition of only a small fraction of random shortcuts to a large underlying lattice can yield a small-world effect.

**5.** Newman [Ne10] estimated the clustering coefficient of the N-W model in the limit of large $n$:

$$C(G_{c,p}^{\text{NW}}) \approx \frac{3(c-2)}{4(c-1) + 8cp + 4cp^2}.$$

Therefore, $C(G_{c,p}^{\text{NW}}) \approx 3/(4(1+p)^2)$ for sufficiently large $c$. In other words, the N-W model with reasonably small values of $p$ exhibits high clustering.

**6.** *Asymptotic bounds on the diameter:* Bollobás and Chung [BoCh88] derived a strong asymptotic result for the diameter of a cycle plus a random matching, which can be considered as a rough approximation of the N-W model with parameters $c = 2$ and $p = 1/2$ (for such parameters the size of a random matching and an estimated number of added shortcuts are both equal to $n/2$). If $G$ is a cycle on $n$ vertices plus a random matching, then

$$\log_2 n - \theta < \text{diam}(G) \le \log_2 n + \log_2 \ln n + \theta$$

asymptotically almost surely (that is, with probability tending to 1 as $n \to \infty$), where $\theta$ is a small constant (at most 10).

**7.** Dorogovtsev and Mendes [DoMe00] obtained an asymptotic expression for $\bar{l}$ as $p \to 0$ and $n \to \infty$ in the D-M model:

$$\frac{\bar{l}(G_{n,p}^{\text{DM}})}{n} \sim \frac{2\rho - 3 + (\rho + 3)e^{-\rho}}{\rho^2},$$

where $\rho = np$ is fixed. Furthermore, consider any function $\rho(n) \to \infty$, which implies $\bar{l}(G_{n,p}^{\text{DM}}) \sim 2n/\rho$. Thus, if one chooses any $p(n) = \Theta(1/\ln n)$, then

$$\bar{l}(G_{n,p}^{\text{DM}}) = \Theta(\ln n).$$

Hence, with a proper choice of $p(n)$, the D-M model exhibits small-world distances as $n \to \infty$.

**Examples:**

**1.** The following figure illustrates a W-S network generated with $n = 16$ vertices, placed in a ring with $c = 2$. The four shaded vertices are selected for rewiring; one end of an edge incident to each such vertex is randomly linked to another vertex, giving the "middle" edges shown in the picture.

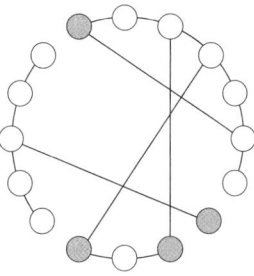

**2.** The following figure illustrates an N-W network generated with $n = 16$ vertices, placed in a ring with $c = 2$. Four new edges (shown highlighted) are randomly generated and added to this ring.

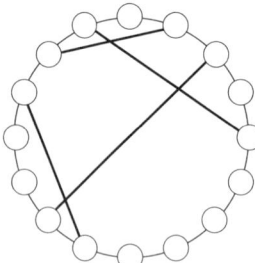

**3.** The following figure illustrates a D-M network generated with $n = 16$ vertices, placed in a ring with $c = 2$. An extra vertex is added to the network and connected to some other randomly selected vertices in the ring. The added edges are shown highlighted.

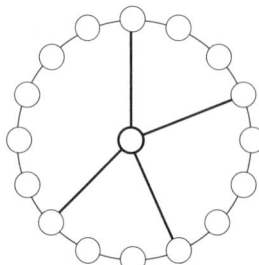

## 10.8.4   KLEINBERG'S GEOGRAPHICAL SMALL-WORLD MODEL

This subsection considers a natural generalization of the W-S small-world network model, referred to as the geographical small-world model, for which there exists a decentralized algorithm capable of finding short paths in logarithmic time with high probability.

**Definitions:**

In an $r$-dimensional lattice, $l_1$-distances are measured by $d(u, v) = \sum_{k=1}^{r} |u_k - v_k|$, where $u = (u_1, \ldots, u_r)$ and $v = (v_1, \ldots, v_r)$.

***Kleinberg's small-world model***, which is a modified version of the N-W model, is obtained by adding random directed edges to an $r$-dimensional lattice, based on $l_1$-distance. In the underlying lattice, each vertex has a directed link to any other vertex within lattice $l_1$-distance $\xi$, for a user-defined constant $\xi \geq 1$; such neighbors are called *local contacts*. Also $q$ long-range connections $(u, v)$ for each vertex $u$ are generated with probability proportional to $d(u, v)^{-\alpha}$ for each vertex $v$; here $\alpha$ is a user-defined parameter.

***Local search*** (or, equivalently, ***decentralized search***) $\mathcal{L}$ in Kleinberg's model is designed to imitate Milgram's small-world experiment (see §10.8.7). Suppose that a starting vertex $i$ is given a message that it must forward to a target vertex $j$. At each iteration, a current vertex should choose which of its neighbors to send the message to. Meanwhile, it uses only local information, i.e., it knows only geographical distances between its neighbors and the target vertex $j$. In Kleinberg's decentralized algorithm every current vertex connects to a neighbor that is geographically closest to vertex $j$. These consecutive choices determine a path from $i$ to $j$ via local search.

The **expected delivery time** $T_A$ of a search algorithm $A$ is the expected number of steps taken by the algorithm to deliver the message over a generated network.

A particular version of Kleinberg's model is based upon probabilities determined by rank rather than by physical distances. The **rank** of a vertex $v$ with respect to a vertex $u$ is defined as the number of other vertices that are geographically closer to $u$ than to $v$: i.e.,

$$\text{rank}_u(v) = |\{w \mid d(w, u) < d(v, u)\}|.$$

**Facts:**

**1.** Kleinberg's geographical small-world model [Kl00] was originally constructed on a two-dimensional lattice ($r = 2$). If no specific information about $r$ is given, this original model with $r = 2$ will be assumed and denoted $G^K_{\alpha,q,\xi}$.

**2.** If $\alpha$ is large, then long-range edges have a small chance of being created, whereas if $\alpha$ is close to 0, then long-range edges may appear in the network with a reasonably high probability. When $\alpha = 0$, Kleinberg's model generates additional edges according to a uniform distribution, which is equivalent to the directed N-W model on the two-dimensional lattice (we treat every undirected edge $(u, v)$ in the underlying lattice of the N-W model as two directed edges $(u, v)$ and $(v, u)$).

**3.** For every graph $G$ and algorithm $A$, the inequality $T_A(G) \geq \bar{l}(G)$ holds.

**4.** In large Kleinberg networks, local search can be carried out most efficiently when $\alpha = 2$, as supported by the following theoretical results [Kl00]:

- For Kleinberg's model with $0 \leq \alpha < 2$, arbitrary $q \geq 1$, $\xi \geq 1$, and any decentralized algorithm $A$

$$T_A(G^K_{\alpha,q,\xi}) = \Omega(n^{(2-\alpha)/3}).$$

  As a consequence, despite the fact that there exist paths between every pair of vertices whose lengths are $O(\ln n)$ for the two-dimensional N-W model (the case $\alpha = 0$), there is no decentralized algorithm capable of finding those chains in logarithmic time.

- For Kleinberg's model with $\alpha = 2$, $q = 1$, $\xi = 1$, and the decentralized algorithm $\mathcal{L}$ (Kleinberg's local search)

$$T_{\mathcal{L}}(G^K_{\alpha,q,\xi}) = O(\ln^2 n).$$

- For Kleinberg's model with $\alpha > 2$, arbitrary $q \geq 1$, $\xi \geq 1$, and any decentralized algorithm $A$

$$T_A(G^K_{\alpha,q,\xi}) = \Omega(n^{(\alpha-2)/(\alpha-1)}).$$

In computational experiments on Kleinberg networks with several hundred million vertices, local search performed most efficiently for values of $\alpha$ in the range $[1.5, 2]$. Moreover, as the network size increases, the most "efficient" value for $\alpha$ approaches 2 [Kl00].

**5.** *Rank-based friendship* is a concept based on the assumption that a directed edge $(u, v)$ is randomly generated with probability $p_{uv}$ proportional to $\text{rank}_u(v)^{-\beta}$, where $\beta$ is a predetermined parameter. If vertices are uniformly distributed in a Euclidean two-dimensional space according to the $l_1$-norm, then $\text{rank}_u(v)$ is approximately $2d(u, v)^2$. Thus, a rank-based friendship network with $\beta = 1$ is approximately the same as Kleinberg's network with $\alpha = 2$, since $p_{uv} \propto d(u, v)^{-2}$ for both models.

**6.** *Blog network*: Liben-Nowell et al. [LiEtal05] employed the rank-based friendship principle to analyze roughly 500,000 users of the blogging site *LiveJournal*. In their case, the rank of a target user $v$ with respect to a primary user $u$ indicates how far $v$

lives from $u$ in comparison with other users connected with $u$ via *LiveJournal*. Liben-Nowell et al. investigated the following question: if we consider all vertex pairs $(u, v)$ with $\text{rank}_u(v) = x$ for some fixed integer $x \geq 1$, what fraction $f(x)$ of these pairs are actual friends? They computationally found that $f(x)$ follows a power law (see §10.8.5) with exponent in the range $(-1.15, -1.2)$. This experimental finding supports the statement (see Facts 4–5) that $\beta = 1$ is an adequate choice for the rank-based friendship model.

**7.** *Facebook network:* Backstrom et al. [BaSuMa10] also considered rank-based friendship. They analyzed the Facebook network at that time and estimated the exponent $\beta$ to be very close to 1. In their case, the bulk of the distribution was approximated by $x^{-0.95}$. These results are consistent with those of Liben-Nowell et al. and further support the formula $p_{uv} \propto \text{rank}_u(v)^{-1}$ as the best choice for the rank-based friendship model.

**8.** Regarding Kleinberg's model based upon an $r$-dimensional lattice with $r > 2$, Easley and Kleinberg [EaKl10] pointed out that local search is efficient for networks built by adding long-range contacts to lattices in $r > 2$ dimensions, when the exponent $\alpha$ is equal to $r$.

**Example:**

**1.** The following figure illustrates an undirected version of Kleinberg's small-world model generated with $n = 16$ vertices, placed in a ring with $\xi = 1$. (Note that any instance of the directed Kleinberg model can be easily transformed into an instance of the undirected Kleinberg model by drawing an edge $(u, v)$ if either the edge $(u, v)$ or $(v, u)$ exists.) New edges were randomly generated with probability $p_{uv} \propto d(u, v)^{-2}$ for each vertex pair $(u, v)$. As a result, four new edges (shown highlighted) were added to this ring. The corresponding distances for the four pairs are 2, 2, 2, 3, which are shorter than the ones of the N-W model instance described in §10.8.3, Example 2 (which are 3, 4, 5, 8).

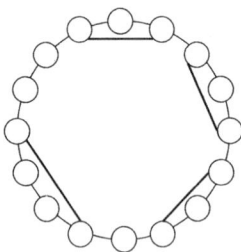

---

## 10.8.5  POWER-LAW RANDOM NETWORKS AND THEIR SMALL-WORLD PROPERTIES

In contrast to the Erdős-Rényi model, most real-world networks have degree distributions substantially different from the Poisson distribution that uniform graphs exhibit for large $n$. Several empirical studies suggest that many real-world networks follow a power-law degree distribution [AiChLu02], [BaAl99], [Ne03]. These observations motivated work on developing more realistic random network models that mitigate the shortcomings of the Erdős-Rényi model. This subsection reviews the facts associated with random network models having general degree distributions, as well as the important special cases of power-law random networks.

**Definitions:**

Given the number of vertices $n$ and degree sequence $\{d_1, d_2, \ldots, d_n\}$, the **configuration model** generates networks using the following steps:

1. For each vertex $i \in V$, attach $d_i$ "half-edges" (also called "stubs").
2. Randomly connect the ends of the half-edges together, one pair at a time.

The **generalized random graph model** generates graphs in which vertices $i, j \in V$ are connected with edge probability

$$p_{ij} = \frac{w_i w_j}{2m},$$

where $\{w_1, w_2, \ldots, w_n\}$ is a given sequence of expected degrees chosen from a distribution of interest.

The **power-law degree distribution** is defined by

$$p_k = Ck^{-\beta} \text{ for } k = k_{\min}, \ldots, k_{\max},$$

where $p_k$ is the fraction of vertices with degree $k$, $C$ is a normalization factor, $\beta$ is the decay exponent, and $k_{\min}$, $k_{\max}$ are (respective) lower and upper limits on the possible values of $k$.

The **pure power-law degree distribution** is given by

$$p_k = Ck^{-\beta} \text{ for } k \geq 1,$$

where $C = 1/\zeta(\beta)$ (the Riemann zeta function).

The **power-law degree distribution with exponential cutoff** is given by

$$p_k = Ck^{-\beta}e^{-k/\kappa} \text{ for } k \geq 1,$$

where $C$ is a normalization constant and $\kappa$ is an exponential cutoff parameter.

*Remark:* The three defined versions of power laws are discrete distributions and should not be confused with continuous power-law distributions.

*Notation:* $G_\beta^{\text{gen}}$ denotes a graph produced by the generalized random graph model with an expected degree sequence $\{w_i\}$ sampled from the pure power-law distribution having parameter $\beta$. $G_\beta^{\text{pl}}$ denotes a theoretical power-law graph with parameter $\beta$, that is, a graph with a power-law degree distribution with an exponent $\beta$ that is assumed to exist; it is used only in theoretical derivations. $G_\beta^{\text{conf}}$ denotes a graph produced by the configuration model with a degree sequence that follows the pure power-law distribution with parameter $\beta$. $G_{\beta,\kappa}^{\text{conf}}$ denotes a graph generated by the configuration model with an expected degree sequence drawn from the power-law distribution with exponential cutoff and parameters $\beta, \kappa$.

**Facts:**

**1.** The configuration model is one of the most popular generalized random network models. This process generates every possible topology having the given parameters with equal probability [Ne10].

**2.** Graphs generated by the configuration model can contain loops and multi-edges. The common way to transform such graphs into simple ones is to delete any loops and multi-edges. As $n$ becomes large, the density of loops and multi-edges tends to 0 for the configuration model, and the degree distribution of the resulting simple graphs is close to the given one. Since we focus on the behavior of graphs for large $n$, such self-loops and multi-edges will be disregarded.

**3.** The paramount property of the configuration model is that the probability $p_{ij}$ of edge $(i, j)$ appearing is given by

$$p_{ij} = \frac{d_i d_j}{2m} \text{ as } n \to \infty.$$

**4.** The property given in Fact 3 makes the configuration model similar to the generalized random graph model, which was proposed by Chung and Lu [ChLu04].

**5.** Despite the fact that the final degree sequence of the generalized random graph model is not exactly equal to the desired degree sequence in general, this model has notable calculation advantages. As in the case of the configuration model, we ignore self-loops and multi-edges produced by the generalized random graph model.

**6.** For the power-law degree distribution with exponential cutoff, $\kappa$ is typically a fixed large positive constant, but it can also be represented as a function of $n$.

**7.** The normalization constant $C$ for the power-law degree distribution with exponential cutoff turns out to be $[\text{Li}_\beta(e^{-1/\kappa})]^{-1}$, where $\text{Li}_n(x)$ is the $n$th polylogarithm of $x$.

**8.** *Small distances in $G_\beta^{\text{gen}}$*:  Chung and Lu [ChLu04] showed that distances in power-law graphs generated via the generalized random graph model are small (or even "ultra-small").

- For a power-law random graph $G_\beta^{\text{gen}}$ with exponent $\beta > 3$ and average degree $\bar{d} > 1$, both the average distance and diameter are $\Theta(\ln n)$ almost surely. More precisely, when $n$ becomes very large

$$\bar{l}(G_\beta^{\text{gen}}) = (1 + o(1))(\ln n / \ln \tilde{d}),$$
$$\text{diam}(G_\beta^{\text{gen}}) = \Theta(\ln n),$$

  hold almost surely, where $\tilde{d} = \sum_{i \in V} w_i^2 / \sum_{i \in V} w_i$ denotes the second-order average degree. These equalities are corroborated by the approximation results obtained by Cohen and Havlin [CoHa03]: $\bar{l}(G_\beta^{\text{pl}}) \propto \ln n$ for $\beta > 3$ and $n \gg 1$.

- For $\beta = 3$ a power-law random graph $G_\beta^{\text{gen}}$ has

$$\text{diam}(G_\beta^{\text{gen}}) = \Theta(\ln n)$$

  almost surely and

$$\bar{l}(G_\beta^{\text{gen}}) = \Theta(\ln n / \ln \ln n)$$

  as $n \to \infty$. The respective approximation results by Cohen and Havlin support the latter formula: $\bar{l}(G_\beta^{\text{pl}}) \approx \ln n / \ln \ln n$ for $\beta = 3$ assuming $\ln \ln n \gg 1$.

- Consider a power-law random graph $G_\beta^{\text{gen}}$ with exponent $2 < \beta < 3$, average degree $\bar{d} > 1$, and maximum degree $\Delta(G_\beta^{\text{gen}})$ satisfying $\ln \Delta \gg \ln n / \ln \ln n$. Then almost surely

$$\text{diam}(G_\beta^{\text{gen}}) = \Theta(\ln n)$$

  and

$$\bar{l}(G_\beta^{\text{gen}}) \leq (2 + o(1))[\ln \ln n / \ln(1/(\beta - 2))]$$

  as $n \to \infty$.

**9.** *Small distances in $G_\beta^{\text{conf}}$*:  Newman et al. [NeStWa01] derived approximation results for the pure power-law distribution realized via the configuration model as $n \to \infty$. Assuming that $G_\beta^{\text{conf}}$ is connected and $\beta > 3$

$$\bar{l}(G_\beta^{\text{conf}}) \approx \frac{\ln n + \ln[\zeta(\beta)/\zeta(\beta - 1)]}{\ln[\zeta(\beta - 2)/\zeta(\beta - 1) - 1] + 1}.$$

More precisely, for sufficiently large $n$, the average distance in pure power-law graphs grows logarithmically: i.e., $\bar{l}(G_\beta^{\text{conf}}) \approx \Theta(\ln n)$.

**10.** *Small distances in $G_{\beta,\kappa}^{\text{conf}}$*: Dorogovtsev et al. [DoMeSa03] obtained the following approximation results for power-law graphs with exponential cutoff generated by the configuration model.

- For a graph $G_{\beta,\kappa}^{\text{conf}}$ with exponent $2 < \beta < 3$ and large but finite cutoff parameter $\kappa$, the average distance is $\Theta(\ln n)$. More specifically, for large $n$

$$\bar{l}(G_{\beta,\kappa}^{\text{conf}}) \sim \frac{e^{1/\kappa}}{3 - \beta} \ln n.$$

- For a graph $G_{\beta,\kappa}^{\text{conf}}$ with exponent $2 < \beta < 3$ and cutoff $\kappa \propto n^\epsilon$, the average distance is ultrasmall: i.e., $\bar{l}(G_{\beta,\kappa}^{\text{conf}}) \sim \Theta(\ln \ln n)$. Specifically, as $n \to \infty$

$$\bar{l}(G_{\beta,\kappa}^{\text{conf}}) \sim \frac{\ln \ln n}{\ln(1/(\beta - 2))},$$

  which matches the results of Chung and Lu for $G_\beta^{\text{gen}}$ with $2 < \beta < 3$.

- Graphs $G_{\beta,\kappa}^{\text{conf}}$ with exponent $\beta = 3$ and a cutoff parameter that is polynomial in $n$ (i.e., $\kappa \propto n^\epsilon$) have average distance

$$\bar{l}(G_{\beta,\kappa}^{\text{conf}}) \sim (1 + \epsilon)\frac{\ln n}{\ln \ln n}$$

  for large $n$. This coincides with the results of Chung and Lu for $G_\beta^{\text{gen}}$ with $\beta = 3$.

**11.** *Clustering in power-law graphs*: For any graph generated by the configuration model using a given degree distribution, the clustering coefficient can be directly estimated as follows [Ne10]:

$$C = \frac{1}{n}\frac{[\bar{d^2} - \bar{d}]^2}{(\bar{d})^3}.$$

Here $\bar{d}$ and $\bar{d^2}$ are the first (average degree) and second moments of the degree distribution. If both of these moments are bounded, then the clustering coefficient converges to zero as $n \to \infty$. This is not the case for power-law graphs and even for graphs with a degree distribution having a power-law tail. Usually, their second moment diverges while the average degree is fixed or bounded.

**Example:**

**1.** To illustrate Fact 11, let $G_\beta^{\text{pl}}$ be a graph having a power-law distribution with parameter $\beta$ for $k \geq k_{\min}$ and an arbitrary distribution for $k < k_{\min}$. If one cuts it off (in a simple graph) at $k = n$, this results in the following approximation of the second moment for $2 < \beta < 3$:

$$\bar{d^2}(G_\beta^{\text{pl}}) \sim n^{3-\beta}.$$

Therefore, as $n$ becomes very large, the clustering coefficient grows as

$$C(G_\beta^{\text{pl}}) \sim 1/n^{\beta-2},$$

which still converges to zero, but at a much slower rate (especially for values of $\beta$ close to 2) than the clustering coefficient of a uniform random network.

---

## 10.8.6    EVOLVING RANDOM NETWORKS WITH SMALL-WORLD PROPERTIES

Models of network growth (formation) are an important class of random network models whose primary goal is to mimic growth processes that may take place in modern dynamically changing real-world networks, such as the World Wide Web, the Internet, and social networks. Such growth models include Price's model [Pr76], uniform attachment [BaAlJe99], preferential attachment [BaAl99] (the well-known Barabási-Albert model) and its generalizations [Ne03], a variety of hybrid models of uniform and preferential attachment [DoMe01], [JaRo07], [Vá03], and several nonlinear preferential attachment models [JeNéBa03], [KrReLe00]. In general, it is difficult to theoretically calculate diameters and shortest path lengths for growing network models; however, some of these models have been shown to generate networks that exhibit small-world properties. These models are reviewed in this subsection.

### Definitions:

***Price's model*** generates directed graphs. A new vertex is added at each time period and its out-degree is defined according to a fixed distribution with mean $\mu$. That newly appearing vertex is connected to already existing vertices chosen at random with probability proportional to their current degree plus a positive parameter $\mu_0$.

In the ***Barabási-Albert model***, a new vertex is added to the growing network at each time period; it is connected to $\mu$ already existing vertices, selected with probability proportional to their current degree. The initial state of the Barabási-Albert model is a clique of cardinality $\mu$ at time 0. Note that the Barabási-Albert model is similar to Price's model, but it generates undirected graphs.

The ***Klemm-Eguíluz model*** has as its initial state a clique of cardinality $\mu$ at time 0. Each vertex of the network can be either in the *active* state or in the *inactive* state. There are always $\mu$ active vertices in the network at the end of every time period. At each step, a new vertex joins the growing network and becomes active. This new vertex is connected to the existing network with $\mu$ edges. Each of these edges is connected either to a randomly chosen vertex (with probability $p$) or to one of the $\mu$ active vertices (with probability $1 - p$). In the former case, random vertices are chosen with probability proportional to their vertex degree at time $t$. At the end of each iteration, one of the active vertices is deactivated. Vertex $i$ is selected for deactivation with probability inversely proportional to its degree at time $t$: i.e., $p_i \propto d_i(t)^{-1}$.

The ***Jackson-Rogers model*** adds, at each step, a new vertex to the existing network. This new vertex identifies $\mu_1$ "parent vertices" uniformly and forms a directed (undirected) link with each of them with probability $p_1$. Furthermore, $\mu_2$ vertices are uniformly selected from the out-neighborhoods (neighborhoods) of the "parent vertices" and directed (undirected) links from the new vertex to the chosen $\mu_2$ vertices are formed with probability $p_2$ each. It is worth noting that $\mu_2$ vertices are selected independently of whether a certain "parent vertex" has been linked to the new vertex. The initial state of the Jackson-Rogers model is a clique on $(\mu_1 + \mu_2 + 1)$ vertices. Henceforth, we will concentrate on the undirected version of the Jackson-Rogers model.

*Notation:*  $G_{\mu}^{\mathrm{BA}}(t)$ denotes a graph generated at time $t$ by the Barabási-Albert model with parameter $\mu$. $G_{\mu,p}^{\mathrm{KE}}(t)$ denotes a graph generated at time $t$ by the Klemm-Eguíluz model with parameters $\mu$ and $p$. $G_{\mu,\rho}^{\mathrm{JR}}(t)$ denotes an undirected graph generated at time $t$ by the undirected version of the Jackson-Rogers model, where $\mu = p_1\mu_1 + p_2\mu_2$ and $\rho = p_1\mu_1/p_2\mu_2$.

**Facts:**

**1.** Price's model [Pr76] is a network formation model aiming at simulating citation networks, which are naturally directed. Price called the respective formation mechanism *cumulative advantage*. He was one of the first to discover a power law governing the out-degree distribution. Indeed, Price's model with parameters $\mu_0$ and $\mu$ exhibits a power-law out-degree distribution with the exponent $\beta = 2 + \mu_0/\mu$ in the limit of large $n$.

**2.** The Barabási-Albert model [BaAl99], also known as the preferential attachment model, is related to Price's model. The main difference is that the Barabási-Albert model considers undirected edges, while the Price model considers directed edges.

**3.** The Klemm-Eguíluz model [KlEg02b] represents a hybrid between the highly clustered dynamic model [KlEg02a] and the Barabási-Albert model. The cases $p = 0$ and $p = 1$ correspond to the highly clustered model and the Barabási-Albert model, respectively. The Klemm-Eguíluz model is often considered for the values of $0 < p \ll 1$.

**4.** The Jackson-Rogers model [JaRo07] is based on the idea of "meeting-based formation", and it represents a crossover between uniform attachment ($p_1 = 1$, $p_2 = 0$) and preferential attachment ($p_1 = 0$, $p_2 = 1$, $\mu_1 = \mu_2 \geq 1$) mechanisms. In this model $\mu$ is the expected number of edges formed at each iteration and $\rho$ is the ratio of the numbers of edges formed using these two attachment mechanisms.

**5.** *Small distances in* $G_\mu^{\mathrm{BA}}$: Bollobás and Riordan [BoRi04] discovered the most significant result about the diameter of the preferential attachment model for a slight modification of the Barabási-Albert model. Here the network formation process is allowed to add multiple edges between pairs of vertices and self-loops for single vertices. This model is a very close approximation to the preferential attachment model.

- If a graph $G_\mu^{\mathrm{BA}}$ is formed using preferential attachment with $\mu = 1$, then the diameter of its largest component $S$ is

$$\mathrm{diam}(G_\mu^{\mathrm{BA}}[S]) = \Theta(\ln n)$$

  asymptotically almost surely, as $n \to \infty$. More precisely,

$$Pr[(1/\gamma - \epsilon)\ln n \leq \mathrm{diam}(G_\mu^{\mathrm{BA}}[S]) \leq (1/\gamma + \epsilon)\ln n] \to 1$$

  as $n \to \infty$, where $\epsilon$ is some positive real number and $\gamma$ is the solution of $\gamma e^{1+\gamma} = 1$.

- If a graph $G_\mu^{\mathrm{BA}}$ is formed via preferential attachment with $\mu \geq 2$, then asymptotically almost surely $G_\mu^{\mathrm{BA}}$ is connected and has diameter smaller than $\ln n$:

$$\mathrm{diam}(G_\mu^{\mathrm{BA}}) = \Theta\left(\frac{\ln n}{\ln \ln n}\right)$$

  as $n \to \infty$. Specifically,

$$Pr[(1 - \epsilon)\ln n/\ln \ln n \leq \mathrm{diam}(G_\mu^{\mathrm{BA}}) \leq (1 + \epsilon)\ln n/\ln \ln n] \to 1$$

  as $n \to \infty$, where $\epsilon$ is a small positive real number.

These results show that the Barabási-Albert model generates graphs with small diameter.

**6.** *Clustering in* $G_\mu^{\mathrm{BA}}$: Klemm and Eguíluz [KlEg02b] derived an approximate expression for the average local clustering of the preferential attachment model. Namely, based on estimations for the Barabási-Albert model, they showed

$$C_{\mathrm{WS}}(G_\mu^{\mathrm{BA}}) = \Theta((\ln^2 n)/n)$$

as $n \to \infty$. More rigorously, $C_{\text{WS}}(G_{\mu}^{\text{BA}}) \sim \mu(\ln^2 n)/(8n)$, which still converges to zero for large $n$ but is larger than the average local clustering of a random network $C_{\text{WS}}(G_{n,p}) \sim \bar{d}/n$ by a factor of $O(\ln^2 n)$. Thus, the Barabási-Albert model exhibits some transitivity, which is higher than that of random networks, but might still not be sufficient to model real-life small-world networks.

**7.** *Small distances in $G_{\mu,p}^{\text{KE}}$:*  Klemm and Eguíluz only managed to empirically estimate average distances in their model for different values of $p \in [0,1]$. They simulated 100 independent realizations of graphs having order $n = 10^4$ and average degree $\bar{d} = 20$ (with $\mu = 10$) for each value of $p$.

- Computational experiments suggested that the average path length in the simulated graphs $G_{\mu,p}^{\text{KE}}$ with $p = 0$ grows linearly, that is

$$\bar{l}(G_{10,0}^{\text{KE}}) \propto n,$$

  which is similar to the behavior of regular one-dimensional lattices.

- Klemm-Eguíluz graphs generated for intermediate values of $p$ are expected to produce average distances that grow logarithmically, since $\bar{l}(G_{\mu,1}^{\text{KE}}) \propto \ln n / \ln \ln n$ for $\mu \geq 2$. Indeed, computational experiments confirmed that

$$\bar{l}(G_{10,0.1}^{\text{KE}}) \propto \ln n$$

  for sufficiently large $n$. Thus, empirical evidence suggests that crossover Klemm-Eguíluz graphs exhibit small average distances.

**8.** *Clustering in $G_{\mu,p}^{\text{KE}}$:*  The average local clustering coefficient of the Klemm-Eguíluz model with $p = 1$ is proportional to $(\ln^2 n)/n$ for large $n$ because $G_{\mu,1}^{\text{KE}} = G_{\mu}^{\text{BA}}$. By contrast, the average local clustering coefficient for the highly clustered model ($p = 0$) reaches the asymptotic value $5/6$. More precisely,

$$C_{\text{WS}}(G_{\mu,0}^{\text{KE}}) \approx \frac{5}{6} - \frac{7}{30\mu} + O\left(\frac{1}{\mu^2}\right),$$

i.e., the Klemm-Eguíluz model has very high transitivity for relatively large values of the parameter $\mu$ [KlEg02b]. Computational experiments also showed that for the crossover model with $p = 0.1$ the average local clustering coefficient is lower than $5/6$ but still reaches an asymptotic value greater than zero. Therefore, the crossover model with values $0 < p \ll 1$ is believed to exhibit not only short average distances but also high transitivity. This conclusion is supported by the empirical evidence obtained by Klemm and Eguíluz.

**9.** *Small distances in $G_{\mu,\rho}^{\text{JR}}$:*  Jackson and Rogers [JaRo07] conducted computational experiments on several real-world networks (fitted to undirected Jackson-Rogers graphs) supporting their conjecture that

$$\text{diam}(G_{\mu,\rho}^{\text{JR}}) = O(\ln n)$$

for various combinations of the model parameters.

**10.** *Clustering in $G_{\mu,\rho}^{\text{JR}}$:*  The Jackson-Rogers model exhibits high clustering under the following conditions: $\mu$ must be an integer, $\rho > 1$, and if $p_1 \geq \rho$, then $\mu_1$ and $p_2\mu_2/\mu_1$ both must be positive integers. Under such assumptions the average local clustering coefficient of the model converges to the fixed number

$$C_{\text{WS}}(G_{\mu,\rho}^{\text{JR}}) = \frac{6p_1}{(1+\rho)[(3\mu - 2)(\rho - 1) + 2\mu\rho]}$$

as $n \to \infty$, thus confirming that the Jackson-Rogers model has high transitivity, which is consistent with small-world characteristics.

**Example:**

**1.** The following six graphs illustrate the growing procedure of the Barabási-Albert model with $\mu = 2$. The process starts with a clique of cardinality $\mu = 2$. Then the model sequentially adds a new vertex (shown highlighted) and connects it to two other vertices which are selected with probability proportional to their current degree. The diagrams show the succession of graphs for $n = 2, 3, 4, 5, 10, 20$ vertices.

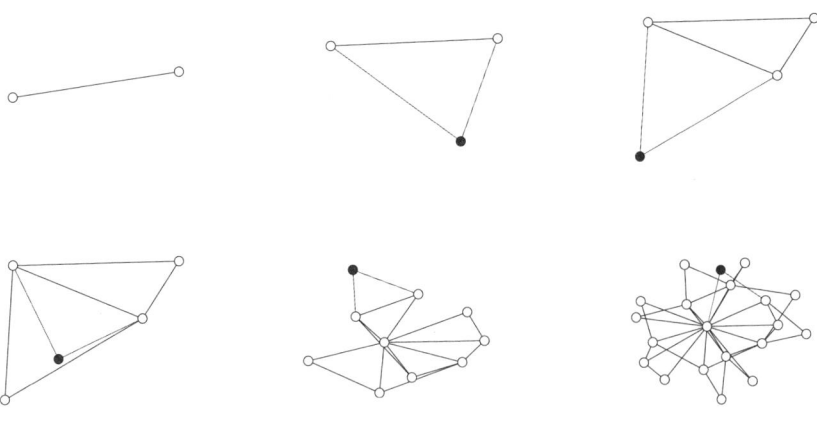

### 10.8.7  APPLICATIONS

Small-world networks arise in a variety of applications, ranging from manmade technological networks to naturally occurring networks in biology, economics, and society. This subsection presents several examples of small-world networks.

**1.** *Milgram's small-world experiment*:  The social psychologist Stanley Milgram pioneered the study of path lengths in real-world social networks through several sets of small-world experiments. Probably the most well-known of these experiments are the 1967 *Kansas study* [Mi67] and the 1969 *Nebraska study* [TrMi69]. In both experiments, a target person in the Boston area was chosen, and a set of randomly selected individuals in Nebraska (or Kansas) were asked to mail a letter to someone they know who would be most likely to have some relation to the target person. The mean number of intermediaries in the network paths between "starters" and the target turned out to be 5.2, which transforms into the average distance 6 if rounded up. Over 20 years later, this result was popularized by John Guare's 1990 play "Six Degrees of Separation" [Gu90] and the movie with the same title. Since then, the concepts of small-world networks and six degrees of separation have been well appreciated both in the scientific literature and in popular culture.

**2.** *Dodds' small-world experiment*:  After Milgram's work, small-world experiments were repeated and extended in several studies. One such study was conducted by Dodds et al. [DoMuWa03] using email messages. The researchers selected more than 60,000 participants who were given instructions in order to reach one of 18 target persons from 13 different countries. Every "starter" was provided some personal information about the assigned target. This experiment was conducted on a significantly larger scale than

Milgram's study as a result of the power of modern internet technology (24,000 email chains were initiated). The results of Dodds' experiment corroborated the fundamental conclusions of Milgram's studies. The average path length of successfully completed chains was 4.05, smaller than Milgram's value 5.2. Besides, Dodds' experiment accounted for uncompleted chains and it estimated that starters could reach their targets in a median of five to seven steps.

**3.** *Collaboration/coauthorship networks:* Collaboration/coauthorship networks represent prominent examples of real-life small-world networks. Perhaps the most well-known such networks are the collaboration network of mathematics researchers (where two researchers are connected by a link if they have ever written a joint paper) and the network of Hollywood actors (where two actors are connected if they have appeared in the same movie). In these networks, the role of a central vertex is traditionally given to Paul Erdős and to Kevin Bacon, respectively. The shortest distance from each vertex to the central vertex is traditionally referred to as that individual's *Erdős number* or *Bacon number*. It turns out that the values of the Erdős number and Bacon number for all vertices in the respective collaboration networks are small (many mathematicians have an Erdős number of at most 5, and the average Bacon number for all actors in the database is about 3), which confirms the small-world structure of these networks.

Coauthorship networks for other scientific disciplines can be similarly analyzed. The following table summarizes characteristics of such networks, demonstrating that these networks exhibit small average distances and diameters, as well as high clustering coefficients (compared to what would be observed if edges in these networks were added uniformly and randomly). This confirms that coauthorship networks do exhibit small-world properties.

| | Biology | Physics | Computer Science | Mathematics | Economics |
|---|---|---|---|---|---|
| number of authors ($n$) | 1,520,251 | 52,909 | 11,994 | 253,339 | 81,217 |
| log number of authors ($\ln n$) | 14.2 | 10.9 | 9.4 | 12.4 | 11.3 |
| mean # authors per paper | 3.75 | 2.53 | 2.22 | 1.45 | 1.56 |
| average # collaborators ($\bar{d}$) | 18.10 | 9.70 | 3.59 | 3.90 | 1.67 |
| size of largest component ($n_S$) | 1,395,693 | 44,337 | 6,396 | 208,200 | 33,027 |
| average distance ($\bar{l}_S$) | 4.6 | 5.9 | 9.7 | 7.6 | 9.5 |
| diameter ($\text{diam}_S$) | 24 | 20 | 31 | 27 | 29 |
| average edge probability ($p$) | $1 \times 10^{-5}$ | $2 \times 10^{-4}$ | $3 \times 10^{-4}$ | $2 \times 10^{-5}$ | $2 \times 10^{-5}$ |
| clustering coefficient ($C$) | 0.066 | 0.430 | 0.496 | 0.150 | 0.136 |

4. *Large-scale real-world networks*: Adamic [Ad99] showed that the World Wide Web (WWW) network has a small-world topology. She considered a graph that was a representation of the WWW at the site level, based on data collected by Jim Pitkow at Xerox PARC. The graph contained around 50 million documents and 259,794 sites. After the deletion of leaf vertices, it was reduced to an undirected graph of order $n = 153,127$. Adamic estimated the average path length $\bar{l}$ to be 3.1 and the average local clustering coefficient $C_{WS}$ to be 0.1078, much larger than the average local clustering coefficient of the respective random graph (that would be $2.3 \times 10^{-4}$). In a second experiment, Adamic considered the respective directed graph or, more precisely, its largest strongly connected component on $n_S = 64,826$ vertices. Here $\bar{l}(G[S]) \approx 4.3$ (average directed distance) and $C_{WS} \approx 0.081$, in contrast to $1.05 \times 10^{-3}$ for the corresponding random graph. Hence, this study provided strong evidence supporting the conjecture about the small-world nature of the WWW.

In a related study, Albert et al. [AlJeBa99] also investigated the average shortest path length in the World Wide Web, in which web pages (URLs) are vertices and hyperlinks are directed edges. Using local connectivity measurements, they constructed a topological model of the WWW. They employed the finite size scaling procedure to assess the average distance in the WWW by measuring distances in samples of increasing size. They calculated $\bar{l} \approx 0.35 + 0.89 \ln n$, which implies that the average path length in the WWW is around 18.59, since the size of the WWW was estimated to be $8 \times 10^8$ at that time. Currently its size is believed to be $\approx 10^{12}$, and so the average distance can be predicted to be close to 25. These results do not contradict Adamic's conclusions because they dealt with different objects in the WWW. Since every site includes many web pages, the order of Albert's graph was substantially greater than the order of Adamic's graph, thus making the estimated $\bar{l}$ much smaller for the latter graph. Unlike Adamic's study, Albert et al. did not study transitivity properties of the WWW.

Leskovec and Horwitz [LeHo08] constructed a network based on one month (June 2006) of communication activities from the Microsoft Instant Messenger network. The initial dataset contained 240 million users, but they extracted 180 million users who had participated in at least one conversation during the selected time period. The undirected communication network contained $1.3 \times 10^9$ edges among active users, where an edge was constructed for each pair of individuals who communicated. The largest connected component appeared to include 99.9% of all the vertices and, more importantly, it was estimated to have $\bar{l}(G[S]) \approx 6.6$. In addition, Leskovec and Horwitz found that the graph was well clustered with $C_{WS} \approx 0.137$, which was rather large for such a massive network.

One of the most extensive studies of the small-world nature of social networks was conducted by Backstrom et al. [BaEtal12]. They analyzed the dataset of all Facebook users active as of May 2011. The resulting graph, with $n \approx 7.21 \times 10^8$ and $m \approx 6.9 \times 10^{10}$ (symmetric friendship links were considered as its edges), was shown to have $O(\ln n)$ average distance. In particular, they found that $\bar{l} \approx 4.74$, utilizing the diffusion-based algorithm HyperANF to do estimations. Moreover, they coined the term "four degrees of separation", referring to the 3.74 intermediaries used on average. Recently, researchers at Facebook updated the estimation of the average shortest path length obtained by Backstrom et al. Edunov et al. [EdEtal16] analyzed the Facebook network, which had became larger (containing approximately twice as many users compared to May 2011) and more interconnected. Their estimation, done via the Flajolet-Martin algorithm, showed that the average distance in the Facebook graph was about 4.57, meaning 3.57 intermediaries on average (this led to the term "three and a half degrees of separation").

# 10.9   NETWORK REPRESENTATIONS AND DATA STRUCTURES

To carry out network optimization algorithms efficiently, careful attention needs to paid to the design of the data structures supporting these algorithms. There are alternative ways to represent a network—differing in their storage requirements and their efficacy in executing certain fundamental operations. These representations need to incorporate both the topology of the underlying graph and also any quantitative information present in the network (such as cost, length, capacity, demand, or supply). Standard representations of networks, and trees in particular, are discussed in this section.

## 10.9.1   NETWORK REPRESENTATIONS

There are various ways to represent networks, just as there are various ways to represent graphs (§8.1.4, §8.3.1). In addition it is necessary to incorporate quantitative information associated with the vertices and edges (or arcs) of the network. While the description here concentrates on directed networks, extensions to undirected networks are also indicated.

**Definitions:**

Let $G = (V, E)$ be a directed graph (§8.3.1) with vertex set $V = \{1, 2, \ldots, n\}$ and arc set $E$. Define $m = |E|$ to be the number of arcs in $G$.

The **adjacency set** $A(i) = \{(i, j) \mid (i, j) \in E\}$ for vertex $i$ is the set of arcs emanating from $i$. (See §10.3.1.)

The **adjacency matrix** for $G$ is the 0-1 matrix $A_G = (a_{ij})$ having $a_{ij} = 1$ if $(i, j) \in E$ and $a_{ij} = 0$ if $(i, j) \notin E$. (See also §6.6.2 and §8.3.1.)

The **arc list** for $G$ (see §8.3.1) can be implemented using two arc-length arrays FROM and TO:

- For each arc $(i, j) \in E$ there is a unique $1 \le k \le m$ satisfying $\text{FROM}(k) = i$ and $\text{TO}(k) = j$.
- Arcs are listed in the parallel FROM and TO arrays in no particular order.

The **linked adjacency list** for $G$ is given by a vertex-length array START and a singly-linked list ARCLIST of arc records:

- $\text{START}(i)$ points to the first record for vertex $i$ in this list, corresponding to a specified first element of $A(i)$.
- Each arc $(i, j) \in A(i)$ has an associated arc record, which contains the fields TO and NEXT. Specifically, ARCLIST.TO gives the adjacent vertex $j$, and ARCLIST.NEXT points to the next arc record in $A(i)$. If there is no such following record, ARCLIST.NEXT = **null**.

The **forward star** for $G$ is given by a vertex-length array START and an arc-length array TO, with the latter in one-to-one correspondence with arcs $(i, j) \in E$:

- $\text{START}(i)$ gives the position in array TO of the first arc leaving vertex $i$.
- The arcs of $A(i)$ are found in the consecutive positions $\text{START}(i)$, $\text{START}(i) + 1$, $\ldots$, $\text{START}(i + 1) - 1$ of array TO. If arc $(i, j)$ corresponds to position $k$ of TO, then $\text{TO}(k) = j$.

- By convention, an additional dummy vertex $n+1$ is added, with $\text{START}(n+1) = m+1$.

**Facts:**

**1.** An undirected graph can be represented by replacing each undirected edge $(i,j)$ by two oppositely directed arcs $(i,j)$ and $(j,i)$.

**2.** The adjacency matrix, the arc list, the linked adjacency list, and the forward star are four standard representations of a directed (or undirected) graph.

**3.** The linked adjacency list and forward star structures are commonly used implementations of the *lists-of-neighbors* representation (§8.3.1).

**4.** The following table shows the (worst-case) computational effort required to carry out certain fundamental operations on $G$: finding an arc, deleting an arc (once found), adding an arc, and scanning the adjacency set of an arbitrary vertex $i$. Here $\alpha_i = |A(i)| \le n$.

| representation | find arc | delete arc | add arc | scan $A(i)$ |
|---|---|---|---|---|
| *adjacency matrix* | $O(1)$ | $O(1)$ | $O(1)$ | $O(n)$ |
| *arc list* | $O(m)$ | $O(1)$ | $O(1)$ | $O(m)$ |
| *linked adjacency list* | $O(\alpha_i)$ | $O(1)$ | $O(1)$ | $O(\alpha_i)$ |
| *forward star* | $O(\alpha_i)$ | $O(n+m)$ | $O(n+m)$ | $O(\alpha_i)$ |

**5.** The storage requirements of the four representations are given in the following table for both directed and undirected graphs. For the last two representations, each undirected edge appears twice: once in each direction.

| representation | storage (directed) | storage (undirected) | exploit sparsity? |
|---|---|---|---|
| *adjacency matrix* | $n^2$ | $\frac{n^2}{2}$ | *no* |
| *arc list* | $2m$ | $2m$ | *yes* |
| *linked adjacency list* | $n+2m$ | $n+4m$ | *yes* |
| *forward star* | $n+m$ | $n+2m$ | *yes* |

**6.** As seen in the table of Example 5, all representations other than the adjacency matrix representation can exploit sparsity in the graph $G$. That is, the storage requirements are sensitive to the actual number of arcs and the computations will generally proceed more rapidly when $G$ has relatively few arcs.

**7.** Quantitative data for network vertices (such as supply and demand) can be stored in an associated vertex-length array, thus supplementing the standard graph representations.

**8.** Quantitative data for network arcs (such as cost, length, capacity, and flow) can be accommodated as follows:

- For the adjacency matrix representation, costs (or lengths) $c_{ij}$ can be imbedded in the matrix $A_G$ itself. Namely, redefine $A_G = (a_{ij})$ so that $a_{ij} = c_{ij}$ if $(i,j) \in E$, whereas $a_{ij}$ is an appropriate special value if $(i,j) \notin E$. For instance, in the shortest path problem (§10.3.1), $a_{ij} = \infty$ can be used to signify that $(i,j) \notin E$. Additional $n \times n$ arrays would be needed, however, to represent more than one type of arc data.

- For the arc list representation, additional arrays parallel to the arrays FROM and TO can be used to store quantitative arc data.

- For the linked adjacency list representation, additional fields within the arc record can be used to store quantitative arc data.

- For the forward star representation, additional arrays parallel to the array TO can be used to store quantitative arc data.

**9.** The arc list representation is best suited for arc-based processing of a network, such as occurs in Kruskal's minimum spanning tree algorithm (§10.1.2).

**10.** The arc list representation is a convenient form for the input of a network to an optimization algorithm. Often this external representation is converted within the algorithm to a more suitable internal representation (linked adjacency list or forward star) before executing the steps of the optimization algorithm.

**11.** The linked adjacency list and forward star representations are best suited to carrying out vertex-based explorations of a graph, such as a breadth-first search or a depth-first search (§9.2.1). It is also ideal for carrying out Prim's minimum spanning tree algorithm (§10.1.2) as well as most shortest path algorithms (§10.3.2).

**12.** Especially in the case of undirected graphs, the linked adjacency list and forward star representations can be enhanced by use of an additional arc-length array MIRROR. The array MIRROR allows one to move from the location of arc $(i, j)$ to the location of arc $(j, i)$ in constant time.

**13.** The linked adjacency list is typically used when the structure of the graph can dynamically change (as by addition/deletion of arcs or vertices). On the other hand, the forward star representation is appropriate for static graphs, in which the graph structure does not change.

**Examples:**

**1.** A directed graph $G$ with five vertices and eight arcs is shown in the following figure.

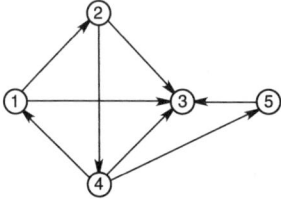

The $5 \times 5$ adjacency matrix for $G$ is given by

$$A_G = \begin{array}{c} \\ 1 \\ 2 \\ 3 \\ 4 \\ 5 \end{array} \begin{array}{c} \begin{array}{ccccc} 1 & 2 & 3 & 4 & 5 \end{array} \\ \left( \begin{array}{ccccc} 0 & 1 & 1 & 0 & 0 \\ 0 & 0 & 1 & 1 & 0 \\ 0 & 0 & 0 & 0 & 0 \\ 1 & 0 & 1 & 0 & 1 \\ 0 & 0 & 1 & 0 & 0 \end{array} \right) \end{array}$$

**2.** An arc list representation of the directed graph in the figure of Example 1 is given in the following table.

| FROM | 1 | 2 | 1 | 4 | 2 | 4 | 4 | 5 |
|------|---|---|---|---|---|---|---|---|
| TO | 2 | 4 | 3 | 1 | 3 | 5 | 3 | 3 |

**3.** The following figure shows a linked adjacency list representation of the directed graph of Example 1. The symbol $\odot$ is used to indicate a **null** pointer.

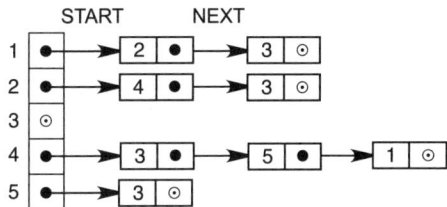

**4.** The following figure shows a forward star representation of the directed graph in Example 1. Since $A(3) = \emptyset$, it is necessary to set $\text{START}(3) = \text{START}(4) = 5$. For example, the arcs in $A(4)$ are associated with positions $[\text{START}(4), \ldots, \text{START}(5) - 1] = [5, 6, 7]$ of the TO array. Similarly, the single arc emanating from vertex 5 is associated with position $[\text{START}(5), \ldots, \text{START}(6) - 1] = [8]$ of the TO array.

| START | | TO | |
|---|---|---|---|
| 1 | 1 | 1 | 2 |
| 2 | 3 | 2 | 3 |
| 3 | 5 | 3 | 4 |
| 4 | 5 | 4 | 3 |
| 5 | 8 | 5 | 3 |
| 6 | 9 | 6 | 5 |
| | | 7 | 1 |
| | | 8 | 3 |

## 10.9.2  TREE DATA STRUCTURES

Since trees are important objects in optimization problems, as well as useful data structures in their own right (see §9.1), additional representations and features of trees are given here.

**Definitions:**

If $T$ is a rooted tree with root $r$ (§9.1.2), then the **predecessor function** $pred: V \rightarrow V$ is defined by $pred(r) = 0$, and $pred(j) = i$ if vertex $i$ is the parent of $j$ in $T$. (See §10.3.1.)

The **principal subtree** $T_j$ rooted at vertex $j$ is the subgraph of $T$ induced by all descendants of $j$ (including $j$). (See §9.1.2.)

The **cardinality** $card(j)$ of vertex $j$ in a rooted tree $T$ is the number of vertices in its principal subtree $T_j$.

The **least common ancestor** $least(i, j)$ of vertices $i$ and $j$ in a rooted tree is the vertex of largest depth (§9.1.2) that is an ancestor of both $i$ and $j$.

**Facts:**

**1.** A rooted tree is uniquely specified by the mapping $pred(\cdot)$.

**2.** In a rooted tree with root $r$, $depth(r) = 0$ and $depth(j) = 1 + depth(pred(j))$ for $j \neq r$.

**3.** In a rooted tree, $height(i) = 0$ if $i$ is a leaf. If $i$ is not a leaf, then $height(i) = 1 + \max\{height(j) \mid j \text{ a child of } i\}$.

**4.** In a rooted tree, $card(i) = 1$ if $i$ is a leaf. If $i$ is not a leaf, then $card(i) = 1 + \sum\{card(j) \mid j \text{ a child of } i\}$.

**5.** The predecessor, depth, height, and cardinality of a rooted tree $T$ can all be calculated while carrying out a preorder or postorder traversal (§9.1.3) of $T$:

- The predecessor and depth can be calculated while advancing from the current vertex to an unvisited vertex.

- The height and cardinality can be updated when retreating from a vertex all of whose children have been visited.

**6.** The depth of a vertex is a monotone increasing function on each path from the root.

**7.** The height and cardinality are monotone decreasing functions on each path from the root.

**Examples:**

**1.** The following figure shows a tree $T$ rooted at vertex 1. Here the vertices have been numbered according to a preorder traversal of $T$. The accompanying table gives the predecessor, depth, height, and cardinality of each vertex of $T$.

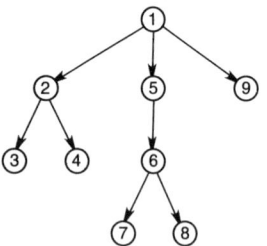

| vertex | 1 | 2 | 3 | 4 | 5 | 6 | 7 | 8 | 9 |
|--------|---|---|---|---|---|---|---|---|---|
| pred   | 0 | 1 | 2 | 2 | 1 | 5 | 6 | 6 | 1 |
| depth  | 0 | 1 | 2 | 2 | 1 | 2 | 3 | 3 | 1 |
| height | 3 | 1 | 0 | 0 | 2 | 1 | 0 | 0 | 0 |
| card   | 9 | 3 | 1 | 1 | 4 | 3 | 1 | 1 | 1 |

**2.** Certain applications (such as cycle detection in the network simplex algorithm, §10.5.2) require finding the least common ancestor $least(i, j)$ of vertices $i$ and $j$ in a rooted tree.

**3.** The calculation of $least(i, j)$ can be carried out efficiently, in $O(n)$ time, by using Algorithm 1.

**4.** Algorithm 1 is based on Fact 6 and employs two auxiliary data structures, the predecessor and depth functions. It repeatedly backs up from a vertex of larger depth until the least common ancestor is found.

---

**Algorithm 1:  Least common ancestor.**

input: rooted tree $T$, vertices $i$ and $j$

output: least common ancestor of $i$ and $j$

```
procedure least(i, j)
while i ≠ j
  if depth(i) > depth(j) then i := pred(i)
  else if depth(i) < depth(j) then j := pred(j)
  else i := pred(i), j := pred(j)
return i
```

# REFERENCES

*Printed Resources*:

[Ad99] L. A. Adamic, "The small world web" in *Research and Advanced Technology for Digital Libraries*, S. Abiteboul and A-M. Vercoustre (eds.), Springer, 1999, 443–452.

[AdEtal07] W. Adams, M. Guignard, P. Hahn, and W. Hightower, "A level-2 reformulation-linearization technique bound for the quadratic assignment problem", *European Journal of Operational Research* 180 (2007), 983–996.

[AgEtal00] R. Agarwala, D. L. Applegate, D. Maglott, G. D. Schuler, and A. A. Schaffler, "A fast and scalable radiation hybrid map construction and integration strategy", *Genome Research* 10 (2000), 350–364.

[AhMaOr93] R. K. Ahuja, T. L. Magnanti, and J. B. Orlin, *Network Flows: Theory, Algorithms, and Applications*, Prentice-Hall, 1993.

[AhEtal95] R. K. Ahuja, T. L. Magnanti, J. B. Orlin, and M. R. Reddy, "Applications of network optimization" in *Network Models*, M. Ball, T. Magnanti, C. Monma, and G. Nemhauser (eds.), North-Holland, 1995, 1–83.

[AhEtal90] R. K. Ahuja, K. Mehlhorn, J. B. Orlin, and R. E. Tarjan, "Faster algorithms for the shortest path problem", *Journal of the ACM* 37 (1990), 213–223.

[AhOrSh01] R. K. Ahuja, J. B. Orlin, and D. Sharma, "Multi-exchange neighborhood structures for the capacitated minimum spanning tree problem", *Mathematical Programming* 91 (2001), 71–97.

[AhOrSh03] R. K. Ahuja, J. B. Orlin, and D. Sharma, "A composite very large-scale neighborhood structure for the capacitated minimum spanning tree problem", *Operations Research Letters* 31 (2003), 185–194.

[AhOrTa89] R. K. Ahuja, J. B. Orlin, and R. E. Tarjan, "Improved time bounds for the maximum flow problem", *SIAM Journal on Computing* 18 (1989), 939–954.

[AiChLu02] W. Aiello, F. Chung, and L. Lu, "Random evolution in massive graphs" in *Handbook of Massive Data Sets*, J. Abello, P. M. Pardalos, and M. G. C. Resende (eds.), Springer, 2002, 97–122.

[AlBa00] R. Albert and A-L. Barabási, "Topology of evolving networks: local events and universality", *Physical Review Letters* 85 (2000), 5234.

[AlJeBa99] R. Albert, H. Jeong, and A-L. Barabási, "Internet: diameter of the world-wide web", *Nature* 401/6749 (1999), 130–131.

[AlGa88] K. Altinkemer and B. Gavish, "Heuristics with constant error guarantees for topological design of local access tree networks", *Management Science* 34 (1988), 331–341.

[AmDoVo96] A. Amberg, W. Domeschke, and S. Voss, "Capacitated minimum spanning trees: algorithms using intelligent search", *Combinatorial Optimization: Theory and Practice* 1 (1996), 9–33.

[AnBr01] K. M. Anstreicher and N. W. Brixius, "A new bound for the quadratic assignment problem based on convex quadratic programming", *Mathematical Programming* 89 (2001), 341–357.

[ApEtal07] D. L. Applegate, R. E. Bixby, V. Chvatal, and W. J. Cook, eds., *The Traveling Salesman Problem: A Computational Study*, 2nd ed., Princeton University Press, 2007.

[ArHaMa90] G. J. R. Araque, L. Hall, and T. Magnanti, "Capacitated trees, capacitated routing and associated polyhedra", CORE discussion paper 1990061, 1990.

[BaEtal12] L. Backstrom, P. Boldi, M. Rosa, J. Ugander, and S. Vigna, "Four degrees of separation", *Proceedings of the 4th Annual ACM Web Science Conference*, ACM, 2012, 33–42.

[BaSuMa10] L. Backstrom, E. Sun, and C. Marlow, "Find me if you can: improving geographical prediction with social and spatial proximity", *Proceedings of the 19th International Conference on World Wide Web*, ACM, 2010, 61–70.

[BaMcBe00] C. A. Bailey, T. W. McLain, and R. W. Beard, "Fuel saving strategies for separated spacecraft interferometry", *AIAA Guidance, Navigation, & Control Conference*, American Institute of Aeronautics & Astronautics, 2000, A00-37143.

[Ba16] A-L. Barabási, *Network Science*, Cambridge University Press, 2016.

[BaAl99] A-L. Barabási and R. Albert, "Emergence of scaling in random networks", *Science* 286 (1999), 509–512.

[BaAlJe99] A-L. Barabási, R. Albert, and H. Jeong, "Mean-field theory for scale-free random networks", *Physica A* 272 (1999), 173–187.

[BaRa10] M. G. Bardossy and S. Raghavan, "Dual-based local search for the connected facility location and related problems", *INFORMS Journal on Computing* 22 (2010), 584–602.

[BiDeSi94] D. Bienstock, Q. Deng, and D. Simchi-Levi, "A branch-and-cut algorithm for the capacitated minimum spanning tree problem", working paper, Columbia University, 1994.

[BiGü95] D. Bienstock and O. Günlük, "Computational experience with a difficult mixed-integer multicommodity flow problem", *Mathematical Programming* 68 (1995), 213–237.

[Bo98] B. Bollobás, *Random Graphs*, Springer, 1998.

[BoCh88] B. Bollobás and F. R. K. Chung, "The diameter of a cycle plus a random matching", *SIAM Journal on Discrete Mathematics* 1 (1988), 328–333.

[BoRi04] B. Bollobás and O. Riordan, "The diameter of a scale-free random graph", *Combinatorica* 24 (2004), 5–34.

[BrLa02] G. Bruno and G. Laporte, "A simple enhancement of the Esau-Williams heuristic for the capacitated minimum spanning tree problem", *Journal of the Operational Research Society* 53 (2002), 583–586.

[BuVa06] S. Burer and D. Vandenbusssche, "Solving lift-and-project relaxations of binary integer programs", *SIAM Journal on Optimization* 16 (2006), 726–750.

[BuSt78] R. Burkard and K. Stratmann, "Numerical investigations on quadratic assignment problems", *Naval Research Logistics Quarterly* 25 (1978), 129–148.

[Ch00] B. Chazelle, "A minimum spanning tree algorithm with inverse-Ackermann type complexity", *Journal of the ACM* 47 (2000), 1028–1047.

[ChLjRa10] S. Chen, I. Ljubic, and S. Raghavan, "The regenerator location problem", *Networks* 55 (2010), 205–220.

[ChLjRa15] S. Chen, I. Ljubic, and S. Raghavan, "The generalized regenerator location problem", *INFORMS Journal on Computing* 27 (2015), 204–220.

[ChMe99] J. Cheriyan and K. Mehlhorn, "An analysis of the highest-level selection rule in the preflow-push max-flow algorithm", *Information Processing Letters* 69 (1999), 239–242.

[ChEtal09] B. V. Cherkassky, L. Georgiadis, A. Goldberg, R. E. Tarjan, and R. F. Werneck, "Shortest-path feasibility algorithms: an experimental evaluation", *Journal of Experimental Algorithmics* 14 (2009), Section 2, Article 7.

[ChGoRa96] B. V. Cherkassky, A. V. Goldberg, and T. Radzik, "Shortest paths algorithms: theory and experimental evaluation", *Mathematical Programming, Series A* 73 (1996), 129–174.

[ChHaLa01] K. W. Chong, Y. Han, and T. W. Lam, "Concurrent threads and optimal parallel minimum spanning trees algorithm", *Journal of the ACM* 48 (2001), 297–323.

[ChEtal11] P. Christiano, J. A. Kelner, A. Madry, D. A. Spielman, and S-H. Teng, "Electrical flows, Laplacian systems, and faster approximation of maximum flow in undirected graphs", *Proceedings of the 43rd ACM Symposium on the Theory of Computing*, ACM, 2011, 273–282.

[ChLu04] F. R. K. Chung and L. Lu, "The average distance in a random graph with given expected degrees", *Internet Mathematics* 1 (2004), 91–113.

[CoHa03] R. Cohen and S. Havlin, "Scale-free networks are ultrasmall", *Physical Review Letters* 90 (2003), 058701.

[CoKeTa94] R. Cole, P. Kelvin, and R. E. Tarjan, "A linear-work parallel algorithm for finding minimum spanning trees", *Proceedings of the 6th ACM Symposium on Parallel Algorithms and Architectures*, 1994, 11–15.

[Co12] W. J. Cook, *In Pursuit of the Traveling Salesman: Mathematics at the Limits of Computation*, Princeton University Press, 2012.

[CoEtal01] T. H. Cormen, C. E. Leiserson, R. L. Rivest, and C. Stein, *Introduction to Algorithms*, 2nd ed., MIT Press and McGraw-Hill, 2001.

[DaFu00] H. K. Dai and S. Fujino, "On designing constrained local access networks", *Proceedings of the 2000 International Symposium on Parallel Architectures, Algorithms, and Networks (ISPAN '00)*, IEEE Computer Society, 2000, 167–176.

[Di69] R. Dial, "Algorithm 360: shortest path forest with topological ordering", *Communications of the ACM* 12 (1969), 632–633.

[DoMuWa03] P. S. Dodds, R. Muhamad, and D. J. Watts, "An experimental study of search in global social networks", *Science* 301 (2003), 827–829.

[DoMe00] S. N. Dorogovtsev and J. F. F. Mendes, "Exactly solvable small-world network", *Europhysics Letters* 50 (2000), 1–7.

[DoMe01] S. N. Dorogovtsev and J. F. F. Mendes, "Scaling properties of scale-free evolving networks: continuous approach", *Physical Review E* 63 (2001), 056125.

[DoMeSa03] S. N. Dorogovtsev, J. F. F. Mendes, and A. N. Samukhin, "Metric structure of random networks", *Nuclear Physics B* 653 (2003), 307–338.

[Dr15] Z. Drezner, "The quadratic assignment problem" in *Location Science*, G. Laporte, S. Nickel, and F. Saldanha da Gama (eds.), Springer, 2015, 345–363.

[EaKl10] D. Easley and J. Kleinberg, *Networks, Crowds, and Markets*, Cambridge University Press, 2010.

[Ed65a] J. Edmonds, "Paths, trees, and flowers", *Canadian Journal of Mathematics* 17 (1965), 449–467.

[Ed65b] J. Edmonds, "Maximum matching and a polyhedron with 0, 1-vertices", *Journal of Research of the National Bureau of Standards* B-69 (1965), 125–130.

[EdEtal16] S. Edunov, C. Diuk, I. O. Filiz, S. Bhagat, and M. Burke, "Three and a half degrees of separation", available at `https://research.facebook.com/blog/three-and-a-half-degrees-of-separation/`, 2016.

[ErRé60] P. Erdős and A. Rényi, "On the evolution of random graphs", *Publication of the Mathematical Institute of the Hungarian Academy of Sciences* 5 (1960), 17–61.

[EsWi66] L. R. Esau and K. C. Williams, "On teleprocessing system design", *IBM Systems Journal* 5 (1966), 142–147.

[EvMi92] J. R. Evans and E. Minieka, *Optimization Algorithms for Networks and Graphs*, Marcel Dekker, 1992.

[FeHu04] P. F. Felzenszwalb and D. P. Huttenlocher, "Efficient graph-based image segmentation", *International Journal of Computer Vision* 59 (2014), 167–181.

[FeRa07] D. Fernholz and V. Ramachandran, "The diameter of sparse random graphs", *Random Structures & Algorithms* 31 (2007), 482–516.

[FoFu62] L. R. Ford and D. R. Fulkerson, *Flows in Networks*, Princeton University Press, 1962.

[FrTa84] M. L. Fredman and R. E. Tarjan, "Fibonacci heaps and their uses in improved network optimization algorithms", *Proceedings of the 25th IEEE Symposium on Foundations of Computer Science*, 1984, 338–346. Full paper in *Journal of the ACM* 34 (1987), 596–615.

[Ga76] H. N. Gabow, "An efficient implementation of Edmonds' algorithm for maximum matching on graphs", *Journal of the ACM* 23 (1976), 221–234.

[GaEtal86] H. N. Gabow, Z. Galil, T. Spencer, and R. E. Tarjan, "Efficient algorithms for finding minimum spanning trees in undirected and directed graphs", *Combinatorica* 6 (1986), 109–122.

[GaSh62] D. Gale and L. S. Shapley, "College admissions and the stability of marriage", *American Mathematical Monthly* 69 (1962), 9–15.

[Ga85] B. Gavish, "Augmented Lagrangian based algorithms for centralized network design", *IEEE Transactions on Communications* COM-33 (1985), 1247–1257.

[Ga91] B. Gavish, "Topological design of telecommunication networks—local access design methods", *Annals of Operations Research* 33 (1991), 17–71.

[GeHeLa94] M. Gendreau, A. Hertz, and G. Laporte, "A tabu search heuristic for the vehicle routing problem", *Management Science* 40 (1994), 1276–1290.

[Ge95] A. M. H. Gerards, "Matching" in *Network Models*, M. Ball, T. Magnanti, C. Monma, and G. Nemhauser (eds.), North-Holland, 1995, 135–224.

[Gi59] E. N. Gilbert, "Random graphs", *The Annals of Mathematical Statistics* 30 (1959), 1141–1144.

[GlKl75] F. Glover and D. Klingman, "Finding minimum spanning trees with a fixed number of links at a node", in *Combinatorial Programming: Methods and Applications*, B. Roy et al. (eds.), 1975, 191–201.

[GoBe93] M. X. Goemans and D. J. Bertsimas, "Survivable networks, linear programming relaxations and the parsimonious property", *Mathematical Programming* 60 (1993), 145–166.

[Go95] A. V. Goldberg, "Scaling algorithms for the shortest paths problem", *SIAM Journal on Computing* 24 (1995), 494–504.

[Go01] A. V. Goldberg, "A simple shortest path algorithm with linear average time", *Algorithms–ESA 2001*, Lecture Notes in Computer Science 2161, Springer, 2001, 230–241.

[GoRa98] A. V. Goldberg and S. Rao, "Beyond the flow decomposition barrier", *Journal of the ACM* 45 (1998), 783–797.

[GoTa86] A. V. Goldberg and R. E. Tarjan, "A new approach to the maximum flow problem", *Proceedings of the 18th ACM Symposium on the Theory of Computing*, 1986, 136–146. Full paper appears in *Journal of the ACM* 35 (1988), 921–940.

[GoTa88] A. V. Goldberg and R. E. Tarjan, "Finding minimum-cost circulations by canceling negative cycles", *Proceedings of the 20th ACM Symposium on the Theory of Computing*, 1988, 7–18. Full paper appears in *Journal of the ACM* 36 (1989), 873–886.

[GoTa14] A. V. Goldberg and R. E. Tarjan, "Efficient maximum flow algorithms", *Communications of the ACM* 57/8 (2014), 82–89.

[GoLj11] S. Gollowitzer and I. Ljubic, "MIP models for connected facility location: a theoretical and computational study", *Computers & Operations Research* 38 (2011), 435–449.

[GoMa99] L. Gouveia and P. Martins, "The Capacitated Minimal Spanning Tree Problem: an experiment with a hop-indexed model", *Annals of Operations Research* 86 (1999), 271–294.

[GoMa00] L. Gouveia and P. Martins, "A hierarchy of hop-indexed models for the Capacitated Minimum Spanning Tree Problem", *Networks* 35 (2000), 1–16.

[GoMa05] L. Gouveia and P. Martins, "The capacitated minimum spanning tree problem: revisiting hop-indexed formulations", *Computers & Operations Research* 32 (2005), 2435–2452.

[GrHe85] R. L. Graham and P. Hell, "On the history of the minimum spanning tree problem", *Annals of the History of Computing* 7 (1985), 43–57.

[GrWh70] G. Graves and A. Whinston, "An algorithm for the quadratic assignment problem", *Management Science* 17 (1970), 453–471.

[GrGoWa11] C. Groer, B. Golden, and E. Wasil, "A parallel algorithm for the vehicle routing problem", *INFORMS Journal on Computing* 23 (2011), 315–330.

[GrMoSt92] M. Grötschel, C. L. Monma, and M. Stoer, "Computational results with a cutting plane algorithm for designing communication networks with low-connectivity constraints", *Operations Research* 40 (1992), 309–330.

[Gu90] J. Guare, *Six Degrees of Separation: A Play*, Vintage Books, 1990.

[GuSp14] G. Guastaroba and M. G. Speranza, "A heuristic for BILP problems: The Single Source Capacitated Facility Location Problem", *European Journal of Operational Research* 238 (2014), 438–450.

[GuIr89] D. Gusfield and R. W. Irving, *The Stable Marriage Problem: Structure and Algorithms*, MIT Press, 1989.

[HaGr98] P. Hahn and T. Grant, "Lower bounds for the quadratic assignment problem based upon a dual formulation", *Operations Research* 46 (1998), 912–922.

[HaEtal12] P. M. Hahn, Y-R. Zhu, M. Guignard, W. L. Hightower, and M. J. Saltzman, "A level-3 reformulation-linearization technique-based bound for the quadratic assignment problem", *INFORMS Journal on Computing* 24 (2012), 202–209.

[Ha96] L. Hall, "Experience with a cutting plane algorithm for the capacitated spanning tree problem", *INFORMS Journal on Computing* 8 (1996), 219–234.

[HaPa16] R. Hall and J. Partyka, "Higher expectations drive performance", *OR/MS Today* 43/1 (2016), 40–47.

[HaAg15] F. Hamid and Y. K. Agarwal, "Solving the two-facility network design problem with 3-partition facets", *Networks* 66 (2015), 11–32.

[HaRa03] B. T. Han and V. T. Raja, "A GRASP heuristic for solving an extended capacitated concentrator location problem", *International Journal of Information Technology and Decision Making* 2 (2003), 597–617.

[HaRo55] T. E. Harris, and F. S. Ross, "Fundamentals of a method for evaluating rail net capacities", *Research Memorandum RM-1573*, The RAND Corporation, 1955.

[Ho98] D. S. Hochbaum, "The pseudoflow algorithm and the pseudoflow-based simplex for the maximum flow problem", in *Integer Programming and Combinatorial Optimization*, Lecture Notes in Computer Science 1412, Springer, 1998, 325–337.

[HoPaRi13] K. L. Hoffman, M. Padberg, and G. Rinaldi, "Traveling salesman problem" in *Encyclopedia of Operations Research and Management Science*, S. I. Gass and M. C. Fu (eds.), Springer, 2013, 1573–1578.

[JaRo07] M. O. Jackson and B. W. Rogers, "Meeting strangers and friends of friends: how random are social networks?", *The American Economic Review* 97 (2007), 890–915.

[JaLuRu00] S. Janson, T. Luczak, and A. Rucinski, *Random Graphs*, John Wiley & Sons, 2000.

[JeNéBa03] H. Jeong, Z. Néda, and A-L. Barabási, "Measuring preferential attachment in evolving networks", *Europhysics Letters* 61 (2003), 567–572.

[JüReRi95] M. Jünger, G. Reinelt, and G. Rinaldi, "The traveling salesman problem" in *Network Models*, M. Ball, T. Magnanti, C. Monma, and G. Nemhauser (eds.), North-Holland, 1995, 225–330.

[KaKlTa95] D. R. Karger, P. N. Klein, and R. E. Tarjan, "A randomized linear-time algorithm to find minimum spanning trees", *Journal of the ACM* 42 (1995), 321–328.

[KeEtal14] J. A. Kelner, Y. T. Lee, L. Orecchia, and A. Sidford, "An almost-linear-time algorithm for approximate max flow in undirected graphs, and its multicommodity generalizations", *Proceedings of the 25th ACM-SIAM Symposium on Discrete Algorithms*, SIAM, 2014, 217–226.

[KiHe03] J. Kim and J. Hespanha, "Discrete approximations to continuous shortest-path: application to minimum-risk path planning for groups of UAVs", *Proceedings of the 42nd IEEE Conference on Decision and Control*, IEEE, 2003, 1734–1740.

[KiLe88] G. A. P. Kindervator and J. K. Lenstra, "Parallel computing in combinatorial optimization", *Annals of Operations Research* 14 (1988), 245–289.

[KiRaTa94] V. King, S. Rao, and R. Tarjan, "A faster deterministic maximum flow algorithm", *Journal of Algorithms* 17 (1994), 447–474.

[Kl00] J. Kleinberg, "The small-world phenomenon: an algorithmic perspective", *Proceedings of the 32nd Annual ACM Symposium on the Theory of Computing*, ACM, 2000, 163–170.

[KlTa13] J. Kleinberg and E. Tardos, *Algorithm Design*, Pearson Education Limited, 2013.

[KlEg02a] K. Klemm and V. M. Eguíluz, "Highly clustered scale-free networks", *Physical Review E* 65 (2002), 036123.

[KlEg02b] K. Klemm and V. M. Eguíluz, "Growing scale-free networks with small-world behavior", *Physical Review E* 65 (2002), 057102.

[KnGo97] D. E. Knuth and M. Goldstein, *Stable Marriage and Its Relation to Other Combinatorial Problems: An Introduction to the Mathematical Analysis of Algorithms*, American Mathematical Society, 1997.

[KrReLe00] P. L. Krapivsky, S. Redner, and F. Leyvraz, "Connectivity of growing random networks", *Physical Review Letters* 85 (2000), 4629.

[KuHe87] A. Kusiak and S. Heragu, "The facility layout problem", *European Journal of Operational Research* 29 (1987), 229–251.

[La10] G. Laporte, "A concise guide to the traveling salesman problem", *Journal of the Operational Research Society* 61 (2010), 35–40.

[LaRoVi14] G. Laporte, S. Ropke, and T. Vidal, "Heuristics for the vehicle routing problem" in *Vehicle Routing: Problems, Methods, and Applications*, 2nd ed., P. Toth and D. Vigo (eds.), SIAM, 2014, 87–116.

[LaEtal85] E. L. Lawler, J. K. Lenstra, A. H. G. Rinnooy Kan, and D. B. Shmoys, eds., *The Traveling Salesman Problem: A Guided Tour of Combinatorial Optimization*, Wiley, 1985.

[Le92] F. T. Leighton, *Introduction to Parallel Algorithms and Architectures*, Morgan Kaufmann, 1992.

[LeHo08] J. Leskovec and E. Horvitz, "Planetary-scale views on a large instant-messaging network", *Proceedings of the 17th International Conference on World Wide Web*, ACM, 2008, 915–924.

[LiEtal05] D. Liben-Nowell, J. Novak, R. Kumar, P. Raghavan, and A. Tomkins, "Geographic routing in social networks", *Proceedings of the National Academy of Sciences USA* 102 (2005), 11623–11628.

[Li81] R. Liggett, "The quadratic assignment problem: an experimental evaluation of solution strategies", *Management Science* 27 (1981), 442–458.

[LoPu07] A. Lodi and A. P. Punnen, "TSP software" in *The Traveling Salesman Problem and Its Variations*, G. Gutin and A. P. Punnen (eds.), Springer, 2007, 737–749.

[LoPu86] L. Lovász and M. D. Plummer, *Matching Theory*, North-Holland, 1986.

[MaMiVa95] T. L. Magnanti, P. Mirchandani, and R. Vachani, "Modeling and solving the two-facility capacitated network loading problem", *Operations Research* 43 (1995), 142–157.

[MiVa80] S. Micali and V. V. Vazirani, "An $O(\sqrt{|V|} \cdot |E|)$ algorithm for finding maximum matching in general graphs", *Proceedings of the 21st Symposium on Foundations of Computer Science*, 1980, 17–27.

[Mi67] S. Milgram, "The small world problem", *Psychology Today* 1 (1967), 61–67.

[NeWo99] G. L. Nemhauser and L. A. Wolsey, *Integer and Combinatorial Optimization*, Wiley, 1999.

[Ne01] M. E. J. Newman, "Scientific collaboration networks. II. Shortest paths, weighted networks, and centrality", *Physical Review E* 64 (2001), 016132.

[Ne03] M. E. J. Newman, "The structure and function of complex networks", *SIAM Review* 45 (2003), 167–256.

[Ne10] M. E. J. Newman, *Networks: An Introduction*, Oxford University Press, 2010.

[NeMoWa00] M. E. J. Newman, C. Moore, and D. J. Watts, "Mean-field solution of the small-world network model", *Physical Review Letters* 84 (2000), 3201.

[NeStWa01] M. E. J. Newman, S. H. Strogatz, and D. J. Watts, "Random graphs with arbitrary degree distributions and their applications", *Physical Review E* 64 (2001), 026118.

[NeWa99] M. E. J. Newman and D. J. Watts, "Scaling and percolation in the small-world network model", *Physical Review E* 60 (1999), 7332.

[ÖnAl09] T. Öncan and I. K. Altinel, "Parametric enhancements of the Esau-Williams heuristic for the capacitated minimum spanning tree problem", *Journal of the Operational Research Society* 60 (2009), 259–267.

[Or88] J. B. Orlin, "A faster strongly polynomial minimum cost flow algorithm", *Proceedings of the 20th ACM Symposium on the Theory of Computing*, ACM, 1988, 377–387. Full paper in *Operations Research* 41 (1993), 338–350.

[Or97] J. B. Orlin, "A polynomial time primal network simplex algorithm for minimum cost flows", *Mathematical Programming* 78 (1997), 109–129.

[Or13] J. B. Orlin, "Max flows in $O(nm)$ time, or better", *Proceedings of the 45th ACM Symposium on the Theory of Computing*, ACM, 2013, 765–774.

[OsKe95] I. Osman and J. Kelly, eds., *Meta-Heuristics: Theory and Applications*, Kluwer, 1995.

[PeRa02] S. Pettie and V. Ramachandran, "An optimal minimum spanning tree algorithm", *Journal of the ACM* 49 (2002), 16–34.

[Pl92] M. D. Plummer, "Matching theory—a sampler: from Dénes König to the present", *Discrete Mathematics* 100 (1992), 177–219.

[Pr76] D. De Solla Price, "A general theory of bibliometric and other cumulative advantage processes", *Journal of the American Society for Information Science* 27 (1976), 292–306.

[RaEtal02] K. G. Ramakrishnan, M. Resende, B. Ramachandran, and J. Pekny, "Tight QAP bounds via linear programming" in *Combinatorial and Global Optimization*, P. M. Pardalos, A. Migdalas, and R. Burkard (eds.), World Scientific Publishing, 2002, 297–303.

[Re93] C. Reeves, ed., *Modern Heuristic Techniques for Combinatorial Problems*, Wiley, 1993.

[ReMa11] C. Rego and F. Mathew, "A filter-and-fan algorithm for the capacitated minimum spanning tree problem", *Computers and Industrial Engineering* 60 (2011), 187–194.

[ReLa06] M. Reimann and M. Laumanns, "Savings based ant colony optimization for the capacitated minimum spanning tree problem", *Computers & Operations Research* 33 (2006), 1794–1822.

[ReSo07] F. Rendl and R. Sotirov, "Bounds for the quadratic assignment problem using the bundle method", *Mathematical Programming* 109 (2007), 505–524.

[ReRaDr95] M. Resende, K. Ramakrishnan, and Z. Drezner, "Computational experiments with the lower bound for the quadratic assignment problem based on linear programming", *Operations Research* 43 (1995), 781–791.

[RoPe99] A. E. Roth and E. Peranson, "The redesign of the matching market for American physicians: some engineering aspects of economic design", *American Economic Review* 89 (1999), 748–780.

[RoSo91] A. E. Roth and M. Sotomayor, *Two-Sided Matching: A Study in Game-Theoretic Modeling and Analysis*, Cambridge University Press, 1991.

[Ro04] F. Roupin, "From linear to semidefinite programming: an algorithm to obtain semidefinite relaxations for bivalent quadratic problems", *Journal of Combinatorial Optimization* 8 (2004), 469–493.

[Ro99] S. Roy, "Stereo without epipolar lines: a maximum-flow formulation", *International Journal of Computer Vision* 34 (1999), 147–161.

[RuEtal15] E. Ruiz, M. Albareda-Sambola, E. Fernández, and M. G. C. Resende, "A biased random-key genetic algorithm for the capacitated minimum spanning tree problem", *Computers & Operations Research* 57 (2015), 95–108.

[ShEtal97] Y. M. Sharaiha, M. Gendreau, G. Laporte, and I. H. Osman, "A tabu search algorithm for the capacitated shortest spanning tree problem", *Networks* 29 (1997), 161–171.

[ShAd99] H. D. Sherali and W. P. Adams, *A Reformulation-Linearization Technique for Solving Discrete and Continuous Nonconvex Problems*, Springer, 1999.

[SlMaKa97] T. H. Sloane, F. Mann, and H. Kaveh, "Powering the last mile: an alternative to powering FITL", *19th International Telecommunications Energy Conference*, 1997, 536–543.

[SoDuRi03] M. C. de Souza, C. Duhamel, and C. C. Ribeiro, "A GRASP heuristic for the capacitated minimum spanning tree problem using a memory-based local search strategy", *Applied Optimization* 86 (2003), 627–658.

[TaRo04] E. Tapia and R. Rojas, "Recognition of on-line handwritten mathematical expressions using a minimum spanning tree construction and symbol dominance", in *Graphics Recognition. Recent Advances and Perspectives*, Springer, 2004, 329–340.

[Ta83] R. E. Tarjan, *Data Structures and Network Algorithms*, SIAM, 1983.

[Th99] M. Thorup, "Undirected single-source shortest paths with positive integer weights in linear time", *Journal of the ACM* 46 (1999), 362–394.

[TrMi69] J. Travers and S. Milgram, "An experimental study of the small world problem", *Sociometry* 32 (1969), 425–443.

[UcEtal08] E. Uchoa, R. Fukasawa, J. Lysgaard, A. Pessoa, M. P. de Aragão, and D. Andrade, "Robust branch-cut-and-price for the Capacitated Minimum Spanning Tree problem over a large extended formulation", *Mathematical Programming* 112 (2008), 443–472.

[Vá03] A. Vázquez, "Growing network with local rules: preferential attachment, clustering hierarchy, and degree correlations", *Physical Review E* 67 (2003), 056104.

[Vo89] A. Volgenant, "A Lagrangean approach to the degree-constrained minimum spanning tree problem", *European Journal of Operational Research* 39 (1989), 325–331.

[WaEtal16] X. Wang, M. Battarra, B. Golden, and E. Wasil, "Vehicle routing and scheduling" in *Routledge Handbook of Transportation*, D. Teodorovic (ed.), Taylor & Francis, 2016, 238–256.

[WaSt98] D. J. Watts and S. H. Strogatz, "Collective dynamics of 'small-world' networks", *Nature* 393 (1998), 440–442.

[XuOlXu02] Y. Xu, V. Olman, and D. Xu, "Clustering gene expression data using a graph-theoretic approach: an application of minimum spanning trees", *Bioinformatics* 18 (2002), 536–545.

[YaChCh12] Z. Yang, F. Chu, and H. Chen, "A cut-and-solve based algorithm for the single-source capacitated facility location problem", *European Journal of Operational Research* 221 (2012), 521–532.

[Ya75] A. Yao, "An $O(|E| \log \log |V|)$ algorithm for finding minimum spanning trees", *Information Processing Letters* 4 (1975), 21–23.

[Zw02] U. Zwick, "All pairs shortest paths using bridging sets and rectangular matrix multiplication", *Journal of the ACM* 49 (2002), 289–317.

**Web Resources**:

`ftp://dimacs.rutgers.edu/pub/netflow/matching/` (Computer code in C for solving weighted matching problems; computer codes in C, Pascal, and Fortran for finding maximum size matchings in nonbipartite networks.)

`ftp://dimacs.rutgers.edu/pub/netflow/maxflow/` (Computer codes for solving maximum flow and minimum cut problems.)

`ftp://dimacs.rutgers.edu/pub/netflow/mincost/` (Computer codes for solving minimum cost flow problems.)

`http://anjos.mgi.polymtl.ca/qaplib/` (Problem instances, solutions, software, and research papers for the QAP.)

`http://comopt.ifi.uni-heidelberg.de/software/TSPLIB95/` (A library of sample problems related to the traveling salesman problem, with their best known solutions.)

`http://lemon.cs.elte.hu/trac/lemon` (Computer codes for solving minimum cost flow problems.)

`http://neo.lcc.uma.es/vrp/` (Datasets and solution methods for vehicle routing problems and variants.)

`http://oracleofbacon.org` (Site for determining the Bacon number for a specified actor.)

`http://riot.ieor.berkeley.edu/Applications/Pseudoflow/maxflow.html` (Computer codes for solving maximum flow and minimum cut problems.)

`http://www.ams.org/mathscinet/collaborationDistance.html` (Website for calculating the minimum distance between mathematical researchers.)

`http://www.avglab.com/andrew/soft.html` (Computer codes to solve single-source shortest path, maximum flow, minimum cut, and minimum cost flow problems.)

`http://www.cs.sunysb.edu/~algorith/` (The Stony Brook Algorithm Repository; see listed Sections 1.4 and 1.5 on Graph Problems.)

`http://www.dis.uniroma1.it/challenge9/` (Computer codes to solve single-source shortest path problems.)

`http://www.mat.uc.pt/~eqvm/cientificos/fortran/codigos.html` (Fortran code for implementing Kruskal's algorithm and Prim's algorithm for minimum spanning trees; Fortran code for implementing the label-correcting algorithm and Dijkstra's algorithm for shortest paths.)

`http://www.netlib.org/toms/479` (Fortran code for implementing Prim's algorithm.)

`http://www.netlib.org/toms/562` (Fortran code for implementing the label-correcting algorithm for shortest paths.)

`http://www.netlib.org/toms/613` (Fortran code for implementing Prim's algorithm.)

`http://www.oakland.edu/enp/` (Website for the Erdős Number Project.)

`http://www3.cs.stonybrook.edu/~algorith/files/minimum-spanning-tree.shtml` (Computer codes for solving minimum spanning tree problems.)

`https://research.facebook.com/blog/three-and-a-half-degrees-of-separation` (Article about three and a half degrees of separation in the Facebook graph.)

国外优秀数学著作
原 版 系 列

离
散
与
组
合
数
学
手
册

Handbook of Discrete and Combinatorial Mathematics, 2e

● ［美］ 肯尼斯·H. 罗森 （Kenneth H. Rosen） 主编

（下册·第二版）

（英文）

哈尔滨工业大学出版社
HARBIN INSTITUTE OF TECHNOLOGY PRESS

黑版贸审字 08－2020－201 号

**图书在版编目(CIP)数据**

离散与组合数学手册：第二版＝Handbook of Discrete and Combinatorial Mathematics，2e. 下册：英文/(美)肯尼斯·H. 罗森(Kenneth H. Rosen)主编. —哈尔滨：哈尔滨工业大学出版社，2023.1
ISBN 978-7-5767-0651-2

Ⅰ.①离…　Ⅱ.①肯…　Ⅲ.①离散数学－英文 ②组合数学－英文　Ⅳ.①O158 ②O157

中国国家版本馆 CIP 数据核字(2023)第 030261 号

LISAN YU ZUHE SHUXUE SHOUCE：DI-ER BAN(XIACE)

策划编辑　刘培杰　杜莹雪
责任编辑　刘立娟
封面设计　孙茵艾
出版发行　哈尔滨工业大学出版社
社　　址　哈尔滨市南岗区复华四道街 10 号　邮编 150006
传　　真　0451－86414749
网　　址　http://hitpress. hit. edu. cn
印　　刷　哈尔滨市石桥印务有限公司
开　　本　787 mm×1 092 mm　1/16　印张 102.25　字数 1 959 千字
版　　次　2023 年 1 月第 1 版　2023 年 1 月第 1 次印刷
书　　号　ISBN 978-7-5767-0651-2
定　　价　248.00 元(全 2 册)

(如因印装质量问题影响阅读，我社负责调换)

# 11

# PARTIALLY ORDERED SETS

## INTRODUCTION

Partially ordered sets play important roles in a wide variety of applications, including the design of sorting and searching methods, the scheduling of tasks, the study of social choice, and the study of lattices. This chapter covers the basic concepts involving partially ordered sets, the various types of partially ordered sets, the fundamental properties of these sets, and their important applications.

A table of notation used in the study of posets is given following the glossary.

## GLOSSARY

**antichain**: a subset of a poset in which no two distinct elements are comparable.

**atom**: in a poset, an element of height 1.

**atomic lattice**: a lattice such that every element is a join of atoms (or equivalently, such that the atoms are the only join-irreducible elements).

**auxiliary graph** (of a simple graph $G$): the graph $G'$ whose vertices are the edges of $G$, with vertex $e_1$ adjacent to vertex $e_2$ in $G'$ if and only if $e_1$ and $e_2$ are adjacent edges in $G$, but do not lie on a 3-cycle in $G$.

**biorder representation** (on a digraph $D$): a pair of real-valued functions $f, g$ on the vertex set $V_D$ such that $u \to v$ is an arc if and only if $f(u) > g(v)$.

**bipartite poset**: a poset of height at most 2.

**Boolean algebra**: the poset whose domain is all subsets of a given set, partially ordered by inclusion.

**Borda consensus function** (on a set of social choice profiles): the consensus function that ranks the alternatives by their Borda count.

**Borda count** (of an alternative social choice $x$): the sum, over all individual rankings, of the number of alternatives $x$ "beats".

**bounded poset**: a poset with both a unique minimal element and a unique maximal element.

**$u$,$v$-bypass** (in a directed graph): a $u$,$v$-path of length at least two such that there is also an arc from $u$ to $v$.

**Cartesian product** (of two posets $P = (X, R)$ and $P' = (X', R')$): the poset $P \times P' = (X \times X', S)$, such that $(x, x')S(y, y')$ if and only if $xRy$ and $x'R'y'$.

**chain**: a subset of a poset in which every two elements are comparable.

**$k$-chain**: a chain of size $k$, i.e., a chain on $k$ elements.

**chain-product**: the Cartesian product of a collection of chains.

**comparability digraph** (of a poset $(X, R)$): the simple digraph whose vertex set is the domain $X$ and which has an arc from $x$ to $y$ if and only if $x \le y$.

**comparability graph** (of a poset $(X, R)$): the simple graph whose vertex set is the domain $X$ and which has an edge joining distinct vertices $x$ and $y$ if and only if $x \le y$.

***comparability invariant*** (for posets): an invariant $f$ such that $f(P) = f(Q)$ whenever posets $P$ and $Q$ have the same comparability graph.

***comparable elements*** (in a poset $(X, R)$): elements $x$ and $y$ such that either $(x, y) \in R$ or $(y, x) \in R$.

***consecutive chain*** (in a ranked poset): a chain whose elements belong to consecutive ranks.

***consensus function*** (on a set of social choice profiles): a function that assigns to each possible profile $P = \{P_i \mid i \in I\}$ on a set of alternatives a linear ordering (ties allowed) of those alternatives.

***consensus ranking*** (on a set of social choice profiles): the linear ordering of the alternatives assigned by the consensus function.

***cover graph*** (of a poset $(X, R)$): the graph with vertex set $X$ and edge set consisting of the pairs satisfying the cover relation.

***cover relation*** (of a poset $(X, R)$): the relation on $X$ consisting of the pairs $(x, y)$ such that $x > y$ in $R$ and such that there is no "intermediate" element $z$ with $x > z > y$.

***cover diagram***: a synonym for the Hasse diagram.

***critical pair*** (in a poset): an ordered incomparable pair that cannot be made comparable by adding any other single incomparable pair as a relation.

***dependent edge*** (in an acyclic directed graph): an arc from $u$ to $v$ such that the graph contains a $u,v$-bypass.

***dimension*** (of a poset): the minimum number of chains in a realizer of the poset.

***distributive lattice***: a lattice in which the meet operator distributes over the join operator, so that $x \wedge (y \vee z) = (x \wedge y) \vee (x \wedge z)$ for all $x, y, z$.

***divisor lattice***: the poset $D(n)$ of divisors of $n$, in which $x \leq y$ means that $y$ is an integer multiple of $x$.

***downset*** (in a poset): a subposet $I$ such that if $x \in I$ and if $y < x$, then $y \in I$; also called an ideal.

***dual*** (of a poset $P = (X, R)$): the poset $P^* = (X, S)$ such that $x \leq y$ in $S$ if and only if $y \leq x$ in $R$.

***extension*** (of a poset $P = (X, R)$): a poset $Q = (X, S)$ such that $R \subseteq S$; meaning that $xRy$ implies $xSy$.

***k-family*** (in a poset): a subposet containing no chain of size $k + 1$.

***Ferrers digraph***: a digraph having a biorder representation.

***filter*** (generated by an element $x$ in a poset $P$): the upset $U[x] = \{y \in P \mid y \geq x\}$.

***filter*** (generated by a subset in a poset $P$): given a subset $A$ of $P$, the upset $U[A] = \bigcup_{x \in A} U[x]$.

***filter*** (in a poset): a subposet whose domain is the set-theoretic complement of the domain of an ideal.

***forbidden subposet description*** (of a class of posets): a characterization of the class as the class of all posets that does not contain any of the posets in a specified collection.

***geometric lattice***: an atomic, upper semimodular lattice of finite height.

**greatest lower bound** (**glb**) (of elements $x$ and $y$ in a poset):  a common lower bound $z$ such that every other common lower bound $z'$ satisfies the inequality $z \geq z'$. Such an element, if it exists, is denoted $x \wedge y$.

**graded poset**:  a poset in which all maximal chains have the same length.

**Hasse diagram** (of a poset):  a straight-line drawing of the cover graph in the plane so that the lesser element of each adjacent pair is below the greater.

**height** (of a poset):  the maximum size of a chain in that poset.

**height** (of an element $x$ of a poset):  the maximum length $h(x)$ of a chain that has $x$ as its maximal element.

**ideal** (generated by an element $x$ in a poset $P$):  the downset $D[x] = \{y \in P \mid y \leq x\}$.

**ideal** (generated by a subset $A$ in a poset $P$):  the downset $D[A] = \bigcup_{x \in A} D[x]$.

**ideal** (in a poset):  a subposet $I$ such that if $x \in I$ and if $y < x$, then $y \in I$.

**incomparability graph** (of a poset $P$):  the edge-complement of the associated comparability graph $G(P)$.

**incomparable pair** (in a poset $(X, R)$):  a pair $x, y \in X$ such that neither $x \leq y$ nor $y \leq x$ in $R$.

**integer partition**:  a nonincreasing nonnegative integer sequence having finitely many nonzero terms, with trailing zeros added as needed for comparison.

**intersecting family**:  a collection of subsets of a set such that every pair of members has nonempty intersection.

**intersection** (of partial orderings $P = (X, R)$ and $Q = (X, S)$ on the set $X$):  the poset $(X, R \cap S)$ that includes the comparisons present in both.

**intersection** (of posets $(X, R)$ and $(X, S)$):  the poset $(X, R \cap S)$.

**interval** (in a poset):  the subposet which contains all elements $z$ such that $x \leq z \leq y$.

**interval order**:  a poset in which there is an assignment to its members of real intervals, such that $x < y$ if and only if the interval for $y$ is totally to the right of the interval for $x$.

**interval representation** (of a poset $P$):  a collection of real intervals corresponding to an interval order for $P$.

**isomorphic** (posets):  posets $P = (X, R)$ and $Q = (Y, S)$ such that there is a poset isomorphism $P \rightarrow Q$.

**isomorphism** (of lattices):  an order-preserving bijection from one lattice to another that also preserves greatest lower bounds and least upper bounds of pairs.

**isomorphism** (of posets):  a bijection from one poset to another that preserves the order relation.

**join**:  given $\{x, y\}$, another name for the least upper bound $x \vee y$.

**join-irreducible element** (in a lattice):  a nonzero element that cannot be expressed as the join of two other elements.

**Jordan-Dedekind chain condition**:  the condition for a poset that every interval has finite length.

**lattice**:  a poset in which every pair of elements has both a greatest lower bound and a least upper bound.

***lattice*** (of bounded sequences): the set $L(m, n)$ of length-$m$ real sequences $a_1, \ldots, a_m$ such that $0 \le a_1 \le \cdots \le a_m \le n$.

***lattice*** (of order ideals in a poset $P = (X, R)$): the set $J(P)$ of order ideals of $P$, ordered by inclusion.

***least upper bound*** (***lub***) (of elements $x$ and $y$ in a poset): a common upper bound $z$ such that every other common upper bound $z'$ satisfies the inequality $z' \ge z$. Such an element, if it exists, is denoted $x \vee y$.

***length*** (of a chain): the number of cover relations in the chain; in other words, one less than the number of elements in the chain.

***length*** (of a poset): the length of a longest chain, which is one less than the height of that poset. (Sometimes *height* is used synonymously with length.)

***lexicographic ordering*** (of the Cartesian product of posets): the ordering for the Cartesian product of the domains in which $(x_1, x_2) \le (y_1, y_2)$ if and only if $x_1 < y_1$ or $x_1 = y_1$ and $x_2 \le y_2$; this is not the usual ordering of the Cartesian product of posets.

***linear extension*** (of a poset): an extension of the poset that is a chain.

***linear order***: See *total order*.

***linearly ordered set***: a poset in which every pair of elements is comparable.

***linear sum*** (of two disjoint posets $P$ and $P'$): the poset in which all the elements of poset $P$ lie "below" all those of poset $P'$.

***locally finite poset***: a poset in which every interval is finite.

***lower bound*** (of elements $x$ and $y$ in a poset): an element $z$ such that $x \ge z$ and $y \ge z$.

***lower semimodular lattice***: a lattice whose dual is upper semimodular.

***majority rule property*** (for a consensus function): the property that $x$ will be preferred to $y$ if and only if a majority of the individuals prefer $x$ to $y$.

***maximal element*** (in a poset): an element such that no other element is greater.

***meet*** (of elements $x$ and $y$): another name for the greatest lower bound $x \wedge y$.

***meet-irreducible element*** (of a lattice): a nonzero element that cannot be expressed as the meet of two other elements.

***minimal element*** (in a poset): an element such that no other element is less.

***minimum realizer encoding*** (of a poset): a poset that lists for each element its position on each extension in a minimum realizer.

***modular lattice***: a lattice in which $x \wedge (y \vee z) = (x \wedge y) \vee z$ for all $x$, $y$, $z$ such that $z \le x$.

***module*** (in a graph $G$): a vertex subset $U \subseteq V_G$ such that each vertex outside $U$ is adjacent to all or none of the vertices in $U$.

***$k$-norm*** (of a sequence $a = \{a_i\}$): the sum $\sum_i \min\{k, a_i\}$, whose value is commonly denoted $m_k(a)$.

***$k$-norm of a chain partition***: the $k$-norm of its sequence of chain sizes.

***normalized matching property*** (for a graded poset): the property stating that for every rank $k$ and every subset $A$ of rank $P_k$, the set $A^*$ of elements in the rank $P_{k+1}$ that are comparable to at least one element of $A$ satisfies the inequality $\frac{|A^*|}{N_{k+1}} \ge \frac{|A|}{N_k}$.

**order module in a poset**: a set $S$ of elements such that every element outside $S$ is above all of $S$, below all of $S$, or incomparable to all of $S$.

**order-preserving mapping** (from poset $P = (X, R)$ to poset $Q = (Y, S)$): a function $f\colon X \to Y$ such that $f(x) \leq f(y)$ whenever $x \leq y$ in $P$.

**order relation** (on a set $X$): a relation $R$ such that $(X, R)$ is a partially ordered set.

**partially ordered set**: a pair $P = (X, R)$ consisting of a set $X$ and a relation $R$ that is reflexive, antisymmetric, and transitive.

**partition lattice**: the poset $\Pi_n$ of partitions of the set $[n] = \{1, \ldots, n\}$, where $\pi < \sigma$ if $\pi$ is a refinement of $\sigma$.

**permutation graph**: a graph whose vertices can be placed in 1-1 correspondence with the elements of a permutation of $[n] = \{1, \ldots, n\}$, such that $v_i$ is adjacent to $v_j$ if and only if the larger of $\{i, j\}$ comes first in the permutation.

**planar poset**: a poset with a Hasse diagram that has no edge-crossings.

**plurality consensus function** (on a set of social choice profiles): the consensus function in which the winner(s) is (are) the alternative(s) appearing in the greatest number of top ranks, after which the winner(s) is (are) deleted and the procedure is repeated to select the next rank of the consensus ranking, etc.

**poset**: a partially ordered set.

**profile** (on a set of alternative social choices): a set $P = \{P_i \mid i \in I\}$ of linear rankings (ties allowed) of the alternatives, one for each member of a set $I$ of "individuals" participating in the decision process.

**quasi-transitive orientation** (on a simple graph $G$): an assignment of directions to the edges of $G$ so that whenever there is an $xy$-arc and a $yz$-arc, there is also an arc between $x$ and $z$.

**rank** (of a graded poset $P$): the length $r(P)$ of any maximal chain in $P$.

**rank function** (on a poset): an integer-valued function $r$ on the elements of the poset so that "$y$ covers $x$" implies that $r(y) = r(x) + 1$.

**ranked poset**: a poset having a rank function.

**kth rank of a ranked poset**: the subset $P_k$ of elements for which $r(x) = k$.

**rank parameters** (of a subset $F$ of elements in a ranked poset $P$): the numbers $f_k = |F \cap P_k|$.

**ranking**: a poset $P$ whose elements are partitioned into ranks $P_1, \ldots, P_k$ such that two elements are incomparable in the poset if and only if they belong to the same rank.

**realizer** (of a poset $P$): a set of linear extensions of $P$ whose intersection is $P$.

**refinement** (of a set partition $\sigma$): replacement of each block $B \in \sigma$ by some partition of $B$.

**regular covering** (of a poset by chains): a multiset of maximal chains such that for each element $x$ the fraction of the chains containing $x$ is $\frac{1}{N_{r(x)}}$, where $N_{r(x)}$ is a Whitney number.

**self-dual poset**: a poset isomorphic to its dual.

**semimodular lattice**: an upper semimodular lattice.

**semiorder**: a poset on which there is a real-valued function $f$ and a real number $\delta > 0$ such that $x < y$ if and only if $f(y) - f(x) > \delta$.

**shadow** (of a family of sets **F**): the collection of sets containing every set that is obtainable by selecting a set in **F** and deleting one of its elements.

**size** (of a finite poset): the number of elements.

**Sperner property** (for a graded poset): the property that some single rank is a maximum antichain.

**k-Sperner property** (for a graded poset): the property that the poset has a maximum $k$-family consisting of $k$ ranks.

**standard k-chain**: the poset $\{1, \ldots, k\}$, under the usual ordering of the integers, written $\underline{k}$.

**standard example of an n-dimensional poset**: the subposet $S_n$ of the Boolean algebra $\underline{2}^n$ induced by the singletons and their complements.

**strict Sperner property**: the property of a graded poset that all maximum antichains are single ranks.

**strong Sperner property**: the property that a graded poset $P$ is $k$-Sperner for all $k \leq r(P)$.

**Steinitz exchange axiom**: for a closure operator $\sigma \colon 2^E \to 2^E$, the rule that $p \notin \sigma(A)$ and $p \in \sigma(A \cup q)$ imply $q \in \sigma(A \cup p)$.

**sublattice** (of a lattice): a subposet that contains the meet and join of every pair of its elements.

**submodular height function** (in a lattice): a height function $h$ such that $h(x \wedge y) + h(x \vee y) \leq h(x) + h(y)$ for all $x$, $y$.

**subposet** (of a poset $(X, R)$): a poset $(Y, S)$ such that $Y \subseteq X$ and $S = R \cap (Y \times Y)$.

**subset lattice**: the Boolean algebra $\underline{2}^n$, that is, the Cartesian product of $n$ copies of the standard 2-chain.

**subspace lattice**: the set $L_n(q)$ of subspaces of an $n$-dimensional vector space over a $q$-element field, partially ordered by set inclusion.

**symmetric chain** (in a ranked poset $P$): a chain that has an element of rank $r(P) - k$ whenever it has an element of rank $k$.

**symmetric chain decomposition** (of a ranked poset): a partition of that poset into symmetric consecutive chains.

**symmetric chain order**: a poset with a symmetric chain decomposition.

**topological ordering** (of an acyclic digraph): a linear extension of the poset it represents.

**topological sort**: an algorithm that arranges the elements of a partially ordered set into a total ordering that is compatible with the original partial ordering.

**total order** (of a set): an order relation in which each pair of distinct elements is comparable.

**transitive orientation** (on a simple graph): an assignment of directions to the edges of a simple graph $G$ so that whenever there is an $xy$-arc and a $yz$-arc, there is also an $xz$-arc.

**triangular chord** (for a walk $x_1, \ldots, x_k$ in an undirected graph): an edge between vertices $x_{i-1}$ and $x_{i+1}$, two apart on the walk.

**upper bound** (of elements $x$ and $y$ in a poset): an element $z$ with $x \leq z$ and $y \leq z$.

**upper semimodular lattice**: a lattice in which whenever $x$ covers $x \wedge y$, it is also true that $x \vee y$ covers $y$.

**upset** (in a poset): a filter.

**weak order**: a ranking, i.e., a poset $P$ whose elements are partitioned into ranks $P_1, \ldots, P_k$ such that two elements are incomparable if and only if they belong to the same rank.

**kth Whitney number** (of a ranked poset $P$): the cardinality $|P_k|$ of the $k$th rank; written $N_k(P)$.

**width** (of a poset): the maximum size of an antichain in the poset.

**Young lattice**: the lattice of integer partitions under component-wise ordering.

| *poset notation* | |
|---|---|
| *notation* | *meaning* |
| $y \geq x$ | $x \leq y$ |
| $x < y$ | $x \leq y$ and $x \neq y$ |
| $x \| y$ | $x \not\leq y$ and $y \not\leq x$ |
| $0$ | minimal element in a bounded poset |
| $1$ | maximal element in a bounded poset |
| $[x, y]$ | the interval $\{z \mid x \leq z \leq y\}$ |
| $\underline{k}$ | standard $k$-chain |
| $P_1 + P_2$ | disjoint union of posets |
| $P_1 \oplus P_2$ | linear sum of two posets |
| $P_1 \times P_2$ | Cartesian product of two posets |
| $P^n$ | iterated Cartesian product of $n$ copies of $P$ |
| $P^*$ | dual of poset $P$ |
| $D(n)$ | divisibility poset of the integer $n$ |
| $r(P)$ | rank of a graded poset $P$ |
| $N_k(P)$ | $k$th Whitney number (= cardinality of $k$th rank) of $P$ |
| $w(P)$ | width of $P$ (= maximum size of an antichain) |
| $D[x]$ | downset (ideal) $\{y \mid y \leq x\}$ |
| $D(x)$ | downset (ideal) $\{y \mid y < x\}$ |
| $U[x]$ | upset (filter) $\{y \mid y \geq x\}$ |
| $U(x)$ | upset (filter) $\{y \mid y > x\}$ |
| $x \vee y$ | lub of $x$ and $y$ |
| $x \wedge y$ | glb of $x$ and $y$ |

# 11.1   BASIC POSET CONCEPTS

## 11.1.1   COMPARABILITY

The integers and the real numbers are totally ordered sets, since every pair of distinct elements can be compared. In a partially ordered set, some pairs of elements may be incomparable. For example, under the containment relation, the sets $\{1, 2\}$ and $\{1, 3\}$ are incomparable.

**Definitions:**

A **partial ordering** (or **order relation**) $R$ on a set $X$ is a binary relation that is

- *reflexive:* for all $x \in S$, $xRx$;
- *antisymmetric:* for all $x, y \in S$, if $xRy$ and $yRx$, then $x = y$;
- *transitive:* for all $x, y, z \in S$, if $xRy$ and $yRz$, then $xRz$.

*Note:* $x \leq y$ or $x \leq_P y$ are often written in place of $xRy$ or $(x, y) \in R$. Also, $y \geq x$ means $x \leq y$. The notation $\preceq$ is sometimes used in place of $\leq$. See the table following the glossary for further poset notation.

A **partially ordered set** (or **poset**) $P = (X, R)$ is a pair consisting of a set $X$, called the **domain**, and a partial ordering $R$ on $X$. Writing $x \in P$ means that $x \in X$. The notation $(X, \leq)$ is often used instead of $(X, R)$ to designate a poset.

The **size** of a finite poset $P$ is the number of elements in the domain.

The elements $x$ and $y$ are **comparable** (**related**) in $P$ if either $x \leq y$ or $y \leq x$ (or both, in which case $x = y$).

A **totally ordered set** (or **linearly ordered set**) is a poset in which every element is comparable to every other element.

The elements $x$ and $y$ are **incomparable** (**unrelated**) if they are not comparable. Writing $x \parallel y$ indicates incomparability.

Element $x$ is **less than** element $y$, written $x < y$, if $x \leq y$ and $x \neq y$. (The notation $\prec$ is sometimes used in place of $<$.)

Element $x$ is **greater than** element $y$, written $x > y$, if $x \geq y$ and $x \neq y$.

An element $x$ of a poset is **minimal** if the poset has no element less than $x$.

An element $x$ of a poset is **maximal** if the poset has no element greater than $x$.

A poset is **bounded** if it has both a unique minimal element (denoted "0") and a unique maximal element (denoted "1").

The **comparability digraph** $D(P)$ of a poset $P = (X, R)$ is the digraph with vertex set $X$, such that there is an arc from $x$ to $y$ if and only if $x \leq y$.

The **comparability graph** $G(P)$ of a poset $P = (X, R)$ is the simple graph with vertex set $X$, such that $xy \in E_G$ if and only if $x$ and $y$ are comparable in $P$, where $x \neq y$.

The **incomparability graph** of a poset $P$ is the edge-complement of the comparability graph $G(P)$.

The **induced poset** of an acyclic digraph $D$ is the poset whose elements are the vertices of $D$ and such that $x \leq y$ if and only if there is a directed path from $x$ to $y$.

The element $y$ **covers** the element $x$ in a poset if $x < y$ and there is no intermediate element $z$ such that $x < z < y$.

The **cover graph** of poset $P$ is the graph with vertex set $X$ such that $x$ and $y$ are adjacent if and only if one of them covers the other in $P$.

A **Hasse diagram** (or **cover diagram** or **diagram**) of poset $P$ is a straight-line drawing of the cover graph in the plane such that the lesser element of each pair satisfying the cover relation is lower in the drawing.

A poset is **planar** if it has a Hasse diagram without edge-crossings.

A **subposet** of $P = (X, \leq)$ is a subset $Y \subseteq X$ with the relation $x \leq y$ in $Y$ if and only if $x \leq y$ in $X$.

The **interval** $[x, y]$ in poset $P$ is the subposet that contains all elements $z$ such that $x \leq z \leq y$.

A poset $P$ is **locally finite** if every interval in $P$ has finitely many elements.

An **order-preserving mapping** from poset $P = (X, \leq_P)$ to poset $Q = (Y, \leq_Q)$ is a function $f \colon X \to Y$ such that $x \leq_P x'$ implies $f(x) \leq_Q f(x')$.

An **isomorphism of posets** $P = (X, R)$ and $Q = (Y, S)$ is a bijection $f \colon X \to Y$ that preserves the order relation: whenever $x_1 \leq_P x_2$, then $f(x_1) \leq_Q f(x_2)$.

**Isomorphic posets** are posets $P = (X, R)$ and $Q = (Y, S)$ such that there is a poset isomorphism $P \to Q$. This is sometimes indicated informally by writing $P = Q$.

Poset $Q = (Y, S)$ **is contained in** (or **imbeds in**) poset $P = (X, R)$ if $Q$ is isomorphic to a subposet of $P$.

A poset $P$ is $Q$**-free** if $P$ does not contain a poset isomorphic to $Q$.

## Facts:

**1.** Every finite nonempty poset has a minimal element and a maximal element.

**2.** The comparability digraph $D(P)$ of a poset $P$ is an acyclic digraph.

**3.** The minimal elements of a poset $P$ induced by a digraph $D$ are the *sources* of $D$; that is, they are the vertices at which every arc points outward.

**4.** The maximal elements of a poset $P$ induced by a digraph $D$ are the *sinks* of $D$; that is, they are the vertices at which every arc points inward.

**5.** The element $y$ covers the element $x$ in a poset $P$ induced by a digraph $D$ if and only if there is an arc in digraph $D$ from $x$ to $y$ and there is no other directed path from $x$ to $y$.

**6.** Suppose that the poset $P$ is induced from an acyclic digraph $D$. Then the comparability digraph of $P$ is the transitive closure of $D$.

**7.** Two different posets cannot have the same Hasse diagram, but they may have the same cover graph or the same comparability graph.

**8.** There is a polynomial-time algorithm to check whether a graph $G$ is a comparability graph, but the problem of deciding whether there exists a poset for which $G$ is the cover graph is NP-complete.

## Examples:

**1.** Any collection of subsets of the same set forms a poset when the subsets are partially ordered by the usual inclusion relation $X \subseteq Y$.

**2.** *Boolean algebra:* The Boolean algebra on a set $X$ is the poset consisting of all the subsets of $X$, ordered by inclusion.

**3.** The Boolean algebra on the set $\{a, b, c\}$ has the following Hasse diagram. The only maximal element is $\{a, b, c\}$. The only minimal element is $\emptyset$.

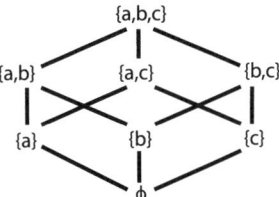

**4.** There are five different isomorphism types of posets of size three, whose Hasse diagrams are as follows.

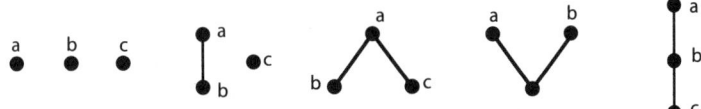

**5.** There are 16 different isomorphism types of posets of size four, whose Hasse diagrams are as follows.

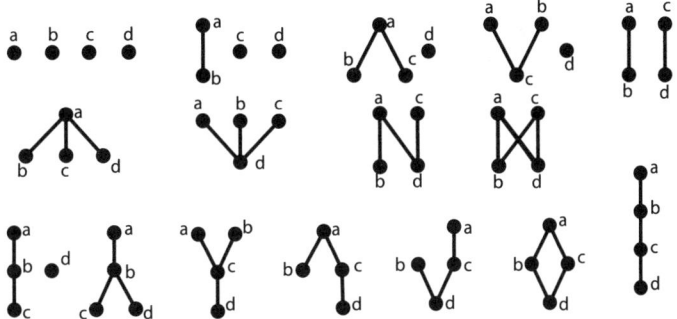

**6.** *Divisibility poset:* The divisibility poset on the set $I$ of positive integers, denoted $D(I)$, has the relation $x \leq y$ if $y$ is an integer multiple of $x$. A number $y$ covers a number $x$ if and only if the quotient $\frac{y}{x}$ is prime.

**7.** The set $D(n)$ of divisors of $n$ forms a subposet of $D(I)$, for any positive integer $n$. The set $D(n)$ is identical to the interval $[1, n]$ in $D(I)$. For instance, the following figure is the Hasse diagram of $D(24) = [1, 24]$.

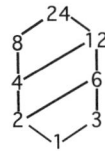

**8.** The interval $[3, 30]$ in $D(I)$ has domain $\{3, 6, 12, 15, 30\}$. The interval $[2, 24]$ has domain $\{2, 4, 6, 8, 12, 24\}$.

**9.** The poset $D(I)$ is infinite, but locally finite.

**10.** The Boolean algebra of all subsets of an infinite set is not a finite poset. Nor is it locally finite, since each interval from a finite set to an infinite superset is infinite.

**11.** The poset $D(6)$ is isomorphic to the poset of subsets of $\{a, b\}$, as seen next.

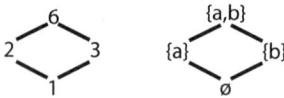

**12.** To generalize Example 11, if $p_1, \ldots, p_n$ are distinct primes, then the divisibility poset $D(p_1 \ldots p_n)$ is isomorphic to the poset of subsets of a set of $n$ objects.

**13.** The partitions of a set form a poset under refinement, as illustrated for the set $\{a, b, c, d\}$. A notation such as $ab\text{-}c\text{-}d$ means a partition of the set $\{a, b, c, d\}$ into the subsets $\{a, b\}$, $\{c\}$, $\{d\}$.

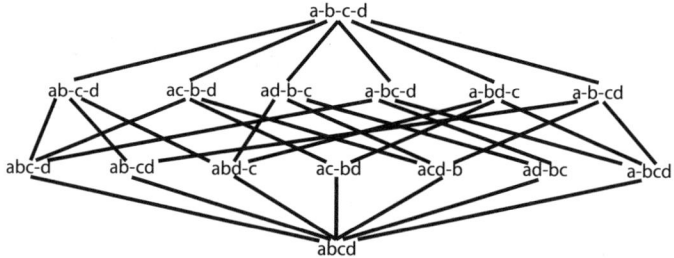

**14.** The 6-cycle is the comparability graph of exactly one (isomorphism type of) poset, which has the following Hasse diagram.

**15.** The 6-cycle is the cover graph of seven posets, all of which are planar. They have the following Hasse diagrams.

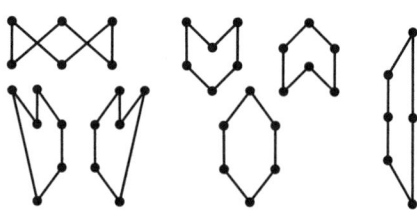

## 11.1.2   CHAINS, ANTICHAINS, AND POSET OPERATIONS

**Definitions:**

A *chain* is a subset $S$ of mutually comparable elements of a poset $P$, or sometimes the subposet of $P$ formed by such a subset.

The **length** of a finite chain $C$ is $|C| - 1$, i.e., the number of edges in the Hasse diagram of that chain, regarded as a poset.

A **k-chain** is a chain of size $k$, i.e., a chain on $k$ elements.

The **standard k-chain** $\underline{k}$ is a fixed $k$-chain, presumed to be disjoint from other objects in the universe of discourse.

The **height of a poset** $P$ is the maximum size of a chain in $P$.

The **height of an element** $x$ in a poset $P$ is the maximum length $h(x)$ of a chain in $P$ that has $x$ as its maximal element.

A **bipartite poset** is a poset of height at most 2.

A **chain-product** (or **grid**) is the Cartesian product of a collection of chains.

An **antichain** (or **clutter** or **Sperner family**) is a subset $S$ of pairwise incomparable elements of a poset $P$, or sometimes the subposet of $P$ formed by such a subset.

A chain or antichain is **maximal** if it is contained in no other chain or antichain.

A chain or antichain in a finite poset is a **maximum** chain or antichain if it is one of maximum size.

The **disjoint union** of two posets $P = (X, R)$ and $P' = (X', R')$ with $X \cap X' = \emptyset$ is the poset $(X \cup X', R \cup R')$, denoted $P + P'$.

The **linear sum** of two posets $P = (X, R)$ and $P' = (X', R')$ with $X \cap X' = \emptyset$ is the poset $(X \cup X', R \cup R' \cup (X \times X'))$, denoted $P \oplus P'$. (This puts all of poset $P$ "below" poset $P'$.)

The **Cartesian product** $P \times P'$ (or **direct product** or **product**) of two posets $P = (X, R)$ and $P' = (X', R')$ is the poset $(X \times X', S)$, such that $(x, x')S(y, y')$ if and only if $xRy$ and $x'R'y'$.

The **iterated Cartesian product** of $n$ copies of a poset $P = (X, \leq)$, written $P^n$, is the set of $n$-tuples in $P$, such that $(x_1, \ldots, x_n) \leq (y_1, \ldots, y_n)$ if and only if $x_j \leq y_j$ for all $j = 1, \ldots, n$.

The **lexicographic ordering** of the Cartesian product $P_1 \times P_2$ of the domains of two posets is the partial ordering in which $(x_1, x_2) \leq (y_1, y_2)$ if and only if $x_1 < y_1$, or $x_1 = y_1$ and $x_2 \leq y_2$.

The **dual** of a poset $P$, denoted $P^*$, is the poset on the elements of $P$ defined by the relation $y \leq_{P^*} x$ if and only if $x \leq_P y$.

A **self-dual poset** is a poset that is isomorphic to its dual.

**Facts:**

1. Every $k$-chain is isomorphic to the linear sum $\underline{1} \oplus \underline{1} \oplus \cdots \oplus \underline{1}$ of $k$ copies of $\underline{1}$.

2. Every antichain of size $k$ is isomorphic to the disjoint union $\underline{1} + \underline{1} + \cdots + \underline{1}$ of $k$ copies of $\underline{1}$.

3. The chains are characterizable as the class of $(\underline{1} + \underline{1})$-free posets.

4. The cover graph of a chain is a path.

5. The comparability graph of a chain of size $n$ is the complete graph $K_n$.

6. The antichains are the class of $\underline{2}$-free posets.

**7.** The comparability graph of an antichain has no edges.

**8.** The maximum size of a chain in a finite poset $P$ equals the minimum number of antichains needed to cover the elements of $P$, that is, the minimum number of antichains whose union equals the domain of poset $P$.

**9.** The bipartite posets are precisely the $\underline{3}$-free posets.

**10.** The bipartite posets are the posets whose comparability graph and cover graph are the same.

**11.** Every maximal chain of a finite poset $P$ extends from a minimal element of $P$ to a maximal element of $P$, and successive pairs on a maximal chain satisfy the cover relation of $P$.

**12.** The Cartesian product of two posets is a poset.

**13.** A poset and its dual have the same comparability graph and the same cover graph.

**14.** The Hasse diagram of the dual of a poset $P$ can be obtained from the Hasse diagram of $P$ either by reflecting through the horizontal axis or by rotating 180 degrees.

**15.** The set of order-preserving maps from a poset $P$ to a poset $Q$ forms a poset, denoted by $Q^P$, under "coordinate-wise ordering": $f \leq g$ in $Q^P$ if and only if $f(x) \leq_Q g(x)$ for all $x \in P$.

**Examples:**

**1.** The following figure shows: (a) a 3-chain, (b) an antichain of width 4, and (c) a bipartite poset.

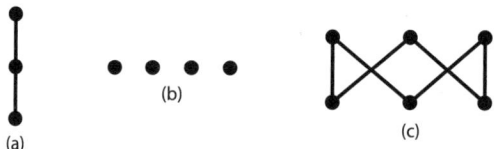

**2.** The poset $\underline{2}^3$ is not planar, even though it has a planar cover graph. However, deleting its minimal element or maximal element leaves a planar subposet.

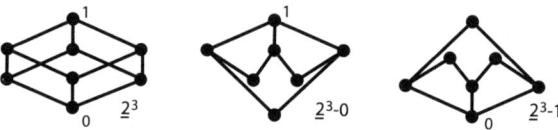

**3.** The cover graph of the poset $\underline{2}^n$ is isomorphic to the $n$-dimensional cube, whose vertices are the bitstrings of length $n$, with bitstrings adjacent if they differ in one position. Each bit encodes the possible presence of an element of the set of which $B_n$ is the Boolean algebra (§5.8.1).

**4.** The interval in the Boolean algebra $\underline{2}^n$ between an element of rank $k$ and an element of rank $l \geq k$ is isomorphic to the poset $\underline{2}^{l-k}$.

**5.** Every maximal chain in the Boolean algebra $\underline{2}^n$ has size $n+1$ and length $n$, and there are $n!$ such chains. There are maximal antichains as small as one element.

**6.** In general, the poset $D(n)$ is isomorphic to a chain product, one factor for each prime divisor of $n$. The elements of $D(n)$ can be encoded as integer vectors $\{a_1, \ldots, a_n \mid 0 \leq a_i < e_i\}$, where $n$ is a product of distinct primes with powers $e_1, \ldots, e_n$, and $a \leq b$ if and only if $a_i \leq b_i$ for all $i$.

**7.** The Hasse diagrams for two possible partial orderings on the Cartesian product of the domains of two posets are shown in the following figure.

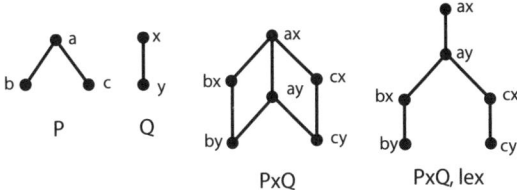

**8.** The two posets $M_5 = \underline{1} \oplus (\underline{1} + \underline{1} + \underline{1}) \oplus \underline{1}$ and $N_5 = \underline{1} \oplus (\underline{2} + \underline{1}) \oplus \underline{1}$ are used in §11.1.5 in a forbidden subposet description.

**9.** The following posets are self-dual.

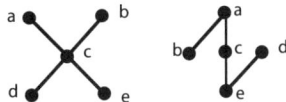

---

## 11.1.3 RANK, IDEALS, AND FILTERS

**Definitions:**

A **graded poset** is a poset in which all maximal chains have the same length.

The **rank** $r(P)$ of a graded poset $P$ is the length of any maximal chain.

A **rank function** $r$ on a poset $P$ is an assignment of integers to the elements so that the relation $y$ covers $x$ implies that $r(y) = r(x) + 1$.

A **ranked poset** is a poset having a rank function.

The **kth rank** of a ranked poset $P$ is the subset $P_k$ of elements $x$ for which $r(x) = k$.

The **kth rank parameter** of a subset of elements $F$ in a ranked poset $P$ is the cardinality $|F \cap P_k|$ of the number of elements of $F$ in the $k$th rank of $P$.

The **kth Whitney number** $N_k(P)$ of a ranked poset $P$ is the cardinality $|P_k|$ of the $k$th rank.

The **length of a poset** $P$ is the length of a longest chain in $P$, which is one less than the height of $P$. *Note:* Sometimes "height" is used synonymously with length.

The **Jordan-Dedekind chain condition** for a poset is that every interval has finite length.

The **width of a poset** $P$, denoted $w(P)$, is the maximum size of an antichain in $P$.

An *ideal* (or *downset*, *order ideal*, or *hereditary family*) in a poset $P$ is a subposet $I$ such that if $x \in I$ and $y < x$, then $y \in I$.

A *filter* (or *upset* or *dual ideal*) in a poset $P$ is a subposet $F$ whose domain is the set-theoretic complement of the domain of an ideal.

The *ideal generated by an element* $x$ in a poset $P$ is the downset $D[x] = \{y \in P \mid y \le x\}$. The related notation $D(x)$ means the downset $\{y \in P \mid y < x\}$.

The *ideal generated by a subset* $A$ in a poset $P$ is the downset $D[A] = \bigcup_{x \in A} D[x]$. The related notation $D(A)$ means the downset $\bigcup_{x \in A} D(x)$.

The *filter generated by an element* $x$ in a poset $P$ is the upset $U[x] = \{y \in P \mid y \ge x\}$. The related notation $U(x)$ means the upset $\{y \in P \mid y > x\}$.

The *filter generated by a subset* $A$ in a poset $P$ is the upset $U[A] = \bigcup_{x \in A} U[x]$. The related notation $U(A)$ means the upset $\bigcup_{x \in A} U(x)$.

A *forbidden subposet description* of a class of posets is a characterization of that class as the class of all posets that does not contain any of the posets in a specified collection. (This generalizes the concept of $Q$-free.)

**Facts:**

**1.** A graded poset has a rank function, in which the rank of each element is defined to be its height.

**2.** If posets $P_1, P_2$ have rank functions $r_1, r_2$, then the Cartesian product $P = P_1 \times P_2$ is ranked, so that the element $x = (x_1, x_2)$ has rank $r(x) = r_1(x_1) + r_2(x_2)$.

**3.** In a Cartesian product of finite ranked posets $P_1$ and $P_2$, the Whitney numbers for the Cartesian product $P = P_1 \times P_2$ satisfy the equation $N_k(P) = \sum_i N_i(P_1) N_{k-i}(P_2)$.

**4.** The Boolean algebra on a set $X$ of cardinality $n$ is isomorphic to $\underline{2}^n$, the Cartesian product of $n$ copies of $\underline{2}$. This poset isomorphism type is often denoted $B_n$.

**5.** The Boolean algebra on a set $X$ of cardinality $n$ is a graded poset, with rank function $r(S) = |S|$, and with Whitney numbers $N_k(\underline{2}^n) = \binom{n}{k}$.

**6.** The sequence of Whitney numbers on the Boolean algebra on a set $X$ of cardinality $n$ is symmetric, since $\binom{n}{k} = \binom{n}{n-k}$. It is also unimodal, since the sequence rises monotonically to the maximum and then falls monotonically.

**7.** *Sperner's theorem:* The only maximum antichains in the Boolean algebra $\underline{2}^n$ are the middle ranks (one such rank if $n$ is even, two if $n$ is odd). Thus the width of $\underline{2}^n$ is $\binom{n}{\lfloor n/2 \rfloor}$.

**8.** The maximal elements of an ideal form an antichain, as do the minimal elements of a dual ideal; these yield natural bijections between the set of antichains in a poset $P$ and the sets of ideals or dual ideals of $P$.

**9.** The divisibility poset $D(I)$ on the integers satisfies the Jordan-Dedekind chain condition.

**Examples:**

**1.** In the poset $P$ of partitions of $\{a, b, c, d\}$ under inverse refinement, illustrated next, the Whitney numbers are $N_1(P) = 1$, $N_2(P) = 6$, $N_3(P) = 7$, and $N_4(P) = 1$.

**2.** In the poset of partitions of $\{a, b, c, d\}$ under inverse refinement, the ideal $D[ac\text{-}bd]$ is the set $\{ac\text{-}bd, a\text{-}bd\text{-}c, ac\text{-}b\text{-}d, a\text{-}b\text{-}c\text{-}d\}$, and the ideal $D(ac\text{-}bd)$ is the set $\{a\text{-}bd\text{-}c, ac\text{-}b\text{-}d, a\text{-}b\text{-}c\text{-}d\}$.

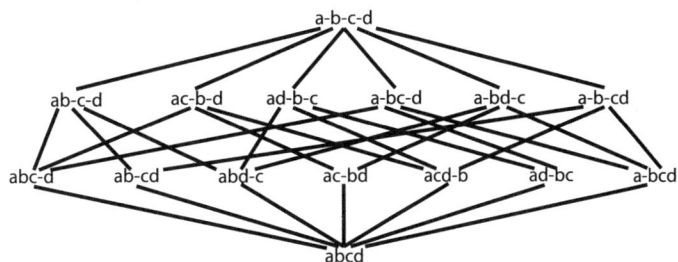

**3.** In the poset of partitions of $\{a, b, c, d\}$ under inverse refinement, the filter $U[a\text{-}bd\text{-}c]$ is the set $\{a\text{-}bd\text{-}c,\ abd\text{-}c,\ ac\text{-}bd,\ a\text{-}bcd,\ abcd\}$, and the filter $U(a\text{-}bd\text{-}c)$ is the set $\{abd\text{-}c,\ ac\text{-}bd,\ a\text{-}bcd,\ abcd\}$.

**4.** In the graded poset $D(n)$ of divisors of $n$, the rank $r(x)$ is the sum of the exponents in the prime power factorization of $x$. The Whitney numbers of $D(n)$ are symmetric because the divisors $x$ and $\frac{n}{x}$ have "complementary" ranks. If $n$ is a product of $k$ distinct primes, then $D(n) \cong \underline{2}^k$.

**5.** The following Hasse diagram corresponds to an ungraded poset, because the lengths of its maximum chains differ, i.e., they are length 2 and length 3.

**6.** In the divisibility poset $D(I)$, the subposet $D(n)$ of divisors of $n$ is a finite ideal, and the non-multiples of $n$ form an infinite ideal, whose complement is the infinite filter $U(n)$ of numbers that are divisible by $n$.

---

### 11.1.4  LATTICES

Lattices are posets with additional properties that capture some aspects of the intersection and the union of sets and (more generally) of the greatest common divisor and least common multiple of positive integers. (See also §5.7.)

**Definitions:**

A (**common**) **upper bound** for elements $x$, $y$ in a poset is an element $z$ such that $x \le z$ and $y \le z$.

A **least upper bound** (or **lub**, pronounced "lub") for elements $x$, $y$ in a poset is a common upper bound $z$ such that every other common upper bound $z'$ satisfies the inequality $z \le z'$. Such an element, if it exists, is denoted $x \vee y$.

The **join** of $x$ and $y$ is the lub $x \vee y$.

A (**common**) **lower bound** for elements $x$, $y$ in a poset is an element $z$ such that $x \ge z$ and $y \ge z$.

A **greatest lower bound** (or **glb**, pronounced "glub") for elements $x$, $y$ in a poset is a common lower bound $z$ such that every other common lower bound $z'$ satisfies the inequality $z \ge z'$. Such an element, if it exists, is denoted $x \wedge y$.

The **meet** of $x$ and $y$ is the glb $x \wedge y$.

A **lattice** is a poset in which every pair of elements has both a lub and a glb.

A lattice is **bounded** if it has both a unique minimal element (denoted "0") and a unique maximal element (denoted "1").

A nonzero element of a lattice $L$ is **join-irreducible** (or simply **irreducible**) if it cannot be expressed as the lub of two other elements. The subposet formed by the join-irreducible elements of $L$ is denoted by $P(L)$.

A nonzero element of a lattice $L$ is **meet-irreducible** if it cannot be expressed as a glb of two other elements. The subposet formed by the meet-irreducible elements of $L$ is denoted by $Q(L)$.

A **complement** of an element $x$ of a lattice is an element $\overline{x}$ such that $x \vee \overline{x} = 1$ and $x \wedge \overline{x} = 0$.

A **complemented lattice** is a lattice in which every element has a complement.

A **lattice isomorphism** is an order-preserving bijection from one lattice to another that also preserves glbs and lubs.

An **atom** of a poset is an element of height 1.

A lattice is **atomic** if every element is a lub of atoms (or equivalently, if the atoms are the only join-irreducible elements).

A **sublattice** of a lattice $L$ is a subposet $P$ such that $x \wedge y$ and $x \vee y$ are in $P$ for all $x$ and $y \in P$.

The **divisor lattice** is the poset $D(n)$ of the positive integer divisors of $n$, in which $x \leq y$ means that $y$ is an integer multiple of $x$.

The **subset lattice** is the Boolean algebra $\mathbf{2}^n$, that is, the Cartesian product of $n$ copies of the standard 2-chain.

The **subspace lattice** $L_n(q)$ is the set of subspaces of an $n$-dimensional vector space over a $q$-element field, partially ordered by set inclusion.

The **lattice of (order) ideals** $J(P)$, for any poset $P = (X, R)$, is the set of order ideals of $P$, ordered by inclusion.

The **lattice of bounded sequences** $L(m, n)$ has as members the length-$m$ real sequences $a_1, \ldots, a_m$ such that $0 \leq a_1 \leq \cdots \leq a_m \leq n$.

An **integer partition** is a nonincreasing nonnegative integer sequence having finitely many nonzero terms, with trailing zeros added as needed for comparison.

The **Young lattice** is the lattice of integer partitions under component-wise ordering.

A **refinement** of a set partition $\sigma$ replaces each block $B \in \sigma$ by some partition of $B$.

The **partition lattice** is the poset $\Pi_n$ of partitions of the set $[n] = \{1, \ldots, n\}$, where $\pi < \sigma$ if $\pi$ is a refinement of $\sigma$.

**Facts:**

1. If $z \leq x$ in a lattice, then $x \wedge (y \vee z) \geq (x \wedge y) \vee z$.

2. **4-point lemma:** If each of the elements $z$, $w$ is less than or equal to each of the elements $x$, $y$ in a lattice, then $z \vee w \leq x \wedge y$.

**3.** An element $z$ is a least upper bound for $x$ and $y$ if and only if it is a unique minimal element among their common upper bounds.

**4.** Every finite lattice is bounded.

**5.** Every chain-product is a lattice.

**6.** If a locally finite poset $P$ with a unique maximal element 1 also has a well-defined glb operation, then $P$ is a lattice.

**7.** Not all lattices are ranked. In particular, the lattice of integer partitions under dominance ordering is unranked.

**8.** Every interval in a lattice is a sublattice, but not every sublattice is an interval.

**9.** In the subspace lattice $L_n(q)$, the Whitney numbers (§11.1.3) satisfy the equation

$$N_k(L_n(q)) = \frac{(q^n - 1)(q^{n-1} - 1)\dots(q^{n-k+1} - 1)}{(q^k - 1)(q^{k-1} - 1)\dots(q^1 - 1)}.$$

**10.** In the subspace lattice $L_n(q)$, the Whitney number $N_k(L_n(q))$ equals the *Gaussian coefficient* $\begin{bmatrix} n \\ k \end{bmatrix}_q$ (§2.3.2), which appears in algebraic identities and in analogues of results on subsets.

**11.** The Whitney number $N_k(\Pi_n)$ of partitions of the set $[n]$ into $n - k$ blocks is the *Stirling subset number* $\left\{ {n \atop n-k} \right\}$ (§2.5.2). This has no closed formula, but the inclusion-exclusion principle yields $\left\{ {n \atop t} \right\} = \sum_{i=0}^{t}(-1)^i \frac{(t-i)^n}{t!}$.

**Examples:**

**1.** The poset specified by the following Hasse diagram is a lattice.

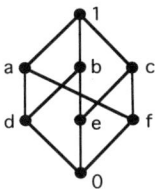

**2.** The poset specified by the following Hasse diagram is *not* a lattice. Although every pair of elements has a common upper bound, none of the three common upper bounds for the elements $c$ and $d$ is a least upper bound.

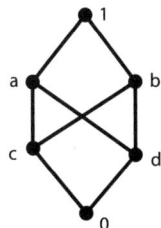

**3.** Two 5-element lattices that occur as subposets of $\underline{2}^3$ but not as sublattices of $\underline{2}^3$ are $M_5 = \underline{1} \oplus (\underline{1} + \underline{1} + \underline{1}) \oplus \underline{1}$ and $N_5 = \underline{1} \oplus (\underline{2} + \underline{1}) \oplus \underline{1}$.

**4.** In the divisor lattice $D(n)$, $a \wedge b = \gcd(a, b)$ and $a \vee b = \text{lcm}(a, b)$.

**5.** The join-irreducible elements of the divisibility lattice $D(I)$ are the powers of primes.

**6.** In the subset lattice $\underline{2}^n$, $a \wedge b = a \cap b$ and $a \vee b = a \cup b$.

**7.** The join-irreducible elements of the subset lattice $\underline{2}^n$ are the singleton sets.

**8.** The subspace lattice $L_n(q)$ is a graded lattice, with rank $r(U) = \dim U$.

**9.** In the subspace lattice $L_n(q)$, the meet of subspaces $U$ and $V$ is their intersection $U \cap V$, and the join is the unique minimal subspace containing their union.

**10.** In the lattice of order ideals $J(P)$, glb and lub are given by intersection and union. Hence $J(P)$ is a sublattice of the Boolean lattice $\underline{2}^{|P|}$; equality holds if and only if $P$ is an antichain.

**11.** The lattice $J(P)$ of ideals of a poset $P = (X, R)$ is finite, with $1_{J(P)} = X$ and $0_{J(P)} = \emptyset$. It is graded, with rank function $r(I) = |I|$.

**12.** By the correspondence between ideals of a poset $P = (X, R)$ and their antichains of maximal elements, the lattice $J(P)$ of ideals is also a lattice on the antichains of $P$. The corresponding ordering on antichains is $A \leq B$ if every element of $A$ is less than or equal to some element of $B$.

**13.** The lattice $L(m, n)$ of bounded sequences is a sublattice of $\underline{n+1}^m$, and $L(m, n) = J(\underline{m} \times \underline{n}) = L(n, m)$. The natural isomorphism maps a sequence $a \in L(m, n)$ to the order ideal of $\underline{m} \times \underline{n}$ generated by $\{(m + 1 - i, a_i) \mid a_i > 0\}$.

**14.** The lattice $L(m, n)$ is a sublattice of the Young lattice.

**15.** In the partition lattice $\Pi_n$, $13|4|2|5 < 123|45$; the order of the blocks and the order of elements within each block are irrelevant.

**16.** The partition lattice $\Pi_n$ is a graded poset, with $1_{\Pi_n} = [n]$ and $0_{\Pi_n} = 1|2|\cdots|n$. The common refinement of $\pi$ and $\sigma$ with the fewest blocks is the greatest lower bound (meet) of $\pi$ and $\sigma$.

**17.** The lattice $\Pi_3$ is isomorphic to the lattice $M_5$.

**18.** In the ordering on antichains of a poset $P$ defined by $A \leq B$ if every element of $A$ is less than or equal to some element of $B$, the maximum antichains of $P$ induce a sublattice.

---

## 11.1.5   DISTRIBUTIVE AND MODULAR LATTICES

### Definitions:

A lattice $L$ is **distributive** if glb distributes over lub in $L$, that is, if $x \wedge (y \vee z) = (x \wedge y) \vee (x \wedge z)$ for all $x, y, z \in L$.

A lattice $L$ is **modular** if $x \wedge (y \vee z) = (x \wedge y) \vee z$ for all $x, y, z \in L$ such that $z \leq x$.

A lattice $L$ is (**upper**) **semimodular** if for all $x, y \in L$, "$x$ covers $x \wedge y$" implies "$x \vee y$ covers $y$".

A lattice $L$ is **lower semimodular** if the reverse implication holds (equivalently, if the dual lattice $L^*$ is semimodular).

The height function $h$ of a lattice $L$ is a **submodular height function** if $h(x \wedge y) + h(x \vee y) \leq h(x) + h(y)$ for all $x, y \in L$.

A lattice is **geometric** if it is atomic, semimodular, and has finite height.

A **closure operator** on the subsets of a set $E$ is a function $\sigma \colon 2^E \to 2^E$ that maps each set to a superset of itself, is order-preserving with respect to set inclusion, and is "idempotent": $\sigma(\sigma(A)) = \sigma(A)$.

The **closed subsets** of a set $E$, with respect to a closure operator $\sigma\colon 2^E \to 2^E$, are the sets with $\sigma(A) = A$.

The **Steinitz exchange axiom** for a closure operator $\sigma\colon 2^E \to 2^E$ is the rule that $p \notin \sigma(A)$ and $p \in \sigma(A \cup q)$ imply $q \in \sigma(A \cup p)$.

## Facts:

**1.** The smallest nondistributive lattices are $M_5 = \underline{1} \oplus (\underline{1} + \underline{1} + \underline{1}) \oplus \underline{1}$ and $N_5 = \underline{1} \oplus (\underline{2} + \underline{1}) \oplus \underline{1}$, which are illustrated in §11.1.2, Example 8.

**2.** A lattice is distributive if and only if it occurs as a sublattice of $\underline{2}^n$ for some $n$.

**3.** Every sublattice of a distributive lattice is a distributive lattice.

**4.** The product of distributive lattices $L_1$ and $L_2$ is a distributive lattice, with $(x_1, x_2) \wedge (y_1, y_2) = (x_1 \wedge y_1, x_2 \wedge y_2)$ and $(x_1, x_2) \vee (y_1, y_2) = (x_1 \vee y_1, x_2 \vee y_2)$.

**5.** In a lattice $L$, distributivity and the dual property that $x \vee (y \wedge z) = (x \vee y) \wedge (x \vee z)$ for all $x, y, z \in L$ are equivalent. Hence the dual of a distributive lattice is a distributive lattice.

**6.** A lattice $L$ is modular if and only if $c \in [a \wedge b, a]$ implies $a \wedge (b \vee c) = c$ for all $a, b \in L$ (equivalently, if $c \in [b, b \vee d]$ implies $b \vee (c \wedge d) = d$ for all $b, d \in L$).

**7.** Let $\mu_a\colon L \to L$ be the operation "take the glb with $a$", and let $\nu_b\colon L \to L$ be the operation "take the lub with $b$". A lattice $L$ is modular if and only if for all $a, b \in L$, the intervals $[a \wedge b, a]$ and $[b, a \vee b]$ are isomorphic sublattices of $L$, with lattice isomorphisms given by $\nu_b$ and $\mu_a$.

**8.** If $y$ covers $x$ in a semimodular lattice $L$, then for all $z \in L$, $x \vee z = y \vee z$ or $x \vee z$ is covered by $y \vee z$.

**9.** A lattice $L$ with a lower bound is semimodular if and only if the following is true: the height function of $L$ is submodular and in each interval the maximal chains all have the same length.

**10.** A lattice is modular if and only if it does not have $N_5$ as a sublattice.

**11.** Every distributive lattice is modular, because in a distributive lattice $x \wedge (y \vee z) = (x \wedge y) \vee (x \wedge z) = (x \wedge y) \vee z$ if $z \leq x$.

**12.** A modular lattice is distributive if and only if it does not have $M_5$ as a sublattice.

**13.** Given a closure operator, the closed sets form a lattice under inclusion with meet and lub given by intersection and closure of the union, respectively.

**14.** If a closure operator $\sigma$ satisfies the Steinitz exchange axiom, then the lattice of closed sets is semimodular.

**15.** The lattice $L_n(q)$ is semimodular. (This follows from the previous fact.)

**16.** A poset is a geometric lattice if and only if it is the lattice of closed sets of a matroid, ordered by inclusion (§12.4). (The span operator in a matroid, which adds to $X$ every element whose addition to $X$ does not increase the rank, is a closure operator that satisfies the Steinitz exchange axiom.)

**17.** A geometric lattice is distributive if and only if it has the form $\underline{2}^n$, and the corresponding matroid is the free matroid, in which all subsets of the elements are independent.

**18.** A complemented distributive lattice is a Boolean algebra.

## Examples:

**1.** Among nondistributive lattices, the lattice $M_5$ is modular, and the lattice $N_5$ is not (which explains the notation).

**2.** The subspace lattices $L_n(q)$ are not distributive.

**3.** The partition lattice $\Pi_n$ is semimodular but not modular for $n > 3$. The lattice $L_n(q)$ is semimodular.

**4.** The partition lattice $\Pi_n$ is geometric, and it is the lattice of closed sets of the cycle matroid of the complete graph $K_n$.

**5.** For $n \geq 3$, the lattice $\Pi(n)$ is not distributive.

**6.** The Boolean lattice $\mathbf{2}^n$, the divisor lattice $D(N)$, the lattice $J(P)$ of order ideals of a poset, and the bounded sequence lattice $L(m, n)$ are distributive.

## 11.2   POSET PROPERTIES

### 11.2.1   POSET PARTITIONS

**Definitions:**

A **chain partition** of a poset is a partition of the domain of that poset into chains.

The **$k$-norm of a sequence** $x = \{x_i\}$ of real numbers is the sum $\sum_i \min\{k, x_i\}$, whose value is commonly denoted $m_k(x)$.

The **$k$-norm of a chain partition** $\mathbf{C}$ of a poset, denoted $m_k(\mathbf{C})$, means the $k$-norm of the sequence of sizes of the chains in the partition.

A **$k$-family** in a poset $P$ is a subposet containing no chain of size $k + 1$. The size of a maximum $k$-family in $P$ is denoted by $d_k(P)$.

A partition of a poset $P$ into chains is **$k$-saturated** if $m_k(\mathbf{C}) = d_k(P)$.

A chain in a ranked poset is **symmetric** if it has an element of rank $r(P) - k$ whenever it has an element of rank $k$.

A chain is **consecutive** if its elements belong to consecutive ranks.

A **symmetric chain decomposition** of $P$ is a partition of $P$ into symmetric consecutive chains.

A **symmetric chain order** is a poset with a symmetric chain decomposition.

A graded poset has the **Sperner property** if some single rank is a maximum antichain.

A graded poset has the **strict Sperner property** if all maximum antichains are single ranks.

A poset $P$ has the **$k$-Sperner property** if it has a maximum $k$-family consisting of $k$ ranks.

A poset has the **strong Sperner property** if it is $k$-Sperner for all $k \leq r(P)$.

A graded poset $P$ satisfies the **normalized matching property** if for every $k$ and every subset $A$ of $P_k$, the set $A^*$ of elements in $P_{k+1}$ that are comparable to at least one element of $A$ satisfies the inequality $\frac{|A^*|}{N_{k+1}} \geq \frac{|A|}{N_k}$, where $N_k$ and $N_{k+1}$ are Whitney numbers.

A **regular covering by chains** is a multiset of maximal chains such that for each $x \in P$ the fraction of the chains containing $x$ is $\frac{1}{N_{r(x)}}$.

To obtain the **bracket representation** of a subset $S$ of $[n] = \{1, \ldots, n\}$, first represent the subset $S$ as a length-$n$ "parenthesis-vector", in which the $j$th bit is a right parenthesis if $j \in S$ and a left parenthesis if $j \notin S$. Then wherever possible, recursively, match a left parenthesis to the nearest unmatched right parenthesis that is separated from it only by previously matched entries.

An **on-line partitioning algorithm** processes the elements of a poset as they are "revealed". Once an element is assigned to a cell, it remains there; there is no backtracking to change earlier decisions.

**Facts:**

**1.** *Dilworth's theorem:* If $P$ is a finite poset, then the width of $P$ equals the minimum number of chains needed to cover the elements of $P$.

**2.** Dilworth's theorem also holds for infinite posets of finite width.

**3.** The 1-families are the antichains.

**4.** Every $k$-family is a union of $k$ antichains.

**5.** A $k$-family in $P$ can be transformed into an antichain in $P \times \underline{k}$ of the same size, and vice versa, and hence $d_k(P) = w(P \times \underline{k})$.

**6.** The discussion of saturated partitions is generally restricted to finite posets.

**7.** If $\alpha_k$ is the number of chains of size at least $k$ in a $k$-saturated chain partition of $P$, then $\Delta_k(P) \geq \alpha_k \geq \Delta_{k+1}(P)$, where $\Delta_k(P) = d_k(P) - d_{k-1}(P)$ for $k \geq 1$.

**8.** *Littlewood-Offord problem:* Let $A = \{a_1, \ldots, a_n\}$ be a set of vectors in $\mathcal{R}^d$, with each vector having length at least 1. Let $R_1, \ldots, R_k$ be regions in $\mathcal{R}^d$ of diameter at most 1. Then of the $2^n$ subsets of $A$, the number whose sum lies in $\bigcup_i R_i$ is at most $d_k(\underline{2}^n)$.

**9.** *Greene-Kleitman (GK) theorem:* For every finite poset $P$ and every $k \geq 0$, there is a chain partition of $P$ that is both $k$-saturated and $(k+1)$-saturated.

**10.** The GK theorem is best possible, since there are infinitely many posets for which no chain partition is both $k$-saturated and $l$-saturated for any nonconsecutive nontrivial values for $k, l$; the smallest has six elements (see the following illustration). The GK theorem extends in various ways to directed graphs.

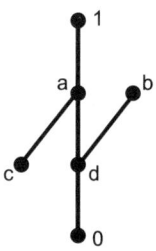

**11.** Dilworth's theorem is the special case of the GK theorem for $k = 0$ (every chain partition is 0-saturated).

**12.** Every product of symmetric chain orders is a symmetric chain order.

**13.** The lattice of bounded sequences $L(m, n)$ (§11.1.4) has a symmetric chain decomposition if $\min\{m, n\} \leq 4$. It is not known whether $L(m, n)$ in general has a symmetric chain decomposition.

**14.** The lattice $L(m, n)$ has the Sperner property.

**15.** The partition lattice $\Pi(n)$ fails to satisfy the Sperner property if $n$ is sufficiently large.

**16.** The Boolean lattice $\underline{2}^n$ and the subspace lattice $L_n(q)$ satisfy the strict Sperner property.

**17.** Every symmetric chain order has the strong Sperner property, and a symmetric chain decomposition is $k$-saturated for all $k$.

**18.** The class of graded posets that have the strong Sperner property and a symmetric unimodal sequence of Whitney numbers is closed under Cartesian product.

**19.** When $N_k \le N_{k+1}$, the normalized matching property implies Hall's condition for the existence of a matching saturating $P_k$ in the bipartite graph of the relations between the two levels.

**20.** Two subsets of the Boolean lattice in $[n]$ are on the same chain of the "bracketing decomposition" if and only if they have the same bracketing representation. This provides an explicit symmetric chain decomposition of $\underline{2}^n$. This generalizes for multisets $(D(N))$.

**21.** *Dedekind's problem:* This is the problem of computing the total number of antichains in the Boolean algebra $\underline{2}^n$. By using the bracketing decomposition, this number is calculated to be at most $3^{\binom{n}{\lfloor n/2 \rfloor}}$. Asymptotically, for even $n$, the number is

$$2^{\binom{n}{n/2}} e^{\binom{n}{n/2-1}} [2^{-n/2} + n^2 2^{-n-5} - n2^{-n-4}(1 + o(1))].$$

The exact values for $n \le 7$ are 3; 6; 20; 168; 7,581; 7,828,354; and 2,414,682,040,998. The stated estimate gives 7,996,118 for $n = 6$.

**22.** *Universal set sequences:* A universal set sequence on a set $S$ is a sequence that contains every subset of $S$ as a consecutive subsequence. The bracketing decomposition yields a universal set sequence on $[n]$ of length asymptotic to $\frac{2}{\pi} 2^n$.

**23.** If two sets, $x$ and $y$, are chosen independently according to a probability distribution on the Boolean lattice $\underline{2}^n$, then the probability that $x$ is contained in $y$ is at least $\binom{n}{\lfloor n/2 \rfloor}^{-1}$.

**24.** There is an on-line algorithm that partitions posets of height $k$ into $\binom{k+1}{2}$ antichains. This is best possible, even for 2-dimensional posets.

**25.** There is an on-line algorithm that partitions posets of width $k$ into $\frac{5^k - 1}{4}$ chains.

**26.** It is impossible to design an on-line algorithm that partitions every poset of width $k$ into fewer than $\binom{k+1}{2}$ chains.

**27.** There is an on-line algorithm to partition every poset of width 2 into five chains, and this is best possible.

### Examples:

**1.** A symmetric chain decomposition of the Boolean lattice $\underline{2}^3$ is shown next.

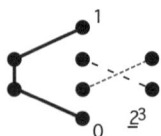

**2.** A regular chain covering of the Boolean lattice $\underline{2}^3$ is shown next.

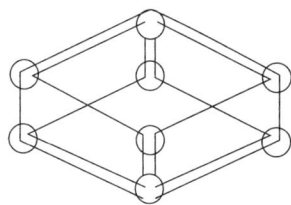

**3.** The poset $\underline{3} \times \underline{4}$ has a regular covering by six chains, using two chains twice and two other chains once each.

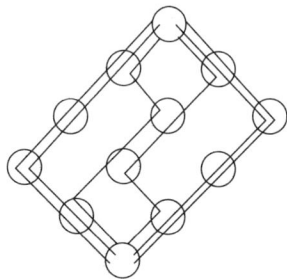

**4.** The poset $\underline{3} \times \underline{4}$ satisfies the Sperner property but not the strict Sperner property, since it has a maximum antichain (size three) that is not confined to a single rank.

**5.** Dilworth's theorem can be used to demonstrate the following result: Any sequence of $n^2 + 1$ real numbers contains a monotone subsequence of length $n + 1$ (Erdős and Szekeres, 1935). Let the given sequence be $a_1, a_2, \ldots, a_{n^2+1}$. Define the poset $P = \{(k, a_k) \mid k = 1, \ldots, n^2 + 1\}$, with $(i, a_i) \le (j, a_j) \Leftrightarrow i < j$ and $a_i \le a_j$. Assume then that the given sequence does not contain a monotone nondecreasing subsequence of length $n + 1$. Since a chain in $P$ corresponds to a monotone nondecreasing subsequence, each chain can contain at most $n$ elements, and so we require at least $n + 1$ chains to cover $P$. Since an antichain corresponds to a monotone nonincreasing subsequence, Dilworth's theorem assures us that there exists an antichain of size at least $n + 1$ and so a monotone nonincreasing subsequence of length at least $n + 1$.

## 11.2.2   LYM PROPERTY

### Definitions:

The **LYM inequality** for a family $F$ in a ranked poset $P$ is the inequality $\sum_{x \in F} \frac{1}{N_{r(x)}} \le 1$, where $N_{r(x)}$ is a Whitney number.

A poset $P$ is an **LYM order** (or satisfies the **LYM property**) if every antichain $F \subseteq P$ satisfies the LYM inequality.

### Facts:

**1.** The LYM property was discovered independently for $\underline{2}^n$ by Lubell, Yamamoto, and Meshalkin.

**2.** The LYM property, the normalized matching property, and the existence of a regular covering by chains are equivalent.

**3.** The LYM property implies the Sperner property and also implies the strong Sperner property (but not the strict Sperner property).

**4.** Every LYM order that has symmetric unimodal Whitney numbers has a symmetric chain decomposition. In particular, $L_n(q)$ is a symmetric chain order.

**5.** It is not known whether every LYM poset has a chain decomposition that is $k$-saturated for all $k$.

**6.** A product of LYM orders may fail the LYM property.

**7.** A product of LYM orders whose sequence of Whitney numbers is log-concave is an LYM order with a log-concave sequence of Whitney numbers. (A sequence $\{a_n\}$ is *log-concave* if $a_n^2 \geq a_{n-1}a_{n+1}$ for all $n$.)

**8.** The divisor lattice $D(N)$ is an LYM order, which follows from the previous fact.

**9.** The partition lattice $\Pi(n)$ is an LYM order if and only if $n < 20$.

**10.** The Boolean lattice $\mathbf{2}^n$ and the subspace lattice $L_n(q)$ have regular coverings by chains and hence are LYM orders.

**11.** If $\{\lambda_x\}$ is an assignment of real-valued weights to the elements of an LYM poset $P$, then for every subset $G \subset P$ and every regular covering $\mathbf{C}$ of $P$,

$$\sum_{x \in G} \frac{\lambda_x}{N_{r(x)}} \leq \max_{C \in \mathbf{C}} \left\{ \sum_{y \in C \cap G} \lambda_y \right\}.$$

**Example:**

**1.** The lattice $L(m, n)$ of bounded sequences is an LYM order if and only if $\min\{m, n\} \leq 2$ or $(m, n) = (3, 3)$.

## 11.2.3   RANKINGS, SEMIORDERS, AND INTERVAL ORDERS

A chain names the "better" of any pair according to a single scale. Realistically, some comparisons may yield indifference. Several families of "chain-like" partial orders successively relax the requirements on indifference.

**Definitions:**

A poset $P$ is a **ranking** or **weak order** if its elements are partitioned into ranks $P_1, \ldots, P_k$ such that two elements are incomparable if and only if they belong to the same rank.

A poset $P$ is a **semiorder** if there is a real-valued function $f$ and a fixed real number $\delta > 0$ ($\delta$ may be taken to be 1) such that $x < y$ if and only if $f(y) - f(x) > \delta$. The pair $(f, \delta)$ is a **semiorder representation** of the poset $P$.

A poset $P$ is an **interval order** if there is an assignment of real intervals to its members such that $x < y$ if and only if the interval for $y$ is totally to the right of the interval for $x$. The collection of intervals is called an **interval representation** of the poset $P$.

A **biorder representation** on a digraph $D$ is a pair of real-valued functions $f$, $g$ on the vertex set $V_D$ such that $u \to v$ is an arc if and only if $f(u) > g(v)$.

A **Ferrers digraph** (or **Ferrers relation** or **biorder**) is a digraph having a biorder representation. (Also see §2.5.1.)

**Facts:**

**1.** Rankings model a single criterion of comparison with "ties" allowed, as in voting.

**2.** A poset is a ranking if and only if its comparability graph is a complete multipartite graph.

**3.** A ranking assigns a score $f(z)$ to each element $z$ such that $x < y$ if and only if $f(x) < f(y)$.

**4.** The forbidden subposet characterization of a ranking is $\underline{1} + \underline{2}$.

**5.** Semiorders were introduced to model intransitivity of indifference; a difference of a few grains of sugar in a coffee cup or a few dollars in the price of a house is not likely to affect one's attitude, but pounds of sugar or thousands of dollars will. The threshold $\delta$ in a semiorder representation indicates a "just-noticeable difference".

**6.** A poset is a semiorder if and only if its incomparability graph is a unit interval graph, that is, an interval graph (§8.1.3) such that all intervals are of unit length.

**7.** An interval representation of a semiorder $P$ with semiorder representation $(f, \delta)$ is obtained by setting $I_x = [f(x) - \frac{\delta}{2} + \epsilon,\ f(x) + \frac{\delta}{2} - \epsilon]$.

**8.** *Scott-Suppes theorem*: The forbidden subposet characterization of a semiorder is $\{\underline{1} + \underline{3}, \underline{2} + \underline{2}\}$.

**9.** The number of nonisomorphic semiorders on an $n$-element set is the Catalan number $C_n = \frac{1}{n+1}\binom{2n}{n}$ (§3.1.3).

**10.** Interval orders model a situation where the value assigned to an element is imprecise.

**11.** The incomparability graph of an interval order is an interval graph.

**12.** Every poset whose incomparability graph is an interval graph is an interval order. This follows from the forbidden subposet characterization of interval orders.

**13.** The forbidden subposet characterization of an interval order is $\underline{2} + \underline{2}$.

**14.** A poset $P$ is an interval order if and only if both the collections of "upper holdings" $U(x) = \{y \in P \mid y > x\}$ and "lower holdings" $D(x) = \{y \in P \mid y < x\}$ form chains under inclusion, in which case the number of distinct nonempty upper holding sets and distinct nonempty lower holding sets is the same. Construction of these chains yields a fast algorithm to compute a representation for an interval order or semiorder.

**15.** The strict comparability digraph of an interval order is a Ferrers digraph, with $f(x)$ and $g(x)$ denoting the left and right endpoints of the interval assigned to $x$. This is the strict comparability digraph of a poset because $f(x) \leq g(x)$ for all $x$. The "upper holdings" and "lower holdings" for an interval order become predecessor and successor sets for a Ferrers digraph.

**16.** For a digraph $D$ with adjacency matrix $A(D)$, the following are equivalent:

- $D$ has a biorder representation (and is a Ferrers digraph);
- $A(D)$ has no 2-by-2 submatrix that is a permutation matrix;
- the successor sets of $D$ are ordered by inclusion;
- the predecessor sets of $D$ are ordered by inclusion;
- the rows and columns of $A(D)$ can be permuted independently so that to the left of a 1 is a 1.

**17.** The greedy algorithm is an optimal on-line algorithm for partitioning an interval order into the minimum number of antichains. It uses at most $2h - 1$ antichains to recursively partition an interval order of height $h$, and this is best possible.

**18.** There is an on-line algorithm to partition every interval order of width $k$ into $3k - 2$ chains, and this is best possible. Equivalently, the maximum number of colors needed for on-line coloring of interval graphs with clique size $k$ is $3k - 2$.

**19.** No on-line partitioning algorithm colors all trees with a bounded number of colors.

**20.** *"Universal" interval orders:* Since the ordering of the interval endpoints is all that matters, interval representations may be restricted to have integer endpoints. The poset $I[0, n]$ or $\mathbf{I}_n$ denotes the interval order whose interval representation consists of all intervals with integer endpoints in $\{0, \ldots, n\}$.

**21.** Every finite interval order is a subposet of some $I[0, n]$.

**Examples:**

**1.** The following Hasse diagram represents a poset that is a ranking. Its three ranks are indicated by the levels in the diagram.

**2.** The following Hasse diagram represents a poset that is a semiorder: for instance, with $\delta = 1$, define $f(a) = 2$, $f(b) = 1.3$, $f(c) = 0.8$, and $f(d) = 0$. It is not a ranking, by Fact 4, because $\underline{1} + \underline{2}$ is a subposet. The interval representation of its incomparability graph is at the right.

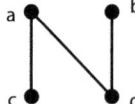
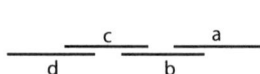

**3.** The following Hasse diagram represents a poset that is an interval order. The interval representation of its incomparability graph is at the right. By Fact 8, it is not a semiorder.

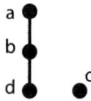

**4.** The skill of a tennis player may vary from day to day, leading to use of an interval $[a_x, b_x]$ to represent player $x$. In this case player $x$ *always* beats player $y$ if $a_x > b_y$.

**5.** The interval order $I[0, 3]$ is not a semiorder.

## 11.2.4   APPLICATION TO SOCIAL CHOICE

When there are more than two candidates for a public office, it is not obvious what is the "best" way to select a winner. Any rule has its pluses and minuses, from the standpoint of public policy. Social choice theory analyzes the effect of various rules for deciding the outcomes of preferential rankings.

## Definitions:

A **profile** on a set $A$ of "alternatives" (e.g., candidates for a public office) is a set $P = \{P_i \mid i \in I\}$ of linear rankings (ties allowed) of $A$, one for each member of a set $I$ of "individuals" (e.g., voters).

A **consensus function** (or **social choice function**) is a function $\phi$ that assigns to each possible profile $P = \{P_i \mid i \in I\}$ on a set $A$ of alternatives a linear order (ties allowed) of $A$ called the **consensus ranking** for $P$.

A consensus function upholds **majority rule** provided it ensures that $x$ is preferred to $y$ if and only if a majority of the individuals prefer $x$ to $y$.

**Plurality** is the consensus function in which the winner(s) is (are) the alternative(s) appearing in the greatest number of top ranks, after which the winner(s) is (are) deleted and the procedure is repeated to select the next rank of the consensus ranking, etc.

The **Borda count** of an alternative $x$ is the sum, over individual rankings, of the number of alternatives $x$ "beats". The resulting **Borda consensus function** ranks the alternatives by their Borda count.

## Facts:

**1.** Plurality can elect someone ranked last by a majority.

**2.** *Condorcet's paradox*: Some profiles have no decisive consensus (i.e., producing a single winner) that upholds majority rule.

**3.** The Borda count is subject to abuse.

**4.** *Arrow's impossibility theorem*: No consensus function exists that satisfies the following four axioms, which were formulated in an attempt to develop a consensus function $\phi$ that avoids the difficulties cited in the facts above:

- *monotonicity*: If $a > b$ in $\phi(P)$ and if profile $P'$ agrees with profile $P$ except for moving alternative $a$ upward in some or all rankings, then $a > b$ in $\phi(P')$.

- *independence of irrelevant alternatives*: If profiles $P$ and $P'$ agree within a set $A' \subseteq A$, then $\phi(P)$ and $\phi(P')$ have the same restriction to $A'$. This axiom implies that votes for extraneous alternatives do not affect the determination of the consensus ranking among the alternatives within the subset $A'$.

- *nondegeneracy*: Given $a, b \in A$, there is a profile $P$ such that $a > b$ in $\phi(P)$. This axiom implies that the structure of the outcome is independent of renaming the alternatives.

- *nondictatorship*: There is no $i \in I$ such that $a > b$ in $P_i$ implies $a > b$ in $\phi(P)$.

## Examples:

**1.** Suppose that $A = \{a, b, c\}$ is the set of alternatives, and suppose that the profile consists of the three rankings $a > b > c$, $c > a > b$, and $b > c > a$. Then for each alternative, there is another alternative that is preferred by $\frac{2}{3}$ of the population.

**2.** The U.S. presidential election of 1912 had three candidates: Wilson (W), Roosevelt (R), and Taft (T). It is estimated that 45% of the voters ranked $W > R > T$, that 30% ranked $R > T > W$, and that 25% ranked $T > R > W$. Wilson won the election, garnering a plurality of the popular vote, but a majority of the population preferred Roosevelt to Wilson. Moreover, 55% regarded Wilson as the least desirable candidate.

**3.** Consider a close election, with four individuals preferring $x$ to $y$ to all other alternatives. A fifth individual prefers $y$ to $x$. If there are enough other alternatives, the fifth individual can throw a Borda-count election to $y$ by placing $x$ at the bottom.

## 11.2.5   LINEAR EXTENSIONS AND DIMENSION

By adding additional comparison pairs to a partial ordering on a set, ultimately a total ordering is obtained. Each of the many ways to do this is called an extension of the original partial ordering.

**Definitions:**

An **extension** of a poset $P = (X, R)$ is a poset $Q = (X, S)$ such that $R \subseteq S$ (i.e., $xRy$ implies $xSy$).

A **linear extension** of a poset $P$ is an extension of $P$ that is a chain.

A **topological sort** is an algorithm that accepts a finite poset as input and produces a linear extension of that poset as output.

A **topological ordering** of an acyclic digraph is a linear extension of the poset arising from it.

The **intersection of two partial orderings** $P = (X, R)$ and $Q = (X, S)$ on the same set $X$ is the poset $(X, R \cap S)$ that includes the relations common to both.

A **realizer** of a poset $P$ is a set of linear extensions of $P$ whose intersection is $P$.

The **order dimension** (or **dimension**) of $P$, written $\dim(P)$, is the minimum cardinality of a realizer of $P$.

The **standard example** $S_n$ of an $n$-dimensional poset is the subposet of $\underline{2}^n$ induced by the singletons and their complements.

An **alternating $k$-cycle** in a poset $P$ is a sequence of ordered incomparable pairs $\{(x_i, y_i)\}_{i=1}^k$ such that $y_i \leq x_{i+1}$, where subscripts are taken modulo $k$.

A **critical pair** (or **unforced pair**) in a poset $P$ is an ordered incomparable pair that cannot be made comparable by adding any other single incomparable pair as a relation.

A linear extension $L$ of a poset $P$ **puts $Y$ over $X$**, where $X$ and $Y$ are disjoint subposets, if $y$ is above $x$ in $L$ whenever $(x, y)$ is an incomparable pair with $x \in X$, $y \in Y$.

Given a subposet $Q \subseteq P$, an **upper extension** of $Q$ is a linear extension of $P$ that puts $P - Q$ over $Q$.

Given a subposet $Q \subseteq P$, a **lower extension** of $Q$ is a linear extension of $P$ that puts $P - Q$ below $Q$.

The **minimum realizer encoding** of a poset lists for each element its position on each extension in a minimum realizer.

The **probability space** on the set of all linear extensions of a (finite) poset $P$ is obtained by taking each linear extension to be equally likely. The notation $Pr(x < y)$ denotes the proportion of linear extensions in which element $x$ comes below element $y$.

**Facts:**

**1.** Every poset is the intersection of all its linear extensions, from which it follows that the concept of dimension is well defined.

**2.** Given incomparable elements $x$ and $y$ in a poset $P$, there is a linear extension of $P$ in which $x$ appears above $y$.

**3.** The chains are the only posets of dimension 1.

**4.** Every antichain has dimension 2, because the intersection of a linear order and its dual is an antichain.

**5.** Topological sort is used to organize activities having a precedence ordering into a sequential schedule.

**6.** The list of minimal forbidden subposets for dimension 2 consists of ten isolated examples and seven one-parameter families.

**7.** If $Q$ is a subposet of $P$, then $\dim(Q) \leq \dim(P)$.

**8.** The dimension of a product of $k$ chains (each of size at least 2) is $k$.

**9.** A poset has dimension at most $k$ if and only if it imbeds in a product of $k$ chains.

**10.** The dimension of a poset $P$ equals the minimum integer $n$ such that $P$ is a subposet of $\mathcal{R}^n$.

**11.** The standard example $S_n$ is a bipartite poset whose comparability graph is obtained from the complete bipartite graph $K_{n,n}$ by deleting a complete matching.

**12.** The minimum realizer encoding of an $n$-element poset of dimension $k$ takes only $O(kn \log n)$ bits, instead of the $O(n^2)$ bits of the order relation. Thus, posets of small dimension have concise representations.

**13.** In the sense of Fact 12, the dimension of a poset may be regarded as a measure of its "space complexity".

**14.** The dimension of a poset $P$ equals the minimum number of linear extensions containing all the critical pairs of $P$.

**15.** The dimension of a poset $P$ is equal to the chromatic number of the hypergraph whose vertex set is the set of critical pairs and whose edges are the sets of critical pairs forming minimal alternating cycles.

**16.** The cover graph of the standard example $S_n$ (of an $n$-dimensional poset) is $K_{n,n}$-(1-factor).

**17.** If $X$ and $Y$ are disjoint subposets of a poset $P$, then $P$ has a linear extension $L$ putting $Y$ over $X$ if and only if $P$ contains no $\underline{2} + \underline{2}$ with minimal elements in $Y$ and maximal elements in $X$.

**18.** $\dim(P) \leq w(P)$. A realizer of size $w(P)$ can be formed by taking upper extensions of the chains in a partition of $P$ into $w(P)$ chains.

**19.** $\dim(P) \leq \frac{|P|}{2}$. The standard example $S_n$ shows that this bound is the best possible.

**20.** *One-point removal theorem*:  For every $x \in P$, $\dim(P) \leq 1 + \dim(P - x)$.

**21.** For every poset $P$, there exist four elements $\{x, y, z, w\}$ such that $\dim(P) \leq 2 + \dim(P - \{x, y, z, w\})$. It is conjectured that, for every poset $P$, there exist two elements $\{x, y\}$ such that $\dim(P) \leq 1 + \dim(P - \{x, y\})$.

**22.** A poset has dimension 2 if and only if the complement of its comparability graph is also a comparability graph; thus there is a polynomial-time algorithm to decide whether a poset has dimension 2. However, recognizing posets of dimension $k$ is NP-complete for every fixed $k$ at least 3.

**23.** If $P$ is a finite poset that is not a chain, then $P$ has a pair of elements $x, y$ such that

$$0.2764 \simeq \frac{1}{2} - \frac{1}{2\sqrt{5}} \leq \Pr(x < y) \leq \frac{1}{2} + \frac{1}{2\sqrt{5}} \simeq 0.7236.$$

**24.** *The $\frac{1}{3}$-$\frac{2}{3}$ conjecture*:  This conjecture states that there is always a pair of elements, $x$ and $y$, such that $\frac{1}{3} \leq \Pr(x < y) \leq \frac{2}{3}$.

**25.** The traditional name *topological sort* is commonly used in applications. However, a topological sort is *not* a sort in the standard meaning of that word. Nor is it directly related to what mathematicians call *topology*.

**26.** Algorithm 1 carries out a topological sort on a given poset; also see §8.3.4.

---

**Algorithm 1**:   **Topological sort.**

input: a finite poset $(X = \{x_1, \ldots, x_n\}, \leq)$

output: a compatible total ordering $A = x_1 \leq x_2 \leq \cdots \leq x_n$ of the elements of $X$

for $j := 1$ to $n$

$\quad x_j :=$ a minimal element of $X$

$\quad X := X - \{x_j\}$

---

**Examples:**

**1.** The following poset has the linear extensions *abc* and *acb* and it is the intersection of these extensions. Thus its dimension is 2.

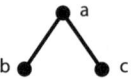

**2.** The following poset has six linear extensions: *abcd*, *acbd*, *acdb*, *cdab*, *cadb*, and *cabd*. Since it is the intersection of *abcd* and *cdab*, its dimension is 2.

**3.** The bipartite poset $S_3$ whose comparability graph and cover graph is the 6-cycle $1, \bar{2}, 3, \bar{1}, 2, \bar{3}$ has dimension 3. The realizer $\{23\bar{1}1\bar{2}3, 13\bar{2}2\bar{1}3, 12\bar{3}3\bar{1}2\}$ establishes the upper bound. Every realizer must have an extension with $\bar{1}$ below 1, one with $\bar{2}$ below 2, and one with $\bar{3}$ below 3. No two of these can occur in the same linear extension, so the dimension is at least 3.

**4.** More generally, for the elements $i \in [n]$ of the standard example $S_n$, a realizer must include distinct linear extensions in which the singleton $\{i\}$ appears above its complement, and any $n$ such extensions suffice.

**5.** For the standard example $S_n$ of dimension $n$, the critical pairs are $\{\bar{i}, i\}$; this reflects the fact that, in a realizer, the extensions need to put $i$ above $\bar{i}$, for each $i$. Each pair of critical pairs forms a minimal alternating cycle. Viewing the minimal alternating cycles as edges creates a hypergraph, namely the complete graph $K_n$, with chromatic number $n$.

**6.** Let $N$ be the bipartite poset with minimal elements $a$ and $b$ and maximal elements $c$ and $d$, in which $a$ lies below $c$, and $b$ lies below $c$ and $d$. This poset has five linear extensions, namely $a < b < c < d$, $a < b < d < c$, $b < a < c < d$, $b < a < d < c$, and $b < d < a < c$. Thus $Pr(a < b) = \frac{2}{5}$.

**7.** *Application of posets to sorting:* The objective of a sort is to arrange the elements of a set $X$ into a sequence by posing sequential queries of the form: "is $x < y$ true?". At any time, the state of cumulative knowledge is representable by a poset $P = (X, R)$, such that the linear extensions of $P$ are remaining candidates for the final sequence order. A desirable query substantially reduces the number of candidates for extensions no matter whether the answer is yes or no, most especially finding a pair, $x$ and $y$, such that $Pr(x < y)$ is close to $\frac{1}{2}$. Thus, Fact 23 shows that the worst case time to sort, in the presence of partial information given by a poset $P$, is $\Omega(\log |P|)$.

**8.** *Application of posets to searching*:   The objective of searching a poset $P$ in which item $s(x)$ is stored at location $x$ is to determine whether a target item $\alpha$ is present in $P$. Each step of the search probes a location and compares its value against the target item. The worst case requires determining for each $x \in P$ whether the item at location $x$ is greater or less than $\alpha$, so the searching problem is the problem of identifying the downset $D_\alpha = \{x \in P \mid s(x) < \alpha\}$. A probe of location $x$ splits the remaining possible downsets into those that contain $x$ and those that do not. The former remain as candidates if $s(x) < \alpha$; the latter remain if $s(x) > \alpha$. A hypothetical adversary would arrange the value $s(x)$ so that the response would leave the larger portion of the ideals. Thus, the number $c(P)$ of probes required in the worst case is at least $\lceil \log_2 i(P) \rceil$, where $i(P)$ denotes the number of ideals in poset $P$.

## 11.2.6   POSETS AND GRAPHS

From the graph-theoretic viewpoint, a comparability graph is by definition a simple graph (§8.1.1) that has a transitive orientation. Comparability graphs are perfect graphs, which motivates most study of comparability graphs.

### Definitions:

A **transitive orientation** on a simple graph $G$ is an assignment of directions to the edges so that whenever there is an $xy$-arc and a $yz$-arc, there is also an $xz$-arc.

A **quasi-transitive orientation** on a simple graph $G$ is an assignment of directions to the edges so that whenever there is an $xy$-arc and a $yz$-arc, there is also an arc between $x$ and $z$.

A **triangular chord** for a walk $x_1, \ldots, x_k$ in an undirected graph $G$ is an edge between vertices $x_{i-1}$ and $x_{i+1}$, two apart on the walk.

The **auxiliary graph** for a simple graph $G$ is the graph $G'$ whose vertices are the edges of $G$, with vertex $e_1$ adjacent to vertex $e_2$ in $G'$ if and only if edges $e_1$ and $e_2$ are adjacent in graph $G$ but do not lie on a 3-cycle.

A **module** in a graph $G$ is a vertex subset $U$ such that each vertex outside $U$ is adjacent to all or none of the vertices in $U$.

An **order module** in a poset $P$ (or **autonomous set**) is a set $S$ of elements such that every element outside $S$ is above all of $S$, below all of $S$, or incomparable to all of $S$.

A **comparability invariant** for posets is an invariant $f$ such that $f(P) = f(Q)$ whenever posets $P$ and $Q$ have the same comparability graph.

A **permutation graph** is a graph whose vertices can be placed in 1-1 correspondence with the elements of a permutation of $[n] = \{1, \ldots, n\}$ such that $v_i$ is adjacent to $v_j$ if and only if the larger of $i$ and $j$ comes first in the permutation.

A **$u,v$-bypass** in a directed graph is a $u,v$-path of length at least two such that there is also an arc from $u$ to $v$.

A **dependent edge** in an acyclic directed graph is an arc from $u$ to $v$ such that the graph contains a $u,v$-bypass.

**Facts:**

**1.** For a simple graph $G$, the following are equivalent:

- $G$ has a transitive orientation;
- $G$ has a quasi-transitive orientation;
- every closed odd walk of $G$ has a triangular chord;
- the auxiliary graph $G'$ is bipartite.

The implications from top to bottom are straightforward, as is the proof that if the auxiliary graph $G'$ is bipartite then $G$ has a quasi-transitive orientation. The proof that if $G$ has a quasi-transitive orientation then $G$ has a transitive orientation takes more work. The last characterization gives an algorithm to decide whether $G$ is a comparability graph in $O(n^3)$ time, where $n$ is the number of vertices. The proof is constructive, so a transitive orientation can also be obtained.

**2.** In any graph, the set of all vertices, the singleton sets of vertices, and the empty set are always modules.

**3.** Modules yield a forbidden subgraph characterization of comparability graphs. The minimal forbidden induced subgraphs consist of eight infinite families and ten special examples.

**4.** If two partial orders have the same comparability graph, then one can be transformed into the other by a sequence of moves involving reversing all the relations inside an order module $S$, i.e., by replacing the partial order induced on $S$ by its dual, and preserving all relations between $S$ and its complement.

**5.** Let $f$ be a poset invariant such that $f(P) = f(P^*)$ for all posets $P$, and such that, if poset $Q$ is obtained from $P$ by replacing a module in $P$ with another module having the same value of $f$, then $f(Q) = f(P)$. Then the invariant $f$ is a comparability invariant.

**6.** Height, width, dimension, and number of linear extensions are all comparability invariants.

**7.** A graph is the complement of a comparability graph if and only if it is the intersection graph of the curves representing a collection of continuous real-valued functions on $[0, 1]$.

**8.** The following conditions are equivalent for a graph $G$:

- $G$ is a permutation graph (adjacency representing the inversions of a permutation);
- $\overline{G}$ is the comparability graph of a 2-dimensional partial order;
- $G$ and $\overline{G}$ are comparability graphs.

**9.** Isomorphism of permutation graphs can be tested in $O(n^2)$ time. Some NP-complete scheduling problems become polynomial when the poset of precedence constraints is 2-dimensional.

**10.** A directed graph corresponds to the diagram of some partial order if and only if it contains no cycles or bypasses.

**11.** Every graph that is the cover graph of some poset is triangle-free.

**12.** If a graph has chromatic number less than its girth, then it is the cover graph of some poset. In particular, a 3-chromatic graph is a cover graph if and only if it is triangle-free.

**13.** It is NP-complete to decide whether a 4-chromatic graph is a covering graph.

**14.** The smallest triangle-free graph that is not a cover graph is the 4-chromatic Grötzsch graph with 11 vertices.

**15.** The maximum number of dependent edges among the orientations of a graph $G$ is equal to the cycle rank $\beta(G) = |E| - |V| + 1$.

**16.** If a graph $G$ has chromatic number less than its girth (§8.7.2), then for all $i$ such that $0 \le i \le \beta(G)$, the graph has an acyclic orientation with exactly $i$ dependent edges.

**17.** The cover graph of a modular lattice is bipartite.

**18.** A modular lattice is distributive if and only if its cover graph does not contain the complete bipartite graph $K_{2,3}$.

**19.** The subgraph of the cover graph of $\underline{2}^{2k+1}$ induced by the $k$-sets and $(k+1)$-sets is a vertex-transitive $(k+1)$-regular bipartite graph. The graph is known to contain cycles using more than 80% of its vertices. The *Erdős revolving door conjecture* asserts that this graph is Hamiltonian.

**20.** *Gallai-Milgram theorem:* The vertices of a digraph $D$ can be covered using at most $\alpha(D)$ disjoint paths, where $\alpha(D)$ is the maximum size of an independent set in $D$.

**21.** Dilworth's theorem (§11.2.1) is the special case of the Gallai-Milgram theorem for comparability digraphs.

**Examples:**

**1.** A transitive orientation for a bipartite graph can be obtained by assigning all the edge directions from one part to the other, as shown here for $K_{3,3}$:

**2.** An odd cycle of length $\ge 5$ has no quasi-transitive orientation (see Fact 1).

**3.** Inserting a triangular chord into a 5-cycle permits the resulting graph to have a transitive orientation, as shown in the following figure.

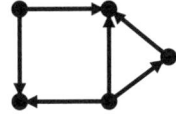

**4.** The following figure shows a graph and its auxiliary graph.

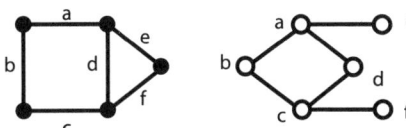

**5.** Any subset of either part of a complete bipartite graph is a module, since the other vertices in its part are not adjacent to any vertex in the module, and the vertices in the other part of the bipartition are each adjacent to all the vertices in the module.

**6.** Deleting a 1-factor from $K_{n,n}$, for $n \ge 3$, yields a graph with no module other than the complete set of vertices, the singletons, and the empty set.

# REFERENCES

***Printed Resources***:

[An87] I. Anderson, *Combinatorics of Finite Sets*, Oxford University Press, 1987.

[Bi95] G. Birkhoff, *Lattice Theory*, 3rd ed., American Mathematical Society, 1995.

[Bo98] B. Bollobás, *Combinatorics: Set Systems, Hypergraphs, Families of Vectors, and Combinatorial Probability*, Cambridge University Press, 1998.

[BoMo94] V. Bouchitté and M. Morvan, eds., *Orders, Algorithms and Applications*, Springer, 1994.

[DaPr02] B. A. Davey and H. A. Priestley, *Introduction to Lattices and Order*, 2nd ed., Cambridge University Press, 2002.

[Fi85] P. C. Fishburn, *Interval Orders and Interval Graphs: A Study of Partially Ordered Sets*, Wiley, 1985.

[Ri82] I. Rival, ed., *Proceedings of the Symposium on Ordered Sets*, Reidel Publishing, 1982.

[Ri85] I. Rival, ed., *Graphs and Order*, Reidel Publishing, 1985.

[Ri86] I. Rival, ed., *Combinatorics and Ordered Sets*, American Mathematical Society, 1986.

[Ri89] I. Rival, ed., *Algorithms and Order*, Kluwer Academic Press, 1989.

[Ro12] K. H. Rosen, *Discrete Mathematics and Its Applications*, 7th ed., McGraw-Hill, 2012.

[Sc03] B. Schröder, *Ordered Sets: An Introduction*, Birkhäuser, 2003.

[Tr92] W. T. Trotter, *Combinatorics and Partially Ordered Sets: Dimension Theory*, The Johns Hopkins University Press, 1992.

# 12

## COMBINATORIAL DESIGNS

## INTRODUCTION

In broad terms, the study of combinatorial designs is the study of the structure of collections of subsets of a finite set when these collections of subsets satisfy certain prescribed properties. In particular, a block design has the property that every one of these subsets has the same size $k$ and every *pair* of points in the set is in exactly the same number of these subsets. Latin squares are also fundamental in this area and can be thought of in this context, but they are commonly thought of as $n \times n$ arrays with the property that each cell contains one element from an $n$-set and each row and each column contain each element exactly once. Some of the questions of general interest include: existence of designs, enumeration of nonisomorphic designs, and the study of subdesigns of designs.

Matroids generalize a variety of combinatorial objects, such as matrices and graphs. These structures arise naturally in a variety of combinatorial contexts and provide a framework for the study of many problems in combinatorial optimization and graph theory.

Much of the information in §12.1–12.3 is condensed from [CoDi07], which provides a comprehensive treatment of combinatorial designs. The main source for the material in §12.4 is [Ox11].

## GLOSSARY

**affine plane**:  a set of points and a set of subsets of points (called lines) such that every two points lie on exactly one line, if a point does not lie on a line $L$ there is exactly one line through the point that does not intersect $L$, and there are three points that are not collinear.

**affine space** (of dimension $n$):  the set $\mathrm{AG}(n, q)$ of all cosets of subspaces of an $n$-dimensional vector space over a field of order $q$.

**automorphism** (a design $D$):  an isomorphism from $D$ onto $D$.

**balanced incomplete block design (BIBD)**:  given a finite set $X$ (of *points*), a collection of subsets (called *blocks*) of $X$ of the same size such that every point belongs to the same number of blocks, and each pair of points belongs to the same number of blocks. The BIBD is described by five parameters: size of $X$, number of blocks, number of blocks to which every element of $X$ belongs, size of each block, and number of blocks to which each pair of distinct points belongs.

**basis** (for a matroid):  a maximal independent set in the matroid.

**basis axioms**:  a set of axioms that specifies the set of bases of a matroid.

**binary matroid**:  a matroid that is isomorphic to a vector matroid of a matrix over the field $GF(2)$.

**biplane**:  symmetric design in which every pair of distinct points belongs to exactly two blocks.

**block**:  each of the subsets in a design.

**circuit**:  a minimal dependent set in a matroid.

**circuit axioms**:  a set of axioms that specifies the set of circuits of a matroid.

**closed set**:  in a matroid, a subset of its ground set that is equal to its closure.

**closed under duality**: the property of a class of matroids that the dual of a matroid in the class is also in the class.

**closure** (of a subset of the ground set in a matroid): given a subset $X$ of the ground set $E$ in a matroid, the set of all points $x \in E$ such that the rank of $X \cup \{x\}$ is equal to the rank of $X$.

**closure axioms**: a set of axioms that specifies the properties that a closure operator of a matroid must satisfy.

**closure operation**: the mapping $K \to \mathbf{B}(K)$, where $K$ is a set of positive integers and $\mathbf{B}(K)$ is the set of positive integers $v$ for which there exists a $(v, K)$-PBD.

**cobasis** (of a matroid): a basis of the dual of a matroid.

**cocircuit** (of a matroid): a circuit of the dual of a matroid.

**cographic matroid**: a matroid isomorphic to the cocyle matroid of a graph.

**coindependent set** (of a matroid): an independent set of the dual of a matroid.

**coloop** (of a matroid): a loop of the dual of a matroid.

**combinatorial geometry**: a simple matroid.

**complete set of mutually orthogonal Latin squares**: a set of $n - 1$ mutually orthogonal Latin squares of side $n$.

**conjugate**: Let $L$ be an $n \times n$ Latin square on symbol set $E_3$, with rows indexed by the elements of the $n$-set $E_1$ and columns indexed by the elements of the $n$-set $E_2$. Let $\mathcal{T} = \{(x_1, x_2, x_3) \mid L(x_1, x_2) = x_3\}$. Let $\{a, b, c\} = \{1, 2, 3\}$. The $(a, b, c)$-conjugate of $L$, $L_{(a,b,c)}$, has rows indexed by $E_a$, columns by $E_b$, and symbols by $E_c$, and is defined by $L_{(a,b,c)}(x_a, x_b) = x_c$ for each $(x_1, x_2, x_3) \in \mathcal{T}$.

**connected**: the property of a matroid that it cannot be written as the direct sum of two nonempty matroids.

**cycle matroid** (of a graph): the matroid on the edge-set of the graph whose circuits are the cycles of the graph.

**t-design**: a $t$-$(v, k, \lambda)$ design.

**t-(v, k, λ) design**: a pair $(X, \mathcal{A})$ where $X$ is a set of $v$ elements (points), $\mathcal{A}$ is a family of $k$-subsets (blocks) of $X$, and every $t$-subset of $X$ occurs in exactly $\lambda$ blocks.

**development** (of a difference set $D$): the incidence structure $\mathrm{dev}(D)$ whose points are the elements of $G$ and whose blocks are the translates $D + g = \{d + g \mid d \in D\}$.

**dual** (of an incidence structure): the incidence structure obtained by interchanging the roles of points and lines.

**dual** (of a matroid): given a matroid $M$, the matroid on the same set as $M$ whose bases are the complements of the bases of $M$.

**equivalent** (Latin squares): two Latin squares $L$ and $L'$ of side $n$ are equivalent if there are three bijections, from the rows, columns, and symbols of $L$ to the rows, columns, and symbols, respectively, of $L'$, that map $L$ to $L'$.

**Fano plane** (or **projective plane of order 2**): the $(7, 7, 3, 3, 1)$ design with point set $X = \{0, \ldots, 6\}$ and the block set $\mathcal{A} = \{013, 124, 235, 346, 450, 561, 602\}$.

**flat**: closed set.

**t-flat**: a subspace of projective dimension $t$ of a projective space; a coset of a subspace of affine dimension $t$ of an affine space.

**k-GDD**:  a group divisible design with $\lambda = 1$ and $K = \{k\}$.

**graphic matroid**:  a matroid that is isomorphic to the cycle matroid of some graph.

**ground set**:  the set of points of a matroid.

**group divisible design** (or **$(K, \lambda)$-GDD**):  given an integer $\lambda$ and a set of positive integers $K$, a triple $(X, \mathcal{G}, \mathcal{A})$ where $X$ is a set (of *points*), $\mathcal{G}$ is a partition of $X$ into at least two subsets (called *groups*), $\mathcal{A}$ is a family of subsets of $X$ (called *blocks*) such that: if $A$ in $\mathcal{A}$, then $|A| \in K$, a group and a block contain at most one common point, and every pair of points from distinct groups occurs in exactly $\lambda$ blocks.

**group-type** (or **type**):  for a group divisible design, the multiset $\{|G| : G \in \mathcal{G}\}$.

**Hadamard design**:  a symmetric $(4n - 1, 2n - 1, n - 1)$ design.

**Hadamard matrix**:  an $n \times n$ matrix $H$ with all entries $\pm 1$ that satisfies $H^{\mathrm{T}} H = nI$.

**hyperplane**:  a subspace of projective dimension $n-1$ of a projective space of projective dimension $n$; a coset of a subspace of affine dimension $n-1$ of an affine space of affine dimension $n$; a maximal nonspanning set of a matroid.

**idempotent**:  the property of a Latin square (or partial Latin square) that for all $i$, cell $(i, i)$ is occupied by $i$.

**imbedded Latin square**:  An $n \times n$ partial Latin square $P$ is imbedded in a Latin square $L$ if the upper $n \times n$ left corner of $L$ agrees with $P$.

**incidence matrix** (of a $(v, b, r, k, \lambda)$ design):  the $b \times v$ matrix with $(i, j)$-entry equal to 1 if the $i$th block contains the $j$th element, and 0 otherwise.

**incidence structure**:  the structure $(V, B, \mathcal{I})$ consisting of a finite set $V$ of *points*, a finite set $B$ of *lines*, and an incidence relation $\mathcal{I}$ between them.

**independent set**:  any set in a special collection of subsets of the ground set in a matroid.

**index**:  the number of blocks to which each pair of distinct points in a design belongs.

**isomorphism** (of block designs $(V, \mathcal{B})$ and $(W, \mathcal{C})$):  a bijection $\psi \colon (V, \mathcal{B}) \to (W, \mathcal{C})$ under which $\psi(B)$ occurs as a block in $\mathcal{C}$ the same number of times that $B$ occurs as a block in $\mathcal{B}$.

**isomorphism** (of matroids):  a bijection between the ground sets of two matroids that preserves independence.

**isotopic**:  equivalent.

**Kirkman's schoolgirl problem**:  the problem of arranging fifteen schoolgirls in five subsets of size 3 for a walk on each of seven days so that every pair of girls walks together exactly once.

**Kirkman triple system**:  a $(v, 3, 1)$ resolvable BIBD, together with a resolution of it.

**Kronecker product**:  for $m \times p$ matrix $M = (m_{ij})$ and $n \times q$ matrix $N = (n_{ij})$, the $mn \times pq$ matrix given by

$$
M \times N = \begin{pmatrix} m_{11}N & m_{12}N & \cdots & m_{1p}N \\ \vdots & \vdots & & \vdots \\ m_{m1}N & m_{m2}N & \cdots & m_{mp}N \end{pmatrix}.
$$

**Latin rectangle**: a $k \times n$ $(k < n)$ array in which each cell contains a single element from an $n$-set such that each element occurs exactly once in each row and at most once in each column.

**Latin square**: a Latin square of side $n$ is an $n \times n$ array in which each entry contains a single element from a set $S$ of size $n$ such that each element occurs exactly once in each row and exactly once in each column.

**line**: a subspace of projective dimension 1 of a projective space; a coset of a subspace of affine dimension 1 of an affine space.

**loop** (in a matroid): an element $e$ of the matroid such that $\{e\}$ is a circuit.

**matroid**: an ordered pair $M = (E(M), \mathcal{I}(M))$ where $E$ (the ground set) is a finite set and $\mathcal{I}$ is a collection of subsets (independent sets) of $E$ such that: the empty set is independent; every subset of an independent set is independent; and if $X$ and $Y$ are independent and $|X| < |Y|$, then there is an element $e$ in $Y - X$ such that $X \cup \{e\}$ is independent.

**matroid representable over a field**: a matroid that is isomorphic to the vector matroid of some matrix over the field $F$.

**multiplier** (of a difference set $D$ in a group $G$): an automorphism $\varphi$ of $G$ such that $\varphi(D) = D + g$ for some $g \in G$.

**mutually orthogonal**: the property of a set of Latin squares that every two are orthogonal.

**orthogonal**: property of two $n \times n$ Latin squares $A = (a_{ij})$ and $B = (b_{ij})$ that all $n^2$ ordered pairs $(a_{ij}, b_{ij})$ are distinct.

**orthogonal array** (of size $N$ with $k$ constraints, $s$ levels, and strength $t$): a $k \times N$ array with entries from a set of $s \geq 2$ symbols, having the property that in every $t \times N$ submatrix every $t \times 1$ column vector appears the same number of times.

**pairwise balanced design** (**PBD**): for a set $K$ of positive integers, a design $(v, K, \lambda)$ consisting of an ordered pair $(X, \mathcal{A})$ where $X$ is a set of size $v$ and $\mathcal{A}$ is a collection of subsets of $X$ with the property that every pair of elements of $X$ occurs in exactly $\lambda$ blocks, and for every block $A \in \mathcal{A}$, $|A| \in K$; a pairwise balanced design is called a $(v, K)$-PBD when $\lambda = 1$.

**parallel class**: a collection of blocks that partition the point set of a design.

**parallel elements**: in a matroid, two elements that form a circuit.

**partial Latin square**: an $n \times n$ array with cells, each of which is either empty or else contains exactly one symbol, such that no symbol occurs more than once in any row or column.

**partial transversal** (of length $k$): in a Latin square, a set of $k$ cells, each from a different row and each from a different column, such that no two contain the same symbol.

**paving matroid**: a matroid such that the number of elements in every circuit is at least as large as the rank of the matroid.

**PBD-closure**: for a set $K$ of positive integers, the set $\mathbf{B}(K) = \{v \mid$ there exists a $(v, K)$-PBD$\}$.

**planar**: the property of a matroid that it is isomorphic to the cycle matroid of a planar graph.

**plane**:  a subspace of projective dimension 2 of a projective space; a coset of a subspace of affine dimension 2 of an affine space.

**projective plane**:  a finite set (of points) and a set of subsets of points (called lines) such that every two points lie on exactly one line, every two lines intersect in exactly one point, and there are four points with no three collinear; equivalently, a symmetric $(n^2 + n + 1, n + 1, 1)$ design.

**projective space** (of dimension $n$):  for a field $F$ of order $q$ and an $(n+1)$-dimensional vector space $S$ over $F$, the set $\mathrm{PG}(n, q)$ of all subspaces of $S$.

**rank** (of a matroid):  the rank of the ground set of the matroid.

**rank** (of a set in a matroid):  the cardinality of every maximal independent subset of the set.

**rank axioms**:  a set of axioms that specifies the properties that a rank function on a matroid must satisfy.

**reduced Latin square**:  a Latin square such that the elements in the first row and the elements in the first column occur in natural order.

**regular matroid**:  a matroid that is representable over all fields.

**replication number**:  the number of blocks to which each point in a design belongs.

**representable over a field**:  the property of a matroid that it is isomorphic to a vector matroid of some matrix over the field.

**resolution**:  a partition of the family of blocks of a balanced incomplete block design into parallel classes.

**resolvable**:  the property of a balanced incomplete block design that it has at least one resolution.

**simple matroid**:  a matroid that has no loops or parallel elements.

**simple (t-design)**:  a $t$-design that contains no repeated blocks.

**spanning set** (of a matroid):  for matroid $M$, a subset of the ground set $E$ of rank $r(M)$.

**Steiner triple system**:  a balanced incomplete block design in which each block has three elements and each pair of points occurs in exactly one block; that is, a $(v, 3, 1)$ design.

**subdesign**:  a collection of points and blocks in a block design that is itself a block design.

**subsquare**:  for $k < n$, a Latin square of side $k$ whose rows and columns are chosen from a Latin square of side $n$.

**symmetric block design**:  a $(v, b, r, k, \lambda)$ design where the number of points $v$ equals the number of blocks $b$.

**ternary matroid**:  a matroid isomorphic to a vector matroid of a matrix over $GF(3)$.

**transversal**:  in a Latin square of side $n$, a set of $n$ cells, one from each row and column, containing each of the $n$ symbols exactly once.

**transversal design**:  a $k$-GDD having $k$ groups of size $n$ and uniform block size $k$.

**transversal matroid**:  given a family of sets, the matroid whose independent sets are partial transversals of this family.

**type**:  See *group-type*.

**uniform matroid**: the matroid with $1, 2, \ldots, n$ as ground set, and all subsets of size less than a specified number as independent sets.

**$(v, k, \lambda)$ design**: a BIBD with parameters $(v, b, r, k, \lambda)$.

**$(v, k, \lambda; n)$ difference set** (of order $n = k - \lambda$): a $k$-subset $D$ of a group $G$ (of order $v$) where every nonzero element of $G$ has exactly $\lambda$ differences $d - d'$ with elements from $D$.

**vector matroid**: the matroid on the columns of a matrix whose independent sets are the linearly independent sets of columns.

**void design**: a BIBD with at most one element.

# 12.1   BLOCK DESIGNS

## 12.1.1   BALANCED INCOMPLETE BLOCK DESIGNS

### Definitions:

A **balanced incomplete block design** (**BIBD**) with parameters $(v, b, r, k, \lambda)$ is a pair $(X, \mathcal{A})$, where $X$ is a set, $\mathcal{A}$ is a collection of subsets of $X$, the five parameters are non-negative integers, either $v \in \{0, 1\}$ (the *void* designs) or $v > k > 0$, and the parameters represent the following:

- $v$ (**order**): the size of $X$ (elements of $X$ are **points**, **varieties**, or **treatments**);
- $b$ (**block number**): the number of elements of $\mathcal{A}$ (elements of $\mathcal{A}$ are **blocks**);
- $r$ (**replication number**): the number of blocks to which every point belongs;
- $k$ (**block size**): the common size of each block;
- $\lambda$ (**index**): the number of blocks to which every pair of distinct points belongs.

*Note*: A BIBD is often referred to as a *design*. Different notations are used for balanced incomplete block designs: $(v, b, r, k, \lambda)$ BIBD, $(v, k, \lambda)$ BIBD and $S_\lambda(2, k, v)$. In this chapter $(v, k, \lambda)$ design will be used. See Fact 6.

A **Steiner triple system** is a $(v, \frac{v(v-1)}{6}, \frac{v-1}{2}, 3, 1)$ design, i.e., a BIBD in which each block has size 3 and each pair of points occurs in exactly one block. A Steiner triple system is denoted $\text{STS}(v)$ or $S(2, 3, v)$. (Jakob Steiner, 1796–1863)

The **incidence matrix** of a $(v, b, r, k, \lambda)$ design is the $b \times v$ matrix $A = (a_{ij})$ defined by

$$a_{ij} = \begin{cases} 1 & \text{if the } i\text{th block contains the } j\text{th point} \\ 0 & \text{otherwise.} \end{cases}$$

### Facts:

**1.** Balanced incomplete block designs are used in the design of experiments when the total number $(v)$ of objects to be tested is greater than the number $(k)$ that can be tested at any one time. They are used to design experiments where the subjects must be divided into subsets (blocks) of the same size to receive different treatments, such that each subject is tested the same number of times and every pair of subjects appears in the same number of subsets.

**2.** Designs are useful in many areas, such as coding theory, cryptography, group testing, and tournament scheduling. Detailed coverage of these and other applications of designs can be found in Chapter V of [CoDi07].

**3.** The word "balanced" refers to the fact that $\lambda$ remains constant. If $\lambda$ changes depending on the pair of points chosen, the design is not balanced.

**4.** The word "incomplete" refers to the fact that $k < v$, that is, the size of each block is less than the number of varieties.

**5.** *Necessary conditions for existence:* If there is a $(v, b, r, k, \lambda)$ design for particular $v$, $b$, $r$, $k$, and $\lambda$, then the parameters must satisfy

- $vr = bk$;
- $\lambda(v - 1) = r(k - 1)$;
- $b \geq v$. (Fisher's inequality, 1940) (Ronald A. Fisher, 1890–1962)

**6.** If a $(v, b, r, k, \lambda)$ design exists, $r = \frac{\lambda(v-1)}{k-1}$ and $b = \frac{\lambda v(v-1)}{k(k-1)}$. In view of these two relationships, $(v, b, r, k, \lambda)$ designs are commonly referred to simply by the three parameters — $v$, $k$, $\lambda$ — as a $(v, k, \lambda)$ design.

**7.** *Necessary conditions for existence:* If there is a $(v, k, \lambda)$ design for particular $v$, $k$, and $\lambda$, then

- $\lambda(v - 1) \equiv 0 \pmod{k - 1}$;
- $\lambda v(v - 1) \equiv 0 \pmod{k(k - 1)}$.

**8.** *Existence of $(v, k, \lambda)$ designs:*

- $(v, 3, \lambda)$ design: exists for all $v$ satisfying the necessary conditions given in Fact 5, namely
  - ⋄ if $\lambda \equiv 2$ or $4 \pmod 6$ and $v \equiv 0$ or $1 \pmod 3$
  - ⋄ if $\lambda \equiv 1$ or $5 \pmod 6$ and $v \equiv 1$ or $3 \pmod 6$
  - ⋄ if $\lambda \equiv 3 \pmod 6$ and $v \equiv 1 \pmod 2$
  - ⋄ if $\lambda \equiv 0 \pmod 6$ and $v \neq 2$;
- $(v, 4, \lambda)$ design: exists for all $v$ and $\lambda$ satisfying the necessary conditions given in Fact 5;
- $(v, 5, \lambda)$ design: exists for all $v$ satisfying the necessary conditions given in Fact 5 except for the case $v = 15, \lambda = 2$;
- $(v, 6, \lambda)$ design: exists for all $v$ satisfying the necessary conditions given in Fact 5, if $\lambda > 1$;
- $(v, 6, 1)$ design: exists for all $v \equiv 1$ or $6 \pmod{15}$, $v \geq 31$, $v \notin \{36, 46\}$, with 29 possible exceptions, the largest being 801. (The first few open cases are 51, 61, 81, and 166.)

**9.** *Existence of Steiner triple systems:* A Steiner triple system with $v$ points exists if and only if $v \equiv 1$ or $3 \pmod 6$. (Kirkman)

**10.** *Wilson's asymptotic existence theorem:* Given $k$ and $\lambda$, there exists a $v_0(k, \lambda)$ such that a $(v, k, \lambda)$ design exists for all $v \geq v_0(k, \lambda)$ that satisfy the necessary conditions given in Fact 5 and make $b$ and $r$ integral. It is known that $v_0(k, \lambda) < \exp(\exp(k^{k^2}))$.

**11.** Assume that $(V, \mathcal{B})$ is a $(v, k, \lambda)$ design. Let $\overline{\mathcal{B}} = \{V - B \mid B \in \mathcal{B}\}$. Then $(V, \overline{\mathcal{B}})$ is a $\left(v, v - k, \lambda\frac{(v-k)(v-k-1)}{k(k-1)}\right)$ design, the *complement* of $(V, \mathcal{B})$.

**12.** Given two Steiner triple systems with $v_1$ and $v_2$ points, respectively, a Steiner triple system with $v_1 v_2$ points can be constructed as follows: Let an STS($v_1$) be defined on the point set $\{x_1, \ldots, x_{v_1}\}$ and an STS($v_2$) be defined on the point set $\{y_1, \ldots, y_{v_2}\}$. Define an STS($v_1 v_2$) on the point set $\{z_{ij} \mid 1 \leq i \leq v_1, 1 \leq j \leq v_2\}$ where $z_{mn} z_{pq} z_{rs}$ is a block in STS($v_1 v_2$) if and only if one of the following holds:

- $m = p = r$ and $y_n y_q y_s$ is a block in STS($v_2$);
- $n = q = s$ and $x_m x_p x_r$ is a block in STS($v_1$);
- $x_m x_p x_r$ is a block in STS($v_1$) and $y_n y_q y_s$ is a block in STS($v_2$).

**13.** The following table lists different types of block designs and their features.

| name | block size | size of subset covered | # times covered | other properties |
|---|---|---|---|---|
| balanced incomplete block design BIBD §12.1.1 | $k$ | 2 | $\lambda$ | |
| pairwise balanced design PBD §12.1.6 | various | 2 | $\lambda$ | also called a linear space if $\lambda = 1$ |
| Steiner triple system STS §12.1.1, §12.1.5 | 3 | 2 | 1 | |
| Kirkman triple system KTS §12.1.4 | 3 | 2 | 1 | resolvable |
| resolvable balanced incomplete block design RBIBD §12.1.4 | $k$ | 2 | $\lambda$ | resolvable |
| projective plane PG$(2, q)$ §12.2.3 | $q + 1$ | 2 | 1 | #points = #blocks $= q^2 + q + 1$ |
| affine plane AG$(2, q)$ §12.2.3 | $q$ | 2 | 1 | resolvable |
| symmetric design SBIBD §12.2.2 | $k$ | 2 | $\lambda$ | #points = #blocks |
| $t$-design $t$-$(v, k, \lambda)$ §12.1.5 | $k$ | $t \geq 2$ | $\lambda$ | |
| Steiner system $S(t, v, k)$ §12.1.5 | $k$ | $t \geq 2$ | 1 | |

**Examples:**

**1.** The following is a $(4, 4, 3, 3, 2)$ design: $X = \{a, b, c, d\}$, blocks $\{abc, abd, acd, bcd\}$.

**2.** *Affine plane of order 3* [a $(9, 3, 1)$ *design*]: Here the point set is $X = \{0, \ldots, 8\}$ and the block set is $\mathcal{A} = \{012, 345, 678, 036, 147, 258, 048, 156, 237, 057, 138, 246\}$. Also see §12.2.3. This design is known as $AG(2, 3)$. This is a Steiner triple system.

**3.** Each of the following is a Steiner triple system (a $(v, 3, 1)$ design). In each of the following a set of *base blocks* $B_i = \{b_{i1}, b_{i2}, b_{i3}\}$ in the group $\mathcal{Z}_v$ is given. To get all the blocks of the design, take all distinct translates $B_i + g = \{b_{i1} + g, b_{i2} + g, b_{i3} + g\}$, for all $g \in \mathcal{Z}_v$, for each of the base blocks $B_i$.

$v = 7$: $\{0, 1, 3\}$ (mod 7)  [Fano plane]

$v = 15$: $\{0, 1, 4\}$ $\{0, 2, 8\}$ $\{0, 5, 10\}$ (mod 15) [The last base block has only 5 ($= \frac{v}{3}$) distinct translates. This is a *short orbit* and occurs for all orders $v \equiv 3$ (mod 6).]

$v = 19$: $\{0, 1, 4\}$ $\{0, 2, 9\}$ $\{0, 5, 11\}$ (mod 19)

$v = 21$: $\{0, 1, 3\}$ $\{0, 4, 12\}$ $\{0, 5, 11\}$ $\{0, 7, 14\}$ (mod 21)
$v = 25$: $\{0, 1, 3\}$ $\{0, 4, 11\}$ $\{0, 5, 13\}$ $\{0, 6, 15\}$ (mod 25)
$v = 27$: $\{0, 1, 3\}$ $\{0, 4, 11\}$ $\{0, 5, 15\}$ $\{0, 6, 14\}$ $\{0, 9, 18\}$ (mod 27)
$v = 31$: $\{0, 1, 3\}$ $\{0, 4, 11\}$ $\{0, 5, 15\}$ $\{0, 6, 18\}$ $\{0, 8, 17\}$ (mod 31)
$v = 33$: $\{0, 1, 3\}$ $\{0, 4, 10\}$ $\{0, 5, 18\}$ $\{0, 7, 19\}$ $\{0, 8, 17\}$ $\{0, 11, 22\}$ (mod 33)
$v = 37$: $\{0, 1, 3\}$ $\{0, 4, 9\}$ $\{0, 6, 21\}$ $\{0, 7, 18\}$ $\{0, 8, 25\}$ $\{0, 10, 24\}$ (mod 37)
$v = 39$: $\{0, 1, 3\}$ $\{0, 4, 9\}$ $\{0, 6, 20\}$ $\{0, 7, 18\}$ $\{0, 8, 23\}$ $\{0, 10, 22\}$ $\{0, 13, 26\}$ (mod 39).

**4.** *Fano plane* or *projective plane of order* 2, PG(2, 2): A $(7, 7, 3, 3, 1)$ design with point set $X = \{0, \ldots, 6\}$ and block set $\mathcal{A} = \{013, 124, 235, 346, 450, 561, 602\}$, shown in the following figure. (Often, as here, a block $\{a, b, c\}$ is written as $abc$.) Also see §12.2.3. (Gino Fano, 1871–1952)

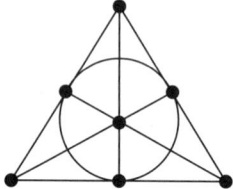

The incidence matrix of the Fano plane is

$$
\begin{pmatrix}
1 & 1 & 0 & 1 & 0 & 0 & 0 \\
0 & 1 & 1 & 0 & 1 & 0 & 0 \\
0 & 0 & 1 & 1 & 0 & 1 & 0 \\
0 & 0 & 0 & 1 & 1 & 0 & 1 \\
1 & 0 & 0 & 0 & 1 & 1 & 0 \\
0 & 1 & 0 & 0 & 0 & 1 & 1 \\
1 & 0 & 1 & 0 & 0 & 0 & 1
\end{pmatrix}
$$

## 12.1.2    ISOMORPHISM AND AUTOMORPHISM

**Definitions:**

Two designs $(V, \mathcal{B})$ and $(W, \mathcal{C})$ are **isomorphic** if there is a bijection $\psi \colon V \to W$ under which $\psi(B) = \{\psi(x) \mid x \in B\}$ occurs as a block in $\mathcal{C}$ the same number of times that $B$ occurs as a block in $\mathcal{B}$. Such a bijection is an **isomorphism**.

An **automorphism** of a design $D$ is an isomorphism from $D$ onto $D$.

The **automorphism group** of a design $D$ is the set of all automorphisms for $D$ with composition as the group operation.

**Facts:**

**1.** Nonisomorphic Steiner triple systems of order $v$ have been enumerated for $v \leq 19$. Up to isomorphism, there are unique designs of order 3, 7, and 9; there are precisely two nonisomorphic designs of order 13, and 80 of order 15. At that point, an explosion occurs: the number of nonisomorphic STS(19) is 11,084,874,829.

**2.** The number of nonisomorphic STS($v$) is at least $(e^{-5}v)^{v^2/6}$ for large $v$. (Wilson)

**3.** The following table lists the parameter sets $(v, k, \lambda)$ that satisfy the necessary conditions for the existence of a block design, with $r \leq 15$ and $3 \leq k \leq \frac{v}{2}$. The parameter sets are ordered lexicographically across the rows of the table by $r$, $k$ and $\lambda$ (in this order). The column $N$ contains the number of pairwise nonisomorphic $(v, k, \lambda)$ designs or the best known lower bound for this number. The notation "?" indicates that no design with these parameters is known to exist, but that existence has not been ruled out.

| $v$ | $k$ | $\lambda$ | $N$ | $v$ | $k$ | $\lambda$ | $N$ | $v$ | $k$ | $\lambda$ | $N$ |
|---|---|---|---|---|---|---|---|---|---|---|---|
| 7 | 3 | 1 | 1 | 9 | 3 | 1 | 1 | 13 | 4 | 1 | 1 |
| 6 | 3 | 2 | 1 | 16 | 4 | 1 | 1 | 21 | 5 | 1 | 1 |
| 11 | 5 | 2 | 1 | 13 | 3 | 1 | 2 | 7 | 3 | 2 | 4 |
| 10 | 4 | 2 | 3 | 25 | 5 | 1 | 1 | 31 | 6 | 1 | 1 |
| 16 | 6 | 2 | 3 | 15 | 3 | 1 | 80 | 8 | 4 | 3 | 4 |
| 15 | 5 | 2 | 0 | 36 | 6 | 1 | 0 | 43 | 7 | 1 | 0 |
| 22 | 7 | 2 | 0 | 15 | 7 | 3 | 5 | 9 | 3 | 2 | 36 |
| 25 | 4 | 1 | 18 | 13 | 4 | 2 | 2,461 | 9 | 4 | 3 | 11 |
| 21 | 6 | 2 | 0 | 49 | 7 | 1 | 1 | 57 | 8 | 1 | 1 |
| 29 | 8 | 2 | 0 | 19 | 3 | 1 | 11,084,874,829 | 10 | 3 | 2 | 960 |
| 7 | 3 | 3 | 10 | 28 | 4 | 1 | $\geq 4{,}747$ | 10 | 5 | 4 | 21 |
| 46 | 6 | 1 | 0 | 16 | 6 | 3 | 18,920 | 28 | 7 | 2 | 8 |
| 64 | 8 | 1 | 1 | 73 | 9 | 1 | 1 | 37 | 9 | 2 | 4 |
| 25 | 9 | 3 | 78 | 19 | 9 | 4 | 6 | 21 | 3 | 1 | $\geq 6.2\times10^7$ |
| 6 | 3 | 4 | 4 | 16 | 4 | 2 | $\geq 2.2\times10^6$ | 41 | 5 | 1 | $\geq 15$ |
| 21 | 5 | 2 | $\geq 22{,}998$ | 11 | 5 | 4 | 4,393 | 51 | 6 | 1 | ? |
| 21 | 7 | 3 | 3,809 | 36 | 8 | 2 | 0 | 81 | 9 | 1 | 7 |
| 91 | 10 | 1 | 4 | 46 | 10 | 2 | 0 | 31 | 10 | 3 | 151 |
| 12 | 3 | 2 | 242,995,846 | 12 | 4 | 3 | $\geq 17{,}172{,}470$ | 45 | 5 | 1 | $\geq 16$ |
| 12 | 6 | 5 | 11,603 | 45 | 9 | 2 | $\geq 16$ | 100 | 10 | 1 | 0 |
| 111 | 11 | 1 | 0 | 56 | 11 | 2 | $\geq 5$ | 23 | 11 | 5 | 1,106 |
| 25 | 3 | 1 | $\geq 10^{14}$ | 13 | 3 | 2 | $\geq 1{,}897{,}376$ | 9 | 3 | 3 | 22,521 |
| 7 | 3 | 4 | 35 | 37 | 4 | 1 | $\geq 51{,}402$ | 19 | 4 | 2 | $\geq 423$ |
| 13 | 4 | 3 | $\geq 3{,}702$ | 10 | 4 | 4 | 13,769,944 | 25 | 5 | 2 | $\geq 118{,}884$ |
| 61 | 6 | 1 | ? | 31 | 6 | 2 | $\geq 72$ | 21 | 6 | 3 | $\geq 236$ |
| 16 | 6 | 4 | $\geq 111$ | 13 | 6 | 5 | 19,072,802 | 22 | 8 | 4 | 0 |
| 33 | 9 | 3 | $\geq 3{,}375$ | 55 | 10 | 2 | 0 | 121 | 11 | 1 | $\geq 1$ |
| 133 | 12 | 1 | $\geq 1$ | 67 | 12 | 2 | 0 | 45 | 12 | 3 | $\geq 3{,}752$ |
| 34 | 12 | 4 | 0 | 27 | 3 | 1 | $\geq 10^{11}$ | 40 | 4 | 1 | $\geq 10^6$ |
| 66 | 6 | 1 | $\geq 1$ | 14 | 7 | 6 | 15,111,019 | 27 | 9 | 4 | $\geq 2.45\times10^8$ |
| 40 | 10 | 3 | ? | 66 | 11 | 2 | $\geq 2$ | 144 | 12 | 1 | ? |
| 157 | 13 | 1 | ? | 79 | 13 | 2 | $\geq 2$ | 53 | 13 | 3 | 0 |
| 40 | 13 | 4 | $\geq 1{,}108{,}800$ | 27 | 13 | 6 | 208,310 | 15 | 3 | 2 | $\geq 685{,}521$ |
| 22 | 4 | 2 | $\geq 7{,}921$ | 8 | 4 | 6 | 2,310 | 15 | 5 | 4 | $\geq 896$ |
| 36 | 6 | 2 | $\geq 5$ | 15 | 6 | 5 | $\geq 117$ | 85 | 7 | 1 | ? |
| 43 | 7 | 2 | $\geq 4$ | 29 | 7 | 3 | $\geq 1$ | 22 | 7 | 4 | $\geq 3{,}393$ |
| 15 | 7 | 6 | $\geq 57{,}810$ | 78 | 12 | 2 | 0 | 169 | 13 | 1 | $\geq 1$ |
| 183 | 14 | 1 | $\geq 1$ | 92 | 14 | 2 | 0 | 31 | 3 | 1 | $\geq 6\times10^{16}$ |
| 16 | 3 | 2 | $\geq 10^{13}$ | 11 | 3 | 3 | $\geq 436{,}800$ | 7 | 3 | 5 | 109 |
| 6 | 3 | 6 | 6 | 16 | 4 | 3 | $\geq 6\times10^5$ | 61 | 5 | 1 | $\geq 10$ |
| 31 | 5 | 2 | $\geq 1$ | 21 | 5 | 3 | $\geq 10^9$ | 16 | 5 | 4 | $\geq 294$ |
| 13 | 5 | 5 | $\geq 76$ | 11 | 5 | 6 | $\geq 127$ | 76 | 6 | 1 | $\geq 1$ |
| 26 | 6 | 3 | $\geq 1$ | 16 | 6 | 5 | $\geq 25$ | 91 | 7 | 1 | $\geq 2$ |
| 16 | 8 | 7 | $\geq 9\times10^7$ | 21 | 9 | 6 | $\geq 10^4$ | 136 | 10 | 1 | ? |
| 46 | 10 | 3 | ? | 28 | 10 | 5 | $\geq 3$ | 56 | 12 | 3 | $\geq 4$ |
| 91 | 13 | 2 | 0 | 196 | 14 | 1 | 0 | 211 | 15 | 1 | 0 |
| 106 | 15 | 2 | 0 | 71 | 15 | 3 | $\geq 72$ | 43 | 15 | 5 | 0 |
| 36 | 15 | 6 | $\geq 25{,}634$ | | | | | | | | |

### 12.1.3  SUBDESIGNS

**Definition:**
Let $Y$ be a subset of $w$ points in a $(v, k, \lambda)$ design. If every block of the BIBD contains 0, 1, or $k$ of the points in $Y$, then a $(w, k, \lambda)$ design is obtained by taking those blocks that contain $k$ points from $Y$. This BIBD on $w$ points is a **subdesign**, called a $(w, k, \lambda)$ subdesign.

**Facts:**
**1.** If there is a $(v, k, 1)$ design containing a $(w, k, 1)$ subdesign, then $v \geq (k - 1)w + 1$. (The parameter lists $(v, k, 1)$ and $(w, k, 1)$ must satisfy the necessary conditions of §12.1.1, Fact 6.)

**2.** In the cases $k = 3$ and $k = 4$, the necessary conditions of §12.1.1, Fact 5 for the presence of a subdesign are sufficient. That is, in the case of $k = 3$, for all $v \geq 2w + 1$, with both $v, w \equiv 1$ or 3 mod 6), there exists a $(v, 3, 1)$ design that contains a $(w, 3, 1)$ subdesign. In the case $k = 4$, for all $v \geq 3w + 1$, with both $v, w \equiv 1$ or 4 (mod 12) there exists a $(v, 4, 1)$ design that contains a $(w, 4, 1)$ subdesign.

**Example:**
**1.** *A construction for a Steiner triple system of order $2v + 1$ given a Steiner triple system of order $v$:*  A variant of this construction dates back at least to Thomas P. Kirkman in 1847. The original STS($v$) is a subdesign of the resulting STS($2v + 1$).

Let $(X, \mathcal{A})$ be an STS($v$) with $X = \{x_0, x_1, \ldots, x_{v-1}\}$. For each $i = 0, 1, \ldots, v - 1$, let $F_i = \{\{x + i, -x + i\} \mid x \in \mathcal{Z}_v, x \neq 0\} \cup \{i, \infty\}$. Then for each $i = 0, , \ldots, v - 1$, construct the triples $\{a, b, x_i\}$ where $\{a, b\} \in F_i$. The set of all such triples in addition to the original triples in $\mathcal{A}$ is the desired STS($2v + 1$) on the point set $X \cup \{0, 1, \ldots, v, \infty\}$.

For $v = 7$, the following STS(15) is obtained. The last row of triples is an STS(7).

$$\{0, \infty, x_0\} \ \{1, 6, x_0\} \ \{2, 5, x_0\} \ \{3, 4, x_0\}$$
$$\{1, \infty, x_1\} \ \{2, 0, x_1\} \ \{3, 6, x_1\} \ \{4, 5, x_1\}$$
$$\{2, \infty, x_2\} \ \{3, 1, x_2\} \ \{4, 0, x_2\} \ \{5, 6, x_2\}$$
$$\{3, \infty, x_3\} \ \{4, 2, x_3\} \ \{5, 1, x_3\} \ \{6, 0, x_3\}$$
$$\{4, \infty, x_4\} \ \{5, 3, x_4\} \ \{6, 2, x_4\} \ \{0, 1, x_4\}$$
$$\{5, \infty, x_5\} \ \{6, 4, x_5\} \ \{0, 3, x_5\} \ \{1, 2, x_5\}$$
$$\{6, \infty, x_6\} \ \{0, 5, x_6\} \ \{1, 4, x_6\} \ \{2, 3, x_6\}$$
$$\{x_0, x_1, x_3\} \ \{x_1, x_2, x_4\} \ \{x_2, x_3, x_5\} \ \{x_3, x_4, x_6\} \ \{x_4, x_5, x_0\} \ \{x_5, x_6, x_1\} \ \{x_6, x_0, x_2\}.$$

### 12.1.4  RESOLVABLE DESIGNS

**Definitions:**
A **parallel class** is a collection of blocks that partition the point set.

A **resolution** of a BIBD is a partition of the family of blocks into parallel classes. A resolution contains exactly $r$ parallel classes.

A BIBD is **resolvable**, denoted RBIBD, if it has at least one resolution.

A $(v, 3, 1)$ RBIBD, together with a resolution of it, is a **Kirkman triple system**, written KTS($v$).

**Facts:**

**1.** Necessary conditions for existence of a $(v, k, \lambda)$ RBIBD are

- $\lambda(v - 1) \equiv 0 \pmod{(k - 1)}$;

- $v \equiv 0 \pmod{k}$.

**2.** If a $(v, k, \lambda)$ RBIBD exists, then $b \geq v + r - 1$ where $b$ is the number of blocks. When $b = v + r - 1$ (or equivalently, $r = k + \lambda$) the RBIBD has the property that two nonparallel lines intersect in exactly $\frac{k^2}{v}$ points. (R. C. Bose, 1901–1987)

**3.** A KTS($v$) exists if and only if $v \equiv 3 \pmod{6}$.

**4.** The following table summarizes the current state of knowledge concerning the existence of resolvable designs.

For the values of $k$ and $\lambda$ given, the number of parameter sets $(v, k, \lambda)$ satisfying all necessary conditions for the existence of a resolvable $(v, k, \lambda)$ design for which the existence of a resolvable $(v, k, \lambda)$ design is not known is given under the column headed "exceptions". The column headed "largest possible exception" gives the largest $v$ satisfying the necessary conditions for the existence of a resolvable $(v, k, \lambda)$ design for which a resolvable $(v, k, \lambda)$ design is not known.

| $k$ | $\lambda$ | exceptions | largest possible exception |
|-----|-----------|------------|----------------------------|
| 3 | 1 | none | |
| 3 | 2 | 6 | |
| 4 | 1 | none | |
| 4 | 3 | none | |
| 5 | 1 | | 645 |
| 5 | 2 | 15 | 395 |
| 5 | 4 | 10,15 | 195 |
| 6 | 5 | none | |
| 6 | 10 | none | |
| 7 | 6 | 14 | 462 |
| 8 | 1 | | 24,480 |
| 8 | 7 | | 1,488 |

**Example:**

**1.** *Kirkman's schoolgirl problem*: In 1850, Kirkman posed the following: fifteen young ladies in a school walk out three abreast for seven days in succession; it is required to arrange them daily, so that no two walk twice abreast. (Thomas P. Kirkman, 1806–1895)

This is equivalent to finding a resolution of some $(15, 3, 1)$ design (or a KTS(15)). Seven nonisomorphic solutions exist; the following is one solution to Kirkman's schoolgirl problem:

| Monday | Tuesday | Wednesday | Thursday | Friday | Saturday | Sunday |
|--------|---------|-----------|----------|--------|----------|--------|
| 9,10,12 | 10,11,13 | 11,12,14 | 12,13,15 | 13,14,9 | 14,15,10 | 15,9,11 |
| 15,8,1 | 9,8,2 | 10,8,3 | 11,8,4 | 12,8,5 | 13,8,6 | 14,8,7 |
| 13,2,7 | 14,3,1 | 15,4,2 | 9,5,3 | 10,6,4 | 11,7,5 | 12,1,6 |
| 11,3,6 | 12,4,7 | 13,5,1 | 14,6,2 | 15,7,3 | 9,1,4 | 10,2,5 |
| 14,4,5 | 15,5,6 | 9,6,7 | 10,7,1 | 11,1,2 | 12,2,3 | 13,3,4 |

## 12.1.5   *t*-DESIGNS AND STEINER SYSTEMS

### Definitions:

A **$t$-$(v, k, \lambda)$ design** (also denoted $S_\lambda(t, k, v)$ and written **$t$-design**) is a pair $(X, \mathcal{A})$ that satisfies the following properties:

- $X$ is a set of $v$ elements (called *points*);
- $\mathcal{A}$ is a family of subsets (*blocks*) of $X$, each of cardinality $k$;
- every $t$-subset of distinct points occurs in exactly $\lambda$ blocks.

A $t$-design is **simple** if it contains no repeated blocks.

A **Steiner system** is a $t$-$(v, k, 1)$ design.

A **Steiner triple system**, denoted STS($v$), is a $(v, 3, 1)$ design. (See §12.1.1.)

A **Steiner quadruple system**, denoted SQS($v$), is a 3-$(v, 4, 1)$ design.

### Facts:

1. A $(v, k, \lambda)$ design (a BIBD) is a 2-$(v, k, \lambda)$ design.

2. If $s < t$, then a $t$-$(v, k, \lambda)$ design is also an $s$-$(v, k, \mu)$ design, where $\mu = \lambda \frac{\binom{v-s}{t-s}}{\binom{k-s}{t-s}}$.

3. As a consequence of Fact 2, a $t$-$(v, k, \lambda)$ design only exists if $\lambda\binom{v-s}{t-s} \equiv 0 \pmod{\binom{k-s}{t-s}}$ for every $0 \le s \le t$; these are the *divisibility conditions*.

4. $t$-$(v, k, \lambda)$ designs exist for all $t$. A $t$-$(v, t + 1, ((t + 1)!)^{2t+1})$ design exists if $v \ge t + 1$ and $v \equiv t \pmod{[(t + 1)!]^{2t+1}}$. (Teirlinck)

5. If a $t$-$(v, k, \lambda)$ design exists, where $t = 2s$ is even, then the number of blocks $b \ge \binom{v}{s}$. (This generalizes Fisher's inequality, §12.1.1, Fact 5.)

6. When $\lambda = 1$, until 2014 $t$-designs were known only for $t \le 5$, and the construction of a 6-$(v, k, 1)$ design was one of the outstanding open problems in the theory of combinatorial designs. In 2014, Keevash established that for every $t$ and $k$, $t$-$(v, k, 1)$ designs exist when $v$ satifies the divisibility conditions (from Fact 3) and $v$ is sufficiently large.

7. Much less is known about the existence of $t$-$(v, k, \lambda)$ designs with $t \ge 3$ compared to BIBDs:

- For $t = 3$, several infinite families are known.
- For every prime power $q$ and $d \ge 2$, there exists a 3-$(q^d + 1, q + 1, 1)$ design, known as an *inversive geometry*. When $d = 2$, these designs are known as *inversive planes*.
- A 3-$(v, 4, 1)$ design (*Steiner quadruple system*) exists if and only if $v \equiv 2$ or 4 (mod 6).

**Examples:**

**1.** The following is a 3-$(8, 4, 1)$ design:

$$X = \{\infty, 0, 1, 2, 3, 4, 5, 6\}$$
$$\mathcal{A} = \{\{0, 1, 3, \infty\}, \ \{1, 2, 4, \infty\}, \ \{2, 3, 5, \infty\}, \ \{3, 4, 6, \infty\}, \ \{4, 5, 0, \infty\},$$
$$\{5, 6, 1, \infty\}, \ \{6, 0, 2, \infty\}, \ \{2, 4, 5, 6\}, \ \{3, 5, 6, 0\}, \ \{4, 6, 0, 1\},$$
$$\{5, 0, 1, 2\}, \ \{6, 1, 2, 3\}, \ \{0, 2, 3, 4\}, \ \{1, 3, 4, 5\}\}$$

**2.** *Simple t-designs (t = 4, 5):* For $t = 4$ or $5$ and $v \leq 30$, the only $t$-$(v, k, 1)$ designs known to exist are those having the following parameters:

$$\begin{array}{ccc} 4\text{-}(11,\ 5,\ 1) & 5\text{-}(12,\ 6,\ 1) & 4\text{-}(23,\ 7,\ 1) \\ 5\text{-}(24,\ 8,\ 1) & 4\text{-}(27,\ 6,\ 1) & 5\text{-}(28,\ 7,\ 1). \end{array}$$

**3.** *Simple t-designs (t = 6):* For $t = 6$ and $v \leq 20$, the only $t$-$(v, k, \lambda)$ designs known to exist are those having the following parameters:

$$\begin{array}{ccccc} 6\text{-}(14,\ 7,\ 4) & 6\text{-}(19,\ 7,\ 4) & 6\text{-}(19,\ 7,\ 6) & 6\text{-}(20,\ 9,\ 112) & 6\text{-}(22,\ 7,\ 8) \end{array}$$
$$6\text{-}(20,10,7m),\ m \in \{48, 58, 62, 63, 67, 68\}.$$

---

## 12.1.6   PAIRWISE BALANCED DESIGNS

**Definitions:**

Given a set $K$ of positive integers and a positive integer $\lambda$, a **pairwise balanced design**, written $(v, K, \lambda)$-PBD, is an ordered pair $(X, \mathcal{A})$ where $X$ is a set (of *points*) of size $v$ and $\mathcal{A}$ is a collection of subsets (*blocks*) of $X$ such that

- every pair of elements of $X$ occurs together in exactly $\lambda$ blocks;
- for every block $A \in \mathcal{A}$, $|A| \in K$.

When $\lambda = 1$, $\lambda$ can be omitted from the notation and the design is called a $(v, K)$-PBD or a **finite linear space**.

Given a set $K$ of positive integers, let $\mathbf{B}(K)$ denote the set of positive integers $v$ for which there exists a $(v, K)$-PBD. The mapping $K \to \mathbf{B}(K)$ is a **closure operation** on the set of subsets of the positive integers, as it satisfies the following properties:

- $K \subseteq \mathbf{B}(K)$;
- $K_1 \subseteq K_2 \Rightarrow \mathbf{B}(K_1) \subseteq \mathbf{B}(K_2)$;
- $\mathbf{B}(\mathbf{B}(K)) = \mathbf{B}(K)$.

The set $\mathbf{B}(K)$ is the **closure** of the set $K$.

If $K$ is any set of positive integers, then $K$ is **PBD-closed** (or **closed**) if $\mathbf{B}(K) = K$.

If $K$ is a closed set, then there exists a finite subset $J \subseteq K$ such that $K = \mathbf{B}(J)$. This set $J$ is a **generating set** for the PBD-closed set $K$.

If $J$ is a generating set for $K$ and if $s \in J$ is such that $J - \{s\}$ is also a generating set for $K$, then $s$ is **inessential** in $K$; otherwise $s$ is **essential**.

A **basis** is a generating set consisting of essential elements.

**Facts:**

**1.** A $(v, k, \lambda)$ design is a special case of a PBD in which the blocks are only permitted to be of one size, $k$.

**2.** *Necessary conditions for existence:* The existence of a $(v, K)$-PBD (with $v > 0$) implies

- $v \equiv 1 \pmod{\alpha(K)}$
- $v(v - 1) \equiv 0 \pmod{\beta(K)}$

where $\alpha(K)$ is the greatest common divisor of the integers $\{k - 1 \mid k \in K\}$ and $\beta(K)$ is the greatest common divisor of the integers $\{k(k - 1) \mid k \in K\}$.

**3.** *Asymptotic existence:* Given $K$, there exists a constant $c_k$ such that a $(v, K)$-PBD exists for all $v \geq c_k$ that satisfy the necessary conditions of Fact 2. The constant $c_k$ is, in general, unspecified. In practice, considerable further work is usually required to obtain a concrete upper bound on $c_k$.

**Examples:**

**1.** The following is a $(10, \{3, 4\})$-PBD:

$$\{1, 2, 3, 4\}, \{1, 5, 6, 7\}, \{1, 8, 9, 10\}, \{2, 5, 8\}, \{2, 6, 9\}, \{2, 7, 10\}$$

$$\{3, 5, 10\}, \{3, 6, 8\}, \{3, 7, 9\}, \{4, 5, 9\}, \{4, 6, 10\}, \{4, 7, 8\}$$

**2.** The following table lists closures of some subsets of $\{3, 4, \ldots, 8\}$. From Fact 3, for a given set $K$ there are only a finite number of values of $v$ (satisfying the necessary conditions) for which there does not exist a $(v, K)$-PBD. These exceptional cases are listed in this table for some small sets $K$. Since $7 \in B(3)$, it is not necessary to include 7 in the list of sets whose closures are given, when 3 is present. Genuine exceptions (values of $v$ satisfying the necessary conditions for the existence of a $(v, K)$-PBD for which it has been *proven* that no such design can exist) are shown in boldface, while possible exceptions (neither existence or nonexistence of a $(v, K)$-PBD is known) are shown in normal type.

| subset $K$ | necessary conditions | exceptions |
|---|---|---|
| 3 | 1, 3 **mod** 6 | — |
| 3,4 | 0, 1 **mod** 3 | — |
| 3,5 | 1 **mod** 2 | — |
| 3,6 | 0, 1 **mod** 3 | **4, 10, 12, 22** |
| 3,8 | $\mathcal{N}$ (natural numbers) | **4, 5, 6, 10, 11, 12, 14, 16, 17, 18, 20, 23, 26, 28, 29, 30, 34, 35, 36, 38** |
| 3,4,5 | $\mathcal{N}$ | **6, 8** |
| 3,4,6 | 0, 1 **mod** 3 | — |
| 3,4,8 | $\mathcal{N}$ | **5, 6, 11, 14, 17** |
| 3,5,6 | $\mathcal{N}$ | **4, 8, 10, 12, 14, 20, 22** |
| 3,5,8 | $\mathcal{N}$ | **4, 6, 10, 12, 14, 16, 18, 20, 26, 28, 30, 34** |
| 3,6,8 | $\mathcal{N}$ | **4, 5, 10, 11, 12, 14, 17, 20, 23** |
| 3,4,5,6 | $\mathcal{N}$ | **8** |
| 3,4,5,8 | $\mathcal{N}$ | **6** |
| 3,4,6,8 | $\mathcal{N}$ | **5, 11, 14, 17** |

| subset $K$ | necessary conditions | exceptions |
|---|---|---|
| 3,5,6,8 | $\mathcal{N}$ | **4, 10, 14, 20** |
| 3,4,5,6,8 | $\mathcal{N}$ | – |
| 4 | $1, 4 \bmod 12$ | – |
| 4,5 | $0, 1 \bmod 4$ | **8, 9, 12** |
| 4,6 | $0, 1 \bmod 3$ | **7, 9, 10, 12, 15, 18, 19, 22, 24, 27**, 33, 34, 39, 45, 46, 51, 87 |
| 4,7 | $1 \bmod 3$ | **10, 19** |
| 4,8 | $0, 1 \bmod 4$ | **5, 9, 12, 17, 20, 21, 24, 33, 41, 44, 45**, 48, 53, 60, 65, 69, 77, 89 |
| 4,5,6 | $\mathcal{N}$ | **7, 8, 9, 10, 11, 12, 14, 15, 18, 19, 23** |
| 4,5,7 | $\mathcal{N}$ | **6, 8, 9, 10, 11, 12, 14, 15, 18, 19, 23, 26, 27**, 30, 39, 42, 51, 54 |
| 4,5,8 | $0, 1 \bmod 4$ | **9, 12** |
| 4,6,7 | $0, 1 \bmod 3$ | **5, 9, 10, 12, 15, 19, 24, 27**, 33, 45, 87 |
| 4,6,8 | $\mathcal{N}$ | **5, 7, 9, 10, 11, 12, 14, 15, 17, 18, 19, 20, 22, 23, 24, 26, 27**, 33, 34, 35, 39, 41, 47, 50, 51, 53, 59, 62, 65, 71, 77, 87, 95, 110, 131, 170 |
| 4,7,8 | $\mathcal{N}$ | **5, 6, 9, 10, 11, 12, 14, 15, 17, 18, 19, 20, 21, 23, 24, 26, 27, 30, 33, 35, 38, 39, 41, 42, 44, 45, 47**, 48, 51, 54, 59, 62, 65, 66, 69, 74, 75, 77, 78, 83, 87, 89, 90, 102, 110, 114, 126, 131, 138, 143, 150, 162, 167, 174, 186 |
| 4,5,6,7 | $\mathcal{N}$ | **8, 9, 10, 11, 12, 14, 15, 18, 19, 23** |
| 4,5,6,8 | $\mathcal{N}$ | **7, 9, 10, 11, 12, 14, 15, 18, 19, 23** |
| 4,5,7,8 | $\mathcal{N}$ | **6, 9, 10, 11, 12, 14, 15, 18, 19, 23, 26, 27**, 30, 42, 51 |
| 4,6,7,8 | $\mathcal{N}$ | **5, 9, 10, 11, 12, 14, 15, 17, 18, 19, 20, 23, 24, 26, 27**, 33, 35, 41, 65, 77, 131 |
| 4,5,6,7,8 | $\mathcal{N}$ | **9, 10, 11, 12, 14, 15, 18, 19, 23** |
| 5,6 | $0, 1 \bmod 5$ | **10, 11, 15, 16, 20, 35**, 40, 50, 51, 80 |

## 12.1.7   GROUP DIVISIBLE DESIGNS AND TRANSVERSAL DESIGNS

**Definitions:**

A **group divisible design** (or $(K, \lambda)$-GDD) is a triple $(X, \mathcal{G}, \mathcal{A})$ where $X$ is a set (of points), $\mathcal{G}$ is a partition of $X$ into at least two subsets (called *groups*), $\mathcal{A}$ is a family of subsets of $X$ (called *blocks*) such that

- if $A$ in $\mathcal{A}$, then $|A| \in K$;
- a group and a block contain at most one common point;
- every pair of points from distinct groups occurs in exactly $\lambda$ blocks.

If $\lambda = 1$, a $(K, \lambda)$-GDD is often denoted by $K$-GDD. If $K = \{k\}$, a $K$-GDD is written $k$-GDD.

The **group-type** (or **type**) of a GDD is the multiset $\{|G| \mid G \in \mathcal{G}\}$. Usually an "exponential notation" is used to describe the type of a GDD: a GDD of type $t_1^{u_1} t_2^{u_2} \ldots t_k^{u_k}$ is a GDD where there are $u_i$ groups of size $t_i$ for $1 \le i \le k$.

A **transversal design** $\mathrm{TD}(k, n)$ is a $k$-GDD of type $n^k$ (that is, one having $k$ groups of size $n$ and uniform block size $k$).

**Fact:**

**1.** The existence of a $\mathrm{TD}(k, n)$ is equivalent to the existence of $k-2$ mutually orthogonal Latin squares of side $n$. (§12.3.2.)

## 12.2 SYMMETRIC DESIGNS AND FINITE GEOMETRIES

### 12.2.1 FINITE GEOMETRIES

**Definitions:**

A finite **incidence structure** $(V, \mathcal{B}, \mathcal{I})$ consists of a finite set $V$ of *points*, a finite set $\mathcal{B}$ of *lines*, and an *incidence relation* $\mathcal{I}$ between them. (Equivalently, a finite incidence structure is a pair $(V, \mathcal{B})$, where $\mathcal{B} = \{\, \{v \mid (v, b) \in \mathcal{I}\} \mid b \in \mathcal{B} \,\}$. In this case, lines are sets of points.)

The **dual** incidence structure is obtained by interchanging the roles of points and lines.

Let $F$ be a finite field, and let $S$ be an $(n+1)$-dimensional vector space over $F$. The set of all subspaces of $S$ is the **projective space of projective dimension $n$** over $F$. When $F$ is the Galois field $GF(q)$ (see §5.6.3), the projective space of projective dimension $n$ is denoted $\mathrm{PG}(n, q)$.

Subspaces of projective dimensions 0, 1, 2, and $n$ are **points**, **lines**, **planes**, and **hyperplanes**, respectively; in general, subspaces of projective dimension $t$ are **$t$-flats**. $\mathrm{PG}_t(n, q)$ denotes the incidence structure of points and $t$-flats in $\mathrm{PG}(n, q)$ (incidence is just containment as subspaces). Often, $\mathrm{PG}_1(n, q)$ is denoted $\mathrm{PG}(n, q)$ (taking the structure of points and lines as the natural geometry of the underlying space).

Let $S$ be an $n$-dimensional vector space over a finite field $F$. The set of all cosets of subspaces of $S$ is the **affine space of affine dimension $n$** over $F$. When $F$ is the Galois field $GF(q)$, the affine space of affine dimension $n$ is denoted $\mathrm{AG}(n, q)$.

Cosets of subspaces of (affine) dimension 0, 1, 2, and $n-1$ of an affine space of affine dimension $n$ are **points**, **lines**, **planes**, and **hyperplanes**, respectively. In general, cosets of subspaces of affine dimension $t$ are **$t$-flats**. $\mathrm{AG}_t(n, q)$ denotes the incidence structure of points and $t$-flats in $\mathrm{AG}(n, q)$ (incidence is containment). Often, $\mathrm{AG}_1(n, q)$ is denoted $\mathrm{AG}(n, q)$ (taking the structure of points and lines as the natural geometry of the underlying space).

*Note:* The term *finite geometry* often just means finite incidence structure. However, incidence structures are often too unstructured to be of much (geometric) interest. Hence the term is sometimes reserved to cover only incidence structures satisfying additional axioms such as those given in §12.2.3, Fact 3 for projective planes.

**Facts:**

**1.** *Projective geometries:* For $q$ a prime power and $1 \le t < n$, $\text{PG}_t(n, q)$ is a $\left( \frac{q^{n+1}-1}{q-1}, \frac{q^{t+1}-1}{q-1}, \frac{(q^{n-1}-1)(q^{n-2}-1)...(q^{n-t-1}-1)}{(q^{t-1}-1)(q^{t-2}-1)...(q-1)} \right)$ design.

**2.** *Affine geometries:* For $q$ a prime power and $1 \le t < n$, $\text{AG}_t(n, q)$ is a $\left( q^n, q^t, \frac{(q^{n-1}-1)(q^{n-2}-1)...(q^{n-t-1}-1)}{(q^{t-1}-1)(q^{t-2}-1)...(q-1)} \right)$ design.

---

## 12.2.2   SYMMETRIC DESIGNS

**Definitions:**

A $(v, b, r, k, \lambda)$ block design is **symmetric** if the number of points equals the number of blocks, that is, $v = b$.

A symmetric design with $\lambda = 2$ is a **biplane**. The parameters of a biplane are $v = \binom{k}{2} + 1$, $k = k$, $\lambda = 2$.

**Facts:**

**1.** In a symmetric $(v, k, \lambda)$ design, $r = k$.

**2.** If a symmetric $(v, k, \lambda)$ design exists, then

- if $v$ is even, then $k - \lambda$ is a perfect square;
- if $v$ is odd, then the Diophantine equation $x^2 = (k - \lambda)y^2 + (-1)^{(v-1)/2} \lambda z^2$ has a solution in integers, not all of which are zero. (Bruck-Ryser-Chowla; the theorem is often referred to as BRC.)

**3.** For every positive integer $k$ there is a symmetric $(2^{k+2}-1, 2^{k+1}-1, 2^k-1)$ block design.

**4.** If $p$ is prime and $k$ is a positive integer, there is a symmetric $(p^{2k} + p^k + 1, p^k + 1, 1)$ block design.

**5.** In a symmetric design any two blocks intersect in exactly $\lambda$ points.

**6.** The *dual* incidence structure obtained by interchanging the roles of points and blocks is also a BIBD with the same parameters (hence the term symmetric).

**7.** The dual of a symmetric design need not be isomorphic to the original design.

**8.** Given a symmetric $(v, k, \lambda)$ design, and a block $A$ of this design, if the points not in $A$ are deleted from all blocks which intersect $A$, the design obtained is the *derived design*. Its parameters are $(k, v - 1, k - 1, \lambda, \lambda - 1)$.

**9.** Given a symmetric $(v, k, \lambda)$ design, and given a block $A$ of this design, delete the block $A$, and delete all points in $A$ from all other blocks. The resulting design is the *residual design*, and has parameters $(v - k, v - 1, k, k - \lambda, \lambda)$. (See Example 1.)

**10.** Any $(v - k, v - 1, k, k - \lambda, \lambda)$ design is a *quasi-residual design*.

**11.** Any quasi-residual BIBD with $\lambda = 1$ or 2 is residual (Hall-Connor); but for $\lambda = 3$, there are examples of quasi-residual BIBDs that are not residual.

**12.** If there is a symmetric $(v, k, \lambda)$ design with $n = k - \lambda$, then $4n - 1 \le v \le n^2 + n + 1$. When $v = 4n - 1$ it is a *Hadamard design*; when $v = n^2 + n + 1$ it is a *projective plane*.

**13.** The only known biplanes have parameters $(7, 4, 2)$, $(11, 5, 2)$, $(16, 6, 2)$, $(37, 9, 2)$, $(56, 11, 2)$, and $(79, 13, 2)$.

**14.** The only known symmetric designs with $\lambda = 3$ have parameters $(11, 6, 3)$, $(15, 7, 3)$, $(25, 9, 3)$, $(31, 10, 3)$, $(45, 12, 3)$, and $(71, 15, 3)$.

**15.** Although infinitely many symmetric designs with $\lambda = 1$ are known, there is no other value of $\lambda$ for which this is known to be true.

### Example:

**1.** In the symmetric $(15, 7, 3)$ design in the following table, if the block $b_0$ is removed and if all the points in $b_0$ are removed from the blocks $b_1, \ldots, b_{14}$, the resulting design is the residual design. It has parameters $(8, 4, 3)$ and its blocks are given on the right.

| | | | | | | | |
|---|---|---|---|---|---|---|---|
| $b_0:$ | 0 | 1 | 2 | 3 | 4 | 5 | 6 |
| $b_1:$ | 0 | 1 | 2 | 7 | 8 | 9 | 10 |
| $b_2:$ | 0 | 1 | 2 | 11 | 12 | 13 | 14 |
| $b_3:$ | 0 | 3 | 4 | 7 | 8 | 11 | 12 |
| $b_4:$ | 0 | 3 | 4 | 9 | 10 | 13 | 14 |
| $b_5:$ | 0 | 5 | 6 | 7 | 8 | 13 | 14 |
| $b_6:$ | 0 | 5 | 6 | 9 | 10 | 11 | 12 |
| $b_7:$ | 1 | 3 | 5 | 7 | 9 | 11 | 13 |
| $b_8:$ | 1 | 3 | 6 | 7 | 10 | 12 | 14 |
| $b_9:$ | 1 | 4 | 5 | 8 | 10 | 11 | 14 |
| $b_{10}:$ | 1 | 4 | 6 | 8 | 9 | 12 | 13 |
| $b_{11}:$ | 2 | 3 | 5 | 8 | 10 | 12 | 13 |
| $b_{12}:$ | 2 | 3 | 6 | 8 | 9 | 11 | 14 |
| $b_{13}:$ | 2 | 4 | 5 | 7 | 9 | 12 | 14 |
| $b_{14}:$ | 2 | 4 | 6 | 7 | 10 | 11 | 13 |

## 12.2.3   PROJECTIVE AND AFFINE PLANES

### Definitions:

A **projective plane** is a finite set of *points* and a set of subsets of points (called *lines*) such that

- every two points lie on exactly one line;
- every two lines intersect in exactly one point;
- there are four points with no three collinear.

An **affine plane** is a set of *points* and a set of subsets of points (called *lines*) such that

- every two points lie on exactly one line;
- if a point does not lie on a line $L$, there is exactly one line through the point that does not intersect $L$;
- there are three points that are not collinear.

### Facts:

**1.** A finite projective plane is a symmetric $(n^2 + n + 1, n + 1, 1)$ design, for some positive integer $n$, called the *order* of the projective plane. The projective plane has $n^2 + n + 1$ points and $n^2 + n + 1$ lines. Each point lies on $n + 1$ lines and every line contains exactly $n + 1$ points.

**2.** *Principle of duality*:   Given any statement about finite projective planes that is a theorem, the dual statement (obtained by interchanging "point" and "line" and interchanging "point lying on a line" with "line passing through a point") is a theorem.

**3.** Any symmetric design with $\lambda = 1$ is a projective plane.

**4.** The existence of a projective plane of order $n$ is equivalent to the existence of a set of $n-1$ mutually orthogonal Latin squares (MOLS) of side $n$.

**5.** Existence of projective planes:   Very little is known about the existence of projective planes:

- There exists a projective plane of order $p^k$ whenever $p$ is prime and $k$ is a positive integer. (See Fact 10.)

- There is no projective plane known for any order $n$ that is not a power of a prime. The smallest open order is 12.

- There is no projective plane of order 10 or any $n \equiv 6 \pmod 8$.

- There exist projective planes for every order $q^2$ and $q^3$ when $q$ is a prime power that cannot be constructed by the method of Fact 9; such planes are called nondesarguesian.

- There are four nonisomorphic projective planes of order 9, three of which are nondesarguesian.

- The following table summarizes the known facts about the existence and number of projective planes of order $n$, for $1 \le n \le 12$:

| order | 2 | 3 | 4 | 5 | 6 | 7 | 8 | 9 | 10 | 11 | 12 |
|---|---|---|---|---|---|---|---|---|---|---|---|
| number of projective planes | 1 | 1 | 1 | 1 | 0 | 1 | 1 | 4 | 0 | $\ge 1$ | ? |

**6.** The proof by Lam, Thiel, and Swiercz in 1989 that there is no projective plane of order 10 involved great amounts of computer power and time. For details, see [CoDi07].

**7.** The existence of a projective plane of order $n$ is equivalent to the existence of an affine plane of order $n$.

**8.** A finite affine plane is a $(n^2, n, 1)$ design, for some positive integer $n$. The affine plane has $n^2$ points and $n^2 + n$ lines. Each point lies on $n + 1$ lines and every line contains exactly $n$ points. The integer $n$ is the *order* of the affine plane.

**9.** Any affine plane of order $n$ has the property that the lines can be partitioned into $n + 1$ parallel classes each containing $n$ lines and hence is a resolvable block design.

**10.** *A direct construction of a projective plane of every order $q = p^k$, when $p$ is prime and $k$ is a positive integer*:   Consider the three-dimensional vector space $\mathcal{F}_q^3$ over GF$(q)$. This vector space contains $\frac{q^3-1}{q-1} = q^2 + q + 1$ 1-dimensional subspaces (lines through the origin $(0,0,0)$) and an equal number of 2-dimensional subspaces (planes through the origin). Now construct an incidence structure where the points are the 1-dimensional subspaces, the lines are the 2-dimensional subspaces, and a point is on a line if the 1-dimensional subspace (associated with the point) is contained in the 2-dimensional subspace (associated with the line). This structure satisfies the axioms and thus is a projective plane (of order $q$). Projective planes such as this, coming from finite fields via the construction in this example, are *desarguesian planes*.

**11.** *A construction of a projective plane of order $n$ from an affine plane of order $n$*:   To construct the projective plane of order $n$ from the affine plane of order $n$, use the fact that the lines of the affine plane can be partitioned into $n+1$ parallel classes each containing $n$ lines. To each line in the $i$th parallel class adjoin the new symbol $\infty_i$. Add one new line, namely $\{\infty_1, \infty_2, \ldots, \infty_{n+1}\}$. Now each line contains $n + 1$ points, there are $n^2 + n + 1$ total points, and each pair of points is on a unique line—this is the projective plane of order $n$.

**12.**  *A construction of an affine plane of order $n$ from a projective plane of order $n$:* The affine plane of order $n$ is the residual design of the projective plane of order $n$. See §12.2.2, Fact 9 for the construction.

**Examples:**

**1.**  The Fano plane (§12.1.1, Example 3) is the projective plane of order 2.

**2.**  The affine plane of order 2 is given in part (a) of the following figure. The set of points is $\{1, 2, 3, 4\}$. The six lines are

$$\{1,2\} \quad \{3,4\} \quad \{1,3\} \quad \{2,4\} \quad \{1,4\} \quad \{2,3\}.$$

The three parallel classes are

$$\{\{1,2\}, \{3,4\}\} \quad \{\{1,3\}, \{2,4\}\} \quad \{\{1,4\}, \{2,3\}\}.$$

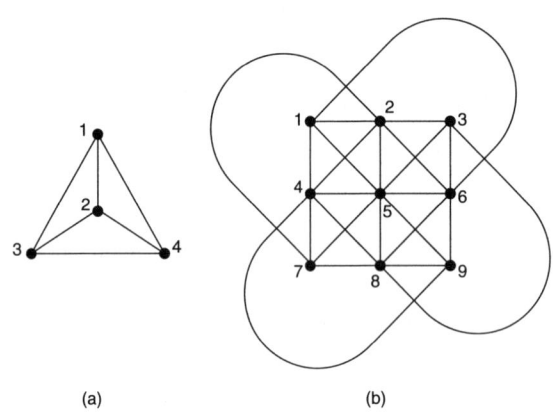

(a)                                    (b)

**3.**  The affine plane of order 3 is given in part (b) of the figure of Example 2. The set of points is $\{1, 2, 3, \ldots, 9\}$. The twelve lines (listed in order in four parallel classes of three lines each) are

$$\{1,2,3\}, \{4,5,6\}, \{7,8,9\} \quad \{1,4,7\}, \{2,5,8\}, \{3,6,9\}$$
$$\{1,5,9\}, \{6,2,7\}, \{4,8,3\} \quad \{3,5,7\}, \{2,4,9\}, \{8,6,1\}.$$

**4.**  Using the construction in Fact 11 on the affine plane of order 3 (Example 3) yields the projective plane of order 3 with thirteen points $\{1, 2, 3, 4, 5, 6, 7, 8, 9, \infty_1, \infty_2, \infty_3, \infty_4\}$ and thirteen lines:

$$\{1,2,3,\infty_1\} \quad \{4,5,6,\infty_1\} \quad \{7,8,9,\infty_1\}$$
$$\{1,4,7,\infty_2\} \quad \{2,5,8,\infty_2\} \quad \{3,6,9,\infty_2\}$$
$$\{1,5,9,\infty_3\} \quad \{6,2,7,\infty_3\} \quad \{4,8,3,\infty_3\}$$
$$\{3,5,7,\infty_4\} \quad \{2,4,9,\infty_4\} \quad \{8,6,1,\infty_4\}$$
$$\{\infty_1,,\infty_2,\infty_3,\infty_4\}.$$

## 12.2.4  HADAMARD DESIGNS AND MATRICES

### Definitions:

A **Hadamard matrix** $H$ of order $n$ is a square $n \times n$ matrix all of whose entries are $\pm 1$ that satisfies the property that $H^{\mathrm{T}} H = nI$ where $I$ is the $n \times n$ identity matrix and $H^{\mathrm{T}}$ is the transpose of $H$.

If $M = (m_{ij})$ is an $m \times p$ matrix and $N = (n_{ij})$ is an $n \times q$ matrix, the **Kronecker product** is the $mn \times pq$ matrix $M \times N$ given by

$$
M \times N = \begin{pmatrix}
m_{11}N & m_{12}N & \cdots & m_{1p}N \\
m_{21}N & m_{22}N & \cdots & m_{2p}N \\
\vdots & \vdots & & \vdots \\
m_{m1}N & m_{m2}N & \cdots & m_{mp}N
\end{pmatrix}.
$$

A **Hadamard design** of order $4n$ is a symmetric $(4n - 1, 2n - 1, n - 1)$ design. The **dimension** of the Hadamard design is $n$.

### Facts:

**1.** A necessary condition for the existence of a Hadamard matrix of order $n$ is that $n = 1$, $n = 2$, or $n \equiv 0 \pmod 4$.

**2.** $HH^{\mathrm{T}} = nI = H^{\mathrm{T}}H$.

**3.** The rows [columns] of a Hadamard matrix are pairwise orthogonal (when considered as vectors of length $n$).

**4.** If a row of a Hadamard matrix is multiplied by $-1$, the result is a Hadamard matrix. Similarly, if a column of a Hadamard matrix is multiplied by $-1$, the result is a Hadamard matrix.

**5.** By multiplying rows and columns of a Hadamard matrix by $-1$, a Hadamard matrix can be obtained where the first row and column consist entirely of $+1$s. A Hadamard matrix of this type is *normalized*.

**6.** In a normalized Hadamard matrix of order $4n$, every row and column (except the first) contains $+1$ and $-1$ exactly $2n$ times each.

**7.** Of all $n \times n$ matrices with entries from $\{-1, +1\}$, a Hadamard matrix $H$ has the maximal determinant. $|\det H| = n^{n/2}$.

**8.** The Kronecker product of two Hadamard matrices is a Hadamard matrix. Thus, if there are Hadamard matrices of order $m$ and $n$, then there is a Hadamard matrix of order $mn$.

**9.** If there exist Hadamard matrices of orders $4m$, $4n$, $4p$, $4q$, then there exists a Hadamard matrix of order $16mnpq$. (Craigen, Seberry, and Zhang)

**10.** Let $q$ be a positive integer. Then there exists a Hadamard matrix of order $2^s q$ for every $s \geq \lfloor 2 \log_2 (q - 3) \rfloor$. (Seberry)

**11.** A Hadamard design of order $4n$ exists if and only if a Hadamard matrix of order $4n$ exists. See the construction in Fact 15.

**12.** *Hadamard's conjecture:* The fundamental question concerning Hadamard matrices remains the *existence question*. The *Hadamard conjecture* is that there exist Hadamard matrices of order $4n$ for all $n \geq 1$. This remains unproved.

**13.** Currently the smallest order for which the existence of a Hadamard matrix is open is 668. Because of Fact 8 and the existence of a Hadamard matrix of order 2, if $q$ is an odd number and there exists a Hadamard matrix of order $2^s q$, then there exists a Hadamard matrix of order $2^t q$ for all $t \geq s$.

The following table gives values of $t$ such that a Hadamard matrix of order $2^t n$ is known, for odd $n < 999$ ($n$ is obtained by adding the three indices of the entry). The notation "." indicates $t = 2$.

|    | 000 | 100 | 200 | 300 | 400 | 500 | 600 | 700 | 800 | 900 |
|----|-----|-----|-----|-----|-----|-----|-----|-----|-----|-----|
| 00 | . . . . . | . . . . . | . . . . . | . . . . . | . . . . . | . . . . . | . . . . . | . . . . . | . . . . . | . . . 3 . |
| 10 | . . . . . | . . . . . | . . . . . | 3 . . . . | . . . . 3 | . . 3 . . | . . . . . | . . . . 4 | . . . . . | . . . 3 3 |
| 20 | . . . . . | . . . . . | . 3 . . . | . . . . . | . . . . . | . . 3 . . | . . . . . | 3 . . . . | . . 3 . . | . . . . . |
| 30 | . . . . . | . . . . . | . . . . . | . . . . . | . . . . . | . . . 3 . | 3 . . . . | . . . . 5 | . . . . 3 | . 4 . . . |
| 40 | . . . . . | . . . . . | . . . . . | . . 3 . . | . 3 . . . | . . . . . | . 3 . 3 . | . . . . . | . . . . . | . . . 3 . |
| 50 | . . . . . | . . . . . | 3 . . . . | . . . . . | . . 3 . . | . . . . . | . . . . 4 | 3 . . . . | . . . 3 . | . . 3 . . |
| 60 | . . . . . | . . . 3 . | . . . . . | . . . . . | . . . . . | . . . . . | . . . 3 . | . . . . . | . . . 3 . | . . . . . |
| 70 | . . . . . | . . . 3 . | . . . . . | . . . . . | . . . . 3 | 3 3 . . . | . . . . . | . . . . . | 5 . . . . | . . . . . |
| 80 | . . . . . | . . . . . | . 3 . . . | . . . . . | . . 3 . . | . . . . . | . . . . . | . . . 3 3 | . . 3 . . | . . . . . |
| 90 | . . . . . | 3 . . . . | . . . . . | . . . . . | 3 . . . . | . . . . 5 | . . . . . | . . . . . | . . . . . | 3 . . . . |
|    | 1 3 5 7 9 | 1 3 5 7 9 | 1 3 5 7 9 | 1 3 5 7 9 | 1 3 5 7 9 | 1 3 5 7 9 | 1 3 5 7 9 | 1 3 5 7 9 | 1 3 5 7 9 | 1 3 5 7 9 |

**14.** *A general construction for Hadamard matrices of order $q+1$ when $q$ is an odd prime power and $q \equiv 3$ (mod 4):*

- construct a $q \times q$ matrix $C = (c_{ij})$ indexed by the elements of the field $\mathrm{GF}(q)$ by letting

$$c_{ij} = \begin{cases} 1, & \text{if } i - j \text{ is a square in } \mathrm{GF}(q) \\ -1, & \text{if } i - j \text{ is not a square in } \mathrm{GF}(q); \end{cases}$$

- construct a Hadamard matrix $H$ of order $q + 1$ from $C$ by adding a first column of all $-1$s and then a top row of all 1s.

This method is used in Example 1 to construct the Hadamard matrix of order 12.

**15.** *Constructing Hadamard designs from Hadamard matrices, and vice versa:* Assume that there exists a Hadamard matrix of order $4n$. Let $H$ be a normalized Hadamard matrix of this order. Remove the first row and column of $H$ and replace every $-1$ in the resulting matrix by a 0. The final $(4n - 1) \times (4n - 1)$ matrix can be shown to be the incidence matrix of a $(4n - 1, 2n - 1, n - 1)$ Hadamard design.

This process can be reversed to construct a Hadamard matrix from a Hadamard design.

**Example:**

**1.** The smallest examples of Hadamard matrices are the following:

$$(1), \begin{pmatrix} 1 & 1 \\ 1 & -1 \end{pmatrix},$$

$$\begin{pmatrix} -1 & 1 & 1 & 1 \\ 1 & -1 & 1 & 1 \\ 1 & 1 & -1 & 1 \\ 1 & 1 & 1 & -1 \end{pmatrix}, \quad \begin{pmatrix} 1 & 1 & 1 & 1 & 1 & 1 & 1 & 1 \\ 1 & -1 & 1 & -1 & 1 & -1 & 1 & -1 \\ 1 & 1 & -1 & -1 & 1 & 1 & -1 & -1 \\ 1 & -1 & -1 & 1 & 1 & -1 & -1 & 1 \\ 1 & 1 & 1 & 1 & -1 & -1 & -1 & -1 \\ 1 & -1 & 1 & -1 & -1 & 1 & -1 & 1 \\ 1 & 1 & -1 & -1 & -1 & -1 & 1 & 1 \\ 1 & -1 & -1 & 1 & -1 & 1 & 1 & -1 \end{pmatrix},$$

$$\begin{pmatrix} 1 & 1 & 1 & 1 & 1 & 1 & 1 & 1 & 1 & 1 & 1 & 1 \\ -1 & 1 & 1 & -1 & 1 & 1 & 1 & -1 & -1 & -1 & 1 & -1 \\ -1 & -1 & 1 & 1 & -1 & 1 & 1 & 1 & -1 & -1 & -1 & 1 \\ -1 & 1 & -1 & 1 & 1 & -1 & 1 & 1 & 1 & -1 & -1 & -1 \\ -1 & -1 & 1 & -1 & 1 & 1 & -1 & 1 & 1 & 1 & -1 & -1 \\ -1 & -1 & -1 & 1 & -1 & 1 & 1 & -1 & 1 & 1 & 1 & -1 \\ -1 & -1 & -1 & -1 & 1 & -1 & 1 & 1 & 1 & 1 & 1 & 1 \\ -1 & 1 & -1 & -1 & -1 & 1 & -1 & 1 & 1 & -1 & 1 & 1 \\ -1 & 1 & 1 & -1 & -1 & -1 & 1 & -1 & 1 & 1 & -1 & 1 \\ -1 & 1 & 1 & 1 & -1 & -1 & -1 & 1 & -1 & 1 & 1 & 1 \\ -1 & -1 & 1 & 1 & 1 & -1 & -1 & -1 & 1 & -1 & 1 & 1 \\ -1 & 1 & -1 & 1 & 1 & 1 & -1 & -1 & -1 & 1 & -1 & 1 \end{pmatrix}.$$

---

## 12.2.5  DIFFERENCE SETS

*Note:* In this section, only difference sets in abelian groups are considered.

### Definitions:

Let $G$ be an additively written group of order $v$. A $k$-subset $D$ of $G$ is a $(v, k, \lambda; n)$ **difference set of order** $n = k - \lambda$ if every nonzero element of $G$ has exactly $\lambda$ representations as a difference $d - d'$ $(d, d' \in D)$. The difference set is **abelian, cyclic,** etc., if the group $G$ has the respective property.

The **development** of a difference set $D$ is the incidence structure $\text{dev}(D)$ whose points are the elements of the group $G$ and whose blocks are the *translates* $D + g = \{d + g \mid d \in D\}$, $g \in G$.

A **multiplier** of a difference set $D$ in a group $G$ is an automorphism $\varphi$ of $G$ such that $\varphi(D) = D + g$ for some $g \in G$. If $\varphi$ is a multiplier and $\varphi(h) = th$ for all $h \in G$, then $t$ is a **numerical** multiplier.

### Facts:

**1.** Both the group $G$ itself and $G - \{g\}$ (for an arbitrary $g \in G$) are $(v, v, v; 0)$ and $(v, v - 1, v - 2; 1)$ difference sets. In the table of Example 2, these *trivial* difference sets are excluded.

**2.** The complement of a $(v, k, \lambda; n)$ difference set is again a difference set with parameters $(v, v - k, v - 2k + \lambda; n)$. Therefore only $k \leq \frac{v}{2}$ (the case $k = \frac{v}{2}$ is actually impossible) is considered.

**3.** The existence of a $(v, k, \lambda; n)$ difference set is equivalent to the existence of a symmetric $(v, k, \lambda)$ design $\mathcal{D}$ admitting $G$ as a point regular automorphism group; that is, for any two points $p$ and $q$, there is a unique group element $g$ which maps $p$ to $q$. The design $\mathcal{D}$ is isomorphic with $\mathrm{dev}(D)$.

**4.** There are many symmetric designs which do not have difference set representations.

**5.** Since difference sets can yield symmetric designs, the parameters $v$, $k$, and $\lambda$ must satisfy the trivial necessary conditions for the existence of a symmetric design ($\lambda(v-1) = k(k-1)$) and must also satisfy the Bruck-Ryser-Chowla condition (§12.2.2, part 2 of Fact 2).

**6.** If $\varphi$ is a multiplier of the difference set $D$, then there is at least one translate $D + g$ of $D$ which is fixed by $\varphi$.

**7.** *The multiplier theorem:* Let $D$ be an abelian $(v, k, \lambda; n)$ difference set. If $p$ is a prime that satisfies $(p, v) = 1$, $p|n$, and $p > \lambda$, then $p$ is a numerical multiplier.

**8.** *The multiplier conjecture:* Every prime divisor $p$ of $n$ that is relatively prime to $v$ is a multiplier of a $(v, k, \lambda; n)$ difference set; that is, the condition $p > \lambda$ in Fact 7 is unnecessary.

**Examples:**

**1.** A $(11, 5, 2; 3)$ difference set in the group $\mathcal{Z}_{11}$ is $\{1, 3, 4, 5, 9\}$.

**2.** The following table lists abelian difference sets of order $n \leq 15$. (See Fact 1.) One difference set for each abelian group is listed. In general, there will be many more examples. There are no other groups or parameters with $n \leq 15$ for which the existence of a difference set is undecided.

In the column "group" the decomposition of the group as a product of cyclic subgroups is given. If the group is cyclic, the integers modulo the group order are used to describe the difference set.

| $n$ | $v$ | $k$ | $\lambda$ | group | difference set |
|---|---|---|---|---|---|
| 2 | 7 | 3 | 1 | (7) | 1 2 4 |
| 3 | 13 | 4 | 1 | (13) | 0 1 3 9 |
| 3 | 11 | 5 | 2 | (11) | 1 3 4 5 9 |
| 4 | 21 | 5 | 1 | (21) | 3 6 7 12 14 |
| 4 | 16 | 6 | 2 | (8)(2) | (00) (10) (11) (20) (40) (61) |
| | | | | $(4)^2$ | (00) (01) (10) (12) (20) (23) |
| | | | | $(4)(2)^2$ | (000) (010) (100) (101) (200) (211) |
| | | | | $(2)^4$ | (0000) (0010) (1000) (1001) (1100) (1111) |
| 4 | 15 | 7 | 3 | (3)(5) | 0 1 2 4 5 8 10 |
| 5 | 31 | 6 | 1 | (31) | 1 5 11 24 25 27 |
| 5 | 19 | 9 | 4 | (19) | 1 4 5 6 7 9 11 16 17 |
| 6 | 23 | 11 | 5 | (23) | 1 2 3 4 6 8 9 12 13 16 18 |
| 7 | 57 | 8 | 1 | (57) | 1 6 7 9 19 38 42 49 |
| 7 | 27 | 13 | 6 | $(3)^3$ | (001) (011) (021) (111) (020) (100) (112) (120) (121) (122) (201) (202) (220) |
| 8 | 73 | 9 | 1 | (73) | 1 2 4 8 16 32 37 55 64 |
| 8 | 31 | 15 | 7 | (31) | 1 2 3 4 6 8 12 15 16 17 23 24 27 29 30 |
| 9 | 91 | 10 | 1 | (91) | 0 1 3 9 27 49 56 61 77 81 |
| 9 | 45 | 12 | 3 | $(3)^2(5)$ | (000) (001) (002) (003) (010) (020) (101) (112) (123) (201) (213) (222) |

| $n$ | $v$ | $k$ | $\lambda$ | group | difference set |
|---|---|---|---|---|---|
| 9 | 40 | 13 | 4 | (40) | 1 2 3 5 6 9 14 15 18 20 25 27 35 |
| 9 | 36 | 15 | 6 | $(4)(3)^2$ | (010) (011) (012) (020) (021) (022) (100) (110) (120) |
| | | | | | (200) (211) (222) (300) (312) (321) |
| | | | | $(2)^2(3)^2$ | (0010) (0011) (0012) (0020) (0021) (0022) (0100) (0110) |
| | | | | | (0120) (1000) (1011) (1022) (1100) (1112) (1121) |
| 9 | 35 | 17 | 8 | (35) | 0 1 3 4 7 9 11 12 13 14 16 17 21 27 28 29 33 |
| 11 | 133 | 12 | 1 | (133) | 1 11 16 40 41 43 52 60 74 78 121 128 |
| 11 | 43 | 21 | 10 | (43) | 1 4 6 9 10 11 13 14 15 16 17 21 23 24 25 31 35 36 38 40 |
| | | | | | 41 |
| 12 | 47 | 23 | 11 | (47) | 1 2 3 4 6 7 8 9 12 14 16 17 18 21 24 25 27 28 32 34 36 37 |
| | | | | | 42 |
| 13 | 183 | 14 | 1 | (183) | 0 2 3 10 26 39 43 61 109 121 130 136 141 155 |
| 15 | 59 | 29 | 14 | (59) | 1 3 4 5 7 9 12 15 16 17 19 20 21 22 25 26 27 28 29 35 36 |
| | | | | | 41 45 46 48 49 51 53 57 |

# 12.3  LATIN SQUARES AND ORTHOGONAL ARRAYS

## 12.3.1  LATIN SQUARES

### Definitions:

A **Latin square** of *side n* is an $n \times n$ array in which each entry contains a single element from an $n$-set $S$, such that each element of $S$ occurs exactly once in each row and exactly once in each column.

A Latin square of side $n$ (on the set $\{1, 2, \ldots, n\}$ or on the set $\{0, 1, \ldots, n-1\}$) is **reduced** or in **standard form** if in the first row and column the elements occur in increasing order.

Let $L$ be an $n \times n$ Latin square on symbol set $E_3$, with rows indexed by the elements of the $n$-set $E_1$ and columns indexed by the elements of the $n$-set $E_2$. Let $\mathcal{T} = \{(x_1, x_2, x_3) \mid L(x_1, x_2) = x_3\}$ and $\{a, b, c\} = \{1, 2, 3\}$. The $(a, b, c)$-**conjugate** of $L$, $L_{(a,b,c)}$, has rows indexed by $E_a$, columns by $E_b$, and symbols by $E_c$, and is defined by $L_{(a,b,c)}(x_a, x_b) = x_c$ for each $(x_1, x_2, x_3) \in \mathcal{T}$.

The **transpose** of a Latin square $L$, denoted $L^T$, is the Latin square which results from $L$ when the roles of rows and columns are exchanged; that is, $L^T(i, j) = L(j, i)$.

A Latin square $L$ of side $n$ is **symmetric** if $L(i, j) = L(j, i)$ for all $1 \le i \le n$, $1 \le j \le n$.

A Latin square $L$ of side $n$ is **idempotent** if $L(i, i) = i$ for all $1 \le i \le n$.

A **transversal** in a Latin square of side $n$ is a set of $n$ cells, one from each row and column, containing each of the $n$ symbols exactly once.

A **partial transversal of length k** in a Latin square of side $n$ is a set of $k$ cells, each from a different row and each from a different column, such that no two contain the same symbol.

Two Latin squares $L$ and $L'$ of side $n$ are **equivalent** (or **isotopic**) if there are three bijections, from the rows, columns, and symbols of $L$ to the rows, columns, and symbols, respectively, of $L'$, that map $L$ to $L'$.

Two Latin squares $L$ and $L'$ of side $n$ are **main class isotopic** if $L$ is isotopic to some conjugate of $L'$.

Let $k < n$. If in a Latin square $L$ of side $n$ the $k^2$ cells defined by $k$ rows and $k$ columns form a Latin square of side $k$, then the cells are a Latin **subsquare** of $L$.

An $n \times n$ array $L$ with cells that are either empty or contain exactly one symbol is a **partial** Latin square if no symbol occurs more than once in any row or column.

A partial Latin square is **symmetric** (or **commutative**) if whenever cell $(i, j)$ is occupied by $x$, cell $(j, i)$ is also occupied by $x$, for every $1 \leq i \leq n$, $1 \leq j \leq n$.

A partial Latin square is **idempotent** if cell $(i, i)$ is occupied by $i$, for all $i$.

A **Latin rectangle** is a $k \times n$ $(k < n)$ array in which each cell contains a single element from an $n$-set such that each element occurs exactly once in each row and at most once in each column.

An $n \times n$ partial Latin square $P$ is said to be **imbedded** in a Latin square $L$ if the upper $n \times n$ left corner of $L$ agrees with $P$.

**Facts:**

**1.** The multiplication table of any (multiplicative) group is a Latin square.

**2.** For each positive integer $k$ a reduced Latin square can be constructed using the following format:

$$
\begin{array}{cccccc}
1 & 2 & 3 & \cdots & k-1 & k \\
2 & 3 & 4 & \cdots & k & 1 \\
3 & 4 & 5 & \cdots & 1 & 2 \\
\vdots & \vdots & \vdots & \ddots & \vdots & \vdots \\
k-1 & k & 1 & \cdots & k-3 & k-2 \\
k & 1 & 2 & \cdots & k-2 & k-1
\end{array}
$$

**3.** Every Latin square has 1, 2, 3, or 6 distinct conjugates.

**4.** A symmetric Latin square of even side can never be idempotent.

**5.** A symmetric idempotent Latin square of side $n$ is equivalent to a 1-factorization of the complete graph $K_{n+1}$ on $n + 1$ points. A Latin square of side $n$ is equivalent to a 1-factorization of the complete graph $K_{n,n}$.

**6.** Every idempotent Latin square has a transversal (the main diagonal).

**7.** Some Latin squares have no transversals. One such class of Latin squares is composed of the addition tables of the integers modulo $2n$ for every $n \geq 1$, or in general the addition table of any group that has a unique element of order 2.

**8.** Every Latin square of side $n$ has a partial transversal of length $k$ where

$$k \geq \max\{n - \sqrt{n}, n - 15(\log n)^2\}. \quad \text{(P. W. Shor)}$$

**9.** A Latin square of side $n$ with a proper subsquare of side $k$ exists if and only if $k \leq \lfloor \frac{n}{2} \rfloor$.

**10.** There exists a Latin square of side $n$ with *no* proper subsquares if $n \neq 2^a 3^b$ or if $n = 3, 9, 12, 16, 18, 27, 81, 243$.

**11.** A partial Latin square of side $n$ with at most $n-1$ filled cells can always be completed to a Latin square of side $n$.

**12.** A $k \times n$ $(k < n)$ Latin rectangle can always be completed to a Latin square of side $n$.

**13.** Let $L$ be a partial Latin square of order $n$ in which cell $(i, j)$ is filled if and only if $i \leq r$ and $j \leq s$. Then $L$ can be completed to a Latin square of order $n$ if and only if $N(i) \geq r + s - n$ for $i = 1, 2, \ldots, n$, where $N(i)$ denotes the number of elements in $L$ that are equal to $i$. (H. Ryser)

**14.** A partial $n \times n$ Latin square can be imbedded in a $t \times t$ Latin square for every $t \geq 2n$.

**15.** An $n \times n$ partial symmetric Latin square can be imbedded in a $t \times t$ symmetric Latin square for every even $t \geq 2n$.

**16.** The number of distinct Latin squares, the number of main classes, and the number of equivalence classes of Latin squares of side $n$ go to infinity as $n \to \infty$. For $1 \leq n \leq 3$, there is one main class and one equivalence class. The number of main classes and equivalence classes of Latin squares of side $4 \leq n \leq 10$ are given in the following table:

| $n$ | 4 | 5 | 6 | 7 | 8 | 9 | 10 |
|---|---|---|---|---|---|---|---|
| *main* | 2 | 2 | 12 | 147 | 283,657 | 19,270,853,541 | 34,817,397,894,749,939 |
| *equiv.* | 2 | 2 | 22 | 563 | 1,676,257 | 115,618,721,533 | 208,904,371,354,363,006 |

**Examples:**

**1.** A $4 \times 4$ Latin square on $\{1, 2, 3, 4\}$, where $1, 2, 3, 4$ represent four brands of tires, gives a way to test each brand of tire on each of the four wheel positions on each of four cars: the $i, j$-entry of the Latin square is the brand of tire to be tested on wheel position $i$ of car $j$.

**2.** A Latin square of side 8 on the symbols $0, 1, \ldots, 7$:

$$
\begin{array}{cccccccc}
0 & 1 & 2 & 3 & 4 & 5 & 6 & 7 \\
1 & 0 & 3 & 4 & 5 & 6 & 7 & 2 \\
2 & 3 & 5 & 0 & 6 & 7 & 4 & 1 \\
3 & 4 & 0 & 7 & 1 & 2 & 5 & 6 \\
4 & 5 & 6 & 1 & 7 & 0 & 2 & 3 \\
5 & 6 & 7 & 2 & 0 & 3 & 1 & 4 \\
6 & 7 & 4 & 5 & 2 & 1 & 3 & 0 \\
7 & 2 & 1 & 6 & 3 & 4 & 0 & 5
\end{array}
$$

**3.** A Latin square of side 4 and its six conjugates:

$$
\begin{array}{ccccccccc}
1 & 4 & 2 & 3 & \quad & 1 & 2 & 4 & 3 & \quad & 1 & 3 & 2 & 4 \\
2 & 3 & 1 & 4 & \quad & 4 & 3 & 1 & 2 & \quad & 2 & 4 & 1 & 3 \\
4 & 1 & 3 & 2 & \quad & 2 & 1 & 3 & 4 & \quad & 4 & 2 & 3 & 1 \\
3 & 2 & 4 & 1 & \quad & 3 & 4 & 2 & 1 & \quad & 3 & 1 & 4 & 2
\end{array}
$$

$\qquad$ (1,2,3)-conjugate $\qquad\qquad$ (2,1,3)-conjugate $\qquad\qquad$ (3,2,1)-conjugate

```
1  2  4  3              1  3  4  2              1  3  2  4
3  4  2  1              3  1  2  4              3  1  4  2
2  1  3  4              2  4  3  1              4  2  3  1
4  3  1  2              4  2  1  3              2  4  1  3
(2,3,1)-conjugate      (1,3,2)-conjugate      (3,1,2)-conjugate
```

**4.** The following gives the main classes of Latin squares of sides 4, 5, and 6. No two Latin squares listed are main class isotopic.

```
                                           0  1  2  3  4      0  1  2  3  4
             0  1  2  3     0  1  2  3      1  2  3  4  0      1  0  3  4  2
             1  0  3  2     1  0  3  2      2  3  4  0  1      2  3  4  0  1
   n = 4:    2  3  0  1     2  3  1  0   n = 5:  3  4  0  1  2      3  4  1  2  0
             3  2  1  0     3  2  0  1      4  0  1  2  3      4  2  0  1  3
```

```
             0  1  2  3  4  5     0  1  2  3  4  5     0  1  2  3  4  5
             1  0  3  2  5  4     1  0  3  2  5  4     1  0  3  4  5  2
             2  3  4  5  0  1     2  3  4  5  0  1     2  3  0  5  1  4
   n = 6:    3  2  5  4  1  0     3  2  5  4  1  0     3  4  5  0  2  1
             4  5  0  1  2  3     4  5  0  1  3  2     4  5  1  2  0  3
             5  4  1  0  3  2     5  4  1  0  2  3     5  2  4  1  3  0
```

```
             0  1  2  3  4  5     0  1  2  3  4  5     0  1  2  3  4  5
             1  0  3  4  5  2     1  0  3  4  5  2     1  0  3  4  5  2
             2  3  0  5  1  4     2  3  1  5  0  4     2  3  4  5  0  1
             3  4  5  0  2  1     3  4  5  1  2  0     3  4  5  2  1  0
             4  5  1  2  3  0     4  5  0  2  3  1     4  5  0  1  2  3
             5  2  4  1  0  3     5  2  4  0  1  3     5  2  1  0  3  4
```

```
             0  1  2  3  4  5     0  1  2  3  4  5     0  1  2  3  4  5
             1  0  3  2  5  4     1  0  3  2  5  4     1  0  3  2  5  4
             2  4  0  5  1  3     2  4  0  5  1  3     2  4  0  5  1  3
             3  5  1  4  0  2     3  5  1  4  0  2     3  5  1  4  2  0
             4  2  5  0  3  1     4  2  5  1  3  0     4  3  5  1  0  2
             5  3  4  1  2  0     5  3  4  0  2  1     5  2  4  0  3  1
```

```
             0  1  2  3  4  5     0  1  2  3  4  5     0  1  2  3  4  5
             1  0  3  2  5  4     1  0  3  4  5  2     1  2  0  4  5  3
             2  4  0  5  3  1     2  3  1  5  0  4     2  0  1  5  3  4
             3  5  4  0  1  2     3  5  4  1  2  0     3  5  4  1  0  2
             4  2  5  1  0  3     4  2  5  0  1  3     4  3  5  2  1  0
             5  3  1  4  2  0     5  4  0  2  3  1     5  4  3  0  2  1
```

**5.** The following are a Latin square of side 7 with a subsquare of side 3 (3 × 3 square in upper left corner) and a Latin square of order 12 with no proper subsquares.

|   |   |   |   |   |   |   |   |   |   |   |   |
|---|---|---|---|---|---|---|---|---|---|---|---|
| 1 | 2 | 3 | 4 | 5 | 6 | 7 | 8 | 9 | a | b | c |
| 2 | 3 | 4 | 5 | 6 | 1 | 8 | 9 | a | b | c | 7 |
| 3 | 1 | 5 | 2 | 7 | 8 | 4 | a | 6 | c | 9 | b |
| 4 | 5 | 6 | 7 | 1 | 9 | b | c | 8 | 3 | 2 | a |
| 5 | 6 | 2 | 8 | a | 7 | 9 | b | c | 4 | 1 | 3 |
| 6 | c | 8 | 1 | 3 | a | 2 | 7 | b | 9 | 4 | 5 |
| 7 | 8 | 1 | a | c | b | 5 | 4 | 2 | 6 | 3 | 9 |
| 8 | 9 | b | 3 | 4 | c | a | 6 | 5 | 1 | 7 | 2 |
| 9 | b | 7 | c | 2 | 5 | 1 | 3 | 4 | 8 | a | 6 |
| a | 7 | c | b | 9 | 4 | 6 | 1 | 3 | 2 | 5 | 8 |
| b | 4 | a | 9 | 8 | 3 | c | 2 | 7 | 5 | 6 | 1 |
| c | a | 9 | 6 | b | 2 | 3 | 5 | 1 | 7 | 8 | 4 |

|   |   |   |   |   |   |   |
|---|---|---|---|---|---|---|
| 1 | 2 | 3 | 4 | 5 | 6 | 7 |
| 2 | 3 | 1 | 6 | 4 | 7 | 5 |
| 3 | 1 | 2 | 7 | 6 | 5 | 4 |
| 4 | 7 | 5 | 1 | 3 | 2 | 6 |
| 7 | 5 | 6 | 3 | 2 | 4 | 1 |
| 6 | 4 | 7 | 5 | 1 | 3 | 2 |
| 5 | 6 | 4 | 2 | 7 | 1 | 3 |

**6.** The partial Latin square $\begin{smallmatrix} 1 & \cdot & 2 \\ \cdot & 3 & 1 \\ 4 & 1 & \cdot \end{smallmatrix}$ is imbedded in the Latin square $\begin{smallmatrix} 1 & 4 & 2 & 3 \\ 2 & 3 & 1 & 4 \\ 4 & 1 & 3 & 2 \\ 3 & 2 & 4 & 1 \end{smallmatrix}$.

---

## 12.3.2  MUTUALLY ORTHOGONAL LATIN SQUARES

**Definitions:**

Two Latin squares $A = (a_{ij})$ and $B = (b_{ij})$ of order $n$ are **orthogonal** if the $n^2$ ordered pairs $(a_{ij}, b_{ij})$ $(1 \leq i, j \leq n)$ are distinct. (The relation of orthogonality is symmetric.)

A set of Latin squares $\{A_1, \ldots, A_k\}$ is a set of **mutually orthogonal Latin squares** (MOLS) if $A_i$ and $A_j$ are orthogonal for all $i, j \in \{1, \ldots, k\}$ $(i \neq j)$. The maximum number of MOLS of order $n$ is written $N(n)$. It is customary to define $N(0) = N(1) = \infty$.

A set of $n-1$ MOLS of side $n$ is a **complete set** of MOLS.

**Facts:**

**1.** If $n \geq 2$, then $N(n) \leq n-1$.

**2.** If $n$ is a prime power, then $N(n) = n-1$.

**3.** $N(n) \geq 2$ for all $n \geq 3$, except $n = 6$. (Bose-Parker-Shrikhande)

**4.** $N(n) \geq 3$ for all $n \geq 4$ except for $n = 6$ and possibly $n = 10$. The table that follows gives the best known lower bounds for $N(n)$ for $0 \leq n \leq 499$. Add the row and column indices to obtain the order.

|    | 0 | 1 | 2 | 3 | 4 | 5 | 6 | 7 | 8 | 9 | 10 | 11 | 12 | 13 | 14 | 15 | 16 | 17 | 18 | 19 |
|----|---|---|---|---|---|---|---|---|---|---|----|----|----|----|----|----|----|----|----|----|
| 0  | $\infty$ | $\infty$ | 1 | 2 | 3 | 4 | 1 | 6 | 7 | 8 | 2 | 10 | 5 | 12 | 4 | 4 | 15 | 16 | 5 | 18 |
| 20 | 4 | 5 | 3 | 22 | 7 | 24 | 4 | 26 | 5 | 28 | 4 | 30 | 31 | 5 | 4 | 5 | 8 | 36 | 4 | 5 |
| 40 | 7 | 40 | 5 | 42 | 5 | 6 | 4 | 46 | 8 | 48 | 6 | 5 | 5 | 52 | 5 | 6 | 7 | 7 | 5 | 58 |
| 60 | 5 | 60 | 5 | 6 | 63 | 7 | 5 | 66 | 5 | 6 | 6 | 70 | 7 | 72 | 5 | 7 | 6 | 6 | 6 | 78 |
| 80 | 9 | 80 | 8 | 82 | 6 | 6 | 6 | 6 | 7 | 88 | 6 | 7 | 6 | 6 | 6 | 6 | 7 | 96 | 6 | 8 |

|      | 0  | 1   | 2 | 3   | 4  | 5  | 6 | 7   | 8   | 9   | 10 | 11  | 12 | 13  | 14 | 15 | 16  | 17  | 18 | 19  |
|------|----|-----|---|-----|----|----|---|-----|-----|-----|----|-----|----|-----|----|----|-----|-----|----|-----|
| 100  | 8  | 100 | 6 | 102 | 7  | 7  | 6 | 106 | 6   | 108 | 6  | 6   | 13 | 112 | 6  | 7  | 6   | 8   | 6  | 6   |
| 120  | 7  | 120 | 6 | 6   | 6  | 124| 6 | 126 | 127 | 7   | 6  | 130 | 6  | 7   | 6  | 7  | 7   | 136 | 6  | 138 |
| 140  | 6  | 7   | 6 | 10  | 10 | 7  | 6 | 7   | 6   | 148 | 6  | 150 | 7  | 8   | 8  | 7  | 6   | 156 | 7  | 6   |
| 160  | 9  | 7   | 6 | 162 | 6  | 7  | 6 | 166 | 7   | 168 | 6  | 8   | 6  | 172 | 6  | 6  | 14  | 9   | 6  | 178 |
| 180  | 6  | 180 | 6 | 6   | 7  | 9  | 6 | 10  | 6   | 8   | 6  | 190 | 7  | 192 | 6  | 7  | 6   | 196 | 6  | 198 |
| 200  | 7  | 8   | 6 | 7   | 6  | 8  | 6 | 8   | 14  | 11  | 10 | 210 | 6  | 7   | 6  | 7  | 7   | 8   | 6  | 10  |
| 220  | 6  | 12  | 6 | 222 | 13 | 8  | 6 | 226 | 6   | 228 | 6  | 7   | 7  | 232 | 6  | 7  | 6   | 7   | 6  | 238 |
| 240  | 7  | 240 | 6 | 242 | 6  | 7  | 6 | 12  | 7   | 7   | 6  | 250 | 6  | 12  | 9  | 7  | 255 | 256 | 6  | 12  |
| 260  | 6  | 8   | 8 | 262 | 7  | 8  | 7 | 10  | 7   | 268 | 7  | 270 | 15 | 16  | 6  | 13 | 10  | 276 | 6  | 9   |
| 280  | 7  | 280 | 6 | 282 | 6  | 12 | 6 | 7   | 15  | 288 | 6  | 6   | 6  | 292 | 6  | 6  | 7   | 10  | 10 | 12  |
| 300  | 7  | 7   | 7 | 7   | 15 | 15 | 6 | 306 | 7   | 7   | 7  | 310 | 7  | 312 | 7  | 10 | 7   | 316 | 7  | 10  |
| 320  | 15 | 15  | 6 | 16  | 8  | 12 | 6 | 7   | 7   | 9   | 6  | 330 | 7  | 8   | 7  | 6  | 8   | 336 | 6  | 7   |
| 340  | 6  | 10  | 10| 342 | 7  | 7  | 6 | 346 | 6   | 348 | 8  | 12  | 18 | 352 | 6  | 9  | 7   | 9   | 6  | 358 |
| 360  | 8  | 360 | 6 | 7   | 7  | 10 | 6 | 366 | 15  | 15  | 7  | 15  | 7  | 372 | 7  | 15 | 7   | 13  | 7  | 378 |
| 380  | 7  | 12  | 7 | 382 | 15 | 15 | 7 | 15  | 7   | 388 | 7  | 16  | 7  | 8   | 7  | 7  | 8   | 396 | 7  | 7   |
| 400  | 15 | 400 | 7 | 15  | 11 | 8  | 7 | 15  | 8   | 408 | 7  | 13  | 8  | 12  | 10 | 9  | 18  | 15  | 7  | 418 |
| 420  | 7  | 420 | 7 | 15  | 7  | 16 | 6 | 7   | 7   | 10  | 6  | 430 | 15 | 432 | 6  | 15 | 6   | 18  | 7  | 438 |
| 440  | 7  | 15  | 7 | 442 | 7  | 13 | 7 | 11  | 15  | 448 | 7  | 15  | 7  | 7   | 7  | 15 | 7   | 456 | 7  | 16  |
| 460  | 7  | 460 | 7 | 462 | 15 | 15 | 7 | 466 | 8   | 8   | 7  | 15  | 7  | 15  | 10 | 18 | 7   | 15  | 6  | 478 |
| 480  | 15 | 15  | 6 | 15  | 8  | 7  | 6 | 486 | 7   | 15  | 6  | 490 | 6  | 16  | 6  | 7  | 15  | 15  | 6  | 498 |

**5.** $N(n) \to \infty$ as $n \to \infty$. (Chowla, Erdős, Straus)

**6.** $N(n \times m) \geq \min\{N(n), N(m)\}$. (MacNeish)

**7.** The existence of $n-1$ MOLS of order $n$ is equivalent to the existence of a projective plane of order $n$ (an $(n^2 + n + 1, n + 1, 1)$ design) and an affine plane of order $n$ (an $(n^2, n, 1)$ design). (§12.2.3)

**8.** The existence of a set of $k-2$ mutually orthogonal Latin squares of order $n$ is equivalent to the existence of a transversal design TD$(k, n)$. (§12.1.7)

**9.** A set of $k-2$ MOLS of order $n$ is equivalent to an $OA(n, k)$ (§12.3.3).

**10.** *Constructing a complete set of MOLS of order $q$ for $q$ a prime power:*  A complete set of MOLS of order $q$ for $q$ a prime power can be constructed as follows:

- for each $\alpha \in GF(q) - \{0\}$, define the Latin square $L_\alpha(i, j) = i + \alpha j$, where $i, j \in GF(q)$ and the algebra is performed in $GF(q)$.

The set of Latin squares $\{L_\alpha \mid \alpha \in GF(q) - \{0\}\}$ is a set of $q - 1$ MOLS of side $q$.

**11.** Let $n_k$ be the largest order for which the existence of $k$ MOLS is unknown. So if $n > n_k$, then there exist at least $k$ MOLS of order $n$. See the following table:

| $k$ | $n_k$ | $k$ | $n_k$ | $k$ | $n_k$ | $k$  | $n_k$ | $k$  | $n_k$ |
|-----|-------|-----|-------|-----|-------|------|-------|------|-------|
| 2   | 6     | 5   | 60    | 8   | 2,766 | 11   | 7,222 | 14   | 7,874 |
| 3   | 10    | 6   | 74    | 9   | 3,678 | 12   | 7,286 | 15   | 8,360 |
| 4   | 22    | 7   | 570   | 10  | 5,804 | 13   | 7,288 |      |       |

**12.** *Constructing a set of r MOLS of size mn × mn from a set of r MOLS of size m × m and a set of r MOLS of size n × n:* Let $A_1, \ldots, A_r$ and $B_1, \ldots, B_r$ be two sets of MOLS, where each $A_i = (a_{xy}^{(i)})$ is of size $m \times m$ and each $B_i = (b_{xy}^{(i)})$ is of size $n \times n$. Construct a set $C_1, \ldots, C_r$ of $mn \times mn$ MOLS as follows: for each $k = 1, \ldots, r$, let

$$C_k = \begin{pmatrix} D_{11}^{(k)} & D_{12}^{(k)} & \cdots & D_{1m}^{(k)} \\ D_{21}^{(k)} & D_{22}^{(k)} & \cdots & D_{2m}^{(k)} \\ \vdots & \vdots & & \vdots \\ D_{m1}^{(k)} & D_{m2}^{(k)} & \cdots & D_{mm}^{(k)} \end{pmatrix}$$

where

$$D_{ij}^{(k)} = \begin{pmatrix} (a_{ij}^{(k)}, b_{11}^{(k)}) & (a_{ij}^{(k)}, b_{12}^{(k)}) & \cdots & (a_{ij}^{(k)}, b_{1n}^{(k)}) \\ (a_{ij}^{(k)}, b_{21}^{(k)}) & (a_{ij}^{(k)}, b_{22}^{(k)}) & \cdots & (a_{ij}^{(k)}, b_{2n}^{(k)}) \\ \vdots & \vdots & & \vdots \\ (a_{ij}^{(k)}, b_{n1}^{(k)}) & (a_{ij}^{(k)}, b_{n2}^{(k)}) & \cdots & (a_{ij}^{(k)}, b_{nn}^{(k)}) \end{pmatrix}.$$

*Note:* In 1782 Leonhard Euler considered the following problem:

> A very curious question, which has exercised for some time the ingenuity of many people, has involved me in the following studies, which seem to open a new field of analysis, in particular in the study of combinations. The question revolves around arranging 36 officers to be drawn from six different ranks and at the same time from six different regiments so that they are also arranged in a square so that in each line (both horizontal and vertical) there are six officers of different ranks and different regiments.

A solution to Euler's problem would be equivalent to a pair of orthogonal Latin squares of order 6, the symbol set of the first consisting of the six ranks and the symbol set of the second consisting of the six regiments. Euler convinced himself that his problem was incapable of solution and goes even further:

> I have examined a very great number of tables ... and I do not hesitate to conclude that one cannot produce an orthogonal pair of order 6 and that the same impossibility extends to $10, 14, \ldots$ and in general to all the orders which are unevenly even.

Euler was proven correct in his claim that an orthogonal pair of order 6 does not exist [G. Tarry, 1900]; however, in 1960 Euler was shown to be wrong for all orders greater than 6. (See the Bose-Parker-Shrikhande theorem, Fact 3.)

**Examples:**

**1.** Two mutually orthogonal Latin squares of side 3:

| | | | | | | |
|---|---|---|---|---|---|---|
| 1 | 2 | 3 | | 1 | 2 | 3 |
| 2 | 3 | 1 | | 3 | 1 | 2 |
| 3 | 1 | 2 | | 2 | 3 | 1 |

**2.** Three mutually orthogonal Latin squares of side 4:

| | | | | | | | | | | | | | |
|---|---|---|---|---|---|---|---|---|---|---|---|---|---|
| 1 | 2 | 3 | 4 | | 1 | 2 | 3 | 4 | | 1 | 2 | 3 | 4 |
| 4 | 3 | 2 | 1 | | 3 | 4 | 1 | 2 | | 2 | 1 | 4 | 3 |
| 2 | 1 | 4 | 3 | | 4 | 3 | 2 | 1 | | 3 | 4 | 1 | 2 |
| 3 | 4 | 1 | 2 | | 2 | 1 | 4 | 3 | | 4 | 3 | 2 | 1 |

**3.** Two MOLS of order 10:

```
0 4 1 7 2 9 8 3 6 5      0 7 8 6 9 3 5 4 1 2
8 1 5 2 7 3 9 4 0 6      6 1 7 8 0 9 4 5 2 3
9 8 2 6 3 7 4 5 1 0      5 0 2 7 8 1 9 6 3 4
5 9 8 3 0 4 7 6 2 1      9 6 1 3 7 8 2 0 4 5
7 6 9 8 4 1 5 0 3 2      3 9 0 2 4 7 8 1 5 6
6 7 0 9 8 5 2 1 4 3      8 4 9 1 3 5 7 2 6 0
3 0 7 1 9 8 6 2 5 4      7 8 5 9 2 4 6 3 0 1
1 2 3 4 5 6 0 7 8 9      4 5 6 0 1 2 3 7 8 9
2 3 4 5 6 0 1 8 9 7      1 2 3 4 5 6 0 9 7 8
4 5 6 0 1 2 3 9 7 8      2 3 4 5 6 0 1 8 9 7
```

## 12.3.3  ORTHOGONAL ARRAYS

**Definition:**

An **orthogonal array of size** $N$, with $k$ constraints (or of degree $k$), $s$ levels (or of order $s$), and *strength* $t$, denoted OA$(N, k, s, t)$, is a $k \times N$ array with entries from a set of $s \geq 2$ symbols, having the property that in every $t \times N$ submatrix, every $t \times 1$ column vector appears the same number $\lambda = \frac{N}{s^t}$ of times. The parameter $\lambda$ is the *index* of the orthogonal array.

*Note:* An OA$(N, k, s, t)$ is also denoted by OA$_\lambda(t, k, s)$; in this notation, if $t$ is omitted it is understood to be 2, and if $\lambda$ is omitted it is understood to be 1.

**Facts:**

**1.** An OA$_\lambda(k, v)$ is equivalent to a transversal design TD$_\lambda(k, v)$.

**2.** OA$_1(t, k, s)$ are known as *MDS codes* in coding theory.

**3.** An OA$_\lambda(k, n)$ exists only if $k \leq \left\lfloor \frac{\lambda v^2 - 1}{v - 1} \right\rfloor$ (Bose-Bush bound). Generally one is interested in finding the largest $k$ for which there exists an OA$_\lambda(k, n)$ (for a given $\lambda$ and $n$).

The following table gives the best known upper bounds and lower bounds for the largest $k$ for which there exists a OA$_\lambda(k, n)$, for $1 \leq n \leq 18$ and $2 \leq \lambda \leq 10$. Entries for which the upper and lower bounds match are shown in boldface.

| $\lambda \backslash n$ | 1 | 2 | 3 | 4 | 5 | 6 | 7 | 8 | 9 | 10 | 11 | 12 | 13 | 14 | 15 | 16 | 17 | 18 |
|---|---|---|---|---|---|---|---|---|---|---|---|---|---|---|---|---|---|---|
| 2 | 3 | 7 | 11 | 15 | 19 | 23 | 27 | 31 | 35 | 39 | 43 | 47 | 51 | 55 | 59 | 63 | 67 | 71 |
| 3 | 4 | 7 | 13 | 16 | 22 | 25 | 31 | 34 | 40 | 43 | 49 | 52 | 58 | 61 | 67 | 70 | 76 | 79 |
|   |   |   | 13 | 10 |   | 13 | 25 |   | 31 | 13 | 49 | 14 | 25 | 37 | 49 | 25 |   |   |
| 4 | 5 | 9 | 14 | 21 | 25 | 30 | 37 | 41 | 46 | 53 | 57 | 62 | 69 | 73 | 78 | 85 | 89 | 94 |
|   |   |   | 13 |   | 10 | 17 | 13 |   | 37 | 21 | 13 | 61 | 21 | 57 | 22 |   | 37 | 37 |
| 5 | 6 | 11 | 17 | 23 | 31 | 36 | 42 | 48 | 56 | 61 | 67 | 73 | 81 | 86 | 92 | 98 | 106 | 111 |
|   |   |   | 8 | 21 |   | 16 | 18 | 26 | 21 |   | 18 | 26 | 26 | 21 | 43 | 81 | 36 | 91 |
| 6 | 3 | 13 | 20 | 27 | 34 | 43 | 49 | 56 | 63 | 70 | 79 | 85 | 92 | 99 | 106 | 115 | 121 | 128 |
|   |   | 7 | 9 | 13 | 8 | 12 | 9 | 13 | 11 | 12 | 9 | 19 | 11 | 17 | 13 | 25 | 11 | 19 |
| 7 | 8 | 15 | 23 | 30 | 38 | 46 | 57 | 64 | 72 | 79 | 87 | 95 | 106 | 113 | 121 | 128 | 136 | 144 |
|   |   |   | 10 | 29 | 12 | 19 |   | 29 | 29 | 19 | 29 | 50 | 29 |   | 36 | 38 | 29 | 64 |
| 8 | 9 | 17 | 26 | 34 | 43 | 52 | 61 | 73 | 81 | 90 | 98 | 107 | 116 | 125 | 137 | 145 | 154 | 162 |
|   |   |   | 10 | 33 | 10 | 17 | 57 |   | 22 | 41 | 33 | 33 | 26 | 65 | 57 |   | 41 | 73 |
| 9 | 10 | 19 | 29 | 38 | 48 | 58 | 68 | 78 | 91 | 100 | 110 | 119 | 129 | 139 | 149 | 159 | 172 | 181 |
|   |   |   | 28 | 37 | 19 | 55 | 28 | 73 |   | 37 | 37 | 109 | 50 | 55 | 55 | 82 | 73 |   |
| 10 | 9 | 21 | 32 | 42 | 53 | 64 | 75 | 86 | 97 | 111 | 121 | 132 | 142 | 153 | 164 | 175 | 186 | 197 |
|   | 4 | 10 | 12 | 10 | 10 | 20 | 12 | 11 | 12 | 12 | 11 | 28 | 12 | 12 | 12 | 19 | 12 | 30 |

# 12.4  MATROIDS

Linearly independent sets of columns in a matrix and acyclic sets of edges in a graph share many similar properties. Hassler Whitney (1907–1989) aimed to capture these similarities when he defined matroids in 1935. These structures arise naturally in a variety of combinatorial contexts. Moreover, they are precisely the hereditary families of sets for which a greedy strategy produces an optimal set.

## 12.4.1  BASIC DEFINITIONS AND EXAMPLES

**Definitions:**
A *matroid* $M$ (also written $(E, \mathcal{I})$ or $(E(M), \mathcal{I}(M))$) is a finite set $E$ (the **ground set** of $M$) and a collection $\mathcal{I}$ of subsets of $E$ (**independent sets**) such that

- the empty set is independent;
- every subset of an independent set is independent ($\mathcal{I}$ is *hereditary*);
- if $X$ and $Y$ are independent and $|X| < |Y|$, then there is $e \in Y - X$ such that $X \cup \{e\}$ is independent.

Subsets of $E$ that are not in $\mathcal{I}$ are **dependent**.

A *basis* of a matroid is a maximal independent set. The collection of bases of $M$ is denoted $\mathcal{B}(M)$.

A **circuit** of a matroid is a minimal dependent set. The collection of circuits of $M$ is denoted $\mathcal{C}(M)$.

Matroids $M_1$ and $M_2$ are **isomorphic** ($M_1 \cong M_2$) if there is a one-to-one function $\varphi$ from $E(M_1)$ onto $E(M_2)$ that preserves independence; that is, a subset $X$ of $E(M_1)$ is in $\mathcal{I}(M_1)$ if and only if $\varphi(X)$ is in $\mathcal{I}(M_2)$.

For a matroid $M$ with ground set $E$ and $A \subseteq E$, all maximal independent subsets of $A$ have the same cardinality, called the **rank** of $A$, written $r(A)$ or $r_M(A)$. The rank $r(M)$ of $M$ is $r(E)$.

A **spanning set** of a matroid $M$ is a subset of the ground set $E$ of rank $r(M)$.

A **hyperplane** of a matroid $M$ is a maximal nonspanning set.

The **closure** $\mathrm{cl}(X)$ (or $\sigma(X)$) of $X$ is $\{x \in E \mid r(X \cup \{x\}) = r(X)\}$.

A set $X$ is a **closed set** or **flat** if $\mathrm{cl}(X) = X$.

A **loop** of $M$ is an element $e$ such that $\{e\}$ is a circuit.

If $\{f, g\}$ is a circuit, then $f$ and $g$ are **parallel elements**.

Matroid $M$ is a **simple matroid** (or **combinatorial geometry**) if it has no loops and no parallel elements.

A **paving matroid** is a matroid $M$ in which all circuits have at least $r(M)$ elements.

Various classes of matroids are defined in the following table.

| matroid $M$ | ground set $E(M)$ | independent sets $\mathcal{I}(M)$ | bases $\mathcal{B}(M)$ | circuits $\mathcal{C}(M)$ |
|---|---|---|---|---|
| **uniform matroid**, $U_{m,n}$ ($0 \leq m \leq n$) | $\{1, 2, \ldots, n\}$ | $\{I \subseteq E : |I| \leq m\}$ | $\{B \subseteq E : |B| = m\}$ | $\{C \subseteq E : |C| = m+1\}$ |
| $M(G)$, **cycle matroid** of graph $G$ | $E(G)$, the edges of $G$ | $\{I \subseteq E(G) \mid I$ contains no cycle$\}$ | for connected $G$: edge-sets of spanning trees | edge-sets of cycles |
| $M[A]$, **vector matroid** of matrix $A$ over field $F$ | column labels of $A$ | $\{I \subseteq E \mid I$ labels a linearly inde-pendent multiset of columns$\}$ | labels of max-imal linearly independent sets of columns | labels of min-imal linearly dependent multi-sets of columns |
| **transversal matroid**, $M(\mathcal{A})$, of family $\mathcal{A} = (A_1, A_2, \ldots, A_m)$ where $A_j \subseteq E$ | $E$ | partial transvers-als of $\mathcal{A}$: sets $\{x_{i_1}, \ldots, x_{i_k}\}$, $i_1 < \ldots < i_k$ and $x_{i_j} \in A_{i_j}$ | maximal partial transversals of $\mathcal{A}$ | minimal sets that are not par-tial transversals |

Matroid $M$ is in the specified class if $M$ satisfies the indicated condition:

- **graphic**: $M \cong M(G)$ for some graph $G$;
- **planar**: $M \cong M(G)$ for some planar graph $G$;
- **representable over** $F$: $M \cong M[A]$ for some matrix $A$ over the field $F$;
- **binary**: representable over $GF(2)$, the 2-element field;

- **ternary**: representable over $GF(3)$;
- **regular**: representable over all fields.

**Facts:**

**1.** If a matroid $M$ is graphic, then $M \cong M(G)$ for some connected graph $G$.

**2.** *Whitney's 2-isomorphism theorem*:   Two graphs have isomorphic cycle matroids if and only if one can be obtained from the other by performing a sequence of the following operations:

- choose one vertex from each of two components and identify the chosen vertices;
- produce a new graph from which the original can be recovered by applying the previous operation;
- in a graph that can be obtained from the disjoint union of two graphs $G_1$ and $G_2$ by identifying vertices $u_1$ and $v_1$ of $G_1$ with vertices $u_2$ and $v_2$ of $G_2$, twist the graph by identifying, instead, $u_1$ with $v_2$ and $u_2$ with $v_1$.

**3.** If $A'$ is obtained from the matrix $A$ over the field $F$ by elementary row operations, deleting or adjoining zero rows, permuting columns, and multiplying columns by nonzero scalars, then $M[A'] \cong M[A]$. The converse of this holds if and only if $F$ is $GF(2)$ or $GF(3)$.

**4.** If a matroid $M$ is representable over $F$ and $r(M) \geq 1$, then $M \cong M[I_{r(M)}|D]$, where $I_{r(M)}|D$ consists of an $r(M) \times r(M)$ identity matrix followed by some other matrix $D$ over $F$.

**5.** A matroid $M$ is regular if and only if $M$ can be represented over the real numbers by a *totally unimodular matrix* (a matrix for which all subdeterminants are 0, 1, or $-1$).

**6.** A matroid $M$ is regular if and only if $M$ is both binary and ternary.

**7.** The smallest matroids not representable over any field have eight elements.

**8.** *Conjecture*:  For all $n \geq 9$, more than half of all matroids on $\{1, 2, \ldots, n\}$ are paving.

**9.** The following table lists the numbers of nonisomorphic matroids, simple matroids, and binary matroids with up to nine elements:

| $|E(M)|$ | 0 | 1 | 2 | 3 | 4 | 5 | 6 | 7 | 8 | 9 |
|---|---|---|---|---|---|---|---|---|---|---|
| matroids | 1 | 2 | 4 | 8 | 17 | 38 | 98 | 306 | 1,724 | 383,172 |
| simple | 1 | 1 | 1 | 2 | 4 | 9 | 26 | 101 | 950 | 376,467 |
| binary | 1 | 2 | 4 | 8 | 16 | 32 | 68 | 148 | 342 | 848 |

**Examples:**

**1.** Let $M$ be the matroid with $E(M) = \{1, 2, \ldots, 6\}$ and $\mathcal{C}(M) = \{\{1\}, \{5, 6\}, \{3, 4, 5\}, \{3, 4, 6\}\}$. Then $\mathcal{B} = \{\{2, 3, 4\}, \{2, 3, 5\}, \{2, 3, 6\}, \{2, 4, 5\}, \{2, 4, 6\}\}$. The following figure shows that $M$ is graphic and binary since $M = M(G_1) = M(G_2)$ and $M = M[A]$ with $A$ being interpreted over $GF(2)$; $M$ is regular since $M = M[A]$ when $A$ is interpreted over any field $F$. Also $M$ is transversal since $M = M(\mathcal{A})$ where $\mathcal{A} = (\{2\}, \{3, 4\}, \{4, 5, 6\})$.

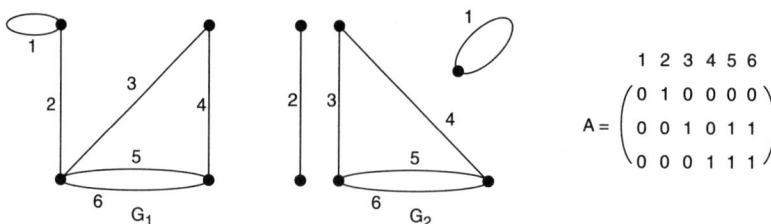

**2. Fano and non-Fano matroids:**   Given a finite set $E$ of points in the plane and a collection of lines (subsets of $E$ with at least three elements), no two of which share more than one common point, there is a matroid with ground set $E$ whose circuits are all sets of three collinear points and all sets of four points no three of which are collinear. Two such matroids are shown in the following figure. Each has ground set $\{1, 2, \ldots, 7\}$. On the right is the non-Fano matroid $F_7^-$. It differs from the Fano matroid $F_7$ on the left by the collinearity of 4, 5, and 6 in the latter.

The matrix in this figure represents $F_7$ over all fields of characteristic 2, and represents $F_7^-$ over all other fields. $F_7$ is binary but non-ternary; $F_7^-$ is ternary but non-binary. Both are non-uniform, non-regular, non-graphic, and non-transversal.

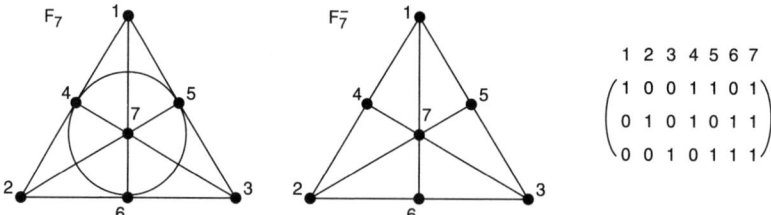

---

## 12.4.2   ALTERNATIVE AXIOM SYSTEMS

Matroids can be characterized by many different axiom systems. Some examples of these systems follow. Throughout, $E$ is assumed to be a finite set and $2^E$ stands for the set of subsets of $E$.

### Definitions:

***Circuit axioms:***   A subset $\mathcal{C}$ of $2^E$ is the set of circuits of a matroid on $E$ if and only if $\mathcal{C}$ satisfies

- $\emptyset \notin \mathcal{C}$;
- no member of $\mathcal{C}$ is a proper subset of another;
- *circuit elimination:*   if $C_1$, $C_2$ are distinct members of $\mathcal{C}$ and $e \in C_1 \cap C_2$, then $\mathcal{C}$ has a member $C_3$ such that $C_3 \subseteq (C_1 \cup C_2) - \{e\}$.

*Note:* The circuit elimination axiom can be strengthened to the following:

- *strong circuit elimination:*   if $C_1, C_2 \in \mathcal{C}$, $f \in C_1 - C_2$, and $e \in C_1 \cap C_2$, then $\mathcal{C}$ has a member $C_3$ such that $f \in C_3 \subseteq (C_1 \cup C_2) - \{e\}$.

***Basis axioms:***   A subset $\mathcal{B}$ of $2^E$ is the set of bases of a matroid on $E$ if and only if

- $\mathcal{B}$ is nonempty;
- if $B_1, B_2 \in \mathcal{B}$ and $x \in B_1 - B_2$, then there is an element $y \in B_2 - B_1$ such that $(B_1 - \{x\}) \cup \{y\} \in \mathcal{B}$.

**Rank axioms**:  A function $r$ from $2^E$ into the nonnegative integers is the rank function of a matroid on $E$ if and only if, for all subsets $X$, $Y$, $Z$ of $E$

- $0 \le r(X) \le |X|$;
- if $Y \subseteq Z$, then $r(Y) \le r(Z)$;
- *submodularity*:  $r(X \cup Y) + r(X \cap Y) \le r(X) + r(Y)$.

**Closure axioms**:  A function cl from $2^E$ into $2^E$ is the closure operator of a matroid on $E$ if and only if, for all subsets $X$ and $Y$ of $E$

- $X \subseteq \mathrm{cl}(X)$;
- if $X \subseteq Y$, then $\mathrm{cl}(X) \subseteq \mathrm{cl}(Y)$;
- $\mathrm{cl}(\mathrm{cl}(X)) = \mathrm{cl}(X)$;
- *MacLane-Steinitz exchange*:  if $x \in E$ and $y \in \mathrm{cl}(X \cup \{x\}) - \mathrm{cl}(X)$, then $x \in \mathrm{cl}(X \cup \{y\})$.

**Fact:**

**1.** If $M$ is a matroid with ground set $E$ and $I \subseteq E$, the following statements are equivalent:

- $I$ is an independent set of $M$;
- no circuit of $M$ contains $I$;
- some basis of $M$ contains $I$;
- $r(I) = |I|$;
- for every element $e$ of $I$, $e \notin \mathrm{cl}(I - \{e\})$.

---

## 12.4.3  DUALITY

**Definitions:**

For a matroid $M$, let $\mathcal{B}^*(M) = \{E(M) - B \mid B \in \mathcal{B}(M)\}$. Then $\mathcal{B}^*(M)$ is the set of bases of a matroid $M^*$, called the **dual** of $M$, whose ground set is also $E(M)$.

Bases, circuits, loops, and independent sets of $M^*$ are called **cobases**, **cocircuits**, **coloops**, and **coindependent sets** of $M$.

For a graph $G$, the **cocycle matroid** (or **bond matroid**) of $G$ is the dual of $M(G)$ and is denoted by $M^*(G)$.

A matroid $M$ is **cographic** if $M \cong M^*(G)$ for some graph $G$.

A class of matroids is **closed under duality** if the dual of every member of the class is also in the class.

**Facts:**

**1.** For all matroids $M$, $(M^*)^* = M$.

**2.** For all matroids $M$, the rank function of $M^*$ is given by $r^*(X) = |X| - r(M) + r(E - X)$.

**3.** The cocircuits of every matroid $M$ are the minimal sets having nonempty intersection with every basis of $M$.

**4.** The cocircuits of every matroid $M$ are the minimal nonempty sets $C^*$ such that $|C^* \cap C| \neq 1$ for every circuit $C$ of $M$.

**5.** For every graph $G$, the circuits of $M^*(G)$ are the minimal edge cuts of $G$.

**6.** A graphic matroid is cographic if and only if it is planar.

**7.** The following classes of matroids are closed under duality: uniform matroids, matroids representable over a fixed field $F$, planar matroids, and regular matroids. The classes of graphic and transversal matroids are not closed under duality.

**8.** The following are special sets and their complements in a matroid $M$ and $M^*$:

| $X$ | basis of $M$ | independent set of $M$ | circuit of $M$ |
|---|---|---|---|
| $E - X$ | basis of $M^*$ | spanning set of $M^*$ | hyperplane of $M^*$ |

**Example:**

**1.** The following are duals of some basic examples:

| matroid | dual |
|---|---|
| $U_{m,n}$ | $U_{n-m,n}$ |
| $M(G)$  ($G$ planar) | $M(G^*)$, where $G^*$ is any dual of $G$ |
| $M[I_r|D]$ ($[I_r|D]$ an $r \times n$ matrix) | $M[-D^T|I_{n-r}]$, same order of column labels as $[I_r|D]$ |

## 12.4.4  FUNDAMENTAL OPERATIONS

**Definitions:**

Three basic constructions for matroids $M$, $M_1$, and $M_2$ are defined in the following table. $M \backslash T$ and $M/T$ are also written as $M|(E - T)$ and $M.(E - T)$ and are called the **restriction** and **contraction** of $M$ to $E - T$; $M\backslash\{e\}$ and $M/\{e\}$ are written as $M\backslash e$ and $M/e$.

| matroid | $\mathcal{I}$ | $\mathcal{C}$ | rank |
|---|---|---|---|
| $M\backslash T$ (**deletion** of $T$ from $M$) | $\{I \subseteq E(M) - T \mid I \in \mathcal{I}(M)\}$ | $\{C \subseteq E(M) - T \mid C \in \mathcal{C}(M)\}$ | $r_{M\backslash T}(X) = r_M(X)$ |
| $M/T$ (**contraction** of $T$ from $M$) | $\{I \subseteq E(M) - T \mid I \cup B_T \in \mathcal{I}(M)$ for some $B_T$ in $\mathcal{B}(M|T)\}$ | minimal nonempty members of $\{C - T \mid C \in \mathcal{C}(M)\}$ | $r_{M/T}(X) = r_M(X \cup T) - r_M(T)$ |
| $M_1 \oplus M_2$ (**direct sum** of $M_1$ and $M_2$) | $\{I_1 \cup I_2 \mid I_j \in \mathcal{I}(M_j)\}$ | $\mathcal{C}(M_1) \cup \mathcal{C}(M_2)$ | $r_{M_1 \oplus M_2}(X) = r_1(X \cap E(M_1)) + r_2(X \cap E(M_2))$ |

Matroid $N$ is a **minor** of matroid $M$ if $N$ can be obtained from $M$ by a sequence of deletions and contractions. The minor $N$ is **proper** if $N \neq M$.

A matroid is **connected** if it cannot be written as the direct sum of two *nonempty matroids* (matroids with nonempty ground sets).

**Facts:**

In each of the following, $M$, $M_1$, and $M_2$ are matroids.

**1.** $M \backslash X \backslash Y = M \backslash (X \cup Y) = M \backslash Y \backslash X$; $M/X/Y = M/(X \cup Y) = M/Y/X$; and $M \backslash X/Y = M/Y \backslash X$.

**2.** $M_1 \oplus M_2 = M_2 \oplus M_1$.

**3.** $(M/T)^* = M^* \backslash T$; and $(M \backslash T)^* = M^*/T$. (Deletion and contraction are dual operations.)

**4.** *The scum theorem:* Every minor of $M$ can be written as $M \backslash X/Y$ for some independent set $Y$ and coindependent set $X$. (The name derives from the fact that an isomorphic copy of every simple minor of a matroid occurs at (that is, floats to) the top of the lattice.) (D. A. Higgs) [CrRo70]

**5.** The following are equivalent:

- $M$ is connected;
- $M^*$ is connected;
- every two distinct elements of $M$ are in a circuit;
- there is no proper nonempty subset $T$ of $E(M)$ such that $M \backslash T = M/T$;
- there is no proper nonempty subset $T$ of $E(M)$ such that $r(T) + r(E(M) - T) = r(M)$;
- there is no proper nonempty subset $T$ of $E(M)$ such that $r(T) + r^*(T) = |T|$.

**6.** If $M$ is connected, then $M$ is uniquely determined by the set of circuits containing some fixed element of $E(M)$.

**7.** If $M$ is connected and $e \in E(M)$, then $M \backslash e$ or $M/e$ is connected.

**8.** $F_7 \oplus F_7^-$ is not representable over any field.

**Examples:**

**1.** $U_{m,n} \backslash e = U_{m,n-1}$ unless $m = n$ when $U_{m,n} \backslash e = U_{m-1,n-1}$.

**2.** $U_{m,n}/e = U_{m-1,n-1}$ unless $m = 0$ when $U_{m,n}/e = U_{m,n-1}$.

**3.** $M(G) \backslash e = M(G \backslash e)$ where $G \backslash e$ is obtained from $G$ by deleting the edge $e$.

**4.** $M(G)/e = M(G/e)$ where $G/e$ is obtained from $G$ by contracting the edge $e$.

**5.** $M[A] \backslash e$ is the vector matroid of the matrix obtained by deleting column $e$ from $A$.

**6.** If $e$ corresponds to a standard basis vector in $A$, then $M[A]/e$ is the vector matroid of the matrix obtained by deleting both the column $e$ and the row containing the one of $e$.

## 12.4.5 CHARACTERIZATIONS

Many matroid results characterize various classes of matroids. Some examples of such results appear below. The Venn diagram in the following figure indicates the relationship between certain matroid classes.

Matroids

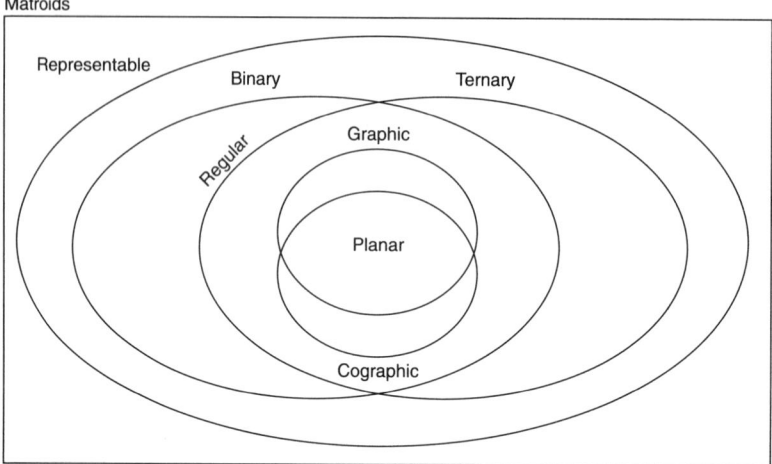

## Definitions:

Let $M_1$ and $M_2$ be two binary matroids such that $E(M_1) \cap E(M_2) = T$, $M_1|T = M_2|T$, and no cocircuit of $M_1$ or $M_2$ is contained in $T$. The **2-sum** and **3-sum** of $M_1$ and $M_2$ are matroids on $(E(M_1) \cup E(M_2)) - T$ whose flats are those sets $F - T$ such that $F \cap E(M_i)$ is a flat of $M_i$ for $i = 1, 2$. The 2-sum occurs when $|T| = 1$, $|E(M_i)| \geq 3$, and $T$ is not a loop of $M_i$, and the 3-sum occurs when $|E(M_i)| \geq 7$ and $T$ is a 3-element circuit of $M_i$.

## Facts:

1. The following are equivalent for a matroid $M$:
   - $M$ is uniform;
   - every circuit of $M$ has $r(M) + 1$ elements;
   - every circuit of $M$ meets every cocircuit of $M$.

2. The following are equivalent for a matroid $M$:
   - $M$ is binary;
   - for every circuit $C$ and every cocircuit $C^*$, $|C \cap C^*|$ is even;
   - for every circuit $C$ and every cocircuit $C^*$, $|C \cap C^*| \neq 3$;
   - for all $C_1, C_2 \in \mathcal{C}$, $(C_1 - C_2) \cup (C_2 - C_1)$ is a disjoint union of circuits.

3. The class of regular matroids is the class of matroids that can be constructed by direct sums, 2-sums, and 3-sums from graphic matroids, cographic matroids, and copies of $R_{10}$ (the matroid that is represented over $GF(2)$ by the ten 5-tuples with exactly three ones). (This last fact is the basis of a polynomial-time algorithm to determine whether a real matrix is totally unimodular.)

4. *Excluded-minor theorems:* Many classes of matroids are *minor-closed*; that is, every minor of a member of the class is also in the class. Such classes can be characterized by listing their *excluded minors* (those matroids that are not in the class but have all their proper minors in the class). Some important examples of such results are given in the following table. The class of transversal matroids is not minor-closed since a contraction of a transversal matroid need not be transversal.

| class | excluded minors | class | excluded minors |
|---|---|---|---|
| binary | $U_{2,4}$ | ternary | $U_{2,5}, U_{3,5}, F_7, F_7^*$ |
| uniform | $U_{0,1} \oplus U_{1,1}$ | graphic | $U_{2,4}, F_7, F_7^*, M^*(K_5), M^*(K_{3,3})$ |
| paving | $U_{0,1} \oplus U_{2,2}$ | regular | $U_{2,4}, F_7, F_7^*$ |

## 12.4.6  THE GREEDY ALGORITHM

**Definitions:**
For a finite set $E$, let $\mathcal{I}$ be a subset of $2^E$ satisfying the first two axioms for independent sets in the definition of matroid (§12.4.1). Let $w$ be a real-valued function on $E$. For $X \subseteq E$, define the **weight** of $X$ by $w(X) = \sum_{x \in X} w(x)$, and let $w(\emptyset) = 0$.

**Facts:**
**1.** Matroids have an important relationship to the greedy algorithm (Algorithm 1) that makes them important in optimization problems.

---

**Algorithm 1**:  **The greedy algorithm for $(\mathcal{I}, w)$.**
$X_0 := \emptyset;\ j := 0$
**while** $E - X_j$ contains an element $e$ such that $X_j \cup \{e\} \in \mathcal{I}$
$\quad e_{j+1} :=$ an element $e$ of maximum weight such that $X_j \cup \{e\} \in \mathcal{I}$
$\quad X_{j+1} := X_j \cup \{e_{j+1}\}$
$\quad j := j + 1$
$B_G := X_j$

---

**2.** $\mathcal{I}$ (a subset of $2^E$ satisfying the first two axioms for independent sets in the definition of matroid) is the set of independent sets of a matroid on $E$ if and only if, for all real-valued weight functions $w$ on $E$, the set $B_G$ produced by the greedy algorithm is a maximal member of $\mathcal{I}$ of maximum weight.

**Example:**
**1.** Let $G$ be a connected graph with each edge $e$ having a cost $c(e)$. Define $w(e) = -c(e)$. Then the greedy algorithm is just Kruskal's algorithm (§10.1.2) and $B_G$ is the edge-set of a spanning tree of minimum cost.

---

## REFERENCES

*Printed Resources*:

[An93] I. Anderson, *Combinatorial Designs: Construction Methods*, Ellis Horwood, 1993.

[BeJuLe86] T. Beth, D. Jungnickel, and H. Lenz, *Design Theory*, Cambridge University Press, 1986.

[CoDi07] C. J. Colbourn and J. H. Dinitz, eds., *Handbook of Combinatorial Designs*, 2nd ed., Chapman and Hall/CRC Press, 2007. (A comprehensive source of information on combinatorial designs.)

[CrRo70] H. H. Crapo and G.-C. Rota, *On the Foundations of Combinatorial Theory: Combinatorial Geometries*, MIT Press, 1970.

[DiSt92] J. H. Dinitz and D. R. Stinson, eds., *Contemporary Design Theory: A Collection of Surveys*, John Wiley & Sons, 1992.

[KeDé15] A. D. Keedwell and J. Dénes, *Latin Squares and Their Applications*, North Holland, 2015.

[Ox11] J. G. Oxley, *Matroid Theory*, 2nd ed., Oxford University Press, 2011. (The main source for §12.4; contains references for all stated results.)

[Re89] A. Recski, *Matroid Theory and Its Applications in Electric Network Theory and in Statics*, Springer-Verlag, 1989.

[St04] D. R. Stinson, *Combinatorial Designs: Constructions and Analysis*, Springer, 2004.

[We10] D. J. A. Welsh, *Matroid Theory*, Dover, 2010.

[Wh08] N. White, ed., *Theory of Matroids*, Cambridge University Press, 2008.

**Web Resources**:

http://designtheory.org (Information on combinatorial, computational, and statistical aspects of design theory.)

http://gams.nist.gov/ (Guide to Available Mathematical Software at the National Institute of Science and Technology: a cross-index and virtual repository of mathematical and statistical software components of use in computational science and engineering; contains a program for finding $t$-designs.)

http://lib.stat.cmu.edu/designs/ (The Designs Archive at Statlib: has some very useful programs for making orthogonal arrays.)

http://link.springer.com/journal/10623 (*Designs, Codes and Cryptography* has a site with its table of contents.)

http://neilsloane.com/gosset/ (GOSSET: A general purpose program for designing experiments.)

http://neilsloane.com/hadamard/index.html (Contains an extensive collection of Hadamard matrices.)

http://neilsloane.com/oadir/index.html (Contains an extensive collection of orthogonal arrays.)

http://onlinelibrary.wiley.com/journal/10.1002/(ISSN)1520-6610 (*Journal of Combinatorial Designs* has a site with its table of contents.)

http://userhome.brooklyn.cuny.edu/skingan/matroids/software.html (Contains an open source, interactive, extensible software system for experimenting with matroids.)

http://www-m10.ma.tum.de/foswiki/pub/Lehrstuhl/PublikationenJRG/21_Orient edMatroids.pdf (Source information on oriented matroids.)

https://www.ccrwest.org/cover.html and https://www.ccrwest.org/cover/table .html (La Jolla Covering Repository: contains coverings $C(v,k,t)$ with $v \leq 32$, $k \leq 16$, $t \leq 8$, and less than 5,000 blocks.)

http://www.cecm.sfu.ca/organics/papers/lam/paper/html/paper.html (Information on the search for a finite projective plane of order 10.)

http://www.cems.uvm.edu/~dinitz/hcd.html (*Handbook of Combinatorial Designs* has a website that lists new results in design theory that have been discovered since its 2007 publication.)

http://www.gap-system.org/Packages/design.html (DESIGN is a package for constructing, classifying, partitioning and studying block designs. )

https://www.math.LSU.edu/~oxley/survey4.pdf (Contains the 45-page paper "What is a Matroid?" by James Oxley.)

`http://www.matroidunion.org/` (A blog for and by the matroid community.)

`http://www.netlib.org/` (Netlib Repository: a collection of mathematical software, papers, and databases.)

`http://www.utu.fi/fi/yksikot/sci/yksikot/mattil/opiskelu/kurssit/Document s/comb2.pdf` (Notes from Ian Anderson and Iiro Honkala for a "Short Course in Combinatorial Designs"; 39 pages, revised 2012.)

# 13

# *DISCRETE AND COMPUTATIONAL GEOMETRY*

## INTRODUCTION

This chapter outlines the theory and applications of various concepts arising in two rapidly growing, interrelated areas of geometry: discrete geometry (which deals with topics such as space filling, arrangements of geometric objects, and related combinatorial problems) and computational geometry (which deals with the many aspects of the design and analysis of geometric algorithms). A more extensive treatment of discrete and computational geometry can be found in [TóORGo17].

## GLOSSARY

**anti-aliasing**: the filtering out of high-frequency spatial components of a signal, to prevent artifacts, or aliases, from appearing in the output image.

**aperiodic** (prototile): a prototile in $d$-dimensional Euclidean space such that the prototile admits a tiling of the space, yet all such tilings are nonperiodic.

**arrangement** (of lines in the plane): the planar straight-line graph whose vertices are the intersection points of the lines and whose edges connect consecutive intersection points on each line (it is assumed that all lines intersect at a common point at infinity).

**arrangement graph**: a graph associated with a Euclidean or projective line arrangement, or a big circle arrangement.

**aspect ratio** (of a simplex): the ratio of the radius of the circumscribing sphere to the radius of the inscribing sphere of the simplex; for a triangulation, the largest aspect ratio of a simplex in the triangulation.

**basis** (of a point configuration in $\mathcal{R}^d$): a subset of the point configuration that is a simplex of the ambient space $\mathcal{R}^d$.

**basis** (of a vector configuration in $\mathcal{R}^d$): a subset of the vector configuration that is a basis of the ambient space $\mathcal{R}^d$.

**big-circle arrangement**: the intersection of a central plane arrangement with the unit sphere in $\mathcal{R}^3$.

**boundary** (of a polyhedron): the vertices, edges, and higher-dimensional facets of the polyhedron.

***cell*** (of a line arrangement): a connected component of the complement in $\mathcal{R}^2$ of the union of the points on the lines.

***centerpoint*** (of a point configuration $P$ of size $n$): a point $q$, not necessarily in $P$, such that for any hyperplane containing $q$ there are at least $\lceil \frac{n}{d+1} \rceil$ points in each semi-space induced by the hyperplane.

***central hyperplane arrangement*** (in $\mathcal{R}^d$): a finite set of central hyperplanes, not all of them going through the same point.

***central plane arrangement*** (in $\mathcal{R}^3$): a finite set of central planes.

***chain***: a planar straight-line graph with vertices $v_1, \ldots, v_n$ and edges $\{v_1, v_2\}$, $\{v_2, v_3\}$, $\ldots$, $\{v_{n-1}, v_n\}$.

***chirotope***: for an ordered point configuration, the set of all signed bases of the configuration; for an ordered vector configuration, the set of all signed bases of the configuration.

***circuit*** (of a set of labeled vectors $V = \{v_1, \ldots, v_n\}$): the signed set $C = (C^+, C^-)$, where $C^+ = \{j \mid \alpha_j > 0\}$ and $C^- = \{j \mid \alpha_j < 0\}$, of indices of the non-null coefficients $\alpha_j$ in a minimal linear dependency $V' = \{v_{i_1}, \ldots, v_{i_k}\}$ of $V$ with $\sum_{j=1}^{k} \alpha_j v_{i_j} = \underline{0}$.

***class library***: in an object-oriented computer language, a set of new data types and operations on them, activated by sending a data item a message.

***closed half-space***: the set of all points on a hyperplane and the points on one side of the same hyperplane.

***cluster of rank 3 hyperline sequences*** (associated with a vector configuration): the ordered set of stars, one for each point in the configuration.

***cluster of stars*** (associated with a point configuration): the ordered set of stars, one for each point in the configuration.

***cocircuit*** (of a labeled vector configuration $V$): a signed set $C = (C^+, C^-)$ of the set $\{1, 2, \ldots, n\}$, induced by a subset of $d - 1$ vectors spanning a central hyperplane $h$. For an arbitrary orientation of $h$, $C^+$ is the set of indices of elements in $V$ lying in $h^+$ and $C^-$ is the set of indices of elements in $V$ lying in $h^-$.

***computational convexity***: the study of high-dimensional convex bodies.

***contraction*** (on element $i$ in a rank $d$ central plane arrangement): the arrangement obtained by identifying $h_i$ with $\mathcal{R}^{d-1}$ and intersecting it with all the other hyperplanes to obtain a rank $d-1$ arrangement with one fewer element.

***convex***: the property of a subset of a Euclidean space that for every pair of points in the set the linear segment joining them is contained in the set.

***convex body***: a closed and bounded convex set with nonempty interior.

***convex d-polyhedron***: the intersection of a finite number of closed half-spaces in $\mathcal{R}^d$.

***convex decomposition*** (of a polyhedron): its partition into interior disjoint convex pieces.

***convex hull*** (of a set of points): the smallest convex set containing the given set of points.

***convex polygon***: a polytope in the plane.

***convex polyhedron***: the intersection of a finite number of half-spaces.

**convex polytope**: a bounded convex polyhedron.

**convex position**: the property of a set of points that it is the vertex set of a polytope.

**convex set**: a subset of $d$-dimensional Euclidean space such that for every pair of distinct points in the set, the segment with these two points as endpoints is contained in the set.

**covering**: a family of convex bodies in $d$-dimensional Euclidean space such that each point belongs to at least one of the convex bodies.

**cyclic $d$-polytope**: the convex hull of a set of $n \geq d + 1$ points on the moment curve in $\mathcal{R}^d$. The moment curve in $\mathcal{R}^d$ is defined parametrically by $x(t) = (t, t^2, \ldots, t^d)$.

**Davenport-Schinzel sequence** (of order $s$): a sequence of characters over an alphabet of size $n$ such that no two consecutive characters are the same, and for any pair of characters, $a$ and $b$, there is no alternating subsequence of length $s + 2$ of the form $\ldots a \ldots b \ldots a \ldots b \ldots$.

**deletion**: the removal of a point (vector, line, etc.) from a configuration and recording the oriented matroid (chirotope, circuits, cluster of stars, etc.) only for the remaining points.

**density** (of a covering): the common value (if it exists) of the lower density and upper density of the covering.

**density** (of a packing): the common value (if it exists) of the lower density and upper density of the packing.

**dual polytopes**: two polytopes $P$ and $Q$ such that there exists a one-to-one correspondence $\delta$ between the set of faces of $P$ and $Q$, where two faces $f_1, f_2 \in P$ satisfy $f_1 \subset f_2$ if and only if $\delta(f_1) \supset \delta(f_2)$ in $Q$.

**duality transformation**: a mapping of points to lines and lines to points that preserves incidences.

**Euclidean hyperplane arrangement** (in $\mathcal{R}^d$): a finite set of affine hyperplanes, not all of them going through the same point.

**Euclidean line arrangement**: a finite set of planar lines, not all of them going through the same point.

**Euclidean pseudoconfiguration of points**: a pair consisting of a planar set of points and a pseudoline arrangement, such that for every pair of distinct points there exists a unique pseudoline incident with them.

**$k$-face**: an open set of dimension $k$ that is part of the boundary of a polyhedron; 0-faces, 1-faces, and $(d-1)$-faces of a $d$-polyhedron are called *vertices*, *edges*, and *facets*.

**face vector** (of a $d$-polyhedron): the $d$-dimensional vector $(f_0, f_1, \ldots, f_{d-1})$, where $f_i$ is the number of $i$-dimensional faces of the $d$-polyhedron.

**face-to-face** (tiling): a tiling of $d$-dimensional Euclidean space by convex $d$-polytopes such that the intersection of any two tiles is a face of each tile, possibly the (improper) empty face.

**general position**: the property of a set of vectors in $\mathcal{R}^d$ that every subset of $d$ elements is a basis; property of a set of points in $\mathcal{R}^d$ that every subset of $d$ elements is a simplex.

**genus** (of a manifold 3-polyhedron): the genus number of its boundary, if the boundary is a 2-manifold.

**geographic information system (GIS)**: an information system designed to capture, store, manipulate, analyze, and display spatial or geographically-referenced data.

**geometric constraint solving**: the problem of locating a set of geometric elements given a set of constraints between them.

**graphical user interface (GUI)**: a mechanism that allows a user to interactively control a computer program with a bitmapped display by using a mouse or pointer to select menu items, move sliders or valuators, etc.

**Grassmann-Plücker relations (rank 3)**: the identities $[123][145] - [124][135] + [125][134] = 0$ satisfied by the determinants $[ijk] = \det(v_i, v_j, v_k)$, for any five vectors $v_i, 1 \leq i \leq 5$.

**half-space**: one of the two connected components of the complement of a hyperplane.

**ham-sandwich cut**: a hyperplane that simultaneously bisects $d$ point configurations in $d$-dimensional Euclidean space.

**hyperplane**: in $d$ dimensions the set of all points on a $(d-1)$-dimensional plane.

**hyperplane arrangement**: the partitioning of the Euclidean space $\mathcal{R}^d$ into connected regions of different dimensions (vertices, edges, etc.) by a finite set of hyperplanes.

**isogonal** (tiling): a tiling such that the group of symmetries acts transitively on the vertices of the tiles.

**isomorphic** (vector or point configurations): configurations having the same order type, after possibly relabeling their elements.

**isotoxal** (tiling): a tiling such that the group of symmetries acts transitively on the edges of the tiles.

**lattice** (tiling): a tiling of $d$-dimensional Euclidean space by translates of a tile such that the corresponding translation vectors form a $d$-dimensional lattice.

**k-level** (in a given nonvertical arrangement of $n$ lines): the lower boundary of the set of points in $\mathcal{R}^2$ having exactly $k$ lines above and $n-k$ below.

**line arrangement**: the partitioning of the plane into connected regions (cells, edges, and vertices) induced by a finite set of lines.

**lower density** (of a covering $\mathcal{C}$): $\underline{\nu}(\mathcal{C}) = \liminf\limits_{R \to +\infty} \dfrac{\sum_{K_i \cap B_R \neq \emptyset} \mathrm{Vol}(K_i)}{\mathrm{Vol}(B_R)}$, where each $K_i$ is a convex body in the covering $\mathcal{C}$ of $d$-dimensional Euclidean space and $B_R$ is the closed ball of radius $R$ centered at the origin.

**lower density** (of a packing $\mathcal{P}$): $\underline{\delta}(\mathcal{P}) = \liminf\limits_{R \to +\infty} \dfrac{\sum_{K_i \subset B_R} \mathrm{Vol}(K_i)}{\mathrm{Vol}(B_R)}$, where each $K_i$ is a convex body in the packing $\mathcal{P}$ of $d$-dimensional Euclidean space and $B_R$ is the closed ball of radius $R$ centered at the origin.

**lower envelope** (of a nonvertical line arrangement): the half-plane intersection of the half-planes below the lines of the arrangement.

**manifold d-polyhedron**: a polyhedron whose boundary is topologically the same as a $(d-1)$-manifold; i.e., every point on the boundary of a manifold $d$-polyhedron has a small neighborhood that looks like an open $d$-ball.

**mathematical programming**: the large-scale optimization of an objective function of many variables subject to constraints.

**minor** (of an oriented matroid given by hyperline sequences): an oriented matroid obtained by a sequence of deletions and/or contractions.

**monohedral** (tiling):  a tiling $\mathcal{T}$ of $d$-dimensional Euclidean space in which all tiles are congruent to one fixed set $T$, the (metrical) prototile of $\mathcal{T}$.

**nonconvex polyhedron**:  the union of a set of convex polyhedra such that the underlying space is connected and nonconvex.

**non-manifold $d$-polyhedron**:  a $d$-polyhedron that does not have a manifold boundary.

**nonperiodic** (tiling):  a tiling such that its group of symmetries contains no translation other than the identity.

**normal** (tiling):  a tiling of $d$-dimensional Euclidean space by convex polytopes such that there exist positive real numbers $r$ and $R$ such that each tile contains a Euclidean ball of radius $r$ and is contained in a Euclidean ball of radius $R$.

**oracle**:  an algorithm that gives information about a convex body.

**order type** (of a vector or point configuration):  the collection of all semi-spaces of the configuration.

**oriented matroid**:  a pair $\mathcal{M} = (n, \mathcal{L})$, where $\mathcal{L}$, the set of covectors of $\mathcal{M}$, is a subset of $\{+, -, 0\}^n$ and satisfies the properties: $0 \in \mathcal{L}$; if $X \in \mathcal{L}$, then $-X \in \mathcal{L}$; if $X, Y \in \mathcal{L}$, then $X \circ Y \in \mathcal{L}$; if $X, Y \in \mathcal{L}$ and $i \in S(X, Y) = \{i \mid X_i = -Y_i \neq 0\}$, then there is $Z \in \mathcal{L}$ such that $Z_i = 0$; for each $j \notin S(X, Y)$, $Z_j = (X \circ Y)_j = (Y \circ X)_j$.

**oriented matroid given by a chirotope**:  an abstract set of points labeled $\{1, \ldots, n\}$, together with a function satisfying the chirotope axioms.

**packing**:  a family of convex bodies in $d$-dimensional Euclidean space such that no two have an interior point in common.

**parallel algorithm**:  an algorithm that concurrently uses more than one processing element during its execution.

**parallel random access machine** (**PRAM**):  a synchronous machine in which each processor is a sequential RAM, and processors communicate using a shared memory.

**parametric search**:  an algorithmic technique for solving optimization problems.

**periodic** (tiling):  a tiling of $d$-dimensional Euclidean space such that the group of all symmetries of the tiling contains translations in $d$ linearly independent directions.

**planar straight-line graph**:  a planar graph such that each edge is a straight line.

**point configuration** (of dimension $d$):  a finite set of points affinely spanning $\mathcal{R}^d$.

**point location problem**:  the problem of determining which region of a given subdivision of $\mathcal{R}^d$ contains a given point.

**polar-duality of vectors and central planes** (in $\mathcal{R}^3$):  a mapping associating with a vector $v$ in $\mathcal{R}^3$ an oriented central plane $h$ having $v$ as its normal vector, and vice versa.

**polyhedron**:  the intersection of a finite number of closed half-spaces in $d$-dimensional Euclidean space.

**polytope**:  a bounded polyhedron.

**$d$-polytope**:  a convex $d$-polyhedron for which there exists a $d$-dimensional cube containing it inside; that is, a bounded convex $d$-polyhedron.

**$\mathcal{H}$-polytope**:  a polytope defined as the intersection of $d$ half-spaces in $d$-dimensional Euclidean space.

*V-polytope*:  a polytope defined as the convex hull of $d$ points in $d$-dimensional Euclidean space.

*projective line arrangement*:  a finite set of projective lines in the projective plane.

*prototile*:  the single tile used repeatedly in a monohedral tiling.

*pseudoline arrangement*:  a finite collection of simple planar curves that intersect pairwise in exactly one point, where they cross.

*Radon partition* (of a set of labeled points $P$):  a signed set $C = (C^+, C^-)$ of points of $P$ such that the convex hull of the points in $C^+$ intersects the convex hull of the points in $C^-$.

*randomized algorithm*:  an algorithm that makes random choices during its execution.

*range counting problem*:  the problem of counting the number of points of a given set of points that lie in a query range.

*range emptiness problem*:  the problem of determining if a query range contains any points of a given set of points.

*range reporting problem*:  the problem of determining all the points of a given set of points that lie in a query range.

*rank 3 hyperline sequence* (associated with some vector $v \in V \subseteq \mathcal{R}^3$):  an alternating circular sequence of subsets of indices, obtained by rotating an oriented central plane in counterclockwise order around the line through $v$.

*ray*:  a half-line that is directed away from its endpoint.

*ray shooting problem*:  the problem of determining the first object in a set of geometric objects that is hit by a query ray.

*real random access machine* (*real RAM*):  a model of computation in which values can be arbitrarily long real numbers, and all standard operations such as $+, -, \times$, and $\div$ can be performed in unit time regardless of operand length.

*realizable* (pseudoline arrangement):  a pseudoline arrangement isomorphic to a line arrangement.

*reflex edges*:  edges of a nonconvex 3-polyhedron that subtend an inner dihedral angle greater than $180°$.

*regular* (polygon):  a polygon with all sides congruent and all interior angles equal.

*regular* (polytope):  a $d$-polytope ($d > 0$) with all its facets regular $(d-1)$-polytopes that are combinatorially equivalent; a regular 0-polytope is a vertex.

*regular* (tiling):  a monohedral tiling of the plane with a regular polygon as prototile.

*reorientation* (of a vector configuration $V = \{v_1, \ldots, v_n\}$):  a vector configuration $V' = \{v'_1, \ldots, v'_n\}$ such that each $v'_i$ is equal to $v_i$ or $-v_i$.

*semiregular* (polyhedron):  a convex polyhedron with each face a regular polygon, but where more than one regular polygon can be used as a face.

*semiregular* (tiling):  a tiling of the plane using $n$ prototiles with the same numbers of polygons around each vertex.

*semi-space* (of a configuration induced by a hyperplane):  the set of indices of the configuration lying on one side of the hyperplane.

*semi-space* (of a vector or point configuration):  a semi-space induced by some hyperplane.

***k-set*** (of a point configuration): a semi-space of the configuration of size $k$.

***d-dimensional simplex*** (or ***d-simplex***): a $d$-polytope with $d + 1$ vertices.

***simplicial complex***: a triangulation of a polyhedron such that for any two simplices in the triangulation, either the intersection of the simplices is empty or is a face of both simplices.

***simplicial polytope***: a polytope in which all faces are simplices.

(***standard affine***) ***pseudo polar-duality***: the association between an $x$-monotone pseudoline arrangement $L = \{l_1, \ldots, l_n\}$ given in slope order and a pseudo configuration of points $(P, L')$, $P = \{p_1, \ldots, p_n\}$, given in increasing order of the $x$-coordinates and with $L'$ being $x$-monotone, satisfying the property that the cluster of stars associated with $L$ and to $P$ are the same.

***straight-line dual***: given the Voronoi diagram of a set $\{p_1, \ldots, p_n\}$ of points in the plane, the planar straight-line graph whose vertices are the points in the set, with two vertices $p_i$ and $p_j$ adjacent if and only if the regions $V(p_i)$ and $V(p_j)$ share a common edge.

***strictly convex***: the property of a convex set that its boundary contains no line segment.

***symmetry*** (of a tiling): a Euclidean motion that maps each tile of the tiling onto a tile of the tiling.

***tile***: an element of a tiling.

***tiling*** (of Euclidean $d$-space): a countable family $\mathcal{T}$ of closed topological $d$-cells of $\mathcal{R}^d$ that cover $\mathcal{R}^d$ without gaps and overlaps.

***triangulation*** (of a $d$-polyhedron): a convex decomposition in which each convex piece of the decomposition is a $d$-simplex.

***triangulation*** (of a simple polygon): an augmentation of the polygon with nonintersecting diagonal edges connecting vertices of the polygon such that in the resulting planar straight-line graph every bounded face is a triangle.

***uniform chirotope***: a chirotope function that takes nonzero values on all $d$-tuples.

***upper density*** (of a covering $\mathcal{C}$): $\overline{\nu}(\mathcal{C}) = \limsup\limits_{R \to +\infty} \frac{\sum_{K_i \cap B_R \neq \emptyset} \text{Vol}(K_i)}{\text{Vol}(B_R)}$, where each $K_i$ is a convex body in the covering $\mathcal{C}$ of $d$-dimensional Euclidean space and $B_R$ is the closed ball of radius $R$ centered at the origin.

***upper density*** (of a packing $\mathcal{P}$): $\overline{\delta}(\mathcal{P}) = \limsup\limits_{R \to +\infty} \frac{\sum_{K_i \subset B_R} \text{Vol}(K_i)}{\text{Vol}(B_R)}$, where each $K_i$ is a convex body in the packing $\mathcal{P}$ of $d$-dimensional Euclidean space and $B_R$ is the closed ball of radius $R$ centered at the origin.

***upper envelope*** (of a nonvertical line arrangement): the half-plane intersection of the half-planes above the lines of the arrangement.

***vector configuration*** (of dimension $d$): a finite set of vectors spanning $\mathcal{R}^d$.

***visibility graph***: given $n$ nonintersecting line segments in the plane, the graph whose vertices are the endpoints of the line segments, with two vertices adjacent if and only if they are visible from each other.

***visibility problem***: the problem of finding what is visible, given a configuration of objects and a viewpoint.

**Voronoi cell** (with center $c_i$):  the convex polyhedral set $V_i = \{\, x \in \mathcal{R}^d : |x - c_i| = \min_j |x - c_j| \,\}$, where $c_1, c_2, \ldots$ are centers of unit balls in a packing of $d$-dimensional Euclidean space.

**Voronoi diagram** (of the set of points $\{p_1, \ldots, p_n\}$ in $d$-dimensional Euclidean space): the partition of $d$-dimensional Euclidean space into convex polytopes $V(p_i)$ such that $V(p_i)$ is the locus of points that are closer to $p_i$ than to any other point in $p_j$.

**zone** (of a line in an arrangement):  the set of cells of the arrangement intersected by the line.

**zonotope**:  the vector (Minkowski) sum of a finite number of line segments.

# 13.1   ARRANGEMENTS OF GEOMETRIC OBJECTS

A wide range of applied fields (statistics, computer graphics, robotics, geographical databases) depend on solutions to geometric problems: polygon intersection, visibility computations, range searching, shortest paths among obstacles, just to name a few. These problems typically start with "consider a finite set of points (or lines, segments, curves, hyperplanes, polygons, polyhedra, etc.)". The *combinatorial* properties of these sets, or arrangements, of objects (incidence, order, partitioning, separation, convexity) set the foundations for the *algorithms* developed in the field of computational geometry.

In this chapter attention is focused on the most studied and best understood arrangements of geometric objects: points, lines, and hyperplanes. Introducing the concepts relies on linear algebra. The combinatorial properties studied belong, however, to a relatively new field, the theory of oriented matroids, which has sometimes been described as *linear algebra without coordinates.*

Several fundamental types of questions are asked about these arrangements. The most basic is the classification problem, whose goal is to find combinatorial parameters allowing the partitioning of the (uncountable) set of all possible arrangements of $n$ objects into a finite number of equivalence classes. Examples of such structures for point and line arrangements include semi-spaces, Radon partitions, chirotopes, hyperline sequences, etc. They satisfy simple properties known as axiomatic systems for oriented matroids, which lead to the definition of an abstract class of objects generalizing finite point and vector sets. In dimension 2, oriented matroids can be visualized topologically as pseudoline arrangements. The numerous definitions needed to introduce arrangements and oriented matroids will be complemented in this section by the most important facts, such as counting the number of finite point, line, and pseudoline arrangements, deciding when a pseudoline arrangement is equivalent to a line arrangement, and basic algorithmic results.

## 13.1.1   POINT CONFIGURATIONS

The simplest geometric objects are points in some $d$-dimensional space. Most of the other objects of interest for applications of computational geometry (sets of segments, polygons, polyhedra) are built on top of, and inherit, geometric structure from sets of points.

The setting for computational geometry problems is the Euclidean (affine) space $\mathcal{R}^d$ and most of its fundamental concepts (convexity, proximity) belong here in a natural way. However, some standard techniques, such as polarity and duality, as well as the abstraction to oriented matroids, are better explained in the context of vector spaces.

Several categories of concepts are introduced in this section, and these are developed and used in the subsequent subsections: vector and point configurations, hyperplanes and half-spaces, convexity, and some combinatorial parameters associated with vector or point configurations, relevant to applications (in statistics, pattern recognition or computational geometry): signed bases, semi-spaces, $k$-sets, centerpoints.

### Definitions:

The (**standard**) **real vector space** of dimension $d$ is the vector space $\mathcal{R}^d = \{\underline{x} \mid \underline{x} = (x_1, \ldots, x_d), x_i \in \mathcal{R}\}$, with vector addition $\underline{x} + \underline{y} = \{x_1 + y_1, \ldots, x_d + y_d\}$ and scalar multiplication $\alpha \underline{x} = \{\alpha x_1, \ldots, \alpha x_d\}$. A vector in $\mathcal{R}^d$ is a **$d$-dimensional vector**.

A **linear combination** of a set of vectors $\{v_1, \ldots, v_n\}$ is a vector of the form $\sum_{i=1}^n \alpha_i v_i$, for coefficients $\alpha_1, \ldots, \alpha_n \in \mathcal{R}$.

A **linearly independent set** of vectors is a set of vectors $\{v_1, \ldots, v_k\}$ such that a linear combination of them equals the zero vector ($\sum_{i=1}^k \alpha_i v_i = \underline{0}$) if and only if $\alpha_i = 0$ for all $i = 1, \ldots, k$.

A **basis** of $\mathcal{R}^d$ is a maximal set of linearly independent vectors, i.e., one that is no longer independent if a new element is added.

A basis is an **ordered basis** if it is given as an ordered set.

The **sign** of an ordered basis is the sign of the determinant of the $d \times d$ matrix with columns given in order by the vectors of the ordered basis.

A **linearly dependent set** of vectors $V = \{v_1, \ldots, v_k\}$ is a set of vectors for which there exists a linear combination with at least one nonzero coefficient yielding the $\underline{0}$ vector; i.e., $\sum_{i=1}^k \alpha_i v_i = \underline{0}$ with some $\alpha_i \neq 0$.

The **linear space** spanned by a set of vectors $V = \{v_1, \ldots, v_k\}, v_i \in \mathcal{R}^d$, is the set of all linear combinations of vectors of $V$.

A **linear $k$-dimensional subspace** of $\mathcal{R}^d$ ($k \leq d$) is the set of all linear combinations of $k$ linearly independent vectors $v_1, \ldots, v_k$ in $\mathcal{R}^d$.

A **line** through $v \in \mathcal{R}^d$ is the 1-dimensional linear subspace of $\mathcal{R}^d$ induced by $v \neq \underline{0}$.

**Euclidean space of dimension $d$** is $\mathcal{R}^d$ seen as an affine space. It is sometimes identified with the $d$-dimensional affine hyperplane $x_{d+1} = 1$ in $\mathcal{R}^{d+1}$.

A (**$d$-dimensional**) **point** is an element of $\mathcal{R}^d$ seen as a Euclidean space.

An **affine combination** of a set of points $\{p_1, \ldots, p_n\}$ is a point of the form $\sum_{i=1}^n \alpha_i p_i$, with $\alpha_i \in \mathcal{R}$ and $\sum_{i=1}^n \alpha_i = 0$.

An **affinely independent set of points** is a set of points $\{p_1, \ldots, p_k\}$ such that no point is an affine combination of the others.

A **simplex** of $\mathcal{R}^d$ is a maximal set of affinely independent vectors. It is an **ordered simplex** if it is given as an ordered set.

The **extended matrix** of an ordered simplex $\{p_1, \ldots, p_{d+1}\}$ is the $(d+1) \times (d+1)$ matrix with its $i$th column equal to $(p_i, 1)$.

The **sign** of an ordered simplex is the sign of the determinant of the extended matrix of the simplex.

An **affinely dependent** set of points $P = \{p_1, \ldots, p_k\}$ is a set of points such that one of the points is an affine combination of the others.

The **affine space** spanned by a set of points $P = \{p_1, \ldots, p_k\}$, with $p_i \in \mathcal{R}^d$, is the set of all affine combinations of points of $P$. It is an affine subspace of $\mathcal{R}^d$.

The **affine $k$-dimensional subspace** of $\mathcal{R}^d$ ($k \leq d$) is the set of all affine combinations of $k$ affinely independent points $p_1, \ldots, p_k$ in $\mathcal{R}^d$.

A **linear function** is a function $h \colon \mathcal{R}^d \to \mathcal{R}$ such that $h(x_1, \ldots, x_d) = \sum_{i=1}^{d} a_i x_i + a_{d+1}$.

A linear function is **homogeneous** if $a_{d+1} = 0$.

The **affine hyperplane** induced by a linear function $h$ is the set $h^0 = \{x \in \mathcal{R}^d \mid h(x) = 0\}$.

An affine hyperplane is a **central hyperplane** if $h$ is a homogeneous linear function.

An **oriented hyperplane** is a hyperplane, together with a choice of a positive side for the hyperplane. This amounts to choosing a (homogeneous or affine) linear function $h$ to generate it, together with all those of the form $\alpha h$, $\alpha > 0$.

A **reorientation** of an oriented hyperplane is a swapping of the negative and positive sides of the hyperplane (or, changing the generating linear function $h$ to $-h$).

The **open half-spaces** induced by an oriented hyperplane $h$ are $h^+ = \{\underline{x} \mid h(x) > 0\}$ (the **positive side** of $h^0$) and $h^- = \{x \mid h(x) < 0\}$ (the **negative side** of $h^0$). The sets $h^+$, $h^0$, and $h^-$ form a partition of $\mathcal{R}^d$: $\mathcal{R}^d = h^+ \cup h^- \cup h^0$, and $h^+$, $h^-$, and $h^0$ are pairwise disjoint.

The **closed half-spaces** induced by $h$ are $h^+ \cup h^0$ and $h^- \cup h^0$.

A **convex combination** of a set of points $\{p_1, \ldots, p_n\}$ is a point of the form $\sum_{i=1}^{n} \alpha_i p_i$ with $\alpha_i \in \mathcal{R}$, $\alpha_i > 0$, and $\sum_{i=1}^{n} \alpha_i = 1$.

The **segment** with endpoints $p_1 \neq p_2$ is the set of all convex combinations of $p_1$ and $p_2$.

A set of points $\{p_1, \ldots, p_k\}$ is **convexly independent** if no point is a convex combination of the others. The points are also said to be in **convex position**.

A **convex set** in $\mathcal{R}^d$ is a set $S \subseteq \mathcal{R}^d$ such that if $p_1$ and $p_2$ are distinct points in $S$, then the segment with endpoints $p_1$ and $p_2$ is contained in $S$.

The **convex hull** of a finite set of points $P$ is the set of all convex combinations of points of $P$.

A **convex polytope** is the convex hull of a finite set of points. Its boundary consists of faces of dimension 0 (*vertices*), 1 (*edges*), ..., $d - 1$ (*facets*).

The **face description** of a convex polytope is a data structure storing information about all the faces and their incidences.

A **convex polygon** is the convex hull of a finite set of points in $\mathcal{R}^2$.

A **vector configuration** [**point configuration**] of dimension $d$ is a finite set of $n$ vectors $\{v_1, \ldots, v_n\}$ ($v_i \in \mathcal{R}^d$) spanning $\mathcal{R}^d$ [points $\{p_1, \ldots, p_n\}$ ($p_i \in \mathcal{R}^d$) affinely spanning $\mathcal{R}^d$].

A configuration is **labeled** if its elements are given as an ordered set. (It may be given as the set of columns of a $d \times n$ matrix with real entries.)

The **rank** of a vector configuration [point configuration] in $\mathcal{R}^d$ is the number $d$ $[d+1]$.

A set of vectors [points] in $\mathcal{R}^d$ is in **general position** if every subset of $d$ elements is a basis [simplex].

An **affine configuration** (or **acyclic vector configuration**) is a configuration with a central hyperplane containing all the vectors of the configuration on one side.

A **reorientation** of a vector configuration $V = \{v_1, \ldots, v_n\}$ is a vector configuration $V' = \{v'_1, \ldots, v'_n\}$ with each $v'_i$ equal to either $v_i$ or $-v_i$.

A **reorientation class** is the set of all labeled vector configurations which are reorientation equivalent.

A **point configuration** $P \subset \mathcal{R}^d$ **induced by an acyclic vector configuration** $V \subset \mathcal{R}^{d+1}$ contained in a half-space $h^+$ is the set of all points $p$ obtained as follows: take the affine plane $h'$ in $h^+$ parallel to $h$ and tangent to the unit sphere $S^d$ in $\mathcal{R}^{d+1}$; the intersection of the line through vector $v \in V$ with the plane $h'$ is a point $p \in h'$. $\mathcal{R}^d$ is identified with the affine plane $h'$.

A **semi-space** of a vector configuration [point configuration] $V$ induced by an oriented central hyperplane [affine hyperplane] $h$ is the set of indices of the elements in $V$ lying on one side of the hyperplane.

A **semi-space of a vector configuration** [**point configuration**] $V$ is a semi-space induced by some hyperplane $h$.

The **order type** of a vector or point configuration $V$ is the collection of all semi-spaces of $V$.

**Isomorphic** vector or point configurations are configurations having the same order type, after possibly relabeling their elements.

A **k-set** of a point configuration $P$ is a semi-space of $P$ of size $k$ $(0 \leq k \leq n)$.

A **centerpoint** of a point configuration $P$ of size $n$ is a point $q$, not necessarily in $P$, such that if $h$ is a hyperplane containing $q$ there are at least $\lceil \frac{n}{d+1} \rceil$ points in each semi-space induced by $h$.

A **ham-sandwich cut** is a hyperplane that simultaneously bisects $d$ point configurations $P_1, P_2, \ldots, P_d$ in $\mathcal{R}^d$.

## Facts:

**1.** A basis of the vector space $\mathcal{R}^d$ has $d$ elements and a simplex of the affine space $\mathcal{R}^d$ has $d+1$ elements.

**2.** The rank of the $d \times n$ matrix associated with an $n$-vector configuration in $\mathcal{R}^d$ is $d$; the rank of the matrix associated with a point configuration in $\mathcal{R}^d$, extended with a row of 1s, is $d+1$.

**3.** The determinant of a $d \times d$ matrix whose columns are a basis of $\mathcal{R}^d$ and the determinant of the $(d+1) \times (d+1)$ extended matrix of a simplex in $\mathcal{R}^d$ are nonnull.

**4.** If $\{v_1, \ldots, v_d\}$ is a basis, then for any vector $v_{d+1}$, $\{v_1, \ldots, v_d, v_{d+1}\}$ is linearly dependent.

**5.** The intersection of linear subspaces [affine subspaces, convex subspaces] of $\mathcal{R}^d$ is linear [affine, convex].

**6.** The intersection of $k$ central hyperplanes [affine hyperplanes] in $\mathcal{R}^d$ $(k \leq d)$ is a linear subspace [affine subspace]. Its dimension is at least $d-k$.

**7.** Every affine subspace of $\mathcal{R}^d$ is a convex set.

**8.** *Carathéodory's theorem*:   Each point in the convex hull of a set of points $P \subset \mathcal{R}^d$ lies in the convex hull of a subset of $P$ with at most $d+1$ points.

**9.** *Radon's theorem*:   Each set of at least $d+2$ points in $\mathcal{R}^d$ can be partitioned into two disjoint sets whose convex hulls intersect in a nonempty set.

**10.** *Helly's theorem*: Let $\mathcal{S}$ be a finite family of $n$ convex sets in $\mathcal{R}^d$ ($n \geq d+1$). If every subfamily of $d+1$ sets in $\mathcal{S}$ has a nonempty intersection, then there is a point common to all the sets in $\mathcal{S}$.

**11.** Every point configuration admits a centerpoint.

**12.** For every $d$ configurations of points in $\mathcal{R}^d$, there exists a ham-sandwich cut.

**13.** *Upper bound theorem*: The number of facets of the convex hull of an $n$-point configuration in $\mathcal{R}^d$ is $O(n^{\lfloor d/2 \rfloor})$. This bound is obtained for configurations of points on the moment curve $P = \{p(t) \mid t \in \{t_1, \ldots, t_n\} \subseteq \mathcal{R}\}$, where $p(t) = (t, t^2, \ldots, t^d) \in \mathcal{R}^d$.

**14.** The number of semi-spaces of a rank $d+1$ vector or point configuration of $n$ elements is $O(n^d)$. The maximum is attained for points in general position.

**15.** For $d = 2$, let $e_k(n)$ denote the maximum number of $k$-sets of any planar $n$-point configuration. Then $\Omega(n \log k) \leq e_k(n) \leq O(nk^{\frac{1}{3}})$.

**16.** *Erdős-Szekeres problem*:   If $c(k)$ is the maximum number of planar points in general position such that no $k$ are in convex position, then $2^{k-2} \leq c(k) \leq \binom{2n-4}{n-2}$.

**17.** The face description of the convex hull of a point set of size $n$ in $\mathcal{R}^d$ can be computed optimally in $O(n \log n)$ time if $d = 2$ or $d = 3$, and $O(n^{\lfloor \frac{d}{2} \rfloor})$ time for $d > 3$.

**18.** A ham-sandwich cut in dimension 2 can be found in linear time.

## Examples:

**1.** The configuration of points in the following figure is given by the columns of the matrix $\begin{pmatrix} 0 & -1 & 3 & 1 & 2 \\ 0 & 1 & 0 & 1 & 2 \end{pmatrix}$. It is not in general position: the three points 1, 4, and 5 are collinear. The extended matrix is $\begin{pmatrix} 0 & -1 & 3 & 1 & 2 \\ 0 & 1 & 0 & 1 & 2 \\ 1 & 1 & 1 & 1 & 1 \end{pmatrix}$. Because $\det \begin{pmatrix} 0 & -1 & 3 \\ 0 & 1 & 0 \\ 1 & 1 & 1 \end{pmatrix} < 0$, the simplex 123 is negative. Some semi-sets are: $\{3\}$ (1-set), $\{2,5\}$ (2-set), $\{1,3,4\}$ (3-set), etc. The convex hull is $\{1, 3, 5, 2\}$.

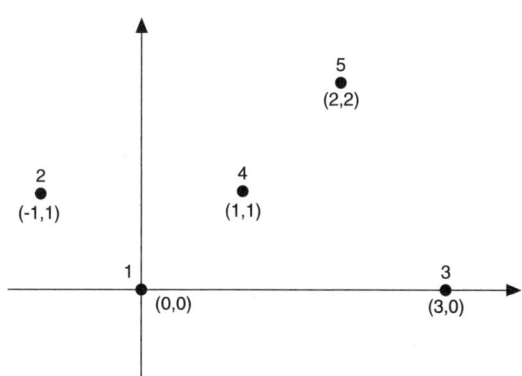

**2.** The two configurations of points from the figure of Example 1 and part (a) of the following figure are isomorphic, but those in parts (a) and (b) of the following figure are not. This can be seen because, for example, they have different numbers of points on their convex hulls.

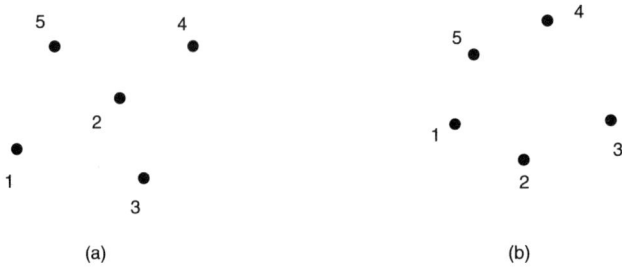

(a)                                    (b)

**3.** The grey point in the following figure is a centerpoint of the point configuration of black points. Some of the separating lines have been shown: they have at least $\frac{1}{3}$ of the points on each side.

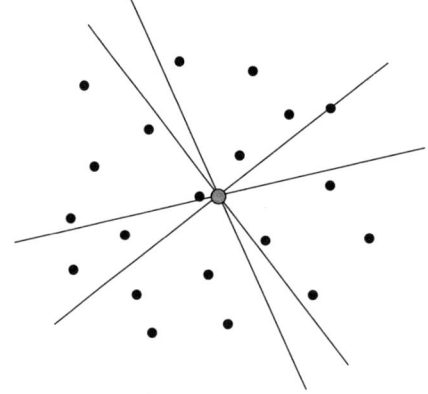

**4.** The line in the following figure is a ham-sandwich cut: it simultaneously bisects the black and the white points.

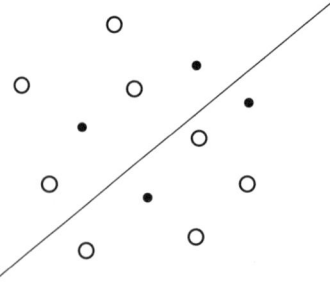

### 13.1.2   LINE AND HYPERPLANE ARRANGEMENTS

Line arrangements and affine point configurations in the plane are related via polar-duality, a transformation which is better understood in terms of 3-dimensional vectors and central planes or using the projective and spherical models. Several types of combinatorial data can be directly translated from the primal setting to the polar. As a consequence, theorems and algorithms on line arrangements follow directly from their counterparts on point configurations, and vice versa. This powerful tool has been used successfully in computational geometry for the design of efficient algorithms. It also generalizes to higher dimensions, where hyperplane arrangements are polar-dual to point configurations in $\mathcal{R}^d$.

**Definitions:**

A (**Euclidean**) **line in** $\mathcal{R}^2$ is an affine subspace of dimension 1. A line is induced by a linear function $l(x, y) = ax + by + c$ and any of its multiples of the form $\alpha l$. A line is **oriented** if a direction has been chosen for it. Its induced half-spaces are called half-planes and denoted by $l^+$ and $l^-$.

A **nonvertical line** is a line given by an equation of the form $y = ax + b$, where $a$ is called the **slope** and $b$ the $y$-intercept of the line. It is oriented in increasing order of the $x$-coordinates of its points and its induced half-planes are above/below it.

A (**Euclidean**) **line arrangement** is a set $\mathcal{L} = \{l_1, \ldots, l_n\}$ of planar lines, not all of them going through the same point; that is, $\bigcap_{i=1}^{n} l_i = \emptyset$. If the lines are oriented, this is an **arrangement of oriented lines**. It is **labeled** if it is given as an ordered set.

A line arrangement is in **general position** if no three lines have a point in common.

An **x-monotone curve** is a curve intersecting each vertical line in exactly one point.

A **half-plane intersection** is the planar region lying in the intersection of a finite set of half-planes. It is described by the (circular) list of the lines incident to its boundary.

An **upper envelope** [**lower envelope**] of a nonvertical line arrangement is the boundary of the half-plane intersection of the half-planes above [below] the lines of the arrangement.

The **k-level** in a nonvertical arrangement of $n$ lines ($1 \le k \le n$) is the lower boundary of the set of points in $\mathcal{R}^2$ having exactly $k$ lines above and $n - k$ below.

The **cell of a line arrangement** is a connected component of the complement in $\mathcal{R}^2$ of the union of the points on the lines.

The **zone of a line** $l$ **in an arrangement** $\mathcal{L}$ ($l \notin \mathcal{L}$) is the set of cells of the arrangement $\mathcal{L}$ intersected by $l$.

A **central plane arrangement** in $\mathcal{R}^3$ is a finite set of central planes. The arrangement is **oriented** if the planes are oriented.

An **acyclic** (or **affine**) **central plane arrangement** in $\mathcal{R}^3$ is a central plane arrangement such that there is a point in $\mathcal{R}^3$ that lies on the positive side of all these planes.

The (**standard**) **line arrangement** induced by a central plane arrangement is the arrangement of the lines of intersection of the central planes with an affine plane $h$ in $\mathcal{R}^3$ that is not parallel with any plane of the arrangement. If the central planes are oriented, an orientation is induced on the lines by keeping the positive side of a line within the positive side of the corresponding plane.

A **big circle** is the intersection of the unit sphere $S^2$ in $\mathcal{R}^3$ with a central plane. If the plane is oriented, the circle is given an orientation so that the positive side of the plane lies on the left of the circle.

A **big-circle arrangement** is the intersection of a central plane arrangement with the unit sphere $S^2$ in $\mathcal{R}^3$. It is oriented if the planes are oriented.

The **big-circle arrangement induced by a central plane arrangement** is the arrangement of the big circles of intersection of the central planes with the sphere $S^2$ in $\mathcal{R}^3$. It is oriented if the planes are oriented.

The **projective plane $P^2$** is the sphere $S^2$ in $\mathcal{R}^3$ with the antipodal points identified.

A **projective line** is the projective curve induced by identifying the antipodal points of a big circle on $S^2$.

A **projective line arrangement** is a finite set of projective lines in the projective plane $P^2$.

The **projective line arrangement induced by a central plane arrangement** is the projective arrangement obtained by the antipodal point identification of the big-circle arrangement on $S^2$ induced by the central plane arrangement.

An **arrangement graph** is a graph associated with a Euclidean or projective line arrangement, or a big-circle arrangement. Its **vertices** correspond to intersection points of lines (or circles) and its **edges** correspond to line (or arc) segments between two intersection points.

*Note:* For the Euclidean case, typically only the bounded line segments are considered as edges (but by adding extra dummy vertices "at infinity", the infinite extremities of each line among the edges can be included). For the Euclidean or spherical case, if the lines are oriented, the arrangement graph is directed, with the edges oriented to be compatible with the orientation of the lines or circles.

**Isomorphic arrangements** are arrangements having isomorphic arrangement graphs. (This applies to Euclidean lines, big-circles (oriented or not), and projective lines.)

A **polar-duality of vectors and central planes** in $\mathcal{R}^3$ is a mapping $\mathcal{D}$ associating a vector $v \in \mathcal{R}^3$ with an oriented central plane having $v$ as its normal vector, and vice versa.

A **polar-duality of points and lines** in the affine space $\mathcal{R}^2$ is any mapping $\mathcal{D}$ associating a point $p \in \mathcal{R}^2$ with an oriented line $l$ in $\mathcal{R}^2$ and vice versa, by the following general procedure: map the points to vectors via some imbedding of $\mathcal{R}^2$ as an affine plane in $\mathcal{R}^3$, apply the polar-duality of vectors and central planes, and then intersect the polar central planes with some affine plane (identified with $\mathcal{R}^2$) to get lines.

A (**standard**) **affine polar-duality** is a mapping $\mathcal{D}$ between nonvertical lines and points in $\mathcal{R}^2$, associating the point $(a, -b)$ with the line $y = ax + b$, and vice versa.

A **Euclidean hyperplane [central hyperplane] arrangement** in $\mathcal{R}^d$ is a finite set $\mathcal{H} = \{h_1, \ldots, h_n\}$ of affine hyperplanes [central hyperplanes], not all of them going through the same point. If the hyperplanes are oriented, the arrangement is oriented.

The following are generalizations to an arbitrary affine space $\mathcal{R}^d$ [vector space $\mathcal{R}^{d+1}$] of previously defined concepts in affine dimension 2 [vector space $\mathcal{R}^3$]:

- **arrangements of big $(d-1)$-spheres** on $S^d$ generalize big-circle arrangements on $S^2$;

- **projective arrangements** of hyperplanes in $P^d$ generalize projective arrangements of lines in $P^2$;

- the **polar-duality** between vectors and central hyperplanes in $\mathcal{R}^{d+1}$ associates with a vector the hyperplane normal to it;

- the **face lattice** of a hyperplane (central, affine, projective) or sphere arrangement, a data structure storing information on faces and their incidences, generalizes the arrangement graph, and is used to define isomorphism of arrangements.

- the **k-level** in an affine arrangement of nonvertical hyperplanes is the lower boundary of the set of points having exactly $k$ hyperplanes above them.

**Facts:**

**1.** A bounded cell in a Euclidean line arrangement is a convex polygon.

**2.** The $k$-level of a nonvertical line arrangement is an $x$-monotone piecewise linear curve incident with vertices and lines of the arrangement.

**3.** The upper envelope is the 0-level of an arrangement of nonvertical lines.

**4.** In a simple big-circle arrangement, every pair of big circles intersects in exactly two points, which are antipodal on the sphere.

**5.** The arrangement graphs of planar line arrangements or spherical big-circle arrangements are planar imbedded graphs. The arrangement graph of a projective line arrangement is projective-planar. The *faces* or *cells* of these graphs are the connected components of the complement of the union of lines or circles.

**6.** The association among central plane arrangements, big-circle arrangements, and projective arrangements preserves isomorphisms. The standard association of an affine line arrangement and big-circle arrangement to an acyclic plane arrangement preserves isomorphisms.

**7.** In a simple Euclidean arrangement of $n$ lines, the number of vertices is $\binom{n}{2}$, the number of segments (bounded or unbounded) is $n^2$, and the number of cells (bounded or unbounded) is $\binom{n}{2} + n + 1$. No nonsimple arrangement exceeds these values.

**8.** *Zone theorem:* The total number of edges (bounded or unbounded) in a zone of an arrangement of $n$ lines is at most $6n$.

**9.** $\mathcal{D}(\mathcal{D}(v)) = v$ and $\mathcal{D}(\mathcal{D}(h)) = h$, for every vector $v$ and hyperplane $h$.

**10.** *Incidence preserving:* If $v \in h$, then $\mathcal{D}(h) \in \mathcal{D}(v)$, for every vector $v$ and hyperplane $h$.

**11.** *Orientation preserving:* If $v \in h^+$, then $\mathcal{D}(h) \in \mathcal{D}(v)^+$, for every vector $v$ and hyperplane $h$.

**12.** *Basic properties of the standard affine polar-dual transformation:*

- *polar-duality preserves above/below properties:* if a point $p$ is above line $l$, then the polar line $\mathcal{D}(p)$ is below the polar point $\mathcal{D}(l)$;

- the polar-dual of a configuration of points $P$ is an arrangement of nonvertical lines $\mathcal{L} = \mathcal{D}(P)$, and vice versa;

- the polar-dual of a set of points given in increasing order of their $x$-coordinates is a set of lines given in increasing order of their slopes;

- the polar-dual of the set of points on the convex hull of $P$ is the set of lines on the upper and lower envelopes of the polar arrangement $\mathcal{D}(P)$; convex hull computation dualizes to half-plane intersection;

- semi-spaces of $P$ dualize to vertices, edges, and cells of the polar arrangement $\mathcal{D}(P)$;

- isomorphic arrangements of lines dualize to isomorphic configurations of points;

- the polar-duals of lines $p_i p_j$ inducing $(k-1)$-sets and $k$-sets in a point configuration $P = \{p_1, \ldots, p_n\}$ are the vertices $l_i \cap l_j$ on levels $k$ and $n-k$ of the polar-dual arrangement $L = \{l_1, \ldots, l_n\}$.

**13.** The polar-dual of an acyclic vector configuration in $\mathcal{R}^3$ is an acyclic central plane arrangement.

**14.** The upper envelope of a line arrangement can be computed optimally in $O(n \log n)$ time.

**15.** The arrangement graph of a line arrangement can be computed in $O(n^2)$ time and space.

**16.** The face incidence lattice of a hyperplane arrangement in $\mathcal{R}^d$ can be computed in $O(n^d)$ time and space.

**17.** The standard polar-duality is ubiquitously used in computational geometry. For example, it is used to derive algorithms for half-plane intersection from convex hull algorithms, to translate between line slope and point $x$-coordinate selection, and to compute the visibility graph in $O(n^2)$ time using the polar-dual arrangement graph. See [Ed87] (Chapter 12) for a collection of such problems.

**18.** In computational geometry, $k$-levels are related to the furthest $k$-neighbors Voronoi diagrams via a lifting transformation that reduces the computation of dimension $d$ Voronoi diagrams to dimension $d+1$ arrangements of hyperplanes.

**19.** $k$-levels in arrangements, as polar-duals to $k$-sets, have an abundance of applications in statistics.

### Examples:

**1.** The following figure shows a line arrangement in general position. The arrangement graph has 10 vertices corresponding to the black points (which could be labeled with pairs of indices of lines such as $12, 13$, etc.), 15 edges corresponding to the bounded line segments such as $(12, 13)$, and $2 \times 5 = 10$ unbounded edges. The upper envelope (0-level), whose list of lines is $\{1, 2, 3, 5\}$, is shown highlighted. The dashed piecewise linear curve is the 2-level.

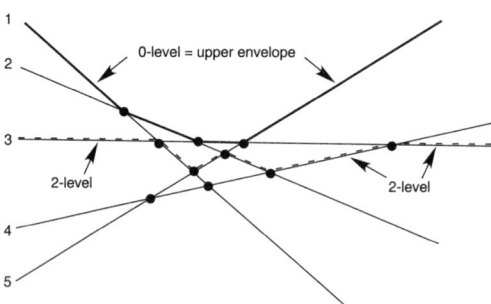

**2.** The zone of line 3 in the line arrangement $\{1, 2, 4, 5\}$ from the figure of Example 1 is depicted in the following figure. It has five cells (one bounded, four unbounded), whose boundaries sum up to 12 segments.

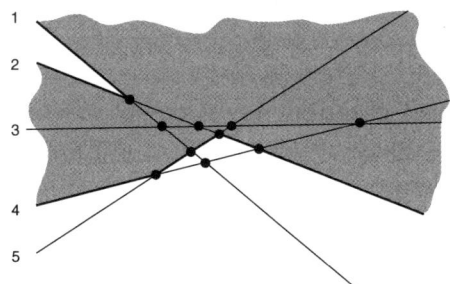

**3.** The line arrangements in the figure for Example 1 and part (a) of the following figure are isomorphic. Those in parts (a) and (b) of the following figure are not isomorphic.

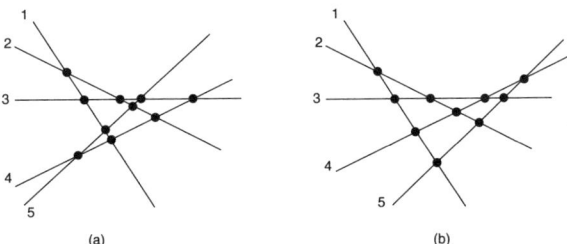

(a)                                                                                   (b)

**4.** The following figure illustrates the standard polar-duality. The arrangement is polar-dual to the configuration of points in the figure of §13.1.1, Example 1; hence the lines are given by the equations 1: $y = 0$, 2: $y = -x - 1$, 3: $y = 3x$, 4: $y = x - 1$ and 5: $y = 2x - 2$. In the primal configuration, point 2 is above line 13. In the polar-dual, line 2 is below the intersection point of lines 1 and 3.

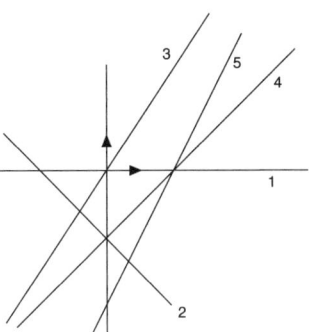

## 13.1.3   PSEUDOLINE ARRANGEMENTS

Pseudoline arrangements represent a natural generalization of line arrangements, retaining incidence and orientation properties, but not straightness. They provide a topological representation for rank 3 oriented matroids (see §13.1.4), which in turn abstract combinatorial properties of vector configurations and oriented line arrangements.

### Definitions:

A **pseudoline** is a planar curve (which may be given an orientation). It is **x-monotone** if the curve is $x$-monotone.

A **pseudoline arrangement** is a finite collection of simple planar curves that intersect pairwise in exactly one point, where they cross, and not all of which have a point in common. It is **labeled** if the pseudolines are given in a fixed order $\{l_1, \ldots, l_n\}$ and **oriented** if the pseudolines are oriented.

A pseudoline arrangement is **realizable** or **stretchable** if it is isomorphic to a line arrangement.

*Note:* The following terms, defined in §13.1.1 and §13.1.2 for line arrangements, have straightforward generalizations to pseudoline arrangements: open half-planes, general position, cell, arrangement graph, isomorphism, upper/lower envelope, $k$-level, zone.

**Facts:**

**1.** Every arrangement of pseudolines is isomorphic to an arrangement of $x$-monotone piecewise linear pseudolines (a wiring diagram such as shown in the figure of Example 2).

**2.** The arrangement graph of a pseudoline arrangement in general position has $\binom{n}{2}$ vertices, $n^2$ edges, and $\binom{n}{2} + n + 1$ faces.

**3.** The number of edges in a zone of a pseudoline in a pseudoline arrangement is at most $6n$.

**4.** Let $e_k(n)$ be the number of edges on the $k$-level of a pseudoline arrangement. Then $\Omega(n \log k) \le e_k(n) \le O(nk^{\frac{1}{3}})$.

**5.** The logarithm of the number of isomorphism classes of pseudoline arrangements is $\Theta(n^2)$. The same number for line arrangements is $\Theta(n \log n)$.

**6.** There exist nonstretchable pseudoline arrangements.

**7.** It is NP-hard to decide whether a pseudoline arrangement is stretchable.

**8.** Pseudoline stretchability is decidable in PSPACE.

**9.** Assume that a predicate is given for deciding when the intersection point of two pseudolines is above or below a third pseudoline. Then the algorithms for computing the upper envelopes, half-space intersection, or the arrangement graph of a line arrangement can be adapted to work for pseudoline arrangements.

**10.** In computational geometry, some algorithmic solutions can be found by reducing the problem to structures behaving like pseudolines — for example, computing the boundary of a union or intersection of unit circles in $\mathcal{R}^2$, all having at least one point in common.

**11.** It is an open problem whether better algorithms can be devised by making explicit use of the straightness of the lines in geometric problems. So far, the only problem where an explicit gap in efficiency between lines and pseudolines has been displayed is in the number of comparisons needed to sort the $x$-coordinates of the vertices of line versus $x$-monotone pseudoline arrangements.

**Examples:**

**1.** The pseudoline arrangement in the following figure is stretchable, because it is isomorphic to the line arrangement in the figure of §13.1.2, Example 1.

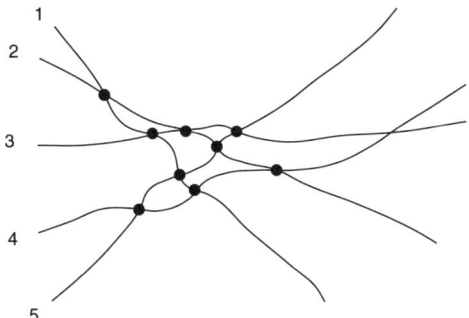

**2.** The following figure shows a standard way of representing a pseudoline arrangement as an $x$-monotone piecewise linear curve arrangement called a *wiring diagram*. The arrangement is the same as in the figure of Example 1.

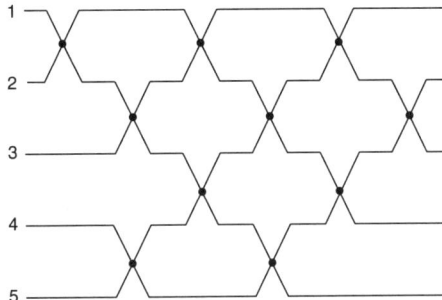

**3.** *A nonstretchable pseudoline arrangement*: The theorem of Pappus in plane geometry states that if the points 1, 2, 3 and 4, 5, 6 are collinear and in this order on two lines, then the three intersection points of the pairs of lines $7 = (15, 24)$, $8 = (16, 34)$, and $9 = (26, 35)$ are also collinear. See the following figure. The perturbed arrangement obtained by replacing the line through $7, 8, 9$ with the dashed pseudoline is not stretchable, since it violates the theorem of Pappus.

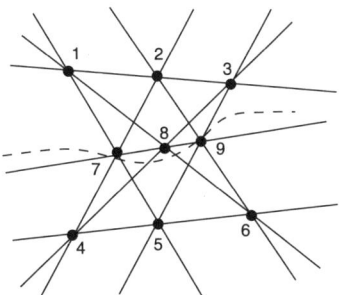

## 13.1.4  ORIENTED MATROIDS

General oriented matroids are abstractions of vector configurations and oriented hyperplane arrangements. Affine oriented matroids model the corresponding situation in affine spaces. They capture in various types of data structures (semi-spaces, chirotopes, Radon partitions, arrangement graphs, hyperline sequences) combinatorial information about $n$-element configurations. This forms the basis of the classification of all $n$-point sets into a finite number of equivalence classes. Each data structure satisfies a set of simple properties, or axioms, which characterize a wider class of objects collectively referred to as oriented matroids.

To simplify the exposition, in some cases only the axiomatization corresponding to points in general position will be presented. Not all oriented matroids arise geometrically from vector sets, but they do have a topological representation via pseudohyperplane arrangements. In rank 3, affine oriented matroids are modeled by pseudoline arrangements. Many geometric algorithms working with line arrangements or, by polarity, point configurations, make use of no more than oriented matroid properties and can be extended to pseudolines.

As potential applications, oriented matroids lay the foundations for a rigorous theory of geometric program verification and testing.

**Definitions:**

Let $E_n = \{1, \ldots, n\}$ and $\overline{E}_n = \{1, \ldots, n\} \cup \{\overline{1}, \ldots, \overline{n}\}$. Triplets $(i, j, k) \in \overline{E}_n^3$ are denoted $ijk$.

A **signed set** $X = (X^+, X^-)$ is a partition of a finite set $X$ into a **positive part** $X^+$ and a **negative part** $X^-$. That is, $X = X^+ \cup X^-$ and $X^+ \cap X^- = \emptyset$. In $E_n$ a signed set may be denoted as a signed sequence of indices, such as $1\overline{2}34$ for $(\{1, 3, 4\}, \{2\})$.

The **complement of a signed set** $X = (X^+, X^-)$ is the set $-X = (X^-, X^+)$.

The **support of a signed set** $X = (X^+, X^-)$ is the unsigned set $X$.

The **size of a signed set** $X = (X^+, X^-)$ is the size of the support of $X$.

The **signed double covering** of a finite set $X$ is the set $\overline{X} = X^+ \cup X^-$, where $X^+ = X$ and $X^- = \{\overline{x} \mid x \in X\}$ is a signed distinct copy of $X$ (its elements called *negated elements*), $X^+ \cap X^- = \emptyset$. If $x \in X^-$, then $\overline{x}$ is the corresponding nonnegated element in $X^+$.

A **basis** of a vector configuration [point configuration] $V \subset \mathcal{R}^d$ is a subset of $V$, identified by a $d$-set of indices, which is a basis [simplex] of the ambient space $\mathcal{R}^d$. A **signed basis** is an ordered basis together with its sign.

The **chirotope** of an ordered vector configuration [point configuration] is the set of all signed bases of $V$.

An **alternating function** is a function $f: E_n^d \to R$ such that the sign of $f(i_1, \ldots, i_d)$ is preserved under an even permutation and negated under an odd permutation of the $d$-tuple $(i_1, \ldots, i_d)$.

An **antisymmetric function** is a function $f: \overline{E}_n^d \to \mathcal{R}$ such that its sign changes when one of the parameters is negated. [For example, $f(\overline{i}_1, i_2, \ldots, i_d) = -f(i_1, i_2, \ldots, i_d)$.]

The (**rank 3**) **Grassmann-Plücker relations** are the identities

$$[1\,2\,3][1\,4\,5] - [1\,2\,4][1\,3\,5] + [1\,2\,5][1\,3\,4] = 0$$

satisfied by the determinants $[i\,j\,k] = \det(v_i, v_j, v_k)$, for any five vectors $v_i, 1 \le i \le 5$.

The (**rank $d$**) **Grassmann-Plücker relations** are the identities

$$[i_1 \ldots i_{d-2}1\,2][i_1 \ldots i_{d-2}3\,4] - [i_1 \ldots i_{d-2}1\,3][i_1 \ldots i_{d-2}2\,4] +$$
$$[i_1 \ldots i_{d-2}1\,4][i_1 \ldots i_{d-2}2\,3] = 0$$

satisfied by the determinants $[i_1 \ldots i_{d-2} j\, k] = \det(v_{i_1}, \ldots, v_{i_{d-2}}, v_j, v_k)$, for any $d+2$ vectors $v_{i_j}$ $(1 \le j \le d{-}2)$ and $v_i$ $(1 \le i \le 4)$.

**Chirotope axioms (rank $d$):**   A function $\chi \colon \overline{E}_n^d \to \{-1, 0, +1\}$ is a chirotope of rank $d$ if it satisfies the following conditions:

- $\chi$ is alternating and antisymmetric;

- for any $d+2$ generic points $i_1 \ldots i_{d-2} 1\,2\,3\,4$, the signs $\chi(i_1 \ldots i_{d-2} j\, k)$ of the six triplets involved in the Grassmann-Plücker relations are such that equality is possible.

A **uniform chirotope** is a chirotope function that takes nonzero values on all $d$-tuples.

**Chirotope axioms (uniform, rank 3):**   A function $\chi \colon \overline{E}_n^3 \to \{-1, +1\}$ is a uniform chirotope of rank 3 if it satisfies the following conditions:

- $\chi$ is alternating and antisymmetric;

- for any five generic points $a, b, i, j, k$, if $\chi(a\, b\, i) = \chi(a\, b\, j) = \chi(a\, b\, k) = +1$ and $\chi(a\, i\, j) = \chi(a\, j\, k) = +1$, then $\chi(a\, i\, k) = +1$.

The chirotope $\chi$ is **affine** if, in addition, it satisfies the following axiom:

- for any four points $i, j, k, l$, if $\chi(i\, j\, k) = \chi(i\, k\, l) = \chi(i\, l\, j) = +1$, then $\chi(j\, k\, l) = +1$.

An **oriented matroid given by a chirotope** is an abstract set of points labeled $\{1, \ldots, n\}$, together with a function $\chi$ satisfying the chirotope axioms.

The **circuit** of a set of labeled vectors $V = \{v_1, \ldots, v_n\}$ is the signed set $C = (C^+, C^-)$, where $C^+ = \{j \mid \alpha_j > 0\}$ and $C^- = \{j \mid \alpha_j < 0\}$, of indices of the nonnull coefficients $\alpha_j$ in a *minimal* linear dependency $V' = \{v_{i_1}, \ldots, v_{i_k}\}$ of $V$ with $\sum_{j=1}^{k} \alpha_j v_{i_j} = \underline{0}$. If $C$ is a circuit, its complement $-C$ is also a circuit.

A **Radon partition** of a set of labeled points $P$ is a signed set $C = (C^+, C^-)$ of points of $P$ such that the convex hull of the points in $C^+$ intersects the convex hull of the points in $C^-$.

A **minimal Radon partition** (or **circuit**) is a Radon partition whose support is minimal with respect to set inclusion.

An **oriented matroid** of an ordered vector [point] configuration given by its circuits is the set of all circuits of $V$.

A set $\mathcal{C}$ of signed subsets of $E_n$ satisfies the **circuit axioms** if

- $\emptyset \notin \mathcal{C}$;

- if $C \in \mathcal{C}$ then $-C \in \mathcal{C}$;

- (*minimality*): if $C = (C^+, C^-)$ is a circuit, then no subset of the support of $\mathcal{C}$ is the support of another circuit;

- (*exchange*): if $C_1$ and $C_2$ are two circuits such that $C_1 \ne -C_2$ and $e \in C_1^+ \cap C_2^-$, then there exists another circuit $D$ such that $e \notin D$, $D^+ \subset C_1^+ \cup C_2^+$, and $D^- \subset C_1^- \cup C_2^-$.

The **oriented matroid given by its circuits** is an abstract set $E_n$ together with a set of signed sets $\mathcal{C}$ satisfying the circuit axioms.

A **cocircuit** of a labeled vector configuration $V$ is a signed set $C = (C^+, C^-)$ of $E_n$, induced by a subset of $d-1$ vectors spanning a central hyperplane $h$. For an arbitrary orientation of $h$, $C^+$ is the set of indices of elements in $V$ lying in $h^+$ and $C^-$ is the set of indices of elements in $V$ lying in $h^-$.

The **cocircuit axioms** are the conditions obtained from the circuit axioms by replacing "circuit" with "cocircuit".

An **oriented matroid given by its cocircuits** is an abstract set $E_n$, together with a set of signed sets $\mathcal{C}$ satisfying the cocircuit axioms.

A **circular sequence** of period $k$ is a doubly infinite sequence $(q_i)_{i \in \mathcal{Z}}$ with $q_i = q_{i+k}$ for all $i \in \mathcal{Z}$.

A **signed permutation** of a set $S$ is a permutation of $S$ whose elements are also assigned a sign; for example, $1\bar{3}4\,2$, where 3 is negative and 1, 2, and 4 are positive.

An **alternating circular sequence** is a circular sequence $(q_i)_{i \in \mathcal{Z}}$ with half-period $k$, defined with elements from a signed double covering $q_i \in \overline{X}$ and satisfying the property $q_i = \bar{q}_{i+k}$ for all $i \in \mathcal{Z}$.

A **representation** of an alternating circular sequence can be obtained from any of its subsequences of $k$ consecutive elements (half period) $\{q_1, \ldots, q_k\}$.

A **star** (or **rank 3 hyperline sequence**) associated with a point $p_i \in P \subseteq \mathcal{R}^2$ [vector $v_i \in V \subseteq \mathcal{R}^3$] is an alternating circular sequence of subsets of indices in $\overline{E}_n$ obtained by rotating an oriented line [oriented central plane] in counterclockwise order around $p_i$ [the line through vector $v_i$] and recording the successive positions where it coincides with lines [central planes] defined by pairs of points $(p_i, p_j)$ with $p_j \in P \setminus \{p_i\}$ [vectors $(v_i, v_j)$, with $v_j \in V \setminus \{v_i\}$]. If a point $p_j$ is encountered by the rotating line in the positive direction from $p_i$, it will be recorded as a positive index, otherwise it will be recorded as a negative index. When the points are not in general position, several may become simultaneously collinear with the rotating line, and they are recorded as one subset $L_j^i$. The sequence is denoted by a half-period $s_I = (L_1^i, L_2^i, \ldots, L_{k_i}^i)$, where $L_j^i \subset \overline{E}_n \setminus \{i, \bar{i}\}$.

A **cluster of stars** (or **rank 3 hyperline sequences**) **associated with a point** (or **vector**) **configuration** $P$ is the ordered set of $n$ stars $s_1, \ldots, s_n$, one for each point $p_i \in P$.

A **uniform cluster of stars** is a cluster of stars corresponding to a set of points in general position. (Each star is a sequence of individual indices.)

An **oriented matroid of a vector** (or **point**) **set** $V$ **given by its cluster of stars** is the cluster of stars associated with $V$.

A **star** (or **rank 3 hyperline sequence**) **associated with an element** $c_i$ **of a big-circle arrangement** $C = \{c_1, \ldots, c_n\}$ on $S^2$ is an alternating circular sequence of subsets of indices in $\overline{E}_n$ obtained by traversing the oriented big circle in its given direction and recording in order the intersections of $c_i$ with the other big circles $c_j$ ($j \neq i$). Each intersection is recorded as a signed index $j$: positive if $c_j$ crosses $c_i$ from left to right, negative otherwise.

The **cluster of stars** (or **rank 3 hyperline sequences**) **associated with a big-circle arrangement** is the set of $n$ stars $s_1, \ldots, s_n$, one for each circle $c_i \in C$.

The **cluster of stars associated with an oriented central plane arrangement** in $\mathcal{R}^3$ [**line arrangement in** $\mathcal{R}^2$] is the cluster of stars of the big-circle arrangement associated with the central plane arrangement [to the central plane arrangement induced by the line arrangement via the imbedding of $\mathcal{R}^2$ as the plane $z = 1$ in $\mathcal{R}^3$].

The **cluster of stars associated with a pseudoline arrangement** [**a pseudoconfiguration of points**] is the generalization from straight lines to pseudolines obtained by recording the order of the vertices of the arrangement along a pseudoline (positive or negative according to whether the line crossing at that vertex comes from right or left) [the circular counterclockwise order of the pseudolines incident with a point].

A **cluster of stars permutation** is an ordered set of alternating circular sequences $s_1, \ldots, s_n$ with the property that the representative half-period of sequence $s_i$ is a signed permutation of the set $E_n \setminus \{i\}$.

A **chirotope function associated with a set of cluster of stars permutations** $s_1, \ldots, s_n$ is a function $\chi \colon E_n^3 \to \{-1, +1\}$ defined by $\chi(i\,j\,k) = +1$ if, in the $i$th sequence $s_i$ and in a half period of it where both $j$ and $k$ occur positively, $j$ occurs before $k$. Otherwise $\chi(i\,j\,k) = -1$.

A set $E_n$, together with an ordered set of alternating circular sequences $s_1, \ldots, s_n$, satisfies the **cluster of stars axioms** (**uniform, rank 3**) if the set of sequences are cluster of star permutations whose associated chirotope function is alternating.

A (**uniform, rank 3**) **oriented matroid given by its cluster of stars** is a set $E_n$ together with $n$ alternating sequences satisfying the cluster of stars axioms.

An abstract set $\overline{E}_n$, together with a set of $n^{d-2}$ (uniform) alternating sequences (indexed by $(d-2)$-tuples $(i_1, \ldots, i_{d-2})$), is an **oriented matroid given by its hyperline sequences** (**uniform, rank $d$**) if the chirotope function $\chi \colon E_n^d \to \{-1, +1\}$ associated with it is alternating. [The function $\chi \colon E_n^d \to \{-1, +1\}$ is defined by $\chi(i_1 \ldots i_{d-2}j\,k) = +1$ if in the star indexed by $i_1, \ldots, i_{d-2}$ and in a half period where both $j$ and $k$ occur positively, $j$ occurs before $k$. Otherwise $\chi(i_1 \ldots i_{d-2}j\,k) = -1$.]

**Deletion** is the removal of a point (vector, line, etc.) from a configuration and recording the oriented matroid (chirotope, circuits, cluster of stars, etc.) only for the remaining points.

In a rank $d$ central plane arrangement, the **contraction** on element $i$ is obtained by identifying $h_i$ with $\mathcal{R}^{d-1}$ and intersecting it with all the other hyperplanes to obtain a rank $d-1$ arrangement with one less element.

The **oriented matroid obtained by a one-element deletion** in the hypersequence representation is the matroid obtained by removing the element from all the hyperline sequences of the original oriented matroid, and discarding all hyperline sequences whose labels contain that element.

The **oriented matroid obtained by a one-element contraction** in the hypersequence representation is the matroid obtained by retaining only the hyperline sequences whose labels contain the element, and dropping it from the labels.

A **rank 2 contraction** of a cluster of stars is one of the stars.

A **minor** of an oriented matroid given by hyperline sequences is an oriented matroid obtained by a sequence of deletions and/or contractions.

A (**Euclidean**) **pseudoconfiguration of points** is a pair $(P, L)$ with $P = \{p_1, \ldots, p_n\}$ a planar set of points and $L = \{l_1, \ldots, l_m\}$ a pseudoline arrangement, such that for every pair of distinct points $(p_i, p_j)$ there exists a unique pseudoline $l_{ij} \in L$ incident with them.

If a pseudoline arrangement is intersected with a vertical line $l_v$ $(x = -M)$, for some very large constant $M$, and all the vertices of the arrangement lie to the right of $l_v$, then the order in which the pseudolines in $L$ cross $v_h$ (decreasing by the $y$-coordinates of the crossings) is the (increasing) **slope order of the pseudolines**.

The (**standard affine**) **pseudo polar-duality** is the association between an $x$-monotone pseudoline arrangement $L = \{l_1, \ldots, l_n\}$ given in slope order and a pseudo configuration of points $(P, L')$, $P = \{p_1, \ldots, p_n\}$, given in increasing order of the $x$-coordinates and with $L'$ being $x$-monotone, satisfying the property that the cluster of stars associated with $L$ and to $P$ is the same.

## Facts:

**1.** Cocircuits correspond to semi-spaces, when the defining hyperplane is incident with $d - 1$ independent elements of the configuration.

**2.** In the rank $d$ uniform oriented matroid associated with a vector configuration in general position in $\mathcal{R}^d$, all the $d$-tuples are bases, all the $(d+1)$-tuples are supports of circuits, and all the $(n-d+1)$-tuples are supports of cocircuits.

**3.** The oriented matroid associated with an affine vector configuration $V$ and the affine oriented matroid associated with the affine point configuration induced by $V$ are the same.

**4.** The chirotope function $\chi$ associated with the cluster of stars of a vector or point configuration is an alternating and antisymmetric function.

**5.** The two given systems of chirotope axioms are equivalent (for the uniform case).

**6.** The hyperline sequences of a contraction by one element of a central plane arrangement are the induced rank $(d-1)$ contraction by that element of the set of rank $d$ hyperline sequences of the original arrangement.

**7.** The induced rank $(d-1)$ contraction of a set of rank $d$ hyperline sequences is a rank $(d-1)$ set of hyperline sequences.

**8.** A minor of an oriented matroid (given by its hyperline sequences) is an oriented matroid.

**9.** For two labeled vector (or point) configurations $V_1$ and $V_2$, the following statements are equivalent:

- $V_1$ and $V_2$ have the same chirotope;
- $V_1$ and $V_2$ have the same order type;
- $V_1$ and $V_2$ have the same hyperline sequences;
- $V_1$ and $V_2$ have the same minors.

Moreover, for any reorientation of $V_1$ or $V_2$:

- $V_1$ and $V_2$ have the same oriented matroid given by circuits;
- $V_1$ and $V_2$ have the same oriented matroid given by cocircuits.

This justifies the unique name *oriented matroid* for the equivalence class of vector configurations with the same chirotope (or clusters, order type, etc.), and for a reorientation class of an oriented matroid.

**10.** For a labeled vector configuration $V$ in $\mathcal{R}^{d+1}$, the following statements are equivalent:

- $V$ is acyclic (affine);

- there is a labeled point configuration $P$ in $\mathcal{R}^d$ whose oriented matroid (or order type or chirotope) is the same as the oriented matroid of $V$.

**11.** Multiplying the elements of a vector configuration by a positive scalar yields a vector configuration with the same oriented matroid. In particular, for any vector configuration $V$ in $\mathcal{R}^{d+1}$, there exists an equivalent vector configuration on the $d$-sphere.

**12.** For any vector configuration, there exists a reorientation of some of its vectors which makes it affine.

**13.** The set of circuits of an affine (acyclic) vector configuration does not contain the *positive* cycle $(E_n, \emptyset)$.

**14.** Two projective line arrangements with the same arrangement graph have polar-dual configurations of points in the same reorientation class (and this can be generalized to arbitrary dimension $d$).

**15.** If a labeled vector configuration in $\mathcal{R}^3$ and a labeled oriented arrangement of central planes are polar-dual, then they have the same hyperline sequences (and this can be generalized to arbitrary dimension $d$).

**16.** The number of oriented matroids of rank $d$ and size $n$ is $2^{O(n^{d-1})}$.

**17.** The number of realizable oriented matroids of rank $d$ and size $n$ is $2^{O(n \log n)}$.

**18.** *Folkman-Lawrence topological representation theorem:* Every oriented matroid of rank $d$ can be represented as a $(d-1)$-pseudosphere arrangement on $S^d$.

**19.** Every affine oriented matroid of rank 3 can be represented as a pseudoline arrangement and as a (polar-dual) pseudoconfiguration of points.

**20.** There exist nonrealizable oriented matroids of any rank.

**21.** Realizable oriented matroids cannot be characterized by a finite set of excluded minors.

**Examples:**

**1.** The function $\chi(i, j, k) = \text{sign} \det(v_i, v_j, v_k)$, for $v_l \in V = \{v_1, \ldots, v_n\} \subset \mathcal{R}^3$, is alternating antisymmetric.

**2.** The following figure shows an example of a point configuration in general position, together with the connecting lines.

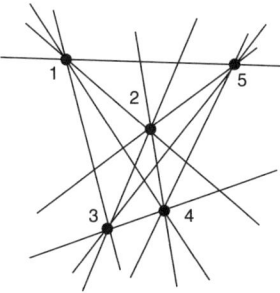

Its oriented matroid is given by

- chirotope: $1\,2\,3-$, $1\,2\,4-$, $1\,2\,5+$, $1\,3\,4+$, $1\,3\,5+$, $2\,3\,4+$, $2\,3\,5+$, $3\,4\,5+$. The other signed triplets are computed by antisymmetry and alternation.
- minimal Radon partitions: $1\,\overline{2}\,\overline{3}\,4$, $1\,\overline{2}\,3\,5$, $1\,\overline{2}\,4\,5$, $\overline{1}\,3\,\overline{4}\,5$, $\overline{2}\,3\,\overline{4}\,5$ and their complements.

- cocircuits: $34\overline{5}$, $245$, $2\overline{3}5$, $234$, $\overline{1}45$, $13\overline{5}$, $\overline{1}34$, $125$, $12\overline{4}$, $123$ and their complements.
- cluster of stars: 1: $3425$, 2: $15\overline{3}4$, 3: $1\overline{4}\,\overline{5}2$, 4: $5213$, and 5: $1234$. (An arbitrary half-period was chosen for the circular sequences.)

**3.** The oriented matroid given as a cluster of stars for the line arrangement in the figure of §13.1.2, Example 1 (or for the pseudoline arrangement in the figure of §13.1.3, Example 1) is

$$1: 23\overline{5}4, \quad 2: \overline{1}354, \quad 3: \overline{1}2\overline{5}4, \quad 4: 5\overline{1}2\overline{3}, \quad 5: \overline{4}\,\overline{1}2\overline{3}.$$

The orientation of the lines is assumed to be in increasing order of the $x$-coordinates. Its minor by the deletion of element 3 is 1: $254$, 2: $\overline{1}54$, 4: $5\overline{1}2$, and 5: $\overline{4}\,\overline{1}2$. Its contraction on point 3 is $\overline{1}\,\overline{2}54$.

# 13.2  SPACE FILLING

Discrete geometry is the study of discrete arrangements of objects in Euclidean and non-Euclidean spaces. This subject dates back to the early 1610s when J. Kepler studied packings of Euclidean space with equally-sized spheres, formulating a conjecture about the greatest possible density of such a packing that was unsettled until the late 1990s. The modern systematic development of the subject began in the late 1940s with the work of L. Fejes Tóth. The Hungarian school he founded focused mainly on packing and covering problems. Two areas from discrete geometry, dense sphere packings and tilings, occupy a substantial part of this section on space filling. Both areas have been active areas of research for the past 75 years.

## 13.2.1  PACKING

The most important results on packings are centered around density and contact numbers, including kissing numbers, foam problems, and soft packings.

**Definitions:**

A **convex body** in $\mathcal{R}^d$ is a compact convex subset of $\mathcal{R}^d$ with nonempty interior.

A family $\mathcal{P} = \{K_1, K_2, \ldots\}$ of the convex bodies $K_1, K_2, \ldots$ in $\mathcal{R}^d$ forms a **packing** of $\mathcal{R}^d$ if no two of the convex bodies $K_1, K_2, \ldots$ have an interior point in common.

Let $\mathcal{P} = \{K_1, K_2, \ldots\}$ be a packing of $\mathcal{R}^d$ and let $B_R$ be the closed ball of radius $R$ centered at the origin in $\mathcal{R}^d$. The **lower density** and **upper density** of $\mathcal{P}$ are defined by

$$\underline{\delta}(\mathcal{P}) = \liminf_{R \to +\infty} \frac{\sum_{K_i \subset B_R} \mathrm{Vol}(K_i)}{\mathrm{Vol}(B_R)} \quad \text{and} \quad \overline{\delta}(\mathcal{P}) = \limsup_{R \to +\infty} \frac{\sum_{K_i \subset B_R} \mathrm{Vol}(K_i)}{\mathrm{Vol}(B_R)}.$$

If $\underline{\delta}(\mathcal{P}) = \overline{\delta}(\mathcal{P}) = \delta(\mathcal{P})$, then $\delta(\mathcal{P})$ is called the **density** of $\mathcal{P}$.

For a convex body $K \subset \mathcal{R}^d$, let $\delta(K)$ denote the largest (upper) density of packings by congruent copies of $K$ in $\mathcal{R}^d$. In particular,

$\delta_T(K)$ = the largest (upper) density of packings by translates of $K$ in $\mathcal{R}^d$;

$\delta_L(K)$ = the largest (upper) density of packings by lattice translates of $K$ in $\mathcal{R}^d$.

Let $\mathcal{P} = \{B_1^d, B_2^d, \ldots\}$ be a packing of unit balls in $\mathcal{R}^d$ with centers $c_1, c_2, \ldots$. The **Voronoi cell** with center $c_i$ is the convex polyhedral set $V_i = \{x \in \mathcal{R}^d \mid |x - c_i| = \min_j |x - c_j|\}$.

The **kissing number** problem asks for the maximum number $k(d)$ of nonoverlapping unit balls that can touch a unit ball in the $d$-dimensional Euclidean space $\mathcal{R}^d$.

The **Hadwiger number** (or **translative kissing number**) of the convex body $K$ in $\mathcal{R}^d$, denoted by $H(K)$, is the largest number of nonoverlapping translates of the convex body $K$ that can all touch $K$.

The **isoperimetric quotient** of the convex body $K$ in $\mathcal{R}^d$ is $\mathrm{iq}(K) = \frac{(\mathrm{Svol}(\mathrm{bd}(K)))^d}{(\mathrm{Vol}(K))^{d-1}}$, where $\mathrm{Svol}(\mathrm{bd}(K))$ denotes the $(d-1)$-dimensional surface volume of the boundary $\mathrm{bd}(K)$ of $K$ and $\mathrm{Vol}(K)$ stands for the $d$-dimensional volume of $K$.

The **contact graph** of a finite unit ball packing in $\mathcal{R}^d$ is the (simple) graph with vertices corresponding to the packing elements and where two vertices are connected by an edge if and only if the corresponding two packing elements touch each other.

The number of edges of a contact graph is called the **contact number** of the underlying unit ball packing.

The **contact number problem** asks for the largest contact number, that is, for the maximum number $c(n, d)$ of edges that a contact graph of $n$ nonoverlapping unit balls can have in $\mathcal{R}^d$. Analogously, the largest contact number of packings by $n$ translates of a convex body $K$ in $\mathcal{R}^d$ is denoted by $c(K, n, d)$.

Let $\mathcal{T}$ be a tiling of $\mathcal{R}^3$ into convex polyhedra $P_i, i = 1, 2, \ldots$, each containing a unit ball $P_i$ containing the closed 3-dimensional ball $B_i$ having radius 1 for $i = 1, 2, \ldots$. Assuming that there is a finite upper bound for the diameters of the convex cells in $\mathcal{T}$, we say that $\mathcal{T}$ is a **normal tiling** of $\mathcal{R}^3$ with the underlying packing $\mathcal{P}$ of the unit balls $B_i, i = 1, 2, \ldots$.

The (**lower**) **average surface area** $\underline{s}(\mathcal{T})$ of the cells in $\mathcal{T}$ is defined by

$$\underline{s}(\mathcal{T}) = \liminf_{L \to \infty} \frac{\sum_{\{i | B_i \subset C_L\}} \mathrm{sarea}(P_i \cap C_L)}{\mathrm{card}\{i \mid B_i \subset C_L\}},$$

where $C_L$ denotes the cube centered at the origin $o$ with edges parallel to the coordinate axes of $\mathcal{R}^3$ and with edge length $L$; here $\mathrm{sarea}(S)$ and $\mathrm{card}(S)$ denote the surface area and cardinality of the set $S$. (Note that $\underline{s}(\mathcal{T})$ is independent of the choice of the coordinate system of $\mathcal{R}^3$.)

Let $P = \{c_i + B^d \mid i = 1, 2, \ldots$ with $|c_j - c_k| \geq 2$ for all $1 \leq j < k\}$ be an arbitrary infinite packing of unit balls in $\mathcal{R}^d$. For any $d \geq 2$ and $\lambda \geq 0$ let $P_\lambda = \bigcup_{i=1}^{+\infty}(c_i + (1 + \lambda)B^d)$ denote the outer parallel domain of $P = \bigcup_{i=1}^{+\infty}(c_i + B^d)$ having outer radius $\lambda$ and let

$$\bar{\delta}_d(P_\lambda) = \limsup_{R \to +\infty} \frac{\mathrm{Vol}(P_\lambda \cap RB^d)}{\mathrm{Vol}(RB^d)}$$

be the (upper) density of the outer parallel domain $P_\lambda$ assigned to the unit ball packing $P$ in $\mathcal{R}^d$. Finally, let

$$\bar{\delta}_d(\lambda) = \sup_P \bar{\delta}_d(P_\lambda)$$

be the largest density of the outer parallel domains of unit ball packings having outer radius $\lambda$ in $\mathcal{R}^d$. Then we say that the family $\{c_i + (1+\lambda)B^d \mid i = 1, 2, \dots\}$ of closed balls of radii $1 + \lambda$ is a **packing of soft balls** with penetrating constant $\lambda$ if $P = \{c_i + B^d \mid i = 1, 2, \dots\}$ is a unit ball packing of $\mathcal{R}^d$.

In particular, $\bar{\delta}_d(P_\lambda)$ is the (upper) **density of the soft ball packing** $\{c_i + (1+\lambda)B^d \mid i = 1, 2, \dots\}$ with $\bar{\delta}_d(\lambda)$ the largest density of packings of soft balls of radii $1 + \lambda$ with penetrating constant $\lambda$.

**Facts:**

1. There are two major problems concerning density:

   - Given a convex body $K \subseteq \mathcal{R}^d$, find efficient packings with congruent copies of $K$; i.e., find packings with congruent copies of $K$ in $\mathcal{R}^d$ having "relatively high" density.

   - Find a "good" upper bound for $\delta(K)$.

2. Sphere packing in $n$-dimensional space, for $n$ large, is important in designing codes that are efficient and unlikely to contain errors when data is transmitted.

3. If $K$ is a convex body in $\mathcal{R}^d$, $\dfrac{2\zeta(d)}{\binom{2d}{d}} \leq \delta_L(K) \leq \delta_T(K) \leq \delta(K)$, where $\zeta(d) = 1 + \frac{1}{2^d} + \frac{1}{3^d} + \cdots$ denotes Riemann's zeta function.

4. If $K$ is a centrally symmetric convex body in $\mathcal{R}^d$, $\dfrac{\zeta(d)}{2^{d-1}} \leq \delta_L(K) \leq \delta_T(K) \leq \delta(K)$.

5. Facts 3 and 4 have been improved slightly for different classes of convex bodies and subclasses of centrally symmetric convex bodies. See [DaRo47], [ElOdRu91], and [Sc63].

6. For each $d$, $(d-1)\dfrac{\zeta(d)}{2^{d-1}} \leq \delta_L(B^d) \leq \delta_T(B^d) = \delta(B^d)$.

7. For every convex domain $D$, $\delta_L(D) = \delta_T(D)$.

8. For every centrally symmetric convex domain $D$, $\delta_L(D) = \delta_T(D) = \delta(D)$.

9. There exists an ellipsoid $C$ in $\mathcal{R}^3$ for which $\delta_L(C) < \delta(C)$.

10. The class of convex bodies $C \subseteq \mathcal{R}^d$ for which $\delta_L(C)$, $\delta_T(C)$, and $\delta(C)$ can be determined (for a given $d$) is very small.

11. The densest packing of unit balls in $\mathcal{R}^3$ has density $\delta(B^3) = \frac{\pi}{\sqrt{18}} = 0.7404...$, which is attained by the "cannonball packing" [Ha05], [Ha12]. This fact was conjectured by Johannes Kepler in the 17th century. It was proved by Thomas Hales in 1998. His proof was based on examining a large number of cases, following an approach first conceived by Fejes Tóth in 1953. Hales's proof by exhaustion relied on complex computer calculations, and, after extensive examination over many years, was generally believed to be complete. To remove any doubt, Hales and a large team embarked on an effort to produce a formal proof of the conjecture. After more than 11 years, Hales and 21 collaborators succeeded in completing a formal proof using the proof assistants Isabelle and HOL Light.

12. No packing of unit balls in $\mathcal{R}^8$ has density greater than $\frac{\pi^4}{384} = 0.2536...$, the density of the $E_8$-lattice packing [Vi16].

13. The Leech lattice achieves the optimal sphere packing density in $\mathcal{R}^{24}$ with $\delta(B^{24}) = \frac{\pi^{12}}{12!} = 0.001929...$ and it is the only periodic packing in $\mathcal{R}^{24}$ with that density [CoEtal16].

**14.** The Cohn-Elkies linear programming bound [CoEl03] improves on all upper bounds for $\delta(B^d)$ obtained earlier in dimensions $4 \le d \le 128$ [CoZh14]. On the other hand, for $d \ge 129$ the best bounds are $2^{-(1+o(1))d} \le \delta(B^d) \le 2^{-(0.5990...+o(1))d}$. See [KaLe78], [Va11], and [Ve13].

**15.** The known values of $k(d)$ are $k(2) = 6$ (trivial), $k(3) = 12$ ([ScVa53]), $k(4) = 24$ ([Mu08]), $k(8) = 240$ ([OdSl79]), and $k(24) = 196{,}560$ ([OdSl79]). In general, $k(d) \le 2^{0.401d(1+o(1))}$ [KaLe78].

**16.** $c(n,2) = \lfloor 3n - \sqrt{12n - 3} \rfloor$ holds for all $n \ge 2$ [Ha74]. On the other hand, one has $6n - 7.863n^{\frac{2}{3}} < c(n,3) < 6n - 0.926n^{\frac{2}{3}}$ for all $n \ge 2$, and $c(n,d) < \frac{1}{2}k(d)\, n - \frac{1}{2^d}\delta(B^d)^{-\frac{d-1}{d}}n^{\frac{d-1}{d}} < \frac{1}{2}2^{0.401d(1+o(1))}n - \frac{1}{2^d}2^{0.599(1+o(1))(d-1)}n^{\frac{d-1}{d}}$ for all $d \ge 4, n \ge 2$ [Be13].

**17.** Let $K$ be a convex domain different from a parallelogram in $\mathcal{R}^2$. Then for all $n \ge 2$, one has $c(K, n, 2) = \lfloor 3n - \sqrt{12n - 3} \rfloor$. If $K$ is a parallelogram, then $c(K, n, 2) = \lfloor 4n - \sqrt{28n - 12} \rfloor$ holds for all $n \ge 2$ [Br96]. On the other hand, if $K$ is a convex body in $\mathcal{R}^d$, $d \ge 3$, then

$$c(K, n, d) \le \frac{H(K)}{2}\, n - \frac{1}{2^d \delta_T(K)^{\frac{d-1}{d}}} \sqrt[d]{\frac{\mathrm{iq}(B^d)}{\mathrm{iq}(K)}}\, n^{\frac{d-1}{d}} \le \frac{3^d - 1}{2}\, n - \frac{\sqrt[d]{\omega_d}}{2^{d+1}}\, n^{\frac{d-1}{d}},$$

where $\omega_d = \frac{\pi^{\frac{d}{2}}}{\Gamma(\frac{d}{2}+1)}$ is the volume of the unit ball $B^d$ [Be10].

**18.** Let $\mathcal{T}$ be an arbitrary normal tiling of $\mathcal{R}^3$. Then the average surface area of the cells in $\mathcal{T}$ is always at least $\frac{24}{\sqrt{3}}$: i.e., $\underline{s}(\mathcal{T}) \ge \frac{24}{\sqrt{3}} = 13.8564...$ [Be13].

**19.** Let $T^d = \mathrm{conv}\{t_1, t_2, \ldots, t_{d+1}\}$ be a regular $d$-simplex of edge length 2 in $\mathcal{R}^d, d \ge 2$, and let $0 \le \lambda < \sqrt{\frac{2d}{d+1}} - 1$. Then $\bar{\delta}_d(\lambda) \le \frac{\mathrm{Vol}\left(T^d \cap \left(\cup_{i=1}^{d+1} t_i + (1+\lambda)B^d\right)\right)}{\mathrm{Vol}(T^d)} < 1$ holds [BeLá15].

**Open Questions and Conjectures:**

**1.** Determine the asymptotic behavior of $\delta(B^d)$ as $d \to +\infty$.

**2.** Determine the asymptotic behavior of $c(n,3)$ as $n \to +\infty$.

**3.** *Foam problem of unit ball packings:* If the Euclidean 3-space is partitioned into convex cells each containing a unit ball, how should the shapes of the cells be designed to minimize the average surface area of the cells? In particular, if $\mathcal{T}$ denotes the Voronoi tiling of an arbitrary unit ball packing in $\mathcal{R}^3$, then prove or disprove that $\underline{s}(\mathcal{T}) \ge 12\sqrt{2} = 16.9705....$

**4.** *Soft dodecahedral conjecture:* Let $D \subset \mathcal{R}^3$ be a regular dodecahedron circumscribed by the unit ball $B^3$. Then, for $\lambda$ with $0 < \lambda < \sqrt{3}\tan\frac{\pi}{5} - 1 = 0.258408\ldots$, we set $\bar{\tau}_3(\lambda) = \frac{\mathrm{Vol}(D \cap (1+\lambda)B^3)}{\mathrm{Vol}(D)}$, where $\sqrt{3}\tan\frac{\pi}{5}$ is the circumradius of $D$. The *soft dodecahedral conjecture* asserts that if $0 < \lambda < \sqrt{3}\tan\frac{\pi}{5} - 1 = 0.258408\ldots$, then $\bar{\delta}_3(\lambda) \le \bar{\tau}_3(\lambda)$.

## 13.2.2   COVERING

**Definitions:**

A family $\mathcal{C} = \{K_i \mid i \in I\}$ of convex bodies in $\mathcal{R}^d$ forms a **covering** of $\mathcal{R}^d$ (that is, **covers** $\mathcal{R}^d$) if each point of $\mathcal{R}^d$ belongs to at least one convex body of $\mathcal{C}$.

The **lower density** and **upper density** of a covering $\mathcal{C}$ are

$$\underline{\nu}(\mathcal{C}) = \liminf_{R \to +\infty} \frac{\sum_{K_i \cap B_R \neq \emptyset} \mathrm{Vol}(K_i)}{\mathrm{Vol}(B_R)} \qquad \text{and} \qquad \overline{\nu}(\mathcal{C}) = \limsup_{R \to +\infty} \frac{\sum_{K_i \cap B_R \neq \emptyset} \mathrm{Vol}(K_i)}{\mathrm{Vol}(B_R)},$$

where $B_R$ denotes the closed ball of radius $R$ centered at the origin in $\mathcal{R}^d$.

If $\underline{\nu}(\mathcal{C}) = \overline{\nu}(\mathcal{C}) = \nu(\mathcal{C})$, then $\nu(\mathcal{C})$ is called the **density** of $\mathcal{C}$.

For a convex body $K \subseteq \mathcal{R}^d$, we define these quantities:

$\nu(K)$ = the smallest (lower) density of coverings of $\mathcal{R}^d$ by congruent copies of $K$;

$\nu_T(K)$ = the smallest (lower) density of coverings of $\mathcal{R}^d$ by translates of $K$;

$\nu_L(K)$ = the smallest (lower) density of coverings of $\mathcal{R}^d$ by lattice translates of $K$.

A **homothety** is an affine transformation of $\mathcal{R}^d$ of the form $x \mapsto t + \lambda x$, where $t \in \mathcal{R}^d$ and $\lambda$ is a nonzero real number.

The image $t + \lambda K$ of a convex body $K$ under a homothety is said to be its **homothetic copy** (or simply a **homothet**). A homothetic copy is *positive* if $\lambda > 0$ and *negative* otherwise. Furthermore, a homothetic copy with $0 < \lambda < 1$ is called a **smaller positive homothet**. Let $\alpha(K)$ denote the **covering number** of $K$, the minimum number of smaller positive homothets of $K$ with a union containing $K$.

Given $0 < k < d$ define a **$k$-codimensional cylinder** $C$ in $\mathcal{R}^d$ as a set which can be presented in the form $C = H + B$, where $H$ is a $k$-dimensional linear subspace of $\mathcal{R}^d$ and $B$ is a measurable set (called the base) in the orthogonal complement $H^\perp$ of $H$.

For a given convex body $K$ and a $k$-codimensional cylinder $C = H + B$, we define the **cross-sectional volume** $\mathrm{crv}_K(C)$ of $C$ with respect to $K$ by

$$\mathrm{crv}_K(C) = \frac{\mathrm{Vol}_{d-k}(C \cap H^\perp)}{\mathrm{Vol}_{d-k}(P_{H^\perp} K)} = \frac{\mathrm{Vol}_{d-k}(P_{H^\perp} C)}{\mathrm{Vol}_{d-k}(P_{H^\perp} K)} = \frac{\mathrm{Vol}_{d-k}(B)}{\mathrm{Vol}_{d-k}(P_{H^\perp} K)},$$

where $P_{H^\perp} : \mathcal{R}^d \to H^\perp$ denotes the orthogonal projection of $\mathcal{R}^d$ onto $H^\perp$. Notice that for every invertible affine map $T : \mathcal{R}^d \to \mathcal{R}^d$ one has $\mathrm{crv}_K(C) = \mathrm{crv}_{TK}(TC)$.

**Facts:**

1. There are two major problems concerning covering:

   - Given a convex body $K \subseteq \mathcal{R}^d$, find efficient coverings of $\mathcal{R}^d$ with congruent copies of $K$; that is, find coverings of $\mathcal{R}^d$ by congruent copies of $K$ having "relatively small" density.

   - Find a "good" lower bound for $\nu(K)$. (This is a highly nontrivial task for most of the convex bodies $K \subseteq \mathcal{R}^d$.)

2. If $K$ is a convex body in $\mathcal{R}^d$, then

$$\nu(K) \leq \nu_T(K) \leq d(\ln d) + d(\ln \ln d) + 4d$$

and

$$\nu_T(K) \leq \nu_L(K) \leq d^{(\log_2 \log_2 d) + c} \quad \text{for some constant } c.$$

3. If $B^d$ denotes the $d$-dimensional closed unit ball in $\mathcal{R}^d$, then

$$\nu(B^d) = \nu_T(B^d) \leq \nu_L(B^d) \leq cd(\ln d)^{\frac{1}{2} \log_2(2\pi e)}$$

for some constant $c$.

**4.** Take a regular simplex inscribed in a unit ball in $\mathcal{R}^d$ and draw unit balls around each vertex. Let $\tau_d$ be the ratio of the sum of the volumes of the portions of these balls lying in the regular simplex to the volume of the regular simplex. Then $\tau_d \leq \nu(B^d)$.

**5.** $\tau_d \sim \frac{d}{e^{3/2}}$. (Thus, Facts 2 and 3 give strong estimates for $\nu(B^d)$. Moreover, for $d = 1$ and 2, the lower bound $\tau_d$ is sharp.)

**6.** The thinnest lattice covering of $\mathcal{R}^d$ by unit balls has been determined up to dimension 5 only. The following table lists the optimal lattices. See [CoSl93].

| dimension | 1 | 2 | 3 | 4 | 5 |
|---|---|---|---|---|---|
| thinnest lattice covering | $\mathcal{Z}$ | $A_2$ | $A_3^*$ | $A_4^*$ | $A_5^*$ |

**7.** If $K$ is a convex domain in $\mathcal{R}^2$, then $\alpha(K) \leq 4$. If $K$ is a convex body in $\mathcal{R}^3$, then $\alpha(K) \leq 16$. Furthermore, if $P$ is a convex polyhedron with affine symmetry in $\mathcal{R}^3$, then $\alpha(P) \leq 8$ [Be10].

**8.** If $K$ is a convex body in $\mathcal{R}^d, d \geq 2$, then $\alpha(K) \leq \frac{\text{Vol}(K-K)}{\text{Vol}(K)}(d \ln d + d \ln \ln d + 5d + 1) \leq \binom{2d}{d}(d \ln d + d \ln \ln d + 5d + 1)$. (Moreover, for sufficiently large $d$, $5d$ can be replaced by $4d$.) Furthermore, if $B[X] \subset \mathcal{R}^d, d \geq 3$ is an intersection of $d$-dimensional unit balls centered at the points of $X \subset \mathcal{R}^d$ having diameter $\text{diam}(X) \leq 1$, then $\alpha(B[X]) < 5d^{\frac{3}{2}}(4 + \ln d)\left(\frac{3}{2}\right)^{\frac{d}{2}}$ [Be13].

**9.** Let $K$ be a convex body in $\mathcal{R}^d$. Let $C_1, \ldots, C_N$ be $k$-codimensional cylinders in $\mathcal{R}^d, 0 < k < d$, such that $K \subset \bigcup_{i=1}^{N} C_i$. Then $\sum_{i=1}^{N} \text{crv}_K(C_i) \geq \frac{1}{\binom{d}{k}}$. Moreover, if $K$ is an ellipsoid and $C_1, \ldots, C_N$ are 1-codimensional cylinders in $\mathcal{R}^d$ such that $K \subset \bigcup_{i=1}^{N} C_i$, then $\sum_{i=1}^{N} \text{crv}_K(C_i) \geq 1$ [Be10].

**10.** The sum of the base areas of finitely many (1-codimensional) cylinders covering a 3-dimensional convex body is always at least one third of the minimum area 2-dimensional projection of the body [Be10].

**11.** If $K$ is a centrally symmetric convex body and $C_1, \ldots, C_N$ are $(d-1)$-codimensional cylinders (i.e., planks) in $\mathcal{R}^d$ such that $K \subset \bigcup_{i=1}^{N} C_i$, then $\sum_{i=1}^{N} \text{crv}_K(C_i) \geq 1$ [Ba91].

**Open Questions and Conjectures:**

**1.** *Hadwiger-Levi covering conjecture (1955):* If $K$ is an arbitrary convex body in $\mathcal{R}^d, d \geq 3$, then $\alpha(K) \leq 2^d$.

**2.** *Bang's affine plank conjecture (1951):* Let $K$ be a convex body in $\mathcal{R}^d, d \geq 2$, and let $C_1, \ldots, C_N$ be $(d-1)$-codimensional cylinders (i.e., planks) in $\mathcal{R}^d$ such that $K \subset \bigcup_{i=1}^{N} C_i$. Then prove that $\sum_{i=1}^{N} \text{crv}_K(C_i) \geq 1$.

**3.** *Bang's cylinder covering problem (1951):* Prove or disprove that the sum of the base areas of finitely many (1-codimensional) cylinders covering a 3-dimensional convex body is at least half of the minimum area 2-dimensional projection of the body.

## 13.2.3  TILING

Only a "diagonal" view of the basic definitions and theorems of this area will be given. See [GrSh90] for additional material.

**Definitions:**

A *tiling* $\mathcal{T}$ of Euclidean $d$-space $\mathcal{R}^d$ is a countable family of closed topological $d$-cells of $\mathcal{R}^d$, the *tiles* of $\mathcal{T}$, which cover $\mathcal{R}^d$ without gaps and overlaps.

A *monohedral tiling* is a tiling $\mathcal{T}$ of $\mathcal{R}^d$ in which all tiles are congruent to one fixed tile $T$, the (metrical) *prototile* of $\mathcal{T}$. In this case, $T$ *admits* the tiling $\mathcal{T}$.

A *regular polygon* is a polygon with all sides congruent and all interior angles equal.

A *regular tiling* is a monohedral tiling of the plane ($\mathcal{R}^2$) with a regular polygon as prototile.

A *semiregular tiling* is a tiling of the plane using $n$ prototiles with the same numbers of polygons around each vertex.

A *semiregular polyhedron* is a convex polyhedron with each face a regular polygon, but where more than one regular polygon can be used as a face.

A tiling $\mathcal{T}$ of $\mathcal{R}^d$ by convex polytopes is *normal* if there exist positive real numbers $r$ and $R$ such that each tile contains a Euclidean ball of radius $r$ and is contained in a Euclidean ball of radius $R$.

A *face-to-face* tiling is a tiling $\mathcal{T}$ by convex $d$-polytopes such that the intersection of any two tiles is a face of each tile, possibly the (improper) empty face. When $d = 2$, such a tiling is an *edge-to-edge* tiling.

A *lattice tiling*, with lattice $L$, is a tiling $\mathcal{T}$ by translates of a single tile $T$ such that the corresponding translation vectors form a $d$-dimensional lattice $L$ in $\mathcal{R}^d$.

A Euclidean motion $\sigma$ of $\mathcal{R}^d$ is a *symmetry* of a tiling $\mathcal{T}$ if $\sigma$ maps (each tile of) $\mathcal{T}$ onto (a tile of) $\mathcal{T}$. The set of all symmetries of $\mathcal{T}$ (a group under composition) is the *symmetry group* $S(\mathcal{T})$ of $\mathcal{T}$.

A *periodic tiling* is a tiling $\mathcal{T}$ of $\mathcal{R}^d$ such that $S(\mathcal{T})$ contains translations in $d$ linearly independent directions. A tiling $\mathcal{T}$ is *nonperiodic* if $S(\mathcal{T})$ contains no translation other than the identity.

An *isohedral tiling* is a tiling $\mathcal{T}$ such that $S(\mathcal{T})$ acts transitively on the tiles of $\mathcal{T}$.

An *isogonal tiling* is a tiling $\mathcal{T}$ such that $S(\mathcal{T})$ acts transitively on the vertices of $\mathcal{T}$.

An *isotoxal tiling* is a tiling $\mathcal{T}$ such that $S(\mathcal{T})$ acts transitively on the edges of $\mathcal{T}$.

Let $\mathcal{T}$ and $\mathcal{T}'$ be tilings of $\mathcal{R}^d$ with symmetry groups $S(\mathcal{T})$ and $S(\mathcal{T}')$. Let $\Phi\colon \mathcal{R}^d \to \mathcal{R}^d$ be a homeomorphism that maps $\mathcal{T}$ onto $\mathcal{T}'$. $\Phi$ is *compatible with a symmetry* $\sigma$ of $\mathcal{T}$ if there exists a symmetry $\sigma'$ of $\mathcal{T}'$ such that $\sigma'\Phi = \Phi\sigma$. $\Phi$ is *compatible with $S(\mathcal{T})$* if $\Phi$ is compatible with each $\sigma$ in $S(\mathcal{T})$. The tilings $\mathcal{T}$ and $\mathcal{T}'$ of $\mathcal{R}^d$ are *homeomeric*, or of the *same homeomeric type*, if there exists a homeomorphism $\Phi\colon \mathcal{R}^d \to \mathcal{R}^d$ that maps $\mathcal{T}$ onto $\mathcal{T}'$ such that $\Phi$ is compatible with $S(\mathcal{T})$ and $\Phi^{-1}$ is compatible with $S(\mathcal{T}')$.

A prototile $T$ in $\mathcal{R}^d$ is *aperiodic* if $T$ admits a tiling of $\mathcal{R}^d$, yet all such tilings are nonperiodic. In general, a set $S$ of prototiles in $\mathcal{R}^d$ is said to be *aperiodic* if $S$ admits a tiling of $\mathcal{R}^d$, yet all such tilings are nonperiodic.

**Facts:**

**1.** There are three monohedral edge-to-edge tilings of $\mathcal{R}^d$ with regular polygons; the prototile must be a triangle, a square, or a hexagon. These are illustrated in the following figure.

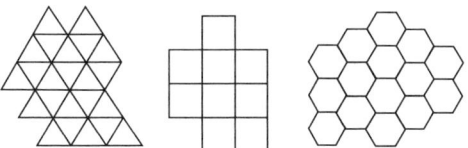

**2.** There are eight semiregular tilings of $\mathcal{R}^d$. These tilings use two or three prototiles.

**3.** Shapes that are not regular polygons [polyhedra] can be used in monohedral tilings of $\mathcal{R}^2$ [$\mathcal{R}^3$].

**4.** Any triangle can be used in a monohedral tiling of the plane. (Join two to form a parallelogram and tile a strip using these parallelograms. Repeat this process with parallel strips to tile the plane.) See the following figure.

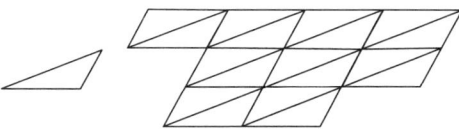

**5.** Any quadrilateral can be used in a monohedral tiling of the plane. (Take a second copy of the quadrilateral and rotate it 180 degrees. Join the two to form a hexagon. Use the hexagons to tile the plane.) See the following figure.

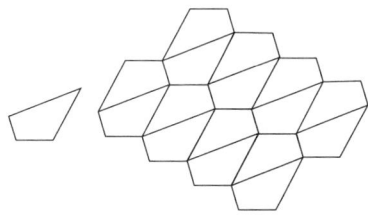

**6.** Any pentagon with a pair of parallel sides can be used in a monohedral tiling of the plane.

**7.** There are at least fourteen types of convex pentagons that can be used in a monohedral tiling of the plane. It is not known if there are more.

**8.** There are three types of convex hexagons that can be used in a monohedral tiling of the plane. Assume that the hexagon has vertices $a, b, c, d, e, f$ in clockwise order, as shown in the following figure. The prototile must be of one of the following forms:

- sum of angles at $a, b, c$ is $360°$; length of $\{a, f\}$ = length of $\{c, d\}$;
- sum of angles at $a, b, e$ is $360°$; length of $\{a, f\}$ = length of $\{d, e\}$ and length of $\{b, c\}$ = length of $\{e, f\}$;
- angles at $a$, $b$, and $c$ are each equal to $120°$; length of $\{a, b\}$ = length of $\{a, f\}$, length of $\{c, b\}$ = length of $\{c, d\}$, and length of $\{e, d\}$ = length of $\{e, f\}$.

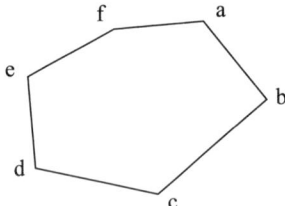

**9.** No convex polygon with more than six sides can be used as prototile in a monohedral tiling of $\mathcal{R}^2$.

**10.** Of the five regular polyhedra (tetrahedron, hexahedron (cube), octahedron, dodecahedron, icosahedron), only the tetrahedron and cube can be used as a prototile in a regular tiling of $\mathcal{R}^3$.

**11.** If $\mathcal{T}$ is a tiling of $\mathcal{R}^d$ with convex tiles, then each tile in $\mathcal{T}$ is a convex $d$-polyhedron.

**12.** If $\mathcal{T}$ is a tiling of $\mathcal{R}^d$ with compact convex tiles, then each tile in $\mathcal{T}$ is a convex $d$-polytope.

**13.** The following classification results have a long history. See [GrSh90].

- There exist precisely 11 distance edge-to-edge isogonal plane tilings, the tiles of which are convex regular polygons (called Archimedean tilings).

- There exist precisely 81 homeomeric types of normal isohedral plane tilings. Precisely 47 of these can be realized by a normal isohedral edge-to-edge tiling with convex polygonal tiles.

- There exist precisely 91 homeomeric types of normal isogonal plane tilings. Precisely 63 types can be realized by normal isogonal edge-to-edge tilings with convex polygonal tiles.

- There exist precisely 26 homeomeric types of normal isotoxal plane tilings. Precisely six types can be realized by a normal isotoxal edge-to-edge tiling with convex polygonal tiles.

**14.** Let $T$ be a convex $d$-polytope. If $T$ tiles $\mathcal{E}^d$ by translation, then $T$ admits (uniquely) a face-to-face lattice tiling of $\mathcal{R}^d$. Such a tile $T$ is called a *parallelotope*. This result is not true for nonconvex polytopes.

**15.** Several aperiodic sets have been found in the plane. Some of them, such as the Wang tiles and Penrose tiles, possess several highly interesting properties [GrSh90].

**16.** Progress has been made for aperiodic tilings in higher dimensions via dynamical systems [Ra95a].

**Open Questions:**

**1.** Extend the classification problems to higher dimensions. (At present, this looks hopeless.)

**2.** Classify all convex $d$-polytopes which are prototiles of monohedral tilings of $\mathcal{R}^d$. (This problem is not even solved for the plane.) However, under suitable restrictions the complexity of the problem changes. (See Fact 4.)

**3.** For $d \geq 5$, determine whether each $d$-parallelotope is a Voronoi cell (see §13.2.1) for some $d$-lattice. (This is known to be true for $1 \leq d \leq 4$.)

# 13.3   COMBINATORIAL GEOMETRY

This section studies geometric results involving combinatorics in the areas of convexity, incidences, distances, and colorings. In some cases the problems themselves have a combinatorial flavor, while in other cases their solution requires combinatorial tools.

## 13.3.1   CONVEXITY

In this subsection, questions of two different kinds are studied. Most of them belong to geometric transversal theory, a subject originating in Helly's theorem. Another group of problems grew out of the Erdős-Szekeres theorem, which turned out to be a starting point of Ramsey theory.

**Definitions:**

A subset $C$ of $d$-dimensional Euclidean space ($d$-space) $\mathcal{R}^d$ is **convex** if the following is true: for any pair of points in $C$, the straight-line segment connecting them is entirely contained in $C$.

A convex set is **strictly convex** if its boundary contains no line segment.

A **convex body** is a compact (i.e., bounded and closed) convex set with nonempty interior.

A **polytope** is a bounded convex body that can be obtained as the intersection of finitely many closed half-spaces. (§13.1.1.)

A **convex polygon** is a polytope in the plane.

A **vertex** of a polytope $P$ is a point $q \in P$, for which there is a hyperplane (§13.1.1) $H$ such that $H \cap P = \{q\}$.

A point set is in **convex position** if it is the vertex set of a polytope.

The **convex hull** of a set $S \subseteq \mathcal{R}^d$ is the smallest convex set containing $S$.

A family $\mathcal{C} = \{C_1, C_2, \ldots\}$ of sets in $d$-space is said to be **intersecting** if all members of $\mathcal{C}$ have a point in common.

A set $T \subseteq \mathcal{R}^d$ is a **transversal** of a family $\mathcal{C}$ of sets if $T \cap C_i$ is nonempty for every $i$. If $\mathcal{C}$ has a $k$-element transversal ($|T| = k$), its members can be **pierced** by $k$ points.

Two sequences $P = \{p_1, \ldots, p_n\}$ and $Q = \{q_1, \ldots, q_n\}$ of points in $\mathcal{R}^k$ have the same **order type** if, for all $1 \le i_1 < i_2 < \cdots < i_{k+1} \le n$, the orientations of the simplices induced by $\{p_{i_1}, \ldots, p_{i_{k+1}}\}$ and $\{q_{i_1}, \ldots, q_{i_{k+1}}\}$ are the same. This order type is **nontrivial** if $P$ and $Q$ are not contained in any hyperplane of $\mathcal{R}^k$.

A **$k$-flat** (an oriented $k$-dimensional plane) $F$ intersects a sequence of $d$-dimensional convex bodies $\mathcal{C} = \{C_1, \ldots, C_n\}$ **consistently** with the above order type if there are $x_i \in F \cap C_i$ such that the sequences $X = \{x_1, \ldots, x_n\}$ and $P$ have the same order type.

**Facts:**

**1.** The convex hull of a set $S$ is the intersection of all convex sets containing $S$.

**2.** For any set $S \subset \mathcal{R}^d$ of finitely many points, not all of which lie in the same hyperplane, the convex hull of $S$ is a polytope. In particular, if $S$ has $d+1$ points, then its convex hull is a simplex whose vertices are the elements of $S$.

**3.** The convex hull of the vertex set of any convex polytope $P$ is identical with $P$.

**4.** *Helly's theorem:* If a family $\mathcal{C}$ of at least $d+1$ convex bodies in $\mathcal{R}^d$ has the property that every $d+1$ of its members have a point in common, then $\mathcal{C}$ is intersecting (i.e., *all* its members have a point in common) [He23].

**5.** *Carathéodory's theorem:* If the convex hull of a set $S \subseteq \mathcal{R}^d$ contains a point $p$, then there exists a subset of $S$ with at most $d+1$ elements whose convex hull contains $p$.

**6.** Let $S$ be a compact set in $\mathcal{R}^d$ with the property that for every $(d+1)$-element subset $T \subset S$, there is a point $s \in S$ such that each segment connecting $s$ to an element of $T$ lies in $S$. Then $S$ has a point such that every segment connecting it to an element of $S$ is entirely contained in $S$ [Kr46].

**7.** Any set of $(k-1)(d+1)+1$ points in $\mathcal{R}^d$ can be partitioned into $k$ parts whose convex hulls have a point in common [Ra21], [Tv66].

**8.** Let $\mathcal{C}$ be any family of convex bodies in $\mathcal{R}^d$ with the property that the volume of the intersection of any $2d$ of them is at least 1. Then the volume of the intersection of all members of $\mathcal{C}$ is at least a positive constant depending only on $d$ [BáKaPa82].

**9.** For any $\epsilon > 0$ and for any $d$ there is a $\delta > 0$ satisfying the following condition: if $\mathcal{C}$ is a family of $n$ $(> d+1)$ convex bodies in $\mathcal{R}^d$ having at least $\epsilon \binom{n}{d+1}$ intersecting $(d+1)$-tuples, then $\mathcal{C}$ has at least $\delta n$ members with a point in common [Ka84].

**10.** For any $d < q \leq p$, there exists $k = k(p,q,d)$ satisfying the following condition: if $\mathcal{C}$ is a family of convex bodies in $\mathcal{R}^d$ such that every subfamily of $\mathcal{C}$ of size $p$ contains $q$ members with a point in common, then $\mathcal{C}$ can be pierced by $k$ points [AlKl92].

**11.** A sequence $\mathcal{C} = \{C_1, \ldots, C_n\}$ of convex bodies in $\mathcal{R}^d$ has a hyperplane transversal if and only if for some $0 \leq k \leq d-1$, there is a nontrivial $k$-dimensional order type of $n$ points such that every $(k+2)$-member subfamily of $\mathcal{C}$ can be met by a suitable $k$-flat consistently with that order type [PoWe90].

**12.** If $\mathcal{S}_k$ is any family of $k$-dimensional linear subspaces of $\mathcal{R}^d$ with the property that any $\binom{k+l}{l}$ of them can be intersected by an $l$-dimensional subspace, then all members of $\mathcal{S}_k$ can be intersected by an $l$-dimensional subspace. (Two subspaces intersect each other if they have at least one point in common, different from the origin.)

**13.** Any set of five points in the plane, no three of which are on a line, has four elements in convex position.

**14.** *Erdős-Szekeres theorem:* For every $k > 2$, there exists a smallest integer $n(k)$ with the property that every set of at least $n(k)$ points in the plane, no three of which are on a line, contains $k$ points in convex position. If $k = 3, 4, 5, 6$, then $n(k) = 2^{k-2} + 1$. (These are the only known values.) Furthermore, $2^{k-2} + 1 \leq n(k) \leq \binom{2k-5}{k-2}$ [ErSz35], [TóVa98], and, asymptotically, $n(k) \leq 2^{(1+o(1))k}$ [Su17].

**Examples:**

**1.** Let $S = \{(1,0,0), (0,1,0), (0,0,1)\}$. The convex hull of $S \subset \mathcal{R}^3$ is a triangular region, which is *not* a convex body in 3-space because its interior is empty.

**2.** Let $S = \{(0,0), (1,0), (0,1)\}$. The convex hull of $S \subset \mathcal{R}^2$ is a triangular region, which *is* a convex body (polygon) in the plane.

**3.** Let $S = \{(x,y) \mid 0 \leq x \leq 2,\ 0 \leq y \leq 2\}$ and $S' = \{(x,y) \mid 0 \leq x \leq 3,\ 0 \leq y \leq 3\}$. The family of all axis-parallel unit squares lying in $S$ is intersecting because each of them contains the point $(1, 1)$. The family of axis-parallel unit squares in $S'$ can be pierced by four points: $(1, 1), (1, 2), (2, 1), (2, 2)$.

**4.** In the line, $\{1, 3, 4, 2\}$ and $\{0, 4, 25, 3\}$ have the same (1-dimensional) order type. The 3-dimensional closed unit balls centered at $(0, 1, 5), (0, 0, 9.6), (0, 0, 9.4), (1, 0, 7)$ are met by the $z$-axis consistently with the above order type, because these balls contain the points $(0, 0, 5)$, $(0, 0, 9)$, $(0, 0, 10)$, and $(0, 0, 7)$, respectively, and the order type of this sequence along the $z$-axis is the same as the 1-dimensional order type of $\{1, 3, 4, 2\}$.

## 13.3.2   INCIDENCES

This subsection studies the structure (and number) of incidences between a set of points and a set of lines (or planes, spheres, etc.). The starting point of many investigations in this field was the Sylvester-Gallai theorem.

**Definitions:**

Given a point set $P$ and a set $L$ of lines (or $k$-flats, spheres, etc.) in Euclidean $d$-space $\mathcal{R}^d$, a point $p \in P$ and a line $l \in L$ are **incident** with each other, if $p \in l$.

Given a set $L$ of lines in the plane, a point incident with precisely two elements of $L$ is called an **ordinary crossing**. Given a set of points $P \subseteq \mathcal{R}^d$, a hyperplane passing through precisely $d$ elements of $P$ is called an **ordinary hyperplane** (for $d = 2$, an **ordinary line**).

Given a set of points $P$, a **Motzkin hyperplane** is a hyperplane $h$ such that all but one element of $h \cap P$ lie in a $(d-2)$-flat.

A family $\Gamma$ of curves in the plane has $d$ **degrees of freedom** if there exists an integer $s$ such that

- no two curves in $\Gamma$ have more than $s$ points in common;
- for any $d$ points, there are at most $d$ curves in $\Gamma$ passing through all the points.

A family of **pseudolines** is a family of simple curves in the plane with the property that every two of them meet precisely once.

A family of **pseudocircles** is a family of simple closed curves in the plane with the property that every two of them meet in at most two points.

**Facts:**

**1.** *Sylvester-Gallai theorem:*   Every finite set of points in the plane, not all of which are on a line, determines an ordinary line. In dual version: every finite set of straight lines in the plane, not all of which pass through the same point, determines an ordinary crossing.

**2.** For every finite set of points in Euclidean $d$-space, not all of which lie on a hyperplane, there exists a Motzkin hyperplane [Ha65], [Ha80].

**3.** Every set of $n$ points in $d$-space, not all of which lie on a hyperplane, determine at least $n$ distinct hyperplanes.

**4.** In 3-space, every set of $n$ noncoplanar points determines at least $\frac{2n}{5}$ Motzkin hyperplanes.

**5.** If $n$ is sufficiently large, then every set of $n$ noncocircular points in the plane determines at least $\binom{n-1}{2}$ distinct circles, and this bound is best possible [El67].

**6.** Every set of $n$ ($>7$) noncollinear points in the plane determines at least $\frac{6n}{13}$ ordinary lines. This bound is sharp for $n = 13$ and false for $n = 7$ [CsSa93].

**7.** If $n$ is sufficiently large, then the number of ordinary lines is at least $n/2$ if $n$ is even, and at least $3\lfloor n/4 \rfloor$ if $n$ is odd [GrTa13]. This proves an old conjecture of Dirac and Motzkin.

**8.** There is a positive constant $c$ such that every set of $n$ points in the plane, not all on a line, has an element incident with at least $cn$ connecting lines. Moreover, any set of $n$ points in the plane, no more than $n - k$ of which are on the same line, determines at least $c'kn$ distinct connecting lines, for a suitable constant $c' > 0$. According to the $d = 2$ special case of Fact 4, due to de Bruijn-Erdős, for $k = 1$ the number of distinct connecting lines is at least $n$. For $k = 2$, the corresponding bound is $2n - 4$ (for $n \geq 10$) [Be83], [SzTr83].

**9.** Every set of $n$ noncollinear points in the plane always determines at least $2\lfloor \frac{n}{2} \rfloor$ lines of different slopes. Furthermore, every set of $n$ points in the plane, not all on a line, permits a spanning tree, all of whose $n - 1$ edges have different slopes [Un82], [Ja87].

**10.** The number of incidences between a set $P$ of points and a set $L$ of lines can be obtained by summing over all $l \in L$ the number of points in $l$ belonging to $P$, or, equivalently, by summing over all $p \in P$ the number of lines in $L$ passing through $p$.

**11.** Let $\Gamma$ be a family of curves in the plane with $d$ degrees of freedom. Then the maximum number of incidences between $n$ points in the plane and $m$ elements of $\Gamma$ is

$$O(n^{d/(2d-1)}m^{(2d-2)/(2d-1)} + n + m)$$

(see [PaSh98]). From the most important special case, when $\Gamma$ is the family of all straight lines ($d = 2$), it follows that for any set $P$ of $n$ points in the plane, the number of distinct straight lines containing at least $k$ elements of $P$ is $O(\frac{n^2}{k^3} + \frac{n}{k})$ [SzTr83]. This bound is asymptotically tight. The same result holds for pseudolines.

For $d > 2$, the above bound is probably not tight: slightly better results are known for the number of incidences between $n$ points and a family $\Gamma$ of $m$ *algebraic* curves with bounded degree [ShShSo16].

**12.** The maximum number of incidences between $n$ points and $m$ spheres in $\mathcal{R}^3$ is

$$O(n^{\frac{4}{7}}m^{\frac{9}{7}}\beta(n,m) + n^2),$$

where $\beta(n,m) = o(\log(nm))$ is an extremely slowly growing function.

If no three spheres contain the same circle, then the following better bound is obtained:

$$O(n^{\frac{3}{4}}m^{\frac{3}{4}} + n + m).$$

Neither of these estimates is known to be asymptotically tight [ClEtal90], [Za13].

**13.** The maximum number of collinear triples determined by $n$ points in the plane, no four of which are on a line, is at most $\frac{n^2}{6} - O(n)$ [BuGrSl79]. More precisely, if $n$ is sufficiently large, the number of collinear triples is at most $\lfloor \frac{1}{6}n(n-3) \rfloor + 1$, and this bound is tight [GrTa13].

**14.** If $M(n)$ denotes the minimum number of different midpoints of the $\binom{n}{2}$ line segments determined by $n$ points in convex position in the plane, then [ErFiFü91] showed

$$\binom{n}{2} - \left\lfloor \frac{n(n+1)(1-e^{-1/2})}{4} \right\rfloor \leq M(n) \leq \binom{n}{2} - \left\lfloor \frac{n^2-2n+12}{20} \right\rfloor.$$

**Examples:**

**1.** Let $P$ be a set of seven points in the plane, consisting of the vertices, the centroid (the point of intersection of the medians), and the midpoints of all sides of an equilateral triangle. Then $P$ determines three ordinary lines (the lines connecting the midpoints of two sides).

**2.** Let $P$ be a $4k$-element set in the plane that can be obtained from the vertex set $\{v_1, v_2, \ldots, v_{2k}\}$ of a regular $2k$-gon by adding the intersection of the line at infinity with every line $v_i v_j$. Then the set $P$ determines precisely $2k$ ordinary lines: every line connecting some $v_i$ to the intersection point of $v_{i-1}v_{i+1}$ and the line at infinity ($1 \leq i \leq 2k$, the indices are taken modulo $2k$). (It can be achieved by a suitable projective transformation that no point of $P$ is at infinity, and the number of ordinary lines remains $\frac{|P|}{2} = 2k$.)

**3.** Let $P$ be a set of $n \geq 4$ points lying on two noncoplanar lines in 3-space so that there are at least two points on each line. Not all points of $P$ are coplanar, but $P$ does not determine any ordinary plane.

**4.** The family of all straight lines in the plane and the family of all unit circles both have two degrees of freedom. The family of all circles with arbitrary radii has three degrees of freedom. The family of the graphs of all polynomials of one variable and degree $d$ has $d$ degrees of freedom.

**5.** Let $P$ be an $n^{\frac{1}{2}} \times n^{\frac{1}{2}}$ part of the integer grid; i.e.,

$$P = \{(i,j) \mid 1 \leq i \leq n^{\frac{1}{2}}, 1 \leq j \leq n^{\frac{1}{2}}\}.$$

Let $k = \left(\frac{cm}{n^{1/2}}\right)^{1/3} > 2$, where $c > 0$ is a sufficiently small constant. For every $1 \leq s < r \leq k$ and for every $1 \leq i \leq r, 1 \leq j \leq \frac{n^{1/2}}{2}$, consider the line passing through $(i,j)$ and $(i+r, j+s)$. If $c$ is sufficiently small, then the number of these lines is at most $m$. There is a constant $c' > 0$ such that the total number of incidences between these lines and the elements of $P$ is at least $c'n^{\frac{2}{3}}m^{\frac{2}{3}}$. (See the case $d=2$ of Fact 11.)

---

## 13.3.3 DISTANCES

The systematic study of the distribution of the $\binom{n}{2}$ distances determined by $n$ points was initiated by Erdős. Given a set of $n$ points $P = \{p_1, p_2, \ldots, p_n\}$, let $g(P)$ denote the number of distinct distances determined by $P$, and let $f(P)$ denote the number of times that the unit distance occurs between two elements of $P$. That is, $f(P)$ is the number of pairs $p_i p_j, i < j$, such that $|p_i - p_j| = 1$. In [Er46], Erdős raised the following general questions: What is the minimum of $g(P)$ and what is the maximum of $f(P)$ over all $n$-element subsets of Euclidean $d$-space or of any other fixed metric space?

**Definitions:**

For any point set $P$ in a metric space, the **unit distance graph** of $P$ is the graph $G(P)$ whose vertex set is $P$ and two points (vertices) are connected by an edge if and only if their distance is 1.

Let $P$ be a finite set of points in a metric space. If the distance between two points $p, q \in P$ is minimum, then $p$ and $q$ form a **closest pair**.

A point $q \in P$ is a **nearest neighbor** of $p \in P$ if no point of $P$ is closer to $p$ than $q$.

A set $P$ in a metric space is a **separated set** if the minimum distance between the points of $P$ is at least 1.

The **diameter** of a finite set of points in a metric space is the maximum distance between two points of the set.

A point $q \in P$ is a **farthest neighbor** of $p \in P$ if no point of $P$ is farther from $p$ than $q$.

A set of points in the plane is said to be in **general position** if no three are on a line and no four on a circle.

**Facts:**

**1.** $f(P)$ is equal to the number of edges of $G(P)$.

**2.** If $p$ and $q$ form a closest pair in $P$, then $q$ is a nearest neighbor of $p$ and $p$ is a nearest neighbor of $q$.

**3.** If the distance between $p$ and $q$ is equal to the diameter of $P$, then $q$ is a farthest neighbor of $p$ and $p$ is a farthest neighbor of $q$.

**4.** The maximum number of times that the unit distance can occur among $n$ points in the plane is $O(n^{4/3})$. *Conjecture:* the asymptotically best bound is $O(n^{1+c/\log\log n})$ [SpSzTr84].

**5.** The maximum number of times that the unit distance can occur in a separated set of $n \leq 3$ points is $\lfloor 3n - \sqrt{12n-3} \rfloor$ [Ha74].

**6.** The maximum number of times that the unit distance can occur in a set of $n$ points in the plane with unit diameter is $n$ [HoPa34].

**7.** For any set of $n > 3$ points in the plane, the total number of farthest neighbors of all elements is at most $3n - 3$ if $n$ is even, and at most $3n - 4$ if $n$ is odd. These bounds cannot be improved [EdSk89].

**8.** The maximum number of times that the unit distance can occur among $n$ points in convex position in the plane is $O(n \log n)$. For $n > 15$, the best known lower bound is $2n - 7$ [Fü90], [EdHa91].

**9.** In 2015 L. Guth and N. Katz showed that the minimum number of distinct distances determined by $n$ points in the plane is $\Omega(n/\log n)$ (see [GuKa15]). It was conjectured by Erdős [Er46] that the best bound is $\Omega\left(\frac{n}{\sqrt{\log n}}\right)$.

**10.** The minimum number of distinct distances determined by $n > 3$ points in convex position in the plane is $\lfloor \frac{n}{2} \rfloor$ [Al63].

**11.** The minimum number of distinct distances determined by $n > 3$ points in the plane, no three of which are on a line, is at least $\lceil \frac{n-1}{3} \rceil$. *Conjecture:* the best possible bound is $\lfloor \frac{n}{2} \rfloor$.

**12.** The minimum number of distinct distances determined by $n$ points in general position in the plane is $O(n^{1+c/\sqrt{\log n}})$, for some positive constant $c$. However, it is not known whether this function is superlinear in $n$ [ErEtal93].

**13.** There are arbitrarily large noncollinear finite point sets in the plane such that all distances determined by them are integers, but there exists no infinite set with this property.

**14.** In an $n$-element planar point set, the maximum number of noncollinear triples that determine the same angle is $O(n^2 \log n)$, and this bound is asymptotically tight [PaSh90].

**15.** Let $f_3(n)$ denote the maximum number of times that the unit distance can occur among $n$ points in $\mathcal{R}^3$. Then

$$\Omega(n^{\frac{4}{3}} \log\log n) \leq f_3(n) \leq n^{\frac{3}{2}},$$

where the lower bound was proved in [Er46] and the upper bound is a slight improvement of a result in [ClEtal90], obtained in [KaEtal12] and [Za13].

**16.** The maximum number of times that the unit distance can occur in a set of $n \geq 4$ points in $\mathcal{R}^3$ with unit diameter is $2n - 2$ [Gr56].

**17.** If $n$ is sufficiently large, then for any set of $n$ points in $\mathcal{R}^3$, the total number of farthest neighbors of all elements is at most $\frac{n^2}{4} + \frac{3n}{2} + 3$ if $n$ is even, at most $\frac{n^2}{4} + \frac{3n}{2} + \frac{9}{4}$ if $n \equiv 1 \pmod 4$, and at most $\frac{n^2}{4} + \frac{3n}{2} + \frac{13}{4}$ if $n \equiv 3 \pmod 4$. These bounds cannot be improved [Cs96].

**18.** Let $f_d(n)$ denote the maximum number of times that the unit distance can occur among $n$ points in $\mathcal{R}^d$. If $d \geq 4$ is even, then

$$f_d(n) = \frac{n^2}{2}\left(1 - \frac{1}{\lfloor \frac{d}{2} \rfloor}\right) + n - O(d).$$

If $d \geq 5$ is odd, then

$$f_d(n) = \frac{n^2}{2}\left(1 - \frac{1}{\lfloor \frac{d}{2} \rfloor}\right) + \Theta(n^{\frac{4}{3}}) \text{ [Er60], [ErPa90]}.$$

**19.** Let $\Phi_d(n)$ denote the maximum of the total number of farthest neighbors of all points over all $n$-element sets in $\mathcal{R}^d$. For every $d \geq 4$,

$$\Phi_d(n) = n^2\left(1 - \frac{1}{\lfloor \frac{d}{2} \rfloor} + o(1)\right) \text{ [ErPa90]}.$$

## Examples:

**1.** Let $P$ be the vertex set of a regular $n$-gon ($n > 3$) in the plane. Then $g(P)$, the number of distinct distances determined by $P$, is equal to $\lfloor \frac{n}{2} \rfloor$. The number of times that the diameter of $P$ is realized is equal to $n$ if $n$ is odd, and $\frac{n}{2}$ if $n$ is even.

**2.** Take a regular hexagon of side length $k$ and partition it into $6k^2$ equilateral triangles with unit sides. Let $P$ denote the union of the vertex sets of these triangles. Then $P$ is a separated set, and $|P| = n = 3k^2 + 3k + 1$. The number of times that the minimum (unit) distance occurs between two elements of $P$ is $9k^2 + 3k = 3n - \sqrt{12n - 3}$.

**3.** Let $P$ denote an $n^{\frac{1}{2}} \times n^{\frac{1}{2}}$ part of the integer grid; i.e., let $P = \{(x, y) \mid 1 \leq x, y \leq n^{\frac{1}{2}}\}$. It follows from classical number-theoretic results that there exists an integer $k$ ($\frac{n}{16} \leq k \leq \frac{n}{8}$) that can be written as the sum of two squares in $2n^{\frac{c}{\log\log n}}$ different ways, for a constant $c > 0$. Thus, for every $(x, y) \in P$, the number of points $(x', y') \in P$ satisfying $(x - x')^2 + (y - y')^2 = k$ is at least $2n^{\frac{c}{\log\log n}}$. In other words, the distance $k^{\frac{1}{2}}$ occurs $n^{1+\frac{c}{\log\log n}}$ times among the elements. By proper scaling, an $n$-element point set $P'$ is obtained in which the unit distance occurs $n^{1+\frac{c}{\log\log n}}$ times. That is, $f(P') = n^{1+\frac{c}{\log\log n}}$. It can also be shown that the number of distinct distances determined by $P'$ satisfies $g(P') = g(P) = \frac{c'n}{\sqrt{\log n}}$ for a suitable positive constant $c'$.

**4.** *Lenz' construction:* Let $C_1, \ldots, C_{\lfloor \frac{d}{2} \rfloor}$ be circles of radius $\frac{1}{\sqrt{2}}$ centered at the origin of $\mathcal{R}^d$, and assume that the supporting planes of these circles are mutually orthogonal. Choose $n_i$ points on $C_i$, where $n_i = \lfloor n/\lfloor \frac{d}{2} \rfloor \rfloor$ or $n_i = \lceil n/\lfloor \frac{d}{2} \rfloor \rceil$, so that $\sum_i n_i = n$. It is clear that any pair of points belonging to different circles $C_i$ are at unit distance from each other. Hence, this point system determines at least

$$\frac{n^2}{2}\left(1 - \frac{1}{\lfloor \frac{d}{2} \rfloor}\right) + n - O(d)$$

unit distances.

**5.** Let $p_1, p_2, p_3, p_4$ be the vertices of a regular tetrahedron with side length 1 in $\mathcal{R}^3$. The locus of points in 3-space lying at unit distance from both $p_1$ and $p_2$ is a circle passing through $p_3$ and $p_4$. Choose distinct points $p_5, p_6, \ldots, p_n$ on the shorter arc of this circle between $p_3$ and $p_4$. An $n$-element point set in $\mathcal{R}^3$ is obtained with diameter 1 and in which the diameter occurs $2n - 2$ times.

### 13.3.4   COLORING

One of the oldest problems in graph theory is the Four Color Problem (§8.6.4). This problem has attracted much interest among professional and amateur mathematicians, and inspired a lot of research about colorings, including Ramsey theory [GrRoSp90] and the study of chromatic numbers, polynomials, etc. In this section, some coloring problems are discussed in a geometric setting.

**Definitions:**

A *coloring* of a set with $k$ colors is a partition of the set into $k$ parts. Two points that belong to the same part are said to have the same color.

The *chromatic number of a graph* $G$ is the minimum number of colors $\chi(G)$ needed to color the vertices of $G$ so that no two adjacent vertices have the same color.

The *chromatic number of a metric space* is the chromatic number of the unit distance graph of the space; that is, the minimum number of colors needed to color all points of the space so that no two points of the same color are at unit distance.

The *polychromatic number of a metric space* is the minimum number of colors $\chi$ needed to color all points of the space so that for each color class $C_i$ $(1 \leq i \leq \chi)$ there is a distance $d_i$ with the property that no two points of this color are at distance $d_i$ from each other.

A point set $P$ in $\mathcal{R}^d$ is *k-Ramsey* if for any coloring of $\mathcal{R}^d$ with $k$ colors, at least one of the color classes has a subset congruent to $P$. If for every $k$, there exists $d(k)$ such that $P$ is $k$-Ramsey in $\mathcal{R}^{d(k)}$, then $P$ is called **Ramsey**.

A point set $P'$ is called a **homothetic copy** (or a **homothet**) of $P$, if $P$ and $P'$ are similar to each other and they are in parallel position.

**Facts:**

**1.** The minimum number of colors needed for coloring the plane so that no two points at unit distance receive the same color is at least 4 and at most 7. That is, the chromatic number of the plane is between 4 and 7 [JeTo95], [So09].

**2.** The following table contains the best known upper and lower bounds on the chromatic numbers of various metric spaces. $(S^{d-1}(r)$ denotes the sphere of radius $r$ in $d$-space, where the distance between two points is the length of the chord connecting them.) See [So09].

| space | lower bound | upper bound |
|---|---|---|
| line | 2 | 2 |
| plane | 4 | 7 |
| rational points of plane | 2 | 2 |
| 3-space | 6 | 15 |
| rational points of $\mathcal{R}^3$ | 2 | 2 |
| $S^2(r),\ \frac{1}{2} \leq r \leq \frac{\sqrt{3-\sqrt{3}}}{2}$ | 3 | 4 |
| $S^2(r),\ \frac{\sqrt{3-\sqrt{3}}}{2} \leq r \leq \frac{1}{\sqrt{3}}$ | 3 | 5 |
| $S^2(r),\ r \geq \frac{1}{\sqrt{3}}$ | 4 | 7 |
| $S^2\left(\frac{1}{\sqrt{2}}\right)$ | 4 | 4 |
| rational points of $\mathcal{R}^4$ | 4 | 4 |
| rational points of $\mathcal{R}^5$ | 8 | $< \infty$ |
| $\mathcal{R}^d$ | $(1+o(1))(1.2)^d$ | $(3+o(1))^d$ |
| $S^{d-1}(r),\ r \geq \frac{1}{2}$ | $d$ | $< \infty$ |

**3.** *Lower bound for the chromatic numbers of $n$-space:*  The chromatic number of $n$-space is at least $n+2$ for $n \geq 2$ (D. Raiskii) [So09].

**4.** *Improved lower bounds for specific $n$:*  The chromatic number of 3-space is at least 6 (O. Nechushtan), the chromatic number of 4-space is at least 7 (K. Cantwell), the chromatic number of 5-space is at least 9 (K. Cantwell), and the chromatic number of 6-space is at least 11 (J. Cibulka). See [So09].

**5.** *Lower bounds for chromatic numbers of $Q_n$ for small $n$:*  In 2008 J. Cibulka showed that $\chi(Q_5) \geq 8$ and $\chi(Q_7) \geq 15$. M. Mann and D. Cherkashin, A. Kulikov, and A. Raigorodskii [ChKuRa15] demonstrated that $\chi(Q_6) \geq 10$, $\chi(Q_8) \geq 16$, and $\chi(Q_9) \geq 22$, respectively.

**6.** The polychromatic number of the plane is at least 4 and at most 6 [So94].

**7.** For any finite $d$-dimensional point configuration $P$ and for any coloring of $d$-space with finitely many colors, at least one of the color classes will contain a homothetic copy of $P$. The corresponding statement is false if "homothetic copy of $P$" is replaced by "translate of $P$".

**8.** A necessary condition for a finite set $P$ to be Ramsey is that it be spherical; i.e., all its points lie on a sphere [GrRoSp90].

**9.** The following conditions are sufficient for a finite set $P$ to be Ramsey:

- $P$ is the vertex set of a right parallelepiped;

- $P$ is the set of points in $d$-space with exactly $k$ ($k < d$) nonzero coordinates having values $x_1, \ldots, x_k$ in this order, where $x_1, \ldots, x_k$ is an arbitrary sequence of nonzero reals;

- $P$ is the vertex set of a regular $n$-gon;

- $P$ is a subset of a Ramsey set;

- $P$ is the Cartesian product of two Ramsey sets [FrRö86], [FrRö90].

**10.** It follows from the first two and the last two conditions of Fact 9 that all "triangles" are Ramsey. Moreover, given any nondegenerate point configuration ("simplex") $S$, there is a constant $c(S) > 1$ such that for every $k < c^d(S)$, $S$ is $k$-Ramsey in $d$-space.

**Examples:**

**1.** Let $G$ be a graph on the vertex set $\{v_1, \ldots, v_7\}$, whose edges are $v_1v_2$, $v_1v_3$, $v_1v_4$, $v_1v_5$, $v_2v_3$, $v_2v_6$, $v_3v_6$, $v_4v_5$, $v_4v_7$, $v_5v_7$, and $v_6v_7$. The chromatic number of $G$ is 4.

**2.** The graph $G$ of Example 1 can be embedded in the plane so that if two of its vertices are connected by an edge, then the corresponding points in the plane are at unit distance. In other words, $G$ is a *subgraph* of the *unit distance graph* of the plane. (In every such imbedding, the points corresponding to $\{v_1, v_2, v_3, v_6\}$ and $\{v_1, v_4, v_5, v_7\}$ form two rhombi of side length 1 that share a vertex.) Hence, the chromatic number of the plane is at least 4.

**3.** Let $P$ be a 2-element point set in Euclidean space. For every positive integer $k$, $P$ is $k$-Ramsey in $k$-space. (To see this, consider a regular simplex in $k$-space, whose side length is equal to the distance between the elements of $P$. Any coloring of $\mathcal{R}^k$ induces a coloring of the vertices of this simplex and, by the pigeonhole principle, one can always find two vertices that get the same color. They form a 2-element set congruent to $P$. Thus, $P$ is Ramsey.)

# 13.4   POLYHEDRA

This section presents basic properties of polyhedra, commonly known as (planar) solids. Any application such as geometric modeling that models the 3-dimensional world of objects must deal with polyhedra. Basic geometric and combinatorial properties of polyhedra as well as their convex decompositions and triangulations are discussed.

## 13.4.1   GEOMETRIC PROPERTIES OF POLYHEDRA

**Definitions:**

A **(d-1)-dimensional plane** is the solution set of the linear equation $a_1x_1 + a_2x_2 + \cdots + a_dx_d = a_{d+1}$, where $a_1, a_2, \ldots, a_{d+1}$ are constants and $x_1, \ldots, x_d$ are $d$ variables.

A **hyperplane** in $d$-dimensional Euclidean space $\mathcal{R}^d$ is the set of all points on a $(d-1)$-dimensional plane.

A **closed half-space** in $\mathcal{R}^d$ is the set of all points on the hyperplane together with the points on one side of the same hyperplane.

A **convex d-polyhedron** is the intersection of a finite number of closed half-spaces in $\mathcal{R}^d$.

A **nonconvex polyhedron** is the union of a set of convex polyhedra such that the underlying space is connected and nonconvex.

A **k-face**, part of the boundary of a polyhedron, lies on at least $d - k$ hyperplanes forming the boundary. In particular, 0-faces, 1-faces, and $(d-1)$-faces of a $d$-polyhedron are **vertices**, **edges**, and **facets**, respectively.

A **polytope (d-polytope)** is a convex $d$-polyhedron that is contained in the interior of some $d$-dimensional cube: that is, a *bounded* convex $d$-polyhedron.

A $d$-polytope is **regular** if all its facets are regular $(d{-}1)$-polytopes that are combinatorially equivalent. A vertex is a regular 0-polytope.

Two polytopes $P$ and $Q$ are **dual polytopes** if there exists a one-to-one correspondence $\delta$ between the set of faces of $P$ and $Q$ such that two faces $f_1, f_2 \in P$ satisfy $f_1 \subset f_2$ if and only if $\delta(f_1) \supset \delta(f_2)$ in $Q$.

A **manifold** (**$d$-manifold**) is a topological space that is locally homeomorphic to $\mathcal{R}^d$ everywhere.

A **manifold $d$-polyhedron** is a polyhedron whose boundary is topologically the same as a $(d{-}1)$-manifold. That is, every point on the boundary of a manifold $d$-polyhedron has a small neighborhood that looks like $\mathcal{R}^d$.

A **non-manifold $d$-polyhedron** is a $d$-polyhedron whose boundary is not a manifold.

A manifold 3-polyhedron has **genus** $g$ if its boundary is a 2-manifold with genus $g$. A 2-manifold surface has genus $g$ if every set of $g+1$ circular cuts separate the surface, but not all sets of $g$ circular cuts do.

Edges in a 3-polyhedron are **reflex edges** if the inner angle subtended by two faces meeting at that edge is greater than $180°$.

**Facts:**

1. Every polytope has a dual polytope.

2. Every polytope is the convex hull of its vertices.

3. A $k$-face is an open set of dimension $k$.

4. *Curvature:*  The curvature $\kappa_v$ of a manifold 3-polyhedron at a vertex $v$ is

$$\kappa_v = \frac{2\pi - \sum_i \theta_i}{2\pi},$$

where $\theta_i$ is the angle between two consecutive edges incident with $v$. Intuitively, curvature at a vertex measures its "sharpness".

5. *Gauss-Bonnet theorem:*  $\sum_v \kappa_v = 2 - 2g$.

6. *Angle sums:*  Let $f$ be a face of a polytope $P$ and let $p$ be an interior point of $f$. The angle at $f$ is measured as the fraction of $P$ covered by a sufficiently small $(d{-}1)$-dimensional sphere centered at $p$. If $\alpha_k$ is the sum of angles at all $k$-dimensional faces, then

$$\sum_{k=0}^{d-1} (-1)^k \alpha_k = (-1)^{d-1} \ (Gram's\ formula,\ [Gr67]).$$

7. A 3-polyhedron is convex if and only if it does not have any reflex edges.

**Examples:**

1. Tetrahedra and cubes are manifold 3-polyhedra with genus 0. These are displayed next.

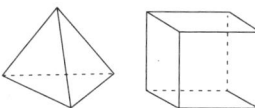

2. Two cubes meeting at a single edge, or two tetrahedra meeting at a single vertex, form non-manifold 3-polyhedra. See the following figure.

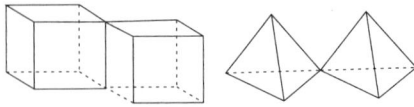

**3.**  A cube has genus 0, but a cube with a cubical through-hole is a manifold 3-polyhedron with genus 1. This is illustrated in the following figure.

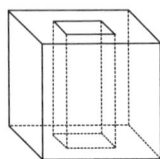

**4.**  A cube and a octahedron (bipyramid) are dual to each other; a tetrahedron is dual to itself.

**5.**  There are five regular polytopes in three dimensions: tetrahedron, cube (hexahedron), octahedron, dodecahedron, icosahedron. They are also called *Platonic solids*. See the following figure.

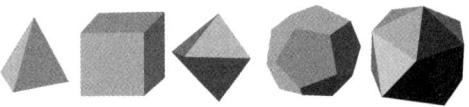

**6.**  There is a circular cut for a toroidal surface that does not separate it, though any two circular cuts always separate it.

### 13.4.2   TRIANGULATIONS

A complex domain is decomposed into simple parts for computational simplicity in many applications. For example, in finite element methods often a domain is triangulated into simplices.

**Definitions:**

A *simplex* (*d-dimensional simplex* or *d-simplex*) is a $d$-polytope with $d+1$ vertices.

A *triangulation* of a $d$-polyhedron is a convex decomposition in which each convex piece of the decomposition is a $d$-simplex.

A polyhedron is *triangulated with Steiner points* if the vertex set of simplices in the triangulation is strictly a superset of the set of vertices of the polyhedron. This type of triangulation uses extra points (other than the vertices of the polyhedron) as vertices.

A triangulation of a polyhedron is a *simplicial complex* if for every two simplices $\sigma_1, \sigma_2$ in the triangulation, $\sigma_1 \cap \sigma_2$ is either empty or a face of both simplices.

A *convex decomposition* of a polyhedron is its partition into convex pieces that have disjoint interiors.

The *aspect ratio of a simplex* is the ratio of the radius of the circumscribing sphere to the radius of the inscribing sphere of the simplex.

The *aspect ratio of a triangulation* is the largest aspect ratio of a simplex in the triangulation.

**Facts:**

**1.** Every $d$-polytope can be triangulated without Steiner points.

**2.** Every $d$-polytope with $n$ faces can be triangulated into $O(n)$ simplices in $O(n)$ time and space.

**3.** There are nonconvex 3-polyhedra that can't be triangulated without Steiner points.

**4.** The problem of deciding if a nonconvex 3-polyhedron can be triangulated without Steiner points or not is NP-complete [RuSe92].

**5.** The problem of decomposing a polyhedron into the minimum number of convex pieces is NP-hard.

**6.** Every polyhedron can be decomposed into disjoint convex pieces by repeatedly slicing the polyhedron through reflex edges [BaDe92], [Ch84].

**7.** There is a class of polyhedra with $n$ edges, of which $r$ are reflex, that requires at least $\Omega(n + r^2)$ convex pieces for its decomposition. These polyhedra have two sets of parallel edges which are created as reflex edges. Two such sets are placed on two hyperbolic paraboloids with an angle of almost $90°$ between them. These polyhedra require at least $\Omega(n + r^2)$ convex pieces for their decomposition [Ch84].

**8.** Every manifold 3-polyhedron can be triangulated into $O(n+r^2)$ tetrahedra in $O((n+r^2)\log r)$ time [ChPa90].

**9.** There exists a polynomial-time algorithm that produces a triangulation of any 3-polyhedron with an aspect ratio and size that are within a constant factor of the optimal [MiVa92].

**Examples:**

**1.** A triangle is a 2-simplex; a tetrahedron is a 3-simplex.

**2.** Part (a) of the following figure shows a tetrahedron with a bad aspect ratio; part (b) shows a tetrahedron with a good aspect ratio.

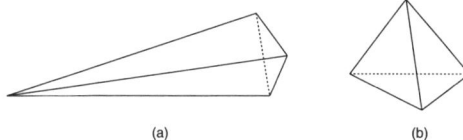

(a)          (b)

**3.** The Schönhardt polyhedron is a nonconvex 3-polyhedron that cannot be triangulated without Steiner points. This polyhedron can be constructed out of a prism whose base and top facets are equilateral triangles. Twist the top triangle, keeping the base fixed. This destroys the planarity of vertical facets. To maintain the planarity, triangulate these facets appropriately [RuSe92].

## 13.4.3 FACE NUMBERS

In many cases the complexity of algorithms dealing with polyhedra depends on the number of faces they contain. Therefore, combinatorial bounds on these numbers play a significant role in analyzing these algorithms.

**Definitions:**

A *cyclic d-polytope* is the convex hull of a set of $n$ $(n \geq d+1)$ points on the moment curve in $\mathcal{R}^d$, $x(t) = (t, t^2, \ldots, t^d)$.

A *face vector* of a $d$-polyhedron $P$ is the $d$-dimensional vector $(f_0, f_1, \ldots, f_{d-1})$, where $f_i = f_i(P)$ is the number of $i$-dimensional faces of $P$.

A *simplicial polytope* is a polytope in which all faces are simplices.

**Facts:**

1.  For $2k \leq d$, every $k$ vertices of a cyclic polytope define a $(k-1)$-face.

2.  *Euler's relation:* For any $d$-polytope, $\sum_{i=0}^{d-1} (-1)^i f_i = 1 - (-1)^d$.

3.  For a manifold 3-polyhedron with genus $g$, $\sum_{i=0}^{2} (-1)^i f_i = 2 - 2g$.

4.  The edges on the boundary of a manifold 3-polyhedron with genus 0 form a planar graph. By the property of planarity, the number of vertices, edges, and facets of such polyhedra are within a constant factor of each other.

5.  *Dehn-Sommerville equations:* The face vectors of simplicial polytopes satisfy the following equations for $-1 \leq k \leq d-2$ with $f_{-1} = 1$:

$$E_d^k : \sum_{j=k}^{d-1} (-1)^j \binom{j+1}{k+1} f_j = (-1)^{d-1} f_k.$$

In particular, $E_d^{-1}$ is Euler's relation.

6.  *Upper bound theorem:* For any $d$-polytope $P$ with $n$ vertices, $f_i(P) = O(n^{\lfloor \frac{d}{2} \rfloor})$ for $1 \leq i \leq d-1$.

7.  *Optimality of cyclic polytopes:* Cyclic polytopes achieve the upper bound since they have $\binom{n}{k} = \Omega(n^k)$ $(k-1)$-faces for $2k \leq d$. This implies that they have $\Omega(n^{\lfloor d/2 \rfloor})$ $\lfloor \frac{d}{2} \rfloor$-faces.

**Examples:**

1.  The 3-dimensional cube $(d = 3)$ has $f_0 = 8, f_1 = 12$, and $f_2 = 6$; thus, by Fact 2, $f_0 - f_1 + f_2 = 2$.

2.  A 3-dimensional cube with a cubical through-hole $(g = 1)$ has $f_0 = 16, f_1 = 24$, and $f_2 = 10$; thus, by Fact 3, $f_0 - f_1 + f_2 = 0$.

## 13.5   ALGORITHMS AND COMPLEXITY IN COMPUTATIONAL GEOMETRY

Computational geometry studies efficient algorithms for solving geometric problems and has applications in computer graphics, robotics, VLSI design, computer-aided design, pattern recognition, statistics, and other fields. The study of computational geometry uses concepts and results from classical geometry, topology, combinatorics, as well as standard techniques from design and analysis of computer algorithms. See [PrSh85] and [TóORGo17].

## 13.5.1   CONVEX HULLS

Finding efficient algorithms for the construction of convex hulls has been a central topic in computational geometry. Several efficient algorithms for constructing boundaries of convex hulls of sets of points in the plane have been developed.

**Definition:**

The **convex hull** of a set of points in $\mathcal{R}^d$ is the smallest convex set containing the points.

**Algorithms:**

1. *Finding boundaries of convex hulls by rotational sweeping*:

   - *Graham Scan*:   Given a set $S$ of $n$ points in the plane, Algorithm 1 scans the points rotationally around a fixed point and eliminates those that are not hull vertices. The remaining points are the vertices of the boundary of the convex hull of $S$. The running time of Graham Scan is $O(n \log n)$, which is dominated by the sorting of the points. The remaining steps take only linear time.

---

**Algorithm 1**:   **Graham Scan.**

input: a finite set $S$ of points in the plane

output: the vertices of the boundary of the convex hull of $S$

$p_0 :=$ the point in $S$ with the minimum $y$-coordinate

sort remaining points by polar angle around $p_0$; append the point $p_0$ to the
   end of the sorted list; let the resulting list be $(p_1, p_2, \ldots, p_n)$, where $p_n = p_0$.

$H[1] := p_n$;  $H[2] := p_1$;  $j := 2$

**for** $i := 2$ **to** $n$

   **while** the path $\{H[j-1], H[j], p_i\}$ does not form a left turn

      $j := j - 1$

   $j := j + 1$

   $H[j] := p_i$

$\{H[1], H[2], \ldots, H[j]$ is the boundary of the convex hull$\}$

---

   - *Jarvis' March*:   Given a set $S$ of $n$ points in the plane, Jarvis' March algorithm constructs the boundary of the convex hull by "marching around" the outer perimeter of $S$. This method is also called "gift-wrapping". Jarvis' March runs in time $O(hn)$, where $h$ is the number of vertices of the convex hull, which in the worst case is $n$.

2. *Divide-and-conquer algorithms*:

   - *QuickHull*:   This algorithm recursively constructs a chain on the boundary of the convex hull, connecting two hull vertices $u$ and $v$. It first finds a hull vertex $w$ on the chain (for example, $w$ is the farthest point from the line $\overline{uv}$). Then the subchains connecting $u$ and $w$, $w$ and $v$, respectively, are constructed recursively and are concatenated [PrSh85]. QuickHull runs practically fast, but in the worst case the running time of QuickHull is $O(n^2)$.

- *MergeHull*:   This algorithm first partitions the set $S$ of points into two subsets $S_1$ and $S_2$ of equal size and then recursively constructs the boundaries of the convex hulls $CH(S_1)$ and $CH(S_2)$. Finally, $CH(S_1)$ and $CH(S_2)$ are "merged" into the convex hull of the set $S$ [PrSh85].

  The boundary of the convex hull for $S$ is the same as the boundary of the convex hull for the hull vertices of $CH(S_1)$ and $CH(S_2)$. Thus, to construct the boundary of $CH(S)$, first sort the hull vertices of $CH(S_1)$ and $CH(S_2)$ (this sorting can be done in linear time), then apply the linear scan of Graham Scan to construct $CH(S)$. Therefore, the boundary of the convex hull $CH(S)$ can be constructed from $CH(S_1)$ and $CH(S_2)$ in linear time. The running time of MergeHull is $O(n \log n)$.

3.  *Other methods*:

   - *Incremental method*:   The *incremental method* for constructing the boundary of the convex hull of a set of points in the plane adds one point at a time to an already constructed boundary of a convex hull. This method has time complexity $O(n \log n)$ [PrSh85]. An advantage of this method is that it can be generalized to construct boundaries of convex hulls in higher dimensions [Ed87].

   - *Algorithm by Kirkpatrick and Siedel based on the prune-and-search method*:   This algorithm partitions a given set of points in the plane into two linearly separable subsets of equal size, finds the two edges of the boundary of the convex hull that "bridge" these two subsets, and recursively constructs the subchains on the boundary of the convex hull between these two bridges. This method has time complexity $O(n \log h)$, where $h$ is the number of vertices on the boundary of the convex hull [Ya90].

**Facts:**

**1.** The problem of finding convex hulls is at least as hard as sorting. The lower bound $\Omega(n \log n)$ of sorting on comparison decision trees also applies to the convex hull problem. This lower bound $\Omega(n \log n)$ can be extended to a more general computation model, the bounded-degree algebraic decision trees.

**2.** An $O(n \log n)$ time algorithm for constructing the boundary of the convex hull of a set of points in $\mathcal{R}^3$ has been developed, which is a generalization of the MergeHull algorithm.

**3.** For dimension $d > 3$, the convex hull of $n$ points in $\mathcal{R}^d$ can have up to $O(n^{\lfloor \frac{d}{2} \rfloor})$ faces. An algorithm based on the incremental method has been proposed to construct the convex hull for a set of $n$ points in $\mathcal{R}^d$ in time $O(n^{\lfloor \frac{d+1}{2} \rfloor})$ [Ya90]. An optimal algorithm of time $O(n^{\lfloor \frac{d}{2} \rfloor})$ has been developed recently by Chazelle. (See the bibliography of [Mu94] for a reference.)

**Examples:**

**1.** The following figure shows a set $S$ of 7 points in the plane and the convex hull of $S$.

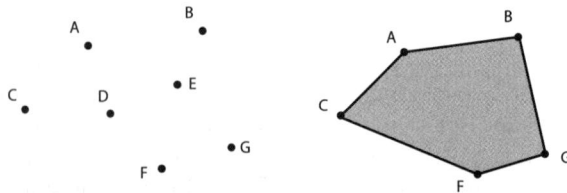

**2.** The convex hull of the set $\{(0,0,0),\ (0,0,1),\ (0,1,0),\ (0,1,1),\ (1,0,0),\ (1,0,1),$ $(1,1,0),\ (1,1,1)\}$ in $\mathcal{R}^3$ is the surface of the unit cube together with its interior.

---

### 13.5.2   TRIANGULATION ALGORITHMS

Triangulation plays an important role in many applications. On a triangulated planar straight-line graph, many problems can be solved more efficiently. Triangulation of a set of points arises in numerical interpolation of bivariate data and in the finite element method.

**Definitions:**

A **planar straight-line graph** (**PSLG**) is a planar graph such that each edge is a straight line.

A **triangulation of a simple polygon** $P$ is an augmentation of $P$ with nonintersecting diagonal edges connecting vertices of $P$ such that in the resulting PSLG, every bounded face is a triangle.

A **triangulation of a PSLG** $G$ is an augmentation of $G$ with nonintersecting edges connecting vertices of $G$ so that every point in the interior of the convex hull of $G$ is contained in a face that is a triangle. In particular, the PSLG to be triangulated can be simply $n$ discrete points in the plane.

A **chain** is a PSLG with vertices $v_1, \ldots, v_n$ and edges $\{v_1, v_2\}, \{v_2, v_3\}, \ldots, \{v_{n-1}, v_n\}$.

A chain is **monotone** if there is a straight line $L$ such that every line perpendicular to $L$ intersects the chain in at most one point.

A simple polygon is **monotone** if its boundary can be decomposed into two monotone chains.

Two vertices $v$ and $u$ in a polygon are **visible** from each other if the open line segment $\overline{uv}$ is entirely in the interior of the polygon.

**Facts:**

**1.** Every simple polygon can be triangulated.

**2.** Every triangulation of a simple polygon with $n$ vertices has $n - 2$ triangles and $n - 3$ diagonals.

**3.** Given a simple polygon with $n$ vertices, there is a diagonal that divides the polygon into two polygons that have at most $\left\lceil \frac{2n}{3} \right\rceil + 1$ vertices.

**4.** For a history of triangulations, see [OR87].

**5.** Simple polygons can be triangulated in $O(n)$ time using an algorithm developed by Chazelle [Ch91].

**6.** PSLGs can be triangulated in $O(n \log n)$ time. (See the *triangulation of a general PSLG* algorithm — item 3 in the following list of algorithms.) This is optimal because a lower bound $\Omega(n \log n)$ has been derived for the time complexity of triangulation of a PSLG.

**Algorithms:**

**1.** *Triangulation of a monotone polygon*: A monotone polygon $P$ can be triangulated in linear time based on the following greedy method. Observe that the monotone polygon $P$

is triangulated if nonintersecting edges are added so that no two vertices are visible from each other.

If necessary, rotate the polygon so that it is monotone with respect to the $y$-axis. Sort the vertices of $P$ by decreasing $y$-coordinate. (This sorting can be done in linear time by merging the two monotone chains of $P$.) Move through the sorted list, and for a vertex $v$, examine each vertex $u$ lower than $v$, in the sorted order, and add an edge between vertices $v$ and $u$ as long as $u$ is visible from $v$. The edge addition process for the vertex $v$ stops at a lower vertex that is not visible from $v$. Then move to the next vertex and perform the edge addition process. Note that once an edge is added between vertices $v$ and $u$, then no vertices between $v$ and $u$ in the sorted list are visible from a vertex that is lower than $u$. Therefore, such vertices can be ignored in the later edge addition process. The edge addition process for all vertices can be performed in linear time if a stack is used to hold the sorted list.

**2.** *Triangulation of a simple polygon:*  Given a general simple polygon $P$, partition $P$ in time $O(n \log n)$ into monotone polygons, then apply the previous linear-time algorithm to triangulate each monotone polygon. This gives a triangulation of a simple polygon in time $O(n \log n)$.

**3.** *Triangulation of a general PSLG:*  To triangulate a general PSLG $G$, first add edges to $G$ so that each face is a simple polygon (no nonconsecutive edges intersect), then apply Chazelle's linear-time algorithm (Fact 5) to triangulate each face.

The complexity of Chazelle's algorithm can be avoided since there is an efficient algorithm that adds edges to a PSLG so that each face is a monotone polygon. To do this, observe that in a PSLG $G$ every face is a monotone polygon if and only if each vertex (except the highest one) has a higher neighbor and each vertex (except the lowest one) has a lower neighbor. Thus, to make each face of $G$ a monotone polygon, check each vertex of $G$ and for those that do not have desired neighbors, add proper edges to them. This process can be accomplished in time $O(n \log n)$ using the plane sweeping method [PrSh85]. Now, the simpler linear-time algorithm for triangulating a monotone polygon (see item 1) is applied to triangulate each face.

**Examples:**
**1.** The following figure illustrates a simple polygon and two of its triangulations.

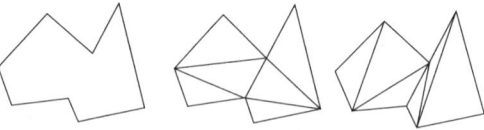

**2.** In part (a) of the following figure the chain is monotone (with respect to any horizontal line); the chain in part (b) is not monotone.

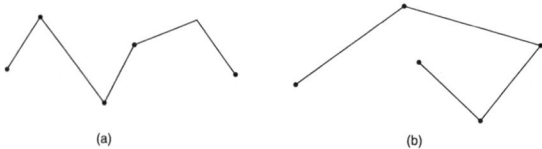

(a)                    (b)

## 13.5.3  VORONOI DIAGRAMS AND DELAUNAY TRIANGULATIONS

### Definitions:

Given a set $S = \{p_1, \ldots, p_n\}$ in $\mathcal{R}^d$, the **Voronoi diagram** Vor($S$) of $S$ is a partition of $\mathcal{R}^d$ into $n$ convex polytopes (**Voronoi cells** or **Dirichlet cells**) $V(p_1), \ldots, V(p_n)$ such that the region $V(p_i)$ is the locus of points that are closer to $p_i$ than to any other point in $S$.

Given the Voronoi diagram Vor($S$) of a set $S = \{p_1, \ldots, p_n\}$ of points in the plane, the **straight-line dual** $D(S)$ of Vor($S$) is a PSLG whose vertices are the points in $S$ and in which two vertices $p_i$ and $p_j$ in $D(S)$ are connected if and only if the regions $V(p_i)$ and $V(p_j)$ share a common edge.

The PSLG $D(S)$ is a triangulation of the set $S$, called the **Delaunay triangulation** of $S$.

### Facts:

**1.** The Voronoi diagram of a set of $n$ points in the plane can be constructed in time $O(n \log n)$.

**2.** The Delaunay triangulation of a set $S$ of points in the plane has the property that the circuit with the three vertices of a triangle of the triangulation on its boundary contains no other point of the set $S$. This property makes the Delaunay triangulation useful in interpolation applications.

**3.** The convex hull problem in the plane can be reduced in linear time to the Voronoi diagram problem in the plane: a point $p$ in a set $S$ is a hull vertex if and only if $V(p)$ is unbounded, and two hull vertices $p_i$ and $p_j$ are adjacent if and only if the two unbounded regions $V(p_i)$ and $V(p_j)$ share a common edge. Thus, the $O(n \log n)$ time algorithm for constructing Voronoi diagrams (Algorithm 2) is optimal.

**4.** The Voronoi diagram problem for $n$ points in $\mathcal{R}^d$ can be reduced in linear time to the convex hull problem for $n$ points in $\mathcal{R}^{d+1}$ [Ed87]. Thus, the Voronoi diagram of a set of $n$ points in $\mathcal{R}^d$ can be constructed in time $O(n^{\lfloor (d+1)/2 \rfloor})$ based on the optimal algorithm for constructing the convex hull of a set of $n$ points in $\mathcal{R}^{d+1}$.

### Algorithms:

**1.** *Construction of Voronoi diagrams in the plane:*  The Voronoi diagram of a set of points in $\mathcal{R}^2$ can be constructed using the divide-and-conquer method of Algorithm 2.

To efficiently partition a set $S$ into a left subset and a right subset of equal size in each recursive construction, pre-sort the set $S$ by $x$-coordinate. To merge the Voronoi diagrams Vor($S_L$) and Vor($S_R$) into the Voronoi diagram Vor($S$), add the part of Vor($S$) that is missing in Vor($S_L$) and Vor($S_R$) and then delete the part of Vor($S_L$) and Vor($S_R$) that does not appear in Vor($S$).

---

**Algorithm 2**:  **Construction of Voronoi diagrams.**

input: a set $S$ of points in the plane

output: the Voronoi diagram of $S$

**if** $|S| < 4$ **then** construct Vor($S$) directly and stop

**else**

> partition $S$ into two equal size subsets $S_L$ (left subset) and $S_R$ (right subset)
>    separated by a vertical line
> construct $\text{Vor}(S_L)$ and $\text{Vor}(S_R)$ recursively;
> merge $\text{Vor}(S_L)$ and $\text{Vor}(S_R)$ into $\text{Vor}(S)$;

**2.** *Voronoi diagrams and geometric optimization problems:* An $O(n \log n)$ time optimal algorithm can be derived via the Voronoi diagram for the problem of finding for each point in a set $S$ of $n$ points in the plane the nearest point in $S$. This is so because each point $p$ in $S$ and its nearest neighbor correspond to two regions in $\text{Vor}(S)$ that share a common edge. This also implies an $O(n \log n)$ time optimal algorithm for the problem of finding the closest pair in a set of $n$ points in the plane.

The Voronoi diagram can be used to design an $O(n \log n)$ time optimal algorithm for constructing a Euclidean minimum spanning tree for a set $S$ of $n$ points in the plane because edges of any Euclidean minimum spanning tree must be contained in the Delaunay triangulation $D(S)$ of $S$. This algorithm implies an $O(n \log n)$ time approximation algorithm for the Euclidean traveling salesman problem which produces a traveling salesman tour of length at most twice the optimum.

**Example:**

**1.** The left half of the following figure illustrates a set of six points in the plane and the Voronoi diagram for the set. The right half shows the Delaunay triangulation of the set.

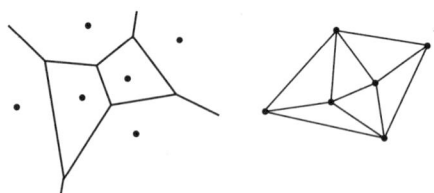

## 13.5.4   ARRANGEMENTS

**Definition:**
Given $n$ lines in the plane, the **arrangement** of the lines in the plane is the PSLG whose vertices are the intersection points of the lines and whose edges connect consecutive intersection points on each line (it is assumed that all lines intersect at a common point at infinity).

**Facts:**

**1.** The arrangement of $n$ lines in the plane partitions the plane into a collection of $O(n^2)$ faces, edges, and vertices.

**2.** The arrangement of $n$ lines can be constructed in $O(n^2)$ time (Algorithm 3), which is optimal.

**3.** An arrangement can be represented by a *doubly-connected-edge-list* in which the edges incident with a vertex can be traversed in clockwise order in constant time per edge [PrSh85].

**4.** Applications of arrangements include finding the smallest-area triangle among $n$ points, constructing Voronoi diagrams, and half-plane range query.

**5.** The arrangement of $n$ hyperplanes in $\mathcal{R}^d$ can be defined similarly, which partitions $\mathcal{R}^d$ into $O(n^d)$ faces of dimension at most $d$.

**6.** Algorithm 3 can be generalized to construct the arrangement of $n$ hyperplanes in $\mathcal{R}^d$ in $O(n^d)$ time, which is optimal [Ed87].

## Algorithm:

**1.** *Constructing the arrangement of a set of lines:* Algorithm 3 constructs the arrangement $\mathcal{A}$ of a set $H$ of $n$ lines $L_1, \ldots, L_n$ in the plane by the incremental method.

To traverse the faces of $\mathcal{A}$ that intersect the line $L_i$, start from a face $F$ that has the point $p_i$ on its boundary, and traverse the boundary of $F$ until an edge $e$ is encountered such that $e$ intersects $L_i$ at a point $q$. A new vertex $q$ is introduced in $\mathcal{A}$ and the adjacencies of the two ends of the edge $e$ are updated. Then reverse the traversing direction on the edge $e$ and start traversing the face that shares the edge $e$ with $F$, and so on. The total number of edges traversed in this process in order to insert the line $L_i$ is bounded by $O(i)$ [Mu94].

---

**Algorithm 3:   Incremental method for constructing the arrangement of a set $H$ of $n$ lines.**

   input: a set $H$ of $n$ lines $L_1, L_2, \ldots, L_n$ in the plane

   output: the arrangement $\mathcal{A}$ of the set $H$

   $\mathcal{A} := L_1$;

   **for** $i := 2$ **to** $n$

      find the intersection point $p_i$ of $L_i$ and $L_1$

      starting from $p_i$, traverse the faces of $\mathcal{A}$ that intersect $L_i$ and update the vertex
         set and edge set of $\mathcal{A}$

---

## Example:

**1.** The following figure shows an arrangement of four lines in the plane. The graph has seven vertices (including the vertex at infinity), 16 edges (of which eight are unbounded), and 11 regions (of which eight are unbounded).

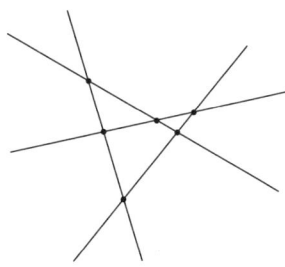

---

## 13.5.5   VISIBILITY

Visibility problems are concerned with determining what can be viewed from a given point (or points) in the plane or 3-dimensional space. See [OR87], [OR93], [Sh92].

## Definitions:

The **visibility problem** is the problem of finding what is visible, given a configuration of objects and a viewpoint.

Given $n$ nonintersecting line segments in the plane, the **visibility graph** is the graph whose vertices are the endpoints of the line segments, with two vertices adjacent if and only if they are visible from each other; i.e., there is an edge joining $a$ and $b$ if and only if the open line segment $ab$ does not intersect any other line segments.

A **star polygon** is a polygon with an interior point $p$ such that each point on the polygon is visible from $p$.

## Facts:

**1.** Visibility problems have important applications in computer graphics and robotics and have served as motivation for research in computational geometry.

**2.** Constructing the visibility graph for a set of $n$ nonintersecting line segments is a critical component of the shortest path problem in the plane. The visibility graph problem can be solved in optimal time $O(n^2)$ [Ya90].

**3.** *Art gallery theorems*: Given a simple polygon with $n$ vertices,

- there is a set $S$ of $\lfloor \frac{n}{3} \rfloor$ vertices of the polygon such that each point on or inside the polygon is visible from a point in $S$;
- there is a set $S$ of $\lceil \frac{n}{3} \rceil$ points on the polygon such that each point on or outside the polygon is visible from a point in $S$;
- there is a set $S$ of $\lceil \frac{n}{2} \rceil$ vertices of the polygon such that each point on, inside, or outside the polygon is visible from a point in $S$.

In each case the number given is the best possible.

## Algorithms:

**1.** Given $n$ line segments in the plane, compute the sequence of subsegments that are visible from the point $y = -\infty$, that is, by using parallel rays. The problem can be solved by a modified version of the plane sweeping algorithm for computing all intersection points of the line segments. The algorithm has worst-case time complexity $O(n^2 \log n)$.

**2.** An alternative algorithm is based on the divide-and-conquer approach: arbitrarily partition the set of the $n$ line segments into two equal size halves, solve both subproblems, and merge the results.

Note that the merging step amounts to computing the minimum of two piecewise (not necessarily continuous) linear functions, which can be easily done in time linear in the number of pieces if it is recursively assumed that the pieces are sorted by $x$-coordinate. For a set of $n$ arbitrary line segments in the plane, in the worst case, the number of subsegments visible from the point $y = -\infty$ is bounded by $O(n\alpha(n))$ [Ya90], where $\alpha(n)$, the inverse of Ackermann's function (§1.3.2), is a monotonically increasing function that grows so slowly that for practical purposes it can be treated as a constant. Therefore, the merging step runs in time $O(n\alpha(n))$. Consequently, the time complexity of the algorithm based on the divide-and-conquer method is $O(n\alpha(n) \log n)$.

**3.** *Three-dimensional visibility*: Given a set of disjoint opaque polyhedra in $\mathcal{R}^3$, find the part of the set that is visible from the viewpoint $z = -\infty$ (that is, with parallel rays). The problem can be solved in time $O(n^2 \log n)$ by a modified plane sweeping algorithm for computing the intersection points of the line segments that are the projections of the edges of the polyhedra on the $xy$-plane. Optimal algorithms of time complexity $O(n^2)$ have been developed based on line arrangements [Do94], [Ya90].

**Examples:**

**1.** The following figure shows three line segments and the visibility graph with six vertices determined by the line segments. The edges of the visibility graph are shown as dotted lines.

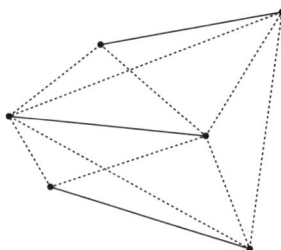

**2.** *Surveillance problems*: A variety of problems require that "guards" be posted at points of a polygon so that corners and/or edges are visible. (See the art gallery theorems of Fact 3.)

**3.** *Hidden surface removal*: An important problem in computer graphics is the problem of finding (and removing) the portions of shapes in 3-dimensional space that are hidden from view, when the object is viewed from a given point. (See the three-dimensional visibility algorithm.)

# 13.6 GEOMETRIC DATA STRUCTURES AND SEARCHING

This section describes the use of data structures for searching, or querying, among a set $S$ of geometric objects. For each of the following problems, there are algorithms that perform a single search in time proportional to $n$, the total complexity of all the geometric objects in $S$. These single-search algorithms use minimal data structures and minimal preprocessing time. When the application searches multiple times among the elements of the same set $S$, it becomes more efficient to preprocess the objects in $S$ into a data structure that would allow a faster searching procedure.

This section presents four fundamental searching problems in computational geometry for which clever data structures reduce the search time to $O(\log_2^k n)$, where $k$ is a small constant (often equal to 1 for problems in the plane and 3-dimensional space). This section covers only the static versions of these problems; that is, $S$ never changes. The dynamic versions allow deletions from $S$ and/or insertions into $S$ in between queries. In the dynamic versions of the problem, in addition to polylogarithmic query time, the goal is to keep the update time polylogarithmic. The dynamic versions are, as a rule, much more difficult.

## 13.6.1 POINT LOCATION

**Definition:**
Let $p$ be a point and $S$ a subdivision of $\mathcal{R}^d$. ($S$ can be a single geometric object, such as a polytope, or can be a general subdivision of $\mathcal{R}^d$.) The **point location problem** is the problem of determining which region of $S$ contains $p$.

**Examples:**

**1.** Locate a point in the subdivision of the space induced by an *arrangement* of a set of hyperplanes. (See §13.5.4.)

**2.** Locate a point in a subdivision all of whose regions are convex (a *convex subdivision*). For example, an arrangement of hyperplanes is a convex subdivision.

**3.** Search for nearest neighbors in Voronoi diagrams. (See §13.5.3.)

**4.** Range searching. (See §13.6.2.)

**5.** Ray shooting. (See §13.6.3.)

**Facts:**

**1.** *Point location in straight-line subdivision using triangulation hierarchy*: Given an $n$-vertex triangulation $R' = (V', T')$, where $V'$ is the set of vertices and $T'$ is the set of triangles, an $(n{+}3)$-vertex *enclosed* triangulation $R = (V, T)$ is the triangulation $R'$ together with the triangulation of the region between $U$ and the convex hull of $R'$, where $U = (V_U, T_U)$ is a triangle that contains $R'$ in its interior. The *triangulation-hierarchy* of [Ki83] consists of a sequence of triangulations $\mathcal{R} = \langle R_1, R_2, \ldots, R_{c \log_2 n} \rangle$, where $R_1 = R$, $R_{c \log_2 n} = U$, and $R_i$ is created from $R_{i-1}$ as follows (see Example 1):

- remove from $V_{i-1} - V_U$ a set $X$ of independent (that is, nonadjacent) vertices and remove from $T_{i-1}$ the set $Z$ of all triangles incident with any vertex in $X$: $V_i = V_{i-1} - X$, $T_i = T_{i-1} - Z$;

- retriangulate any polygons in $R_i = (V_i, T_i)$.

**2.** Algorithm 1 produces a hierarchy for planar subdivision. With minor modifications (for example, "triangles" become tetrahedrons), it can be used for subdivisions in $\mathcal{R}^3$. It can be proven that $|T_{c \log_2 n}| = 1$ for some constant $c$, and that $|\tau(t)|$ is a constant for every $t$.

**3.** Algorithm 1 runs in $O(n)$ time and produces a triangulation hierarchy that takes $O(n)$ space.

---

**Algorithm 1: Computing the triangulation hierarchy.**

input: planar straight-line subdivision $S$

output: triangulation hierarchy of $S$

compute triangulation $R'$ of $S$; compute enclosed triangulation $R$ of $R'$; choose a
     small constant $k$; $R_1 := R$; $i := 1$

**while** $|T_i| > 1$

    $i := i + 1$; $R_i := R_{i-1}$; mark all vertices in $V_i$ having degree $< k$

    **while** there exists some marked vertex $v$

        $P := (V_P, E_P)$ (polygon consisting of vertices adjacent to $v$; that is,

            $V_P := \{v_j \mid (v, v_j, v_k) \in T_i\}$; $E_P := \{(v_j, v_k) \mid (v, v_j, v_k) \in T_i\}$)

        remove $v$ from $V_i$; remove all the triangles incident with $v$ from $T_i$; that is,

            $T_{rem} := \{(v, v_j, v_k) \mid v_j \in V_P\}$; $T_i := T_i - T_{rem}$

        compute the triangulation $R_P$ of $P$

        **for** each triangle $t$ in $R_P$

            $\tau(t) :=$ the set of triangles in $T_{rem}$ that overlap with $t$

            create a pointer from $t$ to every triangle in $\tau(t)$

        unmark $v$ and any marked vertices in $V_P$

**4.** Algorithm 2 carries out point location, given a triangulation hierarchy, and takes $O(\log_2 n)$ time.

---

**Algorithm 2: Performing point location.**

input: a point $q$ and a triangulation hierarchy $\mathcal{R}$

output: triangle that contains $q$

check if $R_{c \log_2 n}$ contains $q$

$i := c \log_2 n - 1;\ t := U$

**while** $i \geq 1$

    determine the triangle $t'$ in $\tau(t)$ that contain $q$ using pointers from $t$ to $\tau(t)$

    $t := t';\ i := i - 1$

---

**5.** The following table shows the complexity of various point location algorithms. The number $m$ denotes the number of regions (or cells) in the subdivision $S$; $n$ denotes the total combinatorial complexity of $S$.

| dimension | subdivision type | query time | space | preprocessing time |
|---|---|---|---|---|
| 2 | convex subdivision | $O(\log_2 n)$ | $O(n)$ | $O(n)$ |
| 3 | simple polytope | $O(\log_2 n)$ | $O(n)$ | $O(n)$ |
| 3 | convex subdivision | $O(\log_2^2 n)$ | $O(n \log_2^2 n)$ | $O(n \log_2^2 n)$ |
| $d$ | arrangement of $n$ hyperplanes | $O(\log_2 n)$ | $O(n^d)$ | $O(n^d)$ |
| $d$ | subdivision of $m$ $(d-1)$-simplices with a total of $n$ faces, $\epsilon > 0$ | $O(\log_2^3 m)$ | $O(m^{d-1+\epsilon} + n)$ | $O(m^{d-1+\epsilon} + n \log_2 m)$ |

**Example:**

**1.** Part (a) of the figure seen on the next page shows triangulation $R_1$ of Fact 1 (the vertices that are removed from $R_1$ are circled). Part (b) shows triangulation $R_2$ (dotted lines are edges in retriangulation). Part (c) gives a list of the pointers from triangles in $R_2$ to triangles in $R_1$.

---

## 13.6.2 RANGE SEARCHING

**Definitions:**

The **range counting problem** is the problem of counting the number of points in a given set $S \subseteq \mathcal{R}^d$ that lie in a given query range $q$.

The **range reporting problem** is the problem of determining all points in a given set $S \subseteq \mathcal{R}^d$ that lie in a given query range $q$.

The **range emptiness problem** is the problem of determining if a given query range $q$ contains any points from a given set $S \subseteq \mathcal{R}^d$.

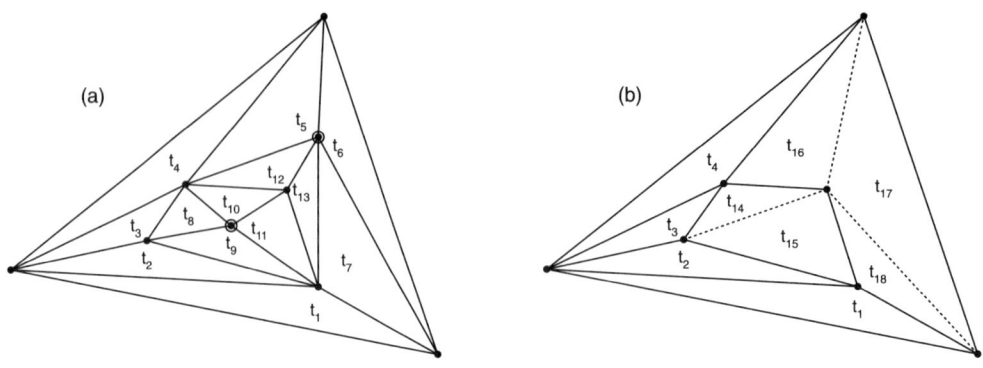

**Facts:**

**1.** The following table gives information on various range searching algorithms. The integer $n$ is the number of points in $S$; $\epsilon$ is an arbitrarily small positive constant. When the query is reporting, the query time has an additive factor of $k$, which is the size of the output.

| dim | range type | query time | space | preprocessing time |
|-----|------------|------------|-------|--------------------|
| 2 | orthogonal | $O(\log_2 n)$ | $O(n \log_2^{2+\epsilon} n)$ | $O(n \log_2 n)$ |
| 2 | convex polygon | $O(\sqrt{n} \log_2 n)$ | $O(n)$ | $O(n^{1+\epsilon})$ |
| 2 | convex polytope | $O(n^{2/3} \log_2^2 n)$ | $O(n \log_2 n)$ | $O(n^{1+\epsilon})$ |
| $d$ | convex polytope for $n \leq m \leq n^d$ | $O(\log_2^{d+1} n)$ $O(n^{1-1/d})$ $O((n/m^{1/d}) \log_2^{d+1} n)$ | $O(n^d)$ $O(n)$ $O(m^{1+\epsilon})$ | $O(n^d (\log_2 n)^\epsilon)$ $O(n^{1+\epsilon})$ $O(m^{1+\epsilon})$ |
| $d$ | half-space for $n \leq m \leq n^d$ | $O(\log_2 n)$ $O(n/m^{1/d})$ | $O(n^d / \log_2^d n)$ $O(m)$ | $O(n^d / \log_2^{d-\epsilon} n)$ $O(n^{1+\epsilon} + m(\log_2 n)^\epsilon)$ |

**2.** The following table gives information on various range reporting algorithms.

| dim | range type | query time | space | preprocessing time |
|---|---|---|---|---|
| 2 | half-plane fixed-radius circle | $O(\log_2 n + k)$ | $O(n)$ | $O(n \log_2 n)$ |
| 2 | orthogonal | $O(\log_2 n + k)$ | $O(n \log_2^\epsilon n)$ | $O(n \log_2 n)$ |
| 3 | half-space | $O(\log_2 n + k)$ | $O(n \log_2 n)$ | $O(n \log_2^3 n \log_2 \log_2 n)$ |
| $d$ | half-space $n \le m \le n^{\lfloor \frac{d}{2} \rfloor}$ | $O(\log_2 n + k)$ $O(n^{1-1/\lfloor \frac{d}{2} \rfloor + \epsilon} + k)$ $O(\frac{n}{m^{1/d}} \log_2 n + k)$ | $O(n^{\lfloor \frac{d}{2} \rfloor + \epsilon})$ $O(n)$ $O(m^{1+\epsilon})$ | $O(n^{\lfloor \frac{d}{2} \rfloor + \epsilon})$ $O(n)$ $O(m^{1+\epsilon})$ |
| $d$ | orthogonal | $O(\log_2^{d-1} n + k)$ $O(dn^{1-\frac{1}{d}} + k)$ | $O(\frac{n \log_2^{d-1} n}{\log_2 \log_2 n})$ $O(dn)$ | $O(n \log_2^{d-1} n)$ $O(dn \log_2 n)$ |

**3.** Orthogonal range searching in $\mathcal{R}^2$ can be carried out using range trees.

**4.** The *range tree* is defined recursively by Algorithm 3. Each node stores a subset of points organized into a threaded binary search tree by the $y$-coordinates of the points. The left child contains half the parent's points, in particular those with lesser $x$-coordinates; the right child contains the other half of the parents' points with greater $x$-coordinates.

Each node also stores the range of $x$-coordinates of its points. For simplicity, all coordinates of all points are assumed to be distinct. It is also assumed that all points of $S = \{(x_1, y_1), (x_2, y_2), \ldots, (x_n, y_n)\}$ have been presorted by their $x$-coordinate so that $x_1 < x_2 < \cdots < x_n$.

**5.** The running time of Algorithm 3 is $O(n \log_2^2 n)$ and the space taken by the range tree is $O(n \log_2 n)$. The running time can be improved by a $\log_2 n$ factor. Essentially the same procedure can be used to build range trees in any dimension.

---

**Algorithm 3:   Computing the range tree.**

**procedure** $RangeTree(S = \{(x_1, y_1), (x_2, y_2), \ldots, (x_n, y_n)\}$: set of points,
        $T$: pointer to root of a range tree)
**if** $S = \emptyset$ **then** return
**else** store the interval $[x_1, x_n]$ in $T.int$
    store $BinarySearchTree(S)$ in $T.y$
    $RangeTree(\{(x_1, y_1), \ldots, (x_{\frac{n}{2}}, y_{\frac{n}{2}})\}, T.left\_child)$
    $RangeTree(\{(x_{\frac{n}{2}+1}, y_{\frac{n}{2}+1}), \ldots, (x_n, y_n)\}, T.right\_child)$

**procedure** $BinarySearchTree(S' = \{(x_1, y_1), \ldots, (x_n, y_n)\}$: set of points)
sort the points of $S'$ by $y$-coordinate so that $y_1 < y_2 < \cdots < y_n$
create a threaded balanced binary search tree $B$ for $S'$:
    store point $(x_i, y_i)$ in the $i$th leftmost leaf $\ell_i$
    $\ell_i.next := \ell_{i+1}$  {connect the leaves into a linked list}
    $\ell_i.key := y_i$
    for each node $v$, $v.key := \min\{\ell_i.key \mid \ell_i \in subtree(v.right\_child)\}$

---

**Algorithm 4**:   **Orthogonal range reporting using range trees**

  **procedure** $OrthoRangeSearching(q = [x_1, x_2] \times [y_1, y_2]$: rectangle in the plane,
      $T$: pointer to root of range tree)

  **if** $T = NIL$ **then** return
  **else if** $T.int \subseteq [x_1, x_2]$ **then** $SearchAll(T.y, [y_1, y_2])$
  **if** $[x_1, x_2] \cap \in T.left\_child.int \neq \emptyset$ **then**
    $OrthoRangeSearching(q, T.left\_child)$
  **if** $[x_1, x_2] \cap \in T.right\_child.int \neq \emptyset$ **then**
    $OrthoRangeSearching(q, T.right\_child)$

  **procedure** $SearchAll(v$: pointer to root of binary tree, $[y_1, y_2]$: query interval)
  **while** $v$ is not a leaf
    **if** $y_1 < v.key$ **then** $v = v.left\_child$
    **else** $v := v.right\_child$
  **if** $v.key < y_1$ **then** $v = v.next$
  **while** $v \neq NIL$ and $v.key < y_2$
    output point stored at $v$
    $v := v.next$

---

**6.** Orthogonal range reporting proceeds as follows down the range tree. If the range of the current node $x$ is a subset of the $x$ range of the query, then all the points in the node's binary search tree with $y$-coordinate in the $y$ range of the query are output. If the $x$ range of the query overlaps the $x$ range of the left child, then the algorithm proceeds recursively to the left child. If the $x$ range of the query overlaps the $x$ range of the right child, then the algorithm proceeds recursively to the right child.

**7.** Algorithm 4 takes $O(\log_2^2 n + k)$ time, where $k$ is the number of reported points. This running time can be improved to $O(\log_2 n + k)$.

**Examples:**

**1.** *Orthogonal range search*:   The query range $q$ is a Cartesian product of intervals on different coordinate axes.

**2.** *Bounded distance search*:   The query range $q$ is a sphere in $\mathcal{R}^d$.

**3.** Other typical search domains are half-spaces and simplices.

**4.** *Machine learning*:   Points are labeled as positive or negative examples of a concept, and range query determines the relative number of positive and negative examples in the range (thus enabling the range to be classified as either a positive or negative example of the concept).

**5.** *Multikey searching in databases*:   Records identified by a $d$-tuple of keys can be viewed as a point in $\mathcal{R}^d$, and the range query on records corresponds to an orthogonal range query.

**6.** Part (a) of the following figure shows a set of eight points in $\mathcal{R}^2$: $a, b, \ldots, h$ and a query range $q$. Part (b) shows the associated range tree.

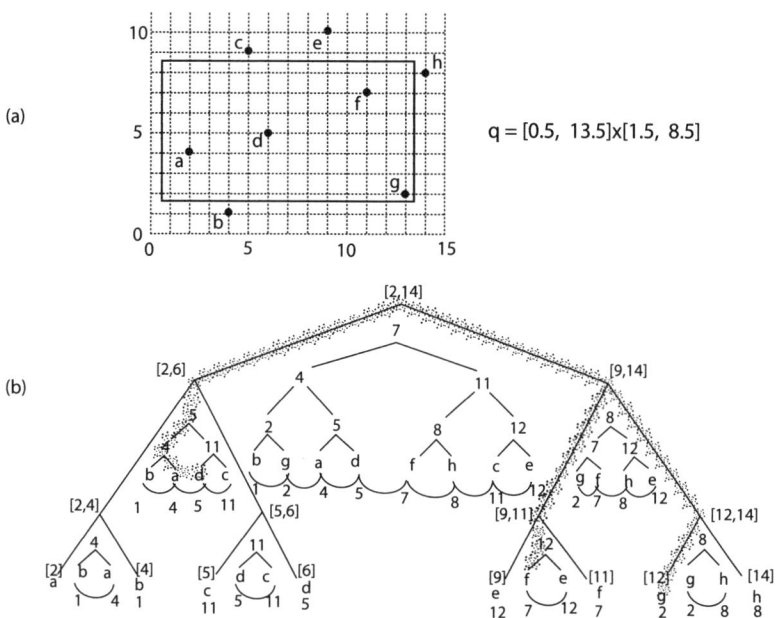

### 13.6.3   RAY SHOOTING AND LINES IN SPACE

#### Definitions:

A **ray** $r$ is a half-line that is directed away from its endpoint; that is, it satisfies the equation $r = p + \lambda \vec{v}$, $\lambda \geq 0$, where $p$ is the **starting point** of $r$ and $\vec{v}$ is the **direction** of $r$.

Given a set $S$ of geometric objects in $\mathcal{R}^d$ and a query ray $r$, the **ray shooting problem** is the problem of determining the first object in $S$ that is hit by $r$, that is, the object $s \in S$ whose intersection with $r$ is closer to $p$ than the intersection between $r$ and any other object in $S$.

A polyhedron is **axis-parallel** if each of its edges is parallel to a coordinate axis.

#### Facts:

**1.** Applications of ray shooting include hidden surface removal, visibility questions and ray tracing in computer graphics, and computing shortest paths in the presence of obstacles in robotics.

**2.** The following table gives information on various ray shooting algorithms.

| dim | subdivision type | query time | space | preprocess- ing time |
|---|---|---|---|---|
| 2 | simple polygon | $O(\log_2 n)$ | $O(n)$ | $O(n)$ |
| 2 | line segments | $O(\log_2 n)$ | $O(n^2\alpha^2(n))$ | $O(n^2\alpha^2(n))$ |
| | | $O(\sqrt{n}\log_2 n)$ | $O(n\log_2^2 n)$ | $O(n\log_2^2 n)$ |
| 3, fix $p$ | axis-parallel polyhedra | $O(\log_2 n)$ | $O(n\log_2 n)$ | $O(n\log_2 n)$ |
| 3, fix $p$ | polyhedra | $O(\log_2 n)$ | $O(n^2\alpha(n))$ | $O(n^2\alpha(n))$ |
| | for any $\epsilon > 0$, $n \leq m \leq n^2$ | $O(n^{1+\epsilon}/\sqrt{m})$ | $O(m^{1+\epsilon})$ | $O(m^{1+\epsilon})$ |
| 3, fix $\vec{v}$ | axis-parallel polyhedra | $O(\log_2 n \times (\log_2\log_2 n)^2)$ | $O(n\log_2 n)$ | $\tilde{O}(n\log_2^2 n)$ |
| | for any $\epsilon > 0$ | $O(\log_2 n)$ | $O(n^{1+\epsilon})$ | $O(n^{1+\epsilon})$ |
| 3, fix $\vec{v}$ | polyhedra | $O(\log_2 n)$ | $O(n^{3+\epsilon})$ | $O(n^{3+\epsilon})$ |
| | for any $\epsilon > 0$, $n \leq m \leq n^3$ | $O(n^{1+\epsilon}/m^{1/3})$ | $O(m^{1+\epsilon})$ | $O(m^{1+\epsilon})$ |
| 3 | axis-parallel polyhedra | $O(\log_2 n)$ | $O(n^{2+\epsilon})$ | $O(n^{2+\epsilon})$ |
| 3 | polyhedra | $O(\log_2 n)$ | $O(n^{4+\epsilon})$ | $\tilde{O}(n^{4+\epsilon})$ |

## Algorithm:

**1.** *Ray shooting from a fixed point among planar nonintersecting segments:* For simplicity, assume that the fixed point $p$ in the plane is at the origin. Define two relations using the same notation: for two points $q_j$ and $q_k$, $q_j \prec q_k$ if $q_j$ makes a smaller polar angle with respect to the origin than does $q_k$; for two nonintersecting segments $s_j$ and $s_k$, $s_j \prec s_k$ if for every ray $r$ that starts at the origin and crosses both $s_j$ and $s_k$, $r$ crosses $s_j$ before crossing $s_k$. Segment $(q_j, q_k)$ starts at $q_j$ and ends at $q_k$ if $q_j \prec q_k$. A *null* segment is denoted $s_\infty$; that is, a query ray hitting $s_\infty$ does not intersect any of the given segments.

Algorithm 5, *VisibilityMap*, creates an array $\mathcal{I}$ of nonoverlapping angle intervals, sorted by their polar angle, with the property that consecutive entries in $\mathcal{I}$ have different "smallest" segments according to the "$\prec$" relation.

This algorithm uses a technique called sweep-plane: the algorithm sweeps the polar coordinates originating at $p$ with a ray, stopping the sweep-ray at all the angles where the sweep-ray intersects a segment endpoint. The set $S'$ stores all the segments intersected by the current sweep-ray; $S'$ is organized as a binary search tree ordered by the "$\prec$" relation on the segments. When the sweep-ray encounters a segment endpoint that starts a segment, the segment is added to $S'$; when the sweep-ray encounters a segment endpoint that ends a segment, the segment is removed from $S'$. At every stop of the sweep-ray, if the smallest (under the "$\prec$" relation) segment of $S'$ is different from the sweep-ray's last stop, a new interval is added to $\mathcal{I}$. See Example 3.

Algorithm 5, *VisibilityMap*, takes $O(n\log_2 n)$ time and can be used for ray shooting among simple polygons. The visibility map consists of array $\mathcal{I}$ and takes $O(n)$ space. The problem is harder if the segments are allowed to intersect.

Algorithm 6, *RayShoot*, takes $O(\log_2 n)$ time.

---

**Algorithm 5**:   **Computing visibility map.**

**procedure** $VisibilityMap(p$: fixed origin point, $S$: set of $n$ segments)

sort endpoints of all segments by their polar angles so that $q_1 \prec q_2 \prec \cdots \prec q_{2n}$

$no\_of\_intervals := 0$; $S' := \langle s_\infty \rangle$

$\{S'$ is a binary search tree containing segments ordered by the "$\prec$" relation$\}$

**for** $i = 1$ **to** $2n$

    $first :=$ the "smallest" (under the "$\prec$" relation) segment in $S'$

    **if** $q_i$ starts a segment $s_j$ **then** insert $s_j$ into $S'$

    **if** $q_i$ ends a segment $s_j$ **then** remove $s_j$ from $S'$

    **if** $first \neq$ smallest segment in $S'$ **then**

        $no\_of\_intervals := no\_of\_intervals + 1$

        $\mathcal{I}[no\_of\_intervals].angle :=$ polar angle of $q_i$

        $\mathcal{I}[no\_of\_intervals].name :=$ the smallest segment in $S'$

---

**Algorithm 6**:   **Ray shooting using the visibility map.**

**procedure** $RayShoot(r = (p, \vec{v})$: query ray, $\mathcal{I}$: visibility map)

consider $\vec{v}$ as a polar angle

do a binary search among $\mathcal{I}[*].angle$ to find the segment $\mathcal{I}[k].name$ such that

    $\mathcal{I}[k].angle \leq \vec{v}$ and $\mathcal{I}[k+1].angle > \vec{v}$

---

**Examples:**

**1.** $S$ can be a single object, such as a simple polygon in the plane.

**2.** $S$ can be a collection of objects, such as a set of polyhedra in 3-dimensional space.

**3.** The following figure illustrates the sweep-plane technique, with rays originating from point $p$. The thick lines in this figure are the segments in $S$ and the thin lines are the boundaries between intervals in the visibility map for $S$. The intervals are labeled by their names.

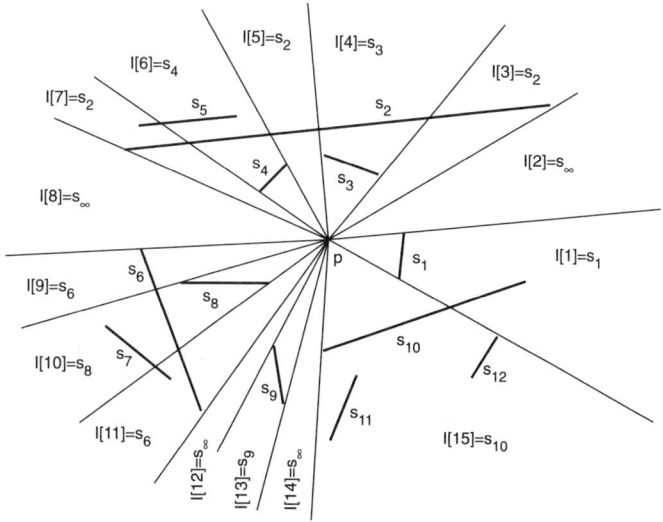

## 13.7  COMPUTATIONAL TECHNIQUES

This section describes some techniques used in the design of geometric algorithms.

### 13.7.1  PARALLEL ALGORITHMS

The goal of parallel computing is to solve problems faster than would be possible on a sequential machine — through the use of parallel algorithms. The complexity of a parallel algorithm is given in terms of its *time* and the number of *processors* used [Já92]. The *work* of a parallel algorithm is bounded above by the *processor-time product*, the product of the number of processors and the time.

**Definitions:**

A *parallel algorithm* is an algorithm that concurrently uses more than one processing element during its execution.

A *parallel machine* is a computer that can execute multiple operations concurrently.

A *parallel random access machine* (**PRAM**) is a synchronous machine in which each processor is a sequential RAM and processors communicate using a shared memory. Depending upon whether concurrent accesses are allowed to the shared memory cells, a PRAM is either *exclusive read* (**ER**) or *concurrent read* (**CR**), and either *exclusive write* (**EW**) or *concurrent write* (**CW**).

**Facts:**

**1.** *Parallel divide-and-conquer:* Parallel algorithms can be obtained using the *divide-and-conquer* paradigm. The subproblems resulting from the divide phase are solved in parallel, and the combining phase of the algorithm is parallelized. In traditional divide-and-conquer algorithms, the problem is partitioned into two subproblems.

**2.** *Many-way divide-and-conquer:* Sometimes faster parallel algorithms can be obtained by partitioning the original problem into multiple, smaller subproblems, often referred to as *many-way divide-and-conquer* [Já92]. The solution to the original problem is obtained from the solutions to the subproblems.

**3.** *Cascading divide-and-conquer:* Some divide-and-conquer algorithms can be speeded up by pipelining (cascading) the work performed in the recursive applications as follows. Consider a binary tree representing the solutions to the recursive computations of the original divide-and-conquer algorithm, where leaves represent terminal subproblems. To obtain a faster algorithm, information is passed from a child to its parent before the solution to the child's subproblem is completely known. The parent then does some precomputation so that the solution to its problem can be computed as soon as the solutions of both of its children's problems are available. Typically, the information passed from a child to its parent is a constant sized sample of the child's current state. Often, such algorithms run in time proportional to the height of the recursion tree.

**4.** Cole first used pipelining to design a work-optimal $O(\log n)$ time parallel version of merge sort. Atallah, Cole, and Goodrich used this strategy to solve several geometric problems including the 3-dimensional maxima problem and computing the visibility of a polygon from a point [AtGo93].

**Algorithms:**

**1.** *Using parallel divide-and-conquer to compute the convex hull $CH(S)$ of a set $S$ of $n$ points in the plane:* In Algorithm 1 the points in $S$ are presorted by $x$-coordinate so that the division in the second step can be accomplished in constant time. The sorting takes $O(\log n)$ time using $O(n \log n)$ work [Já92]. The tangents needed in the third step can be computed in constant time using $|S_L| + |S_R|$ processors on a CREW PRAM [AtGo93]. Thus, as there are $O(\log n)$ recursive calls, Algorithm 1 runs in $O(\log n)$ time using $O(n \log n)$ work, which is both worst-case work-optimal and time-optimal for the CREW PRAM.

---

**Algorithm 1**: **ConvexHull**($S$: a set presorted by $x$-coordinate).
**if** $|S| \leq 3$ **then** construct $CH(S)$ directly
**else**
  partition $S$ into two subsets of equal size, $S_L$ (the left subset) and $S_R$ (the
    right subset), which are separated by a vertical line
  in parallel, recursively construct $CH(S_L)$ and $CH(S_R)$
  construct $CH(S)$ from $CH(S_L)$, $CH(S_R)$, and common tangents between them

---

**2.** *Computing the convex hull of a 3-dimensional point set:* The convex hull of a 3-dimensional point set can be computed using a two-way divide-and-conquer algorithm similar to Algorithm 1. The running time of this algorithm is $O(\log^2 n)$ because the combining in the last step of Algorithm 1 takes $O(\log n)$ time since it is more complex than in the planar case. However, a faster algorithm can be designed using many-way divide-and-conquer. The point set is partitioned into $O(n^{\frac{1}{2}})$ groups, each of size $O(n^{\frac{1}{2}})$, and then, even though the combining still takes $O(\log n)$ time, a total time of $O(\log n)$ can be obtained.

---

## 13.7.2 RANDOMIZED ALGORITHMS

Randomization is a powerful technique that has been used in algorithms for many geometric problems.

**Definition:**
A **randomized algorithm** is an algorithm that makes random choices during its execution.

**Facts:**

**1.** Randomized algorithms are often faster, simpler, and easier to generalize to higher dimensions than deterministic algorithms.

**2.** For many problems, efficient algorithms can be obtained by processing the input objects in a particular order or by grouping them into equal-sized subsets. Although significant computation may be required to exactly determine an appropriate order or a good partition into subsets, in many cases simple random choices can be used instead.

**3.** *Randomized incremental methods:* One of the simplest ways to construct a geometric structure is incrementally. In an incremental construction algorithm, the input objects are inserted one at a time and the current structure is updated after each addition. The desired structure is obtained after all input objects have been inserted. Although the cost of some updates could be very large, in many cases it can be shown that if the objects are inserted in *random order*, then the amount of work for each update is expected to be small and the expected running time of the algorithm will be small as well. Thus, the expectation of the running time depends not on the input distribution but on the ratio of good to bad insertion sequences.

**4.** The power of randomization in incremental algorithms was first noted by Clarkson and Shor, and by Mulmuley. Randomized incremental algorithms have been proposed for constructing many geometric structures including convex hulls, Delaunay triangulations, trapezoidal decompositions, and Voronoi diagrams [Mu94].

**5.** *Randomized divide-and-conquer:* In randomized divide-and-conquer a random subset of the input is used to partition the original problem into subproblems. Then, as in any divide-and-conquer algorithm, the subproblems are solved, perhaps recursively, and their solutions are combined to obtain the solution to the original problem. Ideally, the partition produces subproblems of nearly equal size and the sum of the subproblem sizes roughly equals the input size. This ideal can almost be achieved for many geometric problems. For example, with probability at least $\frac{1}{2}$, $O(r)$ subproblems of size $O(\frac{n \log r}{r})$ can often be obtained, where $n$ is the number of input objects and $r$ is the size of the random subset.

**6.** The fact that a random sample of the input items can often be used to produce subproblems of almost equal size was first shown by Clarkson, and by Haussler and Welzl. Randomized divide-and-conquer algorithms are known for many geometric problems including answering closest-point queries, constructing arrangements, triangulating point sets, and constructing convex hulls [Mu94].

**Examples:**

**1.** *Incrementally constructing the intersection of a set $H$ of $n$ halfplanes in the plane:* First, a random permutation $(h_1, h_2, \ldots, h_n)$ of $H$ is formed. Let $I_i$ denote the intersection of halfplanes $h_1$ through $h_i$. During the $i$th iteration of the algorithm, $I_i$ is computed from $I_{i-1}$ by removing the portion of $I_{i-1}$ that is not contained in $h_i$.

To make this update easier, a vertex of the current intersection $I_{i-1}$ that is not contained in $h_j$ is maintained, for all $j \geq i$; such a vertex is said to *conflict* with $h_j$. Given a conflicting vertex $v$ for $h_i$, the portion of $I_{i-1}$ that must be removed is determined by traversing its boundary in both directions from $v$ until reaching vertices that are contained in $h_i$. Since $I_n$ has size $O(n)$, the amortized cost of each update is $O(1)$ and the algorithm spends a total of $O(n)$ time updating the intersection. After $I_i$ is computed, the conflicting vertex for $h_j$ is updated, for all $j > i$. It can be shown that the total cost of maintaining the conflicting vertices in the algorithm is $O(n \log n)$.

**2.** *Using randomized divide-and-conquer to construct the intersection $I(S)$ of a set $S$ of $n$ halfplanes, each of which contains the origin:* See Algorithm 2. In Step 2, the intersection of the halfplanes in the sample is used to create $O(r)$ triangles, each of which corresponds to a subproblem. For technical reasons, the region corresponding to each subproblem should have constant descriptive complexity.

Often, this condition is achieved by triangulating the resulting structure. Next, the input is distributed to the subproblems. Usually, this is done by finding the objects that intersect, or *conflict* with, a subproblem's region.

Then, the subproblems are solved, recursively if desired, and their solutions are combined to form the solution to the original problem. In the intersection algorithm the final solution is obtained by "gluing" the subproblem solutions together.

If $r$ is some suitably chosen constant, then Algorithm 2 runs in expected time $O(n \log n)$.

---

**Algorithm 2**:   **Intersect**($S$: a set of $n$ halfplanes).

$R :=$ a random sample of size $r$ chosen from $S$.

compute the intersection $I(R)$ and triangulate it by connecting each vertex of $I(R)$ to the origin to obtain $T(R)$

for each triangle $t \in T(R)$, determine its *conflict list* $c(t)$ (the set of halfplanes in $S$ whose bounding lines intersect $t$)

for each triangle $t \in T(R)$, compute the intersection of the halfplanes in $c(t)$ restricted to $t$ {This may be done recursively}

---

## 13.7.3   PARAMETRIC SEARCH

### Definition:

**Parametric search** is an algorithmic technique for solving optimization problems by using an algorithm for solving an associated decision problem as a subroutine.

### Algorithm:

**1.** Parametric search is a powerful algorithmic technique that can be applied to a diverse range of optimization problems. It is best explained in terms of an abstract optimization problem. Consider two versions of this problem:

- a *search problem P*, which can be parametrized by some real, nonnegative value $t$ and whose solution is a value $t^*$ that satisfies the optimization criteria;

- a *decision problem D(t)*, whose input includes a real, nonnegative value $t$ and whose solution is either YES or NO.

In parametric search, the search problem $P$ is solved using an algorithm $A_s$ for the decision problem $D(t)$. In order to apply parametric search, the points corresponding to YES answers for $D(t)$ must form a connected, possibly unbounded, interval in $[0, \infty)$. Then, assuming that $D(0) = $ YES, the search problem $P$ is to find the largest value of $t$ for which $D(t) = $ YES.

The basic idea of parametric search is to use $A_s$ to find $t^*$. This is done by simulating $A_s$, but using a variable (parameter) instead of a value for $t$. Assume that $A_s$ can detect if $t = t^*$, and that the computation in $A_s$ consists of comparisons, each of which tests the sign of a bounded degree polynomial in $t$ and in the original input. During each comparison in the simulation:

- the roots of the appropriate polynomial are computed,

- $A_s$ is run on each root,

- $t^*$ is located between two consecutive roots.

Thus, each comparison in the simulation reduces the interval known to contain $t^*$, which is originally $[0, \infty)$.

**Facts:**

**1.** In most geometric applications it can be shown that at some point the simulation will test a root that is equal to $t^*$.

**2.** If $T_s$ denotes the worst-case running time of $A_s$, the total cost of the parametric search is $O(T_s^2)$ since there are $O(T_s)$ operations in the simulation of $A_s$, and each comparison operation in this simulation takes $O(T_s)$ time.

**3.** With a parallel algorithm $A_p$ to solve the decision problem $D(t)$, the (sequential) parametric search can be done faster. Suppose that $A_p$ runs in $T_p$ parallel time steps using $p$ processors. Then, $A_p$ performs at most $p$ *independent* comparison operations during each parallel time step. The parametric search simulates $A_p$ sequentially as follows. To simulate a parallel time step of $A_p$:

- compute the $O(p)$ roots of the at most $p$ bounded degree polynomials corresponding to independent comparisons in the parallel time step;
- sort these roots;
- use a binary search to locate $t^*$ between two consecutive roots. (The sequential algorithm $A_s$ is used to evaluate the roots during the binary search.)

The cost of simulating each parallel time step is $O(p \log p + T_s \log p)$, and the total cost of the parametric search is $O(T_p(p \log p + T_s \log p))$.

**4.** Parametric search was originally proposed by Megiddo. It has been used for many geometric problems including ray shooting, slope selection, and computing the diameter of a 3-dimensional point set [Mu94].

**Example:**

**1.** *Answering ray shooting queries in a convex polytope Q:* In the search problem $P$, a ray $r$ is given originating at a point $p$ contained in $Q$, and the face of $Q$ hit by $r$ needs to be found. The decision problem $D(t)$ is the *polytope membership problem*: given a point $t$, determine if $t$ lies in $Q$. The value $t^*$ sought by the parametric search is the point where $r$ intersects $Q$. The ray shooting problem can be solved in $O(\log^2 n)$ time by parametric search [Mu94] (since, given an appropriate data structure, the polytope membership problem can be solved sequentially in $O(\log n)$ time).

---

## 13.7.4   FINITE PRECISION

Geometric algorithms are usually designed assuming the *real random access machine* model of computation. In this model, values can be arbitrarily long real numbers, and all standard operations such as $+$, $-$, $\times$, and $\div$ can be performed in unit time regardless of operand length. In reality, however, computers have finite precision and can only approximate real numbers. Several techniques have been suggested for dealing with this problem [Fo93].

**Facts:**

**1.** *Algorithms directly designed for a discrete domain:* This approach can dramatically increase the complexity of the algorithm and has not gained wide use.

**2.** *Floating point numbers:* Floating point numbers provide a convenient and efficient way to approximate real numbers. Unfortunately, naively rounding numbers in geometric algorithms can create serious problems such as topological inversions. There are cases when floating point arithmetic can safely be used. For example, for certain inputs the result of the floating point operation will be unambiguous. Some algorithms have been shown to be sufficiently stable using floating point arithmetic. However, no general method is known for designing stable algorithms.

**3.** *Exact arithmetic*: In exact arithmetic, numbers are represented by vectors of integers and all primitive operations are guaranteed to give correct answers. Integer arithmetic is sufficient for many geometric algorithms since symbolic or algebraic numbers are rarely needed, and in many cases homogeneous coordinates can remove the need for rational numbers. However, since exact representations can have large bit complexity, exact arithmetic can be expensive — typically increasing the cost of arithmetic operations by an order of magnitude. This cost can be decreased somewhat by optimizing the expressions and computations involving exact arithmetic.

**4.** *A combination of floating point and exact arithmetic*: First, the operation is performed using floating point arithmetic. Then, if the result is ambiguous, an exact computation is performed.

**5.** *Adaptive-precision arithmetic*: In adaptive-precision arithmetic each number is approximated by an interval whose endpoints require lower precision. If the exact computation using the approximation is ambiguous, the method iterates using smaller and smaller intervals with higher precision endpoints.

**6.** Although no clear consensus has been reached, a combination of the above strategies may yield the best results.

---

## 13.7.5   DEGENERACY AVOIDANCE

To simplify exposition by reducing the number of cases that must be considered, many geometric algorithms assume that the input is in *general position*. The general position assumption depends on the problem. For example, in problems involving planar point sets, the assumption might be that no two points have the same $x$-coordinate, that no three points lie on the same line, or that no four points lie on the same circle.

**Definitions:**

A set of objects is in **general position** if the objects satisfy certain specified conditions.

A set of input objects that violates the general position assumption is **degenerate**.

**Facts:**

**1.** *Perturbation*: Several schemes have been proposed that apply small perturbations to transform the input so that it does not contain degeneracies [Fo93]. The object of perturbation schemes is to allow the design of simpler algorithms which may validly assume the input is in general position.

Algorithms using perturbation schemes may not always produce correct output. For example, in a convex hull algorithm, points on the boundary could potentially be perturbed into the interior of the polytope and vice versa. Also, perturbation schemes can affect, perhaps adversely, *output-sensitive* algorithms whose running times depend on the size of the output they produce.

**2.** *Deal with degeneracies directly*: Algorithms that deal with degeneracies in a problem specific manner have been designed for problems such as triangulating point sets. Although this approach is not as general and usually leads to more complex algorithms than those employing perturbation schemes, it can provide superior performance.

**3.** *Symbolically perturbing the coordinates of each input point (Edelsbrunner, Mücke):* This is done by adding to each coordinate a suitable power of a small, positive real number represented by a symbolic parameter $\epsilon$. Then, since values are now polynomials in $\epsilon$, the arithmetic operations in the algorithm are replaced by polynomial arithmetic operations. Geometric primitives are implemented symbolically, typically by evaluating a sequence of determinants. Assuming that the dimension of the problem is fixed, this scheme increases the running time of the algorithm by at most a constant factor. However, the overhead incurred can in fact be quite large.

# 13.8   APPLICATIONS OF GEOMETRY

Geometry overlays the entire computing spectrum, having applications in almost every area of science and engineering, including astrophysics, molecular biology, mechanical design, fluid mechanics, computer graphics, computer vision, geographic information systems, robotics, multimedia, and mechanical engineering.

The growing availability of large geometric databases is a major driver of the increase in applications. GIS mapping databases containing most of the world's roads enable route planning applications. Airborne LIDAR creates terrain elevation databases. That enables observer visibility computation, with applications ranging from radio tower siting to visual nuisance mitigation. Laser scanners produce 3D point clouds recording the surfaces of objects from buildings down to sculptures and even people. The ensuing applications include deducing the structure of those objects' surfaces.

## 13.8.1   MATHEMATICAL PROGRAMMING

**Definition:**

**Mathematical programming** is the large-scale optimization of an objective function (such as cost) of many variables subject to constraints, such as supplies and capacities.

**Facts:**

**1.** Mathematical programming includes both *linear programming* (continuous, integer, and network) and *nonlinear programming* (quadratic, convex, general continuous, and general integer).

**2.** Applications include transportation planning and transshipment, factory production scheduling, and even determining a least-cost but adequate diet.

**3.** A *modeling language*, such as AMPL, is often used in applications [FoGaKe93].

**Example:**

**1.** *Linear programming in low dimensions:* This is a special case of linear programming since there are algorithms whose time is linear in the number of constraints but exponential in the number of dimensions.

**Application:**

**1.** Find the *smallest enclosing ball* of a set of points or a set of balls in arbitrary dimension [We91]. This uses a randomized incremental algorithm employing the move-to-front heuristic.

## 13.8.2 POLYHEDRAL COMBINATORICS

Polyhedra have been important to geometry since the classification of regular polyhedra in classical times.

**Fact:**

**1.** The following are some of the many possible operations that can be performed on polygons and polyhedra:

- *Boolean operations*, such as intersection, union, and difference;

- *point location* of new query points in a preprocessed set of polygons, which may partition a larger region;

- *range search* of a preprocessed set of points to find those inside a new query polygon;

- *decomposition* (or *triangulation*) of a polygon into triangles or a polyhedron into tetrahedra (§13.4.2). A simple $n$-gon is always decomposable into $n-2$ triangles, in linear time, and all triangulations have exactly $n-2$ triangles. However, some polyhedra in $\mathcal{R}^3$ cannot be partitioned intro tetrahedra without additional Steiner points. Also, different triangulations of the same polyhedron may have different numbers of tetrahedra.

**Applications:**

**1.** *Aperiodic tilings and quasicrystals*: Tilings (in the plane) and crystallography (in 3-dimensional space) are classic applications of polygons and polyhedra. A recent development is the study of aperiodic (Penrose) tilings [Ga77] and quasicrystals [Ap94]. These are locally but not globally symmetric under 5-fold rotations, quasi-periodic with respect to translations, and self-similar. The 1991 Nobel Prize in Chemistry was awarded for the discovery of actual quasicrystals:

- http://www.nobelprize.org/nobel_prizes/chemistry/laureates/2011/

Large-scale symmetries, such as 5-fold, that are impossible in traditional crystallography, can be visible with X-ray diffraction. Tilings can be constructed by projecting simple objects from, say, $\mathcal{R}^5$. One application is the surface reinforcement of soft metals.

**2.** *Error-correcting codes*: Some error-correcting codes can be visualized with polytopes as follows. Assume that the goal is $k$-bit symbols, where any error of up to $b$ bits can be detected. The possible symbols are some of the $2^k$ vertices of the hypercube in $k$-dimensional space. The set of symbols must contain no two symbols less than $b+1$ distance apart, where the metric is the number of different bits. If errors of up to $c$ bits are to be correctable, then no two symbols can be closer than $2c+1$.

Similarly, in quantum computing, a *quantum error-correcting code* can be designed using Clifford groups and binary orthogonal geometry [CaEtal97].

## 13.8.3 COMPUTATIONAL CONVEXITY

**Definitions:**

An $\mathcal{H}$-*polytope* is a polytope defined as the intersection of $m$ half-spaces in $\mathcal{R}^n$.

A *V-polytope* is a polytope defined as the convex hull of $m$ points in $\mathcal{R}^n$.

A *zonotope* is the vector (Minkowski) sum of a finite number of line segments.

*Computational convexity* is the study of high-dimensional convex bodies.

An *oracle* is an algorithm that gives information about a convex body.

**Facts:**

**1.** Computational complexity is related to linear programming, polyhedral combinatorics, and the algorithmetic theory of polytopes and convex bodies.

**2.** In contrast to computational geometry, computational complexity considers convex structures in normed vector spaces of finite but not restricted dimension. If the body under consideration is more complex than a polytope or zonotope, it may be represented as an oracle. Here the body is a black-box, and all information about it, such as membership, is supplied by calls to the oracle function. Typical algorithms involve *volume computation*, either deterministically, or by *Monte Carlo* methods, perhaps after decomposition into simpler bodies, such as simplices.

**3.** When the dimension $n$ is fixed, the volume of $\mathcal{V}$-polytopes and $\mathcal{H}$-polytopes can be computed in polynomial time.

**4.** There does not exist a polynomial-space algorithm for the exact computation of the volume of $\mathcal{H}$-polytopes (where $n$ is part of the input).

**5.** Additional information on computational convexity can be found at

- http://dimacs.rutgers.edu/TechnicalReports/TechReports/1994/94-31.ps

## 13.8.4   MOTION PLANNING IN ROBOTICS

In Computer Assisted Manufacturing, both the tools and the parts being assembled must often be moved around each other in a cluttered environment. Their motion should be planned to avoid collisions, and then to minimize cost.

**Definition:**

A *Davenport-Schinzel sequence* of order $s$ over an alphabet of size $n$, or $DS(n, s)$, is a sequence of characters such that:

- no two consecutive characters are the same;

- for any pair of characters, $a$ and $b$, there is no alternating subsequence of length $s + 2$ of the form $\ldots a \ldots b \ldots a \ldots b \ldots$.

**Facts:**

**1.** Practical general motion, *path planning*, is solvable with Davenport-Schinzel sequences [ShAg95]. Upper bounds on $\lambda_s(n)$, the length of the longest $DS(n, s)$, determine upper bounds on the complexity of the lower envelopes of certain functions.

For example, given $n$ points in the plane that are moving with positions that are polynomials of degree $s$ in time, the number of times that the closest pair of points can change is $\lambda_{2s}(\binom{n}{2})$.

**2.** Visibility graphs (§13.5.5) are useful in finding a shortest path between two points in the plane, in the presence of obstacles.

**3.** The problem of moving a finite object in the presence of obstacles may also be mapped into a *configuration space* (or *C-space*) problem of moving a corresponding point in a higher dimension. If translational and rotational motion in 2-dimensional (respectively 3-dimensional) is allowed, then the *C*-space is 3-dimensional (respectively 6-dimensional).

**4.** Articulated objects, multiple simultaneous motion, and robot hands also increase the number of degrees of freedom.

**5.** Current problems:

- *representation* of objects, since although planar, faceted models are simpler, the objects should be algebraic surfaces, and even if they are planar, in *C*-space their corresponding versions will be curved;

- *grasping*, or placing a minimal number of fingers to constrain the object's motion;

- *sequence planning* of the assembly of a collection of parts;

- *autonomous navigation* of robots in unstructured environments.

---

### 13.8.5  CONVEX HULL APPLICATIONS

**Facts:**

**1.** The convex hull (§13.5.1) is related to the Voronoi diagram (§13.5.3) since a convex hull problem in $\mathcal{R}^k$ is trivially reducible to a Voronoi diagram problem in $\mathcal{R}^k$, and a Voronoi diagram problem in $\mathcal{R}^k$ is reducible to a convex hull problem in $\mathcal{R}^{k+1}$.

**2.** The definition of convex hull is not constructive, in that it does not lead to a method for finding the convex hull. Nevertheless, there are many constructive algorithms and implementations. One common implementation is *QuickHull* (§13.5.1), a general dimension code for computing convex hulls, Delaunay triangulations, Voronoi vertices, furthest-site Voronoi vertices, and half-space intersections [BaDoHu96].

**3.** *Alpha-shapes* are a useful generalization of convex hulls [Ed05]. An intuitive definition is as follows. For a given $\alpha$, form a surface by rolling around the given point set with a sphere of radius $\alpha$, then smooth out the curved regions. The specification $\alpha = \infty$ gives the convex hull of the point set. The smaller the value of $\alpha$, the more closely the original points are followed. One application is inferring a surface from a 3D point set. A good introduction to the topic is found at

- http://graphics.stanford.edu/courses/cs268-16-fall/Handouts/
  AlphaShapes/as_fisher.pdf

**Examples:**

**1.** *Mathematics*:

- determining the principal components of spectral data;

- studying circuits of matroids that form a Hilbert base;

- studying the neighbors of the origin in the $\mathcal{R}^8$ lattice.

**2.** *Biology and medicine*:

- classifying molecules by their biological activity;

- determining the shapes of left ventricles for electrical analysis of the heart.

**3.** *Engineering*:

- computing support structures for objects in layered manufacturing in rapid prototyping [StBrEa95]. By supporting overhanging material, these structures prevent the object from toppling while partially built.
- designing nonlinear controllers for controlling vibration;
- finding invariant sets for delta-sigma modulators;
- classifying handwritten digits;
- analyzing the training sets for a multilayer perceptron model;
- determining the operating characteristics of process equipment;
- navigating robots;
- creating 6-dimensional wrench spaces to measure the stability of robot grasps;
- building micromagnetic models with irregular grain structures;
- building geographical information systems;
- simulating a spatial database system to evaluate spatial tesselations for indexing;
- producing virtual reality systems;
- performing discrete simulations of incompressible viscous fluids using vortex methods;
- modeling subduction zones of tectonic plates and studying fluid flow and crystal deformation;
- computing 3-dimensional unstructured meshes for computational fluid dynamics.

---

## 13.8.6  NEAREST NEIGHBOR

Variants of the problem of finding the nearest pair of a set of points have applications in fields from handwriting recognition to astrophysics.

**Facts:**

**1.** *Fixed search set, varying query point:* A fixed set $\mathcal{P}$ of $n$ points in $\mathcal{R}^d$ is preprocessed so that the closest point $p \in \mathcal{P}$ can be found for each query point $q$. The search time per query can range from $\log n$ (if $d = 2$) to $n^{\frac{d}{2}}$ (for large $d$). The Voronoi diagram is commonly used in low dimensions. However, because of the Voronoi diagram's complexity in higher dimensions, hierarchical search structures, bucketing, and probabilistic methods perhaps returning approximate answers are common.

**2.** *Moving points:* The points in $\mathcal{P}$ may be moving and the close pairs of points over time is of interest.

**Examples:**

**1.** Fixed search sets, varying query point:

- *Character recognition in document processing:* Each representative character is defined by a vector of features. Each new, unknown character must be mapped to the closest representative character in feature space.
- *Color map optimization in computer graphics:* Many frame buffers allow only the 256 colors in the current color map to be displayed simultaneously, from a palette of $2^{24}$ possible colors. Thus, each color in a new image must be mapped to the closest color in the color map. A related problem is the problem of determining what colors to use in the color map.

- *Clustering algorithms for speech and image compression in multimedia systems*: As in the color map problem, a large number of points must be quantized down to a smaller set.

2. Moving points:

- *Simulation of star motion in astrophysics*: Calculating the gravitational attraction between every pair of stars is too costly, so only close pairs are individually calculated. Otherwise the stars are grouped, and the attraction between close groups is calculated. The groups may themselves be grouped hierarchically.

- *Molecular modeling*: In molecular modeling, close pairs of atoms will be subject to van der Waals forces.

- *Air traffic control*: Air traffic controllers wish to know about pairs of aircraft closer than a minimum safe distance. Here the metric is nonuniform; small vertical separations are more tolerable than small horizontal separations.

- During path planning in robotics and numerically controlled machining, unintended close pairs of objects must also be avoided.

## 13.8.7  COMPUTER GRAPHICS

Computer graphics may be divided into *modeling* of surfaces, and *simulation* of the models. The latter includes *rendering* a scene and its light sources to generate synthetic imagery with respect to some viewpoint. Rendering involves *visibility*, or determining which parts of the surfaces are visible, and *shading* them according to some lighting model.

### Definitions:
**Anti-aliasing** refers to filtering out high-frequency spatial components of a signal, to prevent artifacts, or aliases, from appearing in the output image. In graphics, a high frequency may be an object whose image is smaller than one pixel or a sharp edge of an object.

A **GUI (graphical user interface)** is a mechanism that allows a user to interactively control a computer program with a bitmapped display by using a mouse or pointer to select menu items, move sliders or valuators, and so on. The keyboard is only occasionally used. A GUI contrasts with typing the program name followed by options on a command line, or by preparing a text file of commands for the program. A GUI is easier and more intuitive to use, but can slow down an expert user.

### Examples:
**1.** *Visibility*: Visibility algorithms may be *object-space*, where the visible parts of each object are determined, or *image-space*, where the color of each pixel in the frame buffer is determined. The latter is often simpler, but the output has less meaning, since it is not referred back to the original objects. Techniques include *ray tracing* and *radiosity*.

- *Ray tracing*: Ray tracing extends a line from viewpoint through each pixel of the frame buffer until the first intersecting object. If that surface is a mirror, then the line is reflected from the surface and continues in a different direction until it hits another object (or leaves the scene). If the object is glass, then both a reflecting and a refracting line are continued, with their colors to be combined according to Fresnel's law.

One geometry problem here is that of *sampling for subpixel averaging*. The goal is to color a square pixel of a frame buffer according to the fraction of its area occupied by each visible object. Given a line diagonally crossing a pixel, the fraction of the pixel covered by that face must be obtained for anti-aliasing. If the edges of two faces intersect in this pixel, each face cannot be handled independently, for example with an anti-aliased Bresehnam algorithm. If this is done badly, then it is very obvious in the final image as a possible fringe of a different color around the border of the object [Mi96].

The solution is to pick a small set of points in the pixel (typically 9, 16, or 64 points), determine which visible object projects to each point, and combine those colors. The problem is then to select a set of sampling points in the pixel, such that given a subset region, the number of points in it approximates its area. Four possible methods, from worst to best, are to

⋄ pick the points independently and uniform randomly;

⋄ use a nonrandom uniform distribution;

⋄ start with the above distribution, then jitter the points, or perturb each one slightly;

⋄ use simulated annealing to improve the point distribution.

- *Radiosity*: Radiosity partitions the scene into facets, computes a *form factor* of how much light from each facet will impinge on each other, and solves a system of linear equations to determine each facet's brightness. This models diffuse lighting particularly well.

- *Windowing systems*: Another visibility problem is designing the appropriate data structure for representing the windows in a GUI, so that the window that is in front at any particular pixel location can be determined, in order to receive the input focus.

- *Radio wave propagation*: The transmission of radio waves, as from cellular telephones, which are reflected and absorbed by building contents, is another application of visibility [Fo96].

**2.** *Computer vision*: Applications of geometry to vision include *model-based recognition* (or *pattern matching*), and *reconstruction or recovery of 3-D structure* from 2-D images, such as stereopsis, and structure from motion. In recognition, a model of an object is transformed into a sensor-based coordinate system and the transformation must be recovered. In reconstruction, the object must be determined from multiple projections.

**3.** *Medical image shape reconstruction*: Various medical imaging methods, such as computer tomography, produce data in the form of successive parallel slices through the body. The basic step in reconstructing the 3-dimensional object from these slices in order to view it involves joining the corresponding vertices and edges of two polygons in parallel planes by triangles to form a simple polyhedron. However, there exists a pair of polygons that cannot be so joined [GiORSu96].

---

## 13.8.8  MECHANICAL ENGINEERING DESIGN AND MANUFACTURING

Geometry is very applicable in CAD/CAM, such as in the design and manufacture of automobile bodies and parts, aircraft fuselages and parts such as turbine blades, and ship hulls and propellers. The current leading edge of CAD/CAM is **additive manufacturing**, also known as **3D printing**.

**Examples:**

**1.** *Representations*:   How should mechanical parts be represented? One problem is that geometric descriptions are verbose compared to 2-dimensional descriptions, such as draftings, since those assume certain things that the users will fill in as needed, but which must be explicit in the geometric description. Currently, it is possible to additively manufacture objects that cannot be represented so that their properties can be analyzed. The problem is objects containing lattices with billions of cells and nonhomogeneous (graded) materials. See George Allen's comments in

- http://www.3dcadworld.com/why-cad-is-hard-geometric-problems/
- http://www.cs.technion.ac.il/gdm2014/Presentations/GDM2014_allen.pdf

Possible representation methods include the following:

- *constructive solid geometry*:   Primitive objects, such as cylinders and blocks, are combined with the regularized Boolean operators union, intersection, and difference.
- *faceted boundary representation*:  The object is a polyhedron with a boundary of planar faces.
- *exact boundary representation*:  The object is defined by boundary "faces", but now each face can be curved, such as a NURBS (Non-Uniform Rational B-Spline), or an implicit piecewise quadric, Dupin cyclide (a quartic surface that is good for blending two quadric surfaces), or supercyclide.

The possible methods can be evaluated with the following criteria:

- *robustness* against numerical errors;
- *elegance*;
- *accuracy* in representing complex, curved, shapes, especially blends between the two surfaces at the intersection of two components;
- *ease* of explicitly obtaining geometry such as the boundary;
- *efficiency* when implemented on GPUs, which prefer simple, regular, representations.

**2.** *Mesh generation*:   A *mesh* is the partition of a polyhedron into, typically, tetrahedra or hexahedra to facilitate finite element modeling. A good mesher conforms to constraints, can change scale over a short distance, has no unnecessary long thin elements, and has fewer elements when possible. In some applications, periodic *remeshing* is required.

If the elements are tetrahedra, then a Delaunay criterion that the circumsphere of each tetrahedron contains no other vertices may be used. However, this is inappropriate in certain cases, such as just exterior to an airfoil, where a (hexahedral) element may have an aspect ratio of 100,000:1. This raises numerical computation issues.

Applications of meshing outside mechanical design include *computational fluid dynamics*, *contouring* in GIS, *terrain databases* for real-time simulations, and Delaunay applications in general. Some mesh generation and additive manufacturing programs include the following.

- *Triangle*, a 2-dimensional quality mesh generator and Delaunay triangulator. Triangle generates exact Delaunay triangulations, constrained Delaunay triangulations, conforming Delaunay triangulations, Voronoi diagrams, and triangular meshes. The latter can be generated with no small or large angles, and are thus suitable for finite element analysis.

```
https://www.cs.cmu.edu/~quake/triangle.html
```

- *TetGen*, a quality tetrahedral mesh generator and a 3D Delaunay triangulator.

```
http://wias-berlin.de/software/tetgen/
```

- *MeshLab* is an open-source system for processing and editing 3D triangular meshes. It provides a set of tools for editing, cleaning, healing, inspecting, rendering, texturing, and converting meshes. It offers features for processing raw data produced by 3D digitization tools/devices and for preparing models for 3D printing.

```
http://www.meshlab.net/
```

- *ImplicitCAD* is an open-source, programmatic CAD environment.

```
http://www.implicitcad.org/
```

- *Autodesk Netfabb* is an excellent commercial product with a comprehensive range of connected tools, including design optimization and printing.

```
https://www.netfabb.com/
```

- *Slic3r* is a widely used free open and flexible toolchain for 3D printers. It introduced many features such as multiple extruders, brim, microlayering, bridge detection, command line slicing, variable layer heights, sequential printing (one object at time), honeycomb infill, mesh cutting, object splitting into parts, AMF support, avoiding crossing perimeters, and distinct extrusion widths.

```
http://slic3r.org/
```

**3.** *Minimizing workpiece setup in numerically controlled (NC) machining*: In 4- and 5-axis numerically controlled machining, in order to machine all the faces, the workpiece must be repeatedly dismounted, recalibrated, and remounted. This setup can take much more time than the actual machining. Minimizing the number of setups by maximizing the number of faces that can be machined in one setup is a visibility problem harder than finding an optimal set of observers to cover some geographic terrain. Exact solutions are NP-hard; approximate solutions use geometric duality, topological sweeping, and efficient construction and searching of polygon arrangements on a sphere.

**4.** *Dimensional tolerancing*: *Tolerancing* refers to formally modeling the relationships between mechanical function and geometric form while assigning and analyzing dimensional tolerances to ensure that parts assemble interchangeably [SrVo93]. A tolerance may be specified *parametrically*, as a variation in a parameter, such as the width of a rectangle, or as a *zone* that the object's boundary must remain in. The latter is more general but must be restricted to prohibit pathologies, such as the object's boundary being not connected.

*Tolerance synthesis* attempts to optimize the tolerances to minimize the manufacturing cost of an object, considering that, while large tolerances are cheaper to manufacture, the resulting product may function poorly [Sk96].

## Unsolved Problems:

The following problems are perpetually on the list of those needing more research.

**1.** *Blending* between two surfaces in mechanical design, especially at the ends of the blend, where these surfaces meet others. (A *blending surface* smooths the intersection of two surfaces by being tangent to them, each along a curve.)

**2.** *Variational design* of a class of objects subject to constraints. Well-designed constraint systems may have multiple solutions; the space must be searched for the correct one. Labeling derivative entities, such as the edge resulting from the intersection of two inputs, is an issue partly because this edge may not exist for some parameter values.

**3.** Generally *formalizing the semantics* of solid modeling [Ho96].

**4.** Updating *simplifying assumptions*, such as the linearity of random access memory, and points being in general position, which were useful in the past, but which cause problems now.

**5.** *Accounting for dependencies* between geometric primitives, and *maintaining topological consistency*.

**6.** Designing *robust algorithms* when not only is there numerical roundoff during the computation, but also the input data are imprecise, for example, with faces not meeting properly.

**7.** Better *3-dimensional anti-aliasing* to remove crevices and similar database errors before *rapid prototyping*.

**8.** There still remains a need for many features in geometry implementations, such as more *geometric primitives* at all levels, *default visualization or animation* easily callable for each data structure, more *rapid prototyping* with visualization, a *visual debugger* for geometric software, including changing objects online, and generally *more interactivity*, not just data-driven programs.

## 13.8.9 LAYOUT PROBLEMS

The efficient layout of objects has wide-ranging applications in geometry.

### Examples:

**1.** *Textile part layout*: The clothing industry cuts parts from stock material after performing a tight, nonoverlapping, layout of the parts, in order to minimize the costs of expensive material. Often, because the cloth is not rotationally symmetric, the parts may be translated, but not rotated. Therefore, geometric algorithms for minimizing the overlap of translating polygons are necessary. Since this problem is PSPACE-hard, heuristics must be used [Da95], [LiMi95].

**2.** *VSLI layout*: Both laying out circuits and analyzing the layouts represent important problems. The masks and materials of a VLSI integrated circuit design are typically represented as rectangles, mostly isothetic, although 45 degrees or more general angles of inclination for the edges are becoming common. The rectangles of different layers may overlap. One integrated circuit may be 50MB of data before its hierarchical data structure is flattened, or 2GB after. For further details, see [WeHa11].

Geometry problems include the following.

- *design rule verification*: It is necessary to check that objects are separated by the proper distances and that average metal densities are appropriate for the fabrication process.

- *polygon simplification*: A design described by a complex set of polygons may perhaps be optimized into a smaller set of isothetic polygons (with only horizontal and vertical sides), such that the symmetric difference from the original design is as small as possible.

- *logic verification*: The electrical *circuit* is determined by the graph extracted from the adjacency information of the rectangles, and whether it matches the original logic design is determined. A subproblem is determining *devices* (*transistors*), which occur when rectangles of two particular different layers overlap.

- *capacitance*: This depends on the closeness of the component rectangles, which might be overlapping or separated, representing two conductors.

- *PPC* (Process Proximity Correction): This means to correct the effect that, when *etching* a circuit, a rectangle's edges are displaced outward, possibly causing it to come too close to another rectangle, and change the circuit.

## 13.8.10   GRAPH DRAWING AND VISUALIZATION

The classic field of *graph drawing and visualization* (see [KaWa01], [Ta14], and [TaTo95]) aims automatically to display a graph, emphasizing fundamental properties such as symmetry while minimizing the ratio between longest and shortest edges, number of edge crossings, etc. Applications include advanced GUIs, visualization systems, databases, showing the interactions of individuals and groups in sociology, illustrating connections between components in software engineering, and visualization of biological networks, including phylogenetic trees, metabolic networks, protein-protein interaction networks, and gene regulatory networks.

**Facts:**

**1.** Graph $G$ can be drawn as the 1-skeleton of a convex polytope in $\mathcal{R}^3$ if and only if $G$ is planar and 3-connected (Steinitz). See [Gr67].

**2.** Given a 3-connected planar graph, the graph can be drawn as a convex polyhedron in $\mathcal{R}^3$ using $O(n)$ volume while requiring the vertices to be at least unit distance apart, which allows them to be visually distinguished. This can be done in $O(n^{1.5})$ time [ChGoTa96].

**3.** A special language GraphML is available for representing graphs and drawings of these graphs.

**4.** Three of the most widely used software systems for constructing graph drawings are the Open Graph Drawing Framework (OGDF), the GDTookit, and the Public Implementation of a Graph Algorithm Library and Editor PIGALE, available at

- http://www.ogdf.net
- http://www.dia.uniroma3.it/~gdt
- http://pigale.sourceforge.net

**5.** There are a variety of hardware and software systems available for visualization of 3-dimensional graph drawings. See [La01] and the conference sites

- http://algo.math.ntua.gr/~gd2016/
- http://www.diagrams-conference.org/2016/

## 13.8.11   GEOGRAPHIC INFORMATION SYSTEMS

A *map* (§8.6.4) is a planar graph. Minimally, it contains *vertices*, *edges*, and *polygons*. However, a sequence of consecutive edges and 2-vertices is often called a *chain* (or *polyline*), and its interior vertices *points*. For example, if each polygon is one nation, then the southern border of Canada with the USA is one chain.

**Definition:**

A **geographic information system** (**GIS**) is an information system designed to capture, store, manipulate, analyze, and display spatial or geographically-referenced data.

**Facts:**

Typical simple geometric operations are given in Facts 1–6. More complex ones are given in Facts 7–9.

**1.** *Projecting data from one map projection to another, and determining the appropriate projection*: Since the earth is not a developable surface, no projection meets all the following criteria simultaneously: *equal-area*, *equidistant* (preserving distances from one central point to every other point), *conformal* (preserving all angles), and *azimuthal* (correctly showing the compass angle from one central point to every other point). Since a projection that meets any one criterion exactly is quite bad in the others, the most useful projections tend to be compromises, such as the recent *Robinson projection* [Da95].

**2.** *Rubber-sheeting*, or nonlinear stretching, to align a map with calibration points, and for *edge joining* of adjacent map sheets or databases, which may have slightly different coordinate systems.

**3.** *Generalizing* or reducing the number of points in a chain while preserving certain error properties.

**4.** *Topological cleanup* so that edges that are supposed to meet at one vertex do so, the boundary of each polygon is a closed sequence of vertices and polylines, adjacency information is correct, and so on.

**5.** Choice of the correct *data structure*. Should elevation data be represented in a *gridded* form (as an array of elevations) or should a *triangulated irregular network* (TIN) be used (the surface is partitioned into triangles)?

**6.** *Zone of influence calculation*: For example, find all the national monuments within ten miles of the highway.

**7.** *Overlaying*: Overlaying two maps to produce a third, where one polygon of the overlay map will be those points that are all from the same two polygons of the two input maps is one of the most complex operations in a GIS. If only the *area* or other mass property of the overlay polygons is desired, then it is not necessary completely to find the overlay polygons first; it is sufficient to find the set of vertices and their neighborhoods of each overlay polygon [FrEtal94].

**8.** *Name placement*: Consider a cartographic map containing point features such as cities, line features such as rivers, and area features such as states. The *name placement* problem involves locating the features' names so as to maximize readability and aesthetics [FrAh84]. Efficient solutions become more important as various mapping packages now produce maps on demand. The techniques also extend to *labelling CAD drawings*, such as piping layouts and wiring diagrams.

**9.** *Viewsheds and visibility indices*: Consider a terrain database, and an observer and target, both of which may be some distance above the terrain. The observer can see the target if and only if a line between them does not intersect the terrain. Note that if they are at different heights above the terrain, then this relation is not necessarily commutative.

The (not necessarily connected) polygon of possible targets visible by a particular observer is his *viewshed*. The viewshed's area is the observer's *visibility index*. In order to site observers optimally, the visibility index for each possible observer in the database may be required. Calculating this exactly for an $n \times n$ gridded database takes time $O(n^5)$ so sampling techniques are used [FrRa94].

### Implementations:

**1.** *GRASS GIS*, also known as GRASS (Geographic Resources Analysis Support System), is a free and open-source Geographic Information System (GIS) software suite widely used used for geospatial data management and analysis, image processing, graphics and maps production, spatial modeling, and visualization.

- https://grass.osgeo.org/

**2.** *The Open Source Geospatial Foundation* (OSGeo) was created to support the collaborative development of open-source geospatial software and to promote its widespread use.

- http://www.osgeo.org/

---

## 13.8.12  GEOMETRIC CONSTRAINT SOLVING

Applications of geometric constraint solving include mechanical engineering, molecular modeling, geometric theorem proving, and surveying.

### Definition:

**Geometric constraint solving** is the problem of locating a set of geometric elements given a set of constraints among them.

### Fact:

**1.** The problem may be under-constrained, with an infinite number of solutions, or over-constrained, with no solutions without some relaxation.

### Examples:

**1.** A *receptor* is a rigid cavity in a protein, which is the center of activity for some reaction. A *ligand* is a small molecule that may bind at a receptor. The activity level of a drug may depend on how the ligand fits the receptor, which is made more complicated by the protein molecule's bending.

**2.** In CAD/CAM, there may be constraints such as the specification that opposite sides of a feature must be parallel. For example, commercial systems like Pro/Engineer allow the user to freehand-sketch a part, and then apply constraints such as right angles, to snap the drawing to fit. Then the user is required to add more constraints until the part is well constrained.

**3.** *Molecular modeling*:  There is often a lock-and-key relationship between a flexible protein molecule's receptor and the ligand that it binds. In addition to geometrically matching the fitted shapes, the surface potentials of the molecules is also important. This fitting problem, called *molecular docking*, is important in *computer-aided drug design*. Generally, a heuristic strategy is used to move the molecules to achieve no overlap between the two molecules while maximizing their contact area [IePa95].

## 13.8.13  IMPLEMENTATIONS

One major application of geometry is in implementations of geometric software packages, either as *class libraries* and *subroutine packages* callable from user programs, or as *stand-alone systems*, which the user prepares input data files for and directs with either input command files or a GUI. These implementations are useful both for creating or modeling geometric objects and for visualizing them. This section describes some implementations of general usefulness across several applications.

### Definitions:

In an object-oriented computer language, a **class library** is a set of new data types (classes) and operations on them, activated by sending a message to an **object**, or data item. (For example, a plane object may respond to a message to rotate itself. The internal representation of an object is private, and it may be accessed only by sending it a message.)

In C++, class libraries often use **template metaprogramming**. Here, class templates are expanded at compile time into code that often runs much more efficiently than traditional subroutine libraries. This trades off compilation time for execution time, especially for complicated classes with efficient specialized cases. For example, consider a matrix that is known at compile time to be of size $3 \times 3$, which is common in geometric applications. Here, specialized code is generated for any loop iterating over the rows of the matrix, probably using loop unrolling. Even better, expressions with several matrix operations chained together become code with no temporary matrix storage allocations, and with all the code located together in one routine and available for global optimization.

### Example libraries:

**1.** *Boost.Geometry* (also known as the Generic Geometry Library, GGL) is a part of the collection of Boost C++ Libraries. It defines concepts, primitives, and algorithms for solving geometry problems.

- http://www.boost.org/doc/libs/1_63_0/libs/geometry/doc/html/geometry/introduction.html

Boost.Geometry contains a dimension-agnostic, coordinate-system-agnostic and scalable kernel, based on concepts, meta-functions, and tag dispatching. Supported algorithms include area, length, perimeter, centroid, convex hull, intersection (clipping), within (point in polygon), distance, envelope (bounding box), simplify, and transform. Boost.Geometry can be used wherever geometry plays a role, such as in mapping and GIS, game development, computer graphics and widgets, robotics, and astronomy. However, the development has been mostly GIS-oriented.

**2.** *Computational Geometry Algorithms Library* (CGAL) is a software project that provides access to efficient and reliable geometric algorithms in the form of a C++ library. CGAL is used in various areas needing geometric computation, such as geographic information systems, computer aided design, molecular biology, medical imaging, computer graphics, and robotics.

The library offers data structures and algorithms like triangulations, Voronoi diagrams, Boolean operations on polygons and polyhedra, point set processing, arrangements of curves, surface and volume mesh generation, geometry processing, alpha shapes, convex hull algorithms, shape analysis, AABB and KD trees, among others.

CGAL is a massive project that started in 1996 as a consortium of several existing projects. It was originally funded by the European Union's information technologies program ESPRIT.

- http://www.cgal.org/

**3.** *Geogram* is a C++ programming library of geometric algorithms, with efficient graphics for surfacic and volumetric meshes.

- http://homepages.loria.fr/BLevy/GEOGRAM/

**Example stand-alone systems:**

**1.** *Wolfram|Alpha* is an engine for carrying out dynamic computations, searching for answers, and providing knowledge, based on the *Mathematica* system. Alpha can answer questions such as "How many baseballs fit in a Boeing 747?", where the answer makes a reasonable assumption about the packing density, and states it. More technical questions, such as intersecting lines, computing properties of polytopes, and drawing fractals, are also possible.

- https://www.wolframalpha.com/about.html
- https://www.wolfram.com/mathematica/

**2.** *Blender* is a free and open-source 3D creation suite. It supports modeling, rigging, animation, simulation, rendering, compositing and motion tracking, video editing, and game creation.

- https://www.blender.org/

**3.** Several very efficient geometry and GIS programs for processing large datasets are available at

- www.ecse.rpi.edu/Homepages/wrf/Software

Here is a sampling of the supported software:

- *PinMesh* is a very fast algorithm with the facility to preprocess a polyhedral mesh in order to perform 3D point location queries [MaEtal16].
- *TiledVS* allows one to efficiently compute terrain viewsheds on large raster terrains using small amounts of real memory with a custom virtual memory manager [FeEtal16].
- *UPLAN* is an efficient algorithm for path planning on road networks having polygonal constraints [MaEtal15b].
- *EMFlow* and *RWFlood* efficiently compute hydrography (water flow) on very large ($50000 \times 50000$) terrains containing basins [GoEtal15]; the novel component was to raise the ocean and compute how it flowed over sills to flood interior basins.
- *EPUG-Overlay* allows one to compute the overlay of two GIS maps (plane graphs) in parallel using rational numbers, with the ability to process maps with tens of millions of vertices and hundreds of thousands of polygons in a few minutes [MaEtal15a].
- *NearptD* is a very fast parallel nearest neighbor algorithm and implementation, which has processed $10^7$ points in $E^6$ and $184 \cdot 10^6$ points in $E^3$.

# REFERENCES

*Printed Resources*:

[AgEtal96] P. K. Agarwal, B. Aronov, J. Pach, R. Pollack, and M. Sharir, "Quasi-planar graphs have a linear number of edges", in *Graph Drawing '95*, Lecture Notes in Computer Science 1027, Springer-Verlag, 1996, 1–7.

[AjEtal82] M. Ajtai, V. Chvátal, M. M. Newborn, and E. Szemerédi, "Crossing-free subgraphs", *Annals of Discrete Mathematics* 12 (1982), 9–12.

[AkAl89] J. Akiyama and N. Alon, "Disjoint simplices and geometric hypergraphs", *Combinatorial Mathematics*, Annals of the New York Academy of Sciences 555 (1989), 1–3.

[AlKl92] N. Alon and D. Kleitman, "Piercing convex sets and the Hadwiger-Debrunner $(p, q)$-problem", *Advances in Mathematics* 96 (1992), 103–112.

[Al63] E. Altman, "On a problem of Erdős", *American Mathematical Monthly* 70 (1963), 148–157.

[Ap94] *Aperiodic 1994, International Conference on Aperiodic Crystals*, Les Diablerets, Switzerland, 1994.

[ArEtal91] B. Aronov, B. Chazelle, H. Edelsbrunner, L. Guibas, M. Sharir, and R. Wenger, "Points and triangles in the plane and halving planes in space", *Discrete & Computational Geometry* 6 (1991), 435–442.

[AtGo93] M. J. Atallah and M. T. Goodrich, "Deterministic parallel computational geometry", in *Synthesis of Parallel Algorithms*, J. H. Reif (ed.), Morgan Kaufmann, 1993, 497–536.

[BaKe92] A. Bachem and W. Kerns, *Linear Programming Duality, An Introduction to Oriented Matroids*, Springer-Verlag, 1992.

[BaDe92] C. Bajaj and T. K. Dey, "Convex decomposition of polyhedra and robustness", *SIAM Journal on Computing* 21 (1992), 339-364.

[Ba91] K. Ball, "The plank problem for symmetric bodies", *Inventiones Mathematicae* 104 (1991), 535–543.

[Bá82] I. Bárány, "A generalization of Carathéodory's theorem", *Discrete Mathematics* 40 (1982), 141–152.

[BáFüLo90] I. Bárány, Z. Füredi, and L. Lovász, "On the number of halving planes", *Combinatorica* 10 (1990), 175–183.

[BáKaPa82] I. Bárány, M. Katchalski, and J. Pach, "Quantitative Helly-type theorems", *Proceedings of the American Mathematical Society* 86 (1982), 109–114.

[BaDoHu96] C. B. Barber, D. P. Dobkin, and H. T. Huhdanpaa, "The Quickhull algorithm for convex hulls", *ACM Transactions on Mathematical Software* 22 (1996), 469–483.

[Be83] J. Beck, "On the lattice property of the plane and some problems of Dirac, Motzkin and Erdős in combinatorial geometry", *Combinatorica* 3 (1983), 281–297.

[Be10] K. Bezdek, *Classical Topics in Discrete Geometry*, CMS Books in Mathematics, Springer, 2010.

[Be13] K. Bezdek, *Lectures on Sphere Arrangements – the Discrete Geometric Side*, Fields Institute Monographs, Volume 32, Springer, 2013.

[BeKu90] A. Bezdek and W. Kuperberg, "Examples of space-tiling polyhedra related to Hilbert's Problem 18, Question 2", in *Topics in Combinatorics and Graph Theory*, R. Bodendiek and R. Henn (eds.), Physica-Verlag, 1990, 87–92.

[BeLá15] K. Bezdek and Zs. Lángi, "Density bounds for outer parallel domains of unit ball packings", *Proceedings of the Steklov Institute of Mathematics* 288 (2015), 209–225.

[BjEtal93] A. Björner, M. Las Vergnas, B. Sturmfels, N. White, and G. Ziegler, *Oriented Matroids*, Cambridge University Press, 1993. (A comprehensive monograph on the theory of oriented matroids.)

[Bo93] J. Bokowski, "Oriented matroids", in *Handbook of Convex Geometry, Vol. A*, P. M. Gruber and J. M. Wills (eds.), North Holland, 1993, 555–602.

[BoSt89] J. Bokowski and B. Sturmfels, *Computational Synthetic Geometry*, Lecture Notes in Mathematics 1355, Springer-Verlag, 1989.

[BoFü84] E. Boros and Z. Füredi, "The number of triangles covering the center of an $n$-set", *Geometria Dedicata* 17 (1984), 69–77.

[Br96] P. Brass, "Erdős distance problems in normed spaces", *Computational Geometry* 6 (1996), 195–214.

[BuGrSl79] S. Burr, B. Grünbaum, and N. Sloane, "The orchard problem", *Geometria Dedicata* 2 (1979), 397–424.

[CaEtal97] A. R. Calderbank, E. Rains, P. W. Shor, and N. J. A. Sloane, "Quantum error correction and orthogonal geometry", *Physical Review Letters* 78 (1997), 405–408.

[Ch84] B. Chazelle, "Convex partitions of polyhedra: a lower bound and worst-case optimal algorithm", *SIAM Journal on Computing* 13 (1984), 488–507.

[Ch91] B. Chazelle, "Triangulating a simple polygon in linear time", *Discrete & Computational Geometry* 6 (1991), 485–524.

[ChPa90] B. Chazelle and L. Palios, "Triangulating a nonconvex polytope", *Discrete & Computational Geometry* 5 (1990), 505–526.

[ChKuRa] D. Cherkashin, A. Kulikov, and A. Raigorodskii, "On the chromatic numbers of small-dimensional Euclidean spaces", *arXiv*:1512.03472.

[ChGoTa95] M. Chrobak, M. T. Goodrich, and R. Tamassia, "On the volume and resolution of 3-dimensional convex graph drawing", in *Electronic Proceedings of the 5th MSI-Stony Brook Workshop on Computational Geometry*, 1995.

[ChGoTa96] M. Chrobak, M. T. Goodrich, and R. Tamassia, "Convex drawings of graphs in two and three dimensions (preliminary version)", *Proceedings of the 12th Annual Symposium on Computational Geometry*, ACM, 1996, 319–328.

[ClEtal90] K. Clarkson, H. Edelsbrunner, L. Guibas, M. Sharir, and E. Welzl, "Combinatorial complexity bounds for arrangements of curves and spheres", *Discrete & Computational Geometry* 5 (1990), 99–160.

[CoEl03] H. Cohn and N. Elkies, "New upper bounds on sphere packings I", *Annals of Mathematics* 157 (2003), 689–714.

[CoEtal16] H. Cohn, A. Kumar, S. D. Miller, D. Radchenko, and M. Viazovska, "The sphere packing problem in dimension 24", preprint, 2016, arXiv:1603.06518.

[CoZh14] H. Cohn and Y. Zhao, "Sphere packing bounds via spherical codes," *Duke Mathematics Journal* 163 (2014), 1965–2002.

[CoSl93] J. H. Conway and N. J. A. Sloane, *Sphere Packings, Lattices and Groups*, Springer-Verlag, 1993.

[CsSa93] J. Csima and E. Sawyer, "There exist $6n/13$ ordinary points", *Discrete & Computational Geometry* 9 (1993), 187–202.

[Cs96] G. Csizmadia, "Furthest neighbors in space", *Discrete Mathematics* 150 (1996), 81–88.

[Da95] K. Daniels, "Containment algorithms for nonconvex polygons with applications to layout", Ph.D. thesis, Harvard University, 1995.

[de93] M. de Berg, *Ray Shooting, Depth Orders and Hidden Surface Removal*, Springer-Verlag, 1993.

[DeEd94] T. K. Dey and H. Edelsbrunner, "Counting triangle crossings and halving planes", *Discrete & Computational Geometry* 12 (1994), 281–289.

[Do94] S. E. Dorward, "A survey of object-space hidden surface removal", *International Journal of Computational Geometry and Its Applications* 4 (1994), 325–362.

[Ed87] H. Edelsbrunner, *Algorithms in Combinatorial Geometry*, Springer-Verlag, 1987. (Monograph on combinatorial and algorithmic aspects of point configurations and hyperplane arrangements.)

[Ed05] H. Edelsbrunner, "Smooth surfaces for multi-scale shape representation", in *Foundations of Software Technology and Theoretical Computer Science*, P. S. Thiagarajan (ed.), Lecture Notes in Computer Science 1026, Springer, 1996, 391–412.

[EdHa91] H. Edelsbrunner and P. Hajnal, "A lower bound on the number of unit distances between the points of a convex polygon", *Journal of Combinatorial Theory A* 56 (1991), 312–316.

[EdSk89] H. Edelsbrunner and S. Skiena, "On the number of furthest neighbor pairs in a point set", *American Mathematical Monthly* 96 (1989), 614–618.

[El67] P. D. T. A. Elliott, "On the number of circles determined by $n$ points", *Acta Mathematica Academiae Scientiarum Hungaricae* 18 (1967), 181–188.

[Er46] P. Erdős, "On sets of distances of $n$ points", *American Mathematical Monthly* 53 (1946), 248–250.

[Er60] P. Erdős, "On sets of distances of $n$ points in Euclidean space", *Magyar Tudományos Akadmia Közleményei* 5 (1960), 165–169.

[ErFiFü91] P. Erdős, P. Fishburn, and Z. Füredi, "Midpoints of diagonals of convex $n$-gons", *SIAM Journal on Discrete Mathematics* 4 (1991), 329–341.

[ErEtal93] P. Erdős, Z. Füredi, J. Pach, and Z. Ruzsa, "The grid revisited", *Discrete Mathematics* 111 (1993), 189–196.

[ErPa90] P. Erdős and J. Pach, "Variations on the theme of repeated distances", *Combinatorica* 10 (1990), 261–269.

[ErSz35] P. Erdős and G. Szekeres, "A combinatorial problem in geometry", *Compositio Mathematica* 2 (1935), 463–470.

[FeKu93] G. Fejes Tóth and W. Kuperberg, "Packing and covering with convex sets", in *Handbook of Convex Geometry*, P. M. Gruber and J. M. Wills (eds.), North-Holland, 1993, 799–860.

[FeGo17] S. Felsner and J. E. Goodman, *Pseudoline arrangements*, in C. D. Tóth, J. O'Rourke, and J. E. Goodman (eds.), *Handbook of Discrete and Computational Geometry*, 3rd ed., CRC Press, 2017, Chapter 5.

[FeEtal16] C. R. Ferreira, M. V. A. Andrade, S. V. G. Magalhães, and W. R. Franklin, "An efficient external memory algorithm for terrain viewshed computation", *ACM Transactions on Spatial Algorithms and Systems* 2 (2016), article 6.

[Fo93] S. Fortune, "Progress in computational geometry", in *Directions in Geometric Computing*, R. Martin (ed.), Information Geometers Ltd., 1993, 81–128.

[Fo96] S. Fortune, "A beam tracing algorithm for prediction of indoor radio propagation", in *Applied Computational Geometry Towards Geometric Engineering*, M. C. Lin and D. Manocha (eds.), Lecture Notes in Computer Science 1148, Springer, 1996, 37–40.

[FoGaKe93] R. Fourer, D. M. Gay, and B. W. Kernighan, *AMPL: A Modeling Language for Mathematical Programming*, Duxbury Press/Wadsworth Publishing Co., 1993.

[FrRö86] P. Frankl and V. Rödl, "All triangles are Ramsey", *Transactions of the American Mathematical Society* 297 (1986), 777–779.

[FrRö90] P. Frankl and V. Rödl, "A partition property of simplices in Euclidean space", *Journal of the American Mathematical Society* 3 (1990), 1–7.

[FrRa94] W. R. Franklin and C. Ray, "Higher isn't necessarily better: visibility algorithms and experiments", in *Advances in GIS Research: 6th International Symposium on Spatial Data Handling*, T. C. Waugh and R. G. Healey (eds.), 1994, 751–770.

[FrEtal94] W. R. Franklin, V. Sivaswami, D. Sun, M. Kankanhalli, and C. Narayanaswami, "Calculating the area of overlaid polygons without constructing the overlay", *Cartography and Geographic Information Systems* 21 (1994), 81–89.

[FrAh84] H. Freeman and J. Ahn, "A system for automatic name placement", *4th Jerusalem Conference on Information Technology (JCIT); Next Decade in Information Technology*, IEEE Computer Society Press, 1984, 134–143.

[Fü90] Z. Füredi, "The maximum number of unit distances in a convex *n*-gon", *Journal of Combinatorial Theory A* 55 (1990), 316–320.

[Ga77] M. Gardner, "Mathematical recreations", *Scientific American* 236 (Jan. 1977), 110–121.

[GiORSu96] C. Gitlin, J. O'Rourke, and V. Subramanian, "On reconstructing polyhedra from parallel slices", *International Journal of Computational Geometry & Applications* 6 (1996), 103–122.

[GoEtal15] T. L. Gomes, S. V. G. Magalhães, M. V. A. Andrade, W. R. Franklin, and G. C. Pena, "Efficiently computing the drainage network on massive terrains using external memory flooding process", *Geoinformatica* 19 (2015), 671–692.

[GrRoSp90] R. Graham, B. Rothschild, and J. Spencer, *Ramsey Theory*, 2nd ed., Wiley, 1990.

[GrTa13] B. Green and T. Tao, "On sets defining few ordinary lines", *Discrete & Computational Geometry* 50 (2013), 409–468.

[GrWi93] P. M. Gruber and J. M. Wills, eds., *Handbook of Convex Geometry, Vols. A and B*, North Holland, 1993.

[Gr56] B. Grünbaum, "A proof of Vázsonyi's conjecture", *Bulletin of the Research Council of Israel, Section A* 6 (1956), 77–78.

[Gr67] B. Grünbaum, *Convex Polytopes*, Wiley, 1967.

[Gr72] B. Grünbaum, *Arrangements and Spreads*, CBMS Regional Conference Series in Mathematics 10, American Mathematical Society, 1972.

[GrSh90] B. Grünbaum and G. C. Shephard, *Tilings and Patterns*, Freeman, 1990.

[GuEtal95] P. Gupta, R. Janardan, J. Majhi, and T. Woo, "Efficient geometric algorithms for workpiece orientation in 4- and 5-axis NC-machining", *Electronic Proceedings of the 5th MSI-Stony Brook Workshop on Computational Geometry*, 1995.

[GuKa15] L. Guth and N. H. Katz "On the Erdős distinct distances problem in the plane", *Annals of Mathematics* 181 (2015), 155–190.

[Ha94] T. C. Hales, "The status of the Kepler conjecture", *The Mathematical Intelligencer* 16 (1994), 47–58.

[Ha05] T. C. Hales, "A proof of the Kepler conjecture", *Annals of Mathematics* 162 (2005), 1065–1185.

[Ha12] T. C. Hales, *Dense Sphere Packings: A Blueprint for Formal Proofs*, Cambridge University Press, 2012.

[Ha65] S. Hansen, "A generalization of a theorem of Sylvester on lines determined by a finite set", *Mathematica Scandinavica* 16 (1965), 175–180.

[Ha80] S. Hansen, "On configurations in 3-space without elementary planes and on the number of ordinary planes", *Mathematica Scandinavica* 47 (1980), 181–194.

[Ha74] H. Harborth, "Lösung zu problem 664A", *Elemente der Mathematik* 29 (1974), 14–15.

[He23] E. Helly, "Über Mengen konvexer Körper mit gemeinschaftlichen Punkten", *Jahresbericht der Deutschen Mathematiker-Vereinigung* 32 (1923), 175–176.

[Ho96] C. M. Hoffmann, "How solid is solid modeling?" in *Applied Computational Geometry Towards Geometric Engineering*, M. C. Lin and D. Manocha (eds.), Lecture Notes in Computer Science 1148, Springer, 1996, 1–8.

[HoPa34] H. Hopf and E. Pannwitz, "Aufgabe Nr. 167", *Jahresbericht der Deutschen Mathematiker-Vereinigung* 43 (1934), 114.

[IePa95] D. Ierardi and S. Park, "Rigid molecular docking by surface registration at multiple resolutions", *Electronic Proceedings of the 5th MSI-Stony Brook Workshop on Computational Geometry*, 1995.

[Já92] J. JáJá, *An Introduction to Parallel Algorithms*, Addison-Wesley, 1992.

[Ja87] R. Jamison, "Direction trees", *Discrete & Computational Geometry* 2 (1987), 249–254.

[JeTo95] T. R. Jensen and B. Toft, *Graph Coloring Problems*, Wiley-Interscience, 1995.

[KaLe78] G. A. Kabatiansky and V. I. Levenshtein, "On bounds for packings on a sphere and in space" (in Russian), *Problemy Peredachi Informacii* 14 (1978), 3–25; English translation in *Problems of Information Transmission* 14 (1978), 1–17.

[Ka84] G. Kalai, "Intersection patterns of convex sets", *Israel Journal of Mathematics* 48 (1984), 161–174.

[KaEtal12] H. Kaplan, J. Matoušek, Z. Safernová, and M. Sharir, "Unit distances in three dimensions", *Combinatorics, Probability and Computing* 21 (2012), 597–610.

[KáPaTó98] G. Károlyi, J. Pach, and G. Tóth, "Ramsey-type results for geometric graphs", *Discrete & Computational Geometry* 20 (1998), 375–388.

[KaWa01] M. Kaufmann and D. Wagner, eds., *Drawing Graphs: Methods and Models*, Lecture Notes in Computer Science 2025, Springer-Verlag, 2001.

[Ki83] D. Kirkpatrick, "Optimal search in planar subdivisions", *SIAM Journal on Computing* 12 (1983), 28–35.

[Kn92] D. E. Knuth, *Axioms and Hulls*, Lecture Notes in Computer Science 606, Springer-Verlag, 1992.

[Ko93] P. Komjáth, "Set theoretic constructions in Euclidean spaces", in *New Trends in Discrete and Computational Geometry*, J. Pach (ed.), Springer-Verlag, 1993.

[Kr46] M. A. Krasnoselskiĭ, "Sur un critère pour qu'un domain soit étoilé", (Russian, with French summary), *Matematicheskiĭ Sbornik, N. S.* 19 (1946), 309–310.

[Ku79] Y. Kupitz, "Extremal Problems in Combinatorial Geometry", *Aarhus University Lecture Notes Series* 53, Aarhus University, 1979.

[La01] B. Landgraf, "3D graph drawing", in *Drawing Graphs: Methods and Models*, M. Kaufmann and D. Wagner (eds.), Lecture Notes in Computer Science 2025, Springer-Verlag, 2001, 172–192.

[Le03] F. T. Leighton, *Complexity Issues in VLSI: Optimal Layouts for the Shuffle-Exchange Graph and Other Networks*, The MIT Press, 2003.

[LiMi95] Z. Li and V. Milenkovic, "Compaction and separation algorithms for nonconvex polygons and their applications", *European Journal of Operational Research* 84 (1995), 539–561.

[LiMa96] M. C. Lin and D. Manocha, eds., *Applied Computational Geometry Towards Geometric Engineering*, Lecture Notes in Computer Science 1148, Springer, 1996.

[Lo71] L. Lovász, "On the number of halving lines", *Annales Universitatis Scientarium Budapest, Eötvös, Sectio Mathematica* 14 (1971), 107–108.

[MaEtal15a] S. V. G. Magalhães, M. V. A. Andrade, W. R. Franklin, and W. Li, "Fast exact parallel map overlay using a two-level uniform grid", *Proceedings of the 4th International ACM SIGSPATIAL Workshop on Analytics for Big Geospatial Data*, ACM, 2015, 45–54.

[MaEtal15b] S. V. G. Magalhães, M. V. A. Andrade, W. R. Franklin, and W. Li, "Fast path planning under polygonal obstacle constraints", *4th GIS-focused Algorithm Competition, GISCUP*, 2015, Winner (2nd place).

[MaEtal16] S. V. G. Magalhães, M. V. A. Andrade, W. R. Franklin, and W. Li, "PinMesh – Fast and exact 3D point location queries using a uniform grid", *Computers & Graphics* 58 (2016), 1–11.

[Ma93] J. Matoušek, "Geometric Range Searching", Technical Report, *FB Mathematik und Informatik*, Freie Universität Berlin, 1993.

[Mc80] P. McMullen, "Convex bodies which tile space by translation", *Mathematika* 27 (1980), 113–121.

[MeNä95] K. Mehlhorn and S. Näher, "LEDA: a platform for combinatorial and geometric computing", *Communications of the ACM* 38 (1995), 96–102.

[Mi96] J. S. B. Mitchell, "On some applications of computational geometry in manufacturing and virtual environments", in *Applied Computational Geometry Towards*

*Geometric Engineering*, M. C. Lin and D. Manocha (eds.), Lecture Notes in Computer Science 1148, Springer, 1996, 37–40.

[MiVa92] S. A. Mitchell and S. A. Vavasis, "Quality mesh generation in three dimensions", *Proceedings of the 8th Annual Symposium on Computational Geometry*, ACM, 1992, 212–221.

[Mu94] K. Mulmuley, *Computational Geometry: An Introduction Through Randomized Algorithms*, Prentice Hall, 1994.

[Mu08] O. R. Musin, "The kissing number in four dimensions", *Annals of Mathematics* 168 (2008), 1–32.

[OdSl79] A. M. Odlyzko and N. J. A. Sloane, "New bounds on the number of unit spheres that can touch a unit sphere in $n$-dimensions", *Journal of Combinatorial Theory A* 26 (1979), 210–214.

[OR87] J. O'Rourke, *Art Gallery Theorems and Algorithms*, Oxford University Press, 1987.

[OR93] J. O'Rourke, "Computational Geometry Column 18", *International Journal of Computational Geometry and Its Applications* 3 (1993), 107–113.

[OR94] J. O'Rourke, *Computational Geometry in C*, Cambridge University Press, 1994.

[Pa93] J. Pach, ed., *New Trends in Discrete and Computational Geometry*, Springer-Verlag, 1993.

[PaAg95] J. Pach and P. K. Agarwal, *Combinatorial Geometry*, Wiley, 1995.

[PaShSz94] J. Pach, F. Shahrokhi, and M. Szegedy, "Applications of crossing numbers", *10th ACM Symposium on Computational Geometry*, 1994, 198–202.

[PaSh90] J. Pach and M. Sharir, "Repeated angles in the plane and related problems", *Journal of Combinatorial Theory A* 59 (1990), 12–22.

[PaSh98] J. Pach and M. Sharir, "On the number of incidences between points and curves", *Combinatorics, Probability and Computing* 7 (1998), 121–127.

[PaStSz92] J. Pach, W. Steiger, and M. Szemerédi, "An upper bound on the number of planar $k$-sets", *Discrete & Computational Geometry* 7 (1992), 109–123.

[PaTö94] J. Pach and J. Töröcsik, "Some geometric applications of Dilworth's theorem", *Discrete & Computational Geometry* 12 (1994), 1–7.

[PoWe90] R. Pollack and R. Wenger, "Necessary and sufficient conditions for hyperplane transversals", *Combinatorica* 10 (1990), 307–311.

[PrSh85] F. P. Preparata and M. I. Shamos, *Computational Geometry: An Introduction*, Springer-Verlag, 1985.

[Ra95a] C. Radin, "Aperiodic tilings in higher dimensions", *Proceedings of the American Mathematical Society* 123 (1995), 3543–3548.

[Ra21] J. Radon, "Mengen konvexer Körper, die einen gemeinsamen Punkt enthalten", *Mathematische Annalen* 83 (1921), 113–115.

[Ra95b] V. T. Rajan, "Computational geometry problems in an integrated circuit design and layout tool", *Electronic Proceedings of the 5th MSI-Stony Brook Workshop on Computational Geometry*, 1995.

[Ri96] J. Richter-Gebert, *Realization Spaces of Polytopes*, Lecture Notes in Mathematics 1643, Springer-Verlag, 1996.

[RiZi97] J. Richter-Gebert and G. M. Ziegler, "Oriented matroids", in *Handbook of Discrete and Computational Geometry*, 3rd ed., C. D. Tóth, J. O'Rourke, and J. E. Goodman (eds.), CRC Press, 2017, Chapter 6.

[RuSe92] J. Ruppert and R. Seidel, "On the difficulty of triangulating three-dimensional nonconvex polyhedra", *Discrete & Computational Geometry* 7 (1992), 227–253.

[Sc91] P. Schorn, "Implementing the XYZ GeoBench: a programming environment for geometric algorithms", in *Computational Geometry – Methods, Algorithms and Applications*, Lecture Notes in Computer Science 553, Springer-Verlag, 1991, 187–202.

[ScVa53] K. Schütte and B. L. Van Der Waerden, "Das Problem der dreizehn Kugeln", *Mathematische Annalen* 125 (1953), 325–334.

[ShAg95] M. Sharir and P. K. Agarwal, *Davenport-Schinzel Sequences and Their Geometric Applications*, Cambridge University Press, 1995.

[ShShSo16] M. Sharir, A. Sheffer, and N. Solomon, "Incidences with curves in $\mathcal{R}^d$", *Electronic Journal of Combinatorics* 23 (2016), #P4.16.

[Sh92] T. C. Shermer, "Recent results in art galleries", *Proceedings of the IEEE* 80 (1992), 1384–1399.

[Sk96] V. Skowronski, "Synthesizing tolerances for optimal design using the Taguchi quality loss function", Ph.D. thesis, Rensselaer Polytechnic Institute, 1996.

[So94] A. Soifer, "Six-realizable set $x_6$", *Geombinatorics* 3 (1994), 140–145.

[So09] A. Soifer, *The Mathematical Coloring Book*, Springer, 2009.

[SpSzTr84] J. Spencer, E. Szemerédi, and W. T. Trotter, "Unit distances in the Euclidean plane", in *Graph Theory and Combinatorics*, B. Bollobás (ed.), Academic Press, 1984, 293–303.

[SrVo93] V. Srinivasan and H. B. Voelcker, eds., *Proceedings of the 1993 International Forum on Dimensional Tolerancing and Metrology*, American Society of Mechanical Engineers, Center for Research and Technology Development, and Council on Codes and Standards, 1993.

[StBrEa95] P. Stucki, J. Bresenham, and R. Earnshaw, eds., "Rapid prototyping technology", *IEEE Computer Graphics and Applications* 15 (1995), 17–55.

[Su17] A. Suk, "On the Erdős-Szekeres convex polygon problem", *Journal of the American Mathematical Society*, 2017.

[Sz97] L. A. Székely, "Crossing numbers and hard Erdős problems in discrete geometry", *Combinatorics, Probability and Computing* 7 (1997), 353–358.

[SzTr83] E. Szemerédi and W. T. Trotter, "Extremal problems in discrete geometry", *Combinatorica* 3 (1983), 381–392.

[Ta14] R. Tamassia, ed., *Handbook of Graph Drawing and Visualization*, CRC Press, 2014.

[TaTo95] R. Tamassia and I. G. Tollis, eds., *Graph Drawing*, Lecture Notes in Computer Science 894, Springer-Verlag, 1995.

[TóORGo17] C. D. Tóth, J. O'Rourke, and J. E. Goodman, eds., *Handbook of Discrete and Computational Geometry*, 3rd ed., CRC Press, 2017. (An extensive, comprehensive reference source in discrete and computational geometry.)

[TóVa98] G. Tóth and P. Valtr, "Note on the Erdős-Szekeres theorem", *Discrete & Computational Geometry* 19 (1998), 457–459.

[Tv66] H. Tverberg, "A generalization of Radon's theorem", *Journal of the London Mathematical Society* 41 (1966), 123–128.

[Un82] P. Ungar, "$2N$ noncollinear points determine at least $2N$ directions", *Journal of Combinatorial Theory A* 33 (1982), 343–347.

[Va11] S. Vance, "Improved sphere packing lower bounds from Hurwitz lattices", *Advances in Mathematics* 227 (2011), 2144–2156.

[Ve13] A. Venkatesh, "A note on sphere packings in high dimension," *International Mathematics Research Notices* 7 (2013), 1628–1642.

[Vi16] M. S. Viazovska, "The sphere packing problem in dimension 8," preprint, 2016, arXiv:1603.04246.

[We91] Emo Weltz, "Smallest enclosing disks (balls and ellipsoids)", in *New Results and New Trends in Computer Science*, Lecture Notes in Computer Science 555, Springer-Verlag, 1991, 359–370.

[WeHa11] N. Weste and D. Harris, *CMOS VLSI Design: A Circuits and Systems Perspective*, 4th ed., Pearson, 2011.

[Ya90] F. F. Yao, "Computational geometry", in *Handbook of Theoretical Computer Science, Vol. A*, J. van Leeuwen (ed.), The MIT Press, Elsevier, 1990, Chapter 7.

[Za13] J. Zahl, "An improved bound on the number of point-surface incidences in three dimensions", *Contributions to Discrete Mathematics* 8 (2013), 100–121.

[Zi94] G. M. Ziegler, *Lectures on Polytopes*, Springer-Verlag, 1994.

[Zi98] G. M. Ziegler, "Oriented matroids today", *Electronic Journal of Combinatorics*, Dynamic Survey DS#4, 1998.

[ŽiVr92] R. Živaljević and S. Vrećica, "The colored Tverberg's problem and complexes of injective functions", *Journal of Combinatorial Theory A* 61 (1992), 309–318.

**Web Resources**:

http://cs.brown.edu/cgc/ (Center for Geometric Computing, Brown University.)

http://cs.brown.edu/stc/ (NSF Graphics and Visualization Center.)

http://dimacs.rutgers.edu/TechnicalReports/TechReports/1994/94-31.ps (*On the Complexity of Some Basic Problems in Computational Convexity: II. Volume and Mixed Volumes*, DIMACS Technical Report 94-31.)

http://graphdrawing.org/index.html (Graph drawing resources.)

http://jeffe.cs.illinois.edu/compgeom/compgeom.html (Jeff Erickson's Computational Geometry pages.)

http://www.algorithmic-solutions.com/leda/index.htm (LEDA C++ algorithms for geometric computations.)

http://www-cgrl.cs.mcgill.ca/%7Egodfried/teaching/cg-web.html (Computational geometry resources.)

http://www.cs.cmu.edu/~quake/triangle.html (Triangle: A Two-dimensional Quality Mesh Generator and Delaunay Triangulator.)

http://www.cs.mcgill.ca/~fukuda/soft/polyfaq/polyfaq.html (FAQs in polyhedral computation.)

`http://www3.cs.stonybrook.edu/~algorith/major_section/1.6.shtml` (The Stony Brook Algorithm Repository, which includes Section 1.6 on Computational Geometry.)

`http://www.geom.uiuc.edu/apps/quasitiler/about.html` (QuasiTiler 3.0.)

`http://www.ics.uci.edu/~eppstein/geom.html` (David Eppstein's Geometry in Action pages, a collection of many applications of discrete and computational geometry.)

`https://polymake.org/doku.php` (Polymake, open-source software for research in polyhedral geometry.)

`https://www.math.uci.edu/research/mathematical-visualization` (Mathematical visualization site.)

# 14

# CODING THEORY

*Alfred J. Menezes*
*Paul C. van Oorschot*
*David Joyner and*
*Tony Shaska*

14.8 Quantum Error-Correcting Codes
   14.8.1  Quantum Codes
   14.8.2  Quantum Algebraic Geometry Codes

## INTRODUCTION

This chapter deals with techniques for the efficient and reliable transmission of data over communications channels that may be subject to non-malicious errors. The general topic areas related to these techniques are information theory and coding theory. Information theory is concerned with the mathematical theory of communication, and includes the study of redundancy and the underlying limits of communications channels. Coding theory, in its broadest sense, deals with the translation between source data representations and the corresponding representative symbols used to transmit source data over a communications channel, or store this data. Error-correcting coding is the part of coding theory that adds systematic redundancy to messages to allow transmission errors not only to be detected, but also to be corrected.

## GLOSSARY

**AG code** (**algebraic geometry code**):  a code defined in terms of the divisors of an algebraic curve.

**analog channel**:  a channel that is continuous in amplitude and time.

**BCH code**:  a code arising from a special family of cyclic codes.

**binary symmetric channel** (**BSC**):  a memoryless channel with binary input and output alphabets, and fixed probability $p$ that a symbol is transmitted incorrectly.

**burst error**:  a vector whose only nonzero entries are among a string of successive components, the first and last of which are nonzero.

**capacity of a channel**:  a measure of the ability of a channel to transmit information reliably.

**check symbols**:  the positions in a codeword that provide redundancy.

**code**:  a map from the set of words to the set of all finite strings of elements in a designated alphabet.

**codeword**:  a string produced when a code is applied to a word.

**coding theory**:  the subject concerned with the translation between source data representations and the corresponding representative symbols used to transmit source data over a communications channel.

**complete maximum likelihood decoding** (**CMLD**):  the decoding scheme that decodes a received $n$-tuple to the unique codeword of minimum distance from this $n$-tuple, if such a codeword exists. Otherwise, the scheme arbitrarily decodes the $n$-tuple to one of the codewords closest to this $n$-tuple.

**convolutional code**:  a code in which the encoder has memory, so that an $n$-tuple produced by the encoder not only depends on the message $k$-tuple $u$, but also on some message $k$-tuples produced prior to $u$.

**coset**:  the set $C + x = \{c + x \mid c \in C\}$ determined by a word $x$, given a code $C$.

**coset leader**: a coset member of smallest Hamming weight.

**cyclic code**: a linear code in which every cyclic shift of a codeword is also a codeword.

**data compression**: the transformation of data into a representation which is more compact yet maintains the information content of the original data.

**dual code** (of a code): the orthogonal complement of the code.

**entropy**: a measure of the amount of information provided by an observation of a random variable.

**equivalent codes**: codes for which there is a fixed permutation of the coordinate positions which transform one code to the other.

**erasure**: a transmission error whose position is known, but whose transmission symbol for that position is not.

**error-correction coding**: coding that adds systematic redundancy to messages to allow transmission errors to be detected and corrected.

**error-detection coding**: coding that adds systematic redundancy to messages to allow transmission errors to be detected (but not necessarily corrected).

**extended code**: the code obtained by adding a parity check symbol to each codeword of a code.

**generator matrix for a code**: a matrix whose rows form a basis for that code.

**generator polynomial**: a monic polynomial of least degree in a cyclic code.

**Golay code**: a particular perfect code.

**Hamming code**: a perfect single-error correcting code.

**Hamming distance between two $n$-tuples**: the number of coordinate positions in which they differ.

**Hamming distance of a code**: the smallest Hamming distance over all pairs of distinct codewords in that code.

**Hamming weight of an $n$-tuple**: the number of nonzero coordinates.

**incomplete maximum likelihood decoding (IMLD)**: the decoding scheme that decodes a received $n$-tuple to a unique codeword such that the distance between the $n$-tuple and the codeword is minimum if such a codeword exists. If no such codeword exists, then the scheme reports that errors have been detected, but no correction is possible.

**information symbols**: the positions within a codeword that provide information on the message sent.

**information theory**: the mathematical theory of communication concerned with both the study of redundancy and the underlying limits of communication channels.

**linear code**: a subspace of the set of $n$-tuples with entries from a finite field.

**list decoding**: a decoding algorithm which outputs a list of possible codewords "near" the received word, one of which is correct.

**low-density parity check code**: a linear code whose parity check matrix contains relatively few nonzero elements.

**memoryless source**: a source for which the probability of a particular word being emitted at any point in time is fixed.

**message**: a finite string of source words.

**minimum error probability decoding** (**MED**): the decoding scheme that decodes a received $n$-tuple $r$ to a codeword $c$ for which the conditional probability $Pr(c$ is sent $\mid r$ is received), $c \in C$, is largest.

**nonlinear code**: a code which lacks the structure of a vector space over a finite field.

**Nordstrom-Robinson code**: a special nonlinear code.

**parity check bit**: a bit added to a bit string so that the total number of 1s in the extended string is even.

**parity check matrix** (for a code): a generator matrix for the dual code of the code.

**perfect code**: a code of distance $d$ for which every word is within distance $t = \lfloor \frac{d-1}{2} \rfloor$ of some codeword.

**Preparata code**: a code from an infinite family of nonlinear codes that have efficient encoding and decoding algorithms.

**punctured code**: the code obtained by removing any column of a generator matrix of a linear code.

**qubit** (**quantum bit**): the analog in quantum computation of a bit in classical computation.

**Reed-Muller code**: a code from a particular family of linear codes.

**Reed-Solomon code**: a linear code from a special family of BCH codes.

**self-dual code**: a linear code that is equal to its dual code.

**self-orthogonal code**: a linear code that is contained in its dual code.

**shortened code**: the set of all codewords in a linear code which are 0 in a fixed coordinate position with that position deleted.

**stabilizer code**: a device used to help construct quantum codes from some classical codes.

**syndrome** (of a word $x$): the vector $xH^T$, where $H$ is a parity check matrix for a linear code $C$.

**systematic code**: a linear code that has a generator matrix of the form $[I_k \mid A]$ (alternatively, of the form $[A \mid I_k]$).

**turbo code**: a special type of code built using convolutional codes and an interleaver which permutes the original bits before sending them to the second encoder.

**uniquely decodable code**: a code for which every string of symbols is the image of at most one message.

# 14.1 COMMUNICATION SYSTEMS AND INFORMATION THEORY

## 14.1.1 BASIC CONCEPTS

### Definitions:

A **communication system**, as illustrated in the following figure, is modeled as a data source providing either continuous or discrete output, a **source encoder** transforming source data into binary digits (**bits**), a **channel encoder**, and a channel.

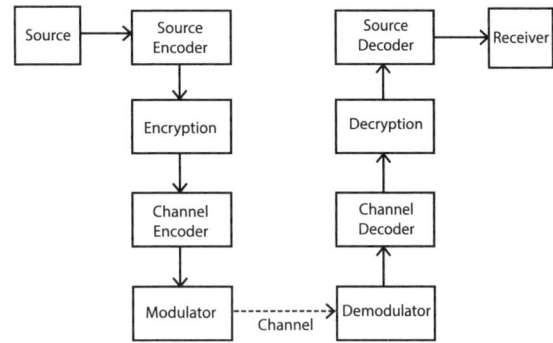

In many communication systems the channel is **analog**, that is, continuous in amplitude and time, in which case a modulator/demodulator (**modem**) is required to transform between analog channel data and discrete encoder/decoder data.

The **source encoder**, the aim of which is to minimize the number of bits required to represent source data while still allowing subsequent reconstruction, typically includes **data compression** to remove unnecessary redundancy.

The objective of the **channel encoder** is to maximize the rate at which information can be reliably conveyed by the channel, in the presence of disruptive channel noise.

**Coding theory** is the study of the translation between source data representations and the corresponding representative symbols (coded data) used to transmit source data over a communication channel.

**Error-correction coding**, located in the channel encoder, adds systematic redundancy to messages to allow transmission errors not only to be detected but also to be corrected.

## 14.1.2 ENTROPY

### Definitions:

**Information theory** is concerned with a mathematical theory of communication and includes the study of redundancy and the underlying limits of communication channels.

Let $X$ be a random variable that takes on a finite set of values $x_1, x_2, \ldots, x_n$ with probability $Pr(X = x_i) = p_i$, where $0 \le p_i \le 1$ for each $i$, $1 \le i \le n$, and $\sum_{i=1}^{n} p_i = 1$. Also, let $Y$ be a random variable that takes on a finite set of values.

The **entropy** (or **uncertainty**) of $X$ is defined to be $H(X) = -\sum_{i=1}^{n} p_i \log_2 p_i$, where $p_i \log_2 p_i = 0$ if $p_i = 0$.

The **joint entropy of $X$ and $Y$** is defined to be

$$H(X,Y) = -\sum_{x,y} Pr(X=x, Y=y) \log_2 Pr(X=x, Y=y).$$

If $X$ and $Y$ are random variables, the **conditional entropy of $X$ given $Y = y$** is

$$H(X \mid Y=y) = -\sum_x Pr(X=x \mid Y=y) \log_2 Pr(X=x \mid Y=y).$$

The **conditional entropy of $X$ given $Y$** (or the **equivocation of $Y$ about $X$**) is $H(X \mid Y) = -\sum_y Pr(Y=y) H(X \mid Y=y)$. (The summation indices $x$ and $y$ range over all values of $X$ and $Y$, respectively.)

**Facts:**

**1.** Useful books that cover information theory include [Ha80], [HaHaJo97], [Mc77], and [We98].

**2.** Information theory provides a theoretical basis for many results in error-correcting codes and cryptography, and provides theoretical bounds useful as metrics for evaluating conjectures in both areas.

**3.** The entropy of $X$ is a measure of the amount of information provided by an observation of $X$.

**4.** The entropy of $X$ is also useful for approximating the number of bits required to encode the elements of $X$.

**5.** If $X$ and $Y$ are random variables, then

- $0 \le H(X) \le \log_2 n$;
- $H(X) = 0$ if and only if $p_i = 1$ for some $i$, and $p_j = 0$ for all $j \ne i$ (that is, there is no uncertainty of the result);
- $H(X) = \log_2 n$ if and only if $p_i = \frac{1}{n}$ for each $i$, $1 \le i \le n$ (that is, all outcomes are equally likely);
- $H(X,Y) \le H(X) + H(Y)$;
- $H(X,Y) = H(X) + H(Y)$ if and only if $X$ and $Y$ are independent.

**6.** The quantity $H(X \mid Y)$ measures the amount of uncertainty remaining about $X$ after $Y$ has been observed.

**7.** If $X$ and $Y$ are random variables, then

- $H(X \mid Y) \ge 0$;
- $H(X \mid X) = 0$;
- $H(X,Y) = H(Y) + H(X \mid Y)$;
- $H(X \mid Y) \le H(X)$;
- $H(X \mid Y) = H(X)$ if and only if $X$ and $Y$ are independent.

**Examples:**

**1.** If $X$ is the random variable on the set $\{x_1, x_2, x_3, x_4\}$ with $p_1 = 0.4$, $p_2 = 0.3$, $p_3 = 0.2$, and $p_4 = 0.1$, then the entropy of $X$ is

$$H(X) = -(0.4 \log_2 0.4 + 0.3 \log_2 0.3 + 0.2 \log_2 0.2 + 0.1 \log_2 0.1) \approx 1.84644.$$

**2.** If $X$ is the random variable on the same set $\{x_1, x_2, x_3, x_4\}$ with $p_1 = p_2 = p_3 = p_4 = \frac{1}{4}$, then the entropy of $X$ is

$$H(X) = -4(\tfrac{1}{4}\log_2 \tfrac{1}{4}) = 2.$$

This situation, with equal probability outcomes, achieves the maximum entropy $\log_2 n = \log_2 4 = 2$.

---

## 14.1.3   THE NOISELESS CODING THEOREM

**Definitions:**

A **source** is a stream of **words** from a set $W = \{w_1, w_2, \dots, w_M\}$.

Let $X_i$ denote the $i$th word produced by a source. The source is said to be **memoryless** if for each word $w_j \in W$, the probability $Pr(X_i = w_j) = p_j$ is independent of $i$, that is, the $X_i$ are independent and identically distributed random variables.

The **entropy** of a memoryless source is $H = -\sum_{j=1}^{M} p_j \log_2 p_j$.

A **code** is a map $f$ from $W$ to $A^*$, the set of all finite strings of elements of $A$ where $A$ is a finite set called the **alphabet**.

For each source word $w_j \in W$, the string $f(w_j)$ is a **codeword**.

The **length** of the codeword $f(w_j)$, denoted $|f(w_j)|$, is the number of symbols in the string.

A **message** is any finite string of source words. If $m = v_1 v_2 \dots v_r$ is a message, then its encoding is obtained by concatenation: $f(m) = f(v_1)f(v_2)\dots f(v_r)$.

The **average length** of a code $f$ is $\sum_{j=1}^{M} p_j |f(w_j)|$.

A code is **uniquely decodable** if every string from $A^*$ is the image of at most one message.

A **prefix code** is a code such that there do not exist distinct words $w_i$ and $w_j$ such that $f(w_i)$ is an initial segment, or prefix, of $f(w_j)$.

**Facts:**

**1.** Prefix codes are uniquely decodable.

**2.** Prefix codes have the advantage of being *instantaneous*. That is, they can be decoded online without looking at future codewords.

**3.** *Kraft's inequality:* A prefix code $f : W \to A^*$ with codeword length $|f(w_i)| = l_i$ for $i = 1, 2, \dots, M$ exists if and only if $\sum_{j=1}^{M} n^{-l_j} \leq 1$, where $n$ is the size of the alphabet $A$.

**4.** *Macmillan's inequality:* If a uniquely decodable code $f : W \to A^*$ with codeword lengths $l_1, l_2, \dots, l_M$ exists, then $\sum_{j=1}^{M} n^{-l_j} \leq 1$, where $n$ is the size of the alphabet $A$.

**5.** A uniquely decodable code with prescribed word lengths exists if and only if a prefix code with the same word lengths exists. As a result, attention can be restricted to prefix codes.

**6.** *Shannon's noiseless coding theorem:* For a memoryless source of entropy $H$, any uniquely decodable code for the source into an alphabet of size $n$ must have average length at least $\frac{H}{\log_2 n}$. Moreover, there exists such a code having average length less than $1 + \frac{H}{\log_2 n}$.

**7.** For a memoryless source, a prefix code with smallest possible average length can be constructed by the Huffman coding algorithm. (See §9.1.2.)

### Examples:

**1.** The code that maps the letters A, B, C, D to 1, 01, 001, 0001, respectively, is a prefix code on this set of four letters.

**2.** The code that maps the letters A, B, C, D to 11, 111, 11111, 111111, respectively, is not a prefix code since the code for A forms the first part of the code for B (and for the codes for C and D as well). It is also not uniquely decodable since a bit string can correspond to more than one string of the letters A, B, C, D. For example, 11111 corresponds to AB, BA, and C.

### 14.1.4  CHANNELS AND CHANNEL CAPACITY

#### Definitions:

A **channel** is a medium that accepts strings of symbols from a finite alphabet $A = \{a_1, \ldots, a_n\}$ and produces strings of symbols from a finite alphabet $B = \{b_1, \ldots, b_m\}$.

Let $X_i$ denote the $i$th input symbol and let $Y_i$ denote the $i$th output symbol. The channel is said to be **memoryless** if the probability $Pr(Y_i = b_j \mid X_i = a_k) = p_{jk}$ (for $1 \leq j \leq m$ and $1 \leq k \leq n$) is independent of $i$.

A **binary symmetric channel** (**BSC**) is a memoryless channel with input and output alphabets $\{0, 1\}$, and probability $p$ that a symbol is transmitted incorrectly. The probability $p$ is called the *symbol error probability* of the channel. See the following figure.

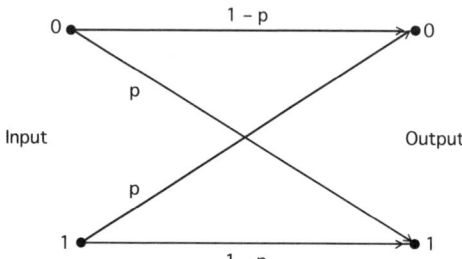

A **q-ary symmetric channel** is a memoryless channel with input and output alphabets each of size $q$ and such that the probability that an error occurs on symbol transmission is a constant $p$. Furthermore, if an error does occur then each of the $q-1$ symbols different from the correct symbol is equally likely to be received.

The **capacity** of a binary symmetric channel with symbol error probability $p$ is $C(p) = 1 + p \log_2 p + (1 - p) \log_2(1 - p)$.

#### Facts:

**1.** The capacity of a communications channel is a (unitless) measure of its ability to transmit information reliably.

**2.** The capacity of a BSC with symbol error probability $p$ is a monotone decreasing function of $p$ for $0 \leq p \leq \frac{1}{2}$, with $1 \geq C(p) \geq 0$. Moreover, $C(0) = 1$ and $C(\frac{1}{2}) = 0$.

**Example:**

**1.** The capacity of a BSC with symbol error probability 0.01 is given by

$$C(0.01) = 1 + 0.01 \log_2(0.01) + 0.99 \log_2(0.99) \approx 0.92.$$

## 14.2  BASICS OF CODING THEORY

Coding theory is the subject devoted to the theory of error-correcting codes. Error-correcting codes were invented to correct errors over unreliable transmission links. With digital communications and digital storage media ubiquitous in the modern world, error-correcting codes have grown in importance. Advances in error-correcting codes have made it possible to transmit information across the solar system using weak transmitters and to store data robustly on storage media so that it is resistant to damage, such as scratches on a compact disk.

Error-correcting codes work by encoding data as strings of symbols, such as bit strings, that contain redundant information that helps identify which codeword may have been sent when a string of symbols, potentially different from the string sent, is received. Coding theory is an active research area, with new and better codes being devised at a steady pace.

### 14.2.1  FUNDAMENTAL CONCEPTS

**Definitions:**

Let $A$ be any finite set (called an *alphabet*), and let $A^n$ denote the set of all $n$-tuples with entries in $A$. A **block code** of length $n$ containing $M$ **codewords** over the alphabet $A$ is a subset of $A^n$ of size $M$. Such a block code is called an **$(n, M)$-code** over $A$.

The **Hamming distance** $d(x, y)$ between two $n$-tuples $x$ and $y \in A^n$ is the number of entries in which they differ.

Let $C$ be an $(n, M)$-code over $A$. The **Hamming distance of $C$** is the smallest Hamming distance over all pairs of distinct codewords in $C$. If $C$ has Hamming distance $d$, then $C$ is sometimes referred to as an **$(n, M, d)$-code**.

The **information rate** (or **rate**) of an $(n, M)$-code over an alphabet of size $q$ is $R = \frac{\log_q M}{n}$.

Suppose that a codeword $c$ from a block code is transmitted and $r$ is received. The **error vector** is $e = r - c$ (formed by componentwise subtraction). The **number of errors** is the number of nonzero components in $e$.

A code is said to **detect** $t$ errors if the decoder is capable of detecting any pattern of $t$ or fewer errors per codeword that may be introduced by the channel.

A code is said to **correct** $t$ errors if the decoder is capable of correcting any pattern of $t$ or fewer errors per codeword that may be introduced by the channel.

Suppose that a codeword $c$ from a block code is transmitted and $r$ is received. An **erasure** is a coordinate within $r$ which is unknown.

A code is said to be ***t-erasure correcting*** if the decoder is capable of correcting any pattern which contains at most $t$ erasures.

A ***burst error*** of length $t$ is a vector whose only nonzero entries are among a string of successive entries, considered cyclically or non-cyclically, the first and last of which are nonzero.

A code is said to be ***t-cyclic burst error correcting*** if the decoder is capable of correcting any pattern which is a cyclic burst of length $t$.

If $C_1$ and $C_2$ are two $(n, M)$-codes over an alphabet $A$, then $C_1$ and $C_2$ are said to be **equivalent codes** if there is a fixed permutation of the coordinate positions which transform one code to the other.

The ***parity check bit*** of a bit string is 0 if there is an even number of bits in the string and is 1 if there is an odd number of bits in the string.

**Facts:**

**1.** Richard Hamming wrote one of the pioneering papers of coding theory in 1948, published in [Ha50].

**2.** Lester Hill, around the time of his paper [Hi27] from the 1920s, wrote an unpublished paper entitled "The checking of the accuracy of transmittal of telegraphic communications by means of operations in finite algebraic fields" [ToChJo12]. In this paper, Hill introduced a very special class of BCH codes for the purpose of error detection.

**3.** Some of the many introductory-level books in coding theory are [Ba97], [Bi05], [GuRuSu15], [JoKi11], [Mo05], [Pr92], [Ro06], [Ro96], and [Ye08]. For more extensive treatments, see [Be15], [Bl03], [JoZi15], [LiCo04], [Ma03], [NeRaSl06], [PeWe72], [PlHuBr98], [TsVl91], [TsVlNo07], [vL99], and especially [HuPl03] and [MaSl77], which contain extensive bibliographies.

**4.** The Error Correcting Codes (ECC) home page provides free software implementing several important error-correcting codes:

- http://www.eccpage.com

**5.** The main objective of coding theory is the design of codes such that

- an efficient algorithm is known for encoding messages;
- an efficient algorithm is known for decoding;
- the error-correcting capability of the code is high;
- the information rate of the code is high.

**6.** For applications in which a two-way communications channel is available (for example, a telephone circuit), it is sometimes economical to use error detection and retransmission upon error, in a so-called *automatic repeat request* (ARQ) strategy, rather than so-called *forward error correction* (FEC) techniques capable of actually correcting errors at the cost of more complex decoding equipment. This is not an option when the communications channel is effectively one-way or unperturbed source data is not available for retransmission (for example in CD-ROM storage and deep-space communications systems).

**7.** For any $n$-tuples $x, y, z \in A^n$, the Hamming distance satisfies

- $d(x, y) \geq 0$ with equality if and only if $x = y$;
- $d(x, y) = d(y, x)$;

- $d(x,y) + d(y,z) \geq d(x,z)$.

**8.** The information rate $R$ of a block code measures the fraction of information of the code which is non-redundant; the information rate $R$ satisfies the inequality $0 < R \leq 1$.

**9.** When a word $r$ is received, the decoder must make some decision. This decision may be one of the following:

- no errors have occurred; accept $r$ as a codeword;
- errors have occurred; correct $r$ to a codeword $c$;
- errors have occurred; no correction is possible.

**10.** Let $C$ be an $(n, M, d)$-code.

- If used only for error detection, $C$ can detect $d-1$ errors.
- If used for error correction, $C$ can correct $\lfloor \frac{d-1}{2} \rfloor$ errors.
- If used for erasure correction, $C$ can correct $d-1$ erasures.

**11.** Let $C$ be an $(n, M, d)$-code, and let $s, t, u \geq 0$.

- If $d \geq 2s + t + 1$, then $C$ can simultaneously correct $s$ errors and detect $s + t$ errors.
- If $d \geq 2s + t + 1$, then $C$ can simultaneously correct $s$ errors and $t$ erasures.
- If $d \geq 2s + t + u + 1$, then $C$ can simultaneously correct $s$ errors and $t$ erasures and detect $s + u$ errors.

**12.** Equivalent codes have the same distance, and hence the same error-correcting capabilities.

**13.** Adding a parity check bit to a bit string of length $n$ produces a bit string of length $n + 1$ with an even number of 0s.

**14.** Different families of error-correcting codes have been, and continue to be, designed to meet the particular requirements of applications. One type of requirement is the ability to correct specific types of errors. For many applications, it is sufficient to assume that errors in different positions occur independently of each other. This assumption follows immediately when using a memoryless channel.

**15.** In other applications, errors may not be independent of each other. For example, when signals are sent over radio channels, including those from deep space, interference can produce errors in a run of bits. Similarly, damage to storage media, such as a compact disk, can produce errors that come in clusters. These types of errors may be better modeled using burst errors and are better handled by cyclic codes (§14.4.1), in particular Reed-Solomon codes (§14.4.4), interleaved Reed-Solomon codes (see [HuPl03] for more information), and fire codes (see [Bl03]).

**Examples:**

**1.** The code produced by adding a parity check bit to each bit string of length $n$ can detect a single error. (It detects an odd number of errors, but not an even number of errors; no error correction is possible using this code.) For example, suppose the bit string 0111 is received where the code word sent is a bit string of length three with a parity check bit added. Since 0111 contains three 1s, it cannot be a codeword. Hence, an error was made in transmission. This error cannot be corrected. To see this, note that if exactly one bit error was made in the transmission, any of the codewords 0110, 0101, 0011, and 1111 could have been sent.

**2.** $C = \{0100011, 1010101, 1101111\}$ is a $(7, 3, 3)$-code over the binary alphabet. Note that $d = 3$ is the minimum Hamming distance between codewords (namely, the first and third codewords). The information rate of $C$ is $R = \frac{\log_2 3}{7} \approx 0.226$.

**3.** The *binary repetition code* of length $n$ is the code $C = \{00\ldots0, 11\ldots1\}$. The code has distance $n$, and so can correct $\lfloor \frac{n-1}{2} \rfloor$ errors. If used only for error detection, then $C$ can detect $n-1$ errors. Although the error-correcting capabilities of $C$ are very good, its information rate $R = \frac{1}{n}$ is very poor.

---

### 14.2.2   MAXIMUM LIKELIHOOD DECODING

**Definitions:**

Suppose $C$ is an $(n, M, d)$-code. Different decoding schemes can be used to recover a codeword from a transmitted bit string received with possible errors. These schemes include

- **Minimum Error Probability Decoding (MED):** If an $n$-tuple $r$ is received, then correct $r$ to a codeword $c$ for which the conditional probability $Pr(c$ is sent $\mid r$ is received$)$, $c \in C$, is largest.

- **Incomplete Maximum Likelihood Decoding (IMLD):** If an $n$-tuple $r$ is received, and there is a unique codeword $c \in C$ such that $d(r, c)$ is a minimum, then correct $r$ to $c$. If no such $c$ exists, then report that errors have been detected, but no correction is possible.

- **Complete Maximum Likelihood Decoding (CMLD):** If an $n$-tuple $r$ is received, and there is a unique codeword $c \in C$ such that $d(r, c)$ is a minimum, then correct $r$ to $c$. Otherwise, arbitrarily select one of the codewords $c \in C$ that is the closest to $r$, and correct $r$ to $c$.

**Facts:**

**1.** For any fixed probability distribution of the source messages, the probability of a decoding error, given that an $n$-tuple $r$ is received, is minimized by MED among all decoding schemes.

**2.** MED has the disadvantage that the decoding algorithm depends on the probability distribution of the source messages. The decoding strategy that is used in practice is CMLD.

**3.** Suppose that the probability that a symbol is transmitted incorrectly in a $q$-ary symmetric channel is $p$, where $0 < p < \frac{q-1}{q}$. Let $r$ be a received word and $c_1, c_2 \in C$ with $d(c_1, r) = d_1$ and $d(c_2, r) = d_2$. Let $Pr(r \mid c)$ denote the probability that $r$ is received, given that $c$ was sent. Then $Pr(r \mid c_1) \leq Pr(r \mid c_2)$ if and only if $d_1 \geq d_2$.

**4.** CMLD chooses a codeword $c$ for which the conditional probability $Pr(r$ is received $\mid c$ is sent$)$, $c \in C$, is largest.

**5.** If all source messages are equally likely, then CMLD performs in exactly the same way as MED.

**Examples:**

**1.** For the binary symmetric channel, Maximum Likelihood Decoding agrees with Nearest Neighbor Decoding (where a received word $v$ is corrected to the codeword $c$ closest to it in the Hamming distance).

**2.** $C = \{010101, 101010, 111111\}$ is a $(6,3,3)$-code over the binary alphabet. The code is transmitted over a binary symmetric channel. The received word $v = 011111$ has respective distances $2, 4, 1$ to the three codewords of $C$ and so is corrected to the codeword $c = 111111$ using Maximum Likelihood Decoding.

### 14.2.3   THE NOISY CHANNEL CODING THEOREM

**Definitions:**
Let $C$ be an $(n, M)$-code, where each word occurs with equal probability. Let $r_i$ be the probability of making an incorrect decision using complete maximum likelihood decoding given that the $i$th codeword was transmitted. The **error probability** of the code $C$ is $P_C = \frac{1}{M} \sum_{j=1}^{M} r_j$.

Let parameters $n$ and $M$ be fixed. Define $P^*(n, M, p)$ to be the smallest error probability $P_C$ of any $(n, M)$-code using a BSC with symbol error probability $p$.

**Facts:**
**1.** *Shannon's noisy channel coding theorem:*   Let $C(p)$ denote the capacity of a BSC with symbol error probability $p$, and define the quantity $M_n = 2^{\lfloor Rn \rfloor}$. If $0 < R < C(p)$, then $P^*(n, M_n, p) \to 0$ as $n \to \infty$.

**2.** By Shannon's noisy channel coding theorem, arbitrarily reliable communication with a fixed information rate is possible on a channel provided that the information rate is less than the channel capacity. Unfortunately, all known proofs of the theorem are non-constructive and do not give bounds on word lengths. So the good codes promised by the theorem may have extremely large word lengths.

## 14.3   LINEAR CODES

Linear codes are an important type of codes with a particular type of structure. In particular, a linear code is a code that is a subspace of a finite-dimensional vector space over a finite field. The main advantages of using linear codes arise from the efficient procedures for correcting errors. These procedures are based on matrix computations that can be carried out easily and rapidly.

### 14.3.1   INTRODUCTION

**Definitions:**
Let $GF(q)^n$ denote the vector space of all $n$-tuples having components from the finite field $GF(q)$ (§5.6.3). The elements of $GF(q)^n$ are called *vectors* or *words*.

An $[n, k]$-**linear code** $C$ over $GF(q)$ is a $k$-dimensional subspace of $GF(q)^n$ over $GF(q)$. More precisely, $C$ is a **linear block code**, but the qualification "block" is generally omitted. The code $C$ is referred to as an $[n, k, d]$-**code**, where $n$ is the **length** of the code, $k$ is the **dimension** of the subspace, and $d$ is the **distance**.

The **Hamming weight** of a word $v \in GF(q)^n$ is the number of nonzero coordinates in $v$.

Let $C$ be an $[n, k]$-code over $GF(q)$. A **generator matrix** $G$ for $C$ is a $k \times n$ matrix with entries from $GF(q)$ whose rows form a basis for $C$.

If an $[n, k]$-code $C$ has a generator matrix of the form $G = [I_k \,|\, A]$, then $C$ is called a **systematic code**, and the generator matrix $G$ is said to be in **standard form**.

The first $k$ positions within a systematic $[n, k]$-code are referred to as the **information symbols** for the code. The last $n - k$ positions are referred to as the **check symbols**.

Let $x = (x_1, x_2, \ldots, x_n)$ and $y = (y_1, y_2, \ldots, y_n)$ be two vectors in $GF(q)^n$. The **inner product** of $x$ and $y$ is the field element $x \circ y = \sum_{i=1}^{n} x_i y_i$. If $x \circ y = 0$, $x$ and $y$ are **orthogonal**.

Let $C$ be an $[n, k]$-code over $GF(q)$. The **orthogonal complement** of $C$, denoted $C^{\perp}$ (read "$C$ perp"), is the set of vectors orthogonal to every vector in $C$:

$$C^{\perp} = \{x \in GF(q)^n \mid x \circ y = 0 \text{ for all } y \in C\}.$$

$C^{\perp}$ is usually called the **dual code** of $C$.

A **parity check matrix** for an $[n, k]$-code $C$ is a generator matrix for $C^{\perp}$.

A linear code $C$ is **self-orthogonal** if $C \subseteq C^{\perp}$. It is **self-dual** if $C = C^{\perp}$.

**Facts:**

**1.** Both [HuPl03] and [MaSl77] are excellent books on linear codes.

**2.** Square parentheses (used to denote an $[n, k]$-code or an $[n, k, d]$-code) denote that a code is linear, while round brackets (used to denote an $(n, M)$-code or an $(n, M, d)$-code as defined in §14.2.1) are used for all codes, linear or not.

**3.** An $[n, k]$-code over $GF(q)$, the finite field of $q$ elements, is an $(n, q^k)$-block code.

**4.** The information rate of an $[n, k]$-code is $R = \frac{k}{n}$.

**5.** The distance of a linear code $C$ is the minimum Hamming weight of a nonzero vector in $C$.

**6.** A linear code is often described by a generator matrix.

**7.** If $G$ is a generator matrix for a code, then any matrix obtained from $G$ by applying a sequence of elementary row operations is also generator matrix for that code.

**8.** Let $C$ be an $[n, k]$-code over $GF(q)$. Then there exists an equivalent systematic code $C'$ with generator matrix $[I_k \,|\, A]$, where $I_k$ is the $k \times k$ identity matrix, and $A$ is a $k \times (n - k)$ matrix with entries from $GF(q)$.

**9.** If $G$ is a generator matrix for an $[n, k]$-code $C$, then $C = \{mG \mid m \in GF(q)^k\}$. The source messages can be taken to be the elements of $GF(q)^k$, and hence encoding is simply multiplication by $G$. Systematic codes are advantageous because if $G$ is in standard form and $c = mG$ is the codeword corresponding to a message $m$, then the first $k$ components of $c$ are identically $m$.

**10.** Within a systematic code, the information symbols provide information about the message while the check symbols provide the redundancy which makes error correction possible.

**11.** If $C$ is an $[n, k]$-code over $GF(q)$, then $C^{\perp}$ is an $[n, n - k]$-code over $GF(q)$.

**12.** If $C$ is an $[n, k]$-code over $GF(q)$, then the dual code of $C^{\perp}$ is $C$ itself.

**13.** There are many important special types and families of linear codes, including Hamming codes (§14.3.4), Golay codes (§14.5.2), Reed-Muller codes (§14.3.5), and cyclic codes (§14.4.1). Among cyclic codes, BCH codes form an important class (§14.4.3) and among BCH codes there is an important class of codes known as Reed-Solomon codes (§14.4.4).

**14.** Reed-Muller codes were used by the Mariner 9 spacecraft on its mission to Mars. A Golay code was used by the Voyager 2 on its mission to Jupiter and Saturn. A Reed-Solomon code was used by the Voyager 2 on its mission to Uranus. (See §15.9 in [HuP03] for more details on these applications.)

**15.** Algorithm 1 uses linear algebra to construct a parity check matrix for a linear code from a generator matrix.

---

**Algorithm 1:   Constructing a parity check matrix $H$ from a generator matrix $G$.**

$G' :=$ the reduced row echelon form of $G$ {use elementary row operations}

$A :=$ the $k \times (n-k)$ matrix obtained from $G'$ by deleting the (pivot) leading columns of $G'$

$H :=$ the $(n-k) \times n$ matrix $H$ obtained by placing, in order, the rows of $-A$ in the columns of $H$ which correspond to the leading columns of $G'$, and placing in the remaining $n-k$ columns of $H$, in order, the columns of the $(n-k) \times (n-k)$ identity matrix $I_{n-k}$

---

**16.** *Parity check matrices:*   Let $C$ be an $[n, k]$-code over $GF(q)$ with a generator matrix $G$, and let $H$ be a parity check matrix for $C$.

- A vector $x \in GF(q)^n$ belongs to $C$ if and only if $xH^T = 0$; it follows that $GH^T = 0$.

- If $G = [I_k \mid A]$ is a generator matrix for $C$, then $H = [-A^T \mid I_{n-k}]$ is a parity check matrix for $C$.

- $C$ has distance at least $s$ if and only if every set of $s-1$ columns of $H$ are linearly independent over $GF(q)$; in other words, the distance of $C$ is equal to the smallest number of columns of $H$ that are linearly dependent over $GF(q)$.

**17.** Let $C$ be an $[n, k]$-code with generator matrix $G$. $C$ is self-orthogonal if and only if $GG^T = 0$.

**18.** Let $C$ be an $[n, k]$-code with generator matrix $G$. $C$ is self-dual if and only if it is self-orthogonal and $k = \frac{n}{2}$ (and hence $n$ is even).

**19.** A *low-density parity check matrix (LDPC)* is a linear code whose parity check matrix has relatively few nonzero elements.

**Examples:**

**1.** Let $C$ be a binary $[7, 4]$-code with generator matrix

$$G = \begin{pmatrix} 0 & 0 & 1 & 0 & 1 & 0 & 1 \\ 1 & 1 & 0 & 0 & 1 & 0 & 1 \\ 0 & 0 & 1 & 0 & 0 & 1 & 1 \\ 1 & 1 & 1 & 0 & 1 & 1 & 1 \end{pmatrix}.$$

Elementary row operations yield the reduced row echelon form of $G$:

$$G' = \begin{pmatrix} 1 & 1 & 0 & 0 & 0 & 1 & 0 \\ 0 & 0 & 1 & 0 & 0 & 1 & 0 \\ 0 & 0 & 0 & 0 & 1 & 1 & 0 \\ 0 & 0 & 0 & 0 & 0 & 0 & 1 \end{pmatrix}.$$

The leading columns of $G'$ are columns 1, 3, 5, and 7, giving

$$A = \begin{pmatrix} 1 & 0 & 1 \\ 0 & 0 & 1 \\ 0 & 0 & 1 \\ 0 & 0 & 0 \end{pmatrix}.$$

Hence, the following parity check matrix is obtained

$$H = \begin{pmatrix} 1 & 1 & 0 & 0 & 0 & 0 & 0 \\ 0 & 0 & 0 & 1 & 0 & 0 & 0 \\ 1 & 0 & 1 & 0 & 1 & 1 & 0 \end{pmatrix}.$$

2. The *extended Hamming code* of order 3 is a binary $[8, 4, 4]$-code with generator matrix

$$G = \begin{pmatrix} 1 & 0 & 0 & 0 & 1 & 1 & 0 & 1 \\ 0 & 1 & 0 & 0 & 1 & 0 & 1 & 1 \\ 0 & 0 & 1 & 0 & 0 & 1 & 1 & 1 \\ 0 & 0 & 0 & 1 & 1 & 1 & 1 & 0 \end{pmatrix}.$$

The code is self-dual since $GG^T = 0$.

## 14.3.2  SYNDROME DECODING

Syndrome decoding is a general decoding technique for linear codes that is useful if the information rate of the code is high. Let $C$ be an $[n, k, d]$-code over $GF(q)$ with parity check matrix $H$.

**Definitions:**
For any $x \in GF(q)^n$, the **coset** of $C$ determined by $x$ is the set $C + x = \{c + x \mid c \in C\}$.

For any $x \in GF(q)^n$, the **syndrome** of $x$ is the vector $xH^T$.

A **coset leader** of a coset of $C$ is one of the coset members of smallest weight.

**Facts:**
1. The coset determined by 0 is $C$.
2. For all $x \in GF(q)^n$, $x \in C + x$.
3. For all $x, y \in GF(q)^n$, if $y \in C + x$, then $C + y = C + x$, that is, each word in a coset determines that coset.
4. The cosets of $C$ partition $GF(q)^n$ into $q^{n-k}$ cosets, each of size $q^k$.
5. A syndrome is a vector of length $n - k$.

**6.** Two vectors $x_1$ and $x_2 \in GF(q)^n$ are in the same coset of $C$ if and only if they have the same syndrome, that is, $x_1 H^T = x_2 H^T$.

**7.** A vector $x \in GF(q)^n$ is a codeword if and only if its syndrome is 0.

**8.** Suppose that a codeword $c$ is transmitted and $r$ is received. If $e = r - c$, then $rH^T = eH^T$, which means that the error vector is in the same coset as the received word. By maximum likelihood decoding, the decoder should choose a vector of smallest weight in this coset as the error vector.

**9.** The fact that there is a one-to-one correspondence between syndromes and coset leaders leads to *syndrome decoding*, a decoding algorithm for linear codes, which is described as Algorithm 2.

---

**Algorithm 2**:  **Syndrome decoding for linear codes.**

> *precomputation*: set up a one-to-one correspondence between coset leaders and
> syndromes; let $r$ be a received word and $H$ the parity check matrix
> compute the syndrome $s = rH^T$ of $r$
> find the coset leader $e$ associated with $s$
> correct $r$ to $r - e$

---

**Example:**

**1.** Consider the binary $[5, 2]$-code $C$ with generator matrix

$$G = \begin{pmatrix} 1 & 0 & 0 & 0 & 1 \\ 0 & 1 & 1 & 1 & 1 \end{pmatrix}$$

and parity check matrix

$$H = \begin{pmatrix} 0 & 1 & 1 & 0 & 0 \\ 0 & 1 & 0 & 1 & 0 \\ 1 & 1 & 0 & 0 & 1 \end{pmatrix}.$$

The eight cosets of $C$ are

$$\{00000, 10001, 01111, 11110\} \quad \{10000, 00001, 11111, 01110\}$$
$$\{01000, 11001, 00111, 10110\} \quad \{00100, 10101, 01011, 11010\}$$
$$\{00010, 10011, 01101, 11100\} \quad \{11000, 01001, 10111, 00110\}$$
$$\{10100, 00101, 11011, 01010\} \quad \{01100, 11101, 00011, 10010\}.$$

The following is a list of coset leaders and their syndromes:

| coset leader | 00000 | 10000 | 01000 | 00100 | 00010 | 11000 | 10100 | 01100 |
|---|---|---|---|---|---|---|---|---|
| syndrome | 000 | 001 | 111 | 100 | 010 | 110 | 101 | 011 |

If the word $r = 01101$ is received, compute the syndrome $01101 \cdot H^T = 010$, which corresponds to a coset leader $e = 00010$. Hence, $r$ is corrected to $r - e = 01111$.

---

## 14.3.3   CONSTRUCTING NEW CODES FROM OLD

There are several methods for modifying a linear code to produce a new linear code. Some of these methods are extending a code, puncturing a code, and shortening a code.

**Definitions:**

If $C$ is a linear code of length $n$ over the field $GF(q)$, then the **extended code** $\overline{C}$ of $C$ is $\overline{C} = \{(c_1, c_2, \ldots, c_n, c_{n+1}) \mid (c_1, c_2, \ldots, c_n) \in C, \sum_{i=1}^{n+1} c_i = 0\}$. The symbol $c_{n+1}$ is called the **overall parity check symbol**.

If $C$ is a linear code over $GF(q)$, the code obtained by removing any column of a generator matrix of $C$ is called a **punctured** $C$, denoted $C^*$.

If $C$ is a linear code of length $n$, a **shortened code** $C'$ of $C$ is a linear code of length $n-1$ which equals the set of all codewords in $C$ having 0 in a fixed coordinate position, with that position deleted.

**Facts:**

**1.** If $C$ is an $[n, k, d]$-code over $GF(q)$ with generator matrix $G$ and parity check matrix $H$, then

- $\overline{C}$ is an $[n+1, k, \overline{d}]$-code over $GF(q)$;

- if $C$ is a binary code, then $\overline{d} = \begin{cases} d, & \text{if } d \text{ is even} \\ d+1, & \text{if } d \text{ is odd}; \end{cases}$

- a generator matrix for $\overline{C}$ is $\overline{G}$, which is obtained by adding a column to $G$ in such a way that the sum of the elements of each row of $\overline{G}$ is 0;

- a parity check matrix for $\overline{C}$ is $\overline{H}$, where $\overline{H} = \begin{pmatrix} 1 & 1 & 1 & 1 & \cdots & 1 \\ & & & & & 0 \\ & & H & & & 0 \\ & & & & & \vdots \\ & & & & & 0 \end{pmatrix}$.

**2.** Puncturing a code is the reverse process to extending a code.

**3.** If $C$ is an $[n, k, d]$-code over $GF(q)$, then $C^*$ is a linear code over $GF(q)$ of length $n-1$, dimension $k$ or $k-1$, and distance $d$ or $d-1$.

**4.** If $C$ is an $[n, k, d]$-code over $GF(q)$, $k \geq 2$, and $C$ has at least one codeword for which the deleted position has a nonzero entry, then $C'$ is an $[n-1, k-1, d']$-code over $GF(q)$, with $d' \geq d$.

### 14.3.4  HAMMING CODES

**Definition:**

A **Hamming code** of order $r$ over $F_q$, denoted $H_r(q)$, is an $[n, k]$-code where $n = \frac{q^r - 1}{q-1}$ and $k = n - r$, with a parity check matrix whose columns are nonzero and such that no two columns are scalar multiples of each other.

**Facts:**

**1.** A decoding algorithm for Hamming codes is shown in Algorithm 3.

**2.** In the binary case $(q = 2)$, the Hamming code $H_r(2)$ has a parity check matrix whose columns consist of all nonzero binary vectors of length $r$, each used exactly once.

**3.** $H_r(q)$ has distance 3, and so is a 1-error correcting code.

**4.** Any two binary Hamming codes of order $r$ are equivalent.

**5.** $H_r(q)$ is a perfect code (§14.5.2).

---

**Algorithm 3**:  **Decoding algorithm for Hamming codes.**

$H :=$ a parity check matrix for a Hamming code $H_r(q)$

$r :=$ a received word

compute the syndrome $s = rH^T$ of $r$

**if** $s = 0$ **then** accept $r$ as the transmitted word

**else**

    compare $s^T$ with the columns of $H$

    **if** $s^T = \alpha h_i$ (where $h_i$ is the $i$th column of $H$) **and** $\alpha \in F_q$ **then**

        the error vector $e$ is the vector with $\alpha$ in position $i$ and 0s elsewhere

        correct $r$ to $c = r - e$

---

**Example:**

**1.** Consider $H_3(2)$, the binary Hamming code of order 3. The code has length $n = 7$ and dimension $k = 4$, and a parity check matrix is

$$H = \begin{pmatrix} 1 & 0 & 0 & 1 & 1 & 0 & 1 \\ 0 & 1 & 0 & 1 & 0 & 1 & 1 \\ 0 & 0 & 1 & 0 & 1 & 1 & 1 \end{pmatrix}.$$

If the received word is $r = 1011101$, compute the syndrome $s = 1011101 \cdot H^T = 001$, which is the third column of $H$. Hence $e = 0010000$, and correct $r$ to 1001101.

---

### 14.3.5  REED-MULLER CODES

**Definition:**

A **first-order Reed-Muller code**, denoted $R(1, m)$, is the binary $[2^m, m + 1]$-code whose generator matrix is formed by adding a single row entirely of 1s to the $m \times 2^m$ submatrix whose columns consist of all binary $m$-tuples.

A **$k$th-order Reed-Muller code**, denoted $R(k, m)$, is the binary $[2^m, s]$-code where $s = \sum_{i=0}^{k} \binom{m}{k}$ whose generator matrix consists of a single row entirely of 1s and the rows of the submatrices $B_1, \ldots, B_k$ which have the following form. $B_1$ is the $m \times 2^m$ matrix whose columns are all distinct binary $m$-tuples and $B_i$ for $i = 2, \ldots, k$ is the $\binom{m}{i} \times 2^m$ matrix where each row is formed from an $i$-subset of the rows of $B_1$ by making an entry 1 if and only if the corresponding entries in each member of the $i$-subset is 1.

**Facts:**

**1.** A first-order Reed-Muller code $R(1, m)$ has weight $2^{m-1}$.

**2.** In practice, the columns in the generator matrix of a first-order Reed-Muller code are listed according to a *proper ordering*. This may be seen as normal lexigraphic order of 0-1 strings, but from right to left. Alternatively, it may be seen as the natural order of the integers in little-endian binary representation. The matrix in Example 1 shows this ordering.

**3.** First-order Reed-Muller codes can be decoded using Algorithm 4.

---

**Algorithm 4**:  **Decoding algorithm for first-order Reed-Muller codes.**

$H_{2^m} :=$ a Hadamard matrix of order $2^m$ in standard form

$G :=$ a generator matrix for the Reed-Muller code

$r := (r_0, \ldots, r_{2^m-1})$ a received word

compute the vector $R := (R_0, \ldots, R_{2^m-1})$ where $R_i = (-1)^{r_i}$

compute the vector $\hat{R} := RH_{2^m}$

let $\hat{R} = (\hat{R}_0, \ldots, \hat{R}_{2^m-1})$; find a component $\hat{R}_j$ with maximum magnitude

suppose $j$ has the binary representation $j = \sum_{i=1}^{m} j_i 2^{i-1}$ with $j_i \in \{0,1\}$

**if** $\hat{R}_j > 0$ **then** decode $r$ to $\sum_{i=1}^{m} j_i \vec{v}_i$ where $\vec{v}_i$ is the $i$th row of $G$ (discounting the row of 1s)

**if** $\hat{R}_j \leq 0$ **then** decode $r$ to $\vec{1} + \sum_{i=1}^{m} j_i \vec{v}_i$ where $\vec{v}_i$ is the $i$th row of $G$ (discounting the row of 1s)

---

**4.** In Algorithm 4, the computation of $RH_{2^m}$ can be speeded up by using a *Fast Hadamard Transform*.

**5.** The first-order Reed-Muller code $R(1,5)$ was used to transmit photographs from the Mariner 9 spacecraft. It had good error-correcting capabilities of correcting 7 errors and detecting 8. However, it had a very low information rate and relatively few codewords.

**6.** The $k$th order Reed-Muller code may also be viewed as a subspace of the vector space $V$ of linear functionals $f : GF(2)^m \to GF(2)$. By considering linear functionals as polynomials in $x_1, \ldots, x_n$, the set of the characteristic functionals $\chi_u = \prod_{i=1}^{m}(x_i + 1 + u_i)$ for all $u = (u_1, \ldots, u_m)$ forms a basis for $V$. Note that $\chi_u$ has the property that $\chi_u(u) = 1$ but $\chi_u(v) = 0$ for all $v \neq u$. The $k$th-order Reed-Muller code is the subspace of functionals of total degree at most $k$. A bijection can be explicitly defined between this representation and the ordinary representation by taking $f \mapsto (c_0, \ldots, c_{2^m-1})$ if $f = \sum_{i=0}^{2^m-1} c_i \chi_{u_i}$ where $u_i$ is the $m$-tuple representing the binary value of $i$.

**7.** The $k$th-order Reed-Muller code can also be represented as the code whose generator matrix is the incidence matrix between the points and the $(m-k)$-flats of an affine geometry of dimension $m$ over $GF(2)$.

**Examples:**

**1.** Consider the first-order Reed-Muller code $R(1,3)$. This code has length $n = 8$, dimension $k = 4$, and distance $d = 4$. Using the proper ordering, the generator matrix for $R(1,3)$ is

$$G = \begin{pmatrix} 1 & 1 & 1 & 1 & 1 & 1 & 1 & 1 \\ 0 & 1 & 0 & 1 & 0 & 1 & 0 & 1 \\ 0 & 0 & 1 & 1 & 0 & 0 & 1 & 1 \\ 0 & 0 & 0 & 0 & 1 & 1 & 1 & 1 \end{pmatrix}.$$

**2.** Suppose the vector $r = (1,0,1,1,0,1,1,0)$ is received. Then we can compute $R = (-1,1,-1,-1,1,-1,-1,1)$. Multiplying $R$ by the Hadamard matrix

$$H_8 = \begin{pmatrix} 1 & 1 & 1 & 1 & 1 & 1 & 1 & 1 \\ 1 & -1 & 1 & -1 & 1 & -1 & 1 & -1 \\ 1 & 1 & -1 & -1 & 1 & 1 & -1 & -1 \\ 1 & -1 & -1 & 1 & 1 & -1 & -1 & 1 \\ 1 & 1 & 1 & 1 & -1 & -1 & -1 & -1 \\ 1 & -1 & 1 & -1 & -1 & 1 & -1 & 1 \\ 1 & 1 & -1 & -1 & -1 & -1 & 1 & 1 \\ 1 & -1 & -1 & 1 & -1 & 1 & 1 & -1 \end{pmatrix}$$

gives us the vector $\hat{R} = (-2, -2, 2, 2, -2, -2, 2, -6)$. The component $\hat{R}_7 = -6$ has largest magnitude. Since $7 = 1 + (1)2^1 + 1(2^2)$, and $\hat{R}_7 < 0$, we decode $r$ to $c = (1, 0, 0, 1, 0, 1, 1, 0)$.

**3.** The second-order Reed-Muller code $R(2,3)$ would have $n = 8$, $k = 7$, and $d = 2$. It would have generator matrix

$$G = \begin{pmatrix} 1 & 1 & 1 & 1 & 1 & 1 & 1 & 1 \\ 0 & 1 & 0 & 1 & 0 & 1 & 0 & 1 \\ 0 & 0 & 1 & 1 & 0 & 0 & 1 & 1 \\ 0 & 0 & 0 & 0 & 1 & 1 & 1 & 1 \\ 0 & 0 & 0 & 1 & 0 & 0 & 0 & 1 \\ 0 & 0 & 0 & 0 & 0 & 1 & 0 & 1 \\ 0 & 0 & 0 & 0 & 0 & 0 & 1 & 1 \end{pmatrix} .$$

## 14.3.6    WEIGHT ENUMERATORS

### Definitions:
Let $C$ be an $(n, M)$-code and let $A_i$ be the number of codewords of weight $i$ in $C$, for $i = 0, 1, \ldots, n$. The vector $(A_0, A_1, \ldots, A_n)$ is called the **weight distribution** of $C$.

Let $C$ be an $[n, k]$-code over $GF(q)$ with weight distribution $(A_0, A_1, \ldots, A_n)$. The **weight enumerator** of $C$ is defined to be the polynomial $W_C(z) = \sum_{i=0}^{n} A_i z^i$.

### Facts:
**1.** Let $C$ be an $[n, k]$-code over $GF(q)$, and let the symbol error probability on the $q$-ary symmetric channel be $p$. If $C$ is used only for error detection, then the probability of an error going undetected is $\sum_{i=0}^{n} A_i \left(\frac{p}{q-1}\right)^i (1 - p)^{n-i}$.

**2.** *MacWilliams' identity:*   Let $C$ be an $[n, k]$-code over $GF(q)$ with dual code $C^\perp$. Then

$$W_{C^\perp}(z) = \tfrac{1}{q^k} [1 + (q-1)z]^n W_C\left(\tfrac{1-z}{1+(q-1)z}\right).$$

### Examples:
**1.** The weight distribution of a binary Hamming code of length $n$ satisfies the recurrence $A_0 = 1$, $A_1 = 0$,

$$(i + 1)A_{i+1} + A_i + (n - i + 1)A_{i-1} = \binom{n}{i}, \quad i \geq 1.$$

**2.** The weight enumerator of the Golay code (§14.5.2) is

$$1 + 253z^7 + 506z^8 + 1288z^{11} + 1288z^{12} + 506z^{15} + 253z^{16} + z^{23}.$$

# 14.4  CYCLIC CODES

Cyclic codes are an important type of linear code which have an additional algebraic structure imposed upon them. They are not only subspaces of finite-dimensional vector spaces, but ideals within a special type of polynomial ring. This additional structure enables the use of shifting and polynomial multiplication in addition to ordinary vector addition. One of the main advantages of cyclic codes is their ability to correct errors which occur in bursts.

## 14.4.1  INTRODUCTION

### Definitions:

A linear code $C$ of length $n$ is **cyclic** if whenever $(a_0, a_1, a_2, \ldots, a_{n-1})$ is a codeword in $C$, then the cyclic shift $(a_{n-1}, a_0, a_1, \ldots, a_{n-2})$ is also a codeword in $C$.

Let $g(x)$ be a polynomial in $GF(q)[x]/(x^n - 1)$. The ideal generated by $g(x)$, namely $\{a(x)g(x) \mid a(x) \in GF(q)[x]/(x^n - 1)\}$, is called the **code generated** by $g(x)$ and is denoted $\langle g(x) \rangle$.

Let $C$ be a nonzero cyclic code in $GF(q)[x]/(x^n - 1)$. A monic polynomial $g(x)$ of least degree in $C$ is called a **generator polynomial** of $C$.

The polynomial $h(x) = \dfrac{x^n - 1}{g(x)}$ is called the **check polynomial** of $C$.

Let $H$ be a parity check matrix for a cyclic code. If $r$ is a received word, the **syndrome polynomial** of $r$ is the polynomial $s(x)$ corresponding to the syndrome $s = rH^T$.

### Facts:

**1.** The study of cyclic codes is facilitated by the attachment of some additional algebraic structure to the vector space $GF(q)^n$.

**2.** If the vector $(a_0, a_1, a_2, \ldots, a_{n-1})$ in $GF(q)^n$ is identified with the polynomial $a_0 + a_1 x + a_2 x^2 + \cdots + a_{n-1} x^{n-1}$, then

- the ring $GF(q)[x]/(x^n - 1)$ can be viewed as a vector space over $GF(q)$;
- the vector spaces $GF(q)^n$ and $GF(q)[x]/(x^n - 1)$ are isomorphic;
- multiplication of a polynomial in $GF(q)[x]/(x^n - 1)$ by $x$ corresponds to a cyclic shift of the corresponding vector;
- a linear code $C$ in the vector space $GF(q)^n$ is cyclic if and only if $C$ is an ideal in the ring $GF(q)[x]/(x^n - 1)$.

**3.** An ideal may contain many elements which will generate the ideal. One of these generators is singled out as *the* generator.

**4.** If $g(x)$ is a generator polynomial of a cyclic code $C$, then $g(x)$ generates $C$; that is, $\langle g(x) \rangle = C$.

**5.** The following are consequences of the fact that the ring $GF(q)[x]/(x^n - 1)$ is a principal ideal domain (i.e., each ideal is generated by a monic polynomial, called the generator polynomial). Here $C$ is a nonzero cyclic code in $GF(q)[x]/(x^n - 1)$ with generator polynomial $g(x)$.

- The generator polynomial of $C$ is unique.

- $g(x)$ divides $x^n - 1$ in $GF(q)[x]$.

- If the degree of $g(x)$ is $n-k$, that is, $g(x) = g_0 + g_1 x + g_2 x^2 + \cdots + g_{n-k} x^{n-k}$ (and $g_{n-k} = 1$), then a basis for $C$ is $\{g(x), xg(x), x^2 g(x), \ldots, x^{k-1} g(x)\}$; hence $C$ has dimension $k$ and a generator matrix for $C$ is

$$
\begin{pmatrix}
g_0 & g_1 & g_2 & \cdots & \cdots & g_{n-k} & 0 & 0 & \cdots & 0 \\
0 & g_0 & g_1 & \cdots & \cdots & g_{n-k-1} & g_{n-k} & 0 & \cdots & 0 \\
0 & 0 & g_0 & \cdots & \cdots & g_{n-k-2} & g_{n-k-1} & g_{n-k} & \cdots & 0 \\
\vdots & \vdots & & \ddots & & & \ddots & \ddots & \ddots & \vdots \\
0 & 0 & \cdots & 0 & g_0 & & \cdots & & & g_{n-k}
\end{pmatrix}.
$$

**6.** Any $c(x) \in C$ can be written uniquely as $c(x) = f(x)g(x)$ in the ring $GF(q)[x]$, where $f(x) \in GF(q)[x]$ has degree less than $k$. Hence, encoding a message polynomial $f(x)$ consists simply of polynomial multiplication by $g(x)$.

**7.** The dual code $C^\perp$ is also cyclic.

**8.** Let $h(x) = h_0 + h_1 x + h_2 x^2 + \cdots + h_k x^k = \dfrac{x^n - 1}{g(x)}$ be in $GF(q)[x]$. Then the reciprocal polynomial $h^*(x) = x^k h(\frac{1}{x})$ of $h(x)$ is a generator of $C^\perp$. (In fact, $(\frac{1}{h_0})h^*(x)$ is the generator polynomial of $C^\perp$.) Hence, a parity check matrix for $C$ is

$$
\begin{pmatrix}
h_k & h_{k-1} & h_{k-2} & \cdots & \cdots & h_0 & 0 & 0 & \cdots & 0 \\
0 & h_k & h_{k-1} & \cdots & \cdots & h_1 & h_0 & 0 & \cdots & 0 \\
0 & 0 & h_k & \cdots & \cdots & h_2 & h_1 & h_0 & \cdots & 0 \\
\vdots & \vdots & & \ddots & & & \ddots & \ddots & \ddots & \\
0 & 0 & \cdots & 0 & h_k & & \cdots & & & h_0
\end{pmatrix}.
$$

**9.** A cyclic code of length $n$ over $GF(q)$ is characterized by its generator polynomial.

**10.** There is a one-to-one correspondence between cyclic codes in $GF(q)^n$ and monic polynomials in $GF(q)[x]$ which divide $x^n - 1$.

**11.** The following table gives the complete factorization of $x^n - 1$ over $GF(2)$ for some small values of odd $n$, $1 \le n \le 31$.

| $n$ | factorization of $x^n - 1$ over $GF(2)$ |
|---|---|
| 1 | $1 + x$ |
| 3 | $(1+x)(1+x+x^2)$ |
| 5 | $(1+x)(1+x+x^2+x^3+x^4)$ |
| 7 | $(1+x)(1+x+x^3)(1+x^2+x^3)$ |
| 9 | $(1+x)(1+x+x^2)(1+x^3+x^6)$ |
| 11 | $(1+x)(1+x+x^2+\cdots+x^{10})$ |
| 13 | $(1+x)(1+x+x^2+\cdots+x^{12})$ |
| 15 | $(1+x)(1+x+x^2)(1+x+x^2+x^3+x^4)(1+x+x^4)(1+x^3+x^4)$ |
| 17 | $(1+x)(1+x+x^2+x^4+x^6+x^7+x^8)(1+x^3+x^4+x^5+x^8)$ |
| 19 | $(1+x)(1+x+x^2+\cdots+x^{18})$ |
| 21 | $(1+x)(1+x+x^2)(1+x^2+x^3)(1+x+x^3)(1+x^2+x^4+x^5+x^6)\cdot$ $\cdot(1+x+x^2+x^4+x^6)$ |
| 23 | $(1+x)(1+x+x^5+x^6+x^7+x^9+x^{11})(1+x^2+x^4+x^5+x^6+x^{10}+x^{11})$ |
| 25 | $(1+x)(1+x+x^2+x^3+x^4)(1+x^5+x^{10}+x^{15}+x^{20})$ |
| 27 | $(1+x)(1+x+x^2)(1+x^3+x^6)(1+x^9+x^{18})$ |
| 29 | $(1+x)(1+x+x^2+\cdots+x^{28})$ |
| 31 | $(1+x)(1+x^2+x^5)(1+x^3+x^5)(1+x+x^2+x^3+x^5)(1+x+x^2+x^4+x^5)\cdot$ $\cdot(1+x+x^3+x^4+x^5)(1+x^2+x^3+x^4+x^5)$ |

**12.** If $C$ is an $[n, k]$-cyclic code generated by $g(x)$, then another parity check matrix for $C$ is the matrix $H$ whose $i$th column is $x^i \bmod g(x)$, for $i = 0, 1, \ldots, n-1$.

**13.** If $r(x)$ is the polynomial corresponding to the received word $r$, then the syndrome polynomial of $r$ is simply $s(x) = r(x) \bmod g(x)$.

**14.** A cyclic code can correct cyclic burst errors of length $t$ if and only if all cyclic burst errors of length $t$ or less have different syndrome polynomials. Algorithm 1 shows how burst errors can be corrected. In the algorithm, $g(x)$ is the generator polynomial for an $[n, k]$-cyclic code which can correct burst errors of length $t$.

---

**Algorithm 1:  Decoding algorithm using burst error correction in cyclic codes.**

suppose a codeword $c$ is transmitted and $r$ is received
**for** $i = 0$ **to** $n - 1$
    compute the syndrome polynomial $s_i(x) = x^i r(x) \bmod g(x)$
    **if** $s_i$ has a (non-cyclic) burst of length $t$ or less **then**
        compute the error polynomial $e(x) = x^{n-i} s_i(x)$
        correct $r(x)$ to $c(x) = r(x) - e(x)$
        break
received word cannot be decoded

---

## Examples:

**1.** Over $GF(2)$, the factorization of $x^7 - 1$ is $x^7 - 1 = (1+x)(1+x+x^3)(1+x^2+x^3)$. The monic divisors of $x^7 - 1$ are

$$g_1(x) = 1$$
$$g_2(x) = 1 + x$$
$$g_3(x) = 1 + x + x^3$$
$$g_4(x) = 1 + x^2 + x^3$$
$$g_5(x) = (1 + x)(1 + x + x^3) = 1 + x^2 + x^3 + x^4$$
$$g_6(x) = (1 + x)(1 + x^2 + x^3) = 1 + x + x^2 + x^4$$
$$g_7(x) = (1 + x + x^3)(1 + x^2 + x^3) = 1 + x + x^2 + x^3 + x^4 + x^5 + x^6$$
$$g_8(x) = 1 + x^7$$

The polynomial $g_5(x)$ generates the binary $[7, 3]$-cyclic code

$$C = \{0000000, 1011100, 0101110, 0010111, 1001011, 1100101, 1110010, 0111001\}.$$

A generator matrix for $C$ is

$$G = \begin{pmatrix} 1 & 0 & 1 & 1 & 1 & 0 & 0 \\ 0 & 1 & 0 & 1 & 1 & 1 & 0 \\ 0 & 0 & 1 & 0 & 1 & 1 & 1 \end{pmatrix}.$$

A parity check matrix for $C$ is

$$H = \begin{pmatrix} 1 & 1 & 0 & 1 & 0 & 0 & 0 \\ 0 & 1 & 1 & 0 & 1 & 0 & 0 \\ 0 & 0 & 1 & 1 & 0 & 1 & 0 \\ 0 & 0 & 0 & 1 & 1 & 0 & 1 \end{pmatrix}.$$

**2.** The cyclic code in Example 1 has distance $d = 4$ and so can correct 1 error, but is also a 2-cyclic burst error correcting code. Suppose the the word $r = 1110100$ is received. Running Algorithm 1 yields the results

| $i$ | $s_i(x)$ | $s_i$ |
|---|---|---|
| 0 | $x + x^3$ | 0101 |
| 1 | $1 + x^3$ | 1001 |
| 2 | $1 + x + x^2 + x^3$ | 1111 |
| 3 | $1 + x$ | 1100 |

The syndrome is a (non-cyclic) burst of length 2. The error polynomial is $e(x) = x^{7-3}s_3(x) = x^4 + x^5$. The received word $r$ is decoded to the codeword $c = r - e = 1110100 - 0000110 = 1110010$.

## 14.4.2  INTERLEAVING

### Definitions:

Let $C$ be an $[n, k]$-code and consider $s$ codewords $c_i = (c_{i,1}, c_{i,2}, \ldots, c_{i,n}) \in C$. Place these codewords in a matrix as follows.

$$\begin{pmatrix} c_{1,1} & c_{1,2} & \cdots & c_{1,n} \\ c_{2,1} & c_{2,2} & \cdots & c_{2,n} \\ \vdots & \vdots & \ddots & \vdots \\ c_{s,1} & c_{s,2} & \cdots & c_{s,n} \end{pmatrix}.$$

The word $(c_{1,1}, c_{2,1}, \ldots, c_{s,1}, c_{1,2}, c_{2,2}, \ldots, c_{s,2}, \ldots, c_{1,n}, c_{2,n}, \ldots, c_{s,n})$ obtained by expanding the matrix entry-by-entry column-wise is said to be made by **interleaving** the $s$ codewords together.

The $[ns, ks]$-code $C^*$ obtained by interleaving all possible $s$-tuples of codewords from $C$ is said to have been created by **interleaving $C$ to a depth of $s$**.

Suppose instead of interleaving the codewords, we use the entries in each column of the above matrix as information symbols within an $[m, s]$-code $\hat{C}$. The result would be a sequence of codewords from $\hat{C}$. This process is referred to as **cross-interleaving $C$ and $\hat{C}$**.

### Facts:

**1.** If an $[n, k]$-code $C$ has distance $d$, then the code $C^*$ obtained by interleaving $C$ to a depth of $s$ has distance $ds$.

**2.** If an $[n, k]$-code $C$ can correct cyclic burst errors of length $t$, then the code $C^*$ obtained by interleaving $C$ to a depth of $s$ can correct cyclic bursts of length $st$.

**3.** Suppose $G$ is a generator matrix for an $[n, k]$-code $C$ and let $C^*$ be the code obtained by interleaving $C$ to a depth of $s$. A generator matrix $G^*$ for $C^*$ can be constructed by interleaving each of the $k$ rows of $G$ separately with $s - 1$ copies of the zero codeword in each of the $s$ possible ways. The resulting $ks$ vectors would form the rows of $G^*$.

**4.** If $C$ is a cyclic code with generator polynomial $g(x)$, then $C^*$, the code obtained interleaving $C$ to a depth of $s$, is also cyclic and its generator polynomial is $g(x^s)$.

**5.** In practice, after cross-interleaving $C$ and $\hat{C}$, the resulting codewords in $\hat{C}$ are often interleaved themselves to some specified depth.

### Examples:

**1.** Let $C$ be the binary $[7, 4]$-code with generator matrix

$$G = \begin{pmatrix} 1 & 0 & 0 & 0 & 1 & 1 & 1 \\ 0 & 1 & 0 & 0 & 1 & 0 & 1 \\ 0 & 0 & 1 & 0 & 0 & 1 & 1 \\ 0 & 0 & 0 & 1 & 1 & 1 & 0 \end{pmatrix}.$$

Interleaving the three codewords 1000111, 1010100, and 0111000 would produce the interleaved codeword 110 001 011 001 110 100 100.

**2.** Suppose that we wish to cross-interleave the three codewords from Example 1 with the binary $[6,3]$-code with generator matrix

$$G = \begin{pmatrix} 1 & 0 & 0 & 1 & 1 & 1 \\ 0 & 1 & 0 & 0 & 1 & 1 \\ 0 & 0 & 1 & 1 & 1 & 0 \end{pmatrix}.$$

Using each group of three symbols as information symbols, we get the seven new codewords 110100, 001110, 011101, 001110, 110100, 100111, and 100111. In practice, we can now interleave these codewords to a desired depth.

## 14.4.3  BCH CODES

### Definitions:

Let $\beta$ be a primitive $n$th root of unity in an extension field of $GF(q)$. Let $g(x)$ be the least common multiple of the minimal polynomials over $GF(q)$ of $\beta^a, \beta^{a+1}, \ldots, \beta^{a+\delta-2}$ where $a$ is an integer. The cyclic code of length $n$ over $GF(q)$ with generator polynomial $g(x)$ is called a **BCH code** (after its discoverers: R. C. Bose, D. Ray-Chaudhuri, and A. Hocquenghem) with **designed distance** $\delta$.

If $a = 1$ in the definition of a BCH code, the code is called **narrow-sense**. If $n = q^m - 1$ for some positive integer $m$ (that is, $\beta$ is primitive in $GF(q^m)$), the code is **primitive**.

### Facts:

**1.** BCH codes are special types of cyclic codes, discovered by A. Hocquenghem in 1959 and independently by R. C. Bose and D. K. Ray-Chaudhuri in 1960.

**2.** *BCH bound:* Let $C$ be a BCH code over $GF(q)$ with designed distance $\delta$. Then $C$ has distance at least $\delta$.

**3.** Algorithm 2 is one method for decoding BCH codes. In the algorithm, let $g(x)$ be a generator polynomial for a BCH code over $GF(q)$ of designed distance $\delta$ and length $n$. Hence $g(x) = \mathrm{lcm}\{m_i(x) \mid a \le i \le a + \delta - 2\}$, where $m_i(x)$ is the minimal polynomial of $\beta^i$ over $GF(q)$, and $\beta$ is a primitive $n$th root of unity in an extension field $GF(q^m)$.

---

**Algorithm 2**:  Decoding algorithm for BCH codes.

suppose a codeword $c$ is transmitted and $r$ is received

compute $S_j = r(\beta^{a+j})$ for $j = 0, 1, \ldots, \delta - 2$ and form the polynomial $S(z) = \sum_{j=0}^{\delta-2} S_j z^j$

use the extended Euclidean algorithm to calculate the greatest common divisor of $S(z)$ and $z^{\delta-1}$ in the ring $GF(q^m)[z]$; stop as soon as the remainder $r_i(z)$ has degree $< \frac{\delta-1}{2}$; this yields polynomials $s_i(z)$ and $t_i(z)$ such that $s_i(z)z^{\delta-1} + t_i(z)S(z) = r_i(z)$; $\sigma(z) := t_i(z)$; $w(z) := r_i(z)$

find $B$, the set of roots of $\sigma(z)$ in $GF(q^m)$

for each $\gamma \in B$, set $E_\gamma = \frac{-\gamma^{-1}w(\gamma)}{\sigma'(\gamma)}$, where $\sigma'(z)$ denotes the formal derivative of $\sigma(z)$

the error vector is $e = (e_0, e_1, \ldots, e_{n-1})$, where $e_i = \begin{cases} 0, & \text{if } \beta^{-i} \notin B, \\ E_\gamma, & \text{if } \beta^{-i} = \gamma \in B \end{cases}$

decode $r$ to $r - e$

---

**4.** In Algorithm 2 the roots comprising $B$ will lie in the subgroup of $GF(q^m)^*$ generated by $\beta$.

**5.** If it is assumed that the number of errors is $l \leq \lfloor \frac{\delta-1}{2} \rfloor$, then the decoding in Algorithm 2 is correct.

**6.** There are more efficient ways of obtaining $\sigma(z)$ and $w(z)$ than by using the Euclidean algorithm; for example, by using the Berlekamp-Massey algorithm (see §5.4.2 in [HuPl03]).

**7.** There exists an analogue to Algorithm 2 for decoding BCH codes in which both errors and erasures occur. See Chapter 5 of [HuPl03] for a description and references.

**Examples:**

**1.** Consider the finite field $GF(3^3)$ generated by a root $\alpha$ of the primitive polynomial $f(x) = 1 + 2x^2 + x^3 \in GF(3)[x]$. A table of powers of $\alpha$ is given in the following table.

| $i$ | $\alpha^i$ | $i$ | $\alpha^i$ | $i$ | $\alpha^i$ |
|---|---|---|---|---|---|
| 0 | 1 | 9 | $2 + 2\alpha + 2\alpha^2$ | 18 | $1 + \alpha$ |
| 1 | $\alpha$ | 10 | $1 + 2\alpha + \alpha^2$ | 19 | $\alpha + \alpha^2$ |
| 2 | $\alpha^2$ | 11 | $2 + \alpha$ | 20 | $2 + 2\alpha^2$ |
| 3 | $2 + \alpha^2$ | 12 | $2\alpha + \alpha^2$ | 21 | $1 + 2\alpha + 2\alpha^2$ |
| 4 | $2 + 2\alpha + \alpha^2$ | 13 | 2 | 22 | $1 + \alpha + \alpha^2$ |
| 5 | $2 + 2\alpha$ | 14 | $2\alpha$ | 23 | $2 + \alpha + 2\alpha^2$ |
| 6 | $2\alpha + 2\alpha^2$ | 15 | $2\alpha^2$ | 24 | $1 + 2\alpha$ |
| 7 | $1 + \alpha^2$ | 16 | $1 + 2\alpha^2$ | 25 | $\alpha + 2\alpha^2$ |
| 8 | $2 + \alpha + \alpha^2$ | 17 | $1 + \alpha + 2\alpha^2$ | | |

The element $\beta = \alpha^2$ is a primitive 13th root of unity in $GF(3^3)$. If $m_i(x)$ denotes the minimal polynomial of $\beta^i$ over $GF(3)$, then

$$m_0(x) = 2 + x$$
$$m_1(x) = 2 + 2x + 2x^2 + x^3$$
$$m_2(x) = 2 + 2x + x^3$$
$$m_4(x) = 2 + x + 2x^2 + x^3$$
$$m_7(x) = 2 + x^2 + x^3.$$

Since $m_1(x) = m_3(x)$, the polynomial

$$g(x) = \operatorname{lcm}(m_0(x), m_1(x), m_2(x), m_3(x)) = m_0(x)m_1(x)m_2(x)$$

has among its roots the elements $\beta^0$, $\beta^1$, $\beta^2$, and $\beta^3$. Hence $g(x)$ is a generator polynomial for a BCH code over $F_3$ of designed distance $\delta = 5$ and length $n = 13$.

**2.** Using the BCH code in Example 1, suppose that the decoder received the word $r = (220\ 021\ 110\ 2110)$. The following steps follow Algorithm 2 to decode $r$:

- Compute $S_0 = r(\beta^0) = 1$, $S_1 = r(\beta^1) = \alpha^{14}$, $S_2 = r(\beta^2) = \alpha^{23}$, and $S_3 = r(\beta^3) = \alpha^{16}$. This gives $S(z) = 1 + \alpha^{14}z + \alpha^{23}z^2 + \alpha^{16}z^3$.

- Applying the extended Euclidean algorithm in $GF(3^3)[z]$ to $S(z)$ and $z^4$ yields

| $i$ | $s_i(z)$ | $t_i(z)$ | $r_i(z)$ | $\deg r_i(z)$ |
|---|---|---|---|---|
| $-1$ | $1$ | $0$ | $z^4$ | $4$ |
| $0$ | $0$ | $1$ | $1 + \alpha^{14}z + \alpha^{23}z^2 + \alpha^{16}z^3$ | $3$ |
| $1$ | $1$ | $\alpha^{17} + \alpha^{23}z$ | $\alpha^{17} + \alpha^{16}z + \alpha^{13}z^2$ | $2$ |
| $2$ | $\alpha^3 + \alpha^{16}z$ | $\alpha^{15} + \alpha^3 z + \alpha^{13}z^2$ | $\alpha^{15} + \alpha^{16}z$ | $1$ |

Stop, since $\deg(r_2(z)) < \frac{\delta-1}{2} = 2$. Hence, $\sigma(z) = \alpha^{15} + \alpha^3 z + \alpha^{13}z^2$ and $w(z) = \alpha^{15} + \alpha^{16}z$.

- By trying all possibilities, obtain the set $B = \{\beta^5, \beta^9\}$ of roots of $\sigma(z)$.

- Compute $E_{\beta^5} = -\beta^{-5}\frac{w(\beta^5)}{\sigma'(\beta^5)} = 2$ and $E_{\beta^9} = -\beta^{-9}\frac{w(\beta^9)}{\sigma'(\beta^9)} = 2$.

- Hence, the error vector is $e = (000\ 020\ 002\ 0000)$ and the word $r$ is decoded to $(220\ 001\ 111\ 2110)$.

## 14.4.4   REED-SOLOMON CODES

**Definition:**

A **Reed-Solomon (RS)** code $C$ is a primitive BCH code of length $n = q-1$ over $GF(q)$. If $C$ has dimension $k$ then we call $C$ an $[n, k]$-RS code.

**Facts:**

**1.** Reed-Solomon codes are special types of BCH codes, and hence they have the same encoding and decoding algorithms.

**2.** Reed-Solomon codes are important because, for a fixed $n$ and $k$, no linear code can have greater distance.

**3.** Reed-Solomon codes are useful for correcting burst errors. (A *binary burst* of length $b$ is a bit string whose only nonzero entries are among $b$ successive components, the first and last of which are nonzero.)

**4.** A Reed-Solomon code was used to encode the data transmissions from the Voyager 2 spacecraft during its encounter with Uranus in January, 1986.

**5.** If $C$ is an $[n, k]$-RS code over $GF(q)$ with designed distance $\delta$, then the generator polynomial for $C$ has the form $g(x) = (x - \beta^a)(x - \beta^{a+1}) \dots (x - \beta^{a+\delta-2})$, where $\beta$ is a primitive element of $GF(q)$.

**6.** If $C$ is an $[n, k]$-RS code over $GF(q)$ with designed distance $\delta$, then the distance of $C$ is exactly $\delta = n - k + 1$.

**7.** A *generalized $[n, k]$-Reed-Solomon code* consists of all the codewords having the form $(f(\alpha_1), \dots, f(\alpha_n))$ for all polynomials $f \in GF(q)[x]$ of degree $< k$ where $\alpha_1, \dots, \alpha_n$ are fixed distinct elements from $GF(q)$. Codewords are sometimes represented as polynomials in $GF(q)[x]$ of degree $< k$.

**8.** Generalized Reed-Solomon codes can be decoded by a *curve-fitting* algorithm proposed by Sudan. This algorithm gives a way of decoding generalized RS-codes beyond the normal error-correction capabilities of an RS-code. It uses a method called *list decoding*, which produces a list of possible candidates to which the transmitted codeword should correspond. An excellent introduction to list decoding can be found in [Gu04].

**9.** It is possible to associate codewords within an RS-code to their polynomial representatives in a generalized RS-code. Suppose $c = (c_1, \ldots, c_n)$ is a codeword in an RS-code such that the $c_i$s are distinct. Setting $\alpha_i = c_i$ gives the association. A polynomial $f$ in the general representation can be converted to an $n$-tuple by evaluating it at the $n$ elements $c_i$. An $n$-tuple $(\hat{c}_1, \ldots, \hat{c}_n)$ can be converted to a polynomial $f$ of degree $< k$ by applying Lagrange interpolation given the relations $\hat{c}_i = f(c_i)$.

**10.** Error correction in compact disks (developed by Philips and Sony) uses a code known as the Cross-Interleaved Reed-Solomon Code (CIRC). The CIRC code is obtained by cross-interleaving two Reed-Solomon codes, one a $[28, 24]$-RS code and the other a $[32, 28]$-RS code.

**Examples:**

**1.** Consider the finite field $GF(5)$ generated by $\beta = 2$. Then $g(x) = (x - \beta)(x - \beta^2) = (x - 2)(x - 4) = x^2 + 4x + 3$ generates a $[4, 2]$-RS code over $GF(5)$ with distance $\delta = 4$.

**2.** The code in Example 1 may be considered as a generalized RS-code by considering all polynomials of degree $< 2$ over $GF(q)$ and using the fixed elements $(1, 2, 4, 3)$. Using this representation, the codeword $(1, 2, 4, 3)$ is associated with the polynomial $x$ while the codeword $(0, 4, 2, 3)$ is associated with the polynomial $1 + 4x$.

---

## 14.4.5 CODES FROM INCIDENCE MATRICES OF GRAPHS

**Definition:**

For any prime $p$, consider the row span $C$ over $GF(p)$ of the $|V| \times |E|$ incidence matrix of a connected graph $\Gamma = (V, E)$. This class of codes arising from incidence matrices of graphs has been investigated by a number of researchers, for example, Dankelmann, Key, and Rodrigues [DaKeRo13].

Recall an incidence structure $D = (P, B, J)$, with point set $P$, block set $B$, and incidence $J$, is a **$t$-$(v, k, \lambda)$ design** if $|P| = v$, every block $b \in B$ is incident with precisely $k$ points, and every $t$ distinct points are together incident with precisely $\lambda$ blocks. The **code $C_F(D)$ of the design $D$ over the finite field $F$** is the space spanned by the incidence vectors of the blocks over $F$.

An **edge cut** of a connected graph $\Gamma$ is a set $S \subset E$ such that $\Gamma - S = (V, E - S)$ is disconnected. The **edge connectivity** $\lambda(\Gamma)$ is the minimum cardinality of an edge cut. The minimum and maximum degrees of the vertices of $\Gamma$ are denoted by $\delta(\Gamma)$ and $\Delta(\Gamma)$.

If $\lambda(\Gamma) = \delta(\Gamma)$ and, in addition, the only edge sets of cardinality $\lambda(\Gamma)$ whose removal disconnects $\Gamma$ are the sets of edges incident with a vertex of degree $\delta(\Gamma)$, then $\Gamma$ is called **super-$\lambda$**.

Let $\Gamma = (V, E)$ be a connected graph and let $G$ be an incidence matrix for $\Gamma$. Denote by $C_p(G)$ the row-span of $G$ over $GF(p)$. Note that $C_2(G)$ is also known as the **cut space** of $\Gamma$.

**Facts:**

**1.** We have $\dim(C_2(G)) = |V| - 1$. For odd $p$, $\dim(C_p(G)) = |V|$ if $\Gamma$ has a closed path of odd length (i.e., if $\Gamma$ is not bipartite), and $\dim(C_p(G)) = |V| - 1$ if $\Gamma$ has no closed path of odd length (i.e., if $\Gamma$ is bipartite).

**2.** Let $\Gamma = (V, E)$ be a connected graph and let $G$ be a $|V| \times |E|$ incidence matrix for $\Gamma$. Then

- $C_2(G)$ is a $[|E|, |V| - 1, \lambda(\Gamma)]_2$-code;
- if $\Gamma$ is super-$\lambda$, then $C_2(G)$ is a $[|E|, |V| - 1, \delta(\Gamma)]_2$-code, and the minimum words are the rows of $G$ of weight $\delta(\Gamma)$.

**3.** Let $\Gamma = (V, E)$ be a connected bipartite graph and let $G$ be a $|V| \times |E|$ incidence matrix for $\Gamma$. Then

1. $C_p(G)$ is a $[|E|, |V| - 1, \lambda(\Gamma)]_p$-code;

2. if $\Gamma$ is super-$\lambda$, then $C_p(G)$ is a $[|E|, |V| - 1, \delta(\Gamma)]_p$-code, and the minimum words are the nonzero scalar multiples of the rows of $G$ of weight $\delta(\Gamma)$.

**Examples:**

**1.** Let $\Gamma = K_5$ denote the complete graph on five vertices and let $G$ be an incidence matrix for $\Gamma$. Then $C_2(G)$ is a $[10, 4, 4]$-code and its dual is a $[10, 6, 3]$-code.

**2.** Let $\Gamma = K_7$ denote the complete graph on seven vertices and let $G$ be an incidence matrix for $\Gamma$. Then $C_2(G)$ is a $[21, 6, 6]$-code and its dual is a $[21, 15, 3]$-code.

# 14.5   BOUNDS FOR CODES

How many codewords can a code have if its codewords are $n$-tuples of elements of $GF(q)$ and it has distance $d$? Although this is a difficult question for all but special sets of values of $n$, $q$, and $d$, there are several different useful bounds on $M$, the number of codewords in the code. There are also special types of codes, called perfect codes, that achieve the maximum number of codewords possible, given values of $n$, $q$, and $d$.

## 14.5.1   CONSTRAINTS ON CODE PARAMETERS

**Definitions:**

Let $A_q(n, d)$ be the maximum $M$ for which there exists an $(n, M, d)$-code over $GF(q)$. A code that achieves this bound is called **optimal**.

Let $V_q(n, d)$ be the number of words in $GF(q)^n$ that have distance at most $d$ from a fixed word.

An $[n, k, d]$-code for which $k = n - d + 1$ is called a **maximum distance separable** (**MDS** code), or an **$[n, k]$-MDS code**.

**Facts:**

**1.** Little is known about $A_q(n, d)$ except for some specific values of $q$, $n$, and $d$.

**2.** For all $n \geq 1$, $A_q(n, 1) = q^n$ and $A_q(n, n) = q$.

**3.** For all $n \geq 2$, $A_q(n, d) \leq qA_q(n - 1, d)$.

**4.** If $d$ is even, then $A_2(n, d) = A_2(n - 1, d - 1)$.

**5.** $V_q(n, d) = \sum_{i=0}^{d} \binom{n}{i}(q - 1)^i$.

**6.** *Hamming bound (or sphere-packing bound):* If $t = \lfloor \frac{d-1}{2} \rfloor$, then $A_q(n, d) \leq \frac{q^n}{V_q(n, t)}$.

**7.** *Singleton bound:* $A_q(n, d) \leq q^{n-d+1}$. Hence, for any $[n, k, d]$-code over $GF(q)$, $k \leq n - d + 1$.

8. *Manin result:* There exists a continuous decreasing function $\alpha_q : [0, 1] \to [0, 1]$ such that

- $\alpha_q$ is strictly decreasing on $[0, \frac{q-1}{q}]$;
- $\alpha_q(0) = 1$;
- if $\frac{q-1}{q} \leq x \leq 1$, then $\alpha_q(x) = 0$;
- The set $\Sigma_q$ of all accumulation points of sequences $(d_C/n_C, k_C/n_C)$, with $C$ a linear code over $GF(q)$, is given by

$$\Sigma_q = \{(\delta, R) \in [0, 1]^2 \mid 0 \leq R \leq \alpha_q(\delta)\}.$$

The Singleton bound, for example, gives an upper estimate for $\alpha_q(\delta)$.

9. *Gilbert-Varshamov bound:*

- $A_q(n, d) \geq \frac{q^n}{V_q(n, d-1)}$;
- if $V_q(n-1, d-2) < q^{n-k}$, then there exists an $[n, k, d]$-linear code over $GF(q)$; thus, if $k$ is the largest integer for which this inequality holds, then $A_q(n, d) \geq q^k$.

10. *Asymptotic Gilbert-Varshamov bound:*

$$\alpha_q(x) \geq 1 - x \log_q(q - 1) - x \log_q(x) - (1 - x) \log_q(1 - x).$$

In other words, for each fixed $\epsilon > 0$, there exists an $[n, k, d]$-code $C$ (which may depend on $\epsilon$) with

$$R(C) \geq 1 - \delta(C) \log_q(q - 1) - \delta(C) \log_q(\delta(C)) - (1 - \delta(C)) \log_q(1 - \delta(C)) - \epsilon.$$

11. For thirty years, the asymptotic version of the Gilbert-Varshamov bound was believed to be the best possible lower bound for good codes. In 1982, using some sophisticated ideas from algebraic geometry, it was proved that the Gilbert-Varshamov bound can be improved. A good survey of these results appears in [TsVl91], [TsVlNo07].

12. Let $C$ be an $[n, k]$-MDS code. If $G$ is a generator matrix for $C$, then any $k$ columns of $G$ are linearly independent.

13. If $C$ is an $[n, k]$-MDS code, then $C^{\perp}$ is also an MDS code.

14. *Johnson bound:* If $d = 2t + 1$, then

$$A_2(n, d) \leq \frac{2^n}{\sum_{i=0}^{t} \binom{n}{i} + \frac{1}{\lfloor \frac{n}{t+1} \rfloor} \binom{n}{t} \left( \frac{n-t}{t+1} - \lfloor \frac{n-t}{t+1} \rfloor \right)}.$$

This is an improvement of the Hamming bound (Fact 6) for binary codes.

**Examples:**

1. The $[13, 10, 3]$ Hamming code over $GF(3)$ meets the Singleton bound.

2. The Reed-Solomon codes are MDS codes.

## 14.5.2 PERFECT CODES

**Definitions:**

An $(n, M, d)$-code over $GF(q)$ is said to be **perfect** if it meets the Hamming bound, that is, $M = \frac{q^n}{V_q(n,t)}$, where $t = \lfloor \frac{d-1}{2} \rfloor$.

The **binary Golay code** is a $[23, 12, 7]$-code over $GF(2)$ with generator matrix $G = [I_{12} \,|\, A]$, where

$$
A = \begin{pmatrix}
1 & 1 & 0 & 1 & 1 & 1 & 0 & 0 & 0 & 1 & 0 \\
1 & 0 & 1 & 1 & 1 & 0 & 0 & 0 & 1 & 0 & 1 \\
0 & 1 & 1 & 1 & 0 & 0 & 0 & 1 & 0 & 1 & 1 \\
1 & 1 & 1 & 0 & 0 & 0 & 1 & 0 & 1 & 1 & 0 \\
1 & 1 & 0 & 0 & 0 & 1 & 0 & 1 & 1 & 0 & 1 \\
1 & 0 & 0 & 0 & 1 & 0 & 1 & 1 & 0 & 1 & 1 \\
0 & 0 & 0 & 1 & 0 & 1 & 1 & 0 & 1 & 1 & 1 \\
0 & 0 & 1 & 0 & 1 & 1 & 0 & 1 & 1 & 1 & 0 \\
0 & 1 & 0 & 1 & 1 & 0 & 1 & 1 & 1 & 0 & 0 \\
1 & 0 & 1 & 1 & 0 & 1 & 1 & 1 & 0 & 0 & 0 \\
0 & 1 & 1 & 0 & 1 & 1 & 1 & 0 & 0 & 0 & 1 \\
1 & 1 & 1 & 1 & 1 & 1 & 1 & 1 & 1 & 1 & 1
\end{pmatrix}.
$$

The **ternary Golay code** is an $[11, 6, 5]$-code over $F_3 = \{0, 1, 2\}$ with generator matrix $G = [I_6 \,|\, B]$, where

$$
B = \begin{pmatrix}
1 & 1 & 1 & 1 & 1 \\
0 & 1 & 2 & 2 & 1 \\
1 & 0 & 1 & 2 & 2 \\
2 & 1 & 0 & 1 & 2 \\
2 & 2 & 1 & 0 & 1 \\
1 & 2 & 2 & 1 & 0
\end{pmatrix}.
$$

**Facts:**

1. A necessary condition for a code to be perfect is that $d$ be odd.

2. The binary Golay code is a perfect code.

3. The extended binary Golay code is a $[24, 12, 8]$-code that is self-dual.

4. The ternary Golay code is a perfect code.

5. The set of all perfect codes over $GF(q)$, determined in 1973 by A. Tietäväinen, consists of the following:

  - the linear code consisting of all words in $GF(q)^n$;
  - the binary repetition codes of odd lengths;
  - the Hamming codes and all codes with the same parameters as them;
  - the binary Golay code and all codes equivalent to it;
  - the ternary Golay code and all codes equivalent to it.

6. There do exist perfect codes with the same parameters as the Hamming codes, but which are not equivalent to them.

**Examples:**

1. The extended binary Golay code is a $[24, 12, 8]$-code that is self-dual.

2. The binary Hamming $[7, 4, 3]$ code is perfect.

# 14.6   NONLINEAR CODES

Although linear codes are studied and used extensively, there are several important types of nonlinear codes. In particular, there are nonlinear codes with efficient encoding and decoding algorithms, as well as nonlinear codes that are important for theoretical reasons.

## 14.6.1   NORDSTROM-ROBINSON CODE

**Definitions:**

Permute the coordinates of the extended binary Golay code so that one of the weight 8 codewords is $1111111100\ldots0$, and call this new code $C'$. For each of the 8-bit words $00000000, 10000001, 01000001, 00100001, 00010001, 00001001, 00000101, 00000011$, there are exactly 32 codewords in $C'$ that begin with that word. The **extended Nordstrom-Robinson code** is the code whose codewords are obtained from these 256 words by deleting the first 8 coordinate positions. The **Nordstrom-Robinson code** is obtained by puncturing the last digit of the extended Nordstrom-Robinson code.

**Facts:**

1. The extended Nordstrom-Robinson code is a binary $(16, 256, 6)$-nonlinear code.

2. The Nordstrom-Robinson code is a binary $(15, 256, 5)$-nonlinear code.

3. The Johnson bound (§14.5.1, Fact 14) yields $A_2(15, 5) \leq 256$, and hence it follows that $A_2(15, 5) = 256$. On the other hand, it has been proved that no linear code of length 15 and distance 5 has more codewords than the binary 2-error correcting BCH code, which has 128.

## 14.6.2   PREPARATA CODES

**Definitions:**

The **Preparata codes** are an infinite family of nonlinear codes that have efficient encoding and decoding algorithms. Let $\beta$ be a primitive element of $GF(2)^m$, and label the elements of $GF(2)^m$ as $\alpha_i = \beta^i$, $0 \leq i \leq 2^m - 2$, and $\alpha_{2^m-1} = 0$. For a subset $X \subseteq GF(2)^m$, let $\chi(X)$ denote the characteristic vector of $X$; that is, $\chi(X)$ is a binary vector of length $2^m$ whose $i$th coordinate is 1 if $\alpha_i \in X$ and is 0 otherwise, for each $0 \leq i \leq 2^m - 1$.

If $m \geq 3$ is odd, the **extended Preparata code** $\overline{P}(m)$ is the set of words of the form $(\chi(X), \chi(Y))$, where $X$ and $Y$ are subsets of $GF(2)^m$ such that

- $|X|$ and $|Y|$ are even;
- $\sum_{x \in X} x = \sum_{y \in Y} y$;

- $\sum_{x \in X} x^3 + \left( \sum_{x \in X} x \right)^3 = \sum_{y \in Y} y^3.$

The **Preparata code** $P(m)$ is obtained from $\overline{P}(m)$ by puncturing the coordinate corresponding to the field element 0 in the first half of each codeword.

**Facts:**

**1.** If $m \geq 3$ is odd, then $\overline{P}(m)$ is a binary nonlinear code with parameters $n = 2^{m+1}$, $M = 2^{2^{m+1}-2m-2}$, $d = 6$.

**2.** If $m \geq 3$ is odd, then $P(m)$ is a binary nonlinear code with parameters $n = 2^{m+1} - 1$, $M = 2^{2^{m+1}-2m-2}$, $d = 5$.

**3.** $P(3)$ is the same as the Nordstrom-Robinson code.

**4.** The Preparata codes can be viewed as linear codes over $\mathcal{Z}_4$.

# 14.7   CONVOLUTIONAL CODES

Convolutional codes are a powerful class of error-correcting codes. They work differently than block codes do. Instead of grouping message symbols into blocks for encoding, check digits are interwoven within streams of information symbols. Convolutional codes can be considered to have memory, since $n$ symbols of information are encoded using these $n$ symbols and previous information symbols.

## 14.7.1   BACKGROUND

**Definitions:**
The figure in §14.1.1 can be used to distinguish two approaches to decoding. For a **hard decision decoder**, the demodulator maps received coded data symbols into the set of transmitted data symbols (for example, 0 and 1). In contrast, the demodulator of **soft decision decoders** may pass extra information to the decoder (for example, three bits of information for each received channel data symbol, indicating the degree of confidence in its being a 0 or a 1).

**Facts:**

**1.** Convolutional codes were introduced by P. Elias in 1955, and are widely used in practice today.

**2.** Convolutional codes are used extensively in radio and satellite links and have been used by NASA for deep-space missions since the late 1970s.

**3.** There are linear codes that differ from block codes in that the codewords do not have constant length.

**4.** Convolutional codes also differ from block codes in that the $n$-tuple produced by an encoder depends not only on the message $k$-tuple $u$, but also on some message $k$-tuples produced prior to $u$; that is, the encoder has memory.

**5.** Soft decision decoding typically allows performance improvements.

**6.** Hard and soft decision techniques can be used in both block and convolutional codes, although soft decision techniques can typically be used to greater advantage in convolutional codes.

**7.** Theoretical results, particularly with respect to BCH codes, position block codes as superior to convolutional codes.

**8.** The minimum distances of BCH codes are typically much larger than the corresponding free distances (§14.7.3) of comparable convolutional codes.

**9.** Decoding techniques for block codes are generally applicable only to $q$-ary (or binary) symmetric channels, which are an appropriate model for only a relatively small fraction of channels that arise in practice.

**10.** Efficient decoding of BCH codes requires hard-decision decoding, which suffers information loss relative to soft-decision strategies, precipitating a performance penalty. The resulting performance of the BCH decoder is significantly inferior to that for a comparable convolutional code, despite the BCH codes being inherently more powerful. Consequently, convolutional codes are used in a majority of practical applications, due to their relative simplicity and performance, and the large number of communication channels which benefit from soft decoding techniques.

**11.** A recently developed class of codes, known as *turbo codes*, are built using convolutional codes. The basic idea behind a turbo encoder is to combine two simple convolutional encoders. Input to the encoder is a block of bits. The two constituent encoders generate parity bits and the information bits are sent unchanged. The key innovation is an interleaver, which permutes the original information bits before they are provided as input to the second encoder. The permutation causes input sequences which produce low-weight codewords for one encoder to generally produce high-weight codewords for the other encoder. See [HeWi99] for information on turbo codes.

**Example:**
**1.** In the simplest version of soft decision decoding, known as the *binary erasure channel* (and usually classified as a hard-decision technique), the demodulator output is one of three values: 0, 1, or "erasure" (indicating that neither a 0 nor a 1 was clearly recognized).

## 14.7.2  SHIFT REGISTERS

**Definitions:**
An **$m$-stage shift register** is a hardware device that consists of $m$ **delay elements** (or **flip-flops**), each having one input and one output, and a clock which controls the movement of data. During each unit of time, the following operations are performed:

- a new input bit and the contents of some of the delay elements are added modulo 2 to form the output bit;

- the content of each delay element (with the exception of the last delay element) is shifted one position to the right;

- the new input bit is fed into the first delay element.

The **generator** of an $m$-stage shift register is a polynomial $g(x) = 1 + g_1 x + g_2 x^2 + \cdots + g_m x^m \in GF(2)[x]$, where $g_i = 1$ if the contents of the $i$th delay element are involved in the modulo 2 sum that produces the output, and $g_i = 0$ otherwise.

**Facts:**
**1.** Assume that the initial contents of a shift register are all 0s. Suppose that a shift register has generator $g(x)$. Let the input stream $u_0, u_1, u_2, \ldots$ be described by the formal

power series $u(x) = u_0 + u_1 x + u_2 x^2 + \cdots$ over $GF(2)$. (If the input stream is finite of length $t$, let $u_i = 0$ for $i \geq t$.) Similarly, let the output stream $c_0, c_1, c_2, \ldots$ be described by the formal power series $c(x) = c_0 + c_1 x + c_2 x^2 + \cdots$ over $GF(2)$. Then $c(x) = u(x)g(x)$.

**2.** In some applications, some of the delay elements are added modulo 2 to the input element. This type of register is referred to as a *linear feedback shift register* (LFSR).

### Examples:

**1.** *Shift-example*: Suppose that the delay elements of the 4-stage shift register in the following figure initially contain all 0s.

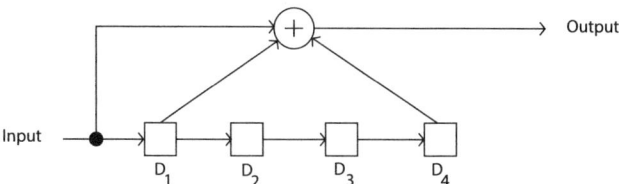

If the input stream to the register is 11011010 (from left to right), the updated contents of the delay elements and the output bits are summarized in the following table:

| time | input | $D_1$ | $D_2$ | $D_3$ | $D_4$ | output |
|------|-------|-------|-------|-------|-------|--------|
| 0 | – | 0 | 0 | 0 | 0 | – |
| 1 | 1 | 1 | 0 | 0 | 0 | 1 |
| 2 | 1 | 1 | 1 | 0 | 0 | 0 |
| 3 | 0 | 0 | 1 | 1 | 0 | 1 |
| 4 | 1 | 1 | 0 | 1 | 1 | 1 |
| 5 | 1 | 1 | 1 | 0 | 1 | 1 |
| 6 | 0 | 0 | 1 | 1 | 0 | 0 |
| 7 | 1 | 1 | 0 | 1 | 1 | 1 |
| 8 | 0 | 0 | 0 | 1 | 0 | 1 | 0 |

**2.** The generator of the shift register in Example 1 is $g(x) = 1 + x + x^4$.

## 14.7.3   ENCODING

*Note*: Throughout this subsection assume that the initial contents of a shift register are all 0s.

### Definitions:

An $(n, 1, m)$-**convolutional code** with generators $g_1(x), g_2(x), \ldots, g_n(x) \in GF(2)[x]$ ($m = \max(\deg g_i(x))$) contains all codewords of the form $c(x) = (c_1(x), c_2(x), \ldots, c_n(x))$, where $c_i(x) = u(x)g_i(x)$, and $u(x) = u_0 + u_1 x + u_2 x^2 + \cdots$ represents the input stream. The **system memory** of the code is $m$.

A convolutional code is **catastrophic** if a finite number of channel errors can cause an infinite number of decoding errors.

The **rate** of an $(n, k, m)$-convolutional code is $\frac{k}{n}$.

The **free distance** $d_{\text{free}}$ of a convolutional code is the minimum weight of all nonzero output streams.

A **Recursive Systematic Convolutional (RSC) code** is a convolutional code which allows some of the output bits to to be fed back into the input to make the code *systematic* in the sense that the data bits within a codeword come before the redundancy bits.

**Facts:**

**1.** A convolutional code is linear.

**2.** Convolutional codes are not block codes since the codewords have infinite length. They are, however, similar to block codes, and in fact can be viewed as block codes over certain infinite fields.

**3.** An $(n, 1, m)$-convolutional code can be described by a single shift register with $n$ outputs, where $c_i(x)$ is the output of the single-output shift register with generator $g_i(x)$ when $u(x)$ is the input. In practice, $c_1(x), c_2(x), \ldots, c_n(x)$ are interleaved to produce one output stream.

**4.** Let $C$ be an $(n, 1, m)$-convolutional code with generators $g_1(x), g_2(x), \ldots, g_n(x)$. Let $G(x) = \sum_{i=1}^{n} x^{i-1} g_i(x^n)$. If the message is $u(x)$, then the corresponding interleaved codeword is $\overline{c}(x) = G(x) u(x^n)$.

**5.** The *Viterbi algorithm* is a maximum likelihood decoding algorithm for convolutional codes [LiCo04]. For an algebraic treatment of convolutional codes, see [Pi88].

**6.** If $\gcd(g_1(x), g_2(x), \ldots, g_n(x)) = 1$ in $GF(2)[x]$, then $C$ is not catastrophic.

**7.** An $(n, k, m)$-convolutional code can be described by $k$ multi-output shift registers, each of maximum length $m$. The message is divided into $k$ streams, each stream being the input to one of the $k$ shift registers. There are $n$ output streams, each formed using some or all of the shift registers.

**8.** The free distance of a convolutional code is a measure of the error-correcting capability of the code, and is a concept analogous to the distance of a block code.

**9.** In contrast to block codes, there are few algebraic constructions known for convolutional codes.

**10.** The convolutional codes used in practice are usually those found by a computer search designed to maximize the free distance among all encoders with fixed parameters $n$, $k$, and $m$. The following table lists the best codes with a rate of $\frac{1}{2}$ ($n = 2$, $k = 1$). The polynomials $g_1(x)$ and $g_2(x)$ are represented by their coefficients, listed from low order to high order.

| $m$ | $g_1(x)$ | $g_2(x)$ | $d_{\text{free}}$ |
|---|---|---|---|
| 2 | 101 | 111 | 5 |
| 3 | 1101 | 1111 | 6 |
| 4 | 10011 | 11101 | 7 |
| 5 | 110101 | 101111 | 8 |
| 6 | 1011011 | 1111001 | 10 |
| 7 | 11100101 | 10011111 | 10 |
| 8 | 101110001 | 111101011 | 12 |
| 9 | 1001110111 | 1101100101 | 12 |
| 10 | 10011011101 | 11110110001 | 14 |
| 11 | 100011011101 | 101111010011 | 15 |
| 12 | 1000101011011 | 1111110110001 | 16 |

**11.** While a convolutional code is constructed with a shift register, an RSC is constructed with an LFSR.

**12.** An RSC code can be shown to be equivalent to a convolutional code. But because they are systematic, data bits can be sent directly to the output. This is generally referred to as the *data stream*. The redundancy bits, meanwhile, are processed through the LFSR. This is generally referred to as the *parity stream*.

### Examples:

**1.** Consider the $(2, 1, 3)$-convolutional code with generators $g_1(x) = 1 + x^3$ and $g_2(x) = 1 + x + x^3$. The code can be described by the shift register of the following figure. The message $u(x) = 1 + x^3 + x^4$, corresponding to the bit string 10011, gets encoded to $c(x) = (u(x)g_1(x), u(x)g_2(x)) = (1 + x^4 + x^6 + x^7, 1 + x + x^5 + x^6 + x^7)$, or in interleaved form to $\bar{c} = 11\ 01\ 00\ 00\ 10\ 01\ 11\ 11\ 00\ 00\ 00\ \ldots$.

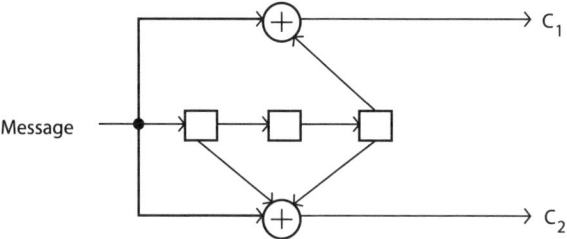

**2.** Suppose that the input stream contains an infinite number of 1s, and the output stream has only finitely many 1s. If the channel introduces errors precisely in the positions of these 1s, then the resulting all-zero output stream will be decoded by the receiver to $m(x) = 0$.

**3.** The following figure is a shift register encoder for a $(3, 2, 3)$-convolutional code $C$.

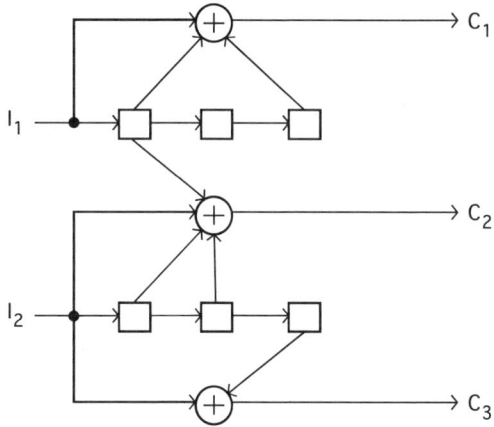

If the input stream is $u = 1011101101$, it is first divided using alternating bits into two streams $I_1 = 11110$ and $I_2 = 01011$. The three output streams are $c_1 = 10010$, $c_2 = 00011$ and $c_3 = 01010$, and the interleaved output is $\bar{c} = 100\ 001\ 000\ 111\ 010$.

## 14.7.4 TURBO CODES

**Definitions:**

A *turbo code* is the code formed by *parallel concatenation* of two RSC codes whose output consists of the following three streams:

- the data stream;
- the parity stream from the first RSC;
- the parity stream from the second RSC where the input data stream, within each input block of a given length $m$, has been permuted.

**Facts:**

**1.** The permutation involved in the input stream for the second RSC code is usually taken to be pseudorandom in practice. When this is the case, any nonzero input will produce a parity stream with large Hamming weight in at least one of the two RSCs.

**2.** Turbo codes (along with LDPC) are the codes which are closest to achieving Shannon's noiseless coding theorem.

**3.** If the input stream consists of blocks of length $m$ and the parity streams output blocks of length $n_1$ and $n_2$ respectively, then turbo codes have a rate of $m/(m+n_1+n_2)$.

**4.** In order to increase the rate of a turbo code, some of the elements in the output stream may be deleted. This is known as *puncturing* the code. When decoding, these symbols are reconstructed using a soft-decision decoder.

**5.** Turbo codes generally suffer from low rates as well as high encoding and decoding complexity.

**6.** Turbo codes were initially used in deep-space applications, 3G telephony, and wireless standards.

# 14.8 QUANTUM ERROR-CORRECTING CODES

## 14.8.1 QUANTUM CODES

Let $q = p^n$, for some prime $p$. Let $\mathcal{H}$ be the $q$-dimensional Hilbert space (i.e., a $q$-dimensional complex linear space).

**Definitions:**

A *quantum system* is given by the tensor product space $V := \mathcal{H}^{\otimes n} = \mathcal{H} \otimes \mathcal{H} \otimes \cdots \otimes \mathcal{H}$. An element of $V$ is called a *quantum state* and denoted by $|\psi\rangle$ and a *quantum code* $Q$ is a $q^k$-dimensional subspace of $V$. A quantum state $|\psi\rangle \in Q$ of $k$ particles is transmitted over a noisy channel and we then receive a quantum state $|\bar{\psi}\rangle \in V$ of $n$ particles.

Let $q = 2$. A *qubit* is the short form for quantum bit and the analog of a bit in classical computation. The possible states of a quantum bit are the states of the vector space $\mathcal{C}^2$. The standard basis of this state space is denoted $|0\rangle = \binom{1}{0}$, $|1\rangle = \binom{0}{1}$. The general state of a qubit is a linear combination $|\psi\rangle = \alpha |0\rangle + \beta |1\rangle$, with $\alpha$, $\beta \in \mathcal{C}$ and $|\alpha|^2 + |\beta|^2 = 1$. For $q > 2$ the analogue of qubits is **qudits**.

The **Pauli group** for $\mathcal{H}$ is defined as

$$G = \langle X^i Z^j \mid i, j \in \mathcal{F}_q \rangle,$$

where

$$X = \begin{pmatrix} 0 & 0 & \cdots & 0 & 1 \\ 1 & 0 & \cdots & 0 & 0 \\ 0 & 1 & \cdots & 0 & 0 \\ \vdots & \vdots & \ddots & \vdots & \vdots \\ 0 & 0 & \cdots & 1 & 0 \end{pmatrix}, \quad Z = \begin{pmatrix} 1 & 0 & \cdots & 0 \\ 0 & \omega_p & \cdots & 0 \\ \vdots & \vdots & \ddots & \vdots \\ 0 & 0 & \cdots & \omega_p^{p-1} \end{pmatrix}.$$

The **Pauli group** on $V$ of length $n$ is the tensor product of the Pauli group on single qubits:

$$G_n = \langle U_1 \otimes \cdots \otimes U_n \mid U_i \in G \text{ for } i = 1, \ldots, n \rangle.$$

As usual, we will call the elements of $G_n$ **operators** in $G_n$.

**Stabilizer codes** are helpful devices that make possible construction of quantum codes from classical codes. Let $V$ denote the qubit state space and let $G_n$ be the Pauli group on $n$ qubits. Let $H \leq G_n$ and denote by $V_H$ the subspace of $V$ fixed by $H$:

$$V_H = \{v \in V \mid gv = v, \forall g \in G_n\}.$$

Subsequently we assume that $H$ is abelian and $-I \notin H$. The subspace $V_H$ is called the **stabilizer code** $C(H)$ of $H$.

Since it suffices to look at a set of generators, we can represent a stabilizer code in an easier way. A **generator matrix** $\mathcal{G}$ of a stabilizer code is an $l \times 2n$ matrix $\mathcal{G}(X|Z)$, where the first $n$ components represent the $X$ errors and the second $n$ components represent the $Z$ errors.

For an operator $E \in G_n$, we denote its adjoint by $E^\dagger$ and for any subgroup $H \leq G_n$, its normalizer will be denoted $N(H)$; see [Sh08] for details.

The **weight** $wt$ of an operator $U_1 \otimes \cdots \otimes U_n$ is the number of elements $U_i$ that are not equal to the identity.

**Facts:**

**1.** $V_H$ is non-trivial if and only if $H$ is abelian and $-I \notin H$.

**2.** Let $\{E_i\}$ be a set of operators in $G_n$ such that $E_j^\dagger E_k \notin N(H) \setminus H$ for all $j$ and $k$. Then $\{E_j\}$ is a correctable set of errors for the code $C(H)$. For a proof, see [AsKn01].

**3.** Let $H$ be an abelian subgroup of $G_n$. If a state $|\psi\rangle$ is in the $+1$ eigenspace of a set of generators $\{g_1, \ldots, g_l\}$ of $H$, it is an eigenstate of all elements in $H$.

**4.** The generator matrix defines an $[[n, k, d]]$ quantum error-correcting code with $k = n - l$.

**5.** Let $E \in G_n$ be an error operator. Then

- if $E \in H$, then $E$ is a codeword;
- if $E \in G_n \setminus N(H)$, then the error is detectable and can be corrected with the most likely error principle;

- if $E \in N(H) \backslash H$, then the error cannot be detected and therefore is not correctable.

**6.** Fact 5 allows us to introduce the distance of a stabilizer code. The *distance d* of a quantum stabilizer code $C$ is the minimum weight of all normalizer elements that are not in the stabilizer:

$$d = \min\{wt(x) \mid x \in N(H) \backslash H\}.$$

**7.** The stabilizer code of the normalizer $N(H)$ is equal to the dual code $C^{\perp_s}$ with respect to the symplectic inner product $\langle , \rangle$. Consequently

$$d = \min\{wt(x) \mid x \in C^{\perp_s} \backslash C\}.$$

**8.** Quantum codes can be constructed from classical codes [AsKn01]. Specifically, let $C \subset F_q^{2n}$ be an $(n+k)$-dimensional subspace such that $C^{\perp_s} \subset C$. Then there exists a quantum code $Q \subset \mathcal{H}^{\otimes n}$ of dimension $q^k$ and minimum distance $d = \dim C \backslash C^{\perp_s}$.

## 14.8.2   QUANTUM ALGEBRAIC GEOMETRY CODES

### Definitions:

Let $F/\mathcal{F}_q$ be an algebraic function field in one variable. Let $P_1, \ldots, P_n$ be places of degree 1 and let $D = P_1 + \cdots + P_n$. Furthermore let $G$ be a divisor with $\mathrm{supp}(G) \cap \mathrm{supp}(D) = \emptyset$. Then the **Goppa code** (respectively **AG code**) $C_{\mathcal{L}} \subseteq F_q^n$ is defined by

$$C_{\mathcal{L}}(D, G) = \{(f(P_1), \ldots, f(P_n)) \mid f \in \mathcal{L}(G)\} \subseteq F_q^n.$$

Define the following linear **evaluation map**

$$\varphi: \mathcal{L}(G) \to \mathcal{F}_q^n$$

$$f \mapsto (f(P_1), \ldots, f(P_n)).$$

### Facts:

**1.** The Goppa code is given by $C_{\mathcal{L}}(D, G) = \varphi(\mathcal{L}(G))$.

**2.** The code $C_{\mathcal{L}}(D, G)$ is a linear $[n, k, d]$-code with parameters

$$k = \dim G - \dim(G - D), \quad d \geq n - \deg G =: d_{des}.$$

The parameter $d_{des}$ is called the *designed distance* of the Goppa code. See [ElSh11], [Sh08], [TsVl91], and [TsVlNo07] for details.

**3.** Assume $\deg G < n$ and let $g$ be the genus of $F/\mathcal{F}_q$. Then the following hold [St93].

- $\varphi: \mathcal{L}(G) \to C_{\mathcal{L}}(D, G)$ is injective and $C_{\mathcal{L}}(D, G)$ is an $[n, k, d]$-code with

$$\begin{aligned} k &= \dim G \geq \deg G + 1 - g \\ d &\geq n - \deg G. \end{aligned}$$

- If in addition $2g - 2 < \deg G < n$, then

$$k = \deg G + 1 - g.$$

- If $(f_1, \ldots, f_k)$ is a basis of $\mathcal{L}(G)$, then a generator matrix for $C_{\mathcal{L}}(D, G)$ is

$$
M = \begin{pmatrix} f_1(P_1) & \cdots & f_1(P_n) \\ \vdots & & \vdots \\ f_k(P_1) & \cdots & f_k(P_n) \end{pmatrix}.
$$

**4.** Let $D = P_1 + \cdots + P_n$ be a divisor, where the $P_i$s are places of degree 1 of an algebraic function field $F/\mathcal{F}_q$. Furthermore let $G$ be a divisor with $\mathrm{supp}(G) \cap \mathrm{supp}(D) = \emptyset$ and define the code $C_{\Omega}(D, G)$ by

$$
C_{\Omega}(D, G) = \{(\mathrm{res}_{P_1}(\omega), \ldots, \mathrm{res}_{P_n}(\omega)) \mid \omega \in \Omega_F(G - D)\} \subseteq \mathcal{F}_q^n,
$$

where $\mathrm{res}_{P_i}(w)$ is the residue of $w$ at $P_i$. In other words, if $t \in F$ is a $P_i$-prime element and $w \in F$ has a $P_i$-adic expansion $w = \sum_{j=n}^{\infty} a_j t^j$ with $n \in \mathcal{Z}$ and $a_j \in \mathcal{F}$, then $\mathrm{res}_{P_i}(w) = a_{-1}$. Then the code $C_{\Omega}(D, G)$, where $D$ and $G$ are as above, has the following properties:

- $C_{\mathcal{L}}(D, G)^{\perp} = C_{\Omega}(D, G)$;
- $C_{\Omega}(D, G) = a \cdot C_{\mathcal{L}}(D, H)$ with $H = D - G + (\eta)$ where $\eta$ is a differential, $v_{P_i}(\eta) = -1$ for $i = 1, \ldots, n$, and $a = (\mathrm{res}_{P_1}(\eta), \ldots, \mathrm{res}_{P_n}(\eta))$;
- $C_{\mathcal{L}}(D, G)^{\perp} = a \cdot C_{\mathcal{L}}(D, H)$.

**5.** One can use differentials with special properties to help construct a self-orthogonal code [St93, Prop. VII.1.2]. Namely, let $x$ and $y$ be elements of $F$ such that $v_{P_i}(y) = 1$, $v_{P_i}(x) = 0$, and $x(P_i) = 1$ for $i = 1, \ldots, n$. Then the differential $\eta = x \cdot \frac{dy}{y}$ satisfies $v_{P_i}(\eta) = -1$ and $\mathrm{res}_{P_i}(\eta) = 1$ for $i = 1, \ldots, n$.

## REFERENCES

### *Printed Resources*:

[AsKn01] A. Ashikhmin and E. Knill, "Nonbinary quantum stabilizer codes", *IEEE Transactions on Information Theory* 47 (2001), 3065–3072.

[Ba97] J. Baylis, *Error-Correcting Codes: A Mathematical Introduction*, Chapman & Hall, 1997.

[Be15] E. R. Berlekamp, *Algebraic Coding Theory*, revised ed., World Scientific Press, 2015.

[Bi05] J. Bierbrauer, *Introduction to Coding Theory*, Chapman & Hall/CRC Press, 2005.

[Bl03] R. E. Blahut, *Algebraic Codes for Data Transmission*, Cambridge University Press, 2003.

[DaKeRo13] P. Dankelmann, J. D. Key, and B. G. Rodrigues, "Codes from incidence matrices of graphs", *Designs, Codes and Cryptography* 68 (2013), 373–393.

[Eb13] W. Ebeling, *Lattices and Codes*, 3rd ed., Springer Spektrum, 2013.

[ElSh11] A. Elezi and T. Shaska, "Quantum codes from superelliptic curves", *Albanian Journal of Mathematics* 5 (2011), 175–191.

[ElSh15] A. Elezi and T. Shaska, "Weight distributions, zeta functions and Riemann hypothesis for linear and algebraic geometry codes", in *Advances on Superelliptic Curves and Their Applications*, L. Beshaj, T. Shaska, and E. Zhupa (eds.), IOS Press, 2015, 328–359.

[Gu04] V. Guruswami, *List Decoding of Error-Correcting Codes*, Lecture Notes in Computer Science 3282, Springer, 2004.

[GuRuSu15] V. Guruswami, A. Rudra, and M. Sudan, *Essential Coding Theory*, available at http://www.cse.buffalo.edu/faculty/atri/courses/coding-theory/book/, 2015.

[Ha50] R. W. Hamming, "Error detecting and error correcting codes", *Bell System Technical Journal* 29 (1950), 147–160.

[Ha80] R. W. Hamming, *Coding and Information Theory*, Prentice-Hall, 1980.

[HaHaJo97] D. Hankerson, G. A. Harris, and P. D. Johnson, Jr., *Introduction to Information Theory and Data Compression*, CRC Press, 1997.

[HeWi99] C. Heegard and S. B. Wicker, *Turbo Coding*, Kluwer, 1999.

[Hi27] L. S. Hill, "The role of prime numbers in the checking of telegraphic communications", *Telegraph and Telephone Age* (April 1 and July 16, 1927) 151–154, 323–324.

[HuPl03] W. C. Huffman and V. Pless, *Fundamentals of Error-Correcting Codes*, Cambridge University Press, 2003.

[JoZi15] R. Johannesson and K. Zigangirov, *Fundamentals of Convolutional Coding*, 2nd ed., Wiley, 2015.

[JoKi11] D. Joyner and J-L. Kim, *Selected Unsolved Problems in Coding Theory*, Birkhäuser, 2011.

[LiCo04] S. Lin and D. J. Costello, Jr., *Error Control Coding*, 2nd ed., Pearson, 2004.

[Ma03] D. MacKay, *Information Theory, Inference, and Learning Algorithms*, Cambridge University Press, 2003.

[MaSl77] F. J. MacWilliams and N. J. A. Sloane, *The Theory of Error-Correcting Codes*, North-Holland, 1977.

[Mc77] R. J. McEliece, *The Theory of Information and Coding*, Addison-Wesley, 1977.

[Mo05] T. Moon, *Error Correction Coding: Mathematical Methods and Algorithms*, Wiley, 2005.

[NeRaSl06] G. Nebe, E. M. Rains, and N. J. A. Sloane, *Self-Dual Codes and Invariant Theory*, Springer-Verlag, 2006.

[PeWe72] W. W. Peterson and E. J. Weldon, Jr., *Error-Correcting Codes*, MIT Press, 1972.

[Pi88] P. Piret, *Convolutional Codes*, MIT Press, 1988.

[PlHuBr98] V. Pless, W. C. Huffman, and R. A. Brualdi, *Handbook of Coding Theory*, Elsevier, 1998.

[Pr92] O. Pretzel, *Error-Correcting Codes and Finite Fields*, Clarendon Press, 1992.

[Ro96] S. Roman, *Introduction to Coding and Information Theory*, Springer-Verlag, 1996.

[Ro06] R. Roth, *An Introduction to Coding Theory*, Cambridge University Press, 2006.

[Sh08] T. Shaska, "Quantum codes from algebraic curves with automorphisms", *Condensed Matter Physics* 11 (2008), 383–396.

[St93] H. Stichtenoth, *Algebraic Function Fields and Codes*, Springer-Verlag, 1993.

[ToChJo12] J. Torres, C. Christensen, and D. Joyner, "Hill's error-detection codes", *Cryptologia* 36 (2012), 88–103.

[TsVl91] M. A. Tsfasman and S. G. Vlăduţ, *Algebraic-Geometric Codes*, Mathematics and its Applications, Kluwer Academic Publishers, 1991.

[TVN07] M. Tsfasman, S. Vlăduţ, and D. Nogin, *Algebraic Geometric Codes: Basic Notions*, American Mathematical Society, 2007.

[vL99] J. H. van Lint, *Introduction to Coding Theory*, third ed., Springer-Verlag, 1999.

[VavO89] S. A. Vanstone and P. C. van Oorschot, *An Introduction to Error Correcting Codes with Applications*, Kluwer Academic Publishers, 1989.

[We98] R. B. Wells, *Applied Coding and Information Theory for Engineers*, Pearson, 1998.

[Ye08] R. Yeung, *Information Theory and Network Coding*, Springer-Verlag, 2008.

**Web Resources**:

http://www.codetables.de/ (Provides bounds on the parameters of various types of codes.)

http://www.eccpage.com (The Error Correcting Codes (ECC) Home Page.)

# 15

---

# *CRYPTOGRAPHY*

*Charles C. Y. Lam, Chapter Editor*

This chapter is a major extension and update of the sections on cryptography in the first edition written by Alfred J. Menezes and Paul C. van Oorshot.

## INTRODUCTION

The history of protecting the secrecy of information goes back to ancient times. For example, the ancient Romans used a secret code to send messages that could not be read by their enemies. The evolution of secret codes ties with the historical development of human intelligence and improvements in technology.

In modern times, there is a constant need to protect information from unauthorized access and from malicious actions. The field of cryptography is devoted to methods that offer such protection. Sending secret messages, authenticating messages, distributing secret keys, sharing secrets, electronic cash, and electronic voting are only some of the applications addressed by modern cryptography.

Symmetric key cryptography and public key cryptography are the two major classes of cryptography. Since prehistoric times, encryption conducted between two parties can be seen as a communication mechanism with a secret key shared between two communicating parties, where the knowledge of the encryption key implies the knowledge of the corresponding decryption key. In the 1970s, through the demonstration of a practical method by Whitfield Diffie and Martin Hellman, and an independent classified discovery within Britain's GCHQ (Government Communications Headquarters), public key cryptography came to life. The discovery led to constructions of many new products that are used in the digital age.

The study of cryptography spans various fields from mathematics, computer science, and engineering, to non-science areas such as business and law. The goal of mathematical cryptography is to ensure the security of the most fundamental cryptographic components.

## GLOSSARY

**active adversary**:  an adversary that threatens the integrity or authentication of stored or transmitted data.

**adaptive chosen-ciphertext attack security (CCA2)**:  given the challenge ciphertext $C$ in CCA1 security, the attacker can continue the decryption requests adaptively, except for $C$ itself, and is unable to recover the plaintext.

**advanced encryption standard (AES)**:  a 128-bit block cipher adopted as a standard in the United States and widely used for commercial applications.

**affine cipher**:  a cipher that encrypts $x \in \mathcal{Z}_{26}$ by sending it to $ax + b$ mod 26, using the key $(a, b)$, where $a, b \in Z_{26}$ and $\gcd(a, b) = 1$.

**attribute-based encryption**:  a public key encryption scheme where the user's secret key and the ciphertext both depend on attributes such as membership and position title.

*authentication*: corroboration that a party, or the origin of data, is as claimed.

*bitcoin*: a widely-used cryptocurrency.

*bitcoin mining*: a process in ensuring data integrity in bitcoin transactions that requires a search of a nonce that produces a desired hash result.

*block cipher*: a cipher that breaks up the plaintext into blocks of a fixed length and encrypts each block via a secret bijective transformation.

*Caesar cipher*: a cipher that shifts each letter forward three positions in the alphabet, wrapping the letters at the end of the alphabet around to the beginning letters. (Sometimes, all shift ciphers are called Caesar ciphers.)

*certification authority* (*CA*): a trusted authority that verifies the identity and RSA public key of a party, and signs the data.

*chosen-ciphertext attack*: an attack in which the adversary has some chosen ciphertext and its corresponding plaintext.

*chosen-plaintext attack*: an attack in which the adversary has some chosen plaintext and its corresponding ciphertext.

*cipher*: an encryption scheme.

*cipher block chaining* (*CBC*) *mode*: a mode of operation of an $n$-bit block cipher in which plaintext is processed $n$ bits at a time, an initialization block is used, and to encrypt each successive $n$-bit block the bitwise XOR of the block with the encrypted version of the previous block is formed and the resulting $n$-bit block is encrypted by the block cipher.

*cipher feedback* (*CFB*) *mode*: a mode of operation of an $n$-bit block cipher in which plaintext is processed $t$ bits at a time, $1 \leq t \leq n$, and in which the ciphertext depends on the current block and previous blocks.

*ciphertext*: encrypted plaintext.

*ciphertext-only attack*: an attack in which the adversary has possession of some ciphertext and nothing else.

*ciphertext space*: the set of all possible ciphertexts.

*cloud computing*: an internet-based model for shared computing resources and data.

*computational security*: the amount of computational effort required by the best currently-known attacks to defeat a system.

*conference key distribution*: a mechanism whereby a trusted authority assigns a piece of secret data to each user so that any desired group of users can compute a common secret key.

*confidentiality*: preventing protected data from becoming available in an intelligible form to unauthorized parties.

*counter mode*: a mode of operation of a $n$-bit block cipher in which the secret keystream is formed by encrypting an incremental counter initialized by an initial value (IV).

*Cramer-Shoup cryptosystem*: a public key cryptosystem based on the decisional Diffie-Hellman problem.

*cryptanalysis*: the use of mathematical, statistical, and other techniques to defeat cryptographic protection mechanisms, including plaintext from ciphertext without knowledge of the key.

**cryptocurrency**:  a class of digital cash for which cryptography is used in secure transactions and in the creation of additional currency.

**cryptographic primitive**:  a basic algorithm used to build cryptographic protocols.

**cryptographic protocol**:  a protocol that performs security functions and applies cryptographic primitives.

**cryptography**:  the science and study of protecting data from malicious or unauthorized actions, including access to, manipulation, impersonation, and forgery.

**cryptology**:  the study of both cryptography and cryptanalysis. (Note that "cryptography" is often used in place of "cryptology".)

**cryptosystem** (or **cryptographic system**):  a system composed of a space of plaintext messages, a space of ciphertext messages, a key space, and families of enciphering and deciphering functions indexed with the keys.

**data encryption standard** (**DES**):  a 64-bit block cipher adopted as a standard in the United States and which is widely used for commercial applications.

**data integrity**:  the assurance that data has not been subjected to unauthorized manipulation, possibly including assurances regarding uniqueness and timeliness.

**data origin authentication**:  corroboration that the origin of data is as claimed.

**decryption**:  the process of recovering plaintext from ciphertext with knowledge of the key.

**differential analysis**:  an attack on symmetric key ciphers that distinguishes the right key from wrong guesses by observing the statistical distribution of differences in the input plaintext and differences in the output ciphertext.

**digital cash**:  a digital equivalent of cash that achieves all the functionality and security (and more) of physical cash.

**digital signature**:  the digital analogue of a handwritten signature, in the form of a value dependent on some secret known only to the signer and on the message being signed.

**digital signature algorithm** (**DSA**):  a digital signature scheme adopted as a United States standard and which is widely used for commercial applications.

**digram cipher**:  a cipher that replaces pairs of plaintext characters (*digrams*) with other digrams.

**electronic code book** (**ECB**) **mode**:  a mode of operation of an $n$-bit block cipher in which plaintext is encrypted $n$ bits at a time.

**electronic voting**:  a mechanism to provide accurate, auditable, fast, and secure voting over a communications network.

**ElGamal cryptosystem**:  a public key cryptosystem based on the discrete logarithm problem.

**elliptic curve cryptosystem**:  a public key cryptosystem based on the elliptic curve discrete logarithm problem.

**encryption**:  the process of mapping plaintext to ciphertext designed to render data unintelligible to all but the intended recipient.

**Enigma machine**:  a rotor-based encryption machine developed before World War II and used extensively by the German military during the war with modifications.

**entity authentication**: corroboration that a party's identity is as claimed.

**Feistel cipher**: a block cipher structure introduced by Horst Feistel of IBM in the 1970s and used in the construction of many ciphers.

**frequency analysis**: a technique to analyze a classical cipher by studying the frequency of letters or groups of letters in a ciphertext.

**Goldwasser-Micali cryptosystem**: the first provably secure public key cryptosystem.

**hash function**: a function that maps arbitrary length bit strings to small fixed length outputs that is easy to compute and has preimage resistance, weak collision resistance, and/or strong collision resistance.

**Hill cipher**: a matrix-based cipher using an invertible $n \times n$ matrix $A$ over $\mathcal{Z}_{26}$ as its key, which encrypts plaintext by splitting it into blocks of size $n$ and then sending the plaintext block $m = (m_1, m_2, \ldots, m_n)'$ as the $m$-tuple $Am$.

**homomorphic encryption**: an encryption mechanism that allows certain computations to be carried out on ciphertext in such a way that decryption of the result yields the same quantity as obtained by performing the computations on the original plaintext.

**identity-based encryption**: a public key cryptosystem that uses an arbitrary string, such as email address or name, as the public key.

**Kerckhoff's assumption**: the statement that the security of a system should rest entirely in the secret key—the adversary is assumed to have complete knowledge of the rest of the cryptographic mechanism(s).

**key**: a secret number or other significant information which parametrizes an encryption or decryption algorithm.

**key exchange** (or **key agreement**): a mechanism that allows two parties to securely share a secret key so that no one else can obtain a copy.

**key schedule**: an algorithm that derives a series of subkeys from an input key used for encryption rounds in a symmetric key encryption scheme.

**keyspace**: the set of all possible keys of a cryptosystem.

**keystream generator**: an algorithm to generate an infinitely long stream of characters to be used as keys starting with a secret seed value.

**known-plaintext attack**: an attack in which the adversary has some plaintext and its corresponding ciphertext.

**linear cryptanalysis**: an attack on symmetric key ciphers that distinguishes the right key from wrong key guesses by observing the linear approximation of input plaintext bits and output ciphertext bits.

**linear feedback shift register** (**LFSR**): a shift register whose input is characterized by a linear recursive function of the output.

**message authentication code** (**MAC**): a keyed function $F_K(\cdot)$ acting on messages $m$ such that when $m$ and $F_K(m)$ are transmitted, the intended recipient will be assured of the authenticity of $m$.

**message space**: the set of all possible plaintexts.

**non-adaptive chosen-ciphertext attack security** (**CCA1**): an attacker that can inquire multiple times the plaintext for any adaptively chosen ciphertext, yet is unable to recover the plaintext for a new challenge ciphertext $C$.

*nonce*:  a pseudorandom number generated for one-time use.

*non-repudiation*:  the ability to ensure that a party to a communication cannot deny the authenticity of their signature or the sending of a message that they originated.

**NTRU cryptosystem**:  a lattice-based public key cryptosystem.

*oblivious transfer*:  a protocol in which one party has a set of bit strings as input and the other party can choose a string to be transferred but the first party cannot determine which string this is.

*one-time pad*:  a stream cipher in which the secret key has length equal to that of the plaintext and is used only once for encryption.

*output feedback* (**OFB**) *mode*:  a mode of operation of an $n$-bit block cipher in which a message is split into blocks of $t$ bits for processing, $1 \leq t \leq n$, and in which error propagation is avoided.

**Paillier cryptosystem**:  a public key cryptosystem based on the intractability of computing $n$-th residue classes.

*passive adversary*:  an adversary that does not interrupt, alter, or insert any transmitted or stored data.

*perfect secrecy* (or *unconditional security*):  a security criteria in which an adversary is assumed to have unlimited computational resources.

*plaintext*:  a message to be encrypted.

*polyalphabetic substitution cipher*:  a family of monoalphabetic substitution ciphers using different substitution mappings in various locations of the plaintext.

*polygram substitution cipher*:  a cipher that substitutes groups of characters with other groups of characters.

*probabilistic encryption*:  an encryption scheme that involves the use of randomness in the encryption algorithm, so that the same message is encrypted into different ciphertexts when different randomness parameters are chosen.

*product cipher*:  a combination of two or more simpler ciphers.

*provable security*:  security in which the difficulty of defeating the method is essentially as difficult as solving a well-known, supposedly difficult problem.

*public key certificate*:  data that binds together a party's identification and public key.

*public key system*:  a cryptosystem in which each user has a pair of encryption (public) and decryption (private) keys.

*quantum attack*:  an attack on cryptographic primitives using quantum computing methods.

*quantum cryptography*:  cryptographic methods employing the science of quantum mechanical properties.

*round function*:  a subroutine in a symmetric key encryption scheme that is run multiple times to obtain the ciphertext.

**RSA cryptosystem**:  a public key cryptosystem in which encryption is based on modular exponentiation with a modulus that is the product of two large primes, and is based on the difficulty of factoring the product of two large primes and the ease of generating large primes.

**scytale**: an ancient encryption device that writes on a piece of parchment wrapped around a cylinder.

**secret sharing**: a mechanism that allows a group of users to share a secret such that any group with more than a threshold number of them can recover the secret, whereas fewer members are unable to obtain any information about the secret.

**secure hash algorithm (SHA)**: a family of cryptographic hash functions adopted as a standard in the United States and widely used for commercial applications.

**secure multiparty computation**: a protocol $\pi$ such that, given $n$ parties $P_1, \ldots, P_n$ where each $P_i$ has a private input $x_i$, $\pi$ realizes a function $f(x_1, \ldots, x_n) = (o_1, \ldots, o_n)$ such that $P_i$ obtains the output $o_i$, $i = 1, \ldots, n$.

**self-synchronizing stream cipher**: a stream cipher capable of re-establishing proper decryption automatically after loss of synchronization, with at most a fixed number of plaintexts unrecoverable.

**shift cipher**: a cipher that replaces each plaintext letter by the letter shifted a fixed number of positions in the alphabet, with letters at the end of the alphabet shifted to the beginning of the alphabet.

**steganography**: the art of concealing the existence of data without the transformation of data.

**substitution cipher**: a cipher that replaces each plaintext character by a fixed substitute according to a permutation of the alphabet.

**stream cipher**: a symmetric key cipher that encrypts individual characters of a plaintext message using a keystream generator and a family of encryption functions indexed by keys.

**symmetric key system**: a cryptosystem in which knowing the encryption function corresponding to a key allows a user to easily determine the corresponding decryption function.

**synchronous stream cipher**: a stream cipher in which the keystream is generated independently of the message.

**transposition cipher**: a cipher that divides plaintext into blocks of a fixed size and rearranges the characters in each block according to a fixed permutation.

**unconditional security** (or **perfect secrecy**): a security criteria for which an adversary is assumed to have unlimited computational resources.

**Vernam cipher**: a one-time pad.

**Vigenère cipher**: a cipher with a $t$-tuple $(k_1, \ldots, k_t)$ as its key that encrypts plaintext messages in blocks of size $t$ so that the $i$th letter in a block is shifted $k_i$ positions in the alphabet, modulo 26.

# 15.1  BASICS OF CRYPTOGRAPHY

The word cryptography comes from Greek words that translate as "hidden writing". Until the mid 20th century, these Greek words completely described the subject, when all activities were devoted to making messages secret, to recovering secret messages, and to preventing unauthorized attempts to find original messages from secret messages. However, during the last fifty years, cryptography has expanded to cover all aspects of protecting data from a wide range of malicious activities. A particular vocabulary now describes different parts of cryptography, the processes of making messages secret and recovering these messages, and attacks on secret messages.

## 15.1.1  BASIC CONCEPTS

### Definitions:

**Cryptography** is the science and study of protecting data from malicious or unauthorized actions, including access to, manipulation, impersonation, and forgery.

**Cryptanalysis** is the use of mathematical, statistical, and other techniques to defeat cryptographic protection mechanisms.

**Cryptology** is the study of both cryptography and cryptanalysis, although "cryptography" is often used in place of "cryptology".

**Steganography** is the art of concealing the existence of data without transforming the data.

A **cipher** is a method whereby a message in some source language (the **plaintext**) is transformed by a mapping, called an **encryption** algorithm, to yield an output, called the **ciphertext**, which is unintelligible to all but an authorized recipient.

A recipient of an encrypted message is able to recover the plaintext from the ciphertext by use of a corresponding **decryption** algorithm.

A **key** is a secret number or other significant information which parametrizes an encryption or decryption algorithm.

The **message space** (or **plaintext space**) $\mathcal{M}$ is the set of all possible plaintexts, the **ciphertext space** $\mathcal{C}$ consists of all possible ciphertexts, and the **keyspace** $\mathcal{K}$ consists of all possible keys.

An **encryption algorithm** $E$ is a family of mappings parametrized by a key $k \in \mathcal{K}$, such that each value $k$ defines a mapping $E_k \in \mathcal{E}$, where $\mathcal{E}$ is the set of all invertible mappings from $\mathcal{M}$ to $\mathcal{C}$. A specific plaintext message $m$ is mapped by $E_k$ to a ciphertext $c = E_k(m)$.

The set $\mathcal{D}$ of **decryption algorithms** consists of all invertible mappings from $\mathcal{C}$ back to $\mathcal{M}$, such that for each encryption key $k \in \mathcal{K}$, there is some mapping $D \in \mathcal{D}$ such that $D_{f(k)}(E_k(m)) = m$ for all $m \in \mathcal{M}$, where $f(k)$ is some key dependent on $k$ and $D_{f(k)}$ is the decryption algorithm corresponding to the decryption key $f(k)$. For so-called **symmetric key** systems, this decryption key $f(k)$ is equal to $k$ itself.

A **cryptographic primitive** is a basic algorithm used to build cryptographic protocols. Hash functions and encryption schemes are examples of cryptographic primitives.

A **cryptosystem** is a five-tuple $(\mathcal{M}, \mathcal{C}, \mathcal{K}, \mathcal{E}, \mathcal{D})$ consisting of the following:

- a set of possible plaintexts $\mathcal{M}$;
- a set of possible ciphertexts $\mathcal{C}$;
- a set of possible keys $\mathcal{K}$;

- a family of encryption algorithms $\mathcal{E}$ and the corresponding family of decryption rules $\mathcal{D}$, parametrized by the keyspace $\mathcal{K}$.

***Probabilistic encryption*** involves the use of randomness in the encryption algorithm, so that the same message is encrypted into different ciphertexts when different randomness parameters are chosen.

***Confidentiality*** means preventing confidential data from being available in an intelligible form to unauthorized parties.

Detection of data manipulation by unauthorized parties (including alteration, insertion, deletion, substitution, delay, and replay) is called ***data integrity***; it should be noted that encryption alone does not guarantee data integrity.

***Authentication*** is corroboration that a party's identity is as claimed (***entity authentication***), or that the origin of data is as claimed (***data origin authentication***); related to this is the assurance that data has not been subjected to unauthorized manipulation (cf. data integrity), possibly including assurances regarding uniqueness and timeliness.

The provision for the resolution of disputes related to digital signatures is called ***non-repudiation***. Digital signatures can be used as the basis of authorization of certain actions. Disputes may occasionally arise subsequently due to either false denials (***repudiated*** signatures) or fraudulent claims (***forged*** signatures).

A ***digital signature*** is intended to be the digital analogue of a handwritten signature; it should be a number dependent on some secret known only to the signer, and, additionally, on the content of the message being signed.

**Facts:**

**1.** Many excellent books on cryptography are available. Those new to the subject can consult [Bu13], [DeKn15], [Ma03], [PaPe09], and [TrWa06]. Books with more historical content include [MevOVa10] and [Sc07]. Textbooks with comprehensive coverage at a more advanced level are [KaLi14] and [St05].

**2.** Cryptography differs from steganography in that while the former involves use of techniques to secure data (for example, codes and ciphers), the latter involves the use of techniques which obscure the existence of data itself (for example, invisible ink, secret compartments, and the use of subliminal channels).

**3.** Cryptographic mechanisms can be used to support a number of fundamental security services, including confidentiality, data integrity, authentication, and non-repudiation.

**4.** The traditional objectives of confidentiality and authentication (although not both required in all cases) lead to the following requirements for a cryptosystem:

- *Fundamental requirement.* (To maintain secrecy of key) it should be infeasible for an adversary to deduce the key $k$ given one or more plaintext-ciphertext pairs $(m, c)$;
- *Confidentiality requirement.* (To maintain confidentiality) it should be infeasible for an adversary to deduce the plaintext $m$ corresponding to any given ciphertext $c$;
- *Authentication requirement.* (To prevent forgery or substitution) it should be infeasible for an adversary to deduce a ciphertext $c'$ corresponding to any message $m'$ of his choosing, or corresponding to any other (meaningful) message $m$.

## 15.1.2 SECURITY OF CRYPTOSYSTEMS

### Definitions:

Adversaries are either **passive** or **active**. Passive adversaries are a threat to confidentiality; they do not interrupt, alter, or insert any transmitted or stored data. Active adversaries additionally threaten integrity and authentication.

There are many models under which one can assume a cryptanalyst is able to attack a cryptographic system. The following types of attack can be hypothesized for increasingly powerful adversaries:

- **ciphertext-only**: the adversary has possession only of some ciphertext;
- **known-plaintext**: the adversary has some plaintext and its corresponding ciphertext;
- **chosen-plaintext**: the adversary has some plaintext of his choosing and its corresponding ciphertext;
- **chosen-ciphertext**: the adversary has some ciphertext of his choosing and its corresponding plaintext.

The most stringent measure of the security of a cryptographic algorithm is **unconditional security**, where an adversary is assumed to have unlimited computational resources, and the question is whether there is enough information available to defeat the system. Unconditional security for encryption systems is called **perfect secrecy**.

To measure an adversary's uncertainty in the key after observing $n$ ciphertext characters $C$, Shannon defined the **key equivocation function** $Q(n) = H(K|C_1C_2 \ldots C_n)$, the conditional entropy (§14.1.2) of the key $K$ given the ciphertext messages $C_1, C_2, \ldots, C_n$; this measures the amount of uncertainty that remains in the key $K$ after observing the ciphertexts $C_1, C_2, \ldots, C_n$. Shannon defined the **unicity distance** of the cipher to be the first value $n = n_0$ for which $Q(n) \approx 0$.

A cryptographic method is said to be **provably secure** if the difficulty of defeating it can be shown to be essentially as difficult as (that is, polynomially equivalent to) solving a well-known and *supposedly* difficult (typically number-theoretic) problem, such as integer factorization or the computation of discrete logarithms. (Thus, "provable" here means provable subject to as yet unproved assumptions.)

A proposed technique is said to be **computationally secure** if the (perceived) level of computation required to defeat it exceeds, by a comfortable margin, the computational resources of the hypothesized adversary.

### Facts:

**1.** It is a standard cryptographic assumption that an adversary will have access to ciphertext.

**2.** *Kerckhoff's assumption*: The security of a system should rest entirely in the secret key — the adversary is assumed to have complete knowledge of the rest of the cryptographic mechanism(s).

**3.** In determining whether the security of a particular cryptosystem is adequate for a particular application, the powers and resources of the anticipated adversary must be taken into account. Potential adversaries may have powers ranging from minimal to unlimited.

**4.** The security of a cryptographic algorithm can be measured according to several different metrics, including unconditional security, provable security, and computational security.

**5.** Let $M$, $C$, and $K$ be random variables ranging over the message space $\mathcal{M}$, ciphertext space $\mathcal{C}$, and keyspace $\mathcal{K}$. Unconditional security for encryption systems can be specified by the condition $H(M|C) = H(M)$; that is, the uncertainty in the plaintext, after observing the ciphertext, is equal to the a priori uncertainty about the plaintext — observation of the ciphertext provides no information (whatsoever) to an adversary.

**6.** A necessary condition for an encryption scheme to be unconditionally secure is that the key should be at least as long as the plaintext. The one-time pad (§15.3.5) is an example of an unconditionally secure encryption algorithm.

**7.** In general, encryption schemes do not offer perfect secrecy, and each ciphertext character observed decreases the uncertainty in the encryption key $k$ used.

**8.** Let $n_0$ be the unicity distance of a cipher. After observing $n_0$ characters, the key uncertainty is zero, meaning an information-theoretic adversary can narrow the set of possible keys down to a single candidate, thus defeating the cipher.

**9.** The computational security measures the amount of computational effort required, by the best currently-known attacks, to defeat a system.

## 15.2  CLASSICAL CRYPTOGRAPHY

The ancient Greeks and Romans were known to use cryptography. Some of the earliest uses of ciphers include the *Scytale* and the *Caesar cipher*. While all of the systems are now considered insecure, they provide insights into the evolution of cryptography.

The approaches in cryptography and cryptanalysis became more systematic in the 19th century. With the development of electromechanical devices, encryption machines were introduced after World War I. The Enigma machine was one of the most famous encryption machines used in World War II. The ciphers up to this point were all restricted to the encryption of individual characters, or groups of characters in the plaintext.

The invention of transistors after World War II led to the computer age where plaintext messages were no longer encoded in pure characters in a specific language. Encryption methods have taken on a much more mathematical approach since then.

**Definitions:**
Let $\mathcal{P} = \mathcal{C} = \mathcal{Z}_n$ be the plaintext and the ciphertext character set for some integer $n$. Let $K \in \mathcal{Z}_n$. The encryption-decryption pair $e_K, d_K$ defined by

$$
\begin{aligned}
e_K(x) &= x + K \bmod n, \\
d_K(y) &= y - K \bmod n
\end{aligned}
$$

is a **shift cipher**.

Suppose the plaintext and the ciphertext character set $\mathcal{A}$ are the same. A **simple** (**monoalphabetic**) **substitution cipher** is a permutation $\pi$ over $\mathcal{A}$, where the message is mapped one character at a time. Decryption is done by applying $\pi^{-1}$ to the ciphertext.

Given the alphabet of size $n$, the **affine cipher** uses the encryption function $e_{a,b}(x) = ax + b \bmod n$, where $\gcd(a,n) = 1$.

A **digram cipher** replaces pairs of plaintext characters (digrams) with other digrams.

A **polygram substitution cipher** involves substituting groups of characters with other groups of characters.

Let $t$ be a positive integer. Let $\pi$ be a permutation on $\{1, 2, \ldots, t\}$. For a plaintext $m = (m_1, m_2, \ldots, m_t)$, the **permutation cipher** maps $m$ to $e_\pi(m_1, m_2, \ldots, m_t) = (m_{\pi(1)}, m_{\pi(2)}, \ldots, m_{\pi(t)})$.

A **polyalphabetic substitution cipher** is a family of monoalphabetic substitution ciphers using different substitution mappings at various locations of the plaintext.

A **simple Vigenère cipher** of period $t$ over an $n$-character alphabet has a key $k_1 k_2 \ldots k_t$, where encryption of plaintext $m = m_1 m_2 m_3 \ldots$ to ciphertext $c = c_1 c_2 c_3 \ldots$ is specified by $c_i = m_i + k_i \bmod n$. The key is reused every $t$ characters (that is, the index $i$ of $k$ is taken modulo $t$). The cipher is an example of a polyalphabetic substitution cipher.

Let $m$ be an $n$-character message in vector form $m = (m_1, m_2, \ldots, m_n)'$. The **Hill cipher** (1929) is an $n$-gram substitution cipher defined by an invertible $n \times n$ matrix $A$ over $\mathcal{Z}_{26}$ as the key, with encryption $c = Am$. Decryption is therefore $m = A^{-1}c$.

**Facts:**

**1.** For a plaintext character set $\mathcal{Z}_n$, there are $n$ distinct shift ciphers. That is, the size of the keyspace is $n$.

**2.** The shift cipher can be easily compromised using *exhaustive key search*, by trying all keys $K$, $0 \le K \le n-1$, until meaningful text appears.

**3.** The *Caesar Cipher*, purportedly used by Julius Caesar, is a shift cipher with $K = 3$.

**4.** In the English language with 26 characters, there are $26! \approx 4 \times 10^{26}$ possible monoalphabetic substitution ciphers, making exhaustive search infeasible.

**5.** The monoalphabetic substitution cipher can be broken easily by *frequency analysis*, which makes use of statistical properties of letters in the English language.

**6.** The affine cipher is a polygram substitution cipher.

**7.** Al-Kindi, an Arab mathematician, published around AD 800 the book *Risalah fi Istikhraj al-Mu'amma* on cryptography. The book was the first publication that described cryptanalytic techniques, including frequency analysis.

**8.** Ibn al-Durayhim (AD 1312–1361) gave detailed descriptions of eight ciphers, including substitution and transposition ciphers (see *Scytale*, Fact 15).

**9.** The first polyalphabetic substitution cipher was invented by Leon Battista Alberti circa 1467.

**10.** Cryptography was widely employed across Europe during and after the Renaissance. One of the most famous stories involved the use of cryptanalysis by Sir Francis Walsingham, leading to the execution of Mary, Queen of Scots.

**11.** In the English alphabet of 26 characters, there are $26^2$ digrams, and thus there are $(26^2)!$ possible keys in the digram cipher.

**12.** The *Playfair Cipher* (1854) is a digram cipher where an alphabet of 25 letters is arranged in a $5 \times 5$ array. In the English alphabet, this can be done by either equating I and J, or omitting Q or X to reduce the number of letters from 26 to 25. Encryption is done by encrypting pairs of plaintext letters as follows:

- If the two letters are in different rows and columns in the table, then the two letters form two corners of a rectangle, and are replaced by the letters representing the other two corners of the rectangle, where the first encrypted letter is in the same column as the first letter of the plaintext.

- If the two letters are in the same row, then each letter is replaced by the letter to the immediate right in the table, where a wrap around to the first column occurs when needed.

- If the two letters are in the same column, then each letter is replaced by the letter immediately below it. The cipher defeats basic frequency analysis, but is easily broken by analysis of frequencies of pairs of letters.

**13.** The Hill Cipher can be easily broken by a known-plaintext attack through solving a system of linear equations.

**14.** The permutation cipher keeps the plaintext characters unchanged, and uses a permutation to alter the positions of the characters in the plaintext.

**15.** The *Scytale* was used by the Greeks, especially by the Spartans, during military campaigns. To operate, a piece of parchment is wrapped around a cylinder. The message is then written on the parchment. The recipient wraps the parchment on a cylinder of the exact specification to retrieve the message. The ciphertext is a result of a special case of permutation cipher called *transposition cipher* due to the regular reordering of plaintext characters.

**16.** Two approaches for the cryptanalysis of the Vigenère cipher are the *Kasiski test*, based on finding repeated strings in cipher text, and the *index of coincidence*, which examines the probability that two letters in the ciphertext are the same.

- Kasiski (1863), and independently Babbage in 1846 who kept it as a military secret, discovered that repeated strings in the plaintext that are apart by multiples of the length of the key are encrypted in the same way. Therefore, by examining repeated patterns, candidate key lengths $k$ can be deduced. Suppose $k$ is the actual key length, let $c = c_1 c_2 c_3 \ldots c_{nk}$ be the ciphertext, and let

$$\mathbf{c}_1 = c_1 c_{k+1} c_{2k+1} \cdots c_{(n-1)k+1},$$
$$\mathbf{c}_2 = c_2 c_{k+2} c_{2k+2} \cdots c_{(n-1)k+2},$$
$$\vdots$$
$$\mathbf{c}_k = c_k c_{2k} c_{3k} \cdots c_{nk}.$$

  Since every $k$th character is encrypted using the same method, each $\mathbf{c}_i$ should have a letter frequency distribution similar to regular English text.

- *Index of coincidence* (Friedman, 1920). Suppose $\mathbf{c} = c_1 c_2 \ldots c_n$ is a ciphertext of $n$ characters. The index of coincidence $E_I(\mathbf{c})$ is the probability that two random elements of $\mathbf{c}$ are identical. In regular English text, the probability is approximately 0.065, whereas for a completely random string, the probability is approximately 0.038. The method involves organizing the ciphertext in a manner similar to the Kasiski test and measures the index of coincidence for each string $\mathbf{c}_i$. The correct key length guess should reveal index of coincidence values similar to those of English text.

**17.** The *Jefferson disk*, a polyalphabetic substitution cipher invented by Thomas Jefferson in 1795, uses rotating disks. A solid cylinder of length 6 inches is sliced into 36 disks. Each disk is numbered, where the 26 letters of the alphabet are inscribed on the disk around the edge in a distinct and random ordering. Each disk has a hole in the middle that allows them to be stacked and mounted on an axle. Encryption is done in blocks of 36 letters. To encrypt, each of the 36 wheels is rotated individually to align the message in one row. Then, any of the other 25 rows can be used as the ciphertext. To decrypt, the recipient arranges the cylinders in the agreed-upon order, aligns the ciphertext in one row, then checks the other 25 rows until meaningful message appears. Since there are 36 disks, the total number of possible orderings is $36! \approx 3.72 \times 10^{41}$.

**18.** A *rotor machine* is a design for mechanical automation of polyalphabetic substitution ciphers. It consists of a number of wired wheels called rotors. Each rotor has connectors on each face of the wheel, wired internally so that an individual rotor represents a fixed monoalphabetic substitution cipher by mapping a character at its input face to another one at its output face. A plaintext character input to the first rotor generates an output that serves as an input to the second rotor. This process continues until the ciphertext character outputs from the last rotor.

- When the rotors are fixed, the combination of rotors produces only a monoalphabetic substitution cipher.
- The rotors are designed to move with each plaintext character encipherment. This results in a polyalphabetic substitution cipher. A basic design mechanism involves rotating one step of the first rotor until it completes a single revolution, then the second rotor will move one step, and so on.
- The encryption key includes the order of the rotors and their respective positions.
- The first rotor machine was built by Dutch naval officers Theo A. van Hengel and R. P. C. Spengler in 1915. Independent inventions have also been attributed to E. H. Hebern, A. G. Damm, H. A. Koch, and A. Scherbius.
- E. H. Hebern founded the Hebern Electric Code, Inc. in 1921 to become the first U.S. cipher machine company. The company went bankrupt in 1926.
- A. Scherbius filed a rotor machine patent in 1918. Scherbius later designed and marketed the *Enigma machine*.
- Scherbius developed the Enigma machine in successive versions. The company Chiffriermaschinen AG was set up in 1923 to produce and market these machines. The initial Model A (1923) was replaced by Model B (1925). In 1926, a reflector and indicator lamps were introduced in Model C. In 1927, Enigma D, with improvements that allowed wheels to be interchangeable, received wide commercial success.
- The Enigma machine as used in World War II consists of versions of similar design with a choice of wheels, and other enhancements such as plugboards to increase the complexity of cryptanalysis.
- Alan Turing, who worked at the British Government Code and Cypher School (GC&CS) at Bletchley Park during the war, provided many of the ideas that led to the construction of the Bombe machine for the cryptanalysis of the Enigma.

**19.** Many other codes were used in World War II. Some notable ones include the American *Codetalkers* using Native American languages as a basis for a substitution cipher, the British *Typex* which is an adaptation of the Enigma machine, and the German *Fish cipher* [Tu00], similar to the Enigma, but only used between the German High Command and Army Group commanders in the field.

**20.** Some references on the history of classical cryptography include *The Code Book* [Si00], *The Codebreakers* [Ka96], *Handbook of Applied Cryptography* [MevOVa10], and *Introduction to Cryptography with Coding Theory* [TrWa06].

**Examples:**

**1.** The shift cipher can be used on any alphabet by a proper mapping from the alphabet to some $\mathcal{Z}_n$. In English, we can map $A \leftrightarrow 0$, $B \leftrightarrow 1$, ..., $Z \leftrightarrow 25$ over $\mathcal{Z}_{26}$. For example, with key $K = 4$, the word "QUIZ" is encoded to $(16, 20, 8, 25)$, and is encrypted to $(20, 24, 12, 3)$, or "UYMD".

**2.** A permutation cipher of the English alphabet such as

```
plaintext   abcdefghijklmnopqrstuvwxyz
ciphertext  UXJAGRDEYLNKOPBCTWVQMIHFZS
```

maps the word "king" to "NYPD".

**3.** Let $A = 0$, $B = 1$, $C = 2, \ldots$ Consider the affine cipher $e_{3,14}(x) = 3x + 14 \bmod 26$. Then the word "QUEEN" is encrypted to "KWAAB".

**4.** Consider the permutation cipher on nine symbols $\pi = (135)(86)(2497)$, where the symbols indicate the positions of characters in the message. The word "XYLOPHONE" is then encrypted to "POXYLNEHO".

**5.** Using the mapping $A \leftrightarrow 0, B \leftrightarrow 1, C \leftrightarrow 2, \ldots, Z \leftrightarrow 25$, the Vigenère cipher with key "MIKE" encrypts the message "SENDMORETROOPS" to "EMXHYWBIFZYSBA".

---

## 15.3   MODERN PRIVATE KEY CRYPTOSYSTEMS

Classical encryption can be performed by hand (e.g., Caesar cipher) or by mechanical means (e.g., Enigma cipher). Modern ciphers refer to ciphers invented since the 1970s which rely upon the power of modern computers. For the most commonly used modern private key cryptosystems, plaintext messages and ciphertext messages are bit strings of a specified length. These cryptosystems are designed so that encryption and decryption can be carried out rapidly using complicated functions that are designed to be resistant to attack. This section covers the basic principles of such cryptosystems and provides information about the most widely used private key systems of today and of the past three decades.

---

### 15.3.1   BLOCK CIPHERS

**Definitions:**

A **block cipher** breaks up the plaintext into blocks of a fixed length and encrypts each block via a secret bijective transformation.

The secret bijective transform of a block cipher is defined by a **key schedule** and a **round function**. The key schedule expands the secret key $K$ into $r$ distinct roundkeys $K_1, K_2, ..., K_r$. To perform encryption, the round function (denoted by $\phi$) is iteratively applied $r$ times on the plaintext $x_0$ and the roundkeys $K_i$ to produce the ciphertext $x_r$.

$$
\begin{aligned}
Keyschedule(K) &\mapsto K_1, K_2, K_3, \ldots, K_r \\
x_0 &= Plaintext \\
x_1 &= \phi(K_1, x_0) \\
x_2 &= \phi(K_2, x_1) \\
x_3 &= \phi(K_3, x_2) \\
&\cdots \\
x_r &= \phi(K_r, x_{r-1}) \\
Ciphertext &= x_r.
\end{aligned}
$$

$r$ is called the **number of rounds of the block cipher** and is carefully chosen to ensure sufficient security without sacrificing throughput.

The round function $\phi$ mixes the intermediate text with the roundkeys by performing the following transformations in sequence:

- **Substitution/Confusion:** To provide (local) nonlinearity. This is usually realized by several 4-bit or 8-bit nonlinear substitution tables (also called **S-boxes**) acting in parallel.
- **Permutation/Diffusion:** To spread the nonlinear effect across the whole block. This is usually realized by a bit permutation or a linear matrix product over $GF(2)$.

A **product cipher** combines two or more simpler ciphers (transformations) such that the resulting cipher is more secure than its components. Thus a block cipher can be seen as a product cipher which combines many simpler transformations (where confusion and diffusion are alternately applied over many rounds) to perform encryption.

**Facts:**

1. The *substitution cipher* is an example of a block cipher where the plaintext is broken up into individual English alphabets and each alphabet is encrypted through a secret substitution table.

2. Modern block ciphers encrypt block sizes of 64 bits or 128 bits.

3. It is not possible to represent the secret bijection of modern block sizes by a single substitution table. Thus modern block ciphers are instantiated by a sequence of simpler substitution tables and permutation transforms which mixes the plaintext with the secret key to produce a ciphertext.

4. Some prominent examples of block ciphers include the Data Encryption Standard (DES, §15.3.2) [Fi77] and the Advanced Encryption Standard (AES, §15.3.3) [DaRi13].

5. The product cipher was introduced by Claude Shannon in his paper "Communication theory of secrecy systems" in 1949.

---

### 15.3.2 FEISTEL CIPHERS

**Definitions:**

The **Feistel cipher** is a block cipher structure introduced by Horst Feistel of IBM in the 1970s. It is used to construct many ciphers including the Data Encryption Standard (DES).

The **round function of a Feistel cipher** (denoted by $\phi_K$) divides the input into two equal halves $(L, R)$ and is defined by

$$\phi_K(L, R) = (R, L \oplus F(K, R)),$$

where $K$ is the roundkey and $F$ is a function mapping the right half $R$ and the roundkey $K$ to be XORed with the left half $L$. For an $r$-round Feistel cipher, a swap is performed on the left and right halves of the final output to produce the ciphertext, that is,

$$E_K(P) = swap(\phi_{K_r} \circ \cdots \circ \phi_{K_2} \circ \phi_{K_1}(P)).$$

where $swap(L, R) = (R, L)$ for equal halves $(L, R)$.

**Facts:**

**1.** One advantage of the Feistel cipher is that it is an *involution*. That is, the encryption circuit can be re-used for decryption when the roundkeys are fed in reverse order. Thus it saves implementation resources as only one copy of the cipher needs to be implemented for both encryption and decryption.

**2.** One of the earliest and most well-known Feistel ciphers is the *Data Encryption Standard* (DES), proposed in 1977 [Fi77]. DES resulted from an IBM submission to a 1974 request by the U.S. National Bureau of Standards (NBS) (which has now become the National Institute of Standards and Technology, NIST) soliciting encryption algorithms for the protection of computer data.

**3.** The early success of DES made Feistel a well-researched structure with many more secure Feistel ciphers being designed later. Some examples include the ciphers KASUMI [Ca05], Camellia [AoEtal00], its parallelizable variant $p$-Camellia [YaKhPo13], and the finalists MARS and Twofish of the AES competition [Ae97] (see §15.3.3). These ciphers still make use of the basic Feistel structure but use more secure components, larger keylength and blocklength, or use a larger number of rounds.

**4.** The basic Feistel cipher with two sub-blocks has also been extended to generalized Feistel structures with four sub-blocks. They include the ciphers Skipjack [Sk98], CLE-FIA [ShEtal07], SMS4 [DiLe08], its parallelizable variant p-SMS4 [YaKhPo13], and the finalist RC6 of the AES competition [Ae97] (see §15.3.3). For illustrative purposes, we can illustrate the round function of SMS4 as an example of a generalized Feistel structure:

$$\phi_K(A, B, C, D) = (B, C, D, A \oplus F(K, B \oplus C \oplus D)).$$

**5.** DES was used by the U.S. government to protect binary coded data during transmission and storage in computer systems and networks. It was also used by the banking industry and businesses to protect financial transactions for commercial data security. DES and its more secure variant Triple-DES (which encrypts the plaintext three times using a longer key) were also part of the SSH/TLS suite of ciphers to protect internet traffic.

**6.** A high-level description of DES is the following:

- DES processes plaintext blocks of 64 bits and produces ciphertext blocks of 64 bits. The encryption mapping $E_K$ is parameterized by a secret 56-bit key $K$. Since decryption requires that the mapping be invertible, $E_K$ is a bijection.

- Encryption of each 64-bit block proceeds in sixteen stages or rounds. The 56-bit key $K$ is used to create sixteen 48-bit subkeys $K_i$, one for each round.

- Within each round, eight fixed, carefully selected 6-to-4 bit substitution mappings (S-boxes) $S_i$, collectively denoted $S$, are used.

- The initial 64-bit plaintext is divided into two 32-bit halves, $L_0$ and $R_0$. Each round is functionally equivalent, taking 32-bit inputs $L_{i-1}$ and $R_{i-1}$ from the previous round and producing outputs $L_i$ and $R_i$ for $1 \le i \le 16$, according to

$$L_i = R_{i-1}, \ R_i = L_{i-1} \oplus F(R_{i-1}, K_i), \ F(R_{i-1}, K_i) = P(S(E(R_{i-1} \oplus K_i))).$$

  Here $E$ is a fixed expansion permutation mapping 32 bits to 48 bits (all bits are used once, some are taken twice), and $P$ is another fixed permutation on 32-bits. An initial permutation (IP) precedes the first round, and its inverse is applied following the last round.

- Decryption makes use of the same algorithm and same key, except the subkeys are applied to the internal rounds in the reverse order.

**7.** DES was used by financial institutions through the 1990s, but in the late 1990s it became apparent that it is not sufficiently secure. With advancements in computing technologies, brute force search of the DES keyspace became cheaper and easier (as outlined in Facts 8–10). Furthermore, several methods, known as Differential Cryptanalysis and Linear Cryptanalysis (see Facts 11–12), were discovered and were shown to be more efficient than brute force.

**8.** The keylength of DES (56 bits) is too short. The original proposal for DES had a keylength of 64 bits. But later, eight bits of the secret key were converted to parity bits making the equivalent keylength 56 bits. The complexity of exhaustive search is $2^{56}$ and this was already criticized early on by researchers to be insufficient protection against brute force search. Diffie and Hellman postulated in 1977 that a custom machine consisting of $10^6$ custom VLSI chips costing \$20,000,000 could search the entire DES keyspace in about one day.

**9.** In 1998, the Electronic Frontier Foundation built a dedicated DES Cracker machine (DEEP CRACK) costing \$250,000 which could find a DES key in 56 hours [De98].

**10.** In 2006, re-configurable parallelized hardware called COPACOBANA was built to break ciphers with keylengths up to 64 bits, including DES [GüEtal08]. It takes longer to find a DES key (on average 8.7 days) but is much cheaper at \$10,000. With the advancement in computing technologies, brute force search on DES became faster, cheaper, and easier.

**11.** *Differential Cryptanalysis* (DC) is an attack on DES that distinguishes the right key from wrong key guesses by observing the statistical distribution of differences in the input plaintext and differences in the output ciphertext. It can recover the secret key with $2^{47}$ chosen plaintext-ciphertext pairs [BiSh91], which is faster than exhaustive search. However, the enormous volume of chosen plaintext-ciphertext pairs required diminishes the value of this attack.

**12.** *Linear Cryptanalysis* (LC) on DES distinguishes the right key from wrong key guesses by observing the linear approximation of input plaintext bits and output ciphertext bits. It can recover the key with $2^{43}$ known plaintext-ciphertext pairs [Ma93], which is faster than exhaustive search. Similar to DC, even though plaintext-ciphertext pairs are not required to be chosen, the enormous volume of pairs required still diminishes the value of this attack.

**13.** Differential and linear cryptanalysis on DES can be thwarted by changing keys before the required number of plaintext-ciphertext pairs needed for cryptanalysis is transmitted. For example, we can change the key after every $2^{32}$ ciphertexts are transmitted.

**14.** The short keylength of DES allows for practical key recovery with just a few plaintext-ciphertext pairs. One countermeasure to the short keylength of DES is to use

Triple-DES which encrypts the plaintext three times with longer keys. In the following description, $K_1, K_2, K_3$ are 56-bit keys, $PT$ is the plaintext, and $CT$ is the ciphertext.

- Two-Key Triple-DES (112-bit secret key):

$$CT = DES_{K_1}(DES_{K_2}^{-1}(DES_{K_1}(PT))).$$

- Three-Key Triple-DES (168-bit secret key):

$$CT = DES_{K_3}(DES_{K_2}^{-1}(DES_{K_1}(PT))).$$

Note that the decryption function $DES^{-1}$ is used for the second pass to allow for backward compatibility with single-key DES. This is because DES with single-key $K$ can be defined as two-key Triple-DES with key $(K_1, K_2) = (K, K)$ or three-key Triple-DES with key $(K_1, K_2, K_3) = (K, K, K)$.

**15.** Another way to counter the short keylength of DES is to use DES-X. It is defined as

$$CT = DES_{K_2}(PT \oplus K_1) \oplus K_3,$$

where $PT$ and $CT$ are the plaintext and ciphertext, $K_2$ is a 56-bit DES key, and $K_1$ and $K_3$ are 64-bit keys to mask the input and output of DES. The total keylength for DES-X is 184 bits.

**16.** *Meet-in-the-middle* (MITM) attacks can be applied to Triple-DES and DES-X to reduce the complexity of brute force search, but their security is still better than DES. Two-key triple DES can be attacked with $2^{32}$ known plaintexts, $2^{88}$ time complexity, and $2^{56}$ memory [vOWi90]. Three-key Triple-DES can be attacked with three known plaintexts, $2^{112}$ time complexity, and $2^{56}$ memory [MevOVa10]. The equivalent keylength of DES-X is shown to be 88 bits under a distinguishing attack with $2^{30}$ known plaintexts [KiRo01]. See [MevOVa10] for a summary of various MITM attacks on Triple-DES and [KiRo01] for a proof of the security of DES-X against MITM.

**17.** Although Triple-DES and DES-X are considered to be secure, neither has sufficiently high throughput to be used for modern applications. Because of this, in 1997 NIST launched a competition for a new encryption standard, to be named the Advanced Encryption Standard (AES).

## 15.3.3   THE ADVANCED ENCRYPTION STANDARD (AES)

In 1997, NIST organized a competition to choose a replacement for DES called the *Advanced Encryption Standard* (AES) [Ae97]. The competition called for ciphers with a 128-bit blocklength and keylengths of 128, 192, and 256 bits, secure against known attacks including differential and linear cryptanalysis, and which provided excellent performance when implemented in hardware and software.

The process of determining which of the 15 submissions provided the best trade-off between security and performance took three years. Finally, in 2000 the winner was declared to be *Rijndael*, the submission of the Belgian cryptographers Joan Daemen and Vincent Rijmen. After winning the competition, the name AES replaced their original name Rijndael.

## Definitions:

The **AES cipher** takes in a 128-bit plaintext and three possible keylengths of 128, 192, and 256 bits. The **internal state** can be viewed as a 4-by-4 array of bytes which is first filled with the 128-bit plaintext. A **key schedule** expands the secret key of 128, 192, or 256 bits into 11, 13, and 15 roundkeys, respectively. A **prewhitening roundkey** is XORed to the plaintext and then the round function is repeatedly applied for $R = $ 10, 12, or 14 times for AES-128, AES-192, or AES-256. Each round of AES uses these operations:

- **SubBytes:** Every byte is nonlinearly transformed by a **Substitution Box** (**S-Box**). The S-Box is affine equivalent to the power function $x^{254}$ on GF$(2^8)$; $x^{254}$ is the multiplicative inverse of $x$ when $x \neq 0$, and 0 when $x = 0$. The finite field representation for GF$(2^8)$ is given by GF$(2)[a]/(f(a))$, where $f(a) = a^8 + a^4 + a^3 + a + 1$ is an irreducible polynomial.

- **ShiftRows:** The ShiftRows operation shifts row $r$ of the AES state array by $r-1$ bytes to the left.

- **MixColumns:** The MixColumns operation multiplies each column of the state array by a right circulant matrix over GF$(2^8)$. The first row of the matrix is given by $(a, a+1, 1, 1)$ where GF$(2^8)$ is also defined by GF$(2)[a]/(f(a))$, where $f(a) = a^8 + a^4 + a^3 + a + 1$.

- **AddRoundKey:** This operation XORs the $i$th subkey $k_i$ to the $i$th round.

## Facts:

1. A complete description of DES can be found in [DaRi13].

2. The AES cipher execution framework is shown in Algorithm 1.

---

**Algorithm 1**:   **AES cipher algorithm.**

input: $R$, plaintext
output: ciphertext
state := plaintext
AddRoundKey(state, $k_0$)
**for** $i = 1$ **to** $R - 1$
    SubBytes(state)
    ShiftRows(state)
    MixColumns(state)
    AddRoundKey(state, $k_i$)
SubBytes(state)
ShiftRows(state)
AddRoundKey(state, $k_R$)
ciphertext := state

---

3. The final round transformation does not have the MixColumns operation.

4. After adoption of AES as the new international block cipher standard by NIST, many banks and financial institutions migrated their encryption algorithm from DES to AES.

5. AES is also used by the IPSec (Internet Protocol Security) standard to protect IP communications. Other applications include TLS/SSH encryption for internet security and WPA/WPA2 encryption for Wifi Protection Access.

**6.** The NSA also stipulated AES as part of its suite B algorithms to protect both unclassified and most classified information for the U.S. government.

**7.** AES is based on the concept of a Substitution Permutation Network (SPN). An SPN starts with a roundkey XOR layer followed by an S-Box substitution layer that provides confusion, followed by a diffusion layer designed to spread the local S-box effect to the entire block.

**8.** The SubBytes operation in AES has the lowest differential and linear probabilities which defend AES against differential [BiSh91] and linear [Ma93] cryptanalysis (two attacks which broke DES). The AES SubBytes is a special case of a power function $x^k$ on GF($2^n$); see [Ca10b] for other power functions with good cryptographic properties.

**9.** The MixColumns matrix is a $4 \times 4$ Maximal Distance Separable (MDS) matrix over GF($2^8$). These matrices have optimal branch number 5. The *branch number* of a matrix is the minimum number of nonzero bytes in the input and output, when ranged among all nonzero inputs. The maximal possible branch number for an $n \times n$ matrix is $n + 1$, in which case the matrix will be called an *MDS matrix*. An MDS matrix ensures a high number of active S-boxes in a block cipher to prevent differential and linear cryptanalysis.

**10.** In general, an $n \times n$ MDS matrix over GF($2^k$) can be formed by reducing the generator matrix of a $[2n, n, n+1]$-MDS code over GF($2^k$) to standard form $[I|A]$. Then $A$ will be an MDS matrix.

**11.** An equivalent formulation for an $n \times n$ MDS matrix $A$ is that every submatrix of $A$ must be nonsingular. In particular, this means that all entries of an MDS matrix have to be nonzero.

**12.** The AES ShiftRows is an example of an *optimal diffusion* structure [DaRi13]. It spreads the effect of MixColumns across the whole 128-bit state and amplifies the diffusion effect of MixColumns from five active S-boxes every two rounds to 25 active S-boxes every four rounds. This gives very strong protection against DC and LC, where the differential and linear characteristic probability over four rounds of AES is already $2^{-150}$, rendering both attacks infeasible [DaRi13].

**13.** Due to the algebraic nature of the components used by AES, it is possible to embed the AES cipher into a bigger cipher BES (Big Encryption System) such that BES can be expressed as a simple set of linear and quadratic equations over $GF(2^8)$ [MuRo02].

**14.** It was claimed that the XSL (eXtreme Sparse Linearization) attack [CoPi02] can then be applied on the BES equations to break AES with complexity $2^{100}$. However, it was shown in [LiKh07] that there were too many linearly dependent equations in the XSL attack on BES for it to work. Despite this, there may be some other way to solve the simple set of equations produced by the BES embedding.

**15.** As of 2016, the most successful attacks on AES are the related-key differential cryptanalysis [BiKhNi09] and biclique attacks [BoKhRe11]. However, they are still not yet practical.

- Related-key differential cryptanalysis (RK-DC) is a differential attack where not just the plaintext, but also the secret key has a nonzero input difference. If the key difference is chosen carefully, then there is a higher chance to find a good differential path such that during the AddRoundKey operation, the roundkey differences cancel the intermediate state differences to result in fewer active bytes. RK-DC can reduce the attack complexity on AES-256 from $2^{256}$ to $2^{130}$ but needs $2^{35}$ related keys, which are extremely unlikely to be available in most practical applications, except perhaps when AES-256 is used in a hash function mode of operation.

- The biclique attack is a variant of the meet-in-the-middle attack (MITM) that uses a biclique structure to extend the number of rounds that can be attacked by MITM. The attack on AES-128, 192, and 256 has complexity $2^{126.1}$, $2^{189.7}$, and $2^{254.4}$, respectively. Although they are the best single-key attacks on AES, their complexity is too close to exhaustive search to be practical.

**16.** For applications that use AES-256 in a hash function mode of operation, the key schedule needs to be strengthened to resist related-key differential attacks. Such work has been carried out in [Ni10] and [ChEtal11], where the main cipher remains intact but the key schedule is modified so that high probability related-key differential paths, crucial for a successful attack, cannot be found.

## 15.3.4   MODES OF OPERATION

With large messages, it is necessary to break up the message into smaller blocks for encryption. Therefore, to use block ciphers, there is a need to employ an encryption mode of operation such that messages can be broken up in blocks, and encrypted in sequence. Various modes are introduced with distinct benefits.

**Definitions:**

An **initial vector** (**IV**) is a random value generated and sent in the clear with the ciphertext in some block cipher modes.

The **Electronic Code Book** (**ECB**) **mode** encrypts the plaintext block by block:

$$c_r = E_K(m_r),$$

for $1 \leq r \leq l$.

In the **Cipher Block Chaining** (**CBC**) **mode**, the previous ciphertext block is XORed to the current plaintext block before encryption:

$$c_1 = E_K(m_1 \oplus IV), \ c_r = E_K(m_r \oplus c_{r-1}),$$

for $2 \leq r \leq l$.

The **Output FeedBack** (**OFB**) **mode** is a stream cipher mode where the IV is repeatedly encrypted, and the most significant $t$ bits of the output blocks form the secret keystream to be XORed to the plaintext $m_r$ to form the ciphertext $c_r$:

$$s_0 = E_K(IV), \ s_r = E_K(s_{r-1}), \ c_r = m_r \oplus MSB_t(s_{r-1}),$$

for $1 \leq r \leq l$.

The **Cipher FeedBack** (**CFB**) **mode** is a variant of the OFB mode. The ciphertext is still formed by XORing the plaintext to the secret internal state. However, the secret internal state evolution is formed by repeatedly encrypting a concatenation of the previous secret state shifted left by $t$ bits and the ciphertext:

$$s_0 = IV, \ s_r = (s_{r-1} \ll t) \oplus c_{r-1}, \ c_r = m_r \oplus MSB_t(E_K(s_{r-1})),$$

for $1 \leq r \leq l$.

Like the OFB mode, the **Counter Mode** is one where the secret keystream is formed by encrypting an incremental counter initialized by IV:

$$c_r = m_r \oplus E_K(IV + r - 1),$$

for $1 \leq r \leq l$.

## Facts:

**1.** ECB is not secure because the same plaintext blocks are mapped to the same ciphertext blocks. For example, if we want to encrypt the pixels of a secret treasure map, it only maps the natural color of a map to different secret colors, but the outline and secret locations would still be visible. That is why we need other modes of operation that conceal the relationship between plaintext blocks.

**2.** *Error propagation.* Counter mode and OFB have 1-bit propagation, i.e., a 1-bit error in the ciphertext during transmission causes a 1-bit error in the plaintext after decryption. CBC and CFB have 1-bit plus 1-block propagation, i.e., a 1-bit error in the ciphertext causes a 1-bit error in the corresponding plaintext block and a 1-block error in the subsequent block after decryption [Dw01].

**3.** *Performance.* CBC needs more resources to implement because it needs both encryption and decryption, whereas OFB, CFB, and counter mode just need encryption. Counter mode is faster than the other three modes because it can be parallelized. CFB allows for $t$-bit block slip recovery.

**4.** *Security against IV misuse.* If IV is repeated, counter mode and OFB are totally broken because the same keystream $KS$ would be repeated for two different encryptions and we have

$$CT = PT \oplus KS \text{ and } CT' = PT' \oplus KS \Rightarrow CT \oplus CT' = PT \oplus PT'.$$

Thus the plaintext XOR relation can be deduced from the ciphertext. This weakness only affects the first block of CFB. CBC is not affected as different plaintexts will give different ciphertexts even when the IV is repeated. Thus many software implementations use CBC mode because software random number generation is less reliable and IV could repeat. However, it is important to protect the integrity of the IV in CBC mode, or else an adversary can selectively manipulate the first message block by manipulating the bits of the IV. This situation may be avoided by encrypting the IV.

**5.** Counter mode is combined with message authentication functions CMAC and GMAC to form the CCM and GCM authenticated encryption modes, respectively [Dw04], [Dw07].

## 15.3.5   STREAM CIPHERS

### Definitions:

A **keystream** is a stream of secret random characters that are combined with the plaintext to form the ciphertext.

A **stream cipher** is a symmetric key cipher that encrypts individual characters of a plaintext message, or small units. It is defined by a keystream generator and an encryption rule.

The **keystream generator** takes in a short secret seed (usually the secret key $K$) and expands it into an infinitely long stream of keystream characters, i.e., $Key\_Gen(K) = z_1, z_2, z_3, \ldots$.

The **encryption rule** $e_z(x)$ defines how the plaintext character $x_i$ is combined with the keystream character $z_i$. There is a corresponding decryption rule $d_z(x)$ such that $d_z(e_z(x)) = x$.

The **one-time pad** is a stream cipher where the secret key is of bitlength equal to the plaintext and never re-used for encryption. Thus there is no need for a keystream generator as the keystream $Z$ is just the secret key $K$. The encryption rule is $e_z(x) = x \oplus z$. The decryption rule is $d_z(y) = y \oplus z$.

A **synchronous stream cipher** is a stream cipher in which the keystream is generated independently of the message.

A **self-synchronizing stream cipher** is a stream cipher capable of re-establishing proper decryption automatically after loss of synchronization, with only a fixed number of plaintexts unrecoverable.

Many stream ciphers use **Linear Feedback Shift Registers** (**LFSR**) in their construction (see §14.7.2). An $n$-word LFSR is defined over a finite field $F$ with an initial state $(s_0, s_1, \ldots, s_{n-1})$, $s_i \in F$, and a feedback relation

$$s_{i+n} = c_{i+n-1} s_{i+n-1} + c_{i+n-2} s_{i+n-2} + \cdots + c_i s_i, \; i \geq 0.$$

In stream cipher applications, the LFSR is usually defined over the finite field $F = GF(2)$ or $GF(2^m)$.

## Facts:

**1.** The one-time pad can be proven to be unconditionally secure but it suffers the drawback that a large quantity of keying material has to be pre-shared to encrypt long messages. Thus in practice, stream ciphers usually generate a long unpredictable keystream from a short key to be XORed with the plaintext for encryption.

**2.** An important advantage of stream ciphers is that they are lighter and faster than block ciphers in hardware implementations. This is because stream ciphers process smaller chunks of data, and thus need fewer resources to store and process the intermediate states. They also have smaller error propagation (than certain block cipher modes such as CFB, CBC) where a one-character error in the ciphertext only causes a one-character error in the plaintext.

**3.** When the encryption rule is $e_z(x) = x \oplus z$, it is insecure to use the same keystream $z_i$ to encrypt two different plaintext streams $x_i$ and $x'_i$. This is because the ciphertext streams will be $c_i = x_i \oplus z_i$ and $c'_i = x'_i \oplus z_i$. And this allows the adversary to deduce the plaintext relation $x_i \oplus x'_i$ from the ciphertext relation $c_i \oplus c'_i$ because they are equal.

**4.** To prevent the repeated keystream weakness, an initialization vector $IV$ is mixed into the key generation and sent in the clear to the receiver for decryption. Now even when the sender uses the same secret key $K$ to encrypt two sessions, he can mix it with two distinct initial vectors $IV, IV'$ to produce distinct keystreams $Key\_Gen(K, IV) = z_1, z_2, z_3, \ldots$ and $Key\_Gen(K, IV') = z'_1, z'_2, z'_3, \ldots$.

**5.** Care must be taken not to repeat the initial vector $IV$ when using stream ciphers or else it is completely broken. It is also not that easy to convert a stream cipher to an authenticated encryption (AE) scheme, whereas numerous AE modes of operation for block ciphers such as CCM, GCM, OCB exist.

**6.** The Vigenère cipher can be seen as a stream cipher. The keystream generator is a cyclical repetition of the secret key:

$$Key\_Gen(k_1, k_2, \ldots, k_s) = k_1, k_2, \ldots, k_s, k_1, k_2, \ldots, k_s, \ldots.$$

The encryption rule is $e_z(x) = x + z \pmod{26}$, where the English alphabet $A, B, \ldots, Z$ is mapped to $0, 1, \ldots, 25$.

**7.** Another stream cipher that works on the English alphabet is the autokey cipher. The keystream generator only uses the secret key once in the beginning and then the plaintext is appended to it. If the plaintext stream is $m_1, m_2, \ldots, m_n$, then the keystream is

$$Key\_Gen(k_1, k_2, \ldots, k_s) = k_1, k_2, \ldots, k_s, m_1, m_2, \ldots, m_{n-s}.$$

The English alphabet is encoded as in the Vigenère cipher and the encryption rule is $e_z(x) = x + z \pmod{26}$.

**8.** On modern computers, the plaintext characters used are binary bits and the encryption rule is usually the XOR operation $e_z(x) = x \oplus z$. Most of the complexity and security of the cipher is provided by the keystream generation.

**9.** Early stream ciphers from the 1950s were based on Linear Feedback Shift Registers (LFSRs). A straightforward way to encrypt is to use an LFSR to expand a short key into a long keystream and XOR it to the plaintext.

**10.** The LFSR structure, however, is insecure as the cryptosystem is linear and can be easily broken by Gaussian elimination, or more efficiently by the Berlekamp-Massey algorithm (even when the LFSR feedback taps are secret) [Ma69]. The Berlekamp-Massey algorithm is derived from the Berlekamp-Welch algorithm for decoding BCH codes.

**11.** To be more secure, the LFSR output can be filtered by a nonlinear Boolean function in one of two configurations [Ru12]:

- *Combinatorial Generator.* This stream cipher consists of $n$ LFSRs and a nonlinear Boolean function $f(x)$ mapping $n$ bits to 1 bit. At each clock cycle, 1 bit is extracted from each LFSR to make up an $n$-bit input vector, which is passed through the Boolean function $f(x)$ to produce 1 output keystream bit. The keystream is XORed with the plaintext to form the ciphertext.

- *Filter Function Generator.* This stream cipher consists of one LFSR and one Boolean function $f(x)$ mapping $n$ bits to 1 bit. At each clock cycle, $n$ bits are extracted from the LFSR and passed as input to the Boolean function $f(x)$ to produce 1 output keystream bit. The keystream is XORed with the plaintext to form the ciphertext.

**12.** The combinatorial and filter function generator stream ciphers are secure against traditional attacks such as correlation attack, divide and conquer attack, and linear span attack if the filter function $f(x)$ is chosen to have the following good cryptographic properties:

- *Balance.* The Boolean function $f(x)$ has an equal number of 0s and 1s in its output. This is to prevent the keystream from having any bias which might leak information about the plaintext.

- *Low Linear Bias.* We want the linear bias of $f(x)$ given by

$$\epsilon = |\, Pr(f(x) = l(x)) - \tfrac{1}{2}|$$

    to be low for all linear functions $l(x)$. This is to defend against fast correlation attacks which make use of soft decoding techniques from LDPC (low-density parity-check), convolutional, and turbo codes [AgEtal12] to recover the stream cipher secret state. See §14.3 and §14.7.

- *High Correlation Immunity.* This property requires $f(x)$ to have linear bias $\epsilon = 0$ for all linear approximations with few linear terms. This is to defend against divide and conquer correlation attacks on combinatorial generators [Ca10a].

- *High Algebraic Degree.* We need $f(x)$ to have high algebraic degree, i.e., its Boolean expression involves monomials that are products of many variables. This is to defend against algebraic attacks and the Berlekamp-Massey attack on keystreams with low linear complexity. Furthermore, for additional resistance to algebraic attacks, we also require $f(x)$ to have high algebraic immunity (explained below).

**13.** It was believed that stream ciphers with high algebraic degree are not susceptible to algebraic attacks. However, the fast algebraic attack invented in 2003 managed to break such ciphers [Co03], [CoMe03].

**14.** *Algebraic Attack on Stream Ciphers.* The algebraic attack from [CoMe03] works on the filter function $f(x)$ in stream ciphers. The attack finds low degree Boolean functions $g(x)$ such that $f(x)g(x)$ also has low degree. Then the keystream equation $f(L_t(s)) = z_t$, which may have high degree, will be transformed to the low degree equation

$$f(L_t(s))g(L_t(s)) = z_t g(L_t(s)).$$

This low degree equation can be solved by algebraic methods such as linearization, eXtreme Linearization (XL), or Gröbner basis finding algorithms.

**15.** *Algebraic Immunity.* To resist the algebraic attacks from [CoMe03] and [Co03] on stream ciphers, we need $f(x)$ to have *high algebraic immunity*. That means we want all Boolean functions $g(x)$ such that $g(x)f(x) = 0$ for all $x$ and all functions $g(x)$ such that $g(x)(f(x) + 1) = 0$ for all $x$ to have high algebraic degree.

**16.** Some practical applications of stream ciphers include the following:

- *SNOW 3G* stream cipher for UEA2/UIA2 telecommunications encryption. The SNOW cipher takes in a 128-bit key and a 128-bit initialization vector $IV$. It is a filter function generator that maintains its secret state in a 16-word LFSR defined over $GF(2^{32})$. The nonlinear combiner is a finite state machine that consists of three 32-bit registers that make use of the AES SubBytes and MixColumns operations for computation. It builds on the success of AES by retaining some of the good cryptographic properties of AES components but aims to be faster by computing each 32-bit word with fewer clock cycles. As of 2015, there is one published analysis on SNOW 3G which shows some related key distinguishing properties [KiYo11], but is not a break of the cipher.

- *RC4* stream cipher for WEP encryption to protect WiFi traffic. The RC4 cipher takes in variable length keys between 40 to 2048 bits. It maintains a key-dependent secret permutation table on 256 bytes, which is iteratively updated by a pseudorandom generation algorithm to produce a keystream of secret bytes. This is XORed with the plaintext to produce the ciphertext. The original specification of RC4 is only defined based on a secret key with no initialization vector $IV$. However, the usage in WEP requires a fresh $IV$ for every packet transmitted. The WEP designers used RC4 with a 128-bit key, where a 24-bit $IV$ is concatenated with a 104-bit secret key. However, the weak design of this key schedule enabled Fluhrer, Mantin, and Shamir to launch a related-key attack (same 104-bit secret key but different 24-bit chosen $IV$s) on WEP encryption to recover the secret key in practical time [FlMaSh01].

- *E0* stream cipher for Bluetooth encryption. The E0 stream cipher takes in a 128-bit key, 48-bit Bluetooth address, and a 26-bit master counter. These are used to initialize four LFSRs of lengths 25, 31, 33, 39 bits and a finite state machine (FSM) that consists of two pairs of 2-bit delay elements. At each clock cycle, 1 bit is taken from the FSM and each of the four LFSRs to be XORed to form a keystream bit. This is then XORed with the plaintext to form the ciphertext. The best attack on E0 to date is a conditional correlation attack by Lu, Meier, and Vaudenay [LuMeVa05]. It uses the first 24 bits of $2^{23.8}$ frames (each frame is 2745 bits long) to recover the secret key with $2^{38}$ complexity.

- *A5/1* stream cipher for GSM encryption. The A5/1 stream cipher takes in a 64-bit key and a 22-bit frame number, which are used to initialize three LFSRs of lengths 19, 22, and 23 bits. The LFSRs are clocked in a stop-and-go fashion based on majority logic. One clocking bit is taken from each LFSR and a majority vote is computed. If the clocking bit of an LFSR agrees with the majority bit, that LFSR is clocked. Thus at each clock cycle, either two or three LFSRs are clocked. The short 64-bit key length of A5/1 makes it vulnerable to brute force attack based on just a few frames of known plaintext, e.g., by using the COPACOBANA search engine [GüEtal08]. With more known plaintext, there are faster attacks such as the correlation attack by Maximov et al. [MaJoBa04], where A5/1 can be broken in less than one minute based on a few seconds of conversation.

**17.** The emergence of algebraic attacks and other prominent breaks on stream cipher standards such as A5/1 and A5/2 (used in GSM encryption) prompted the call for better design/standardization of stream ciphers [Es04]. This prompted the call for a stream cipher competition ESTREAM, organized by EU-ECRYPT, to develop secure stream cipher standards.

**18.** The ESTREAM competition was held by EU-ECRYPT from 2005 to 2008 [RoBi08]. The stream cipher entries fall under two profiles: profile 1 for high throughput software (faster than AES) and profile 2 for lightweight hardware implementations in constrained devices. There were 34 submissions of which seven were chosen to be stream cipher standards after scrutiny by the international cryptographic community.

**19.** The OFB and counter modes of a block cipher expand the secret key into a long keystream (independently of the plaintext) by repeated applications of the block cipher. Thus they are examples of synchronous stream ciphers.

**20.** The CFB mode of a block cipher is an example of a self-synchronizing stream cipher. CFB mode with $t$-bit feedback can recover from a bit slip of $t$ bits. That is, when $t$ consecutive bits of data are lost during transmission, the ciphertext will decrypt incorrectly for $b/t$ iterations, where $b$ is the blocklength, and then resume correct decryption after the misaligned content is completely flushed from the block cipher input register.

**Examples:**

**1.** Consider an autokey cipher with secret key "iloveyou"$= \{8, 11, 14, 21, 4, 24, 14, 20\}$. It maps the plaintext "todayisagoodday"$= \{19, 14, 3, 0, 24, 8, 18, 0, 6, 14, 14, 3, 3, 0, 24\}$ as follows:

$$
\begin{array}{lll}
\text{plaintext} & : & 19, 14, 3, 0, 24, 8, 18, 0, 6, 14, 14, 3, 3, 0, 24 \\
\text{keystream} & : & 8, 11, 14, 21, 4, 24, 14, 20, 19, 14, 3, 0, 24, 8, 18 \\
\text{ciphertext} & : & 1, 25, 17, 21, 2, 6, 6, 20, 25, 2, 17, 3, 1, 8, 16.
\end{array}
$$

The keystream is formed by appending the plaintext after the key and the ciphertext is formed by adding the plaintext to the keystream modulo 26. The ciphertext converted to the alphabet is "bzrvcgguzcrdbiq".

**2.** Consider a simple 9-bit LFSR cipher over $GF(2)$ defined by the feedback relation $s_{i+9} = s_{i+4} + s_i$. Suppose the secret key is 011001010 and the plaintext is the 16-bit string 1100111010110111. To perform encryption, the secret key is expanded to a 16-bit keystream 0110010100011010. It is then XORed to the plaintext to form the ciphertext 1010101110101101.

**3.** One example of the algebraic attack of [CoMe03] is on the *Toyocrypt* cipher. The cipher is a nonlinear filter function generator where a 128-bit LFSR is filtered by a 128-bit Boolean function defined by

$$f(s_0, s_1, \ldots, s_{127}) = s_{127} + \sum_{i=0}^{62} s_i s_{\alpha(i)} + s_{10} s_{23} s_{32} s_{42} +$$

$$s_2 s_9 s_{12} s_{18} s_{20} s_{23} s_{25} s_{26} s_{28} s_{33} s_{38} s_{41} s_{42} s_{51} s_{53} s_{59} + \prod_{i=0}^{62} s_i,$$

where $\alpha(i)$ is a permutation of the indices $i \in \{63, 64, \ldots, 125\}$. This filter function is not easy to attack directly with an algebraic attack because its degree is 63, which is high.

However, when $f(s)$ is multiplied by the linear polynomial $g(s) = s_{23} + 1$, we see that $f(s)g(s)$ becomes a cubic polynomial

$$f(s)g(s) = s_{127}(s_{23} + 1) + \sum_{i=0}^{62} s_i s_{\alpha(i)}(s_{23} + 1).$$

This is because the last three monomials of $f(s)$ disappear when multiplied by $(s_{23} + 1)$, as $s_{23}$ appears in them and $s_{23}^2 = s_{23}$ for Boolean variables. In this way, the degree of the keystream equation is reduced from 63 to 3. By performing a linearization attack (treating each monomial as a new variable), one can solve the resulting linear system for the key in $2^{51.6}$ time using $2^{17.4}$ keystream bits. In [Co03], an improvement is made to this attack where the keystream independent part of the linearized equations is preprocessed by the Berlekamp-Massey algorithm and stored in memory, leading to a faster online attack. There was a further enhancement [ZhEtal09] to the algebraic attacks of [CoMe03] and [Co03] by considering the attack in the re-synchronization scenario, which results in further reduction of the attack complexity to a few seconds on a PC to recover the secret key under four resynchronizations, each producing 128 keystream bits.

## 15.4 HASH FUNCTIONS

Conventional hash functions in non-cryptographic applications are commonly used to map data of arbitrary size to a fixed size. Cryptographic hash functions require additional properties and are used for the purposes of data integrity and message authentication.

### 15.4.1 BASIC CONCEPTS

**Definitions:**

A **hash function** $h$ maps arbitrary length bit strings to small fixed length (for example, 128-bit or 512-bit) outputs called **hash-values**.

A **collision** is a pair of distinct bit strings mapped to the same output by a hash function.

A hash function is **preimage-resistant** if given any $y$ in the range of $h$ (for which a corresponding input is not known), it is computationally infeasible to find any preimage $x^*$ such that $h(x^*) = y$.

A hash function is **weak collision-resistant** if given any one input $x$, it is computationally infeasible to find a second preimage $x^* \neq x$ such that $h(x) = h(x^*)$.

A hash function is **strong collision-resistant** if it is computationally infeasible to find any two distinct inputs $x$ and $x^*$ such that $h(x) = h(x^*)$.

A **one-way hash function** (**OWHF**) is a function $h$ that maps arbitrary length inputs to fixed length outputs, and has the properties of preimage-resistance and weak collision-resistance.

A **collision-resistant hash function** (**CRHF**) is a function $h$ that maps arbitrary length inputs to fixed length outputs, and has the property of strong collision-resistance.

An $n$-bit hash function is said to have **ideal security** if the following properties hold:

- Given a hash output, producing both a preimage and a second preimage given a first, requires approximately $2^n$ operations;
- Producing a collision requires approximately $2^{n/2}$ operations.

**Facts:**

**1.** The basic idea is that a hash-value serves as a compact representative image (sometimes called a *digital fingerprint* or *message digest*) of the input string, and can be used as if it were uniquely identifiable with that string.

**2.** The problem of checking the integrity of the potentially large original input is reduced to verifying that of a small, fixed-size hash-value.

**3.** A hash-value should be uniquely identifiable with a single input *in practice*, and collisions should be *computationally difficult* to find.

**4.** While the utility of hash functions is widespread, the most common cryptographic uses are with digital signatures and for data integrity.

**5.** Regarding digital signatures, long messages are typically hashed first, and then the hash-value is signed rather than signing individual blocks of the original message. Advantages of this oversigning the individual blocks of the original message directly include efficiency with respect to both time and space.

**6.** Regarding data integrity, hash functions together with appropriate additional techniques can be used to verify the integrity of data. Specific integrity applications include virus protection and software distribution.

**7.** MACs (§15.6.2) are a special class of hash functions which take in addition to message input a secret key as a second input, allowing for the verification of both data integrity and data origin authentication.

**8.** Given a hash function $h$ and an input $x$, $h(x)$ should be easy to compute.

**9.** The complete specification of $h$ is usually assumed to be publicly available.

**10.** Collision-resistance is required for applications such as digital signatures and data integrity. Otherwise an adversary might find two messages $x$ and $x'$ that have the same hash-value, obtain a signature on $x$, and claim it as a signature on $x'$.

**11.** Depending on the intended application and the susceptibility of the environment to certain attacks, weak or strong collision-resistance may be required.

**12.** There are no known instances of functions that have been proven to be one-way, that is, for which it can be *proven* (without assumptions) that finding a preimage is difficult. However, it would be most surprising if such functions indeed did not exist. All instances of "one-way functions" given to date should thus properly be qualified as "conjectured" or "candidate" one-way functions.

**13.** Particular hash functions differ in the nature of the compression function and pre-processing of the input.

**14.** A typical usage for data integrity is as follows:

- The hash-value corresponding to a particular input is computed at some point in time;

- The integrity of this hash-value is then protected in some manner;

- At a subsequent point in time, to verify the input data has not been altered, the hash-value is recomputed, using purportedly the same input, and compared for equality with the original hash-value.

## 15.4.2  CONSTRUCTION AND METHODS

In general, hash functions are built by processing small blocks of data iteratively. Two common construction methods account for the majority of existing secure hash functions.

**Facts:**

**1.** *Merkle-Damgård construction:* Most hash functions process fixed-size blocks of the input iteratively as follows:

- A prespecified starting value or *initializing value* (IV) is defined.

- The hash input $x = x_1 x_2 \ldots x_t$ of arbitrary finite length is divided into fixed-length $n$-bit blocks $x_i$. This preprocessing typically involves appending extra bits (*padding*) as necessary to extend the input to an overall bit length that is a multiple of the blocklength $n$. The padding also often includes a partial block indicating the bit length of the unpadded input.

- Each block $x_i$ is then used as input to a simpler hash function $f$, called an $m$-bit *compression function*, which computes a new intermediate result of some fixed bit length $m$ as a function of the previous $m$-bit intermediate result (initially the IV) and the block $x_i$. Letting $H_i$ denote the partial result after the $i$th stage, the hash $h(x)$ of an input $x = x_1 x_2 \ldots x_t$ is defined as follows:

$$H_0 = IV; \qquad H_i = f(H_{i-1}, x_i),\ 1 \le i \le t; \qquad h(x) = H_t.$$

- $H_{i-1}$ serves as the *chaining variable* between stages $i-1$ and $i$.

**2.** *Matyas-Meyer-Oseas construction:* This construction works with any block cipher such as AES. Let $E$ be an $n$-bit block cipher parameterized by a symmetric key $k$. Let $g$ be a function that maps an $n$-bit string to a key $k$ suitable for $E$. Fix an initial value $IV$. The following algorithm is then an $n$-bit hash function which, given any input string $x$, outputs an $n$-bit hash $h(x)$:

- Divide $x$ into $n$-bit blocks and pad if necessary by some method such that all blocks are complete, yielding a padded message of $t$ $n$-bit blocks $x_1 x_2 \ldots x_t$;

- Define $h(x) = H_t$, where

$$H_0 = IV; \qquad H_i = E_{g(H_{i-1})}(x_i) \oplus x_i,\ 1 \le i \le t.$$

This is believed to be a one-way hash function requiring $2^n$ operations to find a preimage, and $2^{n/2}$ operations to find a collision. For underlying ciphers, such as DES, which have relatively small blocklength (for example, with blocks of no more than 64 bits), this is not a collision-resistant hash function since $2^{32}$ operations is well within current computational capability.

## 15.4.3   FAMILIES OF HASH FUNCTIONS

Families of unkeyed hash functions were introduced over the years as cryptographic primitives. These functions were updated in response to increases in computational power, new attacks, as well as industrial needs.

**Facts:**

**1.** The *MD Message-Digest Algorithm Family* was proposed by Rivest et al. *MD2* was proposed in 1989 [Ka92]. The input is in blocks of 128 bits with an output size of the same length. MD2 is considered historic. In CVE-2009-2409 [Cv09], security updates were issued to disable MD2 in Firefox, GnuTLS before 2.6.4 and 2.7.4, OpenSSL 0.9.8 through 0.9.8k, and other products that support MD2 with X.509 certificates.

**2.** *MD4* was an update in 1990 [Ri92]. The input block size is 512 bits with an output size of 128 bits. As of 2011, MD4 was considered "historic" [TuCh11].

**3.** *MD5* was proposed in 1992 as a replacement of MD4. The algorithm follows a Merkle-Damgård construction with the same input and output parameters as MD4. In 2013, Xie, Liu, and Feng demonstrated a collision attack in $2^{18}$ operations [XiLiFe13]. It is advised that MD5 not be used [Vu08].

**4.** *MD6* was first published in 2008 [Ri08] by Rivest et al. as a proposal for the NIST SHA-3 competition. It uses a Merkle tree-like structure to make use of parallel computation of hashes for long inputs. MD6 did not advance to the second round of the competition.

**5.** The *Secure Hash Algorithm* (SHA) is a family of hash functions published by NIST.

**6.** *SHA-0*, published in 1993, produces a hash of 160 bits with inputs in 512-bit blocks. The algorithm follows a Merkle-Damgård construction. It was quickly withdrawn and replaced by SHA-1 in 1995. In 2008, Manuel and Peyrin [MaPe08] showed that a SHA-0 collision can be found in $2^{33.6}$ operations.

**7.** *SHA-1* has the same input and output parameters as SHA-0, and follows a Merkle-Damgård construction. In 2005, Wang et al. [WaYu05] presented a collision attack with complexity $2^{63}$. In 2011, NIST [BaRo11] declared that "SHA-1 shall not be used for digital signature generation after December 31, 2013."

**8.** A SHA-1 collision was demonstrated in 2017 with a complexity of $2^{63.1}$, or 100 GPU years, at a cost of \$110,000 by researchers at CWI Amsterdam and Google Research [StEtal17].

**9.** *SHA-2*, published in 2001, is itself a family of six hash functions with various input/output sizes. These functions follow a Merkle-Damgård construction with a Davies-Meyer compression function. SHA-224 and SHA-256 have block sizes of 512 bits and output a digest (hash value) of 224 and 256 bits, respectively, in 64 rounds. SHA-384, SHA-512, SHA-512/224, and SHA-512/256 have block sizes of 1024 bits and output digests of size 384, 512, 224, and 256 bits, respectively in 80 rounds. Reduced-rounds attacks were found in 2012 [BiEtal11], [KhReSa12].

**10.** *SHA-3* is a result of a NIST call for an alternative to SHA-2. Submissions were accepted until the end of 2008. The standard was ultimately released in August 2015 [Fi15]. The algorithm uses a sponge construction [BeEtal07]. SHA3-224, SHA3-256, SHA3-384, SHA3-512 have block sizes of 1152, 1088, 832, 576 and output digests of size 224, 256, 384, and 512, respectively.

## 15.5    PUBLIC KEY CRYPTOGRAPHY

The invention of public key cryptography in the 1970s significantly changed the field of cryptography. As a result, many security services can now be realized. In a private key system, the encryption and decryption keys are the same, or the knowledge of the encryption key implies an easy derivation of the decryption key. On the other hand, in a public key cryptosystem there is a *public key* portion and a *private key* portion. The public key is not kept secret and allows everyone to use it to encrypt messages to the owner of the private key. The private key is known only to the receiver and is used for decryption of the ciphertext.

A number of public key cryptosystems were proposed in the past 40 years. Soon after their appearance, cryptanalysts worked hard to break them. Most systems were broken quickly. Those that have withstood sustained attacks are generally believed to meet the security requirements of a public key cryptosystem. However, researchers still continue to look for ways to break them and sometimes succeed quite a while after the systems were first proposed. In the following, we introduce the significant systems that have survived existing attacks and appeared hard to break, and that are also reasonably efficient.

### 15.5.1    INTRODUCTION TO PUBLIC KEY CRYPTOSYSTEMS

**Definitions:**

Let $E_p$ and $D_s$ be the encryption and decryption algorithms in a public key cryptosystem with a **public key** $p$ and a **private key** $s$, respectively. To encrypt plaintext $M$, one can generate the ciphertext $C = E_p(M)$. Upon receiving ciphertext $C$, the receiver, using the private key $s$, decrypts by evaluating $M' = D_s(C)$. Given the public key $p$, it should be hard to compute the private key $s$.

**Non-adaptive chosen-ciphertext attack security** (**CCA1**): an attacker that can inquire multiple times the plaintext for any adaptively chosen ciphertext is unable to recover the plaintext for a new challenge ciphertext $C$. Here non-adaptivity means that no query is based on $C$ (i.e., no decryption query is allowed after $C$ is given).

**Adaptive chosen-ciphertext attack security** (**CCA2**): given the challenge ciphertext $C$ in CCA1 security, the attacker can continue the decryption requests adaptively, except for $C$ itself, and is unable to recover the plaintext.

**Facts:**

**1.** The invention of public key cryptography is credited to Diffie and Hellman [DiHe76]. However, it was first conceived in 1970 by James H. Ellis in secret work at the U.K.'s Government Communications Headquarters (GCHQ). This earlier discovery was not made public until 1997.

**2.** In 1978 Ron Rivest, Adi Shamir, and Leonard Adleman invented the first public key cryptosystem that has withstood attack, the RSA cryptosystem (see §15.5.2). However, in 1997 it was revealed that in 1973 Clifford Cocks discovered the equivalent of the RSA cryptosystem in secret work at the GCHQ.

**3.** For a cryptosystem to be useful, it is necessary that $M' = M$. In addition, given the ciphertext $C$ and public key $p$, it should be hard to derive the plaintext $M$. That is, the cryptosystem should be resistant to *ciphertext-only attack*.

**4.** CCA2 is a desired property, but can be difficult to achieve in practice.

**5.** Since the late 1970s many different public key cryptosystems have been developed. The mostly widely used are the RSA cryptosystem (§15.5.2), the ElGamal cryptosystem (§15.5.3), and the elliptic curve cryptosystem (§15.5.4). Other important public key cryptosystems are the NTRU cryptosystem (§15.5.5), the Cramer-Shoup cryptosystem (§15.5.6), the Paillier cryptosystem (§15.5.7), the Goldwasser-Micali probabilistic cryptosystem (§15.5.8), identity-based encryption (§15.5.9), and attribute-based encryption (§15.5.10).

**6.** It will be possible to quickly break many, but not all, public key cryptosystems once quantum computing becomes practical. This is noted in the coverage of the public key cryptosystems in this section.

## 15.5.2    RSA CRYPTOSYSTEM

The RSA cryptosystem is the best known and most widely used public key cryptosystem for secure data transmission. Its suitability as a public key cryptosystem is based on the difficulty of factoring large integers compared with the ease of finding large primes.

**Facts:**

**1.** *Key Generation.*

- Randomly select two large and distinct primes $p$ and $q$. Compute $n = pq$ and $\phi(n) = (p-1)(q-1)$.
- Select an arbitrary value $e$ with $1 < e < \phi(n)$ so that $\gcd(e, \phi(n)) = 1$.
- Compute $d$ with $1 < d < \phi(n)$ such that $ed \equiv 1 \pmod{\phi(n)}$.
- The public key is $(n, e)$ and the private key is $d$.

**2.** *Encryption.* To encrypt a plaintext $M \in \{0, 1, \ldots, n-1\}$, compute $C = M^e \bmod n$.

**3.** *Decryption.* To decrypt $M$ from $C$, compute $M = C^d \bmod n$.

**4.** The correctness of RSA follows from elementary number theory, in particular from Euler's theorem (see §4.3.1).

**5.** Some computational notes:

- The public exponent $e$ should be coprime with $\phi(n) = (p-1)(q-1)$ and can be checked using the Euclidean algorithm (see §4.2.2).
- The secret exponent $d$ can be computed from $(e, \phi(n))$ using the extended Euclidean algorithm (see §4.2.2).
- To compute $M^e \bmod n$ and $C^d \bmod n$, many fast modular exponentiation algorithms such as the square-and-multiply algorithm are available [MevOVa10].

**6.** The hardness of factoring is the security basis of RSA. Some particular concerns with the choice of parameters are the following:

- When $|p - q|$ is small, $n$ will suffer from Fermat factoring attack.
- When $|p - q| < n^{1/4}$, Coppersmith [Co97] presented an attack to factor $n$.
- If $p - 1$ or $q - 1$ has small factors only, then $n$ can be factored using Pollard's $p - 1$ algorithm [Po74]; when $p + 1$ or $q + 1$ has small factors only, then $n$ can be factored using Williams' $p + 1$ algorithm [Wi82].

**7.** One might attempt to use a small $d$ for better decryption efficiency. However, this usage is vulnerable to the factoring attacks by Wiener [Wi90], Boneh and Durfee [BoDu99], and Howgrave-Graham and Seifert [HoSe99]. Hinek and Lam [HiLa10] showed that in practice, three instances of RSA encryption are sufficient to factor $n$ when $d < n^{0.33}$ with at least 93% success rate. Therefore, it is advisable not to use small $d$.

**8.** Due to the nature of factoring algorithms, $p$ and $q$ need to be of similar size to keep the system secure.

**9.** *RSA assumption and factoring assumption.*

- Security against ciphertext-only attack requires that it be hard to derive $M$ from $C$ and $(n, e)$. This is known as the *RSA assumption*.

- Given $n$, if one can find $p$ and $q$, then one can easily compute $d$ from $(n, e)$. So the security of RSA needs the assumption that it is hard to factor $n$. This is known as the *factoring assumption*, and is stronger than the RSA assumption.

**10.** Under a generic computation model over the ring $\mathcal{Z}_n$, the factoring assumption is equivalent to the RSA assumption [AgMa09]. This gives evidence for the widely accepted belief that they are equivalent in the normal computation model.

**11.** Under the RSA assumption, RSA is secure against ciphertext-only attack. However, it is not CCA2 secure due to its multiplicative property.

**12.** To generate the keys $e$ and $d$, one must first generate two primes $p$ and $q$ of similar bitlengths. There are several efficient probabilistic primality tests available, such as the Miller-Rabin test and the Solovay-Strassen test (§4.4).

**13.** There are several general purpose factoring algorithms. See §4.5 for details.

**14.** In 1991, RSA Laboratories began the *RSA Factoring Challenge* to stimulate research in factoring products of two large primes. One goal of this challenge has been to provide guidance as to the minimum recommended size of an *RSA modulus* $n = pq$.

**15.** As of 2016, the largest factored RSA modulus $n$ has bitlength 768 (RSA-768). Therefore, the choice of $n$ needs to be significantly larger. Currently, a 2048-bit (617 decimal digits) modulus is recommended for general applications and a 3072-bit (925 decimal digits) modulus for high-security applications.

**16.** If quantum computers are developed, RSA will be completely broken, since the factoring problem can be efficiently solved on quantum computers. However, building such computers does not seem realistic in the near future.

**17.** An important application of RSA is the *digital envelope*. A digital envelope is a framework for data encryption, in which the data is encrypted under a secret key using a symmetric key cryptosystem such as AES, while this secret key is encrypted using a public key cryptosystem, such as RSA. The details are described in PKCS #7 [Ka98]. A well-known application of the digital envelope is Pretty Good Privacy (PGP), a system for email security.

**18.** An important use of the RSA system is to guarantee the authenticity of a message using a *digital signature*. See §15.6.1 for details.

**Examples:**

**1.** Suppose that $p = 53$ and $q = 83$, so that $n = 53 \cdot 83 = 4399$ and $\phi(n) = (53 - 1)(83 - 1) = 4624$, and further suppose that $e = 23$. Then $(n, e) = (4399, 23)$ is a valid public key for RSA because $\gcd(\phi(n), e) = \gcd(4624, 23) = 1$, as can be verified using the Euclidean algorithm. Using the extended Euclidean algorithm, it can be shown that $23^{-1} \bmod 4624 = 927$. Hence, the exponent in the private key corresponding to the public key $(4399, 23)$ is $d = 927$. (Note that for practical use, the primes $p$ and $q$ need to be hundreds of digits long.)

**2.** In Example 1, to encrypt $M = 19$, compute $C = M^e \bmod n = 2556$. In order to decrypt $C = 2556$, compute $M = C^d \bmod n = 19$.

## 15.5.3  ELGAMAL CRYPTOSYSTEM

The ElGamal cryptosystem is developed from the Diffie-Hellman key-exchange protocol. The system can be defined over any cyclic group. Its security is based on both the discrete logarithm problem and the computational Diffie-Hellman problem.

**Facts:**

**1.** The ElGamal cryptosystem was invented by Taher ElGamal [El85] in 1985.

**2.** *Key Generation.*

- Let $p$ be a large prime. Let $g$ be a generator of the multiplicative group $\mathcal{Z}_p^*$.
- Randomly select $d$ with $1 \le d \le p - 2$, and compute $h = g^d \bmod p$.
- The public key is $(p, g, h)$ and the private key is $d$.

**3.** *Encryption.* To encrypt a plaintext $M \in \langle g \rangle$, randomly select $k$ with $1 \le k \le p - 2$ and then compute $X = g^k \bmod p$ and $Y = Mh^k \bmod p$. The ciphertext is $C = (X, Y)$.

**4.** *Decryption.* To decrypt $(X, Y)$, the receiver computes $M' = YX^{-d} \bmod p$.

**5.** The correctness of the algorithm follows from the fact that $YM^{-1} = h^k = g^{kd} = X^d$.

**6.** Given $p$, $g$, and $h$, if one can determine the private key $d$, then the ciphertext can be decrypted. Thus, the security of ElGamal encryption depends on the hardness of finding $d$ from $(p, g, h)$. This is called the *Discrete Logarithm Problem* (DLP).

**7.** The ciphertext-only security of ElGamal encryption requires that it be hard to determine $Z = h^k \bmod p$ from $(p, g, h, X)$. This is called the *Computational Diffie-Hellman Problem* (CDHP). It is widely believed that the CDHP can only be solved by first solving the DLP.

**8.** The assumption that the DLP is hard does not hold in every finite cyclic group. For instance, the DLP is easy in the additive group $\mathcal{Z}_n$.

**9.** The DLP is insecure under a quantum attack.

**Examples:**

**1.** Let $p = 3767$ and choose $g = 7$. Let $d = 15$. Then $h = 7^{15} \bmod 3767 = 3467$. The public key is $(p, g, h) = (3767, 7, 3467)$ and the private key is $d = 15$. Note that in practice, the parameters should be larger.

**2.** In Example 1, to encrypt message $M = 125$, choose a random value $k = 87$, compute $X = 7^{87} \bmod 3767 = 3113$ and $Y = Mh^k \bmod 3767 = 125 \cdot 3467^{87} \bmod 3767 = 2379$. Thus, the ciphertext is $(3113, 2379)$. To decrypt the ciphertext $(3113, 2379)$, compute $M' = YX^{-d} \bmod 3767 = 2379 \cdot 3113^{-15} \bmod 3767 = 2379 \cdot 464 \bmod 3767 = 125$.

## 15.5.4  ELLIPTIC CURVE CRYPTOSYSTEM

The elliptic curve cryptosystem is a adaptation of the ElGamal cryptosystem over a group of elliptic curves over a finite field. A significant advantage of this system is that it requires a significantly smaller key size compared to other systems in order to achieve the same level of computational security.

**Facts:**

**1.** Elliptic curve cryptography (ECC) is the elliptic curve analogue of discrete logarithm cryptography. It was invented by Neal Koblitz [Ko87] and Victor Miller [Mi85] in 1985. (See §4.11 for the basics of elliptic curve group operations.)

**2.** *Key Generation.*

- Select a large prime $p$ and elliptic curve group $E(\mathcal{F}_p)$.
- Let $P$ be a point of prime order $n$ in $E(\mathcal{F}_p)$ and randomly select an integer $d$, $1 \leq d \leq n-1$. Compute $Q = dP$.
- The public key is $(p, E, n, P, Q)$ and the secret key is $d$.

**3.** *Encryption.* To encrypt message $M \in \langle P \rangle$, select a random integer $k$ from $[1, n-1]$; the ciphertext is $C = (C_1, C_2) = (kP, M + kQ)$.

**4.** *Decryption.* To decrypt $(C_1, C_2)$, compute $M = C_2 - dC_1$.

**5.** Decryption is correct because $C_2 - dC_1 = M + kQ - dkP = M + k(dP) - dkP = M$.

**6.** There are a number of standards on the use of elliptic curve cryptography, covering the choice of curves and various implementation issues.

- ANSI X9.62 (2005) specifies Elliptic Curve Digital Signature Algorithm (ECDSA). The standard includes methods and criteria for public and private key generation, and procedural controls required for the secure use of the algorithm with these keys.
- ANSI X9.63 (2011) specifies key agreement and key transport schemes using elliptic curves for use in symmetric schemes.
- IEEE P1363 (2000, 2004) includes key agreement, signature, and encryption schemes.
- NIST FIPS 186-4 (2013), the Digital Signature Standard (DSS) specifies DSA, RSA, and ECDSA.
- NSA Suite B (2005) includes a set of cryptographic algorithms proposed by the National Security Agency (NSA). Included are ECDSA and Elliptic Curve Diffie-Hellman Key Agreement (ECDH). In 2015, NSA announced preliminary plans for a future transition to quantum-resistant algorithms.

**7.** The security of ECC is based on the difficulty of the *Elliptic Curve Discrete Logarithm Problem* (ECDLP): namely, given point $P \in E(\mathcal{F}_q)$ of prime order $n$ and a point $Q \in \langle P \rangle$, find an integer $\ell$ such that $Q = \ell P$.

- Using Pollard's rho attack. the ECDLP can be solved in $\sqrt{n}$ time.
- The index-calculus attack, which normally applies to the discrete logarithm problem over a finite field, is not generally applicable to the ECDLP.
- In special cases, one can compute an isomorphism from $\langle P \rangle$ to a cyclic group $G$, where the discrete logarithm problem can be solved by algorithms running in subexponential or faster time. The following attacks have been discovered:

- *Prime-field-anomalous curve.* A curve $E$ of prime order $p$ defined over $\mathcal{F}_p$ is isomorphic to the additive group $\mathcal{F}_p^+$. The attack was found in 1998 [Se98], [Sm99], [TaEtal98].
- *Weil and Tate pairing attacks.* When $E$ is defined over $\mathcal{F}_q$ and $\gcd(n, q) = 1$, there is an isomorphism between $E$ and the multiplicative subgroup of order $n$ of $\mathcal{F}_{q^k}^*$ of some extension field $\mathcal{F}_{q^k}$. However, $k$ is very large in general and so the attack is easily circumvented.
- *Weil descent.* This attack reduces the ECDLP in an elliptic curve defined over a binary field $\mathcal{F}_{2^m}$ to the discrete logarithm problem in the Jacobian of a hyperelliptic curve defined over a proper subfield of $\mathcal{F}_{2^m}$. Thus far, the attack is effective only in certain elliptic curves over $\mathcal{F}_{2^m}$ where $m$ is composite. Therefore, for cryptographic purposes, curves defined over $\mathcal{F}_{2^m}$ where $m$ is composite should be avoided [Fr01], [GaHeSm02], [JaMeSt01], [MaMeTe02], [MeQu01], [MeTe06], [MeTeWe04].

8. Since ECC is based on the elliptic curve discrete logarithm problem, ECC is insecure against quantum attacks.

**Examples:**

1. Let $p = 23$ and consider $E(\mathcal{F}_{23}) : y^2 = x^3 + x + 1$. Let $P = (3, 10)$ and $d = 3$. Then $Q = 2P + P = (7, 12) + (3, 10) = (-4, 5)$.

2. In Example 1, to encrypt $M = (9, 7)$, choose $k = 2$, whence $C_1 = kP = (7, 12)$ and $C_2 = kQ + M = 2(7, 12) + (9, 7) = (-6, 3) + (9, 7) = (3, 13)$. To decrypt $(C_1, C_2) = ((7, 12), (3, 13))$, compute $dC_1 = 3(7, 12) = (-6, 3)$. Thus, $C_2 - dC_1 = (3, 13) - (-6, 3) = (3, 13) + (-6, 20) = (9, 7) = M$.

---

### 15.5.5   NTRU CRYPTOSYSTEM

NTRU is a cryptosystem based on lattices. It is one of the very few cryptosystems that is currently resistant to quantum attacks.

**Facts:**

1. NTRU is a lattice-based public key cryptosystem proposed by J. Hoffstein, J. Pipher, and J. Silverman in 1996.

2. For a prime $N$, let $R = \mathcal{Z}[x]/(x^N - 1)$. Let $F(x), G(x) \in R$ with $F(x) = \sum_{i=0}^{N-1} F_i x^i$ and $G(x) = \sum_{j=0}^{N-1} G_j x^j$. Then multiplication $\odot$ in $R$ can be implemented as $F(x) \odot G(x) = H(x) = \sum_{k=0}^{N-1} H_k x^k$, where $H_k = \sum_{i+j=k \bmod N} F_i G_j$.

3. *Key Generation.*

   - Select a large prime $N$, and coprime integers $p$ and $q$ with $q \gg p$.
   - Choose random polynomials $f(x), g(x) \in R$ with coefficients in $\{-1, 0, 1\}$. Compute $f_p(x), f_q(x)$ such that $f(x) \odot f_p(x) = 1 \bmod p$ and $f(x) \odot f_q(x) = 1 \bmod q$.
   - Define $h(x) = pf_q(x) \odot g(x) \bmod q$. Then the public key is $(N, p, q, h)$ and the private key is $(f(x), f_p(x))$.

4. *Encryption.* Consider message $m(x) \in R$ with coefficients in $\{-1, 0, 1\}$. Generate a random polynomial $r(x) \in R$ with small coefficients (relative to $q$) and compute $c(x) = r(x) \odot h(x) + m(x) \bmod q$. The ciphertext is $c(x)$.

**5.** *Decryption.* To decrypt ciphertext $c(x)$, the receiver computes $A(x) = f(x) \odot c(x) \bmod q$ so that the coefficients of $A(x)$ are in $[-q/2, q/2]$. Recover the plaintext $m'(x) = f_p(x) \odot A(x) \bmod p$.

**6.** The correctness of the decryption can be demonstrated as follows. Notice that

$$A(x) = f(x) \odot r(x) \odot (pf_q \odot g(x)) + f(x) \odot m(x) \bmod q.$$

Simplifying, we have $A(x) = pr(x) \odot g(x) + f(x) \odot m(x) \bmod q$. Since $p$ and the coefficients of $r(x), f(x), g(x), m(x)$ are small relative to $q$, it holds that $A(x) = pr(x) \odot g(x) + f(x) \odot m(x)$. Hence, $m'(x) = f_p(x) \cdot A(x) \bmod p = f_p(x) \odot f(x) \odot m(x) \bmod p = m(x)$.

**7.** The security of NTRU relies on the difficulty of finding $f_q(x)$ and $g(x)$ from $h(x)$.

**8.** NTRU security is based on the hardness of the *Shortest Vector Problem* (SVP), finding the shortest vector in a lattice.

**9.** Currently, the most powerful attack on NTRU is the Lenstra-Lenstra-Lovász (LLL) algorithm. However, the attack can be prevented when the security parameters are sufficiently large.

**10.** NTRU is not known to be compromised by any quantum algorithm. This is an advantage over the RSA, ElGamal, and ECC schemes.

**Example:**

**1.** Let $N = 11$, $p = 3$, $q = 31$, $f(x) = x^8 - x^7 + x^3 + x - 1$, $g(x) = x^3 - x + 1$. We have $f_q(x) = 8x^{10} + 11x^9 + 20x^8 + 10x^7 + 28x^6 + 26x^5 + 30x^3 + 17x^2 + 11x + 26$ and $f_p(x) = 2x^{10} + 2x^9 + 2x^8 + x^6 + 2x^4 + x^3 + x^2 + 2$. Thus, $h(x) = pf_q(x) \odot g(x) \bmod q = 21x^{10} + 26x^9 + 15x^8 + 8x^7 + 3x^6 + 5x^5 + 5x^4 + 24x^3 + 11x^2 + 19x + 21$. The public key is $(N, p, q, h)$ and the private key is $(f(x), f_p(x))$.

**2.** In Example 1, to encrypt the message $m(x) = x^{10} - x^9 + x^8 + x + 1$, choose a random polynomial $r(x) = x^9 + x^8 + x + 1$ and compute the ciphertext $c(x) = r(x) \odot h(x) + m(x) \bmod q = 16x^{10} + 18x^9 + 4x^8 + 27x^7 + 18x^6 + 2x^5 + 9x^4 + 12x^3 + 9x^2 + 8x + 16$. To decrypt $c(x)$, compute $A(x) = f(x) \odot c(x) \bmod q = -5x^{10} + 3x^9 + 2x^8 - 2x^6 + 2x^5 + 3x^4 + 4x^3 - x^2 + 2x + 7$. Thus, $m'(x) = f_p(x) \odot A(x) \bmod p = x^{10} - x^9 + x^8 + x + 1 = m(x)$.

---

## 15.5.6  CRAMER-SHOUP CRYPTOSYSTEM

The Cramer-Shoup cryptosystem is an extension of the ElGamal cryptosystem. It is the first practical scheme that is proven to be CCA2 secure without random oracles.

**Facts:**

**1.** The first practical CCA2-secure public key cryptosystem was proposed by R. Cramer and V. Shoup [CrSh98] in 1998.

**2.** The security is based on the *Decisional Diffie-Hellman* (DDH) assumption: it is hard to distinguish $(g, h, g^x, h^x)$ from $(g, h, g^x, h^y)$, where $g, h \in \mathcal{Z}_p^*$ have large prime order $q$ and $x, y$ are randomly selected from $\mathcal{Z}_q$.

**3.** *Key Generation.*

- Let $p$ be a large prime. Let $\mathcal{G}$ be a subgroup of $\mathcal{Z}_p^*$ of large prime order $q$. (All arithmetic in $\mathcal{G}$ is performed modulo $p$.) Randomly select generators $g_1, g_2 \in \mathcal{G}$. Let $H$ be a hash function.

- Randomly select $x_1, x_2, y_1, y_2, z \in \mathcal{Z}_q$ and compute $c = g_1^{x_1} g_2^{x_2}$, $d = g_1^{y_1} g_2^{y_2}$, $h = g_1^z$.

- The public key is $(p, g_1, g_2, c, d, h, H)$ and the private key is $(x_1, x_2, y_1, y_2, z)$.

**4.** *Encryption.* To encrypt the plaintext $m \in \mathcal{G}$, choose a random $r \in \mathcal{Z}_q$ and compute $u_1 = g_1^r$, $u_2 = g_2^r$, $e = mh^r$, $\alpha = H(u_1, u_2, e)$, $v = c^r d^{r\alpha}$. The ciphertext is $(u_1, u_2, e, v)$.

**5.** *Decryption.* To decrypt the ciphertext $(u_1, u_2, e, v)$, the receiver first computes $\alpha = H(u_1, u_2, e)$ and verifies whether $v = u_1^{x_1 + y_1 \alpha} u_2^{x_2 + y_2 \alpha}$. If the verification fails, he rejects; otherwise, he outputs $m' = e/u_1^z$.

**6.** The verification of the decryption step follows, since $u_1^{x_1 + y_1 \alpha} u_2^{x_2 + y_2 \alpha} = u_1^{x_1} u_2^{x_2} \cdot (u_1^{y_1} u_2^{y_2})^\alpha = c^r d^{r\alpha} = v$ and $e = mh^r = mg_1^{zr} = mu_1^z$.

**7.** CCA2 security is attained because to pass the verification the encrypter has to construct $u_1$, $u_2$, $v$ that are consistent with a single $r$. To achieve this, he needs to know $r$ (hence the plaintext $e/h^r$). Thus, the decryption requests in a CCA2 attack are not useful.

**8.** *Plaintext Awareness* (PA) refers to the property that one must know the underlying plaintext when creating a ciphertext. This property is useful to guarantee some privacy properties such as deniability. The plaintext awareness of this scheme was shown in [De06]. The PA property of more general schemes appeared in [JiWa10].

**9.** The Cramer-Shoup cryptosystem has CCA2 security, but it assumes DDH. This assumption does not hold under quantum attacks.

**10.** It is important to verify that $u_1$, $u_2$, $e$, $v$ have prime order $q$. Otherwise, the scheme may be vulnerable to attacks in some circumstances.

**Examples:**

**1.** Let $p = 23$, $g_1 = 4$, $g_2 = 9$. Let $x_1 = 2$, $x_2 = 3$, $y_1 = 5$, $y_2 = 6$, $z = 7$. Compute $c = g_1^{x_1} g_2^{x_2} = 3$, $d = g_1^{y_1} g_2^{y_2} = 13$, $h = g_1^z = 8$. Then the public key is $(p, g_1, g_2, c, d, h) = (23, 4, 9, 3, 13, 8)$ and the private key is $(x_1, x_2, y_1, y_2, z) = (2, 3, 5, 6, 7)$.

**2.** In Example 1, to encrypt the message $M = 5$, select the random value $r = 8$ and compute $u_1 = g_1^r = 9$, $u_2 = g_2^r = 13$, $e = mh^r = 20$. Suppose $\alpha = H(u_1, u_2, e) = 9$ and so $v = c^r d^{r\alpha} = 13$. Thus, the ciphertext is $(u_1, u_2, e, v) = (9, 13, 20, 13)$. To decrypt the ciphertext $(9, 13, 20, 13)$, note that $u_1^{x_1 + y_1 \alpha} u_2^{x_2 + y_2 \alpha} = 13$, which is consistent with $v$. Decryption proceeds by evaluating $m = e/u_1^z = 20/9^7 = 5$.

---

## 15.5.7   PAILLIER CRYPTOSYSTEM

The Paillier cryptosystem is an encryption system that has been proposed as a primitive in electronic voting and cloud computing.

**Facts:**

**1.** The system was proposed by P. Paillier in 1999 [Pa99].

**2.** The security is based on the intractability of computing $n$th residue classes.

**3.** *Key Generation.*

- Randomly select two large primes $p, q$ of similar size so that $\gcd(pq, (p-1)(q-1)) = 1$. Let $n = pq$ and $\lambda = (p - 1)(q - 1)$.

- Randomly select $g \in \mathcal{Z}_{n^2}^*$ such that the order of $g$ is a multiple of $n$. This can be checked by determining if $\gcd(L(g^\lambda \mod n^2), n) = 1$, where $L(x) = \frac{x-1}{n}$.

- The public key is $(n, g)$ and the private key is $(p, q)$.

**4.** *Encryption.* To encrypt the message $m \in \mathcal{Z}_n$, randomly select $r \in \mathcal{Z}_n^*$. Compute the ciphertext $c = g^m \cdot r^n \bmod n^2$.

**5.** *Decryption.* To decrypt the ciphertext $c \in \mathcal{Z}_{n^2}$, compute

$$m' = \frac{L(c^\lambda \bmod n^2)}{L(g^\lambda \bmod n^2)} \bmod n.$$

**6.** The correctness of the algorithm makes use of the fact that the order of $g$ is a multiple of $n$ in $\mathcal{Z}_{n^2}^*$, together with using the binomial theorem on $(1 + n)^x \bmod n^2$. Details can be found in [Pa99].

**7.** The ciphertext is double the size of the message. This is a considerable practical disadvantage.

**8.** The system is *additively homomorphic* (§15.6.4).

**9.** The Paillier cryptosystem is not secure under a quantum attack.

**Examples:**

**1.** Let $p = 149, q = 179$. Then $n = pq = 26671$, $\lambda = (p-1)(q-1) = 26344$. Choose $g = 5$. Note that $\gcd(L(g^\lambda \bmod n^2), n) = \gcd(1213, 26671) = 1$. The public key is therefore $(n, g) = (26671, 5)$ and the private key is $(p, q) = (149, 179)$.

**2.** In Example 1, let the message be $m = 67$. Randomly choose $r = 81$. The ciphertext is $c = g^m \cdot r^n \bmod n^2 = 5^{67} \cdot 81^{26671} \bmod 26671^2 = 577656191$.

**3.** To decrypt the ciphertext 577656191, compute

$$m' = \frac{L(577656191^{26344} \bmod 26671^2)}{L(5^{26344} \bmod 26671^2)} \bmod 26671 = \frac{1258}{1213} \bmod 26671 = 67.$$

Note that the denominator above was calculated in Example 1 during the key setup stage.

---

### 15.5.8  GOLDWASSER-MICALI CRYPTOSYSTEM

The Goldwasser-Micali scheme is a probabilistic public key cryptosystem that is also the first provably secure system.

**Facts:**

**1.** The system was proposed by S. Goldwasser and S. Micali in 1982 [GoMi82].

**2.** The security is based on the intractability of the *Quadratic Residuosity Problem*. Namely, let $n \geq 3$ be an odd composite integer and let $J_n$ be the set of all $a \in \mathcal{Z}_n^*$ having Jacobi symbol 1. Given $b \in J_n$, determine whether or not $b$ is a quadratic residue modulo $n$.

**3.** *Key Generation.*

- Randomly select two large primes $p$ and $q$ of roughly the same size. Compute $n = pq$.
- Select $y \in \mathcal{Z}_n$ such that the *Legendre symbol* $\left(\frac{y}{p}\right) = \left(\frac{y}{q}\right) = -1$, so that the Jacobi symbol $\left(\frac{y}{n}\right) = 1$.
- The public key is $(n, y)$ and the private key is $(p, q)$.

**4.** *Encryption.* Let the message $m$ be represented as a binary string $m = m_1 m_2 \ldots m_t$.

- For each bit $m_i$, randomly select $x_i \in \mathcal{Z}_n^*$. If $m_i = 1$, then compute $c_i = yx_i^2 \bmod n$; otherwise, compute $c_i = x_i^2 \bmod n$.
- The ciphertext is $c = (c_1, c_2, \ldots, c_t)$.

**5.** *Decryption.* To decrypt the ciphertext $c = (c_1, c_2, \ldots, c_t)$,

- For each $c_i$, decide whether it is a quadratic residue modulo $n$ by evaluating the Legendre symbol $e_i = \left(\frac{c_i}{p}\right)$. If $e_i = 1$, then set $m_i' = 0$. Otherwise, set $m_i' = 1$.
- The decrypted message is $m' = m_1' m_2' \ldots m_t'$.

**6.** *Correctness.* When $m_i = 0$, $c_i$ is a quadratic residue modulo $n$. However, when $m_i = 1$, $c_i$ is a pseudosquare modulo $n$ and is revealed when computing the Legendre symbol modulo $p$.

**7.** The system is insecure under a quantum attack due to the ability to factorize $n$.

**8.** The system was developed to demonstrate a provably secure system. Similar to the Paillier system in §15.5.7, message expansion is a practical drawback, in this case with expansion by a factor of $\log_2 n$.

## 15.5.9  IDENTITY-BASED ENCRYPTION

Identity-based encryption uses some unique information about the identity of a user as the public key, thus eliminating the need to manage public key certificates. It uses some central authority to distribute the secret key to the user.

**Facts:**

**1.** Identity-based encryption (IBE) is a public key cryptosystem that uses an arbitrary string (e.g., email address or name) as the public key. It was first proposed by A. Shamir [Sh84b] in 1984. The motivation of this system is to remove the task of managing public key certificates. The first fully functional construction was proposed by D. Boneh and M. Franklin [BoFr01] in 2001.

**2.** Let $G_1$ and $G_2$ be multiplicatively-written cyclic groups of prime order $p$. A bilinear pairing $\hat{e}$ on $(G_1, G_2)$ is an efficiently computable map $\hat{e} : G_1 \times G_1 \to G_2$ satisfying

- *Bilinearity:* $\forall g, h \in G_1, \forall a, b \in \mathcal{Z}_p, \hat{e}(g^a, h^b) = \hat{e}(g, h)^{ab}$; and
- *Non-degeneracy:* $\hat{e}(g, h) = 1$ implies that $g = 1$ or $h = 1$.

**3.** *Key Generation.* The system generator determines $p, G_1, G_2, \hat{e}$, chooses a generator $g$ of $G_1$, chooses two hash functions $G : \{0,1\}^* \to \{0,1\}^n$ and $H : \{0,1\}^* \to G_1$, a random $s \in \mathcal{Z}_p$, and sets $P_{pub} = g^s$. The system parameters are

$$(p, g, G_1, G_2, \hat{e}, G, H, P_{pub}).$$

The system master key is $s$.

**4.** *Extract.* For any identity ID, the system generates its private key as $d_{ID} = H(ID)^s$.

**5.** *Encryption.* To encrypt $M \in \{0,1\}^n$ under identity ID, randomly select $r \in \mathcal{Z}_p$ and compute the ciphertext as

$$(C_1, C_2) = \left(g^r,\ M \oplus G[\hat{e}(H(ID)^r, P_{pub})]\right).$$

**6.** *Decryption.* To decrypt $(C_1, C_2)$ for identity ID, compute

$$M' = C_2 \oplus G[\hat{e}(d_{\mathsf{ID}}, C_1)].$$

**7.** The correctness of this scheme can be easily seen. In the encryption step,

$$\hat{e}(H(\mathsf{ID})^r, P_{pub}) = \hat{e}(H(\mathsf{ID}), g)^{rs} = \hat{e}(H(\mathsf{ID})^s, g^r) = \hat{e}(d_{\mathsf{ID}}, C_1),$$

where the last value is the quantity evaluated in the decryption step.

**8.** Security relies on the *Computational Bilinear Diffie-Hellman* (CBDH) assumption: given $(g^a, g^b, g^c)$, it is hard to compute $\hat{e}(g, g)^{abc}$. If this assumption does not hold, then from $g^r$, $P_{pub} = g^s$ and $H(\mathsf{ID}) = g^w$ (for unknown $w$), an attacker can compute $\hat{e}(g, g)^{rsw}$ which is $\hat{e}(H(\mathsf{ID})^r, P_{pub})$. This can then be used to recover $M$.

**9.** The Boneh-Franklin scheme assumes that the output of hash functions $G$ and $H$ is random so that no information about the input can be deduced from the output. Such idealized hash functions are called *random oracles*.

**10.** Although this assumption is reasonable in practice, it is known that security in the random oracle model does not necessarily hold when the random oracle is replaced by a concrete hash function.

**11.** CCA2-secure IBE without random oracles was constructed by Boneh and Boyen [BoBo04].

**12.** The Boneh-Franklin scheme is not secure against an adaptive chosen ciphertext (CCA2) attack. Given the ciphertext $(C_1, C_2)$, one can easily deduce the ciphertext for the message $M \oplus 1$. However, the scheme can be converted to a CCA2-secure scheme by a Fujisaki-Okamoto transformation [FuOk99].

**13.** Although IBE can eliminate the need to certify public keys, it still does not remove the need for public key management. For instance, if a user's private key is leaked, an official revocation is still needed.

**14.** Just as with RSA, ElGamal, and elliptic curve schemes, the Boneh-Franklin scheme is broken quickly using quantum computers.

---

## 15.5.10    ATTRIBUTE-BASED ENCRYPTION

While identity-based encryption uses one unique informational item about a user as the public key, attribute-based encryption uses multiple attributes of a user to determine access to decryption.

**Facts:**

**1.** Attribute-based encryption is a type of public key encryption where the user's secret key and the ciphertext both depend on attributes.

- *Attribute examples:* membership, position title, age, color, height.

- *Motivation:* one can decrypt a ciphertext only if the attributes for the secret key satisfy the attribute constraints embedded in the ciphertext.

- *Example:* to decrypt a ciphertext for the female professors of a mathematics department, the decryption key needs to contain attributes of "female", "professor", and "mathematics department".

- *Collusion-resistant security:* if an attacker collects many keys, then he can decrypt a ciphertext only if one of these keys satisfies the attribute restriction of the ciphertext. In the previous example, a female student and a male mathematics professor cannot jointly decrypt the ciphertext.

- *Application:* in cloud access control, encrypted data is stored in a cloud server. A user who retrieves a ciphertext can decrypt it if and only if he has a key satisfying the encryption policy.

**2.** The concept of attribute-based encryption was first proposed by Sahai and Waters [SaWa05] and by Goyal et al. [GoEtal06]. In the former, a ciphertext can be decrypted if and only if the number of common attributes embedded in the decryption key and the ciphertext is greater than or equal to a threshold. The latter generalized the scheme: a ciphertext can be decrypted if and only if the attributes in the ciphertext satisfy the access structure embedded in the key.

**3.** The following are some basic notions:

- Let $e : G_1 \times G_1 \to G_2$ be a bilinear pairing, where $G_1$ and $G_2$ are multiplicatively-written groups of prime order $p$.

- The universe of attributes is the set $\{1, \ldots, n\}$. Each user $u$ has a subset of attributes $\mathcal{A}_u \subseteq \{1, \ldots, n\}$.

- A ciphertext $C$ is prepared using an attribute set $\mathcal{B}$; user $u$ can decrypt $C$ only if $|\mathcal{A}_u \cap \mathcal{B}| \geq d$ for a fixed threshold $d$.

- *Lagrange interpolation.* For a polynomial $q(x) \in \mathcal{Z}_p[x]$ of degree $d-1$ and a set $\mathcal{S} \subseteq \mathcal{Z}_p - \{0\}$ of size $d$, we have $q(0) = \sum_{i \in \mathcal{S}} q(i)\lambda_i$, where $\lambda_i = \prod_{j \in \mathcal{S} \setminus \{i\}} \frac{j}{j-i}$.

**4.** The Sahai-Waters attribute-based encryption scheme [SaWa05] is described in the following facts.

**5.** *System Setup.* The system manager selects $p, G_1, G_2, \hat{e}$ and a generator $g$ of $G_1$. Then he selects $t_1, \ldots, t_n, y$ randomly from $\mathcal{Z}_p$ and defines $h_1 = g^{t_1}, h_2 = g^{t_2}, \ldots, h_n = g^{t_n}$, $Y = e(g, g)^y$. The system parameters are $(p, G_1, G_2, \hat{e}, h_1, \ldots, h_n, Y)$ and the master key is $(t_1, \ldots, t_n, y)$.

**6.** *User Key Generation.* For user $u$, randomly select a polynomial $q(x) \in \mathcal{Z}_p[x]$ such that $q(0) = y$. For each $i \in \mathcal{A}_u$, compute $D_i = g^{\frac{q(i)}{t_i}}$. Then the private key for $u$ is $\{D_i\}_{i \in \mathcal{A}_u}$.

**7.** *Encryption.* To encrypt $m \in G_2$ with attribute set $\mathcal{B}$, the sender randomly selects $s \in \mathcal{Z}_p$ and computes the ciphertext

$$(\mathcal{B}, C_0, \{C_i\}_{i \in \mathcal{B}}) = (\mathcal{B}, mY^s, \{h_i^s\}_{i \in \mathcal{B}}).$$

**8.** *Decryption.* To decrypt a ciphertext $(\mathcal{B}, C_0, \{C_i\}_{i \in \mathcal{B}})$ for user $u$ where $|\mathcal{A}_u \cap \mathcal{B}| \geq d$, user $u$ finds a subset $\mathcal{S} \subseteq \mathcal{A}_u \cap \mathcal{B}$ of size $d$ and computes

$$m' = \frac{C_0}{\prod_{i \in \mathcal{S}} \hat{e}(D_i, C_i)^{\lambda_i}}.$$

**9.** The correctness of the decryption is immediate. Note that $C_0 = mY^s$. The denominator in the decryption step can be simplified as follows:

$$\prod_{i \in \mathcal{S}} \hat{e}(D_i, C_i)^{\lambda_i} = \prod_{i \in \mathcal{S}} \hat{e}(g^{\frac{q(i)}{t_i}}, g^{st_i})^{\lambda_i} = \hat{e}(g, g)^{s \sum_{i \in \mathcal{S}} q(i)\lambda_i} = \hat{e}(g, g)^{sq(0)} = Y^s.$$

**10.** The idea behind security is the following. To decrypt $m$, one needs to compute $Y^s$. If $|\mathcal{A}_u \cap \mathcal{B}| < d$, user $u$ does not have sufficiently many shares to recover $q(0) = y$ in the exponentiation $[e^s(g,g)]^{q(0)} = Y^s$ (due to $s$ in $Y^s$, the recovery $Y^{s'}$ in other ciphertexts will be useless). Hence, decryption will be infeasible. Any two users $u_1, u_2$ will have different polynomials $q_1(x), q_2(x)$ when generating their private keys. Hence, it is not productive for them to collude.

**11.** Security relies on the *Decisional Bilinear Diffie-Hellman* (DBDH) assumption, which states that it is hard to distinguish $(g, g^a, g^b, g^c, \hat{e}(g,g)^{abc})$ from $(g, g^a, g^b, g^c, \hat{e}(g,g)^r)$, where $a, b, c, r$ are uniformly random in $\mathcal{Z}_p$. This assumption does not hold under quantum attacks.

# 15.6  CRYPTOGRAPHIC MECHANISMS

The cryptosystems described in the previous sections work with several core cryptographic applications that provide immediate, basic services in communications security. These core products are often components in high-level applications.

## 15.6.1  DIGITAL SIGNATURE

The notion of digital signature was invented by W. Diffie and M. Hellman in 1976. The authors conjectured about the existence of the mechanism. The use of a digital signature is to demonstrate the authenticity of an electronic document. The first practical algorithm was demonstrated through the RSA digital signature scheme.

**Definitions:**

A **digital signature** is an electronic analogue of traditional handwritten signatures.

When using public key methods, a digital signature has a private **signing key** $s$ for a signer and a public **verification key** $v$ for a verifier. The signing procedure is described by a signing algorithm $S$ and the verification procedure is described by a verification algorithm $V$.

A signature scheme is secure against a **public key only attack** if no attacker can forge a signature only with $V$.

A signature scheme is secure against a **known-message attack** if an attacker cannot forge a signature of a new message after given some messages and their corresponding signatures.

A signature scheme is secure against a **chosen-message attack** if an attacker cannot forge a signature of a new message when he adaptively chooses arbitrary messages and receives their corresponding signatures.

**Facts:**

**1.** Signatures must be verifiable in the sense that, should a dispute arise as to whether a party signed a document (caused by either a lying signer trying to *repudiate* a signature it did create, or a fraudulent claimant), an unbiased third party can resolve the matter equitably, without requiring access to the signer's secret information (private key).

**2.** Most of today's digital signature algorithms employ techniques from public key cryptography. It is possible to use symmetric key cryptography to achieve this goal, but it requires the use of a trusted third party, or new keying materials for each signature.

**3.** A secure digital signature should be resistant to the three types of attacks: public key only attack, known-message attack, and chosen-message attack.

**4.** There is a digital signature scheme for each of the public key schemes given in §15.5. There are also other approaches as described in this section.

**5.** *RSA signature scheme.* The plain RSA signature scheme is dual to the RSA public key cryptosystem.

- *Key generation.* The signing key is $d$, and the verification key is $(n, e)$, which are generated in the same manner as in RSA encryption.

- *Signing.* To sign a message $m \in \mathcal{Z}_n$, the signer computes the signature as $sig = m^d \bmod n$.

- *Verifying.* To verify the signature $sig$, check if $sig^e \equiv m \pmod{n}$.

**6.** Given an arbitrary message $m$, since $d$ is unknown, it is hard to forge the signature $m^d \bmod n$. However, it is easy to forge a signature for some messages. For instance, $sig$ is a valid signature of message $sig^e \bmod n$ since $(sig^e)^d \equiv sig \pmod{n}$.

**7.** To avoid this forging attack, one should first hash the message $m$. That is, if $H : \{0,1\}^* \to \{0, \ldots, n-1\}$ is a hash function, the actual signature of $m$ is defined as $sig = H(m)^d \bmod n$. The verification is modified accordingly. Assuming that $H$ is a random oracle and that the RSA assumption holds, the hashed RSA signature is proven secure against chosen-message attacks.

**8.** The *ElGamal digital signature scheme* is a randomized signature algorithm that generates digital signatures on binary messages with arbitraty length. It requires a hash function $h : \{0,1\}^* \to \mathcal{Z}_p$, where $p$ is a large prime number.

- *Key Generation.* As in the ElGamal cryptosystem, the public key is $(p, g, h)$ and the private key is $d$.

- *Signing.* To sign a message $m \in \{0,1\}^*$, the signer randomly selects $k$ with $1 \leq k \leq p-2$ with $\gcd(k, p-1) = 1$, and then computes $X = g^k \bmod p$ and $s = k^{-1}(h(m) - dX) \bmod p - 1$. The signature is $(X, s)$.

- *Verifying.* To verify the signature $(X, s)$, compute $v_1 = h^X \cdot X^s \bmod p$ and $v_2 = g^{h(m)} \bmod p$. Accept the signature if and only if $v_1 = v_2$.

**9.** The correctness of the ElGamal signature algorithm can be easily seen since $ks \equiv h(m) - dX \pmod{p-1}$. That is, $h(m) \equiv dX + ks \pmod{p-1}$ so that $v_2 \equiv g^{h(m)} \equiv g^{dX+ks} \equiv (g^d)^X \cdot (g^k)^s \equiv h^X \cdot X^s \equiv v_1 \pmod{p-1}$.

**10.** The security of the ElGamal scheme is based on the discrete logarithm assumption in $\langle g \rangle$.

**11.** The *Digital Signature Algorithm* (DSA) was adopted in 1994 by the U.S. government as a signature standard [Fi13]. It is a variant of the ElGamal signature scheme and is described as follows:

- *Setup.* The system generator selects a large prime $p$ of bitlength at least 3072 such that $p-1$ has a 256-bit prime factor $q$. It also selects an element $g \in \mathcal{Z}_p^*$ of order $q$.

- *Key generation.* Select $x$ from $\mathcal{Z}_q$ randomly and compute $h = g^x \bmod p$. The signing key is $x$ and the verification key is $(p, q, g, h)$.

- *Signing.* To sign a message $m$ with signing key $x$, the signer randomly selects $k$ from $\mathcal{Z}_q$ and computes

$$r = (g^k \bmod p) \bmod q \ \ \text{and} \ \ s = k^{-1}(H(m) + xr) \bmod q.$$

  The signature of $m$ is $(r, s)$.

- *Verifying.* To verify a signed message $(m, (r, s))$, let $w = s^{-1} \bmod q$. Check that

$$r \stackrel{?}{=} [(g^{H(m)}h^r)^w \bmod p] \bmod q.$$

  If the equality holds, the signature is accepted; otherwise, it is rejected.

**12.** The correctness of the signature can be easily seen since

$$(g^{H(m)}h^r)^w \equiv g^{(H(m)+rx)w} \equiv g^k \pmod{p}.$$

**13.** The security of DSA is based on the discrete logarithm assumption in $\langle g \rangle$.

**14.** An adaptation of elliptic curve cryptography to DSA is described in ANSI X9.62 as the *Elliptic Curve Digital Signature Algorithm* (ECDSA).

**15.** The short signature scheme known as *BLS signature* was proposed by D. Boneh, B. Lynn, and H. Shacham [BoLySh04]. The scheme is built using a bilinear pairing.

Namely, let $G_1$ and $G_2$ be multiplicatively-written cyclic groups of prime order $p$. Let $g$ be a generator of $G_1$ and let $\hat{e} : G_1 \times G_1 \to G_2$ be a bilinear pairing (see §15.5.9). The BLS scheme is specified as follows:

- *Setup.* The system generator selects $p, G_1, G_2, g$ and a hash function $H : \{0, 1\}^* \to G_1$.

- *Key generation.* Select $x \in \mathcal{Z}_q$ randomly and compute $h = g^x$. The verification key is $(G_1, G_2, p, g, h)$ and the signing key is $x$.

- *Signing.* To sign a message $m$, the signer computes $\sigma = H(m)^x$.

- *Verifying.* To verify the signed message $(m, \sigma)$, a verifier checks if $\hat{e}(\sigma, g) \stackrel{?}{=} \hat{e}(H(m), h)$. If equality holds, the signature is accepted; otherwise, it is rejected.

**16.** The correctness of the BLS signature follows from bilinearity of $\hat{e}$.

**17.** It was proven in [BoLySh04] that BLS is secure against chosen-message attack, assuming that $H$ is a random function and that the CDH assumption in $G_1$ holds.

## Examples:

**1.** *RSA Signature.* Using the RSA key pair $(n, e) = (4399, 23)$ and $d = 927$, to sign the message $m = 19$, compute $sig = m^d \bmod n = 19^{927} \bmod 4399 = 4155$. To verify the signature $sig = 4155$, compute $sig^e \equiv 4155^{23} \equiv 19 \equiv m \pmod{4399}$.

**2.** *ElGamal Signature.* Let the ElGamal parameters be $(p, g, h) = (3767, 7, 3467)$ and $d = 15$. Suppose we need to sign the message $m$ where $h(m) = 125$. Randomly choose the value $k = 87$. Note that $\gcd(87, 3766) = 1$. Compute $X = 7^{87} \bmod 3767 = 3113$ and $s = k^{-1}(h(m) - dX) \bmod p - 1 = 87^{-1}(125 - 15 \cdot 3113) \bmod 3766 = 3274$. The signature is $(3113, 3274)$. To verify, compute $v_1 = h^X \cdot X^s \bmod p = 3467^{3113} \cdot 3113^{3274} \bmod 3767 = 3030$, and $v_2 = g^{h(m)} \bmod p = 7^{125} \bmod 3767 = 3030$. Note that $v_1 = v_2$.

**3.** *DSA Key Setup.* Let $p = 15464731$. Since $p - 1 = 15464730 = 2 \cdot 3 \cdot 5 \cdot 17 \cdot 30323$, let $q = 30323$. Note that 2 is a generator of $\mathcal{Z}_p^*$. Then $g = 2^{2 \cdot 3 \cdot 5 \cdot 17} = 2241624$ has order $q$. Consider $x = 5$. Then $h = g^x \bmod p = 6922386$. The signing key is $x = 5$ and the verification key is $(p, q, g, h) = (15464731, 30323, 2241624, 6922386)$. Note that DSA parameters have much larger specified bitlength.

**4.** *DSA Signing.* Suppose the message $m$ has hash value $H(m) = 10628$. The signer randomly selects $k = 14 \in \mathcal{Z}_q$ and computes $r = (g^k \bmod p) \bmod q = 5120962 \bmod q = 26698$ and $s = k^{-1}(H(m) + xr) \bmod q = 14626$. The signature for $m$ is $(r, s) = (26698, 14626)$.

**5.** *DSA Verification.* To verify the signed message $m$ having the signature $(r, s) = (26698, 14626)$, evaluate $w = s^{-1} \bmod q = 18375$. Then $[(g^{H(m)}h^r)^w \bmod p] \bmod q = 5120962 \bmod q = 26698 = r$.

## 15.6.2  MESSAGE AUTHENTICATION CODE

Message authentication code, also called keyed hash function, is a cryptographic primitive that protects the authenticity and integrity of a message.

**Definitions:**

A **message authentication code** (MAC) scheme is a keyed function $F_K(\cdot)$ acting on messages $m$ such that when $m$ and $F_K(m)$ are transmitted, the intended recipient will be assured of the authenticity of $m$.

If a computationally bounded attacker has obtained the MACs of $n$ messages of its choice but still cannot forge the MAC of a new message with probability $\epsilon$, the system is called an **$(n, \epsilon)$-secure MAC**.

If $\epsilon$ is negligible, and $n$ and the runtime of an attacker are polynomially bounded, then the MAC scheme is **existentially unforgeable**.

A **universal$_2$ hash function** from $S$ to $V$ is a keyed hash function $H_K$ such that for any pair of distinct inputs $x, y$, $H_K(x) = H_K(y)$ holds for at most $\frac{1}{|V|}$ of all possible keys.

**Facts:**

**1.** In a MAC scheme, the sender and the receiver share the secret key $K$. This primitive was introduced by E. Gilbert, F. MacWilliams, and N. Sloane [GiMaSl74].

**2.** The security notion for MAC schemes is similar to that of digital signature schemes. An attacker may request the MACs of some messages of its choosing, and its objective is to forge the MAC of some new message.

**3.** *Onetime MAC.* Let $p$ be a large prime and let the key $K = (a, b)$ be chosen randomly from $\mathcal{Z}_p^2$. Consider the MAC scheme $F_{a,b}(x) = ax + b \bmod p$, where $a, b \in \mathcal{Z}_p$.

  - It is immediate that this scheme is $(1, 1/p)$-secure.
  - This is a special case of a universal$_2$ hash function used as a MAC by Wegman and Carter [WeCa81].

**4.** *HMAC.* This is a message authentication code proposed in 1996 by M. Bellare, R. Canetti, and H. Krawczyk [BeCaKr96]. It is standardized as FIPS PUB 198-1.

Let $\mathrm{HMAC}_K(m) = H\big((K \oplus opad)||H((K \oplus ipad)||m)\big)$, with $H$ an iterated hash function (e.g., MD5, SHA-1, SHA-256), $K$ is the secret key, *opad* is the constant 0x5c5c5c...5c5c, and *ipad* is the constant 0x363636...3636 such that *opad* and *ipad* have the same length as $K$.

**5.** *CBC-MAC.* This is a message authentication code obtained from the CBC encryption mode. The construction is as follows. Let $K = (K_1, K_2)$ be the secret key and let $(E, D)$ be a block cipher of blocklength $\ell$.

- Suppose $m$ is the message to be authenticated. The sender divides $m$ into $\ell$-bit blocks $m_1||m_2||\dots||m_t$.
- Set $V_0 = 0$. For $i = 0,\dots,t$, compute $V_i = E_{K_1}(V_{i-1} \oplus m_i)$.
- Define the MAC of $m$ to be $E_{K_2}(V_t)$.

**6.** The plain CBC-MAC is simply $V_t$. However, this construction is not secure. Indeed, given the MAC $\sigma$ of $m_1||m_2||\dots||m_t$ and the MAC $\sigma'$ of $m_1'||m_2'|||\dots||m_s'$, one can compute $m_1'' = V_t \oplus m_1'$. Then the MAC of

$$m_1||\dots||m_t||m_1''||m_2'||m_3'||\dots||m_s'$$

is $\sigma'$, which follows from the fact $m_1'' \oplus \sigma = 0 \oplus m_1'$.

**7.** An alternative patch to the plain CBC-MAC is to prepend the length of $m$ to the input. The problem with this variant is that the message length must be known before the processing begins.

## 15.6.3   KEY EXCHANGE

Two communicating parties wishing to exchange messages using symmetric key encryption must first establish an encryption key. Key exchange is a mechanism that allows two parties to securely share a secret key so that no one else can obtain a copy.

**Facts:**

**1.** A shared key is required whenever a symmetric key system is used. When a key is used for just one communcations session, such a key is called a *session key*.

**2.** The idea of key exchange in the modern era originated in the landmark Diffie-Hellman paper [DiHe76].

**3.** Generally, key exchange is assumed to take place over a public channel, therefore, any key exchange protocol must assume that all messages are available to any person including the adversary. The ultimate goal of the adversary is to compromise the secret key shared between communicating parties.

**4.** The security model describes what information the adversary can obtain and what actions it can take. Two types of security are considered: *passive security* and *active security*. In *passive security*, an adversary can observe messages transmitted over the channel but cannot modify them. In *active security*, an adversary can modify, delete, and replay messages transmitted over the channel. In both cases, the goal is to compute the key shared between the legitimate communicating parties.

**5.** *Diffie-Hellman key exchange.* Malcolm J. Williamson of GCHQ came up with the Diffie-Hellman key exchange in 1974. His work was unknown to the public until its declassification in 1997.

Sender (A) and Receiver (B) agree on large primes $p$ and $q$, where $q$ divides $p - 1$, and an element $g$ of order $q$ in $\mathcal{Z}_p^*$. The steps of this protocol are as follows:

- A selects $x$ randomly, where $1 \le x \le q - 1$, and sends $g^x$ to B.
- Upon receiving $g^x$, B selects a random value $y$, $1 \le y \le q - 1$, and sends $g^y$ to A. Then B computes the shared key as $sk = (g^x)^y = g^{xy}$.
- Upon receiving $g^y$, A computes the shared key as $sk = (g^y)^x = g^{xy}$.

**6.** The protocol is passively secure if the Computational Diffie-Hellman (CDH) assumption holds. That is, it is computationally infeasible to compute $g^{xy}$ from $g^x$ and $g^y$. However, the protocol is not actively secure.

**7.** Nonetheless, the Diffie-Hellman key agreement protocol serves as a basic mechanism for constructing actively-secure key establishment protocols.

**8.** The *Internet Key Exchange* (IKE) protocol [HaCa98] is used in IPsec standards. Its security against active attacks has been proven by R. Canetti and H. Krawczyk [CaKr02].

Let $p, q$ be large primes with $q \mid p - 1$ and let $g$ be an element of order $q$ in $\mathcal{Z}_p^*$. Suppose $(Sig, Ver)$ is a signature scheme and let MAC be a message authentication code. Let $F_K(\cdot) : \{0,1\}^* \to \{0,1\}^{2\lambda}$ be a keyed function (called a *pseudorandom function*) whose outputs appear random to an attacker who does not know $K$. The steps of this protocol are the following:

- *Setup.* Each user $i$ generates a signing key $s_i$ and verification key $v_i$ of $(Sig, Ver)$. Then $s_i$ is the private key and $v_i$ is the public key for user $i$.

- *Protocol.* The key exchange between user $i$ and user $r$ proceeds as follows:
    - ⋄ User $i$ selects a random $x, 1 \leq x \leq q - 1$, and sends $g^x$ and a session-id $s$ to user $r$.
    - ⋄ User $r$ selects a random $y, 1 \leq y \leq q - 1$, computes $(k_0, k_1) = F_{g^{xy}}(0)$, and sends
    
    $$(s, g^y, Sig_{s_r}(1||s||g^x||g^y), \mathrm{MAC}_{k_1}(1||s||r)) = (s, \beta, \sigma, \tau)$$
    
    to user $i$.
    - ⋄ Upon receiving $(s, \beta, \sigma, \tau)$, user $i$ computes $(k_0', k_1') = F_{\beta^x}(0)$ and verifies that $Ver_{v_r}(\sigma) = 1$ and $\tau = \mathrm{MAC}_{k_1'}(1||s||r)$. If the verification passes, user $i$ defines $k_0'$ as the shared key and sends
    
    $$(s, Sig_{s_i}(0||s||g^y||g^x), \mathrm{MAC}_{k_1'}(0||s||i)) = (s, \sigma', \tau')$$
    
    to user $r$; otherwise, user $i$ rejects.
    - ⋄ Upon receiving $(s, \sigma', \tau')$, user $r$ verifies that $\sigma'$ and $\tau'$ are valid. If so, user $r$ accepts $k_0$ as the shared key; otherwise, user $r$ rejects.

**9.** The IKE protocol is essentially an authenticated variant of the Diffie-Hellman key exchange.

**10.** The protocol guarantees that $g^y$ and $g^x$ originated from the legitimate parties. $g^y$ and $g^x$ are cryptographically bound in $\sigma'$ and $\sigma$ to guarantee that both parties see the same $(g^y, g^x)$.

**11.** The purpose of the session id $s$ is to prevent attacks that combine data from different sessions.

**12.** The MAC is used to ensure that the other party has indeed computed the session key. This is called *explicit authentication*.

**13.** *Password-based key exchange.* Passwords are probably the most popular mechanism for access control. Naturally, key exchange using a password as the user's long-term secret is useful.

**14.** M. Bellare, D. Pointcheval, and P. Rogaway [BePoRo00] proposed a provably secure password-based key exchange protocol that is based on the encrypted key exchange (EKE) technique of Bellovin and Merritt [BeMe92].

Let $\pi$ be a password shared between Alice and Bob. Let $p, q$ be large primes with $q \mid p - 1$, and let $g$ be an element of order $q$ in $\mathcal{Z}_p^*$. Let $E_\pi(\cdot) : G \to G$ be a symmetric key encryption scheme with decryption algorithm $D_\pi(\cdot)$, and let $H$ be a hash function. The steps for the password-based key exchange protocol are as follows:

- Party A selects $x$ randomly from $\mathcal{Z}_q$ and sends $(A, E_\pi(g^x)) = (A, \beta)$ to party B.
- On receiving $(A, \beta)$, B recovers $g^x = D_\pi(\beta)$, randomly selects $y \in \mathcal{Z}_q$, defines $k = H(A, B, g^x, g^y, g^{xy})$, and sends $(B, E_\pi(g^y), H(k, 1)) = (B, \gamma, \tau_1)$ to A.
- On receiving $(B, \gamma, \tau_1)$, A recovers $g^y = D_\pi(\gamma)$, computes $k = H(A, B, g^x, g^y, g^{xy})$, and checks if $\tau_1 = H(k, 1)$. If yes, then A sends $\tau_2 = H(k, 2)$ to B and defines the session key to be $sk = H(k, 0)$; otherwise, A rejects.
- On receiving $\tau_2$, B verifies that $\tau_2 = H(k, 2)$. If yes, A defines the session key to be $sk = H(k, 0)$; otherwise, B rejects.

**15.** This protocol is actively secure. The security proof relies on the assumption that $H$ is a random oracle and $E$ is an ideal cipher. Researchers have made progress towards removing such assumptions (see [GeLi03], [GrKa10], [JiGo04], [KaOsYu01]).

---

## 15.6.4  HOMOMORPHIC ENCRYPTION

Homomorphic encryption is an encryption mechanism that allows certain computations to be carried out on ciphertext in such a way that decryption of the result yields the same quantity as would be obtained if the computations were performed on the original plaintext.

**Definition:**
An encryption scheme is **additively homomorphic** if the ciphertext of $m_1 + m_2$ can be obtained from ciphertexts of $m_1$ and $m_2$.

An encryption scheme is **multiplicatively homomorphic** if the ciphertext of $m_1 m_2$ can be obtained from ciphertexts of $m_1$ and $m_2$.

An encryption scheme is **fully homomorphic** if it is both multiplicatively and additively homomorphic.

**Facts:**
**1.** *RSA cryptosystem is multiplicatively homomorphic.* Given ciphertexts $c_1 = m_1^e \bmod n$ and $c_2 = m_2^e \bmod n$, the ciphertext of $m = m_1 \cdot m_2 \bmod n$ is $c = c_1 \cdot c_2 \bmod n$. However, RSA is not additively homomorphic.

**2.** *ElGamal cryptosystem is multiplicatively homomorphic.* Given ciphertexts $c_1 = (g^r, m_1 h^r)$ and $c_2 = (g^s, m_2 h^s)$, the ciphertext of $m = m_1 \cdot m_2$ is $c = (g^{r+s}, m_1 m_2 h^{r+s})$. ElGamal is not additively homomorphic.

**3.** *Paillier cryptosystem is additively homomorphic.* Given ciphertexts $c_1 = g^{m_1} r_1^n$ and $c_2 = g^{m_2} r_2^n$, then $c_1 c_2 = g^{m_1 + m_2} (r_1 r_2)^n$.

**4.** It had long been an open question whether a fully homomorphic public key encryption cryptosystem existed. The existence of such a public key cryptosystem would be important because when such a cryptosystem is used, any computation can be formulated as a circuit of addition and multiplication operations, whereby a third party can compute any function of arbitrary ciphertexts.

**5.** In 2009, C. Gentry solved this open question by constructing the first fully homomorphic public key cryptosystem using lattice-based cryptography [Ge09]. Although this solved the open question, his system is far too slow to be practical.

**6.** An application of fully homomorphic encryption is *cloud computing* (§15.7.3).

**7.** All currently studied fully homomorphic cryptosystems are public key based.

**8.** In 2010, a second fully homomorphic cryptosystem using integer-based cryptography was developed by M. van Dijk, C. Gentry, S. Halevi, and V. Vaikuntanathan [vDEtal10]. Like Gentry's original cryptosystem, this approach is also impractical.

**9.** After the development of the first two fully homomorphic cryptosystems, the search for a practical fully homomorphic cryptosystem continued. Much progress has been made with the development of a second generation of fully homomorphic cryptosystems, but as of 2017 work remains before there is a fully homomorphic cryptosystem that can be used in real-world applications.

**10.** In commercial applications, it may be desirable to chain certain services together while keeping the secrecy of the data intact. Homomorphic encryption can be used to achieve privacy where information is processed in the encrypted form.

**Example:**

**1.** Homomorphic encryption can be used in health care where digital patient data from multiple sources can be stored securely while computations are being performed without violating a patient's right to confidentiality.

## 15.6.5   SECRET SHARING

Secret sharing is a mechanism that allows a group of users to share a secret such that a sufficient number of them can recover the secret whereas an insufficient number of users is unable to obtain any information about the secret.

**Definition:**

In a $(t, n)$-*secret sharing* scheme, each of $n$ users receives a **share** of a secret $s$ so that any $t$ or more of them can recover $s$, but any group of fewer than $t$ parties cannot determine any information about $s$.

**Facts:**

**1.** Secret sharing was independently invented by Shamir [Sh79] and Blakley [Bl79].

**2.** *Shamir's secret sharing scheme.* Let $p$ be a prime; the secret is $s \in \mathcal{F}_p$.

- *Sharing.* The dealer chooses a random polynomial $f(x)$ of degree $t - 1$ with constant term $s$. That is, $f(x) = a_0 + a_1 x + \cdots + a_{t-1} x^{t-1}$ where $a_i$ for $i \geq 1$ is uniformly random over $\mathcal{F}_p$ and $a_0 = s$. The share for party $i$ is $s_i = f(i)$.

- *Recovery.* When at least $t$ parties wish to recover the secret $s$, they jointly do the following

  ◇ Choose some arbitrary $t$ parties $i_1, \ldots, i_t$ from the set of colluders.
  ◇ Since $s_{i_j} = f(i_j)$ for $j = 1, \ldots, t$, they use Lagrange interpolation to obtain

$$f(x) = \sum_{j=1}^{t} s_j \frac{\prod_{1 \leq k \leq t, k \neq j}(x - i_k)}{\prod_{1 \leq k \leq t, k \neq j}(i_j - i_k)}.$$

They then recover

$$s = f(0) = \sum_{j=1}^{t} s_j \frac{\prod_{1 \leq k \leq t, k \neq j} i_k}{\prod_{1 \leq k \leq t, k \neq j}(i_k - i_j)}.$$

- *Security.* If fewer than $t$ parties (say, parties $1, \ldots, t-1$) wish to recover $s$, they can form
$$s_j = a_0 + a_1 j + \cdots + a_{t-1} j^{t-1}, \ j = 1, \ldots, t-1.$$
Notice that $(s_1, \ldots, s_t)$ uniquely recovers $(a_0, \ldots, a_{t-1})$ and vice versa. Hence, $(s_1, \ldots, s_t)$ is in 1-1 correspondence with $(a_0, \ldots, a_{t-1})$. Since the latter is uniformly random over $\mathcal{F}_p^t$, so too is the former. Hence, given $(s_1, \ldots, s_{t-1})$, $a_0$ is uniformly random over $\mathcal{F}_p$ due to the randomness of $s_t$, and thus $(s_1, \ldots, s_{t-1})$ reveals nothing about $s$.

**3.** When $t = n$, an $(n, n)$-secret sharing is one where any $n$ parties can recover the secret while the absence of one or more parties will leak no information about $s$. In this case, a simpler scheme exists whereby the dealer chooses $n-1$ random shares $s_1, \ldots, s_{n-1}$ and defines the share $s_n = s - s_1 - \cdots - s_{n-1} \bmod p$.

**4.** *Blakley's secret sharing scheme.* The scheme uses a $t$-dimensional projective space $PG(t, \mathcal{F}_q)$. Let the secret be a random point $P \in PG(t, \mathcal{F}_q)$. The scheme is described as follows:

- *Sharing.* To share a secret $P \in PG(t, \mathcal{F}_q)$ among $n$ users, the dealer randomly chooses $n$ hyperplanes that contain $P$. Each user receives one of these hyperplanes as his share.

- *Recovery.* If any $v \geq t$ users want to recover $P$, $t$ of the users are selected to combine their shares to form $t$ linear equations, the solution of which determines $P$.

- *Correctness.* In the recovery, the $t$ equations must be linearly independent in order to recover $P$. This can be satisfied with high probability when $n$ is not too large while $q$ is large.

- *Security.* Any set of fewer than $t$ users can only form a linear system with fewer than $t$ equations. This determines a proper subspace of $PG(t, \mathcal{F}_q)$, but cannot determine $P$.

**5.** Compared to Shamir's scheme, the Blakley sharing scheme is not perfectly secure. That is, fewer than $t$ users can still obtain some information about the secret $P$ compared to randomly guessing $P$. The correctness also does not hold with probability 1 as there is a nonzero probability that some subset of fewer than $t$ shares (i.e., hyperplanes) can linearly generate another particular share.

**Examples:**

**1.** Shares of a missile launching code can be given to a list of 12 generals, where at least five of them need to collaborate to reveal the launch code.

**2.** A private company can use secret sharing to protect unauthorized fund transfers over a certain threshold by giving shares of the authorization code to multiple members of management.

## 15.6.6   CONFERENCE KEY DISTRIBUTION

Among the members of a group of users, there may be a need for a subgroup to perform secure communications. Conference key distribution was invented to achieve this purpose.

## Definitions:

A *conference key distribution* scheme is a mechanism whereby a trusted authority assigns a piece of secret data to each user so that any desired group of users can compute a common secret key (called a *group key*).

If a group of users compute their group key interactively, then the scheme is called an *interactive key distribution* (*key agreement* or *group key exchange*); otherwise it is called a *non-interactive key distribution* (or *key pre-distribution*).

If the total number of users is $n$, then a *non-interactive conference distribution scheme* (see Fact 2) secure against $t$ colluding users is called $(n, t)$-*secure*.

If the scheme is secure against computationally unbounded attackers, then it is called *unconditionally secure*; otherwise, it is called *computationally secure*.

## Facts:

**1.** Diffie-Hellman key exchange is an interactive key distribution scheme for a group of size two.

**2.** In a non-interactive scenario, security requires that colluding users should not be able to compute a group key when none of the colluding users is a group member.

- To ensure that an efficient construction is possible, the number of colluding users is usually bounded by a threshold $t$.

**3.** In an interactive scenario, security is more complex, and considers both passive attack and active attacks. An attacker can corrupt a number of users, and thereby obtain the secret data of these users.

- In a passive attack, an attacker can only observe the transmitted messages.

- In an active attack, an attacker can impersonate legitimate users and modify messages sent by users.

- In both cases, the attacker's objective is to learn a conference key that he is not supposed to know.

**4.** *Blom non-interactive key distribution.* Blom [Bl84] proposed a non-interactive conference key distribution scheme without public key components. His scheme is for subgroups of size two and is secure against a collusion of size $k$. Suppose there are $n$ users.

- *Key setup.* The key setup for users is based on the $(k+1)$-by-$n$ generator matrix $G$ of an $(n, k+1)$-MDS code over a finite field $\mathcal{F}_q$. Let $D$ be a $(k+1) \times (k+1)$ purely random but symmetric matrix over $\mathcal{F}_q$. Then the secret key $K_i$ for user $i$ is defined as the $i$th column of $DG$.

- *Conference key computation.* The conference key $k_{ij}$ for users $i$ and $j$ is the entry $(i, j)$ of the square matrix $(DG)^T G = G^T DG$. This is the inner product of the $i$th column of $DG$ and the $j$th column of $G$. User $i$ can compute $k_{ij}$ since $i$ possesses $K_i$ and $G$ is public. On the other hand, user $j$ can compute $k_{ij}$ as the inner product of the $i$th column of $G$ and the $j$th column of $DG$.

- *Security.* Blom's scheme is secure against a collusion of size $k$. The colluders possess $k$ columns of $DG$. However, since $G$ is the generator matrix of an $(n, k+1)$-MDS code, any $k+1$ columns of $G$ are linearly independent. This independence does not change by left multiplying with a full-ranked square matrix $D$ (ignoring the negligible probability that $\text{rank}(D) < k+1$). Hence, $K_i$ is independent of any collusion of size $k$.

**5.** *Polynomial interpolation non-interactive key distribution.* Blundo et al. [BlEtal98] extended Blom's scheme for conferences of arbitrary size $t$. Their scheme is based on polynomial representations. For security against collusions of size $k$, their scheme is described as follows:

- *Key setup.* Let $\sigma$ denote an arbitrary permutation over $\{1, \ldots, t\}$. Then define

$$P(x_1, \ldots, x_t) = \sum_{0 \leq j_1 \leq \cdots \leq j_t \leq k} a_{j_1 \cdots j_t} \sum_{\sigma} x_{\sigma(1)}^{j_1} \cdots x_{\sigma(t)}^{j_t},$$

  where $a_{j_1 \cdots j_t}$ is uniformly random over the finite field $\mathcal{F}_q$. Note that $P(x_1, \ldots, x_t)$ is symmetric. The secret key for user $i$ is defined as $K_i = P(i, x_2, \ldots, x_t)$ for $i = 1, \ldots, n$.

- *Conference key computation.* The conference key for a group $G = \{i_1, \ldots, i_t\}$ is $k_G = P(i_1, \ldots, i_t)$. User $i_\ell$ can compute $k_G = P(i_\ell, i_1, \ldots, i_{\ell-1}, i_{\ell+1}, \ldots, i_t)$. The scheme works because $(\ell, 1, \ldots, \ell-1, \ell+1, \ldots, t)$ is a permutation of $(1, \ldots, t)$ and $P(x_1, \ldots, x_t)$ is symmetric.

- *Security.* $(n, k)$-security requires that any collusion group $\{i_1^*, \ldots, i_k^*\} \subset \{1, \ldots, n\}$ is unable to compute a conference key $k_G$, when no member in $G$ colluded.

  ◇ Suppose $G = \{i_1, \ldots, i_t\}$. Let $P_0(x_1, \ldots, x_t)$ be any symmetric polynomial of degree $\leq k$ such that $P_0(i_u^*, x_2, \ldots, x_t) = K_{i_u^*}$ for $u = 1, \ldots, k$.

  ◇ In addition, note that $A(x_1, \ldots, x_t) = \prod_{j=1}^{k} \prod_{v=1}^{t} (x_v - i_j^*)$ is symmetric with degree $k$ and satisfies $A(i_j^*, x_2, \ldots, x_t) = 0$.

  ◇ Then, for any $\alpha \in \mathcal{F}_q$, $P_0(x_1, \ldots, x_t) + \alpha A(x_1, \ldots, x_t)$ can serve as a candidate of $P(x_1, \ldots, x_t)$ (in view of colluders $\{i_1^*, \ldots, i_k^*\}$), because when evaluated at $i_j^*$, it yields the result $K_{i_j^*}$.

  ◇ However, it has a conference key $P_0(i_1, \ldots, i_t) + \alpha \prod_{j=1}^{k} \prod_{v=1}^{t} (i_v - i_j^*)$ for $G$, which is uniformly random over $\mathcal{F}_q$ (due to the randomness of $\alpha$) as $i_v \neq i_j^*$. Thus, the colluders have no information about $k_G$.

**6.** *Fiat-Naor scheme.* Fiat and Naor [FiNa93] proposed a non-interactive key distribution scheme. This scheme is designed for $n$ users $\{1, \ldots, n\}$ and is unconditionally secure against $w$ colluders.

- *Key setup.* Let $P_1, \ldots, P_\nu$ be all the subsets of $\{1, \ldots, n\}$ of size at least $n-w$. Note that $\nu = \sum_{i=0}^{w} \binom{n}{i}$. Associate each $P_i$ with a random key $k_{P_i}$. The secret key for user $j$ is $K_j = \{k_{P_i} \mid j \in P_i, i = 1, \ldots, \nu\}$. Thus, $K_j$ has size $\sum_{i=0}^{w} \binom{n-1}{i}$.

- *Conference key computation.* For any group $G \subseteq \{1, \ldots, n\}$, the conference key is $k_G = \sum_{i: G \subseteq P_i} k_{P_i}$. Since every member $j \in G$ knows $k_{P_i}$ for any $P_i$ containing $G$, user $j$ can compute $k_G$.

- *Security.* The Fiat-Naor scheme is secure against $w$ colluders.

  ◇ Suppose the colluders are $\{i_1, \ldots, i_w\}$. Then, for any $G$ not containing any colluders, $k_G = \sum_{j: G \subseteq P_j} k_{P_j}$ is independent of $\{K_{i_1}, \ldots, K_{i_w}\}$.

  ◇ Notice that there exists $t$ such that $P_t = \{1, \ldots, n\} - \{i_1, \ldots, i_w\}$ and hence $k_{P_t}$ is independent of the keys of colluders. Since $i_j \notin G$ for any $j$, $G \subseteq P_t$. Thus, $k_{P_t}$ appears in the sum of $k_G$. So $k_G$ is independent of the keys of colluders.

**7.** *Tripartite key distribution from pairing.* Joux [Jo00] presented a one-round computationally secure key distribution scheme for groups of size three. It is based on a bilinear pairing $\hat{e} : G_1 \times G_1 \rightarrow G_2$, where $G_1, G_2$ are groups of prime order $p$.

- *Key setup.* Let $g \in G_1$ be a generator. Then each user $i$ randomly selects $a_i$, $1 \le a_i \le q - 1$, as its private key and publishes $Q_i = g^{a_i}$ as its public key.
- *Conference key computation.* For any group $\{i, j, \ell\}$, the conference key is $k_{\{i,j,\ell\}} = \hat{e}(g, g)^{a_i a_j a_\ell}$. User $i$ can compute this key using $a_i$ as $\hat{e}(Q_j, Q_\ell)^{a_i} = \hat{e}(g, g)^{a_i a_j a_\ell}$.
- *Security.* The security of Joux's scheme requires the *Computational Bilinear Diffie-Hellman* (CBDH) assumption.

**8.** *Burmester-Desmedt interactive key distribution scheme.* Burmester and Desmedt [BuDe94] proposed an interactive key distribution scheme. Their scheme is a generalization of the Diffie-Hellman key exchange to the multiparty setting. It provides a way to compute the conference key for an entire group of users. However, it is insecure when used to compute a group key for a proper subgroup using the same key setup.

Let $p$ and $q$ be large primes with $q \mid p - 1$. Let $g \in \mathcal{Z}_p^*$ be an element of order $q$.

- *Key setup.* For $i = 0, \ldots, n - 1$, user $i$ randomly selects a secret key $a_i$, $1 \le a_i \le q - 1$, and publishes the public key $A_i = g^{a_i}$.
- *Conference key computation.* User $i$ computes $X_i = (A_{i+1}/A_{i-1})^{a_i}$ and broadcasts $X_i$ to group members, where index arithmetic is performed modulo $n$ when necessary. The group key is defined as $k = g^{a_0 a_1 + \cdots + a_{n-2} a_{n-1} + a_{n-1} a_0}$. For any user $j$, notice that $A_{i-1}^{\ell a_i} X_i^{\ell-1} = g^{a_{i-1} a_i} \times A_{i+1}^{(\ell-1) a_i}$ for each $i$. Therefore, user $j$ can compute $k$ by computing

$$A_{j-1}^{a_j n} X_{j+1}^{n-1} X_{j+2}^{n-2} \cdots X_{j-2}^1 \tag{6}$$

$$= g^{a_{j-1} a_j} \times A_j^{a_{j+1}(n-1)} X_{j+2}^{n-2} \cdots X_{j-2}^1$$

$$= g^{a_{j-1} a_j + a_j a_{j+1}} \times A_{j+1}^{a_{j+2}(n-2)} X_{j+3}^{n-3} \cdots X_{j-2}^1$$

$$\vdots$$

$$= g^{a_{j-1} a_j + a_j a_{j+1} + \cdots + a_{j-2} a_{j-1}} = g^{a_0 a_1 + \cdots + a_{n-1} a_0}.$$

- *Security.* The Burmester-Desmedt scheme is secure against a passive outsider attack. The attacker knows all of $X_1, A_1, \ldots, X_{n-1}, A_{n-1}$.

  ◇ In a passive attack, the attacker cannot compute $g^{a_0 a_1 + \cdots + a_{n-1} a_0}$. Otherwise, from Eq. (6), the attacker can compute $A_{j-1}^{a_j} = g^{a_{j-1} a_j}$. Since $A_1, \ldots, A_{n-1}$ is computationally independent of $g^{a_0 a_1}, \ldots, g^{a_{n-1} a_0}$ and hence is computationally independent of $g^{a_{j-1} a_j}, X_1, \ldots, X_{n-1}$, the attacker must compute $g^{a_{j-1} a_j}$ solely based on $X_1, \ldots, X_{n-1}$. However, this is impossible because they are statistically independent.

  ◇ Since conference key distribution is usually used to help a subgroup derive a conference key, one might think that the Burmester-Desmedt scheme can also do the same as key derivation is completely compatible with any subgroup. However, once a user $i$ is involved in a group key computation for a group (say, $\{1, \ldots, n\}$), user $i$ can compute $A_{j-1}^{a_j}$ using the group key and Eq. (6). Then user $i$ can compute the conference key of any group that involves users $j - 1$ and $j$ (and not including user $i$).

## 15.6.7  OBLIVIOUS TRANSFER

There are a variety of reasons why someone may wish to send one of many pieces of information to a receiver without knowing which piece has been transferred. This task can be done using what is known as oblivious transfer.

**Definitions:**

*Oblivious transfer* (*OT*) is a protocol in which Alice has a set of bit strings as her input and Bob can choose only one string to be transferred to him and Alice cannot determine which string this is.

When Alice's set has $k$ elements, the protocol is called $OT_1^k$.

In a *k-out-of-n oblivious transfer*, Alice has a set of $n$ bit strings as her input, where Bob chooses $k$ strings to be transferred.

**Facts:**

**1.** Oblivious transfer originated in Rabin's work [Ra81], where he described a method to send a message $m$ from Alice to Bob with a success probability of $\frac{1}{2}$, while Alice does not know if Bob received the message or not.

- Alice generates RSA parameters $p, q, n, e, d$, and sends $n, e, m^e$ to Bob.

- Bob chooses a random value $x \in \mathcal{Z}_n^*$ and evaluates $c = x^2 \bmod n$. Note that since $x \in \mathcal{Z}_n^*$, there are four square roots of $c \bmod n$. He sends $c$ to Alice.

- With the knowledge of $p, q$, Alice finds a square root $y$ of $c$ easily, and sends $y$ to Bob.

- If $y$ is not $x$ or $-x$, then Bob can factor $n$ by evaluating $\gcd(x - y, n)$, and thus solve for $d$ to recover $m$. Otherwise, Bob cannot recover $m$ from $m^e$. Since there are four square roots, Bob recovers $m$ with probability $\frac{1}{2}$.

**2.** A key requirement of oblivious transfer is that Alice cannot learn which element Bob has chosen while Bob cannot learn any information about the remaining elements in the set.

**3.** Three constructions of Naor et al. [NaPi01], [NaPi05] are the simplest among all provably secure OT schemes.

**4.** The *one-out-of-two oblivious transfer protocol* ($OT_1^2$) is due to Naor and Pinkas [NaPi01].

**5.** Alice has two messages $M_0, M_1$. Bob has a bit $\sigma$ and wishes to learn $M_\sigma$. The protocol proceeds as follows:

- Let $p, q$ be large primes with $q \mid p - 1$ and let $g \in \mathcal{Z}_p^*$ be an element of order $q$. Bob randomly selects $a, b, r \in \mathcal{Z}_q$ and sends $A = g^a$, $B = g^b$, $C_\sigma = g^{ab}$, and $C_{1-\sigma} = g^r$ to Alice.

- Alice verifies that $C_0 \neq C_1$. Then she randomly selects $s_0, s_1, t_0, t_1 \in \mathcal{Z}_q$, and computes and sends

$$W_0 = (A^{s_0} g^{t_0}, M_0 C_0^{s_0} B^{t_0}), \ W_1 = (A^{s_1} g^{t_1}, M_1 C_1^{s_1} B^{t_1})$$

to Bob.

- Upon receiving $W_0 = (\alpha_0, \beta_0)$ and $W_1 = (\alpha_1, \beta_1)$, Bob computes $M_\sigma = \beta_\sigma / \alpha_\sigma^b$.

**6.** *Correctness.* Notice that $C_\sigma = A^b$. Thus, $(A^{s_\sigma} g^{t_\sigma})^b = C_\sigma^{s_\sigma} B^{t_\sigma}$ and $M_\sigma = \beta_\sigma / \alpha_\sigma^b$.

**7.** *Alice's privacy.* Since $C_0 \neq C_1$, $C_{\bar{\sigma}}$ is not the Diffie-Hellman value corresponding to $(A, B)$, where $\bar{\sigma} = 1 - \sigma$. Therefore, $\frac{r}{a} \neq \frac{b}{1}$. Thus, $C_{\bar{\sigma}}^{s_{\bar{\sigma}}} B^{t_{\bar{\sigma}}}$ is independent of $A^{t_{\bar{\sigma}}} g^{t_{\bar{\sigma}}}$. That is, $M_{\bar{\sigma}}$ is independent of Alice's knowledge.

**8.** The Decisional Diffie-Hellman (DDH) assumption states that it is hard to distinguish $(g^a, g^b, g^{ab})$ from $(g^a, g^b, g^r)$, when $a, b, r$ are randomly selected.

**9.** *Bob's privacy.* $\sigma$ is computationally independent of Alice's knowledge $(A, B, C_0, C_1)$ since distinguishing $C_0$ from $C_1$ given $(A, B)$ is the DDH problem.

**10.** *One-out-of-many oblivious transfer* ($\mathrm{OT}_1^k$). The $\mathrm{OT}_1^k$ extension of the $\mathrm{OT}_1^2$ protocol originated in [NaPi01]. Suppose Alice has messages $M_0, \ldots, M_{k-1}$, and Bob has an index $I \in \{0, \ldots, k-1\}$ and wishes to learn $M_I$. The protocol proceeds as follows:

- Let $p, q$ be large primes with $q \mid p-1$ and let $g \in \mathcal{Z}_p^*$ be an element of order $q$. Bob randomly selects $a, b, r_j \in \mathcal{Z}_q$ for $j \neq I$ and computes $A = g^a, B = g^b, C_I = g^{ab}, C_j = g^{r_j}, \forall j \neq I$. Finally, he sends $(A, B, C_0, \ldots, C_{k-1})$ to Alice.

- Alice verifies $C_i \neq C_j$ for $i \neq j$. Then she computes and sends

$$W_i = (A^{s_i} g^{t_i}, M_i C_i^{s_i} B^{t_i}), \ \forall i = 0, \ldots, k-1$$

  to Bob, where $s_i, t_i$ are randomly selected from $\mathcal{Z}_q$.

- Upon receiving $W_i = (\alpha_i, \beta_i), \ i = 0, \ldots, k-1$, Bob computes $M_I = \beta_I / \alpha_I^b$.

**11.** Correctness and privacy analyses for Alice and Bob are similar to $\mathrm{OT}_1^2$ in Fact 10.

**12.** *$\mathrm{OT}_1^k$ from a general setup assumption.* This construction of $\mathrm{OT}_1^k$ is due to Naor and Pinkas [NaPi05]. The scheme, built from an $\mathrm{OT}_1^2$ protocol and a one-way function, is described in the following facts.

**13.** Let $F_K(\cdot)$ be a *pseudorandom function*, i.e., a keyed function such that without the knowledge of $K$, one cannot distinguish outputs of $F_K(\cdot)$ from random values, even if one can adaptively learn the outputs of many other inputs. Pseudorandom functions can be constructed from one-way functions.

**14.** Suppose Alice has input $(M_0, \ldots, M_{2^\ell - 1})$ and Bob has an index $I$ and wishes to learn $M_I$. Let $a_0 \cdots a_{\ell-1}$ be the binary representation of $I$. The protocol proceeds as follows:

- Alice prepares $\ell$ pairs of secret keys for function $F$: $(K_i^0, K_i^1), \ i = 0, \ldots, \ell - 1$. For each $0 \leq J \leq 2^\ell - 1$, let the binary representation of $J$ be $j_0 \cdots j_{\ell-1}$. Alice computes $C_J = M_J \oplus \left( \oplus_{i=0}^{\ell-1} F_{K_i^{j_i}}(J) \right)$ and sends $(C_0, \ldots, C_{2^\ell - 1})$ to Bob.

- For each $i = 0, \ldots, \ell - 1$, Alice and Bob jointly execute $\mathrm{OT}_1^2$: Alice has input $(K_i^0, K_i^1)$ and Bob has $a_i$ as input. As a result, for each $i$, Bob learns $K_i^{a_i}$.

- Bob computes $M_I = C_I \oplus F_{K_i^{a_i}}(I)$.

**15.** *Correctness.* As $I = a_0 \cdots a_{\ell-1}$, it follows that $C_I = M_I \oplus \left( \oplus_{i=0}^{\ell} F_{K_i^{a_i}}(I) \right)$ by definition of $C_I$.

**16.** *Alice's privacy.* Notice that $F_{K_i^b}(0), \ldots, F_{K_i^b}(2^\ell - 1)$ are computationally independent for any $i$ and bit $b$ when $K_i^b$ is not given. In the $\mathrm{OT}_1^2$ protocol, for each $i$, Bob can learn at most one of $K_i^0, K_i^1$ (assume it is $a_i$). Then, for any $J \neq I = a_0 \cdots a_{2^\ell - 1}$, there exists at least one $i$, $j_i \neq a_i$, and hence $K_i^{j_i}$ is not learned by Bob. Thus $C_J$ is computationally independent of Bob's view, and therefore Bob cannot learn $M_J$ for any $J \neq I$.

**17.** *Bob's privacy.* Bob uses $I$ only in the execution of $\text{OT}_1^2$ for each $i$. That is, he inputs $a_i$. However, by Bob's privacy in $\text{OT}_1^2$, Alice learns nothing about $a_i$. Thus, Bob's privacy is guaranteed.

---

## 15.6.8   SECURE MULTIPARTY COMPUTATION

### Definition:

A **secure multiparty computation** is a protocol $\pi$ such that given $n$ parties $P_1, \ldots, P_n$, where each $P_i$ has a private input $x_i$, $\pi$ realizes a function $f(x_1, \ldots, x_n) = (o_1, \ldots, o_n)$ such that $P_i$ obtains the output $o_i$, $i = 1, \ldots, n$.

### Facts:

**1.** The requirements of secure multiparty computation include

- *Correctness.* After jointly executing $\pi$, each $P_i$ should obtain $o_i$ in $f(x_1, \ldots, x_n)$.
- *Security.* After executing $\pi$, $P_i$ cannot obtain any information about $(x_1, \ldots, x_n)$ beyond $(o_i, x_i)$. That is, besides the protocol output $o_i$ and input $x_i$, $P_i$ cannot learn anything from the messages he sees during the execution of $\pi$.

**2.** A set of parties $A \subseteq \{P_1, \ldots, P_n\}$ might collude and are called *corrupted parties.* Security requires that they cannot obtain anything from the execution of $\pi$ beyond $\{(o_i, x_i)\}_{P_i \in A}$.

**3.** *Semi-honest vs. malicious security.* In *semi-honest security*, corrupted parties faithfully follow the specification of $\pi$ during the execution and their goal is to determine some information about the inputs of honest parties. In *malicious security*, corrupted parties may arbitrarily deviate from the specification of $\pi$ and might send inconsistent messages in order to obtain information about the inputs of honest parties.

**4.** *Randomized function vs. deterministic function.* The function value is a random variable for any randomized function $f(x_1, \ldots, x_n)$. However, the randomness parameter of each party can be included in the input. That is, if $P_i$ uses randomness parameter $r_i$, then his input can be considered to be $x_i | r_i$. The function $f$ then becomes deterministic. Thus, we only need to consider deterministic functions.

**5.** *Function represented by a circuit.* Any (vector) Boolean function $f(x_1, \ldots, x_n)$ can be represented using a set of basic logic gates. For example, $\{\text{AND, OR, NOT}\}$ is a complete set for this purpose, and so is $\{\text{NAND}\}$.

**6.** The origins of secure multiparty computation stem from *Yao's millionaires' problem* [Ya82], in which two millionaires Alice and Bob wish to find out who is richer without revealing their wealth.

**7.** *Garbled circuit.* Let $C$ be a Boolean circuit realizing the function to be computed. For simplicity, assume the circuit consists of XOR and AND gates. Assume the circuit inputs are $z_1, \ldots, z_m$. A gate has two inputs and an output. The input of a gate comes from a circuit input or an output of another gate. A circuit input or a gate is called a *node*. We label all nodes from 1 to $N$ for some integer $N$. For each node $i$, select a pair of random keys $(k_i^0, k_i^1)$. Let $(E, D)$ be a symmetric encryption scheme. For a gate node $s$, assume its inputs come from nodes $i$ and $j$. Then we associate this gate with the ciphertext table as follows.

- Let $g : \{0, 1\} \times \{0, 1\} \to \{0, 1\}$ be a gate function for gate $s$. Define

$$c^s_{0,0} = E_{k_i^0}(E_{k_j^0}(k_s^{g(0,0)})), \qquad c^s_{0,1} = E_{k_i^0}(E_{k_j^1}(k_s^{g(0,1)}))$$
$$c^s_{1,0} = E_{k_i^1}(E_{k_j^0}(k_s^{g(1,0)})), \qquad c^s_{1,1} = E_{k_i^1}(E_{k_j^1}(k_s^{g(1,1)})).$$

A circuit with a ciphertext table at each gate is called a *garbled circuit*. The following figure provides an example.

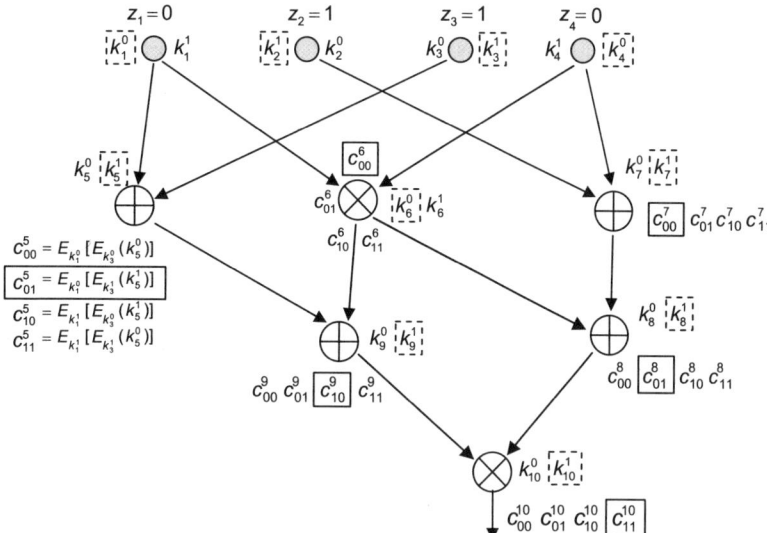

Note that for any input $(\alpha, \beta)$ of gate $s$, $c^s_{\alpha,\beta}$ is a double encryption of one of the random keys at this gate that corresponds to its output bit $g(\alpha, \beta)$. Then, given $(k_i^\alpha, k_j^\beta)$ and $\{c^s_{00}, c^s_{01}, c^s_{10}, c^s_{11}\}$ (in a random order), we can compute $k_s^{g(\alpha,\beta)}$ using the fact that $D_{k_j^\beta}(D_{k_i^\alpha}(c_{\alpha',\beta'}))$ equals $k_s^{g(\alpha,\beta)}$ if $(\alpha', \beta') = (\alpha, \beta)$ and equals $\perp$ (meaning a decryption error) otherwise.

- Anyone who is given $k_1^{z_1}, \ldots, k_m^{z_m}$ (but not the $z_i$s) can evaluate the garbled circuit in a gate-by-gate manner and finally obtain the key $k_N^{C(z_1, \ldots, z_m)}$ (assume that node $N$ is the last gate to be evaluated in $C$).

- From $k_N^{C(z_1, \ldots, z_m)}$, one cannot obtain $C(z_1, \ldots, z_m)$. But given $k_N^0$ and $k_N^1$, one can determine it.

**8.** *Yao's two-party computation protocol.* Suppose that the parties 1 and 2 have inputs $(x_1, \ldots, x_n)$ and $(y_1, \ldots, y_n)$, respectively, where $x_i, y_i \in \{0, 1\}$. The protocol proceeds as follows:

- Party 1 takes a pair of secret keys $(k_i^0, k_i^1)$ for each node $i$ in circuit $C$ for $i = 1, \ldots, N$, where $N$ is the total number of nodes in $C$.

- For each gate $s$, party 1 computes a table $c^s_{00}, c^s_{01}, c^s_{10}, c^s_{11}$ and shuffles them randomly. This generates a garbled circuit of $C$.

- Party 1 sends $G(C)$ and $k_1^{x_1}, \ldots, k_n^{x_n}$ (assuming the input nodes of party 1 are $1, \ldots, n$) to party 2.

- At this moment, party 2 needs to further obtain $k_{n+1}^{y_1}, \ldots, k_{2n}^{y_n}$ in order to decrypt the ciphertext table for each gate and finally obtain $k_N^{C(x_1, \ldots, x_n, y_1, \ldots, y_n)}$.

- Toward this end, let the input nodes for party 2 be $n+1, \ldots, 2n$. For each $i = 1, \ldots, n$, party 1 with input $(k_{n+i}^0, k_{n+i}^1)$ and party 2 with input $y_i$ jointly execute a 1-out-of-2 oblivious transfer to send $k_{n+i}^{y_i}$ to party 2.

- Upon receiving $k_i^{x_i}$ and $k_{n+i}^{y_i}$ for $i = 1, \ldots, n$, party 2 evaluates the garbled circuit of $C$ and obtains $k_N^{C(x_1, \ldots, x_n, y_1, \ldots, y_n)}$ (assuming gate $N$ is the final gate to be evaluated in $C$), and sends it to party 1.

- Party 1 obtains $C(x_1, \ldots, x_n, y_1, \ldots, y_n)$ by comparing $k_N^{C(x_1, \ldots, x_n, y_1, \ldots, y_n)}$ (from party 2) with $k_N^0$ and $k_N^1$, and sends the output to party 2.

**9.** *Passive security of party 1.* Notice that $k_i^{x_i}$ is independent of $x_i$. Thus, prior to the final output from party 1, the input $x_1, \ldots, x_N$ is independent of the messages seen by party 2. Thus, during the entire protocol, party 2 can only obtain information about $(x_1, \ldots, x_n)$ through $C(x_1, \ldots, x_n, y_1, \ldots, y_n)$.

**10.** *Passive security of party 2.* Party 1 obtains $y_1, \ldots, y_n$ only through the 1-out-of-2 oblivious transfers and $C(x_1, \ldots, x_n, y_1, \ldots, y_n)$ in the final step. However, by the security of oblivious transfers, party 1 obtains no knowledge about $y_1, \ldots, y_n$. Hence, the obtained information is covered by $C(x_1, \ldots, x_n, y_1, \ldots, y_n)$.

**11.** *GMW multiparty computation protocol* was invented by Goldreich, Micali, and Wigderson [GoMiWi87].

Consider circuit $C$ with NOT and AND gates. For $i = 1, \ldots, m$, suppose party $i$ has input $\mathbf{x}_i = (x_{i1}, \ldots, x_{in}) \in \{0,1\}^n$. The protocol proceeds as follows:

- Each party $i$ shares input bit $x_{i\ell}$ with all parties. Specifically, he generates uniformly random bits $x_{i\ell}^j$ so that $\sum_{j=1}^m x_{i\ell}^j = x_{i\ell}$ and sends $\{x_{i\ell}^j\}_{\ell=1}^n$ to party $j$ *in private*. Thus, each party $j$ obtains a share $x_{i\ell}^j$ for each input bit $x_{i\ell}$.

- All parties jointly evaluate the gates in $C$ in a top-down manner, where only the top gates are attached to the input wires of $C$. If the input(s) of a gate $g$ is shared among all parties, then $g$ is evaluated, with the result shared among all parties as follows:

  ◇ if $g$ is a NOT gate with input $z$, then by assumption, $z$ is shared among all parties. Assume party $i$ obtains a share $z_i$. Then, party 1 flips $z_1$ and all other parties keep their shares unchanged. Since $\bar{z}_1 + z_2 + \cdots + z_m = \bar{z}$, the NOT gate is evaluated with sharing.

  ◇ if $g$ is an AND gate with inputs $x, y$, then by assumption, $x, y$ are shared and party $i$ holds the share $x_i, y_i$ for $x, y$, respectively. The parties want to jointly compute $z = x \cdot y$ with sharing $z = z_1 + \cdots + z_m$. Notice that

$$\left(\sum_{i=1}^m x_i\right)\left(\sum_{i=1}^m y_i\right) = \sum_{i=1}^m x_i y_i + \sum_{1 \le i < j \le m}(x_i y_j + x_j y_i). \qquad (7)$$

Since party $i$ owns $x_i, y_i$, he can compute $x_i y_i$ locally and define it as $z_{ii}$. Then $(x_i y_j + x_j y_i)$ can be jointly computed by parties $i$ and $j$ with sharing $z_{ij} + z_{ji} = (x_i y_j + x_j y_i)$ via $\mathrm{OT}_1^4$ as follows:

Party $i$ selects $z_{ij}$ uniformly randomly and tries to help party $j$ derive $z_{ji}$. Notice that $z_{ji} = z_{ij}$ for $x_j y_j = 00$; $z_{ji} = z_{ij} + y_i$ for $x_j y_j = 10$; $z_{ji} = z_{ij} + x_i$ for $x_j y_j = 01$; $z_{ji} = z_{ij} + x_i + y_i$ for $x_j y_j = 11$. Then party 1 with input $(z_{ij}, z_{ij}+y_i, z_{ij}+x_i, z_{ij}+y_i+x_i)$ and party 2 with input $(x_j, y_j)$ execute $\mathrm{OT}_1^4$, thereby allowing party 2 to obliviously obtain $z_{ji} = z_{ij} + (x_i y_j + x_j y_i)$.

Finally, each party $i$ defines $z_i = \sum_{j=1}^m z_{ij}$. This yields a shared evaluation of $x \cdot y$.

- After evaluating all gates of $C$, each party $i$ obtains a share $o_i^j$ for party $j$'s output gate. Then, all parties send their shares $o_i^j, i = 1, \ldots, m$ to party $j$. Party $j$ obtains his circuit output $o^j = \sum_{i=1}^m o_i^j$.

**12.** *Correctness.* For each gate $g$, the sharing for its output is achieved. As seen in the previous fact, each gate is computed correctly. Consequently, $C$ is properly evaluated.

**13.** *Privacy of party i.* We need to guarantee that $x_i$ is not leaked beyond $C(x_1, \ldots, x_m)$.

- Evaluating each gate does not leak any information. Namely, if $g$ is a NOT gate, this is true because each output share is computed locally only. If $g$ is an AND gate, $x_i$ is not leaked in any sense. Recall that the inputs are used only in $\mathrm{OT}_1^4$, whereas $\mathrm{OT}_1^4$ does not leak any information beyond $z_{ji} = z_{ij} + (x_i y_j + x_j y_i)$. Further, since $z_{ij}$ is selected by party $i$ in a uniform random manner, $x_i$ is independent of $z_{ji}$.
- In the output stage, party $i$ obtains $o_j^i$ from every party $j$, whereas the $o_j^i, j = 1, \ldots, m$, are jointly uniformly random subject to $o^i = \sum_{j=1}^m o_j^i$. Thus, it does not leak more than $o^i$. The privacy of party $i$ is achieved.

## 15.7  HIGH-LEVEL APPLICATIONS OF CRYPTOGRAPHY

The cryptosystems and cryptographic mechanisms described earlier in this chapter are often components of high-level applications. In this section, three important high-level applications are described: digital cash, electronic voting, and secure cloud computing.

### 15.7.1  DIGITAL CASH

Digital cash aims to achieve all the functionality and security (and more) of paper cash in an electronic world.

**Definitions:**
According to the European Banking Authority, **virtual currency** is "a digital representation of value that is neither issued by a central bank or a public authority, nor necessarily attached to a fiat currency, but is accepted by natural or legal persons as a means of payment and can be transferred, stored, or traded electronically". Cryptography is not necessarily used in the framework.

**Cryptocurrency** is a class of unregulated digital cash in which cryptography is used in secure transactions and in the creation of additional currency.

A **nonce** is a pseudorandom number generated for one-time use.

**Facts:**
**1.** The following are some desirable properties of digital cash suggested in the literature [Ch83], [ChFiNa90], [Wa97]:

- *Anonymity.* Transactions should be anonymous to third parties.

- *Peer-to-peer transactions.* Transactions do not need to involve a trusted third party, nor a central clearance system.

- *Offline.* Transactions can be conducted offline.

- *Proof of payment.* Payment schemes should provide proof of payment.

- *Fraud detection.* There is the ability to determine fraud, such as double spending.

- *Protection from theft.* A stop payment occurs if cash is stolen.

**2.** Much of the research in digital cash involves two issues: maintaining anonymity and preventing double spending. Bitcoin is one recent and high-profile digital cash system.

**3.** *Bitcoin* is a cryptocurrency invented in 2008 by an unknown person under the alias of Satoshi Nakamoto. In 2009, the system was released as open-source software.

**4.** The legal status of bitcoin varies from country to country, from being purely illegal to unregulated, and to being considered as a commodity.

**5.** General information on bitcoin:

- The main cryptographic primitives used are ECDSA and SHA-256.

- Bitcoin is a peer-to-peer version of electronic cash. No financial institution is involved in any transaction.

- An electronic coin is defined to be a chain of digital signatures.

- A timestamp server mechanism is used to verify a chain of transactions. A *block* is a chain of transactions together with information from a previously verified block and a nonce with predefined properties.

- As the chain of transactions grows, the electronic coin takes up memory. A Merkle tree-like structure is used to reduce the space required.

- The hash function SHA-256 is used extensively both for signing the hash of transactions and for providing proof-of-work (see Fact 8) to prevent fraud. ECDSA is used for the signing of the hash.

- Bitcoin does not provide anonymity since all transactions are recorded in a public chain. However, anonymity can be partially achieved by keeping public keys anonymous.

- As of November 2016, bitcoin had a market capitalization of US $11 billion.

**6.** The operations of the bitcoin network follow six steps:

- Every new transaction is broadcast to all nodes.

- Each node then collects new transactions into a block by concatenating the transactions.

- Each node competes to find a proof-of-work (see Fact 8) for its block.

- A node that finds a proof-of-work broadcasts the block to all nodes.

- Nodes accept the block when all transactions are verified and not already spent.

- Nodes accept the block by working on creating the next block in the chain, using the hash of the previous block to begin a new block.

**7.** In a transaction from user A to user B, user A combines the previous transaction and B's public key, and signs its hash using A's signing key. The result is then transmitted to B and publicly announced. Any other user can verify the transaction by verifying A's signature.

**8.** To further reinforce integrity of data, a timestamp mechanism is used to hash a chain of transactions as proof of existence of data. The timestamp server uses a *proof-of-work* system: the timestamp requires the hashing of the previous timestamp hash, together with a nonce and a chain of transactions, resulting in a *block*; the choice of nonce is restricted such that when hashed, the result must begin with a required number of zero bits. This process is called *mining*.

**9.** It is possible that two timestamps on the same set of transactions can be produced at similar times. In this case, the longest block chain is recognized. In the case that there is a tie, then it is broken when the next proof-of-work is found for one of the chains.

**10.** Since proof-of-work requires computing power from contributors, there is an incentive in the form of new bitcoins that are offered to those who provided the proof-of-work service.

**11.** *Bitcoin security.*

- The proof-of-work system is used to prevent fraud. As with any cryptographically secure hash function, the amount of time (work) required to find a nonce doubles with each additional required zero bit.

- Double spending is avoided by recording the series of transactions which is available to all users.

- For a malicious user to nullify a past transaction, the attacker must re-do the proof-of-work of the block, all subsequent blocks, and surpass the work of the honest nodes. The probability of being able to accomplish this diminishes exponentially as more blocks are added.

- Bitcoin can be proven to be secure when there are more honest nodes than malicious nodes.

- The bitcoin wallet, which is a file that contains a collection of private signing keys for use in transactions, is not encrypted by default, and is vulnerable to theft.

## 15.7.2 ELECTRONIC VOTING

Electronic voting was developed to fulfill a need for an accurate, auditable, fast and secure voting system. Suggested voting systems include polling place electronic voting and remote internet voting [Jo], [Mo00].

**Definitions:**

***Electronic voting*** is voting using electronic means to assist in vote casting or counting votes.

***Blinding*** is a technique wherein a supplier can provide a service to a client without knowing the actual content of the input or the output. In the case of voting, a voter uses the technique to conceal the vote, which is then submitted to an administrator for signing. The voter reveals the vote anonymously at a later stage.

**Facts:**

**1.** Some cryptographically-relevant criteria for an electronic voting system include the following [Mo00]:

- *Eligibility and Authentication.* Only registered voters can vote.
- *Uniqueness.* Participants can vote at most once.

- *Accuracy.* Votes should be recorded correctly.
- *Integrity.* The system prevents votes from being modified, forged, or deleted.
- *Verifiability and Auditability.* The system has the ability to verify that votes are accounted for, with reliable and authentic election records.
- *Reliability.* The system should work despite numerous failures.
- *Secrecy and Non-Coercibility.* No one should be able to determine how an individual voted. Voters should not be able to prove how they voted.

**2.** *Blind Signatures Based Voting* (Fujioka-Okamoto-Ohta). This voting scheme was published in 1993 [FuOkOh92]. It uses *blind signatures*, where an administrator signs the concealed vote submitted by the voter. The scheme consists of six major steps: preparation, administration, voting, collecting, opening, and counting.

- *Preparation.* A voter fills in a ballot using a key $k$, attaches a random ID string to the ballot, and blinds the vote using a blinding technique. The voter then sends the blinded vote to the administrator.
- *Administration.* The administrator, after verifying the validity of the voter, signs the blinded vote and sends it back to the voter.
- *Voting.* The voter unblinds the signature and sends the administrator-signed anonymous ballot to the collector.
- *Collecting.* The collector checks the administrator's signature and enters the encrypted vote into a list.
- *Opening.* The voter checks if the ballot is in the list and sends key $k$ through an anonymous channel.
- *Counting.* The collector opens the vote using key $k$.

**3.** The scheme is proven to satisfy eligibility and authentication, uniqueness, accuracy, integrity, verifiability, and secrecy.

**4.** *Cryptographic Counter Based Voting* (Katz-Myers-Ostrovsky). A cryptographic $n$-counter is a triple of algorithms $(G, D, T)$, with $n$ different counting states, such that using any public key cryptosystem, we have

- $G$ is a probabilistic key generation algorithm, where a $k$-bit input generates a public key-private key pair $(pk, sk)$ and a string $s_0$. The secret key implicitly defines an associated set of states $S_{sk}$, where $s_0 \in S_{sk}$. The states represent the count.
- $D$ is the deterministic decryption algorithm that takes in secret key $sk$ and a string $s$, and outputs an integer $i \in \mathcal{Z}_n$ if $s \in S_{sk}$. Otherwise, $D$ outputs *contradiction*.
- $T$ is a probabilistic algorithm known as the *transition algorithm*. It takes the public key $pk$, string $s$, and an integer $i \in \mathcal{Z}_n$, and outputs string $s'$.

**5.** A high-level voting protocol is described in the following steps. Assume that there are $n$ voters, where voters are voting in sequence.

- *Setup.* The authorities run the key generation algorithm, and then announce the public key $pk$ and the initial string $s_0$ to all voters.
- *Voting.* The collector carries the current total. Let $s_i$ be the counter after the $i$th vote. For the $(i + 1)$st vote, the voter takes the current counter $s_i$, public key $pk$, the desired vote, and updates the state to $s_{i+1}$ using the transition function.

- *Tally.* The authorities decrypt the last counter.

**6.** A simple voting protocol where a voter's choice is limited to $\{0, 1\}$ is illustrated using quadratic residuosity in [KaMyOs01]. The scheme is proven to satisfy secrecy, reliability, and verifiability.

**7.** In a *mix-net voting scheme* [BoGo02], [ChPe92], [JaJuRi02], voters submit signed, encrypted votes. A sequence of independent authorities called *mixers* shuffle the votes and perform further encryption. After a chain of mixers has performed the operations, the shuffled votes are completely decrypted. Due to the secret shuffling, the votes become anonymous, thus protecting secrecy. A method of providing shuffle-and-decrypt proofs also provides verifiability.

**8.** The Paillier cryptosystem [Pa99], a homomorphic encryption system (§15.6.4) that satisfies $E(m_1)E(m_2) = E(m_1 + m_2)$, can be used in voting where encrypted votes are multiplied and then decrypted to reveal the total tally [BaEtal01].

---

## 15.7.3   SECURE CLOUD COMPUTING

Cloud computing emerged from the development of high-speed, high-capacity networks and low-cost computers. The proliferation of computer technology has enabled sharing of computing and data resources across the internet. This shared resources model enables businesses and researchers to avoid expensive initial infrastructure costs, and it maximizes the use of computational resources across a network.

### Definition:
**Cloud computing** is an internet-based model that allows shared computing resources and data to electronic devices on demand. The goal is to enable ubiquitous, convenient, on-demand network access to a shared pool of configurable computing resources that can be rapidly provisioned and released with minimal management effort or service provider interaction [MeGr11].

### Facts:
**1.** Since resources are shared, it is essential that confidentiality and auditability are provided throughout the service [ArEtal10].

**2.** In the interest of data privacy, cryptographic separation is needed in which computations and data are concealed in such a way that they appear intangible to the outsider [Ry13], [SuKa11], [ZiLe12]. One can store the encrypted data on a cloud server and the cloud server can run operations such as keyword search, or perform arithmetic on the data, without decrypting the data. This mechanism eliminates the need to download data from the cloud, thereby reducing security risk.

**3.** One solution to protect data privacy in cloud computing is through the use of homomorphic encryption [TeHaGh12], [TeHa14]; see §15.6.4. Paillier [Pa99] and Goldwasser-Micali [GoMi82] cryptosystems can be used in applications where the operation is addition. However, both systems involve message expansion and thus have a practical disadvantage. The RSA and ElGamal systems can be used for multiplication requests.

# REFERENCES

***Printed Resources***:

[AgMa09] D. Aggarwal and U. Maurer, "Breaking RSA generically is equivalent to factoring" in *Advances in Cryptology–Eurocrypt 2009*, Lecture Notes in Computer Science 5479, Springer, 2009, 36–53.

[AgEtal12] M. Ågren et al., "A survey on fast correlation attacks", *Cryptography and Communications* 4 (2012), 173–202.

[AoEtal00] K. Aoki et al., "Camellia: a 128-bit block cipher suitable for multiple platforms –design and analysis" in *Selected Areas in Cryptography–SAC 2000*, Lecture Notes in Computer Science 2012, Springer, 2000, 39–56.

[ArEtal10] M. Armbrust et al., "A view of cloud computing", *Communications of the ACM* 53 (2010), 50–58.

[BaRo11] E. Barker and A. Roginsky, "Transitions: recommendation for transitioning the use of cryptographic algorithms and key lengths", NIST Special Publication 800-131A, 2011.

[BaEtal01] O. Baudron, P. Fouque, D. Pointcheval, J. Stern, and G. Poupard, "Practical multi-candidate election system", *Proceedings of the 20th Annual ACM Symposium on Principles of Distributed Computing*, ACM, 2001, 274–283.

[BeCaKr96] M. Bellare, R. Canetti, and H. Krawczyk, "Keying hash functions for message authentication" in *Advances in Cryptology–Crypto '96*, Lecture Notes in Computer Science 1109, Springer, 1996, 1–15.

[BePoRo00] M. Bellare, D. Pointcheval, and P. Rogaway, "Authenticated key exchange secure against dictionary attacks" in *Advances in Cryptology–Eurocrypt 2000*, Lecture Notes in Computer Science 1807, Springer, 2000, 139–155.

[BeMe92] S. Bellovin and M. Merritt, "Encrypted key exchange: password-based protocols secure against dictionary attacks" in *Symposium on Research in Security and Privacy*, IEEE, 1992, 72–84.

[BeEtal07] G. Bertoni et al., "Sponge functions" in *ECRYPT Hash Workshop*, 2007.

[BiSh91] E. Biham and A. Shamir, "Differential cryptanalysis of DES-like cryptosystems", *Journal of Cryptology* 4 (1991), 3–72.

[BiEtal11] A. Biryukov et al., "Second-order differential collisions for reduced SHA-256" in *Advances in Cryptology–Asiacrypt 2011*, Lecture Notes in Computer Science 7073, Springer, 2011, 270–287.

[BiKhNi09] A. Biryukov, D. Khovratovich, and I. Nikolić, "Distinguisher and related-key attack on the full AES-256" in *Advances in Cryptology–Crypto 2009*, Lecture Notes in Computer Science 5677, Springer, 2009, 231–249.

[Bl79] G. Blakley, "Safeguarding cryptographic keys", *Proceedings of the AFIPS 1979 National Computer Conference* 48 (1979), 313–317.

[Bl84] R. Blom, "An optimal class of symmetric key generation systems" in *Advances in Cryptology–Eurocrypt '84*, Lecture Notes in Computer Science 209, Springer, 1984, 335–338.

[BlEtal98] C. Blundo et al., "Perfectly secure key distribution for dynamic conferences", *Information and Computation* 146 (1998), 1–23.

[BoKhRe11] A. Bogdanov, D. Khovratovich, and C. Rechberger, "Biclique cryptanalysis of the full AES" in *Advances in Cryptology–Asiacrypt 2011*, Lecture Notes in Computer Science 7073, Springer, 2011, 344–371.

[BoBo04] D. Boneh and X. Boyen, "Secure identity based encryption without random oracles" in *Advances in Cryptology–Crypto 2004*, Lecture Notes in Computer Science 3152, Springer, 2004, 443–459.

[BoDu99] D. Boneh and G. Durfee, "Cryptanalysis of RSA with private key $d$ less than $N^{0.292}$" in *Advances in Cryptology–Eurocrypt '99*, Lecture Notes in Computer Science 1592, Springer, 1999, 1–11.

[BoFr01] D. Boneh and M. Franklin, "Identity-based encryption from the Weil pairing" in *Advances in Cryptology–Crypto 2001*, Lecture Notes in Computer Science 2139, Springer, 2001, 213–229.

[BoGo02] D. Boneh and P. Golle, "Almost entirely correct mixing with applications to voting" in *Proceedings of the 9th ACM Conference on Computer and Communications Security*, ACM, 2002, 68–77.

[BoLySh04] D. Boneh, B. Lynn, and H. Shacham, "Short signatures from the Weil pairing", *Journal of Cryptology* 17 (2004), 297–319.

[Bu13] J. Buchmann, *Introduction to Cryptography*, Springer Science & Business Media, 2013.

[BuDe94] M. Burmester and Y. Desmedt, "A secure and efficient conference key distribution system" in *Advances in Cryptology–Eurocrypt '94*, Lecture Notes in Computer Science 950, Springer, 1994, 275–286.

[CaKr02] R. Canetti and H. Krawczyk, "Security analysis of IKE's signature based key-exchange protocol" in *Advances in Cryptology–Crypto 2002*, Lecture Notes in Computer Science 2442, Springer, 2002, 143–161.

[Ca05] C. De Cannière, "Kasumi/Misty1" in *Encyclopedia of Cryptography and Security*, H. C. A. van Tilborg (ed.), Springer, 2005, 322–323.

[Ca10a] C. Carlet, "Boolean functions for cryptography and error correcting codes" in *Boolean Models and Methods in Mathematics, Computer Science, and Engineering*, Y. Crama and P. L. Hammer (eds.), Cambridge University Press, 2010, 257–397.

[Ca10b] C. Carlet, "Vectorial Boolean functions for cryptography" in *Boolean Models and Methods in Mathematics, Computer Science, and Engineering*, Y. Crama and P. L. Hammer (eds.), Cambridge University Press, 2010, 398–469.

[Ch83] D. Chaum, "Blind signatures for untraceable payments" in *Advances in Cryptology*, D. Chaum (ed.), Springer, 1983, 199–203.

[ChFiNa90] D. Chaum, A. Fiat, and M. Naor, "Untraceable electronic cash" in *Advances in Cryptology–Crypto '88*, Lecture Notes in Computer Science 403, Springer, 1990, 319–327.

[ChPe92] D. Chaum and T. Pedersen, "Wallet databases with observers" in *Advances in Cryptology–Crypto '92*, Lecture Notes in Computer Science 740, Springer, 1992, 89–105.

[ChEtal11] J. Choy et al., "AES variants secure against related-key differential and boomerang attacks" in *Information Security Theory and Practice*, Lecture Notes in Computer Science 6633, Springer, 2011, 191–207.

[Co97] D. Coppersmith, "Small solutions to polynomial equations, and low exponent RSA vulnerabilities", *Journal of Cryptology* 10 (1997), 233–260.

[Co03] N. Courtois, "Fast algebraic attacks on stream ciphers with linear feedback" in *Advances in Cryptology–Crypto 2003*, Lecture Notes in Computer Science 2729, Springer, 2003, 176–194.

[CoMe03] N. Courtois and W. Meier, "Algebraic attacks on stream ciphers with linear feedback" in *Advances in Cryptology–Eurocrypt 2003*, Lecture Notes in Computer Science 2656, Springer, 2003, 345–359.

[CoPi02] N. Courtois and J. Pieprzyk, "Cryptanalysis of block ciphers with overdefined systems of equations" in *Advances in Cryptology–Asiacrypt 2002*, Lecture Notes in Computer Science 2501, Springer, 2002, 267–287.

[CrSh98] R. Cramer and V. Shoup, "A practical public key cryptosystem provably secure against adaptive chosen ciphertext attack" in *Advances in Cryptology–Crypto '98*, Lecture Notes in Computer Science 1462, Springer, 1998, 13–25.

[DaRi13] J. Daemen and V. Rijmen, *The Design of Rijndael: AES–The Advanced Encryption Standard*, Springer Science & Business Media, 2013.

[DeKn15] H. Delfs and H. Knebl, *Introduction to Cryptography*, 3rd ed., Springer, 2015.

[De06] A. Dent, "The Cramer-Shoup encryption scheme is plaintext aware in the standard model" in *Advances in Cryptology–Eurocrypt 2006*, Lecture Notes in Computer Science 4004, Springer, 2006, 289–307.

[DiHe76] W. Diffie and M. Hellman, "New directions in cryptography", *IEEE Transactions on Information Theory* 22 (1976), 644–654.

[DiLe08] W. Diffie and G. Ledin, "SMS4 encryption algorithm for wireless networks" in *IACR Cryptology ePrint Archive* Report 2008/329.

[Dw01] M. Dworkin, "Recommendation for block cipher modes of operation: methods and techniques", NIST Special Publication 800-38A, 2001.

[Dw04] M. Dworkin, "Recommendation for block cipher modes of operation: the CCM mode for authentication and confidentiality", NIST Special Publication 800-38C, 2004.

[Dw07] M. Dworkin, "Recommendation for block cipher modes of operation: Galois/counter mode (GCM) and GMAC", NIST Special Publication 800-38D, 2007.

[El85] T. ElGamal, "A public key cryptosystem and a signature scheme based on discrete logarithms", *IEEE Transactions on Information Theory* 31 (1985), 469–472.

[FiNa93] A. Fiat and M. Naor, "Broadcast encryption" in *Advances in Cryptology–Crypto '93*, Lecture Notes in Computer Science 773, Springer, 1993, 480–491.

[Fi77] FIPS PUB 46, "Data Encryption Standard," *Federal Information Processing Standards Publication*, 1977.

[Fi13] FIPS PUB 186-4, "Digital Signature Standard (DSS)", *Federal Information Processing Standards Publication*, 2013.

[Fi15] FIPS PUB 202, "SHA-3 Standard: Permutation-Based Hash and Extendable-Output Functions", *Federal Information Processing Standards Publication*, 2015.

[FlMaSh01] S. Fluhrer, I. Mantin, and A. Shamir, "Weaknesses in the key scheduling algorithm of RC4" in *Selected Areas of Cryptography–SAC 2001*, Lecture Notes in Computer Science 2259, Springer, 2001, 1–24.

[Fr01] G. Frey, "Applications of arithmetical geometry to cryptographic constructions" in *Finite Fields and Applications*, D. Jungnickel and H. Niederreiter (eds.), Springer, 2001, 128–161.

[FuOk99] E. Fujisaki and T. Okamoto, "Secure integration of asymmetric and symmetric encryption schemes" in *Advances in Cryptology–Crypto '99*, Lecture Notes in Computer Science 1666, Springer, 1999, 537–554.

[FuOkOh92] A. Fujioka, T. Okamoto, and K. Ohta, "A practical secret voting scheme for large scale elections" in *Advances in Cryptology–Auscrypt '92*, Lecture Notes in Computer Science 718, Springer, 1992, 244–251.

[GaHeSm02] P. Gaudry, F. Hess, and N. Smart, "Constructive and destructive facets of Weil descent on elliptic curves", *Journal of Cryptology* 15 (2002), 19–46.

[GeLi03] R. Gennaro and Y. Lindell, "A framework for password-based authenticated key exchange" in *Advances in Cryptology–Eurocrypt 2003*, Lecture Notes in Computer Science 2656, Springer, 2003, 524–543.

[Ge09] C. Gentry, "Fully homomorphic encryption using ideal lattices", *Proceedings of the 41st Annual ACM Symposium on Theory of Computing*, ACM, 2009, 169–178.

[GiMaSl74] E. N. Gilbert, F. J. MacWilliams, and N. J. A. Sloane, "Codes which detect deception", *Bell System Technical Journal* 53 (1974), 405–424.

[GoMiWi87] O. Goldreich, S. Micali, and A. Wigderson, "How to play any mental game or a completeness theorem for protocols with honest majority", *Proceedings of the 19th Annual ACM Symposium on Theory of Computing*, ACM, 1987, 218–229.

[GoMi82] S. Goldwasser and S. Micali, "Probabilistic encryption & how to play mental poker keeping secret all partial information", *Proceedings of the 14th Annual ACM Symposium on Theory of Computing*, ACM, 1982, 365–377.

[GoEtal06] V. Goyal et al., "Attribute-based encryption for fine-grained access control of encrypted data", *Proceedings of the 13th ACM Conference on Computer and Communications Security*, ACM, 2006, 89–98.

[GrKa10] A. Groce and J. Katz, "A new framework for efficient password based authenticated key exchange", *Proceedings of the 17th ACM Conference on Computer and Communications Security*, ACM, 2010, 516–525.

[GüEtal08] T. Güneysu et al., "Cryptanalysis with COPACOBANA", *IEEE Transactions on Computers* 57 (2008), 1498–1513.

[HaCa98] D. Harkins and D. Carrel, *The Internet Key Exchange (IKE)*, RFC 2409, Internet Engineering Task Force, 1998.

[HiLa10] M. J. Hinek and C. Lam, "Common modulus attacks on small private exponent RSA and some fast variants (in practice)", *Journal of Mathematical Cryptology* 4 (2010), 58–93.

[HoSe99] N. Howgrave-Graham and J. P. Seifert, "Extending Wiener's attack in the presence of many decrypting exponents" in *Secure Networking–CQRE [Secure] '99*, Lecture Notes in Computer Science 1740, Springer, 1999, 153–166.

[JaMeSt01] M. Jacobson, A. Menezes and A. Stein, "Solving elliptic curve discrete logarithm problems using Weil descent", *Journal of the Ramanujan Mathematical Society* 16 (2001), 231–260.

[JaJuRi02] M. Jakobsson, A. Juels, and R. Rivest, "Making mix nets robust for electronic voting by randomized partial checking" in *USENIX Security Symposium*, 2002, 339–353.

[JiGo04] S. Jiang and G. Gong, "Password based key exchange with mutual authentication" in *Selected Areas in Cryptography–SAC 2004*, Lecture Notes in Computer Science 3357, Springer, 2004, 267–279.

[JiWa10] S. Jiang and H. Wang, "Plaintext-awareness of hybrid encryption" in *Topics in Cryptology–CT-RSA 2010*, Lecture Notes in Computer Science 5985, Springer, 2010, 57–72.

[Jo00] A. Joux, "A one round protocol for tripartite Diffie-Hellman" in *International Algorithmic Number Theory Symposium–ANTS-IV 2000*, Lecture Notes in Computer Science 1838, Springer, 2000, 385–393.

[Ka96] D. Kahn, *The Codebreakers: The Comprehensive History of Secret Communication from Ancient Times to the Internet*, Simon and Schuster, 1996.

[Ka92] B. Kaliski, "The MD2 message-digest algorithm", RFC 1319, Network Working Group, 1992.

[Ka98] B. Kaliski, "PKCS #7: Cryptographic message syntax version 1.5", RFC 2315, Network Working Group, 1998.

[KaLi14] J. Katz and Y. Lindell, *Introduction to Modern Cryptography*, CRC Press, 2014.

[KaMyOs01] J. Katz, S. Myers, and R. Ostrovsky, "Cryptographic counters and applications to electronic voting" in *Advances in Cryptology–Eurocrypt 2001*, Lecture Notes in Computer Science 2045, Springer, 2001, 78–92.

[KaOsYu01] J. Katz, R. Ostrovsky, and M. Yung, "Efficient password authenticated key exchange using human-memorable passwords" in *Advances in Cryptology–Eurocrypt 2001*, Lecture Notes in Computer Science 2045, Springer, 2001, 475–494.

[KhReSa12] D. Khovratovich, C. Rechberger, and A. Savelieva, "Bicliques for preimages: attacks on Skein-512 and the SHA-2 family" in *Fast Software Encryption–FSE 2012*, Lecture Notes in Computer Science 7549, Springer, 2012, 244–263.

[KiRo01] J. Kilian and P. Rogaway, "How to protect DES against exhaustive key search (an analysis of DESX)", *Journal of Cryptology* 14 (2001), 17–35.

[KiYo11] A. Kircanski and A. Youssef, "On the sliding property of SNOW 3G and SNOW 2.0", *IET Information Security* 5 (2011), 199–206.

[Ko87] N. Koblitz, "Elliptic curve cryptosystems", *Mathematics of Computation* 48 (1987), 203–209.

[LiWa14] Y. Li and M. Wang, "Constructing advanced encryption S-boxes for lightweight cryptography with Feistel Structure" in *Cryptographic Hardware and Embedded Systems–CHES 2014*, Lecture Notes in Computer Science 8731, Springer, 2014, 127–146.

[LiKh07] C. Lim and K. Khoo, "An analysis of XSL applied to BES" in *Fast Software Encryption–FSE 2007*, Lecture Notes in Computer Science 4593, Springer, 2007, 242–253.

[LuMeVa05] Y. Lu, W. Meier, and S. Vaudenay, "The conditional correlation attack: a practical attack on Bluetooth encryption" in *Advances in Cryptology–Crypto 2005*, Lecture Notes in Computer Science 3621, Springer, 2005, 97–117.

[MaPe08] S. Manuel and T. Peyrin, "Collisions on SHA-0 in one hour" in *Fast Software Encryption–FSE 2008*, Lecture Notes in Computer Science 5086, Springer, 2008, 16–35.

[Ma03] W. Mao, *Modern Cryptography: Theory and Practice*, Prentice Hall Professional Technical Reference, 2003.

[Ma69] J. Massey, "Shift-register synthesis and BCH decoding", *IEEE Transactions on Information Theory* 15 (1969), 122–127.

[Ma93] M. Matsui, "Linear cryptanalysis method for DES cipher" in *Advances in Cryptology–Eurocrypt '93*, Lecture Notes in Computer Science 765, Springer, 1993, 386–397.

[MaMeTe02] M. Maurer, A. Menezes, and E. Teske, "Analysis of the GHS Weil descent attack on the ECDLP over characteristic two finite fields of composite degree", *LMS Journal of Computation and Mathematics* 5 (2002), 127–174.

[MaJoBal04] A. Maximov, T. Johansson, and S. Babbage, "An improved correlation attack on A5/1" in *Selected Areas in Cryptography–SAC 2004*, Lecture Notes in Computer Science 3357, Springer, 2004, 1–18.

[MeQu01] A. Menezes and M. Qu, "Analysis of the Weil descent attack of Gaudry, Hess and Smart" in *Topics in Cryptology–CT-RSA 2001*, Lecture Notes in Computer Science 2020, Springer, 2001, 308–318.

[MeTe06] A. Menezes and E. Teske, "Cryptographic implications of Hess's generalized GHS attack", *Applicable Algebra in Engineering, Communication and Computing* 16 (2006), 439–460.

[MeTeWe04] A. Menezes, E. Teske, and A. Weng, "Weak fields for ECC" in *Topics in Cryptology–CT-RSA 2004*, Lecture Notes in Computer Science 2964, Springer, 2004, 366–386.

[MevOVa10] A. Menezes, P. van Oorschot, and S. Vanstone. *Handbook of Applied Cryptography*, CRC Press, 2010.

[MeHe78] R. Merkle and M. Hellman, "Hiding information and signatures in trapdoor knapsacks", *IEEE Transactions on Information Theory* 24 (1978), 525–530.

[Mi85] V. Miller, "Uses of elliptic curves in cryptography" in *Advances in Cryptology–Crypto '85*, Lecture Notes in Computer Science 218, Springer, 1985, 417–462.

[Mo00] C. D. Mote, Jr., "Report of the national workshop on internet voting: issues and research agenda" in *Proceedings of the 2000 Annual National Conference on Digital Government Research*, Digital Government Society of North America, 2000, 1–59.

[MuRo02] S. Murphy and M. Robshaw, "Essential algebraic structure within the AES" in *Advances in Cryptology–Crypto 2002*, Lecture Notes in Computer Science 2442, Springer, 2002, 1–16.

[NaPi01] M. Naor and B. Pinkas, "Efficient oblivious transfer protocols", *Proceedings of the 12th Annual ACM-SIAM Symposium on Discrete Algorithms*, SIAM, 2001, 448–457.

[NaPi05] M. Naor and B. Pinkas, "Computationally secure oblivious transfer", *Journal of Cryptology* 18 (2005), 1–35.

[Ni10] I. Nikolic, "Tweaking AES" in *Selected Areas in Cryptography–SAC 2010*, Lecture Notes in Computer Science 6544, Springer, 2010, 198–210.

[PaPe09] C. Paar and J. Pelzl, *Understanding Cryptography: A Textbook for Students and Practitioners*, Springer Science & Business Media, 2009.

[Pa99] P. Paillier, "Public-key cryptosystems based on composite degree residuosity classes" in *Advances in Cryptology–Crypto '99*, Lecture Notes in Computer Science 1592, Springer, 1999, 223–238.

[Po74] J. M. Pollard, "Theorems of factorization and primality testing", *Proceedings of the Cambridge Philosophical Society* 76 (1974), 521–528.

[Ra81] M. O. Rabin, "How to exchange secrets by oblivious transfer", Technical Report TR-81, Aiken Computation Laboratory, 1981.

[Ri08] R. Rivest, "The MD6 hash function", Invited talk at *Advances in Cryptology–Crypto 2008*; available at `http://people.csail.mit.edu/rivest/`.

[RiShAd78] R. Rivest, A. Shamir, and L. Adleman, "A method for obtaining digital signatures and public-key cryptosystems", *Communications of the ACM* 21 (1978), 120–126.

[RoBi08] M. Robshaw and O. Billet, eds., *New Stream Cipher Designs–The eSTREAM Finalists*, Lecture Notes in Computer Science 4986, Springer, 2008.

[Ru12] R. Rueppel, *Analysis and Design of Stream Ciphers*, Springer Science & Business Media, 2012.

[Ry13] M. Ryan, "Cloud computing security: the scientific challenge, and a survey of solutions", *Journal of Systems and Software* 86 (2013), 2263–2268.

[SaWa05] A. Sahai and B. Waters, "Fuzzy identity-based encryption" in *Advances in Cryptology–Eurocrypt 2005*, Lecture Notes in Computer Science 3494, Springer, 2005, 457–473.

[Sc07] B. Schneier, *Applied Cryptography: Protocols, Algorithms, and Source Code in C*, John Wiley & Sons, 2007.

[Se98] I. Semaev, "Evaluation of discrete logarithms in a group of $p$-torsion points of an elliptic curve in characteristic $p$", *Mathematics of Computation* 67 (1998), 353–356.

[Sh79] A. Shamir, "How to share a secret", *Communications of the ACM* 22 (1979), 612–613.

[Sh84a] A. Shamir, "A polynomial-time algorithm for breaking the basic Merkle-Hellman cryptosystem", *IEEE Transactions on Information Theory* 30 (1984), 699–704.

[Sh84b] A. Shamir, "Identity-based cryptosystems and signature schemes" in *Advances in Cryptology–Crypto '84*, Lecture Notes in Computer Science 196, Springer, 1984, 47–53.

[ShEtal07] T. Shirai et al., "The 128-bit blockcipher CLEFIA (extended abstract)" in *Fast Software Encryption–FSE 2007*, Lecture Notes in Computer Science 4593, Springer, 2007, 181–195.

[Si00] S. Singh, *The Code Book: The Science of Secrecy from Ancient Egypt to Quantum Cryptography*, Anchor, 2000.

[Sm99] N. Smart, "The discrete logarithm problem on elliptic curves of trace one", *Journal of Cryptology* 12 (1999), 193–196.

[StEtal17] M. Stevens, E. Bursztein, P. Karpman, A. Albertini, and Y. Markov, "The first collision of full SHA-1"; available `http://shattered.io`.

[St05] D. Stinson, *Cryptography: Theory and Practice*, 3rd ed., CRC Press, 2005.

[SuKa11] S. Subashini and V. Kavitha, "A survey on security issues in service delivery models of cloud computing", *Journal of Network and Computer Applications* 34 (2011), 1–11.

[TaEtal98] S. Takakazu et al., "Fermat quotients and the polynomial time discrete log algorithm for anomalous elliptic curves", *Commentarii Mathematici Universitatis Sancti Pauli, Rikkyo Daigaku Sugaku Zasshi* 47 (1998), 81–92.

[TeHa14] M. Tebaa and S. El Hajji, "Secure cloud computing through homomorphic encryption", preprint, 2014; available at `http://arxiv.org/abs/1409.0829`.

[TeHaGh12] M. Tebaa, S. El Hajji, and A. El Ghazi, "Homomorphic encryption applied to the cloud computing security", *Proceedings of the World Congress on Engineering, Vol. I*, 2012, 536–539.

[TrWa06] W. Trappe and L. Washington, *Introduction to Cryptography with Coding Theory*, 2nd ed., Pearson Education, 2006.

[Tu00] W. T. Tutte, "Fish and I" in *Coding Theory and Cryptography*, D. Joyner (ed.), Springer, 2000, 9–17.

[vDEtal10] M. van Dijk, C. Gentry, S. Halevi, and V. Vaikuntanathan, "Fully homomorphic encryption over the integers" in *Advances in Cryptology–Eurocrypt 2010*, Lecture Notes in Computer Science 6110, Springer, 2010, 24–43.

[vOWi90] P. van Oorschot and M. Wiener, "A known plaintext attack on two-key triple encryption" in *Advances in Cryptology–Eurocrypt '90*, Lecture Notes in Computer Science 473, Springer, 1990, 318–325.

[WaYu05] X. Wang and H. Yu, "How to break MD5 and other hash functions" in *Advances in Cryptology–Eurocrypt 2005*, Lecture Notes in Computer Science 3494, Springer, 2005, 19–35.

[Wa97] P. Wayner, *Digital Cash: Commerce on the Net*, Academic Press Professional, Inc., 1997.

[WeCa81] M. Wegman and J. L. Carter, "New hash functions and their use in authentication and set equality", *Journal of Computer and System Sciences* 22 (1981), 265–279.

[Wi90] M. Wiener, "Cryptanalysis of short RSA secret exponents", *IEEE Transactions on Information Theory* 36 (1990), 553–558.

[Wi82] H. C. Williams, "A $p+1$ method of factoring", *Mathematics of Computation* 39 (1982), 225–234.

[XiLiFe13] T. Xie, F. Liu, and D. Feng, "Fast collision attack on MD5", *IACR Cryptology ePrint Archive*, Report 2013/170.

[Ya82] A. Yao, "Protocols for secure computations", *Proceedings of the 23rd Annual Symposium on Foundations of Computer Science*, IEEE, 1982, 160–164.

[YaKhPo13] H. Yap, K. Khoo, and A. Poschmann, "Parallelisable variants of Camellia and SMS4 block cipher: $p$-Camellia and $p$-SMS4", *International Journal of Applied Cryptography* 3 (2013), 1–20.

[ZhEtal09] A. Zhang et al., "Extensions of the cube attack based on low degree annihilators" in *Cryptology and Network Security*, Lecture Notes in Computer Science 5888, Springer, 2009, 87–102.

[ZiLe12] D. Zissis and D. Lekkas, "Addressing cloud computing security issues", *Future Generation Computer Systems* 28 (2012), 583–592.

***Web Resources***:

[Ae97] `http://csrc.nist.gov/archive/aes/pre-round1/aes_9709.htm` (Request for candidate algorithm nominations for the Advanced Encryption Standard.)

[Cv09] `http://cve.mitre.org/cgi-bin/cvename.cgi?name=cve-2009-2409` (Common Vulnerabilities and Exposures, CVE-2009-2409.)

[De98] `https://w2.eff.org/Privacy/Crypto/Crypto_misc/DESCracker/HTML/199807` `16_eff_des_faq.html` (Frequently Asked Questions (FAQ) about the Electronic Frontier Foundation's *DES Cracker* machine.)

[Es04] `http://www.ecrypt.eu.org/stream/` (eSTREAM, the ECRYPT Stream Cipher Project.)

[Ja] `http://jablon.org/passwordlinks.html` (Site for research papers on password-based cryptography.)

[Jo] `http://www.umic.pt/images/stories/publicacoes1/final_report.pdf` (*A Report on the Feasibility of Internet Voting*, California Internet Voting Task Force, January, 2000.)

[Li] `http://kodu.ut.ee/~lipmaa/crypto/` (Helger Limpmaa's cryptology pointers.)

[MeGr11] `http://nvlpubs.nist.gov/nistpubs/Legacy/SP/nistspecialpublication` `800-145.pdf` (Peter Mell and Timothy Grance, *The NIST Definition of Cloud Computing.*)

[P1363] `http://grouper.ieee.org/groups/1363/` (IEEE P1363 Standard Specifications for Public Key Cryptography.)

[PGP] `http://openpgp.org/` (OpenPGP mail encryption standard.)

[Ri92] `http://tools.ietf.org/html/rfc1320` (R. Rivest, *The MD4 Message-Digest Algorithm.*)

[Sk98] `http://cryptome.org/jya/skipjack-spec.htm` (Specifications for the SKIP-JACK and KEA algorithms.)

[TuCh11] `http://tools.ietf.org/html/rfc6150` (Sean Turner and Lily Chen, *MD4 to Historic Status.*)

[Vu08] `http://www.kb.cert.org/vuls/id/836068` (Note documenting MD5 vulnerability to collision attacks.)

# 16

# DISCRETE OPTIMIZATION

# INTRODUCTION

This chapter discusses various topics in discrete optimization, especially those that arise in applying operations research techniques to applied problems. A fundamental tool is linear programming and its extensions to problems in which certain variables must assume integer values. These techniques are useful in devising solution techniques for a variety of problems in which a given resource must be optimally utilized subject to constraints. The chapter also presents several metaheuristic approaches for approximate solution of difficult discrete optimization problems.

# GLOSSARY

**active constraint**: an inequality satisfied with equality by a given vector.

**balanced matrix**: a 0-1 matrix having no square submatrix of odd order with exactly two 1s in each row and column.

**basic feasible solution** (of an LP): a basic solution that is also a feasible solution.

**basic solution** (of an LP): a solution obtained by setting certain *nonbasic* variables to zero and solving for the remaining *basic* variables.

**bid language**: a format for communicating lists of combinatorial bids.

**bin packing problem**: an optimization problem in which a given set of items is to be packed using the fewest number of bins.

**bounded LP**: a linear programming problem having a finite optimal solution.

**capacitated location problem**: a location problem in which bounds are placed on the amount of demand that can be handled by individual facilities.

**p-center**: a set of $p$ locations for facilities that minimizes the maximum distance from any demand point to its closest facility.

**characteristic function**: a mapping from the set of all coalitions to the nonnegative real numbers.

**characteristic-function game**: a model for distributing a cooperative benefit fairly among players when the concept of fairness is based on the bargaining strengths of coalitions that could form if the players had not already agreed to cooperate.

**coalition**: any subset of the players in a game.

**combinatorial auction**: a situation in which multiple items are auctioned, and each bidder can submit several bids (each of which may contain several items).

***complete information***: a situation arising when a game's structure is known to all players.

***convex hull***: the smallest convex set containing a given set of points.

***convex set***: a set containing the line segment joining any two of its points.

***CPM model***: a deterministic activity net with strict precedence among the activities.

***critical path***: a sequence of activities that determines the completion time of a project.

***criticality index***: the probability that a given path (activity) is (lies on) a critical path of a project.

***cutting plane***: a constraint that can be added to an existing set of constraints without excluding any feasible integer solution.

***decision variables***: the unknowns in an optimization problem.

***demand point***: a point in a metric space that is a source of demand for the service provided by the facilities.

***deterministic activity net***: a directed network in which all the parameters (such as duration, resource requirements, precedence) are known deterministically.

***dual LP***: a minimization LP problem associated with a given maximization LP problem.

***equilibrium***: a strategy combination from which no player has a unilateral incentive to depart.

***facility***: a place where a service (or product) is provided.

***facility location***: a point in a metric space where a facility is located.

***feasible direction***: a direction that preserves feasibility in a sufficiently small neighborhood of a given feasible solution.

***feasible LP***: an LP with a nonempty feasible region.

***feasible region***: the set of all feasible solutions to a given LP.

***feasible solution***: a vector that satisfies the given set of constraints.

***fixed point*** (of a function $f$): a point $x$ such that $f(x) = x$.

***float***: in a deterministic activity net, a measure of the flexibility available in scheduling an activity without delaying the project completion time.

***forward auction***: an auction with several buyers and one seller.

***frequency-based memory***: used in TS to expand the current neighborhood and examine unvisited regions.

***c-game***: a characteristic-function game.

***GAN model***: a probabilistic activity net with conditional progress and probabilistic realization of activities.

***general position***: a set of points $x_1, x_2, \ldots, x_{p+1} \in \mathcal{R}^n$ such that the vectors $x_2 - x_1$, $x_3 - x_1, \ldots, x_{p+1} - x_1$ are linearly independent.

***GERT model***: a probabilistic activity net with exclusive-or branching.

***goal programming (GP) problem***: an LP having multiple objective functions.

***improving direction***: a feasible direction that improves the objective function value.

**imputation**: a distribution among players of the cooperative benefit in a $c$-game.

**infeasible LP**: an LP with an empty feasible region.

**integer programming (IP) problem**: a linear programming problem in which some of the decision variables are required to be integers.

**interior point method**: a technique for solving an LP that iteratively moves through the interior of the feasible region.

**knapsack problem**: an optimization problem in which items are to be selected to maximize the total benefit without exceeding the capacity of the knapsack.

**linear programming (LP) problem**: an optimization problem involving the selection of decision variables that optimize a given linear function and that satisfy linear inequality constraints.

**location problem**: an optimization problem in which $p$ facilities are to be established to minimize the cost of meeting known demands arising from $n$ locations.

**LP relaxation**: the linear programming problem obtained by dropping the integrality requirements of an IP.

**$p$-median**: a set of $p$ locations for facilities that minimizes the total (transportation) cost of satisfying all demands.

**metric space**: a set of points on which a distance function has been defined.

**mixed strategy**: a probability distribution over a set of pure strategies.

**move**: an operation that modifies a given solution $x$ to obtain a "nearby" solution $x'$ to an optimization problem.

**neighbor** (of a solution $x$): another solution $x'$ to an optimization problem reached by a valid move from $x$.

**neighborhood** (of a solution $x$): the set of all neighbors of $x$.

**neighborhood search algorithm**: an iterative algorithm that begins with a solution to an optimization problem and iteratively moves to an improving solution if one exists.

**noncooperative game**: a mathematical model of strategic behavior in the absence of binding agreements.

**normalized characteristic function**: a mapping from the set of all coalitions to $[0, 1]$.

**nucleolus**: a $c$-game solution concept based on minimizing the dissatisfaction of the most dissatisfied coalitions.

**objective function**: the function associated with a given optimization problem that is to be maximized or minimized.

**optimal solution**: a feasible solution to an optimization problem achieving the largest (or perhaps smallest) value of the objective function.

**packing**: a subset of items from a given list that can be placed in a bin of specified capacity.

**path relinking**: a search strategy in TS that chooses moves from initiating solutions that incorporate attributes of guiding solutions.

**payoff function**: a mapping from the set of feasible strategy combinations to $\mathcal{R}^n$, where $n$ is the number of players.

**perfect information**:  a situation arising when the history of a game is known to all players.

**PERT model**:  a probabilistic activity net with strict precedence and activity durations that are known only in probability.

**pivot**:  a move from a given basic solution of an LP to one differing in only one active constraint.

**players**:  a collection of interacting decisionmakers.

**polyhedron**:  the set of points satisfying a given finite set of linear inequalities.

**probabilistic activity net**:  a directed network in which some or all of the parameters, including the realization of the activities, are probabilistically known.

**pure strategy**:  a plan of action available to a player.

**recency-based memory**:  used in TS to obtain a modified neighborhood that excludes (makes tabu) recently visited solutions.

**reduced cost**:  the unit change in the objective function incurred by increasing the value of a given decision variable.

**redundant constraint**:  a constraint that can be removed from a given set of constraints without changing the set of feasible solutions.

**set cover**:  a family of subsets such that each of a specified list of elements is contained in at least one subset.

**set covering problem**:  an optimization problem in which a minimum cost set cover is needed.

**set partition**:  a family of subsets such that each of a specified list of elements is contained in exactly one subset.

**set partitioning problem**:  an optimization problem in which a minimum cost set partition is needed.

**Shapley value**:  a $c$-game solution concept based on players' marginal worths to coalitions on joining, assuming all orders of formation are equally likely.

**$p$-simplex**:  the convex hull of a collection of $p + 1$ points in general position.

**simplex method**:  a technique for solving an LP that moves from vertex to neighboring vertex along the boundary of the feasible region.

**simplicial subdivision** (of a simplex):  a decomposition of the simplex into a collection of simplices that intersect only along entire common faces.

**slack variables**:  the components of $b - Ax^*$ where $x^*$ is a feasible solution to an LP with constraints $Ax \leq b$, $x \geq 0$.

**solution**:  an equilibrium or set of equilibria in a noncooperative game, or an imputation or set of imputations in a $c$-game.

**strategic behavior**:  behavior such that the outcome of an individual's actions depends on actions yet to be taken by others.

**strategic oscillation**:  a search strategy in TS that repeatedly approaches and crosses the feasibility boundary from different directions.

**strategy combination**:  a vector of strategies, one for each player.

**tableau**:  a table storing all information pertinent to a given basic solution for an LP.

**tabu search (TS):**  a metaheuristic solution methodology that employs adaptive memory and responsive exploration of the solution space.

**tabu tenure:**  the number of iterations that an attribute remains tabu-active in TS, thus preventing the revisiting of recently examined solutions.

**totally unimodular matrix:**  a 0-1 matrix such that every square submatrix has determinant 0, +1, or −1.

**unbounded LP:**  a linear programming problem that is not bounded.

**variable-depth neighborhood search algorithm:**  a neighborhood search algorithm that moves to a new solution in a neighborhood of variable depth at each iteration.

**very large-scale neighborhood:**  a neighborhood that has very large size with respect to the problem input.

**very large-scale neighborhood search algorithm:**  a neighborhood search algorithm that selects from a very large-scale neighborhood at each iteration.

**vertex** (of a feasible region):  given a feasible region $S$, a point $x \in S \subseteq \mathcal{R}^n$ defined by the intersection of exactly $n$ linearly independent constraints.

**Vickrey-Clarke-Groves (VCG) mechanism:**  an efficient auction mechanism in which each bidder has an incentive to be honest about his/her valuation of the items.

**winner determination problem:**  the problem of finding a set of winning bids to maximize total monetary value in an auction.

# 16.1  LINEAR PROGRAMMING

Linear programming (LP) involves the optimization of a linear function under linear inequality constraints. Applications of this model are widespread, including problems arising in marketing, finance, inventory, capital budgeting, computer science, transportation, and production. Algorithms are available that, in practice, solve LP problems efficiently.

## 16.1.1  BASIC CONCEPTS

**Definitions:**

A **linear programming (LP) problem** is an optimization problem that can be written

$$
\begin{aligned}
\text{maximize:} \quad & cx \\
\text{subject to:} \quad & \widetilde{A}x \leq \widetilde{b}
\end{aligned}
\tag{1}
$$

where $\widetilde{A}$ is a given $q \times n$ matrix, $c$ is a given row vector of length $n$, and $\widetilde{b}$ is a given column vector of length $q$. The **decision variables** of problem (1) are represented by the column vector $x$ of length $n$.

A **feasible solution** is a vector $x$ satisfying $\widetilde{A}x \leq \widetilde{b}$. The **feasible region** is the subset of all feasible solutions in $\mathcal{R}^n$. If no feasible solution exists (so that the feasible region is empty), the LP problem is **infeasible**; otherwise it is **feasible**.

Each of the $q$ inequalities in $\widetilde{A}x \le \widetilde{b}$ is a **constraint**. A constraint is **redundant** if removing it from (1) doesn't change the feasible region.

For a feasible solution $x$, the function $z = cx$ is the **objective function**, with $cx$ the **objective value** of $x$. When the objective value $z^* = cx^*$ is also maximum, then the feasible $x^*$ is an **optimal** solution. If the objective value can be made arbitrarily large over the feasible region, the LP problem is **unbounded**. Otherwise it is **bounded**.

A vector $y$ is a **feasible direction** at $x^*$ if there is some $\tau > 0$ such that $\widetilde{A}(x^* + \lambda y) \le \widetilde{b}$ for all $0 \le \lambda \le \tau$. If $cy > 0$ also holds, then $y$ is an **improving direction**.

A constraint of the system $\widetilde{A}x \le \widetilde{b}$ that is satisfied with equality by a feasible solution $x^*$ is **active** at $x^*$.

A set of constraints $\{a_i x \le b_i \mid i = 1, 2, \ldots, k\}$ is **linearly independent** if the vectors $\{a_1, a_2, \ldots, a_k\}$ are linearly independent (see §6.1.3).

A **vertex** is a feasible solution with $n$ linearly independent active constraints. A vertex with more than $n$ active constraints is **degenerate**. An LP problem with a degenerate vertex is **degenerate**.

A set $S$ is **convex** if the line segment joining any two of its points is contained in $S$: i.e., $\lambda x + (1 - \lambda)y \in S$ holds for all $x, y \in S$ and $0 \le \lambda \le 1$.

Let $L$ be the line segment connecting the two vertices $x^1$ and $x^2$. Then $x^1$ and $x^2$ are **adjacent** if for all points $y \ne x^1, x^2$ on $L$ and all feasible $y^1$ and $y^2$, the only way $y$ can equal $\frac{1}{2}y^1 + \frac{1}{2}y^2$ is if $y^1$ and $y^2$ are also on $L$. In this case, $L$ is an **edge**.

**Facts:**

**1.** Linear programming models arise in a wide variety of applications, which typically involve the allocation of scarce resources in the best possible way. A sample of such application areas, with reference sources, is given in the following table.

| application | references |
|---|---|
| production scheduling and inventory control | [Ch83], [Ga85] |
| tanker scheduling | [BaJaSh90] |
| airline scheduling | [Ga85] |
| cutting stock problems | [BaJaSh90], [Ch83] |
| workforce planning | [Ga85] |
| approximation of data | [Ch83] |
| matrix games | [Ch83] |
| blending problems | [BaJaSh90] |
| petroleum refining | [Ga85] |
| capital budgeting | [BaJaSh90] |
| military operations | [Ga85] |
| land use planning | [Ga85] |
| agriculture | [Ga85] |
| banking and finance | [Ga85] |
| environmental economics | [Ga85] |
| health care | [Ga85] |
| marketing | [Ga85] |
| public policy | [Ga85] |

**2.** The general concepts of linear programming were first developed by G. B. Dantzig in 1947 in connection with military planning problems for the U.S. Air Force. Earlier, in 1939, L. V. Kantorovich formulated and solved a particular type of LP problem in production planning.

**3.** The term "linear programming" conveys its historical origins and purpose: it is a mathematical model involving *linear* constraints and a *linear* objective function, used for the optimal planning *(programming)* of operations.

**4.** Form (1) of an LP naturally occurs in the selection of levels for production activities that maximize profit subject to constraints on the utilization of the given resources.

**5.** These transformations on an LP do not change feasible (or optimal) solutions:

- *constraints*: change the sense of an inequality by multiplying both sides by $-1$; or replace $a_i x = b_i$ with $a_i x \leq b_i$ and $-a_i x \leq -b_i$; or replace $a_i x \leq b_i$ with $a_i x + s_i = b_i$ and $s_i \geq 0$;

- *variables*: for $x_j$ unrestricted, set $x_j := x'_j - x''_j$ with $x'_j, x''_j \geq 0$; or for $x_j \leq 0$, set $x_j := -x'_j$ with $x'_j \geq 0$;

- *objective function*: change a minimization (maximization) problem to a maximization (minimization) problem by setting $c := -c$.

**6.** *Farkas' lemma*: Suppose $\widetilde{A}$ is a $q \times n$ matrix and $c$ is an $n$-row vector. Then the following are equivalent:

- $cy \geq 0$ for all $y \in \mathcal{R}^n$ such that $\widetilde{A}y \geq 0$;

- there exists some $u \in \mathcal{R}^q$ such that $u \geq 0$, $c = u\widetilde{A}$.

This result is important in establishing the optimality conditions for linear programming problems; it can also be applied to show the existence (and uniqueness) of solutions to linear models of economic exchange and stationary distributions in finite Markov chains (§7.7). (J. Farkas, 1847–1930)

**7.** A feasible solution with an improving direction cannot be optimal for (1).

**8.** A feasible solution with no improving direction is always optimal for (1).

**9.** If a feasible solution to (1) has an improving direction $y$ and if $\widetilde{A}y \leq 0$ then the LP problem is unbounded.

**10.** Each LP problem is either infeasible, unbounded, or has an optimal solution. This need not be the case for nonlinear optimization problems.

**11.** Form (1) of an LP is helpful for understanding the geometric properties of an LP.

**12.** For algorithmic purposes the following form, form (2), of an LP is preferred:

$$\text{maximize:}\quad cx$$
$$\text{subject to:}\quad Ax \leq b \tag{2}$$
$$x \geq 0.$$

Here $A$ is an $m \times n$ matrix.

**13.** The most general form of an LP problem is

$$\text{maximize (or minimize):}\quad dx_1 + ex_2 + fx_3$$
$$\text{subject to:}\quad Ax_1 + Bx_2 + Cx_3 \leq a$$
$$Dx_1 + Ex_2 + Fx_3 \geq b$$
$$Gx_1 + Hx_2 + Kx_3 = c$$
$$x_1 \geq 0,\ x_2 \leq 0,\ x_3 \text{ unrestricted.}$$

In this formulation $A$, $B$, $C$, $D$, $E$, $F$, $G$, $H$, and $K$ are matrices; $a$, $b$, and $c$ are column vectors; and $d$, $e$, and $f$ are row vectors.

**14.** The feasible region of an LP problem is a convex set.

**15.** *Equivalence of forms:*    The general form in Fact 13 is equivalent to both form (1) and form (2) in the following sense: any of these three forms can be transformed into another using the operations of Fact 5. Each form possesses the same set of feasible (or optimal) solutions.

**16.** An excellent glossary of linear programming terms, as well as concepts in general mathematical optimization, can be found at

- http://glossary.computing.society.informs.org/ver2/mpgwiki/

**17.** A comprehensive introduction to building linear programming models and a discussion of various applications are found in Williams [Wi13].

### Examples:

**1.** *Feed mix:*    A manufacturer produces a special feed for farm animals. To ensure that the feed is nutritionally balanced, each bag of feed must supply at least 1250 mg of Vitamin A, 250 mg of Vitamin B, 900 mg of Vitamin C, and 232.5 mg of Vitamin D. Three different grains $(1, 2, 3)$ are blended to create the final product. Each ounce of Grain 1 supplies 2, 1, 5, 0.6 mg of Vitamins A, B, C, D, respectively. Each ounce of Grain 2 provides 3, 1, 3, 0.25 mg of Vitamins A, B, C, D, while each ounce of Grain 3 provides 7, 1 mg of Vitamins A, D. The costs (per ounce) of the constituent grains are 41, 35, and 96 cents for Grains 1, 2, and 3, respectively.

The manufacturer wants to determine the minimum cost mix of grains that satisfies all four nutritional requirements. If $x_i$ is the number of ounces of Grain $i$ that are blended in the final product, then the manufacturer's problem is modeled by the following LP:

$$
\begin{aligned}
\text{minimize:} \quad & 0.41x_1 + 0.35x_2 + 0.96x_3 \\
\text{subject to:} \quad & 2x_1 + 3x_2 + 7x_3 \geq 1250 \\
& x_1 + x_2 \geq 250 \\
& 5x_1 + 3x_2 \geq 900 \\
& 0.6x_1 + 0.25x_2 + x_3 \geq 232.5 \\
& x_1, \ x_2, \ x_3 \geq 0.
\end{aligned}
$$

Each constraint in this LP corresponds to a nutritional requirement. It turns out that the optimal solution to the LP is $x_1^* = 200.1$, $x_2^* = 49.9$, $x_3^* = 100.01$ with $z^* = 195.5$. Note that the amount of Vitamin C supplied by this solution is in excess of 900 mg, while the other vitamins are supplied in exactly the minimum amounts.

**2.** The LP in Example 1 is not in either form (1) or form (2). However, using Fact 5 it can be transformed into form (2), giving the equivalent representation

$$
\begin{aligned}
\text{maximize:} \quad & -0.41x_1 - 0.35x_2 - 0.96x_3 \\
\text{subject to:} \quad & -2x_1 - 3x_2 - 7x_3 \leq -1250 \\
& -x_1 - x_2 \leq -250 \\
& -5x_1 - 3x_2 \leq -900 \\
& -0.6x_1 - 0.25x_2 - x_3 \leq -232.5 \\
& x_1, \ x_2, \ x_3 \geq 0.
\end{aligned}
$$

**3.** The following figure shows the feasible region of the LP problem

$$\begin{aligned}
\text{maximize:} \quad & -x_1 \\
\text{subject to:} \quad & -x_2 \le 0 & \text{(A)} \\
& -x_1 - x_2 \le -4 & \text{(B)} \\
& -x_1 + x_2 \le 4 & \text{(C)} \\
& -3x_1 + 5x_2 \le 30 & \text{(D)} \\
& -x_1 + 3x_2 \le 22 & \text{(E)}
\end{aligned}$$

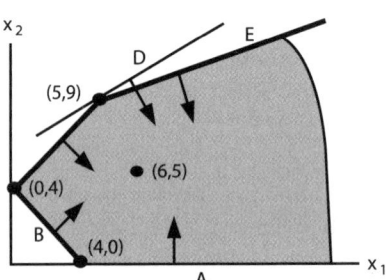

This LP has $n = 2$ decision variables $x_1, x_2$ and it has the vertices $(4, 0)$, $(0, 4)$, and $(5, 9)$. Vertex $(x_1, x_2) = (0, 4)$ is the optimal solution, achieving the maximum objective value $z = 0$. Thus, the LP is bounded, even though its feasible region is not bounded. Constraint (D) is redundant, since dropping it doesn't change the feasible region. Vertex $(5, 9)$ is degenerate, since $3 > n$ constraints are active at this vertex. All vectors are feasible directions at $(6, 5)$. At vertex $(5, 9)$, the direction $(1, -1)$ is feasible, but the direction $(1, 1)$ is not. Vertices $(0, 4)$ and $(5, 9)$ are adjacent, as are $(4, 0)$ and $(0, 4)$.

**4.** *Farkas' lemma:* The row vectors $a^1, a^2$ of $\widetilde{A} = \begin{pmatrix} 2 & 4 \\ 5 & 2 \end{pmatrix}$ are shown in the next figure.

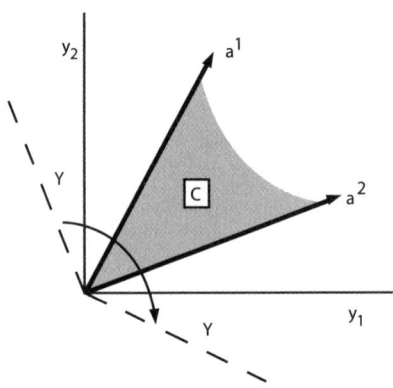

The set $Y = \{y \mid \widetilde{A}y \ge 0\}$ is the region bounded by the two dashed lines. Notice that if $c = u\widetilde{A}$ for some $u \ge 0$ then $c$ must lie in the cone $C$ bounded by the vectors $a^1$ and $a^2$. Geometrically, any $c \in C$ makes an acute angle with every $y \in Y$, hence $cy \ge 0$. Conversely, any $c$ making an acute angle with every $y \in Y$ must be in $C$.

**5.** Fact 10 is illustrated using the following LP problem:

$$\begin{aligned}
\text{maximize:} \quad & -x_1 - x_2 \\
\text{subject to:} \quad & -2x_1 + x_2 \le -1 \\
& -x_1 - 2x_2 \le -2 \\
& x_1, \ x_2 \ge 0.
\end{aligned}$$

This LP has the optimal solution $(x_1^*, x_2^*) = (\frac{4}{5}, \frac{3}{5})$. Suppose the objective function is changed to $z = x_1 - x_2$. Then $(x_1, x_2) = (a, 0)$ is feasible for $a \ge 2$ with objective value $a$. Thus $z$ can be made arbitrarily large and the LP is unbounded. On the other hand, if the second constraint is changed to $4x_1 - x_2 \le -1$, the feasible region is empty and the LP is infeasible.

**6.** *Product mix:* A company manufactures $n$ types of a product, using $m$ shops. Type $j$ requires $a_{ij}$ machine-hours in shop $i$. There is a limitation of $b_i$ machine-hours for shop $i$ and the sale of each type $j$ unit brings the company a profit $c_j$. The optimization problem facing the company is given by an LP problem in form (2). Namely, if $x_j$ is the number of units produced of type $j$, then the optimization problem is

$$\begin{aligned}
\text{maximize:} \quad & \sum_{j=1}^{n} c_j x_j \\
\text{subject to:} \quad & \sum_{j=1}^{n} a_{ij} x_j \le b_i, \qquad i = 1, \ldots, m \\
& x_j \ge 0, \qquad j = 1, \ldots, n.
\end{aligned}$$

**7.** *Transportation:* A product stored at $m$ warehouses needs to be shipped to satisfy demands at $n$ markets. Warehouse $i$ has a supply of $s_i$ units of the product, and market $j$ has a demand of $d_j$ units. The cost of shipping a unit of product from warehouse $i$ to market $j$ is $c_{ij}$. The problem is to determine the number of units $x_{ij}$ to ship from warehouse $i$ to market $j$ in order to satisfy all demands while minimizing cost:

$$\begin{aligned}
\text{maximize:} \quad & \sum_{i=1}^{m} \sum_{j=1}^{n} c_{ij} x_{ij} \\
\text{subject to:} \quad & \sum_{j=1}^{n} x_{ij} \le s_i, \qquad i = 1, \ldots, m \\
& \sum_{i=1}^{m} x_{ij} = d_j, \qquad j = 1, \ldots, n \\
& x_{ij} \ge 0, \qquad i = 1, \ldots, m, \ j = 1, \ldots, n.
\end{aligned}$$

This is an LP in the form specified by Fact 13. Using the transformations in Fact 5, this optimization problem can alternatively be expressed as an LP in the form (1).

---

## 16.1.2   TABLEAUS

### Definitions:

Suppose that an LP is expressed in form (2) of §16.1.1, with $A$ an $m \times n$ matrix. A *tableau* is any table $\dfrac{u \mid z}{D \mid f}$ with the following properties:

- $D$ is an $m \times (m+n)$ matrix with entries $d_{tj}$; $z$ is a real number; $u$ is an $(m+n)$-row vector; and $f$ is an $m$-column vector.

- Associated with the tableau is a partition $\Sigma_B, \Sigma_N$ of the integers $1, \ldots, m + n$. The set $\Sigma_B$, with cardinality $m$, is the **basic set** and $\Sigma_N$ is the **nonbasic set**.

- For every row index $t = 1, \ldots, m$, there is a column of $D$ equal to zero in all coordinates except for the $t$th coordinate, which equals 1. The index of this column is $\varphi(t)$ where $\varphi$ is a function from $\{1, \ldots, m\}$ to $\Sigma_B$ associated with the tableau.

- $\dfrac{u \mid z}{D \mid f}$ can be obtained from $\dfrac{-c \quad 0 \mid 0}{A \quad I \mid b}$ (where $\Sigma_B = \{n + 1, \ldots, m + n\}$ and $\varphi(t) = n + t$, $t = 1, \ldots, m$) by performing the following **pivot** operation a finite number of times:

  P1.  choose a row index $t^* \in \{1, \ldots, m\}$ and a column index $j^* \in \Sigma_N$ with $d_{t^* j^*} \neq 0$;

  P2.  multiply row $t^*$ by $1/(d_{t^* j^*})$;

  P3.  add appropriate multiples of row $t^*$ to all other rows to make $u_{j^*} = 0$ and to make $d_{t j^*} = 0$ for all $t \neq t^*$;

  P4.  remove $j^*$ from $\Sigma_N$ and place it in $\Sigma_B$; remove $\varphi(t^*)$ from $\Sigma_B$ and place it in $\Sigma_N$; set $\varphi(t^*) = j^*$.

In the pivot operation, $\varphi(t^*)$, before replacement, is the index of the **leaving variable** and $j^*$ is the index of the **entering variable**.

The set of variables $\{x_i \mid i \in \Sigma_B\}$ are the **basic variables** and the remaining variables are the **nonbasic variables**.

A **basic solution** is a vector $x^*$ with its basic variables defined by $x_i^* = f_t$ where $t = \varphi^{-1}(i)$; its nonbasic variables have $x_i^* = 0$. If $f \geq 0$ then $x^*$ is a **basic feasible solution (BFS)**.

The **basis matrix** $B$ is the $m \times m$ matrix consisting of the columns of $[A \ I]$ corresponding to the basic variables; the **nonbasis matrix** $N$ is the $m \times n$ matrix corresponding to the nonbasic variables. Let $c_B$ $[c_N]$ denote the vector of basic [nonbasic] components of $c$. Let $x_B$ $[x_N]$ denote the vector of basic [nonbasic] components of $x$.

The **reduced cost** of nonbasic variable $x_j$ is the negative of $u_j$ in the associated tableau.

The **slack variables** are given by $(x_{n+1}, x_{n+2}, \ldots, x_{n+m}) = b - Ax$.

**Facts:**

**1.** Every BFS of (2) corresponds to a vertex of (1), with $q = m + n$, $\widetilde{A} = \begin{bmatrix} A \\ -I \end{bmatrix}$, $\widetilde{b} = \begin{bmatrix} b \\ 0 \end{bmatrix}$.

**2.** In the absence of degeneracy the correspondence in Fact 1 is one-to-one; otherwise it is many-to-one.

**3.** Every LP problem (2) with an optimal solution has an optimal solution that is a vertex. Since the number of vertices is finite, LP problems are combinatorial in nature; that is, an LP can be solved in theory by enumerating its vertices and then selecting one with maximum objective value.

**4.** Let $x^*$ be a BFS of (2). All information pertinent to $x^*$ is contained in its tableau which (after possibly permuting the first $m + n$ columns) is

$$\begin{array}{cc|c} 0 & c_B B^{-1} N - c_N & c_B B^{-1} b \\ \hline I & B^{-1} N & B^{-1} b \end{array}.$$

Here $c_B B^{-1} b$ is the objective value $z$ of $x^*$. The value of the basic variable $x_i^*$ is the $t$th component of $B^{-1}b$, where $i = \varphi(t)$. Every nonbasic variable has value 0.

**5.** A tableau expresses the set of equations below, called a *dictionary* [Ch83]:

$$x_B = B^{-1}b - B^{-1}Nx_N$$
$$z = c_B B^{-1}b + (c_N - c_B B^{-1}N)x_N.$$

**6.** The reduced costs of the nonbasic variables are given by the vector $c_N - c_B B^{-1}N$. Basic variables have zero reduced cost.

**7.** Column $i$ of $B^{-1}$ is identical to $d^{m+i}$, the column of $D$ associated with the slack variable $x_{n+i}$.

**Examples:**

**1.** When slack variables $x_4, x_5, x_6$ are added to the LP

$$\begin{aligned}
\text{maximize:} \quad & 3x_1 + 4x_2 + 4x_3 \\
\text{subject to:} \quad & 3x_1 - x_3 \le 5 \\
& -9x_1 + 4x_2 + 3x_3 \le 12 \\
& -6x_1 + 2x_2 + 4x_3 \le 2 \\
& x_1, x_2, x_3 \ge 0
\end{aligned}$$

the following equivalent LP is formed:

$$\begin{aligned}
\text{maximize:} \quad & 3x_1 + 4x_2 + 4x_3 + 0x_4 + 0x_5 + 0x_6 \\
\text{subject to:} \quad & 3x_1 - x_3 + x_4 = 5 \\
& -9x_1 + 4x_2 + 3x_3 + x_5 = 12 \\
& -6x_1 + 2x_2 + 4x_3 + x_6 = 2 \\
& x_1, x_2, \ldots, x_6 \ge 0.
\end{aligned}$$

The associated tableau, with $\Sigma_B = \{4, 5, 6\}$ and $\Sigma_N = \{1, 2, 3\}$, is then

| $-3$ | $-4$ | $-4$ | 0 | 0 | 0 | 0 |
|---|---|---|---|---|---|---|
| 3 | 0 | $-1$ | 1 | 0 | 0 | 5 |
| $-9$ | 4 | 3 | 0 | 1 | 0 | 12 |
| $-6$ | 2 | 4 | 0 | 0 | 1 | 2 |

Here $\varphi(1) = 4$, $\varphi(2) = 5$, $\varphi(3) = 6$. The basic variables are $x_4 = 5$, $x_5 = 12$, $x_6 = 2$ and the nonbasic variables are $x_1 = 0$, $x_2 = 0$, $x_3 = 0$. The basic feasible solution associated with this tableau is $x = (0, 0, 0, 5, 12, 2)^T$ with objective value $z = 0$. The nonbasic variables $x_1, x_2, x_3$ have reduced costs 3, 4, 4, respectively.

**2.** A pivot is now performed on the tableau in Example 1 using $t^* = 3$ and $j^* = 2$, so the entering variable is $x_2$ and the leaving variable is $x_6$. The resulting tableau (a) follows, where $\Sigma_B = \{2, 4, 5\}$ and $\varphi(1) = 4$, $\varphi(2) = 5$, $\varphi(3) = 2$. The corresponding BFS is $x = (0, 1, 0, 5, 8, 0)^T$ with objective value $z = 4$. If a pivot is performed on (a) using $t^* = 1$ and $j^* = 1$, then tableau (b) results. Here $\varphi(1) = 1$, $\varphi(2) = 5$, $\varphi(3) = 2$ and the new BFS is $x = (\frac{5}{3}, 6, 0, 0, 3, 0)^T$ with objective value $z = 29$.

tableau (a)

| $-15$ | 0 | 4 | 0 | 0 | 2 | 4 |
|---|---|---|---|---|---|---|
| 3 | 0 | $-1$ | 1 | 0 | 0 | 75 |
| 3 | 0 | $-5$ | 0 | 1 | $-2$ | 8 |
| $-3$ | 1 | 2 | 0 | 0 | $\frac{1}{2}$ | 1 |

tableau (b)

| 0 | 0 | $-1$ | 5 | 0 | 2 | 29 |
|---|---|---|---|---|---|---|
| 1 | 0 | $-\frac{1}{3}$ | $\frac{1}{3}$ | 0 | 0 | $\frac{5}{3}$ |
| 0 | 0 | $-4$ | $-1$ | 1 | $-2$ | 3 |
| 0 | 1 | 1 | 1 | 0 | $\frac{1}{2}$ | 6 |

For tableau (b), the basis matrix $B$ corresponds to columns $1, 5, 2$ of $[A\ I]$, namely

$$B = \begin{pmatrix} 3 & 0 & 0 \\ -9 & 1 & 4 \\ -6 & 0 & 2 \end{pmatrix}. \text{ From Fact 7, the inverse matrix } B^{-1} = \begin{pmatrix} \frac{1}{3} & 0 & 0 \\ -1 & 1 & -2 \\ 1 & 0 & \frac{1}{2} \end{pmatrix} \text{ consists of}$$

columns $4, 5, 6$ in tableau (b).

## 16.1.3  SIMPLEX METHOD

The simplex method is in practice remarkably efficient and it is widely used for solving LP problems. The solution idea dates back to J. B. J. Fourier (1768–1830); it was developed and popularized in 1947 by G. B. Dantzig (1914–2005). This section presents two descriptions of the same algorithm: the first is geometrically intuitive, the second is closer to its actual implementation.

**Facts:**

**1.** *Simplex algorithm I*: This method (Algorithm 1) solves a linear programming problem in form (1) of §16.1.1. Assuming that an initial vertex is known, this algorithm travels from vertex to vertex along improving edges until an optimal vertex is reached or an unboundedness condition is detected. In Algorithm 1, the rows of $\tilde{A}$ are denoted $a_1, a_2, \ldots, a_q$ and $\tilde{b}$ has the corresponding components $b_1, b_2, \ldots, b_q$.

---

**Algorithm 1:  Simplex algorithm — form (1).**

input: LP in form (1), initial vertex $x_0$

output: an optimal vertex or an indication of unboundedness

$k := 0$

find a subsystem $Bx \leq r$ of (1) consisting of $n$ linearly independent constraints
      active at $x_k$

$S :=$ list containing the indices of these active constraints

{Main loop}

**if** $u \equiv cB^{-1} \geq 0$ **then** $x_k$ is an optimal solution — stop

**else** {an improving direction}

    $i^* :=$ the smallest index such that $u_{i^*} < 0$

    $y :=$ column $i^*$ of $-B^{-1}$

    **if** $\tilde{A}y \leq 0$ **then** the LP problem is unbounded — stop

    **else** {move to next vertex (possibly same as last)}

        $j^* :=$ smallest index $j$ attaining minimum $\lambda \equiv \min\left\{\frac{b_j - a_j x_k}{a_j y} \mid j \notin S, a_j y > 0\right\}$

        $x_{k+1} := x_k + \lambda y$

        $S[i^*] := j^*$; update $B$

        $k := k + 1$

{Continue with next iteration of main loop}

---

**2.** *Simplex algorithm II*: This method (Algorithm 2) solves a linear programming problem in form (2) of §16.1.1, assuming that $b \geq 0$. It proceeds by successively identifying nonbasic variables having positive reduced cost and pivoting them into the current basis in a way that maintains a basic feasible solution (BFS).

---

**Algorithm 2**:   **Simplex algorithm — form (2).**

input: LP in form (2), with $b \geq 0$

output: an optimal BFS or an indication of unboundedness

begin with the initial tableau: $\dfrac{-c \quad 0 \;\big|\; 0}{A \quad I \;\big|\; b}$, where $\Sigma_B = \{n+1, \ldots, m+n\}$

and $\varphi(t) = n + t, \ t = 1, \ldots, m$

$x_0 := (x_B, x_N)$ where $x_B = b \geq 0$ and $x_N = 0$

$k := 0$

{Main loop}

**if** $u_j \geq 0$ for all $j \in \Sigma_N$ **then** $x_k$ is an optimal solution — stop

**else** {select entering variable}

    $j^* :=$ the smallest index with $u_{j^*} < 0$

    **if** $d_{tj^*} \leq 0$ for $t = 1, \ldots, m$ **then** the LP is unbounded — stop

    **else**

        $t^* :=$ an index $t$ achieving the minimum

            $\min\{\frac{f_t}{d_{tj^*}} \mid t = 1, \ldots, m; \ d_{tj^*} > 0\}$

        (if there are several such $t^*$, make $\varphi(t^*)$ as small as possible)

        do a pivot with entering index $j^*$, leaving index $\varphi(t^*)$

        set component $\varphi(t)$ of $x_{k+1}$ to $f_t$ for $t = 1, \ldots, m$ and the remaining components of $x_{k+1}$ to zero

    $k := k + 1$

{Continue with next iteration of main loop}

---

**3.** There are examples for which Algorithm 1 requires exponential running time, and similarly for Algorithm 2.

**4.** In practice the number of iterations of Algorithms 1 and 2 is proportional to the number of constraints $m$ and grows slowly with the number of variables $n$.

**5.** There is a one-to-one correspondence between the vertex $x_k$ of Algorithm 1 and the BFS $x_k$ of Algorithm 2, when $\widetilde{A}$ is set to $\begin{bmatrix} A \\ -I \end{bmatrix}$ and $\widetilde{b}$ is set to $\begin{bmatrix} b \\ 0 \end{bmatrix}$.

**6.** Interchanging a basic and a nonbasic variable in Algorithm 2 corresponds to interchanging a nonactive and an active constraint in Algorithm 1.

**7.** In the absence of degeneracy, the objective value strictly increases at each step (in both algorithms). The method of breaking ties by choosing the smallest index prevents *cycling* and ensures termination in finite time. In practice, though, cycling is rare and other rules are used.

**8.** When a vertex is not known in Algorithm 1 (when $b \not\geq 0$ in Algorithm 2) a preliminary LP problem, *Phase I*, can be solved to get an initial vertex (a starting tableau).

**9.** The *revised simplex method* is a variation of Algorithm 2. Instead of maintaining the entire tableau at each step only $B^{-1}$ is kept. Columns of $[A \ I]$ are brought in from storage as needed to find $j^*$ and $t^*$. This method is good for sparse matrices $A$ with many columns.

**10.** LP problems can be submitted online for solution using the NEOS server at

- http://www.neos-server.org/neos

**11.** A survey of software packages for solving linear programming problems is described in [Fo15]. Virtually all of these products run on PCs. In many cases, the LP solvers are linked to more general modeling packages that provide a single environment for carrying out the formulation, solution, and analysis of LP problems.

**12.** Many software packages are available to model and to solve LP problems.

–The AMPL and GAMS modeling languages are described at

- http://www.ampl.com
- http://www.gams.com

–Solver software (LINDO API and MATLAB) can be found at

- http://www.lindo.com
- http://www.mathworks.com/products/optimization

–Integrated solver and modeling environments are described at

- http://www.aimms.com
- http://optimization-suite.coin-or.org
- http://www.cplex.com
- http://www.gurobi.com
- http://www.lindo.com

## Examples:

**1.** The LP in Example 1 of §16.1.2 can be placed in the form (1) with

$$\widetilde{A} = \begin{pmatrix} 3 & 0 & -1 \\ -9 & 4 & 3 \\ -6 & 2 & 4 \\ -1 & 0 & 0 \\ 0 & -1 & 0 \\ 0 & 0 & -1 \end{pmatrix}, \quad \widetilde{b} = \begin{pmatrix} 5 \\ 12 \\ 2 \\ 0 \\ 0 \\ 0 \end{pmatrix}, \quad c = (3 \ 4 \ 4).$$

If $x_0 = (0, 1, 0)^T$ then constraints $3, 4, 6$ are active at $x_0$ and $S = [3, 4, 6]$. Thus

$$B = \begin{pmatrix} -6 & 2 & 4 \\ -1 & 0 & 0 \\ 0 & 0 & -1 \end{pmatrix}, \quad B^{-1} = \begin{pmatrix} 0 & -1 & 0 \\ \frac{1}{2} & -3 & 2 \\ 0 & 0 & -1 \end{pmatrix}, \quad u = cB^{-1} = (2 \ -15 \ 4).$$

Here $i^* = 2$, $y = (1, 3, 0)^T$, $\widetilde{A}y = (3, 3, 0, -1, -3, 0)^T$, and $\widetilde{b} - \widetilde{A}x_0 = (5, 8, 0, 0, 1, 0)$. Then $\lambda = \min\{\frac{5}{3}, \frac{8}{3}\} = \frac{5}{3}$ and $j^* = 1$. The new vertex is $x_1 = x_0 + \lambda y = (\frac{5}{3}, 6, 0)^T$ and $S$ is updated to $S = [3, 1, 6]$, so $B$ now contains rows $3, 1, 6$ of $\widetilde{A}$. Additional iterations of Algorithm 1 can then be carried out using the updated $S$ and $B$.

**2.** The same LP can alternatively be solved using Algorithm 2. For illustration, suppose that the tableau (a) from Example 2 (§16.1.2) is given, corresponding to the BFS $x_1 = (0, 1, 0, 5, 8, 0)^T$ and $\Sigma_B = \{2, 4, 5\}$. Here $u = (-15, 0, 4, 0, 0, 2)$ and $j^* = 1$ is chosen. The minimum ratio test gives $\min\{\frac{5}{3}, \frac{8}{3}\} = \frac{5}{3}$ and $t^* = 1$. The next pivot produces tableau (b) in Example 2 (§16.1.2), with $\Sigma_B = \{1, 2, 5\}$ and $x_2 = (\frac{5}{3}, 6, 0, 0, 3, 0)^T$. Here $u = (0, 0, -1, 5, 0, 2)$ so a further pivot is performed using $j^* = 3$ and $t^* = 3$, giving the tableau below. Since $u \geq 0$ the BFS $x_3 = (\frac{11}{3}, 0, 6, 0, 27, 0)^T$ is an optimal solution to the LP, with optimal objective value $z^* = 35$.

| 0 | 1 | 0 | 6 | 0 | $\frac{5}{2}$ | 35 |
|---|---|---|---|---|---|---|
| 1 | $\frac{1}{3}$ | 0 | $\frac{2}{3}$ | 0 | $\frac{1}{6}$ | $\frac{11}{3}$ |
| 0 | 4 | 0 | 3 | 1 | 0 | 27 |
| 0 | 1 | 1 | 1 | 0 | $\frac{1}{2}$ | 6 |

## 16.1.4  INTERIOR POINT METHODS

There are numerous interior point methods for solving LP problems. In contrast to the simplex method, which proceeds from vertex to vertex along edges of the feasible region, these methods move through the interior of the feasible region. In particular this section discusses N. Karmarkar's "projective scaling" algorithm (1984).

**Definitions:**
The **norm** of $x \in \mathcal{R}^n$ is given by $\|x\| = \sqrt{x_1^2 + x_2^2 + \cdots + x_n^2}$. (See §6.1.4.)

Let $e$ denote the row vector of $n$ 1s.

The LP problem

$$\begin{aligned} \text{minimize:} \quad & z = cx \\ \text{subject to:} \quad & Ax = 0 \\ & ex = 1 \\ & x \geq 0 \end{aligned} \tag{3}$$

is in **standard form** for Karmarkar's method if $\frac{1}{n} e$ is a feasible vector and if the optimal objective value is $z^* = 0$.

The $n \times n$ diagonal matrix $\text{diag}(x_1, x_2, \ldots, x_n)$ has diagonal entries $x_1, x_2, \ldots, x_n$. (See §6.3.1.)

The **unit simplex** in $n$ dimensions is $S_n = \{x \in \mathcal{R}^n \mid ex = 1, x \geq 0\}$.

If $\bar{x}$ is feasible to (3), Karmarkar's **centering transformation** $T_{\bar{x}} : S_n \to S_n$ is

$$T_{\bar{x}}(x) = \frac{\text{diag}(\bar{x})^{-1} x}{e \, \text{diag}(\bar{x})^{-1} x}.$$

The **projection** of a vector $v$ onto the subspace $X \equiv \{x \in \mathcal{R}^n \mid \tilde{A}x = 0\}$ is the unique vector $p \in X$ for which $(v - p)^T x = 0$ for all $x \in X$. (See §6.1.4.)

Karmarkar's **potential function** for (3) is $f(x) = \sum_{j=1}^{n} \ln \left(\frac{cx}{x_j}\right)$.

**Facts:**
1. Any LP problem can be transformed into form (3); see [BaJaSh90], [Sc86] for details.
2. The centering transformation $T_{\bar{x}}$ is one-to-one and onto.
3. The inverse of the centering transformation is

$$T_{\bar{x}}^{-1}(y) = \frac{\text{diag}(\bar{x}) \, y}{e \, \text{diag}(\bar{x}) \, y}.$$

**4.** The transformation $T_{\bar{x}}$ places $\bar{x}$ at the center of the transformed unit simplex: $T_{\bar{x}}(\bar{x}) = \frac{1}{n} e$.

**5.** The transformation $T_{\bar{x}}$ maps the feasible region of (3) to

$$Y = \{y \in \mathcal{R}^n \mid A \operatorname{diag}(\bar{x})\, y = 0,\ ey = 1,\ y \geq 0\}.$$

**6.** $W = \{w \in \mathcal{R}^n \mid A \operatorname{diag}(\bar{x})\, w = 0,\ ew = 0,\ w \geq 0\}$ is the set of all feasible directions for $Y$.

**7.** The projection of $v$ onto $W$ is $[I - P^T(PP^T)^{-1}P]v$, where $P = \begin{pmatrix} A \operatorname{diag}(\bar{x}) \\ 1\ 1\ \cdots\ 1 \end{pmatrix}$.

**8.** *Karmarkar's algorithm:*  This method (Algorithm 3) moves through the interior of the feasible region of (3), transforming the problem at each iteration to place the current point at the "center" of the transformed region.

---

**Algorithm 3**:  **Karmarkar's method.**

input: LP in form (3)

output: an optimal solution to (3)

$x_0 := \frac{e}{n}$

$k := 0$

{Main loop}

{Test for optimality within $\epsilon$}

**if** $cx_k < \epsilon$ **then** stop

**else** {find new point $y$ in transformed unit simplex}

$\quad P := \begin{pmatrix} A \operatorname{diag}(x_k) \\ 1\ 1\ \cdots\ 1 \end{pmatrix}$

$\quad c_P := [I - P^T(PP^T)^{-1}P] \operatorname{diag}(x_k)\, c^T$

$\quad y_k := \frac{e}{n} - \left( \frac{\theta}{\sqrt{n(n-1)}} \right) \frac{c_P}{\|c_P\|}$

{find new feasible point in the original space}

$\quad x_{k+1} := T_{x_k}^{-1}(y_k)$

$\quad k := k + 1$

{Continue with next iteration of main loop}

---

**9.** In Algorithm 3, $\epsilon > 0$ is a fixed tolerance chosen arbitrarily small. The parameter $\theta$ is a constant, $0 < \theta < 1$, associated with convergence of the algorithm. The value $\theta = \frac{1}{4}$ ensures the convergence of Algorithm 3.

**10.** There is a positive constant $\delta$ with $f(x_k) - f(x_{k+1}) \geq \delta$ for all iterations $k$ of Karmarkar's method. To ensure this inequality $\operatorname{diag}(x_k)\, c$, rather than $c$, is projected onto the space of feasible directions $W$.

**11.** For large problems, Karmarkar's method requires many fewer iterations than does the simplex method.

**12.** Letting $L$ be the maximum number of bits needed to represent any number associated with the LP problem, the running time of Karmarkar's algorithm is *polynomial*, namely $O(n^{3.5}L^2)$.

**13.** The earliest polynomial-time algorithm for LP problems is the ellipsoid method, proposed by L. G. Khachian in 1979. (See [Ch83], [Sc86].)

**14.** The ellipsoid method has worst-case complexity $O(n^6 L^2)$, where $L$ is defined in Fact 12. Because its calculations require high precision, this method is very inefficient in practice.

**15.** Karmarkar's polynomial-time algorithm was announced in 1984 and it has proven to be seriously competitive with the simplex method. Typically, Karmarkar's algorithm reduces the objective function by fairly significant amounts at the early iterations, often converging within 50 iterations regardless of the problem size.

**16.** More efficient interior point methods have been developed and have proven to be competitive with, and often superior to, the best simplex packages, especially for large LP problems [RoTeVi10], [Wr87].

**17.** LP problems can be submitted online for solution by several solvers that implement interior-point methods using the NEOS server at

- http://www.neos-server.org/neos

**18.** The LP solver packages mentioned in Fact 12 of §16.1.3 all support solution by interior point algorithms.

**Example:**

**1.** In the following LP the vector $x = (\frac{1}{3}, \frac{1}{3}, \frac{1}{3})^T$ is feasible and the problem has the optimal objective value $z^* = 0$, achieved for $x^* = (0, \frac{2}{3}, \frac{1}{3})^T$.

$$\begin{aligned} \text{minimize:} \quad & x_1 \\ \text{subject to:} \quad & x_1 + x_2 - 2x_3 = 0 \\ & x_1 + x_2 + x_3 = 1 \\ & x_1, x_2, x_3 \geq 0. \end{aligned}$$

Karmarkar's algorithm is started with $x_0 = (\frac{1}{3}, \frac{1}{3}, \frac{1}{3})^T$, yielding $cx_0 = \frac{1}{3}$. For illustrative purposes the value $\theta = 0.9$ is used throughout. Since $A = (1 \ 1 \ -2)$ the matrix $P = \begin{pmatrix} \frac{1}{3} & \frac{1}{3} & -\frac{2}{3} \\ 1 & 1 & 1 \end{pmatrix}$, giving $c_P = (\frac{1}{6}, -\frac{1}{6}, 0)$ and $y_0 = (0.0735, 0.5931, 0.3333)^T = x_1$. The new objective value is $cx_1 = 0.0735$. Additional iterations of Algorithm 3 are tabulated in the following table, showing convergence to the optimal $x^* = (0, \frac{2}{3}, \frac{1}{3})^T$ after just a few iterations.

| $k$ | 0 | 1 | 2 | 3 | 4 |
|---|---|---|---|---|---|
| $x_k$ | $\begin{pmatrix} 0.3333 \\ 0.3333 \\ 0.3333 \end{pmatrix}$ | $\begin{pmatrix} 0.0735 \\ 0.5931 \\ 0.3333 \end{pmatrix}$ | $\begin{pmatrix} 0.0056 \\ 0.6611 \\ 0.3333 \end{pmatrix}$ | $\begin{pmatrix} 0.0004 \\ 0.6663 \\ 0.3333 \end{pmatrix}$ | $\begin{pmatrix} 0.0000 \\ 0.6666 \\ 0.3333 \end{pmatrix}$ |
| $y_k$ | $\begin{pmatrix} 0.0735 \\ 0.5931 \\ 0.3333 \end{pmatrix}$ | $\begin{pmatrix} 0.0349 \\ 0.5087 \\ 0.4564 \end{pmatrix}$ | $\begin{pmatrix} 0.0333 \\ 0.4852 \\ 0.4814 \end{pmatrix}$ | $\begin{pmatrix} 0.0333 \\ 0.4835 \\ 0.4832 \end{pmatrix}$ | $\begin{pmatrix} 0.0333 \\ 0.4833 \\ 0.4833 \end{pmatrix}$ |
| $cx_k$ | 0.3333 | 0.0735 | 0.0056 | 0.0004 | 0.0000 |

## 16.1.5   DUALITY

Associated with every LP problem is its dual problem, which is important in devising alternative solution procedures for the original LP. The dual also provides useful information for conducting postoptimality analyses on the given LP.

**Definitions:**

Associated with every LP problem is another LP problem, its **dual**. The original problem is called the **primal**.

The dual of an LP in form (2)

$$\begin{aligned} \text{maximize:} \quad & cx \\ \text{subject to:} \quad & Ax \leq b \\ & x \geq 0 \end{aligned}$$

is defined to be the LP

$$\begin{aligned} \text{minimize:} \quad & ub \\ \text{subject to:} \quad & uA \geq c \\ & u \geq 0. \end{aligned} \tag{4}$$

The components $u_1, u_2, \ldots, u_m$ of $u$ are the **dual variables**.

**Facts:**

**1.** To find the dual of an arbitrary LP problem either transform it (§16.1.1, Fact 5) into form (2) or use the following table:

| primal | dual |
|---|---|
| maximization problem | minimization problem |
| unrestricted variable | equality constraint |
| nonnegative variable | $\geq$ constraint |
| nonpositive variable | $\leq$ constraint |
| equality constraint | unrestricted variable |
| $\leq$ constraint | nonnegative variable |
| $\geq$ constraint | nonpositive variable |

**2.** The dual of the dual LP is the primal LP.

**3.** *Weak duality theorem:* For any feasible solution $x$ to the primal and any feasible solution $u$ to the dual $cx \leq ub$.

**4.** *Strong duality theorem:* If $x^*$ is an optimal solution to (2) then there exists an optimal solution $u^*$ for (4) and $cx^* = u^*b$.

**5.** A given primal LP and its associated dual LP can only produce certain combinations of outcomes, as specified in the following table. For example, if one problem is unbounded then the other must be infeasible.

| primal | dual |
|---|---|
| optimal | optimal |
| infeasible | unbounded |
| unbounded | infeasible |
| infeasible | infeasible |

**6.** Let $x^*$ be an optimal BFS of the primal LP (2), with the corresponding tableau $\dfrac{u \mid z}{D \mid f}$. Then $u$ is an optimal BFS of the dual LP (4).

**7.** *Complementary slackness:* An optimal dual (primal) variable $u_i^*$ ($x_j^*$) can be nonzero only if it corresponds to a primal (dual) constraint active at $x^*$ ($u^*$).

**8.** *Economic interpretation*:  Suppose in the LP (2) that $b_i$ is the amount of resource $i$ available to a firm maximizing its profit. Then the optimal dual variable $u_i^*$ is the price the firm should be willing to pay (over and above its market price) for an extra unit of resource $i$.

**9.** *Dual simplex algorithm*:  This approach (Algorithm 4) can be used when a basic solution for (2) is known that is not necessarily feasible but which has nonnegative reduced costs (i.e., it is a *dual feasible* basic solution). The main idea of the algorithm is to start with the dual feasible basic solution and to maintain dual feasibility at each pivot. An optimal BFS is found once primal feasibility is achieved.

---

**Algorithm 4**:  **Dual simplex algorithm.**

input: LP in form (2), dual feasible basic solution $x_0$

output: an optimal BFS or an indication of infeasibility

associate with $x_0 = (x_B, x_N) = (f, 0)$ the tableau $\dfrac{u \mid z}{D \mid f}$, where $u \geq 0$

$k := 0$

{Main loop}

{Optimality test}

**if** $f \geq 0$ **then** $x_k$ is an optimal solution — stop

**else**

    $t^* :=$ the smallest index with $f_{t^*} < 0$

    **if** $d_{t^* j} \geq 0$ for all $j$ **then** the LP is infeasible — stop

    **else**

        $j^* :=$ the smallest index attaining the maximum

            $\max\left\{\dfrac{u_j}{d_{t^* j}} \mid j = 1, \ldots, m+n; d_{t^* j} < 0\right\}$

        do a pivot with entering index $j^*$, leaving index $\varphi(t^*)$

        set component $\varphi(t)$ of $x_{k+1}$ to $f_t$ for $t = 1, \ldots, m$ and the remaining

            components of $x_{k+1}$ to zero

    $k := k + 1$

{Continue with next iteration of main loop}

---

**10.**  The dual simplex method was devised in 1954 by C. E. Lemke (1920–2004).

**11.**  C++ code that implements the dual simplex algorithm can be found at

    • http://www.codeforge.com/article/214515

**Examples:**

**1.**  Using the table of Fact 1, the dual of

$$\begin{aligned}
\text{maximize:} \quad & 5x_1 - 7x_2 \\
\text{subject to:} \quad & x_1 + 3x_2 - x_3 + x_4 \leq -1 \\
& 2x_1 + x_2 - 4x_3 - x_4 \geq 3 \\
& x_1 + x_2 - 3x_3 + 2x_4 = 2 \\
& x_2 \geq 0, \ x_4 \leq 0, \ x_1, x_3 \text{ unrestricted}
\end{aligned}$$

is

$$\begin{aligned}
\text{minimize:} \quad & -u_1 + 3u_2 + 2u_3 \\
\text{subject to:} \quad & u_1 + 2u_2 + u_3 = 5 \\
& 3u_1 + u_2 + u_3 \geq -7 \\
& -u_1 - 4u_2 - 3u_3 = 0 \\
& u_1 - u_2 + 2u_3 \leq 0 \\
& u_1 \geq 0, \ u_2 \leq 0, \ u_3 \text{ unrestricted.}
\end{aligned}$$

**2.** The LP of §16.1.2, Example 1 has the dual

$$\begin{aligned}
\text{minimize:} \quad & 5u_1 + 12u_2 + 2u_3 \\
\text{subject to:} \quad & 3u_1 - 9u_2 - 6u_3 \geq 3 \\
& 4u_2 + 2u_3 \geq 4 \\
& -u_1 + 3u_2 + 4u_3 \geq 4 \\
& u_1, u_2, u_3 \geq 0 .
\end{aligned}$$

The optimal solution to the primal LP (see §16.1.3, Example 2) is $x^* = (\frac{11}{3}, 0, 6)^T$ with optimal objective value $z^* = 35$. The associated tableau has $u = (0, 1, 0, 6, 0, \frac{5}{2})$. The optimal dual variables for (4) are recovered from the reduced costs of the slack variables $x_4$, $x_5$, and $x_6$, so that $u^* = (6, 0, \frac{5}{2})$. As guaranteed by Fact 4, the optimal dual objective value $5u_1^* + 12u_2^* + 2u_3^* = 30 + 5 = 35 = z^*$. The complementary slackness conditions in Fact 7 hold here: the second primal constraint holds with strict inequality $(x_5^* = 27 > 0)$, so the second dual variable $u_2^* = 0$; also, the second dual constraint holds with strict inequality $(u_5^* = 1 > 0)$, so the second primal variable $x_2^* = 0$.

**3.** Using the transformations in Fact 5 of §16.1.1, the LP problem

$$\begin{aligned}
\text{minimize:} \quad & 2x_1 + 3x_2 + 4x_3 \\
\text{subject to:} \quad & 2x_1 - x_2 + 3x_3 \geq 4 \\
& x_1 + 2x_2 + x_3 \geq 3 \\
& x_1, x_2, x_3 \geq 0
\end{aligned}$$

can be written in form (2), with the corresponding tableau (a) below. Since $u \geq 0$ the current solution $x_4 = -4$, $x_5 = -3$ is dual feasible but not primal feasible. Algorithm 4 can then be applied, giving $t^* = 1$ and $j^* = 1$. The variable $x_4$ leaves the basis and the variable $x_1$ enters, giving tableau (b) and the new basic but not feasible solution $x = (2, 0, 0, 0, -1)^T$ with $z = 4$. One additional dual simplex pivot achieves primal feasibility and produces the optimal solution $x^* = (\frac{11}{5}, \frac{2}{5}, 0, 0, 0)^T$ with $z^* = \frac{28}{5}$.

tableau (a)

| 2 | 3 | 4 | 0 | 0 | 0 |
|---|---|---|---|---|---|
| −2 | 1 | −3 | 1 | 0 | −4 |
| −1 | −2 | −1 | 0 | 1 | −3 |

tableau (b)

| 0 | 4 | 1 | 1 | 0 | −4 |
|---|---|---|---|---|---|
| 1 | $-\frac{1}{2}$ | $\frac{3}{2}$ | $-\frac{1}{2}$ | 0 | 2 |
| 0 | $-\frac{5}{2}$ | $\frac{1}{2}$ | $-\frac{1}{2}$ | 1 | −1 |

## 16.1.6  SENSITIVITY ANALYSIS

Since the data to an LP are often estimates or can vary over time, the analysis of many problems requires studying the behavior of the optimal LP solution to changes in the input data. This form of sensitivity analysis typically uses the solution of the original LP as a starting point for solving the altered LP.

## Definitions:

The **original tableau** for the LP problem (2) is $\dfrac{-c \quad 0 \;\big|\; 0}{A \quad I \;\big|\; b}$.

The **final tableau** for the optimal basic solution $x^*$ (possibly after a permutation of the columns $1, \ldots, m+n$) is

$$\frac{u \;\big|\; z}{D \;\big|\; f} = \frac{0 \quad c_B B^{-1} N - c_N \;\big|\; c_B B^{-1} b}{I \quad B^{-1} N \;\;\big|\; B^{-1} b}.$$

**Row 0** of a tableau refers to the row $u$ of associated dual variables.

A tableau is **suboptimal** if some entries of row 0 are negative. A tableau is **infeasible** if some entries of column $f$ are negative.

Let $a^j$ be the column of $[A \; I]$ associated with variable $x_j$ and let $d^j$ be the column of $D$ associated with variable $x_j$.

## Facts:

**1.** The formulas in Table 1 show how to construct an updated tableau $\mathcal{T}'$ from the final tableau $\mathcal{T}$ of an LP problem:

- if $\mathcal{T}'$ is suboptimal, reoptimize using the simplex method starting with $\mathcal{T}'$;
- if $\mathcal{T}'$ is infeasible, reoptimize using the dual simplex method starting with $\mathcal{T}'$;
- otherwise, $\mathcal{T}'$ corresponds to an optimal BFS for the altered problem.

**Table 1: Formulas for constructing the updated tableau $\mathcal{T}'$.**

| change in LP data | possible changes in tableau | tableau updates |
|---|---|---|
| change in $c_s$, $s$ nonbasic: $c_s := c_s + \Delta$ | only entry $s$ of row 0 can change | $u_s := u_s - \Delta$ |
| change in $c_s$, $s$ basic: $c_s := c_s + \Delta$ | row 0 and $z$ can change | update $c_B$ $u_j := c_B B^{-1} a^j - c_j$, $j$ nonbasic $z := c_B B^{-1} b$ |
| change in $b_r$: $b_r := b_r + \Delta$ | decision variables and $z$ can change | $f := f + \Delta(d^{n+r})$ update $b$ $z := c_B B^{-1} b$ |
| change nonbasic column $s$: $a^s := \tilde{a}^s$ $c_s := \tilde{c}_s$ | tableau column $s$ and $u_s$ can change | update $a^s$ and $c_s$ $u_s := c_B B^{-1} a^s - c_s$ $d^s := B^{-1} a^s$ |
| add a new column $a^\ell$ with cost $c_\ell$ | new tableau column $\ell$ and new $u_\ell$ | $u_\ell := c_B B^{-1} a^\ell - c_\ell$ $d^\ell := B^{-1} a^\ell$ |

**2.** *Ranging:* Table 2 shows how to calculate the (maximal) ranges over which the current basis $B$ remains optimal. In the "range" column of Table 2, $b$ and $c_B$ refer to entries of $\mathcal{T}$, rather than $\mathcal{T}'$.

**3.** When $b_i$ is changed within the allowable range (Table 2), the change in the objective value is $-\Delta$ times the reduced cost of the slack variable associated with row $i$.

**4.** To add a new constraint $a_\ell x \le b_\ell$ to the original LP do the following:

- add a new (identity) column to the tableau corresponding to the slack variable of the new constraint;

- add a new row $\ell$ to the tableau corresponding to the new constraint;

- for each basic $j$ with $d_{\ell j} \neq 0$, multiply row $i = \varphi^{-1}(j)$ by $-d_{\ell j}$ and add to row $\ell$;

- if the updated $f_\ell < 0$ use the dual simplex method to reoptimize.

**5.** For changes in more than one component of $c$, or in more than one right-hand side $b$, use the "100% rule":

- *objective function changes*: If all changes occur in variables $j$ with $u_j > 0$, the current solution remains optimal as long as each $c_j$ is within its allowable range (Table 2). Otherwise, let $\Delta c_j$ be the change to $c_j$. If $\Delta c_j \geq 0$ set $r_j := \frac{\Delta c_j}{\Delta_U}$, else set $r_j := -\frac{\Delta c_j}{\Delta_L}$, where $\Delta_U, \Delta_L$ are computed from Table 2. If $\sum r_j \leq 1$, the current solution remains optimal (if not, the rule tells nothing).

- *right-hand side changes*: If all changes are in constraints not active at $x^*$, the current basis remains optimal as long as each $b_i$ is within its allowable range (Table 2). Otherwise, let $\Delta b_i$ be the change to $b_i$. If $\Delta b_i \geq 0$ set $r_i := \frac{\Delta b_i}{\Delta_U}$, else set $r_i := -\frac{\Delta b_i}{\Delta_L}$, where $\Delta_U, \Delta_L$ are computed from Table 2. If $\sum r_i \leq 1$, the current solution remains optimal (if not, the rule tells nothing).

**Table 2: Ranges over which current basis is optimal.**

| change in LP data | range |
|---|---|
| change in $c_s$, $s$ nonbasic: $c_s := c_s + \Delta$ | $\Delta \leq u_s$ |
| change in $c_s$, $s$ basic: $c_s := c_s + \Delta$ | $\Delta_L \leq \Delta \leq \Delta_U$, where $p = s$th row of $B^{-1}$ $\Delta_L = \max\{\frac{c_j - c_B B^{-1} a^j}{pa^j} \mid pa^j > 0, j \text{ nonbasic}\}$ $\Delta_U = \min\{\frac{c_j - c_B B^{-1} a^j}{pa^j} \mid pa^j < 0, j \text{ nonbasic}\}$ |
| change in $b_r$: $b_r := b_r + \Delta$ | $\Delta_L \leq \Delta \leq \Delta_U$, where $q = r$th column of $B^{-1}$ $\Delta_L = \max\{\frac{-(B^{-1}b)_i}{q_i} \mid q_i > 0\}$ $\Delta_U = \min\{\frac{-(B^{-1}b)_i}{q_i} \mid q_i < 0\}$ |

**Examples:**

**1.** The LP problem

$$\begin{aligned}
\text{maximize:} \quad & 3x_1 + 4x_2 + 4x_3 \\
\text{subject to:} \quad & 3x_1 - x_3 \leq 5 \\
& -9x_1 + 4x_2 + 3x_3 \leq 12 \\
& -6x_1 + 2x_2 + 4x_3 \leq 2 \\
& x_1, x_2, x_3 \geq 0
\end{aligned}$$

has the final tableau $\mathcal{T}$

$$\begin{array}{ccccccc|c}
0 & 1 & 0 & 6 & 0 & \frac{5}{2} & & 35 \\
\hline
1 & \frac{1}{3} & 0 & \frac{2}{3} & 0 & \frac{1}{6} & & \frac{11}{3} \\
0 & 4 & 0 & 3 & 1 & 0 & & 27 \\
0 & 1 & 1 & 1 & 0 & \frac{1}{2} & & 6
\end{array}$$

corresponding to the optimal BFS $x^* = (\frac{11}{3}, 0, 6, 0, 27, 0)^T$ with $z^* = 35$. The associated basis matrix $B$ contains columns $1, 5, 3$ and the inverse basis matrix is $B^{-1}$, where

$$B = \begin{pmatrix} 3 & 0 & -1 \\ -9 & 1 & 3 \\ -6 & 0 & 4 \end{pmatrix}, \qquad B^{-1} = \begin{pmatrix} \frac{2}{3} & 0 & \frac{1}{6} \\ 3 & 1 & 0 \\ 1 & 0 & \frac{1}{2} \end{pmatrix}.$$

If the nonbasic objective coefficient $c_2$ is changed to $4 + \Delta$, the current BFS remains optimal for $\Delta \leq u_2 = 1$, that is, for $c_2 \leq 5$. If the basic objective coefficient $c_1$ is changed to $3 + \Delta$, then $p = (\frac{2}{3}, 0, \frac{1}{6})$ and $\Delta_L = \max\{\frac{-1}{1/3}, \frac{-6}{2/3}, \frac{-5/2}{1/6}\} = -3$. This gives $-3 \leq \Delta$, so the current BFS remains optimal over the range $c_1 \geq 0$. If, however, $c_3$ is changed to the value 2, meaning $\Delta = -2$, the current basis with $\Sigma_B = \{1, 5, 3\}$ will no longer be optimal. Using Table 2 the vector $c_B$ is updated to $c_B = (3, 0, 2)$ and the nonbasic $u_j$ are computed as $u_2 = -1$, $u_4 = 4$, $u_6 = \frac{3}{2}$. The updated $u = (0, -1, 0, 4, 0, \frac{3}{2})$ and $z = 23$ are inserted in tableau $\mathcal{T}$. Since $u_2 < 0$ a simplex pivot with $j^* = 2$ and $t^* = 3$ is performed, leading to the new optimal solution $x^* = (\frac{5}{3}, 6, 0, 0, 3, 0)^T$ with $z^* = 29$.

**2.** Suppose that the right-hand side $b_1$ in the original LP of Example 1 is changed to $b_1 = 3$, corresponding to the change $\Delta = -2$. From Table 1 $f$ is updated to $(\frac{11}{3}, 27, 6)^T - 2(\frac{2}{3}, 3, 1)^T = (\frac{7}{3}, 21, 4)^T$, giving the optimal BFS $x^* = (\frac{7}{3}, 0, 4, 0, 21, 0)^T$. Since $b = (3, 12, 2)^T$ the objective value found from Table 1 is $z = 23$. Notice that the change in objective value is $\Delta z = 23 - 35 = -12$, which is the same as $-\Delta$ times the reduced cost $-u_4$ of $x_4$: namely, $-12 = 2 \cdot (-6)$. To determine the range of variation of $b_1$ so that the basis defined by $\Sigma_B = \{1, 5, 3\}$ remains unchanged, Table 2 is used. Here $q = (\frac{2}{3}, 3, 1)^T$ and $\Delta_L = \max\{-\frac{11/3}{2/3}, -\frac{27}{3}, -\frac{6}{1}\} = -\frac{11}{2}$. Thus $-\frac{11}{2} \leq \Delta$ so that the current basis is optimal for $b_1 \geq -\frac{1}{2}$.

**3.** If the new constraint $3x_1 + 2x_2 - x_3 \leq 4$ is added to the LP in Example 1, the (previous) optimal solution $x^* = (\frac{11}{3}, 0, 6, 0, 27, 0)^T$ is no longer feasible. Using Fact 4, a new row and column are added to the tableau $\mathcal{T}$, giving tableau (a) below. By adding $(-3)$ times row 1 and $+1$ times row 3 to the last row, a new tableau (b) is produced corresponding to the basic set $\Sigma_B = \{1, 5, 3, 7\}$. Since $b_4 < 0$ the dual simplex algorithm is then used with $t^* = 4$ and $j^* = 4$, producing a new tableau that is primal feasible, with the new optimal BFS $(3, 0, 5, 1, 24, 0, 0)^T$ and objective value 29.

tableau (a)

| | | | | | | | | |
|---|---|---|---|---|---|---|---|---|
| 0 | 1 | 0 | 6 | 0 | $\frac{5}{2}$ | 0 | 35 |
| 1 | $\frac{1}{3}$ | 0 | $\frac{2}{3}$ | 0 | $\frac{1}{6}$ | 0 | $\frac{11}{3}$ |
| 0 | 4 | 0 | 3 | 1 | 0 | 0 | 27 |
| 0 | 1 | 1 | 1 | 0 | $\frac{1}{2}$ | 0 | 6 |
| 3 | 2 | -1 | 0 | 0 | 0 | 1 | 4 |

tableau (b)

| | | | | | | | | |
|---|---|---|---|---|---|---|---|---|
| 0 | 1 | 0 | 6 | 0 | $\frac{5}{2}$ | 0 | 35 |
| 1 | $\frac{1}{3}$ | 0 | $\frac{2}{3}$ | 0 | $\frac{1}{6}$ | 0 | $\frac{11}{3}$ |
| 0 | 4 | 0 | 3 | 1 | 0 | 0 | 27 |
| 0 | 1 | 1 | 1 | 0 | $\frac{1}{2}$ | 0 | 6 |
| 0 | 2 | 0 | -1 | 0 | 0 | 1 | -1 |

**4.** One example of the practical use of sensitivity analysis occurred in the airline industry. When the price of aviation fuel was relatively high, and varied by airport location, a linear programming model was successfully used to determine an optimal strategy for refueling aircraft. The key idea is that it might be more economical to take on extra fuel at an enroute stop if the fuel cost savings for the remainder of the flight are greater than the extra fuel burned because of the excess weight of additional fuel. A linear programming model of this situation ended up saving millions of dollars annually. An important feature was providing pilots with *ranges* of fuel prices for each airport location, with associated optimal policies for taking on extra fuel based on the cost range.

**5.** Another example of the beneficial use of sensitivity analysis occurred in a 1997 study to assess the effectiveness of mandatory minimum-length sentences for reducing drug use. One finding of the study was that the longer sentences become more effective than conventional enforcement only when it costs more than \$30,000 to arrest a drug dealer. Thus, rather than producing a single optimal policy, this study identified conditions (parameter ranges) under which each alternative policy is to be preferred.

## 16.1.7   GOAL PROGRAMMING

Goal programming refers to a multicriteria decision-making problem in which a given LP problem can have multiple objectives or goals. This technique is useful when it is impossible to satisfy all goals simultaneously. For example, a model for optimizing the operation of an oil refinery might seek not only to minimize production cost, but also to reduce the amount of imported crude oil and the amount of oil having a high sulfur content. In another instance, the routing of hazardous waste might consider minimizing not only the total distance traveled but also the number of residents living within ten miles of the selected route.

**Definitions:**

A *goal programming* (**GP**) *problem* has linear constraints that can be written

$$Ax \leq b$$
$$Hx + \widetilde{x} - \overline{x} = h$$
$$x \geq 0, \ \widetilde{x} \geq 0, \ \overline{x} \geq 0$$

and objective functions

$$G_1: \quad \text{minimize } z_1 = c_1 \widetilde{x}_1 + d_1 \overline{x}_1$$
$$G_2: \quad \text{minimize } z_2 = c_2 \widetilde{x}_2 + d_2 \overline{x}_2$$
$$\vdots$$
$$G_\ell: \quad \text{minimize } z_\ell = c_\ell \widetilde{x}_\ell + d_\ell \overline{x}_\ell$$

where $A$ is an $m \times n$ matrix and $H$ is an $\ell \times n$ matrix.

The value $h_k$ is the **target value** of the $k$th goal. Goal $k$ is **satisfied** if $(Hx)_k = h_k$ holds for a given vector $x$ of **decision variables**.

The variables $\widetilde{x}$ are the **underachievement variables** while the variables $\overline{x}$ are the **overachievement variables**.

**Facts:**

**1.** In a GP problem, the aim is to find decision variables that approximately satisfy the given goals, which is achieved by jointly minimizing the magnitudes of the underachievement and overachievement variables.

**2.** Assuming $c_k > 0$ and $d_k > 0$, then goal $k$ is satisfied by making $z_k = 0$.

**3.** If all $c_t$ and $d_t$ are positive then for each $k = 1, \ldots, \ell$ at most one of $\widetilde{x}_k, \overline{x}_k$ will be positive in an optimal solution.

**4.** One important case of a GP problem has $c_k = d_k = 1$ for $k = 1, \ldots, \ell$, making the objective to (approximately) satisfy all constraints $Hx = h$.

**5.** When the relative importance of $G_1, \ldots, G_\ell$ is known precisely, an ordinary LP can be used with the objective function being a weighted sum of $z_1, \ldots, z_\ell$.

**6.** *Preemptive goal programming*:  Here the goals are prioritized $G_1 \gg G_2 \gg \cdots \gg G_\ell$, meaning that goal $G_1$ is the most important and goal $G_\ell$ is the least important. Solutions are sought that satisfy the most important goal. Among all such solutions, those are retained that best satisfy the second highest goal, and so forth.

**7.** *Goal programming simplex method*:   The simplex method (§16.1.3, Algorithm 2) can be extended to preemptive GP (minimization) problems, with the following modifications:

- $\ell$ "objective rows" are maintained in the tableau instead of just one.

- Let $i^*$ be the highest-priority index with $z_{i^*} > 0$ for which there exists a nonbasic $j^*$ with $u_{j^*}^{i^*} > 0$ and with $u_{j^*}^i \geq 0$ for any (higher-priority) objective row $i < i^*$. If there is no such $i^*$, stop; else pick $j^*$ corresponding to the most positive $u_{j^*}^{i^*}$.

- All $\ell$ objective rows are updated when a pivot is performed.

- At completion, if the solution fails to satisfy all goals, then every nonbasic variable that would decrease the objective value $z_i$ if it entered the basis, would increase $z_{i'}$ for some higher-priority goal $G_{i'}$, $i' < i$.

**8.** An optimization package (LiPS) that implements goal programming can be found at

- http://sourceforge.net/projects/lipside/

## 16.1.8  INTEGER PROGRAMMING

Integer programming problems are LPs in which some of the variables are constrained to be integers. Such problems more accurately model a wide range of application areas, including capital budgeting, facility location, manufacturing, scheduling, logical inference, physics, engineering design, environmental economics, and VLSI circuit design. However, integer programming problems are much more difficult to solve than LPs.

**Definitions:**

Let $\mathcal{Z}^n$ $[\mathcal{Z}_+^n]$ denote the set of all $n$-vectors with all components integers [nonnegative integers], and let $\mathcal{R}^n$ $[\mathcal{R}_+^n]$ denote the set of all $n$-vectors with all components real numbers [nonnegative real numbers].

A **pure integer programming (IP) problem** is an optimization problem of the form

$$\begin{aligned} \text{maximize:} \quad & z_{IP} = cx \\ \text{subject to:} \quad & Ax \leq b \\ & x \in \mathcal{Z}_+^n \end{aligned} \qquad (5)$$

where $A$ is an $m \times n$ matrix, $b$ is an $m$-column vector, and $c$ is an $n$-row vector.

A **0-1 IP problem** is an IP with each $x_j \in \{0, 1\}$.

A **mixed integer programming (MIP) problem** is of the form

$$\begin{aligned} \text{maximize:} \quad & z_{MIP} = cx + hy \\ \text{subject to:} \quad & Ax + Gy \leq b \\ & x \in \mathcal{Z}_+^n, \ y \in \mathcal{R}_+^p \end{aligned}$$

where $A$ is an $m \times n$ matrix, $G$ is an $m \times p$ matrix, $b$ is an $m$-column vector, $c$ is an $n$-row vector, and $h$ is a $p$-row vector.

For IP problem (5), the **feasible region** is $S \equiv \{x \in \mathcal{Z}_+^n \mid Ax \le b\}$.

A **polyhedron** is a set of points in $\mathcal{R}^n$ satisfying a finite set of linear inequalities.

If $X$ is a finite set of points in $\mathcal{R}^n$, the **convex hull** of $X$ is

$$conv\,(X) \equiv \{\textstyle\sum \lambda_i x_i \mid x_i \in X, \sum \lambda_i = 1, \lambda_i \ge 0\}.$$

The **LP relaxation** of (5) is the linear programming problem

$$
\begin{aligned}
\text{maximize:} \quad & z_{LP} = cx \\
\text{subject to:} \quad & Ax \le b \\
& x \ge 0.
\end{aligned}
\tag{6}
$$

More generally, a **relaxation** of (5) is any problem $\max\{cx \mid x \in T\}$, where $S \subset T$.

The problem $\max\{cx \mid \bar{A}x \le \bar{b},\, x \in \mathcal{Z}_+^n\}$ is a **formulation** for (5) if it contains exactly the same set of feasible integral points as (5). If the feasible region of the LP relaxation of one formulation is strictly contained in another, the first is a **tighter** formulation.

Suppose $\widetilde{x}$ is a feasible solution to (6) but not to (5). A **cutting plane** is any inequality $\pi x \le \pi_0$ satisfied by all points in $conv\,(S)$ but not by $\widetilde{x}$.

A family $\mathcal{S}$ of subsets of $S$ is a **separation** of $S$ if $\bigcup_{S_k \in \mathcal{S}} S_k = S$; a separation is usually a partition of the set $S$.

A **lower bound** $\underline{z}$ for $z_{IP}$ is an underestimate of $z_{IP}$.

**Facts:**

**1.** $z_{IP} \le z_{LP}$. More generally, any relaxation of (5) has an optimal objective value at least as large as $z_{IP}$.

**2.** If $x'$ is feasible to (5) then $z' = cx'$ satisfies $z' \le z_{IP}$.

**3.** The feasible region of an LP problem is a polyhedron, and every polyhedron is the feasible region of some LP problem.

**4.** The set $conv\,(S)$ is a polyhedron, so there is an LP problem $\max\{cx \mid \widetilde{A}x \le \widetilde{b},\, x \ge 0\}$ with the feasible region $conv\,(S)$.

**5.** An optimal solution to the LP in Fact 4 is an optimal solution to (5). However, finding all necessary constraints, called *facets*, of this LP is extremely difficult.

**6.** IP is an NP-hard optimization problem (§17.5.2). Consequently, such problems are harder to solve in practice than LPs. The inherent complexity of solving IPs stems from the nonconvexity of their feasible region, which makes it difficult to verify the optimality of a proposed optimal solution in an efficient manner.

**7.** Formulation of an IP is critical: achieving problem tightness is more important than reducing the number of constraints or variables appearing in the formulation.

**8.** An excellent introduction to building integer linear programming models is found in Williams [Wi13].

**9.** Solution techniques for (5) usually involve some preliminary operations that improve the formulation, called *preprocessing*, followed by an iterative use of heuristics to quickly find feasible solutions.

**10.** Popular solution techniques for solving (5) include cutting plane methods (Fact 11), branch-and-bound techniques (Fact 13), and (hybrid) branch-and-cut methods.

**11.** *Cutting plane method*:   This approach (Algorithm 5) proceeds by first finding an optimal solution $\bar{x}$ to a relaxation $R$ of the original problem (5). If $\bar{x}$ is not optimal, a cutting plane is added to the constraints of the current relaxation and the new LP is then solved. This process is repeated until an optimal solution is found.

---

**Algorithm 5**:   **Cutting plane algorithm for (5).**

input: IP in form (5)

output: an optimal solution $x^*$ with objective value $z^*$

let $R$ be the LP relaxation: $\max\{cx \mid Ax \le b,\ x \ge 0\}$

{Main loop}

optimally solve problem $R$, obtaining $\bar{x}$

**if** $\bar{x} \in \mathcal{Z}_+^n$ **then** stop with $x^* := \bar{x}$ and $z^* := c\bar{x}$

**else**

    find a cutting plane $\pi x \le \pi_0$ with $\pi\bar{x} > \pi_0$ and $\pi x \le \pi_0$ for all feasible

       solutions of (5)

    modify $R$ by adding the constraint $\pi x \le \pi_0$

{Continue with next iteration of main loop}

---

**12.** General methods for finding cutting planes for IP or MIP problems are relatively slow. Cutting plane algorithms using facets for specific classes of IP problems are better, since facets make the "deepest" cuts.

**13.** *Branch-and-bound method*:   This approach (Algorithm 6) decomposes the original problem $P$ into subproblems or *nodes* by breaking $S$ into subsets. Each subproblem $P_j$ is implicitly investigated (and possibly discarded) until an optimal one is found. In this algorithm $z_j^*$ is the optimal value of problem $P_j$, $\overline{z_j}$ is the optimal value of the relaxation $R_j$ of $P_j$, and $\underline{z_j}$ is the best known lower bound for $z_j^*$.

---

**Algorithm 6**:   **Branch-and-bound algorithm for (5).**

input: IP in form (5)

output: an optimal solution $x_0$ with objective value $\underline{z_0}$

let $P$ be the problem: $\max\{cx \mid x \in S\}$

$P_0 := P$; $S_0 := S$; $\underline{z_0} := -\infty$; $\overline{z_0} := +\infty$

put $P_0$ on the list of *live* nodes

{Branching}

    **if** no live node exists **then** go to {termination}

    **else** select a live node $P_j$

{Bounding}

    solve a relaxation $R_j$ of $P_j$

    **if** $R_j$ is infeasible **then** discard $P_j$ and go to {branching}

    **if** $\overline{z_j} = +\infty$ **then** go to {separation}

    {$\overline{z_j}$ is finite}

    **if** $\overline{z_j} \le \underline{z_0}$ **then** discard node $P_j$ and go to {branching}

> **if** $\overline{z_j} = z_j$ **then** update $\underline{z_0} := \max\{\underline{z_0}, z_j\}$ and discard any node $P_i$ for
> which $\overline{z_i} \le \underline{z_0}$
> {Separation}
>     choose a separation $\mathcal{S}_j^*$ of $S_j$ forming new live nodes and go to {branching}
> {Termination}
>     **if** $\underline{z_0} = -\infty$ **then** problem (5) is infeasible
>     **if** $\underline{z_0}$ is finite **then** $\underline{z_0}$ is the optimal objective value and the associated $x_0$ is
>         an optimal solution

**14.** In Algorithm 6 the optimal value $\overline{z_j}$ of relaxation $R_j$ is an upper bound for $z_j^*$. Also $\underline{z_0}$ is the objective value for the best known feasible solution to (5).

**15.** LP relaxations are often used in the bounding portion of Algorithm 6.

**16.** There are specializations of Algorithm 6 for 0-1 IP problems and for MIP problems.

**17.** Branch-and-bound tends to be a computationally expensive solution method. Usually it is applied only when other methods appear to be stalling.

**18.** In the survey [Fo15] of linear programming software, many of the packages listed will handle IP problems as well. When available, these extensions to binary and/or integer-valued variables are indicated by the survey.

**19.** Several software packages that solve IP and MIP problems can be found at

- http://www.aimms.com
- http://optimization-suite.coin-or.org
- http://www.cplex.com
- http://www.gurobi.com
- http://www.lindo.com
- www.mathworks.com/products/optimization

## Examples:

**1.** The next figure shows the convex hull of feasible solutions to the IP

$$
\begin{aligned}
\text{maximize:} \quad & -x_1 + \tfrac{1}{2}x_2 \\
\text{subject to:} \quad & x_2 \le 4 \\
& -x_1 - x_2 \le -\tfrac{5}{2} \\
& 8x_1 + x_2 \le 24 \\
& -3x_1 + 4x_2 \le 10 \\
& x_1,\ x_2 \ge 0 \\
& x_1,\ x_2 \text{ integers.}
\end{aligned}
$$

Here $S = \{(1,2),(1,3),(2,1),(2,2),(2,3),(2,4),(3,0)\}$. The optimal solution occurs at $(1,3)$, with $z_{IP} = \tfrac{1}{2}$. The feasible region of the LP relaxation is also shown, with the optimal LP value $z_{LP} = \tfrac{5}{4}$ attained at $(0, \tfrac{5}{2})$; a cutting plane is depicted by the dashed line.

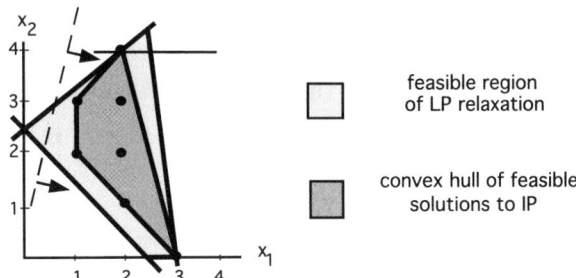

For this problem, $conv\,(S)$ is defined by the following constraints:

$$-x_1 - x_2 \le -3$$
$$-x_1 + x_2 \le 2$$
$$-x_1 + \le -1$$
$$4x_1 + x_2 \le 12$$
$$x_2 \le 4$$
$$x_1,\ x_2 \ge 0.$$

All of these constraints are facets except for $x_2 \le 4$ and the nonnegativity constraints, which are redundant.

**2.** The following IP has the feasible region $S = \{(0,0), (1,1), (2,2), (3,3), (4,4)\}$ and the optimal solution occurs at $(4,4)$ with $z_{IP} = 8$:

$$\begin{aligned}
\text{maximize:} \quad & x_1 + x_2 \\
\text{subject to:} \quad & 2x_1 - 2x_2 \le 1 \\
& -7x_1 + 8x_2 \le 4 \\
& x_1,\ x_2 \ge 0 \\
& x_1,\ x_2 \text{ integers.}
\end{aligned}$$

The LP relaxation has a feasible region defined by vertices $(0,0)$, $(\frac{1}{2}, 0)$, $(0, \frac{1}{2})$, $(8, \frac{15}{2})$, so its optimal solution occurs at $(8, \frac{15}{2})$ with $z_{LP} = \frac{31}{2}$. Consequently, the LP solution is a poor approximation to the optimal IP solution. Moreover, simply rounding the LP solution gives either $(8,7)$ or $(8,8)$, both of which are infeasible to the given IP problem.

**3.** The following IP can be solved using Algorithm 6:

$$\begin{aligned}
\text{maximize:} \quad & 3x_1 + 3x_2 - 8x_3 \\
\text{subject to:} \quad & -3x_1 + 6x_2 + 7x_3 \le 8 \\
& 6x_1 - 3x_2 + 7x_3 \le 8 \\
& x_1,\ x_2,\ x_3 \ge 0 \\
& x_1,\ x_2,\ x_3 \text{ integers.}
\end{aligned}$$

The initial problem $P_0$ has an LP relaxation $R_0$ that is obtained by removing the integer restrictions; solving this LP gives $x = (2.667, 2.667, 0)$ with $z = 16$. A separation is achieved by creating the two subproblems $P_1$ and $P_2$; the constraint $x_1 \le 2$ is appended to $P_0$, creating $P_1$, while the constraint $x_1 \ge 3$ is appended to $P_0$, creating $P_2$. Now the live nodes are $P_1$ and $P_2$. Solving the LP relaxation $R_1$ gives $x = (2, 2.333, 0)$ with $z = 13$. New subproblems $P_3$ and $P_4$ are obtained from $P_1$ by appending the constraints $x_2 \le 2$ and $x_2 \ge 3$, respectively. Now the live nodes are subproblems $P_2, P_3, P_4$. The

LP relaxation $R_2$ of $P_2$ is infeasible, as is the LP relaxation $R_4$ of $P_4$. Solving the LP relaxation $R_3$ gives the feasible integer solution $x = (2, 2, 0)$ with $z = 12$. Since there are no more live nodes, this represents the optimal solution to the stated problem.

**4.** *Fixed-charge problem*:   Find optimal levels of $n$ activities to satisfy $m$ constraints while minimizing total cost. Each activity $j$ has per unit cost $c_j$. In addition, there is a startup cost $d_j$ for certain undertaken activities $j$.

This problem can be modeled as a MIP problem, with a real variable $x_j$ for the level of each activity $j$. If activity $j$ has a startup cost, introduce the additional 0-1 variable $y_j$, equal to 1 when $x_j > 0$ and 0 otherwise. For example, this condition can be enforced by imposing the constraints $M_j y_j \geq x_j$, $y_j \in \{0, 1\}$, where $M_j$ is a known upper bound on the value of $x_j$. The objective is then to minimize $z = cx + dy$.

**5.** *Queens problem*:   On an $n \times n$ chessboard, the task is to place as many nontaking queens as possible.

This problem can be formulated as a 0-1 IP problem, having binary variables $x_{ij}$. Here $x_{ij} = 1$ if and only if a queen is placed in row $i$ and column $j$ of the chessboard. The objective function is to maximize $z = \sum_i \sum_j x_{ij}$ and there is a constraint for each row, column, and diagonal of the chessboard. Such a constraint has the form $\sum_{(i,j) \in S} x_{ij} \leq 1$, where $S$ is the set of entries in the row, column, or diagonal. For example, one optimal solution of this IP for the $7 \times 7$ chessboard is the assignment $x_{16} = x_{24} = x_{37} = x_{41} = x_{53} = x_{65} = x_{72} = 1$, with all other $x_{ij} = 0$.

# 16.2   LOCATION THEORY

Location theory is concerned with locating a fixed number of facilities at points in some space. The facilities provide a service (or product) to the customers whose locations and levels of demand (for the service) are known. The object is to find locations for the facilities to optimize some specified criterion, e.g., the cost of providing the service. Interest in location theory has grown very rapidly because of its variety of applications to such fields as operations research, city planning, geography, economics, electrical engineering, and computer science.

## 16.2.1   *p*-MEDIAN AND *p*-CENTER PROBLEMS

### Definitions:

A **metric space** is a space $S$ consisting of a set of points with a real-valued function $d(x, y)$ defined on all pairs of points $x, y \in S$ with the following properties (§6.1.4):

- $d(x, y) = d(y, x) \geq 0$ for all $x, y \in S$;
- $d(x, y) = 0$ if and only if $x = y$;
- $d(x, z) \leq d(x, y) + d(y, z)$ for all $x, y, z \in S$.

The value $d(x, y)$ is called the **distance** between points $x, y \in S$.

There are $p$ facilities that are to be located at some set $X_p = \{x_1, x_2, \ldots, x_p\}$ of $p$ points in the (metric) space $S$. The elements of $X_p$ are the **facility locations**.

The facilities are to provide a service to the customers whose positions are given by a set $V = \{v_1, v_2, \ldots, v_n\}$ of $n$ points in $S$. The points in $V$ are the **demand points** and the level of **demand** at $v_i \in V$ is given by $w(v_i) \geq 0$.

For $x \in S$ and $X_p \subseteq S$, let $d(x, X_p)$ be the minimum distance from $x$ to a point of $X_p$:
$$d(x, X_p) = \min_{x_i \in X_p} \{d(x, x_i)\}.$$

Suppose $X_p$ is a candidate set of points in $S$ for locating the $p$ facilities. The following two objective functions are defined on $X_p \subseteq S$:

- $F(X_p) = \sum_{i=1}^{n} w(v_i)d(v_i, X_p);$

- $H(X_p) = \max_{1 \leq i \leq n} \{w(v_i)d(v_i, X_p)\}.$

The set $X_p^m \subseteq S$ is a **$p$-median** if $F(X_p^m) \leq F(X_p)$ for all possible $X_p \subseteq S$.

The set $X_p^c \subseteq S$ is a **$p$-center** if $H(X_p^c) \leq H(X_p)$ for all possible $X_p \subseteq S$.

**Facts:**

**1.** It is customary to assume that the demand at $v_i$ is satisfied by its closest facility. Then $w(v_i)d(v_i, X_p)$ indicates the total (transportation) cost associated with having the demand at $v_i \in S$ satisfied by its closest facility in the candidate set $X_p$.

**2.** $F(X_p)$ represents the total transportation cost of satisfying the demands if the facilities are located at $X_p$.

**3.** $H(X_p)$ represents the cost (or unfairness) associated with a farthest demand point not being in close proximity to any facility.

**4.** A $p$-median formulation is designed for locating $p$ facilities in a manner that minimizes the average cost of serving the customers.

**5.** A $p$-center formulation is designed for locating $p$ emergency facilities (e.g., police, fire, and ambulance services), in which the maximum time to respond to an emergency is to be made as small as possible.

**6.** The $p$-median and $p$-center problems are only interesting if $p < n$; otherwise, it is possible to locate at least one facility at each demand point, thereby reducing $F(X_p)$ or $H(X_p)$ to 0.

**Examples:**

**1.** Suppose that a single warehouse is to be located in a way to service $n$ retail outlets at minimum cost. Here $w(v_i)$ is the number of shipments made per week to the outlet at location $v_i$. This can be modeled as a 1-median problem, since the objective is to locate the single warehouse to minimize the total distance traveled by delivery vehicles.

**2.** A new police station is to be located within a portion of a city to serve residents of that area. Neighborhoods in that area can be taken as the demand points, and locating the police station can be formulated as a 1-center problem. Here, the maximum distance from the source of an emergency is critical so the police station should be located to minimize the maximum distance from a neighborhood. The weights at each demand point might be taken to be equal, or in some situations differing weights could signify conversion factors that translate distance into some other measure such as the value of residents' time.

**3.** *Statistics:* Suppose that $n$ given data values $x_1, x_2, \ldots, x_n$ are viewed as points placed along the real line. If the distance between points $x_i$ and $x_j$ is their absolute difference $|x_i - x_j|$, then a 1-median of this set of points (unweighted customer locations) is a point (facility) $\hat{x}$ that minimizes $\sum_{i=1}^{n} |x_i - \hat{x}|$. In fact, $\hat{x}$ corresponds to a *median* of the $n$ data values. If the distance between points $x_i$ and $x_j$ is their squared difference $(x_i - x_j)^2$, then a 1-median is a point $\bar{x}$ minimizing $\sum_{i=1}^{n} (x_i - \bar{x})^2$, which is precisely the *mean* of the $n$ data values. Alternatively, for either distance measure the 1-center of this set of points turns out to correspond to the *midrange* of the dataset: namely, the 1-center is the point located halfway between the largest and the smallest data values.

## 16.2.2    $p$-MEDIANS AND $p$-CENTERS ON NETWORKS

### Definitions:

A **network** is a weighted graph $G = (V, E)$ with vertex set $V = \{v_1, v_2, \ldots, v_n\}$ and edge set $E$, where $m = |E|$; see §8.1.1.

The **weight** of vertex $v \in V$ represents the demand at $v$ and is denoted by $w(v) \geq 0$.

The **length** of edge $e \in E$ represents the cost of travel (or distance) across $e$ and is denoted by $\ell(e) > 0$. Each edge is assumed to be a line segment joining its end vertices.

A **point on a network** $G$ is any point along any edge of $G$. The precise location of the point $x$ on edge $e = (u, v)$ is indicated by the distance of $x$ from $u$ or $v$.

If $x$ and $y$ are any two points on $G$, the **distance** $d(x, y)$ is the length of a shortest path between $x$ and $y$ in $G$, where the length of a path is the sum of the lengths of the edges (or partial edges) in the path.

### Facts:

1. Network $G$ with the above definition of distance constitutes a metric space (§16.2.1).

2. *$p$-median theorem:*  Given a positive integer $p$ and a network $G = (V, E)$, there exists a set of $p$ vertices $V_p \subseteq V$ such that $F(V_p) \leq F(X_p)$ for all possible sets $X_p$ of $p$ points on $G$. That is, a $p$-median can always be found that consists entirely of vertices.

3. A $p$-median of a network $G$ can be found by a finite search through all possible $\binom{n}{p}$ choices of $p$ vertices out of $n$. This is still a formidable task, but if $p$ is a small number (say $p < 5$) it is certainly manageable.

4. The $p$-median theorem also holds if the cost of satisfying the demand at $v_i \in V$ is $f_i(d(v_i, X_p))$, instead of $w(v_i) \cdot d(v_i, X_p)$, provided that $f_i \colon \mathcal{R}^+ \to \mathcal{R}^+$ is a concave nondecreasing function for all $i = 1, 2, \ldots, n$. ($\mathcal{R}^+$ denotes the set of nonnegative real numbers.) In this case, the objective function for the $p$-median problem becomes $F(X_p) = \sum_{i=1}^{n} f_i(d(v_i, X_p))$.

5. Each point $x$ of a $p$-center $X_p^c$ of network $G = (V, E)$ is a point on some edge $e$ such that for some pair of distinct vertices $u$ and $v \in V$, $w(u)d(u, x) = w(v)d(v, x)$; i.e., the point $x$ is the "center" of a shortest path from $u$ to $v$ in $G$ that passes through edge $e$.

6. There are at most $n^2$ predetermined choices of points on each edge of $G$ that could be potential points in $X_p^c$; thus there are $n^2 m$ predetermined choices for points in $X_p^c$.

7. A $p$-center $X_p^c$ of network $G$ can be found by examining all possible $\binom{n^2 m}{p}$ choices of $p$ points out of $n^2 m$. Even for small values of $p$, this is a formidable task.

**Examples:**

**1.** In the following network the levels of demand are given at the vertices and the lengths of the edges are shown on the edges. The 1-median of this network is at the vertex labeled $x_1$. The total transportation cost is $F(\{x_1\}) = 35$.

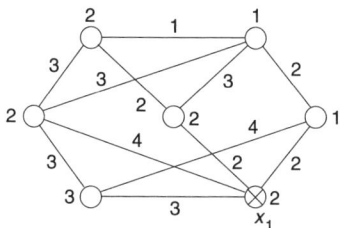

**2.** A tree network $T$ is shown in the following figure, with the vertex demands and edge lengths displayed. A 2-median of $T$ is the set of vertices $X_2 = \{x_1, x_2\}$, with total transportation cost $F(X_2) = 25$. If $x_1$ is kept fixed and $t$ is an arbitrary point along edge $e$, then $\{x_1, t\}$ also constitutes a 2-median of $T$. This is consistent with Fact 2, which only states that there is a $p$-median that is a subset of $V$, not that every $p$-median occurs in this way.

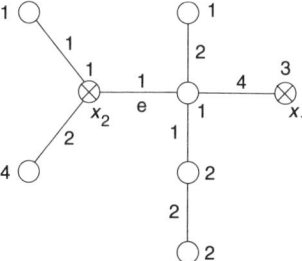

**3.** A 1-center $X_1^C$ is found for the network shown in the left portion of the following figure. For illustration, suppose that the 1-center is along the edge $(u_1, u_2)$, thereby limiting the search to candidate points on this edge. Let $X(x)$ be an arbitrary point along edge $(u_1, u_2)$, parametrized by the scalar $x = d(u_1, X(x))$. Note that $x \in [0, 3]$ with $X(0) = u_1$ and $X(3) = u_2$.

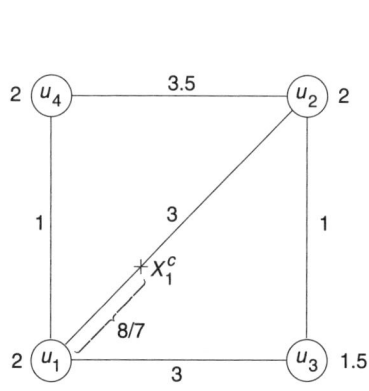
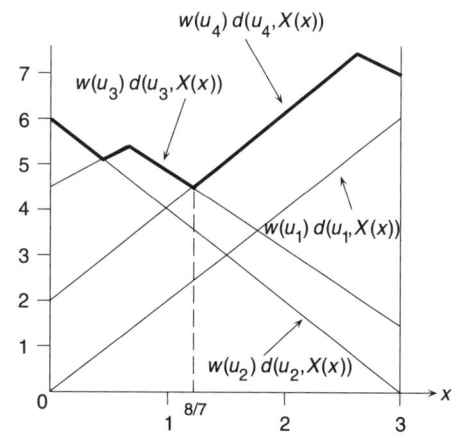

The right portion of the figure shows plots of $w(u_i)d(u_i, X(x))$ as a function of $x$ for $i = 1, 2, \ldots, 4$. The plot of $D(x) \equiv \max_{1 \leq i \leq 4} w(u_i)d(u_i, X(x))$ is highlighted in bold and $D(x)$ assumes its minimum value when $x = \frac{8}{7}$. The 1-center of the network in part (a) is then located along edge $(u_1, u_2)$ a distance of $\frac{8}{7}$ from $u_1$. Note that for this value of $x$, $w(u_4)d(u_4, X(x)) = w(u_3)d(u_3, X(x))$, consistent with Fact 5. In general, $X_1^c$ of a network is not necessarily a unique point; however, here $X_1^c$ is unique and $H(X_1^c) = \frac{30}{7}$.

4. Transportation planners are trying to decide where to locate a single school bus stop along a major highway. Situated along a one-mile stretch of the highway are eight communities. The table below gives the number of school-age students in each community who ride the bus on a daily basis. The distance of each community from the westernmost edge of the one-mile stretch of highway is also shown.

| community | 1 | 2 | 3 | 4 | 5 | 6 | 7 | 8 |
|---|---|---|---|---|---|---|---|---|
| # students | 9 | 4 | 8 | 11 | 5 | 3 | 5 | 11 |
| distance (mi) | 0.0 | 0.2 | 0.3 | 0.4 | 0.6 | 0.7 | 0.9 | 1.0 |

The data of this problem can be represented by an undirected path (§8.1.3) with each vertex $v$ corresponding to a community and $w(v)$ being the number of students from that community riding the bus. Edges join adjacent communities and have a length given by the difference in distance entries from the table. To minimize the total (weighted) distance traveled by the students, a 1-median is sought. By Fact 2, only vertex locations need to be considered. For example, situating the bus stop at vertex 3 incurs a cost of $F(\{3\}) = 9(0.3) + 4(0.1) + 8(0) + 11(0.1) + 5(0.3) + 3(0.4) + 5(0.6) + 11(0.7) = 17.6$. The minimum cost is incurred for vertex 4, with $F(\{4\}) = 16.2$, so that the bus stop should be located at community 4.

---

### 16.2.3   ALGORITHMS FOR LOCATION ON NETWORKS

Algorithms for finding $p$-medians and $p$-centers of a network $G$ can be devised that are feasible for small values of $p$. For general $p$, however, there are no efficient methods known for arbitrary networks $G$. Specialized (and efficient) algorithms are available when $G$ is a tree.

**Definitions:**

Let $T$ be a tree (§9.1.1) with vertex weights $w(v)$.

If $T'$ is a subnetwork of $T$, define the **total weight** of $T'$ by $W(T') = \sum_{v \in V(T')} w(v)$.

For $v \in V(T)$, let $T_{v1}, T_{v2}, \ldots, T_{vd(v)}$ be the components of $T - v$, where $d(v)$ is the degree (§8.1.1) of vertex $v$. Define $M_v = \max_{1 \leq i \leq d(v)} \{W(T_{vi})\}$.

A **leaf vertex** of $T$ is a vertex of degree 1.

**Facts:**

1. For a network $G = (V, E)$ with $n$ vertices and $m$ edges, the $p$-median and $p$-center problems can be solved in $O(n^{p+1})$ and $O(m^p n^p \log^2 n)$ time, respectively [Ta88].

2. Bhattacharya and Shi [BhSh14] provide an improved algorithm for solving the $p$-center problem with worst-case complexity $O(m^p n^{p/2} 2^{\log^* n} \log n)$ when $p \geq 3$, where $\log^* n$ denotes the iterated logarithm of $n$. When $p = 2$, the algorithm's running time is improved to $O(m^2 n \log^2 n)$.

**3.** If $p$ is an independent input variable (i.e., $p$ could grow with $n$), then both the $p$-center and $p$-median problems are NP-hard [KaHa79a], [KaHa79b]. Thus it is highly unlikely that an algorithm will be found with running time polynomial in $n$, $m$, and $p$.

**4.** Considerable success has been reported in solving large $p$-median problems using heuristic methods that do not necessarily guarantee optimal solutions. A survey of various heuristic approaches as well as multiobjective extensions is presented in [DaMa15].

**5.** Exact and heuristic algorithms for variants of the $p$-center problem are reviewed in [CaLaYa15].

**6.** Algorithms of complexities $O(n^2 p)$ and $O(n \log n)$ for the $p$-median and $p$-center problems on tree networks have been reported by Tamir [Ta96] and by Frederickson and Johnson [MiFr90, Chapter 7].

**7.** Vertex $u$ is a 1-median of a tree network $T$ if and only if $M_u \leq \frac{1}{2} W(T)$.

**8.** *1-median of a tree:* This algorithm (Algorithm 1) is based on Fact 7. The main idea is to repeatedly remove a leaf vertex, confining the problem to a smaller tree $T'$.

---

**Algorithm 1: 1-median of a tree.**

input: tree $T$

output: 1-median $\widetilde{v}$

$T' := T$; $W_0 := \sum_{v \in V(T')} w(v)$; $W(v) := w(v)$ for each $v \in V(T')$

{Main loop}

**if** $T'$ consists of a single vertex $\widetilde{v}$ **then** stop

**else**

$\quad$ $\widetilde{v} :=$ a leaf vertex of $T'$

$\quad$ **if** $W(\widetilde{v}) \geq \frac{1}{2} W_0$ **then** stop

$\quad$ **else**

$\qquad$ $u :=$ the vertex adjacent to $\widetilde{v}$ in $T'$

$\qquad$ $W(u) := W(u) + W(\widetilde{v})$

$\qquad$ $T' := T' - \widetilde{v}$

{Continue with next iteration of main loop}

---

**9.** Algorithm 1 can be implemented to run in $O(n)$ time.

**10.** Let $T$ be a tree network with $w(v) = c$ for all $v \in V(T)$. Then the 1-center of $T$ is the unique middle point of a longest path in $T$.

**11.** Select any vertex $v_0$ in a tree network $T$. Let $v_1$ be a farthest vertex from $v_0$, and let $v_2$ be a farthest vertex from $v_1$. Then the path from $v_1$ to $v_2$ is a longest path in $T$.

**12.** *1-center of a tree:* This algorithm (Algorithm 2) applies to "unweighted" trees, in which there are identical weights at each vertex. It is based on Facts 10 and 11.

---

**Algorithm 2: 1-center of an unweighted tree.**

input: tree $T$ with $w(v) = c$ for all $v \in V(T)$

output: 1-center $x$

find a longest path $P$ in $T$ (using Fact 11)

let $u_1$ and $u_2$ be the end vertices of $P$

find the middle point of this path: i.e., the point $x$ such that $d(x, u_1) = d(x, u_2)$

**13.** Algorithm 2 can be implemented to run in $O(n)$ time.

**Examples:**

**1.** Suppose that the vertices of the tree $T$ in Example 2 of §16.2.2 are labeled $v_1, v_2, \ldots, v_8$ in order from top to bottom and left to right at each height. Algorithm 1 can be applied to find the 1-median of $T$. First, the leaf vertex $v_1$ is selected and since $W(v_1) = 1$ is less than $\frac{1}{2}W_0 = \frac{15}{2}$, its weight is added to vertex $v_3$. The following table shows the progress of the algorithm, which eventually identifies vertex $v_4$ as the 1-median of $T$. As guaranteed by Fact 7, $M_{v_4} = \max\{6, 1, 3, 4\} \leq \frac{15}{2}$.

| | | $W(v_i)$ | | | | | | | |
|---|---|---|---|---|---|---|---|---|---|
| *iteration* | $\widetilde{v}$ | 1 | 2 | 3 | 4 | 5 | 6 | 7 | 8 |
| 0 | | 1 | 1 | 1 | 1 | 3 | 4 | 2 | 2 |
| 1 | $v_1$ | − | 1 | 2 | 1 | 3 | 4 | 2 | 2 |
| 2 | $v_2$ | − | − | 2 | 2 | 3 | 4 | 2 | 2 |
| 3 | $v_5$ | − | − | 2 | 5 | − | 4 | 2 | 2 |
| 4 | $v_6$ | − | − | 6 | 5 | − | − | 2 | 2 |
| 5 | $v_3$ | − | − | − | 11 | − | − | 2 | 2 |

**2.** Consider the tree $T$ used in Example 1 whose vertices are labeled as in that example. Algorithm 2 can be applied to find the 1-center of $T$, with all vertex weights being 1. First, select $v_1$ and find a farthest vertex from it, namely $v_5$. A farthest vertex from $v_5$ is then $v_8$, giving a longest path $P = [v_5, v_4, v_7, v_8]$ in $T$. The midpoint $x$ of $P$, located $\frac{1}{2}$ unit from $v_4$ along edge $(v_4, v_5)$, is then the 1-center of $T$. If instead the longest path $Q = [v_6, v_3, v_4, v_5]$ in $T$ had been identified, then the same midpoint $x$ would be found.

## 16.2.4   CAPACITATED LOCATION PROBLEMS

**Definitions:**

Let $X_p = \{x_1, \ldots, x_p\}$ be a set of locations for $p$ facilities in the metric space $S$ with $n$ demand points $V = \{v_1, \ldots, v_n\} \subseteq S$ where $w(v_i) \geq 0$ is the demand at $v_i \in V$.

For each $v_i \in V$ and $x_j \in X_p$, let $w(v_i, x_j)$ be the portion of the demand at $v_i$ satisfied by the facility at $x_j$.

Let $W(x_j)$ be the sum of the demands satisfied by (or allocated to) the facility at $x_j$. In a **capacitated location problem**, upper (and/or lower) bounds are placed on $W(x_j)$.

Given the positive integer $p$ and positive constant $\alpha$, two versions of the **capacitated p-median (CPM) problem** in network $G = (V, E)$ are defined:

(a) Find a set of locations $X_p$ such that $F(X_p) = \sum_{v_i \in V} w(v_i)d(v_i, X_p)$ is minimized subject to $W(x_j) \leq \alpha$ for all $x_j \in X_p$. Here it is assumed that the demands are satisfied by their closest facility, and in the case of ties, a demand, say at $v$, may be allocated in an arbitrary way among the closest facilities to $v$.

(b) Find $X_p$ and $\{w(v_i, x_j) \mid v_i \in V \text{ and } x_j \in X_p\}$ to minimize

$$\sum_{j=1}^{p} \sum_{i=1}^{n} w(v_i, x_j)d(v_i, x_j)$$

subject to

$$\sum_{i=1}^{n} w(v_i, x_j) \le \alpha, \quad j = 1, 2, \ldots, p$$

$$\sum_{j=1}^{p} w(v_i, x_j) = w(v_i), \quad i = 1, 2, \ldots, n.$$

**Facts:**

**1.** Capacitated facility location problems occur in several applied settings, including

- the location of manufacturing plants (with limited output) to serve customers;
- the location of landfills (with limited capacity), which receive solid waste from the members of a community;
- the location of concentrators in a telecommunication network, where each concentrator bundles messages received from individual users and can handle only a certain amount of total message traffic.

**2.** Version (a) of CPM may not have a solution if $\alpha$ is too small.

**3.** If $\alpha$ is sufficiently large in version (a) and CPM has a solution, there may not exist a solution consisting entirely of vertices of $G$. See Example 1.

**4.** Version (b) of the CPM has a solution consisting entirely of vertices of $G$. This was shown by J. Levy; see [HaMi79].

**Examples:**

**1.** Suppose $p = 2$ and $\alpha = 3$ in the following network $G$. A solution to version (a) of the CPM problem consists of the points $X_2 = \{x_1, x_2\}$ with $W(x_1) = W(x_2) = 3$ and $F(X_2) = 9$. It is easy to see that the choice of any two vertices for $X_2$ would violate the allocation constraint to one facility.

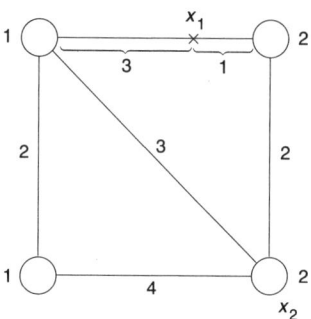

**2.** If $p = 2$ and $\alpha = 3$ for the network $G$ given in Example 1, then version (b) of the CPM problem has a solution containing only vertices of $G$. Suppose that the top two vertices in the figure are $v_1$ and $v_2$ (from left to right) and the bottom two vertices are $v_3$ and $v_4$. Then $X_2 = \{v_1, v_2\}$ is an optimal solution, where all the demand $w(v_3) = 1$ is allocated to $v_1$ whereas the demand $w(v_4) = 2$ is equally split between $v_1$ and $v_2$. (Here, not all demand from $v_4$ is sent to its closest facility $v_2$.) In this solution, $W(v_1) = W(v_2) = 3$ and $F(X_2) = 7$.

## 16.2.5   FACILITIES IN THE PLANE

The $p$-median and $p$-center problems can be defined in the plane $\mathcal{R}^2$. Several measures of distance are commonly considered for these location problems in the plane.

**Definitions:**

Let $x = (x_1, x_2)$ and $y = (y_1, y_2)$ be points of $S = \mathcal{R}^2$.

The **Euclidean** ($\ell_2$) **distance** between $x$ and $y$ is $d(x, y) = [(x_1 - y_1)^2 + (x_2 - y_2)^2]^{1/2}$.

The **rectilinear** ($\ell_1$) **distance** between $x$ and $y$ is $d(x, y) = |x_1 - y_1| + |x_2 - y_2|$.

The (**generalized**) **Weber problem** is the $p$-median problem in $\mathcal{R}^2$ with the $\ell_2$ measure of distance.

The **unweighted Euclidean 1-center problem** is the 1-center problem in $\mathcal{R}^2$ with the $\ell_2$ measure of distance and with $w(v_i) = c$ for all $v_i \in V$.

**Facts:**

**1.** No polynomial-time algorithm for the Weber problem, even when $p = 1$, has been discovered. The history of this problem, and generalizations, are discussed in [We93].

**2.** In practice, an iterative method due to Weiszfeld [We37], [WePl09] has been shown to be highly successful for the Weber problem with $p = 1$.

**3.** Further information on the Weiszfeld algorithm and extensions can be found in [Pl11]. Recent algorithmic improvements are discussed in [Dr15] and [RoVa13].

**4.** The $p$-median and $p$-center problems in $\mathcal{R}^2$ with either $\ell_1$ or $\ell_2$ as the measure of distance have been proven to be NP-hard if $p$ is an input variable [MeSu84].

**5.** The unweighted Euclidean 1-center problem is equivalent to finding the center of the smallest (radius) circle that encloses all points in $V$.

**6.** The following table provides a summary of time complexity results of the best known algorithms for location problems in the plane [Me83], [MeSu84].

|  | *p arbitrary* | *p = 1* |
|---|---|---|
| *p-median* | NP-hard if $p$ is an input variable under $\ell_1$ or $\ell_2$ <br> unknown complexity if $p$ is fixed | unknown complexity under $\ell_1$ or $\ell_2$ |
| *p-center* | NP-hard if $p$ is an input variable under $\ell_1$ or $\ell_2$ <br> unknown complexity for fixed $p > 1$ | $O(n \log^2 n)$ under $\ell_2$ <br> $O(n)$ under $\ell_2$ in the unweighted case <br> $O(n)$ under $\ell_1$ for both the weighted and unweighted cases |

**7.** A unified treatment of location problems in Euclidean space studies the *ordered median location problem*, which generalizes both the $p$-median and $p$-center problems [BlPuEl14], [NiPu05].

**Examples:**

**1.** The floor plan of a factory contains existing machines A, B, C at the coordinate locations $a = (0, 4)$, $b = (2, 0)$, $c = (5, 2)$. A new central storeroom, to house materials needed by the machines, is to be placed at some location $x = (x_1, x_2)$ on the factory floor. Because the aisles of the factory floor run north-south and east-west, transportation between the storeroom and the machines must take place along these perpendicular directions. For example, the distance between the storeroom and the machine C is $|x_1 - 5| + |x_2 - 2|$. Management wants to locate the storeroom so that the weighted sum of distances between the new storeroom and each machine is minimized, taking into account that the demand for material by machine A is twice the demand by machine B, and demand for material by machine C is three times that by machine B. This is a weighted 1-median problem in the plane with the $\ell_1$ measure of distance. The point $(3, 2)$ is an optimal location for the storeroom. In fact, for any $2 \leq u \leq 5$ the point $(u, 2)$ is also an optimal location.

**2.** Suppose that the $\ell_2$ distance measure is used instead in Example 1. Then a weighted 1-median is a location $x = (x_1, x_2)$ that minimizes

$$2\sqrt{x_1^2 + (x_2 - 4)^2} + \sqrt{(x_1 - 2)^2 + x_2^2} + 3\sqrt{(x_1 - 5)^2 + (x_2 - 2)^2}.$$

The minimizing point in this case is $(5, 2)$, which is the *unique* optimal location for the storeroom. On the other hand, if the demands for material are the same for all three machines, then the unweighted 1-median occurs at the unique location $(2.427, 1.403)$.

---

## 16.2.6 OBNOXIOUS FACILITIES

In the preceding subsections, it has been assumed that the consumers at $v_i$ wish to be as close as possible to a facility. That is, the facilities are desirable. In contrast, this subsection discusses location problems where the facilities are undesirable or obnoxious.

### Definitions:
For $v_i \in V$, $w(v_i)d(v_i, X_p)$ represents the **utility** (in contrast to cost) associated with having an obnoxious facility located at distance $d(v_i, X_p)$ from $v_i$.

The following two **obnoxious facility location problems** are defined:

(a) find $X_p \subseteq S$ to maximize $F(X_p) = \sum_{i=1}^{n} w(v_i)d(v_i, X_p)$;

(b) find $X_p \subseteq S$ to maximize $G(X_p) = \min_{1 \leq i \leq n} w(v_i)d(v_i, X_p)$.

If space $S$ is a network $G = (V, E)$, for each edge $e = (u, v) \in E$, let $x(e) = x$ be the point on $e$ such that $w(u)d(u, x) = w(v)d(v, x) \equiv w(e)$.

### Facts:
**1.** If $S$ is a network, problem (a) may not have a solution that is a subset of vertices (see Example 2).

**2.** Suppose $S$ is a network $G$ with $w(v_i) > 0$ for all $v_i \in V$; further assume that at most one point of $X_p$ can be on any particular edge. Renumber the $m$ edges of $G$ so that $w(e_1) \geq w(e_2) \geq \cdots \geq w(e_m)$. Then $x(e_1), x(e_2), \ldots, x(e_p)$ is a solution to problem (b).

**3.** Additional results on this subject can be found in [BrCh89].

### Examples:
**1.** In the location of obnoxious facilities, the distance to a closest facility is to be made as large as possible. This type of problem arises in siting nuclear power plants, sewage treatment facilities, and landfills, for example.

**2.** In the following network, an optimal solution $X_1 = \{x_1\}$ to problem (a) when $p = 1$ is the midpoint of any edge, with $F(X_1) = 5$. However, if the facility is located at any vertex $v$ then $F(\{v\}) = 4$.

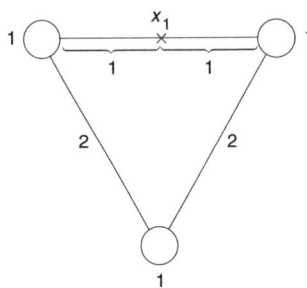

---

## 16.2.7  EQUITABLE LOCATIONS

The $p$-median problem is a widely used model for locating public or private facilities. However, it may leave some demand points (communities) too far from their closest facility and thus might be perceived as inequitable. To remedy this situation, the $p$-median problem can be modified in several ways.

**Definitions:**
Suppose $S$ is a metric space, $V \subseteq S$, $w(v_i) \geq 0$ for all $v_i \in V$, and $p$ is a positive integer.

Let $w'(v_i) = w(v_i)/\sum_{j=1}^{n} w(v_j)$ and define $F'(X_p) = \sum_{i=1}^{n} w'(v_i)d(v_i, X_p)$, $Z(X_p) = \sum_{i=1}^{n} w'(v_i)[d(v_i, X_p) - F'(X_p)]^2$.

The following three **equitable facility location problems** are defined:

    (a) given a constant $\beta$, find a set of $p$ points $X_p \subseteq S$ to minimize

$$F(X_p) = \sum_{i=1}^{n} w(v_i)d(v_i, X_p)$$

        subject to

$$d(v_i, X_p) \leq \beta, \quad \text{for all } v_i \in V;$$

    (b) given a constant $\alpha$, $0 < \alpha < 1$, find a set of $p$ points $X_p \subseteq S$ to minimize $\alpha F(X_p) + (1 - \alpha)H(X_p)$;

    (c) find $X_p \subseteq S$ to minimize $Z(X_p)$.

**Facts:**
**1.** Since the objective function in (b) is a linear combination of the objective functions for the $p$-median and $p$-center problems, the solution $X_p^*$ is called a *centdian*.

**2.** $F'(X_p) = F(X_p)/\sum_{j=1}^{n} w(v_j)$ is the mean distance to the consumers given that the facilities are located at $X_p$.

**3.** $Z(X_p)$ is the variance of the distance to the consumers given that the facilities are located at $X_p$.

**4.** Additional results are discussed in [HaMi79], [Ma86].

**5.** Other models that introduce equity criteria for facility location models are discussed in [Og09].

**Example:**

**1.** The following figure shows a tree network $T$ on vertices $a, b, \ldots, g$, with the edge lengths displayed. Suppose that all vertex weights are 1. Then the 1-median of $T$ is located at vertex $c$, while the 1-center of $T$ is located at the point $x$ one unit from vertex $d$ along $(d, g)$. These locations can be calculated using Algorithms 1 and 2 from §16.2.3. It can be verified that the centdian of $T$ is at point $x$ for $0 \leq \alpha \leq \frac{1}{6}$, at vertex $d$ for $\frac{1}{6} \leq \alpha \leq \frac{1}{2}$, and at vertex $c$ for $\frac{1}{2} \leq \alpha \leq 1$.

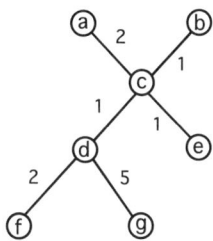

# 16.3  PACKING AND COVERING

Many practical problems can be formulated as either packing or covering problems. In packing problems, known activities are given each of which requires several resources for its completion. The problem is to select a most valuable set of activities to undertake without using any common resources. Such a problem arises, for example, in scheduling as many computational activities as possible on a set of machines (resources) that cannot be used simultaneously for more than one activity. In covering problems, a specified set of tasks must be performed, and the objective is to minimize the resources required to perform the tasks. For example, a number of delivery trucks (resources) operating on overlapping geographical routes need to be dispatched to pick up items at customer locations (tasks). The fewest number of trucks are to be sent so that each customer location is "covered" by at least one of the dispatched trucks. Both exact and heuristic solution algorithms for packing and covering problems are discussed in this section.

## 16.3.1  KNAPSACKS

The knapsack problem arises when a single critical resource is to be optimally allocated among a variety of options. Specifically, there are available a number of items, each of which consumes a known amount of the resource and contributes a known benefit. Items are to be selected to maximize the total benefit without exceeding the given amount of the resource. Knapsack problems arise in many practical situations involving cutting stock, cargo loading, and capital budgeting.

**Definitions:**

Let $N = \{1, 2, \ldots, n\}$ be a given set of $n$ **items**. Utilizing item $j$ consumes (requires) $a_j > 0$ units of the given resource and confers the benefit $c_j > 0$.

The **knapsack problem** (**KP**) is the following 0-1 integer linear programming problem:

$$\text{maximize:} \quad \sum_{j \in N} c_j x_j$$

$$\text{subject to:} \quad \sum_{j \in N} a_j x_j \leq b \tag{1}$$

$$x_j \in \{0,1\}.$$

It is assumed that $a_j \leq b$ for all $j \in N$. Let $z(b)$ denote the optimal objective value in (1) for a given integer $b$.

The **LP relaxation** (§16.1.8) of (1) is the linear programming problem:

$$\text{maximize:} \quad \sum_{j \in N} c_j x_j$$

$$\text{subject to:} \quad \sum_{j \in N} a_j x_j \leq b \tag{2}$$

$$0 \leq x_j \leq 1.$$

Let $z_{LP}$ denote the optimal objective value to the LP relaxation (2).

Let $N^1$ and $N^0$ be the set of variables taking values 1 and 0, respectively, in the optimal solution to (2). Let $\lambda^*$ be the dual variable (§16.1.5) associated with the knapsack inequality in the optimal solution to (2).

A **cover** is a set $S \subseteq N$ such that $\sum_{j \in S} a_j > b$. The cover $S$ is **minimal** if no proper subset of $S$ is a cover.

A **branch-and-bound tree** for KP is a tree $T$ whose nodes correspond to subproblems obtained by fixing certain variables of (1) to either 0 or 1. The **root** of $T$ corresponds to the original problem (1).

A node $t$ of $T$ is specified by its **level** $k = 0, 1, \ldots, n$ and the index set $N_t \subseteq \{1, \ldots, k\}$ of variables currently fixed to 1.

Associated with the set $N_t$ is the current benefit $z_t = \sum_{j \in N_t} c_j$ and the available amount of resource $b_t = b - \sum_{j \in N_t} a_j$.

## Facts:

**1.** Formulation (1) is a 0-1 integer linear programming problem with a single (knapsack) constraint. It expresses the optimization problem in which a subset of the $n$ items is to be selected to maximize the total benefit without exceeding the available amount of the given resource (the capacity of the knapsack).

**2.** KP is an NP-hard optimization problem (§17.5.2).

**3.** KP can be solved in polynomial time for fixed $b$.

**4.** Given a rational $\epsilon > 0$, a 0-1 vector $x^*$ can be found satisfying $\sum_{j \in N} a_j x_j^* \leq b$ and $\sum_{j \in N} c_j x_j^* \geq (1 - \epsilon) z(b)$ in time polynomially bounded by $\frac{1}{\epsilon}$ and by the sizes of $a = (a_1, \ldots, a_n)$, $c = (c_1, \ldots, c_n)$, and $b$.

**5.** If the coefficients $a_j$ can be ordered such that each coefficient is an integer multiple of the previous one, then KP can be solved in polynomial time.

**6.** If $a_{j-1} \geq a_j + \cdots + a_n$ holds for $j = 2, \ldots, n$ then KP can be solved in polynomial time.

**7.** *Greedy heuristic:* This heuristic (Algorithm 1) for the KP processes the variables $x_j$ in nonincreasing order of $\frac{c_j}{a_j}$, making each variable equal to 1 if possible.

---

**Algorithm 1:  Greedy heuristic for the KP.**

input: KP with $\frac{c_1}{a_1} \geq \frac{c_2}{a_2} \geq \cdots \geq \frac{c_n}{a_n}$

output: feasible solution $x = (x_1, x_2, \ldots, x_n)$

**for** $k := 1$ **to** $n$

$x_k := 1$

**if** $\sum_{j=1}^{k} a_j x_j > b$ **then** $x_k := 0$

---

**8.** Suppose $z_H$ is the objective value for the solution $x$ produced by Algorithm 1. Then $z(b) \geq z_H \geq \frac{1}{2} z(b)$.

**9.** Algorithm 1 is most effective if the coefficients $a_j$ are small relative to $b$.

**10.** The LP relaxation (2) can be solved explicitly by filling the knapsack in turn with items $j$ in order of nonincreasing $\frac{c_j}{a_j}$, ignoring the integer restriction. The solution $x$ obtained has at most one fractional component.

**11.** *Core heuristic*: This heuristic (Algorithm 2) for the KP first solves the LP relaxation (2) as in Fact 10, in which at most one variable $x_k$ can be fractional. A smaller knapsack problem is then solved, by setting to 0 any variable $x_j$ with index $j$ sufficiently greater than $k$ and by setting to 1 any variable $x_j$ with index $j$ sufficiently smaller than $k$.

---

**Algorithm 2:  Core heuristic for the KP.**

input: KP with $\frac{c_1}{a_1} \geq \frac{c_2}{a_2} \geq \cdots \geq \frac{c_n}{a_n}$

output: feasible solution $x = (x_1, x_2, \ldots, x_n)$

{Solve LP relaxation}

find the smallest value $k$ such that $\sum_{j=1}^{k} a_j \geq b$

{Solve restricted KP}

select any $r > 0$

$x_j := 1$ for $j \leq k - r$

$x_j := 0$ for $j \geq k + r$

solve to optimality the smaller knapsack problem:

maximize $\sum_{j=k-r+1}^{k+r-1} c_j x_j$

subject to $\sum_{j=k-r+1}^{k+r-1} a_j x_j \leq b - \sum_{j=1}^{k-r} a_j$

$x_j \in \{0, 1\}$

---

**12.** Algorithm 2 is effective if the number of variables $n$ is large since values of $r$ between 10 and 25 give very good approximations in most cases. For further details see [BaZe80].

**13.** $z(b) \leq z_{LP}$.

**14.** If $z^l$ is the objective value of a feasible solution to (1), then $z^l \leq z(b)$.

**15.** Node $t$ of the branch-and-bound tree $T$ corresponds to a subproblem having a nonempty set of feasible solutions if and only if $b_t \geq 0$. When this holds, $z_t$ is a lower bound for $z(b)$.

**16.** An upper bound on the objective value over the subproblem corresponding to node $t$ is $z_t^u = \lfloor z_t^* \rfloor$, where $z_t^* = z_t + \max\{\sum_{j=k+1}^{n} c_j x_j \mid \sum_{j=k+1}^{n} a_j x_j \leq b_t, \, 0 \leq x_j \leq 1\}$.

**17.** *Implicit enumeration*: This is an exact technique based on the branch-and-bound method (§16.1.8). It is implemented using a branch-and-bound tree $T$, with the following specifications:

- The initial tree $T$ consists of the root $t$, with lower bound $z^l$ on $z(b)$ obtained using the greedy heuristic. An upper bound for node $t$ is $z_t^u = z_{LP}$.

- If $z_t^u \leq z^l$, then node $t$ is discarded since it cannot provide a better solution.

- If $z_t^u > z^l$, there are three cases (where node $t$ is at level $k$ of $T$):

  ◇ $a_{k+1} < b_t$: If $k+1 < n$, create a new node with $x_{k+1} = 1$. If $k+1 = n$ an optimal solution for node $t$ has $x_n = 1$. Since this solution is feasible for KP, set $z^l = z_t^u$ and discard node $t$.

  ◇ $a_{k+1} = b_t$: An optimal solution for node $t$ (and a feasible solution for KP) is obtained by setting $x_{k+1} = 1$ and $x_j = 0$ for $j > k+1$. Set $z^l = z_t^u$ and discard node $t$.

  ◇ $a_{k+1} > b_t$: Discard the (infeasible) node with $x_{k+1} = 1$ and create a new node with $x_{k+1} = 0$.

- To backtrack from node $t$ let $N_t = \{j_1, \ldots, j_r\} \subseteq \{1, \ldots, k\}$ with $j_1 < \cdots < j_r$. If $k \notin N_t$ retreat to level $j_r$ and set $x_{j_r} = 0$. If $k = j_r$ retreat to level $j_{r-1}$ and set $x_{j_{r-1}} = 0$.

**18.** *Variable fixing:* Given $z_{LP}$, $z^l$, $N^1$, $N^0$, and $\lambda^*$, the following tests can be used to fix variables and reduce the size of the knapsack problem:

- If $k \in N^1$ and $z_{LP} - (c_k - \lambda^* a_k) \leq z^l$, then fix $x_k = 1$.
- If $k \in N^0$ and $z_{LP} + (c_k - \lambda^* a_k) \leq z^l$, then fix $x_k = 0$.
- Given $k \in N^1$ define

$$z_{LP}^k = \sum_{j \in N^1 - \{k\}} c_j + \max\left\{ \sum_{j \in N - N^1} c_j x_j \,\middle|\, \sum_{j \in N - N^1} a_j x_j \leq b,\ 0 \leq x_j \leq 1 \right\}.$$

  If $z_{LP}^k \leq z^l$, then $x_k$ can be fixed to 1.

- Given $k \in N^0$ define

$$z_{LP}^k = c_k + \max\left\{ \sum_{j \in N - N^0} c_j x_j \,\middle|\, \sum_{j \in N - N^0} a_j x_j \leq b - a_k,\ 0 \leq x_j \leq 1 \right\}.$$

  If $z_{LP}^k \leq z^l$, then $x_k$ can be fixed to 0.

**19.** *Minimal cover inequality:* If $S$ is a minimal cover then each feasible solution $x$ to KP satisfies $\sum_{j \in S} x_j \leq |S| - 1$.

**20.** *Lifted minimal cover inequality:* The minimal cover inequality can be further strengthened. Without loss of generality, assume that $a_1 \geq a_2 \geq \cdots \geq a_n$ and $S = \{j_1 < j_2 < \cdots < j_r\}$. Let $\mu_h = \sum_{k=1}^{h} a_{j_k}$ for $h = 1, \ldots, r$ and define $\lambda = \mu_r - b \geq 1$. Then each feasible solution $x$ to KP satisfies $\sum_{j \in N - S} \alpha_j x_j + \sum_{j \in S} x_j \leq |S| - 1$, where

- if $\mu_h \leq a_j \leq \mu_{h+1} - \lambda$ then $\alpha_j = h$;
- if $\mu_{h+1} - \lambda + 1 \leq a_j \leq \mu_{h+1} - 1$ then (a) $\alpha_j \in \{h, h+1\}$, and (b) there is at least one lifted minimal cover inequality with $\alpha_j = h+1$.

**21.** Computer codes for solving knapsack problems can be found at

  - http://www.diku.dk/~pisinger/codes.html

**22.** Further details on the material in this section are available in [KePfPi04], [NeWo88], and [Sc86].

**23.** Algorithms for the multidimensional 0-1 knapsack problem are discussed in [FrHa05].

## Examples:

**1.** *Investment problem:* An investor has $50,000 to place in any combination of five available investments $(1, 2, 3, 4, 5)$. All investments have the same maturity but are issued in different denominations and have different (one-year) yields, as shown in the table.

| investment | 1 | 2 | 3 | 4 | 5 |
|---|---|---|---|---|---|
| denomination ($) | 10,000 | 20,000 | 30,000 | 10,000 | 20,000 |
| yield (%) | 20 | 14 | 18 | 9 | 13 |

Let variable $x_j = 1$ if Investment $j$ is selected and $x_j = 0$ if it is not. The interest earned for Investment 1 is $(0.20)10,000 = 2,000$; the values of the other investments are found similarly. Then the investor's problem is the knapsack problem

maximize:    $2,000x_1 + 2,800x_2 + 5,400x_3 + 900x_4 + 2,600x_5$

subject to:    $10,000x_1 + 20,000x_2 + 30,000x_3 + 10,000x_4 + 20,000x_5 \leq 50,000$

which has the optimal solution $x_1 = x_3 = x_4 = 1$, $x_2 = x_5 = 0$ with maximum interest of $8,300.

**2.** Consider the knapsack problem in 0-1 variables $x$

maximize:    $30x_1 + 8x_2 + 16x_3 + 20x_4 + 12x_5 + 9x_6 + 5x_7 + 3x_8$

subject to:    $10x_1 + 3x_2 + 7x_3 + 9x_4 + 6x_5 + 5x_6 + 3x_7 + 2x_8 \leq 27.$

Here the variables $x_j$ are indexed in nonincreasing order of $\frac{c_j}{a_j}$. The optimal solution to the LP relaxation (2) is $x_1 = x_2 = x_3 = 1$, $x_4 = \frac{7}{9}$, $x_j = 0$ otherwise, with $z_{LP} = 69\frac{5}{9}$. The greedy heuristic (Algorithm 1) gives the feasible solution $x_1 = x_2 = x_3 = x_5 = 1$, $x_j = 0$ otherwise, with $z_H = 66$. Using $r = 3$, the core heuristic gives $x_1 = x_2 = x_4 = x_6 = 1$, $x_j = 0$ otherwise. This solution is optimal, with objective value 67.

**3.** Consider the knapsack problem in 0-1 variables $x$

maximize:    $x_1 + x_2 + x_3 + x_4 + x_5 + x_6$

subject to:    $10x_1 + 8x_2 + 4x_3 + 3x_4 + 3x_5 + 2x_6 \leq 11.$

The set $S = \{3, 4, 5, 6\}$ is a minimal cover which gives the lifted minimal cover inequality $3x_1 + 2x_2 + x_3 + x_4 + x_5 + x_6 \leq 3$. Adding this inequality and solving the resulting linear program gives $x_4 = x_5 = x_6 = 1$, $x_j = 0$ otherwise. This solution is optimal.

**4.** *General knapsack problem:* The general (or unbounded) knapsack problem allows the decision variables $x_j$ to be any nonnegative integers, not just 0 and 1. For example, an individual can select any number of a given item to pack in a knapsack, subject to the given weight restriction.

## 16.3.2 BIN PACKING

Minimizing the number of copies of a resource required to perform a specified set of tasks can be formulated as a bin packing problem. It is assumed that no such task can be split between two different units of the resource. For example, this type of problem arises in allocating a set of customer loads to (identical) trucks, with no load being split between two trucks. Also, the scheduling of heterogeneous tasks on identical machines can be viewed as a bin packing problem. Namely, find the fewest number of machines of capacity $C$ such that each task is executed on one of the machines and the total capacity of jobs assigned to any machine does not exceed $C$.

**Definitions:**

The positive integer $C$ denotes the **bin capacity**.

Let $L = (p_1, p_2, \ldots, p_n)$ be a list of $n$ **items**, where item $p_i$ has an integer **size** $s(p_i) \leq C$.

A subset $P \subseteq L$ is a **packing** if $\sum_{p_i \in P} s(p_i) \leq C$.

The **gap** of a packing $P$ is given by the quantity $C - \sum_{p_i \in P} s(p_i)$.

The **bin packing problem** is the problem of finding the minimum number of bins (each of capacity $C$) needed to pack all items so that the gap in each bin is nonnegative. The minimum number of bins needed for the list $L$ is denoted $b^*(L)$.

**Facts:**

**1.** The bin packing problem is an NP-hard optimization problem (§17.5.2).

**2.** *First fit (FF) method:* In this heuristic algorithm, item $p_i$ $(i = 1, 2, \ldots, n)$ is placed in the first bin into which it fits. A new bin is started only when $p_i$ will not fit into any nonempty bin.

**3.** Let $b^{FF}(L)$ denote the number of bins produced by the FF algorithm for a list $L$. Then $b^{FF}(L) \leq \min\{\lceil \frac{17}{10}b^*(L) \rceil, 1.75b^*(L)\}$.

**4.** *First fit decreasing (FFD) method:* In this heuristic algorithm, the items are first ordered by decreasing size so that $s(p_1) \geq s(p_2) \geq \cdots \geq s(p_n)$. Then the FF algorithm is applied to the reordered list.

**5.** Let $b^{FFD}(L)$ denote the number of bins produced by the FFD algorithm for a list $L$. Then $b^{FFD}(L) \leq \min\{\frac{11}{9}b^*(L) + 3, 1.5b^*(L)\}$.

**6.** If all item sizes are of the form $C(\frac{1}{k})^j$, $j \geq 0$, for some fixed positive integer $k$, then $b^{FFD}(L) = b^*(L)$.

**7.** If the item sizes are uniformly distributed on $[0, a]$ with $0 < a \leq \frac{C}{2}$, then asymptotically $\frac{b^{FFD}(L)}{b^*(L)} \to 1$.

**8.** *Modified first fit decreasing (MFFD) method:* This heuristic method (Algorithm 3) produces a packing using relatively few bins. After the initial phase of packing the largest sized items $L_A$, let an "A-bin" denote one containing only a single item from $L_A$.

---

**Algorithm 3:**  **MFFD algorithm for bin packing.**

input: list $L$, bin capacity $C$

output: a packing of $L$

partition $L$ into the three sublists $L_A = \{p_i \mid s(p_i) \in (\frac{1}{3}C, C]\}$,

$\qquad L_D = \{p_i \mid s(p_i) \in (\frac{11}{71}C, \frac{1}{3}C]\}$, $L_X = \{p_i \mid s(p_i) \in (0, \frac{11}{71}C]\}$

pack the sublist $L_A$ using the FFD algorithm.

{Pack as much of $L_D$ into A-bins as possible}

    1. let bin $B_j$ be the A-bin with the currently largest gap; if the two smallest unpacked items in $L_D$ will not fit together in $B_j$, go to 4

    2. place the smallest unpacked item $p_i$ from $L_D$ in $B_j$

    3. let $p_k$ be the largest unpacked item in $L_D$ that will now fit in $B_j$; place $p_k$ in $B_j$ and go to 1

    4. combine the unpacked portion of $L_D$ with $L_X$ and add these items to the packing using FFD

---

**9.** Let $b^{MFFD}(L)$ denote the number of bins produced by the MFFD algorithm for a list $L$. Then asymptotically, as $b^*(L)$ gets large, $b^{MFFD}(L) \leq 1.183 b^*(L)$.

**10.** *Best fit (BF) method*: In this heuristic algorithm, item $p_i$ is placed in the bin into which it will fit with the smallest gap left over. Ties are broken in favor of the lowest indexed bin.

**11.** *Best fit decreasing (BFD) method*: In this heuristic algorithm, the items are first ordered so that $s(p_1) \geq s(p_2) \geq \cdots \geq s(p_n)$. Then the BF algorithm is applied to the reordered list.

**12.** Asymptotic worst-case bounds for BF [BFD] are the same as those for FF [FFD]. In practice the BF version performs somewhat better.

**13.** For a fixed number of $d$ different item sizes, the bin packing problem can be solved in polynomial time [GoRo14].

**14.** Further details on the material covered in this section are found in [CoEtal13] and [CoGaJo84].

**Examples:**

**1.** Television commercials are to be assigned to station breaks. This is a bin packing problem where the duration of each station break is $C$ and the duration of each commercial is $s(p_i)$.

**2.** Material such as cable, lumber, or pipe is supplied in a standard length $C$. Demands for pieces of the material are for arbitrary lengths $s(p_i)$ not exceeding $C$. The objective is to use the minimum number of standard lengths to supply a given list of required pieces. This is also a bin packing problem.

**3.** A set of independent tasks with known execution times $s(p_i)$ are to be executed on a collection of identical processors. Determining the minimum number of processors needed to complete all tasks by the deadline $C$ is a bin packing problem.

**4.** Consider the list $L = (4, \ldots, 4, 7, \ldots, 7, 8, \ldots, 8, 13, \ldots, 13)$, consisting of twelve 4s and six each of 7s, 8s, and 13s. Each bin has capacity $C = 24$. Either FF (or BF) when applied to $L$ results in a packing with twelve bins: two bins are packed as $(4, 4, 4, 4, 4, 4)$, two as $(7, 7, 7)$, two as $(8, 8, 8)$, and six as $(13)$.

**5.** If FFD (or BFD) is applied to the list in Example 4, a packing with ten bins results: six bins are packed as $(13, 8)$, two as $(7, 7, 7)$, and two as $(4, 4, 4, 4, 4, 4)$.

**6.** If MFFD is applied to the list in Example 4, then $L_A$ contains the six 13s and $L_D$ contains the remaining items. Packing $L_A$ using FFD results in six A-bins, each containing a single 13 and having gap 11. Steps 1–3 of Algorithm 3 result in six bins packed as $(13, 7, 4)$, and Step 4 yields two bins packed as $(8, 8, 8)$ and one bin packed as $(4, 4, 4, 4, 4, 4)$. This is an optimal solution since all nine bins are completely packed.

---

## 16.3.3 SET COVERING AND PARTITIONING

Set covering or set partitioning problems arise when a specified set of tasks must be performed while minimizing the cost of resources used. Such problems arise in scheduling fleets of vehicles or aircraft, locating fire stations in an urban area, political redistricting, and fault testing of electronic circuits.

**Definitions:**

Let $e$ denote the column vector of all 1s.

Let $A = (a_{ij})$ be a 0-1 **incidence matrix** and let $c = (c_j)$ be a row vector of **costs**.

The set $A_j = \{i \mid a_{ij} = 1\}$ contains all rows **covered** by column $j$.

The **set covering (SC) problem** is the 0-1 integer linear programming problem

$$\begin{aligned} \text{minimize:} \quad & cx \\ \text{subject to:} \quad & Ax \geq e \\ & x_j \in \{0, 1\}. \end{aligned}$$

Let $v^*$ be the optimal objective value to this problem.

The **set partitioning (SP) problem** has the same form as the set covering problem except the constraints are $Ax = e$.

The **LP relaxation** of SC or SP is obtained by replacing the constraints $x_j \in \{0, 1\}$ by $0 \leq x_j \leq 1$. Let $v_{LP}$ be the optimal objective value to the LP relaxation.

The matrix $A$ is **totally unimodular** if the determinant of every square submatrix of $A$ is 0, +1, or −1.

The matrix $A$ is **balanced** if $A$ has no square submatrix of odd order, containing exactly two 1s in each row and column.

The matrix $A$ is in **canonical block form** if, by reordering, its columns can be partitioned into $t$ nonempty subsets $B_1, \ldots, B_t$ such that for each block $B_j$ there is some row $i$ of $A$ with $a_{ik} = 1$ for all $k \in B_j$ and $a_{ik} = 0$ for $k \in \cup_{l=j+1}^{t} B_l$. The rows of $A$ are then ordered so that the row defining $B_j$ becomes the $j$th row for $j = 1, \ldots, t$.

**Facts:**

**1.** Formulation SC expresses the problem of selecting a set of columns (sets) that together cover all rows (elements) at minimum cost. In formulation SP, the covering sets are required to be disjoint.

**2.** Both SC and SP are NP-hard optimization problems (§17.5.2).

**3.** Checking whether a set partitioning problem is feasible is NP-hard.

**4.** In many instances (including bin packing, graph partitioning, and vehicle routing) the LP relaxation of the set covering (partitioning) formulation of the problem is known to give solutions very close to optimality.

**5.** For the bin packing and vehicle routing problems (see Examples 2–3) $v^* \leq \frac{4}{3}\lceil v_{LP} \rceil$.

**6.** If $A$ is totally unimodular or balanced, then the polyhedra $\{x \mid Ax \geq e, \ 0 \leq x_j \leq 1\}$ and $\{x \mid Ax = e, \ 0 \leq x_j \leq 1\}$ have only integer extreme points (vertices). In this case, SC and SP can be solved in polynomial time using linear programming.

**7.** Checking whether a given matrix $A$ is totally unimodular or balanced can be done in polynomial time.

**8.** Every 0-1 matrix that is totally unimodular is also balanced. The converse, however, is not true (see Example 4).

**9.** The matrix $A$ is totally unimodular if and only if each collection of columns of $A$ can be split into two parts so that the sum of the columns in one part minus the sum of the columns in the other part is a vector with entries 0, +1, −1.

**10.** *Greedy heuristic:* This heuristic (Algorithm 4) for the set covering problem successively chooses columns that have smallest cost per covered row.

---

**Algorithm 4**:  **Greedy heuristic for the set covering problem.**

input: 0-1 $m \times n$ matrix $A$, costs $c$

output: feasible set cover $x$

$M^1 := \{1, 2, \ldots, m\}$; $N^1 := \{1, 2, \ldots, n\}$; $k := 1$

{Main loop}

select $j^k \in N^k$ to minimize $\frac{c_j}{|A_j \cap M^k|}$

$N^{k+1} := N^k - \{j^k\}$

obtain $M^{k+1}$ from $M^k$ by deleting all rows containing a 1 in column $j^k$

**if** $M^{k+1} = \emptyset$ **then** $x_j := 1$ for $j \notin N^{k+1}$ and $x_j := 0$ otherwise

**else** $k := k + 1$

{Continue with next iteration of main loop}

---

**11.**  *Randomized greedy heuristic*:  This heuristic for the set covering problem is similar to Algorithm 4 except that at iteration $k$ the column $j^k \in N^k$ is selected at random from among columns $j$ satisfying $\frac{c_j}{|A_j \cap M^k|} \leq (1 + \alpha) \min\{\frac{c_r}{|A_r \cap M^k|} \mid r \in N^k\}$, where $\alpha \geq 0$.

**12.**  Whereas Algorithm 4 is run only once, the randomized greedy heuristic is repeated several times and the best solution is selected.

**13.**  *Implicit enumeration*:  This exact approach (Algorithm 5) for SP works well for dense matrices. In this algorithm, $S$ is the index set of the variables fixed at 1, $z$ is the associated objective value, and $R$ is the set of rows satisfied by $S$. Also $z^*$ denotes the objective value of the best feasible solution found so far.

---

**Algorithm 5**:  **Implicit enumeration method for SP.**

input: 0-1 matrix $A$, costs $c$

output: optimal set of columns $S$ (if any)

place $A$ in canonical block form with blocks $B_j$

order the columns within $B_j$ by nondecreasing $c_t / \sum_i a_{it}$

$S := \emptyset$; $R := \emptyset$; $z := 0$, $z^* := \infty$

1. $r := \min\{i \mid i \notin R\}$; set a marker in the first column of $B_r$

2. examine all columns of $B_r$ in order starting from the marked column

   **if** column $j$ is found with $a_{ij} = 0$ for all $i \in R$ and $z + c_j < z^*$ **then** go to 3

   **if** $B_r$ is exhausted **then** go to 4

3. $S := S \cup \{j\}$; $R := R \cup \{i \mid a_{ij} = 1\}$; $z := z + c_j$

   **if** all rows are included in $R$ **then** $z^* := z$ and go to 4 **else** go to 1

4. **if** $S = \emptyset$ **then** terminate with the best solution found (if any)

   **else** let $k :=$ the last index included in $S$

   $\qquad S := S - \{k\}$; update $z$ and $R$

   $\qquad B_r :=$ the block to which column $k$ belongs

   $\qquad$ move the marker in $B_r$ forward by one column and go to 2

---

**14.**  Other implicit enumeration approaches to set partitioning and set covering are discussed in [BaPa76].

**15.** *Cutting plane methods*: Cutting plane methods (§16.1.8) have been used success-fully to solve large set partitioning and set covering problems. For details regarding an implementation used to solve crew scheduling problems see [HoPa93].

**16.** Further details on the material in this section can be found in [GaNe72], [Ho97], [KoVy12], [NeWo88], and [Sc86].

### Examples:

**1.** *Crew scheduling problem*: An airline must cover a given set of flight segments with crews. There are specified work rules that restrict the assignment of crews to flights. The objective is to cover all flights at minimum total cost. The rows of the matrix $A$ correspond to the flights that an airline has to cover. The columns of $A$ are the incidence vectors of flight "rotations": sequences of flight segments for each flight that begin and end at individual base locations and that conform to all applicable work rules. The objective is to minimize crew costs. This problem can be formulated as either a set covering or set partitioning problem.

**2.** *Bin packing*: The bin packing problem (§16.3.2) can be formulated as a set partitioning problem. The rows of the matrix $A$ correspond to the items and the columns are incidence vectors of any feasible packing of items to a bin. The cost of each variable is 1 if the number of bins is to be minimized. In general, a weighted version can also be formulated where different bins have different costs.

**3.** *Vehicle routing*: Given are a set of customers and the quantity that is to be supplied to each from a warehouse. A fleet of trucks of a specified capacity is available. The objective is to service all the customers at minimum cost. The rows of the matrix $A$ correspond to the customers and the columns are incidence vectors of feasible assignments of customers to trucks (a bin packing problem). The cost of each variable is the cost of the corresponding assignment of customers to the truck. This problem can be formulated as either a set covering or set partitioning problem.

**4.** The following matrix $A$ is not totally unimodular, since $\det(A) = -2$. This can also be seen using Fact 9. If $A$ has columns $C_j$ then $(C_1 + C_2) - (C_3 + C_4) = (0, 2, 0, 0)^T$ has an entry greater than one in absolute value. However, $A$ is a balanced matrix.

$$A = \begin{pmatrix} 1 & 1 & 1 & 1 \\ 1 & 1 & 0 & 0 \\ 1 & 0 & 1 & 0 \\ 1 & 0 & 0 & 1 \end{pmatrix}$$

**5.** There are four requests $R_1, R_2, R_3, R_4$ for information stored in a database, which is composed of five large files $\{1, 2, 3, 4, 5\}$. Request $R_1$ can be fulfilled by retrieving files 1, 3, or 4; request $R_2$ by retrieving files 2 or 3; request $R_3$ by retrieving files 1 or 5; and request $R_4$ by retrieving files 4 or 5. The lengths of the files are $7, 3, 12, 7, 6$ (gigabytes), respectively, and the time to retrieve each file is proportional to its length. Filling all requests in the minimum amount of time is then a set covering problem, with costs $c = (7, 3, 12, 7, 6)$ and incidence matrix

$$A = \begin{pmatrix} 1 & 0 & 1 & 1 & 0 \\ 0 & 1 & 1 & 0 & 0 \\ 1 & 0 & 0 & 0 & 1 \\ 0 & 0 & 0 & 1 & 1 \end{pmatrix}.$$

Applying the greedy heuristic (Algorithm 4) produces $j^1 = 2$, $j^2 = 5$, and $j^3 = 1$, giving $x = (1, 1, 0, 0, 1)$ with total cost 16. This is an optimal solution to the SC problem.

# 16.4  ACTIVITY NETS

Activity nets are important tools in the planning, scheduling, and control of projects. In particular, the CPM (Critical Path Method) and PERT (Program Evaluation and Review Technique) models are widely used in the management of large projects, such as those occurring in construction, shipbuilding, aerospace, computer system design, urban planning, marketing, and accounting.

## 16.4.1  DETERMINISTIC ACTIVITY NETS

The scheduling of large complex projects can be aided by modeling as a directed network of activities having known durations and resource requirements, with the network structure defining the activity precedences. The commonly used critical path method is described as well as extensions that address constrained resources, financial considerations, and project compression.

### Definitions:

A **project** is defined by a set of **activities** that are related by **precedence** relations. An **activity** consumes time and resources to accomplish, whereas a **dummy activity** consumes neither.

Activity $u$ (strictly) **precedes** activity $v$, written $u \prec v$, if activity $u$ must be completed before activity $v$ can be initiated.

A project can be represented using a directed acyclic network $G$ (§8.3.4).

In the **activity-on-node (AoN)** representation of a project, the network $G$ contains a **node** for each activity and the **arcs** of $G$ represent the precedence relations between nodes (activities).

In the **activity-on-arc (AoA)** representation of a project, the network $G$ contains an **arc** for each activity and the **nodes** of $G$ represent certain **events**. Precedence relations are described by the network arcs, possibly requiring the use of **dummy arcs** (dummy activities). In the AoA representation, the network is assumed to have no multiple arcs joining the same pair of nodes, so an activity can be unambiguously referred to by $(i, j)$ for some nodes $i$ and $j$, with the corresponding activity **duration** being $a_{ij}$.

Network $G$ is a **deterministic activity net** if the precedence relations and the parameters associated with the activities are known deterministically. Such a network is also referred to as a **Critical Path Method (CPM) model**.

An **initial node** of $G$ has no entering directed arcs; a **terminal node** has no exiting directed arcs.

**Generalized precedence relations (GPRs)** relax the necessity of a strict precedence between activities. They can be specified in the form of certain lead or lags between a pair of activities, commonly by **start-to-start**, **finish-to-finish**, **start-to-finish**, and **finish-to-start** relations.

The **optimal project compression problem** is that of achieving a target project completion time with least cost, or alternatively minimizing the duration of the project subject to a specified budget constraint.

The complex interaction between the required resources and the duration of an activity is assumed to be given by the functional relationship $c_a = \phi(y_a)$, where $y_a$ is the duration of activity $a$, $\ell_a \leq y_a \leq u_a$, and $c_a$ is its cost. The upper limit $u_a$ is the **normal** duration and the lower limit $\ell_a$ is the **crash** duration of activity $a$.

**Facts:**

**1.** The CPM model arose out of the need to solve industrial scheduling problems; the original work was jointly sponsored by Dupont and Sperry-Rand in the late 1950s.

**2.** In the AoA representation, the network can be assumed to have a single initial node and a single terminal node. These conditions can in general be guaranteed, possibly through the introduction of dummy arcs.

**3.** Suppose that the AoA representation of a network has $n$ nodes, with initial node 1 and terminal node $n$. Then the nodes can always be renumbered (*topologically sorted*) such that each arc leads from a smaller numbered node to a larger numbered one. (See §8.3.4.)

**4.** In the AoA representation, the *earliest time* of realization of node $j$, written $t_j(E)$, is determined recursively from $t_j(E) = \max_{i \in B(j)} \{t_i(E) + a_{ij}\}$ and $t_1(E) = 0$, where $B(j)$ is the set of nodes immediately preceding node $j$.

**5.** Suppose the time of realization of node $n$ is specified as $t_n(L) \geq t_n(E)$. The *latest time* of realization of node $i$, written $t_i(L)$, is determined recursively from $t_i(L) = \min_{j \in A(i)} \{t_j(L) - a_{ij}\}$, where $A(i)$ is the set of nodes immediately succeeding node $i$.

**6.** $t_j(L) \geq t_j(E)$ holds for any node $j$. The difference $t_j(L) - t_j(E) \geq 0$ is called the *node slack* for $j$.

**7.** For each activity $(i, j)$ there are four *activity floats* corresponding to the differences $t_j(X) - t_i(Y) - a_{ij}$, where $X, Y \in \{E, L\}$:

- total float: $TF(i, j) = t_j(L) - t_i(E) - a_{ij}$;
- safety float: $SF(i, j) = t_j(L) - t_i(L) - a_{ij}$;
- free float: $FF(i, j) = t_j(E) - t_i(E) - a_{ij}$;
- interference float: $IF(i, j) = t_j(E) - t_i(L) - a_{ij}$.

**8.** $TF(i, j)$, $SF(i, j)$, and $FF(i, j)$ are always nonnegative, whereas $IF(i, j)$ can be negative, indicating infeasibility of realization under the specified conditions (all activities succeeding node $j$ are accomplished as early as possible and all activities preceding node $i$ are accomplished as late as possible).

**9.** A *critical activity* $(i, j)$ has total float $TF(i, j) = 0$. If $t_n(L) = t_n(E)$, then the set of critical activities contains at least one path from node 1 to node $n$, which represents a longest path in the network from node 1 to node $n$. Such a path is called a *critical path* (CP).

**10.** Floats play an important role in both resource allocation and activity scheduling, since floats give a measure of the flexibility in scheduling activities during project execution without delaying the project completion time.

**11.** The problems of optimal resource allocation and activity scheduling subject to the known precedence constraints are NP-hard optimization problems (§17.5.2).

**12.** Practical solutions to optimal resource allocation and activity scheduling problems are based on heuristics. Virtually all heuristics used in practice rely on ranking the activities according to their float $(TF, SF, FF, IF)$.

**13.** The measure $TF$ is the only float in the AoA mode that is representation-invariant; this measure is the same in both modes of representation and in all AoA models of the same project.

**14.** The $SF, FF, IF$ measures are representation-dependent: they do indeed depend on the structure of the AoA and they may also vary from their AoN values.

**15.** A simple redefinition of $t_i(E)$ for nodes $i$ with all outgoing arcs dummy and $t_j(L)$ for nodes $j$ with all incoming arcs dummy reestablishes the invariance of the activity floats to the mode of representation [ElKa90].

**16.** A plethora of "off-the-shelf" project planning and control software packages for PCs are currently available [DeHe90], [KoPa01]. The review [DeHe90] also outlines criteria against which a software package should be judged.

**17.** An extensive listing of commercial and noncommercial software for project planning can be found at

- `https://wikipedia.org/wiki/Comparison_of_project_management_software`

**18.** GPRs afford the flexibility of modeling relations that are present among activities in many practical situations, gained at a price in computational effort and interpretation of results. The concepts of criticality and float of an activity take on a new meaning, since activities may be "compressed" (speeded up) or "expanded" (slowed down) from their "normal" durations [ElKa92].

**19.** Considerations of resource availabilities are important in project planning and its dynamic control. The most common planning criteria are

- minimization of the project duration;
- smoothing of resource usage;
- minimization of the maximum resource utilization;
- minimization of the cost of resource usage;
- maximization of the present value of the project.

**20.** In the presence of limited resources, the "critical path" may no longer be a "path", in the sense of a connected chain of activities. What emerges is the concept of a *critical sequence* of activities [El77], which need not form a connected chain in the network. (See Example 4.)

**21.** The scheduling of activities related by arbitrary precedence relations subject to resource availabilities is an NP-hard problem. Consequently, such problems are typically approached by integer programming techniques (§16.1.8) or heuristics (e.g., simulated annealing, tabu search, genetic algorithms, neural nets).

**22.** The book [SlWe89] discusses project scheduling under constrained resources; in particular, Chapter 5 of Part I evaluates various heuristics that have been proposed. Also [HeDeDe98] gives a review of recent contributions to this area.

**23.** Typically, resources are available in one or several units or may be acquired at a cost. Mathematical models (large-scale integer linear programs) abound for the minimization of the project duration [El77]. Various branch-and-bound approaches have been proposed for these models [DeHe92], [Sp94].

**24.** Heuristic procedures, let alone optimization algorithms, for activity scheduling under the other criteria mentioned in Fact 19 are not generally available.

**25.** In the CPM model of activity networks there is little problem in defining the cost of an activity, and subsequently the cost of the project.

**26.** Generally, there are two streams of cash flow (from the contractor's point of view): an *in-stream* representing payments to the contractor by the owner, and an *out-stream* representing payments by the contractor in the execution of the activities.

**27.** From the owner's point of view there is only one stream of cash flows: namely, payments to the contractor for work accomplished. Given a particular schedule for the activities, the two streams of cash flow can be easily obtained. The problem is then scheduling activities to maximize the net present value (NPV) of the project.

**28.** Issues concerning the NPV of a project are equally important to those interested in bidding on a proposed project and those who are committed to carry out an already agreed-upon project. Succinctly stated, the problem is to determine the dates of the deliverables in order to maximize the NPV.

**29.** Suppose that the function $\phi(y_a)$ is nonincreasing over the interval $[\ell_a, u_a]$. Reference [El77] gives a treatment of linear, convex, concave, and discrete functions, while [ElKa92] discusses the case in which $\phi(y_a)$ is piecewise linear and convex over the interval $[\ell_a, u_a]$.

**Examples:**

**1.** *Construction planning*: The construction of a house involves carrying out the nine activities listed in the following table. Their durations and (immediate) predecessor activities are also indicated.

| activity | duration (days) | predecessors |
|---|---|---|
| foundation/frame | 12 | — |
| wiring/plumbing | 4 | foundation/frame |
| sheetrock | 7 | wiring/plumbing |
| interior paint | 2 | sheetrock, windows |
| carpet | 3 | interior paint |
| roof | 3 | foundation/frame |
| siding | 7 | roof |
| windows | 2 | siding |
| exterior paint | 2 | windows |

An AoA representation of this project is shown in the following figure. It is necessary to use a dummy activity to ensure that the given precedences are faithfully depicted. The nodes have been numbered in topological order, with node 1 the initial node and node 9 the terminal node. The longest path from node 1 to node 9 is $[1, 2, 4, 5, 6, 7, 8, 9]$ with length 29, corresponding to a project completion time of 29 days.

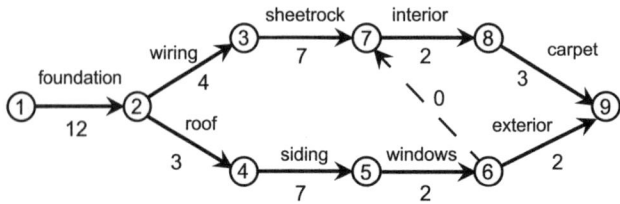

**2.** A project is composed of the four activities $a, b, c, d$ with precedence relations $a \prec c$, $a \prec d$, and $b \prec d$. The AoN representation of this project is shown in part (a) of the following figure. The AoA representation is shown in part (b) of the figure, where the nodes have been numbered in topological order (Fact 3). The dummy activity joining nodes 2 and 3 is needed to maintain the integrity of the precedence relations. Activity durations are indicated on the arcs of part (b) of the figure.

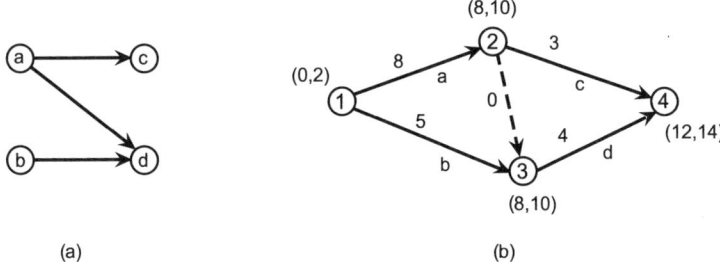

(a)                                                         (b)

**3.** The earliest and latest event times $t_j(E), t_j(L)$ are shown next to each node $j$ in part (b) of the figure of Example 2, where $t_4(L) = 14$, which is 2 units more than $t_4(E) = 12$. Here all nodes have the same slack of 2, which provides no information on the various activity floats, given in the following table. Since $TF(i,j) > 0$ for all activities $(i,j)$, there are no critical activities and no critical path, since a small delay in any activity will not delay the completion time of the project. The critical path can be determined if instead $t_4(L) = t_4(E) = 12$. Then the critical path is given by $[1, 2, 3, 4]$.

| activity | $TF$ | $SF$ | $FF$ | $IF$ |
|----------|------|------|------|------|
| $(1, 2)$ | 2 | 0 | 0 | $-2$ |
| $(1, 3)$ | 5 | 3 | 3 | 1 |
| $(2, 4)$ | 3 | 1 | 1 | $-1$ |
| $(3, 4)$ | 2 | 0 | 0 | $-2$ |

**4.** The following figure gives a project with six activities in AoN representation. There is a single resource, with availability of 6 units. The duration of each activity and the required quantity of the resource are indicated next to each activity (node).

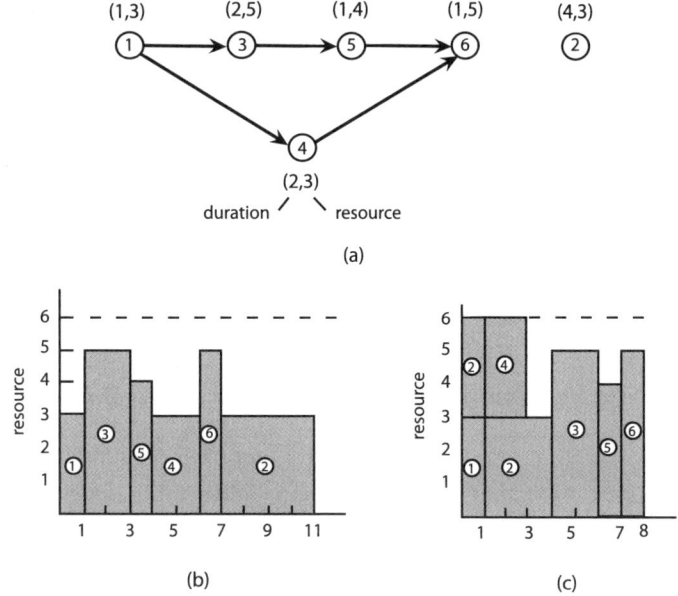

The CP (based solely on durations) is $[1, 3, 5, 6]$ of duration 5. If the integrity of the CP is maintained as long as possible, then activity 4 must be inserted before activity 6 (thus breaking the continuity of the CP), which is then followed by activity 2, as shown in part

1200 Chapter 16 DISCRETE OPTIMIZATION

(b) of the figure. The total duration of the project under this schedule is 11 time units. Now consider the schedule shown in part (c) of the figure, in which the CP is split after activity 1; the total duration of the project is thereby reduced to only 8 time units.

**5.** The project of the following figure is shown in AoA mode, with the duration of each activity written beside each arc. The payment shown next to a node is the income accrued (if positive) or expense incurred (if negative) at the time of realization of that event (node). The CP is $[1, 3, 4]$ with duration 11. Ignoring the time value of money (i.e., assuming a discount factor $\beta = 0$) gives 1,000 as the estimate of project profit. Assuming a discount factor $\beta = 0.99$ and that activities are done as early as possible to maintain the CP, the estimate of project profitability shrinks to $-5,000(.99)^2 + 3,000(.99)^8 + 3,000(.99)^{11} = 553.75$.

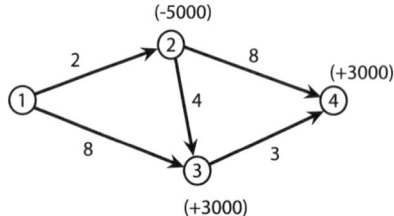

Now suppose that the schedule of activities is modified as follows: delay activity $(1, 2)$ to complete at time $t_2 = 4$ (instead of 2); do activity $(1, 3)$ as early as possible to complete at time $t_3 = 8$; and do activity $(2, 4)$ as early as possible (after the realization of node 2) to complete at time $t_4 = 12$. Then the project profitability increases to $-5,000(.99)^4 + 3,000(.99)^8 + 3,000(.99)^{12} = 624.41$. Note that the increase in project profitability comes as a consequence of ignoring the CP, and in fact delaying the project beyond its normal duration.

**6.** A project involving five activities is shown in the following figure in AoA mode. Each activity (arc) $a$ is labeled with $(u_a, \ell_a, k_a)$ where $k_a$ is the marginal cost of reducing duration from the normal time $u_a$. Next to each node $j$ is its earliest time of realization $t_j(E)$ under normal activity durations.

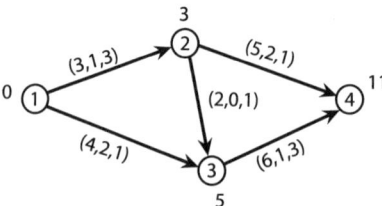

The next table summarizes the breakpoints of the resulting piecewise linear cost function.

| breakpoint | duration $(t_4)$ | marginal cost | cumulative cost |
|:---:|:---:|:---:|:---:|
| 1 | 11 | 1 | 0 |
| 2 | 10 | 2 | 1 |
| 3 | 9 | 3 | 3 |
| 4 | 8 | 4 | 6 |
| 5 | 4 | 5 | 22 |
| 6 | 3 | $\infty$ | 27 |

The function itself is shown in the following figure. With the complete cost function in hand it is easy to answer various questions. For example, the least additional cost required to reduce the project duration from its normal value 11 to 7 is seen to be 10. Alternatively, if 6 additional units of money are available, then the maximum reduction achievable in the project duration is 3 units of time (from 11 to 8).

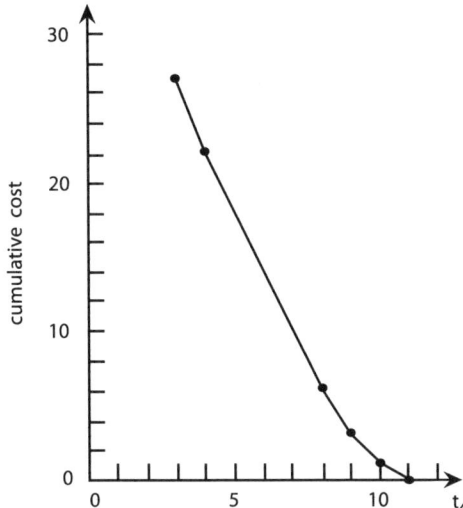

### 16.4.2  PROBABILISTIC ACTIVITY NETS

The CPM model can be extended to incorporate uncertainty or randomness. If the durations of activities are random variables, then the network is a PERT (Program Evaluation and Review Technique) model. Alternatively, the very undertaking of an activity may be determined by chance and this consideration has led to the development of GAN (Generalized Activity Network) models.

### Definitions:

A **probabilistic activity net** is a directed network in which some or all of the parameters, including the realization of the activities, are probabilistically known.

In a **PERT model**, activity durations are random variables. The duration of activity $a$ has expected value $\mu_a$ and variance $\sigma_a^2$.

Let $P(\tau)$ be the probability that the project is completed by time $\tau$.

The **criticality index** of a path $Q$ in the network is the probability that $Q$ is a critical path in any realization of the project.

The **criticality index** of an activity $a$ is the probability that $a$ lies on a critical path in any realization of the project.

A **GAN model** is a probabilistic activity net with conditional progress and probabilistic realization of activities.

If $X$ is a standard normal deviate (§7.3.1), then its (cumulative) distribution function is denoted by $\Phi(x) = Pr(X \leq x)$.

**Facts:**

**1.** The original PERT model evolved in the late 1950s from the U.S. Navy's efforts to plan and accelerate the Polaris submarine missile project.

**2.** A detailed account of the original PERT model, its analysis, and the criticisms levied against it is found in [El77, Chapter 3].

**3.** Estimation of the exact probability distribution function (pdf) of the project duration is an extremely difficult problem due to the nonindependence of the paths leading from the initial node to the terminal node.

**4.** The original PERT model suggested substituting $\mu_a$ for each activity duration and then proceeding with the standard CPM calculations to determine a critical path $Q^*$ in the resulting deterministic network.

**5.** The pdf of the duration of the project can then be approximated using a normal distribution having mean $\widehat{\mu}_{Q^*} = \sum_{a \in Q^*} \mu_a$ and variance $\widehat{\sigma}^2_{Q^*} = \sum_{a \in Q^*} \sigma^2_a$. The normal approximation increases in validity as the number of activities in the path $Q^*$ increases.

**6.** The probability $P(\tau)$ of project completion by time $\tau$ can be approximated using $\widehat{P}(\tau) = \Phi((\tau - \widehat{\mu}_{Q^*})/\widehat{\sigma}_{Q^*})$.

**7.** The value $\widehat{\mu}_{Q^*}$ always underestimates the exact mean project duration (often, seriously). No equivalent statement can be made about the variance estimate $\widehat{\sigma}^2_{Q^*}$ except that it is often a gross approximation of the exact variance.

**8.** PERT analysis goes one step further and uses an approximation to the expected value and the variance of each activity, based on the assumption that each activity duration follows a beta distribution (§7.3.1). In particular, the variance is approximated by $\frac{1}{36}(\text{range})^2$. These additional assumptions render the procedure even more suspect.

**9.** An immediate consequence of randomness in the activity durations is that (virtually) any path can be the CP in some realization of the project. Thus, the criticality index of a path and the criticality index of an activity are more meaningful concepts. See [Wi92] for a critique of the latter.

**10.** In general, it is extremely difficult to determine the exact values of the criticality indices analytically. Monte Carlo sampling is typically used to estimate these values [BaCoPr95].

**11.** Since the early days of PERT, significant strides have been made in estimating the various parameters in the PERT model. The approaches can be classified into the categories of *exact*, *approximating*, and *bounding* procedures. See [El89], [Ka92].

**12.** The concept of a *uniformly directed cutset* has been used to evaluate some common network performance criteria under the assumption of exponentially distributed activity durations [KuAd86]. Attempts to extend the concept to applications in optimal resource allocation have had limited success thus far.

**13.** The restriction of GANs to "exclusive-or" type nodes renders the network a graphical representation of a *semi-Markov process*. The resulting *GERT* (Graphical Evaluation and Review Technique) model has been expanded into SLAM II, an extremely powerful discrete event simulation language.

**14.** The analysis of stochastic activity nets with exclusive-or type nodes (*STEOR-nets*) is thoroughly discussed in [Ne90].

**Examples:**

**1.** The following figure shows a project with six activities whose durations are random variables that assume discrete values with equal probabilities. For example, activity $(1, 2)$ has duration 1, 2, or 5 with probability $\frac{1}{3}$ each.

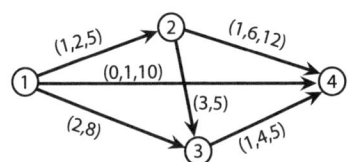

The exact distribution of project completion time (secured by complete enumeration of the 324 realizations) is shown in the following table, from which it is seen that the true mean project duration is $\mu = 12.315$ and the true standard deviation is $\sigma = 2.5735$. The probability that the project duration is no more than 12 time units is $P(12) = 0.4815$. The PERT estimates of these same parameters, based on the deterministic critical path $[1, 2, 3, 4]$, are $\widehat{\mu}_{Q^*} = 10$, $\widehat{\sigma}_{Q^*} = 1$, and $\widehat{P}(12) = 0.9772$.

| duration $(t_4)$ | frequency | relative frequency |
|:---:|:---:|:---:|
| 17 | 36 | 0.1111 |
| 15 | 12 | 0.0370 |
| 14 | 48 | 0.1481 |
| 13 | 72 | 0.2222 |
| 12 | 42 | 0.1296 |
| 11 | 30 | 0.0926 |
| 10 | 36 | 0.1111 |
| 9 | 28 | 0.0864 |
| 8 | 10 | 0.0309 |
| 7 | 6 | 0.0185 |
| 6 | 2 | 0.0062 |
| 5 | 2 | 0.0062 |
| | 324 | 1.0000 |

**2.** The paths from initial node 1 to terminal node 4 for the project in the figure of Example 1 are $Q_1 = [1, 2, 4]$, $Q_2 = [1, 2, 3, 4]$, $Q_3 = [1, 4]$, $Q_4 = [1, 3, 4]$. The following table lists the frequency and relative frequency that each path $Q_i$, or combination of paths, is a critical path. The path criticality indices are then easily determined from this table. For example, the criticality index of $Q_1$ is 0.3951 ($= 0.3457 + 0.0309 + 0.0185$), and of $Q_4$ is 0.2592 ($= 0.1975 + 0.0185 + 0.0432$). The criticality index of each activity can be easily determined from the criticality indices of the paths. For instance, the criticality index of activity $(1, 2)$, which lies on paths $Q_1$ and $Q_2$, is 0.7284 ($= 1 - 0.0741 - 0.1975$).

| $t_4$ | $Q_1$ | $Q_2$ | $Q_3$ | $Q_4$ | $Q_1, Q_2$ | $Q_1, Q_4$ | $Q_2, Q_3$ | $Q_2, Q_4$ |
|---|---|---|---|---|---|---|---|---|
| 17 | 36 | | | | | | | |
| 15 | | 12 | | | | | | |
| 14 | 36 | 12 | | | | | | |
| 13 | 30 | 6 | | 24 | | 6 | | 6 |
| 12 | | 12 | | 24 | | | | 6 |
| 11 | 6 | 18 | | | 6 | | | |
| 10 | | 8 | 24 | | | | 4 | |
| 9 | | 10 | | 16 | | | | 2 |
| 8 | 2 | 6 | | | 2 | | | |
| 7 | 2 | 2 | | | 2 | | | |
| 6 | | 2 | | | | | | |
| 5 | | 2 | | | | | | |
| freq. | 112 | 90 | 24 | 64 | 10 | 6 | 4 | 14 |
| rel. freq. | .3457 | .2778 | .0741 | .1975 | .0309 | .0185 | .0123 | .0432 |

## 16.4.3   COMPLEXITY ISSUES

### Facts:

**1.** The AoN representation of a project is essentially unique.

**2.** The AoA representation is not unique because of the necessity to introduce dummy activities (e.g., to maintain the integrity of the precedence relations).

**3.** Construction of the AoA representation can be carried out with different objectives in mind: to minimize the number of nodes, to minimize the number of dummy activities, or to minimize the *complexity index* of the resulting AoA network [MiKaSt93].

**4.** Analytical solutions to optimization problems for project networks often proceed by conditioning upon certain activities, and then removing the conditioning through either enumeration or multiple integration. Minimizing the computing effort then involves minimizing the number of activities on which such conditioning takes place.

**5.** If the network is series-parallel then no conditioning is required and its analysis is straightforward, though it may be computationally demanding.

**6.** If the network is not series-parallel, then the minimum number of activities for conditioning can be secured by the optimal node reduction procedure of [BeKaSt92], which has polynomial complexity.

**7.** Patterson [Pa83] collected a set of 110 standard test problems, useful for comparing alternative solution procedures. These problems have been supplanted by a more recent sets of test problems [KaSpDr92], [KoSp97]. Benchmark problems for resource-constrained project scheduling problems are available at

- http://www.om-db.wi.tum.de/psplib/library.html

**8.** Several measures of the complexity of a project network were proposed in the 1960s, with questionable validity. The significance of the complexity index [BeKaSt92] in accounting for the difficulty in analysis is discussed in [DeHe96].

# 16.5  GAME THEORY

Games, mathematical models of conflict or bargaining, can be classified in three ways: by mood of play (noncooperative or cooperative), by field of application (e.g., biology or economics), and by mathematical structure (e.g., discrete, continuous, or differential). Correspondingly, game theory is a vast and diverse subject with different traditions in each of many specialties. This section discusses discrete games, in which finitely many strategies are available to finitely many players. Combinatorial and other games form largely separate disciplines to which appropriate references appear in §16.5.4.

## 16.5.1  NONCOOPERATIVE GAMES

This section discusses noncooperative games involving a finite number of players. Collusion among the players is not allowed in these types of games. Such games can model a wide variety of situations, as indicated in §16.5.4.

### Definitions:

An $n$-player game $\Gamma$ in **extensive form** consists of

- a set $\{1, \ldots, n\} \cup \{0\}$ of $n$ decisionmakers (or **players**) augmented by a fictitious player, called 0 (or **chance**), whose actions are random;
- a tree, in which each nonterminal vertex represents a decision point for some player, whose possible actions correspond to arcs emanating from the vertex;
- a **payoff function** that assigns an $n$-vector to each terminal vertex;
- a partition of the nonterminal vertices into $n + 1$ vertex sets, one for each player and for chance;
- a subpartition of each player's vertex set into subsets (**information sets**), such that no vertex follows another in the same subset and all vertices in a subset are followed by the same number of arcs;
- a probability distribution on arcs emanating from any chance vertex.

A **subgame** of $\Gamma$ is a game whose tree is a subtree of the tree for $\Gamma$. A subgame is **proper** if the information set that contains its root contains no other vertices.

A game is **finite** if its tree is finite.

A game has **perfect** information if all information sets contain a single vertex; otherwise, it has **imperfect** information.

A game has **complete** information if all players know the entire extensive form including all terminal payoffs; otherwise it has **incomplete** information.

A **pure strategy** is a function that maps each of a player's information sets to an emanating arc.

An $n$-person game in **normal** (or **strategic**) **form** consists of a set $N = \{1, 2, \ldots, n\}$ of players, a set $S_k$ of possible pure strategies for each $k \in N$, and a payoff function $f = (f_1, f_2, \ldots, f_n)$ that assigns $f_k(w)$ to Player $k$ for every pure strategy combination $w = (w^1, w^2, \ldots, w^n)$, where $w^k \in S_k$. Payoffs are computed by taking expected values over distributions associated with chance vertices in the corresponding extensive form.

Let $D \subseteq S_1 \times S_2 \times \cdots \times S_n$ be the set of all possible pure strategy combinations $w$.

Let $w \,||\, \overline{w}^k$ denote the joint pure strategy combination that is identical to $w$ except for the strategy of Player $k$:

$$w \,||\, \overline{w}^k = (w^1, \ldots, w^{k-1}, \overline{w}^k, w^{k+1}, \ldots, w^n).$$

$w^* \in D$ is a **Nash equilibrium pure strategy combination** (or simply **equilibrium**) if, for every $k \in N$, $f_k(w^*) \geq f_k(w^* \,||\, \overline{w}^k)$ holds for all $\overline{w}^k \in S_k$. (John F. Nash, 1928–2015) Let $E$ denote the set of all such equilibria.

For $k \in N$ define the function $m_k$ that minimizes $f_k(w)$ over components of $w$ that $k$ does not control:

$$m_k(w^k) = \min_{\{\overline{w} \,|\, \overline{w}^k = w^k\}} f_k(\overline{w}).$$

If $\widetilde{w}^k$ maximizes $m_k(w^k)$, then $\widetilde{w}^k$ is a **max-min strategy** for $k$ and $\widetilde{f}_k = m_k(\widetilde{w}^k)$ is the corresponding **max-min payoff**.

Let $D^* = \{w \in D \mid f_k(w) \geq \widetilde{f}_k \text{ for all } k \in N\}$.

The strategy combination $w$ is **individually rational** for all players if $w \in D^*$.

The combination $w \in D$ is **group rational** (or **Pareto-optimal**) if no $\overline{w} \in D$ exists such that $f_k(\overline{w}) \geq f_k(w)$ for all $k \in N$ and $f_i(\overline{w}) > f_i(w)$ for some $i \in N$.

Let $P$ denote the set of all Pareto-optimal $w$. The set $P^* = P \cap D^*$ is the **bargaining set** and each $w \in P^*$ is a **cooperative** strategy combination.

An equilibrium is **subgame perfect** if its restriction to any proper subgame is also an equilibrium. Let $E_S$ denote the set of subgame perfect equilibria.

## Facts:

**1.** Information sets are constructed so that in making a decision a player knows the identity of the information set, but not the particular vertex of the set at which the decision is being made.

**2.** At an equilibrium $w^* \in D$, no $k \in N$ has a unilateral incentive to depart from $(w^*)^k$ if each $j \in N$, $j \neq k$, holds fast to $(w^*)^j$.

**3.** $w \,||\, w^k = w$.

**4.** Different equilibria can yield identical outcomes.

**5.** The bargaining set can also be defined with "threat" strategies in lieu of max-min (or "security") strategies as criteria of individual rationality. Context determines which definition is apt.

**6.** If $E$ is a singleton, or if all elements of $E$ yield the same outcome (see Example 7), then the game is usually regarded as solved.

**7.** In general, however, $E$ may either be empty or yield a multiplicity of outcomes (see Example 8).

**8.** A sufficient condition for $E \neq \emptyset$ in a finite game is that information be perfect (although $E$ need not be computable by all players unless information is also complete). The above condition is not necessary; see Examples 7 and 8.

**9.** If $E$ yields a multiplicity of outcomes, then an equilibrium selection criterion is necessary. One criterion is to reduce $E$ to $E \cap P^*$, thus preferring cooperative equilibria (of a noncooperative game) to noncooperative equilibria. Another criterion is to reduce $E$ to $E \cap E_S$.

**10.** Rationales for the above criteria are discussed in [Me93]. Other equilibrium selection criteria are discussed in [Fr90] and [My91].

**11.** The equilibrium selection problem is one of the important unsolved problems of game theory; see [BiKiTa93].

**Examples:**

**1.** A university (Player 3) must offer a faculty position to either or both of two individuals, a distinguished researcher (Player 1) and a younger colleague in the same area (Player 2), each of whom can say either YES or NO to an offer but cannot communicate with the other. The payoff to Player $i = 1, 2$ (in well-being) is $\sigma_i$ ($> 0$) for an offer, $b_i$ ($> \sigma_i$) for an appointment, and $B_i$ ($> b_i$) if both are appointed. To the university, hiring Player 1 alone is worth 4 (in prestige); but hiring both merits 3, hiring neither is worth 2, and hiring Player 2 alone merits zero, because appointing Player 2 prevents the appointment of another distinguished researcher. The university hides from each candidate whether it has made an offer to the other. The extensive form of this game is shown in the following figure. Each player has a single information set (denoted by a rectangle). There are no chance vertices and no proper subgames. The payoffs to Players 1, 2, and 3 are indicated by the 3-vector at each terminal vertex of the tree.

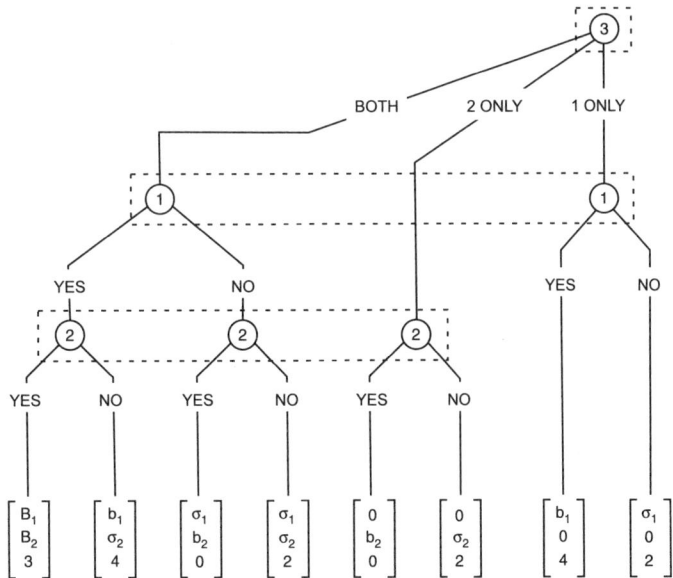

**2.** Suppose in Example 1 that the university now reveals to whom it has made an offer. Also, the university need not offer the position to either candidate this year, in which case a single individual is appointed next year and chance decides with equal probability which current candidate the appointee matches in caliber, giving the university a payoff of $0.5(4) + 0.5(0) = 2$. The extensive form of this game is shown in the following figure. Player 1 has information sets $I$, $J$; Player 2 has information sets $K$, $L$. There is a single chance vertex. Information sets $I$, $J$, $K$ each contain the root of a proper subgame.

**3.** The figures of Examples 1 and 2 are finite games of imperfect information since in both cases Player 2 has an information set with more than one vertex. Each game has incomplete information if players know only their own terminal payoffs.

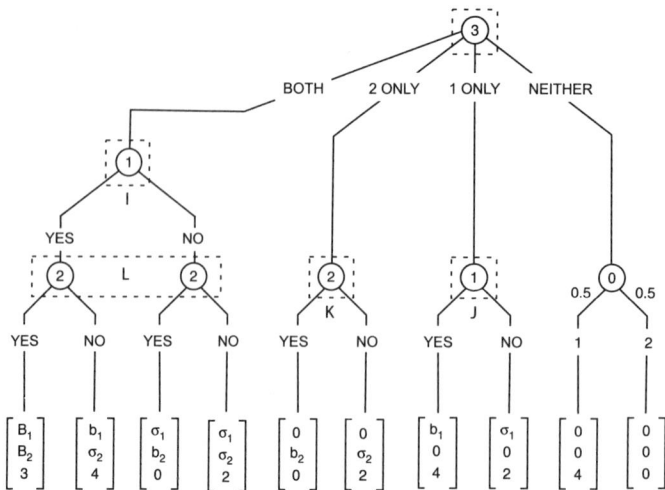

**4.** In the figure of Example 2, Player 2 can say YES or NO at each of $K$ or $L$. Hence Player 2 has four possible strategies: YKNL (yes if $K$, no if $L$), NKYL, an unconditional YES, and an unconditional NO. Likewise, Player 1 has four strategies: YINJ, NIYJ, YES, and NO.

**5.** The following table depicts the strategic form of the figure of Example 1 as a 3-dimensional array. The strategy sets are $S_1 = \{\text{YES, NO}\}$, $S_2 = \{\text{YES, NO}\}$, and $S_3 = \{\text{BOTH, 1 ONLY, 2 ONLY}\}$. The payoff function is defined by $f_1(\text{YES, NO, 1 ONLY}) = b_1$, $f_2(\text{NO, NO, 2 ONLY}) = \sigma_2$, $f_3(\text{YES, YES, BOTH}) = 3$, etc. Player 1's strategies correspond to rows, Player 2's strategies correspond to columns, and Player 3's strategies correspond to arrays.

|       | YES | NO |   |       | YES | NO |   |       | YES | NO |
|-------|-----|-----|--|-------|-----|-----|--|-------|-----|-----|
| YES | $\begin{bmatrix} B_1 \\ B_2 \\ 3 \end{bmatrix}$ | $\begin{bmatrix} b_1 \\ \sigma_2 \\ 4 \end{bmatrix}$ | | YES | $\begin{bmatrix} b_1 \\ 0 \\ 4 \end{bmatrix}$ | $\begin{bmatrix} b_1 \\ 0 \\ 4 \end{bmatrix}$ | | YES | $\begin{bmatrix} 0 \\ b_2 \\ 0 \end{bmatrix}$ | $\begin{bmatrix} 0 \\ \sigma_2 \\ 2 \end{bmatrix}$ |
| NO | $\begin{bmatrix} \sigma_1 \\ b_2 \\ 0 \end{bmatrix}$ | $\begin{bmatrix} \sigma_1 \\ \sigma_2 \\ 2 \end{bmatrix}$ | | NO | $\begin{bmatrix} \sigma_1 \\ 0 \\ 2 \end{bmatrix}$ | $\begin{bmatrix} \sigma_1 \\ 0 \\ 2 \end{bmatrix}$ | | NO | $\begin{bmatrix} 0 \\ b_2 \\ 0 \end{bmatrix}$ | $\begin{bmatrix} 0 \\ \sigma_2 \\ 2 \end{bmatrix}$ |
|       | BOTH | |   |       | 1 ONLY | |  |       | 2 ONLY | |

**6.** The following table depicts the strategic form of the figure of Example 2 as a 3-dimensional array. Player 1's strategies correspond to rows, Player 2's strategies correspond to columns, and Player 3's strategies correspond to arrays. The strategy sets now are $S_1 = \{\text{YES, YINJ, NIYJ, NO}\}$, $S_2 = \{\text{YES, YKNL, NKYL, NO}\}$, and $S_3 = \{\text{BOTH, 1 ONLY, 2 ONLY, NEITHER}\}$. The sets $S_1$, $S_2$ contain more strategies than in Example 5 because Players 1 and 2 have better information: the game is less imperfect. The payoff to Player 3 from NEITHER is an expectation over arcs emanating from the game's single chance vertex.

| | YES | YKNL | NKYL | NO |
|---|---|---|---|---|
| YES | $\begin{bmatrix} B_1 \\ B_2 \\ 3 \end{bmatrix}$ | $\begin{bmatrix} b_1 \\ \sigma_2 \\ 4 \end{bmatrix}$ | $\begin{bmatrix} B_1 \\ B_2 \\ 3 \end{bmatrix}$ | $\begin{bmatrix} b_1 \\ \sigma_2 \\ 4 \end{bmatrix}$ |
| YINJ | $\begin{bmatrix} B_1 \\ B_2 \\ 3 \end{bmatrix}$ | $\begin{bmatrix} b_1 \\ \sigma_2 \\ 4 \end{bmatrix}$ | $\begin{bmatrix} B_1 \\ B_2 \\ 3 \end{bmatrix}$ | $\begin{bmatrix} b_1 \\ \sigma_2 \\ 4 \end{bmatrix}$ |
| NIYJ | $\begin{bmatrix} \sigma_1 \\ b_2 \\ 0 \end{bmatrix}$ | $\begin{bmatrix} \sigma_1 \\ \sigma_2 \\ 2 \end{bmatrix}$ | $\begin{bmatrix} \sigma_1 \\ b_2 \\ 0 \end{bmatrix}$ | $\begin{bmatrix} \sigma_1 \\ \sigma_2 \\ 2 \end{bmatrix}$ |
| NO | $\begin{bmatrix} \sigma_1 \\ b_2 \\ 0 \end{bmatrix}$ | $\begin{bmatrix} \sigma_1 \\ \sigma_2 \\ 2 \end{bmatrix}$ | $\begin{bmatrix} \sigma_1 \\ b_2 \\ 0 \end{bmatrix}$ | $\begin{bmatrix} \sigma_1 \\ \sigma_2 \\ 2 \end{bmatrix}$ |

BOTH

| | YES | YKNL | NKYL | NO |
|---|---|---|---|---|
| YES | $\begin{bmatrix} b_1 \\ 0 \\ 4 \end{bmatrix}$ | $\begin{bmatrix} b_1 \\ 0 \\ 4 \end{bmatrix}$ | $\begin{bmatrix} b_1 \\ 0 \\ 4 \end{bmatrix}$ | $\begin{bmatrix} b_1 \\ 0 \\ 4 \end{bmatrix}$ |
| YINJ | $\begin{bmatrix} \sigma_1 \\ 0 \\ 2 \end{bmatrix}$ | $\begin{bmatrix} \sigma_1 \\ 0 \\ 2 \end{bmatrix}$ | $\begin{bmatrix} \sigma_1 \\ 0 \\ 2 \end{bmatrix}$ | $\begin{bmatrix} \sigma_1 \\ 0 \\ 2 \end{bmatrix}$ |
| NIYJ | $\begin{bmatrix} b_1 \\ 0 \\ 4 \end{bmatrix}$ | $\begin{bmatrix} b_1 \\ 0 \\ 4 \end{bmatrix}$ | $\begin{bmatrix} b_1 \\ 0 \\ 4 \end{bmatrix}$ | $\begin{bmatrix} b_1 \\ 0 \\ 4 \end{bmatrix}$ |
| NO | $\begin{bmatrix} \sigma_1 \\ 0 \\ 2 \end{bmatrix}$ | $\begin{bmatrix} \sigma_1 \\ 0 \\ 2 \end{bmatrix}$ | $\begin{bmatrix} \sigma_1 \\ 0 \\ 2 \end{bmatrix}$ | $\begin{bmatrix} \sigma_1 \\ 0 \\ 2 \end{bmatrix}$ |

1 ONLY

| | YES | YKNL | NKYL | NO |
|---|---|---|---|---|
| YES | $\begin{bmatrix} 0 \\ b_2 \\ 0 \end{bmatrix}$ | $\begin{bmatrix} 0 \\ b_2 \\ 0 \end{bmatrix}$ | $\begin{bmatrix} 0 \\ \sigma_2 \\ 2 \end{bmatrix}$ | $\begin{bmatrix} 0 \\ \sigma_2 \\ 2 \end{bmatrix}$ |
| YINJ | $\begin{bmatrix} 0 \\ b_2 \\ 0 \end{bmatrix}$ | $\begin{bmatrix} 0 \\ b_2 \\ 0 \end{bmatrix}$ | $\begin{bmatrix} 0 \\ \sigma_2 \\ 2 \end{bmatrix}$ | $\begin{bmatrix} 0 \\ \sigma_2 \\ 2 \end{bmatrix}$ |
| NIYJ | $\begin{bmatrix} 0 \\ b_2 \\ 0 \end{bmatrix}$ | $\begin{bmatrix} 0 \\ b_2 \\ 0 \end{bmatrix}$ | $\begin{bmatrix} 0 \\ \sigma_2 \\ 2 \end{bmatrix}$ | $\begin{bmatrix} 0 \\ \sigma_2 \\ 2 \end{bmatrix}$ |
| NO | $\begin{bmatrix} 0 \\ b_2 \\ 0 \end{bmatrix}$ | $\begin{bmatrix} 0 \\ b_2 \\ 0 \end{bmatrix}$ | $\begin{bmatrix} 0 \\ \sigma_2 \\ 2 \end{bmatrix}$ | $\begin{bmatrix} 0 \\ \sigma_2 \\ 2 \end{bmatrix}$ |

2 ONLY

| | YES | YKNL | NKYL | NO |
|---|---|---|---|---|
| YES | $\begin{bmatrix} 0 \\ 0 \\ 2 \end{bmatrix}$ | $\begin{bmatrix} 0 \\ 0 \\ 2 \end{bmatrix}$ | $\begin{bmatrix} 0 \\ 0 \\ 2 \end{bmatrix}$ | $\begin{bmatrix} 0 \\ 0 \\ 2 \end{bmatrix}$ |
| YINJ | $\begin{bmatrix} 0 \\ 0 \\ 2 \end{bmatrix}$ | $\begin{bmatrix} 0 \\ 0 \\ 2 \end{bmatrix}$ | $\begin{bmatrix} 0 \\ 0 \\ 2 \end{bmatrix}$ | $\begin{bmatrix} 0 \\ 0 \\ 2 \end{bmatrix}$ |
| NIYJ | $\begin{bmatrix} 0 \\ 0 \\ 2 \end{bmatrix}$ | $\begin{bmatrix} 0 \\ 0 \\ 2 \end{bmatrix}$ | $\begin{bmatrix} 0 \\ 0 \\ 2 \end{bmatrix}$ | $\begin{bmatrix} 0 \\ 0 \\ 2 \end{bmatrix}$ |
| NO | $\begin{bmatrix} 0 \\ 0 \\ 2 \end{bmatrix}$ | $\begin{bmatrix} 0 \\ 0 \\ 2 \end{bmatrix}$ | $\begin{bmatrix} 0 \\ 0 \\ 2 \end{bmatrix}$ | $\begin{bmatrix} 0 \\ 0 \\ 2 \end{bmatrix}$ |

NEITHER

**7.** {YES, YES, 1 ONLY} and {YES, NO, 1 ONLY} are the equilibria of Example 1; both yield the same outcome, namely, Player 1 is hired without an offer to Player 2.

**8.** Example 2 has 14 equilibria: namely, (YINJ, YES, BOTH), (YINJ, NKYL, BOTH), and all strategy combinations of the form (YES, ·, 1 ONLY), (NIYJ, ·, 1 ONLY), or (NO, ·, NEITHER), where · denotes any of the four strategies of Player 2. Eight of these 14 equilibria correspond to the equilibrium outcome of Example 1, whereas the other six correspond to two different outcomes.

**9.** For Example 1, $m_3(\text{BOTH}) = 0 = m_3(2\ \text{ONLY})$ and $m_3(1\ \text{ONLY}) = 2$, implying $\widetilde{w}^3 = 1\ \text{ONLY}$ and $\widetilde{f}_3 = 2$. For $k \leq 2$, $m_k(\text{YES}) = 0 = m_k(\text{NO})$, implying $\widetilde{f}_k = 0$. So $D^* = D - \{(\text{YES, YES, 2 ONLY}), (\text{NO, YES, 2 ONLY}), (\text{NO, YES, BOTH})\}$.

**10.** For Example 1, $P = \{(\text{YES, YES, BOTH}), (\text{YES, NO, BOTH})\} = P^*$.

**11.** In Example 2, the equilibria (YINJ, YES, BOTH) and (YINJ, NKYL, BOTH) are not subgame perfect because in the subgame beginning at $J$ they would require Player 1 to say NO, which would be irrational. (Player 1's threat to say NO, unless Player 3 makes an offer to BOTH, is not credible because Player 3 has a first mover advantage.)

**12.** While reducing $E$ to $E \cap P^*$ eliminates equilibria of the form (NO, $\cdot$, NEITHER) in Example 2, it is also possible that $E$ and $P$ are disjoint (as in Example 1).

**13.** While reducing $E$ to $E \cap E_S$ eliminates equilibria of the form (YINJ, YES, BOTH) and (YINJ, NKYL, BOTH) in Example 2, it is also possible that $E = E_S$ (as in Example 1, where there are no proper subgames).

## 16.5.2  MATRIX AND BIMATRIX GAMES

This subsection discusses two-player noncooperative games. Such games can be represented in normal form by a pair of matrices.

**Definitions:**

Suppose $S_1 = \{1, \ldots, r\}$ and $S_2 = \{1, \ldots, s\}$.

The $r \times s$ **payoff matrices** $A = (a_{ij})$ and $B = (b_{ij})$, with $a_{ij} = f_1(i,j)$ and $b_{ij} = f_2(i,j)$, define a **bimatrix game**.

The game is **zero-sum** if $a_{ij} + b_{ij} = 0$ for all $i \in S_1$, $j \in S_2$. The game is **symmetric** if $r = s$ and $B = A^T$. In either case, the game is completely determined by $A$ and is called a **matrix game**.

For Player 1, $i \in S_1$ is **dominated** by $i' \in S_1$ if $a_{i'j} \geq a_{ij}$ for all $j \in S_2$, with strict inequality for at least one $j$. For Player 2, $j \in S_2$ is dominated by $j' \in S_2$ if $b_{ij'} \geq b_{ij}$ for all $i \in S_1$ with strict inequality for at least one $i$.

Let $e_k$ denote the $k$-dimensional vector in which every entry is 1, and let $X_k$ denote the $(k-1)$-dimensional **unit simplex** $X_k = \{(x_1, \ldots, x_k) \mid xe_k = 1, x \geq 0\}$.

A **mixed strategy** for Player 1 is a vector $p = (p_1, \ldots, p_r) \in X_r$, where $p_i$ is the probability that Player 1 selects $i \in S_1$. Similarly, a mixed strategy for Player 2 is $q = (q_1, \ldots, q_s) \in X_s$, where $q_j$ is the probability that Player 2 selects $j \in S_2$.

In a **mixed strategy combination** $(p, q) \in X_r \times X_s$, the expected payoffs to Players 1 and 2, respectively, are given by $\phi_1(p, q) = pAq^T$ and $\phi_2(p, q) = pBq^T$.

The pair $(p^*, q^*) \in X_r \times X_s$ is a **Nash equilibrium mixed strategy combination**, or simply an **equilibrium in mixed strategies**, if $\phi_1(p^*, q^*) \geq \phi_1(p, q^*)$ for all $p \in X_r$ and $\phi_2(p^*, q^*) \geq \phi_2(p^*, q)$ for all $q \in X_s$. If the game is zero-sum, then $p^*$ is called an **optimal strategy** for Player 1 and $q^*$ is called an optimal strategy for Player 2.

**Facts:**

**1.** Every bimatrix game has at least one equilibrium in mixed strategies.

**2.** All equilibria in mixed strategies of a zero-sum game yield the same expected payoffs, $v$ to Player 1 and $-v$ to Player 2; $v$ is known as the *value* of the game.

**3.** The value $v$ of a zero-sum game and a pair $(p^*, q^*)$ of optimal strategies can always be computed by solving a dual pair of linear programming (LP) problems (§16.1). The primal LP problem finds $p$ to maximize $v$ subject to $A^T p \geq v1_s$, $p \in X_r$, whereas the dual LP problem finds $q$ to minimize $v$ subject to $Aq \leq v1_r$, $q \in X_s$.

**4.** Player 1 can achieve the value $v$ of a zero-sum game with a mixed strategy that attaches zero probability to any dominated pure strategy. Likewise, Player 2 can achieve value $-v$ by playing dominated pure strategies with zero probability.

**5.** Graphical methods can be used to compute efficiently all equilibria of zero-sum games where $r = 2$ or $s = 2$, or of matrix games (of either type) where $r = s = 3$; see [Dr81], [Ow13], and [Pe08]. There is no general method for computing all equilibria.

**6.** The definition of mixed strategy and the existence of equilibria are readily extended to $n$-player games. This result was one of the fundamental contributions to game theory for which John Nash was awarded the 1994 Nobel Prize in Economic Science.

**Examples:**

**1.** Two advertising agencies are involved in a campaign to promote competing beverages. The payoffs of various promotional strategies are shown in the following table.

| $i$ | $j$ | 1 old | 2 new |
|---|---|---|---|
| 1 | old | 0 | −2 |
| 2 | new | −2 | −1 |
| 3 | diet | 3 | −3 |

The promotional strategies for the first agency are: stress the old formula, advertise a new formula, or advertise a diet drink. The second agency has the possible strategies: stress the old formula, or advertise a new formula. The payoffs in this case indicate the net change in millions of sales gained (by Advertiser 1). For example, if the first agency promotes a diet drink while the other agency promotes the old formula, three million more drinks will be sold. On the other hand, if the other agency happens to promote the new formula, then the first agency will end up losing three million unit sales to the second agency.

**2.** An investor has just taken possession of jewels worth \$45,000 and must store them for the night in one of two locations (A, B). The safe in location A is relatively secure, with a probability $\frac{1}{15}$ of being opened by a thief. The safe at location B is not as secure as the safe at location A, and has a probability $\frac{1}{5}$ of being opened. A notorious thief is aware of the jewels, but doesn't know where they will be stored. Nor is it possible for the thief to visit both locations in one evening. This is a (symmetric) zero-sum game between the investor (Player 1), who selects where to keep the jewels and the thief (Player 2), who decides which safe to try. If the investor puts the jewels in the most secure location (A) and the jewel thief goes to this location, the expected loss in this case is $\frac{1}{15}(-45{,}000) + \frac{14}{15}(0) = -3{,}000$. The other entries of the payoff matrix in the following table are computed similarly, and are expressed in thousands of dollars (to the investor).

| $i$ | $j$ | 1 old A | 2 new B |
|---|---|---|---|
| 1 | A | −3 | 0 |
| 2 | B | 0 | −9 |

No pure strategy combination is a Nash equilibrium, since it is always tempting for one player to defect from the current strategy. However, there is a Nash equilibrium mixed strategy combination: $p^* = (\frac{3}{4}, \frac{1}{4}) = q^*$, with value $v = -\$2,250$ to the investor. The mixed strategy $p^*$ is found by solving the following linear program, in which Player 1 wants to find the largest value of $v$ so that he is guaranteed to receive at least $v$ (regardless of what Player 2 does). The associated optimal dual LP solution gives $q^*$.

$$\text{maximize:}\quad v$$
$$\text{subject to:}\quad -3p_1 + 0p_2 \geq v$$
$$0p_1 - 9p_2 \geq v$$
$$p_1 + p_2 = 1$$
$$p_1, p_2 \geq 0$$

**3.** The zero-sum game of *chump* is played between two camels, a dromedary (Player 1) and a bactrian (Player 2). Player $k$ must simultaneously flash $F_k$ humps and guess that its opponent will flash $G_k$. Possible strategies $(F_k, G_k)$ satisfy $0 \leq F_1, G_2 \leq 1$ and $0 \leq F_2, G_1 \leq 2$. If both players are right or wrong, then the game is a draw; if one is wrong and the other is right, then the first pays $F_1 + F_2$ piasters to the second. The following table shows the strategy sets and corresponding payoffs $a_{ij}$ to Player 1.

|   | $j$ | 1 | 2 | 3 | 4 | 5 |
|---|---|---|---|---|---|---|
| $i$ |   | $(0,0)$ | $(0,1)$ | $(1,0)$ | $(1,1)$ | $(2,0)$ | $(2,1)$ |
|   | $(0,0)$ | 0 | 0 | $-1$ | 0 | $-2$ | 0 |
| 1 | $(0,1)$ | 0 | 0 | 0 | 1 | $-2$ | 0 |
| 2 | $(0,2)$ | 0 | 0 | $-1$ | 0 | 0 | 2 |
| 3 | $(1,0)$ | 1 | 0 | 0 | $-2$ | 0 | $-3$ |
| 4 | $(1,1)$ | 0 | $-1$ | 2 | 0 | 0 | $-3$ |
| 5 | $(1,2)$ | 0 | $-1$ | 0 | $-2$ | 3 | 0 |

The first row and column can be deleted from the full payoff matrix because $(0,0)$ is dominated by $(0,1)$ for both players (Fact 4). Thus it suffices to analyze the reduced payoff matrix in which $r = s = 5$. The value of the game is $-\frac{4}{21}$ for Player 1 ($\frac{4}{21}$ for Player 2). Optimal strategies $p^* = (\frac{2}{21}, \frac{4}{7}, \frac{1}{7}, \frac{4}{21}, 0)$ and $q^* = (\frac{4}{7}, \frac{4}{21}, \frac{2}{21}, \frac{1}{7}, 0)$ are found by linear programming (Fact 3). Note that strategies $i = 5$ and $j = 5$ have zero probability at this equilibrium, despite being undominated.

**4.** The symmetric game of *four ways* [Me01] is played by two left-turning motorists who arrive simultaneously from opposite directions at a 4-way junction. Each has three pure strategies: the first is to go, the second to wait, and the third a conditional strategy of going only if the other appears to be waiting. It takes 2 seconds for one motorist to cross the junction while the other waits. If initially both either go or wait, then both motorists incur an extra "posturing" delay of either 3 or 2 seconds, respectively. Also, the one who ultimately waits is equally likely to be either player. For example, $a_{11} = 0.5 \times (-3 - 2) + 0.5 \times (-3) = -4$ and $a_{22} = 0.5 \times (-2 - 2) + 0.5 \times (-2) = -3$. This game has the payoff matrix

$$\begin{pmatrix} -4 & 0 & 0 \\ -2 & -3 & -2 \\ -2 & 0 & -4 \end{pmatrix}.$$

There are infinitely many equilibria in mixed strategies; these are described in the following table, where $0 \leq a \leq 1$ and $\frac{1}{2} \leq b \leq 1$.

| $p^*$ | $q^*$ |
|---|---|
| $(1,0,0)$ | $(0,a,1-a)$ |
| $(0,a,1-a)$ | $(1,0,0)$ |
| $(0,1,0)$ | $(b,0,1-b)$ |
| $(b,0,1-b)$ | $(0,1,0)$ |
| $\frac{1}{11}(6,2,3)$ | $\frac{1}{11}(6,2,3)$ |

## 16.5.3   CHARACTERISTIC-FUNCTION GAMES

When there exists a binding agreement among all players to cooperate, attention shifts from strategies to the bargaining strengths of coalitions. These strengths are assumed to be measured in terms of a freely transferable benefit (e.g., money or time) and players are assumed to seek a fair distribution of the total benefit available. Also, without loss of generality, the benefit of cooperation will be taken as the savings in costs.

### Definitions:

A **coalition** is a subset $S$ of $N = \{1,\dots,n\}$; equivalently, $S \in 2^N$.

The **cost** associated with coalition $S$ is denoted $c(S)$.

Let $\mathcal{R}^+$ denote the set of nonnegative reals. The **characteristic function** $\bar{v}\colon 2^N \to \mathcal{R}^+$ assigns to each $S$ its cooperative benefit, using $\bar{v}(S) = \max\{0, \sum_{i\in S} c(\{i\}) - c(S)\}$.

A **characteristic-function game**, or **c-game**, is the pair $\Gamma = (N,\bar{v})$.

The game $\Gamma$ is **inessential** if $\bar{v}(N) = 0$. If $\bar{v}(N) > 0$ then the game is **essential**, with **normalized characteristic function** $v\colon 2^N \to [0,1]$ defined by $v(S) = \frac{\bar{v}(S)}{\bar{v}(N)}$.

The game $\Gamma$ is **convex** if $v(S \cup T) \geq v(S) + v(T) - v(S \cap T)$ for all $S,T \in 2^N$.

Let $X = X_n$ be the $(n-1)$-dimensional unit simplex (§16.5.2). Any $x \in X$ is called an **imputation**; it allocates $x_i$ of the total normalized benefit $v(N) = 1$ to Player $i$. An imputation is **unreasonable** if it allocates more to some $i \in N$ than the maximum that $i$ could contribute to any coalition $T - \{i\}$ by joining it.

The **reasonable set** is $X_{RS} = \{x \in X \mid x_i \leq \max_T[v(T) - v(T - \{i\})] \text{ for all } i \in N\}$.

For any $x \in X$ and $S \in 2^N$, the **excess** of coalition $S$ at $x$ is $e(S,x) = v(S) - \sum_{i\in S} x_i$.

The **core** of $\Gamma$ is $C = \{x \in X \mid e(S,x) \leq 0 \text{ for all } S \in 2^N\}$.

The **marginal worth** of Player $i$ to the coalition $T - \{i\}$ is $v(T) - v(T - \{i\})$.

The **Shapley value** of a c-game is the imputation $x^S = (x_1^S, x_2^S, \dots, x_n^S)$ defined by $x_i^S = \frac{1}{n!} \sum_{T\in\Pi^i} (|T| - 1)!\,(n - |T|)!\,(v(T) - v(T - \{i\}))$, where $\Pi^i = \{T \in 2^N \mid T \supseteq \{i\}\}$.

### Facts:
1. An imputation is both individually rational and group rational (see §16.5.1).
2. Convexity is a sufficient (but not necessary) condition for the core to exist.
3. If $C \neq \emptyset$ then $C \subseteq X_{RS}$.

**4.** If $C$ contains a single imputation, then the $c$-game is usually regarded as solved.

**5.** In general, $C$ may either be empty (see Example 1) or contain infinitely many imputations (see Example 2).

**6.** If $C$ contains infinitely many imputations, then there are several ways to single one out as the solution to the $c$-game. One approach is to define a "center" of $C$, which leads to the important concept of the *nucleolus* [Me01].

**7.** Every $c$-game solution concept assumes that players have agreed to enter coalition $N$. If its order of formation were known, players could be allocated their marginal worths; in general, however, this order of formation (and hence marginal worth) is a random variable.

**8.** If all orders of formation of $N$ are equally likely, then the probability that Player $i$ enters $N$ by joining the coalition $T - \{i\}$ is $\frac{(|T|-1)!(n-|T|)!}{n!}$.

**9.** The Shapley value distinguishes a single imputation as the solution of a $c$-game by allocating to players the expected values of their marginal worths, based on the assumption that all orders of formation of $N$ are equally likely.

**10.** $x^S \in X_{RS}$.

**11.** $x^S \in C$ if $\Gamma$ is convex.

**Examples:**

**1.** In the $c$-game *log-hauling* [Me01], three lone drivers of pickup trucks discover a pile of 150 logs too heavy for any one to lift. Players 1, 2, and 3 can haul up to 45, 60, and 75 logs, respectively. Thus $\bar{\nu}(\{1,2\}) = 105$, $\bar{\nu}(\{1,3\}) = 120$, $\bar{\nu}(\{2,3\}) = 135$, and $\bar{\nu}(\{1,2,3\}) = 150$ so that $\nu(\{1,2\}) = \frac{7}{10}$, $\nu(\{1,3\}) = \frac{4}{5}$, and $\nu(\{2,3\}) = \frac{9}{10}$. This $c$-game is not convex; for example, if $S = \{1,2\}$ and $T = \{2,3\}$, then $1 = \nu(S \cup T) < \nu(S) + \nu(T) - \nu(S \cap T) = \frac{8}{5}$. Also, $C = \emptyset$.

**2.** The $c$-game *car pool* [Me01] is played by three co-workers whose office is $d$ miles from their residential neighborhood, shown in the following figure.

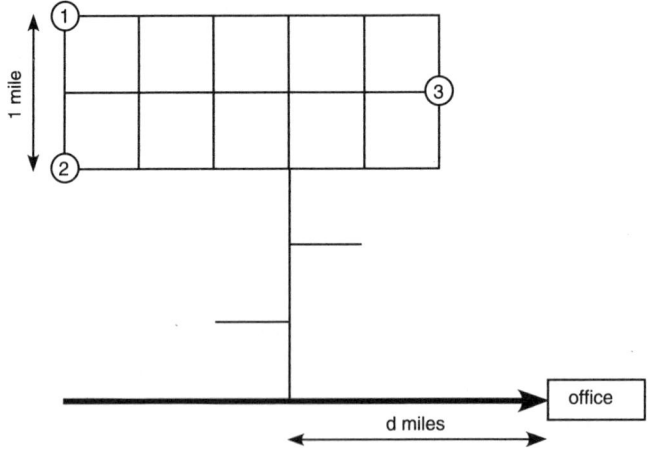

Driving to work costs $\$k$ per mile, and the shortest route is always used. The benefit of cooperation is car pool savings, giving the characteristic function displayed in the following table.

| $S$ | $c(S)$ | $\overline{\nu}(S)$ | $\nu(S)$ | $e(S, x)$ for $d = 1$ |
|---|---|---|---|---|
| $\emptyset$ | $0$ | $0$ | $0$ | $0$ |
| $\{1\}$ | $(4+d)k$ | $0$ | $0$ | $-x_1$ |
| $\{2\}$ | $(3+d)k$ | $0$ | $0$ | $-x_2$ |
| $\{3\}$ | $(3+d)k$ | $0$ | $0$ | $x_1 + x_2 - 1$ |
| $\{1,2\}$ | $(4+d)k$ | $(3+d)k$ | $\frac{3+d}{3+2d}$ | $\frac{4}{5} - x_1 - x_2$ |
| $\{1,3\}$ | $(6+d)k$ | $(1+d)k$ | $\frac{1+d}{3+2d}$ | $x_2 - \frac{3}{5}$ |
| $\{2,3\}$ | $(6+d)k$ | $dk$ | $\frac{d}{3+2d}$ | $x_1 - \frac{4}{5}$ |
| $\{1,2,3\}$ | $(7+d)k$ | $(3+2d)k$ | $1$ | $0$ |

Because $x_3 = 1 - x_1 - x_2$ ($\geq 0$), a set of imputations is determined by its projection onto $x_3 = 0$. In these terms, for $d = 1$, $X$ is the largest triangle in the following figure, $X_{RS}$ is the shaded hexagon, and $C$ is the shaded quadrilateral. Here $C \subset X_{RS} \subset X$ because the $c$-game is convex (for all $d \geq 0$).

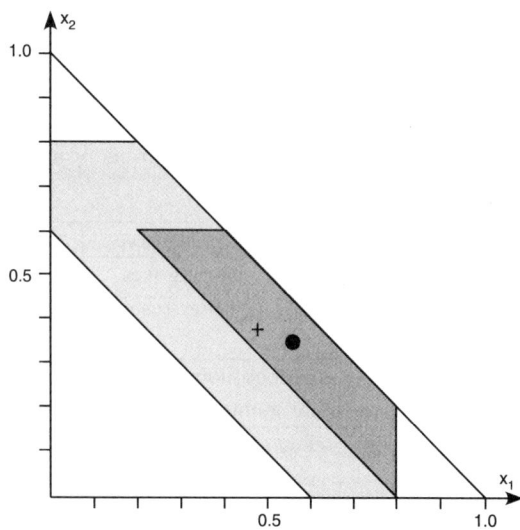

**3.** For the $c$-game in Example 2, it is easy enough to locate a center for $C$; see the figure for Example 2, where the nucleolus is marked by a dot.

**4.** In Example 1, the six possible formation orders of $N$ are 123, 132, 213, 231, 312, 321. Thus, the Shapley value is the imputation $x^S = (\frac{17}{60}, \frac{1}{3}, \frac{23}{60})$; see the following table.

| $i$ | $T \in \Pi^i$ | $\nu(T) - \nu(T - \{i\})$ | probability $i$ enters $N$ by joining $T - \{i\}$ | $x_i^S$ |
|---|---|---|---|---|
| 1 | $\{1\}$ | $0$ | $\frac{1}{3}$ | $\frac{17}{60}$ |
|   | $\{1,2\}$ | $\frac{7}{10}$ | $\frac{1}{6}$ | |
|   | $\{1,3\}$ | $\frac{4}{5}$ | $\frac{1}{6}$ | |
|   | $\{1,2,3\}$ | $\frac{1}{10}$ | $\frac{1}{3}$ | |
| 2 | $\{2\}$ | $0$ | $\frac{1}{3}$ | $\frac{1}{3}$ |
|   | $\{1,2\}$ | $\frac{7}{10}$ | $\frac{1}{6}$ | |
|   | $\{2,3\}$ | $\frac{9}{10}$ | $\frac{1}{6}$ | |
|   | $\{1,2,3\}$ | $\frac{1}{5}$ | $\frac{1}{3}$ | |
| 3 | $\{3\}$ | $0$ | $\frac{1}{3}$ | $\frac{23}{60}$ |
|   | $\{1,3\}$ | $\frac{4}{5}$ | $\frac{1}{6}$ | |
|   | $\{2,3\}$ | $\frac{9}{10}$ | $\frac{1}{6}$ | |
|   | $\{1,2,3\}$ | $\frac{3}{10}$ | $\frac{1}{3}$ | |

**5.** By a calculation very similar to that laid out in the table of Example 4, the Shapley value for Example 2 is the imputation $x^S = (\frac{7}{15}, \frac{11}{30}, \frac{1}{6})$. Because the $c$-game is convex, $x^S \in C$. This is illustrated in the previous figure, where $x^S$ is marked by a cross.

## 16.5.4    APPLICATIONS

Discrete (noncooperative or characteristic-function) games have numerous applications and merge with other categories of games not examined here. The references in the following table provide sources for the definitions, concepts, and applications of such games. This table also lists some representative areas of application of game theory.

| category and references | selected applications | remarks |
|---|---|---|
| characteristic function games [GoGaFi10], [Me93], [Me01], [Ow13], [Pe08], [Wa88] | airport landing fees, voting, water resources | utility is usually assumed to be transferable: in essence, players value benefits identically |
| classical game theory [LuRa57], [vNMo53] | microeconomics, parlor games | economic (as opposed to evolutionary) game theory |
| combinatorial games [AlNoWo07], [Gu91] | chess, go, nim, other parlor games | two players; complete, perfect information; no chance moves; zero-sum |
| continuous games [Dr81], [Fr90] | duels, military combat, oligopoly theory | a discrete game with mixed strategies is a special case of a continuous game |
| cooperative games in strategic form (as opposed to $c$-games) [Fr90], [Me01], [Pe08] | wage bargaining, motoring behavior | agreements among players are binding |
| differential games [BaHa94], [DoEtal00], [Me93] | resource management | extension of optimal control theory |

| category and references | selected applications | remarks |
|---|---|---|
| economic game theory [Fr90], [GoGaFi10], [My91] | microeconomics | equilibria are the result of rational thought processes |
| evolutionary game theory [BrRy13], [Cr03], [Ma82], [Me01], [Sa10], [Si10] | animal behavior | equilibria are the result of natural selection or equivalent populational processes |
| iterated games [Fr90], [Me01], [Si10] | rationality of cooperation | often infinitely many iterations |
| resource games [Me93] | fisheries, forestry, water resources | discrete, continuous, and differential games are all used |
| symmetric matrix games [BrRy13], [Cr03], [Ma82], [Me01], [Sa10], [Si10] | evolutionary game theory | dynamical systems theory provides a rationale for strategic equilibrium |
| zero-sum matrix games [Dr81], [Pe08] | military science | |

# 16.6   SPERNER'S LEMMA AND FIXED POINTS

A fixed point of a function from a set $X$ to itself is a point of $X$ that is mapped into itself. Brouwer (1912) proved that every continuous mapping $f$ on the unit ball has a fixed point. Sperner (1928) gave an elegant proof of Brouwer's fixed-point theorem using a combinatorial lemma known today as Sperner's lemma. This lemma has a number of applications to economics, nonlinear programming, and game theory.

## 16.6.1   SPERNER'S LEMMA

Sperner's lemma is a combinatorial result applicable to certain triangulations of a $p$-dimensional convex set, in which the vertices of the triangulation are given labels from $\{1, 2, \ldots, p+1\}$.

**Definitions:**

The $p+1$ points $x_1, x_2, \ldots, x_{p+1} \in \mathcal{R}^n$ are said to be in **general position** if the vectors $x_2 - x_1, x_3 - x_1, \ldots, x_{p+1} - x_1$ are linearly independent (§6.1.3).

The set $C \subseteq \mathcal{R}^n$ is **convex** if for all $x, y \in C$ and $0 \le \lambda \le 1$, $\lambda x + (1 - \lambda)y \in C$.

The **convex hull** of a finite set of points $v_1, \ldots, v_{p+1} \in \mathcal{R}^n$ is the set $\langle v_1, \ldots, v_{p+1} \rangle = \{\sum_{i=1}^{p+1} \lambda_i v_i \mid \sum_{i=1}^{p+1} \lambda_i = 1, \lambda_i \ge 0 \}$.

A **$p$-simplex** $\sigma$ is the convex hull of $p+1$ points $x_1, \ldots, x_{p+1} \in \mathcal{R}^n$ in general position.

The **vertices** of the $p$-simplex $\sigma = \langle x_1, \ldots, x_{p+1} \rangle$ are the points $x_1, \ldots, x_{p+1}$. The **face** $\tau = \langle x_{j_1}, \ldots, x_{j_k} \rangle$ of $\sigma$ is the simplex spanned by the subset $\{x_{j_1}, \ldots, x_{j_k}\}$ of $\{x_1, \ldots, x_{p+1}\}$. Write $\tau \prec \sigma$ when $\tau$ is a face of $\sigma$.

A *simplicial complex* $K$ is a collection of simplices satisfying

- if $\sigma \in K$ and $\tau \prec \sigma$ then $\tau \in K$;
- if $\sigma, \tau \in K$ intersect, their intersection is a face of each.

The *p-skeleton* of a simplicial complex $K$ is the set of all simplices of dimension $p$ or less. The 0-skeleton is the *vertex set*, denoted $V(K)$.

A *simplicial subdivision* $\mathcal{F}$ of a simplex $\sigma$ is a collection of simplices $\{\tau_j \mid 1 \le j \le m\}$ satisfying

- $\sigma = \bigcup_{j=1}^{m} \tau_j$;
- the intersection of any two $\tau_j$ is either empty or a face of each.

A simplicial subdivision $\mathcal{F}'$ of a simplicial complex $\mathcal{K}$ is a **refinement** of the simplicial subdivision $\mathcal{F}$ of $\mathcal{K}$ if every simplex of $\mathcal{F}$ is a union of simplices of $\mathcal{F}'$.

Given a simplicial subdivision $\mathcal{F}$ of the $p$-simplex $\sigma = \langle x_1, \ldots, x_{p+1} \rangle$, a **proper labeling** of $\mathcal{F}$ is a mapping $\ell \colon V(\mathcal{F}) \to \{1, 2, \ldots, p+1\}$ satisfying

- $\ell(x_m) = m$ for $m = 1, \ldots, p+1$;
- if vertex $v$ lies on a face $\langle x_{k_1}, \ldots, x_{k_q} \rangle$ of $\sigma$, then $\ell(v) \in \{k_1, \ldots, k_q\}$.

Here $\{1, 2, \ldots, p+1\}$ is the **label set**, and if $\ell(v) = k$ then $v$ *receives the label* $k$.

A *distinguished simplex* is a $p$-simplex that receives all $p + 1$ labels 1 through $p + 1$.

**Facts:**

**1.** The convex hull $\langle v_1, \ldots, v_{p+1} \rangle$ is the intersection of all convex sets containing the points $v_1, \ldots, v_{p+1}$.

**2.** The dimension of any $p$-simplex is $p$.

**3.** A $p$-simplex contains $2^{p+1} - 1$ simplices of dimension $p$ or less.

**4.** *Sperner's lemma* (1928): Every properly labeled subdivision of a simplex $\sigma$ has an odd number of distinguished simplices. (E. Sperner, 1906–1980)

**5.** Algorithm 1 gives a method for finding a distinguished triangle in a properly labeled subdivision of a triangle $T$. Each iteration of the outer loop starts at a distinguished 1-simplex and traces out a path, terminating either at a distinguished 2-simplex or at an outer edge of $T$.

---

**Algorithm 1**:  **Distinguished simplex of a 2-simplex.**

  input: properly labeled subdivision of triangle $T$

  output: a distinguished triangle of $T$

  {Outer loop}

    find a distinguished 1-simplex $\tau$ along the bottom of $T$

    {Inner loop}

    **repeat**

      **if** the unique triangle containing $\tau$ is distinguished **then** stop

      **else** proceed to a neighboring triangle whose common edge is distinguished

    **until** either a distinguished triangle is found or the search leads to the bottom

        edge of $T$

  continue outer loop with a new distinguished 1-simplex $\tau$

---

**6.** Since there is an odd number of distinguished 1-simplices along the bottom of $T$ (Fact 4) and since each "failed" outer loop iteration produces a path joining two such distinguished 1-simplices, Algorithm 1 must eventually produce a path terminating at a distinguished 2-simplex.

**Examples:**

**1.** A 0-simplex is a point, a 1-simplex is a line segment, and a 2-simplex is a triangle (interior included). A 3-simplex includes the vertices, edges, faces, and interior of a tetrahedron. See the following figure.

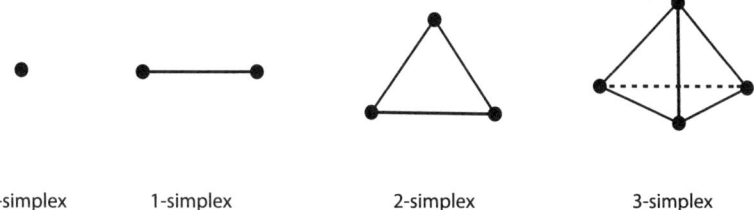

0-simplex      1-simplex          2-simplex          3-simplex

**2.** The 0-skeleton of a simplex $\sigma$ is its vertex set; the 1-skeleton of $\sigma$ is the edge set of $\sigma$ including their endpoints; if $\sigma$ is a 3-simplex, the 2-skeleton is the union of the faces of $\sigma$.

**3.** Part (a) of the following figure shows a simplicial subdivision of a 2-simplex. The subdivision in part (b) of the figure is not simplicial because $\tau_1 \cap \tau_3$ is not a face of the simplex $\tau_3$.

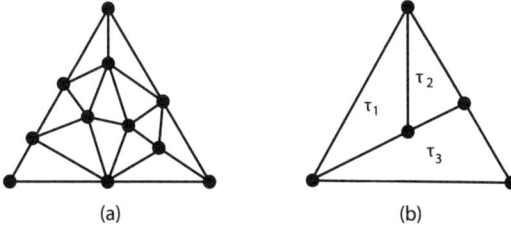

(a)                    (b)

**4.** The following figure shows a proper labeling of a simplicial subdivision of a 1-simplex. A distinguished 1-simplex is a subinterval that receives both the labels 1 and 2. In this example, there are five such 1-simplices, an odd number (as guaranteed by Fact 4).

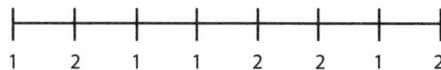

1    2    1    1    2    2    1    2

**5.** The following figure shows a proper labeling of a simplicial subdivision of a 2-simplex. There is one distinguished 2-simplex, receiving all three labels, which is shown shaded in the figure. If the vertex in the interior of the triangle is instead labeled 3, then there will be three distinguished 2-simplices, still an odd number.

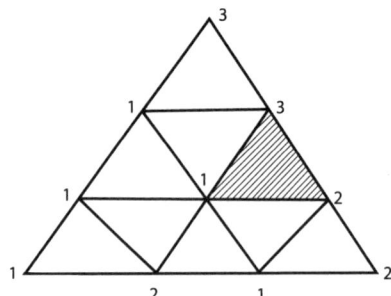

**6.** Several possible paths from executing Algorithm 1 are displayed in the following figure. The rightmost path terminates in a bottom edge, while the leftmost path leads to a distinguished triangle. Note that there are three distinguished triangles in this example, an odd number as required by Sperner's lemma.

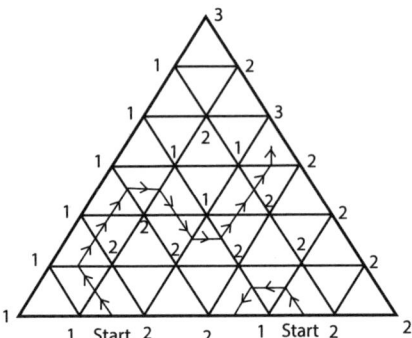

## 16.6.2   FIXED-POINT THEOREMS

Fixed-point theorems have applicability to a number of problems in economics, as well as to game theory and optimization.

### Definitions:

The point $x \in B$ is a **fixed point** of the mapping $f: B \to B$ if $f(x) = x$.

The mapping $f$ defined on a subset $X$ of a normed space $B$ is a **contraction** if there is some $0 \leq \beta < 1$ such that $\|f(x) - f(y)\| \leq \beta \|x - y\|$ for all $x, y \in X$.

The function $F$ is a **set mapping** on $X$ if $F(x)$ is a nonempty subset of $X$ for all $x \in X$.

The set mapping $F$ is **convex** if $F(x)$ is a convex subset of $X$ for all $x \in X$.

### Facts:

**1.** Fixed-point theorems can be used to demonstrate the existence of economic equilibria, solutions to a system of nonlinear equations, and Nash equilibria in two-person nonzero-sum games.

**2.** *Brouwer's fixed-point theorem* I: Every continuous mapping $f: \sigma \to \sigma$ where $\sigma$ is a $p$-simplex has a fixed point. (L. E. J. Brouwer, 1881–1966)

**3.** For a simplex $\sigma = \langle x_1, x_2, \ldots, x_{p+1} \rangle$ and a continuous mapping $f \colon \sigma \to \sigma$, let $f(\sum_{k=1}^{p+1} \lambda_k x_k) = \sum_{k=1}^{p+1} \mu_k x_k$. Then $\sum_{k=1}^{p+1} \mu_k = \sum_{k=1}^{p+1} \lambda_k = 1$.

**4.** Relative to the mapping $f \colon \sigma \to \sigma$, define $T_j = \{\sum_{k=1}^{p+1} \lambda_k x_k \mid \mu_j \le \lambda_j\}$. Then a fixed point of $f$ is any point belonging to $\bigcap_{j=1}^{p+1} T_j$.

**5.** Suppose an interior vertex $v$ of a subdivision $\mathcal{F}$ of $\sigma$ is labeled with $j$ provided that $v \in T_j$, and suppose a vertex $v$ belonging to a face $\langle x_{k_1}, x_{k_2}, \ldots, x_{k_t} \rangle$ is labeled with any one of the labels $k_1, k_2 \ldots, k_t$. Then a fixed point of $f$ occurs in a distinguished simplex of $\sigma$.

**6.** Algorithm 2, based on Sperner's lemma (§16.6.1), produces a sequence of points converging to a fixed point of a $p$-simplex $\sigma$.

---

**Algorithm 2**:  **Fixed point of a $p$-simplex.**

input: function $f$ defined on a $p$-simplex $\sigma$

output: fixed point $x^* \in \sigma$

construct a sequence of subdivisions $\{\mathcal{F}_n \mid n \ge 1\}$ such that $\mathcal{F}_{n+1}$ refines $\mathcal{F}_n$

label the vertex set of $\mathcal{F}_n$ as in Fact 5

for each subdivision $\mathcal{F}_n$ find a distinguished simplex $\tau_n$

$\bigcap \tau_n$ contains the desired fixed point $x^*$

---

**7.**  *Brouwer's fixed-point theorem* II: Every continuous mapping from a convex compact set $B \subseteq \mathcal{R}^n$ into itself has a fixed point.

**8.**  *Contraction mapping theorem:* Every contraction $f \colon X \to X$ has a fixed point. The fixed point is the limit of the sequence $\{f(x_n) \mid n \ge 0\}$, where $x_0$ is an arbitrary element of $X$ and $x_{n+1} = f(x_n)$.

**9.**  *Kakutani's fixed-point theorem:*  Let $X \subseteq \mathcal{R}^n$ be a convex compact set and suppose that $F$ is a convex mapping on $X$. If the graph $\{(x, y) \mid y \in F(x)\} \subseteq \mathcal{R}^{2n}$ is closed, then there exists a point $x^* \in X$ such that $x^* \in F(x^*)$.

**10.**  *Schauder's fixed-point theorem:*  Every continuous mapping $f$ on a convex compact subset $X$ in a normed space $B$ has a fixed point.

**11.**  Reference [Bo89] gives applications of fixed-point theorems to determining market equilibria, maximal elements of binary relations, solutions to complementarity problems, as well as solutions to various types of games (cooperative and noncooperative).

**Examples:**

**1.**  The real-valued function $f(x) = 1 - x$ is a mapping from the 1-simplex $\sigma = [0, 1]$ to itself. It is not a contraction since $|f(x) - f(y)| = |(1 - x) - (1 - y)| = 1 \cdot |x - y|$ holds for all $x, y \in \sigma$ so $\beta \ge 1$. The function $f$ has a fixed point at $x = \frac{1}{2}$. However, the iterative procedure in Fact 8 will not generally locate this fixed point. For example, using $x_0 = \frac{1}{4}$ produces the sequence $x_1 = f(x_0) = \frac{3}{4}$, $x_2 = f(x_1) = \frac{1}{4}$, $x_3 = f(x_2) = \frac{3}{4}$, and so forth, with no limiting value.

**2.**  The following figure shows a real-valued function $f \colon \sigma \to \sigma$ defined over the 1-simplex $\sigma = [0, 1]$. This function has three fixed points, identified by the intersection of the graph of $f$ with the dashed line $y = x$. The sets $T_j$ of Fact 4 relative to $x_1 = 0$, $x_2 = 1$ are also indicated in the figure, and it is verified that $T_1 \cap T_2$ contains the three fixed points of $f$. A subdivision of $\sigma$ into five subintervals is shown in the figure. Using Fact 5, the associated vertices (at $x = 0.0, 0.2, 0.4, 0.6, 0.8, 1.0$) receive the labels $1, 2, 1, 1, 2, 2$ respectively, and so there are three distinguished simplices (each containing a fixed point).

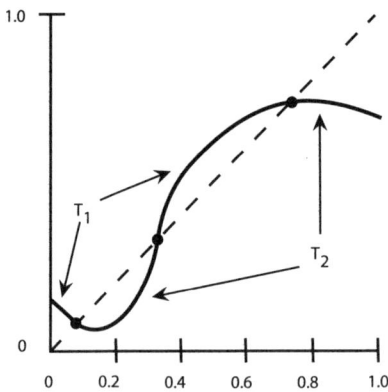

**3.** The real-valued function $f(x) = \frac{1}{4+x^2}$ is a mapping from $\mathcal{R}$ to itself. It can also be shown to be a contraction mapping with $\beta = \frac{1}{16} < 1$. If the iterative procedure in Fact 8 is applied using $x_0 = 1$, then $x_1 = 0.2$, $x_2 = 0.24752$, $x_3 = 0.24623$, $x_4 = 0.24627$, and $x_5 = 0.24627$, yielding the (approximate) fixed point $x^* = 0.24627$.

**4.** *Perron's theorem*:  This theorem (§6.5.5), which assures that every positive matrix has a positive eigenvalue-eigenvector pair, can be proved using the fixed-point theorem in Fact 2. Let $A = (a_{ij})$ be an $n \times n$ matrix, with all $a_{ij} > 0$. The set $\sigma = \{x \in \mathcal{R}^n \mid \sum_{k=1}^{n} x_k = 1,\ x_k \geq 0\}$ is an $(n-1)$-simplex, and the continuous function defined by $f(x) = \frac{Ax}{\|Ax\|_1}$ maps $\sigma$ into itself. (Here $\|w\|_1$ is the $l_1$ norm of vector $w$; see §6.4.5.) By Fact 2, $f$ has a fixed point $\bar{x}$, so that $A\bar{x} = \|A\bar{x}\|_1 \bar{x}$. Since at least one component of $\bar{x}$ is positive and $A$ is positive, the vector $A\bar{x}$ has positive components. It then follows that the eigenvalue $\|A\bar{x}\|_1$ is positive and that the corresponding eigenvector $\bar{x}$ has all positive components.

## 16.7   COMBINATORIAL AUCTIONS

A combinatorial auction (CA) describes a market with several items for sale in which bidders make bids on combinations of items rather than individual items. The specified combination is either won in its entirety or the bid is not accepted. Although initially proposed in 1982 by Rassenti, Smith, and Bulfin [RaSmBu82], the widespread use of CAs in a variety of real-world settings began only later, in the early 21st century. The most prominent CAs have occurred in industrial procurement and in the sale of radio spectrum licenses to telecommunication companies (with individual auctions in some instances generating billions of dollars in revenue). The following subsections discuss basic concepts as well as four issues arising in combinatorial auctions: winner determination, bid language, incentive compatibility, and demand revelation. A few suggested exercises are included sporadically.

## 16.7.1 BASIC CONCEPTS

### Definitions:

A **combinatorial auction** (**CA**) describes a situation in which multiple items are auctioned, and each bidder can submit several bids (each may contain several items). The set of **bidders** is $I = \{1, 2, \ldots, i, \ldots, |I|\}$, the set of **bids** is $J = \{1, 2, \ldots, j, \ldots, |J|\}$, and the set of **items** for sale is $K = \{1, 2, \ldots, k, \ldots, |K|\}$. In addition, there is a fixed **supply** $c_k \geq 1$ of identical, indivisible copies of item $k$ in the auction.

Each bid $j$ consists of a monetary bid amount $b_j$ and an associated (unique) bidder $i_j$. An integer quantity $a_{kj}$ of each item $k$ must be received by bidder $i_j$ at the conclusion of the auction for bid $j$ to be considered **accepted**.

Bidder $i$'s **valuation** for a set of items $S$ is denoted $v_i(S)$.

Bidder $i$'s bid on a set of items $S$ is denoted $b_i(S)$.

A **forward auction** is one with several buyers and one seller, in contrast to a **reverse auction** having several sellers and one buyer.

Except where explicitly noted, this section will focus on forward CAs.

### Facts:

**1.** CAs are also sometimes referred to as *bundle auctions* or *package auctions*.

**2.** A market using combinatorial bids among several buyers and several sellers is called a *combinatorial exchange* (CE), or a *combinatorial double auction*. These two-way markets have different theoretical properties and are less frequently discussed in the literature or used in practice.

**3.** It is convenient to write the (row) vector of all bid amounts as $b$ and the (column) vector of available supplies as $c$. Also the $|K| \times |J|$ matrix $A = (a_{kj})$, indexed by items and bids, indicates the quantity of item $k$ specified in bid $j$.

**4.** For small numerical examples with unique items (i.e., $c_k = 1$ for all $k$), capital letters are often used as item names. For convenience, set braces and commas can be dropped in the expression for a bidder's valuation; e.g., bidder $i$'s valuation for items $A$ and $B$ will be written $v_i(AB)$ rather than as $v_i(\{A, B\})$.

**5.** The valuation function is typically assumed to be *weakly monotonic*: i.e., $v(T) \geq v(S)$, for all $S \subseteq T$. This condition is usually called *free disposal*.

**6.** Bidders are often assumed to have an overall utility that is *quasi-linear* in money: i.e., receiving the set of items $S$ and paying $P_i$ dollars provides the net utility $u_i(S) = v_i(S) - P_i$ to bidder $i$.

### Examples:

**1.** *Government procurement:* The Chilean government procures half a billion dollars in meal services annually using thousands of package bids [KiOlWe14]. Bus route services in London have been bought via a CA with a first-price rule (i.e., pay-as-bid) [CaPe06].

**2.** *Industrial procurement:* Using *expressive commerce* software platforms based on combinatorial procurement auctions, the firm CombineNet conducted \$35 billion in revenue in its first 447 business-to-business auctions over five years. Auction items included direct goods (e.g., food ingredients, steel, chemicals, fibers), indirect goods (e.g., office supplies, furnishing, computers), and services (e.g., transportation, consulting, advertising) [Sa13].

**3.** *Spectrum allocation:*   Using a *combinatorial clock auction* (CCA) format, which combines iterative bidding with a final sealed-bid auction, governments have auctioned the rights to use telecommunications frequencies around the world, including those in the United Kingdom, the Netherlands, Denmark, Ireland, Switzerland, Australia, Austria, Slovenia, Canada, and Slovakia. Revenues were around \$19.5 billion for just ten auctions conducted in 2012 through 2014.

**4.** *Airport arrival and departure slots:*  The FAA adopted a core-selecting combinatorial auction to allocate airport landing slots in 2008, but the auction was canceled for political reasons [CoCoOt09].

**5.** *Electricity auctions:*  In the U.S. and Europe, regional wholesale markets for electrical power generation are governed by a complex sequence of contracts sold at auction, including long-term futures contracts for generation, "day-ahead" markets, and spot markets, some of which include CA concepts in their designs [Sh12].

**6.** *Off-shore wind energy tracts:*   To facilitate the development of wind farms on the U.S. Outer Continental Shelf, CAs were developed that combine both price and non-price attributes into a multi-phase CA design [AuCr11].

**7.** *Real estate auctions:*  A new multistory building in Amsterdam used a CA to sell a mixture of residential, office, dining (etc.) space, subject to municipal constraints, in order to maximize rent [GoEtAl14].

**8.** Suppose we have a set of $|K| = 3$ unique items $\{A, B, C\}$ for sale, with $|J| = 5$ bids from $|I| = 5$ bidders. (When a bidder bids on only one bundle, she is called *single-minded.*) Bidders submit bids $b_1(AB) = 15$, $b_2(C) = 7$, $b_3(AC) = 12$, $b_4(B) = 8$, and $b_5(ABC) = 16$. Since there are unique items, $c = (1, 1, 1)^T$. Bid amounts are collected as $b = (15, 7, 12, 8, 16)$ with the bundle structure yielding

$$A = \begin{pmatrix} 1 & 0 & 1 & 0 & 1 \\ 1 & 0 & 0 & 1 & 1 \\ 0 & 1 & 1 & 0 & 1 \end{pmatrix}.$$

## 16.7.2   WINNER DETERMINATION AND BID LANGUAGE

This subsection studies the winner determination problem (WDP), namely, that of finding a set of winning bids to maximize total monetary value. In order to formulate appropriate optimization problems, it is necessary to define the underlying bid language used to communicate bid information.

**Definitions:**
The **OR bid language** is a format for communicating bids in which a bidder submits a list of combinatorial bids, and the auctioneer is free to accept any subset of them (Bid$_1$ OR Bid$_2$ OR Bid$_3$ ...), subject to supply constraints over all accepted bids. By contrast, in the **XOR bid language** the auctioneer is constrained to accept at most one of the bids made by the same bidder (Bid$_1$ XOR Bid$_2$ XOR Bid$_3$ ...).

The $|I| \times |J|$ binary matrix $B = (b_{ij})$ consists of entries in which $b_{ij} = 1$ if and only if $i_j = i$. Thus, the unit entries of row $i$ of $B$ identify those bids made by bidder $i$.

**Facts:**
**1.** The winner determination problem for the OR bid language (WDP-OR) can be formulated as the following 0-1 integer programming problem (see §16.1.8) in the binary

decision variables $x$: namely, $x_j = 1$ if and only if bid $j$ is accepted by the auctioneer in the final allocation of auction items to bidders. Here it is assumed that all $a_{kj}$ and $c_k$ values are nonnegative.

$$
\begin{aligned}
\text{maximize:} \quad & bx \\
\text{subject to:} \quad & Ax \leq c \\
& x \in \{0,1\}^{|J|}.
\end{aligned} \tag{1}
$$

**2.** The (feasibility) constraints in (1) state that no more than $c_k$ copies of item $k$ can be allocated; i.e., each row enforces $\sum_{j \in J} a_{kj} x_j \leq c_k$ for each item $k$.

**3.** Under the OR bid language, a bidder cannot express preferences for substitute items. (See Example 1.)

**4.** WDP-OR is an NP-hard optimization problem for an arbitrary set of bids. This remains true if $b_j = 1$ for all submitted bids and if each bid is restricted to be on sets of three or fewer items [RoPeHa98].

**5.** Assume $c_k = 1$ for all $k$. Suppose the *permissible* subsets to be bid on form a tree structure: if $S_1$ and $S_2$ are both allowed to be bid on, then either one set contains the other or they are disjoint. Then WDP-OR is solvable in $O(n^2)$ time [RoPeHa98].

**6.** Assume $c_k = 1$ for all $k$, and that permissible packages include only contiguous intervals from the list of items $K$. The resulting matrix $A$ will exhibit the consecutive ones property and WDP-OR is solvable in $O(n^2)$ time. This describes the auctioning of real estate along a coastline with bids only on contiguous land. The case of a closed loop of coastline around an island can be solved in $O(n^3)$ time [RoPeHa98].

**7.** Since the OR bid language cannot express preferences for substitute items (e.g., I wish to have exactly one of these items and would refuse/throw away any additional items), the more expressive XOR bid language may be preferred, which can communicate any value function over subsets.

**8.** The winner determination problem for the XOR bid language (WDP-XOR) can be formulated as the following 0-1 integer programming problem (see §16.1.8) in the binary decision variables $x$: namely, $x_j = 1$ if and only if bid $j$ is accepted by the auctioneer in the final allocation of auction items to bidders.

$$
\begin{aligned}
\text{maximize:} \quad & bx \\
\text{subject to:} \quad & Ax \leq c \\
& Bx \leq 1 \\
& x \in \{0,1\}^{|J|}.
\end{aligned} \tag{2}
$$

**9.** Each of the second set of constraints from (2) is indexed by a unique bidder $i$ and can be written as $\sum_{j \in J | i_j = i} x_j \leq 1$.

**10.** A solution to WDP-XOR is referred to as *efficient*, based on its connection to the First Fundamental Welfare Theorem of Economics. When all bids are truthful, a solution to WDP-XOR corresponds exactly to the classical notion of *allocative efficiency*.

**11.** WDP-XOR is an NP-hard optimization problem for an arbitrary set of bids. This remains true if $b_j = 1$ for all submitted bids and each bid is restricted to be on sets of two or fewer items [vHMu01].

**12.** If each bidder submits a bid on every subset of $K$ (communicating at least $2^{|K|} - 1$ nonempty subsets), then both WDP-OR and WDP-XOR can be solved in polynomial time by dynamic programming. However, even for relatively small $|K|$ the burden of communicating so many bids becomes prohibitive [RoPeHa98].

**13.** Both WDP-OR and WDP-XOR do not admit a polynomial-time approximation scheme (unless $P = NP$) [BeFu99].

**14.** Among the most-studied special cases is that of *unit demand*, in which each bidder wants to consume at most one item, and so is willing to bid only on single-item bundles. This setting is often called the *assignment game*. When bidders are restricted to bid for each item individually, formulation (2) becomes totally unimodular and WDP is hence polynomially solvable. (See also §16.7.3, Fact 3 and §16.7.4, Fact 3.)

**15.** Early work in CAs looked at forming larger bids through the combination of atomic bids on items or subsets using logical connectives like OR, XOR, AND, etc. Various combinations such as OR-of-XOR and XOR-of-OR have been studied.

**16.** The desire to pay \$500 for any $n$ of the items from a particular subset cannot be written efficiently (as the size of the subset grows) with simple logical connectives, unless an explicit "$n$-of-the-set{}" operator is introduced [BoHo01].

**Examples:**

**1.** To illustrate Fact 3, suppose that $v_i(A) = v_i(B) = v_i(AB) = 10$. In the OR bidding language, the bidder cannot express this preference to the auctioneer. If the bidder submits $b_i(A) = b_i(B) = 10$, WDP-OR always allows the auctioneer to award both bids at a total value of 20, ignoring any bid $b_i(AB)$ below 20.

**2.** Returning to Example 8 of §16.7.1, the solutions to both WDP-OR and WD-XOR are identical (because bidders are all single-minded) with winning bids $b_1(AB) = 15$ and $b_2(C) = 7$. Now suppose instead that the first two bids come from one bidder and the remaining from a second bidder (i.e., change the problem to bids $b_1(AB) = 15$, $b_1(C) = 7$, $b_2(AC) = 12$, $b_2(B) = 8$, and $b_2(ABC) = 16$). While WDP-OR remains essentially unchanged (except that the two winning bids are from the same bidder), for WDP-XOR we must employ the matrix

$$B = \begin{pmatrix} 1 & 1 & 0 & 0 & 0 \\ 0 & 0 & 1 & 1 & 1 \end{pmatrix}$$

as in (2), to enforce a single winning bid per bidder. The WDP-XOR solution is to award $b_2(ABC) = 16$, as the auction can no longer accept the first and second bids together, or the third and fourth bids together, both of which are better solutions under WDP-OR.

**3.** The XOR-language has been used in practice in several spectrum license CCAs in Europe. Adjacent licenses with similar transmission properties were grouped into categories and sold in a first CA to allocate the number of licenses won in each of a few categories, with a second "assignment stage" CA used to determine a particular contiguous assignment of bandwidth won in the first auction. This two-stage design helped to set prices on similar items, and to reduce both the communication burden (number of bundles to bid on) and the computational burden of winner determination.

**4.** Combining the flexibility/communication benefits of OR and the richer expression as in XOR, an important bid language introduced in the early CA literature is OR*, which consists of OR bids in which a bidder could place fictitious items into her bids, or equivalently, allow bidders to insert XOR-type constraints [Ni06].

**5.** For many practical auction applications, devising a manageable domain-specific bid language (format to succinctly express relevant packages without an explosion of communication requirements) can be a nontrivial task. Several papers take this approach for a specific application area. For example, [GoEtAl15] look at packages of TV ad slots.

**6.** The 2014 Canadian spectrum license auction used an "XOR of base + OR" bidding language in a CCA format.

- http://www.ic.gc.ca/eic/site/smt-gst.nsf/eng/sf10726.html

For each base + OR collection of bids conveyed, if a bidder receives the base package, then the auctioneer may also accept any of the following OR package bids. At most one base + OR collection could be accepted from each bidder (an XOR structure). This allowed for many more packages to be covered by bids, important because the number of licenses in the auction (spanning several regions and frequency categories) made the number of packages to consider quite large.

## 16.7.3   PRICE DETERMINATION AND INCENTIVE COMPATIBILITY

For simplicity, this subsection focuses on the XOR language.

**Definitions:**

For any auction, $P_i$ will denote the total final **payment** of bidder $i$.

Where convenient, $p_k$ will represent the **unit price** of item $k$.

Bidder $i$'s **surplus** is her observed or inferred profit, given by $s_i = b_i(S_i) - P_i$, where $S_i$ is her winning package, emphasizing that this is not her true payoff or profit if she does not bid **truthfully**, i.e., does not bid $b_i(S) = v_i(S)$. Typically, CAs satisfy **individual rationality**, in which $s_i \geq 0$ for all $i$.

An allocation and bundle-pricing scheme are in **competitive equilibrium** if and only if (i) the allocation is efficient when bids are replaced by prices, and (ii) a bidder prefers the bundle and price assigned to her relative to any other bundle at its prescribed price. The bundle she receives, possibly empty, is said to be in her **demand set** at these prices. The classical concept of a Walrasian equilibrium, also applied to continuous rather than discrete commodities, demands that all items with a positive price are sold.

An auction is **dominant-strategy incentive compatible** (**DSIC**) if the strategy to set $b_i(S) = v_i(S)$ for all $S$ provides at least as much utility to each bidder $i$ as any other bidding strategy she could choose.

A subset of bidders is called a **coalition**. The optimal objective value of WDP solved over coalition $C \subseteq I$ (by ignoring the bids of others) is written as $WDP(C)$. When bids are true values, this gives the value of the **coalitional welfare function**, written $w(C)$.

The **Vickrey-Clarke-Groves** (**VCG**) **mechanism** is the efficient auction mechanism in which each bidder pays $P_i = b_i(S_i) - WDP(I) + WDP(I \backslash i)$. The quantity $WDP(I) - WDP(I \backslash i)$ is called the **VCG discount**.

A welfare function exhibits the **buyers-are-substitutes** condition if and only if $w(I) - w(I \backslash C) \geq \sum_{i \in C} w(I) - w(I \backslash i)$ for all $C \subset I$.

If the buyers-are-substitutes property holds for every subset of $I$, the welfare function is said to exhibit **buyer-submodularity** on $I$.

The outcome of an auction is **blocked** if there is some coalition of bidders who would each agree to an alternative outcome that would also be preferred by the seller (because of greater total payments). As in cooperative game theory, an unblocked outcome is said to be in the **core**.

An efficient auction that selects payments $P_i$ in the core when bids are treated as truthful is called **core-selecting** [DaMi08].

**Facts:**

**1.** An allocation is supported by competitive equilibrium prices if and only if it is efficient, but the prices may be nonlinear in items and non-anonymous (i.e., different bidders may see different prices for the same bundle). See [MiPa07] for the computation of universal competitive equilibrium prices.

**2.** Whenever the LP relaxation of formulation (2) yields an integral solution, the LP dual provides Walrasian equilibrium prices. In particular, the dual variables associated with item capacities yield linear prices for each item that can be added to form anonymous bundle prices. When the LP relaxation solution value strictly exceeds the integer solution value, competitive equilibrium linear item prices do not exist. (Exercise: Prove these statements using LP duality theory and complementary slackness. Hint: first show that dual variables associated with matrix $B$ equal the surplus amounts $s_i$.)

**3.** In the assignment game context (see §16.7.2, Fact 14), [Le83] showed that among all optimal solutions to the LP dual of (2), those that minimize payments are the VCG prices.

**4.** When item prices $p_k$ exist that form a competitive equilibrium, the auction adding those prices to form bundle payments is core-selecting. (Exercise: Prove this fact. Use Fact 2 and Fact 11.)

**5.** The VCG mechanism is the unique efficient DSIC individually rational payment rule [GrLa77].

**6.** Bidders-are-substitutes is a necessary and sufficient condition for VCG to be supported in competitive equilibrium [BiOs02].

**7.** When the VCG outcome is in the core, it is the unique bidder-dominant point (strictly preferred by all winners) in the core [AuMi02].

**8.** When the VCG outcome is not in the core, there is no bidder-dominant outcome. Payments may be as low as zero despite large counter-bids, and VCG is vulnerable to collusive manipulation and false-name bidding [AuMi02]. (See Example 2.)

**9.** Buyer-submodularity of the welfare function guarantees that the VCG outcome is in the core [AuMi02].

**10.** If the *substitutes* (also called *goods-are-substitutes* or *gross substitutes*) condition holds for all bidders (a property discussed in the economics literature in which a price increase on one good cannot cause a necessary drop in demand of a different good), then buyer-submodularity holds [AuMi02]. (See Example 1.)

**11.** Core-selecting payments can be found via constraint generation. Given a candidate set of payments for a particular efficient solution, solving WDP with bids lowered by the current surplus will find the most violated core constraint [DaRa07]. With winning bid amounts $b_i^*$ for winning bidders $i \in W$, the core constraint for any coalition $C$ can be written as $\sum_{i \in W \backslash C} P_i \geq WDP(C) - \sum_{i \in C} b_i^*$.

**12.** When the VCG outcome is in the core, a bidder-optimal core-selecting algorithm arrives at the VCG outcome.

**13.** In a CA, pay-as-bid is always a core-selecting outcome (so the core is always nonempty.) The core in a CE may be empty.

**14.** In a CE, VCG may violate budget balance: VCG payments exceed revenue, requiring outside funds.

**Examples:**

**1.** Let $b_1(A) = 4$, $b_1(B) = 3$, and $b_1(AB) = 6$; $b_2(A) = 3$, $b_2(B) = 4$, and $b_2(AB) = 5$. Here the bidders treat the items as substitutes (Fact 10), and so the VCG outcome is in the core (Fact 9). (Exercise: Determine winning bids and compute VCG payments. Attempt a few non-truthful strategies by a single bidder to see that there is no benefit.)

**2.** Let $b_1(A) = b_2(B) = b_3(AB) = 20$. VCG yields zero payments and is not in the core. (See Fact 8.) If $v_1(A) < 20$ this illustrates successful collusive or false-name bidding for VCG [AuMi02]. Any individually rational outcome with $P_1 + P_2 = 20$ is a bidder-optimal core-selecting outcome.

**3.** Let $b_1(AB) = b_2(AC) = b_3(BC) = 10$ and $b_4(ABC) = 12$. Linear item prices that form a competitive equilibrium do not exist (assuming individual rationality). VCG and bidder-optimal core-selection both have $P_4 = 10$, but this cannot be decomposed into equilibrium item prices. (Exercise: Prove the non-existence of item prices for this numerical example.)

**4.** Let $b_1(A) = b_2(B) = b_3(C) = 20$, $b_4(AB) = 28$, $b_5(AC) = 26$, and $b_7(A) = b_8(B) = b_9(C) = 10$. Minimum-revenue core-selecting payments are given by $P_1 = 16$, $P_2 = 12$, and $P_3 = 10$, which if taken as item prices form a Walrasian equilibrium. However, VCG payments $P_1 = P_2 = P_3 = 10$ are not in the core [DaRa07]. (Exercise: Demonstrate that VCG is not in the core for this example.)

---

## 16.7.4  DEMAND REVELATION

Among the most important properties of traditional auctions is their ability to elicit information about the demand in a market through an iterative process (e.g., calling successively higher prices for a single good). With CAs, however, it is impossible in many cases to guarantee truthful demand revelation through an iterative process (i.e., to implement VCG by calling out market prices and asking for demand reports). Here we mention some facts and heuristic techniques concerning iterative auction procedures in the CA setting.

### Definitions:

An iterative CA consisting of simultaneous single-item auctions, in which item prices are announced for each item in successive rounds, with price increases on items experiencing excess demand, is called a **simultaneous ascending auction (SAA)** or **simultaneous multi-round auction (SMRA)**.

The **exposure problem** occurs in a SAA when a bidder feels the need to keep bidding on a set of complementary items, exposing her to the risk of getting an incomplete (less valuable) subset. It may also refer to the situation in which a bidder must actively bid on substitute items, not knowing which one she might get, exposing her to the risk of winning more than one when only one is desired.

Similar to the SAA process, ascending linear-item prices over several rounds have also been used as the **clock phase** of a multi-phase CCA auction. In this case the bids in each round are translated into XOR package bids to be used in a subsequent sealed-bid WDP called the **supplementary bidding phase**. Bidders can place additional supplementary bids beyond those revealed in the clock phase.

Because a SAA or clock phase is intended to reveal demand information for the market as a whole, **activity rules** are often used to assure that bidders reveal their demand in early rounds, rather than withholding information about the packages of interest, or their value, until later rounds.

Among the simplest activity rules applied in practice is an **eligibility points** system. Each item is assigned a number of eligibility points prior to the auction, and bidding is restricted such that the total of eligibility points for the package bid on by a bidder is not allowed to go up from one round to the next successive auction rounds.

Under a **revealed preference** activity rule, each bidder is forced to bid self-consistently according to an assumed utility functional form.

**Straightforward bidding** occurs in an iterative auction when a bidder always truthfully reports her packages of highest surplus at the current prices. This may also be called **myopic** bidding, as a bidder bids without regard to any potential future rounds, as if the current round will be the last.

Iterative CAs with nonlinear prices have been proposed based on real-time computation of a **winning level** and **deadness level** for any package of interest. If a bidder submits a package bid just above the current winning level, then it is guaranteed to become a winning bid (at least until another later bid might knock it out). Any bid below the deadness level can never become winning (no matter what bids are submitted in the future). A bid between the winning level and deadness level of a bundle will not become winning immediately, but may become winning with the help of other bids in future rounds.

**Facts:**

**1.** There is no ascending-price auction using linear and anonymous item prices that is guaranteed to implement the VCG outcome, even when attention is restricted to bidder preferences satisfying the substitutes property [GuSt00].

**2.** If the substitutes condition holds (see §16.7.3, Fact 10), then the SAA with straightforward bidding closely approximates a competitive equilibrium (dependent on the bid increment). With at least three bidders and one bidder who violates the substitutes property, no competitive linear item prices exist [Mi00]. Efficiency can be quite low for an iterative linear-price auction in the worst case. For example, [BiShZi13] give a $\frac{2}{|K|+1}$ upper bound.

**3.** In the assignment game context (see §16.7.2, Fact 14), [DeGaSo86] among others provide a primal-dual algorithm related to the Hungarian algorithm that implements the VCG auction iteratively. The auctioneer announces prices and bidders report which items maximize their utility at the stated prices (myopic bidding). An assignment is attempted at each round with prices rising on over-demanded items until an assignment is found.

**4.** With $|K| = 1$ and diminishing marginal values for the single item, buyers-are-substitutes holds and VCG can be implemented with an ascending auction [Au04].

**5.** Assuming buyer-submodularity and straightforward bidding, the ascending proxy auction [AuMi02] and $i$Bundle(3) [PaUn00] terminate with minimal competitive equilibrium payments corresponding to the VCG outcome. (This does not guarantee a linear item price decomposition of VCG, however.)

**6.** Assuming concave utility functions that are additively separable in money and items, [BiEtAl11] consider a seller selling bundles corresponding to bases of a matroid, with applications in scheduling and network formation markets. They provide an ascending auction that terminates at the VCG outcome in such a setting.

**7.** Eligibility point activity rules have been used in several spectrum license auctions, but are considered somewhat easy to manipulate. Consequently, there has been a move towards revealed-preference activity rules.

**8.** For the standard (most simple) revealed-preference activity rule, assume quasi-linear preferences and no budget constraints, and consider any two auction rounds $t_1$ and $t_2$, with respective item-price vectors $p^{t_1}$ and $p^{t_2}$. If a given bidder bids on packages defined by vectors $a^{t_1}$ and $a^{t_2}$, then regardless of her hidden values for these bundles, self-consistency demands that $(p^{t_2} - p^{t_1})(a^{t_2} - a^{t_1}) \leq 0$. Bids violating this inequality will be rejected according to the activity rule [AuCtMi06]. (Exercise: Prove that a truthful bidder who knows her valuation function will never violate this activity rule under these assumptions.)

**9.** The previously mentioned revealed-preference rule may be too restrictive in practice, as budget constraints may be important and/or learning about preferences (due to imperfect knowledge of one's own valuation function) may be desired. So relaxed versions of this concept have been explored [AuBa14].

**10.** [AdGu05] introduced the concepts of winning levels and deadness levels for the OR language, which were later extended by [PeZiBi13] to the XOR setting, where the computation of deadness levels is shown to be $\Pi_2^P$-complete.

**Examples:**

**1.** Let bidder 1's bids be given by [(8 for $A$) XOR (8 for $B$)] OR [(8 for $C$) XOR (8 for $D$)]; bidder 2's bids are [(6 for $A$) XOR (2 for $C$)] OR (6 for $B$); bidder 3's bids are [(2 for $B$) XOR (6 for $D$)] OR (6 for $C$). In one efficient allocation, bidder 1 gets $\{A, D\}$, bidder 2 gets $\{B\}$, and bidder 3 gets $\{C\}$, with VCG payments of 12, 2, and 2, respectively. However, the minimal Walrasian equilibrium prices are given by $(6, 6, 6, 6)$. This example demonstrates preferences that satisfy the substitutes property yet cannot achieve the VCG outcome in anonymous linear prices. (See Fact 1.)

**2.** Assume quasi-linear preferences and no budget constraints. Suppose at auction round 1 we have $p^1 = (1, 5)$ with submitted bid $a^1 = (1, 1)^T$. Then suppose that at some later auction round 2 we have $p^2 = (3, 5)$ with $a^2 = (2, 0)^T$. This violates the standard revealed-preference activity rule (Fact 8) since $(p^2 - p^1)(a^2 - a^1) > 0$. In other words, from round 1 to 2 the price of bundle $(1, 1)^T$ increased by 2, while the price of $(2, 0)^T$ increased by 4; if you prefer the former bundle in round 1, you will continue to prefer it after it experiences a smaller price increase.

**3.** Suppose that in the previous example, the bidder has $v((1, 1)^T) = 12$ and $v((2, 0)^T) = 7$, but now assume he has a budget of 7. He would prefer the first bundle under both $p^1$ and $p^2$, but cannot pay 8 for it under $p^2$, so instead demands the latter bundle. This shows a weakness of revealed-preference rules when budgets are taken into consideration.

**4.** Consider an iterative CA in which bidders submit package bids in turn, after querying the auctioneer for the current winning level and deadness level of any bundle(s) of interest (see Fact 10). Suppose the following bids have already been made by five different bidders: $b_1(B) = 5$, $b_2(C) = 5$, $b_3(AD) = 50$, $b_4(AB) = 30$, and $b_5(CD) = 40$. Bidders 4 and 5 are currently winning. If a sixth bidder now queries the auctioneer about her bundle of interest, $\{B, C\}$, a winning level of 20 is reported; bidding at least one bid increment above 20 will assure winning, because added to bidder 3's bid it will exceed the current winning bids of bidders 4 and 5, i.e., $b_3(AD) + b_6(BC) > b_4(AB) + b_5(CD)$. The reported deadness level for $\{B, C\}$ would be 10; any amount not exceeding 10 results in the bid being rejected in favor of $b_1(B) + b_2(C) = 10$.

# 16.8   VERY LARGE-SCALE NEIGHBORHOOD SEARCH

There are numerous optimization problems of interest that are too difficult to solve to optimality in a reasonable amount of time. A practical remedy is to employ heuristics that can quickly compute good quality solutions. Neighborhood search algorithms, also called local search algorithms, represent a wide class of heuristics where at each iteration the algorithm searches for a better quality solution in a "neighborhood" of the current solution. A critical issue in neighborhood search is defining the neighborhood itself. As a rule of thumb, the larger the neighborhood, the better the quality of a locally optimal solution. However, the larger the neighborhood, the longer will be the time required to search the neighborhood explicitly. A large neighborhood will not produce an effective heuristic unless one can search the large neighborhood efficiently. This section concentrates on neighborhood search algorithms where the size of the neighborhood is "very large" with respect to the size of the input data, yet the neighborhood can be implicitly searched in an efficient manner.

## 16.8.1   BASIC CONCEPTS

### Definitions:

A *combinatorial optimization problem* (*COP*) is an optimization problem of the form: minimize $\{f(S) \mid S \in \mathcal{F}\}$ where $\mathcal{F} \subseteq 2^E$ is a finite set of *feasible solutions*. In this context, $E = \{e_1, e_2, \ldots, e_n\}$ is a finite set of elements called the *ground set*, $f : \mathcal{F} \to \mathcal{R}$ is the *objective function*, and $2^E$ denotes the set of all subsets of $E$. For example, in the Traveling Salesman Problem (TSP) of §10.7.1, $E$ corresponds to the edges in the graph, $\mathcal{F}$ corresponds to the set of feasible TSP tours, and $f(S)$ corresponds to the total length of the tour $S \in \mathcal{F}$.

A *neighborhood function* $\mathcal{N}$ is a point-to-set map $\mathcal{N} : \mathcal{F} \to 2^{\mathcal{F}}$. Given a feasible solution $S \in \mathcal{F}$, the set $\mathcal{N}(S)$ is called the *neighborhood* of the solution $S$. The *size* of a neighborhood is the cardinality of the set $\mathcal{N}(S)$.

An operation that modifies a solution $S_i$ to obtain a solution $S_{i+1} \in \mathcal{N}(S_i)$ is called a *move*. If $f(S_{i+1}) < f(S_i)$, then the move is *improving*.

A solution $S^* \in \mathcal{F}$ is *locally optimal* with respect to a neighborhood function $\mathcal{N}$ if and only if $f(S^*) \leq f(S)$ for all $S \in \mathcal{N}(S^*)$.

A *neighborhood search algorithm*, also called a *local search algorithm*, is an iterative algorithm that begins with a solution $S_{i-1}$ at the $i$th *iteration*, and moves to an improving solution $S_i \in \mathcal{N}(S_{i-1})$ if one exists. Otherwise, the algorithm terminates with $S_{i-1}$.

A *very large-scale neighborhood* is a neighborhood that has "very large" size with respect to the problem input. Typically, these neighborhoods are exponentially large, but this term is often used informally to describe neighborhoods that are too large to explicitly search in practice.

A *very large-scale neighborhood* (*VLSN*) *search algorithm* is a neighborhood search algorithm that searches a very large-scale neighborhood at each iteration.

## Facts:

**1.** Algorithm 1 describes a generic neighborhood search algorithm, in which the initial solution $S_0$ is input. The subroutine $\texttt{BestNeighbor}(S_{i-1}, \mathcal{N}, f)$ returns the best solution, according to the objective function $f$, from the neighborhood $\mathcal{N}(S_{i-1})$.

---

**Algorithm 1**:   **Generic neighborhood search.**

input: $S_0$

output: best solution found by neighborhood search

$i \leftarrow 0$

**repeat**

   $i \leftarrow i + 1$

   $S_i \leftarrow \texttt{BestNeighbor}(S_{i-1}, \mathcal{N}, f)$

**until** $f(S_i) \geq f(S_{i-1})$

**return** $S_{i-1}$

---

**2.** The term VLSN search was originally coined in the survey paper by Ahuja et al. [AhEtal02].

**3.** Surveys of VLSN search techniques can be found in [AhEtal02], [AhEtal07], and [PiRo10]. A survey of VLSN search techniques, designed specifically for the Traveling Salesman Problem, can be found in Deĭneko and Woeginger [DeWo00].

**4.** VLSN search algorithms have been used to solve a wide range of well-structured and complex real-world problems: vehicle routing problems [ErOrSt06], facility location problems [AhEtal04b], generalized knapsack problems [CuAh05], and parallel machine scheduling problems [FrNeSc04].

**5.** VLSN search algorithms have also been developed for many industrial applications: airline fleet assignment [AhEtal07a], [AhEtal04a], railroad freight blocking [AhJhLi07], intermodal load planning [NeJhAh09], and steel manufacturing [DaEtal07].

## Examples:

Although the focus is on VLSN search algorithms for NP-hard COPs, the next three illustrations of VLSN search are for polynomially solvable optimization problems.

**1.** *Column generation for linear programming:*   The simplex method (see §16.1) can be viewed as a neighborhood search algorithm. A basic feasible solution is determined from the set of basic columns $B$. A move (or pivot) exchanges one column from $B$ with a column from $N$, the set of nonbasic columns. The neighborhood of a basic feasible solution is the set of all basic feasible solutions that may be obtained by a single pivot. The simplex method can be interpreted as a VLSN search when the set of columns is too large to be represented explicitly and where column generation is used to determine the entering variable.

**2.** *Cycle canceling for the minimum cost flow problem:*   Consider the minimum cost flow problem (see §10.5) for a network $G = (V, A)$, where each vertex $v \in V$ has a supply/demand $b(v)$ and where the unit cost of sending flow on edge $(i, j)$ is $c_{ij}$. For a given feasible flow $x$, the neighborhood of $x$ consists of all feasible flows that can be obtained from $x$ by sending flow around a negative cost cycle in the residual network induced by $x$ [AhMaOr93].

**3.** *Augmenting path algorithm for the maximum cardinality matching problem:*   Consider the maximum cardinality matching problem (see §10.2) for an undirected graph

$G = (V, A)$. A matching is a collection of edges with no two edges incident with a common vertex. Relative to the path $P = [v_1, \ldots, v_{2k}]$ in $G$, let $A_1(P) = \{(v_{2j-1}, v_{2j}) \mid 1 \leq j \leq k\}$ and $A_2(P) = \{(v_{2j}, v_{2j+1}) \mid 1 \leq j \leq k-1\}$. The path $P$ is augmenting with respect to the matching $M$ if $M \cap A_1(P) = \emptyset$ and $A_2(P) \subseteq M$. To augment along the path $P$ is to replace $M$ by $(M \backslash A_2(P)) \cup A_1(P)$. The neighborhood of $M$ consists of all matchings that can be obtained from $M$ via augmenting paths. Berge [Be57] showed that if $M$ is not a maximum matching, then there is an augmenting path with respect to $M$. Edmonds [Ed65] developed the blossom algorithm (see §10.2.3) for determining an augmenting path if one exists. The blossom algorithm is widely considered to be one of the most elegant algorithms in all of combinatorial optimization.

## 16.8.2   VARIABLE-DEPTH NEIGHBORHOOD SEARCH ALGORITHMS

Neighborhoods based on exchanging two or three elements can often be explicitly searched in a reasonable amount of time. Exchanging a larger number of elements may lead to neighborhoods containing better local optimal solutions, but they typically cannot be searched efficiently. Variable-depth search algorithms are local search algorithms that (in principle) consider neighborhoods where a variable number of elements can be exchanged; to make the search practical, heuristics are used. There are variable-depth search algorithms for a large number of applications that attain near optimal solutions efficiently. Such algorithms are the focus of this subsection.

### Definitions:

Let $S_i$ and $S_{i+1}$ be solutions of a COP with ground set $E$. The move from solution $S_i$ to solution $S_{i+1}$ is a **k-exchange** if $|S_i \backslash S_{i+1}| = |S_{i+1} \backslash S_i| = k$ for a positive integer $k$. A neighborhood where $|S_i \backslash S_{i+1}| = |S_{i+1} \backslash S_i| \leq k$ for all $S_{i+1} \in \mathcal{N}(S_i)$ is a **k-exchange neighborhood** of $S_i$. A $k$-exchange neighborhood is said to have **depth $k$**.

A **variable-depth neighborhood search algorithm** begins with a solution $S_{i-1}$ at the $i$th iteration, and moves to an improving solution $S_i$ in the $|E|$-exchange neighborhood of $S_{i-1}$ if an improvement can be determined. Otherwise, the algorithm terminates with $S_{i-1}$. Typically, at each iteration, a variable-depth neighborhood search algorithm only partially explores the $k$-exchange neighborhood for each $k \in \{1, 2, \ldots, |E|\}$, thus avoiding the $\Omega(|E|^k)$ lower bound from explicitly searching the entire neighborhood.

An **ejection chain** is a variable-depth neighborhood search algorithm that iteratively generates a structured alternating sequence of additions/deletions $S_1, S_2, \ldots, S_k$ where

- $|S_1| = |S_j|$ for all odd values of $j$, and
- $|S_2| = |S_j|$ for all even values of $j$.

### Facts:

**1.** The first variable-depth neighborhood search algorithm was developed by Kernighan and Lin [KeLi70] for a set of graph partitioning problems.

**2.** Ejections chains were introduced by Glover [Gl96] for solving the TSP. Glover originally based ejection chains by extending and generalizing the ideas of Lin and Kernighan [LiKe73].

**3.** Many variable-depth search techniques, particularly those for vehicle routing problems, rely on ejection chains at each iteration in order to search the neighborhood.

## Examples:

**1.** A traveling salesman tour $[1, 2, 3, 4, 5, 1]$ is shown in part (a) of the following figure. Part (b) shows a 2-exchange of this tour in which edges $(2, 3)$ and $(4, 5)$ are replaced by edges $(2, 4)$ and $(3, 5)$, forming the new tour $[1, 2, 4, 3, 5, 1]$.

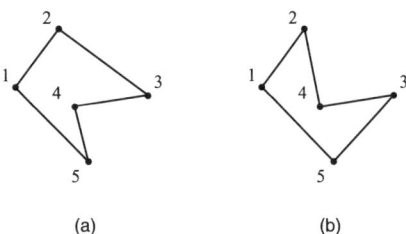

(a)                          (b)

**2.** In the context of the vehicle routing problem (see §10.7.2), the following figure shows in part (a) a feasible assignment of two vehicles that service nine customers from a central depot at location 0. Part (b) indicates the result of a 1-exchange in which customers 5 and 9 are interchanged, again resulting in a feasible solution.

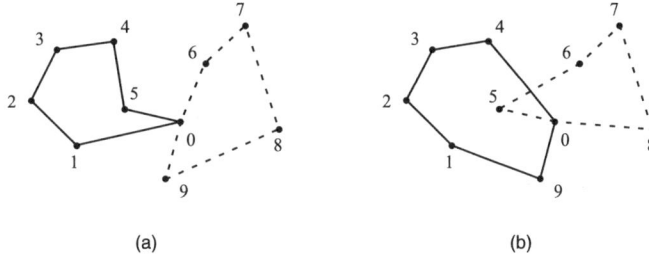

(a)                          (b)

**3.** The maximum cut problem (MAX CUT) is defined as follows. Given an undirected graph $G = (V, A)$ where each edge $e \in A$ has weight $w_e$, partition the vertices $V$ into two subsets $X$ and $Y$ to maximize the sum of the weights of the edges in the cut $(X, Y) = \{(i, j) \mid i \in X \text{ and } j \in Y\}$. Let $f(X, Y)$ denote the sum of the weights in the cut $(X, Y)$.

Algorithm 2 describes a generic variable-depth search algorithm for MAX CUT, given the initial solution $S_0$ as input. The $i$th iteration of the algorithm starts with a solution $S_{i-1}$ and either moves to a new solution $S_i$ where $f(S_i) > f(S_{i-1})$ or maintains the solution $S_{i-1}$ if no better solution is discovered within the $i$th iteration. Within each iteration of the variable-depth search, there are as many as $|V|$ rounds. Here $M_j$ denotes the set of vertices that may not be moved during round $j + 1$, and $(X_j, Y_j)$ denotes the cut constructed during round $j$.

The subroutine $\texttt{flip}(X_{j-1}, Y_{j-1}, v)$ returns a new cut $(X_j, Y_j)$, where vertex $v$ has been moved from the subset containing it to the other subset. For example, if $v \in X_{j-1}$, then $X_j = X_{j-1} \backslash v$ and $Y_j = Y_{j-1} \cup \{v\}$.

---

**Algorithm 2:** **Generic variable-depth search for MAX CUT**

input: $S_0$

output: best solution found by variable-depth search

$i \leftarrow 0$

**repeat**

  $i \leftarrow i + 1$

---

$$M_0 \leftarrow \emptyset$$
$$(X_0, Y_0) \leftarrow S_{i-1}$$
**for** $j = 1$ to $|V|$ **do**
$\qquad v_j^* \leftarrow \mathrm{argmax}_{v \in V \setminus M_{j-1}} \{f(\mathtt{flip}(X_{j-1}, Y_{j-1}, v))\}$
$\qquad M_j \leftarrow M_{j-1} \cup \{v_j^*\}$
$\qquad (X_j, Y_j) \leftarrow \mathtt{flip}(X_{j-1}, Y_{j-1}, v_j^*)$
**end for**
$\quad S_i \leftarrow \{(X_k, Y_k) \mid k = \mathrm{argmax}_{j \in \{1, \dots, |V|\}} f(X_j, Y_j)\}$
**until** $f(S_i) \leq f(S_{i-1})$
**return** $S_{i-1}$

4. Additional applications, with reference sources, are given in the following table.

| application | references |
|---|---|
| clustering | [DoPe94] |
| generalized assignment problem | [YaIbGl04], [YaYaIb99] |
| graph partitioning | [FiMa82], [KeLi70] |
| machine scheduling | [AgEtal07], [BrEtal11] |
| staff scheduling | [Do98] |
| traveling salesman problem | [He00], [JoMc97], [LiKe73] |
| traveling umpire problem | [TrYi07] |
| vehicle routing problem | [Gl96], [NaBr09], [Re98], [XuKe96] |

## 16.8.3  CYCLIC EXCHANGE NEIGHBORHOOD SEARCH ALGORITHMS

The cyclic exchange neighborhood was developed for solving minimum cost partition problems. Suppose that $S = \{S_1, S_2, \dots, S_K\}$ is a partition of $E$. Roughly speaking, a move in a cyclic exchange neighborhood permits one element to be transferred into $S_j$ and permits another element to be transferred out of $S_j$, for several of the subsets $S_j$ of the partition $S$. We discuss cyclic exchange neighborhoods and reference applications to a wide range of settings.

**Definitions:**
Let $E$ be the ground set for a COP. The collection of subsets $\{S_1, S_2, \dots, S_K\}$ is a **K-partition** of $E$ if no subset is empty, the sets are pairwise disjoint, and the union of the sets is $E$.

Suppose that $c$ is a function defined on the subsets of $E$; that is, $c(E')$ is the cost of subset $E'$ of $E$.

A **partitioning problem** is a COP where the objective is to find a partition $T = \{T_1, T_2, \dots, T_K\}$ of $E$ that minimizes $\sum_{j=1}^{K} c(T_j)$.

Let $S = \{S_1, S_2, \dots, S_K\}$ and $T = \{T_1, T_2, \dots, T_K\}$ be two $K$-partitions of $E$. We say that $T$ is a **1-neighbor** of $S$ if $|T_j \setminus S_j| \leq 1$ and if $|S_j \setminus T_j| \leq 1$ for each $j = 1, \dots, K$. If $T$ is a 1-neighbor of $S$, then $T$ is obtained from $S$ via a **1-transfer**.

Suppose that $S = \{S_1, S_2, \ldots, S_K\}$ is a $K$-partition and that $T = \{T_1, T_2, \ldots, T_K\}$ is obtained from $S$ via a 1-transfer. From $S$ and $T$, we can construct the **($S, T$)-transfer graph**, having vertex set $E$. For each $k = 1, \ldots, K$, if $T_k$ is obtained from $S_k$ by inserting element $i$ and deleting element $j$ (that is, $T_k = (S_k \backslash i) \cup \{j\}$), then $(i, j)$ is an edge of the ($S, T$)-transfer graph.

If the ($S, T$)-transfer graph consists of a single cycle, we say that $T$ is obtained from $S$ via a **cyclic exchange**. Specifically, if the cycle is $[i_1, \ldots, i_k, i_1]$, then $T$ is obtained from $S$ as follows: for each $j = 1, \ldots, k$, one inserts $i_j$ into the subset of $S$ containing $i_{j+1}$ and then deletes $i_{j+1}$ from that subset. (Here, $i_{k+1}$ is equivalent to $i_1$.) Similarly, if the ($S, T$)-transfer graph consists of a single path, we say that $T$ is obtained from $S$ via a **path exchange**. The **cyclic exchange neighborhood** (resp., **path exchange neighborhood**) of a $K$-partition $S$ consists of all $K$-partitions that can be obtained from $S$ by a single cyclic (resp., path) exchange.

To find an improving cyclic exchange starting with the $K$-partition $S = \{S_1, S_2, \ldots, S_K\}$ we create a new graph called the **improvement graph**. The vertex set of the improvement graph is $E$. For every pair $i, j$ of elements of $E$ that are in different subsets of the partition $S$, there is an edge $(i, j)$ in the improvement graph. Suppose $S_k$ is the part of $S$ that contains element $j$. Then the edge $(i, j)$ is interpreted as a transfer of element $i$ into $S_k$ and a transfer of element $j$ out of $S_k$. Let $S'_k = (S_k \backslash \{j\}) \cup \{i\}$ be the resulting subset. The cost of the edge $(i, j)$ in the improvement graph is $c_{ij} = c(S'_k) - c(S_k)$, which is the net change in cost of transferring element $i$ into $S_k$ and transferring element $j$ out.

A cycle $C = [i_1, \ldots, i_k, i_1]$ in the improvement graph is **subset-disjoint** with respect to a $K$-partition $S$ if no two elements of $C$ belong to the same subset of $S$.

**Facts:**

**1.** Cyclic exchange neighborhoods were introduced by Thompson and Orlin [ThOr89] and Thompson and Psaraftis [ThPs93].

**2.** Finding a minimum cost subset-disjoint cycle is NP-hard [ThOr89], but it is one of those NP-hard problems that is relatively easy to solve in practice. Thus, heuristics are used to search the improvement graph in practice. Effective heuristics for constructing negative cost subset-disjoint cycles are presented in Thompson and Psaraftis [ThPs93] and in Ahuja et al. [AhOrSh01].

**3.** The improvement graphs and the transfer graphs are closely related. Suppose that $C$ is a subset-disjoint cycle in the improvement graph for partition $S$, and that $T$ is the partition obtained by performing the cyclic exchange induced by $C$. Then the ($S, T$)-transfer graph consists of the cycle $C$.

**4.** *Symmetry*: if $T$ is a 1-neighbor of $S$, then $S$ is a 1-neighbor of $T$.

**5.** If $T$ is a 1-neighbor of $S$, then the ($S, T$)-transfer graph is a disjoint union of cycles and paths because each vertex of the graph has at most one incoming edge and at most one outgoing edge.

**6.** Suppose $S$ is a $K$-partition, and let $G$ be the resulting improvement graph. There is a one-to-one correspondence between cyclic exchanges from $S$ and subset-disjoint directed cycles in $G$.

**7.** Suppose $S$ is a $K$-partition, and let $G$ be the resulting improvement graph. The cost of a subset-disjoint cycle $C$ is the net change in objective value obtained after performing the induced cyclic transfer.

**8.** In practice, it is often much faster to obtain the current improvement graph by updating the improvement graph from the previous iteration rather than constructing a new improvement graph from scratch.

**9.** A path exchange may be transformed into a cyclic exchange by adding dummy vertices. This allows one to use a single unified search algorithm to identify improving cyclic exchanges and path exchanges.

**10.** For $K$-partitions, the cyclic exchange neighborhood is of size $O(n^K)$, where $n = |E|$. When $K$ is permitted to grow with $n$, the size of the neighborhood is typically exponential.

### Examples:

**1.** The following figure illustrates a cyclic exchange. Here, vertex $e_1$ will be transferred from subset $T_1$ to subset $T_2$. Vertex $e_4$ will be transferred from subset $T_2$ to subset $T_4$. Vertex $e_{10}$ will be transferred from subset $T_4$ to $T_5$. Finally, the cyclic exchange will be completed by transferring vertex $e_{13}$ from subset $T_5$ to subset $T_1$.

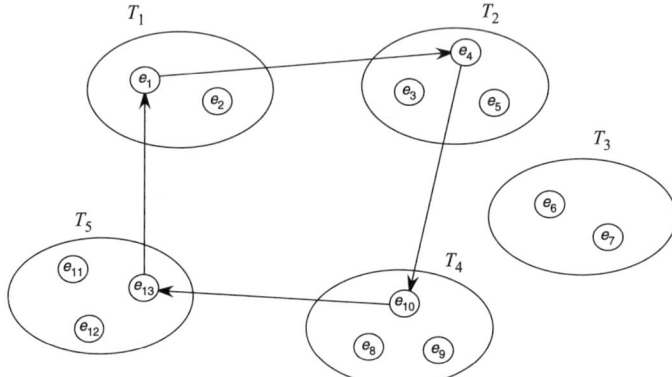

**2.** The following figure illustrates a cyclic exchange for a VRP [AgEtal03]. In general, a cyclic exchange for VRP transfers customer demands in a circular manner among several routes. This example illustrates a "3-cyclic, 2-exchange" scheme where two customers from each of three routes are transferred to the next route in the cycle. Here there are

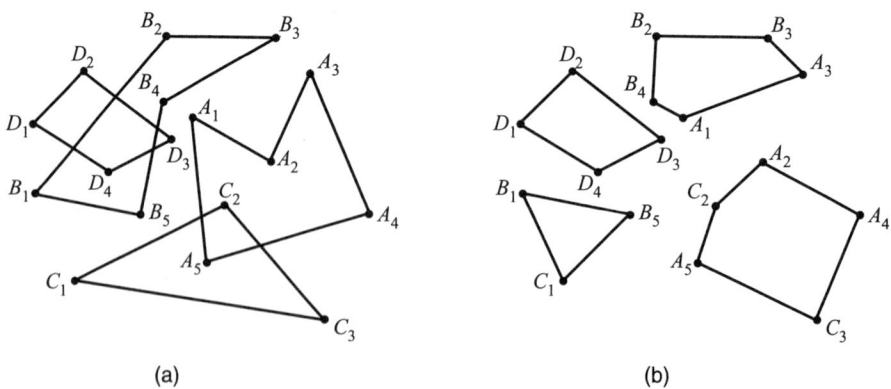

(a)                                        (b)

four routes $I_1, \ldots, I_4$ and the customers being served by each route are $I_1 = \{A_1, \ldots, A_5\}$, $I_2 = \{B_1, \ldots, B_5\}$, $I_3 = \{C_1, \ldots, C_3\}$, and $I_4 = \{D_1, \ldots, D_4\}$. This 3-cyclic 2-exchange involves only routes $I_1$, $I_2$, and $I_3$. Two customers $\{A_1, A_3\}$ from route $I_1$ are transferred to route $I_2$; two customers $\{B_1, B_5\}$ from route $I_2$ are moved to route $I_3$; two customers $\{C_2, C_3\}$ from $I_3$ are transferred to $I_1$; route $I_4$ is left unchanged. Let $S = \{I_1, \ldots, I_4\}$ and $S' = \{I'_1, \ldots, I'_4\}$ represent the set of routes before and after a cyclic exchange occurs, and let $f(I)$ represent the optimal cost of route $I$. The cost of this cyclic exchange is $\sum_{i=1}^{4}[f(I_i') - f(I_i)]$.

**3.** *Capacitated minimum spanning tree problem (CMST):* The CMST (see §10.6.1) is the problem of choosing a minimum weight spanning tree subject to the following constraint: if one deletes the root vertex and its incident edges, then each remaining connected component has at most $K$ vertices, where $K$ is part of the input. This problem models applications in which the goal is to develop a minimum cost computer network connecting $n$ computers to a single hub (or server), where bandwidth is limited. Ahuja et al. [AhOrSh01], [AhOrSh03] used cyclic exchanges to obtain solutions to a set of benchmark instances for the CMST. In each benchmark instance, they either obtained or improved upon the best previously reported solution.

**4.** *Facility location problems:* In facility location problems (see §16.2), the goal is to determine the location of a collection of facilities, and also to assign customers to facilities, to minimize the sum of the distances from customers to facilities. Ahuja et al. [AhEtal04b] developed a cyclic exchange VLSN search algorithm for a capacitated facility location problem. Scaparra et al. [ScPaSc04] created an algorithm based on the cyclic exchange neighborhood for solving the closely related capacitated vertex $p$-center problem. In both cases, the algorithms based on cyclic exchanges produced nearly optimal solutions and were fast on benchmark instances.

**5.** Additional applications, with reference sources, are given in the following table.

| application | references |
|---|---|
| generalized assignment problem | [YaEtal04] |
| intermodal load planning | [NeJhAh09] |
| inventory routing | [SiEtal05] |
| machine scheduling | [FrNeSc04] |
| multiple knapsack problems | [CuAh05] |
| network design | [AmScSc09] |
| production planning | [AhEtal07b] |
| quadratic assignment problem | [AhEtal07c] |
| timetabling | [AbEtal06] |
| vehicle routing problem | [AgEtal03], [ThPs93] |
| weapons target assignment | [AhEtal08] |

### 16.8.4   OTHER VLSN SEARCH ALGORITHMS

In designing very large-scale neighborhood search algorithms, researchers have drawn inspiration and techniques from more than 50 years of research and algorithmic development in mathematical programming and computer science algorithms. In this subsection, we describe a diverse collection of approaches for developing very large-scale neighborhood search algorithms.

**Examples:**

**1.** *Neighborhoods based on combining independent moves:* Compounded independent move (CIM) neighborhoods are neighborhoods based on simultaneously executing two or more moves whose effects on the objective value are independent of one another; that is, the change in the objective value due to the compound move is guaranteed to be the sum of the changes due to the individual moves. CIM neighborhoods were introduced by Congram et al. [CoPoVe02] for a machine scheduling problem and by Potts and van de Velde [PoVe95] for the TSP. Their techniques searched a CIM neighborhood with dynamic programming, and they referred to their approach as *dynasearch*. Ergun et al. [ErOrSt06] and Gendreau et al. [GeEtal06a] searched CIM neighborhoods to solve vehicle routing problems. Taillard [Ta03] used CIM neighborhoods for solving clustering problems. Dror and Levy [DrLe86] used CIM neighborhoods for solving the inventory routing problem.

**2.** *Neighborhoods based on context-free grammars:*   Bompadre and Orlin [BoOr05] showed how to use context-free grammars (CFGs) to generate exponentially large neighborhoods for sequencing problems. Moreover, for some sequencing problems such as the TSP and the linear ordering problem (see §16.9.1), they developed an algorithm that takes the grammar and the problem data as input and then finds the minimum cost neighbor in polynomial time. In addition to presenting an innovative use of CFGs, this paper demonstrates how a single algorithm can be used to search a wide range of different exponentially sized neighborhoods.

**3.** *Neighborhoods based on mathematical programming:* Suppose that $x$ is a solution to a mathematical programming problem. For a given subset $I$ of indices, we say that another feasible solution $y$ is $I$-adjacent to $x$ if $y_i = x_i$ for $i \in I$, and we say that $y$ is in the $I$-neighborhood of $x$. $I$-neighborhoods are conveniently searched using mathematical programming software. This neighborhood can be used in constraint programming (e.g., Pesant and Gendreau [PeGe99] and Davenport et al. [DaEtal07]) as well as in integer programming (e.g., Danna et al. [DaRoLe05]). For example, the relaxation induced neighborhood search (RINS) algorithm of Danna et al. has been implemented in the CPLEX Mixed Integer Programming (MIP) solver.

**4.** *Neighborhoods based on polynomially solvable special cases:* Researchers have developed a substantial literature on polynomial-time algorithms for special cases of NP-hard COPs. Many of the special cases are neighborhood-inducing, and lead to very large-scale neighborhoods that can be searched in polynomial time. For example, Carlier and Villon [CaVi90] presented a polynomial-time algorithm for finding an optimal *pyramidal tour*, that is, a sequence $i_1, \ldots, i_n$ such that for some value $k$, the indices in the subsequence $i_1, \ldots, i_k$ are in increasing order, and the indices in the subsequence $i_k, \ldots, i_n$ are in decreasing order. For a given tour $T$, its pyramidal neighborhood is the set of pyramidal tours that would result if one first relabeled the vertices (or cities) so that $T = [1, \ldots, n]$. Therefore, any algorithm for finding a minimum cost pyramidal tour can be used for finding a minimum cost pyramidal neighbor.

As another example, for a given fixed value $K$, there is a polynomial-time dynamic program for finding an optimal tour $[i_1, \ldots, i_n]$ subject to the constraint that $i_j < j + K$ for each $j$. Simonetti and Balas [SiBa96] developed a very large-scale neighborhood that uses this dynamic program as the search engine.

**5.** *Neighborhoods that exploit local structure:* Ahuja et al. [AhJhLi07] created a neighborhood tailored to a railroad blocking problem. Blocking problems, which also arise in parcel delivery, trucking and air transportation, model how to consolidate a large number of shipments into *blocks* to reduce the total handling costs. The VLSN search algorithm devised by Ahuja et al. iteratively improves solutions by permitting changes to the blocking at one rail station at a time.

# 16.9   TABU SEARCH

Tabu search is a metaheuristic solution methodology applicable to general optimization problems. It is distinguished by its use of adaptive (flexible) memory and responsive exploration of the solution space. In this way the search can learn to exploit specific areas of the solution space as well as explore new terrain.

## 16.9.1   BASIC CONCEPTS

Tabu search starts from an initial solution and explores the solution space by making transformations to move from one solution to the next. The use of flexible memory promotes good move combinations and solution attributes by modifying the neighborhood searched at each step.

### Definitions:

**Tabu search** (**TS**) is a search technique that proceeds iteratively from one (feasible) **solution** to another until a chosen termination criterion is satisfied. It can be applied to **optimization problems** of the form: minimize $\{f(x) \mid x \in X\}$, where the set $X$ summarizes the constraints on the decision variables $x$.

Each solution $x \in X$ has an associated **neighborhood** $N(x) \subseteq X$. The solution $x' \in N(x)$ is reached from $x$ by a **move** operation; in this case, $x'$ is a **neighbor** of $x$.

The **move value** $v$ is the change in the objective function value $f$ from the current solution to the solution after the move: i.e., $v = f(x') - f(x)$. An **improving move** has $v < 0$ while a **nonimproving move** has $v \geq 0$. A **best neighbor** of $x$ (according to the objective function value) is a neighbor $x' \in N(x)$ having the smallest value $f(x')$ over all neighbors of $x$.

**Recency-based** (**short-term**) **memory** is used in TS to obtain a modified neighborhood $N^*(x) \subseteq N(x)$ by eliminating (making tabu) recently visited solutions.

### Facts:

**1.** Tabu search was first proposed by Fred Glover [Gl86]. The most cited reference is the book *Tabu Search* by Fred Glover and Manuel Laguna [GlLa97].

**2.** The word *tabu* (or *taboo*) comes from the Polynesian language Tongan, in which it indicates objects that cannot be touched because they are sacred.

**3.** Whereas the steepest descent procedure only makes improving moves, TS allows the choice of a best neighbor $x'$ of $x$ even when the move is nonimproving.

**4.** By using short-term strategies, it is possible that a solution $x$ can be visited more than once, but it is likely that the corresponding reduced neighborhood $N^*(x)$ will be different each time.

**5.** TS memory structures are based on recording *attributes* as well as the most recent history of the search trajectory (recency-based memory).

**6.** Selected attributes that occur in solutions recently visited are *tabu-active*, and solutions that contain tabu-active elements, or particular combinations of these attributes, are classified as *tabu*. This prevents certain solutions from the recent past from belonging to $N^*(x)$ and hence from being revisited.

**7.** The number of iterations that an attribute remains tabu-active (the *tabu tenure*) can be static (fixed throughout the search) or dynamic (vary either systematically or probabilistically).

**8.** *Aspiration criteria* can be used to override tabu activation rules. The most common aspiration criterion consists of removing a tabu classification from a trial move when the move yields a solution better than the best obtained so far.

**9.** The overall structure of tabu search, employing short-term memory, is displayed in Algorithm 1. The solutions $x^{\text{next}}$ chosen during the Intensification Phase are either not classified tabu (as shown in Algorithm 1) or their tabu status has been overridden by the aspiration criteria (not shown in Algorithm 1).

---

**Algorithm 1: Short-term memory tabu search.**

input: initial solution $x^0 \in X$

output: best solution $x^{\text{best}}$ found during the search

$x^{\text{now}} = x^{\text{best}} = x^0$

**repeat** Intensification Phase

    choose $x^{\text{next}} \in N^*(x^{\text{now}})$

    $x^{\text{now}} \leftarrow x^{\text{next}}$

    if $f(x^{\text{now}}) < f(x^{\text{best}})$ then $x^{\text{best}} \leftarrow x^{\text{now}}$

    update recency-based memory and tabu status

**until** termination criteria satisfied

---

### Examples:

**1.** *Linear ordering problem (LOP):* Given a matrix of weights, the LOP consists of finding a permutation $x$ of the columns (and rows) of the matrix in order to maximize the sum of the weights in the upper triangle. Let $x = (1, 2, 3, 4)$ be a solution to a LOP of size 4, defined by the following table.

|   | 1 | 2 | 3 | 4 |
|---|---|---|---|---|
| 1 | 0 | 12 | −5 | 3 |
| 2 | 6 | 0 | 3 | 4 |
| 3 | −8 | 7 | 0 | −2 |
| 4 | 4 | 0 | 3 | 0 |

The objective function value $f(x) = 12 - 5 + 3 + 3 + 4 - 2 = 15$. If a move is defined as the swap of two elements in the permutation, where each element represents the index of the same row and column, then the swap $(1, 2)$ results in the following table.

|   | 2 | 1 | 3 | 4 |
|---|---|---|---|---|
| 2 | 0 | 6 | 3 | 4 |
| 1 | 12 | 0 | −5 | 3 |
| 3 | 7 | −8 | 0 | −2 |
| 4 | 0 | 4 | 3 | 0 |

The neighborhood of $x$ consists of the following solutions $x'$.

| move | $x'$ | $f(x')$ |
|------|------|---------|
| $(1,2)$ | $(2,1,3,4)$ | 9 |
| $(1,3)$ | $(3,2,1,4)$ | 10 |
| $(1,4)$ | $(4,2,3,1)$ | 8 |
| $(2,3)$ | $(1,3,2,4)$ | 19 |
| $(2,4)$ | $(1,4,3,2)$ | 20 |
| $(3,4)$ | $(1,2,4,3)$ | 20 |

**2.** *Knapsack problem:* Here we seek to maximize the overall profit obtained by placing a subset of items in a knapsack with total weight capacity 100. Each item has a known profit and weight. A formulation in terms of 0-1 variables $x_j$ is

$$\max \;\; 67x_1 + 500x_2 + 98x_3 + 200x_4 + 120x_5 + 312x_6 + 100x_7 + 200x_8 + 180x_9 + 100x_{10}$$
$$\text{s.t.} \;\; 5x_1 + 45x_2 + 9x_3 + 19x_4 + 12x_5 + 32x_6 + 11x_7 + 23x_8 + 21x_9 + 14x_{10} \le 100$$
$$x_j \in \{0,1\}$$

An initial solution $x^0$ can be found using a greedy heuristic: namely, successively select items in decreasing order of their bang-for-buck ratio (profit/weight), continuing until adding one more item exceeds the capacity of the knapsack. The following table shows these ratios; notice that the variables $x_1, \ldots, x_{10}$ are already indexed by decreasing ratios. This greedy heuristic successively chooses items $1, 2, 3, 4, 5$ for the initial solution, giving $x^0 = (1,1,1,1,1,0,0,0,0,0)$ with total profit 985 and total weight 90.

| item | 1 | 2 | 3 | 4 | 5 | 6 | 7 | 8 | 9 | 10 |
|------|-----|-----|------|------|------|------|------|------|------|------|
| profit | 67 | 500 | 98 | 200 | 120 | 312 | 100 | 200 | 180 | 100 |
| weight | 5 | 45 | 9 | 19 | 12 | 32 | 11 | 23 | 21 | 14 |
| ratio | 13.40 | 11.11 | 10.89 | 10.53 | 10.00 | 9.75 | 9.09 | 8.70 | 8.57 | 7.14 |

**3.** We continue with the previous example. Suppose that a move is defined as a "flip" move: i.e., the value of a single variable $x_j$ is changed from either zero to one or from one to zero. The neighborhood $N(x^0)$ of the current solution $x^0$ is then shown in the following table. The *move* column indicates the new value of the flipped variable, while the *neighbor* column shows the indices of the variables set to 1 as a result of the move.

| $j$ | move | neighbor | profit | weight |
|-----|------|----------|--------|--------|
| 1 | $x_1 = 0$ | $\{2,3,4,5\}$ | 918 | 85 |
| 2 | $x_2 = 0$ | $\{1,3,4,5\}$ | 485 | 45 |
| 3 | $x_3 = 0$ | $\{1,2,4,5\}$ | 887 | 81 |
| 4 | $x_4 = 0$ | $\{1,2,3,5\}$ | 785 | 71 |
| 5 | $x_5 = 0$ | $\{1,2,3,4\}$ | 865 | 78 |
| 6 | $x_6 = 1$ | $\{1,2,3,4,5,6\}$ | 1297 | 122 |
| 7 | $x_7 = 1$ | $\{1,2,3,4,5,7\}$ | 1085 | 101 |
| 8 | $x_8 = 1$ | $\{1,2,3,4,5,8\}$ | 1185 | 113 |
| 9 | $x_9 = 1$ | $\{1,2,3,4,5,9\}$ | 1165 | 111 |
| 10 | $x_{10} = 1$ | $\{1,2,3,4,5,10\}$ | 1085 | 104 |

Suppose the search is not allowed to visit infeasible solutions. Then there is no improving move in the neighborhood of the current solution (recall that we have a maximization problem, not a minimization problem). If we adopt the best-improving strategy, then the first move in the table, changing the value of $x_1$ to zero, is the best one in $N(x^0)$, giving $x^{\text{next}} = (0, 1, 1, 1, 1, 0, 0, 0, 0, 0)$ with a profit of 918.

**4.** We show how short-term memory can be incorporated into the previous 0-1 knapsack example. Let the attribute of interest be the index of the variable chosen to change values. Any move that includes such an index is classified as tabu, and we set the tabu tenure to be three iterations. Using this memory structure, we determine the next solution $x^{\text{next}}$. Subject to the constraint that $x_1 = 0$ and considering only a single "flip" move, a best neighbor of $x^{\text{now}} = (0, 1, 1, 1, 1, 0, 0, 0, 0, 0)$ turns out to be $x^{\text{next}} = (0, 1, 1, 1, 1, 0, 1, 0, 0, 0)$, achieved by changing the value of $x_7$. The corresponding profit is 1018 with a weight of 96. The following table shows shows ten iterations of the procedure using this short-term memory structure. The *tabu active* column lists the indices of the tabu active variables; each index remains tabu for three iterations. The *move* column indicates the flipped variable, while the *solution* column shows the solution after the move has been executed. The resulting profit and weight are shown in the last two columns. The best solution found after ten iterations occurs in the second iteration and has a profit of 1018.

| iteration | tabu active | move | solution | profit | weight |
|:---:|:---:|:---:|:---:|:---:|:---:|
| 0 | | | $\{1, 2, 3, 4, 5\}$ | 985 | 90 |
| 1 | | 1 | $\{2, 3, 4, 5\}$ | 918 | 85 |
| 2 | 1 | 7 | $\{2, 3, 4, 5, 7\}$ | 1018 | 96 |
| 3 | 7 1 | 3 | $\{2, 4, 5, 7\}$ | 920 | 87 |
| 4 | 3 7 1 | 5 | $\{2, 4, 7\}$ | 800 | 75 |
| 5 | 5 3 7 | 8 | $\{2, 4, 7, 8\}$ | 1000 | 98 |
| 6 | 8 5 3 | 7 | $\{2, 4, 8\}$ | 900 | 87 |
| 7 | 7 8 5 | 3 | $\{2, 3, 4, 8\}$ | 998 | 96 |
| 8 | 3 7 8 | 4 | $\{2, 3, 8\}$ | 798 | 77 |
| 9 | 4 3 7 | 9 | $\{2, 3, 8, 9\}$ | 978 | 98 |
| 10 | 9 4 3 | 8 | $\{2, 3, 9\}$ | 778 | 75 |

**5.** An alternative neighborhood definition for the 0-1 knapsack problem could be one in which the values of two variables are flipped. Instead of 10 neighbors, this move generates 45 neighbors, corresponding to all combinations of two variables. A potential short-term memory structure for this neighborhood could be as follows.

- attributes: indices of the variables chosen to change values
- tabu move: one that includes at least one tabu active variable
- tabu tenure: three iterations for a variable flipping from 1 to 0, and two iterations for a variable flipping from 0 to 1

The tabu tenure recognizes that in a typical knapsack problem there will be more variables outside the knapsack (i.e., with their values set to zero) than those inside the knapsack (i.e., with their values set to one). Therefore, if a variable leaves the knapsack, the tabu tenure keeps this item outside longer than when an item enters the knapsack.

**6.** The LOP in Example 1 provides a few choices for a short-term memory design in a tabu search that employs swap moves (i.e., exchanges of the positions of two rows/columns). One of those designs is as follows.

- attributes: indices of the elements (rows/columns) chosen to exchange positions
- tabu move: one that includes at least one tabu active variable
- tabu tenure: two iterations for the element exchanging to the lower position, and three iterations for the element moving to the higher position in the permutation

This tabu tenure structure identifies that elements moving to the earlier position in the permutation make a larger contribution to the objective function value. Arguably, these elements have moved to their "preferred" position and therefore there is no need to force them to stay there.

**7.** Additional applications, with reference sources, are given in the following table.

| application | references |
|---|---|
| maritime transportation | [KoFa10], [KoFaLa10] |
| chemical industry | [LiEtal05], [LiMi04a], [LiMi04b] |
| satellite range scheduling | [ZuAmGi08] |
| conservation area network design | [CiBaSa10] |
| high level synthesis | [SeSiRo11] |
| real-world routing | [ArHeSp06], [GeEtal06b], [GeEtal08], [OpLø08], [PaEtal07] |
| forestry | [CaEtal03] |
| scheduling | [AlEtal04], [HoLe10] |
| DNA sequencing | [BlEtal00], [BlEtal04], [BlGlKa04] |
| logistics | [ScSø09], [Su12] |

## 16.9.2  LONG-TERM MEMORY

Long-term memory plays an important role in creating the right balance between intensification and diversification of a tabu search. Intensification strategies are based on modifying choice rules to encourage move combinations and solution features historically found good. They may also initiate a return to attractive regions to search them more thoroughly. Diversification, on the other hand, encourages the search process to examine unvisited regions and the generation of solutions that differ in significant ways from those previously seen.

**Definitions:**

***Frequency-based (long-term) memory*** is used in TS to obtain a modified neighborhood $N^*(x) \supseteq N(x)$ in order to expand the neighborhood and examine unvisited regions.

***Frequencies*** consist of ratios, whose numerators represent either transition counts or residence counts. A ***transition count*** is the number of iterations where an attribute changes (enters or leaves) the solutions visited. A ***residence count*** is the number of iterations where an attribute belongs to the solutions visited. The denominators generally represent one of three types of quantities: (1) the total number of occurrences of all events represented by the numerators (such as the total number of iterations), (2) the sum (or average) of the numerators, or (3) the maximum numerator value.

**Facts:**

**1.** The use of long-term memory does not require long solution runs before its benefits become visible.

**2.** The chance of finding still better solutions as time grows—in the case where an optimal solution has not already been found—is enhanced by using long-term TS memory in addition to short-term memory.

**3.** Residence frequencies and transition frequencies sometimes convey related information, but in general they carry different implications.

**4.** A high residence frequency may indicate that an attribute is highly attractive if the domain consists of high quality solutions, or may indicate the opposite if the domain consists of low quality solutions.

**5.** Frequency-based memory is used to define penalty and incentive values to modify the evaluation of moves and therefore determine which moves are selected. In a minimization problem, the modified move value function has the following mathematical form:

$$v' = v(1 + pq).$$

Here $v$ is the original move value, $p$ is the penalty factor, $q$ is the frequency ratio, and $v'$ is the modified move value. When $p = 0$ the original move value is not modified. A value $p > 0$ penalizes the move and $v'$ makes it less attractive. A value $p < 0$ provides an incentive and $v'$ makes the move more attractive.

**6.** Penalty/incentive functions based on frequency ratios are also used in restarting procedures that employ greedy choices. Penalty/incentive functions bias the selections and induce diversification.

**7.** Algorithm 2 shows how a restarting procedure based on long-term memory may be superimposed on the short-term memory tabu search in Algorithm 1.

---

**Algorithm 2**:  **Tabu search with long-term memory.**
input: initial solution $x^0 \in X$
output: best solution $x^{\text{best}}$ found during the search
$x^{\text{now}} = x^{\text{best}} = x^0$
**repeat** Diversification Phase
   reset recency-based memory
   if $x^{\text{now}} = \emptyset$ then $x^{\text{now}} \leftarrow$ diversified starting solution
   **repeat** Intensification Phase
      choose $x^{\text{next}} \in N^*(x^{\text{now}})$
      $x^{\text{now}} \leftarrow x^{\text{next}}$
      if $f(x^{\text{now}}) < f(x^{\text{best}})$ then $x^{\text{best}} \leftarrow x^{\text{now}}$
      update recency-based memory and tabu status
      update frequency-based memory
   **until** termination criteria satisfied
**until** termination criteria satisfied

---

**Example:**

**1.** Suppose that in the knapsack problem described in §16.9.1, Example 2 we define the residency ratio $q$ as the number of times an item was in the knapsack divided by the current number of visited solutions. We then restart the search by erasing the short-term memory and generating a solution with the following function:

$$r' = r(1 - q),$$

where $r$ is the original bang-for-buck ratio. This function reduces the attractiveness of elements with large $q$ values. The new starting solution is built by giving priority to the elements with large $r'$ and thus inducing diversification.

## 16.9.3  STRATEGIC OSCILLATION

Strategic oscillation provides a means to achieve an effective interplay between intensification and diversification over the intermediate to long term. Strategic oscillation operates by orienting moves in relation to the feasibility boundary, which represents a point where a search would normally stop. Instead of stopping when this boundary is reached, however, the rules for selecting moves are modified to permit the region defined by the feasibility boundary to be crossed. The approach then proceeds for a specified depth beyond the oscillation boundary, and turns around. The oscillation boundary again is approached and crossed, this time from the opposite direction, and the method proceeds to a new turning point.

### Definitions:

The process of repeatedly approaching and crossing the feasibility boundary from different directions creates an oscillatory behavior, and so characterizes the **strategic oscillation** method.

The implementation of strategic oscillation entails the selection of a proximity measure and the oscillation amplitude. The **proximity measure** assigns a numerical value to moving toward or away from the feasibility boundary. Proximity measures are defined in reference to the problem constraints. The **oscillation amplitude** is controlled by a specified number of moves $m$. If a move is made that causes the search to cross the feasibility boundary, then the search is allowed to make $m - 1$ additional moves in the same direction (i.e., moving away from the feasibility boundary) before turning around.

### Facts:

**1.** In combinatorial optimization problems that require the selection of $k$ elements, the rule to delete elements from the solution will typically be different in character from the one used for adding elements. In other words, one rule is not simply the inverse of the other.

**2.** Rule differences are features of strategic oscillation that provide an enhanced heuristic vitality. The application of different rules may be accompanied by crossing a boundary to different depths on different sides. An option is to approach and retreat from the boundary while remaining on a single side, without crossing (i.e., electing a crossing of "zero depth").

**3.** In both one-sided and two-sided oscillation approaches it is frequently important to spend additional search time in regions close to the feasibility boundary, and especially to spend time at the boundary itself.

**4.** In problems for which feasibility is determined by a set of constraints, vector-valued functions can be used to control the oscillation. In this case, controlling the search by bounding this function can be viewed as manipulating a parameterization of the constraint set.

**Examples:**

**1.** In a combinatorial optimization problem consisting of selecting $k$ out of a given number of elements, a proximity measure $t$ may be the difference between the number of elements that have been selected and $k$. Feasible solutions are those for which $t = 0$. Solutions with $t > 0$ have too many elements and solutions with $t < 0$ have too few. Moves can be designed to keep the search at $t = 0$ or to create an oscillation pattern around $t = 0$.

**2.** The feasibility boundary in the knapsack problem is defined by the capacity of the knapsack, which is 100 in §16.9.1, Example 2. To create simple oscillation around the feasible boundary, the search is allowed to cross to the infeasible region and $m$ is set to 1. The proximity measure $t$ is defined as the difference between the total weight of the items in the knapsack and the knapsack capacity. Therefore, $t > 0$ indicates an infeasible solution while $t \leq 0$ corresponds to a feasible solution. The procedure operates with the following rules:

- When approaching the feasibility boundary from the feasible region, select the best non-tabu feasible move. If no feasible move is available, then select the non-tabu move that improves profit the most.

- When approaching the feasibility boundary from the infeasible region, select the non-tabu move that removes the variable with the smallest bang-for-buck ratio.

The short-term memory structure is the same as the one defined in §16.9.1. Therefore, the move selections have to be done by considering the tabu classifications. The following table shows 10 iterations of TS with strategic oscillation. The initial solution is given by the bang-for-buck heuristic.

| iteration | tabu active | move | solution | profit | weight |
|:---:|:---:|:---:|:---:|:---:|:---:|
| 0 | | | $\{1, 2, 3, 4, 5\}$ | 985 | 90 |
| 1 | | 6 | $\{1, 2, 3, 4, 5, 6\}$ | 1297 | 122 |
| 2 | 6 | 5 | $\{1, 2, 3, 4, 6\}$ | 1177 | 110 |
| 3 | 5 6 | 4 | $\{1, 2, 3, 6\}$ | 977 | 91 |
| 4 | 4 5 6 | 8 | $\{1, 2, 3, 6, 8\}$ | 1177 | 114 |
| 5 | 8 4 5 | 6 | $\{1, 2, 3, 8\}$ | 865 | 82 |
| 6 | 6 8 4 | 9 | $\{1, 2, 3, 8, 9\}$ | 1045 | 103 |
| 7 | 9 6 8 | 3 | $\{1, 2, 8, 9\}$ | 947 | 94 |
| 8 | 3 9 6 | 4 | $\{1, 2, 4, 8, 9\}$ | 1147 | 113 |
| 9 | 4 3 9 | 8 | $\{1, 2, 4, 9\}$ | 947 | 90 |
| 10 | 8 4 3 | 3 | $\{1, 2, 3, 4, 9\}$ | 1045 | 99 |

The oscillation can be observed in the *weight* column. It is noteworthy to discuss the move made in the last iteration. The move is selected even though the move is classified tabu (note that 3 appears in the *tabu active* list). The initial solution is the best-known feasible solution at that point and it has an objective function value of 985. Therefore, an aspiration criterion that overrides the tabu status of a move can be invoked (see §16.9.1, Fact 8). The criterion states that the tabu classification can be overridden if the move leads to a solution that is better than the best solution found so far.

## 16.9.4   PATH RELINKING

A useful integration of intensification and diversification strategies occurs in the approach called path relinking. This approach generates new solutions by exploring trajectories that connect reference solutions—by starting from one of these solutions, called an initiating solution, and generating a path in the neighborhood space that leads toward the other solutions, called guiding solutions. This is accomplished by selecting moves that introduce attributes contained in the guiding solutions.

### Definition:

**Path relinking** is a strategy that seeks to incorporate attributes of high quality solutions, by creating inducements to favor these attributes in the moves selected. However, instead of using an inducement that merely encourages the inclusion of such attributes, the path relinking approach subordinates all other considerations to the goal of choosing moves that introduce the attributes of the guiding solution, in order to create a "good attribute composition" in the current solution.

### Facts:

**1.**  At each step, path relinking chooses the best move, according to the change in the objective function value, from the restricted set of moves that incorporate a maximum number of the attributes in the guiding solutions.

**2.**  Path relinking can be used to generate intensification strategies by choosing reference solutions that lie in a common region or that share common features.

**3.**  Diversification strategies based on path relinking characteristically select reference solutions that come from different regions or that exhibit contrasting features.

**4.**  Membership in the reference set is determined by setting a threshold that is connected to the objective function value of the best solution found during the search.

**5.**  The initiating solution can be used to give a beginning partial construction, by specifying particular attributes as a basis for the remaining constructive steps. Constructive neighborhoods can be viewed as a feasibility-restoring mechanism since a null or partially constructed solution does not satisfy all conditions to qualify as feasible.

**6.**  Path relinking can make use of destructive neighborhoods, where an initial solution is "overloaded" with attributes donated by the guiding solutions, and such attributes are progressively stripped away or modified until reaching a set with an appropriate composition.

**7.**  Path relinking consists of the following steps:

- Identify the neighborhood structure and associated solution attributes for path relinking (possibly different from those of other TS strategies applied to the problem).

- Select a collection of two or more reference solutions, and identify which members will serve as the initiating solution and the guiding solution(s). (Reference solutions can be infeasible, such as "incomplete" or "overloaded" solution components treated by constructive or destructive neighborhoods.)

- Move from the initiating solution toward the guiding solution(s), generating one or more intermediate solutions.

**Example:**

1. Consider the following two reference solutions for the knapsack problem in §16.9.1, Example 2.

| solution | 1 | 2 | 3 | 4 | 5 | 6 | 7 | 8 | 9 | 10 | profit | weight |
|---|---|---|---|---|---|---|---|---|---|---|---|---|
| 1 | 0 | 1 | 1 | 0 | 1 | 0 | 1 | 0 | 1 | 0 | 998 | 98 |
| 2 | 1 | 1 | 1 | 1 | 1 | 0 | 0 | 0 | 0 | 0 | 985 | 90 |

These solutions were found with a TS that used simple moves consisting of flipping one variable at a time. Both of these solutions are feasible. Suppose that solution 2 is chosen as the initiating solution and that solution 1 is the guiding solution. The relinking process starts by identifying the values that are common to both solutions. These correspond to variables 2, 3, and 5. Swap moves are used during the path relinking process in order to keep the intermediate solutions close to the feasibility boundary. A swap in this context corresponds to adding one item to the knapsack while taking another item out of the knapsack. This means that a variable that is set to 1 in the initiating solution and to 0 in the guiding solution must be paired with a variable that is set to 0 in the initiating solution and to 1 in the guiding solution. The pairs in our example are $(1,7), (1,9), (4,7)$, and $(4,9)$. The next table shows the intermediate solutions that result from these swaps.

| swap | 1 | 2 | 3 | 4 | 5 | 6 | 7 | 8 | 9 | 10 | profit | weight |
|---|---|---|---|---|---|---|---|---|---|---|---|---|
| $(1,7)$ | 0 | 1 | 1 | 1 | 1 | 0 | 1 | 0 | 0 | 0 | 1018 | 96 |
| $(1,9)$ | 0 | 1 | 1 | 1 | 1 | 0 | 0 | 0 | 1 | 0 | 1098 | 106 |
| $(4,7)$ | 1 | 1 | 1 | 0 | 1 | 0 | 1 | 0 | 0 | 0 | 885 | 82 |
| $(4,9)$ | 1 | 1 | 1 | 0 | 1 | 0 | 0 | 0 | 1 | 0 | 965 | 92 |

The best feasible swap is $(1,7)$ and the search moves to the intermediate solution with profit 1018 and weight 96. After the move, there is only one swap left to reach the guiding solution, namely $(4,9)$. In this case, the process produced only one intermediate solution; however, in general, path relinking visits more solutions during the transformation of the initiating solution into the guiding solution.

# REFERENCES

*Printed Resources*:

[AbEtal06] S. Abdullah, S. Ahmadi, E. K. Burke, and M. Dror, "Investigating Ahuja-Orlin's large neighborhood search approach for examination timetabling", *OR Spectrum* 29 (2006), 351–372.

[AdGu05] G. Adomavicius and A. Gupta, "Toward comprehensive real-time bidder support in iterative combinatorial auctions", *Information Systems Research* 16 (2005), 169–185.

[AgEtal03] R. Agarwal, R. K. Ahuja, G. Laporte, and Z. J. Shen, "A composite very-large-scale neighborhood search algorithm for the vehicle routing problem", in *Handbook of Scheduling: Algorithms, Models and Performance Analysis*, J. Y-T. Leung (ed.), CRC Press, 2003, Chapter 49.

[AgEtal07] R. Agarwal, Ö. Ergun, J. B. Orlin, and C. N. Potts, "Solving parallel machine scheduling problems with very large-scale neighborhood search", working paper, Industrial and Systems Engineering School, Georgia Institute of Technology, 2007.

[AhEtal02] R. K. Ahuja, Ö. Ergun, J. B. Orlin, and A. P. Punnen, "A survey of very large-scale neighborhood search techniques", *Discrete Applied Mathematics* 123 (2002), 75–102.

[AhEtal07] R. K. Ahuja, Ö. Ergun, J. B. Orlin, and A. P. Punnen, "Very large-scale neighborhood search: theory, algorithms, and applications", in *Approximation Algorithms and Metaheuristics*, T. F. Gonzalez (ed.), Chapman & Hall/CRC, 2007, Chapter 20.

[AhEtal04a] R. K. Ahuja, J. Goodstein, J. Liu, A. Mukherjee, and J. B. Orlin, "A neighborhood search algorithm for the combined through fleet assignment model with time windows", *Networks* 44 (2004), 160–171.

[AhEtal07a] R. K. Ahuja, J. Goodstein, A. Mukherjee, J. B. Orlin, and D. Sharma, "A very large-scale neighborhood search algorithm for the combined through-fleet assignment model", *INFORMS Journal on Computing* 19 (2007), 416–428.

[AhEtal07b] R. K. Ahuja, W. Huang, H. E. Romeijn, and D. Romero Morales, "A heuristic approach to the multi-period single-sourcing problem with production and inventory capacities and perishability constraints", *INFORMS Journal on Computing* 19 (2007), 14–26.

[AhJhLi07] R. K. Ahuja, K. C. Jha, and J. Liu, "Solving real-life railroad blocking problems", *Interfaces* 37 (2007), 404–419.

[AhEtal07c] R. K. Ahuja, K. C. Jha, J. B. Orlin, and D. Sharma, "A very large-scale neighborhood search algorithm for the quadratic assignment problem", *INFORMS Journal on Computing* 19 (2007), 646–657.

[AhEtal08] R. K. Ahuja, A. Kumar, K. C. Jha, and J. B. Orlin, "Exact and heuristic algorithms for the weapon-target assignment problem", *Operations Research* 55 (2008), 1136–1146.

[AhMaOr93] R. K. Ahuja, T. L. Magnanti, and J. B. Orlin, *Network Flows: Theory, Algorithms and Applications*, Prentice Hall, 1993.

[AhEtal04b] R. K. Ahuja, J. B. Orlin, S. Pallottino, M. P. Scaparra, and M.G. Scutellà, "A multi-exchange heuristic for the single-source capacitated facility location problem", *Management Science* 50 (2004), 749–760.

[AhOrSh01] R. K. Ahuja, J. B. Orlin, and D. Sharma, "Multi-exchange neighborhood structures for the capacitated minimum spanning tree problem", *Mathematical Programming* 91 (2001), 71–97.

[AhOrSh03] R. K. Ahuja, J. B. Orlin, and D. Sharma, "A composite very large-scale neighborhood structure for the capacitated minimum spanning tree problem", *Operations Research Letters* 31 (2003), 185–194.

[AlNoWo07] M. H. Albert, R. J. Nowakowski, and D. Wolfe, *Lessons in Play: An Introduction to Combinatorial Game Theory*, A K Peters/CRC Press, 2007. (An introduction to combinatorial game theory.)

[AlEtal09] D. Aloise, A. Deshpande, P. Hansen, and P. Popat, "NP-hardness of Euclidean sum-of-squares clustering", *Machine Learning* 75 (2009), 245–248.

[AlEtal04] A. C. F. Alvim, C. C. Ribeiro, F. Glover, and D. J. Aloise, "A hybrid improvement heuristic for the one-dimensional bin packing problem", *Journal of Heuristics* 10 (2004), 205–229.

[AmScSc09] D. Ambrosino, A. Sciomachen, and M. G. Scutellà, "A heuristic based on multi-exchange techniques for a regional fleet assignment location-routing problem", *Computers & Operations Research* 36 (2009), 442–460.

[ArHeSp06] C. Archetti, A. Hertz, and M. G. Speranza, "A tabu search algorithm for the split delivery vehicle routing problem", *Transportation Science* 40 (2006), 64–73.

[Au04] L. M. Ausubel, "An efficient ascending-bid auction for multiple objects", *American Economic Review* 94 (2004), 1452–1475.

[AuBa14] L. M. Ausubel and O. V. Baranov, "Market design and the evolution of the combinatorial clock auction", *American Economic Review* 104 (2014), 446–451.

[AuCr11] L. M. Ausubel and P. Cramton, "Multiple factor auction design for wind rights", Report to Bureau of Ocean Energy Management, Regulation and Enforcement, 2011.

[AuCtMi06] L. Ausubel, P. Cramton, and P. Milgrom, "The clock-proxy auction: a practical combinatorial auction design", in *Combinatorial Auctions*, P. Cramton, Y. Shoham, and R. Steinberg (eds.), MIT Press, 2006, Chapter 5.

[AuMi02] L. M. Ausubel and P. R. Milgrom, "Ascending auctions with package bidding", *Frontiers of Theoretical Economics* 1 (2002), 1–42.

[BaPa76] E. Balas and M. W. Padberg, "Set partitioning: a survey", *SIAM Review* 18 (1976), 710–760.

[BaZe80] E. Balas and E. Zemel, "An algorithm for large zero-one knapsack problems", *Operations Research* 28 (1980), 1130–1154.

[BaCoPr95] M. O. Ball, C. J. Colbourn, and J. S. Provan, "Network reliability" in *Network Models*, M. Ball, T. Magnanti, C. Monma, and G. Nemhauser (eds.), North-Holland, 1995, 673–762.

[BaHa94] T. Basar and A. Haurie (eds.), *Advances in Dynamic Games and Applications*, Birkhäuser, 1994.

[BaJaSh90] M. S. Bazaraa, J. J. Jarvis, and H. D. Sherali, *Linear Programming and Network Flows*, 2nd ed., Wiley, 1990.

[BeKaSt92] W. W. Bein, J. Kamburowski, and M. F. M. Stallmann, "Optimal reduction of two-terminal directed acyclic graphs", *SIAM Journal on Computing* 21 (1992), 1112–1129.

[Be57] C. Berge, "Two theorems in graph theory", *Proceedings of the National Academy of Sciences USA* 43 (1957), 842–844.

[BeFu99] P. Berman and T. Fujito, "On approximation properties of the independent set problem for low degree graphs", *Theory of Computing Systems* 32 (1999), 115–132.

[BhSh14] B. Bhattacharya and Q. Shi, "Improved algorithms to network $p$-center location problems", *Computational Geometry* 47 (2014), 307–315.

[BiShZi13] M. Bichler, P. Shabalin, and G. Ziegler, "Efficiency with linear prices? A game-theoretical and computational analysis of the combinatorial clock auction", *Information Systems Research* 24 (2013), 394–417.

[BiEtAl11] S. Bikhchandani, S. de Vries, J. Schummer, and R. Vohra, "An ascending Vickrey auction for selling bases of a matroid", *Operations Research* 59 (2011), 400–413.

[BiOs02] S. Bikhchandani and J. M. Ostroy, "The package assignment model", *Journal of Economic Theory* 107 (2002), 377–406.

[BiKiTa93] K. Binmore, A. Kirman, and P. Tani (eds.), *Frontiers of Game Theory*, MIT Press, 1993.

[BlPuEl14] V. Blanco, J. Puerto, and S. El Haj Ben Ali, "Revisiting several problems and algorithms in continuous location with $\ell_\tau$ norms", *Computational Optimization and Applications* 58 (2014), 563–595.

[BlEtal04] J. Blazewicz, P. Formanowicz, M. Kasprzak, W. T. Markiewicz, and A. Swierc, "Tabu search algorithm for DNA sequencing by hybridization with isothermic libraries", *Computational Biology and Chemistry* 28 (2004), 11–19.

[BlEtal00] J. Blazewicz, P. Formanowicz, M. Kasprzak, W. T. Markiewicz, and J. Weglarz, "Tabu search for DNA sequencing with false negatives and false positives", *European Journal of Operational Research* 125 (2000), 257–265.

[BlGlKa04] J. Blazewicz, F. Glover, and M. Kasprzak, "DNA sequencing—tabu and scatter search combined", *INFORMS Journal on Computing* 16 (2004), 232–240.

[BoOr05] A. Bompadre and J. B. Orlin, "Using grammars to generate very large-scale neighborhoods for the traveling salesman problem and other sequencing problems", *Proceedings of the Eleventh International Conference on Integer Programming and Combinatorial Optimization (IPCO)*, 2005, 437–451.

[Bo89] K. C. Border, *Fixed Point Theorems with Applications to Economics and Game Theory*, Cambridge University Press, 1989.

[BoHo01] C. Boutilier and H. H. Hoos, "Bidding languages for combinatorial auctions", *Proceedings of the Seventeenth International Joint Conference on Artificial Intelligence*, 2001, 1211–1217.

[BrCh89] M. L. Brandeau and S. S. Chiu, "An overview of representative problems in location research", *Management Science* 35 (1989), 645–674.

[BrRy13] M. Broom and J. Rychtář, *Game-Theoretical Models in Biology*, CRC Press, 2013.

[BrEtal11] T. Brueggemann, J. Hurink, T. Vredeveld, and G. Woeginger, "Exponential size neighborhoods for makespan minimization scheduling", *Naval Research Logistics* 58 (2011), 795–803.

[CaLaYa15] H. Calik, M. Labbé, and H. Yaman, "$p$-Center Problems", in *Location Science*, G. Laporte, S. Nickel, and F. Saldanha da Gama (eds.), Springer, 2015, 79–92.

[CaPe06] E. Cantillon and M. Pesendorfer, "Auctioning bus routes: the London experience", in *Combinatorial Auctions*, P. Cramton, Y. Shoham, and R. Steinberg (eds.), MIT Press, 2006, Chapter 22.

[CaVi90] J. Carlier and P. Villon, "A new heuristic for the traveling salesman problem", *RAIRO - Operations Research* 24 (1990), 245–253.

[CaEtal03] F. Caro, M. Constantino, I. Martins, and A. Weintraub, "A 2-opt tabu search procedure for the multiperiod forest harvesting problem with adjacency, green-up, old growth and even flow constraints", *Forest Science* 49 (2003), 738–751.

[Ch83] V. Chvátal, *Linear Programming*, Freeman, 1983.

[CiBaSa10] M. Ciarleglio, J. W. Barnes, and S. Sarkar, "ConsNet—A tabu search approach to the spatially coherent conservation area network design problem", *Journal of Heuristics* 16 (2010), 537–557.

[CoEtal13] E. G. Coffman, J. Csirik, G. Galambos, S. Martello, and D. Vigo, "Bin packing approximation algorithms: survey and classification", in *Handbook of Combinatorial Optimization*, P. M. Pardalos, D. Z. Du, and R. L. Graham (eds.), Springer, 2013, 455–531.

[CoGaJo84] E. G. Coffman, M. R. Garey, and D. S. Johnson, "Approximation algorithms for bin-packing: an updated survey", in *Algorithm Design for Computer System Design*, G. Ausiello, M. Lucertini, and P. Serafini (eds.), Springer-Verlag, 1984, 49–106.

[CoCoOt09] J. P. Cohen, C. C. Coughlin, and L. S. Ott, "Auctions as a vehicle to reduce airport delays and achieve value capture", *Federal Reserve Bank of St. Louis Review* 91 (2009), 569–587.

[CoPoVe02] R. K. Congram, C. N. Potts, and S. L. van de Velde, "An iterated dynasearch algorithm for the single machine total weighted tardiness scheduling problem", *INFORMS Journal on Computing* 14 (2002), 52–67.

[Cr03] R. Cressman, *Evolutionary Dynamics and Extensive Form Games*, MIT Press, 2003.

[CuAh05] C. B. Cunha and R. K. Ahuja, "Very large-scale neighborhood search for the $K$-constrained multiple knapsack problem", *Journal of Heuristics* 11 (2005), 465–481.

[DaRoLe05] E. Danna, E. Rothberg, and C. Le Pape, "Exploring relaxation induced neighborhoods to improve MIP solutions", *Mathematical Programming* 102 (2005), 71–90.

[DaMa15] M. S. Daskin and K. L. Maass, "The $p$-median problem", in *Location Science*, G. Laporte, S. Nickel, and F. Saldanha da Gama (eds.), Springer, 2015, 21–45.

[DaEtal07] A. Davenport, J. Kalagnanam, C. Reddy, S. Siegel, and J. Hou, "An application of constraint programming to generating detailed operations schedules for steel manufacturing", *Proceedings of Constraint Programming*, 2007, 64–75.

[DaMi08] R. Day and P. Milgrom, "Core-selecting package auctions", *International Journal of Game Theory* 36 (2008), 393–407.

[DaRa07] R. W. Day and S. Raghavan, "Fair payments for efficient allocations in public sector combinatorial auctions", *Management Science* 53 (2007), 1389–1406.

[DeWo00] V. G. Deĭneko and G. J. Woeginger, "A study of exponential neighborhoods for the Travelling Salesman Problem and for the Quadratic Assignment Problem", *Mathematical Programming* 87 (2000), 519–542.

[DeGaSo86] G. Demange, D. Gale, and M. Sotomayor, "Multi-item auctions", *Journal of Political Economy* 94 (1986), 863–872.

[DeHe92] E. Demeulemeester and W. S. Herroelen, "A branch-and-bound procedure for the multiple resource-constrained project scheduling problem", *Management Science* 38 (1992), 1803–1818.

[DeHe96] B. De Reyck and W. S. Herroelen, "On the use of the complexity index as a measure of complexity in activity networks", *European Journal of Operational Research* 91 (1996), 347–366.

[DeHe90] J. De Wit and W. S. Herroelen, "An evaluation of microcomputer-based software packages for project management", *European Journal of Operational Research* 49 (1990), 102–139.

[DoEtal00] E. Dockner, S. Jørgensen, N. V. Long, and G. Sorger, *Differential Games in Economics and Management Science*, Cambridge University Press, 2000.

[DoPe94] U. Dorndorf and E. Pesch, "Fast clustering algorithms", *ORSA Journal of Computing* 6 (1994), 141–153.

[Do98] K. A. Dowsland, "Nurse scheduling with tabu search and strategic oscillation", *European Journal of Operational Research* 106 (1998), 393–407.

[Dr81] M. Dresher, *The Mathematics of Games of Strategy: Theory and Applications*, Dover, 1981. (First published in 1961.)

[Dr15] Z. Drezner, "The fortified Weiszfeld algorithm for solving the Weber problem", *IMA Journal of Management Mathematics* 26 (2015), 1–9.

[DrLe86] M. Dror and L. Levy, "A vehicle routing improvement algorithm comparison of a greedy and a matching implementation for inventory routing", *Computers & Operations Research* 13 (1986), 33–45.

[Ed65] J. Edmonds, "Paths, trees and flowers", *Canadian Journal of Mathematics* 17 (1965), 449–467.

[El77] S. E. Elmaghraby, *Activity Networks: Project Planning and Control by Network Models*, Wiley, 1977.

[El89] S. E. Elmaghraby, "The estimation of some network parameters in the PERT model of activity networks: review and critique", in R. Słowinski and J. Weglarz (eds.), *Advances in Project Scheduling*, Elsevier, 1989, 371–432.

[ElKa90] S. E. Elmaghraby and J. Kamburowski, "On project representation and activity floats", *Arabian Journal of Science and Engineering* 15 (1990), 627–637.

[ElKa92] S. E. Elmaghraby and J. Kamburowski, "The analysis of activity networks under generalized precedence relations (GPRs)", *Management Science* 38 (1992), 1245–1263.

[ErOrSt06] Ö. Ergun, J. B. Orlin, and A. Steele-Feldman, "Creating very large-scale neighborhoods out of smaller ones by compounding moves", *Journal of Heuristics* 12 (2006), 115–140.

[FiMa82] C. M. Fiduccia and R. M. Mattheyses, "A linear time heuristic for improving network partitions", *Proceedings of ACM IEEE Nineteenth Design Automation Conference*, IEEE Computer Society, 1982, 175–181.

[Fo15] R. Fourer, "Linear programming: software survey", *OR/MS Today* 42/3 (2015), 52–63.

[FrMcWh92] R. L. Francis, L. F. McGinnis, Jr., and J. A. White, *Facility Layout and Location: An Analytical Approach*, Prentice-Hall, 1992.

[FrNeSc04] A. Frangioni, E. Necciari, and M. G. Scutellà, "Multi-exchange algorithms for minimum makespan machine scheduling problems", *Journal of Combinatorial Optimization* 8 (2004), 195–220.

[FrHa05] A. Fréville and S. Hanafi, "The multidimensional 0-1 knapsack problem—bounds and computational aspects", *Annals of Operations Research* 139 (2005), 195–227.

[Fr90] J. W. Friedman, *Game Theory with Applications to Economics*, 2nd ed., Oxford University Press, 1990.

[GaNe72] R. S. Garfinkel and G. L. Nemhauser, *Integer Programming*, Wiley, 1972.

[Ga85] S. I. Gass, *Linear Programming: Methods and Applications*, 5th ed., Boyd & Fraser, 1985.

[GeEtal06a] M. Gendreau, F. Guertin, J-Y. Potvin, and R. Seguin, "Neighborhood search heuristics for a dynamic vehicle dispatching problem with pick-ups and deliveries", *Transportation Research* Part C 14 (2006), 157–174.

[GeEtal06b] M. Gendreau, M. Iori, G. Laporte, and S. Martello, "A tabu search algorithm for a routing and container loading problem", *Transportation Science* 40 (2006), 342–350.

[GeEtal08] M. Gendreau, M. Iori, G. Laporte, and S. Martello, "A tabu search heuristic for the vehicle routing problem with two-dimensional loading constraints", *Networks* 51 (2008), 4–18.

[Gl86] F. Glover, "Future paths for integer programming and links to artificial intelligence", *Computers & Operations Research* 13 (1986), 533–549.

[Gl90] F. Glover, "Tabu search: a tutorial", *Interfaces* 20 (1990), 74–94.

[Gl96] F. Glover, "Ejection chains, reference structures, and alternating path algorithms for traveling salesman problems", *Discrete Applied Mathematics* 65 (1996), 223–253.

[GlLa97] F. Glover and M. Laguna, *Tabu Search*, Springer, 1997.

[GlLa02] F. Glover and M. Laguna, "Tabu search", in *Handbook of Applied Optimization*, P. M. Pardalos and M. G. C. Resende (eds.), Oxford University Press, 2002, 194–208.

[GlLa13] F. Glover and M. Laguna, "Tabu search", in *Handbook of Combinatorial Optimization*, 2nd ed., D. Z. Du and P. M. Pardalos (eds.), Springer, 2013, 3261–3362.

[GlLaMa07] F. Glover, M. Laguna, and R. Martí, "Principles of tabu search", in *Approximation Algorithms and Metaheuristics*, T. Gonzalez (ed.), Chapman & Hall/CRC, 2007, Chapter 23.

[GoRo14] M. X. Goemans and T. Rothvoß, "Polynomiality for bin-packing with a constant number of item types", *Proceedings of the 25th Annual ACM-SIAM Symposium on Discrete Mathematics*, 2014, 830–839.

[GoEtAl15] A. Goetzendorff, M. Bichler, P. Shabalin, and R. W. Day, "Compact bid languages and core pricing in large multi-item auctions", *Management Science* 61 (2015), 1684–1703.

[GoGaFi10] J. González-Díaz, I. García-Jurado, and M. G. Fiestras-Janeiro, *An Introductory Course on Mathematical Game Theory*, American Mathematical Society, 2010.

[GoEtAl14] D. R. Goossens, S. Onderstal, J. Pijnacker, and F. C. R. Spieksma, "Solids: a combinatorial auction for real estate", *Interfaces* 44 (2014), 351–363.

[GrLa77] J. Green and J-J. Laffont, "Characterization of satisfactory mechanisms for the revelation of preferences for public goods", *Econometrica* 45 (1977), 427–438.

[GuSt00] F. Gul and E. Stacchetti, "The English auction with differentiated commodities", *Journal of Economic Theory* 92 (2000), 66–95.

[Gu91] R. Guy, ed., *Combinatorial Games*, Proceedings of Symposia in Applied Mathematics, American Mathematical Society, Volume 43, 1991.

[HaMi79] G. Y. Handler and P. B. Mirchandani, *Location on Networks: Theory and Algorithms*, MIT Press, 1979.

[He00] K. Helsgaun, "An effective implementation of the Lin-Kernighan traveling salesman heuristic", *European Journal of Operational Research* 126 (2000), 106–130.

[HeDeDe98] W. S. Herroelen, B. De Reyck, and E. Demeulemeester, "Resource-constrained project scheduling: a survey of recent developments", *Computers & Operations Research* 25 (1998), 279–302.

[HoLe10] S. C. Ho and J. M. Y. Leung, "Solving a manpower scheduling problem for airline catering using metaheuristics", *European Journal of Operational Research* 202 (2010), 903–921.

[Ho97] D. S. Hochbaum, *Approximation Algorithms for NP-Hard Problems*, PWS Publishing Company, 1997.

[HoPa93] K. Hoffman and M. W. Padberg, "Solving airline crew scheduling problems by branch-and-cut", *Management Science* 39 (1993), 657–682.

[JoMc97] D. S. Johnson and L. A. McGeoch, "The travelling salesman problem: a case study in local optimization", in *Local Search in Combinatorial Optimization*, E. H. L. Aarts and J. K. Lenstra (eds.), Wiley, 1997, 311–336.

[Ka92] J. Kamburowski, "Bounding the distribution of project duration in PERT networks", *Operations Research Letters* 12 (1992), 17–22.

[KaSpDr92] J. Kamburowski, A. Sprecher, and A. Drexl, "Characterization and generation of a general class of resource-constrained project scheduling problems: easy and hard instances", *Research Report N0301*, Institut für Betriebswirtschaftslehere, Christian-Albrechts Universität zu Kiel, 1992.

[KaHa79a] O. Kariv and S. L. Hakimi, "An algorithmic approach to network location problems. I: the $p$-centers", *SIAM Journal on Applied Mathematics* 37 (1979), 513–538.

[KaHa79b] O. Kariv and S. L. Hakimi, "An algorithmic approach to network location problems. II: the $p$-medians", *SIAM Journal on Applied Mathematics* 37 (1979), 539–560.

[KePfPi04] H. Kellerer, U. Pferschy, and D. Pisinger, *Knapsack Problems*, Springer Science & Business Media, 2004.

[KeLi70] B. Kernighan and S. Lin, "An efficient heuristic procedure for partitioning graphs", *Bell System Technical Journal* 49 (1970), 291–307.

[KiOlWe14] S. W. Kim, M. Olivares, and G. Y. Weintraub, "Measuring the performance of large-scale combinatorial auctions: a structural estimation approach", *Management Science* 60 (2014), 1180–1201.

[KoPa01] R. Kolisch and R. Padman, "An integrated survey of deterministic project scheduling", *Omega* 29 (2001), 249–272.

[KoSp97] R. Kolisch and A. Sprecher, "PSPLIB – A project scheduling problem library", *European Journal of Operational Research* 96 (1997), 205–216.

[KoFa10] J. E. Korsvik and K. Fagerholt, "A tabu search heuristic for ship routing and scheduling with flexible cargo quantities", *Journal of Heuristics* 16 (2010), 117–137.

[KoFaLa10] J. E. Korsvik, K. Fagerholt, and G. Laporte, "A tabu search heuristic for ship routing and scheduling", *Journal of the Operational Research Society* 61 (2010), 594–603.

[KoVy12] B. Korte and J. Vygen, *Combinatorial Optimization*, 3rd edition, Springer, 2012.

[KuAd86] V. G. Kulkarni and V. G. Adlakha, "Markov and Markov-regenerative PERT networks", *Operations Research* 34 (1986), 769–781.

[Le83] H. B. Leonard, "Elicitation of honest preferences for the assignment of individuals to positions", *Journal of Political Economy* 91 (1983), 461–479.

[LiEtal05] B. Lin, S. Chavali, K. Camarda, and D. C. Miller, "Computer-aided molecular design using tabu search", *Computers & Chemical Engineering* 29 (2005), 337–347.

[LiKe73] S. Lin and B. Kernighan, "An effective heuristic algorithm for the traveling salesman problem", *Operations Research* 21 (1973), 498–516.

[LiMi04a] B. Lin and D. C. Miller, "Solving heat exchanger network synthesis problems with tabu search", *Computers & Chemical Engineering* 28 (2004), 1451–1464.

[LiMi04b] B. Lin and D. C. Miller, "Tabu search algorithm for chemical process optimization", *Computers & Chemical Engineering* 28 (2004), 2287–2306.

[LuRa57] R. D. Luce and H. Raiffa, *Games and Decisions*, Wiley, 1957. (Covers much the same ground as [vNMo53] but at a more elementary level.)

[Ma86] O. Maimon, "The variance equity measures in location decisions on trees", *Annals of Operations Research* 6 (1986), 147–160.

[MaPiTo99] S. Martello, D. Pisinger, and P. Toth, "Dynamic programming and strong bounds for the 0-1 knapsack problem", *Management Science* 45 (1999), 414–424.

[MaTo90] S. Martello and P. Toth, *Knapsack Problems: Algorithms and Computer Implementations*, Wiley, 1990.

[Ma82] J. Maynard Smith, *Evolution and the Theory of Games*, Cambridge University Press, 1982. (The founding treatise on evolutionary game theory.)

[Me83] N. Megiddo, "Linear-time algorithms for linear programming in $\mathcal{R}^3$ and related problems", *SIAM Journal on Computing* 12 (1983), 759–776.

[MeSu84] N. Megiddo and K. J. Supowit, "On the complexity of some common geometric location problems", *SIAM Journal on Computing* 13 (1984), 182–196.

[Me93] M. Mesterton-Gibbons, "Game-theoretic resource modeling", *Natural Resource Modeling* 7 (1993), 93–147. (A survey of game theory and applications accessible to the general reader.)

[Me01] M. Mesterton-Gibbons, *An Introduction to Game-Theoretic Modelling*, 2nd ed., American Mathematical Society, 2001. (A unified introduction to classical and evolutionary game theory.)

[MiKaSt93] D. J. Michael, J. Kamburowski, and M. Stallmann, "On the minimum dummy-arc problem", *RAIRO Recherche Opérationnelle* 27 (1993), 153–168.

[Mi00] P. Milgrom, "Putting auction theory to work: the simultaneous ascending auction", *Journal of Political Economy* 108 (2000), 245–272.

[MiFr90] P. B. Mirchandani and R. L. Francis (eds.), *Discrete Location Theory*, Wiley, 1990.

[MiPa07] D. Mishra and D. Parkes, "Ascending price Vickrey auctions for general valuations", *Journal of Economic Theory* 132 (2007), 335–366.

[My91] R. B. Myerson, *Game Theory: Analysis of Conflict*, Harvard University Press, 1991. (General introduction to economic game theory.)

[NaBr09] Y. Nagata and O. Bräysy, "A powerful route minimization heuristic for the vehicle routing problem with time windows", *Operations Research Letters* 37 (2009), 333–338.

[NeJhAh09] A. K. Nemani, K. C. Jha, and R. K. Ahuja, "The load planning problem at an intermodal railroad terminal", working paper, Industrial and Systems Engineering, University of Florida, 2009.

[NeWo88] G. L. Nemhauser and L. A. Wolsey, *Integer and Combinatorial Optimization*, Wiley, 1988.

[Ne90] K. Neumann, *Stochastic Project Networks: Temporal Analysis, Scheduling, and Cost Minimization*, Lecture Notes in Economics and Mathematical Systems 344, Springer-Verlag, 1990.

[NiPu05] S. Nickel and J. Puerto, *Location Theory: A Unified Approach*, Springer, 2005.

[Ni06] N. Nisan, "Bidding languages", in *Combinatorial Auctions*, P. Cramton, Y. Shoham, and R. Steinberg (eds.), MIT Press, 2006, Chapter 9.

[Og09] W. Ogryczak, "Inequality measures and equitable locations", *Annals of Operations Research* 167 (2009), 61–86.

[OpLø08] J. Oppen and A. Løkketangen, "A tabu search approach for the livestock collection problem", *Computers & Operations Research* 35 (2008), 3213–3229.

[Ow13] G. Owen, *Game Theory*, 4th ed., Emerald Group Publishing Limited, 2013. (An overview of the mathematical theory of games.)

[PaEtal07] D. C. Paraskevopoulos, P. P. Repoussis, C. D. Tarantilis, G. Ioannou, and G. P. Prastacos, "A reactive variable neighborhood tabu search for the heterogeneous fleet vehicle routing problem with time windows", *Journal of Heuristics* 14 (2007), 247–254.

[PaUn00] D. C. Parkes and L. H. Ungar, "Iterative combinatorial auctions: theory and practice", *Proceedings of the 17th National Conference on Artificial Intelligence*, 2000, 74–81.

[Pa83] J. H. Patterson, "Exact and heuristic solution procedures for the constrained-resource, project scheduling problem: Volumes I, II, and III", *Research Monograph*, privately circulated, 1983.

[PeGe99] G. Pesant and M. Gendreau, "A constraint programming framework for local search methods", *Journal of Heuristics* 5 (1999), 255–279

[Pe08] H. Peters, *Game Theory: A Multi-Leveled Approach*, Springer, 2008.

[PeZiBi13] I. Petrakis, G. Ziegler, and M. Bichler, "Ascending combinatorial auctions with allocation constraints: on game theoretical and computational properties of generic pricing rules", *Information Systems Research* 24 (2013), 768–786.

[PiRo10] D. Pisinger and S. Ropke, "Large neighborhood search", in *Handbook of Metaheuristics*, M. Gendreau and J-Y. Potvin (eds.), Springer, 2010, 399–419.

[Pl11] F. Plastria, "The Weiszfeld algorithm: proof, amendments, and extensions", in *Foundations of Location Analysis*, H. A. Eiselt and V. Marianov. (eds.), Springer, 2011, 357–389.

[PoVe95] C. N. Potts and S. L. van de Velde, "Dynasearch: iterative local improvement by dynamic programming: Part I, the traveling salesman problem", Technical Report, University of Twente, 1995.

[RaSmBu82] S. J. Rassenti, V. L. Smith, and R. L. Bulfin, "A combinatorial auction mechanism for airport time slot allocation", *The Bell Journal of Economics* 13 (1982), 402–417.

[Re98] C. Rego, "A subpath ejection method for the vehicle routing problem", *Management Science* 44 (1998), 1447–1459.

[RoVa13] A. M. Rodríguez-Chía and C. Valero-Franco, "On the global convergence of a generalized iterative procedure for the minisum location problem with $\ell_p$ distances for $p > 2$", *Mathematical Programming* 137 (2013), 477–502.

[RoTeVi10] C. Roos, T. Terlaky, and J-Ph. Vial, *Interior Point Methods for Linear Optimization*, 2nd ed., Springer, 2010.

[RoPeHa98] M. H. Rothkopf, A. Pekeč, and R. M. Harstad, "Computationally manageable combinational auctions", *Management Science* 44 (1998), 1131–1147.

[Sa13] T. Sandholm, "Very-large-scale generalized combinatorial multi-attribute auctions: lessons from conducting $60 billion of sourcing", in *The Handbook of Market Design*, N. Vulkan, A. E. Roth, and Z. Neeman (eds.), Oxford University Press, 2013, Chapter 16.

[Sa10] W. H. Sandholm, *Population Games and Evolutionary Dynamics*, MIT Press, 2010.

[ScPaSc04] M. P. Scaparra, S. Pallottino, and M. G. Scutellà, "Large-scale local search heuristics for the capacitated vertex $p$-center problem", *Networks* 43 (2004), 241–255.

[ScSø09] P. Schittekat and K. Sørensen, "Supporting 3PL decisions in the automotive industry by generating diverse solutions to a large-scale location-routing problem", *Operations Research* 57 (2009), 1058–1067.

[Sc10] U. Schneider, "A tabu search tutorial based on a real-world scheduling problem", *Central European Journal of Operations Research* 19 (2010), 467–493.

[Sc86] A. Schrijver, *Theory of Linear and Integer Programming*, Wiley, 1986.

[SeSiRo11] M. Sevaux, A. Singh, and A. Rossi, "Tabu search for multiprocessor scheduling: application to high level synthesis", *Asia-Pacific Journal of Operational Research* 28 (2011), 201–212.

[Sh12] G. B. Sheblé, *Computational Auction Mechanisms for Restructured Power Industry Operation*, Springer, 2012.

[Si10] K. Sigmund, *The Calculus of Selfishness*, Princeton University Press, 2010.

[SiBa96] N. Simonetti and E. Balas, "Implementation of a linear time algorithm for certain generalized traveling salesman problems", *Proceedings of the 5th International Conference on Integer Programming and Combinatorial Optimization (IPCO)*, 1996, 316–329.

[SiEtal05] S. Sindhuchao, H. E. Romeijn, E. Akçali, and R. Boondiskulchok, "An integrated inventory-routing system for multi-item joint replenishment with limited vehicle capacity", *Journal of Global Optimization* 32 (2005), 93–118.

[SlWe89] R. Słowinski and J. Weglarz, eds., *Advances in Project Scheduling*, Elsevier, 1989.

[Sp94] A. Sprecher, *Resource-Constrained Project Scheduling: Exact Methods for the Multi-Mode Case*, Lecture Notes in Economics and Mathematical Systems 409, Springer-Verlag, 1994.

[Su12] M. Sun, "A tabu search heuristic procedure for the capacitated facility location problem", *Journal of Heuristics* 18 (2012), 91–118.

[Ta03] E. D. Taillard, "Heuristic methods for large centroid clustering problems", *Journal of Heuristics* 9 (2003), 51–73.

[Ta88] A. Tamir, "Improved complexity bounds for center location problems on networks using dynamic data structures", *SIAM Journal on Discrete Mathematics* 1 (1988), 377–396.

[Ta96] A. Tamir, "An $O(pn^2)$ algorithm for the $p$-median and related problems on tree graphs", *Operations Research Letters* 19 (1996), 59–64.

[ThOr89] P. M. Thompson and J. B. Orlin, "The theory of cyclic transfers", Operations Research Center Working Paper, MIT, 1989.

[ThPs93] P. M. Thompson and H. N. Psaraftis, "Cyclic transfer algorithms for multivehicle routing and scheduling problems", *Operations Research* 41 (1993), 935–946.

[TrYi07] M. A. Trick and H. Yildiz, "Bender's cuts guided large neighborhood search for the traveling umpire problem", in *Integration of AI and OR Techniques in Constraint Programming for Combinatorial Optimization Problems*, P. Van Hentenryck and L. Wolsey (eds.), Springer, 2007, 332–345.

[vHMu01] S. van Hoesel and R. Müller, "Optimization in electronic markets: examples in combinatorial auctions", *Netnomics* 3 (2001), 23–33.

[vNMo53] J. von Neumann and O. Morgenstern, *Theory of Games and Economic Behavior*, 3rd ed., Princeton University Press, 1953. (The founding treatise on economic game theory.)

[Wa88] J. Wang, *The Theory of Games*, Oxford University Press, 1988. (An overview of the mathematical theory of games emphasizing bimatrix games and $c$-games.)

[We37] E. Weiszfeld, "Sur le point pour lequel la somme des distances de $n$ points donnés est minimum", *Tohoku Mathematical Journal* 43 (1937), 355–386.

[WePl09] E. Weiszfeld and F. Plastria, "On the point for which the sum of the distances to $n$ given points is minimum", *Annals of Operations Research* 167 (2009), 7–41.

[We93] G. Wesolowsky, "The Weber problem: history and perspective", *Location Science* 1 (1993), 5–23.

[Wi13] H. P. Williams, *Model Building in Mathematical Programming*, 5th ed., Wiley, 2013.

[Wi92] T. M. Williams, "Criticality of stochastic networks", *Operations Research* 43 (1992), 353–357.

[Wr87] S. J. Wright, *Primal-Dual Interior-Point Methods*, SIAM, 1987.

[XuKe96] J. Xu and J. P. Kelly, "A network flow-based tabu search heuristic for the vehicle routing problem", *Transportation Science* 30 (1996), 379–393.

[YaIbGl04] M. Yagiura, T. Ibaraki, and F. Glover, "An ejection chain approach for the generalized assignment problem", *INFORMS Journal on Computing* 16 (2004), 133–151.

[YaEtal04] M. Yagiura, S. Iwasaki, T. Ibaraki, and F. Glover, "A very large-scale neighborhood search algorithm for the multi-resource generalized assignment problem", *Discrete Optimization* 1 (2004), 87–98.

[YaYaIb99] M. Yagiura, T. Yamaguchi, and T. Ibaraki, "A variable-depth search algorithm for the generalized assignment problem", in *Metaheuristics: Advances and Trends in Local Search Paradigms for Optimization*, S. Voss, S. Martello, I. H. Osman, and C. Roucairol (eds.), Kluwer, 1999, 459–471.

[ZuAmGi08] N. Zufferey, P. Amstutz, and P. Giaccari, "Graph colouring approaches for a satellite range scheduling problem", *Journal of Scheduling* 11 (2008), 263–277.

**Web Resources**:

`http://akira.ruc.dk/~keld/research/LKH/` (Helsgaun's implementation of the Lin-Kernighan heuristic (LKH) for solving the traveling salesman problem. Helsgaun's LKH has produced optimal solutions for all TSPLIB instances that have been solved to optimality. Furthermore, Helsgaun's LKH has improved the best known solutions for a series of large-scale instances with unknown optima, among these a 1,904,711-city instance that is known as the "World TSP".)

`http://crcv.ucf.edu/source/K-Means` (MATLAB and C codes for the $k$-means algorithm.)

`http://glossary.computing.society.informs.org/ver2/mpgwiki/` (Excellent glossary of linear programming terms, as well as concepts in general mathematical optimization.)

`http://optimization-suite.coin-or.org` (COIN-OR open source software for optimization problems.)

`http://people.brunel.ac.uk/~mastjjb/jeb/info.html` (OR-Library, a collection of test datasets for a wide variety of operations research problems, including those for the generalized assignment problem, the traveling salesman problem, and various kinds of knapsack problems and vehicle routing problems.)

`http://sourceforge.net/projects/lipside` (LiPS package for linear, integer, and goal programming problems.)

`http://wikipedia.org/wiki/Comparison_of_project_management_software` (A comprehensive survey of models, data, and algorithms for project scheduling problems.)

`http://www.aimms.com` (AIMMS integrated solver and modeling environment.)

`http://www.ampl.com` (Modeling language for optimization problems.)

`http://www.codeforge.com/article/214515` (C++ code for the dual simplex method.)

`http://www.cplex.com/` (Software package CPLEX to solve LP and IP/MIP problems.)

`http://www.cs.nott.ac.uk/~rxq/data.htm` (Test instances of different timetabling problems maintained by the Department of Computer Science and Information Technology at the University of Nottingham.)

`http://www.di.unipi.it/optimize` (Test instances for different optimization problems produced and maintained by the Operations Research Group in the Department of Computer Science at the University of Pisa.)

`http://www.diku.dk/~pisinger/codes.html` (Optimization codes in C for knapsack problems.)

`http://www.gams.com` (High-level modeling platform for optimization problems.)

`http://www.gurobi.com` (Integrated solver and modeling environment for LP and IP/MIP problems.)

http://www.ic.gc.ca/eic/site/smt-gst.nsf/eng/sf10726.html (Licensing framework for broadband radio service in Canada, setting out details of the auction format.)

http://comopt.ifi.uni-heidelberg.de/software/TSPLIB95/ (TSPLIB, a widely referenced library of benchmark instances of the symmetric traveling salesman problem, the Hamiltonian cycle problem, the asymmetric traveling salesman problem, the sequential ordering problem, and the capacitated vehicle routing problem.)

http://www.lindo.com/ (Software packages LINDO/LINGO to solve LP problems and IP/MIP problems.)

http://www.mathworks.com/products/optimization (MATLAB toolbox for solving optimization problems.)

http://www.neos-server.org/neos (Online solver for LPs using simplex and interior point methods.)

http://www.netlib.org/toms/632 (Fortran code for solving knapsack problems.)

http://www.om-db.wi.tum.de/psplib/library.html (Contains benchmark problem sets for various types of resource-constrained project scheduling problems.)

# 17

# *THEORETICAL COMPUTER SCIENCE*

---

## INTRODUCTION

Theoretical computer science is concerned with modeling computational problems and solving them algorithmically. It strives to distinguish what can be computed from what cannot. If a problem can be solved by an algorithm, it is important to know the amount of space and time needed.

---

## GLOSSARY

**abelian square**: a word having the pattern $xx^p$, where $x^p$ is a permutation of $x$.

**acceptance probability** (of input word by a probabilistic TM): the sum of the probabilities over all acceptance paths of computation.

**Ackermann function**: a very rapidly growing function that is recursive, but not primitive recursive.

**algorithm**: a finite list of instructions designed to accomplish a specified computation or other task.

**alphabet**: a finite set of *symbols*.

**ambiguous context-free grammar**: a grammar whose language has a string having two different leftmost derivations.

**analysis of an algorithm**: an estimation of its cost of execution, especially of its running time.

**antecedent of a production** $\alpha \to \beta$: the string $\alpha$ that precedes the arrow.

**average-case running time**: the expected running time of an algorithm, usually expressed asymptotically in terms of the input size.

**Backus-Naur** (or **Backus normal**) **form** (**BNF**): a metalanguage used to specify computer language syntax.

**busy beaver function**: the function $BB(n)$ whose value is the maximum number of 1s that an $n$-state Turing machine can print and still halt.

**busy beaver machine, $n$-state**: an $n$-state Turing machine on the alphabet $\Sigma = \{\#, 1\}$ that accepts an input tape filled with blanks (#s) and halts after placing a maximum number of 1s on the tape.

**cellular automaton** ($n$-**dimensional**): an interconnection network in which there is a processor at each integer lattice point of $n$-dimensional Euclidean space, and each processor communicates with its immediate neighbors.

**characteristic function** (of a language): the function on strings that has value *yes* for elements in the language, and *no* otherwise.

**characteristic function** (of a set): the function whose value is 1 for elements of the set, and 0 otherwise.

**Chomsky hierarchy**: four classes of grammars, with increasing restrictions.

**Chomsky normal form** (for a production rule): the form $A \to BC$ where $B$ and $C$ are nonterminals or the form $A \to a$ where $a$ is a terminal.

***Church's thesis*** (or the ***Church-Turing thesis***): the premise that the intuitive notion of what is computable or partially computable should be formally defined as computable by a Turing machine.

***code*** (for an alphabet): a language $C$ of nonempty words, such that whenever a word $w$ can be written as a concatenation of words in $C$, the write-up is always unique. That is, if $w = x_1 \ldots x_m = y_1 \ldots y_n$ where $x_i, y_j \in C$, then $m = n$ and $x_i = y_i$ for $i = 1, \ldots, m$.

***code indicator*** (of a language): the sum of the code indicators of all words in the language.

***code indicator*** (of a word $w \in V^*$): the number $ci(w) = |V|^{-|w|}$.

***collapse*** (of the polynomial hierarchy to the $i$th rank): the circumstance that PH $= \Sigma_i^p$, for some $i \geq 0$.

***common PRAM*** (or ***CRCW$^{com}$***): a CRCW PRAM model in which concurrent writes to the same location are permitted if all processors write the same data.

***comparison sort***: a sorting algorithm that uses only comparisons between keys to determine the sorted order.

***complement*** (of language $L$ over alphabet $V$): the language $\overline{L}$, where complementation is taken with respect to $V^*$.

***$\mathcal{C}$-complete language*** (where $\mathcal{C}$ is a class of languages): a language $A$ such that $A$ is $\mathcal{C}$-hard and $A \in \mathcal{C}$.

***complexity*** (of an algorithm): an asymptotic measure of the number of operations or the running time needed for a complete execution; sometimes, a measure of the total amount of computational space needed.

***complexity*** (of a function): usually, the minimum complexity of any algorithm representing the function; sometimes, the length or complicatedness of the list of instructions.

***complexity*** (of a function), ***Kolmogorov-Chaitin type***: a measure of the minimum complicatedness of any algorithm representing the function, usually according to the number of instructions in the algorithm (and *not* related to its running time).

***coNP***: the complexity class $\Pi_1^p$, which contains every language $A$ such that $\overline{A} \in \Sigma_1^p$.

***concatenation*** (of two languages $L_1$ and $L_2$): the set $\{xy \mid x \in L_1, \ y \in L_2\}$, denoted $L_1 L_2$.

***concatenation*** (of two strings): the result of appending the second string to the right end of the first.

***consequent*** (of a production $\alpha \to \beta$): the string $\beta$ that follows the arrow.

***context-free*** (or ***type 2***) ***grammar***: a grammar in which the antecedent of each production is a single nonterminal.

***context-sensitive*** (or ***type 1***) ***grammar***: a grammar where every production $\alpha \to \beta$ (except possibly $S \to \lambda$) has the form $\alpha = uAv$ and $\beta = uxv$.

***CRCW concurrent read concurrent write***: a PRAM model in which both concurrent reads from and concurrent writes to the same location are allowed.

***CREW concurrent read exclusive write***: a PRAM model where concurrent reads are allowed, but not concurrent writes to the same location.

**cube** (over an alphabet): a word having the pattern $xxx$.

**derivation** (of the string $y$ from the string $x$): a sequence of substitutions, according to the production rules, that transforms string $y$ into string $z$. The notation $x \Longrightarrow^* y$ means that such a derivation exists.

**deterministic finite automaton (DFA)**: a model of a computer for deciding membership in a set.

**emptiness problem for grammars**: deciding whether the language generated by a grammar is empty.

**empty string**: the string of length zero, that is, the string with no symbols; often written $\lambda$.

**equivalence problem for grammars**: deciding whether two grammars are equivalent.

**equivalent automata**: automata that accept the same language.

**equivalent grammars**: grammars that generate the same language.

**EREW exclusive read exclusive write**: a PRAM model in which neither concurrent reads from nor concurrent writes to the same location are allowed.

**existential lower bound** (for an algorithm): a lower bound for its number of execution steps that holds for at least one input.

**existential lower bound** (for a problem): a lower bound for every algorithm that can solve that problem.

**finite automaton**: a DFA or an NFA.

**finite-state machine**: a finite automaton or finite transducer.

**finite-state machine with output**: another name for a finite transducer.

**finiteness problem** (for a grammar): deciding whether the language generated by that grammar is finite.

**finite transducer**: a model of a computer for calculating a function, like a finite automaton, except that it also produces an output string each time it reads an input symbol.

**free monoid** (generated by an alphabet): the set of all strings composable from symbols in the alphabet, with the operation of string concatenation.

**frequency** (of a symbol in a string): the number of occurrences of the symbol in the string.

**Game of Life**: a 2-dimensional cellular automaton designed by John H. Conway.

**Gödel numbering** (of a set): a method for encoding Turing machines as products of prime powers; more generally, a similar one-to-one recursive function on an arbitrary set whose image in $\mathcal{N}$ is a recursive set.

**grammar**: a set of *production* rules of the form $\alpha \rightarrow \beta$ that allow one to derive strings by repeated substitution.

**halting problem**: the problem of designing an algorithm capable of deciding which computations $P(x)$ halt and which do not, where $P$ is a computer program (or Turing machine) and $x$ is a possible input to $P$.

**$\mathcal{C}$-hard language** ($\mathcal{C}$ a class of languages): a language $A$ such that every language in class $\mathcal{C}$ is polynomial-time reducible to $A$.

**Hilbert's tenth problem**:  the (undecidable) problem of determining for an arbitrary multivariate polynomial equation $p(x_1, \ldots, x_n) = 0$ whether there exists a solution consisting of integers.

**inclusion problem for languages**:  deciding whether one language is included in another.

**inherently ambiguous context-free language**:  a context-free language such that every context-free grammar for the language is ambiguous.

**input size**:  the quantity of data supplied as input to a computation.

**interconnection network model**:  a parallel computation model as a digraph in which each vertex represents a processor, and in each phase, each processor communicates with its neighbors and makes a computation.

**inverse of a morphism**:  for morphism $h\colon V^* \to U^*$, the mapping $h^{-1}\colon U^* \to 2^{V^*}$ defined for $x \in U^*$ by $h^{-1}(x) = \{\, y \in V^* \mid h(y) = x \,\}$.

**Kleene closure** (or **Kleene star**):  of a language $L$, the set of all iterated concatenations of zero or more words in $L$, denoted $L^*$.

**language** (accepted by a machine, e.g., finite automaton, PDA, or Turing):  the set of all accepted strings.

**language** (generated by the grammar $G$):  the language $L(G)$ of words consisting of terminal symbols derivable from the start symbol.

**language** (over an alphabet $V$):  a subset of the free monoid $V^*$; that is, a set of strings.

**Las Vegas algorithm**:  an algorithm that always produces correct output, whose running time is a random variable.

**Las Vegas to Monte Carlo transformation**:  the Monte Carlo algorithm obtained by running the Las Vegas scheme for $kE[T]$ steps and halting, where $E[T]$ is the expected Las Vegas running time.

**leftmost derivation**:  a derivation in which at each step the leftmost nonterminal is replaced.

**leftmost language** (generated by the grammar $G$):  the language of strings of terminals with leftmost derivations from the start symbol.

**length set** (of a language $L$):  the set $\{\, |x| \mid x \in L \,\}$.

**length-increasing (or type 1) grammar**:  a grammar in which the consequent $\beta$ of each production $\alpha \to \beta$ (except $S \to \lambda$, if present) is at least as long as its antecedent $\alpha$.

**linear grammar**:  a context-free grammar where each consequent contains at most one nonterminal.

**Mealy machine**:  a finite transducer whose output function always produces a single symbol.

**membership problem** (for a grammar $G$):  given an input string $x$, deciding whether $x \in L(G)$.

**mirror image** (of a language $L$):  the language obtained by reversing every string in $L$.

**Monte Carlo algorithm**:  an algorithm that has a bounded number of computational steps and produces incorrect output with some low probability.

**Monte Carlo to Las Vegas transformation**:  the Las Vegas algorithm of repeatedly running that Monte Carlo algorithm until correct output occurs.

**Moore machine**:  a Mealy machine such that for every state $k$ and every pair of input symbols $s_1$ and $s_2$, the outputs $\tau(k, s_1)$ and $\tau(k, s_2)$ are the same.

**morphism** (from the alphabet $V$ to the alphabet $U$):  a function $s\colon V \to U$.

**nondeterministic finite automaton (NFA)**:  a model like a DFA, but there may be several different states to which a transition is possible, instead of only one.

**nondeterministic polynomial-time computation on a TM**:  a computation such that there exists a polynomial function $p(n)$ such that for any input of size $n$ there is a computational path on the TM whose length is at most $p(n)$ steps.

**nondeterministic Turing machine**:  a model like a deterministic Turing machine, except that the transition function $\Delta$ maps each state-symbol pair to a set of state-symbol-direction triples.

**nonterminal** (in a grammar):  a symbol that may be replaced when a production is applied.

**nontrivial family of languages**:  a family that contains at least one language different from $\emptyset$ and $\{\lambda\}$.

**NP**:  the complexity class of languages nondeterministically TM-decidable in polynomial time.

**NP-complete language**:  a language $A$ such that $A$ is NP-hard and $A \in \text{NP}$.

**NP-complete problem**:  a decision problem equivalent to deciding membership in an NP-complete language.

**NP-hard language**:  a language $A$ such that every language in complexity class NP is polynomial-time reducible to $A$.

**oracle** (for a language):  a machine state that decides whether or not a given string is in that language.

**oracle Turing machine**:  like a Turing machine, except equipped with a special second tape on which it can write a string to query an oracle for a fixed language $L$.

**P**:  the complexity class of languages deterministically TM-decidable in polynomial time.

**palindrome**:  a string that is identical to its reverse.

**parallel computation model**:  a computational model that permits more than one instruction to be executed simultaneously, instead of requiring that they be executed sequentially.

**parsing a string**:  in theoretical computer science, a *derivation*.

**partial function**:  an incomplete rule that assigns values to some elements in its domain but not necessarily to all of them.

**partial function $\phi_M$ induced by a TM** $M$:  the rule that associates to each input $v$ for which $M$'s computation halts the output $\phi_M(v)$, and is otherwise undefined.

**partial recursive function**:  a partial function derivable from the constant zero functions $\zeta_n(x_1, \ldots, x_n) = 0$, the successor function $\sigma(n) = n + 1$, and the projection functions $\pi_i^n(x_1, \ldots, x_n) = x_i$, using multivariate composition, multivariate primitive recursion, and unbounded minimalization.

**pattern** (over an alphabet $V$):  a string of variables over that alphabet; regarded as present in a particular word $w \in V^*$ if there exists an assignment of strings from $V^+$ to the variables in that pattern such that the word formed thereby is a substring of $w$.

**polynomial hierarchy (PH)**: the union of the complexity classes $\Sigma_n^p$, for $n \geq 0$.

**polynomial-space computation** (of a function by a TM $M$): a computation by $M$ of that function such that there exists a polynomial function $p(n)$ such that for every input of size $n$, the calculation workspace takes at most $p(n)$ positions on the tape.

**polynomial-time computation** (of a function by a TM $M$): a computation by $M$ of that function such that there exists a polynomial function $p(n)$ such that for every input of size $n$, the calculation takes at most $p(n)$ steps.

**positive closure (or Kleene plus)**: Of a language $L$, the set of all iterated concatenations of words in $L$ excluding the empty word, denoted $L^+$.

**nth power of a language**: the set of all iterated concatenations $w_1 w_2 \ldots w_n$ where each $w_i$ is a word in the language.

**PRAM memory conflict**: the conflict that occurs when more than one processor attempts concurrently to write to or read from the same global memory register.

**PRAM parallel random access machine**: a model of parallel computation as a set of global memory registers and a set of processors, each with access to its own local registers.

**primitive recursion**: a restricted way of defining $f(n+1)$ in terms of $f(n)$.

**primitive recursive function**: any function derivable from the constant zero functions $\zeta_k(x_1, \ldots, x_k) = 0$, the successor function $\sigma(n) = n + 1$, and the projection functions $\pi_i^n(x_1, \ldots, x_n) = x_i$, using multivariate composition and multivariate primitive recursion.

**probabilistic Turing machine**: a nondeterministic Turing machine $M$ with exactly two choices of a next state at each step, both with probability $\frac{1}{2}$ and independent of all previous choices.

**production rule** (in a grammar): a rule for making a substitution in a string.

**projection function, n-place**: a function $\pi_i^n(x_1, \ldots, x_n) = x_i$ that maps an $n$-tuple to its $i$th coordinate.

**PSPACE**: the complexity class of languages TM-decidable in polynomial space.

**pumping lemma**: any one of several results in formal language theory concerned with rewriting strings; used in impossibility proofs.

**pushdown automaton (PDA)**: a (possibly nondeterministic) finite automaton equipped with a stack.

**random access machine (RAM)**: a computation model with several arithmetic registers and an infinite number of memory registers.

**randomized algorithm**: an algorithm that makes random choices during its execution, guided by the output of a random (or pseudo-random) number generator.

**recursive language**: a language with a decidable membership question.

**recursive set**: a set whose characteristic function is recursive.

**recursively enumerable set**: a set that is either empty or the image of a recursive function.

**reducibility in polynomial-time** (of language $A$ to language $B$): the existence of a polynomial-time computable function $f$ such that $x \in A$ if and only if $f(x) \in B$, for each string $x$ in the alphabet of language $A$; denoted by $A \leq_m^p B$.

**reduction**: a strategy for solving a problem by transforming its natural form of input into the input form for another problem, solving that other problem on the transformed input, and transforming the answer back into the original problem domain.

**regular expression** (over an alphabet $V$): a string $w$ in the symbols of $V$ and the special set $\{\epsilon, ), (, +, *\}$ such that $w \in V$ or $w = \epsilon$, or (continuing recursively) $w = (\alpha\beta), (\alpha + \beta)$, or $\alpha^*$, where $\alpha$ and $\beta$ are regular expressions.

**regular** (or **type 3**) **grammar**: a grammar such that every production $\alpha \to \beta$ has antecedent $\alpha \in N$ and consequent $\beta \in T \cup TN \cup \{\lambda\}$.

**regular language**: a language that can be obtained from elements of its alphabet using finitely many times the operations of union, concatenation, and Kleene star.

**regularity problem** (for grammars): deciding whether $L(G)$ is a regular language.

**reverse** (of the string $x$): the string $x^R$ obtained by writing $x$ backwards.

**running time**: the number of primitive operation steps executed by an algorithm, usually expressed in big-$O$ asymptotic notation (or sometimes $\Theta$-notation) as a formula based on the input size.

**space complexity** (of an algorithm): a measure of the amount of computational space needed in the execution, relative to the size of the input.

**sparse language**: a language $A$ for which there is a polynomial function $p(n)$ such that for every $n \in \mathcal{N}$, there are at most $p(n)$ elements of length $n$ in $A$.

**square** (over an alphabet): the pattern $xx$, or any word having that pattern.

**square-free word**: a word having no subwords with the pattern $xx$.

**start symbol** (in a grammar): a designated nonterminal from which every word of the language is generated.

**state diagram** (for a finite automaton): a labeled digraph whose vertices represent the states and whose arcs represent the transitions.

**string** (accepted by a finite automaton): a string that can cause the automaton to end up in an accepting state, immediately after the last transition.

**string** (accepted by a PDA): an input string that can lead to the stack being empty and the PDA being in an accepting state after the last transition.

**string** (accepted by a TM $M$): a string $w$ such that $M$ halts on input $w$.

**string**: a finite sequence of symbols from some alphabet.

**substitution** (for the alphabet $V$ in the alphabet $U$): a mapping $s: V \longrightarrow 2^{U^*}$, which means that each symbol $b \in V$ may be replaced by any of the strings in the set $s(b)$; extends to strings of $V^*$.

**terminal** (in a grammar): a symbol that cannot be replaced by other symbols.

**time complexity** (of an algorithm): a function representing the number of operations or the running time needed, using the size of the input as its argument.

**total function**: a partial function defined on all of its domain, i.e., a function.

**tractable problem**: a problem that can be solved by an algorithm with polynomial-time complexity.

**$\lambda$-transition** (in an NFA): a transition that could occur without reading any symbols of the input string.

**transition table** (for a DFA): a table whose rows are indexed by the states and whose columns are indexed by the symbols, such that the entry in row $r$ and column $c$ is the state to which the DFA moves if it reads symbol $c$ while in state $r$.

**trapping state** (of a finite automaton): a non-accepting state from which every outward arc is a self-loop.

**trio**: a nontrivial family of languages closed under $\lambda$-free morphisms, inverse morphisms, and intersection with regular languages.

**Turing-acceptable language**: a language that has a TM $M$ that accepts it.

**Turing-computable function**: a function $f$ such that there is a TM $M$ with $f = \phi_M$.

**Turing-decidable language**: a language whose characteristic function is Turing-computable.

**Turing machine (TM)**: an automaton whose head can move one character in either direction and that can replace the symbol it reads by a different symbol.

**Turing-p-reducibility** (of language $A$ to language $B$): the existence of a deterministic oracle TM $M^B$ that decides language $A$ in polynomial time. Notation: $A \leq_T^p B$.

**Turing's test** (of whether a given computer can think): are its responses to a series of written questions indistinguishable from human responses (by a person who does not know who gave the responses)?

**type 0 grammar**: a grammar with no restrictions.

**type 1 grammar**: a length-increasing grammar, or equivalently, a context-sensitive grammar.

**type 2 grammar**: a context-free grammar.

**type 3 grammar**: a regular grammar.

**unambiguous context-free language**: a context-free language that has a context-free grammar that is not ambiguous.

**unbounded minimalization**: a way of using a function or partial function to define a new function or partial function.

**uncomputable function**: a function whose values cannot be calculated by a Turing machine (or by a computer program).

**undecidable problem**: a decision problem whose answers cannot be given by a Turing machine (or by a computer program).

**universal Turing machine**: a TM that can simulate every TM.

**unsolvable problem**: a problem that cannot be decided by a recursive function. Equivalently, an undecidable problem.

**variable** (over an alphabet $V$): a symbol not in $V$ whose values range over $V^*$.

**word**: usually a finite sequence of symbols (same as *string*), sometimes a countably infinite sequence.

**word equation** (over an alphabet $V$): an expression $\alpha = \beta$, such that $\alpha$ and $\beta$ are words containing letters from $V$ and some variables over $V$.

**word inequality**: the negation of a word equation, commonly written as $\alpha \neq \beta$.

**worst-case running time**: the maximum number of execution steps of an algorithm, usually expressed in big-$O$ asymptotic notation (or sometimes $\Theta$-notation) as a formula based on the input size.

# 17.1  COMPUTATIONAL MODELS

The objective of a computer, no matter what input/output or memory devices are attached, is ultimately to make logical decisions and to calculate the values of functions. Any decision problem can be represented as recognizing whether an input string is in a specified subset; calculating a function amounts to accepting an input string and producing an output string. At this fundamental level, the fundamental models in Table 1 can serve as the theoretical basis for all sequential computers.

**Table 1: Fundamental computational models.**

| model | description | comment |
|---|---|---|
| DFA $= (K, s, F, \Sigma, \delta)$ <br> $K =$ set of states, start at <br> $s \in K$, accepting states $F \subseteq K$, <br> input alphabet $\Sigma$, <br> transition fn $\delta \colon K \times \Sigma \to K$ | deterministic finite automaton: scans a tape once; decides whether to accept | recognizes regular languages |
| NFA $= (K, s, F, \Sigma, \Delta)$ <br> $K, s, F, \Sigma$ like DFA, <br> transition fn $\Delta \colon K \times \Sigma^* \to \mathcal{P}(K)$ | nondeterministic finite automaton | equivalent power to DFA |
| $(K, s, \Sigma_I, \Sigma_O, \delta, \tau)$ <br> $K =$ set of states, start at <br> $s \in K$, in-alphabet $\Sigma_I$, <br> out-alphabet $\Sigma_O$, <br> transition fn $\delta \colon K \times \Sigma_I \to K$, <br> output fn $\delta \colon K \times \Sigma_I \to \Sigma_O^*$ | Finite transducer: also called "finite-state machine with output" | Special cases: Mealy machine writes one output symbol per input symbol; in Moore machine, output symbol depends only on state before transition |
| PDA $= (K, s, F, \Sigma, \Gamma, \Delta)$ <br> $K, s, F, \Sigma$ as DFA, stack <br> alphabet $\Gamma$, transition fn <br> $\Delta \colon K \times \Sigma^* \times \Gamma^* \to \mathcal{P}(K \times \Gamma^*)$ | Pushdown automaton: uses stack for storage, nondeterministic | recognizes context-free languages |
| TM $= (K, s, h, \Sigma, \delta)$ <br> $K =$ set of states, start at <br> $s \in K$, halting state $h \notin K$, <br> alphabet $\Sigma$, transition fn <br> $\delta \colon K \times \Sigma \to K \times \Sigma \times \{L, R\} \cup \{h\}$ | Turing machine: has two-way tape with rewritable symbols | decides membership in recursive sets; equivalent power to nondeterministic version |

## 17.1.1  FINITE-STATE MACHINES

### Definitions:

If $\Sigma$ is a set of symbols, then $\Sigma^*$ is the set of all (finite) strings made from symbols of $\Sigma$.

A **deterministic finite automaton**, abbreviated **DFA**, models a computer for decision-making as a 5-tuple $(K, s, F, \Sigma, \delta)$ such that

- $K$ is a finite set of **states**;
- $s \in K$ (**starting state**);
- $F \subseteq K$ (**accepting states**);

- $\Sigma$ is a finite **alphabet** of **symbols**;
- $\delta : K \times \Sigma \to K$ (**transition function**).

A DFA is also known as a **(deterministic) finite-state recognizer**.

The **computer model** for a DFA consists of a logic box, programmed by the transition function $\delta$. It has a read-only head that examines an input tape while moving in only one direction. Whenever it reads symbol $c$ on the tape while in state $q$, the automaton switches into state $\delta(q, c)$ and moves to the next symbol. The string is considered to be accepted if the automaton is in an accepting state after the last transition.

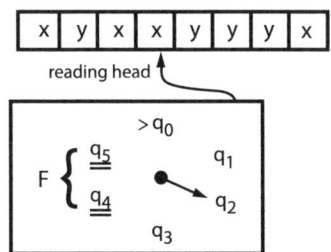

A **nondeterministic finite automaton**, or **NFA**, is a 5-tuple $(K, s, F, \Sigma, \Delta)$ like a DFA, except that $\Delta$ is a function from $K \times \Sigma^*$ to $\mathcal{P}(K)$ (the set of all subsets of $K$).

The **computer model** for an NFA is like the computer model for a DFA. However, if it reads string $u$ on the input string while in state $q$, the automaton may switch to any of the states in the set $\Delta(q, u)$ and move to the symbol after $u$. The string is considered to be accepted if *it is possible for* the automaton to be in an accepting state after the last transition.

A **finite automaton** is either a DFA or an NFA.

A **state diagram** for a DFA is a digraph with vertex set $K$ such that for each state $q$ and each symbol $c$ there is an arc from vertex $q$ to vertex $\delta(q, c)$, labeled with $c$. (Sometimes a single arc is labeled with multiple symbols, instead of drawing multiple arcs between the same pair of states.) The starting state is designated by an entering arrow and the accepting states by double circles. A state diagram for an NFA is similar, except that for each state $p$ in $\Delta(q, u)$ there is an arc from vertex $q$ to vertex $p$ labeled with the string $u$, which may be the empty string $\lambda$.

A **transition table** for a DFA is a table with rows indexed by the states in $K$ and columns indexed by the symbols in $\Sigma$, such that the entry in row $q$ and column $c$ is $\delta(q, c)$. The starting-state row label is marked with a ">" and the accepting-state row labels are underscored.

A **configuration** for a finite automaton is a pair $(q, w)$ with $q \in K$ and $w \in \Sigma^*$. It signifies that the automaton is in state $q$ with the head at the first symbol of string $w$. (Since the head moves in only one direction, a common assumption is that it consumes each symbol as it reads.)

The DFA configuration $(q, w)$ **yields the configuration** $(q', w')$ **in one step** if deleting the initial symbol, call it $c$, of the string $w$ gives the string $w'$, and $\delta(q, c) = q'$. Notation: $(q, w) \vdash_M (q', w')$.

The NFA configuration $(q, w)$ **yields the configuration** $(q', w')$ **in one step** if there is an initial prefix $u$ on the string $w$ whose deletion gives the string $w'$, and $q' \in \Delta(q, u)$.

The configuration $(q, w)$ **yields** the configuration $(q', w')$ if there is a sequence of configurations $(q, w) = (q_0, w_0), (q_1, w_1), \ldots, (q_n, w_n) = (q', w')$ such that $(q_{i-1}, w_{i-1}) \vdash_M (q_i, w_i)$, for $i = 1, \ldots, n$. Notation: $(q, w) \vdash_M^* (q', w')$.

A string $w$ is **accepted by finite automaton** $M$ if there is an accepting state $q$ such that $(s, w) \vdash_M^* (q, \lambda)$. That is, machine $M$ accepts string $w$ if, starting in state $s$ at the first symbol, its transition sequence can ultimately lead to an accepting state, immediately after its last transition.

The **language accepted by a finite automaton** $M$ is the set of all strings accepted by $M$. It is denoted $L(M)$.

Finite automata $M_1$ and $M_2$ are **equivalent** if $L(M_1) = L(M_2)$; that is, if they accept the same language.

A **trapping state** of a finite automaton is a non-accepting state where every outward arc is a self-loop. To simplify state diagrams, trapping states are often not explicitly drawn.

A (**deterministic**) **finite transducer**, also known as a **finite-state machine with output**, models a function-calculating computer as a 6-tuple $(K, s, \Sigma_I, \Sigma_O, \delta, \tau)$ such that

- $K$ is a finite set of **states**;
- $s \in K$ (**starting state**);
- $\Sigma_I$ is a finite alphabet of **input symbols**;
- $\Sigma_O$ is a finite alphabet of **output symbols**;
- $\delta \colon K \times \Sigma_I \to K$ (**transition function**);
- $\tau \colon K \times \Sigma_I \to \Sigma_O^*$ (**output function**).

A **Mealy machine** is a finite transducer whose output function always produces a single symbol.

A **Moore machine** is a Mealy machine such that for every state $k$ and every pair of input symbols $s_1$ and $s_2$, the outputs $\tau(k, s_1)$ and $\tau(k, s_2)$ are the same.

A **finite-state machine** is a finite automaton or a finite transducer.

**Facts:**

**1.** Finite-state machines are the design plan of many practical electronic control devices, for instance in wristwatches or automobiles.

**2.** The phrase "finite-state machine" refers to a finite-state model that may or may not have output capacity and that may or may not be nondeterministic.

**3.** For every NFA, there is an equivalent DFA. (M. Rabin and D. Scott, 1959)

**4.** In software design, NFAs are commonly used in preference to DFAs because they often achieve the same task with fewer states.

**5.** The nondeterminism of an NFA is that possibly $u = \lambda$, or that possibly there are two states $p$ and $p'$ in $\Delta(q, u)$, so that in state $q$ the NFA might read substring $u$ and switch into either state $p$ or state $p'$.

**6.** NFAs are often defined so that $\lambda$-transitions are the only possible instances of nondeterminism. In this seemingly more restrictive kind of NFA, the transition function is a map from $K \times (\Sigma \cup \{\lambda\})$ to $K$.

**7.** The class of languages accepted by finite automata is closed under all of the following operations:

- union;
- concatenation;
- Kleene star (see §17.3.2);
- complementation;
- intersection.

**8.** *Kleene's theorem*:  A language is regular (§17.3.4) if and only if it is the language accepted by some finite automaton. (S. Kleene, 1956)

**9.** Some lexical scanning processes of compilers use finite automata.

**10.** The relation "yields" for finite automata is the reflexive, transitive closure of the relation $\vdash_M$.

**11.** Both Moore machines and Mealy machines have the computational capability of an unrestricted finite transducer.

**12.** More comprehensive coverage of finite-state machines is provided by many textbooks, including [Si14] and [LePa97].

**Examples:**

**1.** The DFA specified by the following transition table and state diagram decides whether a binary string has an even number of 1s. Formally, $M = (K, s, F, \Sigma, \delta)$ with $K = \{\text{Even, Odd}\}$, $s = \text{Even}$, $F = \{\text{Even}\}$, and $\Sigma = \{0, 1\}$.

**2.** In one early form of the language BASIC, an identifier could be a letter or a letter followed by a digit. The following state diagram specifies a DFA that accepts this:

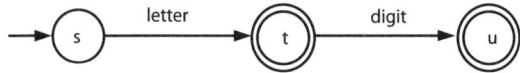

**3.** A "proper fixed-point numeral" is a nonempty string of decimal digits, followed by a decimal point, and then another nonempty string of digits. For instance, the number zero would be represented as "0.0". The following DFA decides whether the input string is such a numeral.

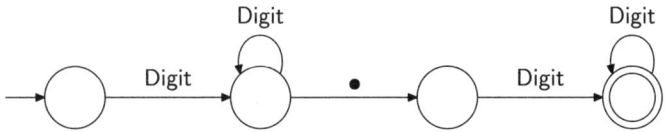

**4.** An "integer" in some programming languages is a nonempty string of decimal digits, possibly preceded by a sign + or −. The following NFA recognizes such an integer.

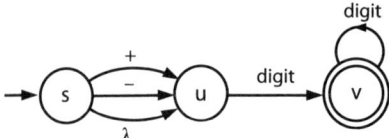

**5.** The following finite-state transducer has $\{0, 1, \$\}$ and $\{0, 1\}$ for its input and output alphabets, respectively, where "$\$$" serves as an end-of-string marker. It reads a binary numeral, starting at the units digit, and prints a binary numeral whose value is double the input numeral.

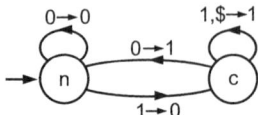

**6.** The following finite-state machine models a vending machine for a 20-cent toy. The possible inputs are nickels and dimes, and a push of a button releases the toy if enough change has been deposited. Each state indicates the amount that has been deposited. This machine may be regarded as a transducer that produces symbol T (toy) if it receives input B while in state 20.

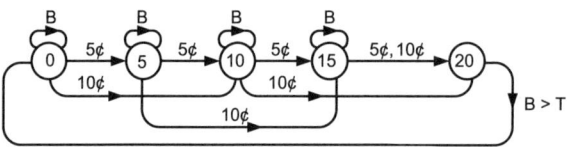

**7.** NFAs can be used to model various solitaire games and puzzles, such as making a complete knight's tour of a chessboard.

---

### 17.1.2  PUSHDOWN AUTOMATA

**Definitions:**

A **pushdown automaton** (**PDA**) is essentially a (possibly nondeterministic) finite automaton equipped with a stack. It is given by a 6-tuple $(K, s, F, \Sigma, \Gamma, \Delta)$ such that

- $K$ is a finite set of **states**;
- $s \in K$ (**starting state**);
- $F \subseteq K$ (**accepting states**);
- $\Sigma$ is a finite **alphabet of input symbols**;
- $\Gamma$ is a finite **alphabet of stack symbols**;
- $\Delta \colon K \times \Sigma^* \times \Gamma^* \to \mathcal{P}(K \times \Gamma^*)$ (**transition function**).

A **transition** of a PDA is an element $(q, \gamma)$ in some $\Delta(p, u, \beta)$. The idea is that from state $p$, the PDA may read the substring $u$ and the stack substring $\beta$, and transfer into state $q$ while popping $\beta$ and pushing $\gamma$, thereby replacing $\beta$ by $\gamma$.

*Note*: A PDA is frequently defined so that the only strings that can be read or written or pushed or popped are single characters and the empty string. This has no effect on the computational generality.

The **computer model** for a PDA consists of a logic box, programmed by $\Delta$, equipped with a read-only head that examines an input tape while moving in only one direction, and also equipped with a stack. When it reads substring $u$ from input while in state $p$ with substring $\beta$ at the top of the stack, the automaton selects a corresponding entry from $\Delta$ and makes that transition.

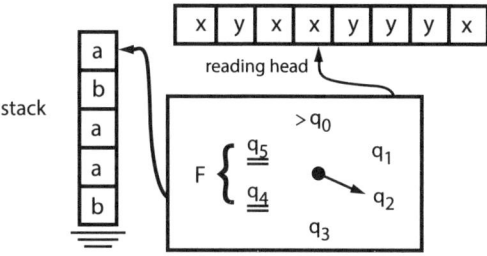

A **configuration** for a PDA is a triple $(p, u, \beta)$ such that $p \in K$, $u \in \Sigma^*$, and $\beta \in \Gamma^*$.

The PDA configuration $(p, ux, \beta\alpha)$ **yields the configuration** $(q, x, \gamma\alpha)$ **in one step** if there is a transition $((p, u, \beta), (q, \gamma))$. Notation: $(p, ux, \beta\alpha) \vdash_M (q, x, \gamma\alpha)$.

The PDA configuration $C$ **yields** the configuration $C'$, denoted $C \vdash_M^* C'$, if there is a sequence of configurations $C = C_0, C_1, \ldots, C_n = C$ such that $C_{i-1} \vdash_M C_i$, for $i = 1, \ldots, n$.

A string $w \in \Sigma^*$ is **accepted** by a PDA $M$ if there is an accepting state $q$ such that $(s, w) \vdash_M^* (q, \lambda, \lambda)$. That is, it accepts string $w$ if, starting in state $s$ at the first symbol, its transition sequence can ultimately lead to an accepting state and an empty stack after it has read the last symbol.

The **language accepted** by a PDA $M$ is the set of all strings accepted by $M$. It is denoted $L(M)$.

A **state diagram** for a PDA is a digraph with vertex set $K$ such that for each pair $(q, \gamma)$ in $\Delta(p, u, \beta)$ there is an arc from vertex $p$ to vertex $q$, labeled $u, \beta \to \gamma$. (Sometimes a single arc is labeled with multiple symbols, instead of drawing multiple arcs between the same pair of states.) The starting state is designated by an entering arrow, and the accepting states by a double circle.

## Facts:

**1.** The PDA model was invented by A. G. Oettinger in 1961.

**2.** A language $L$ is context-free (see §17.3.3) if and only if there is a pushdown automaton $M$ such that $L$ is the language accepted by $M$. (M. Schützenberger 1963, and independently by N. Chomsky and by J. Evey)

**3.** A PDA can test whether a string is a palindrome, or whether all the left and right parentheses are matched, but a DFA cannot.

**4.** The class of languages accepted by deterministic PDAs is smaller than the class accepted by nondeterministic PDAs.

**5.** More comprehensive coverage of pushdown automata is provided by many textbooks, including [Si14] and [LePa97].

**Examples:**

**1.** The following PDA decides whether a sequence of left and right parentheses is well-nested, in the sense that every left parenthesis is uniquely matched to a right parenthesis, and vice versa. It is necessary and sufficient that, in counting left and right parentheses while reading from left to right, the number of right parentheses never exceeds the number of left parentheses and that the total counts are the same.

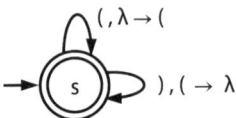

**2.** The following PDA decides whether a string from the alphabet $\{0, 1, m\}$ has the form $bmb^r$ where $b$ is a bitstring and $b^r$ its reverse. The middle $m$ signals when to switch from pushing symbols onto the stack to popping them off.

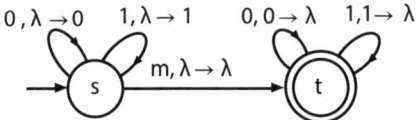

**3.** The following nondeterministic PDA decides whether a binary string is of the form $bb^r$, that is, an even-length palindromic bitstring. In effect, it considers every character interspace as the possible middle.

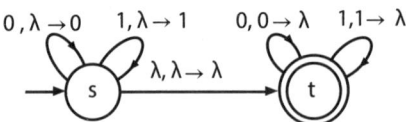

---

### 17.1.3   TURING MACHINES

**Definitions:**

A **Turing machine** (**TM**) models a computer as a 5-tuple $(K, s, h, \Sigma, \delta)$ such that

- $K$ is a finite set of **states**;
- $s \in K$ (**starting state**);
- $h \notin K$ (**halting state**);
- $\Sigma$ is a finite set of symbols, including the **blank** #; (**alphabet**);
- $\delta \colon K \times \Sigma \to (K \times \Sigma \times \{L, R\}) \cup \{h\}$ (**transition function**).

The **computer model** for a Turing machine consists of a logic box, programmed by $\delta$, equipped with a read-write head that examines an input tape with a left end but no right end. To start a computation, the input string is written at the left end of the tape, and the rest of the tape filled with blanks. The head starts at the leftmost symbol. Whenever the Turing machine reads symbol $b$ on the tape while in state $q$, its internal logic produces the triple $\delta(q, b) = (p, c, D)$, and the machine switches into state $p$, replaces $b$ by $c$, and moves one space in direction $D$, that is, either to the left ($L$) or to the right ($R$).

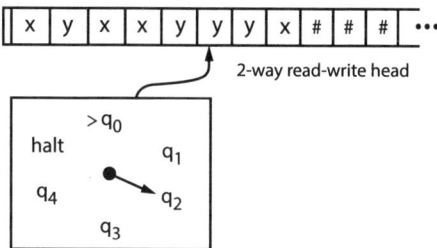

A **transition table** for a Turing machine is a table whose rows are labeled with the states and whose columns are labeled with the symbols, such that the entry in row $q$ and column $b$ is $\delta(q, b)$. Thus, a typical table entry is a triple indicating to which state to switch from $q$, the symbol to replace $b$, and whether to move one square to the right or to the left. However, another possibility is that a table entry is the halt state $h$.

A **configuration** for a Turing machine is a quadruple $(q, u, b, v)$, such that $q \in K \cup \{h\}$, $u, v \in \Sigma^*$, and $b \in \Sigma$, commonly written as $(q, u\underline{b}v)$. This means that the Turing machine is in state $q$, that the present value of the tape is $ubv$, that the head is presently at the indicated $b$, and that the rest of the tape to the right is filled with blanks.

The Turing machine configuration $(p, u\underline{b}v)$ **yields the configuration** $(q, x\underline{c}y)$ **in one step** if the transition $\delta(p, b)$ changes configuration $(p, u\underline{b}v)$ to configuration $(q, x\underline{c}y)$. This is denoted by $(p, u\underline{b}v) \vdash_M (q, x\underline{c}y)$.

The Turing machine **accepts** string $w$ if it enters the halting state $h$ after starting with $w$ on the tape.

An **infinite loop** for a Turing machine is an infinite sequence of configurations $C_0$, $C_1$, $C_2, \ldots$ such that $C_{i-1} \vdash_M C_i$, for $i = 1, 2, \ldots$.

A Turing machine **hangs** if it moves left off the left end of the tape.

The **output** of a Turing machine is the string from the left end of the tape up to the last non-blank character if the computation halts, and undefined otherwise.

The **partial function $\phi_M$ induced by a Turing machine** $M$ maps each input $v$ for which the computation halts to the output $\phi_M(v)$, and is otherwise undefined.

A function $f: \Sigma^* \to \Sigma^*$ is **Turing-computable** if there is a Turing machine $M$ that halts for all inputs, and computes the function $f$, that is, $f(v) = \phi_M(v)$ for all $v \in \Sigma^*$.

A Turing machine $M = (K, s, h, \Sigma, \delta)$ **simulates** another Turing machine $M' = (K', s', h, \Sigma', \delta')$ if there exists a Turing-computable function $\beta: \Sigma'^* \to \Sigma^*$ such that $\phi_M(\beta(w)) = \phi_{M'}(w)$ for all $w \in domain(\phi_{M'})$ and $\phi_M(\beta(w))$ is undefined for all $w \notin domain(\phi_{M'})$.

A **universal Turing machine** is a Turing machine $U = (K_U, s_U, h_U, \Sigma_U, \delta_U)$ that can simulate every Turing machine, in the following sense. There is a rule $\alpha_U$ for encoding any Turing machine $M$ and a rule $\beta_U$ for encoding any input $w$ to $M$, such that $\phi_U(\alpha_U(M)\#\beta_U(w))$ is defined and equals $\phi_M(w)$ whenever $\phi_M(w)$ is defined, and is undefined otherwise.

A language $L \subseteq \Sigma^*$ is **Turing-acceptable** if there exists a Turing machine $M$ that accepts it. That is, $L = \{w \in \Sigma^* \mid M$ accepts $w\}$.

The **characteristic function** $\chi_L: \Sigma^* \to \{yes, no\}$ of a language $L \subseteq \Sigma^*$ is given by the rule $\chi_L(w) = yes$ if $w \in L$, and $no$ otherwise.

A language is **Turing-decidable** if its characteristic function is Turing-computable.

A (subset-membership) decision problem is **undecidable** if it does not correspond to a Turing-decidable language.

An **$n$-state busy beaver machine** is an $n$-state Turing machine on the alphabet $\Sigma = \{\#, 1\}$ that accepts a two-way infinite input tape filled with #s and halts after placing a maximum number of 1s on the tape. (The name *busy beaver* is from an analogy between the machine piling up 1s and a beaver piling up logs.)

The **busy beaver function** $BB(n)$ has as its value the number of 1s on the output tape of an $n$-state busy beaver machine.

A **nondeterministic Turing machine** is defined like a Turing machine, except that instead of a transition function that assigns a unique change of symbol and direction of motion for the head, there is a transition function that permits more than one possibility.

A **linear bounded automaton** (or **LBA**) is representable as a (possibly nondeterministic) Turing machine that is fed only the finite stretch of tape containing the input word, rather than an infinite tape.

**Facts:**

**1.** A Turing machine is commonly regarded as a program to compute the partial function $\phi_M$.

**2.** Every Turing-decidable language is Turing-acceptable.

**3.** If a language $L \subseteq \Sigma^*$ is Turing-decidable, then its complement $\overline{L}$ is also Turing-decidable.

**4.** Every Turing-acceptable language $L \subseteq \Sigma^*$ whose complement $\overline{L}$ is also Turing-acceptable is a Turing-decidable language.

**5.** The following problems about Turing machines are undecidable:

    (a) Given a TM $M$ and an input string $w$, does $M$ halt on input $w$?

    (b) Given a TM $M$, does $M$ halt on the empty tape?

    (c) Given a TM $M$, does there exist an input $w$ for which $M$ halts?

    (d) Given a TM $M$, does $M$ halt on every input string?

    (e) Given two TMs $M_1$ and $M_2$, do they accept the same input?

    (f) Given two numbers $n$ and $k$, is $BB(n) > k$?

**6.** In view of part (d) of the preceding fact, there is no way to tell whether an arbitrary computer program in a general language always halts, much less whether it calculates what it is supposed to calculate.

**7.** It is possible to construct a universal Turing machine. A universal Turing machine with six states and four symbols was constructed in 1982 by Y. Rogozhin.

**8.** The busy beaver problem was invented by Tibor Radó in 1962.

**9.** Turing machines have been extended in several ways, including: infiniteness in two directions, more work tapes, and two-dimensional tapes.

**10.** Some of the extensions of a Turing machine can perform computations more quickly and are easier to program.

**11.** Any function that can be computed by a Turing machine with (a) a two-way infinite tape, (b) $k$ tapes, and/or (c) two-dimensional tapes, can also be computed by some standard Turing machine.

**12.** Any language that can be decided by a nondeterministic Turing machine can also be decided by some standard Turing machine.

**13.** The following table gives known values and lower bounds for the busy beaver function:

| $n$ | 1 | 2 | 3 | 4 | 5 | 6 |
|------|---|---|---|---|------|------|
| $BB(n)$ | 1 | 4 | 6 | 13 | $\geq 4098$ | $> 10^{18267}$ |

**14.** A linear bounded automaton is less powerful than a Turing machine. However, it is more powerful than a pushdown automaton.

**15.** Alan M. Turing (1912–1954) was a British mathematician whose cryptanalytic work during World War II led to the decryption of ciphertext from the German cipher machine called the Enigma. This work was dramatized in the 2014 film "The Imitation Game".

**16.** Turing proposed that a machine be regarded as "thinking" if its responses to written questions cannot be distinguished from those of a person. This criterion is called *Turing's test*.

**17.** More comprehensive coverage of Turing machines is provided by many textbooks, including [Si14] and [LePa97].

**Examples:**

**1.** This is a 1-state Turing machine with alphabet $\Sigma = \{0, 1, \#\}$ that changes every character preceding the first blank into a blank. It accepts any string over its alphabet.

|  | 0 | 1 | # |
|------|------|------|---|
| $\rightarrow a$ | $\#aR$ | $\#aR$ | $h$ |

**2.** This 3-state Turing machine with alphabet $\Sigma = \{0, 1, \#\}$ doesn't change its input tape at all. It halts whenever it encounters the third '1'. Thus, it accepts any tape with at least three 1s but accepts no other strings.

|  | 0 | 1 | # |
|------|------|------|------|
| $\rightarrow a$ | $0aR$ | $1bR$ | $\#aR$ |
| $b$ | $0bR$ | $1cR$ | $\#bR$ |
| $c$ | $0cR$ | $h$ | $\#cR$ |

**3.** This 3-state Turing machine with alphabet $\Sigma = \{1, \#\}$ adds two positive integers, each represented as a string of 1s. For instance, the tape $111\#11\#\#\#\cdots$ becomes $11111\#\#\#\cdots$.

|  | 1 | # |
|------|------|------|
| $\rightarrow a$ | $1aR$ | $1bR$ |
| $b$ | $1bR$ | $\#cL$ |
| $c$ | $\#cR$ | $h$ |

**4.** This 2-state Turing machine shows that $BB(2)$ is at least 4.

|  | # | 1 |
|------|------|------|
| $\rightarrow a$ | $1bL$ | $1bR$ |
| $b$ | $1aR$ | $h$ |

**5.** This 3-state Turing machine shows that $BB(3)$ is at least 6.

|              | #      | 1      |
| -----------: | :----- | :----- |
| $\rightarrow a$ | $1bR$  | $1cL$  |
| $b$          | $1cR$  | $h$    |
| $c$          | $1aL$  | $\#bL$ |

## 17.1.4   PARALLEL COMPUTATIONAL MODELS

### Definitions:

A *parallel computation model* permits more than one instruction to be executed simultaneously, instead of requiring that they be executed sequentially.

An *interconnection network* models parallel computation as a digraph in which each vertex represents a processor. In each phase a processor communicates with its neighbors and makes a computation.

An $n$-dimensional *cellular automaton* is an interconnection network in which there is a processor at each integer lattice point of $n$-dimensional Euclidean space, and each processor communicates with its immediate neighbors.

A *random access machine* (**RAM**) has several arithmetic registers and an infinite number of memory registers (often modeled as an infinite array), each of which can be accessed immediately via its address (the index in the array).

A *parallel random access machine* (**PRAM**) models parallel computation as a set of *global memory registers* $\{ M_j \mid j = 1, 2, \ldots \}$ and a set of *processors* $\{ P_j \mid j = 1, 2, \ldots \}$. Each processor $P_j$ also has sole access to an infinite sequence of *local registers*.

In a PRAM, a *processor* performs *read* and *write* instructions involving global memory, and other instructions involving only its local memory. All processors of a PRAM perform the same program in synchrony, so that all processors that are not idle are performing their task under the same instruction of the program.

In a PRAM, the concurrent construct $\mathbf{par}[a \leq j \leq b]P_j : S_j$ means that each processor $P_j$ for $a \leq j \leq b$ is performing the operation $S_j$. The operation could be a calculation, a read from some global register, or a write to some global register.

In a PRAM, a *computation* starts when all the processors execute the first instruction. It stops when all processors halt. The contents of the global memory are regarded as the output.

In a PRAM, a *memory conflict* occurs when more than one processor attempts concurrently to write to or read from the same global memory register.

- In *exclusive read exclusive write* (**EREW**), neither concurrent reads from nor concurrent writes to the same location are allowed.

- In *concurrent read exclusive write* (**CREW**), concurrent reads are allowed, but not concurrent writes to the same location.

- In *concurrent read concurrent write* (**CRCW**), both concurrent reads from and concurrent writes to the same location are allowed.

In the *common PRAM* (**CRCW$^{com}$ PRAM**) model, concurrent writes to the same location are permitted provided all processors are trying to write the same data.

**Facts:**

**1.** Some commercially available parallel computers have an array of elements in which a single broadcast instruction applying to every element is executed simultaneously for all the elements.

**2.** Random access machines are considered theoretical models of real sequential computers.

**3.** PRAM programs are often described using high-level programming language constructs for array processing that are similar to sequential array processing, except that the PRAM array locations are processed in parallel.

**4.** Whereas $O(n)$ is regarded as fast for sequential processing, a parallel algorithm tends to be regarded as fast if it runs in $O(\log n)$ time, where $n$ is the size of the input.

**Examples:**

**1.** A parallel computer for *sorting* up to $n$ items can be modeled as a row of processors $P_1, P_1, \ldots, P_n$ each joined to its immediate predecessor $P_{j-1}$ and its immediate successor $P_{j+1}$. On the first phase and on all subsequent odd-numbered phases, each processor pair $(P_{2j+1}, P_{2j+2})$ compares items and swaps, if necessary, so that the smaller item ends up in the lower-indexed processor. On the second phase and on all subsequent even-numbered phases, each processor pair $(P_{2j}, P_{2j+1})$ compares items and swaps, if necessary, so that the smaller item ends up in the lower-indexed processor. After $n$ phases, the items are sorted in ascending order.

**2.** *EREW PRAM: finding the maximum:*   Given $n$ numbers, with $n = 2^r$, store the numbers in global registers $M_n, \ldots, M_{2n-1}$. Then execute the following:

> **for** $i = r - 1$ **downto** $0$
>      **par**$[2^i \le j \le 2^{i+1}] \, P_j : M_j := \max\{M_{2j}, M_{2j+1}\}$

The maximum ends up in global register $M_1$.

**3.** *$CRCW^{com}$ PRAM: finding the maximum:*   Given $n$ numbers, store the numbers in global registers $M_1, \ldots, M_n$. Use $n^2$ processors $P_{i,j}, 1 \le i, j \le n$ as follows:

> **par**$[1 \le i, j \le n] \, P_{i,j} : M_{i+n} := 0$
> **par**$[1 \le i, j \le n] \, P_{i,j} : $ **if** $M_i < M_j$ **then** $M_{i+n} := 1$
>      $\{M_{i+n} = 0$ if and only if $M_i = \max\{M_1, \ldots, M_n\}\}$
> **par**$[1 \le i, j \le n] \, P_{i,j} : $ **if** $M_{n+i} = 0$ **then** $M_0 := M_i$

This program is much faster than the EREW program, because all pairs are compared simultaneously in a single parallel step.

**4.** *Game of Life:*   The "Game of Life", invented by mathematician John H. Conway, is played on an infinite checkerboard. The *neighbors* of a square are the eight squares that touch it, including at its corners. In the initial configuration $c_0$ of the game, some squares are regarded as *alive* and all others *dead*. Each configuration $c_k$ gives birth to a new configuration $c_{k+1}$, according to the following rules:

- a live cell remains alive if it has either exactly two or exactly three live neighbors;

- a dead cell comes alive if and only if it has exactly three live neighbors.

The following sequence of configurations illustrates these rules.

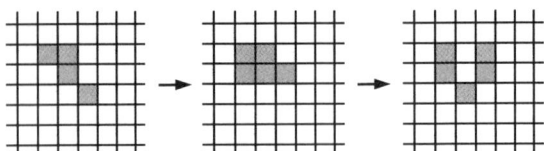

The "Game of Life" can be regarded as a cellular automaton in which the squares are the processors and each processor is joined to its eight neighbors.

**5.** A configuration in the "Game of Life" has *periodicity* $n$ if every $n$th configuration is the same, and $n$ is the smallest such number. Here are three periodic configurations.

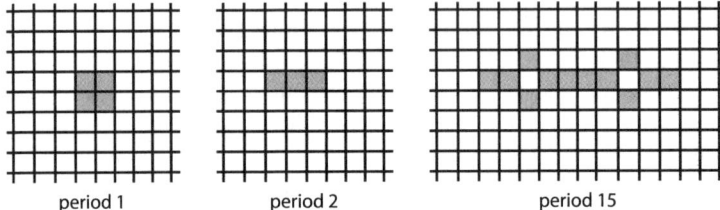

period 1                  period 2                       period 15

# 17.2  COMPUTABILITY

The theory of computability is concerned with distinguishing what can be computed from what cannot. This is not a question of skill at performing calculations. The remarkable truth is that the impossibility of computing certain functions can be proved from the definition of what it means to compute a function.

## 17.2.1  RECURSIVE FUNCTIONS AND CHURCH'S THESIS

The implicit domain for the theory of computability is the set $\mathcal{N}$ of natural numbers. The encoding of problems concerned with arbitrary objects into terms of natural numbers permits general application of this theory.

**Definitions:**

The $n$-place **constant zero function** is the function $\zeta_n(x_1, \ldots, x_n) = 0$.

The **successor function** is the function $\sigma(n) = n + 1$.

The $i$th $n$-place **projection function** is the function $\pi_i^n(x_1, \ldots, x_n) = x_i$.

The (**multivariate**) **composition** of the $n$-place function $f(x_1, \ldots, x_n)$ and the $n$ $m$-place functions $g_1(x_1, \ldots, x_m), \ldots, g_n(x_1, \ldots, x_m)$ is the $m$-place function

$$h(x_1, \ldots, x_m) = f(g_1(x_1, \ldots, x_m), \ldots, g_n(x_1, \ldots, x_m)).$$

(**Multivariate**) **primitive recursion** uses a previously defined $(n+2)$-place function $f(x_1, \ldots, x_{n+2})$ and a previously defined $n$-place function $g(x_1, \ldots, x_n)$ to define the following new $(n+1)$-place function:

$$h(x_1, \ldots, x_{n+1}) = \begin{cases} g(x_1, \ldots, x_n) & \text{if } x_{n+1} = 0; \\ f(x_1, \ldots, x_{n+1}, h(x_1, \ldots, x_n, x_{n+1} - 1)) & \text{otherwise.} \end{cases}$$

**Unbounded minimalization** uses an $(n+1)$-place function $f(x_1, \ldots, x_{n+1})$ to define the following new $n$-place function, which is denoted $\mu_m[f(x_1, \ldots, x_n, m) = 0]$:

$$g(x_1, \ldots, x_n) = \begin{cases} \text{the least } y \text{ such that } f(x_1, \ldots, x_n, y) = 0 & \text{if it exists;} \\ 0 & \text{otherwise.} \end{cases}$$

An $(n+1)$-place function $f(x_1, \ldots, x_{n+1})$ is **regular** if for every $n$-tuple $(x_1, \ldots, x_n)$ there is a $y \in \mathcal{N}$ such that $f(x_1, \ldots, x_n, y) = 0$.

The class $\mathcal{P}$ of **primitive recursive functions** is the smallest class of functions that contains

- the constant zero functions $\zeta_n(x_1, \ldots, x_n) = 0$, for all $n \in \mathcal{N}$;
- the successor function $\sigma(n) = n + 1$;
- the projection functions $\pi_i^n(x_1, \ldots, x_n) = x_i$, for all $n \in \mathcal{N}$ and $1 \leq i \leq n$;

and is closed under multivariate composition and multivariate primitive recursion.

The class $\mathcal{RF}$ of **recursive functions** is the smallest class of functions that contains

- the constant zero functions $\zeta_n(x_1, \ldots, x_n) = 0$, for all $n \in \mathcal{N}$;
- the successor function $\sigma(n) = n + 1$;
- the projection functions $\pi_i^n(x_1, \ldots, x_n) = x_i$, for all $n \in \mathcal{N}$ and $1 \leq i \leq n$;

and is closed under multivariate composition, multivariate primitive recursion, and the application of unbounded minimalization to regular functions.

A **recursive function** is a function in $\mathcal{RF}$.

**Church's thesis**, or the **Church-Turing thesis**, is the premise that recursive functions and Turing machines are capable of representing every function that is computable or partially computable.

A **partial function** on $\mathcal{N}$ is a function whose values are possibly undefined for certain natural numbers.

A partial function on $\mathcal{N}$ is called **total** if it is defined on every natural number.

The class $\mathcal{PR}$ of **partial recursive functions** is the smallest class of partial functions that contains the constant zero functions $\zeta$, the successor function $\sigma$, and the projection functions $\pi_i^n$, and is closed under multivariate composition, multivariate primitive recursion, and the arbitrary application of unbounded minimalization.

A **partial recursive function** is a function in $\mathcal{PR}$.

A partial recursive function $f$ is **represented by the Turing machine** $M$ if machine $M$ calculates the value $f(n)$ for every number $n$ on which $f$ is defined and fails to halt for every number on which $f$ is undefined.

The **Ackermann function** $A\colon \mathcal{N} \times \mathcal{N} \to \mathcal{N}$ is defined by

$$A(0, j) = j + 1;$$
$$A(i + 1, 0) = A(i, 1);$$
$$A(i + 1, j + 1) = A(i, A(i + 1, j)).$$

**Facts:**

**1.** The standard integer functions of arithmetic, including addition, subtraction, multiplication, division, and exponentiation, are all primitive recursive functions.

**2.** A function is a partial recursive function if and only if it can be represented by a Turing machine.

**3.** There are several other models of computation that are equivalent to partial recursive functions and to Turing machines, including labeled Markov algorithms and Post production systems.

**4.** Church's thesis identifies formal concepts (recursive functions and Turing machines) with the intuitive concept of what is computable; so it is not something that is subject to proof.

**5.** Church's thesis is often invoked in the proof of theorems about computable functions to avoid dealing with low-level details of the model of computation.

**6.** The Ackermann function is recursive but not primitive recursive.

**7.** The Ackermann function grows faster than any primitive recursive function, in the following sense. For every primitive recursive function $f(n)$, there is an integer $n_0$ such that $f(n) < A(n, n)$ for all $n > n_0$.

**Examples:**

**1.** Addition is primitive recursive. Define
$$a(x, 0) = \pi_1^1(x);$$
$$a(x, y + 1) = \sigma(\pi_3^3(x, y + 1, a(x, y))).$$
Then $a(x, y) = x + y$.

**2.** Multiplication is primitive recursive. Define
$$m(x, 0) = 0;$$
$$m(x, y + 1) = a(m(x, y), \pi_2^2(x, y)), \text{ where } a(x, y) \text{ is addition.}$$
Then $m(x, y) = x \cdot y$.

**3.** Predecessor is primitive recursive. Define
$$p(0) = 0;$$
$$p(x + 1) = \pi_1^1(x).$$
Then $p(x)$ is the predecessor function (where the predecessor of 0 is 0).

**4.** Nonnegative subtraction, denoted $x \dot{-} y$, is defined as $x - y$ if $x \geq y$ and 0 otherwise. Define
$$s(x, 0) = \pi_1^1(x);$$
$$s(x, y + 1) = p(s(x, y)), \text{ where } p(x) \text{ is predecessor.}$$
Then $s(x, y) = x \dot{-} y$.

**5.** The function $p(n) = $ the $n$th prime number is a primitive recursive function.

## 17.2.2  RECURSIVE SETS AND DECIDABLE PROBLEMS

### Definitions:

The **characteristic function** of a set $A$ is the function

$$f(x) = \begin{cases} 1 & \text{if } x \in A; \\ 0 & \text{if } x \notin A. \end{cases}$$

A set $A$ is **recursive** if its characteristic function is recursive.

A problem is (**computationally**) **decidable** (or solvable) if it can be represented as a membership problem that can be decided by a recursive function.

A set $A \subseteq \mathcal{N}$ is **recursively enumerable** (**r.e.**) if $A = \emptyset$ or $A$ is the image of a recursive function.

A **Gödel numbering** of a set $S$ is a one-to-one recursive function $g \colon S \to \mathcal{N}$ whose image in $\mathcal{N}$ is a recursive set.

### Facts:

**1.** If a set $A$ and its complement $\overline{A}$ are both recursively enumerable, then $A$ is recursive.

**2.** If a recursively enumerable set $A$ is the image of a nondecreasing function, then $A$ is recursive.

**3.** A set is recursively enumerable if and only if it is the image of a partial recursive function.

**4.** A set is recursively enumerable if and only if it is the domain of a partial recursive function.

**5.** The set of Turing machines has a Gödel numbering.

### Examples:

**1.** Every finite set of numbers is recursive.

**2.** The prime numbers are a recursive set.

**3.** The problem of deciding which Turing machines halt on all inputs is undecidable. The set of Gödel numbers for these Turing machines is neither recursive nor recursively enumerable.

**4.** For any fixed $c \in \mathcal{N}$, the problem of deciding which Turing machines halt when the number $c$ is supplied as input is undecidable. The set of Gödel numbers for these Turing machines is recursively enumerable, but not recursive.

**5.** The problem of deciding which Turing machines halt when their own Gödel number is supplied as input is undecidable. The set of Gödel numbers for these Turing machines is recursively enumerable, but not recursive.

**6.** *Hilbert's tenth problem*: Hilbert's tenth problem (posed in 1900) was the problem of devising an algorithm to determine, given a polynomial $p(x_1, \ldots, x_n)$ with integer coefficients, whether there exists an integer root. Y. Matiyasevich proved in 1970 that no such algorithm exists. That is, the set of polynomials with integer coefficients that have an integer solution is not recursive. Hilbert's tenth problem is called a "natural example" of an undecidable problem, since the concepts used to define it are not from within computability theory (i.e., unlike problems concerned with the behavior of Turing machines). [Ma93]

# 17.3   LANGUAGES AND GRAMMARS

Strings of symbols are a general way to represent information, both in written text and in a computer. A language is a set of strings that are used within some domain of discourse, and a grammar is a system for generating a language. A grammar is what enables a compiler to determine whether a body of code is syntactically correct in a given computer language. Formal language theory is concerned with languages, grammars, and rudiments of combinatorics on strings. Applications of formal language theory range from natural and programming languages, developmental biology, and computer graphics, to semiotics, artificial intelligence, and artificial life.

## 17.3.1   ALPHABETS

### Definitions:

An **alphabet** is a finite nonempty set.

A **symbol** is an element of an alphabet.

A **string** in an alphabet is a finite sequence of symbols over that alphabet.

A **word** in an alphabet is a finite or countably infinite sequence of symbols over that alphabet.

The **length of a string** $w$ is the number of symbols in $w$, denoted $|w|$.

The **empty string** $\lambda$ (or sometimes $\epsilon$) is the string of length zero, that is, the string with no symbols.

The **frequency** $|w|_a$ of a symbol $a$ in a string $w$ is the number of occurrences of $a$ in string $w$.

A **substring** of a string $w$ is a sequence of consecutive symbols that occurs in $w$.

A **subword** of a word $w$ is a sequence of consecutive symbols that occurs in $w$.

A **prefix** of a string $w$ is a substring that starts at the leftmost symbol.

A **suffix** of a string $w$ is a substring that ends at the rightmost symbol.

The **reverse** or **mirror image** $x^R$ of the string $x = a_1 a_2 \ldots a_n$ is the string $a_n \ldots a_2 a_1$.

A **palindrome** is a string that is identical to its reverse.

A **pseudopalindrome** in an alphabet (such as English) that includes punctuation symbols (such as comma, hyphen, or blank) is a word that becomes a palindrome when all of its punctuation symbols are deleted.

The **concatenation** $xy$ of two strings $x = a_1 a_2 \ldots a_m$ and $y = b_1 b_2 \ldots b_n$ is the string $a_1 a_2 \ldots a_m b_1 b_2 \ldots b_n$ obtained by appending string $y$ to the right of string $x$.

The **nth power of a string** $w$, denoted $w^n$, is the concatenation of $n$ copies of $w$.

The **shuffle** $x \shuffle y$ of two strings $x = x_1 \ldots x_n$ and $y = y_1 \ldots y_n$ is the string $x_1 y_1 \ldots x_n y_n$.

**Facts:**

**1.** A symbol of an alphabet is usually conceptualized as something that can be represented by a single byte or by a written character.

**2.** A finite word is a string.

**3.** The sum of the frequencies $|w|_a$ taken over all the symbols $a$ in the alphabet equals the length of the string $w$.

**4.** The length $|xy|$ of a concatenation equals the sum $|x|+|y|$ of the lengths of the strings $x$ and $y$ from which it is formed.

**5.** The length of the $n$th power of a string $w$ is $n \cdot |w|$.

**6.** When a pseudopalindrome occurs in a natural language, it is commonly called a palindrome.

**Examples:**

**1.** The English alphabet includes lower and upper case English letters, the blank symbol, the digits $0, 1, \ldots, 9$, and various punctuation symbols.

**2.** ASCII (*American Standard Code for Information Interchange*) is an alphabet of size 128 for many common computer languages. See Table 1.

**3.** ABLE WAS I ERE I SAW ELBA is a palindrome.

**4.** The names EVE, HANNAH, and OTTO are palindromes.

**5.** MADAM I'M ADAM and SIX AT-NOON TAXIS are pseudopalindromes.

**6.** The third power of the string 011 is 011011011.

**7.** The concatenation of BOOK and KEEPER is BOOKKEEPER.

**8.** The shuffle of FLOOD and RIVER is FRLIOVOEDR.

## 17.3.2 LANGUAGES

**Definitions:**

The *free monoid* $V^*$ generated by the alphabet $V$ is the structure whose domain is the set of all strings composable from symbols over $V$, with the operation of string concatenation.

A (*formal*) *language* on the alphabet $V$ is a subset of the free monoid $V^*$.

The $\lambda$-*free semigroup* $V^+$ on an alphabet $V$ is the set $V^* - \{\lambda\}$, with the concatenation operation.

A $\lambda$-*free language* on the alphabet $V$ is a subset of the $\lambda$-free semigroup $V^+$.

The *length set* of a language $L$ is the set $length(L) = \{|x| \mid x \in L\}$.

The *concatenation* $L_1 L_2$ of two languages is the set $\{xy \mid x \in L_1, y \in L_2\}$.

The *ith power* of a language $L$ is the language $L^i$ defined recursively by the rule $L^0 = \{\lambda\}$ and $L^{i+1} = L^i L$, $i \geq 0$.

The *Kleene closure* (or *Kleene star*) $L^*$ of a language $L$ is the union $\bigcup_{i \geq 0} L^i$ of all its powers.

The *positive closure* (or *Kleene plus*) $L^+$ of a language $L$ is the union $\bigcup_{i \geq 1} L^i$ of all its powers excluding the zeroth power.

## Table 1: ASCII codes and binary.

| | | | | | | | |
|---|---|---|---|---|---|---|---|
| 000 0000 | NUL | 010 0000 | SP | 100 0000 | @ | 110 0000 | ' |
| 000 0001 | SOH | 010 0001 | ! | 100 0001 | A | 110 0001 | a |
| 000 0010 | STX | 010 0010 | " | 100 0010 | B | 110 0010 | b |
| 000 0011 | ETX | 010 0011 | # | 100 0011 | C | 110 0011 | c |
| 000 0100 | EOT | 010 0100 | $ | 100 0100 | D | 110 0100 | d |
| 000 0101 | ENQ | 010 0101 | % | 100 0101 | E | 110 0101 | e |
| 000 0110 | ACK | 010 0110 | & | 100 0110 | F | 110 0110 | f |
| 000 0111 | BEL | 010 0111 | ' | 100 0111 | G | 110 0111 | g |
| 000 1000 | BS | 010 1000 | ( | 100 1000 | H | 110 1000 | h |
| 000 1001 | HT | 010 1001 | ) | 100 1001 | I | 110 1001 | i |
| 000 1010 | LF | 010 1010 | * | 100 1010 | J | 110 1010 | j |
| 000 1011 | VT | 010 1011 | + | 100 1011 | K | 110 1011 | k |
| 000 1100 | FF | 010 1100 | , | 100 1100 | L | 110 1100 | l |
| 000 1101 | CR | 010 1101 | - | 100 1101 | M | 110 1101 | m |
| 000 1110 | SO | 010 1110 | . | 100 1110 | N | 110 1110 | n |
| 000 1111 | SI | 010 1111 | / | 100 1111 | O | 110 1111 | o |
| 001 0000 | DLE | 011 0000 | 0 | 101 0000 | P | 111 0000 | p |
| 001 0001 | DC1 | 011 0001 | 1 | 101 0001 | Q | 111 0001 | q |
| 001 0010 | DC2 | 011 0010 | 2 | 101 0010 | R | 111 0010 | r |
| 001 0011 | DC3 | 011 0011 | 3 | 101 0011 | S | 111 0011 | s |
| 001 0100 | DC4 | 011 0100 | 4 | 101 0100 | T | 111 0100 | t |
| 001 0101 | NAK | 011 0101 | 5 | 101 0101 | U | 111 0101 | u |
| 001 0110 | SYN | 011 0110 | 6 | 101 0110 | V | 111 0110 | v |
| 001 0111 | ETB | 011 0111 | 7 | 101 0111 | W | 111 0111 | w |
| 001 1000 | CAN | 011 1000 | 8 | 101 1000 | X | 111 1000 | x |
| 001 1001 | EM | 011 1001 | 9 | 101 1001 | Y | 111 1001 | y |
| 001 1010 | SUB | 011 1010 | : | 101 1010 | Z | 111 1010 | z |
| 001 1011 | ESC | 011 1011 | ; | 101 1011 | [ | 111 1011 | } |
| 001 1100 | FS | 011 1100 | < | 101 1100 | \ | 111 1100 | | |
| 001 1101 | GS | 011 1101 | = | 101 1101 | ] | 111 1101 | } |
| 001 1110 | RS | 011 1110 | > | 101 1110 | ˆ | 111 1110 | ~ |
| 001 1111 | US | 011 1111 | ? | 101 1111 | − | 111 1111 | DEL |

*Control codes:* ACK: acknowledge, BEL: bell, BS: backspace, CAN: cancel, CR: carriage return, DC1–4: device controls, DEL: delete, DLE: data link escape, EM: end of medium, ENQ: enquiry, EOT: end of transmission, ESC: escape, ETB: end of transmission block, ETX: end of text, FF: form feed, FS: file separator, GS: group separator, HT: horizontal tab, LF: line feed, NAK: negative acknowledgment, NUL: null, RS: record separator, SI: shift in, SO: shift out, SOH: start of heading, SP: space, STX: start of text, SUB: substitute, SYN: synchronous/idle, US: united separator, VT: vertical tab.

The **union** of two languages $L_1$ and $L_2$ is $L_1 \cup L_2$, using the usual set operation.

The **intersection** of two languages $L_1$ and $L_2$ is $L_1 \cap L_2$, using the usual set operation.

The **complement** of a language $L$ over an alphabet $V$ is the language $\overline{L}$, where complementation is taken with $V^*$ as the universe.

A language is **regular** if it is any of the languages $\emptyset$, $\{\lambda\}$, or $\{b\}$, where $b$ is a symbol of its alphabet, or if it can be obtained by applying the operations of union, concatenation, and Kleene star finitely many times to one of those languages.

The **shuffle** $L_1 \text{⧢} L_2$ of two languages $L_1, L_2$ is the language $\{w \in V^* \mid w = x \text{⧢} y, \text{ for some } x \in L_1, y \in L_2\}$.

The **mirror image** $mi(L)$ of the language $L$ is the language $\{x^R \mid x \in L\}$. It is also called the **reverse** of the language $L$.

The **left quotient** of the language $L_1$ with respect to the language $L_2$ on the same alphabet $V$, is the language $L_2 \backslash L_1$ containing every string of $V^*$ that can be obtained from a string in $L_1$ by erasing a prefix from $L_2$. That is, $L_2 \backslash L_1 = \{w \in V^* \mid \text{there is } x \in L_2 \text{ such that } xw \in L_1\}$.

The **left derivative** of language $L$ with respect to the string $x$ over the same alphabet $V$ is the language $\partial_x(L) = \{x\} \backslash L$.

The **right quotient** is the notion symmetric to left quotient.

The **right derivative** is the notion symmetric to left derivative.

A **substitution** for the alphabet $V$ in the alphabet $U$ is a mapping $s \colon V \longrightarrow 2^{U^*}$. This means that each symbol $b \in V$ may be replaced by any of the strings in the set $s(b)$.

A **finite substitution** is a substitution such that the replacement set $s(a)$ for each symbol $a \in V$ is finite.

The **extension of a substitution** $s \colon V \longrightarrow 2^{U^*}$ from its domain alphabet $V$ to the set $V^*$ of strings over $V$ is given by the rules $s(\lambda) = \{\lambda\}$ and $s(ax) = s(a)s(x)$, for $a \in V, x \in V^*$.

A **morphism from the alphabet** $V$ to the alphabet $U$ is a substitution $s \colon V \longrightarrow 2^{U^*}$ such that the replacement set $s(a)$ for every symbol $a \in V$ is a singleton set.

A **$\lambda$-free substitution** is a substitution such that $\lambda$ is never substituted for a symbol. That is, $\lambda \notin s(a)$, for every symbol $a \in V$.

A **$\lambda$-free morphism** is a morphism such that $s(a) \neq \{\lambda\}$, for every symbol $a \in V$.

The **extension of a substitution** $s \colon V \longrightarrow 2^{U^*}$ to the language $L \subseteq V^*$ is the language $s(L) = \bigcup_{x \in L} s(x)$ that contains every string in $U^*$ obtainable from a string in $L$ by making replacements permissible under substitution $s$.

The **inverse of a morphism** $h \colon V^* \longrightarrow U^*$ is the mapping $h^{-1} \colon U^* \longrightarrow 2^{V^*}$ defined by $h^{-1}(x) = \{y \in V^* \mid h(y) = x\}$, $x \in U^*$.

A family of languages is **nontrivial** if it contains at least one language different from $\emptyset$ and $\{\lambda\}$.

**Facts:**

**1.** The set of all binary strings with at least as many 1s as 0s is a language.

**2.** The set of all binary strings in which no two occurrences of 1 are consecutive is a language.

**3.** Some strings of a natural language such as English are categorized as nouns, verbs, and adjectives. Other more complicated strings are categorized as sentences.

**4.** Some strings of common computer languages are categorized as identifiers and arithmetic expressions. Other more complicated strings are categorized as statements, with subcategories such as assignment statements, if-statements, and while-statements.

**Examples:**

**1.** Natural languages and computer languages are formal languages.

**2.** The Kleene closure of the language $\{00, 01, 10, 11\}$ is the language of all strings of even length.

**3.** The left derivative $\{\text{bee}\}\backslash\text{English}$ includes the following strings: f, n, p, r, s, t, tle, ts, keeper, swax, feater, ping.

**4.** The substitution $0 \mapsto \{00, 01\}, 1 \mapsto \{10, 11\}$ over the free monoid $\{0, 1\}^*$ is the language of all strings of even length.

**5.** Given the alphabet $\{a, b\}$, define the morphism $\phi: \{a, b\} \longrightarrow \{a, b\}^*$ by the replacements $\phi(a) = ab$ and $\phi(b) = ba$, and define the string $w_n$ by the recursion $w_0 = a$ and $w_{n+1} = \phi(w_n)$. Then

$$w_1 = ab, \quad w_2 = abba, \quad w_3 = abbabaab, \quad w_4 = abbabaabbaababba, \dots .$$

**6.** In Example 5 each word $w_n$ is a prefix of the next word $w_{n+1}$. The *Thue $\omega$-word* is the infinite word $\lim_{n \to \infty} w_n$.

**7.** Given the alphabet $\{a, b\}$, define the morphism $\rho: \{a, b\} \longrightarrow \{a, b\}^*$ by the replacements $\rho(a) = ab$ and $\rho(b) = a$, and define the string $w_n$ by the recursion $w_0 = a$ and $w_{n+1} = \phi(w_n)$. Then

$$w_1 = ab, \quad w_2 = aba, \quad w_3 = abaab, \quad w_4 = abaababa, \dots .$$

**8.** In Example 7 each word $w_n$ is a prefix of the next word $w_{n+1}$. The *Fibonacci $\omega$-word* is the infinite word $\lim_{n \to \infty} w_n$.

**9.** A language is regular if and only if it is the language of strings accepted by some finite-state recognizer.

## 17.3.3  GRAMMARS AND THE CHOMSKY HIERARCHY

**Definitions:**

A **phrase-structure grammar** (or **unrestricted grammar** or **type 0 grammar**) is a quadruple $G = (N, T, S, P)$ such that

- $N$ is a finite alphabet of symbols called **nonterminals**;
- $T$ is a finite nonempty alphabet, disjoint from $N$, of symbols called **terminals**;
- $S$ is a nonterminal called the **start symbol**;

- $P$ is a finite set of **production rules** of the form $\alpha \to \beta$, where $\alpha$ is a string in $N \cup T$ that contains at least one nonterminal and $\beta$ is a string in $N \cup T$.

The **antecedent** of a production $\alpha \to \beta$ is $\alpha$. The **consequent** of a production $\alpha \to \beta$ is $\beta$.

The string $y$ is **directly derivable** from the string $x$ with respect to the grammar $G$ if there is a production rule $u \to v \in P$ and if there are strings $w_1, w_2 \in (N \cup T)^*$ such that $x = w_1 u w_2$ and $y = w_1 v w_2$.

The **direct derivability relation** $x \Longrightarrow_G y$ (or $x \Longrightarrow y$, when the grammar $G$ is understood) means that $y$ is directly derivable from string $x$.

A **derivation** of the string $y$ from the string $x$ is a sequence of direct derivations $x \Longrightarrow z_1$, $z_1 \Longrightarrow z_2, \dots, z_n \Longrightarrow y$. This is sometimes called **parsing**.

The string $y$ is **derivable** from the string $x$ with respect to the grammar $G$ if there is a derivation of $y$ from $x$. Notation: $x \Longrightarrow^* y$.

The **Chomsky normal form for a production rule** is $A \to BC$, where $B$ and $C$ are nonterminals or the form $A \to a$ where $a$ is a terminal.

The **language generated by the grammar** $G$ is the language $L(G) = \{x \in T^* \mid S \Longrightarrow^* x\}$.

Grammars $G_1$ and $G_2$ are **equivalent** if $L(G_1) = L(G_2)$.

A **leftmost derivation** $x \Longrightarrow_{left} y$ is a derivation $x \Longrightarrow y$ in which at each step the leftmost nonterminal is replaced.

The **leftmost language generated by the grammar** $G$ is the language $L_{left}(G)$ of strings of terminals with leftmost derivations from the start symbol $S$.

A grammar $(N, T, S, P)$ is **length-increasing** (or **monotonic**) if $|u| \le |v|$ for all productions $u \to v$. (However, the production $S \to \lambda$ is allowed, provided that $S$ does not appear in any consequent.)

A grammar $(N, T, S, P)$ is **context-sensitive** (or of **type 1**) if for each production $u \to v$, the antecedent and consequent have the form $u = u_1 A u_2$ and $v = u_1 x u_2$, for $u_1, u_2 \in (N \cup T)^*$, $A \in N$, $x \in (N \cup T)^+$. (The production $S \to \lambda$ is allowed, provided that $S$ does not appear in any consequent.)

A grammar $(N, T, S, P)$ is **context-free** (or of **type 2**) if the antecedent of each production $u \to v$ is a nonterminal.

An **L-system** is a production-based model for growth and life development.

A grammar $(N, T, S, P)$ is **linear** if each production $u \to v$ has its antecedent $u \in N$ and its consequent $v \in T^* \cup T^* N T^*$.

A grammar $(N, T, S, P)$ is **right-linear** if each production $u \to v$ has $u \in N$ and $v \in T^* \cup T^* N$.

A grammar $(N, T, S, P)$ is **left-linear** if each production $u \to v$ has $u \in N$ and $v \in T^* \cup N T^*$.

A grammar $(N, T, S, P)$ is **regular** (or of **type 3**) if each production $u \to v$ has $u \in N$ and $v \in T \cup TN \cup \{\lambda\}$.

Given a class of grammars, there are some basic decision problems about arbitrary grammars $G_1, G_2$ in the class:

> **equivalence**:  are the grammars $G_1$ and $G_2$ equivalent?
>
> **inclusion**:  is the language $L(G_1)$ contained in the language $L(G_2)$?
>
> **membership**:  given an arbitrary string $x$, is $x$ an element of $L(G_1)$?
>
> **emptiness**:  is the language $L(G_1)$ empty?
>
> **finiteness**:  is the language $L(G_1)$ finite?
>
> **regularity**:  is $L(G_1)$ a regular language? (see §17.3.2)

The **recursive** languages are the languages with a decidable membership question.

The various classes of languages are denoted as follows:

> **RE** (type 0):  the class of all unrestricted languages;
>
> **CS** (type 1):  the class of all context-sensitive languages;
>
> **CF** (type 2):  the class of all context-free languages;
>
> **LIN**:  the class of all linear languages;
>
> **REG** (type 3):  the class of all regular languages.

**Facts:**

1.  *Chomsky hierarchy*:  The following strict inclusions hold.

$$\textbf{REG} \subset \textbf{LIN} \subset \textbf{CF} \subset \textbf{CS} \subset \textbf{RE}.$$

2.  The language of an unrestricted grammar is recursively enumerable (RE).

3.  **CS** (context sensitive) $\subset$ {recursive languages} $\subset$ **RE** (unrestricted).

4.  The class of languages generated by monotonic grammars is identical to the class of languages generated by context-sensitive grammars.

5.  L-systems were introduced in 1968 by Aristid Lindenmayer (1922–1990), a Dutch biologist, to model the development of some plant systems. (See [Li16].)

6.  The classes of languages generated by right-linear or by left-linear grammars coincide. This class is identical to the family of languages generated by regular grammars, as well as to the class of regular languages (see §17.3.2).

7.  $L_{left}(G) \in \textbf{CF}$ (context-free) for each type-0 grammar $G$.

8.  If $G$ is a context-free grammar, then $L_{left}(G) = L(G)$.

9.  Let $G$ be a context-free grammar. Then there is an equivalent grammar $G'$, with every rule in Chomsky normal form. Moreover, there is constructive method for transforming grammar $G$ into the grammar $G'$.

10.  The following table gives closure properties for Chomsky hierarchy classes.

| | RE | CS | CF | LIN | REG |
|---|---|---|---|---|---|
| union | yes | yes | yes | yes | yes |
| intersection | yes | yes | no | no | yes |
| complement | no | yes | no | no | yes |
| concatenation | yes | yes | yes | no | yes |
| Kleene star | yes | yes | yes | no | yes |
| intersection with | | | | | |
|      regular languages | yes | yes | yes | yes | yes |
| substitution | yes | no | yes | no | yes |
| $\lambda$-free substitution | yes | yes | yes | no | yes |
| morphisms | yes | no | yes | yes | yes |
| $\lambda$-free morphisms | yes | yes | yes | yes | yes |
| inverse morphisms | yes | yes | yes | yes | yes |
| left/right quotient | yes | no | no | no | yes |
| left/right quotients with | | | | | |
|      regular languages | yes | no | yes | yes | yes |
| left/right derivative | yes | yes | yes | yes | yes |
| shuffle | yes | yes | no | no | yes |
| mirror image | yes | yes | yes | yes | yes |

**11.** *Rice's theorem*: Let $P$ be a nontrivial property of recursively enumerable languages (i.e., at least one grammar has property $P$ and at least one grammar does not have property $P$). Then property $P$ is undecidable.

**12.** A language is context-free if and only if it is the language accepted by some (possibly nondeterministic) pushdown automaton.

**13.** The following table summarizes the decidability properties of the grammar classes in the Chomsky hierarchy. In this table U stands for *undecidable*, D for *decidable*, and T for *trivial*.

| | RE | CS | CF | LIN | REG |
|---|---|---|---|---|---|
| equivalence | U | U | U | U | D |
| inclusion | U | U | U | U | D |
| membership | U | D | D | D | D |
| emptiness | U | U | D | D | D |
| finiteness | U | U | D | D | D |
| regularity | U | U | U | U | T |

**Examples:**

**1.** In the grammar $G = (N, T, S, P)$, where $N = \{S\}$, $T = \{0, 1\}$, and $P = \{S \to 0S1, S \to \lambda\}$, a derivation of the string 0011 is $S \Longrightarrow 0S1 \Longrightarrow 00S11 \Longrightarrow 0011$.

**2.** The following are examples of languages generated by the grammar $G = (N, T, S, P)$ with $N = \{S, x, y, z\}$, $T = \{0, 1, 2\}$:

| production set $P$ | language $L(G)$ | class |
|---|---|---|
| $S \to 0x,\ x \to 1y,\ y \to 0x,\ x \to 1,\ y \to \lambda$ | $\{01, 0101, 010101, \ldots\}$ | regular |
| $S \to \lambda, S \to 0x, S \to 01, x \to S1$ | $\{0^n 1^n \mid n \geq 0\}$ | linear |
| $S \to \lambda,\ S \to 0Sx2,\ 2x \to x2,\ 0x \to 01,\ 1x \to 11$ | $\{0^n 1^n 2^n \mid n \geq 0\}$ | unrestricted |

## 17.3.4   REGULAR AND CONTEXT-FREE LANGUAGES

### Definitions:

Given an alphabet $V$, a **regular expression** over $V$ is a string $w$ over the alphabet $V \cup \{\epsilon, ), (, +, *\}$ that has one of the following forms:

- $w \in V$ or $w = \epsilon$;
- $w = (\alpha\beta)$, where $\alpha$ and $\beta$ are regular expressions;
- $w = (\alpha + \beta)$, where $\alpha$ and $\beta$ are regular expressions;
- $w = \alpha^*$, where $\alpha$ is a regular expression.

The set of all regular expressions over alphabet $V$ is denoted $\mathbf{rex}_V$.

The function $L$ maps $\mathbf{rex}_V$ to the set of all languages over the alphabet $V$, using the following rules:

- $L(\epsilon) = \emptyset$, and $L(a) = \{a\}$ for all $a \in V$;
- $L((\alpha\beta)) = L(\alpha)L(\beta), L((\alpha + \beta)) = L(\alpha) \cup L(\beta)$, and $L(\alpha^*) = (L(\alpha))^*$.

A context-free grammar $G$ is **ambiguous** if there is a string $x \in L(G)$ having two different leftmost derivations in $G$.

A context-free language $L$ is **inherently ambiguous** if every context-free grammar of $L$ is ambiguous; otherwise, language $L$ is called **unambiguous**.

### Facts:

**1.** *Kleene's theorem*: A language $L$ is regular if and only if there is a regular expression $e$ such that $L = L(e)$.

**2.** Every context-free language over a one-letter alphabet is regular.

**3.** Every regular language $L$ can be represented in the form $L = h_4(h_3^{-1}(h_2(h_1^{-1}(a^*b))))$, where $h_1, h_2, h_3, h_4$ are morphisms.

**4.** Each regular language is unambiguous.

**5.** There are inherently ambiguous linear languages.

**6.** The ambiguity problem for context-free grammars is undecidable.

**7.** The length set of a context-free language is a finite union of arithmetical progressions.

**8.** Every language $L$ can be represented in the form $L = h(L_1 \cap L_2)$, as well as in the form $L = L_3 \backslash L_4$, where $h$ is a morphism and $L_1, L_2, L_3, L_4$ are linear languages.

**9.** *Pumping lemma for regular languages*: If $L$ is a regular language over the alphabet $V$, then there are numbers $p$ and $q$ such that every string $z \in L$ with length $|z| > p$ can be written in the form $z = uvw$, with $u, v, w \in V^*$, where $|uv| \le q, v \ne \lambda$, so that $uv^i w \in L$ for all $i \ge 0$.

**10.** *Pumping lemma for linear languages*: If $L$ is a linear language on the alphabet $V$, then there are numbers $p$ and $q$ such that every string $z \in L$ with length $|z| > p$ can be written in the form $z = uvwxy$, with $u, v, w, x, y \in V^*$, where $|uvxy| \le q$ and $vx \ne \lambda$, so that $uv^i wx^i y \in L$ for all $i \ge 0$.

**11.** *Bar-Hillel ($uvwxy$, pumping) lemma for context-free languages*: If $L$ is a context-free language over the alphabet $V$, then there are numbers $p$ and $q$ such that every string $z \in L$ with length $|z| > p$ can be written in the form $z = uvwxy$, with $u, v, w, x, y \in V^*$, where $|vwx| \le q$ and $vx \ne \lambda$, so that $uv^i wx^i y \in L$ for all $i \ge 0$.

**12.** *Ogden's pumping lemma (pumping with marked positions):* If $L$ is a context-free language on the alphabet $V$, then there is a number $p$ such that for every string $z \in L$ and for every set of at least $p$ marked occurrences of symbols in $z$, we can write $z = uvwxy$, where (a) either each of $u, v, w$ or each of $w, x, y$ contains at least one marked symbol; (b) $vwx$ contains at most $p$ marked symbols; and (c) $uv^i wx^i y \in L$ for all $i \geq 0$.

**13.** Let $G$ be a context-free grammar $G$. Then there is a grammar $G' = (N, T, S, P)$, with every rule of the form $A \to a\alpha$, for $A \in N, a \in T, \alpha \in (N \cup T)^*$, such that $L(G') = L(G) - \{\lambda\}$. Moreover, there is a constructive method for transforming grammar $G$ into the grammar $G'$, which is said to be in the *Greibach normal form*.

**14.** Let $G$ be a context-free grammar $G$ and $(k, l, m)$ a triple of nonnegative integers. Then an equivalent grammar $G' = (N, T, S, P)$ can be effectively constructed whose every rule is in one of the following two forms:

- $A \to xByCz$, with $A, B, C \in N, x, y, z \in T^*$, and $|x| = k, |y| = l, |z| = m$;
- $A \to x$, with $A \in N, x \in T^*, |x| \in length(L(G))$.

Such a grammar $G'$ is said to be in *super normal form*.

**Examples:**

**1.** The following are some regular expressions over $\{0, 1\}$ and the languages they represent.

| | |
|---|---|
| $1^*$ | all strings with no 0s |
| $1^*01^*$ | all strings with exactly one 0 |
| $1^*(0 + \epsilon)1^*$ | all strings with one or no 0s |
| $(0 + 1)(0 + 1)$ | all strings of length 2 |
| $(0 + 1)(0 + 1 + \epsilon)$ | all strings of length 1 or 2. |

**2.** *Backus-Naur form (BNF)* (or *Backus normal form*) for specifying computer language syntax uses context-free production rules. Nonterminals are enclosed in brackets; the symbol ::= is used in place of $\to$; and all the consequents of the same antecedent are written in the same statement with the alternative consequents separated by vertical bars. For instance, in some programming languages, the following might be the BNF for the lexical token called an identifier.

⟨identifier⟩ ::= ⟨letter⟩|⟨letter⟩⟨alphameric string⟩
⟨letter⟩ ::= a|b|c|d|e|f|g|h|i|j|k|l|m|n|o|p|q|r|s|t|u|v|w|x|y|z
⟨alphameric string⟩ ::= ⟨alphameric⟩|⟨alphameric string⟩⟨alphameric⟩
⟨alphameric⟩ ::= ⟨letter⟩|⟨digit⟩
⟨digit⟩ ::= 0|1|2|3|4|5|6|7|8|9

## 17.3.5   COMBINATORICS ON WORDS

*Note:* In this subsection, a *word* is taken to be finite.

**Definitions:**

A (*word*) *variable* over an alphabet $V$ is a symbol (such as $x$ or $y$) not in $V$ whose values range over $V^*$.

A *pattern* in a word is a string of word variables.

A pattern is **present** in a word $w \in V^*$ if there exists an assignment of strings from $V^+$ to the variables in that pattern such that the word formed thereby is a substring of $w$.

A **square** is a word of the pattern "$xx$".

A **square-free** word is a word with no subwords of the pattern "$xx$".

A **cube** is a word of the pattern "$xxx$".

An **abelian square** is a word of the form $xx^p$, where $x^p$ is some permutation of the word $x$.

A **word equation** over an alphabet $V$ is an expression of the form $\alpha = \beta$ such that $\alpha$ and $\beta$ are words containing letters of an alphabet $V$ and some variables over $V$.

A **word inequality** is the negation of a word equation, which is commonly written in the form $\alpha \neq \beta$.

A **solution** to a system $S$ of (finitely many) word equations and word inequalities is a list of words whose substitutions for their respective variables converts every word equation and word inequality in the system into a true proposition.

A **code** is a nonempty language $C \subseteq V^+$ such that whenever a word $w$ in $V$ can be written as a concatenation of words in $C$, the write-up is always unique. That is, if $w = x_1 \ldots x_m = y_1 \ldots y_n$, where $m, n \geq 1$, and $x_i, y_j \in C$, then $m = n$ and $x_i = y_i$ for $i = 1, \ldots, m$. This property is called **unique decodability**.

The **code indicator of a word** $w \in V^*$ is the number $ci(w) = |V|^{-|w|}$.

The **code indicator of a language** is the sum of the code indicators of all words in the language.

**Facts:**

**1.** Certain patterns are *unavoidable* in sufficiently long words.

**2.** Squares are avoidable in alphabets with three or more letters; that is, there are arbitrarily long square-free words.

**3.** Cubes are avoidable over two-letter alphabets.

**4.** Although squares are avoidable in three-letter alphabets, abelian squares are unavoidable. Every word of length $\geq 8$ over $V = \{a, b, c\}$ contains a subword of the form $xx^p, x \in V^+$, where $x^p$ is a permutation of $x$.

**5.** Abelian squares are avoidable in alphabets with four or more letters.

**6.** It is *decidable* (by *Makanin's algorithm*) whether or not a system of word equations and inequalities has a solution.

**7.** It is decidable whether or not a given finite language is a code.

**8.** Every code $C$ satisfies the inequality $ci(C) \leq 1$.

**9.** If a language $C = \{w_1, \ldots, w_n\}$ over $V$ is not a code then, by the so-called *defect theorem*, the algebraic structure of $C^*$ can be simulated by an alphabet with at most $n-1$ letters: the smallest free submonoid of $V^*$ containing $C$ is generated by at most $n-1$ words.

**10.** The following three conditions are equivalent for any two words $u$ and $v$:

- $\{u, v\}$ is not a code;
- $u$ and $v$ are powers of the same word;
- $uv = vu$.

(This is a corollary to Fact 9.)

**11.** For every word $w \in V^+$, there are a unique shortest word $\rho(w)$ and an integer $n \geq 1$ such that $w = (\rho(w))^n$. (The word $\rho(w)$ is called the *primitive root* of $w$.)

**12.** *Lyndon's theorem:* If $uv = vw$ with $u, v, w \in V^*$, then there exist words $x, y \in V^*$ and a number $n \geq 0$ such that $u = xy, w = yx$ and $v = (xy)^n x = x(yx)^n$.

**13.** If $uv = vu$ with $u, v \in V^+$, then $\rho(u) = \rho(v)$ and, consequently, $u$ and $v$ are powers of the same word. (This is a corollary to Lyndon's theorem.)

**14.** Assume that words $u^m$ and $v^n$ have a common prefix or suffix of length $|u| + |v| - d$, where $u, v \in V^+, m, n \geq 1$ and $d = \gcd(|u|, |v|)$. Then $\rho(u) = \rho(v)$ and $|\rho(u)| \leq d$. Thus, if $d = 1$ then $u$ and $v$ are powers of the same letter.

**15.** If $u^m = v^n$, where $m, n \geq 1$, then $u$ and $v$ are powers of the same word. (This is a corollary to Fact 14.)

**16.** If $u^m v^n = w^p$, where $m, n, p \geq 2$, then $\rho(u) = \rho(v) = \rho(w)$.

**Examples:**

**1.** In the alphabet $V = \{a, b\}$, the only square-free three-letter words are *aba* and *bab*. The two possible extensions of *aba* by one letter are *abaa*, which contains the square *aa*, and *abab*, which is a square. Similarly, both extensions of *bab* by one letter contain a square. Thus, squares are unavoidable in words of length $\geq 4$ over two-letter alphabets.

**2.** All solutions for the system $xaba = abax, xx \neq x, x \neq aba$, over the alphabet $V = \{a, b\}$ are (by the corollary to Lyndon's theorem) of the form $x = (aba)^n, n \geq 2$.

---

# 17.4   ALGORITHMIC COMPLEXITY

The "complexity of an algorithm" usually means a measure of the computational effort or cost of execution, relative to the "size" of the problem. Other factors that may affect this kind of complexity are the characteristics of the particular input and the values returned by random number generators. The most common complexity measure is *running time*, but other measures, such as *space utilized* and *number of comparisons*, are sometimes used. Another view of complexity focuses on the complicatedness of the algorithm, rather than on the effort needed to execute it.

---

## 17.4.1   OVERVIEW OF COMPLEXITY

To simplify discussion, it is assumed that every function and algorithm under consideration here has one argument. (Everything is easily generalized to multivariate functions by regarding the list of arguments as an $n$-tuple.)

**Definitions:**

A function $f: \mathcal{N} \to \mathcal{N}$ is **asymptotic** to a function $g: \mathcal{N} \to \mathcal{N}$ if $\lim_{n \to \infty} f(n)/g(n) = 1$. *Notation:* $f(n) \sim g(n)$. (See §1.3.3.)

The **input size** of the argument of an algorithm is the number of bits required to specify a value of that argument (or sometimes the argument's numeric value).

A (*cost-based*) *complexity measure* for an algorithm is any of several different asymptotic measures of cost or difficulty in running that algorithm, relative to the input size. It is given in big-$O$ notation (or sometimes in $\Theta$-notation). (See §1.3.3.)

A *time-complexity* measure of an algorithm is a big-$O$ expression for the number of operations or the running time needed for a complete execution of that algorithm, represented as a function of the size of the input.

A *space-complexity* measure of an algorithm is a big-$O$ expression for the amount of computational space needed in the execution of that algorithm, represented as a function of the size of the input.

An algorithm runs in *polynomial time* if its time-complexity is dominated by a polynomial.

An algorithm runs in *polynomial space* if its space-complexity is dominated by a polynomial.

A *Kolmogorov-Chaitin complexity measure* of an algorithm is a measure based on the number of instructions of the algorithm, which is taken as an estimate of the logical complicatedness.

The *time-complexity* of a computable function is the minimum time-complexity taken over all algorithms that compute the function.

The *parallel time-complexity* of a computable function is the minimum time-complexity taken over all parallel algorithms that compute the function.

The *space-complexity* of a computable function is the minimum space-complexity taken over all algorithms that compute the function.

A *decision function* is a function on a countably infinite domain that decides whether an object is in some specified subset of that domain.

A computable decision function is in *class* **P** (*polynomial*) if its time-complexity is polynomial.

A computable decision function is in *class* **NP** (*nondeterministic polynomial*) if its parallel time-complexity is polynomial.

A function $g$ *reduces* a decision function $h$ to a decision function $f$ if $h = f \circ g$.

A computable decision function $f$ is **NP-hard** if every decision function in class NP can be reduced to $f$ by a polynomial-time function.

A computable decision function is **NP-complete** if it is NP-hard and is in class NP.

A *tractable problem* is a set-membership problem with a decision function in class P.

**Facts:**

1. The previous definitions can be rephrased in terms of problems and algorithms.

   - A problem is in *class P* (or *tractable*) if it can be solved by an algorithm that runs in polynomial time;
   - A problem is in *class NP* if, given a tentative solution (obtained by any means), it is possible to check that the solution is correct in polynomial time;
   - A problem is *NP-complete* if it is in class NP and is NP-hard.

**2.** When considering whether a given problem belongs to P or NP, and whether it might be NP-complete, it is helpful to rewrite the problem, or an associated problem, as a decision problem (which has a yes/no answer) because decision problems have been easier to characterize and classify than general problems. For example, see the description of the traveling salesman problem in Example 3 in this section.

**3.** Time-complexity of sorting algorithms is typically measured by the number of comparisons needed.

**4.** The words *good*, *efficient*, and *feasible* are commonly used interchangeably to mean polynomial-time.

**5.** Additive and multiplicative constants that are ignored in big-$O$ analysis of an algorithm can sometimes be too large for practical application.

**6.** That a problem belongs to P does not necessarily imply that it can be solved in a practical amount of time, since the polynomial bound of its complexity can be of high degree. Fortunately, however, for most problems in P arising in practical applications, the polynomial bound is of relatively small degree.

**7.** Belonging to class NP means that a solution can be *checked* in polynomial time, but not necessarily *found* in polynomial time.

**8.** When a problem is in class NP, it may be possible to solve the problem for cases arising in practical applications in a reasonable amount of time, even though there are other cases for which this is not true. Also, such problems can often be attacked using *approximation algorithms* that do not produce the exact solution, but instead produce a solution guaranteed to be close in some precise sense to the actual solution.

**9.** Every problem in class P is in class NP.

**10.** It often requires only a small change to transform a problem in class P to one in class NP. For example, the first four problems in Example 2 (Euler graph, edge cover, linear Diophantine equation, 2-satisfiability) are in class P, but the similar first four problems in Example 3 (Hamilton graph, vertex cover, quadratic Diophantine equation, 3-satisfiability), each of which results from seemingly small changes in the respective problem from class P, are in class NP.

**11.** To show a problem is NP-complete, the problem can be transformed (in a specific way) to a problem already known to be NP-complete. This is often much easier than showing directly that the problem is NP-complete. See [GaJo79] for details.

**12.** If there is an NP-hard problem that belongs to P, then P = NP.

**13.** Not all NP problems are known to be NP-complete. (See Example 4 for such a problem.)

**14.** Deciding whether P = NP is the outstanding problem in the theory of computational complexity. It is the common belief that P $\neq$ NP, based on an extensive search for polynomial-time solutions to various NP problems.

**15.** The first problem to be shown to be NP-complete was the *satisfiability problem* (Example 3). That the satisfiability problem is NP-complete is called *Cook's theorem*, after Steven A. Cook, who discovered it in 1971. [Co71]

**16.** In 1972 Richard Karp proved that the traveling salesman problem (TSP) (and many others) were NP-complete. [Ka72]

**17.** Hundreds of thousands of problems (in many areas, including mathematics, computer science, operations research, physics, biology) are known to be NP-complete.

**18.** Extensive information on NP-completeness (methods of proof, examples, etc.) can be found in [AtBl09a], [GaJo79], [GoDíTu14], and [vL90].

**19.** A more formal approach to complexity, given in terms of Turing machines, appears in §17.5.

**Examples:**

**1.** The following table gives some different input size variables for different problem types.

| *problem type* | *typical input size parameters* |
|---|---|
| database sorting | number of records |
| graph algorithms | number of vertices and/or number of edges |
| arithmetic computation | numbers of digits in the numerals |
| convex hull construction | number of points |

**2.** The following problems are in class P.

- *Euler graph*: given a graph, determine whether the graph has an Euler circuit;

- *edge cover*: given a graph $G$ and positive integer $n$, determine whether there is a subset $E$ of edges of $G$ with $|E| \leq n$ and every vertex of $G$ an endpoint of an edge in $E$;

- *linear Diophantine equation*: given positive integers $a$, $b$, $c$, determine whether $ax + by = c$ has a solution in positive integers $x$ and $y$;

- *2-satisfiability*: given a Boolean expression in conjunctive normal form in which each sum contains only two variables, determine whether the expression is "satisfiable" (i.e., there is an assignment of 0 and 1 to the variables such that the expression has value 1);

- *cycle rank*: given a graph $G$ and positive integer $n$, determine whether there is a subset $E$ of edges of $G$ with $|E| \leq n$ such that each cycle in $G$ contains an edge in $E$;

- *linear programming*: maximize $cx$ subject to $Ax \leq b$ where $A$ is a given $q \times n$ matrix, $c$ is a given row vector of length $n$, and $b$ is a given column vector of length $q$ (see §16.1.1).

**3.** The following problems are NP-complete.

- *Hamilton graph*: given a graph, determine whether the graph has a Hamilton circuit;

- *vertex cover*: given a graph $G$ and positive integer $n$, determine whether there is a subset $V$ of vertices of $G$ with $|V| \leq n$ with every edge of $G$ having an endpoint in $V$;

- *quadratic Diophantine equation*: given positive integers $a$, $b$, $c$, determine whether the equation $ax^2 + by = c$ has a solution in positive integers $x$ and $y$;

- *3-satisfiability*: given a Boolean expression in conjunctive normal form in which each sum contains only three variables, determine whether the expression is "satisfiable" (i.e., there is an assignment of 0 and 1 to the variables such that the expression has value 1);

- *satisfiability*: given a Boolean expression in conjunctive normal form, determine whether the expression is "satisfiable" (i.e., there is an assignment of 0 and 1 to the variables such that the expression has value 1) (see Fact 15);

- *traveling salesman problem*: given a weighted graph and positive number $k$, determine whether there is a Hamilton circuit of weight at most $k$ (see §10.7.1);

- *independent vertex set*: given a graph $G$ and a positive integer $n$, determine whether $G$ contains an independent vertex set of size at least $n$;
- *knapsack problem*: given a set $S$, values $a_i$ and $b_i$ for each $i \in S$, and numbers $a$ and $b$, determine whether there is a subset $T \subseteq S$ such that $\sum_{i \in T} a_i \leq a$ and $\sum_{i \in T} b_i \geq b$ (see §16.3.1);
- *bin packing problem*: given $k$ bins (each of capacity $c$) and a collection of weights, determine whether the weights can be placed in the bins so that no bin has its capacity exceeded (see §16.3.2);
- *3-coloring*: given a graph $G$, determine whether its vertices can be colored with 3 colors;
- *clique problem*: given a graph $G$ and positive integer $n$, determine whether $G$ has a clique of size at least $n$;
- *dominating set*: given a graph $G$ and positive integer $n$, determine whether $G$ has a dominating set of size at most $n$;

**4.** The following problem is an NP problem, but not known to be NP-complete nor known to be in P.

- *graph isomorphism*: given two graphs, determine whether they are isomorphic.

## 17.4.2  WORST-CASE AND AVERAGE-CASE ANALYSIS

### Definitions:

A **worst-case complexity measure** of an algorithm is based on the maximum computational cost for any input of that size. It is usually expressed in big-$O$ asymptotic notation (or sometimes $\Theta$-notation) as a formula based on the input size variables.

An **average-case complexity measure** of an algorithm is based on the expected computational cost over a random distribution of its inputs of a given size.

### Facts:

**1.** Algorithmic analysis of deterministic algorithms often assumes a uniform random distribution of the possible inputs, when the actual distribution is unknown.

**2.** For sorting algorithms, an average-case analysis may assume that all input permutations of the keys to be sorted are equally likely. In practice, however, some permutations may be far more likely than others, e.g., already sorted, almost sorted, or reverse sorted.

**3.** The input size measures for average-case analysis are usually the same as for worst-case analysis.

### Examples:

**1.** The following table gives the worst-case running times of some sorting algorithms [CoEtal09], where the input size parameter $n$ is the number of records.

| sorting method | worst-case complexity |
|---|---|
| insertion sort | $\Theta(n^2)$ |
| selection sort | $\Theta(n^2)$ |
| bubble sort | $\Theta(n^2)$ |
| heapsort | $\Theta(n \log n)$ |
| quicksort | $\Theta(n^2)$ |
| mergesort | $\Theta(n \log n)$ |

**2.** The following table gives the worst-case running times of some graph algorithms, based on input size parameters $|V|$ and $|E|$, which are the numbers of vertices and edges.

| graph algorithm | worst-case complexity |
|---|---|
| Kruskal's MST algorithm | $\Theta(|E| \log |V|)$ |
| Dijkstra's shortest-path algorithm with linked-list priority queue | $O(|V|^2)$ |
| Dijkstra's shortest-path algorithm with heap-based priority queue | $O(|E| \log |V|)$ [CoEtal09] |
| Dijkstra's shortest-path algorithm with Fibonacci-heap priority queue | $O(|E| + |V| \log |V|)$ [CoEtal09] |
| Edmonds-Karp max-flow algorithm | $O(|V| \cdot |E|^2)$ |

**3.** The following table gives the worst-case running times of some plane convex hull algorithms (see §13.5.1), based on the number $n$ of points supplied as input.

| convex hull algorithm | worst-case complexity |
|---|---|
| Graham Scan | $\Theta(n \log n)$ |
| Jarvis March ("gift-wrapping") | $\Theta(nh)$, $h = \#$ corners (convex hull) |
| QuickHull | $O(n^2)$ |
| MergeHull | $O(n \log n)$ |

**4.** The following table gives the average-case running times of some sorting algorithms, where the input size parameter $n$ is the number of records.

| sorting method | average-case complexity |
|---|---|
| insertion sort | $O(n^2)$ |
| selection sort | $O(n^2)$ |
| bubble sort | $O(n^2)$ |
| heapsort | $O(n \log n)$ |
| quicksort | $O(n \log n)$ |
| mergesort | $O(n \log n)$ |

**5.** *Randomized quicksort* (Algorithm 1) [CoEtal09]: A subarray from index $p$ to index $r$ of an array $A$ is sorted, using an external subroutine $random(p, r)$ that generates a number in the set $\{p, \ldots, r\}$ in $O(1)$ time. Another external subroutine $partition(A, p, r)$ rearranges the subarray $A[p..r]$ and returns an index $q$, $p \leq q < r$, such that for $i = p, \ldots, q$, $A[i] \leq A[q]$ and such that for $i = q + 1, \ldots, r$, $A[i] > A[q]$; this subroutine runs in $\Theta(r - p)$ worst-case time.

---

**Algorithm 1:  Randomized quicksort.**

**procedure** *randomized-quicksort*$(A, p, r)$

**if** $p < r$ **then**

　　$i := random(p, r)$

　　exchange $A[p]$ and $A[i]$

　　$q := partition(A, p, r)$

　　*randomized-quicksort*$(A, p, q)$

　　*randomized-quicksort*$(A, q + 1, r)$

To sort $n$ keys, randomized quicksort takes $\Theta(n^2)$ time in the worst case (when unlucky enough to have partition sizes always unbalanced), but only $\Theta(n \log n)$ time in the average case (partition sizes are usually at least a constant fraction of the total).

**6.** *Convex hull*: For some distributions of $n$ points in the plane, the expected value $E[h]$ of the number of vertices on the convex hull is known. This bound implies that the average-case running time of Jarvis March is an additional factor of $n$ greater:

| distribution | $E[h]$ | average-case running time |
|---|---|---|
| uniform in convex polygon | $O(\log n)$ | $O(n \log n)$ |
| uniform in circle | $O(n^{\frac{1}{3}})$ | $O(n^{\frac{4}{3}})$ |
| normal in plane | $O(\sqrt{\log n})$ | $O(n\sqrt{\log n})$ |

## 17.4.3 LOWER BOUNDS

Lower bounds on running times of algorithms are typically given as functions of input size using $\Omega$-notation (see §1.3.3).

**Definitions:**

An **existential lower bound** for an algorithm is a lower bound for its running time that holds for at least one input.

An **existential lower bound** for a problem is a lower bound for every algorithm that could solve that problem.

A **comparison sort** is a sorting method that rearranges records based only on comparisons between keys.

The **Euclidean minimum spanning tree** (or **Euclidean MST**) **problem** has as input vertices a set of $n$ points in the plane and as output a spanning tree of minimum total edge length.

A **reduction of a problem** $A$ to another problem $B$ is the following sequence of steps:

- the input to problem $A$ is transformed into an input to problem $B$;
- problem $B$ is solved on the transformed input;
- the output of problem $B$ is transformed back into a solution to problem $A$ for the original input.

An $f(n)$ **time reduction** of problem $A$ to problem $B$ is a reduction such that the time for the three steps together is $f(n)$.

**Facts:**

**1.** For a given model of computation, if problem $A$ has a lower bound of $T(n)$ and it reduces in $f(n)$ time to problem $B$, then problem $B$ has a lower bound of $T(n) - f(n)$.

**2.** Every comparison sort on $n$ records requires $\Omega(n \log n)$ comparisons in the worst case. This follows since there are $n!$ possible solutions, and each comparison can reduce this by at most a factor of 2. (Note that $\log_2(n!)$ is $\Omega(n \log n)$.)

**3.** Computing the Euclidean minimum spanning tree on $n$ points takes $\Omega(n \log n)$ time in the worst case.

**4.** Unlike the Euclidean MST problem, most graph *problems* have no known nontrivial lower bound. Some graph *algorithms*, however, have lower bounds on their implementation.

**5.** Running Dijkstra's algorithm (see §10.3.2) on a directed graph with $n$ vertices takes $\Omega(n \log n)$ time in the worst case.

**6.** Finding the vertices for the convex hull of $n$ points in the plane, in *any* order, takes $\Omega(n \log n)$ time in the worst case.

**7.** Constructing the Voronoi diagram (see §13.5.3) on $n$ points in the plane takes in the worst case $\Omega(n \log n)$ time.

**Examples:**

**1.** *An $O(n)$-time reduction of sorting to a gift-wrap of a convex hull:* Given a set of $n$ positive numbers $\{x_1, \ldots, x_n\}$, first produce in $\Theta(n)$ time their respective squares $\{x_1^2, \ldots, x_n^2\}$. Since each point $(x_j, x_j^2)$ lies on the parabola given by $y = x^2$, the Jarvis march on the convex hull of the points $(x_j, x_j^2)$ is a list of points, ordered by abscissa. Sequentially read off the first coordinate of every point of the convex hull in $\Theta(n)$ time, thereby producing the sorted list of numbers. This implies that finding the gift-wrapped convex hull of $n$ points requires at least $\Omega(n \log n) - \Theta(n) = \Omega(n \log n)$ time.

**2.** *An $O(n)$-time reduction of sorting numbers to Euclidean MST:* To sort $n$ numbers $\{x_1, \ldots, x_n\}$, create $n$ points $\{(x_i, 0) \mid 1 \leq i \leq n\}$ in the Euclidean plane. The Euclidean MST of this set contains an edge between points $(x_i, 0)$ and $(x_j, 0)$ if and only if the numbers $x_i$ and $x_j$ are consecutive in the sorted list of numbers. The Euclidean MST is easily converted back to a sorted list of numbers in $O(n)$ time. This implies that Euclidean MST requires at least $\Omega(n \log n) - \Theta(n) = \Omega(n \log n)$ time.

# 17.5  COMPLEXITY CLASSES

From a formal viewpoint, complexity theory is concerned with classifying the difficulty of testing for membership in various languages. This means deciding whether any given string is in the language. The general application of complexity theory is achieved by encoding decision problems on natural topics such as graph coloring and finding integer solutions to equations as set membership problems. A more comprehensive coverage of complexity theory is provided by many textbooks, including [DuKo00] and [HeOg02].

## 17.5.1  ORACLES AND THE POLYNOMIAL HIERARCHY

Throughout this section, whenever the *alphabet* is unspecified, it may be assumed to be the binary set $\{0, 1\}$. Also, throughout this section a *Turing machine* (see §17.1.3) is assumed to have among its states a unique *acceptance state* $q_A$ and a unique *rejection state* $q_R$. All other states continue the computation.

**Definitions:**

A *language over an alphabet* is a set of strings on that alphabet (see §17.3.2).

A *nondeterministic Turing machine* is a 5-tuple $(K, s, h, \Sigma, \Delta)$ otherwise like a deterministic Turing machine, except that the transition function $\Delta$ maps each state-symbol pair $(q, b)$ to a set of state-symbol-direction triples.

An **oracle for a language** $L$ is a special computational state to which a machine presents a string $w$, which switches to special state $Y$ ("yes") if $w \in L$ and to special state $N$ ("no") if $w \notin L$.

An **oracle Turing machine** is a 6-tuple $(K, s, h, \Sigma, \delta \text{ or } \Delta, L)$, equipped with an oracle for language $L$ and with a special second tape on which it can write a string over the alphabet of language $L$ (which might be different from $\Sigma$). Aside from oracle steps, it is a Turing machine.

A Turing machine $M$ **accepts** string $w$ if there exists a computational path from the starting configuration with input $w$ to the acceptance state $q_A$.

A Turing machine $M$ **rejects** string $w$ if it does not accept $w$. (Either $M$ halts in rejection state $q_R$ or does not halt.)

The **language accepted by a Turing machine** $M$ is the set of all the strings it accepts. It is denoted $\mathcal{L}(M)$.

The Turing machine $M$ **decides** the language $\mathcal{L}(M)$ if it always halts, even for input strings not in $\mathcal{L}(M)$.

A (possibly nondeterministic) Turing machine $M$ is said to be of **time complexity $T(n)$** if for each $n$, each word $x$ of length $n$, and (if $M$ is nondeterministic) each sequence of nondeterministic choices made by $M$ on input $x$, $M$ halts within $T(|x|)$ steps.

A (possibly nondeterministic) Turing machine $M$ is said to be of **space complexity $S(n)$** if for each $n$, each word $x$ of length $n$, and (if $M$ is nondeterministic) each sequence of nondeterministic choices made by $M$ on input $x$, $M$ scans at most $S(|x|)$ cells of any storage tape.

A Turing machine $M$ has **polynomial time complexity** if there exists a polynomial $p(n)$ such that $M$ is of time complexity $T(n)$.

A Turing machine $M$ has **polynomial space complexity** if there exists a polynomial $p(n)$ such that $M$ is of space complexity $T(n)$.

The complexity class **P** contains every language that can be decided by a deterministic TM with polynomial time complexity.

The complexity class **PSPACE** contains every language that can be decided by a deterministic TM with polynomial space complexity.

The complexity class **NP** contains every language that can be decided by a nondeterministic TM with polynomial time complexity.

For any language $L$, the complexity class **$P^L$** contains every language that is decided in polynomial time by a deterministic TM with oracle $L$.

For any language $L$, the complexity class **$NP^L$** contains every language that is decided in polynomial time by a nondeterministic TM with oracle $L$.

For any class $\mathcal{C}$ of languages, the complexity class **$P^{\mathcal{C}}$** contains every language that is decidable in polynomial time by a deterministic TM with oracle $L \in \mathcal{C}$. Equivalently, notated in terms of reductions (see §17.5.2), $P^{\mathcal{C}} = \{L \mid (\exists B \in \mathcal{C})[L \leq_T^p B]\}$.

For any class $\mathcal{C}$ of languages, the complexity class **$NP^{\mathcal{C}}$** contains every language that is decidable in polynomial time by a nondeterministic TM with oracle $L \in \mathcal{C}$.

The complexity classes $\Sigma_n^p$ are defined inductively by

$$\Sigma_k^p = \begin{cases} \text{P} & \text{if } k = 0 \\ \text{NP}^{\Sigma_{k-1}^p} & \text{if } k \geq 1. \end{cases}$$

The **polynomial hierarchy PH** is the collection comprising every language $A$ for which there exists an $n$ such that $A \in \Sigma_n^p$.

For $n \geq 0$, the polynomial hierarchy is said to **collapse to its nth level** if $\text{PH} = \Sigma_n^p$. The polynomial hierarchy is said to **collapse** if for some $n \geq 0$ it collapses to its $n$th level.

For $n \geq 0$, the complexity class $\mathbf{\Pi_n^p}$ contains every language $A$ such that $\overline{A} \in \Sigma_n^p$.

The complexity class **coNP** is $\Pi_1^p$.

The complexity class $\mathbf{\Delta_0^p}$ is P. For $n \in \mathcal{Z}^+$, the complexity class $\mathbf{\Delta_n^p}$ is $\text{P}^{\Sigma_{n-1}^p}$.

The complexity class **P/poly** contains each set $A$ such that there exist a set $B \in \text{P}$, a function $h$, and a polynomial $f(\cdot)$ such that $(\forall n)[|h(n)| \leq f(n)]$ and $(\forall x)[x \in A \iff \langle x, h(|x|) \rangle \in B]$.

A language $B$ is **sparse** if there exists a polynomial $p(n)$ such that for every $n \in \mathcal{N}$, there are at most $p(n)$ elements of length $n$ in $B$.

### Facts:

**1.** The following equalities hold: $\Sigma_0^p = \Pi_0^p = \Delta_0^p = \Delta_1^p = \text{P}$.

**2.** For $n \geq 0$, the following relationships hold:

$$\Delta_n^p \subseteq \Sigma_n^p \cap \Pi_n^p \begin{smallmatrix} \subseteq \Sigma_n^p \subseteq \\ \subseteq \Pi_n^p \subseteq \end{smallmatrix} \Sigma_n^p \cup \Pi_n^p \subseteq \Delta_{n+1}^p.$$

**3.** $\text{PH} \subseteq \text{PSPACE}$.

**4.** If $\text{PH} = \text{PSPACE}$, then the polynomial hierarchy collapses.

**5.** *Downward separation:* For each $n \geq 0$ it holds that: if $\Sigma_n^p = \Sigma_{n+1}^p$, then $\text{PH} = \Sigma_n^p$. In particular, $\text{P} = \text{NP}$ if and only if $\text{P} = \text{PH}$.

**6.** *Downward separation:* For each $n \geq 1$ it holds that: if $\Sigma_n^p = \Pi_n^p$, then $\text{PH} = \Sigma_n^p$.

**7.** $\text{NP} = \Sigma_1^p$.

**8.** The complexity class $\Sigma_n^p$ is closed under union and intersection, for all $n \geq 0$.

**9.** $\text{P}^{\text{NP} \cap \text{coNP}} = \text{NP} \cap \text{coNP}$. More generally, $\text{P}^{\Sigma_n^p \cap \Pi_n^p} = \Sigma_n^p \cap \Pi_n^p$ and $\text{P}^{\Delta_n^p} = \Delta_n^p$, for all $n \geq 0$.

**10.** *Upward separation:* Nondeterministic exponential time $\left( \bigcup_{c>0} \text{NTIME}[2^{cn}] \right)$ is equal to deterministic exponential time $\left( \bigcup_{c>0} \text{DTIME}[2^{cn}] \right)$ if and only if $\text{NP} - \text{P}$ contains no sparse sets.

**11.** *Succinct certificates:* For every language in NP there is a proof scheme in which each member (and only members) has a polynomial-size "proof" of membership that can be checked in deterministic polynomial time. Such a short membership proof is sometimes called a *succinct certificate.*

**12.** Sets in P/poly are said to "have small circuits." Many circuit classes less inclusive than P/poly play an important role in complexity theory; see [Vo99].

**13.** $\text{P/poly} = \text{P}^{\{S \mid S \text{ is sparse}\}}$.

**Examples:**

**1.** *Logical proposition problems*: The problem of deciding whether a particular assignment of TRUE-FALSE values to the variables satisfies a logical proposition is in P. Deciding whether a proposition has an assignment that satisfies it is in NP. Deciding whether all assignments satisfy it (i.e., whether the proposition is a tautology) is in coNP.

**2.** *Graph isomorphism problems*: Deciding whether a given vertex bijection between two graphs realizes a graph isomorphism is in P. Deciding whether two graphs are isomorphic is in NP.

**3.** *Graph coloring problems*: Deciding whether an assignment of colors from a set of three colors to the vertices of a graph is a proper coloring is in P. Deciding whether a graph has a proper 3-coloring is in NP.

**4.** *Unique maximum clique problem*: Define UMC to be the set of graphs $G$ with a clique $U \subseteq V_G$ such that every other clique is strictly smaller than $U$. Then UMC is in the class $\Delta_2^p = \mathrm{P}^{\mathrm{NP}}$.

**5.** To prove by succinct certificate that a given graph has some clique of size at least $k$, one can provide a list of $k$ distinct vertices that are mutually adjacent. (The mutual adjacency condition for the $k$ vertices can be verified in polynomial time.)

**6.** *P-selectivity*: A set $L$ is said to be *semi-feasible* (or, equivalently, *P-selective*) if there is a polynomial-time computable function $f$ such that for all $x$ and $y$, (a) $f(x,y) = x$ or $f(x,y) = y$, and (b) $\{x,y\} \cap L \neq \emptyset \implies f(x,y) \in L$. All semi-feasible sets are in P/poly. (This can be seen by a divide and conquer argument.)

**7.** *The primality problem*: The language PRIMES, which consists of the bitstrings that represent prime numbers when interpreted as binary numerals, is in P.

---

## 17.5.2  REDUCIBILITY AND NP-COMPLETENESS

**Definitions:**

A language $A$ over alphabet $\Sigma$ is **polynomial-time reducible** (or **m-p-reducible**) to a language $B$, denoted $A \leq_m^p B$, if there exists a polynomial-time computable function $f$ such that, for each $x \in \Sigma^*$, it holds that $x \in A$ if and only if $f(x) \in B$.

A language $A$ is **NP-hard** if every language in NP is polynomial-time reducible to $A$.

A language $A$ is **NP-complete** if $A$ is NP-hard and $A \in$ NP.

For a class of languages $\mathcal{C}$, a language $A$ is **$\mathcal{C}$-hard** if every language in $\mathcal{C}$ is polynomial-time reducible to $A$.

A language $A$ is **$\mathcal{C}$-complete** if $A$ is $\mathcal{C}$-hard and $A \in \mathcal{C}$.

A language $A$ is **Turing-p-reducible** to the language $B$, denoted $A \leq_T^p B$, if there is a deterministic oracle TM $M$ such that $M^B$ decides $A$ in polynomial time.

A language $A$ is **$\mathcal{C}$-Turing-p-hard** if every language in $\mathcal{C}$ is Turing-p-reducible to $A$.

A language $A$ is **$\mathcal{C}$-Turing-p-complete** if $A$ is $\mathcal{C}$-Turing-p-hard and $A \in \mathcal{C}$.

**Facts:**

**1.** For most NP-complete problems, showing membership in NP is easy.

**2.** For integer linear programming, however, it is easy to show NP-hardness, but showing membership in NP is nontrivial.

**3.** Polynomial-time reducibility is also called *Karp reducibility* after Richard Karp.

**4.** Turing-p-reducibility is also called *Cook reducibility* after Stephen Cook.

**5.** The complement of any NP-complete problem is coNP-complete.

**6.** If $A$ is polynomial-time reducible to $B$, then $A$ is Turing-p-reducible to $B$.

**7.** If $A \leq_m^p B$ and $B \leq_m^p C$, then $A \leq_m^p C$.

**8.** If $A \leq_T^p B$ and $B \leq_T^p C$, then $A \leq_T^p C$.

**9.** *Downward closure:* For $n \geq 0$, if $A \in \Sigma_n^p$ and $B \leq_m^p A$ then $B \in \Sigma_n^p$. In particular, if any NP-complete set is in P, then P = NP.

**10.** *Karp-Lipton theorem:* If there is a sparse NP-$\leq_T^p$-hard set, then PH $= \Sigma_2^p$.

**11.** If there is a sparse NP-$\leq_T^p$-complete set, then PH $= \Delta_2^p$.

**12.** *Mahaney's theorem:* If there is a sparse NP-hard (or NP-complete) set, then P = NP.

**13.** *Ladner's theorem:* If P $\neq$ NP, then there exists a set in NP $-$ P that is not NP-complete.

**14.** A large catalog of NP-complete problems appears in [GaJo79]. A few of the most commonly cited appear in §17.4.1. A similar catalog for higher levels of the polynomial hierarchy appears in [ScUm02].

**Examples:**

**1.** For examples of NP-complete problems, see §17.4.1, Example 3.

**2.** *Quantified Boolean formulas:* Let QBF be the class of true statements of the form

$$(\exists x_1)\,(\forall x_2)\,(\exists x_3)\,(\forall x_4)\,\ldots\,(Q_z x_z)\,[F(x_1, x_2, \ldots, x_z)],$$

where $F$ is a quantifier-free formula over the Boolean variables $x_1, \ldots, x_z$ and where $Q_z$ is $\exists$ if $z$ is odd and $\forall$ if $z$ is even. QBF is PSPACE-complete.

**3.** *Tautologies problem:* The classic coNP-complete language is the set TAUTOLOGY of all logical propositions that are satisfied by every assignment of logical values to the variables.

**4.** *Graph isomorphism problem:* It is not known whether the set GI of isomorphic graph pairs is in coNP or whether GI is NP-complete, though it is known that GI is NP-complete only if the polynomial hierarchy collapses.

---

## 17.5.3   PROBABILISTIC TURING MACHINES

**Definitions:**

A **probabilistic Turing machine** is a nondeterministic Turing machine $M$ with exactly two choices at each step. Each such choice occurs with probability $\frac{1}{2}$, and is independent of all previous choices.

The **acceptance probability** $p_M(w)$ that a probabilistic Turing machine $M$ accepts input word $w$ is the sum of the probabilities over all accepting computation paths on input $w$.

A probabilistic Turing machine $M$ **accepts language** $L$ **with one-sided error** if $p_M(w) > \frac{1}{2}$ if $w \in L$, and $p_M(w) = 0$ if $w \notin L$.

A probabilistic Turing machine $M$ **accepts language** $L$ **with two-sided error** if $p_M(w) > \frac{1}{2}$ if $w \in L$, and $p_M(w) \leq \frac{1}{2}$ if $w \notin L$.

A probabilistic Turing machine $M$ **accepts language** $L$ **with bounded two-sided error** if there exists some $\epsilon > 0$ such that for all $w$, $p_M(w) > \frac{1}{2} + \epsilon$ if $w \in L$ and $p_M(w) < \frac{1}{2} - \epsilon$ if $w \notin L$.

The complexity class **RP** (the random polynomial-time languages) is the class of languages that are decided by Turing machines with one-sided error in polynomial time.

The complexity class **coRP** contains the language $A$ if $\overline{A} \in$ RP.

The complexity class **ZPP** (the zero-error probabilistic polynomial-time languages) is the intersection RP ∩ coRP.

The complexity class **PP** (the probabilistic polynomial-time languages) is the class of languages that are decided by Turing machines with two-sided error in polynomial time.

The complexity class **BPP** (the bounded-error probabilistic polynomial-time languages) is decided by Turing machines with bounded two-sided error in polynomial time.

The complexity class **#P** is the class of all functions $f$ such that there exists a nondeterministic polynomial-time Turing machine $M$ such that, for each $x$, $f(x) = \#\mathrm{acc}_M(x)$, where $\#\mathrm{acc}_M(x)$ denotes the number of accepting computation paths of machine $M$ on input $x$.

The major types of polynomial-time reductions defined for functions include Turing reductions, metric reductions, many-one reductions, and parsimonious reductions. The latter three of these are defined as follows. A function $e : \Sigma^* \to \mathcal{N}$ **polynomial-time metric reduces** to a function $h : \Sigma^* \to \mathcal{N}$ if there exist two polynomial-time computable functions $\varphi$ and $\psi$ such that $(\forall x \in \Sigma^*)[e(x) = \psi(x, h(\varphi(x)))]$. A function $e : \Sigma^* \to \mathcal{N}$ **polynomial-time many-one reduces** to a function $h : \Sigma^* \to \mathcal{N}$ if there exist two polynomial-time computable functions $\varphi$ and $\psi$ such that $(\forall x \in \Sigma^*)[e(x) = \psi(h(\varphi(x)))]$. A function $e : \Sigma^* \to \mathcal{N}$ **parsimoniously reduces** to a function $h : \Sigma^* \to \mathcal{N}$ if there is a polynomial-time computable function $\varphi$ such that $(\forall x \in \Sigma^*)[e(x) = h(\varphi(x))]$.

Let $\leq_\alpha$ be a reduction that is defined for functions. A function $f$ is said to be **#P-complete with respect to** $\leq_\alpha$ **reductions** if $f \in$ #P and for every $g \in$ #P it holds that $g \leq_\alpha f$.

**Facts:**

**1.** ZPP is exactly the class of languages accepted by error-free probabilistic Turing machines running in expected polynomial time.

**2.** ZPP = RP ∩ coRP $\overset{\subseteq \mathrm{RP} \subseteq}{\subseteq \mathrm{coRP} \subseteq}$ RP ∪ coRP ⊆ BPP ⊆ PP ⊆ PSPACE.

**3.** RP ⊆ NP ⊆ PP.

**4.** $\mathrm{ZPP}^{\mathrm{ZPP}} = \mathrm{P}^{\mathrm{ZPP}} = \mathrm{ZPP}$; $\mathrm{BPP}^{\mathrm{BPP}} = \mathrm{P}^{\mathrm{BPP}} = \mathrm{BPP}$.

**5.** BPP ⊆ P/poly.

**6.** $\mathrm{BPP} \subseteq \mathrm{NP}^{\mathrm{BPP}} \subseteq \mathrm{ZPP}^{\mathrm{NP}} \subseteq \Sigma_2^p \cap \Pi_2^p$.

**7.** If there is a sparse NP-$\leq_T^p$-hard set, then PH = $\mathrm{ZPP}^{\mathrm{NP}}$. (This extends the Karp-Lipton theorem of §17.5.2.)

**8.** $\mathrm{PH} \subseteq \mathrm{PP}^{\mathrm{PH}} \subseteq \mathrm{P}^{\mathrm{PP}} = \mathrm{P}^{\#\mathrm{P}}$.

**9.** PP is closed under all Boolean operations.

**10.** If NP ⊆ BPP then BPP = PH and RP = NP.

**11.** It remains an open question whether BPP, RP, coRP, or ZPP have complete languages.

**Examples:**

**1.** $SAT \in PP$:  Consider a probabilistic polynomial-time Turing machine $M$ that, given a proposition $F$, immediately flips its coin. If the result is "heads", then proposition $F$ is accepted and machine $M$ halts. If "tails", then the machine, by a series of coin flips, randomly assigns each variable to be either true or false, and (on that computation path) accepts $F$ if the resulting assignment satisfies the proposition. Thus $F$ is accepted with probability exactly $\frac{1}{2}$ if $F$ is unsatisfiable, and is accepted with probability at least $\frac{1}{2} + \frac{1}{2^{k+1}}$ if $F$ is satisfiable, where $k$ is the number of logical variables in $F$. Thus SAT $\in$ PP. This implies that NP $\subseteq$ PP, since the language SAT is NP-complete.

**2.** *Equality of polynomial products*:  Given two lists of rational-coefficient polynomials, where each polynomial in the lists has been specified by a list of (coefficient, degree) pairs, the problem of deciding whether the product of the polynomials in the first list yields the same polynomial as the product of the polynomials in the second list is in the class coRP. This is because if the products are equal, then they will evaluate to the same value on any argument, yet if the products are unequal then it can be argued that the products will differ with sufficient probability on a randomly chosen argument.

**3.** *MAJORITY-SAT is PP-complete*:  The language *MAJORITY-SAT* is the set of (quantifier-free) Boolean formulas $F$ such that $F$ is satisfied by more than half of the possible variable assignments.

**4.** *#SAT*:  #SAT, the function that maps from Boolean formulas $F$ to the number of solutions of $F$, is #P-complete with respect to parsimonious reductions (and thus with respect to many-one, metric, and Turing reductions).

**5.** *#SAT Is as Hard to Enumeratively Approximate (aka List Approximate) as It Is to Compute Exactly*:  If there exists a polynomial-time function $f$ mapping from Boolean formulas to lists of integers such that for each Boolean formula $F$ it holds that the number of solutions of $F$ (i.e., $\#SAT(F)$) is one of the integers in the list $f(F)$, then #SAT can be computed in polynomial time.

**6.** *The Power Index*:  Power indices are very important in cooperative game theory and political science. Given a list of states and for each a positive natural number indicating the number of votes that state has, the raw Shapley-Shubik power index of a state $s$ is the number of permutations of the list in which the sum of the weights of all the states coming before $s$ in the permutation is less than or equal to half the total weight but the sum of the weights of all the states up to and including $s$ is strictly greater than half the total weight. Informally, the index is a measure of how often $s$ is the player who puts a coalition over the finish line if players join the coalition in random order. The raw Shapley-Shubik power index is #P-complete with respect to many-one reductions (and thus with respect to metric and Turing reductions), but is not #P-complete with respect to parsimonious reductions.

# 17.6 RANDOMIZED ALGORITHMS

General randomization principles for algorithms have many specific applications. In particular, random algorithms from number theory have applications in cryptography and fingerprinting, Also, randomized algorithms for partitioning, for searching and sorting, and for graph problems such as mincut and matching, including some heuristics for NP-complete problems, have applications in testing and applications for parallel or distributed environments.

## 17.6.1 OVERVIEW AND GENERAL PARADIGMS

Most randomized algorithms follow a few general paradigms. For many further topics not covered here, see the excellent survey papers [Ka91,We83] and also the textbook [MoRa95].

**Definition:**

A *randomized algorithm* is an algorithm that makes random choices during its execution. Such random choices are guided by the output of a random (or, in practice, pseudo-random) number generator.

**Facts:**

**1.** Many problems have no known deterministic algorithms to match the efficiency of randomized algorithms. Even for problems for which efficient deterministic algorithms are known, randomized algorithms are often remarkably easier to understand and implement.

**2.** Worst-case instances of a randomized algorithm occur when the algorithm performs badly for the overwhelming majority of its probabilistic choices.

**3.** *Abundance of witnesses paradigm:* Deciding whether a given input has a certain property sometimes reduces to finding a combinatorial object "witnessing" the property. When the space of all potential witnesses is too large to be searched exhaustively, it sometimes suffices to inspect a small random sample, selected so that one of the elements of the sample will be a suitable witness with very high probability.

**4.** *Random sampling:* Sometimes a small random sample is indicative of the population as a whole.

**5.** Intuitively, the power of randomization is analogous to the standard game-theoretic fact that probabilistic game strategies are substantially more effective than deterministic ones. That is, an algorithm can be regarded as a player, and the problem to be solved can be regarded as an adversary trying to present the player with input instances on which the algorithm exhibits worst-case performance.

**6.** If an algorithm is deterministic, then the game-theoretic adversary knows in advance the entire strategy of the player. Thus, the worst-case instances are well defined and can be presented as input to the algorithm. If an algorithm is probabilistic, then the game-theoretic adversary does not know in advance the output of the random number generator. In particular, worst-case instances under deterministic strategies may be smoothed out by randomization.

**Examples:**

**1.** *Cryptography*:   Some public key cryptography is based on the sharp dichotomy between the efficiency of deciding whether a number is prime or composite, and the apparent hardness of actually factoring composite numbers.

**2.** *Fingerprinting*:   A large data object is represented by a much smaller "fingerprint" such that, with very high probability, distinct objects map to distinct fingerprints. A similar strategy is used for "hashing" (see §18.4) where large objects are mapped to much smaller keys with very low probabilities of collisions.

**3.** *Testing identities*:   It is often possible to check if an algebraic expression is identically equal to zero by substituting random values for the variables and checking whether the expression evaluates to zero.

**4.** *Symmetry breaking*:   It is often necessary for a set of processes to come collectively to an arbitrary but consistent decision among a set of indistinguishable possibilities. There is a method to break such symmetries using randomization: this yields an efficient parallel perfect matching algorithm, as well as protocols for distributed environments, computation in the presence of errors, and Byzantine agreements.

**5.** *Load balancing*:   For problems involving choice among resources, such as processors or communication links, randomization can be useful in spreading out the load.

**6.** *The probabilistic method*:   The probabilistic method is to demonstrate that a combinatorial object of interest occurs with nonzero probability in a suitably defined probability space. Sometimes the probabilistic method yields efficient algorithmic constructions rather than mere existential arguments. (See §7.11.)

---

**17.6.2   LAS VEGAS AND MONTE CARLO ALGORITHMS**

Randomized algorithms are classified into two types: Monte Carlo algorithms and Las Vegas algorithms.

**Definitions:**

A **Monte Carlo algorithm** has bounded running time and produces correct output with probability bounded away from zero.

The **success amplification method for a Monte Carlo algorithm** is to perform $k$ independent runs of the algorithm.

A **Las Vegas algorithm** always produces correct output. However, its running time is a random variable, whose expectation and variance are quantified in the analysis of the algorithm.

The **success amplification method for a Las Vegas algorithm** is to perform $k/2$ independent Las Vegas runs of $2E[T]$ steps each, where $E[T]$ is the expected Las Vegas running time.

The **Monte Carlo to Las Vegas transformation**, starting from a Monte Carlo algorithm, is the Las Vegas algorithm of repeatedly running the Monte Carlo algorithm until a success occurs.

The **Las Vegas to Monte Carlo transformation**, starting from a Las Vegas algorithm, is the Monte Carlo algorithm obtained by running the Las Vegas scheme for $kE[T]$ steps and halting.

**Facts:**

**1.** If the probability of success of a single run is $p$, then the probability under the success amplification method that $k$ independent runs fail is $(1-p)^k$. Thus, the probability of success becomes $1-(1-p)^k$.

**2.** If $p$ is the probability of success of a Monte Carlo algorithm, then the expected number of Las Vegas trials before a success occurs is $1/p$.

**3.** *Markov's inequality*: The probability that a positive random variable exceeds $k$ times its expectation is at most $\frac{1}{k}$.

**4.** Markov's inequality yields a general method to bound variances of Las Vegas algorithms: If $T$ is the running time of a Las Vegas algorithm, then $Pr[T > kE[T]] < 1/k$.

**5.** The probability that a transformed Las Vegas to Monte Carlo algorithm is successful is at least $1-1/k$.

**6.** If the expected running time of a Las Vegas algorithm is $E(T)$, then the running time of the amplified algorithm is $kE[T]$. However, the probability of success becomes $1-(\frac{1}{2})^{\frac{k}{2}}$.

**Examples:**

**1.** *A database problem*: In a large database whose keys are stored in no particular order, find a key that is not contained in that database, within time $O(N)$, where $N$ is the size of the database. This would match the natural lower bound of $\Omega(N)$, the time required just to read the entire database. The deterministic strategy of sorting and checking for the first missing key would take time $O(N \log N)$. Assume $N = 2^{25}$ and that the keys are 32 digits long.

- a *Monte Carlo randomized strategy*: Pick a random 32-digit key and then scan the database! There are $2^{32}$ potential keys and only $N$ keys in the database. Thus, the probability that a randomly chosen key is not in the database is at least $1-2^{-7}$, which is greater than 99%. The running time is dominated by a single scan of the database to check whether the randomly chosen key is suitable. Thus, it completes in $O(N)$ time.

- *success amplification*: The probability that among $k$ independently chosen random keys none is found suitable is less than $0.01^k = 10^{-2k}$; this quantity becomes negligible, even for very small values of $k$. The running time of this amplified algorithm is $O(kN)$.

- *from Monte Carlo to Las Vegas*: Repeatedly pick random keys until a suitable key is found. The expected number of trials before a suitable key is found is $\frac{1}{p} < \frac{100}{99}$. Thus, the expected running time is $O(\frac{100N}{99}) = O(N)$.

**2.** *Finding the median*: Among $n$ keys find the $m$th in increasing order. The deterministic strategy of sorting would take time $O(n \log n)$. Algorithm 1 does this within time $O(n)$ as follows.

- a *Las Vegas randomized strategy*: Pick a random key $s$ from the database and consider the sets $X$ and $Y$ of keys in the database that are smaller and larger, respectively, than $s$. If $|X| \geq m$, then the problem reduces to finding the $m$th key in $X$. If $n - |Y| \leq m$, then the problem reduces to finding the $(m-(n-|Y|))$th key in $Y$. Finally, if $|X| < m < n - |Y|$, then $s$ is the $m$th key.

---

**Algorithm 1:**   In a set $A$ of $n$ distinct keys, find the $m$th smallest.

input: a set $A$ with $n = |A|$, and an integer $m$ with $1 \leq m \leq n$

**FIND** $(A, m)$

**if** $A = \{s\}$ **then** return $s$

**else**

  pick $s$ uniformly at random from $A$

  compute $X = \{a \in A \mid a < s\}$ and $Y = \{a \in A \mid a > s\}$

  **if** $|X| \geq m$ **then** call FIND$(X, m)$

  **if** $n - |Y| \leq m$ **then** call FIND$(Y, m - (n - |Y|))$

  **if** $|X| < m < n - |Y|$ **then** return $s$

---

- *expected running time*: The randomly chosen key $s$ splits the database into pieces $X$ and $Y$ which are, on average, of size $\frac{n}{2}$, and in most cases substantially smaller than $n$. Thus, the problem of looking for a key in a set of size $n$ reduces to a problem of looking for a key in a set of size "approximately" $\frac{n}{2}$ and a running time of the type $T(n) \approx T(\frac{n}{2}) + O(n) = O(n)$ can intuitively be expected. More precisely, let $T(n, m)$ denote the running time to find the $m$th key. Since any of the keys could equally likely be picked as the splitter $s$, one can formulate a recurrence for the expectation $E[T(n, m)]$, which solves to $E[T(n, m)] = O(n)$, for all $m$.

- *variance*: Markov's inequality bounds the variance of the running time by $Pr[T(n, m) > kE[T(n, m)]] < \frac{1}{k}$.

**3.** *Primality testing*: Algorithm 2 produces correct output with probability at least $1 - (\frac{1}{2})^k$. Thus, after $\log n$ trials of selecting a random integer less than $n$ and testing, the likelihood is very high for reasonably large $n$, that a prime number will be obtained. This follows from the prime number theorem and Markov's inequality.

---

**Algorithm 2:**  Test primality of $n$ (with $k$ witnesses).

input: positive integers $n$ and $k$ with $n$ odd

pick $a_1, \ldots, a_k$, each independently and uniformly at random from $\{2, 3, \ldots, n-1\}$

compute $\gcd(n, a_i)$ for all $1 \leq i \leq k$

  $\{\gcd(n, a_i)$ can be computed efficiently using the Euclidean algorithm$\}$

**if** there exists an $a_i$ with $\gcd(n, a_i) \neq 1$ **then** output "*composite*" and halt

**else:** compute $z_i = a_i^{(n-1)/2} \pmod{n}$ for all $a_i$ with $1 \leq i \leq k$

  $\{z_i$ can be computed efficiently by repeated squaring$\}$

**if** for some $i$, $z_i \not\equiv \pm 1 \pmod{n}$ **then** output "*composite*"

**else if** for some $i$, $z_i \equiv -1 \pmod{n}$ **then** output "*probably prime*"

**else** output "*probably composite*"

## REFERENCES

***Printed Resources***:

[AhEtal06] A. V. Aho, M. S. Lam, R. Sethi, and J. D. Ullman, *Compilers: Principles, Techniques, and Tools*, 2nd ed., Addison Wesley, 2006.

[AhMaOr93] R. K. Ahuja, T. L. Magnanti, and J. B. Orlin, *Network Flows: Theory, Algorithms, and Applications*, Prentice Hall, 1993.

[AtBl09a] M. J. Atallah and M. Blanton, *Algorithms and Theory of Computation Handbook, Volume 1: General Concepts and Techniques*, 2nd ed., Chapman & Hall/CRC, 2009.

[AtBl09b] M. J. Atallah and M. Blanton, *Algorithms and Theory of Computation Handbook, Volume 2: Special Topics and Techniques*, 2nd ed., Chapman & Hall/CRC, 2009.

[Co71] S. A. Cook, "The complexity of theorem-proving procedures", *Proceedings of the Third Annual ACM Symposium on the Theory of Computing*, 1971, 151–158.

[CoEtal09] T. H. Cormen, C. E. Leiserson, R. L. Rivest, and C. S. Stein, *Introduction to Algorithms*, 3rd ed., MIT Press, 2009.

[DuKo00] D-Z. Du and K-I Ko, *Theory of Computational Complexity*, Wiley, 2000.

[Ed65] J. Edmonds, "Paths, trees and flowers", *Canadian Journal of Mathematics* 17 (1965), 449–467.

[FlBe94] R. W. Floyd and R. Beigel, *The Language of Machines: An Introduction to Computability and Formal Languages*, Computer Science Press, 1994.

[Fr77] R. Frievalds, "Probabilistic machines can use less running time", B. Gilchrist (ed.), *Information Processing 77, Proceedings of IFIP 77*, North-Holland, 1977, 839–842.

[GaJo79] M. R. Garey and D. S. Johnson, *Computers and Intractability, A Guide to the Theory of NP-Completeness*, W. H. Freeman, 1979.

[GoDíTu14] T. Gonzalez, J. Díaz-Herrera, and A. Tucker, eds., *Computing Handbook: Computer Science and Software Engineering*, 3rd ed., CRC Press, 2014.

[GrHoRu95] R. Greenlaw, H. J. Hoover, and W. L. Ruzzo, *Limits to Parallel Computation: P-Completeness Theory*, Oxford University Press, 1995.

[HeOg02] L. A. Hemaspaandra and M. Ogihara, *The Complexity Theory Companion*, Springer-Verlag, 2002.

[Jo81] D. S. Johnson, "The NP-completeness column: an ongoing guide (1st edition)", *Journal of Algorithms* 2 (1981), 393–405.

[Jo90] D. S. Johnson, "A catalog of complexity classes", *Handbook of Theoretical Computer Science*, J. van Leeuwen (ed.), MIT Press/Elsevier, 1990, 67–161.

[Ka93] D. R. Karger, "Global mincuts in RNC, and other ramifications of a simple mincut algorithm", *Proceedings of the 4th Annual ACM-SIAM Symposium on Discrete Algorithms*, 1993, 21–30.

[Ka72] R. M. Karp, "Reducibility among combinatorial problems", *Complexity of Computer Computations*, R. E. Miller and J. W. Thatcher (eds.), Plenum Press, 1972, 85–103.

[Ka91] R. M. Karp, "An introduction to randomized algorithms", *Discrete Applied Mathematics* 34 (1991), 165–201.

[Kn97] D. E. Knuth, *Seminumerical Algorithms*, Volume 2 of *The Art of Computer Programming*, 3rd ed., Addison-Wesley, 1997.

[LePa97] H. R. Lewis and C. H. Papadimitriou, *Elements of the Theory of Computation*, 2nd ed., Prentice-Hall, 1997.

[Li16] P. Linz, *An Introduction to Formal Languages and Automata*, 6th ed., Jones and Bartlett, 2016.

[Lo83] M. Lothaire, *Combinatorics on Words*, Addison-Wesley, 1983.

[Lo79] L. Lovász, "On determinants, matchings, and random algorithms", L. Budach (ed.), *Fundamentals of Computing Theory*, Akademia-Verlag, 1979.

[Ma93] Y. Matiyasevich, *Hilbert's Tenth Problem*, MIT Press, 1993.

[MoRa95] R. Motwani and P. Raghavan, *Randomized Algorithms*, Cambridge University Press, 1995.

[MuVaVa87] K. Mulmeley, U. V. Vazirani, and V. V. Vazirani, "Matching is as easy as matrix inversion", *Combinatorica* 7 (1987), 105–113.

[Pa94] C. Papadimitriou, *Computational Complexity*, Addison-Wesley, 1994.

[PaYa91] C. H. Papadimitriou and M. Yannakakis, "Optimization, approximation, and complexity classes", *Journal of Computer and Systems Sciences* 43 (1991), 425–440.

[PrSh85] F. P. Preparata and M. I. Shamos, *Computational Geometry: An Introduction*, Springer-Verlag, 1985.

[Ra80] M. O. Rabin, "Probabilistic algorithms for testing primality", *Journal of Number Theory* 12 (1980), 128–138.

[Ré12] G. E. Révész, *Introduction to Formal Languages*, Dover, 2012.

[RiShAd78] R. L. Rivest, A. Shamir, and L. Adleman, "A method for obtaining digital signatures and public key cryptosystems", *Communications of the ACM* 21 (1978), 120–126.

[Sa85] A. Salomaa, *Computation and Automata*, Cambridge University Press, 1985.

[ScUm02] M. Schaefer and C. Umans, "Completeness in the polynomial-time hierarchy: Part I: A compendium", *SIGACT News* 33 (2002), 32–49.

[SeWa11] R. Sedgewick and K. Wayne, *Algorithms*, 4th ed., Addison-Wesley, 2011.

[Si14] M. Sipser, *Introduction to the Theory of Computation*, 3rd ed., Cengage, 2014.

[So87] R. I. Soare, *Recursively Enumerable Sets and Degrees*, Perspectives in Mathematical Logic, Springer-Verlag, 1987.

[SoSt77] R. Solovay and V. Strassen, "A fast Monte Carlo test for primality", *SIAM Journal on Computing* 6 (1977), 84–85.

[Ta83] R. E. Tarjan, *Data Structures and Network Algorithms*, SIAM, 1983.

[vL90] J. van Leeuwen, ed., *Handbook of Theoretical Computer Science, Vol. A: Algorithms and Complexity*, Elsevier, 1990.

[Va82] L. G. Valiant, "A scheme for fast parallel communication", *SIAM Journal on Computing* 11 (1982), 350–361.

[Vo99] H. Vollmer, *Introduction to Circuit Complexity: A Uniform Approach*, Springer-Verlag, 1999.

[We83] D. J. A. Welsh, "Randomized algorithms", *Discrete Applied Mathematics* 5 (1983), 133-145.

**Web Resources**:

`cafaq.com/lifefaq` (Frequently Asked Questions About Cellular Automata.)

`www3.cs.stonybrook.edu/~algorith/` (The Stony Brook Algorithm Repository.)

`www.drb.insel.de/~heiner/BB/` (Busy Beaver Turing Machines.)

`wwwhomes.uni-bielefeld.de/achim/gol.html` (Achim's Game of Life page.)

# 18

# *INFORMATION STRUCTURES*

# INTRODUCTION

Information structures are groupings of related information into records and organization of the records into databases. The mathematical structure of a record is specified as an abstract datatype and represented concretely as a linkage of segments of computer memory. General chapter references are [AhHoUl83], [Kn97], and [Kn98].

# GLOSSARY

**abstract datatype (ADT)**: a mathematically specified datatype equipped with operations that can be performed on its data objects.

**adaptive bubblesort**: a bubblesort that stops the first time a scan produces no transpositions.

**ADT-constructor**: any of the three operations *string of*, *set of*, or *tuple of* used to build more complex ADTs from simpler ADTs.

**alphabetic datatype**: an elementary datatype whose domain is a finite set of symbols, and whose only primary operation is a total ordering query.

**ambivalent data structure**: a structure that keeps track of several alternatives at many of its vertices, even though a global examination of the structure would determine which of these alternatives is optimal.

**array data structure**: an indexed sequence of cells $\langle a_j \mid j = d, \dots, u \rangle$ of fixed size, with consecutive indices.

**AVL tree**: a binary search tree with the property that the two subtrees of each node differ by at most one in height.

**binary search**: a recursive search method that proceeds by comparing the target key to the key in the middle of the list, in order to determine which half of the list could contain the target item, if it is present.

**binary search tree**: a binary tree in which the key at each node is larger than all the keys in its left subtree, but smaller than all the keys in its right subtree.

**binary search tree structure**: a binary-tree structure in which for every cell, all cells accessible through the left child have lower keys, and all cells accessible through the right child have higher keys.

**binary tree structure**: a tree structure such that each cell has two pointers.

**bubblesort**: a sort that repeatedly scans an array from the highest index to the lowest, on each iteration swapping every out-of-order pair of consecutive items that is encountered.

**cell** (in a concrete data structure): a storage unit within the data structure that may contain data and pointers to other cells.

**certificate** (for a property of a graph $G$): another graph that has the specified property if and only if the graph $G$ has the property.

**chaining method** (for hash tables): a hashing method that resolves collisions by placing all the records whose keys map to the same location in the main array into a linked list (chain), which is rooted at that location, but stored in the secondary array.

**circular linked list**: a set of cells, each with two pointers, one designated as its *forward pointer* and the other as its *backward pointer*, plus a header with one or more pointers to *current cells*, such that these conditions hold: (1) the sequence of cells formed by following the forward pointers, starting from any cell, traverses the entire set and returns to the starting cell; and (2) the sequence of cells formed by following the backward pointers, starting from any cell, traverses the entire set and returns to the starting cell.

**closed hash table**: a hash table in which collisions are resolved without the use of secondary storage space, that is, by probing in the main array to find available locations.

**cluster** (in a spanning tree): a set of vertices whose induced subgraph is connected.

**clustering property** (of a probe function): the undesirable possibility that parts of the probe sequences generated for two different keys are identical.

**collision instance** (of a hash function): a pair of different keys for which the value of the hash function is the same.

**collision resolution** (of a hashing process): a procedure within the hashing process used to define a sequence of alternative locations for storage of a record whose key collides with the key of an existing record in the table.

**comparison sort**: a sorting method in which the final sorted order is based solely on comparisons between elements in the input sequence.

**concrete data structure**: a mathematical model for storing the current value of a structured variable in computer memory.

**database**: a set of *records*, stored in a computer.

**datatype**: a set of objects, called the domain, and a set of mappings, called primary operations, from the domain to itself or to the domain of some other datatype.

**deheaping**: removing the highest priority entry from a heap and patching the result so that the heap properties are restored.

**dictionary**: an abstract datatype whose domain is a set of keyed pairs, in which arbitrary pairs may be accessed directly.

**domain** (of a datatype): the set of objects within that datatype.

**dyadic graph property**: a property defined with respect to pairs of vertices.

**dynamic structure** (for a database): an information structure for the database whose configuration may be changed, for instance, by the insertion or deletion of elements.

**dynamic update operation**: (on a graph) an operation that changes the graph and keeps track of whether the graph has some designated property.

**edge-incidence table** (for a graph): a dictionary whose keys are the vertices of a graph or digraph. The data component for each key vertex is a list of all the edges that are incident on that vertex. Each self-loop occurs twice in the list.

**elementary datatype**: an alphabetic datatype or a numeric datatype, usually intended for direct representation in the hardware of a computer.

**endpoint table** (for a graph): a dictionary whose keys are the edges. The data component for each key edge is the set of endpoints for that edge. If an edge is directed, then its endpoints are marked as head and tail.

**enheaping**: placing a new entry into its correctly prioritized position in a heap.

**entry** (in a database): a 2-tuple, whose first component is a key, and whose second component is some data; also called a *record*.

**external sorting method**: a method that uses external storage, such as hard disk or tape, outside the main memory during the sorting process.

**far end** (of a one-way linked list): the cell that contains a null pointer.

**Fibonacci heap**: a modification of a heap, using the Fibonacci sequence, that permits more efficient implementation of a priority queue than a heap based on a left-complete binary tree.

**FIFO property** (of a database): the property that the item retrieved is always the item inserted the longest ago. FIFO means "first-in-first-out".

**flat notation** (in a postcondition of a primary operation specification): the value $X^\flat$ of the variable $X$ before the specified operation is executed.

**fullness** (of a closed hash table): the ratio of the number of records presently in the table to the size of the table.

**generic datatype**: a specification in an ADT-template that means that there are no restrictions whatsoever on that datatype.

**hash function** (for storing records in a table): a function that maps each key to a location in the table.

**hash table**: an array of locations for records (entries) in which each record is identified by a unique key, and in which a hash function is used to perform the table-access operations (of insertion, deletion, and search), possibly involving the use of a secondary array to resolve competition for locations.

**hashing**: storage-retrieval in a large table in which the table location is computed from the key of each data entry.

**header** (of a concrete data structure): a special memory unit (but not a cell) that contains current information about the entire configuration and pointers to some critical cells in the structure.

**heap**: a concrete data structure that represents a priority tree as an array.

**heapsort**: sorting a set of entries by first enheaping all items and then successively deheaping them.

**incidence matrix** (for a graph): a 0-1 matrix that specifies the incidence relation, where rows are indexed by the vertices and columns by the edges. The entry in the row corresponding to vertex $v$ and edge $e$ is 1, if $v$ is an endpoint of $e$, and 0 otherwise.

**in-place realization** (of a sorting method): a method that uses, beyond the space needed for one copy of each data entry, only a constant amount of additional space, regardless of the size of the list to be sorted.

**insertion sort**: a sort that transforms an unsorted list into a sorted list by iteratively transferring the next item from the remaining items in the unsorted input list and inserting it into correct position in the sorted output list.

**internal sorting method**: any method that keeps all the entries in the primary memory of the computer during the process of rearrangement.

**key** (in a database entry): a value from an ordered set, used to store and retrieve data.

**key domain**: the ordered set from which values of keys are drawn.

**key entry** (of a record in a table): a value from an ordered set (e.g., integer identification codes or alphabetic strings) used to store records in the table.

**key randomization**: a "preliminary" procedure within the hashing process for mapping non-numeric keys (or keys with poor distribution) into (more uniformly) random distributed integers in the domain of the hash function.

**keyed pair**: a 2-tuple whose first component, called a *key*, is used to locate the data in the second component.

**left child** (of a cell in a binary tree structure): the cell to which the first pointer points.

**left-complete binary tree**: either a binary tree that is complete, or a balanced binary tree (§9.1.2) such that at depth one less than the maximum, the following hold: (1) all nodes with two children are to the left of all nodes with one or no children; (2) all nodes with no children are to the right of all nodes with one or two children; and (3) there is at most one node with only one child, which must be a left child.

**LIFO property** (of a database): the property that the item retrieved is always the item most recently inserted. LIFO means "last-in-first-out".

**linear search**: the technique of scanning the entries of a list in sequence, until either some stopping condition occurs or the entire list has been scanned.

**mergesort**: a sort that partitions an unsorted list into lists of length one and then iteratively merges them until a single sorted list is obtained.

**near end** (of a one-way linked list): the cell that is pointed to by the header and by no other cell.

**nearly complete** (property of a binary tree): the possible property that the binary tree is complete at every level except possibly at the bottom level. At the bottom level, all the missing leaves are to the right of all the present leaves.

**null pointer**: a pointer that points to an artificial location, which serves as a signal to an algorithm to react somewhat differently than to a pointer to an actual location.

**numeric datatype**: an elementary datatype whose domain is a set of *numbers* and whose primary operations are a total ordering query and the arithmetic operators $+$ (addition), $\times$ (multiplication), and $-$ (change of sign).

**one-way linked list**: a set of cells, each with one pointer, such that: (1) exactly one of these cells is pointed to by the header but by no cell; (2) exactly one cell contains a null pointer; and (3) the sequence of cells formed by following the pointers, starting from the header, traverses the entire set, ending with the cell containing the null pointer.

**open hash table**: a hash table that uses a secondary array to resolve collisions.

**ordered datatype**: a datatype with an order relation such that any two elements can be compared.

**pivot** (in a quicksort): an entry at which the sequence is split.

**plane graph**: a planar graph, together with a particular imbedding in the plane.

**pointer** (to a cell): a representation of that cell's location in computer memory.

**postcondition** (of a primary operation): a list of conditions that must hold after the operation is executed, if the precondition is satisfied when the operation commences.

**precondition** (of a primary operation): a list of conditions that must hold immediately before the operation is executed, for the operation to execute as specified.

***primary key***: the key component of highest precedence, when the key has more than one component.

***primary operation*** (for a datatype): a basic operation that retrieves information from an object in the domain or modifies the object.

***priority queue***: an abstract datatype whose domain is a set of records, in which only the entry with the largest key is immediately accessible.

***priority tree***: a nearly complete binary tree whose nodes are assigned data entries from an ordered set of "priorities", such that there is no node whose priority supersedes the priority of its parent node.

***probe function***: a function used iteratively to calculate an alternative location in a closed hash table when the initial location calculated from the key or the previous probe location is already occupied.

***probe sequence*** (for a hash table location): the sequence of locations calculated by the probe function in its effort to find an unoccupied place in the table.

***query*** (to a datatype): a primary or secondary operation that changes nothing and returns a logical value, i.e., *true* or *false*.

***queue***: an abstract datatype that organizes the records into a sequence, such that records are inserted at one end (called the *back* of the queue) and extractions are made from the other end (called the *front*).

***quicksort***: sorting by recursively partitioning a list around an entry (the pivot) so that all smaller items precede the pivot and all larger items follow it.

***radix sort***: a sort using iterative partitioning into queues and recombining by concatenation, in which the partitioning is based on a digit in a numeral.

***random access list***: an abstract datatype whose domain is a set of records such that the values of the key field range within an interval of integers $a \le k \le b$; this permits implementations that execute primary operations faster than a general table.

***rank*** (of an element of a finite ordered set): the number of elements that it exceeds or equals.

***rank-counting sort***: sorting by calculating the rank for each element, and then assigning each element to its correct position according to its rank.

***record***: a 2-tuple, whose first component is a key, and whose second component is some data; also called an *entry*.

***record in a table***: a table entry containing a key and some data.

***right child*** (of a cell in a binary-tree structure): the cell to which the second pointer points.

***root cell*** (of a tree structure): the cell to which the header points.

***scanning a database*** (or a portion of a database): examining every record in that database (or portion).

***searching*** (a database): seeking either a target entry with a specific key or a target entry whose key has some specified property.

***secondary key***: the key component of next highest precedence, when the key has more than one component.

***secondary operation*** (for a datatype): an operation constructed from primary operations and from previously defined secondary operations.

**selection sort**: a sort that transforms an unsorted list into a sorted list by iteratively finding the item with smallest key from the remaining items in the unsorted input list and appending it to the end of the sorted output list.

**sequence of**: an ADT-constructor that converts a datatype with domain $D$ into a new datatype, whose domain is the set of all finite sequences of objects from domain $D$, and whose primary operations are some sequence operations.

**set of**: an ADT-constructor that converts a datatype with domain $D$ into a new datatype, whose domain is the set of all subsets of objects from domain $D$, and whose primary operations are set operations.

**shakersort**: a bubblesort variation that alternates between bubbling upward and sinking downward on alternate scans.

**Shellsort**: a sorting method that involves partitioning a list into sublists and insertion sorting each of the sublists.

**sinking sort**: a "reverse bubblesort" that scans an array repeatedly from the lowest index to the highest, each time swapping every out-of-order pair of consecutive items that is encountered.

**size** (of a cell in a data structure): the number of bytes of computer memory that the cell occupies.

**size** (of a hash table): the number of locations in the main array in which the records are stored. (If chaining is used to resolve collision, the total number of records stored may exceed the size of the main array.)

**sorting algorithm**: a method for arranging the entries of a database into a sequence that conforms to the order of their keys.

**sparse certificate**: a strong certificate (for a property of a graph $G$) in which the number of edges is $O(|V_G|)$.

**sparse sequence**: a sequence in which nearly all the entries are zeros.

**stable certificate**: a certificate produced by a stable function.

**stable (certificate) function**: a function $A$ that maps graphs to strong certificates such that: (1) $A(G \cup H) = A(A(G) \cup H)$; and (2) $A(G - e)$ differs from $A(G)$ by $O(1)$ edges, where $e$ is an edge in $G$.

**stack**: an abstract datatype that organizes the records into a sequence, in which insertion and extraction are made at the same end (called the *top* of the stack).

**static structure** (for a database): an information structure for the database whose configuration does not change during an algorithmic process.

**strong certificate** (for a property of a graph $G$): a certificate graph $G'$ for $G$ with the same vertex set as $G$ such that, for every graph $H$, the graph $G \cup H$ has property $\mathcal{P}$ if and only if $G' \cup H$ has property $\mathcal{P}$.

**table**: a set of keyed pairs, in which arbitrary pairs may be accessed directly; used as the domain of a dictionary.

**target** (of a database search): an entry whose key has been designated as the objective of the search.

**2-3 tree**: a tree in which each non-leaf node has 2 or 3 children, and in which every path from the root to a leaf is of the same length.

***tree structure***: a concrete data structure such that the header points to a single cell, and such that from that cell to each other cell, there is a single chain of pointers.

***k-tuple of***: an ADT-constructor that converts a list of $k$ datatypes into a new datatype, whose domain is the Cartesian product of the domains of the datatypes in that list, and whose primary operations are the projection functions from a $k$-tuple to each of its coordinates.

***two-way incidence structure*** (for a graph): a pair consisting of an edge-incidence table and an endpoint table.

***two-way linked list***: a set of cells, each with two pointers, one designated as its *forward pointer* and the other as its *backward pointer*, plus a header with a forward pointer and a backward pointer, such that these conditions hold: (1) considering only the forward pointers, it is a one-way linked list; and (2) following the sequence of backward pointers yields the reverse of the sequence obtained by following the forward pointers.

***two-way sequential list***: an ADT-template whose domain is strings, in which an entry is reached by applying the access operations *forward* and *backward*. Insertions are made before or after the *current* location.

***union-find datatype***: an abstract datatype whose records are mutually disjoint sets, in which there is a primary operation to locate the set containing a specified target element and a primary operation of merging two sets.

---

# 18.1   ABSTRACT DATATYPES

Organizing numbers and symbols into various kinds of records is a principal activity of information engineering. The organizational structure of a record is called a datatype. Abstractly, a datatype is characterized by a formal description of its domain and of the intrinsic operations by which information is entered, modified, and retrieved. Providing the specification at this abstract level ensures that the datatype is independent of the underlying types of information elements stored within the structure, and independent also of the hardware and software used to implement this organization. See [AhHoUl83] and [Kn97].

---

## 18.1.1   ABSTRACT SPECIFICATION OF RECORDS AND DATABASES

Information engineering uses discrete mathematics as a source of models for various kinds of records and databases. The language of abstract mathematics is used to specify a complex structure in terms of its elements. Constructors and templates are used to create new kinds of data from old kinds.

### Definitions:

A ***datatype*** consists of a set of objects, called the *domain*, and a set of mappings, called *primary operations*, from the domain to itself or to the domain of some other datatype.

The ***domain*** of a datatype is its set of objects.

A **primary operation** for a datatype is a basic operation that retrieves information from an object in the domain or modifies the object.

A **secondary operation** on the domain of a datatype is an operation constructed from primary operations and previously defined secondary operations.

A **query** is a primary or secondary operation on a datatype domain that preserves the values of all its arguments and returns a logical value, i.e., *true* or *false*.

An **alphabetic datatype** is a datatype whose domain is a finite set of *symbols*. Its only primary operation is a total ordering query.

A **numeric datatype** is a datatype whose domain is a set of *numbers* and whose primary operations are a total ordering query and the arithmetic operators $+$ (addition), $\times$ (multiplication), and $-$ (change of sign).

An **elementary datatype** is an alphabetic datatype or a numeric datatype, usually intended for direct representation in the hardware of a computer.

An **abstract datatype** (**ADT**) is a mathematically specified datatype equipped with operations that can be performed on its data objects.

An **ADT-constructor** is a template that converts a datatype into a new ADT.

The constructor **sequence of** transforms a datatype $X$-type with domain $D$ into a new datatype "sequence of $X$-type" whose domain is the set $Seq_D$ of all finite sequences of elements of $D$. The primary operations of the resulting datatype are

- **header**($s$), which yields a singleton sequence whose only element is the first object in the sequence $s$ (or the empty sequence, if the sequence $s$ is empty);
- **trailer**($s$), which deletes the first entry of sequence $s$ (or yields the empty sequence, if the sequence $s$ is empty);
- **concat**($s, t$), which concatenates the two sequences;
- **first**($s$), which gives the value of the first entry of a nonempty sequence $s$;
- **append**($s, d$), which appends to sequence $s \in Seq_D$ an entry $d \in D$;
- **nullseq**( ), whose value is the null sequence $\lambda$.

The constructor **set of** converts a datatype $X$-type with domain $D$ into a new datatype "set of $X$-type" whose domain is the set of all subsets of $D$. The primary operations are

- **inclusion**($S, T$), a query whose value is true if $S \subseteq T$;
- **union**($S, T$), whose value is $S \cup T$;
- **intersection**($S, T$), whose value is $S \cap T$;
- **difference**($S, T$), whose value is $S - T$;
- **choose**($S$), whose value is an arbitrary element of a nonempty subset $S$;
- **singleton**($d$), which transforms an element of $D$ into the singleton set whose only entry is $d$;
- **emptyset**( ), whose value is the empty set $\emptyset$;
- **universe**( ), whose value is the underlying domain $D$.

The constructor **k-tuple of** converts a list of $k$ datatypes

$$X_1\text{-type}, X_2\text{-type}, \ldots, X_k\text{-type}$$

into a new datatype "$k$-tuple $(X_1, \ldots, X_k)$" whose domain is the Cartesian product $D_1 \times D_2 \times \cdots \times D_k$ of the domains of the respective datatypes in that list. The primary operations are the projection functions

- **coord**$_j(s)$, which gives the value of the $j$th coordinate of the $k$-tuple $s$;
- **entuple** $(d_1, \ldots, d_k)$, whose value is the $k$-tuple whose $j$th coordinate is the element $d_j$ of domain $D_j$.

An **elementary ADT-constructor** is any of the three operations *sequence of*, *set of*, or *tuple of* used to build more complex ADTs from simpler ADTs.

The **Iverson truth function** assigns to a proposition $p$ the integer value $(p)$ such that

$$(p) = \begin{cases} 1 & \text{if } p \text{ is true;} \\ 0 & \text{otherwise.} \end{cases}$$

A **datatype specification** uses a combination of elementary datatypes and ADT-constructors to specify the domain and the primary operations. It may also use the following mathematical notation:

- $\emptyset$ denotes the empty set;
- $\lambda$ denotes the empty sequence;
- $\cdot$ denotes the operation of appending one element to a sequence;
- $\circ$ denotes the sequence concatenation operation.

Moreover, every primary operation is either a query or a procedure.

Specifying a datatype as **generic** in an ADT-template means that any datatype can be used in that part of the template as a block in the construction of the new datatype.

Specifying a datatype as **ordered** in an ADT-template means that any ordered datatype can be used in that part of the template as a block in the construction of the new datatype.

The **precondition** of a primary operation is a list of conditions that must hold immediately before the operation is executed, for it to execute as described.

The **postcondition** of a primary operation is a specification of conditions that must hold after the operation is executed, if the precondition is satisfied when the operation commences.

The **flat notation** $X^\flat$ in a postcondition of a primary operation specification means the value of the variable $X$ before that operation is executed. Unadorned $X$ (without the $\flat$) means the value of $X$ after the operation.

## Facts:

**1.** The domain of an ADT is specified as a mathematical model, without saying how its elements are to be represented.

**2.** Sometimes the domain of a datatype is specified by roster. Other times it is specified with the use of set-theoretic operations.

**3.** A primary operation of an ADT is specified functionally. That is, its value on every element of the domain is declared, but the choice of an algorithm to be used in its implementation is omitted.

**4.** A primary operation may be implemented so that it has direct access to the data representing the value of the computational variable to which it is applied.

**5.** A primary operation can modify the information within a variable in its datatype or retrieve information from a variable.

**6.** A secondary function is implemented through calls to the primary operations from which it is ultimately composed.

**7.** There is no set of standard conventions for writing ADTs.

**8.** Software designers frequently specify a particular concrete information structure (see §18.2), instead of writing an ADT.

**9.** The advantage of writing an ADT, rather than a concrete datatype, is that it leaves the implementer room to find a new (and possibly improved) way to meet the requirements of the task.

**10.** In a datatype specification, a functional subprogram is represented by a procedure that produces a non-boolean value, and a variable to receive that value is specified as its last parameter.

**Examples:**

**1.** Elementary numeric datatypes include the integers and the reals.

**2.** Elementary alphabetic datatypes include the ASCII set and the decimal digits.

**3.** *Complex_number* is a datatype that represents complex numbers and their addition and multiplication.

> ADT **complex_number**:
>
> Domain
>
>> 2-tuple ($re$: real, $im$: real)
>
> Primary Operations
>
>> sum ($w$: complex_number, $z$: complex_number)
>>
>> Comment: add two complex numbers.
>>
>>> {pre: *none*}
>>>
>>> {post: $sum(w, z) = entuple(re(w) + re(z),\ im(w) + im(z))$}
>>
>> prod ($w$: complex_number, $z$: complex_number)
>>
>> Comment: multiply two complex numbers.
>>
>>> {pre: *none*}
>>>
>>> {post: $prod(w, z) = entuple$
>>>
>>>> $(re(w) \cdot re(z) - im(w) \cdot im(z),\ re(w) \cdot im(z) + im(w) \cdot re(z))$}

**4.** *Baseten_digit* is a datatype that might be used in the construction of base-ten numerals representing arbitrarily large integers and their addition.

> ADT **baseten_digit**:
>
> Domain
>
>> $\{0, 1, 2, 3, 4, 5, 6, 7, 8, 9\}$: integers
>
> Primary Operations
>
>> add_digits ($x$: baseten_digit, $y$: baseten_digit)
>>
>>> {pre: *none*}
>>>
>>> {post: $add\_digits(x, y) = x + y \textbf{ mod } 10$}
>>
>> addcarry: ($x$: baseten_digit, $y$: baseten_digit)
>>
>>> {pre: *none*}
>>>
>>> {post: $addcarry(x, y) = (x + y \geq 10)$}

**5.** The datatype *alphastring* represents sequences of lowercase English letters.

> ADT **alphastring**:
>
> Domain
>
>   sequence of $\{a, b, c, d, e, f, g, h, i, j, k, l, m, n, o, p, q, r, s, t, u, v, w, x, y, z\}$
>
> Primary Operations
>
>   none except from the *constructor* sequence of

**6.** The **union-find** constructor transforms a datatype on domain set $S$ into a datatype whose objects are of three kinds: elements of $S$, subsets of $S$, and partitions of $S$. There is a primary operation to merge two cells of a partition, and a primary operation to locate the cell of a partition that contains a specified target element.

---

## 18.1.2   STACKS AND QUEUES

Access to entries in the interior of a list is unnecessary much of the time. Restricting access to the first and last entries is a precaution to prevent mistakes.

### Definitions:

A **stack** is an ADT whose domain is a sequence, one end of which is called the *top*, and the other the *bottom*. One primary operation, called *pushing*, appends a new entry to the top; the other, called *popping*, removes the entry at the top and returns its value. No entry may be examined, added to the stack, or deleted from the stack except by an iterated composition of these operations.

The **top of a stack** is the end of that *stack* that can be accessed directly.

**Pushing an entry onto a stack** means appending it to the top of the stack.

**Popping an entry from a stack** means deleting it from the top of the stack and possibly examining the data it contains.

The **LIFO property** of a database is that the item retrieved is always the item most recently inserted. LIFO means "last-in-first-out".

A **queue** is an ADT whose domain is a sequence, one end of which is called the *front*, and the other the *back*. One primary operation, called *enqueueing*, appends a new entry to the back; the other, called *dequeueing*, removes the entry at the front and returns its value. No entry may be examined, added to the queue, or deleted from the queue except by an iterated composition of these operations.

The **back of a queue** is the end to which entries may be appended.

The **front of a queue** is the end from which entries may be deleted and possibly examined.

**Enqueueing an entry into a queue** means appending it to the back of the queue.

**Dequeueing an entry from a queue** means deleting it from the front of the queue, and possibly examining the data it contains.

The **FIFO property** of a database is that the item retrieved is always the item inserted the longest ago. FIFO means "first-in-first-out".

**Facts:**

**1.** Abstract specification of stacks and queues mentions only the behavior of those datatypes and totally avoids all details of implementation. This permits a skillful implementer to innovate with efficient concrete structures (see §18.2) that meet the behavioral specification.

**2.** Abstract specification of stacks and queues is consistent with the principles of object-oriented programming, in which details of implementation are hidden inside the data objects, so that the rest of the program perceives only the specified functional behavior.

**3.** All stacks have the LIFO property. For any stack $S$ and for any element $b$, after executing the sequence of instructions

$$\mathrm{push}(S, b), \ \mathrm{pop}(S, x)$$

the resulting value of the stack $S$ is whatever it was before the operations.

**4.** A stack is most commonly implemented as a linked list (see §18.2.2).

**5.** After changing the value of the variable $y$ to that of the top entry on the stack $S$, the sequence of instructions

$$\mathrm{pop}(S, x), \ y := x, \ \mathrm{push}(x)$$

restores $S$ to its previous state.

**6.** All queues have the FIFO property. Given an empty queue $Q$ and two elements $b_1$ and $b_2$, the sequence of operations

$$\mathrm{enqueue}(Q, b_1), \ \mathrm{enqueue}(Q, b_2), \ \mathrm{dequeue}(Q, x_1), \ \mathrm{dequeue}(Q, x_2),$$

yields $x_1 = b_1$, $x_2 = b_2$, and $Q = \lambda$.

**7.** A queue is most commonly implemented as a linked list (see §18.2.2).

**Examples:**

**1.** The following pseudocode specifies the ADT *stack of D*, where $D$ is an arbitrary datatype.

> Domain
> > sequence of $D$: generic
> Primary Operations
> > create_stack ($S$: stack)
> > Comment: Initialize variable $S$ as an empty stack.
> > > {pre: *none*}
> > > {post: $S = \lambda$}
> > push ($S$: stack, $x$: element of $D$)
> > Comment: Put value of $x$ at *top* of stack $S$
> > > {pre: *none*}
> > > {post: $S = x \cdot S^b$}
> > pop ($S$: stack, $x$: element of $D$)
> > Comment: Remove top item of stack $S$; return it as value of variable $x$.
> > > {pre: $S \neq \lambda$}
> > > {post: $x \cdot S = S^b$}
> > query_empty_stack ($S$: stack)
> > Comment: Decide whether stack $S$ is empty.
> > > {pre: *none*}
> > > {post: query_empty_stack $= (S = \lambda)$}

**2.** The following pseudocode specifies the ADT **queue of D**, where $D$ is an arbitrary datatype.

> Domain
>> sequence of $D$: generic
>
> Primary Operations
>> create_queue ($Q$: queue)
>> Comment: Initialize $Q$ as an empty queue.
>>> {pre: *none*}
>>> {post: $Q = \lambda$ }
>>
>> enqueue ($Q$: queue, $x$: element of $D$)
>> Comment: Put $x$ at the **back** of queue $Q$.
>>> {pre: *none*}
>>> {post: $Q = Q^b \cdot x$}
>>
>> dequeue ($Q$: queue, $x$: element of $D$)
>> Comment: Delete **front** of $Q$; return as $x$.
>>> {pre: $Q \neq \lambda$}
>>> {post: $Q^b = x \cdot Q$}
>>
>> query_empty_queue ( $Q$: queue)
>> Comment: Decide whether queue $Q$ is empty.
>>> {pre: *none*}
>>> {post: query_empty_queue $= (Q = \lambda)$}

**3.** The following figure illustrates the difference between stacking (last-in-first-out) and queueing (first-in-first out).

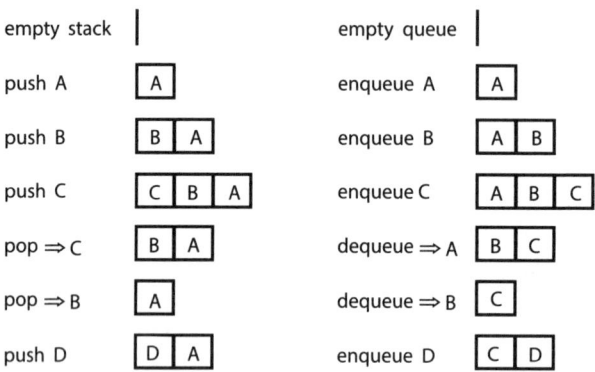

---

## 18.1.3   TWO-WAY SEQUENTIAL LISTS

A two-way sequential list conceptualizes a linear list as having a current location, so that entries may be inserted or deleted only at the current location.

### Definitions:

A **two-way sequential list** is a list with a designated location at which access is permitted.

The **current location** of a two-way sequential list is the location at which access is permitted.

The **forepart** of a two-way sequential list is the part preceding the current location, which is empty when the current location is at the start of the list.

The **aftpart** of a two-way sequential list is the part following the current location, which is empty when the current location is at the finish of the list.

**Facts:**

**1.** A two-way sequential list does not maintain place-in-list numbers for the entries. The result of such an additional requirement would force the insert operation to renumber the part of the list following a newly inserted entry. This would slow the performance.

**2.** A two-way sequential list is easily implemented as a pair of stacks.

**Example:**

**1.** The following pseudocode specifies the ADT **seq_list of D**.

Domain
  2-tuple (*fore*: sequence of $D$, *aft*: sequence of $D$)
  type $D$: generic
Primary Operations
  create_list ($L$: seq_list)
  Comment: Initialize an empty list $L$.
    {pre: *none*}
    {post: $fore(L) = \lambda \ \wedge \ aft(L) = \lambda$}
  reset_to_start ($L$: seq_list)
  Comment: Reset to start of list.
    {pre: *none*}
    {post: $fore(L) = \lambda \ \wedge \ aft(L) \ = \ fore(L^\flat) \circ \ aft(L^\flat)$}
  advance ($L$: seq_list)
  Comment: Advance current position by one element.
    {pre: $aft(L) \neq \lambda$}
    {post: $(\exists x : D)[fore(L) = fore(L^\flat) \cdot x \wedge \ aft(L^\flat) = x \cdot aft(L)]$}
  query_atstart ($L$: seq_list)
    {pre: *none*}
    {post: $query\_atstart = (fore(L) = \lambda)$}
  query_atfinish ($L$: seq_list)
    {pre: *none*}
    {post: $query\_atfinish = (aft(L) = \lambda)$}
  insert ($L$: seq_list, $x$: element of $D$)
    {pre: *none*}
    {post: $aft(L) = x \cdot aft(L^\flat) \wedge fore(L) = fore(L^\flat)$}
  remove ($L$: seq_list, $x$: element of $D$)
    {pre: $aft(L) \neq \lambda$}
    {post: $aft(L^\flat) = x \cdot aft(L) \wedge fore(L) = fore(L^\flat)$}
  swap_right ($L$: seq_list, $M$: seq_list)
    {pre: *none*}
    {post: $fore(L) = fore(L^\flat) \wedge fore(M) = fore(M^\flat)$
      $\wedge aft(L) = aft(M^\flat) \wedge aft(M) = aft(L^\flat)$}

## 18.1.4  DICTIONARIES AND RANDOM ACCESS LISTS

**Definitions:**

A **keyed pair** is a 2-tuple whose first entry, which is called a *key*, is from an ordered datatype and is used to access data in the second entry.

A **table** is a set of keyed pairs such that no two keys are identical.

A **random access list** is a table whose keys are consecutive integers.

A **dictionary** is another name for a table.

**Facts:**

**1.** A static table (whose size $n$ does not change) can be implemented as an array.

**2.** A dynamic table (which permits inserts and deletes) is often implemented as a binary search tree (see §18.2.3).

**3.** Specifying a datatype as a dictionary means that its primary retrieval operation can execute in $\Theta(n)$ time.

**4.** Specifying a datatype as a random access list means that its primary retrieval operation can execute in $\Theta(1)$ time.

**Examples:**

**1.** The following pseudocode specifies the ADT **table**.

>  Domain
>>  set of table_entry
>>  type table_entry: 2-tuple (*key*: ordered, *data*: generic)
>
>  Primary Operations
>>  create_table ($T$: table)
>>>  {pre: *none*}
>>>  {post: $T = \lambda$}
>>
>>  insert_entry ($T$: table, $e$: table_entry)
>>>  {pre: $(\forall e' \in T)[\ \text{key}(e') \neq \text{key}(e)]$}
>>>  {post: $T = T^{\flat} \cup \{e\}$}
>>
>>  remove_entry ($T$: table, $e$: table_entry)
>>>  {pre: $e \in T$}
>>>  {post: $T = T^{\flat} - \{e\}$}
>>
>>  find_entry ($T$: table, $k$: key, found: boolean, $e$: table_entry)
>>>  {pre: *none*}
>>>  {post: $\big((\exists e' \in T)[e'.key = k]\big) \wedge (found = true) \wedge (e = e'))$
>>>  $\vee\big(\neg(\exists e' \in T)[e'.key = k] \wedge found = false\big)$}

**2.** The following pseudocode specifies the ADT **Random access list**.

>  Domain
>>  set of table_entry
>>  type table_entry: 2-tuple (*key*: subrange of integers, *data*: generic)
>
>  Primary Operations
>>  Exactly the same as for the ADT table.

### 18.1.5  PRIORITY QUEUES

A priority queue is an "unfair queue", in which entries are not dequeued on a first-enqueued basis. Instead, each entry has a priority, and is dequeued on a highest priority basis.

**Definition:**
A *priority queue* is a set of keyed pairs, such that the key of the entry returned by a dequeue operation is not exceeded by the key of any other entry currently in the queue.

**Facts:**
**1.** A priority queue is usually implemented as a heap (see §18.2.4).

**2.** Two different entries in a priority queue may have the same key.

**3.** The operating system for a multi-user programming environment places computational tasks into a priority queue.

**Example:**
**1.** The following pseudocode specifies the ADT *P_queue*.

> Domain
> > set of Pq_entry
> > type Pq_entry: 2-tuple (*key*: ordered, *data*: generic)
>
> Primary Operations
> > create_Pq (*PQ*: P_queue)
> > > { pre: *none* }
> > > { post: $PQ = \lambda$ }
> >
> > enPqueue(*PQ*: P_queue, *e*: Pq_entry)
> > > { pre: *none* }
> > > { post: $PQ = PQ^\flat \cup \{e\}$ }
> >
> > dePqueue (*PQ*: P_queue, *e*: Pq_entry)
> > > { pre: $PQ \neq \emptyset$ }
> > > { post: $(e \in PQ^\flat) \wedge (\forall e' \in PQ)[key(e) \leq key(e')] \wedge PQ = PQ^\flat - \{e\}$ }
> >
> > query_empty_Pqueue(*PQ*: P_queue)
> > > { pre: *none* }
> > > { post: query_empty_Pqueue = (PQ = $\lambda$) }

---

## 18.2  CONCRETE DATA STRUCTURES

Concrete data structures configure computer memory into containers of related information. They are used to implement abstract datatypes. Contiguous stretches of memory are regarded as arrays, and noncontiguous portions are linked with pointers.

## 18.2.1   MODELING COMPUTER STORAGE AND RETRIEVAL

There are a few generic concepts common to nearly all concrete data structures.

**Definitions:**

A *concrete data structure* is a mathematical model for storing the current value of a structured variable in computer memory.

A *cell* in a concrete data structure $S$ is a unit within the data structure that may contain data and pointers to other cells.

The *header* of a concrete data structure is a special unit that contains current information about the entire configuration and pointers to some critical cells in the structure. It is *not* a cell.

An *insert operation* insert($S$: structure, $c$: cell, *loc*: location) inserts a new cell $c$ into structure $S$ at location *loc*.

A *delete operation* delete($S$: structure, *loc*: location) deletes from a structure $S$ the cell at location *loc*.

A *target predicate* for a concrete data structure is a predicate that applies to the cells.

A *find operation* find($S$: structure, $t$: target, *loc*: location) searches a structure $S$ for a cell that satisfies target predicate $t$. It returns *false* if there is no such cell. In addition to returning the boolean value *true* if there is such a cell, it also assigns to its location parameter *loc* the location of such a cell.

A *next operation* next($S$: structure, *loc*: location) returns the boolean value *true* if the structure $S$ is nonempty, in which case it also assigns to its location parameter *loc* the location of whatever cell it regards as next; it returns *false* if $S$ is empty.

The *size of a cell* is the number of bytes of computer memory it occupies.

A *pointer to a cell* is a representation of its location in computer memory.

A *null pointer* is a pointer that points to an artificial location. Detecting a null pointer is a signal to an algorithm to react somewhat differently than to a pointer to an actual location.

**Facts:**

**1.** There may be several alternative suitable concrete data structures that can be used to implement a given abstract datatype.

**2.** If the records of a database are all of the same fixed size, then the records themselves may be in the cells of a concrete data structure.

**3.** If the size of records is variable, then the cells of the concrete data structure often contain pointers to the actual data, rather than the data itself. This permits faster execution of operations.

**4.** The most common form of target predicate for a concrete data structure is an assertion that a key component of the cell matches some designated value.

## 18.2.2    ARRAYS AND LINKED LISTS

### Definitions:

An **array** is an indexed sequence of identically structured cells $\langle a_j \mid j = d, \dots, u \rangle$, with consecutive indices.

An array is **zero-based** if its lowest index is zero.

A **one-way linked list** is a set of cells, each with one pointer, such that

- exactly one of these cells is pointed to by the header but by no cell;
- exactly one cell contains a null pointer;
- the sequence of cells formed by following the pointers, starting from the header, traverses the entire set, ending with the cell containing the null pointer.

The **far end of a one-way linked list** is the cell that contains a null pointer.

The **near end of a one-way linked list** is the cell that is pointed to by the header and by no other cell.

A **two-way linked list** is a set of cells, each with two pointers, one designated as its *forward pointer* and the other as its *backward pointer*, plus a header with a forward pointer and a backward pointer, such that

- considering only the forward pointers, it is a one-way linked list;
- following the sequence of backward pointers yields the reverse of the sequence obtained by following the forward pointers.

A **sparse sequence** is a sequence in which nearly all the entries are zeros.

A **circular linked list** is a set of cells, each with two pointers, one designated as its *forward pointer* and the other as its *backward pointer*, plus a header with one or more pointers to *current cells*, such that

- the sequence of cells formed by following the forward pointers, starting from any cell, traverses the entire set and returns to the starting cell;
- the sequence of cells formed by following the backward pointers, starting from any cell, traverses the entire set and returns to the starting cell.

### Facts:

**1.** A random access list (see §18.1.4) can be implemented as an array so that a *find* operation executes in $O(1)$ time.

**2.** A stack (see §18.1.2) can be implemented as a one-way linked list with its top at the near end, so that *push* and *pop* both execute in $O(1)$ time.

**3.** A queue (see §18.1.2) can be implemented as a two-way linked list with its back at the near end of the forward list and its front at the far end, so that *enqueue* and *dequeue* both execute in $O(1)$ time.

**4.** A two-way sequential list (see §18.1.3) can be implemented as a two-way linked list, or as a pair of one-way linked lists.

### Examples:

**1.** The following figure illustrates an array with cells $a_d, \dots, a_u$.

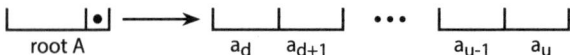

**2.** The following figure illustrates a one-way linked list, with cell $a_d$ at the near end and cell $a_u$ at the far end.

**3.** The following figure illustrates a two-way linked list.

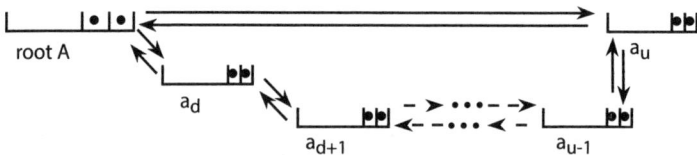

**4.** Representing a sparse finite sequence by a linked list can save space. The cell given to each nonzero entry includes its position in the sequence and points to the cell with the next nonzero entry.

**5.** A queue whose maximum length is bounded can be represented by a circular linked list with two header pointers, one to the back and one to the front of the queue. The number of cells equals the maximum queue length. This eliminates the need for "garbage collection".

---

### 18.2.3   BINARY SEARCH TREES

#### Definitions:

A **tree structure** on $n$ cells is a concrete data structure such that the header points to a single cell, and such that from that cell to each other cell, there is a single chain of pointers.

The **root cell** of a tree structure is the cell to which the header points.

A **binary tree structure** is a tree structure such that each cell has two pointers.

The **left child** of a cell in a binary tree structure is the cell to which the first pointer points.

The **right child** of a cell in a binary tree structure is the cell to which the second pointer points.

A **binary search tree structure** is a binary tree structure in which for every cell, all cells accessible through the left child have lower keys, and all cells accessible through the right child have higher keys.

#### Facts:

**1.** The *ADT table* is commonly implemented as a binary search tree structure.

**2.** The average running time for the *ADT table* operations of *insertion*, *deletion*, and *find* is $O(\log n)$. The time may be worse if relatively few cells have two children.

**3.** Using a *2-3 tree structure* instead of a binary search tree structure for the *ADT table* operations reduces the worst-case running time from $O(n)$ to $O(\log n)$.

**4.** Algorithm 1 can be used in the binary search tree operations of finding, inserting, and deleting.

---

**Algorithm 1: BSTsearch**$(T, t)$**.**

input: a binary-search tree $T$ and a target key $t$

output: if $t \in T$, the address of the node with key $t$, else the address where $t$
       could be inserted

**if** $root(T) = $ NULL **then** return address of *root*

**else if** $t < key(root)$ **then** BSTsearch(*leftsubtree*$(T), t$)

**else if** $t = key(root)$ **then** return address of *root*

**else** BSTsearch(*rightsubtree*$(T), t$)

---

**5.** To find a target key $t$ in the binary search tree $T$, first apply BSTsearch$(T, t)$. If the address of a null pointer is returned, then there is no node with key $t$. Otherwise, the address returned is a node with key $t$.

**6.** To insert a node with target key $t$ into the binary search tree $T$, first apply the algorithm BSTsearch$(T, t)$. Then install the node at the location returned.

**7.** To delete node $t$ from the binary search tree $T$, first apply BSTsearch$(T, t)$. Then replace node $t$ either by the node with the largest key in the left subtree or by the node with the smallest key in the right subtree.

**Examples:**

**1.** The following figure illustrates a binary search tree.

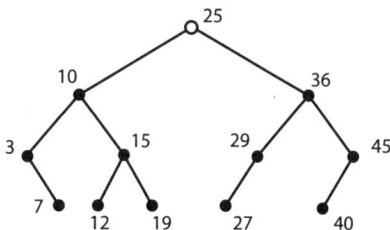

**2.** Inserting 32 into the BST of Example 1 yields the following BST.

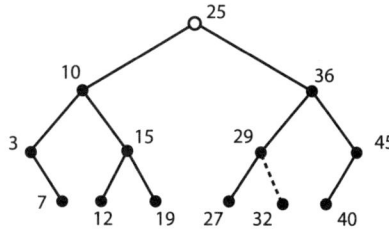

**3.** Deleting node 10 from the BST of Example 1 would yield one of the following two BSTs.

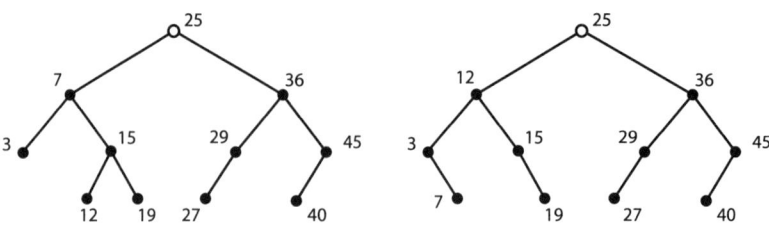

---

## 18.2.4   PRIORITY TREES AND HEAPS

### Definitions:

A binary tree is **left-complete** if it is complete or if it is a balanced binary tree (§9.1.2) such that at depth one less than the maximum, the following conditions hold:

- all nodes with two children are to the left of all nodes with one or no children;
- all nodes with no children are to the right of all nodes with one or two children;
- there is at most one node with only one child, which must be a left-child.

A **priority tree** is a left-complete binary tree, with the following additional structure:

- each node has an attribute called a *key*;
- the values of the keys are drawn from a partially ordered set;
- no node has a higher priority key than its parent.

A **heap** is a representation of a priority tree as a zero-based array, such that each node is represented at the location in the array whose index equals its location in the breadth-first-search order of the tree. Thus

- $index(root) = 0$;
- $index(leftchild(v)) = 2 \times index(v) + 1$;
- $index(rightchild(v)) = 2 \times index(v) + 2$;
- $index(parent(v)) = \left\lfloor \frac{index(v)-1}{2} \right\rfloor$.

**Enheaping an entry** into a heap means placing it into a correctly prioritized position.

**Trickle-up** means enheaping by Algorithm 2.

---

**Algorithm 2**:  **PriorityTreeEnqueue** $(T, x)$.

input: a priority tree $T$ and a new entry $x$

output: tree $T$ with the new node $x$ inserted so that it remains a priority tree

install entry $x$ into the first vacant spot in the left-complete tree $T$

**while** $x \neq root(T)$ **and** $priority(x) > priority(parent(x))$

swap $x$ with $parent(x)$

---

**Deheaping an entry** from a heap means taking the root as the deheaped entry and patching its left subtree and its right subtree back into a single tree.

***Trickle-down*** means deheaping by Algorithm 3.

---

**Algorithm 3:  PriorityTreeDequeue** $(T)$.

input: a priority tree $T$

output: tree $T - root(T)$ with priority-tree shape restored

replace $root(T)$ by rightmost entry $y$ at bottom level of $T$

**while** $y$ is not a leaf **and** $[priority(y) \leq priority(leftchild(y))$ **or**
$\qquad priority(y) \leq priority(rightchild(y))]$

$\quad$ **if** $priority(leftchild(y)) > priority(rightchild(y))$

$\qquad$ **then** swap $y$ with $leftchild(y)$

$\qquad$ **else** swap $y$ with $rightchild(y)$

---

A ***Fibonacci heap*** is a modification of a heap, using the Fibonacci sequence, that permits more efficient implementation of a priority queue than a heap based on a left-complete binary tree.

**Facts:**

**1.** For a tree with $n$ nodes, the worst-case execution time of the priority tree enqueueing algorithm (Algorithm 2) is in the class $\Theta(\log n)$.

**2.** For a tree with $n$ nodes, the worst-case execution time of the priority tree dequeueing algorithm (Algorithm 3) is in the class $\Theta(\log n)$.

**Examples:**

**1.** The following figure shows a left-complete binary tree.

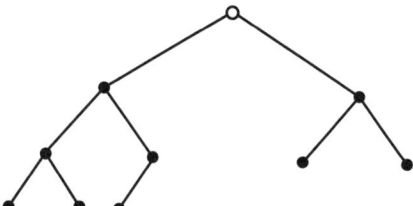

**2.** The following figure shows a priority tree of height 3.

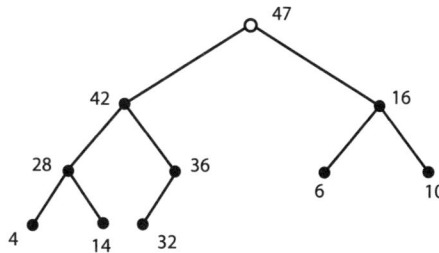

**3.** The following figure illustrates a priority tree insertion. It shows how 45 is inserted into the priority tree of Example 2 in the correct location to maintain the left-compete binary tree shape and then rises until the priority property is restored.

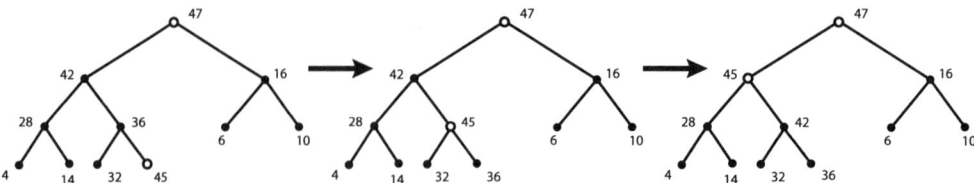

**4.** The following figure illustrates a priority tree deletion. It shows how the left-complete binary tree shape and priority property are restored after the root is removed from the priority tree of Example 2.

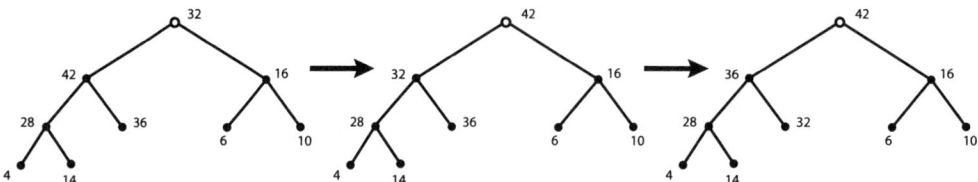

**5.** The following heap corresponds to the priority tree of Example 2. Observe that the keys occur in the array according to the breadth-first-search order of their vertices.

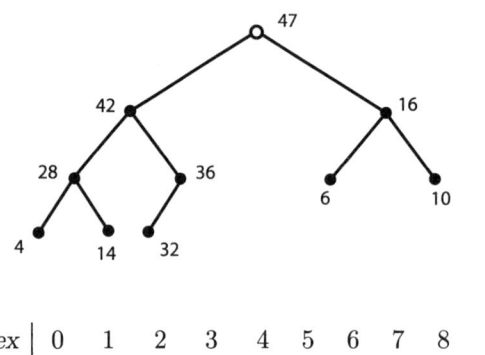

| index | 0 | 1 | 2 | 3 | 4 | 5 | 6 | 7 | 8 | 9 |
|-------|---|---|---|---|---|---|---|---|---|---|
| key   | 47 | 42 | 16 | 28 | 36 | 6 | 10 | 4 | 14 | 32 |

## 18.2.5   NETWORK INCIDENCE STRUCTURES

### Definitions:

An **incidence matrix** for a graph is a 0-1 matrix that specifies the incidence relation. The rows are indexed by the vertices and the columns by the edges. The entry in the row corresponding to vertex $v$ and edge $e$ is 1 if $v$ is an endpoint of $e$, and 0 otherwise.

An **endpoint table** for a graph (§8.1.1) is a *dictionary* whose keys are the edges. The data component for each key edge is the set of endpoints for that edge. If an edge is directed, then its endpoints are marked as *head* and *tail*.

An **edge-incidence table** is a *dictionary* whose keys are the vertices of a graph or digraph. The data component for each key vertex is a list of all the edges that are incident on that vertex. Each self-loop occurs twice in the list.

A **two-way incidence structure** for a graph is a pair consisting of an edge-incidence table and an endpoint table.

**Facts:**

**1.** The time required to insert a new vertex $v$ into a two-way incidence structure for a graph with $n$ vertices and $m$ edges is in $\Theta(\log n)$. By way of contrast, the time for an incidence matrix is in $\Theta(n \cdot m)$.

**2.** The time required to delete a vertex $v$ from a two-way incidence structure for a graph with $n$ vertices and $m$ edges is in $\Theta(\log n + \deg(v))$. By way of contrast, the time for an incidence matrix is in $\Theta(n \cdot m)$.

**3.** The time required to insert a new edge $e$ into a two-way incidence structure for a graph with $n$ vertices and $m$ edges is in $\Theta(\log m)$. By way of contrast, the time for an incidence matrix is in $\Theta(m \cdot n)$.

**4.** An edge-incidence table can represent an imbedding of a graph on a surface as a rotation system (§8.8.3).

**Example:**

**1.** The following graph corresponds to the network incidence structure given below.

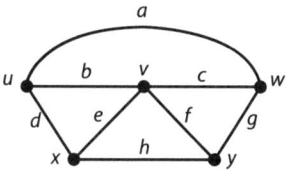

ENDPOINT TABLE

EDGE-INCIDENCE TABLE

| | | | | | | | |
|---|---|---|---|---|---|---|---|
| *u.* | *a* | *b* | *d* | | *a.* | *u* | *w* |
| *v.* | *b* | *c* | *e* | *f* | *b.* | *u* | *v* |
| *w.* | *a* | *c* | *g* | | *c.* | *v* | *w* |
| *x.* | *d* | *e* | *h* | | *d.* | *u* | *x* |
| *y.* | *f* | *g* | *h* | | *e.* | *v* | *x* |
| | | | | | *f.* | *v* | *y* |
| | | | | | *g.* | *w* | *y* |
| | | | | | *h.* | *x* | *y* |

# 18.3  SORTING AND SEARCHING

Since commercial data processing involves frequent sorting and searching of large quantities of data, efficient sorting and searching algorithms are of great practical importance. Sorting and searching strategies are also of fundamental theoretical importance, since sorting and searching steps occur in many algorithms. The following table compares the performance of some of the most common methods for sorting $n$ items.

| sorting method | average time factors | comments |
|---|---|---|
| *expanding a sorted subsequence* | | |
| selection sort | $\approx \frac{n^2}{2}$ comparisons $\approx n$ exchanges | |
| insertion sort | $\approx \frac{n^2}{4}$ comparisons $\approx \frac{n^2}{8}$ exchanges | linear if input file is "almost sorted" |
| Shellsort | $< n^{3/2}$ comparisons | for "good" increments $1, 4, 13, 40, 131, \ldots$ |
| *exchanging out-of-order pairs* | | |
| bubblesort | $\approx \frac{n^2}{2}$ comparisons $\approx \frac{n^2}{4}$ exchanges | one pass if input file is already sorted |
| sinking sort | $\approx \frac{n^2}{2}$ comparisons $\approx \frac{n^2}{4}$ exchanges | one pass if input file is already sorted |
| shakersort | $\approx \frac{n^2}{2}$ comparisons $\approx \frac{n^2}{4}$ exchanges | |
| heapsort | $< 2n \log_2 n$ comparisons | always $\Theta(n \log n)$ |
| *divide-and-conquer* | | |
| mergesort | $\approx n \log_2 n$ comparisons | always $\Theta(n \log n)$ |
| quicksort | $\approx 2n \log_2 n$ comparisons | worst-case $\frac{n^2}{2}$ |
| *sorting by distribution* | | |
| rank-counting | $\Theta(n)$ | |
| radix sort on $k$-digit key | $\approx n \log_2 n$ comparisons | |

## 18.3.1   GENERIC CONCEPTS FOR SORTING AND SEARCHING

### Definitions:

A **database** is a set of *entries*, stored in a computer as an *information structure*.

An **entry** in a database is a 2-tuple whose first component is a *key* and whose second component is some data.

A **key** in a database entry is a value from an ordered set, used to store and retrieve data.

The **key domain** is the ordered set from which keys are drawn.

The **primary key** is the component of highest precedence, when the key for database records has more than one component.

The **secondary key** is the component of next highest precedence after the primary key.

A **record** is another name for a database entry.

**Sorting** is the process of arranging a collection of database entries into a sequence that conforms to the order of their keys.

**Searching a database** means using a systematic procedure to find an entry with a key designated as the objective of the search.

***Scanning a database*** (or a portion of a database) means examining every record in that database (or portion).

The ***target*** of a database search is an entry whose key has been designated as the objective of the search.

A ***comparison sort*** is any sorting method that uses only comparisons of keys.

An ***internal sorting method*** keeps all the entries in the primary memory of the computer during the process of rearrangement.

An ***external sorting method*** uses external storage outside the main memory during the sorting process.

An ***in-place realization*** of a sorting method uses, beyond the space needed for one copy of each data entry, only a constant amount of additional space, regardless of the size of the list to be sorted.

A ***dynamic structure*** for a database is an information structure whose configuration may change during an algorithmic process, for instance, by the insertion or deletion of elements.

A ***static structure*** for a database is a data structure whose configuration does not change during an algorithmic process.

**Facts:**

**1.** Several different general strategies for sorting are given in the following subsections. Each leads to more than one method for sorting.

**2.** Some elementary sorting methods take $O(n^2)$ time. Most practical comparison sorting methods require $O(n \log n)$ time.

**3.** The worst-case running time of any comparison sort is at least $\Omega(n \log n)$.

**Examples:**

**1.** The following are all internal comparison sorts: selection sort (§18.3.2), insertion sort (§18.3.2), Shellsort (§18.3.2), bubblesort (§18.3.3), heapsort (§18.3.3), and quicksort (§18.3.4).

**2.** Mergesort (§18.3.4) is a comparison sort that may be either internal or external.

**3.** *Database model for a telephone directory:*   Each entry has as key the name of a person and as data that person's telephone number. The target of a search is the entry for a person whose number one wishes to call. Names of persons form an ordered key domain under a modified lexicographic ("alphabetic") ordering, in which it is understood that a family name (a "last name" in European-based cultures) has higher precedence than a given name.

**4.** *Database model for a reverse telephone directory:*   In a reverse telephone directory entry, the key is a telephone number and the data is the name of the person with that number. This permits the telephone company to retrieve the name of the person who has a particular phone number, for instance, if someone inquires why some particular telephone number occurs on a long-distance phone bill.

**5.** *Database model for credit card information:*   In a credit card database, the key to each entry is a credit card number, and the data includes the name of the cardholder, the maximum credit limit, and the present balance.

## 18.3.2   SORTING BY EXPANDING A SORTED SUBSEQUENCE

One general strategy for sorting is to iteratively expand a sorted subsequence, most often implemented as an array or a linked list, until the expanded subsequence ultimately contains all the entries of the database.

### Definitions:

A *selection sort* iteratively transforms an unsorted input sequence into a sorted output sequence. At each iteration, it *selects* the item with smallest key from the remaining items in the unsorted input sequence and appends that item at the end of the sorted output sequence.

An *insertion sort* iteratively transforms an unsorted input sequence into a sorted output sequence. At each iteration, it takes the first remaining item from the unsorted input subsequence and *inserts* it into its proper position in the sorted output sequence.

A *Shellsort* of an unsorted sequence $a_1, \ldots, a_n$ is based on a list of *increments* of decreasing size: $h_1 > h_2 > \cdots > h_t = 1$. On the $k$th iteration, the sequence is partitioned into $h_k$ subsequences, such that for $j = 1, \ldots, h_k$, the $j$th subsequence is

$$\langle a_{j+rh_k} \mid 0 \leq r \leq \frac{n-j}{h_k} \rangle$$

and each of these $j$ subsequences is sorted by an insertion sort.

### Facts:

**1.** Selection sorts and insertion sorts both have time-complexity $O(n^2)$ in the worst case.

**2.** The time-complexity of a selection sort is independent of the order of the input sequence, since finding the smallest remaining item requires scanning all the remaining items.

**3.** The running time of an insertion sort can be significantly reduced for "almost sorted" sequences, with time $O(n)$ as the limiting case.

**4.** Optimizing the running time of a Shellsort involves some very difficult mathematical problems, many of which have not yet been solved. In particular, it is not known which choice of increments yields the best result.

**5.** It is known that Shellsort increments should not be multiples of each other, if the objective is to achieve fast execution.

**6.** Evidence supporting the efficiency of the Shellsort increment list $\ldots, 40, 13, 4, 1$ according to the rule $h_{i-1} = 3h_i + 1$ is given by Knuth [Kn98]. The increment list $\ldots, 15, 7, 3, 1$ satisfying the rule $h_{i-1} = 2h_i + 1$ is also recommended.

**7.** Shellsort is a refinement of a straight insertion sort. The motivation for its design in 1959 by D. L. Shell is based on the observation that an insertion sort works very fast for "almost sorted" sequences.

**8.** Shellsort is guaranteed to produce a sorted list, because on the last pass, it applies an insertion sort to the whole sequence.

**9.** An in-place realization of the strategy of expanding a sorted subsequence conceptually partitions the array into a sorted subsequence at the front of the array $A[1..n]$ and an unsorted subsequence of remaining items at the back. Initially, the sorted subsequence is the empty sequence and the unsorted subsequence is the whole list. At each step of the iteration, the sorted front part expands by one item and the unsorted back part contracts by one item.

## Algorithms:

**1.** Algorithm 1 is an in-place realization of a selection sort.

---

**Algorithm 1**: **SelectionSort of array** $A[1..n]$.

  **for** $i := 1$ **to** $n - 1$

    minindex $:= i$; minkey $:= A[i]$

    **for** $j := i + 1$ **to** $n$

      **if** $A[j] <$ minkey **then**

        minindex $:= j$; minkey $:= A[j]$

    swap $A[i]$ with $A[\text{minindex}]$

---

**2.** Algorithm 2 is an in-place realization of an insertion sort.

---

**Algorithm 2**: **InsertionSort of array** $A[1..n]$.

  **for** $i := 2$ **to** $n$

    nextkey $:= A[i]$; $j := i - 1$

    **while** $j > 0$ **and** $A[j] >$ nextkey

      $A[j + 1] := A[j]$; $j := j - 1$

    $A[j + 1] :=$ nextkey

---

## Examples:

In the following examples of single-list implementations of SelectionSort and Insertion-Sort, the symbol " | " separates the sorted subsequence at the front from the remaining unsorted subsequence at the back. The arrows "←" and "→" indicate how far the index $j$ moves during an iteration.

**1.** On the sequence $15, 8, 10, 6, 13, 17$, SelectionSort would progress as follows:

| minkey | minkey | minkey |
|---|---|---|
| 15  8  10  $\widehat{6}$  13  17 | 6 \| $\widehat{8}$  10  15  13  17 | 6  8 \| $\widehat{10}$  15  13  17 |
| $i$  $j$ ———————→ | $i$  $j$ ———————→ | $i$  $j$ ———————→ |

| minkey | minkey |  |
|---|---|---|
| 6  8  10 \| 15  $\widehat{13}$  17 | 6  8  10  13 \| $\widehat{15}$  17 | 6  8  10  13  15  17 \| |
| $i$  $j$ ——→ | $i$  $j$ |  |

**2.** On the sequence $15, 8, 10, 6, 13, 17$, InsertionSort would progress as follows:

| shift | shift | shift  shift  shift |
|---|---|---|
| 15 ↑ 8  10  6  13  17 | 8  15 ↑ 10  6  13  17 | 8→ 10→ 15 ↑ 6  13  17 |
| ← $j$  $i$ | ← $j$  $i$ | ←———— $j$  $i$ |

| shift |  |  |
|---|---|---|
| 6  8  10  15 ↑ 13  17 | 6  8  10  13  15 \| 17 | 6  8  10  13  15  17 \| |
| ← $j$  $i$ | $j$  $i$ |  |

**3.** If $n = 13$ and $h_3 = 4$, then on the third iteration, ShellSort would insertion sort the following subsequences:

$$a_1, a_5, a_9, a_{13}$$
$$a_2, a_6, a_{10}$$
$$a_3, a_7, a_{11}$$
$$a_4, a_8, a_{12}$$

## 18.3.3   SORTING BY EXCHANGING OUT-OF-ORDER PAIRS

A standard measure of the totality of disorder of a sequence of $n$ items is the number of pairs $(a_i, a_j)$ such that $i < j$ but $a_i > a_j$. Thus, the disorder ranges from 0 (i.e., totally ordered) to $\binom{n}{2}$ (i.e., in reverse order). The strategy of exchange sorts is to swap out-of-order pairs until all pairs are in order.

**Definitions:**

A **bubblesort** scans an array $A[1..n]$ repeatedly from the highest index to lower indices, each time swapping every out-of-order pair of consecutive items that is encountered.

A **sinking sort** scans an array $A[1..n]$ repeatedly from the lowest index to higher indices, each time swapping every out-of-order pair of consecutive items that is encountered.

A **shakersort** scans an array $A[1..n]$ repeatedly, and alternates between bubbling upward and sinking downward on alternate scans.

A bubblesort, sinking sort, or shakersort is **adaptive** if it stops the first time a scan produces no transpositions.

**Heapsort** sorts a sequence of entries by iteratively enheaping them all into a heap (§18.2.4) and then iteratively deheaping them all. The order in which they deheap is sorted.

**Facts:**

**1.** The name "bubblesort" suggests imagery in which lighter items (i.e., earlier in the prescribed order of the key domain) bubble to the top of the list.

**2.** The name "sinking sort" suggests that heavier items sink to the bottom.

**3.** The name "shakersort" suggests a salt shaker that is turned upside down.

**4.** Since each swap during an exchange sort reduces the total disorder, it follows that each scan brings the list closer to perfect order. By transitivity of the order relation, it follows that if every consecutive pair in a sequence is in the correct order, then the entire sequence is in order.

**5.** After the first pass of a bubblesort from bottom to top, the smallest element is certain to be in its correct final position at the beginning of the list. After the second pass, the second largest element must be in its correct position, and so on.

**6.** Bubblesort has worst-case time complexity $O(n^2)$.

**7.** For "almost sorted" sequences, an adaptive bubblesort can run much faster than $O(n^2)$ time.

**8.** The priority property implies that the root of a priority tree is assigned the data entry with first precedence.

**9.** Whereas a sequence of length $n$ has $\binom{n}{2}$ pairs that might be out of order, a binary tree of $n$ elements has at most $n \log n$ pairs that could be out of order, if one compares only those pairs such that one node is an ancestor of the other.

**10.** Heapsort improves upon the idea of bubblesort because it bubbles only along tree paths between a bottom node and the root, instead of along the much longer path in a linear sequence from a last item to the first.

**11.** Heapsort runs in $O(n \log n)$ time.

**12.** Heapsort was invented by J. W. J. Williams in 1964.

## Algorithms:

**1.** Algorithm 3 is an adaptive version of bubblesort.

---

**Algorithm 3**:  **BubbleSort of array $A[1..n]$.**
*first* := 1; *last* := $n$; *exchange* := true
**while** *exchange*
   *first* := *first* + 1; *exchange* := false
   **for** $i$ := *last* **downto** *first*
     **if** $A[i] < A[i-1]$ **then**
       swap $A[i]$ and $A[i-1]$; *exchange* := true

---

**2.** Algorithm 4 is a heapsort algorithm.

---

**Algorithm 4**:  **HeapSort of array $A[0..n]$ into array $B[0..n]$.**
**procedure** heapify($i$)
**if** ($A[i]$ is not a leaf) **and** (a child of $A[i]$ is larger than $A[i]$) **then**
   let $A[k]$ be the larger child of $A[i]$
   swap $A[i]$ and $A[k]$
   heapify($k$)

**procedure** buildheap
**for** $i$ := $\lfloor \frac{n}{2} \rfloor$ **downto** 0
   heapify($i$)

**main program** heapsort
buildheap **for** $i = 0$ **to** $n$
   deheap root of $A$ and transfer its value to $B[n-i]$

---

## Examples:

**1.** When cancelled checks are returned to the payer by a bank, they may be in nearly sorted order, since the payees are likely to deposit checks quite soon after they arrive. Thus, they arrive for collection in an order rather close to the order in which they are written. A shakersort might work quite quickly on such a distribution.

**2.** Starting with the unsorted list $L = 15, 8, 10, 6, 17, 13$, bubblesort would produce the following sequence of lists.

$$\begin{aligned}
\text{initial list: } & 15 \quad 8 \quad 10 \quad 6 \quad 17 \quad 13 \\
\text{after one pass: } & 6 \quad 15 \quad 8 \quad 10 \quad 13 \quad 17 \\
\text{after two passes: } & 6 \quad 8 \quad 15 \quad 10 \quad 13 \quad 17 \\
\text{after three passes: } & 6 \quad 8 \quad 10 \quad 15 \quad 13 \quad 17 \\
\text{after four passes: } & 6 \quad 8 \quad 10 \quad 13 \quad 15 \quad 17
\end{aligned}$$

**3.** Starting with the unsorted list $L = 15, 8, 10, 6, 17, 13$, sinking sort would produce the following sequence of lists.

$$\begin{aligned}
\text{initial list: } & 15 \quad 8 \quad 10 \quad 6 \quad 17 \quad 13 \\
\text{after one pass: } & 8 \quad 10 \quad 6 \quad 15 \quad 13 \quad 17 \\
\text{after two passes: } & 8 \quad 6 \quad 10 \quad 13 \quad 15 \quad 17 \\
\text{after three passes: } & 6 \quad 8 \quad 10 \quad 13 \quad 15 \quad 17
\end{aligned}$$

**4.** Starting with the unsorted list $L = 15, 8, 10, 6, 17, 13$, shakersort would produce the following sequence of lists:

$$\begin{aligned}
\text{initial list: } & 15 \quad 8 \quad 10 \quad 6 \quad 17 \quad 13 \\
\text{after one pass: } & 6 \quad 15 \quad 8 \quad 10 \quad 13 \quad 17 \\
\text{after two passes: } & 6 \quad 8 \quad 10 \quad 13 \quad 15 \quad 17
\end{aligned}$$

## 18.3.4  SORTING BY DIVIDE-AND-CONQUER

The strategy of a divide-and-conquer sort is to partition the given sequence into smaller subsequences, to sort the subsequences recursively, and finally to merge the sorted subsequences into a single sorted sequence.

**Definitions:**

A *top-down mergesort* splits the input sequence into two equal (or nearly equal) sized subsequences, recursively mergesorts the two subsequences, and finally merges the two sorted subsequences into a single sorted sequence.

A *bottom-up mergesort* initially regards each entry in its input sequence as a list of length one. It merges two consecutive pairs at a time into lists of length two. Then it merges the lists of length two into lists of length four. Ultimately, all the initial items are merged into a single list.

A *quicksort* selects an element $x$ (called the *pivot*) in the input list and splits the input list into two subsequences $S_1$ and $S_2$ such that every element in $S_1$ is no larger than $x$ and every element in $S_2$ is no smaller than $x$. Next it recursively sorts $S_1$ and $S_2$. Then it concatenates the two sorted subsequences into a single sorted sequence.

The *pivot* in a quicksort iteration is the element $x$ at which the sequence is split.

## Facts:

**1.** A top-down mergesort is usually implemented as an internal sort.

**2.** A bottom-up mergesort is a common form of external sort.

**3.** An outstanding merit of quicksort is that it can be performed quickly within a single array.

**4.** Quicksort was first described by C. A. R. Hoare in 1962.

**5.** The running time of a mergesort is $O(n \log n)$.

**6.** In the worst case, a quicksort takes time $\Omega(n^2)$.

**7.** Choosing the quicksort pivot at random tends to avoid worst-case behavior.

**8.** The average running time for a quicksort is $O(n \log n)$.

**9.** External sorting is used to process very large files, much too large to fit into the primary memory of any computer.

**10.** The emphasis in devising good external sorting algorithms is on decreasing the number of times the data are accessed because the time required to transfer data back and forth between the the primary memory and the tape usually far outweighs the time required to perform comparisons on data in the primary memory.

**11.** Formal algorithms and more detailed discussions of external sorting can be found in [Kn98].

**12.** In an external sort, the number of times each element is read from or written to the external memory is $\log(\frac{n}{m}) + 1$, where $m$ is the available internal memory size. Improvements on the construction of runs as well as on the merging process are possible (see [Kn98]).

## Algorithms:

**1.** Algorithm 5 merges two sorted sequences into a single sorted sequence.

---

**Algorithm 5**:   **Merge two sequences.**

**procedure** merge($A[1..m], B[1..h], C[\ ]$)

{Merge two sorted sequences $A$ and $B$ into a single sorted sequence $C$}

$i_A := 1;\ i_B := 1;\ i_C := 1$

**while** $i_A \leq m$ **and** $i_B \leq h$

  **if** $A[i_A] \leq B[i_B]$ **then**

    $C[i_C] := A[i_A];\ i_A := i_A + 1$

  **else**

    $C[i_C] := B[i_B];\ i_B := i_B + 1$

  $i_C := i_C + 1$

**if** $i_A > m$ **then**

  move the remaining elements in $B$ to $C$

**else**

  move the remaining elements in $A$ to $C$

---

**2.** Algorithm 6 mergesorts a sequence internally.

---

**Algorithm 6**:  MergeSort $S$.

**procedure** mergesort$(S)$

**if** $length(S) \le 1$ **then** return **else**

  split $S$ into two (equal or nearly equal)-sized subsequences $S_1$ and $S_2$

  mergesort $S_1$

  mergesort $S_2$

  merge$(S_1, S_2)$

---

**3.** In a typical external mergesort such as Algorithm 7, there are two input tapes and two output tapes. The entries are initially arranged onto the two input tapes, with half the entries on each tape, and regarded as sublists of length one. A sublist from the first input tape is merged with a sublist from the second input tape and written as a sublist of doubled length onto the first output tape. Then the next sublist from the first input tape is merged with the next sublist from the second input tape and written as a sublist of doubled length onto the second output tape. The alternating process is iterated until the sublists from the input tapes have all been merged into sublists of doubled length onto the two output tapes. Then the two output tapes become input tapes to another iteration of the merging process. This continues until all the original entries are in a single list.

---

**Algorithm 7**:  **External MergeSort sequence $S$ of length $n$.**

  **for** $i := 1$ **to** $\lceil \log n \rceil$

    **for** $j := 1$ **to** $\lceil \frac{\log n}{4i} \rceil$

      merge next sublist from input $A$ with next sublist from input $B$,

        writing merged sublist onto output tape $C$

      merge next sublist from input $A$ with next sublist from input $B$,

        writing merged sublist onto output tape $D$

    reset output tape $C$ as input tape $A$ and vice versa

    reset output tape $D$ as input tape $B$ and vice versa

---

**4.** The generic quicksort algorithm QuickSort (Algorithm 8) does not specify how to select a pivot.

---

**Algorithm 8**:  **QuickSort.**

**procedure** split$(x, S)$

**for** each element $y$ in $S$ **do**

  **if** $x \ge y$ **then**

    put $y$ in $S_1$

  **else**

    put $y$ in $S_2$

**main program**

**if** $length(S) \le 1$ **then** return **else**

  choose an arbitrary element $x$ in sequence $S$

---

> split$(x, S)$ into $S_1$ and $S_2$
>
> recursively sort $S_1$ and $S_2$
>
> concatenate the two sorted subsequences

**Example:**

**1.** The following illustrates MergeSort on the sequence $S = 21, 6, 8, 11, 10, 17, 15, 13$.

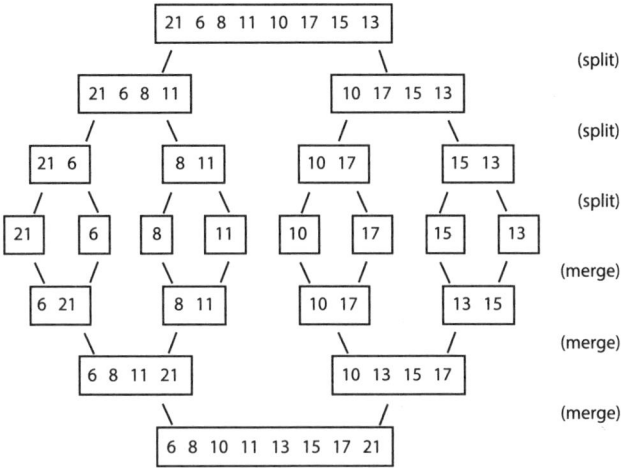

---

### 18.3.5 SORTING BY DISTRIBUTION

Prior knowledge of the distribution of the elements of the input sequence sometimes permits sorting algorithms to break the lower bound of $\Omega(n \log n)$ for running time of comparison sorts.

**Definitions:**

The **rank** of an element of a finite ordered set is the number of elements that it exceeds or equals.

A **rank-counting sort** calculates the "rank" for each element, and then assigns the elements directly to their correct position according to their rank.

In a **base-ten radix sort**, the keys are base-ten integer numerals with at most $k$ digits. Each entry is appended to one of ten queues $Q_0, \ldots, Q_9$, according to the value of its least significant digit, after which the list $Q_0 \circ \cdots \circ Q_9$ is formed by concatenation. The concatenated list is then similarly separated into ten queues, according to the values of the next least significant digit. This process is iterated up to the most significant digit.

A **radix sort** is a sort like the base-ten radix sort, using an arbitrary radix, not necessarily ten.

**Facts:**

**1.** A rank-counting sort gives favorable results when the input keys are $n$ different positive integers, all bounded by $cn$ for some constant $c$.

**2.** The running time of a rank-counting sort is $O(n)$. The RankCountingSort (Algorithm 9) can be modified so that it sorts in linear time even when the input elements are not all distinct [CoEtal09].

**3.** It can be proved that a radix sort correctly sorts the input (see [Kn98]).

**4.** The running time of RadixSort is bounded by $O(kn)$, where $k$ is the maximum number of digits in a key. When $k$ is a constant independent of $n$, RadixSort sorts in linear time. Note, however, that if the input consists of $n$ distinct numbers and the base of the numbers is fixed, then $k$ is of order $\Omega(\log n)$.

### Algorithms:

**1.** In the rank-counting sort (Algorithm 9), the array $A$ contains $n$ input elements, the array $B$ is the output array, and the array $C$ is an auxiliary array of size $cn$ used for counting. Step 3 causes count $C[A[j]]$ to be the rank of entry $A[j]$.

---

**Algorithm 9**:  **RankCountingSort of array $A[1..]$ into array $B[1..n]$.**
$\{\mathrm{pre} : \max(A[i]) \le cn\}$
**for** $i := 1$ **to** $cn$
  $C[i] := 0$
**for** $j := 1$ **to** $n$
  $C[A[j]] := 1$
**for** $i := 2$ **to** $cn$
  $C[i] := C[i] + C[i-1]$
**for** $j := 1$ **to** $n$
  $B[C[A[j]]] := A[j]$

---

**2.** The base-ten radix sort (Algorithm 10) starts with an input list $A$ whose keys have at most $k$ digits.

---

**Algorithm 10**:  **RadixSort of array $A[1..n]$.**
**for** $d := 1$ **to** $k$
  **for** $i := 0$ **to** $9$
    make $Q_i$ an empty queue
  **for** $j := 1$ **to** $n$
    let $h$ be the $j$th digit of $A[j]$
    append $A[j]$ to queue $Q_h$
  $A := Q_0 \circ \cdots \circ Q_9$   {Concatenation}

---

## 18.3.6   SEARCHING

### Definitions:

***Searching a database*** means seeking either a target entry with a specific key or a target entry whose key has some specified property.

***Linear search*** is the technique of scanning the entries of a list in sequence, either until some stopping condition occurs or the entire list has been scanned.

***Binary search*** is a recursive technique for seeking a specific target entry in a list. The target key is compared to the key in the middle of the list, in order to determine which half of the list could contain the target item, if it is present.

***Hashing*** is storage-retrieval in a large table in which the table location is computed from the key of each data entry (see §18.4).

A ***binary search tree*** is a binary tree in which each node has an attribute called its *key*, and the keys are elements of an ordered datatype (e.g., the integers or alphabetic strings). Moreover, at each node $v$ the key is larger than all the keys in its left subtree, but smaller than all the keys in its right subtree.

A ***2-3 tree*** is a tree in which each non-leaf node has 2 or 3 children, and in which every path from the root to a leaf is of the same length.

An ***AVL tree*** is a binary search tree with the property that the two subtrees of each node differ by at most 1 in height.

**Facts:**

**1.** Some common database search objectives are for a specified target entry, for the maximum entry, for the minimum entry, or for the $k$th smallest entry.

**2.** The performance of a dynamic database structure that permits insertions and deletions is measured by the time needed for insertions and deletions, as well as the time needed for searching.

**3.** A binary search runs in average time $O(\log n)$ to search for a specified element $x$ in a sorted list of $n$ elements.

**4.** In the worst case of searching by comparison-based algorithms for a specified target element in a sorted list of length $n$, $\Omega(\log n)$ comparisons are necessary.

**5.** A randomly constructed $n$-node binary search tree has expected height of at most $2 \log n$.

**6.** An AVL tree of $n$ nodes has depth $O(\log n)$.

**7.** Insertion and deletion on an AVL tree, with patching if needed so that the result is an AVL tree, can be performed in $O(\log n)$ worst-case running time [Kn98].

**8.** AVL trees are named for their inventors, G. M. Adelson-Velskii and Y. M. Landis.

**9.** A 2-3 tree for a set $S$ of entries can be constructed by assigning the entries to the leaves of the tree in order of increasing key from left to right. Each non-leaf node $v$ is labeled with two elements $L[v]$ and $M[v]$, which are the largest keys in the subtrees rooted at its left child and middle child, respectively.

**10.** The operations of searching, finding a maximum or minimum, inserting a new entry, and deleting an entry all execute within $O(\log n)$ time.

**Algorithms:**

**1.** To search for a specified target key $x$ in a sorted list $A[1..n]$, a call to the recursive algorithm BinarySearch $(A, 1, n, x)$ in Algorithm 11 can be used. Its technique is to compare the target to the middle entry of the list and to decide thereby in which half the target might occur; then that half remains as the active portion of the list for the next iteration of the search step, while the other half becomes inactive.

---

**Algorithm 11:  BinarySearch $(A, L, U, x)$.**

{Look for $x$ in $A[L..U]$; report its position if found, else report 0}

**if** $L = U$ **then**

    **if** $x = A[L]$ **then** return $L$ **else** return 0

**else**

    $M := \lfloor \frac{L+U}{2} \rfloor$

    **if** $x > A[M]$ **then**

        return BinarySearch $(A, M + 1, U, x)$

    **else** return BinarySearch $(A, L, M, x)$

---

**2.** *Find the maximum [minimum] in an unsorted list:* Scan the list from start to finish and keep track of the largest [smallest] seen so far.

**3.** *Finding the maximum in a binary search tree:* Start at the root and follow the right-child pointers until some node has no right child. That node must contain the maximum.

**4.** *Finding the minimum in a binary search tree:* Start at the root and follow the left-child pointers until some node has no left child. That node must contain the minimum.

**5.** *Searching for a target entry x in a 2-3 tree:* Start at the root, and use the keys at non-leaf nodes to locate the correct leaf, as described by Algorithm 12.

---

**Algorithm 12:   23TSearch$(x, r)$.**

{Case 1: $r$ is a leaf}

    **if** $r$ is labeled with $x$ **then** return "yes" **else** return "no"

{Case 2: $r$ is not a leaf}

    **if** $x \leq L[r]$ **then** return 23TSearch$(x, \text{leftchild}(r))$

    **else if** $x \leq M[r]$ **then** return 23TSearch$(x, \text{midchild}(r))$

    **else if** $x$ has a right child **then** return 23TSearch$(x, \text{rightchild}(r))$

    **else return** "no"

---

**6.** *Finding the maximum in a 2-3 tree:* Starting from the root, follow the right-child pointers to the rightmost leaf, which contains the maximum entry.

**7.** *Finding the minimum in a 2-3 tree:* Starting from the root, follow the left-child pointers to the leftmost leaf, which contains the minimum entry.

**8.** *To insert a new entry x into a 2-3 tree:* First locate the non-leaf node $v$ whose child $x$ "should" be. If $v$ is a 2-child node $v$, then simply install $x$ as a third child of $v$. If $v$ already has three children, then let $v$ keep as children the two smallest of the set comprising its three children and $x$. A new non-leaf node $u$ becomes the parent of the largest member of this set. Now recursively insert node $u$ as a new child to the parent of node $v$. If the process eventually makes the root of the tree a 4-child node, then the last step is to create a new root with two new children, each of which has two of the four children of the former root. Note that the labels of some non-leaf nodes may be updated in this process.

**9.** *To delete an entry x from a 2-3 tree:* Essentially, reverse the manner by which an element is inserted. First find the leaf $v$ containing $x$. If the parent $p$ of $v$ has three children, then the leaf $v$ is simply deleted. If $p$ has only two children $v$ and $v'$, then select an adjacent sibling $p'$ of $p$. If $p'$ has only two children, then make $v'$ a child of $p'$, and recursively delete the node $p$ from the tree. If $p'$ has three children, then make an appropriate child of $p'$ into a new child of $p$ and delete the node $v$ (note that now both $p$ and $p'$ have two children). Again the process may progress recursively to the root of the tree, such that it is necessary to delete one of the only two children of the root. In this case, delete the root and make the remaining child of the root into a new root of the tree. Labels of some non-leaf nodes may need to be updated in this process.

**10.** *Searching in a random list:* Finding the $k$th smallest element, for an arbitrary $k$, in a random list of size $n$ can also be done in linear time. The algorithm, Algorithm 13, is based on the method of "Prune and Search". That is, the process first prunes away in linear time a constant factor of the elements in the input, then recursively searches the rest. A careful analysis shows that each of the two sets $S_1$ and $S_2$ contains at most $\frac{7n}{10}$ elements. Therefore, if $T(n)$ is the running time of Algorithm 13 for $n$ elements, it follows that $T(n) \le T(\frac{n}{5}) + T(\frac{7n}{10}) + O(n)$. This relation gives $T(n) = O(n)$.

---

**Algorithm 13: Finding-the-$k$th-Smallest.**

 divide the $n$ input elements into $\lceil \frac{n}{5} \rceil$ groups of five elements

 find the median for each of the $\lceil \frac{n}{5} \rceil$ groups

 recursively find the median $m^*$ of these group medians

 partition the input into two sets $S_1$ and $S_2$ such that each element in $S_1$ is no

   larger than $m^*$ and each element in $S_2$ is no smaller than $m^*$

 **if** $S_1$ has $\ge k$ elements **then**

  recursively find the $k$th smallest element in $S_1$

 **else**

  recursively find the $(k - |S_1|)$th smallest element in $S_2$

---

**Example:**

**1.** We apply Algorithm 11 to search for the target 64 in the following 16-element list:

$$5, 8, 9, 13, 16, 22, 25, 36, 47, 49, 64, 81, 100, 121, 144, 169.$$

First split it into these two 8-element sublists

$$5, 8, 9, 13, 16, 22, 25, 36 \qquad 47, 49, 64, 81, 100, 121, 144, 169$$

and then compare 64 to the largest item in the first list. Since $64 > 36$, it follows that 64, if present in the original list, would have to be in the second sublist. Next split the active sublist further into these two 4-element sublists

$$47, 49, 64, 81 \qquad 100, 121, 144, 169$$

and then compare 64 to the largest item in the first new sublist. Since $64 \le 81$, it follows that 64, if present, would have to be in the second sublist. Therefore, resplit the active sublist further into these two 2-element sublists

$$47, 49 \qquad 64, 81$$

and then compare the target 64 to the largest item in the first new sublist. Since $64 > 49$, it follows that 64, if present, would have to be in the second sublist. Therefore, resplit the active sublist further into these two 1-element sublists

$$64 \qquad 81$$

and then compare 64 to the largest item in the first new sublist. Since $64 \leq 64$, it follows that 64, if present, would have to be in the first sublist. Since that sublist has only one element, namely 64, the target 64 is compared to that one element. Since they are a match, the target has been located as the 11th item in the original list.

# 18.4  HASHING

Hashing, also known as "address calculation" and as "scatter storage", is a mathematical approach to organizing records within a table. The objective is to reduce the amount of time needed to find a record with a given key. Hashing is best suited for "dynamic" tables, that is, for databases whose use involves interspersed lookups and insertions. Dynamic dictionaries (such as spelling checkers) and compiler-generated symbol tables are examples of applications where hashing may be useful.

## 18.4.1  HASH FUNCTIONS

Hashing is an approach to placing records into a table and retrieving them when needed, in which the location for a record is calculated by an algorithm called a hash function.

### Definitions:

A **record** is a pair of the form ($k$:*key*, $d$:*data*), in which the second component is data and the first component is a *key* used to store it in a table and to retrieve it subsequently from that table.

A **key domain** is an ordered set, usually the integers, whose members serve as keys for the *records* of a table. No two different records have the same key.

A **hash table** is an array, in which the location for storing and retrieving a record are calculated from the key of that record.

A **hash function** $h$ is a function that maps a key domain to an interval of integers $[0..m-1]$. The intent is that a record with key $k$ is to be stored in or retrieved from the location $h(k)$ in the table.

A **collision** occurs when a hash function $h$ assigns the same table location to two different keys $k_1 \neq k_2$, i.e., when $h(k_1) = h(k_2)$.

**Collision resolution** is the process of finding an alternative location for a new record whose key collides with that of a record already in the table.

The **fullness** of a hash table $T$ is the ratio $\alpha(T) = \frac{n}{m}$ of the number $n$ of records in the table to the capacity $m$ of the table.

### Facts:

**1.** Hashing is often used when the set of keys (of the records in the database) is not a consecutive sequence of integers or easily convertible to a consecutive sequence of integers.

**2.** Keys that are non-numeric (or do not have a good random distribution) can be transformed into integers by using or modifying their binary representation. The resulting integers are called *virtual keys*.

**3.** It is desirable for a hash function to have the *simplicity property*, i.e., that it takes only a few simple operations to compute the hash function value.

**4.** It is desirable for a hash function to have the *uniformity property*, i.e., that each possible location in the range $0, \ldots, m-1$ of the hash function $h \colon K \to [0..m-1]$ is generated with equal likelihood, that is, with probability $\frac{1}{m}$.

**Examples:**

**1.** The *division method* $h(k) = k \bmod m$ is a simple hash function that can be used when the keys of the records are integers that extend far beyond a feasible table size. The table size $m$ must be chosen carefully to avoid high instance of collision, without wasting too much storage space. Selecting $m$ to be a prime not close to a power of 2 is typically a good choice.

**2.** The *multiplication method* is another simple hashing rule. First the key (in base 2) is multiplied by a constant value $A$ such that $0 < A < 1$, and then $p = \log m$ bits are selected for the hash function value from somewhere in the fractional part of the resulting base-2 numeral, often required to be away from its low end. (This is similar to some methods for generating random numbers.)

**3.** As a simplified example of a multiplicative hash function, consider table size $m = 16$, address size $p = \log 16 = 4$ bits, and key size $w = 6$ bits. With fractional constant $A = 0.101011_2$, use the first four bits of the fractional part. For instance, given key $k = 011011_2$, first calculate $R = A \cdot k = 010010.001001_2$. Then take $k = 0010_2$, the low-end four bits of the fractional part. Knuth [Kn98] suggests $A = 0.6180339887...$ as a good choice of a multiplier.

---

## 18.4.2  COLLISION RESOLUTION

**Definitions:**

***Collision resolution*** is the process of computing an alternative location for a colliding record. The two basic methods are *chaining* and *rehashing*.

***Chaining*** is a collision resolution method that involves auxiliary storage space for records outside the confines of the main array. Each slot in the main table can be used as the root of a linked list that contains all the records assigned to that location by the hash function. Each additional colliding record is inserted at the front of the linked list. When searching for a record, the list is traversed until the record is found, or until the end of the list is reached.

The ***size of a chained hash table*** is the number of linked list headers (i.e., the size of the main array). Thus, a chained hash table said to be of size $m$ may be used to store a database with more than $m$ records.

***Rehashing*** is a collision resolution method in which there is no auxiliary storage outside the main table, so that a colliding record must be stored elsewhere in the main table, that is, at a location other than that assigned by the hash function to its key $k$. A collision resolution function finds the substitute location.

A ***collision resolution function*** under rehashing generates a *probe sequence* $\langle h_0(k) = h(k), h_1(k), h_2(k), \ldots, h_{m-1}(k) \rangle$. The new record is inserted into the first unoccupied probe location. When searching for a record, the successive probes are tried until the record is found or the probe finds an unoccupied location (i.e., unsuccessful search).

A **probe sequence** for key $k$ is a sequence $\langle h_0(k), h_1(k), h_2(k), \ldots, h_{m-1}(k) \rangle$ of possible storage locations in the table $T$ that runs without repetition through the entire set $\langle 0, 1, 2, \ldots, m-1 \rangle$ of locations in the table, as possible places to store the record with key $k$.

**Clustering** is a hashing phenomenon in which after two keys collide, their probe sequences continue to collide.

**Linear probing** means trying to resolve a collision at location $h(k)$ with a probe sequence of the form $h_i(k) = (h_0(k) + ci) \bmod m$.

**Quadratic probing** means trying to resolve a collision at location $h(k)$ by using a probe sequence of the form $h_i(k) = (h_0(k) + c_1 i + c_2 i^2) \bmod m$.

A **secondary hash function** is a hash function used to generate a probe sequence, once a collision has occurred.

**Double hashing** means using a primary hash function $h$ and a secondary hash function $h'$ to resolve collisions, so that $h_i(k) = (h(k) + ih'(k)) \bmod m$.

## Facts:

**1.** In designing a hash function, an objective is to keep the length of the probe sequences short, so that records are stored and retrieved quickly.

**2.** Under chain hashing, inserting a record always requires $O(1)$ time.

**3.** Under chain hashing, the time to find a record is proportional to the length of the chain from its key, and the average length of a chain in the table equals the fullness $\alpha$ of the table.

**4.** Under chain hashing, if the number of records in the table is proportional to the table capacity, then a *find* operation needs $O(1)$ time, on average.

**5.** Under chain hashing, a *delete* operation consists of a *find* with time proportional to table fullness, followed by removal from a linked list, which requires only $O(1)$ time.

**6.** Analysis of rehashing performance is based on the following assumptions:

- uniform distribution of keys: each possible key in the key domain is equally likely to occur as the key of some record;
- uniform distribution of initial hash locations (see §18.4.1);
- uniform distribution of probe sequences: each possible probe sequence

$$\langle h_0(k), h_1(k), h_2(k), \ldots, h_{m-1}(k) \rangle,$$

regarded as a permutation on the set of all table locations, is equally likely.

**7.** Under rehashing, the expected time to perform an *insert* operation is the same as the expected time for unsuccessful search, and is at most $\frac{1}{1-\alpha}$.

**8.** Under rehashing, the expected time $E(\alpha)$ to perform a successful *find* operation is at most $\frac{1}{\alpha} \ln \frac{1}{1-\alpha} + \frac{1}{\alpha}$. For instance, $E(0.5) = 3.386$ and $E(0.9) = 3.670$. That means that if a table is 90% full, a record will be found, on average, with 3.67 probes.

**9.** Under rehashing, location of the deleted record needs to be marked as *deleted* so that subsequent probe sequences do not terminate prematurely. Moreover, the running time of a *delete* operation is the same as for a successful *find* operation. (It also causes the measure of fullness for searches to be different from that used for insertions, because a new record can be inserted into the location marked as *deleted*). However, in most applications that use hashing, records are never deleted.

**Examples:**

**1.** The following example of linear probing uses prime modulus $m = 1013$ and prime multiplier $c = 367$. The keys are taken to be social security numbers.

| key $k$ | $h_0(k)$ | $h_1(k)$ |
|---------|----------|----------|
| 113548956 | 773 | |
| 146834522 | 172 | |
| 207639313 | 651 | |
| 359487245 | 896 | |
| 378545523 | 592 | |
| 435112760 | 896 | 250 |
| 670149788 | 651 | |
| 721666437 | 172 | 539 |
| 762456748 | 12 | |

**2.** Linear probing suffers from clustering. That is, if $h_i(k_1) = h_j(k_2)$, then $h_{i+p}(k_1) = h_{j+p}(k_2)$ for all $p = 1, 2, \ldots$. All probe sequences follow the same (linear) pattern, from which it follows that long chains of filled locations will cause a large number of probes needed to insert a new record (and to locate it later).

**3.** Quadratic probing suffers from clustering. That is, if $h_0(k_1) = h_0(k_2)$, then $h_i(k_1) = h_i(k_2)$ for $i = 1, 2, \ldots$.

**4.** Pairing the primary hash function $h(k) = k \bmod p$ with the secondary hash function $h'(k) = k \operatorname{\mathbf{div}} p$, where $p$ is a prime, yields the double hash function $h_i(k) = (h_0(k) + ih'(k)) \bmod p$.

# 18.5 DYNAMIC GRAPH ALGORITHMS

Dynamic graph algorithms are algorithms that maintain information with regard to properties of a (possibly edge-weighted) graph while the graph is changing. These algorithms are useful in a number of application areas, including communication networks, VLSI design, distributed computing, and graphics, where the underlying graphs are subject to dynamic changes. Efficient dynamic graph algorithms are also used as subroutines in algorithms that build and modify graphs as part of larger tasks, e.g., the algorithm for constructing Voronoi diagrams by building planar subdivisions.

*Notation:* Throughout this section, $n$ and $m$ denote the number of vertices and the number of edges, respectively, of the graph $G$ that is being maintained and queried dynamically.

## 18.5.1 DYNAMICALLY MAINTAINABLE PROPERTIES

**Definitions:**

A (**dynamic**) **update operation** is an operation on a graph that keeps track of whether the graph has a designated property.

A **query** is a request for information about the designated property.

**Facts:**

**1.** The primitive update operations for most dynamic graph algorithms are edge insertions and deletions and, in the case of edge-weighted graphs, changes in edge weights.

**2.** For most dynamic graph algorithms, insertion or deletion of an isolated vertex can be accomplished by an easy modification of a nondynamic algorithm.

**3.** The insertion or deletion of vertices together with their incident edges is usually harder and has to be done by iterating the associated edge update operation.

**4.** There is a trade-off between the time required for update operations and the time required to respond to queries about the property being maintained. Thus, running times of the update operations depend strongly on the property being maintained.

**5.** Nontrivial dynamic algorithms corresponding to several graph properties are known (see Examples).

**Examples:**

**1.** *Connectivity*: The permitted query is whether two vertices $x$ and $y$ are in the same component. Permitted updates are edge insertions, edge deletions, and isolated vertex insertions. Frederickson [Fr85] provides an algorithm for maintaining minimum spanning forests that can easily be adapted to this problem. Improvements in running times have been achieved by [EpEtal92b] and [EpGaIt93].

**2.** *Bipartiteness*: Update operations are the same as for connectivity (Example 1). A query simply asks whether a graph is bipartite. An algorithm is presented in [EpEtal92b], with an improvement in [EpGaIt93].

**3.** *Minimum spanning forests*: The query is whether an edge is in a minimum spanning forest. The graph is weighted, and the update operations are increments and decrements of weights. (Edge insertion is accomplished by lowering the edge weight from $\infty$ and edge deletion by incrementing the edge weight to $\infty$.) [Fr85] contains the early result, with improvements by [EpGaIt93]. The plane and planar graph cases have been considered by [EpEtal92b] and [EpEtal93].

**4.** *Biconnectivity and 2-edge connectivity*: Update operations are the same as for connectivity (Example 1). Queries ask whether two given vertices lie in the same biconnected (resp., 2-edge connected) component. Efficient algorithms for maintaining biconnectivity are found in [EpEtal92b], [EpGaIt93], [Ra93], and [HeRaSu94]. Efficient algorithms for maintaining 2-edge connectivity are found in [EpEtal92b], [Fr91], and [Fr85]. Any algorithm for dynamically maintaining biconnectivity translates to an algorithm with the same time bounds for 2-edge connectivity [GaIt91].

**5.** *Planarity*: Update operations include edge insertions and deletions. Queries ask whether the graph is currently planar. Variants include queries that would test whether the addition of a particular edge would destroy the current imbedding. Algorithms are described in [EpEtal93] and [Ra93].

---

## 18.5.2   TECHNIQUES

**Definitions:**

A **partially dynamic algorithm** is usually an algorithm that handles only edge insertions and, for edge-weighted graphs, decrements in edge weights. Less commonly, this term can refer to an algorithm that handles only edge deletions or weight increments.

A **cluster** in a spanning tree $T$ for a graph $G$ is a set of vertices such that the subgraph of $T$ induced on these vertices is connected.

An **ambivalent data structure** is a structure that, at many of its vertices, keeps track of several alternatives, despite the fact that a global examination of the structure would determine which of these alternatives is optimal.

A **certificate** for property $\mathcal{P}$ and graph $G$ is a graph $G'$ such that $G$ has property $\mathcal{P}$ if and only if $G'$ has property $\mathcal{P}$.

A **strong certificate** for property $\mathcal{P}$ and graph $G$ is a graph $G'$, on the same vertex set as $G$, such that, for every graph $H$, $G \cup H$ has property $\mathcal{P}$ if and only if $G' \cup H$ has property $\mathcal{P}$.

A **sparse certificate** is a strong certificate in which the number of edges is $O(n)$.

A function $A$ that maps graphs to strong certificates is **stable** if it satisfies

- $A(G \cup H) = A(A(G) \cup H)$;
- $A(G - e)$ differs from $A(G)$ by $O(1)$ edges, where $e$ is an edge in $G$.

A **stable certificate** is one produced by a stable mapping.

A **plane graph** is a planar graph, together with a particular imbedding in the plane.

A **compressed certificate** for a property $\mathcal{P}$ of $G$, where $G = (V, E)$ is a subgraph of a larger graph $F$ and $X \subset V$ separates $G$ from $F - G$, is a small certificate $G' = (V', E')$ with $X \subset V'$ such that, for any graph $H$ that is attached to $G$ only at the vertices of $X$, $H \cup G$ has property $\mathcal{P}$ if and only if $H \cup G'$ does, and $|V'| = O(|X|)$.

A graph property $\mathcal{P}$ is **dyadic** if it is defined with respect to a pair of vertices $(x, y)$.

A graph $C$ is a **certificate of a dyadic property** $\mathcal{P}$ for $X$ in $G$ if and only if, for any $H$ with $V(H) \cap V(G) \subset X$ and every $x$ and $y$ in $V(H)$, $\mathcal{P}$ is true for $(x, y)$ in the graph $G \cup H$ if and only if it is true for $(x, y)$ in the graph $C \cup H$.

## Facts:

**1.** Using the union-find data structure [Ta75], it is possible to maintain connectivity information in $O(\alpha(m, n))$ amortized time per update or query. Here $\alpha(m, n)$ denotes the inverse Ackermann function, and it grows extremely slowly.

**2.** For other graph properties, such as 2-edge connectivity and biconnectivity, a data structure called the *link/condense tree* [WeTa92] maintains information in $O(\alpha(m, n))$ amortized time per update or query.

**3.** The link/condense tree supports the operation of condensing an entire path in the tree into a single vertex. This is important in the applications considered, because the insertion of an edge may cause several biconnected components or 2-edge connected components to be combined into one.

**4.** Link/condense trees are based on *dynamic trees*. [SlTa83]

**5.** The dynamic tree data structure maintains a set of rooted trees. It supports the operations of *linking* the root of one tree as the child of a vertex in another tree, *cutting* a tree at a specified edge, and *everting* a tree to make a specified vertex the root in worst-case $O(\log n)$ time per operation. It also supports other operations based on keys stored at vertices, such as finding the minimum key on the path from a given vertex to the root in $O(\log n)$ time.

**6.** To maintain minimum spanning trees in an edge-weighted, connected graph subject to changes in edge weights, Frederickson [Fr85] uses *clustering* and *topology trees* in Algorithm 1.

---

**Algorithm 1**:  **Maintaining a minimum spanning tree.**

*Preprocessing*:

   find a minimum spanning tree $T$ of the initial graph $G$

   maintain a dynamic tree of $T$, using [SlTa83]

   let $z := n^{2/3}$

   group the vertices of $T$ into *clusters* whose sizes are between $z$ and $3z - 2$
   {There will be $\Theta(n^{1/3})$ such clusters}

   **for** each pair of clusters $i, j$ maintain the set of edges $E_{ij}$ as a min-heap

*Updates*:

   decreases in tree edge weights do not change anything

   increases in non-tree edge weights are handled by a suitable update of the
      appropriate min-heap

   decreases in non-tree edge weights are handled by using the dynamic tree
      appropriately

   if tree edge $e$ increases in weight, remove it, thus partitioning the clusters into
      two sets; then find an edge of minimum cost between clusters on opposite sides
      of the partition

---

**7.** Ambivalence may permit faster updates, possibly at the cost of slower queries.

**8.** Frederickson [Fr91] presents an ambivalent data structure for spanning forests that builds on the ideas of multilevel partitions and (2-dimensional) topology trees developed in [Fr85].

**9.** Let $\mathcal{P}$ be a property for which sparse certificates can be found in time $f(n, m)$. Suppose that there is a data structure that can be built in time $g(n, m)$ and permits static testing of property $\mathcal{P}$ in time $q(n, m)$. Then there is a fully dynamic data structure for testing whether a graph has property $\mathcal{P}$; update time for this structure is $f(n, O(n))O(\log(\frac{m}{n})) + g(n, O(n))$, and query time is $q(n, O(n))$. This "basic sparsification technique" is used to dynamize static algorithms. To use it, one need only be able to compute sparse certificates efficiently.

**10.** The *sparsification* method of [EpEtal92b] partitions the input graph into sparse subgraphs (with $O(n)$ edges) and summarizes the relevant information about each subgraph in an even sparser "certificate". Certificates are merged in pairs, producing larger subgraphs that are themselves sparsified using the certificate technique. The result is a balanced binary tree in which each vertex is a sparse certificate. Each insertion or deletion of an edge in the input graph causes changes in $\log(\frac{m}{n})$ tree vertices. Because these changes occur in graphs with $O(n)$ edges, instead of the $m$ edges in the input graph, time bounds for updates are reduced in most natural problems.

**11.** Let $\mathcal{P}$ be a property for which stable sparse certificates can be maintained in time $f(n, m)$ per update. Suppose that there is a fully dynamic algorithm for $\mathcal{P}$ with update time $g(n, m)$ and query time $q(n, m)$. Then this algorithm can be sped up; specifically, $\mathcal{P}$ can be maintained fully dynamically in time $f(n, O(n))O(\log(\frac{m}{n})) + g(n, O(n))$ per update, with query time $q(n, O(n))$. Because this "stable sparsification technique" is used to speed up existing dynamic algorithms, it often yields better results than the basic sparsification technique described in Fact 10. However, to use it, one must be able to maintain stable sparse certificates efficiently; this is a more stringent requirement than what is needed to apply basic sparsification.

**12.** Eppstein, Galil, and Italiano [EpGaIt93] improve the sparsification technique to remove the $\log(\frac{m}{n})$ factor in these bounds. They achieve this improvement by partitioning the edge set of the original graph more carefully.

**13.** Dynamic algorithms restricted to plane graphs have been considered by several authors. Eppstein et al. [EpEtal93] introduce a variant of sparsification that permits the design of efficient dynamic algorithms for planar graphs in an imbedding-independent way, as long as the updates to the graph preserve planarity. Because these graphs are already sparse, Eppstein et al. design a separator-based sparsification technique.

**14.** The fact that separator sizes are sublinear ($O(\sqrt{n})$ for planar graphs) allows the possibility of maintaining sublinear certificates. Eppstein et al. [EpEtal93] use a separator-based decomposition tree as the sparsification tree and show how to compute it in linear time and maintain it dynamically. They use it to show the following: For a property $\mathcal{P}$ for which compressed certificates can be built in time $T(n)$, a data structure for testing $\mathcal{P}$ built in time $P(n)$, and queries answered in time $Q(n)$, a fully dynamic algorithm for maintaining $\mathcal{P}$ under planarity-preserving insertions and deletions takes amortized time $P(O(n^{1/2})) + T(O(n^{1/2}))$ per update and $Q(O(n^{1/2}))$ per query.

**15.** A dyadic property $\mathcal{P}$, for which compressed certificates can be built in time $T(n)$, a data structure for testing $\mathcal{P}$ built in time $P(n)$, and queries answered in time $Q(n)$, can be maintained with updates taking $T(O(n^{1/2}))$ amortized time and queries taking $P(O(n^{1/2})) + Q(O(n^{1/2})) + T(O(n^{1/2}))$ worst-case time.

**16.** In dealing with plane (as opposed to planar) graphs and allowing only updates that can be performed in a planarity-preserving manner on the existing imbedding, simpler techniques that rely on planar duality can be used [EpEtal92b].

**17.** When maintaining minimum spanning trees under updates that change only edge weights, the most difficult operation to handle is an increase in the weight of an MST edge. However, in the dual graph this can be viewed as a decrease in the weight of a non-MST edge. This idea and the handling of edge insertions and deletions are addressed by the data structures of [GuSt85] and the *edge-ordered tree* data structure of [EpEtal92b]. These data structures help maintain the subdivision and its dual in the face of general updates and also help perform required access operations efficiently. Edge-ordered trees are an adaptation of the dynamic trees of [SlTa83].

**18.** Knowledge of the imbedding allows one to use topology trees in more efficient ways. Specifically, Rauch [Ra94] partitions the non-tree edges into equivalence classes called *bundles*. In the cyclical ordering of edges emanating from a cluster, bundles are carefully chosen, consecutive subsets of edges.

### 18.5.3   APPLICATIONS

**Examples:**

**1.** *Bipartiteness* [EpEtal92a], [EpGaIt93]:   A graph that is not bipartite contains an odd cycle. The graph formed by adding the shortest edge inducing an odd cycle (if any) to the minimum spanning forest is a stable certificate of (non-)bipartiteness. Using the clustering techniques of [Fr85] and the improved sparsification techniques of [EpGaIt93], this certificate can be maintained in $O(n^{1/2})$ time per update. The query time in this example is $O(1)$; one bit is used to indicate whether the operation is maintaining a certificate of bipartiteness or of non-bipartiteness.

**2.** *Minimum spanning forests* [EpEtal92a], [EpGaIt93], [Fr85]: In this example, the goal is not to maintain a data structure that supports efficient testing of a property, but rather to maintain the minimum spanning forest itself as edges are added to and deleted from the input graph. It is shown in [EpEtal92a] how to define a canonical minimum spanning forest that serves as the analogue of a stable sparse certificate. Frederickson [Fr85] uses the topological approach to obtain a fully dynamic algorithm that maintains minimum spanning forests in time $O(m^{1/2})$ per update. Applying the improved stable sparsification technique with $f(n, m) = g(n, m) = O(m^{1/2})$ yields a fully dynamic minimum spanning forest algorithm with update time $O(n^{1/2})$. For plane graphs, [EpEtal92b] show that both updates and queries can be performed in $O(\log n)$ time per operation; for planar graphs, [EpEtal93] show that $O(\log^2 n)$ time per deletion and $O(\log n)$ time per insertion are sufficient.

**3.** *Connectivity* [EpEtal92a], [EpGaIt93], [Fr85]: Simple enhancements to the minimum spanning forest algorithms in [Fr85] yield fully dynamic algorithms for the connectivity problem in which the update times are the same as they are for minimum spanning forests, and the query times are $O(1)$. Thus, as in Example 2, applying improved stable sparsification with $f(n, m) = g(n, m) = O(m^{1/2})$ yields a fully dynamic connectivity algorithm with update time $O(n^{1/2})$ and query time $O(1)$. Similarly, the planar and plane graph algorithms for minimum spanning trees can be generalized to work for minimum spanning forests and adapted to maintain connected components.

**4.** *Biconnectivity* [EpEtal92a], [EpGaIt93], [HeRaSu94], [Ra93]: Cheriyan, Kao, and Thurimella [ChKaTh93] show that $C_2 = C_1 \cup B_2$ is a sparse certificate for biconnectivity, where $C_1$ is a breadth-first spanning forest of the input graph $G$, and $B_2$ is a breadth-first spanning forest of $G - C_1$. Eppstein et al. [EpEtal92a] show that $C_2$ is in fact a strong certificate of biconnectivity. These strong certificates can be found in time $O(m)$, using classical breadth-first search algorithms. Applying improved sparsification with $f(n, m) = g(n, m) = O(m)$ yields a fully dynamic algorithm for maintaining the biconnected components of a graph that has update time $O(n)$.

**5.** The approach to biconnectivity in [Ra94] is to partition the graph $G$ into clusters and decompose a query that asks whether vertices $u$ and $v$ lie in the same biconnected component into a query in the cluster of $u$, a query in the cluster of $v$, and a query between clusters. The 2-dimensional topology tree is adapted in a novel way, and the ambivalent data structures previously defined for connectivity and 2-edge connectivity are extended to test biconnectivity between clusters. To test biconnectivity within a cluster $C$, the entire subgraph induced by $C$ and a compressed certificate of $G - C$ are maintained. Using all these ingredients, [Ra94] obtains amortized $O(m^{1/2})$ time per update and $O(1)$ worst-case time per query.

**6.** Using clever data structures based on topology trees, bundles, and the idea of *recipes*, first introduced in this context in [HeRaSu94], the problem of fully dynamic biconnectivity for plane graphs can be solved in $O(\log^2 n)$ time per update and $O(\log n)$ time per query.

**7.** *2-edge connectivity* [EpEtal92a], [Fr85], [Fr91]: Thurimella [Th89] and Nagamochi and Ibaraki [NaIb92] show that the following structure $U_2$ is a certificate for 2-edge connectivity: $U_2 = U_1 \cup F_2$, where $U_1$ is a spanning forest of $G$, and $F_2$ is a spanning forest of $G - U_1$. Eppstein et al. [EpEtal92a] show that $U_2$ is in fact a stable, sparse certificate. Frederickson's minimum spanning forest algorithm [Fr85] can be adapted to maintain

$U_2$ in time $f(n,m) = O(m^{1/2})$. Frederickson's ambivalent data structure technique [Fr91] can be used to test 2-edge connectivity with update time $g(n,m) = O(m^{1/2})$ and query time $q(n,m) = O(\log n)$. Here a "query" is a pair of vertices, and the answer is "yes" if they are in the same 2-edge connected component and "no" otherwise. Applying improved stable sparsification yields a fully dynamic algorithm with update time $O(n^{1/2})$ and query time $O(\log n)$.

**8.** *Planarity* [EpEtal93], [Ra93]:   Eppstein et al. [EpEtal93] use the separator-based sparsification technique to obtain a fully dynamic planarity-testing algorithm for general graphs that answers queries of the form "is the graph currently planar?" and "would the insertion of this edge preserve planarity?". Their algorithm requires amortized running time $O(n^{1/2})$ per update or query. Italiano, La Poutré, and Rauch [ItLaRa93] use topology trees, bundles, and recipes to obtain a fully dynamic algorithm on plane graphs that tests whether the insertion of a particular edge would destroy the given imbedding. Their algorithm requires time $O(\log^2 n)$ for updates and queries.

---

## 18.5.4   RECENT RESULTS AND OPEN QUESTIONS

**Examples:**

**1.** Alberts and Henzinger [AlHe95] investigate dynamic algorithms on random graphs with $n$ vertices and $m_0$ edges on which a sequence of $k$ arbitrary update operations is performed. They obtain expected update times of $O(k \log n + \sum_{i=1}^{k} \frac{n}{\sqrt{m_i}})$ for minimum spanning forest, connectivity, and bipartiteness and $O(k \log n + \sqrt{\log n} \sum_{i=1}^{k} \frac{n}{\sqrt{m_i}})$ for 2-edge connectivity. The data structures required for these algorithms use linear space, and the preprocessing times match those of the best algorithms for finding a minimum spanning forest.

**2.** Fredman and Henzinger [FrHe98] investigate lower bounds in the cell probe model of computation and obtain good results for $k$-edge connectivity, $k$-vertex connectivity, and planarity-testing of imbedded planar graphs. Both average-case analysis and lower bounds are important topics for future research on dynamic graph algorithms.

**3.** Klein et al. [KlEtal94] give a fully dynamic algorithm for the all-pairs shortest path problem on planar graphs. If the sum of the absolute values of the edge-lengths is $D$, then the time per operation is $O(n^{9/7} \log D)$ (worst case for queries, edge deletion, and length changes, and amortized for edge insertion); the space requirement is $O(n)$. Several types of partially dynamic algorithms for shortest paths appear in [AuEtal91], [EvGa85], [FrMaNa94], and [Ro85].

**4.** King [Ki99] developed a fully dynamic algorithm for maintaining all pairs shortest paths in directed graphs where the weights of the edges are positive integers less than a bound $b$; this algorithm runs in $O(n^{2.5}\sqrt{b \log n})$ amortized time.

**5.** Henzinger and King [HeKi99] obtained fully dynamic, randomized algorithms for connectivity, 2-edge connectivity, bipartiteness, cycle equivalence, and constant-weight minimum spanning trees that have polylogarithmic expected time per operation.

# REFERENCES

*Printed Resources*:

[AhHoUl74] A. V. Aho, J. E. Hopcroft, and J. D. Ullman, *The Design and Analysis of Computer Algorithms*, Addison-Wesley, 1974.

[AhHoUl83] A. V. Aho, J. E. Hopcroft, and J. D. Ullman, *Data Structures and Algorithms*, Addison-Wesley, 1983.

[AlHe95] D. Alberts and M. Rauch Henzinger, "Average case analysis of dynamic graph algorithms", *Proceedings of the 6th Symposium on Discrete Algorithms*, ACM/SIAM, 1995, 312–321.

[AuEtal91] G. Ausiello, G. F. Italiano, A. Marchetti-Spaccamela, and U. Nanni, "Incremental algorithms for minimal length paths", *Journal of Algorithms* 12 (1991), 615–638.

[ChKaTh93] J. Cheriyan, M-Y. Kao, and R. Thurimella, "Scan-first search and sparse certificates: an improved parallel algorithm for $k$-vertex connectivity", *SIAM Journal on Computing* 22 (1993), 157–174.

[CoEtal09] T. H. Cormen, C. E. Leiserson, R. L. Rivest, and C. Stein, *Introduction to Algorithms*, 3rd ed., MIT Press, 2009. (A comprehensive study of algorithms and their analysis.)

[EpGaIt93] D. Eppstein, Z. Galil, and G. Italiano, *Improved Sparsification*, Technical Report 93-20, Department of Information and Computer Science, University of California, Irvine, 1993.

[EpEtal92a] D. Eppstein, Z. Galil, G. Italiano, and A. Nissenzweig, "Sparsification – A technique for speeding up dynamic graph algorithms", *Proceedings of the 33rd Symposium on Foundations of Computer Science*, IEEE Computer Society, 1992, 60–69.

[EpEtal93] D. Eppstein, Z. Galil, G. Italiano, and T. Spencer, "Separator based sparsification for dynamic planar graph algorithms", *Proceedings of the 25th Symposium on Theory of Computing*, ACM, 1993, 208–217.

[EpEtal92b] D. Eppstein, G. Italiano, R. Tamassia, R. Tarjan, J. Westbrook, and M. Yung, "Maintenance of a minimum spanning forest in a dynamic plane graph", *Journal of Algorithms* 13 (1992), 33–54.

[EvGa85] S. Even and H. Gazit, "Updating distances in dynamic graphs", *Methods of Operations Research* 49 (1985), 371–387.

[Fr85] G. N. Frederickson, "Data structures for on-line updating of minimum spanning trees, with applications", *SIAM Journal on Computing* 14 (1985), 781–798.

[Fr91] G. N. Frederickson, "Ambivalent data structures for dynamic 2-edge-connectivity and $k$ smallest spanning trees", *Proceedings of the 32nd Symposium on Foundations of Computer Science*, IEEE Computer Society, 1991, 632–641.

[FrHe98] M. Fredman and M. R. Henzinger, "Lower bounds for fully dynamic connectivity problems in graphs", *Algorithmica* 22 (1998), 351–362.

[FrMaNa94] D. Frigioni, A. Marchetti-Spaccamela, and U. Nanni, "Incremental algorithms for the single-source shortest path problem", in *Foundations of Software Technology and Theoretical Computer Science*, Lecture Notes in Computer Science 880, Springer, 1994, 113–124.

[GaIt91] Z. Galil and G. Italiano, "Reducing edge connectivity to vertex connectivity", *SIGACT News* 22 (1991), 57–61.

[GoBa91] G. H. Gonnett and R. Baeze-Yates, *Handbook of Algorithms and Data Structures in Pascal and C*, 2nd ed., Addison Wesley, 1991. (An extensive collection of algorithms with analysis and implementation; includes mathematical formulas used in analysis.)

[GuSt85] L. J. Guibas and J. Stolfi, "Primitives for the manipulation of general subdivisions and the computation of Voronoi diagrams", *ACM Transactions on Graphics* 4 (1985), 74–123.

[HeKi99] M. R. Henzinger and V. King, "Randomized fully dynamic algorithms with polylogarithmic time per operation", *Journal of the ACM* 46 (1999), 502–516.

[HeRaSu94] J. Hershberger, M. Rauch, and S. Suri, "Fully dynamic 2-edge-connectivity in planar graphs", *Theoretical Computer Science* 130 (1994), 139–161.

[ItLaRa93] G. Italiano, H. La Poutré, and M. Rauch, "Fully dynamic planarity testing in embedded graphs", *Algorithms – ESA '93*, Lecture Notes in Computer Science 726, Springer, 1993, 212–223.

[KlEtal94] P. Klein, S. Rao, M. Rauch, and S. Subramanian, "Faster shortest-path algorithms for planar graphs", *Proceedings of the 26th Symposium on Theory of Computing*, ACM, 1994, 27–37.

[Ki99] V. King, "Fully dynamic algorithms for maintaining all-pairs shortest paths and transitive closure in digraphs," *Proceedings of the 40th Symposium on Foundations of Computer Science*, IEEE Computer Society, 1999, 81–89.

[Kn97] D. E. Knuth, *The Art of Computer Programming. Vol. 1: Fundamental Algorithms*, 3rd ed., Addison-Wesley, 1997. (A seminal reference on the subject of algorithms.)

[Kn98] D. E. Knuth, *The Art of Computer Programming. Vol. 3: Sorting and Searching*, 2nd ed., Addison-Wesley, 1998. (A seminal reference on the subject of algorithms.)

[NaIb92] H. Nagamochi and T. Ibaraki, "Linear time algorithms for finding a sparse $k$-connected spanning subgraph of a $k$-connected graph", *Algorithmica* 7 (1992), 583–596.

[Ra93] M. Rauch, "Fully dynamic graph algorithms and their data structures", PhD thesis, Computer Science Department, Princeton University, 1993.

[Ra94] M. Rauch, "Improved data structures for fully dynamic biconnectivity", *Proceedings of the 26th Symposium on Theory of Computing*, ACM, 1994, 686–695.

[Ro85] H. Rohnert, "A dynamization of the all-pairs least-cost path problem", *Proceedings of the 2nd Symposium on Theoretical Aspects of Computer Science*, Lecture Notes in Computer Science 182, Springer, 1985, 279–286.

[SlTa83] D. Sleator and R. Tarjan, "A data structure for dynamic trees", *Journal of Computer and System Sciences* 26 (1983), 362–391.

[Ta75] R. Tarjan, "Efficiency of a good but not linear set union algorithm", *Journal of the ACM* 22 (1975), 215–225.

[Th89] R. Thurimella, "Techniques for the design of parallel graph algorithms", PhD thesis, Computer Science Department, University of Texas, Austin, 1989.

[WeTa92] J. Westbrook and R. Tarjan, "Maintaining bridge-connected and biconnected components on-line", *Algorithmica* 7 (1992), 433–464.

**Web Resources:**

http://www.cs.stonybrook.edu/~algorith/major_section/1.1.shtml  (The Stony Brook Algorithm Repository: Data Structures, developed by Steven Skiena.)

https://visualgo.net  (Animations of many data structures and algorithms.)

https://www.cs.auckland.ac.nz/~jmor159/PLDS210/ds_ToC.html  (*Data Structures and Algorithms* by John Morris.)

https://www.cs.auckland.ac.nz/~jmor159/PLDS210/niemann/s_man.htm  (*Sorting and Searching Algorithms: A Cookbook* by Thomas Niemann.)

https://www.cs.usfca.edu/~galles/visualization/Algorithms.html  (Interactive animations of data structures and algorithms.)

https://www.hackerearth.com/practice/notes/heaps-and-priority-queues/ (Has notes on heaps and priority queues.)

https://www.toptal.com/developers/sorting-algorithms  (Animations of various sorting algorithms.)

# 19

# *DATA MINING*

## INTRODUCTION

The amount of data generated daily has exploded over the past few decades. This data is found in digital files, databases, and on the Internet. Much of the data is publicly available, but much is privately held by individuals and a wide variety of different types of organizations. With the continuing collection of so much data, attention has grown in using this data to discover new information. The subject of data mining has rapidly evolved to meet this need; it is the process of uncovering relationships between elements in large datasets. Data mining techniques take the input data and output a representation that expresses information contained in the data or knowledge about the data. Data mining plays an important role in many diverse disciplines, including retail business, climate science, bioinformatics, telecommunications, and computer security.

Although data mining is usually considered to be part of computer science, it employs a wide range of mathematical concepts and techniques. This chapter can be used to learn about many of the important tasks in data mining, the key mathematical ideas used, and some of the important algorithms used in data mining.

Data mining has an excellent chance of being useful wherever a large amount of information has been collected. For example, data mining is used in banking, marketing, retail sales, manufacturing, customer relationship management, fraud detection, intrusion detection, healthcare, bioinformatics, criminology, meteorology, and most other disciplines where large information stores are amassed.

*Note.* Different and sometimes conflicting terminology is used in the data mining literature. Consequently, when consulting data mining literature it is important to determine the terminology being used. Here, we present the most commonly used terms and mention some synonymous terminology.

## GLOSSARY

**agglomerative hierarchical clustering**: a technique that starts from single-entity clusters and successively merges clusters until all objects belong to the same cluster.

**anomaly**: a data point that is significantly different from other data points and that is considered abnormal in some way.

**anomaly detection**: a method for finding data points that are anomalies (or outliers).

**association rule**: a rule stating that if some features in a specified set have particular values in an instance, then another feature has a high probability to have a specified value.

**basket**: a collection of data items (synonymous with *itemset* or *transaction*).

**classification**: the process of constructing a classifier from data (also known as *supervised learning*) and using this classifier to predict the class of new data points.

**classifier**: a function or model that takes an element of a test set and predicts the class of that element.

**cluster**: a set of instances that are similar in some way to each other, but distinct from points outside the cluster in this same manner.

**clustering**: the process of partitioning a collection of data points into sets, where the points in each set are similar to one another in some way and separated from other data points in the same way. (In machine learning, it is also known as *unsupervised learning*.)

**count**: the number of transactions in which an item or set of items occurs.

**data mining**: the process of analyzing data using mathematical techniques and algorithms to learn new information.

**decision tree**: a representation as a tree, where each non-leaf node denotes a test on an attribute, each branch represents an outcome of the test, and each leaf node has a class label.

**dimension**: a function on data points. See also *feature*.

**distance measure**: a numerical indication of how far apart two instances (points) are.

**divisive hierarchical clustering**: a technique that starts from a cluster containing all objects and successively refines clusters until all objects belong to different clusters.

**entropy**: a measure of the amount of disorder in a set of examples.

**feature**: a function on data points. See also *dimension*.

**frequent itemset mining**: the process of finding those itemsets that occur in a predefined fraction of the transactions in a specified set of transactions.

**homogeneity** (of a cluster): a measure of similarity of the objects in a cluster.

**instance**: a data point.

**item**: a data element.

**itemset**: a collection of data items.

**knowledge discovery in databases**: the complete process of obtaining useful knowledge from collections of data, of which data mining is a part.

**linear model for classification**: a model in which a line or hyperplane is drawn to divide the dataset into positive and negative (in the binary case) examples. New points are classified depending on which side of the line/hyperplane they fall.

**metric space**: a set of points together with a real-valued distance function $d(x, y)$ called a *metric* that maps pairs of points $x, y$ to the real numbers, where distances cannot be negative, the distance between two points can only be 0 if the points are the same, the distance is symmetric, and the distance satisfies the triangle inequality.

**naive Bayes classifier**: a classifier, predicated on the assumption that features are independent of one another, which predicts that an item belongs to the class producing the largest probability that this item belongs to it.

**nearest neighbor classifier**: a classifier in which training instances are stored and a test example is put into the class of its closest point among the training instances.

**outlier**: a data point significantly different from the other data points and considered to be abnormal in some way.

**outlier detection** (or **anomaly detection**): a method for finding data points that are outliers (or anomalies).

**separation of a cluster**: a measure of how different objects in a cluster are from those in different clusters.

**similarity measure**: a numerical indication of the closeness of two instances (points).

***spectral clustering***: a clustering technique that uses the eigenvalues and eigenvectors of certain graph-based matrices.

***support*** (for a set of items): the fraction of the transactions that contain this set of items as a subset.

***test set***: a set of points with class labels, where each point is described by a vector of feature values.

***training set***: a set of points with class labels, used by a classification method to learn the classifier.

***transaction***: a collection of data items (synonymous with *itemset* and *basket*).

# 19.1   DATA MINING FUNDAMENTALS

Diverse techniques can be used in data mining to derive useful information from large sets of data. Nevertheless, a large percentage of data mining involves solving a few important types of problems. These problems are: finding relationships between variables in databases (solved by constructing association rules, covered in §19.2), assigning categories to data items (solved by classifying items, covered in §19.3), partitioning datasets into meaningful categories (solved by creating clusters of items, covered in §19.4), and finding data points that do not conform to an expected pattern (solved using outlier or anomaly detection, covered in §19.5). Comprehensive coverage of data mining, discussing the five areas listed here as well as many other problems, can be found in [Ag15], [HaKaPe12], [LeRaUl14], and [MaRo10].

## 19.1.1   BASIC CONCEPTS

### Definitions:

An ***association rule*** is a rule stating that if some features have particular values in an instance, then another feature has a high probability to have a specified value.

***Big data*** is the name used to describe datasets that are so large or so complex that conventional software tools for working with them are inadequate.

A ***classifier*** is a function that can be used to distinguish a particular class of data from other data.

***Classification*** is the process of constructing a classifier from the data. (In machine learning, classification is also known as ***supervised learning***.)

***Clustering*** is the process of partitioning the data points into groups of data points, where the points in each group are similar to one another in some way and separated from other data points in the same way. (In machine learning, it is also known as ***unsupervised learning***.)

A ***cluster*** is a group of instances that are similar in some way to each other, but distinct from points outside the cluster in the same way.

***Data mining*** is the discipline that employs mathematical techniques and algorithms to build models and analyze large collections of data in order to obtain useful information.

A *dimension* or *feature* is a function on data points.

An *instance* is a data point.

The process of obtaining useful knowledge from collections of data is called **Knowledge Discovery in Databases** (**KDD**).

An *outlier* (or *anomaly*) is a data point that so significantly differs from the other data points that it is considered to be abnormal in some way.

*Outlier detection* or *anomaly detection* is a method for determining that a data point is an outlier. It can be seen as complementary to clustering.

A *relational database* consists of a collection of tables, each consisting of a set of attributes (or fields) and containing a set of records, where each record is an $n$-tuple for some positive integer $n$.

**Facts:**

**1.** Data mining techniques are used within the larger process known as Knowledge Discovery in Databases (KDD) (see [FaPiSm96] and [MaRo10]), which involves other aspects of computing.

**2.** The goal of KDD is to obtain useful knowledge from stored data. The steps of KDD (as described in [FaPiSm96]) are

- Modeling the application domain, the relevant prior knowledge, and the goals.
- Creating a target dataset by focusing on specific data samples and features.
- Carrying out data cleaning and preprocessing by removing noise and outliers, collecting the necessary information to account for noise, and deciding how to handle missing data fields.
- Performing data reduction and projection by finding useful features to represent the data (depending on the goals of the process) and using dimensionality reduction methods to reduce the number of features.
- Choosing the particular data mining method, such as classification, clustering, constructing association methods, and so on.
- Choosing the particular data mining algorithm(s) and deciding on particular parameters.
- Performing the data mining task and deriving the representations that are output by the chosen method.
- Interpreting the mined patterns.
- Assimilating the discovered knowledge.

**3.** Data mining generally refers to the modeling and analysis steps of KDD, and not the other steps. In particular, data mining focuses on building mathematical models, selecting the appropriate data mining tasks, applying the appropriate algorithms, and generating and interpreting the output of these algorithms.

**4.** Special methods are often needed to handle *big data*, that is, large datasets that cannot be handled in the usual way on computers because of their massive size (ranging from terabytes to petabytes). Also, the number of dimensions or features may be so large that it is difficult to derive knowledge about the data. For discussions of these issues see [LeRaUl14].

**5.** Data mining can be used to analyze large collections of a wide range of data types. In this chapter the focus is on mining data that is stored in flat files (containing numbers or text) or in relational databases.

**6.** Techniques for mining data not stored in flat files or rational data bases, such as data from multimedia databases, spatial databases, time-series databases, and the Web, are not addressed in this chapter. (See [MaRo10] for information about data mining of these and other types of specialized data collections.)

**7.** Following [Ag15], we can abstractly view the data mining problem by considering the data being in the form of an $n \times d$ data matrix $\mathcal{D}$, where $n$ is the number of records and $d$ is the number of columns (or dimensions).

- *Relationships between columns.* Finding frequent relationships between columns is the association rule mining problem. If a particular column is identified as being more important (i.e., a class), then the problem of data classification is that of determining the relationships between the other columns and this column.

- *Relationships between rows.* Finding those subsets of rows in which the values of the columns are similar is clustering. Finding those rows in which the values of the columns are very different from the values of the columns in other rows is outlier analysis, and the row with the different values is an anomaly.

**8.** The Iris dataset, created by the statistician Sir Ronald Fisher in 1936, is commonly used to illustrate methods of data mining and machine learning. This dataset contains 150 instances, encompassing three classes of flowers: *Iris setosa*, *Iris versicolor*, and *Iris virginica*. The data includes four numerical attributes (sepal length, sepal width, petal length, and petal width), measured in centimeters.

**9.** Singular Value Decomposition is used in a wide variety of data mining tasks, including classification and clustering, two of the key tasks of data mining. It is also useful in developing recommender systems. (See §6.7.4 for this and other applications.)

**10.** There are many diverse data mining software packages available, including both public domain and commercial packages. Public domain packages include

- Weka, written in Java and developed at the University of Waikato in New Zealand (`http://www.cs.waikato.ac.nz/ml/weka/`)

- R, a free software platform for statistical computing that can be extended with many different packages (`https://www.r-project.org/`)

- Rapid Miner, an environment for machine learning and data mining processes (`https://rapidminer.com/`)

- ELKI, open-source data mining software in Java that specializes in classification and outlier detection (`https://elki-project.github.io/`)

Commercial software packages for data mining include

- SAS Enterprise Modeler (`http://support.sas.com/software/products/miner/`)

- IBM SPSS Modeler (`https://www.ibm.com/us-en/marketplace/spss-modeler`)

- MATLAB toolboxes with data mining capabilities (`https://www.mathworks.com/products.html`)

Consult the later sections of this chapter for more information about these and other more specialized software tools for data mining tasks.

**Examples:**

**1.** Financial accesses, such as using a credit or debit card, create data that can be mined to learn interesting information about spending and buying behavior.

**2.** Telephone and cell phone usage data records can be mined to obtain many types of useful information, such as finding groups of people who communicate with one another.

**3.** Many on-line retailers mine customer purchase records to recommend other products for customers to buy. For example, two customers can be considered similar if the purchase sets of the two customers have a large *similarity measure* (§19.1.2). Likewise, we can consider two products to be similar if the two sets of customers who have purchased these items have a large similarity measure.

**4.** The *Netflix Challenge* was a competition to develop an algorithm to predict a person's ratings for films based upon his/her ratings for other films. Other similar problems are product recommendations made by on-line retailers such as Amazon.com, as well as the recommendation of news articles to a reader based on articles they have read previously.

**5.** Classification may be used to develop a method of determining which loan applications are acceptable and which are too risky.

**6.** Association rules may be used by a retail establishment to determine which products to suggest to a customer who has just purchased a product. If the customer has just bought a PC and a digital camera, perhaps he/she might buy a memory card [HaKaPe12].

**7.** Outlier analysis can be used by a credit card company to determine if a transaction is likely to be due to fraud. For example, the purchase may be in a city or a type of store not usual for that particular customer.

**8.** Records of which items are commonly bought together can help a large retailer with product placement and advertising.

**9.** The database of adverse drug reactions has been mined to find new signals of adverse reactions to drugs.

**10.** When users access Web services, access logs are created at servers and various cluster profiles are maintained at commercial Web sites. These logs can be mined to understand patterns of user accesses, or unusual patterns indicating criminal behavior.

**11.** Records of meteorological data have been mined to learn predictors of conditions like temperature and rainfall.

---

### 19.1.2    Measures of Proximity

In data mining, models not only include collections of data, but also specific ways to determine how similar different data points are or how far apart they are. We describe here some of the most important of these measures of proximity. They are used in building models and in classification, clustering, and outlier detection.

**Definitions:**

A **distance measure** or (**dissimilarity measure**) is a numerical indication of how far apart two instances (points) are. High values indicate that the instances (points) are different, whereas low values indicate that they are similar.

A **similarity measure** is a numerical indication of the closeness of two instances (points). High values indicate that the two instances (points) are similar, whereas low values indicate that they are different.

A **metric space** consists of a set of points and a real-valued function $d(x, y)$, called a **metric**, mapping pairs of points to the real numbers and satisfying for all points $x, y, z$

- $d(x, y) \geq 0$ (distances cannot be negative);
- $d(x, y) = 0$ if and only if $x = y$ (distance can only be 0 if the two points are the same);
- $d(x, y) = d(y, x)$ (**symmetry**);
- $d(x, y) \leq d(x, z) + d(z, y)$ (**triangle inequality**).

**Euclidean $n$-space** is the set of $n$-tuples of real numbers.

The **Euclidean distance** between $x = (x_1, x_2, \ldots, x_n)$ and $y = (y_1, y_2, \ldots, y_n)$ is given by $d_2(x, y) = \left( \sum_{i=1}^{n} (x_i - y_i)^2 \right)^{1/2}$.

The **Manhattan distance** between $x = (x_1, x_2, \ldots, x_n)$ and $y = (y_1, y_2, \ldots, y_n)$ is given by $d_1(x, y) = \sum_{i=1}^{n} |x_i - y_i|$.

The **Chebyshev distance** (or **supremum distance**) between $x = (x_1, x_2, \ldots, x_n)$ and $y = (y_1, y_2, \ldots, y_n)$ is given by $d_\infty(x, y) = \max_{i=1}^{n} |x_i - y_i|$.

More generally, whenever $r$ is a real number with $r \geq 1$, the distance measure $d_r(x, y) = \left( \sum_{i=1}^{n} |x_i - y_i|^r \right)^{1/r}$ is a metric, known as the **Minkowski distance**.

The **Hamming distance** between two vectors is the number of components in which the vectors differ.

The **cosine similarity** of the $n$-tuples $x = (x_1, x_2, \ldots, x_n)$ and $y = (y_1, y_2, \ldots, y_n)$ is

$$\frac{x \cdot y}{\| x \| \| y \|}.$$

The **cosine distance** of the $n$-tuples $x = (x_1, x_2, \ldots, x_n)$ and $y = (y_1, y_2, \ldots, y_n)$ is the arccosine of their cosine similarity.

The **Jaccard similarity** of the sets $A$ and $B$ equals $|A \cap B| / |A \cup B|$.

The **Jaccard distance** between two sets $A$ and $B$ is defined to be $d(A, B) = 1 - |A \cap B| / |A \cup B|$.

The **edit distance** between the strings $x$ and $y$ is the smallest number of insertions and/or deletions of single characters that can be used to convert string $x$ into string $y$.

The **term vector** of a document containing $n$ different words $w_1, w_2, \ldots, w_n$ is the $n$-tuple $(f_1, f_2, \ldots, f_n)$, where $f_i$ is the number of occurrences of word $w_i$ in the document.

**Facts:**

**1.** In data mining, a wide variety of different distance measures and metrics are used. These measure the distance between real numbers, sets, strings, and other objects.

**2.** A metric is a distance measure, but not all distance measures are metrics.

**3.** Among the proximity measures commonly used in data mining are Euclidean distance, cosine similarity and distance, Manhattan distance, and supremum distance.

**4.** The Euclidean distance and Manhattan distance are special cases of the Minkowski metric with $r = 2$ and $r = 1$, respectively.

**5.** Minkowski distance with parameter $r$ is also called the $\ell_r$ distance (based on the $\ell_r$ norm). Euclidean distance, Chebyshev distance, and Manhattan distance correspond to $\ell_2$, $\ell_\infty$, and $\ell_1$, respectively.

**6.** From real analysis it follows that $\lim_{r\to\infty} d_r(x,y) = \max_{i=1}^{n} |x_i - y_i|$, which is the Chebyshev distance between $x$ and $y$. This implies that Chebyshev distance is a metric, but it is easier to prove this directly from the definition.

**7.** Chebyshev distance is also known as *chessboard distance*, because in chess the minimum number of moves needed by a king to go from one square on a chessboard to another is the Chebyshev distance between the centers of the squares.

**8.** The cosine similarity of two vectors $x$ and $y$ is the cosine of the angle $\theta$ between them. Here, the vectors have real numbers as entries. In data mining, vectors may have some or all entries that only take only a finite number of values. For instance, some entries may only take on the values 0 and 1.

**9.** The cosine distance is used to measure the angle between vectors.

**10.** The Jaccard similarity of two sets is the ratio of the size of their intersection to the size of their union.

**11.** Jaccard distance is a metric on every space of sets [LeRaUl14].

**12.** On-line retailers can use customer purchase records to recommend other products for customers to buy. For example, two customers can be considered similar if their purchase sets have large Jaccard similarity. Likewise, two products can be considered similar if the two sets of customers who have purchased these items have large Jaccard similarity.

**13.** The similarity of two documents can be measured by finding the cosine similarity of their term vectors.

**14.** The Jaccard similarity and Jaccard distance between two documents depends on the particular sets we associate with these documents. The most common choices are the set of all words a document contains or the set of all strings of length $k$ it contains (called *k-shingles* or *k-grams*), where $k$ is a positive integer selected large enough so that the probability that any given shingle appearing in the document is low.

**15.** The edit distance $d(x,y)$, where $x$ and $y$ are two strings, equals $l(x) + l(y) - l(LCS(x,y))$, where $l(s)$ denotes the length of the string $s$ and $LCS(x,y)$ is the longest common subsequence of $x$ and $y$.

**16.** Hamming distance is a metric on every space of vectors.

**17.** The Manhattan distance gets its name because it measures the distance required to walk between two addresses in the part of Manhattan (in New York City) where all streets run either north-south or east-west. It is also called *taxicab distance*.

**Examples:**

**1.** Let $S = \{2, 4, 5, 6, 8\}$ and $T = \{1, 3, 4, 5, 7\}$. Because $S \cap T$ is $\{4, 5\}$ and $S \cup T$ is $\{1, 2, 3, 4, 5, 6, 7, 8\}$, the Jaccard similarity of $S$ and $T$ is $|S \cap T|/|S \cup T| = 2/8 = 0.25$. The Jaccard distance is $1 - S \cap T|/|S \cup T| = 1 - 0.25 = 0.75$.

**2.** Jaccard distance can be used to measure the distance between documents. There are a variety of ways to associate sets with documents. For instance, documents can be viewed as sets of words. They can also be viewed as collections of $k$-shingles or $k$-grams, where $k$ specifies the length of the character sequence.

**3.** Here we find the similarity of and the distance between the vectors $x = (2, 4, -1)$ and $y = (4, 2, 1)$ using a variety of measures and metrics.

- The Hamming distance between $x$ and $y$ is 3, because $x$ and $y$ differ in all three components.

- The Manhattan distance between $x$ and $y$ is $|2-4|+|4-2|+|-1-1| = 2+2+2 = 6$ and the Chebyshev distance between them is 2.
- The cosine similarity of $x$ and $y$ is

$$\frac{8+8-1}{\sqrt{2^2+4^2+(-1)^2}\cdot\sqrt{4^2+2^2+1^2}} = \frac{17}{\sqrt{21}\cdot\sqrt{21}} = \frac{17}{21} \approx 0.8095.$$

- The cosine distance between $x$ and $y$ is $\arccos(\frac{17}{21}) \approx 35.95^o$.

4. The edit distance between the strings $x = dcba$ and $y = fcab$ is 4 because one can convert $x$ to $y$ by first deleting the $d$, then inserting an $f$ before the $c$, then deleting the $b$, and finally inserting a $b$ after the $a$. There is no way to change $x$ into $y$ using only three insertions and/or deletions.

## 19.2  FREQUENT ITEMSET MINING AND ASSOCIATION RULES

The mining of association rules is one of the classic areas of data mining. It is based on one of the more fundamental methods for characterizing data, the finding of frequent itemsets. There are many areas of application besides the original retail application.

### Definitions:

In this area of data mining, an element of the dataset is called an **item**.

A collection of items is known as an **itemset** (or **basket** or **transaction**). The items are objects of some sort from a universe $U$. The term **k-itemset** is used to denote an itemset that contains exactly $k$ items.

A **transaction** is an $n$-tuple of items for some positive integer $n$.

The data to be analyzed consists of a set $T$ of **transactions** (or **itemsets** or **market-baskets**).

The number of transactions in which an item $I$ or set of items occurs is called the **count** of $I$, defined as

$$count(I) = |\{t_i \mid I \subseteq t_i,\ t_i \in T\}|,$$

where $T = \{t_1, t_2, \ldots, t_n\}$ is the set of transactions. [Sometimes the word "support" is used for this quantity, conflicting with the more common usage of this word given in the next definition.]

The **support** for an item or set of items $I$ is the fraction of the transactions that contain $I$ as a subset. That is,

$$support(I) = \frac{count(I)}{|T|}.$$

Note that the support of $I$ is the probability of $I$ occurring in a transaction.

If an item or set of items $I$ occurs in more than $s$ percent of the transactions, where $s$ is called the **support threshold**, $I$ is said to be **frequent**. [For those authors who treat "support" as an absolute number, $s$ is an absolute number as well.]

Suppose $S$ is a set of items. An **association rule** is an implication of the form $I \to J$, where $I$ and $J$ are disjoint subsets of $S$.

Given a set of transactions $T = \{t_1, t_2, \ldots, t_n\}$, **frequent itemset mining** is the process of finding those itemsets that occur in a predefined fraction $m$ of the transactions in $T$.

The predefined fraction $m$ is called the **minimum support** (or **minimum support threshold**).

The **confidence** of an association rule $I \rightarrow J$ is the support for $I \cup J$ divided by the support for $I$. That is,

$$confidence(I \rightarrow J) = \frac{support(I \cup J)}{support(I)}.$$

The **support** of an association rule $I \rightarrow J$ is the support for $I \cup J$. That is,

$$support(I \rightarrow J) = support(I \cup J).$$

The **interest** [LeRaUl14] of an association rule $I \rightarrow J$ is the difference between the confidence of the rule and the support of $J$. That is,

$$interest(I \rightarrow J) = confidence(I \rightarrow J) - support(J).$$

The **lift** of an association rule $I \rightarrow J$ is the confidence of the rule divided by the support of $J$. That is,

$$lift(I \rightarrow J) = \frac{confidence(I \rightarrow J)}{support(J)}.$$

**Facts:**

**1.** The original application of mining the frequency of itemsets to find association rules analyzed grocery or department store data. This is the origin of the *market-basket* model of data, which is based on *baskets* of *items*. Typically the number of baskets is very large and they are stored in a file containing a sequence of baskets. The number of items in a basket is much smaller than the total number of possible items.

**2.** The folklore is that one of of the earliest discoveries made from market-basket data analysis is that people who buy diapers tend to be likely to buy beer. The discovered rule is then *diapers* $\rightarrow$ *beer*.

**3.** The support of every subset $J$ of an itemset $I$ is at least that of the support of itemset $I$. This property is called the *support monotonicity property*.

**4.** If a set $I$ of items is frequent, then so is every subset of $I$. This property is called *monotonicity* or the *downward closure property*.

**5.** Rules with high confidence are useful, but confidence does not require a connection between itemsets $I$ and $J$. The confidence of the rule expresses the conditional probability $Pr(J \mid I)$.

**6.** Interest expresses the type of connection between $I$ and $J$. If the percentage of the transactions with $I$ and also $J$ is the same as the percentage with $J$ alone, then there is no connection and the interest value will be 0. A positive number indicates a positive connection and a negative one indicates that the occurrence of $I$ makes it less likely that $J$ occurs.

**7.** If the lift of the association rule $I \rightarrow J$ is greater than 1, then $I$ and $J$ are positively correlated. If the lift is 1, then $I$ and $J$ are independent and the rule is not interesting. If the lift is below 1, then there is a negative correlation.

**8.** The best-known algorithm for counting frequent subsets is the *apriori algorithm* introduced by C. Agrawal and R. Srikant in 1994. Given a set of transactions and a minimum support threshold *minsup*, this algorithm finds the itemsets that are subsets containing at least *minsup* transactions in the database using a bottom-up approach. Frequent subsets are extended, an item at a time (*candidate generation*), and groups of candidates are tested to determine those transactions with successively larger numbers of elements. The algorithm terminates when no additional extensions exist. Details of this algorithm can be found in [Ag15].

**9.** If there is a large number of items, the space needed to count the itemsets can become prohibitive. For 2-itemsets in a universe of 1000 objects, the number of 2-itemsets is $\binom{1000}{2} = 499,500$. If we want to consider 3-itemsets, then an additional $\binom{1000}{3} = 166,167,000$ possibilities must be included. There are several techniques to reduce the amount of space needed to store the counts of itemsets [LeRaUl14].

**10.** Generally, the size of the file containing the transactions is very large and does not fit in main memory. An important issue is reducing the number of times it is necessary to read the file containing the transactions.

**11.** Another well-known method for obtaining frequent itemsets is called *frequent pattern growth* or *FP-growth*. The first scan of the database counts the frequent pairs as in the apriori algorithm. The second pass builds the *FP-tree*. Further mining is done on the FP-tree and no further scans over the transaction database are needed. Discussion of this method as well as others can be found in [Ag15], [HaKaPe12], and [TaStKu06].

**12.** A transaction can be a document or a web page. The items are the words that occur in the document. Assuming that we ignore frequently occurring words (known as *stop words*), then frequent itemsets would represent words that often occur together in documents.

**13.** Transactions can be sentences and the items can represent the documents in which the sentence occurs. Frequent itemsets are then documents that share common sentences.

**14.** Many measures have been proposed to evaluate whether rules are interesting or useful. Discussion of these various options can be found in [Ag15], [HaKaPe12], and [TaStKu06]. Generally an association rule is considered to be *interesting* if it satisfies both a minimum support threshold and a minimum confidence threshold, where these threshold values are set by the user.

**15.** Following are some free software tools for finding association rules and mining frequent itemsets:

- arules, an R extension package (`https://cran.r-project.org/web/packages/arules/index.html`)
- ARtool, a collection of algorithms and tools for the mining of association rules in binary databases (`http://www.cs.umb.edu/~laur/ARtool/index.html`)
- FIMI (Frequent Item Mining Implementations) site (`http://fimi.ua.ac.be/`)
- opusminer, an R package providing an interface to OPUS Miner for finding key associations in transaction data (`https://cran.r-project.org/web/packages/opusminer/index.html`)

**Examples:**

**1.** Suppose the association rule iPhone 8 → iPad Pro for transactions made at an Apple Store has support 3% and confidence 15%. Then in 3% of all transactions at the store an iPhone 8 and an iPad Pro were purchased (perhaps with other items), and 15% of the customers who bought an iPhone 8 at the Apple Store also bought an iPad Pro at this store.

**2.** Consider the following supermarket transactions from one hour at the Healthy Eating Emporium:

| Transaction ID | Items |
|---|---|
| 1 | {Kefir, Tortillas} |
| 2 | {Cone Filters, Kefir, Kombucha, Tortillas, Omega-3} |
| 3 | {Almonds, Apples, Cone Filters, Kefir, Soy Butter} |
| 4 | {Tortillas, Coconut Milk, Cone Filters, Kefir, Kombucha, Tortillas} |
| 5 | {Cone Filters, Tortillas, Kefir, Kombucha, Omega-3} |

The count for {Tortillas} is 4, and therefore the support is 4/5. The count for {Kefir, Kombucha} is 3, and therefore the support is 3/5. The support for {Kombucha, Kefir} → {Omega-3} is 2/5 and the confidence is $\frac{2/5}{3/5} = 2/3$. Since the support for Omega-3 is 2/5, the rule is somewhat interesting as the interest is $2/3 - 2/5 = 4/15$.

**3.** Frequent itemset analysis can be used for applications that do not involve market-basket analysis. Some examples [LeRaUl14] are the following:

- The items consist of biomarkers (e.g., genes or blood proteins) and diseases. Each basket represents data about a patient, e.g., genome, blood-chemistry analysis, and medical history with regard to diseases. A frequent itemset that contains a disease and one or more biomarkers provides a possible test for the disease.

- The items are documents and the baskets are sentences. An item is in a basket if the sentence represented by the basket occurs in that document. Pairs of items that occur in several baskets provide an indication that those documents share several sentences and can be evidence of plagiarism.

- Documents are represented by baskets and the items are the words that occur in the document. If the most common words (stop words) are ignored, then frequent pairs may indicate a joint concept or related concepts.

# 19.3  CLASSIFICATION METHODS

Classification is the problem of identifying the class or category of a data item given a dataset of examples that are already identified as belonging to particular classes. A new example can be given to the classifier and the classifier will predict the category of the new example.

**Definitions:**
Items that are to be classified are $m$-tuples, where the different dimensions are called **features**. Different features can belong to different sets, such as the set of real numbers, the set of positive integers, or some finite set. These items are also called **instances** or **data points**.

The goal of classification is to create a model that will specify (or predict) the **class label** that identifies the class each new data item belongs to.

The **training set** given to a classification method is a set of items used by the classification method to **learn** (or create) the classifier. Each point or example in the training set is an $m$-tuple $(x_1, x_2, \ldots, x_m)$ of feature values, where each $x_i$ is a value or measurement for the corresponding feature $F_i$, except for one value $x_i$, which is the class of that item.

The **test set** is a set of items, each described by a vector of feature values, given to a classifier that predicts the class of each of these items.

The **classifier** is a model that takes an item $\mathbf{X}_j$ of the test set and predicts the class of that point.

The **instance space** is the $m$-dimensional space containing the items to be classified.

A **nearest neighbor classifier** is a classifier that assigns the class of the nearest item to this item using a specified metric.

A **k-nearest neighbor classifier** finds the classes of the $k$ nearest neighbors to an item (using a specified metric) and assigns to the item the class a majority of these items belong to. Here the parameter $k$ is chosen so that the classifier selects the most appropriate classes to the items.

A **perceptron** is a threshold function applied to the $n$-tuple $(x_1, x_2, \ldots, x_n)$, where each $x_i$ is a real number. A perceptron with weights $w_1, w_2, \ldots, w_n$ and threshold value $\Theta$ outputs 1 if $\sum_{i=1}^{n} w_i x_i \geq \Theta$ and gives $-1$ otherwise.

A dataset is **linearly separable** given a perceptron if there is a hyperplane that separates all the points for which a perceptron has value 1 from those where it has value $-1$.

A **linear model** for classification draws a line or hyperplane dividing the dataset into positive and negative examples (in the binary case). New points are classified depending on which side of the line/hyperplane they fall.

An **decision tree** represents a concept using a tree structure, where each non-leaf node denotes a test on an attribute, each branch represents an outcome of the test, and each leaf node has a class label.

A **naive Bayes classifier** provides the probability that an item is a member of a particular class, assuming that the features occur independently of one another.

Suppose that $D$ is a partition of the classified training items, where there are $v$ different classes $C_1, C_2, \ldots, C_v$. Let $C_{i,D}$ be the set of data items of $C_i$ that belong to $D$. Then the **entropy** or **expected information** required to classify an item in $D$ is

$$Entropy(D) = -\sum_{i=1}^{v} |C_{i,D}|/|D| \log_2\{|C_{i,D}|/|D|\}.$$

Suppose that $D$ is a partition of the classified training items and $D$ is partitioned using an attribute $A$ with $m$ distinct values $a_1, a_2, \ldots, a_m$ into subsets $D_1, D_2, \ldots D_m$, where $D_i$ contains the items in $D$ for which the value of $A$ is $a_i$. Then the **entropy** or **expected information** of $D$ with partition $A$ is

$$Entropy_A(D) = \sum_{i=1}^{m} |D_i| Entropy(D_i)/|D|.$$

The **information gain** obtained by a partition using an attribute $A$ of a set $D$ of classified training items is $Gain(A) = Entropy(D) - Entropy_A(D)$.

## Facts:

**1.** Classification is also called *supervised learning* (in the parlance of machine learning) because the training set includes the classes of the items in the training set and a new item is classified using a classifier built from this training set. This contrasts with clustering (discussed in §19.4), where the classes of the items in the training set are not known in advance.

**2.** Many different classification methods are used, including the nearest neighbor classifier, the $k$-nearest neighbor classifier, perceptrons, linear classifiers, support vector machines, decision tree learning classifiers, naive Bayesian classifiers, and neural network classifiers. However, many other classification methods have been invented and new ones are being developed with some regularity.

**3.** A perceptron can be used as a classifier only if the data is linearly separable.

**4.** Perceptrons can be used for binary classification, that is, when each feature takes on one of two values (such as TRUE and FALSE, 1 and $-1$, and so on). They can be used even when there is an extremely large number of features, such as for the classification of documents where there is a feature for each word in a dictionary that takes the value 1 if the word appears in the document and the value $-1$ if it does not.

**5.** In nearest neighbor and $k$-nearest neighbor classifiers, when all the values of items in all dimensions are numerical, Euclidean distance is usually used. However, for certain other applications alternative metrics might be preferable. When all values are discrete, the Hamming distance can be used.

**6.** The choice of $k$ for a $k$-nearest neighbor classifier depends on the particular type of data. However, the general idea is to make $k$ small enough so that only relatively nearby items are among the closest $k$ items to new examples, but large enough to minimize the effects of data items that may have resulted from measurement error or noise. Some experts suggest that $k$ should equal the square root of the number of items in the training set.

**7.** The nearest neighbor classifier produces a Voronoi tessellation (§13.5.3) of the instance space. The cell of a new example determines its class.

**8.** When a nearest neighbor or a $k$-nearest neighbor classifier is used, it is important to scale or normalize the units in each dimension. Otherwise, dimensions where the values fall in wider ranges will dominate and dimensions where values are closer together will have little effect on the results of the classifier.

**9.** Consider the case of deciding whether or not a point $x = (x_1, x_2, \ldots, x_n)$ is a member of some category $C$. In a linear classifier, prediction is based on a weighted sum, where each feature has a weight $w_i$, giving

$$f(x) = \sum_{i=1}^{n} w_i x_i.$$

If $f(x)$ is positive then the point $x$ is a positive example of $C$, otherwise its class is not $C$. Typically the vectors $x$ are normalized so that $\| x \| = 1$. The vector $w$ can be seen as defining a hyperplane separating the positive and negative examples of $C$.

**10.** A *support vector machine (SVM)* uses a nonlinear mapping (if necessary) to transform the training set into a higher dimension where the data is linearly separable. Here we can search for an optimal separating hyperplane, which operates as a decision boundary that separates the points of the two classes. The optimality of the separating hyperplane is in the sense that it is a *maximal margin hyperplane*, one that separates the points with the largest *margin* so that none of the data points lie within this margin of the hyperplane. The margin of a hyperplane is the distance from this hyperplane to the data points closest to it. We describe part of the process used by SVM in Fact 11. Full details can be found in [Ag15] and [HaKaPe12].

**11.** The goal of the SVM process is to find a hyperplane $w \cdot x + b = 0$ that not only separates the data points, but also has the maximum margin $\gamma$. The following figure illustrates the difference that choice of the decision line or hyperplane can make on the margin. The core of the SVM optimization process ensures that this margin $\gamma$ is as large as possible.

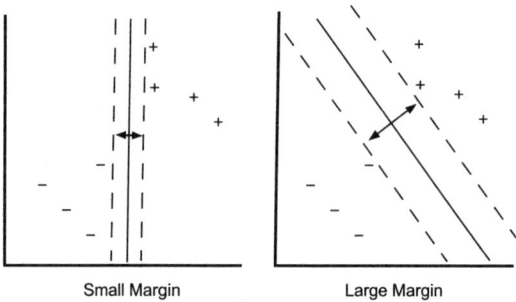

Small Margin          Large Margin

**12.** A decision tree can be built in various ways using a greedy algorithm that uses a top-down recursive divide-and-conquer approach. The training set is recursively partitioned into successively smaller subsets during the construction of the tree. At the end of the process, the tree that has been constructed contains both *decision nodes* and *leaf nodes*. The decision nodes represent steps where sets are partitioned into smaller subsets, while leaf nodes represent classifications of items.

**13.** The ID3 algorithm, developed by J. R. Quinlan, uses the attribute with the highest information gain to drive a greedy algorithm used to build a decision tree. At a splitting step of the algorithm, the attribute with the highest information gain is selected as the splitting node.

**14.** To illustrate the process of creating a decision tree, we consider a simple scenario. The recursive decision tree learning algorithm (Algorithm 1) begins at the root and successively works downwards.

---

**Algorithm 1:  Decision Tree Learning.**

**Decision-Tree-Learning**(*Examples, Attributes*)

   **if** *Examples* is empty, **return** the leaf node labeled with the most frequent
      categorization in the example set

   create a new *node* for the tree

   **if** all examples are positive, return the node as a leaf node labeled with **yes**

   **if** all examples are negative, return the node as a leaf node labeled with **no**

   **if** *Attributes* is empty, return the leaf node labeled with the most frequent
      categorization in the example set

   **else** select the attribute $A$ from *Attributes* that best classifies the examples
      and make $A$ the label for *node*

   **for** each value $v_i$ of $A$

      add a new branch below *node* with the branch label of $v_i$

> set $Examples_{v_i}$ to the portions of *Examples* that have $v_i$ for attribute $A$
> add to this branch the subtree
>    **Decision-Tree-Learning**($Examples_{v_i}$, *Attributes*−$A$)

**15.** Algorithm 1 at each call creates a node (the outermost call creating the root). Its arguments are a set of examples and the attributes by which the examples are categorized (the outermost call being given the complete set of examples and attributes). As the recursion proceeds, the sets *Examples* and *Attributes* become smaller.

At each call, a determination is made with regard to how to label the node created in that call. If all of the examples are of the same category, then the node is a leaf and is labeled with either Yes or No. If the set of examples is empty or if the attributes are empty, the algorithm with this training set has no basis upon which it can categorize instances corresponding to this path in the tree. The node therefore must be a leaf and by default it is given the most frequent label.

Otherwise, at each step the algorithm needs to determine which is the best attribute to use as the label for a node. Once the attribute for a node is chosen, there will then be as many branches below that node as there are values for the chosen attribute. The idea is to prefer smaller trees. Therefore at each step the best attribute is the one that classifies the most examples. This can be given a mathematically precise definition by using the notions of entropy and information gain.

**16.** Locality sensitive hashing can be used to find nearest neighbors in near constant time [LeRaUl14].

**17.** Classification can also be done using *neural networks*, which are loosely based on a biological model. In a neural network there are layers of nodes and weights between the nodes. A perceptron is the simplest type of neural network. More expressive learning methods use networks with hidden layers of nodes and weights between the input and output layers. Backpropagation is used to train the network [LeRaUl14], [HaKaPe12].

**18.** Many software packages, including those mentioned in §19.1.1, support a wide variety of classification algorithms. Some packages that can be used without charge are listed next:

- ELKI is open-source data mining software written in Java with an emphasis on unsupervised methods in cluster analysis (https://elki-project.github.io/)

- SGI MLC++ provides open-source code written in C++ at Stanford University for machine learning (classification), including implementations of the ID3 algorithm for decision trees and algorithms for nearest neighbor algorithms (http://www.sgi.com/tech/mlc/)

- The Scikit-learn open-source toolkit contains many classification methods including those discussed here. They are all implemented in Python. Scikit-learn is included with both the Anaconda Continuum and the Enthought Canopy distributions of Python packages (http://scikit-learn.org/stable/)

**Examples:**

**1.** Classification can be used to determine whether an e-mail is spam, using two classes for e-mails: spam and nonspam. Features that can be used are whether there are punctuation errors in e-mail headers, whether certain suspect keywords are present, whether nonsense strings are found, and so on.

**2.** Classification can help determine whether a patient has a particular disease using as features the symptoms and medical data of this patient.

**3.** Categorizing documents by primary topic using keyword counts as features can be done via classification.

**4.** People can be categorized into the type of vacation they may be interested in based on personal data including age, income, age of children, whether they have visited certain places, and so on.

**5.** Optical character recognition can classify each written character into the letter it represents by looking at various features.

**6.** Classifying videos can be done to determine whether a new example depicts a particular activity.

**7.** We can use two dimensions of data from the Iris database, sepalwidth and petalwidth, to classify new instances using the nearest neighbor classifier. The training set we use includes three instances, with the following known values for (sepalwidth, petalwidth, class).

> Sample 1:    (3.4, 0.4, *Iris setosa*)
> Sample 2:    (2.7, 1.3, *Iris versicolor*)
> Sample 3:    (3.0, 2.1, *Iris virginica*)

We classify a new instance (3.8, 2.0) by computing the Euclidean distance between this new instance and each of the three classified instances (using just the first two dimensions), obtaining 1.65 for Sample 1, 1.30 for Sample 2, and 0.81 for Sample 3. Because the third instance is the closest, the new sample is classified as *Iris virginica*.

**8.** The days when a particular person will play an outdoor sport (such as golf) based on the weather can be classified using a decision tree (as done in [Qu93],[Mi97], and [WiFrHa11]). A possible decision tree for this problem is displayed next. To explore the decision tree, one starts at the root and follows the values for each attribute of the day to be classified until arriving at a leaf. The classification is then the label of this leaf. For example, if the Outlook for today is *Sunny* and the Humidity is *High*, then golf will not be played. But if the Outlook for today is *Cloudy* then golf will be played.

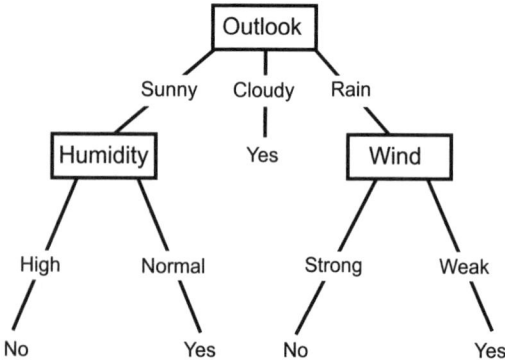

**9.** To illustrate the method of constructing decision trees, consider the training examples shown in the following table. Because there are nine positive examples and five negative examples, it follows that $Entropy(S) = -(9/14) \cdot \log_2(9/14) - (5/14) \cdot \log_2(5/14) = 0.940$. Now, consider the options for choosing the first attribute for the decision tree of whether golf will be played based on the given data. The information gain values for the attributes are computed as follows:

$$Gain(S, \text{Outlook}) = 0.246 \quad Gain(S, \text{Humidity}) = 0.151$$
$$Gain(S, \text{Wind}) = 0.048 \quad\quad Gain(S, \text{Temperature}) = 0.246$$

| Day | Outlook | Temperature | Humidity | Wind | Play |
|-----|---------|-------------|----------|------|------|
| Ex1 | Sunny | Hot | High | Weak | No |
| Ex2 | Sunny | Hot | High | Strong | No |
| Ex3 | Cloudy | Hot | High | Weak | Yes |
| Ex4 | Rain | Mild | High | Weak | Yes |
| Ex5 | Rain | Cool | Normal | Weak | Yes |
| Ex6 | Rain | Cool | Normal | Strong | No |
| Ex7 | Cloudy | Cool | Normal | Strong | Yes |
| Ex8 | Sunny | Mild | High | Weak | No |
| Ex9 | Sunny | Cool | Normal | Weak | Yes |
| Ex10 | Rain | Mild | Normal | Weak | Yes |
| Ex11 | Sunny | Mild | Normal | Strong | Yes |
| Ex12 | Cloudy | Mild | High | Strong | Yes |
| Ex13 | Cloudy | Hot | Normal | Weak | Yes |
| Ex14 | Rain | Mild | High | Strong | No |

We see that Outlook gives us the highest information gain. Splitting on Outlook gives us three branches. The instances where the value equals *Cloudy* are all decided as positive. Of those with the value *Sunny*, 3 out of 5 are negative; of those with the value *Rain*, 2 out of 5 are negative. If we were to choose Wind, then there would be two branches. Those instances under the *Weak* branch give 6 out of 8 positive while those under the *Strong* branch give 3 out of 6 positive. The value of information gain does roughly correspond with our intuitions of how well the choice of attribute serves to categorize the instances.

**10.** Consider the problem of using the naive Bayes classifier to classify a new instance of the weather data above [Mi97], [WiFrHa11]. For this instance **X**, Outlook = *Sunny*, Temperature = *Cool*, Humidity = *High*, and Wind = *Strong*. We can estimate the needed probabilities as follows:

$$Pr(\text{Play} = Yes) = 9/14 = 0.64$$
$$Pr(\text{Play} = No) = 5/14 = 0.36$$
$$Pr(\text{Wind} = Strong \,|\, \text{Play} = Yes) = 3/9 = 0.33$$
$$Pr(\text{Wind} = Strong \,|\, \text{Play} = No) = 3/5 = 0.60$$
$$Pr(\text{Outlook} = Sunny \,|\, \text{Play} = Yes) = 2/9 = 0.22$$
$$Pr(\text{Outlook} = Sunny \,|\, \text{Play} = No) = 3/5 = 0.60$$
$$Pr(\text{Play} = Yes \,|\, \mathbf{X}) = 0.64 \times 0.33 \times 0.22 \times 0.33 \times 0.33 = 0.0051$$
$$Pr(\text{Play} = No \,|\, \mathbf{X}) = 0.36 \times 0.60 \times 0.60 \times 0.20 \times 0.60 = 0.016$$

So the naive Bayes classifier predicts that the specified day should be placed in the *No* category.

## 19.4  CLUSTERING

Clustering is a powerful tool for automated analysis of data. It addresses the following general problem: given a set of entities, find subsets (or clusters) that are homogeneous and/or well separated. Homogeneity means that entities in the same cluster should be similar, while separation means that entities in different clusters should differ from one another. Clustering is ubiquitous, with applications in the natural sciences, psychology, medicine, engineering, economics, marketing, and other fields.

### 19.4.1  BASIC CONCEPTS

**Definitions:**

A **sample** is a set $O = \{o_1, o_2, \ldots, o_n\}$ of $n$ entities/objects among which clusters are to be found.

An $n \times s$ **data matrix** $X$ is obtained by measuring or observing $s$ features of the entities of $O$.

An $n \times n$ **matrix of dissimilarities** $D = (d_{ij})$ between entities of $O$ is computed from the matrix $X$. Typically, these quantities satisfy $d_{ij} = d_{ji} \geq 0$ for $i, j = 1, 2, \ldots, n$ and $d_{ii} = 0$ for $i = 1, 2, \ldots, n$. These values do not need to satisfy the triangle inequalities, i.e., to be distances.

A **criterion** expresses **homogeneity** and/or **separation** of the clusters in the clustering to be found.

Homogeneity of a cluster $C_\ell$ is commonly measured by

- the **diameter** $d(C_\ell)$ of $C_\ell$, or maximum dissimilarity between entities of $C_\ell$:

$$d(C_\ell) = \max_{i,j: o_i, o_j \in C_\ell} d_{ij};$$

- the **radius** $r(C_\ell)$ of $C_\ell$, or minimum among all entities $o_i$ of $C_\ell$ of the maximum dissimilarity between $o_i$ and another entity of $C_\ell$:

$$r(C_\ell) = \min_{i: o_i \in C_\ell} \max_{j: o_j \in C_\ell} d_{ij};$$

- the **star** $st(C_\ell)$ of $C_\ell$, or minimum among all entities $o_i$ of $C_\ell$ of the sum of dissimilarities between $o_i$ and the other entities of $C_\ell$:

$$st(C_\ell) = \min_{i: o_i \in C_\ell} \sum_{j: o_j \in C_\ell} d_{ij};$$

- the **clique** $cl(C_\ell)$ of $C_\ell$, or sum of dissimilarities between entities of $C_\ell$:

$$cl(C_\ell) = \sum_{i<j: o_i, o_j \in C_\ell} d_{ij}.$$

Separation of the cluster $C_\ell$ can be expressed by

- the **split** $s(C_\ell)$ of $C_\ell$, or minimum dissimilarity between an entity of $C_\ell$ and one outside $C_\ell$:

$$s(C_\ell) = \min_{i:o_i \in C_\ell, j:o_j \notin C_\ell} d_{ij};$$

- the **cut** $c(C_\ell)$ of $C_\ell$, or sum of dissimilarities between entities of $C_\ell$ and entities outside $C_\ell$:

$$c(C_\ell) = \sum_{i:o_i \in C_\ell} \sum_{j:o_j \notin C_\ell} d_{ij}.$$

Also used are the **normalized** star, clique, or cut obtained by dividing $st(C_\ell)$ by $|C_\ell| - 1$, $cl(C_\ell)$ by $|C_\ell|(|C_\ell| - 1)$, and $c(C_\ell)$ by $|C_\ell|(n - |C_\ell|)$, respectively.

**Facts:**

**1.** Criteria for assessing the overall effectiveness of a given clustering can be based on the six measures defined earlier, which can be combined by summing or by calculating the threshold (maximum or minimum) of the component cluster measures. The overall objective is to *maximize* the combined separation measure or to *minimize* the combined homogeneity measure.

**2.** The clustering problem is old. It can be traced back to Aristotle and was already much studied by 18th century naturalists such as Buffon, Cuvier, and Linné.

**3.** The set of entities $O$ may be associated with a mixture of distributions, the number and parameters of which are to be found [Mi96, Chapter 2].

**4.** Dissimilarities can be computed from sources other than the data matrix $X$, for instance when comparing biological sequences or partitions.

**5.** Specific distance and dissimilarity for entities represented by binary variables can be derived in an ad hoc manner [An73], [GrRa69].

**6.** Nominal (or categorical) variables with more than two states as well as numerical data can be transformed into binary variables [BoEtal97].

**7.** An axiomatic study of dissimilarities can be found in [Ba89].

**8.** Instead of computing dissimilarities, clustering can be performed directly on the matrix $X$ (e.g., Algorithm 3 of §19.4.4).

**9.** The computational complexity of a clustering problem depends on the criterion used. For instance, split maximization is polynomially solvable [DeHa80] while diameter minimization is NP-hard [Br78].

**10.** Cluster analysis is not the only way to study dissimilarities or distances between entities in the field of data analysis. Another common technique is *principal component analysis* [Jo02].

**11.** A popular online source for benchmark datasets commonly used in cluster analysis is the UCI Machine Learning Repository located at

- `https://protect-us.mimecast.com/s/oXA1BRUZEdqZIQ?domain=archive.ics.` `uci.edu`

**Examples:**

**1.** Consider four entities $o_1, o_2, o_3, o_4$ comprising a cluster $C_\ell$, for which the associated matrix of dissimilarities is given by

$$D = \begin{bmatrix} 0 & 3 & 4 & 5 \\ 3 & 0 & 6 & 8 \\ 4 & 6 & 0 & 1 \\ 5 & 8 & 1 & 0 \end{bmatrix}.$$

Then $d(C_\ell) = 8$ since $d_{24} = 8$, $r(C_\ell) = 5$ since $d_{14} = 5$, $st(C_\ell) = 11$, and $cl(C_\ell) = 27$.

**2.** The next figure illustrates three clusters, consisting of 2, 3, and 4 entities, respectively. The top portion of the figure uses the *diameter* of the clusters to obtain an overall measure: the left-hand side shows the maximum (threshold) cluster diameter, while the right-hand side shows that the sum of the three cluster diameters is considered. The bottom portion of the figure uses the *split* of the clusters to obtain an overall measure: the left-hand side shows the minimum (threshold) cluster split, while the right-hand side shows that the sum of the three splits is considered.

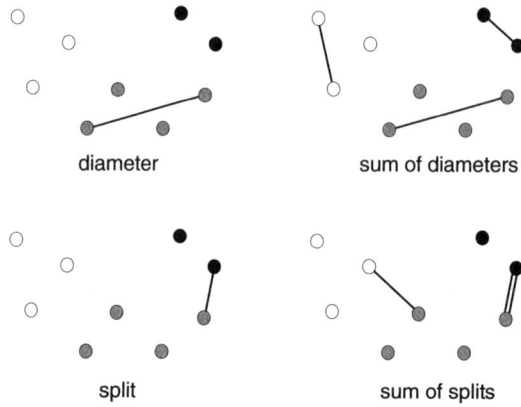

**3.** The Iris dataset [Fi36] is one of the most popular datasets used in cluster analysis. It consists of $n = 150$ samples obtained from each of three species of Iris flowers (*Iris setosa*, *Iris virginica*, and *Iris versicolor*). Each sample has $s = 4$ measured characteristics (in mm): sepal length, sepal width, petal length, and petal width. Clustering is used in this context to determine species of plants or animals from a given set of observations.

**4.** To find out which of 15 well-known politicians are perceived as similar, a representative sample of people is given all 105 possible pairs of these politicians, and asked to rank the pairs in terms of their similarity. By suitable aggregation of the rankings, a symmetric distance matrix can be obtained whose values can be used to find groups of similar politicians, with as much difference as possible between the groups. This could, for example, be relevant in forming election campaign teams [Sp85].

**5.** Recommendation algorithms are best known for their use on e-commerce web sites, where they exploit input about a customer's interests to generate a list of recommended items. Many applications use only the items that customers purchase and explicitly rate to represent their interests, but they can also use other attributes, including items viewed, demographic data, subject interests, and favorite artists. Clustering techniques work by identifying groups of consumers who appear to have similar preferences. Once the clusters are created, predictions for an individual can be made by averaging the opinions of the other consumers in that cluster. Thus, the store radically changes based on customer interests, showing programming titles to a software engineer and baby toys to a new parent [LiSmYo03], [ScKoRi01].

## 19.4.2   TYPES OF CLUSTERING

**Definitions:**

Cluster analysis methods are based on generating certain collections of sets:

(i) **Subset** $C$ of $O$;

(ii) **Partition** $P_k = \{C_1, C_2, \ldots, C_k\}$ of $O$ into $k$ clusters such that

(ii a)  $C_j \neq \emptyset$, for $j = 1, 2, \ldots, k$;

(ii b)  $C_i \cap C_j = \emptyset$, for $i, j = 1, 2, \ldots, k$ and $i \neq j$;

(ii c)  $\bigcup\limits_{j=1}^{k} C_j = O$;

(iii) **Packing** $Pa_k = \{C_1, C_2, \ldots, C_k\}$ of $O$ with $k$ clusters: same as (ii) but without (ii c);

(iv) **Covering** $Co_k = \{C_1, C_2, \ldots, C_k\}$ of $O$ with $k$ clusters: same as (ii) but without (ii b);

(v) **Hierarchy** $H = \{P_1, P_2, \ldots, P_q\}$ of $q \leq n$ partitions of $O$. The set of partitions $P_1, P_2, \ldots, P_q$ of $O$ is such that $C_i \in P_c, C_j \in P_d$ and $c > d$ imply $C_i \subset C_j$ or $C_i \cap C_j = \emptyset$ for all $i, j$, with $i \neq j$, and $c, d = 1, 2, \ldots, q$.

**Facts:**

**1.** By far the most frequently used types of clustering are the partition and the *complete hierarchy of partitions*, i.e., the one containing $n$ partitions.

**2.** Not all clustering criteria are independent. For instance, partitioning with the minimum sum of cliques criterion is equivalent to partitioning with the maximum sum of cuts criterion, since

$$\sum_{\ell=1}^{k} cl(C_\ell) + \frac{\sum_{\ell=1}^{k} c(C_\ell)}{2} = \Delta,$$

with $\Delta$ constant for all partitions $P_k$.

**3.** Relaxations of hierarchies are also studied. They include *hierarchies of packings* [MaBe83], *weak hierarchies* [BaDr89], and *pyramids* [BeDi85].

**4.** In *constrained clustering*, additional requirements are imposed on the clusters. The most frequent ones are bounds on their cardinality, bounds on their weight, assuming entities with weights (or connectedness), and assuming an adjacency matrix between entities is given.

**5.** In *fuzzy clustering*, each entity has a degree of membership in one or several clusters. These values represent the likelihood that the entity is assigned to that cluster.

**Examples:**

**1.** For the graph shown in part (a) of the following figure, we wish to cluster the five entities into two clusters $C_1$ and $C_2$ to minimize the overall radius, i.e., $\max\{r(C_1), r(C_2)\}$. The associated matrix of dissimilarities is given by

$$D = \begin{bmatrix} 0 & 10 & 1 & \infty & \infty \\ 10 & 0 & 1 & \infty & \infty \\ 1 & 1 & 0 & 1 & 1 \\ \infty & \infty & 1 & 0 & 10 \\ \infty & \infty & 1 & 10 & 0 \end{bmatrix}.$$

In an optimal *partition*, one of the clusters must have radius equal to 10 due to the presence of entities $o_1$ and $o_2$, or entities $o_4$ and $o_5$, in the same cluster. In a *covering*, entity $o_3$ is allowed to belong to two clusters at the same time, and so it is possible to construct two clusters with radius equal to 1, as shown in part (b) of the figure.

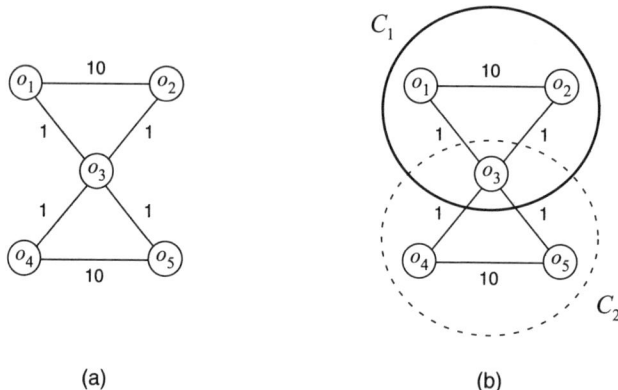

(a)                                      (b)

**2.** A phylogenetic tree (§20.2) is a classical example of a hierarchy of partitions. Fusions of clusters are represented by the moment when two species (or two sets of species) have separated from each other in the course of history. The next figure illustrates the evolution of dogs—the Family Canidae.

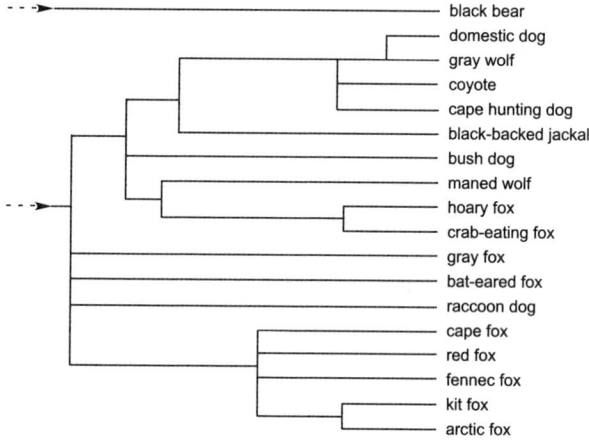

### 19.4.3   HIERARCHICAL CLUSTERING

This subsection discusses algorithms that can be used to carry out hierarchical clustering. Both agglomerative methods as well as divisive methods are discussed.

#### Agglomerative Algorithms:

Agglomerative hierarchical clustering algorithms are among the oldest and still most frequently used methods for cluster analysis. They start from an initial partition into $n$ single-entity clusters, and successively merge clusters according to some specified local criterion until all entities belong to the same cluster.

#### Definitions:

A **local criterion** is a criterion that uses only the information provided by the current partition.

Let $G_c = (V, E)$ denote a complete graph, having a vertex $v_i$ associated with each entitiy

$o_i$, for $i = 1, 2, \ldots, n$, and with edge $(v_i, v_j)$ weighted by the dissimilarity $d_{ij}$, for $i < j$. Let MST denote a ***minimum spanning tree*** of $G_c$; see §10.1.1.

The ***single-linkage*** algorithm merges at each step two clusters achieving the smallest inter-cluster dissimilarity.

The ***complete-linkage*** algorithm merges at each step two clusters for which the resulting cluster has smallest diameter.

**Facts:**

**1.** Algorithm 1 uses a local criterion to successively merge partitions until all entities belong to a single cluster.

---

**Algorithm 1**:   **Agglomerative hierarchical algorithm.**

input: set $O = \{o_1, o_2, \ldots, o_n\}$ of $n$ entities

output: a hierarchy $H = \{P_1, P_2, \ldots, P_n\}$ of partitions of $O$

$C_i := \{o_i\}, i = 1, 2, \ldots, n$

$P_n := \{C_1, C_2, \ldots, C_n\}$

$k := 1$

$\quad$ **while** $n - k > 1$ **do**

$\quad\quad$ select $C_i, C_j \in P_{n-k+1}$ following a local criterion

$\quad\quad$ $C_{n+k} := C_i \cup C_j$

$\quad\quad$ $P_{n-k} := (P_{n-k+1} \cup \{C_{n+k}\}) \setminus \{C_i, C_j\}$

$\quad\quad$ $k := k + 1$

---

**2.** The values of the split for all subsets of entities of $O$, and hence for all partitions of $O$, belong to the set of dissimilarity values associated with the set of edges of any MST of $G_c$ [Ro67].

**3.** The single-linkage algorithm provides maximum split partitions at all levels of the hierarchy [DeHa80]. This is a corollary of Fact 2.

**4.** For other criteria, the partitions obtained after several steps of an agglomerative algorithm are not necessary optimal. For instance, the complete-linkage algorithm does not guarantee that after two or more steps the resulting partition is optimal regarding the minimum diameter.

**5.** Computer code, in Fortran and Javascript, that implements the single-linkage and complete-linkage algorithms can be found at the sites

$\quad$ • https://protect-us.mimecast.com/s/ANrmBzUvd34vSd?domain=pitt.edu
$\quad$ • https://protect-us.mimecast.com/s/3RmXBoh8AQO8uM?domain=
$\quad\quad\quad\quad$ code.google.com

**6.** A parametric formula gives new dissimilarity values between cluster $C_k$ and $C_i, C_j$ when these latter two are merged [LaWi67]:

$$d_{k,i \cup j} = \alpha_i d_{ik} + \alpha_j d_{jk} + \delta |d_{ik} - d_{jk}|$$

The parameters for single-linkage and complete-linkage algorithms are given next.

| method | $\alpha_i$ | $\alpha_j$ | $\delta$ |
|---|---|---|---|
| single-linkage | $\frac{1}{2}$ | $\frac{1}{2}$ | $-\frac{1}{2}$ |
| complete-linkage | $\frac{1}{2}$ | $\frac{1}{2}$ | $\frac{1}{2}$ |

Clusters to be merged at each iteration are those corresponding to the smallest updated dissimilarity. Using a heap yields an $O(n^2 \log n)$ implementation of Algorithm 1.

**7.** Four other hierarchical clustering methods are based on an extension of the previous formula; further extensions are due to [Ja91].

**8.** Better algorithmic complexity can be derived in a few cases. For example, finding the MST of $G_c$, ranking its edges by nondecreasing values, and merging entities at endpoints of successive edges yields a $\theta(n^2)$ implementation of the single-linkage algorithm. At each iteration, clusters correspond to connected components of a graph with the same vertex set as $G_c$ and having as edges those of the MST considered.

**9.** The *reducibility property*

$$d(C_i, C_j) \le \min\{d(C_i, C_k), d(C_j, C_k)\}$$

implies

$$\min\{d(C_i, C_k), d(C_j, C_k)\} \le d((C_i \cup C_j), C_k) \quad \forall i, j, k.$$

This means that by merging two clusters $C_i$ and $C_j$, less dissimilar between themselves than with another cluster $C_k$, cannot make the resultant cluster $C_i \cup C_j$ less dissimilar to $C_k$ than $C_i$ and $C_j$ alone.

The dissimilarities $D = (d_{ij})$ induce a *nearest neighbor* relation, with one or more pairs of reciprocal near neighbors. When the reducibility property holds, each pair of reciprocal near neighbors will be merged before merging with other clusters.

**10.** Updating chains of nearest neighbors yields $\theta(n^2)$ agglomerative hierarchical clustering algorithms [Be82], [Mu83].

**11.** The results of hierarchical clustering can be represented graphically by a *dendrogram* [Co71] or an *espalier* [HaJaSi96]. Vertical lines correspond to entities or clusters, and horizontal lines joining endpoints of vertical lines correspond to mergings of clusters. The height of the horizontal lines corresponds to the value of the updated dissimilarity between the clusters merged, which can be a measure of separation or homogeneity of the clusters obtained. In an espalier the length of the horizontal lines is used to represent a second measure of homogeneity or separation of the clusters.

**Examples:**

**1.** The following figure illustrates a hierarchical clustering of eight entities. The left portion shows a dendrogram and the right portion shows an espalier.

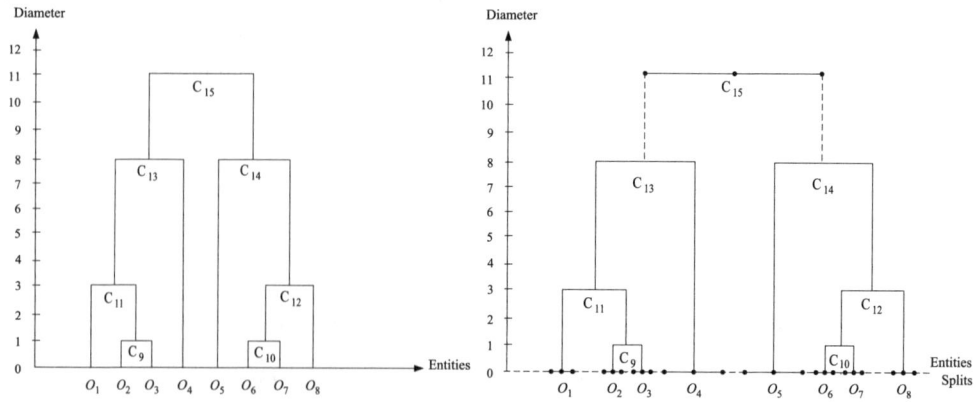

**2.** We apply the single-linkage algorithm to the following matrix of dissimilarities associated with five entities:

|  | $C_1 = \{o_1\}$ | $C_2 = \{o_2\}$ | $C_3 = \{o_3\}$ | $C_4 = \{o_4\}$ | $C_5 = \{o_5\}$ |
|---|---|---|---|---|---|
| $C_1 = \{o_1\}$ | 0 | 1* | 4 | 7 | 5 |
| $C_2 = \{o_2\}$ | 1* | 0 | 7 | 6 | 2 |
| $C_3 = \{o_3\}$ | 4 | 7 | 0 | 3 | 8 |
| $C_4 = \{o_4\}$ | 7 | 6 | 3 | 0 | 5 |
| $C_5 = \{o_5\}$ | 5 | 2 | 8 | 5 | 0 |

The smallest split is between clusters $C_1$ and $C_2$, which are therefore merged into a single cluster $C_6 = C_1 \cup C_2$. By using the updating formula presented in Fact 6, new dissimilarity values are obtained.

|  | $C_6$ | $C_3$ | $C_4$ | $C_5$ |
|---|---|---|---|---|
| $C_6$ | 0 | 4 | 6 | 2* |
| $C_3$ | 4 | 0 | 3 | 8 |
| $C_4$ | 6 | 3 | 0 | 5 |
| $C_5$ | 2* | 8 | 5 | 0 |

Since the current smallest split is found now between clusters $C_5$ and $C_6$, they are merged into cluster $C_7 = C_5 \cup C_6$, and then the matrix of dissimilarities is updated likewise.

|  | $C_7$ | $C_3$ | $C_4$ |
|---|---|---|---|
| $C_7$ | 0 | 4 | 5 |
| $C_3$ | 4 | 0 | 3* |
| $C_4$ | 5 | 3* | 0 |

Clusters $C_3$ and $C_4$ are next merged together to create cluster $C_8 = C_3 \cup C_4$. After that, we reach the last level of the hierarchy of partitions by merging clusters $C_7$ and $C_8$ to form a single cluster $C_9$.

|  | $C_7$ | $C_8$ |
|---|---|---|
| $C_7$ | 0 | 4* |
| $C_8$ | 4* | 0 |

The single-linkage algorithm obtains maximum split partitions at the five levels of the hierarchy. As next shown, this is not a coincidence.

**3.** Consider the complete graph $G_c = (V, E)$ with vertices $v_i$ associated with the entities $o_i$ of Example 2, and with edges $(v_i, v_j)$ weighted by the dissimilarities $d_{ij}$ presented in its first table. The MST of $G_c$ contains the following edges, listed by nondecreasing weight: $(v_1, v_2)$ (weight 1), $(v_2, v_5)$ (weight 2), $(v_3, v_4)$ (weight 3), and $(v_1, v_3)$ (weight 4). Now consider a graph $G'$ with the same vertex set as $G_c$ and where clusters correspond to its different connected components. This is shown in part (a) of the following figure.

The single-linkage algorithm, based on finding the MST of $G_c$ (Fact 8), considers the MST edges in nondecreasing order, and at each iteration connects components of $G'$ using these edges. In this example, it starts by considering edge $(v_1, v_2)$ and connecting these vertices in order to create the connected component $C_6 = \{v_1, v_2\}$; see part (b). The following steps of the clustering are shown in parts (c)–(e), Namely, the next edge $(v_2, v_5)$ connects $v_5$ to $C_6$, creating $C_7 = \{v_1, v_2, v_5\}$. A new connected component $C_8$

is formed by connecting vertices $v_3$ and $v_4$. Finally, the two connected components are merged into a single connected component of $G'$ by considering edge $(v_1, v_3)$. The sequence of clusters obtained corresponds to that found in the previous example (Fact 2).

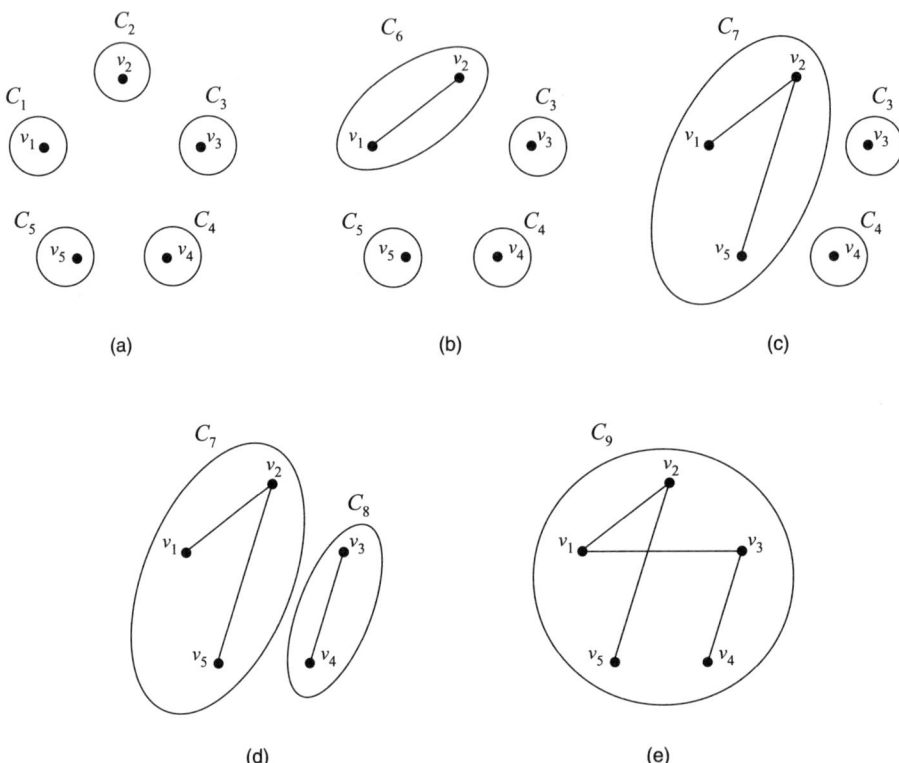

**4.** Consider the following set of five entities as well as its matrix of dissimilarities:

|       | $o_1$ | $o_2$ | $o_3$ | $o_4$ | $o_5$ |
|-------|-------|-------|-------|-------|-------|
| $o_1$ | 0     | 4     | 9     | 7     | 11    |
| $o_2$ | 4     | 0     | 1     | 6     | 3     |
| $o_3$ | 9     | 1     | 0     | 8     | 6     |
| $o_4$ | 7     | 6     | 8     | 0     | 4     |
| $o_5$ | 11    | 3     | 6     | 4     | 0     |

The complete-linkage algorithm applied to this set of entities produces the dendrogram presented next, which represents the successive agglomerations performed by the algorithm with the purpose of minimizing the diameter of each of the partitions obtained in the hierarchy.

The complete-linkage algorithm is not assured of finding minimum diameter partitions at all levels of the hierarchy (Fact 4). For instance, a better bipartition could have been obtained by clustering entities $o_1$ and $o_4$ in one cluster and entities $o_2$, $o_3$, and $o_5$ in another one, yielding a bipartition of diameter 7.

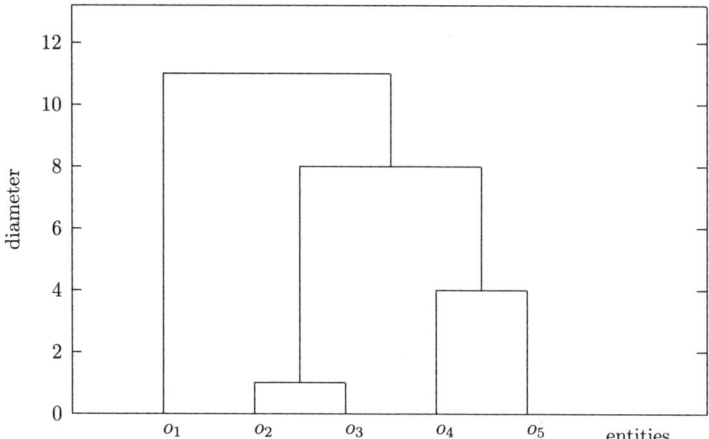

## Divisive Algorithms:

Divisive hierarchical clustering algorithms are less frequently used than agglomerative ones. They proceed from an initial cluster containing all entities and then perform successive bipartitions of one cluster at a time until all entities belong to different clusters.

## Definitions:

Let MST$'$ denote a **maximum spanning tree** of $G_c$; see §10.1.1.

A **coloring** of a graph is an assignment of colors to its vertices such that no two adjacent vertices have the same color; see §8.6.1. A **bicoloring** of a graph is a coloring that uses only two colors.

An **odd cycle** in a graph is one containing an odd number of vertices.

## Facts:

**1.** Algorithm 2 uses two local criteria to successively refine partitions until each entity belongs to a single cluster.

---

**Algorithm 2**:   **Divisive hierarchical algorithm.**
  input: set $O = \{o_1, o_2, \ldots, o_n\}$ of $n$ entities
  output: a hierarchy $H = \{P_1, P_2, \ldots, P_n\}$ of partitions of $O$
  $P_1 := \{C_1\} = \{\{o_1, o_2, \ldots, o_n\}\}$
  $k := 1$
    **while** $k < n$ **do**
      select $C_i \in P_k$ following a first local criterion
      partition $C_i$ into $C_{2k}$ and $C_{2k+1}$ following a second local criterion
      $P_{k+1} := (P_k \cup \{C_{2k}\} \cup \{C_{2k+1}\}) \setminus \{C_i\}$
      $k := k + 1$

---

**2.** The role of the first local criterion in Algorithm 2 is not crucial, as it only determines the order in which clusters will be bipartitioned. The second criterion is more difficult to apply since optimal bipartitioning of the chosen cluster according to a given criterion is a problem which may be NP-hard.

**3.** The unique bicoloring of MST′ defines a minimum diameter bipartition of $O$ [MoSu91]. The diameter of this bipartition is equal to the largest dissimilarity of an edge outside MST′ closing an odd cycle in MST′.

**4.** Fact 3 yields an $O(n^2 \log n)$ divisive hierarchical algorithm for the minimum diameter criterion by building simultaneously maximum spanning trees at all levels of the hierarchy [GuHaJa91].

**Example:**

**5.** Consider the matrix of dissimilarities between five entities from Example 4 in §19.4.3:

|        | $o_1$ | $o_2$ | $o_3$ | $o_4$ | $o_5$ |
|--------|-------|-------|-------|-------|-------|
| $o_1$  | 0     | 4     | 9     | 7     | 11    |
| $o_2$  | 4     | 0     | 1     | 6     | 3     |
| $o_3$  | 9     | 1     | 0     | 8     | 6     |
| $o_4$  | 7     | 6     | 8     | 0     | 4     |
| $o_5$  | 11    | 3     | 6     | 4     | 0     |

The MST′ of the associated complete graph $G_c$ consists of the edges $(v_1, v_5)$ (weight 11), $(v_1, v_3)$ (weight 9), $(v_3, v_4)$ (weight 8), and $(v_2, v_4)$ (weight 6). The largest dissimilarity of an edge outside MST′ closing an odd cycle in MST′ is provided by edge $(v_1, v_4)$ having weight 7; see part (a) of the following figure. This is the value of the diameter of the bipartition obtained from bicoloring the vertices in MST′ (Fact 3), i.e., the partition formed by clusters $C_2 = \{o_1, o_4\}$ and $C_3 = \{o_2, o_3, o_5\}$.

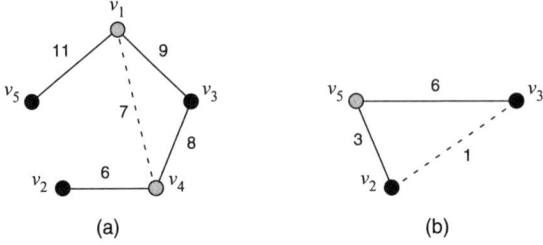

(a)                    (b)

For the next step, since the diameter of the partition results from cluster $C_2 = \{o_1, o_4\}$, this cluster is chosen to be divided by the first local criterion, yielding a new partition $C_4 = \{o_1\}$, $C_5 = \{o_4\}$, and $C_3 = \{o_2, o_3, o_5\}$ of diameter equal to 6 due to the dissimilarity between entities $o_3$ and $o_5$ in $C_3$. Thus, cluster $C_3$ is chosen to be divided by the first local criterion. A new MST′ for the partial complete graph composed of vertices $v_2, v_3, v_5$ has edges $(v_3, v_5)$ (weight 6) and $(v_2, v_5)$ (weight 3); see part (b) of the figure. Then a bipartition of $C_3$ is performed in order to build clusters $C_6 = \{o_2, o_3\}$ and $C_7 = \{o_5\}$, yielding a partition of diameter equal to 1 due to the dissimilarity between entities $o_2$ and $o_3$ in cluster $C_6$. Finally, the last bipartition over $C_6$ is trivial, and the last level of the hierarchy is attained.

---

## 19.4.4   PARTITIONING

**Definitions:**

Let $h(\cdot)$ be a homogeneity (or heterogeneity) function on the subsets of $O$.

An ***optimal substructure*** indicates that optimal solutions of subproblems can be used to find optimal solutions of the overall problem.

A problem has ***overlapping subproblems*** if the problem can be broken down into subproblems which are reused several times.

***Dynamic programming*** is a method of solving problems exhibiting the properties of overlapping subproblems and optimal substructure in a much faster way than complete enumeration of solutions.

***Recursion*** is a method of defining functions in which the function being defined is applied within its own definition.

***Branch-and-bound*** (§16.1.8) consists of an implicit enumeration of all solutions using a tree, where large subsets of solutions are pruned before being explored by using upper and lower bounds on the quantity being optimized.

***Valid inequalities*** are inequalities in a mathematical programming formulation of a problem that are necessarily satisfied by its variables.

### Facts:

**1.** Dynamic programming provides an exact algorithm for partitioning by exploiting the recursion

$$f_k^*(O') = \begin{cases} \underset{O_k \subset O'}{opt} \ (f_{k-1}^*(O' - O_k) + h(O_k)) \text{ for } k > 1 \\ h(O'), \text{ for } k = 1 \end{cases}$$

where $f_k^*(O')$ denotes the optimal value for partitioning the set $O' \subseteq O$ into $k$ clusters, and additivity is assumed. This formula is easily modified if the value of the objective function is equal to the maximum or the minimum value for each of the clusters. Applying this equation takes time exponential in $n$, so only small sets of entities may be considered. Constraints can sometimes accelerate the computations, e.g., if all clusters must be small.

**2.** The single-linkage algorithm provides optimal partitions for the split criterion at all levels of the hierarchy. So it is also a $\theta(n^2)$ algorithm for maximizing the split of a partition of $O$ into $k$ clusters.

**3.** Let $\mathcal{C} = \{C_1, C_2, \ldots, C_{2n-1}\}$ denote the set of clusters obtained when applying the single-linkage algorithm to $O$. Then for all $k$ there exists a partition $P_k^*$ that maximizes the sum of splits and consists solely of clusters of $\mathcal{C}$ [HaJaFr89].

**4.** Let $t$ be the smallest dissimilarity value such that the partial graph $G_t = (V, E_t)$ of $G$ with $E_t = \{(v_i, v_j) \mid d_{ij} \geq t\}$ is $k$-colorable. Then the color classes in any optimal $k$-coloring of $G_t$ define a minimum diameter partition of $O$ into $k$ clusters [Ch75], [HaDe78].

**5.** Partitions obtained with the single-linkage algorithm may suffer from the *chaining effect*: dissimilar entities at the ends of a long chain of pairwise similar entities are assigned to the same cluster. Partitions obtained by the coloring algorithm for minimum diameter may suffer from the *dissection effect*: similar entities may be assigned to different clusters.

**6.** Bicriterion cluster analysis algorithms seek compromise solutions in order to avoid chaining and dissection effects [DeHa80].

**7.** Branch-and-bound algorithms have been applied, with some success, to several partitioning problems in cluster analysis [Di85], [KlAr91], [KoNaFu75]. Their efficiency depends on sharpness of the bounds used, availability of a good heuristic solution, and efficient branching, i.e., rules that improve bounds for all subproblems obtained in a fairly balanced way.

**8.** Brusco and Stahl [BrSt05] proposed a *repetitive branch-and-bound algorithm* (RBBA) that solves subproblems in sequence by repeating a branch-and-bound procedure. In RBBA, the resolution of a given subproblem by branch-and-bound depends on the optimal solutions of smaller subproblems already solved by the algorithm, as a look-ahead component in the bound. For instance, using the minimum diameter criterion, RBBA can solve exactly instances with $n \leq 250$ entities and $k \leq 20$ clusters.

**9.** Good results can also be obtained when bounds result from the solution of a mathematical program combined with heuristic methods. For minimum sum-of-stars partitioning, primal-dual variable neighborhood search [HaEtal07] solved exactly for the first time instances with $n = 15,000$.

**10.** Several valid inequalities are given in [GrWa90] for the sum-of-cliques criterion. These results are used in a cutting plane algorithm [GrWa89] to solve instances with $n \leq 158$. It appears that triangle inequalities (i.e., valid inequalities which state that if the pairs of entities $o_i, o_j$ and $o_j, o_k$ belong to the same cluster then $o_i, o_k$ also belong to the same cluster) suffice in almost all cases.

**11.** Partitioning problems in cluster analysis can be formulated by considering all possible clusters in the following mathematical program:

$$
\begin{aligned}
\text{minimize:} \quad & \sum_{t \in T} h(C_t) y_t \\
\text{subject to:} \quad & \sum_{t \in T} a_{it} y_t = 1, \quad i = 1, \ldots, n \\
& \sum_{t \in T} y_t = k \\
& y_t \in \{0, 1\}, \quad t \in T
\end{aligned}
$$

where $T = \{1, \ldots, 2^n - 1\}$ and $a_{it}$ equals 1 if entity $o_i$ belongs to cluster $C_t$ and is 0 otherwise. Variable $y_t$ equals 1 if cluster $C_t$ is in the optimal partition and is 0 otherwise. This is a large linear partitioning problem with a size constraint, for which the number of variables is exponential in the number of entities $n$.

**12.** The standard way to tackle the mathematical program in Fact 11 is to solve the linear relaxation by means of *column generation* [DeDeSo05], and then apply branch-and-bound. The entering column is obtained by solving the following auxiliary problem, whose unknowns are the coefficients $a_i$ of the column:

$$
\begin{aligned}
\text{minimize:} \quad & h(C_t) - \sum_{i=1}^{n} a_i \lambda_i + \sigma \\
\text{subject to:} \quad & a_i \in \{0, 1\}, \quad i = 1, \ldots, n.
\end{aligned}
$$

Here $\lambda_i$ for $i = 1, \ldots, n$ and $\sigma$ are the dual variables at the current iteration. Difficulty varies depending on the form of $h(C_t)$ as a function of the $a_i$.

Once the entering column is found, the linear program solver proceeds. However, convergence may be slow, particularly if there are few clusters in the partition and hence massive degeneracy of the optimal solution. In fact, even when the optimal solution is found, many more iterations may be needed to prove its optimality. Once the linear relaxation of the problem is solved, one must check for integrality of the solution. If the solution is not integer, branch-and-bound is needed.

**Examples:**

**1.** Consider three entities with the associated matrix of dissimilarities given by

|       | $o_1$ | $o_2$ | $o_3$ |
|-------|-------|-------|-------|
| $o_1$ | 0     | 4     | 9     |
| $o_2$ | 4     | 0     | 1     |
| $o_3$ | 9     | 1     | 0     |

They are easily clustered by simple inspection regardless of the criterion chosen. The goal here is to show how dynamic programming works in solving a partitioning problem (Fact 1). Suppose we want to minimize the diameter of a bipartition over these entities.

The algorithm starts with $f_1^*(o_1) = 0$, $f_1^*(o_2) = 0$, $f_1^*(o_3) = 0$ since $h(o_i) = 0$ for $i = 1, 2, 3$. It is useless to calculate $f_2^*(o_i)$ for $i = 1, 2, 3$ since $k$ is larger than $n$. Then the algorithm proceeds by calculating

$$f_1^*(O_{12} = \{o_1, o_2\}) = h(O_{12}) = 4, \ f_1^*(O_{13} = \{o_1, o_3\}) = h(O_{13}) = 9,$$
$$f_1^*(O_{23} = \{o_2, o_3\}) = h(O_{23}) = 1, \ f_1^*(O_{123} = \{o_1, o_2, o_3\}) = h(O_{123}) = 9.$$

With all the $f_1^*(\cdot)$ now calculated we obtain

$$
\begin{aligned}
f_2^*(O_{123}) &= \min\{f_1^*(O_{123}) + h(\emptyset), f_1^*(O_{12}) + h(o_3), f_1^*(O_{13}) + h(o_2), f_1^*(O_{23}) + h(o_1), \\
&\qquad f_1^*(o_1) + h(O_{23}), f_1^*(o_2) + h(O_{13}), f_1^*(o_3) + h(O_{12})\} \\
&= f_1^*(O_{23}) + h(o_1) = 1,
\end{aligned}
$$

which results in $C_1 = \{o_1\}$ and $C_2 = \{o_2, o_3\}$.

**2.** In Example 2 of §19.4.3, the single-linkage algorithm for $n = 5$ entities obtained $(5 \times 4)/2 - 1 = 9$ different clusters: $C_1 = \{o_1\}, C_2 = \{o_2\}, C_3 = \{o_3\}, C_4 = \{o_4\}, C_5 = \{o_5\}, C_6 = \{o_1, o_2\}, C_7 = \{o_1, o_2, o_5\}, C_8 = \{o_3, o_4\}$, and $C_9 = \{o_1, o_2, o_3, o_4, o_5\}$. If a maximum sum-of-splits partition with four clusters is demanded, from $\binom{5}{2} = 10$ different possibilities, we can restrict ourselves to only two due to Fact 3:

$$P_4^1 = \{C_6, C_3, C_4, C_5\}, \ P_4^2 = \{C_1, C_2, C_8, C_5\},$$

which results in $P_4^* = P_4^1$ with sum of splits equal to 10. Here, the maximum split partition of size 4 obtained by the single-linkage algorithm coincides with the one that maximizes the sum of splits, though this is not always true.

**3.** The minimum diameter partition with $k = 3$ for the five entities with matrix of dissimilarities presented in Example 4 of §19.4.3 is determined by the smallest dissimilarity value $t$ such that the partial graph $G_t = (V, E_t)$ of $G$ with $E_t = \{(v_i, v_j) \mid d_{ij} \geq t\}$ is 3-colorable (Fact 4). For the given example, the smallest $t$ for which $G_t$ is 3-colorable is equal to 5, since the partial graph $G_t$ shown next is no longer 3-colorable if edges $(v_1, v_2)$ and $(v_4, v_5)$ of weight 4 are added to it. The diameter of the partition $P_3^* = \{\{o_1, o_2\}, \{o_3\}, \{o_4, o_5\}\}$ is equal to 4.

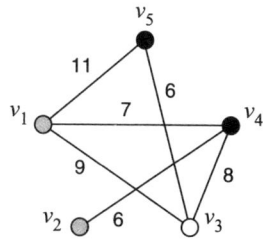

---

▷ **Partitioning in Euclidean Space**

Here we suppose that the entities $o_i$ are points $x_i$ of an $s$-dimensional Euclidean space.

**Definitions:**

The homogeneity of cluster $C_j$ is measured by reference to a **center** of $C_j$, which is not in general a point belonging to the cluster.

A popular criterion for partitioning points in Euclidean space is the **minimum sum-of-squares** criterion $ss(C_j)$ given by

$$ss(C_j) = \sum_{i:o_i \in C_j} (\|x_i - \overline{x}\|_2)^2,$$

where $\|\cdot\|_2$ denotes Euclidean distance and $\overline{x}$ is given by the **centroid** of the points $x_i$ in cluster $C_j$, i.e.,

$$\overline{x} = \frac{1}{|C_j|} \sum_{i:o_i \in C_j} x_i.$$

Other criteria can also be used:

- the **variance** $v(C_j)$ of $C_j$, defined as $ss(C_j)$ divided by $|C_j|$;
- the **continuous radius** $cr(C_j)$ of $C_j$, defined by

$$cr(C_j) = \min_{x \in \mathcal{R}^s} \max_{i:o_i \in C_j} \|x_i - x\|_2;$$

- the **continuous star** $cst(C_j)$ of $C_j$, defined by

$$cst(C_j) = \min_{x \in \mathcal{R}^s} \sum_{i:o_i \in C_j} \|x_i - x\|_2.$$

**Facts:**

1. The minimum sum-of-squares criterion addresses both homogeneity and separation [Sp80]. It is well known as the criterion tackled by the classical $k$-means clustering procedure [Ma67], described by Algorithm 3. From an initial partition, the $k$-means algorithm proceeds by reassigning the entities to their closest centroids and updating the cluster centroids until stability is reached. It produces not only the cluster centroids $\overline{x}_j$ but also the cluster membership function $m$.

---

**Algorithm 3:** $k$-means.

input: set $X = \{x_1, x_2, \ldots, x_n\}$ of $n$ points in $\mathcal{R}^s$, number of clusters $k$

output: cluster centroids $\overline{X} = \{\overline{x}_1, \ldots, \overline{x}_k\} \in \mathcal{R}^s$, $m : X \to \{1, \ldots, k\}$

choose initial values of $\overline{X}$

**for** $i := 1$ **to** $n$

  $m(x_i) := \text{argmin}_{j \in \{1,\ldots,k\}} (\|x_i - \overline{x}_j\|_2)^2$

**repeat**

  **for** $j := 1$ **to** $k$

    calculate the centroid $\overline{x}_j$

  **for** $i := 1$ **to** $n$

    $m(x_i) := \text{argmin}_{j \in \{1,\ldots,k\}} (\|x_i - \overline{x}_j\|_2)^2$

**until** $m$ does not change

---

**2.** The centroids used by criteria $ss(C_j)$ and $v(C_j)$ may be usefully considered as representative of clusters in some applications.

**3.** The sum of squares of cluster $C_j$ is equal to the sum of all squared distances between pairs of entities of this cluster divided by its cardinality:

$$ss(C_j) = \frac{1}{|C_j|} \sum_{i:o_i \in C_j} \sum_{\ell \neq i:o_\ell \in C_j} (\|x_i - x_\ell\|_2)^2.$$

**4.** The entities of clusters $C_1$ and $C_2$ in a bipartition may be separated by a hyperplane of $\mathcal{R}^s$. Moreover, one may choose this hyperplane from among those going through 1 to $s$ entities [Ha67]. This implies that the number of bipartitions is polynomially bounded for fixed dimensions. Using this fact, one can devise an exact polynomial $O(n^{s+1} \log n)$ algorithm for minimum sum-of-squares hierarchical divisive clustering of points in $s$-dimensional space [HaJaMl98].

**5.** Minimum sum-of-squares clustering is NP-hard for general dimensions even when $k = 2$ [AlEtal09]. For general values of $k$, the problem is NP-hard even in the plane [MaNiVa09]. If both $k$ and the dimension $s$ are fixed, the problem can be solved in $O(n^{sk+1})$ time [InKaIm94], which may be very time consuming even for instances in the plane. Exact methods with exponential computational complexity include dynamic programming [vOMe04], branch-and-bound [Br06], column generation [dMEtal00], concave minimization [XiPe05], and semidefinite programming [AlHa09], [PeXi05].

**6.** Implementations of the $k$-means algorithm can be found at the sites

- `https://protect-us.mimecast.com/s/dqYaBRipEOKpFJ?domain=bonsai.` `hgc.jp`
- `https://protect-us.mimecast.com/s/DzxkBmIOQ58OUg?domain=cs.umd.edu`
- `https://protect-us.mimecast.com/s/8JQwBdfQvNrQhK?domain=alglib.net`

**7.** Because reassignments are performed only if profitable and the number of partitions is finite, the $k$-means algorithm always converges to a local minimum. Better local minima are provided with the $j$-means heuristic [HaMl01a] combined with variable-depth neighborhood search [HaMl01b]; see §16.8.2. A detailed comparative study of heuristics for minimum sum-of-squares clustering with many references is provided in [BrSt07].

**8.** Problems in one-dimensional Euclidean space are best solved by dynamic programming. This method works well when optimal clusters possess the *string property*, i.e., consist of consecutive points on the line. Assume $o_1, o_2, \ldots, o_n$ are indexed in order of nondecreasing values of $x_1, x_2, \ldots, x_n$. Let $h(\cdot)$ be a homogeneity function on the subsets of $O$ and let $F_m^k$ be the optimal value of clustering $o_1, o_2, \ldots, o_m$ into $k$ clusters. The dynamic programming recursion may be written

$$F_m^k = \min_{i \in \{k, k+1, \ldots, m\}} \{F_{i-1}^{k-1} + h(C_{i,m})\}$$

where $C_{i,m} = \{o_i, o_{i+1}, \ldots, o_m\}$. Using updating to compute $h(\cdot)$ for all potential clusters yields $O(n^2)$ algorithms for various criteria [Ra71], [Sp80].

**9.** The objective in the classical *Fermat-Weber problem* is to locate a facility in $\mathcal{R}^s$ in order to minimize the sum of its weighted Euclidean distances to the locations of a given set of users. It is a central problem in continuous location theory [We93]; see §16.2.5. Applying the continuous star criterion is equivalent to the Fermat-Weber problem with equally weighted Euclidean distances.

**10.** For $A \in \mathcal{R}^{s \times s}$ and $b \in \mathcal{R}^s$,

$$\|x_1 - x_2\| = \|Ax_1 + b - Ax_2 - b\|$$

if and only if $A$ is orthogonal, i.e., criteria involving Euclidean distances are invariant with respect to orthogonal transformations and translations.

An interesting question is whether the optimal solution would change if scaling of measurements is performed over the data. In general, a scale transformation is defined as a nonsingular matrix $H \in \mathcal{R}^{s \times s}$ with $H = \text{diag}(h_1, \ldots, h_s)$ ($h_\ell \neq 0$ for $\ell = 1, \ldots, s$). From the previous equation, we observe that Euclidean distance criteria are invariant with respect to scale transformations if $h_\ell = \pm 1$ for $\ell = 1, \ldots, s$, i.e., the composition of optimal partitions is unaffected by scale transformations only if they amount to no more than a change of sign. This is very restrictive for practical applications.

### Examples:

**1.** Color image quantization is a data compression technique that reduces the total set of colors in a digital image to a representative subset. In the RGB color model, a true color image is represented by a matrix of pixel colors, each consisting of 24 bits: one byte for red, one byte for green, and one byte for blue components. However, it is often the case that only $k \ll 2^{24} \approx 16.8$ million possible colors are stored.

The problem can be modeled by considering an image $I$ as a function $I : \Omega \to \mathcal{R}^3$, $\Omega \subset \mathcal{Z} \times \mathcal{Z}$, with each dimension of $I(\Omega) = \{I(x) \mid x \in \Omega\}$ representing one of the red, green, and blue components of a pixel color, with values ranging between 0 and 255. Thus, color quantization of $I$ for the RGB color model consists of solving a partitioning problem in the 3-dimensional Euclidean space $I(\Omega)$ with the number of clusters equal to $k$. After clusters $C_1, C_2, \ldots, C_k$ are obtained, the color $I(x)$ of each pixel $x \in \Omega$ is replaced by the centroid of its cluster.

**2.** We provide an example of the $k$-means algorithm applied in $\mathcal{R}^2$ to partitioning nine entities into three clusters. The first figure illustrates the initialization of the $k$-means algorithm, defined by first determining the initial positions for the centroids of each cluster (e.g., by random selection).

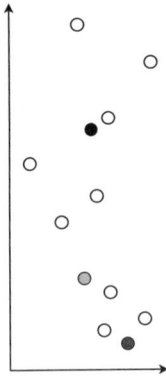

The next figure presents three iterations of the algorithm, performed until stabilization is reached. The left-hand side of each panel shows the assignment step of the algorithm where each point is assigned to its closest centroid, while the right-hand side indicates the updated positions of the cluster centroids.

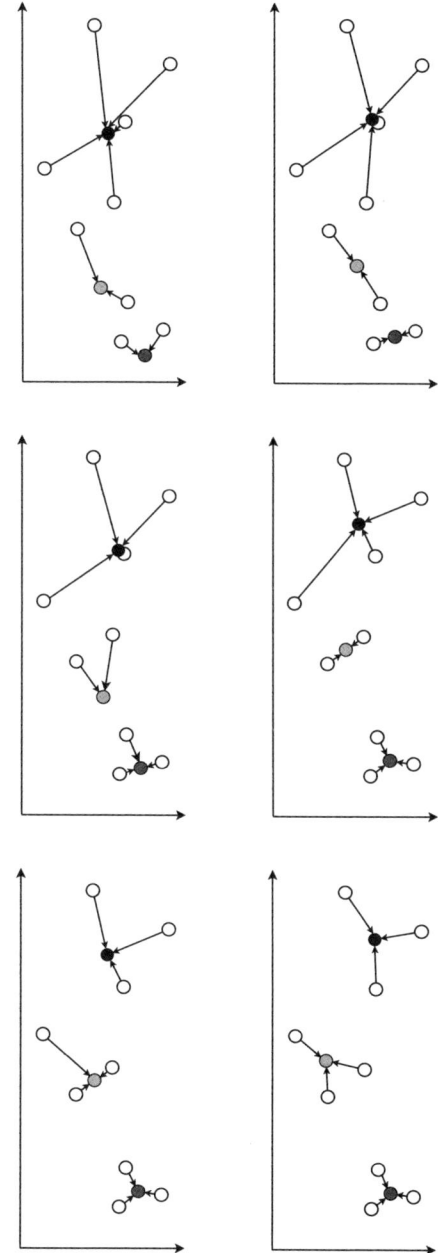

**3.** Consider two data matrices $X$ and $X'$, each corresponding to the coordinates of six points in the plane:

$$X = \begin{pmatrix} 0 & 48 \\ 6 & 48 \\ 0 & 24 \\ 6 & 24 \\ 0 & 0 \\ 6 & 0 \end{pmatrix}, \quad X' = \begin{pmatrix} 0 & 12 \\ 24 & 12 \\ 0 & 6 \\ 24 & 6 \\ 0 & 0 \\ 24 & 0 \end{pmatrix}.$$

It is easy to verify that $X' = XH$ where $H = \begin{pmatrix} 4 & 0 \\ 0 & 1/4 \end{pmatrix}$.

Not surprisingly, since $H$ is not orthogonal, the optimal partitions differ for the data matrices $X$ and $X'$ (Fact 10). For instance, if the minimum sum-of-squares criterion is used with $k = 2$, the optimal bipartition for points in $X$ is given by $\{\{x_1, x_2, x_3, x_4\}, \{x_5, x_6\}\}$, while for points in $X'$ it is given by $\{\{x'_1, x'_3, x'_5\}, \{x'_2, x'_4, x'_6\}\}$.

---

## 19.4.5  SPECTRAL CLUSTERING

There are several other clustering methods found in the literature (e.g., model-based clustering [FrRa02], additive clustering [Mi87], and subspace clustering [PaHaLi04]). Among these methods, spectral clustering algorithms have become very popular in recent years. Their main appeal relies on the fact that they are based on standard linear algebra and are simple to implement. Spectral clustering techniques use the eigenvalues and eigenvectors of certain graph-based matrices in order to cluster data.

### Definitions:

The **unnormalized graph Laplacian** is defined as $L = S - D$, where $D = (d_{ij})$ and $S = \mathrm{diag}(s_1, s_2, \ldots, s_n)$ with diagonal elements defined by $s_i = \sum_{j=1}^{n} d_{ij}$. It is customary to let $D$ represent a matrix of *similarities*, i.e., if $d_{ij}$ is very large (small), then objects $o_i$ and $o_j$ are very similar (different).

The **normalized graph Laplacian** is given by $L_{norm} = S^{-1}L = I - S^{-1}D$.

### Facts:

**1.** A thorough coverage of Laplacian matrices and spectral clustering is found in [Ch97].

**2.** A different normalized graph Laplacian matrix $L_b = S^{-1/2}LS^{-1/2}$ can be used by spectral clustering algorithms [NgJoWe02].

**3.** The term *normalization* means that the largest eigenvalue is less than or equal to 2, with equality only when the graph is bipartite.

**4.** Since $L_{norm}$ and $L_b$ are positive semidefinite all their eigenvalues range between 0 and 2.

**5.** According to [vL07], $L_b$ does not present any computational advantages over $L_{norm}$. In fact, spectral clustering with the eigenvectors of $L_b$ might be problematic if they contain particularly small entries.

**6.** The matrix $L$ satisfies the following properties [Mo91]:

- $f^T L f = \frac{1}{2} \sum_{i,j=1}^{n} d_{ij}(f_i - f_j)^2$ holds for every vector $f \in \mathcal{R}^n$;
- $L$ is symmetric and positive semidefinite;

- The smallest eigenvalue of $L$ is 0, and a corresponding eigenvector is the constant one vector $(1, 1, \ldots, 1)^T$;

- $L$ has $n$ nonnegative, real-valued eigenvalues $0 = \lambda_1 \leq \lambda_2 \leq \cdots \leq \lambda_n$.

**7.** The unnormalized spectral clustering algorithm (Algorithm 4) uses the unnormalized Laplacian matrix $L$ to change the representation of the data matrix $X$ to points $y_i \in \mathcal{R}^k$ for $i = 1, \ldots, n$. Then the $k$-means procedure is executed on these points in order to obtain the output clustering for the original data. Thus, if the $k$-means procedure classifies points $y_i$ and $y_j$ into the same cluster, then the output clustering likewise assigns entities $o_i$ and $o_j$ to the same cluster.

---

**Algorithm 4**:   **Unnormalized spectral clustering.**

input: similarity matrix $D \in \mathcal{R}^{n \times n}$, number of clusters $k$

output: clusters $C_1, \ldots, C_k$

construct a graph $G = (V, E)$ from $D$

compute the unnormalized Laplacian $L$

compute the first $k$ eigenvectors $\overline{f}_1, \ldots, \overline{f}_k$ of $L$ associated with its $k$ smallest
   eigenvalues

let $\overline{V} \in \mathcal{R}^{n \times k}$ be the matrix containing the vectors $\overline{f}_1, \ldots, \overline{f}_k$ as columns

**for** $i := 1$ **to** $n$

   let $y_i \in \mathcal{R}^k$ be the vector corresponding to the $i$th row of $\overline{V}$

cluster the points $(y_i)_{i=1,\ldots,n} \in \mathcal{R}^k$ into $k$ clusters using the $k$-means algorithm

---

**8.** The normalized Laplacian $L_{norm}$ satisfies the following properties [Ch97]:

- $\lambda$ is an eigenvalue of $L_{norm}$ with eigenvector $\overline{f}$ if and only if $\lambda$ and $\overline{f}$ solve the generalized eigenproblem $L\overline{f} = \lambda S \overline{f}$;

- 0 is an eigenvalue of $L_{norm}$, and a corresponding eigenvector is the constant one vector $(1, 1, \ldots, 1)^T$;

- $L_{norm}$ is positive semidefinite and has $n$ real-valued eigenvalues $0 = \lambda_1 \leq \lambda_2 \leq \cdots \leq \lambda_n$.

**9.** The normalized spectral clustering algorithm (Algorithm 5) is similar to the unnormalized version presented in Fact 7, except that it uses the eigenvectors of the normalized Laplacian matrix $L_{norm}$ instead of those of $L$.

---

**Algorithm 5**:   **Normalized spectral clustering.**

input: similarity matrix $D \in \mathcal{R}^{n \times n}$, number of clusters $k$

output: clusters $C_1, \ldots, C_k$

construct a graph $G = (V, E)$ from $D$

compute the unnormalized Laplacian $L$

compute the first $k$ eigenvectors $\overline{f}_1, \ldots, \overline{f}_k$, associated with the $k$ smallest
   eigenvalues of $L_{norm}$ that solve the generalized eigenproblem $L\overline{f} = \lambda S \overline{f}$

let $\overline{V} \in \mathcal{R}^{n \times k}$ be the matrix containing the vectors $\overline{f}_1, \ldots, \overline{f}_k$ as columns

**for** $i := 1$ **to** $n$

   let $y_i \in \mathcal{R}^k$ be the vector corresponding to the $i$th row of $\overline{V}$

cluster the points $(y_i)_{i=1,\ldots,n} \in \mathcal{R}^k$ into $k$ clusters using the $k$-means algorithm

---

**10.** The similarity graphs $G$ used by Algorithm 4 and Algorithm 5 may not be complete. In fact, there are several ways of constructing a graph from a similarity matrix $D$ such that similarity relationships are well represented, e.g., fully connected graphs, $\varepsilon$-neighborhood graphs, or $k$-nearest neighbor graphs [vL07].

**11.** Let $G = (V, E)$ be an undirected graph with nonnegative weights on edges and with $|V| = n$. Then the multiplicity of 0 as an eigenvalue of $L$ and $L_{norm}$ is equal to the number of connected components of $G$ [Mo97]. Moreover, the eigenspace of 0 is spanned by the indicator vectors associated with those components, where the indicator vector $\sigma_A = (t_1, \ldots, t_n)^T \in \mathcal{R}^n$ of a connected component with vertices $A \subseteq V$ is defined such that $t_i = 1$ if $v_i \in A$ and $t_i = 0$ otherwise.

**12.** Unnormalized spectral clustering provides a relaxation to minimizing the *ratio cut* criterion [HaKa92], which is expressed as

$$Rcut(P_k = \{C_1, C_2, \ldots, C_k\}) = \sum_{j=1}^{k} \frac{c(C_j)}{|C_j|}.$$

**13.** Normalized spectral clustering provides a relaxation to minimizing the *normalized cut* criterion [ShMa00], which is defined as

$$Ncut(P_k = \{C_1, C_2, \ldots, C_k\}) = \sum_{j=1}^{k} \frac{c(C_j)}{\sum_{i:o_i \in C_j} s_i}.$$

**14.** An argument in favor of normalized spectral clustering is that $Ncut$ expresses both homogeneity and separation. In fact,

$$\sum_{i,j:o_i,o_j \in C_\ell} d_{ij} = \sum_{i,j:o_i \in C_\ell} d_{ij} - \sum_{i,j:o_i \in C_\ell, o_j \in O \setminus C_\ell} d_{ij} = \sum_{i:o_i \in C_\ell} s_i - c(C_\ell).$$

Hence, homogeneity is maximized for a cluster $C_\ell$ if $c(C_\ell)$ is small and if $\sum_{i:o_i \in C_\ell} s_i$ is large. Both of these conditions are optimized for the $Ncut$ criterion. However, $Rcut$ aims to maximize $|C_\ell|$ of each cluster $\ell = 1, \ldots, k$, though the cardinality of a cluster is not necessarily related to homogeneity as it also depends on the similarities between the entities it contains.

**15.** Separation is maximized for both the $Ncut$ and $Rcut$ criteria since there is a term $c(C_\ell)$ for each cluster $C_\ell$, $\ell = 1, \ldots, k$, which is minimized in the numerator of both formulas (recall that $D$ is now a similarity matrix).

**16.** MATLAB code for spectral clustering is available at the sites

- https://protect-us.mimecast.com/s/QQegB6F2AeL2C2?domain=
      cis.upenn.edu
- https://protect-us.mimecast.com/s/xN7GBqUevdgeU0?domain=
      mathworks.com/44879-spectral-clustering

**17.** Besides its parallel with graph partitioning, spectral clustering is also related to random walks on the similarity graph [MeSh01] and perturbation theory [vL07].

**Example:**

**1.** One hundred points are generated in the plane and plotted in following figure. The similarity matrix is constructed such that $d_{ij}$ for two points $x_i$ and $x_j$ is given by the inverse of their Euclidean distance. Each group of 25 points is generated by adding normal(0,1) perturbations to the points $(1, 1), (1, 10), (10, 1), (10, 10)$. The different types of marker represent the natural clusters for the dataset. A complete graph $G$, weighted by the entries of matrix $D$, is used as the similarity graph. The first four eigenvectors of $L_{norm}$ are displayed in the next figures. The coordinates of an eigenvector $\bar{f} \in \mathcal{R}^{100}$ are plotted consistent with the original 100 data points.

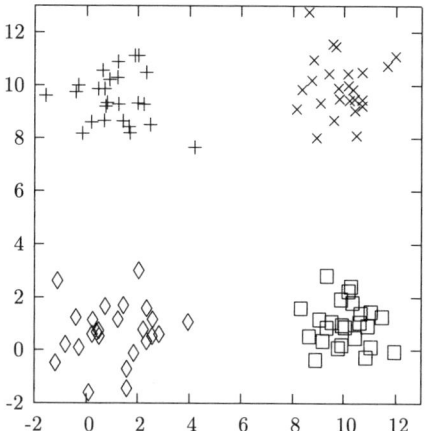

The first eigenvalue of $L_{norm}$ is 0. Since $G$ is fully connected its multiplicity is equal to 1, and the first eigenvector is the constant one vector multiplied by a scalar (note that $G$ is a weighted graph).

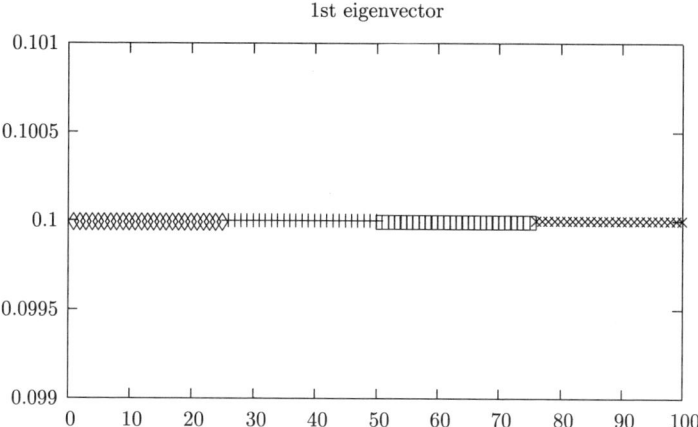

The following eigenvectors carry all the information about the clusters. If clusters in the upper parts of the graphics are distinguished from clusters in the lower parts, we notice that the second eigenvector separates clusters $\Diamond$ and $\Box$ from clusters $+$ and $\times$. Similarly, the third eigenvector separates clusters $\Diamond$ and $+$ from clusters $\Box$ and $\times$, while the fourth eigenvector separates clusters $\Diamond$ and $\times$ from $\Box$ and $+$. If the $k$-means procedure is then applied to the rows of these eigenvectors, it identifies the correct four clusters.

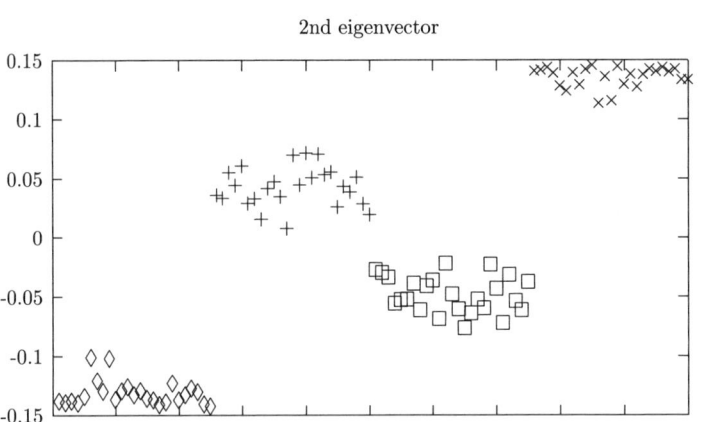

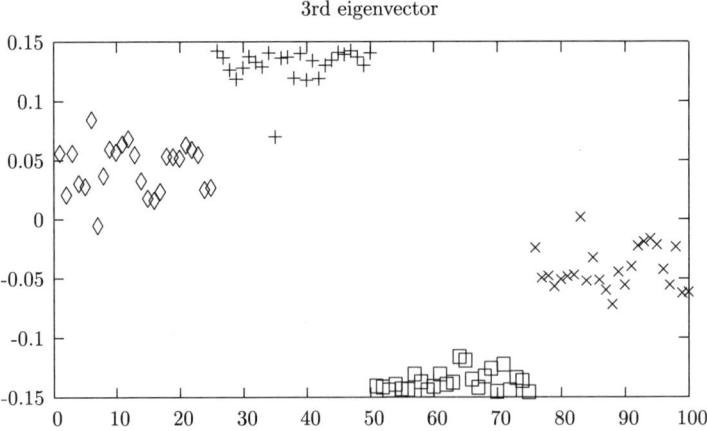

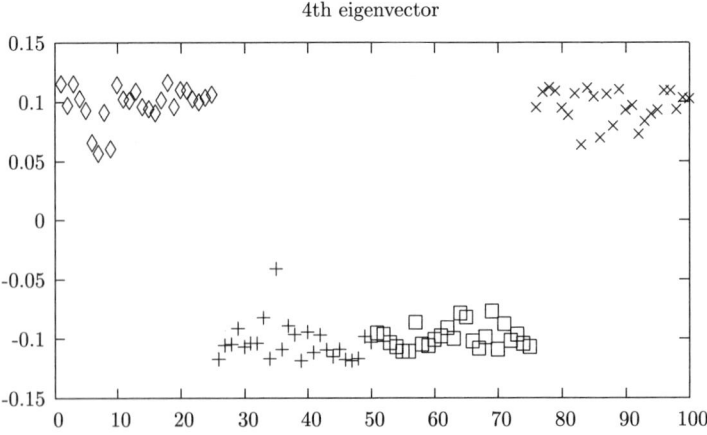

# 19.5   OUTLIER DETECTION

Outlier detection, also known as anomaly detection, involves finding data points that do not fit in with other data points in some important way. Outlier detection has many important applications in diverse areas, including finance, medicine, network security, and image processing. Outlier detection is complementary to that of clustering in that it determines individual data points that are so different from the bulk of the data that they can be classified as anomalies and not included in a cluster.

**Definitions:**

An **outlier** (or anomaly) is a data point that differs so significantly from other data points that it is unlikely to have been generated by the same mechanism.

**Collective outliers** collectively deviate from a dataset, even though individual data points do not.

**Statistical methods** (or **model-based methods**) for outlier detection assume that normal data objects are generated by a statistical model. (However, outliers are data points that do not follow the model.)

Among statistical methods, **parametric methods** for outlier detection assume normal data are objects that are generated by a parametric distribution with parameter $\theta$. The *probability density function* $f(y, \theta)$ gives the probability than an object $y$ is generated by the distribution. If the probability is small, then it is likely that it is an outlier [HaKaPe12].

Among statistical methods, **nonparametric methods** for outlier detection do not assume a prior statistical model, but instead try to determine the model from the data.

**Proximity-based methods** classify as outliers those objects that are significantly farther away than are other objects to points nearby, using some measure of distance or similarity.

The **distance-based outlier score** of an object is its distance to its $k$th nearest neighbor, where the integer $k$ is specified by the user.

Among proximity-based methods, **distance-based outlier detection methods** look at the *neighborhood of a point*. A point is an outlier if its neighborhood does not have a sufficient number of other points.

Among proximity-based methods, **density-based outlier detection methods** classify a point as an outlier if its density is much lower than that of its neighbors.

**Clustering-based methods** classify as outliers objects that either do not belong to a cluster or belong to small or sparse clusters.

**Classification-based approaches** are given a training set of examples of outliers and non-outliers. A classification method is then used to learn to predict whether a point is an outlier.

**Mahalanobis distance** is the distance between a point and the centroid of a cluster, normalized by the standard deviation in each dimension.

## Facts:

**1.** Surveys of methods for outlier detection can be found in [Ag15] and [HoAu04].

**2.** Classification methods can be used to find outliers. One way this might be done successfully is to use an outlier class containing items that deviate significantly from the other data points.

**3.** Clustering methods can be adapted to find outliers by finding clusters and then looking for data points that are far from these. Normal points may belong to large dense clusters, while outliers do not. However, this approach does not work when normal data points do not share strong patterns, while outliers do.

**4.** If the assumption is made that univariate data has a normal distribution, then the *maximum likelihood method* [HaKaPe12] can be used to estimate the mean $\mu$ and the standard deviation $\sigma$ by maximizing the *log-likelihood function*:

$$\ln \mathcal{L}(\mu, \sigma^2) = \sum_{i=1}^{n} \ln f(x_i \mid (\mu, \sigma^2)) =$$

$$-\frac{n}{2} \ln(2\pi) - \frac{n}{2} \ln(\sigma^2) - \frac{1}{2\sigma^2} \sum_{i=1}^{n} (x_i - \mu)^2.$$

Here $n$ is the number of examples and $f(x \mid (\mu, \sigma^2))$ is the normal probability density function. Derivatives with respect to $\mu$ and $\sigma^2$ are taken and one obtains the following *maximum likelihood estimates*:

$$\hat{\mu} = \bar{x} = \frac{1}{n} \sum_{i=1}^{n} x_i, \quad \hat{\sigma}^2 = \frac{1}{n} \sum_{i=1}^{n} (x_i - \bar{x}).$$

Any object more than three standard deviations from the mean of the estimated distribution is considered to be an outlier.

**5.** Another method is to compute the *interquartile range* (IQR). The first quartile $Q1$ is the point dividing the lower 25% of the data from the upper 75%, and the third quartile $Q3$ is the point dividing the lower 75% of the data from the upper 25%. The IQR is defined as $Q3 - Q1$. Any point that is more than $1.5 \times IQR$ below the lower quartile $Q1$ or $1.5 \times IQR$ above the upper quartile $Q3$ is considered to be an outlier [HaKaPe12].

**6.** Grubb's test is a straightforward statistical method for finding univariate outliers [HaKaPe12]. A z-score is computed for each point $x$ in the dataset using

$$z = \frac{\mid x - \bar{x} \mid}{s},$$

where $\bar{x}$ is the sample mean and $s$ is the sample standard deviation of the given data. A point $x$ is considered an outlier if

$$z \geq \frac{N-1}{\sqrt{N}} \sqrt{\frac{t_{\alpha/(2N), N-2}^2}{N - 2 + t_{\alpha/(2N), N-2}^2}},$$

where $t_{\alpha/(2N), N-2}^2$ is the value taken by a $t$-distribution at a significance level of $\alpha/(2N)$, with $N$ being the number of points in the dataset.

**7.** Multivariate outlier detection can be done using the Mahalanobis distance measure [HaKaPe12]. For a multivariate dataset, let $\bar{\mathbf{o}}$ be the mean vector. For a point $\mathbf{o}$ in the dataset, the Mahalanobis distance from $\mathbf{o}$ to $\bar{\mathbf{o}}$ is given by

$$\mathrm{MDist}(\mathbf{o}, \bar{\mathbf{o}}) = (\mathbf{o} - \bar{\mathbf{o}})^T S^{-1} (\mathbf{o} - \bar{\mathbf{o}}),$$

where $S$ is the covariance matrix. The following steps are needed for outlier detection:

- Calculate the mean vector from the multivariate dataset.
- For each point $o$, calculate $\text{MDist}(o, \bar{o})$, the Mahalanobis distance from $o$ to $\bar{o}$.
- Detect outliers using Grubb's test on the set $\{\text{MDist}(o, \bar{o}) \mid o \in D\}$.
- If $\text{MDist}(o, \bar{o})$ represents an outlier then $o$ is an outlier.

**8.** Multivariate outlier detection can also be done using the $\chi^2$ statistic [HaKaPe12].

**9.** A nonparametric method can be used to construct a histogram from the input data. The bins can be of equal width or equal depth. If a point falls within a bin, then it is considered normal. If it falls outside the bins, it is considered an outlier [HaKaPe12].

**10.** A proximity-based approach can be based on the concept of *distance-based outliers*. Given a set $D$ of data items, the user specifies a distance threshold $r$ to define a neighborhood of a point. For each point $o$, the number of points within the $r$-neighborhood is determined. If most of the points in the dataset are far away from $o$, then the point $o$ is an outlier. Specifically, given a *distance threshold* $r \geq 0$, a *fraction threshold* $0 < \pi \leq 1$, and a distance measure $\text{dist}(x, y)$, a point $o$ is a *distance-based outlier* if

$$\frac{\| \{o' \mid \text{dist}(o, o') \leq r\} \|}{|D|} \leq \pi.$$

An algorithm for finding distance-based outliers using this notion is given in [HaKaPe12].

**11.** Most data mining software packages support outlier detection. Some worthy of special mention are the following:

- Scikit-learn, bundled with many Python data science distributions, contains outlier detection methods (`http://scikit-learn.org/stable/modules/outlier_detection.html`)

- Weka, a Java-based data mining kit, has tools to detect and to remove outliers and extreme values (`http://www.cs.waikato.ac.nz/ml/weka/`)

- ELKI is a data mining software written in Java with an emphasis on unsupervised methods in outlier analysis (`https://elki-project.github.io/`)

**12.** The Outlier Detection DataSets (ODDS) collection includes a large collection of datasets from different domains (`http://odds.cs.stonybrook.edu`). For instance, it contains the YelpCHI database of reviews of hotels and restaurants in the Chicago area. Yelp has a filtering algorithm that identifies reviews as fake or suspicious and puts them in a filtered list. Of the 67,395 reviews in this database, 13.23% have been put in the filtered list.

**Examples:**

**1.** Outlier detection can be used to identify credit card fraud. For this application, outliers represent unusual purchases or patterns of purchases in credit card activity.

**2.** Network intrusions can be detected using outlier detection by determining whether TCP connection data is dissimilar from normal traffic.

**3.** Insider stock trading has been detected using techniques from outlier analysis.

**4.** Outlier detection has many applications in image processing, including mammography and video surveillance.

**5.** (Adapted from [HaKaPe12]) Suppose that a city's average July temperatures over ten years are $24.0°, 28.9°, 28.9°, 29.0°, 29.1°, 29.1°, 29.2°, 29.2°, 29.3°, 29.4°$. Assume that the average temperatures follow a normal distribution determined by the mean $\mu$ and the standard deviation $\sigma$. Then the maximum likelihood estimates for the mean and standard deviation are

$$\hat{\mu} = \frac{24.0 + 28.9 + 28.9 + 29.0 + 29.1 + 29.1 + 29.2 + 29.2 + 29.3 + 29.4}{10} = 28.61$$

and

$$\hat{\sigma}^2 = [(24.0-28.61)^2 + (28.9-28.61)^2 + (28.9-28.61)^2 + (29.0-28.61)^2 + (29.1-28.61)^2 +$$

$$(29.1-28.61)^2 + (29.2-28.61)^2 + (29.2-28.61)^2 + (29.3-28.61)^2 + (29.4-28.61)^2]/10 \approx 2.29,$$

giving $\hat{\sigma} = \sqrt{2.29} = 1.51$. Then, following Fact 4, the value $24.0°$ is identified as an outlier because it is $4.61°$ away from the sample mean and $\frac{4.61}{1.51} = 3.04 > 3$.

# REFERENCES

*Printed Resources*:

[Ag14] C. Aggarwal, ed., *Data Classification, Algorithms and Applications*, Chapman and Hall, 2014.

[Ag15] C. Aggarwal, *Data Mining: The Textbook*, Springer, 2015.

[Ag17] C. Aggarwal, *Outlier Analysis*, 2nd ed., Springer, 2017.

[AgRe13] C. Aggarwal and C. Reddy, eds., *Data Clustering: Algorithms and Applications*, Chapman and Hall, 2013

[AlEtal09] D. Aloise, A. Deshpande, P. Hansen, and P. Popat, "NP-hardness of Euclidean sum-of-squares clustering", *Machine Learning* 75 (2009), 245–248.

[AlHa09] D. Aloise and P. Hansen, "A branch-and-cut SDP-based algorithm for minimum sum-of-squares clustering", *Pesquisa Operacional* 29 (2009), 503–516.

[An73] M. R. Anderberg, *Cluster Analysis for Applications*, Academic Press, 1973.

[BaDr89] H. J. Bandelt and A. W. M. Dress, "Weak hierarchies associated with similarity measures: an additive clustering technique", *Bulletin of Mathematical Biology* 51 (1989), 133–166.

[Ba89] F. B. Baulieu, "A classification of presence/absence based dissimilarity coefficients", *Journal of Classification* 6 (1989), 233–246.

[Be82] J. P. Benzecri, "Construction d'une classification ascendante hiérarchique par la recherche en chaîne des voisins réciproques", *Les Cahiers de l'Analyse des Données* VII(2) (1982), 209–218.

[BeDi85] P. Bertrand and E. Diday, "A visual representation of the compatibility between an order and a dissimilarity index: the pyramids", *Computational Statistics Quarterly* 2 (1985), 31–44.

[BoEtal97] E. Boros, P. L. Hammer, T. Ibaraki, and A. Kogan. "Logical analysis of numerical data", *Mathematical Programming* 79 (1997), 163–190.

[Br78] P. Brücker, "On the complexity of clustering problems", *Lecture Notes in Economic and Mathematical Systems* 157 (1978), 45–54.

[Br06] M. J. Brusco, "A repetitive branch-and-bound procedure for minimum within-cluster sums of squares partitioning", *Psychometrika* 71 (2006), 347–363.

[BrSt05] M. J. Brusco and S. Stahl, *Branch-and-Bound Applications in Combinatorial Data Analysis*, Springer-Verlag, 2005.

[BrSt07] M. J. Brusco and D. Steinley, "A comparison of heuristic procedures for minimum within-cluster sums of squares partitioning", *Psychometrika* 72 (2007), 583–600.

[Ch75] N. Christofides, *Graph Theory: An Algorithmic Approach*, Academic Press, 1975.

[Ch97] F. Chung, *Spectral Graph Theory*, American Mathematical Society, 1997.

[Co71] R. M. Cormack, "A review of classification (with discussion)", *Journal of the Royal Statistical Society A* 134 (1971), 321–367.

[DeHa80] M. Delattre and P. Hansen, "Bicriterion cluster analysis", *IEEE Transactions on Pattern Analysis and Machine Intelligence* PAMI-2(4) (1980), 277–291.

[DeDeSo05] G. Desaulniers, J. Desrosiers, and M. M. Solomon, eds., *Column Generation*, Springer, 2005.

[Di85] G. Diehr, "Evaluation of a branch-and-bound algorithm for clustering", *SIAM Journal on Scientific and Statistical Computing* 6 (1985), 268–284.

[dMEtal00] O. du Merle, P. Hansen, B. Jaumard, and N. Mladenović, "An interior point algorithm for minimum sum-of-squares clustering", *SIAM Journal on Scientific Computing* 21 (2000), 1485–1505.

[FaPiSm96] U. Fayyad, G. Piatesky-Shapiro, and P. Smyt, "From data mining to knowledge discovery in databases", *AI Magazine* 1 (1996), 37–54.

[Fi36] R. A. Fisher, "The use of multiple measurements in taxonomic problems", *Annals of Eugenics* VII (1936), 179–188.

[FrRa02] C. Fraley and A. E. Raftery, "Model-based clustering, discriminant analysis, and density estimation", *Journal of the American Statistical Association* 97 (2002), 611–631.

[GrRa69] P. E. Green and V. R. Rao, "A note on proximity measures and cluster analysis", *Journal of Marketing Research* 6 (1969), 359–364.

[GrWa89] M. Grötschel and Y. Wakabayashi, "A cutting plane algorithm for a clustering problem", *Mathematical Programming* 45 (1989), 59–96.

[GrWa90] M. Grötschel and Y. Wakabayashi, "Facets of the clique partitioning polytope", *Mathematical Programming* 47 (1990), 367–387.

[GuHaJa91] A. Guénoche, P. Hansen, and B. Jaumard, "Efficient algorithms for divisive hierarchical clustering with the diameter criterion", *Journal of Classification* 8 (1991), 5–30.

[HaKa92] L. Hagen and A. Kahng, "New spectral methods for ratio cut partitioning and clustering", *IEEE Transactions on Computer-Aided Design of Integrated Circuits and Systems* 11 (1992), 1074–1085.

[HaKaPe12] J. Han, M. Kamber, and J. Pei, *Data Mining: Concepts and Techniques*, 3rd ed., Morgan Kaufmann, 2012.

[HaEtal07] P. Hansen, J. Brimberg, D. Urosević, and N. Mladenović, "Primal-dual variable neighborhood search for the simple plant location problem", *INFORMS Journal on Computing* 19 (2007), 552–564.

[HaDe78] P. Hansen and M. Delattre, "Complete-link cluster analysis by graph coloring", *Journal of the American Statistical Association* 73 (1978), 397–403.

[HaJaFr89] P. Hansen, B. Jaumard, and O. Frank, "Maximum sum-of-splits clustering", *Journal of Classification* 6 (1989), 177–193.

[HaJaMl98] P. Hansen, B. Jaumard, and N. Mladenović, "Minimum sum of squares clustering in a low dimensional space", *Journal of Classification* 15 (1998), 37–55.

[HaJaSi96] P. Hansen, B. Jaumard, and B. Simeone, "Espaliers, a generalization of dendrograms", *Journal of Classification* 13 (1996), 107–127.

[HaMl01a] P. Hansen and N. Mladenović, "J-means: a new local search heuristic for minimum sum of squares clustering", *Pattern Recognition* 34 (2001), 405–413.

[HaMl01b] P. Hansen and N. Mladenović, "Variable neighborhood search: principles and applications", *European Journal of Operational Research* 130 (2001), 449–467.

[Ha67] E. F. Harding, "The number of partitions of a set of $n$ points in $k$ dimensions induced by hyperplanes", *Proceedings of the Edinburgh Mathematical Society* 15 (1967), 285–289.

[Ha80] D. M. Hawkins. *Identification of Outliers*, Chapman and Hall, 1980.

[HoAu04] V. Hodge and J. Austin, "A survey of outlier detection methodologies," *Artificial Intelligence Review* 22 (2004), 85–126.

[InKaIm94] M. Inaba, N. Katoh and H. Imai, "Applications of weighted Voronoi diagrams and randomization to variance-based $k$-clustering", *Proceedings of the 10th ACM Symposium on Computational Geometry*, 1994, 332–339.

[Ja91] M. Jambu, *Exploratory and Multivariate Data Analysis*, Academic Press, 1991.

[Jo02] I. T. Jolliffe, *Principal Component Analysis*, 2nd ed., Springer, 2002.

[KlAr91] G. Klein and J. E. Aronson, "Optimal clustering: a model and method", *Naval Research Logistics* 38 (1991), 447–461.

[KoNaFu75] W. L. G. Koontz, P. M. Narendra, and F. Fukunaga, "A branch and bound clustering algorithm", *IEEE Transactions on Computers* C-24 (1975), 908–915.

[LaWi67] G. N. Lance and W. T. Williams, "A general theory of classificatory sorting strategies. 1. Hierarchical systems", *The Computer Journal* 9 (1967), 373–380.

[LeRaUl14] J. Leskovec, A. Rajaraman, and J. D. Ullman, *Mining of Massive Datasets*, 2nd ed., Cambridge University Press, 2014.

[LiSmYo03] G. Linden, B. Smith, and J. York, "Amazon.com recommendations: item-to-item collaborative filtering", *IEEE Internet Computing* 7 (2003), 76–80.

[Ma67] J. B. MacQueen, "Some methods for classification and analysis of multivariate observations", *Proceedings of 5th Berkeley Symposium on Mathematical Statistics and Probability*, 1967, 281–297.

[MaNiVa09] M. Mahajan, P. Nimbhorkar, and K. Varadarajan, "The planar $k$-means problem is NP-hard", *Lecture Notes in Computer Science* 5431 (2009), 274–285.

[MaRo10] O. Maimon and L. Rokach, *Data Mining and Knowledge Discovery Handbook*, 2nd ed., Springer, 2010.

[MaBe83] D. W. Matula and L. L. Beck, "Smallest-last ordering and clustering and graph-coloring algorithms", *Journal of the ACM* 30 (1983), 417–427.

[MeSh01] M. Meila and J. Shi, "Learning segmentation by random walks" in *Advances in Neural Information Processing Systems 12*, S. A. Solla, T. K. Leen, and K.-R. Müller (eds.), MIT Press, 2001, 873–879.

[Mi87] B. Mirkin, "Additive clustering and qualitative factor analysis methods for similarity matrices", *Journal of Classification* 4 (1987), 7–31.

[Mi96] B. Mirkin, *Mathematical Classification and Clustering*, Kluwer, 1996.

[Mi97] T. M. Mitchell, *Machine Learning*, WCB/McGraw-Hill, 1997.

[Mo91] B. Mohar, "The Laplacian spectrum of graphs" in *Graph Theory, Combinatorics, and Applications*, G. Chartrand, O. R. Oellermann, and A. J. Schwenk (eds.), Wiley, 1991, 871–898.

[Mo97] B. Mohar, "Some applications of Laplace eigenvalues of graphs" in *Graph Symmetry: Algebraic Methods and Applications*, G. Hahn and G. Sabidussi (eds.), Kluwer, 1997, 225–275.

[MoSu91] C. Monma and S. Suri, "Partitioning points and graphs to minimize the maximum or the sum of diameters" in *Graph Theory, Combinatorics, and Applications*, G. Chartrand, O. R. Oellermann, and A. J. Schwenk (eds.), Wiley, 1991, 899–912.

[Mu83] F. Murtagh, "A survey of recent advances in hierarchical clustering algorithms", *The Computer Journal* 26 (1983), 329–340.

[NgJoWe02] A. Ng, M. Jordan, and Y. Weiss, "On spectral clustering: analysis and an algorithm" in *Advances in Neural Information Processing Systems*, T. Dietterich, S. Becker, and Z. Guahramani (eds.), MIT Press, 2002, 849–856.

[PaHaLi04] L. Parsons, E. Haque, and H. Liu, "Subspace clustering for high dimensional data: a review", *SIGKDD Explorations Newsletter* 6 (2004), 90–105.

[PeXi05] J. Peng and Y. Xia, "A new theoretical framework for $k$-means-type clustering", *Studies in Fuzziness and Soft Computing* 180 (2005), 79–96.

[Qu93] J. R. Quinlin, *C4.5: Programs for Machine Learning*, Morgan Kaufmann, 1983.

[Ra71] M. R. Rao, "Cluster analysis and mathematical programming", *Journal of the American Statistical Association* 66 (1971), 622–626.

[Ro67] P. Rosenstiehl, "L'arbre minimum d'un graphe" in *Théorie des Graphes*, P. Rosenstiehl (ed.), Dunod, 1967, 357–368.

RuNo10] S. Russell and P. Norvig, *Artificial Intelligence: A Modern Approach*, 3rd ed., Prentice Hall, 2010.

[ScKoRi01] J. B. Schafer, J. A. Konstan, and J. Riedl, "E-commerce recommendation applications", *Data Mining and Knowledge Discovery* 5 (2001), 115–153.

[ShMa00] J. Shi and J. Malik, "Normalized cuts and image segmentation", *IEEE Transactions on Pattern Recognition and Machine Intelligence* 22 (2000), 888–905.

[SiUp12] K. Singh and S. Upadhaya, "Outlier detection: applications and techniques," *International Journal of Computer Science Issues* 9 (2012), 307–323.

[Sp80] H. Späth, *Cluster Analysis Algorithms for Data Reduction and Classification of Objects*, Wiley, 1980.

[Sp85] H. Späth, *Cluster Dissection and Analysis: Theory, FORTRAN Programs, Examples*, Ellis Horwood, 1985.

[TaStKu06] P. Tan, M. Steinbach, and V. Kumar, *Introduction to Data Mining*, Addison Wesley, 2006.

[vOMe04] B. J. van Os and J. J. Meulman, "Improving dynamic programming strategies for partitioning", *Journal of Classification* 21 (2004), 207–230.

[vL07] U. von Luxburg, "A tutorial on spectral clustering", *Statistics and Computing* 17 (2007), 395–416.

[We93] G. Wesolowsky, "The Weber problem: history and perspective", *Location Science* 1 (1993), 5–23.

[WiFrHall] I. H. Witten, E. Frank, and M. A. Hall, *Data Mining: Practical Machine Learning Tools and Techniques*, 3rd ed., Morgan Kaufmann, 2011.

[WuEtal07] X. Wu et al., "Top 10 algorithms in data mining", *Knowledge and Information Systems* 14 (2008), 1–37.

[XiPe05] Y. Xia and J. Peng, "A cutting algorithm for the minimum sum-of-squared error clustering", *Proceedings of the SIAM International Data Mining Conference*, 2005, 150–160.

**Web Resources**:

`http://archive.ics.uci.edu/ml/` (UCI Machine Learning Repository of benchmark datasets for clustering algorithms.)

`http://id3alg.altervista.org` (Implementation of ID3 algorithm in C.)

`http://www.alglib.net/dataanalysis/clustering.php` (Implementations of the $k$-means algorithm.)

`http://www.cis.upenn.edu/~jshi/software/` (MATLAB code for normalized spectral clustering.)

`http://www.cs.umd.edu/~mount/Projects/KMeans` (C++ code for the $k$-means algorithm.)

`http://www.cs.waikato.ac.nz/ml/weka/` (Machine learning tools for data mining.)

`http://www.mathworks.com/matlabcentral/fileexchange/44879-spectral-clustering` (MATLAB code for spectral clustering.)

`http://www.mmds.org/` (Online site for the textbook *Mining of Massive Datasets* by J. Leskovec, A. Rajaraman, and J. D. Ullman; includes the latest version of the text, lecture slides, and links to courses.)

`https://code.google.com/p/figue/` (Javascript for a variety of clustering algorithms.)

# 20

# DISCRETE BIOMATHEMATICS

\*This author was supported by the Intramural Research Program of the National Institutes of Health, National Library of Medicine.

## INTRODUCTION

Modern biology makes use of a variety of discrete and combinatorial methods in order to better understand biological phenomena at both the microscopic and macroscopic levels. This chapter explains how such methods are useful in understanding the structure and properties of the genome (both DNA and RNA structures), gene regulatory networks, the firings of neurons, the classification and evolution of species, as well as competition between species. A variety of modern algorithms for analyzing biological data are also presented in this chapter.

## GLOSSARY

**alignment** (of two sequences):  a matching in which individual letters/nulls from each sequence are placed into correspondence.

**alignment score**:  the sum of scores for aligned pairs of letters/nulls in an alignment.

**arc diagram**:  a representation of an RNA structure in which the bases are placed along a line and arcs (drawn in the upper-half plane) join base pairs.

**basal species** (in a food web):  species located at the bottom of the food web.

**Boolean network**:  a polynomial dynamical system over $\mathbf{F}_2$.

**boxicity** (of a graph $G$):  the smallest dimension $p$ such that two vertices are connected by an edge in $G$ if and only if their corresponding $p$-dimensional boxes overlap.

**Burrows-Wheeler Transform BWT** (of a string):  a reversible transformation of the string used to perform efficient query searches.

**cellular automaton**:  a lattice of cells, which at each step take on a value based on the values present in its neighboring cells.

**code** (of a cover):  the combinatorial neural code whose codewords correspond to regions defined by intersections of open sets in the cover.

**combinatorial neural code**:  a collection of binary codewords in which each binary digit is interpreted as the state (on/off) of a neuron.

**common enemy graph**:  a graph whose vertices represent the species in a food web and whose edges $(a, b)$ signify that species $a$ and species $b$ have a common predator.

**competition graph**:  a graph whose vertices represent the species in a food web and whose edges $(a, b)$ signify that species $a$ and species $b$ have a common prey.

**de Bruijn graph**: a directed graph whose vertices consist of all length-$k$ sequences over an alphabet and in which the edge $(u, v)$ indicates that the suffix of $u$ equals the prefix of $v$.

**dependency graph** (for a finite dynamical system): a directed graph whose vertices correspond to variables and whose directed edges indicate functional dependencies between variables.

**discrete-time dynamical system**: a system based on variables $x_1, x_2, \ldots, x_n$ and state transition functions for each variable $x_i$.

**DNA segment**: a sequence of bases on a DNA strand, represented as a string over the alphabet {A, C, G, T}.

**DNA target sequence**: a long unknown DNA sequence that is to be decoded.

**Euler path** (in a graph): a path that traverses every edge of the graph exactly once.

**expanded sequence**: a sequence into which null characters may be inserted.

**food web**: a directed graph whose vertices correspond to species and whose arcs correspond to direct predator/prey relationships.

**genetic algorithm**: a heuristic search algorithm that mimics natural biological evolution, whereby a population of candidate solutions evolves over time via mutation and recombination.

**genome assembler**: a computer program that carries out the gene assembly process from given input data.

**global pairwise alignment** (of sequences): an alignment in which all letters and nulls of the two (expanded) sequences must be aligned.

**Hamilton path** (in a graph): a path that visits every vertex of the graph exactly once.

**hybridization** (of a DNA strand): a technique to produce a list of all subsequences of the DNA strand having some fixed length $k$.

**hyperplane code**: a combinatorial code obtained from the output of a one-layer feedforward neural network.

**KMP algorithm**: an efficient algorithm for finding an exact match between a query string and a database.

**Knudsen-Hein grammar**: a probabilistic context-free grammar used to represent RNA secondary structures.

**local pairwise alignment** (of sequences): an alignment of arbitrary-length segments of the two (expanded) sequences.

**maximum agreement subtree**: a subtree induced by the largest set of common taxa.

**nearest neighbor interchange**: an exchange of two subtrees appearing on opposite sides of an internal tree edge.

**nerve** (of a cover): the simplicial complex defined by the nonempty intersections of the given open sets.

**neural ring**: a quotient ring associated with a combinatorial code.

**nucleotide**: an organic molecule (having a distinguished nitrogenous base) that serves as a building block for DNA and RNA.

**overlap graph**: a graph whose vertices represent known DNA strands and whose edges represent overlaps between these DNA strands.

***overlap-layout-consensus method***:  a technique for reassembling a DNA strand from known DNA fragments.

***path graph***:  a representation of possible alignments of two sequences using a rectangular array of nodes joined by directed horizontal, vertical, and diagonal edges.

***phylogenetic tree***:  a tree whose leaves are labeled with taxa (taxonomical units).

***polytomy***:  a vertex in a phylogenetic tree with degree exceeding three.

***predator projection graph***:  a weighted graph based on the competition graph.

***prey projection graph***:  a weighted graph based on the common enemy graph.

***RNA sequence***:  a sequence over the four letter alphabet {A, C, G, U}.

***Robinson-Foulds distance***:  a measure of proximity between two trees, given by the number of edge-induced partitions occurring in one tree but not the other.

***secondary structure***:  an RNA structure whose arc diagram has no crossings.

***seed***:  a relatively short matching subsequence that serves as an anchor point for larger matches between two sequences.

***shortest common superstring***:  a shortest string that contains each of a given set of strings as a consecutive sequence.

***shotgun sequencing***:  a method in which many copies of a DNA target sequence are broken up into pieces, which are then individually sequenced and recombined.

***state-space graph*** (for an $n$-dimensional dynamical system):  a directed graph whose vertices represent $n$-tuple states and whose edges represent transitions between such states.

***substitution matrix***:  a prescription of the score produced when two specified letters are aligned.

***suffix array*** (for a string):  a lexicographically sorted array of all suffixes of the string.

***suffix tree*** (for a string):  a compact representation of all suffixes of the string.

***trophic level*** (of a species $x$):  a positioning measure based on the length of paths to $x$ from basal species that $x$ consumes, either directly or indirectly, in a food web.

***Yule-Harding distribution***:  a model for generating all rooted binary trees based on certain rules for speciation.

# 20.1   SEQUENCE ALIGNMENT

Alignments are a powerful way to compare related DNA or protein sequences. They can be used to capture various facts about the sequences aligned, such as common evolutionary descent or common structural function. We take the general view that the alignment of letters from two or multiple sequences represents the hypothesis that they are descended from a common ancestral sequence.

DNA molecules are composed of chains of nucleotides, and protein molecules are composed of chains of amino acids. The specific order of nucleotides or amino acids within these chains are respectively called DNA and protein sequences. Perhaps chief among the various biological functions of DNA sequences is to encode protein sequences, because proteins are involved in most of the biological functions of living cells.

DNA sequences, and the protein sequences they encode, evolve by mutation followed by natural selection. There are a variety of mechanisms for DNA mutation, but the most common result is the substitution of a single nucleotide for another, or the deletion or insertion of one or several adjacent nucleotides. At the protein level, the most common resulting mutations are the substitution of one amino acid for another, or the insertion or deletion of one or multiple adjacent amino acids. There is no simple biological mechanism for exchanging the order of two letters in a DNA or protein sequence, so an alignment representing the common descent of two DNA or protein sequences is co-linear, with no "crossovers" between corresponding letters.

## 20.1.1  GLOBAL AND LOCAL PAIRWISE ALIGNMENTS

### Definitions:
An **alphabet** is a finite set of letters and a **sequence** $S$ is a finite string of letters, each chosen from the alphabet.

A **null character**, generally represented by the symbol "–", is a character not in the alphabet that signifies an absent letter.

Given a sequence $S$, an **expanded sequence** $S'$ is the sequence $S$ with an arbitrary number of null characters placed at its start, its end, or between any two of its characters.

A **global pairwise alignment** of sequences $S$ and $T$ is a one-to-one co-linear correspondence of expanded sequences $S'$ and $T'$, such that no nulls from $S'$ and $T'$ correspond.

A **local pairwise alignment** of sequences $S$ and $T$ is a one-to-one co-linear correspondence of equal-length segments of the expanded sequences $S'$ and $T'$, such that no nulls from the segments correspond.

### Facts:
**1.** DNA molecules are composed of two strands, each a string of nucleotides, represented by the four letters A, C, G, T.

**2.** The strands of a DNA molecule are directional and run in opposite directions. Each nucleotide in one strand of a DNA molecule is paired, through chemical interaction, with a complementary nucleotide in the other strand: A always pairs with T, and C with G. Because of this complementarity, the nucleotide sequence of one strand determines the nucleotide sequence of the other, and thus a DNA molecule is generally represented by the nucleotide sequence of one of its strands, with that of the other implied.

**3.** Proteins are composed of one, or multiple, chemically interacting amino acid chains, but here we confine our attention to single chains. An amino acid chain is a directional string of amino acids. There are twenty commonly occurring amino acids abbreviated by the letters A, C, D, E, F, G, H, I, K, L, M, N, P, Q, R, S, T, V, W, Y.

**4.** A particular amino acid is encoded by three adjacent DNA nucleotides, called a *codon*; for example, the codon ATG represents the amino acid methionine, or M. Some amino acids are encoded by only one codon, whereas others by as many as six. The table describing the correspondence between all $4^3 = 64$ possible codons and the 20 amino acids they represent is called the *genetic code*; except for minor variations, it is universal to all life on earth. The genetic code contains three *stop codons* that represent no amino acid.

**Examples:**

**1.** One possible global alignment of the protein sequences VHLTPEEKSAVTALWG and VAFTEKQEALVSSSLEAF is

```
VHLT--PEEKSAV-TALWG-
VAFTEKQEA--LVSSSLEAF
```

**2.** One possible local alignment of the DNA sequences GTTACTTTGGACCCTCAA and AAATTGATCTTTTAAC is

```
TTTGGACC
T-T-GATC
```

## 20.1.2   PAIRWISE ALIGNMENT SCORES

In order to select among the many possible global or local alignments of two sequences, it is useful to assign an objective function, or score, to each possible alignment. We then seek an alignment with an optimal score, using the convention that higher scores are better. The simplest way to score an alignment is to specify scores for aligning particular letters to one another, or for aligning letters to nulls, and then to define the score of an alignment as the sum of the scores of all its aligned letters and nulls.

**Definitions:**

A *column* of a pairwise alignment is the one-to-one correspondence of a single letter (or null) in one sequence with a single letter (or null) in the other.

A *substitution* is a column that aligns two letters. A *substitution score* is a score defined for a substitution involving a particular pair of letters.

An *indel* is a column that aligns a letter with a null. An *indel score* is the score defined for aligning a letter with a null.

A *gap of length k* is composed of $k$ adjacent indels, each of which contains a letter from one expanded sequence and a null from the other.

The *alignment score* is the sum of the substitution and the indel scores of an alignment's columns.

An *optimal alignment* is an alignment with maximum score.

**Facts:**

**1.** An alignment of two identical letters is still termed a substitution.

**2.** The term *indel* represents an abbreviation for "insertion or deletion".

**3.** Indel scores are usually chosen to be independent of the particular letter aligned with a null, but this restriction is not necessary.

**4.** The definition of an alignment score will be generalized later (see §20.1.6).

**Example:**

**1.** Consider the following alignment of two DNA sequences:

```
TGA-CG
-GTACC
```

Suppose the substitution scores for all columns aligning identical letters are $+5$ and for all columns aligning mismatching letters are $-1$. Also, suppose all indel scores are $-2$. Then the alignment score for this DNA alignment is $-2 + 5 - 1 - 2 + 5 - 1 = 4$.

## 20.1.3  PATH GRAPHS AND OPTIMAL GLOBAL PAIRWISE ALIGNMENT

The most fruitful algorithms for pairwise sequence alignment are based on the concept of a path graph.

**Definitions:**
Suppose we have two sequences of lengths $m$ and $n$, respectively. The **path graph** consists of a rectangular $(m + 1) \times (n + 1)$ array of **nodes**, with directed horizontal, vertical, and diagonal **edges** between adjacent nodes. The $m$ letters of the first sequence are placed between successive rows of this array and the $n$ letters of the second sequence are placed between successive columns of this array. Diagonal edges of the path graph correspond to substitutions, while horizontal and vertical edges correspond to indels. Scores for the directed edges derive from the corresponding substitution and indel scores.

**Facts:**
**1.** The rectangular array underlying the path graph is composed of $m \times n$ *cells*. Each cell is associated with a letter of the first sequence and a letter of the second sequence. The diagonal edge ($\searrow$) in this cell has weight corresponding to the substitution score for these two letters. The lower horizontal edge ($\rightarrow$) in a cell corresponds to a null in the first expanded sequence, and the rightmost vertical edge ($\downarrow$) in a cell corresponds to a null in the second expanded sequence; in each case, the null follows the letter that the diagonal edge signifies aligning.

**2.** There is a one-to-one correspondence between alignments and directed paths beginning at the upper left node $(0, 0)$ of the path graph and ending at the lower right node $(m, n)$. The score of an alignment equals the sum of the edge scores along its corresponding path.

**3.** The problem of finding an optimal global alignment is then transformed into finding an optimal path through the path graph.

**4.** Although the number of possible paths through the path graph grows exponentially with $m$ and $n$, optimal alignments can be found without examining all paths.

**5.** Algorithm 1 uses a dynamic programming approach to find optimal paths in an efficient way. In brief, there are at most three nodes at locations $(i - 1, j), (i, j - 1), (i - 1, j - 1)$ with edges leading into a particular node at location $(i, j)$. Therefore, if one knows the score of an optimal path from the origin to each of these three nodes, one can easily calculate the score of an optimal path through each of these nodes into $(i, j)$. The highest score among these paths is then the score of an optimal path to $(i, j)$. Optimal alignment scores for each node in a path graph may thus be calculated by simply filling in scores for the nodes sequentially, beginning at the origin, and proceeding from left to right along each row in turn [NeWu70], [Sa72], [Se74].

---

**Algorithm 1**:  **Global pairwise alignment algorithm.**
  input: two sequences of lengths $m$ and $n$, respectively
  output: the optimal global alignment score as well as optimal alignments

create the corresponding path graph $G$ with $(m+1) \times (n+1)$ nodes, indexed from 0 to $m$ and from 0 to $n$

label the upper left node of $G$ with 0

label the rest of the first row and the first column of $G$, based on the indel scores

for each remaining node in location $(i,j)$ of the array compute

Vscore $:=$ label$(i-1,j)$ + score of edge from $(i-1,j)$ to $(i,j)$

Hscore $:=$ label$(i,j-1)$ + score of edge from $(i,j-1)$ to $(i,j)$

Dscore $:=$ label$(i-1,j-1)$ + score of edge from $(i-1,j-1)$ to $(i,j)$

label$(i,j) :=$ max$\{$Vscore, Hscore, Dscore$\}$; mark any edge achieving this maximum value

label$(m,n)$ represents the optimal global alignment score and the marked edges can be used to retrieve alignments achieving this optimal score

**6.** In Algorithm 1, the values Vscore, Hscore, and Dscore represent the scores of paths using (respectively) the vertical, horizontal, and diagonal edges entering the node located at $(i,j)$. By tracing back along the marked edges, starting from the lower right corner of $G$, one can obtain optimal alignments for the given pair of sequences.

**7.** To align sequences of lengths $m$ and $n$, the time and space complexity of Algorithm 1 are each $O(mn)$. The space complexity can be reduced to $O(\min\{m,n\})$ [MyMi88].

**8.** There are various methods for speeding up the algorithm by avoiding the consideration of nodes through which an optimal alignment cannot possibly pass [Fi84], [Sp89].

**9.** Optimal alignments are unaltered by multiplying all substitution and indel scores by a positive constant. Optimal alignments are unaltered by adding a constant $a$ to all substitution scores, and $a/2$ to all indel scores.

**Examples:**

**1.** We wish to align the sequences $S_1 =$ ACGC and $S_2 =$ GACTAC. The corresponding path graph $G$ has nodes arranged in an $(m+1) \times (n+1) = 5 \times 7$ rectangular array. The four letters of the first sequence are placed between successive rows of $G$ and the six letters of the second sequence are placed between successive columns of $G$. The shaded cell corresponds to aligning the first letter C of the first sequence and the second letter A of the second sequence.

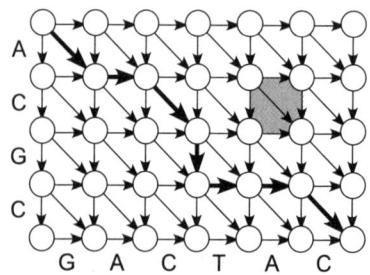

The highlighted path (from the upper left node to the lower right node in $G$) corresponds to the alignment

```
A-CG--C
GAC-TAC
```

**2.** Suppose that substitution scores are $+1$ for matches and $0$ for mismatches, and that indel scores are $-1$. Then the alignment in Example 1 has score $0-1+1-1-1-1+1 = -2$. This is not, however, an optimal global alignment of $S_1$ and $S_2$.

**3.** Optimal alignments, based on the substitution and indel scores specified in Example 2, can be found using Algorithm 1. After labeling the upper left node with 0, we label nodes in the first row and first column using $-1, -2, -3, \ldots$ since the indel scores are $-1$. Each remaining node is labeled with the maximum sum of node label plus edge score, taken over the three nodes adjacent to and leading into the given node.

For example, to label the node at location $(2, 1)$ we consider its three incident edges:

$(1, 1) \to (2, 1)$: label of $(1, 1)$ + edge score (indel) $= 0 - 1 = -1$

$(2, 0) \to (2, 1)$: label of $(2, 0)$ + edge score (indel) $= -2 - 1 = -3$

$(1, 0) \to (2, 1)$: label of $(1, 0)$ + edge score (mismatch C, G) $= -1 + 0 = -1$

Node $(2, 1)$ thus receives as its label the maximum sum $-1$. The following figure shows all node labels obtained by following Algorithm 1. Since the label of the lower right node $(4, 6)$ is 1, this represents the optimal global alignment score.

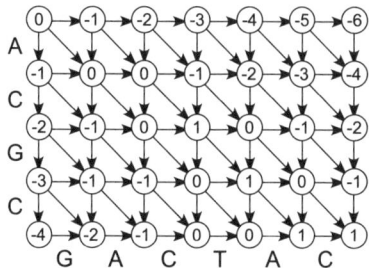

**4.** In order to reconstruct the alignment or alignments corresponding to this optimal score, the edge or edges corresponding to the best path into each node have been marked. In Example 3, the edges $(1, 1)$ and $(1, 0)$ both give the maximum label $-1$ and so are marked. All edges marked during the execution of Algorithm 1 are shown in the following path graph.

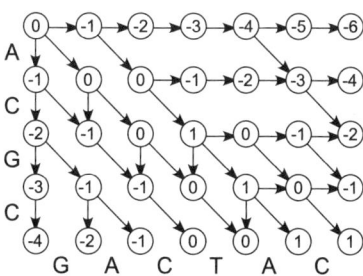

**5.** We can then trace back along marked edges from the terminal node $(5, 7)$ to the origin node $(1, 1)$, giving the highlighted edges shown in the following figure. From these two paths, we obtain the two optimal alignments of $S_1$ and $S_2$, each with (maximum) score $= +1$:

```
    -ACG-C        -AC-GC
    GACTAC        GACTAC
```

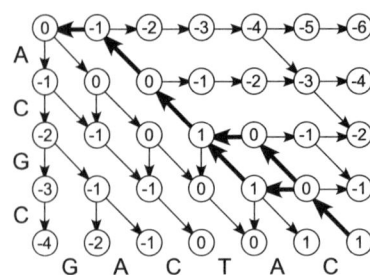

## 20.1.4  OPTIMAL LOCAL PAIRWISE ALIGNMENT

Many protein or DNA sequences are related only across subsequences, but not over their entire lengths. To recognize these relationships, it is useful to introduce the concept of a local alignment (§20.1.1) and to describe algorithms capable of finding optimal local alignments.

**Facts:**

**1.** Algorithm 1 from the previous section can be modified [SmWa81] to find optimal local alignment(s) of two sequences by making a few minor modifications.

- Allow a path to start, with score 0, at any node within the path graph, not just at the origin (top left) node. Thus, if all paths into a given a node have negative score, none of them is chosen as optimal.

- Record which node within the path graph receives the highest score, and start tracing back from there, rather than from the terminal node.

- Terminate the traceback when a node with score 0 is reached.

**2.** To align sequences of lengths $m$ and $n$, the time and space complexity for this local alignment algorithm are each $O(mn)$. The space complexity can be reduced to $O(\min\{m,n\})$ [MyMi88].

**3.** Optimal local alignments are unaltered by multiplying all substitution and indel scores by a positive constant.

**4.** In addition to optimal local alignment(s), one may wish to find other local alignments that align regions of the sequences in question that are not implicated in the optimal local alignment. One may define "locally optimal" local alignments as those whose path within the path graph has a score greater than or equal to the score for the path of any local alignment that shares a common edge or node [Se84]. All such locally optimal local alignments can also be found in $O(mn)$ time [AlEr86b].

**Examples:**

**1.** We wish to align the sequences $S_1 = \text{ACGC}$ and $S_2 = \text{GATTGA}$. We construct a path graph with $(m+1) \times (n+1) = 5 \times 7$ nodes. Suppose that substitution scores are $+4$ for matches and $-1$ for mismatches, and that indel scores are $-2$. The following figure shows the resulting labels for locally aligning these sequences, with only marked edges shown.

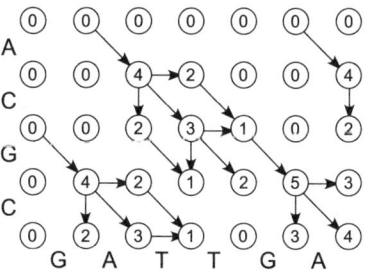

**2.** The largest score obtained is 5 so this is the optimal local alignment score. Tracing back from this location $(4, 6)$ yields the highlighted edges shown next.

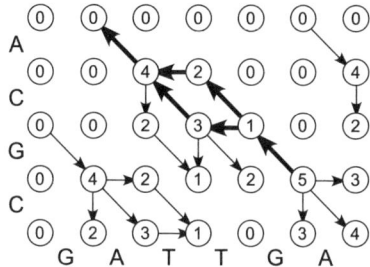

As a result, we find two optimal local alignments, each with score 5:

```
AC-G        A-CG
ATTG        ATTG
```

## 20.1.5   SUBSTITUTION MATRICES

Although the algorithms described in §20.1.3 and §20.1.4 are unaffected by the choice of substitution scores, the quality of the alignments they return depends strongly on which scores are used. In particular, this subsection discusses the use of certain substitution matrices to specify scoring values.

**Definitions:**

A **substitution matrix** $S = (s_{ij})$ prescribes the score value when letters $i$ and $j$ are aligned.

A **log-odds matrix** $S$ is defined by $s_{ij} = \log \frac{q_{ij}}{p_i p_j}$. Here $q_{ij}$ is the **target frequency** with which the letters $i$ and $j$ are estimated to appear in columns from accurate alignments of related sequences; $p_i$ and $p_j$ are the **background probabilities** with which the letters $i$ and $j$ would appear by chance.

**Facts:**

**1.** Many different substitution matrices have been proposed, but theory and practice suggest that the best substitution matrices are log-odds matrices [Al91], [DaScOr78], [KaAl90].

**2.** The target frequencies should depend upon the degree of evolutionary divergence separating the sequences compared, and various methods of estimating these frequencies for proteins have been described, giving rise to the commonly used PAM [ScDa78] and BLOSUM [HeHe92] series of substitution matrices.

## 20.1.6   AFFINE GAP SCORES

A single mutational event can insert or delete multiple nucleotides into or from a DNA sequence, and this can result in multiple contiguous amino acids being inserted into or deleted from the encoded protein sequence. Thus, an alignment in which the several insertions or deletions are adjacent to one another makes more biological sense than one in which these insertions and deletions are separated. The scoring systems considered earlier, however, do not reflect this biological fact. One remedy is to modify the alignment scoring system by defining appropriate scores for gaps of various lengths in place of scores for individual indels.

**Definition:**

**Affine gap scores** are given by $g(k) = -(a + bk)$, where $k$ is the gap length and $a, b$ are (generally nonnegative) constants.

**Facts:**

**1.** It is evident that when $a = 0$, the affine gap score $g(k)$ is equivalent to indel scores of $-b$. However, when $a > 0$, the score for $k$ adjacent indels is better than that for $k$ indels separated by substitution columns.

**2.** For either global or local alignments, the basic algorithms described in §20.1.3 and in §20.1.4 must be modified to accommodate affine gap costs. Fortunately, it is possible to do this while retaining $O(mn)$ time complexity by remembering not just the score of the best path into each node, but also the best score for a path that arrives through a horizontal, vertical, or diagonal edge [AlEr86a], [Go82]. For example, the incremental score for a path leading out of a node along a horizontal edge is $-b$ for a path that arrived along a horizontal edge, but is $-(a + b)$ for a path that arrived along any other type of edge.

**3.** It is possible to construct $O(mn)$ algorithms based on $g(k)$ that are more involved than affine gap scores [MiMy88], but these algorithms are fairly complex and in practice almost never used.

**4.** The elements of sequence alignment considered here been generalized in a great variety of ways. These include rapid heuristic database search algorithms [AlEtal90], [PeLi88]; protein profile construction and comparison [AlEtal97], [GrMcEi87]; and multiple sequence alignment, for which there is a vast and growing literature.

**Example:**

**1.** The practical difference in using affine gap costs in place of simple indel costs can be illustrated in the optimal local alignments resulting from the comparison of two protein sequences. Both alignments shown next employ BLOSUM-62 substitution scores [HeHe92], but the first alignment uses indel scores of $-6$, while the second uses gap scores of $g(k) = -(11 + k)$. Identical letters are echoed on the central lines of these alignments, and the sequence positions of the letters at the start and end of each line are provided. The second alignment contains many fewer gaps, but also a smaller number of identical aligned letters.

Alignment 1:

```
 49 CERTLKYFLGIAGGKWVVSYFWVTQSIKERKMLNEHDFEVRGDVVNGRNHQGPKRARESQDRK-IFRGLEICCYG 122
    C RT KYFL  A G   VS  WV  S       N            G     R    QR   F L
865 C-RTRKYFLCLASGIPCVSHVWVHDSCHANQLQNYRNY-L---LPAGYSLE-EQRILDWQPRENPFQNLKVLLVS 933
```

```
123 PFTNMPTDQLEWM-VQLC-GASVVKE-LSS-FT--LGTGVHPIVVVQPDAWTEDNGFHAIGQMCEAPVVTREWVL 191
         L W     GA VK  SS     GV  VV  P                PVV   EWV
934 D-QQQNFLEL-WSEILMTGGAASVKQhHSSAHNKDIALGVFDVVVTDPSC-PA-SVLKC-AEALQLPVVSQEWVI 1003
```

Alignment 2:

```
 51 RTLKYFLGIAGGKWVVSYFWVTQSIKERKMLNEHDFEVRGDVVNGRNHQGPKRARESQDRK-IFRGLEICCYG 122
    RT KYFL  A G   VS  WV  S       N                 R    QR   F L
866 RTRKYFLCLASGIPCVSHVWVHDSCHANQLQNYRNY-----LLPAGYSLEEQRILDWQPRENPFQNLKVLLVS 933
```

```
123 PFTNMPTDQLEWMVQLCGASVVKELSSFT----LGTGVHPIVVVQPDAWTEDNGFHAIGQMCEAPVVTREWVL 191
                  GA VK  S       GV  VV  P                PVV   EWV
934 DQQQNFLELWSEILMTGGAASVKQHHSSAHNKDIALGVFDVVVTDPSC---PASVLKCAEALQLPVVSQEWVI 1003
```

## 20.1.7   SEQUENCE ALIGNMENT HEURISTICS

Sequence alignment, as described so far, is computationally intensive, with the algorithms requiring an execution time quadratic in the size of the sequences being aligned. In the general case this run time is unavoidable. However, faster execution times can be obtained when the sequences being aligned have high levels of similarity. Here we discuss banded methods, exclusion methods, and the use of seeds to initiate searches.

**Definitions:**

Suppose that $G$ is the path graph for the alignment of two sequences. The **diagonal** of $G$ consists of all nodes at locations $(0,0), (1,1), \ldots, (m,n)$. A **band of width $d$** consists of all nodes within (Manhattan) distance $d$ of the diagonal of $G$.

A **$k$-mer** is a subsequence of length $k$ found within a given sequence (see §20.4.1).

A **seed** is a relatively short matching subsequence that serves as an anchor point for larger matches between sequences. A **spaced seed** is a $k$-mer within which a number $t$ of fixed positions (denoted by $*$) are allowed to match any letter; the spaced seed then has **width** $k$ and **weight** $k - t$.

**Facts:**

**1.** Suppose that two given sequences differ by at most $d$ edits (at most $d$ letters are changed, inserted, or deleted between the two sequences). Then an optimal path in the path graph $G$ cannot pass through nodes that are further than distance $d$ from the diagonal of $G$ as each "move" away from the diagonal implies an additional difference being incorporated in the alignment.

**2.** By Fact 1 it is sufficient to compute the alignment scores for just a subset of the nodes in $G$, specifically for nodes in the band of width $d$. Using this banded alignment algorithm, an optimal alignment can then be determined in $O(\min\{m,n\}d)$ time [ChPeMi92].

**3.** The value $d$ does not need to be known a priori. Rather, an algorithm that adapts dynamically to the actual similarity between the sequences can be devised by performing a binary search for the band that encompasses the optimal path through the path graph. Specifically, one starts with an initial guess for $d$, e.g., by setting $d = 1$. The banded alignment algorithm is then executed to find an alignment that occurs within the band of width $d$. If the resulting alignment contains fewer than $d$ edits, then this is also the optimal alignment. If the alignment contains more than $d$ edits, then $d$ is doubled, and the process is repeated until an optimal alignment is found.

**4.** The dynamic algorithm presented in Fact 3 also takes $O(\min\{m,n\}d)$ time, though it can be up to four times slower than an algorithm that starts with an initial correct guess for the value of $d$.

**5.** A special case of alignment occurs when one sequence is much longer than the other one, e.g., when aligning one short sequence against an entire genome, or when aligning a sequence against a database. In such situations, a run time proportional to the product of the lengths of the two sequences can be prohibitive. Heuristics have been developed that can focus on the regions of the alignment table that contain high-quality alignments.

**6.** Suppose we are searching for an alignment of a (query) sequence of length $m$ within a longer sequence (database) of length $n$, where $m \ll n$. Specifically, we are looking for alignments with up to $d$ differences (indels or substitutions). Then any inexact alignment with $d$ or fewer differences must contain an exact match between the query sequence and the database that is of length at least $\lfloor m/(d+1) \rfloor$. Indeed, the worst-case scenario has the $d$ differences equally separated along the sequence, breaking it up into $d+1$ segments.

**7.** *Exclusion methods*: These approaches, based on Fact 6, use exact matching to speed up inexact alignment [WuMa92]. First we identify all exact matches of length $\lfloor m/(d+1) \rfloor$ between the two strings, corresponding to diagonal segments within the path graph. Then we perform the banded alignment procedure to find the highest scoring alignment within the neighborhood of the exact match. This procedure is guaranteed to find an optimal alignment with fewer than $d$ edits, if one exists, and is faster than the full alignment algorithm as only a small fraction of the entire path graph needs to be explored. Regions of the path graph that cannot contain good alignments are excluded from consideration.

**8.** A key factor determining the speed-up we can expect is the length of the exact match being sought: $k = \lfloor m/(d+1) \rfloor$. For small values of $k$ (i.e., when the error being tolerated is high with respect to the size of the query sequence), the expected number of $k$-mers that yield a match between the sequence and the database can be very high, making this approach inefficient.

**9.** The size of $k$ specified in Fact 8 is unnecessarily conservative as it derives from a worst-case scenario where errors are equally distributed throughout the sequence. A random distribution of errors within the sequence is unlikely to be uniform and so contains exact segments that can be much longer than $\lfloor m/(d+1) \rfloor$.

**10.** The value of $k$ could be determined from theoretical principles, e.g., by setting $k$ to be the expected length of the longest edit-free segment within an alignment of length $m$ that contains $d$ randomly distributed edits. In practice, the exact value of $d$ is usually not known, and $k$ is set empirically to a value that minimizes the number of alignments that need to be performed without missing too many alignments.

**11.** To increase sensitivity without sacrificing speed, Ma et al. [MaTrLi02] suggested the use of spaced seeds. The intuition behind the use of spaced seeds is that the "don't care" symbols allow the seed to match even if a difference occurs in the middle of the seed.

**12.** Spaced seeds have a higher sensitivity than the exact $k$-mer seeds of the same weight without a corresponding increase in the number of matches that need to be explored. In practice, algorithms relying on spaced seeds use a combination of multiple seeds in order to maximize sensitivity. Finding the optimal combination of seeds is NP-hard [MaLi07]; however simple heuristic algorithms are highly effective in practice [IIIl07].

**13.** Spaced seeds can be generalized by extending the original observation underlying all exclusion methods—that any alignment of high enough fidelity must contain a long enough exact match between the query and the database. It can be shown [GhPo09], [KaNa07] that any alignment of high enough fidelity must contain a long enough inexact alignment of higher fidelity than the original one. Specifically, if strings $L_1$ and $L_2$ of length $\ell$ match with an edit distance $k$, there exist substrings $E_1$ in $L_1$ and $E_2$ in $L_2$ of length $e$ such that the two substrings match with an edit distance no greater than $k/(\ell - e)$ [GhPo09].

**14.** Using Fact 13, one could search for inexact seeds that match between a query and a database using a fast inexact alignment algorithm that can tolerate a limited number of edits, and then use the resulting anchors to search for the full alignment. This generalization provides a higher sensitivity than approaches based on spaced seeds.

**Examples:**

**1.** The following figure illustrates a path graph with a band of width $d = 1$ indicated. An optimal alignment with at most $d$ edits must occur within a band of width $d$ on either side of the diagonal. Paths traversing the grayed nodes involve more than $d$ edits.

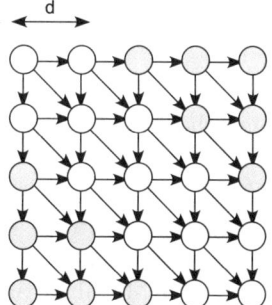

**2.** In the worst-case scenario, the $d$ edits are equally spaced along the sequence, implying that exact matches between the two strings are at least $\lfloor m/(d+1) \rfloor$ characters long:

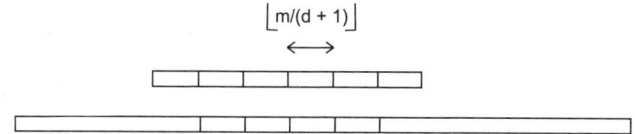

**3.** The spaced seed $111 * 111$ denotes the set of $k$-mers of width 7 and weight 6. This seed allows the middle nucleotide to differ between the query sequence and the database sequence. In the following two strings, a spaced seed of width 7 is shared by two strings. Only the characters aligned to 1s in the spaced seed need to match between the two strings.

```
TATT CGTCTGA TCAT
     111*111
GTCT CGTATGA TGCT
```

## 20.1.8   EXACT STRING MATCHING

Underlying the exclusion methods described in §20.1.7 is the assumption that finding exact matches between two strings is computationally more efficient than finding inexact alignments. Exact string matching is of independent interest as well in many applications (e.g., web search engines largely rely on exact matching of query strings). As such, the problem has been extensively researched and many variants and algorithms have been described in the literature. Here we highlight several key results.

### Definitions:

Given a string $S$, a **suffix tree** is a compact representation of all suffixes in $S$. It is a tree that contains a leaf for each suffix in $S$, and where each edge is labeled with a string of characters so that the path from the root to each leaf spells the corresponding suffix. The string spelled by the path from the root of a suffix tree to an internal node is termed the **string label** of the node.

The suffix tree also contains a collection of **suffix links**, special edges connecting the node in the tree with string label $c\alpha$ (where $c$ is a single character and $\alpha$ is a string) to the node with string label $\alpha$.

The **suffix array** is a lexicographically sorted array of all suffixes of the string, together with information necessary for speeding up the search process.

### Facts:

**1.** A naive search for exact matches between a query string of length $m$ and a database of length $n$ requires $O(mn)$ run time, by checking whether all $m$ letters in the query match, starting at each of the $n - m + 1$ positions within the database.

**2.** *KMP algorithm*: Knuth, Morris, and Pratt [KnMoPr77] reduced the run time for exact matching to $O(m + n)$ by simple preprocessing of the query string. Specifically, at each position $i$ in the query, the KMP algorithm defines $sp_i$ to be the length of the longest nontrivial prefix of the query that matches a suffix of the string that ends at position $i$.

**3.** This information can be computed in $O(m)$ time and can be used to speed up alignment as follows. The query is compared to the database, one letter at a time, until a mismatch is detected at position $i + 1$ in the query. If $sp_i$ is defined, the query is shifted to the right so that the prefix of the query aligns to the corresponding suffix of the string ending at position $i$ prior to the shift. The comparison to the database continues from position $sp_i + 1$, as the properties of the $sp_i$ values guarantee that all prior positions match the database sequence (these were already matched prior to the shift). If $sp_i = 0$, the query is shifted such that its first letter is aligned to the character mismatched prior to the shift, and the comparison between the query and the database starts from the first letter of the pattern.

**4.** Two observations establish an $O(n)$ run time for the search procedure, giving an $O(m + n)$ run time once the cost of preprocessing is taken into account.

- Once a character in the database is successfully matched to the query it is never again compared to the query, thereby bounding the total number of exact matches performed during the execution of the algorithm to $n$, the size of the database.

- Every mismatch triggers a shift of the query by at least one position; thus the number of mismatches is also bounded by $n$, yielding an overall $O(n)$ run time.

**5.** The KMP algorithm can be used to find full-length matches of a query against a database, as well as matches that involve just a prefix of the query (e.g., finding the longest prefix match in case a full-length match cannot be found). To address more complex exact matching problems, Weiner [We73] and McCreight [Mc76] independently developed suffix trees.

**6.** The strings represented by the path from the root to any node in the suffix tree are unique. In other words, the shared prefixes of strings represented in the tree are represented by the same path from the root to some internal node in the tree.

**7.** To ensure that the suffixes of $S$ reach the leaves of the tree (i.e., no suffix is a prefix of some other suffix in the tree), a special character $ not found in the alphabet is appended to the end of each string stored in the suffix tree.

**8.** Suffix links are useful in the efficient construction of suffix trees as well as for solving certain exact matching problems, such as identifying the longest substring that matches between the query and the database.

**9.** When appropriately structured, suffix trees require $O(m)$ space to store all the suffixes of a given string of length $m$. To achieve such space efficiency, each edge is labeled not with its string label, but with the coordinates within the string where the string label occurs (see Example 2). Each edge in the tree thus requires constant storage space, and the total size of the tree is determined by the number of leaves, equal to the number of suffixes $m$.

**10.** If a query is represented within a suffix tree, it can be aligned against a database as follows. Starting with the root of the tree, the matching process follows the edges of the tree that match the database sequence until either reaching a leaf (in which case the corresponding suffix represents a partial match between the query and the database) or a mismatch occurs. The algorithm then follows the suffix link starting from the node immediately above the location of the mismatch and the matching process continues from the same position in the database, starting from the new location in the tree. Following the suffix link implicitly discards the first character in the database and is equivalent to the shift procedure described in the KMP algorithm.

**11.** The search procedure requires $O(m+n)$ time, matching the time complexity of the KMP algorithm.

**12.** Despite its asymptotically linear space complexity, the suffix tree data structure requires a substantial amount of memory, estimated at over 20 bytes per letter. Manber and Myers [MaMy93] described an alternative approach for storing all the suffixes of a string in a compact way that uses substantially less space at the cost of a small additional increase in run time.

**13.** In a suffix array, the suffixes themselves are stored as simply an index within the original string, ensuring that the space usage is proportional to the size of the string even though the total length of all suffixes combined is proportional to the length of the string.

**14.** In addition to the suffixes, the data structure for a suffix array comprises an array that stores the *longest common prefix* (LCP) between adjacent suffixes. This information is necessary to efficiently search within the suffix array using binary search with just an additive overhead of $O(\log m)$. A trivial implementation of binary search within an array of $m$ strings of length $O(m)$ requires $O(m \log m)$ time. Since the LCP values are only used for the binary search process, it is sufficient to record just the longest common prefixes of the pairs of strings that will be compared during the binary search, or $O(m)$ values. Suffix arrays thus occupy $O(m)$ space, and a sequence of length $n$ can be searched against a suffix array in $O(n + \log m)$ time.

**15.** Constructing a suffix array involves sorting the set of suffixes of a string, a process that can be performed in $O(m \log m)$ time using an algorithm that takes advantage of the relationship between the suffixes of a same string [MaMy93].

**Examples:**

**1.** The following diagram illustrates the $sp_i$ values used by the KMP algorithm and a shift after a mismatch occurs at position $i$ in the query. The shaded regions correspond to identical sequences. After the shift, matching continues from the position marked X.

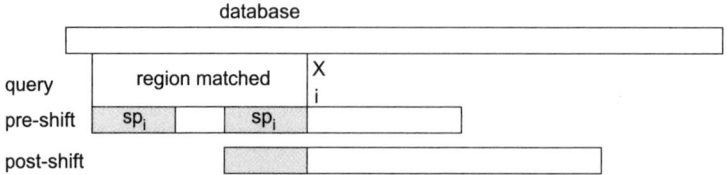

**2.** The following figure shows a suffix tree for the string BABBAB. The dashed line is a suffix link connecting the node labeled BAB to the node labeled AB. Edge labels are provided for clarity, but an efficient implementation would compress them to two integers. For example, the edge BAB$ can be represented as $(4, 7)$, the range within the original string that spells this string.

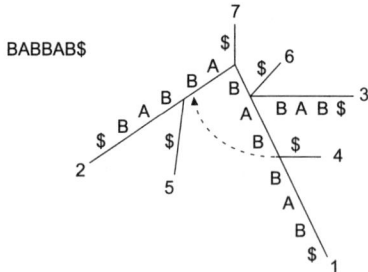

## 20.1.9   INDEXING METHODS FOR STRING MATCHING

When searching for multiple query sequences against the same database, search efficiency can be increased by preprocessing or indexing the database. For the following, assume we are searching for exact matches between multiple sequences of length $m$ and a database of length $n$.

**Facts:**

**1.** Storing the database in a suffix tree (§20.1.8) enables exact matching queries to be executed in time independent of the database size. For each query sequence, the matching process traverses the suffix tree from the root, following the path that matches the query until either a match is found, or a mismatch indicates the query does not match within the database, yielding a run time of $O(m)$ per query.

**2.** As discussed in §20.1.8, the main drawback of this strategy is the large memory necessary to store the suffix tree.

**3.** *Inverted index:* Finding short exact matches of length $k$ between one or more query sequences and a database, a component of the exclusion methods described in §20.1.7, can be performed with a simple (inverted) index linking all strings of length $k$ within the database to their location in the database. Lookups within the index can be performed efficiently if the entire index is stored in memory; however, the memory space necessary to store all $k$-mers within a database can be very large. Furthermore, $k$-mers that occur frequently within the database impact both memory usage and the time needed to search the index, and are usually excluded from the index in practical implementations.

**4.** *Burrows-Wheeler transform (BWT)*: A reversible permutation of the indexed string can be used to construct a memory-efficient indexing structure called the *FM index* [FeMa00]. The Burrows-Wheeler transform [BuWh94] of a string $S$ can be constructed using Algorithm 2. It is assumed here that the added character $ is lexicographically smallest.

---

**Algorithm 2**:   **Burrows-Wheeler transform.**

input: string $S$

output: Burrows-Wheeler transform of $S$

augment string $S$ by appending to its end a character $ not found in the alphabet

construct a table $T$ comprising all circular rotations of $S$ as rows

sort the rows of $T$ into lexicographic order

return the last column in $T$ as the Burrows-Wheeler transform of $S$

---

**5.** The BWT and the suffix array of $S$ are conceptually linked. The BWT represents the characters that precede the lexicographically ordered suffixes of $S$, and the BWT can be trivially constructed from the suffix array of $S$. See Example 2.

**6.** The BWT can be reversed, obtaining the original string, with the help of a simple observation: the order in which multiple instances of the same character occur in the first column of the sorted BWT table is the same as the order in which these characters occur in the last column of the table. Specifically, the first A in the last column corresponds to the same character in the original string as the first A in the first column, the third C in the last column corresponds to the same character in the original string as the third C in the first column, etc. This *last-to-first* (LF) mapping guides the reversal of the BWT; see Algorithm 3.

---

**Algorithm 3**:   **BWT reversal.**

input: BWT table $T$ for $S$

output: original string $S$

start with the first row of the BWT table and output the character $c$ in the last column; this is the last character in the original string

**repeat until** the entire string is reconstructed

use the LF property to identify the corresponding instance of $c$ in the first column

prepend the character at the end of the respective row to the string reconstructed so far

---

**7.** In Algorithm 3 only the first and the last columns of the BWT matrix are necessary to reverse the transformation. The last column is the BWT itself, while the first column is simply the lexicographically sorted list of characters in the original string.

**8.** To efficiently perform the LF mapping it is necessary to know the index of each character within the last column. That is, for a character $c$ within the $i$th row of the matrix we need to know the number of characters equal to c that occur within the rows above. Storing this information for each row of the matrix allows the LF lookup to occur in constant time; however, it requires a large amount of memory $O(n \log n)$ for a string of length $n$. A trade-off between run time and memory usage can be obtained by storing the index information at every $b$th row, yielding a memory usage of $O((n \log n)/b)$. Each LF mapping, however, requires $O(b)$ time as the index values need to be computed on the fly within the block between the selected rows. See Example 5.

**9.** The same procedure used for reversing the BWT can be used to match a query $Q$ against a database stored in the BWT. This process is similar to a binary search in that the transitions between the first and last columns of the BWT matrix repeatedly shrink the range within which the query string may be found. See Algorithm 4.

---

**Algorithm 4**:  **BWT query match.**

input: query $Q$, BWT matrix $T$

output: matching string in the database

identify the block of rows in $T$ starting with the last character in $Q$

**repeat until** the query is fully matched

within the last column of these rows, find those rows containing the rightmost unmatched character in $Q$; if no rows end in this character, no match has been found

within the first columns of the BWT matrix, identify the rows starting with the characters found above

---

**Examples:**

**1.** We illustrate the Burrows-Wheeler transform of the string BABBAB. The character $ is added to this string, giving the first row of the table shown on the left; the rows of this table represent successive circular rotations. After lexicographically sorting the rows, we obtain the BWT matrix on the right, whose last column is the Burrows-Wheeler transform of the original string.

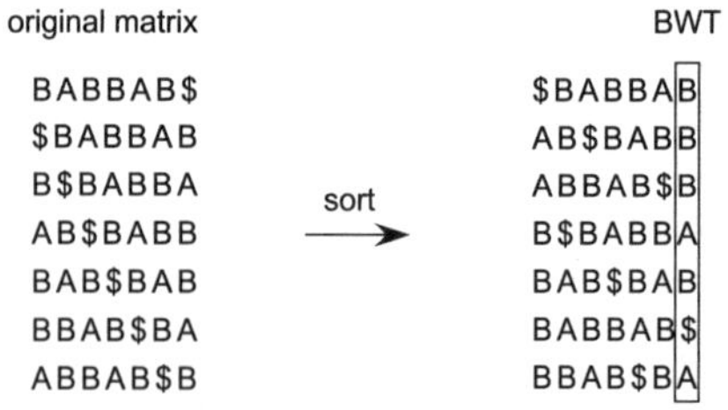

|  original matrix | | BWT |
|---|---|---|
| BABBAB$ | | $BABBA**B** |
| $BABBAB | | AB$BAB**B** |
| B$BABBA | sort | ABBAB$**B** |
| AB$BABB | $\longrightarrow$ | B$BABB**A** |
| BAB$BAB | | BAB$BA**B** |
| BBAB$BA | | BABBAB**$** |
| ABBAB$B | | BBAB$B**A** |

**2.** In the following figure, the suffix array of the string in Example 1 is shown highlighted in bold within the Burrows-Wheeler matrix.

BWT

$BABBA**B**
AB$BAB**B**
ABBAB$**B**
B$BABB**A**
BAB$BA**B**
BABBAB**$**
BBAB$B**A**

**3.** We illustrate the LF mapping for the BWT matrix shown in Example 1. The first B in the last column is the fourth B occurring in the original string BABBAB; the first B in the first column also corresponds to the fourth B in the original string. The second B in the last column is the third B occurring in the original string BABBAB; the second B in the first column also corresponds to the third B in the original string.

**4.** Here we show how to reverse the BWT in Example 1 to generate the original string BABBAB, starting from the rightmost character. The arrows indicate the LF mapping. The string being reconstructed is shown above the table with the latest character added shown in bold. Note that only the first and last columns of the table are needed for this operation.

| **B** | **AB** | **BAB** | **BBAB** | **ABBAB** | **BABBAB** |
|---|---|---|---|---|---|
| $BABBA**B** | $BABBAB | $BABBAB | $BABBAB | $BABBAB | $BABBAB |
| AB$BABB | AB$BABB | **AB**$BAB**B** | AB$BABB | AB$BABB | AB$BABB |
| ABBAB$B | ABBAB$B | ABBAB$B | ABBAB$B | ABBAB$B | **ABBAB$B** |
| B$BABBA | **B**$BABB**A** | B$BABBA | B$BABBA | B$BABBA | B$BABBA |
| BAB$BAB | BAB$BAB | BAB$BAB | **BAB$BAB** | BAB$BAB | BAB$BAB |
| BABBAB$ | BABBAB$ | BABBAB$ | BABBAB$ | BABBAB$ | BABBAB$ |
| BBAB$BA | BBAB$BA | BBAB$BA | BBAB$BA | **BBAB$BA** | BBAB$BA |

**5.** A character index is needed for performing the LF mapping. The full index is shown on the left. On the right is a sparse index that stores the information every $b$ rows.

|  | A | B |  |  | A | B |
|---|---|---|---|---|---|---|
| $BABBAB | 0 | 1 |  | $BABBAB | 0 | 1 |
| AB$BABB | 0 | 2 |  | AB$BABB |  |  |
| ABBAB$B | 0 | 3 |  | ABBAB$B |  |  |
| B$BABBA | 1 | 3 |  | B$BABBA | 1 | 3 |
| BAB$BAB | 1 | 4 |  | BAB$BAB |  |  |
| BABBAB$ | 1 | 4 |  | BABBAB$ |  |  |
| BBAB$BA | 2 | 4 |  | BBAB$BA | 2 | 4 |

**6.** We illustrate the search process using a BWT/FM-index. The search range is iteratively refined as more characters in the query string ABB are being matched, starting with the rightmost character. The matched characters are shown in bold.

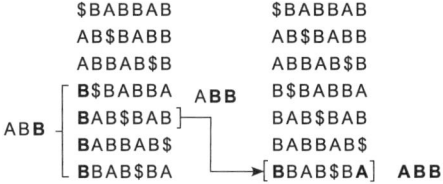

# 20.2   PHYLOGENETICS

Phylogenetics is the study of the evolutionary relationships among a collection of organisms. These relationships are displayed in the form of a tree whose leaves are labeled by the organisms. Internal vertices of the tree represent common ancestors. Biologists apply statistical and combinatorial techniques to construct trees from data such as DNA sequences. Phylogenetic trees are used in epidemiology, systematics, ecology, and linguistics. We review the basic combinatorial principles of the discipline. Most of the results can be found in [SeSt03] and [DrEtal11]. We refer the reader to [Fe04] for a biological perspective on phylogenetics which is grounded in sound mathematics. Discrete mathematicians may enjoy the recent textbook [HuRuSc10] on phylogenetic networks, while [PaSt05] provides a viewpoint on the subject from the perspective of algebraic geometry.

## 20.2.1   Phylogenetic Trees

### Definitions:
The **leaves** of a tree are the vertices incident with a single edge.

The leaves of a **phylogenetic tree** are labeled with a set of operational taxonomical units or **taxa**. A phylogenetic tree may be rooted or unrooted.

An **internal edge** is not incident with any of the taxa.

A vertex of a phylogenetic tree with degree greater than three is called a **polytomy**. A **binary tree** has no polytomies.

Two leaves form a **cherry** if they are the only leaves adjacent to a common vertex.

**Phylogenetic reconstruction** is the process of transforming data about the relationships among a particular set of taxa into a phylogenetic tree which represents the underlying relationships.

### Facts:
**1.** Evolution is thought of as a binary process. Therefore most phylogenetic reconstructions aim to return binary trees.

**2.** Operational taxonomical units typically refer to a set of organisms or a set of genes. However, they may also refer to non-living objects such as languages.

**3.** The input to a phylogenetic reconstruction algorithm is typically an aligned collection of nucleotide sequences, amino acid sequences, or discrete morphological data.

**4.** An unrooted binary tree with $n \geq 3$ taxa has $n - 3$ internal edges. A rooted binary tree has a single additional internal edge.

**5.** For $n$ taxa the number of rooted binary phylogenetic trees is

$$(2n - 3)!! = 1 \cdot 3 \cdot 5 \cdot 7 \ldots (2n - 3).$$

This is also the number of unrooted binary trees with $n + 1$ taxa. As there are more 57-taxa trees than there are atoms in the visible universe, phylogenetic reconstruction algorithms are often restricted from considering all possible trees as models.

**Example:**

**1.** The following figure displays a phylogenetic tree with 11 internal edges and 15 taxa $\{a, b, \ldots, o\}$. This tree has four cherries $(ab, ef, gh$ and $kl)$ and a single polytomy adjacent to taxa $m, n$ and $o$.

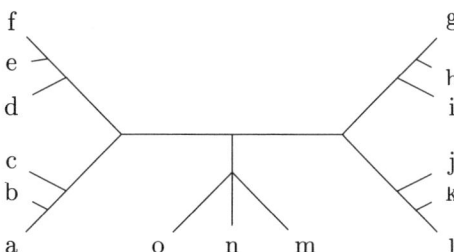

---

## 20.2.2   Tree Comparisons

When testing phylogenetic reconstruction algorithms, one generates data assuming a fixed tree and model of evolution. Then one reconstructs trees using this simulated data. To test the performance of the reconstruction algorithm, the reconstructed trees are then compared to the fixed tree using a tree metric.

**Definitions:**

Every edge in a phylogenetic tree induces a **split** or partition of the set of taxa into two disjoint subsets.

The **Robinson-Foulds distance** between two trees is the number of splits which occur in one tree but not the other.

Each internal edge divides a binary tree into four subtrees. A **nearest neighbor interchange** exchanges two of the subtrees on opposite sides of the internal edge.

The **nearest neighbor interchange distance** between two binary trees is the minimum number of nearest neighbor interchanges required to transform one tree into the other.

**Facts:**

**1.** Both the Robinson-Foulds distance and the nearest neighbor interchange distance define metrics on the set of $n$-taxa trees.

**2.** The Robinson-Foulds distance can be computed in linear time [Da85], while computing the nearest neighbor interchange distance is an NP-complete problem [LiTrZh96].

**3.** Asymptotically, the proportion of trees that share $s$ splits with a tree $T$ follows a Poisson distribution with mean $\lambda = \frac{c}{2n}$, where $n$ is the number of taxa and $c$ is the number of cherries of $T$ [BrSt09].

**4.** There exists a Hamilton walk connecting all binary $n$-taxa trees through a sequence of nearest neighbor interchanges [GoFoStJ13].

**5.** For an $n$-taxa tree, the nearest neighbor distance between any two trees is at most $n \log n + O(n)$ [LiTrZh96].

**6.** Maximum likelihood reconstruction assigns a likelihood score to each tree. Biologists select the tree which maximizes the likelihood of observing the data given the particular tree. In practice it is not feasible to compute a likelihood score for all possible trees. To address this issue a *guide tree* is computed using a fast algorithm. Likelihood scores for trees near the guide tree under the nearest neighbor interchange distance are compared to estimate the maximum likelihood tree [TaEtal11].

**Examples:**

**1.** Three trees differing by a single nearest neighbor interchange are shown in the following figure. Here $S_1, S_2, S_3, S_4$ represent subtrees connected by an internal edge $e$.

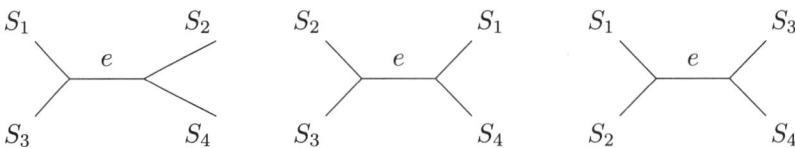

**2.** The tree shown on the left of the next figure has splits $\mathcal{S} = \{ab|cdef, abc|def, abcd|ef\}$. The tree pictured on the right has splits $\mathcal{S}' = \{ab|cdef, abd|cef, abdf|ce\}$.

Since there are four splits which appear in one tree but not the other (all but $abcd|ef$) the Robinson-Foulds distance between these trees is four.

**3.** The nearest neighbor interchange distance between the trees above is two as is demonstrated by the sequence of nearest neighbor interchanges shown in the following figure.

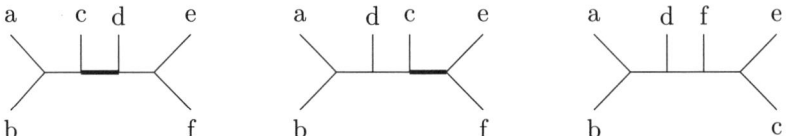

The nearest neighbor interchange distance cannot be smaller than two, since the trees differ by a Robinson-Foulds distance of four, and each interchange can only change a single split, so the difference in the number of splits can change by a maximum of two.

---

## 20.2.3   Tree Agreement

There are a variety of algorithms for combining trees on a shared set of taxa. These methods may seek to reconcile the tree differences or highlight their similarities.

**Definitions:**

Let $T$ be a phylogenetic tree on a set of taxa $\mathcal{X}$. For every subset $\mathcal{S} \subseteq \mathcal{X}$, the **induced subtree** $T|\mathcal{S}$ is the tree constructed by taking the unique minimal connected subgraph of $T$ containing the leaves in $\mathcal{S}$, and then removing all non-root vertices of degree two.

Given trees $T_1, T_2, \ldots, T_k$, a **maximum agreement subtree** is the induced subtree on the largest set $\mathcal{S} \subseteq \mathcal{X}$ such that $T_1|\mathcal{S} = T_2|\mathcal{S} = \cdots = T_k|\mathcal{S}$. The maximum agreement subtree need not be unique.

Given trees $T_1, T_2, \ldots, T_k$ on a set of taxa $\mathcal{X}$, the **majority rules consensus tree** is the tree $T$ containing exactly the splits which occur in the majority of the trees $T_1, T_2, \ldots, T_k$.

### Facts:

**1.** For any collection of trees on a set of taxa, both the majority rules consensus tree and the maximum agreement subtree exist. The majority rules consensus tree contains all of the taxa, while the maximum agreement subtree typically contains a proper subset of the taxa.

**2.** The majority rules consensus tree can be computed in linear time [AmClStJ03], while the maximum agreement subtree can be computed in $O(n \log n)$ time [CoEtal00].

**3.** Two random binary phylogenetic trees on $n$ taxa share a maximum agreement subtree on $\Omega(\log n)$ leaves [BrMcKSt03].

### Examples:

**1.** The induced subtree on the set $\{a, b, c, d\}$ for the 15-taxa figure shown next is the 4-taxa tree on the right.

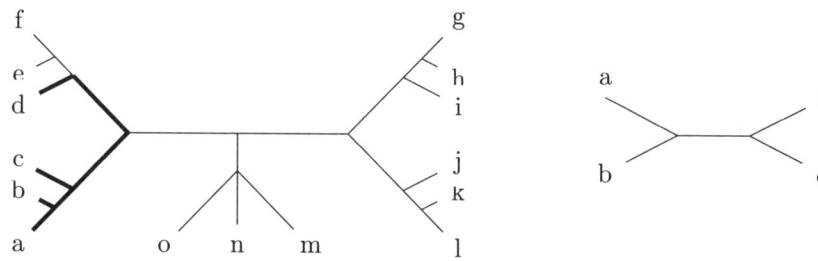

**2.** In the next figure, a maximum agreement subtree of the two 6-taxa trees pictured on the left is the 4-taxa tree pictured on the right.

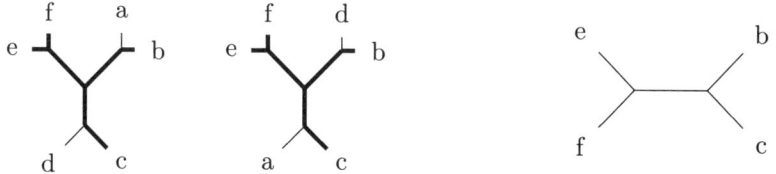

**3.** The majority rules consensus tree will contain all of the taxa. The two trees from the preceding figure share only the split $(abcd|ef)$ so their majority rules consensus tree is forced to contain a polytomy, as pictured on the following page.

---

### 20.2.4   Tree Reconstruction

Distance methods of phylogenetic reconstruction algorithms were among the first applied to DNA sequence data and remain in use due to their speed and simplicity.

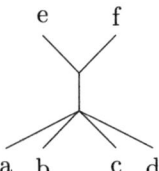

## Definitions:

One may assign a positive **branch length** to each edge of a phylogenetic tree.

Given a collection of $n$ taxa, a **dissimilarity matrix** is a nonnegative, symmetric $n \times n$ matrix where entry $a_{i,j}$ indicates the difference between taxa $i$ and taxa $j$. We assume $a_{i,i} = 0$ for all $1 \leq i \leq n$.

A dissimilarity matrix is a **metric** if $a_{i,k} \leq a_{i,j} + a_{j,k}$ for all $i, j, k$.

A dissimilarity matrix is a **tree metric** if there exists a phylogenetic tree $T$ such that for each pair $i,j$ the matrix entry $a_{i,j}$ is the sum of the branch lengths along the unique path from leaf $i$ to leaf $j$.

A dissimilarity matrix is **ultra-metric** if it is a tree metric for a rooted tree $T$ in which every leaf is equidistant from the root.

## Facts:

**1.** A dissimilarity matrix is frequently computed using a modified Hamming distance on a pair of DNA sequences. The number of evolutionary changes is under-counted by the Hamming distance. Various distance functions such as the Jukes-Cantor distance [JuCa69] have been developed to account for these missing changes.

**2.** Every metric on three or fewer taxa is a tree metric.

**3.** When there are at least four taxa, a dissimilarity matrix is a tree metric if and only if for all $i, j, k, l$ we have $a_{i,j} + a_{k,l} \leq \max\{a_{i,k} + a_{j,l}, a_{i,l} + a_{j,k}\}$. This condition is known as the *four point condition*.

**4.** The set of tree metrics can be described by a *tropical variety* [SpSt04], while the set of ultra-metric trees can be described as a *polyhedral fan* [ArKl06].

**5.** With over 40,000 citations, neighbor joining [SaNe87], described in Algorithm 1, is the most popular distance-based algorithm.

---

**Algorithm 1: Neighbor joining.**
input: dissimilarity matrix $M = (m_{i,j})$
output: phylogenetic tree $T$
1. construct the matrix $S = (s_{i,j})$ with
$$s_{i,j} = (n-2)m_{i,j} - \sum_{k=1}^{n}(m_{i,k} + m_{j,k}).$$
2. find a pair of taxa $i$ and $j$ which minimize $s_{i,j}$; construct a vertex $A$ on the tree which joins $i$ and $j$.
3. calculate the distance from the taxa $i$ and $j$ to the new vertex $A$ using the formula $d_{i,A} = \frac{1}{2}m_{i,j} + \frac{1}{2(n-2)}\sum_{k=1}^{n}(m_{i,k} - m_{j,k})$.
4. calculate the distance from all other taxa $x$ to the new vertex $A$ using the formula $d_{x,A} = \frac{1}{2}(m_{x,i} + m_{x,j} - m_{i,j})$.
5. repeat Step 1 using the dissimilarity matrix constructed by replacing the taxa $i$ and $j$ with the single newly constructed vertex $A$.

---

**Examples:**

1. Given the dissimilarity matrix $M = \begin{bmatrix} 0 & 3 & 8 & 9 \\ 3 & 0 & 9 & 10 \\ 8 & 9 & 0 & 9 \\ 9 & 10 & 9 & 0 \end{bmatrix}$ for the set $\mathcal{X} = \{a, b, c, d\}$,

we note that $m_{a,b} + m_{c,d} = 12, m_{a,c} + m_{b,d} = 18$, and $m_{a,d} + m_{b,c} = 18$. The four point condition is satisfied, as each individual sum is less than or equal to the maximum of the other two. The four point condition can be reformulated to state that the maximum value of the three sums is obtained by more than one of the three pairs of taxa (here $m_{a,c} + m_{b,d}$ and $m_{a,d} + m_{b,c}$). This viewpoint leads to the interpretation of the space of trees as a *Tropical Grassmanian* (see [PaSt05] for details).

2. Since $M$ satisfies the four point condition, we can use the neighbor joining algorithm to compute a tree whose pairwise distances match the dissimilarity matrix. First we

compute the associated matrix $S = \begin{bmatrix} 0 & -36 & -30 & -30 \\ -36 & 0 & -30 & -30 \\ -30 & -30 & 0 & -36 \\ -30 & -30 & -36 & 0 \end{bmatrix}.$

We choose $s_{a,b}$ as the minimum entry and construct a new vertex $A$ connecting the first two taxa. Using Steps 3 and 4 from Algorithm 1 we compute $d_{a,A} = 1, d_{b,A} = 2, d_{c,A} = 7, d_{d,A} = 8$. This gives the following tree.

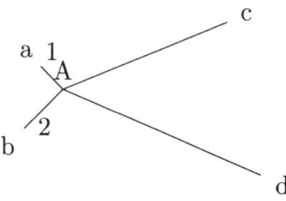

We repeat the process on the new matrix where the first row and column now represent

the newly constructed vertex $A$. First we compute $M = \begin{bmatrix} 0 & 7 & 8 \\ 7 & 0 & 9 \\ 8 & 9 & 0 \end{bmatrix}$. Then calculating

the $S$ matrix yields $S = \begin{bmatrix} 0 & -24 & -24 \\ -24 & 0 & -24 \\ -24 & -24 & 0 \end{bmatrix}$. Since there is a tie between all scores it

doesn't matter which pair we join. For convenience we select $m_{A,c}$ to construct a new vertex $B$ which connects $A$ and the taxa $c$. Then we compute $d_{A,B} = 3, d_{c,B} = 4$ and $d_{d,B} = 5$. Neighbor joining returns the phylogenetic tree shown in the following figure. Notice that the pairwise distances among the taxa on this tree exactly match those in the dissimilarity matrix.

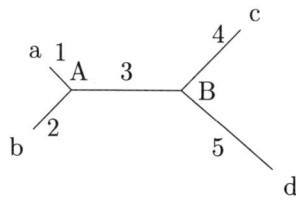

### 20.2.5   Tree Distributions

We would like to compare trees reconstructed from biological data to random trees. To make this notion precise requires an underlying distribution on the space of trees.

**Definitions:**

The **uniform distribution** on the space of rooted or unrooted trees assumes that each tree occurs with equal likelihood.

A **ranked** phylogenetic tree is a rooted binary tree where each internal vertex is assigned a number from $1, 2, \ldots, n-2$ such that if $v_i$ is contained in the path from the root to a vertex $v_j$, then the number assigned to $v_i$ must be less than the number assigned to $v_j$.

The **Yule-Harding distribution** on rooted binary phylogenetic trees is constructed by taking the uniform distribution on ranked phylogenetic trees and ignoring the rank.

**Facts:**

**1.** There are $\dfrac{n!(n-1)!}{2^{n-1}}$ ranked binary trees on a set of $n$ taxa.

**2.** The probability of selecting a rooted binary tree $T$ under the uniform model is $\dfrac{1}{(2n-3)!!}$. Under the Yule-Harding model this probability is $\dfrac{2^{n-1}}{n! \prod\limits_{v \in T} (Des(v) - 1)}$, where

$Des(v)$ denotes the number of leaves which are descendants of an internal vertex $v$.

**3.** In general, trees generated under Yule-Harding are more balanced about the root than trees generated under the uniform distribution.

**4.** Trees generated under a broad class of evolutionary assumptions follow the Yule-Harding distribution [LaSt13].

**Examples:**

**1.** The probability of observing the tree in the following figure is $\frac{1}{15}$ under the uniform distribution as it is one of the fifteen rooted 4-taxa trees.

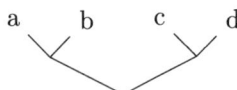

**2.** The root of the tree above has four leaf descendants, and each of the other two internal vertices have two descendants. This allows us to compute the probability of this tree under the Yule-Harding model as

$$\frac{2^{n-1}}{n! \prod\limits_{v \in T} (Des(v) - 1)} = \frac{8}{24(4-1)(2-1)(2-1)} = \frac{1}{9}.$$

**3.** The probability of a tree under the Yule-Harding model is completely determined by the tree topology. The three 4-taxa trees with the balanced topology all occur with probability $\frac{1}{9}$. The twelve 4-taxa trees which have unbalanced topology around the root occur with probability $\frac{1}{18}$. Under the uniform distribution $\frac{1}{5}$ of the rooted 4-taxa trees sampled are balanced, while under the Yule-Harding model $\frac{1}{3}$ are balanced.

## 20.2.6 Subtrees and Supertrees

Biologists seek to reconstruct trees of ever increasing size with the ultimate goal of building a tree containing every living organism. To reconstruct large trees, it is important to determine which collection of subtrees best reflects the structure of the tree itself. These subtrees can then be used as inputs into a supertree algorithm designed to reconstruct large trees from a collection of smaller input trees.

### Definitions:

A **quartet tree** is a binary tree with four leaves. An unrooted quartet tree $T$ is denoted $ab|cd$ if $a$ and $b$ form a cherry of $T$.

The **support of a tree**, denoted $supp(T)$, is the collection of taxa at the leaves of $T$.

A tree $T_1$ **displays** a tree $T_2$ if $T_1|supp(T_2) = T_2$.

A collection of trees $T_1, T_2, \ldots, T_n$ is called **definitive** if there exists a unique tree which displays all of the trees in the collection.

A collection of subsets of taxa $\mathcal{S}_1, \mathcal{S}_2, \ldots, \mathcal{S}_k \subseteq \mathcal{X}$ is called **decisive** if for any tree $T$ with taxa $\mathcal{X}$ the subtrees $T|\mathcal{S}_1, T|\mathcal{S}_2, \ldots, T|\mathcal{S}_k$ are definitive.

A collection of subsets $\mathcal{S}_1, \mathcal{S}_2, \ldots, \mathcal{S}_k \subseteq \mathcal{X}$ satisfies the **four-way partition property** if for all partitions of $\mathcal{X}$ into four nonempty sets $\mathcal{X}_1, \mathcal{X}_2, \mathcal{X}_3$ and $\mathcal{X}_4$ there exists a subset $\mathcal{S}_i$ which has a nonempty intersection with each of $\mathcal{X}_1, \mathcal{X}_2, \mathcal{X}_3, \mathcal{X}_4$.

### Facts:

**1.** The collection of all 4-element subsets of a set of taxa is decisive.

**2.** For any unrooted binary $n$-taxa tree, there exists a collection of $n - 3$ quartet trees which are definitive.

**3.** A collection of subsets of taxa $\mathcal{S}_1, \mathcal{S}_2, \ldots, \mathcal{S}_k \subseteq \mathcal{X}$ is decisive if and only if it satisfies the four-way partition property.

### Examples:

**1.** A quartet tree $q = ab|cd$ is shown in the following figure. The support of $q$ is $\{a, b, c, d\}$.

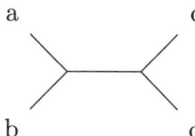

**2.** The quartet tree $q = ab|cd$ is displayed by the following 5-taxa tree since the induced subtree on the support of $q$ is isomorphic to $q$.

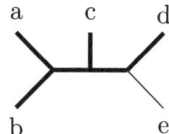

**3.** For the taxon set $X = \{a, b, c, d, e\}$ the quartet trees $ab|cd$ and $ac|de$ are definitive as they are only displayed by the tree in the next figure.

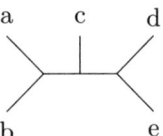

**4.** The quartets $ab|cd$ and $ab|ce$ are not definitive as they are displayed by both trees in the following figure.

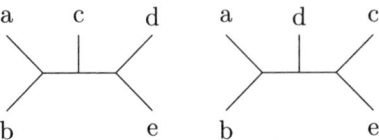

**5.** The previous example also shows that the collection $\mathcal{T} = \{\{a, b, c, d\}, \{a, b, c, e\}\}$ is not decisive. However, the collection of subsets

$$\mathcal{S} = \{\{a, b, c, d\}, \{a, b, c, e\}, \{a, b, d, e\}, \{a, c, d, e\}\}$$

satisfies the four-way partition property and is thus decisive.

## 20.2.7 Applications

### Facts:

**1.** Most phylogenetic reconstruction algorithms return an unrooted tree. For that reason, biologists include a distantly related organism or *outgroup* in their sample to help place the root of the tree.

**2.** Phylogenetic trees provide a tool for understanding biodiversity. The simplest *measure of the biodiversity* of a set of taxa is the sum of the branch lengths in the subtree they induce. This metric is used in conservation biology when making decisions about which organisms to protect.

**3.** It is possible that trees reconstructed from different genes from the same collection of species do not agree. A *coalescent theory* describes how different gene trees can result from a single species tree. In fact the species tree may not even match the most prevalent gene tree [DeRo06].

### Examples:

**1.** The following figure shows how a tiger could be used as an outgroup when reconstructing the evolutionary history of primates.

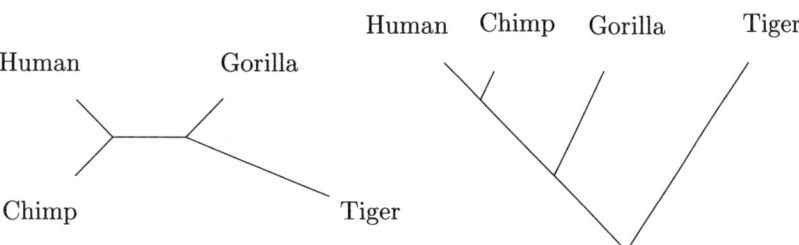

**2.** In the following figure, the taxa $\{a, f, g, l, n\}$ have the maximum biodiversity of any 5-taxa set for this tree.

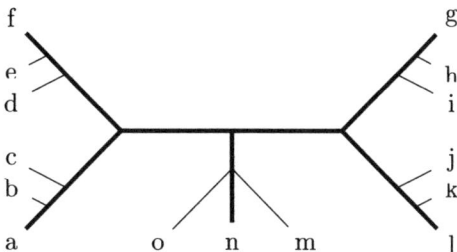

**3.** In the next figure two gene trees evolve under the coalescent model within a species tree which is represented as a band-like tree enclosed in the thick lines. In this figure $a, b$ and $c$ represent distinct species. The dotted gene tree exhibits discordance with the species tree topology. The solid gene tree matches the species tree topology.

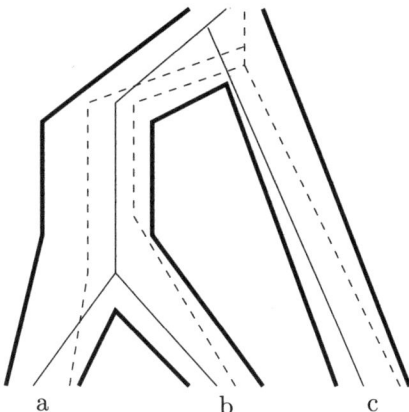

## 20.3  DISCRETE-TIME DYNAMICAL SYSTEMS

Discrete-time dynamical systems are used in mathematical biology to model and simulate interaction-based networks, such as gene regulatory networks. While the behavior of a biological system may be seen as continuous, in that it moves continuously from one state to another, the technology to record observations of the system, such as microarray chips, is not continuous and the available experimental data consist of collections of discrete instances of continuous processes.

### 20.3.1  BASIC CONCEPTS

**Definitions:**

A *discrete-time dynamical system* on $n$ vertices is a triple $(N, \mathbf{X}, F)$ with the following properties:

**1.** $N$ is a set of $n$ variables $x_1, x_2, \ldots, x_n$;

**2.** $\mathbf{X} = X_1 \times \cdots \times X_n$ is the Cartesian product of *state sets*, in which each $X_i$ is the set of values, called *states*, that variable $i$ can take; and

**3.** $F = (f_1, \ldots, f_n) : \mathbf{X} \to \mathbf{X}$ is a function of the variables in $N$, called the *global function*, and each $f_i : \mathbf{X} \to X_i$ is a *transition function*, associated with vertex $x_i$.

The first *iterate* of point $\mathbf{x} = (x_1, \ldots, x_n) \in \mathbf{X}$ is $F(\mathbf{x})$, the second iterate is $F \circ F(\mathbf{x})$, and so on.

A sequence of iterates $\mathbf{x} \mapsto F(\mathbf{x}) \mapsto F \circ F(\mathbf{x}) \mapsto \cdots \mapsto F \circ \cdots \circ F(\mathbf{x})$ is called a *trajectory*. The set of all iterates is called the *orbit* of $\mathbf{x} \in X$. A trajectory that starts and ends at the same point is called a (limit) *cycle*. A point $\mathbf{x} \in \mathbf{X}$ is a *fixed point* (cycle of length one) when $F(\mathbf{x}) = \mathbf{x}$.

A dynamical system is a *finite dynamical system* if each state set $X_i$ is a finite set.

A finite dynamical system is a *polynomial dynamical system* if $\mathbf{X} = k^n$ where $k$ is a finite field.

Let $R$ be a commutative ring. A *polynomial ring* $R[x_1, \ldots, x_n]$ in $n$ variables is the set of all linear combinations of monomials of the form $\mathbf{x}^\alpha = x_1^{\alpha_1} \ldots x_n^{\alpha_n}$ over $R$, where $\alpha$ is the $n$-tuple exponent $\alpha = (\alpha_1, \ldots, \alpha_n) \in \mathcal{Z}_{\geq 0}^n$.

The *support* of $f \in R[x_1, \ldots, x_n]$, denoted $supp(f)$, is a subset $\{x_{i_1}, \ldots, x_{i_m}\}$ of $\{x_1, \ldots, x_n\}$ such that $m$ is the smallest integer with $f \in R[x_{i_1}, \ldots, x_{i_m}]$.

Let $F$ be an $n$-dimensional finite dynamical system. The *dependency graph* of $F$, denoted $\mathcal{D}(F)$, is a directed graph $(V, E)$ where $V = \{x_1, \ldots, x_n\}$ and $E = \{(x_i, x_j) \mid x_i \in supp(f_j)\}$.

The *state-space graph* of an $n$-dimensional polynomial dynamical system $F : k^n \to k^n$, denoted $\mathcal{S}(F)$, is a directed graph $(V, E)$ where $V = k^n$ and $E = \{(a, b) \mid a, b \in V \text{ and } F(a) = b\}$.

**Facts:**

**1.** For simplicity, it is often assumed that all vertices have the same state set, that is, $X_i = X$ for all $i = 1, \ldots, n$ and $\mathbf{X} = X^n$.

**2.** Let $k$ be a finite field. Then every function $f : k^n \to k$ is a polynomial of degree at most $n$.

**3.** The structure of a finite field can be imposed on $X$ by requiring that $: X :$ is a power of a prime number.

**4.** Each transition function $f_i : X^n \to X$ is local in the sense it only depends on those variables that correspond to vertices to which $i$ is connected by an edge in $\mathcal{D}(F)$.

**5.** Any polynomial dynamical system gives rise to a directed graph on $n$ vertices through the construction of a dependency graph; the converse is also true: any directed graph on $n$ vertices can be the dependency graph for a polynomial dynamical system (not unique).

**6.** The dynamics of a polynomial dynamical system $F : k^n \to k^n$ are uniquely represented by its state-space graph $\mathcal{S}(F)$ which has $k^n$ vertices.

**7.** For visualization of state-space and dependency graphs of polynomial dynamical systems, the web-based software package ADAM is available at

- http://adam.plantsimlab.org

**8.** Boolean networks (see §20.3.2) can be viewed as polynomial dynamical systems over $k = \mathbf{F}_2$, the field of two elements.

**Examples:**
**1.** Let $D_1 = (N, \mathbf{X}, F)$ be a discrete-time dynamical system with $N = \{x_1, x_2, x_3\}$, $\mathbf{X} = Q^3$, and $F = \{f_1, f_2, f_3\}$ where

$$
\begin{aligned}
f_1 = f_1(x_1, x_2, x_3) &= 2x_1 + x_3 \\
f_2 = f_2(x_1, x_2, x_3) &= \frac{5.1x_3}{x_2 + 1} \\
f_3 = f_3(x_1, x_2, x_3) &= 3^{x_2}.
\end{aligned}
$$

We notice that $supp(f_1) = \{x_1, x_3\}$, $supp(f_2) = \{x_2, x_3\}$, and $supp(f_3) = \{x_2\}$. We find the first iterate of point $(0, 0, 1)$ by $F(0, 0, 1) = (f_1(0, 0, 1), f_2(0, 0, 1), f_3(0, 0, 1)) = (1, 5.1, 1)$. In this example the system iterations are determined through a synchronous update of the vertices; when the vertices operate on different time scales, a *sequential* updating scheme is needed. See [MoRe07] for an introduction to sequential dynamical systems.

**2.** Let $D_2 = (N, \mathbf{X}, F)$ be a finite (polynomial) dynamical system such that $N = \{x_1, x_2\}$, $\mathbf{X} = \mathbf{F}_3^2$, and $F : \mathbf{F}_3^2 \to \mathbf{F}_3^2$ with transition functions

$$
\begin{aligned}
f_1 = f_1(x_1, x_2) &= 2x_2 \\
f_2 = f_2(x_1, x_2) &= x_1 + x_1^2.
\end{aligned}
$$

The dependency graph of $D_2$ is shown in part (a) of the following figure. The state-space graph in part (b) consists of $3^2 = 9$ vertices arranged into three connected components. The system has two fixed points, $(00)$ and $(12)$, and a cycle of length 2.

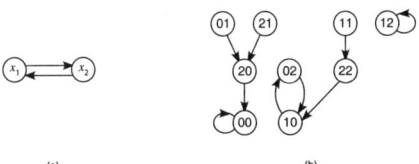

(a)                                (b)

**3.** Let $D_3 = (N, \mathbf{X}, F)$ be a polynomial dynamical system with $N = \{x_1, x_2\}$, $\mathbf{X} = \mathbf{F}_3^2$, and $F = \{f_1, f_2\}$ where

$$
\begin{aligned}
f_1 &= x_1^2 x_2 \\
f_2 &= x_2.
\end{aligned}
$$

$D_3$ is an example of a *monomial* dynamical system: each transition polynomial consists of a single monomial, i.e., $f_i = \mathbf{x}^\alpha$ for $i = 1, \ldots, n$. While little can be said about the length and shape of the cycles of a general finite dynamical system without explicitly calculating its state-space graph, for the case of monomial dynamical systems more is known [CoJaLa06].

**4.** Let $D_3$ be as in the previous example. It is also a representative of another important class of dynamical systems—*fixed point systems*. Such systems have only fixed points as cycles. As seen in the following figure, the system $D_3$ has the five fixed points $(00), (01), (02), (11), (22)$ and no other, longer cycles. Fixed point systems are important in biochemical modeling since biochemical networks often exhibit such steady-state dynamics [Lu12].

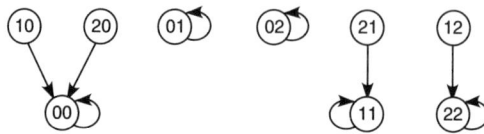

**5.** Laubenbacher and Stigler [LaSt04] address the reverse engineering problem of gene regulatory networks in the context of polynomial dynamical systems. They propose an algorithm that takes a discrete time series dataset $D = \{s_1, \ldots, s_m \mid s_i \in k^n, 1 \leq i \leq m\}$, where $k$ is a finite field, and generates a *minimal* polynomial dynamical system $F = (f_1, \ldots, f_n)$ such that $F(s_i) = s_{i+1}$ for each $i < m$. By minimal it is meant that each $f_j$ is reduced with respect to the ideal $\mathbf{I}(\{s_1, \ldots, s_m\}) = \cap_i \mathbf{I}(s_i)$ of polynomials vanishing on all input data points, i.e., all polynomial dynamical systems that fit the data differ by elements of $\mathbf{I}(\{s_1, \ldots, s_m\})$. Their algorithm is based on the Buchberger-Möller algorithm for Gröbner basis computation. It has been applied to the reverse engineering of a gene regulatory network responsible for the segment polarity development in *D. melanogaster*.

**6.** Simão et al. [SiEtal05] develop a qualitative model of tryptophan biosynthesis in *E. coli*. While the other vertices in the network model can be represented by binary variables, a ternary variable (taking the values 0, 1, 2) is needed for the vertex representing tryptophan.

## 20.3.2  BOOLEAN NETWORKS

Boolean networks were first used in the life sciences in the 1960s, when Stuart Kauffman introduced them to model gene regulatory networks as switching networks [Ka69]. They have the advantage of being computationally simple and intuitive, and they have been successfully used to model regulatory networks, preserving the qualitative, generic properties of global network dynamics.

**Definitions:**

A ***Boolean function*** is a function of the form $f : \mathcal{B}^n \to \mathcal{B}$ where $\mathcal{B} = \{0, 1\}$.

A ***Boolean network*** $G(V, F)$ is a set $V = \{x_1, \ldots, x_n\}$ of vertices representing network nodes, together with a global function $F = (f_1, \ldots, f_n)$, where $f_i : \mathcal{B}^n \to \mathcal{B}$ is the transition function associated with vertex $x_i$.

The ***maximum degree*** (sometimes called ***maximum connectivity***) $\Delta(G)$ of a Boolean network $G(V, F)$ refers to the maximum number of vertices in $V$ that are adjacent to a single vertex.

A **probabilistic Boolean network** [ShEtal02] is a Boolean network in which to every vertex $x_i$, there corresponds a set $F_i = \{f_j^{(i)}\}_{j=1,\ldots,l(i)}$ where each $f_j^{(i)}$ (called a **predictor**) is a possible transition function determining the value of $x_i$ and $l(i)$ is the number of possible transition functions for $x_i$. If $l(i) = 1$ for all $i = 1, \ldots, n$, then $\prod_{i=1}^{n} l(i) = 1$ and the probabilistic Boolean network reduces to a standard Boolean network.

Let $\mathbf{f} = (f^{(1)}, \ldots, f^{(n)})$ be a random vector taking values in $F_1 \times \cdots \times F_n$. If $f^{(1)}, \ldots, f^{(n)}$ are independent, i.e., if the selection of the Boolean functions comprising a specific network is independent, the probabilistic Boolean network is said to be **independent** (see Fact 10).

A **random Boolean network** [Ka93] is another variation of a Boolean network in which the state of each vertex is determined by the connections coming from other (or the same) vertices. The connections are wired randomly, but remain fixed during the dynamics of the network. The way in which the vertices affect each other is not only determined by their connections, but also by logic functions, which are generated randomly using lookup tables for each vertex, taking the states of the connecting vertices as inputs and producing the state of the vertex as output. These also remain fixed during the dynamics of the network.

**Facts:**

**1.** A Boolean network on $n$ vertices can be seen as a polynomial dynamical system over $\mathbf{F}_2$, the field of two elements, i.e., $F = (f_1, \ldots, f_n) : \mathbf{F}_2^n \to \mathbf{F}_2^n$ with $f_i : \mathbf{F}_2^n \to \mathbf{F}_2$. Its state-space graph $\mathcal{S}(F)$ consists of $2^n$ vertices and its dependency graph is $\mathcal{D}(F) = (V, E)$ where $E = \{(x_i, x_j) \mid x_i \in supp(f_j)\}$.

**2.** There are $2^{n2^n}$ different Boolean networks on $n$ vertices.

**3.** The dynamics over a Boolean network flow according to the updating functions and scheme. Since the state space is finite (size $2^n$, where $n$ is the number of vertices), eventually a state will be repeated and it is said to have reached an *attractor*. If the attractor consists of one state, it is called a *steady state* or *fixed point*, whereas if it consists of two or more states, it is called a *cycle*. The set of states that flow towards an attractor is called the *attractor basin*.

**4.** Boolean functions are traditionally represented in terms of logical operations (AND ($\wedge$), OR ($\vee$), NOT ($\neg$)). They can also be expressed as polynomials over $\mathbf{F}_2$. Namely, any two logical variables $a$ and $b$ can be considered as values in $\mathbf{F}_2$ so $a \wedge b$ can be expressed as $ab$, $a \vee b$ as $a + b + ab$, and $\neg a$ as $a + 1$.

**5.** When Boolean networks are used to model gene regulatory networks (see Examples 1 and 2), typically $x_i = 1$ represents the fact that gene $i$ is expressed and $x_i = 0$ means it is not expressed. The list of Boolean functions $F = (f_1, \ldots, f_n)$ represents the rules of regulatory interactions between genes. All network genes are assumed to update synchronously in accordance with the functions assigned to them and this process is then repeated. The artificial synchrony simplifies computation while preserving the qualitative, generic properties of global network dynamics [Hu99].

**6.** The maximum connectivity $\Delta$ in a Boolean network that models a gene regulatory network is interpreted as the maximum number of genes that regulate some single gene.

**7.** A low connectivity $\Delta$ is more realistic for a real gene regulatory network [ThEtal98].

**8.** A Boolean network model is often a suitable representation of genetic networks because genetic manipulation often involves either overexpression or deletion of a gene [Hu99].

**9.** A Boolean network model is often able to retain sufficient biological information to realistically model genetic regulatory networks. For example, a Boolean network model is able to provide a clear distinction between different classes of sarcomas and different subclasses of gliomas [ShZh02].

**10.** A realization of a probabilistic Boolean network at a given instant of time is determined by a vector of Boolean functions. If there are $N$ possible realizations, then there are $N$ vector functions $\mathbf{f}_1, \ldots, \mathbf{f}_N$ of the form $\mathbf{f}_k = (f_{k_1}^{(1)}, \ldots, f_{k_n}^{(n)})$, for $k = 1, \ldots, N$, $1 \le k_i \le l(i)$ and where $f_{k_i}^{(i)} \in F_i$, $i = 1, \ldots, n$.

**11.** Let $\mathbf{f} = (f^{(1)}, \ldots, f^{(n)})$ be a random vector taking values in $F_1 \times \cdots \times F_n$; the $f^{(i)}$ are not necessarily independent. Then the probability that a predictor $f_j^{(i)}$ is used to predict gene $i$ is $c_j^{(i)} = \Pr\{f^{(i)} = f_j^{(i)}\} = \sum_{k: f_{k_i}^{(i)} = f_j^{(i)}} \Pr\{\mathbf{f} = \mathbf{f}_k\}$ and since the $c_j^{(i)}$ are probabilities, they must satisfy $\sum_{j=1}^{l(i)} c_j^{(i)} = 1$.

**12.** Kauffman was the first to propose random Boolean networks, supporting the hypothesis that living organisms could be constructed from random elements, without the need of precisely programmed elements [Ka93]. Random Boolean networks are generalizations of Boolean cellular automata (see §20.3.3).

**13.** The updating of the vertices in classic random Boolean networks is synchronous: the states of vertices at time $t+1$ depend on the states of vertices at time $t$. There can be drastic differences if the updating scheme is changed [HaBo97].

**14.** There are three phases that can be distinguished in a random Boolean network: *ordered*, *chaotic*, and *critical*. For definitions and methods for identifying the phases, see [Ka00].

**Examples:**

**1.** Let $G(V, F)$ be a Boolean network with $V = \{x_1, x_2, x_3\}$ and $F = (f_1, f_2, f_3)$ where

$$
\begin{aligned}
f_1 &= x_1 \wedge x_2 \wedge x_3 \\
f_2 &= (x_1 \vee x_2) \wedge x_3 \\
f_3 &= \neg x_1 \wedge x_2.
\end{aligned}
$$

As seen in the following figure, the state space contains two types of attractors: two fixed points, (000) and (011), with basins of attraction of sizes 2 and 1, respectively, and one cycle of length 2, with basin of attraction of size 5.

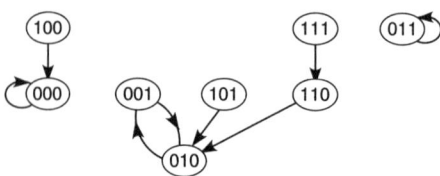

**2.** A simple model of the lactose metabolism in *E. coli* [StVe11], constructed under the assumptions that molecular synthesis and degradation require one time unit and extracellular lactose is always available, can be represented as

$$
\begin{aligned}
H_M &= \neg R \wedge C \\
H_P &= M & H_B &= M \\
H_C &= \neg G_e & H_R &= \neg A \wedge \neg A_l \\
H_A &= L \wedge B & H_{A_l} &= A \wedge L \wedge L_l \\
H_L &= \neg G_e \wedge P \wedge L_e & H_{L_l} &= \neg G_e \wedge (L \vee L_e)
\end{aligned}
$$

where

$M = lac$ mRNA

$P, B = $ lac permease and $\beta$-galactosidase, respectively

$C = $ catabolite activator protein CAP

$R = $ repressor protein LacI

$L, A = $ lactose and allolactose (inducer), respectively

$L_l, A_l = $ (at least) low concentration of lactose and allolactose, respectively.

Here $L_e$ and $G_e$ represent extracellular lactose and glucose, respectively, and are considered as parameters in the model.

**3.** Albert and Othmer [AlOt03] constructed a Boolean model of the network of segment polarity genes in the fruitfly *D. melanogaster*. The model was built from inferences given gene and protein expression data. The following figure depicts the graph of connections in the Boolean model. The shape of the nodes indicates whether the corresponding substances are mRNAs (ellipses), proteins (rectangles), or protein complexes (hexagons). The edges of the network signify either biochemical reactions or regulatory interactions. Terminating arrows generally indicate activation, while terminating segments indicate inhibition.

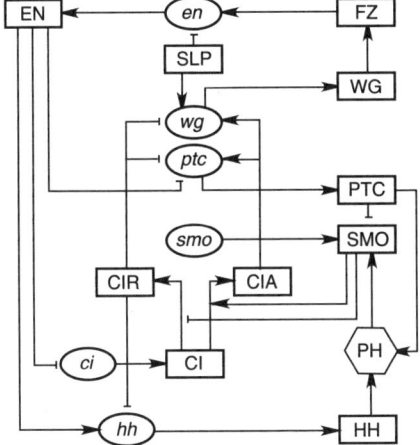

**4.** Elementary cellular automata, discussed in §20.3.3, can also be viewed as Boolean networks whose dependency graph is circular and in which all vertices have the same local Boolean function.

### 20.3.3  CELLULAR AUTOMATA

Cellular automata are simple mathematical idealizations of natural systems. They were introduced by John von Neumann and Stanislaw Ulam as simple models in which to study biological processes such as self-reproduction [vN66, Ch. 5]. A cellular automaton is a discrete model that consists of a finite or infinite regular grid of cells, each in one of a finite number of states. Among other applications in systems biology, cellular automata are used to model reaction-diffusion systems [We97] and biological cells [MaDe06].

Largely inspired by cellular automata, scientists of various backgrounds, such as Arthur Burks (philosopher) [Bu71] and Christopher Langton (computer scientist) [La89], started seeking to mimic life mathematically and generate known features of life from basic principles under the premise that life can be manifested outside of biochemistry.

**Definitions:**

A **cellular automaton** is a lattice of identical sites (**cells**), each taking on a finite set of values. The lattice can be in any finite number of dimensions. The state of a cell at time $t$ (time is discrete) is a function of the states of a finite number of cells (called its **neighborhood**) at time $t - 1$. These **neighbors** are a selection of cells relative to the specified cell, and do not change. Every cell has the same **rule** for updating, based on the values in this neighborhood. Each time the rules are applied to the whole lattice, a new **generation** is produced.

A **one-dimensional cellular automaton** of length $n$ can be defined as a network on $n$ vertices, where its dependency graph is a circle.

An **elementary cellular automaton** is a one-dimensional cellular automaton whose cells can only assume two values, e.g., 0 or 1.

**Facts:**

**1.** In the 1970s, John Conway's "Game of Life" [Ga70] (also see §17.1.4), a two-state, two-dimensional cellular automaton, became widely known. It is one of the simplest examples demonstrating that elaborate patterns and behaviors can emerge from very simple rules. A Java implementation can be found at

- http://www.ibiblio.org/lifepatterns/

**2.** An elementary cellular automaton is a ring consisting of a finite number of squares (cells). Each square has two neighbors: a square to its left and a square to its right. Each square can assume one of two colors. The color of each square in the next step depends on its current color and the colors of its two neighbors. All squares use the same rule to update their colors. Squares can be initialized to different colors.

**3.** In 1982 Stephen Wolfram published the first of a series of papers [Wo82] systematically investigating elementary cellular automata. The unexpected complexity of the behavior of these simple rules led Wolfram to suspect that complexity in nature may be due to similar mechanisms.

**Examples:**

**1.** The following figure presents two generations of an elementary cellular automaton of length 8, with initial state given in part (a). Update rule: at time $t$ each cell assumes whichever color had the majority among itself and its neighborhood at time $t - 1$. An end cell's neighbors are its adjacent cell and the cell at the other end (the array "wraps around"). Part (b) shows the automaton's state after one update.

(a)                              (b)

**2.** The next figure shows a pattern obtained with a simple one-dimensional cellular automaton. Starting from a "seed" containing a single nonzero cell, subsequent lines are obtained by successive applications of the following update rule: $a_i^{(t+1)} = a_{i-1}^{(t)} + a_{i+1}^{(t)}$ (mod 2), where $a_i^{(t)}$ is the value of the cell at position $i$ on time step $t$. The pattern obtained is Pascal's triangle of binomial coefficients, reduced modulo 2 [Wo82].

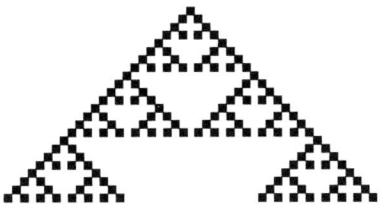

**3.** Some organisms, such as *Conus textile*, grow complicated forms by repeated application of simple local rules, closely resembling cellular automata [DeDo05].

**4.** Some researchers adopt a broader notion of artificial life and include synthetic biochemistry and robotics [Ke09], [ItSa14]. Other interesting examples of attempts at creating artificial life include the Digital Organism Simulation Environment (DOSE) [CaLi14], an executable DNA driven digital organism simulator, and OpenWorm [PaEtal12], an international open science project to simulate the roundworm *Caenorhabditis elegans* at the cellular level.

## 20.3.4  GENETIC ALGORITHMS

A genetic algorithm (GA) is a heuristic search algorithm that mimics the metaphor of natural biological evolution. GAs were originally developed as a means of studying the natural evolution process, and have been adapted for use in optimization since the 1960s [Go89], [Ho75], [Ko91]. They operate on a population of potential solutions, applying the principle of survival of the fittest to produce better and better approximations to a solution. GAs often outperform classical methods of optimization when applied to problems in fields as diverse as biology, genetics, engineering, art, economics, marketing, operations research, robotics, social sciences, physics, politics, and chemistry. Currently three are three main avenues of research in simulated evolution: genetic algorithms, evolution strategies, and evolutionary programming, each method emphasizing a different facet of natural evolution. A good introductory survey can be found in [Fo94].

### Definitions:

Since GAs are inspired by natural evolution, terminology from genetics is often used. The definitions below are valid only within the context of GAs and may even sometimes disagree with their counterpart concepts from evolutionary biology [EiSm03].

A **chromosome** is a set of parameters that define a proposed solution to the given optimization problem. The chromosome is often represented as a vector, with elements being described as **genes** and varying values at specific positions called **alleles**. Chromosomes are points in the **genotype space** where the search takes place.

The objects forming possible solutions within the original problem context are referred to as **phenotypes** and their encoding, the **individuals** within the GA, are called **genotypes**.

A **population** is a collection of genotypes (**individuals**), representing a possible solution.

**Crossover**, also called **recombination**, is an operator applied to two or more selected individuals (**parents**) and results in one or more individuals (**children**). **Mutation** is applied to one individual and results in one new individual. Recombination and mutation, collectively called **variation operators**, produce a set of new individuals that compete (based on their fitness) with the old ones for a place in the next generation.

A **fitness function** is a particular type of objective function that quantifies the optimality of a chromosome so that a particular chromosome may be ranked against all the other chromosomes. The chosen chromosomes are allowed to "breed" and mix their datasets, thus producing a new generation that will hopefully be even closer to the optimal.

**Fitness landscapes** are used to visualize the relationship between genotypes and reproductive success. Genotype **fitness** corresponds to the height of the landscape. The set of all possible genotypes, their degree of similarity, and their fitness values constitute the fitness landscape. In the following figure, points A and C are local optima, while point B is the global optimum. The trajectory of the filled circle traces a population that moves from a very low fitness value to the top of a peak.

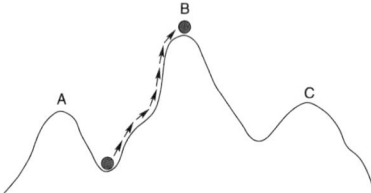

**Facts:**

**1.** GAs consistently perform well in approximating solutions to all types of problems because they do not make any assumption about the underlying fitness landscape.

**2.** While there are variations in different GAs, the common idea behind all of them is that given a population of individuals, the environmental pressure causes natural selection that results in a rise in the fitness of the population. The environmental pressure is mimicked through the application of a fitness function that is to be maximized. The process starts by randomly creating a set of candidate solutions and applying the fitness function to identify the best candidates for creating the next generation by means of recombination and mutation. The process is repeated until an individual of sufficiently high fitness is found.

**3.** There is no guarantee that the best solution found by a GA is actually the optimal solution. In the context of GAs, "optimal solution" means the best solution we are able to find. The advantage of using a GA is that there is inherently some level of stochasticity; that is, there is always the possibility of mutating to a better solution, as long as the mutation rate is nonzero [LaHiOr15].

**4.** *Uniform crossover* is a typical method of crossover in which there is an equally likely chance that the child will take each particular gene from either parent (see Example 1). *One-point crossover* is another popular option (see Example 7).

**5.** During selection, fitter individuals have a higher chance to be selected but even weaker ones have a chance to survive and become parents. Combining recombination and mutation with selection generally leads to improved fitness.

**6.** Binary representation is commonly used for candidate solutions, encoding them as binary strings (see Example 5). Other representations, such as floating-point, can be found in the GA literature as well.

**Examples:**

**1.** In the "Rabbits and Grass" example discussed in [LaHiOr15], two randomly chosen solutions are selected, $p_1 = (0101010101)$ and $p_2 = (1110110010)$. A child solution is created using uniform crossover by going through each gene, i.e., each entry in the parent solutions, and randomly selecting one of the values. An example of a child solution might be $p_{new} = (0111110111)$. The child solution is then subjected to mutation, e.g., the second and the last entry of $p_{new}$ are "mutated" through *bit-flipping* to obtain $p_{new}^{mut} = (0011110110)$. The algorithm continues for a predetermined number of steps, or until a certain condition is met. For example, one might run the algorithm for 50 generations. Another method is to repeat the process until no better solution has been found for some specified number of consecutive generations. When the algorithm terminates, the best current solution is selected as the candidate for an optimal solution [LaHiOr15].

**2.** GAs are population-based stochastic algorithms that repeatedly test individuals in a population against a fitness criterion. Algorithm 1 outlines the general principles of GAs [EiSm03].

---

**Algorithm 1**:  **General scheme of a GA.**

  input: population with random candidate solutions

  output: suitable solution (candidate with sufficient quality)

  EVALUATE each candidate in the population

  **while** TERMINATION CONDITION not satisfied **do**

    SELECT parents

    CROSSOVER pairs of parents

    MUTATE resulting offspring

    EVALUATE new candidates

    SELECT individuals for next generation

  **return** an individual (that meets TERMINATION CONDITION)

---

**3.** If the problem has a known optimal solution value, then reaching the optimal fitness level within a given precision can be used as a TERMINATION CONDITION in Algorithm 1. However, because of the stochastic nature of the GA, there is no guarantee of convergence. To ensure algorithm termination, other criteria are often applied, such as restricting the CPU time allowed, fixing the number of fitness evaluations permitted, terminating when the population diversity drops below a certain level, and setting a minimum fitness improvement for a number of generations or fitness evaluations.

**4.** There are many GA variations. Algorithm 1 can be modified, for example, so that the initial chromosomes are selected in a certain way, e.g., at random or so that they are very different from one another (i.e., solutions that come from different regions of the solution space). The crossover process can also be modified so that one parent is favored over another, or one can forego the mutation step altogether. The likelihood of mutation is another area where user input is important: a high level of mutation will result in more variation of child chromosomes, and thus will not incorporate the relative fitness of the parent chromosomes as much. On the other hand, if the mutation step is not included then there is a greater risk of the solutions converging to some solution that is only locally minimal [LaHiOr15].

**5.** Suppose we want to find an integer value $x$ that maximizes $F(x) = -x^2$. Then the set of integers $\mathcal{Z}$ will be the set of phenotypes. Thus the integer 9 is a phenotype and, if binary coding is used, its genotype representation is 1001. The fitness function of this genotype could be defined as (the negative of) the square of its phenotype, $9^2 = 81$.

**6.** Suppose the goal is to find a vector of 100 bits, $\mathbf{x} \in \{0, 1\}^{100}$, such that the sum of all of the bits in the vector is maximized. The fitness function could be written as

$F(\mathbf{x}) = \sum_{i=1}^{100} x_i$, where $\mathbf{x} = (x_1, x_2, \ldots, x_{100})$. A candidate solution is evaluated using $F$ and its fitness is identified in the range between 0 and 100. Let an initial population of 100 parents be selected at random and subjected to selection using $F(\mathbf{x})$ with the probabilities of recombination and bit mutation being 0.8 and 0.01, respectively. The following figure shows the rate of improvement of the best vector in the population, and the average of all parents, at each generation. The process rapidly converges to the vector of all 1s, with optimum fitness value 100 [Fo94].

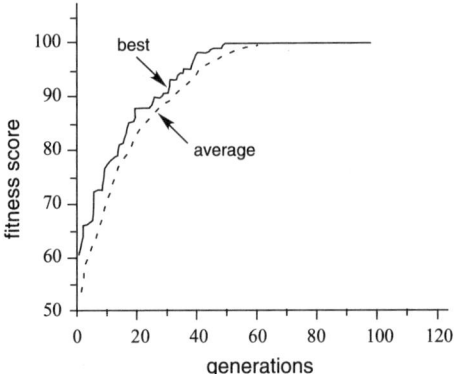

**7.** The *0-1 knapsack problem* (§16.3.1), an optimization problem that arises in many industrial problems, can be briefly described as follows. Given a set of $n$ items, each of which has some value $v_i$ attached to it and some cost $c_i$, how do we select a subset of those items with cost at most $C_{max}$ that maximizes the overall value. A candidate solution for this problem can be represented as a binary string of length $n$, where a 1 in a given position indicates that an item is included and a 0 indicates that it is omitted. The corresponding genotype space $G$ is the set of all such strings, whose size $2^n$ increases exponentially with the number of items considered.

While it might seem natural to define the phenotype space $P$ to be identical to $G$, a one-to-one correspondence between the genotype and phenotype spaces leads to problems [EiSm03]. Instead, let the fitness of a given solution $p$, represented by a binary genotype $g$, be determined by summing the values of the included items: $Q_p = \sum_{i=1}^{n} v_i g_i$. In order to break the one-to-one correspondence between $G$ and $P$, when creating a solution proceed as follows. Read from left to right along the binary string, and keep a running tally of the cost of included items. When a value 1 is encountered, first check to see if including the item would violate the capacity constraint $Q_p = \sum_{i=1}^{n} v_i g_i \le C_{max}$; i.e., rather than interpreting a value 1 as meaning to include this item, interpret it as meaning to include this item *if* it does not violate the cost constraint.

For this problem, [EiSm03] chose as a recombination operator the one-point crossover, which aligns two parents and picks a random point along their length. The two offspring are created by exchanging the tails of the parents at that point. This operator is applied with 0.7 probability. That is, for each pair of parents select a random value uniformly between 0 and 1; if it is below 0.7 then create two offspring by crossover, otherwise make copies of the parents. Bit-flipping is used as a mutation operator: in each position invert the value with a small probability $p_m \in [0, 1)$, for example $p_m = 1/n$. A termination condition might be, for example, when no improvement in the fitness of the best member of the population has been observed for twenty five generations. Such a termination criterion is appropriate since the maximum value that can be achieved is not known.

# 20.4   GENOME ASSEMBLY

Every living organism has a genome that contains its complete hereditary information, encoded by long DNA sequences. Abstractly, the genome can be viewed as a set of strings over the alphabet $\{A, T, C, G\}$. The task of decoding, or sequencing a genome, is of considerable interest. However, it is a computationally daunting task because these sequences can be hundreds of billions of base pairs in length, and even the most advanced technology can only read sequences of a few thousand base pairs. As a result, sequencing a genome is done by breaking up numerous copies into small pieces which can be individually read. These fragments can then be reassembled into the original sequence by analyzing the overlaps, or by recording the length-$k$ subsequences that they contain. This section provides an overview of approaches and algorithms developed to solve this massive biological jigsaw puzzle.

## 20.4.1   BASIC CONCEPTS

### Definitions:

A strand of DNA consists of a sequence of four types of **nucleotides** distinguished by a nitrogenous **base**: adenine (A), thymine (T), cytosine (C), or guanine (G).

A **DNA segment** is a consecutive subsequence of bases on a DNA strand, which one can represent as a string or sequence over the alphabet $\{A, T, C, G\}$.

A **DNA target sequence** is a long unknown DNA sequence that one wishes to decode, but is typically much too long to do so all at once, even using the state-of-the-art tools.

Long DNA strands need to be broken up into smaller strands called **clones**, which are short enough to be individually sequenced using modern technology.

DNA clones that are sequenced are called **reads**, or **fragments**.

A **k-mer** of a string is a (consecutive) length-$k$ substring. The **prefix** (respectively, **suffix**) of a $k$-mer is its initial (respectively, terminal) $(k-1)$-mer.

The technique of **sequencing by hybridization** (SBH) uses a DNA microarray, or biochip, to determine all $k$-mers in the reads. This information is then used to reconstruct the target sequence.

**Shotgun sequencing** is a method where many copies of a long DNA strand are broken up randomly into clones, which are sequenced into reads. The original target sequence is reconstructed by analyzing the overlaps of the reads, or by the set of $k$-mers in the reads.

A **genome assembler** is a computer program that performs the algorithmic gene assembly process from given input data.

### Facts:

**1.** Each base can chemically bond with a particular *complement* base, forming a *base pair* (bp). Specifically, A bonds with T, and C bonds with G.

**2.** The bases in a DNA strand are strung along a sugar-phosphate backbone. The carbon atoms in the sugar (ribose) are numbered $1'$ through $5'$. The $3'$ carbon in each sugar molecule is connected through a phosphate group to a $5'$ carbon on the next one, giving each DNA strand a directionality.

**3.** A DNA molecule consists of two DNA strands that bond to each other and twist into a double-helix structure. These two strands bond in an anti-parallel fashion: namely, if the sequence of a DNA strand is $\mathbf{s} = (s_1, \ldots, s_n)$, then the complementary sequence is $\mathbf{t} = (t_1, \ldots, t_n)$, where $t_i = \overline{s}_{n-i+1}$, $i = 1, \ldots, n$, with the bar denoting the complement base.

**4.** DNA sequencing was invented in the late 1970s, using an in vitro chain termination method known as *Sanger sequencing* [SaNiCo77]. This led to the "first generation" of sequencers, which were the most widely used methods for about 25 years and which produced reads of 300–900 bp in length.

**5.** First-generation sequencers have been supplanted (but not replaced) by the development of high-throughput parallelized *next-generation* (Next-Gen) sequencing technologies.

**6.** Sequencing by hybridization using $k$-mers was developed in 1988 [DrEtal89]. The range $8 \leq k \leq 25$ was common.

**7.** DNA hybridization for genome assembly proved to be very limited due to both its accuracy and low values of $k$, but microarray technology that it utilized has since been widely used to measure gene expression.

**8.** The bacterium *Carsonella ruddii* has the smallest known genome—about 160,000 bp [NaEtal06]. The largest known genome belongs to the ameboid *Polychaos dubium*, containing approximately 670 billion bp. In contrast, the human genome has about 3 billion bp. The Animal Genome Size Database [Gr16] contains genomic data for over 5,600 animal species and is available online at

- http://www.genomesize.com

**9.** The first successful genome sequence was completed in 1995 by a team at The Institute for Genomic Research (TIGR) and Johns Hopkins University. They sequenced the entire 1.83 million base pair genome of the bacterium *Haemophilus influenzae* [FlEtal95].

**10.** A breakthrough whole-genome sequence assembly project was completed in 2000 by Eugene Myers et al., using the Celera assembler [MyEtal00]. They sequenced the 135 million base pair genome of the fruit fly *Drosophila melanogaster*, which was over 25 times larger than any other genome assembled.

**11.** As of 2016, the single molecule real-time (SMRT) sequencer [RoCaSc13] from Pacific Biosciences can generate reads with an average length of 10,000–15,000 bp and a maximum of about 60,000 bp.

**12.** Other popular Next-Gen sequencers produce shorter but more accurate reads, such as those developed by the companies Illumina [BeEtal08] (up to 300 bp/read) and 454 Life Sciences [MaEtal05] (up to 600 bp/read).

**Examples:**

**1.** The following diagram shows the bases appearing in a short DNA strand CGACTTGC, oriented in the $5'$ to $3'$ direction. The complementary DNA strand is GCAAGTCG, also oriented in the $5'$ to $3'$ direction.

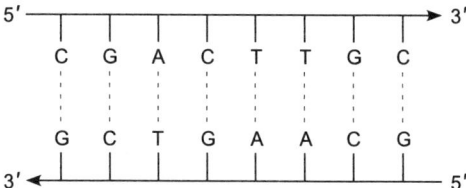

**2.** The following diagram illustrates a microarray that can be used to identify $k$-mers for $k = 3$. In particular, the DNA fragment ACTAGC contains the 3-mers ACT, CTA, TAG, and AGC, which are detected by the microarray shown.

| AAA | AAC | AAG | AAT | ACA | ACC | ACG | ACT |
|-----|-----|-----|-----|-----|-----|-----|-----|
| AGA | AGC | AGG | AGT | ATA | ATC | ATG | ATT |
| CAA | CAC | CAG | CAT | CCA | CCC | CCG | CCT |
| CGA | CGC | CGG | CGT | CTA | CTC | CTG | CTT |
| GAA | GAC | GAG | GAT | GCA | GCC | GCG | GCT |
| GGA | GGC | GGG | GGT | GTA | GTC | GTG | GTT |
| TAA | TAC | TAG | TAT | TCA | TCC | TCG | TCT |
| TGA | TGC | TGG | TGT | TTA | TTC | TTG | TTT |

## 20.4.2  OVERLAP GRAPH APPROACHES

The fragment assembly problem takes in a set of reads (strings) and asks for a larger string **s** that "best explains" the reads. Originally, this was approached as a modified shortest common superstring (SCS) problem with added constraints. The SCS problem can be solved by constructing a graph whose vertices are the reads and whose edges represent overlaps. Due to repeated substrings, it is well known that the fragment assembly problem is more complicated than SCS. Despite this, many assemblers still construct some type of "overlap graph" from the reads and then seek a Hamilton path. Such assemblers often break this into three distinct phases: overlap, layout, and consensus. This approach was introduced in 1984 [PeSöUk84]. It was the first widespread method for sequence assembly and was widely used for over 20 years, in popular assemblers such as PHRAP, PCAP [HuWa03], TIGR, and Celera [MyEtal00].

**Definitions:**

Given a collection of strings (reads) $S = \{\mathbf{s}_1, \ldots, \mathbf{s}_n\}$, a solution to the **shortest common superstring (SCS) problem** is a string **s** of minimum length that contains each $\mathbf{s}_i$ as a (consecutive) substring.

The **overlap** of $\mathbf{s}_i$ with $\mathbf{s}_j$ is the length of the longest suffix of $\mathbf{s}_i$ that is a prefix of $\mathbf{s}_j$.

Given $S$ and a threshold $\ell > 0$, the **overlap graph** is a weighted directed graph constructed as follows:

- the vertices are the strings $\mathbf{s}_i \in S$;
- there is a directed edge $\mathbf{s}_i \to \mathbf{s}_j$ if the overlap of $\mathbf{s}_i$ with $\mathbf{s}_j$ is at least $\ell$;
- the cost (or weight) of an edge is negative of the overlap between $\mathbf{s}_i$ and $\mathbf{s}_j$.

A **Hamilton path** in a graph is a path that visits every vertex exactly once.

The three phases in the overlap-layout-consensus approach are

- *overlap*: find significant overlaps between reads, taking into account difficulties such as unknown orientations and possible errors;
- *layout*: construct an appropriate graph and find a Hamilton path corresponding to the optimal layout of overlapping reads;
- *consensus*: fix errors and resolve discrepancies in the layout.

If the orientation of overlapping reads is known, then this information can be tracked by using annotated directed edges called **dovetail edges**, as shown next:

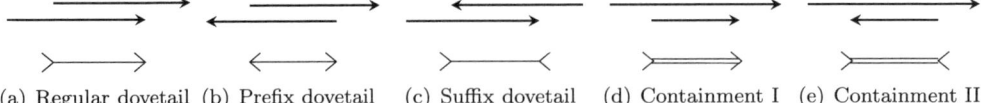

(a) Regular dovetail (b) Prefix dovetail   (c) Suffix dovetail   (d) Containment I  (e) Containment II

A **dovetail path** is a path in the overlap graph with the added requirement that it respects the orientation of the tails; i.e., two consecutive edges common to a vertex correspond to different ends of that fragment.

**Facts:**

**1.** An SCS corresponds to a minimum cost Hamilton path in the overlap graph.

**2.** Finding a minimum cost Hamilton path is an NP-hard problem, even if all edge weights are identical.

**3.** Some genomes are circular and others are linear. If the genome is circular, then an SCS corresponds to a minimum cost *Hamilton tour* (or cycle) instead of a minimum cost Hamilton path.

**4.** Non-optimal superstrings for the SCS problem can be found with a simple greedy heuristic [TaUk88]: at each step, merge two strings with maximum overlap. This is known to be a $c$-approximation algorithm for small $c$, with $c = 2$ conjectured. This heuristic is used by popular genome assemblers such as TIGR (see [SuEtal95], [PoKo04]), PHRAP, and CAP.

**5.** Due to the nature of DNA sequencing, the fragment assembly problem is harder than the standard SCS, due to the following difficulties:

- the orientation of the strings may not be known;
- there could be sequencing errors;
- there are usually long repeated substrings in the actual genome;
- there could be differences between inherited copies of a genome (for example, in humans, one copy is from the mother and the other is from the father).

**6.** Due to high computational costs, look-up tables are often used in the overlap phase. Dynamic programming is also used to find optimal alignments.

**7.** The biggest challenge with the fragment assembly problem is dealing with the abundance of (frequently long) repeated subsequences. This is usually part of the layout phase. As a result, the real genome is usually not the actual solution to the SCS problem, but much longer.

**8.** One way to handle repeats using an SCS approach is to assume that the sampling coverage is uniform, so that maximum likelihood methods can give evidence for repeated substrings.

**9.** Majority voting, Bayesian models, and dynamic programming have all been used within the consensus phase to determine an optimal multiple alignment.

**Examples:**

**1.** Given reads $S = \{\text{AAT, CGTA, TAA, TCGT}\}$, the overlap graph with threshold $\ell = 1$ is shown next.

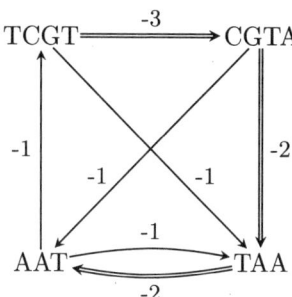

A minimum cost Hamilton path in this graph is given by TCGT $\rightarrow$ CGTA $\rightarrow$ TAA $\rightarrow$ AAT with cost $-7$. This sequence yields the shortest common superstring TCGTAAT, of length 7.

**2.** Suppose that the greedy heuristic is applied to the reads in Example 1. The strings TCGT and CGTA have maximum overlap (3) and are merged, giving the string TCGTA. Next, strings TAA and AAT have maximum overlap (2) and so are merged, giving the string TAAT. Finally, the strings TCGTA and TAAT are merged, producing the string TCGTAAT with length 7, which is optimal (see Example 1).

**3.** To see that the greedy heuristic need not produce a minimum length superstring, consider reads $S = \{\text{ATATATATC, GATATATAT, TATATATA}\}$. The strings GATATATAT and ATATATATC have maximum overlap (8); these two strings are merged, giving the new string GATATATATC. Since TATATATA and GATATATATC have no overlap, they can be concatenated in either order, say giving the string TATATATAGATATATATC of length 18. However, the SCS has length 12, obtained by overlapping GATATATAT, TATATATA, and ATATATATC, producing the optimal superstring GATATATATATC. This example can be generalized to show that the ratio between the heuristic length and the optimal length approaches 2 in these circumstances.

**4.** Consider a short circular genome CACTACGTAA, and suppose that shotgun sequencing produces the reads $\{\text{ACAC, ACGTA, CACTAC, CGTAACA, CTACG}\}$. The overlap graph is shown in the following figure, using an overlap threshold of $\ell = 2$ for the edges. The target sequence is represented in the overlap graph by a Hamilton tour, denoted by the thick edges.

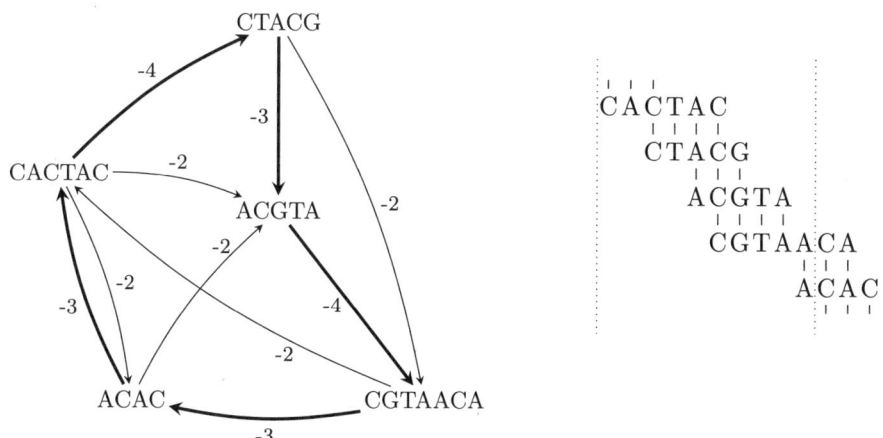

**5.** Simplifications of the overlap graph, such as the following [My95], are frequently implemented in the layout phase:

- require minimum overlap for edges;
- contained read removal: if **s** contains **s′**, then remove **s′** and all incident edges;
- topologically smooth vertices: replace edges $a \to b$ and $b \to c$ with a single edge $a \to c$ if $b$ has no other incident edges.

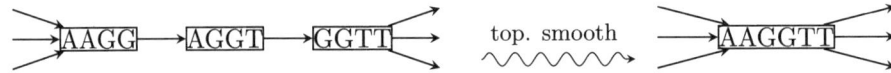

- transitive edge removal: given edges $a \to b$, $b \to c$, and $a \to c$, remove $a \to c$.

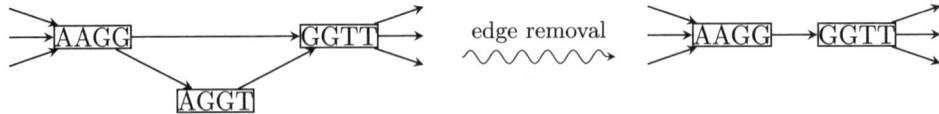

**6.** The consensus phase uses a multiple alignment to correct errors, as shown in the following example with four overlapping reads.

```
                        GACGGATTAAAATAAGC
Alignment:      TTAATGACGGATTAATA
                          ATTAAAATAAGCAGT
                ATGTCGGACTAAAATAA

Consensus:      TTAATGACGGATTAAAATAAGCAGT
```

## 20.4.3   DE BRUIJN GRAPH METHODS

The standard de Bruijn graph on $m$ symbols (for genome assembly, $m = 4$) is a directed graph whose vertices consist of all length-$k$ sequences and in which the edge $(u, v)$ indicates that the suffix of $u$ equals the prefix of $v$ (see §3.1.7). A number of genome assemblers record all of the $k$-mers present in the reads, and then use a "modified de Bruijn graph" to reconstruct the target sequence. There are two main approaches: (i) represent the vertices as $k$-mers and edges as (prefix, suffix) pairs of $(k-1)$-mers, or (ii) represent each $k$-mer as an edge, whose vertices are its (prefix, suffix) $(k-1)$-mers. In the first approach, the solution is encoded as a Hamilton path (visiting every vertex exactly once), and in the second approach, as an Euler path (visiting every edge exactly once). There is no consensus name for these graphs in the literature, so we will call them Hamilton and Euler de Bruijn graphs, respectively.

### Definitions:

Given a set of reads, construct the **Hamilton de Bruijn graph** as follows:

- the vertices are the $k$-mers that appear in some read;
- there is a directed edge $u \to v$ if the suffix of $u$ is the prefix of $v$.

An **Euler path** in a graph is a path that traverses every edge exactly once.

Given a set of reads, construct the **Euler de Bruijn graph** as follows:

- the vertices are the $(k-1)$-mers that appear in some read;
- there is a directed edge $u \rightarrow v$ if some $k$-mer contains $u$ as a prefix and $v$ as a suffix.

**Facts:**

**1.** The early uses of sequencing by hybridization (SBH) used an overlap-layout-consensus approach. An Euler de Bruijn graph approach was proposed in 1989 [Pe89]. Another early SBH approach is described in [DrEtal93], with $k = 8$.

**2.** A Hamilton path in the Hamilton de Bruijn graph represents a feasible solution to the fragment assembly problem.

**3.** An Euler path in the Euler de Bruijn graph represents a feasible solution to the fragment assembly problem [IdWa95], [PeTaWa01].

**4.** If the genome is circular, then solutions to the fragment assembly problem correspond to Hamilton (or Euler) cycles instead of paths.

**5.** A directed graph has an Euler cycle (or tour) if and only if the in-degree of each vertex equals its out-degree. This property holds for both the standard and the Euler de Bruijn graphs.

**6.** Algorithm 1 (Hierholzer's algorithm) describes a simple approach for finding an Euler cycle in a directed graph satisfying the property given in Fact 5. It can be implemented to run in linear time. Also see §8.4.3.

---

**Algorithm 1**:   **Euler cycle algorithm.**

pick a vertex $v$, and travel along unused edges until returning to $v$
**while** the cycle is not Eulerian
    pick a vertex $w$ on the cycle that has unused outgoing edges
    travel along unused edges until returning back at $w$
    merge cycles

---

**7.** Algorithm 1 can be easily adapted to find an Euler path in a directed graph that possesses such a path (but not an Euler cycle).

**8.** Further modifications or generalizations of de Bruijn graphs, such as weighted edges (counting how many times the $k$-mers occur), paired de Bruijn graphs [MeEtal11], and pathset graphs [PhEtal13], attempt to improve the process of resolving repeats.

**9.** Other modified de Bruijn graphs are used in different problems in bioinformatics, such as RNA assembly [GrEtal11], protein sequencing [BaEtal08], and synteny block reconstructions [PhPe10].

**10.** Unlike the overlap graphs, these modified de Bruijn graphs are not *read coherent*, meaning that some paths through the graph are not consistent with the reads [My05].

**11.** de Bruijn graph assemblers perform well when the read length is short. For longer reads (e.g., over 100 bp), overlap graph assemblers perform well [ScDeSa10].

**12.** When constructing a modified de Bruijn graph, the value of $k$ should be small enough so that nearly all of the $k$-mers in the genome will be found in the reads. The range $30 \le k \le 60$ is common, and these $k$-mers are found directly from the reads (rather than from hybridization).

**13.** Overlap graph approaches were used in most first-generation sequencing projects which assembled the the first microbial genome in 1995 [FlEtal95] and the human genome in 2001. However, most Next-Gen assemblers favor de Bruijn graph methods.

**Examples:**

**1.** Recall Example 4 from §20.4.2, in which shotgun sequencing produced the reads {ACAC, ACGTA, CACTAC, CGTAACA, CTACG} from the (unknown) circular genome CACTACGTAA. The Hamilton de Bruijn graph for $k = 3$ is shown in the following figure.

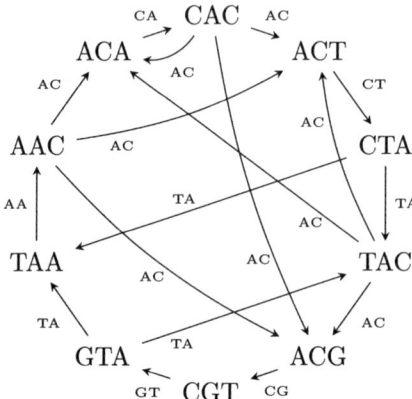

The target sequence CACTACGTAA is represented in this directed graph by the Hamilton cycle that traverses the vertices in a clockwise order. Note that this directed graph contains other Hamilton cycles such as CAC → ACG → CGT → GTA → TAC → ACT → CTA → TAA → AAC → ACA → CAC, which encodes the alternative target sequence CACGTACTAA. Further sequencing data would then be needed to identify the correct target sequence.

In this example, the coverage of the reads is enough to recover all 3-mers in the original genome. However, removing any read destroys this property, which would change the de Bruijn graph. This highlights the importance of having a sufficiently high coverage so that all (or almost all) of the $k$-mers appear in at least one of the reads.

**2.** The next figure displays the Euler de Bruijn graph for the 3-mers present in the reads given in Example 1. The target sequence is represented in this graph by the Euler cycle CA → AC → CT → TA → AC → CG → GT → TA → AA → AC → CA.

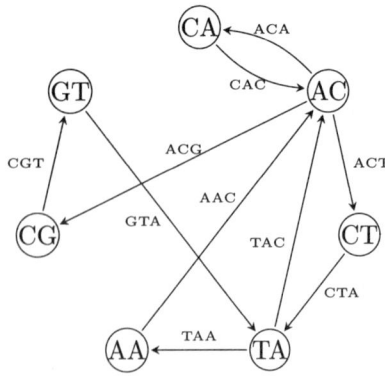

Note that this graph supports several Euler cycles representing different target sequences. For example, the Euler cycle CA → AC → CG → GT → TA → AA → AC → CT → TA → AC → CA represents the different circular target sequence CACGTAACTA. Another difficulty arises from repeats: this same Euler de Bruijn graph would also be generated from the circular genome CACACTACGTAA in which CAC and ACA are repeated.

# 20.5   RNA FOLDING

Deoxyribonucleic acid (DNA) is a double-stranded macromolecule built from four types of nucleotides linked together along a sugar-phosphate backbone that twists into the familiar double-helix shape. Its cousin ribonucleic acid (RNA) is chemically similar but with several important differences: it is single-stranded, and the nucleobase uracil (U) takes the place of thymine (T). Since RNA has only one strand, it can fold and bond to itself. The specific structure into which RNA folds often plays an important role in its function. A central problem in computational biology is to predict how an RNA strand will fold given only its raw sequence of nucleotides. There are two main approaches to the folding prediction problem that both employ dynamic programming: the first attempts to minimize the structure's free energy, and the second generates structures using a context-free grammar and then selects the most likely structure.

## 20.5.1   BASIC CONCEPTS

### Definitions:

A strand of RNA consists of a sequence of four types of **nucleotides** distinguished by a nitrogenous **base** (adenine (A), cytosine (C), guanine (G), uracil (U)) and connected along a sugar-phosphate backbone.

The carbon atoms in the sugar (ribose) are numbered $1'$ through $5'$ and the $3'$ carbon in each sugar molecule is connected through a phosphate group to a $5'$ carbon of the next sugar molecule. This gives directionality to the RNA chain: a "front" end called the **$5'$-end** and a "back" end called the **$3'$-end**.

Adenine and guanine are **purines**, and cytosine and uracil are **pyrimidines**.

The RNA strand folds onto itself by a formation of hydrogen bonds between the bases from different nucleotides. Two bases that share a chemical bond are called a **base pair**.

### Facts:

**1.** The large majority of observed base pairs are either one of the two *Watson-Crick* pairs, AU (two hydrogen bonds) and CG (three hydrogen bonds), or the weaker *wobble pair*, GU (two hydrogen bonds). Other base pairs (e.g., purine-purine and pyrimidine-pyrimidine pairs) are thermodynamically unstable and thus rare.

**2.** The length of an RNA strand can vary from a few tens to a few thousand nucleotides.

**3.** The structure of functional RNA has been evolutionary preserved. Therefore, when given a set of homologous RNA sequences (the ones with shared ancestry), a common structure can be found using covariation analysis. Even though highly accurate, this is a manual process that cannot be applied to a single sequence and an important problem in computational biology is the design of an equally accurate prediction method of low computational complexity.

**4.** There exist viruses whose genetic material consists of RNA instead of DNA (e.g., SARS, polio, ebola, measles, and HIV).

### Example:

**1.** The following figure illustrates the structure of RNA. The four chemical bases (A, U, G, C) are attached to sugar groups (ribose R) on the sugar-phosphate (P) backbone.

The 5′ end is at the upper left and the 3′ end is at the lower right.

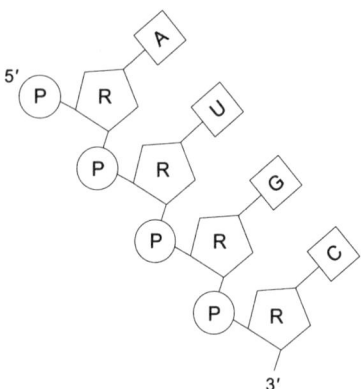

---

## 20.5.2   COMBINATORIAL MODELS

An RNA structure can be represented by a sequence over a four-letter alphabet with additional information encoding which bases form pairs. These are typically represented by graphs called arc diagrams.

**Definitions:**

An **RNA sequence** is a sequence $\mathbf{b} = b_1 b_2 \ldots b_n$, where $b_i \in \{A, C, G, U\}$.

A **partial matching** of $V = \{1, \ldots, n\}$ is a collection $\{(i_1, j_1), \ldots, (i_k, j_k)\}$ of disjoint size-2 subsets of $V$ called **arcs**.

An **RNA structure** is an RNA sequence $\mathbf{b} = b_1 b_2 \ldots b_n$ along with a partial matching of $V = \{1, \ldots, n\}$ which describes the base pairs. Elements in $V$ are called **nodes** or **vertices**.

A canonical way to view an RNA structure is by its **arc diagram**: draw either $1, \ldots, n$ or the sequence of bases horizontally, and then draw each arc $(i, j)$ in the partial matching in the upper-half plane.

The **length** of the arc $(i, j)$ is $j - i$. Two arcs $(i_1, j_1)$ and $(i_2, j_2)$ with $i_1 < i_2$ are **crossing** if $i_1 < i_2 < j_1 < j_2$. Otherwise, they are **noncrossing**. These correspond to geometric crossings and noncrossings of arcs in the arc diagram representation of RNA.

If the RNA structure has no crossings then it is called a **secondary structure**. Otherwise, it is called a **pseudoknot structure** [Re10].

A pseudoknot structure is **$k$-noncrossing** if it has no $k$ arcs that are mutually crossing.

A **stack** or **helix** of size $\sigma$ is a maximal sequence of **nested arcs**

$$(i, j), (i+1, j-1), \ldots, (i + (\sigma - 1), j - (\sigma - 1)).$$

A node $v$ is **accessible** from a base pair $(i, j)$ if $i < v < j$ and there is no base pair $(i', j')$ such that $i < i' < v < j' < j$. A base pair $(v, w)$ is **accessible** from $(i, j)$ if both $v$ and $w$ are accessible from it [SaEtal83].

The ***null loop***, or ***0-loop***, of an RNA structure is the set of nodes not accessible from any base pair.

The ***k-loop*** $\mathbf{L}_{i,j}$ closed by $(i,j)$ is the subset of $V$ formed from the $k-1$ base pairs and the isolated bases that are accessible from $(i,j)$. The ***size*** of a loop is the number of isolated bases in it.

1-loops are also called ***hairpin loops*** and $k$-loops for $k \geq 3$ are termed ***multibranch loops***. There are three types of 2-loops. If $(i',j')$ is the (unique) base pair accessible from $(i,j)$, then the 2-loop closed by $(i,j)$ is a

- ***stacked pair*** if $i' - i = j - j' = 1$ (i.e., it has size 0),
- ***bulge loop*** if exactly one of $i' - i$ and $j - j'$ is $> 1$,
- ***internal loop*** if both $i' - i$ and $j - j'$ are $> 1$.

**Facts:**

**1.** Base pairs of length 1 are biophysically infeasible. Base pairs of length 2 are thermodynamically unstable and thus very rare. Most combinatorial models require a minimum arc length $\lambda \geq 2$, usually $\lambda = 3$ or $\lambda = 4$.

**2.** Let $T^\lambda(n)$ denote the number of length-$n$ RNA secondary structures with minimum arc length at least $\lambda$. There is no known closed formula for $T^\lambda(n)$, but it satisfies the following recurrence [Wa78]:

$$T^\lambda(n) = T^\lambda(n-1) + \textstyle\sum_{k=0}^{n-(\lambda+1)} T^\lambda(n-2-k)T^\lambda(k).$$

**3.** The numbers $T^\lambda(n)$ grow exponentially in $n$ [StWa79]. For example,

$$T^2(n) \sim \sqrt{\tfrac{15+7\sqrt{5}}{8\pi}}\, n^{-3/2} \left(\tfrac{3+\sqrt{5}}{2}\right)^n.$$

**4.** There is a bijection between secondary structures (noncrossing partial matchings) of length $n$ and *Motzkin paths* of length $n$: namely, lattice paths in the plane from $(0,0)$ to $(n,0)$ consisting of three types of steps: $\nearrow$, $\searrow$, and $\longrightarrow$, where each has width 1.

**5.** The *point-bracket notation* of a size-$n$ RNA secondary structure is a length-$n$ string over the 3-element alphabet $\{\,(, \bullet, )\,\}$, where the three characters designate left arc endpoint, isolated vertex, and right arc endpoint, respectively. Such a string corresponds to a secondary structure if and only if any initial segment contains at least as many open parentheses as closed ones.

**6.** RNA secondary structures can also be represented as plane trees in different ways, capturing various degrees of detail of the structure.

**7.** There is a bijection between $k$-noncrossing pseudoknot structures and the following combinatorial structures (not defined here, see [ChEtal07]):

- integer lattice walks in $\mathcal{Z}^{k-1}$ that start and end at $(k-1,\dots,2,1)$ in the fundamental Weyl chamber of type $B_{k-1}$;
- vacillating standard Young tableaux of length $n$ and height less than $k$.

**8.** Every RNA secondary structure has a well-defined *loop decomposition* [SaEtal83] of the vertex set into loops:

$$\mathbf{b} = \mathbf{L}_0 \cup \left( \textstyle\bigcup_{(i,j)} \mathbf{L}_{i,j} \right),$$

where $\mathbf{L}_0$ is the null loop and the $\mathbf{L}_{i,j}$ are distinct.

**9.** RNA pseudoknots were not discovered until 1982, when they were observed in the RNA of the turnip yellow mosaic virus [RiEtal82].

**10.** A database of hundreds of pseudoknots can be found at [TaEtal09]. Most observed pseudoknots are 3-noncrossing [TaEtal09].

**11.** The 3-noncrossing pseudoknots are precisely those that can be drawn as noncrossing diagrams, if base pairs are additionally allowed to be drawn as arcs in the lower half-plane.

**12.** An RNA pseudoknot can be naturally associated a topological surface via a *ribbon graph* or *fatgraph* [BoEtal08], [PeEtal10]. This allows pseudoknots to be classified by their genus [AnEtal13].

### Examples:

**1.** Two folds of the RNA strand **b** = AAAGUUCCUUUUUUGGAAAAAAA are shown in the following diagram. The RNA structure on the left is a secondary structure because its arc diagram is noncrossing. The RNA structure on the right is a 3-noncrossing pseudoknot.

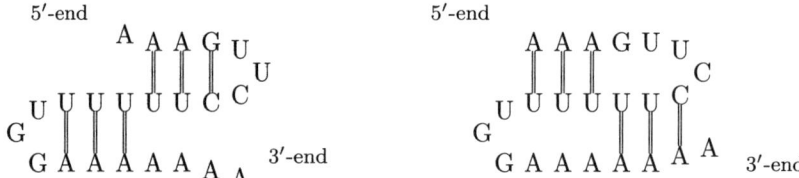

**2.** The arc diagrams corresponding to the RNA structures in Example 1 are shown next.

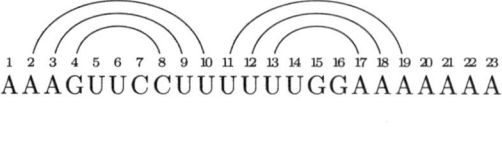

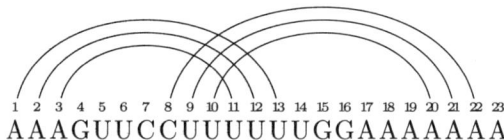

**3.** The Motzkin path and point-bracket notation of the secondary structure in Example 1 are displayed next.

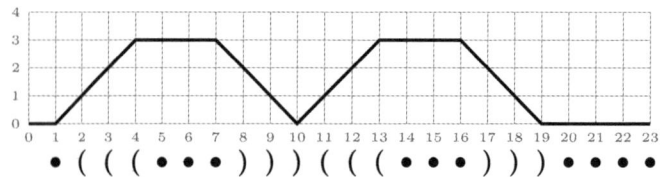

**4.** The following figure shows an RNA structure with different types of loops together with several of its tree representations, capturing different levels of structural detail.

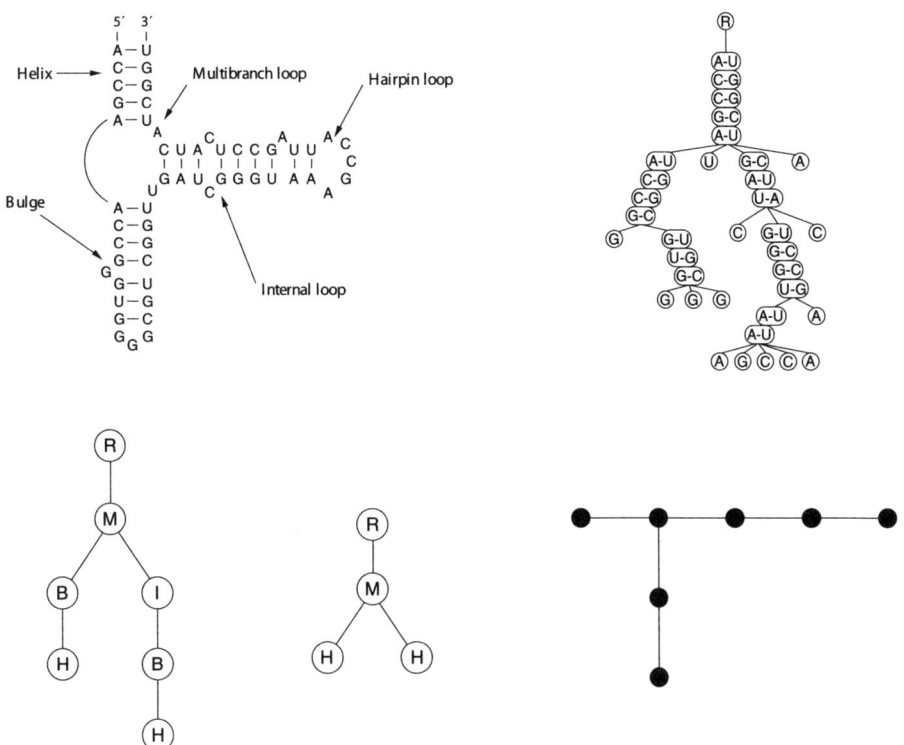

## 20.5.3  MINIMAL FREE ENERGY FOLDING ALGORITHMS

In thermodynamics, the *free energy* of a chemical structure is the amount of energy needed to maintain its structural integrity. The lower the free energy the more stable the structure is, whereas a positive free energy means that it would require energy to maintain it. In most thermodynamics models for RNA folding, energy parameters or "scores" are assigned to *substructures* (base pairs, loops, etc.), and the free energy is determined by summing over all of its substructures. Computing the minimum free energy structure of an RNA sequence is a discrete optimization problem.

### Definitions:

The **dynamic programming** approach to finding the minimal free energy of a secondary structure of $\mathbf{b} = b_1 b_2 \ldots b_n$ has two steps:

- **recursive solution** of the problem on substrings to find the overall minimal free energy;
- **traceback**, which recovers a structure that achieves this minimum.

Let $e(i, j)$ be the free energy contribution if $i$ and $j$ were to form a base pair, and let $E(i, j)$ be the minimal energy of the subsequence $\mathbf{b}_{i,j} = b_i b_{i+1} \ldots b_j$.

### Facts:

**1.** A simple energy-based RNA folding model that maximizes the number of base pairs was proposed by Nussinov et al. [NuEtal78].

**2.** A weighted variant of the Nussinov model specifies the free energies $e(i,j)$ using

$$e(i,j) = \begin{cases} -3 & \{b_i, b_j\} = \{C, G\}, & i \leq j - 4 \\ -2 & \{b_i, b_j\} = \{A, U\}, & i \leq j - 4 \\ -1 & \{b_i, b_j\} = \{G, U\}, & i \leq j - 4 \\ 0 & \text{otherwise.} \end{cases}$$

**3.** In the model specified by Fact 2, the energy scores of $-3$ and $-2$ represent the number of hydrogen bonds in the respective base pairs. The energy score $-1$ represents the two hydrogen bonds appearing in the (less stable) wobble pairs.

**4.** The $E(i,j)$ can be computed recursively via dynamic programming using

$$E(i,j) = \min \begin{cases} E(i, j-1) \\ \min_{i \leq k \leq j-4} E(i, k-1) + E(k+1, j-1) + e(k, j). \end{cases}$$

**5.** The recursion in Fact 4 can be justified as follows. If $i > j - 4$ then $E(i,j) = 0$. Otherwise, there are two possibilities for the structure that minimizes the energy: either $j$ is unpaired, in which case there is a secondary structure on $\mathbf{b}_{i,j-1}$, or $j$ is paired with some $k$, where $i \leq k \leq j - 4$, in which case the noncrossing condition implies that the secondary structure splits into substructures on $\mathbf{b}_{i,k-1}$ and $\mathbf{b}_{k+1,j-1}$. The optimal energy score $E(i,j)$ is the minimum value resulting from these two cases.

**6.** The value $E(1, n)$ is the minimal free energy of $\mathbf{b} = \mathbf{b}_{1,n}$.

**7.** The main assumption in the energy-based structure prediction methods is that the RNA sequence folds to a structure that minimizes the free energy.

**8.** Nussinov's model is overly simplistic. A more complex model for the free energy of secondary structures is proposed in [TiEtal73]. In this model the free energy of a secondary structure is the sum of independent energies for each loop in the structure. The model has evolved substantially over the years and since it assumes that the thermodynamic stability of a base pair is dependent on the identity of the adjacent base pairs, it's known as the *nearest-neighbor thermodynamic model* (NNTM).

**9.** The free energy of an RNA structure depends primarily on three factors: (i) the helices, or stacks, which are generally stabilizing and contribute negative free energy; (ii) other loops, which are generally destabilizing and contribute positive free energy; and (iii) the surrounding temperature.

**10.** State-of-the-art models have thousands of experimentally and computationally determined parameters [TuMa10] and require computational resources to run, such as the publicly available UNAFold web server [MaZu08]. Another popular RNA folding program is the Vienna RNA Package [HoEtal94]. These programs, which exclude pseudoknots, can be accessed at

- `http://unafold.rna.albany.edu`
- `http://www.tbi.univie.ac.at/RNA`

**11.** The theromodynamics of multi-branch loops and pseudoknots is still poorly understood [DiTuMa01], [LyPe00].

**12.** The accuracy of the NNTM predictions varies widely [RoHe14].

**13.** Recent work in loop energy estimation ranks substructures more accurately but in ways that are not handled well by dynamic programming [AaNa10], [ZhEtal08].

**14.** Predicting RNA secondary structures containing pseudoknots of arbitrary types is NP complete for a large class of reasonable free-energy functions [LyPe00]. Energy-based, polynomial time algorithms for prediction of RNA secondary structures have been developed in [Ak00], [RiEd99], [UeEtal99] but each of them allows a restricted class of pseudoknots.

**Examples:**
**1.** Consider the sequence $\mathbf{b} = $ GGGACCUUCC. The $e(i, j)$ can be calculated using Fact 2. Namely, $e(1, 5) = -3$, $e(1, 6) = -3$, $e(2, 6) = -3$, $e(1, 7) = -1$, $e(2, 7) = -1, \ldots$, $e(1, 10) = -3$, $e(2, 10) = -3$, $e(3, 10) = -3$, $e(4, 10) = 0$, $e(5, 10) = 0$, $e(6, 10) = 0$.

**2.** Using the sequence in Example 1, we can recursively calculate the $E(i, j)$ values. For example, $E(1, 5) = \min\{E(1, 4), E(1, 0) + E(2, 4) + e(1, 5)\} = \min\{0, -3\} = -3$. In a similar way, we determine $E(1, 6) = -3$ and $E(2, 6) = -3$. Consequently, $E(1, 7) = \min\{E(1, 6), E(1, 0) + E(2, 6) + e(1, 7), E(1, 1) + E(3, 6) + e(2, 7), E(1, 2) + E(4, 6) + e(3, 7)\} = \min\{-3, -4, -1, -1\} = -4$.

**3.** The values $E(i, j)$ can be conveniently stored in a table. The following table shows the values of $E(i, j)$ for the sequence $\mathbf{b} = $ GGGACCUUCC.

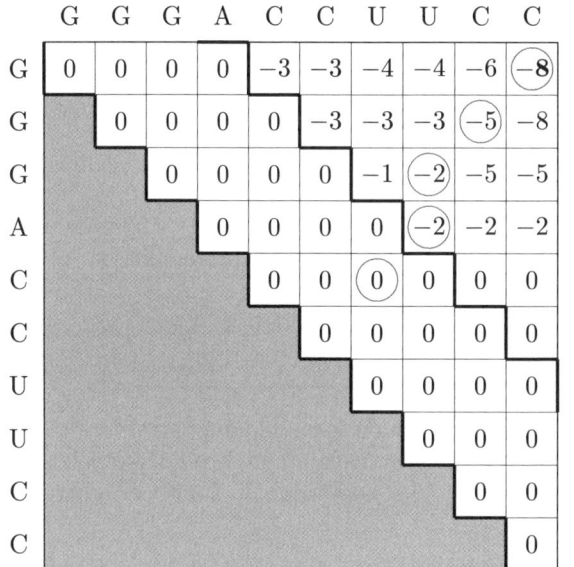

The minimal free energy is $E(1, 10)$, which is the upper-right entry in the table. It is calculated as

$$
\begin{aligned}
E(1, 10) \;=\; & \min\{E(1, 9),\, E(1, 0) + E(2, 9) + e(1, 10),\, E(1, 1) + E(3, 9) + e(2, 10),\\
& E(1, 2) + E(4, 9) + e(3, 10),\, E(1, 3) + E(5, 9) + e(4, 10),\\
& E(1, 4) + E(6, 9) + e(5, 10),\, E(1, 5) + E(7, 9) + e(6, 10)\}\\
\;=\; & \min\{-6, -8, -8, -5, 0, 0, -3\} = -8.
\end{aligned}
$$

The actual RNA structure can be recovered by the dynamic programming traceback step: start by circling the $E(1, 10)$ entry, then circle a value that could have preceded it in the recursive step, and repeat until a value of 0 is reached. The base pairs can be recovered from the "jumps" in the circled values. One RNA structure achieving this minimum energy is shown in the preceding figure.

## 20.5.4  LANGUAGE-THEORETIC METHODS

Formal language theory arose in the field of linguistics, in an attempt to understand the syntactic structure and evolution of natural languages. Since the 1950s, it has been applied to genetics, because DNA and RNA sequences can be viewed as languages over a four-letter alphabet.

**Definitions:**

A *language* is a set of (finite) strings over an alphabet $\Sigma$ of **terminal symbols**.

A *grammar* is a collection of production rules for strings in a given language that dictates how temporary **nonterminal symbols** can be rewritten into strings of terminal symbols, along with a distinguished **start symbol** $S$.

A *derivation* of a string $\alpha$ is a sequence of production rules transforming $S$ into $\alpha$.

A *context-free grammar* (**CFG**) is a grammar in which every production rule is of the form $V \to \alpha$, where $V$ is a nonterminal symbol and $\alpha$ is a string of terminal and/or nonterminal symbols.

A *stochastic context-free grammar* (**SCFG**) assigns probabilities to the production rules, so that for each nonterminal symbol $V$, the probabilities $Pr(V \to \alpha_i)$ taken over all $\alpha_i$ sum to 1.

The **Knudsen-Hein grammar** [KnHe99] is an SCFG for RNA folding defined as follows, where $p_i + q_i = 1$ for each $i = 1, 2, 3$.

| production | probability | production | probability |
|---|---|---|---|
| $S \to LS$ | $p_1$ | $S \to L$ | $q_1$ |
| $L \to dFd'$ | $p_2$ | $L \to s$ | $q_2$ |
| $F \to dFd'$ | $p_3$ | $F \to LS$ | $q_3$ |

Each string in the resulting language corresponds to a secondary structure with minimal arc length 3. The terminals $s, d, d'$ represent an isolated nucleotide, a left base pair end, and a right base pair end, respectively. The nonterminals $L$ and $F$ create loops and helices, respectively.

The **probability of a derivation** is the product of the probabilities of each individual rule used and the **probability of a structure** is the sum of all left-to-right derivations that yield that structure.

**Facts:**

**1.** The most restrictive class of grammars in the Chomsky hierarchy is the class of *regular grammars*. These can be used to generate RNA sequences but not RNA structures.

**2.** The location of terminal symbols in regular languages is uncorrelated. Long distance pairwise correlations between nucleotides in a sequence can be modeled by CFGs but not by regular grammars. Therefore, CFGs are generally used for RNA structure prediction algorithms.

**3.** Probabilistic modeling approaches for RNA secondary structure prediction using SCFGs were first introduced in the 1990s [SaEtal94].

**4.** The unknown probability parameters in an SCFG can be estimated using "training algorithms" such as the *inside-outside algorithm* [Ba79], [LaYo90].

**5.** The Cocke-Younger-Kasami algorithm finds the probability of the most likely secondary structure for a given RNA sequence. A traceback procedure can be used to recover a structure that achieves this maximum probability [DuEtal98].

**6.** The Knudsen-Hein grammar is used in the secondary structure prediction software Pfold [KnHe03]. This method predicts a structure for a set of aligned sequences. Predictions can be done on a web server at

- `http://www.daimi.au.dk/~compbio/pfold`

**7.** The Knudsen-Hein grammar was compared to eight other grammars in [DoEd04], and was shown to perform at least as well as more complicated grammars.

**8.** It has been shown that the number of different motifs (base pairs, helices, hairpin loops, internal loops, etc.) is normally distributed among structures generated by the Knudsen-Hein grammar [HePo14].

**9.** A topological approach to fold RNA pseudoknots with an SCFG has been proposed [ReEtal11]; citations within this work refer to other language-theoretic approaches that include pseudoknots.

**Example:**

**1.** The unique left-to-right derivation of the structure $\mathcal{S}$ for $\mathbf{b} = $ GGACUGC is

$$S \xrightarrow{q_1} L \xrightarrow{p_2} dFd' \xrightarrow{q_3} dLSd' \xrightarrow{p_2} ddFd'Sd' \xrightarrow{q_3} ddLSd'Sd' \xrightarrow{q_2} ddsSd'Sd'$$
$$\xrightarrow{q_1} ddsLd'Sd' \xrightarrow{q_2} ddssd'Sd' \xrightarrow{q_1} ddssd'Ld' \xrightarrow{q_2} ddssd'sd'$$

and therefore $Pr(\mathcal{S}) = p_2^2 q_1^3 q_2^3 q_3^2$.

From the string $\alpha = ddssd'sd'$ we infer the following RNA structure

# 20.6   COMBINATORIAL NEURAL CODES

Neural codes are the brain's way of representing, transmitting, and storing information about the world. Combinatorial neural codes are based on binary patterns of neural activity, as opposed to the precise timing or rate of neural activity. The structure of a combinatorial code may reflect important aspects of the represented stimuli or network architecture. Combinatorial codes can be analyzed using an algebraic object called the neural ring.

## 20.6.1   BASIC CONCEPTS

From simultaneous recordings of neurons in the brain we can infer which subsets of neurons tend to fire together. This information is captured by a combinatorial code.

### Definitions:

The set of neurons is denoted by $[n] = \{1, \ldots, n\}$.

An **action potential**, or **spike**, is an electrical event in a single neuron. This is the fundamental unit of neural activity. We say that a neuron "fires" action potentials, or spikes.

A **spike train** is a sequence of spike times for a single neuron. This captures the electrical activity of the neuron over time.

A **codeword** is a string of 0s and 1s, with a 1 for each active neuron and a 0 denoting silence; equivalently, it is a subset $\sigma \subseteq [n]$ of (active) neurons firing together. For example, if $n = 6$ the subset $\sigma = \{145\} \subseteq [6]$ is also denoted 100110.

A **combinatorial neural code** is a collection of codewords $\mathcal{C} \subseteq 2^{[n]}$. In other words, it is a binary code of length $n$, where each binary digit is interpreted as the "on" or "off" state of a neuron.

A **maximal codeword** is a codeword that is maximal in the code under inclusion. If $\sigma \in \mathcal{C}$ is maximal, then there is no $\tau \in \mathcal{C}$ such that $\tau \supsetneq \sigma$.

An **abstract simplicial complex** $\Delta \subseteq 2^{[n]}$ is a collection of subsets of $[n]$ that is closed under inclusion (see §16.6.1). That is, if $\sigma \in \Delta$ and $\tau \subset \sigma$, then $\tau \in \Delta$. A **facet** of $\Delta$ is an element of $\Delta$ that is maximal under inclusion.

The **simplicial complex of a code**, $\Delta(\mathcal{C})$, is the smallest abstract simplicial complex on $[n]$ that contains all elements of $\mathcal{C}$:

$$\Delta(\mathcal{C}) = \{\sigma \subseteq [n] \mid \sigma \subseteq \tau \text{ for some } \tau \in \mathcal{C}\}.$$

### Facts:

**1.** Spikes (action potentials) are all-or-none electrical events. It thus suffices to keep track only of the spike times, as in a spike train.

**2.** Most combinatorial neural codes appear ill-suited for error correction [CuEtal13a].

**3.** Simplicial complexes are heavily-studied objects in topology and algebraic combinatorics.

**4.** Each facet of $\Delta(\mathcal{C})$ corresponds to a maximal codeword of $\mathcal{C}$.

**5.** The simplicial complex $\Delta(\mathcal{C})$ is useful for analyzing a code, but discards important information. All codes with the same maximal codewords have the same simplicial complex.

**6.** Manin [Ma15] provides an historical overview contrasting neural codes with error-correcting codes and cryptography.

### Example:

**1.** Combinatorial codes can be obtained from neural data by temporally binning the spikes into patterns of 0s and 1s. The following figure depicts a set of binned spike trains and the resulting codewords. The set of unique codewords is the code $\mathcal{C}$. The simplicial complex $\Delta(\mathcal{C})$ has facets corresponding to the two maximal codewords, 1110 and 1101.

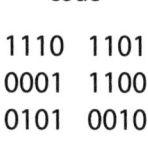

spike trains

code

1110  1101
0001  1100
0101  0010

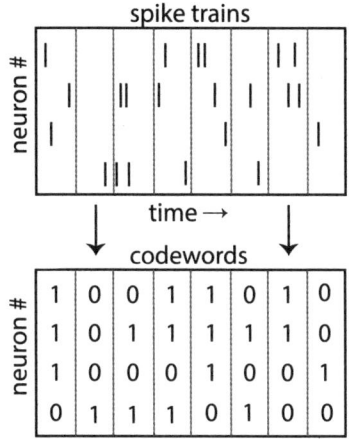

time →

codewords

| 1 | 0 | 0 | 1 | 1 | 0 | 1 | 0 |
|---|---|---|---|---|---|---|---|
| 1 | 0 | 1 | 1 | 1 | 1 | 1 | 0 |
| 1 | 0 | 0 | 0 | 1 | 0 | 0 | 1 |
| 0 | 1 | 1 | 1 | 0 | 1 | 0 | 0 |

simplicial complex

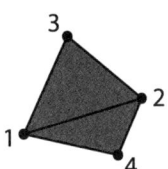

## 20.6.2  THE CODE OF A COVER

An important type of combinatorial neural code is one defined by an arrangement of open sets in Euclidean space. The open sets correspond to receptive fields.

**Definitions:**

A **stimulus space** $X$ is a parametric space of stimuli. The stimuli could be sensory, such as visual, auditory, or olfactory signals, or higher-level, such as an animal's position in space. Typically, a stimulus space is modeled as a subset of Euclidean space, $X \subseteq \mathcal{R}^d$.

A **receptive field** is a subset $U_i \subseteq X$ of the stimulus space corresponding to a single neuron $i$. The stimuli in $U_i$ induce neuron $i$ to fire.

A subset $V \subseteq \mathcal{R}^n$ is **convex** if, given any pair of points $x, y \in V$, the point $z = tx+(1-t)y$ is contained in $V$ for any $t \in [0,1]$.

**Convex receptive fields** are convex subsets $U_i \subseteq X$.

A collection of open sets $\mathcal{U} = \{U_1, \ldots, U_n\}$ is an **open cover** of their union $\bigcup_{i=1}^{n} U_i$.

$\mathcal{U}$ is a **good cover** if every nonempty intersection $\bigcap_{i\in\sigma} U_i$ is contractible (that is, if it can be continuously shrunk to a point).

The **nerve** of an open cover $\mathcal{U}$ is the simplicial complex

$$\mathcal{N}(\mathcal{U}) = \{\sigma \subseteq [n] \mid \bigcap_{i\in\sigma} U_i \neq \emptyset\}.$$

Given an open cover $\mathcal{U}$, the **code of the cover** is the combinatorial neural code

$$\mathcal{C}(\mathcal{U}) = \{\sigma \subseteq [n] \mid \bigcap_{i\in\sigma} U_i \setminus \bigcup_{j\in[n]\setminus\sigma} U_j \neq \emptyset\}.$$

**Facts:**

**1.** Neurons in many brain areas, such as sensory cortices and the hippocampus, have activity patterns that can be characterized by receptive fields.

**2.** Receptive fields are computed experimentally by correlating neural responses to independently measured external stimuli.

**3.** Intersections of convex sets are always convex, and all convex sets are contractible. Thus, any open cover consisting of convex sets is a good cover.

**4.** Each codeword in $\mathcal{C}(\mathcal{U})$ corresponds to a region that is defined by the intersections of the open sets in $\mathcal{U}$ [CuEtal13b].

**5.** If $\mathcal{U}$ is an open cover, then $\mathcal{C}(\mathcal{U}) \subseteq \mathcal{N}(\mathcal{U})$ and $\Delta(\mathcal{C}(\mathcal{U})) = \mathcal{N}(U)$. The nerve of the cover can thus be recovered from the code by completing it to a simplicial complex, but the code contains additional information about $\mathcal{U}$ that is not captured by the nerve alone.

**6.** *Nerve lemma:* If $\mathcal{U}$ is a good cover, then the covered space $Y = \bigcup_{i=1}^{n} U_i$ is homotopy-equivalent to $\mathcal{N}(\mathcal{U})$. In particular, $Y$ and $\mathcal{N}(\mathcal{U})$ have exactly the same homology groups.

**7.** *Helly's theorem:* Consider $k$ convex subsets $U_1, \ldots, U_k \subseteq \mathcal{R}^d$, for $d < k$. If the intersection of every $d+1$ of these sets is nonempty, then the full intersection $\bigcap_{i=1}^{k} U_i$ is also nonempty.

**8.** In addition to Helly's theorem and the Nerve lemma, there is a great deal known about $\mathcal{N}(\mathcal{U})$ for collections of convex sets in $\mathcal{R}^d$. In particular, the $f$-vectors of such simplicial complexes have been completely characterized by G. Kalai [Ka84], [Ka86].

**9.** The Nerve lemma has been exploited in the context of two-dimensional place field codes to show that topological features of an animal's environment could be inferred from neural codes representing position in the hippocampus [CuIt08].

**Example:**

**1.** The following figure, adapted from [CuEtal15], depicts an open cover consisting of four convex sets (A) as well as the corresponding code (B). The nerve of the cover (C) is identical to the simplicial complex $\Delta(\mathcal{C})$.

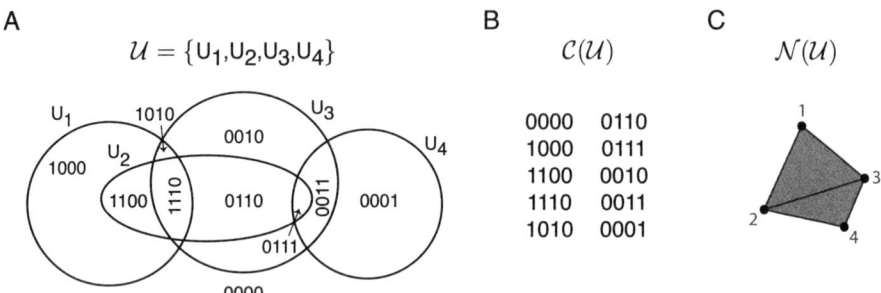

### 20.6.3   THE NEURAL RING AND IDEAL

The structure of a combinatorial code can be analyzed using the neural ring and ideal. These are algebraic objects that keep track of the combinatorics of the code, much as the Stanley-Reisner ring and ideal encode a simplicial complex [MiSt05]. For more details, see [CuEtal13b].

**Definitions:**

$\mathcal{F}_2$ is the field with two elements $\{0,1\}$. We can regard a codeword on $n$ neurons as an element of $\mathcal{F}_2^n$ and a combinatorial neural code as a subset $\mathcal{C} \subseteq \mathcal{F}_2^n$.

$\mathcal{F}_2[x_1, \ldots, x_n]$ is a polynomial ring with coefficients in $\mathcal{F}_2$.

The **ideal $I_{\mathcal{C}}$** is the set of polynomials that vanish on all codewords in $\mathcal{C}$:

$$I_{\mathcal{C}} = I(\mathcal{C}) = \{f \in \mathcal{F}_2[x_1, \ldots, x_n] \mid f(c) = 0 \text{ for all } c \in \mathcal{C}\}.$$

The **neural ring $R_{\mathcal{C}}$** is the quotient ring

$$R_{\mathcal{C}} = \mathcal{F}_2[x_1, \ldots, x_n]/I_{\mathcal{C}}.$$

A **pseudo-monomial** is a polynomial $f \in \mathcal{F}_2[x_1, \ldots, x_n]$ that can be written as

$$f = \prod_{i \in \sigma} x_i \prod_{j \in \tau}(1 - x_j),$$

where $\sigma, \tau \subseteq [n]$ satisfy $\sigma \cap \tau = \emptyset$.

For any binary string $v \in \mathcal{F}_2^n$, the **indicator function**

$$\chi_v = \prod_{\{i \mid v_i = 1\}} x_i \prod_{\{j \mid v_j = 0\}} (1 - x_j)$$

is a pseudo-monomial with the property that $\chi_v(v) = 1$ and $\chi_v(c) = 0$ for any $c \neq v$.

The **neural ideal $J_{\mathcal{C}}$** is generated by the indicator functions of all non-codewords:

$$J_{\mathcal{C}} = \langle \chi_\nu \mid \nu \in \mathcal{F}_2^n \setminus \mathcal{C} \rangle.$$

A pseudo-monomial $f \in J_{\mathcal{C}}$ is called **minimal** if there does not exist another pseudo-monomial $g \in J_{\mathcal{C}}$ with $\deg(g) < \deg(f)$ such that $f = hg$ for some $h \in \mathcal{F}_2[x_1, \ldots, x_n]$.

The **canonical form** of $J_{\mathcal{C}}$ is the set of all minimal pseudo-monomials:

$$\mathrm{CF}(J_{\mathcal{C}}) = \{f \in J_{\mathcal{C}} \mid f \text{ is a minimal pseudo-monomial}\}.$$

**Facts:**

**1.** A polynomial $f \in \mathcal{F}_2[x_1, \ldots, x_n]$ can be evaluated on a binary string of length $n$ (such as a codeword) by simply replacing each indeterminate $x_i$ with the 0/1 value of the $i^{\text{th}}$ position in the string. For example, if $f = x_1 x_3(1 - x_2) \in \mathcal{F}_2[x_1, \ldots, x_4]$, then $f(1011) = 1$ and $f(1100) = 0$.

**2.** Irrespective of $\mathcal{C}$, the ideal $I_{\mathcal{C}}$ always contains the relations $\mathcal{B} = \langle x_1^2 - x_1, \ldots, x_n^2 - x_n \rangle$, due to the binary nature of codewords.

**3.** The ideals $I_{\mathcal{C}}$ and $J_{\mathcal{C}}$ carry all the combinatorial information about the code $\mathcal{C}$. They are closely related: $I_{\mathcal{C}} = J_{\mathcal{C}} + \mathcal{B}$.

**4.** *Fundamental lemma:* Let $\mathcal{C} \subseteq \{0, 1\}^n$ be a neural code, and let $\mathcal{U} = \{U_1, \ldots, U_n\}$ be any collection of open sets (not necessarily convex) such that $\mathcal{C} = \mathcal{C}(\mathcal{U})$. Then, for any pair of subsets $\sigma, \tau \subseteq [n]$,

$$\prod_{i \in \sigma} x_i \prod_{j \in \tau}(1 - x_j) \in I_{\mathcal{C}} \iff \bigcap_{i \in \sigma} U_i \subseteq \bigcup_{j \in \tau} U_j.$$

**5.** The canonical form is a special basis, similar to a Grobner basis but tailored to a different purpose. From the canonical form one can read off minimal relationships between receptive fields.

**6.** The canonical form $\mathrm{CF}(J_{\mathcal{C}})$ can be computed algorithmically, starting from the code $\mathcal{C}$. In [CuEtal13b, Section 4.5], one such algorithm is described that uses the primary decomposition of pseudo-monomial ideals. This algorithm has since been improved [CuYo15], and software for computing $\mathrm{CF}(J_{\mathcal{C}})$ is publicly available at

- `https://github.com/nebneuron/neural-ideal`

**Examples:**

**1.** The code $\mathcal{C}(\mathcal{U})$ shown in panel (B) of §20.6.2, Example 1 has ten codewords and six non-codewords: $0100, 1001, 0101, 1101, 1011$, and $1111$. The neural ideal is

$$\begin{aligned} J_{\mathcal{C}} \;=\; & \langle x_2(1-x_1)(1-x_3)(1-x_4),\; x_1 x_4(1-x_2)(1-x_3),\; x_2 x_4(1-x_1)(1-x_3), \\ & x_1 x_2 x_4(1-x_3),\; x_1 x_3 x_4(1-x_2),\; x_1 x_2 x_3 x_4 \rangle. \end{aligned}$$

The canonical form is

$$\mathrm{CF}(J_{\mathcal{C}}) = \{x_1 x_4, x_2(1-x_1)(1-x_3), x_2 x_4(1-x_3)\}.$$

Using the fundamental lemma, we can read off the following receptive field relationships: $U_1 \cap U_4 = \emptyset$, $U_2 \subseteq U_1 \cup U_3$, and $U_2 \cap U_4 \subseteq U_3$. This is consistent with the original arrangement of open sets shown in panel (A) of §20.6.2, Example 1.

**2.** The code $\mathcal{C} = \{111, 011, 001, 000\}$ on three neurons has the canonical form $\mathrm{CF}(J_{\mathcal{C}}) = \{x_1(1-x_2), x_1(1-x_3), x_2(1-x_3)\}$. This indicates that $U_1 \subseteq U_2$, $U_1 \subseteq U_3$, and $U_2 \subseteq U_3$.

---

## 20.6.4  CONVEX CODES

**Definitions:**

Let $\mathcal{C}$ be a combinatorial neural code on $n$ neurons.

If there exists an open cover $\mathcal{U} = \{U_1, \ldots, U_n\}$ such that $\mathcal{C} = \mathcal{C}(\mathcal{U})$ and each $U_i$ is a convex open subset of $\mathcal{R}^d$, then $\mathcal{C}$ is a **convex code**.

The **minimum embedding dimension** $d(\mathcal{C})$ of a convex code is the minimum dimension such that $\mathcal{C}$ admits a convex representation.

A code $\mathcal{C} = \mathcal{C}(\mathcal{U})$ has a **local obstruction** if there exists a nonempty intersection $U_\sigma = \bigcap_{i \in \sigma} U_i$ such that $U_\sigma \subseteq \bigcup_{i \in \tau} U_i$, but the nerve of the cover $\{U_i \cap U_\sigma\}_{i \in \tau}$ is *not* contractible.

**Facts:**

**1.** Convex codes have been observed in several brain areas. Orientation-selective neurons in the visual cortex [BeBaSo95] have convex receptive fields that reflect a neuron's preference for a particular angle. Hippocampal place cells [McNEtal06], [OKDo71] are neurons that have spatial receptive fields, called *place fields*, that are typically convex.

**2.** If $\mathcal{C}$ has a local obstruction, then $\mathcal{C}$ is not a convex code.

**3.** All codes on $n \leq 2$ neurons are convex.

**4.** For $n \leq 3$ there are six non-convex codes (up to permutation-equivalence, including the all-zeros word): namely, $\{000, 010, 001, 110, 101\}$, $\{000, 010, 110, 101\}$, $\{000, 110, 101\}$, $\{000, 100, 010, 110, 101, 011\}$, $\{000, 100, 110, 101, 011\}$, $\{000, 110, 101, 011\}$.

**Examples:**

**1.** The code $\mathcal{C} = \mathcal{C}(\mathcal{U})$ shown in panel (B) of §20.6.2, Example 1 is convex by construction. Panel (A) shows a two-dimensional convex realization. The minimum embedding dimension for this code is $d(\mathcal{C}) = 2$.

**2.** Consider the code $\hat{\mathcal{C}} = \mathcal{C} \setminus \{0110\}$, where $\mathcal{C}$ is the code in Example 1. Code $\hat{\mathcal{C}}$ differs from $\mathcal{C}$ by only one codeword and has the same simplicial complex $\Delta(\hat{\mathcal{C}}) = \Delta(\mathcal{C})$. However, $\hat{\mathcal{C}}$ is not a convex code. It has a local obstruction because $U_2 \cap U_3 \subseteq U_1 \cup U_4$, yet the nerve of the cover of $U_\sigma = U_2 \cap U_3$ by $U_1 \cap U_\sigma$ and $U_4 \cap U_\sigma$ is disconnected, and thus not contractible.

**3.** The codes $\mathcal{C}_1 = \{111, 011, 001\}$ and $\mathcal{C}_2 = \{111, 101, 011, 110, 100, 010\}$ are both convex and have the same simplicial complex, but possess different embedding dimensions: $d(\mathcal{C}_1) = 1$, while $d(\mathcal{C}_2) = 2$.

**Open Questions:**

**1.** How do we determine, in general, whether or not a code is convex?

**2.** Are there other obstructions to convexity beyond local obstructions?

**3.** If a code is convex, what is the minimum embedding dimension?

---

### 20.6.5  FEEDFORWARD AND HYPERPLANE CODES

*Hyperplane codes* are an important class of combinatorial codes. These are codes that arise as an output of a one-layer feedforward neural network, and they are sometimes referred to as *feedforward codes*.

**Definitions:**

A **hyperplane code** is a convex code, where the underlying open convex cover $\mathcal{U} = \{U_i\}_{i=1}^n$ can be obtained as $U_i = X \cap H_i^+$, where $X \subseteq \mathcal{R}^m$ is an open convex set and the

$$H_i^+ = \{y \in \mathcal{R}^m \mid \sum_{a=1}^m w_{ia}y_a - \theta_i > 0\} \tag{3}$$

are open half-spaces.

A **one-layer feedforward neural network** is a network with input and output layers connected as shown in the following figure.

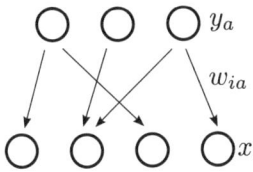

The network inputs nonnegative numbers $y_a \geq 0$ and outputs nonnegative numbers $x_i \geq 0$ according to the rule

$$x_i(y) = \phi\left(\sum_{a=1}^m w_{ia}y_a - \theta_i\right), \qquad i \in [n]. \tag{4}$$

Here $\theta_i \in \mathcal{R}$ are the neuronal thresholds, $w_{ia} \in \mathcal{R}$ are the effective strengths of the feedforward connections, and the **transfer function** $\phi\colon \mathcal{R} \to \mathcal{R}_{>0}$ satisfies the condition $\phi(t) = 0$ if $t \leq 0$ and $\phi(t) > 0$ if $t > 0$.

A **feedforward code** is a hyperplane code, where the underlying convex set $X$ can be chosen to be the positive orthant $\mathcal{R}_+^m$. This class of codes arises as the output of a one-layer feedforward neural network (4), where positivity of each row of (4) corresponds to the halfspace $H_i^+$ in (3). Specifically the code of the network (4) is

$$\mathcal{C}(w, \theta) = \{\sigma \subseteq [n] \mid \exists y \in \mathcal{R}_+^m \text{ such that } x_i(y) > 0 \Leftrightarrow i \in \sigma\}.$$

**Facts:**

**1.** Hyperplane codes (and thus feedforward codes) are convex.

**2.** Not every convex code is a hyperplane code. Perhaps the smallest example is the code $C = \{\varnothing, 2, 3, 4, 12, 13, 14, 123, 124\}$, which can be easily seen to possess a 2-dimensional convex realization. However, it can be proved to be not realizable as a hyperplane code.

**3.** *Theorem* [GiIt14]: For every simplicial complex $K$ with $n$ vertices, there exists a feedforward network $(w, \theta)$ described by (4) so that $K$ is the simplicial complex of the appropriate feedforward code $K = \Delta\left(C\left(w, \theta\right)\right)$.

**Example:**

**1.** The following figure, adapted from [GiIt14], displays a feedforward code for a network with two neurons in the input layer, corresponding to the axes $y_1$ and $y_2$, and three neurons in the output layer, corresponding to the (oriented) hyperplanes $H_1$, $H_2$, and $H_3$. For each output neuron $x_i$, the inputs $y = (y_1, y_2)$ that yield $x_i(y) > 0$ lie in the positive halfspace $H_i^+$. The resulting code $C$ consists of combinations of output neurons that can be simultaneously activated by at least one choice of nonnegative inputs.

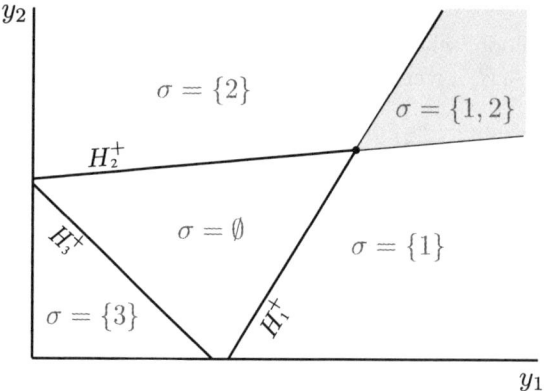

$$C = \{\varnothing, \{1\}, \{2\}, \{3\}, \{1,2\}\}$$
$$= \{000, 100, 010, 001, 110\}$$

# 20.7  FOOD WEBS AND GRAPHS

The study of food webs has occurred over the last fifty years, primarily undertaken by ecologists working in natural habitats. Mathematicians subsequently became interested in the graph-theoretical properties of food webs and their corresponding competition graphs. This section introduces food webs, together with associated graphs and parameters that play an important role in linking mathematics and ecology.

## 20.7.1  MODELING PREDATOR-PREY RELATIONSHIPS WITH FOOD WEBS

Food webs, through both direct and indirect interactions, describe the flow of energy through an ecosystem, moving from one organism to another. Understanding food webs can help to predict how important any given species is, and how ecosystems change with the addition of a new species or removal of an existing species.

**Definitions:**

A **food web** is represented by a directed graph (digraph) $D = (V, A)$ with vertex set $V$ and arc set $A$. Each vertex represents a species in the ecosystem and the arc $(x, y)$ is directed from a prey species $x$ to a predator $y$ of that prey.

A **basal species** is one that does not depend for food on any other organism in the ecosystem. That is, these are species located at the bottom of the food web.

**Facts:**

**1.** In the early 1960s, when food webs were first used to model predator-prey relationships, arcs were directed from predator to prey. The current usage, which tracks the flow of energy from prey to predator, reverses this earlier convention.

**2.** The interactions of species as they attempt to acquire food determine much of the structure of a community. Food webs represent these feeding relationships within a community.

**3.** Basal species correspond to vertices with no incoming arcs: vertices with *indegree* 0. Species at the top of the food web correspond to vertices with *outdegree* 0.

**4.** The digraph of a food web contains no directed cycles (since a species does not prey upon itself, either directly or indirectly).

**5.** Various online tools can be used to construct food webs from ecological data, such as

- http://bioquest.org/esteem

**Examples:**

**1.** The following figure depicts a food web in which sharks eat sea otters, sea otters eat sea urchins and large crabs, large crabs eat small fishes, and sea urchins and small fishes eat kelp. Equivalently, sea urchins and large crabs are eaten by sea otters (both are prey for sea otters) and sea otters are prey for sharks. Kelp is the only basal species, whereas sharks are at the top of this food web.

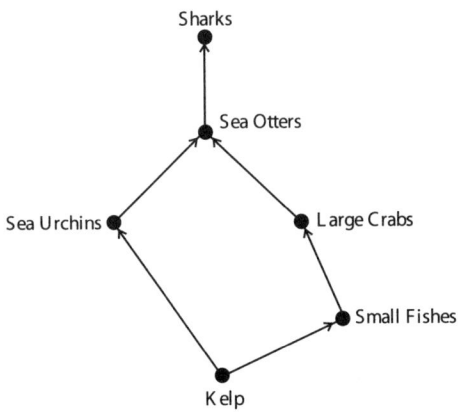

**2.** A larger food web is defined by the following predator-prey relationships.

| species | species they feed on |
|---------|----------------------|
| sharks | sea otters |
| sea otters | sea stars, sea urchins, large fish & octopus, large crabs, abalone |
| sea stars | abalone, small herbivorous fishes, sea urchins, organic debris |
| sea urchins | kelp, sessile invertebrates, organic debris |
| abalone | organic debris |
| large crabs | sea stars, small predatory fishes, organic debris, small herbivorous fishes, kelp |
| small predatory fishes | sessile invertebrates, planktonic invertebrates |
| small herbivorous fishes | kelp |
| large fish & octopus | large crabs, small predatory fishes |
| sessile invertebrates | microscopic planktonic algae, planktonic invertebrates |
| planktonic invertebrates | microscopic planktonic algae |

From these relationships, the following food web can be constructed. The basal species are seen to be kelp, organic debris, and microscopic planktonic algae. Sharks are the only species at the top of this food web.

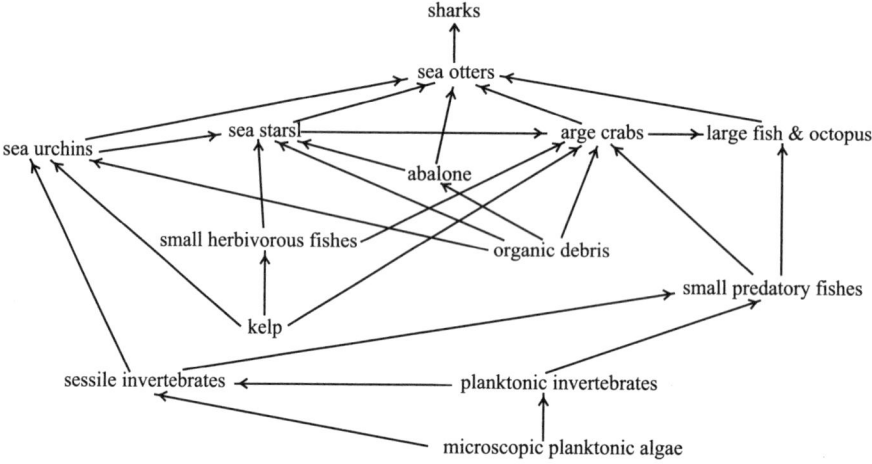

## 20.7.2   TROPHIC LEVEL AND TROPHIC STATUS

Trophic levels in food webs provide a way of organizing species in a community food web into feeding groups. Scientists have used various measures to classify species in a food web into these various feeding groups, typically based on the positioning of species in the food web.

### Definitions:

The **length** of a path in a directed graph (digraph) $D$ is the number of arcs in the path.

A **shortest path** between vertices $x$ and $y$ in $D$ has the smallest length of any path from $x$ to $y$; this shortest path length is denoted $d(x, y)$.

A *longest path* between vertices $x$ and $y$ in $D$ has the largest length of any path from $x$ to $y$; this longest path length is denoted $D(x, y)$.

*Notation*: We write $u \prec v$ if species $v$ consumes species $u$, either directly or indirectly.

The *trophic level* (option 1) $TL_1(x)$ of species $x$ is defined to be 0 if $x$ is a basal species. Otherwise, it is the smallest length of a path to $x$ from any basal species that $x$ consumes, either directly or indirectly: $TL_1(x) = \min\{d(y, x) : y \prec x, \text{and } y \text{ is basal}\}$.

The *trophic level* (option 2) $TL_2(x)$ of species $x$ is defined to be 0 if $x$ is a basal species. Otherwise, it is the largest length of a path to $x$ from any basal species that $x$ consumes, either directly or indirectly: $TL_2(x) = \max\{D(y, x) : y \prec x, \text{and } y \text{ is basal}\}$.

The *trophic status* $TS(x)$ of species $x$ is defined to be 0 if $x$ is a basal species. Otherwise, it is given by $TS(x) = \sum\{D(y, x) : y \prec x\}$.

Species $v$ is called **dominant** if there exists an arc $(u, v)$, whose removal from the food web allows $u$ to have uncontrolled growth, and thus become a new "dominant" species.

A species is **trophic status dominant** if its trophic status is greater than the number of non-basal species in the food web.

## Facts:

**1.** Since the digraph $D$ corresponding to a food web is acyclic, the longest path length $D(x, y)$ between any two vertices is well defined and finite.

**2.** The trophic level $TL_1(x)$ can be calculated recursively as follows:

$$TL_1(x) = 1 + \min\{TL_1(y) : (y, x) \in A\}, \text{ where } TL_1(x) = 0 \text{ if } x \text{ is a basal species.}$$

**3.** The trophic level $TL_2(x)$ can be calculated recursively as follows:

$$TL_2(x) = 1 + \max\{TL_2(y) : (y, x) \in A\}, \text{ where } TL_2(x) = 0 \text{ if } x \text{ is a basal species.}$$

**4.** $TL_2(x) \geq TL_1(x)$ holds for any species $x$ in a food web.

**5.** Neither of the trophic level definitions $TL_1$ or $TL_2$ is entirely satisfactory in determining the hierarchical structure of a food web. For instance, neither definition reflects the number of species that are direct or indirect prey of a species.

**6.** A reasonable property is that if species $x$ is a predator of species $y$, then the trophic level of species $x$ is greater than the trophic level of species $y$. This property is satisfied by $TL_2$ and $TS$, but not by $TL_1$.

**7.** High trophic status of a species means that there are many paths of various lengths that reach that species in the food web, so the loss of any one path is inconsequential.

**8.** If vertex $u$ has only one outgoing arc $(u, v)$, then $v$ is dominant. Such a vertex $u$ therefore has outdegree 1.

**9.** The concept of trophic status dominance incorporates both the number of species that are direct or indirect prey and the extent of energy transfer, based on the trophic status of the species. This definition resembles the definition of status for people in a community or social network. In a social network, the status of person $A$ is determined by the number of people who can reach $A$ and the (worst-case) lengths of paths to $A$.

**Examples:**

**1.** The trophic level (option 1) values for Example 1 of §20.7.1 are computed as follows: $TL_1(\text{kelp}) = 0$; $TL_1(\text{sea urchins}) = TL_1(\text{small fishes}) = 1$; $TL_1(\text{large crabs}) = TL_1(\text{sea otters}) = 2$; $TL_1(\text{sharks}) = 3$. Notice that from Fact 2, we have $TL_1(\text{sea otters}) = 1 + \min\{TL_1(\text{sea urchins}), TL_1(\text{large crabs})\} = 1 + \min\{1, 2\} = 2$.

**2.** The trophic level (option 1) values for Example 2 of §20.7.1 are as follows: $TL_1(\text{kelp}) = TL_1(\text{organic debris}) = TL_1(\text{microscopic planktonic algae}) = 0$; $TL_1(\text{abalone}) = TL_1(\text{sea stars}) = TL_1(\text{sea urchins}) = TL_1(\text{large crabs}) = TL_1(\text{small herbivorous fishes}) = TL_1(\text{sessile invertebrates}) = TL_1(\text{planktonic invertebrates}) = 1$; $TL_1(\text{sea otters}) = TL_1(\text{small predatory fishes}) = TL_1(\text{large fish \& octopus}) = 2$; $TL_1(\text{sharks}) = 3$. Fact 2 verifies that $TL_1(\text{sea otters}) = 1 + \min\{TL_1(\text{sea urchins}), TL_1(\text{sea stars}), TL_1(\text{large crabs}), TL_1(\text{large fish \& octopus})\} = 1 + \min\{1, 1, 1, 2\} = 2$.

**3.** The trophic level (option 2) values for Example 1 of §20.7.1 are computed as follows: $TL_2(\text{kelp}) = 0$; $TL_2(\text{sea urchins}) = TL_2(\text{small fishes}) = 1$; $TL_2(\text{large crabs}) = 2$; $TL_2(\text{sea otters}) = 3$; $TL_2(\text{sharks}) = 4$. Notice that from Fact 3, we have $TL_2(\text{sea otters}) = 1 + \max\{TL_2(\text{sea urchins}), TL_1(\text{large crabs})\} = 1 + \max\{1, 2\} = 3$.

**4.** The trophic level (option 2) values for Example 2 of §20.7.1 are computed as follows: $TL_2(\text{kelp}) = TL_2(\text{organic debris}) = TL_2(\text{microscopic planktonic algae}) = 0$; $TL_2(\text{small herbivorous fishes}) = TL_2(\text{abalone}) = TL_2(\text{planktonic invertebrates}) = 1$; $TL_2(\text{sessile invertebrates}) = 2$; $TL_2(\text{sea urchins}) = TL_2(\text{small predatory fishes}) = 3$; $TL_2(\text{sea stars}) = 4$; $TL_2(\text{large crabs}) = 5$; $TL_2(\text{large fish \& octopus}) = 6$; $TL_2(\text{sea otters}) = 7$; $TL_2(\text{sharks}) = 8$. Notice that $TL_2(x) \geq TL_1(x)$ holds for each species $x$, as guaranteed by Fact 4.

**5.** The trophic status of sea otters in Example 1 of §20.7.1 can be calculated by first determining longest path lengths to sea otters from all species that sea otters feed upon (directly or indirectly): namely, $D(\text{sea urchins, sea otters}) = D(\text{large crabs, sea otters}) = 1$, $D(\text{small fishes, sea otters}) = 2$, $D(\text{kelp, sea otters}) = 3$. This gives $TS(\text{sea otters} = 1 + 1 + 2 + 3 = 7$. Similar calculations produce $TS(\text{kelp}) = 0$, $TS(\text{sea urchins}) = TS(\text{small fishes}) = 1$, $TS(\text{large crabs}) = 3$, and $TS(\text{sharks}) = 12$.

**6.** Trophic status for the species in Example 2 of §20.7.1 are found as $TS(\text{kelp}) = TS(\text{organic debris}) = TS(\text{microscopic planktonic algae}) = 0$; $TS(\text{planktonic invertebrates}) = TS(\text{abalone}) = TS(\text{small herbivorous fishes}) = 1$; $TS(\text{sessile invertebrates}) = 3$; $TS(\text{small predatory fishes}) = 6$; $TS(\text{sea urchins}) = 8$; $TS(\text{sea stars}) = 16$; $TS(\text{large crabs}) = 26$; $TS(\text{large fish \& octopus}) = 37$; $TS(\text{sea otters}) = 49$; $TS(\text{sharks}) = 62$.

**7.** If the arc (sea urchins, sea otters) is removed from the food web in Example 1 of §20.7.1, then sea urchins will have uncontrolled growth and become a new "dominant" species in the food web. Therefore sea otters are considered a dominant species. Using Fact 8, since (small fishes, large crabs) is the only arc leaving small fishes, large crabs are a dominant species. Similarly, (sea otters, sharks) is the only arc leaving sea otters, so sharks are dominant.

**8.** In the food web for Example 2 of §20.7.1, both sharks and sea otters are dominant.

**9.** In the food web for Example 2 of §20.7.1, sharks, sea otters, sea stars, large crabs, and large fish & octopus are trophic status dominant since each has trophic status exceeding $14 - 3 = 11$. By contrast, only sharks and sea otters are dominant using the arc removal definition (see Example 8).

### 20.7.3   WEIGHTED FOOD WEBS

Not all ecological relationships have the same strength. Species may consume much more of one prey species than another. To model this, we can assign weights to the arcs of a food web to indicate food preferences.

**Definitions:**

The **weight** of arc $(x, y)$ in a food web, denoted $w_{xy}$, is the proportional food contribution of species $x$ to species $y$ in the food web.

The **flow-based trophic level** $TL_F(x)$ of species $x$ is defined to be 1 if $x$ is a basal species. Otherwise, it is based on the weighted sum of flow-based trophic levels of the species $y$ that it directly consumes: $TL_F(x) = 1 + \sum\{w_{yx}TL_F(y) \mid (y, x) \in A\}$.

**Fact:**

**1.** The sum of the weights of the incoming arcs to a species is 1, since the set of incoming arcs represents the full diet of the species.

**Examples:**

**1.** The following figure depicts a weighted food web. For example, the weight 0.6 on the arc from rodents to snakes indicates that snakes eat rodents more frequently than other lizards in the ratio of 6 to 4. Specifically, 60% of a snake's diet comes from rodents, while 40% of its diet comes from other lizards.

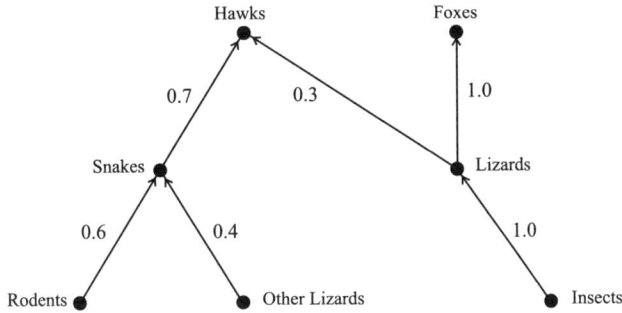

Here $TL_F(\text{snakes}) = 1 + 0.6(1) + 0.4(1) = 2$ and $TL_F(\text{lizards}) = 1 + 1(1) = 2$, so that $TL_F(\text{hawks}) = 1 + 0.7(2) + 0.3(2) = 3$.

**2.** If prairie dogs are removed from the following weighted food web, then black-footed ferrets have no food source and so will die off. If jackrabbits & small rodents are removed, and species can survive on 50% of their normal diet, then coyotes will survive, but golden eagles will have to increase their consumption of antelopes, prairie dogs, and black-footed ferrets, a challenge given their limited numbers.

Flow-based trophic levels for species in this weighted food web are given by $TL_F(\text{grasses }$ & sedges$) = 1, TL_F(\text{prairie dogs}) = TL_F(\text{jackrabbits \& small rodents}) = TL_F(\text{prong-}$horn antelopes$) = 2, TL_F(\text{black-footed ferrets}) = 3, TL_F(\text{golden eagles}) = TL_F(\text{coyotes})$ $= 3.05$.

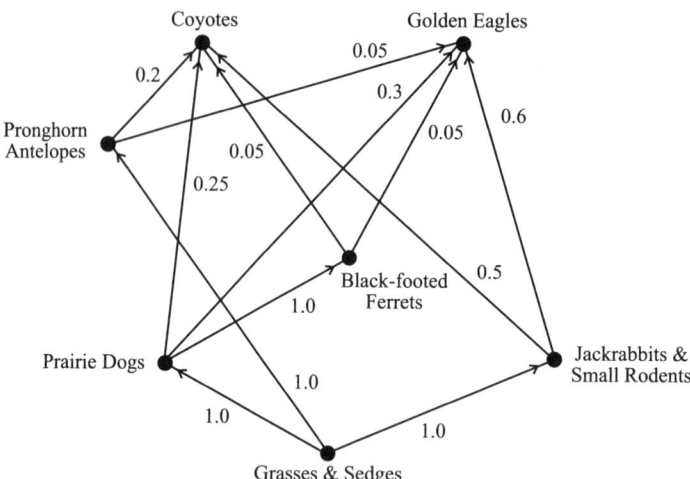

---

## 20.7.4  COMPETITION GRAPHS

There has been considerable attention paid lately to creating graphical models for better understanding predator-prey relationships, especially to inform conservation policy makers. This subsection introduces several undirected graphs and parameters that are useful in understanding the competition of species.

### Definitions:

Suppose a food web is represented by the directed acyclic graph $D$ with $n$ vertices and $m$ arcs.

The **directed connectance** (**density**) of the digraph $D$ is defined to be $C = \frac{m}{n^2}$.

The **competition graph** associated with $D$ is an undirected graph $G$ whose vertices are the species in $D$. There is an edge in $G$ between species $a$ and species $b$ if and only if $a$ and $b$ have a common prey: i.e., there is some vertex $x$ such that there exist arcs $(x, a)$ and $(x, b)$ in $D$.

The **competition number** $k(G)$ for a graph $G$ is the fewest number of isolated vertices that need to be added to $G$ so that $G$ is the competition graph for some directed acyclic graph.

The **common enemy graph** associated with $D$ is an undirected graph $G$ whose vertices are the species in $D$. There is an edge in $G$ between species $a$ and species $b$ if and only if $a$ and $b$ have a common predator: i.e., there is some vertex $x$ such that there exist arcs $(a, x)$ and $(b, x)$ in $D$.

A **clique** of the graph $G$ is a subgraph of $G$ in which every pair of distinct vertices is connected by an edge in the subgraph. A **maximal clique** is a clique that cannot be extended by including additional vertices.

### Facts:

**1.** If all possible arcs exist in the digraph $D$, there would be $n(n-1)$ arcs, so the maximum connectance is $1 - \frac{1}{n} < 1$; the minimum connectance is $0$ if there are no arcs. Since species at the top of the food web have no outgoing arcs and the food web has no cycles, the maximum is much less than $1 - \frac{1}{n}$.

**2.** It was long believed that the higher the connectance the more stable the food web. In fact, using real or simulated models and population dynamics on the food web and competition graph, the steady state (stability) is achieved only for small $n$ and $C$, particularly when the product $nC = \frac{m}{n} < 2$. These results are robust under the change of initial conditions and ecological parameters [DuWiMa02].

**3.** Competition graphs are also known as *niche overlap graphs* and *predator graphs*.

**4.** Common enemy graphs are also known as *prey graphs*.

**5.** If $D$ is a directed acyclic graph, then there must exist an isolated vertex in its corresponding competition graph $G$. One such vertex would be a vertex having no incoming arcs in $D$. (Every directed acyclic graph contains at least one vertex of indegree 0.)

**6.** Any graph $G$ can be the competition graph for some directed acyclic graph $D$ by adding a sufficient number of isolated vertices to $G$.

**7.** If $G$ has exactly one hole (induced cycle of length at least 4), then $k(G) \leq 2$ [ChKi05].

**8.** $k(G) \leq$ number of holes $+ 1$ [MaScSc14].

**9.** Determining the competition number is an NP-hard problem [Op82].

**Examples:**

**1.** In the food web of §20.7.1, Example 1 sea urchins and small fishes both have kelp as a common prey, so (sea urchins, small fishes) is an edge of the competition graph. This simple competition graph, shown next, has one edge and four independent vertices.

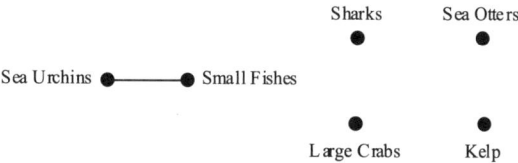

**2.** The following figure shows a food web $D$ for the polar bear. Here we use the abbreviations PB (Polar Bear), AB (Arctic Birds), P (Phytoplankton), RS (Ringed Seal), ACd (Arctic Cod), HZ (Herbivorous Zooplankton), KW (Killer Whale), HS (Harbour Seal), ACh (Arctic Char), HpS (Harp Seal), CZ (Carnivorous Zooplankton), and C (Capelin). Note that Phytoplankton (P) is the only basal species of this food web.

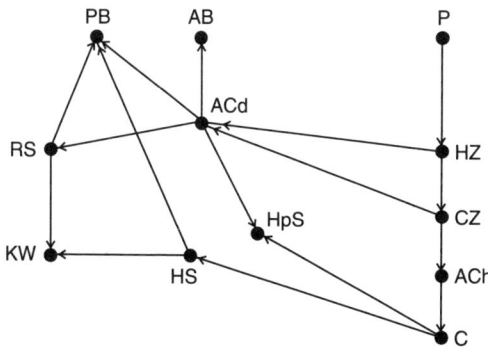

The corresponding competition graph $G$ is shown next. The competition graph $G$ contains the isolated vertices P, C, and HZ. Since there are arcs in $D$ from ACd to PB, AB, RS, and HpS, those four vertices form a clique, in fact a maximal clique, in $G$.

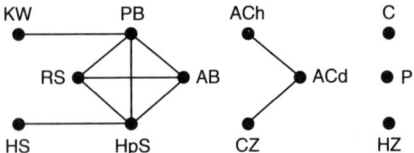

**3.** The following graph $G$ contains a single hole, the induced 4-cycle $abcd$. Therefore by Fact 7 it is possible to add at most two isolated vertices to $G$ so that the result is a competition graph for some digraph $D$.

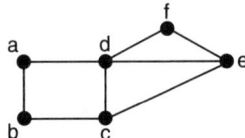

The digraph $D$ shown next generates a competition graph that consists of $G$ plus a single isolated vertex $x$. Thus $k(G) = 1$.

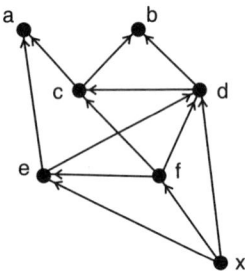

---

## 20.7.5   INTERVAL GRAPHS AND BOXICITY

This subsection discusses interval graphs and related concepts that are useful in assessing the dimension of community habitats.

**Definitions:**

A graph is an **interval graph** if we can find a set of intervals on the real line so that each vertex is assigned an interval and two vertices are joined by an edge if and only if their corresponding intervals overlap.

The **boxicity** $box(G)$ of graph $G$ is the smallest positive integer $p$ having the property that we can assign a box in Euclidean $p$-space to each vertex of $G$ so that two vertices are connected by an edge of $G$ if and only if their corresponding boxes overlap.

The **clique matrix** for a graph $G$ is the binary matrix $M = (m_{ij})$, whose rows correspond to the maximal cliques of $G$ and whose columns correspond to the vertices of $G$; $m_{ij}$ is 1 if and only if vertex $j$ is in maximal clique $i$, and is 0 otherwise.

A binary matrix has the **consecutive ones property for columns** if its rows can be permuted so that the 1s in each column occur consecutively.

The **sink food web** for a subset of species $X$ in a food web is the induced subgraph defined by all species that are prey of species in $X$.

The **source food web** for a subset of species $W$ in a food web is the induced subgraph defined by all species that are predators of species in $W$.

**Facts:**

**1.** If $G$ is the competition graph corresponding to a real community food web and $G$ is an interval graph, then the species in the food web have one-dimensional habitats or niches. That is, each species can be mapped to the real line with overlapping intervals if they have common prey, and this single dimension applies to each species in the web. This single dimension might be determined by temperature, moisture, pH, or a number of other factors.

**2.** The maximal cliques of an interval graph can be ordered in a sequence such that for any vertex $v$, the maximal cliques containing $v$ occur consecutively in the sequence [FuGr65]. This is one way of determining if a graph is an interval graph—does such an ordering of maximal cliques exist?

**3.** A graph is an interval graph if and only if its clique matrix $M$ has the consecutive ones property for columns [FuGr65].

**4.** A graph is an interval graph if and only if it does not contain one of the following "forbidden" structures as a subgraph: the cycle $C_4$, the complete bipartite graph $K_{3,3}$, or the graph $H$ having a pendant edge attached to each vertex of the complete graph $K_3$ [GiHo64].

**5.** A graph is an interval graph if and only if it is a *chordal graph* (contains no holes) and its complement is a *comparability graph* (has a transitive orientation) [LeBo62].

**6.** Interval graphs have boxicity 1.

**7.** If the boxicity of the competition graph is 2, then the species in the associated food web have 2-dimensional habitats or niches.

**8.** Every graph can be represented as the intersection graph of boxes in some dimension, so boxicity is well defined in general. However, it is hard to compute [Co81].

**9.** There are fast algorithms to test if a graph is an interval graph. However, there are no fast ways known for computing the boxicity of a general graph [Co81], [CoRo83].

**10.** The parameter $box(G)$ is bounded above by the minimum size of a maximal matching (§10.2.1) in the complement of $G$. When $G$ is dense and its complement is sparse, the boxicity tends to be small [Ro69].

**11.** In the 1960s, Joel Cohen found that food webs arising from "single habitat ecosystems" (homogeneous ecosystems) generally have competition graphs that are interval graphs. This remarkable empirical observation of Cohen, that real-world competition graphs are usually interval graphs, has led to a great deal of research on the structure of competition graphs and on the relation between the structure of digraphs and their corresponding competition graphs. It has also led to a great deal of research in ecology to determine just why this might be the case [Co78].

**12.** Using randomly generated food webs (digraphs), Cohen et al. showed that the probability that a competition graph is an interval graph goes to 0 as the number of species increases. In other words, it should be highly unlikely that competition graphs corresponding to food webs are interval graphs [CoKoMu79].

**13.** Cohen showed that a food web has a competition graph that is an interval graph if and only if each sink food web contained in it is an interval graph. However, the same is not true for source food webs [Co78].

**14.** A food web can have a competition graph that is an interval graph while some source food web contained in it has a competition graph that is not an interval graph. The following site provides an example of a food web whose competition graph is an interval graph, together with a subset $W$ of vertices whose source food web has a competition graph of boxicity 2.

• http://dimacs.rutgers.edu/IMB/TalksEtc/Foodwebs-and-Biodiversity-
       7-29-13.pdf

**Examples:**

**1.** The competition graph $G$ shown next is an interval graph, since we can associate intervals on the real line with its vertices so that the intervals corresponding to two vertices overlap if and only if there is an edge between the two vertices in $G$. One such representation is shown. Notice that the maximal cliques of $G$ are $cd, bde, ef$. In this ordering, the maximal cliques containing each vertex occur consecutively (Fact 2).

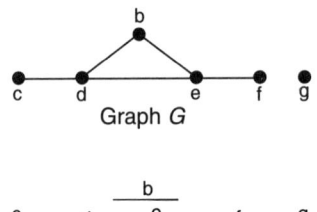

Using the above ordering of cliques and the alphabetical ordering of the vertices produces the following clique matrix, which has the consecutive ones property for columns (Fact 3):

$$M = \begin{pmatrix} 0 & 1 & 1 & 0 & 0 & 0 \\ 1 & 0 & 1 & 1 & 0 & 0 \\ 0 & 0 & 0 & 1 & 1 & 0 \end{pmatrix}.$$

**2.** The competition graph $G$ shown next is not an interval graph. An attempt to add an interval for vertex $a$ to the interval representation given in Example 1 is not successful. There is no place for such an interval that intersects only the interval for $b$ and no others. Since $G$ contains the forbidden subgraph $H$ from Fact 4, we are assured that $G$ is not an interval graph. In fact, $G$ has boxicity 2.

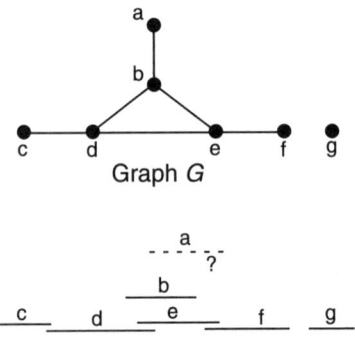

**3.** Fact 4 guarantees that the cycle $C_4$ is not an interval graph. A representation using 2-dimensional boxes is shown below, so $box(C_4) = 2$.

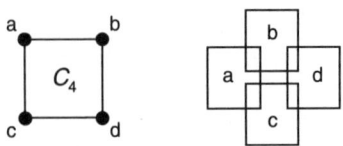

## 20.7.6  PROJECTION GRAPHS

Projection graphs are weighted graphs based on competition graphs and common enemy graphs.

**Definitions:**

Suppose $D$ is a food web with $n$ vertices. Let $T$ denote the set of top species (no outgoing arcs), $B$ the set of basal species (no incoming arcs), and $I$ the remaining (intermediate) species. Let $x_{ij}$ be a 0-1 variable, with $x_{ij} = 1$ if arc $(i, j)$ is in $D$ and $x_{ij} = 0$ otherwise.

The **predator projection graph** (**PP**) associated with $D$ has a vertex for each species in $D$ and contains the edge $(i, j)$ if $i$ and $j$ have a common prey. The weight $A(i, j)$ of edge $(i, j)$ reflects the number of common prey:

$$A(i,j) = \frac{1}{n(|B| + |I|)} \sum \{x_{ki}x_{kj} \mid k \in B \cup I\}.$$

The **prey projection graph** (**EP**) associated with $D$ has a vertex for each species in $D$ and contains the edge $(i, j)$ if $i$ and $j$ have a common predator. The weight $B(i, j)$ of edge $(i, j)$ reflects the number of common predators:

$$B(i,j) = \frac{1}{n(|T| + |I|)} \sum \{x_{ik}x_{jk} \mid k \in T \cup I\}.$$

**Facts:**

**1.** The PP graph is a weighted competition graph minus the basal species, since they have no prey. We can use the competition graph to construct the PP graph, by removing the basal species and then labelling edge $(i, j)$ with weight $A(i, j)$.

**2.** The EP graph is a weighted common enemy graph minus the top species, since they have no predators. We can use the common enemy graph to construct the EP graph, by removing the top species and then labelling edge $(i, j)$ with weight $B(i, j)$.

**3.** The matrix of edge weights in the PP graph is given by $A = \alpha X^T X$, where $X = (x_{ij})$ and $\alpha = (n(n - |T|))^{-1}$.

**4.** The matrix of edge weights in the EP graph is given by $B = \beta X X^T$, where $X = (x_{ij})$ and $\beta = (n(n - |B|))^{-1}$.

**Examples:**

**1.** For the polar bear food web (§20.7.4, Example 2), Phytoplankton (P) is the only basal species so P does not appear in the corresponding PP graph, shown next.

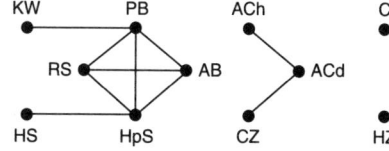

The matrix of edge weights in the PP graph, normalized by the divisor $96 = 12 \times 8$, is given by

|     | PB | AB | RS | ACd | KW | HS | HpS | HZ | CZ | ACh | C |
|-----|----|----|----|-----|----|----|-----|----|----|-----|---|
| PB  | 0 | .0104 | .0104 | 0 | .0208 | 0 | .0104 | 0 | 0 | 0 | 0 |
| AB  | .0104 | 0 | .0104 | 0 | 0 | 0 | .0104 | 0 | 0 | 0 | 0 |
| RS  | .0104 | .0104 | 0 | 0 | 0 | 0 | .0104 | 0 | 0 | 0 | 0 |
| ACd | 0 | 0 | 0 | 0 | 0 | 0 | 0 | 0 | .0104 | .0104 | 0 |
| KW  | .0208 | 0 | 0 | 0 | 0 | 0 | 0 | 0 | 0 | 0 | 0 |
| HS  | 0 | 0 | 0 | 0 | 0 | 0 | .0104 | 0 | 0 | 0 | 0 |
| HpS | .0104 | .0104 | .0104 | 0 | 0 | .0104 | 0 | 0 | 0 | 0 | 0 |
| HZ  | 0 | 0 | 0 | 0 | 0 | 0 | 0 | 0 | 0 | 0 | 0 |
| CZ  | 0 | 0 | 0 | .0104 | 0 | 0 | 0 | 0 | 0 | 0 | 0 |
| ACh | 0 | 0 | 0 | .0104 | 0 | 0 | 0 | 0 | 0 | 0 | 0 |
| C   | 0 | 0 | 0 | 0 | 0 | 0 | 0 | 0 | 0 | 0 | 0 |

**2.** For the polar bear food web (§20.7.4, Example 2), the top species are PB, AB, KW, and HpS; they do not appear in the corresponding EP graph, shown next.

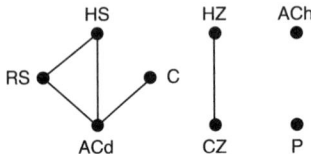

The matrix of edge weights in the EP graph, normalized by the divisor $132 = 12 \times 11$, is given by

|     | RS | ACd | HS | P | HZ | CZ | ACh | C |
|-----|----|----|----|---|----|----|-----|---|
| RS  | 0 | .00758 | .01515 | 0 | 0 | 0 | 0 | 0 |
| ACd | .00758 | 0 | .00758 | 0 | 0 | 0 | 0 | .00758 |
| HS  | .01515 | .00758 | 0 | 0 | 0 | 0 | 0 | 0 |
| P   | 0 | 0 | 0 | 0 | 0 | 0 | 0 | 0 |
| HZ  | 0 | 0 | 0 | 0 | 0 | .00758 | 0 | 0 |
| CZ  | 0 | 0 | 0 | 0 | .00758 | 0 | 0 | 0 |
| ACh | 0 | 0 | 0 | 0 | 0 | 0 | 0 | 0 |
| C   | 0 | .00758 | 0 | 0 | 0 | 0 | 0 | 0 |

## 20.7.7   OPEN QUESTIONS

There remain a number of open questions that suggest further study of graphs and that may inform conservationists and policy makers. Several are mentioned here.

**1.** Can we characterize the directed graphs whose corresponding competition graphs are interval graphs? This is a fundamental open question in applied graph theory. Indeed there is no forbidden list of digraphs (finite or infinite) such that when these digraphs are excluded, one automatically has a competition graph that is an interval graph.

**2.** What are the ecological characteristics of food webs that seem to lead to their competition graphs being interval graphs? Most directed graphs do not have interval graph competition graphs, yet statistically most actual food webs do have interval competition graphs. (This is an important unsolved problem described by Cohen, with no answers to date.)

**3.** What is the relationship between the boxicity of the competition graph of a food web and the boxicities of the competition graphs of its source food webs (note for every subset $W$ there is a source food web)? These can indeed be different; see §20.7.5, Fact 14.

**4.** It has been shown recently that it is possible to determine if a graph has boxicity 2, yet it is difficult to determine if a graph has boxicity $k$ for $k > 2$ (this is an NP-complete problem). There are no nice characterizations known for graphs with boxicity 2, in contrast to the case for interval graphs. Find a forbidden subgraph characterization for graphs of boxicity 2.

**5.** Is there any relationship between the indegrees and outdegrees of vertices in a food web, or the maximum indegree or outdegree over all vertices, and the competition number of the graph? If you limit the indegree or outdegree (or both) of all vertices in a digraph, is it more likely to have a competition graph that is an interval graph?

**6.** Is there any relationship between the connectance of a food web and the corresponding competition graph?

**7.** Are there mathematical results obtainable from using the PP and EP graphs that are not achievable from simply considering the competition graph, or its counterpart the common enemy graph? Do these have ecological consequences?

---

# REFERENCES

*Printed Resources*:

[AaNa10] D. P. Aalberts and N. Nandagopal, "A two-length-scale polymer theory for RNA loop free energies and helix stacking", *RNA* 16 (2010), 1350–1355.

[Ak00] T. Akutsu, "Dynamic programming algorithms for RNA secondary structure prediction with pseudoknots", *Discrete Applied Mathematics* 104 (2000), 45–62.

[AlOt03] R. Albert and H. Othmer, "The topology of the regulatory interactions predicts the expression pattern of the segment polarity genes in *Drosophila melanogaster*", *Journal of Theoretical Biology* 223 (2003), 1–18.

[Al91] S. F. Altschul, "Amino acid substitution matrices from an information theoretic perspective", *Journal of Molecular Biology* 219 (1991), 555–565.

[AlEr86a] S. F. Altschul and B. W. Erickson, "Optimal sequence alignment using affine gap costs", *Bulletin of Mathematical Biology* 48 (1986), 603–616.

[AlEr86b] S. F. Altschul and B. W. Erickson, "Locally optimal subalignments using nonlinear similarity functions", *Bulletin of Mathematical Biology* 48 (1986), 633–660.

[AlEtal90] S. F. Altschul, W. Gish, W. Miller, E. W. Myers, and D. J. Lipman, "Basic local alignment search tool", *Journal of Molecular Biology* 215 (1990), 403–410.

[AlEtal97] S. F. Altschul, T. L. Madden, A. A. Schäffer, J. Zhang, Z. Zhang, W. Miller, and D. J. Lipman, "Gapped BLAST and PSI-BLAST: a new generation of protein database search programs", *Nucleic Acids Research* 25 (1997), 3389–3402.

[AmClStJ03] N. Amenta, F. Clarke, and K. St. John, "A linear-time majority tree algorithm", *Algorithms in Bioinformatics* 2812 (2003), 216–227.

[AnEtal13] J. E. Andersen, R. C. Penner, C. M. Reidys, and M. S. Waterman, "Topological classification and enumeration of RNA structures by genus", *Journal of Mathematical Biology* 67 (2013), 1261–1278.

[ArKl06] F. Ardila and C. Klivans, "The Bergman complex of a matroid and phylogenetic trees", *Journal of Combinatorial Theory B* 96 (2006), 38–49.

[Ba79] J. K. Baker, "Trainable grammars for speech recognition", *Proceeding of the Spring Conference of the Acoustical Society of America*, Boston, 1979, 547–550.

[BaEtal08] N. Bandeira, V. Pham, P. Pevzner, D. Arnott, and J. R. Lill, "Automated *de novo* protein sequencing of monoclonal antibodies", *Nature Biotechnology* 26 (2008), 1336–1338.

[BeEtal08] D. R. Bentley et al., "Accurate whole human genome sequencing using reversible terminator chemistry", *Nature* 456 (2008), 53–59.

[BeBaSo95] R. Ben-Yishai, R. L. Bar-Or, and H. Sompolinsky, "Theory of orientation tuning in visual cortex", *Proceedings of the National Academy of Sciences USA* 92 (1995), 3844–3848.

[BoEtal08] M. Bon, G. Vernizzi, H. Orland, and A. Zee, "Topological classification of RNA structures", *Journal of Molecular Biology* 379 (2008), 900–911.

[BrMcKSt03] D. Bryant, A. McKenzie, and M. Steel, "The size of a maximum agreement subtree for random binary trees", *DIMACS Series in Discrete Mathematics and Theoretical Computer Science* 61 (2003), 55–66.

[BrSt09] D. Bryant and M. Steel, "Computing the distribution of a tree metric", *IEEE/ACM Transactions on Computational Biology and Bioinformatics* 6 (2009), 420–426.

[Bu71] A. W. Burks, *Essays on Cellular Automata*, University of Illinois Press, 1971.

[BuWh94] M. Burrows and D. Wheeler, "A block sorting lossless data compression algorithm", *Technical Report* 124, Digital Equipment Corporation, 1994.

[CaLi14] C. F. G. Castillo and M. H. T. Ling, "Digital Organism Simulation Environment (DOSE): a library for ecologically-based *in silico* experimental evolution", *ACSIJ Advances in Computer Science: An International Journal* 3 (2014), 44–50.

[ChFrSi10] L. S. Chandran, M. C. Francis, and N. Sivadasan, "Geometric representation of graphs in low dimension using axis parallel boxes", *Algorithmica* 56 (2010), 129–140.

[ChEtal07] W. Y. C. Chen, E. Y. P. Deng, R. R. X. Du, R. P. Stanley, and C. H. Yan, "Crossing and nesting of matchings and partitions", *Transactions of the American Mathematical Society* 359 (2007), 1555–1575.

[ChPeMi92] K-M. Chao, W. R. Pearson, and W. Miller, "Aligning two sequences within a specified diagonal band", *Computer Applications in the Biosciences* 8 (1992), 481–487.

[ChKi05] H. H. Cho and S. R. Kim, "A class of acyclic digraphs with interval competition graphs", *Discrete Applied Mathematics* 148 (2005), 171–180.

[Co78] J. E. Cohen, *Food Webs and Niche Space*, Princeton University Press, 1978.

[CoKoMu79] J. E. Cohen, J. Komlós, and M. Mueller, "The probability of an interval graph, and why it matters", *Proceeding of Symposia in Pure Mathematics* 34 (1979), 97–115.

[CoEtal00] R. Cole, M. Farach-Colton, R. Hariharan, T. Przytycka, and M. Thorup, "An $O(n \log n)$ algorithm for the maximum agreement subtree problem for binary trees", *SIAM Journal on Computing* 30 (2000), 1385–1404.

[CoJaLa06] O. Colon-Reyes, A. Jarrah, R. Laubenbacher, and B. Sturmfels, "Monomial dynamical systems over finite fields", *Complex Systems* 4 (2006), 333–342.

[CoPeTe11] P. E. C. Compeau, P. A. Pevzner, and G. Tesler, "How to apply de Bruijn graphs to genome assembly", *Nature Biotechnology* 29 (2011) 987–991.

[Co81] M. B. Cozzens, "Higher and multi-dimensional analogues of interval graphs", Ph.D. thesis, Department of Mathematics, Rutgers University, 1981.

[Co11] M. B. Cozzens, "Food webs, competition graphs, and habitat formation", *Mathematical Modeling of Natural Phenomena* 6 (2011), 22–38.

[Co15] M. B. Cozzens, "Food webs and graphs" in *Algebraic and Discrete Mathematical Methods for Modern Biology*, R. Robeva (ed.), Elsevier/Academic Press, 2015, 29–49.

[CoRo83] M. B. Cozzens and F. S. Roberts, "Computing the boxicity of a graph by covering its complement by cointerval graphs", *Discrete Applied Mathematics* 6 (1983), 217–228.

[CuEtal15] C. Curto, E. Gross, J. Jeffries, K. Morrison, M. Omar, Z. Rosen, A. Shiu, and N. Youngs, "What makes a neural code convex?", preprint, Nov. 2015; available at `http://arxiv.org/abs/1508.00150`.

[CuIt08] C. Curto and V. Itskov, "Cell groups reveal structure of stimulus space", *PLoS Computational Biology* 4 (2008), e1000205.

[CuEtal13a] C. Curto, V. Itskov, K. Morrison, Z. Roth, and J. L. Walker, "Combinatorial neural codes from a mathematical coding theory perspective", *Neural Computation* 25 (2013), 1891–1925.

[CuEtal13b] C. Curto, V. Itskov, A. Veliz-Cuba, and N. Youngs, "The neural ring: an algebraic tool for analyzing the intrinsic structure of neural codes, *Bulletin of Mathematical Biology* 75 (2013), 1571–1611.

[CuYo15] C. Curto and N. Youngs, "Neural ring homomorphisms and maps between neural codes", preprint, Nov. 2015; available at `http://arxiv.org/abs/1511.00255`.

[Da85] W. Day, "Optimal algorithms for comparing trees with labeled leaves", *Journal of Classification* 2 (1985), 7–28.

[DaScOr78] M. O. Dayhoff, R. M. Schwartz, and B. C. Orcutt, "A model of evolutionary change in proteins", in *Atlas of Protein Sequence and Structure*, Volume 5, Supplement 3, M. O. Dayhoff (ed.), National Biomedical Research Foundation, Washington, DC, 1978, pp. 345–352.

[DeRo06] J. Degnan and N. Rosenberg, "Discordance of species trees with their most likely gene trees", *PLoS Genetics* 2 (2006), e68.

[DeDo05] A. Deutsch and S. Dormann, *Cellular Automaton Modelling of Biological Pattern Formation*, Birkhauser, 2005.

[DoEd04] R. D. Dowell and S. R. Eddy, "Evaluation of several lightweight stochastic context-free grammars for RNA secondary structure prediction", *BMC Bioinformatics* 5:71 (2004).

[DiTuMa01] J. M. Diamond, D. H. Turner, and D. H. Mathews, "Thermodynamics of three-way multibranch loops in RNA", *Biochemistry* 40 (2001), 6971–6981.

[DrEtal11] A. Dress, K. Huber, J. Koolen, V. Moulton, and A. Spillner, *Basic Phyloge-netic Combinatorics*, Cambridge University Press, 2011.

[DrEtal89] R. Drmanac, I. Labat, I. Brukner, and R. Crkvenjakov, "Sequencing of mega-base plus DNA by hybridization: theory of the method", *Genomics* 4 (1989), 114–128.

[DrEtal93] R. Drmanac et al., "DNA sequence determination by hybridization: a strategy for efficient large-scale sequencing", *Science* 260 (1993), 1649–1652.

[DuWiMa02] J. A. Dunne, R. J. Williams, and N. D. Martinez, "Network structure and biodiversity loss in food webs: robustness increases with connectance", *Ecology Letters* 5 (2002), 558–567.

[DuEtal98] R. Durbin, S. R. Eddy, A. Krogh, and G. Mitchison, *Biological Sequence Analysis: Probabilistic Models of Proteins and Nucleic Acids*, Cambridge University Press, 1998.

[EiSm03] A. E. Eiben and J. E. Smith, *Introduction to Evolutionary Computing*, Springer, Natural Computing Series, 2003.

[Fe04] J. Felsenstein, *Inferring Phylogenies*, Sinauer Associates, 2004.

[FeMa00] P. Ferragina and G. Manzini, "Opportunistic data structures with applica-tions", *Proceedings of the 41st Annual Symposium on Foundations of Computer Science*, 2000, 390–398.

[Fi84] J. W. Fickett, "Fast optimal alignment", *Nucleic Acids Research* 12 (1984), 175–180.

[FlEtal95] R. D. Fleischmann et al., "Whole-genome random sequencing and assembly of *Haemophilus Influenzae* Rd", *Science* 269 (1995), 496–512.

[Fo94] D. B. Fogel, "An introduction to simulated evolutionary optimization", *IEEE Transactions on Neural Networks: Special Issue on Evolutionary Computation* 5 (1994), 3–14.

[FuGr65] D. R. Fulkerson and O. A. Gross, "Incidence matrices and interval graphs", *Pacific Journal of Mathematics* 15 (1965), 835–855.

[Ga74] H. Gabai, "Bounds for the boxicity of a graph", York College, City University of New York, 1974.

[Ga70] M. Gardner, "The fantastic combinations of John Conway's new solitaire game 'life'", *Scientific American* 223 (1970), 120–123.

[GhPo09] M. Ghodsi and M. Pop, "Inexact local alignment search over suffix arrays", *Pro-ceedings of the IEEE International Conference on Bioinformatics and Biomedicine (BIBM '09)*, 2009, 83–87.

[GiHo64] P. C. Gilmore and A. J. Hoffman, "A characterization of comparability graphs and of interval graphs", *Canadian Journal of Mathematics* 16 (1964), 539–548.

[GiIt14] C. Giusti and V. Itskov, "A no-go theorem for one-layer feedforward networks", *Neural Computation* 26 (2014), 2527–2540.

[Go89] D. E. Goldberg, *Genetic Algorithms in Search, Optimization, and Machine Learn-ing*, Addison-Wesley, 1989.

[GoFoStJ13] K. Gordan, E. Ford, and K. St. John, "Hamiltonian walks of phylogenetic treespaces", *IEEE/ACM Transactions on Computational Biology and Bioinformatics* 10 (2013), 1076–1079.

[Go82] O. Gotoh, "An improved algorithm for matching biological sequences", *Journal of Molecular Biology* 162 (1982), 705–708.

[GrEtal11] M. G. Grabherr et al., "Full-length transcriptome assembly from RNA-Seq data without a reference genome", *Nature Biotechnology* 29 (2011), 644–652.

[Gr16] T. R. Gregory, "Animal genome size database", http://www.genomesize.com, 2016.

[GrMcEi87] M. Gribskov, A. D. McLachlin, and D. Eisenberg, "Profile analysis: detection of distantly related proteins", *Proceedings of the National Academy of Sciences USA* 84 (1987), 4355–4358.

[HaBo97] I. Harvey and T. Bossomaier, "Time out of joint: attractors in asynchronous random Boolean networks", in *Proceedings of the Fourth European Conference on Artificial Life (ECAL97)*, P. Husbands and I. Harvey (eds.), MIT Press, 1997, 67–75.

[HeMaDa15] Q. He, N. Macauley, and R. Davies, "RNA secondary structures: combinatorial models and folding algorithms" in *Algebraic and Discrete Mathematical Methods for Modern Biology*, R. Robeva (ed.), Elsevier/Academic Press, 2015, 321–345.

[HePo14] C. Heitsch and S. Poznanović, "Asymptotic distribution of motifs in a stochastic context-free grammar model of RNA folding", *Journal of Mathematical Biology* 69 (2014), 1743–1772.

[HeHe92] S. Henikoff and J. G. Henikoff, "Amino acid substitution matrices from protein blocks", *Proceedings of the National Academy of Sciences USA* 89 (1992), 10915–10919.

[HoEtal94] I. L. Hofacker, W. Fontana, P. F. Stadler, L. S. Bonhoeffer, M. Tacker, and P. Schuster, "Fast folding and comparison of RNA secondary structures", *Monatshefte für Chemie* 125 (1994), 167–188.

[Ho75] J. H. Holland, *Adaptation in Natural and Artificial Systems*, University of Michigan Press, 1975.

[Hu99] S. Huang, "Gene expression profiling, genetic networks, and cellular state: an integrating concept for tumorigenesis and drug discovery", *Journal of Molecular Medicine* 77 (1999), 469–480.

[HuWa03] X. Huang and J. Wang, "PCAP: a whole-genome assembly program", *Genome Research* 13 (2003), 2164–2170.

[HuRuSc10] D. Huson, R. Rupp, and C. Scornavacca, *Phylogenetic Networks: Concepts, Algorithms and Applications*, Cambridge University Press, 2010.

[IdWa95] R. M. Idury and W. S. Waterman, "A new algorithm for DNA sequence assembly", *Journal of Computational Biology* 2 (1995), 291–306.

[IlIl07] L. Ilie and S. Ilie, "Multiple spaced seeds for homology search", *Bioinformatics* 23 (2007), 2969–2977.

[ItSa14] K. Ito and N. Sakuraba, "Importance of real-world properties in chasing task: simulation and analysis of dragonfly's behavior", *Artificial Life and Robotics* 19 (2014), 370–374.

[JuCa69] T. Jukes and C. Cantor, "Evolution of protein molecules", *Mammalian Protein Metabolism* 3 (1969), 21–132.

[Ka84] G. Kalai, "Characterization of $f$-vectors of families of convex sets in $\mathcal{R}^d$. I. Necessity of Eckhoff's conditions", *Israel Journal of Mathematics* 48 (1984), 175–195.

[Ka86] G. Kalai, Characterization of $f$-vectors of families of convex sets in $\mathcal{R}^d$. II. Sufficiency of Eckhoff's conditions", *Journal of Combinatorial Theory A* 41 (1986), 167–188.

[KäNa07] J. Kärkkäinen and J. C. Na, "Faster filters for approximate string matching", *Proceedings of the 9th Workshop on Algorithm Engineering and Experiments (ALENEX' 07)*, 2007, 84–90.

[KaAl90] S. Karlin and S. F. Altschul, "Methods for assessing the statistical significance of molecular sequence features by using general scoring schemes", *Proceedings of the National Academy of Sciences USA* 87 (1990), 2264–2268.

[Ka69] S. A. Kauffman, "Metabolic stability and epigenesis in randomly constructed genetic nets", *Journal of Theoretical Biology* 22 (1969), 437–467.

[Ka93] S. A. Kauffman, *The Origins of Order*, Oxford University Press, 1993.

[Ka00] S. A. Kauffman, *Investigations*, Oxford University Press, 2000.

[Ke09] E. F. Keller, *Making Sense of Life: Explaining Biological Development with Models, Metaphors, and Machines*, Harvard University Press, 2009.

[KnHe99] B. Knudsen and J. Hein, "RNA secondary structure prediction using stochastic context-free grammars and evolutionary history", *Bioinformatics* 15 (1999), 446–454.

[KnHe03] B. Knudsen and J. Hein, "Pfold: RNA secondary structure prediction using stochastic context-free grammars", *Nucleic Acids Research* 31 (2003), 3423–3428.

[KnMoPr77] D. Knuth, J. H. Morris, and V. Pratt, "Fast pattern matching in strings", *SIAM Journal on Computing* 6 (1977), 323–350.

[Ko91] J. R. Koza, "Evolution and co-evolution of computer programs to control independently-acting agents", in *Proceedings of the First International Conference on Simulation of Adaptive Behavior*, MIT Press, 1991, 366–375.

[LaSt13] A. Lambert and T. Stadler, "Birth–death models and coalescent point processes: the shape and probability of reconstructed phylogenies", *Theoretical Population Biology* 90 (2013), 113–128.

[La89] C. G. Langton, ed., *Artificial Life*, Volume 6 of Santa Fe Institute Studies in the Sciences of Complexity, Addison-Wesley, 1989.

[LaYo90] K. Lari and S. J. Young, "The estimation of stochastic context-free grammars using the inside-outside algorithm", *Computer Speech and Language* 4 (1990), 35–56.

[LaHiOr15] R. Laubenbacher, F. Hinkelmann, and M. Oremland, "Agent-based models and optimal control in biology: a discrete approach" in *Algebraic and Discrete Mathematical Methods for Modern Biology*, R. Robeva (ed.), Elsevier/Academic Press, 2015, 143–178.

[LaSt04] R. Laubenbacher and B. Stigler, "A computational algebra approach to the reverse engineering of gene regulatory networks", *Journal of Theoretical Biology* 229 (2004), 523–537.

[LeBo62] C. B. Lekkerkerker and J. C. Boland, "Representation of a finite graph by a set of intervals on the real line", *Fundamenta Mathematicae* 51 (1962), 45–64.

[LiTrZh96] M. Li, J. Tromp, and L. Zhang, "On the nearest neighbor interchange distance between evolutionary trees", *Journal of Theoretical Biology* 182 (1996), 463–467.

[Lu12] A. C. J. Luo, *Regularity and Complexity in Dynamical Systems*, Springer, 2012.

[LyPe00] R. B. Lyngsö and C. N. S. Pedersen, "RNA pseudoknot prediction in energy-based models", *Journal of Computational Biology* 7 (2000), 409–427.

[MaLi07] B. Ma and M. Li, "On the complexity of the spaced seeds", *Journal of Computer and System Sciences* 73 (2007), 1024–1034.

[MaTrLi02] B. Ma, J. Tromp, and M. Li, "PatternHunter: faster and more sensitive homology search", *Bioinformatics* 18 (2002), 440–445.

[Ma55] R. H. MacArthur, "Fluctuations of animal populations and a measure of community stability", *Ecology* 36 (1955), 533–536.

[MaScSc14] B. MacKay, P. Schweitzer, and P. Schweitzer, "Competition numbers, quasi-line graphs, and holes", *SIAM Journal on Discrete Mathematics* 28 (2014), 77–91.

[MaDe06] D. G. Malleta and L. G. De Pillis, "A cellular automata model of tumor-immune system interactions", *Journal of Theoretical Biology* 239 (2006), 334–350.

[MaMy93] U. Manber and G. Myers, "Suffix arrays: a new method for on-line string searches", *SIAM Journal on Computing* 22 (1993), 935–948.

[Ma15] Y. I. Manin, "Neural codes and homotopy types: mathematical models of place field recognition", *Moscow Mathematical Journal* 15 (2015), 741–748.

[MaEtal05] M. Margulies et al., "Genome sequencing in microfabricated high-density picolitre reactors", *Nature* 437 (2005), 376–380.

[MaZu08] N. R. Markham and M. Zuker, "Unafold: software for nucleic acid folding and hybridization", in *Bioinformatics, Volume II. Structure, Function and Applications*, J. M. Keith (ed.), Springer, 2008, 3–31.

[Ma09] R. May, "Food web assembly and collapse: mathematical models and implications for conservation", *Transactions of the Royal Society B: Biological Sciences* 364 (2009), 1643–1646.

[Mc76] E. M. McCreight, "A space-economical suffix tree construction algorithm", *Journal of the ACM* 23 (1976), 262–272.

[McNEtal06] B. L. McNaughton, F. P. Battaglia, O. Jensen, E. I. Moser, and M. B. Moser, "Path integration and the neural basis of the 'cognitive map'", *Nature Reviews Neuroscience* 7 (2006), 663–678.

[MeEtal11] P. Medvedev, S. Pham, M. Chaisson, G. Tesler, and P. Pevzner, "Paired de Bruijn graphs: a novel approach for incorporating mate pair information into genome assemblers", *Journal of Computational Biology* 18 (2011), 1625–1634.

[Me09] J. Memmott, "Food webs: a ladder for picking strawberries or a practical tool for practical problems?", *Transactions of the Royal Society B: Biological Sciences* 364 (2009), 1693–1699.

[MiMy88] W. Miller and E. W. Myers, "Sequence comparison with concave weighting functions", *Bulletin of Mathematical Biology* 50 (1988), 97–120.

[MiSt05] E. Miller and B. Sturmfels, *Combinatorial Commutative Algebra*, Springer, 2005.

[MoRe07] H. S. Mortveit and C. M. Reidys, *An Introduction to Sequential Dynamical Systems*, Springer-Verlag, 2007.

[My95] E. W. Myers, "Toward simplifying and accurately formulating fragment assembly", *Journal of Computational Biology* 2 (1995), 275–290.

[My05] E. W. Myers, "The fragment assembly string graph", *Bioinformatics* 21 Suppl. 2 (2005), 79–85.

[MyEtal00] E. W. Myers et al., "A whole-genome assembly of *Drosophila*", *Science* 287 (2000), 2196–2204.

[MyMi88] E. W. Myers and W. Miller, "Optimal alignments in linear space", *CABIOS* 4 (1988), 11–17.

[NaEtal06] A. Nakabachi, A. Yamashita, H. Toh, H. Ishikawa, H. E. Dunbar, N. A. Moran, and M. Hattori, "The 160-kilobase genome of the bacterial endosymbiont *Carsonella*", *Science* 314 (2006), 267.

[NeWu70] S. B. Needleman and C. D. Wunsch, "A general method applicable to the search for similarities in the amino acid sequences of two proteins", *Journal of Molecular Biology* 48 (1970), 443–453.

[NuEtal78] R. Nussinov, G. Pieczenik, J. R. Griggs, and D. J. Kleitman, "Algorithms for loop matchings", *SIAM Journal on Applied Mathematics* 35 (1978), 68–82.

[OKDo71] J. O'Keefe and J. Dostrovsky, "The hippocampus as a spatial map. Preliminary evidence from unit activity in the freely-moving rat", *Brain Research* 34 (1971), 171–175.

[Op82] R. Opsut, "On the competition number of a graph", *SIAM Journal on Algebraic and Discrete Methods* 3 (1982), 470–478.

[PaSt05] L. Pachter and B. Sturmfels, *Algebraic Statistics for Computational Biology*, Cambridge University Press, 2005.

[PaEtal11] G. M. Palamara, V. Zlatić, A. Scala, and G. Caldarelli, "Population dynamics on complex food webs", *Advances in Complex Systems* 16 (2011), 635–647.

[PaEtal12] A. Palyanov, S. Khayrulin, S. D. Larson, and A. Dibert, "Towards a virtual *C. elegans*: a framework for simulation and visualization of the neuromuscular system in a 3D physical environment", *In Silico Biology* 11 (2012), 137–147.

[PeLi88] W. R. Pearson and D. J. Lipman, "Improved tools for biological sequence comparison", *Proceedings of the National Academy of Sciences USA* 85 (1988), 2444–2448.

[PeEtal83] H. Peltola, H. Söderlund, J. Tarhio, and E. Ukkonen, "Algorithms for some string matching problems arising in molecular genetics", *Proceedings of the 9th IFIP World Computer Congress*, Elsevier, 1983, 59–64.

[PeSöUk84] H. Peltola, H. Söderlund, and E. Ukkonen, "SEQUAID: a DNA sequence assembly program based on a mathematical model", *Nucleic Acids Research* 12 (1984), 307–321.

[PeEtal10] R. C. Penner, M. Knudsen, C. Wiuf, and J. E. Andersen, "Fatgraph models of proteins", *Communications on Pure and Applied Mathematics* 63 (2010), 1249–1297.

[Pe89] P. Pevzner, "L-tuple DNA sequencing: computer analysis", *Journal of Biomolecular Structure and Dynamics* 7 (1989), 63–73.

[PeTaWa01] P. A. Pevzner, H. Tang, and M. S. Waterman, "An Eulerian path approach to DNA fragment assembly", *Proceedings of the National Academy of Sciences USA* 14 (2001), 9748–9753.

[PhEtal13] S. Pham, D. Antipov, A. Sirotkin, G. Tesler, P. Pevzner, and M. Alekseyev, "Pathset graphs: a novel approach for comprehensive utilization of paired reads in genome assembly", *Journal of Computational Biology* 20 (2013), 359–371.

[PhPe10] S. K. Pham and P. A. Pevzner, "DRIMM-Synteny: decomposing genomes into evolutionary conserved segments", *Bioinformatics* 26 (2010), 2509–2516.

[PoKo04] M. Pop and D. Kosack, "Using the TIGR assembler in shotgun sequencing projects", *Methods in Molecular Biology* 255 (2004), 279–294.

[Re10] C. M. Reidys, *Combinatorial Computational Biology of RNA: Pseudoknots and Neutral Networks*, Springer, 2010.

[ReEtal11] C. M. Reidys, F. W. D. Huang, J. E. Andersen, R. C. Penner, P. F. Stadler, and M. E. Nebel, "Topology and prediction of RNA pseudoknots", *Bioinformatics* 27 (2011), 1076–1085.

[RiEtal82] K. Rietveld, R. Van Poelgeest, C. W. Pleij, J. H. Van Boom, and L. Bosch, "The tRNA-like structure at the $3'$ terminus of turnip yellow mosaic virus RNA. Differences and similarities with canonical tRNA", *Nucleic Acids Research* 10 (1982), 1929–1946.

[RiEd99] E. Rivas and S. R. Eddy, "A dynamic programming algorithm for RNA structure prediction including pseudoknots", *Journal of Molecular Biology* 285 (1999), 2053–2068.

[Ro69] F. S. Roberts, "On the boxicity and cubicity of a graph" in *Recent Progress in Combinatorics*, W. T. Tutte (ed.), Academic Press, 1969, 301–310.

[Ro78] F. S. Roberts, "Food webs, competition graphs, and the boxicity of ecological phase space" in *Theory and Applications of Graphs*, Y. Alavi and D. Lick (eds.), Springer-Verlag, 1978, 477–490.

[RoCaSc13] R. J. Roberts, M. O. Carneiro, and M. C. Schatz, "The advantages of SMRT sequencing", *Genome Biology* 14 (2013), 405.

[RoHe14] E. Rogers and C. E. Heitsch, Profiling small RNA reveals multimodal substructural signals in a Boltzmann ensemble", *Nucleic Acids Research* 42 (2014), e171.

[SaNe87] N. Saitou and M. Nei, "The neighbor-joining method: a new method for reconstructing phylogenetic trees", *Molecular Biology and Evolution* 4 (1987), 406–425.

[SaEtal94] Y. Sakakibara, M. Brown, R. Hughey, I. S. Mian, K. Sjölander, R. C. Underwood, and D. Haussler, "Stochastic context-free grammars for tRNA modeling", *Nucleic Acids Research* 22 (1994), 5112–5120.

[SaNiCo77] F. Sanger, S. Nicklen, and A. R. Coulson, "DNA sequencing with chain-terminating inhibitors", *Proceedings of the National Academy of Sciences USA* 74 (1977), 5463–5467.

[Sa72] D. Sankoff, "Matching sequences under deletion-insertion constraints", *Proceedings of the National Academy of Sciences USA* 69 (1972), 4–6.

[SaEtal83] D. Sankoff, J. B. Kruskal, S. Mainville, and R. J. Cedergren, "Fast algorithms to determine RNA secondary structures containing multiple loops", in *Time Warps, String Edits, and Macromolecules: The Theory and Practice of Sequence Comparison*, D. Sankoff and J. B. Kruskal (eds.), Addison-Wesley, 1983, 93–120.

[ScDeSa10] M. C. Schatz, A. L. Delcher, and S. L. Salzberg, "Assembly of large genomes using second-generation sequencing." *Genome Research* 20 (2010), 1165–1173.

[ScDa78] R. M. Schwartz and M. O. Dayhoff, "Matrices for detecting distant relationships", in *Atlas of Protein Sequence and Structure*, Volume 5, Supplement 3, M. O. Dayhoff (ed.), National Biomedical Research Foundation, Washington, DC, 1978, pp. 353–358.

[Se74] P. H. Sellers, "On the theory and computation of evolutionary distances", *SIAM Journal on Applied Mathematics* 26 (1974), 787–793.

[Se84] P. H. Sellers, "Pattern recognition in genetic sequences by mismatch density", *Bulletin of Mathematical Biology* 46 (1984), 501–514.

[SeSt03] C. Semple and M. Steel, *Phylogenetics*, Oxford University Press, 2003.

[ShEtal02] I. Shmulevich, E. R. Dougherty, S. Kim, and W. Zhang, "Probabilistic Boolean networks: a rule-based uncertainty model for gene regulatory networks", *Bioinformatics* 18 (2002), 261–274.

[ShZh02] I. Shmulevich and W. Zhang, "Binary analysis and optimization-based normalization of gene expression data", *Bioinformatics* 18 (2002), 555–565.

[SiEtal05] E. Simão, E. Remy, D. Thieffry, and C. Chaouiya, "Qualitative modelling of regulated metabolic pathways: application to the tryptophan biosynthesis in *E. coli*", *Bioinformatics* 21 (2005) Suppl. 2, 190–196.

[SiPo15] J. T. Simpson and M. Pop, "The theory and practice of genome sequence assembly", *Annual Review of Genomics and Human Genetics* 16 (2015), 153–172.

[SmWa81] T. F. Smith and M. S. Waterman, "Identification of common molecular subsequences", *Journal of Molecular Biology* 147 (1981), 195–197.

[SpSt04] D. Speyer and B. Sturmfels, "The tropical Grassmanian", *Advances in Geometry* 4 (2004), 389–411.

[Sp89] J. L. Spouge, "Speeding up dynamic programming algorithms for finding optimal lattice paths", *SIAM Journal on Applied Mathematics* 49 (1989), 1552–1566.

[StWa79] P. R. Stein and M. S. Waterman, "On some new sequences generalizing the Catalan and Motzkin numbers", *Discrete Mathematics* 26 (1979), 261–272.

[StVe11] B. Stigler and A. Veliz-Cuba, "Boolean models can explain bistability in the *lac* operon", *Journal of Computational Biology* 18 (2011), 783–794.

[SuEtal95] G. Sutton, O. White, M. Adams, and A. Kerlavage, "TIGR assembler: a new tool for assembling large shotgun sequencing projects", *Genome Science and Technology* 1 (1995), 9–19.

[TaEtal11] K. Tamura, D. Peterson, N. Peterson, G. Stecher, M. Nei, and S. Kumar, "MEGA5: molecular evolutionary genetics analysis using maximum likelihood, evolutionary distance, and maximum parsimony methods", *Molecular Biology and Evolution* 28 (2011), 2731–2739.

[TaUk88] J. Tarhio and E. Ukkonen, "A greedy approximation algorithm for constructing shortest common superstrings", *Theoretical Computer Science* 51 (1988), 131–145.

[TaEtal09] M. Taufer, A. Licon, R. Araiza, D. Mireles, F. H. D. Van Batenburg, A. P. Gultyaev, and M-Y. Leung, "Pseudobase++: an extension of pseudobase for easy searching, formatting and visualization of pseudoknots", *Nucleic Acids Research* 37 (2009), D127–D135.

[ThEtal98] D. Thieffry, A. M. Huerta, E. Perez-Rueda, and J. Collado Vides, "From specific gene regulation to genomic networks: a global analysis of transcriptional regulation in *Escherichia coli*", *BioEssays* 20 (1998), 433–440.

[TiEtal73] I. Tinoco, P. N. Borer, B. Dengler, M. D. Levine, O. C. Uhlenbeck, D. M. Crothers, and J. Gralla, "Improved estimation of secondary structure in ribonucleic acids", *Nature* 246 (1973), 40–41.

[TrSa12] T. J. Treangen and S. L. Salzberg, "Repetitive DNA and next-generation sequencing: computational challenges and solutions", *Nature Reviews Genetics* 13 (2012), 36–46.

[Tr79] W. T. Trotter, Jr., "A forbidden subgraph characterization of Roberts' inequality for boxicity", *Discrete Mathematics* 28 (1979), 303–314.

[TuMa10] D. H. Turner and D. H. Mathews, "NNDB: the nearest neighbor parameter database for predicting stability of nucleic acid secondary structure", *Nucleic Acids Research* 38 (2010), D280–D282.

[UeEtal99] Y. Uemura, A. Hasegawa, S. Kobayashi, and T. Yokomori, "Tree adjoining grammars for RNA structure prediction", *Theoretical Computer Science* 210 (1999), 277–303.

[vN66] J. von Neumann, *The Theory of Self-Reproducing Automata*, A. Burks (ed.), Univ. of Illinois Press, 1966.

[Wa78] M. S. Waterman, "Secondary structure of single-stranded nucleic acid", *Advances in Mathematics Supplementary Studies* 1 (1978), 167–212.

[We97] J. Weimar, "Cellular automata for reaction-diffusion systems", *Parallel Computing* 23 (1997), 1699–1715.

[We73] P. Weiner, "Linear pattern matching algorithms", *14th Annual IEEE Symposium on Switching and Automata Theory*, 1973, 1–11.

[Wo82] S. Wolfram, "Cellular automata as simple self-organizing systems", Caltech report CALT-68-938, 1982.

[WuMa92] S. Wu and U. Manber, "Fast text searching: allowing errors", *Communications of the ACM* 35 (1992), 83–91.

[ZhEtal08] J. Zhang, M. Lin, R. Chen, W. Wang, and J. Liang, "Discrete state model and accurate estimation of loop entropy of RNA secondary structures", *Journal of Chemical Physics* 128 (2008), 125107.

**Web Resources**:

`http://adam.plantsimlab.org/` (Web-based software ADAM to analyze the dynamics of discrete biological systems.)

`http://bioquest.org/esteem` (Contains Excel worksheets and Java scripts for analyzing food webs.)

`http://dimacs.rutgers.edu/IMB/TalksEtc/Foodwebs-and-Biodiversity-7-29-13.pdf` (Powerpoint slides for the presentation "Graphs, Food Webs and Biodiversity".)

`http://evolution.genetics.washington.edu/phylip/software.html` (Website that is maintained by J. Felsenstein; it lists available software for phylogenetic reconstruction. This page contains links to almost 400 phylogenetics software packages.)

`http://mbcf149.dfci.harvard.edu/cmsmbr/biotools/biotools16.html` (List of sequence alignment servers and databases.)

`http://rosalind.info/problems/locations/` (Online platform for learning bioinformatics algorithms through coding. Includes an extensive collection of exercises and problems related to sequence alignment.)

`http://seqanswers.com/wiki/Software` (Detailed listing of software for bioinformatics applications.)

`http://treebase.org/treebase-web/` (The Treebase website contains an open access database of recently published phylogenetic trees.)

`http://unafold.rna.albany.edu` (UNAfold web server for nucleic acid folding and hybridization.)

`http://www.arn.org/docs/newman/rn_artificiallife.htm` (Artificial Life and Cellular Automata by Robert C. Newman.)

`http://www.daimi.au.dk/~compbio/pfold` (Pfold RNA folding server that uses the Knudsen-Hein grammar.)

`http://www.ebi.ac.uk/Tools/emboss/align/` (Local and global tools for pairwise sequence alignment.)

`http://www.genomesize.com` (Comprehensive catalog of animal genome size data.)

`http://www.ibiblio.org/lifepatterns/` (Java applet for Conway's "Game of Life".)

`http://www-igm.univ-mlv.fr/~lecroq/string/` (Exact string matching algorithms in C.)

`http://www.langmead-lab.org/teaching-materials/` (Videos and lecture slides for string matching algorithms from Ben Langmead, including Python code.)

`http://www.phrap.com` (PHRAP package for sequence assembly.)

`http://www.tbi.univie.ac.at/RNA` (RNA folding package to calculate minimal free energy for secondary structures.)

`https://blast.ncbi.nlm.nih.gov/Blast.cgi/` (Web server for running BLAST.)

`https://github.com/nebneuron/neural-ideal` (Software for computing with the neural ideal.)

`https://www.cbcb.umd.edu/software` (Open-source software for genome assembly.)

# BIOGRAPHIES

*Victor J. Katz*

**Niels Henrik Abel** (1802–1829), born in Norway, was self-taught and studied the works of many mathematicians. When he was nineteen years old, he proved that there is no closed formula for solving the general fifth degree equation. He also worked in the areas of infinite series and elliptic functions and integrals. The term *abelian* group was coined in Abel's honor in 1870 by Camille Jordan.

**Leonard Adleman** (born 1945) is credited with coining the term "computer virus". He was one of the inventors (with Adi Shamir and Ronald Rivest) of the RSA encryption algorithm, currently used in secure data transmission. He received his Ph.D. from the University of California, Berkeley in 1976. Adleman is also the creator of the field of DNA computing, beginning with a 1994 paper in which he used DNA to solve an instance of the Hamiltonian Graph problem. He is currently a Professor of Computer Science and Molecular Biology at the University of Southern California

**Kenneth Appel** (1932–2013) is best known for his solution of the four-color problem (along with Wolfgang Haken). He received his Ph.D. from the University of Michigan in 1959 and spent most of his career at the University of Illinois. The proof of the four-color theorem broke new ground, since it rests on hundreds of hours of computer calculation. Although the proof was initially disdained by many in the mathematical community, it actually provided a start to a change in the attitudes of mathematicians to computers, and has led to much further use of computers, even in theoretical mathematics.

**Aristotle** (384–322 B.C.E.) was the most famous student at Plato's academy in Athens. After Plato's death in 347 B.C.E., he was invited to the court of Philip II of Macedon to educate Philip's son Alexander, who soon thereafter began his successful conquest of the Mediterranean world. Aristotle himself returned to Athens, where he founded his own school, the Lyceum, and spent the remainder of his life writing and lecturing. He wrote on numerous subjects, but is perhaps best known for his works on logic, including the *Prior Analytics* and the *Posterior Analytics*. In these works, Aristotle developed the notion of logical argument, based on several explicit principles. In particular, he built his arguments out of syllogisms and concluded that demonstrations using his procedures were the only certain way of attaining scientific knowledge.

**Emil Artin** (1898–1962) was born in Vienna and in 1921 received a Ph.D. from the University of Leipzig. He held a professorship at the University of Hamburg until 1937, when he came to the United States. In the U.S. he taught at the University of Notre Dame, Indiana University, and Princeton. In 1958 he returned to the University of Hamburg. Artin's mathematical contributions were in number theory, algebraic topology, linear algebra, and especially in many areas of abstract algebra.

**Charles Babbage** (1792–1871) was an English mathematician best known for his invention of two of the earliest computing machines, the *Difference Engine*, designed to calculate polynomial functions, and the *Analytical Engine*, a general purpose calculating machine. The Difference Engine was designed to use the idea that the $n$th order differences in $n$th degree polynomials were always constant and then to work backwards from those differences to the original polynomial values. Although Babbage received a grant from the British government to help in building the Engine, he never was able to complete one because of various difficulties in developing machine parts of sufficient accuracy. In addition, Babbage became interested in his more advanced Analytical Engine. This latter device was to consist of a *store*, in which the numerical variables were kept, and a *mill*, in which the operations were performed. The entire machine was to be controlled by instructions on punched cards. Unfortunately, although Babbage made numerous engineering drawings of sections of the Analytical Engine and gave a series of seminars in 1840 on its workings, he was never able to build a working model.

**Paul Gustav Heinrich Bachmann** (1837–1920) studied mathematics at the University of Berlin and at Göttingen. In 1862 he received a doctorate in group theory and held positions at the universities at Breslau and Münster. He wrote several volumes on number theory, introducing the big-$O$ notation in his 1892 book.

**John Backus** (1924–2007) received bachelor's and master's degrees in mathematics from Columbia University. He led the group at IBM that developed FORTRAN in 1954. Later in the 1950s, he was an influential member of the international team that developed ALGOL, using the Backus-Naur form for the syntax of the language. He received the National Medal of Science in 1974 and the Turing Award in 1977 "for profound, influential, and lasting contributions to the design of practical high-level programming systems."

**Abu-l-'Abbas Ahmad ibn Muhammad ibn al-Banna al-Marrakushi** (1256–1321) was an Islamic mathematician who lived in Marrakech in what is now Morocco. Ibn al-Banna developed the first known proof of the basic combinatorial formulas, beginning by showing that the number of permutations of a set of $n$ elements was $n!$ and then developing in a careful manner the multiplicative formula to compute the values for the number of combinations of $k$ objects in a set of $n$. Using these two results, he also showed how to calculate the number of permutations of $k$ objects from a set of $n$. The formulas themselves had been known in the Islamic world for many years, in connection with specific problems like calculating the number of words of a given length which could be formed from the letters of the Arabic alphabet. Ibn al-Banna's main contribution, then, was to abstract the general idea of permutations and combinations out of the various specific problem situations considered earlier.

**Thomas Bayes** (1702–1761) an English Nonconformist, wrote an *Introduction to the Doctrine of Fluxions* in 1736 as a response to Berkeley's *Analyst* with its severe criticism of the foundations of the calculus. He is best known, however, for attempting to answer the basic question of statistical inference in his *An Essay Towards Solving a Problem in the Doctrine of Chances*, published three years after his death. That basic question is to determine the probability of an event, given empirical evidence that it has occurred a certain number of times in a certain number of trials. To do this, Bayes gave a straightforward definition of probability and then proved that for two events $E$ and $F$, the probability of $E$ given that $F$ has happened is the quotient of the probability of both $E$ and $F$ happening divided by the probability of $F$ alone. By using areas to model probability, he was then able to show that, if $x$ is the probability of an event happening in a single trial, if the event has happened $p$ times in $n$ trials, and if $0 < r < s < 1$, then the probability that $x$ is between $r$ and $s$ is given by the quotient of two integrals. Although in principle these integrals can be calculated, there has been a great debate since Bayes' time about the circumstances under which his formula gives an appropriate answer.

**James Bernoulli** (**Jakob I**) (1654–1705) was one of eight mathematicians in three generations of his family. He was born in Basel, Switzerland, studied theology in addition to mathematics and astronomy, and entered the ministry. In 1682 be began to lecture at the University of Basel in natural philosophy and mechanics. He became professor at the University of Basel in 1687, and remained there until his death. His research included the areas of the calculus of variations, probability, and analytic geometry. His most well-known work is *Ars Conjectandi*, in which he described results in combinatorics and probability, including applications to gambling and the law of large numbers; this work also contained a reprint of the first formal treatise in probability, written in 1657 by Christiaan Huygens.

**Bhaskara** (1114–1185), the most famous of medieval Indian mathematicians, gave a complete algorithmic solution to the *Pell equation* $Dx^2 \pm 1 = y^2$. That equation had been studied by several earlier Indian mathematicians as well. Bhaskara served much of his adult life as the head of the astronomical observatory at Ujjain, some 300 miles northeast of Bombay, and became widely respected for his skills in astronomy and the mechanical arts, as well as mathematics. Bhaskara's mathematical contributions are chiefly found in two chapters, the *Lilavati* and the *Bijaganita*, of a major astronomical work, the *Siddhāntasiromani*. These include techniques of solving systems of linear equations with more unknowns than equations as well as the basic combinatorial formulas, although without any proofs.

**Belá Bollobás** (born 1943) was on the Hungarian team in the first three International Mathematical Olympiads, beginning in 1959, and won two gold medals, thereby coming to the attention of Paul Erdős, who encouraged him in his studies. Although he received his Ph.D. from the University of Budapest in 1967, he soon was able to leave Hungary and vowed not to return while it was still under communist control. He received a fellowship to Trinity College, Cambridge, where he remained until 1996, then accepted a professorship at the University of Memphis. He has made major contributions to fields of extremal graph theory and the theory of random graphs and has published several research monographs in his field as well as introductory textbooks in combinatorics and graph theory.

**George Boole** (1815–1864) was an English mathematician most famous for his work in logic. Born the son of a cobbler, he had to struggle to educate himself while supporting his family. But he was so successful in his self-education that he was able to set up his own school before he was 20 and was asked to give lectures on the work of Isaac Newton. In 1849 he applied for and was appointed to the professorship in mathematics at Queen's College, Cork, despite having no university degree. In 1847, Boole published a small book, *The Mathematical Analysis of Logic*, and seven years later expanded it into *An Investigation of the Laws of Thought*. In these books, Boole introduced what is now called Boolean algebra as part of his aim to "investigate the fundamental laws of those operations of the mind by which reasoning is performed; to give expression to them in the symbolical language of a Calculus, and upon this foundation to establish the science of Logic and construct its method." In addition to his work on logic, Boole wrote texts on differential equations and on difference equations that were used in Great Britain until the end of the nineteenth century.

**William Burnside** (1852–1927), was born in London, graduated from Cambridge in 1875, and remained there as lecturer until 1885. He then went to the Royal Naval College at Greenwich, where he stayed until he retired. Although he published much in applied mathematics, probability, and elliptic functions, he is best known for his extensive work in group theory (including the classic book *Theory of Groups*). His conjecture that groups of odd order are solvable was proved by Walter Feit and John Thompson and published in 1963.

**Georg Ferdinand Ludwig Philip Cantor** (1845–1918) was born in Russia to Danish parents, received a Ph.D. in number theory in 1867 at the University of Berlin, and in 1869 took a position at Halle University, where he remained until his retirement. He is regarded as a founder of set theory. He was interested in theology and the nature of the infinite. His work on the convergence of Fourier series led to his study of certain types of infinite sets of real numbers, and ultimately to an investigation of transfinite numbers.

**Augustin-Louis Cauchy** (1789–1857), the most prolific mathematician of the nineteenth century, is most famous for his textbooks in analysis written in the 1820s for use at the École Polytechnique, textbooks which became the model for calculus texts for the next hundred years. Although born in the year the French Revolution began, Cauchy was a staunch conservative. When the July Revolution of 1830 led to the overthrow of the last Bourbon king, Cauchy refused to take the oath of allegiance to the new king and went into a self-imposed exile in Italy and then in Prague. He did not return to his teaching posts until the Revolution of 1848 led to the removal of the requirement of an oath of allegiance. Among the many mathematical subjects to which he contributed besides calculus were the theory of matrices, in which he demonstrated that every symmetric matrix can be diagonalized by use of an orthogonal substitution, and the theory of permutations, in which he was the earliest to consider these from a functional point of view. In fact, he used a single letter, say $S$, to denote a permutation and $S^{-1}$ to denote its inverse and then noted that the powers $S$, $S^2$, $S^3$, ... of a given permutation on a finite set must ultimately result in the identity. He also introduced the current notation $(a_1 a_2 \ldots a_n)$ to denote the cyclic permutation on the letters $a_1, a_2, \ldots, a_n$.

**Arthur Cayley** (1821–1895), although graduating from Trinity College, Cambridge as Senior Wrangler, became a lawyer because there were no suitable mathematics positions available at that time in England. He produced nearly 300 mathematical papers during his fourteen years as a lawyer, and in 1863 was named Sadlerian professor of mathematics at Cambridge. Among his numerous mathematical achievements are the earliest abstract definition of a group in 1854, out of which he was able to calculate all possible groups of order up to eight, and the basic rules for operating with matrices, including a statement (without proof) of the Cayley-Hamilton theorem that every matrix satisfies its characteristic equation. Cayley also developed the mathematical theory of *trees* in an article in 1857. In particular, he dealt with the notion of a *rooted tree*, a tree with a designated vertex called a root, and developed a recursive formula for determining the number of different rooted trees in terms of its *branches* (edges). In 1874, Cayley applied his results on trees to the study of chemical isomers.

**Pafnuty Lvovich Chebyshev** (1821–1894) was a Russian mathematician who received his master's degree in 1846 from Moscow University. From 1860 until 1882 he was a professor at the University of St. Petersburg. His mathematical research in number theory dealt with congruences and the distribution of primes; he also studied the approximation of functions by polynomials.

**Avram Noam Chomsky** (born 1928) received a Ph.D. in linguistics at the University of Pennsylvania. For many years he was a professor of foreign languages and linguistics at M.I.T. He has made many significant contributions to the study of linguistics and the study of grammars, but is probably better known for his left-wing political activism.

**Chrysippus** (280–206 B.C.E.) was a Stoic philosopher who developed some of the basic principles of the propositional logic, which ultimately replaced Aristotle's logic of syllogisms. He was born in Cilicia, in what is now Turkey, but spent most of his life in Athens, and is said to have authored more than 700 treatises. Among his other achievements, Chrysippus analyzed the rules of inference in the propositional calculus, including the rules of *modus ponens*, *modus tollens*, the hypothetical syllogism, and the alternative syllogism.

**Fan Chung** (born 1949) was born in Taiwan and received her Ph.D. from the University of Pennsylvania in 1974. She became an expert in Ramsey theory and has also made numerous contributions to graph theory, in particular to the applications of graph theory in Internet computing, communication networks, and software reliability. She spent many years at Bell Labs, where she collaborated with Ron Graham and eventually married him. She is currently a Distinguished Professor of Mathematics at the University of California, San Diego.

**Alonzo Church** (1903–1995) studied under Hilbert at Göttingen, was on the faculty at Princeton from 1927 until 1967, and then held a faculty position at UCLA. He is a founding member of the Association for Symbolic Logic. He made many contributions in various areas of logic and the theory of algorithms, and stated the Church-Turing thesis (if a problem can be solved with an effective algorithm, then the problem can be solved by a Turing machine).

**Clifford Cocks** (born 1950) is a British mathematician and cryptographer who invented the public key algorithm in 1973 now known as the RSA algorithm (after its independent invention four years later by Ronald Rivest, Adi Shamir, and Len Adleman). Cocks had invented the algorithm, using prime factorization, while he was working for the Communications-Electronics Security Group in Britain, an arm of the United Kingdom Government Communications Headquarters (GCHQ). But although GCHQ did not use the algorithm, they treated it as classified information and did not release the research until 1997. In 2001, Cocks developed one of the first secure identity based encryption schemes.

**George Dantzig** (1914–2005) is an American mathematician who formulated the general linear programming problem of maximizing a linear objective function subject to several linear constraints and developed the simplex method of solution in 1947. His study of linear programming grew out of his World War II service as a member of Air Force Project SCOOP (Scientific Computation of Optimum Programs), a project chiefly concerned with resource allocation problems. After the war, linear programming was applied to numerous problems, especially military and economic ones, but it was not until such problems could be solved on a computer that the real impact of their solution could be felt. The first successful solution of a major linear programming problem on a computer took place in 1952 at the National Bureau of Standards. After he left the Air Force, Dantzig worked for the RAND Corporation and then served as a professor of operations research at Stanford University.

**Richard Dedekind** (1831–1916) was born in Brunswick, in northern Germany, and received a doctorate in mathematics at Göttingen under Gauss. He held positions at Göttingen and in Zurich before returning to the Polytechnikum in Brunswick. Although at various times he could have received an appointment to a major German university, he chose to remain in his home town where he felt he had sufficient freedom to pursue his mathematical research. Among his many contributions was his invention of the concept of ideals to resolve the problem of the lack of unique factorization in rings of algebraic integers. Even though the rings of integers themselves did not possess unique factorization, Dedekind showed that every ideal is either prime or uniquely expressible as the product of prime ideals. Dedekind published this theory as a supplement to the second edition (1871) of Dirichlet's *Vorlesungen über Zahlentheorie*, of which he was the editor. In the supplement, he also gave one of the first definitions of a field, confining this concept to subsets of the complex numbers.

**Abraham de Moivre** (1667–1754) was born into a Protestant family in Vitry, France, a town about 100 miles east of Paris, and studied in Protestant schools up to the age of 14. Soon after the revocation of the Edict of Nantes in 1685 made life very difficult for Protestants in France, however, he was imprisoned for two years. He then left France for England, never to return. Although he was elected to the Royal Society in 1697, in recognition of a paper on "A method of raising an infinite Multinomial to any given Power or extracting any given Root of the same", he never achieved a university position. He made his living by tutoring and by solving problems arising from games of chance and annuities for gamblers and speculators. De Moivre's major mathematical work was *The Doctrine of Chances* (1718, 1736, 1756), in which he devised methods for calculating probabilities by use of binomial coefficients. In particular, he derived the normal approximation to the binomial distribution and, in essence, invented the notion of the standard deviation.

**Augustus De Morgan** (1806–1871) graduated from Trinity College, Cambridge in 1827. He was the first mathematics professor at University College in London, where he remained on the faculty for 30 years. He founded the London Mathematical Society. He wrote over 1000 articles and textbooks in probability, calculus, algebra, set theory, and logic (including DeMorgan's laws, an abstraction of the duality principle for sets). He gave a precise definition of limit, developed tests for convergence of infinite series, and gave a clear explanation of the Principle of Mathematical Induction.

**René Descartes** (1596–1650) left school at 16 and went to Paris, where he studied mathematics for two years. In 1616 he earned a law degree at the University of Poitiers. In 1617 he enlisted in the army and traveled through Europe until 1629, when he settled in Holland for the next 20 years. During this productive period of his life he wrote on mathematics and philosophy, attempting to reduce the sciences to mathematics. In 1637 his *Discours* was published; this book contained the development of analytic geometry. In 1649 he has invited to tutor the Queen Christina of Sweden in philosophy. There he soon died of pneumonia.

**Leonard Eugene Dickson** (1874–1954) was born in Iowa and in 1896 received the first Ph.D. in mathematics given by the University of Chicago, where he spent much of his faculty career. His research interests included abstract algebra (including the study of matrix groups and finite fields) and number theory.

**Edsger W. Dijkstra** (1930–2002) was a Dutch computer scientist who helped shape the emerging field of computer science from both an engineering and theoretical perspective. In particular, he turned computer programming into a scientific discipline. He made contributions in many areas of the emerging domain of computer science. For example, he originated the phrase "structured programming", and his ideas in this area helped lay the foundation for the new discipline of software engineering. He is known for Dijkstra's algorithm, which found the shortest path in a graph between any two given nodes. He was also on the team that implemented the first compiler for the language ALGOL 60. After twenty years of work in the Netherlands, he joined the Burroughs Corporation in 1973 as a research fellow, moving eleven years later to the University of Texas at Austin, where he remained until his retirement.

**Diophantus** (c. 250) was an Alexandrian mathematician about whose life little is known except what is reported in an epigram of the *Greek Anthology* (c. 500), from which it can be calculated that he lived to the age of 84. His major work, however, the *Arithmetica*, has been extremely influential. Despite its title, this is a book on algebra, consisting mostly of an organized collection of problems translatable into what are today called indeterminate equations, all to be solved in rational numbers. Diophantus introduced the use of symbolism into algebra and outlined the basic rules for operating with algebraic expressions, including those involving subtraction. It was in a note appended to Problem II-8 of the 1621 Latin edition of the *Arithmetica* — to divide a given square number into two squares — that Pierre de Fermat first asserted the impossibility of dividing an $n$th power ($n > 2$) into the sum of two $n$th powers. This result, now known as Fermat's Last Theorem, was finally proved in 1994 by Andrew Wiles.

**Charles Lutwidge Dodgson** (1832–1898) is more familiarly known as Lewis Carroll, the pseudonym he used in writing his famous children's works *Alice in Wonderland* and *Through the Looking Glass*. Dodgson graduated from Oxford University in 1854 and the next year was appointed a lecturer in mathematics at Christ Church College, Oxford. Although he was not successful as a lecturer, he did contribute to four areas of mathematics: determinants, geometry, the mathematics of tournaments and elections, and recreational logic. In geometry, he wrote a five-act comedy, "Euclid and His Modern Rivals", about a mathematics lecturer Minos in whose dreams Euclid debates his *Elements* with various modernizers but always manages to demolish the opposition. He is better known, however, for his two books on logic, *Symbolic Logic* and *The Game of Logic*. In the first, he developed a symbolical calculus for analyzing logical arguments and wrote many humorous exercises designed to teach his methods, while in the second, he demonstrated a game which featured various forms of the syllogism.

**Eratosthenes** (276–194 B.C.E) was born in Cyrene (North Africa) and studied at Plato's Academy in Athens. He was tutor of the son of King Ptolemy III Euergetes in Alexandria and became chief librarian at Alexandria. He is recognized as the foremost scholar of his time and wrote in many areas, including number theory (his sieve for obtaining primes) and geometry. He introduced the concepts of meridians of longitude and parallels of latitude and used these to measure distances, including an estimation of the circumference of the earth.

**Paul Erdős** (1913–1996) was born in Budapest. At 21 he received a Ph.D. in mathematics from Eőtvős University. After leaving Hungary in 1934, he traveled extensively throughout the world, with very few possessions and no permanent home, working with other mathematicians in combinatorics, graph theory, number theory, and many other areas. He was author or coauthor of approximately 1500 papers with 500 coauthors.

**Euclid** (c. 300 B.C.E.) is responsible for the most famous mathematics text of all time, the *Elements*. Not only does this work deal with the standard results of plane geometry, but it also contains three chapters on number theory, one long chapter on irrational quantities, and three chapters on solid geometry, culminating with the construction of the five regular solids. The axiom-definition-theorem-proof style of Euclid's work has become the standard for formal mathematical writing up to the present day. But about Euclid's life virtually nothing is known. It is, however, generally assumed that he was among the first mathematicians at the Museum and Library of Alexandria, which was founded around 300 B.C.E by Ptolemy I Soter, the Macedonian general of Alexander the Great who became ruler of Egypt after Alexander's death in 323 B.C.E.

**Leonhard Euler** (1707–1783) was born in Basel, Switzerland and became one of the earliest members of the St. Petersburg Academy of Sciences. He was the most prolific mathematician of all time, making contributions to virtually every area of the subject. His series of analysis texts established many of the notations and methods still in use today. He created the calculus of variations and established the theory of surfaces in differential geometry. His study of the Königsberg bridge problem led to the formulation and solution of one of the first problems in graph theory. He made numerous discoveries in number theory, including a detailed study of the properties of residues of powers and the first statement of the quadratic reciprocity theorem. He developed an algebraic formula for determining the number of partitions of an integer $n$ into $m$ distinct parts, each of which is in a given set $A$ of distinct positive integers. And in a paper of 1782, he even posed the problem of the existence of a pair of orthogonal Latin squares: If there are 36 officers, one of each of six ranks from each of six different regiments, can they be arranged in a square in such a way that each row and column contains exactly one officer of each rank and one from each regiment?

**Abraham ibn Ezra** (1089–1164) was a Spanish poet, philosopher, astrologer, and biblical commentator who was born in Tudela, but spent the latter part of his life as a wandering scholar in Italy, France, England, and Palestine. It was in an astrological text that ibn Ezra developed a method for calculating numbers of combinations, in connection with determining the number of possible conjunctions of the seven "planets" (including the sun and the moon). He gave a detailed argument for the cases $n = 7$, $k = 2$ to 7, of a rule which can easily be generalized to the modern formula $C(n, k) = \sum_{i=k-1}^{n-1} C(i, k - 1)$. Ibn Ezra also wrote a work on arithmetic in which he introduced the Hebrew-speaking community to the decimal place-value system. He used the first nine letters of the Hebrew alphabet to represent the first nine numbers, used a circle to represent zero, and demonstrated various algorithms for calculation in this system.

**Kamāl al-Dīn al-Fārisī** (died 1320) was a Persian mathematician most famous for his work in optics. In fact, he wrote a detailed commentary on the great optical work of Ibn al-Haytham. But al-Fārisī also made major contributions to number theory. He produced a detailed study of the properties of *amicable numbers* (pairs of numbers in which the sum of the proper divisors of each is equal to the other). As part of this study, al-Fārisī developed and applied various combinatorial principles. He showed that the classical figurate numbers (triangular, pyramidal, etc.) could be interpreted as numbers of combinations and thus helped to found the theory of combinatorics on a more abstract basis.

**Pierre de Fermat** (1607–1665) was a lawyer and magistrate for whom mathematics was a pastime that led to contributions in many areas: calculus, number theory, analytic geometry, and probability theory. He received a bachelor's degree in civil law in 1631, and from 1648 until 1665 was King's Counsellor. He suffered an attack of the plague in 1652, and from then on he began to devote time to the study of mathematics. He helped give a mathematical basis to probability theory when, together with Blaise Pascal, he solved Méré's paradox: why is it less likely to roll a 6 at least once in four tosses of one die than to roll a double 6 in 24 tosses of two dice? He was a discoverer of analytic geometry and used infinitesimals to find tangent lines and determine maximum and minimum values of curves. In 1657 he published a series of mathematical challenges, including the conjecture that $x^n + y^n = z^n$ has no solution in positive integers if $n$ is an integer greater than 2. He wrote in the margin of a book that he had a proof, but the proof would not fit in the margin. His conjecture was finally proved by Andrew Wiles in 1994.

**Fibonacci (Leonardo of Pisa)** (c. 1175–c. 1250) was the son of a Mediterranean merchant and government worker named Bonaccio (hence his name *filius Bonaccio*, "son of Bonaccio"). Fibonacci, born in Pisa and educated in Bougie (on the north coast of

Africa where his father was administrator of Pisa's trading post), traveled extensively around the Mediterranean. He is regarded as the greatest mathematician of the Middle Ages. In 1202 he wrote the book *Liber Abaci*, an extensive treatment of topics in arithmetic and algebra, and emphasized the benefits of Arabic numerals (which he knew about as a result of his travels). In this book he also discussed the rabbit problem that led to the sequence that bears his name: $1, 1, 2, 3, 5, 8, 13, \ldots$. In 1225 he wrote *Liber Quadratorum* (*Book of Squares*) as a response to a challenge from Master John of Palermo to "find a square number from which when five is added or subtracted, always arises a square number."

**Lester R. Ford, Jr.** (born 1927) is a pioneer in the field of network flows. He received his doctorate from the University of Illinois in 1953. Ford worked as a researcher for the Council for Economic and Industry Research and the RAND Corporation. While working at RAND, he collaborated with D. R. Fulkerson on seminal work on network flows. Their Ford-Fulkerson algorithm, first published in a technical report in 1954, finds the maximum flow in a capacitated network. His work on graph algorithms includes the widely used Bellman-Ford algorithm. This algorithm, invented with Richard Bellman, finds the shortest path in a weighted digraph in which edge weights can be negative. His father, Lester R. Ford, Sr., was a well-known mathematician who worked in number theory and was esteemed for his mathematical exposition.

**Joseph Fourier** (1768–1830), orphaned at the age of 9, was educated in the military school of his home town of Auxerre, 90 miles southeast of Paris. Although he hoped to become an army engineer, such a career was not available to him at the time because he was not of noble birth. He therefore took up a teaching position. During the Revolution, he was outspoken in defense of victims of the Terror of 1794. Although he was arrested, he was released after the death of Robespierre and was appointed in 1795 to a position at the École Polytechnique. After serving in various administrative posts under Napoleon, he was elected to the Académie des Sciences and from 1822 until his death served as its perpetual secretary. It was in connection with his work on heat diffusion, detailed in his *Analytic Theory of Heat* of 1822, and, in particular, with his solution of the *heat equation* $\frac{\partial v}{\partial t} = \frac{\partial^2 v}{\partial x^2} + \frac{\partial^2 v}{\partial y^2}$, that he developed the concept of a Fourier series. Fourier also analyzed the relationship between the series solution of a partial differential equation and an appropriate integral representation and thereby initiated the study of Fourier integrals and Fourier transforms.

**Georg Frobenius** (1849–1917) organized and analyzed the central ideas of the theory of matrices in his 1878 memoir "On linear substitutions and bilinear forms". Frobenius there defined the general notion of *equivalent* matrices. He also dealt with the special cases of *congruent* and *similar* matrices. Frobenius showed that when two symmetric matrices were similar, the transforming matrix could be taken to be *orthogonal*, one whose inverse equaled its transpose. He then made a detailed study of orthogonal matrices and showed that their eigenvalues were complex numbers of absolute value 1. He also gave the first complete proof of the Cayley-Hamilton theorem that a matrix satisfies its characteristic equation. Frobenius, a full professor in Zurich and later in Berlin, made his major mathematical contribution in the area of group theory. He was instrumental in developing the concept of an abstract group, as well as in investigating the theory of finite matrix groups and group characters.

**Delbert Ray Fulkerson** (1924–1976) received his Ph.D. at the University of Wisconsin, Madison in 1951. His most important work was the development (with Lester Ford, Jr.) of the Ford-Fulkerson algorithm, used to solve the maximum flow problem in capacitated networks. Such problems occur in many real-world situations such as airline crew scheduling and railway traffic flow. Fulkerson began his career at the

RAND Corporation and later served as Professor of Engineering at Cornell University. In his honor, the Fulkerson Prize for outstanding papers in discrete mathematics is awarded every three years jointly by the Mathematical Optimization Society and the American Mathematical Society.

**David Gale** (1921–2008) was a game theorist who received his Ph.D. in 1949 from Princeton University. He made major contributions to mathematical economics, including solving the $n$-dimensional Ramsey Problem and his theory of optimal economic growth. In fact, for the last half of his career, he held a joint appointment in mathematics, operations research, and economics at the University of California, Berkeley. He was always fascinated with various kinds of puzzles and math games and wrote the Mathematical Entertainments column of *The Mathematical Intelligencer* from 1991 to 1997.

**Évariste Galois** (1811–1832) led a brief, tragic life which ended in a duel fought under mysterious circumstances. He was born in Bourg-la-Reine, a town near Paris. He developed his mathematical talents early and submitted a memoir on the solvability of equations of prime degree to the French Academy in 1829. Unfortunately, the referees were never able to understand this memoir nor his revised version submitted in 1831. Meanwhile, Galois became involved in the revolutionary activities surrounding the July revolution of 1830 and was arrested for threatening the life of King Louis-Phillipe and then for wearing the uniform of a National Guard division which had been dissolved because of its perceived threat to the throne. His mathematics was not fully understood until fifteen years after his death when his manuscripts were finally published by Liouville in the *Journal des mathématique*. But Galois had in fact shown the relationship between subgroups of the group of permutations of the roots of a polynomial equation and the various extension fields generated by these roots, the relationship at the basis of what is now known as *Galois theory*. Galois also developed the notion of a finite field in connection with solving the problem of finding solutions to congruences $F(x) \equiv 0 \pmod{p}$, where $F(x)$ is a polynomial of degree $n$ and no residue modulo the prime $p$ is itself a solution.

**Carl Friedrich Gauss** (1777–1855), often referred to as the greatest mathematician who ever lived, was born in Brunswick, Germany. He received a Ph.D. from the University of Helmstedt in 1799, proving the Fundamental Theorem of Algebra as part of his dissertation. At age 24 Gauss published his important work on number theory, the *Disquisitiones Arithmeticae*, a work containing not only an extensive discussion of the theory of congruences, culminating in the quadratic reciprocity theorem, but also a detailed treatment of cyclotomic equations in which he showed how to construct regular $n$-gons by Euclidean techniques whenever $n$ is prime and $n-1$ is a power of 2. Gauss also made fundamental contributions to the differential geometry of surfaces as well as to complex analysis, astronomy, geodesy, and statistics during his long tenure as a professor at the University of Göttingen. It was in connection with using the method of least squares to solve an astronomical problem that Gauss devised the systematic procedure for solving a system of linear equations today known as Gaussian elimination. (Unknown to Gauss, the method had appeared in Chinese mathematics texts 1800 years earlier.) Gauss' notebooks, discovered after his death, contained investigations in numerous areas of mathematics in which he did not publish, including the basics of non-Euclidean geometry.

**Sophie Germain** (1776–1831) was forced to study in private due to the turmoil of the French Revolution and the opposition of her parents. She nevertheless mastered mathematics through calculus and wanted to continue her study in the École Polytechnique when it opened in 1794. But because women were not admitted as students, she diligently collected and studied the lecture notes from various mathematics classes and, a few years later, began a correspondence with Gauss (under the pseudonym Monsieur LeBlanc, fearing that Gauss would not be willing to recognize the work of a woman) on ideas in number theory. She was, in fact, responsible for suggesting to the French general leading the army occupying Brunswick in 1807 that he insure Gauss' safety. Germain's chief mathematical contribution was in connection with Fermat's Last Theorem. She showed that $x^n + y^n = z^n$ has no positive integer solution where $xyz$ is not divisible by $n$ for any odd prime $n$ less than 100. She also made contributions in the theory of elasticity and won a prize from the French Academy in 1815 for an essay in this field.

**Kurt Gödel** (1906–1978) was an Austrian mathematician who spent most of his life at the Institute for Advanced Study in Princeton. He made several surprising contributions to set theory, demonstrating that Hilbert's goal of showing that a reasonable axiomatic system for set theory could be proven to be complete and consistent was in fact impossible. In several seminal papers published in the 1930s, Gödel proved that it was impossible to prove internally the consistency of the axioms of any reasonable system of set theory containing the axioms for the natural numbers. Furthermore, he showed that any such system was inherently incomplete, that is, that there are propositions expressible in the system for which neither they nor their negations are provable. Gödel's investigations were stimulated by the problems surrounding the axiom of choice, the axiom that for any set $S$ of nonempty disjoint sets, there is a subset $T$ of the union of $S$ that has exactly one element in common with each member of $S$. Since that axiom led to many counterintuitive results, it was important to show that the axiom could not lead to contradictions. But given his initial results, the best Gödel could do was to show that the axiom of choice was relatively consistent, that its addition to the Zermelo-Fraenkel axiom set did not lead to any contradictions that would not already have been implied without it.

**Ronald Graham** (born 1935) is a central figure in the development of discrete mathematics. He received his Ph.D. in 1962 from the University of California, Berkeley and spent most of his professional life at Bell Labs, although he was appointed to an endowed chair at the University of California, San Diego in 1999. His work has been seminal in the origin of three new branches of mathematics: Ramsey theory, computational geometry, and worst-case analysis of multiprocessing algorithms. He was a long-time collaborator with Paul Erdős, frequently housing that wandering mathematician during the last decades of his life. Graham served as president of the American Mathematical Society in 1993-1994 and the Mathematical Association of America in 2003-2004 and received the Steele Award for Lifetime Achievement from the former in 2003. He is also a skilled juggler and has served as president of the International Jugglers' Association.

**Wolfgang Haken** (born 1928), along with Kenneth Appel, produced the first proof of the four-color problem with the help of a computer, announcing it to the world on July 22, 1976. He then presented a paper describing the proof at the August, 1976 Summer Meeting of the American Mathematical Society in Toronto, to a very mixed reaction. Haken received his education in Germany, but spent most of his academic career at the University of Illinois. In addition to his work on the four-color problem, he has contributed numerous ideas to the subject of algorithmic topology.

**Marshall Hall** (1910–1990) was a group theorist whose 1959 textbook on the subject is still being used today. Hall received his Ph.D. from Yale in 1936 and later taught at Ohio State, the California Institute of Technology, and Emory University. During World War II, he spent time at Bletchley Park, working on breaking the German

Enigma codes. Among his important results in group theory was his solution of the Burnside problem of exponent 6, in which he showed that a finitely generated group in which the order of every element divides 6 must be finite. Later, he worked out (with James Senior) detailed descriptions of every group of order 32 and 64, publishing the results in an organized fashion in *The Groups of Order $2^n$ $n \leq 6$*. Hall is also famous for some fundamental work in the theory of projective planes as well as other areas of combinatorics.

**William Rowan Hamilton** (1805–1865), born in Dublin, was a child prodigy who became the Astronomer Royal of Ireland in 1827 in recognition of original work in optics accomplished during his undergraduate years at Trinity College, Dublin. In 1837, he showed how to introduce complex numbers into algebra axiomatically by considering $a + ib$ as a pair $(a, b)$ of real numbers with appropriate computational rules. After many years of seeking an appropriate definition for multiplication rules for triples of numbers which could be applied to vector analysis in 3-dimensional space, he discovered that it was in fact necessary to consider quadruplets of numbers, which Hamilton named quaternions. Although quaternions never had the influence Hamilton forecast for them in physics, their noncommutative multiplication provided the first significant example of a mathematical system which did not obey one of the standard arithmetical laws of operation and thus opened the way for more "freedom" in the creation of mathematical systems. Among Hamilton's other contributions was the development of the *Icosian* game, a graph with 20 vertices on which pieces were to be placed in accordance with various conditions, the overriding one being that a piece was always placed at the second vertex of an edge on which the previous piece had been placed. One of the problems Hamilton set for the game was, in essence, to discover a cyclic path on his game board which passed through each vertex exactly once. Such a path in a more general setting is today called a Hamilton circuit.

**Richard W. Hamming** (1915–1998) was born in Chicago and received a Ph.D. in mathematics from the University of Illinois in 1942. He was the author of the first major paper on error-correcting and detecting codes (1950). His work on this problem had been stimulated in 1947 when he was using an early Bell System relay computer on weekends only. During the weekends the machine was unattended and would dump any work in which it discovered an error and proceed to the next problem. Hamming realized that it would be worthwhile for the machine to be able not only to detect an error but also to correct it, so that his jobs would in fact be completed. In his paper, Hamming used a geometric model by considering an $n$-digit code word to be a vertex in the unit cube in the $n$-dimensional vector space over the field of two elements. He was then able to show that the relationship between the word length $n$ and the number $m$ of digits which carry the information was $2^m \leq \frac{2^n}{n+1}$. (The remaining $k = n - m$ digits are check digits which enable errors to be detected and corrected.) In particular, Hamming presented a particular type of code, today known as a Hamming code, with $n = 7$ and $m = 4$. In this code, the set of actual code words of 4 digits was a 4-dimensional vector subspace of the 7-dimensional space of all 7-digit binary strings.

**Godfrey Harold Hardy** (1877–1947) graduated from Trinity College, Cambridge in 1899. From 1906 until 1919 he was lecturer at Trinity College, and, recognizing the genius of Ramanujan, invited Ramanujan to Cambridge in 1914. Hardy held the Sullivan chair of geometry at Oxford from 1919 until 1931, when he returned to Cambridge, where he was Sadlerian professor of pure mathematics until 1942. He developed the Hardy-Weinberg law which predicts patterns of inheritance. His main areas of mathematical research were analysis and number theory, and he published over 100 joint papers with Cambridge colleague John Littlewood. Hardy's book *A Course in Pure Mathematics* revolutionized mathematics teaching, and his book *A Mathematician's Apology* gives his view of what mathematics is and the value of its study.

**Abū 'Alī al-Hasan ibn al-Haytham (Alhazen)** (965–1039) was one of the most influential Islamic scientists. He was born in Basra (now in Iraq) but spent most of his life in Egypt, after he was invited to work on a Nile control project. Although the project, an early version of the Aswan dam project, never came to fruition, ibn al-Haytham did produce in Egypt his most important scientific work, the *Optics*. This work was translated into Latin in the early thirteenth century and was studied and commented on in Europe for several centuries thereafter. Although there was much mathematics in the *Optics*, ibn al-Haytham's most interesting mathematical work was the development of a recursive procedure for producing formulas for the sum of any integral powers of the integers. Formulas for the sums of the integers, squares, and cubes had long been known, but ibn al-Haytham gave a consistent method for deriving these and used this to develop the formula for the sum of fourth powers. Although his method was easily generalizable to the discovery of formulas for fifth and higher powers, he gave none, probably because he only needed the fourth power rule in his computation of the volume of a paraboloid of revolution.

**David Hilbert** (1862–1943) was one of the last of the universal mathematicians, who contributed greatly to many areas of mathematics. Hilbert spent the first 33 years of his life in and around Königsberg, then capital of East Prussia, now in Russia. He attended the university there and, after receiving his doctorate, joined the faculty in 1885. He only rose to prominence, however, after he was called by Felix Klein to Göttingen, where he soon became one of the major reasons for that university's taking over from Berlin as the preeminent university for mathematics in Germany, and probably the world, through the first third of the twentieth century. Hilbert began his career with the study of algebraic forms, then turned to algebraic number theory, the foundations of geometry, integral equations, theoretical physics, and finally the foundations of mathematics. He is probably most famous for his lecture at the International Congress of Mathematicians in Paris in 1900, where he presented a list of 23 problems which he felt would be of central importance for mathematics in the twentieth century. Hilbert firmly believed that it was problems that drove mathematical progress and was always confident that, "*wir mussen wissen, wir werden wissen* (we must know, we will know)." After the Nazi seizure of power, Hilbert was forced to witness the demise of the Göttingen he knew and loved and died a lonely man during the Second World War.

**Hypatia** (c. 370–415), the first woman mathematician on record, lived in Alexandria. She was given a very thorough education in mathematics and philosophy by her father Theon and became a popular and respected teacher. She was responsible for detailed commentaries on several important Greek works, including Ptolemy's *Almagest*, Apollonius' *Conics*, and Diophantus' *Arithmetica*. Unfortunately, Hypatia was caught up in the pagan-Christian turmoil of her times and was murdered by an enraged mob.

**Leonid Kantorovich** (1912–1986) was a Soviet economist responsible for the development of linear optimization techniques in relation to planning in the Soviet economy. The starting point of this development was a set of problems posed by the Leningrad timber trust at the beginning of 1938 to the Mathematics Faculty at the University of Leningrad. Kantorovich explored these problems in his 1939 book *Mathematical Methods in the Organization and Planning of Production*. He believed that one way to increase productivity in a factory or an entire industrial organization was to improve the distribution of the work among individual machines, the orders to various suppliers, the different kinds of raw materials, the different types of fuels, and so on. He was the first to recognize that these problems could all be put into the same mathematical language and that the resulting mathematical problems could be solved numerically, but for various reasons his work was not pursued by Soviet economists or mathematicians.

**Abū Bakr al-Karajī** (died 1019) was an Islamic mathematician who worked in Baghdad. In the first decade of the eleventh century he composed a major work on algebra entitled *al-Fakhrī* (*The Marvelous*), in which he developed many algebraic techniques, including the laws of exponents and the algebra of polynomials, with the aim of systematizing methods for solving equations. He was also one of the early originators of a form of mathematical induction, which was best expressed in his proof of the formula for the sum of integral cubes.

**Muhammad ibn Muhammad al-Fullāni al-Kishnāwī** (died 1741) was a native of northern Nigeria and one of the few African black scholars known to have made contributions to "pure" mathematics before the modern era. Muhammad's most important work, available in an incomplete manuscript in the library of the School of Oriental and African Studies in London, deals with the theory of magic squares. He gave a clear treatment of the "standard" construction of magic squares and also studied several other constructions — using knight's moves, borders added to a magic square of lower order, and the formation of a square from a square number of smaller magic squares.

**Stephen Cole Kleene** (1909–1994) studied under Alonzo Church and received his Ph.D. from Princeton in 1934. His research has included the study of recursive functions, computability, decidability, and automata theory. In 1956 he proved Kleene's Theorem, in which he characterized the sets that can be recognized by finite-state automata.

**Felix Klein** (1849–1925) received his doctorate at the University of Bonn in 1868. In 1872 he was appointed to a position at the University of Erlangen, and in his opening address laid out the *Erlanger Programm* for the study of geometry based on the structure of groups. He described different geometries in terms of the properties of a set that are invariant under a group of transformations on the set and gave a program of study using this definition. From 1875 until 1880 he taught at the Technische Hochschule in Munich, and from 1880 until 1886 in Leipzig. In 1886 Klein became head of the mathematics department at Göttingen and during his tenure raised the prestige of the institution greatly.

**Donald E. Knuth** (born 1938) received a Ph.D. in 1963 from the California Institute of Technology and held faculty positions at the California Institute of Technology (1963–1968) and Stanford (1968–1992). He made contributions in many areas, including the study of compilers and computational complexity. He is the designer of the mathematical typesetting system TeX, and is the author of *The Art of Computer Programming*, the first comprehensive manual in the subject, published in many installments beginning in 1968. He received the Turing Award in 1974 and the National Medal of Technology in 1979.

**Kazimierz Kuratowski** (1896–1980) was the son of a famous Warsaw lawyer who became an active member of the Warsaw School of Mathematics after World War I. He taught both at Lwów Polytechnical University and at Warsaw University until the outbreak of World War II. During that war, because of the persecution of educated Poles, he went into hiding under an assumed name and taught at the clandestine

Warsaw University. After the war, he helped to revive Polish mathematics, serving as director of the Polish National Mathematics Institute. His major mathematical contributions were in topology. In particular, he formulated a version of a maximal principle equivalent to the axiom of choice. This principle is today known as Zorn's lemma. Kuratowski also contributed to the theory of graphs by proving in 1930 that any non-planar graph must contain a copy of one of two particularly simple non-planar graphs.

***Joseph-Louis Lagrange*** (1736–1813) was born in Turin into a family of French descent. He was attracted to mathematics in school and at the age of 19 became a mathematics professor at the Royal Artillery School in Turin. At about the same time, having read a paper of Euler's on the calculus of variations, he wrote to Euler explaining a better method he had recently discovered. Euler praised Lagrange and arranged to present his paper to the Berlin Academy, to which he was later appointed when Euler returned to Russia. Although most famous for his *Analytical Mechanics*, a work which demonstrated how problems in mechanics can generally be reduced to solutions of ordinary or partial differential equations, and for his *Theory of Analytic Functions*, which attempted to reduce the ideas of calculus to those of algebraic analysis, he also made contributions in other areas. For example, he undertook a detailed review of solutions to quadratic, cubic, and quartic polynomials to see how these methods might generalize to higher degree polynomials. He was led to consider permutations on the roots of the equations and functions on the roots left unchanged by such permutations. As part of this work, he discovered a version of Lagrange's Theorem to the effect that the order of any subgroup of a group divides the order of the group. Although he did not complete his program and produce a method of solving higher degree polynomial equations, his methods were applied by others early in the nineteenth century to show that such solutions were impossible.

***Gabriel Lamé*** (1795–1870) was educated at the École Polytechnique and the École des Mines before going to Russia to direct the School of Highways and Transportation in St. Petersburg. After his return to France in 1832, he taught at the École Polytechnique while also working as an engineering consultant. Lamé contributed original work to number theory, applied mathematics, and thermodynamics. His best-known work is his proof of the case $n = 5$ of Fermat's Last Theorem in 1839. Eight years later, he announced that he had found a general proof of the theorem, which began with the factorization of the expression $x^n + y^n$ over the complex numbers as $(x + y)(x + \alpha y)(x + \alpha^2 y) \ldots (x + \alpha^{n-1} y)$, where $\alpha$ is a primitive root of $x^n - 1 = 0$. He planned to show that the factors in this expression are all relatively prime and therefore that if $x^n + y^n = z^n$, then each of the factors would itself be an $n$th power. He would then use the technique of infinite descent to find a solution in smaller numbers. Unfortunately Lamé's idea required that the ring of integers in the cyclotomic field of the $n$th roots of unity be a unique factorization domain. And, as Kummer had already proved three years earlier, unique factorization in fact fails in many such domains.

***Edmund Landau*** (1877–1938) received a doctorate under Frobenius and taught at the University of Berlin and at Göttingen. His research areas were analysis and analytic number theory, including the distribution of primes. He used the big-$O$ notation (also called a Landau symbol) in his work to estimate the growth of various functions.

***Pierre-Simon de Laplace*** (1749–1827) entered the University of Caen in 1766 to begin preparation for a career in the church. He soon discovered his mathematical talents, however, and in 1768 left for Paris to continue his studies. He later taught mathematics at the École Militaire to aspiring cadets. Legend has it that he examined, and passed, Napoleon there in 1785. He was later honored by both Napoleon and King Louis XVIII. Laplace is best known for his contributions to celestial mechanics, but he was also one of the founders of probability theory and made many contributions to mathematical statistics. In fact, he was one of the first to apply his theoretical results in statistics to a genuine problem in statistical inference, when he showed from the surplus of male to female births in Paris over a 25-year period that it was "morally certain" that the probability of a male birth was in fact greater than $\frac{1}{2}$.

**Gottfried Wilhelm Leibniz** (1646–1716), born in Leipzig, developed his version of the calculus some ten years after Isaac Newton, but published it much earlier. He based his calculus on the inverse relationship of sums and differences, generalized to infinitesimal quantities called differentials. Leibniz hoped that his most original contribution to philosophy would be the development of an alphabet of human thought, a way of representing all fundamental concepts symbolically and a method of combining these symbols to represent more complex thoughts. Although he never completed this project, his interest in finding appropriate symbols ultimately led him to the $d$ and $\int$ symbols for the calculus that are used today. Leibniz spent much of his life in the diplomatic service of the Elector of Mainz and later was a Counsellor to the Duke of Hanover. But he always found time to pursue his mathematical ideas and to carry on a lively correspondence on the subject with colleagues all over Europe.

**Levi ben Gerson** (1288–1344) was a rabbi as well as an astronomer, philosopher, biblical commentator, and mathematician. He lived in Orange, in southern France, but little is known of his life. His most famous mathematical work is the *Maasei Hoshev* (The Art of the Calculator) (1321), which contains detailed proofs of the standard combinatorial formulas, some of which use the principle of mathematical induction. About a dozen copies of this medieval manuscript are extant, but it is not known whether the work had any direct influence elsewhere in Europe. Also, at the request of a French music theorist, Levi produced an elegant proof of the theorem that a power of two and a power of three cannot be consecutive numbers, except for the obvious cases of 2 and 3, 3 and 4, and 8 and 9.

**László Lovász** (born 1948) was so inspired as a high school student by Paul Erdős that he competed in the International Mathematical Olympiad in 1964, 1965, and 1966, winning a gold medal each time. He published his first paper, on graph theory, when he was only seventeen, and has since gone on to a remarkable career in which he has split time between Hungary and the United States. He has numerous results in discrete mathematics to his credit, results that have applications in other areas of mathematics as well as in computer science. He has developed new methods for solving combinatorial problems, including those relying on geometric polyhedral and topological techniques, thus showing that combinatorics deserves to be considered as one of the major branches of mathematics. Besides authoring several important research monographs, he has also written an elementary textbook introducing students to the basic ideas of discrete mathematics. His work has won him numerous prizes, and he served as president of the International Mathematical Union from 2007 to 2010.

**Augusta Ada Byron King Lovelace** (1815–1852) was the child of the famous poet George Gordon, the sixth Lord Byron, who left England five weeks after his daughter's birth and never saw her again. She was raised by her mother, Anna Isabella Millbanke, a student of mathematics herself, so she received considerably more mathematics education than was usual for girls of her time. She was tutored privately by well-known mathematicians, including William Frend and Augustus DeMorgan. Her husband, the Earl of Lovelace, was made a Fellow of the Royal Society in 1840, and through this connection, Ada was able to gain access to the books and papers she needed to continue her mathematical studies and, in particular, to understand the workings of Babbage's Analytical Engine. Her major mathematical work is a heavily annotated translation of a paper by the Italian mathematician L. F. Menabrea dealing with the Engine, in which she gave explicit descriptions of how it would solve specific problems and described, for the first time in print, what would today be called a computer program, in this case a program for computing the Bernoulli numbers. Interestingly, only her initials, A.A.L., were used in the published version of the paper. It was evidently not considered proper in mid-nineteenth century England for a woman of her class to publish a mathematical work.

**Jan Łukasiewicz** (1878–1956) studied at the University of Lwów and taught at the University of Lwów, the University of Warsaw, and the Royal Irish Academy. A logician, he worked in the area of many-valued logic, writing papers on three-valued and $m$-valued logics, He is best known for the parenthesis-free notation he developed for propositions, called Polish notation.

**Percy Alexander MacMahon** (1854–1929) was born into a British army family and joined the army himself in 1871, reaching the rank of major in 1889. Much of his army service was spent as an instructor at the Royal Military Academy. His early mathematical work dealt with invariants, following on the work of Cayley and Sylvester, but a study of symmetric functions eventually led to his interest in partitions and to his extension of the idea of a partition to higher dimensions. MacMahon's two volume treatise *Combinatorial Analysis* (1915–16) is a classic in the field. It identified and clarified the basic results of combinatorics and showed the way toward numerous applications.

**Mahāvīra** (ninth century) was an Indian mathematician of the medieval period whose major work, the *Ganitasārasaṅgraha*, was a compilation of problems solvable by various algebraic techniques. For example, the work included a version of the hundred fowls problem: "Doves are sold at the rate of 5 for 3 coins, cranes at the rate of 7 for 5, swans at the rate of 9 for 7, and peacocks at the rate of 3 for 9. A certain man was told to bring at these rates 100 birds for 100 coins for the amusement of the king's son and was sent to do so. What amount does he give for each?" Mahāvīra also presented, without proof and in words, the rule for calculating the number of combinations of $r$ objects out of a set of $n$. His algorithm can be easily translated into the standard formula. Mahāvīra then applied the rule to two problems, one about combinations of tastes and another about combinations of jewels on a necklace.

**Andrei Markov** (1856–1922) was a Russian mathematician who first defined what are now called Markov chains in a paper of 1906 dealing with the Law of Large Numbers and subsequently proved many of the standard results about them. His interest in these chains stemmed from the needs of probability theory. Markov never dealt with their application to the sciences, only considering examples from literary texts, where the two possible states in the chain were vowels and consonants. Markov taught at St. Petersburg University from 1880 to 1905 and contributed to such fields as number theory, continued fractions, and approximation theory. He was an active participant in the liberal movement in pre-World War I Russia and often criticized publicly the actions of state authorities. In 1913, when as a member of the Academy of Sciences he was asked to participate in the pompous ceremonies celebrating the 300th anniversary of the Romanov dynasty, he instead organized a celebration of the 200th anniversary of Jacob Bernoulli's publication of the Law of Large Numbers.

**Marin Mersenne** (1588–1648) was educated in Jesuit schools and in 1611 joined the Order of Minims. From 1619 he lived in the Minim Convent de l'Annonciade near the

Place Royale in Paris and there held regular meetings of a group of mathematicians and scientists to discuss the latest ideas. Mersenne also served as the unofficial "secretary" of the republic of scientific letters in Europe. As such, he received material from various sources, copied it, and distributed it widely, thus serving as a "walking scientific journal". His own contributions were primarily in the area of music theory as detailed in his two great works on the subject, the *Harmonie universelle* and the *Harmonicorum libri*, both of which appeared in 1636. As part of his study of music, he developed the basic combinatorial formulas by considering the possible tunes one could create out of a given number of notes. Mersenne was also greatly interested in the relationship of theology to science. He was quite concerned when he learned that Galileo could not publish one of his works because of the Inquisition and, in fact, offered his assistance in this matter.

**Hermann Minkowski** (1864–1909) was a German mathematician who received his doctorate at the University of Königsberg. He became a lifelong friend of David Hilbert and, on Hilbert's suggestion, was called to Göttingen in 1902. In 1883, he shared the prize of the Paris Academy of Sciences for his essay on the topic of the representations of an integer as a sum of squares. In his essay, he reconstructed the entire theory of quadratic forms in $n$ variables with integral coefficients. In further work on number theory, he brought to bear geometric ideas beginning with the realization that a symmetric convex body in $n$-space defines a notion of distance and hence a geometry in that space. The connection with number theory depends on the representation of forms by lattice points in space.

**Aḥmad ibn Mun'im al-'Abdari** (died 1228) was born in Andalusia, but then lived and taught in Marrakesh, probably at the Almohade court during the reign of Muḥammad al-Nāṣir. He is best known for his work *On the Science of Calculation*, which contains a long section titled "On denumerating the Words Such That a Person Cannot Express Himself without One of Them." That is, he attempted to count all possible words that can be formed out of the letters, including vowels, of the Arabic alphabet. He used various counting techniques, including what is now called the Pascal triangle, to show how this problem can be attacked. Although the question as posed cannot be answered in full, ibn Mun'im did give methods for finding partial answers. For example, he showed that the number of possible words of nine letters of which two are not repeated, two are repeated twice, and one is repeated three times is 5,968,924,232,544,000. In another section of his work, he discussed figured numbers, showing how they can be generated recursively and giving a rule for calculating the sum of consecutive figured numbers of a given type.

**John F. Nash, Jr.** (1928–2015) shared the 1994 Nobel Prize in Economics (together with Reinhard Selten and John Harsanyi) for fundamental contributions to game theory and its applications to economics. Nash earned his Ph.D. at Princeton University with a dissertation defining and explaining what became known as the Nash equilibrium in non-cooperative games. A group of players are in the Nash equilibrium if each one has chosen a strategy and no one can benefit by changing strategies, provided that all the other players keep their strategies unchanged. Nash proved that such an equilibrium exists under many game-theoretic conditions. As detailed in Sylvia Nasar's biography of Nash, "A Beautiful Mind", he suffered from paranoid schizophrenia for much of his adult life. He still managed to make major contributions, not only to game theory but to differential geometry and partial differential equations. In fact, in 2015 he shared the Abel Prize with Louis Nirenberg for his work on nonlinear partial differential equations. He was killed in an automobile accident on his way home from Norway, where he had received the award.

**Peter Naur** (1928–2016), born in Denmark, was originally an astronomer, using computers to calculate planetary motion, but in 1959 he became a full-time computer scientist. He was a developer of the programming language ALGOL and worked on compilers for ALGOL and COBOL. In 1969 he took a computer science faculty position at the University of Copenhagen, from which he retired in 1998. He won the Turing award in 2005 "for fundamental contributions to programming language design and the definition of ALGOL 60, to compiler design, and to the art and practice of computer programming."

**Amalie Emmy Noether** (1882–1935) received her doctorate from the University of Erlangen in 1908 and a few years later moved to Göttingen to assist Hilbert in the study of general relativity. During her eighteen years there, she was extremely influential in stimulating a new style of thinking in algebra by always emphasizing its structural rather than computational aspects. She was forced to leave Germany in the 1930s by the Nazis, but in 1934, she became a professor at Bryn Mawr College and a member of the Institute for Advanced Study. She is most famous for her work on Noetherian rings, and her influence is still evident in today's textbooks in abstract algebra.

**Blaise Pascal** (1623–1662) showed his mathematical precocity with his *Essay on Conics* of 1640, in which he stated his theorem that the opposite sides of a hexagon inscribed in a conic section always intersect in three collinear points. Pascal is better known, however, for his detailed study of what is now called Pascal's triangle of binomial coefficients. In that study Pascal gave an explicit description of mathematical induction and used that method, although not quite in the modern sense, to prove various properties of the numbers in the triangle, including a method of determining the appropriate division of stakes in a game interrupted before its conclusion. Pascal had earlier discussed this matter, along with various other ideas in the theory of probability, in correspondence with Fermat in the 1650s. These letters, in fact, can be considered the beginning of the mathematization of probability.

**Giuseppe Peano** (1858–1932) studied at the University of Turin and then spent the remainder of his life there as a professor of mathematics. He was originally known as an inspiring teacher, but as his studies turned to symbolic logic and the foundations of mathematics and he attempted to introduce some of these notions in his elementary classes, his teaching reputation changed for the worse. Peano is best known for his axioms for the natural numbers, first proposed in the *Arithmetices principia, nova methodo exposita* of 1889. One of these axioms describes the principle of mathematical induction. Peano was also among the first to present an axiomatic description of a (finite-dimensional) vector space. In his *Calcolo geometrico* of 1888, Peano described what he called a *linear system*, a set of quantities provided with the operations of addition and scalar multiplication which satisfy the standard properties. He was then able to give a coherent definition of the *dimension* of a linear system as the maximum number of linearly independent quantities in the system.

**Charles Sanders Peirce** (1839–1914) was born in Massachusetts, the son of a Harvard mathematics professor. He received a master's degree from Harvard in 1862 and an advanced degree in chemistry from the Lawrence Scientific School in 1863. He made contributions to many areas of the foundations and philosophy of mathematics. He was a prolific writer, leaving over 100,000 pages of unpublished manuscript at his death.

**George Pólya** (1887–1985) was a Hungarian mathematician who received his doctorate at Budapest in 1912. From 1914 to 1940 he taught in Zurich, then emigrated to the United States where he spent most of the rest of his professional life at Stanford

University. Pólya developed some influential enumeration ideas in several papers in the 1930s, in particular dealing with the counting of certain configurations that are not equivalent under the action of a particular permutation group. For example, there are 16 ways in which one can color the vertices of a square using two colors, but only six are non-equivalent under the various symmetries of the square. In 1937, Pólya published a major article in the field, "Combinatorial Enumeration of Groups, Graphs and Chemical Compounds", in which he discussed many mathematical aspects of the theory of enumeration and applied it to various problems. Pólya's work on problem solving and heuristics, summarized in his two volume work *Mathematics and Plausible Reasoning*, insured his fame as a mathematics educator; his ideas are at the forefront of recent reforms in mathematics education at all levels.

**Qin Jiushao** (1202–1261), born in Sichuan, published a general procedure for solving systems of linear congruences — the Chinese remainder theorem — in his *Shushu jiuzhang* (*Mathematical Treatise in Nine Sections*) in 1247, a procedure which makes essential use of the Euclidean algorithm. He also gave a complete description of a method for numerically solving polynomial equations of any degree. Qin's method had been developed in China over a period of more than a thousand years; it is similar to a method used in the Islamic world and is closely related to what is now called the Horner method of solution, published by William Horner in 1819. Qin studied mathematics at the Board of Astronomy, the Chinese agency responsible for calendrical computations. He later served the government in several offices, but because he was "extravagant and boastful", he was several times relieved of his duties because of corruption. These firings notwithstanding, Qin became a wealthy man and developed an impressive reputation in love affairs.

**Willard Van Orman Quine** (1908–2000) was a logician and philosopher who received his Ph.D. from Harvard in 1932 and remained with the university in various positions for the remainder of his life. He was a chief proponent of the view that philosophy is not conceptual analysis but is an abstract branch of the empirical sciences. While he wrote many expository works in logic, his most innovative work was in set theory, where at different times in his life he proposed three variants of axiomatic set theory. He also made many contributions to symbolic logic, where some of his results found applications in computer science.

**Richard Rado** (1906–1989) left Germany for Great Britain in 1933, once the racial laws were promulgated by the Nazis. He had already received a Ph.D. from the University of Berlin, but then received a second from Cambridge University two years later. Rado's most important contributions were in combinatorics, and he wrote many joint papers in the field with Paul Erdős. In particular, the two of them developed the partition calculus, which generalizes the following example: If six people meet by chance, then either at least three all know each other or there are at least three with no two knowing each other. Rado spent his academic career at three universities in England: the University of Sheffield, King's College, London, and the University of Reading, from which he retired.

**Srinivasa Ramanujan** (1887–1920) was born near Madras into the family of a book-keeper. He studied mathematics on his own and soon began producing results in combinatorial analysis, some already known and others previously unknown. At the urging of friends, he sent some of his results to G. H. Hardy in England, who quickly recognized Ramanujan's genius and invited him to England to develop his untrained mathematical talent. During the war years from 1914 to 1917, Hardy and Ramanujan collaborated on a number of papers, including several dealing with the theory of partitions. Unfortunately, Ramanujan fell ill during his years in the unfamiliar climate of England and died at age 32 soon after returning to India. Ramanujan left behind several notebooks containing statements of thousands of results, enough work to keep many mathematicians occupied for years in understanding and proving them. The 2015 film "The Man Who Knew Infinity" depicts his time at Trinity College, Cambridge, and his mathematical collaborations with Hardy.

**Frank Ramsey** (1903–1930), son of the president of Magdalene College, Cambridge, was educated at Winchester and Trinity Colleges. He was then elected a fellow of King's College, where he spent the remainder of his life. Ramsey made important contributions to mathematical logic. What is now called Ramsey theory began with his clever combinatorial arguments to prove a generalization of the pigeonhole principle, published in the paper "On a Problem of Formal Logic". The problem of that paper was the *Entscheidungsproblem* (the decision problem), the problem of searching for a general method of determining the consistency of a logical formula. Ramsey also made contributions to the mathematical theory of economics and introduced the subjective interpretation to probability. In that interpretation, Ramsey argues that different people when presented with the same evidence, will have different degrees of belief. And the way to measure a person's belief is to propose a bet and see what are the lowest odds the person will accept. Ramsey's death at the age of 26 deprived the mathematical community of a brilliant young scholar.

**Ronald Rivest** (born 1947) is a cryptographer and professor at MIT. He received his Ph.D. in computer science from Stanford University in 1974 and three years later, along with Adi Shamir and Len Adleman, invented the RSA public key algorithm, now commonly used for secure data transmission. The algorithm is based on the difficulty of factoring the product of two large prime numbers. (The algorithm had been earlier invented by Clifford Cocks, but had not been published because it was classified.) According to legend, the algorithm was invented after the three had consumed much wine during a Passover seder at the home of a student. Rivest has invented other symmetric key encryption algorithms as well.

**Alvin Roth** (born 1951) won the Nobel Prize in Economics in 2012 (together with Lloyd Shapley) for his work in market design. One of his major achievements was demonstrating, in 1984, that the National Resident Matching Program, which matches new medical doctors with hospital residency programs, was stable and strategy-proof. His work was based on the theoretical foundations introduced by David Gale and Lloyd Shapley in 1962. He later worked with the New York public school system to redesign its matching programs for students entering high school, as well as with the Boston public school system for its matching of primary school students. More recently, he applied the ideas of stable matchings to the problem of paired kidney exchanges. Roth received his Ph.D. from Stanford University and is currently a faculty member at the same institution.

**Bertrand Arthur William Russell** (1872–1970) was born in Wales and studied at Trinity College, Cambridge. A philosopher/mathematician, he is one of the founders of modern logic and wrote over 40 books in different areas. In his most famous work, *Principia Mathematica*, published in 1910–13 with Alfred North Whitehead, he attempted to deduce the entire body of mathematics from a single set of primitive axioms. A pacifist, he fought for progressive causes, including women's suffrage in Great Britain and nuclear disarmament. In 1950 he won a Nobel Prize for literature.

**al-Samaw'al ibn Yahyā ibn Yahūda al-Maghribī** (1125–1180) was born in Baghdad to well-educated Jewish parents. Besides giving him a religious education, they encouraged him to study medicine and mathematics. He wrote his major mathematical work, *Al-Bāhir* (*The Shining*), an algebra text that dealt extensively with the algebra of polynomials, at the age of 19. In it, al-Samaw'al worked out the laws of exponents, both positive and negative, and showed how to divide polynomials even when the division was not exact. He also used a form of mathematical induction to prove the binomial theorem, that $(a+b)^n = \sum_{k=0}^{n} C(n,k) a^{n-k} b^k$, where the $C(n,k)$ are the entries in the Pascal triangle, for $n \le 12$. In fact, he showed why each entry in the triangle can be formed by adding two numbers in the previous row. When al-Samaw'al was about 40, he decided to convert to Islam. To justify his conversion to the world, he wrote an autobiography in 1167 stating his arguments against Judaism, a work which became famous as a source of Islamic polemics against the Jews.

**Issai Schur** (1875–1941) was born in the Russian Empire but spent most of his academic life in Germany. He received his doctorate from the University of Berlin in 1901 and was a professor there from 1919 until he was forced to retire by the Nazis in 1935. In fact, he led a well-known school of mathematicians at Berlin working in group representations as well as some related ones. He is best known for his work in that field, but he also contributed significantly to combinatorics and number theory. In particular, when he realized in the 1920s that group representations were important for theoretical physics, he returned to their study with renewed energy and eventually gave a complete description of the rational representations of the general linear group. Unfortunately, after he was dismissed from his professorship, he declined some invitations to go abroad. Eventually, all avenues of work were closed to him in Germany. He was forced to relinquish membership in the Prussian Academy of Science. Finally, in 1939, he left Germany for Palestine where he died two years later.

**Adi Shamir** (born 1952) is an Israeli cryptographer who received his Ph.D. in computer science from the Weizmann Institute in 1977. He is the inventor (along with Ronald Rivest and Len Adleman) of the RSA public key algorithm, now used in secure data transmission. He has made numerous other contributions to cryptography and computer science, including the independent discovery of differential cryptanalysis (previously known to the U.S. National Security Agency). He is currently on the faculty both of the Weizmann Institute and the École Normale Supérieure in Paris.

**Claude Elwood Shannon** (1916–2001) applied Boolean algebra to switching circuits in his master's thesis at M.I.T in 1938. Shannon realized that a circuit can be represented by a set of equations and that the calculus necessary for manipulating these equations is precisely the Boolean algebra of logic. Simplifying these equations for a circuit would yield a simpler, equivalent circuit. Switches in Shannon's calculus were either open (represented by 1) or closed (represented by 0); placing switches in parallel was represented by the Boolean operation "+", while placing them in parallel was represented by "·". Using the basic rules of Boolean algebra, Shannon was, for example, able to construct a circuit which would add two numbers given in binary representation. He received his Ph.D. in mathematics from M.I.T. in 1940 and spent much of his professional life at Bell Laboratories, where he worked on methods of transmitting data efficiently and made many fundamental contributions to information theory.

**Lloyd Shapley** (1923–2016) received the Nobel Prize in Economics (together with Alvin Roth) in 2012 after a long career in game theory and mathematical economics. He served in the U.S. Army Air Corps in World War II, after which he finished his education with a Ph.D. from Princeton in 1953. He was a professor at the University of California, Los Angeles from 1981, where he had a joint appointment in mathematics and economics. His Nobel Prize was awarded "for the theory of stable allocations and the practice of market design." In particular, he used Cooperative Game Theory to study methods to pair different players in a game in a stable matching, one in which no two players would prefer one another over their current counterparts.

**Emanuel Sperner** (1905–1980) was a German mathematician who received his doctorate in 1928 from the University of Hamburg. He is famous for several results in set theory discovered when he was a young man. Sperner's Theorem, published in 1928, describes the largest possible collection of subsets of a given set with the property that no subset in the collection contains any other subset in the collection. The result known as "Sperner's Lemma," proved about the same time, deals with the labelings of vertices in a special type of graph. Surprisingly, the lemma is crucial to an elegant proof of the Brouwer Fixed Point Theorem and has numerous other applications in topology. Sperner is also famous for his 1951 textbook *Introduction to Modern Algebra and Matrix Theory* (with Otto Schreier), in which fundamental concepts in algebra and in affine and projective geometry are treated simultaneously. Sperner became secretary of the German Mathematical Society in 1935, then resigned that position to become editor of the *Jahresbericht* of the Society until its closure at the end of 1943. He was a member of the board of the Society which, in 1938, wrote to all remaining Jewish members asking them to resign.

**James Stirling** (1692–1770) studied at Glasgow University and at Balliol College, Oxford and spent much of his life as a successful administrator of a mining company in Scotland. His mathematical work included an exposition of Newton's theory of cubic curves and a 1730 book entitled *Methodus Differentialis* which dealt with summation and interpolation formulas. In dealing with the convergence of series, Stirling found it useful to convert factorials into powers. By considering tables of factorials, he was able to derive the formula for $\log n!$, which leads to what is now known as Stirling's approximation: $n! \approx (\frac{n}{e})^n \sqrt{2\pi n}$. Stirling also developed the Stirling numbers of the first and second kinds, sequences of numbers important in enumeration.

**Sun Zi** (4th century) is the author of *Sunzi suanjing* (*Master Sun's Mathematical Manual*), a manual on arithmetical operations which eventually became part of the required course of study for Chinese civil servants. The most famous problem in the work is one of the first examples of what is today called the Chinese remainder problem: "We have things of which we do not know the number; if we count them by threes, the remainder is 2; if we count them by fives, the remainder is 3; if we count them by sevens, the remainder is 2. How many things are there?" Sun Zi gives the answer, 23, along with some explanation of how the problem should be solved. But since this is the only problem of its type in the book, it is not known whether Sun Zi had developed a general method of solving simultaneous linear congruences.

**James Joseph Sylvester** (1814–1897), who was born into a Jewish family in London and studied for several years at Cambridge, was not permitted to take his degree there for religious reasons. Therefore, he received his degree from Trinity College, Dublin and soon thereafter accepted a professorship at the University of Virginia. His horror of slavery, however, and an altercation with a student who did not show him the respect he felt he deserved led to his resignation after only a brief tenure. After his return to England, he spent 10 years as an attorney and 15 years as professor of mathematics at the Royal Military Academy at Woolwich. Sylvester returned to the United States in 1871 to accept the chair of mathematics at the newly opened Johns Hopkins University in Baltimore, where he founded the *American Journal of Mathematics* and helped initiate a tradition of graduate education in mathematics in the United States. Sylvester's primary mathematical contributions are in the fields of invariant theory and the theory of partitions.

**Terence Tao** (born 1975) is the youngest participant to date in the International Mathematical Olympiad, where he first competed on the Australian team at the age of ten and where he won a gold medal at thirteen. He received his Ph.D. from Princeton University at the age of 20 and then joined the faculty at UCLA, where he was promoted to full professor at the age of 24. In 2006, he won the Fields Medal "for his contributions to partial differential equations, combinatorics, harmonic analysis and additive number theory." Among his many accomplishments are his proof (with Ben Green) that it is always possible to find a progression of prime numbers of equal spacing and any length, and his new method of attack, in 2014, on the Navier-Stokes Millennium Problem.

**Thābit ibn Qurra** (836–901) was born in the south of what is now Turkey. His mathematical abilities were discovered by scholars from Baghdad, and so he was invited to the capital of the caliphate where he eventually became court astronomer. He translated many Greek mathematical works into Arabic and also produced an algebra text in which justifications of the algorithms for solving quadratic equations were based on theorems of Euclid's Book II. His major mathematical work, however, was the statement and proof of a result producing amicable numbers, a result that was transmitted to Europe via several Hebrew mathematical texts.

**John Wilder Tukey** (1915–2000) received a Ph.D. in topology from Princeton in 1939. After World War II he returned to Princeton as professor of statistics, where he founded the Department of Statistics in 1966. His work in statistics included the areas of spectra of time series and analysis of variance. He invented (with J. W. Cooley) the fast Fourier transform. He was awarded the National Medal of Science and served on the President's Science Advisory Committee. He also coined the word "bit" for a binary digit.

**Alan Turing** (1912–1954) studied mathematics at King's College, Cambridge and in 1936 invented the concept of a Turing machine to answer the questions of what a computation is and whether a given computation can in fact be carried out. This notion today lies at the basis of the modern all-purpose computer, a machine which can be programmed to do any desired computation. At the outbreak of World War II, Turing was called to serve at the Government Code and Cypher School in Bletchley Park in Buckinghamshire. It was there, during the next few years, that he led the successful effort to crack the German "Enigma" code, an effort which turned out to be central to the defeat of Nazi Germany. After the war, Turing continued his interest in automatic computing machines and so joined the National Physical Laboratory to work on the design of a computer, continuing this work after 1948 at the University of Manchester. Turing's promising career came to a grinding halt, however, when he was arrested in 1952 for homosexual acts. The penalty for this "crime" was submission to psychoanalysis and hormone treatments to "cure" the disease. Unfortunately, the cure proved worse than the disease, and, in a fit of depression, Turing committed suicide in June, 1954.

**Alexandre-Théophile Vandermonde** (1735–1796) was directed by his physician father to a career in music. However, he later developed a brief but intense interest in mathematics and wrote four important papers published in 1771 and 1772. These papers include fundamental contributions to the theory of the roots of equations, the theory of determinants, and the knight's tour problem. In the first paper, he showed that any symmetric function of the roots of a polynomial equation can be expressed in terms of the coefficients of the equation. His paper on determinants was the first logical, connected exposition of the subject, so he can be thought of as the founder of the theory. Toward the end of his life, he joined the cause of the French Revolution and held several different positions in government.

**François Viète** (1540–1603), a lawyer and advisor to two kings of France, was one of the earliest cryptanalysts and successfully decoded intercepted messages for his patrons. In fact, he was so successful in this endeavor that he was denounced by some who thought that the decipherment could only have been made by sorcery. Although a mathematician only by avocation, he made important contributions to the development of algebra. In particular, he introduced letters to stand for numerical constants, thus enabling him to break away from the style of verbal algorithms of his predecessors and treat general examples by formulas rather than by giving rules for specific problems.

**Edward Waring** (1734–1798) graduated from Magdalen College, Cambridge in 1757 with highest honors and shortly thereafter was named a Fellow of the University. In 1760, despite opposition because of his youth, he was named Lucasian Professor of Mathematics at Cambridge, a position he held until his death. To help solidify his position, then, he published the first chapter of his major work, *Miscellanea analytica*, which in later editions was renamed *Meditationes algebraicae*. Waring is best remembered for his conjecture that every integer is the sum of at most four squares, at most nine cubes, at most 19 fourth powers, and, in general, at most $r$ $k$th powers, where $r$ depends on $k$. The general theorem that there is a finite $r$ for each $k$ was proved by Hilbert in 1909. Although the result for squares was proved by Lagrange, the specific results for cubes and fourth powers were not proved until the twentieth century.

**Andrew Viterbi** (born 1935) is an electrical engineer who invented the Viterbi algorithm. This is a dynamic programming algorithm for finding the most likely sequence of hidden states that results in a sequence of observed events. The algorithm is used in cell phones, speech recognition, DNA analysis and other applications. Viterbi immigrated to the United States with his parents in 1939 because of the Italian racial laws restricting the civil rights of Jews. He received his Ph.D. from the University of Southern California. Besides serving as a professor of electrical engineering at both the University of California, Los Angeles and the University of California, San Diego, he was the co-founder of Qualcomm Inc., a semiconductor and telecommunications equipment company.

**Hassler Whitney** (1907–1989) received bachelor's degrees in both physics and music from Yale; in 1932 he received a doctorate in mathematics from Harvard. After a brief stay in Princeton, he returned to Harvard, where he taught until 1952, when he moved to the Institute for Advanced Study. Whitney produced more than a dozen papers on graph theory in the 1930s, after his interest was aroused by the four color problem. In particular, he defined the notion of the *dual graph* of a map. It was then possible to apply many of the results of the theory of graphs to gain insight into the four color problem. During the last twenty years of his life, Whitney devoted his energy to improving mathematical education, particularly at the elementary school level. He emphasized that young children should be encouraged to solve problems using their intuition, rather than only be taught techniques and results which have no connection to their experience.

**Andrew Wiles** (born 1953) is best known for his proof of Fermat's Last Theorem. Born in England, he first learned of the theorem when he was ten and vowed to be the first person to prove it. After receiving his Ph.D. from Cambridge, he spent time at the Institute for Advanced Study in Princeton and then joined the Princeton mathematics faculty, where he remained until he took a research professorship at Oxford in 2011. After he announced his proof in 1993 at a series of lectures at Cambridge University, the review process found a flaw in the proof. Fortunately, with the help of Richard Taylor, Wiles corrected the flaw the following year. The 1995 issue of the *Annals* *of Mathematics* contained the complete proof. The many new ideas in Wiles's work has led to many other new results in number theory. Wiles himself received the Abel prize in 2016 in recognition of his achievement.

## REFERENCES

*Printed Resources*:

*Dictionary of Scientific Biography*, Macmillan, 1998.

D. M. Burton, *The History of Mathematics, An Introduction*, 3rd ed., McGraw-Hill, 1996.

H. Eves, *An Introduction to the History of Mathematics*, 6th ed., Saunders, 1990.

H. Eves, *Great Moments in Mathematics (Before 1650)*, Dolciani Mathematical Expositions, No. 5, Mathematical Association of America, 1983.

H. Eves, *Great Moments in Mathematics (After 1650)*, Dolciani Mathematical Expositions, No. 7, Mathematical Association of America, 1983.

V. J. Katz, *History of Mathematics, an Introduction*, 3rd ed., Addison-Wesley, 2008.

V. J. Katz, M. Folkerts, B. Hughes, R. Wagner, and J. L. Berggren, eds., *Sourcebook in the Mathematics of Medieval Europe and North Africa*, Princeton University Press, 2016.

V. J. Katz, A. Imhausen, E. Robson, J. W. Dauben, K. Plofker, and J. L. Berggren, eds., *The Mathematics of Egypt, Mesopotamia, China, India, and Islam: A Sourcebook*, Princeton University Press, 2007.

**Web Resources**:

http://turnbull.mcs.st-and.ac.uk/~history/ (The MacTutor History of Mathematics archive.)

# INDEX

世界著名数学家 B. Thwaites 曾指出:

必须比过去更加多地强调离散数学. 在较初级阶段应该详尽地介绍有限集及其上的封闭二元运算, 而不是无限集的概念. 无论在哪一个水平上, 再也不能在学生面前把数学描述成由被天赐规律决定的证明、论证和争辩组成的系统. 必须有意识地, 及时地对我们的学生进行算法方面的教育, 并辅以启发方式的教育. 必须明确地教导他们, 任何数学方法及其成功的机会必定依赖于可运用的手段及所得结果的用处.

过去我们过多的关注像函数论、微分方程这样的所谓以连续函数为研究对象的分支, 而对以研究离散对象的组合数学重视不够. 新世纪来临, 组合数学终于迎来了黄金时代.

本书是一部英文版的权威的数学工具书, 中文书名或可译为《离散与组合数学手册 (第二版)》.

本书的主编为肯尼斯·H. 罗森, 美国数学家, 麻省理工学院博士, AT&T 贝尔实验室的研究员. 他发表过很多有关数论与数学建模的研究性论文, 是 CRC/Chapman & Hall 的教科书系列的系列编辑, 也是备受推崇的离散数学及其应用系列的创始编辑. 他还是《离散数学》期刊的副主编. 他曾在四所大学任职, 现在还继续任教. 他在 AT&T 贝尔实验室的工作获得了 70 项专利.

正如本书作者在前言中所述:

在过去的几十年里,离散和组合数学的重要性急剧增加.编写第二版是为了更新第一版的内容并扩大其覆盖范围.我们对本书第一版的成功感到欣慰,也希望许多要求再版的读者会发现第二版是值得等待的.

《离散与组合数学手册》出版的目的是为需要离散与组合数学信息的计算机科学家、工程师、数学家和学生、物理与社会科学家,以及相关的图书管理员提供一本全方位的参考书.

本书的第一版是以准参考资料的形式,为在工作或学习中用到与本书主题相关知识的人准备的展现这类信息的第一手资料.第二版是对第一版的重大修订.它包括了大量的添加和更新,这些添加与更新在本书前言的后半部分进行了总结.本书的范围包括通常被认为是离散数学的一部分的许多领域,集中于被认为在计算机科学、工程和其他学科的应用中至关重要的信息.本版本涵盖的一些基本主题领域包括:

| | |
|---|---|
| 逻辑与集合论 | 图论 |
| 枚举 | 树 |
| 整数序列 | 网络流 |
| 递归关系 | 组合设计 |
| 生成函数 | 计算几何 |
| 数论 | 编码理论 |
| 抽象代数 | 密码学 |
| 线性代数 | 离散最优化 |
| 离散概率论 | 自动机理论 |
| 数据挖掘 | 数据结构与算法 |
| 离散生物信息学 | |

本书是一部大型的工具书.本书呈现材料的方式可以让读者快速、轻松地找到和使用关键信息.每章都包含一个词汇表,用来对该章节中最重要的术语提供简洁的定义.单独的主题包含在每章的各个部分和小节中,每个部分都是清晰可辨的:定义、事实和示例.事实清单包括:

- 有关材料的运用方式与重要性的信息.
- 历史信息.
- 关键定理.
- 最新结果.
- 开放性问题的情况.
- 通常不容易计算的数值表.
- 汇总表.
- 简单伪代码中的关键算法.
- 关于算法的信息,比如它们的复杂性.
- 主要应用.
- 网络与印刷材料中的额外资料指向.

事实被简明扼要地列出来,以便于读者查找和理解.本书还提供了连接各部分的交叉引用参考资料.希望进一步研究某个主题的读者可以查阅列出的资料.

本书中材料的选择主要是因为它们的重要性与实用性.为了确保全面性,本书还添加了额外的资料,以便读者在探索中遇到离散数学中的新术语和概念时能够从本书中获得帮助.

本书提供了一些例子用以说明一些关键的定义、事实和算法.读者可能会发现一些有趣的事实和谜题也包括在内.读者还将在主要章节之后找到大量传记,重点介绍了离散数学的许多重要贡献者的生平.

本书的每一章都包含了一个分为印刷资源和相关网站的参考清单.

本书是靠团队的力量完成的.

本书第一版的组织和结构由包括 1 名主编、3 名副主编、1 名项目主编和编辑在内的 CRC 出版社的一个团队制定.该团队整理了一份拟议的目录,然后由一组顾问编辑进行分析,每个顾问编辑都是离散数学的一个或多个方面的专家.这些顾问编辑提出了修改建议,包括是否涵盖其他重要主题.目录完全编排完成后,本书的各个部分由来自工业界和学术界的 70 多名参与者来准备,他们了解如何使用这些材料以及这些材料重要的原因.参与者在副主编和主编的指导下工作,这些编辑确保风格的一致性以及材料呈现的清晰度和全面性.作者和我们的编辑团队仔细审查了材料,以确保风格的准确性和一致性.

在第二版中,我们组建了一个新团队.这个团队的第一个目标是整理一个新的目录.这涉及确定新章节和新段落的机会,以扩大第二版的内容范围,提高第二版的吸引力.在以前和新的参与者的帮助下,本书添加了额外的材料,更新和扩展了现有的材料,遵循了一贯风格并保持或改进了第一版中的陈述.

本书第二版的出版是付出了多年努力的.自 1999 年第一版问世以来,已有许多新发现出现,新领域的重要性日益凸显.重要的变化包括:

- 从 17 章增加到 20 章,增加了 360 多页.
- 关于离散生物信息学和数据挖掘的新章节.
- 关于编码理论和密码学的个别章节涵盖的范围更广(以前只在一个章节中介绍过).
- 关于许多主题的新章节,包括:

| | |
|---|---|
| 代数数论 | 椭圆曲线 |
| 奇异值分解 | 隐马尔科夫模型 |
| 概率方法 | 完满图 |
| 扩展图 | 小世界网络 |
| 组合拍卖(问题) | 超大规模邻域搜索 |
| 禁忌搜索 | 量子纠错码 |
| 经典密码学 | 密码散列函数 |
| 密码机制 | 现代私钥密码学 |

| | |
|---|---|
| 密码应用 | 联想法 |
| 分类 | 聚类 |
| 离群值分析 | 序列对比 |
| 进化树 | 离散 — 时间动力系统 |
| RNA 折叠 | 食物网与竞争图 |
| 神经编码 | 基因组组装 |

• 对现有部分进行了数以千计的更新和添加,重大更改或新的小节包括:

| | |
|---|---|
| 素数 | 组合矩阵理论 |
| 循环码 | 公钥密码学 |
| 划分 | 序列的渐近性 |
| 因式分解 | 距离、连通性、遍历性和匹配性 |
| 扩展图 | 图形着色、标签和相关参数 |
| 仿真 | 通信网络 |
| 区位论 | 困难的路由与分配问题 |

• 30 多篇传记.

• 数百个新的网络资源,为了验证它们的有效性,我们已在 2017 年年中访问过这些网站.


本书的内容极其广泛,从目录中可窥一斑.


1. 基础知识
   1.1  命题与谓词逻辑
   1.2  集合论
   1.3  函数
   1.4  关系
   1.5  证明方法
   1.6  公理程序验证
   1.7  基于逻辑的计算机编程范式
2. 计数方法
   2.1  计数问题概要
   2.2  基本计数技巧
   2.3  排列与组合
   2.4  包含 / 排除
   2.5  划分
   2.6  伯恩赛德 / 波利亚计数公式
   2.7  默比乌斯反演计数
   2.8  杨氏表
3. 序列

　　虽然笔者并不是搞组合数学的学者，但是由于多年从事数学竞赛的教学工作，所以养成了对组合数学的热爱与关注.

　　不久前笔者读到一篇第 51 届 IMO 金牌得主、现普林斯顿大学的聂子佩先生的短文，讲述了若干竞赛试题的背景，讲得相当深入浅出，其正是本书的部分内容.

　　这是一个历史悠久的故事，从头讲起至少应该追溯到六十年前，但我们先从幂级数的定义开始说.

一

　　幂级数是多项式的推广形式，指的是形如 $\sum_{n=0}^{\infty} a_n t^n$ 的式子，我们可以认为

这里的 $t$ 和求和号都只是符号,那么幂级数的意义就只在于一个数列 $\{a_n\}$.不过我们不会满足于此.当幂级数作为一个级数在某些意义下收敛时,它又可以代表一个函数,我们可以对函数做各种运算,而泰勒公式提示我们,在一些时候运算得到的函数又可以写成幂级数的形式.如此一来,我们有时可以直接定义幂级数上的运算.比如加法,我们有

$$\sum_{n=0}^{\infty} a_n t^n + \sum_{n=0}^{\infty} b_n t^n = \sum_{n=0}^{\infty} (a_n + b_n) t^n$$

看上去天经地义.

那么,乘法呢?

美国人发明了一个词"freshman's dream",指代的是 $(x+y)^n = x^n + y^n$ 这个式子.在实践中,大一学生做得更多的事情是将矩阵中对应项相乘得到的矩阵当作矩阵的乘法.类似地,我们也可以定义幂级数的"乘法"为

$$\sum_{n=0}^{\infty} a_n t^n \circ \sum_{n=0}^{\infty} b_n t^n = \sum_{n=0}^{\infty} a_n b_n t^n$$

简洁明了.然而这样的乘法却和幂级数对应的函数之间的乘法是不相容的.如果要相容,我们只能定义

$$\sum_{n=0}^{\infty} a_n t^n \cdot \sum_{n=0}^{\infty} b_n t^n = \sum_{n=0}^{\infty} \left( \sum_{k=0}^{n} a_k b_{n-k} \right) t^n$$

为了区分这两种乘法,我们把第一种称为幂级数的阿达玛乘法,第二种称为幂级数的乘法.

那么幂级数的阿达玛乘法与幂级数对应的函数之间就没有多大联系了吗?也不完全是.至少,我们还可以有这样的命题:

**命题 1**    如果两个幂级数对应的函数是有理函数,那么它们的阿达玛乘积对应的函数也是有理函数.

有理函数指的是多项式的商,这与这些多项式的系数是不是有理数无关,请注意不要混淆.

设有理函数 $\dfrac{P(t)}{Q(t)}$ 对应的幂级数为 $\sum_{n=0}^{\infty} a_n t^n$,由于满足

$$Q(t) \cdot \sum_{n=0}^{\infty} a_n t^n = P(t)$$

由幂级数的乘法可知,数列 $\{a_n\}$ 自某项起形成一个常系数线性递推数列,反之亦然.由特征方程法或者考虑部分分式,我们可以推得,有理函数对应的幂级数系数自某项起有通项公式

$$a_n = \sum_{i=1}^{s} p_i(n) \alpha_i^n$$

这里 $\alpha_i$ 是一些两两不同的复数,而 $p_i$ 是一些非零复系数多项式.

注意:即使 $a_n$ 全是实数,甚至全是正整数,我们也无法保证 $p_i$ 的系数是实数或者 $\alpha_i$ 是实数,事实上,$\alpha_i$ 是 $Q(t)$ 的某些根.

因为两个有理函数对应的阿达玛乘积的幂级数系数的通项公式也有这样的形式,所以我们可以反其道而行之,推得其幂级数系数自某项起也形成

一个常系数线性递推数列,因而阿达玛乘积对应的函数也是有理函数,即命题 1 成立.

那么,除法呢?

我们定义阿达玛除法为阿达玛乘法的逆运算,即当 $b_n$ 全不为零时,定义 $\sum_{n=0}^{\infty} a_n t^n$ 和 $\sum_{n=0}^{\infty} b_n t^n$ 的阿达玛商为 $\sum_{n=0}^{\infty} \frac{a_n}{b_n} t^n$.

那么两个有理函数对应的幂级数的阿达玛商对应的函数也是有理函数吗? 答案是否定的.

类似地,我们还可以定义 $\sum_{n=0}^{\infty} a_n t^n$ 的阿达玛 $k$ 次方根为满足按阿达玛乘法自乘 $k$ 次等于 $\sum_{n=0}^{\infty} a_n t^n$ 的所有幂级数.然而,有理函数对应的幂级数也未必有一个阿达玛 $k$ 次方根,使得其对应的函数是有理函数.

一切就到此为止了吗? 不,让我们回到六十多年前.

## 二

1959 年,法国数学家皮索在前人的特例的启发下提出了如下的猜想:

**命题 2**  如果两个整系数幂级数对应的函数是有理函数,且它们的阿达玛商也是整系数幂级数,则它们的阿达玛商对应的函数也是有理函数.

**命题 3**  如果一个幂级数对应的函数是有理函数,且它的系数都是整数的 $k$ 次方,则它有一个阿达玛 $k$ 次方根,使得其对应的函数也是有理函数.

命题 2 解决于 20 世纪 80 年代,命题 3 解决于 2000 年,对于这些证明,我们先放一放,我们且来看看皮索是怎么处理这两个命题的.

皮索用了一种现在我们称为"最大根方法"的技巧.他首先做了一些额外的假设:在命题 2 中,设除数 $\sum_{n=0}^{\infty} b_n t^n$ 的系数自某项起的通项公式

$$b_n = \sum_{i=1}^{s'} q_i(n) \beta_i^n$$

中,$\beta_1$ 的模长大于任何 $\beta_i (2 \leqslant i \leqslant s')$ 的模长.在命题 3 中,设幂级数 $\sum_{n=0}^{\infty} a^n t^n$ 的系数自某项起的通项公式

$$a_n = \sum_{i=1}^{s} p_i(n) \alpha_i^n$$

中,$\alpha_1$ 的模长大于任何 $\alpha_i (2 \leqslant i \leqslant s)$ 的模长,这种条件被称为最大根条件.

我们首先来看命题 2.

设被除数 $\sum_{n=0}^{\infty} a_n t^n$ 的系数自某项起的通项公式为

$$a_n = \sum_{i=1}^{s} p_i(n) \alpha_i^n$$

除数 $\sum_{n=0}^{\infty} b_n t^n$ 的系数自某项起的通项公式为

$$b_n = \sum_{i=1}^{s'} q_i(n) \beta_i^n$$

那么阿达玛商的系数自某项起的通项公式则为

$$\frac{a_n}{b_n} = \frac{\displaystyle\sum_{i=1}^{s} p_i(n)\alpha_i^n}{\displaystyle\sum_{i=1}^{s'} q_i(n)\beta_i^n} = \frac{\displaystyle\sum_{i=1}^{s} \frac{p_i(n)}{q_1(n)}\left(\frac{\alpha_i}{\beta_1}\right)^n}{1 + \displaystyle\sum_{i=2}^{s'} \frac{q_i(n)}{q_1(n)}\left(\frac{\beta_i}{\beta_1}\right)^n}$$

$$= \left(\sum_{i=1}^{s} \frac{p_i(n)}{q_1(n)}\left(\frac{\alpha_i}{\beta_1}\right)^n\right) \sum_{j=0}^{\infty} \left(-\sum_{i=2}^{s'} \frac{q_i(n)}{q_1(n)}\left(\frac{\beta_i}{\beta_1}\right)^n\right)^j$$

展开后可以写作

$$\frac{a_n}{b_n} = \sum_{i=1}^{\infty} r_i(n)\gamma_i^n$$

这里 $\{r_i(n)\}$ 是一列非零有理函数,而 $\{\gamma_i\}$ 是模长不增的复数列.估计每一项的大小,我们得到,存在正整数 $l$ 和实数 $0 < \gamma < 1$,使得 $|r_i(n)\gamma_i^n| \ll n^{li}\gamma^{ni}$.

设正整数 $M$ 满足 $|\gamma_{M+1}| < 1$,并将 $\frac{a_n}{b_n}$ 写作

$$\frac{a_n}{b_n} = \sum_{i=1}^{M} r_i(n)\gamma_i^n + \sum_{i=M+1}^{\infty} r_i(n)\gamma_i^n$$

则后一半当 $n$ 趋于无穷时趋于零.由条件 $a_n, b_n$ 均为整数,故 $p_i, q_j$ 的所有系数以及 $\alpha_i, \beta_j$ 均为代数数,故 $r_i$ 的所有系数以及 $\gamma_i$ 也均为代数数.因为 $r_i$ 的所有系数都是代数数,所以存在非零整系数多项式 $R$,使得所有 $R(n)r_i(n)$ $(1 \leqslant i \leqslant M)$ 均为多项式.因为 $\gamma_i$ 都是代数数,所以

$$\sum_{i=1}^{M} R(n)r_i(n)\gamma_i^n$$

形成一个整系数线性递推数列,即存在不全为零的整数 $c_1, \cdots, c_h$,使得

$$\sum_{k=1}^{h} c_k\left(\sum_{i=1}^{M} R(n+k)r_i(n+k)\gamma_i^{n+k}\right) = 0$$

于是,我们有

$$\sum_{k=1}^{h} \frac{a_{n+k}}{b_{n+k}}c_k R(n+k) = \sum_{k=1}^{h} c_k\left(\sum_{i=M+1}^{\infty} R(n+k)r_i(n+k)\gamma_i^{n+k}\right)$$

由于左边是整数,右边当 $n$ 趋于无穷时趋于零,我们知道当 $n$ 足够大时,左右均等于零,即幂级数 $\displaystyle\sum_{n=0}^{\infty} \frac{a_n}{b_n}R(n)t^n$ 的系数自某项起形成常系数线性递推数列,换句话说,幂级数 $\displaystyle\sum_{n=0}^{\infty} \frac{a_n}{b_n}R(n)t^n$ 对应的函数是有理函数.

如果我们额外假设了 $q_1(n)$ 是常数,那么 $R(n)$ 也是常数,此时我们已经不需要再做什么了.而在一般情况下,我们再回过头来看看命题 2 就会发现,我们已经把最大根条件下的命题 2 化归成了在除数 $\displaystyle\sum_{n=0}^{\infty} b_n t^n$ 的系数是整值多项式的条件下的命题 2.换句话说,我们只需证明如下命题,这个命题的证明应该归功于皮索的后继者康托尔:

**命题 4**    如果整系数幂级数 $\sum\limits_{n=0}^{\infty} a_n t^n$ 对应的函数是有理函数,且存在一个整值多项式 $R$,使得 $\dfrac{a_n}{R(n)}$ 总是整数,则幂级数 $\sum\limits_{n=0}^{\infty} \dfrac{a_n}{R(n)} t^n$ 对应的函数也是有理函数.

与之前一样,我们设 $\sum\limits_{n=0}^{\infty} a_n t^n$ 的系数自某项起的通项公式为

$$a_n = \sum_{i=1}^{s} p_i(n) \alpha_i^n$$

这时

$$\frac{a_n}{R(n)} = \sum_{i=1}^{s} \frac{p_i(n) \alpha_i^n}{R(n)}$$

其分母总是整数,而其分子在一般情况下却只是个代数数,看来我们在这里无论如何也得用一些代数数论的知识才能处理下去了. 世界就是如此,有的问题高等而肤浅,比如如何将一个初等数论的证明平行推广到代数数论中去,而有的问题却初等而深刻,比如生命的意义是什么. 为了让本文更可读,我决定只证明其初等的特例,而将完全版的证明留给了解一些代数数论或者对代数数论感兴趣的读者们. 我们假设,这里 $p_i$ 的系数和 $\alpha_i$ 都是有理数.

我们不妨设 $R(n)$ 模任何素数 $p$ 都不恒同余于零,不然我们用 $\dfrac{R(n)}{p}$ 代替 $R(n)$,而这样的操作只能做有限次,否则某个不为零的整值就会变成绝对值小于 1 的数,取 $N$ 为一大于所有 $p_i$ 的系数的分母的绝对值,所有 $\alpha_i$ 的分子和分母的绝对值,所有 $\alpha_i - \alpha_j (i \neq j)$ 的分子和分母的绝对值,以及所有 $R(n)$ 的系数的分母的绝对值的正整数. 由中国剩余定理,存在无穷多个正整数 $d$,使得 $R(d)$ 的所有素因子都不小于 $N$.

取任意这样的 $d$,对 $R(d)$ 的每个素因子 $p$,假设 $p^h$ 恰好整除 $R(d)$,那么 $p^h$ 整除 $R(d + kp^h)$,这里 $k = 0, \cdots, s - 1$. 由条件

$$0 \equiv a_{d+kp^h} = \sum_{i=1}^{s} p_i(d + kp^h) \alpha_i^{d+kp^h} \equiv \sum_{i=1}^{s} p_i(d) \alpha_i^{d+kp^h} \pmod{p^h}$$

对每个 $k = 0, \cdots, s - 1$,把这 $s$ 个式子想成关于 $p_1(d), \cdots, p_s(d)$ 的 $s$ 元一次方程组. 由范德蒙行列式及费马小定理,系数行列式等于

$$\prod_{i=1}^{s} \alpha_i^d \prod_{i>j} (\alpha_i^{p^h} - \alpha_j^{p^h}) \equiv \prod_{i=1}^{s} \alpha_i^d \prod_{i>j} (\alpha_i - \alpha_j) \pmod{p}$$

不是 $p$ 的倍数,由克莱姆法则,系数矩阵可逆,所以这个方程组在模 $p^h$ 意义下没有非零解,故 $p^h$ 整除每个 $p_i(d)$. 由 $p$ 的任意性,我们知道每个 $p_i(d)$ 的分子都是 $R(d)$ 的倍数. 由于这样的 $d$ 可以任意大,我们知道 $R$ 作为多项式整除每个 $p_i$,我们便得到了命题 4 的结论.

皮索对于命题 3 的处理是类似的,只是把幂级数展开式

$$\frac{1}{1-x} = \sum_{i=0}^{\infty} x^i$$

换作

$$(1+x)^{\frac{1}{k}} = \sum_{i=0}^{\infty} \begin{pmatrix} \dfrac{1}{k} \\ i \end{pmatrix} x^i$$

如此他得到的结论是:若最大根条件成立,且 $p_1(n)$ 是常数,则命题 4 成立. 如果想要摆脱 $p_1(n)$ 是常数这个条件,我们需要通过分析素因子来区分多项式里的 $n$ 和指数上的 $n$ 带来的影响.

也许我们应该在陷入更多对细枝末节的探讨前停止,再回头看看什么是最大根方法,为什么我们需要最大根来处理这些问题.

一言以蔽之,最大根方法就是以渐进的手段分析整数序列,由于绝对值小于 1 的整数只有 0,或者,等价地,依据代数数论的刘维尔不等式,在不等式中得到等式.至于为什么需要最大根,这是为了让得到的级数收敛.

如果最大根不唯一,这个方法就失效了吗? 也不是.

我们至少还有两种普遍的方法在最大根不唯一的时候依然用最大根方法处理问题.

第一,是从最大根到最小根的转变.如果 $\{a_n\}(n \in \mathbf{N})$ 是我们的常系数线性递推数列,那么我们可以依样画葫芦把其定义拓展到 $\{a_n\}(n \in \mathbf{Z})$,把数列倒过来看,最大根就变成了最小根,最小根则变成了最大根,这时我们需要考虑两个问题:(1) $\{a_n\}(n \in \mathbf{Z})$ 这时未必是整数数列了,我们需要对我们的命题做一点推广;(2)我们需要证明对于 $n \in \mathbf{N}$ 时题目给出的整除条件或者 $k$ 次方数条件可以推广到对于所有 $n \in \mathbf{Z}$ 成立,这通常需要分别考虑每个素因子得到.

第二,是改变所在的度量空间.级数 $\sum_{i=0}^{\infty} 2^i$ 在通常意义下发散,却在 2 进度量中收敛于 $-1$.级数收敛条件的改变意味着最大根条件的改变.

## 三

从方法到结论,我们都可以利用本文中讲述的内容更深入地了解一些竞赛题的背景.比如下面这道题是 2000 年全国高中数学联赛加试题.证明一个常系数线性递推式总是完全平方数,这种题目甚至已经成为初中数学竞赛的套路.

**问题 1**　设数列 $\{a_n\}$ 和 $\{b_n\}$ 满足 $a_0 = 1, b_0 = 0$,且 $\begin{cases} a_{n+1} = 7a_n + 6b_n - 3 \\ b_{n+1} = 8a_n + 7b_n - 4 \end{cases}$. 求证: $a_n$ 是完全平方数.

如果你遇到类似的考题,记住本文的命题 3,求通项 —— 开方 / 幂级数展开 —— 算结果的线性递推式,必然能得到证明.

以下三道题直接按最大根方法作幂级数展开就可以得到:

**问题 2**　设 $a,b$ 为整数,使得对所有 $n \in \mathbf{N}$ 都有 $2^n a + b$ 是完全平方数.证明: $a = 0$.

(2001 年波兰数学奥林匹克第三轮)

**问题 3**　设 $a,b$ 为大于 1 的整数,使得对所有 $n \in \mathbf{N}$ 都有 $a^n - 1$ 整除 $b^n - 1$.证明:存在正整数 $k$ 使得 $b = a^k$.

(美国数学月刊问题 10674)

**问题 4**　设 $a,b$ 为正整数,使得对所有 $n \in \mathbf{N}$ 都有 $b^n + n$ 是 $a^n + n$ 的倍数.证明:$a = b$.

(2005 年 IMO 预选题 N6)

最后这个问题可以通过按最大根方法和系数比较得到,其中的多项式部分不是常数,这使得比较系数的过程大为简化:

**问题 5**　设 $a_1, a_2, a_3, b_1, b_2, b_3$ 是两两不同的正整数,使得对所有 $n \in \mathbf{N}$ 都有

$$(n+1)a_1^n + na_2^n + (n-1)a_3^n \mid (n+1)b_1^n + nb_2^n + (n-1)b_3^n$$

证明:存在正整数 $k$,使得 $b_i = ka_i (i = 1,2,3)$.

(2010 年中国数学奥林匹克)

本工作室成立已有十七个年头了,承各位读者不弃,使得我们有了长期发展的动力.当然囿于笔者的学识与眼界难免会有个别格调与品味不高的图书混迹其中.

几年前,笔者曾读过一本优秀的科学家传记,是写苏联天才物理学家朗道的,文中有这样一段:

一位听众请朗道谈谈对一本新出版的科普书籍的看法,这本书由于谬误百出而引来一片哗然.大家期待着暴风骤雨般的批判.

然而朗道的回答却出乎意料:

"不要过于悲观地看待这本荒诞的书籍,它没有危害任何人."朗道说:"这本书给大家带来了震惊和欢乐,但是出版十本有缺陷的书,也好过一本好书都不出."

(摘自《朗道传》,迈娅·比萨拉比著,李雪莹译.北京:高等教育出版社,2018:221.)

是的,用中文说就是聊胜于无!

刘培杰

2022.12.11

于哈工大

# 刘培杰数学工作室
## 已出版(即将出版)图书目录——原版影印

| 书　　名 | 出版时间 | 定　价 | 编号 |
|---|---|---|---|
| 数学物理大百科全书.第1卷(英文) | 2016－01 | 418.00 | 508 |
| 数学物理大百科全书.第2卷(英文) | 2016－01 | 408.00 | 509 |
| 数学物理大百科全书.第3卷(英文) | 2016－01 | 396.00 | 510 |
| 数学物理大百科全书.第4卷(英文) | 2016－01 | 408.00 | 511 |
| 数学物理大百科全书.第5卷(英文) | 2016－01 | 368.00 | 512 |
| zeta函数,q-zeta函数,相伴级数与积分(英文) | 2015－08 | 88.00 | 513 |
| 微分形式:理论与练习(英文) | 2015－08 | 58.00 | 514 |
| 离散与微分包含的逼近和优化(英文) | 2015－08 | 58.00 | 515 |
| 艾伦·图灵:他的工作与影响(英文) | 2016－01 | 98.00 | 560 |
| 测度理论概率导论,第2版(英文) | 2016－01 | 88.00 | 561 |
| 带有潜在故障恢复系统的半马尔柯夫模型控制(英文) | 2016－01 | 98.00 | 562 |
| 数学分析原理(英文) | 2016－01 | 88.00 | 563 |
| 随机偏微分方程的有效动力学(英文) | 2016－01 | 88.00 | 564 |
| 图的谱半径(英文) | 2016－01 | 58.00 | 565 |
| 量子机器学习中数据挖掘的量子计算方法(英文) | 2016－01 | 98.00 | 566 |
| 量子物理的非常规方法(英文) | 2016－01 | 118.00 | 567 |
| 运输过程的统一非局部理论:广义波尔兹曼物理动力学,第2版(英文) | 2016－01 | 198.00 | 568 |
| 量子力学与经典力学之间的联系在原子、分子及电动力学系统建模中的应用(英文) | 2016－01 | 58.00 | 569 |
| 算术域(英文) | 2018－01 | 158.00 | 821 |
| 高等数学竞赛:1962—1991年的米洛克斯·史怀哲竞赛(英文) | 2018－01 | 128.00 | 822 |
| 用数学奥林匹克精神解决数论问题(英文) | 2018－01 | 108.00 | 823 |
| 代数几何(德文) | 2018－04 | 68.00 | 824 |
| 丢番图逼近论(英文) | 2018－01 | 78.00 | 825 |
| 代数几何学基础教程(英文) | 2018－01 | 98.00 | 826 |
| 解析数论入门课程(英文) | 2018－01 | 78.00 | 827 |
| 数论中的丢番图问题(英文) | 2018－01 | 78.00 | 829 |
| 数论(梦幻之旅):第五届中日数论研讨会演讲集(英文) | 2018－01 | 68.00 | 830 |
| 数论新应用(英文) | 2018－01 | 68.00 | 831 |
| 数论(英文) | 2018－01 | 78.00 | 832 |

# 刘培杰数学工作室

## 已出版(即将出版)图书目录——原版影印

| 书　　名 | 出版时间 | 定　价 | 编号 |
|---|---|---|---|
| 湍流十讲(英文) | 2018—04 | 108.00 | 886 |
| 无穷维李代数:第3版(英文) | 2018—04 | 98.00 | 887 |
| 等值、不变量和对称性(英文) | 2018—04 | 78.00 | 888 |
| 解析数论(英文) | 2018—09 | 78.00 | 889 |
| 《数学原理》的演化:伯特兰·罗素撰写第二版时的<br>手稿与笔记(英文) | 2018—04 | 108.00 | 890 |
| 哈密尔顿数学论文集(第4卷):几何学、分析学、天文学、<br>概率和有限差分等(英文) | 2019—05 | 108.00 | 891 |
| 偏微分方程全局吸引子的特性(英文) | 2018—09 | 108.00 | 979 |
| 整函数与下调和函数(英文) | 2018—09 | 118.00 | 980 |
| 幂等分析(英文) | 2018—09 | 118.00 | 981 |
| 李群,离散子群与不变量理论(英文) | 2018—09 | 108.00 | 982 |
| 动力系统与统计力学(英文) | 2018—09 | 118.00 | 983 |
| 表示论与动力系统(英文) | 2018—09 | 118.00 | 984 |
| 分析学练习.第1部分(英文) | 2021—01 | 88.00 | 1247 |
| 分析学练习.第2部分,非线性分析(英文) | 2021—01 | 88.00 | 1248 |
| 初级统计学:循序渐进的方法:第10版(英文) | 2019—05 | 68.00 | 1067 |
| 工程师与科学家微分方程用书:第4版(英文) | 2019—07 | 58.00 | 1068 |
| 大学代数与三角学(英文) | 2019—06 | 78.00 | 1069 |
| 培养数学能力的途径(英文) | 2019—07 | 38.00 | 1070 |
| 工程师与科学家统计学:第4版(英文) | 2019—06 | 58.00 | 1071 |
| 贸易与经济中的应用统计学:第6版(英文) | 2019—06 | 58.00 | 1072 |
| 傅立叶级数和边值问题:第8版(英文) | 2019—05 | 48.00 | 1073 |
| 通往天文学的途径:第5版(英文) | 2019—05 | 58.00 | 1074 |
| 拉马努金笔记.第1卷(英文) | 2019—06 | 165.00 | 1078 |
| 拉马努金笔记.第2卷(英文) | 2019—06 | 165.00 | 1079 |
| 拉马努金笔记.第3卷(英文) | 2019—06 | 165.00 | 1080 |
| 拉马努金笔记.第4卷(英文) | 2019—06 | 165.00 | 1081 |
| 拉马努金笔记.第5卷(英文) | 2019—06 | 165.00 | 1082 |
| 拉马努金遗失笔记.第1卷(英文) | 2019—06 | 109.00 | 1083 |
| 拉马努金遗失笔记.第2卷(英文) | 2019—06 | 109.00 | 1084 |
| 拉马努金遗失笔记.第3卷(英文) | 2019—06 | 109.00 | 1085 |
| 拉马努金遗失笔记.第4卷(英文) | 2019—06 | 109.00 | 1086 |
| 数论:1976年纽约洛克菲勒大学数论会议记录(英文) | 2020—06 | 68.00 | 1145 |
| 数论:卡本代尔1979:1979年在南伊利诺伊卡本代尔大学<br>举行的数论会议记录(英文) | 2020—06 | 78.00 | 1146 |
| 数论:诺德韦克豪特1983:1983年在诺德韦克豪特举行的<br>Journees Arithmetiques数论大会会议记录(英文) | 2020—06 | 68.00 | 1147 |
| 数论:1985—1988年在纽约城市大学研究生院和大学中心<br>举办的研讨会(英文) | 2020—06 | 68.00 | 1148 |

# 刘培杰数学工作室
## 已出版(即将出版)图书目录——原版影印

| 书 名 | 出 版 时 间 | 定 价 | 编号 |
|---|---|---|---|
| 数论:1987年在乌尔姆举行的Journees Arithmetiques数论大会会议记录(英文) | 2020—06 | 68.00 | 1149 |
| 数论:马德拉斯1987:1987年在马德拉斯安娜大学举行的国际拉马努金百年纪念大会会议记录(英文) | 2020—06 | 68.00 | 1150 |
| 解析数论:1988年在东京举行的日法研讨会会议记录(英文) | 2020—06 | 68.00 | 1151 |
| 解析数论:2002年在意大利切特拉罗举行的C.I.M.E.暑期班演讲集(英文) | 2020—06 | 68.00 | 1152 |
| 量子世界中的蝴蝶:最迷人的量子分形故事(英文) | 2020—06 | 118.00 | 1157 |
| 走进量子力学(英文) | 2020—06 | 118.00 | 1158 |
| 计算物理学概论(英文) | 2020—06 | 48.00 | 1159 |
| 物质,空间和时间的理论:量子理论(英文) | 2020—10 | 48.00 | 1160 |
| 物质,空间和时间的理论:经典理论(英文) | 2020—10 | 48.00 | 1161 |
| 量子场理论:解释世界的神秘背景(英文) | 2020—07 | 38.00 | 1162 |
| 计算物理学概论(英文) | 2020—06 | 48.00 | 1163 |
| 行星状星云(英文) | 2020—10 | 38.00 | 1164 |
| 基本宇宙学:从亚里士多德的宇宙到大爆炸(英文) | 2020—08 | 58.00 | 1165 |
| 数学磁流体力学(英文) | 2020—07 | 58.00 | 1166 |
| 计算科学:第1卷,计算的科学(日文) | 2020—07 | 88.00 | 1167 |
| 计算科学:第2卷,计算与宇宙(日文) | 2020—07 | 88.00 | 1168 |
| 计算科学:第3卷,计算与物质(日文) | 2020—07 | 88.00 | 1169 |
| 计算科学:第4卷,计算与生命(日文) | 2020—07 | 88.00 | 1170 |
| 计算科学:第5卷,计算与地球环境(日文) | 2020—07 | 88.00 | 1171 |
| 计算科学:第6卷,计算与社会(日文) | 2020—07 | 88.00 | 1172 |
| 计算科学.别卷,超级计算机(日文) | 2020—07 | 88.00 | 1173 |
| 多复变函数论(日文) | 2022—06 | 78.00 | 1518 |
| 复变函数入门(日文) | 2022—06 | 78.00 | 1523 |
| 代数与数论:综合方法(英文) | 2020—10 | 78.00 | 1185 |
| 复分析:现代函数理论第一课(英文) | 2020—07 | 58.00 | 1186 |
| 斐波那契数列和卡特兰数:导论(英文) | 2020—10 | 68.00 | 1187 |
| 组合推理:计数艺术介绍(英文) | 2020—07 | 88.00 | 1188 |
| 二次互反律的傅里叶分析证明(英文) | 2020—07 | 48.00 | 1189 |
| 旋瓦兹分布的希尔伯特变换与应用(英文) | 2020—07 | 58.00 | 1190 |
| 泛函分析:巴拿赫空间理论入门(英文) | 2020—07 | 48.00 | 1191 |
| 卡塔兰数入门(英文) | 2019—05 | 68.00 | 1060 |
| 测度与积分(英文) | 2019—04 | 68.00 | 1059 |
| 组合学手册.第一卷(英文) | 2020—06 | 128.00 | 1153 |
| *—代数、局部紧群和巴拿赫*—代数丛的表示.第一卷,群和代数的基本表示理论(英文) | 2020—05 | 148.00 | 1154 |
| 电磁理论(英文) | 2020—08 | 48.00 | 1193 |
| 连续介质力学中的非线性问题(英文) | 2020—09 | 78.00 | 1195 |
| 多变量数学入门(英文) | 2021—05 | 68.00 | 1317 |
| 偏微分方程入门(英文) | 2021—05 | 88.00 | 1318 |
| 若尔当典范性:理论与实践(英文) | 2021—07 | 68.00 | 1366 |
| 伽罗瓦理论.第4版(英文) | 2021—08 | 88.00 | 1408 |

# 刘培杰数学工作室
# 已出版(即将出版)图书目录——原版影印

| 书　　名 | 出版时间 | 定　价 | 编号 |
|---|---|---|---|
| 典型群,错排与素数(英文) | 2020—11 | 58.00 | 1204 |
| 李代数的表示:通过 gln 进行介绍(英文) | 2020—10 | 38.00 | 1205 |
| 实分析演讲集(英文) | 2020—10 | 38.00 | 1206 |
| 现代分析及其应用的课程(英文) | 2020—10 | 58.00 | 1207 |
| 运动中的抛射物数学(英文) | 2020—10 | 38.00 | 1208 |
| 2—纽结与它们的群(英文) | 2020—10 | 38.00 | 1209 |
| 概率,策略和选择:博弈与选举中的数学(英文) | 2020—11 | 58.00 | 1210 |
| 分析学引论(英文) | 2020—11 | 58.00 | 1211 |
| 量子群:通往流代数的路径(英文) | 2020—11 | 38.00 | 1212 |
| 集合论入门(英文) | 2020—10 | 48.00 | 1213 |
| 酉反射群(英文) | 2020—11 | 58.00 | 1214 |
| 探索数学:吸引人的证明方式(英文) | 2020—11 | 58.00 | 1215 |
| 微分拓扑短期课程(英文) | 2020—10 | 48.00 | 1216 |
| 抽象凸分析(英文) | 2020—11 | 68.00 | 1222 |
| 费马大定理笔记(英文) | 2021—03 | 48.00 | 1223 |
| 高斯与雅可比和(英文) | 2021—03 | 78.00 | 1224 |
| π 与算术几何平均:关于解析数论和计算复杂性的研究(英文) | 2021—01 | 58.00 | 1225 |
| 复分析入门(英文) | 2021—03 | 48.00 | 1226 |
| 爱德华·卢卡斯与素性测定(英文) | 2021—03 | 78.00 | 1227 |
| 通往凸分析及其应用的简单路径(英文) | 2021—01 | 68.00 | 1229 |
| 微分几何的各个方面.第一卷(英文) | 2021—01 | 58.00 | 1230 |
| 微分几何的各个方面.第二卷(英文) | 2020—12 | 58.00 | 1231 |
| 微分几何的各个方面.第三卷(英文) | 2020—12 | 58.00 | 1232 |
| 沃克流形几何学(英文) | 2020—11 | 58.00 | 1233 |
| 彷射和韦尔几何应用(英文) | 2020—12 | 58.00 | 1234 |
| 双曲几何学的旋转向量空间方法(英文) | 2021—02 | 58.00 | 1235 |
| 积分:分析学的关键(英文) | 2020—12 | 48.00 | 1236 |
| 为有天分的新生准备的分析学基础教材(英文) | 2020—11 | 48.00 | 1237 |
| 数学不等式.第一卷.对称多项式不等式(英文) | 2021—03 | 108.00 | 1273 |
| 数学不等式.第二卷.对称有理不等式与对称无理不等式(英文) | 2021—03 | 108.00 | 1274 |
| 数学不等式.第三卷.循环不等式与非循环不等式(英文) | 2021—03 | 108.00 | 1275 |
| 数学不等式.第四卷.Jensen 不等式的扩展与加细(英文) | 2021—03 | 108.00 | 1276 |
| 数学不等式.第五卷.创建不等式与解不等式的其他方法(英文) | 2021—04 | 108.00 | 1277 |

# 刘培杰数学工作室
# 已出版(即将出版)图书目录——原版影印

| 书　名 | 出版时间 | 定　价 | 编号 |
|---|---|---|---|
| 冯·诺侬曼代数中的谱位移函数:半有限冯·诺侬曼代数中的谱位移函数与谱流(英文) | 2021—06 | 98.00 | 1308 |
| 链接结构:关于嵌入完全图的直线中链接单形的组合结构(英文) | 2021—05 | 58.00 | 1309 |
| 代数几何方法.第1卷(英文) | 2021—06 | 68.00 | 1310 |
| 代数几何方法.第2卷(英文) | 2021—06 | 68.00 | 1311 |
| 代数几何方法.第3卷(英文) | 2021—06 | 58.00 | 1312 |

| 书　名 | 出版时间 | 定　价 | 编号 |
|---|---|---|---|
| 代数、生物信息和机器人技术的算法问题.第四卷,独立恒等式系统(俄文) | 2020—08 | 118.00 | 1199 |
| 代数、生物信息和机器人技术的算法问题.第五卷,相对覆盖性和独立可拆分恒等式系统(俄文) | 2020—08 | 118.00 | 1200 |
| 代数、生物信息和机器人技术的算法问题.第六卷,恒等式和准恒等式的相等 问题、可推导性和可实现性(俄文) | 2020—08 | 128.00 | 1201 |
| 分数阶微积分的应用:非局部动态过程,分数阶导热系数(俄文) | 2021—01 | 68.00 | 1241 |
| 泛函分析问题与练习:第2版(俄文) | 2021—01 | 98.00 | 1242 |
| 集合论、数学逻辑和算法论问题:第5版(俄文) | 2021—01 | 98.00 | 1243 |
| 微分几何和拓扑短期课程(俄文) | 2021—01 | 98.00 | 1244 |
| 素数规律(俄文) | 2021—01 | 88.00 | 1245 |
| 无穷边值问题解的递减:无界域中的拟线性椭圆和抛物方程(俄文) | 2021—01 | 48.00 | 1246 |
| 微分几何讲义(俄文) | 2020—12 | 98.00 | 1253 |
| 二次型和矩阵(俄文) | 2021—01 | 98.00 | 1255 |
| 积分和级数.第2卷,特殊函数(俄文) | 2021—01 | 168.00 | 1258 |
| 积分和级数.第3卷,特殊函数补充:第2版(俄文) | 2021—01 | 178.00 | 1264 |
| 几何图上的微分方程(俄文) | 2021—01 | 138.00 | 1259 |
| 数论教程:第2版(俄文) | 2021—01 | 98.00 | 1260 |
| 非阿基米德分析及其应用(俄文) | 2021—03 | 98.00 | 1261 |
| 古典群和量子群的压缩(俄文) | 2021—03 | 98.00 | 1263 |
| 数学分析习题集.第3卷,多元函数:第3版(俄文) | 2021—03 | 98.00 | 1266 |
| 数学习题:乌拉尔国立大学数学力学系大学生奥林匹克(俄文) | 2021—03 | 98.00 | 1267 |
| 柯西定理和微分方程的特解(俄文) | 2021—03 | 98.00 | 1268 |
| 组合极值问题及其应用:第3版(俄文) | 2021—03 | 98.00 | 1269 |
| 数学词典(俄文) | 2021—01 | 98.00 | 1271 |
| 确定性混沌分析模型(俄文) | 2021—06 | 168.00 | 1307 |
| 精选初等数学习题和定理.立体几何.第3版(俄文) | 2021—03 | 68.00 | 1316 |
| 微分几何习题:第3版(俄文) | 2021—05 | 98.00 | 1336 |
| 精选初等数学习题和定理.平面几何.第4版(俄文) | 2021—05 | 68.00 | 1335 |
| 曲面理论在欧氏空间 $E_n$ 中的直接表示(俄文) | 2022—01 | 68.00 | 1444 |
| 维纳一霍普夫离散算子和托普利兹算子:某些可数赋范空间中的诺特性和可逆性(俄文) | 2022—03 | 108.00 | 1496 |
| Maple 中的数论:数论中的计算机计算(俄文) | 2022—03 | 88.00 | 1497 |
| 贝尔曼和克努特问题及其概括:加法运算的复杂性(俄文) | 2022—03 | 138.00 | 1498 |

| 书　　名 | 出版时间 | 定　价 | 编号 |
|---|---|---|---|
| 复分析:共形映射(俄文) | 2022—07 | 48.00 | 1542 |
| 微积分代数样条和多项式及其在数值方法中的应用(俄文) | 2022—08 | 128.00 | 1543 |
| 蒙特卡罗方法中的随机过程和场模型:算法和应用(俄文) | 2022—08 | 88.00 | 1544 |
| 线性椭圆型方程组:论二阶椭圆型方程的迪利克雷问题(俄文) | 2022—08 | 98.00 | 1561 |
| 动态系统解的增长特性:估值、稳定性、应用(俄文) | 2022—08 | 118.00 | 1565 |
| 群的自由积分解:建立和应用(俄文) | 2022—08 | 78.00 | 1570 |
| 混合方程和偏差自变数方程问题:解的存在和唯一性(俄文) | 2023—01 | 78.00 | 1582 |
| 拟度量空间分析:存在和逼近定理(俄文) | 2023—01 | 108.00 | 1583 |
| 二维和三维流形上函数的拓扑性质:函数的拓扑分类(俄文) | 2023—03 | 68.00 | 1584 |
| 齐次马尔科夫过程建模的矩阵方法:此类方法能够用于不同目上的的复杂系统研究、设计和完善(俄文) | 2023—03 | 68.00 | 1594 |
|  |  |  |  |
| 狭义相对论与广义相对论:时空与引力导论(英文) | 2021—07 | 88.00 | 1319 |
| 束流物理学和粒子加速器的实践介绍:第2版(英文) | 2021—07 | 88.00 | 1320 |
| 凝聚态物理中的拓扑和微分几何简介(英文) | 2021—05 | 88.00 | 1321 |
| 混沌映射:动力学、分形学和快速涨落(英文) | 2021—05 | 128.00 | 1322 |
| 广义相对论:黑洞、引力波和宇宙学介绍(英文) | 2021—06 | 68.00 | 1323 |
| 现代分析电磁均质化(英文) | 2021—06 | 68.00 | 1324 |
| 为科学家提供的基本流体动力学(英文) | 2021—06 | 88.00 | 1325 |
| 视觉天文学:理解夜空的指南(英文) | 2021—06 | 68.00 | 1326 |
| 物理学中的计算方法(英文) | 2021—06 | 68.00 | 1327 |
| 单星的结构与演化:导论(英文) | 2021—06 | 108.00 | 1328 |
| 超越居里:1903年至1963年物理界四位女性及其著名发现(英文) | 2021—06 | 68.00 | 1329 |
| 范德瓦尔斯流体热力学的进展(英文) | 2021—06 | 68.00 | 1330 |
| 先进的托卡马克稳定性理论(英文) | 2021—06 | 88.00 | 1331 |
| 经典场论导论:基本相互作用的过程(英文) | 2021—07 | 88.00 | 1332 |
| 光致电离量子动力学方法原理(英文) | 2021—07 | 108.00 | 1333 |
| 经典域论和应力:能量张量(英文) | 2021—05 | 88.00 | 1334 |
| 非线性太赫兹光谱的概念与应用(英文) | 2021—06 | 68.00 | 1337 |
| 电磁学中的无穷空间并矢格林函数(英文) | 2021—06 | 88.00 | 1338 |
| 物理科学基础数学.第1卷,齐次边值问题、傅里叶方法和特殊函数(英文) | 2021—07 | 108.00 | 1339 |
| 离散量子力学(英文) | 2021—07 | 68.00 | 1340 |
| 核磁共振的物理学和数学(英文) | 2021—07 | 108.00 | 1341 |
| 分子水平的静电学(英文) | 2021—08 | 68.00 | 1342 |
| 非线性波:理论、计算机模拟、实验(英文) | 2021—06 | 108.00 | 1343 |
| 石墨烯光学:经典问题的电解解决方案(英文) | 2021—06 | 68.00 | 1344 |
| 超材料多元宇宙(英文) | 2021—07 | 68.00 | 1345 |
| 银河系外的天体物理学(英文) | 2021—07 | 68.00 | 1346 |
| 原子物理学(英文) | 2021—07 | 68.00 | 1347 |
| 将光打结:将拓扑学应用于光学(英文) | 2021—07 | 68.00 | 1348 |
| 电磁学:问题与解法(英文) | 2021—07 | 88.00 | 1364 |
| 海浪的原理:介绍量子力学的技巧与应用(英文) | 2021—07 | 108.00 | 1365 |
| 多孔介质中的流体:输运与相变(英文) | 2021—07 | 68.00 | 1372 |
| 洛伦兹群的物理学(英文) | 2021—08 | 68.00 | 1373 |
| 物理导论的数学方法和解决方法手册(英文) | 2021—08 | 68.00 | 1374 |

# 刘培杰数学工作室
## 已出版(即将出版)图书目录——原版影印

| 书　　名 | 出版时间 | 定　价 | 编号 |
|---|---|---|---|
| 非线性波数学物理学入门(英文) | 2021—08 | 88.00 | 1376 |
| 波:基本原理和动力学(英文) | 2021—07 | 68.00 | 1377 |
| 光电子量子计量学.第1卷,基础(英文) | 2021—07 | 88.00 | 1383 |
| 光电子量子计量学.第2卷,应用与进展(英文) | 2021—07 | 68.00 | 1384 |
| 复杂流的格子玻尔兹曼建模的工程应用(英文) | 2021—08 | 68.00 | 1393 |
| 电偶极矩挑战(英文) | 2021—08 | 108.00 | 1394 |
| 电动力学:问题与解法(英文) | 2021—09 | 68.00 | 1395 |
| 自由电子激光的经典理论(英文) | 2021—08 | 68.00 | 1397 |
| 曼哈顿计划——核武器物理学简介(英文) | 2021—09 | 68.00 | 1401 |
| 粒子物理学(英文) | 2021—09 | 68.00 | 1402 |
| 引力场中的量子信息(英文) | 2021—09 | 128.00 | 1403 |
| 器件物理学的基本经典力学(英文) | 2021—09 | 68.00 | 1404 |
| 等离子体物理及其空间应用导论.第1卷,基本原理和初步过程(英文) | 2021—09 | 68.00 | 1405 |

| 书　　名 | 出版时间 | 定　价 | 编号 |
|---|---|---|---|
| 拓扑与超弦理论焦点问题(英文) | 2021—07 | 58.00 | 1349 |
| 应用数学:理论、方法与实践(英文) | 2021—07 | 78.00 | 1350 |
| 非线性特征值问题:牛顿型方法与非线性瑞利函数(英文) | 2021—07 | 58.00 | 1351 |
| 广义膨胀和齐性:利用齐性构造齐次系统的李雅普诺夫函数和控制律(英文) | 2021—06 | 48.00 | 1352 |
| 解析数论焦点问题(英文) | 2021—07 | 58.00 | 1353 |
| 随机微分方程:动态系统方法(英文) | 2021—07 | 58.00 | 1354 |
| 经典力学与微分几何(英文) | 2021—07 | 58.00 | 1355 |
| 负定相交形式流形上的瞬子模空间几何(英文) | 2021—07 | 68.00 | 1356 |

| 书　　名 | 出版时间 | 定　价 | 编号 |
|---|---|---|---|
| 广义卡塔兰轨道分析:广义卡塔兰轨道计算数字的方法(英文) | 2021—07 | 48.00 | 1367 |
| 洛伦兹方法的变分:二维与三维洛伦兹方法(英文) | 2021—08 | 38.00 | 1378 |
| 几何、分析和数论精编(英文) | 2021—08 | 68.00 | 1380 |
| 从一个新角度看数论:通过遗传方法引入现实的概念(英文) | 2021—07 | 58.00 | 1387 |
| 动力系统:短期课程(英文) | 2021—08 | 68.00 | 1382 |
| 几何路径:理论与实践(英文) | 2021—08 | 48.00 | 1385 |
| 论天体力学中某些问题的不可积性(英文) | 2021—07 | 88.00 | 1396 |
| 广义斐波那契数列及其性质(英文) | 2021—08 | 38.00 | 1386 |
| 对称函数和麦克唐纳多项式:余代数结构与Kawanaka恒等式(英文) | 2021—09 | 38.00 | 1400 |

| 书　　名 | 出版时间 | 定　价 | 编号 |
|---|---|---|---|
| 杰弗里·英格拉姆·泰勒科学论文集:第1卷.固体力学(英文) | 2021—05 | 78.00 | 1360 |
| 杰弗里·英格拉姆·泰勒科学论文集:第2卷.气象学、海洋学和湍流(英文) | 2021—05 | 68.00 | 1361 |
| 杰弗里·英格拉姆·泰勒科学论文集:第3卷.空气动力学以及落弹数和爆炸的力学(英文) | 2021—05 | 68.00 | 1362 |
| 杰弗里·英格拉姆·泰勒科学论文集:第4卷.有关流体力学(英文) | 2021—05 | 58.00 | 1363 |

# 刘培杰数学工作室
# 已出版(即将出版)图书目录——原版影印

| 书　名 | 出版时间 | 定　价 | 编号 |
|---|---|---|---|
| 非局域泛函演化方程:积分与分数阶(英文) | 2021—08 | 48.00 | 1390 |
| 理论工作者的高等微分几何:纤维丛、射流流形和拉格朗日理论(英文) | 2021—08 | 68.00 | 1391 |
| 半线性退化椭圆微分方程:局部定理与整体定理(英文) | 2021—07 | 48.00 | 1392 |
| 非交换几何、规范理论和重整化:一般简介与非交换量子场论的重整化(英文) | 2021—09 | 78.00 | 1406 |
| 数论论文集:拉普拉斯变换和带有数论系数的幂级数(俄文) | 2021—09 | 48.00 | 1407 |
| 挠理论专题:相对极大值,单射与扩充模(英文) | 2021—09 | 88.00 | 1410 |
| 强正则图与欧几里得若尔当代数:非通常关系中的启示(英文) | 2021—10 | 48.00 | 1411 |
| 拉格朗日几何和哈密顿几何:力学的应用(英文) | 2021—10 | 48.00 | 1412 |

| 书　名 | 出版时间 | 定　价 | 编号 |
|---|---|---|---|
| 时滞微分方程与差分方程的振动理论:二阶与三阶(英文) | 2021—10 | 98.00 | 1417 |
| 卷积结构与几何函数理论:用以研究特定几何函数理论方向的分数阶微积分算子与卷积结构(英文) | 2021—10 | 48.00 | 1418 |
| 经典数学物理的历史发展(英文) | 2021—10 | 78.00 | 1419 |
| 扩展线性丢番图问题(英文) | 2021—10 | 38.00 | 1420 |
| 一类混沌动力系统的分歧分析与控制:分歧分析与控制(英文) | 2021—11 | 38.00 | 1421 |
| 伽利略空间和伪伽利略空间中一些特殊曲线的几何性质(英文) | 2022—01 | 68.00 | 1422 |
| 一阶偏微分方程:哈密尔顿—雅可比理论(英文) | 2021—11 | 48.00 | 1424 |
| 各向异性黎曼多面体的反问题:分段光滑的各向异性黎曼多面体反边界谱问题:唯一性(英文) | 2021—11 | 38.00 | 1425 |

| 书　名 | 出版时间 | 定　价 | 编号 |
|---|---|---|---|
| 项目反应理论手册.第一卷,模型(英文) | 2021—11 | 138.00 | 1431 |
| 项目反应理论手册.第二卷,统计工具(英文) | 2021—11 | 118.00 | 1432 |
| 项目反应理论手册.第三卷,应用(英文) | 2021—11 | 138.00 | 1433 |
| 二次无理数:经典数论入门(英文) | 2022—05 | 138.00 | 1434 |
| 数,形与对称性:数论,几何和群论导论(英文) | 2022—05 | 128.00 | 1435 |
| 有限域手册(英文) | 2021—11 | 178.00 | 1436 |
| 计算数论(英文) | 2021—11 | 148.00 | 1437 |
| 拟群与其表示简介(英文) | 2021—11 | 88.00 | 1438 |
| 数论与密码学导论:第二版(英文) | 2022—01 | 148.00 | 1423 |

# 刘培杰数学工作室
# 已出版(即将出版)图书目录——原版影印

| 书　名 | 出版时间 | 定　价 | 编号 |
|---|---|---|---|
| 几何分析中的柯西变换与黎兹变换:解析调和容量和李普希兹调和容量、变化和振荡以及一致可求长性(英文) | 2021—12 | 38.00 | 1465 |
| 近似不动点定理及其应用(英文) | 2022—05 | 28.00 | 1466 |
| 局部域的相关内容解析:对局部域的扩展及其伽罗瓦群的研究(英文) | 2022—01 | 38.00 | 1467 |
| 反问题的二进制恢复方法(英文) | 2022—03 | 28.00 | 1468 |
| 对几何函数中某些类的各个方面的研究:复变量理论(英文) | 2022—01 | 38.00 | 1469 |
| 覆盖、对应和非交换几何(英文) | 2022—01 | 28.00 | 1470 |
| 最优控制理论中的随机线性调节器问题:随机最优线性调节器问题(英文) | 2022—01 | 38.00 | 1473 |
| 正交分解法:涡流流体动力学应用的正交分解法(英文) | 2022—01 | 38.00 | 1475 |
| 芬斯勒几何的某些问题(英文) | 2022—03 | 38.00 | 1476 |
| 受限三体问题(英文) | 2022—05 | 38.00 | 1477 |
| 利用马利亚万微积分进行 Greeks 的计算:连续过程、跳跃过程中的马利亚万微积分和金融领域中的 Greeks(英文) | 2022—05 | 48.00 | 1478 |
| 经典分析和泛函分析的应用:分析学的应用(英文) | 2022—03 | 38.00 | 1479 |
| 特殊芬斯勒空间的探究(英文) | 2022—03 | 48.00 | 1480 |
| 某些图形的施泰纳距离的细谷多项式:细谷多项式与图的维纳指数(英文) | 2022—05 | 38.00 | 1481 |
| 图论问题的遗传算法:在新鲜与模糊的环境中(英文) | 2022—05 | 48.00 | 1482 |
| 多项式映射的渐近簇(英文) | 2022—05 | 38.00 | 1483 |
| 一维系统中的混沌:符号动力学,映射序列,一致收敛和沙可夫斯基定理(英文) | 2022—05 | 38.00 | 1509 |
| 多维边界层流动与传热分析:粘性流体流动的数学建模与分析(英文) | 2022—05 | 38.00 | 1510 |
| 演绎理论物理学的原理:一种基于量子力学波函数的逐次置信估计的一般理论的提议(英文) | 2022—05 | 38.00 | 1511 |
| $R^2$ 和 $R^3$ 中的仿射弹性曲线:概念和方法(英文) | 2022—08 | 38.00 | 1512 |
| 算术数列中除数函数的分布:基本内容、调查、方法、第二矩、新结果(英文) | 2022—05 | 28.00 | 1513 |
| 抛物型狄拉克算子和薛定谔方程:不定常薛定谔方程的抛物型狄拉克算子及其应用(英文) | 2022—07 | 28.00 | 1514 |
| 黎曼-希尔伯特问题与量子场论:可积重正化、戴森-施温格方程(英文) | 2022—08 | 38.00 | 1515 |
| 代数结构和几何结构的形变理论(英文) | 2022—08 | 48.00 | 1516 |
| 概率结构和模糊结构上的不动点:概率结构和直觉模糊度量空间的不动点定理(英文) | 2022—08 | 38.00 | 1517 |

# 刘培杰数学工作室
# 已出版(即将出版)图书目录——原版影印

| 书　名 | 出版时间 | 定　价 | 编号 |
|---|---|---|---|
| 反若尔当对:简单反若尔当对的自同构(英文) | 2022—07 | 28.00 | 1533 |
| 对某些黎曼－芬斯勒空间变换的研究:芬斯勒几何中的某些变换(英文) | 2022—07 | 38.00 | 1534 |
| 内诣零流形映射的尼尔森数的阿诺索夫关系(英文) | 2023—01 | 38.00 | 1535 |
| 与广义积分变换有关的分数次演算:对分数次演算的研究(英文) | 2023—01 | 48.00 | 1536 |
| 强子的芬斯勒几何和吕拉几何(宇宙学方面):强子结构的芬斯勒几何和吕拉几何(拓扑缺陷)(英文) | 2022—08 | 38.00 | 1537 |
| 一种基于混沌的非线性最优化问题:作业调度问题(英文) | 即将出版 | | 1538 |
| 广义概率论发展前景:关于趣味数学与置信函数实际应用的一些原创观点(英文) | 即将出版 | | 1539 |
| 纽结与物理学:第二版(英文) | 2022—09 | 118.00 | 1547 |
| 正交多项式和q－级数的前沿(英文) | 2022—09 | 98.00 | 1548 |
| 算子理论问题集(英文) | 2022—09 | 108.00 | 1549 |
| 抽象代数:群、环与域的应用导论:第二版(英文) | 即将出版 | | 1550 |
| 菲尔兹奖得主演讲集:第三版(英文) | 2023—01 | 138.00 | 1551 |
| 多元实函数教程(英文) | 2022—09 | 118.00 | 1552 |
| 球面空间形式群的几何学:第二版(英文) | 2022—09 | 98.00 | 1566 |
| 对称群的表示论(英文) | 2023—01 | 98.00 | 1585 |
| 纽结理论:第二版(英文) | 2023—01 | 88.00 | 1586 |
| 拟群理论的基础与应用(英文) | 2023—01 | 88.00 | 1587 |
| 组合学:第二版(英文) | 2023—01 | 98.00 | 1588 |
| 加性组合学:研究问题手册(英文) | 2023—01 | 68.00 | 1589 |
| 扭曲、平铺与镶嵌:几何折纸中的数学方法(英文) | 2023—01 | 98.00 | 1590 |
| 离散与计算几何手册:第三版(英文) | 2023—01 | 248.00 | 1591 |
| 离散与组合数学手册:第二版(英文) | 2023—01 | 248.00 | 1592 |
| 分析学教程.第1卷,一元实变量函数的微积分分析学介绍(英文) | 2023—01 | 118.00 | 1595 |
| 分析学教程.第2卷,多元函数的微分和积分,向量微积分(英文) | 2023—01 | 118.00 | 1596 |
| 分析学教程.第3卷,测度与积分理论,复变量的复值函数(英文) | 2023—01 | 118.00 | 1597 |
| 分析学教程.第4卷,傅里叶分析,常微分方程,变分法(英文) | 2023—01 | 118.00 | 1598 |

**联系地址:**哈尔滨市南岗区复华四道街10号　哈尔滨工业大学出版社刘培杰数学工作室
**网　址:**http://lpj.hit.edu.cn/
**邮　编:**150006
**联系电话:**0451－86281378　　13904613167
**E-mail:**lpj1378@163.com